Radiation Detection

Radiation Detection

Concepts, Methods, and Devices

Douglas S. McGregor
J. Kenneth Shultis

CRC Press
Taylor & Francis Group
Boca Raton London New York

CRC Press is an imprint of the
Taylor & Francis Group, an **informa** business

First edition published 2020
by CRC Press
6000 Broken Sound Parkway NW, Suite 300, Boca Raton, FL 33487-2742

and by CRC Press
2 Park Square, Milton Park, Abingdon, Oxon, OX14 4RN

CRC Press is an imprint of Taylor & Francis Group, an Informa business

Library of Congress Cataloging-in-Publication Data
[Insert LoC Data here when available]

ISBN: 978-1-4398-1939-5 (hbk)
ISBN: 978-1-4398-1940-1 (ebk)

Typeset in CMR
by Nova Techset Private Limited, Bengaluru & Chennai, India

Visit the Taylor & Francis Web site at
http://www.taylorandfrancis.com

and the CRC Press Web site at
http://www.crcpress.com

For Product Safety Concerns and Information please contact our EU representative: GPSR@taylorandfrancis.com
Taylor & Francis Verlag GmbH, Kaufingerstraße 24, 80331 München, Germany

To the countless pioneers, inventors and scholars
whose seminal contributions have developed
radiation detection into a mature discipline

About the Cover Picture

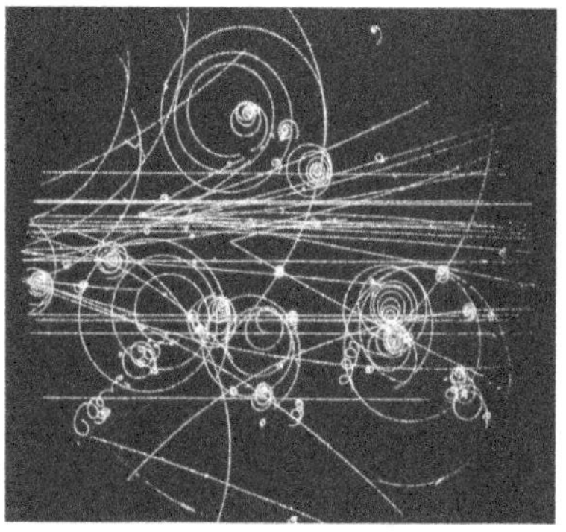

A 1960 image from CERN's first liquid hydrogen bubble chamber, which was 32 cm in diameter. Here 16-GeV negative pions enter from the left and are seen as the horizontal bubble trails. One pion interacts with a proton in the liquid hydrogen producing a spray of new particles including a neutral Λ^0 that quickly decays into a proton and negative pion to produce the sideway "V" (slightly left of center). Lower energy charged particles produced in the original and subsequent interactions create spiral tracks in the magnetic field of the chamber. Photograph courtesy of CERN.

Contents

Preface

Ever since the discovery of x rays by Röntgen almost 125 years ago, who used fluorescing barium-platinocyanide coated plates and later photographic film as the first radiation detectors, there has been increasing activity in the design and development of detectors to measure ionizing radiation. Early radiation detector development was driven by both the discovery of new types of ionizing radiation and a better understanding of how radiation interacts with matter. In the last half of the twentieth century detector technology saw major advances with the introduction of supporting electronics, notably the photomultiplier tube, and the introduction of new detector materials such as semiconductors and new scintillator materials. In this era, the fundamental principles of radiation detection and measurement were firmly established. Then in the last two decades, a vast number of researchers introduced an ever increasing number of new detector designs, almost all based on twentieth century technology. In the past decade the introduction of new variations in basic detector technology has become a growth industry, fueled largely by national security concerns and government sponsorship.

At the same time that radiation detector technology was advancing, many books on the subject were introduced. Many were limited to a single aspect or type of detector technology and were intended for a specialist in radiation detection and, consequently, were not suitable as texts in courses on radiation technology. Other books designed primarily as text books attempted to present a broad coverage of the many facets of radiation detections but were somewhat superficial because of their reliance on old analyses, rules of thumb, simple derivations, and a limited number of detector designs. Yet other more ambitious text books became encyclopedic by trying to cover the many different types of detectors while necessarily scrimping on analytic details or simply quoting analytical results. Also lacking in most of these treatises was a review of the history of radiation detector development and a presentation of the important milestones achieved by the vast research on radiation detection.

When we embarked on this book project over ten years ago, we envisioned a book that could serve both as a text for different courses in radiation detection and as a reference work for practitioners in radiation detection. Four goals guided us: (1) to include information about the historical development and the major achievements in detector technology, (2) to provide fundamental information about radiation interactions, nuclear physics, and electrostatics, (3) to present a broad and detailed coverage of the many types of radiation detectors that have been introduced, and (4) to present important mathematical models and analyses used in detector designs and applications. With these goals in mind, we have attempted to develop the fundamentals of radiation detection and detector technology so that a researcher or user of these devices can understand radiation interactions in materials, physics of the radiation detectors, commercially available and conceptual detectors, detector designs, characterization techniques, and special electronics used for various detector configurations and applications. Many of the topics covered are interrelated and, hence, the presentation of these topics could be arranged in a number of different ways. The sequence chosen by us was done so as to reduce redundancy while, at the same time, to develop a logical and clear path of presentation.

Overall, this book is designed to give the reader more than a glimpse at radiation detection devices and a few packaged equations. Rather it seeks to provide an understanding that allows the reader to choose the appropriate detection technology for a particular application, to design detectors, and to competently perform radiation measurements. For this reason, there are many equations derived from first principles, derivations which may seem tedious to some readers. Yet, we believe that a firm understanding of the assumptions used to derive many frequently encountered equations used in the discipline helps a radiation detector user to decide when, and also when not, to apply them, and how to arrive at an appropriate alternative, if needed. Many of the later chapters on specific aspects of radiation detectors attempt to provide a comprehensive review of the historical development and current state of the topic. Such a review necessarily entails citations to many of the important discoveries and the many references provided allow a reader to quickly find additional and more detailed information.

This book has five main themes: General Concepts for Radiation Detection, Common Radiation Detector Properties, Types of Radiation Detectors, Radiation Measurements, and finally a single chapter on Introductory Electronics. Although these themes are represented by various chapters, there is some redundancy between chapters both because the themes are closely related and because some overlap assists with clarity. For instance, semiconductor photodiodes are used as light collecting devices for some scintillators, hence there is a discussion on photodiodes in the chapter on light collecting devices. Yet, the physics of semiconductor diode detectors is detailed in those chapters covering semiconductor detectors, which comes after the chapter on light collecting devices. This order may seem out of sequence, yet is unavoidable because the chapters on scintillator detectors precede the chapters on semiconductor detectors, and the authors deem it best to include the chapter on light gathering devices next to the chapters on the scintillation detectors because they are used together as a single unit. Neutron detectors are configured as gas-filled, scintillator, and semiconductor devices, as well as a few special detector designs. As a result, there are chapters dedicated specifically to neutron detectors. Yet many of the principles of these neutron detectors borrow heavily from those concepts outlined in the chapters dedicated to gas-filled, scintillator, and semiconductor detectors.

Acknowledgments

Although we have included a few original contributions, the vast amount of material presented is the result of the work by thousands of researchers in radiation detection and its related fields. For their efforts and for what they have taught us we are extremely grateful. Although it is impossible to give credit to all of these specialists, we have attempted to cite many of the key papers, reports and books that have guided the development of radiation technology. Because of the breadth of the goals for this book, the time required to write it and its resulting size are unusually long. Special thanks go our publisher's forbearance and patience in this book project.

The authors acknowledge and give their appreciation for the support and assistance of numerous individuals who made this project possible through their contributions, guidance, and encouragement. Great appreciation is extended to those current and former students who took time to proofread this text and, in doing so, made valuable contributions and suggestions. These include Elsa Ariesanti, Steven Bellinger, Brian Cooper, Benjamin Damm, Nathaniel Edwards, Ryan Fronk, Priyarshini Ghosh, Nikolas Hinson, Alireza Kargar, Michael Meier, Benjamin Montag, Kyle Nelson, Taylor Ochs, and Michael Reichenberger. Many of these same students contributed to text material in the form of radiation detection data and instrument photographs. We thank the library staff at Kansas State University for acquiring the several hundreds of citations, many rare and difficult to locate, used in the references listed at the end of each chapter. Special gratitude is extended to Rick Brake (retired Los Alamos Nat. Lab.), Anthony Caruso (UMKC), Stephan Friedrich (LLNL), Kimberlee Kearfott (U. Michigan), Mike Kusner (Saint-Gobain), Michael Mayhugh (retired Saint-Gobain), Tim Sobering (KSU), and Jack Thiesen (U. Michigan) for reviewing certain chapters of this book and providing valuable comments, corrections, and improvements.

The authors also extend appreciation to their colleagues and friends who contributed to the knowledge and material presented in the text. These include Anthony Caruso (UMKC), Elias Chavez (Ludlum), Paul Davison (ADIT), William Dunn (Kansas State U.), Peggy Groff (Photonis), Daniel Herr (Saint-Gobain Crystals), Kimberly Kearfott (U. Michigan), Ben Kennedy (Ortec Ametek-AMT), Mike Kusner (Saint-Gobain Crystals), Bill Lehnert (LND, Inc.), Michael Mayhugh (retired Saint-Gobain Crystals), Jeremy Roberts (Kansas State U.), Ronald Rojeski, and Ron Stubberfield (ET Enterprises).

A special acknowledgement is extended to Richard (Rick) Brake, whose encouragement and financial support allowed one author (McGregor) to build his first radiation detectors while a graduate student intern at Los Alamos National Laboratory. Along with Rick Brake were those who took time to teach the author fundamentals of semiconductor device fabrication and detector construction, including Victor Swenson and Bill Gibler of Texas A&M University, and Alan Gibbs and David Brown of Los Alamos National Laboratory. It is the initial encouragement from these mentors that eventually led to the compilation of this text. Also acknowledged is former advisor and instructor Mark Weichold of Texas A&M University, whose constructive and detailed explanations of solid state and semiconductor physics have served one author (McGregor) well.

Above all we have valued the support and encouragement of our wives, Vilelmina and Sue, whose forbearance of too many evenings of neglect because of this tome will forever be remembered and appreciated.

Manhattan, Kansas
June 2019

Douglas S. McGregor
J. Kenneth Shultis

About the Authors

Douglas S. McGregor, born in Dallas, Texas in 1957, received BA (1985) and MS (1989) degrees, both in electrical engineering, at Texas A&M University. He then received MS (1992) and PhD (1993) degrees, both in nuclear engineering, from the University of Michigan. From 1994 to 1996 he held a post doctoral position at Sandia National Laboratory in Livermore, California before returning to the University of Michigan where he was an assistant research professor from 1997 to 2002. In 2002, he joined the Nuclear Engineering faculty at Kansas State University (KSU) where he now is a University Distinguished Professor and holds the Boyd D. Brainard Chair in Mechanical and Nuclear Engineering. Professor McGregor serves as director of the Semiconductor Materials and Radiological Technologies Laboratory at KSU, a 9500 sq ft laboratory dedicated to radiation detector research.

Professor McGregor teaches and conducts research on the design, development, and deployment of radiation detectors and detector systems. In particular, he develops systems for measuring various ionizing and non-ionizing radiations based on semiconductor, scintillator, and gas-filled detectors. He specializes in semiconductor device physics, semiconductor device designs, radiation detector physics, neutron detector designs, and radiation detection and measurement. He has published over 200 research articles and reports, is co-inventor on over 20 radiation detector patents, and his research group has received five R&D-100 Awards for radiation detector innovations. Prof. McGregor is also the recipient of various other honors, including the KSU College of Engineering (CoE) Frankenhoff Outstanding Research Award (2006) and the CoE Engineering Distinguished Researcher Award (2016).

J. Kenneth Shultis, born in Toronto Canada in 1941, graduated from the University of Toronto with a BASc degree in Engineering Physics. He gained his MS (1965) and PhD (1968) degrees in Nuclear Science and Engineering from the University of Michigan. After a postdoctoral year at the Mathematics Institute of the University of Groningen, the Netherlands, he joined the Nuclear Engineering faculty at Kansas State University in 1969 and where he presently holds the Black and Veatch Distinguished Professorship and is the Ike and Letty Conerstone teaching scholar.

Professor Shultis teaches and conducts research in neutron and radiation transport, radiation shielding, reactor physics, numerical analysis, particle combustion, remote sensing, and utility energy and economic analyses. He has had a rich collaboration in research and scholarship. Besides being coauthor of this book he has coauthored the books *Fundamentals of Nuclear Science and Engineering*, *Radiation Shielding*, *Radiological Assessment*, *Principles of Radiation Shielding*, and *Exploring Monte Carlo Methods*. In addition, he has produced over 150 research papers and reports and served as a consultant to many private and governmental organizations. He is a Fellow of the American Nuclear Society (ANS), and has received many awards for his teaching and research, including the infrequently awarded ANS Rockwell Lifetime Achievement Award for his contributions over 50 years to the practice of radiation shielding.

Chapter 1

Origins

Great discoveries are made accidentally less often than the populace likes to think.

Wilhelm Röntgen

1.1 A Brief History of Radiation Discovery

In 1895, within the town of Strasbourg (then located in Germany, but now in France), Wilhelm Conrad Röntgen (see Fig. 1.1) made the first observation of mysterious penetrating rays. For many years, these mysterious rays were called "Röntgen rays"; however, Röntgen actually named them "x rays," as we know them today, after the common algebraic symbol for an unknown. In the famous experiment, Röntgen was operating a type of Crookes tube (see Fig. 1.2) while performing experiments with "cathode rays." With the Crookes tube covered with black paper, he noticed that a plate coated with barium-platinocyanide ($BaPt(CN)_4$), located approximately six feet away in the darkened room, was glowing. By placing objects with varying densities and thicknesses between the Crookes tube and the plate, he deduced that the mysterious emanations originated from the Crookes tube and that they were attenuated according to mass density and thickness of the absorber. A short while later, it was discovered that x rays are actually a form of electromagnetic radiation emitted from accelerated charged particles (electrons in Röntgen's experiment). Röntgen also used photographic plates for his studies after he learned that film emulsions were exposed by the x rays. After experimenting with many x-ray image exposures of inanimate objects, perhaps the most famous x-ray photograph he developed is traditionally believed to be an image of his wife's hand, Bertha Röntgen (see Fig. 1.3). Indeed photographic film is a type of radiation detector, and the x-ray image of Bertha Röntgen's hand marks the beginning of medical imaging. For his discovery of x rays, Röntgen became the recipient in 1901 of the first Nobel Prize in Physics. These discoveries were made with two types of *radiation detectors*, namely the scintillating $BaPt(CN)_4$-coated plate and the photographic plates.

Figure 1.1. Wilhelm Conrad Röntgen (1845–1923).

In 1896, shortly after the announcement of the discovery of x rays, Henri Becquerel (Fig. 1.4), while conducting experiments with his father's geology samples, discovered natural radiation emissions. It was already known that many substances could be made to fluoresce by exposing them to visible light, and Becquerel was studying the fluorescence of sulphates of uranium and potassium by exposing his samples

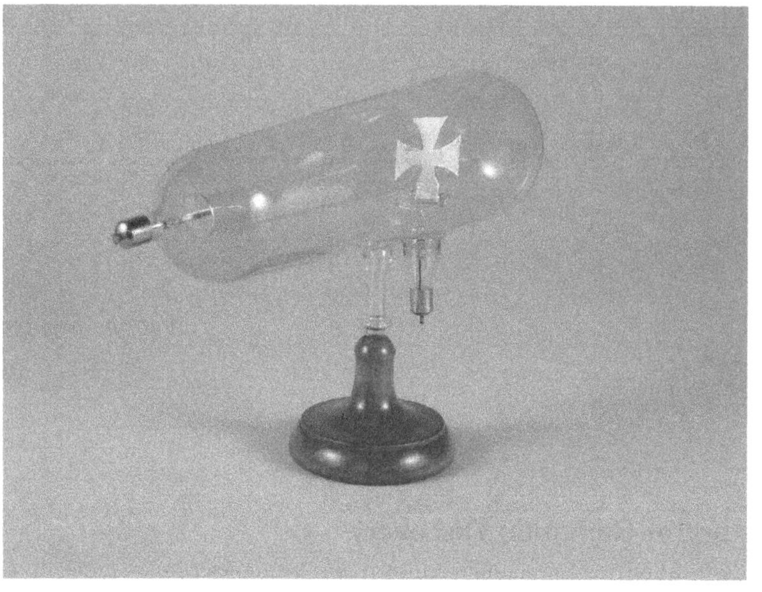

Figure 1.2. A Crookes tube was used as the x-ray source in Röntgen's experiments. The Maltese cross in the vacuum tube is connected to a hinge that allows it to be flipped up or down.

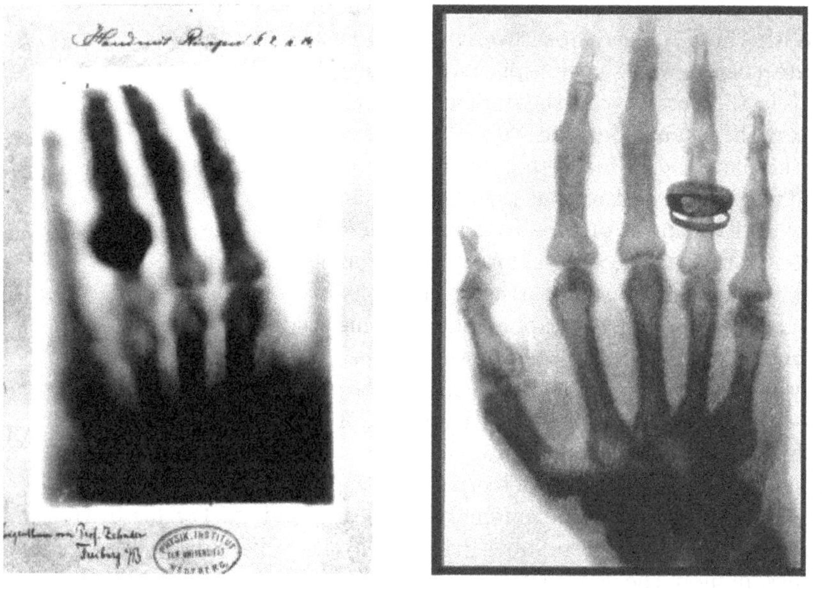

Figure 1.3. (Left) Traditionally labeled as Bertha Röntgen's hand, and probably the case, it actually was not labeled as such at the time of publication. (Right) Often mistakenly labeled as Bertha Röntgen's hand, this image taken by Röntgen is actually that of Alfred von Kölliker's hand. The discovery that x rays can be used to produce images of bone structure marks the beginning of the modern medical imaging profession.

Figure 1.4. Antoine Henri Becquerel (1852–1908).

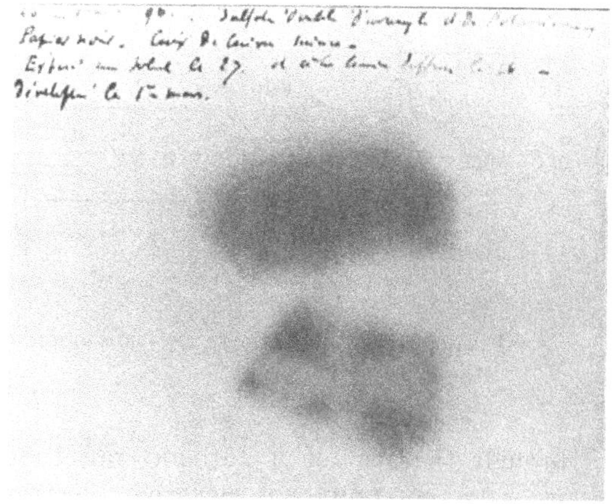

Figure 1.5. The original photographic plate images, taken by Henri Becquerel, of uranium ore samples.

to sunlight and measuring the light emissions with photographic plates. In one instance, he prepared a sample set of uranium salts for an exposure experiment, but decided to delay the experiment that day due to cloudy weather. Instead, he placed the uranium salt samples in a dresser drawer, fortuitously atop a stack of photographic plates. A few days later, he decided to check the quality of his photographic plates before trying the fluorescence experiment again, and developed the top photographic plate on which the samples lay, only to find that an image had formed (see Fig. 1.5). At first, Becquerel actually thought that the uranium salt samples were exposed to enough light to cause fluorescence, which he believed then exposed the film. He continued to conduct the experiments, but instead covered the photographic plates with black paper, upon which he placed the samples. He later discovered uranium salts not yet exposed to sunlight had the same effect on the photographic plates. Finally, several months after the initial discovery, he found that non-fluorescent uranium ore could also expose the plates, and he deduced that unknown emanations from the rock samples themselves caused the exposure. Initially referred to as "Becquerel rays," we now know the emanations he observed were a combination of alpha particles, beta particles, and gamma rays. For the discovery of radioactivity, Becquerel shared in the third Nobel Prize in Physics in 1903[1] along with Pierre and Marie Curie for their work and discovery of polonium and radium. Again, the discovery was made possible with a type of radiation detector (photographic plates). It is interesting to note that in February 1896 an English physicist and mathematician by the name of Silvanus Thompson independently conducted the same experiment as Becquerel, in which he exposed uranium samples to sunlight as they were placed upon a paper-covered photographic plate. He also discovered the mysterious emanations, but when he sent his results in for publication (to the Royal Society in London), he was informed that Becquerel had already reported the same findings only two weeks earlier to the French Academy of Sciences.

During the 1890s, many other significant experiments were being conducted with cathode ray tubes. Cathode ray tubes were built in many configurations, but common features included a negatively charged electrode (cathode) and a positively charged electrode (anode) sealed in an elongated vacuum tube, where the far end of the tube opposite the cathode was coated with a phosphor.

[1] Interestingly enough, there is evidence that natural radiation had been observed from uranium salts much earlier by Claude Félix Abel Niepce de Saint-Victor, in which a uranyl nitrate sample had been brought into close proximity to a photo sensitive paper sheet [Saint-Victor 1858]. The Niepce de Saint-Victor experiments were reported to the French Academy of Sciences in 1858, and evidently Becquerel was aware of the work [Fournier and Fournier 1990]. Regardless, Becquerel was the first to explain and understand the significance of the discovery.

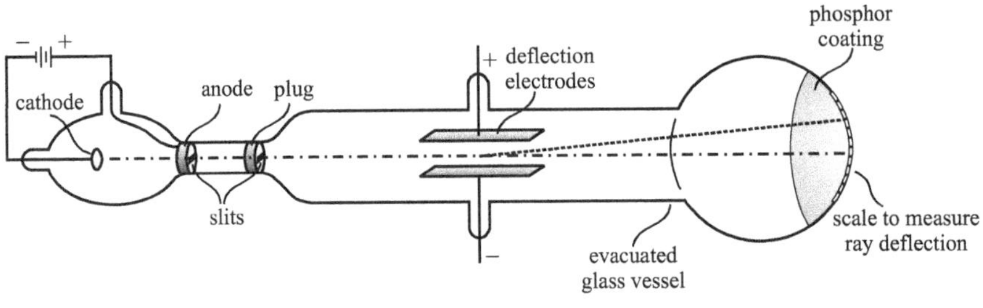

Figure 1.7. The cathode ray tube apparatus used by Thomson to measure the q/m ratio for cathode rays.

In 1895, through the use of a cathode ray tube, Jean Baptiste Perrin made the discovery that cathode rays were composed of negative electricity. Building on this knowledge, Joseph John Thomson (see Fig. 1.6) conducted experiments with different cathode ray tubes (CRTs) to investigate the nature of the cathode rays, where he paid special attention to evacuating the CRT. This simple precaution allowed experiments to be conducted without contaminant gases which could become ionized during the operation of the tube. A schematic of such a tube is shown in Fig. 1.7. With these tubes Thomson made three fundamental experimental observations. First, he noted that magnetic fields and electric fields could deflect the trajectory of the cathode rays.

Figure 1.6. Joseph John Thomson (1856–1940).

Further, given a known magnetic field strength, he noted that the deflection of these cathode rays was far greater than observed for positively charged hydrogen gas. Finally, by comparing the results of the cathode rays and the hydrogen gas, Thomson noted that the charge-to-mass ratio was approximately 1800 times greater for the cathode rays than for hydrogen ions.[2] Because he was confident no gas was inside the cathode ray tube, in 1897 Thomson correctly deduced that cathode rays were in fact negatively charged particles emitted from the cathode surface or, more correctly, emitted from the atoms composing the cathode. Although Thomson named these newly identified negative particles "corpuscles," the eventual name given to them was *electrons* as suggested by George Johnston Stoney, who had actually predicted the presence of these negative particles as early as 1874.

It should be noted that Thomson's discovery was the first recorded indication that atoms consisted of subatomic particles, thereby, earning him the 1906 Nobel Prize in Physics. Thomson's discovery was made possible, in part, with another type of radiation detector, namely the fluorescent screen of the CRT.

The observations and findings of Becquerel caught the interests of Pierre and Marie Curie, both chemists by education (see Fig. 1.8). Marie began studying under Becquerel, and began a systematic investigation of the known elements to determine if other materials emitted energetic rays similar to those from uranium. While studying pitchblende, an uranium ore containing various oxides, thorium, and other rare earth elements, she found that thorium and thorium compounds also emitted energetic rays. Marie is generally

[2]The same observation was made by Arthur Schuster in 1890, where he also measured the charge-to-mass ratio of cathode rays to be over 1000 times greater than positively charged hydrogen gas. However, he believed the measurements to be in error and his observations went largely unnoticed.

credited with recommending the name *radioactivity*[3] to describe these energetic emissions. Marie quickly reported these findings, only to learn that German physicist Gerhard Schmidt had published similar findings on thorium only two months earlier. However, of genuine interest in Curie's work was the observation that the pitchblende, after removing the uranium, was more radioactive than the uranium. The Curies thus deduced correctly that another radioactive element must be present.

Figure 1.8. Marie (1867–1934) and Pierre (1859–1906) Curie discovered polonium and radium.

In 1898, Pierre joined Marie's research effort, and together they worked to separate and identify the other radioactive substances in pitchblende. They used a quartz piezoelectric electrometer (named a Curie electrometer), invented by Pierre and his brother Jacques, to measure the ionization produced in air by the radiation emitted from the purified samples. They separated literally tons of pitchblende material to attain just a few milligrams of the purified radioactive substance, which they named polonium (after Marie's country of birth, Poland) and they reported the new material in July 1898. Five months later in December 1898, they reported a second new element, which they named radium. For their work on radioactive materials, both Pierre and Marie Curie shared the 1903 Nobel Prize in Physics with Henri Becquerel. Marie Curie also received the 1911 Nobel Prize in Chemistry for discovering two new elements[4] (Pierre had been accidentally killed by a horse and wagon in 1906, and did not share in the award).

In 1899 Ernest Rutherford (see Fig. 1.9), who was a former student of J.J. Thomson, took on the task of identifying the natural radiation discovered by Becquerel. Rutherford used a gold foil electroscope as the detector in his experiments. Understanding that the radiation produced free charges in air, he placed a uranium ore sample in the electroscope compartment, charged the electroscope, and then recorded the discharge rate of the gold foils. Rutherford then placed thin layers of aluminum foil (each 5-μm thick), one at a time, over the uranium ore sample. For each foil added, he observed the change in the discharge rate. He noted that a significant change in discharge rate occurred each time for the first three foils used, but the change in discharge rate was dramatically less afterwards. In fact, after the first three foils were in place the discharge rate had dropped to 8% of the initial rate with no attenuators. However, it took an extra 100 aluminum foil layers to reduce the discharge rate to 2% of the initial rate. Rutherford correctly deduced that two types of radiation were present. He named those radiations stopped with only three foils *alpha rays* and those higher penetrating radiations were named *beta rays.* We now know these rays to be particles, with α particles being ^{4}He nuclei and β particles being electrons. For these discoveries, Ernest Rutherford was awarded the 1906 Nobel Prize in Chemistry. Using magnetic fields, Rutherford later showed that beta particles could be easily deflected, while

Figure 1.9. Sir Ernest Rutherford (1871–1937) discovered alpha and beta particles.

[3]The word radioactivity was chosen because of its Latin root word "radius," which means 'ray' or 'spoke of a wheel.'
[4]Marie Curie is the first (and presently only) person to receive Nobel prizes in two different scientific fields.

Figure 1.10. Paul Ulrich Villard (1860–1934) discovered what were later called gamma rays.

Figure 1.11. Arthur Holly Compton (1892–1962) discovered what is known as Compton scattering.

alpha particles were deflected very little, and in the opposite direction of beta particles, indicating different masses and different characteristic charges.

Shortly thereafter, in 1900, Paul Villard (see Fig. 1.10) conducted experiments with radium samples and photographic plates. He collimated the radiation from a radium sample towards a photographic plate that was shielded with a thin layer of lead. The lead stopped the α particles, but allowed the highly penetrating radiation to pass. Using a magnetic field, he noted that two types of penetrating radiation were present, one type that was deflected by a magnetic field, and another that was not. He concluded that the deflected radiation was composed of the β particles discovered by Rutherford, but the undeflected radiation must be a third and different kind. Initially, Rutherford thought these emanations to be nothing more than highly energetic β particles, but eventually came to believe that Villard was correct. Villard did not name the radiation he discovered, but instead, in 1903, it was Rutherford who named the third radiation *gamma rays.* In 1914 it was shown that these highly penetrating rays, which were not affected by magnetic fields, could also be diffracted off crystal surfaces. Hence, Villard was correct, and gamma radiation was something quite different. The nature of gamma rays was explored further in 1918 by Arthur Holly Compton (see Fig. 1.11), who developed the fundamental physics behind what is referred to as Compton scattering. He showed that gamma rays lose energy after scattering with electrons, demonstrating the particle nature of gamma rays, for which he earned the 1927 Nobel Prize in Physics.

Noted in Millikan's book *Electricity, Sound and Light* [1908], there were certain aspects of the atom not clearly understood. For instance, as Millikan deduced, because electrons possess negative electricity and atoms themselves are neutral, then there must be some form of positive charge within the atomic structure to cancel the negative charge of the electron. J.J. Thomson proposed an atomic model, later disproved, in which the atomic nucleus consisted of positive electricity (now known as protons) about which, or within which, the negative electrons were rotating. This model is often referred to as the "plum pudding model," with the pudding being the positive stuff and the electrons the plums. In 1909, Ernest Rutherford performed an important experiment that proved the plum pudding model wrong. In his experiment, Rutherford collimated

Figure 1.12. Charles Thomson Rees Wilson (1869–1959).

Figure 1.13. Robert Andrews Millikan (1868–1953).

alpha particles from a Po sample into an ultra-thin sheet of Au foil, using the alpha particles as a method to probe the atomic structure of Au. If Thomson was correct and the Au atoms resembled the plum pudding model, then the alpha particles should just pass through the Au foil with hardly any deflection. Although Rutherford found that many alpha particles did actually pass directly through the Au foil, he also found that a small fraction were strongly deflected, and some scattered in the backwards direction. He concluded that such strong interactions could only happen if the positively charged alpha particles came into the close vicinity of another small but heavy positive mass, unlike the small deflections expected from the large neutralized atomic core suggested by Thomson. From his findings, Rutherford postulated the following:

(1) Most of the atomic mass and positive charge are contained in a small core called the nucleus.[5]

(2) Most of the volume of the atom is empty space about which the negative electrons are dispersed.

(3) Within the atom, the number of negatively charged electrons and the number of positively charged particles (protons) are equal.

In 1911 Charles Wilson (see Fig. 1.12) devised what is now known as a cloud chamber[6] to visualize the paths of charged particles by trails of condensed droplets in supersaturated water vapor and air mixture. These vapor trails would drift downwards under the influence of gravity. Wilson realized that the ion trails could be forced downwards faster by applying an electric field such that the droplets fell under the influence of both gravity and the electrostatic field [Kaplan 1962]. Wilson managed to obtain an estimate of the charge of an electron by comparing the rates of fall with and without the electric field force applied. However, two significant errors prevented accuracy, namely (a) the droplets could reach terminal velocity in the air and water vapor mixture regardless of the applied field, hence a true measure of the electric field influence was not possible, and (b) a water droplet would begin to evaporate so that the droplet mass changed as the cloud

[5]Traditionally, because of his identification of a distinct positive core within the atom, Rutherford is credited with discovering the proton.

[6]Wilson received the 1919 Nobel Prize in Physics for this interesting detection device.

trail fell downwards. Further, Wilson measured only the upper part of the cloud trails because the lower droplets in the trails were more heavily charged and generally moved faster than the upper portion.

Robert Millikan (see Fig. 1.13) thought that Wilson's method could be improved by applying the electric field force in the opposite direction of gravity so that the tiny charged vapor droplets could be made to remain stationary (balancing electric field and gravitational forces). No assumptions need be made about terminal velocities and what portion of the cloud should be inspected. Specific droplets could be selected, using a microscope, and observed. However, the problem remained that water and alcohols quickly evaporated. As the story goes, Millikan thought of the idea of substituting clock oil for water vapor while on a train trip. When looking at his watch, he realized that clock oil was used as a long lasting lubrication because it did not evaporate. Using an atomizer to inject a fine mist of oil droplets into a chamber, Millikan irradiated the chamber with x rays to produce charges on the oil droplets (see Fig. 1.14). He held a single droplet stationary by balancing the electric field force with the gravitational force. Regardless of the droplet chosen, he observed that the measured electric field (coulombs per unit distance) required to hold an oil droplet stationary were multiples of 1.6×10^{-19} coulombs. A single oil droplet might have an excess or deficiency of one or more electrons, but could not gain or lose a fraction of an electron.[7] This measurement also led to the electron mass being estimated as 9.1×10^{-31} kg, again remarkably close to the presently accepted value of $9.109\,382\,91 \times 10^{-31}$ kg.

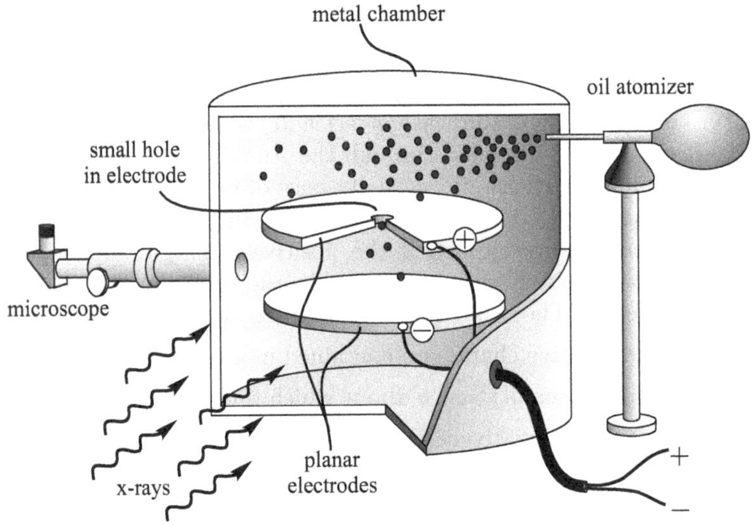

Figure 1.14. The basic components of Millikan's oil drop apparatus.

In 1920, Rutherford noted that there was some problem balancing the atomic masses of elements and their charge. For instance, the mass of hydrogen could be measured, as could other elements. It had also

[7]There is controversy over Millikan's selective use of data, in which over 60% of his recorded measurements were not included in his publications. Note also that a great controversy arose between Millikan and Felix Ehrenhaft, who also tried to measure the fundamental negative charge. Ehrenhaft's data suggests the existence of a smaller fundamental charge, sometimes recorded as low as 1/10 that of 1.6×10^{-19} coulombs. It has been suggested that Ehrenhaft may have observed the effect of elementary particles (such as quarks), although a more likely explanation is that the observed differences were a result of experimental error. Regardless, Millikan's conclusion, reported in 1913, has less than 0.5% error of the presently accepted value of $q_e = 1.602\,176\,565 \times 10^{-19}$ coulombs. Delayed by the "Millikan-Ehrenhaft controversy," Millikan did eventually receive the Nobel Prize in Physics (1923) for his work.

been determined that a proton had an intrinsic charge equal in magnitude to that of an electron, yet opposite in polarity. This all made sense, except that the masses did not scale linearly with the quantity of electrons or protons. Rutherford hypothesized, therefore, that some additional mass, neutral in nature, must be present. Many years later, in 1928, experiments were conducted by Walther Bothe[8] and Herbert Becker where alpha particles from Po were used to bombard Be. The result was the discovery of a new form of highly penetrating neutral radiation. Initially thought to be high energy gamma rays, this idea was quickly discarded as unlikely because of the constraints of conservation of energy and mass requirements in Compton scattering. Irene Curie (daughter of Marie and Pierre Curie) and her husband, Frédérick Joliot, studied the effects further by also bombarding Be with alpha particles emitted from Po, causing the unidentified neutral radiation to be emitted into a paraffin sample. The result was the emission from the paraffin of energetic protons. These protons were measured with a gas-filled counter (a detector developed by Hans Geiger and Ernest Rutherford) to have approximately 5.3 MeV of kinetic energy. From Compton's previous findings with gamma-ray scattering, the observed result was unlikely if the unknown neutral emissions really were gamma rays.

James Chadwick (see Fig. 1.15), who was Rutherford's student, used a similar arrangement as Curie and Joliot with a few changes. A Po alpha particle source was used to produce the neutral particles from a boron sample, and a gas-filled detector similar to early Geiger models was placed near the Be sample such that the neutral emissions could enter the detector [Chadwick 1932]. Separately using N_2 and then H_2 gas in the detector, along with conservation of mass and energy calculations, Chadwick measured and calculated the mass (or energy) of the neutral radiation to be 938 MeV/c^2, nearly the same, but slightly larger than that of protons. Thus, Chadwick found the missing particle that Rutherford had hypothesized, and because they have no intrinsic charge, he named them *neutrons*. Remarkably, Chadwick's calculations were within 0.17% error of the presently accepted value of 939.272046 MeV/c^2. For the discovery of neutrons, Chadwick received the 1935 Nobel Prize in Physics.

Figure 1.15. James Chadwick (1891–1974) discovered neutrons.

Following the discovery of these basic subatomic particles (electrons, protons, and neutrons), numerous other subatomic particles and elementary particles have been discovered. *Quarks* are elementary particles that combine to form *hadrons*, and all quarks follow Fermi-Dirac statistical models in which no quark can occupy the same quantum state as another. Hence, quarks belong to the group of particles classified as *fermions*. Independently postulated as existing by Murray Gell-Mann and George Zweig in 1964, the quark was discovered in 1968 at the Stanford Linear Accelerator Center (SLAC). Gell-Mann received the 1969 Nobel Prize in Physics for his contributions to the theory and his prediction of the quark.

There are six types of quarks, with the fanciful names of up, down, top, bottom, charm and strange, in what is now known as the standard model. Hadrons are subcategorized as *mesons* and *baryons*, where mesons are composed of two quarks and baryons are composed of three quarks. The two most well-known and stable baryons are protons and neutrons. Protons are composed of two up-quarks (each with charge of $+2/3\ q_e$) and one down-quark (charge of $-1/3\,q_e$), yielding a total

[8]Walther Bolthe used Geiger counters to invent a method of coincidence counting in 1924. He used this technique to discover and verify several physical principles about penetrating radiation and light quanta. For his invention of coincidence counting, and his many discoveries, he received the 1954 Nobel Prize in Physics.

charge of $+q_e$. Neutrons are composed of one up-quark and two down-quarks, yielding a total neutral charge. Neutrons not bound in an atomic nucleus are radioactively unstable with a half-life of 10.23 minutes. When a neutron is free from an atomic nucleus, one of the down-quarks decays into an up-quark, causing the neutron to decay into a proton, an electron, and an antineutrino.

The existence of mesons, a particle with binding force capable of holding the atomic nucleus together, was predicted by Hideki Yukawa in 1935. Convincing evidence of Yukawa's theory was found in 1945 with the discovery of the π-meson (pion) at the University of Bristol, a particle that was later proved to have a strong nuclear force that participates in binding of the nucleus. Yakawa received the 1948 Nobel Prize in Physics for his theory on the behavior and properties of mesons, in particular, the pion.

Another set of elementary particles are the *leptons*, also classified as fermions, which are categorized into six *flavors* subdivided as three *generations*. The electronic generation is composed of electrons and electron-neutrinos, the second generation is composed of muons and muon-neutrinos, and the third generation is composed of tauons and tauon-neutrinos. The existence of the various neutrino forms was discovered in 1962 by Leon M. Lederman, Melvin Schwartz, and Jack Steinberger who were awarded the 1988 Nobel Prize in Physics.

All fermions (hadrons and leptons) have corresponding anti-particles. For hadrons the anti-particle is composed of the corresponding anti-quarks producing anti-hadrons with identical mass but opposite charge as their real hadron counterparts. In fact anti-matter seems to interact with the strong and electromagnetic forces identically as does matter. Thus, it is possible to envision anti-atoms and macroscopic masses composed of anti-matter. However, matter and anti-matter cannot coexist. Just as positrons annihilate with electrons, anti-nucleons annihilate with nucleons.

Finally, a third branch of elementary particles contains the *bosons* consisting of photons, gluons, Z and $W^{\pm}$ particles. These particles are governed by Bose-Einstein statistics and, hence, they can occupy the same quantum state at the same time. Photons represent the force component of an electromagnetic field, gluons are the strong binding force among the quarks in the neutron and proton as well as the binding of the nucleons in an atomic nucleus, and the Z and $W^{\pm}$ particles are responsible for the so-called weak nuclear force. A fifth boson, the Higgs boson (H^0) is a theoretical particle predicted by the standard model, but for years it remained elusive. Finally in 2012, researchers at the European Organization for Nuclear Research (known as CERN) announced its discovery, completing the search for all the elementary particles required by the standard model. This last boson is important because it is the particle that gives mass to all the other elementary particles.

Today the search for a better understanding of the sub-atomic realm continues. The ingenious methods used by Thompson, Rutherford, Millikan and their contemporaries have been largely replaced with enormous particle accelerators and sophisticated detection systems. Rather than a single observer performing experiments in a one-room laboratory, teams of physicists and engineers work together at large facilities, such as CERN, to measure and observe the effects of these numerous elementary particles. Yet, even today, it is the innovative thought and creativity of *individuals* that continues to be the driving force behind discovery.

1.2 A Brief History of Radiation Detectors

There are three essential features in any radiation detector. First, the detector must have a material capable of interacting with the radiation of interest. Otherwise, the radiation would simply pass through it as if no detector were present. Second, upon interacting with the radiation, the material must exhibit an observable response. The response can be any of a multitude of possibilities such as temperature changes, color shifts, light emission, or ionization of the material. Third, the observable must be measurable. For instance, radiation interactions may occur in a common stone on the ground and ionization may occur in the stone as a result, but accurate measurement of the ionization is prevented because the stone can not conduct the charges to allow their measurement. A person seeking either to use or to design radiation detectors must keep these three simple, yet important, concepts in mind.

All the important discoveries of subatomic particles and radiations were accomplished with the use of various types of radiation detectors, including photographic plates, electroscopes, scintillating materials, and gas-filled detectors. In the case of photographic film, the absorber is the photo-emulsion, the observable is the chemical change in the Ag halide crystals in the emulsion, and the measurable is the developable Ag left behind as an image. In scintillating film, the absorber is the scintillation material, the observable is the excitation of electrons in the material, which deexcite to give a measurable number of visible photons. In an electrometer, two Au leaves are held apart by positive charges. When radiation is absorbed in the air surrounding the Au leaves, it ionizes the air—the observable. Electrons are attracted to the Au leaves and neutralize some of the charge on the leaves, thereby, causing the Au leaves to move closer together—the measurable result.

An odd form of radiation detector, not recommended by the authors, is the warming or reddening of skin from radiation exposure. During the year 1900, Friedrich Walkoff and Friedrich Giesel both conducted self-exposure experiments with radium. Giesel [1900] reported that he strapped 270 mg of Ra salt to his inner forearm for hours, and he communicated his observations to Henri Becquerel and Pierre Curie. Walkoff [1900] reported a similar experiment, except he performed two exposures, each 20 minutes in duration. He wrote that a skin inflammation formed that had "already lasted over two weeks." Apparently these findings piqued the interest of Pierre Curie, who decided to perform a similar experiment of his own [Becquerel and Curie 1901]. He purposely exposed his arm to Ra for several hours, which "resulted into a lesion that resembled a burn that developed progressively and required several months to heal." The circumstances surrounding Becquerel are not as clear, but apparently he carried a glass vial of Ra salts in his vest pocket and developed a similar radiation burn which took 49 days to heal and left a permanent scar [Becquerel 1903]. In these experiments the skin tissue was the absorber, the observable was a radiation burn, and the measurable was the intensity of red as seen by the eye. Amazingly, technicians demonstrating x-ray machines to potential customers often held one hand near the x-ray tubes until the hand became warm, which indicated to them that the machine was operating. Since those times, much has been learned regarding the health risks involved with radiation exposure, and government regulations are in place to limit the radiation dose to civilians and radiation workers.

As the understanding of radiation interactions in materials increased, more convenient methods of detecting the presence of radiation were sought and developed. For several years after Röntgen's initial discovery of x rays, tedious methods were employed using photographic plates, fluorescing screens, or electroscopes. These early methods of radiation detection relied upon a large amount of radiation exposure and were not capable of recording individual radiation quanta. The first radiation detector capable of distinguishing individual radioactive particles was the *spinthariscope*, invented by Crookes in 1903. It consisted of a ZnS scintillation screen and a magnifying eyepiece as one unit. Within the device, a tiny Ra source was suspended by a wire between the eyepiece and the ZnS screen. In a darkened room, the spinthariscope could be used to visually see individual alpha particle interactions from the radiation source.[9] As a post-doctoral researcher studying under Rutherford, Hans Geiger used the same principle by counting, through a microscope, flashes of light caused by individual alpha particles as they interacted with a scintillating ZnS screen. The process was tedious and Geiger would quickly become fatigued.

With the understanding that radiation would ionize gas, in 1908 Ernest Rutherford and Hans Geiger fashioned a type of gas-filled radiation detector based on the principle of drifting the liberated charges with

[9]Believe it or not, spinthariscope rings (named *Lone Ranger Atomic Bomb Rings*), complete with alpha particle emitting material, were available from General Mills Cereals between 1947 and the early 1950s for the bargain price of 15 cents and a Kix cereal box top.

an electric field [Flakus 1981; Frame 2004; Geiger 1913]. This simple detector, at first, consisted of a thin wire strung coaxially through a metal cylinder. Gas was introduced into the cylinder, and a charged particle interacting in the gas would produce ionization. The opposite charges (electrons and ions) were drifted apart by an electric field applied across the wire and cylinder and the charge movement was measured as a current. The device was redesigned such that the central wire was replaced with a needle point inserted into a metal cavity, known as *Geiger point counters*. Later in 1928, returning to the original coaxial design, Walther Müller (Geiger's student) improved the device by running the central wire through a glass tube and attaching the device to electronic tube circuit for readout [Flakus 1981; Frame 2004]. This was an early version of present day *Geiger-Müller* counters. Various forms of gas-filled detectors were designed specific to experimental needs of the times.

Over the following decades, a variety of improvements in the design and operation of gas-filled detectors were implemented. For instance, it was learned that the proper application and selection of gas components permitted much better operation of Geiger-Müller counters. New electrode designs permitted large detectors to be built, some spanning over a meter in area with multiple anodes. Gas detectors have found uses as gamma-ray spectrometers and neutron detectors and, today, gas-filled detectors continue to be used routinely for radiation detection applications.

Better detectors were also sought that might be more efficient, less cumbersome and easier to use. It was understood that solids absorbed radiation more effectively than gases, and so there was effort to develop a solid-state conducting detector. Röntgen and Joffé [1913] published the first (known) observations of induced currents in certain solids. Röntgen [1921] continued the efforts, as did many other enthusiasts such as Schiller [1926] and Jaffé [1932]. Yet, their choices of materials were unfortunate, being mainly quartz, mica, rocksalt, glass, and other electrically insulating materials. Hence, the results were poor mainly caused by the very small induced currents, inadequate amplification electronics, crude counting methods, and time-dependent material polarization.

Scintillators, materials that emit visible light when irradiated with energetic particles or photons, offered another method to produce solid-state detectors. Scintillation materials were known since Röntgen's initial discovery of x rays with the $BaPt(CN)_4$-coated plate, yet a number of problems prevented the realization of a practical device for decades, mainly (1) that the scintillators used had poor light yield efficiency and (2) that there were then no practical devices available to measure or amplify the scintillation light output. Two major developments allowed scintillators to become practical and useful radiation detectors and spectrometers, namely the development of the photomultiplier tube (PMT) and the discovery of scintillating thallium-doped sodium iodide (NaI:Tl). There is evidence that Leonid A. Kubetsky [Lubsandorzhiev 2006] developed an electron amplifier that utilized secondary emission. Known as "Kubetsky tubes," the devices constitute an early version of the modern PMT. Independently, in 1935, a single-stage PMT was developed by RCA that had a gain of 8 [Burle 1980]. A multistage PMT was reported by Zworykin in 1936 that was operated by adjusting electric and magnetic fields but which proved to be impractical. Meanwhile, in 1937 Kubetsky published the descriptions of numerous photomultiplier designs with gains as high as 10^6. Within a few years, in 1941, RCA developed and commercialized the 931A PMT that had high electron signal gains, thereby providing a light amplification device that could be used with scintillators.

Various known organic scintillators were used in conjunction with the newly developed RCA 931A PMT with encouraging results. However, it is Hofstadter's discovery of NaI:Tl in 1947 that marks the beginning of a truly practical portable gamma-ray scintillation spectrometer [Hofstadter 1948; Hofstadter and McIntyre 1950]. NaI:Tl is a bright scintillator that emits approximately 43,000 visible light photons per MeV of energy absorbed, and the light yield as a function of energy is acceptably linear. Within a few years, commercial NaI:Tl gamma-ray spectrometer units were available.

Despite the success of NaI:Tl as a solid-state gamma-ray spectrometer, effort still continued to develop a solid-state version of an electron conduction device. Van Heerden [1945] demonstrated what is often referred to as the first "practical" semiconductor radiation detector. In his Ph.D. dissertation, "The Crystal Counter," he decries the use of AgCl crystals cooled to low temperatures to detect alpha particles from a ^{210}Po source [see also Van Heerden and Milatz 1950]. Shortly thereafter, CdS crystals were used to observe pulses from gamma rays and beta particles (as reported by Frerichs and Warminsky [1946] and later by Goldsmith and Lark-Horovitz [1949]). In 1947, Wooldringe et al., followed by Jentschke 1948, reported induced electrical pulses from diamond samples.[10]

Although semiconductor detectors remained of interest, the development of NaI:Tl devices kept the status of semiconductor detectors, then referred to as "crystal counters," to little more than interesting gadgets. The main problem with semiconductor devices was the unfortunate high concentrations of defects in the crystals. The defects, which included impurities, dislocations, and intrinsic defects, caused severe problems with charge carrier collection from the devices. With the success of NaI:Tl, it appeared that crystal counters offered little hope for improvement, if any, to radiation spectroscopy.

Nevertheless, a variety of advances in semiconductor detectors began to occur after 1948 [see reviews by Hofstadter 1949a, 1949b; Chynoweth 1952; McGregor and Hermon 1997]. McKay [1949, 1951] developed a radiation detector using a point contact on a Ge semiconductor substrate. The rectifying point contact was irradiated with alpha particles from a Po source, and this detector yielded sizeable electronic pulses on the order of 8-12 millivolts at reverse biases between 2-10 volts. Soon after, various forms of diffused junction and "surface barrier" detectors were developed based on Ge and Si [Dearnaley and Northrop 1966]. However, the development that finally allowed semiconductor detectors to become serious devices for radiation detection and radiation spectroscopy was the introduction of Li ion drifting by Pell [1960]. Li ions were electrically drifted into Si and Ge crystals to nullify, or "compensate for," the effects of residual impurity dopants.

Originally, university and government researchers found it necessary to construct their own custom semiconductor detectors, and many techniques were developed and published to build surface barrier, *pn* junction, and Li-drifted detectors. However, with the success of Li drifting, a variety of semiconductor detectors became commercially available by 1964. Also by this time the traditional name of "crystal counter" was dropped and replaced with the more descriptive name "semiconductor detector." The semiconductor detectors enabled gamma-ray and particle energy identification with outstanding resolution, far better than that achieved with scintillation detectors [Bertolini and Coche 1968; Poenaru and Vilcov 1969; Deme 1971]. Lithium drifted Ge detectors, or Ge(Li) detectors, were of more use for gamma-ray spectroscopy than the Si(Li) detectors due to the greater atomic number of Ge (32) over Si (14), which gives Ge a higher absorption efficiency. Both Ge(Li) and Si(Li) detectors[11] needed to be operated at cryogenic temperatures near 77K to reduce thermal noise and leakage currents. This operating criterion was especially true for Ge(Li) devices for an added reason; if the detector warmed up, the Li would diffuse and redistribute in the crystal and the detector would be ruined [Dearnaley and Northrop 1966; Bertolini and Coche 1968]. Hence, Ge(Li) detectors had to be kept at cryogenic temperatures at all times, which is an obvious disadvantage.

Another method of dealing with the contaminant impurities is to remove them by various methods. Zone refinement, one such method, had already been used to purify and remove background impurity contaminants from a number of substances. Robert Hall at General Electric Co. (GE) suggested in 1968 that this method might be used to purify Ge to such levels that Li drifting would no longer be necessary. Two research groups, one at GE led by Robert Hall and another at Lawrence Berkeley Laboratory led by W.L. Hansen, developed a method of zone refinement for Ge [Yu and Cardona 1996]. Soon high-purity Ge (HPGe) detectors became commercially available, and over the years Ge(Li) detectors have been largely phased out for the preferable

---

[10] Jentschke references a diamond device that was reported by Stetter [1941], thereby predating Van Heerden's work. The device was used to observe pulses from an alpha particle source, yet the article is unclear on many aspects of the experiment, and the work was largely overlooked for many years.

[11] The traditional pronunciation is "jelly detector" for the Ge(Li) detector and "silly detector" for the Si(Li) detector.

HPGe detectors. HPGe detectors must still be cooled to cryogenic temperatures to prevent damage from excessive leakage currents; however, when not operating these detectors are no longer ruined if they are allowed to warm up. As explained in a later chapter, Si(Li) detectors currently remain important as x-ray spectrometers and are still commercially available.

Although Ge and Si have been established as important radiation detectors, the need for cryogenic cooling is inconvenient, especially in situations where a portable device is required. Hence, the search for semiconductor devices that can operate at room temperature remains a priority. Detectors made from gallium arsenide (GaAs), a wide band-gap semiconductor that can be operated at room temperature, were studied and reported as radiation counters in 1960. In 1970, Eberhardt, Ryan, and Tavendale demonstrated the first high energy resolution semiconductor room-temperature-operated gamma-ray spectrometer. The detector was fabricated from a high-purity, epitaxially grown GaAs crystal, and the active region of the device was only 80 microns thick. Since then, many other, and perhaps more promising, semiconductors have been studied. These include mercuric iodide, cadmium telluride, and cadmium zinc telluride. The main impediment to widespread use of compound semiconductors is the difficulty of producing quality detector crystals. It is much harder to produce defect-free compound semiconductors than it is to produce defect-free single-element semiconductors, of which only Si and Ge are practical materials.[12]

A timeline of the major developments in radiation detector inventions is shown in Figs. 1.16 and 1.17. This chart provides a snapshot up through 2010. However, this is not the end of detector history. Detector development actively continues today and the discipline is far from being exhausted, as evidenced by many new devices and designs produced in the past two decades. Novel approaches and designs, such as the gas electron multiplier (GEM), have expanded the use of gas-filled detectors. New scintillation materials, such as various lanthanide halides and elpasolites, have been developed that offer gamma-ray energy resolution that is much better than NaI:Tl. Also, new compact photomultiplier devices based on semiconductors have become available. Clever semiconductor detector designs have significantly reduced the effects of defects previously impeding progress with compound semiconductor detectors. Novel detector approaches using cryogenics have yielded devices with unprecedented energy resolution. Micro-machined semiconductors have produced the first high-efficiency semiconductor neutron detectors. All of these devices, amongst a myriad of other designs and technologies, some still in development, are described in chapters throughout this text.

[12]There are other single element semiconductors, such as C (diamond) and Te, but these materials have intrinsic properties that generally disqualify them. Because of its high melting point, diamond cannot be produced through traditional bulk crystal growth processes and is generally grown as thin chemical vapor deposition (CVD) films. Unfortunately, these diamond materials continue to have issues with defects. Further, with C having such a low Z number (6), they are simply not useful as gamma-ray spectrometers. However, there has been some effort investigating diamond as radiation-hard particle detectors. Te is a semiconductor, but its band gap (0.33 eV) is smaller than Ge (0.72 eV), hence such detectors would need to be operated at cryogenic temperatures, thereby offering no significant advantage over Ge detectors.

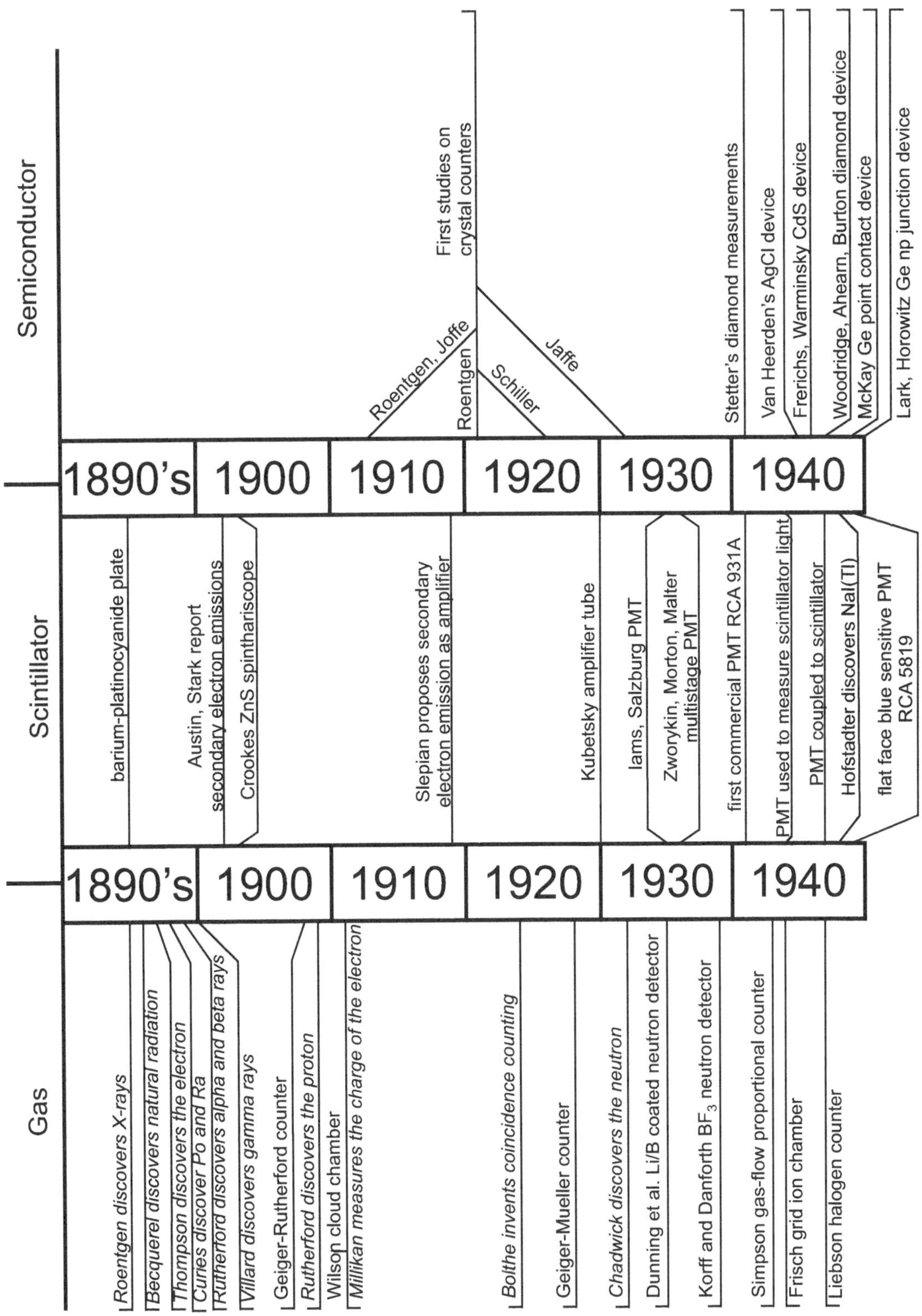

Figure 1.16. Important events in the development of radiation detectors.

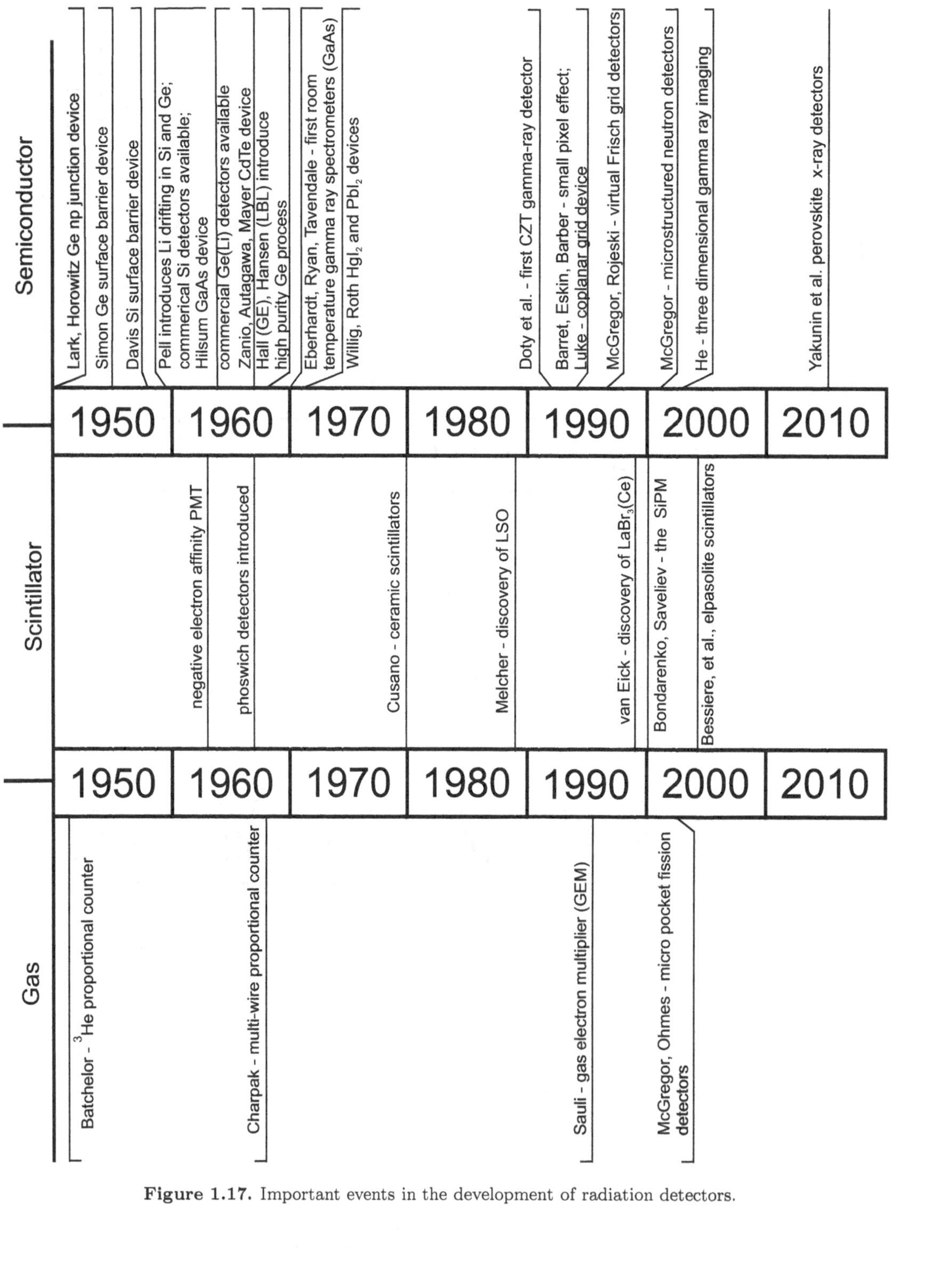

Figure 1.17. Important events in the development of radiation detectors.

REFERENCES

BECQUEREL, H., AND P. CURIE, "Action Physiologiques des Rayons du Radium," *Compte Rendus des Séances l'Acad. Sci.*, **132**, 1289–1291, (1901).

BECQUEREL, H. "Action physiologique du rayonnement du radium sur la peau et sur les graines." *Mémoires de l'Académie des Sciences de lInstitut de France*, Paris: Firmin-Didot et Cie, pp. 262-265, 1903.

BERTOLINI, G., AND A. COCHE, *Semiconductor Detectors*. New York: Wiley, 1968.

BURLE Technologies Inc., *Photomultiplier Handbook*, 1980.

CHADWICK, J., "The Existence of a Neutron," *Proc. R. Soc. Lond. A*, **136**, 692–708, (1932).

CHYNOWETH, A.G., "Conductivity Crystal Counters," *Am. J. Physics*, **20**, 218–226, (1952).

DEARNALEY, G., AND D.C. NORTHROP, *Semiconductor Counters for Nuclear Radiations*, 2nd Ed. New York: Wiley, 1966.

DEME, S., *Semiconductor Detectors for Nuclear Radiation Measurement*, New York: Wiley, 1971.

FLAKUS, F.N., "Detecting and Measuring Radiation—A Short History," IAEA Bulletin, **23**, No. 4, 31–36, (1981).

FOURNIER, P., AND J.A. FOURNIER, "Niepce de Saint-Victor (1805-1870), M.E. Chevreul (1786-1889) et la Découverte de la Radioactivité," *New J. Chem.*, **14**, 785–790, (1990).

FRAME, P., "A History of Radiation Detection Instrumentation," *Health Physics*, **87**, 111–135, (2004).

FRERICHS, R. AND R. WARMINSKY, "The Measurement of $\beta-$ and $\gamma-$Rays by Internal Photo-Effect in Crystal Phosphors." *Naturwissenschaften*, **33**, 251, (1946).

GEIGER, H., "A Method of Counting Alpha and Beta Rays," *Deutsch Phys. Ges.*, **15**, 534–539, (1913).

GIESEL, F., "Über Radioactive Stoffe," *Ber. Dtsche. Chem. Ges.*, **33**, 3569-3571, (1900).

GOLDSMITH, G.J., AND K. LARK-HOROVITZ, "Cadmium Sulphide as a Crystal Counter." *Phys. Rev.*, **75**, 526–527, (1949).

HOFSTADTER, R., "Alkali Halide Scintillation Counters." *Phys. Rev.*, **74**, 100–101, (1948).

HOFSTADTER, R., "Crystal Counters—I." *Nucleonics*, **4**, 2–28, (1949a).

HOFSTADTER, R., "Crystal Counters—II." *Nucleonics*, **4**, 29–43, (1949b).

HOFSTADTER, R., AND J.A. MCINTYRE, "Gamma-Ray Spectroscopy with Crystals of NaI(Tl)," *Nucleonics*, **6**, 32–37, (1950).

JAFFÉ, G. "Effect of Alpha Rays on the Passage of Electricity Through Crystals," *Physikalische Zeitschrift*, **33**, 393–399, (1932).

JENTSCHKE, W., "The Crystal Counter," *Phys. Rev.*, **73**, 77–78, (1948).

KAPLAN, I., *Nuclear Physics*, Reading: Addison-Wesley, 1962.

KUBETSKY, L.A., "Multiple Amplifier," *Proc. IRE*, **25**, 421–433, (1937).

LUBSANDORZHIEV, B.K., "On the History of Photomultiplier Tube Invention," *Nucl. Instrum. and Meth.*, **567A**, 236–238, (2006).

MCGREGOR, D.S. AND H. HERMON, "Room Temperature Compound Semiconductor Radiation Detectors," *Nucl. Instrum. and Meth.*, **A395**, 101–124, (1997).

MCKAY, K.G., "A Germanium Counter," *Phys. Rev.*, **76**, 1537–1537, (1949).

MCKAY, K.G., "Electron-Hole Production in Germanium by Alpha Particles," *Phys. Rev.*, **84**, 829–832, (1951).

MILLIKAN, R.A., AND J. MILLS, *Electricity, Sound, and Light*, Boston: Ginn & Company, 1908.

PELL, E.M., "Ion Drift in an n-p Junction," *J. App. Phys.*, **31**, 291–302, (1960).

POENARU, D.N. AND N. VILCOV, *Measurements of Nuclear Radiations with Semiconductor Detectors*, New York: Chemical Publishing Co., 1969.

RÖNTGEN, W.C., AND A. JOFFÉ, "Über die Elektrizitätsleitung in einigen Kristallen und über den Einfluss der Bestrahlung darauf," *Annalen der Physik*, **346**, 449–498, (1913).

RÖNTGEN, W.C., "Über die Elektrizitätsleitung in einigen Kristallen und über den Einfluss einer Bestrahlung darauf." *Annalen der Physik*, **369**, 1–195, (1921).

SAINT-VICTOR, C.N., "Dèuxieme Mémoire sur Une Nouvelle Action de la Lumière," *Compte Rendus Acad. Sci.*, **46**, 448–452, (1858).

SCHILLER, H., "Untersuchungen über die Elektrizittsleitung in festen Dielektriken bei hohen Feldstärken." *Annalen der Physik*, **386**, 32–90, (1926).

STETTER, G., "Durch Korpuskularstrahlen in Kristallen hervorgerufene Elektronenleitung," *Verhandl. Deut. Physik. Ges.*, **22**, 13–14, (1941).

VAN HEERDEN, P.J., *The Crystal Counter*, University of Utrecht, Dissertation, 1945.

VAN HEERDEN, P.J., AND J.M.W. MILATZ, "The Crystal Counter; a New Apparatus in Nuclear Physics for the Investigation of $\beta-$ and $\gamma-$Rays," *Physica*, **16**, 505–527, (1950).

WALKOFF, F., "Unsichtbare, Photographisch Wirksame Strahlen." *Photographische Rundsch Z Freunde Photographie*, **14**, 189-191, (1900).

WOOLDRIDGE, D.E., A.J. AHEARN, AND J.A. BURTON, "Conductivity Pulses Induced in Diamond by Alpha Particles," *Phys. Rev.*, **71**, 913–913, (1947).

YU, P.Y., AND M. CARDONA, *Fundamentals of Semiconductors*, New York: Springer, 1996.

ZWORYKIN, V.K., G.A. MORTON, AND L. MALTER, "The Secondary Emission Multiplier—A New Electronic Device," *Proc. IRE*, **24**, 351–375, (1936).

Chapter 2

Introduction to Nuclear Instrumentation

"...that a module be developed by the National Laboratories with the intent that the module will become standard in all of the National Laboratories and will be duplicated by many manufacturers."

U.S. National Bureau of Standards, 1963

Why is This Chapter Here?

In many ways this is an "orphan chapter" in that it does not discuss directly the theoretical principles or design of radiation detectors. Consequently, it could appear at the end of the book or even as an appendix. However, traditionally, courses designed to teach the principles of radiation detection and measurement usually have an associated laboratory component in which the actual detectors of various types are used to demonstrate radiation detection principles. The authors have learned that for such courses students must be introduced early to the various electronic devices they are going to use to make different radiation measurements. This early introduction is essential to minimize the many (often quite creative and destructive) ways students try to misuse the electronics to produce observable signals and measurements in the laboratory.

This chapter also serves as a historical view of how the electronics for various radiation detectors has evolved and become standardized over the past many decades. However, the reader may safely skip this chapter in the initial reading because little that follows critically depends on the information in this chapter.

2.1 Introduction

Nuclear instruments are designed to acquire and produce meaningful signals from radiation detectors. Almost all radiation detectors must have electrical power applied to them in some form or fashion and, hence, it is important to have a stable power supply.[1] Detectors that operate by producing electronic signals must have accompanying electronics that can accurately measure the magnitude of induced current or induced charge.

[1]Exceptions are detectors that use radiation to produce measurable chemical changes, e.g., photographic films, or changes in crystalline and chemical structures, e.g., color changes in crystals or material embrittlement.

Often the signal from the detector must be *shaped* and/or *amplified* so that it can be manipulated by other nuclear electronic instruments. Typically, there is also a user interface that allows one to easily interpret the data that is being accumulated or recorded. Such an interface can assume several forms, ranging from a simple analog meter up to a complex computer interface. The basic components for a radiation detection system are found in all such systems ranging from small portable hand-held devices to systems with sizes of hundreds of meters or kilometers that are used in high-energy physics experiments.

2.2 The Detector

The detector is the active element used to absorb radiation particles and subsequently produce a recognizable signal associated with each event. Detectors may rely upon measurable phenomena, such as electronic, thermal, or visual changes. Detector types are commonly categorized as gas-filled detectors, scintillation detectors, semiconductor detectors, and alternate detectors. Gas-filled detectors, as the name implies, are vessels filled with a gas, which may be under pressure or under a slight vacuum (and are covered in Chapters 9–11). Ionizing radiation interactions produce mobile charges in the gas, which, in turn, are usually identified as an induced current or voltage caused by the motion of these mobile charges. Scintillation detectors fluoresce when ionizing radiation interacts in the absorbing medium (see Chapters 13 and 14). These scintillation detectors are fashioned from solid, liquid, or gaseous materials that produce measurable light when stimulated by ionizing radiation particles. Semiconductor detectors are solid-state devices that change conductivity when radiation particles interact in the material, mainly due to the excitation of mobile charges to higher energy levels (see Chapters 15–16). Alternate detectors include devices that indicate the presence of radiation by acoustical changes, color changes, temperature changes, and variety of other phenomena (see Chapter 19).

Electronic detectors in a low-level radiation environment are usually operated in pulse mode, whereas detectors in a high radiation environment are usually operated in current mode. To understand the difference, shown in Fig. 2.1 is a common electronics arrangement for pulse mode counting. Radiation particles interacting in the detector produce an electronic signal which can be sensed and amplified by the preamplifier circuit. The signal from the preamplifier is then shaped and amplified further, where it either is rejected or accepted by discriminating electronics. Afterwards, the count rate or energy spectrum is recorded and displayed. The advantage of this method is that small currents can be converted into large voltage pulses through the pulse shaping and amplification stages. However, the entire process involves some finite amount of time, especially in the pulse processing stage, which defines the resolving (dead) time of the system. Radiation particles entering the detector at a higher rate than can be processed result in dead time losses; hence an alternate measurement method must be used when particles interact in the detector at a rate faster than the pulses can be processed.

In current mode, the current produced in a detector is directly measured with a current meter, usually by measuring the voltage across a load resistor through which the induced current flows. Unfortunately, these currents for a single ionizing event can range from tens of picoamperes to a few nanoamperes. Hence, current mode measurements are usually reserved for high radiation environments in which the signal is a result of many, nearly simultaneous, ionizing events.

Some radiation detectors must be used in pulse mode rather than current mode. For instance, the gas-filled ion chamber can be operated in either pulse mode or current mode, but other gas-filled detectors, such as Geiger-Müller counters can only be operated in pulse mode. There are many other examples that are covered in later chapters.

The detector for any single radiation measurement must be carefully chosen for the specific application. No single device can provide all of the desired aspects of a radiation detector. When selecting a detector, the operator must consider the radiation being measured (gamma-rays, neutrons, charged-particles, etc.), the radiation exposure rate, and the information sought (energy, count rate, etc.). Some detectors are best suited for energy identification, others are best suited for high interaction efficiency, while still others may provide a more economical but satisfying result. Hence, the operator must choose a detector that provides the required information for a measurement, often from many possible sensors.

2.3 Nuclear Instrumentation

Instrumentation for the nuclear industry was specified in a 1964 report TID-20893 to the U.S. Atomic Energy Commission (now, after many reincarnations, the Department of Energy) to what are referred to as Nuclear Instrument Modules (NIM). Several revisions to the standard were submitted over the following years, eventually established in the 4th revision of 1974 (TID-20893-REV-4), with an updated version described in the 1990 report DOE/ER-0457T. The standard defines the dimensions, standard connections, and power requirements for modular nuclear electronic components, as well as the cabinet or bin into which these components are inserted. The NIM standard is a flexible and simple system that can incorporate many useful detection tools, such as amplifiers, analog-to-digital converters (ADCs), counting and timing electronics, discriminators, and power supplies. The NIM standard provides a common footprint for electronic modules (amplifiers, ADCs, discriminators, etc.), all of which plug into a larger chassis (NIM bin). This bin must supply ±12 and ±24 volts DC power to the modules via a backplane; the standard also specifies ±6 V DC and 220 V or 110 V AC pins; however, not all NIM bins provide these other voltages. Shown in Fig. 2.1 is a block diagram of components commonly used in a pulse-mode counting system, and shown in Fig. 2.2 is a NIM bin with common components described in Section 2.5.

2.4 History of NIM Development

Early nuclear instruments were composed of vacuum tube circuits within a standardized 19-inch wide chassis. The components each had their own DC power supply and each connected directly to AC power. These components easily installed in standard 19-inch wide instrument racks. Hence, there was no issue with interchangeability between components, mainly because each unit operated independently.

The invention of the transistor in 1948 later allowed the electronics to be built much more compactly. Further, the overall efficiency and heat load was significantly less for transistorized components. Yet, these early transistorized nuclear instruments were built in a similar fashion as were their vacuum-tube counterparts, designed for large 19-inch wide containers. It was eventually realized, however, that these transistorized electronic components could be built as smaller and more efficient modular units, thereby eliminating the need for a single 19-inch wide chassis. Many independent laboratories and commercial companies began designing and constructing various different modular electronic component systems.

Two such systems, advanced at the time, were the systems developed by the European Organization for Nuclear Research (CERN) and the United Kingdom Atomic Energy Research Establishment. The European Standards community on Nuclear Electronics (ESONE), founded in 1961, also began working on a new standardized modular system. Along with these systems, numerous commercial companies began working on various modular nuclear electronics systems. Further, the independent U.S. national laboratories began developing modular nuclear electronics systems for their own purposes. As a result, numerous different nuclear electronics systems were under development, none of which was compatible with any of the other systems. Overall, research at nuclear laboratories was seriously limited by the compatibility of instruments, and researchers were forced to invest in numerous different modular nuclear components.

To remedy this divergent development of nuclear modules, the U.S. National Bureau of Standards (NBS),[2] in a 1963 report to the U.S. Atomic Energy Commission (AEC), recommended that the U.S. national

[2]The National Bureau of Standards is now the National Institute of Standards and Technology (NIST).

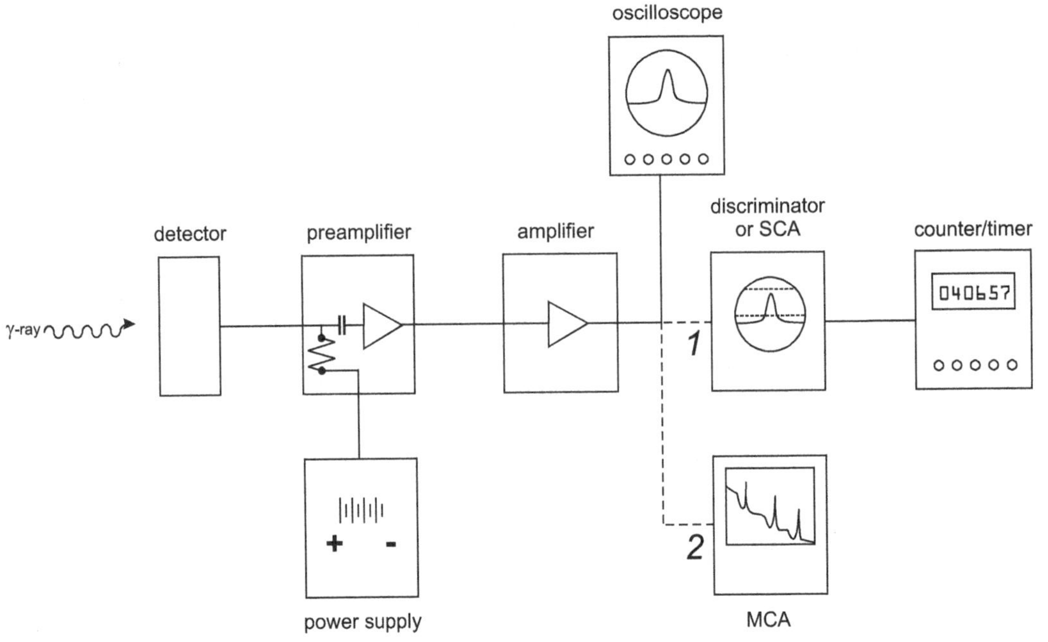

Figure 2.1. The basic components of a pulse-mode electronic detection system. A detector is powered by a bias supply or power supply usually through the preamplifier. The induced electronic signal from the detector is shaped and amplified through the preamplifier and amplifier circuits. The signal can be observed and monitored with an oscilloscope. When used as a basic counter, the shaped signal pulse is routed through a discriminator or single channel analyzer (SCA) and then into a counter/timer (route 1). When used as a spectrometer, the shaped signal pulse is routed into a multichannel analyzer (MCA) in which the pulse amplitudes are sorted and categorized according to the pulse heights (route 2).

laboratories develop a modular system to be duplicated by commercial manufacturers. The U.S. AEC convened a meeting of representatives from several national laboratories on Feb. 26, 1964 to determine if such a task was of interest to the laboratories. In agreement, the AEC Committee on Nuclear Instrument Modules was established, composed of members from all AEC national laboratories and a few other prominent nuclear laboratories.

The first NIM Committee meeting was held on March 17, 1964, with follow-up meetings over several ensuing months. The committee studied the advantages and shortcomings of the existing modular systems, and a system similar to the CERN design was eventually developed. The prototype bins were manufactured by Oak Ridge National Laboratory (ORNL), Lawrence Radiation Laboratory in Berkeley and Lawrence Radiation Laboratory in Livermore.[3] These early prototypes were studied at later meetings of the NIM Committee, eventually leading to a recommended design. Final details were resolved by the NIM Executive Committee with representatives from the U.S. NBS, ORNL, Lawrence Radiation Laboratory in Berkeley, and with later added members from Princeton-Pennsylvania Accelerator and Brookhaven National Laboratory (BNL). The final specifications were published in July 1964, and later updated versions of the specifications were published in 1966, 1968, 1969, and 1974. The standard was well received, with the first commercial NIM components being available by November 1964. During 1965, numerous more NIM instruments became

[3]The Lawrence Radiation Laboratory in Berkeley is now the Lawrence-Berkeley National Laboratory and the Lawrence Radiation Laboratory in Livermore is now the Lawrence-Livermore National Laboratory.

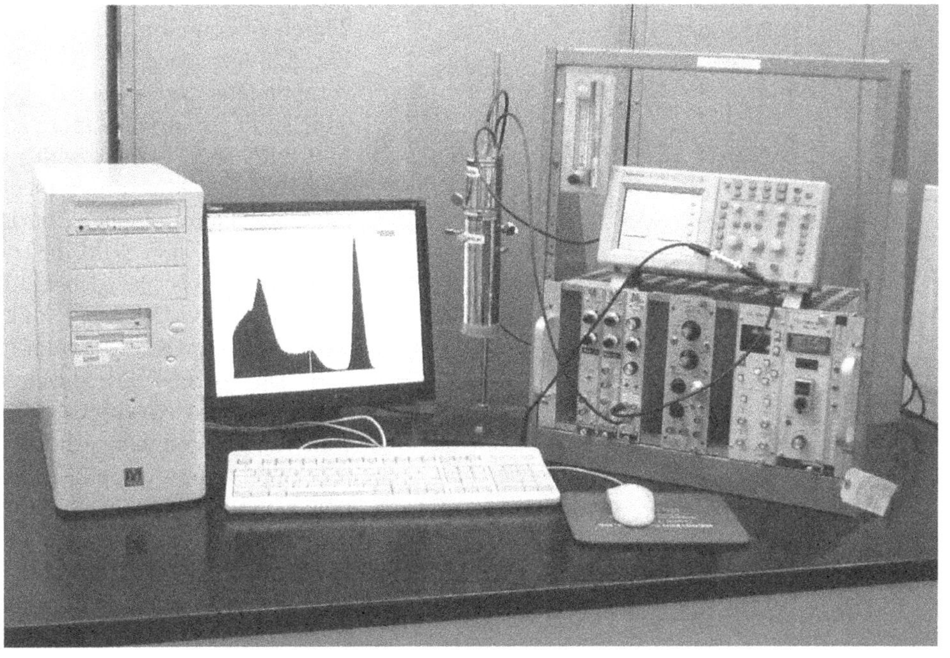

Figure 2.2. A typical measurement system in which various NIM components are installed into a NIM bin shown to the right. This NIM bin contains several nuclear instrument modules, including a power supply, a pulser, an amplifier, an SCA, and a counter/timer. The detector, to the left of the NIM bin, is a NaI:Tl device with the preamplifier attached to the detector back. The system is also attached to an oscilloscope, shown on top of the NIM bin. The detector is also connected to a computer-based MCA, which shows a spectrum of ^{137}Cs gamma rays on the monitor, behind the keyboard. The ^{137}Cs gamma ray check source is located directly beneath the NaI:Tl detector.

available, and NIM components accounted for over 95% of all modular nuclear instruments produced in the U.S. in the immediate years following 1967. The NIM standard was again revised in 1990, to make the standard compatible with existent technology and manufacturing processes, as described in Department of Energy (DOE) document DOE/ER-0457T.

2.5 NIM Components

Commercial manufacturers offer a variety of NIM components for radiation detection applications. Many of these modules are versatile and can be used for general radiation counting applications. There are also special NIM modules optimized for narrowly applied or special applications. Overall, the list of common NIM components include a NIM bin and power supply, a detector voltage supply, preamplifier, amplifier, discriminator, counter, timer, and often a data analyzer. Described in the following sections are the basic properties of these components. Further details on the design and operation of NIM components and other nuclear electronics can be found in Chapter 22.

2.5.1 The NIM Bin

The container for NIM components is traditionally referred to as a NIM bin. The NIM bin is used as the power supply docking port for most NIM components, except for a few high voltage power supplies that

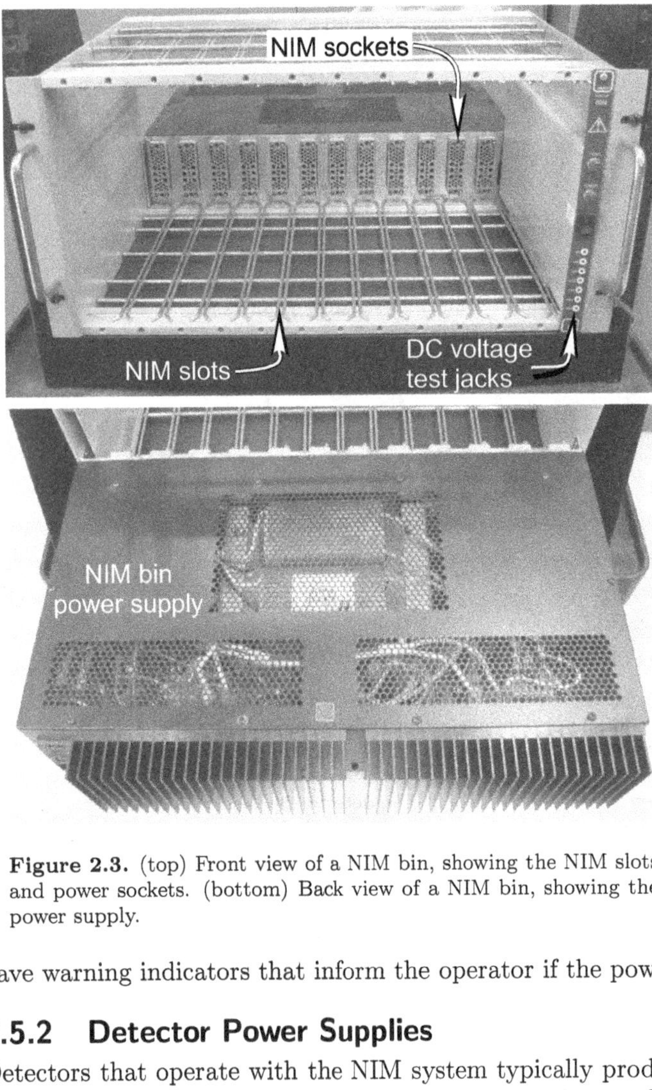

Figure 2.3. (top) Front view of a NIM bin, showing the NIM slots and power sockets. (bottom) Back view of a NIM bin, showing the power supply.

plug directly into a 120 VAC wall socket. The most common NIM bin size, shown in Fig. 2.3, has space for 12 single-width NIM components, as specified by DOE/ER-0457T. There are also more compact NIM bins with only 4 single width slots. These smaller NIM bins are typically used for portable measurement situations that require only a few NIM instruments. A power supply is placed behind the instrument slots in a NIM bin. This unit provides ±6 volts DC, ±12 volts DC, and ±24 volts DC to the NIM sockets and voltage jacks,[4] as well as 117 volts AC. Power is commonly rated between 150–160 W for the larger power supplies, although lower power models rated at ~95 W are available (usually for models that do not offer the ±6 volt DC option). The NIM bin sockets and pin assignments conform to the specifications outlined in DOE/ER-0457T. Test jacks in the front of the NIM bin provide convenient access to monitor the DC voltage, and can also be used to conveniently supply power to custom electronics. Modern power supplies are filtered from electromagnetic interference (EMI), are short-circuit proofed, thermally protected, and provide stable power required for NIM components. Usually NIM bins have warning indicators that inform the operator if the power supply is approaching the temperature limit.

2.5.2 Detector Power Supplies

Detectors that operate with the NIM system typically produce an electronic output and, therefore, require a power supply of their own. Traditionally, the majority of such NIM power supplies for detectors apply a high voltage to the detectors.[5] Stability of the supply voltage is of particular importance in high-resolution measurements systems. High-voltage power supplies are used to operate detectors that do not suffer changes in capacitance as a function of voltage, such as gas-filled and scintillation detectors. High-voltage bias supplies are usually used for semiconductor detectors, in which the operating voltage can affect the detector capacitance.

High Voltage Power Supply

The high voltage power supply, as the name implies, supplies high voltage to the detection device, and can supply it with either positive or negative polarity. Typically, the voltage is applied to the detector

[4]Some models provide only ±12 VDC and ±24 VDC.

[5]Recently, there has been a move toward special detector systems based on semiconductor photon detectors that operate on lower voltages. These detectors, such as avalanche photodiodes, are discussed in Chapter 14.

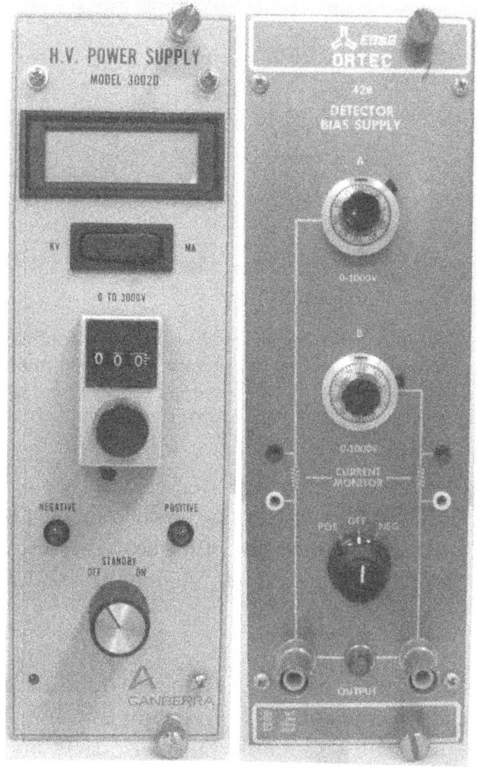

Figure 2.4. (left) A common all-purpose HV power supply used for gas-filled and scintillation detectors. (right) A dual source DC bias supply used for smaller semiconductor detectors, such as silicon surface barrier detectors.

across a load resistor with the connection routed through the preamplifier unit. Also, a power supply often has a meter indicating the applied voltage, which is reasonably accurate because the detector resistance is typically much greater than the preamplifier load resistance. HV power supplies are available with the maximum voltages exceeding 3.5 kV. Because PMTs and gas-filled devices need relatively high power, a HV power supply is plugged directly into a wall socket, although many models have rails that allow the power supply to slip into a NIM bin for convenience. A HV power supply is shown in the left-hand side of Fig. 2.4.

High Voltage Bias Supply

For semiconductor detectors, the power supply is generally referred to as a "bias supply." As with a HV power supply, the bias supply provides a choice of positive or negative high voltage to the detector. However, this power supply does not have a meter, but rather only a dial indicator for the total applied voltage to the system.

Because the system includes load resistances in series with the detector, there is a voltage drop across this resistance, which reduces that applied to the detector. Thus, a meter indicating total voltage would be misleading. Instead jacks are provided that allow determination of the bias current. Because the series load resistance is known, the voltage applied to the detector, which is less than the total, can also be determined. Bias supplies are available for which the maximum allowable voltages can exceed 5 kV. Because typical

Figure 2.5. (left) Front view of a common charge-sensitive preamplifier, showing the safe high voltage (SHV) connector for the detector and the nine pin D-subminiature connector for the preamplifier power. (right) Back view of a the same charge-sensitive preamplifier, showing the connector terminals for the power supply, pulser input, timing input, and the output signal.

detectors are highly resistive and draw little current, the power requirement of the bias supply is low and, consequently, most bias supplies are plugged directly into the NIM bus for power, unlike the HV power supply. A HV bias NIM module is shown in the right-hand side of Fig. 2.4.

2.5.3 Preamplifier

A preamplifier unit has two basic purposes, namely (1) to provide low noise coupling of the detector to the string of amplifier and readout electronics, and (2) to produce a first stage of signal amplification. The preamplifier thus acts as a filter against objectionable line noise that may alter the input signal. Preamplifiers are designed to be either current sensitive, voltage sensitive, or charge sensitive. Optimum detector performance is obtained when the user selects the proper preamplifier matched to the type and characteristics of the detector and the analysis electronics. The general application of these preamplifiers are described here, with details on the design and operation reserved for Chapter 22.

Voltage-Sensitive Preamplifiers

Voltage-sensitive preamplifiers are so named because they tend to respond to and linearly amplify the voltage input from the detector into the preamplifier circuit. As discussed in Chapter 8, the voltage input V_{in} into the preamplifier from the detector is generally an inverse function of the detector capacitance, $V_{in} \approx Q_i/C_d$, where Q_i is the charge induced by the motion of electrons and ions excited in the detector by a radiation interaction event and C_d is the combined capacitance of the detector and coupling cables. The preamplifier output is usually defined by the feedback resistors, where for a typical inverting configuration $V_{out} \approx -(R_f/R_L)V_{in}$, where R_f is the feedback resistance and R_L is the input load resistance.

For many detectors, the capacitance remains constant for different applied operating voltages V_o; hence the preamplifier output V_{out} is mostly linear with respect to the charge excited in the detector regardless of the voltage applied to the detector. Examples include gas-filled detectors and scintillation detectors that are coupled to photo-multiplier tubes.

Charge-Sensitive Preamplifiers

There are some detectors that change capacitance as voltage is applied, particularly semiconductor diode detectors. These semiconductor detectors have active regions that increase with applied operating voltage, which, in turn, causes the capacitance C_d to decrease with increasing voltage. As a result, the preamplifier output changes with a varying operating voltage for identical induced charge Q_i. By redesigning the preamplifier feedback circuit, the capacitive component of the detector C_d can be minimized compared to the feedback capacitance C_f of the preamplifier, thereby, effectively rendering the preamplifier output almost entirely dependent upon the induced charge Q_i. Hence, for semiconductor diode detectors, the voltage output V_{out} is largely dependent only upon the induced charge Q_i and feedback capacitance C_f, where $V_{out} \approx -AV_{in} \approx -Q_i/C_f$, where A is the preamplifier gain. Commercial charge sensitive preamplifiers are designed for optimum performance when matched to specific ranges of detector capacitance. Hence, it is advised that the user consult the preamplifier specification sheets to properly match detectors to preamplifiers. A common commercial charge sensitive preamplifier is shown in Fig. 2.5.

Current-Sensitive Preamplifiers

As the name implies, a current-sensitive preamplifier is designed to measure the instantaneous current flowing from a detector. The configuration is seldom used for most detector applications; however, on rare occasions a preamplifier with a fast rise time for timing applications is needed, and a current sensitive preamplifier is required. The current sensitive configuration requires that the input impedance be low and the preamplifier gain A be large. These conditions can be realized by using the general design of a voltage sensitive preamplifier and reducing the load resistance at the input of the preamplifier. Note that the *Johnson noise*[6] can be high unless the feedback resistor in the preamplifier circuit is $R_f \geq 10^9\ \Omega$. If the preamplifier input impedance is 50 Ω, the current sensitive preamplifier will convert the input current I_{in} to an output voltage V_{out} described by $V_{out} = I_{in}(50\Omega)$ A.

2.5.4 Amplifier

The amplifier unit has two main purposes, namely, (1) pulse shaping and (2) pulse height amplification. The pulse at the output of the preamplifier typically has a fast rise time with a slower falling tail with the timing characteristics determined by the leading and trailing time constants. This pulse is then integrated and shaped by the amplifier to provide a Gaussian-like pulse that is far easier to manipulate with counting electronics that are placed further down in the pulse processing system. The amplifier, depending on its design, may allow the user to change rise times, decay times, and pulse width. Further, the gain on an amplifier has both course and fine adjustments, important features needed when calibrating the system. Lastly, the NIM standard allows for a 10-volt maximum output from the amplifier, i.e., any pulses whose amplitude is larger than 10 volts is "clipped" at 10 volts. Hence, it is wise to adjust the amplifier gain so that the largest expected output pulses are less than 10 volts; otherwise, the clipped signal may not be recognized or counted.

2.5.5 Oscilloscope

Although an oscilloscope is not part of the NIM system of instruments, it is a fundamental and important piece of equipment with which all NIM users should become familiar. The oscilloscope allows the user to determine how the system is manipulating the electronic pulses at every stage of the system by simply hooking into that portion of the pulse processing system with a "tee" connector. It can be used to measure pulse heights, pulse widths, repetition rates, pulse symmetry and variety of other values that allow for trouble-shooting and system calibration.

2.5.6 Pulse Discriminators

A simple discriminator is used to reject pulses outside set pulse-amplitude boundaries, thereby passing only pulses of certain voltages that are of interest. The electronic pulses emerging from the amplifier are indicative of the amount of energy deposited within the detector, typically linear with energy in well-designed systems. Hence, once calibrated, the final signal pulse height is a measure of the energy absorbed in the detector.

[6] Johnson noise is the electronic noise generated by the thermal agitation of the charge carriers inside an electrical conductor.

The discriminator allows the user to select a voltage at a lower level discriminator (LLD) threshold, below which pulses are rejected. In older discriminators, any signal above the LLD is passed. However, some newer systems reject pulses that exceed the 10-volt upper limit. Often a "discriminator" unit has only an LLD, whereas the single channel analyzer (SCA) has both an LLD and an upper level discriminator (ULD). The ULD rejects signal pulses that exceed the set threshold. Hence, the SCA allows the user to pick a voltage region (or energy region) of interest defined by lower and upper energy thresholds, and pass only those signal pulses that fall within those boundaries.

2.5.7 Counter/Timer

Voltage pulses that pass through the discriminator or SCA are usually routed to a counter/timer. The function of the counter/timer is to record the number of radiation induced pulses within a predetermined amount of time. Hence, a count rate from a radiation source can be determined. Most counter/timers can be set to automatically count for a preset amount of time and stop so as to reveal the number of pulses or counts obtained within that time period. Conversely, they can be set to record a certain number of counts and stop when that limit is obtained to reveal the time required for the measurement. Counter/timers are common instruments used for radiation measurement and dosimetry.

2.5.8 Pulse Generator

A valuable piece of equipment that assists with calibration of the radiation counting system is the pulse generator, and is commonly referred to as a *pulser*. Pulsers provide accurate voltage test pulses into the radiation detection system. The pulses can be set to have different shapes (tail pulse being the most common), decay times, pulse frequencies, and amplitudes in order to calibrate the amplifier, MCA, SCA, or counter/timers. The test pulses can also be used to help determine system dead times and system electronic noise. Although not necessary to operate a radiation counting system, they are quite valuable when first configuring the system and calibrating it for radiation measurements. Further, it is typical that a pulser be operating during spectroscopic radiation measurements in order to have a comparison peak for system noise determination. Typically the pulser peak is set to an MCA channel that does not interfere with the accumulated radiation spectrum.

2.5.9 Coincidence Modules

There are many applications that require the measurement of events that occur in two separate detectors within a given time interval Δt. Examples include identifying the simultaneous arrival of photon-photon or photon-particle events, positron lifetime studies, decay scheme studies. Such measurements are commonly referred to as *coincidence measurements*. By contrast there are some measurements that require the opposite condition, i.e., when two radiation events are recorded within Δt are rejected. Such measurements are referred to as *anti-coincidence* measurements. Regardless, the main function of a coincidence module is to discern the arrival of two or more radiation particles in separate detectors within some preset time interval Δt.

In practice it is not possible to analyze coincidence events with 100% confidence due to the uncertainties associated with the statistical nature of radiation emissions. Statistical timing errors may occur from the detection process and uncertainties in the electronics resulting from timing jitter, amplitude walk, and noise, all of which lead to statistically variable time delays between processed events. A simple coincidence circuit solves this problem by essentially summing the two input pulses, passing the resultant sum pulse through a discriminator level, and generating an output pulse when the two input pulses overlap.

The period of time in which the two input pulses can be accepted is defined as the coincidence resolving time, and is determined by the width of the pulses, τ, such that the resolving time is equal to 2τ. A coincidence analyzer produces a logic pulse output when the input pulses, on the active inputs, occur within the resolving time window selected on the front panel. Because detector events occur at random times, accidental coincidences can occur between two pulses which produce background in the coincidence counting. The rate of accidental or random coincidences r_R is given by

$$r_R = r_1 r_2 (2\tau), \tag{2.1}$$

where r_1 and r_2 are the count rates observed by the two detectors.

The number of counts in the detectors depends upon the experiment and the detectors, so the best way to reduce accidental coincidences is to make the resolving time as small as possible. However, the resolving time cannot be reduced below the amount of time jitter in the detector pulses without losing true coincidences, so the type of detector determines the minimum useful resolving time.

2.5.10 Time-to-Amplitude Converters

There are some radiation measurements in which the time interval between two separate events is sought. A time-to-amplitude converter (TAC) measures the time delay between two events Δt and converts it into a voltage output between 0–10 volts. The magnitude of the output voltage is linearly proportional to Δt. These pulses can be displayed as a pulse height spectrum, thereby yielding the average delay and the timing resolution defined as the full width at half maximum (FWHM) of the pulse height spectrum.

2.5.11 Analog-to-Digital Converters

An analog-to-digital converter (ADC) generates a digital "word" proportional to the amplitude of an input pulse. In nuclear applications, ADCs are used to digitize the output signals from spectroscopy amplifiers. Because these amplifiers generate output pulses whose amplitudes are directly proportional to the energies of the incident radiation, the ADC can be used with an amplifier and a multichannel analyzer (MCA) to generate energy distributions (spectra) of the radiation emitted by radioactive samples.

2.5.12 Photomultiplier Tube Base

The photomultiplier tube (PMT) base is a separate electronic unit and not a NIM bin module and is used to supply voltage to a PMT. PMTs are vacuum tubes commonly coupled to a scintillating detection medium, which might be a solid, liquid, or gas. Scintillation light excited by radiation absorbed in the scintillator detector may strike the photosensitive cathode of the PMT and eject an electron into the tube. Beyond the light sensitive cathode, a series of metal plates or meshes, called dynodes, are biased with voltage in order to attract and accelerate electrons. These electrons strike the dynode and knock off more secondary electrons than were incident on the dynode, thereby amplifying the signal that increases with the number of dynodes in the PMT. The PMT base divides the applied voltage evenly among these dynodes, with ten stage dividers being common amongst commercial vendors. The output of the PMT base can be coupled to a preamplifier or, in some models, the preamplifier may be incorporated within the PMT base. Some PMT bases allow adjustment of the gain and applied bias in order to achieve optimal detector performance.

2.5.13 Multichannel Analyzer

The multichannel analyzer (MCA) measures the pulse heights emerging from the amplifier with high precision for many pulses, one after the other, and creates a histogram of small voltage intervals or bins along the abscissa (x-axis) and number of observed pulse heights within each interval along the ordinate (y-axis). In the NIM standard the maximum pulse height is 10 volts so the amplifier gain must be set so that no pulse heights of interest exceed this limit. The voltage measurement is made with a quick analog-to-digital converter and the subsequent manipulation and storage is today done with conventional computer code and memory. Because the underlying analog-to-digital converters and computer hardware are binary, the

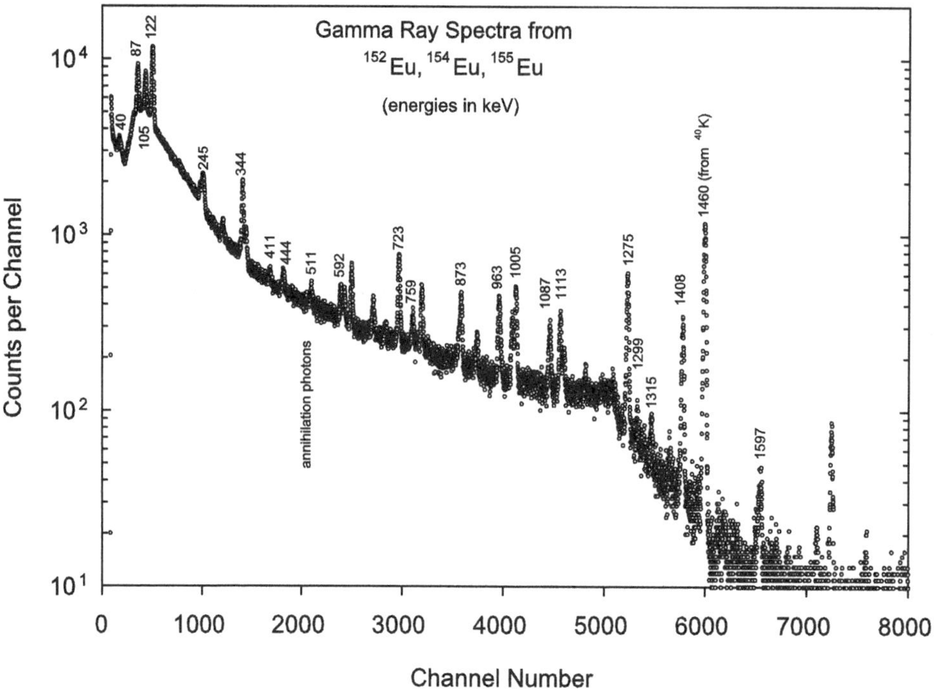

Figure 2.6. Differential pulse height spectrum of gamma rays emitted by the radioactive decay of ^{152}Eu, ^{154}Eu, and ^{155}Eu as measured by a high purity germanium (HPGe) semiconductor detector.

first MCAs divided the ten-volt span into a selectable number of bins that are increasing powers of 2 and this approach is still used. The bins are called channels that range from as low as 128 in number to 4096 or even more. For typical scintillation or gas filled detectors 1024 or 2048 channels are sufficient, but for higher resolution semiconductor detectors pulses are typically collected in 4096 or 8192 bins. The *pulse height spectrum* that is formed from radiation induced pulses forms a histogram on the MCA output display, usually a computer screen. The histogram represents the voltage distribution of pulse height amplitudes provided by the detector, where the voltage pulses are linearly related to the radiation energy absorbed by the detector for each event. Hence, the pulse height spectrum is a relative measure of the spectroscopic energy distribution absorbed in the detector. Overall, the MCA performs the function of numerous SCAs in series as well as a counter/timer. Modern MCAs can be set with over 16000 channels spread over 10 volts, a requirement for radiation spectroscopy applications with ultra-high resolution. Shown in Fig. 2.6 is a MCA differential pulse height spectrum of a mixed Eu source (^{152}Eu, ^{154}Eu,^{155}Eu) as taken with a high purity Ge (HPGe) detector. Details of various types of MCAs are reserved for Chapter 22.

2.5.14 Other NIM Components

Although the main NIM (or CAMAC) components commonly used in a detecting system are listed above, there are many other NIM components that can be used for special counting purposes. Such units include timing modules, coincidence modules, delay lines, time-to-amplitude converters (TAC), and analog-to-digital converters. There are also many variations of those NIM components discussed in the previous sections. The user should consult the specifications regarding the NIM components to determine which modules are best suited for the experiments or measurements to be made.

2.6 CAMAC

For more complex detection systems, as may be encountered at large research facilities and high-energy accelerator centers, there is another international standard for modularized electronics, namely, Computer Automated Measurement And Control (CAMAC), which defines a standard bus for data acquisition and control. The interface system, upon which this standard is based, was developed by the ESONE Committee of European Laboratories with the collaboration of the NIM Committee of the US Department of Energy (DOE) [ESONE, 1964; CAMAC, 1972]. This standard is based on ERDA Reports TID-25875, July 1972 (corresponding to ESONE Report EUR 4100e) and TID-25877. The standard defines the mechanical construction and electrical dataway for the modules. The CAMAC also includes the IEEE standards of 583 (the basic CAMAC design), 595 (serial highway system), 596 (parallel branch highway system), 675 (auxiliary crate controller), 683 (block transfer specifications), 726 (real-time BASIC computer language) and 758 (FORTRAN computer language subroutines for CAMAC).

The container for instrumentation modules is referred to as a 'crate' or 'CAMAC crate.' The CAMAC bus allows data exchange between plug-in modules (up to 24 in a single crate) and a crate controller, which then interfaces to a PC or to a VME-CAMAC interface. The original standard was capable of one 24-bit data transfer every μs. Later a revision to the standard was released to support shorter cycles which allow a datum transfer every 450 ns. A follow-up on upwardly compatible standard Fast CAMAC allows the crate cycle time to be tuned to the capabilities of the modules in each slot. Typically, CAMAC components are significantly more costly than NIM components, and are best used when computer automation is necessary. Regardless of the standard, both NIM and CAMAC systems use basic components to operate and manipulate data from radiation detectors.

2.7 Nuclear Instruments other than NIM or CAMAC

Although NIM electronics are popular and still in widespread use, most of the systems that use NIM electronics are not considered portable. However, there are integrated components available as handheld or totable instruments. Many of these instruments, or survey meters, consist of a battery-operated unit with a detachable detector (see Fig. 2.7). The monitoring unit may have an analog or digital display that conveys information regarding the radiation exposure or the energy signature, and often has a dial selectable radiation exposure range feature.

Portable instrumentation is available for gas-filled, scintillation, and semiconductor detectors, and range in complexity from simple radiation counters to cryogenically cooled high-resolution energy spectrometers. Some relatively new instrumentation that has become available include miniature Geiger-Müller counters the size of key-chain holders, compact solid state VLSI[7] detector packages, to sophisticated 3-dimensional imaging devices capable of identifying radioactive sources by energy and location.

Although these devices may have special operating electronics and features, the same basic pulse processing and signal recognition electronics described available for NIM components are usually a part of the detector package. For instance, a handheld rate meter will typically have a power supply, amplification and shaping, pulse discrimination, and a visual output much like a NIM system, but instead in a compact handheld platform. Some examples of these systems are presented in many of the following chapters on detectors.

2.8 Cables and Connectors

The proper choice of cables and connectors used with nuclear instrumentation is influenced by signal velocity, instrument impedance, operation temperature, radiation environment, and mechanical environment.

[7]Very large scale integration.

Figure 2.7. Two commercial handheld portable gas-filled detectors. The detector on the left is an ion chamber manufactured by Eberline, and the detector on the right is a Geiger-Müller counter attached to a rate meter, both manufactured by Ludlum.

Although numerous variations exist, there are a few standard choices for cables and connectors used in the nuclear industry. Briefly described here are some of those choices. For a more detailed discussion on cables and connectors for nuclear measurements, the reader is referred to Chapter 22.

2.8.1 Cables

Signal transmission of voltage pulses is most often done with coaxial cables. Such cables have a central conductor, around which is an insulating dielectric material. A second conducting layer is wrapped around the dielectric, over which a protective insulator is often applied. Common dielectric materials include polyethylene and Teflon^{TM}. For special radiation environments, cables with mineral dielectrics may be a wise choice.

The propagation speed v of a signal in a cable is dependent upon the dielectric and magnetic properties of the internal insulating materials separating the central conductor from the outer conductor, and is given by

$$v = \frac{1}{\sqrt{\mu\epsilon}} = \frac{1}{\sqrt{\mu_r\epsilon_r\epsilon_0\mu_0}} = \frac{c}{\sqrt{\mu_r\epsilon_r}}, \tag{2.2}$$

where ϵ_r is the relative permittivity, μ_r is the relative permeability and, by definition, $\epsilon_0\mu_0 = c^{-2}$. Most often, $\mu_r \approx 1$, hence the wave propagation speed becomes primarily dependent upon the relative permittivity. Electric waves propagate as transverse electro-magnetic (TEM) waves, meaning that the electrical and magnetic fields are perpendicular to the direction of wave propagation. At some "cutoff" frequency, other modes can propagate, those being transverse magnetic (TM) and transverse electric (TE), which do not necessarily travel at the same wave velocity, thereby potentially causing undesirable interference. Hence, it is best to operate within the specified bandwidth of the cable.

Historically, conductor investigations indicated that wave attenuation is minimum at characteristic impedances of 77 ohms, the breakdown voltage is greatest at 60 ohms and the power-carrying capacity is maximized at 30 ohms. The widespread use of 75 ohm and 50 ohm cables comes from a compromise of these properties. "Radio Guide" (RG) cable is named from the defunct military specification for cable identification, but is commercially still in common use. RG cables have typical characteristic impedances between 50 and 93 ohms. A table of values for various cables can be found in Chapter 22.

Finally, the perils of mixing connectors that are mechanically different is fairly obvious while mixing cables of components of different impedances is more subtle, but can be equally problematic.

2.8.2 Delay Lines

Under some circumstances it is necessary to delay the signal from a detector before it enters the analyzing electronics. For instance, a coincidence measurement may be desired for two different types of detectors, but the signal processing times are different for the two devices. A delay line can be used to retard the progression of the faster signal, thereby allowing coincident events to arrive at the same time. A simple delay line consists of a long coaxial cable where the propagation time per unit length of cable is

$$t_d = \frac{1}{v} = \frac{\sqrt{\mu_r \epsilon_r}}{c} = \sqrt{\frac{LC}{d^2}}, \tag{2.3}$$

where d is the cable length, L is the cable inductance, and C cable capacitance. For typical coaxial cables that have polyethylene as the insulating filling ($\epsilon_r \simeq 2.25$), the delay time is approximately 5 nanoseconds per meter. There are special cables designed as delay lines that have spiral conductors to increase the time delay per unit length, thereby shortening the required cable.

Delay lines not only cause a delay in the arrival time of a signal, but also can attenuate the signal as it travels down the cable. If the delay line is purposely shorted at the output terminal, then the signal can be reflected back, inverted in form, towards the input terminal. As a result, the combined attenuation and reflection of the original signal can be used to alter the final output electronic signal to a desirable outcome, referred to as single delay line pulse shaping. The method is discussed in more detail in Chapter 22.

2.8.3 Connectors

Bayonet Neill-Concelman connectors, most often referred to as BNC connectors, are quick disconnect RF connectors designed for use with coaxial cables (see Fig. 2.8). Cables commonly used with BNC connectors for nuclear measurements are RG-58 and RG-59. BNC connectors can be acquired with a characteristic impedance of 50 ohms or 75 ohms, where it is important to match the impedance of the connector to that of cable impedance to minimize signal reflection. Although typically rated for voltage applications below 500 V, some BNC connectors can be acquired with ratings up to 900 V.

Miniature High Voltage (MHV) connectors are RF connectors for coaxial cable (see Fig. 2.9). They have a similar appearance to BNC connectors, except that the insulation around the male pin protrudes slightly out of the MHV connector, whereas the insulation around the male pin of a BNC connector does not. BNC and MHV connectors do not match. MHV and BNC connectors are not interchangeable, and damage could occur if a BNC connector is mated to a MHV connector.[8]

MHV connectors have potential ratings up to 5000 V DC at 3 A. MHV connectors were at one time popular for detectors requiring applied voltages exceeding that of BNC specifications. Although still in

[8]One author discovered the aftermath of a student's amazing feat of forcing a male MHV connector onto a female BNC connector, a result that rivals spot welding as a process to permanently fuse electrical connectors together.

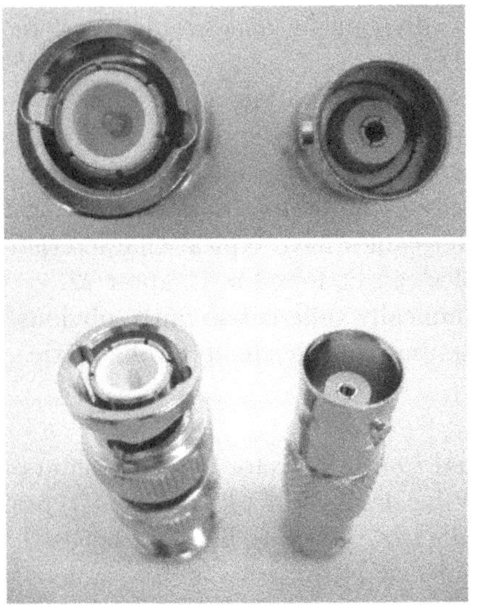

Figure 2.8. Male and Female BNC connectors.

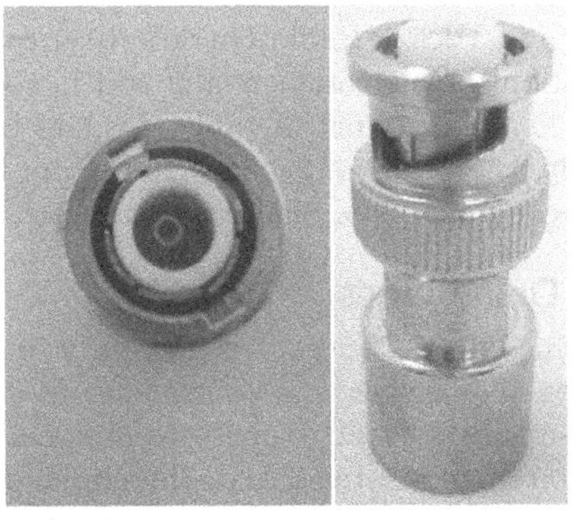

Figure 2.9. Male MHV connectors. Notice the slight insulator protrusion.

use, a MHV connector has design issues, namely, a user can be exposed to ungrounded high voltage during disconnection; hence, SHV connectors (discussed below) have largely displaced the once popular MHV connectors.

Safe High Voltage (SHV) connectors are RF connectors designed for use with coaxial cables. Similar in design to a BNC and/or an MHV connector, the male SHV connector has the distinctive feature of a shielded insulator protruding significantly from the connector body (see Fig. 2.10). The protruding insulator on a male SHV connector remedies the safety issues of an MHV connector by preventing contact with the high voltage conductor. Further, the design eliminates the chance of mismatching a high voltage line to a low voltage connector, as can happen by misidentifying an MHV as a BNC. Most SHV connectors are rated for 5000 V and 5 A.

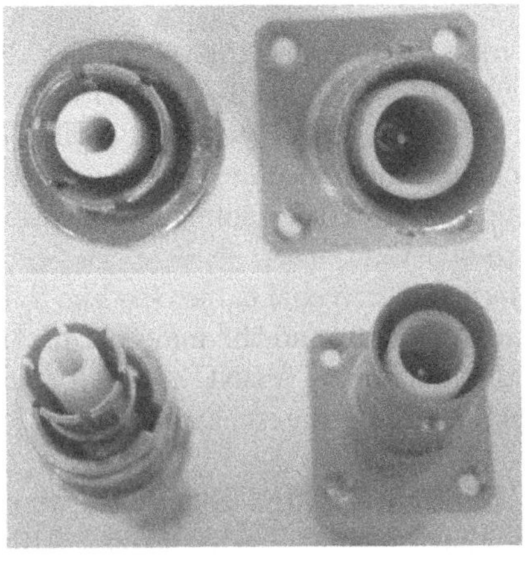

Figure 2.10. Male (left) and female (right) SHV connectors.

Figure 2.11. Female (left) and male (right) N-type connectors, showing (top) 75 Ω and (bottom) 50 Ω variants.

The "N" connector, named after Paul Neill of Bell Laboratories, is a weatherproof RF connector for use with coaxial cables (see Fig. 2.11). The connector is somewhat rugged with a threaded system rather than a bayonet connection system. N connectors can be used with RG-8, RG-58, RG-141, and RG-225 cables. N connectors come in 50 ohm and 75 ohm variants, and are typically rated for 1000–1500 V. In general, 50 ohm and 75 ohm variants are not interchangeable, and damage can occur if a 50 ohm N connector is mated to a 75 ohm N connector.

The "C" connector, presumably named after inventor Carl Concelman of Amphenol Corp., is a bayonet style RF connector that resembles that of a BNC connector, but is considerably larger in size (see Figs. 2.12 and 2.13). C connectors are relatively weatherproof and are often employed with portable survey instrumentation. C connectors have either 50 ohm or 75 ohm impedance, and are typically rated up to 1500 V and 2 A. The actual insulator resistance is greater than 5 GΩ. C-type 50 ohm connectors can be used with 75 ohm cable with little loss at frequencies below 300 MHz. The recommended operating temperature range is between −65°C and 165°C.

The High-Voltage N-Type (HN) connectors are threaded weatherproof coaxial connectors designed to withstand temperature and mechanical stresses. The connectors typically are designed with a characteristic impedance of 50 ohms with a voltage rating of 1500 V. Operational temperature range is −65°C to +165°C. HN connectors are often employed with detector instrumentation used for nuclear reactor power monitoring.

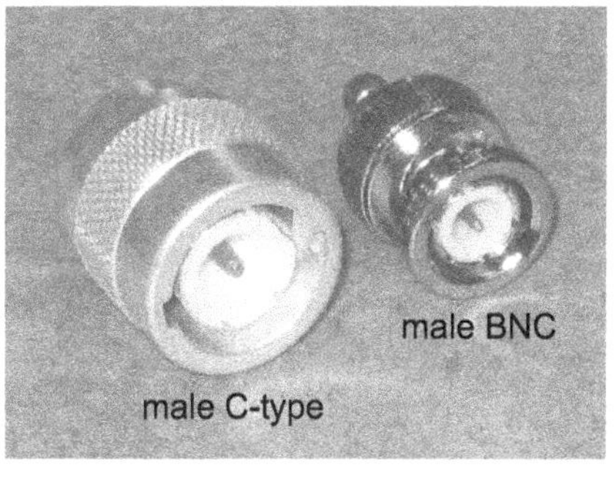

Figure 2.12. Comparison of a C-type connector with a BNC connector.

Figure 2.13. Female and male C-type connectors, showing (top) 75 Ω and (bottom) 50 Ω variants.

LEMO connectors, named after company founder Léon Mouttet, are push-pull miniature coaxial connectors with significantly smaller diameters than BNC or other common coaxial connectors. These compact connectors can be employed for instrumentation where little space is available for multiple connectors. The 50 ohm LEMO 00 series connectors are rated at voltages above 2000 V with limit of 4 A, and they are often used in conjunction with NIM and CAMAC instrumentation.

PROBLEMS

1. Explain the purpose behind implementing the NIM system.

2. Common RG-58/U cable has a characteristic impedance of 50 ohms per meter and characteristic capacitance of 93.5 pF per meter. Calculate the length of cable required to delay a signal by 0.5 microseconds.

3. What is the upper limit for signal voltage in a NIM system? What happens if a signal greater than the limit passes into a NIM component such as a SCA?

4. If the standard for the upper limit for signal voltage were 10 mV, rather than the 10 V of the NIM standard, then only a preamplifier would be necessary to provided any needed signal amplification and, thus, would eliminate the need for an additional amplifier. Why is this low voltage limit not a good idea?

5. What is the voltage width of each bin in a MCA with (a) 128 channels and with (b) 4096 channels?

6. Although Oliver Heaviside invented and patented the coaxial cable in 1880, almost all early work on the development of radiation detectors and the important atomic-level discoveries in the first half of the twentieth century, simply used wires (often bare!) for making electrical connections. So why are coaxial cables so important in today's detector techology?

7. Today with ultra-large-scale integrated circuits, tens of billions of electronic components can be placed on a small chip of silicon. Indeed a complete PC, no larger than the tip of a finger, has been fabricated. Why have electronics for radiation detectors not experienced a similar revolutionary reduction in size?

REFERENCES

American National Standard for Signal Connectors for Nuclear Instruments, ANSI N3.3-1968, New York: American National Standards Institute, 1968.

American National Standard Nomenclature and Dimensions for Panel Mounting Racks, Panels, and Associated Equipment, ANSI N83.9-1968, New York: American National Standards Institute, 1968.

CAMAC, A Modular Instrumentation System for Data Handling - Description and Specification, European Atomic Energy Community, EURATOM Report EUR 4100e, Luxembourg: Office Central de Vente des Publications des Communautes Europeennes, 1972.

Coaxial Cable Connectors Used in Nuclear Instrumentation, International Electrotechnical Commission Publication 313, 1st. Ed., Geneva, Switzerland: International Electrotechnical Commission, 1969.

COSTRELL, L., "NIM Standard," in *Instrumentation in Applied Nuclear Chemistry,* Ch. 5, J. Kruger, Ed., New York: Plenum Press, 1973.

DOE/ER-0457T, Standard NIM Instrumentation System, U.S. NIM Committee, US DOE, 1990.

ESONE System of Nuclear Electronics, European Atomic Energy Community, EURATOM, Report EUR 1831e, Luxembourg: Office Central de Vente des Publications des Communautes Europeennes, 1964.

IEEE STANDARD, *583-1982 IEEE Standard Modular Instrumentation and Digital Interface System (CAMAC),* IEEE, 1982.

IEEE STANDARD, *595-1982 IEEE Standard Serial Highway Interface System (CAMAC),* IEEE, 1982.

IEEE STANDARD, *596-1982 IEEE Standard Parallel Highway Interface System (CAMAC),* IEEE, 1982.

IEEE STANDARD, *675-1976 IEEE Standard Multiple Controllers in a CAMAC Crate,* IEEE, 1982.

IEEE STANDARD, *683-1976 IEEE Recommended Practice for Block Transfers in CAMAC Systems,* IEEE, 1976.

IEEE STANDARD, *726-1982 IEEE Standard Real-Time BASIC for CAMAC,* IEEE, 1982.

IEEE STANDARD, *758-1979 IEEE Standard Subroutines for Computer Automated Measurement and Control (CAMAC),* IEEE, 1979.

SAGNELL, L., *CERN 19 Inch Chassis Systems,* CERN Report 62-29, Geneva, Switzerland: European Organization Nuclear Research, 1962.

Standard Nuclear Instrument Modules, U.S. AEC Report TID-20893, Washington DC: U.S. Gov. Printing Office, 1964; superseded in 1974 by TID-20893, Rev. 4.

Standard NIM Instrumentation System, U.S. DOE Report ER/DOE-0457T, Washington, DC: U.S. Department of Energy, 1990.

United Kingdom Atomic Energy Authority Specifications and Guide to the 2000 Series Unit-Equipment, AERE(R) 11048, Berkshire, England: U.K. Atomic Energy Research Establishment, Harwell, 1962.

Why 50 Ohms?, *Microwaves 101,* Online encyclopedia, 2009-01-13; Retrieved 2013-04-10.

Coax Power Handling, *Microwaves 101,* Online encyclopedia, 2009-01-13; Retrieved 2013-04-10.

Chapter 3

Basic Atomic and Nuclear Physics

I claim that relativity and the rest of modern physics is not complicated. It can be explained very simply. It is only unusual or, put another way, it is contrary to common sense.

Edward Teller

The main topic of this book is radiation detection and measurement; however, in order to understand the principles involved in this discipline it is necessary to comprehend the basic ideas of *modern* physics upon which these principles are based. It is not the intent in this and the next few chapters to present a detailed derivation of the many topics addressed by atomic and nuclear physics. Rather, several topics important for radiation detection are reviewed. Such a review does not go into great detail about the derivation of the results, mainly because there are many excellent and thorough treatments already published such as books by Kaplan [1962], Evans [1955], and Griffiths [2004, 2005]. It should also be noted that much of the material in this and the next chapter have been extracted from Shultis and Faw [2008].

In this chapter, the important concepts of special relativity, wave-particle duality, and quantum mechanics are reviewed. Then brief descriptions of the constituents of matter, atomic and nuclear models and the energetics of nuclear interactions are provided. Finally, a review of radioactivity and its dynamics are discussed.

3.1 Modern Physics Concepts

During the first three decades of the twentieth century, our understanding of the physical universe underwent tremendous changes. Although the results of this revolution in physics are now called "modern" physics, they are now almost a century old. Three of these modern physical concepts are (1) Einstein's theory of special relativity, which extended Newtonian mechanics; (2) wave-particle duality, which says that both electromagnetic waves and atomic particles have dual wave and particle properties; and (3) quantum mechanics, which revealed that the microscopic atomic world is far different from our everyday macroscopic world. The results and insights provided by these three advances in physics are fundamental to an understanding of nuclear science and technology.

3.1.1 The Special Theory of Relativity

The classical laws of dynamics as developed by Newton were believed, for over 200 years, to describe all motion in nature. Students still spend considerable effort mastering the use of these laws of motion. For example, Newton's second law, in the form originally stated by Newton, says the rate of change of a body's momentum $\mathbf{p}$ equals the force $\mathbf{F}$ applied to it, i.e.,

$$\mathbf{F} = \frac{d\mathbf{p}}{dt} = \frac{d(m\mathbf{v})}{dt}. \tag{3.1}$$

For a constant mass m, as assumed by Newton, this equation immediately reduces to the modern form of the second law, $\mathbf{F} = m\mathbf{a}$, where $\mathbf{a} = d\mathbf{v}/dt$, the acceleration of the body.

In 1905 Einstein discovered that classical mechanics could not describe properly the behavior of an object traveling near the speed of light and he found the necessary correction. In his theory of special relativity,[1] Einstein showed that Eq. (3.1) is still correct, but that the mass of a body is not constant, but increases with the body's speed v. The form $\mathbf{F} = m\mathbf{a}$ is thus incorrect. Specifically, Einstein showed that m varies with the body's speed as

$$m = \frac{m_o}{\sqrt{1 - v^2/c^2}}, \tag{3.2}$$

where m_o is the body's "rest mass," i.e., the body's mass when it is at rest, and c is the speed of light ($\simeq 3 \times 10^8$ m/s). The validity of Einstein's correction was immediately confirmed by observing that the electron's mass did indeed increase as its speed increased in precisely the manner predicted by Eq. (3.2).

Most fundamental changes in physics arise in response to experimental results that reveal an old theory to be inadequate. However, Einstein's correction to the laws of motion was produced theoretically before being confirmed experimentally. This is perhaps not too surprising since in our everyday world the difference between m and m_o is incredibly small. For example, a satellite in a circular earth orbit of 7100 km radius moves with a speed of 7.5 km/s. For this satellite the mass correction factor $\sqrt{1 - v^2/c^2} = 1 - 0.31 \times 10^{-9}$, i.e., relativistic effects change the satellite's mass only in the ninth significant figure or by less than one part in a billion! Thus for practical engineering problems in our macroscopic world, relativistic effects can safely be ignored. However, at the atomic and nuclear level, these effects can be very important.

3.1.2 Principle of Relativity

The principle of relativity is older than Newton's laws of motion. In Newton's words (actually translated from Latin) "The motions of bodies included in a given space are the same amongst themselves, whether the space is at rest or moves uniformly forward in a straight line." This means that experiments made in a laboratory in uniform motion (e.g., in a non-accelerating train) produce the same results as when the laboratory is at rest. Indeed, this principle of relativity is widely used to solve problems in mechanics by shifting to moving frames of reference to simplify the equations of motion.

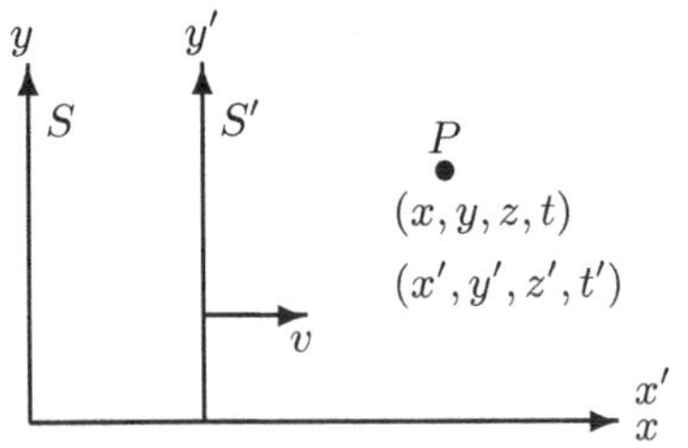

Figure 3.1. Two inertial coordinate systems.

The relativity principle is a simple intuitive and appealing idea. But do all the laws of physics indeed remain the same in all non-accelerating (*inertial*) coordinate systems? Consider the two coordinate systems shown in Fig. 3.1. System S is at rest, while system S' is moving uniformly to the right with speed v. At $t = 0$, the origin of S' is at the origin of S. The coordinates of some point P are (x, y, z) in S and (x', y', z') in S'. Clearly, the primed and unprimed coordinates are related by

$$x' = x - vt; \quad y' = y; \quad z' = z; \quad \text{and} \quad t' = t. \tag{3.3}$$

If these coordinate transformations are substituted into Newton's laws of motion, the equations are unchanged. For example, consider a force in the x-direction, F_x, acting on some mass m. Then the second

[1] In 1915 Einstein published the general theory of relativity, in which he generalized his special theory to include gravitation.

law in the S' moving system is $F_x = m\, d^2x'/dt'^2$. Now transform this law to the stationary S system. For v constant, one finds

$$F_x = m\frac{d^2x'}{dt'^2} = m\frac{d^2(x-vt)}{d(t)^2} = m\frac{d^2x}{dt^2}.$$

Thus the second law has the same form in both systems. Since the laws of motion are the same in all inertial coordinate systems, it follows that it is impossible to tell, from results of mechanical experiments, whether or not the system is moving.

In the 1870s, Maxwell introduced his famous laws of electromagnetism. These laws explained all observed behavior of electricity, magnetism, and light in a uniform system. However, when Eqs. (3.3) are used to transform Maxwell's equations to another inertial system, they assume a different form. Thus from optical experiments in a moving system, one should be able to determine the speed of the system. For many years Maxwell's equations were thought to be somehow incorrect, but 20 years of research only continued to reconfirm them. Eventually, some scientists began to wonder if the problem lay in the Galilean transformation of Eqs. (3.3). Indeed, Lorentz observed in 1904 that if the transformation

$$x' = \frac{x - vt}{\sqrt{1 - v^2/c^2}}; \quad y' = y; \quad z' = z; \quad t' = \frac{t - vx/c^2}{\sqrt{1 - v^2/c^2}} \tag{3.4}$$

is used, Maxwell's equations become the same in all inertial coordinate systems. Poincaré, about this time, even conjectured that *all* laws of physics should remain unchanged under the peculiar looking Lorentz transformation. The Lorentz transformation is indeed strange, since it indicates that space and time are not independent quantities. Time in the S' system, as measured by an observer in the S system, is different from the time in the observer's system.

3.1.3 Results of the Special Theory of Relativity

It was Einstein who, in 1905, showed that the Lorentz transformation was indeed the correct transformation relating all inertial coordinate systems. He also showed how Newton's laws of motion must be modified to make them invariant under this transformation.

Einstein based his analysis on two postulates:

- The laws of physics are expressed by equations that have the same form in all coordinate systems moving at constant velocities relative to each other.
- The speed of light in free space is the same for all observers and is independent of the relative velocity between the source and the observer.

The first postulate is simply the principle of relativity, while the second states that light is observed to move with speed c even if the light source is moving with respect to the observer. From these postulates, Einstein demonstrated several amazing properties of our universe.

1. The laws of motion are correct, as stated by Newton, if the mass of an object is made a function of the object's speed v, i.e.,

$$m(v) = \frac{m_o}{\sqrt{1 - v^2/c^2}}. \tag{3.5}$$

   This result also shows that no material object can travel faster than the speed of light since the relativistic mass $m(v)$ must always be real. Further, an object with a rest mass ($m_o > 0$) cannot even travel at the speed of light; otherwise, its relativistic mass would become infinite and give it an infinite kinetic energy.

2. The length of a moving object in the direction of its motion appears smaller to an observer at rest, namely,
$$L = L_o\sqrt{1 - v^2/c^2}. \tag{3.6}$$
where L_o is the "proper length" or length of the object when at rest.

3. The passage of time appears to slow in a system moving with respect to a stationary observer. The time t required for some physical phenomenon (e.g., the interval between two heart beats) in a moving inertial system appears to be longer (dilated) than the time t_o for the same phenomenon to occur in the stationary system. This phenomenon is called *time dilation*. The relation between t and t_o is
$$t = \frac{t_o}{\sqrt{1 - v^2/c^2}}. \tag{3.7}$$

4. Perhaps the most famous result from special relativity is the demonstration of the equivalence of mass and energy by the well-known equation $E = mc^2$. By using Eqs. (3.1) and (3.2), Einstein showed that the kinetic energy T of a particle is given by $T = mc^2 - m_oc^2$. Thus it is seen that the kinetic energy is associated with the increase in the mass of the particle. Equivalently, this result can be written as $mc^2 = m_oc^2 + T$. The quantity mc^2 can be interpreted as the particle's "total energy" E, which equals its rest-mass energy plus its kinetic energy. If the particle was also in some potential field, for example, an electric field, the total energy would also include the potential energy. Thus it follows that
$$\boxed{E = mc^2.} \tag{3.8}$$
This well-known equation is the cornerstone of nuclear energy analyses. It shows the equivalence of energy and mass. One can be converted into the other in precisely the amount specified by $E = mc^2$.

The derivation of these important results can be found, for example, in Shultis and Faw [2008].

Relation Between Kinetic Energy and Momentum

Both classically and relativistically, the momentum p of a particle is given by
$$p = mv. \tag{3.9}$$
In classical physics, a particle's kinetic energy T is given by
$$T = \frac{mv^2}{2} = \frac{p^2}{2m},$$
which yields
$$p = \sqrt{2mT}. \tag{3.10}$$

For relativistic particles, the relationship between momentum and kinetic energy is not as simple. Square Eq. (3.5) to obtain
$$m^2\frac{c^2 - v^2}{c^2} = m_o^2,$$
or, upon rearrangement,
$$p^2 \equiv (mv)^2 = (mc)^2 - (m_oc)^2 = \frac{1}{c^2}[(mc^2)^2 - (m_oc^2)^2].$$

Because $T = mc^2 - m_oc^2$, the above result reduces to

$$p^2 = \frac{1}{c^2}\left[(T + m_oc^2)^2 - (m_oc^2)^2\right] = \frac{1}{c^2}\left[T^2 + 2Tm_oc^2\right]. \tag{3.11}$$

Thus for relativistic particles

$$\boxed{p = \frac{1}{c}\sqrt{T^2 + 2Tm_oc^2}.} \tag{3.12}$$

Relativistic Particles

For most moving objects encountered in engineering analyses, the classical expression for kinetic energy can be used. Only if an object has a speed near c must relativistic expressions be used. If the changeover from classical to relativistic mechanics is associated with a specific relativistic mass increase, 0.1% for example, the associated speed can be calculated from Eq. (3.5), which, for this example, yields $1.001 = 1/\sqrt{1 - v^2/c^2}$. Solving this result for v gives the relativistic speed threshold as $v = 0.045c$. Listed in Table 3.1 for several important atomic particles are the rest mass energies and the kinetic energies required for a 0.1% mass change.

Table 3.1. Rest mass energies and kinetic energies for a 0.1% relativistic mass increase for four particles.

Particle	rest mass energy m_oc^2	kinetic energy for a 0.1% increase in mass
electron	0.511 MeV	511 eV $\simeq$ 0.5 keV
proton	938 MeV	938 keV $\simeq$ 1 MeV
neutron	940 MeV	940 keV $\simeq$ 1 MeV
α-particle	3751 MeV	3.8 MeV $\simeq$ 4 MeV

3.2 Highlights in the Evolution of Atomic Theory

The concept of the atom is ancient. The Greek philosophers Leucippus and his pupil Democritus in the fifth century BC conjectured that all matter was composed of indivisible particles or *atoms* (lit. "not to be cut"). Unfortunately, Aristotle, whose ideas were more influential far into the Middle Ages, favored the "fire, air, earth, and water" theory of Empedocles. According to the Roman poet Lucretius (100 B.C.), who wrote at length about the nature of the universe in *De Rerun Natura*, the philosopher Epicurus (300 B.C.) tried to revive the atomistic idea but little came from his effort because, of course, the idea was beyond any experimental verification test at that time.

The modern concept of the atom had its origin in the observations of chemical properties made by eighteenth and nineteenth century alchemists. Lavoisier first discovered that mass was conserved in chemical reactions, and shortly after Gay-Lussac found that gases combined in simple ratios by volume. These observations prompted chemist Dalton in the period 1803–1808 to develop an explanation for the observations. Dalton proposed his atomic hypothesis that stated: (1) each element consists of a large number of identical particles (called atoms) that cannot be subdivided and that preserve their identity in chemical reactions; (2) a compound's mass equals the sum of the masses of the constituent atoms; and (3) chemical compounds are formed by the combination of atoms of individual elements in simple proportions (e.g., 1:1, 1:2, etc.). This atomic hypothesis explained chemical reactions and the distinct ratios in which elements combined to form compounds. However, Dalton made no statement about the structure of an atom.

At the beginning of the twentieth century, a wealth of experimental evidence allowed scientists to develop ever more refined models of the atom, until our present model of the atom was essentially established by the 1940s. The structure of the nucleus of an atom has now also been well developed qualitatively and is supported by a wealth of nuclear data. However, work still continues on developing more refined mathematical models to quantify the properties of nuclei and atoms.

In this section, a brief historical summary of the development of atomic models is presented. The emphasis of the presentation is on the novel ideas that were developed, and little discussion is devoted to the many experiments that provided the essential data for model development. Those interested in more detail, especially about the seminal experiments, should refer to any modern physics text.

3.2.1 Radiation as Waves and Particles

For many phenomena, radiant energy can be considered as electromagnetic waves. Indeed Maxwell's equations, which describe very accurately interactions of long wave-length radiation, readily yield a wave equation for the electric and magnetic fields of radiant energy. Phenomena such as diffraction, interference, and other related optical effects can be described only by a wave model for radiation.

However, near the beginning of the twentieth century, several experiments involving light and x rays were performed that indicated that radiation also possessed particle-like properties. Today it is understood, through quantum theory, that matter (e.g., electrons) and radiation (e.g., x rays) both have wave-like and particle properties. This dichotomy, known as the *wave-particle duality principle*, is a cornerstone of modern physics. For some phenomena, a wave description works best; for others, a particle model is appropriate.

3.2.2 Early Observations

In 1887, Heinrich Hertz observed that when ultraviolet light was illuminated upon one contact of a high voltage circuit that an electric spark would issue a longer distance than when the contact was not exposed to the light. J.J. Thomson demonstrated that the effect was caused by a negative charge being produced on the contact surface by the ultraviolet light, referred to as the "photo-electric" effect. Later, in 1897, J.J. Thomson correctly explained that cathode rays, which had been observed for decades, were actually accelerated negative particles, which he referred to as "electrons", as was correctly predicted by Benjamin Franklin in 1746 with his Leyden jar experiments.[2]

3.2.3 The Photoelectric Effect

According to a classical (wave theory) description of light, the light energy was absorbed by the metal surface, and when sufficient energy was absorbed to free a bound electron, a photoelectron would "boil" off the surface. If light were truly a wave, one would expect the following observations:

- Photoelectrons should be produced by light of all frequencies.
- At low intensities, a time lag would be expected between the start of irradiation and the emission of a photoelectron since it takes time for the surface to absorb sufficient energy to eject an electron.
- As the light intensity (i.e., wave amplitude) increases, more energy is absorbed per unit time and, hence, the photoelectron emission rate should increase.
- The kinetic energy of the photoelectron should increase with the light intensity since more energy is absorbed by the surface.

[2]Franklin was first to give the positive and negative designations to electricity, and he was the first to discover the principle of conservation of charge. The cgs unit of electric charge was named after him, where one franklin (Fr) is equal to one statcoulomb.

However, experimental results differed dramatically from these predictions. It was observed:

- For each metal there is a minimum light frequency below which no photoelectrons are emitted, no matter how high the intensity.
- There is no time lag between the start of irradiation and the emission of photoelectrons, no matter how low the intensity.
- The intensity of the light affects only the emission *rate* of photoelectrons.
- The kinetic energy of the photoelectron depends only on the frequency of the light and not on its intensity. The higher the frequency, the more energetic is the photoelectron.

Einstein's Explanation of the Photoelectric Effect

In 1905 Einstein introduced a new light model which explained all these observations.[3] Einstein assumed that light energy consists of *photons* or "quanta of energy," each with an energy $E = h\nu$, where h is Planck's constant (6.626×10^{-34} J s) and ν is the light frequency. He further assumed that the energy associated with each photon interacts as a whole, i.e., either all the energy is absorbed by an atom or none is. With this "particle" model, the maximum kinetic energy of a photoelectron would be

$$E = h\nu - A, \tag{3.13}$$

where A is the amount of energy (the so-called *work function*) required to free an electron from the metal. Thus if $h\nu < A$, no photoelectrons are produced. Increasing the light intensity only increases the number of photons hitting the metal surface per unit time and, thus, the rate of photoelectron emission.

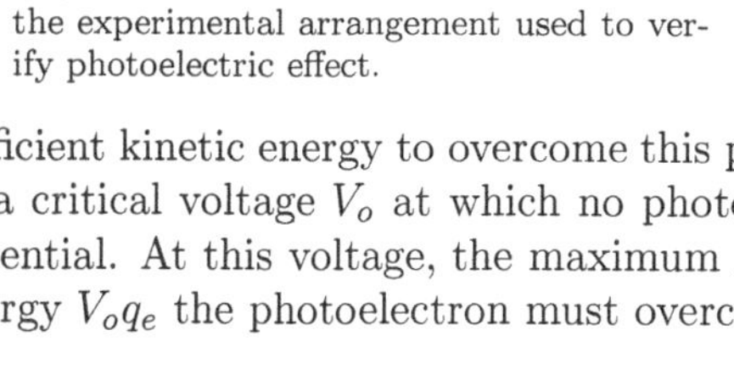

Figure 3.2. A schematic illustration of the experimental arrangement used to verify photoelectric effect.

Although Einstein was able to explain qualitatively the observed characteristics of the photoelectric effect, it was several years later before Einstein's prediction of the maximum energy of a photoelectron, Eq. (3.13), was verified quantitatively using the experiment shown schematically in Fig. 3.2. Photoelectrons emitted from freshly polished metallic surfaces were absorbed by a collector causing a current to flow between the collector and the irradiated metallic surface. As an increasing negative voltage was applied to the collector, fewer photoelectrons had sufficient kinetic energy to overcome this potential difference and the photoelectric current decreased to zero at a critical voltage V_o at which no photoelectrons had sufficient kinetic energy to overcome the opposing potential. At this voltage, the maximum kinetic energy of a photoelectron, Eq. (3.13), equals the potential energy $V_o q_e$ the photoelectron must overcome, i.e.,

$$V_o q_e = h\nu - A,$$

or

$$V_o = \frac{h\nu}{q_e} - \frac{A}{q_e}, \tag{3.14}$$

[3]It is an interesting historical fact that Einstein received the Nobel prize for his photoelectric research and not for his theory of relativity, which he produced in the same year.

where q_e is the electron charge (1.602×10^{-19} C).[4] In 1912 Hughes showed that, for a given metallic surface, V_o was a linear function of the light frequency ν. In 1916 Millikan, who had previously measured the electron charge q_e, verified that plots of V_o versus ν for different metallic surface had a slope of h/q_e, from which h could be evaluated. Millikan's value of h was in excellent agreement with the value determined from measurements of black-body radiation, in whose theoretical description Planck first introduced the constant h.

The prediction by Einstein and its subsequent experimental verification clearly demonstrated the quantum nature of radiant energy. Although the wave theory of light explained diffraction and interference phenomena, scientists saw that the energy of electromagnetic radiation was in the form of individual quanta, each with the same energy $E = h\nu$. A quantum of energy could then be transferred to a single atomic electron. This quantization occurs no matter how weak the radiant energy.

3.2.4 Compton Scattering

Other experimental observations showed that light, besides having quantized energy characteristics, must have another particle-like property, namely, momentum. According to the wave model of electromagnetic radiation, radiation should be scattered from an electron with no change in wavelength. However, in 1922 Compton observed that x rays scattered from electrons had an increase in the wavelength $\Delta\lambda = \lambda' - \lambda$ proportional to $(1 - \cos\theta_s)$ where θ_s was the scattering angle (see Fig. 3.3). To explain this observation, it was necessary to treat x rays as particles with a linear momentum $p = h/\lambda$ and energy $E = h\nu = pc$.

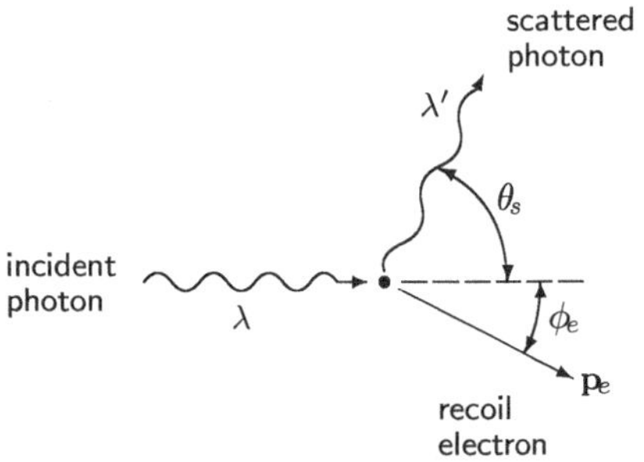

Figure 3.3. A photon with wavelength λ is scattered by an electron. After scattering, the photon has a longer wavelength λ' and the electron recoils with an energy T_e and momentum $\mathbf{p}_e$.

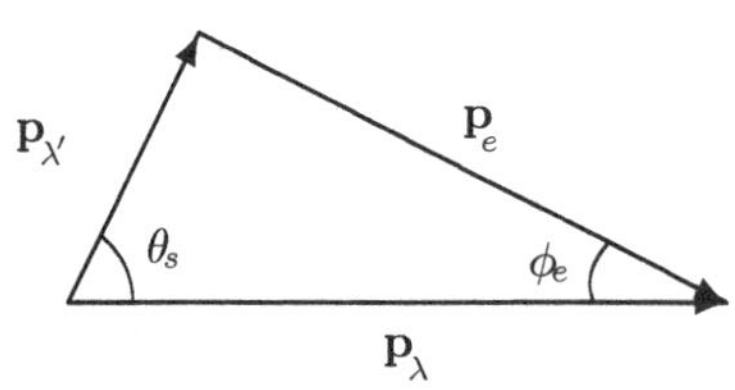

Figure 3.4. Conservation of momentum requires the initial momentum of the photon p_λ equal the vector sum of the momenta of the scattered photon and recoil electron.

In an x-ray scattering interaction, the energy and momentum before scattering must equal the energy and momentum after scattering. Conservation of linear momentum requires the initial momentum of the incident photon (the electron is assumed to be initially at rest) to equal the vector sum of the momenta of the scattered photon and the recoil electron. This requires the momentum vector triangle of Fig. 3.4 to be

[4]Historically, the unit charge of an electron is denoted by 'e', which is also the SI symbol. Unfortunately, 'e' is also used as Euler's number

$$e = \lim_{n\to\infty}\left(1 + \frac{1}{n}\right)^n = \sum_{n=0}^{\infty}\frac{1}{n!}.$$

Hence, to avoid confusion the authors chose to designate the symbol q_e as the magnitude for the unit charge $1.602\,176\,565 \times 10^{-19}$ coulombs.

closed, i.e.,

$$\mathbf{p}_\lambda = \mathbf{p}_{\lambda'} + \mathbf{p}_e \tag{3.15}$$

or from the law of cosines

$$p_e^2 = p_\lambda^2 + p_{\lambda'}^2 - 2p_\lambda p_{\lambda'} \cos\theta_s. \tag{3.16}$$

The conservation of energy requires

$$p_\lambda c + m_e c^2 = p_{\lambda'} c + mc^2 \tag{3.17}$$

where m_e is the rest-mass of the electron before the collision when it has negligible kinetic energy, and m is its relativistic mass after scattering the photon. This result, combined with Eq. (3.11) (in which $m_e \equiv m_o$), can be rewritten as

$$p_\lambda + m_e c - p_{\lambda'} = \sqrt{p_e^2 + (m_e c)^2}. \tag{3.18}$$

Substitute for p_e from Eq. (3.16) into Eq. (3.18), square the result, and simplify to obtain

$$\frac{1}{p_{\lambda'}} - \frac{1}{p_\lambda} = \frac{1}{m_e c}(1 - \cos\theta_s). \tag{3.19}$$

Because $\lambda = h/p$, this result gives the increase in the wavelength of the scattered electron as

$$\lambda' - \lambda = \frac{h}{m_e c}(1 - \cos\theta_s), \tag{3.20}$$

where $h/(m_e c) = 2.431 \times 10^{-6}\mu$m. Thus, Compton was able to predict the wavelength change of scattered x rays by using a particle model for the x rays, a prediction which could not be obtained with a wave model.

This result can be expressed in terms of the incident and scattered photon energies, E and E', respectively. With the photon relations $\lambda = c/\nu$ and $E = h\nu$, Eq. (3.20) gives

$$\boxed{\frac{1}{E'} - \frac{1}{E} = \frac{1}{m_e c^2}(1 - \cos\theta_s).} \tag{3.21}$$

3.2.5 Electromagnetic Radiation: Wave-Particle Duality

Electromagnetic radiation assumes many forms encompassing radio waves, microwaves, visible light, x rays, and gamma rays. Many properties are described by a wave model in which the wave travels at the speed of light c and has a wavelength λ and frequency ν, which are related by the wave speed formula

$$c = \lambda\nu. \tag{3.22}$$

The wave properties account for many phenomena involving light such as diffraction and interference effects.

However, as Einstein and Compton showed, electromagnetic radiation also has particle-like properties, namely, the light energy being carried by discrete quanta or packets of energy called *photons*. Each photon has an energy $E = h\nu$ and interacts with matter (atoms) in particle-like interactions (e.g., in the photoelectric interactions described above).

Thus, light has both wave-like and particle-like properties. The properties or model to use depend on the wavelength of the radiation being considered. If, for example, the electromagnetic radiation is visible or infrared, radar or radio, with wavelengths upwards of $\sim 10^{-6}$ m and thus much greater than atom dimensions, the wave model is usually most useful. However, if the electromagnetic radiation consists of ultraviolet, x rays or gamma rays, with wavelengths $\lesssim 10^{-8}$ m or less, the corpuscular or photon model is usually used. This is the model generally used throughout this book on radiation detection, because the radiation of concern is usually penetrating short-wavelength electromagnetic radiation.

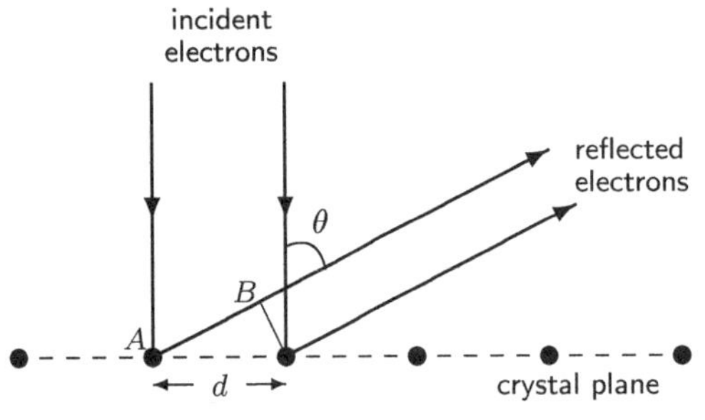

Figure 3.5. Electrons scattering from atoms on a crystalline plane interfere constructively if the distance AB is a multiple of the electron's de Broglie wavelength.

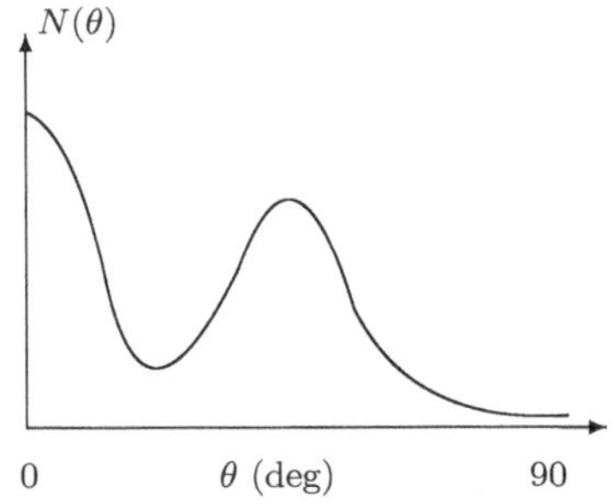

Figure 3.6. Observed number of electrons $N(\theta)$ scattered into a fixed cone or directions about an angle θ by the atoms in a nickel crystal.

Photon Properties

Some particles must always be treated relativistically. For example, photons, by definition, travel with the speed of light c. From Eq. (3.5), one might think that photons have an infinite relativistic mass, and hence, from Eq. (3.12), infinite momentum. This is obviously not true since objects, when irradiated with light, are not observed to jump violently. This apparent paradox can easily be resolved if the rest mass of the photon is exactly zero, although its relativistic mass is finite. In fact, the total energy of a photon, $E = h\nu$, is due strictly to its motion. Equation (3.12) immediately gives the momentum of a photon (with $m_o \equiv 0$) as

$$p = \frac{E}{c} = \frac{h\nu}{c} = \frac{h}{\lambda}. \tag{3.23}$$

From Eq. (3.8), the photon's relativistic mass is $mc^2 = E = h\nu$ or

$$m = \frac{h\nu}{c^2}. \tag{3.24}$$

3.2.6 Electron Scattering

In 1924 de Broglie postulated that, since light had particle properties, then for symmetry (physicists love symmetry!), particles should have wave properties. Because photons had a discrete energy $E = h\nu$ and momentum $p = h/\lambda$, de Broglie suggested that a particle, because of its momentum, should have an associated wavelength $\lambda = h/p$.

Confirmation of Matter Waves

Davisson and Germer in 1927 confirmed that electrons did indeed behave like waves with de Broglie's predicted wavelength. In their experiment, shown schematically in Fig. 3.5, Davisson and Germer illuminated the surface of a Ni crystal by a perpendicular beam of 54-eV electrons and measured the number of electrons $N(\theta)$ reflected at different angles θ from the incident beam. According to the particle model, electrons should be scattered by individual atoms isotropically and $N(\theta)$ should exhibit no structure. However, $N(\theta)$ was observed to have a peak near 50° (see Fig. 3.6). This observation could only be explained by recognizing the peak as a constructive interference peak — a wave phenomenon. Specifically, two reflected electron waves are in phase (constructively interfere) if the difference in their path lengths AB in Fig. 3.5 is an integral number of wavelengths, i.e., if $d \sin\theta = n\lambda$, $n = 1, 2, \ldots$ where d is the distance between atoms of the crystal.

In the following year, G.P. Thomson used a beam of more energetic electrons to pass through a thin foil of a polycrystalline material. Film behind the foil clearly showed diffraction circles as a result of constructive and destructive interference of the electron waves. In 1930 Estermann and Stern [1930] used a similar method

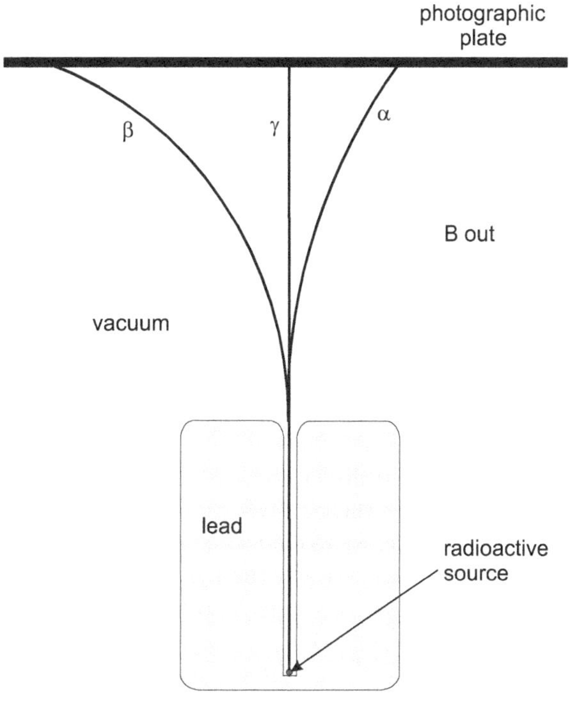

Figure 3.7. Deflection of α, β, and γ rays by a magnetic field out of and perpendicular to the page.

to demonstrate that helium atoms also exhibit wave characteristics. These experiments and many similar ones clearly demonstrated that electrons (and other particles such as atoms) have wave-like properties.

3.3 Development of the Modern Atom Model

In the first half of the twentieth century, the efforts to develop a model of an atom also provided additional evidence that orbital electrons must have wave-like properties. In this section the historical development of atomic models is reviewed.

3.3.1 Discovery of Radioactivity

In 1896, Becquerel discovered that uranium salts emitted rays similar to x rays in that they also fogged photographic plates. Becquerel's discovery was followed by isolation of two other radioactive elements, radium and polonium, by the Curies in 1898. The radiation emission rate of radium was found to be more than a million times that of uranium, for the same mass.

Experiments in magnetic fields showed that three types of radiation could be emitted from naturally occurring radioactive materials. The identification of these radiations was made by the experimental arrangement shown in Fig. 3.7. Radioactive material was placed in a lead enclosure and the emitted radiation collimated in the upward direction by passing through the collimator. The entire chamber was evacuated and a magnetic field was applied perpendicularly and directed outwardly from the plane of the page. With this arrangement, Becquerel found three distinct spots at which radiation struck the photographic plate used as a detector. The three different types of radiation, whose precise nature was then unknown, were called *alpha*, *beta*, and *gamma* rays.

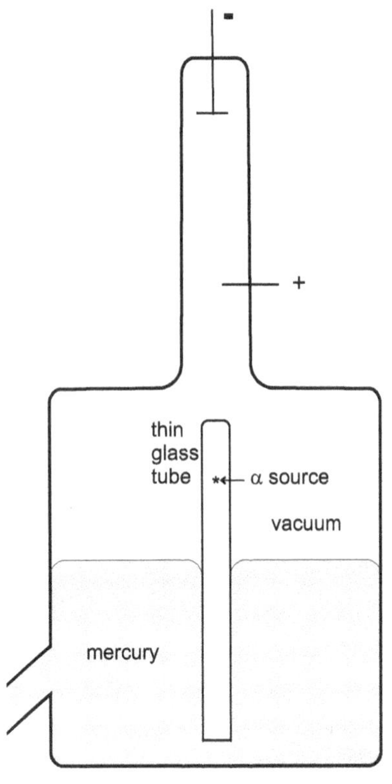

Figure 3.8. Experimental arrangement used to identify the identity of α particles.

The beta rays, which were deflected to the left, were obviously negatively charged particles, and were later found to be the same as the "cathode" rays seen in gas discharge tubes. These rays were identified by J.J. Thomson in 1898 as electrons. The gamma rays were unaffected by the magnetic field and hence had to be uncharged. Today, we know that gamma rays are high frequency electromagnetic radiation whose energy is carried by particles called photons. The alpha particles, being deflected to the right, had positive charge. They were deflected far less than were the beta rays, an indication that the alpha particles have a charge-to-mass ratio q_e/m far less than that of beta particles. Either the positive charge of the alpha particle was far less than the negative charge of beta particles and/or the alpha particle's mass was far greater than that of a beta particle.

A quantitative analysis of the deflection of alpha particles showed that their speeds were of the order of 10^7 m/s. The charge-to-mass ratio was found to be 4.82×10^7 C kg^{-1}. By contrast, the charge-to-mass ratio for the hydrogen ion is twice as large, namely, 9.59×10^7 C kg^{-1}. Thus, if the alpha particle had the same charge as the hydrogen ion, its mass would have to be twice that of the hydrogen ion. If the alpha particle were doubly charged, its mass would be four times as large and would correspond to that of the helium atom. That an alpha particle is an ionized helium atom was demonstrated by Rutherford who used the experimental arrangement of Fig. 3.8. The alpha particles from the radioactive source penetrate the thin-walled glass tube and are collected in the surrounding evacuated chamber. After slowing, the α particles capture ambient electrons to form neutral helium atoms. The accumulated helium gas is then compressed so that an electrical discharge occurs when a high voltage is applied between the electrodes. The emission spectrum from the excited gas atoms was found to have the same wavelengths as that produced by an ordinary helium-filled discharge tube. Therefore, the alpha particle must be a helium ion. The mass of the alpha particle is four times the mass of hydrogen, but the charge on the alpha particle is only twice the charge on a hydrogen ion.

3.3.2 Thomson's Atomic Model: The Plum Pudding Model

The discovery of radioactivity by Becquerel in 1896 and Thomson's discovery of the electron in 1898 provided a basis for the first theories of atomic structure. In radioactive decay, atoms are transformed into different atoms by emitting positively charged or negatively charged particles. This led to the view that atoms are composed of positive and negative charges. If correct, the total negative charge in an atom must be an integral multiple of the electronic charge and, since atoms are electrically neutral, the positive and negative charges must be numerically equal.

The emission of electrons from atoms under widely varying conditions was convincing evidence that electrons exist as such inside atoms. The first theories of atomic structure were based on the idea that atoms were composed of electrons and positive charges. There was no particular assumption concerning the nature of the positive charges because the properties of the positive charges from radioactive decay and from gas discharge tubes did not have the q_e/m consistency that was shown by the negative charges (electrons).

At the time of Thomson's research on atoms, there was no information about the way that the positive and negative charges were distributed in the atom. Thomson proposed a model, simple but fairly accurate, considering the lack of information about atoms at that time. He assumed that an atom consisted of a sphere of positive charge of uniform density, throughout which was distributed an equal and opposite charge in the form of electrons. The atom was like a "plum pudding," with the negative charges dispersed like raisins or plums in a dough of positive electricity (see Fig. 3.9). The diameter of the sphere was of the order of 10^{-10} m, the magnitude found for the size of the atom.

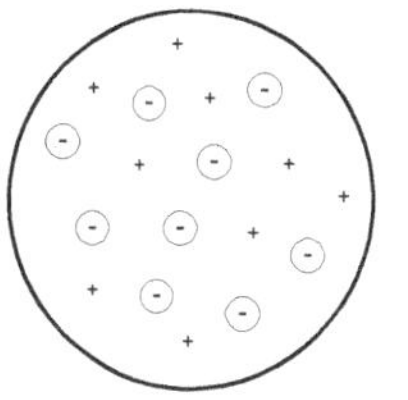

Figure 3.9. Thomson's plum-pudding model of the atom.

This model explained three important experimental observations: (1) an ion is just an atom from which electrons have been lost, (2) the charge on a singly ionized atom equals the negative of the charge of an electron, and (3) the number of electrons in an atom approximately equals one-half of the atomic weight of the atom, i.e., if the atom's mass doubles, the number of electrons double.

Also it was known that the mass of the electron was known to be about one eighteen hundredth the mass of the hydrogen atom. Therefore, the total mass of the electrons in an atom is a very small part of the total mass of the atom, and, hence, practically all of the mass of an atom is associated with the positive charge of the "pudding."

3.3.3 The Rutherford Atomic Model

At the beginning of the twentieth century, Geiger and Marsden used alpha particles from a radioactive source to irradiate a thin gold foil. According to the Thomson plum-pudding model, for a gold foil 4×10^{-5} cm in thickness, the probability an alpha particle scatters at least by an angle ϕ was calculated to be $\exp(-\phi/\phi_m)$, where $\phi_m \simeq 1°$. Thus, the probability that an alpha particle scatters by more than 90° should be about 10^{-40}. In fact, it was observed that one alpha particle in 8000 was scattered by more than 90°!

To explain these observations, Rutherford in 1911 concluded that the positively charged mass of an atom must be concentrated within a sphere of radius about 10^{-12} cm. The electrons, therefore, must revolve about a massive, small, positively charged nucleus in orbits with diameters of the size of atoms, namely, about 10^{-8} cm. With such a model of the nucleus, the predicted deflection of alpha particles by the gold foil fit the experimental data very well.

3.3.4 The Bohr Atomic Model

The Rutherford model of an atom was quickly found to have serious deficiencies. In particular, it violated classical laws of electromagnetism. According to classical theory, an accelerating charge (the electrons in circular orbits around a nucleus) should radiate away their kinetic energy within about 10^{-9} s and spiral into the nucleus.[5]

But obviously atoms do not collapse. Were they to do so, the electromagnetic radiation emitted by the collapsing electrons should be continuous in frequency, since as the electrons collapse, they should spiral ever faster around the nucleus and hence experience increasing central accelerations. When atoms are excited by an electrical discharge, for example, atoms are observed to emit light not in a continuous wavelength spectrum but with very discrete wavelengths characteristic of the element. For example, part of the spectrum of light emitted by hydrogen is shown in Fig. 3.10.

[5]This radiation produced by an accelerated electric charge explains how radio waves are generated. Electrons are accelerated back and forth along a wire (antenna) by an alternating potential with a well-defined frequency, namely that of the radio waves that are emitted from the wire.

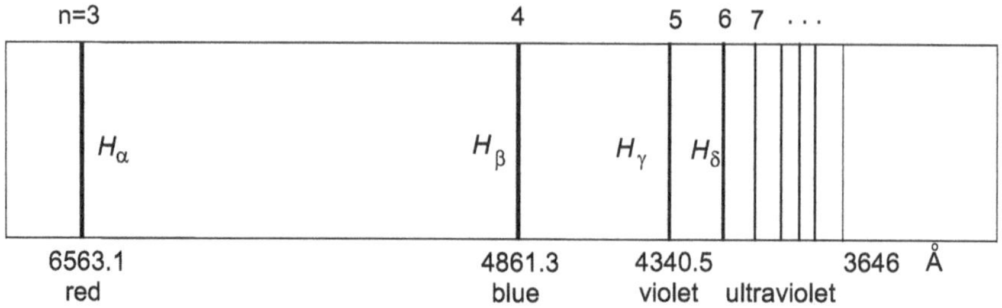

Figure 3.10. Diagram of the lines of the Balmer series of atomic hydrogen. After Kaplan [1963].

Table 3.2. Various hydrogen spectral series.

n_o	Series name	Spectrum location
1	Lyman	ultraviolet
2	Balmer	visible and near ultraviolet
3	Paschen	infrared
4	Brackett	infrared
5	Pfund	infrared

It was found empirically that the wavelength of light emitted by hydrogen (excited by electrical discharges) could be described very accurately by the simple equation

$$\frac{1}{\lambda} = R_H \left[\frac{1}{n_o^2} - \frac{1}{n^2}\right] \tag{3.25}$$

where n_o, n are positive integers with $n > n_o$ and R_H is the Rydberg constant found empirically to have the value $R_H = 10\,967\,758$ m^{-1}. The integer n_o defines a series of spectral lines discovered by different researchers identified in Table 3.2.

The observed discrete-wavelength nature of light emitted by excited atoms was in direct conflict with Rutherford's model of the atom. In a series of papers published between 1913 and 1915, Bohr developed an atomic model which predicted very closely the observed spectral measurements in hydrogen. Bohr visualized an atom much like the Rutherford model with the electrons in orbits around a small dense central nucleus. However, his model included several non-classical constraints. Bohr postulated:

1. An electron moves in a circular orbit about the nucleus obeying the laws of classical mechanics.

2. Instead of an infinity of orbits, only those orbits whose angular momentum L is an integral multiple of $h/2\pi$ are allowed.

3. Electrons radiate energy only when moving from one allowed orbit to another. The energy $E = h\nu$ of the emitted radiation is the difference between characteristic energies associated with the two orbits. These characteristic energies are determined by a balance of centripetal and Coulombic forces.

Consider an electron of mass m_e moving with speed v in a circular orbit of radius r about a central nucleus with charge Zq_e. From postulate (1), the centripetal force on the electron must equal the Coulombic attractive force, i.e.,

$$\frac{m_e v^2}{r} = \frac{Zq_e^2}{4\pi\epsilon_o r^2}, \tag{3.26}$$

where ϵ_o is the *permittivity* of free space and q_e is the charge of the electron. From postulate (2)

$$L \equiv m_e v r = n\frac{h}{2\pi}, \qquad n = 1, 2, 3, \ldots. \tag{3.27}$$

Solution of these relations for r and v yields

$$v_n = \frac{Zq_e^2}{2\epsilon_o n h} \quad \text{and} \quad r_n = \frac{n^2 h^2 \epsilon_o}{\pi m_e Z q_e^2}, \qquad n = 1, 2, 3, \ldots \tag{3.28}$$

For $n = 1$ and $Z = 1$, these expressions yield $r_1 = 5.293 \times 10^{-11}$ m and $v_1 = 2.187 \times 10^6$ m/s $\ll c$, indicating that our use of non-relativistic mechanics is justified.

The electron has both potential energy[6] $V_n = -Zq_e^2/(4\pi\epsilon_o r_n)$ and kinetic energy $T_n = \frac{1}{2}m_e v_n^2$ or, from Eq. (3.26), $T_n = \frac{1}{2}[Zq_e^2/(4\pi\epsilon_o r_n)]$. Thus, the electron's total energy $E_n = T_n + V_n$ or, with Eq. (3.28),

$$E_n = T_n + V_n = \frac{1}{2}\left(\frac{Zq_e^2}{4\pi\epsilon_o r_n}\right) - \left(\frac{Zq_e^2}{4\pi\epsilon_o r_n}\right) = -\frac{1}{2}\frac{Zq_e^2}{4\pi\epsilon_o r_n} = -\frac{m_e(Zq_e^2)^2}{8\epsilon_o^2 n^2 h^2}. \tag{3.29}$$

For the ground state of hydrogen ($n = 1$ and $Z = 1$), this energy is $E_1 = -13.606$ eV, which is the experimentally measured ionization energy for hydrogen.

Finally, by postulate (3), the frequency of radiation emitted as an electron moves from an orbit with "quantum number" n to an orbit denoted by $n_o < n$ is given by

$$h\nu_{n\to n_o} = E_n - E_{n_o} = \frac{m_e Z^2 q_e^4}{8\epsilon_o^2 h^2}\left[\frac{1}{n_o^2} - \frac{1}{n^2}\right]. \tag{3.30}$$

Since $1/\lambda = \nu/c$, the wavelength of the emitted light is thus

$$\frac{1}{\lambda_{n\to n_o}} = \frac{m_e Z^2 q_e^4}{8\epsilon_o^2 c h^3}\left[\frac{1}{n_o^2} - \frac{1}{n^2}\right] = R_\infty\left[\frac{1}{n_o^2} - \frac{1}{n^2}\right], \quad n > n_o. \tag{3.31}$$

Example 3.1: What is the energy (in eV) required to remove the electron in the ground state from singly ionized helium?

Solution:

For the helium nucleus $Z = 2$ and the energy of the ground state (n=1) is, from Eq. (3.29),

$$E_1 = -\frac{m_e(2q_e^2)^2}{8\epsilon_o^2 h^2}.$$

From Table A.1 in Appendix A, one finds $h = 6.626\times10^{-34}$ J s, $m_e = 9.109\times10^{-31}$ kg, $q_e = 1.6022\times10^{-19}$ C, and $\epsilon_o = 8.854 \times 10^{-12}$ F m^{-1} (= C^2 J^{-1} m^{-1}). Substitution of these values into the above expression for E_1 yields

$$E_1 = -8.720 \times 10^{-18}\left(\frac{\text{kg}}{\text{m}^{-2}\ \text{s}^2}\right) = -8.720 \times 10^{-18}\text{J}$$
$$= (-8.720 \times 10^{-18}\text{J})/(1.6022 \times 10^{-19}\text{J/eV}) = -54.43 \text{ eV}.$$

Thus, it takes 54.43 eV of energy to remove the electron from singly ionized helium.

Equation (3.31) has the same form as the empirical Eq. (3.25). However, in the present analysis it has been implicitly assumed that the proton is infinitely heavy compared to the orbiting electron so that the electron moves in a circular orbit about a stationary proton. The Rydberg constant in the above result (with $Z = 1$) is thus denoted by R_∞, and is slightly different from R_H for the hydrogen atom. In reality, the proton and electron revolve about each other. To account for this, the electron mass in

[6] The potential energy is negative because of the attractive Coulombic force, with energy required to extract the electron from the atom. The zero reference for potential energy is associated with infinite orbital radius.

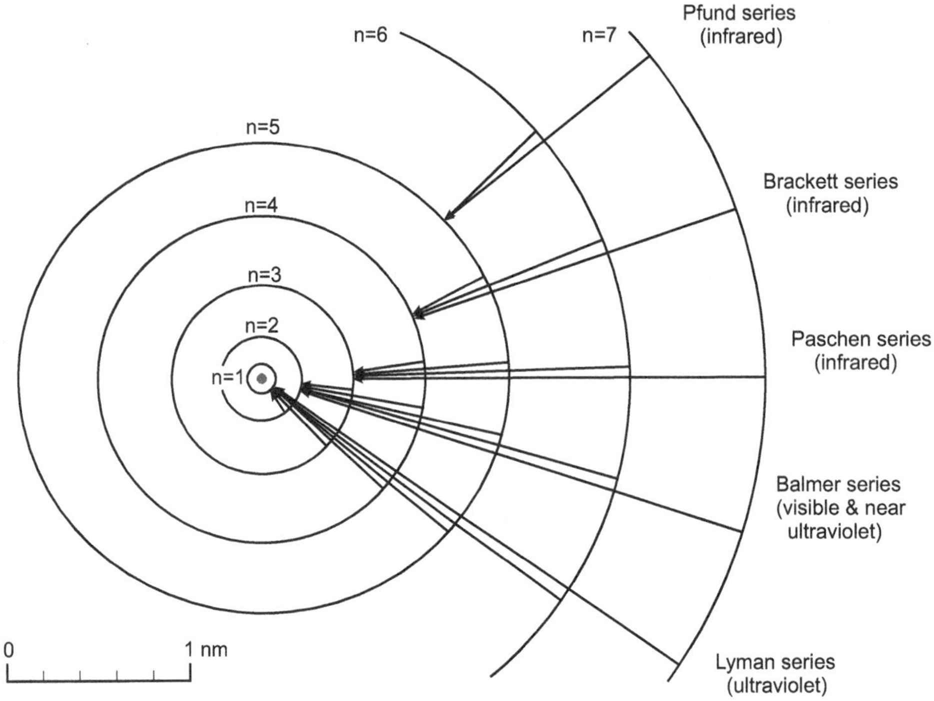

Figure 3.11. Bohr orbits for the hydrogen atom showing the deexcitation transitions that are responsible for the various spectral series observed experimentally. After Kaplan [1963].

Eq. (3.31) should be replaced by the electron's *reduced* mass $\mu_e = m_e m_p/(m_e + m_p)$ [Kaplan 1963]. This is a small correction since $\mu_e = 0.999445568\, m_e$, nevertheless an important one. With this correction, $R_H = \mu_e q_e^4/(8\epsilon_o^2 ch^3) = 10\,967\,758 \text{ m}^{-1}$, which agrees with the observed value to eight significant figures!

The innermost orbits of the electron in the hydrogen atom are shown schematically in Fig. 3.11 to scale. The electronic transitions that give rise to the various emission series of spectral lines are also shown. The Bohr model, with its excellent predictive ability, *quantizes* the allowed electron orbits. Each allowed orbit is defined by the *quantum number* n introduced through Bohr's second postulate, a postulate whose justification is that the resulting predicted emission spectrum for hydrogen is in amazing agreement with the observation.

3.3.5 Extension of the Bohr Theory: Elliptic Orbits

The Bohr theory was very successful in predicting with great accuracy the wavelengths of the spectral lines in hydrogen and singly ionized helium. Shortly after Bohr published his model of the atom, there was a significant improvement in spectroscopic resolution. Refined spectroscopic analysis showed that the spectral lines were not simple, but consisted of a number of lines very close together, a so-called *fine structure*. In terms of energy levels, the existence of a fine structure means that instead of a single electron energy level for each quantum number n, there are actually a number of energy levels lying close to one another. Sommerfeld partly succeeded in explaining some of the lines in hydrogen and singly ionized helium by postulating elliptic orbits as well as circular orbits (see Fig. 3.12). These elliptical orbits required the introduction of another quantum number ℓ to describe the angular momentum of orbits with varying eccentricities.

Sommerfeld showed that, in the case of a one-electron atom, the fine structure could be partially explained. When the predictions of Sommerfeld's model were compared to experimental results on the resolution of the Balmer lines for He^+, the theory predicted more lines than were observed in the fine structure. Agreement

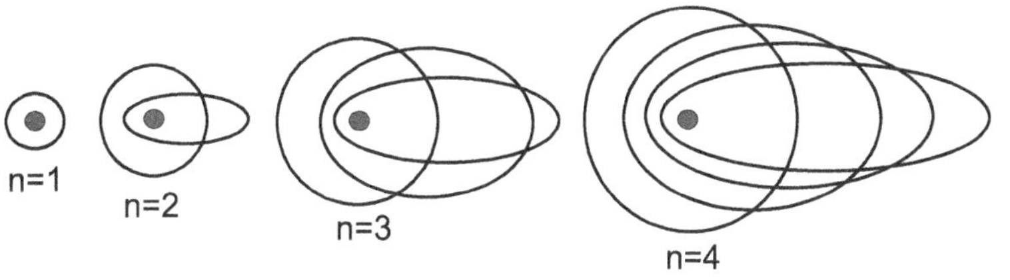

Figure 3.12. Relative positions and dimensions of the elliptical orbits corresponding to the first four values of the quantum number n in Sommerfeld's modification for the hydrogen atom. After Kaplan [1963].

between theory and experiment required the new quantum number ℓ to be constrained by an empirical or *ad hoc* selection rule which limited the number of allowed elliptical orbits and the transitions between the orbits.

For two-electron atoms, difficulties arose in the further refinement of Bohr's theory. To account for the observed splitting of the spectral lines in an applied magnetic field, it was necessary to introduce a third quantum number m, related to the orientation of the elliptical orbits in space. In the presence of a magnetic field, there is a component of the angular momentum in the direction of the field. This modification again led to the prediction of more spectral lines than were experimentally observed, and a second set of selection rules had to be empirically introduced to limit the number of lines.

Further difficulties arose when the model was applied to the spectrum of more complicated atoms. In the spectra of more complicated atoms, multiplet structure is observed. Multiplets differ from fine structure lines in that lines of a multiplet can be widely separated. The lines of sodium with wavelengths of 589.593 nm and 588.996 nm are an example of a doublet. Triplets are observed in the spectrum of magnesium.

These difficulties (and others) could not be resolved by further changes of the Bohr theory. It became apparent that these difficulties were intrinsic to the model. An entirely different approach was necessary to solve the problem of the structure of the atom and to avoid the need to introduce empirical selection rules.

3.3.6 The Quantum Mechanical Model of the Atom

From the *ad hoc* nature of quantum numbers and their selection rules used in refinements of the basic Bohr model of the atom, it was apparent an entirely different approach was needed to explain the details of atomic spectra. This new approach was introduced in 1925 by Schrödinger who brought wave or quantum mechanics to the world. Here, only a brief summary is given. Schrödinger's new theory (or model) showed that there were indeed three quantum numbers (n, ℓ, m) analogous to the three used in the refined Bohr models. Further, the values and constraints on these quantum numbers arose from the theory naturally—no *ad hoc* selection rules were needed.

To explain the multiple fine-line structure in the observed optical spectra from atoms and the splitting of some of these lines in a strong magnetic field (the *anomalous Zeeman effect*), the need for a fourth quantum number m_s was needed. This quantum number accounted for the inherent angular momentum of the electron equal to $\pm h/2\pi$. In 1928, Dirac showed that this fourth quantum number also arises from the wave equation if it is corrected for relativistic effects.

In the quantum mechanical model of the atom, the electrons are no longer point particles revolving around the central nucleus in orbits. Rather, each electron is visualized as a standing wave around the nucleus. The amplitude of this wave (called a *wave function*) at any particular position gives the probability that the electron is in that region of space. Example plots of these wave functions for the hydrogen atom are given in Fig. 3.16. Each electron wave function has a well-defined energy which is specified uniquely by the four quantum numbers defining the wave function.

3.3.7 Wave-Particle Duality

The fact that particles can behave like waves and that electromagnetic waves can behave like particles seems like a paradox. What really is a photon or an electron? Are they waves or particles? The answer is that entities in nature are more complex than is often thought, and they have, simultaneously, both particle and wave properties. Which properties dominate depends on the object's energy and mass. The de Broglie wavelength is given by

$$\lambda = \frac{h}{p} = \frac{hc}{\sqrt{T^2 + 2Tm_oc^2}} \tag{3.32}$$

For a classical object ($T^2 \ll 2Tm_oc^2$), the wavelength is given by $\lambda = h/\sqrt{2m_oT}$. However, as the object's speed increases, its behavior eventually becomes relativistic ($T \gg 2m_oc^2$) and the wavelength varies as $\lambda = hc/T$, the same as that for a photon. When the wavelength of an object is much less than atomic dimensions ($\sim 10^{-10}$ m), it behaves more like a classical particle than a wave. However, for objects with longer wavelengths, wave properties tend to be more apparent than particle properties.

3.4 Quantum Mechanics

The demonstration that particles (point objects) also had wave properties led to another major advance of modern physics. Because a material object such as an electron has wave properties, it should obey some sort of wave equation. Indeed, Schrödinger in 1925 showed that atomic electrons could be well described as standing waves around the nucleus. Further, the electron associated with each wave could have only a discrete energy. The branch of physics devoted to this wave description of particles is called *quantum mechanics* or *wave mechanics.*

3.4.1 Schrödinger's Wave Equation

To illustrate Schrödinger's wave equation, consider an analogy to the standing waves produced by a plucked string of length L anchored at both ends. The wave equation that describes the displacement $\Psi(x,t)$ as a function of position x from one end of the string, which has length L, and at time t is [Riley et al. 2006]

$$\frac{\partial^2\Psi(x,t)}{\partial x^2} = \frac{1}{u^2}\frac{\partial^2\Psi(x,t)}{\partial t^2}. \tag{3.33}$$

Here u is the wave speed. There are infinitely many discrete solutions to this homogeneous partial differential equation, subject to the boundary condition $\Psi(0,t) = \Psi(L,t) = 0$, namely,

$$\Psi(x,t) = A\sin\left(\frac{n\pi x}{L}\right)\sin\left(\frac{n\pi ut}{L}\right), \quad n = 1,2,3\ldots \tag{3.34}$$

That this is the general solution can be verified by substitution of Eq. (3.34) into Eq. (3.33). The frequencies ν of the solutions are also discrete. The time for one cycle is $t_c = 1/\nu$ so that $n\pi ut_c/L = 2\pi$; thus

$$\nu = \frac{nu}{2L}, \quad n = 1,2,3,\ldots.$$

Notice that the solution of the wave equation Eq. (3.34) is *separable*, i.e., it has the form $\Psi(x,t) = \psi(x)T(t) = \psi(x)\sin(2\pi\nu t)$. Substitution of this separable form into Eq. (3.33) yields

$$\frac{d^2\psi(x)}{dx^2} + \frac{4\pi^2\nu^2}{u^2}\psi(x) = 0$$

or, since $u = \lambda\nu$,

$$\frac{d^2\psi(x)}{dx^2} + \frac{4\pi^2}{\lambda^2}\psi(x) = 0. \tag{3.35}$$

To generalize to three-dimensions, the operator d^2/dx^2 is replaced by $\partial^2/\partial x^2 + \partial^2/\partial y^2 + \partial^2/\partial z^2 \equiv \nabla^2$ giving

$$\nabla^2\psi(x,y,z) + \frac{4\pi^2}{\lambda^2}\psi(x,y,z) = 0. \tag{3.36}$$

Now apply this wave equation to an electron bound to an atomic nucleus. The positively charged nucleus produces an electric field and an attractive electric force on the electron. The electron with (rest) mass m has a total energy E, kinetic energy T, and a potential energy U such that $T = E - U$. The wavelength of the electron is $\lambda = h/p = h/\sqrt{2mT} = h/\sqrt{2m(E-U)}$ (assuming the electron is non-relativistic). Substitution for λ into Eq. (3.36) gives

$$\boxed{-\frac{h^2}{8\pi^2 m}\nabla^2\psi(x,y,z) + U(x,y,z)\psi(x,y,z) = E\psi(x,y,z).} \tag{3.37}$$

This equation is known as the steady-state *Schrödinger's wave equation*, and is the fundamental equation of quantum mechanics. This is a homogeneous equation in which everything on the left-hand side is known except, of course, $\psi(x,y,z)$; but on the right-hand side the electron energy E is not known. Such an equation generally has only the trivial null solution ($\psi = 0$); however, non-trivial solutions can be found if E has very precise and discrete values[7] $E = E_n, n = 0, 1, 2, \ldots$. This equation then says that an electron around a nucleus can have only very discrete values of $E = E_n$, a fact well verified by experiment. Moreover, the wave solution of $\psi_n(x,y,z)$ associated with a given energy level E_n describes the amplitude of the electron wave. The interpretation of ψ_n is discussed below.

3.4.2 The Wave Function

The non-trivial solution $\psi_n(x,y,z)$ of Eq. (3.37) when $E = E_n$ (an eigenvalue of the equation) is called a *wave function.* In general, this is a complex quantity which extends over all space, and may be thought of as the relative amplitude of a wave associated with the particle described by Eq. (3.37). Further, because Eq. (3.37) is a homogeneous equation, then, if ψ' is a solution, so is $\psi = A\psi'$, where A is an arbitrary constant. It is usual to choose A so that the integral of $\psi\psi^* = |\psi|^2$ over all space equals unity, i.e.,[8]

$$\iiint \psi(x,y,z)\psi^*(x,y,z)\,dV = 1. \tag{3.38}$$

Just as the square of the amplitude of a classical wave defines the intensity of the wave, the square of the amplitude of the wave function $|\psi|^2$ gives the probability of finding the particle at any position in space. Thus, the probability that the particle is in some small volume dV around the point (x,y,z) is

$$\text{Prob} = |\psi(x,y,z)|^2\,dV = \psi(x,y,z)\psi^*(x,y,z)\,dV.$$

From this interpretation of ψ, the normalization condition of Eq. (3.38) requires that the particle be somewhere in space.

[7] Mathematicians call such an equation (subject to appropriate boundary conditions) an *eigenvalue* problem in which E_n is called the *eigenvalue* and the corresponding solution $\psi_n(x,y,z)$ the *eigenfunction.*

[8] Here ψ^* denotes the complex conjugate of ψ.

3.4.3 The Uncertainty Principle

With quantum/wave mechanics, it is no longer possible to say that a particle is at a particular location; rather, one can say only that the particle has a probability $\psi\psi^* \, dV$ of being in a volume dV. It is possible to construct solutions to the wave equation such that $\psi\psi^*$ is negligibly small except in a very small region of space. Such a wave function thus localizes the particle to the very small region of space. However, such localized *wave packets* spread out very quickly so that the subsequent path and momentum of the particle is known only within very broad limits. This idea of there being uncertainty in a particle's path and its speed or momentum was first considered by Heisenberg in 1927.

If one attempts to measure both a particle's position along the x-axis and its momentum, there will be an uncertainty Δx in the measured position and an uncertainty Δp in the momentum. Heisenberg's uncertainty principle says there is a limit to how small these uncertainties can be, namely,

$$\Delta x \, \Delta p \geq \frac{h}{4\pi}. \tag{3.39}$$

This limitation is a direct consequence of the wave properties of a particle. The uncertainty principle can be derived rigorously from Schrödinger's wave equation; however, a more phenomenological approach is to consider an attempt to measure the location of an electron with very high accuracy. Conceptually, one could use an idealized microscope, which can focus very short wavelength light to resolve points that are about 10^{-11} m apart. To "see" the electron, a photon must scatter from it and enter the microscope. The more accurately the position is to be determined (i.e., the smaller Δx), the smaller must be the light's wavelength (and the greater the photon's energy and momentum). Consequently, the greater is the uncertainty in the electron's momentum Δp since a higher energy photon, upon rebounding from the electron, will change the electron's momentum even more. *By observing a system, the system is necessarily altered.*

The limitation on the accuracies with which both position and momentum (speed) can be known is an important consideration only for systems of atomic dimensions. For example, to locate a mass of 1 g to within 0.1 mm, the minimum uncertainty in the mass's speed, as specified by Eq. (3.39), is about 10^{-26} m/s, far smaller than errors introduced by practical instrumentation. However, at the atomic and nuclear levels, the uncertainty principle provides a very severe restriction on how position and speed of a particle are fundamentally intertwined.

There is a second uncertainty principle (also by Heisenberg) relating the uncertainty ΔE in a particle's energy E and the uncertainty in the time Δt at which the particle had the energy, namely,

$$\Delta E \, \Delta t \geq \frac{h}{4\pi}. \tag{3.40}$$

This restriction on the accuracy of energy and time measurements is a consequence of the time-dependent form of Schrödinger's wave equation (not presented here), and is of practical importance only in the atomic world. In the atomic and subatomic world involving transitions between different energy states, energy need not be rigorously conserved during very short time intervals Δt, provided the amount of energy violation ΔE is limited to $\Delta E \simeq h/(2\pi\Delta t)$. This uncertainty principle is an important relation used to estimate the lifetimes of excited nuclear states.

3.4.4 Particle in a Potential Well

To illustrate a solution of the Schrödinger's wave equation, consider the case in which a particle is confined within a one-dimensional potential well of finite depth and width a, as shown in Fig. 3.13. The potential is constant at $U(x) = 0$ inside the well and is constant at a value labeled U_0 for all regions outside the well. A solution is sought for the three regions $x < 0$, $0 < x < a$, and $a < x$. To simplify the analysis, it is common to write Eq. (3.37) as

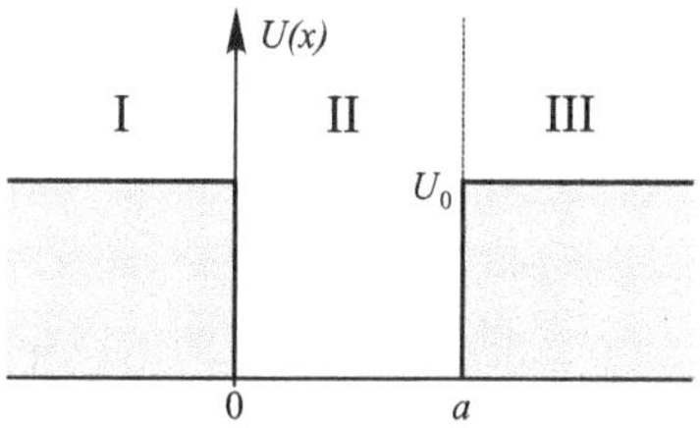

Figure 3.13. Finite square potential well in which the particle energy is less than the well potential.

$$\boxed{\nabla^2\psi(x,y,z) - \frac{2m}{\hbar^2}[U(x,y,z) - E]\psi(x,y,z) = 0,} \tag{3.41}$$

where $\hbar \equiv h/(2\pi)$. Equation (3.41) can then be written for this one-dimensional problem, under the the assumption that $E < U_0$, as

$$\frac{d^2\psi(x)}{dX^2} - k^2\psi(x) = 0 \qquad \text{for } x < 0 \text{ or } x > a, \tag{3.42}$$

and

$$\frac{d^2\psi(x)}{dX^2} + \phi^2\alpha^2\psi(x) = 0 \quad \text{for } 0 < x < a, \tag{3.43}$$

where $k^2 \equiv 2mE/\hbar^2$ and $\phi^2 \equiv 2m(U_0 - E)/\hbar^2$, both positive real numbers. From Eq. (3.42) one infers

$$\psi_1(x) = Ae^{kx} + Be^{-kx} \quad (x < 0) \tag{3.44}$$

and

$$\psi_3(x) = Fe^{kx} + Ge^{-kx} \quad (x > a). \tag{3.45}$$

The negative exponent term in region I and the positive exponent in region III must be excluded because $\psi(x)$ must not become infinite anywhere. Consequently, it is concluded that $B = F = 0$. From Eq. (3.43), one has

$$\psi_2(x) = C\sin(kx) + D\cos(kx), \tag{3.46}$$

the wave function in region II. A condition for the validity of Eq. (3.37) is that $\nabla^2\psi$ must exist and remain finite everywhere. Here this condition requires ψ and $d\psi/dx$ be continuous across all boundaries (and at infinity). From this requirement, values of the arbitrary constants A, G, C, and D can be determined. Beforehand, a special case will be considered.

Particle in a Box

As $U_0 \to \infty$ the parameter $\phi \to \infty$ so that $\psi_1(0)$ and $\psi_3(a)$ must vanish so the only non-zero values of $\psi(x)$ are to be found in region II. In other words, the particle must be confined inside the well. From Eq. (3.46) it is seen that

$$\psi_2(0) = 0 = D \qquad \text{and } \psi_2(x) = C\sin(kx). \tag{3.47}$$

Because $\psi_2(a) = 0 = C\sin(kx)$ to obtain non-trivial solutions the argument ka must assume discrete values (or *eigenvalues*)

$$k = \frac{n\pi}{a}, \quad n = \pm 1,\ \pm 2,\ \pm 3,\ \ldots,. \tag{3.48}$$

so the wavefunctions have the form

$$\psi_n(x) = C\sin\left(\frac{n\pi x}{a}\right), \tag{3.49}$$

and the energy of the particles is restricted to discrete values

$$E_n = \frac{n^2h^2\pi^2}{2m} = \frac{1}{2m}\left(\frac{n\pi\hbar}{a}\right)^2. \tag{3.50}$$

Because the energy of the particle is quantized, so is its momentum. Consequently,

$$E_n = \frac{p_n^2}{2m} = \frac{\hbar^2 k_n^2}{2m} = \frac{n^2\pi^2\hbar^2}{2ma^2}. \tag{3.51}$$

Finally the constant C is determined by the normalization condition

$$\int_{\infty}^{\infty} \psi_n^*(x)\psi_n(x)\, dx = \int_0^a A_n^2 \sin^2\left(\frac{n\pi x}{a}\right) dx = 1. \tag{3.52}$$

Upon evaluation of the integral, it is found $A_n = \sqrt{2/a}$ so the wave functions for a particle in a box are

$$\psi_n(x) = \sqrt{\frac{2}{a}} \sin\left(\frac{n\pi x}{a}\right). \tag{3.53}$$

In Fig. 3.14 the wave function is shown for $n = 3$.

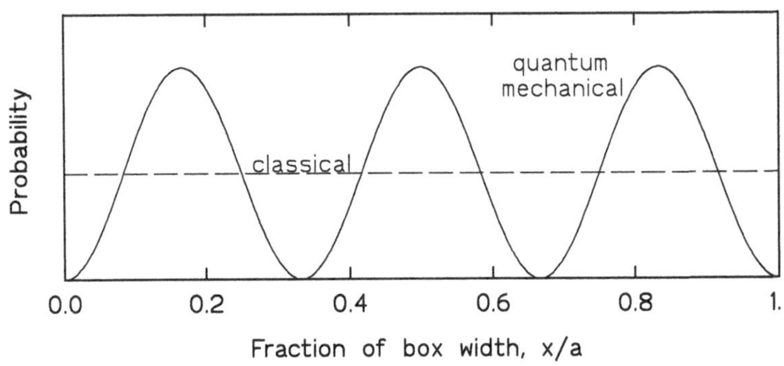

Figure 3.14. The probability distribution for finding a particle in quantum state $n = 3$ with energy E_3 at points in a one-dimensional box.

From this analysis two distinct differences between classical mechanics and wave mechanics are apparent. First, a particle confined to a box with rigid boundaries can have any speed, energy or linear momentum, whereas in wave mechanics these variables have very distinct discrete values. Second, classically the particle can be anywhere in the box with equal probability, whereas in wave mechanics the probability of observing the particle varies with position.

Particle in a Finite Potential Well

Now let us return to a particle in a finite potential well. The most general solution is given by Eqs. (3.44), (3.45), and (3.46), but now the boundary conditions are

$$\psi_1(-\infty) = 0\ ,\ \psi_3(\infty) = 0$$

$$\psi_1(0) = \psi_2(0)\ ,\ \psi_2(a) = \psi_3(a)$$

$$\left.\frac{d\psi_1}{dx}\right|_{x=0} = \left.\frac{d\psi_2}{dx}\right|_{x=0},\ \left.\frac{d\psi_2}{dx}\right|_{x=a} = \left.\frac{d\psi_3}{dx}\right|_{x=a}$$

From the first two boundary conditions, the coefficients B and F must be zero if the wave function disappears at $\pm\infty$. Hence, the wavefunctions reduce to

$$\psi_1(x) = Ae^{\phi x}, \qquad \psi_2(x) = C\sin(kx) + D\cos(kx), \qquad \psi_3(x) = Ge^{-\phi x}. \tag{3.54}$$

From the boundary conditions on $\psi_1(0)$ and $\psi_3(a)$ one finds $A = D$ and $\phi A = kC$ from which it follows that $D = kC/\phi$. Finally, the boundary conditions on the slope of the wave functions give

$$C\sin(ka) + D\cos(ka) = Ge^{-\phi a} \quad \text{and} \quad k(C\cos(ka) - D\sin(ka)) = -\phi Ge^{-\phi a}. \tag{3.55}$$

Substitution of $D = kC/\phi$ into Eqs. (3.55) yields

$$C\sin(ka) + \frac{kC}{\phi}\cos(ka) = Ge^{-\phi a}, \tag{3.56}$$

and

$$\frac{k}{\phi}\left((\frac{kC}{\phi}\sin(ka) - C\cos(ka)\right) = Ge^{-\phi a} \quad . \tag{3.57}$$

Equating these two results and rearranging terms yields

$$C\big[(k^2 - \phi^2)\sin(ka) - 2\phi k\cos(ka)\big] = 0. \tag{3.58}$$

Here it is assumed that the wavefunction ψ and coefficient C are non-zero. Equation (3.58) can then be rewritten as

$$\tan(ka) = \frac{2\phi k}{k^2 - \phi^2}, \tag{3.59}$$

a relation that relates the allowable values of ϕ and k. With the following substitutions,

$$\phi_0 = \sqrt{\frac{2mU_0}{\hbar^2}}, \quad \xi = \frac{E}{U_0}, \quad \phi = \phi_0\sqrt{1-\xi}, \quad k = \phi_0\sqrt{\xi}, \tag{3.60}$$

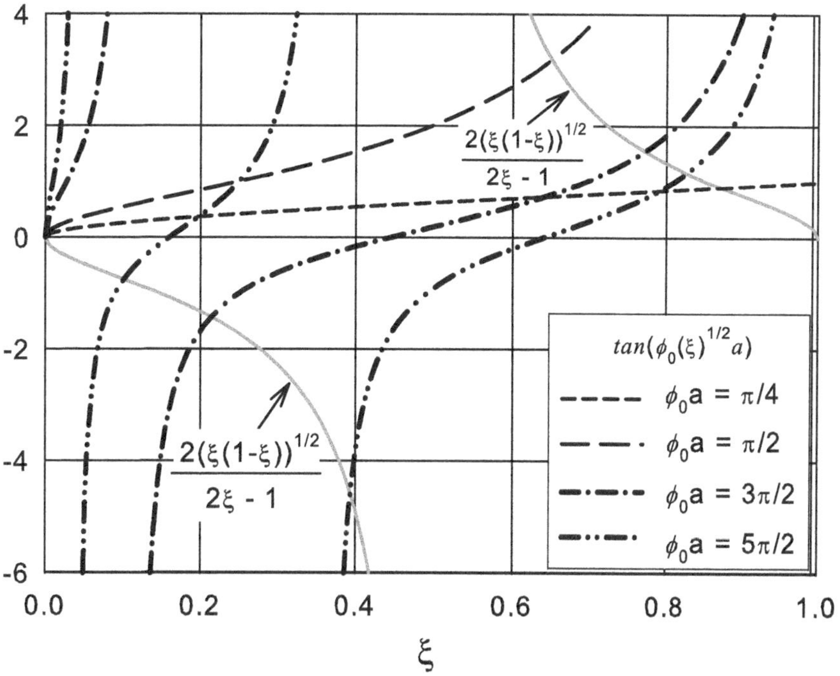

Figure 3.15. Solutions to the finite square well for different values of $\phi_0\alpha$.

Eq. (3.59) is reduced to the following

$$\tan(\phi_0\sqrt{\xi}a) = \frac{2\sqrt{\xi(1-\xi)}}{2\xi - 1} \quad . \tag{3.61}$$

The solutions to Eq. (3.61) can be found by plotting both sides versus ξ and locating the intercepts, the results of which are shown in Fig. 3.15. The allowed energies are found by multiplying the intercept value of ξ by the potential energy U_0. The number of possible solutions, for a given value of $\phi_0 a$ is seen to be

$$\text{If } (n-1)\,\pi \; < \phi_0 a < n\pi \quad \text{there are } n \text{ solutions,} \quad n = 1, 2, 3, \ldots.$$

Each solution defines an allowed energy state in the well. Observe from Fig. 3.15 that more states appear as a is increased. Further, as a is decreased, the general trend is the solution for E (a function of ξ) increases in value. In conclusion, (1) the number of allowed states increases as the well dimension a increases, and (2) the allowed states are pushed upwards towards U_0 as the well dimension a decreases.

3.4.5 The Hydrogen Atom

The hydrogen atom, the simplest atom, can be considered as a system of two interacting point charges, the proton (nucleus) and an electron. The electrostatic attraction between the electron and proton is described by Coulomb's law. The potential energy of a bound electron is given by,

$$U(r) = -\frac{q_e^2}{r}, \tag{3.62}$$

where r is the distance between the electron and the proton.

The three-dimensional Schrödinger wave equation in spherical coordinates for the electron bound to the proton is [Griffiths 2005]

$$\frac{1}{r^2}\frac{\partial}{\partial r}\left(r^2\frac{\partial\psi}{\partial r}\right) + \frac{1}{r^2\sin\theta}\frac{\partial}{\partial\theta}\left(\sin\theta\frac{\partial\psi}{\partial\theta}\right) + \frac{1}{r^2\sin^2\theta}\frac{\partial^2\psi}{\partial\phi^2} + \frac{8\pi^2\mu}{h^2}\left[E - U(r)\right]\psi = 0, \tag{3.63}$$

where $\psi = \psi(r,\theta,\phi)$ and μ is the electron mass (more correctly the reduced mass of the system).

To solve this partial differential equation for the wave function $\psi(r,\theta,\phi)$, first replace it by three equivalent ordinary differential equations involving functions of only a single independent variable. To this end, the "separation of variables" method is used to seek a solution of the form

$$\psi(r,\theta,\phi) = R(r)\Theta(\theta)\Phi(\phi). \tag{3.64}$$

Substitute this form into Eq. (3.63) and multiply the result by $r^2\sin^2\theta/(R\Theta\Phi)$ to obtain

$$\frac{\sin^2\theta}{R(r)}\frac{d}{dr}\left(r^2\frac{dR(r)}{dr}\right) + \frac{1}{\Phi(\phi)}\frac{d^2\Phi(\phi)}{d\phi^2} + \frac{\sin\theta}{\Theta(\theta)}\frac{d}{d\theta}\left(\sin\theta\frac{d\Theta(\theta)}{d\theta}\right) + \frac{8\pi^2\mu}{h^2}r^2\sin^2\theta\left[E - U(r)\right] = 0. \tag{3.65}$$

The second term is only a function of ϕ, while the other terms are independent of ϕ. This term, therefore, must equal a constant, $-m^2$ say. Thus,

$$\frac{d^2\Phi(\phi)}{d\phi^2} = -m^2\Phi(\phi). \tag{3.66}$$

Equation 3.65, upon division by $\sin^2\theta$ and rearrangement becomes

$$\frac{1}{R(r)}\frac{d}{dr}\left(r^2\frac{dR(r)}{dr}\right) + \frac{8\pi^2\mu r^2}{h^2}\left[E - U(r)\right] = -\frac{1}{\sin\theta\Theta(\theta)}\frac{d}{d\theta}\left(\sin\theta\frac{d\Theta(\theta)}{d\theta}\right) + \frac{m^2}{\sin^2\theta} = 0. \tag{3.67}$$

Since the terms on the left are functions only of r and the terms on the right are functions only of θ, both sides of this equation must be equal to the same constant, β say. Thus two ordinary differential equations are obtained, one for $R(r)$ and one for $\Theta(\theta)$, namely,

$$\frac{1}{\sin\theta}\frac{d}{d\theta}\left(\sin\theta\frac{d\Theta(\theta)}{d\theta}\right) - \frac{m^2}{\sin^2\theta}\Theta(\theta) + \beta\Theta(\theta) = 0, \tag{3.68}$$

and

$$\frac{1}{r^2}\frac{d}{dr}\left[r^2\frac{dR(r)}{dr}\right] + \frac{\beta}{r^2}R(r) + \frac{8\pi^2\mu}{h^2}\left[E - U(r)\right]R(r) = 0. \tag{3.69}$$

Thus Eq. (3.63) is reduced to three ordinary, homogeneous, differential equations [Eq. (3.66), Eq. (3.68), and Eq. (3.69)], whose solutions, when combined, give the entire wave function $\psi(r,\theta,\phi) = R(r)\Theta(\theta)\Phi(\phi)$. Each of these equations is an eigenvalue problem which yields "quantum numbers." These quantum numbers are used to describe the possible electron configurations in a hydrogen atom. The solutions also yield relationships between the quantum numbers. The solution details are omitted—see any book on quantum mechanics for the explicit solutions, for example Griffiths [2005]. The present discussion is restricted to the essential features that arise from each equation.

The most general solution of Eq. (3.66) is

$$\Phi(\phi) = A\sin m\phi + B\cos m\phi, \tag{3.70}$$

where A and B are arbitrary. However, it is required $\Phi(0) = \Phi(2\pi)$ since $\phi = 0$ and $\phi = 2\pi$ are the same azimuthal angle. This boundary condition then requires m to be an integer, i.e., $m = 0, \pm1, \pm2, \ldots$. For each *azimuthal* quantum number m, the corresponding solution is denoted by $\Phi_m(\phi)$.

Equation (3.68) in θ has normalizable solutions only if the separation constant has the form $\beta = \ell(\ell+1)$ where ℓ is a positive integer or zero, and is called the *angular momentum* quantum number. Moreover, the azimuthal quantum number m must be restricted to $2\ell+1$ integer values, namely, $m = 0, \pm1, \pm2, \ldots, \pm\ell$. The corresponding solutions of Eq. (3.68) are denoted by $\Theta_{\ell m}(\theta)$ and are known to mathematicians as the associate Legendre functions of the first kind; however, these details are not of concern here.

Finally, the solution of Eq. (3.69) for the radial component of the wave function, with $\beta = \ell(\ell+1)$ and $U(r) = -q_e^2/r$, has normalizable solutions only if the electron's energy has the (eigen)value

$$E_n = -\frac{2\pi^2\mu q_e^2}{h^2n^2}, \quad \text{where } n = 1, 2, 3, \ldots \tag{3.71}$$

The integer n is called the *principal* quantum number. Moreover, to obtain a solution, the angular quantum number ℓ must be no greater than $n-1$, i.e., $\ell = 0, 1, 2, \ldots, (n-1)$. The corresponding radial solution is denoted by $R_{n\ell}(r)$ and mathematically is related to the associated Laguerre function.

Thus, the wave functions for the electron bound in the hydrogen atom have only very discrete forms $\psi_{n\ell m}(r,\theta,\phi) = R_{n\ell}(r)\Theta_{\ell m}(\theta)\Phi_m(\phi)$. For the hydrogen atom, the energy of the electron is given by Eq. (3.71) and is independent of the angular moment or azimuthal quantum numbers m and ℓ (this is not true for multielectron atoms).

Special Notation for Electron States

A widely used, but strange, notation has been of long standing in describing the n and ℓ quantum numbers of particular electron states. The letters s, p, d, f, g, h, and i are used to denote values of the angular momentum quantum number ℓ of 0, 1, 2, 3, 4, 5, and 6, respectively. The value of n is then used as a prefix to the angular moment letter. Thus a bound electron designated as $5f$ refers to an electron with $n = 5$ and $\ell = 3$.

Examples of Wave Functions for Hydrogen

In Fig. 3.16, density plots of $|\psi_{n\ell m}|^2 = \psi^*_{n\ell m}\psi_{n\ell m}$ are shown for different electron states in the hydrogen atom. These plots are slices through the three-dimensional $|\psi|^2$ in a plane perpendicular to the x-axis and through the atom's center. Because $|\psi|^2$ is the probability of finding the electron in a unit volume, the density plots directly show the regions where the electron is most likely to be found. The s states ($\ell = 0$) are spherically symmetric about the nucleus, while all the others have azimuthal and/or polar angle dependence.

Electron Energy Levels in Hydrogen

For hydrogen, the energy of the bound electron is a function of n only (see Eq. (3.71)). Thus, for $n > 1$, the quantum numbers ℓ and m may take various values without changing the electron energy, i.e., the allowed electron configurations fall into sets in which all members of the set have the same energy. Thus, in Fig. 3.16, an electron has the same energy in any of the four $4f$ electron states or the three $4d$ states, even though the distribution of the electron wave function around the nucleus is quite different for each state. Such states with the same electron binding energy are said to be *degenerate.*

The Spin Quantum Number

It was found, first from experiment and later by theory, that each quantum state (specified by values for n, ℓ, and m) can accommodate an electron in either of two spin orientations. The electron, like the proton and neutron, has an inherent angular momentum with a value of $\frac{1}{2}h/(2\pi)$. In a bound state (defined by n, ℓ, and m) the electron can have its spin "up" or "down" with respect to the z-axis used to define angular momentum. Thus, a fourth quantum number $m_s = \pm\frac{1}{2}$ is needed to unambiguously define each possible electron configuration in an atom.

Dirac showed in 1928 that when the Schrödinger wave equation is rewritten to include relativistic effects, the spin quantum number m_s is inherent in the solution along with the quantum numbers n, ℓ, and m, which were also inherent in the wave function solution of the non-relativistic Schrödinger's wave equation [Schiff 1968].

3.4.6 Energy Levels for Multielectron Atoms

The solution of Schrödinger's equation for the hydrogen atom can be obtained analytically, but the solution for a multielectron system cannot. This difficulty arises because of the need to add a repulsive component to the potential energy term to account for the interactions among the electrons. To obtain a solution for a multielectron atom, numerical approximation techniques must be used together with high speed computers. Such a discussion is far beyond the scope of this text.

There is, however, one important result for multielectron atoms that should be described here. The energies of the electronic levels in atoms with more than one electron are functions of both n and ℓ. The energy degeneracy in the angular momentum quantum number disappears. Thus states with the same n but different ℓ values have slightly different energies, and there is a significant reordering of the electron energy levels compared to those in the hydrogen atom.

Electron energy levels with the same value of n and ℓ but different values of m are still degenerate in energy; however, this degeneracy is removed in the presence of a strong external magnetic field (the Zeeman effect). An electron moving about a nucleus creates a magnetic dipole whose strength of interaction with an external magnetic field varies with the quantum numbers ℓ and m.

The Electronic Structure of Atoms

Each electron in an atom can be characterized by its four quantum numbers n, ℓ, m, and m_s. Further, according to Pauli's exclusion principle (1925), no two electrons in an atom can have the same quantum numbers. An assignment of a set of four quantum numbers to each electron in an atom, no sets being alike in all four numbers, then defines a quantum state for an atom as a whole. For ground-state atoms, the

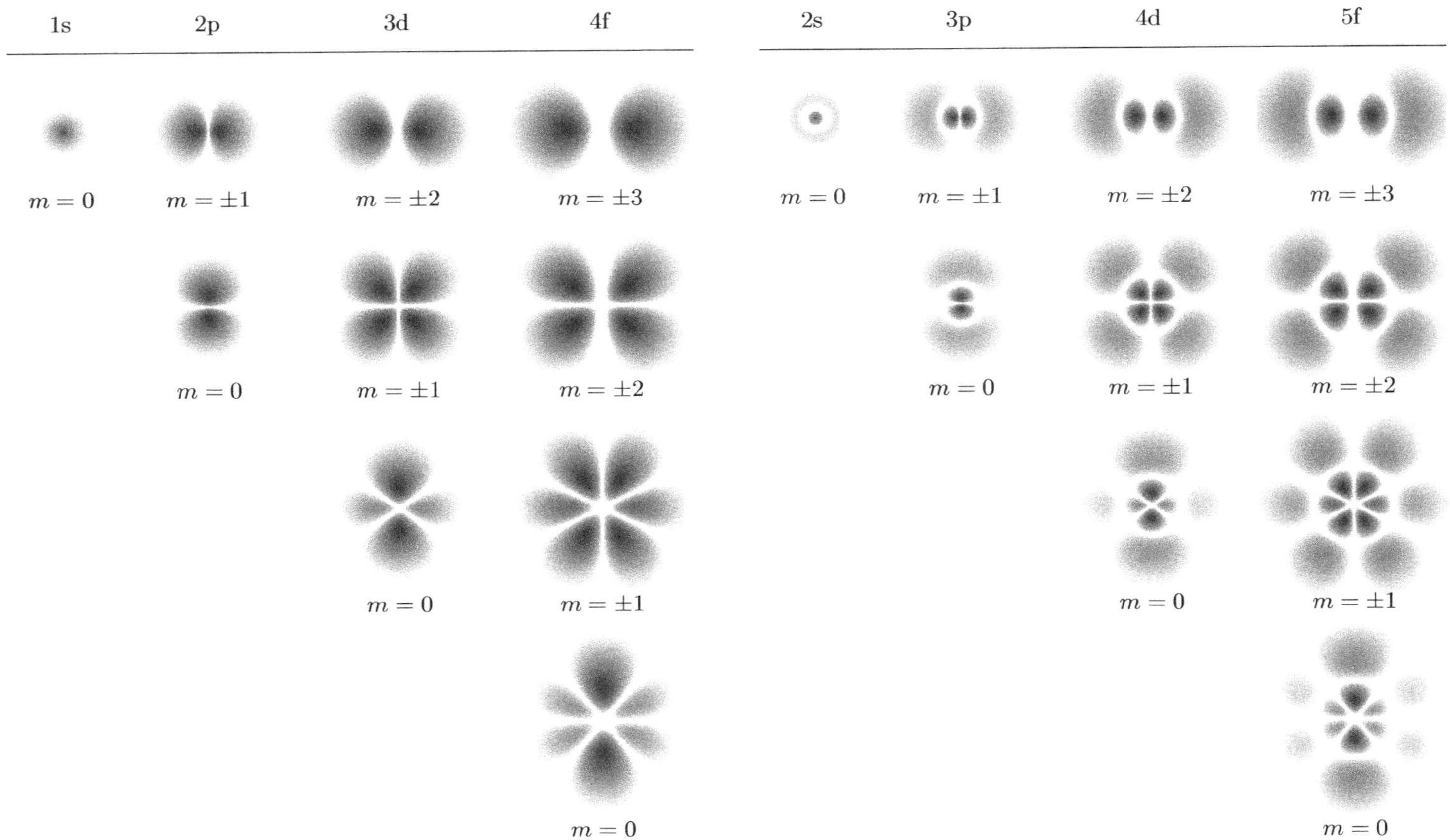

Figure 3.16. Density plots of $\psi^*\psi$ for different electron eigenstates in the hydrogen atom. Plots are sectional views of the probability density in a plane containing the polar axis, which is vertical and in the plane of the paper. Scale of the right group is about 15% bigger than that of the left group. After: R.B. Leighton, *Principles of Modern Physics*, McGraw Hill, New York, 1959.

electrons are in the lowest energy electron states. For an atom in an excited state, one or more electrons are in electron states with energies higher than some vacant states. Electrons in excited states generally drop very rapidly ($\sim 10^{-7}$ s) into vacant lower energy states. During these spontaneous transitions, the difference in energy levels between the two states must be emitted as a photon (*fluorescence* or *x rays*) or be absorbed by other electrons in the atom (thereby causing them to change their energy states).

Crucial to the chemistry of atoms is the arrangement of atomic electrons into various electron *shells*. All electrons with the same n number constitute an *electron shell*. For $n = 1, 2, \ldots, 7$, the shells are designated $K, L, M, \ldots, Q$.

Consider electrons with $n = 1$ (K shell). Since $\ell = 0$ (s state electrons), then $m = 0$ and $m_s = \pm\frac{1}{2}$. Hence, there are only 2 $1s$ electrons, written as $1s^2$, in the K shell. In the L shell ($n = 2$), $\ell = 0$ or 1. For $\ell = 0$, there are two $2s$ electrons (denoted by $2s^2$), and for $\ell = 1$ ($m = -1, 0, 1$), there are six $2p$ electrons (denoted by $2p^6$). Thus in the L shell there are a total of eight electrons ($2s^2 2p^6$). Electrons with the same value of ℓ (and n) are referred to as a *subshell*. For a given subshell, there are $(2\ell + 1)$ m values, each with two m_s values, giving a total of $2(2\ell + 1)$ electrons per subshell, and $2n^2$ electrons per shell. A shell or subshell containing the maximum number of electrons is said to be *closed*.

The Periodic Table of Elements can be described in terms of the possible number of electrons in the various subshells. The number of electrons in an atom equals its atomic number Z and determines its position in the Periodic Table. The chemical properties are determined by the number and arrangement of the electrons. Each element in the table is formed by adding one electron to that of the preceding element in the Periodic Table in such a way that the electron is most tightly bound to the atom. The arrangement of the electrons for the elements with electrons in only the first four shells is shown in Table 3.3.

3.4.7 Success of Quantum Mechanics

Quantum mechanics has been an extremely powerful tool for describing the energy levels and the distributions of atomic electrons around a nucleus. Each energy level and configuration is uniquely defined by four quantum numbers: n the principal quantum number, ℓ the orbital angular momentum quantum number, m_ℓ the z-component of the angular momentum, and $m_s = \pm 1/2$, the electron spin number. These numbers arise naturally from the analytical solution of the wave equation (as modified by Dirac to include special relativity effects) and thus avoid the *ad hoc* introduction of orbital quantum numbers required in earlier atomic models.

Inside the nucleus, quantum mechanics is also thought to govern. However, the nuclear forces holding the neutrons and protons together are much more complicated than the electromagnetic forces binding electrons to the nucleus. Consequently, much work continues in the application of quantum mechanics (and its more general successor quantum electrodynamics) to predicting energy and configuration states of nucleons. Nonetheless, the fact that electronic energy levels of an atom and nuclear excited states are discrete with very specific configurations is a key concept in modern physics. Moreover, when one state changes spontaneously to another state, energy is emitted or absorbed in specific discrete amounts.

3.5 The Fundamental Constituents of Ordinary Matter

Throughout this book, atoms, neutrons, and protons are treated as the only "fundamental" entities of interest. Indeed, for the energies involved in practical nuclear and atomic interactions, this view is well justified. However, at kinetic energies produced in particle accelerators or found in some cosmic rays, the microscopic structure of matter is found to be much more complicated. This fine structure is not of practical importance, however, in understanding radiation detectors; but for completeness, a brief overview of the constituents of ordinary matter is presented here.

Although neutrons and protons are often considered "fundamental" particles, it is now known that they are composed of other smaller particles called *quarks* held together by the exchange of massless particles

Table 3.3. Electron shell arrangement for the lightest elements.

Element	Z	shell and electron designation							
		K	L		M			N	
		1s	2s	2p	3s	3p	3d	4s	4p
H	1	1							
He	2	2							
Li	3	2	1						
Be	4	2	2						
B	5	2	2	1					
C	6	2	2	2					
N	7	2	2	3					
O	8	2	2	4					
F	9	2	2	5					
Ne	10	2	2	6					
Na	11	neon configuration			1				
Mg	12				2				
Al	13				2	1			
Si	14				2	2			
P	15				2	3			
S	16				2	4			
Cl	17				2	5			
Ar	18				2	6			
K	19	argon configuration						1	
Ca	20							2	
Sc	21						1	2	
Ti	22						2	2	
V	23						3	2	
Cr	24						5	1	
Mn	25						5	2	
Fe	26						6	2	
Co	27						7	2	
Ni	28						8	2	
Cu	29						10	1	
Zn	30						10	2	
Ga	31						10	2	1
Ge	32						10	2	2
As	33						10	2	3
Se	34						10	2	4
Br	35						10	2	5
Kr	36						10	2	6

Source: H.D. Bush, *Atomic and Nuclear Physics*, Prentice Hall, Englewood Cliffs, NJ, 1962.

called *gluons*, which are the carriers of the nuclear force. The protons and neutrons are bound together in a nucleus by the residual interactions of the gluons in the neutrons and protons.

From studies of nuclear interactions produced by energetic cosmic rays and by particle accelerators, it is believed that there are six different types or "flavors" of quarks with the fanciful names of up (u), down (d), strange (s), charm (c), top (t), and bottom (b). For each quark there exists an antiquark, denoted by $\bar{u}$, $\bar{d}$, etc. Each quark has a charge that is a fraction of the electron charge, namely the u, d, c, s, t, and b quarks have charge $2/3, -1/3, 2/3, -1/3, 2/3$, and $-1/3$, respectively. An antiquark has the same mass and angular momentum, but opposite charge, as its real quark counterpart.

With the quarks and antiquarks two types of composite particles are formed. *Baryons* are composed of 3 quarks; for example, the neutron is a (ddu) composite (with a net charge of $-1/3 - 1/3 + 2/3 = 0$) and the proton is a (uud) composite (with a net charge of $2/3 + 2/3 - 1/3 = 1$). *Mesons* are composed of a quark-antiquark pair; for example, a π^+ meson is a (u,$\bar{d}$) pair and a π^- meson is a ($\bar{u}$,d) pair.

In addition to the quarks, there is another class of fundamental particles, called *leptons* consisting of the electron, muon, and tau particles, each with an associated neutrino. Each of the leptons also has an antiparticle. The quarks and leptons are exceedingly small with sizes less than about 10^{-19} m and are indeed fundamental particles, i.e., not composite particles.

"Ordinary" matter is mostly composed of neutrons, protons, electrons, and neutrinos, each of which has an angular momentum of $\pm h/(2\pi)$ (or a "spin" of 1/2) and, as such, are called *fermions*. Most of the possible composite baryons and mesons are extremely unstable, with the heavier quarks decaying rapidly to less massive ones by the forces mitigated by *bosons*, such as the photon or $W^\pm$ boson, all of which have spin = 1. For example, a free neutron can decay to a proton when a d quark changes to an u quark by emitting a W^- boson, which almost instantly decays to an electron (e^-) and an electron antineutrino ($\overline{\nu}_e$). The known "fundamental" particles with their mass and charge are shown in Table 3.4. Whether there are additional fundamental particles or even whether these particles are themselves composite particles made from even more complex structures are unknown and are presently topics of great interest to high energy physicists.

However, in this work the electron, neutron, and proton are treated as fundamental indivisible particles, because the composite nature of nucleons becomes apparent only under extreme conditions, such as those encountered during the first second after the creation of the universe (the "big bang") or in high energy particle accelerators. Such gigantic energies are not addressed in this book. Rather, the energy of radiation considered in this work is sufficient only to rearrange or remove the electrons in an atom or the neutrons and protons in a nucleus.

3.5.1 Dark Matter and Energy

In our universe all *ordinary* matter is composed of electrons, neutrinos, and up and down quarks which are bound together in protons and neutrons. However, over the past several decades, it has become apparent that the gravitational forces from ordinary matter are insufficient to explain (1) the motion of stars near the center of galaxies, (2) the filigreed distribution of galaxies in the universe, and (3) the accelerating expansion of the universe itself. To explain these observations, astrophysicists have come to accept that there must be vast amounts of *dark matter* to provide extra gravitational forces on stars in a galaxy and between galaxies. Further, the amount of dark matter must be about five times that of ordinary matter. Although dark matter can exert gravitational forces, it does not interact with light and, hence, is invisible to us. The fundamental particles of dark matter are unknown and are the subject of much current speculation.

Astrophysicists have also had to postulate the existence of something called *dark energy* that must pervade the universe in order to explain the observed accelerated expansion of our universe. Again, what constitutes dark energy is unknown, but the amount of dark energy represents about 70% of the total mass/energy of our universe!

Table 3.4. The fundamental particles that make up "ordinary" matter. Each fermion has a corresponding antiparticle with the same mass but of opposite charge. Data are from Yao et al. [2006]

Fermions (spin = 1/2)					
quarks			**leptons**		
flavor	mass (GeV/c^2)	charge (q_e)	type	mass (GeV/c^2)	charge (q_e)
u	0.0015–0.0030	2/3	ν_e	$< 2 \times 10^{-9}$	0
d	0.003–0.007	−1/3	e^-	0.00051100	−1
c	1.25	2/3	ν_μ	< 0.00019	0
s	0.095 ± 0.025	−1/3	μ^-	0.10566	−1
t	176	2/3	ν_τ	< 0.018	0
b	4.70	−1/3	τ^-	1.7769	−1

Interaction Bosons (spin = 1)			
symbol	force	mass (GeV/c^2)	charge (q_e)
gluon g	strong	0	0
γ	electromagnetic	0	0
W^-	weak	80.4	−1
W^+	weak	80.4	1
Z^0	weak	91.187	0
Higgs H	explains mass	125	0

3.5.2 Atomic and Nuclear Nomenclature

The identity of an atom is uniquely specified by the number of neutrons N and protons Z in its nucleus. For an electrically neutral atom, the number of electrons equals the number of protons Z, which is called the *atomic number.* All atoms of the same *element* have the same atomic number. Thus, all oxygen atoms have 8 protons in the nucleus while all uranium atoms have 92 protons. The nucleus of an atom with a unique N and A numbers is often referred to as a *nuclide.* There are about 3200 known nuclides.

However, atoms of the same element may have different numbers of neutrons in the nucleus. Atoms of the same element, but with different numbers of neutrons, are called *isotopes.* The symbol used to denote a particular isotope is

$$\boxed{{}^A_Z\mathrm{X}}$$

where X is the chemical symbol and $A \equiv Z + N$, which is called the *mass number.* For example, two uranium isotopes are ${}^{235}_{92}U$ and ${}^{238}_{92}U$. The use of both Z and X is redundant because one specifies the other. Consequently, the subscript Z is often omitted, so that one may write, for example, simply ${}^{235}U$ and ${}^{238}U$.[9]

Because isotopes of the same element have the same number and arrangement of electrons around the nucleus, the chemical properties of such isotopes are nearly identical. Only for the lightest isotopes (e.g., hydrogen ${}^1H$, deuterium ${}^2H$, and tritium ${}^3H$) are small differences observed. For example, light water ${}^1H_2O$

[9]To avoid superscripts, which were hard to make on old-fashioned typewriters, the simpler form U-235 and U-238 was often employed. However, with modern word processing, this form should no longer be used.

freezes at 0°C, while heavy water ^2H_2O (or D_2O since deuterium is often given the chemical symbol D) freezes at 3.82°C.

3.5.3 Relative Atomic Masses

The *relative atomic mass* $\mathcal{A}$ of any particular isotope, a dimensionless physical quantity, is the ratio of the mass of one atom of the isotope to 1/12 the mass of an atom of ^{12}C. It is equivalent to the older term *atomic weight*.[10]

The relative atomic mass of an element (typically composed of several isotopes) is slightly more involved. The relative atomic mass of an element is ratio of the *average* mass of atoms in a sample of the element to 1/12 the mass of an atom of ^{12}C. Thus the elemental relative atomic mass can vary depending on the source of the sample. For example, elemental samples from different geological, biological, or extraterrestrial sources often have different relative atomic masses because of their different isotopic compositions. In a similar manner, the relative molecular weight is the average mass of molecules in a sample of the molecules to 1/12 the mass of an atom of ^{12}C.

Most naturally occurring elements are composed of two or more isotopes. The *isotopic abundance* γ_i of the i-th isotope in a given element is the fraction of the atoms in the element that are that isotope. As with the elemental relative atomic mass, the isotopic abundances of a sample depends on the origins of the sample. For a specified element, the *elemental atomic weight* is the weighted average of the atomic weights of all naturally occurring isotopes of the element, weighted by the isotopic abundance of each isotope, i.e.,

$$\mathcal{A} = \sum_i \frac{\gamma_i(\%)}{100} \mathcal{A}_i, \tag{3.72}$$

where the summation is over all the isotopic species comprising the element.

Both the terms relative atomic mass and atomic weight are often loosely used to refer to a *standard* atomic weight. The elemental standard atomic weight is the average mass of the atoms in a *normal* sample of the element to 1/12 the mass of an atom of ^{12}C. A normal sample is one from any reasonable source of the element or its compounds used in commerce for industry and one that has not undergone significant isotopic modification. These values are widely published in books and on wall charts of the Periodic Table.

3.5.4 The Atomic Mass Unit

Closely related to the concept of the relative atomic mass is the *unified atomic mass unit*,[11] abbreviated as u, defined such that the mass of a neutral atom of ^{12}C atom is 12 u. Equivalently, the mass of Avogadro's number of ^{12}C atoms (Avogadro's number = 1 mole) is 0.012 kg. Thus, the SI unit for 1 u equals $(1/12)(0.012 \text{ kg}/N_a) = 1.660\,538\,7 \times 10^{-27}$ kg. An alternative and useful value of 1 u is $931.494\,043/c^2$ MeV. Masses of the different atomic nuclides are available in many compilations, e.g., Shultis and Faw [2008]. It then follows that the mass M of an atom measured in atomic mass units numerically equals the atom's relative atomic mass $\mathcal{A}$. The atomic weight of a nuclide almost equals (within less than one percent) the atomic mass number A of the nuclide. Thus for approximate calculations, one can usually assume $\mathcal{A} \simeq A$.

[10]The continued use of "atomic weight" and "relative atomic mass" has caused much controversy since the 1960s, primarily because of the distinction in physics between "weight" and "mass."

[11]Previously called the *atomic mass unit* and abbreviated *amu*, it was based on ^{16}O rather than ^{12}C. This older metric and terminology are obsolete. In terms of today, *unified atomic mass unit* and *atomic mass unit* are often casually interchanged, but both refer to 'u' based on ^{12}C.

3.5.5 Avogadro's Number

Avogadro's constant is the key to the atomic world since it relates the number of microscopic entities in a sample to a macroscopic measure of the sample. Specifically, Avogadro's constant $N_a \simeq 6.022 \times 10^{23}$ equals the number of atoms in 12 grams of ^{12}C. Few fundamental constants need be memorized, but an approximate value of Avogadro's constant should be.

The importance of Avogadro's constant lies in the concept of the mole. A *mole* (abbreviated mol) of a substance is defined to contain as many "elementary particles" as there are atoms in 12 g of ^{12}C. In older texts, the mole was often called a "gram-mole" but is now called simply a mole. The "elementary particles" can refer to any identifiable unit that can be unambiguously counted. One can, for example, speak of a mole of stars, persons, molecules, or atoms.

Because the atomic weight of ^{12}C is defined to be 12 and because 12 g of ^{12}C is defined to contain 1 mol of atoms, it follows that the mass *in grams* of any atomic species that numerically equals the dimensionless atomic weight of the species must also contain 1 mole of atoms. The mass in grams of a substance that equals the dimensionless atomic or molecular weight is sometimes called the *gram atomic weight* or *gram molecular weight*. Thus, one gram atomic or molecular weight of any substance represents one mole of the substance and contains as many atoms or molecules as there are atoms in one mole of ^{12}C, namely, N_a atoms or molecules. That one mole of any substance contains N_a entities is known as *Avogadro's law* and is the fundamental principle that relates the microscopic world to the everyday macroscopic world. Table 3.5 illustrates how Avogadro's number N_a, together with the atomic weight $\mathcal{A}$, is the key to transforming between macroscopic masses and the number of atoms.

Table 3.5. Use of Avogadro's constant N_a (atoms/mol) and the atomic weight $\mathcal{A}$ (g/mol) to move between the atomic and macroscopic worlds.

Macroscopic World	Transformation $\overset{\longleftarrow}{\longrightarrow}$			**Microscopic World**
m (g)	$\xrightarrow{\div\mathcal{A}}$	$\dfrac{m}{\mathcal{A}}$ (mols)	$\xrightarrow{\times N_a}$	$\dfrac{m}{\mathcal{A}} N_a$ (atoms)
$m = \dfrac{N}{N_a}\mathcal{A}$ (g)	$\xleftarrow{\times\mathcal{A}}$	$\dfrac{N}{N_a}$ (mols)	$\xleftarrow{\div N_a}$	N (atoms)
$\rho \left(\dfrac{\text{g}}{\text{cm}^3}\right)$	$\xrightarrow{\div\mathcal{A}}$	$\dfrac{\rho}{\mathcal{A}} \left(\dfrac{\text{mol}}{\text{cm}^3}\right)$	$\xrightarrow{\times N_a}$	$\dfrac{\rho N_a}{\mathcal{A}} \left(\dfrac{\text{atoms}}{\text{cm}^3}\right)$
$m = \dfrac{\mathcal{A}}{N_a} \left(\dfrac{\text{g}}{\text{atom}}\right)$	$\xleftarrow{\times\mathcal{A}}$	$\dfrac{1}{N_a}$ (mol)	$\xleftarrow{\div N_a}$	1 (atom)

3.5.6 Mass of an Atom

With Avogadro's number, many basic properties of atoms can be inferred. For example, the mass of an individual atom can be found. Since a mole of a group of identical atoms (with a mass of $\mathcal{A}$ grams) contains N_a atoms, the mass of an individual atom is

$$M\,(\text{g/atom}) = \mathcal{A}/N_a \simeq A/N_a. \tag{3.73}$$

The approximation of $\mathcal{A}$ by A is usually quite acceptable for all but the most precise calculations.

3.5.7 Number Density of Atoms and Isotopes

Avogadro's number also determines the atom density (number of atoms per cm^3) of a substance. For a homogeneous substance of a single species and with mass density ρ g/cm^3, 1 cm^3 contains $\rho/\mathcal{A}$ moles of the substance and $\rho N_a/\mathcal{A}$ atoms. The atom density N is thus

$$\boxed{N\ (\text{atoms/cm}^3) = \frac{\rho}{\mathcal{A}} N_a.} \tag{3.74}$$

In radiation detection the number density of a particular isotope of an element is often of interest. To find the atom density N_i of isotope i of an element with atom density N, simply multiply N by the fractional isotopic abundance $\gamma_i/100$ for the isotope, i.e., $N_i = (\gamma_i/100)N$.

The same manner the atom density can be found of a particular element contained in a homogeneous substance composed entirely of a certain molecule. In this case, N in Eq. (3.74) is the molecular density and $\mathcal{A}$ the gram molecular weight. The number of atoms of a particular type, per unit volume, is found by multiplying the molecular density by the number of the atoms of interest per molecule.

The composition of a mixture such as air is often specified by the mass fraction w_i of each constituent. If the mixture has a mass density ρ, the mass density of the ith constituent is $\rho_i = w_i\rho$. The density N_i of the ith component is thus

$$N_i = \frac{\rho_i N_a}{\mathcal{A}_i} = \frac{w_i \rho N_a}{\mathcal{A}_i}. \tag{3.75}$$

If the composition of a substance is specified by a chemical formula, such as X_nY_m, the molecular weight of the mixture is $\mathcal{A} = n\mathcal{A}_X + m\mathcal{A}_Y$ and the mass fraction of component X is

$$w_X = \frac{n\mathcal{A}_X}{n\mathcal{A}_X + m\mathcal{A}_Y}. \tag{3.76}$$

Finally, as a general rule of thumb, it should be remembered that atom densities in solids and liquids are usually between 10^{21} and 10^{23} atoms cm^{-3}. Gases at standard temperature and pressure are typically less by a factor of 1000.

3.5.8 Size of an Atom

For a substance with an atom density of N atoms/cm^3, each atom has an associated volume of $V = 1/N$ cm^3. If this volume is considered a cube, the cube's width is $V^{1/3}$. For ^{238}U, the cubical size of an atom is thus $1/N^{1/3} = 2.7 \times 10^{-8}$ cm. Measurements of the size of atoms reveal a diffuse electron cloud about the nucleus. Although there is no sharp edge to an atom, an effective radius can be defined such that outside this radius, an electron is very unlikely to be found. Almost all atoms with $Z > 10$ have radii between 1×10^{-8} and 2×10^{-8} cm. As Z increases, i.e., as more electrons and protons are added, the size of the electron cloud changes little, but simply becomes more dense. Hydrogen, the lightest element, is also the smallest with a radius of about 0.5×10^{-8} cm.

3.5.9 Nuclear Dimensions

Size of a Nucleus

If each proton and neutron in the nucleus has the same volume, the volume of a nucleus should be proportional to A. This has been confirmed by many measurements that have explored the shape and size of nuclei. Nuclei, to a first approximation, are spherical or very slightly ellipsoidal with a somewhat diffuse surface. In particular, it is found that an effective spherical nuclear radius is

$$R = R_o A^{1/3}, \qquad \text{with } R_o \simeq 1.25 \times 10^{-13} \text{ cm}. \tag{3.77}$$

The associated volume is

$$V_{\text{nucleus}} = \frac{4}{3}\pi R^3 \simeq 7.25 \times 10^{-39} A \quad \text{cm}^3. \tag{3.78}$$

Since the atomic radius of about 2×10^{-8} cm is 10^5 times greater than the nuclear radius, the nucleus occupies only about 10^{-15} of the volume of an atom. If an atom were to be scaled to the size of a large concert hall, then the nucleus would be the size of a very small gnat!

Nuclear Density

Since the mass of a nucleon (neutron or proton) is much greater than the mass of electrons in an atom ($m_n \simeq 1837\ m_e$), the mass density of a nucleus is

$$\rho_{\text{nucleus}} = \frac{m_{\text{nucleus}}}{V_{\text{nucleus}}} = \frac{A/N_a}{(4/3)\pi R^3} = 2.4 \times 10^{14}\ \text{g/cm}^3.$$

This is the density of the earth if it were compressed to a ball 200 m in radius.

3.6 Nuclear Reactions

Nuclear reactions play a very important role in many areas of nuclear science and engineering, because it is through such reactions that various types of radiation are produced. The detection of such radiation is the focus of this book. There are two main categories of nuclear reactions.

In the first category, the initial reactant X is a single atom or nucleus that spontaneously changes by emitting one or more particles, i.e.,

$$X \longrightarrow Y_1 + Y_2 + \cdots.$$

Such a reaction is called *radioactive decay.* Of the over 3200 known nuclides (an atom whose nucleus has a unique number neutron and protons), all but 266 are radioactive. The variation in the number of radioactive atoms over time and the different types of radioactive decay are addressed in the next section. Radioactive decay is a particular type of nuclear reaction that, because it occurs spontaneously, must necessarily be exothermic; i.e., mass must decrease in the decay process and energy must be emitted, usually in the form of the kinetic energy of the reaction products.

The second broad category of nuclear reactions are *binary reactions* in which *two* nuclear particles (nucleons, nuclei or photons) interact to form different nuclear particles.

$$X + x \longrightarrow Y_1 + Y_2 + \cdots.$$

The most common types of such nuclear reactions are those in which some nucleon or nucleus x moves with some kinetic energy and strikes and interacts with a nucleus X to form a pair of product nuclei y and Y, i.e.,

$$x + X \longrightarrow Y + y.$$

As a shorthand notation, such a reaction is often written as $X(x, y)Y$ where x and y are usually the lightest of the reaction pairs.

3.6.1 Q-Value for a Reaction

In any nuclear reaction, energy must be conserved, i.e., the total energy, including rest-mass energy, of the initial particles must equal the total energy of the final products, i.e.,

$$\sum_i [E_i + m_i c^2] = \sum_i [E_i' + m_i' c^2] \tag{3.79}$$

where E_i (E_i') is the kinetic energy of the ith initial (final) particle with a rest mass m_i (m_i').

Any change in the total kinetic energy of particles before and after the reaction must be accompanied by an equivalent change in the total rest mass of the particles before and after the reaction. To quantify this change in kinetic energy or rest-mass change in a reaction, a so-called Q-value is defined as

$$Q = (\text{KE of final particles}) - (\text{KE of initial particles}). \tag{3.80}$$

Thus, the Q-value quantifies the amount of kinetic energy gained in a reaction. Equivalently, from Eq. (3.79), this gain in kinetic energy must come from a decrease in the rest mass, i.e.,

$$Q = (\text{rest mass of initial particles})c^2 - (\text{rest mass of final particles})c^2. \tag{3.81}$$

The Q-value of a nuclear reaction may be either positive or negative. If the rest masses of the reactants exceed the rest masses of the products, the Q-value of the reaction is positive with the decrease in rest mass being converted into a gain in kinetic energy. Such a reaction is exothermic.

Conversely, if Q is negative, the reaction is endothermic. For this case, kinetic energy of the initial particles is converted into rest-mass energy of the reaction products. The kinetic energy decrease equals the rest-mass energy increase. Such reactions cannot occur unless the colliding particles have at least a certain amount of kinetic energy.

Binary Reactions

For the binary reaction $x + X \rightarrow Y + y$, the Q-value is given by

$$Q = (E_y + E_Y) - (E_x + E_X) = [(m_x + m_X) - (m_y + m_Y)]c^2. \tag{3.82}$$

For most binary nuclear reactions, the number of protons is conserved so that the same number of electron masses may be added to both sides of the reactions and, neglecting differences in electron binding energies, the Q-value can be written in terms of atomic masses as

$$Q = (E_y + E_Y) - (E_x + E_X) = [(M_x + M_X) - (M_y + M_Y)]c^2. \tag{3.83}$$

Radioactive Decay Reactions

For a radioactive decay reaction $X \rightarrow Y + y$, there is no particle x and the nucleus of X generally is at rest so $E_X = 0$. In this case

$$Q = (E_y + E_Y) = [m_X - (m_y + m_Y)]c^2 > 0. \tag{3.84}$$

Thus radioactive decay is always exothermic and the mass of the parent nucleus must always be greater than the sum of the product masses. In some types of radioactive decay (e.g., beta decay and electron capture), the number of protons is *not* conserved and care must be exercised in expressing the nuclear masses in terms of atomic masses. This nuance is discussed in Section 3.6.3.

3.6.2 Conservation of Charge and the Calculation of Q-Values

In all nuclear reactions, total charge must be conserved. Sometimes this is not clear when writing the reaction and subtle errors can be made when calculating the Q-value from Eq. (3.81). Consider, for example, the reaction ${}^{16}_{8}\text{O}(n,p){}^{16}_{7}\text{N}$ or

$${}^{1}_{0}\text{n} + {}^{16}_{8}\text{O} \longrightarrow {}^{16}_{7}\text{N} + {}^{1}_{1}\text{p}. \tag{3.85}$$

One might be tempted to calculate the Q-value as

$$Q = \{m_n + M(^{16}_{8}\text{O}) - M(^{16}_{7}\text{N}) - m_p\}c^2.$$

However, this is incorrect because the number of electrons is not conserved on both sides of the reaction. On the left side there are 8 electrons around the $^{16}_{8}$O nucleus, while on the right there are only 7 electrons in the product atom $^{16}_{7}$N. In reality, when the proton is ejected from the $^{16}_{8}$O nucleus by the incident neutron, an orbital electron is also lost from the resulting $^{16}_{7}$N nucleus. Thus, the reaction of Eq. (3.85), to conserve charge, should be written as

$$^{1}_{0}\text{n} + {}^{16}_{8}\text{O} \longrightarrow {}^{16}_{7}\text{N} + {}^{\ 0}_{-1}\text{e} + {}^{1}_{1}\text{p}. \tag{3.86}$$

The released electron can now be conceptually combined with the proton to form a neutral atom of $^{1}_{1}$H, the electron binding energy to the proton being negligible compared to the reaction energy. Thus, the reaction can be approximated by

$$^{1}_{0}\text{n} + {}^{16}_{8}\text{O} \longrightarrow {}^{16}_{7}\text{N} + {}^{1}_{1}\text{H} \tag{3.87}$$

and its Q-value is calculated as

$$Q = \{m_n + M(^{16}_{8}\text{O}) - M(^{16}_{7}\text{N}) - M(^{1}_{1}\text{H})\}c^2. \tag{3.88}$$

In any nuclear reaction, *in which the numbers of neutrons and protons are both conserved*, the Q-value is calculated by replacing any charged particle by its neutral-atom counterpart. Recall that 1 u is $931.494\,043/c^2$ MeV, a value that readily allows reaction Q-values to be calculated in units of MeV when the atomic masses are in atomic mass units.

3.6.3 Special Case for Changes in the Proton Number

The above procedure for calculating Q-values in terms of neutral-atom masses *cannot* be used when the number of protons and neutrons are not conserved, and a more careful analysis is required. Such reactions involve the so-called weak force that is responsible for changing neutrons into protons or vice versa. These reactions are recognized by the presence of a neutrino ν (or antineutrino $\overline{\nu}$) as one of the reactants or products. In such reactions, the conceptual addition of electrons to both sides of the reaction to form neutral atoms often results in the appearance or disappearance of an extra free electron.

3.7 Radioactivity

Radioactive nuclei and their radiations have properties that are the basis of many of the ideas and techniques of atomic and nuclear physics [Shultis and Faw 2008]. The spontaneous emission of alpha and beta particles led to the concept that atoms are composed of smaller fundamental units. Likewise, the scattering of alpha particles led to the idea of the nucleus, which is fundamental to the models used in atomic physics. The discovery of isotopes resulted from the analysis of the chemical relationships among the various radioactive elements. The bombardment of the nucleus by alpha particles caused the disintegration of nuclei and led to the discovery of the neutron and the present model for the composition of the nucleus.

The discovery of artificial, or induced, radioactivity started a new line of nuclear research and hundreds of artificial nuclei have been produced by many different nuclear reactions. The investigation of the emitted radiations from radionuclides has shown the existence of nuclear energy levels similar to the electronic energy levels. The identification and the classification of these levels are important sources of information about the structure of the nucleus.

A number of radioactive nuclides occur naturally on the earth. One of the most important is $^{40}_{19}$K, which has an isotopic abundance of 0.0118% and a half-life of 1.28×10^9 y. Potassium is an essential element needed by plants and animals, and is an important source of human internal and external radiation exposure. Other naturally occurring radionuclides are of cosmogenic origin. Tritium ($^3_1$H) and $^{14}_6$C are produced by cosmic ray interactions in the upper atmosphere, and also can cause measurable human exposures. $^{14}_6$C (half-life 5730 y), which is the result of a neutron reaction with $^{14}_7$N in the atmosphere, is incorporated into plants by photosynthesis. By measuring the decay of ^{14}C in ancient plant material, the age of the material can be determined.

Other sources of terrestrial radiation are uranium, thorium, and their radioactive progeny. All elements with Z $>$ 83 are radioactive. Uranium and thorium decay into daughter radionuclides, forming a series (or chain) of radionuclides that ends with a stable isotope of lead or bismuth. There are four naturally occurring radioactive decay series.

In all nuclear interactions, including radioactive decay, there are several quantities that are always conserved or unchanged by the nuclear transmutation. The most important of these conservation laws include:

- Conservation of charge, i.e., the number of elementary positive and negative charges in the reactants must be the same as in the products.
- Conservation of the number of nucleons, i.e., A is always constant. With the exception of EC and $\beta^{\pm}$ radioactive decay, in which a neutron (proton) transmutes into a proton (neutron), the number of protons and neutrons is also generally conserved.
- Conservation of mass/energy (total energy). Although, neither rest mass nor kinetic energy is generally conserved, the total (rest-mass energy equivalent plus kinetic energy) is conserved.
- Conservation of linear momentum. This quantity must be conserved in all inertial frames of reference.
- Conservation of angular momentum. The total angular momentum (or the *spin*) of the reacting particles must always be conserved.

3.7.1 Types of Radioactive Decay

There are several types of spontaneous changes (or *transmutations*) that can occur in radioactive nuclides. In each transmutation, the nucleus of the parent atom ^A_ZP is altered in some manner and one or more particles are emitted. If the number of protons in the nucleus is changed, then the number of orbital electrons in the daughter atom D must subsequently also be changed, either by releasing an electron to or absorbing an electron from the ambient medium. The most common types of radioactive decay are summarized in Table 3.6.

3.7.2 Radioactive Decay Diagrams

Detailed descriptions of radioactive decay involve a considerable amount of information. One convenient method for presenting such data is through a radioactive decay diagram. In Fig. 3.17 the hypothetical decay of a radionuclide ^A_ZX is shown.

In a radioactive decay diagram, also sometimes called a *nuclear-level diagram*, the y-axis represents the masses (or energy equivalents) of the nuclides involved and the x-axis shows the atomic number of the nuclides. At the highest energy is the radionuclide of interest, the parent, whose ground state energy is denoted by the bold horizontal line. The radioactive parent may decay in several ways to produce different daughter nuclides. Decay that leads to daughters with lower atomic number than that of the parent have daughter states shown to the left of the parent, with thin horizontal lines denoting excited states and a bold line for the ground state. Daughters with increased atomic numbers are shown to the right of the parent.

Because all radioactive decays are exoergic, the mass of the daughter states must always be less than the mass of the parent. Thus, the various decay modes (shown by descending arrows) always proceed from the parent nuclide at the top of the diagram to daughter states below the parent, i.e., to states with less mass.

Table 3.6. Summary of important types of radioactive decay. The parent atom is denoted as P and the product or daughter atom by D.

Decay Type	Reaction	Description
gamma (γ)	${}^{A}_{Z}\mathrm{P}^* \longrightarrow {}^{A}_{Z}\mathrm{P} + \gamma$	An excited nucleus decays to its ground state by the emission of a gamma photon.
alpha (α)	${}^{A}_{Z}\mathrm{P} \longrightarrow {}^{A-4}_{Z-2}\mathrm{D} + \alpha$	An α particle is emitted leaving the daughter with 2 fewer neutrons and 2 fewer protons than the parent.
negatron (β^-)	${}^{A}_{Z}\mathrm{P} \longrightarrow {}_{Z+1}^{A}\mathrm{D}+\beta^-+\overline{\nu}$	A neutron in the nucleus changes to a proton. An electron (β^-) and an antineutrino ($\overline{\nu}$) are emitted.
positron (β^+)	${}^{A}_{Z}\mathrm{P} \longrightarrow {}_{Z-1}^{A}\mathrm{D}+\beta^++\nu$	A proton in the nucleus changes into a neutron. A positron (β^+) and a neutrino (ν) are emitted.
electron capture (EC)	${}^{A}_{Z}\mathrm{P}+\mathrm{e}^- \longrightarrow {}_{Z-1}^{A}\mathrm{D}^*+\nu$	An orbital electron is absorbed by the nucleus, converts a nuclear proton into a neutron and a neutrino (ν), and, generally, leaves the nucleus in an excited state.
proton (p)	${}^{A}_{Z}\mathrm{P} \longrightarrow {}^{A-1}_{Z-1}\mathrm{D} + p$	A nuclear proton is ejected from the nucleus.
neutron (n)	${}^{A}_{Z}\mathrm{P} \longrightarrow {}^{A-1}{}_{Z}\mathrm{D} + n$	A nuclear neutron is ejected from the nucleus.
internal conversion (IC)	${}^{A}_{Z}\mathrm{P}^* \longrightarrow [{}^{A}_{Z}\mathrm{P}]^+ + \mathrm{e}^-$	The excitation energy of a nucleus is used to eject an orbital electron (usually a K-shell electron).

The vertical energy/mass axis is on a relative scale for each daughter with zero energy being the ground state of each daughter. The energy of each excited state is written on or adjacent to the horizontal line representing the state. For an excited state with energy E_i above the ground state, the excited nucleus has an excess mass $\Delta m_i = E_i/c^2$ greater than the mass of the nucleus in its ground state. Also often given is the Q-value for the decay to the ground state. The Q-value for decay to an excited state is the Q-value to the ground state less the nuclear excitation energy.

In the decay diagram of Fig. 3.17, the parent nucleus ${}^{A}_{Z}\mathrm{X}$ can decay to four different daughters: by β^- decay to ${}_{Z+1}^{A}\mathrm{B}$, by neutron decay to ${}^{A-1}{}_{Z}\mathrm{X}$, by positron decay or electron capture to ${}_{Z-1}^{A}\mathrm{C}$, and by α decay to ${}^{A-4}_{Z-2}\mathrm{D}$. For the alpha decay, note the break in the vertical line descending from the parent to indicate the parent and daughters are on different energy/mass scales. Without such a break, the alpha decay daughter would be far (almost 4 GeV) below the other decay daughters. Each decay arrow to a ground or excited state is labeled by the type of decay, e.g., α_1, and the decay probability or *frequency*, e.g., $f(\alpha_1)$. The frequency f_i, usually expressed in percent, gives the probability of that decay mode happening *per decay of the parent.* Thus, all the decay frequencies should sum to 100%.

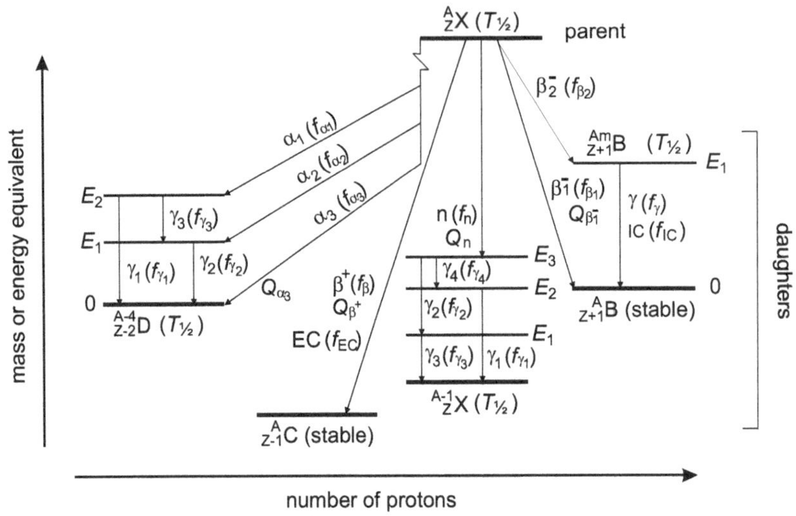

Figure 3.17. A radioactive decay diagram showing four decay modes of the hypothetical parent nuclide ^A_ZX in which four different daughter nuclides are produced.

Excited states of the daughter nucleus usually decay very rapidly to lower excited states or to the ground state either by internal conversion, whereby the excitation energy is transferred to an orbital electron, or by the emission of a photon called a *gamma ray.*[12] The gamma ray energy is essentially equal to the difference between the two energy states, because the kinetic energy of the recoiling atom is negligible. Thus, the unique energies of gamma rays emitted in radioactive decay can be found readily from the energy levels in the decay diagram.

Occasionally, an excited state can persist for an appreciable time. Such excited states are called *metastable* and labeled as such with its half-life (see Section 3.7.5). For example, in Fig. 3.17, the first excited state of the daughter $_{Z+1}^{A}$B is denoted as metastable by $_{Z+1}^{Am}$B. This metastable state then decays to the ground state by emission of a gamma ray or by internal conversion with the frequencies f_γ and f_{IC}, respectively.

3.7.3 Energetics of Radioactive Decay

In this section, the energies involved in the various types of radioactive decay are examined. Of particular interest are the energies of the particles emitted in the decay process.[13]

Gamma Decay

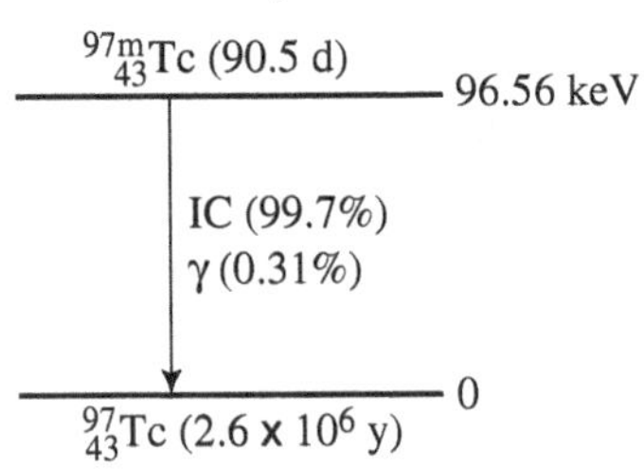

Figure 3.18. Radioactive decay diagram for the decay of ^{97m}Tc.

In many nuclear reactions, a nuclide is produced whose nucleus is left in an excited state. These excited nuclei usually decay very rapidly within 10^{-9} s to the ground state by emitting the excitation energy in the form of a gamma ray. However, a few excited nuclei remain in an excited state for a much longer time before they decay by gamma-ray emission. These long-lived excited nuclei are called *metastable* nuclei or *isomers.* When such an excited nuclide decays by gamma-ray emission, the decay is called an *isomeric transition.* A simple example is shown in Fig. 3.18. The first excited state of $^{97m}_{43}$Tc is a metastable state with a half-life of 90.5 d. It decays to the ground state with the emission of a 96.5 keV gamma photon.

[12] Any photon emitted from a nucleus is termed a *gamma* photon. Photons emitted when atomic electrons change their energy state are called *x rays.* Although x rays generally have lower energies than gamma photons, there are many exceptions.

[13] From now on, kinetic energy is given the usual symbol E instead of T used up to now.

The gamma-decay reaction of an excited isotope of element P can be written as

$$\gamma \text{ decay:} \qquad {}^{A}_{Z}\mathrm{P}^{*} \rightarrow {}^{A}_{Z}\mathrm{P} + \gamma \, . \tag{3.89}$$

Energy conservation for this nuclear reaction requires

$$M({}^{A}_{Z}\mathrm{P}^{*})c^2 \equiv M({}^{A}_{Z}\mathrm{P})c^2 + E^{*} = M({}^{A}_{Z}\mathrm{P})c^2 + E_P + E_\gamma \quad \text{or} \quad E^{*} = E_P + E_\gamma, \tag{3.90}$$

where E_γ is the energy of the gamma photon, E^* is the excitation energy (above the ground state) of the initial parent nucleus, and E_P is the recoil kinetic energy of the resulting ground-state nuclide. If the initial nuclide is at rest, the Q-value of this reaction is simply the sum of the kinetic energies of the products, which, with Eq. (3.90), yields

$$Q_\gamma = E_P + E_\gamma = E^{*}. \tag{3.91}$$

Linear momentum must also be conserved.[14] Again with the parent at rest before the decay (i.e., with zero initial linear momentum), the gamma photon and recoil nucleus must move in opposite directions and have equal magnitudes of linear momentum. Since the photon has momentum $p_\gamma = E_\gamma/c$ and the recoil nuclide has momentum $M_P v_P = \sqrt{2 M_P E_P}$, conservation of momentum requires

$$E_\gamma/c = \sqrt{2 M_P E_P} \qquad \text{or} \qquad E_P = \frac{E_\gamma^2}{2 M_P c^2} \tag{3.92}$$

where $M_P \equiv M({}^{A}_{Z}\mathrm{P})$. Substitution of this result for E_P into Eq. (3.91) yields

$$E_\gamma = Q_\gamma \left[1 + \frac{E_\gamma}{2 M_P c^2} \right]^{-1} \simeq Q_\gamma = E^{*}. \tag{3.93}$$

The approximation in this result follows from the fact that E_γ is at most 10–20 MeV, while $2 M_P c^2 \geq 4000$ MeV. Thus, in gamma decay, the kinetic energy of the recoil nucleus is negligible compared to the energy of the gamma photon and $E_\gamma \simeq Q_\gamma = E^*$.

Alpha-Particle Decay

From the prior discussion of nuclear structure, it was noted that nucleons tend to group themselves into subunits of ${}^{4}_{2}\mathrm{He}$ nuclei, often called alpha particles. Thus, it is not surprising that, for proton-rich heavy nuclei, a possible mode of decay to a more stable state is by alpha-particle emission.

In alpha decay, the nucleus of the parent atom ${}^{A}_{Z}\mathrm{P}$ emits an alpha particle. The resulting nucleus of the daughter atom ${}^{A-4}_{Z-2}\mathrm{D}$ then has two fewer neutrons and two fewer protons. Initially, the daughter still has Z electrons, two too many, and thus is a doubly negative ion $[{}^{A-4}_{Z-2}\mathrm{D}]^{2-}$, but these extra electrons quickly break away from the atom, leaving it in a neutral state. The fast moving doubly charged alpha particle quickly loses its kinetic energy by ionizing and exciting atoms along its path of travel and acquires two orbital electrons to become a neutral ${}^{4}_{2}\mathrm{He}$ atom. Since the atomic number of the daughter is different from that of the parent, the daughter is a different chemical element. The alpha decay reaction is thus represented by

$$\alpha \text{ decay:} \qquad {}^{A}_{Z}\mathrm{P} \longrightarrow [{}^{A-4}_{Z-2}\mathrm{D}]^{2-} + {}^{4}_{2}\alpha \longrightarrow {}^{A-4}_{Z-2}\mathrm{D} + {}^{4}_{2}\mathrm{He}. \tag{3.94}$$

[14]Throughout this section, because of the comparatively large mass and low energy of the recoil atom, momentum of the recoil is computed using laws of classical mechanics.

Decay Energy Recall from the previous chapter, the Q-value of a nuclear reaction is defined as the decrease in the rest mass energy (or increase in kinetic energy) of the product nuclei (see Eq. (3.81)). For radioactive decay, the Q-value is sometimes called the *disintegration energy*. For alpha decay we have

$$\begin{aligned} Q_\alpha/c^2 &= M({}^{A}_{Z}\mathrm{P}) - [M([{}^{A-4}_{Z-2}\mathrm{D}]^{2-}) + m({}^{4}_{2}\alpha)] \\ &\simeq M({}^{A}_{Z}\mathrm{P}) - [M({}^{A-4}_{Z-2}\mathrm{D}) + 2m_e + m({}^{4}_{2}\alpha)] \\ &\simeq M({}^{A}_{Z}\mathrm{P}) - [M({}^{A-4}_{Z-2}\mathrm{D}) + M({}^{4}_{2}\mathrm{He})]. \end{aligned} \tag{3.95}$$

In this reduction to atomic masses, the binding energies of the two electrons in the daughter ion and in the He atom have been neglected since these are small (several eV) compared to the Q-value (several MeV). For alpha decay to occur, Q_α must be positive, or, equivalently, $M({}^{A}_{Z}\mathrm{P}) > M({}^{A-4}_{Z-2}\mathrm{D}) + M({}^{4}_{2}\mathrm{He})$.

Kinetic Energies of the Products The disintegration energy Q_α equals the kinetic energy of the decay products. How this energy is divided between the daughter atom and the α particle is determined from the conservation of momentum. The momentum of the parent nucleus was zero before the decay, and thus, from the conservation of linear momentum, the total momentum of the products must also be zero. The alpha particle and the daughter nucleus must, therefore, leave the reaction site in opposite directions with equal magnitudes of their linear momentum to ensure the vector sum of their momenta is zero.

If we assume neither product particle is relativistic,[15] conservation of energy requires

$$Q_\alpha = E_D + E_\alpha = \frac{1}{2}M_D v_D^2 + \frac{1}{2}M_\alpha v_\alpha^2, \tag{3.96}$$

and conservation of linear momentum requires

$$M_D v_D = M_\alpha v_\alpha, \tag{3.97}$$

where $M_D \equiv M({}^{A-4}_{Z-2}\mathrm{D})$ and $M_\alpha \equiv M({}^{4}_{2}\mathrm{He})$. These two equations in the two unknowns v_D and v_α can be solved to obtain the kinetic energies of the products. Solve Eq. (3.97) for v_D and substitute the result into Eq. (3.96) to obtain

$$Q_\alpha = \frac{1}{2}\frac{M_\alpha^2}{M_D}v_\alpha^2 + \frac{1}{2}M_\alpha v_\alpha^2 = \frac{1}{2}M_\alpha v_\alpha^2\left[\frac{M_\alpha}{M_D} + 1\right]. \tag{3.98}$$

Hence, we find that the kinetic energy of the alpha particle is

$$E_\alpha = Q_\alpha\left[\frac{M_D}{M_D + M_\alpha}\right] \simeq Q_\alpha\left[\frac{A_D}{A_D + A_\alpha}\right]. \tag{3.99}$$

Notice that, in alpha decay, the alpha particle is emitted with a well-defined energy. The recoiling nucleus carries off the remainder of the available kinetic energy. From Eq. (3.96) we see $E_D = Q_\alpha - E_\alpha$, so that from the above result

$$E_D = Q_\alpha\left[\frac{M_\alpha}{M_D + M_\alpha}\right] \simeq Q_\alpha\left[\frac{A_\alpha}{A_D + A_\alpha}\right]. \tag{3.100}$$

[15]Here a non-relativistic analysis is justified since the energy liberated in alpha decay is less than 10 MeV, whereas the rest mass energy of an alpha particle is about 3727 MeV.

Alpha Decay Diagrams The above calculation assumes that the alpha decay proceeds from the ground state of the parent nucleus to the ground state of the daughter nucleus. As discussed earlier, sometimes the daughter nucleus is left in an excited nuclear state (which ultimately relaxes to the ground state by the emission of one or more gamma rays). In these cases, the Q_α-value of the decay is reduced by the excitation energy of the excited state. Often a radionuclide that decays by alpha emission is observed to emit alpha particles with several discrete energies. This is an indication that the daughter nucleus is left in excited states with nuclear masses greater than the ground state mass by the mass equivalent of the excitation energy.

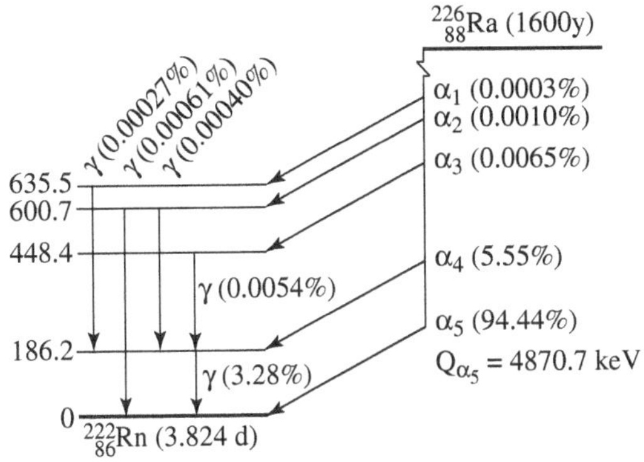

Figure 3.19. Energy levels for α decay of ^{226}Ra.

A simple case is shown in Fig. 3.19 for the decay of $^{226}_{88}$Ra. This nuclide decays to the first excited state of $^{222}_{86}$Rn 5.55% of the time and to the ground state 94.45% of the time. Two alpha particles with kinetic energies of $E_{\alpha 1} = 4.783$ MeV and $E_{\alpha 2} = 4.601$ MeV are thus observed. Also a 0.186-MeV gamma ray is observed, with an energy equal to the energy difference between the two alpha particles or the two lowest nuclear states in $^{222}_{88}$Rn.

Example 3.2: What is the initial kinetic energy of the alpha particle produced in the radioactive decay $^{226}_{88}\text{Ra} \rightarrow {}^{222}_{86}\text{Rn} + {}^{4}_{2}\text{He}$?

Solution:

The Q_α value in mass units (i.e., the mass defect) is, from Eq. (3.95),

$$\begin{aligned} Q_\alpha &= [M(^{226}_{88}\text{Ra}) - M(^{222}_{86}\text{Rn}) - M(^{4}_{2}\text{He})]c^2 \\ &= [226.025402 - 222.017571 - 4.00260325]\text{u} \times 931.5 \text{ MeV/u} \\ &= 0.005228 \text{ u} \times 931.5 \text{ MeV/u} = 4.870 \text{ MeV}. \end{aligned}$$

The kinetic energy of the alpha particle from Eq. (3.99) is

$$E_\alpha = Q_\alpha \left[\frac{A_D}{A_D + A_\alpha}\right] = 4.870 \left[\frac{222}{222+4}\right] = 4.784 \text{ MeV}.$$

The remainder of the Q_α energy is the kinetic energy of the product nucleus, $^{222}_{86}$Rn, namely, 4.870 MeV − 4.783 MeV = 0.087 MeV.

Beta-Particle Decay

Many neutron-rich radioactive nuclides decay by changing a neutron in the parent (P) nucleus into a proton and emitting an energetic electron. Many different names are applied to this decay process: electron decay, beta minus decay, negatron decay, negative electron decay, negative beta decay, or simply beta decay. The ejected electron is called a *beta particle* denoted by β^-. The daughter atom, with one more proton in the nucleus, initially lacks one orbital electron, and thus is a single charged positive ion, denoted by $[_{Z-1}^{\ A}\text{D}]^+$. However, the daughter quickly acquires an extra orbital electron from the surrounding medium. The general β^- decay reaction is thus written as

$$\boxed{\beta^- \text{ decay:} \quad {}^{A}_{Z}\text{P} \longrightarrow [{}_{Z+1}^{\ \ A}\text{D}]^+ + {}_{-1}^{\ 0}\text{e} + \overline{\nu}_e.} \qquad (3.101)$$

Here $\overline{\nu}$ is an antineutrino, a chargeless particle with very little, if any, rest mass.[16] That a third product particle is involved with β^- decay is implied from the observed energy and momentum of the emitted β^- particle. If the decay products were only the daughter nucleus and the β^- particle, then, as in α decay, conservation of energy and linear momentum would require that the decay energy be shared in very definite proportions between them. However, β^- particles are observed to be emitted with a continuous distribution of energies that has a well-defined maximum energy (see Fig. 3.20). Rather than abandon the laws of conservation of energy and momentum, Pauli suggested in 1933 that at least three particles must be produced in a β^- decay. The vector sum of the linear momenta of three products can be zero without any unique division of the decay energy among them. In 1934 Fermi used Pauli's suggestion of a third neutral particle to produce a beta-decay theory which explained well the observed beta-particle energy distributions. This mysterious third particle, which Fermi named the neutrino (lit. "little neutral one"), has since been verified experimentally and today it is extensively studied by physicists trying to develop fundamental theories of our universe. The maximum energy of the β^- spectrum corresponds to a case in which the neutrino obtains zero kinetic energy, and the decay energy is divided between the daughter nucleus and the β^- particle.

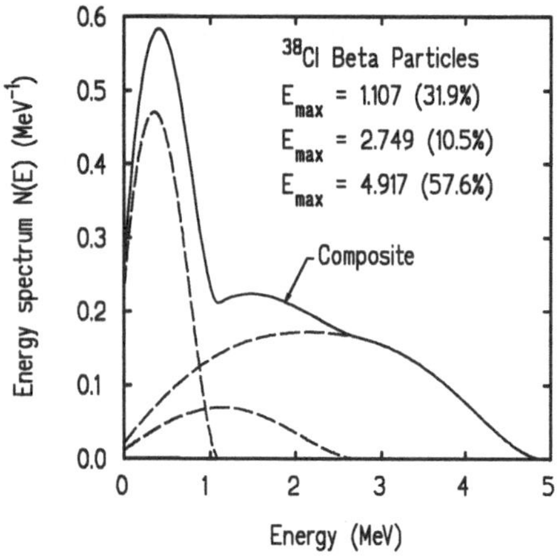

Figure 3.20. Energy spectra of principle ^{38}Cl β^- particles. From Shultis and Faw [2000].

Decay Energy The beta decay energy is readily obtained from the Q-value of the decay reaction. Specifically,

$$Q_{\beta^-}/c^2 = M({}^A_Z\text{P}) - [M([{}_{Z+1}^{\;\;A}\text{D}]^+) + m_{\beta^-} + m_{\overline{\nu}_e}]$$

$$\simeq M({}^A_Z\text{P}) - [\{M({}_{Z+1}^{\;\;A}\text{D}) - m_e\} + m_{\beta^-} + m_{\overline{\nu}_e}] = M({}^A_Z\text{P}) - M({}_{Z+1}^{\;\;A}\text{D}). \qquad (3.102)$$

For β^- decay to occur spontaneously, Q_{β^-} must be positive or, equivalently, the mass of the parent atom must exceed that of the daughter atom, i.e., $M({}^A_Z\text{P}) > M({}_{Z+1}^{\;\;A}\text{D})$.

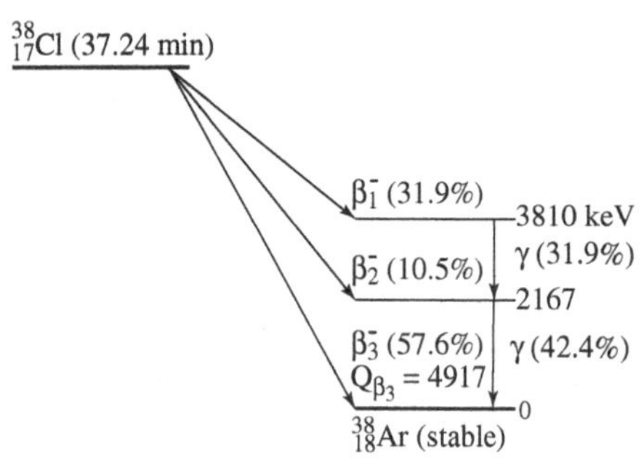

Figure 3.21. Radioactive decay diagram for the decay of ^{38}Cl. Three distinct groups of β^- particles are emitted.

Often in β^- decay, the nucleus of the daughter is left in an excited state. For example, ^{38}Cl decays both to the ground state of the daughter ^{38}Ar as well as to two excited states (see Fig. 3.21). The resulting β^- energy spectrum (Fig. 3.20) is a composite of the β^- particles emitted in the transition to each energy level of the daughter. For a decay to an energy level E^* above the ground level, the mass of the daughter atom in Eq. (3.102) $M({}_{Z+1}^{\;\;A}\text{D})$ must be replaced by the mass of the *excited* daughter $M({}_{Z+1}^{\;\;A}\text{D}^*) \simeq M({}_{Z+1}^{\;\;A}\text{D}) + E^*/c^2$. Thus, Q_{β^-} for β^- decay to an excited level with energy E^* above ground level in the daughter is

$$Q_{\beta^-}/c^2 = M({}^A_Z\text{P}) - M({}_{Z+1}^{\;\;A}\text{D}) - E^*/c^2. \qquad (3.103)$$

[16]The mass of the emitted neutrino is essentially that of the electron neutrino ν_e. The mass of ν_e is currently a subject of great interest in the high-energy physics community. Recent experiments show that, if it has any mass, its mass is exceedingly small and is less than a few eV mass equivalent (see Table 3.4). For the energetics of radioactive beta decay, the neutrino rest mass is completely negligible and can be taken as zero.

Because the kinetic energy of the parent nucleus is zero, the Q_{β^-} decay energy must be divided among the kinetic energies of the products. The maximum kinetic energy of the β^- particle occurs when the antineutrino obtains negligible energy. In this case, since the mass of the β^- particle is much less than that of the daughter nucleus, $Q = E_D + E_{\beta^-} \simeq E_{\beta^-}$ or

$$(E_{\beta^-})_{\max} \simeq Q_{\beta^-}. \tag{3.104}$$

Positron Decay

Nuclei that have too many protons for stability often decay by changing a proton into a neutron. In this decay mechanism, an anti-electron or *positron* β^+ or ${}_{+1}^{0}\text{e}$, and a neutrino ν are emitted. The daughter atom, with one less proton in the nucleus, initially has one too many orbital electrons, and thus is a negative ion, denoted by $[{}_{Z-1}^{\;\;A}\text{D}]^-$. However, the daughter quickly releases the extra orbital electron to the surrounding medium and becomes a neutral atom. The β^+ decay reaction is written as

$$\boxed{\beta^+ \text{ decay:} \qquad {}_{Z}^{A}\text{P} \longrightarrow [{}_{Z-1}^{\;\;A}\text{D}]^- + {}_{+1}^{0}\text{e} + \nu_e.} \tag{3.105}$$

The positron has the same physical properties as an electron, except that it has one unit of positive charge. The positron β^+ is the antiparticle of the electron. The neutrino ν is required (as in β^- decay) to conserve energy and linear momentum since the β^+ particle is observed to be emitted with a continuous spectrum of energies up to some maximum value $(E_{\beta^+})_{\max}$. The neutrino ν in Eq. (3.105) is the antiparticle of the antineutrino $\overline{\nu}$ produced in beta-minus decay.

Decay Energy The decay energy is readily obtained from the Q-value of the decay reaction. Specifically,

$$\begin{aligned} Q_{\beta^+}/c^2 &= M({}_{Z}^{A}\text{P}) - [M([{}_{Z-1}^{\;\;A}\text{D}]^-) + m({}_{+1}^{0}\text{e}) + m_{\nu_e}] \\ &\simeq M({}_{Z}^{A}\text{P}) - [\{M({}_{Z-1}^{\;\;A}\text{D}) + m_e\} + m_{\beta^+} + m_{\nu_e}] \\ &= M({}_{Z}^{A}\text{P}) - M({}_{Z-1}^{\;\;A}\text{D}) - 2m_e \end{aligned} \tag{3.106}$$

where the binding energy of the electron to the daughter ion and the neutrino mass have been neglected. If the daughter nucleus is left in an excited state, the excitation energy E^* must also be included in the Q_{β^+} calculation, namely,

$$Q_{\beta^+}/c^2 = M({}_{Z}^{A}\text{P}) - M({}_{Z-1}^{\;\;A}\text{D}) - 2m_e - E^*/c^2. \tag{3.107}$$

Thus, for β^+ decay to occur spontaneously, Q_{β^+} must be positive, i.e., $M({}_{Z}^{A}\text{P}) > M({}_{Z-1}^{\;\;A}\text{D}) + 2m_e + E^*/c^2$.

The maximum energy of the emitted positron occurs when the neutrino acquires negligible kinetic energy, so that the Q_{β^+} energy is shared by the daughter atom and the positron. Because the daughter atom is so much more massive than the positron (by factors of thousands), almost all the Q_{β^+} energy is transferred as kinetic energy to the positron. Thus

$$(E_{\beta^+})_{\max} \simeq Q_{\beta^+}. \tag{3.108}$$

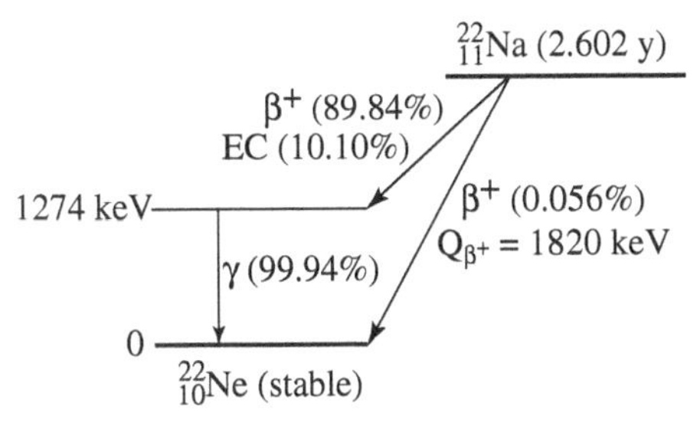

Figure 3.22. Radioactive decay diagram for positron emission from ^{22}Na.

An example of a radionuclide that decays by positron emission is ${}_{11}^{22}\text{Na}$. The decay reaction is

$${}_{11}^{22}\text{Na} \longrightarrow {}_{10}^{22}\text{Ne} + \beta^+ + \nu_e,$$

and the level diagram for this decay is shown in Fig. 3.22. Notice that the daughter is almost always left in its first excited state. This state decays, with a mean lifetime of 3.63 ps, by emitting a 1.274-MeV gamma ray.

The emitted positron loses its kinetic energy by ionizing and exciting atomic electrons as it moves through the surrounding medium. Eventually, it captures an ambient electron, forming for a brief instant a pseudo-atom called *positronium* before they annihilate each other. Their entire rest mass energy $2m_ec^2$ is converted into photon energy (the kinetic energy at the time of annihilation usually being negligible). Before the annihilation, there is zero linear momentum, and there must be no net momentum remaining; thus, two photons traveling in opposite directions must be created, each with energy $E = m_ec^2 = 0.511$ MeV.

Electron Capture

In the quantum mechanical model of the atom, the orbital electrons have a finite (but small) probability of spending some time inside the nucleus, the innermost K-shell electrons having the greatest probability. It is possible for an orbital electron, while inside the nucleus, to be captured by a proton, which is thus transformed into a neutron. Conceptually we can visualize this transformation of the proton as $p + {}_{-1}^{0}\text{e} \rightarrow n + \nu$, where the neutrino is again needed to conserve energy and momentum. The general electron capture (EC) decay reaction is written as

$$\text{EC decay:} \quad {}_{Z}^{A}\text{P} \longrightarrow {}_{Z-1}^{A}\text{D}^* + \nu_e. \tag{3.109}$$

where the daughter is generally left in an excited nuclear state with energy E^* above ground level. Unlike in most other types of radioactive decay, no charged particles are emitted. The only *nuclear* radiations emitted are gamma photons produced when the excited nucleus of the daughter relaxes to its ground state. As the outer electrons cascade down in energy to fill the inner shell vacancy, x rays and Auger electrons are also emitted.

Decay Energy The decay energy is readily obtained from the Q-value of the decay reaction. If we assume the daughter nucleus is left in its ground state

$$\begin{aligned} Q_{EC}/c^2 &= M({}_{Z}^{A}\text{P}) - [M({}_{Z-1}^{A}\text{D}) + m_{\nu_e}] \\ &\simeq M({}_{Z}^{A}\text{P}) - M({}_{Z-1}^{A}\text{D}). \end{aligned} \tag{3.110}$$

If the daughter nucleus is left in an excited state, the excitation energy E^* must also be included in the Q_{EC} calculation, namely,

$$Q_{EC}/c^2 = M({}_{Z}^{A}\text{P}) - M({}_{Z-1}^{A}\text{D}) - E^*/c^2. \tag{3.111}$$

Thus, for EC decay to occur spontaneously, Q_{EC} must be positive, i.e., $M({}_{Z}^{A}\text{P}) > M({}_{Z-1}^{A}\text{D}) + E^*/c^2$.

Notice that both β^+ and EC decay produce the same daughter nuclide. In fact, if the mass of the parent is sufficiently large compared to the daughter, both decay modes can occur for the same radionuclide. From Eq. (3.107) and Eq. (3.111), we see that if the parent's atomic mass is not at least two electron masses greater than the daughter's mass, Q_{β^+} is negative and β^+ decay cannot occur. However, Q_{EC} is positive as long as the parent's mass is even slightly greater than that of the daughter, and EC can still occur.

An example of a radionuclide that decays by electron capture is ${}_{4}^{7}\text{Be}$. The decay reaction is

$${}_{4}^{7}\text{Be} \longrightarrow {}_{3}^{7}\text{Li} + \nu_e.$$

The level diagram for this decay is shown in Fig. 3.23. Notice that in this example, the Q_{EC} of 862 keV is less than $2m_ec^2 = 1022$ keV so that there can be no competing β^+ decay. Finally, in an EC

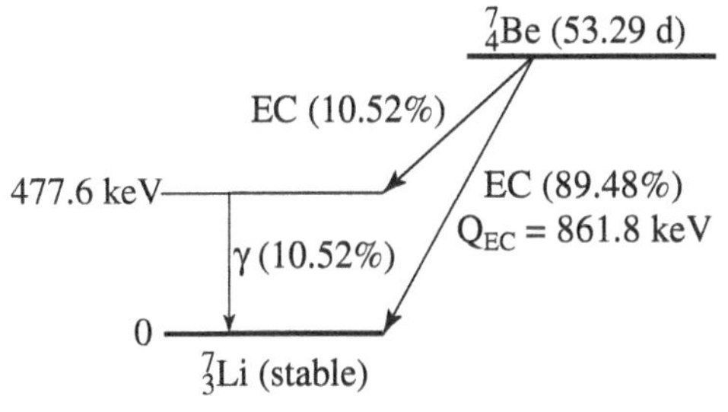

Figure 3.23. Decay of ^{7}Be by electron capture.

transition, an orbital electron (usually from an inner shell) disappears leaving an inner electron vacancy. The remaining atomic electrons cascade to lower orbital energy levels to fill the vacancy, usually emitting x rays as they become more tightly bound. The energy change in an electronic transition, instead of being emitted as an x ray, may also be transferred to an outer orbital electron ejecting it from the atom. These ejected electrons are called *Auger* electrons, named after their discoverer, and appear in any process that leaves a vacancy in an inner electron shell.

Neutron Decay

A few neutron-rich nuclides decay by emitting a neutron producing a different isotope of the same parent element. Generally, the daughter nucleus is left in an excited state which subsequently emits gamma photons as it returns to its ground state. This decay reaction is

$$n \text{ decay:} \qquad {}^{A}_{Z}\text{P} \longrightarrow {}^{A-1}{}_{Z}\text{P}^* + {}^{1}_{0}\text{n}. \tag{3.112}$$

An example of such a neutron decay reaction is ${}^{138}_{54}\text{Xe} \rightarrow {}^{137}_{54}\text{Xe} + n$. Although, neutron decay is rare, it plays a very important role in nuclear reactors. A small fraction of the radioactive atoms produced by fission reactions decay by neutron emission at times up to minutes after the fission event in which they were created. These neutrons contribute to the nuclear chain reaction and thus effectively slow it down, making it possible to control nuclear reactors.

The Q-value for such a decay is

$$Q_n/c^2 = M({}^{A}_{Z}\text{P}) - [M({}^{A-1}{}_{Z}\text{P}^*) + m_n] = M({}^{A}_{Z}\text{P}) - M({}^{A-1}{}_{Z}\text{D}) - m_n - E^*/c^2, \tag{3.113}$$

where E^* is the initial excitation energy of the daughter nucleus. Thus, for neutron decay to occur to even the ground state of the daughter, $M({}^{A}_{Z}\text{P}) > M({}^{A-1}{}_{Z}\text{D}) + m_n$.

Proton Decay

A few proton-rich radionuclides decay by emission of a proton. In such decays, the daughter atom has an extra electron (i.e., it is a singly charged negative ion). This extra electron is subsequently ejected from the atom's electron cloud to the surroundings and the daughter returns to an electrically neutral atom. The proton decay reaction is thus

$$p \text{ decay:} \qquad {}^{A}_{Z}\text{P} \longrightarrow [{}^{A-1}_{Z-1}\text{D}^*]^- + {}^{1}_{1}\text{p}. \tag{3.114}$$

In this reaction P and D refer to atoms of the parent and daughter. Thus, the Q-value for this reaction is

$$\begin{aligned} Q_p/c^2 &= M({}^{A}_{Z}\text{P}) - [M([{}^{A-1}_{Z-1}\text{D}^*]^-) + m_p] \simeq M({}^{A}_{Z}\text{P}) - [\{M({}^{A-1}_{Z-1}\text{D}^*) + m_e\} + m_p] \\ &\simeq M({}^{A}_{Z}\text{P}) - [\{M({}^{A-1}_{Z-1}\text{D}) + E^*/c^2 + m_e\} + m_p] \simeq M({}^{A}_{Z}\text{P}) - M({}^{A-1}_{Z-1}\text{D}) - M({}^{1}_{1}\text{H}) - E^*/c^2. \end{aligned} \tag{3.115}$$

Thus, for proton decay to occur and leave the daughter in the ground state ($E^* = 0$), it is necessary that $M({}^{A}_{Z}\text{P}) > M({}^{A-1}_{Z-1}\text{D}) - M({}^{1}_{1}\text{H})$.

Internal Conversion

Often the daughter nucleus is left in an excited state, which decays (usually within about 10^{-9} s) to the ground state by the emission of one or more gamma photons. However, the excitation may also be transferred to an atomic electron (usually a K-shell electron) causing it to be ejected from the atom leaving the nucleus in the ground state but the atom singly ionized with an inner electron-shell vacancy. Symbolically,

$$\boxed{\text{IC decay:} \qquad {}^{A}_{Z}\text{P}^* \longrightarrow [{}^{A}_{Z}\text{P}]^+ + {}^{\,0}_{-1}\text{e}.} \tag{3.116}$$

The inner electrons are very tightly bound to the nucleus with large binding energies BE_e^K for K-shell electrons in heavy atoms. The amount of kinetic energy shared by the recoil ion and the ejected electron should take this into account. The Q value for the IC decay is calculated as follows:

$$\begin{aligned} Q_{\text{IC}}/c^2 &= M({}^{A}_{Z}\text{P}^*) - [M([{}^{A}_{Z}\text{P}]^+) + m_e] \\ &\simeq \{M({}^{A}_{Z}\text{P}) + E^*/c^2\} - [\{M({}^{A}_{Z}\text{P}) - m_e + \text{BE}_e^K/c^2\} + m_e] \\ &= [E^* - \text{BE}_e^K]/c^2. \end{aligned} \tag{3.117}$$

This decay energy is divided between the ejected electron and the daughter ion. To conserve the zero initial linear momentum, the daughter and IC electron must divide the decay energy as

$$E_e = \left(\frac{M({}^{A}_{Z}\text{P})}{M({}^{A}_{Z}\text{P}) + m_e}\right)[E^* - \text{BE}_e^K] \simeq E^* - \text{BE}_e^K \tag{3.118}$$

and

$$E_D = \left(\frac{m_e}{M({}^{A}_{Z}\text{P}) + m_e}\right)[E^* - \text{BE}_e^K] \simeq 0. \tag{3.119}$$

Besides the monoenergetic IC electron, other radiation is also emitted in the IC process. As the outer electrons cascade down in energy to fill the inner-shell vacancy, x rays and Auger electrons are also emitted.

3.7.4 Exponential Decay

Radioactive decay is a stochastic process, i.e., one cannot predict when a particular radionuclide will decay. All that one can do is make probability statements about when it will decay. For example, as shown in this section, one can, at best, calculate the probability that it will or will not decay in a finite time interval t.

Each radionuclide species has its own characteristic *decay constant* λ which is the probability a radionuclide decays in a unit time for an infinitesimal time interval. The smaller λ, the more slowly the radionuclides decay. For stable nuclides, $\lambda = 0$. The decay constant for a radionuclide is independent of most experimental parameters such as temperature and pressure, since it depends only on the nuclear forces inside the nucleus.

Consider a sample composed of a large number of identical radionuclides with decay constant λ. With a large number of radionuclides ($N >>> 1$), one can use continuous mathematics to describe an inherently discrete process. In other words, $N(t)$ is interpreted as the *average* or *expected* number of radionuclides in the sample at time t, a continuous quantity. Then, the probability any one radionuclide decays in an interval dt is λdt, so the expected number of decays in the sample that occurs in dt at time t is $\lambda\, dt\, N(t)$. This must equal the decrease $-dN$ in the number of radionuclides in the sample, i.e., $-dN = \lambda N(t)dt$, or

$$\frac{dN(t)}{dt} = -\lambda N(t). \tag{3.120}$$

The solution of this differential equation is

$$\boxed{N(t) = N(0)e^{-\lambda t},} \tag{3.121}$$

where $N(0)$ is the number of radionuclides in the sample at $t = 0$. This exponential decay of a radioactive sample is known as the *radioactive decay law.* Such an exponential variation with time not only applies to radionuclides, but to any process governed by a constant rate of change, such as decay of excited electron states of an atoms, the rate of growth of money earning compound interest, and the growth of human populations, to name a few such processes.

3.7.5 The Half-Life

Any dynamic process governed by exponential decay has a remarkable property. The time it takes for it to decay to one-half of the initial value, $T_{1/2}$, is a constant called the *half-life.* From Eq. (3.121)

$$N(T_{1/2}) \equiv \frac{N_o}{2} = N_o e^{-\lambda T_{1/2}}. \tag{3.122}$$

Solving for $T_{1/2}$ yields

$$\boxed{T_{1/2} = \frac{\ln 2}{\lambda} \simeq \frac{0.693}{\lambda}.} \tag{3.123}$$

Notice that the half-life is independent of time t. Thus, after n half-lives, the initial number of radionuclides has decreased by a multiplicative factor of $1/2^n$, i.e.,

$$N(nT_{1/2}) = \frac{1}{2^n} N_o. \tag{3.124}$$

The number of half-lives n needed for a radioactive sample to decay to a fraction ϵ of its initial value is found from

$$\epsilon \equiv \frac{N(nT_{1/2})}{N_o} = \frac{1}{2^n}$$

which, upon solving for n, yields

$$n = -\frac{\ln \epsilon}{\ln 2} \simeq -1.44 \ln \epsilon. \tag{3.125}$$

Alternatively, the radioactive decay law of Eq. (3.121) can be expressed in terms of the half-life as

$$N(t) = N_o \left(\frac{1}{2}\right)^{t/T_{1/2}}. \tag{3.126}$$

3.7.6 Decay Probability for a Finite Time Interval

From the exponential decay law, some useful probabilities and averages can be obtained. If we have N_o identical radionuclides at $t = 0$, we expect to have $N_o e^{-\lambda t}$ atoms at a later time t. Thus, the probability $\overline{P}$ that any one of the atoms *does not* decay in a time interval t is

$$\overline{P}(t) = \frac{N(t)}{N(0)} = e^{-\lambda t}. \tag{3.127}$$

The probability P that a radionuclide *does* decay in a time interval t is

$$P(t) = 1 - \overline{P}(t) = 1 - e^{-\lambda t}. \tag{3.128}$$

As the time interval becomes very small, i.e., $t \to \Delta t \ll 1$, it is seen that

$$P(\Delta t) = 1 - e^{-\lambda \Delta t} = 1 - \left[1 - \lambda \Delta t + \frac{1}{2!}(\lambda \Delta t)^2 - + \ldots\right] \simeq \lambda \Delta t. \tag{3.129}$$

This approximation is in agreement with the earlier interpretation of the decay constant λ as being the decay probability per infinitesimal time interval.

From these results, the probability distribution function for when a radionuclide decays can be obtained. Specifically, let $p(t)dt$ be the probability a radionuclide, which exists at time $t = 0$, decays in the time interval between t and $t + dt$. Clearly,

$$\begin{aligned} p(t)dt &= \{\text{prob. it doesn't decay in } (0,t)\}\,\{\text{prob. it decays in the next } dt \text{ time interval}\} \\ &= \{\overline{P}(t)\}\,\{P(dt)\} = \{e^{-\lambda t}\}\,\{\lambda\, dt\} = \lambda e^{-\lambda t} dt. \end{aligned} \tag{3.130}$$

3.7.7 Mean Lifetime

In a radioactive sample, radionuclides decay at all times. From Eq. (3.121) it is seen that an infinite time is required for all the radioactive atoms to decay. However, as time increases, fewer and fewer atoms decay. The average lifetime of a radionuclide can be obtained by using the decay probability distribution $p(t)dt$ of Eq. (3.130). The average or mean lifetime T_{av} of a radionuclide is thus

$$T_{av} = \int_0^\infty t\,p(t)\,dt = \int_0^\infty t\lambda e^{-\lambda t} dt = -te^{-\lambda t}\Big|_0^\infty + \int_0^\infty e^{-\lambda t} dt = \frac{1}{\lambda}. \tag{3.131}$$

Example 3.3: What is the value of the decay constant and the mean lifetime of ^{40}K (half-life 1.29 Gy)?

Solution:
The decay constant is

$$\lambda = \frac{\ln 2}{T_{1/2}} = \frac{\ln 2}{1.29 \times 10^9 \text{ y}} = 5.37 \times 10^{-10} \text{ y}^{-1}.$$

The average lifetime is $\bar{t} = 1/\lambda = 1.86 \times 10^9$ y.

3.7.8 Activity

For detection and safety purposes, one is generally not really interested in the number of radioactive atoms in a sample; rather one is interested in the number of decays or transmutations per unit of time that occur within the sample. This decay rate, or *activity* $A(t)$, of a sample is given by[17]

$$A(t) \equiv -\frac{dN(t)}{dt} = \lambda N(t) = \lambda N_o e^{-\lambda t} = A_o e^{-\lambda t}, \tag{3.132}$$

where A_o is the activity at $t = 0$. Since the number of radionuclides in a sample decreases exponentially, the activity also decreases exponentially.

[17]Do not confuse the symbol A used here for activity with the same symbol used for the atomic mass number.

The SI unit used for activity is the becquerel (Bq) and is defined as one transformation per second. An older unit of activity, and one that is still sometimes encountered, is the curie (Ci) defined as 3.7×10^{10} Bq. One Ci is the approximate activity of one gram of $^{226}_{88}$Ra (radium). To obtain 1 Ci of tritium ($T_{1/2} = 12.6$ y, $\lambda = 1.74 \times 10^{-9}$ s^{-1}) one needs 1.06×10^{-4} g of tritium. By contrast, one Ci of ^{238}U ($\lambda = 4.88 \times 10^{-18}$ s^{-1}) is 3.00×10^{6} g.

3.7.9 Decay by Competing Processes

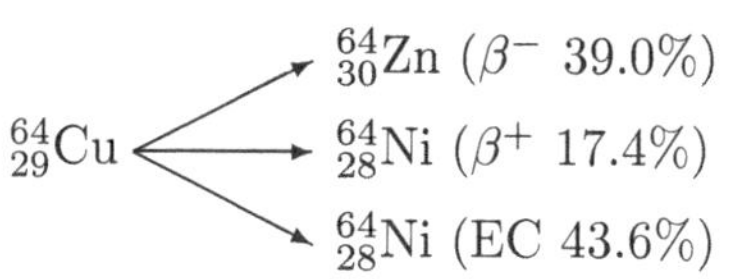

Figure 3.24. $^{62}_{29}$Cu has three radioactive decay modes.

Some radionuclides decay by more than one process. For example, $^{64}_{29}$Cu decays by β^+ emission 17.4% of the time, by β^- emission 39.0% of the time, and by electron capture 43.6% of the time, as shown in Fig. 3.24. Each decay mode is characterized by its own decay constant λ_i. For the present example of $^{64}_{29}$Cu, the decay constants for the three decay modes are $\lambda_{\beta^+} = 0.009497$ h^{-1}, $\lambda_{\beta^-} = 0.02129$ h^{-1}, and $\lambda_{\rm EC} = 0.02380$ h^{-1}.

To find the effective decay constant when the decay process has n competing decay modes, write the differential equation that models the rate of decay. Denote the decay constant for the ith mode by λ_i. Thus, the rate of decay of the parent radionuclide is given by,

$$\frac{dN(t)}{dt} = -\lambda_1 N(t) - \lambda_2 N(t) - \cdots - \lambda_n N(t) = -\sum_{i=1}^{n} \lambda_i N(t) \equiv -\lambda N(t). \tag{3.133}$$

where λ is the overall decay constant, namely,

$$\lambda \equiv \sum_{i=1}^{n} \lambda_i. \tag{3.134}$$

The probability f_i that the nuclide will decay by the ith mode is

$$f_i = \frac{\text{decay rate by } i\text{th mode}}{\text{decay rate by all modes}} = \frac{\lambda_i}{\lambda}. \tag{3.135}$$

3.7.10 Decay Dynamics

The transient behavior of the number of atoms of a particular radionuclide in a sample depends on the nuclide's rates of production and decay, the initial values of it and its parents, and the rate at which it escapes from the sample. In this section, several common decay transients are discussed.

Decay with Production

In many cases the decay of radionuclides is accompanied by the creation of new ones, either from the decay of a parent or from production by nuclear reactions such as cosmic ray interactions in the atmosphere or from neutron interactions in a nuclear reactor. If $Q(t)$ is the rate at which the radionuclide of interest is being created, the rate of change of the number of radionuclides is

$$\frac{dN(t)}{dt} = -\text{rate of decay} + \text{rate of production} \tag{3.136}$$

or

$$\frac{dN(t)}{dt} = -\lambda N(t) + Q(t). \tag{3.137}$$

The most general solution of this differential equation is

$$\boxed{N(t) = N_o e^{-\lambda t} + \int_0^t dt'\, Q(t') e^{-\lambda(t-t')}} \tag{3.138}$$

where again N_o is the number of radionuclides at $t = 0$.

For the special case that $Q(t) = Q_o$ (a constant production rate), the integral in Eq. (3.138) can be evaluated analytically to give

$$N(t) = N_o e^{-\lambda t} + \frac{Q_o}{\lambda}[1 - e^{-\lambda t}]. \tag{3.139}$$

As $t \to \infty$ it is seen that $N(t) \to N_e = Q_o/\lambda$, where N_e is the equilibrium value. This constant equilibrium value N_e can be obtained more directly from Eq. (3.137). Upon setting the left-hand side of Eq. (3.137) to zero (the equilibrium condition), it is seen that $0 = -\lambda N_e + Q_o$, or $N_e = Q_o/\lambda$.

Three Component Decay Chains

Often a radionuclide decays to another radionuclide which in turn decays to yet another. The chain continues until a stable nuclide is reached. For simplicity, first consider a three-component chain. Such three component chains are quite common and an example is

$${}^{90}_{38}\text{Sr} \xrightarrow{T_{1/2}=29.1\text{y}} {}^{90}_{39}\text{Y} \xrightarrow{T_{1/2}=64\text{h}} {}^{90}_{40}\text{Zn}(\text{ stable}),$$

in which both radionuclides decay by β^- emission. Such three-member decay chains can be written schematically as

$$\text{X}_1 \xrightarrow{\lambda_1} \text{X}_2 \xrightarrow{\lambda_2} \text{X}_3(\text{ stable}).$$

At $t = 0$ the number of atoms of each type in the sample under consideration is denoted by $N_i(0)$, $i = 1, 2, 3$. The differential decay equations for each species are (assuming no loss from or production in the sample)

$$\frac{dN_1(t)}{dt} = -\lambda_1 N_1(t). \tag{3.140}$$

$$\frac{dN_2(t)}{dt} = -\lambda_2 N_2(t) + \lambda_1 N_1(t). \tag{3.141}$$

$$\frac{dN_3(t)}{dt} = \lambda_2 N_2(t). \tag{3.142}$$

The solution of Eq. (3.140) is just the exponential decay law of Eq. (3.121), i.e.,

$$N_1(t) = N_1(0)e^{-\lambda_1 t}. \tag{3.143}$$

The number of first daughter atoms $N_2(t)$ is obtained from Eq. (3.138) with the production term $Q(t) = \lambda_1 N_1(t)$. The result is

$$N_2(t) = N_2(0)e^{-\lambda_2 t} + \frac{\lambda_1 N_1(0)}{\lambda_2 - \lambda_1}[e^{-\lambda_1 t} - e^{-\lambda_2 t}]. \tag{3.144}$$

The number of second daughter (granddaughter) atoms is obtained by integrating Eq. (3.142). Thus,

$$\begin{aligned} N_3(t) &= N_3(0) + \lambda_2 \int_0^t dt'\, N_2(t') \\ &= N_3(0) + \lambda_2 N_2(0) \int_0^t dt'\, e^{-\lambda_2 t'} + \frac{N_1(0)\lambda_1\lambda_2}{\lambda_2 - \lambda_1} \int_0^t dt'\, [e^{-\lambda_1 t'} - e^{-\lambda_2 t'}] \\ &= N_3(0) + N_2(0)[1 - e^{-\lambda_2 t}] + \frac{N_1(0)}{\lambda_2 - \lambda_1}[\lambda_2(1 - e^{-\lambda_1 t}) - \lambda_1(1 - e^{-\lambda_2 t})]. \end{aligned} \tag{3.145}$$

General Decay Chain
The general decay chain can be visualized as

$$X_1 \xrightarrow{\lambda_1} X_2 \xrightarrow{\lambda_2} X_3 \xrightarrow{\lambda_3} \cdots X_i \xrightarrow{\lambda_i} \cdots \xrightarrow{\lambda_{n-1}} X_n \text{ (stable)}$$

The decay and buildup equations for each member of the decay chain are

$$\begin{aligned} \frac{dN_1(t)}{dt} &= -\lambda_1 N_1(t) \\ \frac{dN_2(t)}{dt} &= \lambda_1 N_1(t) - \lambda_2 N_2(t) \\ \frac{dN_3(t)}{dt} &= \lambda_2 N_2(t) - \lambda_3 N_3(t) \\ &\vdots \quad \vdots \\ \frac{dN_{n-1}(t)}{dt} &= \lambda_{n-2} N_{n-2}(t) - \lambda_{n-1} N_{n-1}(t) \\ \frac{dN_n(t)}{dt} &= \lambda_{n-1} N_{n-1}(t). \end{aligned} \tag{3.146}$$

For the case when only radionuclides of the parent X_1 are initially present, the initial conditions are $N_1(0) \neq 0$ and $N_i(0) = 0$, $i > 1$. The solution of these so-called Bateman equations with these initial conditions is [Faw and Shultis 1999]

$$A_j(t) = \lambda_j N_j(t) = N_1(0)(C_1 e^{-\lambda_1 t} + C_2 e^{-\lambda_2 t} + \cdots + C_j e^{-\lambda_j t}) = N_1(0) \sum_{m=1}^{j} C_m e^{-\lambda_m t}. \tag{3.147}$$

The coefficients C_m are

$$C_m = \frac{\prod_{i=1}^{j} \lambda_i}{\prod_{\substack{i=1 \\ i \neq m}}^{j} (\lambda_i - \lambda_m)} = \frac{\lambda_1 \lambda_2 \lambda_3 \cdots \lambda_j}{(\lambda_1 - \lambda_m)(\lambda_2 - \lambda_m) \cdots (\lambda_j - \lambda_m)}, \tag{3.148}$$

where the $i = m$ term is excluded from the denominator.

PROBLEMS

1. In 1911, Robert A. Millikan determined the singular charge held by an electron. In the experiment, parallel metal plates were fashioned as depicted in Fig. 1.14 with a voltage applied. Millikan would discharge an atomized cloud of oil droplets into the chamber. Free electrons would attach to oil droplets, and an appropriate adjustment of the voltage would allow the electric force on the oil droplet to equal the gravitational force, thereby suspending such an oil droplet motionless in the chamber. Calculate the change in voltage needed to suspend an oil droplet onto which three electrons have attached. The electron charge is 1.602×10^{-19} C.

2. For slowly moving particles, that is, $v \ll c$, show that Eq. (3.8) yields the usual classical result,

$$T = mc^2 - m_o c^2 \simeq \frac{1}{2} m_o v^2. \tag{3.149}$$

3. An accelerator increases the kinetic energy of electrons uniformly to 10 GeV over a 3000 m path. That means that at 30 m, 300 m, and 3000 m, the kinetic energy is 10^8, 10^9, and 10^{10} eV, respectively. At each of these distances, compute the velocity, relative to light (v/c), and the mass in atomic mass units.

4. Muons are subatomic particles that have the negative charge of an electron but are 206.77 times more massive. They are produced high in the atmosphere by cosmic rays colliding with nuclei of oxygen or nitrogen, and muons are the dominant cosmic-ray contribution to background radiation at the earth's surface. A muon, however, rapidly decays into an energetic electron, existing, from its point of view, for only 2.20 μs, on the average. Cosmic-ray generated muons typically have speeds of about $0.998c$ and thus should travel only a few hundred meters in air before decaying. Yet muons travel through several kilometers of air to reach the earth's surface. Using the results of special relativity, explain how this is possible. HINT: consider the atmospheric travel distance as it appears to a muon, and the muon lifetime as it appears to an observer on the earth's surface.

5. A 1-MeV gamma ray loses 200 keV in a Compton scatter. Calculate the scattering angle.

6. At what energy (in MeV) can a photon lose at most one-half of its energy in Compton scattering?

7. What is the de Broglie wavelength of a water molecule moving at a speed of 2400 m/s? What is the wavelength of a 3-g bullet moving at 400 m/s?

8. If a neutron is confined somewhere inside a nucleus of characteristic dimension $\Delta x \simeq 10^{-14}$ m, what is the uncertainty in its momentum Δp? For a neutron with momentum equal to Δp, what is its total energy and its kinetic energy in MeV? Verify that classical expressions for momentum and kinetic energy may be used.

9. Repeat the previous problem for an electron trapped in the nucleus. HINT: relativistic expressions for momentum and kinetic energy must be used.

10. Estimate the wavelengths of the first three spectral lines in the Lyman spectral series for hydrogen. What energies (eV) do photons with these wavelengths have?

11. Consider an electron in the first Bohr orbit of a hydrogen atom. (a) What is the radius (in meters) of this orbit? (b) What is the total energy (in eV) of the electron in this orbit? (c) How much energy is required to ionize a hydrogen atom when the electron is in the ground state?

12. What photon energy (eV) is required to excite the hydrogen electron in the innermost (ground state) Bohr orbit to the first excited orbit?

13. What is the de Broglie wavelength of the electron in the first Bohr orbit? Compare this wavelength to the circumference of the first Bohr orbit. What does this comparison reveal about the standing wave in the first Bohr orbit?

14. Calculate the limiting (smallest) wavelength of the Lyman, Balmer, and Paschen series for the Bohr model of the hydrogen atom.

15. What is the net energy released (in MeV) for each of the following fusion reactions? (a) ${}^2_1\text{H} + {}^2_1\text{H} \longrightarrow {}^3_2\text{He} + {}^1_0\text{n}$ and (b) ${}^2_1\text{H} + {}^3_1\text{H} \longrightarrow {}^4_2\text{He} + {}^1_0\text{n}$.

Data: $M({}^2_1\text{H}) = 2.0141018$ u; $M({}^3_1\text{H}) = 3.0160493$ u;
$M({}^3_2\text{He}) = 3.0160293$ u; $M({}^4_2\text{He}) = 4.0026032$ u; $m_n = 1.0086649$ u, .

16. Calculate the Q-values for the following two beta radioactive decays.
(a) ${}^{22}_{11}\text{Na} \longrightarrow {}^{22}_{10}\text{Ne} + {}^{0}_{+1}\text{e} + \nu$ and (b) $M{}^{38}_{17}\text{Cl} \longrightarrow {}^{38}_{18}\text{Ar} + {}^{0}_{-1}\text{e} + \overline{\nu}$.

Data: $M({}^{22}_{11}\text{Na}) = 21.9944368$ u; $M({}^{22}_{10}\text{Ne}) = 21.9913855$ u;
$M({}^{38}_{17}\text{Cl}) = 37.9680106$ u; $M({}^{38}_{18}\text{Ar}) = 37.9637322$ u; $m_e = 5.485799 \times 10^{-4}$ u. .

17. The radioisotope ^{224}Ra decays by α emission primarily to the ground state of ^{220}Rn (94% probability) and to the first excited state 0.241 MeV above the ground state (5.5% probability). What are the energies of the two associated α particles?

Data: $M({}^{224}_{88}\text{Ra}) = 224.0202020$ u; $M({}^{220}_{86}\text{Rn}) = 220.0113841$ u;
$M({}^4_2\text{He}) = 4.0026032$ u;

18. The activity of a radioisotope is found to decrease by 30% in one week. What are the values of its (a) decay constant, (b) half-life, and (c) mean life?

19. How many atoms are there in a 1.20 MBq source of (a) ^{24}Na and (b) ^{238}U?

20. The isotope ^{132}I decays by β^- emission to ^{132}Xe with a half-life of 2.3 h. (a) How long will it take for 7/8 of the original number of ^{132}I nuclides to decay? (b) How long will it take for a sample of ^{132}I to lose 95% of its activity?

21. How many grams of ^{32}P are there in a 5 mCi source?

22. The average mass of potassium in the human body is about 140 g. From the abundance and half-life of ^{40}K (see Table 5.1), estimate the average activity (Bq) of ^{40}K in the body.

23. A 6.2 mg sample of ^{90}Sr (half-life 29.12 y) is in secular equilibrium with its daughter ^{90}Y (half-life 64.0 h). (a) How many Bq of ^{90}Sr are present? (b) How many Bq of ^{90}Y are present? (c) What is the mass of ^{90}Y present? (d) What will the activity of ^{90}Y be after 100 y?

24. Consider a sample in which A is a radioactive parent that decays to a radioactive daughter B. If the initial activity of B at $t = 0$ is zero, show that the maximum activity of B is reached at time

$$t_{max} = A_0 \frac{\ln\left(\dfrac{\lambda_B}{\lambda_A}\right)}{\lambda_B - \lambda_A}. \tag{3.150}$$

25. Using a chart of nuclides, start with ^{241}Am and list the daughter products and methods of decay for each daughter product to the end of the chain (stable isotope).

REFERENCES

ESTERMANN, I., AND O. STERN, "Beugung von Molekularstrahlen," *Z. Physik.*, **61**, 95–125, (1930).

EVANS, R.D., *The Atomic Nucleus,* New York: McGraw-Hill, 1955; republished by Melborne, FL: Krieger Publishing Co., 1982.

FAW, R.E. AND J.K. SHULTIS, *Radiological Assessment,* La Grange Park, IL: American Nuclear Society, 1999.

FIRESTONE, R.B., AND V.S. SHIRLEY, Eds., *Table of Isotopes,* Vols. 1 and 2, 8th Ed, New York: Wiley, 1996.

GRIFFITHS, D.J., *Introduction to Elementary Particles,* Strauss: Wiley-VCH, 2004.

GRIFFITHS, D.J., *Introduction of Quantum Mechanics,* 2nd Ed., Upper Saddle River, NJ: Pearson Prentice Hall, 2005.

ICRP, *Radionuclide Transformations,* Publication 38, International Commission on Radiological Protection, *Annals of the ICRP*, **11-13**, 1983.

KAPLAN, I., *Nuclear Physics*, Reading: Addison-Wesley, 1962.

POVH, B., K. RITH, C. SCHOLZ, AND F. ZETSCHE, *Particles and Nuclei,* 5th Ed., Berlin: Springer, 2006.

RILEY, K.F., M.P. HOBSON, AND S.J. BENCE, *Mathematical Methods for Physics and Engineering,* 3rd Ed., Cambridge: Cambridge University Press, 2006.

SCHIFF, L.I., *Quantum Mechanics,* 3rd Ed., New York: McGraw Hill, 1968.

SHULTIS, J.K. AND R.E. FAW, *Fundamentals of Nuclear Science and Engineering,* 3rd Ed., Boca Raton, FL: CRC Press, 2017.

THOMSON, G.P., "Experiments on the Diffraction of Cathode Rays," *Proc. Roy. Soc.*, **117**, 600, (1928).

WEBER, D.A., K.F. ECKERMAN, L.T. DILLMAN, AND J.C. RYMAN, *MIRD: Radionuclide Data and Decay Schemes*, New York: Society of Nuclear Medicine, 1989.

YAO, Y-M, et al., "Review of Particle Physics," *J. Phys. G.*, **33**, 1–1232, (2006).

Chapter 4

Radiation Interactions in Matter

Every great discovery I ever made, I gambled that the truth was there, and then I acted in faith until I could prove its existence.

Arthur H. Compton

4.1 Introduction

It is important that any professional working with or designing radiation detectors and instrumentation be familiar with fundamental interactions of radiation with matter. It is through these interactions that observable phenomena can be manipulated to yield information about the radiation type, intensity, and energy, as well as other possible uses, such as expected results and characteristic signatures. For radiation to be detected, it must first interact with neutral atoms in the detector and produce ionization and/or excitation in the detector material. Ionizing radiation is generally subdivided into two classes: *directly ionizing* radiation whose interactions produce ionization and excitation in a medium, and *indirectly ionizing* radiation that cannot ionize atoms but that can cause interactions whose charged products, known as *secondary* radiation, are directly ionizing. Fast moving charged particles, such as alpha particles, beta particles and fission fragments, can directly ionize and excite matter. Neutral particles, such as neutron and photons, cannot interact Coulombically with the electrons of the matter through which they pass; rather they cause interactions that transfer some of their incident kinetic energy to charged secondary particles.

In this chapter, how these two types of ionizing radiation interact with matter is discussed. Particular emphasis is given to quantifying the rate at which the radiation interacts with the medium and how the kinetic energy of the radiation is transfered to the medium.

4.2 Indirectly Ionizing Radiation

First considered is indirectly ionizing radiation whose particles carry no charge. Of primary concern are the interactions of photons and neutrons, although the general principles discussed here also apply to other neutral particles such as neutrinos and neutral mesons. Later in the chapter the interaction of directly ionizing radiation is considered.

4.2.1 Attenuation of Neutral Particle Beams

The interaction of a photon or neutrons with constituents of matter is dominated by short-range forces. Consequently, unlike charged particles, neutral particles move in straight lines through a medium, punctuated by occasional "point" interactions, in which the neutral particle may be absorbed or scattered or cause some other type of reaction. The interactions are stochastic in nature, i.e., the travel distance between interactions with the medium can be predicted only in some average or expected sense.

The interaction of a given type of neutral radiation with matter may be classified according to the type of interaction and the matter with which the interaction takes place. The interaction may take place with an electron, and in many cases the electron behaves as though it were free. Similarly, the interaction may take place with an atomic nucleus, which in many cases behaves as though it were not bound in a molecule or crystal lattice. However, in some cases, particularly for radiation particles of comparatively low energy, molecular or lattice binding must be taken into account.

The interaction may be a scattering of the incident radiation accompanied by a change in its energy. A scattering interaction may be elastic or inelastic. Consider, for example, the interaction of a gamma photon with an electron in what is called Compton scattering. In the sense that the interaction is with the entire atom within which the electron is bound, the interaction must be considered as inelastic, since some of the incident photon's energy must compensate for the binding energy of the electron in the atom. However, in most practical cases, electron binding energies are orders of magnitude lower than gamma-photon energies, and the interaction may be treated as a purely elastic scattering of the photon by a free electron. Neutron scattering by an atomic nucleus may be elastic, in which case the incident neutron's kinetic energy is shared by that of the scattered neutron and that of the recoil nucleus, or it may be inelastic, in which case, some of the incident neutron's kinetic energy is transformed to internal energy of the nucleus and thence to a gamma ray emitted from the excited nucleus. It is important to note that, for both elastic and inelastic scattering, unique relationships between energy exchanges and angles of scattering arise from conservation of energy and linear momentum.

Other types of interactions are absorptive in nature. The identity of the incident particle is lost, and total relativistic momentum and energy are conserved, some of the energy appearing as nuclear excitation energy, some as translational, vibrational, and rotational energy. The ultimate result may be the emission of particulate radiation, as occurs in the photoelectric effect and in neutron radiative capture.

The discussion in this section of how a beam of radiation is attenuated as it passes through matter applies equally to both neutrons and photons. In later sections, descriptions specific to each type of neutral radiation and the particular radiation-medium interactions involved are given. The concept of the interaction coefficient is first introduced to describe how readily radiation particles interact with matter, and then use it to quantify the attenuation of a beam of neutral particles passing through some material.

4.2.2 The Linear Interaction Coefficient

The interaction of radiation with matter is always statistical in nature, and, therefore, must be described in probabilistic terms. Consider a particle traversing a homogeneous material and let $P_i(\Delta x)$ denote the probability that this particle, while traveling a distance Δx in the material, causes a reaction of type i (e.g., it is scattered). It is found empirically that the probability per unit distance traveled, $P_i(\Delta x)/\Delta x$, approaches a constant as Δx becomes very small, i.e.,

$$\mu_i \equiv \lim_{\Delta x \to 0} \frac{P_i(\Delta x)}{\Delta x}. \tag{4.1}$$

The quantity μ_i is a property of the material for a given incident particle and interaction. In the limit of small path lengths, μ_i is seen to be the probability, per unit differential path length of travel, that a particle undergoes an ith type of interaction.

That μ_i is constant for a given material and for a given type of interaction implies that the probability of interaction, per unit differential path length, is independent of the path length traveled prior to the interaction. In this book, when the interaction coefficient is referred to as the *probability per unit path length* of an interaction, it is understood that this is true only in the limit of very small path lengths.

The constant μ_i is called the *linear coefficient* for reaction i. For each type of reaction, there is a corresponding linear coefficient. For, example, μ_a is the *linear absorption coefficient*, μ_s the *linear scattering coefficient*, and so on. Although this nomenclature is widely used to describe photon interactions, μ_i is often referred to as the *macroscopic cross section* for reactions of type i, and is usually given the symbol Σ_i when describing neutron interactions. In this section, the photon jargon is used, although the present discussion applies equally to neutrons.

The probability, per unit path length, that a neutral particle undergoes some sort of reaction, μ_t, is the sum of the probabilities, per unit path length of travel, for each type of possible reaction, i.e.,

$$\mu_t(E) = \sum_i \mu_i(E). \tag{4.2}$$

Because these coefficients generally depend on the particle's kinetic energy E, this dependence has been shown explicitly. The total interaction probability per unit path length, μ_t, is fundamental in describing how indirectly ionizing radiation interacts with matter and is usually called the *linear attenuation coefficient.* It is perhaps more appropriate to use the words *total linear interaction coefficient* since many interactions do not "attenuate" the particle in the sense of an absorption interaction.

4.2.3 Attenuation of Uncollided Radiation

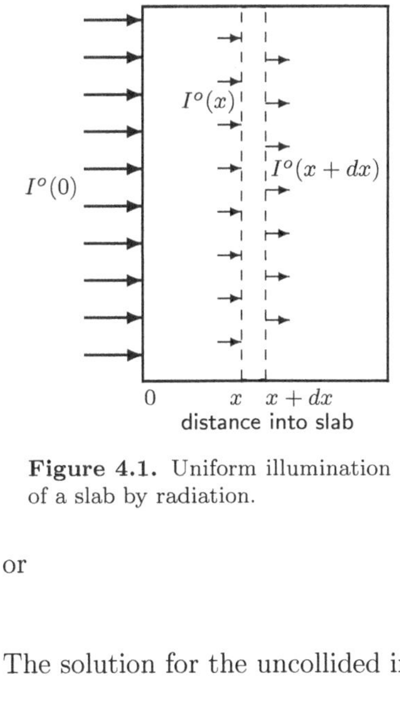

Figure 4.1. Uniform illumination of a slab by radiation.

Consider a plane parallel beam of neutral particles of intensity I_o particles cm^{-2}s^{-1} normally incident on the surface of a thick slab (see Fig. 4.1). As the particles pass into the slab, some interact with the slab material. Of interest in many situations, is the intensity $I^o(x)$ of *uncollided* particles at depth x into the slab. At some distance x into the slab, some uncollided particles undergo interactions for the first time as they cross the next Δx of distance, thereby reducing the *uncollided* beam intensity at x, $I^o(x)$, to some smaller value $I^o(x+\Delta x)$ at $x+\Delta x$. The probability an uncollided particle interacts as it crosses Δx is

$$P(\Delta x) = \frac{I^o(x) - I^o(x+\Delta x)}{I^o(x)}.$$

In the limit as $\Delta x \to 0$, Eq. (4.1) yields

$$\mu_t = \lim_{\Delta x \to 0} \frac{P(\Delta x)}{\Delta x} = \lim_{\Delta x \to 0} \frac{I^o(x) - I^o(x+\Delta x)}{\Delta x} \frac{1}{I^o(x)} \equiv -\frac{dI^o(x)}{dx} \frac{1}{I^o(x)}$$

or

$$\frac{dI^o(x)}{dx} = -\mu_t I^o(x). \tag{4.3}$$

The solution for the uncollided intensity is

$$\boxed{I^o(x) = I^o(0)e^{-\mu_t x}.} \tag{4.4}$$

Uncollided, indirectly ionizing radiation is thus exponentially attenuated as it passes through a medium.

From this result, the interaction probability $P(x)$ that a particle interacts somewhere along a path of length x is

$$P(x) = 1 - \frac{I^o(x)}{I^o(0)} = 1 - e^{-\mu_t x}. \tag{4.5}$$

The probability $\overline{P}(x)$ that a particle does *not* interact while traveling a distance x is

$$\overline{P}(x) = 1 - P(x) = e^{-\mu_t x}. \tag{4.6}$$

As $x \to dx$, it is found that $P(dx) \to \mu_t dx$, which is in agreement with the definition of μ_t.

4.2.4 Average Travel Distance Before an Interaction

From the above results, the probability distribution for how far a neutral particle travels before interacting can be derived. Let $p(x)dx$ be the probability that a particle interacts for the first time between x and $x + dx$. Then

$$\begin{aligned} p(x)dx &= \{\text{prob. particle travels a distance } x \text{ without interaction}\} \times \{\text{prob. it interacts in the next } dx\} \\ &= \left\{\overline{P}(x)\right\}\{P(dx)\} = \left\{e^{-\mu_t x}\right\}\{\mu_t dx\} = \mu_t e^{-\mu_t x} dx. \end{aligned} \tag{4.7}$$

Note that $\int_0^\infty p(x)\,dx = 1$, as is required for a proper probability distribution function.

This probability distribution can be used to find the average distance $\overline{x}$ traveled by a neutral particle to the site of its first interaction, namely, the average distance such a particle travels before it interacts. The average of x is

$$\overline{x} = \int_0^\infty x\,p(x)\,dx = \mu_t \int_0^\infty x\,e^{-\mu_t x}\,dx = \frac{1}{\mu_t}. \tag{4.8}$$

This average travel distance before an interaction, $1/\mu_t$, is called the *mean-free-path length.*

The total linear attenuation coefficient μ_t can be interpreted, equivalently, as (1) the probability, per unit differential path length of travel, that a particle interacts, or (2) the inverse of the average distance traveled before interacting with the medium. Note the analogy to radioactive decay, where the decay constant λ is the inverse of the mean lifetime of a radionuclide.

4.2.5 Half-Thickness

To estimate how incident radiation interacts in a detector medium, a convenient concept is that of the *half-thickness*, $x_{1/2}$, namely, the thickness of a medium required for half of the incident radiation to undergo an interaction. For the uncollided beam intensity to be reduced to one-half of its initial value, $x_{1/2}$ must be such that

$$\frac{1}{2} = \frac{I^o(x_{1/2})}{I^o(0)} = e^{-\mu_t x_{1/2}}$$

from which one finds

$$x_{1/2} = \frac{\ln 2}{\mu_t}. \tag{4.9}$$

Again, note the similarity to the half-life of radioactive decay.

4.2.6 Microscopic Cross Sections

The linear coefficient $\mu_i(E)$ depends on the type and energy E of the incident particle, the type of interaction i, and the composition and density of the interacting medium. One of the more important quantities that determines μ is the density of target atoms or electrons in the material. It seems reasonable to expect that μ_i should be proportional to the "target" atom density N in the material, that is,

$$\boxed{\mu_i = \sigma_i N = \sigma_i \frac{\rho N_a}{A},} \tag{4.10}$$

where σ_i is a constant of proportionality independent of N. Here ρ is the mass density of the medium, N_a is Avogadro's number (mol^{-1}), and A is the atomic weight of the medium.

The proportionality constant σ_i is called the *microscopic cross section* for reaction i, and is seen to have dimensions of area. It is often interpreted as being the effective cross-sectional area presented by the target atom to the incident particle for a given interaction. Indeed, in many cases σ_i has dimensions comparable to those expected from the physical size of the nucleus. However, this simplistic interpretation of the microscopic cross section, although conceptually easy to grasp, leads to philosophical difficulties when it is observed that σ_i generally varies with the energy of the incident particle and, for a crystalline material, the particle direction. The view that σ is the interaction probability per unit differential path length, normalized to one target atom per unit volume, avoids such conceptual difficulties while emphasizing the statistical nature of the interaction process. Cross sections are usually expressed in units of cm^2. A widely used special unit is the *barn*, equal to 10^{-24} cm^2.

Data on cross sections and linear interaction coefficients, especially for photons, are frequently expressed as the ratio of μ_i to the density ρ, called the *mass interaction coefficient* for reaction i. Upon division of Eq. (4.10) by ρ, one obtains

$$\frac{\mu_i}{\rho} = \frac{\sigma_i N}{\rho} = \frac{N_a}{A}\sigma_i. \tag{4.11}$$

From this result, it is seen that μ_i/ρ is an intrinsic property of the interacting medium—independent of its density. This method of data presentation is used much more for photons than for neutrons, in part because, for a wide variety of materials and a wide range of photon energies, μ_i/ρ is only weakly dependent on the nature of the interacting medium.

For compounds or homogeneous mixtures, the linear and mass interaction coefficients for interactions of type i are, respectively,

$$\mu_i = \sum_j \mu_i^j = \sum_j N^j \sigma_i^j \tag{4.12}$$

and

$$\frac{\mu_i}{\rho} = \sum_j w_j \left(\frac{\mu_i}{\rho}\right)^j, \tag{4.13}$$

in which the superscript j refers to the jth component of the material, the subscript i to the type of interaction, and w_j is the weight fraction of component j. In Eq. (4.12), the atomic density N^j and the linear interaction coefficient μ_i^j are values for the jth material *after* mixing.

4.2.7 Calculation of Radiation Interaction Rates

To quantify the number of interactions caused by neutral particles as they pass through matter, one often needs to estimate the number of a specific type of interaction that occurs, in a unit time and in a unit volume, at some point in the matter. This density of reactions per unit time is called the *reaction rate density*, denoted by F_i. This quantity depends on (1) the amount or strength of the radiation at the point of interest, and (2) the readiness with which the radiation can cause the interaction of interest.

Flux Density

At any given location $\mathbf{r}$, the radiation field is generally a function of time t (e.g., a detector can pass by a radiation source) and of the radiation particles' energy E (e.g., radiation sources often emit photons or neutrons with many energies). To quantify the "strength" of a radiation field one could use the particle density $n(\mathbf{r}, E, t)$ where $n(\mathbf{r}, E, t)dV\,dE$ is the expected number of particles in differential volume element

dV about $\mathbf{r}$ with energies in dE about E at time t. Alternatively, one could use the flux density, often just called the flux, which is defined as

$$\phi(\mathbf{r}, E, t) \equiv v(E)\, n(\mathbf{r}, E, t), \tag{4.14}$$

where $v(E)$ is the speed of a radiation particle with kinetic energy E.[1]

Because $v(E)$ is the distance a radiation particle of energy E travels in a unit time, the flux $\phi(\mathbf{r}, E, t)dV\, dE$ has the interpretation of being the distance traveled in a unit time by particles in dV and $\mathbf{r}$ that have energies in dE and E.

Reaction-Rate Density

With the concepts of flux and interaction coefficient, the rate at which radiation interacts with a medium can be quantified. Specifically, let $\widehat{F}_i(\mathbf{r}, E, t)dE$ be the expected number of ith type interactions per unit time that occurs in a unit volume at $\mathbf{r}$ caused by radiation particles with energies in dE about E. Thus,

$$\begin{aligned} \widehat{F}_i(\mathbf{r}) &= \{\text{total path-length traveled by the particles in one cm}^3 \text{ in one second}\}/ \\ &\quad \{\text{average distance particles must travel for an } i\text{th type interaction} \\ &= \{\phi(\mathbf{r}, E, t)dE\}/\{1/\mu_i(\mathbf{r}, E, t)\} \end{aligned}$$

or

$$\boxed{\widehat{F}_i(\mathbf{r}, E, t)dE = \mu_i(\mathbf{r}, E, t)\, \phi(\mathbf{r}, E, t)dE.} \tag{4.15}$$

This simple expression for reaction-rate densities is a key equation for many nuclear calculations.

From the interaction rate density, all sorts of useful information can be calculated about the effects of the radiation. For example, the total number of scattering interactions that occur in some detector volume V between times t_1 and t_2 by radiation particle of all energies is

$$\text{Number of scatters in } (t_1, t_2) = \int_{t_1}^{t_2} dt \iiint_V dV \int_0^{E_{\max}} dE\, \mu_s(\mathbf{r}, E, t)\phi(\mathbf{r}, E, t). \tag{4.16}$$

Finally, to obtain the reaction rate density per *unit mass*, simply replace μ in the above expression by (μ/ρ), where ρ is the mass density of the interacting medium.

Radiation Fluence

In most interaction rate calculations, the interaction properties of the medium do not change appreciably in time. For this usual situation, the example of Eq. (4.16) reduces to

$$\text{Number of fissions in } (t_1, t_2) = \iiint_V dV \int_0^{E_{\max}} dE\, \Sigma_f(\mathbf{r}, E)\, \Phi(\mathbf{r}, E), \tag{4.17}$$

where the *fluence* of radiation between t_1 and t_2 is the time-integrated flux density, namely

$$\boxed{\Phi(\mathbf{r}, E) \equiv \int_{t_1}^{t_2} dt\, \phi(\mathbf{r}, E, t).} \tag{4.18}$$

[1]Photons, of course, always travel with the speed of light so $v(E) = c$. However, neutrons of different energies have different speeds.

For a steady-state radiation field, the fluence is simply $\Phi(\mathbf{r}, E) = (t_2 - t_1)\phi(\mathbf{r}, E)$.

Often the time interval over which the flux density is integrated to obtain the fluence is implicitly assumed to be over all prior time. Thus, the fluence at time t is

$$\Phi(\mathbf{r}, E, t) \equiv \int_{-\infty}^{t} dt' \, \phi(\mathbf{r}, E, t'). \tag{4.19}$$

The rate of change of the fluence with time, or the fluence rate, is simply the flux density. The fluence is thus used to quantify the cumulative effect of radiation interactions, while the flux density is used to quantify the rate of interactions.

4.3 Scattering Interactions

Scattered radiation is of great concern in interpreting the results of radiation detectors. First scattering transfers a spectrum of energies to the scattering medium as a result of the many different scattering angles that can occur. Likewise scattering of radiation from material near a detector cause radiation with a great variety of energies to reach a detector.

4.3.1 Differential Scattering Coefficients

The scattering coefficient μ_s is generally a function of the energy of the incident radiation. However, to describe the many features of a scattering interaction, μ_s may depend on other variables, leading to so-called *differential scattering coefficients.* Such coefficients may be a function of (1) the energy of the particle after scattering, (2) the energy of the recoil atom or electron, (3) the angles of deflection of the scattered radiation and recoil atom or electron,[2] and (4) the angles of emission of secondary particles.

For example, the doubly differential scattering interaction coefficient is defined in such a way that $d^2\mu_s(E, E', \theta_s)/dE' \, d\Omega_s$ is the probability per unit path length for an interaction in which the incident particle of energy E emerges from the interaction with energy between E' and $E' + dE'$ and with scattering angle θ_s, measured with respect to the incident direction, within the differential solid angle $d\Omega_s$.[3] In this form, μ has units such as cm^{-1} MeV^{-1} sr^{-1}. Alternatively, the interaction coefficient could be expressed in terms of energy T_r and angles θ_r for recoil electrons or atoms, or other secondary radiations. Thus, one has an equivalent doubly differential scattering interaction coefficient $d^2\mu_s(E, T_r, \theta_r)/dT_r \, d\Omega_r = d^2\mu_s(E, E', \theta_s)/dE' \, d\Omega_s$. This later description is most useful for radiation detection, because it is the energy of recoil secondary radiation that is of interest, i.e., the energy deposited in the detector.

Often it is also of interest to deal only with, say, the energy dependence or the angular distribution of scattered radiation. In this case, one or the other of the following forms of *singly differential* scattering coefficient may be used:

$$\frac{d\mu_s(E, T_r)}{dT_r} \equiv \int_{4\pi} d\Omega_r \, \frac{d^2\mu_s(E, T_r, \theta_r)}{dT_r \, d\Omega_r} \tag{4.20}$$

or

$$\frac{d\mu_s(E, \theta_r)}{d\Omega_r} \equiv \int_{0}^{\infty} dT_r \, \frac{d^2\mu_s(E, T_r, \theta_r)}{dT_r \, d\Omega_r}. \tag{4.21}$$

Here $d\mu_s(E, T_r)/dT_r$ is the probability, per unit path length of travel, that scattering produces a recoil particle with energy dT_r about T_r, without regard to scattering angle. Likewise, $d\mu_s(E, \theta_r)/\, d\Omega_r$ is the probability

---

[2] This dependence is true for isotropic media, which is assumed to be the case unless otherwise noted specifically. For crystalline and other anisotropic media, μ_s generally depends on the incident radiation direction $\mathbf{\Omega}$ and the exit radiation direction $\mathbf{\Omega}'$.

[3] In many references the differential coefficient $d^2\mu_s(E, E', \theta_s)/dE' \, d\Omega_s$ is simply written as $\mu_s(E, E', \theta_s)$.

for scattering into direction range $d\Omega_r$, without regard to the energy of the exit radiation. Finally,

$$\mu_s(E) \equiv \int_0^\infty dT_r \frac{d\mu_s(E,T_r)}{dT_r} \equiv \int_{4\pi} d\Omega_r \frac{d\mu_s(E,\theta_r)}{d\Omega_r}, \tag{4.22}$$

which is just the total linear interaction coefficient for the scattering of incident radiation of energy E, without regard to the recoil particle's energy or angle of recoil. Associated with the differential scattering coefficients are the differential microscopic scattering cross sections as a generalization of Eq. (4.10). For example

$$\frac{d^2\mu_s(E,E',\theta_s)}{dE'\,d\Omega_s} = N\frac{d^2\sigma_s(E,E',\theta_s)}{dE'\,d\Omega_s} = \frac{\rho N_a}{A}\frac{d^2\sigma_s(E,E',\theta_s)}{dE'\,d\Omega_s}. \tag{4.23}$$

4.3.2 Conservation Laws for Scattering Reactions

Conservation of Momentum

Consider the nuclear reaction illustrated in Fig. 4.2. Reactants are a stationary target of rest mass M and an incident particle of rest mass m, kinetic energy E, and momentum $\mathbf{p}$. Products are one particle of rest mass m', kinetic energy E', and momentum $\mathbf{p}'$, and a second particle of rest mass M', kinetic energy T, and momentum $\mathbf{p}_r$. Energies E, E', and T are kinetic energies in the laboratory system, except for photons for which the adjective *kinetic* is not used.

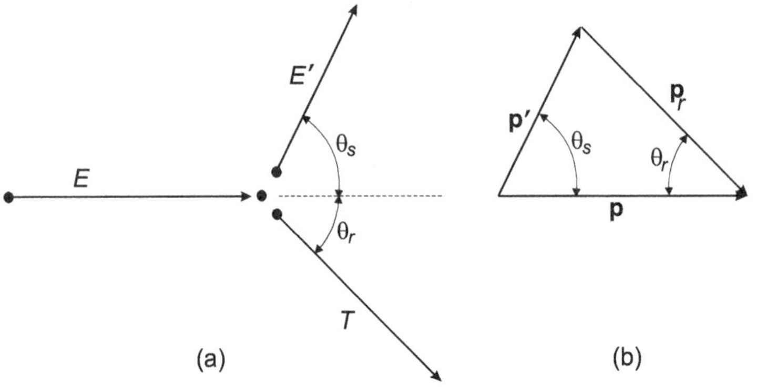

Figure 4.2. Nuclear reaction with one reactant initially stationary and with products emerging at angles θ_s and θ_r in the laboratory system. (a) Energies and scattering angles. (b) Conservation of linear momentum.

Energy and momentum of the incident particle are related by the equation

$$E = \sqrt{p^2c^2 + m^2c^4} - mc^2, \tag{4.24}$$

in which scalar p is the magnitude of the momentum $\mathbf{p}$ and c is the speed of light in vacuum, 2.9979×10^8 m/s. Similar equations relate E' and $\mathbf{p}'$, and T and $\mathbf{p}_r$. Conservation of linear momentum is depicted in the vector diagram Fig. 4.2(b), in which the initial momentum vector $\mathbf{p}$ must equal the sum of the product momentum vectors $\mathbf{p}'$ and $\mathbf{p}_r$. From the law of cosines,

$$(p')^2 = p^2 + p_r^2 - 2pp_r\cos\theta_r \tag{4.25}$$

and

$$p_r^2 = p^2 + (p')^2 - 2pp'\cos\theta_s. \tag{4.26}$$

It is also clear from the figure that from the law of sines,

$$p'/\sin\theta_r = p_r/\sin\theta_s. \tag{4.27}$$

Conservation of Energy

The fact that total energy, the sum of kinetic and rest-mass energy, is conserved in the nuclear reaction is expressed by the equation

$$E = E' + T - Q, \tag{4.28}$$

in which Q is the Q *value* of the reaction. If it is assumed that the reactants are initially in the ground state, i.e., $m = m_o$ and $M = M_o$, the Q value is given by

$$Q = \Delta M_o c^2 - W \equiv (m_o + M_o - m_o' - M_o')c^2 - \sum E_{ex}, \tag{4.29}$$

in which $\Delta M_o c^2$ is energy equivalent of the changes in the ground-state rest masses of the reactants, and W is the sum of the nuclear excitation energies E_{ex} of the reaction products.

Application of the Conservation Laws

If an incident particle is a photon, the rest mass is zero and, from Eq. (4.24), $p = E/c$. If the incident particle is a neutron, classical mechanics may be assumed to apply for energies less than about 15 MeV, and $p = \sqrt{2mE}$. If the incident particle is an electron, then relativistic conservation laws generally should be applied and the following equations are useful. The ratio of the particle speed to the speed of light in vacuum is identified as $\beta = v/c$, and, with $\epsilon \equiv E/mc^2$, then obtains the results

$$\beta^2 = \frac{\epsilon(\epsilon+2)}{(\epsilon+1)^2}, \quad \frac{\beta^2}{1-\beta^2} = \epsilon(\epsilon+2) \tag{4.30}$$

and

$$p^2c^2 = E(E + 2mc^2) = \frac{\beta^2 m^2 c^4}{1-\beta^2}. \tag{4.31}$$

Equations (4.24) through (4.31) provide the framework for treatment of the kinematics of a large class of nuclear reactions. They are drawn upon throughout this book. The topics now to be addressed are scattering reactions, for which $m' = m$ and $M' = M$. Most reactions addressed are elastic scattering, for which the Q value is zero. The exception is inelastic scattering of a neutron by an atomic nucleus. That reaction is *endothermic*, and the Q value is negative and equal in magnitude to the excitation energy of the final state of the nucleus with respect to its ground state.

4.3.3 Scattering of Photons by Free Electrons

This scattering process is known as the *Compton effect*, to recognize its 1923 discovery by A.H. Compton. Since the photon has zero rest mass, $p = E/c$ and $p' = E'/c$. Because the electron is free, the scattering is elastic and $E = E' + T$. Thus, Eqs. (4.25) and (4.26) reduce to

$$E' = \frac{E}{1 + (E/m_e c^2)(1 - \cos\theta_s)}, \quad 0 \le \theta_s \le \pi, \tag{4.32}$$

and

$$T = \frac{2m_e c^2 E^2 \cos^2\theta_r}{(E + m_e c^2)^2 - E^2\cos^2\theta_r}, \quad 0 \le \theta_r \le \pi/2. \tag{4.33}$$

Here m_ec^2 is the rest-mass energy of the electron, 0.51099892 MeV or 8.1871×10^{-14} J. These equations are very much simplified if expressed in terms of the dimensionless variables[4] $\Lambda = m_ec^2/E$ and $\tau = T/m_ec^2$, namely,

$$\Lambda' = 1 + \Lambda - \cos\theta_s, \tag{4.34}$$

and

$$\tau = \frac{2\cos^2\theta_r}{(1+\Lambda)^2 - \cos^2\theta_r}. \tag{4.35}$$

Note that $\Lambda \leq \Lambda' \leq \Lambda + 2$. Note too that the maximum energy transfer to initial kinetic energy of a recoil electron is given by

$$\left[\frac{T}{E}\right]_{\max} = \frac{2}{\Lambda + 2}, \tag{4.36}$$

which approaches unity for incident photons of very high energy. Similarly,

$$\frac{E'}{E} = \frac{\Lambda}{\Lambda + 1 - \cos\theta_s} \tag{4.37}$$

and

$$\left[\frac{E'}{E}\right]_{\min} = \frac{\Lambda}{\Lambda + 2}, \tag{4.38}$$

which approaches unity for incident photons of very low energy. A relationship between scattering angles is given by

$$\cot\theta_r = (1 + \Lambda^{-1})\tan(\theta_s/2). \tag{4.39}$$

4.3.4 Scattering of Neutrons by Atomic Nuclei

The scattering interaction is the most probable interaction of fast neutrons and is the mechanism relied upon to slow these neutrons to thermal energies, at which they can be absorbed through (n, γ) reactions. There are two distinct types of scattering processes, both of importance in fast neutron attenuation. In *capture scattering* the incident neutron is absorbed by the scattering nucleus to form a compound nucleus which subsequently decays by the emission of a neutron. If the residual nucleus is left in the ground state, the scattering is called *elastic.* If the residual nucleus is left in an excited state, the scattering is called *inelastic.* The other type of scattering is referred to as *potential scattering.* In this process, which is always elastic, the incident neutron is scattered by the nucleus as a whole—analogous to the diffraction of the incident neutron wave by the entire nuclear potential. Capture-scattering cross sections generally exhibit resonance behavior, while potential scattering cross sections usually vary slowly with energy.

In all scattering processes the total energy and momentum must be conserved. Except for thermal neutron scattering, for which the thermal motion of the target atoms may be comparable to the neutron speed, one can properly neglect the initial kinetic energy of the scattering nucleus in the laboratory coordinate system. Furthermore, these scattering interactions may be treated by classical mechanics, so that energy and momentum are related by $p^2 = 2mE$. The ratio M/m of the mass of the target nucleus to the mass of

---

[4]In this book, the dimensionless variable Λ is called the *reduced* wavelength. Clearly it does not have units of length. Usually in most texts the symbol for reduced wavelength is λ, but because λ is already used as the decay constant, and later is used as actual wavelength, to avoid confusion the symbol Λ is used as a substitute. It might better be called the *wave number.* It may be construed more properly as the ratio of the actual wavelength hc/E to the *Compton wavelength* $\Lambda_c = hc/m_ec^2 = 2.4263 \times 10^{-12}$ m, in which h is Planck's constant, 6.6261×10^{-34} J s. More correctly, then, Λ is the wavelength in units of the Compton wavelength.

the neutron is so nearly equal to the atomic mass A of the target nucleus that in this book, the difference is neglected. In the general case of inelastic scattering, Eqs. (4.24) to (4.26) reduce to (see Problem 16)

$$\omega_s(E,E') = \frac{1}{2}\left[(A+1)\sqrt{\frac{E'}{E}} - (A-1)\sqrt{\frac{E}{E'}} - \frac{QA}{\sqrt{EE'}}\right], \tag{4.40}$$

in which $\omega_s \equiv \cos\theta_s$, $-1 \le \omega_s \le +1$. It follows that

$$\sqrt{E'} = \frac{1}{(A+1)}\left[\omega_s\sqrt{E} \pm \sqrt{E(\omega_s^2 + A^2 - 1) + A(A+1)Q}\right] \tag{4.41}$$

and

$$T = \frac{AE}{(1+A)^2}\left[\Delta + 2\cos^2\theta_r \pm \sqrt{(\Delta + 2\cos^2\theta_r)^2 - \Delta^2}\right], \tag{4.42}$$

in which

$$\Delta = \frac{Q(1+A)}{AE} \tag{4.43}$$

and $0 \le \theta_r \le \pi/2$.

Because $\sqrt{E'}$ physically must be non-negative, only the plus sign in Eqs. (4.41) and (4.42) gives meaningful results for elastic scattering ($Q = 0$) and for most inelastic scattering. However, when a neutron with energy only slightly greater than $|Q|$ is inelastically scattered, both signs may lead to physically realistic results. This so-called "double value" region is discussed below.

4.3.5 Threshold Energies for Neutron Inelastic Scattering

From Eq. (4.41), it is clear that for E' to be real it is necessary that

$$E \ge -\frac{A(A+1)Q}{\omega_s^2 + A^2 - 1}. \tag{4.44}$$

The least value of E allowing inelastic scatter, the threshold energy E_t, corresponds to $\omega_s = 1$, namely,

$$E_t = -\frac{A+1}{A}Q. \tag{4.45}$$

The energies of the first and second excited nuclear states tend to decrease as the atomic weight A increases. Consequently, the threshold for inelastic scattering tends to decrease as the atomic mass of the scatterer increases. The level spacings of the light elements and the magic number[5] nuclides are relatively large and hence inelastic scattering is generally less significant for such nuclides. Moreover, the odd-even and even-odd nuclides tend to have smaller thresholds than the even-even nuclides.

It is also clear from Eq. (4.41) that for values of E just greater than E_t, the scattered neutron can appear in the forward direction, $\omega_s > 0$, with either of two distinct positive energies. The largest value of E for which this is possible—the cutoff energy E_c—is the largest value of E for which $\omega_s\sqrt{E} - \sqrt{E(\omega_s^2 + A^2 - 1) + A(A+1)Q} \ge 0$, namely,

$$E_c = -\frac{AQ}{A-1}. \tag{4.46}$$

For E between E_t and E_c, there is a maximum scattering angle, or minimum ω_s, permitting real values of E', namely, $\omega_{s,min} = \sqrt{1 - A^2 - QA(A+1)/E}$. Discussion of the implications of dual values of E' may be found in the works of Evans [1955] and Amaldi [1959].

[5] A magic number nucleus is one in which the number of neutrons or protons equals 2, 8, 20, 50, 82, or 126. When the nucleus is magic, a particularly stable configuration of the nucleons in the nucleus is achieved analogous to closed electron shells in atomic physics.

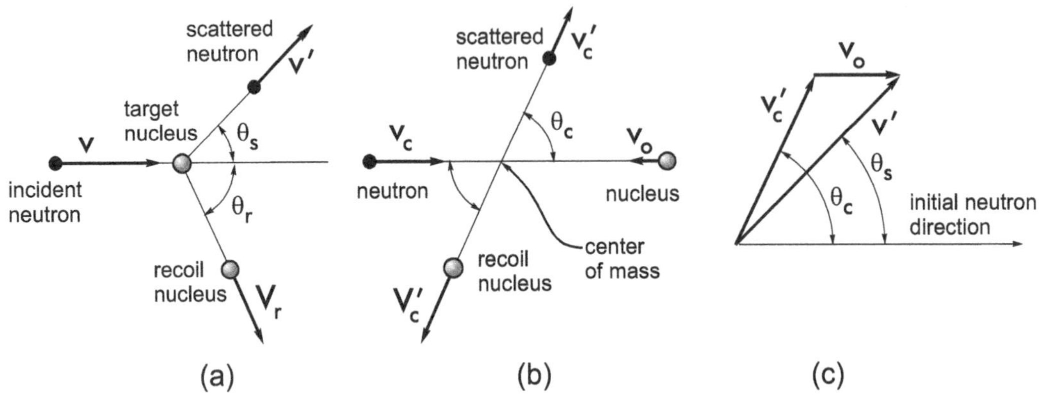

Figure 4.3. Conversion between laboratory and center-of-mass coordinate systems. (a) Velocities in the laboratory system. (b) Velocities in the center-of-mass system. (c) Relationship between scattering angles in the two systems.

4.3.6 Neutron Scattering in the Center-of-Mass System

Particle kinematics may be described and understood completely in the laboratory coordinate system used implicitly to this point. It may therefore seem a needless complication to introduce a different coordinate system—the center-of-mass system. Nevertheless, there are certain aspects of particle kinematics that are greatly simplified when described in the new system. This is especially true for angular characteristics of scattering cross sections and angular distributions of neutrons after scattering. Such distributions are more easily and more precisely described in the center-of-mass system. For example, the angular distribution of neutrons scattered from hydrogen atoms or, indeed, the angular distribution of billiard balls after elastic collision is very anisotropic in the laboratory system but completely isotropic in the center-of-mass system.

Consider the scattering reaction illustrated in Fig. 4.3(a). The target nucleus is initially at rest. Prior to scatter, the center of mass is moving to the right with velocity $\mathbf{v}_o$ defined in such a way that the total linear momentum $m\mathbf{v}$ is equal to $(m+M)\mathbf{v}_o$. In the center-of-mass system, Fig. 4.3(b), the center of mass is stationary. Prior to scatter, the target atom is moving to the left with velocity $\mathbf{v}_o$ and the incident neutron is moving to the right with velocity $\mathbf{v}_c = \mathbf{v} - \mathbf{v}_o$. The total momentum before and after scattering is zero. Thus, products of the scattering must move in opposite directions, the neutron with velocity $\mathbf{v}'_c$ and the recoil nucleus with velocity $\mathbf{V}'_c$. Figure 4.3(c) illustrates the relationship between the scattering angles in the two systems and, as addressed in Problem 18, reveals that in terms of their cosines, where from Fig. 4.3, $\omega_s = \cos\theta_s$, $\omega_r = \cos\theta_r$, and $\omega_c = \cos\theta_c$,

$$\omega_s = \frac{\gamma + \omega_c}{\sqrt{1 + 2\gamma\omega_c + \gamma^2}} \tag{4.47}$$

and[6]

$$\omega_c = -\gamma(1-\omega_s^2) \pm \omega_s\sqrt{1-\gamma^2(1-\omega_s^2)}, \tag{4.48}$$

[6]For elastic scattering, only the positive sign in Eq. (4.48) applies, with one exception. For $A = 1$, $\omega_s \geq 0$, and $\omega_c = -1 + 2\omega_s^2$. For inelastic scattering, the positive sign applies except in the region of dual values of E', and then $\omega_s > 0$ and ω_c is dual valued.

in which (see Problem 20)

$$\gamma \equiv \frac{v_o}{v_c'} = \left[A^2 + \frac{A(A+1)Q}{E}\right]^{-1/2}. \tag{4.49}$$

Energies after scattering are given by (see Problem 21)

$$E' = \frac{1}{2}E(1+\alpha) + \frac{1}{2}(1-\alpha)E\omega_c\sqrt{1+\Delta} + \frac{QA}{A+1} \tag{4.50}$$

and

$$T = E - E' + Q = \frac{1}{2}E(1-\alpha)\left[1 - \omega_c\sqrt{1+\Delta}\right] + \frac{Q}{A+1}, \tag{4.51}$$

in which

$$\alpha \equiv \left[\frac{A-1}{A+1}\right]^2. \tag{4.52}$$

Note that E'/E always lies between the limits

$$\frac{E'_{\rm min}}{E} = \left[\frac{A\sqrt{1+\Delta}-1}{1+A}\right]^2 \text{ and } \frac{E'_{\rm max}}{E} = \left[\frac{A\sqrt{1+\Delta}+1}{1+A}\right]^2. \tag{4.53}$$

4.3.7 Limiting Cases in Classical Mechanics of Elastic Scattering

By setting $Q = 0$ and $E' = E - T$ in Eqs. (4.41) and (4.50) and using Eq. (4.27), one finds that

$$\frac{T}{E} = \frac{2A}{(1+A)^2}(1-\omega_c) = \frac{4A}{(1+A)^2}\omega_r^2, \tag{4.54}$$

where $A \equiv M/m$. Thus,

$$\omega_r = \sqrt{\frac{1-\omega_c}{2}} = \sin\!\left(\frac{\theta_c}{2}\right). \tag{4.55}$$

By substituting $\omega_c = 1 - 2\omega_r^2$ from this equation into Eq. (4.47), one finds that for the classical mechanics of elastic scattering,

$$\omega_s = \frac{1 - [2A\omega_r^2/(A+1)]}{\sqrt{1 - [4A\omega_r^2/(A+1)^2]}}. \tag{4.56}$$

Listed as follows are limiting values of scattering parameters for three cases.

Case 1: $m = M$. This case describes neutron-proton scattering and the scattering by stationary free electrons of low-energy electrons or positrons. All scattering angles are possible in the center-of-mass system, but in the laboratory system, θ_s and θ_r are limited to the range 0 to $\pi/2$.

θ_c	θ_s	θ_r	E'/E	T/E
0	0	$\pi/2$	1	0
$\pi/2$	$\pi/4$	$\pi/4$	1/2	1/2
π	$\pi/2$	0	0	1

Case 2: $m < M$. This case describes electron scattering by nuclei and neutron scattering from heavier nuclei. Scattering is possible for values of θ_s and θ_c between 0 and π but for θ_r only between 0 and $\pi/2$. As A increases, θ_s approaches θ_c.

θ_c	θ_s	θ_r	E'/E	T/E
0	0	$\pi/2$	1	0
$\pi/2$	$\tan^{-1}A$	$\pi/4$	$(1+\alpha)/2$	$(1-\alpha)/2$
π	π	0	α	$1-\alpha$

Case 3: $m > M$. This case approximately describes scattering of heavy charged particles by electrons. Scattering is possible for $0 \le \sin\theta_s \le M/m$. As $M/m \to 0$, $\theta_s \to 0$.

4.3.8 Relativistic Elastic Scattering of Electrons and Heavy Charged Particles

The basic working equation follows directly from the conservation laws, Eqs. (4.24) through (4.29), namely,

$$T = \frac{2Mc^2 p^2c^2 \omega_r^2}{\left[Mc^2 + \sqrt{p^2c^2 + m^2c^4}\right]^2 - p^2c^2 \omega_r^2}. \tag{4.57}$$

The maximum energy is transferred to the recoil target when $\omega_r = 1$. In terms of $\beta \equiv v/c$, and in terms of Eqs. (4.30) and (4.31),

$$T_{\max} = \frac{\beta^2}{1-\beta^2}\left[\frac{2Mm^2c^2}{m^2 + M^2 + [2mM/\sqrt{1-\beta^2}]}\right]. \tag{4.58}$$

Electron-Electron Scattering

Here $M = m = m_e$. The scattered and recoil electrons are indistinguishable, so the convention is adopted that the one with the lesser energy is identified as the recoil:

$$T = \frac{\beta^2}{1-\beta^2}\left[\frac{2m_ec^2\,\omega_r^2}{\left(1 + [1/\sqrt{1-\beta^2}]\right)^2 - \beta^2\omega_r^2/(1-\beta^2)}\right] \tag{4.59}$$

and

$$T_{\max} = \frac{\beta^2}{1-\beta^2}\left[\frac{m_ec^2}{1 + [1/\sqrt{1-\beta^2}]}\right] = E. \tag{4.60}$$

4.4 Cross Sections for Photon Interactions

For details of the mechanisms of photon interactions, the reader is referred to the standard reference works of Heitler [1954] and Evans [1955]. For comprehensive data tabulations, the reader is referred to ANSI [1991], Biggs and Lighthill [1972], Cullen [1994], Cullen et al. [1989], Hubbell [1969, 1982], Hubbell and Seltzer [1995], Hubbell et al. [1975, 1979], Plechaty et al. [1975], Seltzer [1993], and Trubey et al. [1989]. Photon energies between 10 eV and 10 MeV are important in the design and analysis of photon radiation detectors. For this energy range, only the photoelectric effect, pair production, and Compton scattering mechanisms of interaction are significant. Of these three, the photoelectric effect predominates at the lower photon energies. Pair production is important only for higher-energy photons. Compton scattering predominates at intermediate energies. In rare instances one may need to account also for coherent scattering.

4.4.1 Thomson Cross Section for Incoherent Scattering

Incoherent scattering refers to the interaction of a photon with an individual electron, as distinguished from the coherent interaction of a photon with all electrons of an atom. It is assumed in the discussion that follows that the incident radiation is not polarized.

In the limit of zero photon energy, scattering of the photon by a free electron may be treated by the classical theory of radiation. The electron in the electromagnetic field of the incident radiation vibrates with the same frequency as that of the incident radiation, thereby giving rise to the emission of secondary electromagnetic radiation of the same frequency. Named for discoverer of the electron, J.J. Thomson (1856–1940) [see Thomson 1897], the total cross section *per electron* for such scattering is

$$\sigma_T = \frac{8}{3}\pi r_e^2, \tag{4.61}$$

in which r_e is the *classical electron radius*. The value of r_e is given by

$$r_e = \frac{q_e^2}{4\pi\epsilon_o m_e c^2}, \tag{4.62}$$

where q_e is the electronic charge, 1.6022×10^{-19} C, and ϵ_o is the permittivity of free space, 8.8542×10^{-14} F cm^{-1}. Thus, $r_e = 2.8179 \times 10^{-13}$ cm and $\sigma_T = 6.6525 \times 10^{-25}$ cm^2.

For unpolarized incident radiation, the Thomson *electronic* cross section per steradian for scattering at angle θ_s, as in Fig. 4.2, is

$$\frac{d\sigma_T(\theta_s)}{d\Omega_s} = \frac{1}{2} r_e^2 (1 + \cos^2\theta_s). \tag{4.63}$$

Knowledge of the Thomson cross section is important for two reasons. It is the low-energy limit for the incoherent (Compton) scattering cross section. It is also, as is evident from Eq. (4.82), the basis for computing Rayleigh-scattering cross sections for interactions of photons coherently with atomic electrons.

4.4.2 Klein-Nishina Cross Section for Incoherent Scattering

The atomic Compton cross section for incoherent scattering may be approximated as the atomic number times the electronic cross section for scattering by a free electron (i.e., the total Klein-Nishina cross section σ_{KN}), namely

$$\sigma_c = Z\,\sigma_{KN}. \tag{4.64}$$

The differential scattering cross section *per electron* $d\sigma_{KN}(\Lambda, \theta_s)/d\Omega_s$ was calculated by Klein and Nishina [1929] and by Tamm [1930]. This differential scattering cross section for unpolarized photons is[7]

$$\frac{d\sigma_{KN}(\Lambda,\theta_s)}{d\Omega_s} = \frac{1}{2} r_e^2 \frac{\Lambda^2}{(1+\Lambda-\cos\theta_s)^2}\left(\frac{\Lambda}{1+\Lambda-\cos\theta_s} + \frac{1+\Lambda-\cos\theta_s}{\Lambda} - \sin^2\theta_s\right). \tag{4.65}$$

A convenient alternative form is

$$\frac{d\sigma_{KN}(E,\theta_s)}{d\Omega_s} = \frac{1}{2} r_e^2 \zeta[1+\zeta^2 - \zeta(1-\cos^2\theta_s)], \tag{4.66}$$

in which $\zeta \equiv E'/E = \Lambda/\Lambda'$ is given by Eq. (4.37). This cross section is illustrated in Fig. 4.4. Note that as Λ approaches infinity, that is, E approaches zero, ζ approaches unity and Eq. (4.66) reduces to the Thomson formula, Eq. (4.63).

[7] Although both polarized and unpolarized cases were derived, the unpolarized case is generally of more interest for radiation detection.

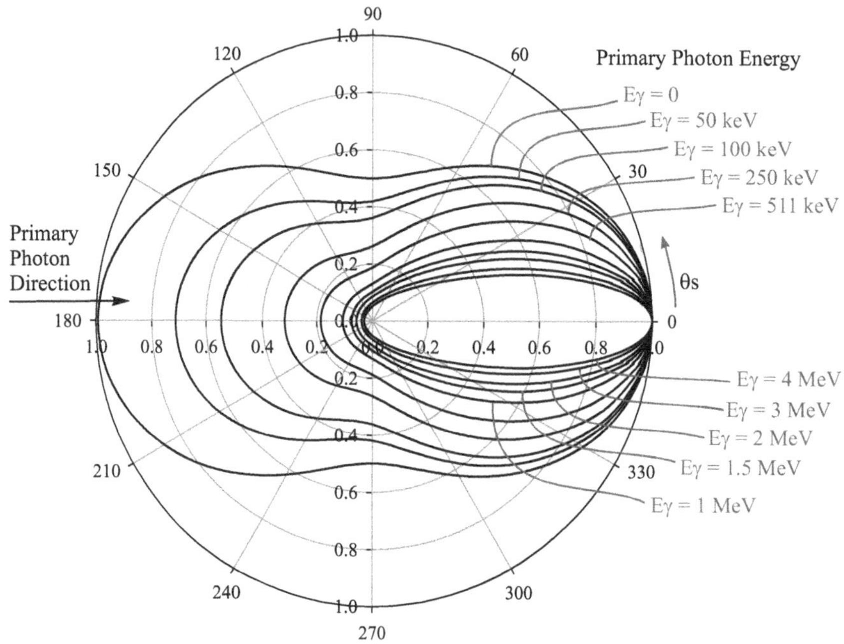

Figure 4.4. Shown is the function $d\sigma_{KN}/d\Omega_s$ divided by r_e^2 as a function of scattering angle θ. Notice that the number of photons scattered into a unit differential solid angle increases in the forward direction as the primary photon energy increases.

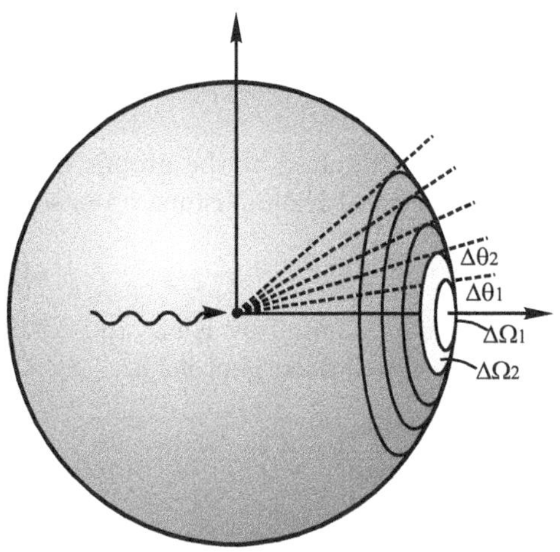

Figure 4.5. For equal $\Delta\theta_i$, the solid angle $\Delta\Omega_i$ increases as θ increases from 0 to $\pi/2$.

One must be wary in interpreting Fig. 4.4. At first glance an observer might mistakenly think that high energy photons are primarily scattered directly forward. Although the term $d\sigma_{KN}/d\Omega_s$ increases with decreasing angle θ as energy increases (as shown in Fig. 4.4), the total differential solid angle defined by $d\Omega_s$ *decreases* as θ decreases (see Fig. 4.5). This change in solid angle with angle θ is defined by

$$\frac{d\Omega_s}{d\theta_s} = 2\pi \sin\theta_s, \tag{4.67}$$

which clearly increases as θ_s approaches 90°. Further, when $\theta_s = 0$, $d\Omega_s/d\theta_s = 0$. Hence, the density of photons scattering into $d\Omega_s$ increases with decreasing θ, but the total probability of photons scattering into the differential solid angle, $d\Omega_s$, is proportional to the product of the differential solid angle and the density of photons scattered into a unit solid angle about angle θ, namely the differential scattering cross section.

The distribution of photons scattered per unit polar angle θ_s is quite different than $d\sigma_{KN}/d\Omega_s$. Multiplying Eq. (4.66) by Eq. (4.67) yields the probability of scattering a photon of initial energy E through angle θ_s, namely

$$\frac{d\sigma_{KN}(E,\theta_s)}{d\theta_s} = \frac{d\sigma_{KN}(E,\theta_s)}{d\Omega_s}\frac{d\Omega_s}{d\theta_s} = \pi r_e^2 \zeta[1+\zeta^2 - \zeta(1-\cos^2\theta_s)]\sin\theta_s. \tag{4.68}$$

This cross section is plotted in Fig. 4.6. Notice that there is still a preference for forward scattering with increasing photon energy; however, the highest cross section is no longer at $\theta_s = 0^o$, and it changes as a function of E. Notice also, quite unlike Fig. 4.4, which shows the density of photons scattering into $d\Omega_s$, that the highest differential cross section in Fig. 4.6 for low energy photon (Thomson) scattering is at scattering angles of 55° and 125°.

A related quantity is the *energy scattering* differential cross section,

$$\frac{d\sigma_{KNe}(E,\theta_s)}{d\Omega_s} \equiv \frac{E'}{E}\frac{d\sigma_{KN}(E,\theta_s)}{d\Omega_s} = \frac{Z}{2}r_e^2\zeta^2[1+\zeta^2-\zeta(1-\cos^2\theta_s)]. \quad (4.69)$$

The total Compton cross section *per atom*, based on the free-electron approximation, is obtained from Eq. (4.65) by integration over all directions.

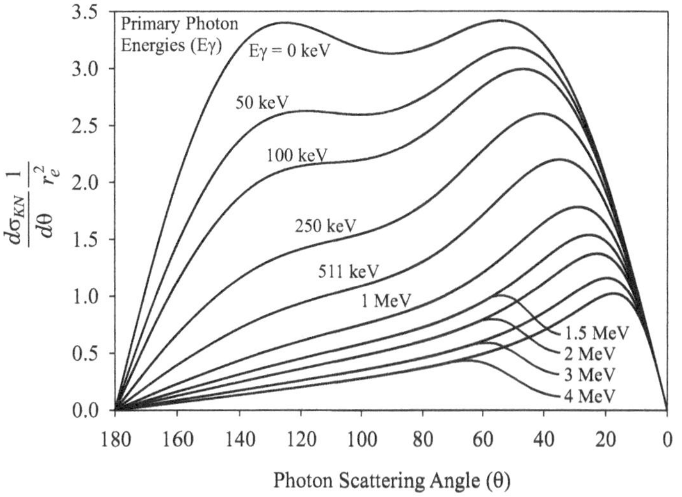

Figure 4.6. Shown is $d\sigma_{KN}/d\theta_s$ divided by r_e^2 as a function of scattering angle θ_s. Notice that the number of photons scattered into angle θ_s increases towards the forward direction (approaching $\theta_s = 0$) as the primary photon energy increases; however, $d\sigma_{KN}/d\theta_s$ approaches zero as θ_s goes to zero for all photon energies.

$$\sigma_C(\Lambda) = Z\sigma_{KN}(\Lambda) = 2\pi Z\int_{-1}^{+1} d(\cos\theta_s)\,\frac{d\sigma_{KN}(\Lambda,\theta_s)}{d\Omega_s}$$
$$= \pi Z r_e^2\Lambda\left[(1-2\Lambda-2\Lambda^2)\ln\left(1+\frac{2}{\Lambda}\right)+\frac{2(1+9\Lambda+8\Lambda^2+2\Lambda^3)}{(\Lambda+2)^2}\right]. \quad (4.70)$$

Other Differential Compton Cross Sections

The cross section, per unit wavelength, for scattering of a photon into wavelength Λ' without regard to angle is, from Eq. (4.69),[8]

$$\frac{d\sigma_C(\Lambda,\Lambda')}{d\Lambda'} = Z\frac{d\sigma_{KN}(\Lambda,\Lambda')}{d\Lambda'} = Z\pi\, r_e^2\left(\frac{\Lambda}{\Lambda'}\right)^2\left[\left(\frac{\Lambda}{\Lambda'}\right)+\left(\frac{\Lambda'}{\Lambda}\right)+(\Lambda'-\Lambda)(\Lambda'-\Lambda-2)\right]. \quad (4.71)$$

A related quantity is the cross section, per unit recoil electron energy, for creating a recoil electron with energy T. Here it is convenient to use the ratio $\tau \equiv T/m_ec^2$. Because $1/\Lambda - 1/\Lambda' = \tau$, $d\Lambda'/d\tau = \Lambda^2(1-\Lambda\tau)^{-2}$. Thus, $\sigma_{KN}(\Lambda,\tau) = \sigma_{KN}(\Lambda,\Lambda')d\Lambda'/d\tau$, or

$$\frac{d\sigma_C(\Lambda,\tau)}{d\tau} = \pi Z\, r_e^2\Lambda^2\left[(1-\Lambda\tau)+(1-\Lambda\tau)^{-1}+\frac{(\Lambda^2\tau)(\Lambda^2\tau+2\Lambda\tau-2)}{(1-\Lambda\tau)^2}\right]. \quad (4.72)$$

Energy Spectrum of Recoil Electrons

Unlike shielding analyses in which the more penetrating scattered photon is of primary interest, for energy spectroscopy detectors it is the energy spectrum of the recoil secondary electron produced in Compton

[8]Note that $\sigma(\Lambda,\theta_s) = -\sigma(\Lambda,\Lambda') \times d\Lambda'/d\Omega_s$ and that $d\Lambda'/d\Omega_s = (1/2\pi)\,d\Lambda'/d\omega_s = -1/2\pi$.

scattering that is of primary interest. The energy of the secondary electron is simply $T = E - E'$ or from Eq. (4.37), in terms of dimensionless quantities,

$$\tau = \frac{1}{\Lambda} - \frac{1}{\Lambda'} = \frac{1 - \cos\theta_s}{\Lambda(1 + \Lambda - \cos\theta_s)}. \tag{4.73}$$

Because $\Lambda = m_e c^2/E$ and $\tau = T/m_e c^2$, this relation reduces to

$$T = \frac{E(1 - \cos\theta_s)}{m_e c^2/E + (1 - \cos\theta_s)}. \tag{4.74}$$

The Compton recoil electron energy function described in Eq. (4.73) is plotted in Fig. 4.7, which clearly shows that the majority of energy is lost to the Compton recoil electrons for angles θ_s greater than 90°. This result means that, although more photons are generally scattered in the forward direction, as shown in Fig. 4.6, the energy transferred to the Compton recoil electrons is distributed over a wider range of energies. Conversely, fewer photons are scattered from 90°–180°; however, the energy transferred to the Compton recoil electrons is distributed over a smaller range of energies as T approaches the maximum energy defined by Eq. (4.36).

Also of importance is the probability the recoil electron has a particular value. Let $g(\Lambda, T)$ be the probability that the recoil electron has an energy, per unit energy about T. This probability density function can be computed as

$$g(\Lambda, T) = g(\Lambda, \tau)\frac{d\tau}{dT} = \frac{1}{m_e c^2}\frac{d\sigma_C(\Lambda, \tau)/d\tau}{\sigma_C(\Lambda)}. \tag{4.75}$$

This expression is readily evaluated from Eqs. (4.70) and (4.72) and is shown in Fig. 4.8.

Compton Energy-Absorption and Energy-Scattering Cross Sections

The mean fraction of the photon energy transferred to the recoil electron is designated as f_C and the Compton energy-absorption cross section[9] per electron is defined as

$$\sigma_{Ca}(\Lambda) \equiv f_C\, \sigma_C(\Lambda). \tag{4.76}$$

The factor f_C is the weighted average of $T/E = \Lambda\tau$, namely,

$$f_C = \frac{1}{\sigma_C(\Lambda)} \int_0^{2/[\Lambda(\Lambda+2)]} d\tau\ \Lambda\tau\, \sigma_C(\Lambda, \tau). \tag{4.77}$$

The energy-scattering cross section is the product of the total cross section and the mean fraction of the photon energy retained by the scattered photon:

$$\sigma_{Ce}(\Lambda) = (1 - f_C)\, \sigma_C(\Lambda). \tag{4.78}$$

It is evident from the definition of f_C that the factor $(1 - f_C)$ can be evaluated as the mean value of $E'/E = \Lambda/\Lambda'$, namely,

$$1 - f_C = \frac{1}{\sigma_C(\Lambda)} \int_\Lambda^{\Lambda+2} d\Lambda' \frac{\Lambda}{\Lambda'}\, \sigma_C(\Lambda, \Lambda'). \tag{4.79}$$

[9]The Compton energy absorption cross section thus defined is not a true cross section for photon absorption since, in the interaction, a scattered photon always results. Rather, it is an effective energy absorption cross section with respect to the incident photon energy such that the product $E(\mu_{Ca}/\rho)\phi$ is the rate per unit mass at which energy is transferred to initial kinetic energy of Compton recoil electrons.

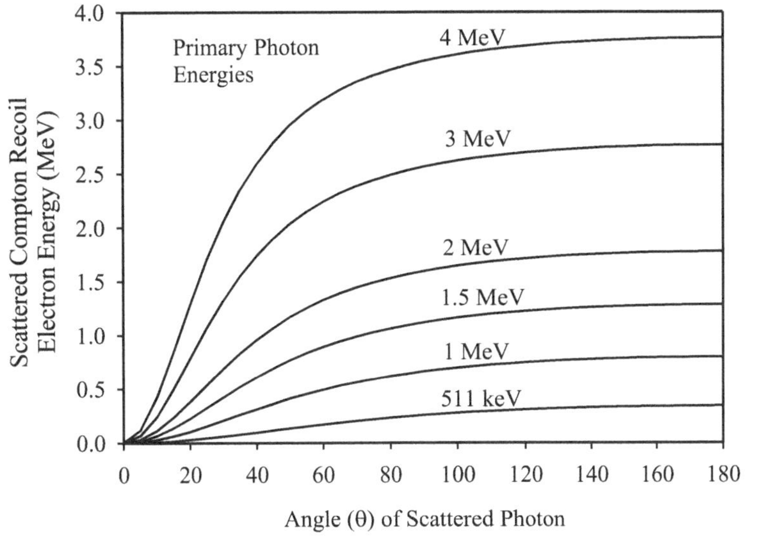

Figure 4.7. The Compton recoil electron energy as a function of primary photon energy and scattered photon angle θ_s. Notice that the change in Compton electron energy is small for photon scattering angles above 90°.

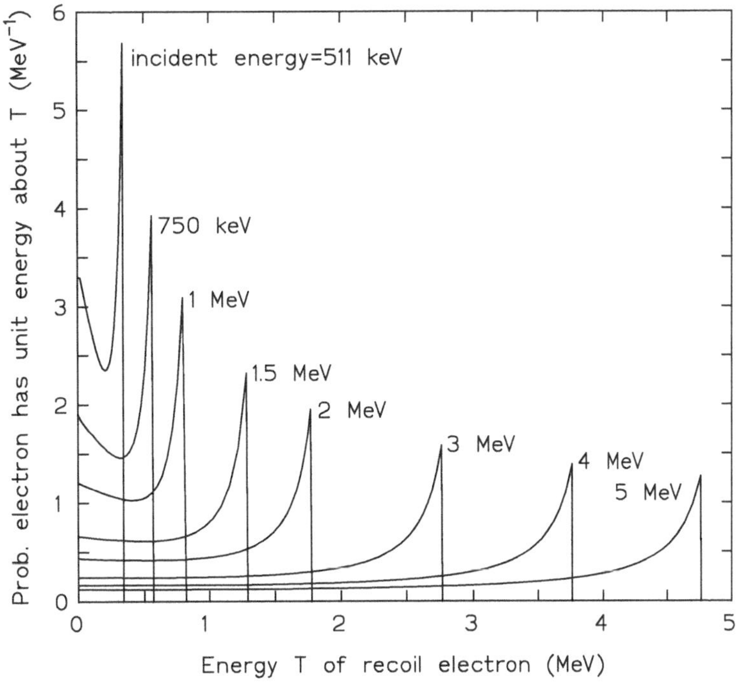

Figure 4.8. The Compton recoil electron spectrum as a function of Compton electron energy.

4.4.3 Incoherent Scattering Cross Sections for Bound Electrons

The equations for scattering from free electrons break down when the kinetic energy of the recoil electron would be comparable to its binding energy in the atom. Thus, binding effects might be thought to be an important consideration for the attenuation of low-energy photons in media of high atomic number. For example, the binding energy of K-shell electrons in lead is 88 keV. However, under these same circumstances, cross sections for the photoelectric interaction of photons greatly exceed incoherent scattering cross sections. Radiation attenuation in this energy region is dominated by photoelectric interactions, and in most attenuation calculations, the neglect of electron binding effects on incoherent scattering causes negligible error. Corrections for electron binding and related data are available in the literature [e.g., Storm and Israel 1967; Biggs and Lighthill 1972; Plechaty et al. 1975; Hubbell 1982; Trubey et al. 1989]. The total atomic cross section for incoherent scattering by bound electrons is given by

$$\sigma_{incoh}(\Lambda) = 2\pi \int_{-1}^{+1} d(\cos\theta_s)\, S(x, Z) \frac{d\sigma_{KN}(\Lambda, \theta_s)}{d\Omega_s}, \tag{4.80}$$

in which $S(x, Z)$ is the *incoherent scattering function* [Hubbell et al. 1975] and x is the *momentum-transfer parameter*, given approximately by

$$x = \frac{E}{hc} \sin(\theta_s/2). \tag{4.81}$$

Figure 4.9 shows the relative importance, in lead, of electron binding effects by comparing photoelectric cross sections with those for incoherent scattering from both free and bound electrons. As is apparent, $S(x, Z)$ is very small for low-energy photons and approaches Z as photon energy increases.

4.4.4 Coherent (Rayleigh) Scattering

In competition with the incoherent scattering of photons by individual electrons is coherent scattering by the electrons of an atom collectively. Since the recoil momentum in the Rayleigh interaction is taken up by the atom as a whole, the energy loss of the gamma photon is slight and the scattering angle small. For example, for 1-MeV photons scattering coherently from iron atoms, 75% of the photons are scattered within a cone of less than 4° half-angle [Hubbell 1969]. As is apparent from Fig. 4.9, coherent scattering cross sections may greatly exceed incoherent scattering cross sections, especially for low-energy photons and high-Z materials. However, because of the minimal effect on photon energy and direction, and because the coherent scattering cross section is so much less than the cross section for the photoelectric effect, it is common practice to ignore Rayleigh scattering in the design of many photon detectors, especially when electron binding effects mentioned in the preceding section are ignored. Named for the 4th Lord Rayleigh, R.J. Strutt (1875–1947), the scattering cross section per atom is

$$\frac{d\sigma_R(E, \theta_s)}{d\Omega_s} = \frac{1}{2} r_e^2 (1 + \cos^2\theta_s)[F(x, Z)]^2, \tag{4.82}$$

in which $F(x, Z)$ is the *atomic form factor*, and the momentum-transfer parameter x is given by Eq. (4.81). Form factors are tabulated by Hubbell and Overbø [1979]. As E approaches zero, $F(x, Z)$ approaches Z and σ_R varies as Z^2. Effects of coherent scattering are addressed in detail by Trubey and Harima [1987] and data are available in an ANSI standard [1991].

4.4.5 Photoelectric Effect

In the photoelectric effect, a photon interacts with an entire atom, resulting in the emission of a photoelectron, usually from the K shell of the atom. Although the difference between the photon energy E and the electron

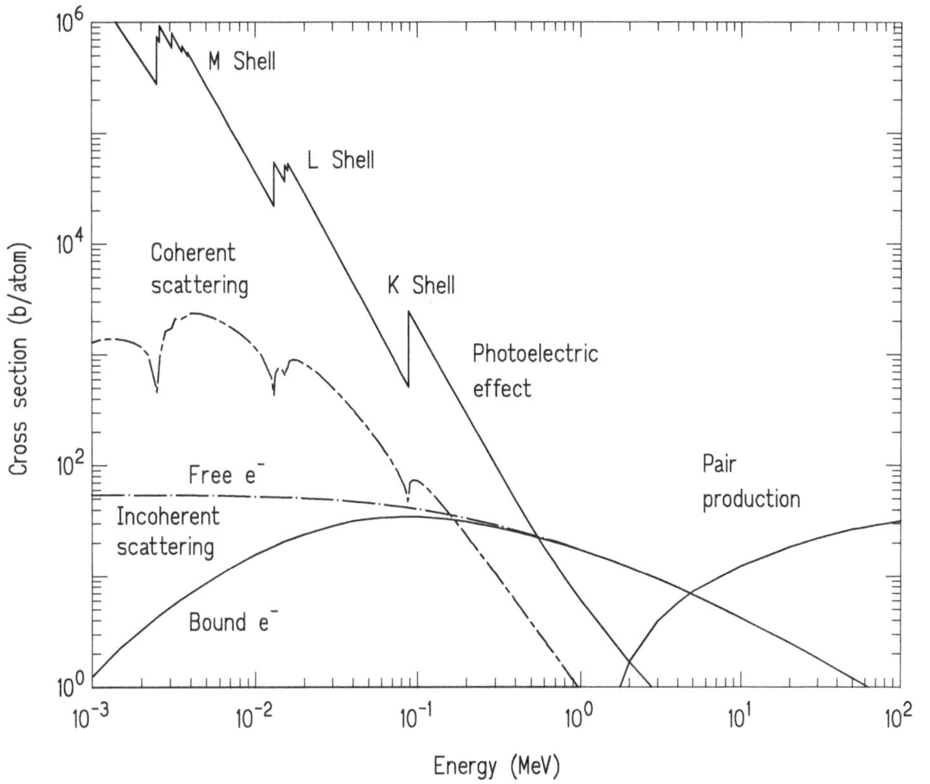

Figure 4.9. Comparison of scattering, photoelectric-effect, and pair-production cross sections for photon interactions in lead.

binding energy E_b is distributed between the electron and the recoil atom, virtually all of that energy is carried as kinetic energy of the photoelectron because of the comparatively small electron mass. Thus, $T = E - E_b$.

K-shell binding energies E_k vary from 13.6 eV for hydrogen to 7.11 keV for iron, 88 keV for lead, and 116 keV for uranium. As the photon energy drops below E_k, the cross section drops discontinuously. As E decreases further, the cross section increases until the first L edge is reached, at which energy the cross section drops again, then rises once more, and so on for the remaining edges. These "edges" for lead are readily apparent in Fig. 4.9. The cross section varies as E^{-n}, where $n \simeq 3$ for energies less than about 150 keV and $n \simeq 1$ for energies greater than about 5 MeV. The atomic cross section varies as Z^m, where m varies from about 4 at $E = 100$ keV to 4.6 at $E = 3$ MeV. As a very crude approximation in the energy region for which the photoelectric effect is dominant,

$$\sigma_{ph}(E) \propto \frac{Z^4}{E^3}. \tag{4.83}$$

Although it is true that for light nuclei, K-shell electrons are responsible for almost all photoelectric interactions, such interactions are normally much less important than incoherent scattering. As a general rule, about 80% of photoelectric interactions with heavy nuclei result in ejection of a K-shell electron. Consequently, the approximation is often made for heavy nuclei that the total photoelectric cross section is 1.25 times the cross section for K-shell electrons.

As the vacancy left by the photoelectron is filled by an electron from an outer shell, either fluorescence x rays or Auger electrons[10] may be emitted. The probability of x-ray emission is given by the *fluorescent yield*. For the K shell, fluorescent yields vary from 0.005 for $Z = 8$ to 0.965 for $Z = 90$. Although x rays of various energies may be emitted, the approximation is often made that only one x ray or Auger electron is emitted, with energy equal to the binding energy of the photoelectron.

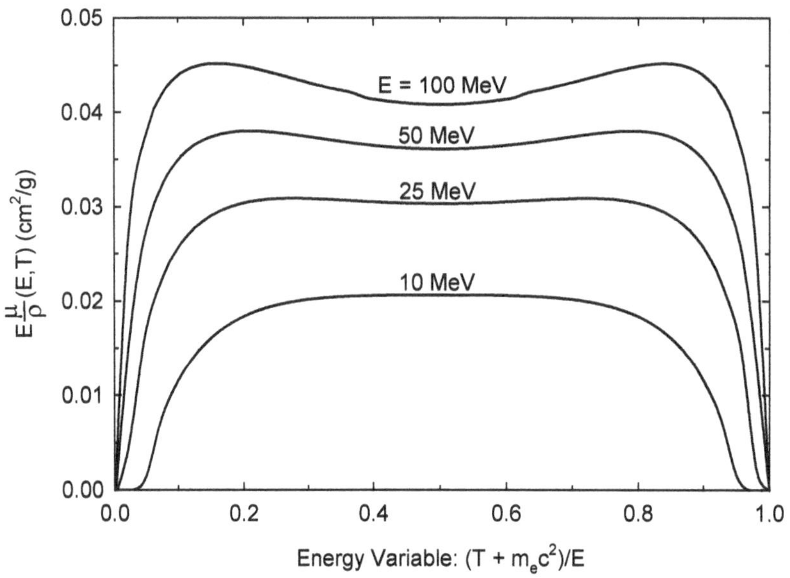

Figure 4.10. Differential pair-production cross section in iron, in the form of the mass interaction coefficient. The function $\mu(E, T)\, dT$ is the probability per unit distance of travel that a photon of energy E experiences a pair-production interaction in which a product electron has kinetic energy between T and $T + dT$. Data derived using the PEGS4 Code [Nelson, Hirayama, and Rogers 1985].

4.4.6 Pair Production

In this process, the incident photon is completely absorbed and in its place appears a positron-electron pair. The phenomenon is induced by the strong electric field in the vicinity of the nucleus and has a photon threshold energy of $2m_ec^2$ (= 1.02 MeV). In the nuclear pair production process, the nucleus acquires indeterminate momentum but negligible kinetic energy. Thus,

$$T_+ + T_- = E - 2m_ec^2, \tag{4.84}$$

in which T_+ and T_- are the kinetic energies of the positron and electron, respectively. To a first approximation, the total atomic pair production cross section varies as Z^2. The cross section increases with photon energy, approaching a constant value at high energy. As illustrated in Fig. 4.10, the resulting electron and

[10]If an electron in an outer shell, say Y, makes a transition to a vacancy in an inner shell, say X, an x ray may be emitted with energy equal to the difference in binding energy between the two shells. Alternatively, an electron in some other shell, say Y′, which may be the same as Y, may be emitted with energy equal to the binding energy of the electron in shell X less the sum of the binding energies of electrons in shells Y and Y′. This electron is called an *Auger electron.* If an electron makes a transition from one subshell to a vacancy in another subshell of the same shell, the small difference in binding energies may be transferred to an outer-shell electron, in this case called a *Coster-Kronig electron.*

positron are widely distributed in energy. Both have directions not far from the original direction of the photon but separated by π radians in azimuth about the photon direction. As an approximation, the angles ϑ with respect to the photon direction are m_ec^2/E radians.

Far less likely than nuclear pair production, a photon of sufficient energy can interact in the electric field of an orbital electron producing an electron-positron pair and a recoil orbital electron. This electron pair production is referred to as *triplet production* and is, except for the lightest nuclei, a small fraction of the total pair production events (see Fig. 4.11). It can be shown that because of the constraints of conservation of energy and momentum the threshold photon energy for triplet production is $4m_ec^2 \simeq 2.044$ MeV, twice that for pair production. The photon energy less the energy equivalent of four electron masses is shared as kinetic energy among the positron and the two electrons, namely

$$T_{e^+} + T_{1e^-} + T_{2e^-} = E - 4m_ec^2. \qquad (4.85)$$

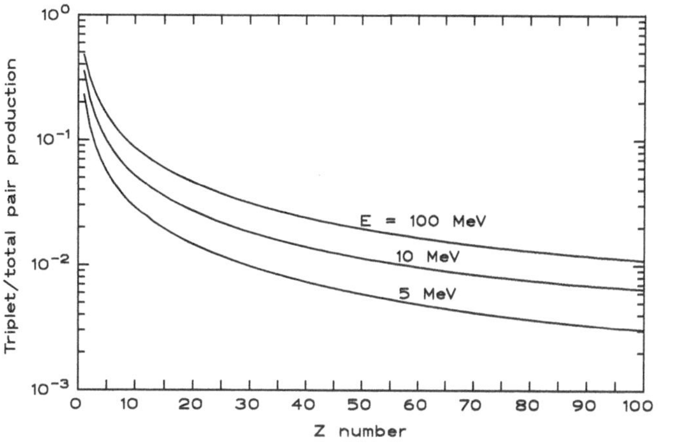

Figure 4.11. Ratio of the triplet to total pair production cross sections as a function of the Z-number of the nuclide for three incident photon energies.

The fate of the positron produced in pair production is annihilation with an ambient electron, generally after slowing to practically zero kinetic energy. This annihilation process results in the creation of two photons moving in opposite directions (to conserve the near zero linear momentum of the slowed electron and positron), each with energy m_ec^2.

4.4.7 Photon Interactions—Minor Effects

Discussed above are the photon interaction mechanisms that are important for radiation detection. Photons can interact with matter in several other ways, but are relatively unimportant for most detector designs.

In 1933 Max Delbrück predicted that high energy photons could be scattered in the Coulombic field near a nucleus as a result of the polarization in the vacuum produced by this field. In this process the incident photon is most often scattered coherently by a virtual electron-positron pair in the field, resulting in little change in direction or loss of energy. This predicted scattering mechanism was finally confirmed experimentally only 20 years later.

Another photon scattering mechanism is that of *nuclear resonance scattering.* In this process an incident photon can be absorbed by a nucleus placing the nucleus in an excited nuclear state. Because these states have very short lifetimes (typically ps), the energy of the incident photon must, consequently, almost exactly match the nuclear energy level (recall the uncertainty principle $\Delta t \Delta E \leq h/(2\pi)$). The excited nuclear state decays, almost always by electric dipole transitions, to lower nuclear states emitting a cascade of gamma rays isotropically, analogous to the fluorescence x-ray cascades produced by the photoelectric effect.

Another very minor scattering process is that of nuclear Rayleigh scattering in which the nucleons scatter an incident photon coherently just as in Rayleigh scattering from all the orbital electrons. However, this is an extremely minor phenomenon, again with little change in direction or energy loss. These minor scattering mechanisms are reviewed by Kane et al. [1986].

At higher energies (5-50 MeV), photons can cause (γ, n) reactions and at even higher energies they can result in meson production. Such phenomena are of importance only in the design of detectors for high energy physics research.

4.4.8 Photon Attenuation Coefficients

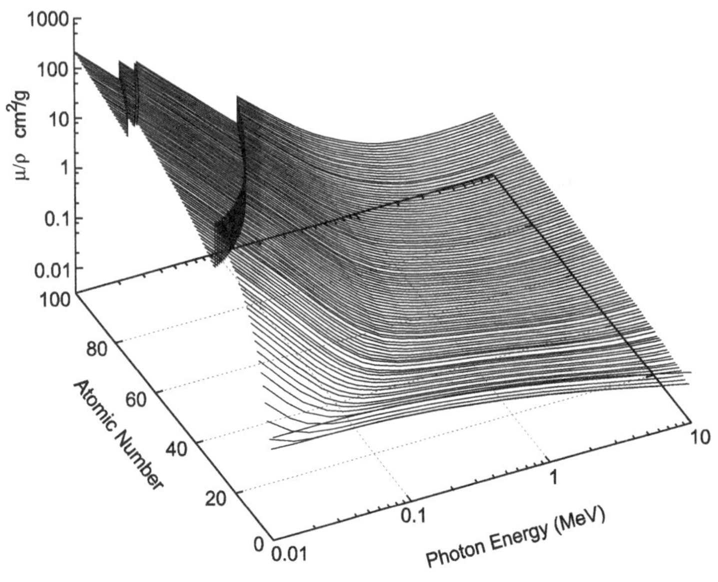

Figure 4.12. Mass attenuation coefficients as a function of photon energy for all the elements. Notice the smooth variation of μ/ρ with energy and the atomic number Z, except for the discontinuities at the edge energies.

The photon linear attenuation coefficient μ is, in the limit of small path lengths, the probability per unit distance of travel that a gamma photon undergoes any significant interaction. Thus, for a specified medium,

$$\mu(E) = N\left[\sigma_c(E) + \sigma_{ph}(E) + \sigma_{pp}(E)\right], \tag{4.86}$$

in which $N = \rho N_a/A$ is the atom density. Note that Rayleigh scattering and other minor effects are specifically excluded from this definition.[11] More common in data presentation is the mass interaction coefficient

$$\begin{aligned}\frac{\mu}{\rho} &= \frac{N_a}{A}\left[\sigma_c(E) + \sigma_{ph}(E) + \sigma_{pp}(E)\right] \\ &= \frac{\mu_c}{\rho} + \frac{\mu_{ph}}{\rho} + \frac{\mu_{pp}}{\rho},\end{aligned} \tag{4.87}$$

in which N_a is Avogadro's number. It is important to note that the mass interaction coefficients are independent of the mass density ρ of the material, and it is for this reason that μ/ρ rather than μ values are usually tabulated. The variation of μ/ρ with energy for all the elements is shown in Fig. 4.12.

4.5 Neutron Interactions

The interaction processes of neutrons with matter are fundamentally different from those for the interactions of photons. Whereas photons interact, more often, with the atomic electrons, neutrons interact essentially only with the atomic nucleus. Although neutron-electron interactions do occur, this type of interaction is highly improbable and therefore negligible when compared to the neutron-nucleus interactions. The cross sections that describe the various neutron interactions are also very unlike those for photons. Neutron cross sections not only can vary rapidly with the incident neutron energy, but also can vary erratically from one element to another and even between isotopes of the same element. The description of the interaction of a neutron with a nucleus involves complex interactions between all the nucleons in the nucleus and the incident neutron, and consequently fundamental theories which can be used to predict neutron cross-section variations in any accurate way are still lacking. As a result, all cross-section data are empirical in nature, with little guidance available for interpolation between different energies or isotopes.

Over the years, many compilations of neutron cross sections have been generated, and the more extensive, such as the Evaluated Nuclear Data Files (ENDF) [Kinsey 1979; Rose and Dunford 1991], contain so much information that digital computers are used almost exclusively to process these cross-section libraries to extract cross sections or data for a particular neutron interaction. Even with the large amount of cross-section information available for neutrons, there are still energy regions and special interactions for which

[11] When referring to data tables in other publications, the reader should be aware that occasionally Rayleigh scattering and incoherent scattering are included.

the cross sections are poorly known. For example, cross sections for interactions which produce energetic secondary photons or charged particles are of concern in radiation detection and often still are not known with an accuracy sufficient to perform satisfactory analyses of neutron detectors that rely on such interactions.

4.5.1 Classification of Types of Interactions

There are many possible neutron-nuclear interactions, only some of which are of concern in radiation-protection calculations. Ultra-high-energy interactions can produce many and varied secondary particles; however, the energies required for such reactions are usually well above the neutron energies commonly encountered, and therefore such interactions can be neglected here. Similarly, for low-energy neutrons many complex neutron interactions are possible—Bragg scattering from crystal planes, phonon excitation in a crystal, coherent scattering from molecules, and so on—none of which is of particular importance in the design of many neutron detectors. The neutron reactions of principal importance are those that produce various secondary radiations that, in turn, deposit energy in the active medium of a radiation detector.

The total cross section, which is the sum of cross sections for all possible interactions, gives a measure of the probability that a neutron of a certain energy interacts in some manner with the medium through which the neutrons are traveling. The components of the total cross section for absorption and scattering interactions are usually of primary concern. Nonetheless, when the total cross section is large, the probability of some type of interaction is great and thus the total cross section, which is the most easily measured and the most widely reported, gives at least an indication of the neutron energy regions over which one must investigate the neutron interactions in greater detail.

The total cross sections, although they vary from nuclide to nuclide and with the incident neutron energy, have certain common features. For the sake of classification, the nuclides are usually divided into three broad categories: light nuclei, with mass number $\lesssim 25$; intermediate nuclei; and heavy nuclei, with mass number $\gtrsim 150$. Example total cross sections for each category are shown in Figs. 4.13 through 4.15.

For light nuclei and some magic number nuclei the cross section at low energies (< 1 keV) often varies as

$$\sigma_t = \sigma_1 + \frac{\sigma_2}{\sqrt{E}}, \tag{4.88}$$

where E is the neutron energy, σ_1 and σ_2 are constants, and the two terms on the right-hand side represent the elastic scattering and the radiative capture (or absorption) reactions, respectively. For solids at energies less than about 0.01 eV, there may be *Bragg cutoffs* representing energies below which coherent scattering from the various crystalline planes of the material is no longer possible. These cutoffs are not shown in Figs. 4.13 through 4.15. At energies greater than about 0.1 eV, the cross sections are usually slowly varying and "smooth" up to the MeV energy range, at which energies fairly wide (keV to MeV) resonances appear.

Of all the nuclides, only hydrogen and its isotope deuterium exhibit no resonances. For both isotopes, the cross sections above 1 eV are almost constant up to the MeV region, above which they decrease with increasing energy.

For heavy nuclei, as illustrated in Fig. 4.15, the total cross section, unless masked by a low-energy resonance, exhibits a $1/\sqrt{E}$ behavior at low energies and usually a Bragg cutoff (not shown). The resonances appear at much lower energies than for the light nuclei (usually in the eV region) and have very narrow widths (1 eV or less) with large peak values. Above a few keV the resonances are so close together and so narrow that they cannot be resolved and the cross sections appear to be smooth except for a few broad resonances. Finally, the intermediate nuclei, as would be expected, are of intermediate character between the light and heavy nuclei, with resonances in the region from 100 eV to several keV. The resonances are not as high or as narrow as for the heavy nuclei.

Because neutrons are invariably born with high energies (100 keV – 10 MeV), neutron radiation fields almost always contain a spectrum of neutrons ranging from thermal energies to high energy. As neutrons

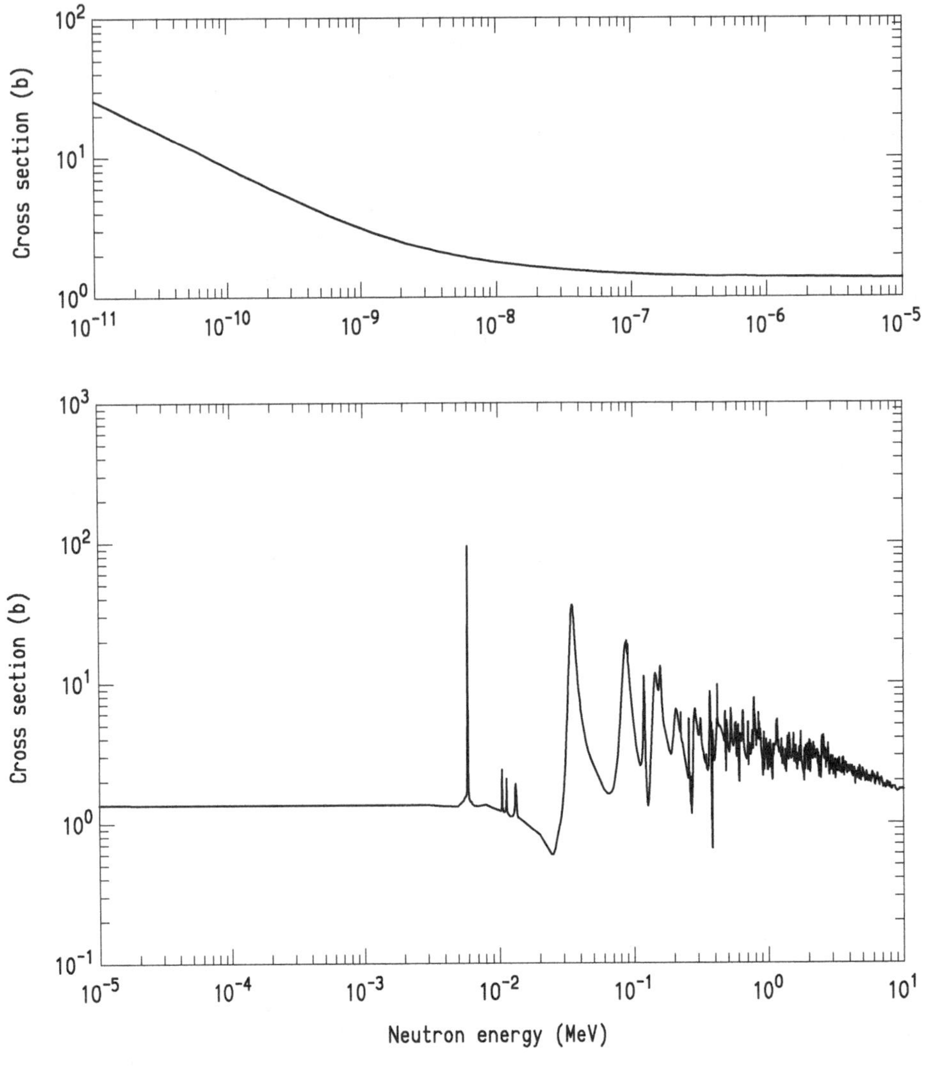

Figure 4.13. Total neutron cross section for aluminum computed using NJOY-processed ENDF/B (version V) data.

scatter they lose energy and, unless they are absorbing while they slow down, they eventually come into thermal equilibrium with the ambient medium. Detector materials that have large thermal absorption cross sections for producing charged reaction products are typically used to detect thermal neutrons. Three commonly used isotopes are ^{3}He, ^{10}B, and ^{6}Li, all having large thermal $\sigma_{(n,p)}$, $\sigma_{(n,\alpha)}$, and $\sigma_{(n,t)}$ cross sections, respectively. These cross sections are shown in Fig. 4.16. Another charged-particle producing reaction that is useful for detecting thermal neutrons is the fission reaction.

Alternatively, one could use (n, γ) absorption reactions and use a photon detector. If the result of such a reaction produces a radioactive isotope, then detectors for the radiation emitted by the isotope's decay can be used to infer the presence of neutrons. However, the absorption cross sections at all energies, except

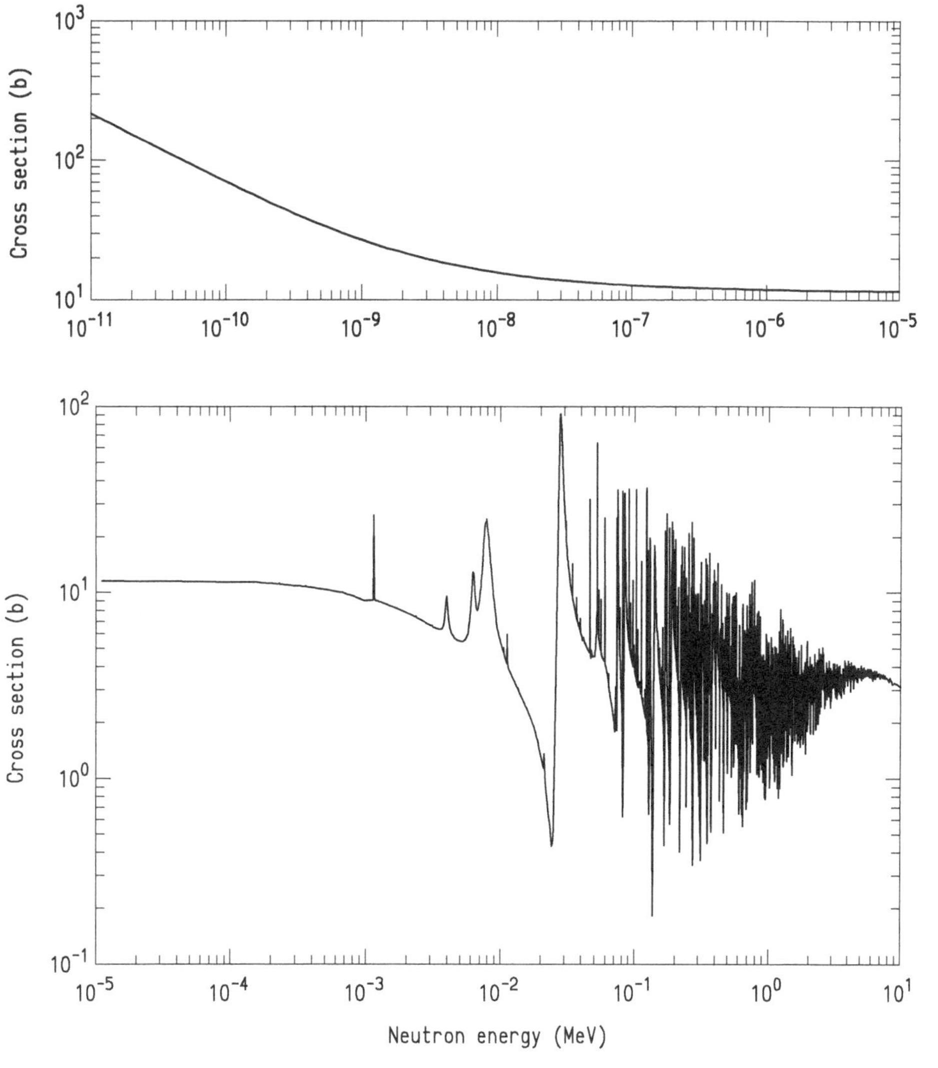

Figure 4.14. Total neutron cross section for iron computed using NJOY-processed ENDF/B (version V) data.

thermal energies, for all nuclides are usually small compared to other reactions. Over the fission-neutron energy spectrum, the (n, γ) reaction cross sections seldom exceed 200 mb for the heavy elements, and for the lighter elements this cross section is considerably smaller. Only for thermal neutrons and a few isolated absorption resonances in the keV region for heavy elements is the (n, γ) reaction important.

In the high energy region, by far the most important neutron interaction is the scattering process. Generally, elastic scattering is more important, although, when the neutron energy somewhat exceeds the energy level of the first excited state of the scattering nucleus, inelastic scattering becomes possible. The recoil energy of the scattering nucleus can be measured and used to infer the energy distribution of incident neutrons. Also in the MeV region the (n, α) reaction cross sections for Be, N, and O are appreciable fractions

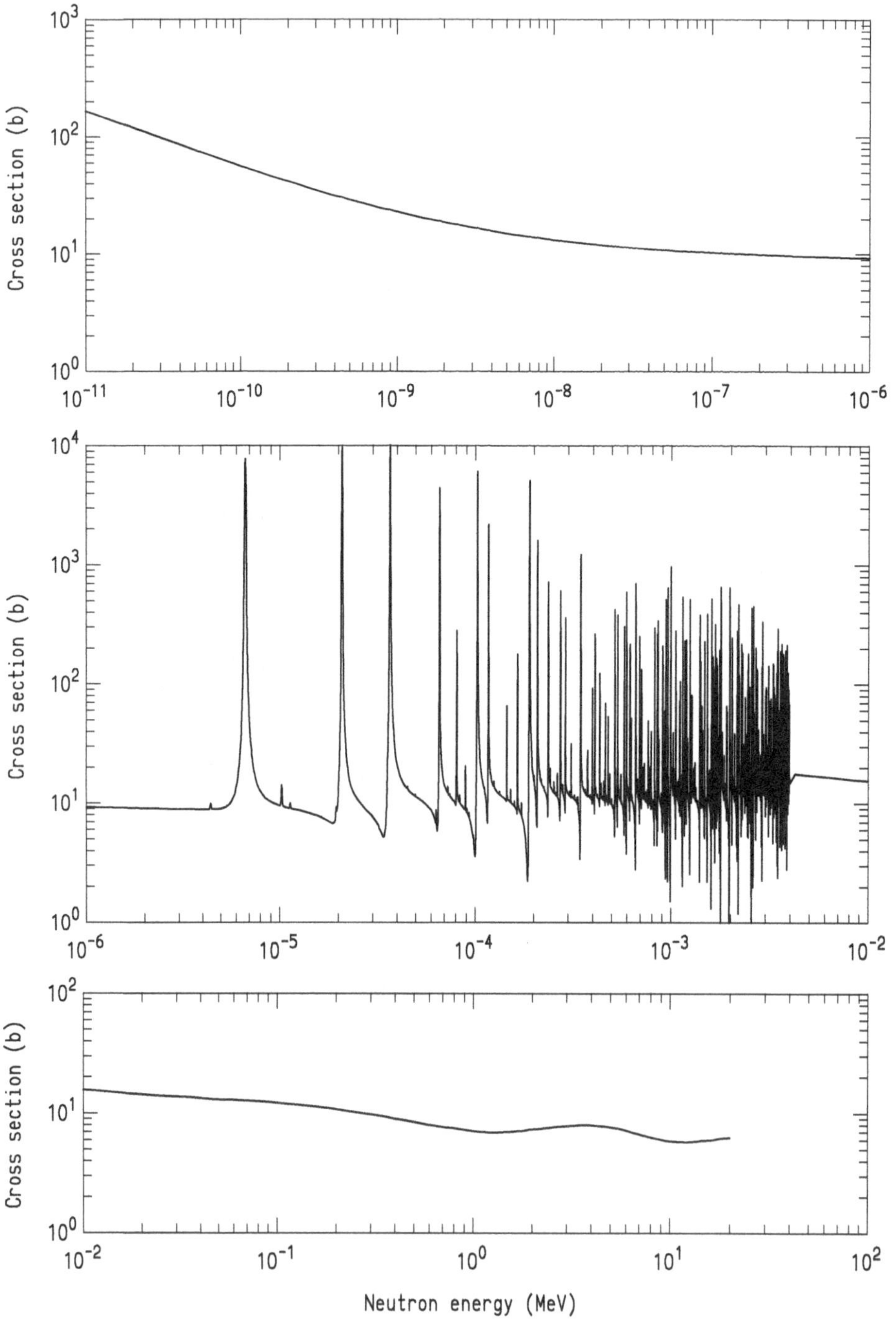

Figure 4.15. Total neutron cross section for uranium computed using NJOY-processed ENDF/B (version V) data. Above 4 keV the resonances are no longer resolved and only the average cross-section behavior is shown.

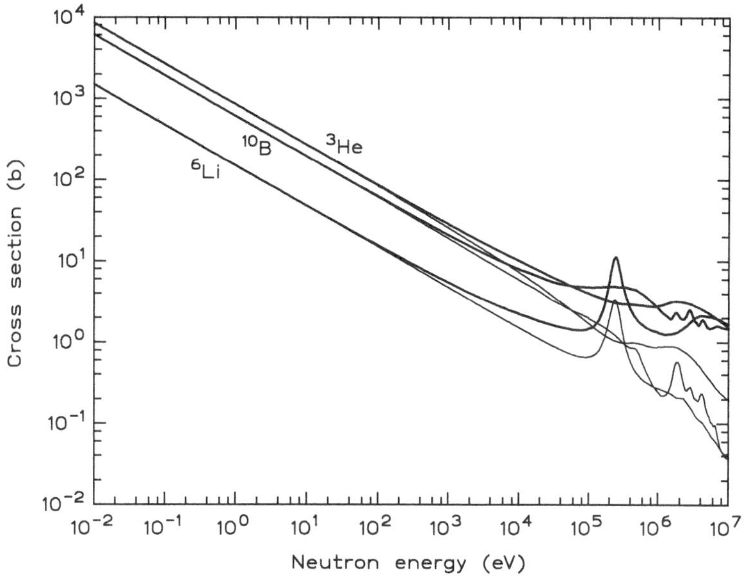

Figure 4.16. Total neutron cross sections for ^{3}He, ^{10}B, and ^{6}Li (heavy lines) and the (n, p), (n, α), and (n, t) cross sections (light lines), respectively.

of the total cross sections and may exceed the inelastic scattering contributions. This situation is probably true for most light elements, although only partial data are available. For heavy and intermediate nuclei the charged-particle emission interactions are at most a few percent of the total inelastic interaction cross section and hence are usually ignored.

4.5.2 Thermal Neutron Interactions

As neutrons slow through scattering, their speeds become comparable to those of the ambient atoms in thermal motion. In this case it is important to account for the motion of *both* the neutrons and the interacting nuclei. In the following treatment, it is assumed that neutrons interact with only a single atom at a time (unlike, for example, Bragg scattering) so that interactions can be described by microscopic cross sections.

Let $n(\mathbf{v})d\mathbf{v}$ and $N(\mathbf{V})d\mathbf{V}$ be the number density at some point of interest of, respectively, the neutrons with laboratory velocities in $d\mathbf{v}$ about $\mathbf{v}$ and the atoms with laboratory velocities in $d\mathbf{V}$ about $\mathbf{V}$. Neutrons and nuclei thus approach each other with a relative velocity $\mathbf{v}_r = \mathbf{v} - \mathbf{V}$. Now consider a coordinate system in which the atomic nuclei are at rest, so that the neutrons approach the nuclei with the relative velocity $\mathbf{v}_r$. Thus, from the nuclei's perspective, the neutrons appear as a beam with intensity $dI = n(\mathbf{v})v_r\, d\mathbf{v}$, where $v_r = |\mathbf{v}_r|$. These neutrons cause interactions of type i at a rate of

$$dF_i = n(\mathbf{v})N(\mathbf{V})\sigma_i(v_r)v_r\, d\mathbf{v}\, d\mathbf{V} \tag{4.89}$$

interactions cm^{-3}s^{-1}. The total interaction rate density for neutrons and nuclei of all speeds is thus [Lamarsh 1966]

$$F_i = \iint n(\mathbf{v})N(\mathbf{V})\sigma_i(v_r)v_r\, d\mathbf{v}\, d\mathbf{V}, \tag{4.90}$$

where the integration is over the six components of $\mathbf{v}$ and $\mathbf{V}$. Below several special cases are considered. More general cases are considered by Meghreblian and Holmes [1960].

$1/v$ Absorption

Consider now the case of thermal neutron absorption (capture or fission). In the thermal energy range $\sigma_a(v)$ varies as $1/v$, so that

$$\sigma_a(v_r) = \frac{v_{ro}}{v_r}\sigma_a(v_{vo}), \tag{4.91}$$

where v_{ro} is an arbitrary reference speed. Substitution of this result into Eq. (4.90) gives

$$F_a = v_{ro}\sigma_a(v_{ro}) \iint n(\mathbf{v})N(\mathbf{V})\, d\mathbf{v}\, d\mathbf{V} = N\sigma_a(v_{ro})nv_{ro}, \tag{4.92}$$

where N and n are, respectively, the atomic and neutron number densities. This astounding result shows that the absorption *rate* of a $1/v$ absorber is a constant that is independent of the velocity distributions of either the neutrons or atoms! Here F_a depends only on a single arbitrary relative speed v_{ro} between the neutron and nucleus. McGregor and Shultis [2011] show the benefit of this result for neutron detector calibrations. One could also view all the nuclei at rest and treat v_{ro} as an arbitrary *laboratory* speed of the neutron. Thus, the absorption rate density can be calculated as

$$F_a = \Sigma_a(E_o)\phi_o \tag{4.93}$$

where $\phi_o = nv_o$ with v_o an arbitrary laboratory speed and E_o is the corresponding energy. Typically, v_o is taken as 2200 m/s so that $E_o = 0.0253$ eV, the most probable energy of a Maxwellian distribution at $T_o = 293.61$ K. Here $\phi_o = nv_o$ is the "2200-m/s flux", the flux that would result if all thermal neutrons were to travel at 2200 m/s.

Non-$1/v$ Absorption

The absorption cross section of some nuclei, particularly those heavy nuclei with resonances just above the thermal region, deviates significantly from the ideal $1/v$ behavior. However, for such heavy nuclei the center-of-mass and laboratory coordinate systems are nearly the same, so that v_r in Eq. (4.90) equals the laboratory speed v of the neutrons. Equation (4.90) becomes

$$F_a = \iint n(\mathbf{v})N(\mathbf{V})\sigma(v)v\, d\mathbf{v}\, d\mathbf{V} = N\int n(\mathbf{v})\sigma(v)v\, d\mathbf{v}. \tag{4.94}$$

Because $n(\mathbf{v})\, d\mathbf{v} = n(E)\, dE$ where E is the laboratory neutron energy, Eq. (4.94) transforms to

$$F_a = N\int n(E)\sigma_a(E)v(E)\, dE = \int \Sigma_a(E)\phi(E)\, dE, \tag{4.95}$$

where the integration is over all thermal energies.

In many instances the energy dependence of thermal neutrons is well approximated by a Maxwellian distribution with an effective neutron temperature T, i.e., by

$$\phi_M(E,T) = \frac{2\pi n}{(\pi kT)^{3/2}}\sqrt{\frac{2}{m_n}}E\exp[-E/kT], \tag{4.96}$$

where n is the neutron density, m_n is the neutron mass, k is Boltzmann's constant ($k = 8.617\,343 \times 10^{-5}$ eV K^{-1}).

In this case Eq. (4.95) can be numerically evaluated to give

$$F_a = g_a(T)\Sigma_a(E_o)\phi_o = \overline{\Sigma}_a\phi_t, \tag{4.97}$$

where the total thermal flux

$$\phi_t = \int_0^{\sim 5kT} \phi(E)\, dE \simeq \int_0^{\infty} \phi_M(E,T)\, dE, \tag{4.98}$$

and the thermal-averaged macroscopic absorption cross section is

$$\overline{\Sigma}_a = \frac{\sqrt{\pi}}{2} g_a(T) \left(\frac{T_o}{T}\right)^{1/2} \Sigma_a(E_o). \tag{4.99}$$

In this result, the Westcott non-$1/v$ factor $g_a(T)$ accounts for deviations from the $1/v$ behavior of absorption cross sections in the thermal energy region.[12] If $\sigma_a(E) \simeq 1/\sqrt{E}$, then $g_a(T) = 1$. Finally, the relation between the thermal flux ϕ_t and the fictitious 2200-m/s flux is obtained from Eqs. (4.97) and (4.99) as

$$\phi_t = \frac{2}{\sqrt{\pi}} \left(\frac{T}{T_o}\right)^{1/2} \phi_o. \tag{4.100}$$

The result for the absorption rate density of Eq. (4.95) is also obtained for light nuclei with non-$1/v$ behavior and for which the center-of-mass and laboratory systems are not equal, *provided* both the neutrons and atoms have Maxwellian distributions of energy with the same temperature [Lamarsh 1966].

Scattering Interaction Rate

Scattering cross sections at thermal energies (always elastic) are usually nearly constant, i.e., $\Sigma_s(E) \simeq \Sigma_s(E_o)$. The scattering interaction rate density is given by

$$F_s = \int \Sigma_s(E)\phi(E)\, dE. \tag{4.101}$$

If $\Sigma_s(E) = \Sigma_s(E_o)$, then clearly $F_s = \Sigma_s(E_o)\phi_t$. Sometimes $\Sigma(E)$ is not quite constant. In this case, Eq. (4.101) is integrated numerically after assuming a Maxwellian for $\phi(E)$ to give, in analogy to Eq. (4.97),

$$F_s = g_s(T)\Sigma_s(E_o)\phi_o = \overline{\Sigma}_s\phi_t, \tag{4.102}$$

where the thermal averaged scattering cross section is

$$\overline{\Sigma}_s = \frac{\sqrt{\pi}}{2} g_s(T) \left(\frac{T_o}{T}\right)^{1/2} \Sigma_s(E_o). \tag{4.103}$$

For the case $\Sigma_s(E) = \Sigma_s(E_o)$, $g_s(T) = 2/\sqrt{\pi} \simeq 1.1283$, a result seen in most of the scattering cross sections listed in Appendix B, Table B.1.

4.5.3 Neutron Differential Scattering Cross Sections

The differential scattering cross section $\sigma_s(E,\theta)$ is sometimes reported in cross-section libraries in terms of ω_c, the cosine of the scattering angle in the center-of-mass coordinate system. This is done because, in the center-of-mass system, scattering of low- and intermediate-energy neutrons is nearly isotropic, that is,

$$\frac{d\sigma_s(E,\omega_c)}{d\Omega_c} \simeq \frac{\sigma_s(E)}{4\pi}. \tag{4.104}$$

[12]Westcott used a more general form to account also for epithermal neutron absorption. See Westcott [1955].

In fact, for hydrogen, the scattering in the center-of-mass system is isotropic for energies up to about 30 MeV. Generally, the heavier the nuclide, the lower is the energy above which elastic scattering becomes anisotropic. The differential scattering cross section is thus well represented by a low-order Legendre polynomial expansion in the form

$$\frac{d\sigma_s(E,\omega_c)}{d\Omega_c} \simeq \frac{\sigma_s(E)}{4\pi}\sum_{n=0}^{N}(2n+1)f_n(E)P_n(\omega_c). \tag{4.105}$$

Cross-section data may be tabulated in terms of the expansion coefficients $f_n(E)$. Because of the way that Eq. (4.105) is formulated, f_0 is always equal to 1. The order of the expansion seldom exceeds $N = 8$. Few data are available on the angular distributions of inelastically scattering neutrons. When only one or two levels are involved, the scattering may be anisotropic; however, it has generally been found that it is a good approximation to assume that the inelastically scattered neutrons are emitted isotropically in the center-of-mass system. This is particularly true when multiple levels are involved in the inelastic process. The potential scattering of fast neutrons is never isotropic but is highly peaked in the forward directions. In fact, the angular distribution can exhibit several maxima as the scattering angle varies.

The relation between the laboratory and center-of-mass scattering angular distributions may be obtained from a formal change of independent variable or from recognition that the probability of scattering into corresponding differential solid angles $d\Omega_s = 2\pi\, d\omega_s$ and $d\Omega_c = 2\pi\, d\omega_c$ must be the same, that is,[13]

$$\frac{d\sigma_s(E,\omega_s)}{d\Omega_s}\,d\omega_s = \frac{d\sigma_s(E,\omega_c)}{d\Omega_c}\,d\omega_c. \tag{4.106}$$

Transformations between the two systems make use of Eqs. (4.47) and (4.48) as well as the interrelationship

$$\frac{d\omega_c^\pm}{d\omega_s} = \pm\frac{(1+2\gamma\omega_c^\pm+\gamma^2)^{3/2}}{1+\gamma\omega_c^\pm} = 2\gamma\omega_s \pm \frac{1-\gamma^2(1-2\omega_s^2)}{[1-\gamma^2(1-\omega_s^2)]^{1/2}}. \tag{4.107}$$

For elastic scattering and for inelastic scattering with single values of E', only the positive sign in this equation applies, and there is only a single value of ω_c. However, for inelastic scattering with dual values of E', there are two values of ω_c, as indicated in Eq. (4.48). The positive signs in Eqs. (4.48) and (4.107) are associated with ω_c^+, and the negative signs with ω_c^-.

4.5.4 Average Energy Transfer in Neutron Scattering

Here it is necessary to be very careful in distinguishing between the kinetic energy lost by the neutron and the kinetic energy gained by the recoil atom, for the two are not the same when the scattering is inelastic. Knowledge of the energy lost by the neutron is important in the examination of neutron moderation. Knowledge of the energy gained by the recoil atom is important in dosimetry. The mean *fraction* of the energy E lost is described as $1-\langle E'\rangle/E$, and the mean fraction of E transferred to the recoil atom as $\langle T\rangle/E$. The two are related by $\langle T\rangle = \langle E - E'\rangle + Q$.

The mean fraction of energy transferred $\langle E'\rangle/E$ may be determined based on $[d\sigma_s(E,E')/dE']\,dE'/\sigma_s(E)$ being the probability that a neutron of energy E leaves a scattering interaction with energy between E' and $E'+dE'$. Thus,

$$\frac{\langle E'\rangle}{E} = \frac{1}{\sigma_s(E)}\int_{E'_{\min}}^{E'_{\max}} dE'\,\frac{E'}{E}\frac{d\sigma_s(E,E')}{dE'}, \tag{4.108}$$

[13]It should be noted that following standard usage, $\sigma_s(E,\omega_c)$ and $\sigma_s(E,\omega_s)$ are different functions even though they carry the same symbol, σ_s.

Table 4.1. Average fraction $1 - \langle E' \rangle / E$ of neutron energy lost in center-of-mass scattering interactions. Note that $\Delta = Q(A+1)/AE$ and that $\langle T' \rangle / E = 1 - \langle E' \rangle / E + Q/E$.

	Isotropic scatter	Anisotropic scatter
Elastic scatter	$\frac{1-\alpha}{2}$	$\frac{1-\alpha}{2}[1 - f_1(E)]$
Inelastic scatter	$\frac{1-\alpha}{2}\left[1 - \frac{A}{2}\Delta\right]$	$\frac{1-\alpha}{2}\left[1 - \frac{A}{2}\Delta - f_1(E)\sqrt{1+\Delta}\right]$

in which the limits are given in Eqs. (4.53). Scattering into a particular range of energies requires scattering within an associated range of directions, that is, $[d\sigma_s(E, E')/dE']\, dE' = 2\pi[d\sigma_s(E, \omega_c)/d\Omega_c]\, d\omega_c$. Thus, from Eq. (4.50), with $\Delta = Q(1 + A)/AE$,

$$\frac{d\sigma_s(E, E')}{dE'} = \frac{4\pi}{(1-\alpha)E\sqrt{1+\Delta}} \frac{d\sigma_s(E, \omega_c)}{d\Omega_c}. \tag{4.109}$$

The determination of $\langle E' \rangle / E$ may be accomplished more simply by using the angular distribution of the cross section directly, that is,

$$\frac{\langle E' \rangle}{E} = \frac{2\pi}{\sigma_s(E)} \int_{-1}^{+1} d\omega_c \, \frac{E'}{E} \frac{d\sigma_s(E, \omega_c)}{d\Omega_c}, \tag{4.110}$$

where the ratio E'/E is given in Eq. (4.50), and by using the Legendre expansion of Eq. (4.105) for $d\sigma_s(E, \omega_c)/d\Omega_c$.

Evaluation of energy transfer for several cases of neutron scattering is summarized in Table 4.1. In this table f_1 is the first order Legendre expansion coefficient of Eq. (4.105). Notice that as A increases, α approaches unity, and the average energy loss in elastic scattering approaches zero. Only by inelastic scattering can appreciable energy losses be realized. Although for hydrogen the average energy loss is one-half of the initial energy, the total scattering cross section $\sigma_s(E)$ declines with increasing energy in the MeV energy region and hydrogen scattering events become relatively improbable. For this reason inelastic scattering by heavy nuclides plays a crucial role in the slowing of fast neutrons.

The energy of the recoil nucleus from a scattering event caused by a fast neutron is quickly dissipated in solids or liquids, and hence for all practical purposes the recoil energy can be assumed to be deposited locally. For example, a 5-MeV proton travels at most 0.5 mm in aluminum. Heavier recoiling nuclei are stopped in much shorter distances. Only for the case of neutron scattering in a gas does one have to be concerned with the travel of the recoil atoms and then only if very detailed calculations are required. The energy distribution of the recoil nuclei is usually of great interest in radiation detection and dosimetry.

4.5.5 Radiative Capture of Neutrons

When a neutron is absorbed by a nucleus any kinetic energy of the neutron plus its binding energy in the resulting compound nucleus (usually 7 to 9 MeV) leaves the compound nucleus in a highly excited state. The excited nucleus usually decays within 1 ps of the capture, often through several intermediate states, thereby emitting one or more energetic gamma photons. These *capture gamma photons* from (n, γ) reactions can be used to identify the isotope that absorbed the neutron.

The cross section for radiative capture is very small for high energy neutrons, typically no more than a few hundred millibarns for neutrons with energies between 20 keV and 10 MeV. For many nuclides, the (n, γ) capture cross section is poorly known in the keV and MeV energy region. Only for certain important nuclides such as fissionable isotopes is the cross section known with a good degree of certainty. Of more importance are the (n, γ) reactions caused by thermal neutrons which have been slowed by scattering and come into equilibrium with the thermal motion of atoms in the shield. The (n, γ) cross section for thermal neutrons may be quite large—up to thousands of barns for nuclides such as cadmium that have capture resonances near the thermal energy region—and for most isotopes it comprises almost the total absorption cross section. In a material at room temperature, the thermal neutrons have an average energy of 0.025 eV, corresponding to a speed of 2200 m s^{-1}. In Appendix B, thermal neutron cross sections, as well as yields and energies of the capture photons, are given for common elements.

4.5.6 Neutron-Induced Fission

When a neutron is absorbed by a heavy nucleus, the resulting compound nucleus is produced in a very excited state. The "nuclear fluid" of such an excited nucleus undergoes large oscillations and deformations in shape. The compound nucleus may, during one of its oscillations in shape, deform into an elongated or dumbbell configuration in which the two ends Coulombically repel each other and the nuclear forces, being very short ranged, are no longer able to hold the two ends together. The two ends then separate (*scission*) within about 10^{-20} s into two nuclear pieces, repelling each other with such tremendous Coulombic force that many of the orbital electrons are left behind. Two highly charged *fission fragments* are thus created.[14] The fission fragments are produced in such highly excited nuclear states that neutrons can "boil" off them. Anywhere from 0 to about 8 *prompt* neutrons evaporate from the primary fission fragments within about 10^{-17} s of the scission. After prompt neutron emission, the fission fragments are still in excited states, but with excitation energies insufficient to cause further neutron emission. They quickly decay to lower energy levels only by the emission of *prompt* gamma rays within about 2×10^{-14} s after the prompt-neutron emission.

The highly charged fission fragments pass through the surrounding medium producing millions of ionizations and excitations of the ambient atoms. As the fission fragments slow, they gradually acquire electrons reducing their ionization charge, until, by the time they are stopped after about 10^{-12} s, they become electrically neutral atoms. After prompt neutron and gamma ray emission, the fission fragments are then termed *fission products.* The fission products are generally radioactive and, thus, start decay chains whose members radioactively decay, usually by isobaric β^- decay, until a stable end-nuclide is reached. Occasionally, short-lived fission products decay by neutron emission producing *delayed neutrons* seconds to minutes after the fission event. The half-lives of the many possible fission product daughters range from fractions of a second to many thousands of years.

Sometimes very heavy radionuclides, e.g., ^{238}U and ^{252}Cf, are already sufficiently unstable that the addition of a neutron is not needed to excite the nucleus to produce oscillations that lead to fission. Such nuclides are said to *spontaneously* fission.

4.6 Charged-Particle Interactions

Knowledge of the ranges of ionizing charged particles and the rates at which energy is dissipated along their paths is of crucial importance in the design of radiation detectors and in radiation dosimetry. A charged particle is slowed as a result of both Coulombic interactions with (atomic) electrons and radiation losses (bremsstrahlung). Energetic charged particles cause thousands of ionizations and excitations of the atoms along their path before they are slowed and become part of the ambient medium.

---

[14] Sometimes three fission fragments are formed in *ternary* fission with the third being a small nucleus. Alpha particles are created in about 0.2% of the fissions, and nuclei of ^{2}H, ^{3}H, and others up to about ^{10}B are formed much less frequently.

The *stopping power*, often denoted as $-dE/dx$, is the expected energy loss per unit distance of travel by the charged particle along its trajectory. As used in this book, the term is synonymous with linear energy transfer (LET), denoted as $L(E)$. During deceleration, the stopping power generally increases until the energy of the particle is so low that charge neutralization or quantum effects bring about a reduction in the rate of energy loss.

Unlike photons and neutrons, charged particles have a finite range in matter, i.e., there is a distance beyond which none can travel. The distance the particle travels before being stopped is called the *range*. However, this definition is not very precise because, for stochastic reasons, not every particle starting with the same energy travels the same distance before stopping. Also the distance used to define the range can be either along the particle's trajectory or a straight-line (crow-flight) distance from the source of the charged particles. The resulting slightly different definitions of range are described more fully later in this section. A charged particle range is often reported as a *mass thickness range* defined as R/ρ, with units such as g cm^{-2}, in order to mostly remove the density dependence of the range on the density ρ of stopping medium. The paths traveled by light particles, like electrons and positrons, are quite different from those of heavy charged particles such as protons and alpha particles, and are considered separately in the following sections.

Heavy charged particles (HCPs) are ions with masses greater than or equal to that of a proton. With kinetic energies much less than their rest-mass energies, HCPs lose energy almost entirely from Coulombic interactions with atomic electrons. A multitude of such interactions take place—so many that the slowing down is virtually continuous along a straight-line path. These interactions, taken individually, may range from ionization processes producing energetic recoil electrons called *delta rays* to weak atomic or molecular excitation, which may not result in ionization at all. The LET resulting from Coulombic interactions, L_{coll}, is called the *collisional LET* or *collisional stopping power*. Only for heavy charged particles of very low energy do collisions with atomic nuclei of the stopping medium become important and can result in large angular deflections from their usual straight-line trajectories.

Another energy-loss mechanism, which is especially important for electrons, is radiative energy loss through bremsstrahlung production and is characterized by the *radiative LET* or *radiative stopping power* L_{rad}. Also, a careful treatment of electron slowing down requires consideration of delta-ray production and the concomitant deflection of the incident electron from its original direction. As a result the trajectory of slowing electrons is tortuous exhibiting many large angle deflections.

The total stopping power is defined as

$$L = -\frac{dE}{dx} = -\left.\frac{dE}{dx}\right|_{coll} - \left.\frac{dE}{dx}\right|_{rad}, \tag{4.111}$$

which is the sum of the collisional and radiative energy losses. The *mass stopping power* is defined as

$$\frac{L}{\rho} = -\frac{dE}{dx}\frac{1}{\rho}, \tag{4.112}$$

where ρ is the mass density of the stopping material, and has units such as MeV cm^2 g^{-1}. Also, the *specific ionization* is defined as

$$\frac{L}{w} = -\frac{dE}{dx}\frac{1}{w}, \tag{4.113}$$

where w is the average energy required to liberate an electron from its host atom, usually described as the production of electron-ion pairs (gases) or electron-hole pairs (solids).

4.6.1 Collisional Energy Loss

Collisional energy loss results from momentum transfer caused by the Coulombic force between the incident particle with speed v and charge number z and the target, of mass M, which is an atomic nucleus with charge number Z equal to the atomic number, or an electron of charge number $Z = 1$. The kinematics of individual interactions were described earlier in this chapter with notation as illustrated in Fig. 4.2. The LET may be computed on the basis of a differential interaction cross section $d\sigma(E,T)/dT$ defined so that $[d\sigma(E,T)/dT]\, dT$ is the cross section for interaction of an incident particle of energy E, resulting in a target recoil with energy between T and $T + dT$. A differential linear interaction coefficient $d\mu(E,T)/dT$ is the product of $d\sigma(E,T)/dT$ and the density N of targets, atomic nuclei, or atomic electrons, as appropriate. The product $T\,[d\mu(E,T)/dT]\, dT$ is the total energy loss per unit differential distance of travel of the primary particle associated with individual losses between T and $T + dT$. Thus, the LET $L(E)$ is given, in principle, by

$$L(E) = \int_0^{T_{\max}} T \frac{d\mu(E,T)}{dT}\, dT. \tag{4.114}$$

The total LET is the sum of the collisional and radiative LETs, $L_{coll}(E) + L_{rad}(E)$ based on integrals over $d\mu_{coll}(E,T)/dT$ and $d\mu_{rad}(E,T)/dT$.

Rutherford Cross Section

The fundamental formula for collisional loss, named for Ernest Rutherford (1871–1937), is based on classical mechanics and was originally derived for scattering of alpha particles by atomic nuclei. Its derivation may be found in most textbooks on atomic and nuclear physics. A rigorous derivation, based on conservation of angular momentum as well as kinetic energy and linear momentum, may be found in Evans [1955]. The differential cross section may be written as

$$\frac{\sigma_{ruth}(E,T)}{dT} = \frac{2\pi z^2 Z^2 q_e^4}{(4\pi\epsilon_o)^2 M v^2 T^2} = \frac{2\pi z^2 Z^2 q_e^4}{(4\pi\epsilon_o)^2 M c^2} \frac{1}{\beta^2 T^2}, \tag{4.115}$$

in which q_e is the unit electronic charge. Note that if targets are atomic electrons, their density is $N_e = \rho(Z/A)N_a$, where Z is the medium atomic number. Then $M = m_e$ and the target charge number is unity. In terms of dimensionless variables $\epsilon = E/m_e c^2$ and $\tau = T/m_e c^2$, and with $d\mu(\epsilon,\tau)/d\tau = m_e c^2 d\mu(E,T)/dT$, the linear coefficient for interactions with atomic electrons is given by

$$\frac{d\mu_{ruth}(\epsilon,\tau)}{d\tau} = \frac{2\pi N_e z^2 r_e^2}{\beta^2}\left[\frac{1}{\tau^2}\right], \tag{4.116}$$

in which r_e is the classical electron radius, given in Eq. (4.62). Equation (4.115) commonly appears in the cgs system of units, within which the factor $4\pi\epsilon_o$ does not appear and $r_e = q_e^2/m_e c^2$. Bhabha [1938] showed that for heavy charged incident particles and electron targets, the Rutherford cross section must be modified to

$$\frac{d\mu_{coll}(\epsilon,\tau)}{d\tau} = \frac{2\pi N_e z^2 r_e^2}{\beta^2}\left[\frac{1}{\tau^2} - \frac{\beta^2}{\tau\tau_{\max}}\right], \tag{4.117}$$

where $\tau_{\max} \equiv T_{\max}/m_e c^2$, with $T_{\max}$ being the maximum energy transferred to the recoil particle (see Section 4.3.8). This equation applies to alpha particles or other particles with spin 0. An additional correction term, important only for highly relativistic conditions, is required for protons or particles with spin 1/2.

Møller and Bhabha Cross Sections

The relativistic collisional energy loss in electron-electron interactions was addressed by Møller [1932], who found that

$$\frac{d\mu_{coll}(\epsilon,\tau)}{d\tau} = \frac{2\pi N_e r_e^2}{\beta^2}\left[\frac{1}{\tau^2} + \frac{1}{(\epsilon-\tau)^2} + \frac{1}{(\epsilon+1)^2} - \frac{2\epsilon+1}{\tau(\epsilon-\tau)(\epsilon+1)^2}\right]. \tag{4.118}$$

This interaction coefficient is illustrated in Fig. 4.17. A similar expression for positron energy losses was derived by Bhabha [1936].

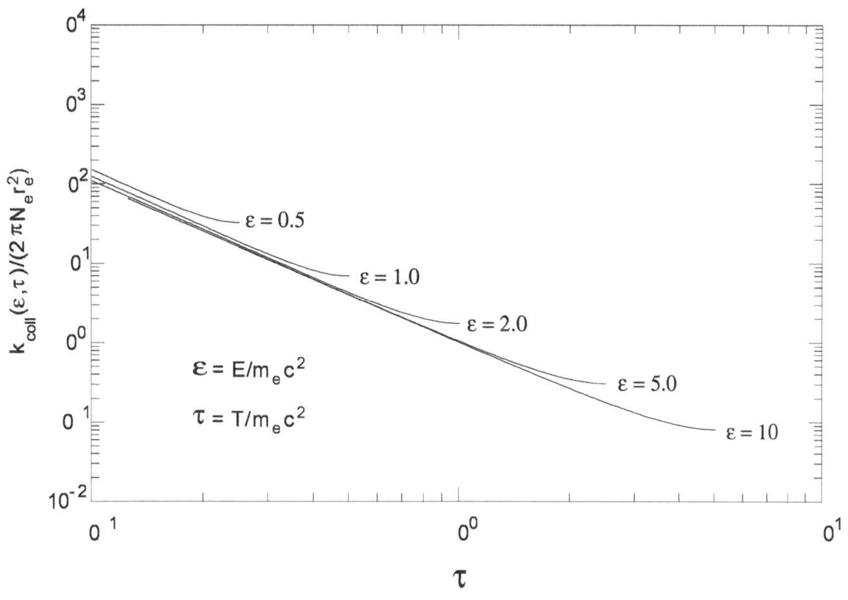

Figure 4.17. Møller cross section for electron-electron interactions.

Collisional LET

Both the classical and the relativistic approaches fail when Eq. (4.116) or (4.118) is substituted into Eq. (4.114). The integral $\int d\tau\, \tau^{-2}$ is infinite for a lower limit of zero. However, the zero limit is not realistic physically. A quantum mechanical approach is necessary because energy losses less than some finite value are prohibited due to quantized states of atomic electrons in the stopping medium. Bethe [1930, 1932] showed that the collisional LET can be written in the form

$$L_{coll}(\epsilon) = 2\pi z^2 N_e m_e c^2 r_e^2 \frac{1}{\beta^2} \int_0^{\tau_{\max}} d\tau\, \tau\, [\tau^{-2} + \cdots] \simeq \frac{\rho Z z^2 \mathcal{C}}{A\beta^2} f(\tau_{max}, \bar{I}), \tag{4.119}$$

in which $\mathcal{C} = 4\pi N_a r_e^2 m_e c^2 = 0.30705$ cm^2 MeV mol^{-1}. In the function f, $\bar{I} \equiv I/m_e c^2$ is a dimensionless *mean excitation energy* and is determined empirically for the stopping medium. Note that when the target is an electron

$$\tau_{\max} = \begin{cases} \epsilon/2 & \text{for electron collisions,} \\ \epsilon & \text{for positron collisions,} \\ 2\beta^2/(1-\beta^2) & \text{for non-relativistic heavy particles.} \end{cases} \tag{4.120}$$

Selected values of I are given in Table 4.2. For $Z \geq 13$, I for elemental substances, in units of eV, is given approximately by the empirical formula [Barkas and Berger 1964]

$$I = 9.76Z + 58.8Z^{-0.19}. \tag{4.121}$$

Janni [1982] gives the following formula, valid within 10% for $Z > 34$:

$$I = 10.3Z - 8.17Z^{1/3}. \tag{4.122}$$

Table 4.2. Selected values of the mean excitation energy for compounds and atomic constituents of compounds in the condensed state.

Material	Form	I (eV)	Material	Form	I (eV)
H	saturated bond	19.0	H_2	gas	19.2
H	unsaturated bond	16.0	N_2	gas	82.0
C	saturated bond	81.1	O_2	gas	95.0
C	unsaturated bond	79.8	CO_2	gas	85.0
C	highly chlorinated	69.0	Air	dry	85.7
N	amines, nitrates, etc.	105.7	Water	gas	71.6
N	in rings	81.9	Water	liquid	75.0
O	-O-	104.6	Tissue	ICRU muscle	74.7
O	=O	94.4	Bone	ICRU compact	91.9

Source: ICRU [1984].

For atomic constituents of compounds, I should be increased by a factor of 1.13 [ICRU 1984]. For mixtures of elements identified by the index j, the average values of Z/A and I are given by the empirical formulas

$$\langle Z/A \rangle = \sum_j w_j \left(\frac{Z_j}{A_j}\right) \tag{4.123}$$

and

$$\ln I = \langle Z/A \rangle^{-1} \sum_j w_j \left(\frac{Z_j}{A_j}\right) \ln I_j. \tag{4.124}$$

in which w_j is the weight fraction of the jth constituent.

4.6.2 Radiative Energy Loss

Classical electromagnetic theory requires that the rate at which electromagnetic energy is radiated from an accelerating charged particle be proportional to the square of the acceleration. In the electric field of a nucleus with atomic number Z, the force on a charged particle with z elemental charges is proportional to the product Zz. The resulting acceleration of the charged particle is proportional to this force and inversely proportional to the particle's mass m. Thus, the rate of radiative energy loss by the charged particle would be expected to be proportional to $(Zz/m)^2$. This inverse-square dependence on particle mass explains why bremsstrahlung from protons or heavier charged particles is negligible in most circumstances compared to that from electrons or positrons. The dependence on the square of Z also explains why, except for the lightest elements, bremsstrahlung in the field of the nucleus far exceeds bremsstrahlung in the fields of atomic electrons.

One may define a differential interaction coefficient for radiative energy loss in such a way that $\mu_{rad}(E, E')\,dE'$ is the probability per unit differential distance of travel that deceleration of an electron of energy E results in emission of a photon with energy between E' and $E' + dE'$. However, it is not possible to write a simple expression for $\mu_{rad}(E, E')$ or for the associated microscopic cross section $\sigma_{rad}(E, E')$. The interaction coefficient for radiative energy loss by an electron is illustrated in Fig. 4.18, which is based on calculations performed by the PEGS4 computer program [Nelson et al. 1985].

For a relativistic heavy charged particle of rest mass M, with $E \gg Mc^2$, it can be shown that the ratio of radiative to ionization losses is approximately [Evans 1955]

$$\frac{(-dE/ds)_{rad}}{(-dE/ds)_{coll}} \simeq \frac{EZ}{700}\left(\frac{m_e}{M}\right)^2, \tag{4.125}$$

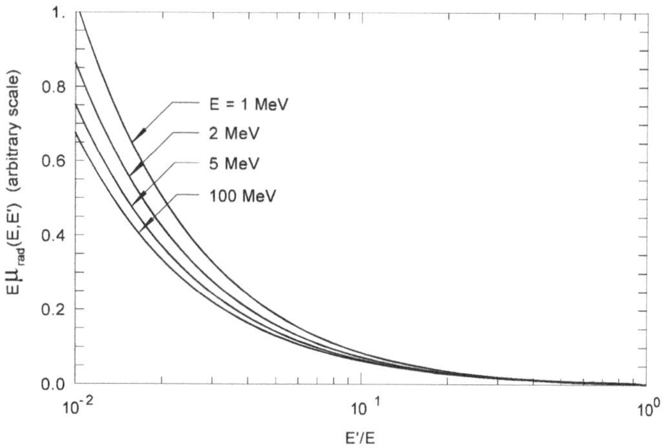

Figure 4.18. Energy spectrum of bremsstrahlung photons released in lead by radiative energy losses of electrons with initial energy E.

where E is in MeV. From this result it is seen that bremsstrahlung is more important for high-energy particles of small mass incident on high-Z material. In most dosimetry situations, only electrons ($m_e/M = 1$) are of importance for their associated bremsstrahlung. All other charged particles are far too massive to produce significant amounts of bremsstrahlung. Bremsstrahlung from electrons is of particular radiological interest for devices that accelerate electrons, such as betatrons and x-ray tubes, or for situations involving radionuclides that emit only beta particles.

4.6.3 Estimating Charged-Particle Ranges

Charged particles as they travel through some medium lose energy through Coulombic interactions with ambient electrons (and, infrequently, with atomic nuclei) and by bremsstrahlung production. As a consequence, they eventually come to rest (discounting thermal motion) after a finite distance of travel. The distance a charged particle travels before coming to rest is called its *range*. However, there are several measures used for the range of a charged particle.

The Continuous Slowing Down Range

It is common to neglect energy-loss fluctuations and assume that particles lose energy continuously along their tracks, with a mean energy loss per unit path length given by the total stopping power. Under this approximation [ICRU 1984], the *continuous slowing down range* R_{CSDA} of a charged particle with initial kinetic energy E_o is given by [ICRU 1984]

$$R_{\text{CSDA}} = \int_0^R ds = \int_0^{E_o} \frac{dE}{(-dE/ds)_{tot}}. \tag{4.126}$$

where $(-dE/ds)_{tot} = (-dE/ds)_{coll} + (-dE/ds)_{rad}$, the total stopping power. This range is the expected length of a particle's trajectory before it comes to rest.

Evaluation of the integral is complicated by difficulties in formulating both the radiative stopping power and the collisional stopping power for low-energy particles, particularly electrons, and it is common to assume that the reciprocal of the stopping power is zero at zero energy and increases linearly to the known value at the least energy. It should be noted that this range cannot be measured experimentally since, to do so, would require a three-dimensional visualization of a particle's track and an evaluation of the length of this track.

Ranges for Penetration Depths

For a charged particle of a specific type traveling in a given medium, there is some average distance the particle travels between where it starts with energy E_0 and where it comes to rest and becomes part of the ambient medium. The *mean forward* or *projected* range R_p is the expected or average maximum depth of penetration in the initial direction of the charged particle. This range can be measured experimentally by finding the distance from a monodirectional source along the emission direction at which the intensity of charged particles is reduced by 50%.

Because charged particles, especially electrons and positrons, can be backscattered as they slow down, another range, which is sometimes encountered, is the average penetration distance R_{end} to where the particle comes to rest, i.e., to where the particle trajectory ends. As in the definition of R_p, this range is defined along a straight line in the initial particle direction. For HCPs $R_p \le R_{end}$ but for electrons R_{end} can be substantially smaller than R_m.

There are other ranges that are sometimes encountered for heavy charged particles. One is the *maximum* range R_{max} that is the crow-flight distance, in the initial particle direction, between the particle source to a point beyond which no particle reaches. Yet another straight-line range is the *extrapolated* range R_e that is based on a linear extrapolation of the distance versus intensity plot from R_p to zero intensity. These two ranges are discussed in more detail later in this section.

The CSDA and projected ranges are non-stochastic quantities and for heavy charged particles are nearly equal. For electrons, however, R_{CSDA} can be significantly greater than R_p and R_{end}. This difference is quantified by the *detour* factor, discussed next.

Detour Factor

The difference between R_{CSDA} and R_p is due to the deviation of a charged particle's trajectory from an ideal straight-line trajectory, a result of the multiple scattering events that occur along the trajectory. This deviation is very pronounced for electrons and positrons but much less so for heavy charged particles. The detour factor is defined as R_p/R_{CSDA} [Andreo et al. 2017]. It should be noted that this definition is subtly different for protons and alpha particles as defined by the ICRU [1993] who, instead of R_p used R_{end}, the depth of penetration to the termination point. Examples of the detour factor for electrons, protons and alpha particles are shown in Fig. 4.19 where it is seen that electrons have smaller detour factors than the heavy charged particles. This difference is a consequence of the much more tortuous path an electron travels while traversing matter compared to the path taken by a heavy charged particle.

4.6.4 Electron Energy Loss and Range

Before modern-day computers, analytical expressions were used to approximate the energy loss of electrons through matter. One widely used expression for relativistic electrons, based on the electron scattering results of Møller [1932], was derived by Bethe for electron stopping power in a substance with electron density N_e [Bethe and Ashkin 1953]

$$L = \frac{2\pi N_e q_e^4}{mv^2}\left\{ \ln\left[\frac{mv^2 E}{2I^2(1-\beta)}\right] - \left(2\sqrt{1-\beta^2} - 1 + \beta^2\right)\ln 2 \right.$$
$$\left. +1 - \beta^2 + \frac{1}{8}(1 - \sqrt{1-\beta^2})^2 \right\}. \tag{4.127}$$

where E is the incident particle energy and I is the average excitation potential. Although generally convenient for a wide variety of materials, Eq. (4.127) becomes inaccurate at low electron velocities. More recently, computer codes with corrections to the Bethe equation render improved results. For *electrons and positrons,* integration of the Møller [1932] or Bhabha [1938] cross section, with $z = 1$, leads to the Rohrlich and Carlson

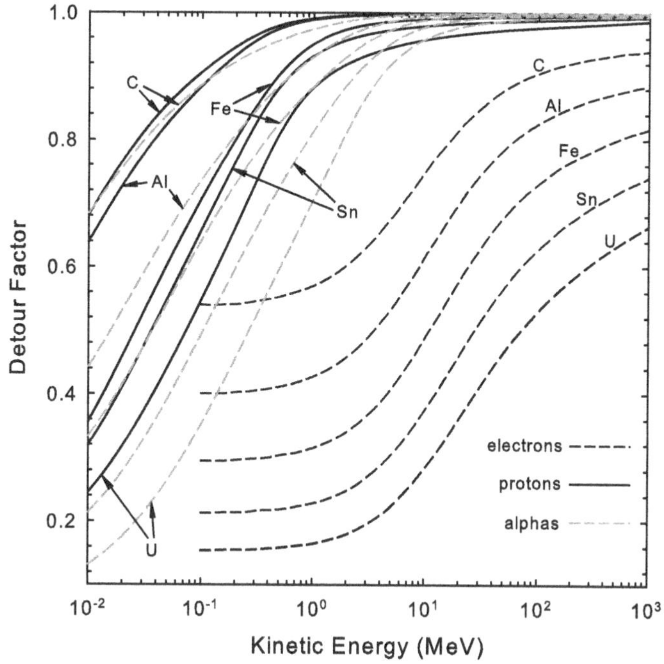

Figure 4.19. Detour factors $R_p/R_{\rm CSDA}$ for electrons (heavy dashed lines), protons (solid lines) and alpha particles (light dashed lines). The data for protons and alpha particles were obtained with the STAR codes [Berger 1992] and for electrons from Andreo et al. [2017].

[1954] formula used by Seltzer and Berger [1985; 1986] in their compilations of stopping powers [ICRU 1984]

$$L_{coll}(E) = \frac{\rho Z C}{2A\beta^2}\left[2\ln(\epsilon/\bar{I}) + \ln(1+\epsilon/2) + F^{\pm}(\epsilon) - \delta\right]. \qquad (4.128)$$

The factor F^- for electrons is given by

$$F^-(\epsilon) = (1-\beta^2)\left[1 + \epsilon^2/8 - (1+2\epsilon)\ln 2\right]. \qquad (4.129)$$

The factor F^+ for positrons is given by

$$F^+(\epsilon) = 2\ln 2 - (\beta^2/12)\left[23 + 14/(\epsilon+2) + 10/(\epsilon+2)^2 + 4/(\epsilon+2)^3\right]. \qquad (4.130)$$

In Eq. (4.128), the *density-effect correction term* δ accounts for polarization of atoms in the stopping medium caused by passage of the electron, the result of which is a reduction of the electric field acting on the moving electron and thus a reduction in the stopping power. Evaluation of δ is quite involved. Procedures are described by the ICRU [1984]. In liquid water, for example, the density effect causes a 1.2% reduction in L_{coll} at $E = 1$ MeV, 3.9% at 2 MeV, and 11.5% at 10 MeV. Selected values of mean excitation energies are given in Table 4.2. Stopping powers are illustrated in Fig. 4.20.

Restricted Electron Stopping Power The rate at which energy is deposited locally along the track of an electron is of great importance in certain radiation detector analyses. This requires establishing a cutoff energy, say E_{cut}, below which an energy loss is considered to be local. A value of E_{cut} on the order of 200

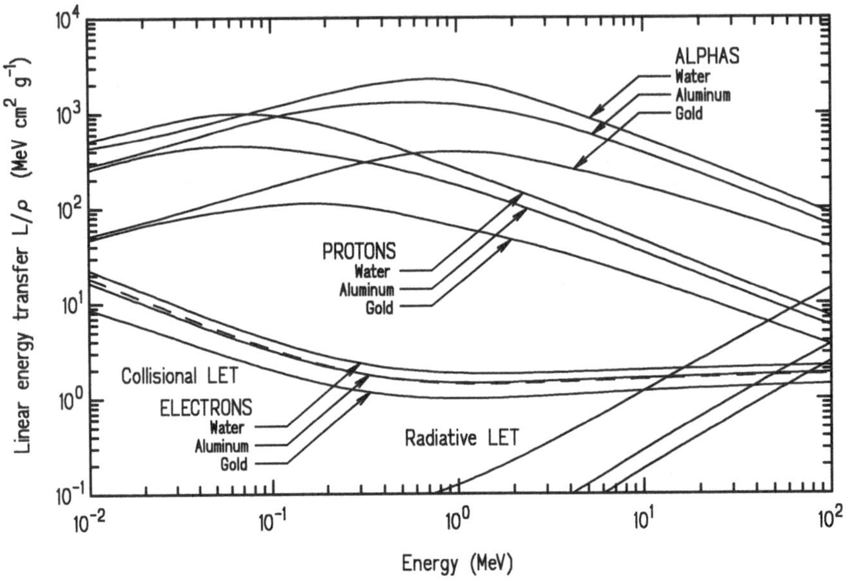

Figure 4.20. Linear energy transfer L/ρ in mass units (MeV cm^2/g) for alpha particles, protons and electrons. The dashed line for aluminum is the collisional LET for positrons. Data is from the STAR series of codes [Berger 1992].

eV is often chosen. The restricted stopping power $L_{coll}(E, E_{cut})$ is defined in analogy with Eq. (4.119), but with $\tau_{\max}$ expressed in terms of E_{cut}. The result is

$$L_{coll}(E, E_{cut}) = \frac{\rho Z C}{2A\beta^2}\left[2\ln(\epsilon/\bar{I}) + \ln(1+\epsilon/2) + G^{\pm}(\epsilon,\eta) - \delta\right], \tag{4.131}$$

in which $\eta \equiv E_{cut}/E$. The factor G^- for electrons is given by

$$G^-(\epsilon,\eta) = (1-\beta^2)\left[\epsilon^2\eta^2/2 + (2\epsilon+1)\ln(1-\eta)\right] - 1 - \beta^2 + \ln\left[4\eta(1-\eta)\right] + (1-\eta)^{-1}. \tag{4.132}$$

The factor G^+ for positrons is given by

$$\begin{aligned} G^+(\epsilon,\eta) = \ln 4\eta - \beta^2[1 &+ (2-\xi^2)\eta - (3+\xi^2)(\xi\epsilon/2)\eta^2 \\ &+ (1+\epsilon\xi)(\xi^2\epsilon^2/3)\eta^3 - (\xi^3\epsilon^3/4)\eta^4], \end{aligned} \tag{4.133}$$

in which $\xi \equiv (\epsilon+2)^{-1}$.

Electron Radiative LET No simple formula can describe the stopping power or LET for electron radiative energy losses (bremsstrahlung). The total stopping power can be written as the sum of stopping powers associated with radiative losses in the force fields of nuclei and electrons, namely, $L_{rad}(E) = L_{rad,n}(E) + L_{rad,e}(E) = L_{rad,n}(E)(1+\xi/z)$, in which ξ is the "correction term" $L_{rad,e}/L_{rad,n}$. According to ICRU [1984], ξ depends hardly at all on the medium and is less than 1.2 in magnitude for all E. It is about 0.5 at $E = 700$ keV and approaches zero at low E. The radiative stopping power may be written as [ICRU 1984]

$$L_{rad}(E) = N r_e^2 (E + m_e c^2) Z^2 \alpha (1 + \xi/Z)\Phi(E, Z), \tag{4.134}$$

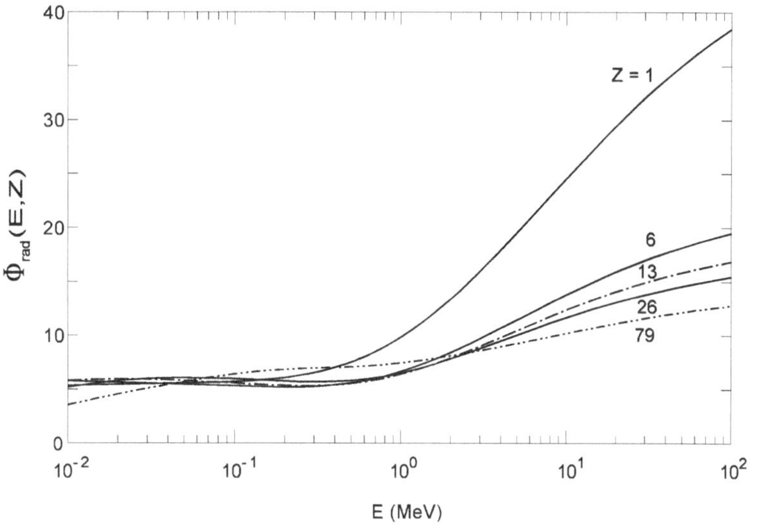

Figure 4.21. Scaled dimensionless radiative energy loss cross sections $\Phi_{rad}(E, Z)$ for electrons in various media. Based on data from Seltzer and Berger [1986].

in which N is the atomic density, $\alpha \simeq 1/137$ is the *fine-structure constant*, and $\Phi(E, Z)$ is the dimensionless scaled radiative energy-loss cross section given by

$$\Phi(E, Z) = \frac{1}{\alpha r_e^2 Z^2 (E + m_e c^2)} \int_0^E dE' \; E' \sigma_{rad}(E, E'). \tag{4.135}$$

The function Φ, too, is not strongly dependent on Z, as is illustrated in Fig. 4.21 for several values of Z. Tables of data are given by Seltzer and Berger [1985, 1986].

Electron Range The mean crow-flight range of electrons in a material with density ρ can be approximated by [Katz and Penfold 1952]

$$\rho R_p \text{ (g cm}^{-2}) = \begin{cases} 0.412 E^{1.265 - 0.0954 \ln E} & 0.1 \text{ MeV } \le E \le 3 \text{ MeV} \\ .530E - 0.106 & 2.5 \text{ MeV } \le E \le 20 \text{ MeV} \end{cases}, \tag{4.136}$$

where E is the energy of the incident electron. Notice that this empirical approximation is independent of the Z number of the attenuating material, a result that follows from the observation that, when the range is expressed in terms of mass thickness, the range is almost the same for all materials (as can be seen from Fig. 4.23).

When determining the maximum range of beta particles, then the maximum particle emission energy should be used. Notably, Eq. (4.136) implies that the electron mass range ρR_p is independent of the target material. The usefulness of this outcome allows easy range estimations between differing materials. Hence, the result of Eq. (4.136) need only be divided by the mass density of the target material to determine the actual range. However, in reality there is a difference in particle range between materials, and modern computer programs can provide better range estimates than the older empirical formulas.

Beta particles (electrons or positrons) produce numerous ionizations and electronic excitations as they move through a medium. Most interactions are small deflections accompanied by small energy losses. However, beta particles, having the same mass as the electrons of the medium, can also undergo large-angle

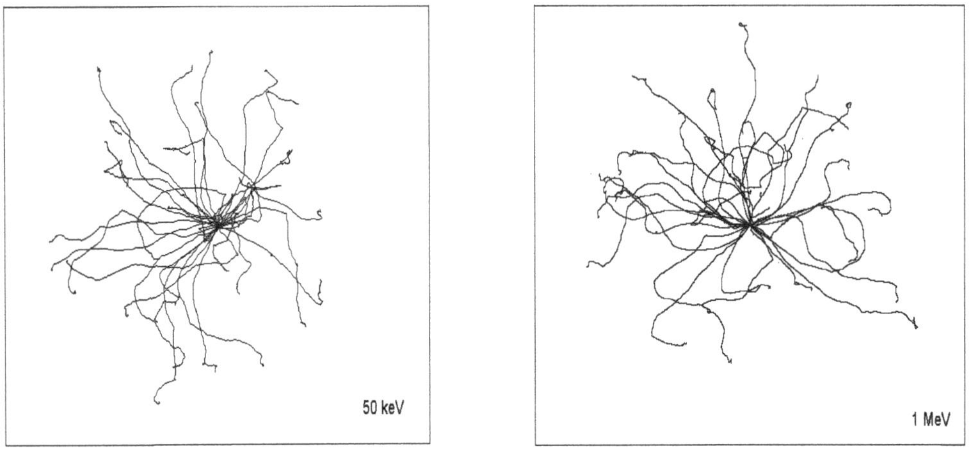

Figure 4.22. Tracks of 30 electrons from a 50-keV (left) and a 1-MeV (right) point isotropic source in water, shown as orthographic projections of tracks into a single plane. The box has dimensions $2r_o \times 2r_o$ where r_o is the CSDA range of the electrons. Calculations performed using the EGS4 code, courtesy of Robert Stewart, Kansas State University.

scatters producing secondary electrons with substantial recoil energy. These energetic secondary electrons (called *delta* rays) in turn pass through the medium causing additional ionization and excitation. Large-angle deflections from atomic nuclei can also occur with negligible energy loss. Consequently, the paths traveled by beta particles as they transfer their energy to the surrounding medium are far from straight. Examples of such paths are shown in Fig. 4.22.

For such paths, it is seen that the distances beta particles travel into a medium vary tremendously. The range of beta particles is defined as the path length of the particles, i.e., the distance they travel along their twisted trajectories. This range is well approximated by the CSDA range and is the maximum distance a beta particle can penetrate, although most, because of the twisting of their paths, seldom penetrate this far (see Section 4.6.3).

It may be inferred from data given by Cross, Freedman, and Wong [1992] that for a beam of electrons, ranging in energy from 0.025 to 4 MeV, normally incident on water, about 80% of the electron energy is deposited within a depth of about 60% of the CSDA range, 90% of the energy within about 70% of the range, and 95% of the energy within about 80% of the range. For a point isotropic source of monoenergetic electrons of the same energies in water, 90% of the energy is deposited within about 80% of the range, and 95% of the energy within about 85% of the range. In either geometry, all of the energy is deposited within about 110% of the CSDA range. Electron CSDA ranges are presented in Fig. 4.23.

4.6.5 Spatial Distribution of the Electron Energy Absorption

In both the design of radiation detectors and radiation dosimetry, it is important to know where the energy of a charged particle is deposited at different distances from the source of the charged particles. Because of the tortuous trajectories of electrons, the spatial distribution of this absorbed energy is not easily calculated from first principles. This section considers point and uniform-beam sources of monoenergetic electrons in infinite homogeneous media. In particular, spatial distributions of the absorbed dose (absorbed energy per unit mass) are expressed in terms of point and plane kernels and associated scaled dimensionless dose distributions.

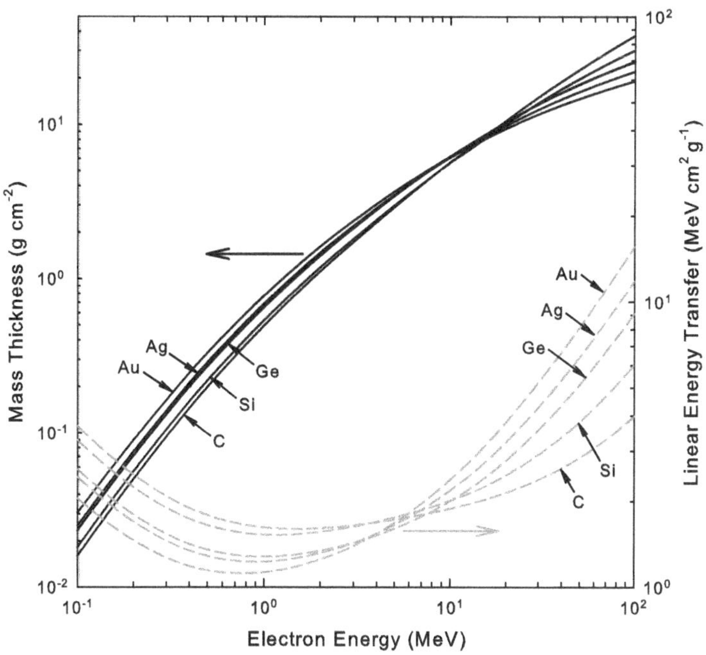

Figure 4.23. The CSDA range or path length ρR, in mass thickness (g/cm^2), for electrons in various materials. The CSDA ranges are shown as solid black lines. Also shown, as gray dotted lines, are linear energy transfer (LET) as functions of energy. Data are from the STAR series of codes [Berger 1992].

Point and Plane Kernels Defined

The expected absorbed dose at distance r from a unit point isotropic source of monoenergetic charged particles of energy E is denoted as $\mathcal{G}(r, E)$. Typical units for $\mathcal{G}$ are MeV/g. The point kernel $\mathcal{G}$ is very conveniently expressed in terms of the dimensionless dose distribution $\mathcal{F}(r/r_o, E)$, with radial distance r scaled by the CSDA range $R_{CSDA} \equiv r_o$, namely,

$$\mathcal{G}(r, E) = \frac{E}{4\pi r^2 \rho r_o} \mathcal{F}(r/r_o, E) = \frac{E}{4\pi r^2 \hat{r}_o} \mathcal{F}(\hat{r}/\hat{r}_o, E), \tag{4.137}$$

in which $\hat{r} = \rho r$ and $\hat{r}_o$ is the CSDA range of Eq. (4.126) in mass units. A useful interpretation of $\mathcal{F}$, which is illustrated in Fig. 4.24, is that the fraction of E deposited between radii r and $r + dr$ is $(dr/r_o)\mathcal{F}(r/r_o, E)$. Note that, without a subscript, $\mathcal{F}$ and $\mathcal{G}$ refer to a point isotropic source.

Similarly, for an infinite plane perpendicular[15] source of monoenergetic particles in an infinite homogeneous medium, the expected absorbed dose at distance r along the beam is denoted as $\mathcal{G}_{pl}^{\perp}(r, E)$. This is conveniently expressed in terms of a scaled dimensionless dose distribution as

$$\mathcal{G}_{pl}^{\perp}(r, E) = \frac{E}{\rho r_o} \mathcal{F}_{pl}^{\perp}(r/r_o, E), \tag{4.138}$$

which is illustrated in Fig. 4.25.

[15]The source is an area source with all particles released in a parallel beam normal to the source area.

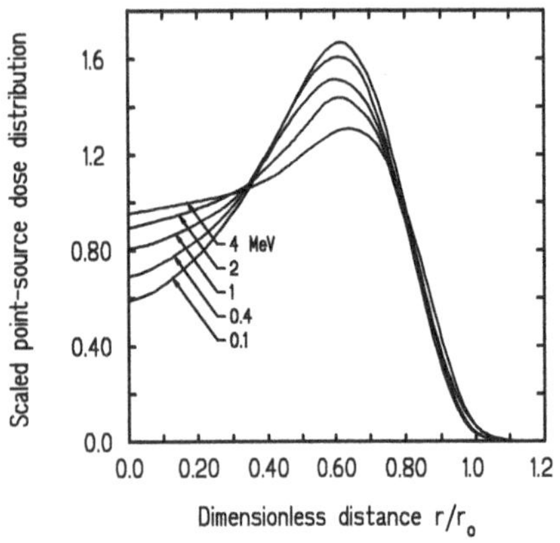

Figure 4.24. Scaled dimensionless dose distributions, $\mathcal{F}(r/r_o, E)$, for point isotropic electron sources in water. From Cross, Freedman, and Wong [1992].

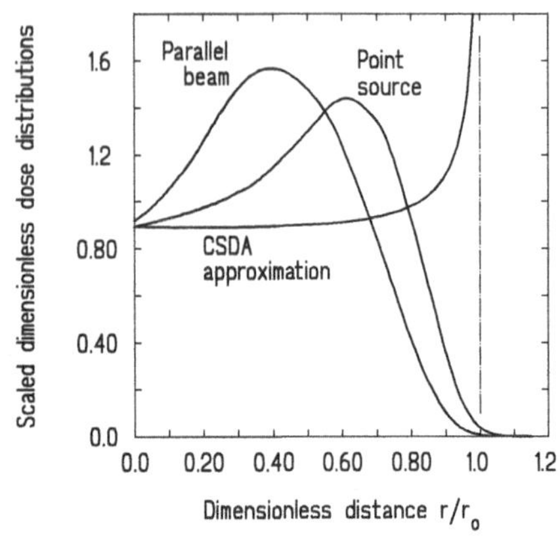

Figure 4.25. Comparison of $\mathcal{F}(r/r_o, E)$ point-source, $\mathcal{F}_{pl}^{\perp}(r/r_o, E)$ parallel-beam-source, and CSDA scaled dimensionless dose distributions for 2-MeV electron sources in water. Data for the point and parallel-beam sources from Cross, Freedman, and Wong [1992].

CSDA Absorbed Energy Distribution

In the continuous slowing-down approximation, all particle tracks are straight and of length r_o. It is easily shown that in this approximation,

$$\left[\mathcal{F}_{pl}^{\perp}(r,E)\right]_{\text{CSDA}} = \left[\mathcal{F}(r,E)\right]_{\text{CSDA}} = \frac{L(r_o - r)}{E/r_o}, \quad r < r_o. \tag{4.139}$$

Note that the linear energy transfer L is evaluated at the energy of the particle after passage through distance r, that is, for a particle with residual range $r_o - r$. The residual range[16] determines both the energy and the LET of a particle. The denominator of Eq. (4.126) is just the average LET along the entire path of the particle. Point and plane dose distributions are compared with the distribution in the CSDA approximation in Fig. 4.25.

Electron Absorbed Energy Distributions

The pioneering moments-method electron-transport calculations of Spencer [1959] were used by Berger [1971] and Cross et al. [1982] to construct scaled dimensionless dose distributions for monoenergetic electrons and beta-particle sources. These were revised by Berger in 1973, based on Monte Carlo calculations. Cross, Freedman, and Wong [1992] also used Monte Carlo methods to prepare comprehensive tables of dose distributions for point, plane parallel, and plane isotropic sources of electrons and beta particles in water. Example dose distributions for point sources are illustrated in Fig. 4.25. Tabulations of $\mathcal{F}(r/r_o, E)$ versus r/r_o are given by Shultis and Faw [2000] and Cross et al. [1992].

Figure 4.25 compares kernels for 2-MeV point and plane parallel electron sources in water. In contrast, the point kernel based on the CSDA approximation is shown to be a very poor approximation for electrons. The reason is apparent from examination of Fig. 4.22, which shows the spatial distributions of energy deposition from point isotropic sources of monoenergetic electrons. Not only is much of the energy deposited well within

[16] A charged particle energetic enough to penetrate a distance x and emerge with energy E' has a residual range $R_{CSDA}(E')$ in the same material. So the range $R_{CSDA}(E)$ of the particle with initial energy E must have been $x + R_{CSDA}(E')$.

distance r_o, but the spatial distribution of the energy distribution scaled by r_o is insensitive to E. The scaled patterns shown in Fig. 4.22 for 50-keV and 1-MeV electrons reinforce the logic for the similarity in shapes of the curves in Fig. 4.24.

Kernels in Non-Aqueous Media

It has long been observed [Cross 1967; Berger 1971] that when distance is measured in mass thickness $\hat{r} = \rho r$, the point kernel in medium m at distance $\hat{r}$ is proportional to that in water, medium w, at a scaled distance $\eta\hat{r}$. The scale factor η was found to depend on medium m but only very weakly on the energy, so that the same scale factor could be applied to beta-particle sources as well as monoenergetic electron sources. Conservation of energy for a point source requires that

$$\int_0^\infty d(\hat{r}/\rho_m)\,(\hat{r}/\rho_m)^2\rho_m\,\mathcal{G}_m(\hat{r},E) = \int_0^\infty d(\eta\hat{r}/\rho_w)\,(\eta\hat{r}/\rho_w)^2\rho_w\,\mathcal{G}_w(\eta\hat{r},E). \tag{4.140}$$

It therefore follows that

$$\mathcal{G}_m(\hat{r},E) = \eta^3\left[\frac{\rho_m}{\rho_w}\right]^2\mathcal{G}_w(\eta\hat{r},E). \tag{4.141}$$

Similarly,

$$\mathcal{G}^{\perp}_{m,pl}(\hat{r},E) = \eta\,\mathcal{G}^{\perp}_{w,pl}(\eta\hat{r},E). \tag{4.142}$$

In terms of the dimensional dose distribution $\mathcal{F}$ the point kernel for medium m is

$$\mathcal{G}_m(\hat{r},E) = \frac{E}{4\pi r^2(\hat{r}_o/\eta)}\mathcal{F}\left(\frac{\hat{r}}{\hat{r}_o/\eta},E\right) = \frac{E}{4\pi r^2\hat{r}_m}\mathcal{F}\left(\frac{\hat{r}}{\hat{r}_m},E\right), \tag{4.143}$$

in which the mass CSDA range $\hat{r}_m(E)$ of electrons in medium m, in terms of that in water, is given by $\hat{r}_m(E) = \hat{r}_o(E)/\eta$. Thus, $\mathcal{F}$ may be interpreted as a universal function of the ratio $\hat{r}/\hat{r}_m = r/r_m$, the distance from the point source in units of the CSDA range.

Cross, Freedman, and Wong [1992] recommend that η be computed using the following formula, which is based on experiments and calculations for beta-particle attenuation in various media relative to that in water:

$$\eta = 0.77(1 + 0.0491\overline{Z} - 0.0009\overline{Z}^2)\frac{(L/\rho)_m}{(L/\rho)_w}, \tag{4.144}$$

in which the mean atomic number is given as follows, based on mass fractions of the atomic constituents in the medium:

$$\overline{Z} = \frac{\sum_i w_i Z_i^2/A_i}{\sum_i w_i Z_i/A_i}. \tag{4.145}$$

The ratio of stopping powers is insensitive to energy, and Berger [1971] recommends that they be evaluated at 0.2 MeV. Cross, Freedman, and Wong [1992] recommend that, instead, $(L/\rho)_m/(L/\rho)_w$ be replaced by $\hat{r}_w/\hat{r}_m$ evaluated at 500 keV. These formulas predict that $\eta = 0.89$ for air. Note that the ratio of the mass CSDA range in water to that in air varies from 0.89 for 0.01-MeV electrons to 0.91 for 3-MeV electrons. For the human body, η is 0.98 for soft tissue and 0.97 for compact bone.

4.6.6 Heavy Charged-Particle Energy Loss

Rossi [1952] describes the stopping power for relativistic *heavy charged particles* as

$$L_{coll}(E) = \frac{\rho Z z^2\mathcal{C}}{A\beta^2}\left[\ln\left(\frac{2\beta^2}{(1-\beta^2)\overline{I}}\right) - \beta^2\right]. \tag{4.146}$$

This approximation is mainly used for protons with energies between about 10 and 1000 MeV or other heavy particles with comparable speeds. Note that L_{coll} is directly proportional to z^2 and that $N_e = \rho(Z/A)N_a$. It does not vary independently with mass and kinetic energy of the incident charged particle. Instead, L_{coll} is a function of the particle velocity or β^2.

At low energies, β becomes small and Eq. (4.146) reduces to

$$L_{coll}(E) = \frac{\rho Z z^2 C}{A\beta^2} \ln\left(\frac{2m_e v^2}{I}\right). \tag{4.147}$$

Note that L_{coll} is only weakly dependent on the mean excitation energy I. For a non-relativistic particle,[17] $v = \sqrt{2T/M}$. Thus, in a given medium, a 1-MeV proton has roughly the same collisional stopping power as a 2-MeV deuteron and roughly one-fourth that of a 4-MeV alpha particle. In this way, stopping-power data for protons may be used to approximate stopping powers for other heavy charged particles. Equation (4.146) overestimates the LET by less than about 10% for proton energies between 1 and 10^4 MeV and for stopping media such as air, water, and elements ranging in atomic number from aluminum to gold. For proton energies much below 1 MeV, the formula fails utterly. For low energies, Anderson and Ziegler [1977] recommend a general relation

$$L_{coll}(E) \simeq A_1 E^{A_2}, \tag{4.148}$$

where A_1 is an experimentally determined constant and $A_2 \simeq 0.5$. The collisional stopping power for protons and alpha particles, as a function of the particle's energy, is shown in Fig. 4.20.

Corrections to Eq. (4.146) account for shell and density effects, among other minor effects. Shell corrections compensate for the fact that the atomic binding of inner-shell electrons reduces their participation in charged-particle slowing down. Density corrections compensate for the polarization of the stopping medium induced by passage of the charged particle and the resulting perturbation of the electronic structure and reduction in stopping power. Stopping of low-energy charged particles is affected by neutralization of the charge of the particle and by collisions with atomic nuclei of the stopping medium. Detailed data on the LET and range of heavy charged particles are given in reports by Janni [1982], Hubert et al. [1990], and the ICRU [1993]. Computer codes are also available for calculation of ranges, stopping powers, and related quantities for electrons, protons, helium nuclei, and heavy ions [Berger 1992; Ziegler et al. 2013].

4.6.7 Heavy Charged-Particle Range

When a heavy charged particle (HCP) with mass M and energy E_M scatters from an electron, the maximum energy loss by the HCP and the maximum energy of the recoil electron is [Shultis and Faw 2017]

$$(E_e)_{\max} = 4\frac{m_e}{M}E_M. \tag{4.149}$$

As an example, consider a 4-MeV alpha particle. From Eq. (4.149), the maximum energy loss of such an alpha particle when it interacts with an electron is found to be 2.2 keV. Most such alpha-electron interactions result in far smaller energy losses, although usually sufficient energy transfer to free the electron from its atom or at least to raise the atom to a high excitation state. Consequently, many thousands of these interactions are required for the alpha particle to lose its kinetic energy.

Moreover, the mass of the alpha particle being over 7000 times that of an electron, the alpha particle experiences negligible deflection from its path of travel. An important characteristic of such heavy charged particles, as they move through a medium, is that they travel in almost straight lines, unless they are

[17] By this is meant a particle whose kinetic energy is very much less than its rest-mass energy: 511 keV for electrons, about 930 MeV for protons or neutrons, and about 3727 MeV for alpha particles.

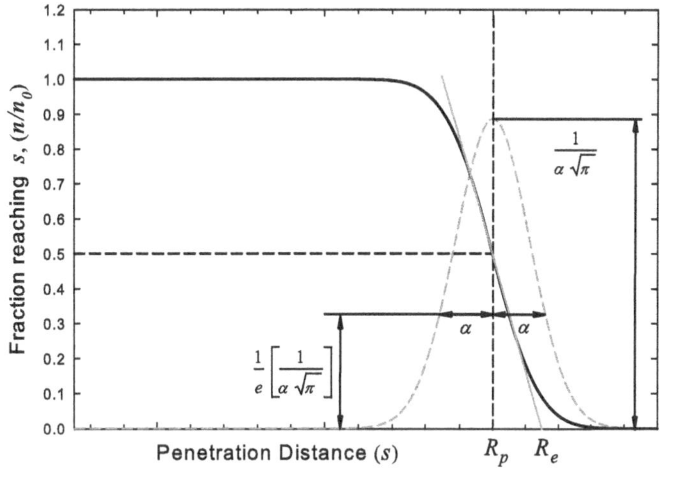

Figure 4.26. Fraction of α-particles that penetrate a distance s. The gray dotted line shows the derivative $dn(s)/ds$ and is termed the straggling curve, a result of the large-angle scattering near the end of the particles' paths and the statistical fluctuations in the number of interactions needed to absorb the particles' initial energy.

deflected by a nucleus. Such nucleus-particle interactions are rare and usually occur only when the charged particles have lost most of their kinetic energy.

Many measurements of the *range* of heavy particles have been made. In Fig. 4.26, the fraction of alpha particles, with the same initial energy, that travel various distances s into a medium is illustrated. We see from this figure, that the number of alpha particles penetrating a distance s into a medium remains constant until near the end of their range. The number penetrating beyond this maximum distance falls rapidly.

The distance a heavy charged particle travels in a medium before it is stopped is not a precise quantity because of the stochastic nature of the number and energy losses of all the particle-medium interactions, and because of possible large-angle deflections near the end of the path. From Fig. 4.26, we see that the range of a heavy charged particle is not an exact quantity. Several measures of the range can be defined. We can define R_p as the *mean* or *projected* range, that is, the distance at which the intensity of heavy particles has been halved. This projected range is at the maximum of the differential energy curve or *straggling curve* $dn(s)/ds$, shown as the dotted curve in Fig. 4.26. Alternatively, the range can be taken as R_e, the *extrapolated* range, that is, the value obtained by drawing the tangent to the curve at its inflection point and extrapolating the tangent until it crosses the s-axis. Finally, if we treat the slowing down of heavy charged particles as a continuous process, a unique range, called the *continuous slowing-down approximation* (CSDA), a unique CSDA range can be calculated. Although slightly different, all three of these definitions of a heavy particle's range are, for practical purposes, nearly the same. The mean and CSDA ranges for protons and alpha particles are presented in Fig. 4.27.

Because of the large number of interactions with electrons of the ambient medium, heavy charged particles, of the same mass and initial kinetic energy, travel in straight lines and penetrate almost the same distance into a medium before they are stopped. Because of the stochastic nature of the particle-medium interactions and because, near the end of their paths, some heavy particles experience large angle deflections from atomic nuclei, the penetration distances exhibit a small variation called *straggling*. The straggling is defined by a range-straggling parameter α_0, where $\alpha_0 \simeq 0.015R_p$ [Evans 1955]. The straggling curve indicates the final

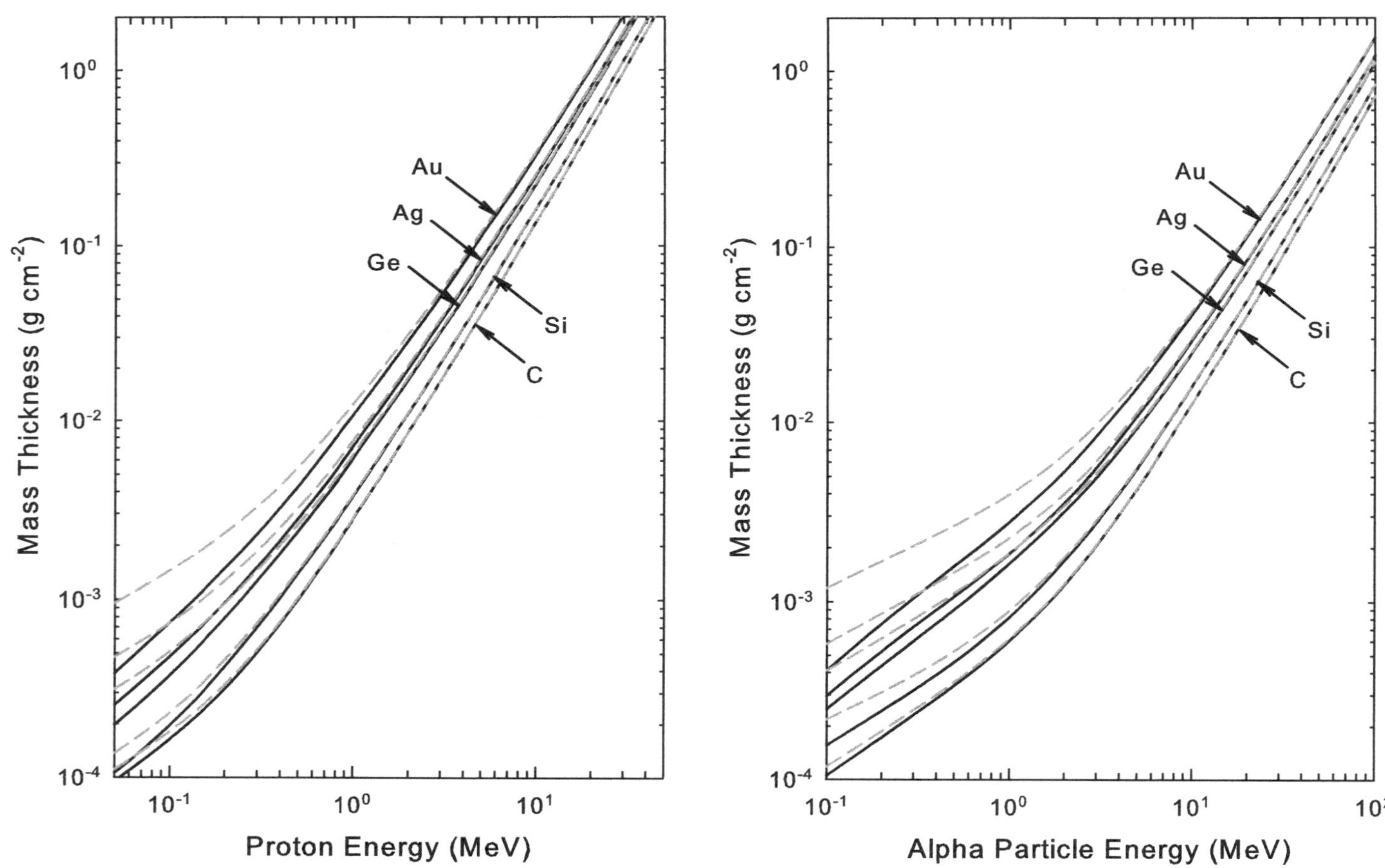

Figure 4.27. Range or path length ρR, in mass thickness (g/cm^2), for protons and alpha particles in several materials. The average (or projected) ranges are shown as black solid lines while the CSDA ranges are shown as gray dotted lines. Data are from the STAR series of codes [Berger 1992].

resting distribution of ions in the target material, and is modeled as a Gaussian distribution

$$\frac{1}{n_0}\frac{dn(s)}{ds} = \frac{1}{\alpha\sqrt{\pi}}e^{-[(s-R_p)/\alpha]^2}, \tag{4.150}$$

where n_0 is the total number of ions and $\alpha \simeq \sqrt{2}\alpha_0$. Notably, the width 2α marks the locations in the number-range curve where the maximum of Eq. (4.150) is reduced by e^{-1}.

For *heavy* charged particles, we may use Eq. (4.126) to obtain the dependence of the range on the particle's mass m and charge z. Because bremsstrahlung is seldom important for heavy charged particles with energies below several tens of MeV, we may take $(-dE/ds)_{tot} \simeq (-dE/ds)_{coll}$. From Eq. (4.146) we see the collisional stopping power is of the form $(-dE/ds)_{coll} = \rho(Z/A)z^2 f(v)$ where $f(v)$ is a function only of v or $\beta = v/c$. From Eq. (4.126), the mass thickness range, ρR_{CSDA} is

$$\rho R_{CSDA} = \rho \int_0^{E_o} \frac{dE}{(-dE/ds)_{coll}} = \rho \int_0^{E_o} \frac{1}{(-dE/ds)_{coll}}\frac{dE}{dv}\,dv. \tag{4.151}$$

Substitution of $dE/dv = mv$ and $(-dE/ds)_{coll} = \rho(Z/A)z^2 f(v)$ then gives

$$\rho R_{CSDA} = \left(\frac{A}{Z}\right)\frac{m}{z^2}\int_0^{v_o}\frac{v\,dv}{f(v)} = \left(\frac{A}{Z}\right)\frac{m}{z^2}g(v_o), \tag{4.152}$$

where $g(v_o)$ is a function only of the initial speed v_o of the particle. From this result, three useful "rules" for relating different ranges follow.

1. The mass-thickness range ρR_{CSDA} is independent of the density of the medium.

2. For particles of the *same initial speed* in the same medium, ρR_{CSDA} is *approximately* proportional to m/z^2, where m and z are for the charged particle.

3. For particles of the *same initial speed*, in different media, ρR is *approximately* proportional to $(m/z^2)(Z/A)$, where Z is the atomic number of the stopping medium and A is its atomic weight.

Thus, from rule 2, in a given medium, a 4-MeV alpha particle has about the same range as a 1-MeV proton. However, these rules fail for particle energies less than about 1 MeV per atomic mass unit. For example, the CSDA range of a 0.4-MeV alpha particle in aluminum is about twice that of a 0.1-MeV proton.

4.6.8 The Bragg Curve

In reporting on his 1904 research on the ionization produced along the track of an alpha-particle, W.H. Bragg plotted a graph of the ion pairs produced per unit distance as a function of the distance penetrated by alpha particles in air. Known as the *Bragg curve*, this type of graph has been used countless times to illustrate the concepts of range and stopping power. A Bragg ionization curve is the average distribution of energy deposited, through ionization, per unit distance over the particle range and is described by

$$I(s) = \int_r^{\infty} \frac{i(r-s)}{\alpha\sqrt{\pi}}e^{-[(r-R)/\alpha]^2}dx, \tag{4.153}$$

where α is the straggling parameter, $i(r-s)$ is the *specific ionization* along the path of an *individual* particle at a distance of $(r-s)$, s is the penetration distance in the medium, r is the range of an *individual* particle, and R_p is the average particle range.

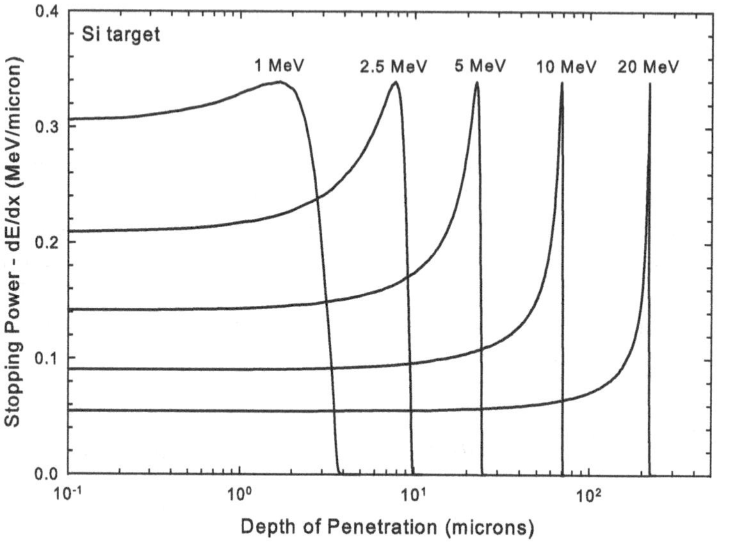

Figure 4.28. "Bragg curves" for alpha particles (helium nuclei) in silicon showing the stopping power versus depth of penetration. Data obtained using the SRIM program [Ziegler et al. 2013].

Closely related to the Bragg ionization curve is a graph of stopping power versus distance penetrated along the track of an alpha particle or other charged particle. Examples are shown in Fig. 4.28, illustrating energy loss along the tracks of alpha particles in silicon. In silicon, the stopping power is maximum for alpha particles of energy about 460 keV. For alpha particles with initial energies above 460 keV, stopping power increases along the particle path, reaches a maximum, and then decreases rapidly as the particle releases most of its kinetic energy near the end of its path.

4.6.9 Approximate Range Formula for Charged Particles

An empirical mass range $\rho R_{\rm CSDA}$ for electrons, protons and alpha particles can be obtained by fitting the equation

$$\rho R_{\rm CSDA}\ ({\rm cm^2 g^{-1}}) = 10^{a+bx+cx^2} \tag{4.154}$$

to calculated CSDA range data. Here $x = \log_{10} E$ with the initial energy E of the particle in units of MeV. The empirical constants a, b, and c depend on the stopping material. This material dependence ameliorates a deficiency of the empirical formula of Eq. (4.136) for the mean electron range. Values of a, b, and c for a few materials are given in Table 4.3 for electrons and Table 4.4 for protons and alpha particles. Values of these parameters for other materials can be obtained from the STAR computer code package [Berger 1992].

If computer codes, empirical data, and/or particles ranges, such as listed in Table 4.4, are not available, then alternative relations do exist that can yield an acceptable result. For instance, from the analysis of Bragg and Kleeman [1905], an expression is available for heavy ions as a ratio of ranges for monoenergetic particles in different media

$$\frac{R_1}{R_2} = \frac{\rho_2}{\rho_1}\sqrt{\frac{A_1}{A_2}}, \tag{4.155}$$

where R_1 and R_2 are the ranges for alpha particles in two materials of density ρ_1 and ρ_2, and atomic masses A_1 and A_2. Eq. (4.155) is commonly called the *Bragg-Kleeman Rule*, and is best used for materials with similar atomic masses. The use of Eq. (4.155) is straightforward for elemental materials, but requires

Table 4.3. Constants for the empirical formula of Eq. (4.154). Valid for energies between 0.1 and 10 MeV.

	Electrons		
Material	a	b	c
C	-0.30434	1.2633	-0.21918
Al	-0.25876	1.2380	-0.22597
Si	-0.27136	1.2360	-0.22740
Ar	-0.22887	1.2209	-0.23682
Ge	-0.18537	1.2069	-0.23702
Ag	-0.16377	1.1903	-0.24781
Xe	-0.14796	1.1738	-0.25395
Pb	-0.10787	1.1458	-0.25628
air	-0.31149	1.2419	-0.23063
H_2O	-0.36173	1.2587	-0.21600
LiF	-0.26944	1.2572	-0.21674

Source: Fit to data from STAR codes [Berger 1992].

Table 4.4. Constants for the empirical formula $y = a+bx+cx^2$ relating charged-particle energy and its projected range, in which y is $\log_{10}$ of $\rho R_{\rm CSDA}$ (g/cm^2) and x is $\log_{10}$ of the initial particle energy E_o (MeV). Valid for energies between 0.1 and 10 MeV.

	Protons			Alpha Particles		
Material	a	b	c	a	b	c
C	-2.5542	1.4997	0.22046	-3.2132	1.0080	0.29815
Al	-2.4108	1.4570	0.19861	-3.0581	.99132	0.27018
Si	-2.4232	1.4802	0.17421	-3.0813	.98208	0.29625
Ar	-2.3756	1.4944	0.16909	-3.0030	.93249	0.32566
Ge	-2.2097	1.4045	0.17980	-2.7867	.96529	0.19442
Ag	-2.1440	1.3815	0.18943	-2.7267	.95425	0.21643
Xe	-2.1265	1.4375	0.13890	-2.7556	.97004	0.26094
Pb	-1.9923	1.3663	0.17444	-2.5803	.98740	0.19178
air	-2.5510	1.5066	0.21732	-3.1854	.94174	0.34192
H_2O	-2.6144	1.4975	0.23219	-3.2357	.93102	0.33442
LiF	-2.4883	1.4501	0.25453	-3.1240	.96343	0.30053

Source: Fit to data from STAR codes [Berger 1992].

a conversion to an *effective atomic mass* when used with mixtures and compounds [Henriksen and Baarli 1957]. There are various expressions for the effective atomic mass, some fairly complicated [see for example, Birgani et al. 2012]. Offered here is a relatively simple variation [Birks 1964]

$$A_{eff} = \left[\sum_{i=1} f_i A_i \Big/ \sum_{i=1} f_i\sqrt{A_i}\right]^2, \tag{4.156}$$

where f_i and A_i are the atomic fraction and atomic mass of element i. Knowledge of alpha particle range data for one material will allow estimations of alpha particle ranges in another material.

Example 4.1: Estimate the range in water of a 6-MeV triton (nucleus of ^{3}H). Data for the empirical range formula of Eq. (4.154) are not provided for a triton. We must, therefore, estimate the triton's range from the observation that, for a given medium, the range R of heavy charged particles with the same speed is proportional to m/z^2, where m and z are the mass and charge of the particle.

Solution:

Step 1: First find the kinetic energy of a proton with the same speed as the 6-MeV triton. Classical mechanics, appropriate for these energies, gives

$$\frac{E_p}{E_T} = \frac{(1/2)m_p v_p^2}{(1/2)m_T v_T^2} = \frac{m_p}{m_T} = \frac{1}{3}.$$

The proton energy with the same speed as a 6-MeV triton is, thus, $E_p = E_T/3 = 2$ MeV.

Step 2: Now we find the range in water of a 2-MeV proton using the empirical formula of Eq. (4.154) and the data in Table 4.4. With $x = \log_{10} 2 = 0.30103$

$$\rho R_p(2 \text{ MeV}) = 10^{-2.6144+1.4975x+0.23219x^2} = 0.007202 \text{ g/cm}^2.$$

Since $\rho = 1$ g/cm^3, $R_p(2 \text{ MeV}) = 0.007202$ cm.

Step 3: From the range "rule" for the same medium (following Eq. (4.152)), we have

$$\frac{R_T(6 \text{ MeV})}{R_p(2 \text{ MeV})} = \frac{m_T}{m_p}\frac{z_p^2}{z_T^2} = \frac{3}{1}\frac{1}{1} = 3.$$

The range of the triton is then $R_T(6 \text{ MeV}) = 3R_p(2 \text{ MeV}) = 0.0216$ cm.

Example 4.2: What is the range of 5.5 MeV alpha particles in silicon dioxide (SiO_2)?

Solution:

Step 1: Determine the effective atomic mass of SiO_2. From Eq. (4.156),

$$\begin{aligned} A_{\mathit{eff}} &= \left[\sum_{i=1} f_i A_i \Big/ \sum_{i=1} f_i\sqrt{A_i}\right]^2 \\ &= \left[\left(f_{\mathrm{Si}}A_{\mathrm{Si}} + f_{\mathrm{O}}A_{\mathrm{O}}\right) \Big/ \left(f_{\mathrm{Si}}\sqrt{A_{\mathrm{Si}}} + f_{\mathrm{O}}\sqrt{A_{\mathrm{O}}}\right)\right]^2 \\ &= \left[\left(0.333(28) + .666(16)\right) \Big/ \left(0.333\sqrt{28} + .666\sqrt{16}\right)\right]^2 = 20.38 \text{ g mol}^{-1} \end{aligned}$$

Step 2: Determine the range of 5.5 MeV alpha particles in a material with similar atomic mass. Here, Si is chosen for the material with a known range. From Eq. (4.154) and Table 4.4, with $x = \log_{10} 5.5 = 0.7404$,

$$\rho R_\alpha(5.5 \text{ MeV}) = 10^{-3.0813+0.98208(0.7404)+0.29625(0.7404)^2} = 0.0064 \text{ g/cm}^2.$$

The density of Si is 2.32 g cm^{-3}, yielding

$$R_\alpha(5.5 \text{ MeV}) = \frac{0.0064}{2.32} = 0.00277 \text{ cm} = 27.7\ \mu\text{m}.$$

Step 3: Use Eq. (4.155) to find the range in SiO_2 (quartz). The density of quartz is 2.65 g cm^{-3}. Therefore,

$$R_{SiO_2} = (0.00277 \text{ cm}) \left(\frac{2.32 \text{ g cm}^{-3}}{2.65 \text{ g cm}^{-3}} \right) \sqrt{\frac{20.38}{28}} = 0.0021 \text{ cm} = 21 \ \mu\text{m}.$$

By comparison to the SRIM code [Ziegler et al. 2013], the answers are $R_{Si} = 28$ microns and $R_{SiO_2} = 22$ microns.

4.6.10 Range of Fission Fragments

The range of a *fission fragment* is difficult to calculate because the fission fragment's charge changes as it slows down. The lighter of the two fragments, having more initial kinetic energy, has a somewhat longer range. An empirical formula with an accuracy of ±10% for the range (in units of mass thickness) of a fission product, with initial kinetic energy E (in MeV), is [Alexander and Gazdik 1960]

$$\rho R \ (\text{mg/cm}^2) \simeq C\, E^{2/3}, \quad \frac{E_{H,L}}{2} \leq E \leq E_{H,L} \tag{4.157}$$

where $E_{H,L}$ refers to the initial energy of either the midpoint heavy or light fission fragment (nominally 67.9 MeV and 99.9 MeV, respectively), and

$$C = \begin{cases} 0.14, & \text{air} \\ 0.19, & \text{aluminum} \\ 0.27, & \text{nickel} \\ 0.50, & \text{gold} \end{cases} \tag{4.158}$$

Measurements by Fulmer [1957] indicate slightly longer ranges than predicted by Eq. (4.157). Figures 4.29 and 4.30 demonstrate the residual energy after a fission fragment passes through a mass thickness ρR. The x-axis intercept is the range of the fragment.

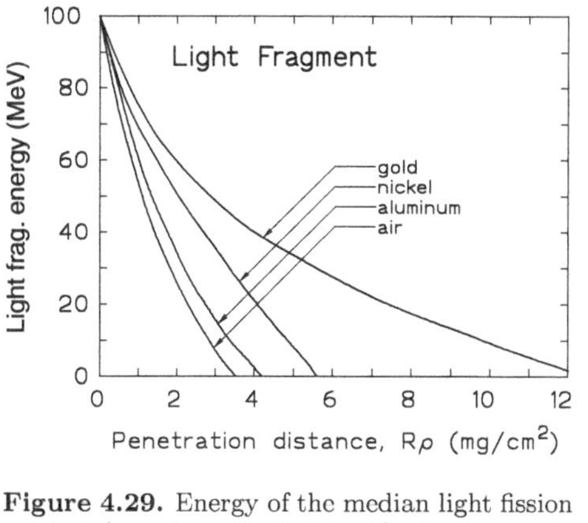

Figure 4.29. Energy of the median light fission product (initial energy 99.9 MeV) after penetrating a distance R into various materials. After Fulmer [1957].

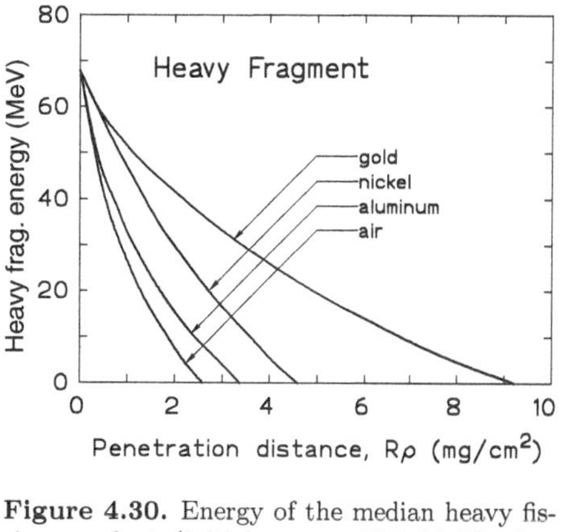

Figure 4.30. Energy of the median heavy fission product (initial energy 67.9 MeV) after penetrating a distance R into various materials. After Fulmer [1957].

PROBLEMS

1. A broad beam of neutrons is normally incident on a homogeneous slab 6-cm thick. The intensity of neutrons transmitted through the slab without interactions is found to be 30% of the incident intensity. (a) What is the total interaction coefficient μ_t for the slab material? (b) What is the average distance a neutron travels in this material before undergoing an interaction?

2. Based on the interaction coefficients tabulated in Appendix B, plot the tenth-thickness (in centimeters) versus photon energy from 0.1 to 10 MeV for water, concrete, iron, and lead.

3. With the data of Table B.5, calculate the half-thickness for 1-MeV photons in (a) water, (b) iron, and (c) lead.

4. A material is found to have a tenth-thickness of 2.3 cm for 1.25-MeV gamma rays. (a) What is the linear attenuation coefficient for this material? (b) What is the half-thickness? (c) What is the mean-free-path length for 1.25-MeV photons in this material?

5. Calculate the linear interaction coefficients in pure air at 20°C and 1 atm pressure for a 1-MeV photon and a thermal neutron (2200 m s^{-1}). Assume that air has the composition 75.3% nitrogen, 23.2% oxygen, and 1.4% argon *by mass*. Use the following data:

	Photon	Neutron
Element	μ/ρ (cm^2 g^{-1})	σ_{tot} (b)
Nitrogen	0.0636	11.9
Oxygen	0.0636	4.2
Argon	0.0574	2.2

6. In natural uranium, 0.720% of the atoms are the isotope ^{235}U, 0.0055% are ^{234}U, and the remainder ^{238}U. From the data in Table B.3, what is the total linear interaction coefficient (macroscopic cross section) for a 2200 m s^{-1} neutron in natural uranium? What is the total macroscopic fission cross section for 2200 m s^{-1} neutrons?

7. At a particular position, the flux density of particles is 2×10^{12} cm^{-2} s^{-1}. (a) If the particles are photons, what is the density of photons at that position? (b) If the particles are thermal neutrons (2200 m/s), what is the density of neutrons?

8. A beam of 2-MeV photons with intensity 10^8 $cm^{-2}s^{-1}$ irradiates a small sample of water. (a) How many photon-water interactions occur in one second in one cm^3 of the water? (b) How many positrons are produced per second in one cm^3 of the water?

9. A small homogeneous sample of mass m (g) with atomic mass A is irradiated uniformly by a constant flux density ϕ (cm^{-2} s^{-1}). If the total atomic cross section for the sample material with the irradiating particles is denoted by σ_t (cm^2), derive an expression for the fraction of the atoms in the sample that interact during a 1-h irradiation. State any assumptions made.

10. A 1-mCi source of ^{60}Co is placed in the center of a cylindrical water-filled tank with an inside diameter of 20 cm and depth of 100 cm. The tank is made of iron with a wall thickness of 1 cm. What is the uncollided flux density at the outer surface of the tank nearest the source?

11. Derive the Compton formula, Eq. (4.32), given that the momentum of a photon of energy E is equal to E/c and the momentum of an electron of kinetic energy T is equal to $(1/c)\sqrt{T(T+2m_ec^2)}$.

12. Verify Eqs. (4.34) to (4.39).

13. Verify Eq. (4.61); that is, given Eq. (4.63), carry out the integration

$$\sigma_T = \int_{4\pi} d\Omega \, \frac{a\sigma_T(\theta_s)}{d\Omega_s}.$$

14. What is the maximum possible kinetic energy (keV) of a Compton electron and the corresponding minimum energy of a scattered photon resulting from scattering of **(a)** a 100-keV photon, **(b)** a 1-MeV photon, and **(c)** a 10-MeV photon? Estimate for each case the range the electron would have in air of 1.2 mg/cm^3 density and in water of 1 g/cm^3 density.

15. Compute and plot the energy spectrum $F(T)$ of Compton recoil electrons for incident photons of 0.5 and 2.0 MeV. Here $F(T)\,dT$ is the fraction of electrons with energies between T and $T + dT$.

16. Consider the classical mechanics of neutron inelastic scattering by a target atom initially at rest. Derive Eq. (4.40), which relates the angle of scattering in the laboratory system and the neutron energies before and after scattering. *Suggestion*: Apply the law of cosines to the triangle of Fig. 4.2(b). Note that $p^2 = 2mE$, $(p')^2 = 2mE'$, $p_r^2 = 2MT$, and $T = E - E' + Q$.

17. From Fig. 4.14, the total microscopic cross section in iron for neutrons with energy of 27 keV is about 0.4 b, and for a neutron with an energy of 28 keV about 90 b. (a) Estimate the fraction of 27-keV neutrons that pass through a 10-cm thick slab without interaction. (b) What is this fraction for 28-keV neutrons?

18. Consider the classical mechanics of neutron inelastic scattering by a target atom initially at rest. Derive Eq. (4.47), which relates the angles of scattering in the laboratory and center-of-mass systems. *Suggestion*: From Fig. 4.3(c) note that $v_c' \sin\theta_c = v' \sin\theta_s$, and $v_o + v_c' \cos\theta_c = v' \cos\theta_s$. Apply trigonometric identities and the definition $\gamma \equiv v_o/v_c'$ to reach the solution.

19. What is the maximum possible kinetic energy (keV) of a Compton electron and the corresponding minimum energy of a scattered photon resulting from scattering of **(a)** a 100-keV photon, **(b)** a 1-MeV photon, and **(c)** a 10-MeV photon? Estimate for each case the range the electron would have in air of 1.2 mg/cm^3 density and in water of 1 g/cm^3 density.

20. Consider the classical mechanics of neutron inelastic scattering by a target atom initially at rest. Derive Eq. (4.49) for the ratio of the speed of the target atom before scatter to the speed of the neutron after scatter. Show that for elastic scatter, $\gamma = m/M = 1/A$. *Suggestion*: Note that the total kinetic energy relative to the center of mass is equal to the kinetic energy of the reduced mass $\mu \equiv mM/(M+m) = mA/(1+A)$ moving at the relative velocity between the neutron and atom. Note, too, that the kinetic energy in the center-of-mass system after scatter differs from that before scatter by the Q value. That is, $\frac{1}{2}\mu(v_c' + V_c')^2 = \frac{1}{2}\mu v^2 + Q$ so that $(v_c')^2 = [2A/m(A+1)][EA/(1+A) + Q]$. Similarly, $(m+M)v_o \equiv mv$, so that $v_o^2 = 2E/m(A+1)^2$.

21. Consider the classical mechanics of neutron inelastic scattering by a target atom initially at rest. Derive Eq. (4.50), which relates the angle of scattering in the center-of-mass system and the neutron energies before and after scattering. *Suggestion*: Apply the law of cosines to Fig. 4.3c, with the result that $v'^2 = (v_c')^2 + v_o^2 + 2v_o v_c' \omega_c$ and apply the results of Problem 20 for v_o and v_c'.

22. Take as given that elastic scattering of neutrons by hydrogen atoms is isotropic in the center-of-mass system. How would you determine the fraction of neutrons scattered at angles less than 45 degrees in the laboratory system?

23. The differential scattering cross section (b/sr) for 1-MeV neutrons on Mg in the laboratory system is

$$\frac{d\sigma_s(E,\omega_s)}{d\Omega_s} = \frac{2.7}{4\pi}[1 + 1.2P_1(\omega_s) + 0.75P_2(\omega_s) + 0.08P_3(\omega_s)].$$

(a) Evaluate the total scattering cross section $\sigma_s(E)$. (b) Evaluate the mean cosine of the laboratory scattering angle $\langle\omega_s\rangle$. (c) Evaluate the mean cosine of the center-of-mass scattering angle $\langle\omega_c\rangle$. (d) Evaluate the average energy lost by such neutrons in scattering. (e) Compute and plot versus ω_c the scattering cross section $\sigma_s(E,\omega_c)$ for such neutrons.

24. When an electron moving through air has 5 MeV of energy, what is the ratio of the rates of energy loss by bremsstrahlung to that by collision? What is this ratio for lead?

25. About what thickness of aluminum is needed to stop a beam of (a) 2.5-MeV electrons, (b) 2.5-MeV protons, and (c) 10-MeV alpha particles? Hint: For parts (a) and (b), use Tables 4.4 and 4.3 and compare your values to ranges shown in Figs. 4.23 and (4.27). For part (c), use the range interpolation rules on page 143.

26. Estimate the range of a 10-MeV tritium nucleus in air.

27. Verify that under the continuous slowing-down approximation for charged particles,

$$\mathcal{F}_{pl}^{\perp}(r,E) = \mathcal{F}(r,E) = \frac{L(r_o - r)}{E/r_o},$$

in which $L(r_o - r)$ is the LET evaluated at the energy of the particle with residual CSDA range $r_o - r$.

REFERENCES

ALEXANDER, J.M., AND M.F. GASZDIK, "Recoil Properties of Fission Products," *Phys. Rev.*, **120**, 874–886, (1960).

AMALDI, E., "The Production and Slowing Down of Neutrons." *Handbuch der Physik,* Vol. 38, Part 2, S. Flügge, Ed., Berlin: Springer-Verlag, 1959.

ANDERSON, H.H., AND J.F. ZIEGLER, *Hydrogen: Stopping Powers and Ranges in All Elements,* Vol. 3, New York: Pergamon, 1977.

ANSI, *Gamma-Ray Attenuation Coefficients and Buildup Factors for Engineering Materials,* ANSI/ANS-6.4.3, New York: American National Standards Institute, 1991.

AREDRO, P, D.T. BURNS, A.E. NAHUM, J. SEUNTJENS, AND F.H. ATTIX, *Fundamentals of Ionizing Radiation Dosimetry,* 2nd Ed., Weinheim, Germany: Wiley-VHC, 2017.

BARKAS, W.H., AND M.J. BERGER, *Tables of Energy Losses and Ranges of Heavy Charged Particles,* Report NASA SP-3013, Washington DC: National Aeronautics and Space Administration, 1964.

BERGER, M.J., *Distribution of Absorbed Dose Around Point Sources of Electrons and Beta Particles in Water and Other Media,* NM/MIRD Pamphlet 7, *J. Nucl. Med.*, **12**, Suppl. 5, (1971).

BERGER, M.J., *ESTAR, PSTAR, and ASTAR: Computer Programs for Calculating Stopping Power and Range Tables for Electrons, Protons, and Helium Ions,* Report NISTIR 4999. Gaithersburg, MD: National Institute of Standards and Technology, 1992. [Distributed as Peripheral Shielding Routine PSR-330 by Radiation Shielding Information Center, Oak Ridge National Laboratory, Oak Ridge, TN.]

BETHE, H.A., "Zur Theorie des Durchgangs schneller Korpuskularstrahlen durch Materie," *Ann. Physik*, **397**, 325–400, (1930).

BETHE, H.A., "Bremsformel für Elektronen relativistischer Geschwindigkeit," *Z. Physik*, **76**, 293–299, (1932).

BETHE, H.A., AND J. ASHKIN, "Passage of Radiation Through Matter," Section II, in *Experimental Nuclear Physics,* Vol. 1, SEGRÉ, E., Ed., New York: Wiley, 1953.

BHABHA, H.J., "The Scattering of Positrons by Electrons with Exchange on Dirac's Theory of the Positron," *Proc. R. Soc. London*, **A154**, 195–206, (1936).

BHABHA, H.J., "On the Penetrating Component of Cosmic Radiation," *Proc. R. Soc. London*, **A164**, 257–294, (1938).

BIGGS, F. AND R. LIGHTHILL, *Analytical Approximations for Photon-Atom Differential Scattering Cross Sections Including Electron Binding Effects,* Report SC-RR-72 0659, Albuquerque, NM: Sandia Laboratories, 1972.

BIRGANI, M.J.T., F. SEIF, N. CHEGENI, AND M.R. BAYATIANI, "Determination of the Effective Atomic and Mass Numbers for Mixture and Compound Materials in High Energy Photon Interactions," *J. Radioanal. Nucl. Chem.*, **292**, 1367–1370, (2012).

BIRKS, J.B., *Theory and Practice of Scintillation Counting,* New York: Pergamon Press, 1964.

BRAGG, W.H., AND R. KLEEMAN, "On the α Particles of Radium, and their Loss of Range in Passing Through Various Atoms and Molecules," *Phil. Mag.*, **10**, 318–340, (1905).

CROSS, W.G., "The Distribution of Absorbed Energy from a Point Beta Source," *Can. J. Phys.*, **45**, 2121–2040, (1967).

CROSS, W.G., H. ING, N.O. FREEDMAN, AND J. MAINVILLE, *Tables of Beta-Ray Dose Distributions in Water, Air, and Other Media,* Report AECL-7617, Chalk River, Ontario: Atomic Energy of Canada Limited, 1982.

CROSS, W.G., N.O. FREEDMAN, AND P.Y. WONG, *Tables of Beta-Ray Dose Distributions in Water,* Report AECL-10521, Chalk River, Ontario: Atomic Energy of Canada Limited, 1992.

CROSS, W.G., N.O. FREEDMAN, AND P.Y. WONG, *Tables of Beta-Ray Dose Distributions in Water*, Report AECL-10521, Chalk River, Ontario: Chalk River Laboratories, Atomic Energy of Canada, Ltd., 1992.

CULLEN, D.E., M.H. CHEN, J.H. HUBBELL,, S.T. PERKINS, E.F. PLECHATY, J.A. RATHKOPF, AND J.H. SCOFIELD, *Tables and Graphs of Photon Interaction Cross Sections from 10 eV to 100 GeV Derived from the LLNL Evaluated Photon Data Library (EPDL),* Report UCRL-50400, Vol. 6, Rev. 4, Part A: $Z = 1$–50, Part B: $Z = 51$–100. Livermore CA: Lawrence Livermore National Laboratory, 1989.

CULLEN, D.E., *Program EPICSHOW: A Computer Program to Allow Interactive Viewing of the EPIC Data Libraries,* Report UCRL-ID-116819, Livermore, CA: Lawrence Livermore National Laboratory, 1994.

EVANS, R.D., *The Atomic Nucleus,* New York: McGraw-Hill, 1955; republished by Melbourne, FL: Krieger Publishing Co., 1982.

FULMER, C.B., "Scintillation Response of CsI(Tl) Crystals to Fission Fragments and Energy vs Range in Various Materials for Light and Heavy Fission Fragments," *Phys. Rev.*, **108**, 1113–1116, (1957).

HEITLER, W., *The Quantum Theory of Radiation,* 3rd Ed., Oxford: Oxford University Press, 1954.

HENRIKSEN, T. AND J. BAARLI, "The Effective Atomic Number," *Rad. Research*, **6**, 415–423, (1957).

HUBBELL, J.H., *Photon Cross Sections, Attenuation Coefficients, and Energy Absorption Coefficients from 10 keV to 100 GeV,* Report NSRDS-NBS 29, Washington, DC: National Bureau of Standards, 1969.

HUBBELL, J.H., W.J. VEIGELE, E.A. BRIGGS, R.T. BROWN, D.T. CROMER, AND R.J. HOWERTON, "Atomic Form Factors, Incoherent Scattering Functions, and Photon Scattering Cross Sections," *J. Phys. Chem. Ref. Data*, **4**, 471–616, (1975).

HUBBELL, J.H., AND I. OVERBØ, "Relativistic Atomic Form Factors and Photon Coherent Scattering Cross Sections," *J. Phys. Chem. Ref. Data*, **8**, 69–106, (1979).

HUBBELL, J.H., "Photon Mass Attenuation and Energy Absorption Coefficients," *Int. J. Appl. Radiat. Isot.*, **33**, 1269–1290, (1982).

HUBBELL, J.H., AND S.M. SELTZER, *Tables of X-Ray Attenuation Coefficients 1 keV to 20 MeV for Elements $Z = 1$ to 92 and 48 Additional Substances of Dosimetric Interest,* Report NISTIR 5632, Gaithersburg, MD: National Institute of Standards and Technology, 1995.

HUBERT, F., AND R. BIMBOT,, AND H. GAUVIN, "Range and Stopping-Power Tables for 2.5–500 MeV/Nucleon Heavy Ions in Solids," *At. Data and Nucl. Data Tables*, **46**, 11–213, (1990).

ICRU, *Stopping Powers for Electrons and Positrons,* Report 37, Washington, DC: International Commission on Radiation Units and Measurements, 1984.

ICRU, *Stopping Powers and Ranges for Protons and Alpha Particles,* Report 49, Washington, DC: International Commission on Radiation Units and Measurements, 1993.

JANNI, J.F., "Proton Range-Energy Tables, 1 keV–10 GeV," *At. Data and Nucl. Data Tables*, **27**, 147–529, (1982).

KANE, P.P., L. KISSEL, R.H. PRATT, AND S.C. ROY, "Elastic Scattering of γ-rays and X-rays by Atoms," *Physics Reports*, **140**, No. 2, 75–159, (1986).

KATZ, L., AND A.S. PENFOLD, "Range-Energy Relations for Electrons and the Determination of Beta-Ray End-Point Energies by Absorption," *Rev. Mod. Phys.*, **24**, 28–44, (1952).

KINSEY, R., *ENDF/B-V Summary Documentation*, Report BNL-NCS-17541 (ENDF-201), 3rd Ed., Upton, NY: Brookhaven National Laboratory, 1979.

KLEIN, O. AND Y. NISHINA, "Über die Streuung von Strahlung durch freie Elektronen nach der neuen relativistischen Quantendynamik von Dirac," *Z. Physik*, **52**, 853–868, (1929).

LAMARSH, J.R., *Introduction to Nuclear Reactor Theory,* Reading, MA: Addison-Wesley, 1966.

MCGREGOR, D.S., AND J.K. SHULTIS, "Reporting Detection Efficiency for Semiconductor Neutron Detectors: A Need for a Standard," *Nucl. Instrum. Meth.*, **A 632**, 167–174, (2011).

MEGHREBLIAN, R.V., AND D.K. HOLMES, *Reactor Theory,* New York: McGraw-Hill, 1960.

MØLLER, C., "Zur Theorie des Durchgangs schneller Elektronen durch Materie," *Ann. Phys.*, **406**, 531–585, (1932).

NELSON, W.R., H. HIRAYAMA, AND D.W.O. ROGERS, *The EGS4 Code System,* Report SLAC-265, Stanford, CA: Stanford Linear Accelerator Center, 1985.

PLECHATY, E.F., D.E. CULLEN, AND R.J. HOWERTON, Report UCRL-50400, Vol. 6, Rev. 1, Springfield, VA: National Technical Information Service, 1975. [Data are available as the DLC-139 code package from the Radiation Shielding Information Center, Oak Ridge, TN: Oak Ridge National Laboratory.]

ROHRLICH, F., AND B.C. CARLSON, "Positron-Electron Differences in Energy Loss and Multiple Scattering," *Phys. Rev.*, **93**, 38–44, (1954).

ROSE, P.F., AND C.L. DUNFORD, Eds., *ENDF-102, Data Formats and Procedures for the Evaluated Nuclear Data File ENDF-6*, Report BNL-NCS 44945 (Rev.), Upton, NY: Brookhaven National Laboratory, 1991.

ROSSI, B., *High-Energy Particles,* Englewood Cliffs, NJ: Prentice-Hall, Inc., 1952.

SELTZER, S.M., "Calculation of Photon Mass Energy-Transfer and Mass Energy Absorption Coefficients," *Rad. Res.*, **136**, 147-179, (1993).

SELTZER, S.M., AND M.J. BERGER, "Bremsstrahlung Spectra from Electron Interactions with Screened Atomic Nuclei and Orbital Electrons," *Nucl. Instrum. Meth.*, **B12**, 95–134, (1985).

SELTZER, S.M., AND M.J. BERGER, "Bremsstrahlung Energy Spectra from Electrons with Kinetic Energy 1 keV–10 GeV Incident on Screened Nuclei and Orbital Electrons of Neutral Atoms with $Z = 1$–100," *At. Data and Nucl. Data Tables*, **35**, 345–418, (1986).

SHULTIS, J.K., AND R.E. FAW, Radiation Shielding, LaGrange Park, IL: Am. Nucl. Soc., 2000.

SHULTIS, J.K., AND R.E. FAW, Fundamentals of Nuclear Science and Engineering, 3rd Ed., Boca Raton, FL: CRC Press, 2017.

STORM, E., AND H.I. ISRAEL, *Photon Cross Sections from 0.001 to 100 MeV for Elements 1 through 100,* Report LA-3753, Los Alamos, NM: Los Alamos Scientific Laboratory, 1967.

TAMM, I., "Über die Wechselwirkung der freien Elektronen mit der Strahlung nach der Diracsehen Theorie des Elektrons und nach der Quantenelektrodynamik," *Z. Physik*, **62**, 545–568, (1930).

THOMSON, J.J., "Cathode Rays," *Phil. Mag.*, 5th Series, **44**, 359–364, (1897).

TRUBEY, D.K., M.J. BERGER, AND J.H. HUBBELL., "Photon Cross Sections for ENDF/B-VI," American Nuclear Society Topical Meeting, *Advances in Nuclear Engineering Computation and Radiation Shielding*, Santa Fe, NM, 1989. [Data are available as the DLC-136/PHOTX code package from the Radiation Shielding Information Center, Oak Ridge National Laboratory, Oak Ridge, TN.]

TRUBEY, D.K., AND Y. HARIMA, "New Buildup Factor Data for Point Kernel Calculations," in *Proc. of a Topical Conf. on Theory and Practices in Radiation Protection and Shielding*, Knoxville, TN, April 1987, Vol. 2, p. 503, American Nuclear Society, La Grange Park, IL, (1987). [Data are available as the DLC-129 code package from the Radiation Shielding Information Center, Oak Ridge National Laboratory, Oak Ridge, TN.]

WESTCOTT, C.H., "The Specification of Neutron Flux and Nuclear Cross-Sections in Reactor Calculations," *J. Nucl. Energy*, **2**, 59–75, (1955).

ZIEGLER, J.F., J.P. BIERSACK, AND M.D. ZIEGLER, *SRIM: The Stopping and Range of Ions in Matter,* Available through the web at www.srim.org, 2013.

Chapter 5

Sources of Radiation

I proposed the word radioactivity which has since become generally adopted; the radioactive elements have been called radio elements.

Marie Curie

5.1 Introduction

In our world ionizing radiation is ever present, though rarely noticed by our human senses. It is with radiation sensors that such radiation is detected and its characteristics measured. However, the type of sensor to use depends on the type of radiation, its energy, and what information is sought about the radiation. In this book many types of radiation detectors are discussed, each designed for specific types of radiation measurements. Key to any radiation measurement is an understanding of the type and characteristics of the radiation being measured. In this chapter, the most common sources of ionizing radiation are reviewed and many of their properties examined. Much of the material presented here has been excerpted from expanded treatments given in Shultis and Faw [2000] and Faw and Shultis [2012].

5.1.1 Origins of Ionizing Radiation

Ionizing radiation is invariably the consequence of physical reactions, involving subatomic particles, at the atomic or nuclear level. The possible radiation-producing reactions are many and, usually, although not always, involve altering the configuration of neutrons and protons in an atomic nucleus or the rearrangement of atomic electrons about a nucleus. These radiation producing reactions can be divided into the following two categories.

Radioactive Decay

In the first type of reaction, the nucleus of an atom spontaneously changes its internal arrangement of neutrons and protons to achieve a more stable configuration. In such spontaneous *radioactive* transmutations, ionizing radiation is almost always emitted. The number of known different atoms, each with a distinct combination of Z and A, is about 3200. Of these, only 266 are stable and 65 are very long-lived radioisotopes still found in nature. The remaining nuclides have been made by humans and are radioactive with lifetimes much shorter than the age of the solar system. Both naturally occurring and manufactured radionuclides are the most commonly encountered sources of ionizing radiation.

Binary Reactions

The second category of radiation-producing interactions involves two impinging atomic or subatomic particles that react to form one or more reaction products. Examples include neutrons interacting with nuclei of atoms, photons interacting with nuclei or atomic electrons, and cosmic rays interacting with nuclei in our atmosphere. Such reactions invariably produce, as reaction products, ionizing radiation.

5.1.2 Physical Characterization of Radiation Sources

The most fundamental type of source is a *point source*. Clearly no real source can have zero size, but a real source can be approximated as a point source provided (1) that the volume is sufficiently small, that is, it has dimensions much smaller than the dimensions of the attenuating medium between source and detector, and (2) that there is negligible interaction of radiation with the matter in the source volume. The second requirement may be relaxed if source characteristics are modified to account for source self-absorption and other source-particle interactions.

In general, a point source may be characterized as depending on energy, direction, and time. In almost all cases, time is not treated as an independent variable because the time delay between a change in the emission rate of the source and the resulting change in the radiation field is usually negligible so that the time dependence of the radiation field is that of the source.[1] Therefore, the most general characterization of a point source used here is in terms of energy and direction. Most radiation sources treated in radiation measurements are isotropic, so that source characterization requires only knowledge of energy dependence. Radioisotope sources are certainly isotropic, as are fission sources and capture gamma-ray sources.

Radiation sources may also be distributed along a line, over an area, or within a volume. Source characterization requires, in general, spatial and energy dependence. Occasionally it is necessary to include angular dependence. This is especially true for effective area sources associated with computed angular flows across certain planes.

Energy dependence may be discrete, such as for radionuclide sources, or continuous, as for bremsstrahlung or fission neutrons and photons. When discrete energies are numerous, an energy multigroup approach is often used. The same multigroup approach may be used to approximately characterize a source whose emissions are continuous in energy.

5.2 Sources of Gamma Rays

Radioactive sources abound in our technological age and are used for a wide variety of purposes in many educational, medical, research, industrial, governmental, and commercial facilities. The radionuclides in these sources almost always leave their progeny in excited nuclear states whose subsequent transitions to lower-energy states usually result in the emission of one or more gamma photons. Consequently, these radioactive sources are frequently encountered in radiation measurements and in verifying the proper operation of a radiation detector.

In characterizing the gamma radiation emitted by a radioactive source, a careful distinction must be made between the activity of a radioisotope and its source strength. *Activity* is precisely defined as the expected number of atoms undergoing *radioactive transformation* per unit time. The SI unit of activity is the becquerel (Bq), equivalent to 1 transformation per second, while the older unit for activity is the curie (Ci) equal to 3.7×10^{10} transformations per second. Activity is *not* defined as the number of particles emitted per unit time, and the decay of two common laboratory radioisotopes, ^{60}Co and ^{137}Cs, can be used to illustrate this point. Each transformation of ^{60}Co results in the emission of *two* gamma rays, one at 1.173 MeV and the other at 1.333 MeV. However, each transformation of ^{137}Cs, accompanied by a transformation of its decay product ^{137m}Ba, results in emission of a 0.662-MeV gamma ray with an 85% probability.

Extensive compilations of the energies and intensities of the gamma and x rays emitted by radionuclides are available, for example [Kocher 1981; ICRP 1983; Weber et al. 1989; Eckerman et al. 1994], or computer programs that can be used to prepare specialized compilations [Dillman 1980]. Especially useful is the on-line decay data base for radionuclides that is found at `https://www.nndc.bnl.gov/nudat2`.

[1]Exceptions include nuclear explosions or supernovae where the radiation field is very far from the source.

5.2.1 Naturally Occurring Radionuclides

Radionuclides existing on earth arise from two sources. First, earth has always been bombarded by cosmic rays coming from the sun and from intergalactic space. For the most part, cosmic rays consist of protons and alpha particles, with heavier nuclei contributing less than 1% of the incident nucleons.[2] These cosmic rays interact with atoms in the atmosphere and produce a variety of light radionuclides.

The second source of naturally occurring radionuclides is the residual radioactivity in the matter from which the earth was formed. These heritage or primordial radionuclides were formed in the stars from whose matter the solar system was formed. Most of these radionuclides have long since decayed away during the 4 billion years since the earth was formed.

When making measurements of a gamma-ray source, there is always a background source of radiation as a consequence of the ubiquitous presence of naturally occurring radionuclides. Here the major sources of such radionuclides are discussed.

Cosmogenic Radionuclides

Cosmic rays interact with constituents of the atmosphere, sea, or earth, but mostly with the atmosphere, leading directly to radioactive products. Capture of secondary neutrons produced in primary interactions of cosmic rays leads to many more. Only those produced from interactions in the atmosphere lead to significant radiation exposure to humans.

The most prominent of the cosmogenic radionuclides are tritium ^{3}H and ^{14}C. The ^{3}H (symbol T or t) is produced mainly from the ^{14}N(n,T)^{12}C and ^{16}O(n,T)^{14}N reactions. Tritium has a half-life of 12.3 years and, upon decay, emits one β^- particle with a maximum energy of 18.6 (average energy 5.7) keV. Tritium exists in nature almost exclusively as HTO.[3]

The nuclide ^{14}C is produced mainly from the ^{14}N$(n,p)^{14}$C reaction. It exists in the atmosphere as CO_2, but the main reservoirs are the oceans. ^{14}C has a half-life of 5730 years and decays by β^- emission with a maximum energy of 157 keV and average energy of 49.5 keV.

Over the past century, combustion of fossil fuels with the emission of CO_2 without any ^{14}C has diluted the cosmogenic content of ^{14}C in the environment. Moreover, since the use of nuclear weapons in World War II, artificial introduction of ^{3}H and ^{14}C (and other radionuclides) by human activity has been significant, especially from atmospheric nuclear-weapons tests. Consequently, these isotopes no longer exist in natural equilibria in the environment.

Singly Occurring Primordial Radionuclides

Of the many radionuclide species present when the solar system was formed 4.57 billion years ago, some 17 very long-lived radionuclides still exist as singly occurring or isolated radionuclides, that is, as radionuclides not belonging to a radioactive decay chain. These radionuclides are listed in Table 5.1, and all but one are seen to have half-lives greater than the age of the solar system. Of these radionuclides, the most significant (from a human exposure perspective) are ^{40}K and ^{87}Rb because they are inherently part of our body tissue.

Decay Series of Primordial Origin

Each naturally occurring radioactive nuclide with $Z > 83$ is a member of one of three long decay chains, or radioactive series, stretching through the upper part of the Chart of the Nuclides. These radionuclides decay by α or β^- emission and they have the property that the number of nucleons (mass number) A for each member of a given decay series can be expressed as $4n + i$, where n is an integer and i is a constant (0, 2, or 3) for each series. The three naturally occurring series are named the thorium ($4n$), uranium ($4n + 2$),

[2]Electrons, primarily from the sun, also are part of the cosmic rays incident on the earth. But they are deflected by earth's magnetosphere and do not produce cosmogenic radionuclides.

[3]H_2O with one hydrogen replaced by tritium.

Table 5.1. The 17 isolated primordial radionuclides. Data taken from GE-NE [1996].

Radionuclide & the Decay Modes		Half-life (years)	% El. Abund.	Radionuclide & the Decay Modes		Half-life (years)	% El. Abund.
$^{40}_{19}$K	β^- EC β^+	1.27×10^9	0.0117	$^{50}_{23}$V	β^- EC	1.4×10^{17}	0.250
$^{87}_{37}$Rb	β^-	4.88×10^{10}	27.84	$^{113}_{48}$Cd	β^-	9×10^{15}	12.22
$^{115}_{49}$In	β^-	4.4×10^{14}	95.71	$^{123}_{52}$Te	EC	$> 1.3 \times 10^{13}$	0.908
$^{138}_{57}$La	EC β^-	1.05×10^{11}	0.090	$^{144}_{60}$Nd	α	2.38×10^{15}	23.80
$^{147}_{62}$Sm	α	1.06×10^{11}	15.0	$^{148}_{62}$Sm	α	7×10^{15}	11.3
$^{152}_{64}$Gd	α	1.1×10^{14}	0.20	$^{176}_{71}$Lu	β^-	3.78×10^{10}	2.59
$^{174}_{72}$Hf	α	2.0×10^{15}	0.162	$^{180}_{73}$Ta	EC β^+	$> 1.2 \times 10^{15}$	0.012
$^{187}_{75}$Re	β^-	4.3×10^{10}	62.60	$^{186}_{76}$Os	α	2×10^{15}	1.58
$^{190}_{78}$Pt	α	6.5×10^{11}	0.01				

Table 5.2. Principal radioisotopes in two naturally occurring primordial decay series. Source: NuDat 2.5, National Nuclear Data Center, Brookhaven National Laboratory.

Thorium Series			Uranium Series		
Nuclide & decay mode		Half-life[a] $T_{1/2}$	Nuclide & decay mode		Half-life[a] $T_{1/2}$
$^{232}_{90}$Th	α	**14.05 Gy**	**$^{238}_{92}$U**	α	**4.468 Gy**
$^{228}_{88}$Ra	β	5.75 y	$^{234}_{90}$Th	β	24.10 d
$^{228}_{89}$Ac	β	6.15 h	$^{234m}_{91}$Pa	β	1.159 m
$^{228}_{90}$Th	α	1.912 y	$^{234}_{92}$U	α	244.5 ky
$^{224}_{88}$Ra	α	3.66 d	$^{230}_{90}$Th	α	75.4 ky
$^{220}_{86}$Rn	α	55.6 s	$^{226}_{88}$Ra	α	1600 y
$^{216}_{84}$Po	α	0.145 s	$^{222}_{86}$Rn	α	3.8235 d
$^{212}_{82}$Pb	β	10.64 h	$^{218}_{84}$Po	α, β	3.098 m
$^{212}_{83}$Bi	α, β	60.55 m	$^{218}_{85}$At	α	1.5 s
$^{212}_{84}$Po	α	0.299 μs	$^{214}_{82}$Pb	β	26.8 m
$^{208}_{81}$Tl	β	3.053 m	$^{214}_{83}$Bi	α, β	19.9 m
$^{208}_{82}$Pb		∞	$^{214}_{84}$Po	α	164 μs
			$^{210}_{81}$Tl	β	1.3 m
			$^{210}_{83}$Bi	α, β	5.012 d
			$^{210}_{82}$Pb	α, β	22.2 y
			$^{210}_{84}$Po	α	138.4 d
			$^{206}_{80}$Hg	β	8.2 m
			$^{206}_{81}$Tl	β	4.2 m
			$^{206}_{82}$Pb		∞

[a] Gy = 10^9 y, ky = 10^3 y, μs = 10^{-6} s.

and actinium ($4n + 3$) series, named after the radionuclide at, or near, the head of the series. The head of each series has a half-life much greater than any of its daughters.

There is no naturally occurring series represented by $4n + 1$. This series was recreated after $^{241}_{94}$Pu was made in nuclear reactors. This series does not occur naturally since the half-life of the longest lived member of the series, $^{237}_{93}$Np, is only 2.14×10^6 y, much shorter than the age of the earth (4.54×10^9 y). Hence, any members of this series that were in the original material of the solar system have long since decayed away.

The principal radioisotopes produced by the two most important naturally occurring series (the uranium and thorium series) are given in Table 5.2. In both of these decay chains, a few members decay by both α and β^- decay causing the decay chain to branch (see Shultis and Faw [2000]). Nevertheless, each decay chain ends in the same stable isotope.

5.2.2 Prompt Fission Gamma Photons

Whenever a heavy nucleus fissions, copious gamma photons are produced either within the first 6×10^{-8} s after the fission event (the *prompt fission gamma photons*) or from the subsequent decay of the fission products. If radiation measurements are being made in the presence of fission sources, for example, these prompt and delayed gamma rays are of great importance.

Most investigations of prompt fission gamma photons have centered on the thermal-neutron-induced fission of ^{235}U. For this nuclide it has been found that the number of prompt fission photons is 8.13 ± 0.35 photons per fission over the energy range 0.1 to 10.5 MeV, and the energy carried by this number of photons is 7.25 ± 0.26 MeV per fission [Keepin 1965]. The energy spectrum of prompt gamma photons from the thermal fission of ^{235}U between 0.1 and 0.6 MeV is approximately constant at 6.6 photons MeV^{-1} fission^{-1}. At higher energies the spectrum falls off sharply with increasing energy. The measured energy distribution of the prompt fission photons can be represented by the following empirical fit over the range 0.1 to 10.5 MeV [Keepin 1965]:

$$N(E) = \begin{cases} 6.6 & 0.1 < E < 0.6 \text{ MeV} \\ 20.2e^{-1.78E} & 0.6 < E < 1.5 \text{ MeV} \\ 7.2e^{-1.09E} & 1.5 < E < 10.5 \text{ MeV}, \end{cases}$$

where E is in MeV and $N(E)$ is in units of photons MeV^{-1} fission^{-1}.

Investigation of ^{233}U, ^{239}Pu, and ^{252}Cf indicates that the prompt fission photon energy spectra for these isotopes resembles very closely that for ^{235}U, and hence for most purposes it is reasonable to use the ^{235}U spectrum for other fissioning isotopes.

5.2.3 Fission-Product Gamma Photons

With the widespread application of nuclear fission, an important concern is the consideration of the very long lasting gamma-ray activity produced by the decay of fission products. In the fission process, most often two fragments are produced (*binary fission*). About 0.3% of the time a third light fragment is produced (*ternary fission*), most often ^{3}H. The mass distribution or *fission-product chain yield* is bi-modal, with many products having atomic mass number around 95 and around 140. Among the former are the important long-lived radionuclide ^{90}Sr, several isotopes of the halogen bromine, and various isotopes of the noble gas krypton. Among the heavy fragments are the important long-lived radionuclide ^{137}Cs, radioisotopes of halogen iodine, notably ^{131}I, and isotopes of the noble gas xenon. The fission-products are neutron-rich and decay almost exclusively by β^- emission, often forming long decay chains. From the range of mass numbers produced, about 100 different decay chains are formed. An example of a short chain is

$$^{140}_{54}\text{Xe} \xrightarrow[16\text{ s}]{\beta^-} {}^{140}_{55}\text{Cs} \xrightarrow[66\text{ s}]{\beta^-} {}^{140}_{56}\text{Ba} \xrightarrow[12.8\text{ d}]{\beta^-} {}^{140}_{57}\text{La} \xrightarrow[40\text{ h}]{\beta^-} {}^{140}_{58}\text{Ce (stable)}.$$

The total gamma-ray energy released by the fission product chains is comparable to that released as prompt fission gamma photons. The gamma-ray energy release rate declines rapidly with time after fission. About three-fourths of the delayed gamma-ray energy is released in the first thousand seconds after fission. In most calculations involving spent nuclear fuel, the gamma activity at several months or even years after removal of fuel from the nuclear reactor is of interest and so only the long-lived fission products need be considered.

It has been found that the gamma energy released from fission products is relatively independent of the energy of the neutrons causing the fissions. However, the gamma-ray energy released and the photon energy spectrum depend significantly on the fissioning isotope, particularly in the first 10 s after fission. Generally, fissioning isotopes, having a greater proportion of neutrons to protons, produce fission-product chains of longer average length, with isotopes richer in neutrons and hence with greater available decay energy. Also, the photon energy spectrum generally becomes less energetic as the time after fission increases.

The delayed gamma photon energy spectrum from the fission of ^{235}U, at times up to about 5 s, may be approximated by the proportionality

$$N(E) \sim e^{-1.1E}, \tag{5.1}$$

where $N(E)$ is the delayed gamma-ray yield (photons MeV^{-1} fission^{-1}) and E is the photon energy in MeV. The time dependence for the total gamma photon energy emission rate $F(t)$ (MeV s^{-1} fission^{-1}) is often described by the simple decay formula

$$F(t) = 1.4t^{-1.2}, \quad 10 \text{ s } < t < 10^7 \text{ s}, \tag{5.2}$$

where t is in seconds. More complicated (and accurate) expressions for $F(t)$ have been obtained from fits to experimental data (see, for example, Shultis and Faw [2000]); but for preliminary calculations the simpler result is usually adequate. It is observed that both ^{235}U and ^{239}Pu have roughly the same total gamma-ray-energy decay characteristics for up to 200 days after fission, at which time ^{235}U products begin to decay more rapidly. At 1 year after fission, the ^{239}Pu gamma-ray emission rate is about 60% greater than that of ^{235}U.

For accurate calculations involving fission products, the variation with time after fission of the energy spectra of the photons must be taken into account. Often the energy spectra are averaged over discrete energy intervals and the energy emission rate in each energy group is considered as a function of time after fission. Computer codes, based on extensive libraries of radionuclide data, have been developed to compute the abundances and decay rates of the hundreds of fission-product radionuclides. One such code is ORIGEN [Hermann and Westfall 1981; RSIC 1991].

5.2.4 Capture Gamma Photons

The compound nucleus formed by neutron absorption is initially created in a highly excited state with excitation energy equal to the kinetic energy of the incident neutron plus the neutron binding energy, which averages about 7 MeV. The decay of this nucleus, usually within 10^{-12} s, and usually by way of intermediate states, typically produces several energetic photons. Generally, the probability a neutron causes an (n, γ) reaction is greatest for slow moving *thermal neutrons*, i.e., neutrons whose speed is in equilibrium with the thermal motion of the atoms in a medium. At high energies, it is more likely that a neutron scatters, thereby losing some of its kinetic energy and, thus, slows towards thermal energies. Table B.4 in Appendix B lists the number of capture gamma photons by 1-MeV energy bin when a thermal neutron is absorbed in various elements

Capture photons may be created intentionally by placing a material with a high thermal-neutron (n, γ) cross section in a thermal neutron beam. The energy spectrum of the resulting capture gamma photons can then be used to identify trace elements in the sample. More often, however, capture gamma photons are an undesired secondary source of radiation.

The absorption of a thermal neutron by an isotope or element typically produces dozens of capture photons, each with a unique energy and probability of emission. Detailed compilations of the energies and frequencies are provided by Révay et al. [2004]. Analysis of the capture gamma photon spectrum emitted by an unknown sample can be used to infer the elemental composition of the sample. Generally, only a few of the most intense capture photons for a particular element are needed. In Table B.5 (Appendix B), the energies of the three capture photons recommended for each element for elemental identification are given.

5.2.5 Inelastic Scattering Gamma Photons

The excited nucleus formed when a neutron is inelastically scattered decays to the ground state usually within about 10^{-14} s, with the excitation energy being released by one or more photons. Because of the constraints imposed by the conservation of energy and momentum in all scattering interactions, inelastic neutron scattering cannot occur unless the incident neutron energy is greater than $(A+1)/A$ times the energy required to excite the scattering nucleus to its first excited state (see Section 4.3.5). Except for the heavy nuclides, neutron energies above about 0.5 MeV are typically required for inelastic scattering.

The detailed calculation of secondary photon source strengths from inelastic neutron scattering requires knowledge of the fast-neutron flux, the inelastic scattering cross sections, and spectra of resultant photons, all as functions of the incident neutron energy. The cross sections and energy spectra of the secondary photons depend strongly on the incident neutron energy and the particular nuclide. Such inelastic scattering data are known only for the more important structural and shielding materials, and even the known data require extensive data libraries. Often inelastic scattered gamma rays are ignored in routine radiation measurements. But in some cases their identification is vital because, from their energies and intensities, isotopic abundances of a sample can be inferred.

5.2.6 Activation Gamma Photons

For many materials, absorption of a neutron or a proton produces a radionuclide with a half-life ranging from a fraction of a second to many years. Many radionuclides encountered in research laboratories, medical facilities, and industry are produced as activation nuclides from neutron or proton absorption in some parent material (see Table 5.3). Such nuclides decay, usually by beta emission, leaving the daughter nucleus in an excited state which usually decays quickly to its ground state with the emission of one or more gamma photons. Thus, the apparent half-life of the photon emitter is that of the parent (or activation nuclide), while the number and energy of the photons are characteristic of the nuclear structure of the decay daughter.

5.2.7 Positron Annihilation Photons

Positrons, generated either from the positron decay of radionuclides or from pair production interactions induced by high energy photons, slow down in matter within about 10^{-10} s and are subsequently annihilated with ambient electrons. With rare exception, the rest-mass energy of the electron and positron is emitted in the form of two annihilation photons, each of energy m_ec^2 (= 0.511 MeV).

5.3 Sources of X Rays

The interaction of photons or charged particles with matter leads inevitably to the production of secondary x-ray photons which have energies $\lesssim$ 100 keV. Because the energies of the x rays are unique to the element that emits them, their energies and intensities can be used to determine the elemental composition of an irradiated sample. For measurements of these x rays it is necessary to understand how the x rays are produced and some characteristics of the production mechanisms. There are two principal methods whereby secondary

Table 5.3. Important radioisotopes produced by reactors and accelerators for use in medical, research, and industrial applications. Those isotopes commercially available for medical use are shown in bold. Sources: References [4] and [5]. Decay data from NUDAT 2.5, National Nuclear Data Center, Brookhaven National Laboratory.

Nuclide	Half-life	Decay modes[a]	Nuclide	Half-life	Decay modes[a]
^{3}H	12.33y	β^- *	^{81m}Kr	13.1s	EC IT
^{11}C	20.39m	β^+ EC	^{85}Kr	10.76y	β^-
^{13}N	9.965m	β^+ EC	**^{82}Sr**[b]	25.6d	EC *
^{14}C	5730y	β^- *	**^{89}Sr**	50.6d	β^-
^{15}O	122.2s	β^+ EC	**^{90}Sr**[c]	28.90y	β^- *
^{18}F	109.8m	β^+ EC	**^{99}Mo**[d]	65.94h	β^-
^{22}Na	2.603y	β^+ EC	^{103}Pd	16.99d	EC
^{26}Al	7.17E5y	β^+ EC	^{110m}Ag	249.8d	β^- IT
^{28}Mg	20.91h	β^-	**^{111}In**	2.80d	EC
^{32}Si	153y	β^- *	^{113m}In	99.48m	IT
^{32}P	14.26d	β^- *	**^{123}I**	13.2h	EC
^{33}P	25.3d	β^- *	**^{125}I**	59.40d	EC
^{35}S	87.51d	β^- *	**^{131}I**	8.052d	β^-
^{36}Cl	3.01E5y	$\beta^\pm$ EC	**^{133}Xe**	5.243d	β^-
^{46}Sc	83.79d	β^-	^{137}Cs[e]	30.0y	β^- *
^{51}Cr	27.70d	EC	^{140}La	1.679d	β^-
^{54}Mn	312.1d	β^- EC	^{148}Gd	74.6y	α *
^{57}Co	271.7d	EC	**^{153}Sm**	46.3h	β^-
^{57}Cu	196ms	β^+ EC	^{159}Gd	18.48h	β^-
^{57}Cr	21.1s	β^-	^{169}Yb	32.02d	EC
^{59}Fe	44.50d	β^-	^{170}Tm	128.6d	β^- EC
^{60}Co	5.271y	β^-	^{186}Re	89.25h	β^- EC
^{64}Cu	12.70h	$\beta^\pm$ EC	^{191}Os	15.4d	β^-
^{65}Zn	244.1d	β^+ EC	^{192}Ir	73.83d	β^- EC
^{67}Ga	3.261d	EC	^{198}Au	2.696d	β^-
^{68}Ga	67.71m	β^+ EC	**^{201}Tl**	73.0h	EC
^{75}Se	119.8d	EC	^{204}Tl	3.78y	β^- EC *
^{81}Rb	4.572h	β^+ EC	^{210}Pb	22.2y	β^-
^{82}Rb	1.273m	β^+ EC	^{241}Am	432.6 y	α

[a] Decays without any gamma photon emission are denoted by *.

[b] In equilibrium with decay product ^{82}Rb (1.273 m, β^+ EC).

[c] In equilibrium with decay product ^{90}Y (64.00 h, β^-).

[d] In equilibrium with decay product ^{99m}Tc (6.01 h, IT).

[e] In equilibrium with decay product ^{137m}Ba (2.552 m, IT).

x-ray photons are generated: the rearrangement of atomic electron configurations leads to characteristic x rays, and the deflection of charged particles in the nuclear electric field results in bremsstrahlung. Both mechanisms are discussed below.

5.3.1 Characteristic X Rays

The electrons around a nucleus are arranged in shells or layers, each of which can hold a maximum number of electrons (see Section 3.4.6). The two electrons in the innermost shell (K shell) are the most tightly bound, the six electrons in the next shell (L shell) are the next most tightly bound, and so on outward for the M, N,... shells. If the normal electron arrangement around a nucleus is altered through ionization of an inner electron or through excitation of electrons to higher energy levels, the electrons begin a complex series of transitions to vacancies in the lower shells (thereby acquiring higher binding energies) until the unexcited state of the atom is achieved. In each electronic transition, the difference in binding energy between the final and initial states is either emitted as a photon, called a *characteristic x ray*, or given up to an outer-shell electron which is ejected from the atom, called an *Auger electron*. The discrete electron energy levels and the transition probabilities between levels vary with the Z number of the atom and, thus, the characteristic x rays provide a unique signature for each element.

The number of x rays with different energies is greatly increased by the multiplicity of electron energy levels available in each shell (1, 3, 5, 7,... distinct energy levels for the K, L, M, N,... shells, respectively). To identify the various characteristic x rays for an element, many different schemes have been proposed. One of the more popular uses the letter of the shell whose vacancy is filled together with a numbered Greek subscript to identify a particular electron transition (e.g., $K_{\alpha 1}$ and $L_{\gamma 5}$). In Fig. 5.1 the electron-energy-level diagram is shown for lead, and the more probable and energetic transitions are indicated together with the nomenclature for the associated characteristic x rays. The notation shown in this example is valid for any element; that is, $K_{\beta 1}$ refers to the characteristic x ray produced when an electron from the third energy level of the M shell fills a vacancy in the K shell. Fortunately, in many radiation measurements such detail is seldom needed, and often only the dominant K series of x rays is considered, with a single representative energy being used for all x rays.

Production of Characteristic X Rays

There are several methods commonly encountered in radiation measurement applications whereby atoms may be excited and characteristic x rays produced. A photoelectric absorption leaves the absorbing atom in an ionized state. If the incident photon energy is sufficiently greater than the binding energy of the K-shell electron, which ranges from 14 eV for hydrogen to 115 keV for uranium, it is most likely (80 to 100%) that a vacancy is created in the K shell and thus that the K series of x rays dominates the subsequent secondary radiation. These x-ray photons produced from photoelectric absorption are often called *fluorescent radiation* and are widely used to identify trace elements in a sample by bombarding the sample with low-energy photons from a radioactive source or with x rays from an x-ray machine and then observing the induced fluorescent radiation.

Characteristic x rays can also arise following the decay of a radionuclide. In the decay process known as *electron capture*, an orbital electron, most likely from the K shell, is absorbed into the nucleus, thereby decreasing the nuclear charge by one unit. The resulting K-shell vacancy then gives rise to the K series of characteristic x rays. A second source of characteristic x rays, which occurs in many radionuclides, is a result of *internal conversion*. Most daughter nuclei formed as a result of any type of nuclear decay are left in excited states. This excitation energy may be either emitted as a gamma photon or transferred to an orbital electron which is ejected from the atom. Again it is most likely that a K-shell electron is involved in this internal conversion process.

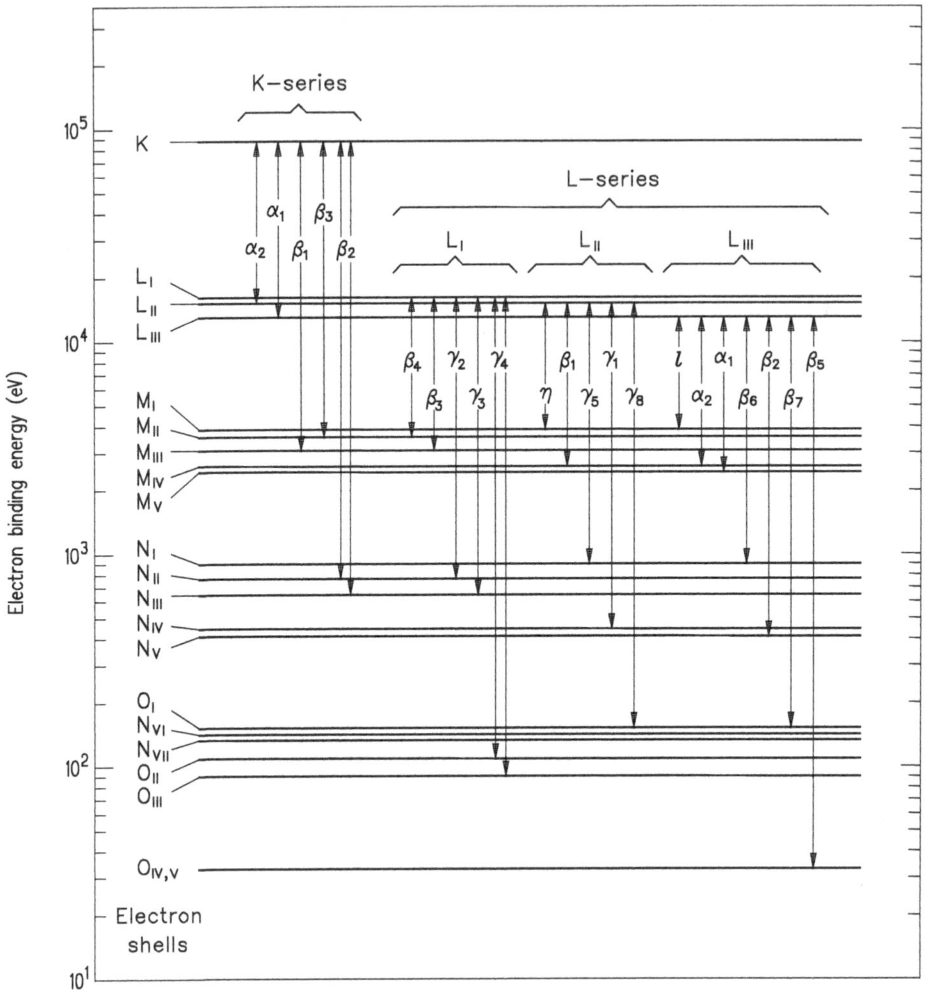

Figure 5.1. X-ray energy-level diagram for lead ($Z = 82$) showing the principal characteristic x-ray transitions and their standard nomenclature. After Flügge [1957].

Energies of Characteristic X Rays

To generate a particular series of characteristic x rays, an electron vacancy must be created in an appropriate electron shell. Such vacancies are created only when sufficient energy is transferred to an electron in that shell so as to allow it to break free of the atom or at least be transferred to an energy level above all the other electrons. In Chapter 4 it was seen that the photoelectric cross section changed abruptly at well-defined energies, called *edge energies*, corresponding to the ionization or binding energy for electrons in the various shells. Below the edge energy for a shell, vacancies can no longer be created in the shell and the photoelectric cross section, therefore, is smaller because fewer electrons are now available to interact with the incident photon. These shell edge energies increase with the atomic number Z of the atom as a result of the electrons becoming more tightly bound in the atom. The edge energies or ionization energies for different shells and even for the different electron states within the same shell can be found in many books on x-ray analysis [e.g., Flügge 1957], and in photon data libraries [e.g., Trubey et al. 1989]. A reasonably accurate approximation

Table 5.4. Values of the constants a b and c used to express principal characteristic x-ray energies and the K-edge energy as $E = aZ^c \exp(-bZ^3)$. Maximum error is less than 100 eV, and mostly less than 30 eV.

X Ray or Edge	a (eV)	b	c	Fit Range
K-edge	6.85198	1.168×10^{-7}	2.13240	$7 < Z < 95$
$K_{\alpha 1}$	7.51002	1.489×10^{-7}	2.07109	$7 < Z < 95$
$K_{\alpha 2}$	7.63540	1.079×10^{-7}	2.06582	$7 < Z < 95$
$K_{\beta 1}$	7.37998	1.208×10^{-7}	2.10694	$7 < Z < 95$
$K_{\beta 2}$	6.95769	1.138×10^{-7}	2.12778	$7 < Z < 95$
$K_{\beta 3}$	7.39760	1.116×10^{-7}	2.10624	$7 < Z < 95$
$L_{\alpha 1}$	0.34810	-8.338×10^{-8}	2.35265	$30 < Z < 95$
$L_{\alpha 2}$	0.35173	-9.667×10^{-8}	2.34975	$30 < Z < 95$
$L_{\beta 1}$	0.31810	1.514×10^{-7}	2.38418	$30 < Z < 95$
$L_{\beta 2}$	0.21302	-1.346×10^{-7}	2.51145	$30 < Z < 95$

for the K-edge energy is

$$E_K \text{ (eV)} = a\, Z^c \exp[-bZ^3], \tag{5.3}$$

where the parameters for a, b, and c are listed in Table 5.4.

The characteristic x rays emitted when electrons fill a vacancy in a shell always have less energy than that required to create the vacancy. Extensive tabulations of characteristic x-ray energies are available [ICRP 1983; Weber et al. 1989; Eckerman et al. 1994]. From these results it is seen that the K_α x-ray energy varies from only 0.52 keV for oxygen ($Z = 8$) to 6.4 keV for iron ($Z = 26$) to 98 keV for uranium ($Z = 92$). By comparison, the L series of x rays for uranium occurs at energies around 15 keV.

Intensity of Characteristic X Rays

The *fluorescent yield* of a material is the fraction of the atoms with a vacancy in an inner electron shell that emit an x ray upon the filling of the vacancy. Values for the fluorescent yield are found in Browne and Firestone [1986] or in nuclear data libraries such as Eckerman et al. [1994]. For vacancies in the K shell, the fluorescent yield can be approximated for $1 < Z < 100$ by [Hubbell 1989]

$$\omega_K = \frac{B^4}{1 + B^4}, \tag{5.4}$$

where the parameter B is given by

$$B = 0.0370 + 0.03112Z + 5.44 \times 10^{-5}Z^2 - 1.250 \times 10^{-6}Z^3. \tag{5.5}$$

From the results of Eqs. (5.4) and (5.5) it is seen that the fluorescent yield increases dramatically with the Z number, varying from 0.0069 for oxygen ($Z = 8$) to 0.97 for uranium ($Z = 92$). Thus, the secondary fluorescent radiation is of more concern for heavy materials.

5.3.2 Bremsstrahlung

A charged particle gives up its kinetic energy either by collisions with electrons along its path or by photon emission as it is deflected, and hence accelerated, by the electric fields of nuclei. The photons produced by the deflection of the charged particle are called *bremsstrahlung* (literally, "braking radiation"). For a given type of charged particle, the radiative stopping power, L_{rad}, increases with the particle energy and with the square of the atomic number (Z) of the absorber [see Eq. (4.134)], while the collisional (ionization) stopping power, L_{coll}, decreases with particle energy and increases only with the first power of Z [see Eq. (4.131)].

For a relativistic particle of rest mass M (i.e., $E \gg Mc^2$) it can be shown that the ratio of radiative to ionization losses is approximately [Evans 1955]

$$\frac{L_{rad}}{L_{coll}} \simeq \frac{EZ}{700}\left(\frac{m_e}{M}\right)^2, \tag{5.6}$$

where E is in MeV. From this result it is seen that bremsstrahlung is more important for high energy particles of small mass incident on high-Z material. In most radiation measurement situations, only electrons ($m_e/M = 1$) are ever of importance for their associated bremsstrahlung. All other charged particles are far too massive to produce significant amounts of bremsstrahlung. Bremsstrahlung from electrons, however, is of particular radiological interest for devices that accelerate electrons, such as betatrons and x-ray tubes, or for situations involving radionuclides that emit only beta particles.

Thick-Target Bremsstrahlung for Monoenergetic Electrons

The energy distribution of the photons produced by the bremsstrahlung mechanism is continuous up to a maximum energy corresponding to the maximum kinetic energy of the incident charged particles. The exact shape of the continuous bremsstrahlung spectrum depends on many factors, including the energy distribution of the incident charged particles, the thickness of the target, and the amount of bremsstrahlung absorbed in the target and other masking material.

For monoenergetic electrons of energy E_o incident on a target thicker than the electron range, the number of bremsstrahlung photons of energy E, per unit energy and per incident electron, emitted as the electron is completely slowed down can be approximated by the distribution [Wyard 1952]

$$N_{br}(E_o, E) = 2kZ\left[\left(\frac{E_o}{E} - 1\right) - \frac{3}{4}\ln\left(\frac{E_o}{E}\right)\right], \qquad E \leq E_o, \tag{5.7}$$

where k is a normalization constant independent of E. The fraction of the incident electron's kinetic energy that is subsequently emitted as bremsstrahlung can then be calculated from this approximation as

$$Y(E_o) = \frac{1}{E_o}\int_0^{E_o} dE\, E N_{br}(E_o, E) = \frac{5}{8}kZE_o, \tag{5.8}$$

which is always a small fraction in most situations. For example, only 4% of the energy of a 0.5-MeV electron, when stopped in lead, is converted into bremsstrahlung. Equation (5.8) can be used to express the normalization constant k in terms of $Y(E_o)$, namely $kZ = 8Y(E_o)/(5E_o)$. With this choice for k, the approximation of Eq. (5.7) agrees quite well with the thick-target bremsstrahlung spectrum calculated by much more elaborate methods, such as the continuous slowing-down model (see Fig. 5.2).

The angular distribution of bremsstrahlung is generally quite anisotropic and varies with the incident electron energy. Bremsstrahlung induced by low-energy electrons ($\lesssim$ 100 keV) is emitted predominantly at 90° to the direction of the incident electron. As the electron energy increases, the direction of the peak intensity shifts increasingly toward the forward direction, until for electrons above a few MeV, the bremsstrahlung is confined to a very narrow forward beam [NCRP 1977]. The angular distribution of radiation leaving a target is very difficult to compute since it depends on the target size and orientation. For thin targets the anisotropy of the bremsstrahlung resembles that for a single electron-nucleus interaction, while for thick targets multiple electron interactions and photon absorption in the target must be considered.

Bremsstrahlung from Beta Particles

The electrons and positrons emitted by radionuclides undergoing beta decay produce bremsstrahlung as they slow down in the source material. In beta decay, the electrons or positrons are emitted with a continuous distribution of energies from zero to a maximum energy $E_{\max}$. The shape of the spectrum is a complicated

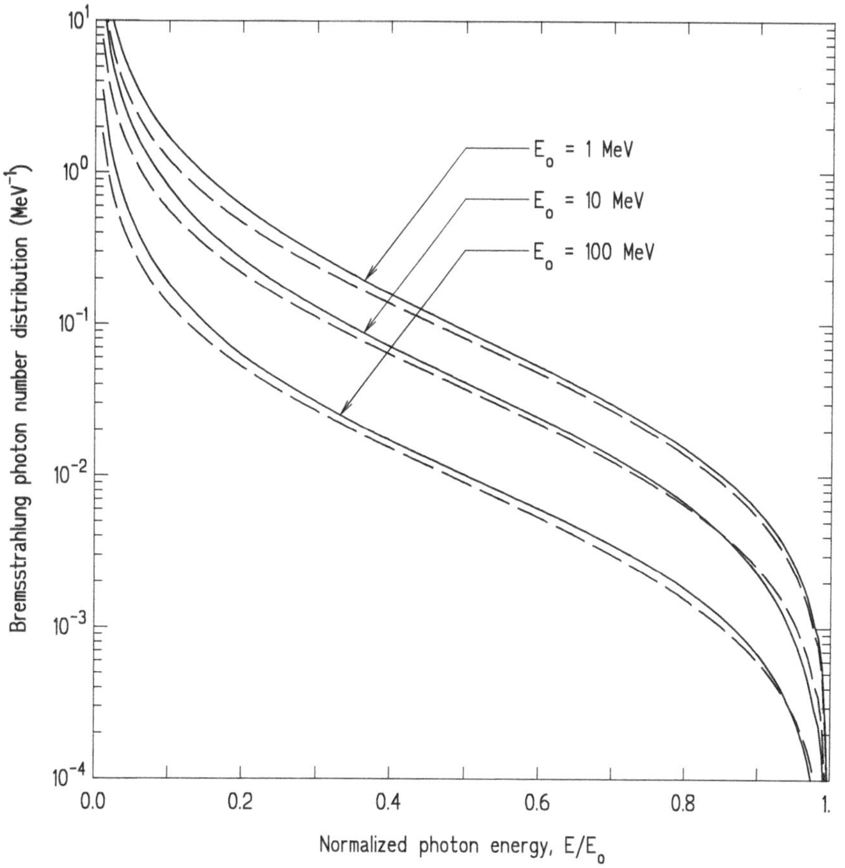

Figure 5.2. Distribution of bremsstrahlung photons, $N_{br}(E_o, E)$, produced by monoenergetic electron beams (energy E_o) incident on a thick lead target. The solid lines are calculations based on the continuous slowing-down approximation (CSDA) and the dashed lines are based on the Wyard approximation of Eq. (5.7).

function of the quantum mechanical properties of the parent nucleus (see Section 5.5.1), although computer codes are available to calculate the beta spectrum for most radionuclides [Dillman 1980]. If the number of beta particles emitted in unit energy about E (normalized to one electron) is denoted by $N_\beta(E)$, the energy distribution $N_{br}(E')$ of the resulting bremsstrahlung intensity produced in a target thicker than the range of the most energetic beta particles can be computed from Eq. (5.7) as

$$N_{br}(E') = \int_E^{E_{\max}} dE\, N_\beta(E) N_{br}(E, E'). \tag{5.9}$$

This energy distribution of bremsstrahlung from beta decay is quite different from that from monoenergetic electrons, the former being much softer and comparatively less intense at higher energies for $E_{\max} = E_o$.

The average fraction of the beta-particle energy that is emitted as bremsstrahlung is

$$f_\beta = \int_0^{E_{\max}} dE\, E\, Y(E) N_\beta(E) \bigg/ \int_0^{E_{\max}} dE\, E\, N_\beta(E). \tag{5.10}$$

Table 5.5. Characteristic X-ray properties of target materials used in X-ray tubes.

Element	X-ray line	Wavelength (10^{-10} m)	Energy (keV)	Excitation voltage (kV)
Tungsten	$K_{\alpha 1}$	0.2090	59.3182	69.525
	$K_{\beta 1}$	0.1844	67.2443	69.525
	$L_{\alpha 1}$	1.4764	8.3976	10.207
	$L_{\beta 1}$	1.2818	9.6724	11.514
Silver	$K_{\alpha 1}$	0.5594	22.1629	25.514
	$L_{\alpha 1}$	4.1544	2.9843	3.351
Molybdenum	$K_{\alpha 1}$	0.7093	17.4793	20.000
	$K_{\beta 1}$	0.6323	19.6083	20.000
	$L_{\alpha 1}$	5.4066	2.2932	2.520
Copper	$K_{\alpha 1}$	1.5406	8.0478	8.979
	$K_{\beta 1}$	1.3922	8.9053	8.979
Nickel	$K_{\alpha 1}$	1.6579	7.4782	8.333
Cobalt	$K_{\alpha 1}$	1.7890	6.9303	7.709
Iron	$K_{\alpha 1}$	1.9360	6.4038	7.112
Chromium	$K_{\alpha 1}$	2.2897	5.4147	5.989
Aluminum	$K_{\alpha 1}$	8.3393	1.4867	1.560

Inner Bremsstrahlung

During the beta-decay process, the beta particle is accelerated by the positive charge of the nucleus, and consequently, a small amount of bremsstrahlung is emitted. These x rays, called "inner" bremsstrahlung, can usually be ignored because only a small fraction of the beta-decay energy, on the average, is emitted as this type of radiation.

5.3.3 X-Ray Machines

The production of x-ray photons as bremsstrahlung and fluorescence occurs in any device that produces high energy electrons. Devices that can produce significant amount of x rays are those in which a high voltage is used to accelerate electrons, which then strike an appropriate target material. Such is the basic principle of all x-ray tubes used in medical diagnosis and therapy, industrial applications, and research laboratories.

Although there are many different designs of x-ray sources for different applications, most designs for low to medium voltage sources ($\lesssim$ 180 kV) place the electron source (cathode) and electron target (anode) in a sealed glass tube. The anode and cathode extend from the glass tube, which acts as an insulator between these electrodes. Further, the glass tube contains the necessary vacuum through which electrons are accelerated by the applied high voltage between the anode and cathode. The anodes of x-ray tubes incorporate a suitable metal upon which the electrons impinge and generate bremsstrahlung and characteristic x rays. Most of the electron energy is deposited in the anode as heat rather than being radiated away as x rays and, thus, heat removal is an important aspect in the design of x-ray tubes. Tungsten is the most commonly used target material because of its high atomic number and because of its high melting point, high thermal conductivity, and low vapor pressure. Occasionally, other target materials are used when different characteristic x-ray energies are desired (see Table 5.5). Generally, the operating conditions of a particular tube (current, voltage, and operating time) are limited by the rate at which heat can be removed from the anode. For most medical and dental diagnostic units, voltages between 40 and 150 kV are used, while medical therapy units may use 6 to 150 kV for superficial treatment or 180 kV to 50 MV for treatment requiring very penetrating radiation.

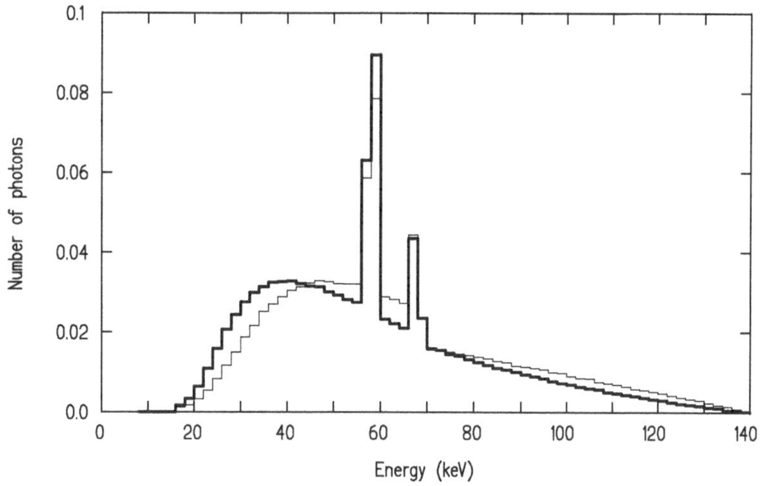

Figure 5.3. Measured photon spectra from a Machlett Aeromax x-ray tube (tungsten anode) operated at a constant 140 kV potential. This tube has an inherent filter thickness of 2.50-mm aluminum equivalent and yields the spectrum shown by the thick line. The addition of an external 6-mm aluminum filter hardens the spectrum (thin line). Both spectra are normalized to unit area. Data are from Fewell, Shuping, and Hawkins [1981].

The energy spectrum of x-ray photons emitted from an x-ray tube has a continuous bremsstrahlung component up to the maximum electron energy (i.e., the maximum voltage applied to the tube). If the applied voltage is sufficiently high as to cause ionization in the target material, there will also be characteristic x-ray lines superimposed on the continuous bremsstrahlung spectrum. In Fig. 5.3 two calculated exposure spectra of x rays are shown for the same operating voltage but for different amounts of beam filtration (i.e., different amounts of material attenuation in the x-ray beam). As the beam filtration is increased, the low-energy x rays are preferentially attenuated and the x-ray spectrum hardens and becomes more penetrating. Also readily apparent in these spectra are the tungsten $K_{\alpha 1}$ and $K_{\beta 1}$ characteristic x rays.

The characteristic x rays may contribute a substantial fraction of the total x-ray emission. For example, the L-shell radiation from a tungsten target is between 20% and 35% of the total energy emission when voltages between 15 kV and 50 kV are used [ICRU 1970]. Above and below this voltage range, the L component rapidly decreases in importance. However, even a small degree of filtering of the x-ray beam effectively eliminates the low-energy portion of the spectrum containing the L-shell x rays. The higher-energy K-series x rays from a tungsten target contribute a maximum of 12% of the total x-ray exposure for operating voltages between 100 and 200 kV [ICRU 1970].

An ideal x-ray spectrum can be estimated from Eq. (5.7). The rate $\dot{N}(E)$ at which bremsstrahlung energy, per unit energy about E, is produced in the target (MeV per second) when bombarded by a current of I amperes of electrons which have been accelerated through a potential of V volts is

$$\dot{N}(E) = \frac{I}{q_e} E N_{br}(E_o, E) \simeq I k' \left[E_o - E - \frac{3}{4} E \ln\left(\frac{E_o}{E}\right)\right], \tag{5.11}$$

in which q_e is the electron charge, k' a constant, and $E_o = 10^{-6} V$. This spectrum, which is proportional to the exposure spectrum, is an ideal *thick target spectrum* in that the incident electron is completely stopped and no absorption of photons is considered. It is accepted practice in x-ray technology to use the dimensional (not Compton) x-ray wavelength λ rather than the photon energy E. Since these two quantities are related by $\lambda = hc/E$, the ideal spectral distribution of bremsstrahlung energy emission rate per unit wavelength,

$\dot{N}(\lambda)$, is

$$\dot{N}(\lambda) = \dot{N}(E)\left|\frac{dE}{d\lambda}\right| = Ik''\frac{1}{\lambda^3}\left[\left(\frac{\lambda}{\lambda_o} - 1\right) - \frac{3}{4}\ln\left(\frac{\lambda}{\lambda_o}\right)\right], \tag{5.12}$$

where the minimum wavelength $\lambda_o = hc/E_o$ and k'' is a constant. As the voltage increases above the critical voltage for ionization of the target atoms, the emission rate of the characteristic x rays also increases. For voltages up to six times the critical voltage for K-shell ionization, V_K, the emission of the K_α x rays increases approximately as $(V - V_K)^2$ [Flügge 1957].

The shape of the x-ray spectrum from a real x-ray machine, such as those of Fig. 5.3, differs considerably from the ideal spectrum discussed above. The high-voltage power supply of most units usually does not provide a constant voltage across the x-ray tube but rather gives a pulsating or fluctuating voltage.[4] This results in fewer high energy bremsstrahlung photons near the cutoff than would be expected on the basis of the peak voltage alone. In effect, electrons with a distribution of energies up to the peak voltage impinge on the anode and cause a shift of the bremsstrahlung spectrum toward lower energies (longer wavelengths). On the other hand, the low-energy photons are preferentially absorbed by the target material, the glass envelope of the x-ray tube, and the air outside the tube. In addition, metal filters are often placed in the beam path to absorb preferentially the low-energy photons and thus to shift the spectrum toward higher energies. The strength of the x-ray beam thus depends on the target material, tube voltage, tube current, and the type and amount of filtering experienced by the x rays.

5.3.4 Synchrotron X Rays

When a charged particle initially moving in a straight line is accelerated by deflecting it in an electromagnetic field, the perturbation in the particle's electric field travels away from the particle at the speed of light and is observed as electromagnetic radiation (photons). Such is the origin of bremsstrahlung produced when fast electrons (beta particles) are deflected by the electric field of a nucleus.

This same mechanism can be used to produce intense photon radiation by deflecting an electron beam by magnetic fields. In a special accelerator called a *synchrotron*, highly relativistic electrons are forced to move in a circular path inside a storage ring by placing *bending magnets* along the ring. Photons are emitted when the beam is accelerated transversely by (1) the bending magnets (used to form the circular electron beam), and by (2) insertion device magnets such as *undulators*, *wigglers*, and *wave-length shifters*.

Because the electrons are highly relativistic, the synchrotron radiation is emitted in a very narrow cone in the direction of electron travel as they are deflected. Undulators cause the beam to be deflected sinusoidally by a weak oscillatory magnetic field, thereby producing nearly monochromatic photons. By contrast, a wiggler uses a strong oscillatory magnetic field which, because of relativistic effects, produces distorted sinusoidal deflections of the electron beam and synchrotron radiation with multiple harmonics, i.e., a line spectrum. If very strong magnetic fields are used, many harmonics are produced that merge to yield a continuous spectrum ranging from the infrared to hard x rays. By placing undulators or wigglers at a specific location in the storage ring, very intense and narrowly collimated beams of photons with energies up to a few keV can be produced, useful in x-ray diffraction analyses.

5.4 Sources of Neutrons

5.4.1 Fission Neutrons

Many heavy nuclides fission after the absorption of a neutron, and some nuclides fission spontaneously, producing several energetic fission neutrons. Almost all of the fast neutrons produced from a fission event

[4]If a constant potential is applied to the x-ray tube, the designation "kVcp" is sometimes used, while the peak voltage for a pulsating applied voltage is often indicated by "kVp."

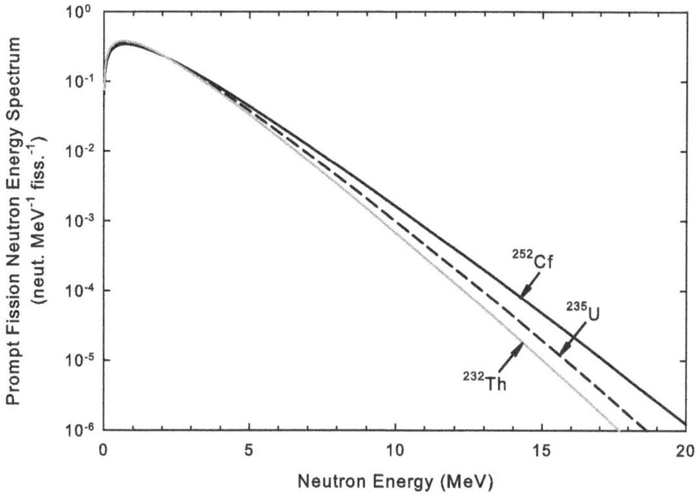

Figure 5.4. Energy spectra of prompt neutrons produced from the fission of fissionable nuclei as calculated by Walsh [1989].

are emitted within 10^{-14} s of the fission event, and are called *prompt neutrons*. Generally less than 1% of the total fission neutrons are emitted as *delayed neutrons*, which are produced by the neutron decay of fission products at times up to many seconds or even minutes after the fission event. As the energy of the neutron which induces the fission in a heavy nucleus increases, the average number of fission neutrons also increases. For example, the fission of ^{235}U by a thermal neutron (average energy 0.025 eV) produces, on the average, 2.43 fission neutrons. A fission caused by a 10-MeV neutron, by contrast, yields 3.8 fission neutrons. For ^{239}Pu, fission by thermal or 10 MeV neutrons yields 2.87 or 4.2 neutrons. The fission of ^{238}U is induced only by fast neutrons, with a 10-MeV neutron yielding 3.9 fission neutrons.

Since the advent of fission reactors, many transuranic isotopes have been produced in significant quantities. Many of these isotopes have appreciable spontaneous fission probabilities, and consequently they can be used as very compact sources of fission neutrons. For example, 1 g of ^{252}Cf releases 2.3×10^{12} neutrons per second, and very intense neutron sources can be made from this isotope, limited in size only by the need to remove the fission heat through the necessary encapsulation. Almost all spontaneously fissioning isotopes decay much more frequently by α emission than by fission.

The energy dependence of the fission neutron spectrum has been investigated extensively, particularly for the important isotope ^{235}U. All fissionable nuclides produce prompt-fission neutrons with energy frequency distributions that go to zero at low and high energies (see Fig. 5.4), reaching a maximum at about 0.7 MeV, and have an average energy of about 2 MeV. The fraction of prompt fission neutrons emitted per unit energy about E, $\chi(E)$, can be described quite accurately by a Watt distribution

$$\chi(E) = ae^{-E/b}\sinh\sqrt{cE}, \tag{5.13}$$

where the parameters a, b and c depend on the fissioning isotope (see Table 5.6). The fission-neutron spectrum for thermal-neutron fission of ^{235}U is often used as an approximation for other fissioning isotopes.

5.4.2 Fusion Neutrons

Neutrons can be produced as products of nuclear reactions in which energetic charged particles hit target atoms. Most such reactions require accelerators to produce the energetic charged particles and hence such neutrons are to be encountered only near accelerator targets.

Table 5.6. Parameters for the Watt approximation of Eq. (5.13).

Nuclide	Type of Fission	a	b	c
^{233}U	thermal	0.6077	1.1080	1.2608
^{235}U	thermal	0.5535	1.0347	1.6214
^{239}Pu	thermal	0.5710	1.1593	1.2292
^{232}Th	fast (2 MeV)	0.5601	0.9711	1.8262
^{238}U	fast (2 MeV)	0.5759	1.0269	1.5776
^{252}Cf	spontaneous	0.6400	1.1750	1.0401

One major exception to the insignificance of charged-particle-induced reactions are those in which light elements fuse exoergically to yield a heavier nucleus and which are accompanied quite often by the release of energetic neutrons. Neutron-producing fusion reactions of most interest in the development of thermonuclear fusion power are

$$^{2}\mathrm{H} + {}^{2}\mathrm{H} \longrightarrow {}^{3}\mathrm{He}\ (0.82\ \mathrm{MeV}) + {}^{1}n\ (2.45\ \mathrm{MeV})$$

$$^{2}\mathrm{H} + {}^{3}\mathrm{H} \longrightarrow {}^{4}\mathrm{He}\ (3.5\ \mathrm{MeV}) + {}^{1}n\ (14.1\ \mathrm{MeV}).$$

When these reactions are produced by accelerating one nuclide toward the other, the velocity of the center of mass must first be added to the center-of-mass neutron velocity before determining the neutron energy in the laboratory coordinate system. In most designs for fusion power, the velocity of the center of mass is negligible, and the concern is with monoenergetic 2.45- or 14.1-MeV fusion neutrons. The 14.1-MeV fusion neutrons are also produced copiously in a thermonuclear explosion.

A beam of relatively low energy deuterons (100 to 300 keV) incident on a deuterium or tritium target can produce a significant number of thermonuclear neutrons. Thus, these D-D or D-T reactions are used in relatively compact accelerators, called *neutron generators*, in which deuterium ions are accelerated through a high voltage (100 to 300 kV) and allowed to fall on a thick deuterium- or tritium-bearing target. Typically in such devices, a 1-mA beam current produces up to 10^9 14-MeV neutrons per second from a thick tritium target.

5.4.3 Photoneutrons

A gamma photon with energy sufficiently large to overcome the neutron binding energy (about 7 MeV in most nuclides) may cause a (γ, n) reaction. Very intense and energetic photoneutron production can be realized in an electron accelerator where the bombardment of an appropriate target material with the energetic electrons produces intense bremsstrahlung (see Section 5.3.2) with a distribution of energies up to that of the incident electrons. The probability a photon will cause a (γ, n) reaction increases with the photon energy, reaching a maximum over a broad energy range of approximately 20 to 23 MeV for light nuclei ($A \lesssim 40$) and 13 to 18 MeV for medium and heavy nuclei. The peak energy of this *giant resonance* can be approximated by $80A^{-1/3}$ MeV for $A > 40$. The width of the resonance varies from about 10 MeV for light nuclei to 3 MeV for heavy nuclei. Consequently, in medical or accelerator facilities that produce photons with energies above about 15 MeV, neutron production in the surrounding walls can lead to significant neutron fields.

The photoneutron mechanism can be used to create laboratory neutron sources by mixing intimately a beryllium or deuterium compound with a radioisotope that decays with the emission of high energy photons.

Alternatively, the encapsulated radioisotope may be surrounded by a beryllium- or deuterium-bearing shell. A common reactor photoneutron source is an antimony-beryllium mixture, which has the advantage of being rejuvenated by exposing the source to the neutrons in a nuclear reactor to transmute the stable ^{123}Sb into the required ^{124}Sb isotope (half-life of 60.2 days). Characteristics of some (γ, n) sources are given in Table 5.7.

One very attractive feature of such (γ, n) sources is the nearly monoenergetic nature of the neutrons if the photons are monoenergetic. However, in large sources, the neutrons may undergo significant scattering in the source material and thereby degrade the nearly monoenergetic nature of their spectrum. These photoneutron sources generally require careful use because of their inherently large photon emission rates. Because nominally only one in a million high energy photons actually interacts with the source material to produce a neutron, these sources generate gamma rays that are of far greater biological concern than are the neutrons. A particularly useful combination is an SbBe neutron source coupled with an ^{56}Fe shield. The ^{56}Fe 24 keV anti-resonance allows the passage of the SbBe 23 keV neutrons while working to shield other emissions.

Table 5.7. Characteristics of some important (γ, n) sources.

Source	Half-life	Principal photons: E (MeV)	Principal photons: Number per decay	Average neutron energy (MeV)	Standard yield[a]
^{24}Na + Be	15.0 h	2.754	1.00	0.967	3.5
^{24}Na + D_2O	15.0 h	2.754	1.00	0.262	7.3
^{28}Al + Be	2.24 m	1.779	1.00	0.100	
^{56}Mn + Be	2.58 h	1.810	0.272	0.128	0.78
		2.113	0.143	0.397	
		2.522	0.010	0.761	
^{56}Mn + D_2O	2.58 h	2.522	0.010	0.146	0.08
		2.657	0.007	0.214	
^{72}Ga + Be	14.1 h	1.861	0.053	0.173	1.4
		2.202	0.259	0.476	
		2.491	0.077	0.733	
		2.508	0.128	0.748	
^{72}Ga + D_2O	14.1 h	2.491	0.077	0.131	1.6
		2.508	0.128	0.139	
^{76}As + Be	26.3 h	1.788	0.003	0.108	1.9
		2.096	0.007	0.382	
^{88}Y + Be	107 d	1.836	0.993	0.151	2.7
		2.734	0.006	0.949	
^{88}Y + D_2O	107 d	2.734	0.006	0.252	0.08
^{116m}In + Be	54.2 m	2.112	0.154	0.396	0.22
^{124}Sb + Be	60.2 d	1.691	0.490	0.023	5.1
		2.091	0.057	0.378	
^{140}La + Be	40.4 h	2.522	0.034	0.761	0.08
^{140}La + D_2O	40.4 h	2.522	0.034	0.146	0.2
^{226}Ra + Be	1622 y	many		0.68 max	0.8
^{226}Ra + D_2O	1622 y	many		0.11 max	0.03
^{228}Ra + Be	6.7 y	2.62		0.848	0.95
		1.80		0.119	
^{228}Ra + D_2O	6.7 y	2.62		0.195	2.6

[a] Number of neutrons emitted from 1 g of Be or D_2O per 10^6 disintegrations of the radionuclide placed 1 cm away. Yield is for all photons emitted.

Sources: Blizard and Abbott [1962], Shleien [1992], and Shultis and Faw [2000]

5.4.4 Alpha-Neutron Sources

Many compact laboratory neutron sources use energetic alpha particles from various radioisotopes (*emitters*) to induce (α, n) reactions in appropriate materials (*converters*). Although a large number of nuclides emit neutrons if bombarded with alpha particles of sufficient energy, the energies of the alpha particles from radioisotopes are capable of penetrating the potential barriers of only the lighter nuclei.

Of particular interest are those light isotopes for which the (α, n) reaction is exoergic ($Q > 0$) or, at least, has a low threshold energy. For endoergic reactions ($Q < 0$), the threshold alpha energy is $-Q(1+4/A)$. Thus, for an (α, n) reaction to occur, the alpha particle must (1) have enough energy to overcome the repulsive Coulombic force field of the nucleus, and (2) exceed the threshold energy for the reaction. Converter materials used to make practical (α, n) sources include lithium, beryllium, boron, carbon, fluorine and sodium.

A neutron source can be fabricated by mixing intimately a light converter element, such as lithium or beryllium, with a radioisotope that emits energetic alpha particles. Most of the practical alpha emitters are actinide elements, which form intermetallic compounds with beryllium. Such a compound, e.g., $PuBe_{13}$, ensures both that the emitted alpha particles immediately encounter converter nuclei, thereby producing a maximum neutron yield, and that the radioactive actinides are bound into the source material, thereby reducing the risk of leakage of the alpha-emitting component.

The neutron yield from an (α, n) source varies strongly with the converter material, the energy of the alpha particle, and the relative concentrations of the emitter and converter elements. The degree of mixing between converter and emitter and the size, geometry, and source encapsulation may also affect the neutron yield. For example, a ^{239}Pu/Be source has an optimum neutron yield of about 60 neutrons per 10^6 primary alpha particles. Characteristics of some (α, n) sources are given in Table 5.8.

Table 5.8. Characteristics of some (α, n) sources.

Source	Half-life	Principal alpha energies (MeV)	Average neutron energy (MeV)	Optimum neutron yield per 10^6 primary alphas^a
^{210}Po / Li	138.4 d	5.305	0.48	1.3
^{239}Pu / Be	24100 y	5.155, 5.143, 5.105	4.6	60
^{210}Po / Be	138.4 d	5.305	4.5	70
^{238}Pu / Be	87.8 y	5.499, 5.457, 5.358	4.5	80
^{241}Am / Be	432 y	5.486, 5.443, 5.388	4.4	75
^{244}Cm / Be	18.1 y	5.805, 5.763	4.3	100^b
^{242}Cm / Be	163 d	6.113, 6.070	4.1	110
^{226}Ra / Be	1600 y	7.687, 6.003, 5.490	3.9	500^c
+ daughters		5.304, 4.785, 4.602		
^{227}Ac / Be	21.8 y	7.386, 6.819, 6.623	3.9	700^c
+ daughters		6.038, 5.960, 5.715		
^{241}Am / B	432 y	5.486, 5.443, 5.388	3	13
^{210}Po / C	138.4 d	5.305		0.10
^{241}Am / F	432 y	5.486, 5.443, 5.388	1.5	4.1
^{210}Po / F	138.4 d	5.305		5
^{210}Po / Na	138.4 d	5.305		1

^{a} Yield for alpha particles incident on a target thicker than the alpha-particle ranges.

^{b} Does not include a 4% contribution from the spontaneous fission of ^{244}Cm.

^{c} Yield is dependent on the proportion of daughters present. Value for ^{226}Ra corresponds to a 22-year-old source (50% contribution for ^{210}Po).

Sources: Jaeger [1968], GPO [1970], and Knoll [1989].

The energy distributions of neutrons emitted from (α, n) sources are continuous below some maximum neutron energy with definite structure at well-defined energies determined by the energy levels of the converter and the excited product nuclei (see Fig. 5.5). The use of the same converter material with different alpha emitters produces similar neutron spectra with different portions of the same basic spectrum accentuated or reduced as a result of the different alpha-particle energies. Average energies of neutrons typically are several MeV. For example, the neutrons produced by a ^{239}Pu/Be source have an average energy of 4.6 MeV.

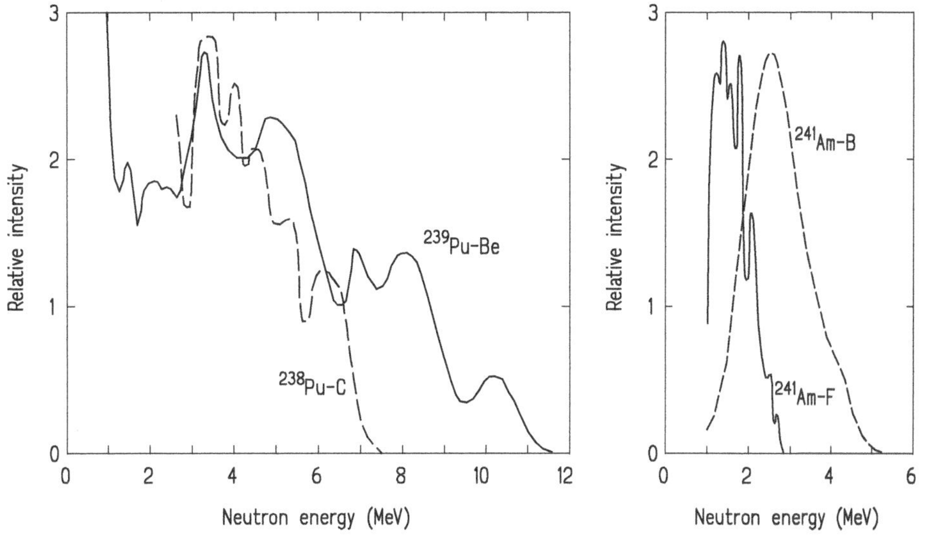

Figure 5.5. Neutron energy spectra produced by four (α, n) sources. The spectral structure is determined by the converter material, while the relative intensity of various portions of the spectrum depends on the energies of the alpha particles emitted by the radioactive component. Based on experiment data from Anderson and Neff [1972] and Lorch [1973].

5.4.5 Activation Neutrons

A few highly unstable nuclides decay by the emission of a neutron. The delayed neutrons associated with fission arise from such decay of the fission products. However, there are nuclides other than those in the fission-product decay chain which also decay by neutron emission. Only one of these nuclides, ^{17}N, is of importance in nuclear reactor situations. This isotope is produced in water-moderated reactors by an (n, p) reaction with ^{17}O (threshold energy, 8.0 MeV). The decay of ^{17}N by beta emission (half-life 4.4 s) produces ^{17}O in a highly excited state, which in turn decays rapidly by neutron emission. Most of the decay neutrons are emitted within $\pm$ 0.2 MeV of the most probable energy of about 1 MeV, although neutrons with energies up to 2 MeV may be produced.

5.4.6 Spallation Neutron Sources

In a spallation neutron source, pulses of very energetic protons (up to 1 GeV), produced by an accelerator, strike a heavy metal target such as mercury or liquid bismuth. Such an energetic proton when it strikes a target nucleus "spalls" or knocks out neutrons. Additional neutrons boil off as the struck nucleus heats up. Typically, 20 to 30 neutrons are produced per spallation reaction. These pulses of neutrons are then slowed down or thermalized by passing them through cells filled with water, or even liquid hydrogen if very slow neutrons are needed.

5.5 Sources of Charged Particles

Charged particles are directly ionizing radiations and, unlike photons and neutrons, have well-defined ranges as they interact with the long-range Coulombic forces of the atoms of the medium through which they pass. Such particles arise from several sources. Radioactive decay products almost always produce charged particles such as alpha and beta particles.[5] Likewise, photon and neutron induced reactions with an atom usually produce secondary charged particles. Indeed it is these secondary charged particles that are used to detect the presence of these indirectly ionizing radiations. Finally in our technological modern world there are many machines that produce energetic charged particles. Indeed accelerators that produce energetic charged particles are used for a myriad of tasks such as industrial processing, production of medical radioisotopes, or numerous research purposes.

5.5.1 Beta Decay

Beta decay of a radionuclide, in effect, involves either (1) a transformation within the nucleus of a neutron into a proton accompanied by the emission of an electron, or beta particle, and an antineutrino, or (2) a nuclear transformation of a proton in the decaying nucleus to a neutron accompanied by emission of a positron and a neutrino. The energy released in the process—the Q value—is shared by the electron and the antineutrino or by the positron and the neutrino. Beta decay of a nucleus may occur by any of several different nuclear transitions. For each there is a unique endpoint energy, the maximum kinetic energy of the electron or positron. Also, for each transition, there is a unique statistical distribution of energy between the electron and antineutrino or positron and neutrino, and to a very minor extent the residual nucleus. The balance of this discussion deals with the more common beta-particle emissions, with the recognition that positron emission may be treated very similarly.

ΔJ	$\Delta\Pi$	Transition classification	n
0,1	-1	Allowed	0
0,1	$+1$	Non-unique, first-forbidden	0
2	$+1$	Unique, first-forbidden	1
2	-1	Non-unique, second-forbidden	1
3	-1	Unique, second-forbidden	2

Although the energy spectrum of the beta particles for any one transition certainly depends on the endpoint energy, the spectrum also depends, in a complicated way, on the nuclear spin and parity quantum numbers of the nucleus before and after beta-particle emission. As shown to the left, these quantum numbers determine whether the transition is referred to as *allowed* or *forbidden*, and if forbidden, whether the transition is *unique* (U) or *non-unique* (non-U). In a transition, the spin quantum number J may remain constant or it may change by one or more units, that is, $\Delta J = 0, 1, 2, \ldots$. The parity quantum number Π may or may not change, that is, $\Delta\Pi = -1$ (no change) or $+1$ (change). To a first approximation, a single classification index n may be used to characterize the energy spectrum. For $\Delta J \geq 2$, $n = \Delta J - 1$. Otherwise, the various combinations of $\Delta\Pi$ and ΔJ establish the following selection rules, which determine the *shape factor*, which, in part, governs the beta-particle energy spectrum for a particular transition.

As an example, consider the decay of ^{38}Cl to ^{38}Ar. Three beta transitions are possible, with the following characteristics:

[5] Only in gamma-ray decay of nuclear isomers, electron capture, and neutron radioactive decay (from fission fragments, for instance) are no charged particles emitted.

Frequency (%)	E_{max} (keV)	E_{av} (keV)	ΔJ	$\Delta \Pi$	n	Classification
32.5	1107	420	1	−1	0	Allowed
11.5	2749	1182	0	+1	0	Non-U, first forbidden
56.0	4917	2244	2	+1	1	Unique, first forbidden

The energy spectrum of beta particles arising from a transition is conveniently expressed in terms of the electron total energy W (including rest-mass energy), that is, $W = E + m_e c^2$, and the momentum $p = \sqrt{E^2 + 2m_e c^2 E}/c$. The energy spectrum $N_i(E)$ for transition i is defined in such a way that $N_i(E)dE$ is the probability that, in a transformation of the radionuclide via transition i, a beta particle is emitted with kinetic energy in the range dE about E. This spectrum can be expressed as

$$N_i(E) = C_i\, p\, W[E_{\max,i} - E]^2 F(Z, A, W)\; S_n(Z, A, E_{\max,i}, W), \tag{5.14}$$

in which $E_{\max,i}$ represents the maximum (or endpoint) energy, Z and A are the charge and mass numbers of the nucleus after the transition, F is the *Fermi function*, S_n is a *shape factor* determined by the selection-rule classification index n, and C_i is a normalization constant.

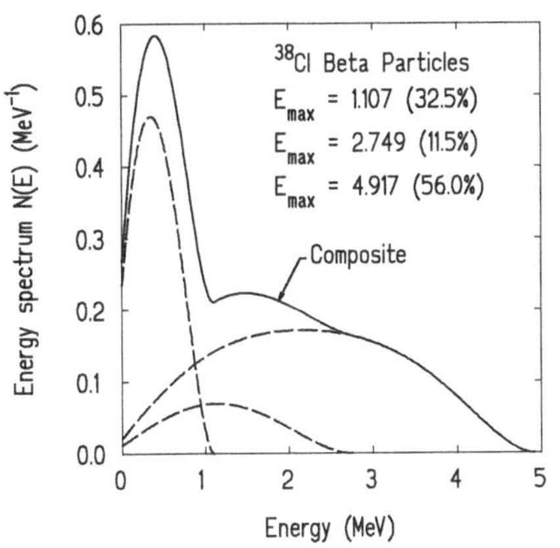

Figure 5.6. Energy spectra of ^{38}Cl beta particles.

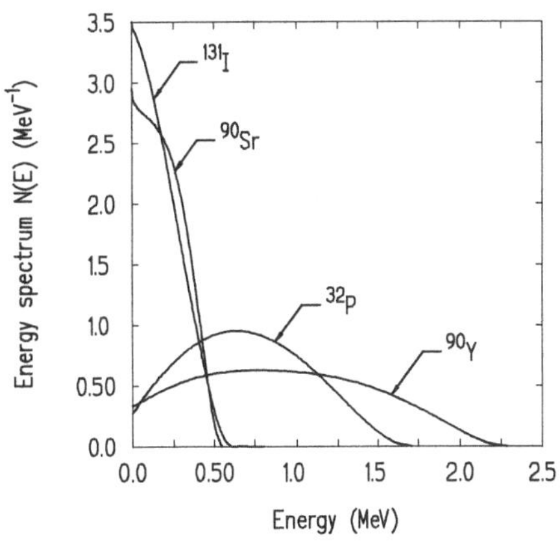

Figure 5.7. Energy spectra of selected beta-particle sources.

The product $p\,W\,[E_{\max,i} - E]^2$ is a statistical factor and is a major determinant of the energy spectrum. The shape factor S_n is unity for allowed and nonunique first-forbidden transitions; for other transitions it is a complicated function. Suffice to say here that it is an extremely tedious task to evaluate the Fermi function and the shape factor. Procedures are clearly described by Dillman and Von der Lage [1975] and by Dillman [1980], as are procedures required to account for screening by atomic electrons, an effect that is especially important for positron decay. Figure 5.6 illustrates individual and composite spectra for the three transitions involved in the beta decay of ^{38}Cl. In Fig. 5.7 the spectra for four important beta-emitting radionuclides are given. Table 5.9 lists commonly encountered radionuclide sources that emit beta particles.

If f_i is the frequency for the ith transition, then the spectrum normalization constant C_i is determined by the requirement that

$$\int_0^{E_{\max,i}} dE\, N_i(E) = f_i. \tag{5.15}$$

Table 5.9. Half-lives of common beta-particle emitters, maximum emission energies, and yields. Positron emission is indicated by (+).

Nuclide	Half-life	E_{max} (MeV)	% yield
^{3}H	12.26 y	0.0186	100
^{63}Ni	100.1 y	0.067	100
^{14}C	5730 y	0.156	100
^{35}S	87.51 d	0.167	100
^{147}Pm	2.6234 y	0.224	99.994
^{33}P	25.34 d	0.249	100
^{45}Ca	163.8 d	0.257	100
^{99}Tc	2.111×10^{5} y	0.294	99.998
^{22}Na	2.602 y	0.545(+)	90.3
		1821(+)	.055
^{131}I	8.02 d	0.304	0.643
		0.334	7.20
		0.606	89.4
		0.807	0.39
^{36}Cl	3.01×10^{5} y	0.710	98.1
^{204}Tl	3.78 y	0.764	97.08
^{210}Pb/^{210}Bi	22.3 y/ 5.01 d	0.017	80.2
		0.064	19.8
		1.161	99.99
^{137}Cs	30.07 y	0.514	94.36
		1.176	5.64
^{60}Co	5.271 y	0.317	99.88
		1.491	0.12
^{32}Si/^{32}P	172 y/ 14.26 d	0.225	100
		1.710	100
^{68}Ge/^{68}Ga	270.8 d/ 67.63 m	0.822(+)	1.2
		1.899(+)	87.68
^{90}Sr/^{90}Y	28.78 y/ 64.1 hr	0.546	100
		2.280	99.99
^{106}Ru/^{106}Rh	1.023 y/ 29.8 s	0.039	100
		1.979	1.77
		2.407	10.0
		3.029	8.1
		3.541	78.6

Sources: Firestone [1996]; NuDat [2019].

The average beta-particle energy for a particular transition is

$$\langle E_i \rangle = \frac{1}{f_i} \int_0^{E_{\text{max},i}} dE\, E N_i(E). \tag{5.16}$$

The average beta-particle energy released per transformation of the radionuclide is given by $\sum_i f_i \langle E_i \rangle$.

5.5.2 Alpha Decay

Radioactive decay by alpha-particle emission is possible only in the heavier elements. One radionuclide may have several transitions involving alpha-particle emission; but for each transition, the particles are monoenergetic and usually of several MeV energy. The Geiger-Nuttall rule [Kaplan 1962] accounts for the strong inverse relationship between the half-life of a radionuclide and the energy of its alpha particle. Table 5.10, which lists several important alpha-particle emitters, illustrates this inverse relationship.

Table 5.10. Half-lives of common alpha-particle emitters. Also listed are the prominent energies and frequencies of particles emitted.

Nuclide	Half-life	E (MeV)	Percent
^{252}Cf†	2.645 y	6.118	84.3
		6.076	15.5
^{244}Cm	18.11 y	5.805	76.4
		5.763	23.6
^{238}Pu	87.74 y	5.499	71.4
		5.456	28.6
^{241}Am	432.2 y	5.486	85.1
		5.443	12.8
		5.388	1.4
^{228}Th	1.913 y	5.423	73.4
		5.340	26.6
(^{228}Th has multiple daughters ranging from 5.449 to 8.784 MeV)			
^{210}Po	138.4 d	5.304	100
^{239}Pu	2.411×10^4 y	5.157	73.3
		5.144	15.1
		5.106	11.5
^{226}Ra	1600 y	4.785	94.4
		4.602	5.5
(^{226}Ra has multiple daughters ranging from 5.489 to 7.687 MeV)			
^{237}Np	2.14×10^6 y	4.789	47.6
		4.774	18.1
		4.769	14.3
		4.644	5.9
^{230}Th	7.538×10^4 y	4.687	76.3
		4.621	23.4
^{235}U	7.038×10^8 y	4.596	5.6
		4.398	57
		4.366	17
		4.214	6.4
^{238}U	4.468×10^9 y	4.198	77
		4.151	23
^{232}Th	1.405×10^{10} y	4.013	77
		3.950	23
^{148}Gd	74.6 y	3.183	100

Source: Firestone [1996].
†Also a spontaneous neutron emitter.

5.5.3 Photon Interactions

In all the principal photon interactions, secondary electrons are produced. The photoelectric interactions lead to photoelectrons and, perhaps, Auger electrons. In a Compton scattering interaction, an orbital electron recoils from the atom, and in pair production a positron-electron pair are created. The energetics of these secondary electrons are discussed in Sections 4.3.3 and 4.4. X- and gamma-ray detectors then measure the ionization and excitation of the ambient atoms produced by these secondary electrons.

5.5.4 Neutron Interactions

Scattering

Often, when neutrons interact with a nucleus, secondary charged particles are produced. When fast neutrons scatter from a nucleus, the recoil nucleus is a charged particle which gives up its kinetic energy through ionization and excitation of the ambient atoms. The energy transfered from the neutrons to the recoil nucleus is described in Section 4.3.4. As the mass of the scattering nucleus decreases ever more energy is transferred to the recoil nucleus. In particular, for scattering from hydrogen, the recoil nucleus (a proton) can receive up to all of the incident neutron's energy. By measuring the energy distribution of the recoil protons, the energy distribution of the fast incident neutrons can be inferred.

Because the neutron has an internal charge distribution that is not uniform throughout the neutron, it is possible for neutrons to interact with electrons. However, such neutron-electron interactions are exceedingly rare and are of little importance in radiation detection.

Charged-Particle Reactions

Neutron detectors are often designed to detect the charged particles produced in neutron reactions that produce secondary charged particles. Most such reactions such as ^{3}He$(n,p)^{3}$H, ^{10}B$(n,\alpha)^{7}$Li, or ^{6}Li$(n,t)^{4}$He, occur predominately at thermal neutron energies. However, for neutrons with energies in the MeV region (n,p), (n,α) and other knockout reactions can also occur. These reactions are especially important when light elements are involved. The (n,α) reaction cross sections for Be, N, and O are appreciable fractions of the total cross sections and may exceed the inelastic scattering contributions. This situation is probably true for most light elements, although only partial data are available. For heavy and intermediate nuclei, the charged-particle emission interactions are at most a few percent of the total inelastic interaction cross sections and, hence, are usually ignored.

Fission

A special case of charged-particle producing reactions is that of neutron-induced fission. Any very heavy nucleus can be made to fission accompanied by a net release of energy if hit sufficiently hard by some incident particle. However, very few nuclides can fission by the absorption of a neutron with negligible incident kinetic energy. The binding energy of this neutron in the compound nucleus provides sufficient excitation energy that the compound nucleus can deform sufficiently to split spontaneously. Nuclides, such as ^{235}U, ^{233}U, and ^{239}Pu, that fission upon the absorption of a slow moving neutron, are called *fissile* nuclei and play an important role in thermal neutron fission detectors. Nuclides that fission only when struck with a neutron with one or more MeV of kinetic energy, such as ^{238}U and ^{240}Pu, are said to be *fissionable*, i.e., they undergo fast fission.

The highly excited compound nucleus almost always splits assymetrically producing a heavy fission fragment Y_H and a lighter fission fragment Y_L.[6] The initial fission fragments are highly charged, missing 10 to 15 electrons. They are also in such highly excited states that, within 10^{-17} s of the scission, several prompt neutrons "boil" away, and, within 2×10^{-14} s, more excitation energy is lost from the emission of prompt fission gamma rays.

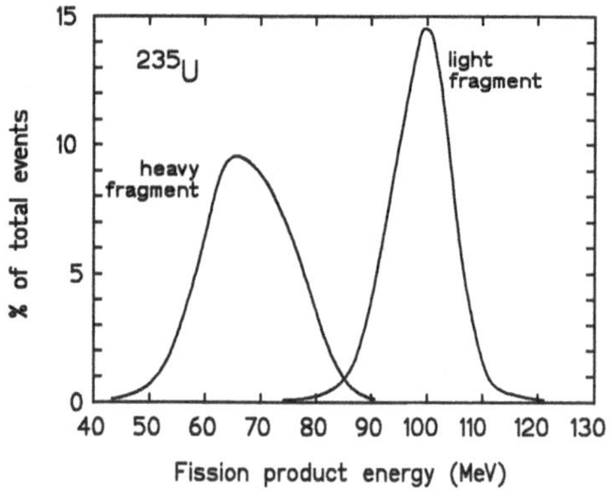

Figure 5.8. Energy distribution of fission products produced by the thermal fission of ^{235}U. The mean energy of the light fragment is 99.2 MeV and that of the heavy fragment is 68.1 MeV. After Keepin [1965].

[6]Sometimes three fission fragments are formed in *ternary* fission with the third being a small nucleus. Alpha particles are created in about 0.2% of the fissions and nuclei of ^{2}H, ^{3}H, and other nuclei up to about ^{10}B with much less frequency.

The highly charged fission fragments pass through the surrounding medium causing millions of Coulombic ionization and excitation interactions with the electrons of the medium. As the fission fragments slow, they gradually acquire electrons that reduce their ionization charge, until, by the time they are stopped, they have become electrically neutral atoms. This transfer of the fission fragments' kinetic energy to the ambient medium takes about 10^{-12} s. It is this enormous energy transfer to the ambient medium that makes neutron fission detectors possible.

Because the mass distribution of the fission fragments is bimodal (two peaks), one would expect the initial kinetic energy of the fragments to also be bimodal. In Fig. 5.8, the kinetic energy distribution of the fission fragments produced in the thermal fission of ^{235}U is shown.

5.5.5 Accelerators

Beginning in the 1930s, machines were developed to accelerate charged subatomic particles such as protons and alpha particles to speeds that could induce nuclear reactions. The first such *particle accelerator* produced protons with 700 keV of energy. Recently an accelerator became operational that produces 7-TeV protons or heavy ions such as lead with 1200 TeV of kinetic energy.

Today the development of ever more energetic particle accelerators is driven by the high energy physics community. With these enormous and costly machines, physicists will perform experiments that will reveal information about the fundamental physics governing the subatomic world. Accelerators with lower energies are also central to other areas of research such as the study of atomic and nuclear physics. A 1-GeV proton accelerator is now used at the Spallation Neutron Source at Oak Ridge National Laboratory to bombard a liquid mercury target. The resulting spallation reactions release copious neutrons which are ideal probes to determine molecular structures. Accelerators can also be used to produce intense x-ray beams that in turn can be used in fundamental research on materials. As in most areas of fundamental research, accelerator technology has spun off many practical applications such as cancer therapy, production of important radionuclides, ion implanting, and food preservation to name a few.

For high energy physics research, special radiation detectors are needed to detect the myriad of subatomic particles that are produced in these machines. Such detectors are huge, very expensive, and usually one of a kind. Consequently, only brief consideration of such detectors is given in this book.

5.6 Cosmic Rays

At the beginning of the twentieth century, the reason why electrometers would discharge for no apparent reason was a mystery. One postulated explanation was that earth itself emitted ubiquitous ionizing radiation. But in 1911 Victor Hess, using electrometers on a balloon, found that ionizing radiation increased in strength with altitude, thereby showing that this radiation came from outer space.

Although much is now known about this *cosmic* radiation, there are many aspects of it that are still mysterious. Cosmic radiation has been measured from a few GeV to over 10^{20} eV. At a few GeVs the composition is 98% nuclei (of which 87% are protons, 12% helium nuclei, and 1% heavier nuclei), $< 2\%$ electrons, and $< 10^{-4}\%$ gamma rays. One unsettled question is how the most energetic cosmic rays are produced. For example, since 1991 more that 20 protons with energies $\geq 3 \times 10^{20}$ eV have been observed. This energy corresponds to 48 J, which is the kinetic energy of a 58-g tennis ball moving at about 150 km/h! The energy spectrum of cosmic rays follows a "power law" $E^{-\alpha}$ where $\alpha \simeq 2.7$ for $10^{10} < E < 3 \times 10^{15}$ eV and, above 3×10^{15} eV, $\alpha \simeq 3$.

At times, some stars throw off their outer sheath, forming a shock wave rapidly moving outward. In 1949 Enrico Fermi suggested that charged particles moving back and forth through a moving and magnetized shock front would statistically gain energy. As a particle gains energy, the shorter is the time that the charged particle stays in the front of the shock wave, and the higher the probability it will escape. In such a case, the charged particles leaving the front would have a power-law distribution in energies.

But where are such cosmic accelerators located? Prime suspects are Active Galactic Nuclei (AGN), which are galaxies that produce more electromagnetic emission than can be deduced from their stellar content, stellar remnants, and their interstellar medium. Such AGNs are assumed to harbor at their centers a supermassive black hole, millions to billions more massive that our sun. Such a black hole accretes matter from the surrounding galaxy, forming an accretion disk and two "jets," each up to several thousand light-years in length, perpendicular to the disk. In these jets it is thought that charged particles can be accelerated to the highest possible energies and, in the process, more radiation is produced than by the entire galaxy.

Another possible source is a supernova remnant. The ejected outer shell of the supernova builds into a spherically shaped nebula (the *supernova remnant*). The inner core of the supernova becomes a black hole, a neutron star, or a white dwarf, depending on the mass of the progenitor star. A shock front is formed when the ejected material flows into the interstellar medium. And in this shock front, Fermi acceleration can occur that produces enormous numbers of relativistic particles.

A neutron star produced by a supernova explosion that spins with a period of less than a few seconds is called a *pulsar*. Because pulsars have very strong corotating magnetic fields ($> 10^{12}$ G), a strong electric field is induced, which in turns accelerates charged particles.

Finally, *microquasars*, are binary systems consisting of a compact neutron star or black hole and a massive companion star. The compact object accretes matter from the star forming an accretion disk that produces a pair of perpendicular relativistic jets in which charged particles are accelerated, as occurs in AGNs.

5.6.1 High Energy Gamma Rays

Another cosmic ray mystery is how cosmic gamma rays with energies above several MeV are produced, the so-called high energy (HE) and very high energy (VHE) gamma rays. The hottest structures in the universe are accretion disks around compact objects. Yet their thermal peak emissions produce x rays with energies up to only a few tens of keVs. Moreover, the energy spectrum of HE and VHE gamma rays is found to follow a power law $E^{-\alpha}$, a distribution not followed by the thermal black body spectrum. Hence, high energy gamma photons cannot be produced by thermal emissions.

The HE and VHE gamma rays undoubtedly arise because of the interaction of high energy charged particles with matter. If a cosmic proton accelerator is surrounded by dense matter, copious numbers of neutral pions π^0 are produced by the highest energy protons. The main decay mode of these pions (98.8%) is into two very energetic gamma photons. Another mechanism involves relativistic electrons. In strong magnetic fields these electrons lose energy through synchrotron and curvature emissions. Alternatively, in inverse Compton scattering, a low energy photon scatters from a high energy electron to produce a high energy photon.

Beginning in the 1960s it was revealed that bursts of gamma rays appeared randomly in the sky. Today it is known that these gamma ray bursts (GRB) occur about once a day lasting from milliseconds to hundreds of seconds and originate far outside our galaxy. The energy release is enormous and the mechanism(s) for the creation of such bursts is not well established. Conjectured possibilities include the merging of two black holes or two neutron stars and the collapse of a hypernova.

PROBLEMS

1. The average mass of potassium in the adult human body is about 140 g of which 0.0117% of the radioisotope ^{40}K. (a) Estimate the average activity (Bq) of ^{40}K in the body. (b) From the data in Ap. C, estimate the energy and number of gamma photons emitted per second from the body.

2. An obsolete unit of activity, although still widely used, is the curie (Ci) which equals 3.7×10^{10} Bq. How many atoms and what mass of the isotope ^{3}H are needed to produce a radioactive source with an activity of 1 Ci? Perform the same calculation for the isotopes ^{60}Co and ^{239}U.

3. Why does coal contain no ^{14}C?

4. How many alpha particles are emitted per second from a 1-gram sample of vanadium? Compare this emission rate to that for one gram of ^{238}U.

5. A 1-g sample of ^{235}U is placed in a thermal flux of 4×10^4 cm^{-2} s^{-1}. How many prompt fission gamma photons are emitted per second from this sample?

6. The global inventory of ^{14}C is about 8.5 EBq. If all that inventory is a result of cosmic ray interactions in the atmosphere, how many kilograms of ^{14}C are produced each year in the atmosphere?

7. Calculate the neutron emission rate from a 20-GBq point source of ^{124}Sb surrounded by 40 g of beryllium. The ^{9}Be(γ, n) cross section is about 1 mb.

8. An isotope that decays by ejecting alpha particles with energies 6.82 MeV (10%) and 4.30 MeV (90%) is mixed intimately with a large amount of beryllium. For an (α, n) source using 40 GBq of this mixture, estimate the neutron emission rate and sketch the energy spectrum of the emitted neutrons.

9. Estimate the rate at which activation product neutrons from the decay of ^{17}N are produced in a water-moderated, uranium-fueled reactor which is operating at a steady thermal power of 3000 MW.

10. Estimate the number of spontaneous-fission neutrons emitted from a 1 kilogram sample of ^{238}U.

11. A thermonuclear device (^{2}H-^{3}H reaction only) is detonated in space with a yield of 1.00 MWd. What is the resulting neutron fluence 1 km from the detonation point?

12. One gram of natural iron is present in a uniform thermal flux density of 10^{12} cm^{-2} s^{-1}. Describe quantitatively the resulting emission of capture gamma rays from the iron.

13. A beam of thermal neutrons (10^{10} neutrons cm^{-2} s^{-1}) irradiates a 2-g sample of calcium for 30 s. What is the flux density (in vacuum) of gamma photons with energies greater than 1 MeV at a point 1 m from the sample 2 min after the irradiation ceases? Consider the decay of both ^{47}Ca and ^{49}Ca.

14. Cadmium is commonly used to attenuate thermal neutrons because of its large (n, γ) cross section. However, capture gamma photons are produced as a result of the neutron capture. As an example, use Table B.4 to sketch the energy-dependent flux density of capture photons above 1 MeV at 10 m (in vacuum) from a point where a thermal neutron beam of 10^8 neutrons s^{-1} is stopped by a cadmium sheet.

15. Electrons accelerated through a potential of 100 kV strike a gold target. Sketch the spectrum of the resulting x rays, first as a function of x-ray energy down to 50 keV, and, then as a function of wavelength. Show both the bremsstrahlung and characteristic x rays. If the electron beam current and voltage were

doubled and the target replaced by an aluminum one, by what percentage would the bremsstrahlung power change?

16. Calculate the flux-density energy spectrum of the bremsstrahlung at 1 m (in vacuum) from a small tungsten target bombarded by 10 mA of electrons accelerated through a constant potential of 500 kV. Neglect x-ray absorption and assume isotropic emission of the x rays.

REFERENCES

Anderson, M.E., and R.A. Neff, "Neutron Energy Spectra of Different Size ^{239}Pu-Be(α,n) Sources," *Nucl. Instrum. Meth.*, **99**, 231–235, (1972).

Blizard, E.P., and L.S. Abbott, Eds., *Reactor Handbook,* Vol. III, Part B: *Shielding,* 2nd Ed., New York: Interscience, 1962.

Browne, E., and R.B. Firestone, *Table of Radioactive Isotopes,* New York: Wiley, 1986.

Dillman,L.T., and F.C. Von der Lage, *Radionuclide Decay Schemes and Nuclear Parameters for Use in Radiation Dose Estimation,* NM/MIRD Pamphlet No. 10, Society of Nuclear Medicine, 1975.

Dillman, L.T., *EDISTR—A Computer Program to Obtain a Nuclear Data Base for Nuclear Dosimetry,* Report ORNL/TM-6689, Oak Ridge, TN: Oak Ridge National Laboratory, 1980.

Eckerman, K.F., R.J. Westfall, J.C. Ryman, and M. Cristy, "Availability of Nuclear Decay Data in Electronic Form, Including Beta Spectra Not Previously Published," *Health Phys.*, **67**, 338–345, (1994). [Available as DLC-172, NUCDECAY Data Library, Radiation Shielding Information Center, Oak Ridge National Laboratory, Oak Ridge, TN.]

Evans, R.D., *The Atomic Nucleus,* New York: McGraw-Hill, 1955; republished by Melbourne, FL: Krieger Publishing Co., 1982.

Faw, R.E., and J.K. Shultis, "Radiation Sources," Ch. 13, in *Encyclopedia of Sustainability Science and Technology,* R.A. Meyers Ed., New York: Springer, ISBN 987-0-387-89469-0, pp 343–387, 2012, .

Fewell, T.R., R.E. Shuping, and K.R. Hawkins, *Handbook of Computed Tomography X-Ray Spectra,* HHS (FDA) 81-8162, Rockville, MD: U.S. Department of Health and Human Services, Food and Drug Administration, 1981.

Firestone, R.B., *Table of Isotopes,* 8th Ed. V.S. Shirley, Ed. New York: Wiley, 1996.

Flügge, S., Ed., *Handbuch der Physik,* Vol. 30, Berlin: Springer-Verlag, 1957.

GE-NE (General Electric Nuclear Energy), *Nuclides and Isotopes: Chart of the Nuclides,* 15th Ed., prepared by J.R. Parrington, H.D. Knox, S.L. Breneman, E.M. Baum, and F. Feiner, San Jose, CA: GE Nuclear Energy, 1996.

GPO, *Radiological Health Handbook,* Washington, DC: U.S. Government Printing Office, 1970.

Hubbell, J.M., "Bibliography and Current Status of K, L and Higher Shell Fluorescence Yields for Computations of Photon Energy-Absorption Coefficients," NISTR 89-4144. Gaithersburg, MD: National Institute of Standards and Technology, 1989.

ICRP, *Radionuclide Transformations: Energy and Intensity of Emissions.* Pub. 38. *Annals of the ICRP,* Vols. 11–13, Elmsford, NY: International Commission on Radiological Protection, Pergamon Press, 1983.

ICRU, *Radiation Dosimetry: X Rays Generated at Potentials of 5 to 150 kV,* Rep. 17, Washington, DC: International Commission on Radiation Units and Measurements, 1970.

Jaeger, R.G., Ed., *Engineering Compendium on Radiation Shielding,* Vol. I: *Shielding Fundamentals and Methods,* New York: Springer-Verlag, 1968.

Kaplan, I., *Nuclear Physics,* 2nd Ed., Reading: Addison-Wesley, 1962.

Keepin, G.R., *Physics of Nuclear Kinetics,* Reading, MA: Addison-Wesley, 1965.

Knoll, G.F., *Radiation Detection and Measurement,* 2nd Ed., New York: Wiley, 1989.

Lorch, E.A., "Neutron Spectra of ^{241}Am/B, ^{241}Am/Be, ^{241}Am/F, ^{242}Cm/Be, ^{238}Pu/^{13}C, and ^{252}Cf Isotopic Neutron Source," *Int. J. Appl. Radiat. Isot.*, **24**, 585–591, (1973).

NCRP, *Radiation Protection Design Guidelines for 0.1–100 MeV Particle Accelerator Facilities.* Rep. 51, Bethesda, MD: National Council on Radiation Protection and Measurements, 1977.

NuDat 2.7, National Nuclear Data Center, Brookhaven Nat. Lab., www.nndc.bnl.gov/ nudat2/, accessed 2019.

Révay, Z., R.B. Firestone, T. Belgya, and G.L. Molnár, "Prompt Gamma-Ray Spectrum Catalogue," in *Handbook of Prompt Gamma Ray Activation Analysis in Neutron Beams,* G. Molnár, Ed. Dordrecht, The Netherlands: Kluwer Acad. Pub., 2004.

Shleien, B., *The Health Physics and Radiological Health Handbook,* Silver Springs, MD: Scinta Inc., 1992.

Shultis, J.K. and R.E. Faw, *Radiation Shielding,* La Grange Park, IL: American Nuclear Society, 2000.

Trubey, D.K., M.J. Berger, and J.H. Hubbell, "Photon Cross Sections for ENDF/B-VI," American Nuclear Society Topical Meeting, *Advances in Nuclear Engineering Computation and Radiation Shielding,* Santa Fe, NM, April 9–13, 1989. [Available as DLC-136, PHOTOX Data Library, Radiation Information and Shielding Center, Oak Ridge National Laboratory, Oak Ridge, TN.]

Walsh, R.L., "Spin-Dependent Calculation of Fission Neutron Spectra and Fission Spectrum Integrals for Six Fissioning Systems," *Nucl. Sci. Eng.*, **102**, 119–113, (1989).

Weber, D.A., K.F. Eckerman, L.T. Dillman, and J.C. Ryman, *MIRD: Radionuclide Data and Decay Schemes,* New York: Society of Nuclear Medicine, 1989.

Wyard, S.J., "Intensity Distribution of Bremsstrahlung from Beta Rays," *Proc. Phys. Soc. London,* **A65**, 377–379, (1952).

Chapter 6

Probability and Statistics for Radiation Counting

To understand God's thoughts we must study statistics,
for these are the measure of His purpose.

Florence Nightingale[1]

6.1 Introduction

The science of probability can be traced to 1654 with a study of gambling, in which Blaise Pascal and Pierre de Fermat were challenged with a die rolling problem. It was Chevalier de Méré who approached Pascal and asked what the remaining chance of winning a game of dice would be between two players at any moment during the progression of the game, with the idea that the two players would split the stakes fairly if they wished to discontinue playing. The correspondence between Pascal and Fermat and their early observations on probability and expected values have developed over time into a valuable science used to interpret data and predict scientific outcomes. Although there are numerous definitions for statistics, an accepted definition is that statistics is the branch of mathematics dealing with the collection, analysis, interpretation, and presentation of numerical data. Presently, the fields of probability and statistics are combined, in which collections of observed data are described by probabilistic models, making possible predictions and categorization of observed phenomena.

The proper analysis of any type of experimental data requires an assessment of the uncertainties associated with each measurement. Without such an estimate, the data have limited value. There are several types of uncertainties or errors associated with any experimental measurement. These errors include stochastic uncertainties, sampling errors and systematic errors. For example, the decay of radioactive atoms occurs randomly or stochastically so that a measurement of the number of decays observed in a fixed time interval has an inherent *stochastic uncertainty.* Repeated measurements would give slightly different results. *Systematic* errors are introduced by some constant bias or error in the measuring system and are often very difficult to assess because they may arise from biases unknown to the experimenter. *Sampling errors* arise from making measurements on a different population from the one desired. These latter two errors are often hard to detect, let alone quantify.

Engineers and scientists must always be aware of the difference between *accuracy* and *precision*, even though popular usage often blurs or ignores the important distinction. Precision refers to the degree of

[1]Lesser known are Florence Nightingale's contributions to the discipline of statistics. She collected much data on hospital conditions and correlated her findings with death rates. She presented these results with her invention of statistical coxcomb graphics, emphasizing the need for cleanliness in hospitals, ultimately leading to reformed hospital practices.

measurement quantification as determined, for example, by the number of significant figures. Accuracy is a measure of how closely the measured value is to the true (and usually unknown) value. A precise measurement may also be very inaccurate. In radiation counting, methods of probability and statistics are used to provide accurate measurements together with an estimate of the precision of the measurement, often termed the error or uncertainty associated with the measurement. In this chapter, the fundamental probability and statistical concepts used for radiation counting are introduced, along with associated probability models and error analyses.

6.1.1 Types of Measurement Uncertainties

Data measured with an ionizing-radiation detection system contain both random uncertainties and systematic errors. Uncertainty assignment requires knowledge of both. Even with a perfect measurement system capable of operating over a long time period without introducing significant systematic error, one must always consider the random nature of the data caused by the stochastic nature of all radiation sources. This chapter deals only with one component of the total uncertainty—the random or statistical uncertainty, which can be estimated through various probability distribution functions (PDFs), most commonly the Gaussian (or normal) distribution.

Radiation counting *is* a stochastic process. Further complications in the data analysis can be introduced if the source emission rate changes in time. For example, the activity of a single radionuclide sample exponentially decays in time or the radiation leaking from a reactor increases in time as the reactor power increases. In this chapter, unless otherwise noted, it is assumed that the strength of any radiation source is constant in time.

6.1.2 Probability and Statistics

The term *probability* refers to the study of randomness and uncertainty in a series of non-deterministic or stochastic measurements, and naturally arises from the study of an expected value of certain possible outcomes as compared to all possible outcomes. One common example would be the toss of a fair coin, where it is assumed that one side or the other (heads or tails) lands horizontally up when the tossed coin comes to rest. Suppose that *heads* is selected as a "success" should it land face up. Because there are only two possible outcomes from a coin toss and only one method by which a heads can be observed, the probability is equal to the one possible method of observing a heads divided by two possible outcomes, or $1/2$. Another example is that of rolling a six-sided die, and each side of the die has a different number ranging from one to six. Suppose that rolling a number *one* is considered a "success". There are six possible outcomes, of which only one outcome results in the observation of the number *one*. The anticipated outcome is divided by the total number of possible outcomes to determine that there is a $1/6$ probability of observing a number *one* after one roll of the die. Chances are that the reader is already familiar with these examples, and the results are intuitive. Yet, what of large sample sets, as expected with the observation of radioactive decay in radioactive materials? Most materials have atomic densities between 10^{21} cm^{-3} and 10^{23} cm^{-3}; hence even tiny radioactive samples have enormous numbers of atoms which may decay within a certain time period (or may not). For radioactive materials, how is the probability of observing an average number of radiation emissions within a set time period predicted, and how is the uncertainty for experimental observations determined and reported? To understand the models and methods used to predict, report, and analyze radiation counting data, it is important to first define certain terms and statistical distributions commonly used in probability theory and statistics.

6.2 Probability and Cumulative Distribution Functions

6.2.1 Continuous Random Variable

The probability distribution function (PDF) describes the expected outcomes for a random variable x over the entire range of possible outcomes. A proper PDF has the following three properties: (1) it is defined on an interval $[a, b]$, where $a < b$, (2) it is non-negative on this interval although it can be zero for some $x \in [a, b]$, and (3) the PDF must be normalized such that $\int_a^b f(x)\,dx = 1$. Here a and b represent real number or infinite limits (i.e., $a \to -\infty$ and/or $b \to \infty$) and the interval can be either closed or open.

A PDF is a *density* function, i.e., it specifies the probability *per unit of* x, so $f(x)$ has units that are the inverse of the units of x. The probability that x falls within the interval $a \le x_1 < x < x_2 \le b$ is

$$\text{Prob}\{x_1 \le x \le x_2\} = \int_{x_1}^{x_2} f(x)\,dx. \tag{6.1}$$

The integral defined by

$$\boxed{F(x) \equiv \int_a^x f(x')\,dx',} \tag{6.2}$$

where $f(x)$ is a PDF over the interval $[a, b]$, is called the *cumulative distribution function* (CDF) of f. Note that, from this definition, a CDF has the following properties: (1) $F(a) = 0$, (2) $F(b) = 1$, and (3) $F(x)$ is monotone increasing, because f is always non-negative.

The CDF is a direct measure of probability. The value $F(x_i)$ represents the probability that a random sample of the stochastic variable x has a value between a and x_i, i.e., $\text{Prob}\{a \le x \le x_i\} = F(x_i)$. More generally,

$$\text{Prob}\{x_1 \le x \le x_2\} = \int_{x_1}^{x_2} f(x)\,dx = F(x_2) - F(x_1). \tag{6.3}$$

6.2.2 Discrete Random Variable

Some PDFs are defined only for discrete values of a random variable x. For example, in rolling a die the number of dots on the upward face has only six possible outcomes, namely $x_1 = 1, x_2 = 2, \ldots, x_6 = 6$. In such cases, $f_i \equiv f(x_i)$ represents the probability of event i, where $i = 1, 2, \ldots, I$. In general, the number of outcomes I can be finite or unbounded. Then for $f_1, f_2, \ldots, f_I$ to be a PDF it is required that (1) $f_i \ge 0$ and (2) $\sum_{i=1}^{I} f_i = 1$. Further, the discrete CDF

$$F_i \equiv \sum_{j=1}^{i} f_j \tag{6.4}$$

can be interpreted as the probability that one of the first i events occurs. This discrete CDF has the properties that (1) $F_1 = f_1$, (2) $F_I = 1$, and (3) $F_i \ge F_j$, if $i \ge j$.

6.3 Mode, Mean and Median

Radiation measurements may consist of a single measurement or a series of measurements of a radioactive source. Consider a set of measurements $x_1, x_2, \ldots, x_N$, where each x_i is an observed value and N represents the number of measurements taken. How are certain properties of this set of data determined, and which properties contain important information to the experimenter? One such property may be a central value of the data set, referred to as a *location index* [Kendall and Stuart 1977]. The location index may be the true value or an experimentally determined value. Further, the location index may be the *mode* (or most probable value), the average (or *mean* value), or it may be the *median* (or middle) value of the data set. An example of these three location indices is shown in Fig. 6.1.

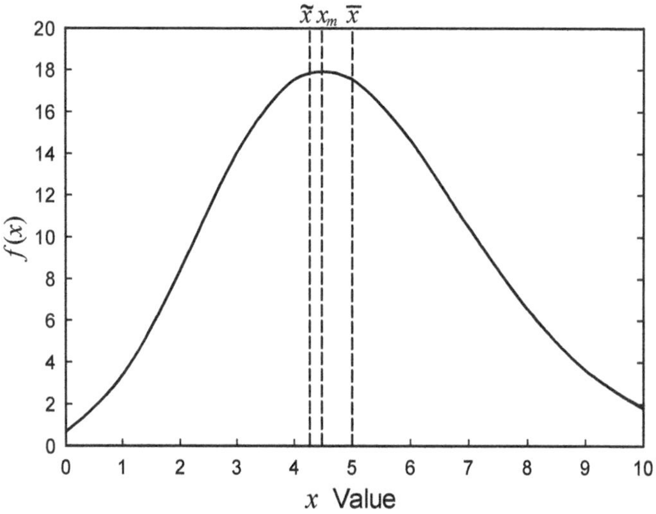

Figure 6.1. Three often quoted location indices of a distribution function, the mode x_m, the median $\tilde{x}$, and the mean $\overline{x}$.

6.3.1 Mode

The mode of a numerical data set is the most probable value, or the most frequently observed value, of that data set. For a continuous function $f(x)$ the mode is the value x_m at which $f(x)$ is maximum and is found from

$$\left.\frac{df(x)}{dx}\right|_{x=x_m} = 0. \tag{6.5}$$

For a discrete data set, the mode can be determined by noting the frequency at which a result is observed, and selecting the most commonly observed result.

Unfortunately, $f(x)$ may actually have several points that yield maxima in the distribution, hence the mode may not be a unique attribute. For such a case, the PDF is said to be multimodal and each maximum can be called a mode. However, in this book, the distribution functions that describe radiation counting always have a single mode.

6.3.2 Mean

The mean of some random variable x, also referred to as the *average* or *expected value* of x, is defined for any continuous function $g(x)$ defined on the interval $[a, b]$ as

$$\mu(x) \equiv \int_a^b xg(x)\,dx \Big/ \int_a^b g(x)\,dx. \tag{6.6}$$

If $g(x)$ is also a PDF, now represented by $f(x)$, then $\int_a^b f(x)\,dx = 1$ and the mean for a continuous PDF is given by

$$\mu(x) = \int_a^b xf(x)\,dx. \tag{6.7}$$

The mean of a PDF is sometimes denoted by $\langle x\rangle$ or by $E(x)$ (for the expected value of x). More generally, the *central moments* of a PDF are defined as

$$\mu(x^n) = \int_{-\infty}^{\infty} x^n f(x)\,dx, \qquad n = 1, 2, \ldots. \tag{6.8}$$

Thus, the mean of a PDF is just the first central moment. The higher central moments give information about the shape of the PDF. In fact, a PDF can be uniquely expressed in terms of its moments so that knowing *all* the moments is tantamount to knowing the PDF itself [Cramèr 1946]. For a *discrete* PDF, the mean is calculated as

$$\mu(x) = \frac{1}{I}\sum_{i=1}^{I} x_i f_i. \tag{6.9}$$

6.3.3 Median

The median and mean of a variable are not the same, yet are often erroneously used synonymously as the average. The word median actually refers to "middle," not average. For a PDF the median $\tilde{x}$ or med(x) is that value of x for which half the area under the distribution lies to the left of $\tilde{x}$ and half to the right of $\tilde{x}$, i.e.,

$$\int_{-\infty}^{\tilde{x}} f(x)\,dx = \int_{\tilde{x}}^{\infty} f(x)\,dx = \frac{1}{2}, \tag{6.10}$$

or, in words, the probability that $x < \tilde{x}$ is exactly equal to the probability that $x > \tilde{x}$.

To find the median of a discrete PDF, the f_i are rearranged in ascending (or descending) order, from smallest to the largest (or largest to smallest). The ordered sequence is denoted by $f_{(1)}, f_{(2)}, \ldots, f_{(I)}$. The median is then defined as

$$\tilde{x} = \begin{cases} \text{middle value of } f_{(i)}, & \text{if } I \text{ is odd} \\ \text{mean value of two middle } f_{(i)}\text{'s}, & \text{if } I \text{ is even} \end{cases}. \tag{6.11}$$

6.4 Variance and Standard Deviation of a PDF

As discussed above, PDFs have three location indices that give a measure of where the central portion of the PDF is located, with the mean value $\overline{x}$ the most commonly used. However, this value gives no information about the spread of the PDF for values above and below $\overline{x}$. Typically, a location index, such as $\overline{x}$, is quoted along with a *dispersion index*, where the dispersion index is a measure of the spread of data relative to the mean. From the definition of the mean one has for any PDF

$$\int_a^b (x - \overline{x}) f(x) dx = 0 \qquad \text{or} \qquad \sum_{i=1}^{I} (f_i - \mu(x)) = 0. \tag{6.12}$$

This result says that the average deviation from the mean is always equal to 0, and does little to assist with a description of the spread of the PDF. Instead, the two most used dispersion indices in radiation counting statistics are the *variance* and the *standard deviation*.

The variance for a continuous PDF $f(x)$ is defined as

$$\sigma^2(x) \equiv \text{var}(x) = \int_a^b (x - \mu(x))^2 f(x) dx, \tag{6.13}$$

where $\sigma(x)$ is the *standard deviation* (or standard error) of the PDF $f(x)$ and $\mu(x)$ is the mean of the PDF given by Eq. (6.7). From Eq. (6.13) the variance is seen to be the average value of $(x - \mu(x))^2$ and the standard deviation is the positive root of that average value. Equation (6.13) can be rearranged as

$$\sigma^2(x) = \int_a^b (x - \mu(x))^2 f(x)\, dx = \int_a^b x^2 f(x) dx - 2\mu(x) \int_a^b x f(x)\, dx + \mu^2(x) \int_a^b f(x)\, dx$$
$$= \mu(x^2) - 2\mu(x)\,\mu(x) + \mu^2(x) = \mu(x^2) - \mu^2(x). \tag{6.14}$$

Thus, the variance is the difference between the second central moment and the square of the first central moment of the PDF. This result is a far more meaningful and useful result compared to Eq. (6.13). The standard deviation $\sigma(x) = \sqrt{\sigma^2(x)}$ relates the spread about the mean value of a distribution function, and hence allows an experimenter to quickly assess the expected error when designing an experiment that closely follows a known PDF. For a *discrete* PDF a similar result is obtained for the variance:

$$\sigma^2(x) \equiv \sum_{i=1}^{I} (x_i - \mu)^2 f_i = \mu(x^2) - \mu^2(x). \tag{6.15}$$

6.5 Probability Data Distribution

Closely related to a discrete PDF is the *data distribution function* $f_{\mathcal{D}}(x_i)$, which is an *experimentally* determined distribution function as distinct from a theoretical model. The data distribution function is defined by a set of observed data. Suppose an outcome x is observed on occasion from a set of N trial experiments. Typically, the number of times n_i the desired outcome x_i is observed is divided by the number of trials conducted N. The only exact method to determine the probability $f(x_i)$ of observing the outcome x_i is to perform an *infinite* number of trials, namely

$$f(x_i) = \lim_{N\to\infty} \frac{n_i}{N}. \tag{6.16}$$

Conducting an infinite number of trials is obviously impractical. In reality, an experimenter must decide upon some practical finite number of N trials that can produce an estimate of $p(x_i)$, within some acceptable margin of error, as

$$f(x_i) \simeq f_{\mathcal{D}}(x_i) = \frac{n_i}{N}. \tag{6.17}$$

In general, accuracy increases for the estimate of $f(x_i)$ given by Eq. (6.17) as the number of trials N increases. However, in some cases, an experimenter may be unable to perform a desired number of trials, because, for example, of time constraints, and consequently must instead rely upon a smaller number of trials. In such a case reporting the uncertainty associated with the estimated value of $f_{\mathcal{D}}(x_i)$ is imperative.

As an example of constructing a data PDF, consider an experiment in which a radiation detector is used to obtain a series $\hat{C}_1, \hat{C}_2, \ldots, \hat{C}_N$ of N measurements or counts (each under identical conditions).[2] In this set of N measurements, the number of resulting counts for some measurements may be equal; others may be unique. The number of times n_i identical counts x_i are observed is tallied. This tallying leads to a tally sequence $n_1, n_2, \ldots, n_M$ of observation of counts $x_1, x_2, \ldots, x_M$, where $M(N) \leq N$. Note $\sum_{i=1}^{M} n_1$ equals the total number N of measurements made. The data distribution function is then determined by dividing each tally by N, i.e.,

[2]Throughout the text, a *measurement* is defined as the process of collecting counting data from a radiation detector, and a *count* is defined as the single observation of a radiation event during a measurement.

$$f_{\mathcal{D}}(x_i) \equiv f_{\mathcal{D}i} = \frac{n_i}{N}, \quad i = 0, 1, 2, \ldots, M. \tag{6.18}$$

Here $f_{\mathcal{D}i}$ is the *estimate* of the true probability $f(x_i)$ of observing x_i. Summation of all these estimates gives

$$\sum_{i=0}^{M} f_{\mathcal{D}}(x_i) = \frac{n_0}{N} + \frac{n_1}{N} + \frac{n_2}{N} + \ldots + \frac{n_M}{N} = 1. \tag{6.19}$$

This result shows that a proper *data* PDF must be normalized to unity.

Example 6.1: Suppose the number of washing loads brought into a laundromat per person between the hours of 1:00 pm and 4:00 pm one day are observed to be 1, 2, 1, 3, 2, 2, 3, 1, 4, 2, 1, 2. What is the data distribution function?

Solution:
From the 12 persons observed, 4 people had 1 load, 5 people had 2 loads, 2 people had 3 loads, and only 1 person had 4 loads. Values for the data distribution function are as follows:

$$f_{\mathcal{D}1} = \frac{n_1}{12} = \frac{4}{12} \qquad f_{\mathcal{D}2} = \frac{n_2}{12} = \frac{5}{12}$$
$$f_{\mathcal{D}3} = \frac{n_3}{12} = \frac{2}{12} \qquad f_{\mathcal{D}4} = \frac{n_4}{12} = \frac{1}{12}$$

and $f_{\mathcal{D}n} = 0$ for $n = 0$ and $n \geq 5$. By summing all possibilities, the data distribution function is shown to be normalized, i.e.,

$$\sum_{i=0}^{\infty} f_{\mathcal{D}}(x_i) = 0 + \frac{4}{12} + \frac{5}{12} + \frac{2}{12} + \frac{1}{12} + 0 + 0 + \ldots = 1. \tag{6.20}$$

6.5.1 Sample Mean

In most experiments, counts are accumulated and an average value is determined from the observations $\widehat{C}_i$. Hence, the data generally are in the form of discrete observations rather than as a continuous function. The *true mean* of an experimentally observed random variable is correctly written as

$$\mu(x) = \sum_{i=1}^{I} x_i f_i = \lim_{N\to\infty} \frac{1}{N} \sum_{i=1}^{N} \widehat{C}_i = \lim_{N\to\infty} \sum_{i=1}^{M(N)} x_i f_{\mathcal{D}i}, \tag{6.21}$$

where as $N \to \infty$ then $M(N) \to I$, the number of possible different observations. The above result indicates that the exact value of an average can be determined only if an infinite number of measurements are recorded. The luxury of conducting infinite measurements for an experiment is rare, hence a more practical method of estimating the mean is to use an arithmetic average. This estimate is called the *sample mean.* The sample mean (or experimental mean) is

$$\boxed{\mu(x) \simeq \overline{x} = \frac{1}{N} \sum_{i=1}^{N} \widehat{C}_i.} \tag{6.22}$$

It should be remembered that the sample mean is only an approximation or estimate of the true mean, As more measurements are made, thereby increasing N, it is expected that the value $\overline{x}$ approaches the true mean $\mu(x)$. Usually, $\overline{x}$ is believed to yield a more accurate result of the expected value if a series of measurements is made rather than any single measurement, all experimental conditions being the same for each measurement. In other words, accuracy increases as data are accumulated. Typically, only one significant digit of precision is added to $\overline{x}$ above that reported for the individual x_i.

The result $\overline{x}$ can be defined in a similar fashion as Eq. (6.9), provided that the data distribution function $f(x_i)$ is known, namely

$$\boxed{\overline{x} = \sum_{i=0}^{N} x_i f_D(x_i)} \quad . \tag{6.23}$$

For instance, a given set of data can be cataloged according to the frequency of observations. The observed frequency becomes a legitimate estimate of the probability distribution for the random variable. This experimentally determined data distribution function changes as more data are accumulated, yet converges on the true probability distribution function as the number of sampled data points approaches infinity.

Example 6.2: Suppose a local farmer sells apples each day at the farmer's market. He sells the following number of apples for each day of the week, starting on Sunday and ending Saturday: 28, 14, 19, 24, 18, 21, 32. What is the sample mean for the week of sales?

Solution:

The sample mean of apples sold per day from these 7 x_i's is found to be

$$\overline{x} = \frac{1}{7}\sum_{i=1}^{7} x_i = \frac{28+14+19+24+18+21+32}{7} = \frac{156}{7} = 22.3.$$

From the limited data set, it is concluded that the farmer sells 22.3 apples per day on average.

Example 6.3: Given two fair dice, what is the probability distribution function for expected numbers resulting from tosses of the die compared to a an experimental distribution function composed of 100 experimental trials?

Solution:

For two dice there are 36 possible number combinations. The probability distribution function representing the summed die values and expected frequencies are shown in Table 6.1 and Fig. 6.2.

Table 6.1. Probabilities associated with throwing two dice.

Value of Thrown x_i	1	2	3	4	5	6	7	8	9	10	11	12
No. of Combinations	0	1	2	3	4	5	6	5	4	3	2	1
$f(x_i)$	0	$\frac{1}{36}$	$\frac{2}{36}$	$\frac{3}{36}$	$\frac{4}{36}$	$\frac{5}{36}$	$\frac{6}{36}$	$\frac{5}{36}$	$\frac{4}{36}$	$\frac{3}{36}$	$\frac{2}{36}$	$\frac{1}{36}$

Because there is no possibility of throwing a 0 or 1, the following is found.

$$\sum_{x=1}^{12} f(x_i) = \frac{0}{36} + \frac{1}{36} + \frac{2}{36} + \frac{3}{36} + \frac{4}{36} + \frac{5}{36} + \frac{6}{36} + \frac{5}{36} + \frac{4}{36} + \frac{3}{36} + \frac{2}{36} + \frac{1}{36} = 1,$$

which is the expected result for a probability distribution function. Because the actual PDF is known, the true mean is calculated to be

$$\mu(x) = \sum_{x=1}^{12} x_i f(x_i) = 1\left(\frac{0}{36}\right) + 2\left(\frac{1}{36}\right) + 3\left(\frac{2}{36}\right) + 4\left(\frac{3}{36}\right) + 5\left(\frac{4}{36}\right) +$$
$$6\left(\frac{5}{36}\right) + 7\left(\frac{6}{36}\right) + 8\left(\frac{5}{36}\right) + 9\left(\frac{4}{36}\right) + 10\left(\frac{3}{36}\right) + 11\left(\frac{2}{36}\right) + 12\left(\frac{1}{36}\right) = 7.$$

That the expectation value is 7 is intuitively correct because the die-tossing PDF is symmetric about the number 7.

By comparison, the data distribution function for an experiment in which 100 tosses of two dice were observed is shown in Table 6.2 and in Fig. 6.2. Equation (6.23) is used to determine the experimental mean

Table 6.2. Observed data for 100 throws of two dice.

Value of Thrown x_i	1	2	3	4	5	6	7	8	9	10	11	12
No. of Combinations	0	1	2	3	4	5	6	5	4	3	2	1
Observed Occurences	0	3	5	11	7	15	16	13	10	10	7	3
$f_\mathcal{D}(x_i)$	.00	.03	.05	.11	.07	.15	.16	.13	.10	.10	.07	.03

of the observed data from the dice tosses, i.e.,

$$\overline{x} = \sum_{x=1}^{12} x_i f_{\mathcal{D}_i} = 1\,(.00) + 2\,(.03) + 3\,(.05) + 4\,(.11) + 5\,(.07) + 6\,(.15)$$
$$7\,(.16) + 8\,(.13) + 9\,(.10) + 10\,(.10) + 11\,(.07) + 12\,(.03) = 7.09$$

Here it is found that the sample mean $\overline{x}$ is close to the true mean $\mu(x)$. Should the experiment be conducted again, the outcome of the experimental mean would once again be near the true mean (although not necessarily the same as the previous result), and the data distribution function would be slightly different. If more data are accumulated in the experiment, thereby increasing the number of trials N, the data distribution function approaches that of the probability distribution function, and the experimental mean approaches the value of the true mean.

6.5.2 Sample Median

Consider the *sample median* $\tilde{x}$ from a set of measurements $x_1, x_2, ..., x_N$, where each x_i is a non-negative number and N represents the number of measurements taken. To find the sample median, the x_i are rearranged in ascending (or descending) order, from smallest to largest (or largest to smallest). The ordered sequence is denoted by $x_{(1)}, x_{(2)}, \ldots, x_{(N)}$. The sample median is defined as

$$\tilde{x} = \begin{cases} \text{middle value of } x_{(i)}, & \text{if } N \text{ is odd} \\ \text{mean value of the two middle } x_{(i)}, & \text{if } N \text{ is even} \end{cases} . \tag{6.24}$$

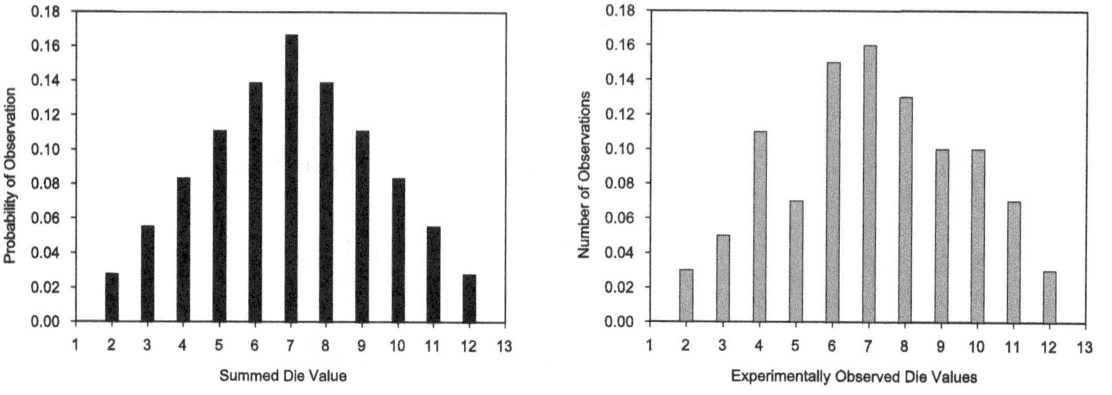

Figure 6.2. The PDFs for the tossing of two dice. Left: theoretical probabilities for obtaining all possible summed values of the dice. Right: experimental probabilities obtained after 100 throws.

Example 6.4: Revisit Example 6.2, in which a farmer sold the following number of apples in consecutive days of a single week: $x_1, x_2, \ldots, x_N =$ 28, 14, 19, 24, 18, 21, 32. Find the experimental median.

Solution:

Reordering the data from lowest to highest values produces the set

$$x_{(1)}, x_{(2)}, \ldots, x_{(N)} = 14, 18, 19, 21, 24, 28, 32.$$

Here $N = 7$ is odd and the sample median (or middle value) is represented by the 4th entry in the ordered list, i.e., $\tilde{x} = x_{(4)} = 21$. Notice the sample median is not equal to the sample mean found in Example 6.2.

Example 6.5: Suppose a well pumping company accumulates data regarding the lifetimes of 1/2 horsepower pumps. Over a period of 12 years, the following failure times, in months, for 20 different pumps were observed,

$$31, 29, 60, 49, 30, 95, 37, 85, 39, 47, 135, 50, 52, 21, 54, 45, 58, 12, 64, 40$$

What are the sample mean and median values for these data?

Solution:

The sample mean is

$$\overline{x} = \frac{1}{20}\sum_{i=1}^{20} x_i = \frac{1}{20}(31 + 60 + 29 + 49 + 30 + 95 + 37 + 85 + 39 + 47 + 135 + 50 + 52 + 21 + 54 + 45 + 58 + 12 + 64 + 40) = \frac{1033}{20} = 51.7.$$

Hence, the average lifetime of the 1/2 horsepower pump is found to be 51.7 months. To find the sample median, the order is arranged from smallest to largest values,

$$12, 21, 29, 30, 31, 37, 39, 40, 45, 47, 49, 50, 52, 54, 58, 60, 64, 85, 95, 135$$

Because $N = 20$ is even, the sample median is the average of the 10th and 11th values in the ordered list,

$$\tilde{x} = \frac{x_{(10)} + x_{(11)}}{2} = \frac{47 + 49}{2} = 48.0.$$

It should be noticed that the sample median is close to, but not equal to, the sample mean in this case.

6.5.3 Trimmed Sample Mean

It is common to encounter experimental data in which some measurements yield values that are quite far from either the sample mean or sample median. Such points are commonly referred to as "outliers," which an analyst may regard as non-representative for the data set. Hence, it is important to notice a major distinction between the sample mean and the sample median. The sample mean takes into account all observed values, whereas the sample median eliminates those values that are not in the middle of the ordered distribution. The sample mean is affected by outlying values, whereas the sample median is not affected by outlying values at all, thereby representing two opposite extremes for interpreting the data distribution.

The sample mean uses all data values, even those from the extreme ends of the distribution, whereas the sample median is insensitive to the extreme values at the two ends of the distribution. One method to eliminate the outliers' influence on the sample mean is to use a *trimmed mean.* As an example, a 5% trimmed mean is computed by eliminating the smallest 5% and the largest 5% of the sample values, and then calculating the sample mean from the remaining data.

There are instances when a standard percentage trimmed mean is prescribed, but the number of measurements does not allow this to be exactly done. For instance, one might have 23 data points, in which case a 10% trimmed mean would need to have 2.3 data points eliminated from the smallest and largest extremes of the ordered distribution. Clearly, this is not possible. Instead, interpolation between the nearest possible points is generally used. For instance, the 10% trimmed mean should be calculated by determining the trimmed mean with only 2 data points removed from upper and lower portions of the distribution, and then repeated with 3 data points removed from upper and lower portions of the distribution. The 10% trimmed mean is determined by appropriate interpolation between the two trimmed means.

Example 6.6: The ordered data in Example 6.5 of the lifetime of pumps, in months, are

$$12, 21, 29, 30, 31, 37, 39, 40, 45, 47, 49, 50, 52, 54, 58, 60, 64, 85, 95, 135$$

It appears that some pumps had very short lifetimes (possibly due to manufacturing defects or ove use) and some had very long lifetimes (possible due to low usage). To avoid these outliers, one should calculate a trimmed mean, for example, a 10% trimmed mean.

Solution:

With this data censoring, the trimmed data are

$$29, 30, 31, 37, 39, 40, 45, 47, 49, 50, 52, 54, 58, 60, 64, 85$$

With only 16 data measurements remaining, the 10% trimmed mean is

$$\overline{x}_{tr(10)} = \frac{1}{16}\sum_{i=1}^{16} x_i = \frac{1}{16}(29 + 30 + 31 + 37 + 39 + 40 + 45 + 47$$

$$+49 + 50 + 52 + 54 + 58 + 60 + 64 + 85) = \frac{770}{16} = 48.1 \text{ months.}$$

Notice that the trimmed mean is now much closer to the uncensored sample median of 48.0 months. This result might indicate to the pump manufacturer that those pumps with lifetimes less than two years were actually defective, and did not represent the true quality of the product.

6.5.4 Sample Variance

Often an experimenter encounters a situation in which data about random variables have been accumulated, yet the true probability distribution function $f(x)$ is unknown. There are several options that might be used to determine an informative measure of the reasonableness of a sample set. The *population* variance and *sample* variance are dispersion indices that deal with the variability of actual data. If the true mean $\mu(x)$ of a distribution is known, then the population variance is denoted $\sigma^2(x)$, and the population standard deviation is $\sigma(x)$, where,

$$\sigma^2(x) = \frac{1}{N}\sum_{i=1}^{N}(x_i - \mu(x))^2. \tag{6.25}$$

However, the actual value of $\mu(x)$ is rarely known. Instead, the experimentally observed average, $\overline{x}$, is usually used as an estimate of $\mu(x)$. Typically, the *sample variance*, denoted s^2, is used to estimate the variance in the data distribution, where

$$\boxed{s^2 = \frac{1}{N-1}\sum_{i=1}^{N}(x_i - \overline{x})^2}\ . \tag{6.26}$$

The sample standard deviation, denoted by s, is the positive square root of the sample variance. That $N-1$ is used in the denominator instead of N, used in the definition of the population variance, is because all of the N x_i are used to calculate $\overline{x}$, so that only $N-1$ of the x_i are independent of $\overline{x}$. In common parlance, "one degree of freedom" is lost by using $\overline{x}$ to approximate $\mu(x)$ [Box et al. 1978]. Another way of understanding this issue is to recall the result of Eq. (6.22). If $N-1$ of the x_i's are known, the final x_i is already determined because the summation of $(x_i - \overline{x})$ must equal zero. Hence, there are only $N-1$ degrees of freedom that can be used to determine the sample variance.

Example 6.7: A series of 30 radiation measurements are performed on a check source, each measurement 6 seconds long, to calibrate a detection system. Given the following counts as recorded for each measurement, determine the sample mean, the sample variance, the sample standard deviation and plot the data distribution function.

Solution:

Measurement data:

$$29, 30, 26, 33, 32, 31, 19, 35, 29, 31, 38, 25, 23, 33, 24,$$
$$28, 37, 26, 29, 41, 30, 27, 23, 34, 40, 27, 21, 30, 29, 31.$$

From Eq. (6.22), the experimental mean is found to be

$$\overline{x} = \frac{1}{N}\sum_{i=1}^{N} x_i = \frac{891}{30} = 29.7.$$

The sample variance is found from Eq. (6.26)

$$s^2 = \frac{1}{N-1}\sum_{i=1}^{N}(x_i - \overline{x})^2 = \frac{1}{30-1}\sum_{i=1}^{30}(x_i - 29.7)^2 = \frac{816.3}{29} = 28.15.$$

The sample standard deviation is

$$s = \sqrt{28.15} = 5.31.$$

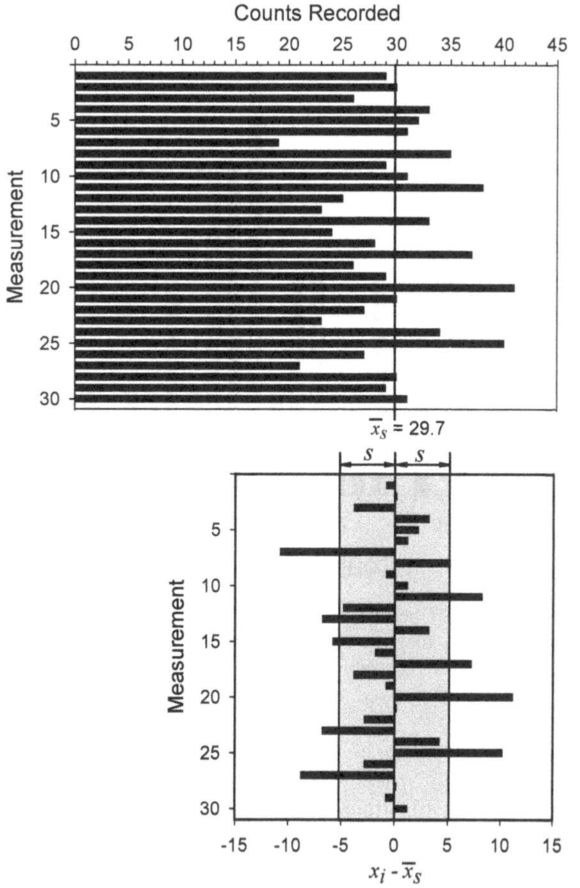

Figure 6.3. The distribution of counts in each of the 30 measurements and the difference, positive and negative, from the mean value ($\overline{x} = 29.7$). Also shown is a comparison with the sample standard deviation ($s = 5.31$).

The sum of the differences, shown in Fig. 6.3, is zero. However, the sample variance and sample standard deviation are non-zero. The sample standard deviation is shown in Fig. 6.3 as compared to the difference $x_i - \overline{x}$ for the 30 measurements.

The data distribution function, $f_{\mathcal{D}}(x_i)$, is determined with Eq. (6.17), where each frequency of observed counts is divided by 30. For instance, four measurements resulted in the observation of $x_i = 29$ counts, yielding

$$f_{\mathcal{D}}(x_i = 29) = \frac{4}{30} = 0.13\overline{3}.$$

The resulting histogram for the 30 measurements is plotted in Fig. 6.4. Finally, it should be noted that as more data are accumulated, the data distribution function begins to take on the appearance of the theoretical PDF, in this case, a Gaussian distribution.

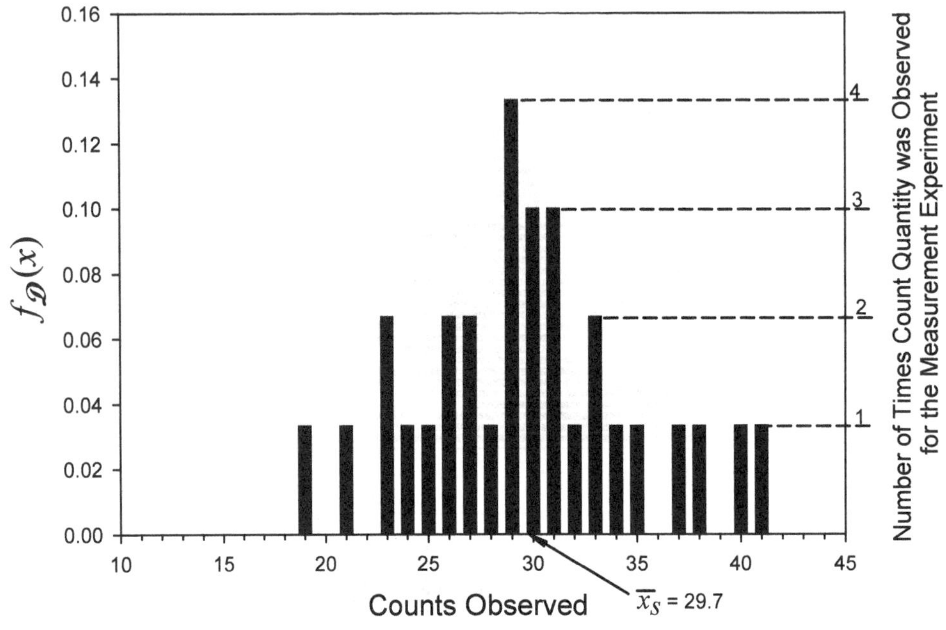

Figure 6.4. The distribution of counts in each of the 30 measurements projected as a data distribution function. The data distribution function describes the experimentally determined probability of observing a randomly chosen number of counts per measurement.

6.6 Binomial Distribution

There are experiments in which only two outcomes are possible, hence the name binomial, which translates as meaning "two terms". Examples of binomial statistical experiments include tossing a coin for "heads" or "tails", picking marbles of two different colors from a box, uncovering shells with or without a prize underneath, and selecting whether or not today is your birthday. If such an experiment, regardless of type, is conducted over N number of trials, then the possibilities of observing either one or the other result can be listed. Typically one of the possible outcomes is listed as a "success" (S) and the other a "failure" (F). For instance, consider a coin toss experiment in which exactly three trials are conducted. In such a case, if the outcome is a "heads" it is listed as a success, and an outcome of "tails" is listed as a failure. For any independent set of 3 trials, there are only 8 possible outcomes, as listed in Table 6.3.

Table 6.3. Possible outcomes of a binomial experiment with $N = 3$.

Outcome	Number of successes	Probability
SSS	3	p^3
SSF	2	$p^2(1-p)$
SFS	2	$p^2(1-p)$
FSS	2	$p^2(1-p)$
SFF	1	$p(1-p)^2$
FFS	1	$p(1-p)^2$
FSF	1	$p(1-p)^2$
FFF	0	$(1-p)^3$

The probability of observing a success is denoted p. Because there are only two possible outcomes from any single trial, the probability of observing a failure must be $(1-p)$. Also listed in Table 6.3 are the probabilities of achieving any of the possible combinations. For instance, the probability of achieving a success 3 times in a row is $ppp = p^3$. The probability of observing an outcome for x number of times from N trials can be determined by dividing the number of times the desired outcome can appear by the total number of possible outcomes. Because only one outcome results in three successes, the probability of throwing three heads in a row out of three trials is $1/2^3 = 0.125$. Unfortunately, for a large number of trials N, listing all possible outcomes becomes an explosively difficult method to use. For example, there are 1024 possible outcome combinations for 10 independent trials.

Fortunately, a Swiss mathematician Jacques Bernoulli (a.k.a. James or Jacob) worked out a general solution to binomial experiments,[3] often referred to as *Bernoulli trials*, hence eliminating the need to list all possible experimental outcomes. Let p denote the probability of observing a success for a particular event. Then the probability that the event is not observed in any trial is $q = 1-p$. In any trial one must either observe the event or not observe the event. Thus, $p+q = 1 = (p+q)^N$. A binomial expansion then gives

$$(p+q)^N = p^N + Np^{(N-1)}q + \frac{N(N-1)}{2!}p^{N-2}q^2 + \frac{N(N-1)(N-2)}{3!}p^{N-3}q^3 + ... + q^N = 1. \quad (6.27)$$

Figure 6.5. Jacques Bernoulli (1654–1705).

Substitution of $(1-p)$ for q yields,

$$(p+q)^N = p^N + Np^{(N-1)}(1-p) + \frac{N(N-1)}{2!}p^{N-2}(1-p)^2 + \frac{N(N-1)(N-2)}{3!}p^{N-3}(1-p)^3 + ... + (1-p)^N = 1, \quad (6.28)$$

which can be rewritten as,

$$(p+q)^N = f(N) + f(N-1) + f(N-2) + ... + f(0) = 1. \quad (6.29)$$

Here the PDF $f(x)$ is the probability of observing the event x (an integer) number of times in N trials. Each individual term in Eqs. (6.28) and (6.29) can be denoted as

$$\boxed{\mathcal{B}(x|p,N) \equiv f_{\mathcal{B}}(x) = \frac{N!}{(N-x)!x!}\, p^x(1-p)^{N-x}, \quad x = 0,1,2,\ldots,N.} \quad (6.30)$$

In this discrete PDF (see Sec. 6.2.2), known as the *binomial* distribution, the random variable x assumes non-negative integer values, i.e., $x_1 = 0, x_2 = 1, \ldots$.

The binomial distribution is the governing distribution whenever the following experimental conditions are fulfilled:

1. The experiment consists of a fixed number of N trials, where N is specified before the experiment begins.

2. Each trial is identical in condition, and each trial can result in one of two possible outcomes. Generally, the outcome is denoted as a success S or a failure F.

[3] Late in the 17th century, Jacques Bernoulli (see Fig. 6.5) derived the binomial PDF and, of even greater importance, the law of large numbers. These results were eventually published posthumously in 1713 by his nephew, Nicolaus Bernoulli, in *Ars Conjectandi* ("The Art of Conjecture").

3. The trials are independent, and the outcome of each trial does not affect the outcomes of other trials.
4. The probability of a success p is the same from trial to trial.

For instance, if a test involves the probability of removing specific balls from a box (for instance, black or white in a mixture of both), for each ball removed from the box the actual probability for the next test has changed because a ball has been removed. For large sample sets, the resulting change in p may be small enough to be of no consequence; however, the experimenter must review the possibility of whether or not replacement does matter. Generally, an experiment can be approximated as being binomial provided that the sample size, N, is less than 5% of the population size. For instance, with a box of 10,000 marbles, extracting 500 or less marbles during the experiment can be approximated as a binomial experiment.[4]

For any given binomial experiment, the average number of successes observed is

$$\mu(x) = \sum_{x=0}^{N} x\mathcal{B}(x|p,N)) = \sum_{x=0}^{N} \frac{xN!}{(N-x)!x!} p^x(1-p)^{N-x}. \tag{6.31}$$

Let $y = x - 1$ and $M = N - 1$; then $x = y + 1$ and $N = M + 1$, and it is easy to show $N - x = M - y$. Thus

$$\begin{aligned}\mu(x) &= \sum_{y=0}^{M} \frac{(y+1)(M+1)!}{(M-y)!(y+1)!} p^{y+1}(1-p)^{M-y} = \sum_{y=0}^{M} \frac{(M+1)(M)!}{y!(M-y)!} p\,p^y(1-p)^{M-y} \\ &= (M+1)p \sum_{y=0}^{M} \frac{(M)!}{y!(M-y)!} p^y(1-p)^{M-y} = Np\sum_{y=0}^{M} f(y|p,M)\end{aligned}$$

Because the summation of a discrete PDF over all possible outcomes is unity, the above simplifies to

$$\mu(x) = Np. \tag{6.32}$$

In a similar manner, the variance of the binomial distribution is found to be

$$\sigma^2(x) = pN(1-p). \tag{6.33}$$

Example 6.8: A simple experiment, with which most readers are familiar, is that of tossing a fair coin. Intuitively it is known that the probability of observing a "head" or a "tail" is 50%. Suppose the observation of a heads is determined to be a success. What is the probability that out of 10 tosses exactly 7 successes are observed? What is the average number of successes expected from 10 coin tosses?

Solution:

For the parameters $p = 0.500$, $x = 7$, and $N = 10$, the probability of throwing exactly 7 heads out of 10 trials is

$$\mathcal{B}(x|0.5,10) = \frac{N!}{(N-x)!x!} p^x(1-p)^{N-x} = \frac{10!}{(10-7)!7!}(0.5)^7(1-0.5)^{10-7} = \frac{10!}{3!7!}(0.5)^7(0.5)^3 = 0.117.$$

---

[4]Note that, while the binomial distribution is a good approximation for large sample populations, its use for small populations without replacement can result in unacceptable error. For such small populations, the *hypergeometric distribution* must be used [Devore 1991]. Because this distribution is rarely used for radiation counting, the hypergeometric distribution is not covered here.

There is, thus, an 11.7% probability that exactly 7 out of 10 tosses result in the observation of heads. The expected number of heads observed from 10 coin tosses is $\mu = pN = (10)(0.5) = 5$. The variance is $\sigma^2 = pN(1-p) = (0.5)(10)(0.5) = 2.50$. Finally, the standard deviation $\sigma = \sqrt{2.50} = 1.58$. Thus, for 10 coin tosses, the expected number of heads that are observed is 5.00 ± 1.58.

The expected number of successes from a binomial distribution for various values of p are shown in Table 6.4 for the condition in which $N = 10$. The binomial distribution becomes symmetric for $p = 0.5$, such as with the coin toss example. The summed probabilities equal unity in each case, meaning that out of 10 trials, at least one of the observations in the column marked x are observed. The condition in which $p = .5$ with $N = 10$, as with the coin toss example, is plotted in Fig. 6.6. For the case of $N = 10$ trials the probabilities of different outcomes for fixed values of p are listed in Table 6.4.

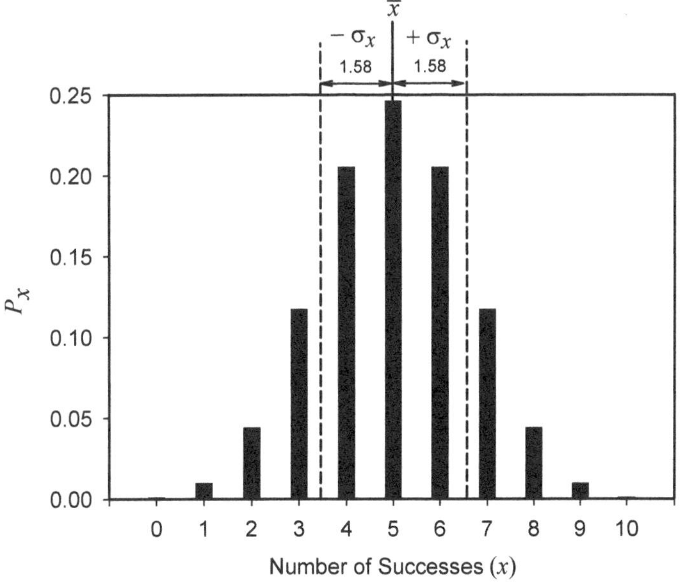

Figure 6.6. The binomial distribution for the coin toss example, in which $N = 10$, showing the probabilities of throwing exact numbers of heads (successes).

6.6.1 Radioactive Decay and the Binomial Distribution

Radionuclides are governed by binomial statistics because a radionuclide either decays in a time interval t or it does not. Consider a radioactive sample initially containing N_o identical radionuclides with a decay constant λ. The probability a radionuclide does not decay in time t is $(1-p) = e^{-\lambda t}$, so the probability it does decay is $p = (1 - e^{-\lambda t})$. From Eq. (6.30) the probability $f_{\mathcal{B}}(x)$ that x atoms decay in time T is

$$f_{\mathcal{B}}(x) = \frac{N_o!}{(N_o - x)!x!}(1 - e^{-\lambda T})^x (e^{-\lambda T})^{N_o - x} \qquad x = 0, 1, 2, \ldots, N_o. \tag{6.34}$$

From Eq. (6.32) the average number decaying in time T is

$$\mu(x) = N_o(1 - e^{-\lambda T}) \tag{6.35}$$

and the variance is

$$\sigma^2(x) = N_o(1 - e^{-\lambda T})e^{-\lambda T} = \mu(x)e^{-\lambda T}. \tag{6.36}$$

Table 6.4. Probabilities for the number of successes for a binomial distribution with $N = 10$ and $p = 0.1, 0.3, 0.5, 0.7$ and 0.9.

	$f_{\mathcal{B}}(x)$				
x	$p = 0.1$	$p = 0.3$	$p = 0.5$	$p = 0.7$	$p = 0.9$
0	0.348678	0.028248	0.000977	5.90×10^{-6}	10^{-10}
1	0.387420	0.121061	0.009766	0.000138	9×10^{-9}
2	0.193710	0.233474	0.043945	0.001447	3.65×10^{-7}
3	0.057396	0.266828	0.117188	0.009002	8.75×10^{-6}
4	0.011160	0.200121	0.205078	0.0036757	0.000138
5	0.001488	0.102919	0.246094	0.102919	0.001488
6	0.000138	0.036757	0.205078	0.200121	0.011160
7	8.75×10^{-6}	0.009002	0.117188	0.266828	0.057396
8	3.65×10^{-7}	0.001447	0.043945	0.233474	0.193710
9	9×10^{-9}	0.000138	0.009766	0.121061	0.387420
10	10^{-10}	5.90×10^{-6}	0.000977	0.028248	0.348678

If $\lambda T \ll 1$, i.e., the measurement time is very much less than the half-life of the radionuclide, $\sigma(x) = \sqrt{\mu(x)}$.

Let c be the probability that a decay produces a count in a radiation detection system. Then the probability a count is observed in time T is $p_c = c(1 - e^{-\lambda T}) \simeq c\lambda T$ if $\lambda T \ll 1$. Thus, the probability of obtaining x counts in time T is

$$f_{\mathcal{B}}(x) = \frac{N_o!}{(N_o - x)!x!}(c\lambda T)^x(1 - c\lambda T)^{N_o - x}, \qquad x = 0, 1, 2, \ldots, N_o. \tag{6.37}$$

Although, this approximation for $f_{\mathcal{B}}(x)$ and the exact result of Eq. (6.34) describe the statistics of radioactive decay, these descriptions have little utility. Because N_o is typically very large (10^5–10^{15}), the terms involving N_o typically are uncalculably huge or ridiculously small. Another difficulty with using the binomial distribution to describe radioactive decays occurs if a radioactive sample has multiple radioactive species. In this case, the probability of obtaining x decays in time T is no longer described by a binomial distribution. However, both of these problems are avoided by using the Poisson distribution discussed next.

6.7 Poisson Distribution

Suppose that an experimenter is observing a radiation source anticipating the emission of a radiation quantum. The source contains m different radioactive species with M_i atoms and a decay constant λ_i for the ith species. Assume there is negligible decay in the observation time T, i.e., $\lambda_i T \ll 1$, so that the M_i do not change appreciably. The observing time T is segmented into N smaller intervals of the same length Δt; hence, $T = N\Delta t$. From Eq. (6.34) the probability that no atoms of species i decays in Δt is

$$f_{\mathcal{B}}(0) = \frac{M_i!}{(M_i - 0)!0!}(1 - e^{-\lambda_i \Delta t})^0(e^{-\lambda_i \Delta t})^{(M_i - 0)} = e^{-\lambda_i M_i \Delta t}. \tag{6.38}$$

The probability none of the radionuclides decay in Δt is, thus, $\prod_{i=1}^{m} \exp(-\lambda_i M_i \Delta t)$ and the probability one (or more) radionuclides decay is

$$p_{\text{decay}} = 1 - \prod_{i=1}^{m} e^{-\lambda_i M_i \Delta t} = 1 - e^{-(\lambda_1 M_1 + \ldots + \lambda_m M_m)\Delta t}. \tag{6.39}$$

As $N \to \infty$, $\Delta t \to 0$ and

$$p_{\text{decay}} \simeq \left(\sum_{i=1}^{m} \lambda_i M_i\right) \Delta t, \tag{6.40}$$

which also vanishes as $N \to \infty$ with the probability of two or more decays vanishing faster than the probability of a single decay in Δt. Now define c_i as the probability of recording a count per decay of radionuclide species i,[5] so the probability of recording a count in a small Δt is

$$p = \left(\sum_{i=1}^{m} c_i \lambda_i M_i\right) \Delta t, \tag{6.41}$$

Again as $N \to \infty$, both $\Delta t \to 0$ and $p \to 0$; but the product Np remains finite, i.e., Eq. (6.41) with $\Delta t = T/N$ yields

$$Np = N\left(\sum_{i=1}^{m} c_i \lambda_i M_i\right) \frac{T}{N} = \left(\sum_{i=1}^{m} c_i \lambda_i M_i\right) T \equiv \mu. \tag{6.42}$$

Thus, in the limit of large N, with $p = \mu/N$ the binomial distribution of Eq. (6.30) becomes

$$f_{\mathcal{B}}(x) = \lim_{N\to\infty} \frac{N!}{(N-x)!x!} \left(\frac{\mu}{N}\right)^x \left(1 - \frac{\mu}{N}\right)^{N-x}, \quad x = 0, 1, 2, ..., N \tag{6.43}$$

When N is large, the evaluation of the binomial distribution of Eq. (6.43) becomes computationally difficult because the various factorials in the binomial coefficient become enormous. But Eq. (6.43) can be considerably simplified as follows:

$$\begin{aligned} \lim_{N\to\infty} f_{\mathcal{B}}(x) &= \lim_{N\to\infty} \frac{N(N-1)\cdots(N-x+1)}{N\ \ (N)\ \ \cdots\ \ (N)} \frac{\mu^x}{x!} \left(1 - \frac{\mu}{N}\right)^N \left(1 - \frac{\mu}{N}\right)^{-x} \\ &= \lim_{N\to\infty} \frac{\left(1 - \frac{1}{N}\right)\cdots\left(1 - \frac{x-1}{N}\right)}{\left(1 - \frac{\mu}{N}\right)^x} \frac{\mu^x}{x!} \left(1 - \frac{\mu}{N}\right)^N. \end{aligned} \tag{6.44}$$

As $N \to \infty$, the terms $(1 - i/N)$ and $(1 - \mu/N)$ approach unity. Also from the properties of the base of the natural logarithms e, it is known that

$$\lim_{N\to\infty} \left(1 - \frac{\mu}{N}\right)^N = e^{-\mu}. \tag{6.45}$$

Hence Eq. (6.44), in the limit as $N \to \infty$, yields

$$\boxed{\mathcal{P}(x|\mu) \equiv f_{\mathcal{P}}(x) = \frac{\mu^x}{x!} e^{-\mu}, \quad x = 0, 1, 2, \ldots}\ . \tag{6.46}$$

[5]Here it assumed that an ideal counting system is being used so that c_i is a constant and unaffected by realities such as dead time.

This PDF with a single parameter μ, is referred to as the Poisson (pronounced /pwä-´sōn/) probability distribution function.[6] Here it is learned that the Poisson distribution can be used provided that the probability of observing any single event is small (by comparison to N), or $p \ll 1$. Unlike the binomial PDF, the Poisson PDF describes the radioactive decay of both a single radioactive species or a sample containing multiple different radionuclides.

Figure 6.7. Siméon-Denis Poisson (1781–1840).

Certain properties of the Poisson distribution should be noted. Because it is a probability distribution function, it must be normalized, i.e.,

$$\sum_{x=0}^{\infty} \frac{(pN)^x}{x!} e^{-pN} = 1. \tag{6.47}$$

For a large number of trials, the average or expected number of successes would be $Np = \mu$. This result can be derived more formally as follows:

$$\langle x \rangle = \sum_{x=0}^{\infty} x \frac{\mu^x}{x!} e^{-\mu} = \sum_{x=1}^{\infty} \frac{\mu^x}{(x-1)!} e^{-\mu} = \mu \sum_{y=0}^{\infty} \frac{\mu^y}{y!} e^{-\mu} = \mu. \tag{6.48}$$

In a similar fashion, the variance of the Poisson distribution is found to be

$$\sigma^2(x) = \mu. \tag{6.49}$$

This is a very important property of Poisson statistics, namely that the variance is equal to its mean, i.e., $\sigma^2 = \mu$. Because the Poisson distribution is the limiting distribution for the binomial distribution, this result is in agreement with the limiting value of the binomial variance for large N, p small, and $Np = \mu$. From Eq. (6.33) one has

$$\sigma_{\mathcal{B}}^2 = Np(1-p) = \mu(1-p) = \lim_{p \to 0} \mu(1-p) = \sigma_{\mathcal{P}}^2. \tag{6.50}$$

Example 6.9: During harvest season, a farmer learns that approximately 1% of the corn in his field has insects that could damage the ears. If he pulls exactly 200 corn ears from his field to inspect them, what is the probability that he finds exactly 3 ears of corn with insects?

Solution:

The PDF governing the number of infected ears found in N tries is described by the binomial PDF because either an ear has insects or it does not. A comparison between the binomial and Poisson methods will be conducted. For the binomial PDF, with $p = 0.01$, $x = 3$, and $N = 200$, the expected probability that three corn ears have insects for a sample set of 200 is

$$f_{\mathcal{B}}(x=3) = \frac{N!}{(N-x)!x!} p^x (1-p)^{N-x} = \frac{200!}{97!\,3!} (0.1)^3 (1-0.1)^{97} = 0.18135$$

Also, the average number of ears that will have insects out of a sample of 200 ears is

$$\mu_{\mathcal{B}} = pN = (0.01)(200) = 2,$$

---

[6]Siméon-Denis Poisson (see Fig. 6.7) published the distribution function, now named after him, in *Recherches sur la probabilité des jugements en matière criminelle et en matière civile (1837)*, an analysis of probability regarding the French criminal justice system, where the Poisson distribution appears for the first and only time in his publications.

with a standard deviation of $\sigma_{\mathcal{B}} = pN(1-p) = \sqrt{(0.01)(200)(1-0.01)} = \pm 1.407$.

The approximation of the binomial PDF by a Poisson PDF yields the expected probability that exactly three ears of corns are infested as

$$f_{\mathcal{P}}(x=3) = \frac{(pN)^x}{x!}e^{-pN} = \frac{[(0.01)(200)]^3}{3!}e^{-(0.01)(200)} = 0.18045.$$

Also, the average number of ears and the standard deviation that have insects out of a sample of 200 ears is

$$\mu_{\mathcal{P}} = pN = (0.01)(200) = 2 \qquad \text{and} \qquad \sigma_{\mathcal{P}} = pN = \sqrt{(0.01)(200)} = \pm 1.414.$$

Notice how well the Poisson result agrees with the true result of the binomial PDF. Fig. 6.8 shows the resulting Poisson distribution for $x = 0, \ldots, 8$.

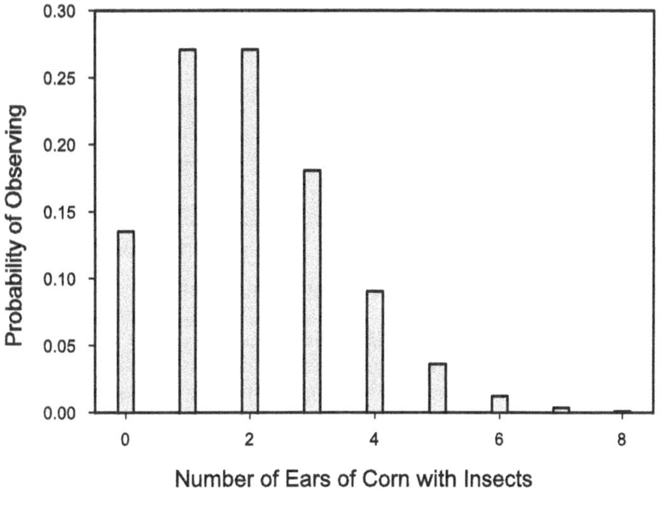

Figure 6.8. The Poisson distribution for Example 6.9, showing the expected probabilities of finding insects in specific numbers of corn ears from a sample set of 200.

Example 6.10: A radioactive sample has a half-life of 10 years. The sample mass has $N = 10^9$ identical radioactive atoms at time $t = 0$. What is the probability that exactly 120 atoms decay within a one-minute observation period of the sample beginning at $t = 0$?

Solution:

Here $\lambda = \ln(2)/10$ yrs $= 1.3188 \times 10^{-7}$ min^{-1}, $x = 120$ and $N = 10^9$. The expected number of decays in one minute is

$$p = \lambda \Delta t = (1.3188 \times 10^{-7} \text{min}^{-1})(1 \text{ min}) = 1.3188 \times 10^{-7}.$$

The average number of decays observed within a one-minute observation period is

$$\mu_{\mathcal{P}} = pN = (1.3188 \times 10^{-7})(10^9) = 131.88 \text{ decays}.$$

Here Poisson statistics must be used because in this case binomial statistics produce fatal overflow problems when attempting to evaluate $f_{\mathcal{B}}(x = 120)$.

$$f_{\mathcal{P}}(x=120) = \frac{(pN)^x}{x!}e^{-pN} = \frac{131.88^{120}}{120!}e^{-131.88} = 0.02096,$$

so there is about a 2% probability that exactly 120 decays are observed in a one-minute period. Notice that the probability distribution shown in Fig. 6.9 is nearly symmetric about the mean value. Indeed, as the probability of observing an event decreases to small values, and the average value for observing the event increases (a large sample set), the Poisson PDF converges on a symmetric distribution about the mean.

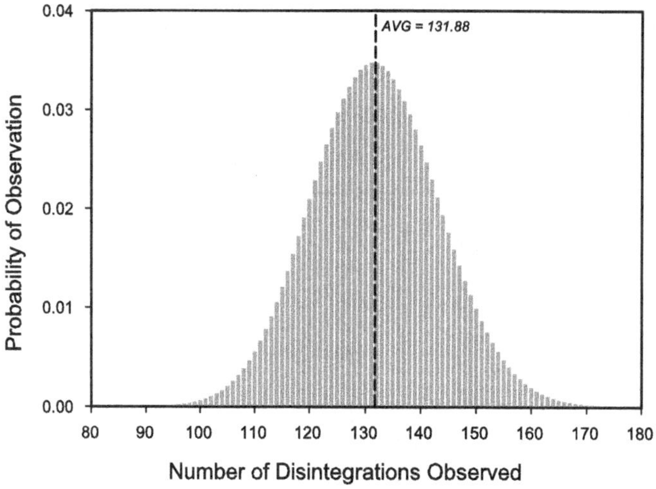

Figure 6.9. The Poisson distribution for Example 6.10, showing the probabilities of observing an exact number of disintegrations per minute from a radioactive sample initially composed of 10^9 radioactive atoms.

6.8 Gaussian or Normal Distribution

For large sample sets and high averages, both the binomial and Poisson distributions are cumbersome to implement numerically. A more manageable probability distribution function is the *Gaussian* PDF, also called the *normal* PDF. With this PDF the probability of observing an event within some dx is[7]

$$\mathcal{N}(x|\mu,\sigma)\,dx = \frac{1}{\sqrt{2\pi}\sigma}\exp\left[-\frac{(x-\mu)^2}{2\sigma^2}\right]dx. \tag{6.51}$$

Here $\mathcal{N}(x|\mu,\sigma) \equiv f_{\mathcal{N}}(x)$ is the PDF that has the following properties: (1) It is properly normalized, i.e.,

$$\int_{-\infty}^{\infty} f_{\mathcal{N}}(x)\,dx = \int_{-\infty}^{\infty}\frac{1}{\sqrt{2\pi}\sigma}\exp\left[-\frac{(x-\mu)^2}{2\sigma^2}\right]dx = 1. \tag{6.52}$$

(2) The average value of x is

$$\int_{-\infty}^{\infty} x f_{\mathcal{N}}(x)\,dx = \int_{-\infty}^{\infty}\frac{x}{\sqrt{2\pi}\sigma}\exp\left[-\frac{(x-\mu)^2}{2\sigma^2}\right]dx = \mu. \tag{6.53}$$

[7]Independently, Adrain in 1808 and Gauss (see Fig. 6.10) a year later developed the formula for the normal distribution and showed that experimental errors were fit well by this distribution. However, almost thirty years earlier this distribution had been discovered by by Laplace in 1778 as he derived the extremely important central limit theorem.

(3) The variance of the Gaussian distribution is

$$\int_{-\infty}^{\infty} (x-\mu)^2 f_{\mathcal{N}}(x)\, dx = \int_{-\infty}^{\infty} \frac{(x-\mu)^2}{\sqrt{2\pi}\sigma} \exp\left[-\frac{(x-\mu)^2}{2\sigma^2}\right] dx = \sigma^2(x) = \text{var}(x). \tag{6.54}$$

Figure 6.10. Carl Friedrich Gauss (1777–1855).

The normal distribution is ubiquitous in nature. The famous "bell-shape" curve keeps popping up everywhere in every discipline. That the normal distribution is so prevalent is explained by the very general and powerful *central limit theorem* (CLT). In words, this theorem states that if N samples z_i are drawn from *any* PDF with mean $\mu(z)$ and variance $\sigma^2(z)$, then the arithmetic average $\overline{z}$ is asymptotically distributed, for large N, as a normal distribution with mean μ and standard deviation $\sigma(z)/\sqrt{N}$. The astonishing feature about the CLT is that nothing is said about the distribution function used to generate the N samples of z, from which the random variable $\overline{z}$ is formed. No matter what the distribution is, provided it has a finite variance, the sample mean $\overline{z}$ has an approximately normal distribution for large samples. The restriction to distributions with finite variance is of little practical consequence because most useful PDFs have a well-defined mean and variance.[8]

In radiation counting, the measured counts are governed by the binomial distribution and well approximated by the Poisson. Because of the CLT, it is not surprising that samples from a Poisson PDF assume a Gaussian distribution for sufficiently large N. A normal distribution in which $\sigma = \mu = \sqrt{pN}$ is, in fact, a very good approximation to the Poison PDF with mean μ, especially for large μ. Shown in Fig. 6.11 is a comparison of the Poisson distribution and the normal distribution for a mean value of 25. Thus, it is seen that the normal distribution is a good approximation for μ as small as 25. Although the Poisson distribution is a discrete PDF and the Gaussian distribution is a continuous PDF, the similarity between the two is unmistakable.

The basic characteristics of the continuous Gaussian distribution are shown in Fig. 6.12, where it has been renormalized by multiplying $f_{\mathcal{N}}(x)$ by $(\sqrt{2\pi})\sigma$. Data from such a distribution are routinely reported as $\overline{x} \pm \sigma$ where $\overline{x}$ is the average number of measured counts during a specified time period and σ is the *standard deviation* of $\overline{x}$. When an error is reported as one standard deviation (or "one sigma") it is called the *standard error*.

Another important feature of the continuous Gaussian distribution is the *full width at half maximum* (FWHM), which is the width of the distribution along the abscissa where it has half the maximum value of the distribution along the ordinate. For instance, the renormalized distribution of Fig. 6.12 has a maximum value of 1, hence the FWHM is the change in x between the points at which the distribution equals 0.5. The relationship between σ and the FWHM is found by evaluating $f_{\mathcal{N}}(x') = 0.5 f_{\mathcal{N}}(\mu)$, where as seen in Fig. 6.12, $|\mu - x'| = 0.5$ FWHM. To find the FWHM write $f_{\mathcal{N}}(x') = 0.5 f_{\mathcal{N}}(\mu)$ explicitly as

$$\frac{1}{\sqrt{2\pi}\sigma} \exp\left[-\frac{(0.5\ \text{FWHM})^2}{2\sigma^2}\right] = \frac{1}{2}\frac{1}{\sqrt{2\pi}\sigma}.$$

[8]One exception is the Cauchy or Lorentzian PDF, which has no mean or variance,

$$f_C(x) = \frac{1}{\pi}\frac{\beta}{(x-a)^2+\beta^2}, \quad -\infty < \alpha < \infty, \quad \beta > 0, \quad -\infty < x < \infty.$$

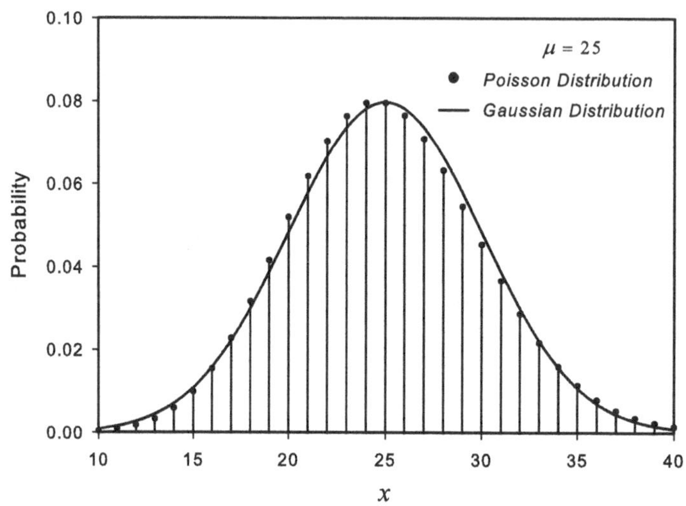

Figure 6.11. A comparison of the Poisson PDF $f_{\mathcal{P}}(x_i)$ and the Gaussian PDF $f_{\mathcal{N}}(x)$ for a mean value of 25. The Poisson PDF is slightly asymmetric, whereas the Gaussian PDF is symmetric about the mean. Yet, the Gaussian distribution is a good approximation for the Poisson PDF, hence is commonly used as a substitute for the Poisson PDF in radiation counting.

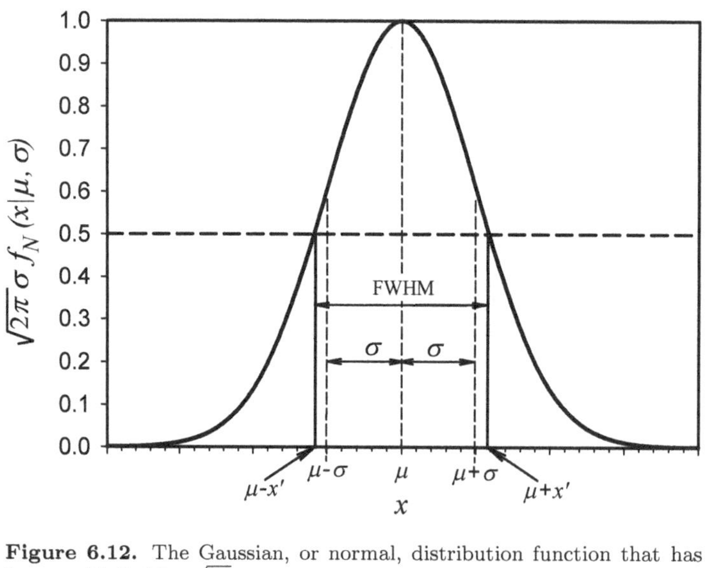

Figure 6.12. The Gaussian, or normal, distribution function that has been multiplied by $\sqrt{2\pi}\sigma$.

Take the log of both sides and solve for FWHM to obtain

$$\boxed{\text{FWHM} = 2\sqrt{2\ln(2)}\sigma \simeq 2.355\sigma.} \tag{6.55}$$

6.8.1 Standard Normal Distribution

The normal distribution of Eq. (6.51) can be changed into a more symmetric one. Let $z = (x - \mu)/\sigma$ so that Eq. (6.51) becomes

$$f_{\mathcal{N}}(z) = \frac{1}{\sqrt{2\pi}} \exp\left[-\frac{z^2}{2}\right] \equiv f_{\mathcal{N}}(z|0,1). \tag{6.56}$$

The mean of this standard normal is

$$\mu = \int_{-\infty}^{\infty} z f_{\mathcal{N}}(z) dz = \int_{-\infty}^{\infty} \frac{z}{\sqrt{2\pi}} \exp\left[-\frac{z^2}{2}\right] dz = 0, \tag{6.57}$$

and the variance is

$$\sigma^2(z) = \int_{-\infty}^{\infty} z^2 f_{\mathcal{N}}(z) dz = \int_{-\infty}^{\infty} \frac{z^2}{\sqrt{2\pi}} \exp\left[-\frac{z^2}{2}\right] dz = 1. \tag{6.58}$$

Here it is found that the average is 0, meaning that the distribution is centered at $z = 0$, and the variance and standard deviations are both equal to unity. This particular form of the Gaussian distribution is referred to as the *standard normal probability distribution* and sometimes is denoted by $f_{\mathcal{N}}(z) \equiv \mathcal{N}(z|0,1)$.

A normal distribution $\mathcal{N}(x|\mu,\sigma)$ with any μ or σ is transformed to the standard normal $\mathcal{N}(z|0,1)$ by letting $z = (x - \mu)/\sigma$. By converting a normal distribution to the standard normal PDF the probability distribution becomes "standardized," the results of which can be found in a number of different mathematical tables for various values of z. Hence, by converting the Gaussian distribution to the standard normal distribution, the probabilities can be easily determined for any values of μ and σ. The standard normal PDF is shown in Fig. 6.13.

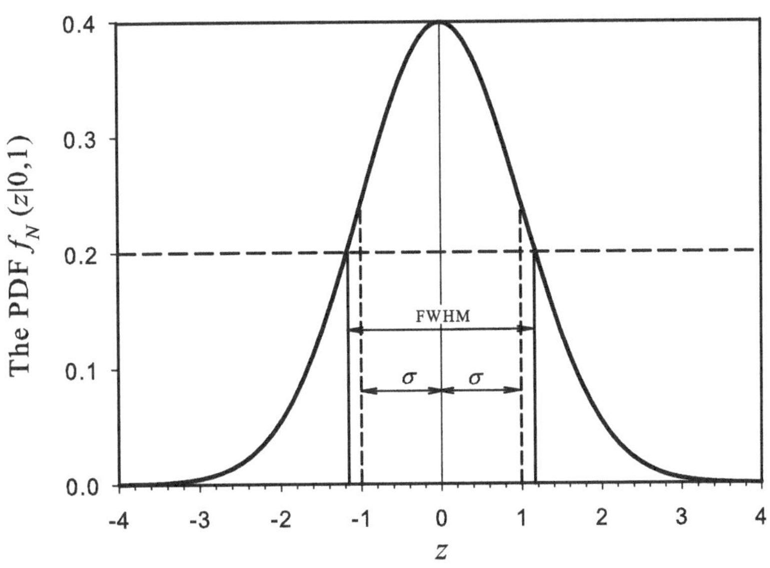

Figure 6.13. The standard normal distribution $f_{\mathcal{N}}(z|0,1)$ is centered at $z = 0$ with both σ^2 and σ equal to 1. The x-axis labels are the number of standard deviations from the mean ($z = 0$).

Example 6.11: A factory produces ball bearings with an average radius of $\mu = 6$ mm and with a standard deviation of $\sigma = 0.04$ mm. The diameters of the ball bearings must not be any less than 5.9 mm nor larger than 6.1 mm in order to fit into the sleeves for which they were designed. Out of a daily manufacture run of 20,000 ball bearings, how many ball bearings must be rejected?

Solution:

Good ball bearings fall between the diameters of 6 mm ± 0.1 mm, hence, the rejected ball bearings have radii in the intervals $(0, 5.9)$ mm and $(6.1, \infty)$. Here it is assumed that a normal PDF describes the distribution of radii. Although negative radii are unphysical, the area under the Gaussian tail of $(-\infty, 0)$ is negligibly small and totally ignorable here. Thus

$$\text{Prob}\{x < 5.9\text{mm}\} + \text{Prob}\{x > 6.1\text{mm}\} =$$

$$\int_{-\infty}^{5.9} \frac{1}{\sqrt{2\pi}} \exp\left[-\frac{(x-6)^2}{2(0.04)^2}\right] dx + \int_{6.1}^{\infty} \frac{1}{\sqrt{2\pi}} \exp\left[-\frac{(x-6)^2}{2(0.04)^2}\right] dx$$

To evaluate these two integrals, convert the normal PDF to the standard normal PDF by substituting

$$z = \frac{x-\mu}{\sigma} = \frac{x-6}{0.04}$$

so that the acceptance limits become

$$z_{low} = \frac{5.9-6}{0.04} = -2.5 \quad \text{and} \quad z_{high} = \frac{6.1-6}{0.04} = 2.5.$$

Because the Gaussian distribution is symmetric about the mean, it is necessary to calculate the probability of finding ball bearings within only one of these intervals, and afterwards simply doubling the answer. Thus, to find probability in the interval $(2.5, \infty)$ integrate the standard normal PDF between $-\infty$ and 2.5 and subtract the results from 1 as shown in Fig. 6.14.

$$\begin{aligned}\text{Prob}\{x < 5.9\text{mm}\} + \text{Prob}\{x > 6.1\text{mm}\} &= 2\left[\int_{2.5}^{\infty} \frac{1}{\sqrt{2\pi}} \exp\left[-\frac{z^2}{2}\right] dz\right] \\ &= 2\left[1 - \int_{-\infty}^{2.5} \frac{1}{\sqrt{2\pi}} \exp\left[-\frac{z^2}{2}\right] dz\right].\end{aligned}$$

The value of the last integral is easily found from a table of the standard normal PDF, where it is found

$$\int_{-\infty}^{2.5} \frac{1}{\sqrt{2\pi}} \exp\left[-\frac{z^2}{2}\right] dz = 0.99379.$$

Hence the probability a ball bearing is rejected is

$$\text{Prob}\{x < 5.9\text{mm}\} + \text{Prob}\{x > 6.1\text{mm}\} = 2\,[1 - 0.99379] = 0.01242.$$

It is, thus, expected that $(20,000)(0.01241) = 248.4$ ball bearings must be rejected each day.

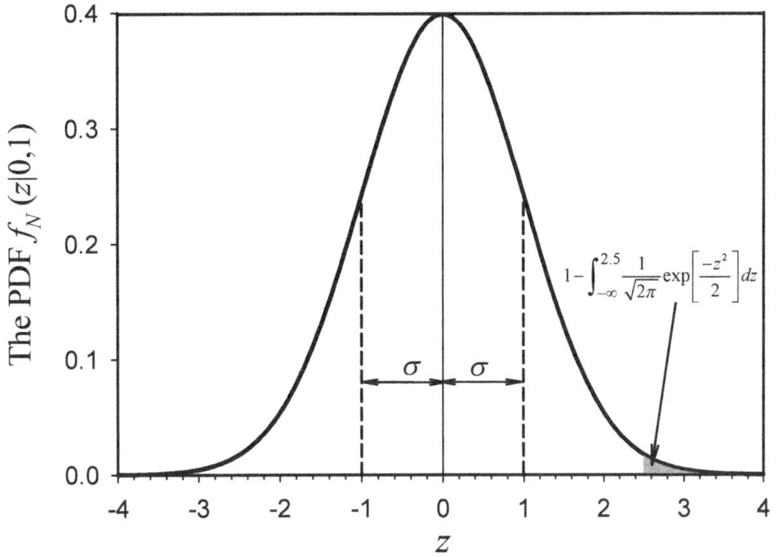

Figure 6.14. The standard normal PDF of Example 6.11, showing the probability region between 2.5 and ∞.

6.8.2 Cumulative Normal Distribution and the Error Function

The cumulative normal distribution has a special importance for a variety of mathematical applications. It is defined as

$$F_{\mathcal{N}}(y) = \int_{-\infty}^{y} \frac{1}{\sqrt{2\pi}\sigma} \exp\left[-\frac{(x-\mu)^2}{2\sigma^2}\right] dx. \tag{6.59}$$

Notice that the $F_{\mathcal{N}}(y)$ is the probability of observing an event between $-\infty$ and some specified value y. This cumulative normal distribution can be converted to the cumulative standard normal distribution by using again the substitution $z = (x-\mu)/\sigma$ to give

$$F_{\mathcal{N}}(u) = \int_{-\infty}^{u} \frac{1}{\sqrt{2\pi}} \exp\left[-\frac{z^2}{2}\right] dz. \tag{6.60}$$

The above integration of the standard normal distribution is shown in Fig. 6.15.

A function closely related to the CDF of the standard normal distribution is the *error function*, defined as

$$\text{erf}(u) = \int_{0}^{u} \frac{2}{\sqrt{\pi}} \exp\left[-z^2\right] dz. \tag{6.61}$$

Because the standard normal PDF is symmetric about the origin, the relationship between the error function and the cumulative standard normal distribution is, for $u > 0$,

$$F_{\mathcal{N}}(u) = \frac{1}{2} + \int_{0}^{u} \frac{1}{\sqrt{2\pi}} \exp\left[-\frac{z^2}{2}\right] dz = \frac{1}{2}\left[1 + \text{erf}\left(\frac{u}{\sqrt{2}}\right)\right]. \tag{6.62}$$

For any positive value of the product $u_\sigma \equiv u\,\sigma$, $\text{erf}(u_\sigma/\sqrt{2}\sigma)$ is the probability that the error of a single measurement lies between $-u_\sigma$ and $+u_\sigma$. Because $\sigma = 1$ for the standard normal distribution, the probability interval between $\pm\sigma$ is $\text{erf}(1/\sqrt{2}) = 0.683$. This is the same probability of observing x from $\mathcal{N}(x|\mu, \sigma)$ in the interval $[\mu - \sigma, \mu + \sigma]$. A comparison of the functions $F_{\mathcal{N}}(u)$ and $\text{erf}(u)$ is shown in Fig. 6.16.

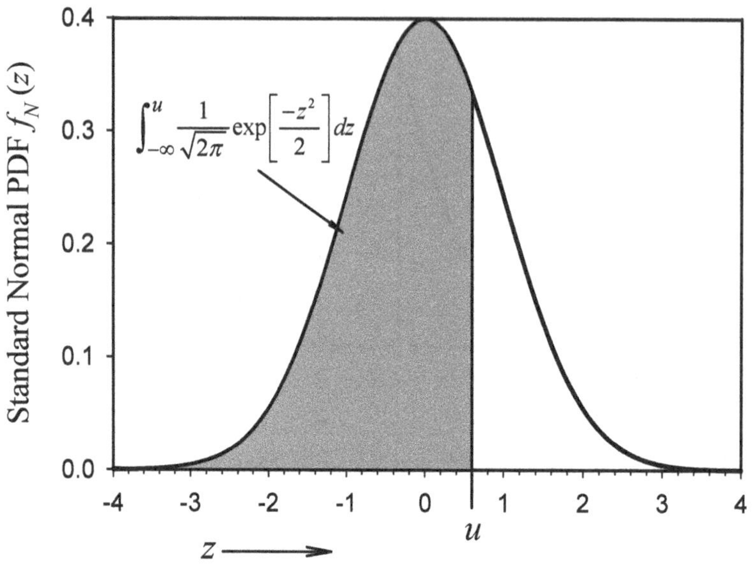

Figure 6.15. The cumulative standard normal distribution showing the probability region up to the value of u (shaded area).

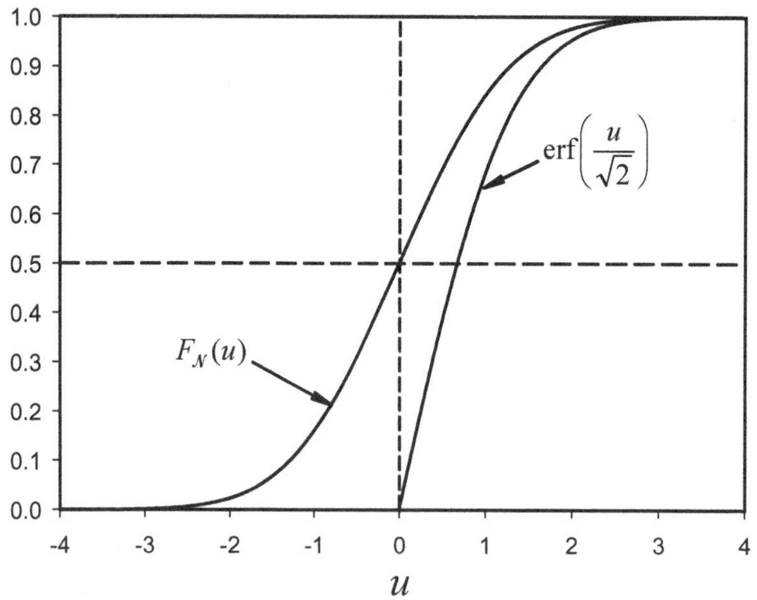

Figure 6.16. A comparison of the cumulative standard normal distribution $F_{\mathcal{N}}(u)$ of Eq. (6.59) and the error function $\text{erf}(u/\sqrt{2})$ given by Eq. (6.61).

Example 6.12: Use the error function to calculate the probability of observing x in the interval $[\mu - m\sigma, \mu + m\sigma]$ for $m = 0.5$, $m = 1.5$, and $m = 2.5$.

Solution:
The probability is

$$\text{Prob}\{\mu - m\sigma \le x \le \mu + m\sigma\} = \text{erf}\left(\frac{\text{u}_\sigma}{\sqrt{2}\sigma}\right) = \text{erf}\left(\frac{\text{u} \cdot \sigma}{\sqrt{2}\sigma}\right) = \text{erf}\left(\frac{\text{u}}{\sqrt{2}}\right).$$

Here u equals 1.5, 2.5 and 2.5. From error function tables,

$$\text{Prob}\{\mu - 0.5\sigma \le x \le \mu + 5\sigma\} = \text{erf}\left(\frac{0.5}{\sqrt{2}}\right) = \text{erf}(0.354) = 0.383$$

$$\text{Prob}\{\mu - 1.5\sigma \le x \le \mu + 1.5\sigma\} = \text{erf}\left(\frac{1.5}{\sqrt{2}}\right) = \text{erf}(1.061) = 0.866$$

$$\text{Prob}\{\mu - 2.5\sigma \le x \le \mu + 2.5\sigma\} = \text{erf}\left(\frac{2.5}{\sqrt{2}}\right) = \text{erf}(1.768) = 0.988.$$

6.8.3 Discrete Gaussian Distribution

The previous sections about the normal PDF have described continuous functions. However, for radiation counting, either the presence of a radiation disintegration (or count) is observed or it is not. In other words, a fraction of an event can never be observed. The Poisson PDF is a discrete distribution that is a better model for describing radiation counting statistics; however, it is numerically difficult to use for large numbers of counts. Thus, one is forced to use the normal PDF. Because only discrete integers of counts can be observed, it, therefore, becomes necessary to use the *discrete* Gaussian distribution to describe sample populations.

The discrete Gaussian distribution is defined, measured and analyzed only at integer values x_i of counts, although the mean μ is not necessarily an integer. This discrete distribution is normalized, i.e.,

$$\sum_{i=1}^{\infty} f_{\mathcal{N}}(x_i) = 1, \tag{6.63}$$

where, because the normal PDF approximating the Poisson has $\sigma^2 = \mu$,

$$f_{\mathcal{N}}(x_i) = \frac{1}{\sqrt{2\pi}\sigma} \exp\left[-\frac{(x_i - \mu)^2}{2\sigma^2}\right] = \frac{1}{\sqrt{2\pi\mu}} \exp\left[-\frac{(x_i - \mu)^2}{2\mu}\right]. \tag{6.64}$$

It is actually the discrete form of the Gaussian distribution that is observed for radiation counting experiments. The observations of counts must be either null or positive discrete numbers, yet the mean, standard deviation, and variance are determined from the accumulated data and may take on any value between 0 and ∞, because they are not necessarily integers.

Suppose an experimenter has a radioactive source used for calibration, such as ^{137}Cs. During a single measurement over some time interval, the experimenter observes 42 counts. Because there is only one measurement, the observed number of counts in the measurement also serves as the mean; hence as a first estimate $\sigma^2 = \mu = 42$. Ignore background counts for the moment, so from this single measurement, the distribution is expected to follow a Gaussian PDF as shown in Fig. 6.17.

The distribution of Fig. 6.17 shows the expected probabilities $f_{\mathcal{N}}(x_i)$ of observing various numbers of counts, based on an average of 42 counts. The summed values of the probabilities $\sum_{i=1}^{\infty} f_{\mathcal{N}}(x_i) = 1$. Should the experimenter take a second identical measurement, according to the results in Fig. 6.17, there is a 6.16% probability of once again observing 42 counts, meaning that there is a 93.84% probability that 42 counts will not be observed. Should the experimenter take another measurement, it would most likely be some other number than 42, and combining the two results would produce a new estimate of the mean $\mu \simeq \overline{x}$.

Suppose the experimenter continues to take measurements under identical conditions, tallying the number of times, or probability, each time a particular number of counts in a measurement is observed. The data average, from Eq. (6.22), would change with each measurement. After numerous measurements (perhaps a hundred or so), the experimenter should notice that the experimentally observed distribution approaches that described by Eq. (6.64) with the observed *new* average $\mu \simeq \overline{x}$, defined by Eq. (6.22), and with $\sigma^2 \simeq s^2$ defined by Eq. (6.26).

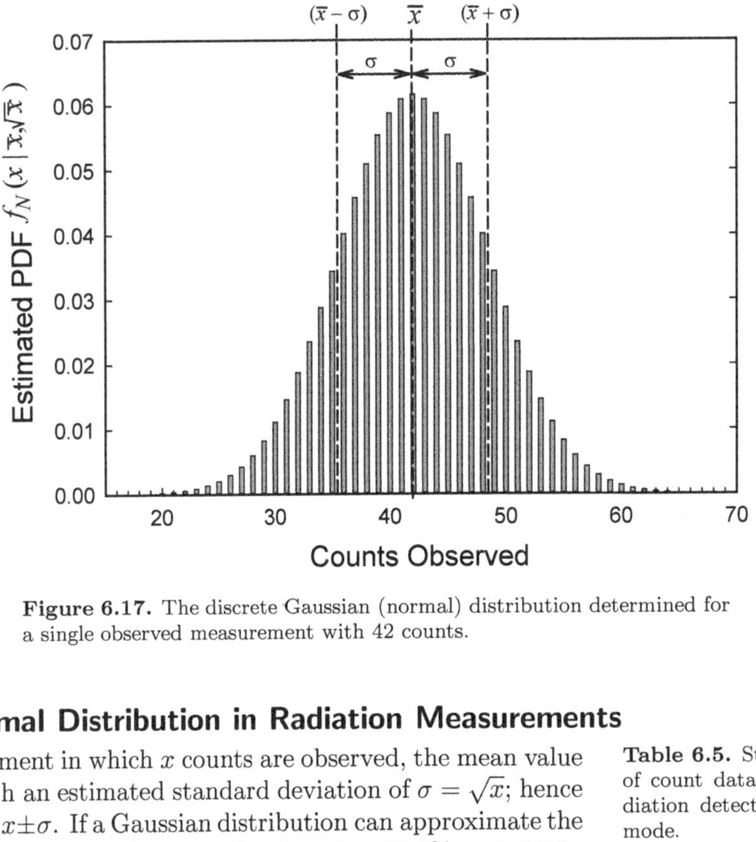

Figure 6.17. The discrete Gaussian (normal) distribution determined for a single observed measurement with 42 counts.

6.8.4 The Normal Distribution in Radiation Measurements

For a single measurement in which x counts are observed, the mean value is estimated as x with an estimated standard deviation of $\sigma = \sqrt{x}$; hence the reported value is $x \pm \sigma$. If a Gaussian distribution can approximate the data, then σ has a specific meaning, namely, there is a 68.3% probability that the next observed measurement has a value that falls within the range $x \pm \sigma$. Table 6.5 lists the relationship between the number of radiation induced counts recorded and the percent standard deviation. Thus to obtain a standard error of 1% or less requires recording at least 10,000 gross counts.

Table 6.5. Standard deviation (%) of count data measured with a radiation detector operating in pulse mode.

Observed Counts	Standard Deviation (1-σ error)
100	10%
400	5%
1100	3%
2500	2%
10000	1%

For a single radiation measurement, the counts observed are a good first estimate of the number of counts that would be observed in a second independent measurement. If the counts observed are greater than approximately 25, a normal PDF can be used to model the expected results. Hence, with x being the gross number of counts observed in the single measurement, the mean and standard deviation are, respectively,

$$\overline{x} = x \quad \text{and} \quad \sigma(\overline{x}) = \sqrt{x}. \tag{6.65}$$

This result would then be reported as "counts $= x \pm \sqrt{x}$." Often it is preferred that the standard deviation σ be reported in terms of a relative to mean (fraction of the mean) or as a percent of the average, respectively given by,

$$\sigma_{\text{rel}} = \frac{\sqrt{x}}{x} = \frac{1}{\sqrt{x}} \quad \text{and} \quad \sigma_{\%} = 100\frac{\sqrt{x}}{x} = \frac{100}{\sqrt{x}}. \tag{6.66}$$

Typically, the measurement is reported as "counts $= x \pm \sigma_{\text{rel}}$" or "counts $= x \pm \sigma_{\%}$."

Example 6.13:

A radiation measurement is conducted in which the gross counts observed are 258. How should the counts be reported in terms of absolute, fractional and percent standard errors?

Solution:

With the above results one has

$$\text{counts} = x \pm \sqrt{x} = 258 \pm \sqrt{258} = 258 \pm 16.1 \qquad \text{absolute.}$$

$$\text{counts} = x \pm \frac{1}{\sqrt{x}} = 258 \pm \frac{1}{\sqrt{258}} = 258 \pm 0.0622 \qquad \text{relative.}$$

$$\text{counts} = x \pm 100\frac{1}{\sqrt{x}} = 258 \pm 100\frac{1}{\sqrt{258}} = 258 \pm 6.22\% \qquad \text{percentage.}$$

Uncertainties are not always reported as one standard deviation. Sometimes a larger uncertainty is reported in order to increase the probability that the mean is included within the range between $x - k\sigma$ and $x + k\sigma$. As shown in Table 6.6, if the error is reported as one standard deviation, the probability is 68.3%. However for an error range of $x - 1.65\sigma$ to $x + 1.65\sigma$ the probability increases to 90%. This new interval would be reported as the 90% *confidence interval.* However, the convention is to report errors as one standard deviation. For anything other than one standard deviation, one should specify the number of standard deviations or the confidence interval.

Table 6.6. Probability intervals relative to the number of standard deviations based upon the Gaussian distribution.

No. of standard deviations $k\sigma$	Probability event observed in $\pm k\sigma$
0.683σ	0.500
1.000σ	0.683
1.645σ	0.900
1.960σ	0.950
2.000σ	0.955
2.576σ	0.990
3.000σ	0.997
3.291σ	0.999
4.000σ	0.99994
5.000σ	0.999999

Lastly, the shape of full-energy peaks observed from gamma-ray *spectra* are typically Gaussian. Figure 6.18 shows a common gamma-ray spectrum from a ^{22}Na source, which emits 511-keV annihilation photons from positron emissions, and 1.28-MeV gamma rays. On comparing Fig. 6.18 to Fig. 6.12, one can observe that the photon energy peaks are Gaussian in shape. The Gaussian shaped peaks arise from statistical fluctuations in the number of free signal carriers (charge carriers) excited per monoenergetic photon interaction event.[9] Typically the resolution of the gamma-ray

[9]The details regarding signal formation in gamma-ray spectrometers are addressed in the chapters on scintillation and semiconductor detectors.

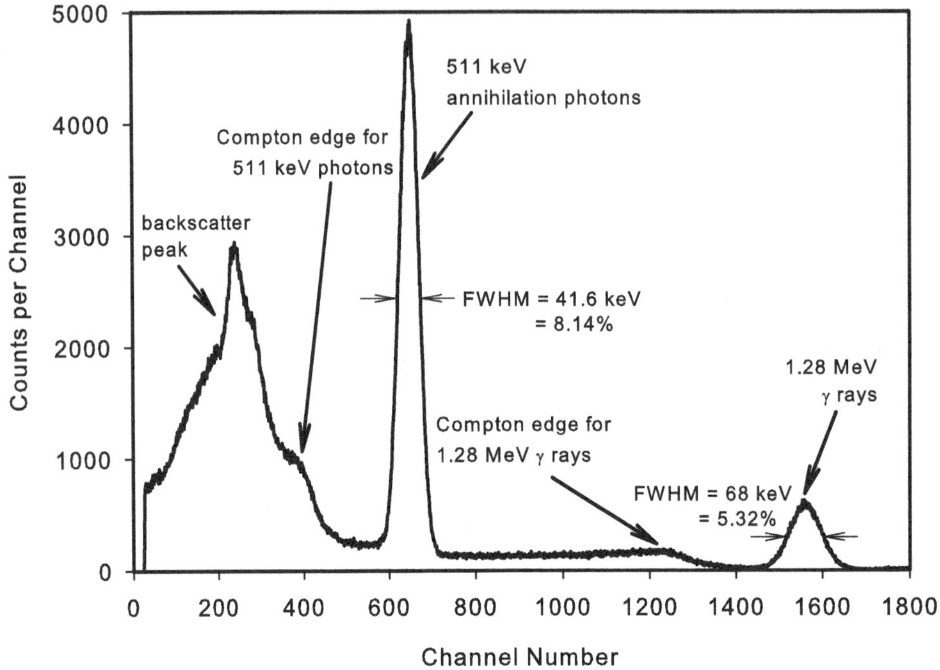

Figure 6.18. Differential pulse height distribution of the gamma rays emitted by the radioactive decay of ^{22}Na as measured by a NaI(Tl) scintillation detector. ^{22}Na is a positron and gamma-ray emitter, hence shown are the 511-keV full-energy peaks from positron annihilation and the 1.28-MeV γ ray emissions. In addition to the two full-energy peaks, several other features are also apparent. The end points of continua for Compton scattered photons are shown, referred to as "Compton edges", one for the 511-keV annihilation photons and one for the 1.28-MeV γ rays. The backscatter peak arises from ^{22}Na emissions scattering in the material surrounding the detector, and then entering and interacting in the NaI detector.

peak from the detector is quoted as a function of the full peak width at half the maximum value (FWHM), which for a Gaussian distribution is 2.355σ, where σ is in units of energy. In terms of percent, the energy resolution is found by dividing 235.5σ by the gamma-ray energy. A detailed discussion on these matters follows in the chapter on spectroscopy.

6.9 Error Propagation

A measurement, generally, may have several sources of error, all of which may be of such importance that they cannot be ignored. It becomes necessary to understand how each random variable affects the measurement error and, moreover, how such measurement error can be calculated from the values of the observed variables [Bevington and Robinson 1992; Taylor 1997]. The usual method to estimate the error of a function of the measured variables is to propagate the errors in each variable.

For a function described by $f(\mathbf{x}) \equiv f(x_1, x_2, x_3, ..., x_N)$, where the $\mathbf{x} = \{x_1, x_2, ..., x_N\}$ are random variables, the variables are usually measured quantities used to determine some average value of the random variable f. For instance, it is common practice to perform a *background* measurement, with no radiation source present, after performing a measurement on a suspected radiation source. The counts observed in the background measurement are subtracted from the counts observed from the radiation source measurement.

However, both measurements have an uncertainty or standard error associated with them. What then is the error for the difference between the counts in these two measurements?

To estimate the standard deviation of f in terms of the standard deviations $\sigma(x_i)$ of the x_i, it is necessary to "propagate" these errors. The usual way to develop such an *error propagation formula* is to begin by expanding $f(\mathbf{x})$ in a Taylor series about the mean values of the x_i, namely about the $\boldsymbol{\mu} = \{\mu_1, \mu_2, \ldots, \mu_N\}$. Such an expansion can be written as[10]

$$f(x_1, x_2, \ldots, x_N) = \sum_{i_1=0}^{\infty} \sum_{i_2=0}^{\infty} \cdots \sum_{i_N=0}^{\infty} \frac{\partial^{i_1}}{\partial x_1^{i_1}} \frac{\partial^{i_2}}{\partial x_2^{i_2}} \cdots \frac{\partial^{i_N}}{\partial x_N^{i_N}} \frac{f(\mathbf{x})}{i_1!, i_2!, \ldots, i_N!}\bigg|_{\mathbf{x}=\boldsymbol{\mu}}$$

$$\times (x_1 - \mu_1)^{i_1} (x_2 - \mu_2)^{i_2} \ldots (x_N - \mu_N)^{i_N}. \tag{6.67}$$

Now assume that higher order terms are negligibly small so that only the first two terms are retained. With this assumption the Taylor series reduces to

$$f(\mathbf{x}) \simeq f(\boldsymbol{\mu}) + \sum_{i=1}^{N} (x_i - \mu_i) \frac{\partial f(\boldsymbol{\mu})}{\partial x_i}, \tag{6.68}$$

where the notation

$$\frac{\partial f(\boldsymbol{\mu})}{\partial x_i} \equiv \frac{\partial f(\mathbf{x})}{\partial x_i}\bigg|_{\mathbf{x}=\boldsymbol{\mu}},$$

is used. Let $a_0 = f(\mu_1, \mu_2, ..., \mu_N)$ and $a_i = \partial f/\partial x_i|_{\mathbf{x}=\boldsymbol{\mu}}$. The average or *expected* value of $f(\mathbf{x})$ is calculated from Eq. (6.68) as

$$E(f) \equiv \langle f \rangle = \left\langle a_0 + \sum_{i=1}^{N} (x_i - \mu_i) a_i \right\rangle = \langle a_0 \rangle + \sum_{i=1}^{N} \langle a_i x_i - a_i \mu_i \rangle$$

$$= a_0 + \sum_{i=1}^{N} \langle a_i x_i - a_i \mu_i \rangle = a_0 + \sum_{i=1}^{N} a_i (\mu_i - \mu_i) = a_0.$$

Thus, it is seen that $\langle f \rangle = f(\boldsymbol{\mu})$.

Now, what of the error? The variance of F is determined as follows:

$$\sigma^2(f) \equiv \langle (f(\mathbf{x}) - f(\boldsymbol{\mu}))^2 \rangle \simeq \left\langle \left\{ \left(a_0 + \sum_{i=1}^{N} (x_i - \mu_i) a_i \right) - a_0 \right\}^2 \right\rangle = \left\langle \left(\sum_{i=1}^{N} (x_i - \mu_i) a_i \right)^2 \right\rangle$$

$$= \left\langle \sum_{i=1}^{N} a_i^2 (x_i - \mu_i)^2 + \sum_{i=1}^{N} \sum_{\substack{j=1 \\ j \neq i}}^{N} a_i a_j (x_i - \mu_i)(x_j - \mu_j) \right\rangle$$

$$= \sum_{i=1}^{N} a_i^2 \langle (x_i - \mu_i)^2 \rangle + 2 \sum_{i=1}^{N} \sum_{\substack{j=1 \\ j < i}}^{N} a_i a_j \langle (x_i - \mu_i)(x_j - \mu_j) \rangle$$

$$= \sum_{i=1}^{N} \left(\frac{\partial f(\boldsymbol{\mu})}{\partial x_i} \right)^2 \sigma_i^2 + 2 \sum_{i=1}^{N} \sum_{\substack{j=1 \\ j < i}}^{N} \frac{\partial f(\boldsymbol{\mu})}{\partial x_i} \frac{\partial f(\boldsymbol{\mu})}{\partial x_j} \sigma_{ij}^2. \tag{6.69}$$

[10]For a Taylor series to exist, it is necessary for f to be continuous and possess derivatives of all orders.

Here σ_{ij} is the covariance defined as

$$\text{cov}(x_i, x_j) \equiv \sigma_{ij}^2 = \langle (x_i - \mu_i)(x_j - \mu_j) \rangle. \tag{6.70}$$

Equivalently, the covariance is given by

$$\text{cov}(x_i, x_j) = \langle x_i x_j \rangle - \mu_i \mu_j. \tag{6.71}$$

Although determination of the covariance is important in cases where the variables are correlated, the result is dependent upon the units of measure. Hence, the same covariance can vary widely by simply changing the units from, for example, mm to cm. To remedy the issue, a scaling factor is used to produce a more meaningful result, referred to as the *correlation coefficient*,

$$\rho_{ij} \equiv \frac{\text{cov}(x_i, x_j)}{\sigma_i \sigma_j}, \tag{6.72}$$

where $\sigma_i \equiv \sigma(x_i)$ and $\sigma_j \equiv \sigma(x_j)$. Substitution of Eq. (6.71) into Eq. (6.72) yields,

$$\rho_{ij} = \frac{\langle x_i x_j \rangle - \mu_i \mu_j}{\sigma_i \sigma_j}. \tag{6.73}$$

and Eq. (6.69) becomes

$$\sigma^2(f) = \sum_{i=1}^{N} \left(\frac{\partial f(\boldsymbol{\mu})}{\partial x_i} \right)^2 \sigma_i^2) + 2 \sum_{i=1}^{N} \sum_{\substack{j=1 \\ j \neq i}}^{N} \frac{\partial f(\boldsymbol{\mu})}{\partial x_i} \frac{\partial f(\boldsymbol{\mu})}{\partial x_j} \rho_{ij} \sigma^2(\hat{x}_i) \sigma^2(\hat{x}_j). \tag{6.74}$$

But how is one to use this formula since the μ_i are generally not known? It is usual to use the measured values $\hat{x}_i$ as an estimate of μ_i. In many situations, such as radiation measurements, it is often assumed, usually justifiably, that the experimentally observed random variables are independent so the covariances are zero. With these two simplifications, Eq. (6.74) reduces to the following error propagation formula:

$$\boxed{\sigma^2(f) = \sum_{i=1}^{N} \left(\frac{\partial f(\mathbf{x})}{\partial x_i} \right)^2 \Bigg|_{x_i = \hat{x}_i} \sigma^2(\hat{x}_i).} \tag{6.75}$$

If the x_i are radiation counts that can be described by Gaussian distributions, then $\sigma^2(x_i) \simeq \sigma^2(\hat{x}_i) = \hat{x}_i$.

The "error propagation formula" of Eq. (6.75) can be used provided that

1. the function can be expressed as a Taylor expansion in which the first two terms adequately approximate the function, and
2. the variables are truly independent of each other so that the covariance between the variables is zero (or, at least, negligibly small).

The second requirement can be usually verified by understanding the details of the physics and the measurement systems used to obtain the measurements $\hat{x}_i$. However, the first requirement is generally more difficult to verify, especially if f is highly non-linear or even if f depends very sensitively on one or more of the $\hat{x}_i$.

There are a few common situations frequently encountered in radiation counting where the error propagation formula is used. These are discussed in the following subsections. For clarity, in the following examples of error propagation a "measurement" refers to a single experiment in which a number of radiation "counts" is observed over a preset period of time.

6.9.1 A Measurement is Scaled by a Constant

There are measurements in which the end result must be scaled by a constant. For instance, consider the case

$$f(x) = Ax,$$

where A is a constant and x is the number of counts in the measurement. With the error propagation formula, Eq. (6.75), and because $\sigma^2(\hat{x}) = \hat{x}$ for a single radiation measurement, one obtains

$$\sigma^2(f) = \sum_{i=1}^{1} \left(\frac{\partial f(x_i)}{\partial \hat{x}_i}\right)^2 \sigma^2(\hat{x}_i) = \left(\frac{\partial (Ax)}{\partial x}\right)^2 \Bigg|_{x=\hat{x}} \sigma^2(\hat{x}) = A^2\sigma^2(\hat{x}). \tag{6.76}$$

Thus, the standard deviation of a random variable multiplied by a constant is

$$\boxed{\sigma(f) = A\sigma(\hat{x}).} \tag{6.77}$$

Hence, it is found that the standard error of the initial measurement is simply scaled by the constant.

Example 6.14: Suppose an experimenter wished to scale the counts obtained in a radiation measurement. A good example is when the counts were taken for a non-standard time period, and the experimenter wishes to scale the counts to a standardized period. For instance, if the measurement lasted 6 minutes and $\hat{x} = 3078$ counts were observed. What number of counts and standard error should the experimenter report if the measurement times had been 60 minutes?

Solution:

The scaling factor A would be 10 in order to change the time from 6 to 60 minutes. Thus the 60-minute measurement would produce

$$f(\hat{x}) = A\,\hat{x} = (10)(3078 \text{ counts}) = 30780 \text{ counts},$$

where $\hat{x}$ is the total gross counts observed in the 6-minute measurement. From the error propagation formula

$$\sigma(f) = A\sigma(\hat{x}) = (10)\sqrt{3078} = 554.8 \text{ counts}.$$

The expected counts and standard error for a 60-minute measurement time are then reported as

$$N_{\text{counts}} = 30780 \text{ counts } \pm \text{ } 554.80 \text{ counts}.$$

Had the experimenter actually counted for one hour and recorded 30780 counts, the standard error would have been $\pm\sqrt{30780} = \pm 175.4$. It is, thus, important to note that the scaled standard error is larger than it would have been had the experimenter actually performed a one-hour measurement. This outcome reinforces the fact that the total number of observed counts determines the statistical error, and not the count rate. Whether the experimenter performed the measurement for one minute or one hour, the average count rate remains the same. Yet, accumulating more observed counts in a measurement reduces the overall error in the measurement.

6.9.2 Random Variables are Added or Subtracted

There are circumstances in which two independent variables must be added or subtracted. Consider a function

$$f(x) = a_1x_1 \pm a_2x_2 \pm ... \pm a_Nx_N,$$

where x_1 through x_N are independent random variables and the a_i are arbitrary constants. With the error propagation formula of Eq. (6.75) and measured values $\hat{x}_i$, the estimate of the variance of f is

$$\sigma^2(f) = \sum_{i=1}^{N}\left(\frac{\partial f(\hat{x}_i)}{\partial \hat{x}_i}\right)^2 \sigma^2(\hat{x}_i) = \sum_{i=1}^{N}\left(\frac{\partial(\pm a_i\hat{x}_i)}{\partial \hat{x}_i}\right)^2 \sigma^2(\hat{x}_i) = \sum_{i=1}^{N}(\pm a_i)^2\sigma^2(\hat{x}_i) = \sum_{i=1}^{N} a_i^2\sigma^2(\hat{x}_i).$$

Therefore, the standard deviation of random variables either added or subtracted is the sum square of the absolute variances, i.e.,

$$\boxed{\sigma^2(f) = \sum_{i=1}^{N} a_i^2\sigma^2(\hat{x}_i).} \tag{6.78}$$

Example 6.15: In most radiation counting environments, there is an ever present background radiation field that can interfere with radiation measurements. To correct for this background, a measurement is first taken with the suspected radiation source present, and then a second measurement is taken with the suspected radiation source removed. If the measurement duration times are the *same* for both measurements, then neither needs be scaled for different measurement times. The results of the second measurement, referred to as a *background* measurement, is subtracted from the first measurement that recorded the summed counts from *both* the suspected radiation source and the background radiation. It is necessary to report the error for the net counts N_{net} observed. Consider the case

$$f(\mathbf{x}) = f(B, G) = G - B,$$

where G is the initial radiation measurement of gross counts and B is the background measurement. The background measurement B is subtracted from the gross measurement G to determine the net measurement of counts from only the radiation source. However, both G and B have error associated with their measurements, and both contribute to the error in N_{net}. Suppose that radiation measurements of a source and background yield $G = 2032$ counts and $B = 256$ counts. What are the net number of counts?

Solution:

The average net counts are $N_{\text{net}} = G - B = 2032$ cts-256 cts $= 1776$ cts. Because $\partial f/\partial G = \partial f/\partial B = 1$, the error propagation formula of Eq. (6.75) yields

$$\sigma(N_{\text{net}}) = \sqrt{\sum_{i=1}^{2}\sigma_i^2} = \sqrt{\sigma^2(G) + \sigma^2(B)} = \sqrt{G + B} = \sqrt{2032 + 256} = 47.83 \text{ cts}.$$

The net counts and standard error are then reported as

$$N_{\text{net}} = 1776 \text{ cts} \pm 47.83 \text{ cts}.$$

6.9.3 Random Variables are Multiplied or Divided

Under some situations, random variables must be either multiplied or divided. First consider the multiplication case, in which

$$f(x) = ax_1x_2 \ldots x_N = a\prod_{i=1}^{N} x_i,$$

where the x_i are independent random variables with measured value $\hat{x}_i$ and a is some constant. Because $\sigma^2(\overline{x}) = \overline{x}$, the error propagation formula, Eq. (6.75), gives

$$\sigma^2(f) = \sum_{i=1}^{N}\left(\frac{\partial f}{\partial x_i}\right)^2 \sigma_i^2 = \left(\frac{\partial}{\partial x_1}(ax_1x_2...x_N)\right)^2 \sigma_1^2 + \ldots + \left(\frac{\partial}{\partial x_N}(ax_1x_2...x_N)\right)^2 \sigma_N^2$$

$$= \left[a\prod_{\substack{i=1\\ i\neq 1}}^{N} x_i\right]^2 \sigma_1^2 + \ldots + \left[a\prod_{\substack{i=1\\ i\neq N}}^{N} x_i\right]^2 \sigma_N^2 = \sum_{j=1}^{N}\left[\left(a\prod_{\substack{i=1\\ i\neq j}}^{N} x_i\right)^2 \sigma_j^2\right].$$

Now divide both sides by $f^2(x)$ to get the sum square of the relative variances,

$$\left(\frac{\sigma(f)}{f(x_1, \ldots, x_N)}\right)^2 = \sum_{i=1}^{N}\left(\frac{\sigma_i^2}{x_i}\right)^2. \tag{6.79}$$

Now consider the division case, in which a product of random variables is divided by another product of random variables,

$$f(x) = \frac{g(x)}{h(x)} \equiv \frac{ax_{g1}x_{g2}\ldots x_{gN}}{x_{h1}x_{h2}\ldots x_{hM}}, \tag{6.80}$$

where x_{gi} and x_{hj} are independent random variables. From Eq. (6.79),

$$\frac{\sigma^2(g)}{g^2(x)} = \sum_{i=1}^{N}\frac{\sigma^2(x_{gi})}{x_{gi}^2} = \sum_{i=1}^{N}\frac{\sigma_{gi}^2}{x_{gi}^2} \quad \text{and} \quad \frac{\sigma^2(h)}{h^2(x)} = \sum_{j=1}^{M}\frac{\sigma^2(x_{hj})}{x_{hj}^2} = \sum_{j=1}^{M}\frac{\sigma_{hj}^2}{x_{hj}^2}. \tag{6.81}$$

Application of the error propagation formula of Eq. (6.75) to Eq. (6.80) yields,

$$\sigma^2(f) = \left(\frac{\partial f}{\partial g(x)}\right)^2 \sigma^2(g) + \left(\frac{\partial f}{\partial h(x)}\right)^2 \sigma^2(h) = \left(\frac{1}{h(x)}\right)^2 \sigma^2(g) + \left(-\frac{g(x)}{h^2(x)}\right)^2 \sigma^2(h). \tag{6.82}$$

Divide both sides by $f^2(x)$ to obtain

$$\frac{\sigma^2(f)}{f^2(x)} = \left(\frac{\sigma(g)}{g(x)}\right)^2 + \left(\frac{\sigma(h)}{h(x)}\right)^2, \tag{6.83}$$

and substitution of the terms from Eqs. (6.81) yields

$$\frac{\sigma^2(f)}{f^2(x)} = \sum_{i=1}^{N}\frac{\sigma_{gi}^2}{x_{gi}^2} + \sum_{j=1}^{M}\frac{\sigma_{hj}^2}{x_{hj}^2}. \tag{6.84}$$

This is essentially the same result as Eq. (6.79). For multiplied or divided random variables, it is thus concluded that the general expression for the variance, based on measured values of the random variable, i.e., $x_i \to \hat{x}_i$, is

$$\boxed{\sigma^2(f) = f^2(\hat{x}_1, \ldots, \hat{x}_N)\left(\sum_{i=1}^{N}\left[\frac{\sigma(\hat{x}_i)}{\hat{x}_i}\right]^2\right)} \quad . \tag{6.85}$$

Example 6.16: Suppose that the activity of a small unknown source is to be determined. One popular method to do this is to compare it to a similar small source of known activity. For example, a calibrated source is measured for a specified time. Afterwards, an uncalibrated source of the same type is placed in the identical location and measured for the same amount of time. A background measurement is then taken for the identical time period. What is the activity and standard error of the uncalibrated source given the following data?

Calibrated source activity (A_c): 0.9 μCi
Calibrated source measurement (G_c): 11280 counts
Unknown source measurement (G_u): 14560 counts
Background measurement (B): 120 counts

Solution:

The average value of the unknown activity is

$$f(x) \equiv A_u = A_c \frac{G_u - B}{G_c - B}$$

where G_u is the uncalibrated source measurement, G_c is the calibrated source measurement, B is the background measurement, A_c is the activity of the known source and A_u is the activity of the unknown source.

$$A_u = A_c \frac{G_u - B}{G_c - B} = 0.9\ \mu\text{Ci}\ \frac{14560 \text{ cts} - 120 \text{ cts}}{11280 \text{ cts} - 120 \text{ cts}} = 1.165\ \mu\text{Ci}.$$

The error propagation formula is used to find the standard error. For a first approximation, assume that the background has a negligible effect on the standard error of A_u. Because the variance of a single measurement is the observed number of counts, the error propagation formula yields

$$\begin{aligned}
\sigma(A_u) &\simeq \left[A_u^2\left(\left[\frac{\sigma(x_u)}{x_u}\right]^2 + \left[\frac{\sigma(x_c)}{x_c}\right]^2\right)\right]^{1/2} = \left[A_u^2\left(\frac{x_u}{x_u^2} + \frac{x_c}{x_c^2}\right)\right]^{1/2} \\
&= \left[A_u^2\left(\frac{1}{x_u} + \frac{1}{x_c}\right)\right]^{1/2} = \left[(1.165\ \mu\text{Ci})^2\left(\frac{1}{14560 \text{ cts}} + \frac{1}{11280 \text{ cts}}\right)\right]^{1/2} \\
&= 0.0146\ \mu\text{Ci}
\end{aligned}$$

The measured activity and standard error are then reported as

$$A_u = 1.165\ \mu\text{Ci} \pm 0.0146\ \mu\text{Ci}.$$

6.9.4 Random Variables in Exponents or Logarithms

The emission rate of a radiative source decreases exponentially, and it is common to encounter situations in which the uncertainty of exponentials and logarithms of random variables must be determined. Consider the case

$$f(x) = e^{ax}, \tag{6.86}$$

where x is a random variable and a is a constant.[11] From the error propagation formula, Eq. (6.75), one has

$$\sigma^2(f) = \sum_{i=1}^{1}\left(\frac{\partial}{\partial \hat{x}_i} f(\hat{x}_i)\right)^2 \sigma^2(\hat{x}_i) = \left(\frac{\partial}{\partial \hat{x}} e^{a\hat{x}}\right)^2 \sigma^2(\hat{x}) = \left(ae^{a\hat{x}}\right)^2 \sigma^2(\hat{x}). \tag{6.87}$$

Therefore, the standard deviation for an exponential function is

$$\boxed{\sigma(f) = a\exp[a\hat{x}]\sigma(\hat{x}).} \tag{6.88}$$

For logarithmic functions,

$$f(x) = \ln(ax),$$

where x is a random variable and a is a constant. With the error propagation formula, Eq. (6.75), one has

$$\sigma^2(f) = \sum_{i=1}^{1}\left(\frac{\partial}{\partial \hat{x}_i} f(\hat{x}_i)\right)^2 \sigma^2(\hat{x}_i) = \left(\frac{\partial}{\partial \hat{x}} \ln(a\hat{x})\right)^2 \sigma^2(\hat{x}) = \hat{x}^{-2}\sigma^2(\hat{x}).$$

Therefore, the standard deviation for a logarithmic function is

$$\boxed{\sigma(f) = \sigma(\hat{x})/\hat{x}.} \tag{6.89}$$

Example 6.17: The total attenuation coefficient can be estimated with a simple, counting experiment. A radiation measurement G_1 is made from a source with no attenuator present, and a second measurement G_2, of equal duration, is repeated with an attenuator of known thickness t placed between the source and detector. Because radiation is exponentially attenuated, the two measurements are related by

$$G_2 = G_1 e^{-xt}$$

where $\overline{x}$ is the attenuation coefficient. Solving for $\overline{x}$, one obtains

$$x = \frac{1}{t}\ln\left(\frac{G_1}{G_2}\right) = \frac{1}{t}\left[\ln(G_1) - \ln(G_2)\right].$$

Suppose the following values are observed: $t = 1.5$ mm, $\sigma(t) = 0.05$ mm, $G_1 = 1500$ cts, and $G_2 = 1235$ cts. Estimate the attenuation coefficient.

[11] Caution should be exercised here, mainly because the exponential function is highly non-linear. As a result, the first two terms of the Taylor expansion may not adequately describe $f(x)$. The reader is advised to refer to Section 6.9.7.

Solution:

The linear attenuation coefficient is thus estimated as

$$\hat{x} = \frac{1}{1.5}\left[\ln(1500) - \ln(1235)\right] = 0.1296 \text{ mm}^{-1}.$$

The variance is found from the error propagation formula as

$$\begin{aligned}\sigma^2(\hat{x}) &= \left(\frac{\partial}{\partial t}\frac{1}{t}\left[\ln(G_1) - \ln(G_2)\right]\right)^2 \sigma^2(t) \\ &\quad + \left(\frac{\partial}{\partial G_1}\frac{1}{t}\left[\ln(G_1) - \ln(G_2 t)\right]\right)^2 \sigma^2(G_1) + \left(\frac{\partial}{\partial G_2}\frac{1}{t}\left[\ln(G_1) - \ln(G_2)\right]\right)^2 \sigma^2(G_2) \\ &= \left(-\frac{1}{t^2}\left[\ln(G_1) - \ln(G_2)\right]\right)^2 \sigma^2(t) + \left(\frac{1}{G_1 t}\right)^2 \sigma^2(G_1) + \left(-\frac{1}{G_2 t}\right)^2 \sigma^2(G_2) \\ &= \left(-(1.5)^{-2}\left[\ln(1500) - \ln(1235)\right]\right)^2 (0.05)^2 + \left(\frac{1}{(1500)(1.5)}\right)^2 (1500) + \left(-\frac{1}{(1235)(1.5)}\right)^2 (1235) \\ &= 6.75 \times 10^{-4} \text{ mm}^{-2}.\end{aligned}$$

The standard error for the measured linear attenuation coefficient is $\sigma_{\hat{x}} = \sqrt{6.75 \times 10^{-4}} = 2.6 \times 10^{-2}$ mm^{-1}. Hence, linear attenuation coefficient is reported to be $x = 0.1296 \pm 0.026$ mm^{-1}.

6.9.5 A Series of Radiation Measurements

Quite often a series of radiation counts is conducted rather than a single measurement. For instance, the variance about the average in a series might reveal the "goodness" of the data.[12] Here it is assumed that a consecutive series of counts $\hat{x}_i$ is taken in which the conditions for each measurement are identical (counting time, source position, etc.). What then is the total standard error in the sum of all the measurements? The sum of the independent random variables x_i, itself another random variable, is

$$f(x_1, \ldots, x_N) \equiv x_{\text{sum}} = x_1 + x_2 + \ldots + x_N = \sum_{i=1}^{N} x_i$$

With the error propagation formula of Eq. (6.75) and the measured values $\hat{x}_i$, the variance of the sum is

$$\sigma^2_{\text{sum}} = \sum_{i=1}^{N}\left(\frac{\partial}{\partial \hat{x}_i} x_i\right)^2 \sigma^2(\hat{x}_i) = (1)^2\sigma^2(\hat{x}_1) + (1)^2\sigma^2(\hat{x}_2) + \ldots + (1)^2\sigma^2(\hat{x}_N) = \hat{x}_1 + \hat{x}_2 + \ldots + \hat{x}_N = \hat{x}_{\text{sum}}.$$

This is an important result because it shows that the variance of the sum is dependent upon only the total number of radiation counts observed and not the number of measurements taken! Hence, the total value of x_{sum} determines the error, whether it is from a single measurement or a series of many measurements. In other words, there is no appreciable difference in the error observed from a single long measurement accumulated over some time t, or a series of N measurements in which each individual measurement is taken for a time t/N.

---

[12] The goodness of data refers to the correlation between the expected data fluctuations and the observed data fluctuations. This concept is covered in Section 6.10.

What then is the error in the average of the N counts? Again, it is assumed that the conditions for each measurement are identical. The average of the counts is

$$\overline{x} = \frac{1}{N}(x_1 + x_2 + ... + x_N) = \sum_{i=1}^{N} \frac{x_i}{N}.$$

Again use the error propagation formula to determine the variance of this average,

$$\sigma^2(\overline{x}) = \sum_{i=1}^{N} \left(\frac{\partial}{\partial \hat{x}_i} \frac{x_i}{N}\right)^2 \sigma^2(\hat{x}_i) = \frac{1}{N^2} \sum_{i=1}^{N} \sigma^2(\hat{x}_i) = \frac{1}{N^2} \sum_{i=1}^{N} \hat{x}_i = \frac{\overline{x}}{N}.$$

Thus, the standard deviation of an average of a set of measurements is

$$\boxed{\sigma(\overline{x}) = \sqrt{\frac{\overline{x}}{N}} = \frac{1}{N}\left[\sum_{i=1}^{N} \hat{x}_i\right]^{1/2}.} \tag{6.90}$$

When $N = 1$, Eq. (6.90) reduces to Eq. (6.65). For N replicate measurements, the error in the average observed count is reduced by $\sqrt{N}$. This result assumes, of course, that the measurements are independent and that no changes occur in the radiation field or with the detector's response during the time that the N data points are recorded. If another series of N measurements were to be taken, a new average value for $\overline{x}$ would be found; yet there would be a 68.3% chance that the new average would fall within the interval ranging from $\overline{x} - \sigma_{\overline{x}}$ and $\overline{x} + \sigma_{\overline{x}}$ from the first series of measurements. The average number of counts observed per measurement and standard error for N measurements is

$$\overline{x} \pm \sigma(\overline{x}) = \overline{x} \pm \sqrt{\frac{\overline{x}}{N}}. \tag{6.91}$$

Example 6.18: An experimenter makes five identical and independent radiation measurements, each two minutes, and records the counts 1207, 1167, 1235, 1193, and 1243. What is the average expected number of counts per minute and the standard error?

Solution:

The average counts observed per measurement is

$$\overline{G} = \sum_{i=1}^{N} \frac{G_i}{N} = \frac{1}{5}(1207 + 1167 + 1235 + 1193 + 1243) = 1209 \text{ counts}.$$

From Eq. (6.90),

$$\sigma(\overline{x}) = \sqrt{\frac{\overline{G}}{N}} = \sqrt{\frac{1209}{5}} = 15.55 \text{ counts}.$$

Division by 2 yields the average expected counts per minute and error as 604.5 ± 7.77 cpm.

6.9.6 Measurements Over Different Time Intervals

Statistical error in radiation counting is determined by the total number of counts observed during a measurement or series of measurements. However, there may be cases in which the measurements are not all taken over the same period of time. What then is the error? It is incorrect to simply use the error of a *count rate*. Instead, the experimenter must first determine the error of the total counts observed and convert that result into a count rate. To illustrate this, consider an experimenter who takes four different measurements over four different measurement times producing the results shown in Table 6.7. All four measurements represent the same count rate, but the error in each is tremendously different.

Table 6.7. Comparison of errors for measurements with different measurement times but with the same count rate.

Observed counts G	Count time t	Standard error σ	Standard error %	Count rate g
100	10 s	± 10	10%	10 cps
1000	100 s	± 31.6	3.16%	10 cps
10000	1000 s	± 100	1%	10 cps
100000	10000 s	± 316	0.316%	10 cps

The count rate for the ith measurement is given by $g_i = G_i/t_i$. The average count rate for N independent measurements of the same radiation source is

$$\overline{f}(\mathbf{x}) \equiv \overline{g}(G_1, \ldots, G_n, t_1, \ldots, t_n) = \sum_{i=1}^{N} G_i \Big/ \sum_{i=1}^{N} t_i. \tag{6.92}$$

To determine the error in the average count rate, write each count rate measurement in terms of total counts observed per unit time. Assume that the error in the measurement time is negligible so that $\sigma_{t_i} \simeq 0$. Then with the error propagation formula, the variance in the count rate is[13]

$$\sigma_{\overline{g}}^2 = \sum_{j=1}^{N} \left(\frac{\partial}{\partial G_i} \sum_{i=1}^{N} G_i \Big/ \sum_{i=1}^{N} t_i \right)^2 \sigma^2(G_j) = \sum_{i=1}^{N} \sigma^2(G_i) \Big/ \left(\sum_{i=1}^{N} t_i \right)^2 = \sum_{i=1}^{N} G_i \Big/ \left(\sum_{i=1}^{N} t_i \right)^2. \tag{6.93}$$

Now suppose that $f(\mathbf{x})$ involves M different radiation source configurations and that multiple measurements are made for each source configuration. Further, suppose $f(\mathbf{x})$ takes sums and differences of measurements for the different source configurations. For example, measurements are often taken with a radioisotopic source present and with the source removed to obtain background counts. Then the background count rate is subtracted from the source count rate to obtain the *net* or background-corrected source count rate. In general one seeks the average

$$\overline{f}(\mathbf{x}) = \sum_{m=1}^{M} (\pm)\overline{g}_m(G_{m1}, \ldots, G_{mN_m}, t_{m1}, \ldots, t_{mN_m}), = \sum_{m=1}^{M} \left[(\pm) \sum_{i=1}^{N_m} G_{mi} \Big/ \sum_{i=1}^{N_m} t_{mi} \right]. \tag{6.94}$$

Here N_m is the number of different measurements taken with the mth source configuration and G_{mi} and t_{mi} are the counts and count time for the ith measurement. From the error propagation formula, the variance of this average is found to be

$$\sigma_{\overline{f}}^2 = \sum_{m=1}^{M} \left[\sum_{i=1}^{N_m} G_{mi} \Big/ \left(\sum_{i=1}^{N_m} t_{mi} \right)^2 \right], \tag{6.95}$$

from which the standard deviation $\sigma_{\overline{f}}$ is readily found.

[13] Alternatively, if N is greater than 10 or so, the sample variance $s^2(\overline{g}) = \frac{1}{\sqrt{N-1}} \sum_{i=1}^{N} (g_i - \overline{g})^2$ can be used to estimate $\sigma_{\overline{g}}^2$.

Example 6.19: A scientist takes a 15-minute background measurement, resulting in $B_1 = 241$. The scientist then takes two radiation measurements of the sample, each over 10 minute periods, and observes $G_1 = 1993$ and $G_2 = 1957$. During a third measurement, he is called to meeting, and stops the third measurement at 6 minutes, recording $G_3 = 1165$ counts. Before leaving, he removes the source and starts a background measurement. When the scientist returns, the background measurement has continued for 67 minutes, yielding $B_2 = 1043$ counts. What is the average net count rate and standard error?

Solution:
The net count rate is found by subtracting the background count rate from the gross count rate. From Eq. (6.94), the net average count rate is

$$\overline{n} = \left[\sum_{i=1}^{N_G} G_i \Big/ \sum_{i=1}^{N_G} t_{Gi}\right] - \left[\sum_{i=1}^{N_B} B_i \Big/ \sum_{i=1}^{N_B} t_{Bi}\right], \quad \text{where } N_G = 3, N_B = 2,$$

$$= \left(\frac{1993 \text{ cts} + 1957 \text{ cts} + 1165 \text{ cts}}{10 \text{ min} + 10 \text{ min} + 6 \text{ min}}\right) - \left(\frac{241 \text{ cts} + 1043 \text{ cts}}{15 \text{ min} + 67 \text{ min}}\right) = 180.1 \text{ cpm}.$$

From Eq. (6.95), the variance is

$$\sigma_{\overline{n}}^2 = \left[\sum_{i=1}^{N_G} G_i \Big/ \left(\sum_{i=1}^{N_G} t_{Gi}\right)^2\right] + \left[\sum_{i=1}^{N_B} B_i \Big/ \left(\sum_{i=1}^{N_B} t_{Bi}\right)^2\right], \quad \text{where } N_G = 3, N_B = 2,$$

$$= \frac{1993 \text{ cts} + 1957 \text{ cts} + 1165 \text{ cts}}{(10 \text{ min} + 10 \text{ min} + 6 \text{ min})^2} + \frac{241 \text{ cts} + 1043 \text{ cts}}{(15 \text{ min} + 67 \text{ min})^2} = 7.758 \text{ min}^{-2}$$

The standard error is found to be $\sigma_{\overline{m}} = \sqrt{7.758} = 2.785 \text{ min}^{-1}$. The net count rate and error are then reported as 180.1 ± 2.79 cpm.

Equal Counting Times

If the count times are identical for the N_m measurements in each source configuration, i.e., $t_{m1} = t_{m2} = \ldots = t_{mN_m} \equiv t_m$, then Eq. (6.92) reduces to,

$$\overline{n} = \sum_{m=1}^{M}\left((\pm)\frac{1}{N_m}\sum_{i=1}^{N_m}\frac{G_{mi}}{t_m}\right) = \sum_{m=1}^{M}\left((\pm)\frac{1}{N_m}\sum_{i=1}^{N_m} g_{mi}\right) = \sum_{m=1}^{M}\left((\pm)\frac{1}{N_m t_m}\sum_{i=1}^{N_m} G_{mi}\right). \tag{6.96}$$

and the variance reduces to

$$\sigma_{\overline{n}}^2 = \sum_{m=1}^{M}\left[\sum_{i=1}^{N_m} G_{mi} \Big/ \left(\sum_{i=1}^{N_m} t_m\right)^2\right] = \sum_{m=1}^{M}\left[\sum_{i=1}^{N_m} G_{mi} \Big/ (N_m t_m)^2\right] = \sum_{m=1}^{M}\left[\frac{1}{N_m^2 t_m}\sum_{i=1}^{N_m} g_{mi}\right]. \tag{6.97}$$

Example 6.20: A researcher takes a single radiation measurement on a sample in a low background counting environment and observes $G = 2310$ counts in a one-minute counting period. Afterwards, the researcher conducts a single one-hour long background measurement and observes $B = 213$ counts. What is the net count rate and standard error?

Solution:
The net count rate is found by subtracting the background count rate from the gross count rate. From Eq. (6.96), the net average count rate is,

$$\overline{n} = \frac{1}{N_G}\sum_{i=1}^{N_G}\frac{G_i}{t_G} - \frac{1}{N_B}\sum_{i=1}^{N_B}\frac{B_i}{t_B}, \text{ where } N_G = 1, N_B = 1,$$

$$= \frac{G}{t_G} - \frac{B}{t_B} = \left(\frac{2310 \text{ cts}}{1 \text{ min}} - \frac{213 \text{ cts}}{60 \text{ min}}\right) = 2306.15 \text{ cpm}.$$

From Eq. (6.97), the standard error is

$$\sigma_{\overline{n}} = \left(\frac{G}{t_G^2} + \frac{B}{t_B^2}\right)^{1/2} = \left(\frac{g}{t_g} + \frac{b}{t_b}\right)^{1/2} = \left(\frac{2310 \text{ cts}}{(1 \text{ min})^2} + \frac{213 \text{ cts}}{(60 \text{ min})^2}\right)^{1/2} = 48.06 \text{ cpm}.$$

The count rate and error are reported to be 2306.15 ± 48.06 cpm.

Example 6.21: Before measuring a radioactive sample, a technician takes a 30-minute background measurement, resulting in $B_1 = 513$. The technician then takes four radiation measurements of the sample, each over 5-minute measurement periods, and observes $G_1 = 2310, G_2 = 2219, G_3 = 2387$, and $G_4 = 2264$ counts. Afterwards, the technician conducts another 30-minute long background measurement and observes $B_2 = 478$ counts. What is the average net count rate and its standard error?

Solution:
The net count rate is found by subtracting the background count rate from the gross count rate. From Eq. (6.96), net average count rate is (with $N_G = 4$ and $N_B = 2$)

$$\overline{n} = \frac{1}{N_G}\sum_{i=1}^{N_G}\frac{G_i}{t_G} - \frac{1}{N_B}\sum_{i=1}^{N_B}\frac{B_i}{t_B}, = \frac{1}{4}\left(\frac{9180 \text{ cts}}{5 \text{ min}}\right) - \frac{1}{2}\left(\frac{991 \text{ cts}}{30 \text{ min}}\right) = 442.48 \text{ cpm}.$$

From Eq. (6.97), the standard error is

$$\sigma_n = \left\{\frac{1}{N_G^2}\sum_{i=1}^{N_G}\frac{G_i}{t_G^2} + \frac{1}{N_B^2}\sum_{i=1}^{N_B}\frac{B_i}{t_B^2}\right\}^{1/2}, \text{ where } N_G = 4, N_B = 2,$$

$$= \left\{\frac{1}{16}\left(\frac{9180 \text{ cts}}{25 \text{ min}^2}\right) + \frac{1}{4}\left(\frac{991 \text{ cts}}{900 \text{ min}^2}\right)\right\}^{1/2} = 4.82 \text{ cpm}.$$

The count rate and error are reported to be 442.48 ± 4.82 cpm.

6.9.7 Caution about Error Propagation

The error propagation formula of Eq. (6.75) is based on the assumption that the first two terms in the Taylor series expansion for a function $f(x_1, x_2, ..., x_N)$, evaluated at measured values $\hat{x}_1, \hat{x}_2, \ldots, \hat{x}_N$, adequately describes its variation with the x_i. The first two terms of the Taylor series expansion describe a linear variation with the x_i, and for radiation counting the function is evaluated about the mean of the data distribution. Hence, for highly non-linear functions, such as high order polynomials and exponential functions, Eq. (6.75) may not provide an adequate description. This discrepancy between the actual function and an estimate from the truncated Taylor series expansion is especially of concern for arguments that are far from the mean values $(\overline{x}_i)$.

Further, data are assumed to be independent and uncorrelated in Eq. (6.75). But there are situations where this assumption is simply incorrect. For example, often a series of data is analyzed with linear regression to determine a specific characteristic (or characteristics) of the underlying distribution. Examples include the determination of a radioactive decay constant or a photon attenuation coefficient of a material. From the *least squares* fit of the data, a slope and intercept are determined, and the data values actually *are* correlated when determining these parameters. Under such situations, the covariance of the data may be an important component of the standard error, hence it is advisable to actually use Eq. (6.74) in such cases.

6.10 Data Interpretation

In any statistical analysis where prediction or interpretation is drawn from a probability distribution function lies the possibility of misinterpreting the data. For instance, in radiation counting, there is a possibility of declaring a sample radioactive when in fact it is not, or the possibility of measuring a radioactive source, but declaring it non-radioactive. Such concerns are especially important if the count rate from the suspected radioactive source is comparable to that from background radiation. Consider a sample suspected of being radioactive, with the hypothesis that the sample is not radioactive. This is termed the *null hypothesis*, denoted by H_0. The other possibility is, of course, that the sample actually is radioactive, is termed the *alternative hypothesis*, denoted H_a. If an experimenter measures the activity of the sample, then two following possible errors in the interpretation of the data can occur [Devore 1991].

Type I error consists of rejecting the null hypothesis in favor of the alternative hypothesis H_a when in fact H_0 is true, meaning in this case declaring the sample radioactive when in fact it is not, termed a *false positive.*

Type II error consists of accepting the null hypothesis H_0 when it is incorrect, meaning the sample is declared non-radioactive when it actually is radioactive, termed a *false negative.*

Such errors are difficult to identify, yet the probability of making either a Type I or Type II error can be determined. The method used for determining these probabilities is referred to as "hypothesis testing."[14] The test procedure is usually defined by some measure called a *test statistic* or a *rejection region.* The test statistic defines a quantity, whether it be the average value or some other statistic, with which the observed data are to be interpreted. The rejection region is a range of values in which a new hypothesis is accepted over the previous hypothesis. For radiation counting, one might use the average counts observed from a calibration with the background radiation as a test statistic and the region greater than $\overline{x} + 2\sigma$ as the rejection region. Hence, samples producing counts greater than $\overline{x} + 2\sigma$ are declared radioactive, and samples producing counts below $\overline{x}+2\sigma$ are deemed non-radioactive. Or, it may be hypothesized that a single measurement is a good first estimate of the average number of counts expected per measurement, and the rejection region is defined as some function of the standard deviation.

[14] Although hypothesis testing is often used in probability and statistics, it is widely considered to be a branch of *decision theory.*

Figure 6.19 shows the standard normal distribution in which the regions outside of $z \pm 1.96\sigma$ are designated rejection regions. The remaining values of z are in the acceptance region. Given any hypothesis about the distribution of Fig. 6.19, for instance the mean value, a measurement falling outside of $\overline{z} \pm 1.96\sigma$ implies that the hypothesis be rejected. The form shown in Fig. 6.19 is referred to as a *two-tailed test* because regions on both sides of the mean value are designated as rejection regions. There are some circumstances in which extreme values on only one side of the distribution is of interest. Such a test is referred to as a *one-tailed test.*

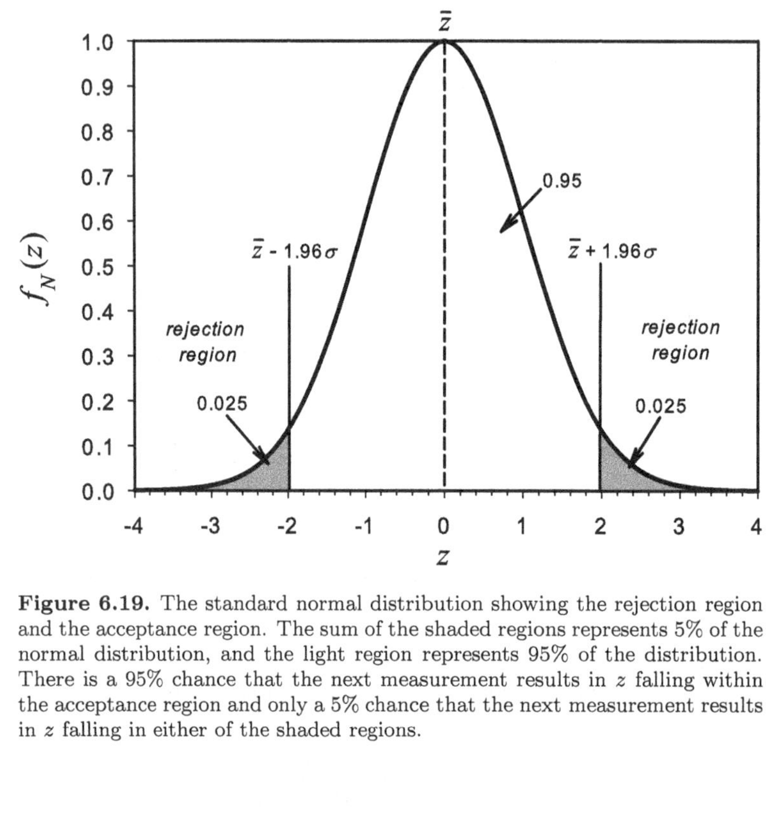

Figure 6.19. The standard normal distribution showing the rejection region and the acceptance region. The sum of the shaded regions represents 5% of the normal distribution, and the light region represents 95% of the distribution. There is a 95% chance that the next measurement results in z falling within the acceptance region and only a 5% chance that the next measurement results in z falling in either of the shaded regions.

Example 6.22: Suppose a single radiation measurement is performed in which $\hat{x} = 194$ counts are observed. Such counts follow a Gaussian distribution. The hypothesis H_0 is declared that 194 is the mean value with the standard deviation $\sigma = \sqrt{\hat{x}} = 13.93$. Hence, the mean value is the test statistic. It is also asserted that the hypothesis is rejected should the next independent measurement produce a count outside of $\pm 1.96\sigma$ (the rejection region—see the shaded area in Fig. 6.20). The rejection region is 5% of the area under the Gaussian PDF, whereas the acceptance region is 95% of the area.

Should the next measurement produce a count that falls in the shaded area of Fig. 6.20, then the hypothesis H_0 is rejected at a 0.05 level of significance. For instance, if another measurement produced 121 counts, then at the 0.05 level, the hypothesis that 194 is the mean value of the underlying Gaussian distribution would be rejected. What this also means is, because 5% of the PDF lies outside of $\pm 1.96\sigma$, then there is a 5% probability that the experimenter can make a Type I error by declaring the mean is *not* 194 when it actually is!

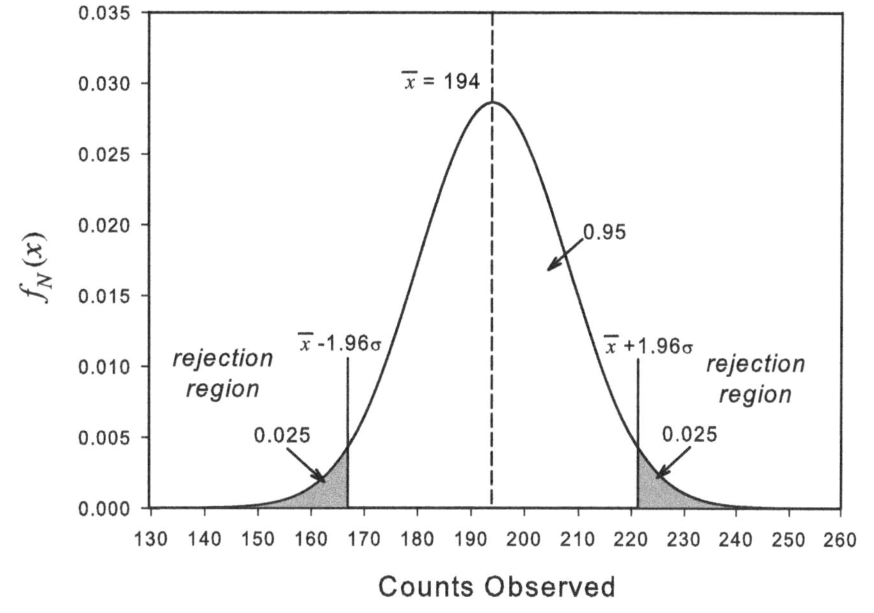

Figure 6.20. The Gaussian distribution $f_{\mathcal{N}}(x|\bar{x}, \bar{x})$ as predicted for the case in which it is assumed that $\bar{x} = \sqrt{\sigma} = 194$. The sum of the shaded regions represent 5% of the Gaussian PDF, and the light region represents 95% of the PDF.

6.10.1 Radiation Detection Limits

Sometimes a suspected radioactive sample produces counts similar to those of a background count for the same counting time. How is it possible to determine if the sample is actually radioactive or not? Clearly such a case involves Types I and II errors.

As shown in Fig. 6.21, the Gaussian background and radioactive sample distributions of counts can overlap. In this plot the expected background counts are subtracted from both the background and sample count distribution, i.e., the Gaussian distributions are plotted as a function of the net counts. Because the two distributions overlap it is possible an experimenter can declare a sample radioactive, when it is not (Type I error), or declare the sample "clean" when it is actually radioactive (Type II error).

The net sample counts are defined the gross sample counts G less the background counts B, i.e., $N_s = G - B$. The standard deviation of the background is σ_B, and the standard deviation of the total counts from the sample is σ_G. If the sample is *not* radioactive ($N_s \simeq 0$), then it is expected that $G - B \simeq B - B \simeq 0$. Under such a condition, the standard deviation is

$$\sigma(N_{\text{net}}) \equiv \sigma_0 = \sqrt{\sigma_B^2 + \sigma_B^2} = \sqrt{2\sigma_B^2} = \sqrt{2B}. \tag{6.98}$$

But should the sample produces net counts comparable to that of the background ($N_s \simeq B$ or $G \simeq 2B$), the standard deviation is approximately,

$$\sigma(N_{\text{net}}) \simeq \sqrt{G + B} = \sqrt{2B + B} = \sqrt{3B}. \tag{6.99}$$

The concepts of Type I and Type II error are important for radiation counting in the presence of significant background radiation. Typically the limiting ability to detect non-background radiation is given by the *minimum detectable activity* (MDA), which defines the minimal amount of radiation that can be detected given a specific confidence limit defined by the rejection region and radiation background [Currie 1968; Rucker 1995].[15]

Related to the MDA is the *critical detection limit* (CDL), which defines the limit by which counts from a sample can be reported with a preset rejection region. Hence, any measurement resulting in counts below the CDL is not worth reporting, based on a one-tailed test at the low end of the sample measurement count distribution. The CDL is defined by the confidence limit defined by some fixed multiple of the standard deviation of the sample distribution.

Shown in Fig. 6.22 are two net count distributions for background and sample measurements in which the tails of the distributions overlap. It should be remembered that the background influences the MDA, and the MDA is determined by the error that an experimenter deems as tolerable. Hence, given some predetermined test and rejection regions, the adjacent shaded regions α and β can be defined, in which α represents the probability of making a Type I error and β represents the probability of making a Type II error. The decision to state that a sample is not radioactive is defined by $1-\alpha$, in which $\alpha = k_\alpha \sigma_0$, is a fraction of the high-end tail of the Gaussian background distribution, where values of $k\sigma$ are listed in Table 6.6. For instance, it is quite common to choose $k_\alpha = 1.645$ so that $\alpha = 0.05$ and that $1-\alpha = 1 - 1.645\sigma_B = 0.95$. Because the background net count distribution is centered about zero, the CDL (denoted as L_C) is

$$\boxed{L_C = k_\alpha \sigma_0 = k_\alpha \sqrt{2B}.} \tag{6.100}$$

Hence, a sample measurement producing net counts below the CDL is declared non-radioactive. The MDA must be larger than CDL and can be defined in terms of the CDL by setting the MDA equal to the CDL plus some confidence interval with an overlapping rejection region defined in Fig. 6.22 as β. Hence, the MDA, denoted by L_D, is

$$L_D = L_C + k_\beta \sigma_N = k_\alpha \sigma_0 + k_\beta \sigma_{N_s}. \tag{6.101}$$

Typically, β is chosen to be 0.05, hence $k_\beta = k_\alpha = 1.645$.

Two limiting cases in which the sample yields low counts are now considered. In the first case, assume that the background and net sample count rates are similar.

CASE 1: $(N_s \simeq B)$.

$$\begin{aligned} L_D &= k_\alpha \sigma_B + k_\beta \sigma_N = k_\alpha \sqrt{2B} + k_\beta \sqrt{G+B} = k_\alpha \sqrt{2B} + k_\beta \sqrt{3B} \\ &= 1.645(\sqrt{2}\sigma_B + \sqrt{3}\sigma_B) = 5.176\sigma_B. \end{aligned} \tag{6.102}$$

Suppose, however, the net sample counts are quite a bit smaller than those of the background radiation. With the same criteria for k_β and k_α, the following is stated,

CASE 2: $(N \ll B)$. The MDA can be rewritten as

$$L_D = L_C + k_\beta \sigma_N = L_C + k_\beta \sqrt{L_D + \sigma_0^2}, \tag{6.103}$$

which, when combined with Eq. (6.100), and solved for L_D yields

$$L_D = L_C + \frac{k_\beta^2}{2}\left[1 + \sqrt{1 + \frac{4L_C}{k_\beta^2} + \frac{4L_C^2}{k_\alpha^2 k_\beta^2}}\right]. \tag{6.104}$$

[15]The use of the word "activity" in the MDA is unfortunate because the MDA does not refer to the true activity of a sample (number of decays per unit time), but rather to the number of counts produced above background.

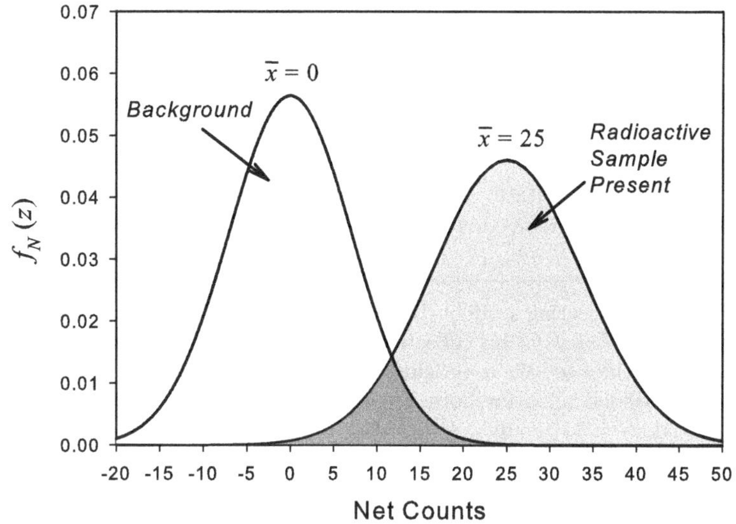

Figure 6.21. Overlapping Gaussian distributions for a case in which a radioactive sample produces comparable counts as does the background radiation. Here the background counts have been subtracted from both distributions and are thus plotted as a function of the net counts, i.e., background subtracted counts or net counts. In this hypothetical case, the background $B = 25$ so that an average net measurement for a non-radioactive sample is 0 counts. The low-level radioactive sample has $N_s = 25$ counts. Notice the significant overlap between the two distributions, hence leaving considerable room for misinterpretation of the radioactive nature of the sample.

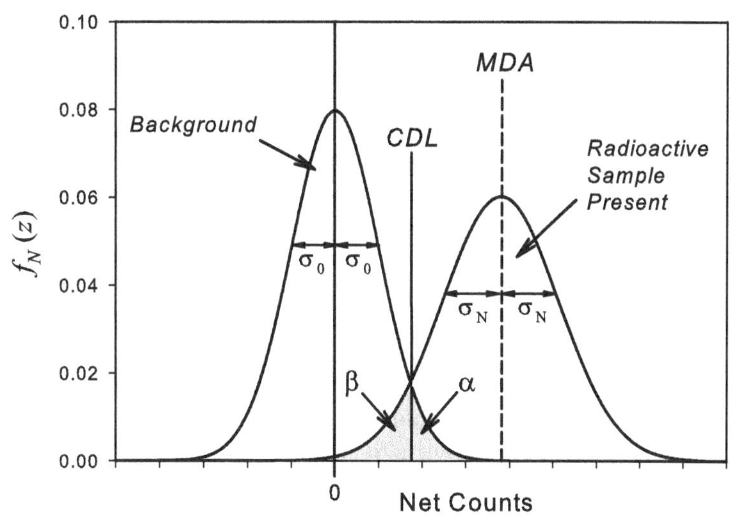

Figure 6.22. Overlapping Gaussian distributions for a case in which a radioactive sample produces a net count comparable to that of the background radiation. The area of the shaded region α is the probability of a Type I error, and the area of the shaded region β is the probability of a Type II error.

If $k_\alpha = k_\beta = k$, then the above reduces to [Currie 1968],

$$\boxed{L_D = k^2 + 2L_C,} \tag{6.105}$$

a commonly used test for low level radiation. Hence, a sample measurement with net counts above the MDA is declared radioactive with 95% confidence.

Example 6.23: A nuclear engineering student discovers an advertisement for "radioactive marbles" on the internet and decides to purchase a few for his collection of atomic memorabilia. Upon receiving the yellowish-green marbles, the student decides to take a measurement to determine if the marbles really are radioactive. A 95% confidence level is decided upon for both Types I and II error, so that $k = k_\alpha = k_\beta = 1.645$. The following data are collected: $G = 217$ counts, $B = 194$ counts, and $N_s = 217 - 194 = 23$ counts. Then

$$\begin{aligned} L_D &= k^2 + 2L_C = k^2 + 2k\sqrt{2B} \\ &= (1.645)^2 + 1.645(2)\sqrt{2}\sqrt{194} = 2.706 + 4.65(13.93) = 67.3. \end{aligned}$$

This result indicates that at least 67.3 net counts should be observed from the sample to reduce the probability of a Type II error to only 5%. With the MDA at 67.3, but $N_s = 23$, the student suspects that the marbles are not radioactive. The CDL is found to be

$$L_C = k_\alpha \sigma_0 = k_\alpha \sqrt{2B} = (1.645)\sqrt{2}(13.93) = 32.4.$$

Since $N_s < L_C$, the result is considered "not worth reporting," and the student determines with 95% confidence that the marbles are not radioactive. (Next time, buy the red Fiesta Ware!)

In recent years, the standard that defined the use of MDA and CDL was updated to address perceived deficiencies [Strom, 2001; ANSI 13.30, 2011]. Measurements and analysis indicate that Eq. (6.100) routinely overestimates the fraction of false positive events [Strom, 2001], becoming large as α is reduced ($\alpha \leq 0.05$). The inaccuracy stems from the fact that a Gaussian distribution is assumed for these low count rates, when in fact the actual average and variance for the observed counts are not well defined.

An alternative, but equivalent, expression for Eq. (6.100) that incorporates varying count rates is [Nicholson 1963],

$$L_C(R_n, \alpha) = k_\alpha \sqrt{\frac{B}{t_B}\left(\frac{1}{t_B} + \frac{1}{t_G}\right)}, \tag{6.106}$$

where t_B is the background count time and t_G is the gross count time, and L_C is given as a count rate. Eq. (6.106) is a function of the net count rate R_n and α, whereas Eq. (6.100) is a function of the observed net counts and α. One suggested method of correction is to substitute $(B+1)$ for B in the analysis [Strom 2001], which implies that the actual average background count observed (or predicted with a Gaussian model) is higher than the measured average. The correction is based on the "belief" that the observations can be described by a uniform Bayesian probability distribution function where the expected value of the observed background counts is $B + 1$. Substitution for B yields,

$$L_C(R_n, \alpha) = k_\alpha \sqrt{\frac{(B+1)}{t_B}\left(\frac{1}{t_B} + \frac{1}{t_G}\right)}, \tag{6.107}$$

where L_C is given as a count rate. Although Eq. (6.107) reduces the frequency of false positives from Eq. (6.100), the estimated false positive rate is still higher than the actual rate.

Alternative treatments for L_C are listed and briefly described in the literature [Strom 2001]. The study compares eight different decision rules in which it was concluded that all of the methods introduce bias into the predicted false positive rate, with Eq. (6.100) yielding the highest false positive rate. Eq. (6.107) adequately suppresses the false positive rate; however, alternative methods appear to yield even lower false positive rates, but apparently risk an increase in false negatives for low counting rates.

6.10.2 Chi-Square Test for "Goodness" of Data

Data accumulated by a counting system may yield results that are affected by systematic errors. For instance, electronic noise may be a source of significant problems, especially if the lower level discriminator (LLD) is set too low. It is often important that an experimenter test the counting system for such possible issues before beginning a radiation measurement experiment.

In Section 6.8 it was shown that counting statistics can be modeled with a Gaussian PDF with mean μ and variance $\sigma^2 = \mu$, provided that the probability p of observing a decay is small and the sample mean $\overline{x}$ is larger than approximately 25. Therefore, given a set of measurements $\hat{x}_1, \hat{x}_2, \ldots, \hat{x}_N$, it should be possible to compare in some way the population variance σ^2 to the sample variance s^2 of Eq. (6.26). One approach is to calculate the χ^2 *statistic* defined as[16]

$$\chi^2 \equiv \frac{\sum_{i=1}^{N}(x_i - \mu)^2}{\sigma^2(x)}, \tag{6.108}$$

where x_i is the ith random sample from $\mathcal{N}(x|\mu,\sigma)$. Here χ^2 is a random variable and is distributed according to the χ^2 PDF [Mood et al. 1974]

$$f(\chi^2|\nu) = \frac{1}{2^{\nu/2}\Gamma(\nu/2)}(\chi^2)^{(\nu/2)-1}e^{-\chi^2/2}, \quad x > 0, \tag{6.109}$$

with $\nu = N$ degrees of freedom. This PDF is illustrated in Fig. 6.23 for $\nu = 1, 2, 3, 4$, and 5. From this figure this distribution is seen to be asymmetrical about the mean, although the χ^2 distribution becomes increasingly symmetric as ν becomes large.

How might χ^2 help one to decide if a counting system is operating properly? If one were to estimate μ by $\overline{x}$ and use the measured $\hat{x}_i$ in place of the x_i then, with Eq. (6.26), χ^2 would be estimated as

$$\hat{\chi}^2 \equiv \frac{\sum_{i=1}^{N}(\hat{x}_i - \overline{x})^2}{\sigma^2} = \frac{(N-1)s^2}{\sigma^2}. \tag{6.110}$$

Divide by $(N-1)$ to obtain the so-called *reduced* chi-square statistic $\chi^2_\nu = s^2/\sigma^2$. Thus, if the data are a good fit to the Gaussian model, then χ^2_ν should be close to unity. If, however, s^2 is greatly different from σ^2, then χ^2_ν would be either very small or much greater than unity, an indication that something is probably amiss with the counting system. Calculation of the reduced χ^2_ν is a very easy and quick way to check a counting system.

Although the above reduced chi-squared procedure is a quick and easy method, it does not yield any quantification of the confidence one has about how well the data fit the Gaussian model. The chi-squared test for "goodness" of data, begins by calculating a *measured* or critical χ^2 value from Eq. (6.108), with the

[16] The square on χ^2 is to remind one that this statistic can never be negative.

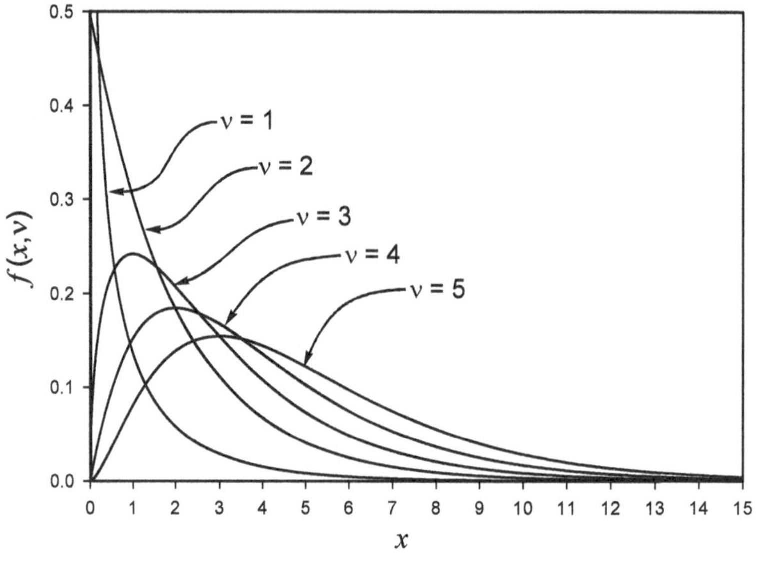

Figure 6.23. The χ^2 probability distribution function for $\nu = 1, 2, 3, 4$, and 5.

observed data, i.e., $\mu \simeq \overline{x}$ and $\sigma^2 \simeq \overline{x}$, as

$$\widehat{\chi}^2 = \frac{1}{\overline{x}} \sum_{i=1}^{N} (\hat{x}_i - \overline{x})^2. \tag{6.111}$$

This statistic is also distributed as the χ^2 PDF but now with $N - 1$ degrees of freedom because μ has been approximated by $\overline{x}$ (see Section 6.5.4).

The probability α of observing a χ^2 greater than χ^2_α is

$$\alpha(\nu) = \int_{\chi^2_\alpha}^{\infty} \frac{1}{2^{\nu/2}\Gamma(\nu/2)} (\chi^2)^{(\nu/2)-1} e^{-\chi^2/2}\, d\chi^2. \tag{6.112}$$

For example, shown in Fig. 6.24 is the χ^2 distribution for $\nu = 20$, in which the shaded region defines the probability of observing a $\widehat{\chi}^2$ value between the limits $\chi^2_\alpha = 27$ and ∞.

Given any χ^2_α and ν, the probability $\alpha(\nu)$ of obtaining a value of $\widehat{\chi}^2$ greater that χ^2_α can be found by numerical evaluation of Eq. (6.112). Conversely, $(1 - \alpha(\nu))$ is the probability of obtaining a $\widehat{\chi}^2$ less than χ^2_α. Figure 6.25 shows values of α for $\nu = 1, 2, 3, 4$, and 5 as a function of the critical value χ^2_α. Shown in Table 6.8 are the values χ^2_α, with the associated probabilities α, for ν ranging from 1 to 30. Thus, the chi-square goodness-of-fit test consists of calculating $\widehat{\chi}^2$ from the measured data. Then from Table 6.8 find the probabilities of obtaining a value larger or smaller than $\widehat{\chi}^2$. If either of these probabilities is much greater or much less than 0.5, then the data are suspect.

Typically, an experimenter should choose a confidence interval to determine whether or not the observed data are unusual. For instance, if a $\widehat{\chi}^2$ value for a set of N measurements yields an α of 0.05, then there is only a 5% probability of observing a higher value should a new data set of N measurements be acquired. Similarly, should the $\widehat{\chi}^2$ value yield α of 0.95, there is a 95% probability that the next set of N measurements produces a higher value. In either case, the result is at the extremes of the χ^2 distribution, a strong indication

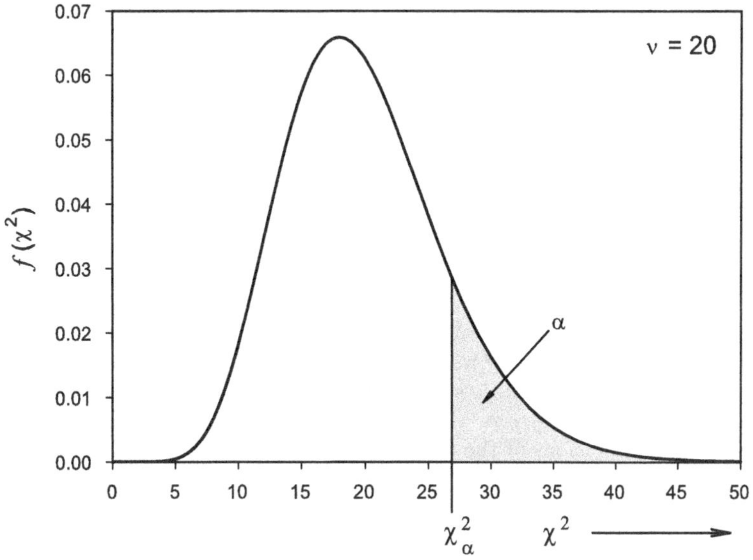

Figure 6.24. The χ^2 probability distribution function for $\nu = 20$, showing a critical value of $\chi^2_\alpha = 27$, where α is the shaded region.

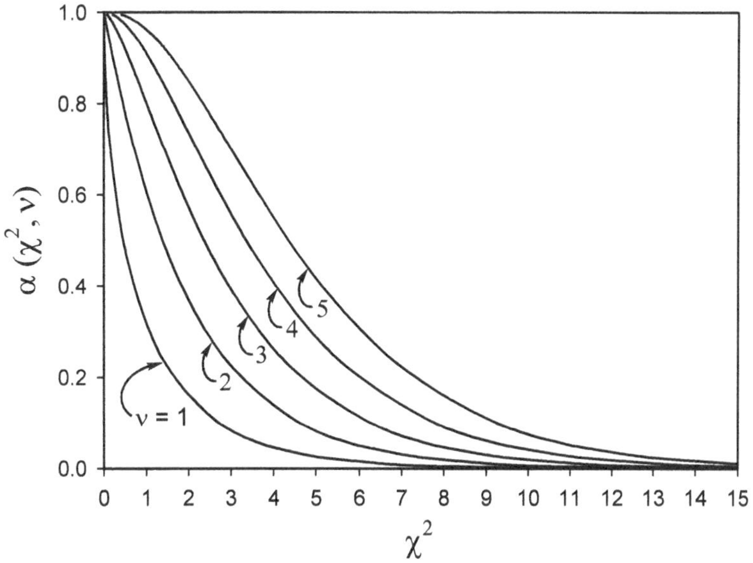

Figure 6.25. The values of α for $\nu = 1, 2, 3, 4$, and 5 as a function of various critical values of χ^2.

that the data are not following the expected Gaussian model. An acceptable confidence interval should be decided upon by the experimenter that minimizes either Type I or Type II error, depending on which is more important in the data analysis.

Example 6.24: An experimenter takes a series of five measurements with a counting system to test the goodness of the data and obtains the following counts: 302, 311, 297, 307, and 299. Find the value of $\widehat{\chi^2}$ and determine the probability of observing a value greater than (or less than) $\widehat{\chi}^2$.

Solution:

From Eq. (6.111),

$$\widehat{\chi}^2 = \frac{\sum_{i=1}^{N} (\overline{x} - x_i)^2}{\overline{x}} = \frac{1.44 + 60.84 + 38.44 + 14.44 + 17.64}{303.2} = 0.438.$$

From Table 6.8 with $\nu = 5 - 1 = 4$, the value of $\widehat{\chi}^2 = 0.438$ falls between 0.297 and 0.484 under the probability columns (α) of 0.99 and 0.975, respectively. Application of linear interpolation between these χ^2_α values, α is found to be 0.9787. This result means that a new series of measurements would have a 97.87% chance of having a $\widehat{\chi}^2$ value greater than 0.438 (or a 2.13% of observing a value less than 0.438). The experimenter should be suspicious of the data!

What do these results mean? Because the distribution yields a low value for χ^2, the variance about the mean is much smaller than should be observed for samples from a Gaussian distribution. Had the $\widehat{\chi}^2$ value been relatively large, for instance near 14 for the case with $\nu = 4$, then the variance about the mean is much larger than should be observed for a Gaussian distribution. Finally, a $\widehat{\chi}^2$ value yielding $\alpha \approx 0.5$ describes the expected variance for a Gaussian distribution.

The χ^2 test is an important tool. If the value of the probability α falls far from 0.5, then something may be wrong with the detector, the electronics, or the counting system arrangement. Possible sources of the problem include:

1. The lower level discriminator (LLD) is set too low, hence the electronic noise is being measured as a significant addition to the real radiation counts.
2. Pickup from ground loops and other electronic noise.
3. Poor preparation of the radioactive sample.
4. Faulty equipment or malfunctioning equipment.

Finally, the experimenter must decide if the result of the chi-squared test is simply a matter of statistics. Because the χ^2 distribution has a range $[0,\infty]$, all positive values of $\widehat{\chi}^2$ are possible. Hence, the observation of a single $\widehat{\chi}^2$ value yielding an extreme value of α is not alarming, and might be considered an outlier. However, if several subsequent tests also produce $\widehat{\chi}^2$ values with extreme values of α, then it is almost certain that a problem exists with the counting system.

6.10.3 Presentation of Data

Presentation of counting data should also give the estimated one-sigma errors. In plots of counting data *error bars* are typically used to show the experimental error associated with each data marker. The length of the error bars are scaled to properly depict the values of $\pm\sigma$. If the error is too small to be properly represented by error bars, then the data are represented only by the data markers. In some cases, error bars should be shown along both the ordinate and abscissa of a Cartesian graph.

Theoretical plots, empirical plots and curve fitting routines are depicted by a continuous smooth curve *without data points.* With plotting software, an experimenter may be tempted to use data points and a

Table 6.8. Critical Values for the χ^2_α Distribution.

ν \ α	0.995	0.99	0.975	0.95	0.90	0.75	0.50	0.25	0.10	0.05	0.025	0.01	0.005
1	0.000	0.000	0.001	0.004	0.016	0.102	0.455	1.323	2.706	3.843	5.025	6.637	7.882
2	0.010	0.020	0.051	0.103	0.211	0.575	1.386	2.773	4.605	5.992	7.378	9.210	10.597
3	0.072	0.115	0.216	0.352	0.584	1.213	2.366	4.108	6.251	7.815	9.348	11.344	12.837
4	0.207	0.297	0.484	0.711	1.064	1.923	3.357	5.385	7.779	9.488	11.143	13.277	14.860
5	0.412	0.554	0.831	1.145	1.610	2.675	4.351	6.626	9.236	11.070	12.832	15.085	16.748
6	0.676	0.872	1.237	1.635	2.204	3.455	5.348	7.841	10.645	12.592	14.440	16.812	18.548
7	0.989	1.239	1.690	2.167	2.833	4.255	6.346	9.037	12.017	14.067	16.012	18.474	20.276
8	1.344	1.646	2.180	2.733	3.490	5.071	7.344	10.219	13.362	15.507	17.534	20.090	21.954
9	1.735	2.088	2.700	3.325	4.168	5.899	8.343	11.389	14.684	16.919	19.022	21.665	23.587
10	2.156	2.558	3.247	3.940	4.865	6.737	9.342	12.549	15.987	18.307	20.483	23.209	25.188
11	2.603	3.053	3.816	4.575	5.578	7.584	10.341	13.701	17.275	19.675	21.920	24.725	26.757
12	3.074	3.571	4.404	5.226	6.304	8.438	11.340	14.845	18.549	21.026	23.337	26.217	28.300
13	3.565	4.107	5.009	5.892	7.042	9.299	12.340	15.984	19.812	22.362	24.736	27.688	29.819
14	4.075	4.660	5.629	6.571	7.790	10.165	13.339	17.117	21.064	23.685	26.119	29.141	31.319
15	4.601	5.229	6.262	7.261	8.547	11.037	14.339	18.245	22.307	24.996	27.488	30.578	32.801
16	5.142	5.812	6.908	7.962	9.312	11.912	15.339	19.369	23.542	26.296	28.845	32.000	34.267
17	5.697	6.408	7.564	8.672	10.085	12.792	16.338	20.489	24.769	27.587	30.191	33.409	35.718
18	6.265	7.015	8.231	9.390	10.865	13.675	17.338	21.605	25.989	28.869	31.526	34.805	37.156
19	6.844	7.633	8.907	10.117	11.651	14.562	18.338	22.718	27.204	30.144	32.852	36.191	38.582
20	7.434	8.260	9.591	10.851	12.443	15.452	19.337	23.828	28.412	31.410	34.170	37.566	39.997
21	8.034	8.897	10.283	11.591	13.240	16.344	20.337	24.935	29.615	32.671	35.479	38.932	41.401
22	8.643	9.542	10.982	12.338	14.041	17.240	21.337	26.039	30.813	33.924	36.781	40.289	42.796
23	9.260	10.196	11.689	13.091	14.848	18.137	22.337	27.141	32.007	35.172	38.076	41.638	44.181
24	9.886	10.856	12.401	13.848	15.659	19.037	23.337	28.241	33.196	36.415	39.364	42.980	45.559
25	10.520	11.524	13.120	14.611	16.473	19.939	24.337	29.339	34.382	37.652	40.646	44.314	46.928
26	11.160	12.198	13.844	15.379	17.292	20.843	25.336	30.435	35.563	38.885	41.923	45.642	48.290
27	11.808	12.879	14.573	16.151	18.114	21.749	26.336	31.528	36.741	40.113	43.195	46.963	49.645
28	12.461	13.565	15.308	16.928	18.939	22.657	27.336	32.620	37.916	41.337	44.461	48.278	50.993
29	13.121	14.256	16.047	17.708	19.768	23.567	28.336	33.711	39.087	42.557	45.722	49.588	52.336
30	13.787	14.953	16.791	18.493	20.599	24.478	29.336	34.800	40.256	43.773	46.979	50.892	53.672
40	20.707	22.164	24.433	26.509	29.051	33.660	39.335	45.616	51.805	55.759	59.342	63.691	66.766
50	27.991	29.707	32.357	34.764	37.689	42.942	49.335	56.334	63.167	67.505	71.420	76.154	79.490
60	35.535	37.485	40.482	43.188	46.459	52.294	59.335	66.981	74.397	79.082	83.298	88.379	91.952
70	43.275	45.442	48.758	51.739	55.329	61.698	69.334	77.577	85.527	90.531	95.023	100.425	104.215
80	51.172	53.540	57.153	60.392	64.278	71.145	79.334	88.130	96.578	101.879	106.629	112.329	116.321
90	59.196	61.754	65.647	69.126	73.291	80.625	89.334	98.650	107.565	113.145	118.136	124.116	128.299
100	67.328	70.065	74.222	77.930	82.358	90.133	99.334	109.141	118.498	124.342	129.561	135.807	140.169

dot-to-dot line linking the calculated or theoretical values; but this is simply incorrect. In plotting data a clear distinction must always be made between experimental data points and a theoretical comparison (or curve fit to the data).

Shown in Fig. 6.26 is a *counting curve* for a proportional gas-filled radiation counter[17] versus the high voltage applied to operate the detector. The curve shows the counts observed and the standard error ($\pm\sigma$) for each measurement. In this example, the uncertainties in the applied voltage were deemed to be smaller than the data markers and, hence, horizontal error bars are omitted. Also in the graph a *curve fitted* line is drawn freehand through the data. Such a curve does not represent a theoretical or empirical equation. Equivalently, bicubic splines, for example, can be used to produce a *smooth* curve through the data [Press et al. 1992]. However, such a line is not necessary, but is usually provided to show the trend of the data for presentation purposes. Notice that the trend line is not drawn through each data point, but rather is a smooth representation, showing an estimated average of the data. Error bars are present on data markers in which the value of $\pm\sigma$ exceeds the marker size.

Fig. 6.27 depicts data and a theoretical comparison for an experiment in which different thicknesses of a Cu attenuator are placed between a detector and a ^{137}Cs source. As the Cu thickness is increased, the gamma rays passing through the Cu sheet and entering the detector are exponentially reduced. The example shows the theoretical variation with Cu thickness, for a specific detector and source strength, to be 300 counts when no attenuator is present. Also shown are the theoretical limits $\pm\sigma$ for the attenuated counts based on Gaussian counting statistics. The actual data observed are presented as data markers with error bars defined by $\pm\sigma$ for each data point. Finally two Gaussian distributions show how data samples at two Cu thicknesses should be distributed.

Fig. 6.27 also reveals another important aspect of data presentation, namely it is statistically expected that only 68.3% of the error bars should actually come into contact with the true mean value. In other words, for data following a Gaussian PDF, 31.7% of the data markers and accompanying error bars should not contact either the theoretical or sample mean! Hence, it is simply incorrect to connect all experimental data points together by lines drawn between each pair of data markers.

6.10.4 Concluding Remarks

Radiation emissions are stochastic (non-deterministic) such that the observation of a radioactive decay event cannot be predicted exactly; rather only probability statements can be made of observing an event within a specified period of time. A true mean value (population mean) of the radioactive decay rate can be calculated from the decay constant of the radionuclide, and the probability of observing a radioactive decay is bounded by the variance about that mean. However, experimenters, who do not know the decay constant, must use measured values to obtain the sample mean and sample variance. Almost always there is a difference between the true and sample means, and a difference between the population sample variances. These differences generally become smaller as more data are accumulated.

Three basic statistical models are developed that can describe radioactive decay and counting statistics with the Gaussian (or normal) distribution most often used for radiation counting. Many examples are included to assist in understanding the use of these models. The reader is warned that the models apply only to stochastic processes, and do not account for systematic and sampling errors. Systematic errors may be introduced by biases unknown to the experimenter, such as malfunctioning equipment. The χ-square test can often expose that such biases exist. Sampling errors arise from performing measurements on a sample population different from the one desired. For example, a radioactive sample containing two types of radionuclides or a background count with the radioactive sample not sufficiently far away from the detector would cause sampling errors.

[17] Proportional counters, a type of gas-filled detector, are covered in Chapter 10.

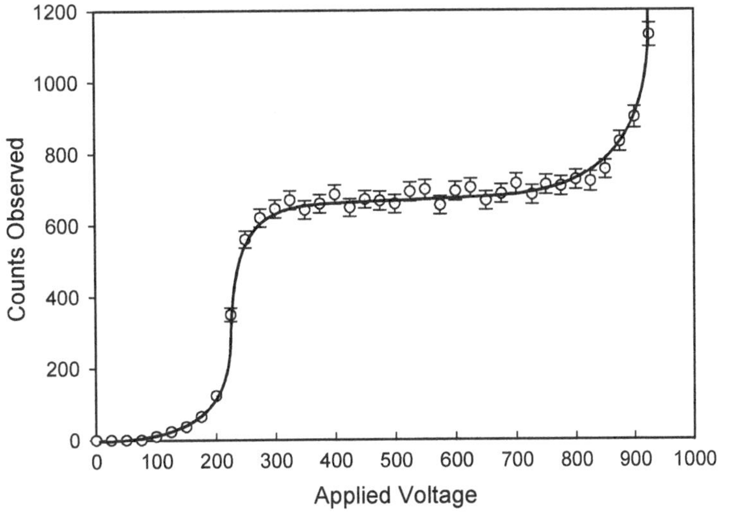

Figure 6.26. Example of data points with error bars, showing a counting curve for a gas-filled proportional counter and a smooth curve fit through the data.

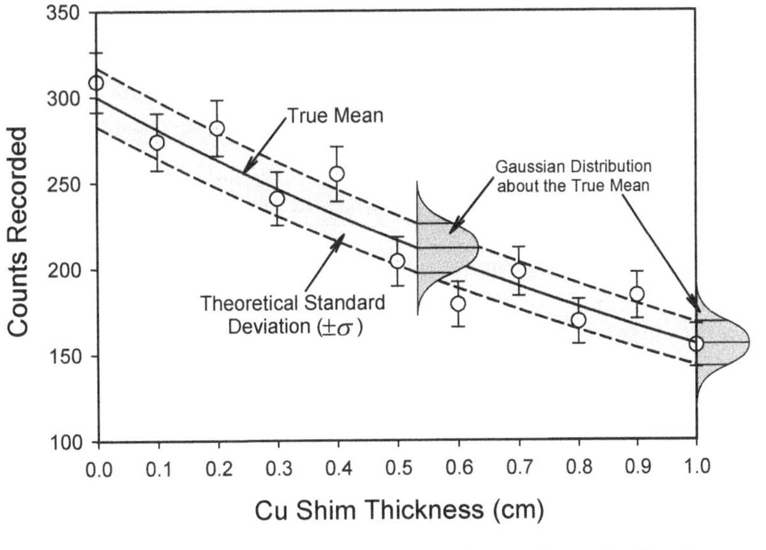

Figure 6.27. Example of data points with error bars, showing the true mean and standard deviation, along with experimental standard deviations for each data point.

The purpose of this chapter is to provide the necessary tools for an experimenter to perform basic radiation measurements. Throughout this book other more specialized statistical models used in radiation detection are introduced. There are also a huge number of other statistical models used in other branches of science and engineering that are not discussed here. For details on these other probability and statistical models for engineering and science, and their proper application, the reader is directed to the reference list at the end of the chapter.

PROBLEMS

1. What is the probability of tossing heads 10 times out of 15 with a "fair" coin?

2. "Introduction to Engineering Concepts 101" is a huge class with 375 students enrolled. Use Poisson statistics to determine the probability that exactly 3 students in the class have a birthday today. Plot the probability distribution function for 0 through 10 students having birthdays.

3. An amateur archer practices once every afternoon, shooting 100 arrows and, generally, hits the target. One time out of 15, the archer hits the bull's eye. After shooting 100 arrows, what is the average number of bull's eyes achieved, and what is the standard deviation? What is the probability that the archer strikes exactly 10 bull's eyes after 100 tries?

4. A set of radiation measurements is made with the following results:

$$54, 89, 64, 102, 90, 78, 56, 74, 88, 42,$$
$$70, 94, 50, 82, 100, 64, 72, 78, 58, 76$$

Find the sample mean, variance and standard deviation.

5. Ten radiation measurements are conducted from which a standard error is determined. If an eleventh measurement is conducted under identical conditions, by what amount does the standard deviation change?

6. ACME Roadrunner Company manufactures alternators with a one-year guarantee. Random sampling of alternators yielded the following number of months to failure for 22 alternators:

$$37, 25, 44, 51, 26, 41, 18, 29, 58, 34, 43$$
$$31, 11, 27, 40, 39, 52, 23, 36, 38, 45, 55$$

Determine the experimental mean, median and variance of the data distribution. Find the 10% trimmed mean and compare it to the median. Use a Gaussian PDF to estimate what percentage of alternators fail and must be returned for a warranty replacement?

7. Show that Eq. (6.95) reduces to

$$\sigma_n = \frac{1}{N}\sqrt{\sum_{i=1}^{N}\left(\frac{G_i}{t_{Gi}^2} + \frac{B_i}{t_{Bi}^2}\right)}$$

if the number of measurements performed N is the same for both gross and background counts, with all t_G equal and all t_B equal.

8. A measurement is conducted with 662 keV gamma rays from a ^{137}Cs source. A detection system records an average counting rate of 125 cpm. A background measurement records an average count rate of 15 cpm. If a 30-minute background measurement is performed, how long must the ^{137}Cs source measurement be in order to reduce the standard error of the net counts below 1%?

9. A radiation measurement is conducted in which a total gross number of counts is 2109, taken over 1 minute. A 10-minute background measurement is taken, resulting in 312 counts. What is the net count rate and standard deviation (in counts and in percent)?

10. The following set of measurements, each for a six-second duration, resulted in the following number of counts:

$$70, 64, 49, 58, 57, 52, 59, 53, 53, 58, 63, 61, 62, 68, 67, 51, 56, 54, 51, 66$$
$$66, 68, 71, 67, 51, 64, 73, 58, 69, 62, 56, 60, 57, 53, 57, 52, 55, 65, 61, 58$$
$$56, 61, 55, 70, 65, 68, 73, 57, 62, 58, 55, 63, 54, 58, 57, 64, 58, 66, 54, 49$$
$$68, 54, 62, 58, 79, 74, 53, 63, 58, 51, 56, 60, 56, 57, 62, 62, 59, 50, 63, 57$$
$$72, 45, 57, 47, 48, 62, 59, 53, 40, 77, 79, 48, 51, 69, 65, 60, 58, 71, 64, 48$$

(a) Calculate the sample mean, sample variance and the sample standard deviation. (b) Plot the data distribution function and indicate the mean and standard deviation. (c) Use the sample mean to plot the expected Gaussian PDF on the same graph. (d) Compare the experimental and theoretical distributions by examining the variances and standard deviations.

11. A measurement is conducted with a 356-keV gamma-ray source that produces 1267 counts in a period of 5 minutes. A 2-mm thick sheet of material is placed between the source and the detector and the detector records 874 counts in 5 minutes? What is the attenuation coefficient and its standard error for the material?

12. The following measurements were taken over five-minute periods each:

$$1288, 1443, 1377, 1341$$

and the following background measurements were taken over ten-minute periods each:

$$351, 436, 379, 414$$

Calculate the net counting rate per minute and the associated percent error.

13. A measurement was taken over a period of five minutes, resulting in 1417 counts. The background was measured for one hour, resulting in 315 counts.

1. What is the average counting rate per minute and its percent standard error?
2. What is the estimated number of counts and its percent standard error for a one-hour measurement?
3. Use the average cpm to find the percent standard error if the measurement was conducted for one hour as opposed to five minutes. How does this result compare to the five-minute measurement?

14. Prove that the covariance $\langle (x_i - \overline{x}_i)(x_j - \overline{x}_j) \rangle$ can be approximated by $\langle x_i x_j \rangle - \overline{x}_i \overline{x}_j$. [Hint: Start by expanding $(x_i - \overline{x}_i)(x_j - \overline{x}_j)$].

15. The half-life of a radiation source is sought. The first measurement, taken over a period of 1 minute, yielded 2350 counts. Ten minutes later, a second measurement was started, again over a 1-minute period, and yielded 1780 counts. What is the half-life of the source and the standard error in the measured half-life?

16. Perform the χ^2 test on the data of Example 6.7. If a new set of measurements is taken under identical conditions, what is the probability that the new $\widehat{\chi}^2$ value is greater? What is your determination of the "goodness" of the data?

17. Use the reduced chi-squared statistic to examine the "goodness" of the data in Example 6.7.

18. Derive Eq. (6.104) using the results of Eq. (6.100) and Eq. (6.101). Show that if $k_\alpha = k_\beta$, then Eq. (6.104) reduces to Eq. (6.105).

19. An unmarked sample was found on the bottom shelf, way in the back, of the neutron activation analysis (NAA) lab. It is unknown how old this sample is, and unfortunately, whether or not it is still radioactive. The background recorded by a counting system in the NAA lab is 35 cpm. For a one-minute measurement, what is the CDL required to produce with 95% confidence that the sample is not radioactive? What is the MDA required for 95% confidence that the sample is still radioactive? Repeat your analysis for a 60-minute measurement.

20. You are a final contestant on the *Let's Make a Deal!* show, and must pick one of three doors, behind which only one door has the grand prize, and behind the other two doors there is a pallet of dog food. After making your initial selection, one of the doors that you did not choose is opened to reveal a pallet of dog food, hence either the door you picked or the remaining closed door has the grand prize. You are then given the opportunity to change your choice to the other closed door. Should you make the change? Determine the statistics of winning if you change as opposed to keeping your original choice.

REFERENCES

ANSI 30.13, "Performance Criteria for Radiobioassay, An American National Standard," Health Physics Society, American National Standards Institute, ANSI/HPS N13.30-2011, 2011.

BEVINGTON, P.R. AND D.K. ROBINSON, *Data Reduction and Error Analysis for the Physical Sciences,* 2nd Ed., New York: McGraw-Hill, 1992.

BOX, G.E.P., W.G. HUNTER, AND J.S. HUNTER, *Statistics for Experimenters,* New York: Wiley, 1978.

CRAMÈR, H., *Mathematical Methods of Statistics,* Princeton, NJ: Princeton Univ. Press, 1946.

CURRIE, L.A., "Limits for Qualitative Detection and Quantitative Determination," *Analytical Chemistry,* **40**, 586–593, 1968.

DEVORE, J.A., *Probability and Statistics for Engineering and the Sciences,* 3rd Ed., Pacific Grove: Brooks-Cole, 1991.

KENDALL, M. AND A. STUART, *The Advanced Theory of Statistics,* 4th ed., Vol. 1, New York: MacMillan Publishing, 1977.

MOOD, A.M., F.A. GRAYBILL, AND D.C. BOES, *Introduction to the Theory of Statistics,* 3rd Ed., New York: McGraw-Hill, 1974.

NICHOLSON, W.L., *Fixed Time Estimation of Counting Rates with Background Corrections,* AEC R and D Report HW-76279, Richland, WA: Hanford Works, 1963.

PRESS, W.H., B.P. FLANNERY, S.A. TEUKOLSKY, AND W.T. VETTERLING, *Numerical Recipes,* Cambridge, UK: Cambridge University Press, 1986.

RUCKER, T.L., "Methodologies for the Practical Determination and Use of Method Detection Limits," *J. Radioanalytical and Nuclear Chemistry, Articles,* **192**, 345–350, 1995.

STROM, D.J. AND J.A. MACLELLAN, "Evaluation of Eight Decision Rules for Low-Level Radioactivity Counting," *Health Physics,* **81**, 27–34, 2001.

TAYLOR, J.R., *An Introduction to Error Analysis,* 2nd Ed., Mill Valley: University Science, 1997.

Chapter 7

Source and Detector Effects in Radiation Measurements

An experiment is a question which science poses to Nature, and a measurement is the recording of Nature's answer.

Max Planck

The purpose of making radiation measurements is usually to infer information about the source of the radiation. In particular, a detector response, such as a count rate, is often used to estimate the strength or intensity of the radiation particles emitted by a source. Typically, the emission rate of a specified type of radiation is sought. Although the detector count rate is *approximately* proportional to the emission rate of the radiation from the source, there are many correction factors (some small, some large) that must be made to the observed count rate to obtain an accurate estimate of the source strength. For example, not all emitted radiation is emitted in a direction that intersects the detector volume. Likewise radiation emitted toward the detector can be scattered by the surrounding medium into directions pointing away from the detector. Thus, to accurately infer information about source strengths it is important to understand how the source strength is related to the detector's response. The several effects that must be accounted for are the topics of this chapter.

To limit the scope of this chapter, the sources of radiation considered in this chapter are assumed to be radioactive sources which emit radiation isotropically. Other sources of radiation such as cosmic rays, accelerator sources, radiation from reactor beam ports, x-ray machines, etc., have their own special phenomena that must be considered to infer accurate source strengths.

7.1 Detector Efficiency

Upon the detection of radiation, a proper calculation of the source activity, or radiation field, requires an understanding of the detector response in the local environment. The ability of a detector to respond to radiation is often reported as "efficiency", yet there are multiple definitions for detector efficiency, which include *absolute* efficiency, *relative* efficiency, and *intrinsic* efficiency [McGregor and Shultis 2011].

Absolute efficiency ϵ_A is defined as the ratio of the measured number of counts recorded by the detector to the total number of emissions from the radiation source during a time Δt, or, equivalently, the ratio of the count rate R_{det} to the source emission rate,

$$\epsilon_A = \frac{R_{det}}{AB\prod_{i=1}^{n} C_i}, \tag{7.1}$$

where A is the source activity, B is the branching ratio or the probability, per radioactive decay, of emission of the radiation of interest and C_i are various correction factors that apply to the measurement. A significant characteristic of absolute efficiency measurements is that they are sensitive to changes in the source/detector geometry. Hence, reporting the absolute efficiency is of little value unless the experimental arrangement is described with adequate specificity to permit accurate reproduction of the measurement. The absolute efficiency depends critically on the detector's performance characteristics and on the geometrical positioning of the detector with respect to the radiation source. There are conditions under which absolute efficiency is a standard for reporting a detector's performance. For instance, the measurement method to report absolute efficiency for high-purity Ge (HPGe) detectors (see Chapter 20) is outlined under ANSI/IEEE 325-1996, which describes a standardized distance, orientation and radiation energy. However, for general use, the characterization of a detector by its absolute efficiency may be difficult to repeat.

Relative efficiency ϵ_R is the absolute efficiency of a test detector compared directly with another reference detector of known absolute efficiency ϵ_A^{ref} under identical conditions, i.e.,

$$\epsilon_R = \frac{\epsilon_A}{\epsilon_A^{ref}} = \frac{R_{det}}{AB\epsilon_A^{ref}\prod_{i=1}^{n} C_i}. \tag{7.2}$$

Typically, the reference detector is an established standard. For instance, using the HPGe example once again, the ANSI/IEEE 325-1996 standard prescribes that the HPGe detector be compared directly to the performance of a 3-in. × 3-in. right-cylindrical NaI:Tl detector (see Chapter 13), where the established absolute efficiency for the NaI:Tl detector is 1.2×10^{-3} [Fairstein et al. 1996]. Commercial vendors commonly quote relative efficiencies for various detectors; hence, it is important that the user understand the characterization method used to determine the efficiency.

Intrinsic efficiency ϵ_I is the ratio of the detector response (counts or count rate) to the particle emission or emission rate further divided by the solid angle Ω subtended by the detector with respect to the (point) source divided by the 4π solid angle of radiation emission, i.e.,

$$\epsilon_I = \frac{4\pi\epsilon_A}{\Omega} = \frac{4\pi R_{det}}{AB\Omega\prod_{i=1}^{n} C_i}. \tag{7.3}$$

Here Ω is proportional to the probability that emitted radiation intersect the detector's entrance aperture. For the case that the radiation source is not a point source, $\Omega/4\pi$ is replaced by a correction factor P_F which is again the probability the emitted radiation is in a direction toward the detector aperture. The intrinsic efficiency is often reported in an attempt to reduce the dependence upon system geometry, by considering only the radiation that is emitted toward the detector. Intrinsic efficiency, can, in principle, be used to scale the expected absolute efficiency for various alternate source/detector geometries.

Although the use of intrinsic efficiency mitigates the geometric dependence on efficiency measurements, it does not completely eliminate this dependence. Other effects such as the source preparation, detector characteristics, and the medium between the source and detector can also affect the measured results. These various effects that alter the measured efficiency can be classified into the following three broad categories.

Source Effects: The source preparation and encapsulation can alter the energy and number of radiation particles arriving at the detector aperture.

Detector Effects: The operational characteristics of a detector can alter the measured energy and number of radiation particles recorded by the detector.

Geometric Effects: The geometry of the source, detector and their configuration with respect to each other affect the energy and number of radiation particles arriving at the detector aperture.

These topics and how to correct for them are discussed in the remainder of this chapter.

7.2 Source Effects

Preparation of the source material and its encapsulation has an effect on the emission rate of radiation from the source assembly. Both alpha and beta particles sources can suffer significant backscatter from the source backing, or can also have considerable energy attenuation if the source material or source encapsulation is too thick. Likewise, the material (such as air) between the source and detector and the detector window reduce the number and energy of particles reaching the sensitive volume of the detector. Measurement of gamma rays and neutrons may produce significant room scatter if the shielding around the detector is inadequate. However, even with shielding to prevent room scatter, scattering from the shielding material itself produces scattered particles than can still reach the detector. Correction for all of these effects in order to determine accurately the source strength from a measured detector response is a formidable task. However, with proper source and detector designs, the needed correction factors can be made small (a few percent) so that approximate estimates of them can suffice.

Consider an idealized source/detector geometry shown in Fig. 7.1 in which the source material, the source substrate or backing, the source encapsulation cover and the detector window consist of parallel planes. As shown, there are many possibilities for source radiation to be prevented from reaching the active volume of the detector and also possibilities for which emitted radiation that would miss the detector can be redirected into the detector. To address all of the possibilities would require a detailed radiation transport calculation. In this section the two most important source effects are considered, namely (1) self-absorption of the radiation by the source and (2) the backscattering of radiation from the source substrate. These effects are addressed using simple and, thus, approximate analyses and are based on several key assumptions such as straight-line radiation trajectories, exponential or finite range attenuation, no source cover, a radionuclide source with isotropic emission, and a detector response that is insensitive to the energy and angle of incidence of the entering radiation. First a few key parameters are defined and then results for different types of radiation are given.

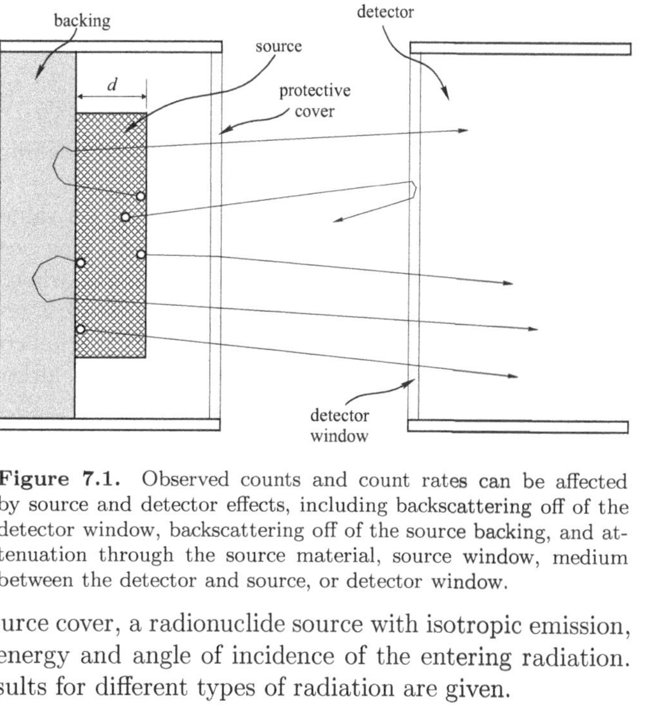

Figure 7.1. Observed counts and count rates can be affected by source and detector effects, including backscattering off of the detector window, backscattering off of the source backing, and attenuation through the source material, source window, medium between the detector and source, or detector window.

Source Self-Absorption Effect

Define the self-absorption factor for a source layer of thickness d

$$f_a(d) \equiv \frac{\text{no. particles exiting the source}}{\text{no. particles exiting without self-absorption}}, \tag{7.4}$$

where only particles emitted towards the front surface (i.e., the surface adjacent to the detector are considered). This factor is the probability $P_a^{esc}(d)$ that a particle born uniformly in the source volume escapes through the front surface. Division of the measured source strength S_{meas}, inferred from the detector count rate, by f_a gives a source strength corrected for source self-absorption that is larger than S_{meas} because f_a is always less than unity.

As $d \to 0$ the self-absorption factor $f_a(d) \to 1$, i.e., there is no self-absorption and all source particles escape the source layer. It should be noted that both the energy and angular distributions of the escaping particles are, generally, different than those particles emitted by the radionuclide. Also the self-absorption factor cannot be measured experimentally since all sources must be composed of matter and, thus, be capable of interacting with the radiation.

Source Backscatter Effect

Radiation particles emitted by the source in directions away from the detector can reach the source backing, of thickness T, in which they can scatter and some are subsequently reflected back towards the detector. To correct S_{meas} for this backscatter effect, a backscatter factor $f_b(T)$ is defined as

$$f_b(T) \equiv \frac{\text{no. particles recorded with source backing}}{\text{no. particles recorded without source backing}}. \tag{7.5}$$

To evaluate $f_b(T)$ it is assumed there is negligible source self-absorption so that half of the source particles S leave towards the detector and half impinge on the source backing. Because the recorded count rate is proportional to the rate at which particles leave the source in the direction of the detector, the backscatter factor is given by

$$f_b(T) = \frac{0.5S + 0.5S\ \eta_b(T)}{0.5S} = 1 + \eta_b(T), \tag{7.6}$$

where η_b is the *reflection coefficient* and is the probability a particle entering the substrate is reflected back to the surface.

This coefficient increases from 0 to a saturation value η_b as the thickness of the source backing increases from 0 (no backscattering) to a value beyond which negligible particles are reflected. For particles with a finite range R_{eff} saturation reflection is achieved when $T = R_{eff}/2$ since no particles can reach a greater depth and return to the backing surface. In practice, $T \simeq 0.3R_{eff}$ achieves saturation reflection. For particles that are exponentially attenuated with an attenuation coefficient μ, $T \simeq 1/\mu$ results in near saturation. The reflection coefficient usually is a measured quantity, although with modern Monte Carlo transport techniques it can be calculated.

7.2.1 Alpha Particles

The preparation of the radioisotope that emits alpha particles can affect the observed count rate and spectroscopic energy resolution. If the radioisotope is deposited as a thick layer, the alpha particles lose energy as they travel through the source material. The energy loss depends on the alpha-particle trajectory and where in the source material the particle was emitted. Likewise particles emitted towards the material upon which the radioisotope was deposited can make multiple Coulombic scatters and be reflected back towards the detector. This backscatter component can cause an overestimation of the source activity by increasing the count rate over that expected from a source with no backing.

In some cases, alpha-particle sources have a protective covering to reduce risk of contamination from radioisotope leakage. These thin windows, as well as the detector window, can cause alpha particles to scatter away from the detector, thereby causing an underestimation in the source activity. Further, thin windows can absorb energy from the particles, thereby reducing the range and measured energy while increasing the uncertainty for spectroscopy measurements. To minimize the effect of the detector window a thin form of boPET[1] is typically used for alpha-particle source windows and detector windows, thereby, reducing both the effects of energy loss and backscattering. These multiple measurement problems are illustrated in Fig. 7.1. Finally, if there is a medium, such as air, between the source and the detector window further degradation of the energy of the alpha particles occurs.

[1]Biaxially oriented polyethylene terephthalate (boPET) is a stretched polyester film commonly known by trade names such as Mylar. Applied as detector windows, these films are generally between 3 to 6 microns thick (or 0.4 to 0.8 mg cm^{-2}), and being primarily composed hydrogen, carbon and oxygen ($C_{10}H_8O_4$), alpha-particle backscattering is minimal.

Attenuation

Consider the arrangement, shown in Fig. 7.2, in which an alpha particle source of thickness d, plated upon a planar backing, is placed inside the chamber of a gas-filled detector, such that half of the emissions are initially emitted towards the gas, the other half being emitted in the direction of the backing. Under ideal conditions, half of the emissions should be detected.[2] However, particles emerging at oblique angles from the normal (perpendicular) trajectory may pass through so much material that they lose too much energy for detection.

Recall that alpha particles have discrete ranges in matter, denoted here as $R_o \simeq R_{\text{CSDA}}$, about which there is a minimal amount of straggle. As alpha particles are emitted in the source material, they lose energy before emerging into the detector. At some critical angle θ the alpha particle has just enough energy to still be detected as it emerges from the source material, a condition that is dependent on the alpha-particle trajectory and the detector electronic lower level discriminator (LLD) threshold ($E_\alpha \geq E_{LLD}$). This effective range R_{eff} is the maximum distance that the alpha particle can travel through the source material before $E_\alpha < E_{LLD}$, i.e.,

$$R_{eff} = R_o - R_z, \tag{7.7}$$

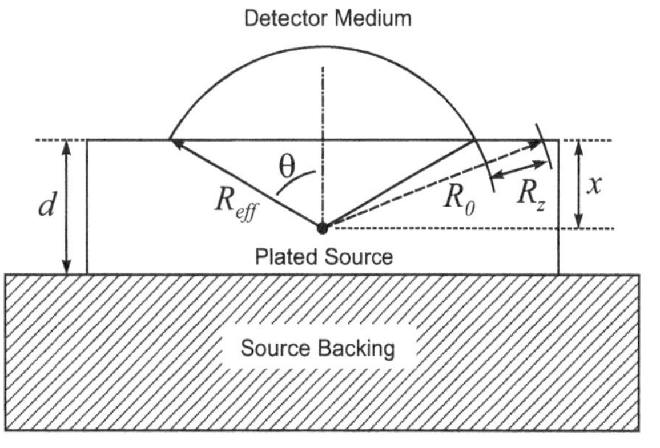

Figure 7.2. Depicted are the average total range R_o and the average effective range R_{eff} of alpha-particle emissions from a plated source of thickness d born at location x.

where R_z is the *residual range*, the range in the source material that an alpha particle with energy E_{LLD} would have.

Consider an alpha particle emitted at depth x in the source material. Because the emission is isotropic, the probability it reaches the source surface with sufficient energy to be detected is $p_{esc}(x) = \Delta\Omega(x)/4\pi$, where $\Delta\Omega(x)$ is the solid angle subtended by the disk on the surface defined by the cone with the half-angle θ in Fig. 7.2. A differential solid angle in spherical coordinates (see Section 7.4) is $d\Omega = \sin\theta\, d\theta\, d\psi$ where θ is the polar angle and ψ is the azimuthal angle. Hence

$$\Delta\Omega(x) = \int d\Omega = \int_0^{2\pi} d\psi \int_0^{\theta} \sin\theta\, d\theta = 2\pi(1-\cos\theta) = 2\pi\left(1 - \frac{x}{R_{eff}}\right). \tag{7.8}$$

Note that if $x \geq R_{eff}$, $p_{esc}(x) = 0$ because either no particles can escape the source material or escaping particles have energies below E_{LLD} and cannot be detected. Then the probability P_a an alpha particle born uniformly within the source material escapes attenuation from the source material is given by

$$P_a^{esc} = \int_0^{\min(d,R_{eff})} \{\text{probability an alpha particle is emitted in } dx \text{ about } x\} \times \{p_{esc}(x)\}. \tag{7.9}$$

Because the radioisotope is distributed uniformly in the source material, the probability an alpha particle is emitted in dx about x is dx/d so that

$$P_a^{esc} = \int_0^{\min(d,R_{eff})} \left\{\frac{dx}{d}\right\}\left\{\frac{\Delta\Omega(x)}{4\pi}\right\} = \frac{1}{2d}\int_0^{\min(d,R_{eff})} \left(1 - \frac{x}{R_{eff}}\right) dx. \tag{7.10}$$

[2]Such a detector is often referred to as a 2π detector, because the source emits particles into the detector chamber over a solid angle of 2π steradians.

Evaluation of the integral yields

$$P_a^{esc} = \begin{cases} \frac{1}{2}\left[1 - \frac{d}{2R_{eff}}\right], & \text{if } d < R_{eff} \\ \frac{R_{eff}}{4d}, & d \geq R_{eff} \end{cases} . \tag{7.11}$$

In the above analysis it was assumed that the source material composition is known so that the ranges R_o and R_{eff} are known. Unfortunately, the chemical form of the source material is seldom known because source manufacturers often regard this information as proprietary.

Attenuation of alpha particles through the intermediate medium between the detector and source, typically air, causes energy loss and range reduction. Alpha particle detectors must be placed close enough to the source such that the particles can enter the detector medium with sufficient energy to be detected. In the prior case of the 2π detector, the alpha-particle source is placed inside the detector cavity. However, operation of common survey meters and thin-window detectors does not permit such a placement, and the sensitive detector volume must be placed within the range of the alpha particles. Further, the thin window of the device must be taken into account as part of the total range reduction.

The total energy attenuation becomes a function of the source thickness, the medium distance between the detector and source, and the detector entrance window thickness. Often it is the "mass thickness" of a material that is quoted for range (see Fig. 7.3), defined as the product of the alpha-particle range and the mass density of the attenuating material, i.e., $R_{\text{mass}} = \rho R_o$ g cm^{-2}.

Backscattering

Consider the backscatter events shown in Fig. 7.1. From the general expression for Rutherford backscattering [Kaplan 1962], it is unlikely that an alpha particle backscatters towards the detector from a nucleus in the source backing material in a single scattering event. Instead, it is multiple small-angle scatters that can cause the alpha particle, originally on a trajectory towards the source backing, to eventually emerge from the source towards the detector. The overall consequence of alpha-particle backscattering is to produce a higher count rate than that without backscattering. Hence a backscattering correction factor should be estimated and used when reporting the measured source activity.

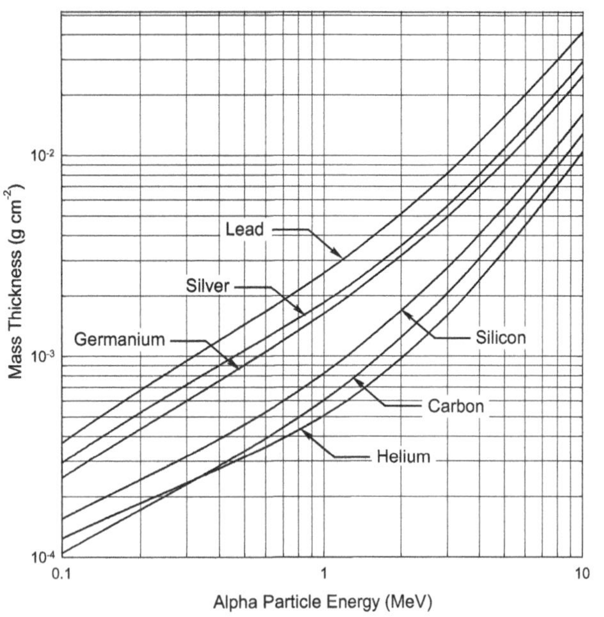

Figure 7.3. Alpha particle mean range R_p as a function of mass thickness and energy (from Ziegler, 2013).

Careful source preparation may allow an experimenter to produce an ultra-thin source backing, which produces minimal backscattering, and, thereby, effectively eliminates the backscattering problem. However, repeatability of source preparation often becomes an issue, compounded by calibration issues. Instead, it is far more common to prepare sources that reach a saturation backscatter limit. Intuitively, to achieve this limit, the source support (backing) should be $\geq R_o/2$ for the backing material; hence, a particle scattering deeper than $R_o/2$ cannot emerge from the backing material.

A general expression for alpha-particle backscattering is [Crawford 1949]

$$\eta_b = K \left[\frac{\sum_{i=1}^{N} f_i A_i^2}{A_r^{3/2} \sum_{i=1}^{N} f_i A_i^{1/2}} \right]^{1/2}, \tag{7.12}$$

where η_b is the probability that a particle is backscattered if it is emitted in the direction of the source backing, K is a constant, A_r is the atomic weight of a reference backing material, A_i are the atomic weights of the constituents in the backing material (or window material), and f_i are the atomic fractions of the elements (see also [Deruytter 1962]). For elemental source backings, Eq. (7.12) reduces to

$$\eta_b = K \left[\frac{A}{A_r} \right]^{3/4}. \tag{7.13}$$

From Eq. (7.13), the amount of backscatter increases as A (or Z) increases, indicating that low Z materials are best used for alpha-particle source backing.

Reported experimental results indicate that the alpha-particle backscatter probability measured with ^{210}Po ($E_\alpha = 5.307$ MeV) is [Hutchinson et al. 1968]

$$\eta_b = \frac{(1.30 \pm 0.08) \times 10^{-3} Z}{C_0}, \tag{7.14}$$

where C_0 is a constant measured to be 2.525 ± 0.002. Variations with alpha-particle energy and the thickness of the backing material were not considered in Hutchinson's work; presumably, saturation backscatter coefficients were investigated and measured. According to Eq. (7.14), a 5.31-MeV alpha-particle source on an Al backing has a 0.67% backscatter coefficient, while the same alpha-particle source applied to a Au backing has a 4.07% backscatter coefficient. Hutchinson et al. [1968] also report that the surface morphology of the source backing affects the backscatter coefficient, η_b being generally higher for polished source backings as opposed to unpolished source backings.

With the source of sufficient thickness to reach the asymptotic backscatter limit, Rossi and Staub [1949] approximate the backscatter probability as

$$\eta_b \simeq 0.201\phi, \tag{7.15}$$

where ϕ is a function of the effective range in the backing material, defined as $R_e = R_o - R_z$. To some degree, Eq. (7.15) takes into account the energy of the alpha particle, in which η_b is dependent upon the necessary effective range R_e. Rossi and Staub [1949] published a short table of values for ϕ that can be used to estimate η_b.

Little, if any, experimental data is available about the energy distribution of the alpha particles that are backscattered from the source backing. However, with modern Monte Carlo methods it is now possible to track alpha particles as they undergo Coulombic scattering and estimate η_b and the energies of backscattered alpha particles from any backing material. Example calculations are shown in Fig. 7.4 for an isotropic source. Notice the energy spectra are peaked towards very lower energies (for alpha particles that penetrated relatively far into the backing and took many small-angle scatters to re-emerge) and at energies close to the source energy (from particles that entered the backing at grazing angles and, hence, needed only a few scatters to re-emerge). The area under each histogram gives η_b. For example, for aluminum $\eta_b = 0.83\%$. However, if the low-energy reflected alpha particles below 0.5 MeV are ignored, then $\eta_b = 0.67\%$, a result in agreement with Hutchinson's measured value quotes from above.

MCNP6 has been used to calculate the total backscattering probability for different backing materials (see Fig. 7.5). The variation of $\eta_b(sat)$ with the Z number of the backing medium is well described, as shown

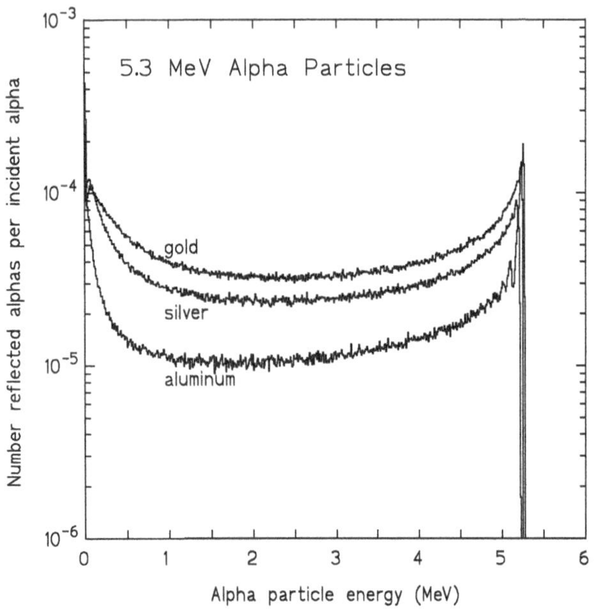

Figure 7.4. A histogram of the number of ^{210}Po alpha particles backscattered from three thick metal surfaces. Energy bin widths are 0.01 MeV. Results obtained with MCNP6 [2013].

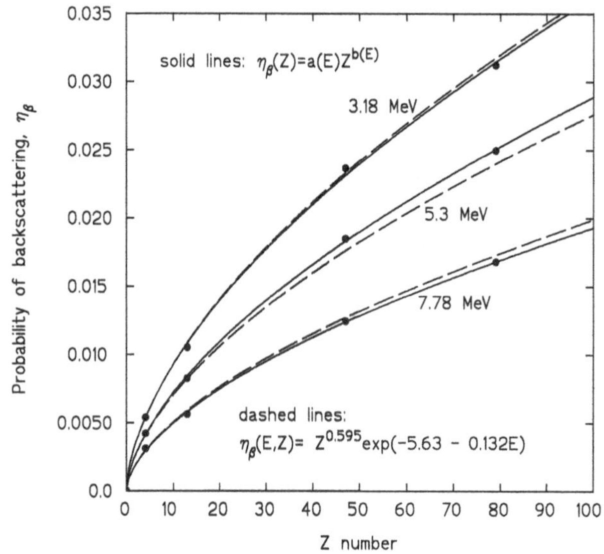

Figure 7.5. The saturation backscatter probability for alpha particles as estimated with MCNP6 [2013] (solid circles) as a function of the backscattering material Z number. The associated uncertainty is less than 0.5% for all points. Here E is the initial energy of the alpha particle and the energies 3.18 MeV, 5.30 MeV, and 8.78 MeV are the energies of the alpha particles emitted by ^{148}Gd, ^{210}Po, and ^{212}Po, respectively.

by the solid lines in Fig. 7.5, by the simple relation $\eta_b(Z) = aZ^b$, where a and b depend on the initial energy E_α of the alpha particle. Based on MCNP6 calculations shown in Fig. 7.5, the probability of backscattering is approximately (see the dashed lines in Fig. 7.5)

$$\eta_b(E_\alpha, Z) \simeq Z^{0.595}\exp(-5.63 - 0.132E_\alpha), \tag{7.16}$$

where E_α is the initial energy of the alpha particle (in MeV).

In general, when performing alpha-particle measurements, the experimenter should correct the acquired count rate g, and hence the measured activity, by using a backscattered corrected count rate $\tilde{g}$

$$\tilde{g} = \frac{g}{1+\eta_b} = \frac{g}{f_b}. \tag{7.17}$$

7.2.2 Beta Particles

As with alpha-particle sources, beta-particle sources can also suffer from attenuation (self-absorption) in the source material and its encapsulation and from backscatter effects. This latter effect can be particularly severe for backscattering from heavy elements. Hence, correction for these effects is important to measurement accuracy.

Attenuation

Consider a volume of source material of thickness d in which beta-emitting radionuclide is uniformly distributed. At any depth x in the source material beta particles are emitted isotropically with a continuum of energies up a maximum energy $E_{\max}$ such that $\chi_\beta(E)\,dE$ is the probability the beta particle (either an electron or positron) is emitted with energy in dE about E. The probability a beta particle reaches the surface of the source material and is not absorbed in the source film can formally be written as

$$P_a^{esc} = \int_0^{E_{\max}} \chi_\beta(E)dE \int_0^{\pi} d\theta \int_0^{d} dx f_\beta(x,\theta,E), \tag{7.18}$$

where $f_\beta(x,\theta,E)$ is the probability a beta particle emitted at depth x with energy E reaches the surface traveling in direction θ with respect to the surface normal. However, there is no analytical solution or even numerical solution of Eq. (7.18), because there is no known expression for $f_\beta(x,\theta,E)$. Unlike the case of alpha particles which travel in straight lines with a well-defined range, beta particles travel along twisted trajectories (see Fig. 4.22) that can terminate at any distance from the emission point up the *continuous slowing down approximation* (CSDA) range of the beta particle. To evaluate P_a^{esc} accurately for beta particles, it is necessary to employ Monte Carlo methods.

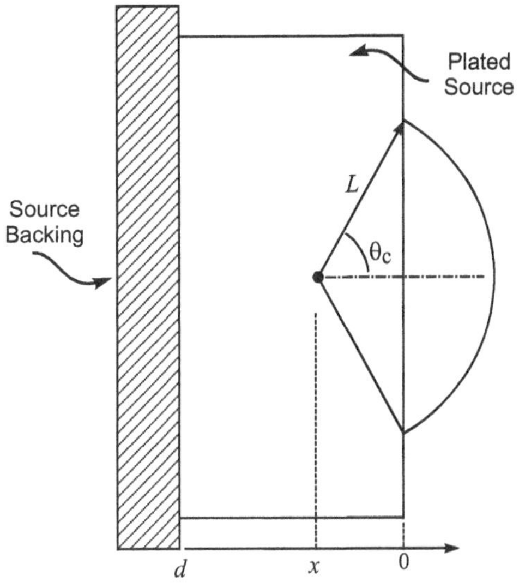

Figure 7.6. Geometry for a beta-particle source.

Although not strictly accurate, the attenuation of beta particles through a material can sometimes be modeled by an exponential function

$$\frac{I(x)}{I_0} = I_0 e^{-\mu x}, \tag{7.19}$$

where $I(x)/I_0$ is the fraction of the initial ionization measured (say) by an ion chamber after the beta particles have passed through an absorber of thickness x and μ is an effective linear attenuation coefficient. Note that Eq. (7.19) is not based on a physical phenomenon, but instead is derived from rather remarkable linear fits of beta-particle transmission data plotted upon semilogarithmic graphs [Baltakmens 1970].[3] An empirical relation that gives the mass absorption coefficient $\rho\mu$ is [Evans 1955]

$$\frac{\mu}{\rho} = \frac{17}{E_{\max}^{1.14}} \qquad \text{cm}^2\text{g}^{-1}, \tag{7.20}$$

where $E_{\max}$ is in MeV. This mass absorption coefficient only weakly depends on the Z number of the absorber increasing slightly with Z.

Consider the beta source shown in Fig. 7.6 with thickness d in which beta particles can be emitted from any depth x in the film. Assume that Eq. (7.19) is an adequate estimate for beta-particle attenuation, thereby removing the energy dependence in Eq. (7.18). Further, assume that the beta particles travel an average distance $\overline{L}$ from their emission point (an average over the beta energy spectrum). Then the escape probability for a beta particle emitted at depth x is the fraction emitted into a cone of directions with half angle $\theta_c = \cos^{-1}(x/\overline{L})$, namely $\Delta\Omega(x)/4\pi$ (see Eq. (7.8)) times the probability the beta particle is not exponentially attenuated on the way to the surface, i.e.,

$$P_a^{esc} = \frac{1}{2}\int_0^{\theta_c(x)} \int_0^{\min(\overline{L},d)} \left(1 - \frac{x}{\overline{L}}\right) \exp\left[-\mu\left(d - \frac{x}{\cos\theta(x)}\right)\right] dx\, d\theta. \tag{7.21}$$

Even numerical evaluation of this result is difficult because average ranges $\overline{L}$ for a beta spectrum are generally unknown. But more important, this approach based on exponential attenuation of beta particles

[3]Exponential attenuation implies an infinite range which is not true for beta particles. Equation (7.19) applies only to the initial portion of the attenuation curve.

along a straight line to the source surface is approximately true only for the initial portion of the intensity versus penetration relation. Moreover, this approach does not address the change of the energy spectrum and angular distribution of the escaping electrons. However, many texts (including this one) continue to propagate this questionable approach, primarily because of its simplicity and the avoidance of a full-blown electron transport calculation.

Backscattering

An empirical formula for the backscatter coefficient for normally-incident monoenergetic electrons was proposed by Tabata et al. [1971], namely

$$\eta_b = \frac{b_1 \exp[-b_2 Z^{-b_3}]}{1 + (b_4 + b_5 Z^{-b_6})\tau^{b_7-(b_8/Z)}}, \tag{7.22}$$

Table 7.1. Constants for the determination of beta-particle backscattering. From [Tabata et al. 1971].

Constant	Value	Constant	Value
b_1	1.15 ± 0.06	b_5	15.7 ± 3.1
b_2	8.35 ± 0.25	b_6	1.59 ± 0.07
b_3	0.525 ± 0.020	b_7	1.56 ± 0.02
b_4	0.0185 ± 0.0019	b_8	4.42 ± 0.18

where $\tau = E_\beta/(m_e c^2)$. Here E_β is the initial energy of the beta particle, Z is the atomic number of the target material, $m_e c^2$ is the rest mass energy equivalent of an electron, and the constants a_1 through a_8 are listed in Table 7.1. For energies from approximately 50 keV up to 22 MeV, and for $Z \geq 6$, Eq. (7.22) has $\simeq$ 7% RMS deviation from the data measured by Tabata et al. [1971]. The results of Eq. (7.22) for several elements are shown in Fig. 7.7. Although Eq. (7.22) is often used for backscattering of beta particles emitted isotropically from radioisotopes, it is based on experimental data in which electrons were normally incident on the substrate. As calculated by Kuzminikh and Vorobiev [1975] the backscatter coefficient varies strongly with the incident angle with respect to the surface normal and η_b increases as the incidence angle increases. Because beta particles are emitted isotropically, an uncollimated source will produce a different average backscatter coefficient than described by Eq. (7.22). Thus, Eq. (7.22) likely underestimates η_b, but its use for beta particles, which are incident at all angles, is continued for lack of an alternative.

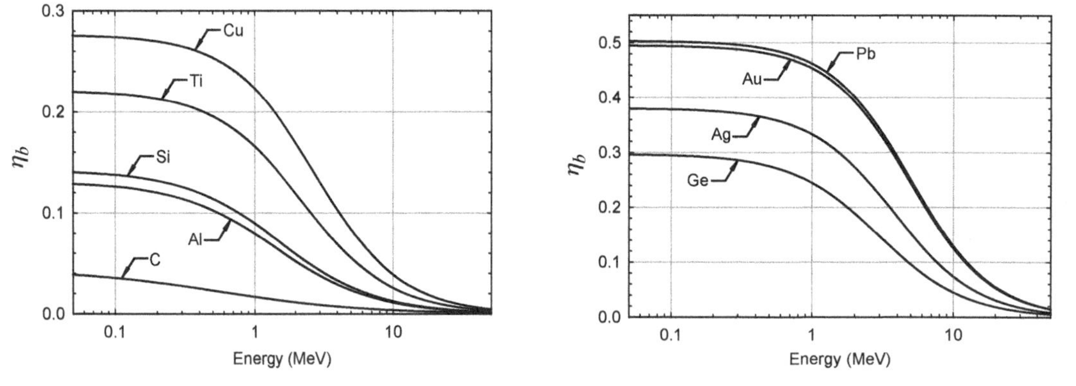

Figure 7.7. Calculated beta-particle backscatter coefficients as functions of target Z and beta-particle energy (after Tabata [1971]).

Monte Carlo Correction Factors

For beta particles, the only accurate way to account for source back scattering and attenuation by the source encapsulation is to use electron transport techniques. Several Monte Carlo codes are available to perform such calculations. Many use condensed-history models, but some are now able to use point-wise

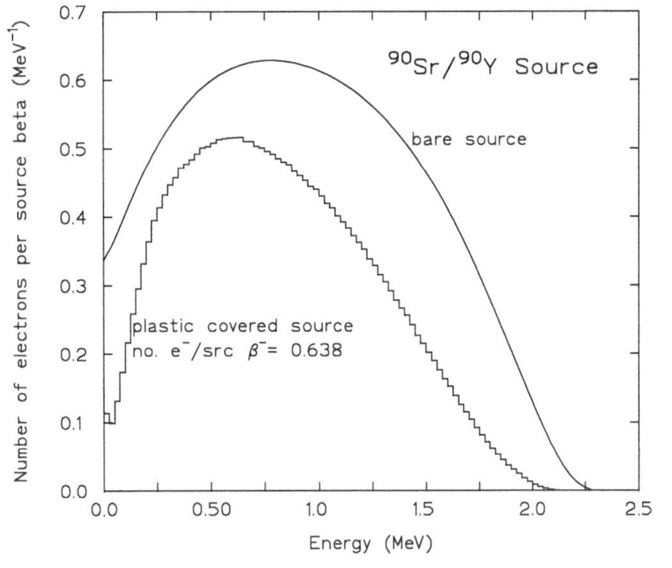

Figure 7.8. The MCNP6 calculated energy spectrum of electrons leaving a 1.55 × 2.54 mm plastic disk containing a $^{90}Sr/^{90}Y$ beta particle source in the center of the disk. In this example, 0.638 electrons are emitted by the disk per source beta particle.

electron interaction models. Another advantage to this approach is that both the backscatter and source encapsulation effects can be determined in a single calculation. Moreover, the energy spectrum of the emitted beta particles and electrons can be found at the same time. In Fig. 7.8 the emitted spectrum from a common $^{90}Sr/^{90}Y$ check source is shown.

7.2.3 Penetrating Radiations (γ rays, x rays, neutrons)

Although penetrating radiations, such as gamma rays, x rays, and neutrons, can all have backscatter and attenuation effects from the source, these effects are generally small and ignored. The exception may be for low-energy photons in heavy encapsulation. However, scattering from walls, floors and other massive objects in the room can make a significant difference in a radiation measurement, especially when dealing with neutron sources.

Often shielding is placed around and in close proximity to the source and detector to reduce roomshine and background from radiation sources outside the counting chamber. This shielding is important if a correct measurement of the source activity is to be obtained. Standard lead bricks are often used to provide such shielding. However, for photon sources, such shielding can emit copious 88-keV fluorescence when source photons photoelectrically interact with the lead. To minimize this secondary radiation, the counting chamber can be lined with metals of decreasing Z values, e.g., copper followed by aluminum. To design proper shielding and to correct for roomshine, transport codes must generally be used.

7.2.4 Source Decay During Measurement

A measurement taken in the presence of a radiation background overestimates the result for any sample under investigation. As covered in Chapter 6, corrections for the background counts or background count rate, and the associated error, must be considered for any radiation measurement.

If background is negligible by comparison to the radioactive source emission rate, but the radioactive source under test suffers appreciable decay during the measurement time t_1, i.e., $\lambda t_1 \gtrsim 0.01$, then the count rate n_0 at starting time t_0 can be determined from the total observed counts recorded between t_0 and t_1. Consider the simple case of a radionuclide sample of N_0 identical radionuclides at time t_0. The decay constant is λ and the radionuclides emit a single particle per decay, i.e., the frequency or *branching ratio* per decay is 1. Further, the detector efficiency for recording this particle is denoted by ϵ. The count rate $n(t)$ at $t \geq t_0$ is

$$n(t) = \epsilon \lambda N_0 e^{-\lambda t} \tag{7.23}$$

The average count rate $\overline{n}$ between $t_0 = 0$ and t_1 is

$$\overline{n} = \frac{\lambda N_0 \epsilon}{t_1} \int_0^{t_1} e^{-\lambda t} dt = -\frac{n_0}{\lambda t_1}(e^{-\lambda t_1} - 1), \tag{7.24}$$

where $n_0 = \lambda N_0 \epsilon$ is the count rate at t_0. Rearrangement gives

$$n_0(t_0) = \overline{n} \frac{\lambda t_1}{1 - e^{-\lambda t_1}}. \tag{7.25}$$

Finally the count rate at any time $t > 0$ is

$$n(t) = \overline{n} \frac{\lambda t_1}{1 - e^{-\lambda t_1}} e^{-\lambda t}. \tag{7.26}$$

This result applies to the more usual case in which the radionuclide emits M particles, each with a different energy. Denoting the emission frequency of the ith particle by f_i and the detector efficiency for the ith particle by ϵ_i yields the initial count rate n_0 as

$$n_0 = \lambda N_0 \sum_{i=1}^{M} \epsilon_i f_i. \tag{7.27}$$

7.2.5 Contamination

As a final word on errors associated with radioactive samples and preparation, there is always the possibility of contaminating the detector work area and/or detector with radioactive material. Radioactive sources that produce airborne daughter products (^{222}Rn for instance) can contaminate the detector aperture, a situation that causes the detector itself to become a background radiation source. Alpha particle sources are sometimes encapsulated with a boPET thin film to reduce leaks and contamination, which unfortunately reduces their use as spectroscopy standards. Gamma-ray and many beta-particle sources can be encapsulated in plastic to reduce the risk of contamination. As a precaution, radioactive sources should be checked periodically for leaks.

7.3 Detector Effects

7.3.1 Scattering and Absorption by the Detector Window

For almost all radiation detection systems the source is located outside the sensitive volume of the detector which is enclosed in some housing. Radiation then must enter the detector through a window or aperture.[4]

[4]There are exceptions. For example, gas flow proportional counters and some liquid scintillators have the source in the detecting medium.

As radiation particles reach the detector window, most pass through into the sensitive detector volume and are recorded. But a small fraction are either absorbed or scattered back by the window material as depicted in Fig. 7.1. Even if a particle passes through the window it may lose sufficient energy in doing so that the signal produced by it falls below the lower level discriminator (LLD) setting and so the particle is not recorded.

Correction for window effects is very difficult because such corrections depend on the type of radiation, its energy, its angle of incidence to the window, the window material, and the window thickness. Because of this complex dependence, there is no direct way to correct for window effects. However, to minimize window effects one should make the window as thin as possible and of a material that is as transparent as possible to the incident radiation. For low-energy photons and changed particles windows of low Z material are generally better than those made from high Z materials. For neutron detectors, window materials with very small neutron cross sections at the neutron energies of interest are preferred. Common window materials are boPET, glass, mica, or low-Z metals. Often the absorption effect of the window is incorporated into the efficiency of a detector. But for an energy spectrometer, corrections for energy loss in the window must be made separately to the spectrum.

7.3.2 Time Interval Distribution between Radioactive Decays

The times at which interactions occur in a detector from particles emitted from a radioactive (or stochastic) source have a Poisson distribution (see Section 6.7). The mean or expected rate at which interactions (or events) occur is denoted by $\bar{r}$. In a time interval t the mean number of events is, thus, $\mu = \bar{r}t$ with a variance of μ. Thus, the probability of having n events in time t is given by

$$\mathcal{P}(n|\mu) = \frac{(\bar{r}t)^n}{n!}e^{-\bar{r}t}. \tag{7.28}$$

In the following section on dead time, information about the time intervals between events is needed. In particular, the probability there is no event in a time t following an event, $P_0(t)$. From the Poisson PDF, the probability that no events are observed within a time interval t is

$$\mathcal{P}(0|\bar{r}t) = \frac{(\bar{r}t)^0}{0!}e^{-\bar{r}t} = e^{-\bar{r}t}. \tag{7.29}$$

The probability that an event occurs within some differential time interval dt about t is expressed as $\bar{r}dt$. From these two results, it follows that the probability of observing no event in the time interval $(0, t)$ and an event in the next dt interval is

$$dP_0(t) = \left(e^{-\bar{r}t}\right)(\bar{r}dt) = \bar{r}e^{-\bar{r}t}dt. \tag{7.30}$$

Therefore, the total probability that there is no event within time t is expressed as the probability that the event occurs anytime time past t, i.e.,

$$P_0(t) = \int_t^\infty dP(t) = \int_t^\infty \bar{r}e^{-\bar{r}t}dt = e^{-\bar{r}t}. \tag{7.31}$$

The probability that there are one or more events within time t is then $[1 - \exp(-\bar{r}t)]$.

7.3.3 Dead Time

All radiation detection systems operating in pulse mode have a limit on the maximum rate at which data can be recorded. The limiting component may be either the time response of the radiation detector, which is usually the case for the Geiger-Müller counter, or may be the resolving capability of the electronics.

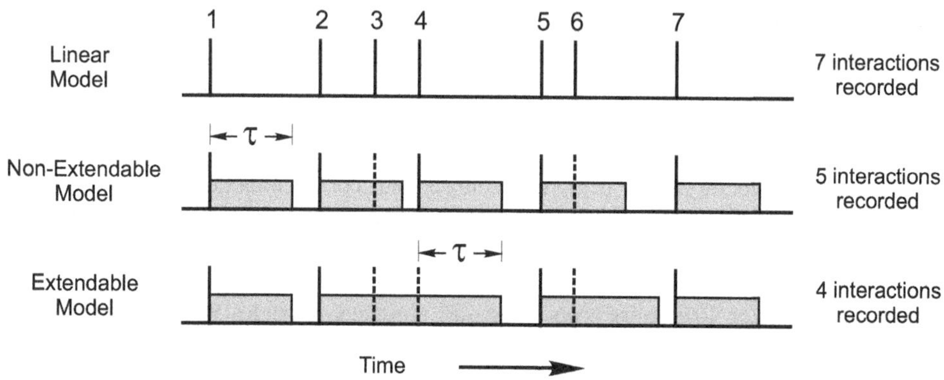

Figure 7.9. The three basic models for detector interval operation. The linear model assumes that the pulse is a delta function and has no processing time and, hence, has no dead time. The non-extendable dead time model assumes that once a pulse is registered, a preset time interval τ must pass, before another event can be processed. The extendable dead-time model assumes that each pulse entering the system restarts the dead-time interval, even during a previously initiated dead-time interval (after NCRP, 1985).

Although a radiation counting system is treated as a system that counts radiation interactions in a detector, it is instead actually recording the number of detector response time intervals initiated by a radiation particle, each time interval being terminated by the interval processing time. In other words, the detector records intervals rather than events. If two radiation particles interact within the detector during a single processing interval, only one of the interactions is recorded.

After a radiation interaction, radiation detectors generally require time to reset before they are fully sensitive to record a subsequent radiation interaction, especially for pulse mode operated electronic detectors. The time during which a detector is insensitive is commonly referred to as *dead time*, although the term is often misused. Dead time actually refers to a time duration in which a detector in insensitive to a radiation interaction. *Resolving time* is the period required before the detector can form a signal pulse large enough to register in the detector system, and is dependent upon the detector characteristics and the preset electronic conditions (lower level discriminator setting, pulse shaping time, etc.). The *recovery time* is the period required after a radiation interaction for the detector to reset to its initial starting condition. Often the dead time and resolving time are interchanged. Although technically different, the overall effect of either dead or resolving times on the radiation counting system is the same. Detector insensitivity and dead-time issues can arise from many sources, such as limitations in pulse processing times, limitations in detector signal formation, and time constants required for the device to return to its full sensitivity. The details of pulse processing are saved for Ch. 22 on nuclear electronics. In the present section, the effect that dead time and resolving time have on a radiation measurement is presented. Also, because both dead time and resolving time have the same effect on a radiation measurement, for simplicity, both are referred to in this chapter as dead time.

When a radiation interaction occurs in a detector and produces a measurable event, the detector and counting system become occupied while processing the event. As a consequence, the counting system does not respond to a subsequent event until some time τ has passed. Because the counting system is unavailable for a portion of time during a measurement, the recorded number of counts is consequently less than the number of interactions that actually occurred in the detector during the measurement time. Three possible ways in which a counting system processes a radiation event are shown in Fig. 7.9.

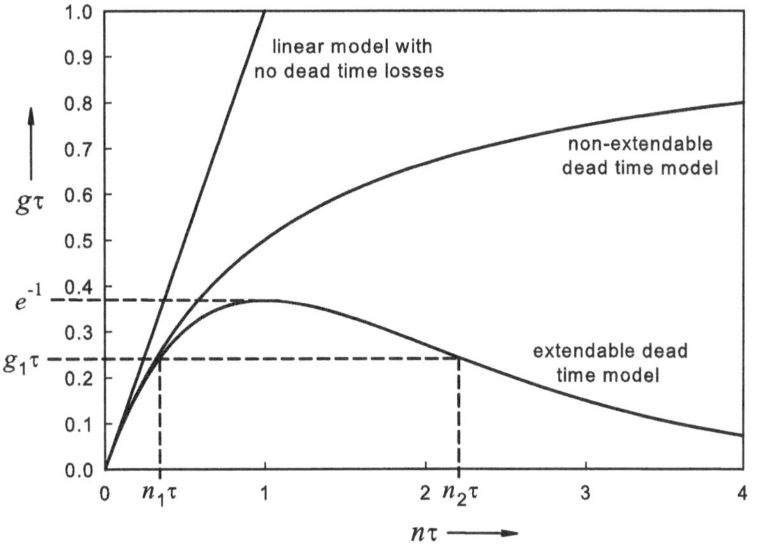

Figure 7.10. The three basic models for detector interval operation. The linear model assumes no processing time, hence has no dead-time losses. The non-extendable dead-time model shows a gradual reduction in the observed events $g\tau$ as the interaction rate $n\tau$ increases. The extendable dead-time response has a maxima in observed events as the interaction rate $n\tau$ increases, beyond which the observed event rate $g\tau$ decreases. Note that the extendable model produces for a single observable event rate, $g_1\tau$, two possible interaction rates (shown as $n_1\tau$ and $n_2\tau$).

7.3.4 Models for Dead Time

There are three basic models regarding pulse intervals produced in a radiation detector. These models, each with different processing times, are depicted in Fig. 7.10. The simplest is the linear model, in which events (pulses) are treated as delta functions and there is no processing time interval associated with a radiation event. The second model, referred to as the non-extendable (or non-paralyzing) model (NCRP, 1985; Evans, 1955), treats the intervals as preset time periods, unalterable in length by additional events arriving during the pulse processing interval. Hence, after the time interval τ begins, it remains the same length regardless if more events interact in the detector during the interval τ. The third model, referred to as the extendable (or paralyzing) model, treats an initial pulse interval τ as being restarted with another time τ for subsequent interactions arriving during the entire pulse interval. Hence, regardless of when in the initial interval a previous radiation interaction occurs, the interval is extended from that time forward by τ. The total processing interval time, whether using the extendable or non-extendable model, is generally referred to as the system *dead time* and denoted by τ.

Typically, the dead-time responses of detector systems consist of a combination of extendable and non-extendable characteristics. However, most detection systems are designed in such a way that the extendable dead-time effect is the least contributor, with the goal of producing a system that performs with non-extendable characteristics.

Extendable (Paralyzable) Dead Time

In an extendable system, only events with time intervals greater than the dead time τ can be registered, because there are no recorded events with time intervals less than the dead time. Suppose n represents the *true* average number of events per unit time. For large n, the number of events g, per unit time, observed

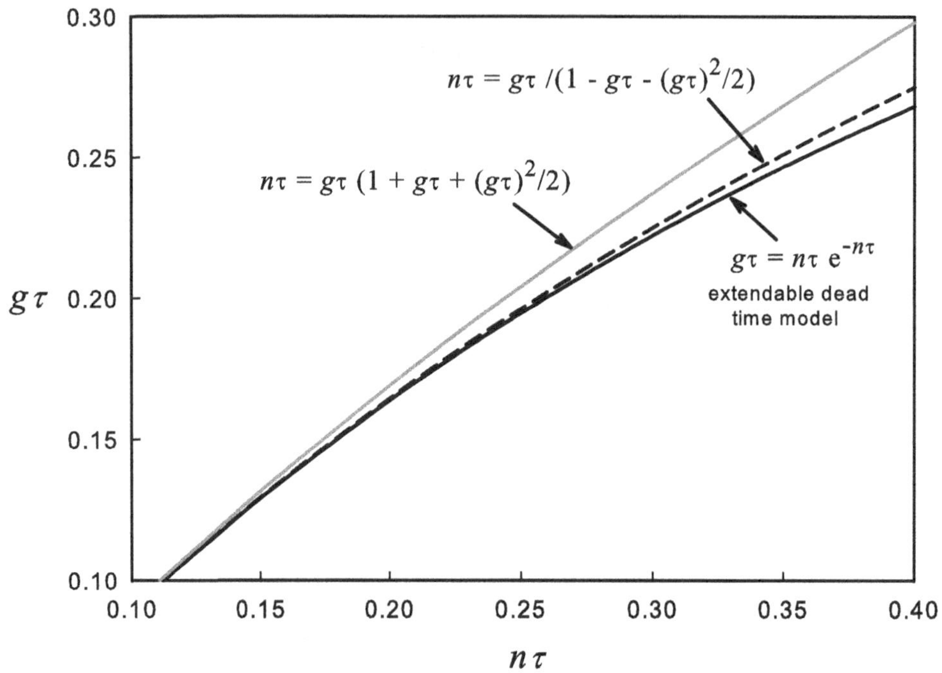

Figure 7.11. Comparison of extendable dead time and approximations from Eq. (7.35) and Eq. (7.37).

with intervals longer than τ is found with Eq. (7.31) to be

$$g = ne^{-n\tau} \qquad \text{or} \qquad g\tau = n\tau e^{-n\tau}. \tag{7.32}$$

If the true event rate n is small, then few events are missed, and $n \gtrsim g$. To find the exact true event rate after observing g, n must be expressed as a function of g. Unfortunately, a closed form solution to Eq. (7.32) cannot be found analytically. For small dead-time losses, $g \simeq n$, and Eq. (7.32) can be approximated as [NCRP 1985]

$$g\tau = n\tau e^{-n\tau} \simeq n\tau e^{-g\tau}, \tag{7.33}$$

which yields

$$n\tau \simeq g\tau e^{g\tau}. \tag{7.34}$$

This result can be expanded for $g\tau \ll 1$ as

$$n\tau \simeq g\tau\left(1 + g\tau + \frac{(g\tau)^2}{2!}\right), \tag{7.35}$$

which, unfortunately, underestimates $n\tau$. However, for $\epsilon = g\tau \ll 1$, the following approximation compensates, partially, for the underestimate

$$\left(1 + g\tau + \frac{(g\tau)^2}{2!}\right) \simeq (1+\epsilon) \simeq \frac{1}{(1-\epsilon)}. \tag{7.36}$$

Substitution of this result into Eq. (7.35) gives another approximation for $n\tau$, namely

$$n\tau \simeq \frac{g\tau}{1 - g\tau - (g\tau)^2/2}. \tag{7.37}$$

Equation (7.37) produces a result with less than 0.1 percent error for dead-time losses of $\leq 10\%$. A comparison of Eq. (7.32) with Eq. (7.35) and Eq. (7.39) is shown in Fig. 7.11.

For high dead-time losses, an exact solution for n can easily be found numerically by iteration. The maximum observable event rate, g_{max}, for an extendable system can be found by differentiating Eq. (7.32) and solving for maxima, i.e.,

$$\frac{dg}{dn} = e^{-n\tau}(1 - n\tau) = 0, \tag{7.38}$$

from which the maximum is seen to occur at $n\tau = 1$. The maximum is thus

$$g_{max} = ne^{-1} = 0.368n. \tag{7.39}$$

This result implies a dead-time loss exceeding 63% at the count rate maxima. The extendable dead-time response function is shown in Fig. 7.10 where it is observed that the observed event rate decreases beyond $n\tau = 1$ even as n is increasing. The extendable dead-time response can be dangerous in high radiation environments. An operator using a radiation detection device with extendable dead time can observe a low count rate, when in fact the radiation environment is actually at highly dangerous levels.[5]

Non-Extendable (Non-Paralyzable) Dead Time

For a system with non-extendable dead-time intervals, a system is insensitive to radiation for time interval τ directly after an event, but no longer than this time interval. Therefore, for an observed count rate g over a measurement interval t, the system is dead for the fraction of time $n\tau$ and, therefore, the system is *live* for the fraction of time $1 - n\tau$. The true count rate n is related to g by

$$\frac{g}{n} = 1 - g\tau \quad \text{or} \quad n = \frac{g}{1 - g\tau}. \tag{7.40}$$

At low count rates, $g\tau \ll 1$, and this last result can be approximated as

$$n \simeq g(1 + g\tau). \tag{7.41}$$

As $n\tau$ increases, the maximum number of events that can be observed is limited to the maximum number of intervals τ that fit within the counting time interval t. Hence, the maximum observed count rate asymptotically approaches $g_{max} = 1/\tau$. The non-extendable dead time response function is shown in Fig. 7.10 where it is seen that the observed event rate continues to increase as $n\tau$ increases, asymptotically approaching g_{max}.

Note that the term $g\tau$ is the fraction of the time that the detector is unable to respond to additional ionization in the active volume of the detector. In the design of a counting system, it is best to minimize these losses by trying, if possible, to ensure that $g\tau \lesssim 0.05$. For example, for a GM counter with a typical dead time of $\tau = 100$ μs, the maximum count rate would be 500 counts/s.

[5]Such an event occurred after the Chernobyl accident, when Geiger-Müller (GM) counters were used as survey meters. Radioactive core debris outside the reactor building, after the explosion, caused such high dead times that the GM counters falsely showed low level radiation fields, when in reality the fields were dangerously high.

7.3.5 Counting Error Associated with Dead Time

The variance of the observed counts is often assumed to be that of a Poisson distribution and is frequently approximated with a normal distribution. Then the observed count rate g over some counting interval t is assumed to be the total number of true events for the measurement, i.e., $N = \bar{r}t$. Such an approach is valid for detector systems with negligible dead times in which $g\tau \simeq n\tau$.

However, for systems with appreciable dead-time losses, those theoretical intervals Δt between events for which $\Delta t < \tau$ are removed; hence, the interval distribution of recorded events does not strictly follow a Poisson PDF and the variance must be corrected. Recall that a Poisson distribution has a variance equal to the total number of observed counts, where $\sigma^2 = gt$ and with standard deviation $\sigma = \sqrt{gt}$. However, a counting system actually records the number of time intervals registered during a measurement, and not the actual number of events occurring in the detector. Because there are no intervals recorded for which $\Delta t < \tau$, the actual standard deviation for nt is reduced by

$$\sigma(gt) = (1 - g\tau)\sqrt{gt}. \tag{7.42}$$

Furthermore, as derived by Vincent [1973], the standard deviation for the corrected number of recorded counts nt is greater than the Poisson prediction by

$$\sigma(nt) = \sqrt{(1 + n\tau)\, nt}. \tag{7.43}$$

From Chapter 6, the net count rate is determined by $n = g - b$, where g is the observed count rate and b is the observed background count rate. However, this result does not correct g and b for the detector dead time. Denote by a subscript c the dead-time corrected rates, and assume a non-paralyzable detector, then from Eq. (7.40) the corrected true event rate becomes

$$n_c = g_c - b_c = \frac{g}{1 - g\tau} - \frac{b}{1 - b\tau} = \left(\frac{G}{t_G}\right)\frac{1}{1 - (G\tau/t_G)} - \left(\frac{B}{t_B}\right)\frac{1}{1 - (B\tau/t_B)}, \tag{7.44}$$

where G is the total number of counts observed over measurement time t_G, and B is the total number of background counts observed over measurement time t_B. Use of error propagation, discussed in Section 6.9, the variance is

$$\begin{aligned}\sigma^2(n_c) = \left(\frac{\partial r}{\partial G}\right)^2 \sigma_G^2 + \left(\frac{\partial r}{\partial B}\right)^2 \sigma_B^2 &= \left[\left(\frac{1}{t_G}\right)\frac{1}{1 - (G\tau/t_G)} + \left(\frac{G}{t_G}\right)\left(\frac{1}{1 - (G\tau/t_G)}\right)^2\left(\frac{\tau}{t_G}\right)\right]^2 \sigma_G^2 \\ &+ \left[\left(\frac{1}{t_B}\right)\frac{1}{1 - (B\tau/t_B)} + \left(\frac{B}{t_B}\right)\left(\frac{1}{1 - (B\tau/t_B)}\right)^2\left(\frac{\tau}{t_B}\right)\right]^2 \sigma_B^2 \\ = \left(\frac{1}{1 - (G\tau/t_G)}\right)^4 \frac{G}{t_G^2} + \left(\frac{1}{1 - (B\tau/t_B)}\right)^4 \frac{B}{t_B^2}.\end{aligned} \tag{7.45}$$

Typically, the background count rate is small so that $b\tau \ll 1$, and Eq. (7.45) reduces to

$$\boxed{\sigma^2(n_c) = \left(\frac{1}{1 - (G\tau/t_G)}\right)^4 \frac{G}{t_G^2} + \frac{B}{t_B^2}}\;. \tag{7.46}$$

7.3.6 Methods for Measuring Dead Time

There are many methods of measuring the dead time of a detector system and some of the more popular methods are outlined here. The measurement of dead time may be dictated by the availability of sources and equipment. The user is advised that these methods vary in accuracy; hence, the choice of a method may also be dictated by how much error is acceptable.

Two Source Method

The determination of dead time by the two source method is a simplification of the multi-source method. Two nearly identical calibrated sources are used in this measurement method. Often these sources are available as "split pairs." For instance, the two sources in a split-pair beta-particle source, identified as sources 1 and 2, are often conveniently shaped as two halves of a disc.

Sources 1 and 2 are measured independently in the system to yield count rates 1 and 2, or g_1 and g_2. Afterwards, both sources are measured together to produce a combined count rate g_{12}. In order to reduce possible geometric anomalies from source placement, a common practice is to place one source in the detection location for measurement, followed by a second source, and finally the removal of the first source. For instance, source 1 is located in the detection region to measure g_1. Next, the second source is carefully placed next to the first source to measure g_{12}. Afterwards, the first source is carefully removed, so as to not disrupt the location of source 2, in order to measure g_2. A background measurement then follows.

Non-Extendable Dead-Time Model First consider a detector with a non-extendable dead time. The true event rates for each measurement are related by

$$n_{c1} - n_{cb} + n_{c2} - n_{cb} = n_{c12} - n_{cb}. \tag{7.47}$$

With the use of Eq. (7.40) and Eq. (7.44), this result yields

$$\frac{g_1}{1 - g_1\tau} + \frac{g_2}{1 - g_2\tau} = \frac{g_{12}}{1 - g_{12}\tau} + \frac{b}{1 - b\tau}. \tag{7.48}$$

An approximate solution for the dead time τ can be found by use of the approximation, valid for $g\tau \ll 1$, namely

$$\frac{g}{1 - g\tau} \simeq g(1 + g\tau). \tag{7.49}$$

Equation (7.48) then becomes

$$g_1(1 + g_1\tau) + g_2(1 + g_2\tau) \simeq g_{12}(1 + g_{12}\tau) + b(1 + b\tau), \tag{7.50}$$

which is readily solved for τ to give

$$\boxed{\tau \simeq \frac{g_{12} + b - g_1 - g_2}{g_1^2 + g_2^2 - g_{12}^2 - b^2}}\;. \tag{7.51}$$

However, Eq. (7.51) can introduce significant error because it is based on a first order expansion. Equation (7.48) can be rearranged to produce the following quadratic equation

$$(g_1 g_2(g_{12} + b) - g_{12}b(g_1 + g_2))\,\tau^2 - 2\,(g_1 g_2 - g_{12}b)\,\tau + (g_1 + g_2 - g_{12} - b) = 0, \tag{7.52}$$

whose solution for the smaller of the two positive roots is

$$\boxed{\tau = \frac{1}{2A}\left[-B - \sqrt{B^2 - 4AC}\right]} \quad , \tag{7.53}$$

where $A = g_1 g_2(g_{12} + b) - g_{12}b(g_1 + g_2)$, $B = 2\,(g_{12}b - g_1 g_2)$, and $C = g_1 + g_2 - g_{12} - b$.

This expression for τ can be simplified by the assumption that the background count rate is insignificant ($b \simeq 0$). The result is

$$\boxed{\tau = \frac{1}{g_{12}}\left[1 - \left[1 - \frac{g_{12}}{g_1 g_2}(g_1 + g_2 - g_{12})\right]^{1/2}\right]} \quad . \tag{7.54}$$

If the background is negligible and if $g_1 \simeq g_2$, then in Eq. (7.48) one can approximate both $(1 - g_1\tau)$ and $(1 - g_2\tau)$ by $[1 - (g_1 + g_2)\tau/2]$. With these approximations Eq. (7.48) yields the result

$$\boxed{\tau = \frac{2(g_1 + g_2 - g_{12})}{(g_1 + g_2)g_{12}}} \quad . \tag{7.55}$$

Example 7.1: A split-pair source is used to determine the dead time of a common Geiger-Müller tube detector, where the following count rates are measured:

$G_1 = 93,600$ counts	$t_{G_1} = 60$ seconds	$G_2 = 106,800$ counts	$t_{G_2} = 60$ seconds
$G_{12} = 173,400$ counts	$t_{G_{12}} = 60$ seconds	$B = 30,600$ counts	$t_B = 30$ minutes

Determine and compare the dead times calculated with Eqs. (7.51), (7.53), (7.54), (7.55), and (7.53).

Solution:
The estimated count rates are:

$$g_1 = \frac{G_1}{t_{G_1}} = \frac{93600 \text{ cts}}{60 \text{ seconds}} = 1560 \text{ cps} \qquad g_2 = \frac{G_2}{t_{G_2}} = \frac{106,800 \text{ cts}}{60 \text{ seconds}} = 1780 \text{ cps}$$

$$g_{12} = \frac{G_{12}}{t_{G_{12}}} = \frac{173,400 \text{ cts}}{60 \text{ seconds}} = 2890 \text{ cps} \qquad b = \frac{B}{t_B} = \frac{30600 \text{ cts}}{(30 \text{ minutes})(60 \text{ sec/min})} = 17 \text{ cps}$$

Solution 1: Substitution of g_1, g_2, g_{12} and b into Eq. (7.51) gives

$$\tau \simeq \frac{g_{12} + b - g_1 - g_2}{g_1^2 + g_2^2 - g_{12}^2 - b^2} = \frac{2890 + 17 - 1560 - 1780}{(1560)^2 + (1780)^2 - (2890)^2 - (17)^2} = 1.574 \times 10^{-4} \text{ seconds} = 157.4\ \mu\text{s}.$$

Solution 2: From Eq. (7.54),

$$\tau = \frac{1}{g_{12}}\left[1 - \left[1 - \frac{g_{12}}{g_1 g_2}(g_1 + g_2 - g_{12})\right]^{1/2}\right] = \frac{1}{2890}\left[1 - \left[1 - \frac{2890}{(1560)(1780)}(1560 + 1780 - 2890)\right]^{1/2}\right]$$

$$= 9.372 \times 10^{-5} \text{ s} = 93.72\ \mu s$$

Solution 3: The dead time is determined from Eq. (7.55) as

$$\tau = \frac{2(g_1 + g_2 - g_{12})}{(g_1 + g_2)g_{12}} = \frac{2(1560 + 1780 - 2890)}{(1560 + 1780)2890} = 9.324 \times 10^{-5} \text{ s} = 93.24 \ \mu s$$

Solution 4: From Eq. (7.53), coefficients A, B and C are determined as

$$\begin{aligned} A &= g_1 g_2 (g_{12} + b) - g_{12} b (g_1 + g_2) = [(1560)(1780)(2890 + 17)] \\ &\quad -[(2890)(17)(1560 + 1780)] = 7,908,063,400 \\ B &= 2\,(g_{12} b - g_1 g_2) = 2\,[(2890)(17) - (1560)(1780)] = -5,455,340 \\ C &= g_1 + g_2 - g_{12} - b = 1560 + 1780 - 2890 - 17 = 433. \end{aligned}$$

Substitution of these values into Eq. (7.53) gives the dead time

$$\tau = \frac{1}{2A}\left[-B - \sqrt{B^2 - 4AC}\right] = 9.151 \times 10^{-5} \text{ s} = 91.51 \ \mu s.$$

The exact solution, $\tau = 91.51$ μs, was found with Eq. (7.53). The approximation calculated by Eq. (7.51), $\tau = 157.4$ μs, overestimates the exact answer by a factor of 1.72, a consequence of the use of a first order expansion approximations with a relatively large value of $g\tau \simeq 0.15$. However, the approximate value found from Eq. (7.54), $\tau = 93.72$ μs, overestimates the exact answer by a factor of only 1.024 (2.4%). The best approximate value was found with Eq. (7.55), yielding 93.24 μs, noting that this approximation works only if $b \ll g_1$, g_2 and g_{12}, and $g_1 \simeq g_2$. It can be concluded that Eq. (7.54) can be used to calculate the dead time provided that the background count rate is much less than the radioactive sample count rate, and Eq. (7.55) can be used with the added stipulation that the count rates of g_1 and g_2 are similar. Otherwise, it is best to use the more tedious operation required by Eq. (7.53).

Extendable Dead-Time Model The extendable dead-time model is transcendental and has no closed form solution for τ. However, there are methods to find an acceptable approximation to the dead time, one of which, reported elsewhere [NCRP 1985] is derived here. Assume an extendable model with negligible background, so the true count rates are related by Eq. (7.47), namely

$$n_1 + n_2 = n_{12}.$$

The measured rates g_i are then related through Eq. (7.32) as

$$g_1 e^{g_1 \tau} + g_2 e^{g_2 \tau} = g_{12} e^{g_{12} \tau}.$$

Assume that the count rates of g_1 and g_2 are similar, i.e., $g_1\tau \simeq g_2\tau$ so the above relation can be written as

$$g_1 e^{g_1 \tau} + g_2 e^{g_1 \tau} \simeq g_1 e^{g_2 \tau} + g_2 e^{g_2 \tau} = g_{12} e^{g_{12} \tau}. \tag{7.56}$$

By using only the $\exp[g_1/\tau]$ terms in this result, one can solve for τ as

$$\frac{g_1 + g_2}{g_{12}} \simeq e^{(g_{12} - g_1)\tau},$$

from which

$$\tau \simeq \frac{1}{g_{12} - g_1} \ln\left(\frac{g_1 + g_2}{g_{12}}\right). \tag{7.57}$$

In a similar fashion, using only the $\exp[g_2/\tau]$ in Eq. (7.56), one obtains

$$\tau \simeq \frac{1}{g_{12} - g_2} \ln\left(\frac{g_1 + g_2}{g_{12}}\right). \tag{7.58}$$

A better estimate of τ should be obtained by averaging these two results, namely

$$\tau \simeq \frac{1}{2}\left[\frac{1}{(g_{12} - g_1)} + \frac{1}{(g_{12} - g_2)}\right] \ln\left(\frac{g_1 + g_2}{g_{12}}\right) \simeq \frac{1}{2}\left[\frac{(g_{12} + g_{12} - (g_1 + g_2))}{(g_{12}^2 + g_1 g_2 - g_{12}(g_1 + g_2))}\right] \ln\left(\frac{g_1 + g_2}{g_{12}}\right).$$

Because $g_1 + g_2 \simeq g_{12}$, this approximation for τ reduces to that stated by the NCRP [1985]

$$\boxed{\tau \simeq \left[\frac{g_{12}}{2g_1 g_2}\right] \ln\left(\frac{g_1 + g_2}{g_{12}}\right)}. \tag{7.59}$$

This approximation for τ of Eq. (7.59) is based on the assumptions that $g\tau$ is a small number, background is negligible, and the count rates g_1 and g_2 are similar. To obtain the exact solution of Eq. (7.3.6) for the dead time, a computer program based, for example, on the bisection method can be used to iteratively arrive at a numerical solution for τ. This numerical method can treat all values of $g\tau$.

Pulser Methods

There may be occasions when an experimenter does not have access to a split-pair source to measure the dead time of a detector system. For such instances, Baerg [1965] derived an approximation for non-extendable dead-time systems which uses a periodic pulse generator (pulser) and a single radiation source. Suppose a pulser inserts test pulses of frequency ν into the test port of the detector's preamplifier. The pulser signal by itself is not stochastic; however, if a radiation source is present, then the probability of observing a count from the pulser is dependent upon the randomness of a count being registered by a radiation event, namely $(1 - g\tau)$. Here g is the observed source count rate without the pulser. Hence, the observed pulser count rate p in the presence of the radiation source is $p = \nu(1 - g\tau)$.

A pulser input activates the electronics and causes the detector system to be inactive for time τ. Consequently, the observed count rate g_1 from the radiation source, with the pulser activated, is

$$g_1 = g(1 - p\tau) = g(1 - \nu(1 - g\tau)\tau). \tag{7.60}$$

The total count rate is determined by adding the probability of observing a count from the pulser with the probability of observing a count from the radioactive source, namely

$$g_{1p} = \nu(1 - g\tau) + g(1 - \nu(1 - g\tau)\tau) = \nu + g - 2\nu g\tau - \nu g^2\tau^2, \tag{7.61}$$

which can be solved for τ to yield

$$\boxed{\tau = \frac{1}{g}\left[1 - \left(\frac{g_{1p} - g}{p}\right)^{1/2}\right]}. \tag{7.62}$$

Given the pulser frequency ν, the experimenter need only take measurements with (g_{1p}) and without (g) the pulser to determine the dead time. Note that the assumptions used to arrive at Eq. (7.62) can lead to inaccuracies in the dead-time measurements, but the accuracy is considered to be acceptable for pulser signals with frequencies $\nu < 1/(3\tau)$ [Müller 1973, 1976].

There is no exact analytical solution for extendable dead time; however, an approximate solution for extendable dead time with the use of a single pulser is offered by NCRP [1985], namely,

$$\tau \simeq \frac{\xi e^{-\xi}}{g}, \quad \text{where} \quad \xi = (q^2 + 4q - 1)^{1/2} - (q+1) \quad \text{and} \quad q = \nu/(g_{1p} - g). \tag{7.63}$$

The NCRP [1985] states Eq. (7.63) as being accurate to better than 0.3% over a wide range of pulser frequencies, provided that $g\tau \leq 0.20$.

A method introduced by Baerg [1965], and described in more detail later by Müller [1973], uses two pulse generators to measure the dead time of an electronic system without the detector. Suppose two pulsers are used with different frequencies ν_1 and ν_2, where the frequency ratio is not an integer, then an exact system dead time is given by

$$\tau = \frac{g_{\nu 1} + g_{\nu 2} - g_{\nu 1 \nu 2}}{2 g_{\nu 1} g_{\nu 2}}, \tag{7.64}$$

where $g_{\nu 1}$ is the count rate from pulser frequency ν_1, $g_{\nu 2}$ is the count rate from pulser frequency ν_2, and $g_{\nu 1 \nu 2}$ is the count rate with both pulsers operating. If the highest pulser frequency has $\nu < 1/(2\tau)$ then the solution to Eq. (7.64) provides the value of τ regardless of whether the dead time is extendable or non-extendable.

However, if $\nu > 1/(2\tau)$ then Eq. (7.64) provides a method to determine if the dead time is extendable or non-extendable. If ν_{av} is the average frequency of the combined pulsers after passing dead time τ, with $T = 1/\nu \equiv \min(T_1, T_2)$, then the response from a *non-extendable* dead-time system is

$$\nu_{av} = \begin{cases} \nu_1 + \nu_2 - 2\tau\nu_1\nu_2, & \text{for } 0 < \tau < T/2 \,, \\ \nu, & \text{for } T/2 < \tau < T \,. \end{cases} \tag{7.65}$$

For an *extendable* dead-time system

$$\nu_{av} = \begin{cases} \nu_1 + \nu_2 - 2\tau\nu_1\nu_2, & \text{for } 0 < \tau < T \,, \\ 0, & \text{for } \tau > T \,. \end{cases} \tag{7.66}$$

The dead time is determined from

$$\tau = \frac{g_{\nu 1} + g_{\nu 2} - g_{\nu_{av}}}{2 g_{\nu 1} g_{\nu 2}}. \tag{7.67}$$

A comparison of results for ν_{av} beyond $\tau > T/2$ reveals the type of dead-time for the system [Müller 1973].

Decaying Source Method

A rapidly decaying source, such as ^{116m}In ($t_{1/2} = 54.2$ min), is measured at incremental times as the sample undergoes radioactive decay. The count rate is monitored, corrected for background, and plotted on a semi-log graph. If there were no dead time, the plot would decrease linearly in time. But at high count rates dead-time effects are important, and a plot such as that of Fig. 7.12 is obtained after the sample decays through several half-lives.

The slope M of the linear portion of the semilogarithmic plot is then found by fitting a straight line to the measured data. Typically, the experimenter determines this slope from a least squares fit to the lower count rate data that yields an acceptable coefficient of determination.[6] The slope M then yields the half-life from

[6] The coefficient of determination, often denoted R^2, is an indicator of linearity for a linear regression curve fit. Values converging on $R^2 = 1$ are considered "perfect" predictors of an unknown variable.

$t_{1/2} = \ln(2)/M$. This experimentally determined half-life can be compared to the literature value, thereby providing another test for accuracy.

At the high count-rate end of the plot, the measured count rates g_i underpredicts the true event rate n_i, which varies exponentially in time, because dead time prevents some events from being recorded. For a non-extendable detector the predicted and observed count rates are related by Eq. (7.40), from which the dead time is found to be

$$\tau = \frac{n_i - g_i}{g_i n_i}. \tag{7.68}$$

Note that in the linear portion of Fig. 7.12, $n_i \simeq g_i$ and Eq. (7.68) yields $\tau \simeq 0$. To find τ use the highest count-rate measurements and the predicted true event rate

$$n_i = n(t_i) = g_n \exp[\lambda(t_n - t_i)], \tag{7.69}$$

where g_n is an observed count rate well into the linear portion of the decay curve. Typically, one uses $n_i = n_0$, the initial observation. Substitution of measured value g_n and predicted value n_i into Eq. (7.68) yields τ (see Fig. 7.12).

An alternative approach is to use the decay relation

$$n(t) = n_0 \exp(-\lambda t), \tag{7.70}$$

where $t_0 = 0$, and substitute it into Eq. (7.40) for a non-extendable dead-time detector to obtain [Martin 1961]

$$g(t) \exp[\lambda t] = -g(t)\, n_0 \tau + n_0. \tag{7.71}$$

This is a linear equation, where $-n_0\tau$ is the slope. Hence, by plotting the left-hand side of Eq. (7.71) versus $g(t)$ a plot similar to that of Fig. 7.13 is obtained, from which the slope S of the resulting straight line can be determined. The dead time is then found as $\tau = -n_0/S$.

For an extendable dead-time detector, the dead time is found from Eq. (7.32), with $t_i = t_0$, from the relation between n_0 and g_0, namely

$$g_0 = n_0 \exp(-n_0 \tau). \tag{7.72}$$

This result must be solved numerically. However, a similar graphical method can be used, as described for the non-extendable dead-time case, where substitution of Eq. (7.70) into Eq. (7.32) yields

$$\lambda t + \ln(g) = -n_0 \tau \exp[-\lambda t] + \ln[n_0]. \tag{7.73}$$

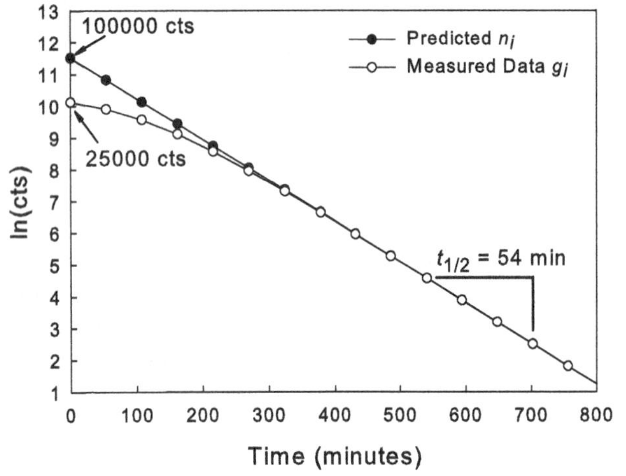

Figure 7.12. The dead time is determined by plotting the logarithm of the measured counts versus the decay time. From the plot the expected values of n_i are determined and the dead time τ if found from either Eq. (7.40) or Eq. (7.32).

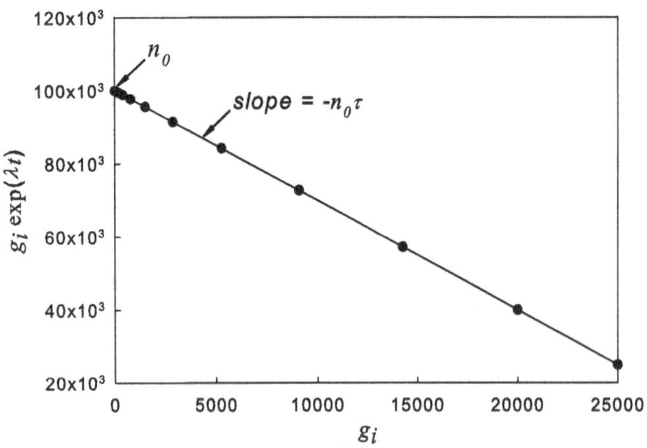

Figure 7.13. The dead time for a non-extendable detector is found by plotting $g_i \exp(\lambda t)$ versus g_i. The slope of straight line fitted to the data is τn_0. Note one must wait through several half-lives until dead-time effects are negligible and an accurate value of n_0 is obtained.

Eq. (7.73) is another linear equation in which $-n_0\tau$ is the slope. Thus, the procedure is to plot the left-hand side of Eq. (7.73) versus $e^{-\lambda t}$, find the slope of the resulting straight line, and evaluate $\tau = -n_0/S$.

The advantage of the graphical methods using Eq. (7.71) and Eq. (7.73) over the methods of Eq. (7.68) and Eq. (7.72) is debatable. Both methods for finding the dead time improve as the sample is allowed to decay through more half-lives. Regardless, these graphical estimations are generally considered inaccurate [NCRP 1985], although accuracy can be improved by subtracting background count rates and collecting data through enough half-lives so that dead-time effects are negligible and a reasonably accurate value of n_0 can be determined.

Increasing Power Method

The increasing power method has some similarity to the decaying source method, and requires a variable field radiation source, such as a nuclear reactor. The detector is placed in the measurement location and a background measurement is acquired. The radiation field is increased and measured systematically at various power levels, preferably increasing in power if a nuclear reactor is used.[7]

The detector response without dead time should be linear with respect to the power level. A simple least squares fit through the low count rate data points yields a linear function for comparison to the measured values at high count rates. For non-extendable dead time, the difference between a linear response plotted through the low radiation field measurements and the high field measurements yields the dead time from Eq. (7.68), where p is the projected correct count rate from the linear fit and g is the measured count rate. For extendable dead time, Eq. (7.72) is used. Accuracy is limited by error in the linear curve fit for low counting rates.

7.4 Geometric Effects: View Factors

Not all radiation emitted by a radioactive source, of strength S_{tot} particles/second, enters the detector aperture. The fraction of emitted radiation that does enter the detector, in the absence of any radiation attenuation between the source and detector, is called here the *view factor* P_v. No account of backscatter or direction of radiation incidence on the aperture is made. The view factor is strictly a geometric probability depending only on the shapes and sizes of the source and aperture and their relative orientation to each other. With this quantity, the *intrinsic efficiency* ϵ_I of the detector is

$$\epsilon_I = \frac{C_{obs}}{S_{tot}P_v}, \tag{7.74}$$

where C_{obs} is the observed count rate (corrected for dead time).

Two general classes of view factors are considered in this section. The first class consists of view factors for point isotropic sources and for plane detector apertures of simple shapes. Such sources are frequently encountered in measurements of radioactive samples. The second class is for plane sources with different shapes.

Finally, in this section it is assumed that source radiation is emitted isotropically, a good assumption for most photon sources. However, the encapsulation of charged particle sources can make the emission quite anisotropic. Such anisotropic sources are not considered here.

[7] Although the experiment can be conducted by starting at a high reactor power and incrementally lowering the power, it takes much longer because a reactor's power can be raised more quickly than it can be decreased.

7.4.1 Point Isotropic Sources

The view factor for an aperture of area A is closely related to the solid angle subtended at the source point by the aperture. The basic equation for the solid angle is

$$\Omega = \int_A \frac{\hat{\mathbf{n}} \bullet d\mathbf{A}}{R^2}, \tag{7.75}$$

where R is the distance between the source and dA, $\hat{\mathbf{n}}$ is a unit vector along R, $\hat{\mathbf{n}} \bullet d\mathbf{A}$ is the area of the projection of dA on the plane perpendicular to R. The view factor is simply $P_v = \Omega/4\pi$, namely, the fraction of the emitted radiation particles into the solid angle Ω.

Solid Angle for dA on a Sphere

Consider a radioactive point source placed at the origin of spherical coordinate system, as shown in Fig. 7.14. Given a sphere of radius r, a direction vector extending from the origin can be described by Cartesian unit vectors

$$\mathbf{r} = r \sin\theta \cos\phi \, \mathbf{i} + r \sin\theta \sin\phi \, \mathbf{j} + \cos\theta \, \mathbf{k}. \tag{7.76}$$

Emissions isotropically radiating from the source uniformly intersect a sphere centered about the origin. A differentially small area dA, defined by the limits $r \sin\theta$ and $r \sin\theta + d\theta$ and also ϕ and $\phi + d\phi$ through which a portion of this radiation passes is

$$dA = r \, \sin\theta \, d\theta \, r \, d\phi = r^2 \, \sin\theta \, d\theta \, d\phi. \tag{7.77}$$

The solid angle is defined as dA/r^2. Integration over the entire sphere surface gives

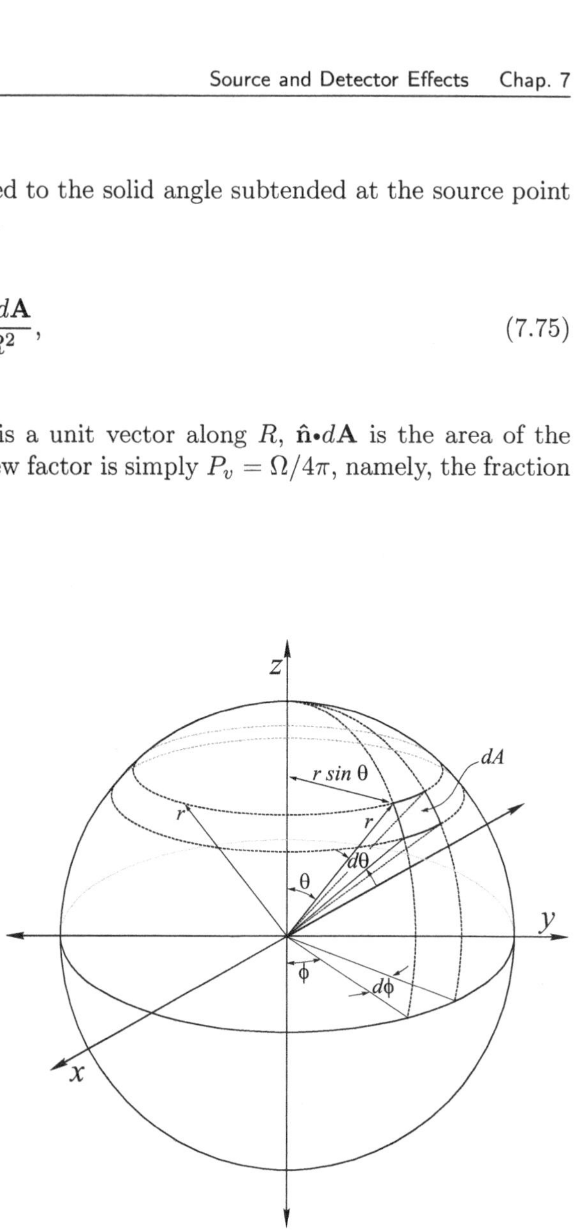

Figure 7.14. Coordinate system for isotropic radiation emissions (at the origin) intersecting differential area dA.

$$\int_{4\pi} d\Omega = \int_{4\pi} \frac{dA}{r^2} = \int_0^{\pi} \sin\theta \, d\theta \int_0^{2\pi} d\phi = 4\pi. \tag{7.78}$$

This result shows that the solid angle for the entire surface of the sphere surface is 4π. The solid angle is actually dimensionless, yet units of steradians (sr) are typically attributed to the solid angle; hence there are 4π steradians subtended by the surface of a sphere, regardless of radius r. The solid angle for a detector of face area A can be determined with the method of Eq. (7.78).

Solid Angle for a Point Source and a Cylindrical Detector

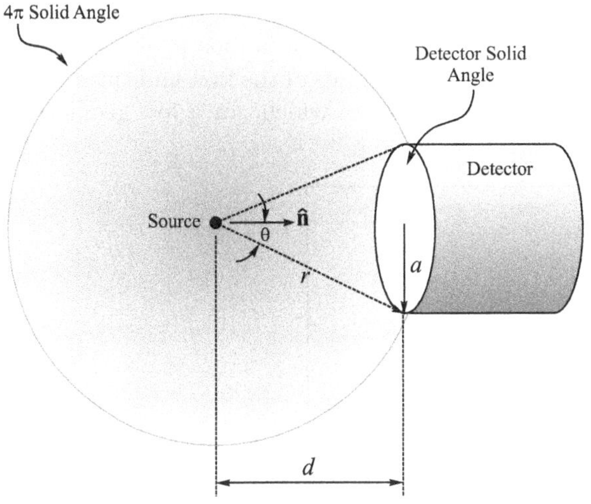

Figure 7.15. Representation of the relationship between a radioactive point source and a cylindrical radiation detector of radius a with the face spaced a distance d away from the source.

A typical geometry for a radiation detector is a cylinder with radius a, as shown in Fig. 7.15. The projected area (or solid angle) of the meniscus formed at the intersection of the cylinder and the sphere is found from

$$\int_{4\pi} d\Omega = \frac{1}{r^2}\int_0^{\theta} r^2 \sin\theta d\theta \int_0^{2\pi} d\phi = 2\pi\left(1-\cos\theta\right), \tag{7.79}$$

where θ is defined by the angle between a normal vector $\hat{\mathbf{n}}$ and $\mathbf{r}$.
Because

$$\cos\theta = \frac{d}{\sqrt{d^2+a^2}}, \tag{7.80}$$

where d is the distance from the radioactive point source to the center of the detector face, Eq. (7.79) becomes

$$\Omega = 2\pi\left(1-\frac{d}{\sqrt{d^2+a^2}}\right). \tag{7.81}$$

The fraction P_v of radiation emissions that pass through the detector face is found by dividing the detector solid angle by the solid angle of the sphere, namely

$$P_v = \frac{\Omega}{4\pi} = \frac{1}{2}\left(1-\frac{d}{\sqrt{d^2+a^2}}\right). \tag{7.82}$$

Example 7.2:

A 3-inch × 3-inch right cylindrical NaI:Tl detector[8] is placed 6 inches away from a ^{60}Co source for a radiation measurement. What are the solid angle and the view factor for the detector? Assume no intermediate absorption of the gamma rays.

Solution:

Here $d = 6$ inches and $a = 1.5$ inches
From Eq. (7.81), the solid angle is

$$\Omega = 2\pi\left(1-\frac{d}{\sqrt{d^2+a^2}}\right) = 2\pi\left(1-\frac{6}{\sqrt{6^2+1.5^2}}\right) = 0.188 \text{ sr}. \tag{7.83}$$

The view factor is thus

$$P_v = \Omega/(4\pi) = 0.015. \tag{7.84}$$

Thus, only 1.5% of radiation emissions from the source cross the detector aperture.

[8]The common notation refers to the diameter and length of a right straight cylinder.

Point Source Offset from Center of Circular Aperture

Consider the case of an isotropic point source set a distance d from the plane of a circular aperture of radius a. The source is also offset by a distance s from the center of the disk. Paxton [1959] obtained analytic solutions for the solid angle Ω subtended by the disk at the source. The view factor is thus $P_v = \Omega/4\pi$. The expressions, however, are quite complicated and involve complete elliptic integrals of the first and third kinds. This problem is more easily obtained by Monte Carlo analyses, the results of which, for a few geometries, are shown in Fig. 7.16.

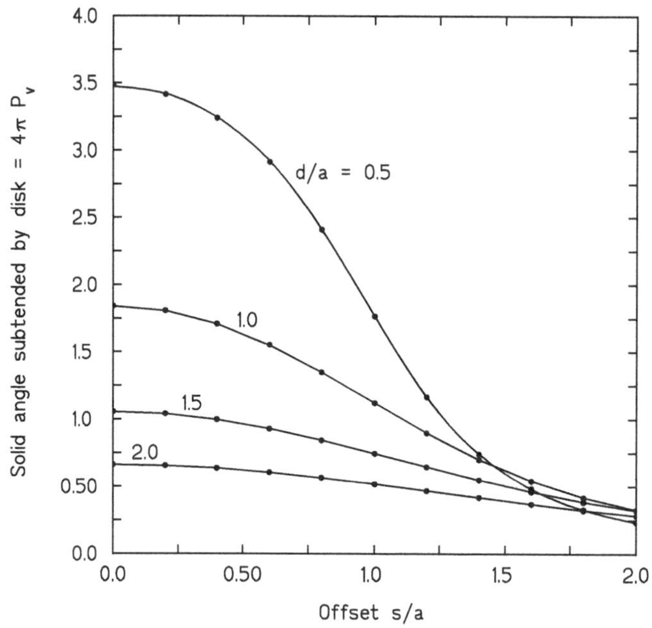

Figure 7.16. The solid angle subtended by a circular aperture at a point source a distance d from the aperture plane and offset from the disk center by s. The circles are Monte Carlo results and the lines are spline fits to data reported by Paxton [1959].

Point Source and a Rectangular Aperture

The view factor for a point source at some distance d above the *center* of a detector with a rectangular aperture with dimensions $a \times b$ is [Gossman et al. 2010, 2011]

$$P_v(a/d, b/d) = \frac{1}{\pi} \arcsin\left[\frac{(a/d)(b/d)}{\sqrt{[(a/d)^2+4][(b/d)^2+4]}}\right]. \tag{7.85}$$

The view factors for rectangular apertures are shown in Fig. 7.17. For case of a square aperture with dimension a, Eq. (7.85) reduces to

$$P_v(a/d) = \frac{1}{\pi} \arcsin\left[\frac{(a/d)^2}{(a/d)^2+4}\right]. \tag{7.86}$$

Of more utility is the case when the point source is a distance d above a corner of a rectangular aperture with dimensions $a \times b$. For this case, the view factor is one-quarter that of Eq. (7.85) with $a \to 2a$ and

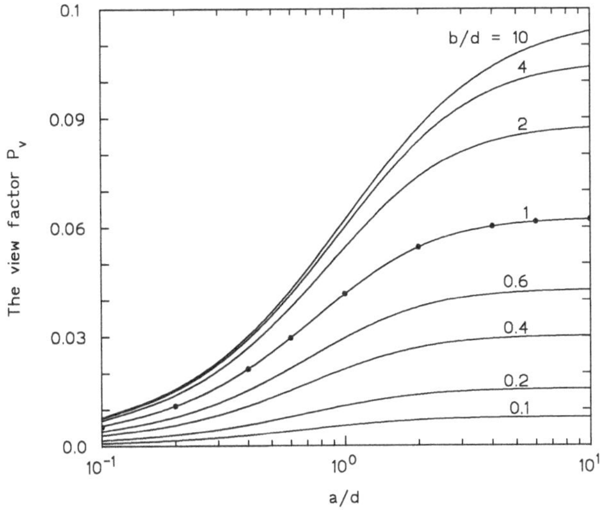

Figure 7.17. The view factor for a point source over the center of a rectangular aperture as calculated by Eq. (7.85). Data points are Monte Carlo calculated values using MCNP6 [2013].

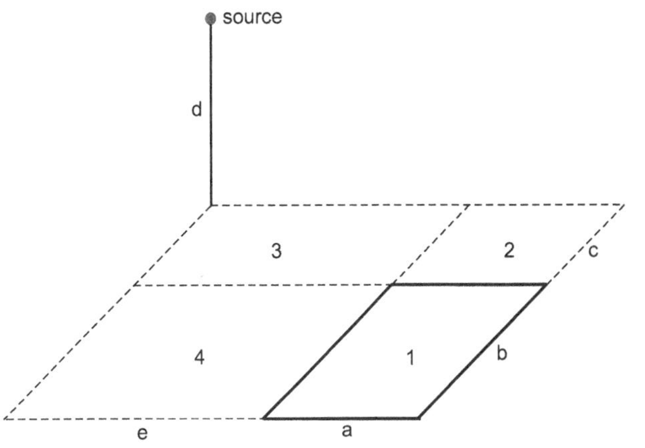

Figure 7.18. Calculation of the view factor for a point source offset from a rectangular detector aperture (solid lines) when the source, a distance d above the aperture plane. With Eq. (7.87) the view factor for the aperture is $P_v^1 = P_v^{1+2+3+4} - P_v^{2+3} - P_v^{3+4} + P_v^3$, where the superscript is the union of indicated rectangular regions.

$b \longrightarrow 2b$. The result is

$$P_v(a/d, b/d) = \frac{1}{\pi} \arcsin\left[\frac{(a/d)(b/d)}{\sqrt{[(a/d)^2+1][(b/d)^2+1]}}\right]. \tag{7.87}$$

With this result the view factor can be calculated for the point source located above any point in the aperture plane! If the source is above the aperture, then a perpendicular line extending from the source to the detector divides the aperture into four smaller rectangular apertures.

Likewise if the source is not above the aperture, the aperture view factor can be found by adding and subtracting the view factors for each rectangular region, each of which has the source over a corner. See Fig. 7.18 for an example. Note, however, that radiations can now enter the side of the detector if the source is offset from the detector aperture, thereby confusing the view factor situation altogether. Hence, an additional calculation is needed to account for radiations entering into the side of the detector.

Approximation Method for the View Factor: Under the condition that the point source is sufficiently far from the detector aperture, and the dimensions of the aperture are similar ($a \simeq b$), the view factor can be approximated by replacing the rectangular aperture by a circular one with the same area. Hence, the aperture area $A = ab = \pi\tilde{a}^2$ from which the equivalent aperture radius $\tilde{a}$ is obtained. Then, from Eq. (7.82) the view factor is

$$P_v^{approx} \simeq \frac{1}{2}\left(1 - \frac{d}{\sqrt{d^2 + A/\pi}}\right) = \frac{1}{2}\left(1 - \frac{d}{\sqrt{d^2 + ab/\pi}}\right), \tag{7.88}$$

where A is the projected area of the detector aperture. Typically, the distance d should be at least equal to the sum of the aperture's vertical and horizontal dimensions. Notice that once again P_v depends only on dimensionless ratios a/d and b/d. Comparison of Eq. (7.85) with Eq. (7.88) shows that if a/d is no greater than $\pm 40\%$ of b/d, then P_v^{approx} overestimates P_v by less than 4%.

Example 7.3:

A gas-filled detector with a 3-inch × 2-inch aperture is placed 3 inches away from a ^{137}Cs source for a radiation measurement. The source is on the detector axis. What is the solid angle subtended by the NaI:Tl detector and the resulting view factor? How does the approximation method compare to the exact results? Assume no intermediate absorption of the gamma rays.

Solution:

From Eq. (7.85)

$$\Omega = 4\pi P_v(a/d, b/d) = 4 \arcsin \frac{(a/d)(b/d)}{\sqrt{[(a/d)^2+4][(b/d)^2+4]}}$$

$$= 4 \arcsin \frac{(3/3)(2/3)}{\sqrt{[(3/3)^2+4][(2/3)^2+4]}} = 0.568 \text{ sr}.$$

The view factor is thus $P_v = \Omega/(4\pi) = 0.0452$.

The approximate method for P_v is given by Eq. (7.88) so

$$\Omega = 4\pi P_v^{approx} \simeq 2\pi \left(1 - \frac{3}{\sqrt{3^2+(2)(3)/\pi}}\right), = 0.576 \text{ sr}.$$

Thus, the approximate method overestimates by about 1.5%.

7.4.2 Isotropic Area Sources

Solid Angle for a Disk Source and a Cylindrical Detector

As shown in Fig. 7.19, a radioactive disk source of radius a_s is centered and placed a distance d from circular aperture or radius a_d. The planes of both disks are parallel. Decay particles can reach the detector aperture from any location on the disk source. The source emits radiation isotropically at a rate of S_0 particles per unit area per unit time. The rate at which radiation emitted from a differential area of the source, $dA_s = r_1 dr_1 d\theta_1$, reaches a differential area on the detector aperture, $dA_d = r_2 dr_2 d\theta_2$, is

$$dN = \frac{S_0\ (\hat{\mathbf{n}}\bullet\mathbf{R})\ dA_s\ dA_d}{4\pi R^3}. \tag{7.89}$$

Here $\mathbf{n}$ is a unit vector perpendicular to the source, and the term $\hat{\mathbf{n}}\bullet\mathbf{R}/R$ is the cosine of the angle between $\mathbf{R}$ and $\mathbf{n}$, namely d/R. Thus,

$$dN = \frac{S_0\ d\ dA_s\ dA_d}{[d^2 + r_1^2 + r_2^2 - 2r_1 r_2 \cos(\phi_1 - \phi_2)]^{3/2}}. \tag{7.90}$$

The total emission rate from the radioactive source is $N_0 = S_0 4\pi a_s^2$. Division of Eq. (7.90) by N_0 then gives differential view factor for radiation from dA_s to dA_d, i.e.,

$$dP_v = \frac{1}{4\pi^2 a_s^2} \frac{d\ dA_s\ dA_d}{[d^2 + r_1^2 + r_2^2 - 2r_1 r_2 \cos(\phi_1 - \phi_2)]^{3/2}}. \tag{7.91}$$

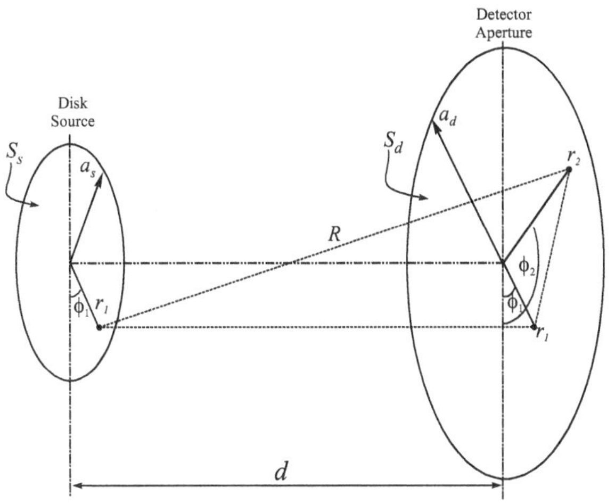

Figure 7.19. Geometry used to determine the solid angle for a disk source and cylindrical detector.

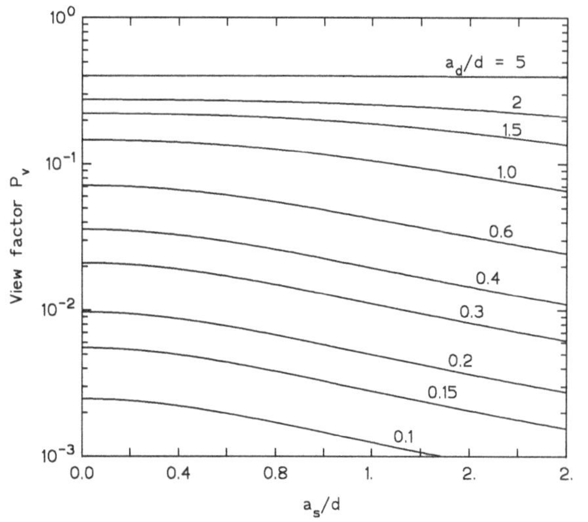

Figure 7.20. The view factor for an axially aligned disk source and disk detector apertures as calculated with Eq. (7.93).

To obtain the total view factor, the above result must be integrated over all dA_s and dA_d. Ruby [1994] expressed the result as a single integral involving Bessel functions of the first kind of order one, namely

$$P_v \frac{a_d}{a_s} \int_0^\infty \frac{dk}{k} \exp[-kd] J_1(ka_s) J_1(ka_d). \tag{7.92}$$

This expression, like previous view factors, can be expressed in terms of dimensionless parameters by defining $kd = k'$. Equation (7.92) becomes

$$P_v(a_s/d, a_d/d) = \frac{(a_d/d)}{(a_s/d)} \int_0^\infty \frac{dk'}{k'} \exp[-k'] J_1(k'\frac{a_s}{d}) J_1(k'\frac{a_d}{d}). \tag{7.93}$$

Although this result cannot be evaluated analytically, it is easily evaluated using numerical integration. Example results are shown in Fig. 7.20.

Ruby [1994], by substituting a truncated series approximation for the Bessel functions in Eq. (7.93), found

$$P_v \simeq \omega^2 \left\{ 1 - \left(\frac{3}{4}\right)(\psi^2 + \omega^2) + \left(\frac{15}{24}\right)(\psi^4 + \omega^4 + 3\psi^2\omega^2) - \left(\frac{35}{64}\right)\left[\psi^6 + \omega^6 + 6\psi^2\omega^2(\psi^2 + \omega^2)\right] \right\}, \tag{7.94}$$

where $\psi = a_s/d$ and $\omega = a_d/d$. Accuracy increases as the distance between the source and detector increases. In other words, the accuracy increases as ψ and ω decrease, where the error is $< 1\%$ for $\psi < 0.2$ and $\omega < 0.5$. Another much more complicated, but perhaps a more accurate approximation to Eq. (7.93) was published later [Vega-Carrillo 1996, 2005]. However, it turns out that numerous solutions and tables for various geometries for the disk source/cylindrical detector geometry have been published [Jaffey 1954; Paxton 1959; Hubbell et al. 1961], offering a simplistic alternative method for arriving at view factors for a disk source.

7.4.3 Monte Carlo Approach to View Factor Angle Calculations

The number of source/detector combinations for which the view factors can be evaluated analytically, or even approximated, is quite limited. However, Monte Carlo techniques can be easily used to numerically determine view factors, from simple to very complex geometries. Sources and detector apertures of any shape and orientation to each other are readily treated. Indeed, the programming effort to numerically evaluate Eq. (7.93) is far greater than that needed for a Monte Carlo evaluation. Monte Carlo results are shown to agree well with analytical and experimental results for several examples shown earlier in this section.

More important, all of the examples given in this section have assumed isotropic emissions from the various sources. However, there are many cases in which the emission is quite anisotropic. For example, radiation emitted from the outer surface of a shielded source is often proportional to some power of the cosine of the emission angle with respect to the outward normal to the surface. Similarly, radiation emitted from a collimated source, such as a beam port of a nuclear reactor, is quite anisotropic. Monte Carlo methods can treat such cases with little extra effort.

7.5 Geometric Corrections: Detector Parallax Effects

Uncollided gamma rays are exponentially attenuated as they travel through a medium. The probability a gamma ray makes, at least, its first interaction after traveling a distance x in the detector is, thus,

$$P_{int}(x) = 1 - e^{-\mu(E)x}, \tag{7.95}$$

where $\mu(E)$ is the total interaction coefficient for gamma rays of energy E in the medium inside the detector.

As a first approximation, Eq. (7.95) describes the expected fraction of photons that have undergone a single interaction while traveling a distance x. For a photoelectric interaction, the photon's full energy is transferred to the detector medium less, perhaps, some losses from x-ray escape events. Compton scattering and pair production lead to partial energy deposition. Equation (7.95) does not take into account small-angle Compton or incoherent scattering, which usually deposits very little energy in the detector with the secondary photon continuing on in nearly the same direction with almost the same energy. This possibility can be accounted for by removing incoherent and small-angle Compton scattering components from the interaction coefficient $\mu(E)$ in Eq. (7.95).

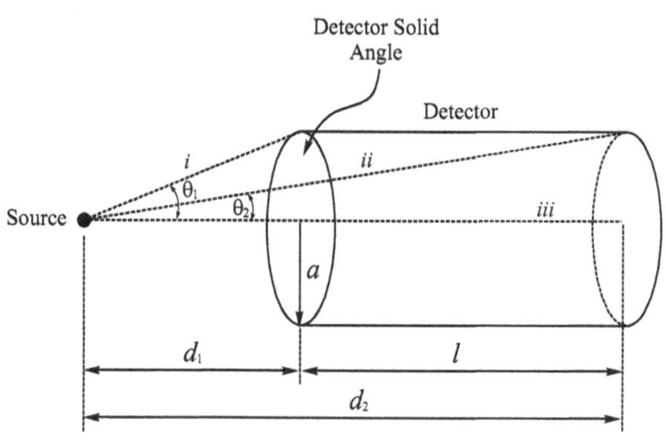

Figure 7.21. A radioactive point source at distance d_1 from the center of the aperture of a cylindrical detector. Particles entering the detector encounter different mass thicknesses of the detector medium depending on their trajectories. Shown by the dotted lines are the three limiting trajectories.

Hence, the intrinsic efficiency, as given by Eq. (7.74), pertains to the efficiency of the detector to have at least one radiation interaction within the detector volume after radiation particles have crossed the plane of the detector aperture. Because some events may deposit energy below the system's electronic threshold, then Eq. (7.74) and Eq. (7.95) give the best possible efficiency that can be achieved by the detector.

Consider the situation of Fig. 7.21, depicting a cylindrical detector of radius a and length l. The radiation source is placed at distance d_1 from the center of the aperture of the cylindrical detector. However, the source is also at distance d_2 from the exit aperture of the detector. The interaction probability of gamma rays entering the aperture at angle θ_1 is zero. Gamma rays entering the detector at angles between θ_1 and

θ_2 have a probability of interaction in the range

$$0 \le p(x) \le 1 - \exp\left[-\frac{\mu(E)l}{\cos\theta_2}\right],$$

while those entering between θ_2 and 0° have an interaction probability in the range

$$1 - \exp\left[-\frac{\mu(E)l}{\cos\theta_2}\right] \ge p(x) \ge \exp[-\mu(E)l].$$

Although Eq. (7.74) indicates that ϵ_I is constant for all values of $\theta \in [0, \theta_1]$ because all source particles emitted in this range reach the detector aperture, parallax effects reduce this efficiency because different angles of incidence lead to different probabilities that the incident photon will interact within the detector and lead to a recorded count. A measured efficiency is only reliable for a single geometry. Parallax effects change the efficiency as the geometry is scaled as by, for example, altering the distance between the source and detector.

The intrinsic efficiency is determining by both the view factor P_v, the probability a source particle is emitted toward the detector aperture, and the probability that such particles interact in the detector upon crossing the aperture. For gamma rays and neutrons, the interaction probability is described by the exponential dependence of Eq. (7.95). The number of particles interacting within the detector divided by the number of particles crossing the detector aperture gives the intrinsic efficiency of the detector. If one assumes a point isotropic source and that any interaction results in a detectable signal, the intrinsic efficiency, modified to account for the interaction probability $p(x)$ in the detector, is given by

$$\widehat{\epsilon}_I = \frac{\dfrac{1}{4\pi}\dfrac{1}{r^2}\displaystyle\int_0^\theta r^2 \left(1 - \exp\left[-\mu(E)l(\theta)\right]\right)\sin\theta\, d\theta \int_0^{2\pi} d\phi}{\dfrac{1}{4\pi}\dfrac{1}{r^2}\displaystyle\int_0^\theta r^2 \sin\theta\, d\theta \int_0^{2\pi} d\phi} = \frac{\displaystyle\int_0^\theta \left(1 - \exp\left[-\mu(E)l(\theta)\right]\right)\sin\theta\, d\theta}{\int_0^\theta \sin\theta\; d\theta}. \tag{7.96}$$

For the cylindrical case shown in Fig. 7.21, the modified intrinsic efficiency is,

$$\widehat{\epsilon}_I = \frac{1}{1-\cos\theta_1}\left[\int_0^{\theta_2}\left(1 - \exp\left[-\frac{\mu(E)l}{\cos\theta}\right]\right)\sin\theta\, d\theta + \int_{\theta_2}^{\theta_1}\left(1 - \exp\left[-\frac{\mu(E)(a\cos\theta - d_1\sin\theta)}{\cos\theta\ \sin\theta}\right]\right)\sin\theta\, d\theta\right]. \tag{7.97}$$

This result cannot be evaluated analytically and numerically methods must be used. Note that most often a detector is operated with a discriminator level set to reduce (or eliminate) electronic noise and background radiations. Consequently, the lower level discriminator (LLD) rejects electronic signals that fall below the set threshold, thereby producing a lower intrinsic detector efficiency than predicted in Eq. (7.97).

The detection efficiency for charged particles is usually adequately described by Eq. (7.74); however, *parallax clipping* can yield a measured efficiency that is lower than the theoretical treatment, another consequence of the signal discrimination level. As shown in Fig. 7.22, an alpha-particle point source subtends a solid angle Ω defined by distance d and aperture radius a. However, alpha particles deposit energy through ionization along their straight-line trajectories, and sufficient energy must be deposited within the detector to produce a signal above the LLD setting. At large angles from the detector axis, alpha particles can strike the detector wall before depositing sufficient energy and hence avoid detection. Suppose the average ionization required to produce a measurable signal requires a transit length in the detector of $\ge h$. Consequently, the solid angle that defines intrinsic efficiency, denoted Ω' in Fig. 7.22, is defined by the virtual aperture $a' = a - h\sin\theta'$.

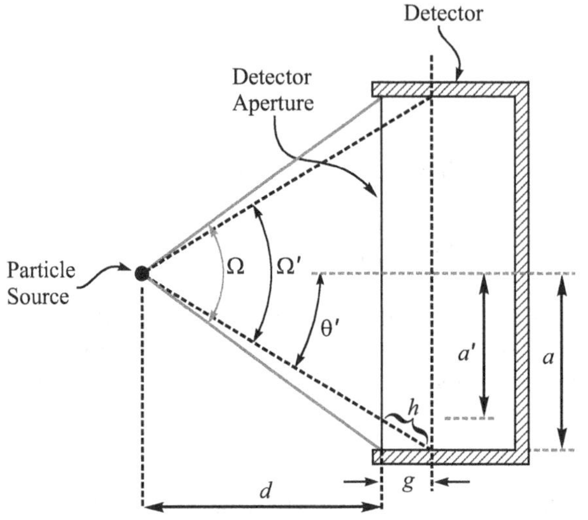

Figure 7.22. A radioactive point source of alpha particles at distance d from the center of the detector aperture. The energy deposited in the detector depends on the emission angle from the axis. Ω is the solid angle determined by d and the aperture radius a from Eq. (7.81). The distance h represents the transit length that a particle must travel to deposit enough energy for detection. Consequently, the actual solid angle for detection, Ω', is determined by $a' = a - h\sin\theta'$ and the distance d.

7.5.1 Attenuation and Scattering Effects Outside the Detector

X and Gamma Rays

The determination of a detector's response is complicated when absorption and attenuation in the environment around the detector are considered. Gamma rays emitted towards a detector may be scattered away from the detector or otherwise interact in the intervening air. For large distances between the source and detector, the air attenuation may be significant, although parallax effects are greatly reduced. The attenuation length of gamma rays, defined as the mean-free-path length, is given by $1/\mu(E)$ and is shown in Fig. 7.23 for different air temperatures at sea level. Such far-field measurements are also made more complex by photons scattered into the detector from the air and other objects around the detector. Corrections for such skyshine, roomshine, groundshine, etc., generally require transport analyses.

Gamma-ray attenuation is most pronounced in air and in the source and detector encapsulations for low-energy photons. Almost all radiation detectors are in sealed packages. Gas detectors are sealed within a gas chamber, scintillators are typically hermetically sealed in a reflecting cannister, and semiconductor detectors are sealed in light-tight encapsulation to reduce electromagnetic noise. Also, many types of semiconductor detectors are chilled to low temperatures, thereby requiring isolation in vacuum cannisters to prevent icing and convective warming. In all of these packaging cases, low-energy x rays and gamma rays can be attenuated as they pass through the container towards the detecting medium.

Alpha Particles

Corrections for alpha-particle (and other heavy charged particle) detectors can be more complicated than just parallax effects. Recall that charged particles deposit energy through Coulombic interactions along their straight-line trajectories. They lose energy as they travel through any material between their emission points and the sensitive volume of the detector.

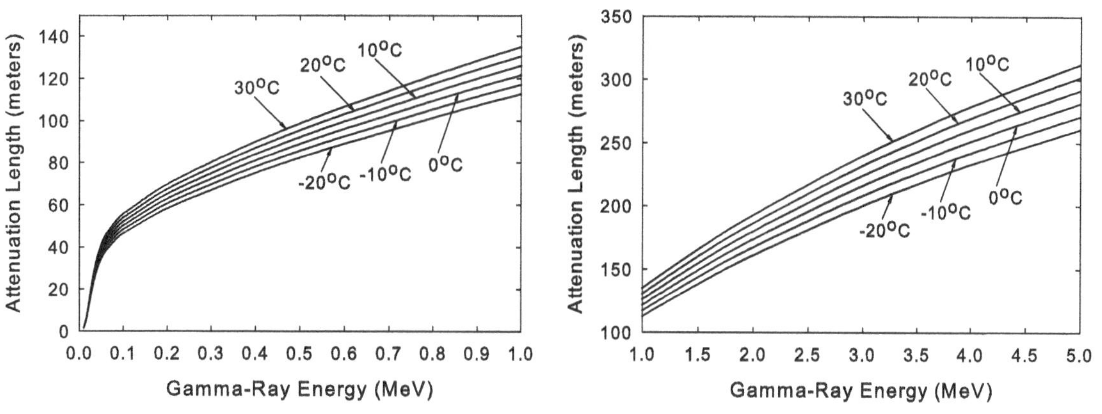

Figure 7.23. Gamma-ray attenuation lengths, $1/\mu(E)$, in air as a function of energy and temperature at sea level. Calculated values based on data in the *Radiological Health Handbook* [1970].

In Fig. 7.24 are shown two cases for a gas-filled detector having a thin window for the aperture. Such windows are commonly fabricated from aluminized boPET, thereby allowing energetic alpha particles to pass through to the detection gas. In the case of Fig. 7.24(a), the alpha-particle point source is placed such that the range of the alpha particles can reach any location on the detector aperture. The view factor is determined by the solid angle Ω subtended at the source by the aperture. As discussed earlier, because of parallax effects, the actual solid angle that allows alpha particles to deposit energy above the discriminator setting is $\Omega' < \Omega$. Note that energy loss per unit distance $(-dE/dx)$ is typically higher in the boPET than the medium between the source and detector, thereby strongly influencing Ω'.

In the case of Fig. 7.24(b), the alpha-particle point source is placed such that the range of the alpha particles is too short to reach all locations on the detector aperture. Consequently, the aperture radius a does not determine the solid angle within which particles can reach the aperture. Instead, the solid angle Ω_s is determined by a smaller effective a_s, where a_s is the greatest radius inside of which the alpha particles have enough energy to reach the detector. Again, due to parallax and energy loss in the window, the actual solid angle that allows alpha particles to deposit energy above the discriminator setting is Ω'.

In both cases shown in Fig. 7.24, use of Ω to correct the count underestimates the source strength. To determine the correct emission rate, the correct solid angle to use is Ω', which accounts for both parallax and window energy-loss effects. The correction is further complicated by the fact that energy deposition by alpha particles is defined by the Bragg ionization curve, in which $-dE/dx$ is not constant. Hence, the length h, the minimum distance the alpha particle must travel within the detector to be detected, is not a constant value.

A common method to reduce solid angle and parallax effects for alpha particle measurements is to highly collimate the alpha-particle source. As a result, the count rate diminishes, thereby requiring significantly longer measurement times. Also, energy dissipation in the intermediate medium (typically air) can be eliminated by placing the alpha particle source and detector in a vacuum chamber, a measurement method usually used for alpha-particle spectroscopy with a semiconductor detector, but not with a gas-filled detector.[9]

[9]The gas pressure inside a gas-filled detector can cause the boPET window to expand like a balloon if operated in vacuum, further complicating the calculation of a correction factor.

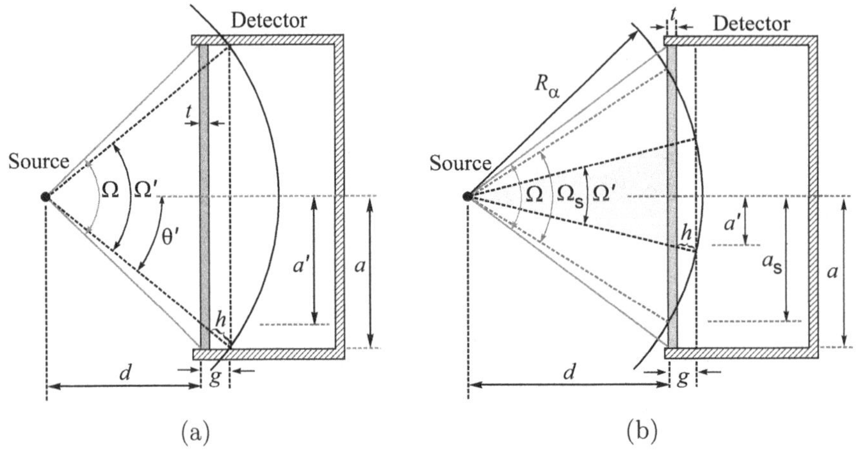

Figure 7.24. In (a) the range of the alpha particle is greater that the longest distance to the outer boundary of the detector aperture, whereas in (b) the range of the alpha particle is shorter than this distance. Note that Ω is defined by the detector aperture, Ω_s is defined by the maximum range that the alpha particles can reach the detector aperture, and Ω' is defined by the required transit length that a particle must pass through to still deposit enough energy for detection.

Beta Particles

Because beta particles, as they give up energy to the ambient medium, travel in very tortuous trajectories, the above discussion for heavy charged particles does not apply. As discussed in Section 7.2.2, parallax effects, along with source reflection, air and aperture attenuation, and source encapsulation effects can all be treated by using a single Monte Carlo simulation. There are no simple methods.

Neutrons

In many neutron experiments, the neutrons travel through air from their source to a detector. Thermal neutrons, in particular, can undergo significant air scattering, primarily by the nitrogen in air. For example, consider thermal neutrons produced in a reactor that travel down a reactor beam port to some experimental area. Such neutrons have a Maxwellian energy distribution characterized by a neutron temperature T and a thermal-averaged total interaction coefficient (or macroscopic cross section), as given by [Lamarsh 1966], namely

$$\overline{\Sigma}_t = \frac{\sqrt{\pi}}{2}\Sigma_t(E_o)\sqrt{\frac{T_o}{T}}. \tag{7.98}$$

Here E_o is a reference energy (usually 0.0253 eV) and T_o is a reference temperature (usually taken as 293 K).[10]

For dry air at STP, $\Sigma(E_o)$ is 5.01×10^{-4} cm^{-1}. Thus, the average distance thermal neutrons, with $T = T_o$, travel before they interact in air is $\lambda_t = 1/\Sigma_t = 2.00$ m. This result shows that for every 2 m of neutron travel in air, the fraction that interact is $1 - e^{-1} = 0.63$. To avoid such losses, sometimes a beam port is fitted with a long polyethylene tube inflated and sealed with a non-interacting gas, such as ^{4}He, or other canister containing a vacuum or ^{4}He.

[10]The most probable neutron energy in a Maxwellian flux distribution with a neutron temperature T_o is $E_o = 0.0253$ eV.

PROBLEMS

1. A thin layer of a radioisotope that emits 4.13-MeV, 5.1-MeV and 6.95-MeV alpha particles is electroplated on an iron substrate opposite and alpha-particle detector. What fraction of the alpha particles of each energy emitted towards the substrate are reflected back towards the detector? How would one distinguish alpha particles emitted directly towards the detector from those that are reflected from the substrate?

2. Plot the backscatter coefficient for monoenergetic electrons from an iron target for energies between 0.1 and 10 MeV.

3. An ^{241}Am alpha-particle source is placed in a 2π counter which has a dead time of 30 μs. The source was prepared with a 4-mm diameter, 50-angstrom thick plated deposit upon a 1-mm Ni backing. The lower level discriminator (LLD) is set to 100 keV. What is the expected count rate, per source particle, in the detector?

4. A ^{137}Cs disk source of 1-cm radius is placed and centered 25 cm away from a 3-in × 3-in NaI:Tl detector. Determine the fraction of gamma rays that intersect the detector aperture.

5. An ^{116m}In foil is activated in a nuclear reactor to use as a source to measure the dead time of a gas-filled proportional counter. Radiation measurements are performed at set time intervals, each measurement being 10 sec long, with the following results:

Data for dead-time determination for an ^{116m}In foil source.

T (min)	0	54	108	163	216	270
Counts	331204	260509	182493	114217	65323	35034
T (min)	324	378	432	486	540	594
Counts	18302	9308	4689	2402	1197	604

(a) Use the dead-time decay measurement method of Eq. (7.68)-Eq. (7.69) to find the detector dead time. (b) Use the dead-time decay measurement method of Eq. (7.71) to find the detector dead time. (c) Compare the results and discuss.

6. A radiation measurement is conducted in air for a ^{137}Cs source with a 5-in × 5-in NaI:Tl detector at a distance of 100 meters. Estimate the fraction of unscattered gamma rays that reach the detector.

7. An ^{241}Am alpha-particle source is centered and placed 3 mm away from the circular aperture of a gas-filled proportional counter. The aperture is 5.04 cm in diameter and is composed of 0.4 mg cm^{-2} boPET. The experiment is conducted at sea level and at 30°C. Assume that the alpha particle must pass through at least 2.5 mm of gas in the detector to be recorded. Radiation measurements are performed for 3 min each. After each measurement, the source is moved backwards by 2.5 mm, followed by a new measurement, resulting in the tabular data shown below.

Data for the determination of alpha-particle range.

D (mm)	3	5.5	8	10.5	13	15.5	18	20.5
Counts	237170	224568	210324	193868	176166	159315	142533	125979
D (mm)	23	25.5	28	30.5	33	35.5	38	40.5
Counts	109079	91118	73509	54969	36693	18363	1283	1

(a) After correcting for equivalent air thickness of the window, plot counts versus range.

(b) Based on the subtended solid angle, correct for counts per measurement and plot atop the previous graph. Determine source strength. Discuss features.

(c) Using the virtual aperture, correct the counts and plot atop the previous graph. Determine source activity. Discuss the features.

REFERENCES

BAERG, A.P., "Variation on the Paired Source Method of Measuring Dead Time," *Metrologia*, **1**, 131–133, (1965).

BALTAKMENS, T., "A Simple Method for Determining the Maximum Energy of Beta Emitters by Absorption Measurements," *Nucl. Instrum. Meth.*, **82**, 264–268, (1970).

CRAWFORD, J.A., "Theoretical Calculations Concerning Back-Scattering of Alpha Particles," in *The Transuranium Elements*, G.T. Seaborg, J.J. Katz, W.M. Manning, Eds., Vol. 2, New York: McGraw-Hill, 1949.

DERUYTTER, A.J., "Evaluation of the Absolute Activity of Alpha Emitters and of the Number of Nuclei in Thin Alpha Active Layers," *Nucl. Instrum. Meth.*, **15**, 164–170, (1962).

EVANS, R.D., *The Atomic Nucleus*, New York: McGraw-Hill, 1955.

FAIRSTEIN, E., ET AL., "IEEE Standard Test Procedures for Germanium Gamma-Ray Detectors," *IEEE Std 325-1996*, NIDC, IEEE, 1996.

GOSSMAN, M.S., A.J. PAHIKKALA, M.B. RISING, AND P.H. MC GINLEY, "Providing Solid Angle Formalism for Skyshine Calculations," *J. Appl. Clin. Med. Phys.*, **11**, (4), 278-282, (2010).

GOSSMAN, M.S., A.J PAHIKKALA,M.B RISING, P.H. MCGINLEY,"Letter to the Editor," *J. Appl. Clin. Med. Phys.*, **12**, (1), 242–243, (2011).

HUBBELL, J.H., R.L. BACH, AND R.J. HERBOLD, "Radiation Field From a Circular Disk Source," *J. Res. National Bureau of Standards*, **65C**, 249–264, (1961).

HUTCHINSON, J.M.R., C.R. NASS, D.H. WALKER, AND W.B. MANN, "Backscattering of Alpha Particles from Thick Metal Backings as a Function of Atomic Weight," *Int. J. Appl. Rad. Iso.*, **19**, 517–522, (1968).

JAFFEY, A.H., "Solid Angle Subtended by a Circular Aperture at Point and Spread Sources," *Rev. Sci. Instrum.*, **25**, 349–354, (1954).

KAPLAN, I., *Nuclear Physics*, Reading: Addison-Wesley, 1962.

KUZMINIKH, V.A. AND, S.A. VORBIEV, "Backscattering Coefficients Calculations of Monoenergetic Electrons and Positrons," *Nucl. Instrum. Meth.*, **129**, 561–563, (1975).

LAMARSH, J.R., *Introduction to Nuclear Reactor Theory*, Reading, MA: Addison-Wesley, 1966.

MARTIN, G.R., "The Estimation of the Resolving Time of a Counting Apparatus," *Nucl. Instrum. Meth.*, **13**, 263, (1961).

MCGREGOR, D.S. AND J.K. SHULTIS, "Reporting Detection Efficiency for Semiconductor Neutron Detectors, A Need for a Standard," *Nucl. Instrum. Meth.*, **632A**, 167–174, (2011).

MCNP6, *MCNP6 User's Manual*, LA-CP-13-00634, Los Alamos National Laboratory, 2013.

MÜLLER, J.W., "Dead-Time Problems," *Nucl. Instrum. Meth.*, **112**, 47–57, (1973).

MÜLLER, J.W., "On the Evaluation of the Correction Factor $\mu(\rho', \tau')$ for the Periodic Pulse Method," Report BIPM-76/3, *Bureau Int. des Poids et Mesures*, Sévres, 1976.

NCRP, "A Handbook of Radioactivity Measurements Procedures," Report 58, 2nd Ed., Bethesda, Maryland: NCRP, 1985.

PAXTON, F., "Solid Angle Calculation for a Circular Disk," *Rev. Sci. Instrum.*, **30**, 254–258, (1959).

Radiological Health Handbook, Rev., Rockville, MD: U.S. Department of Health, Education, and Welfare, 1970.

ROSSI, B.B. AND H.H. STAUB, *Ionization Chambers and Counters*, New York: McGraw-Hill, 1949.

RUBY, L., "Further Comments on the Geometrical Efficiency of a Parallel-Disk Source and Detector System," *Nucl. Instrum. Meth.*, **A337**, 531–533, (1994).

TABATA, T., R. ITO, AND S. OKABE, "An Empirical Equation for the Backscattering Coefficient of Electrons," *Nucl. Instrum. Meth.*, **94**, 509–513, (1971).

VEGA-CARILLO, H.R., "Geometrical Efficiency for a Parallel Disk Source and Detector," *Nucl. Instrum. Meth.*, **A371**, 535–537, (1996).

VEGA-CARILLO, H.R., Erratum to "Geometrical Efficiency for a Parallel Disk Source and Detector," *Nucl. Instrum. Meth.*, **A538**, 814, (2005).

VINCENT, C.H., *Random Pulse Trains, Their Measurement and Statistical Properties*, IEE Monograph Series 13, London: Peter Peregrinus, 1973.

ZEIGLER, J.F., J.P. BIERSACK, AND M.D. ZEIGLER, *SRIM: The Stopping and Range of Ions in Matter*, available through the web at `www.srim.org`, 2013.

Chapter 8

Essential Electrostatics

I happen to have discovered a direct relation between magnetism and light, also electricity and light, and the field it opens is so large and I think rich.

Michael Faraday

Numerous radiation detectors rely on the current produced by the motion of mobile charges created by radiation events. Radiation interactions within the detector volume ionize the absorbing medium, thereby creating mobile point charges, usually referred to as charge carriers (because these particles carry the charge through the detector). The detector may be a solid, liquid or gas, within which two or more electrodes are installed. Generally, a voltage is applied to the detector electrodes to move, or *drift*, the mobile charges across the detector volume to produce the current.

Electronic detectors range from simple radiation counters to complex radiation spectrometers. The former are used to indicate the presence of ionizing radiation, whereas the latter can also identify the energies and types of radiation. In either case, the fundamental physics of current induction formed by mobile charge carriers applies. Introduced in this chapter are the basic concepts governing charge and current induction in electronic radiation detectors.

8.1 Electric Field

Consider a point charge of Q coulombs, as shown in Fig. 8.1. The charge itself has *electric field* lines emanating from it radially outward (or inward). If a point charge is centered within an imaginary spherical surface, the electric field lines intersect the surface at right angles (perpendicular), with the imaginary surface area $4\pi r^2$, where r is the radius of the sphere. Suppose the number of field lines is denoted by N, then the surface density of field lines becomes $N/(4\pi r^2)$ per unit area. It is easy to understand that the electric field line density decreases with r^2 as the spherical surface area increases. The electric field vector (with SI units of volts/meter or, equivalently newtons/coulomb) at the surface of the sphere is

$$\boldsymbol{\mathcal{E}} = \frac{Q}{4\pi\epsilon_o r^2}\hat{\mathbf{r}}, \tag{8.1}$$

where $\hat{\mathbf{r}}$ is a unit vector normal to the spherical surface and ϵ_o is the *permittivity of free space* $\simeq 8.854\times10^{-12}$ farads per meter or, equivalently, A s V^{-1} m^{-1}. The magnitude of $\mathcal{E}$ is

$$\mathcal{E} = \boldsymbol{\mathcal{E}}\bullet\hat{\mathbf{r}} = \frac{Q\cos\theta}{4\pi\epsilon_o r^2}. \tag{8.2}$$

Because the field lines are perpendicular to the spherical surface, then multiplication by the area of the sphere $A = 4\pi r^2$ gives

$$\mathcal{E}A = \frac{Q\cos\theta}{4\pi\epsilon_o r^2}4\pi r^2 = \frac{Q}{\epsilon_o}. \tag{8.3}$$

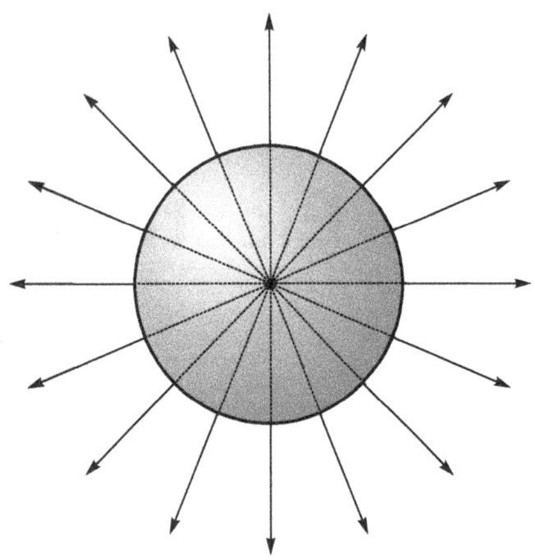

Figure 8.1. An imaginary spherical surface enclosing a point charge at the center.

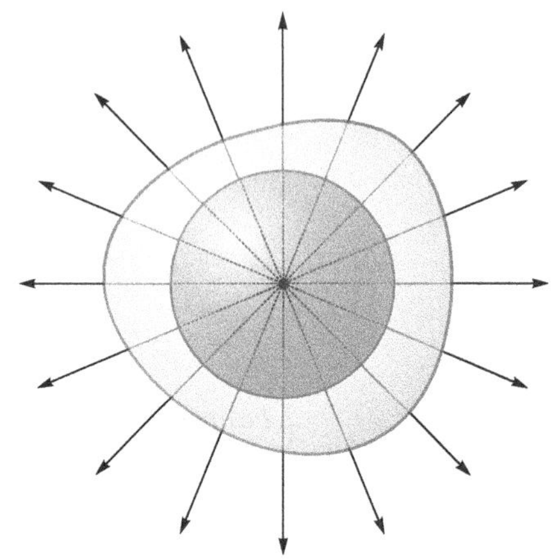

Figure 8.2. An irregularly shaped surface surrounding both the spherical surface and the point charge.

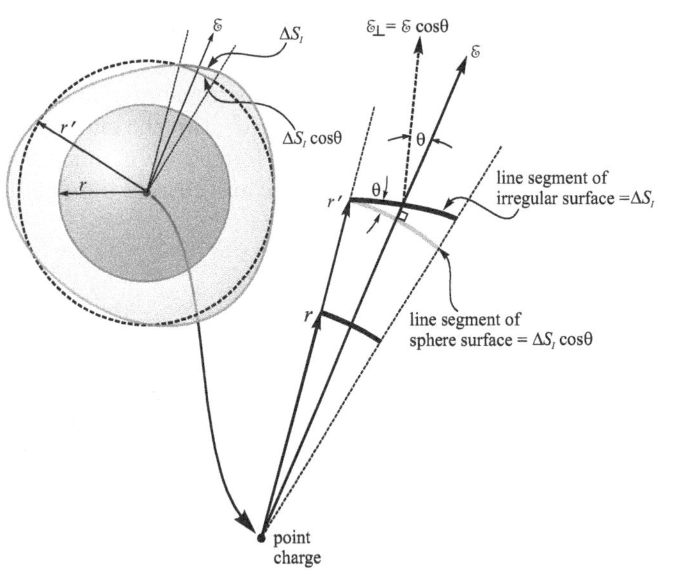

Figure 8.3. An irregularly shaped surface surrounding both the spherical surface and the point charge. The spherical surface is expanded to a new radius of r' that intersects the irregular surface.

Note that if the charge is positive, then the electric field lines point outwards (or positive direction), and if the charge is negative, then the electric field lines point inwards (or negative direction). In either case, Eq. (8.3) is still correct. From this result it is seen Q/ϵ_o is independent of the size of the sphere. If the imaginary spherical surface is then surrounded by an irregularly shaped surface, as shown in Fig. 8.2, the product $\mathcal{E}A$ can be shown to still be the same.

Suppose that the sphere of radius r is expanded to a radius r' such that portions of the new spherical surface lie inside and outside the irregular surface, as shown in Fig. 8.3. The electric field lines are still orthogonal to the sphere of radius r'. Consider a small segment ΔS_I along the irregular surface and a small line segment ΔS_S along the spherical surface. The angle between the two small line segments is designated θ; hence the length of the line segment on the sphere must be $\Delta S_I \cos\theta$. Notice also that the angle between electric field vector and the normal vector to the irregular surface is also θ.

Now apply the same argument to a small area segment ΔA_I on the irregular surface that intersects the spherical surface (see Fig. 8.4). The small area formed by projecting ΔA_I back onto the spherical surface is denoted ΔA_S. From the previous results for the line segments, the product $\mathcal{E}\Delta A_I$ for the irregular surface becomes $\Delta A_S = \mathcal{E}\Delta A_I \cos\theta$ for the spherical surface. In the limit as $\Delta A_I \to dA_I$ and $\Delta A_S \to dA_S$, the integration of $\mathcal{E}\cos\theta dA_I$ over the irregular surface must equal the integration of $\mathcal{E}dA_S$ over the spherical

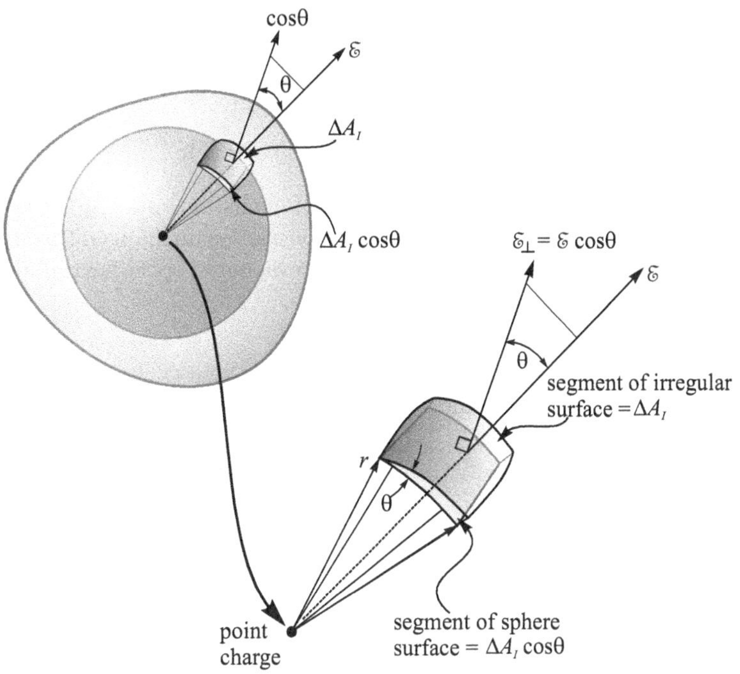

Figure 8.4. It can be shown that the electric field normal to the irregular surface ($E_\perp$) is related to the spherical surface by a factor of $\cos\theta$. Similarly, the small area on the irregular surface, ΔA_I, is related to the small area projected onto the spherical surface, ΔA_S, by a factor of $\Delta A_I \cos\theta$.

surface, which from Eq. (8.3) equals Q/ϵ_o, i.e.,

$$\int_{A_I} \mathcal{E} \cos\theta \, dA_I = \int_{A_S} \mathcal{E} \, dA_S = \frac{Q}{\epsilon_o}. \tag{8.4}$$

Thus, the product of the electric field and total surface area always is Q/ϵ_o regardless of the shape of a *closed* surface surrounding a point charge! Because the irregular surface is arbitrary, this also means the point charge can be anywhere inside this surface. For multiple point charges inside S_I, each charge Q_i contributes Q_i/ϵ_o, meaning that the total charge is the summation of charges within the surface surrounding the charges $Q_{total} = \sum_i Q_i$.

Finally, if $\hat{\mathbf{n}}$ is a unit vector perpendicular to any closed surface S, then $\mathcal{E} \cos\theta \, dA_S = \boldsymbol{\mathcal{E}} \bullet \hat{\mathbf{n}} \, dA_S$ and Eq. (8.4) can be written as

$$\boxed{\int_{A_S} \boldsymbol{\mathcal{E}} \bullet \hat{\mathbf{n}} \, dA_S = \iint_S \boldsymbol{\mathcal{E}} \bullet \hat{\mathbf{n}} \, dS = \frac{\sum_i Q_i}{\epsilon_o} = \frac{Q_{total}}{\epsilon_o}.} \tag{8.5}$$

This is commonly referred to as *Gauss' Law.*[1]

[1]Eq. (8.5), credited to Carl Friedrich Gauss, is actually an application of his Law of Quadratic Reciprocity that he rediscovered in 1795. This result was actually derived by Joseph-Louis Lagrange many years earlier in 1785.

8.1.1 Alternate Derivation of Gauss' Law

A more mathematically direct way, although not so insightful, to derive Gauss' law is to start with the first of Maxwell's four equations of electromagnetism, namely

$$\nabla \cdot \mathcal{D}(\mathbf{r}) = \nabla \cdot [\kappa \epsilon_o \mathcal{E}(\mathbf{r})] = \rho_V(\mathbf{r}), \tag{8.6}$$

where $\mathcal{D}(\mathbf{r})$ is the dielectric displacement, κ is the dielectric constant, and $\rho_V(\mathbf{r})$ is the volumetric charge density (coulombs per cubic meter). Usually the dielectric constant is unchanged through a medium, although not always. Suppose in this example that the charge is in an evacuated space surrounded by a surface. Hence, $\kappa = 1$ and Eq. (8.6) can be rewritten as,

$$\nabla \cdot \mathcal{E}(\mathbf{r}) = \frac{\rho_V(\mathbf{r})}{\epsilon_o}, \tag{8.7}$$

Equation 8.7 is integrated over a closed volume V to obtain

$$\iiint_V \nabla \cdot \mathcal{E}(\mathbf{r})\, dV = \frac{1}{\epsilon_o} \iiint_V \rho_V(\mathbf{r})\, dV = \frac{Q_{total}}{\epsilon_o}. \tag{8.8}$$

Here Q_{total} is the total charge contained in the volume V. The remaining volume integral can be converted to an integral over the surface S of V by using Gauss' divergence theorem [Riley et al. 2006] to obtain

$$\iint_S \mathcal{E} \cdot \hat{\mathbf{n}}\, dS = \frac{Q_{total}}{\epsilon_o}, \tag{8.9}$$

where $\hat{\mathbf{n}}$ is the outward unit normal to the surface S. Equation (8.6), or the alternative form Eq. (8.7), are usually referred to as Gauss' law in differential form, while Eq. (8.9) is Gauss' law in integral form.

8.2 Electrical Potential Energy

An electric field causes mobile charges to move, hence an electric field does *work* on a mobile charge. The work required to move the particle from position $\mathbf{r}_1$ to position $\mathbf{r}_2$ is given by

$$W = \int_{\mathbf{r}_1}^{\mathbf{r}_2} \mathbf{F} \cdot d\mathbf{l} = \int_{\mathbf{r}_1}^{\mathbf{r}_2} F \cos\theta \; dl, \tag{8.10}$$

where $\mathbf{F}$ is the force vector applied to the particle, $d\mathbf{l}$ is a differential length along the direction of travel, and θ is the angle between these two vectors. Coulomb's law states that the force on a particle at $\mathbf{r}'$ with charge q' from a second particle at $\mathbf{r}$ with charge q is

$$\mathbf{F} = \frac{QQ'(\mathbf{r} - \mathbf{r}')}{4\pi\epsilon_o|\mathbf{r} - \mathbf{r}'|^3} \quad \text{with magnitude} \quad F = \frac{QQ'}{4\pi\epsilon_o r^2}, \tag{8.11}$$

where r is the distance between the charges.

Consider the electric field of a point charge Q shown in Fig. 8.1. If another point charge Q' is placed in the electric field of the first charge Q, then the electric field of Q exerts a force on Q', and the electric field of Q' exerts an equal force on Q.

In Fig. 8.5, a single charge Q' is moving in the electric field of a second charge Q from position r_1 to position r_2 across the field lines of Q. In the limit of decreasing angles between field lines, $\Delta\omega \rightarrow d\omega$, $\Delta l \rightarrow dl$, $\Delta r \rightarrow dr$, and $\theta_1 = \theta_2 = \theta$. From Fig. 8.5, the relationship between the differential distance traveled dl and the differential change in distance between the charges dr is

$$\cos\theta \; dl = dr. \tag{8.12}$$

Substitution of Eq. (8.12) into Eq. (8.10) then yields the work done on the point charge Q' as it moves from some point $\mathbf{r}_1$ to $\mathbf{r}_2$,

$$W = \int_{\mathbf{r}_1}^{\mathbf{r}_2} F dr = \int_{\mathbf{r}_1}^{\mathbf{r}_2} \frac{1}{4\pi\epsilon_o} \frac{QQ'}{r^2} dr = \frac{QQ'}{4\pi\epsilon_o} \left(\frac{1}{r_1} - \frac{1}{r_2} \right). \tag{8.13}$$

This result shows that the path taken from $\mathbf{r}_1$ to $\mathbf{r}_2$ is immaterial, but instead only the change in radial distance from Q determines the work done. In other words, it is only the starting and ending locations that determine the total work performed on the charge moving in the electric field.[2]

The potential energy of a particle with charge Q located a distance r away from another particle of charge Q' is represented by

$$U = \frac{1}{4\pi\epsilon_o} \frac{QQ'}{r}. \tag{8.14}$$

Suppose that the particle with charge Q' is surrounded by n particles with charge Q_i at distance r_i from the particle with charge Q'. The total potential energy exerted on the particle with charge Q' from the collective electric fields of the surrounding particles is

$$U = \frac{Q'}{4\pi\epsilon_o} \left(\frac{Q_1}{r_1} + \frac{Q_2}{r_2} + \frac{Q_3}{r_3} + ... + \frac{Q_n}{r_n} \right) = \frac{Q'}{4\pi\epsilon_o} \sum_{i=1}^{n} \frac{Q_i}{r_i}. \tag{8.15}$$

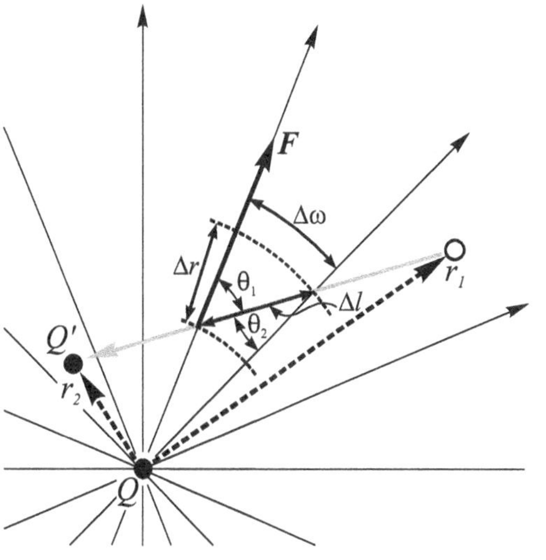

Figure 8.5. Depiction of charge Q' moving through the electric field of charge Q.

The potential energy of an individual charged particle is useful, and is important for understanding signal formation within an electronic radiation detector. However, it is more common to use the general term *potential*, which is defined as the potential energy per unit charge,

$$V = \frac{U}{Q'} = \frac{1}{4\pi\epsilon_o} \sum_{i=1}^{n} \frac{Q_i}{r_i}, \tag{8.16}$$

and is expressed in units of volts (or joules per coulomb). Note that the potential is no longer dependent upon the "test" charge Q'.[3] The force exerted upon Q' may also be expressed in terms of the electric field, produced by one or more point charges, in which $\mathbf{F} = Q'\boldsymbol{\mathcal{E}}$. Substitution of $q'\boldsymbol{\mathcal{E}}$ into Eq. (8.13) and division by Q' gives the potential difference between two points within the electric field. Hence, the potential difference between arbitrary locations a and b is

$$V_{ab} = \Delta V = \int_a^b \boldsymbol{\mathcal{E}} \bullet d\mathbf{l} = \int_a^b \mathcal{E} \cos\theta \, dl. \tag{8.17}$$

[2]The path independence is a consequence that $\mathbf{F}$ of Eq. (8.11) is given by the gradient of a *scalar field potential*, i.e.,

$$\mathbf{F} = \boldsymbol{\nabla} \left(\frac{QQ'}{4\pi\epsilon_o r} \right).$$

As a result $\boldsymbol{\nabla} \bullet \mathbf{F} = 0$. Force fields with this property are called *conservative* force fields.

[3]The normalization unit charge Q' is that of a free electron, q_e, such that $Q' = q_e N'$ where N' is an integer. For a single electron, N' must equal unity. Other charges are therefore integer representations of q_e, namely, $Q_i = q_e N_i$.

In summary, Eq. (8.17) is the voltage that an experimenter would measure between two points (a and b) within an electric field. The work done on a unit test charge moving from some point a to another point b in the electric field is $Q'V_{ab}$.

8.3 Capacitance

Consider the arrangement depicted in Fig. 8.6. Two conductive plates, separated by a distance d, have equal, but opposite, charges. An electric field is produced between the plates by the charges on the plates. The positively charged plate (or terminal) has a voltage V_1 and the negatively charged plate has a voltage V_2. The *capacitance* of the two plates is defined as the ratio of the charge magnitude on either plate to the magnitude of the potential difference between the plates,

$$C = \left| \frac{Q}{\Delta V} \right|. \tag{8.18}$$

If ΔV is taken as the applied voltage V between the electrodes, then the above definition gives the important relation

$$CV = Q \tag{8.19}$$

The SI unit for capacitance is the farad (one coulomb per volt). The reader should understand that the charge stored in a capacitor has a summed positive charge on one terminal and an equal summed negative charge stored on the opposite terminal; hence, a capacitor with stored charge Q actually has $+Q$ on one terminal and $-Q$ on the other terminal.

Many radiation detectors use a voltage applied between two conductive terminals. Of these detectors, the parallel plate configuration, similar to that shown in Fig. 8.6, is the simplest and easiest to understand. The finite area of either plate has a charge of magnitude Q. Hence, there is an average charge magnitude per unit area, defined as a surface charge density σ. Assume that there is a vacuum between the conductive plates and that the conductive plates have equal areas A and are separated by a distance d, then the magnitude of the electric field between the parallel plates is

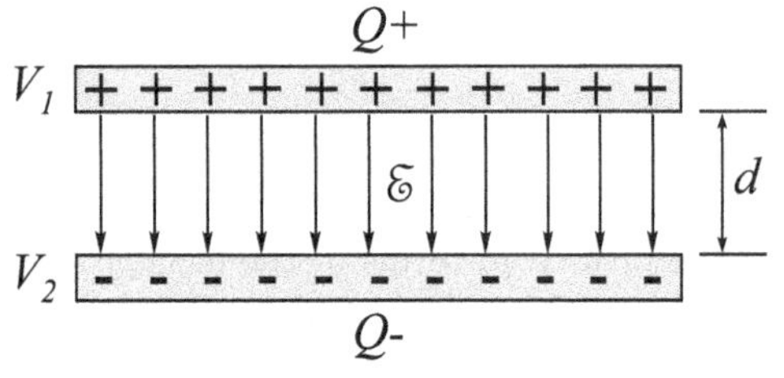

Figure 8.6. Two conductive plates separated by distance d have equal, but opposite, charges stored upon them. The cumulative charges produce an electric field, with a voltage of V_1 at the positive charge terminal and a voltage V_2 at the negative charge terminal.

$$\mathcal{E} = \frac{\sigma}{\epsilon_o} = \frac{Q}{\epsilon_o A}. \tag{8.20}$$

From Eq. (8.17), the voltage difference between the plates is

$$V = \mathcal{E}d = \frac{Qd}{\epsilon_o A}. \tag{8.21}$$

Substitution of Eq. (8.21) into Eq. (8.19) yields the capacitance

$$C = \epsilon_o \frac{A}{d}. \tag{8.22}$$

A radiation detector has a radiation absorbing medium, such as a gas or semiconductor, between the conductive electrodes. The permittivity of the radiation absorbing material must be substituted into Eq. (8.22), which is accomplished by multiplying ϵ_o by the *dielectric correction constant* κ_s of the radiation absorbing material,

$$C = \kappa_s \epsilon_o \frac{A}{d}. \tag{8.23}$$

Note that the electric field, voltage and charge terms are not necessary to define capacitance; hence, the capacitance is dependent only upon the device dimensions and permittivity of the insulating material between the electrodes.

Many advanced detector designs do not have simple parallel plate geometries, and many detectors are designed to reduce capacitive effects. Gas filled detectors are often designed in a coaxial configuration with a single anode wire through the center with a much larger cylindrical outer conductive cathode shell. There are detector designs with numerous anodes (or cathodes), some detectors have concentric spherical electrodes, and many detectors have a variety of different electrode shapes. Because Eq. (8.23) applies only to simple planar geometries, capacitance for these other detector geometries must be found by deriving a fundamental form of Eq. (8.19) for each particular geometry of interest.

8.4 Current and Stored Energy

Capacitors are generally used as charge storage devices, and these charges are transported to a capacitor through an electronic circuit. Consider the movement of charge ΔQ to a capacitor in some time interval Δt. The *current* I is defined as the amount of charge transported divided by the time taken for the charge movement, i.e.,

$$I = \frac{\Delta Q}{\Delta t} = C\frac{\Delta V}{\Delta t}. \tag{8.24}$$

The rate at which charge flows to a capacitor, or other electronic components may not be constant in time; hence, the instantaneous current is defined as

$$I = \frac{dQ}{dt} = C\frac{dV}{dt}. \tag{8.25}$$

Charges flowing to a capacitor build up the charges on one side of the capacitor, which causes charges of the opposite polarity to accumulate on the opposing electrode. Consequently, work is performed to accumulate the change in charge ΔQ in the capacitor. From Eq. (8.10), the work to increase the electrode charge by dq

$$dW = dq\int \mathcal{E}\cos\theta \, dl = V\,dq = \frac{q}{C}dq. \tag{8.26}$$

where V is the magnitude of the voltage on the capacitor at a time when the accumulated charge is q. If no charge is initially stored on the capacitor, Eq. (8.26) is integrated to find the total work performed to accumulate charge Q on the capacitor, i.e.,

$$W = \int_0^W dW = \int_0^Q \frac{q}{C}dq = \frac{Q^2}{2C}. \tag{8.27}$$

With Eq. (8.19), the work performed to store Q, which is also the *energy* stored by the capacitor, can be expressed as

$$W = \frac{Q^2}{2C} = \frac{QV}{2} = \frac{CV^2}{2}. \tag{8.28}$$

This result is important for electronic radiation detectors. Suppose a radiation interaction occurs within an electronic radiation detector, on which a voltage V_0 is applied across the detector terminals. The mobile charges created by the radiation event are caused to move (or *drift*) through the detector. Positive charges drift towards the negatively charged electrode and negative charges drift towards the positively charged electrode. As these charges separate, they form a small, time-dependent electric field of opposite polarity to that of the electric field applied to the detector. As a result, a small change in voltage appears between the

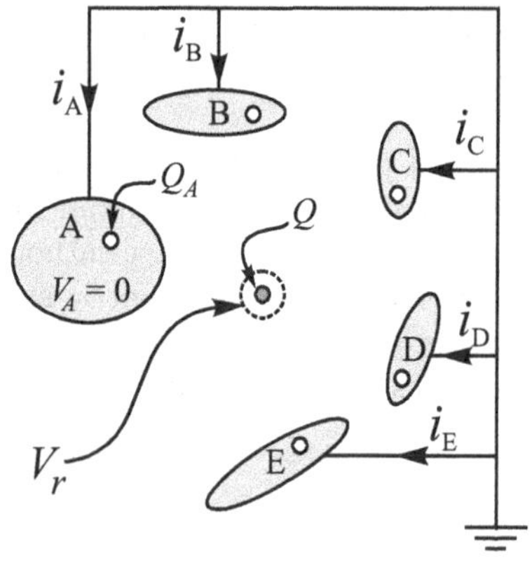

Figure 8.7. A point charge Q is placed in the vicinity of grounded conductive surfaces, and produces image charges on those surfaces.

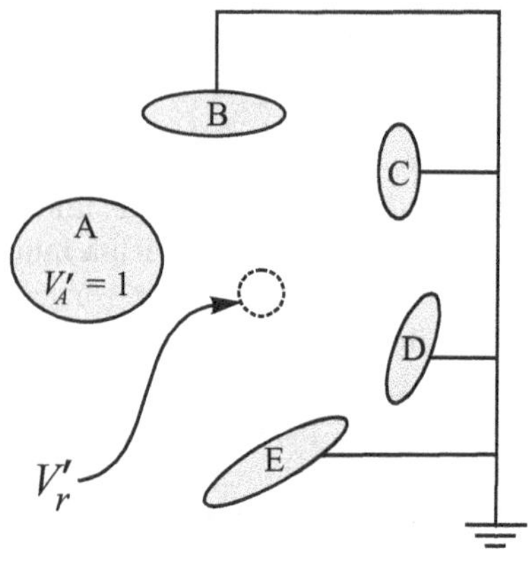

Figure 8.8. The point charge is removed and a potential equal to unity is applied to **A**. All other electrodes remain grounded.

electrodes caused by the electric field of the drifting charges. Hence, stored energy is expended to move the charges, and the energy expended to move the charges can be measured as a change in current or voltage that, typically, is performed by an externally connected circuit.

Charges moving in a detector cause current to flow in an externally connected circuit through a process called *induction*. In other words, mobile charges moving in the detector cause electron current to flow in the circuit externally connected to the detector. The change in electron current can be measured directly, commonly referred to as *current mode operation*, or it can flow to a capacitor in the external circuit and be stored, thereby providing a measure of the voltage change as defined by Eq. (8.19). This change in voltage ΔV, or voltage *pulse*, signals the occurrence of a radiation interaction event in the detector. Such measurements are referred to as *pulse mode operation*.

8.5 Basics of Charge Induction

Many radiation detectors are based on the principle of charge induction, whereby moving point charges work to induce a current in an external circuit. The detector itself is a capacitor, and charges moving in the detector cause current flow to or from the device terminals or electrodes. The charge or current induced on any electrode can be found through Shockley's or Ramo's method (both methods are essentially the same) [Ramo 1939; Shockley 1938].

Consider a single point charge Q situated at position $\mathbf{r}_q$ in a vacuum within which there are an arbitrary number of grounded conducting surfaces, as depicted in Fig. 8.7. The point charge Q will cause electrons on the surrounding conductive surfaces to change in position. If Q is negative, then the surface electrons will be slightly repelled, while if positive, slightly attracted to the surface. The overall effect is to have the apparent appearance of an *image charge* of opposite sign on the conductive surface. Suppose the goal is to determine the induced current and induced charge produced upon electrode **A** by the point charge Q, thereby producing an image charge of Q_A on the electrode. Perhaps the other electrodes are also affected, but it will shortly be seen to be of no consequence to this derivation. Surround the point charge Q by a small equipotential conducting sphere. From Gauss' law one has

$$\int_{sph} \boldsymbol{\mathcal{E}} \cdot \hat{\mathbf{n}}\, ds = \int_{sph} \frac{\partial V}{\partial n}\, ds = \frac{Q}{\epsilon_o}, \tag{8.29}$$

where $\hat{\mathbf{n}}$ is the outward unit normal to the surface and $\partial V/\partial n$ denotes $\nabla V \bullet \hat{\mathbf{n}}$.

Because there are no charges in the vacuum bounding the electrodes and the small sphere, the electrostatic field is given by Laplace's equation

$$\nabla^2 V(\mathbf{r}) = 0, \quad \mathbf{r} \in \text{ all space outside the conducting surfaces.} \tag{8.30}$$

The boundary conditions needed to specify a unique solution of Eq. (8.29) are $V(\mathbf{r}) \to 0$ as $|\mathbf{r}| \to \infty$ and by the voltages on the electrodes. The voltage on the small imaginary sphere is denoted by $V_r = V(\mathbf{r})$.

Now for the same set of electrodes with the charge Q removed, electrode **A** is given a voltage V_A with the other electrodes still grounded, as shown in Fig. 8.8. The new electrostatic field, $V'(\mathbf{r})$, is described by

$$\nabla^2 V'(\mathbf{r}) = 0, \tag{8.31}$$

with boundary conditions specified by the voltage on all surfaces. The voltage at $\mathbf{r}$ at the surface of the imaginary sphere is now $V'_r = V'(\mathbf{r})$. Apply Green's theorem [Riley et al. 2006] to all surfaces, namely

$$\int_{vol} [V'\nabla^2 V - V\nabla^2 V']dv = \int_S \left[V'\frac{\partial V}{\partial n} - V\frac{\partial V'}{\partial n}\right] ds. \tag{8.32}$$

Here the surface integration is over all electrode surfaces, the imaginary surface, and the surface at infinity, and the volume integration is over all space bounded by all the surfaces. From Eq. (8.30) and Eq. (8.31) the left-hand side of Eq. (8.32) is seen to vanish. On the right-hand side the surface integral at infinity vanishes because $V(\mathbf{r} \to \infty) = V'(\mathbf{r} \to \infty) = 0$. The surface integrals around all grounded electrodes also have the potential V and V' equal to zero, hence, the right side of Eq. (8.32) vanishes for these electrodes. Thus, only the surface integrals around the charged electrode **A** and around the small sphere remain, and Eq. (8.32) reduces to [Ramo 1939],

$$V'_A \int_{S_A} \frac{\partial V_A}{\partial n} ds - V_A \int_{S_A} \frac{\partial V'_A}{\partial n} ds + V'_r \int_{sph} \frac{\partial V_r}{\partial n} ds - V_r \int_{sph} \frac{\partial V'_r}{\partial n} ds = 0. \tag{8.33}$$

From Gauss' law the first integral equals Q_A/ϵ_o, where Q_A is the image charge on electrode **A**. Similarly, the third integral equals Q/ϵ_o. The second term vanishes because $V_A = 0$. From Eq. (8.29), the fourth integral vanishes because there is no charge inside the small imaginary sphere. Thus,

$$V'_A \frac{Q_A}{\epsilon_o} + V'_r \frac{Q}{\epsilon_o} = 0 \quad \text{or} \quad Q_A = -Q\frac{V'_r}{V'_A}. \tag{8.34}$$

From this result the induced current produced by the point charge in Fig. 8.7, moving in any direction between the electrodes with vector velocity $\mathbf{v} = d\mathbf{r}/dt$, is

$$I_A = \frac{dQ_A}{dt} = -\frac{Q}{V'_A}\frac{dV'_r}{dt} = -\frac{Q}{V'_A}\nabla V'_r(\mathbf{r}) \bullet \frac{d\mathbf{r}}{dt} = \frac{Q}{V'_A}\boldsymbol{\mathcal{E}}'(\mathbf{r}) \bullet \mathbf{v}, \tag{8.35}$$

where $\boldsymbol{\mathcal{E}}'(\mathbf{r})$ is the electric field at point $\mathbf{r}$ under the condition that the "potential" at electrode **A** is set at V_A with all other electrodes grounded.

It is useful to normalize $V'(\mathbf{r})$ by setting V'_A to one volt, i.e., define $V_w(\mathbf{r}) \equiv V'_Q(\mathbf{r})/V'_A$. This normalized "voltage" V_w is called the *weighting potential* although it is dimensionless. In terms of this normalized potential, Eq. (8.34) and Eq. (8.35) are expressed as

$$\boxed{Q_A = -QV_w(\mathbf{r}) \quad \text{and} \quad I_A = -Q\nabla V_w(\mathbf{r}) \bullet \frac{d\mathbf{r}}{dt} = Q\boldsymbol{\mathcal{E}}_w(\mathbf{r}) \bullet \mathbf{v}} \tag{8.36}$$

where the normalized or weighting electric field is $\boldsymbol{\mathcal{E}}_w = \boldsymbol{\mathcal{E}}'/V_A' = -\boldsymbol{\nabla} V_w$. This result can be applied to any one of the electrodes by setting its potential to one volt and grounding all other electrodes [see also Jen 1941; Beck 1953]. The results of Eq. (8.36) are significant and important, and are often referred to as *Ramo's Theorem.*[4] Hence, for any geometric detector structure, the expected current induction for the electrode tied to the readout circuit can be found from Eq. (8.36).

One can calculate the weighting potential for practically any electrode configuration by solving Laplace's equation (Eq. (8.30)) with the boundary conditions that (1) the contact connected to the readout electronics is set to unity and (2) all other contacts are grounded (set to zero). Weighting potentials for a few select geometries are easily calculated by analytical means, while weighting potentials for more complicated electrode geometries may be calculated by finite difference methods.

8.5.1 Green's Reciprocation Theorem

Suppose there is an array of M conductors in a vacuum and charges Q_i, $i = 1, \ldots, M$ are placed on the conductors to produce potentials V_i, $i = 1, \ldots, M$. Then if these charges are replaced by different charges Q_i', $i = 1, \ldots, M$ a different set of potentials V_i', $i = 1, \ldots, M$ would result. How are the old charges and potentials related to the new charges and potentials?

The potential fields are given by $\nabla^2 V(\mathbf{r}) = 0$ and $\nabla^2 V'(\mathbf{r}) = 0$ with appropriate boundary conditions. Equation (8.32) is Green's theorem and relates V to V'. Because $\nabla^2 V(\mathbf{r}) = 0$ and $\nabla^2 V'(\mathbf{r}) = 0$, the left-hand side of Eq. (8.32) vanishes. Also the integral over the surface at infinity, where both V and V' are zero, vanishes. Thus, Eq. (8.32) reduces to [Smythe 1989]

$$\sum_{i=1}^{M} \int_{S_i} \left[V_i' \frac{\partial V}{\partial n} - V_i \frac{\partial V'}{\partial n} \right] ds_i = 0,$$

where S_i is a closed surface around electrode i. The V and V' are constant on the surface so Eq. (8.5.1) can be written as

$$\sum_{i=1}^{M} V_i' \int_{S_i} \frac{\partial V}{\partial n} ds_i = \sum_{i=1}^{M} V_i \int_{S_i} \frac{\partial V'}{\partial n} ds_i, \tag{8.37}$$

which, from Gauss' law (see Eq. (8.29)), yields the important result

$$\boxed{\sum_{i=1}^{M} Q_i V_i' = \sum_{i=1}^{M} Q_i' V_i.} \tag{8.38}$$

This result is commonly referred to as *Green's reciprocation theorem* [for more details, see Jeans 1943; Jackson 1975; Smythe 1989].

8.6 Charge Induction for a Planar Detector

The simplest geometry for a radiation detector that depends on charge induction is the *planar device* and consists of two parallel conductors or terminals separated by a radiation absorbing medium. This configuration is often used for simple gas-filled or semiconductor radiation detectors. A voltage is typically applied

[4] Although Eq. (8.36) is commonly referred to as "Ramo's theorem," the same result was actually published by Shockley in 1938 a year before Ramo's publication in 1939. Because both V_w and $\boldsymbol{\mathcal{E}}_w$ are dimensionless, Eqs. (8.36) are dimensionally correct, although many forget this fact.

across the terminals to produce an electric field throughout the radiation absorbing medium. When a radiation interaction occurs within the detector volume, N_o electron-ion pairs are created. Because of the electric field across the device, free electrons, with charge $-q$ and positive ions, with charge q drift within the detector medium toward the conductive surfaces, with negative charges moving towards the anode and positive charges moving towards the cathode. Both conductive terminals are connected within the same circuit, as shown in Fig. 8.9.

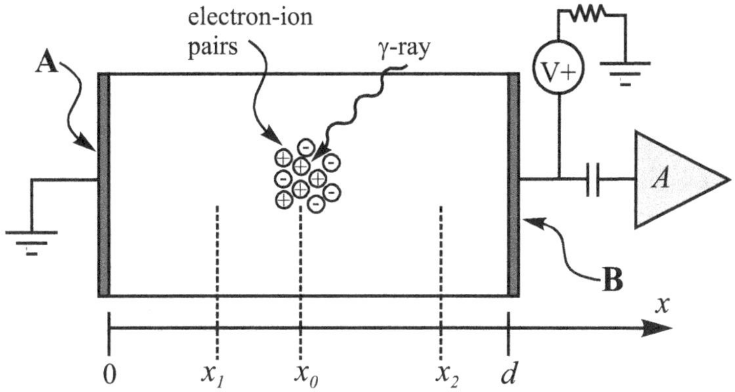

Figure 8.9. Depiction of a planar radiation detector where a gamma ray is absorbed at location x_0. The figure shows a preamplifier attached to the positively biased electrode at $x = d$.

The electrostatic field within the absorbing medium, in the absence of free charges, is given by Laplace's equation $\nabla^2 V(\mathbf{r}) = 0$.[5] By picking the x-axis to be perpendicular to the terminals, with the origin at the grounded terminal, as shown in Fig. 8.9, the potential varies only with x and the potential is given by

$$\nabla^2 V(x) = \frac{d^2 V(x)}{dx^2} = 0, \tag{8.40}$$

subject to the boundary conditions $V(0) = 0$ and $V(d) = V_B$. Integration of Eq. (8.40) twice and application of the boundary conditions to evaluate the two constants of integration yields

$$V(x) = \frac{V_B}{d} x. \tag{8.41}$$

The dimensionless weighting potential $V_w(x)$ is obtained by dividing this result by V_B or, equivalently, by setting $V_B = 1$ volt. In either way, for a planar device $V_w(x) = x/d$. With this weighting potential, the induced charge in the electrodes is readily calculated.[6]

In some circumstances, it is important to separately calculate the induced charge contributions from the electrons and positive ions liberated by radiation. Instances requiring separate treatments may involve

---

[5] Laplace's equation is simply Poisson's equation with ρ equal to zero,

$$\nabla^2 V(\mathbf{r}) = -\nabla \bullet \boldsymbol{\mathcal{E}}(\mathbf{r}) = -\frac{\rho}{\kappa \epsilon_o}. \tag{8.39}$$

[6] Some textbooks describe the output of a planar detector with conservation of energy methods. However, the reader is warned that the conservation of energy approach is generally incorrect, although it may appear to work for special cases. The consequence of using the conservation of energy approach is especially problematic with semiconductor detectors and should be avoided.

different charge carrier velocities or different recombination/trapping times for electrons and ions. Such treatments are covered in later chapters. Here it is assumed that charge carrier velocities and trapping are of no consequence, so the separate contributions of electrons and positive ions to the induced current can be determined in a straightforward manner. Consider the situation depicted in Fig. 8.9, where a radiation interaction event occurs at distance x_0 between two electrodes, labeled **A** and **B**, separated by a distance of d. A positive voltage potential is applied at electrode **B**, such that electrons drift to electrode **B** and positive ions drift to electrode **A**.

The change in the induced charge on electrode **B** resulting from the total displacement of N_o charge pairs moving between the electrodes can be found by using Ramo's Theorem. Upon integration of the induced current given by Eq. (8.36), the induced charge on the electrodes ΔQ after the ions have moved from x_0 to x_1 and the electrons have moved from x_0 to x_2 is

$$\begin{aligned}\Delta Q &= \int_0^{t(x_1)} I_A(t)\,dt\Big|_{ions+} + \int_0^{t(x_2)} I_A(t)\,dt\Big|_{e^-} \\ &= -q_e N_o \int_0^{t(x_1)} \nabla V_w \bullet \frac{d\mathbf{r}}{dt}\,dt\Big|_{ions+} - (-q_e N_o)\int_0^{t(x_2)} \nabla V_w \bullet \frac{d\mathbf{r}}{dt}\,dt\Big|_{e^-},\end{aligned} \tag{8.42}$$

where q_e is the magnitude of the electron charge, $t(x_1)$ is the time required for the ions to move from x_0 to x_1, and $t(x_2)$ is the time required for the free electrons to move from x_0 to x_2. In this geometry $V_w(\mathbf{r}) = V_w(x) = x/d$ so Eq. (8.42) can be evaluated as follows.

$$\begin{aligned}\Delta Q &= -q_e N_o \int_0^{t(x_1)} \frac{dV_w}{dx}\left(\frac{dx}{dt}\right) dt\Big|_{ions+} - (-q_e N_o)\int_0^{t(x_2)} \frac{dV_w}{dx}\frac{dx}{dt}\,dt\Big|_{e^-} \\ &= -q_e N_o \int_0^{t(x_1)} \frac{d}{dx}\left(\frac{x}{d}\right)\frac{dx}{dt}\,dt\Big|_{ions+} + q_e N_o \int_0^{t(x_2)} \frac{d}{dx}\left(\frac{x}{d}\right)\frac{dx}{dt}\,dt\Big|_{e^-} \\ &= -q_e N_o \int_{x_0}^{x_1} \frac{1}{d}\,dx\Big|_{ions+} + q_e N_o \int_{x_0}^{x_2} \frac{1}{d}\,dx\Big|_{e^-} = q_e N_o\left[\frac{x_0 - x_1}{d}\Big|_{ions+} + \frac{x_2 - x_0}{d}\Big|_{e^-}\right],\end{aligned} \tag{8.43}$$

which yields the final result,

$$\boxed{\Delta Q|_{planar} = q_e N_o\left[\frac{x_2 - x_1}{d}\right].} \tag{8.44}$$

Hence, the total displacement of the free carriers divided by the detector active region width yields the change in the induced charge measured by the detector. If all of the electrons drift to the positive contact at $x = d$ and all of the positive ions drift to the negative contact at $x = 0$, then $x_2 - x_1 = d - 0$, and the induced charge becomes $Q = q_e N_o$.

Analysis Based on Green's Reciprocation Theorem

The induced charge can also be found by using Green's reciprocation theorem as stated by Eq. (8.38). Consider the two electrodes in the planar device, both of equal dimensions separated by distance d. Two different situations are now considered. In Case 1, a charge Q_o is placed between the plates on an imaginary infinitely small conductor (labeled conductor 3). Plate **A** has, as a consequence, an induced charge Q_A and plate **B** has an induced charge Q_B. The potentials on the electrodes **A** and **B** are zero (Case 1). The induced charges Q_A and Q_B are to be determined by Green's reciprocation theorem. In Case 2 the charge Q_o is removed from imaginary conductor 3 (so that $Q_o = 0$), and placed on electrode **B**. In both cases the potentials and resulting induced charges are unknown and are summarized in Table 8.1.

Table 8.1. Charge and potential conditions for two electrodes between which a unit mobile charge (Q_o) is present.

	Case 1 (unprimed)			Case 2 (primed)		
Electrode	**A**	**B**	3	**A**	**B**	3
Charge	Q_A	Q_B	Q_o	Q'_A	Q_o	0
Potential	0	0	V_o	V'_A	V'_B	V'_o

From Green's reciprocation theorem of Eq. (8.38) for a general electrode problem

$$Q_1V'_1 + Q_2V'_2 + Q_3V'_3 = Q'_1V_1 + Q'_2V_2 + Q'_3V_3,$$

or rewritten for this specific case,

$$Q_AV'_A + Q_BV'_B + Q_oV'_o = Q'_AV_A + Q'_BV_B + Q'_3V_o.$$

In the present analysis, $V_A = V_B = Q'_3 = 0$ and the above general relation reduces to

$$Q_AV'_A + Q_BV'_B + Q_oV'_o = 0. \tag{8.45}$$

Because the sum of the induced charges and the charge Q_o, causing the induced charges, must sum to zero, one also has

$$Q_A + Q_B + Q_o = 0. \tag{8.46}$$

From Eq. (8.45) and Eq. (8.46) the induced charges in Case 1 are found to be

$$Q_A = -Q_o\frac{V'_o - V'_B}{V'_A - V'_B} \quad \text{and} \quad Q_B = -Q_o\frac{V'_A - V'_o}{V'_A - V'_B}. \tag{8.47}$$

What does this mean? The charge induced by the test charge on a given electrode is proportional to the potential difference between the location where the charge originally appeared (in our case, at x_o) and the opposite electrode. It is a change in the induced charge that causes induced current to flow, and we seek to find the change in the induced charge as the test charge moves through the space between the electrodes.

With these results, the change in induced charges on the terminal electrodes as the test charge Q_o in Case 1 is moved to a new location between the terminals. The potential on the imaginary electrode in Case 2 changes from V'_o to V'_n. The change on electrode **A** is thus

$$\Delta Q_A = \left(-Q_o\frac{V'_n - V'_B}{V'_A - V'_B}\right) - \left(-Q_o\frac{V'_o - V'_B}{V'_A - V'_B}\right) = -Q_o\frac{V'_n - V'_o}{V'_A - V'_B}, \tag{8.48}$$

and the change in the induced charge on electrode **B** is

$$\Delta Q_B = \left(-Q_o\frac{V'_A - V'_n}{V'_A - V'_B}\right) - \left(-Q_o\frac{V'_A - V'_o}{V'_A - V'_B}\right) = Q_o\frac{V'_n - V'_o}{V'_A - V'_B}. \tag{8.49}$$

Notice that Eq. (8.48) and Eq. (8.49) have opposite charge signs, hence ΔQ_A on electrode **A** is opposite in sign for ΔQ_B on electrode **B**. A knowledge of the equipotential lines between the conductors is needed in order to apply the results of Eqs. (8.48) and (8.49), otherwise referred to as the weighting potential. As stated before, the weighting potential can be calculated by solving Laplace's equation with the electrode of interest placed at a unit potential while all other electrodes are grounded. Hence, the difference $V'_A - V'_B$ becomes unity, and the normalized potentials V'_o and V'_n can be determined.

Once again consider the situations shown in Fig. 8.9. Suppose that the electrons move in the positive x direction to point x_2 with potential $V'(x_2) = V_2'$ and the positive ions move in the negative x direction to point x_1 with potential $V'(x_1) = V_1'$. From Eq. (8.49), the change in induced charge on electrode **B**, from both the contribution of the negative electrons and positive ions, is

$$\Delta Q_B = -q_e N_o \left. \frac{V_2' - V_o'}{V_A' - V_B'} \right|_{e^-} \quad \text{and} \quad + q_e N_o \left. \frac{V_1' - V_o'}{V_A' - V_B'} \right|_{ions^+},$$

where N_o is the number of electron-ion pairs in motion. A unit negative charge on electrode **B** would produce a potential profile described by Eq. (8.41), namely

$$V(x) = -V_o \frac{x}{d}. \tag{8.50}$$

A normalized potential difference within the device is

$$\frac{V(x_2) - V(x_1)}{-V_o} = \frac{x_2}{d} - \frac{x_1}{d} = \frac{x_2 - x_1}{d}. \tag{8.51}$$

This result shows that the potential between the two parallel terminals is simply the displaced distance divided by d. With the proper charge sign for the ions and electrons, the two contributions to Eq. (8.6) are

$$-q_e N_o \left. \frac{V_2' - V_o'}{V_A' - V_B'} \right|_{e^-} = -q_e N_o \frac{x_2 - x_0}{d} \quad \text{and} \quad q_e N_o \left. \frac{V_1' - V_o'}{V_A' - V_B'} \right|_{ions^+} = q_e N_o \frac{x_1 - x_0}{d}. \tag{8.52}$$

Addition of these two results and use of Eq. (8.49) yields

$$\Delta Q_B = -q_e N_o \frac{x_1 - x_0}{d} + q_e N_o \frac{x_2 - x_0}{d} = q_e N_o \frac{x_2 - x_1}{d}.$$

This is the same result as Eq. (8.44) obtained by using Ramo's theorem. With the use of the same method, it can be shown that the change in the induced charge on electrode **A** is

$$\Delta Q_A = q_e N_o \frac{x_1 - x_0}{d} - q_e N_o \frac{x_2 - x_0}{d} = -q_e N_o \frac{x_2 - x_1}{d}.$$

Although Ramo's theorem and Green's reciprocation relations give the same result for the induced charge, it is obvious that Ramo's theorem is considerably easier to use than Green's reciprocation theorem.

8.6.1 Planar Detector with Stationary Space Charge

Now consider a planar detector in which a stationary uniform distribution of space charge is present as shown in Fig. 8.10. Such space charge is typical for many *semiconductor* detector devices. In this case the induced charge can be found once again from Ramo's theorem; however, the operating potential must be solved using Poisson's equation for electricity.

First, consider Green's reciprocation theorem for two cases. In Case 1 there is a single mobile charge Q_o placed on a small imaginary electrode and a single stationary charge Q_s also on a small imaginary electrode, both located between electrodes **A** and **B**, with potentials V_o and V_s, respectively. Now in Case 2, each charge is independently placed upon electrode **A** and analyzed. The charges Q_{Ao} and Q_{Bo} are the charges induced by the mobile charge Q_o on electrodes **A** and **B**, and the charges Q_{As} and Q_{Bs} are the charges induced by the stationary charge Q_s on electrodes **A** and **B**, respectively. The electrode **M** is the

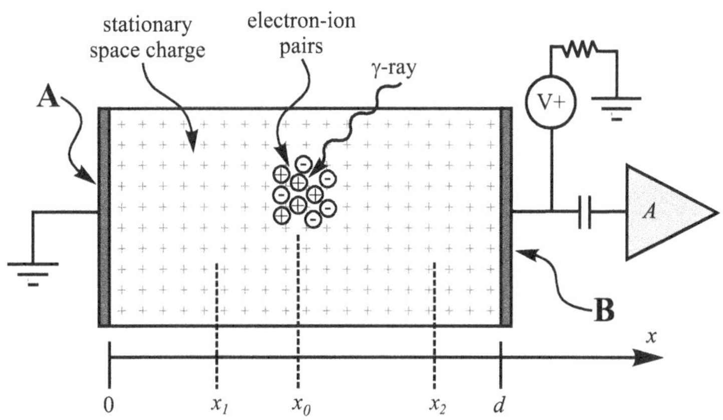

Figure 8.10. Depiction of a planar radiation detector with a uniform distribution of positive stationary space charge. The gamma ray is absorbed at location x_0.

Table 8.2. Charge and potential conditions for two electrodes between which a unit space charge Q_s and a unit mobile charge Q_o are present.

		Case 1 (unprimed)				Case 2 (primed)			
Mobile Charge	Electrode	**A**	**B**	**M**	**S**	**A**	**B**	**M**	**S**
	Charge	Q_{Ao}, Q_{As}	Q_{Bo}, Q_{Bs}	Q_o	Q_s	Q_o, Q_{As}	Q'_{Bo}, Q_{Bs}	0	Q_s
	Potential	0, 0	0, 0	V_o	V_s	$V'_{Ao}, 0$	$V'_{Bo}, 0$	V'_o	V_s
Stationary Charge	Electrode	**A**	**B**	**M**	**S**	**A**	**B**	**M**	**S**
	Charge	Q_{Ao}, Q_{As}	Q_{Bo}, Q_{Bs}	Q_o	Q_s	Q_{Ao}, Q_s	Q_{Bo}, Q'_{Bs}	Q_o	0
	Potential	0, 0	0, 0	V_o	V_s	$0, V'_{As}$	$0, V'_{Bs}$	V_o	V'_s

imaginary electrode at the initial location of the mobile charge and electrode **S** is the imaginary electrode at the location of the stationary charge. The Green's reciprocation conditions for the mobile and stationary charges are summarized in Table 8.2, in which V'_{A0} and V'_{B0} are the potentials produced at electrodes **A** and **B**, respectively, by the mobile point charge Q_o, and V'_{As} and V'_{Bs} are the potentials produced at electrodes **A** and **B**, respectively, by the stationary point charge Q_s. The induced charges on electrodes **A** and **B**, shown in Table 8.2, have been separated for the stationary and mobile charges.

Green's reciprocation theorem of Eq. (8.38) is used for this analysis

$$\sum_{i=1}^{12} Q_i V'_i = \sum_{i=1}^{12} Q'_i V_i,$$

which, upon using values from Table 8.2, expands as

$$Q_{Ao}V'_{Ao} + (Q_{As} \times 0) + Q_{Bo}V'_{Bo} + (Q_{Bs} \times 0) + Q_o V'_o + Q_s V_s + (Q_{Ao} \times 0) + Q_{As}V'_{As} +$$
$$(Q_{Bo} \times 0) + Q_{Bs}V'_{Bs} + Q_o V_o + Q_s V'_s = (Q_o \times 0) + (Q_{As} \times 0) + (Q'_{Bo} \times 0) + (Q_{Bs} \times 0) +$$
$$(V_o \times 0) + Q_s V_s + (Q_s \times 0) + (Q_{Ao} \times 0) + (Q'_{Bs} \times 0) + (Q_{Bo} \times 0) + Q_o V_o + (V_s \times 0).$$

This result simplifies to

$$Q_{Ao}V'_{Ao} + Q_{Bo}V'_{Bo} + Q_oV'_o + Q_{As}V'_{As} + Q_{Bs}V'_{Bs} + Q_sV'_s = 0. \tag{8.53}$$

The sum of the charge and its resulting induced charges must equal zero, i.e.,

$$Q_{Ao} + Q_{Bo} + Q_o = 0 \quad \text{and} \quad Q_{As} + Q_{Bs} + Q_s = 0. \tag{8.54}$$

From Eqs. (8.53) and Eqs. (8.54), the induced charge on electrodes **A** and **B** are

$$Q_A = Q_{Ao} + Q_{As} = -Q_o\frac{V'_o - V'_{Bo}}{V'_{Ao} - V'_{Bo}} - Q_s\frac{V'_s - V'_{Bs}}{V'_{As} - V'_{Bs}}, \tag{8.55}$$

and

$$Q_B = Q_{Bo} + Q_{Bs} = -Q_o\frac{V'_{Ao} - V'_o}{V'_{Ao} - V'_{Bo}} - Q_s\frac{V'_{As} - V'_s}{V'_{As} - V_{Bs}}. \tag{8.56}$$

Now suppose that m space charges are added between the electrodes. The induced charge upon the output electrode **B** becomes

$$Q_B = -Q_o\frac{V'_{Ao} - V'_o}{V'_{Ao} - V'_{Bo}} - \sum_{i=1}^{m} Q_{s_i}\frac{V'_{As_i} - V'_{s_i}}{V'_{As_i} - V'_{Bs_i}}. \tag{8.57}$$

The observed change in the induced charge upon electrode **B** if the mobile charge moves from position x_0 to a new position x_n is

$$\Delta Q_B = \left\{-Q_o\frac{V'_{Ao} - V'_n}{V'_{Ao} - V'_{Bo}} - \sum_{i=1}^{m} Q_{s_i}\frac{V'_{As_i} - V'_{s_i}}{V'_{As_i} - V'_{Bs_i}}\right\}$$
$$- \left\{-Q_o\frac{V'_{Ao} - V'_o}{V'_{Ao} - V'_{Bo}} - \sum_{i=1}^{m} Q_{s_i}\frac{V'_{As_i} - V'_{s_i}}{V'_{As_i} - V'_{Bs_i}}\right\} = Q_o\frac{V'_n - V'_o}{V'_{Ao} - V'_{Bo}}.$$

This is the same answer as Eq. (8.49)! Hence, it is seen that any space charge does not affect the change in the induced charge created by the mobile charge. Also notice that the potential change is with respect to the potential distribution that is created by the mobile charge ($V'_{Ao} - V'_{Bo}$) and *not* the applied voltage or the potential distribution created by the numerous m number of stationary charges. As a result, the change in the induced charge for the planar device, even with stationary space charge present, yields the same result as given by Eq. (8.44), i.e.,

$$\Delta Q|_{planar} = qN_o\left[\frac{x_2 - x_1}{d}\right].$$

The operating potential is determined, in general, from Poisson's equation,

$$\nabla^2 V(\mathbf{r}) = -\frac{\rho(\mathbf{r})}{\epsilon_s}, \tag{8.58}$$

subject to suitable boundary conditions. For the planar detector $V(\mathbf{r}) \rightarrow V(x)$ and ρ is constant. Thus, Poisson's equation becomes

$$\frac{d^2V(x)}{dx} = -\frac{d\mathcal{E}(x)}{dx} = -\frac{\rho}{\epsilon_s}, \tag{8.59}$$

where the $\epsilon_s = \kappa_s \epsilon_o$ is the permittivity of the detector material. Integration of Eq. (8.59) gives the electric field as

$$\mathcal{E}(x) = \frac{\rho}{\epsilon_s} x + C_1,$$

where C_1 is a constant of integration. For simplicity, it is assumed that the device is fully depleted so that the electric field vanishes at $x = 0$ so that $C_1 = 0$.[7] Integration once again gives $V(x)$

$$\int \frac{dV(x)}{dx} dx = -\int \mathcal{E}(x) dx = -\int \frac{\rho}{\epsilon_s} x\, dx \qquad \text{or} \qquad V(x) = -\frac{\rho}{2\epsilon_s} x^2 + C_2.$$

The boundary condition $V(0) = 0$ requires that $C_2 = 0$ so the voltage in this planar detector is

$$V(x) = -\frac{\rho}{2\epsilon_s} x^2, \tag{8.60}$$

and the corresponding electric field is

$$\mathcal{E}(x) = -\frac{dV}{dx} = \frac{\rho x}{\epsilon_s}. \tag{8.61}$$

Notice that the voltage distribution is parabolic, while the electric field is linear.

The voltage normalized to unity at $x = d$ is

$$\frac{V(x)}{V(d)} \equiv V_n(x) = \left(\frac{\rho x^2}{2\epsilon_s}\right) \Big/ \left(\frac{\rho d^2}{2\epsilon_s}\right) = \frac{x^2}{d^2}, \tag{8.62}$$

and is shown in Fig. 8.11. Also notice that the weighting potential is linear, just as in a planar device with no space charge while the operating potential is clearly non-linear.

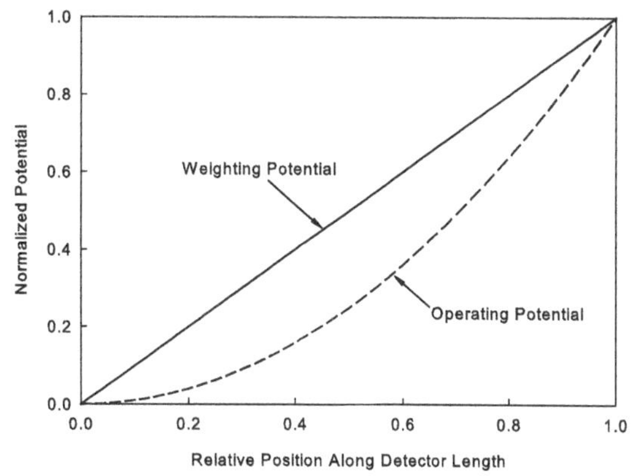

Figure 8.11. A comparison of the weighting potential and the normalized operating potential for a planar detector with uniform space charge.

The above analysis for a planar device with uniform space charge reveals an important point: the space charge in a detector does not affect the weighting potential distribution, but does affect the operating potential. Intuitively, the result should be obvious because Gauss' law shows that the electric field produced by a unit of charge in a volume is independent of the electric fields produced by other charges in the volume. Hence, the induced current produced by the moving charge is independent of the electric field or potential produced by the space charge, and produces the same effect as if there were no space charge. The observed induced charge produced by a charge moving from one point to another depends on the weighting potential between two specific points, and is independent of the trajectory taken from one point in the detector to another.

However, the operating potential does affect the performance of the detector. The operating potential and the accompanying electric field determine the charge carrier velocities and the active volume of a device. The shape of the operating potential and the electric field also determines the charge carrier trajectories as they travel through the detector volume. In gas-filled detectors, the operating potential determines the operation region (ion chamber, proportional, Geiger-Müller). Further, for semiconductor devices the operating potential determines the extent of recombination, trapping and detrapping effects.

---

[7]The variation of the voltage and electric field in semiconductors is explored in greater detail in Chapter 15.

Historical Note

During the decade of the 1960s, there was some disagreement about the application of Ramo's theorem to detectors with space charge, or more precisely, about how the theorem could be applied to semiconductor detectors. The confusion arose from the role of space charge in the depleted or active regions of the semiconductor material, and how the charge induction from mobile free carriers was affected. As it turns out, space charge within any radiation detector alters the electric field and, as a consequence, changes the charge carrier velocities. When analyzing a detector, one must take great caution in analyzing the *operating potential*, which is found by solving Poisson's equation for electric fields with the boundary condition that voltages are set to the actual operating voltages. However, from Gauss' law of electricity, the applied voltages do *not* affect the voltage or electric field resulting from a point charge and, hence, the space charge also does not affect the charge induction of a point charge as it moves in the cavity of the detector [Martini and Ottaviani 1969]. It can be concluded that (1) the analyst can use the normalized Laplace equation (weighting potential) with all contacts grounded except for the collecting electrode to determine the charge or current induction as a function of charge position and displacement, and (2) the analyst must use Poisson's equation with the actual operating voltages on all contacts to determine the charge carrier trajectories and velocities within the detector cavity.

8.6.2 A Planar Detector Composed of Two Materials

In the previous analysis of a planar device, the medium between the two terminals was a single homogeneous material. Now consider a planar device composed of two dielectric materials, as shown in Fig. 8.12. In such a device, the electric field intensity is no longer constant in each material, although the dielectric displacement is constant in each region.

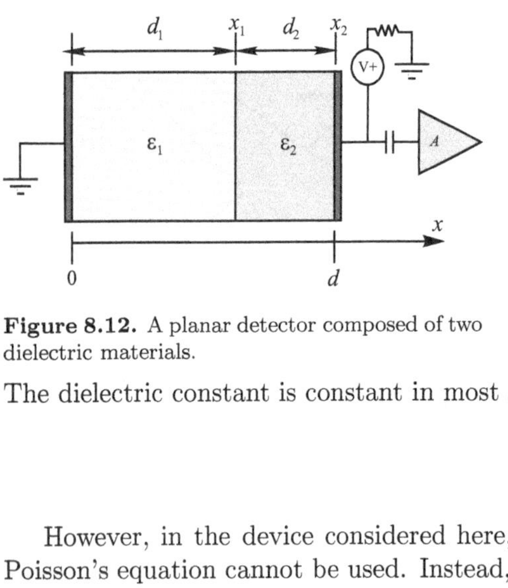

Figure 8.12. A planar detector composed of two dielectric materials.

The dielectric displacement is defined as

$$\mathbf{D} = \epsilon_s \mathbf{E}, \tag{8.63}$$

where $\mathbf{E}$ is the electric field and $\epsilon_s = \kappa_s \epsilon_o$ is the dielectric constant of the material. Recall from Maxwell's equations or, equivalently, Gauss' differential law of electricity that

$$\nabla \bullet \mathbf{D} = \nabla \bullet \epsilon_s \mathcal{E} = \rho, \tag{8.64}$$

where ρ is the space charge density. Note that Eq. (8.64) indicates that $\mathbf{D}$ must be constant if the space charge density is zero. The dielectric constant is constant in most situations, so that Eq. (8.64) can represented as

$$\nabla \bullet \mathcal{E} = \frac{\rho}{\epsilon_s}. \tag{8.65}$$

However, in the device considered here, the dielectric constant is not constant across the device, and Poisson's equation cannot be used. Instead, for this one-dimensional geometry one has

$$\frac{d[\epsilon_s(x)\mathcal{E}(x)]}{dx} = \epsilon_s \frac{d\mathcal{E}(x)}{dx} + \mathcal{E}(x)\frac{d\epsilon_s(x)}{dx} = \rho(x), \tag{8.66}$$

and if no space charge is present, as in the present case,

$$\epsilon_s(x)\frac{d\mathcal{E}(x)}{dx} + \mathcal{E}(x)\frac{d\epsilon_s(x)}{dx} = 0. \tag{8.67}$$

For a detector with two different dielectric materials and no space charge in the device, the dielectric displacement is constant and is given by

$$D = \epsilon_1 \mathcal{E}_1 = \epsilon_2 \mathcal{E}_2.$$

However, the electric field across the device is *not* constant if the dielectric constants of both regions are not the same. The voltage across the region between 0 to x_1, a width of d_1, is

$$V_1 = (x_1 - 0)\mathcal{E}_1 = d_1 \mathcal{E}_1,$$

and across the region x_1 to x_2, of width d_2, is

$$V_2 = (x_2 - x_1)\mathcal{E}_2 = d_2 \mathcal{E}_2,$$

where the total voltage across the entire device is

$$V_T = V_1 + V_2 = d_1 \mathcal{E}_1 + d_2 \mathcal{E}_2 = D\left(\frac{d_1}{\epsilon_1} + \frac{d_2}{\epsilon_2}\right).$$

Here V_1 and V_2 are the voltage drops across region d_1 and d_2, respectively. The capacitance of this detector is

$$C = \frac{Q}{V_T} = \frac{Q}{D}\left(\frac{\epsilon_1 \epsilon_2}{d_1 \epsilon_2 + d_2 \epsilon_1}\right).$$

The voltage $V(x)$ increases in a piecewise linear manner from $V(0) = 0$ to $V(x_2) = V_T$, i.e., as

$$V(x) = \begin{cases} x\mathcal{E}_1 = \dfrac{xD}{\epsilon_1}, & \text{for } x \in [0, x_1] \\ (x - x_2)\mathcal{E}_2 + V_1 = (x - x_2)\dfrac{D}{\epsilon_2} + \dfrac{d_1 D}{\epsilon_1}, & \text{for } x \in [x_1, x_2] \end{cases}. \tag{8.68}$$

The weighting potential is thus $V_w(x) = V(x)/V_T$.

If a gamma ray is absorbed in this planar device and produces N_o ion-electron pairs, the induced charge after the ions and electrons have drifted and are separated by a distance Δx is

$$\Delta Q = \begin{cases} \dfrac{qN_o D}{V_T}\dfrac{\Delta x}{\epsilon_1} = \dfrac{qN_o V_1}{V_T}\dfrac{\Delta x}{d_1}, & 0 \le \Delta x \le x_1 \\ \dfrac{qN_o D}{V_T}\dfrac{\Delta x}{\epsilon_2} = \dfrac{qN_o V_2}{V_T}\dfrac{\Delta x}{d_2}, & x_1 \le \Delta x \le x_2 \end{cases}. \tag{8.69}$$

Notice that charge induction increases as the dielectric constant decreases. Hence, a point charge moving identical distances in both regions induces more charge when moving in the region of lower ϵ. One could conceivably create a device that concentrates most of the charge induction to a small region with a low dielectric constant, which is aligned with a larger volume with a high dielectric constant. In the present case, the weighting potential and the operating potential for the device are identical in shape. The weighting potential is scaled from 1 to 0, whereas the operating potential is scaled from the highest voltage V_{high} at one contact to the lowest voltage V_{low} at the opposite contact.

8.7 Charge Induction for a Cylindrical Detector

Often detectors are formed into cylindrical shapes, in which the active region is between an inner electrode of radius r_1 and an outer electrode of radius r_2. This geometry makes it possible to construct a volumetrically large device while maintaining a relatively low capacitance on the smaller inner electrode. To take advantage of the geometry and low capacitance, the signal is typically read from the smaller radius inner electrode.

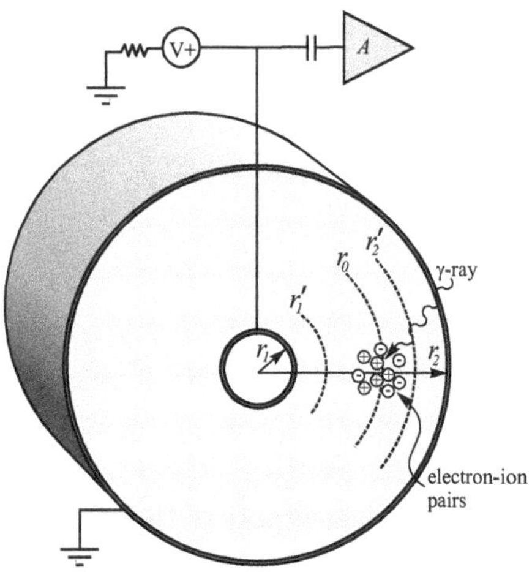

Figure 8.13. Depiction of a cylindrical radiation detector where a gamma ray is absorbed at radius r_0. The figure shows a preamplifier attached to the positively biased inner electrode at $r = r_1$.

The Laplacian operator for cylindrical geometry is

$$\nabla^2 V = \frac{1}{r}\frac{\partial}{\partial r}\left(r\frac{\partial V}{\partial r}\right) + \frac{1}{r^2}\frac{\partial^2 V}{\partial \phi^2} + \frac{\partial^2 V}{\partial z^2} = 0. \qquad (8.70)$$

If the detector is a long coaxial cylinder, then the change in potential with angle θ and length z is negligible, so that $V(r,\phi,z) \to V(r)$ and Eq. (8.70) reduces to

$$\nabla^2 V = \frac{1}{r}\frac{d}{dr}\left(r\frac{dV(r)}{dr}\right) = \frac{dV(r)}{dr} + r\frac{d^2V(r)}{dr^2} = 0. \qquad (8.71)$$

Equation (8.71) can be rewritten as

$$\mathcal{E}(r) + r\frac{d\mathcal{E}(r)}{d} = 0 \qquad \text{or} \qquad \frac{d[\mathcal{E}(r)r]}{dr} = 0.$$

Because the derivative of $\mathcal{E}(r)r$ is zero, then the product must be a constant, i.e., $E(r)r = C_1$ or

$$\mathcal{E}(r) = \frac{C_1}{r}. \qquad (8.72)$$

The potential is then found as

$$V(r) = \int \mathcal{E}(r)dr = \int \frac{C_1}{r}dr = C_1 \ln(r) + C_2. \qquad (8.73)$$

From Fig. 8.13, the preamplifier readout is seen to be attached to the inner electrode, which has a positive bias V^+. To find the potential, the inner electrode potential $V(r_1)$ is set to V^+ volts and the voltage $V(r_2)$ on the outer electrode is set to zero. Use these boundary conditions to evaluate the constants C_1 and C_2. Substitution for C_1 and C_2 in Eq. (8.73) gives

$$V(r) = V^+ \ln\left(\frac{r}{r_2}\right)\left[\ln\left(\frac{r_1}{r_2}\right)\right]^{-1}. \qquad (8.74)$$

The corresponding electric field, from Eq. (8.72), is

$$\mathcal{E}(r) = \frac{V^+}{r}\left[\ln\left(\frac{r_1}{r_2}\right)\right]^{-1}. \qquad (8.75)$$

The weighting potential is found by setting $V^+ = 1$ in Eq. (8.74), namely

$$V_w(r) = \ln\left(\frac{r}{r_2}\right)\left[\ln\left(\frac{r_1}{r_2}\right)\right]^{-1}. \qquad (8.76)$$

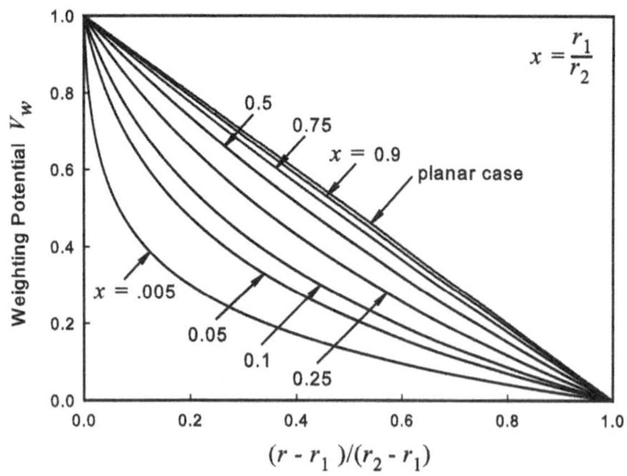

Figure 8.14. Weighting potentials for various values of r_1/r_2 versus the normalized distance from r_1 to r_2 for a cylindrical detector.

The weighting potentials, as a function of the ratio r_1/r_2 for a cylindrical detector, are shown in Fig. 8.14. At small values of r_1/r_2, the weighting potential shows a strong increase near the small anode, r_1. This increase means that charges moving near the inner electrode create more induced charge than those free carriers moving near the large outer electrode. Also it is seen that the weighting potential approaches that of a planar detector as r_1/r_2 approaches unity.

The ΔQ induced by the motion of free carriers as they drift toward the electrodes can be found from Ramo's theorem. After time $t(r_1')$ the electrons have drifted from where they were created at r_0 to r_1' and after $t(r_2')$ the ions have drifted to a radial distance r_2'. From Ramo's theorem one has

$$\begin{aligned}\Delta Q &= \int_0^{t(r_2')} I(t)\,dt\bigg|_{ions+} + \int_0^{t(r_1')} I(t)\,dt\bigg|_{e-} \\ &= -qN_o \int_0^{t(r_2')} \nabla V_w \bullet \frac{d\mathbf{r}}{dt}\,dt\bigg|_{ions+} - (-qN_o)\int_0^{t(r_1')} \nabla V_w \bullet \frac{d\mathbf{r}}{dt}\,dt\bigg|_{e-} . \end{aligned} \tag{8.77}$$

In this geometry V is a function only of r so Eq. (8.77) is evaluated as follows.

$$\Delta Q = -qN_o \int_{r_0}^{r_2'} \frac{dV_w(r)}{dr}\,dr\bigg|_{ions+} - (-q)N_o \int_{r_0}^{r_1'} \frac{dV_w(r)}{dr}\,dr\bigg|_{e-} \tag{8.78}$$

Substitution of Eq. (8.76) for $V_w(r)$ gives

$$\begin{aligned}\Delta Q &= -qN_o \left[\ln\left(\frac{r_1}{r_2}\right)\right]^{-1} \left\{\int_{r_0}^{r_2'} \frac{d}{dr}\ln\left(\frac{r}{r_2}\right)dr - \int_{r_0}^{r_1'} \frac{d}{dr}\ln\left(\frac{r}{r_2}\right)dr\right\} \\ &= qN_o\left[\ln\left(\frac{r_2}{r_1}\right)\right]^{-1}\left\{\ln\left(\frac{r_2'}{r_0}\right) - \ln\left(\frac{r_1'}{r_0}\right)\right\}\end{aligned} \tag{8.79}$$

which yields the final result,

$$\boxed{\Delta Q = qN_o\left[\ln\left(\frac{r_2}{r_1}\right)\right]^{-1}\ln\left(\frac{r_2'}{r_1'}\right).} \tag{8.80}$$

This result shows that the induced charge is a function of only the total charge carrier displacement in r. If the electrons drift to the positive contact at r_1 and the positive ions drift to the negative contact at r_2, then $r_1' = r_1$ and $r_2' = r_2$, and the induced charge becomes $Q = qN_o$. It is important to note that the weighting potential for a cylinder is not linear, and depends strongly upon the values of r_1 and r_2, whereas the weighting potential for the planar device is linear with x.

8.8 Charge Induction for Spherical and Hemispherical Detectors

Another detector configuration that has been successfully demonstrated is the hemispherical device, in which a half sphere has the rounded outermost surface coated as the cathode contact and small spot at the center is the anode contact. Typically, the preamplifier, used to measure the induced charge signal, is connected to the small innermost contact. Fig. 8.15 shows a section of a spherical detector, which has the same characteristics as a hemispherical detector. Ramo's theorem is used to determine ΔQ induced by free charge carriers moving in the device.

But first the weighting potential must be found. In spherical coordinates, the Laplacian for the detector of Fig. 8.15 is

$$\nabla^2 V = \frac{1}{r^2}\frac{\partial}{\partial r}\left(r^2 \frac{\partial V}{\partial r}\right) + \frac{1}{r^2 \sin\theta}\frac{\partial}{\partial \theta}\left(\sin\theta \frac{\partial V}{\partial \theta}\right) + \frac{1}{r^2 \sin^2\theta}\frac{\partial^2 V}{\partial \phi^2} = 0. \tag{8.81}$$

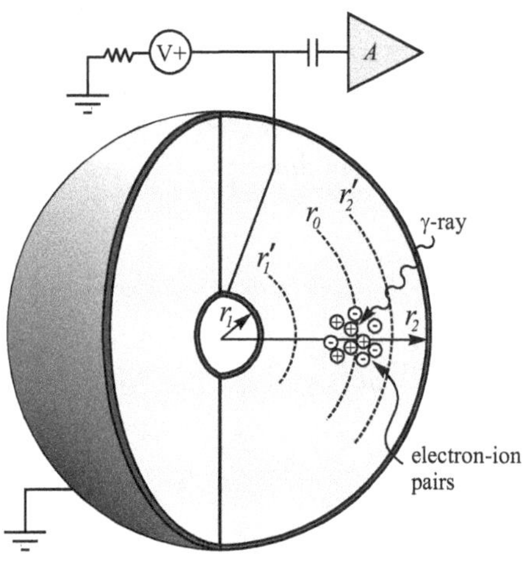

Figure 8.15. Depiction of a spherical radiation detector where a gamma ray is absorbed at radius r_0. The figure shows a preamplifier attached to the positively biased inner electrode at $r = r_1$.

Because of the spherical symmetry in the present geometry, the potential changes with r but not with ϕ or θ. Equation (8.81) then reduces to

$$\nabla^2 V = \frac{1}{r^2}\frac{d}{dr}\left(r^2 \frac{dV(r)}{dr}\right) = 0, \tag{8.82}$$

which can be rewritten as

$$\frac{dV(r)}{dr} + \frac{r}{2}\frac{d^2V(r)}{dr^2} = \frac{d}{dr}\left(\frac{d[rV]}{dr}\right) = 0.$$

Integration over r twice gives

$$rV(r) = C_1 r + C_2, \tag{8.83}$$

where C_1 and C_2 are constants of integration to be found from the boundary conditions. As before, the weighting potential is found from Eq. (8.83) with boundary conditions of $V_w(r_1) = 1$ and $V_w(r_2) = 0$. The result is

$$V_w(r) = \left(\frac{r_1}{r_2 - r_1}\right)\left(\frac{r_2}{r} - 1\right). \tag{8.84}$$

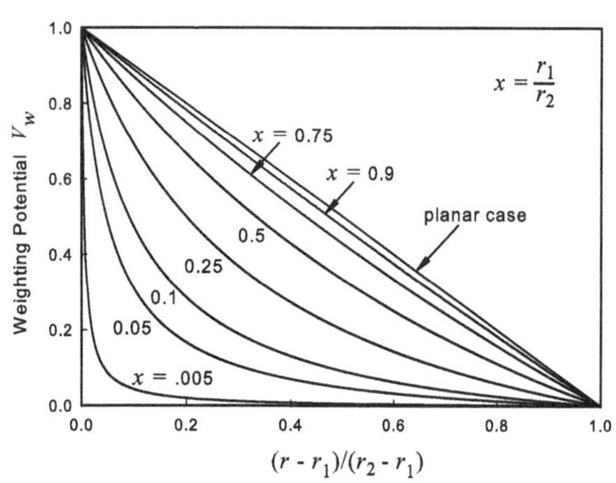

Figure 8.16. Weighting potentials for various values of r_1/r_2 versus the normalized distance from r_1 to r_2 for a spherical detector.

The weighting potentials are shown in Fig. 8.16 as a function of the ratio r_1/r_2 for a spherical (or hemispherical) detector. At small values of r_1/r_2, the weighting potential shows a strong increase near the small anode at r_1, which means that charges moving in the vicinity near the inner contact have more influence on the induced charge than those free carriers moving near the large electrode at r_2. As with the cylindrical detector, the weighting potential approaches that of the planar device as r_1/r_2 approaches unity. One should notice that the weighting potential for a spherical detector is much more skewed than for the cylindrical detector.

The induced charge is now found with Ramo's Theorem with the contributions of the electrons and ions being analyzed separately. Because $V_w(r)$ is a function of only r, as in the previous cylindrical geometry, Eq. (8.78) is applicable to the spherical geometry as well. Substitution of the weighting potential from Eq. (8.84) into Eq. (8.78) yields

$$\Delta Q = -qN_o\left(\frac{r_1}{r_2 - r_1}\right)\left\{\int_{r_0}^{r_2'} \frac{d}{dr}\left(\frac{r_2}{r}\right)dr - \int_{r_0}^{r_2'} \frac{d}{dr}\left(\frac{r_2}{r}\right)dr\right\}$$
$$= qN_o\left(\frac{r_1}{r_1 - r_2}\right)\left\{\left(\frac{r_2}{r_2'} - \frac{r_2}{r_0}\right) - \left(\frac{r_2}{r_1'} - \frac{r_2}{r_0}\right)\right\},$$

which, upon rearrangement, yields the final result,

$$\Delta Q = qN_o \frac{r_1 r_2}{r_1' r_2'} \left[\frac{r_1' - r_2'}{r_1 - r_2} \right]. \tag{8.85}$$

8.9 Concluding Remarks

The topics discussed in this chapter constitute the fundamental physics governing radiation detectors whose signal formation is based on current induction. The reader is referred to the references for sources with more detailed discussions on electricity and magnetism. It should be noted there are circumstances, not covered in the present chapter, that can alter detector performance. For instance, current induction in gas-filled ion chambers may depend mainly upon the motion of electrons, and in other gas-filled detectors the current induction may depend mainly on positive ion motion. Semiconductor detectors, based on current induction, may have imperfections that cause an exponential decrease of the charge carrier density with time, thereby causing the current induction to be a function of time, charge carrier velocity, and interaction position. Such alterations in detector response are described in the chapters on gas-filled and semiconductor detectors. Further, unique detector electrode designs can be used to change the current induction contributions of electrons and positive ions as a function of position, as shown here for the cylindrical and spherical devices. In the absence of space charge, the weighting and operating potentials of two-terminal devices have essentially the same shape, the difference being the former is normalized and the latter is not. In the presence of space charge, the shapes of the weighting and operating potentials can be significantly different.

PROBLEMS

1. A charge of 2.5×10^{-8} C is placed in an upwardly directed electric field with magnitude 5×10^4 N C^{-1}. Determine the work performed for the following three cases: (a) the charge is moved 24 cm to the right, (b) the charge is moved 24 cm downward, and (c) the charge is moved 34 cm at a downward 45° angle.

2. A 5.5 MeV alpha particle has a direct head-on collision with a Au nucleus at rest. What is the minimum distance achieved between the two particles?

3. There are two point charges placed 1 cm apart, where

$$Q_1 = +q_e N_1 = q_e(10^4) \qquad \text{and} \qquad Q_2 = -q_e N_2 = -q_e(5 \times 10^3).$$

Determine: (a) the potential halfway between the two charges, (b) the potential at a point exactly 6 mm from Q_1 and 8 mm from Q_2, and (c) the work performed moving the charge from the location of B to the location of A.

4. Consider an evacuated parallel plate chamber with 1000 volts applied between the contacts spaced 3 cm apart. Simultaneously, an electron is released from the cathode and a He 2+ ion is released from the anode. (a) At what distance from the positive electrode do the particles pass each other? (b) What are the particle speeds when they arrive at their respective collection electrodes? (c) What are the particle energies when they arrive at their respective collection electrodes?

5. For a cylindrical capacitor with dielectric constant ϵ_s, inner conductor radius r_a and outer conductor radius r_b, (a) derive an expression for the capacitance per unit length. (b) Derive an expression for the total capacitance of a hemispherical capacitor.

6. Consider the diagram below (Fig. 8.17). An electron is released from electrode C and travels through electrode B to electrode A at constant speed. (a) Sketch the output pulse from the electron if measured from electrode A. (b) Sketch the output pulse from the electron if measured from electrode C. (c) Sketch the output pulse from the electron if measured from electrode B.

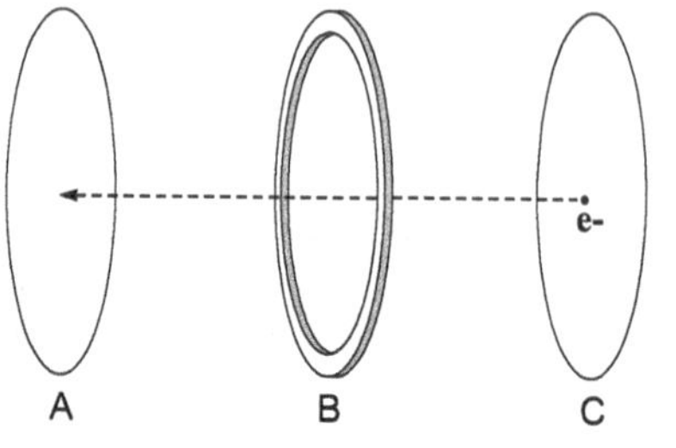

Figure 8.17. Diagram for problem 6.

7. You have a coaxial detector filled with 1 atm Ar and with inner electrode radius of 25 microns and outer electrode radius of 1.25 cm. A potential of 1000 volts is applied. A radiation interaction occurs at radius $r = 0.625$ cm, producing 50,000 electron-ion pairs. Sketch the pulse shape as a function of time for: (a) positive voltage applied to the inner electrode and (b) negative voltage applied to the inner electrode.

REFERENCES

BECK, A.H.W., *Thermionic Valves*, London: Cambridge University Press, 1953.

JACKSON, J.D., *Classical Electrodynamics*, 2nd Ed., New York: Wiley, 1975.

JEANS, J., *The Mathematical Theory of Electricity and Magnetism*, 5th Ed., London: Cambridge University Press, 1943.

JEN, C.K., "On the Induced Current and Energy Balance in Electronics," *Proc. IRE*, **29**, 345–349, (1941).

MARTINI, M. AND G. OTTAVIANI, "Ramo's Theorem and the Energy Balance Equations in Evaluating the Current Pulse from Semiconductor Detectors," *Nucl. Instrum. Meth.*, **67**, 177–178, (1969).

RAMO, S., "Currents Induced by Electron Motion," *Proc. IRE*, **27**, 584–585, (1939).

RILEY K.F., M.P. HOBSON, AND S.J. BENCE, *Mathematical Methods for Physics and Engineering*, 3rd Ed., Cambridge: Cambridge University Press, 2006.

SHOCKLEY, W., "Currents to Conductors Induced by a Moving Point Charge," *J. Appl. Phys.*, **9**, 635–636, (1938).

SMYTHE, W.R., *Static and Dynamic Electricity*, 3rd Ed., Bristol: Taylor & Francis, 1989.

Chapter 9

Gas-Filled Detectors: Ion Chambers

Gases are distinguished from other forms of matter, not only by their power of indefinite expansion so as to fill any vessel, however large, and by the great effect heat has in dilating them, but by the uniformity and simplicity of the laws which regulate these changes.

James Clerk Maxwell

In 1908, Ernest Rutherford and Hans Geiger constructed the first gas-filled radiation detector from a metallic cylinder with a thin axially positioned wire inside [Rutherford and Geiger 1908]. The gas medium in this case was air. When a voltage was applied between the cylinder and wire, a current was measured with an electrometer when alpha particles entered the device. They also noticed that the behavior of the detector changed with increasing voltage, namely, that alpha particles could be detected at much lower applied voltages than beta particles. This ability to discriminate between the two radiations became known as *proportional* counting. Experiments conducted later with the gas-filled detectors clearly showed distinctive regions of operation, as shown in Fig. 9.1.

The physical principle behind a gas-filled detector is quite simple. Radiation interacts in either the chamber gas or the chamber walls, thereby liberating electrons from parent nuclei. These electrons (and ions in the gas) produce additional ionization in the detector gas, which creates a charge "cloud" composed of electrons and positive ions. A voltage placed across electrodes in the gas chamber causes the electrons and ions to drift apart in opposite directions, with electrons drifting towards the anode and the positive ions drifting towards the cathode. In some cases, electron attachment to a neutral atom produces negative ions. As these gas ions and electrons, or *charge carriers*, move through the chamber, they induce a current to flow in an external circuit connected to the chamber. This current, or change in current, can then be measured as an indication that a radiation interaction occurred in the chamber.

9.1 General Operation

Gas detectors can be operated in either a *pulse mode* or *current mode*. Pulse mode is generally used in low to moderate radiation fields. In this case, a single radiation particle, such as an alpha particle, beta particle, or gamma ray, interacts in the chamber volume, producing a charge cloud around the interaction site. The charge carriers then drift apart and induce current between the device terminals. A charging circuit, usually consisting of a preamplifier and feedback circuit, time-integrates the current and records a voltage pulse. This voltage pulse is treated as a *single* event, i.e, a single radiation particle has been detected. The preamplifier circuit is subsequently discharged and reset, so as to allow the device to detect the next radiation-interaction event. Hence, each voltage pulse from the detector indicates a radiation interaction event. Although extremely useful, there are drawbacks to pulse-mode operation. Should another radiation interaction occur while the detector is integrating or discharging the current from a previous interaction

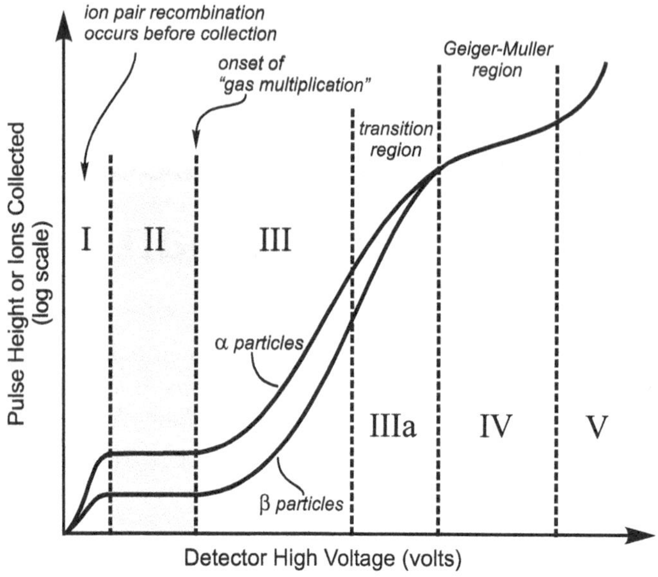

Figure 9.1. The observed output pulse height versus the applied high voltage for a gas-filled detector. Several different regions can be seen: (I) recombination, (II) ion chamber, (III) proportional, (IIIa) transition region, (IV) Geiger-Müller, and (V) continuous discharge.

event, the detector may not, and usually does not, record the new interaction, a condition referred to as *pulse pile up*. The time interval in which a new pulse is not recorded is the detector *recovery time*, also referred to as *dead time* (see Ch. 7). A pulse-mode detector operated in low radiation fields has little problem with dead time; however, a detector operated in high radiation fields may have significant dead time losses, thereby yielding an incorrect measurement of the radiation intensity.

For high radiation fields, gas detectors are commonly operated in current mode, in which the radiation induced current is measured with a current meter. Under such conditions, many interactions can occur in the device in very short periods of time, and the current observed increases with radiation exposure. Hence, current mode can be used to measure high radiation fields in which the magnitude of the current is a measure of the radiation induced ionization rate in the detector. The output current is then proportional to the intensity of the radiation field at the detector location. The drawback to pulse-mode operation is that it does not distinguish between individual radiation interactions.

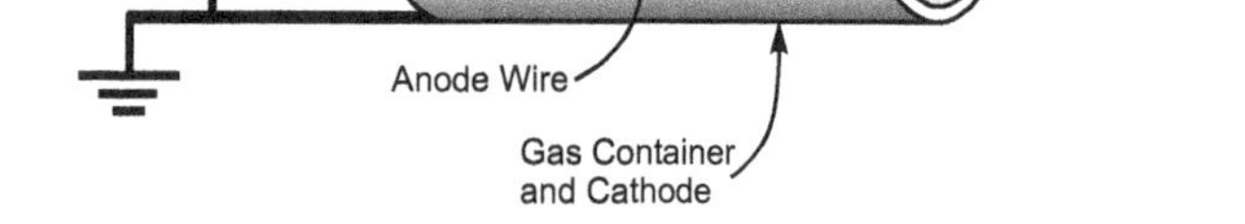

Figure 9.2. Shown is a coaxial gas detector, which is commonly used for Geiger-Müller tubes, and sometimes used for proportional counters. High voltage is applied to the central wire anode, while the outer cylinder wall, the cathode, is held at ground.

Consider a gas-filled detector as shown in Fig. 9.2, similar to the device first devised by Geiger and Rutherford, in which α-particles and β-particles enter through a thin window. These particles can interact in the gas (or tube wall) depositing some or all of their energy in the tube gas and, in the process, generate a miniature cloud of free electrons and positive ions, referred to as *electron-ion pairs*. As required for any radiation detector, there is an absorber (the fill gas), and an observable (the electron-ion pairs). To complete the detector, a method is needed to measure the amount of ionization. Suppose the detector is connected to a simple electrometer so as to measure the current produced by the motion of the electron-ion pairs. Without an applied voltage, the electron-ion pairs diffuse randomly *in all directions* and eventually recombine. As a result, the net current measured by the electrometer is zero. However, if a small positive voltage is applied to the central thin wire (anode) of the device, the free electrons (negative charges) drift towards the anode and the free ions (positive charges) drift towards the detector wall. At low voltages, some measurable current is observed, yet considerable recombination will still occur, the expected result within Region I, the *recombination region*, as shown in Fig. 9.1. As the voltage is increased, electron-ion pair separation is more rapid until practically no recombination occurs. Hence, the current measured is a measure of the total number of electron-ion pairs formed. This is Region II, and is referred to as the *ionization chamber region.*

As the voltage is increased further, the electrons gain enough kinetic energy from the electric field to create more electron-ion pairs through impact ionization, thereby providing a mechanism for signal gain, often referred to as *gas multiplication.* As a result, the observed current increases as the voltage increases, but the amount of ionization is still proportional to the energy deposited in the gas by the original radiation particle. This is Region III, the *proportional region.* Still further increasing the applied voltage causes disproportional current increases to form, labeled in Fig. 9.1 as the transition region IIIa, beyond which, in Region IV, all currents, regardless of origin, radiation species or energies, have the same magnitude. Region IV is the *Geiger-Müller region.* Finally excessive voltage drives the detector into region V where the voltage causes sporadic arcing and other spontaneous electron emissions to occur, causing *continuous discharging* in the detector. Because of possible equipment damage and the loss of meaningful signals, gas detectors should not be operated in the continuous discharge region. In the following chapter, detector operation in Regions I and II is described.

9.2 Electrons and Ions in Gas

Radiation interactions in gas can cause excitation and ionization. Excitation is a process in which the energy transferred to a gas atom or molecule is less than the ionization energy. Hence, electrons are elevated to a higher energy level, but are still bound to the nucleus. Energy transferred to electrons greater than their ionization energy causes ionization, so that the electrons and positive ions become separate charge carriers. An electric field applied to the gas volume separates these charge carriers and, in the process, produces an induced current, as described in Chapter 8. The electrons, with a mass more than a thousand times less than that of the positive ions, also have quite different transport properties than heavy ions. Also, negative ions, formed from electron attachment to neutral atoms, have different transport characteristics than positive ions (although similar). Detector and gas properties that affect the charge transport for electrons and ions include mobility, electric field, ionization potential, charge transfer, electron attachment, and recombination. In the following chapter on gas-filled ion chambers, pertinent information on charge transport and ion behavior in gases is presented. Exhaustive studies dedicated to the complexities of electron and ion motion in gases have been published in far more detail than presented here. See, for example, [Loeb 1939, 1960; Huxley and Crompton 1974; Sitar et al. 1993].

9.2.1 Ionization

Directly ionizing radiation with energy above the minimum ionizing energy I_0 of a gas will produce electron-ion pairs (eip) along their trajectories. Due to competing energy loss mechanisms, such as atomic excitation and bremsstrahlung production, the average energy w required to produce an electron-ion pair is typically much greater than I_0. As the ionizing particle passes through the detector gas, it produces a continuous chain of electron-ion pairs. The *total* number of electron-ion pairs produced per unit distance of travel by the charged particle is

$$n_t \simeq \frac{dE}{dx}\frac{1}{w}. \tag{9.1}$$

Note that, although Eq. (9.1) may be used to estimate the density of ion pairs per unit path length, the stopping power dE/dx for heavy ions (alpha particles, for instance) is not constant. The values for n_t and dE/dx in Table 9.1 are listed for *minimum ionizing particles* (MIPs), i.e., beta particles with energies above

Table 9.1. Useful properties of various gases at 293 K and 760 torr. From [Sauli 1977].

Gas	Z	A	ρ_o (1 atm) g/cm^3	I_o eV	w eV/eip*	$(dE/dx)/\rho_o$ MeV g^{-1} cm^2	dE/dx keV/cm	n_p eip*/cm	n_t eip*/cm
H_2	2	2	8.38×10^{-5}	15.4	37	4.03	0.34	5.2	9.2
He	2	4	1.66×10^{-4}	24.6	41	1.94	0.32	5.9	7.8
N_2	14	28	1.17×10^{-3}	15.5	35	1.68	1.96	(10)	56
O_2	16	32	1.33×10^{-3}	12.2	31	1.69	2.26	22	73
Ne	10	20.2	8.39×10^{-4}	21.6	36	1.68	1.41	12	39
Ar	18	39.9	1.66×10^{-3}	15.8	26	1.47	2.44	29.4	94
Kr	36	83.8	3.49×10^{-3}	14.0	24	1.32	4.60	(22)	192
Xe	54	131.3	5.49×10^{-3}	12.1	22	1.23	6.76	44	307
CO_2	22	44	1.86×10^{-3}	13.7	33	1.62	3.01	(34)	91
CH_4	10	16	6.70×10^{-4}	13.1	28	2.21	1.48	16	53
C_4H_{10}	34	58	2.42×10^{-3}	10.8	23	1.86	4.50	(46)	195

* eip = electron-ion pair

1 MeV.[1] The total number N_T of electron-ion pairs produced by a beta particle can then be estimated with the measured data listed in Table 9.1 as

$$N_T \simeq \frac{\Delta E}{w}, \tag{9.2}$$

where ΔE is the energy deposited in the gas. The initial ionizing particle creates n_p *primary* electron-ion pairs per unit length as it passes through the detector gas and deposits ΔE of energy. If $\Delta E \gg I_0$, the energetic free electrons and positive ions can have sufficient kinetic energy to produce more electron-ion pairs through secondary ionization (delta rays). By summing contributions from primary and secondary ionization, the *total* density of electron-ion pairs per unit pathlength, n_t, can be found. Values of n_p and n_t for several common detector gases are listed in Table 9.1. For mixtures of different gases, a general rule often used to describe the total number of ion pairs created is

$$N_T \simeq \sum_{i=1}^{K} \frac{f_i \Delta E}{w_i}, \tag{9.3}$$

where f_i is the concentration fraction of each gas constituent. The total ionization, per unit path length of travel, for MIPs is

$$n_t \simeq \sum_{i=1}^{K} f_i n_t^i \frac{\rho}{\rho_\circ}, \tag{9.4}$$

where $\rho_\circ$ is the gas density at 1 atm, ρ is the gas density at the operating pressure, and n_t^i is n_t for gas species i. Similarly, the primary ionization per unit path length for MIPs is

$$n_p \simeq \sum_{i=1}^{K} f_i n_p^i \frac{\rho}{\rho_\circ}. \tag{9.5}$$

[1]From Fig. 4.20 the collisional stopping power for electrons is nearly constant for energies above about 1 MeV.

9.2.2 Diffusion Effects

Following the creation of the electron-ion pairs, the ions and electrons begin to diffuse to regions with lower charge densities according to Fick's law of diffusion. Also the initial high concentration of electrons and positive ions promotes recombination losses. Because the electron mass is much less than that of a positive ion, the electrons diffuse much faster than the positive ions. This phenomenon has a significant effect on detector operation and performance.

Diffusion of Electrons in a Gas

After the initial ionization event, the electrons and ions rapidly come into thermal equilibrium with the gas and assume a Maxwellian distribution of speeds ω_e and ω_{ion}. The number density for the electron thermal speed ω_e distribution in a charge cloud is given by

$$n_e(\omega_e) = 4\pi \left(\frac{m_e}{2\pi kT}\right)^{3/2} \omega_e^2 \exp\left(-\frac{m_e\omega_e^2}{2kT}\right), \tag{9.6}$$

where m_e is the electron mass. The average electron thermal speed $\overline{\omega}_e$ is

$$\overline{\omega}_e = \int_0^\infty \omega_e n(\omega_e)\, d\omega_e \Big/ \int_0^\infty n(\omega_e)\, d\omega_e = \sqrt{\frac{8kT}{m_e\pi}}. \tag{9.7}$$

In terms of energy, the number density for the electron energy distribution in a charge cloud is

$$n_e(E) = 2\pi \left(\frac{1}{\pi kT}\right)^{3/2} \sqrt{E} \exp\left(-\frac{E}{kT}\right), \tag{9.8}$$

and the average electron energy is[2]

$$\overline{E}_e = \int_0^\infty E\, n(E)\, dE \Big/ \int_0^\infty n(E) dE = \frac{3kT}{2}. \tag{9.9}$$

The diffusion rate of the electrons and ions, to a first approximation, can be evaluated with Fick's law, namely,

$$\frac{dn_e(r,t)}{dt} = D_e \nabla^2 n_e(r,t), \quad \text{for electrons,} \tag{9.10}$$

$$\frac{dn_{ion}(r,t)}{dt} = D_{ion} \nabla^2 n_{ion}(r,t), \quad \text{for ions.} \tag{9.11}$$

Here D_e and D_{ion} are the diffusion coefficients for electrons and ions with units of cm^2/s, and $n_e(r,t)$ and $n_{ion}(r,t)$ are the charge carrier concentrations. Also, r is the distance from the center of the initial charge cloud whose size is negligible compared to volume of the gas tube. Thus, all charges are assumed to begin to diffuse radially outwards from $r = 0$. In this spherically symmetric coordinate system Eq. (9.10) can be written as

$$\frac{\partial n_e(r,t)}{\partial t} = D_e \frac{1}{r^2}\frac{\partial}{\partial r}\left(r^2 \frac{\partial n_e(r,t)}{\partial r}\right) = D_e \left(\frac{2}{r}\frac{\partial n_e(r,t)}{\partial r} + \frac{\partial^2 n_e(r,t)}{\partial r^2}\right). \tag{9.12}$$

The solution for $n_e(r,t)$ in three dimensions in an infinite gas medium is

$$n_e(r,t) = \frac{n_0}{(4\pi D_e t)^{3/2}} \exp\left(\frac{-r^2}{4D_e t}\right), \tag{9.13}$$

[2]Note that $\frac{1}{2}m_e\overline{\omega}_e^2 \neq \overline{E}_e$.

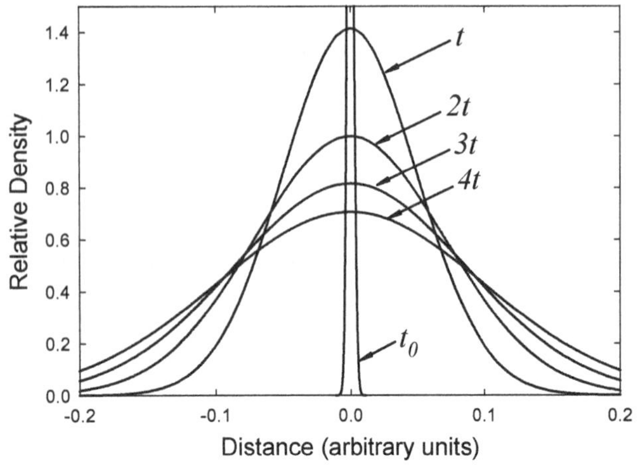

Figure 9.3. Diffusion of an electron cloud based upon Eq. (9.15).

where n_0 is the initial number of electrons. The mean distance squared $\overline{r^2}$ is

$$\overline{r^2} = \int_0^\infty n_e(r,t)\, r^2\, dV \bigg/ \int_0^\infty n_e\, dV = 6D_e t. \tag{9.14}$$

Often a simpler one-dimensional diffusion model is used in which electrons diffuse away from a plane of electrons created at $t = 0$ containing n_0 electrons cm^{-2}. Then in Eq. (9.10) $\nabla^2 \to d^2/dx^2$ where x is the perpendicular distance from the plane source. The solution in an infinite homogeneous gas medium is

$$n_e(x,t) = \frac{n_0}{\sqrt{4\pi D_e t}} \exp\left(\frac{-x^2}{4D_e t}\right), \tag{9.15}$$

which is recognized as a normal (Gaussian) distribution function with mean $x = 0$ and variance $2D_e t$. The diffusion of electrons as a function of time in a gas, according to Eq. (9.15), is illustrated in Fig. 9.3.

The Diffusion Coefficient D_e

The momentum-transfer cross section for electrons is defined as [Buckman and Elford 2000]

$$\sigma_m(E) = 2\pi \int_0^\pi \frac{d\sigma}{d\Omega}\left(1 - \frac{v}{v_0}\cos\theta\right)\sin\theta\, d\theta, \tag{9.16}$$

where $d\sigma/d\Omega$ is the differential scattering cross section, v_0 is the initial electron speed, and v is the electron speed after scattering. The momentum-transfer cross sections for several gases used in radiation detectors are shown in Fig. 9.4. There is a notable decrease in the cross section for Ar, Kr, and Xe gases between the energies of 0.1 eV to 1 eV, known as the Ramsauer-Townsend effect, a consequence of the quantum mechanical wave nature of low-energy electrons and their resonant transmission through the potentials of select gases. The differential scattering cross section is defined as the fraction of electrons of energy E that scatter at angle θ into differential solid angle $d\Omega$, where $d\Omega = 2\pi \sin\theta d\theta$. The electron diffusion coefficient is given by [Rice-Evans 1974]

$$D_e = \frac{\overline{\omega}_e}{3n_a\sigma_m}, \tag{9.17}$$

where n_a is the molecular density at temperature T (absolute) and pressure P (torr) and is found from

$$n_a = n_L \frac{P}{P_0}\frac{T_0}{T}. \tag{9.18}$$

Here n_L is the Loschmidt constant,[3] defined as

$$n_L = \frac{P_0}{kT_0} = \frac{P_0 N_A}{RT_0}, \tag{9.19}$$

[3]The Loschmidt constant is the atomic or molecular density for an ideal gas, quoted as 2.6867774×10^{19} cm^{-3} at 0°C at 760 torr.

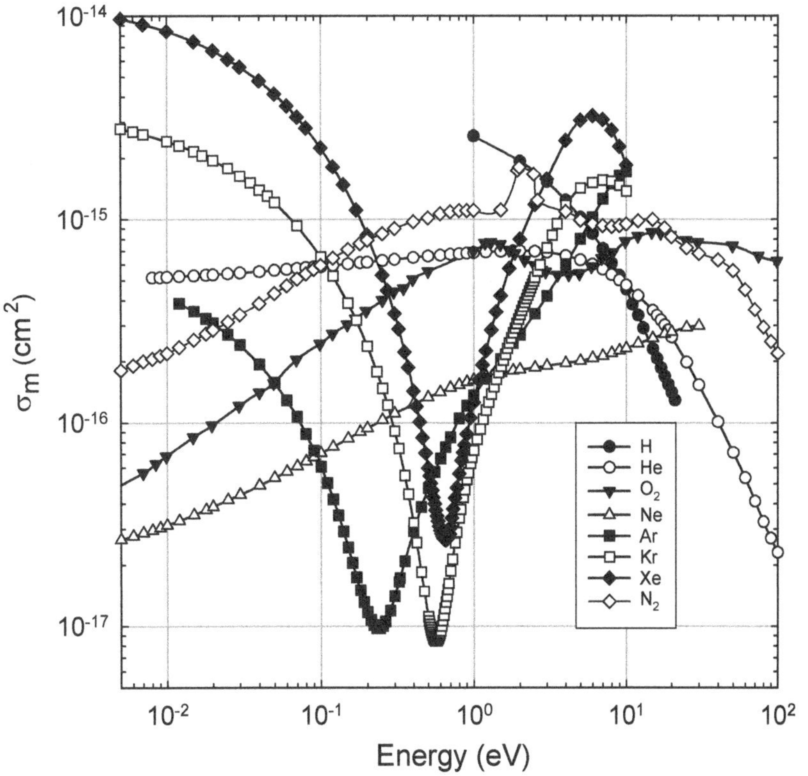

Figure 9.4. The measured momentum transfer cross section of electrons in various gases. From [Phelps 1985; Buckman and Elford 2000; Itikawa 2006].

where $P_0 = 760$ torr, $T_0 = 273$K, N_A is Avogadro's number, R is the gas constant,[4] and k is Boltzmann's constant. The frequency between collisions is [Sitar et al. 1993]

$$\nu_e = n_a \overline{\omega}_e \sigma_e = \tau_e^{-1}, \tag{9.20}$$

where σ_e is the elastic scattering cross section for electrons, $\overline{\omega}_e$ is the average electron thermal speed and τ_e is the average collision time. Note that $\sigma_e = \sigma_m$ for isotropic scattering; hence, the mean free path between collisions is

$$\lambda_e = \frac{1}{n_a \sigma_m} = \frac{kT}{P\sigma_m}. \tag{9.21}$$

Substitution of Eq. (9.7) and Eq. (9.21) into Eq. (9.17), the diffusion coefficient for the average electron energy is

$$D_e = \frac{\overline{\omega}_e \lambda_e}{3} = \frac{(2kT)^{3/2}}{\sqrt{m_e \pi}} \frac{1}{3P\sigma_m}. \tag{9.22}$$

The diffusion of electrons in a gaseous medium is largely governed by the concentration gradient and the diffusion coefficient. However, Coulombic attraction and repulsion between ions and electrons can alter the motion of electrons in the gas from the ideal diffusion model for neutral particles.

---

[4] $R = 8.3144621 \frac{\text{J}}{\text{mol K}}$.

Different mixtures and concentrations of gases can be used to alter the diffusion properties. For instance, adding gases that have low σ_m values increases D_e and, consequently, increases the electron diffusion. Likewise, diffusion can be hindered by adding gases that have large σ_m values. An estimate of the diffusion coefficient can be found with [Sitar et al. 1993] as

$$D_e = \left[\sum_{i=1} \frac{P_i}{D_i}\right]^{-1}, \tag{9.23}$$

where P_i is the relative partial pressure of each gas and D_i is the diffusion coefficient of each gas.

Diffusion of Ions in a Gas

The equations developed for electron diffusion in a gas hold true for ion diffusion. Hence, the one-dimensional solution of Eq. (9.15) gives

$$n_{ion}(x,t) = \frac{n_0}{\sqrt{4\pi D_{ion} t}} \exp\left(\frac{-x^2}{4D_{ion}t}\right), \tag{9.24}$$

where the mean squared distance $\overline{r^2} = 2D_{ion}t$, which is also the variance of the normal distribution of Eq. (9.24). The diffusion coefficient is

$$D_{ion} = \frac{\overline{\omega}_{ion}\lambda_{ion}}{3} = \frac{(2kT)^{3/2}}{\sqrt{m_{ion}\pi}} \frac{1}{3P\sigma_{ion}}, \tag{9.25}$$

where λ_{ion}, $\overline{\omega}_{ion}$, and m_{ion} are the ion counterparts of λ_e, $\overline{\omega}_e$ and m_e, respectively. Note that the diffusion coefficient is inversely proportional to the square root of the ion mass. For this reason the diffusion coefficients for ions are significantly smaller than that for electrons, as can be seen from Table 9.2. Consequently, the diffusion speed of ions can be over 1000 times less than that of electrons.

9.2.3 Electron and Ion Transport

As explained in Chapter 8, it is the motion of charges in some preferred direction that induces current flow in an electronically operated detector. However, if both negative and positive charges are moving equally in all directions, as occurs in the diffusion of both electrons and ions, then the net induction current must be zero. Although there is no induced current, the diffusion process still causes the electrons and ions to separate because the ions diffuse much more slowly than do the electrons.

But with the application of an electric field to the chamber, the electrons and ions further separate and begin to drift in opposite directions. This motion in a preferred direction then induces a current, whose measurement indicates that an ionizing event has occurred in the detector.

Influence of an Electric Field

The average kinetic energy of thermal electrons in a gas at 20°C is given by the Boltzmann distribution

$$\overline{E}_e = \frac{3kT}{2} = 0.038\text{eV}. \tag{9.26}$$

This average energy corresponds to a speed of $\omega_e = 4.2 \times 10^7$ cm/s. Yet, these randomly oriented trajectories produce no induced current in the detector.

The application of an electric field $\mathcal{E}$ exerts a force on the diffusing electrons and ions causing the electrons to drift preferentially towards the anode and the ions to drift preferentially towards the cathode. However, the drift speeds are typically very much less than the thermal motion speeds. For electrons the drift speeds

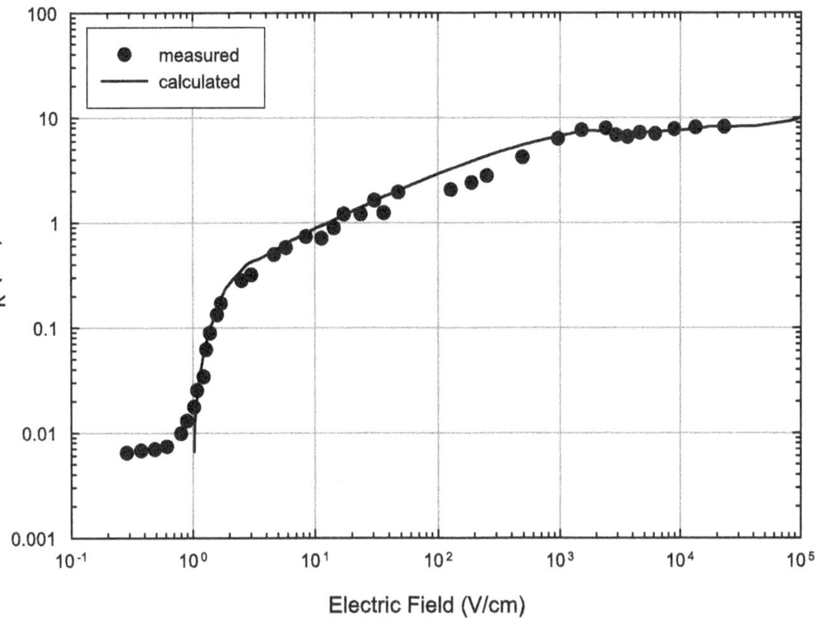

Figure 9.5. The characteristic energy as a function of electric field for Ar gas at 1 ATM. After [Palladino and Sadoulet 1975].

are on the order of 10^5 to 10^6 cm/s. The average electron drift speed v_e of the electrons in the direction of the applied electric field is approximated by Palladino and Sadoulet [1975] as

$$v_e = \frac{q_e \mathcal{E}}{m_e} \frac{\lambda_e}{\overline{\omega}_e}, \tag{9.27}$$

where q_e is the unit electronic charge. A measurable quantity, referred to as the *electronic characteristic energy*, ϵ_K, is a function of the diffusion coefficient and the electric field, namely, [Rice-Evans 1974]

$$\epsilon_K = \frac{q_e D_e \mathcal{E}}{v_e} = \frac{q_e D_e}{\mu_e}. \tag{9.28}$$

and D_e is given by Eq. (9.22). The electronic characteristic energy is the effective kinetic energy of the free electron in an electric field. It is much greater than that for electrons undergoing only thermal diffusion. Measured and calculated values of ϵ_K for pure Ar are shown in Fig. 9.5. Theoretical and measured values of ϵ_K as a function of reduced electric field ($\mathcal{E}/P$) for many gases can be found in the literature [Christophorou 1971; Schultz and Gresser 1978; Peisert and Sauli 1984]. The quantity

$$\mu_e = \frac{q_e \lambda_e}{m_e \overline{\omega}_e}, \tag{9.29}$$

is commonly referred to as the electron *charge carrier mobility.*[5]

[5] Loeb [1960] reviews alternative, but similar, expressions for electron mobility.

From Eq. (9.27) and Eq. (9.29), the drift speed in the presence of an electric field is thus given by the simple relation

$$v_e = \mu_e \mathcal{E}, \tag{9.30}$$

in which the charge carrier mobility is a function of gas pressure, namely,

$$\mu_e = \frac{P_0\, \mu_0}{P}, \tag{9.31}$$

where μ_0 is the mobility in units of $cm^2\ V^{-1}\ s^{-1}$ at 1 atm (760 torr) pressure, $P_0 = 760$ torr, $\mathcal{E}$ is the electric field, and P is the gas pressure in torr.

The mobility is related to the diffusion coefficient by[6]

$$\frac{D_e}{\mu_e} = \frac{kT}{q_e} \tag{9.32}$$

where D_e is the electron diffusion coefficient, q_e is the unit charge of an electron, k is Boltzmann's constant and T is the absolute temperature. Equation (9.30) is valid up to a saturation electric field $\mathcal{E}_{sat}$, beyond which charged particle scattering becomes the determining factor for the drift speed. At low fields, the electron drift speed is linear with respect to the applied field; however, this region of linearity is usually below a reduced electric field value of 0.3 V cm^{-1} $torr^{-1}$ for many important detector gases. As electrons gain energy under higher electric fields, their small mass allows high speeds to be reached with moderately low voltages. Electron collisions with other electrons in the gas can cause more ionization, referred to as *impact ionization*, which increases the number of electron-ion pairs in the gas, but such collisions reduce the mean free path of the electrons. As a result, electrons reach a saturation speed v_{sat} at electric fields approaching $\mathcal{E}_{sat}$.

Consider the ratio ξ between the characteristic energy (for any value D_e and associated μ_e) and the average thermal energy, namely

$$\xi = \frac{\epsilon_K}{\overline{E}_e} = \frac{2q_e D_e}{3\mu_e kT}. \tag{9.33}$$

Substitution of Eq. (9.32) into Eq. (9.33) and rearrangement gives

$$\frac{D_e}{\mu_e} = \xi \frac{3kT}{2q_e}. \tag{9.34}$$

The electron drift speed, for the same reduced electric field, varies among detector gases and gas mixtures. In argon, a common detector gas, electrons reach a fairly constant saturation speed of approximately 0.37 cm μs^{-1} at a reduced electric field of 0.2 V cm^{-1} $torr^{-1}$, as can be seen from Fig. 9.6. However, the small addition of 0.2% nitrogen increases the saturation speed above 1.0 cm μs^{-1} (at 0.6 V cm^{-1} $torr^{-1}$). From the data shown in Fig. 9.6, the mean drift speed for electrons in pure argon gas can be approximated by

$$v_e = \frac{3.64(\mathcal{E}/P) + 114.6(\mathcal{E}/P)^2}{1 + 12.7(\mathcal{E}/P) + 304.33(\mathcal{E}/P)^2}\ \text{cm}/\mu\text{s}, \tag{9.35}$$

[6]Eq. (9.32) is commonly called the Einstein relation because it appeared in his 1905 paper in *Annalen der Physik*. On rare occasion, Eq. (9.32) has been called the Einstein-Smoluchowski relation because it was published later by Smoluchowski [1906] (also in *Annalen der Physik* with reference to Einstein's 1905 paper). However, as pointed out by Huxley [1974], Nernst [1888] published the relation for electrolytic ions in 1888. Townsend [1900] applied a form of Eq. (9.32) to gases and published it in 1900 in *Philosophical Transactions A*, although in a considerably less recognizable form. Regardless, it has been suggested by Huxley [1974] that the proper name for Eq. (9.32) should be the Nernst-Townsend relation due to their prior publications.

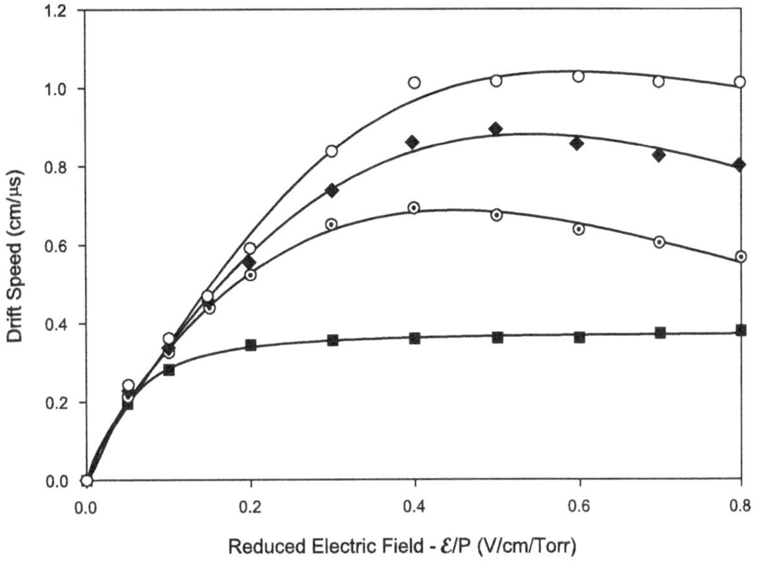

Figure 9.6. The measured drift speed of electrons in argon/nitrogen mixtures as a function of reduced electric field ($\mathcal{E}/\mathcal{P}$) and nitrogen concentration; ■ pure argon; ⊙ 0.05% N_2 ; ♦ 0.1% N_2; ○ 0.2% N_2. After [Bortner et al. 1957].

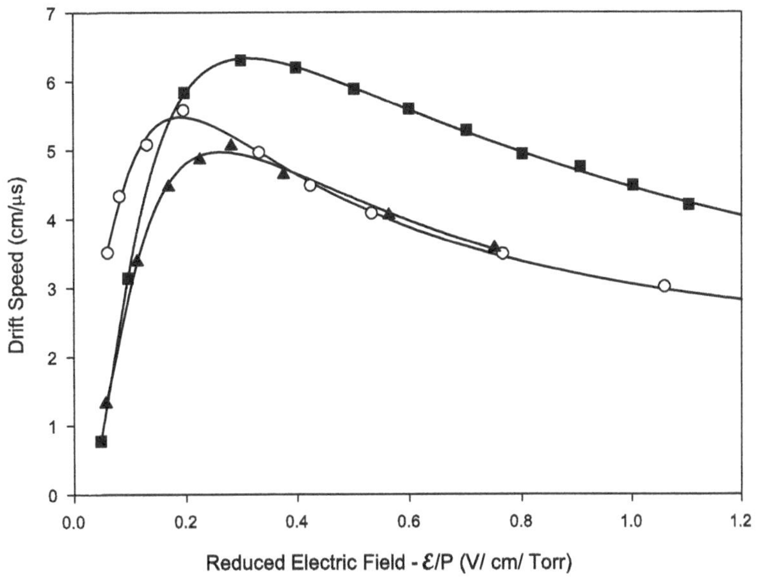

Figure 9.7. The measured drift speed of electrons in P-10 gas (90% argon, 10% methane) as a function of reduced electric field (E/P). Data from ■ [Bortner et al. 1957]; ○ [Mattern 1988]; ▲ [Ferreira et al. 2005]. Discrepancies between measured values may be due to impurity contamination in the gas samples.

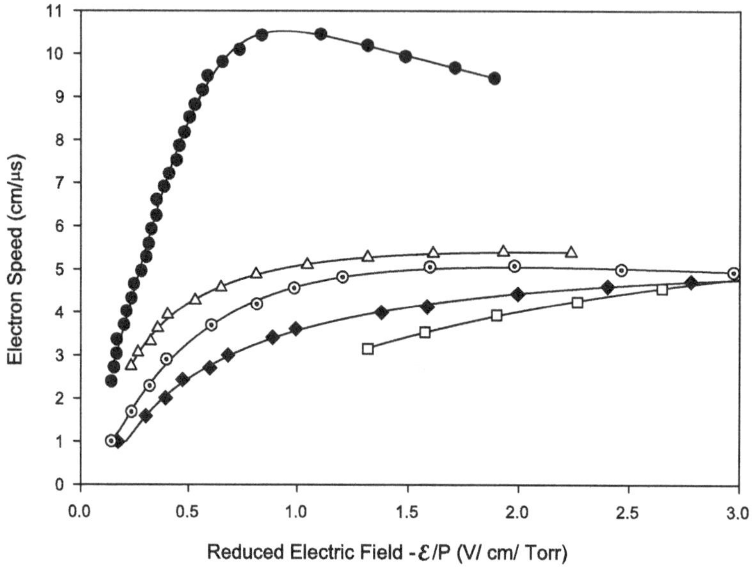

Figure 9.8. The measured drift speed of electrons in various organic polyatomic gases as a function of reduced electric field (E/P); • CH_4; △ C_2H_6 ; ⊙ C_2H_4; ◆ C_3H_8; and □ C_2H_2. After [Cottrell and Walker 1965].

where $\mathcal{E}/P \leq 0.8$ V cm^{-1} torr^{-1}.

Another common detector gas is a mixture of 90% Ar and 10% CH_4, generally referred to as *P-10* gas. Electrons reach a maximum speed at reduced electric fields between 0.2 and 0.3 V cm^{-1} torr^{-1}, beyond which electron scattering causes the speed to drastically decrease as can be seen in Fig. 9.7. With the data of Fig. 9.7, the mean drift speed for electrons in P-10 gas can be approximated by

$$v_e = \frac{6.852\sqrt{\mathcal{E}/P} + 4.626(\mathcal{E}/P)}{1 - 3.306\sqrt{\mathcal{E}/P} + 6.059(\mathcal{E}/P)} \text{ cm}/\mu\text{s}, \tag{9.36}$$

where $\mathcal{E}/P \leq 1.4$ V cm^{-1} torr^{-1}.

Drift speeds of many other organic gases used in gas-filled detectors are shown in Fig. 9.8. Note that CH_4 has a similar electron drift speed characteristic as P-10 gas, with a maximum of approximately 10.5 cm μs^{-1}, whereas the other organic gases shown in Fig. 9.8 do not show this trend, but instead gradually approach a saturation speed.

The drift speed of ions, having much smaller mobilities than electrons because of their much larger masses, typically follow Eq. (9.30) for electric fields commonly used in ion chambers (or other gas-filled detectors). Measured ion mobilities (μ_0) for several gases, as reported by different sources, are listed in Table 9.2.

The motion of charge carriers within an electric field is influenced by the combined effects of diffusion

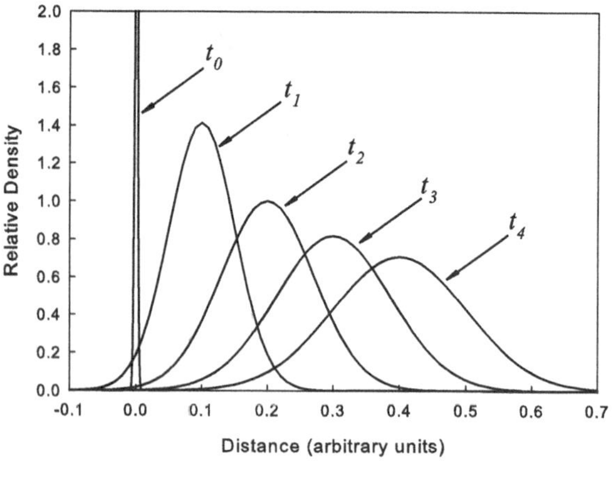

Figure 9.9. Diffusion and drift of an electron cloud based upon Eq. (9.37).

Table 9.2. Transport properties of various ions in the same gas at 1 atm. Data are from [Loeb 1929; Rossi and Staub 1949; Staub 1953; Sharpe, 1964; Huxley 1974; and Sauli 1977].

Gas	λ_{ion} cm	$\overline{\omega}_{ion}$ cm s^{-1}	D^+_{ion} cm^2 s^{-1}	μ^+_{ion} cm^2 V^{-1} s^{-1}	D^-_{ion} cm^2 s^{-1}	μ^-_{ion} cm^2 V^{-1} s^{-1}
Air			0.028	1.36	0.043	1.87
Ar	1.0×10^{-5}	4.4×10^5	0.04	1.7	0.043	1.7
H_2	1.8×10^{-5}	2×10^5	0.123	6.7	.190	7.95
C_2H_2			0.018	0.71	0.021	0.86
C_2H_4					.019	0.75
C_2H_6					0.027	1.07
Cl_2					0.018	0.73
CO			0.027	1.1	0.027	1.07
CO_2	6×10^{-6}	5.5×10^5	0.023	0.76	0.026	0.81
H_2O	1.0×10^{-5}	7.1×10^5	0.02	0.7	0.019	0.75
He	2.8×10^{-5}	1.4×10^5	0.26	10.2	0.157	6.31
Ne				4.4		
N_2	3×10^{-5}	9.0×10^5	0.032	1.27	0.045	1.82
O_2	1.0×10^{-5}	5.0×10^5	0.025	1.36	0.0396	1.80

and drift. The distribution of electrons or ions as a function of electric field and time is

$$n_i(x,t,\mathcal{E}) = \frac{n_0}{(4\pi D_i t)^{1/2}} \exp\left(\frac{-(x-\mu_i \mathcal{E} t)^2}{4D_i t}\right), \tag{9.37}$$

where the subscript i refers to electrons or ions, $\overline{v}_i = \mu_i \mathcal{E} \le v_{sat}$. The diffusion and drift of electrons in a gas as a function of time, according to Eq. (9.37), is illustrated in Fig. 9.9.

9.2.4 Charge Transfer

Heavy positive ions colliding with neutral molecules (or atoms) can exchange charge, i.e., a singly charged positive gas ion accepts an electron from a neutral gas component, thereby becoming neutralized while ionizing the previously neutral molecule. For ions moving in the same host gas, for instance Ar^+ moving through Ar, an electron can be transferred from a neutral gas atom to a positively charged gas ion, referred to as symmetric resonance charge transfer [Sitar et al. 1993],

$$X^+ + X \rightarrow X + X^+, \tag{9.38}$$

with a cross section on the order of 10^{-15} to 10^{-14} cm^2. This form of charge transfer effectively reduces the mobility of positive ions in the same neutral gas, quoted as being on the order of 7.5% reduction in mobility from that of the neutral gas atom [Wilkinson 1950].

In many cases, a second gas is purposely introduced in with the main detector gas to alter the ion transport characteristics. The ability for a positive ion to pick up an electron in a collision is strongly dependent upon the separate ionization potentials. If the ionization potential of the host gas is higher than that of the added neutral gas atoms or molecules, then charge transfer from host gas ions colliding with neutral gas molecules (atoms) of the second component is possible, whereas the reverse electron exchange is not. This type of charge transfer is referred to as asymmetric non-resonance charge transfer [Sitar et al. 1993] and is written as

$$X^+ + Y \rightarrow X + Y^+ + \Delta\epsilon, \tag{9.39}$$

where $\Delta\epsilon$ is the difference in ionization energies $(I_{0X} - I_{0Y})$, and has a charge transfer cross section of about 10^{-15} cm^2. For instance, the introduction of CH_4 (I_0 = 13.1 eV) into Ar (I_0 = 15.8 eV) permits charge transfer to occur between Ar^+ and CH_4, but not the reverse exchange. Hence,

$$Ar^+ + CH_4 \rightarrow Ar + CH_4^+ + 2.7 \text{ eV}. \tag{9.40}$$

In the next chapter on proportional counters, this reaction is seen to be a very important charge transfer reaction. If the detector gas is a mixture of both Ar and CH_4, ionizing radiation creates a cloud of both Ar^+ and CH_4^+ ions. However, within a short period of time, ion exchange causes virtually all Ar atoms to become neutralized, having transferred their charge to the CH_4 molecules.

9.2.5 Electron Attachment

After the initial charge cloud has been formed, the electric field separates the positive ions and electrons. Electrons, having significantly higher mobility than ions, rapidly drift towards the detector anode, while the positive ions slowly drift towards the cathode. There is a possibility of the formation of negative heavy ions in the chamber, a process in which electrons become attached to a neutral gas atoms or molecules. The consequence of such electron attachment is the formation of slow moving negative ions, thereby altering the speed and shape of the signal formation.

The mechanisms of electron attachment, described by Bloch and Bradbury [1935] and Herzenberg [1969], and summarized by Huk et al. [1988], constitute a basic two-step process. Electrons come into contact with an electro-negative molecule to form an excited negative ion

$$Y + e^- \rightarrow Y^{-*}, \tag{9.41}$$

where Y represents the electro-negative gas molecule. The excited electro-negative molecule may then deexcite through re-emission of the electron

$$Y^{-*} \rightarrow Y + e^-, \tag{9.42}$$

or become stabilized by emission of photon

$$Y^{-*} \rightarrow Y^- + \gamma. \tag{9.43}$$

If the excited gas molecule releases the electron, the effect of electron attachment is basically nullified. However, if the gas molecule becomes stabilized, then the electron charge carrier is converted into a slow moving negative ion, with a decreased mobility by three or more orders of magnitude. Often, these parasitic gases are contaminants in the host detection gas X. The negative ion in the excited state may instead collide with the host detection gas such that

$$Y^{-*} + X \rightarrow Y + X + e^-, \tag{9.44}$$

or become stabilized by,

$$Y^{-*} + X \rightarrow Y^- + X^*. \tag{9.45}$$

The average probability of forming negative ions is a function of the electron *attachment coefficient* C_a [Compton and Langmuir 1930], which is the average probability that a single collision causes an electron to attach to an atom or molecule and can be expressed as

$$C_a = \lambda_e f_V N \overline{\sigma}_{at}. \tag{9.46}$$

Here N is the atomic density of the gas for the species being considered, f_V is the volume fraction of that gas in the chamber gas, and $\overline{\sigma}_{at}$ is the average electron capture cross section. Unfortunately, there appears

Table 9.3. Electron attachment coefficients C_a in various gases. From [Korff 1946; Loeb 1929]. Gases that have practically zero values for C_a include He, Ne, Ar, Kr, Xe, H_2, N_2 and CH_4 [Loeb 1929, 1939].

Gas	C_a	Gas	C_a	Gas	C_a
Air	5.0×10^{-6}	C_2H_6	4.0×10^{-7}	H_2O	2.5×10^{-5}
BF_3	1.24×10^{-8}	CO	6.25×10^{-9}	N_2O	1.64×10^{-6}
C_2H_2	1.28×10^{-7}	CO_2	6.2×10^{-9}	NH_3	1.01×10^{-8}
C_2H_4	2.13×10^{-8}	Cl_2	$> 5.0 \times 10^{-4}$	O_2	2.5×10^{-5}

Table 9.4. Electron attachment rate coefficients α_I for O_2 in various Ar/CH_4 gas mixtures as measured by Huk et al. [1988]. The samples had 200 ppm O_2 added and the measurements were conducted at a pressure of 4 bar.

Ar/CH_4 – 90%/10%			Ar/CH_4 – 80%/20%		
$\mathcal{E}/\mathcal{P}$ V cm^{-1} bar^{-1}	v_e cm/μs	α_I μs^{-1}	$\mathcal{E}/\mathcal{P}$ V cm^{-1} bar^{-1}	v_e cm/μs	α_I μs^{-1}
100	5.36	0.048 ± 0.003	100	5.54	0.103 ± 0.006
138	5.45	0.034 ± 0.003	138	6.61	0.098 ± 0.007
163	5.32	0.029 ± 0.003	163	6.91	0.089 ± 0.007
200	5.07	0.024 ± 0.003	200	7.08	0.074 ± 0.007
250	4.70	0.019 ± 0.003	250	7.10	0.069 ± 0.007

to be some discrepancy regarding this simple definition, because the designation of "attachment coefficient" has been assigned to various dimensioned numbers in the literature [Huk et al. 1988; Sitar et al. 1993]. In the present work, the *interaction coefficient* ψ_I is defined as a function of the attachment coefficient, such that [Sitar et al. 1993]

$$\psi_I \equiv \frac{f_V N \overline{\sigma}_{at} \overline{\omega}_e}{v_e} = \frac{C_a \overline{\omega}_e}{\lambda_e v_e} = \frac{f_V N C_a \overline{\omega}_e \sigma_m}{v_e}, \tag{9.47}$$

in which the last result is obtained from Eq. (9.21).

The *attachment rate coefficient* is defined as the rate at which electrons are attached to gas molecules (atoms), per unit time, and is a function of gas pressure and electric field. It is defined as

$$\alpha_I = \psi_I v_e = \frac{C_a \overline{\omega}_e}{\lambda_e}. \tag{9.48}$$

In Table 9.3 values of the attachment coefficient are given, and in Table 9.4 attachment rate coefficients for Ar/CH_4 are listed for several reduced potentials.

Consider a cloud of electron-ion pairs produced with density N_p at some location x from the anode within a planar gas chamber of width d. Under bias, the electrons drift towards the anode and the positive ions towards the cathode. To be collected, the electrons must drift a distance x over the time $t = x/v_e$ and the positive ions must drift a distance $d - x$ over the time $(d - x)/v_{ion}$. As the electrons drift, they are lost to electron attachment described by

$$N_e(x) = N_p e^{-\psi_I x}, \tag{9.49}$$

or, equivalently,

$$N_e(t) = N_p e^{-\alpha_I t}. \tag{9.50}$$

Common detector gases, such as Ar and CH_4, have relatively low values for C_a, and Staub [1953] states that Ar, CO_2, H_2, He and N_2 have attachment coefficients that are very small, a statement supported in part by data [Loeb 1929]. However, contamination in the detection gas from O_2 or water vapor can increase electron attachment and compromise performance [Staub 1953; O'Kelly et al. 1960; Huk et al. 1988]. Overall, it is usually best to avoid using gases, or gas mixtures, in which a constituent has a fairly high value of C_a. Further, care should be taken to ensure that the detector gas does not become contaminated with impurities having large values of C_a, such as O_2 or water vapor.

Example 9.1: In the operation of a Ar-filled ion chamber, operated with an electric field of 300 V cm^{-1}, it was discovered that the chamber is contaminated with 1% O_2. The chamber is operated at 20°C with pressure of 1 atm. What is the expected electron attachment fraction per unit differential travel distance in the chamber?

Solution:

From Fig. 9.5 the interpolated value (from the data) of the characteristic energy ϵ_K is 3.0 eV, and is the effective kinetic energy of free electrons in the electric field. From Eq. (9.35) (or Fig. 9.6) the drift speed is 3.6×10^5 cm s^{-1}, and from Fig. 9.4, the momentum transfer cross section is 4.2×10^{-16} cm^2. Also, from Table 9.3, the capture coefficient for O_2 is C_a is 2.5×10^{-5}, while that for Ar is essentially zero ($C_a \simeq 0$). The average thermal speed is

$$\overline{\omega}_e = \sqrt{\frac{2\epsilon_K}{m_e}} = \sqrt{\frac{2(3\text{ eV})(1.6 \times 10^{-19}\text{ J/eV})}{9.1 \times 10^{-31}\text{ kg}}}$$
$$= 1.027 \times 10^6 \text{ m s}^{-1} = 1.027 \times 10^8 \text{ cm s}^{-1}$$

The molecular density of O_2 in the chamber is $fN_{O_2} = f\rho_{O_2}N_a/A_{O_2}$. From Table 9.1, $\rho_{O_2} = 1.33$ mg/cm^3 at 293 K and 760 torr. Thus,

$$f_V N_{O_2} = (0.01)(1.33 \times 10^{-3}\text{ g cm}^{-3})\frac{(6.022 \times 10^{23}\text{ mol}^{-1})}{32\text{ g/mol}} = 2.50 \times 10^{17}\text{ cm}^{-3}.$$

Then from Eq. (9.47) the interaction coefficient is

$$\psi_I = \frac{f_V N_{O_2} C_a \overline{\omega}_e \sigma_m}{v_e}$$
$$= \frac{(2.50 \times 10^{17}\text{ cm}^{-3})(2.5 \times 10^{-5})(1.027 \times 10^8\text{ cm s}^{-1})(4.2 \times 10^{-16}\text{ cm}^2)}{(3.6 \times 10^5\text{ cm/s})}$$
$$= 0.7497\text{ cm}^{-1}.$$

Hence

$$1 - \frac{N_e(x)}{N_p} = 1 - e^{-\psi_I x} = 1 - e^{-(0.7497\text{ cm}^{-1})(1\text{ cm})} = 0.5275.$$

This results means that for every centimeter traveled by the electrons in the chamber gas, 52.75% of the electrons become attached to the O_2 molecules.

Although the average number of collisions per electron in a gas chamber varies with chamber size, gas pressure, and applied electric field, Korff [1946] quotes the average number of collisions per electron per detection event as approximately 10^5. The average probability for electron attachment can be estimated by multiplying the number of collisions by C_a. Hence, for those gases with $C_a < 10^{-5}$, the probability of electron attachment is minimal. However, for gases with $C_a > 10^{-5}$, electron attachment can occur, thereby slowing and distorting the electrical signal from the detector. The number of collisions expected for a single electron traversing a chamber can be estimated from published values [Staub 1953], with the mean free path generally ranges between 0.5 μm to 3.0 μm. Some halogen gases have relatively high values for C_a, as do O_2 and water vapor.

9.3 Recombination

After the formation of electron-ion pairs, it is possible that some electrons recombine with positive ions before they are separated by drift and diffusion. There are three basic types of recombination described in the literature [Wilkinson 1950], namely *columnar* recombination, *volumetric* recombination, and *preferential* recombination.[7] Columnar recombination occurs from the localized formation of a relatively dense cloud of ion pairs around the trajectory of, for instance, an alpha particle in the detector gas. Electrons recombine with positive ions, but generally not with their original parent atoms. Volumetric recombination is the recombination of positive and negative ions produced by unrelated ionizing interactions. Preferential recombination occurs when an electron attaches to a neutral atom and, subsequently, recombines with the parent positive ion from whence the electron came, or, less probably, the electron recombines immediately with its original parent atom.[8]

Note that recombination occurs only if the electrons and positive ions are locally close enough to enable the process to occur. The product n^-n^+ is the combined localized charge carrier concentration within the gas chamber. The larger this concentration, the greater is the rate of recombination. Even under low electric field conditions, the diffusion of electrons is much higher than that of positive ions, and rapid separation can be expected; hence, the effect of electron-ion recombination is typically minor. Recombination is less prevalent in gases that have small electron interaction coefficients than found with electro-negative gases. However, in gases that form heavy negative ions, through electron attachment for instance, recombination may cause significant charge carrier losses.

9.3.1 Columnar Recombination

Columnar recombination is of most concern for gas-filled detectors operated in pulse mode, because the background or interaction rate under most situations is usually not high enough to cause appreciable volumetric recombination. Columnar recombination is almost entirely caused by the recombination of positive and negative heavy ions; therefore, those detector gases that do not suffer from electron attachment have minimal columnar recombination. However, gases that have components that do have large electron attachment coefficients, such as air, alcohols, or contaminated gases, can have appreciable columnar recombination.

Jaffé [1913] developed a theory for columnar recombination based on gases that have appreciable electron attachment coefficients, in which the fraction of ions escaping recombination is described by

$$F(\mathcal{E}) = \frac{N_\infty}{N_o} = \frac{1}{1 + \dfrac{\beta N_o}{8\pi D_{ion}}\sqrt{\dfrac{\pi}{z}}S(z)}. \tag{9.51}$$

Here β is the recombination coefficient, N_o is the initial number of electron-ion pairs in the ionization path (eip density), N_∞ is the number of ions that escape recombination, D_{ion} is the ion diffusion coefficient, and

$$S(z) = \sqrt{\frac{1}{\pi}}\int_0^\infty \frac{e^{-S}dS}{\sqrt{S\left(1 + \dfrac{S}{z}\right)}}. \tag{9.52}$$

[7] Loeb [1939] suggests as many as five types of recombination.

[8] In the strictest sense, preferential recombination is the only true recombination of the many types, in that an electron stripped from an atomic nucleus recombines back with its original parent atom. In the other two cases mentioned, columnar and volumetric recombinations, electrons combine with any random positive ions, other than their original parents.

The substitution

$$f(z) = \sqrt{\frac{\pi}{z}}S(z) = e^{z/2}\left(\frac{i\pi}{2}\right)H_0^{(1)}(\frac{iz}{2}), \tag{9.53}$$

yields

$$F(\mathcal{E}) = \frac{N_\infty}{N_o} = \frac{1}{1+\frac{\beta N_o}{8\pi D_{ion}}f(z)}. \tag{9.54}$$

where $H_0^{(1)}$ is a Hankel function of the first kind of order zero,[9] listed [Jahnke and Emde 1945] and plotted elsewhere [Zanstra 1935]. The variable z is [Loeb 1939]

$$z = \frac{1}{2}\left(\frac{\mathcal{E}\mu b\sin\theta}{D_{ion}}\right)^2, \tag{9.55}$$

where θ is the angle between the ionization track and the applied electric field, μ is the ion mobility, K_1 is a constant, $\mathcal{E}$ is the applied electric field and P is the gas pressure in atm. With the substitution $z = 2x$ into Eq. (9.53) and Eq. (9.55), Diebner [1931] shows that

$$f(x) = e^x\left(\frac{i\pi}{2}\right)H_0^{(1)}(ix), \tag{9.56}$$

and is approximated by

$$f(x) \approx 2.303\left[1+\left(\frac{x}{2}\right)^2\right]e^x\log_{10}\left(\frac{1.122}{x}\right), \tag{9.57}$$

for $x < 0.1$, thereby providing a straightforward method to evaluate Eq. (9.51) as a function of x. However, these equations, Eqs. (9.51) to (9.57), are all very interesting, but, without knowledge of b, F cannot be directly evaluated.[10] The evaluation method developed by Zanstra [1935] expresses Eq. (9.51) as,

$$F(\mathcal{E}) = \frac{N_\infty}{N_o} = \frac{I}{I_o} = \frac{1}{1+gf(x)}. \tag{9.58}$$

where I is the measured current, I_o is the current without recombination, and

$$g = \frac{\beta N_o}{8\pi D_{ion}}. \tag{9.59}$$

Rearranging terms, Eq. (9.58) can be rewritten as

$$\frac{1}{I} = \frac{1}{I_o} + \frac{gf(x)}{I_o}. \tag{9.60}$$

Zanstra [1931] asserts that, provided g is constant, Eq. (9.60) will yield a linear response. At $f(x) = 0$, the condition $I = I_o$ is satisfied; hence by plotting $1/I$ vs $f(x)$ and extrapolating to the ordinate intercept at $f(x) = 0$ then I_o can be obtained. This method was confirmed with satisfactory results by Clay [1939] and Greening [1964]. For $x > 10$, Eq. (9.56) can be approximated by [Boag 1966]

$$f(x) \approx \sqrt{\frac{\pi}{2x}}. \tag{9.61}$$

[9] Hankel functions are otherwise known as Bessel functions of the third kind (or Weber functions). Here $H_0^{(1)}(z) = J_0(z)+iY_0(z)$, where $J_0(z)$ and $Y_0(z)$ are Bessel functions of the first and second kind, respectively.

[10] An average value of the product $b\sin\theta$ was used for the work [Jaffé 1913].

Substituting Eq. (9.55) and Eq. (9.61) into Eq. (9.58) yields,

$$\frac{N_\infty}{N_o} = \frac{1}{1 + \left(\frac{\beta N_o}{8\pi D_{ion}}\right)\left(\frac{2D_{ion}}{\mathcal{E}\mu b \sin\theta}\right)\left(\frac{\pi}{2}\right)^{1/2}}. \tag{9.62}$$

Rearranging terms yields, and acknowledging that $N_o \propto Q_o$,

$$\frac{1}{N_\infty} = \frac{1}{N_o} + \left(\frac{\beta}{8\pi D_{ion}}\right)\left(\frac{2D_{ion}}{\mathcal{E}\mu b \sin\theta}\right)\left(\frac{\pi}{2}\right)^{1/2} \quad \text{so that} \quad \frac{1}{Q} = \frac{1}{Q_o} + \frac{C_1}{V_o}, \tag{9.63}$$

where Q is the measured charge, Q_o is the actual charge liberated, V_o is the applied voltage, and C_1 is constant, resulting in a linear equation identical to that expressed by Attix [2004]. Hence, for relatively high applied voltages, the inverse of the measured charge Q can be plotted against measured values on the right side of Eq. (9.63), as shown in Fig. 9.10. Extrapolating back to the origin where $C_1/V_o = 0$, the actual value of a Q_o can be found.

There are a few trends, perhaps obvious, that can be seen from Eq. (9.51). First, columnar recombination increases with the density of N_o and the recombination coefficient β. Second, the recombination decreases with larger diffusion coefficients D_{ion}. Finally, perhaps not so obvious from Eq. (9.51), but affirmed by Eq. (9.63), columnar recombination decreases as the applied voltage is increased or gas pressure is decreased.

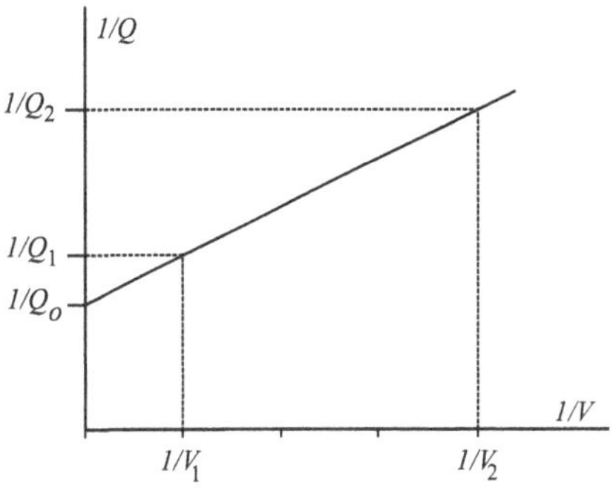

Figure 9.10. Graphic depiction of a method to determine the correct Q_o in the presence of columnar recombination from $1/Q = 1/Q_o + C_1/V_o$.

9.3.2 Volumetric Recombination

Volumetric recombination occurs upon irradiation of a gas medium in high radiation fields, such that negative and positive ions encounter oppositely polarized charge carriers generated from multiple random events. The volumetric rate at which negative charges, with concentration n^-, are recombined with the positive charges, with concentration n^+, thereby reducing both n^- and n^+, is $-\beta n^- n^+$, where β is the recombination coefficient. Loeb [1939] quotes $\beta_e \simeq 2 \times 10^{-10}$ cm^3 s^{-1} for electron-ion pairs in Ar at 1 atm, and Rossi and Staub [1949] makes the claim that β_e is similar for most detector gases. The value of β is much larger if the negative ions are not electrons, but instead negatively charged heavy ions, where β_{ion}^- ranges between 0.9×10^{-6} cm^3 s^{-1} – 1.7×10^{-6} cm^3 s^{-1} at 1 atm [Loeb 1929].

Effects of Diffusion and Recombination in an Ion Chamber

Consider the one-dimensional ion chamber shown in Fig. 9.11 whose gas is uniformly and constantly irradiated such that S_o electron-ion pairs are created per unit volume per unit time. The intensity of the irradiation is sufficiently high that the chamber is operated in the current mode. In this section three cases are considered: (1) no diffusion or recombination effects, (2) only diffusion is important, (3) only recombination is important. In the following analysis the approach taken by Rossi and Staub [1949] is used.

Case (1): No Diffusion or Recombination Effects

The positive and negative current densities $\mathbf{j}^\pm$ formed, respectively, by the positive and negative ion drift velocities $\mathbf{v}^\pm$ caused by the electric field are given by

$$\boldsymbol{\nabla}\boldsymbol{\cdot}\mathbf{j}^\pm = \pm q_e S_o. \tag{9.64}$$

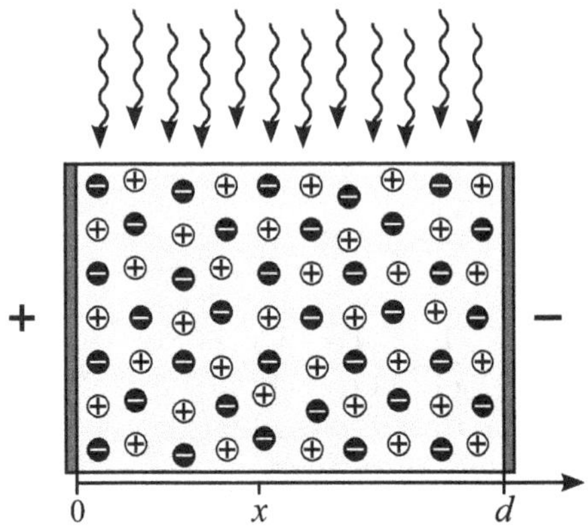

Figure 9.11. Uniform irradiation of a planar configuration gas-filled ion chamber. Volumetric recombination increases as the source term S_o, the production of electron-ion pairs, is increased.

Equivalently, in terms of the ion concentrations

$$\mathbf{j}^{\pm} = \pm q_e n^{\pm} \mathbf{v}^{\pm}. \tag{9.65}$$

For the one-dimensional case of Fig. 9.11, the ion speeds are

$$\mathbf{v}^{\pm} = \pm \mathbf{i} v^{\pm}, \tag{9.66}$$

and the current densities are

$$\mathbf{j}^{\pm} = \mathbf{i} j^{\pm}, \tag{9.67}$$

where $j^{\pm}$ and $v^{\pm}$ are the absolute values, respectively. Substitution of Eq. (9.65) into Eq. (9.64) and use of the magnitudes $v^{\pm}$ yields the following differential equations for the steady-state ion concentrations

$$v^{\pm} \frac{dn^{\pm}(x)}{dx} = \pm S_o. \tag{9.68}$$

The general solution of these equations is

$$n^{\pm}(x) = \pm \frac{S_o}{v^{\pm}} x + C^{\pm}, \tag{9.69}$$

where $C^{\pm}$ are arbitrary constants that can be determined from the boundary conditions $n^{+}(0) = 0$ and $n^{-}(d) = 0$. In this manner, the concentration profiles are found to be

$$n^{+}(x) = \frac{S_o}{v^{+}} x \qquad \text{and} \qquad n^{-}(x) = \frac{S_o}{v^{-}} (d - x). \tag{9.70}$$

In terms of the magnitudes $v^{\pm}$ and $j^{\pm}$ the current densities for the geometry of Fig. 9.11 are

$$j^{\pm} = q_e n^{\pm}(x) v^{\pm}, \tag{9.71}$$

so that

$$j = j^{+} + j^{-} = q_e[v^{+} n^{+}(x) + v^{-} n^{-}(x)] = q_e S_o d \tag{9.72}$$

Finally, multiplication of j by the area A of the electrodes gives the steady-state *saturation* induced current

$$I_s = q_e S_o A d. \tag{9.73}$$

Case (2): Ion Drift with Diffusion

When diffusion caused by thermal agitation becomes appreciable, some ions near the electrode of the same sign reach that electrode and are neutralized. This reduces the number ions that drift towards their collecting electrode and, hence, reduces the induced current.

To quantify this effect, add diffusion to the previous analysis. Equation (9.64) still holds but Eqs. (9.65) become

$$\mathbf{j}^{+} = q_e n^{+} \mathbf{v}^{+} - q_e D^{+} \nabla n^{+} \qquad \text{and} \qquad \mathbf{j}^{-} = -q_e n^{-} \mathbf{v}^{-} + q_e D^{-} \nabla n^{-}. \tag{9.74}$$

For the one-dimensional case of Fig. 9.11 and use of absolute values these equations yields

$$j^{+}(x) = q_e n^{+}(x) v^{+} - q_e D^{+} \frac{dn^{+}(x)}{dx} \quad \text{and} \quad j^{-}(x) = q_e n^{-}(x) v^{-} + q_e D^{-} \frac{dn^{-}(x)}{dx}. \tag{9.75}$$

Equation (9.64), as before, gives

$$\frac{dj^{\pm}(x)}{dx} = \pm q_e S_o \qquad \text{or} \qquad j^{\pm}(x) = \pm q_e S_o x + C_1^{\pm}. \tag{9.76}$$

where $C_1^{\pm}$ are constants of integration. Equating the right-hand sides of Eqs. (9.75) and (9.76) gives, after rearrangement,

$$\frac{dn^+(x)}{dx} = \frac{v^+}{D^+}n^+(x) - \frac{S_o}{D^+}x + \widetilde{C}_1^+, \tag{9.77}$$

$$\frac{dn^-(x)}{dx} = -\frac{v^-}{D^-}n^-(x) - \frac{S_o}{D^-}x + \widetilde{C}_1^-, \tag{9.78}$$

where $\widetilde{C}_1^{\pm} = C_1^{\pm}/D^{\pm}$.

The most general solutions of Eqs. (9.77) and (9.78) are

$$n^+(x) = C_2^+ e^{v^+x/D^+} + \frac{S_o x}{v^+} + C_3^+, \tag{9.79}$$

$$n^-(x) = C_2^- e^{-v^-x/D^-} - \frac{S_o x}{v^-} + C_3^-, \tag{9.80}$$

where $C_2^{\pm}$ and $C_3^{\pm}$ are arbitrary constants. To evaluate these four constants, use the four boundary conditions $n^{\pm}(0) = 0$ and $n^{\pm}(d) = 0$. With these conditions the concentration profiles are found to be

$$n^+(x) = \frac{S_o}{v^+}x - \frac{S_o d}{v^+}\left[\frac{1 - e^{v^+x/D^+}}{1 - e^{v^+d/D^+}}\right], \tag{9.81}$$

$$n^-(x) = -\frac{S_o}{v^-}x + \frac{S_o d}{v^-}\left[\frac{1 - e^{-v^-x/D^-}}{1 - e^{-v^-d/D^-}}\right]. \tag{9.82}$$

For most practical ion chambers $v^{\pm}d/D^{\pm}$ are large numbers so that Eqs. (9.81) and (9.82) can be approximated as

$$n^+(x) \simeq \frac{S_o}{v^+}x - \frac{S_o d}{v^+}e^{-v^+(d-x)/D^+}, \tag{9.83}$$

$$n^-(x) \simeq \frac{S_o}{v^-}(d - x) - \frac{S_o d}{v^-}e^{-v^-x/D^-}. \tag{9.84}$$

To obtain the current densities, substitute Eqs. (9.83) and (9.84) into Eqs. (9.75) to obtain

$$j^+(x) \simeq q_e S_o x - q_e\frac{D^+}{v^+}S_o \qquad \text{and} \qquad j^-(x) \simeq q_e S_o(d - x) - q_e\frac{D^-}{v^-}S_o. \tag{9.85}$$

Finally, the induced current is seen to be

$$I = A(j^+ + j^-) \simeq q_e S_o A d - q_e S_o A\left[\frac{D^+}{v^+} + \frac{D^-}{v^-}\right]. \tag{9.86}$$

Thus, the decrease in the saturation current I_s due to diffusion effects is

$$-\left.\frac{\delta I_s}{I_s}\right|_{\text{diffusion}} \simeq \left[\frac{D^+}{v^+d} + \frac{D^-}{v^-d}\right]. \tag{9.87}$$

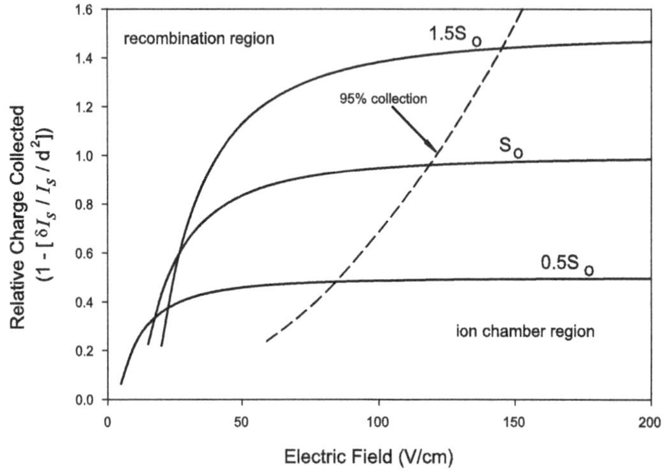

Figure 9.12. The relative charge collection $1 - (\beta R S_o)/(6v^+v^-)$ for air as a function of the electric field. Here the distance d between electrodes is in cm and $q_e S_o = 1$ nA cm^{-3}. Values of the other variables are calculated in Example 9.2.

Substitution of Eq. (9.34) into Eq. (9.87), which also applies to ions, gives

$$-\frac{\delta I_s}{I_s}\bigg|_{\text{diffusion}} \simeq \frac{3kT}{2q_e V}\left(\xi^+ + \xi^-\right), \tag{9.88}$$

where V is the applied voltage. For heavy ions, the difference between the characteristic energy and the thermal energy is small, hence for both negative and positive heavy ions $\xi \simeq 1$, and

$$-\frac{\delta I_s}{I_s}\bigg|_{\text{diffusion}} \simeq \frac{3kT}{q_e V}, \tag{9.89}$$

from which it is seen that diffusion losses are negligible for high voltages. However, for electrons, the value of ξ can be considerably higher, on the order of 200 or more for typical electric fields employed in ion chamber operation. In this case

$$-\frac{\delta I_s}{I_s}\bigg|_{\text{diffusion}} \simeq \frac{3kT}{2q_e V}\left(\xi_e^- + 1\right). \tag{9.90}$$

This result shows that there can be electron losses from diffusion if the operating voltage is on the order of only a few hundred volts. Raising the operating voltage minimizes diffusion losses.

Case (3): Effect of Recombination

Assume that diffusion losses are negligible so that the ion concentration profiles are given by Eqs. (9.70). The induced current, with recombination effects, is thus

$$I = I_s - \delta I_s = q_e S_o A d - q_e A \int_0^d \beta n^+(x) n^-(x) R\, dx, \tag{9.91}$$

where R is the fraction of free electrons that produce heavy negative ions through electron attachment. Substitution of Eqs. (9.70) into this result yields

$$-\frac{\delta I_s}{I_s}\bigg|_{\text{recombination}} = \frac{\beta R}{S_o d}\int_0^d \frac{S_o^2 x(d-x)}{v^+v^-} dx = \frac{\beta R}{6}\frac{S_o d^2}{v^+v^-}. \tag{9.92}$$

From Eq. (9.87) it is seen that losses from diffusion decrease with increasing electrode spacing while, from Eq. (9.92), losses from recombination are seen to increase with electrode spacing. Equation (9.87) also indicates that diffusion losses are independent of the source term S_o, and are inversely proportional to the ion drift speeds, which are proportional to the applied electric field. Equation (9.92) indicates that recombination losses increase proportionally to the source term S_o and inversely to the drift speeds of the charge carriers (or applied voltage). The electric field required to move the operation of the detector from the recombination region into the ion chamber region increases with radiation exposure, i.e., with S_o, as depicted in Fig. 9.12. For gases with negligible electron attachment and, consequently, relatively small recombination losses, the recombination region spans a small range of voltages. Examples of such gases with little to no recombination losses include Ar, CH_4, P-10 and diatomic forms of halogens. However, contamination from electronegative gases, such as O_2, O, NO, OH, and Cl, can cause considerable recombination, thereby extending the range of the recombination region. Equation (9.92) indicates that the recombination region increases as the volumetric ionization increases, as shown in Fig. 9.12. Hence, as S_o increases, the applied electric field must also be increased to transition from the recombination region into the ion chamber region.

Example 9.2: An open air ion chamber is irradiated with 662-keV gamma rays to produce $q_e S_o = 1$ nA cm^{-3}. The ion chamber's aperture is 2.5 cm × 2.5 cm and the electrode spacing is 2.5 cm. The electrodes are designed as squares, each being 2.5 cm × 2.5 cm. The chamber is operated at 300 volts. The saturation speed of electrons in air at 760 torr is $v_e = 6.58 \times 10^5$ cm s^{-1}. What is the fractional loss of saturation current due to recombination?

Solution:
The electric field across the chamber is $\mathcal{E} = 300$ V/2.5 cm $= 120$ V/cm. For energies below 0.2 eV, there is an unfortunate lack of data for the ratio D/μ (or ϵ_K) for oxygen, although it can be assumed from available measurements that $\epsilon_K \lesssim 0.2$ eV [Huxley and Crompton 1974; Peisert and Sauli 1984]. Instead, assume that $\xi = 1$ so that $\epsilon_K = \overline{E}_e$, thereby giving as a first approximation the average thermal electron energy as $3kT/2$. The characteristic energy is then

$$\epsilon_K = \frac{q_e D_e}{\mu_e} = \frac{2\overline{E}_e}{3} = kT \simeq .026 \text{ eV},$$

which can be interpreted as the minimum thermal energy.[11] From Fig. 9.4, the momentum transfer cross section σ_m for O_2 is $\simeq 7 \times 10^{-16}$ cm^2. Also, from Table 9.3, the capture coefficient for O_2 is C_a is 2.5×10^{-5}. The capture coefficient for N_2 is essentially zero ($C_a \simeq 0$).

$$\overline{\omega}_e = \sqrt{\frac{2\epsilon_K}{m_e}} = \sqrt{\frac{2(0.026 \text{ eV})(1.6 \times 10^{-19} \text{ J/eV})}{9.1 \times 10^{-31} \text{ kg}}} = 9.56 \times 10^4 \text{ m s}^{-1} = 9.56 \times 10^6 \text{ cm s}^{-1}$$

The mean free path λ_e between scatters with O_2 is estimated as a function of O_2 density, namely $\lambda_e = 1/(fN\sigma_m)$, where $n_a = fN$. From Eq. (9.47),

$$\psi_I = \frac{C_a \overline{\omega}_e}{\lambda_e v_e} = \frac{C_a \overline{\omega}_e \sigma_m f N}{v_e} = \frac{(2.5 \times 10^{-5})(9.56 \times 10^6 \text{ cm s}^{-1})(7 \times 10^{-16}\text{cm}^2)}{6.58 \times 10^5 \text{ cm s}^{-1}}$$
$$\times (0.21)(2.7 \times 10^{19}\text{cm}^{-3})\frac{273\text{K}}{293\text{K}} = 1.34 \text{ cm}^{-1}. \quad (9.93)$$

If, instead, $\epsilon_K = 0.2$ eV is used, then $\psi_I = 3.72$ cm^{-1}. Hence, it can be assumed that the interaction coefficient lies between the two extremes, i.e., 1.34 cm$^{-1} \le \psi_I \le 3.72$ cm^{-1}. With such a high interaction

[11] Typically, the characteristic energy is much higher than the minimum thermal energy. Use of the minimum thermal energy serves to provide the minimum thermal speed $\overline{\omega}_e$.

coefficient, essentially all electrons become attached to an ion within 1 to 2 cm of air; hence $R \simeq 1$. Now assume that direct electron-ion recombination is negligible, leaving only heavy ion recombination between negative and positive heavy ions. From Table 9.2, the positive ion and negative ion speeds are

$$v^+ = \frac{760\mu^+_{air}\mathcal{E}}{P} = \frac{1.36 \text{ cm}^2}{\text{V s}}\frac{300\text{V}}{2.5\text{cm}} = 163.2 \text{ cm s}^{-1},$$
$$v^- = \frac{760\mu^-_{air}\mathcal{E}}{P} = \frac{1.87 \text{ cm}^2}{\text{V s}}\frac{300\text{V}}{2.5\text{cm}} = 224.4 \text{ cm s}^{-1}$$

For $q_e S_o = 10^{-9}$ A cm^{-3}, the number of electron-ion pairs S_o produced per unit volume per unit time is 6.25×10^9 cm^{-3} s^{-1}. Finally, assume a high recombination coefficient of 1.7×10^{-6} cm^3 ion^{-1} s^{-1} [see data from Loeb 1929]. Thus,

$$-\frac{\delta I_s}{I_s} = \frac{\beta S_o d^2}{6v^-v^+} = \frac{(1.7 \times 10^{-6}\text{cm}^3 \text{ ion}^{-1} \text{ s}^{-1})(6.25 \times 10^9 \text{ ion pairs cm}^{-3}\text{s}^{-1})((2.5 \text{ cm})^2)}{6(224.4 \text{ cm s}^{-1})(163.2 \text{ cm s}^{-1})} = 0.3022.$$

This result means that approximately 30% of the charge carriers are lost to recombination.

Equation (9.92) is at times written as a collection efficiency [Boag 1966]

$$f = \frac{Q}{Q_o} = 1 - \frac{\beta R}{6}\frac{S_o d^2}{v^+v^-} = 1 - \frac{\beta R}{6}\frac{Q_o^* d^4}{q_e\mu^+\mu^- V_o^2} = 1 - \frac{1}{6}\zeta^2, \tag{9.94}$$

where Q_o^* is the true ionization rate (esu cm^{-3} s^{-1}), q_e is the unit electric charge (in esu),[12] and V_o is the applied voltage across electrode separation d, and

$$\zeta = \sqrt{\frac{\beta R}{q_e\mu^-\mu^+}}\left(\frac{d^2\sqrt{Q_o^*}}{V_o}\right) = k^*\left(\frac{d^2\sqrt{Q_o^*}}{V_o}\right), \tag{9.95}$$

where k^* is a constant. Through comparisons of experimental observations, an accepted average value of k^* is 36.7 ± 2.2 V s$^{1/2}$ cm$^{1/2}$ esu$^{1/2}$ for air at STP [Greening 1964]. Other terms in Eq. (9.95) are defined as follows [Attix 2004]: d is the plate separation (cm), $Q_o^* = qN_o/v_l t$ (esu/cm^3 s), q_e the unit electrical charge (esu) $= 4.8032 \times 10^{-10}$ esu, N_o the number of ion pairs produced in the chamber volume, v_l the chamber volume (cm^3), t the irradiation duration (s), V_o the applied potential (volts), β the recombination coefficient (cm^3/ ion s), μ^- the mobility of negative ions (cm^3/V s), μ^+ the mobility of positive ions (cm^3/V s), and R the fraction of negative ions from electron attachment. It is assumed the irradiation time t is long by comparison to the ion-transit time across the detector ($\sim$ 1 ms). The solution Eq. (9.92), and also Eq. (9.94), are based on the charge density distributions described by Eq. (9.70), which ideally form a linear dependence with position. Boag [1966] points out that the solutions Eq. (9.92) and Eq. (9.94) overestimate the recombination in an ion chamber and offer a correction altering Eq. (9.70) with the charge carrier collection efficiencies,

$$n^+(x) = \frac{fS_o}{v^+}x \quad \text{and} \quad n^-(x) = \frac{fS_o}{v^-}(d - x). \tag{9.96}$$

Substitution into Eq. (9.91) yields

$$f = 1 - \frac{f^2\zeta^2}{6}, \tag{9.97}$$

[12] The esu, or statcouloumb, is a nearly defunct unit in the cgs system, with a conversion to the SI system, 1 C = 2.9979×10^9 statC or 1 statC = 3.3357×10^{-10} C.

which actually underestimates the recombination losses. The derivations by Mie [1904] and Greening [1964] produce a geometric average between the two derivations of Eq. (9.92) and Eq. (9.97), namely,

$$f = 1 - \frac{f\zeta^2}{6}, \qquad \text{or} \tag{9.98}$$

$$f = \frac{1}{1 + \dfrac{\zeta^2}{6}}. \tag{9.99}$$

Eq. (9.99) is accurate for d between 0.5 cm up to 8 cm provided that $f \geq 0.7$. Rearranging terms with the substitution $q_e N_o / v_l t = Q_o^*$, Eq. (9.99) reduces to the linear equation

$$\frac{1}{Q} = \frac{1}{q_e N_o} + \frac{k^{*2} d^4}{6 v_l t V_o^2} = \frac{1}{Q_o} + \frac{C_1}{V_o^2}, \tag{9.100}$$

where $C_1 = k^{*2} d^4 / 6 v_l t$ is constant. Eq. (9.99) can be plotted to determine the true charge collection efficiency f, or the true value of Q_o, as shown in Fig. 9.13. Hence, by measuring the current, or charge, at two different operating voltages, a line can be projected through the ordinate at $1/V^2 = 0$ to determine the value $1/Q_o$.

Equation (9.100) also holds true for cylindrical and spherical ion chambers, although the device dimensions, and therefore ζ, must be altered. For cylindrical detectors, in which a is the inner electrode wire radius (usually the anode) and b is the outer wall electrode radius (usually the cathode), one has

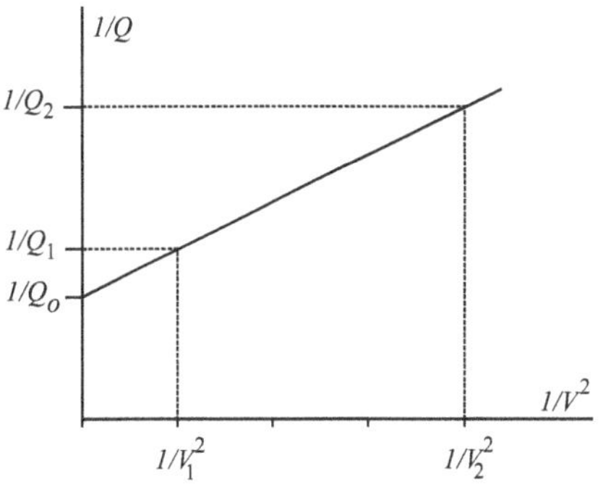

Figure 9.13. Graphic depiction of a method to determine the correct Q_o in the presence of volumetric recombination from $1/Q = 1/Q_o + C_1/V_o^2$.

$$\zeta_{cyl} = k^* \frac{[(b-a)\kappa_{cyl}]^2 \sqrt{Q_o^*}}{V_o} \tag{9.101}$$

where

$$\kappa_{cyl} = \left[\left(\frac{b+a}{2(b-a)} \right) \ln\left(\frac{b}{a} \right) \right]^{1/2}. \tag{9.102}$$

For spherical detectors, in which a is the inner electrode wire radius and b is the outer wall electrode radius, *zeta* is given by

$$\zeta_{sph} = k^* \frac{[(b-a)\kappa_{sph}]^2 \sqrt{Q_o^*}}{V_o} \tag{9.103}$$

where

$$\kappa_{sph} = \left[\frac{1}{3} \left(\frac{b}{a} + 1 + \frac{a}{b} \right) \right]^{1/2}. \tag{9.104}$$

9.3.3 Preferential Recombination

Preferential recombination occurs when a liberated electron recombines with its original parent ion. The process can occur when a heavy negative ion, formed through electron attachment, remains in close proximity of the positive parent ion. The probability that preferential recombination occurs is [Onsager 1938]

$$P_{pr} = 1 - \exp\left[\frac{-q_e^2}{4\pi\epsilon_o d_s k T} \right], \tag{9.105}$$

where d_s is the distance between the negative and positive ions. Preferential recombination occurs when the electrostatic attractive forces of the negative and positive ions overcome the thermal Brownian motion of the ion-pair, i.e., when

$$\frac{q_e^2}{4\pi\epsilon_o d_s} \gtrsim \frac{3}{2}kT. \tag{9.106}$$

Wilkinson (1950) points out that preferential recombination is generally unimportant for ion chambers operated under typical gas pressures. Regardless, the effect of preferential recombination is to remove charge carriers from the signal formation process, the same as columnar and volumetric recombination. Hence, its effect is nearly indistinguishable from other forms of recombination, except, perhaps, at high gas pressures where $D^{\pm}$ becomes small. Loeb [1939] suggests that gases with high attachment coefficients, such as O_2, Cl_2, SO_2, and CO_2, may experience preferential recombination for pressures above 5 to 10 atm.

9.4 Ion Chamber Operation

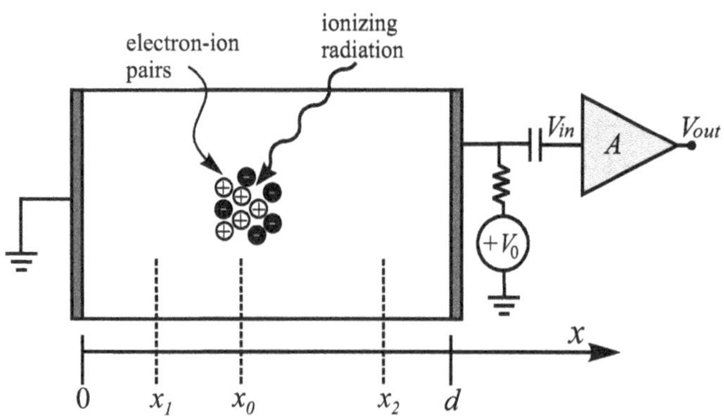

Figure 9.14. Depiction of an ion chamber of planar design.

Perhaps the simplest gas-filled detector is the *ion chamber.* There are many configurations of ion chambers, and they are operated in Region II of the gas curve. The detection method is simple. Ionizing radiation, such as charged particles or energetic photons, enter a region filled with a gas, such as Ar or air. The chamber has electrodes across which a voltage is applied. As the radiation interacts with the gas, electron-ion pairs are created, depicted in Fig. 9.14, whose number is proportional to the radiation energy absorbed. The voltage applied across the electrodes causes the negative electrons to separate from the positive ions and drift across the chamber volume. Electrons or negative ions drift towards the anode, and positive ions drift towards the cathode. Such charge motion *induces* current to flow in the external circuit.

Typically, this induced current is sensed by either directly measuring the current or by first storing the charge in a capacitor and then measuring the resulting voltage across the capacitor. The first case is referred to as *current mode* operation and the second case is *pulse mode* operation. Current mode operation is used in high radiation fields, and the magnitude of the current measured gives a relative measure of the radiation field. Pulse mode is used for lower radiation fields, and allows each radiation particle interacting in the chamber to be counted. The voltage produced by the ion chamber detector is the input voltage to the amplification circuitry, and hence, is referred to as V_{in}. The shaped voltage pulse exiting the accompanying amplification circuits is referred to as the output voltage V_{out}.

Ion chambers come in many forms: they can be used for reactor power measurements, in which the radiation field is very high, and for small personnel dosimeters, in which the radiation levels are typically very low. Although simple in concept, two main problems occur when the ion chamber is operated in pulse mode, namely, (1) the measured signal is small due to relatively few electron-ion pairs produced by a single radiation particle and (2) the time to form a signal can be relatively long due to the slow motion of the heavy positive ions. To ameliorate this second problem, an RC circuit is often connected to an ion chamber to reduce the time constant of the system, thereby reducing the time response. Consequently, the capacitor usually discharges *before* all of the ions are collected.

9.4.1 Planar Ion Chambers

A planar ion chamber of the design shown in Fig. 9.14 is operated with positive potential V_0 applied across the electrodes separated by distance d. A radiation particle deposits energy at location x_0 in the chamber and creates a cloud of electron-ion pairs. The number of electron-ion pairs produced is proportional to the energy deposited within the detector gas by the radiation particle.

Current Mode Operation

A simple current mode arrangement for measuring radiation interactions is shown in Fig. 9.15. Radiation interactions in the chamber create electron-ion pairs, which are drifted across the chamber to their respective collecting electrodes by the applied voltage.

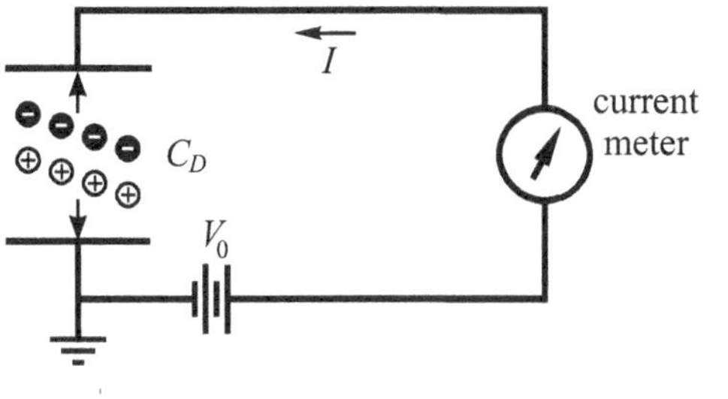

Figure 9.15. Current mode operation of an ion chamber.

One method used to measure the energy deposited by radiation interactions within an ion chamber is to directly measure the current. This action can be accomplished by inserting a current meter into the circuit, as depicted in Fig. 9.15. Because electron speeds are much higher than ions, the current is highest while electrons are in motion. The measured current is defined as

$$I = \frac{Q}{t} = \frac{q_e E v_e}{w \Delta x_e} + \frac{q_e E v_{ion}}{w \Delta x_{ion}} = \frac{q_e E}{w t_e} + \frac{q_e E}{w t_{ion}}, \quad 0 < t \le t_e. \tag{9.107}$$

and

$$I = \frac{q_e E}{w t_{ion}}, \quad t_e < t \le t_{ion}, \tag{9.108}$$

where E is the energy deposited in the detector, w is the average energy the radiation particle must deposit in the gas to produce an electron-ion pair, and t_e and t_{ion} are the electron and ion extraction or sweep-out times, respectively.

Example 9.3: Calculate the initial instantaneous current measured from an energy deposition of 1 MeV by a radiation particle in an ion chamber operated with pure Ar gas at 1 atm. The distance d between the electrodes is 2 cm with an applied bias of 200 volts. The interaction occurs at $x_0 = 0.5d$.

Solution:

From Table 9.1, thc average energy required to produce a electron-ion pair in Ar is 26 eV. For an electric field of 200 V, the reduced electric field at 1 atm (760 torr) is

$$\frac{\mathcal{E}}{P} = \frac{200 \text{ V}}{(2 \text{ cm})(760 \text{ torr})} = 0.132 \text{ V cm}^{-1} \text{ torr}^{-1},$$

From Eq. (9.35), the electron drift speed in argon is

$$\begin{aligned} v_e &= \frac{3.64(\mathcal{E}/P) + 114.6(\mathcal{E}/P)^2}{1 + 12.7(\mathcal{E}/P) + 304.33(\mathcal{E}/P)^2} \\ &= \frac{3.64(0.132 \text{ V cm}^{-1} \text{ torr}^{-1}) + 114.6(0.132 \text{ V cm}^{-1} \text{ torr}^{-1})^2}{1 + 12.7(0.132 \text{ V cm}^{-1} \text{ torr}^{-1}) + 304.33(0.132 \text{ V cm}^{-1} \text{ torr}^{-1})^2} \\ &= 0.364 \text{ cm}/\mu\text{s} = 3.64 \times 10^5 \text{ cm/s}. \end{aligned} \tag{9.109}$$

From Table 9.2, the positive ion mobility is 1.7 cm^2 V^{-1} s^{-1}, and from Eq. (9.30)

$$v_{ion}^{+} = \mu_{ion}^{+} \mathcal{E} = (1.7 \text{ cm}^2 \text{ V}^{-1} \text{ s}^{-1})(100 \text{ V cm}^{-1}) = 170 \text{ cm/s}.$$

The initial current is mainly from electron motion; hence, the extraction time over distance $d - x_0 = 1$ cm is

$$t = \frac{1 \text{ cm}}{3.64 \times 10^5 \text{ cm s}^{-1}} = 2.75\ \mu\text{s}.$$

Thus, the initial current is

$$I = \frac{Q}{t} \simeq \frac{(10^6 \text{ eV})(1.6 \times 10^{-19} \text{ Coulombs})}{(26 \text{ eV/e-ion pair})(2.75 \times 10^{-6} \text{ seconds})} = 2.24 \text{ nA}.$$

After the electrons are collected, the positive ions are collected with an extraction time of (1 cm)/(170 cm s^{-1})= 5.88 ms. Thus the current decreases to

$$I = \frac{(10^6 \text{ eV})(1.6 \times 10^{-19} \text{ Coulombs})}{(26 \text{ eV/e-ion pair})(5.88 \times 10^{-3} \text{ s})} = 1.05 \text{ pA},$$

decreasing to zero after all ions are collected.

Ion chambers are frequently used in high radiation environments in which pulse processing is not possible because of unacceptable pulse pile-up and dead time losses. However, under low irradiation conditions, the induced currents can be very small and difficult to measure. Thus under these conditions it is more practical to employ pulse processing methods and to operate the detector in pulse mode.

Pulse Mode Operation

A planar ion chamber is operated with positive potential V_0 applied across the electrodes separated by distance d as shown in Fig. 9.14. As before, a radiation particle deposits energy at location x_0 in the chamber and creates a cloud of electron-ion pairs. The motion of these mobile charges induces current to flow across a coupling capacitor in which the charge is stored and measured as a voltage $V = Q/C$. The pulse shape provided by the detector V_{in} is determined by several physical parameters, including the type of gas, chamber, line and coupling capacitances, and the detector load resistor. Three operating cases are considered in the following analyses.

Case 1: No Electron Attachment Here it is assumed that the gases chosen for the ion chamber have negligible electron attachment coefficients; hence, the charge carriers are restricted to electrons and positive ions. The solution for current and charge induction for a planar detector was developed in Chapter 8, where it was found (see Eq. (8.43))

$$\Delta Q = q_e N_o \left[\left.\frac{x_0 - x_1}{d}\right|_{ions^+} + \left.\frac{x_2 - x_0}{d}\right|_{e^-} \right]. \tag{9.110}$$

Because the mobility of electrons in gas far surpasses that of the heavy ions, the saturation drift speed of electrons $v_e = \mu_e \mathcal{E}$ is always much higher than the drift speeds of ions $v_{ion} = \mu_{ion}\mathcal{E}$, i.e., $v_e \gg v_{ion}$. Consequently, the electron sweep-out time $t_e = (x_2 - x_0)/v_e$ is much smaller than the ion sweep-out time $t_{ion} = (x_0 - x_1)/v_{ion}$, i.e., $t_e \ll t_{ion}$.

If t is measured from the time of the ionizing event, the induced charge changes in time as the charge carriers are swept out of the chamber. From Eq. (9.111) the induced charge is first seen to vary as

$$\Delta Q(t) = \frac{q_e N_o t}{d} \left[v_{ion} + v_e \right], \qquad 0 < t \le t_e. \tag{9.111}$$

Regardless of the location of x_0, the total charge induced at time t_e is

$$\Delta Q(t_e) = \frac{q_e N_o t_e}{d} \left[v_{ion} + v_e \right], \tag{9.112}$$

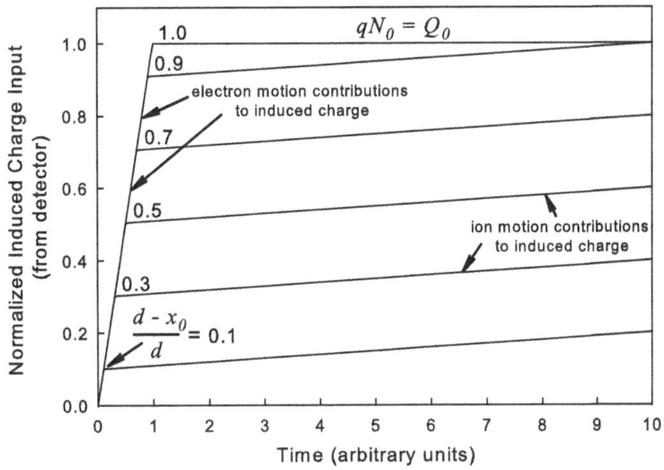

Figure 9.16. Normalized induced charge input from a planar ion chamber as a function of interaction location $d - x_0$. Depicted is the case in which $v_e = 100v_{ion}$.

which, because $v_e \gg v_{ion}$, can be approximated as

$$\Delta Q(t_e) \simeq q_e N_o \frac{v_e t_e}{d} = q_e N_o \frac{d - x_0}{d}. \tag{9.113}$$

After the electrons are rapidly collected, the positive ions continue to drift across the chamber at a much lower speed, so the collected charge is

$$\Delta Q(t) = q_e N_o \left[\frac{v_{ion} t}{d} + \frac{d - x_0}{d}\right], \qquad t_e < t \leq t_{ion}. \tag{9.114}$$

Finally, for $t > t_{ion}$ all the charge has been collected and ΔQ remains constant at

$$\Delta Q(t) = q_e N_o \left[\frac{x_0}{d} + \frac{d - x_0}{d}\right] = q_e N_o, \qquad t > t_{ion}. \tag{9.115}$$

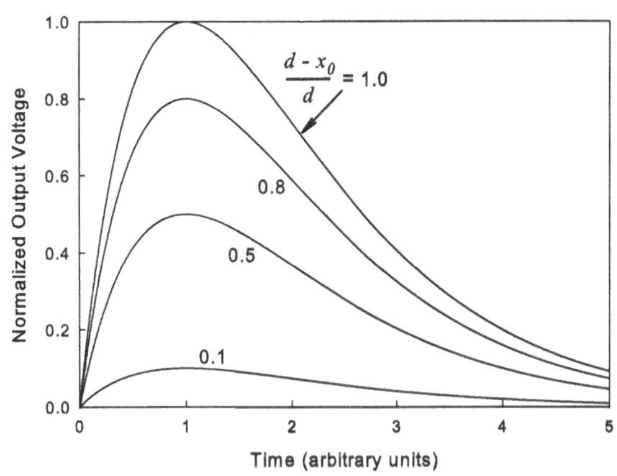

Figure 9.17. Normalized output voltage V_{out} for various interaction locations $d - x_0$ in a planar ion chamber.

The induced charge input for a planar device is shown in Fig. 9.16. Typically, v_{ion} for normal operating voltages of an ion chamber is much too slow for pulse mode operation, and the system RC network discharges the stored charge before all of the ions are collected. Consequently, the induced charge applied to the detector shaping electronics is best described by Eq. (9.113), in which the contribution of positive ions is mostly neglected. Also, the pulse height described by Eq. (9.113) is dependent upon the initial location of the ionization cloud, i.e., the pulses are *position dependent*. The electronic signals from ion chambers operated with gases having low electron attachment coefficients are dominated by electron motion. As a result, ion chambers generate signals that are strongly dependent upon the interaction location within the detector. Recall from Chapter 8 that $V = Q/C$. The time dependent rise in voltage from an ion chamber with intrinsic capacitance C is, thus,

$$V_{in}(t) = \frac{\Delta Q}{C} = \frac{q_e N_o t}{Cd}\left[v_{ion} + v_e\right] \simeq \frac{q_e N_o t}{Cd} v_e. \tag{9.116}$$

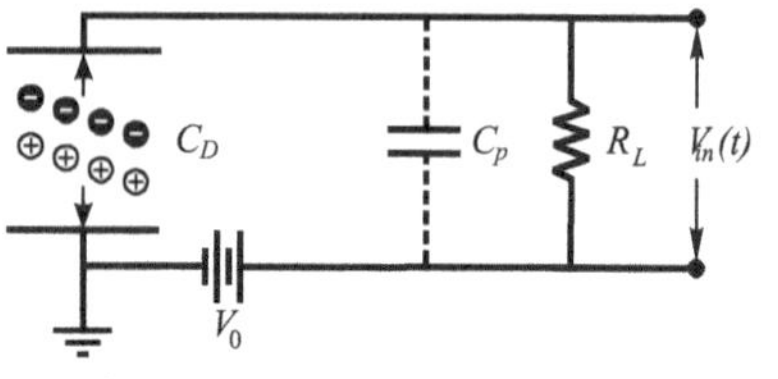

Figure 9.18. Equivalent circuit model for the time variant input voltage from an ion chamber.

The ideal voltage output from a planar ion chamber, for an input voltage V_{in} to the electronic circuitry, is depicted in Fig. 9.17. However, the ion chamber capacitance C_D, line and stray capacitances C_p, and the load resistor R_L connected to the voltage supply form an RC network that introduces a time constant τ intrinsic to the detector. The equivalent circuit for the time dependent input from a planar ion chamber is depicted in Fig. 9.18, that, as discussed in Chapter 22, forms a high pass differentiator circuit. The load resistor provides a leakage path to drain charge from the coupling capacitor between the detector and the preamplifier circuit, without which the coupling capacitor would eventually saturate. If R_L is too small, current from the detector is drained too quickly and the resulting voltage is small. If R_L is too large, τ becomes large, causing pulse pile-up from consecutive radiation interactions and, eventually, causing the coupling capacitor to saturate. The time dependent voltage from an ion chamber with intrinsic capacitance C and load resistance R_L is

$$V_{in}(t) \simeq \frac{q_e N_o}{Cd}\left[v_{ion} + v_e\right] t \exp\left[-\frac{t}{RC}\right], \tag{9.117}$$

Typically, the load resistor is chosen such that the time constant is much greater than the longest possible electron collection time, while remaining shorter than the ion collection time, or $t_{ion} \gg RC \gg t_e$. As a result, the intrinsic differentiator circuit has little discharge effect upon the electron induced current, yet causes the coupling capacitance to discharge before the ions are collected. The effect of the RC time constant upon $V_{in}(t)$ is depicted in Fig. 9.19. The combined effects of position dependent charge collection and the intrinsic system RC time constant lower the voltage pulse height from the ideal maximum. This difference in the expected value is called the *pulse height deficit*, an example of which is shown in Fig. 9.19. The magnitude of the pulse height deficit is a function of the RC time constant and the radiation interaction location in the ion chamber.

Example 9.4: An ion chamber filled with pure argon at 1 atm is operated with the following parameters: $V_0 = 200$ volts; $R_L = 5$ MΩ; $C = 65$ pF; $d = 5$ cm; $x_0 = 2$ cm from the anode; deposition energy $E = 5.5$ MeV. What is the maximum input voltage pulse after the radiation interaction? Compare the answer to that of an interaction occurring at $x_0 = 5$ cm from the anode. Determine the pulse height deficit for both cases.

Solution:
The time constant is $\tau = RC = 300$ μs. From Table 9.1 the average energy to produce an electron-ion pair in Ar is 26 eV, so that the total number of electron-ion pairs produced is

$$N_0 = \frac{5.5 \times 10^6 \text{ eV}}{26 \text{ eV/e-ion pair}} = 211538 \text{ eip}.$$

For $\tau \to \infty$, the expected input voltage is from Eq. (9.115)

$$V_{in} = \frac{\Delta Q}{C} = \frac{q_e N_0}{C} = \frac{(1.6 \times 10^{-19} \text{ C})(211538 \text{ e-ion pairs})}{65 \times 10^{-12} \text{ F}} = 0.5207 \text{ mV}.$$

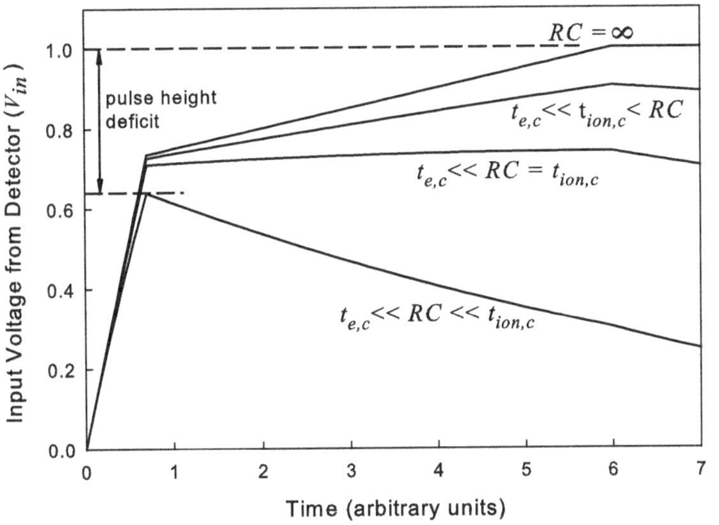

Figure 9.19. Normalized input voltage V_{in} from a planar ion chamber, where $(d - x_0)/d = 0.7$ and $v_e = 10v_{ion}$. Comparison between systems with different RC time constants is shown.

With an electric field of 200 V/ 5 cm = 40 V/cm, the reduced electric field at 1 atm (760 torr) is

$$\frac{\mathcal{E}}{P} = \frac{200 \text{ V}}{(5 \text{ cm})(760 \text{ torr})} = 0.0526 \text{ V cm}^{-1} \text{ torr}^{-1}.$$

From Eq. (9.35), the electron speed in argon is

$$\begin{aligned} v_e &= \frac{3.64(\mathcal{E}/P) + 114.6(\mathcal{E}/P)^2}{1 + 12.7(\mathcal{E}/P) + 304.33(\mathcal{E}/P)^2} \\ &= \frac{3.64(0.0526 \text{ V cm}^{-1} \text{ torr}^{-1}) + 114.6(0.0526 \text{ V cm}^{-1} \text{ torr}^{-1})^2}{1 + 12.7(0.0526 \text{ V cm}^{-1} \text{ torr}^{-1}) + 304.33(0.0526 \text{ V cm}^{-1} \text{ torr}^{-1})^2} = 0.203 \text{ cm}/\mu\text{s} = 2.03 \times 10^5 \text{ cm/s} \end{aligned}$$

From Table 9.2, the positive ion mobility is 1.7 cm^2 V^{-1} s^{-1}, and from Eq. (9.30)

$$v_{ion}^{+} = \mu_{ion}^{+}\mathcal{E} = (1.7 \text{ cm}^2 \text{ V}^{-1} \text{ s}^{-1})(40 \text{ V/cm}) = 68 \text{ cm/s}.$$

The electron extraction time is

$$t_e = \frac{x_0}{v_e} = \frac{2 \text{ cm}}{2.03 \times 10^5 \text{ cm/s}} = 9.85 \times 10^{-6} \text{ s},$$

and the ion extraction time is

$$t_{ion} = \frac{d - x_0}{v_{ion}} = \frac{2 \text{ cm}}{68 \text{ cm/s}} = 44.12 \times 10^{-3} \text{ s}.$$

Because $t_e \ll \tau \ll t_{ion}$, the maximum input voltage is estimated from Eq. (9.117) at t_e as

$$\begin{aligned} V_{in}(t_e) &\simeq -\frac{q_e N_o}{dC}\left[v_{ion} + v_e\right] t_e \exp\left[-\frac{t_e}{RC}\right], \\ &= -\frac{(1.6 \times 10^{-19} \text{ C})(211,538 \text{ e-ion pairs})}{(5 \text{ cm})(65 \times 10^{-12} \text{ F})} \\ &\quad \times \left[68 \text{ cm/s} + 2.03 \times 10^5 \text{ cm/s}\right](9.85 \times 10^{-6} \text{ s}) \exp\left[-\frac{9.85 \times 10^{-6} \text{ s}}{300 \times 10^{-6} \text{ s}}\right] = 0.2016 \text{ mV}. \end{aligned}$$

The pulse height deficit is 0.5207 mV − 0.2016 mV = 0.3191 mV. By comparison, if the interaction occurred at $x_0 = d$, then the induced current is produced entirely from electron motion and $t_e = 2.46 \times 10^{-5}$ s. The input voltage in this case is

$$V_{in}(t_e) \simeq -\frac{q_e N_o}{dC} v_e t_e \exp\left[-\frac{t_e}{RC}\right],$$
$$= -\frac{(1.6 \times 10^{-19}\ \text{C})(211{,}538\ \text{e-ion pairs})}{(65 \times 10^{-12}\ \text{F})} \exp\left[-\frac{2.46 \times 10^{-5}\ \text{s}}{300 \times 10^{-6}\ \text{s}}\right] = 0.4797\ \text{mV}.$$

The pulse height deficit is 0.5207 mV − 0.4797 mV = 0.041 mV.

Although discussions on nuclear electronics are reserved for Chapter 22, to illustrate the consequence of slow positive ion speeds, the following exposition is offered. The input voltage $V_{in}(t)$ from a ion chamber is commonly applied to a *CR-RC* preamplifier. The preamplifier consists of (1) a low-pass integrating/shaping circuit that has an R_iC_i time constant τ_i, and (2) a high-pass differentiating/shaping circuit that has an R_oC_o time constant of τ_o. If it is assumed that the collection time for electrons is much less than the time constant ($t_e \ll \tau$), and the time constant is much less than the collection time for the ions ($\tau \ll t_{ion}$), then Eq. (9.113) can be approximated as a step input to the electronic circuitry. The observed output voltage from the preamplifier circuit for a step input is,

$$V_{out}(t) \simeq V_{in}(t)\frac{\tau_o}{\tau_o - \tau_i}\left(e^{-t/\tau_o} - e^{-t/\tau_i}\right), \tag{9.118}$$

Often, the time constant for input and output are equal, i.e., $\tau_o = \tau_i$. Substitution of Eq. (9.113) and use of L'Hôpital's rule[13] yields

$$V_{out}(t) \simeq \frac{q_e N_o v_e t_e}{Cd}\frac{t}{\tau}e^{-t/\tau} \qquad t > t_e, \tag{9.119}$$

where it is assumed that t_e is relatively small such that the input appears as a step input, and

$$V_{out}(t) \simeq \frac{q_e N_o (d - x_0)}{Cd}\frac{t}{\tau}e^{-t/\tau}. \tag{9.120}$$

The normalized output voltages V_{out} as a function of interaction location $d - x_0$ for monoenergetic ionizing radiation particles are shown in Fig. 9.17, from which the responses are seen to be overwhelmingly position dependent. As a result, a monoenergetic ionizing radiation source produces a continuous spectrum of signals with varying amplitudes, thus, precluding the use of a planar ion chamber as an energy spectrometer.

Case 2: High Electron Attachment Consider the case in which the gas in the ion chamber has high electronegativity; in other words, the electrons become attached to neutral molecules almost immediately after liberation. The induced charge now becomes

$$\Delta Q = -\frac{q_e N_o t}{d}\left[v_{ion}^+ + v_{ion}^-\right] = -\frac{q_e N_o t \mathcal{E}}{d}\left[\mu_{ion}^+ + \mu_{ion}^-\right]. \tag{9.121}$$

The time dependent input voltage from the detector is

$$V_{in}(t) = \frac{\Delta Q}{C} = -\frac{q_e N_o t}{Cd}\left[v_{ion}^+ + v_{ion}^-\right]. \tag{9.122}$$

[13]Although named after Guillaume François Antoine Marquis de L'Hôpital, evidently it was his tutor Johann Bernoulli who discovered the rule (for pay), as indicated by communications between the two mathematicians [Dunn and Shultis 2012].

Because mobilities for positive and negative ions are only slightly different, the pulse mode signals produced are small for typical operating voltages applied to ion chambers; however, they are much more consistent in magnitude than the case in which electrons dominate. Nevertheless, there is some dependence of the pulse magnitude on the energy deposited and the resulting induced signal formed from the drifting ions. Background and electronic noise further degrade the spectroscopic performance. Consequently the ion chamber typically cannot be used for energy spectroscopy. However, these detectors can be used in current mode to measure high radiation fields.

Case 3: Moderate Electron Attachment If the ion chamber is contaminated with an electronegative gas, the electrons are lost by electron attachment according to Eq. (9.49) and Eq. (9.50). In pulse mode operation, the positive ions and electrons are rapidly separated; hence, it is assumed that recombination is minimal. Electrons lost from attachment to neutral molecules become slow moving negative ions. But if electrons are the main contributor to the current, the instantaneous current produced at time t is

$$I = \frac{dQ}{dt} = -\frac{q_e N_o}{d}\left[v_e e^{-\alpha_I t}\right], \qquad t \le \frac{d - x_0}{v_e} = t_e, \tag{9.123}$$

where α_I is the electron interaction rate coefficient. Integration of Eq. (9.123) yields the time dependent input voltage from the detector as

$$V_{in}(t) \simeq -\int_0^t \frac{q_e N_o}{Cd}\left[v_e e^{-\alpha_I t'}\right] dt' = -\frac{q_e N_o}{Cd}\left[\frac{v_e}{\alpha_I}(1 - e^{-\alpha_I t})\right], \quad t \le t_e. \tag{9.124}$$

Within the electron collection time t_e, the electrons have either attached to neutral molecules or have been collected. The induced signal is position dependent, dominated by electron motion (unless α_I is very large). The observed slope of the time dependent signal decreases with t, thereby reducing the output signal below that expressed by Eq. (9.116).

After the electrons have been collected, the positive ions continue to drift and produce an induced voltage described by

$$V_{in}^+(t) = -\frac{q_e N_o t}{Cd}\left[\mu_{ion}^+ \mathcal{E}\right], \qquad t \le \frac{x_0}{\mu_{ion}^+ \mathcal{E}} = t_{ion}. \tag{9.125}$$

The heavy negative ions produced from electron attachment also continue to drift. Assuming that $v_e \gg v_{ion}^-$, the input voltage produced by the heavy negative ions is found to be (see Exercise Problem 9.7)

$$V_{in}^-(t) = -\frac{q_e N_o v_e}{Cd}\left\{\frac{\mu_{ion}^- \mathcal{E}}{v_e} t + \frac{e^{-\alpha_I t_e}}{\alpha_I}\left[1 - \exp\left(\frac{\alpha_I t \mu_{ion}^- \mathcal{E}}{v_e}\right)\right]\right\}, \quad t \le \frac{d - x_0}{\mu_{ion}^- \mathcal{E}}. \tag{9.126}$$

The total input voltage from the detector with moderate electron attachment is found by summing Eq. (9.124), Eq. (9.125), and Eq. (9.126). An example is shown in Fig. 9.20.

9.4.2 Coaxial Ion Chambers

In Chapter 8, it was shown that the charge induction for a coaxial detector was a function of the inner electrode diameter and the outer electrode diameter. The shape of the signal output changes significantly depending upon the bias configuration. Typically, a positive voltage is applied to the inner electrode, as shown in Fig. 9.21. The charge induction for electrons and positive ions is given by Eq. (8.79) which can be written as

$$\Delta Q = q_e N_o \left[\ln\left(\frac{r_2}{r_1}\right)\right]^{-1} \left\{\ln\left(\frac{r_2'}{r_0}\right) - \ln\left(\frac{r_1'}{r_0}\right)\right\}. \tag{9.127}$$

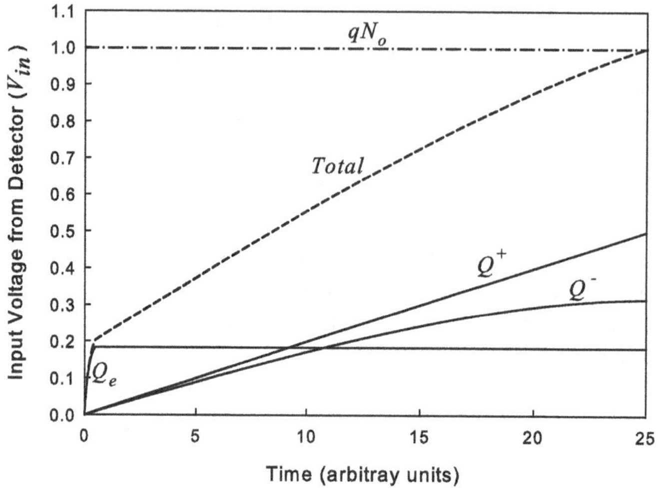

Figure 9.20. Normalized detector input voltage $V_{in}(t)$ for an ion chamber with moderate electron attachment. Shown is the case for $v_e = 50v_{ion}$, $x_o = 0.5d$ and $RC = \infty$. $V(t)$ is the sum of induced voltage from electrons (Q_e/C), positive (Q^+/C), and negative (Q^-/C) ions.

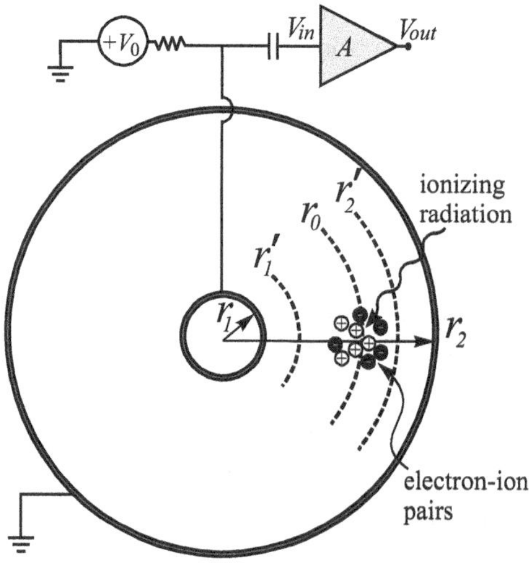

Figure 9.21. Depiction of an ion chamber of cylindrical design.

The electric field for a cylindrical detector is given by Eq. (8.75), namely,

$$\mathcal{E}(r) = \frac{V_0^+}{r} \ln\left[\frac{r_2}{r_1}\right], \tag{9.128}$$

where V_0 is the positive voltage applied to the inner electrode. Because the electric field now varies with r, the drift speeds of the charge carrier are position dependent such that

$$v^{\pm}(r) = \mu^{\pm}\mathcal{E}(r) = \mu^{\pm}\frac{V_0}{r} \ln\left[\frac{r_2}{r_1}\right]. \tag{9.129}$$

The minimum electron speed is at radius r_2, namely,

$$v_{min}^- = \mu_e^- \frac{V_0}{r_2} \ln\left[\frac{r_2}{r_1}\right]. \tag{9.130}$$

If the electron speed is much higher than the positive ion speed, as is usually the case, then the electrons are swept out much more rapidly than the ions, thereby contributing the most to the induced charge, as was found for the planar case. Just after the electrons have been collected $r_1' = r_1$ and the positive ions have moved little, i.e., $r_2' \simeq r_0$, the induced charge is, from Eq. (9.127),

$$\Delta Q(t_e) = q_e N_o \left[\ln\left(\frac{r_2}{r_1}\right)\right]^{-1} \ln\left(\frac{r_0}{r_1}\right) = q_e N_o \left[\ln(R_{21})\right]^{-1} \ln\left(R_{21}\frac{r_0}{r_2}\right), r_0 \geq r_1 \tag{9.131}$$

where R_{21} is the ratio r_2/r_1. Reversal of the bias polarity yields the induced charge just after the electrons

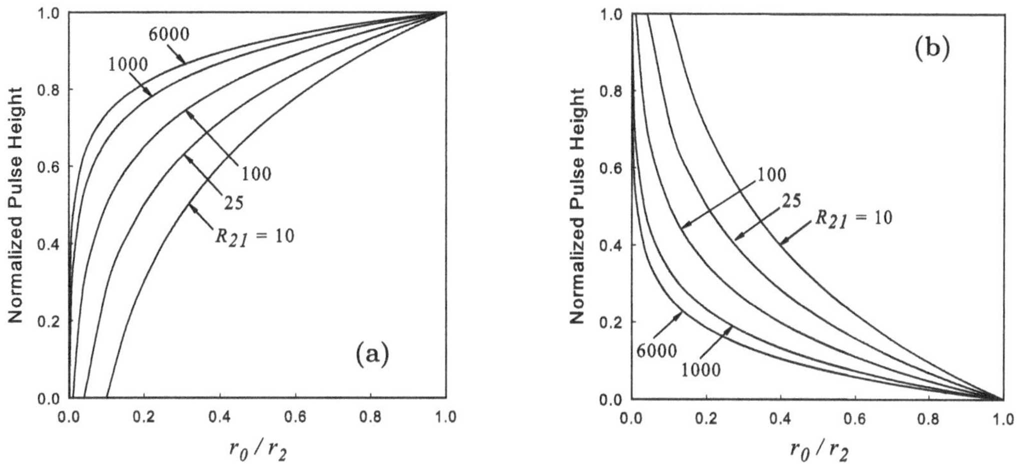

Figure 9.22. The normalized induced charge $\Delta Q/(qN_o)$ just after the electrons have been collected for a cylindrical ion chamber as a function of interaction location r_0 and ratio R_{21}. In (a) positive bias is applied to the inner electrode of radius r_1 and the outer electrode at r_2 is grounded. In (b) negative bias is applied to the inner electrode with outer electrode grounded. Applying positive voltage to the inner electrode has the clear advantage of producing larger signals over much of the detector volume.

have been collected

$$\Delta Q(t_e) = q_e N_o \left[\ln\left(\frac{r_2}{r_1}\right)\right]^{-1} \ln\left(\frac{r_2}{r_0}\right) = -q_e N_o \left[\ln(R_{21})\right]^{-1} \ln\left(\frac{r_0}{r_2}\right), r_0 \geq r_1. \quad (9.132)$$

The dependency of $V_{in}(t)$ on radiation interaction location, the ratio r_2/r_1 and the bias voltage polarity is shown in Fig. 9.22.

The predicted electron drift speed is complicated by the fact that low energy (thermal) electrons, with an average energy of $3/2kT$, lose very little energy when colliding with a heavy ion. Hence, as an electric field is applied to a gas, free electrons gain much higher energies than the average thermal energy. The ratio between the electron characteristic energy (ϵ_K) and the thermal energy is denoted by the ratio ξ. As result, the mobility of the electrons is not constant but is a function of the electric field. To develop the expected time-dependent pulse shape from a coaxial ion chamber, the following semi-empirical relation is adopted for the electron speed [Wilkinson 1950], namely,

$$v_e = C_1 \left(\frac{\mathcal{E}}{P}\right)^{1/2}. \quad (9.133)$$

Here C_1 is an experimentally determined constant. This expression is valid for low values of reduced electric fields.

The input pulse into the coupling electronics is

$$V_{in}(t) = \frac{\Delta Q(t)}{C}, \quad (9.134)$$

where C is the combined coupling and detector capacitance. Because the electron drift speeds are much greater than those of the positive ion, the charge induction between $0 \leq t \leq t_e$ is dominated by electron motion. From Eq. (8.44) for $r_2' \simeq r_0$ one has

$$\Delta Q(t) \simeq q_e N_0 \left[\ln(R_{21})\right]^{-1} \ln\left(\frac{r_0}{r_e(t)}\right), \quad (9.135)$$

where r_0 is the radial location of the initial ionization event and $r_e(t) = r_1'(t)$ is the electron cloud location at time t. The electron drift speed changes with radial location as

$$v_e(r) = -\frac{dr}{dt} = \frac{C_1}{P^{1/2}}\mathcal{E}(r)^{1/2} = \frac{C_1}{P^{1/2}}\left[\frac{V_0}{r\ln(R_{21})}\right]^{1/2}. \tag{9.136}$$

Integration of this result gives

$$-\int_{r_0}^{r-e(t)} r^{1/2}\, dr = \int_0^t \frac{C_1}{P^{1/2}}\left[\frac{V_0}{\ln(R_{21})}\right]^{1/2} dt \tag{9.137}$$

or

$$\frac{2}{3}\left(r_0^{3/2} - r_e^{3/2}(t)\right) = \frac{C_1}{P^{1/2}}\left[\frac{V_0}{\ln(R_{21})}\right]^{1/2} t. \tag{9.138}$$

The electron sweep out time t_e, i.e., when $r_e(t_e) = r_1$, is found to be

$$t_e = \left(r_0^{3/2} - r_1^{3/2}\right) \Big/ \frac{3}{2}\frac{C_1}{P^{1/2}}\left[\frac{V_0}{\ln(R_{21})}\right]^{1/2}. \tag{9.139}$$

Equation (9.138) can be solved for $r_e(t)$ and combining this result with Eqs. (9.135) and (9.134) yields the pulse height $V_{in}(t)$ as a function of time, namely,

$$V(t) = -\frac{qN_0}{C\ln(R_{21})}\left[\ln r_0 - \ln\left\{r_0^{3/2} - \frac{3}{2}\frac{C_1}{P^{1/2}}\left[\frac{V_0}{\ln(R_{21})}\right]^{1/2} t\right\}^{2/3}\right]. \tag{9.140}$$

The general voltage pulse height time response for coaxial ion chambers, with positive voltage applied to the central wire, is shown in Fig. 9.23. Although the coaxial ion chamber pulses are still dominated by electron motion, the coaxial geometry reduces the problem of widely varying position-dependent pulses that occurs in planar ion chambers.

9.5 Ion Chamber Designs

Ion chambers have been used as radiation detectors from the time of Rutherford's and Geiger's invention of the gas-filled radiation detector. Designs have been improved over the past 100 years to reduce leakage current, stabilize output, and reduce the dependence on the position of the ionizing event. A list of specialized ion chambers would be much too long to cover in the present work, but descriptions for many of these devices can be found in the literature [Rossi and Staub 1949; Staub 1953; Price 1964; Boag 1966, 1986; Attix 2004; Sauli 2014]. Presented here are several common designs encountered in radiation detection and measurement.

The profile of energy deposition in an ion chamber depends on the gas container, the type of gas and the gas pressure. Penetrating radiation, such as gamma-rays, interact readily with the chamber walls, producing energetic electrons (photoelectrons, Compton electrons, or, if $E_\gamma > 1.022$ MeV, electrons and positrons from pair-production). These secondary electrons, if sufficiently energetic, can escape the walls and enter the gas, thereby producing a cloud of electron-ion pairs. The response of the chamber to these secondary electrons can be altered by lining the chamber with materials having of varying Z to change the probability a gamma-ray interacts in the wall and the probability a secondary electron escapes into the gas. Gamma rays can also interact directly with the chamber gas. Increasing gas pressure and the Z number of the detection gas both increase the efficiency of the detector. Because ion chambers are usually designed to measure the electrons produced by ionization, except in special cases, the contribution of the ion motion is minor. Hence, the adverse effects that increasing gas pressure has on ion motion are of little consequence, and the electron speeds remain high.

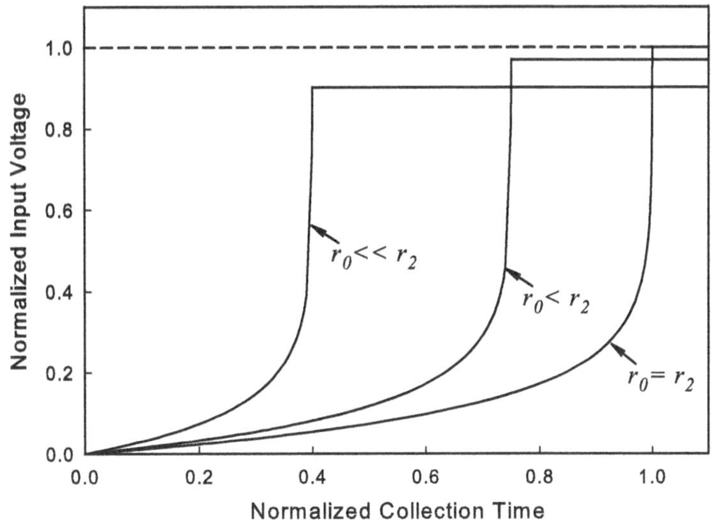

Figure 9.23. Normalized input voltage as a function of normalized time, where $t = 1$ is the drift time between r_2 and r_1 and $R_{21} = 500$. Shown are the cases in which a radiation interaction occurs at the cathode, and at two different locations within the detector where $r_0 < r_2$.

9.5.1 Basic Designs and Characteristics

Although an ion chamber is a simple concept, there are important design details that cannot be overlooked. For instance, the insulators must be selected to reduce leakage current during operation in a radiation field. The resistivity of the feedthroughs must be extremely high to reduce leakage current from moving around the device from anode to cathode.

Planar Ion Chamber Designs

A simple planar ion chamber configuration and its equivalent circuit are shown in Fig. 9.24. A voltage V_0 is applied across the chamber volume and load resistor R_L. Because the actual gas chamber resistance R_C is extremely high, almost all of the voltage is dropped across the chamber and very little across the load resistor. As a result, most of the voltage is also dropped across the insulating feedthroughs and the chamber body. The resistance of the feedthroughs and chamber are represented by R_F, and the leakage current is $I_L = V_0/R_F$. The leakage current can be reduced by using high resistance materials, thereby ensuring that R_F is large and I_L is small.

The detector input voltage becomes contaminated by the leakage current, becoming $V_{in} = (I_C + I_L)R_L$. The leakage current adds white noise and can produce an incorrect assessment of the radiation environment. To illustrate the problem, a common ion chamber operated at the modest voltage of 100 V requires the insulator feedthroughs have a combined resistance of 10^{16} Ω to keep the leakage current below 10^{-14} amp. By comparison, the background current produced by cosmic rays

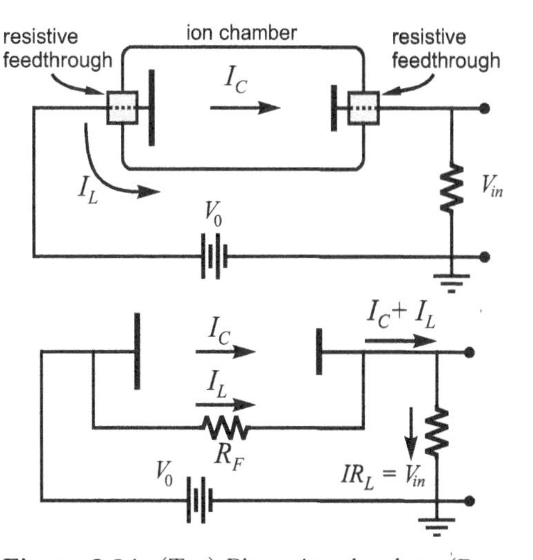

Figure 9.24. (Top) Planar ion chamber. (Bottom) The equivalent circuit model for the planar ion chamber.

is on the order of 10^{-16} amp. Common high resistivity materials used for electrical feedthroughs include Teflon ($\simeq 10^{25}$ Ω cm), quartz ($\simeq 7.5 \times 10^{19}$ Ω cm) and alumina ($\simeq 10^{16}$ Ω cm).

Although the volumetric resistivity of many candidate materials is sufficiently high to serve as feedthroughs, it is usually the surface leakage current that causes problems. The surface leakage is largely dependent on the cleanliness of the surface and the ability of the insulator to absorb moisture. Scratches or other surface damage can also increase surface leakage current. Many attractive high resistivity materials, such as Teflon, undergo resistivity changes as a result of radiation damage. If the ion chamber is to be operated in high radiation fields, materials to be avoided include plastics, Teflon and, under some circumstances, polystyrene. Many ceramic materials, such as alumina and quartz, appear to have much better radiation resistance, and therefore are frequently used as feedthroughs for ion chambers.

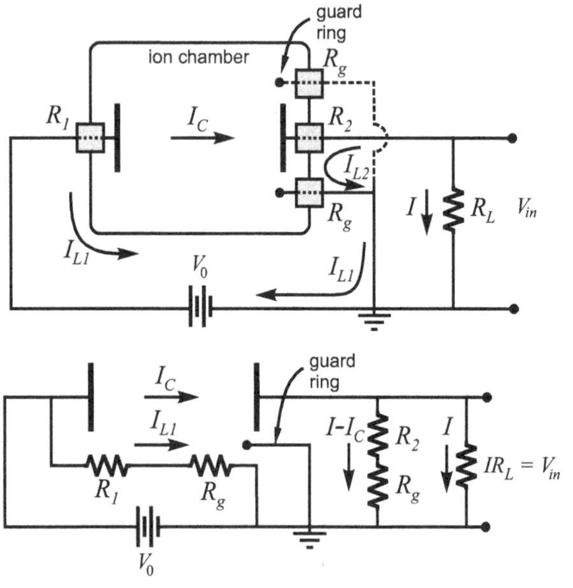

Figure 9.25. (Top) Planar ion chamber with a guard ring. (Bottom) The equivalent circuit model for the planar ion chamber with a guard ring.

An interesting consequence of using plastics and polystyrene, or other soft insulating materials, is the generation of *stress* currents. Stress currents arise from electron migration through the insulator when the material is under high voltage. Although not an issue for systems with detection currents greater than 10^{-13} amp, stress currents can become problematic for chamber currents less than 10^{-14} amp. Thus, the stress current can determine the lower limit of sensitivity for the ion chamber. Dielectric strength is a measure of the electric stress required to abruptly move substantial charge on or through a dielectric. Ceramic materials such as alumina and quartz appear to have low stress currents.

One method to reduce bulk and surface leakage through and around an insulator is to incorporate a guard ring into the detector. A planar ion chamber with a guard ring is depicted in Fig. 9.25, along with the equivalent circuit. The guard ring is placed around the collection electrode and grounded into the detector circuit, as shown in Fig. 9.25. The operating voltage is applied across the chamber, and leakage current I_{L1} flows through (or around) the resistive feedthroughs for the electrode, with resistance R_1, and guard ring, with resistance R_g. This leakage current I_{L1} bypasses the loading resistor R_L and instead returns to ground. Hence, the leakage current I_{L1} does not contribute to V_{in}.

The voltage drop across the guard ring insulator and the collection electrode insulator is determined by leakage current I_{L2}, as depicted in Fig. 9.25. However, because almost all voltage is dropped across the high resistance of the ion chamber, the actual voltage drop between the guard ring and the collection electrode is practically zero; hence, $V_{g-2} = I_{L2}(R_2 + R_g) \simeq 0$, meaning that the leakage current I_{L2} is practically zero as well. As a result, the measured detector current has little contamination from leakage currents.

Guard rings are often employed to define a specific region in the detector that is sensitive to radiation. Electrons and ions collected by the guard ring produce induced current that returns to ground without adding to $IR_L = V_{in}$. As a result, radiation interactions in the electric field region influenced by the guard ring go undetected. The free air ion chamber, discussed in a later section, is one such example of an ion chamber with a sensitive region well defined by a guard ring.

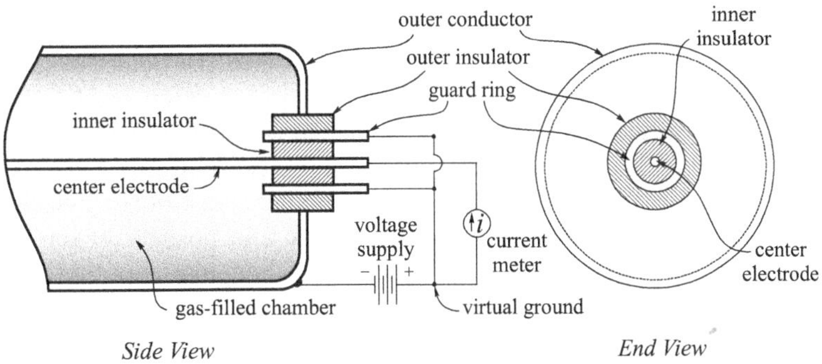

Figure 9.26. Depiction of a coaxial ion chamber with a guard ring.

Coaxial Ion Chamber Designs

Coaxial detectors operate mainly with a positive voltage applied to the central electrode. For different types of detectors, the anode diameter may range between 25 microns to a few hundred microns. Coaxial proportional counters and coaxial Geiger-Müller counters typically have central anodes between 25 to 100 microns in diameter for reasons discussed in Chapters 10 and 11.

Due to the cylindrical geometry, there is a higher probability that gamma rays interact in the larger volume of gas nearest the cathode than in the smaller volume near the anode. This effect is referred to as *geometrical weighting* [McGregor and Rojeski 1999]. As a result the coaxial design reduces the position dependence of the induced signal V_{in} that is transferred to the electronics, as shown in Fig. 9.22(a). Although negative voltage can be applied to the anode, the average voltage signal produced is considerably less, as shown in Fig. 9.22(b).

Guard rings are usually used in coaxial ion chambers as depicted in Fig. 9.26. Just as a guard ring is used in a planar chamber, the guard ring is fashioned so that the leakage current is diverted from adding to the detector signal. Detector leakage currents on the order of only 10^{-16} amp have been successfully realized in designs of coaxial ion chambers. However, variations in the insulator design and other irregularities can blur the edges of the active region in a coaxial ion chamber. Often a second guard ring is placed around the first guard ring, referred to as a *field tube*, to define better the active region of the detector. Because field tubes have a special importance to the operation of proportional counters, a discussion on field tubes is delayed until Chapter 10.

9.5.2 Gamma-Ray Ion Chamber Designs

Gamma-ray ion chambers, operated in the current mode, are stable, have long lives, and they can be fabricated in a variety of sizes and shapes. Large ion chambers are used as area monitors for ionizing radiation, and high-pressure chambers offer a relatively high sensitivity, permitting measurement of exposure rates as low as 1 μR/h. Small chambers with low gas pressures can be operated in radiation fields with exposure rates as great as 10^7 R/h.

Because of the lower energy deposition $(-dE/dx)$ of electrons compared to that of heavy charged particles, electrons produced by gamma-ray interactions in the gas often have ranges greater than the dimensions of an ion chamber. As a result, the energy deposition by electrons is usually less than that for heavy charged particles and, consequently, the resulting input voltage V_{in} is smaller for electrons. Obviously, ion chambers operating at relatively low gas pressures have lower efficiencies than those with higher pressures.

Sensitivity to gamma rays can be enhanced by lining the walls of the chambers with high Z materials to increase the interaction probability. Recall from Chapter 3 that the total interaction coefficient for gamma rays increases with the Z number. However, the electron range decreases with Z, thereby reducing the sensitive wall thickness for which secondary electrons can escape into the detector gas.

Air-filled ionization chambers vented to atmospheric pressure are often used to measure radiation exposure. After the formation of an electron-ion pair, the electrons and ions begin to move through the gas towards their collecting electrodes. However, the electronegative oxygen atoms quickly capture free electrons to produce heavy negative ions. Because of this electron attachment, most of the free electrons disappear and the chamber current is produced by the motion of both negative and positive heavy ions. This condition is not a problem when a detector is operated in the current mode as long as the potential difference is sufficient to prevent significant charge recombination. However, such a chamber is not suited for pulse-mode operation, because the positive and negative ions move slowly compared to electron speeds in the same electric field, consequently, very few and indistinct pulses are produced.

9.5.3 Neutron-Sensitive Ion Chambers

If an ion chamber is coated with a neutron reactive material with a large neutron cross section or filled with a neutron reactive gas, such that ionizing particles are released from the neutron reactions, the chamber can be used as a neutron detector. Commonly used isotopes for neutron detectors are ^{3}He, ^{10}B, ^{6}Li, and ^{235}U. Neutron sensitive ion chambers are usually filled with $^{10}BF_3$ or ^{3}He gas, or the inside walls of the chamber are coated with ^{10}B, ^{6}LiF, or ^{235}U. These gas-filled neutron detectors can be operated as both ion chambers or proportional counters.

Ion chambers that use ^{235}U are often referred to as fission chambers, since it is the fission fragments from the ^{235}U that ionize the chamber gas. Fission chambers are often used in mixed radiation fields containing large numbers of neutrons and gamma rays. Fission fragments deposit as much as 50 times the energy as do gamma rays in a common fission chamber. Hence, when operated in pulse mode, the voltage pulses formed by fission fragments are much larger than those formed by gamma rays, thereby making it possible to discriminate between the two radiations. Due to pulse pile-up, ion and fission chambers are generally not operated in pulse mode when in high radiation fields, although some special pulse mode designs incorporating ^{235}U are used for in-core nuclear reactor monitoring. Gas-filled neutron detectors are described with more detail in Chapters 17 and 18.

9.5.4 Compensated Ion Chambers

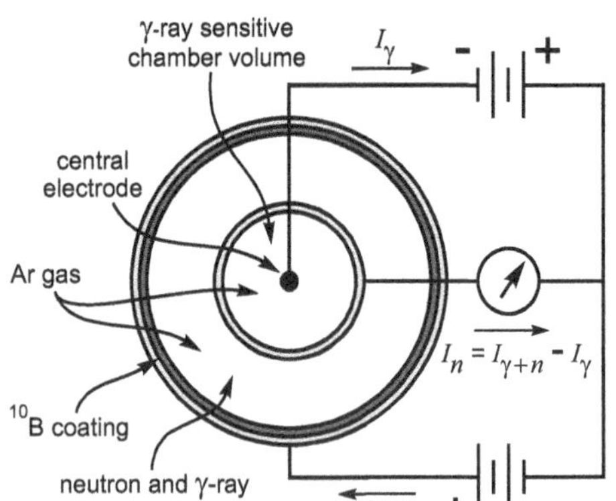

Figure 9.27. Cross section diagram of a concentric compensated ion chamber.

A form of ion chamber frequently used for control of nuclear reactors is known as the *compensated ion chamber.* Ion chambers, when operated in *current mode,* can be used in high radiation environments. If a gas-filled neutron detector is placed near a nuclear reactor, it responds to both neutrons and gamma rays. Yet, current mode operation does not permit pulse-height discrimination between neutron and gamma ray interactions as does pulse mode operation. The compensated ion chamber design is used to distinguish between the two types of radiation in high radiation fields.

Typically the chamber has two concentric electrodes and a central wire electrode. One concentric electrode is coated with a neutron sensitive material such as ^{235}U or a compound containing ^{10}B. The configuration depicted Fig. 9.27 allows both chambers to experience the same radiation field. Differences between the two chambers can be properly calibrated by adjusting the operating voltages. The ^{10}B (or ^{235}U) coated chamber is referred to as the *working chamber* and the uncoated chamber is referred to as the *compensating chamber*. When exposed to both gamma rays and neutrons, the voltage potential for the working chamber causes current to flow that deflects the current meter in one direction. The voltage potential in the compensating chamber, sensitive only to gamma rays, causes current to flow in the opposite direction. The voltage potentials on the chambers can be adjusted such that the two gamma ray signals exactly cancel. As a result, the compensated chamber produces a net current that is directly proportional to the strength of the (thermal) neutron field. Compensated ion chambers are widely used in nuclear reactors because of their ability to respond to neutron fields that vary up to ten orders of magnitude; i.e., these detectors have a very large *dynamic range*.

9.5.5 Frisch Grid Ion Chambers

An ion chamber design attributed to Otto Frisch [Frisch 1942; Bunemann et al. 1949] practically eliminates the dependence of the output pulse height on the position of the ionizing event in ion chambers, thereby allowing the detector to be used as an energy spectrometer. Shown in Fig. 9.28 is the basic arrangement for a Frisch grid or *gridded* ion chamber design. The gridded ion chamber is separated into two compartments: one is a much larger main chamber and the other is the measurement or detector chamber. The two chambers are separated by a conductive screen. Shielding can be arranged to confine radiation interactions to the main chamber. A voltage is placed across both chambers by use of a resistive divider. The electric field is applied such that the grid is positive with respect to the cathode, and the anode (the collecting electrode in the measurement compartment) is positive with respect to the grid.

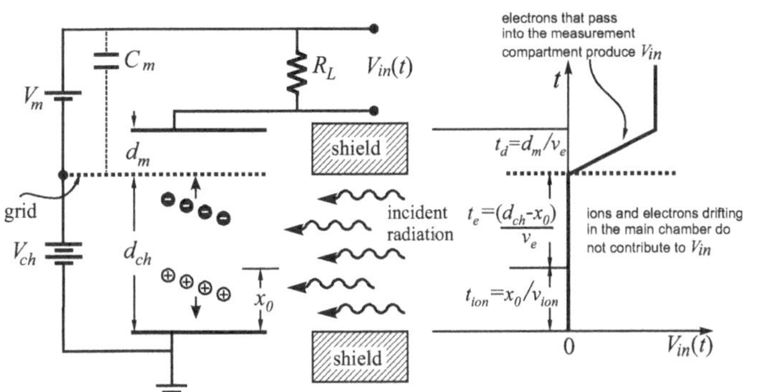

Figure 9.28. Depiction of a Frisch grid ion chamber.

Gamma rays, or other ionizing particles, that interact in the main chamber produce electron-ion pairs just as in any other ion chamber. The positive ions drift toward the cathode and electrons drift toward the grid. However, the main chamber is not connected directly to the input, and the induced current produced in the main chamber does not charge the coupling capacitance. The grid essentially shields the anode from induced current while charges move in the main chamber. As a result, charges moving in the main chamber compartment do not contribute to the voltage signal.

With sufficient speed, the electrons pass through the grid into the measurement compartment, which is connected to the output electronics. As the electrons drift through the measurement compartment, they induce an input voltage $V_{in}(t)$

$$V_{in}(t) = \frac{q_e N_0 v_e t_m}{d_m},$$

where t_m is the drift time across the measurement compartment and d_m is the length of the measurement compartment. If there is no electron attachment, the input voltage reduces to $q_e N_0$ after all electrons are collected. Note that the interaction position dependence has been eliminated from the input voltage. Further, because positive ions do not contribute to the signal, the slow signal component usually observed from ion drift is also eliminated, thereby reducing the effect of pulse height deficit.

9.5.6 Free Air Ion Chambers

A standard instrument used for radiation exposure measurements is the *free air ion chamber.* The traditional radiation exposure unit, the Roentgen, abbreviated R, is defined precisely as the quantity of x or gamma radiation that produces 2.58×10^{-4} coulomb of separated charge of either sign per kilogram of air in the incremental volume where the primary photon interactions occur. The SI exposure unit, the X unit, equals 3881 R, which is defined as the amount of x or gamma radiation that produces 1 Coulomb of charge per kg of air. Hence, by measuring the amount of ionization in air, the radiation exposure field can be determined. This task can be accomplished with a free air ionization chamber.

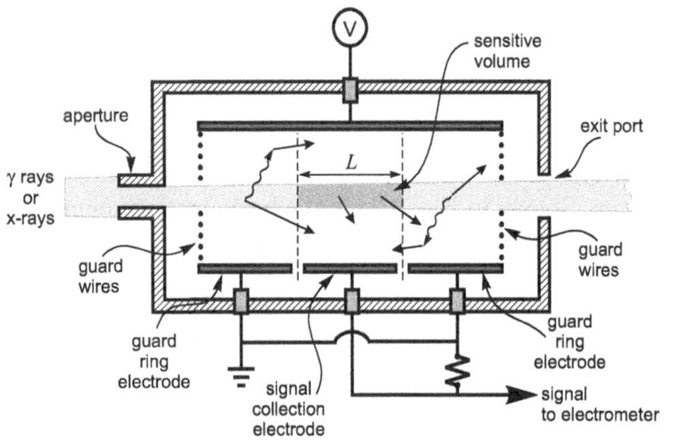

Figure 9.29. Free air ionization chamber configuration.

The chamber is constructed as shown in Fig. 9.29 with one electrode is segmented to form a guard ring structure. The chamber is shielded with lead and is filled only with air at ambient temperature. X or gamma rays enter through an aperture in the box, interact in the air, and whose reaction products then ionize the air. However, only the ionization formed in that region defined by the radiation aperture and the center electrode (of width L) is measured, thereby giving an air volume and the ionization. As a result, the radiation exposure can be determined.

Here the initial assumption is that ion recombination does not significantly affect measured results. However, the high density of O_2 indicates that significant electron attachment will occur in the chamber which can result in recombination losses. To reduce the effect of charge loss to recombination, the ionized volume is kept small, and it is the replenishment of ionized charges in the sensitive volume that indicates the radiation exposure.

Two important design features should be noted. First the interaction rate density along the photon beam inside the chamber must be relatively constant, and second the length of the beam on either side of the central portion of length L must be at least equal to the maximum distance in which ions are produced from the primary photon interaction location. Thus, ions produced outside of L by interactions in L are exactly compensated for by ions produced in L by photon interactions outside of L. This compensation is called *charged particle equilibrium* [Shultis and Faw 2002] and is central to radiation dosimetry.

Because of the need for charge particle equilibrium, the use of the free air ionization chamber is limited to photons with energies less than about 500 keV [Cember 1983]. Although larger units would allow measurements of radiation exposure for higher energy gamma rays, such chambers appear to be impractically large and would require much more extensive shielding.

Example 9.5: A free air ionization chamber is operated at 20°C with air pressure of 750 torr. The chamber aperture diameter is 1 cm and the signal collection electrode is 4 cm in length. At 150 volts operating bias, a current of 12 nA is measured. What is the radiation exposure rate?

Solution:

As discussed earlier, electron attachment is severe in air; hence, almost all electrons undergo electron attachment before collection and, thereby produce heavy negative ions. However, because the exposed volume is small compared to the chamber size, the positive and negative ions are quickly separated so that it is assumed that few ions recombine.

The pressure and temperature must be corrected for the standard conditions of radiation exposure rate. Hence,

$$\frac{dX}{dt} = \frac{I}{\rho_0 V_{ch}} \frac{P_0}{P} \frac{T}{T_0}$$

where V_{ch} is the sensitive volume of the chamber, ρ_0 is the air density, and P_0 and T_0 are the standard pressures (in torr) and absolute temperatures.

The sensitive volume is $V_{ch} = \pi(0.5)^2(4)\ \text{cm}^3 = 3.14\ \text{cm}^3$. The density of air at standard conditions is $1.2922 \times 10^{-6}\ \text{kg/cm}^3$. The resulting exposure rate is

$$\frac{dX}{dt} = \frac{1.2 \times 10^{-8}\text{A}}{(1.2922 \times 10^{-6}\ \text{kg/cm}^3)(3.14\ \text{cm}^3)} \left(\frac{760\ \text{torr}}{750\ \text{torr}}\right)\left(\frac{293.15\ \text{K}}{273.15\ \text{K}}\right)$$

$$= 3.214 \times 10^{-3}\ \text{X-units/sec} = 12.47\ \text{R/sec}$$

9.5.7 Pocket Ion Chambers

Compact ion chambers, often referred to as *pocket ion chambers*, are a convenient method to measure exposure to radiation. There are fundamentally two types of pocket ion chambers, one is a condenser-type chamber and the other an electroscope chamber, both shown in Fig. 9.30. The condenser-type ion chamber is a form of air-wall ion chamber, in which the air volume is relatively small, usually on the order of 2 cm^3, with walls composed of "air-equivalent" material. The condenser pocket ion chamber operation is quite simple. Coaxial in form, the outer portion is composed of electrically conducting plastic with a gamma-ray response equivalent to pressurized air. A central wire is attached inside the chamber through insulating standoffs, and is positively charged with respect to the outer wall. These chambers are typically backfilled with air. As radiation interactions occur within the chamber walls, the resulting ionization discharges the condenser such that the stored voltage decreases. The decrease in voltage is directly correlated to the energy deposited within the chamber, thereby yielding a measure of the radiation exposure. The condenser pocket ion chamber is conveniently simple, but must be read from a charging device. The sensitivity of common condenser pocket ion chambers ranges from 25 mR to 250 R.

The electroscope pocket ion chamber is closely related to the gold leaf electroscope. These devices are commonly used to measure relative radiation exposure for the wearer. The device consists of a tube approximately 10 cm long and 1 cm in diameter. Inside the tube are two small metal-coated quartz fibers, each approximately 4 microns in diameter. One of the quartz fibers is stationary and the other is hinged, and both are inside an air cavity of the tube. The tube also has viewing optics such that when held up to a light, the observer can see the shadow of the hinged quartz fiber against a display scale. The device is inserted into a power supply and charged so that the two fibers, having like charges, are repulsed apart. This charged-up device can now be worn as a dosimeter. Electrons produced by ionization caused by radiation interactions in the chamber are attracted to the quartz fibers, thereby reducing the net charge and causing the quartz fibers to move closer together. The change in location of the hinged fiber yields the change in exposure to the wearer, which is determined by the change in location of the fiber with respect to the display scale in the tube (maximum scale is usually 200 mR).[14] The clear advantage of the electroscope pocket ion chamber is the convenient method of viewing the exposure. However, interpretation of the exposure is visually left to the user rather than a measuring instrument.

[14]The Federal Civil Defense Administration (FCDA) had millions of pocket ion chambers produced during the 1950s and 1960s for civilian use. By far the most popular pocket ion chamber, the V-742, was read in units of R with a maximum range of 200 R. Some of the units ranged up to 600 R (V-746), although the production run was limited to only a few hundred units.

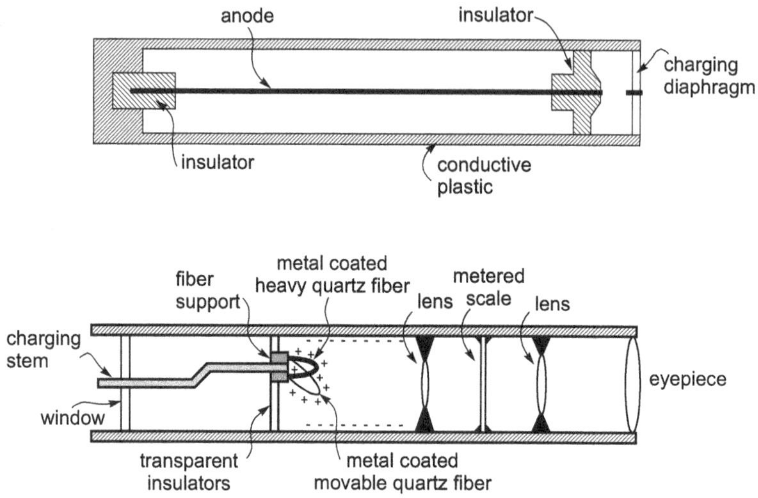

Figure 9.30. The basic components of a pocket ion chambers. (Top) Depiction of a condenser-type pocket ion chamber. (Bottom) Depiction of an electroscope pocket ion chamber.

9.5.8 Cloud Chambers

Cloud chambers are a type of ion chamber, but certainly not in the traditional electronic sense. Developed by Charles Thomson Rees Wilson in 1911 (see [Wilson 1912]), the initial devices operated on the principle of cooling by adiabatic expansion. In the original device, an ionizing radiation source was placed inside an air-filled cylinder outfitted with a tightly sealed plunger. With the plunger in a nearly closed position, the air in the cylinder compartment was saturated with water vapor. When the plunger is rapidly retracted, the rapid decrease in temperature and pressure causes the water vapor to condense, particularly upon ions in the chamber. The condensation upon these ions makes the ionization tracks visible to the eye. The effect is near immediate and quite dramatic. Although the cloud chamber is a simple device, it led to numerous discoveries, including the discovery of the positron (1932), the muon (1936), and the kaon (1947).[15] The cloud chamber works best with electro-negative gas molecules, for which electron attachment is rapid, thereby retarding the diffusion of the negative ions and prolonging the visual effect [Sitar et al. 1993].

Expansion Cloud Chambers

Consider the expansion cloud chamber depicted in Fig. 9.31(a). The chamber is filled with a non-condensible gas that adequately follows the ideal gas law. A vapor, water for instance, is added to the chamber such that it reaches saturation. Initially, the partial pressure of the gas is P_{g1} and the partial pressure of the vapor is P_{v1}. If the total mass of the vapor is m_{v1} at absolute temperature T_1 and chamber volume V_1, then,

$$P_{v1}V_1 = \frac{m_{v1}}{M}RT_1 \tag{9.141}$$

where M is the molecular weight of the vapor molecules and R is a constant. If the piston is rapidly retracted to cause a sudden expansion in the closed system, in which the volume V_1 is expanded to V_2, then the temperature immediately reduces from T_1 to T_2'. For an ideal gas, undergoing a reversible adiabatic

[15]For his invention of the cloud chamber, Charles Thomson Rees Wilson shared the 1927 Nobel prize in physics with Arthur Holly Compton.

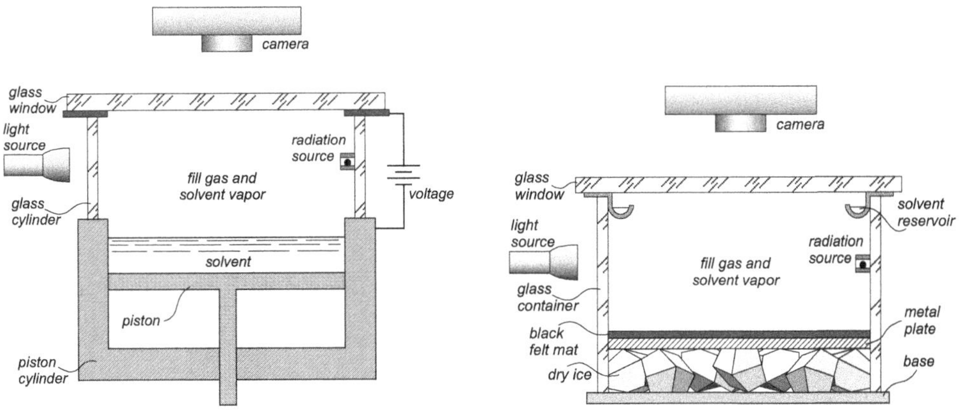

Figure 9.31. (a) Basic components of an expansion cloud chamber and (b) basic components of a diffusion cloud chamber.

process [Das Gupta and Ghosh 1946],

$$\frac{T_1}{T_2'} = \left(\frac{V_2}{V_1}\right)^{\gamma-1}, \tag{9.142}$$

where γ is the adiabatic index[16] for the *gas mixture*

$$\gamma = \frac{C_P}{C_V}. \tag{9.143}$$

Here C_P is the specific heat capacity at constant pressure and C_V is the specific heat capacity at constant volume. Immediately after the expansion, but before condensation takes place, the new vapor partial pressure is

$$P_2'V_2 = \frac{m_{v1}}{M}RT_2', \tag{9.144}$$

where the vapor mass is the same as before, but now distributed over the new volume V_2. Yet, this state is short lived, because the cooling causes part of the water vapor to condense, thereby changing the partial pressure, vapor mass and the temperature (slightly). As equilibrium is reached, the temperature rises slightly from T_2' to T_2, caused by the condensation of vapor. The equilibrium condition is

$$P_2V_2 = \frac{m_{v2}}{M}RT_2, \tag{9.145}$$

where m_{v2} is the equilibrium vapor mass. The density of the vapor immediately after the piston is retracted is represented by

$$\rho_2' = \frac{m_{v1}}{V_2} \tag{9.146}$$

and after the system has reached equilibrium, the density of the vapor is

$$\rho_2 = \frac{m_2}{V_2}. \tag{9.147}$$

[16] Also called the heat capacity ratio or ratio of specific heats.

The amount of supersaturation can be defined as the ratio of the densities of ρ_2' to ρ_2, i.e.,

$$S = \frac{\rho_2'}{\rho_2} = \frac{m_{v1}}{m_{v2}} = \frac{P_1 V_1 T_2}{P_2 V_2 T_1}. \tag{9.148}$$

Dividing Eq. (9.144) by Eq. (9.145) also yields,

$$S = \frac{P_2' T_2}{P_2 T_2'}. \tag{9.149}$$

If the temperature rise from condensation is small, then $T_2' \simeq T_2$, and with the substitution of Eq. (9.142) into Eq. (9.148) one obtains [Das Gupta and Ghosh 1946]

$$S \simeq \frac{P_1}{P_2}\left(\frac{V_1}{V_2}\right)^{\gamma} = \frac{P_1}{P_2}\left(\frac{1}{F_C}\right)^{\gamma}, \tag{9.150}$$

where F_C is the expansion ratio. Note that the supersaturation is strongly dependent upon γ and the partial vapor pressure P_2. The required expansion ratio to produce a specific amount of supersaturation decreases with increasing γ. Smaller expansion ratios are desirable because the lower amount of volumetric disturbance reduces gas turbulence that can blur the ionization tracks.

Figure 9.32. Ionization tracks produced by alpha particles in a cloud chamber [http://klarifi.blogspot.com/search/label/nuclear Dec. 2013]

The rapid decrease in gas temperature causes a supersaturation of the water vapor, which enables the water to condense on select surfaces. If the vapor pressure of small water droplets is lower than the supersaturation, the droplets then grow in size, and continue to increase in size until the vapor pressure of the droplets corresponds to the supersaturation condition. As the drops are forming, thermal conduction is heating the chamber gas and reducing the degree of supersaturation, which rapidly returns the gas mixture to the equilibrium condition. The time duration over which the supersaturation conditions permit the formation of visible tracks is relatively short, usually lasting between a few milliseconds to a few seconds.

Viewing is best performed from the top of the chamber through a clear glass window (see Fig. 9.31(a)). Gravity pulls the vapor tracks downward, and diffusion broadens the tracks. Together these two effects serve to blur images if viewed from the side; however, if viewed from the top, blurring from gravity is kept to a minimum, as shown in Fig. 9.32. There is also the possibility of producing turbulence in the gas when the piston is withdrawn, causing additional distortions in the ionization tracks.

Competing condensation surfaces include the chamber walls and foreign matter in the chamber. It can be shown that small particle surfaces have higher vapor pressures than smooth planar surfaces [Staub 1953]; hence operating the system with optimized expansion ratios reduces (or eliminates) surface condensation while permitting condensation on neutral and charged particles. The system can be cycled a few times to reduce the background contamination, such as dust, from the system. Dust particles and other contaminants appear as a faint fog in the chamber when first operated, but eventually drop into the solvent solution from gravity, thereby being cleared from the gas after a few cycles. Tracks formed from ions can be cleared from the system by temporarily applying a voltage across the chamber volume, which resets the system for the next expansion operation.

Diffusion Cloud Chambers

A simpler method of achieving the needed supersaturation condition is used in the diffusion cloud chamber [Wilson 1951]. Depicted in Fig. 9.31(b), the diffusion cloud chamber operates with a temperature gradient from the top to the bottom of the chamber. The chamber bottom is filled with dry ice and covered with a metal plate. Typically a black felt or velvet mat is placed over the metal plate to improve viewing contrast. A reservoir of solvent is placed near the top of the chamber. The entire chamber is covered with a transparent window, usually glass, although modern cloud chambers have plastic lids. Often the lid is slightly warmed to produce a thermal gradient from the lid to the chilled metal plate. The solvent vaporizes and diffuses throughout the chamber, thereby causing the supersaturation condition to appear in the lower temperature region of the chamber.

Diffusion cloud chambers are commonly operated with air as the gas and with either methanol or ethanol as the solvent. These gases and vapors have good electron attachment, facilitating the formation of heavy negative ions, which helps with the formation of visible particle tracks.

Advantages of the diffusion cloud chamber design include the continuous mode of operation and the lack of moving parts, thereby reducing gas turbulence. However, the radiation source must be placed in a location above the region of condensation or condensate coats the source and attenuates the ionizing radiation. For alpha particle sources, the condensate can completely block the alpha particles from emerging into the chamber. Because of its simplicity, diffusion cloud chambers are often used as demonstration devices for classroom instruction.

9.5.9 Smoke Detector Ionization Chambers

A typical form of smoke detector commonly found in a household environment is actually a small free air ionization chamber with an embedded ^{241}Am source that emits alpha particles. It can detect particles of smoke that are too small to be visible. The alpha particles produce ionization in the tiny ion chamber, which consists of an air-filled space between two electrodes, and permits a small, constant current to flow between the electrodes. Smoke entering the chamber absorbs and neutralizes the alpha particles, thereby reducing the alpha-particle induced ionization and current flow. This reduction in current sets off the alarm. Hot air entering the chamber can change the rate of ionization, thereby altering the current and setting off the alarm.

9.6 Summary

There are five distinct regions of operation for a gas-filled detector, the first two appearing at the lowest voltages, identified as the recombination and ion chamber regions. At low applied voltages, charge pairs may recombine, consequently nullifying, at least partially, the charge current induction effect necessary for signal formation. The recombination probability is a strong function of the electron attachment coefficient, the charge carrier diffusion coefficient, the gas density (pressure), and the applied voltage. At high enough voltages, the free charges can be collected at the detector electrodes with reasonably high efficiency. Such voltages are identified as the ionization chamber region of operation. Ion chambers can be operated in current mode or pulse mode. Current mode operation is usually employed for relatively high radiation fields, while pulse mode operation is usually employed for relatively low count rates ($< 5 \times 10^4$ cps). Under pulse mode operation, the signal from a ion chamber is strongly dependent on the electron motion and less so on the heavy ion motion. Ion chambers can be acquired in a variety of configurations and are often application specific.

PROBLEMS

1. A planar ion chamber with spacing between the electrodes of 3 cm is biased at 600 volts. The backfill gas is P-10 filled at 1 atm. Unfortunately, the P-10 is contaminated with 0.04% O_2. Determine the maximum fractional electron losses due to electron attachment for the detector.

2. Verify the general solution of Eqs. (9.79) and (9.80).

3. Use the boundary conditions to obtain solutions of Eqs. (9.81) and (9.82).

4. Verify the approximations of Eqs. (9.83) and (9.84), and then use these results to obtain Eqs. (9.85).

5. An ion chamber backfilled with air at 1 atm is operated at 200 volts. The electrode spacing is 4 cm. The time constant of the detector is much less than the collection time for ions, but much longer than the collection time for electrons. If a 5.5 MeV alpha particle enters into the chamber center parallel to the electrodes and deposits all energy in the gas, what is the total value of free charge measured?

6. A planar ion chamber with electrodes spaced 2 cm apart is backfilled with 1 atm air at 20°C. Both anode and cathode electrodes are square, with area 4 cm^2 each. The chamber is then fully irradiated with a one curie (Ci) 200-keV gamma-ray source at a distance of 25 cm from the detector. At what voltage is the 90% saturation current reached?

7. Given the case in which $v_e \gg v_{ion}^-$, derive Eq. (9.126).

8. A free air ion chamber is operated at 200 volts.

   The properties of the chamber include the following:
   - Electrode separation = 5 cm
   - Aperture diameter = 1.5 cm
   - Anode/cathode width = 2.5 cm

   The operating conditions include:
   - Temperature = 82°F
   - Pressure = 1.05 atm

   The operating response = 0.0095 μA

   (a) What is the exposure rate from the radiation entering the chamber?

   (b) The operator was in the radiation environment for 15 minutes. What is the integrated exposure in X-units and roentgens?

9. The operating voltage of the free air ion chamber in Example 9.5 is increased to 200 volts with a resulting current of 1.3×10^{-8} A. What is the corrected exposure as compared to the result of Example 9.5?

10. A pressurized condenser type pocket ion chamber has a chamber capacitance of 100 pF. When fully charged, the chamber registers a voltage of 100 volts. The volume of the device is 10 cm^3, backfilled with air at a pressure of 3 atm. After a tour of a radiation facility, the wearer of the ion chamber reads a change in voltage of -4 volts. What was the radiation exposure?

REFERENCES

ATTIX, F.H., "Introduction to Radiological Physics and Radiation Dosimetry," New York: Wiley, 2004.

BLOCH, F. AND N.E. BRADBURY, "On the Mechanism of Unipolar Electron Capture," *Phys. Rev.*, **48**, 689–695, (1935).

BOAG, J.W., "Ionization Chambers," Ch. 9, in *Radiation Dosimetry*, Vol. II, F. ATTIX and W.C. ROESCH, eds., New York: Academic Press, 1966.

BOAG, J.W., "Ionization Chambers," Ch. 3, in *Dosimetry of Ionizing Radiation*, Vol. II, K.R. KASE, B. BJARNGARD, AND F. ATTIX, eds., New York: Academic Press, 1986.

BORTNER, T.E., G.S. HURST, AND W.G. STONE, "Drift Velocities of Electrons in Some Commonly Used Counting Gases," *Rev. Sci. Instrum.*, **28**, 103–108, (1957).

BUCKMAN, S.J. AND M.T. ELFORD, "Momentum Transfer Cross Sections," in *Electron Collisions with Atoms*, Vol. 17A of series *Interaction of Photon and Electrons with Atoms,* Y. ITIKAWA, Ed., Berlin: Springer, 2000.

BUNEMANN, O., T.E. CRANSHAW, AND J.A. HARSHAW, "Design of Grid Ionization Chambers," *Can. J. Research*, **A27**, 191–206, (1949).

CEMBER, H., *Introduction to Health Physics*, New York: Pergamon, 1983.

CHRISTOPHOROU, L.G., *Atomic and Molecular Radiation Physics*, New York: Wiley, 1971.

CLAY, J., "The Absolute Value of Cosmic-Ray Ionization at Sea Level in Different Gases," *Rev. Mod. Phys.*, **11**, 123–127, (1939).

COMPTON, K.T. AND I. LANGMUIR, "Electrical Discharges in Gases; Part 1. Survey of Fundamental Processes," *Rev. Mod. Phys.*, **2**, 123–242, (1930).

COTTRELL, T.L. AND I.C. WALKER, "Drift Velocities of Slow Electrons in Polyatomic Gases," *Trans. Faraday Soc.*, **61**, 1585–1593, (1965).

DIEBNER, K., "Über die Kolonnenionization einzelner α-Strahlen," *Annalen der Physik*, **402**, 947–982, (1931).

DUNN, W.L. AND J.K. SHULTIS, "Exploring Monte Carlo Methods," San Diego, CA: Academic Press, 2012.

EINSTEIN, A., "Über die von der molekularkinetischen Theorie der Wärme geforderte Bewegung von in ruhenden Flüssigkeiten suspendierten Teilchen," *Annalen der Physik*, **322**, 549-560, (1905).

FERREIRA, Ix-B.G., J.G. HERRERA, AND L. VILLASEÑOR, "The Drift Chambers Handbook; Introductory Laboratory Course (Based on, and Adapted from, A.H. Walenta's Course Notes)," *J. Phys. Conf. Ser.*, **18**, 346-361, (2005).

FRISCH, O.R., "Isotope Analysis of Uranium Samples by Means of Their α-Ray Groups," British Atomic Energy Report, BR-49, 1942.

GREENING, J.R., "Saturation Characteristics of Parallel-Plate Ionization Chambers," *Phys. Med. Biol.*, **9**, 143–154, (1964).

DAS GUPTA, N.N. AND S.K. GHOSH, "A Report on the Wilson Cloud Chamber and Its Applications in Physics," *Rev. Mod. Phys*, **18**, 225–294, (1946).

HERZENGBERG, A., "Attachment of Slow Electrons to Oxygen Molecules," *J. Chem. Phys.*, **51**, 4942–4950, (1969).

HUK, M., P. IGO-KEMENES, AND A. WAGNER, "Electron Attachment to Oxygen, Water, and Methanol in Various Drift Chamber Gas Mixtures," *Nucl. Instrum. Meth.*, **A267**, 107–119, (1988).

HUXLEY, L.G.H. AND R.W. CROMPTON, *The Diffusion and Drift of Electrons in Gases*, New York: Wiley, 1974.

ITIKAWA, Y., "Cross Sections for Electron Collisions with Nitrogen Molecules," *J. Phys. Chem. Ref. Data*, **35**, 31–56, (2006).

JAFFÉ, G., "Zur Theorie der Ionisation in Kolonnen," *Annalen der Physik*, **347**, 303–345, (1913).

JAHNKE, E. AND F. EMDE, *Tables of Functions with Formulae and Curves*, New York: Dover, 1945.

KORFF, S.A., *Electron and Nuclear Counters—Theory and Use*, New York: Van Nostrand Co., 1946.

LOEB, L.B., "Properties of Carriers of Free Electricity in Gases," *International Critical Tables of Numerical Data, Physics, Chemistry and Technology*," Vol. VI, New York: McGraw-Hill, 1929.

LOEB, L.B., *Fundamental Processes of Electrical Discharge in Gases*, New York: Wiley, 1939.

LOEB, L.B., *Basic Processes of Gaseious Electronics*, Berkeley, CA: University of California Press, 1960.

MATTERN, D., *Bestimmung des Einflusses von Driftparametern auf die Signalform einer TEC (Time Expansion Chamber)*, PhD Thesis, University of Siegen, 1988.

MCGREGOR, D.S. AND R.A. ROJESKI, "Performance of Geometrically Weighted Semiconductor Frisch Grid Radiation Spectrometers," *IEEE Trans. Nucl. Sci.*, **NS-46**, 250–259, (1999).

MIE, G., "Der Elektrische Strom in Ionisierter Luft in Einem Ebenen Kondensator," *Ann. Physik*, **318**, 857–889, (1904).

MORSE, P.M., W.P. ALLIS, AND E.S. LAMAR, "Velocity Distributions for Elastically Colliding Electrons," *Physical Review*, **48**, 212–219, (1935).

NERNST, W., "zur Kinetik der in Lösung befindlichen Körper 1.Theorie der Diffusion," *Z. Phys. Chem.* **2**, 613–637, (1888).

O'KELLY, L.B., G.S. HURST, AND T.E. BARTNER, "Measurement of Electron Attachment in Oxygen-Methane and Oxygen Carbon-Dioxide Mixtures," ORNL Report 2887, Oak Ridge National Laboratory, Oak Ridge, TN, Jan. 11, 1960.

ONSAGER, L., "Initial Recombination of Ions," *Phy. Rev.*, **54**, 554–557, (1938).

PALLADINO, V. AND B. SADOULET, "Application of Classical Theory of Electrons in Gases to Drift Proportional Chambers," *Nucl. Instrum. Meth.*, **128**, 323–335, (1975).

PEISERT, A., AND F. SAULI, *Drift and Diffusion of Electrons in Gases: A Compilation*, CERN 84-08, Geneva, 1984.

PHELPS, A.V., "Tabulations of Collision Cross Sections and Calculated Transport and Reaction Coefficients for Electron Collisions with O_2," *JILA Information Center Report*, **28**, JILA, Boulder, CO, 1985.

PRICE, W.J., *Nuclear Radiation Detection*, 2nd. Ed., New York: McGraw-Hill, 1964.

RICE-EVANS, P., *Spark, Streamer, Proportional and Drift Chambers*, London: The Richelieu Press, 1974.

ROSSI, B.B. AND H.H. STAUB, *Ionization Chambers and Counters*, New York: McGraw-Hill, 1949.

RUTHERFORD, F.R.S. AND H. GEIGER, "An Electrical Method of Counting the Number of α-Particles from Radio-active Substances," *Proc. of the Royal Society of London A*, **81**, 141–161, (1908).

SAULI, F., *Principles of Operation of Multiwire Proportional and Drift Chambers*, CERN 77-09, Geneva, 1977.

SAULI, F., *Gaseous Radiation Detectors*, Cambridge: Cambridge University Press, 2014.

SCHULTZ, G. AND J. GRESSER, "A Study of Transport Coefficients of Electrons in Some Gases Used in Proportional and Drift Chambers," *Nucl. Instrum. Meth.*, **151**, 413–431, (1978).

SHARPE, J., *Nuclear Radiation Detectors*, London: Methuen & Co., 1964.

SHULTIS, J.K. AND R.E. FAW, *Radiation Shielding*, La Grange Park, IL: American Nuclear Society, 2002.

SITAR, B., G.I. MERSON, V.A. CHECHIN, AND YU.A. BUDAGOV, *Ionization Measurements in High Energy Physics*, Springer Tracts in Modern Physics, Vol. 124, Berlin: Springer-Verlag, 1993.

STAUB, H.H., "Detection Methods," Ch. 1 in *Experimental Nuclear Methods*, Vol. 1, E. SEGRÉ, Ed., New York: Wiley, 1953.

VON SMOLUCHOWSKI, M., "Zur kinetischen Theorie der Brownschen Molekularbewegung und der Suspensionen," *Annalen der Physik*, **326**, 756-780, (1906).

TOWNSEND, J.S., "The Diffusion of Ions in Gases," *Phil. Trans. of the Royal Society A*, **193**, 129–151, (1900).

WILKINSON, D.H., *Ionization Chambers and Counters*, Cambridge: University Press, 1950.

WILSON, C.T.R.,"On an Expansion Apparatus for making Visible the Tracks of Ionising Particles in Gases and some Results Obtained by its Use," *Phil. Trans. Royal Soc.* **A87**, 277-292, (1912) .

WILSON, J.G., *The Principles of Cloud-Chamber Technique*, Cambridge: University Press, 1951.

ZANSTRA, H., "Ein Kurzes Verfahren zur Bestimmung des Sättigungsstromes nach der Jaffé'schen Theorie der Kolonnenionisation," *Physica*, **2**, 817–824, (1935).

Chapter 10

Gas-Filled Detectors: Proportional Counters

I've worked on many detectors, some were very elegant and useless, and didn't have a Nobel Prize, so this one was not the most elegant, but it was useful.

Georges Charpak[1]

10.1 Introduction

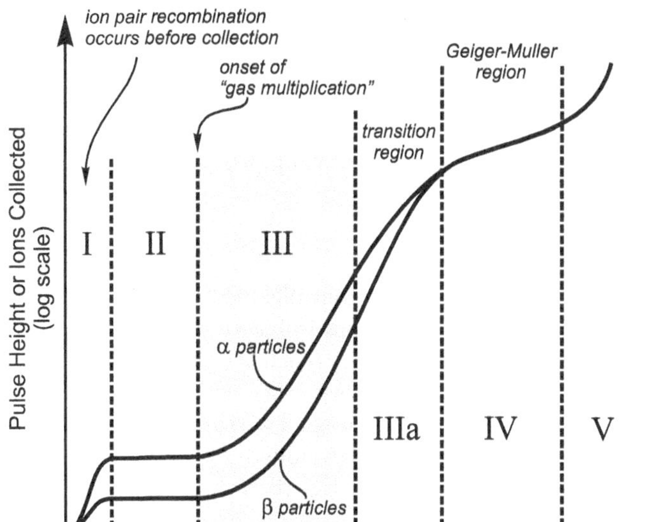

Figure 10.1. Proportional counters are operated in region III of the pulse height curve for gas-filled detectors.

Proportional counters are operated in region III of the pulse height curve for gas detectors (see Fig. 10.1). In this region an electronic pulse produced by ions moving through the detector is proportional to the original energy absorbed in the detector gas by a radiation particle, be they charged particles, neutrons, gamma rays or x rays. Although the gas-flow proportional counter was invented by John A. Simpson, Jr., in the mid 1940s [Simpson 1950a, 1950b], the actual effect of pulse height proportionality was known from those initial experiments conducted by Rutherford and Geiger with their gas-filled chambers. Argon is the most commonly used gas in a proportional counter, although there are many other gases that can be used, such as ^{3}He, Xe, and $^{10}BF_3$.

As with the ion chamber, a radiation particle can interact with the chamber, either directly with the chamber gas or with the chamber walls. If a gamma ray interacts with the chamber wall and a primary or secondary energetic electron escapes into the gas volume, this electron produces a cloud of electron-ion pairs. If the gamma ray interacts directly with the gas, then the energetic electron produced by the interaction also produces a cloud of electron-ion pairs. In either case, a cloud of electron-ion pairs is formed in which the total number of

[1]Received the Nobel Prize in physics in 1992 for inventing the multiwire proportional counter.

electron-ion pairs produced is proportional to the radiation energy deposited in the detector gas. Hence, by measuring the number of electron-ion pairs formed, the amount of energy deposited in the gas can be determined. Application of a voltage across the chamber and measurement of the current produced as the electrons and ions drift through the chamber volume allows the number of electron-ion pairs to be measured. Yet, as with the ion chamber, such a current can be miniscule and hard to measure.

10.2 General Operation

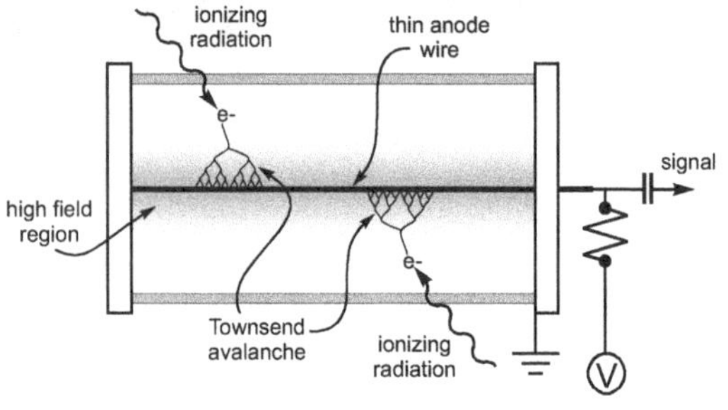

Figure 10.2. With a high electric field near the anode of a gas-filled detector, the signal is amplified by impact or Townsend avalanching, often referred to as "gas multiplication".

At sufficiently high voltages, electrons can gain enough kinetic energy or speed to produce more ionization of the gas, an effect called *impact ionization.* These newly liberated electrons can themselves gain enough energy from the electric field to cause even more ionization (depicted in Fig. 10.2). The process continues until the electrons are collected at the anode. The entire process of generating the impact ionization cloud is called a *Townsend avalanche*, or sometimes *gas multiplication.* There is a threshold electric field, $\mathcal{E}_t$, required for avalanche multiplication, below which the electrons do not gain sufficient energy to cause impact ionization. This threshold defines the difference between Region II and Region III in the gas pulse height curve of Fig. 10.1.

Although parallel plate electrodes work for ion chambers, they are seldom used for proportional counters. The preferred geometry is a coaxial configuration, as depicted in Fig. 10.3. To see why, compare the difference in electric fields between coaxial and parallel plate geometries.

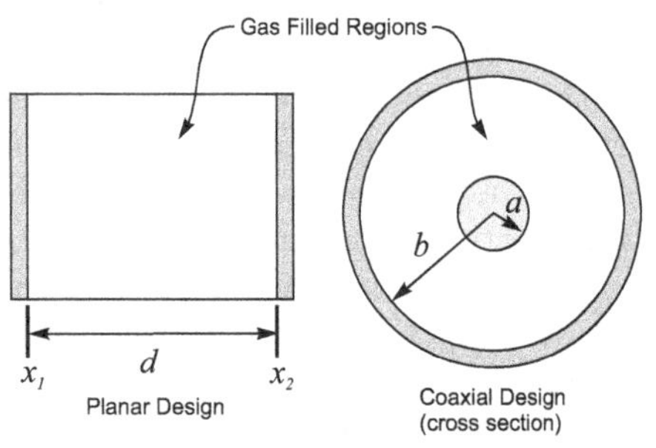

Figure 10.3. Planar and coaxial geometries are often used for gas-filled radiation detectors.

The parallel plate detector configuration is shown in Fig. 10.3. A voltage V_o is applied to the electrode at $x = x_1$ and zero (grounded) at $x = x_2$. The electric field at any position between the electrodes is

$$\mathcal{E}(x) = \frac{V_o}{x_2 - x_1} = \frac{V_o}{d}, \tag{10.1}$$

where d is the width between the parallel electrodes. Notice that the electric field for a planar configuration is constant and, hence, a relatively large voltage is required to reach the threshold $\mathcal{E}_t$ needed to produce avalanching. By contrast consider the coaxial design shown in Fig. 10.3 in which a voltage V_o is applied to the inner anode and the outer electrode is grounded. The electric field at any radius r for $a < r < b$ is, from Eq. (8.75),

$$\mathcal{E}(r) = \frac{V_o}{r\ln(b/a)}, \tag{10.2}$$

where a is the radius of the inner anode and b is the radius of the cathode shell wall. Unlike the planar case, the electric field is not constant for the coaxial case, and the highest electric field is at $r = a$. Suppose the distance between b and a in the cylindrical case is the same as the distance between x_2 and x_1 in the planar case such that $b - a = x_2 - x_1 = d$. The condition is set such that the highest value of the electric field in both cases just barely reaches the threshold electric field $\mathcal{E}_t$,

$$\mathcal{E}_t = \frac{V_o|_{\text{cylindrical}}}{a\ln(b/a)} = \frac{V_o|_{\text{planar}}}{d}. \tag{10.3}$$

or

$$\frac{V_o|_{planar}}{V_o|_{cylindrical}} = \frac{d}{a\ln(b/a)}, \tag{10.4}$$

and, if $a \ll b$ such that $d = b - a \approx b$, then

$$\frac{V_o|_{planar}}{V_o|_{cylindrical}} \approx \frac{b}{a\ln(b/a)}. \tag{10.5}$$

Because $a \ll b$, $b/a > 1$ and the right-hand side of Eq. (10.5) is always greater than unity (usually $\gg 1$). Thus, for similar chamber dimensions it is found that the voltage needed to reach $\mathcal{E}_t$ for the planar device is always greater than that needed for the cylindrical device.

10.3 Townsend Avalanche Multiplication

At a threshold electric field $\mathcal{E}_t$ electrons gain enough kinetic energy to produce impact ionization of neutral gas atoms (or molecules). Although the exact value for $\mathcal{E}_t$ varies with the gas, the threshold electric field required for impact ionization is generally about 10 kV/cm. As the electric field is increased above $\mathcal{E}_t$, the increase dn in the number of electron-ion pairs $n(x)$ over a differential distance dx traveled by the electrons is

$$dn = \alpha_T n(x) dx, \tag{10.6}$$

where α_T is the *first Townsend ionization coefficient* [Townsend 1915]. This coefficient can be interpreted as the probability, per unit differential path length of travel, that an electron causes impact ionization.[2] Also the mean free path between impact ionization events is $1/\alpha_T$.

From Eq. (10.6), the number of new electrons is found to increase exponentially with distance, i.e.,

$$n(x) = n(0)\exp(\alpha_T x), \tag{10.7}$$

where $n(0)$ is the initial number of electron-ion pairs. The *gas gain* or gas multiplication M at distance x from the initial location of the original electron-ion cloud is, therefore,

$$M = \frac{n(x)}{n_0} = e^{\alpha_T x}. \tag{10.8}$$

In Fig. 10.4 the first Townsend ionization coefficient is shown for various gases.

To achieve the necessary electric field threshold for a planar detector, potentials exceeding several thousand volts must be applied across the electrodes. However, coaxial or hemispherical detectors can be operated in region III of the gas pulse height curve with significantly lower voltages, commonly ranging between 200 and 1500 volts.

Consider a coaxial detector of radius b with anode wire of radius a, in which the electric field at the surface of the anode is $\mathcal{E}_t$. Thus, from Eq. (10.2)

$$\mathcal{E}_t(a) \approx 10^4 \text{ volts cm}^{-1} = \frac{V_t}{a\ln(b/a)}, \tag{10.9}$$

or the threshold voltage is approximately,

$$V_t = a\mathcal{E}_t \ln\left(\frac{b}{a}\right) \approx 10^4 a\ln(b/a) \text{ volts}. \tag{10.10}$$

[2]Such a coefficient is analogous to the interaction coefficients for indirectly ionizing radiation discussed in Sec. 4.2.

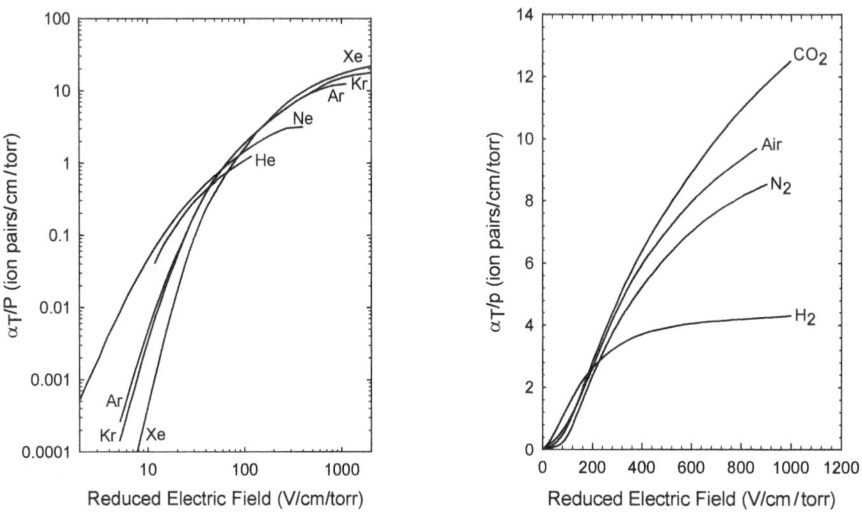

Figure 10.4. The reduced first Townsend coefficient as a function of the reduced electric field for several detector gases. From [Brown 1959].

The critical electric field needed for avalanching can be described in terms of the critical radius r_c at the onset of avalanche multiplication,

$$\mathcal{E}_t = \mathcal{E}(r_c) = \frac{V_0}{r_c \ln(b/a)}. \tag{10.11}$$

Equating Eq. (10.9) with Eq. (10.11) yields,

$$\frac{V_0}{V_t} = \frac{r_c}{a}. \tag{10.12}$$

Example 10.1: Consider a coaxial proportional counter backfilled with argon gas that is 1 inch in diameter with a 25-micron diameter anode wire. Estimate the voltage required to reach the avalanche electric field at the surface of the anode and compare the result to the expected voltage required for a planar detector with a similar electrode spacing.

Solution:
From Eq. (10.9), the threshold voltage for the coaxial detector is,

$$V_t = (10^4)\frac{(25 \times 10^{-4}\text{ cm})}{2} \ln\left(\frac{(2)(2.54\text{ cm})}{(2)(25 \times 10^{-4}\text{ cm})}\right) \text{ volts} = 86.55 \text{ volts.}$$

By comparison, a planar detector with electrode spacing of 1/2 inch would require

$$V_t = (10^4\text{ V/cm})\left(\frac{2.54\text{ cm}}{2}\right) = 12700 \text{ volts.}$$

10.3.1 The Rose-Korff Formula for M

A commonly used, but simplistic, expression for the first Townsend coefficient can be derived by first considering the probability that an electron gains enough energy to produce impact ionization [Rice-Evans

1974]. Given an average distance Δx that an electron must traverse to gain enough energy to cause impact ionization, the condition for impact ionization requires

$$q\mathcal{E}\Delta x \geq qI_0 \tag{10.13}$$

where $\mathcal{E}$ is the electric field and I_0 is the ionization potential of the gas. The fraction of electrons with transit paths $\geq \Delta x$ or, equivalently, have energies above E_t needed to cause impact ionization is

$$\frac{n_e(\Delta x)}{n_e(0)} = e^{-\Delta x/\lambda} = e^{-I_0/\lambda\mathcal{E}}, \tag{10.14}$$

where $n_e(0)$ is the initial number of electrons produced from the radiation interaction event, and λ is electron mean free path (for all types of interaction). The first Townsend coefficient is then the probability the electron has $E \geq E_t$ divided by the probability of an interaction in a unit differential distance of travel, namely[3]

$$\alpha_T = \frac{n_e(\Delta x)}{n_e(0)}\frac{1}{\lambda} = \frac{1}{\lambda}e^{-I_0/\lambda\mathcal{E}}. \tag{10.15}$$

The mean free path λ is a function of the gas pressure and can be approximated by

$$\lambda = \frac{1}{AP}, \tag{10.16}$$

where A is an experimentally determined constant and P is the gas pressure. Substitution of Eq. (10.16) into Eq. (10.15), the first Townsend coefficient is thus found to be [Korff 1946],

$$\frac{\alpha_T}{P} = Ae^{-I_0AP/\mathcal{E}} = Ae^{-BP/\mathcal{E}}. \tag{10.17}$$

For a limited range of the reduced electric field $\mathcal{E}/P$, the microscopic cross section for impact ionization σ_i is proportional to the electron energy E_e [Sauli 1977], i.e.,

$$\sigma_i = \overline{z}E_e, \tag{10.18}$$

where $\overline{z}$ is an experimental constant. The mean free path *between impact ionizations* is

$$\lambda_i = 1/\alpha_T = 1/\sigma_i N \tag{10.19}$$

where N is the atomic or molecular density of the chamber gas. Substitution of Eq. (10.18) into Eq. (10.19) yields,

$$\alpha_T = \overline{z}NE_e. \tag{10.20}$$

Measured values for A, B, and $\overline{z}$ for several gases are listed in Table 10.1. From Eq. (10.17) it is seen that α_T increases as the electric field is raised above the avalanche threshold electric field $\mathcal{E}_t$.

[3]Implicit in this derivation of α_T is the assumption that, once $E \geq E_t$, the next interaction is an impact ionization event and not, for example, a scatter.

Table 10.1. Some parameters used to calculate α_T from Eq. (10.17), Eq. (10.18), and Eq. (10.20). A and B are generally useful between 100 - 600 V cm^{-1} torr^{-1}, except H_2 (150 - 600) and He (20 - 150). Data from [Brown 1959; Korff 1946; Rice-Evans 1974].

Gas	A eip cm^{-1} torr^{-1}	B V cm^{-1} torr^{-1}	$\overline{z}$ 10^{17} cm^2 V^{-1}
H_2	5	130	0.46
He	3	34	0.11
N_2	12	342	0.70
Ne	4	100	0.14
Ar	14	180	1.81

Multiplication in a Coaxial Detector If r_c is the *critical radius* at which the electric field reaches the critical electric field necessary for a Townsend avalanche to occur, then the total gain, or multiplication factor, is given by

$$M = \exp\left[\int_a^{r_c} \alpha_T(r)dr\right]. \tag{10.21}$$

The capacitance, per unit length, of a coaxial chamber is

$$C_l = \frac{2\pi\epsilon_0\kappa}{\ln(b/a)}, \tag{10.22}$$

where κ is the dielectric constant of the gas. Recall that the electric field depends upon the radius r, and substitution of Eq. (10.22) into Eq. (10.2) gives this dependence as[4]

$$\mathcal{E}(r) = \frac{V_o}{r\ln(b/a)} = \frac{C_l V_o}{2\pi r\epsilon_o\kappa}, \tag{10.23}$$

and

$$V(r) = \frac{V_o\ln(r/a)}{\ln(b/a)} = \frac{C_l V_o}{2\pi\epsilon_o\kappa}\ln(r/a). \tag{10.24}$$

The average energy gained by electrons between impact ionizing collisions is, by definition,

$$\overline{E}_e = \lambda_i\mathcal{E}(r) = \frac{\mathcal{E}(r)}{\alpha_T}, \tag{10.25}$$

which, when substituted into Eq. (10.20), gives

$$\alpha_T = \sqrt{\overline{z}N\mathcal{E}(r)}. \tag{10.26}$$

Substitution of Eq. (10.26) into Eq. (10.21) and use of Eq. (10.23) allows the gas multiplication to be evaluated as

$$M = \exp\left[\int_a^{r_c}\sqrt{\overline{z}N\frac{C_l V_o}{2\pi r\epsilon_0\kappa}}dr\right] = \exp\left[\sqrt{2\overline{z}N\frac{C_l V_o}{\pi\epsilon_0\kappa}}\left(r_c^{1/2} - a^{1/2}\right)\right] = \exp\left[\sqrt{2a\overline{z}N\frac{C_l V_o}{\pi\epsilon_0\kappa}}\left(\sqrt{\frac{r_c}{a}} - 1\right)\right]. \tag{10.27}$$

[4]Note that from Poisson's equation, $\nabla^2 V = -\nabla\cdot\vec{\mathcal{E}} = -\rho/\kappa\epsilon_0$. Here the authors assume that the initial space charge is negligible and use the magnitudes of the voltage and electric fields to determine M.

Table 10.2. The reduced first Townsend coefficient α_T/P and the gas multiplication factor M as a function of $\ln M/(aPS_a)$ where $S_a \equiv \mathcal{E}_a/P$. Here $S_a \equiv S(a)$ and $\mathcal{E}_a \equiv \mathcal{E}(a)$. Also C_1 to C_{14}, K_1 to K_{10}, m, and d are empirical constants. The formulas are from [Kowalski 1986].

Citation	α_T/P	$\ln M/(aPS_a)$
Rose and Korff [1941]	$\frac{c_1}{2}\sqrt{S}$	$K_1 - \frac{c_1}{\sqrt{S_a}}$
Hristov [1957]	c_2	$K_1 - \frac{c_2}{S_a}$
Diethorn [1956]	$c_3 S$	$c_3(\ln S_a - \ln K_3)$
Townsend [1915] Williams, Sara [1962]	$c_4 \exp\left(\frac{-c_5}{S}\right)$	$\frac{c_4}{c_5}(\exp\frac{-c_5}{S_a} + K_4)$
Ward [1958] Charles [1972]	$c_6 \exp\left(\frac{-c_7}{\sqrt{S}}\right)$	$K_5 + \frac{2c_6}{c_7^2}\left(\frac{c_7}{\sqrt{S_a}} + 1\right)\exp\left(\frac{-c_7}{\sqrt{S_a}}\right)$
Zastawny [1966]	$c_8(S - S_0)$	$K_6 + c_8\left(\ln\frac{S_a}{c_9} + \frac{c_9}{S_a} - 1\right)$
Kowalski [1985]	$\frac{c_{10}S^{1.5}}{2}$	$c_{10}\sqrt{S_a} + K_7$
Aoyama [1985]	$c_{11}S^m \exp\left(\frac{-c_{12}}{S^{1-m}}\right)$	$\frac{c_{11}}{c_{12}}\frac{1}{1-m}\exp\left(-c_{12}S_a^{m-1}\right) + K_8$
Kowalski [1986]	$c_{13}S^d + c_{14}$	$\frac{c_{13}}{d-1}S_a^{d-1} - \frac{c_{14}}{S_a} + K_9$
Kowalski [1986]	$c_{13}S^d$	$\frac{c_{13}}{d-1}S_a^{d-1} + K_{10}$

Finally, substitution of Eq. (10.12) into Eq. (10.27) and use of Eq. (10.23) gives the following two equivalent results

$$\boxed{\begin{aligned} M &= \exp\left[\sqrt{2a\bar{z}N\frac{C_l V_o}{\pi\epsilon_0\kappa}}\left(\sqrt{\frac{V_0}{V_t}} - 1\right)\right] \\ M &= \exp\left[2\sqrt{\frac{a\bar{z}NV_o}{\ln(b/a)}}\left(\sqrt{\frac{V_0}{V_t}} - 1\right)\right] \end{aligned}} . \tag{10.28}$$

Equation (10.28) represents the derived result of at least one formulation of the avalanche coefficient, traditionally named the "Rose-Korff" formula [Rose and Korff 1941]. In the present case, Eq. (10.28) yields acceptable results for many gases for which $10 \le M \le 10^3$ [Staub 1953]. However, above this range, the formulation of α_T may need adjustment. Further, several experimental studies indicate that Eq. (10.28) has limited use for many detector gas mixtures, and has limited use over a large range of gas detector design parameters [Diethorn 1956; Kiser 1960; Zastawny 1966; Zastawny and Mizeraczyk, 1966]. Consequently, there are many alternative empirical formulas for M listed in the literature [Kowalski 1986]. A summary of these formulas is given in Table 10.2.

10.3.2 The Diethorn Formula for M

Another commonly used expression for M is the Diethorn formula [Diethorn 1956]. At a constant gas pressure P in the detector chamber under a reduced electric field $S(r) \equiv \mathcal{E}(r)/P$, the distance between successive impact ionizations varies as $1/S$. Also, within a *constant* reduced electric field S, the mean free path between ionizing collisions varies with $1/P$. Hence,

$$\lambda_i \propto \frac{1}{SP} \propto \frac{P}{\mathcal{E}P} \propto \frac{1}{\mathcal{E}}. \tag{10.29}$$

This relation indicates that the distance between ionizing collisions decreases as the electric field increases and is independent of the gas pressure P. The ionization mean free path λ_i can be much greater than the collision mean free path λ. The kinetic energy acquired by an electron between successive ionizing events is,

$$q\Delta V = q\lambda_i\mathcal{E} \tag{10.30}$$

where ΔV is the potential difference that an electron must pass through over a distance λ_i in order to obtain sufficiently kinetic energy to produce an ionizing event. From Eq. (10.29), ΔV must be a constant characteristic of the fill gas. It is therefore assumed that the first Townsend coefficient is linearly related to the electric field, or

$$\frac{\alpha_T}{P} = c\frac{\mathcal{E}}{P}, \tag{10.31}$$

where c is an experimentally determined constant. Also, for any gas pressure P, there is a critical threshold electric field for the onset of avalanche multiplication, or

$$\mathcal{E}_t = KP, \tag{10.32}$$

where K is an empirical constant.

Diethorn [1956] estimates the number of electron collisions producing secondary ionization as $n = V_t/\Delta V$ where V_t is the potential at $\mathcal{E}_t$.[5] Because the first ionization event yields 2 electrons both of which cause subsequent ionization events, thereby doubling their number, and so on, then after n impact ionizations there are 2^n electrons for each initial electron. Thus, the multiplication factor is

$$M = 2^n \qquad \text{or} \qquad \ln M = n \ln 2 = \frac{V_t}{\Delta V}\ln 2. \tag{10.33}$$

For a coaxial detector, the reduced electric field can be described by,

$$S(r) = \frac{\mathcal{E}(r)}{P} = \frac{V_0}{rP\ln(b/a)}. \tag{10.34}$$

with reduced potential distribution

$$U(r) = \frac{V(r)}{P} = \frac{V_0\ln(r/a)}{P\ln(b/a)}. \tag{10.35}$$

Solve Eq. (10.34) for r and evaluate the result at the critical radius r_c to obtain

$$r_c = \frac{V_0}{\mathcal{E}_t \ln \dfrac{b}{a}} \tag{10.36}$$

[5] Diethorn's treatment indicates that a negative potential is applied to the cathode of a coaxial detector with the anode grounded, a case rarely used, if ever. However, the treatment is mathematically correct. If instead a positive potential was applied to the anode with the cathode grounded, the estimated number of ionizing collisions would be $n = (V_0 - V_t)/\Delta V$.

Table 10.3. Constants for different gases used to calculate M from Eq. (10.38).

Gas	ΔV V	K V cm^{-1} $torr^{-1}$	Citation
95% Xe/5% CO_2	31.4	48.16	Hendricks [1972]
90% Xe/10% CH_4	33.9	47.63	Hendricks [1972]
P-5 (95% Ar/5% CH_4)	21.8 ± 4.4	59.21 ± 5.26	Wolff [1974]
P-10 (90% Ar/10% CH_4)	23.6 ± 5.4	63.16 ± 3.95	Wolff [1974]
CH_4	36.5 ± 5	90.79 ± 6.58	Wolff [1974]
C_3H_8	29.5 ± 2	131.58 ± 5.26	Wolff [1974]

Substitution of Eq. (10.32) and Eq. (10.36) into Eq. (10.35) yields,

$$V_t = \frac{V_0}{\ln \frac{b}{a}} \ln \left(\frac{V_0}{aKP \ln \frac{b}{a}} \right). \tag{10.37}$$

Finally, substitution of Eq. (10.37) into Eq. (10.33) yields the Diethorn formula [Diethorn 1956],

$$\boxed{\ln M = \frac{V_0}{\ln(b/a)} \frac{\ln 2}{\Delta V} \left(\ln \frac{V_0}{aP \ln(b/a)} - \ln K \right)}, \tag{10.38}$$

or the general form used in Table 10.2,

$$\boxed{\frac{\ln M}{aPS_a} = \frac{\ln 2}{\Delta V} (\ln S_a - \ln K)}, \tag{10.39}$$

where $S_a \equiv S(a)$. Equation (10.38) is particularly useful for CH_4 and many forms of Xe-CH_4 and Ar-CH_4 mixtures. Reported values for constants in Eq. (10.38) are listed in Table 10.3.

10.3.3 The Zastawny Formula for M

However, Eq. (10.38) is less useful for pure Ar, N_2, and CO_2, or for those gas mixtures involving CO_2. To address the problems with Eq. (10.38), Zastawny [1966] developed an alternative solution to the Diethorn equation (also listed in Table 10.2). Differentiation of Eq. (10.34) and evaluation of $S_a = S(a)$ from this equation gives the relation

$$dr = -\frac{aS_a}{S^2(r)} dS. \tag{10.40}$$

With this relation between dr and dS Eq. (10.21) may be expressed as

$$\ln M = aPS_a \int_{S_{r_c}}^{S_a} \frac{\alpha_T(S)}{P} \frac{dS}{S^2(r)}, \tag{10.41}$$

Table 10.4. Constants for different gases used to calculate M from Eq. (10.43).

Gas	K V^{-1}	c_1 V^1	c_2 V cm^{-1} torr^1	Citation
CO_2	0.5×10^{-3}	1.93×10^{-2}	69	Zastawny [1966]
N_2	$(0.655 \pm 0.03) \times 10^{-3}$	$(18.6 \pm 0.9) \times 10^{-2}$	85 ± 4	Zastawny, Mizeraczyk [1966]
Ar	$(6.25 \pm 0.3) \times 10^{-3}$	$(39.6 \pm 2) \times 10^{-2}$	34 ± 2	Zastawny, Mizeraczyk [1966]
P-10	$(0.5 \pm 0.1) \times 10^{-3}$	$(3.0 \pm 0.03) \times 10^{-2}$	25 ± 1	Zastawny [1967]
Xe/5% CO_2	2.49×10^{-3}	3.01×10^{-3}	32.37	Hendricks [1972]
Xe/10% CH_4	5.45×10^{-3}	2.46×10^{-3}	19.47	Hendricks [1972]

where $S_{r_c} = S(r_c)$ is the reduced electric field at the radial location for the onset of avalanching, and $\alpha_T(S)$ is the first Townsend coefficient as a function of the reduced electric field. Zastawny [1966] proposed

$$\frac{\alpha_T}{P} = c_1(S - c_2), \tag{10.42}$$

which, when substituted into Eq. (10.41) yields,

$$\boxed{\frac{\ln M}{aPS_a} = K + c_1\left(\ln\frac{S_a}{c_2} + \frac{c_2}{S_a} - 1\right)}\,, \tag{10.43}$$

where K, c_1, and c_2 are experimentally determined constants. Experimental comparisons indicate that Eq. (10.43) works well for Ar, N_2, CO_2, and mixtures of CO_2 gas [Zastawny 1966, 1967; Zastawny and Mizeraczyk 1966]. Reported values of constants in Eq. (10.43) for several gases are listed in Table 10.4.

10.3.4 The Kowalski Formula for M

Kowalski mentions that equations for M listed in Table 10.2 have diverse results when tested with various gases over wide ranges of S. In an attempt to produce a generalized formula for the gas multiplication factor, Kowalski suggests the use of

$$\frac{\alpha_T}{P} = cS^d \tag{10.44}$$

where c and d are experimentally determined constants. Insertion of Eq. (10.44) into Eq. (10.41) yields

$$\boxed{\frac{\ln M}{aPS_a} = \frac{c}{d-1}S_a^{d-1} + K} \tag{10.45}$$

where K is an experimentally determined constant. When fit to data reported by various researchers, Eq. (10.45) appears to have merit for a variety of complex gas mixtures, many of which are listed in Table 10.5.

10.4 Gas Dependence

The ionization potentials and electron mobilities for many detector gases are significantly different, many of which are listed in Tables 9.1 and 9.2. The first Townsend coefficient should also vary among gases because

Table 10.5. Constants for different gases used to calculate M from Eq. (10.45). Data from [Kowalski 1986].

Gas	c $(\text{mPa})^{d-1}\ \text{V}^{-d}$	d	K V^{-1}
Ar	6.34×10^{-4}	1.85	-6.79×10^{-3}
Ar/5.7% C_5H_{12}	1.46×10^{-3}	1.775	-1.22×10^{-2}
Ar/3.1% C_5H_{12}	2.46×10^{-3}	1.70	-1.52×10^{-2}
Kr/4.11% C_2H_6O	2.14×10^{-3}	1.525	-1.54×10^{-2}
Kr/1.92% C_2H_6O	3.00×10^{-3}	1.45	-1.97×10^{-2}
Xe/6.16% H_2	1.10×10^{-4}	2.225	-3.02×10^{-3}
Xe/16.7% Kr/7% H_2	1.80×10^{-4}	2.125	-4.08×10^{-3}

M is strongly dependent on the mean free path length λ_i between ionizing collisions. As a result, gas multiplication can vary significantly for different detector parameters, gases, and gas mixtures. Examples of experimentally measured multiplication factors as a function of operating voltage are shown in Fig. 10.5 and Fig. 10.6.

Electron attachment, discussed in Sec. 9.2.5, can adversely affect the performance of proportional counters. Hence, gases with small electron attachment coefficients are commonly chosen as the gas in a proportional counter and include CH_4, CO_2, and Ar. Extreme care is required to prevent contaminant gases with relatively high electron attachment coefficients, such as O_2, from entering the detector.

Various gas mixtures are often introduced into the detector chamber to improve performance. Gas additives and mixtures can improve electron velocities (see Fig. 9.6 and Fig. 9.7), improve avalanche gain, decrease the required operating voltage, improve energy resolution, and eliminate spurious pulses. These gas mixtures may include Penning mixtures to increase the number of charge carriers per ionizing event, and may include quench gases to prevent successive avalanches from a single event.

10.4.1 Quenching Gas

After an impact collision, the resulting ion is often left in an excited state which can rapidly decay by emitting a photon (usually in the ultraviolet). This photon may then interact photoelectrically with another gas atom or molecule releasing a photoelectron which then causes another avalanche of electrons.

Let δ be the probability that an electron liberated through impact ionization leads to a tertiary electron through photoionization (typically $\delta \ll 1$). The gas multiplication per initial electron-ion pair is then given by

$$M = f + \delta f^2 + \delta^2 f^3 + ... + \delta^{n-1} f^n = \sum_{i=1}^{n} \delta^{i-1} f^i \quad , \tag{10.46}$$

where f is the total number of electrons liberated per primary electron-ion pair and would equal the multiplication factor if there were no photoionization, i represents the consecutive avalanche number (first, second, third, and so on) up to the final avalanche (n). Because δf is strongly dependent upon the applied operating voltage and is generally less than unity, Eq. (10.46) reduces to

$$M = \frac{f}{1 - \delta f} \quad . \tag{10.47}$$

If, in fact, $\delta f > 1$, the avalanching process becomes uncontrollable and the detector develops a self-sustaining discharge, which may occur with too high of voltage (as in Region V). Continuous waves of

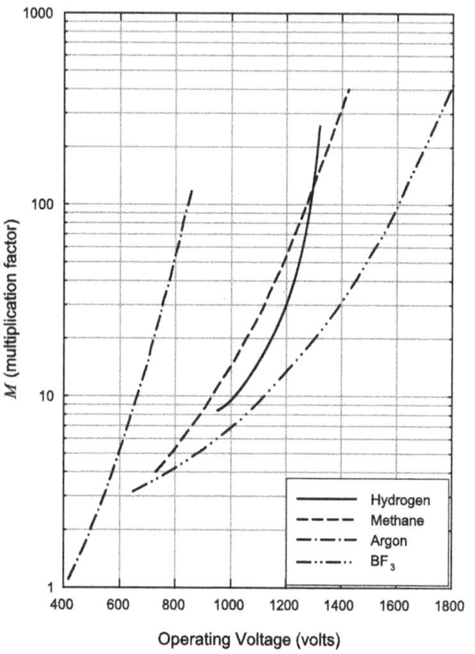

Figure 10.5. The multiplication factor M for several gases each at a pressure of 100 torr. The detectors for H_2, CH_4 and Ar had $a = 127\mu$m and $b = 11.05$ mm. The BF_3 detector had $a = 127\mu$m and $b = 19.05$ mm. From [Staub 1953].

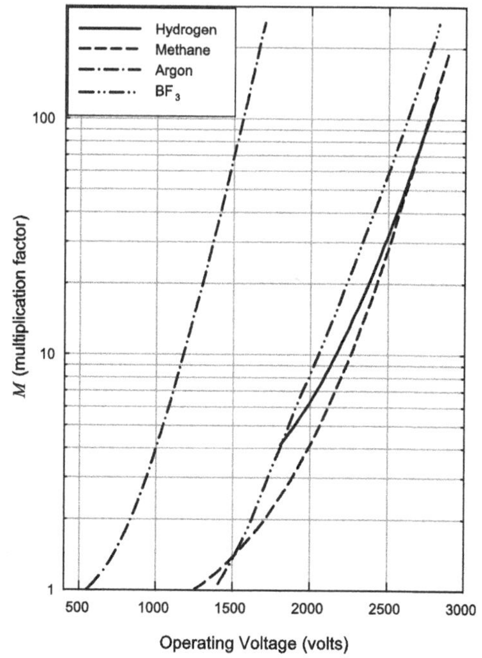

Figure 10.6. The multiplication factor M for several gases and pressures. The detectors for H_2 (550 torr), CH_4 (400 torr) and Ar (400 torr) had $a = 127\mu$m and $b = 11.05$ mm. The BF_3 detector (804 torr) had $a = 127\mu$m and $b = 19.05$ mm. From [Staub 1953].

avalanches can occur if UV light released by excited ions causes photoionization of outer shell electrons in too many more neutral Ar atoms. This is the most probable source of new electrons because of the high opacity of the gas at UV energies. However, UV light striking the cathode may also release photoelectrons from the cathode material. Finally, when a gas ion reaches the cathode wall, more electrons may be released by two different mechanisms. First, the energy difference between the work function of the cathode material and the ionization potential of the ion may be radiated into the tube gas as an ultraviolet photon which may, in turn, interact with a gas atom releasing a photoelectron. Second, the energy difference may liberate an electron directly from the cathode in a radiationless energy transfer. These mechanisms for starting additional avalanches are depicted in Fig. 10.7(a).

To prevent continuous waves of avalanches from occurring in the chamber after a radiation interaction, a small amount of a *quenching gas* is added to the gas mixture, typically a polyatomic molecule. When an ionizing particle enters the detector, it ionizes both the host and the quenching gases. More important, the argon ions, traveling towards the cathode, collide with a quenching gas molecules from which they readily take an electron, thus transferring almost all of the positive charge to quench gas molecules. UV photons can also interact with the quench gas, but the molecules dissociate instead of releasing photoelectrons. These molecules, when reaching the cathode, dissociate rather than liberate electrons, as would an argon ion if it were to strike the wall. In this manner further avalanches are suppressed. These quenching mechanisms are depicted in Fig. 10.7(b).

Quench gases are usually chosen with ionization potentials below that of the host gas, a precaution that allows charge to transfer from the host gas to the quench gas while rendering the reverse process improbable.

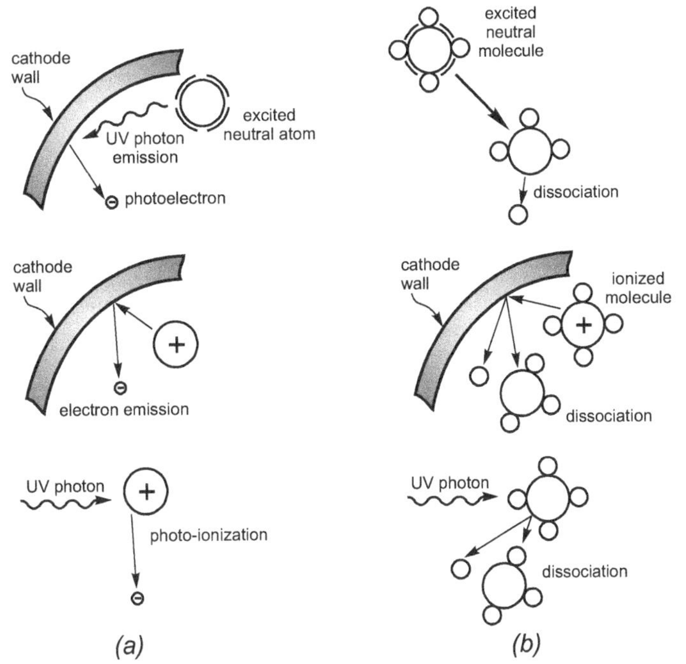

Figure 10.7. A quench gas is used to prevent continuous avalanches in the proportional counter. When a noble gas ion or UV photon strikes the cathode wall or a UV photon is absorbed by a neutral noble gas atom, electrons may be ejected that can start another avalanche, as depicted in *(a)*. The quench gas, usually an organic molecule, dissociates when it strikes the cathode wall or when it absorbs an energetic photon; hence it does not cause the release of an electron that can start a new avalanche, as depicted in *(b)*.

Quench gases should also not be electronegative so they are not prone to electron attachment. Several different organic molecules have been successfully used as quench gases (see Table 10.6). As the host gas ions drift through the chamber, they pass their charge to the quench gas molecules. These charged quench gas ions continue to drift and carry the positive charge to the cathode wall. When a quench gas is either struck by a UV photon or strikes the cathode wall, it dissociates, rather than eject an electron, as illustrated in Fig. 10.7(b). As a result, the quench gas prevents continuous waves of ionization. A common proportional counter gas mixture is *P-10*, which is a mixture of 90% Ar and 10% methane.

10.4.2 Penning Mixtures

Electron collisions in a gas can leave a gas in an excited state, which generally de-excites by the emission of a photon. However, long-lived metastable states can also be formed by electron collisions, in which radiative transitions to the ground state are forbidden. These excited metastable states do not produce ionization and consequently do not add to the induced charge signal. However, if a small amount of gas is added to the gas mixture, in which the added gas has a lower ionization potential than that of the excited metastable state of the host gas, then the energy stored by the metastable state can be transferred through a collision to produce another electron-ion pair,

$$A^* + B \rightarrow A + B^+ + e^-, \tag{10.48}$$

Table 10.6. Common detector host and quench gases with their ionization potentials I_o. From [Christophorou 1971].

Detector Gas	Common Name	I_o (eV)	Detector Gas	Common Name	I_o (eV)
Ar	Argon	15.755	CH_4	Methane	12.98
He	Helium	24.580	CO_2	Carbon Dioxide	13.79
H_2	Hydrogen	15.42	C_2H_2	Acetylene	11.36
O_2	Oxygen	12.08	C_2H_5OH	Ethyl Alcohol	10.50
N_2	Nitrogen	15.6	C_3H_8	Propane	11.08
Ne	Neon	21.559	C_4H_{10}	Isobutane	10.55
Kr	Krypton	13.996	Cl_2	Chlorine	11.48
Xe	Xenon	12.127	Br_2	Bromine	10.55

known as the Penning effect [Penning 1927]. This gas additive is also referred to as a "quench gas", although the effect is quite different than described in the previous subsection. Quite instead, adding a Penning quench gas increases ionization, and thereby increases the charge signal for the primary radiation absorption event. In effect, the average ionization energy of the gas mixture is lower than either individual gas, which works to improve charge collection statistics and improve energy resolution in a proportional counter [Sipilä 1976].

An example of the Penning effect is the small addition of Ar to a host gas of Ne. From Table 10.7, Ne has a metastable state energy of 16.53 eV while Ar has a first ionization energy of 15.76 eV. Hence, the metastable Ne gas atom can transfer sufficient energy to an Ar gas atom to produce an electron-ion pair. Typically, excellent improvement is achieved with only 0.01% Ar added [Druyvesteyn and Penning 1940]. The Penning effect is also observed with numerous polyatomic gases, generally because polyatomic gases have relatively low ionization energies (see Table 10.6). Experimental studies have shown that addition of H_2C_2 (acetylene) to the host gas Ar improves energy resolution for low avalanche gains (≤ 10) [Järvinen and Sipilä 1984; Agrawal et al. 1989]. Studies with gas-filled proportional detectors have shown that $\sim 2\%$ Xe added to Ar does improve performance, providing higher avalanche gains than other mixtures tested in the study [Agrawal et al. 1989], while a gas mixture of $\sim 20\%$ Xe added to Ar produced improved energy resolution ($\approx 7\%$ FWHM at 22 keV) nearly equal to mixtures of Ar/H_2C_2. In a similar study for multiple organic gas additives with Xe as the host gas, a strong Penning effect was observed with trimethylamine ($I_o \approx 8.32$ eV) and dimethylamine ($I_o \approx 8.4$ eV) as additives [Ramsey and Agrawal, 1989], where it is reported that the Penning mixture of Ar + 5% trimethylamine allowed for a 50% reduction in operation voltage while improving energy resolution by $\approx 22\%$.

10.5 Proportional Counter Operation

The creation of a pulse in a proportional counter starts with the initial ionizing event and ends with the collection of the liberated charges at the electrodes (see Fig. 10.8). A radiation particle enters the detector, interacts with the chamber gas, and liberates N_0 electron-ion pairs. The electrons are quickly swept towards the anode at speeds far exceeding those of the positive ions. Hence, the slow moving ions contribute very little to the initial induced charge. Upon entering the avalanche region near the anode, the electrons create numerous electron-ion pairs through impact ionization, equaling approximately $N_0\overline{M}$, where $\overline{M}$ is the average multiplication factor. The dense cloud of electrons is rapidly collected, leaving behind a dense cloud of slow moving positive ions. These positive ions slowly drift towards the cathode. Because of the high weighting potential in the anode region, it is the positive ions drifting from the anode towards the cathode that contribute to nearly all of the input voltage $V_{in}(t)$ to the amplification circuit.

Table 10.7. Ionization and excitation data for several gases. Data from [Christophorou 1971; Rice-Evans 1974].

Gas	First Ionization Energy (eV)	Second Ionization Energy (eV)	First Excited State (eV)	Recombination Coefficient (cm^3 s^{-1})
He	24.58	54.40	20.9 19.8 meta	10^{-8}
Ne	21.559	41.07	16.68 16.53 meta 16.62 meta	2×10^{-7}
Ar	15.755	27.52	11.56 11.49 meta 11.66 meta	9×10^{-7}
Kr	13.996	24.56	9.98 9.86 meta 10.51 meta	10^{-5}
Xe	12.127	21.2	8.39 8.28 meta 9.4 meta	2×10^{-6}
H	13.60		10.2	
N	14.53	29.59	6.3	
O	13.61	35.11	9.1	
H_2	15.42		11.2	2×10^{-6}
N_2	15.6		6.1	2×10^{-6}
O_2	12.08			4×10^{-7}
I_2	9.0		1.9	1.5×10^{-8}

10.5.1 Pulse Shape

The pulse shape of a coaxial gas-filled detector varies greatly depending on which region of the pulse height gas curve of Fig. 10.1 the chamber is operated. The expected pulse shape for electron dominated voltage

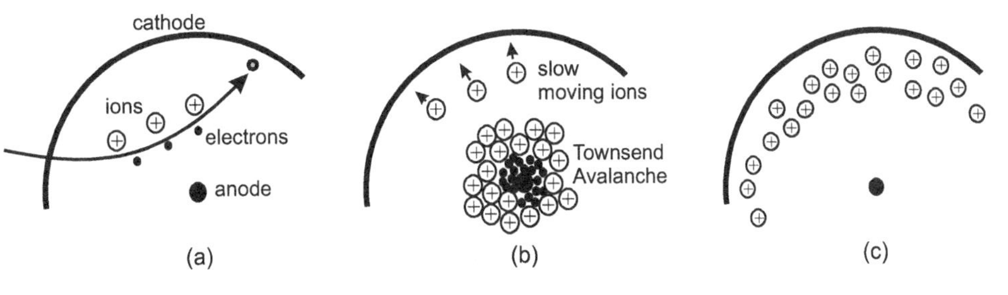

Figure 10.8. Pulse progression for a proportional counter: (a) an ionizing particle enters the detector gas and creates electron-ion pairs, (b) the electrons rapidly drift towards the anode and create an avalanche of electron-ion pairs, (c) the electrons are rapidly collected, whereas the positive ions drift much more slowly towards the cathode. It is the positive ion drift that contributes to most of the input pulse $V(t)$ from a proportional counter.

pulses in a coaxial ion chamber was derived in Sec. 9.4.2. However, operating the same detector in the proportional region causes a significant difference in the input pulse shape $V_{in}(t)$. For simplicity, assume that a radiation interaction creates a localized electron-ion cloud that can be treated as the location of N_0 charged pairs at a single point. Initially, the input voltage induced by drifting electrons is essentially the same as that in an ion chamber. However, once the electrons enter the avalanching region, a huge cloud of electron-ion pairs is produced, mostly within a few microns of the anode. Because the total charge is increased by the gas multiplication factor, most of the electron-ion pairs are produced very close to the anode. Consequently, the drift distance for electron collection is relatively short, and the electron contribution to the input voltage is relatively small. It is the motion of the cloud of positive ions drifting back towards the cathode that is responsible for most the time-dependent input voltage. Hence, the input voltage in proportional counters is *positive ion dominated*, whereas for ion chambers the input voltage is *electron dominated.*

Recall that the time dependent input pulse into the coupling electronics is

$$V_{in}(t) = \frac{\Delta Q(t)}{C} \tag{10.49}$$

where C is the combined coupling and detector capacitance. The expression for $\Delta Q(t)$ for a coaxial detector was derived in Sec. 8.7. Equation (8.80) with the variables used here is

$$\Delta Q(t) = qN_0[\ln(b/a)]^{-1}\ln(r_{ion}(t)/r_e(t)). \tag{10.50}$$

where $r_{ion}(t)$ and $r_e(t)$ are the radial locations of the positive ions and electrons, respectively, at time t. Unlike an ion chamber in which only N_0 electron-ion pairs are collected, in a proportional counter MN_0 pairs are collected The electrons are rapidly collected in time t_e, leaving behind the positive ions to drift toward the cathode. Almost all of the positive ions, save those very few produced in the original ionization event, begin their drift, after all of the electrons are collected, essentially at the anode surface. Thus, for $t > t_e$, Eq. (10.50) becomes

$$\Delta Q(t) \approx qMN_0\ln(b/a)]^{-1}\ln(r_{ion}(t)/a) \qquad t > t_e \simeq 0. \tag{10.51}$$

The drift speed of the positive ions is given by

$$v_{ion} \equiv \frac{dr_{ion}}{dt} = \frac{P_0\ \mu_{ion}\mathcal{E}(r)}{P} = \frac{P_0\ \mu_{ion}V_0}{rP\ln(b/a)}. \tag{10.52}$$

The ion position $r_{ion}(t)$ is found by integrating Eq. (10.52) as

$$\int_a^{r(t)} r\,dr = \frac{P_0\ \mu_{ion}V_0}{P\ln(b/a)}\int_0^t dt \qquad \text{or} \qquad \frac{r_{ion}^2}{2} - \frac{a^2}{2} = \frac{P_0\ \mu_{ion}V_0}{P\ln(b/a)}t.$$

Rearrangement of this last result yields

$$r_{ion}(t) = \sqrt{\frac{2P_0\ \mu_{ion}V_0t}{P\ln(b/a)} + a^2}. \tag{10.53}$$

Upon substitution of Eq. (10.53) into Eq. (10.50) and use of Eq. (10.49) the input voltage is

$$V_{in}(t) \approx \frac{qMN_0}{C\ln(b/a)}\ln\left(\sqrt{\frac{2P_0\ \mu_{ion}V_0t}{a^2P\ln(b/a)} + 1}\right), \qquad t_e < t < t_{ion}, \tag{10.54}$$

for which the collection time for the positive ions is

$$t_{ion} = \frac{(b^2 - a^2)P\ln(b/a)}{2P_0\ \mu_{ion} V_0}. \quad (10.55)$$

The maximum input voltage V_{max} is found by evaluating Eq. (10.54) at $t = t_{ion}$. The normalized input voltage $V(t)/V_{max}$ is, thus

$$\frac{V_{in}(t)}{V_{max}} \approx \frac{1}{\ln(b/a)} \ln\left(\sqrt{\frac{2P_0\ \mu_{ion} V_0 t}{a^2 P\ln(b/a)} + 1}\right). \quad (10.56)$$

The magnitude of $V(t)/V_{max}$ initially rises rapidly, a fortuitous characteristic of the coaxial detector design, and is due to the combined effects of an initially high positive ion speed because of the high electric field near the anode and the enhanced weighting potential near the anode. For example, the time required to reach half the maximum input voltage $V_{max}/2$ is found from Eq. (10.56) as

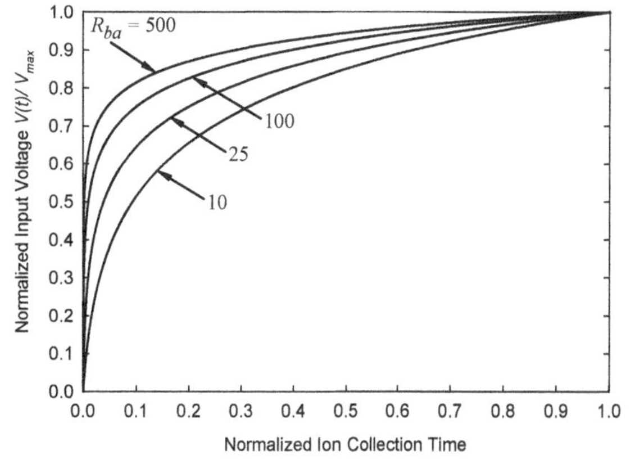

Figure 10.9. The normalized time-dependent input voltage $V(t)/V_{max}$ for a proportional counter as a function of t/t_{ion} for different values of $R_{ba} = b/a$ as given by Eq. (10.56). For the results shown, $b = 1.25$ cm and $a = b/R_{ba}$.

$$t_{V_{max}/2} = \left[\frac{b}{a} - 1\right] \frac{a^2 P\ln(b/a)}{2P_0\ \mu_{ion} V_0}. \quad (10.57)$$

The normalized time-dependent input voltages $V(t)/V_{max}$ as a function of t/t_{ion} are shown in Fig. 10.9 for different values of $R_{ba} \equiv b/a$.

Example 10.2: A coaxial proportional counter backfilled with argon at 1 atm is operated at 1000 volts. The detector has an anode radius of $a = 25$ microns and cathode radius $b = 1.25$ cm. Exclude the possibility of a time delay caused by space charge. At what time past $t = 0$ does the input voltage $V(t)$ reach 1/2 the maximum possible pulse height? How far have the positive ions drifted in the chamber?

Solution:

From Table 9.2, the mobility μ_{ion}^{+} at 1 atm for argon is 1.7 cm^2 V^{-1}s^{-1}. The ratio $b/a = 500$. The maximum speed of the positive ions is found from Eq. (10.52) as

$$v_{max} = \frac{P_0\ \mu_{ion} V_0}{aP\ln(b/a)} = \frac{(1\text{ atm})(1.7\text{ cm}^2\text{ V}^{-1}\text{ s}^{-1})(1000\text{ V})}{(25\times^{-4}\text{ cm})(1\text{ atm})\ln(500)} = 0.109\text{ cm}/\mu\text{s}.$$

The highest positive-ion speed is less than the saturation speed shown in Fig. 9.6, thereby validating the use of the ion-speed approximation. From Eq. (10.57)

$$t_{V_{max}/2} = \left[\frac{b}{a} - 1\right] \frac{a^2 P\ln(b/a)}{2P_0\ \mu_{ion} V_0} = [500 - 1] \frac{(25\times^{-4}\text{ cm})^2(1\text{ atm})\ln(500)}{2(1\text{ atm})(1.7\text{ cm}^2\text{ V}^{-1}\text{ s}^{-1})1000\text{ V}} = 5.7\times 10^{-6}\text{ seconds}.$$

Insertion of this result into Eq. (10.53), the radial location is

$$r(t_{V_{max}/2}) = \sqrt{\frac{2P_0\ \mu_{ion} V_0 t}{P\ln(b/a)} + a^2}$$

$$= \sqrt{\frac{2(1\text{ atm})(1.7\text{ cm}^2\text{ V}^{-1}\text{ s}^{-1})(1000\text{ V})(5.7\times 10^{-6}\text{ s})}{(1\text{ atm})\ln(500)} + (25\times^{-4}\text{ cm})^2} = 0.0559\text{ cm} = 559\ \mu\text{m},$$

which indicates that the positive ions have drifted

$$d = (559 - 25)\ \mu\text{m} = 534\ \mu\text{m}.$$

The input voltage $V_{in}(t)$ from a proportional counter is often applied to a preamplifier *CR-RC* circuit (as is done with an ion chamber). From Eqs. (9.118) and (10.56), the output voltage of the preamplifier is

$$\begin{aligned} V_{out}(t) &= V_{in}(t)\frac{\tau_o}{\tau_o - \tau_i}\left(e^{-t/\tau_o} - e^{-t/\tau_i}\right), \\ &= \frac{qMN_0}{C\ln(b/a)}\ln\left(\sqrt{\frac{2P_0\ \mu_{ion}V_0 t}{a^2 P\ln(b/a)}} + 1\right)\frac{\tau_o}{\tau_o - \tau_i}\left(e^{-t/\tau_o} - e^{-t/\tau_i}\right). \end{aligned} \tag{10.58}$$

With the same analysis used to obtain Eq. (9.120) for the case $\tau_o = \tau_i$, Eq. (10.58) reduces to

$$V_{out}(t) = \frac{qMN_0}{C\ln(b/a)}\ln\left(\sqrt{\frac{2P_0\ \mu_{ion}V_0 t}{a^2 P\ln(b/a)}} + 1\right)\frac{t}{\tau}e^{-t/\tau}. \tag{10.59}$$

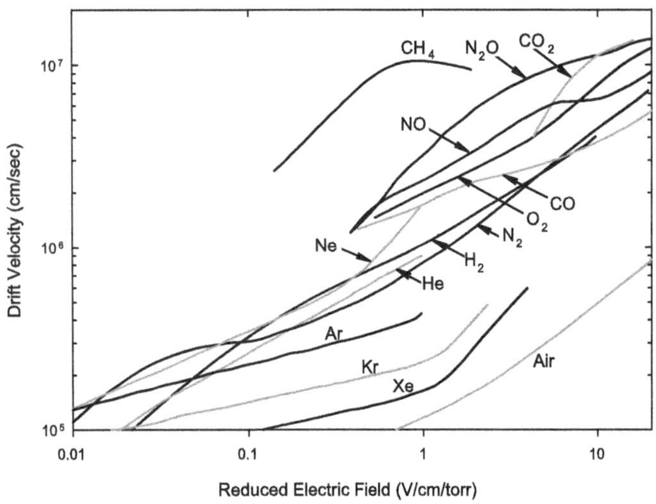

Figure 10.10. Electron drift speeds, measured at room temperature (293–300 K), for several detector gases as a function of reduced electric field. Data from [Bowe 1960; Brown 1959; Cottrell and Walker 1965; Pack and Phelps 1961; Phelps et al. 1960].

Typically the time constant of the circuit is adjusted to be much shorter than the total time required to collect the positive ions; otherwise, the recovery time would result in significant dead time loss. However, because a large portion of the input voltage is induced within the first several microseconds, the time constant can be adjusted to be small while still allowing the induction of large input currents. For monoenergetic events, the proportional counter nearly eliminates the position dependent charge collection issues typical of ion chambers. In principle, proportional counters should be capable of operating as energy spectrometers provided there is strict proportionality, i.e., M does not depend on the primary ionization. This requirement is true if space-charge effects by the positive ions are negligible (see Sec. 10.5.2) and if all the electrons have approximately the same travel distance and produce the same avalanche multiplication. This last requirement is achieved by using a very thin anode wire, typically a few hundred μm in radius. Thus, the high electric fields that cause avalanching are very close to the anode and almost all secondary ionization is in a very small volume around the anode.

Radiation interactions rarely create electron-ion pairs at a point location. Typically, radiation charged particles or charged particle reaction products have paths oriented over all possible directions in the chamber. The situation is further complicated for energetic electrons because they move along very tortuous trajectories (see Fig. 4.22). For instance, a cloud of electron-ion pairs oriented parallel to the central anode wire produces a time-dependent input pulse similar to that described by Eq. (10.54). However, an ion pair track perpendicular to the anode wire produces an input voltage distorted from those shown in Fig. 10.9. Those electrons liberated nearest the anode start an avalanche at time t_0, whereas those electrons liberated furthest from the anode enter the avalanche region at $t_0 + \Delta t$, where Δt is the time required for all the electrons to drift into the avalanche region. As a result, the rise of $V_{in}(t)$ is further extended over Δt causing a distortion in the shape of the input voltage. Consequently, the input pulses can have different maxima for monoenergetic events, thereby compromising the energy resolution of the proportional counter. This problem of varying pulse heights, with examples, is discussed in the literature [Gott and Charles 1969]. To reduce the effect, it is typical to choose proportional gases with high electron mobilities and low electron attachment coefficients. Recall from Chapter 9 that electron drift speeds in Ar are relatively low. The same is true for other noble gases such as Kr and Xe. However, the average electron speed can be significantly increased by adding small portions of CH_4 or CO_2 into the proportional gas (for example, compare Fig. 9.6 and Fig. 9.7). Shown in Fig. 10.10 are the speed characteristics as a function of reduced electric field for several gases often used in proportional counters.

10.5.2 Space Charge Effects

As detailed by Wilkinson [1950] *space charge* accumulation can affect the gain, and ultimately the energy resolution, of a proportional counter. The buildup of space charge has two basic formation mechanisms, namely, *self-induced* space charge and *cumulative* space charge [Hendricks 1969; Mori and Watanabe 1982].

Self-induced space charge appears as an accumulation of positive charge near the anode as a consequence of the avalanching process. This buildup of positive charge around the anode decreases the electric field in the region nearest the anode wire. This space charge causes the local avalanche region around the anode to decrease in size and causes a slight decrease in gain. The higher the radiation energy deposited in the chamber, or rather, the more initial electron-ion pairs produced, the more avalanche electron-ion pairs are produced. Hence, the space charge effect can introduce a small amount of non-proportionality in the detector response at moderate operating voltages. As the operating voltage is increased, the multiplication factor increases (see Fig. 10.5 and Fig. 10.6) and a larger cloud of space charge is formed. Hence, increasing the operating voltage also increases the non-proportionality response to energy deposition and, generally, causes a decrease in the expected pulse height [Fergason 1966]. Eventually, at sufficiently high voltages, the detector is driven into the *transition region* (region IIIa of Fig. 10.1), in which pulses have limited proportionality to the energy deposited in the detector chamber.

Cumulative space charge appears as a consequence of multiple events contributing to the total number of drifting positive ions. Recall that, for those positive ions originating near the anode, the drifting times required to reach the cathode can be relatively long compared to the electron drift time. During that drifting period of the ions, additional radiation interaction events may occur that create additional clouds of positive ions near the anode and which drift, in sequence, towards the cathode. Although these ions are drifting in a region of low electrical field, the cumulative effect of multiple ion cloud formations causes the total electric field to drop, including that in the avalanching region. Consequently, the induced charge is observed to decrease from that expected from a single event [Burkhalter 1966]. This space charge effect becomes worse with higher count rate, in which the uncertainty in pulse height for monoenergetic radiations increases, and the average spectral shift in pulse height increases [Hendricks 1969]. To reduce the overall effect of space charge within the chamber, it has been recommended that the gas gain M be kept low [Wilkinson 1950; Hendricks 1969].

10.5.3 Counting Curve

It is always a good practice to calibrate gas-filled detectors before conducting measurements. For survey instruments, such a calibration may be performed once per year with sufficient confidence in the measurement results. However, laboratory instrumentation coupled to modular counting equipment, such as NIM or CAMAC, requires calibration much more often. The reason why, of course, is because of the possible frequent changes in equipment settings made by other users, such as the amplifier gain or the lower level discriminator (LLD) setting, changes which alter the overall detector response. Consequently, before any radiation measurements are conducted, a *counting curve* is usually first created to locate region III of the pulse height curve for the chamber being used and also for the type of radiation particle being measured in the samples.

Ideally, one should use a known source that emits the energy and type of particle to be measured in the samples. For example, suppose the samples were thought to contain a radionuclide that emits an alpha particle of energy E_o. Then with a check source that emits the same alpha particle energy, the lower level discriminator (LLD) is adjusted by observing, with an oscilloscope, the amplitudes of the pulses being sent to the counter. The LLD is set to reject pulses produced by background and other lower amplitude pulses produced by other radiations. In this fashion only the higher amplitude pulses formed by the alpha particle of energy E_o are counted.

Table 10.8. Data for the counting curve of Fig. 10.11.

HV	Counts	HV	Counts	HV	Counts	HV	Counts
0	0	250	560	500	658	750	711
25	0	275	620	525	692	775	707
50	0	300	645	550	698	800	725
75	0	325	669	575	654	825	720
100	10	350	642	600	693	850	754
125	23	375	659	625	703	875	832
150	37	400	685	650	667	900	900
175	65	425	648	675	686	925	1130
200	123	450	671	700	714	950	1342
225	349	475	667	727	683	975	1800

Once the LLD is set, a series of measurements or counts of the radiation source are recorded at incrementally increasing voltages beginning with a very low or zero voltage. Each measurement has the same time duration. As the output pulses triggered by the radiation sources become larger, a direct result of increasing M with applied voltage, they eventually pass through the LLD and are counted. The counts for each measurement are plotted versus the operating voltage used. For a monoenergetic α particle source, as the operating voltage is increased, eventually almost every output pulse from the radiation source exceeds the LLD, and the number of counts per measurement reaches a near constant value over a finite voltage region as shown in Fig. 10.11. As the voltage is increased further, the avalanche gain in the chamber continues to increase until eventually lower energy radiations from the source or background radiations create pulses that also pass through the LLD, at which point the count rate rapidly increases. Usually, the operator selects an operating voltage in the center of the perceived *counting plateau* to conduct the sample measurements.

A measure of the performance of a proportional counter is the slope of counting plateau. To obtain the slope first draw a smooth line through the data points and, somewhat subjectively, determine the voltage V_a at the start of the plateau and the voltage V_b and the end of the plateau. The fractional plateau slope is given by

$$\text{slope} = \left(\frac{\text{Cts}_b - \text{Cts}_a}{V_b - V_a}\right)\left(\frac{2}{\text{Cts}_b + \text{Cts}_a}\right) \tag{10.60}$$

where Cts_a and Cts_b are the measured counts at operating voltages V_a and V_b, respectively.[6]

[6]An alternative expression is

$$\text{slope} = \left(\frac{\text{Cts}_b - \text{Cts}_a}{V_b - V_a}\right)\left(\frac{1}{\text{Cts}_a}\right),$$

where ΔCts is divided by Cts_a. But, because proportional counters are operated in the counting plateau center, and because of the subjective nature of choosing the counting plateau limits, the authors believe that dividing by the average counts offers a better representation of the fractional slope.

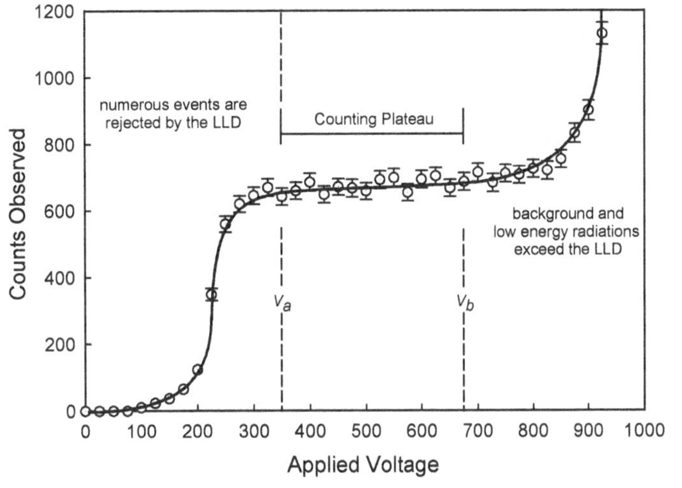

Figure 10.11. A counting curve for a gas-filled proportional counter in which a monoenergetic alpha particle source was used to calibrate the detector counting plateau.

The slope can also be expressed in terms of percent change per 100 volts of operating potential,

$$\% \text{ slope}|_{100V} = 100 \left(\frac{100(\text{Cts}_b - \text{Cts}_a)}{V_b - V_a} \right) \left(\frac{2}{\text{Cts}_b + \text{Cts}_a} \right). \tag{10.61}$$

Example 10.3: Measurements were conducted with a proportional counter with a monoenergetic α-particle source for many operating voltages and yielded the data of Table 10.8. These data were used to construct Fig. 10.11. Determine the % slope of the detector.

Solution:
The limits of the counting plateau are determined to be $V_a = 350$ volts and $V_b = 675$ volts, yielding

$$\begin{aligned} \text{slope} &= \left(\frac{\text{cnts}_b - \text{cnts}_a}{V_b - V_a} \right) \left(\frac{2}{\text{cnts}_b + \text{cnts}_a} \right) \\ &= \left(\frac{686 \text{ cnts} - 642 \text{ cnts}}{675 \text{ volts} - 350 \text{ volts}} \right) \left(\frac{2}{686 \text{ cnts} + 642 \text{ cnts}} \right) = 0.000204 \end{aligned}$$

$$\begin{aligned} \% \text{ slope}|_{100V} &= 100 \left(\frac{100(\text{cnts}_b - \text{cnts}_a)}{V_b - V_a} \right) \left(\frac{2}{\text{cnts}_b + \text{cnts}_a} \right) \\ &= 2.04\% \text{ per 100 volts.} \end{aligned}$$

Source and Detector Effects

Quite often a source may emit several different radiations. Such mixed sources may emit alpha particles, beta particles and gamma-rays of different energies. If performed correctly, the proportional counting can be used to distinguish among the different radiations and their energies. Consider the pulse height histogram of Fig. 10.12, showing the spectral features of a mixed source that emits alpha and beta particles.

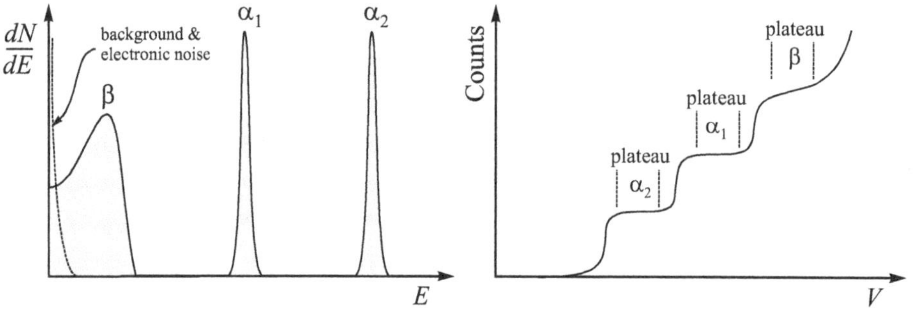

Figure 10.12. Depiction of the differential pulse height spectrum from a mixed α and β particle source and the resulting counting curve for a relatively large volume proportional counter.

For many proportional counter designs, the ranges of the α or β particles are longer than the chamber dimensions. In other words, the particles do not deposit all of their initial energy in the detector. Because the differential energy deposition, or stopping power $-dE/dx$, is much greater for α particles than that for β particles, the density of the ionized charge cloud produced by the α particles is greater than that produced by β particles. Consequently, the operating voltage required to produce pulses exceeding the LLD is lower for α particles than β particles.

As shown in Fig. 10.12, the highest energy α particle yields a counting plateau at the lowest operating voltage. The α particle with the lower energy, α_1, requires a higher operating voltage to produce a counting plateau. Note that the counting plateau for α_1 includes the sum of counts from both α_2 and α_1. At a higher voltage, the β particle source produces its counting plateau.

Because β particles are emitted with a continuum of energies, ranging from zero to E_{max}, the counting plateau for β particles has a much higher increasing slope with increasing operating voltage than do the slopes for the alpha particles. Consider Fig. 10.12, in which a beta particle spectrum is depicted. As the operating voltage increases, the highest energy β particles eventually are able to produce pulses above the LLD. As the operating voltage is increased further, more β particles with lower energies produce pulses above the LLD, a trend that continues with increasing voltage. At high enough voltages, background counts are included in the measurements, as well as the possibility of including spurious "counts" from the high electric field, thereby causing a rapid increase in counts unrelated to the radiation source.

Electrons ejected by γ-ray interactions, β particles and positrons all have similar stopping powers $-dE/dx$ in the detector gas. Usually, proportional counter chambers are not large enough to capture all of the energy from β particle emissions before they collide with the chamber wall. Consequently, the counting plateau usually appears at a similar location in the counting curve regardless of the energy of the electron. Distinction among energies becomes difficult unless the detector chamber is large enough to absorb all (or most) of the energy. However, such chambers are very much larger than normal proportional counters. Note that large gas-filled chambers do exist, many used for high energy physics experiments that require the capture of all or much of the energy from weakly ionizing particles.

For some proportional counters, particularly pancake designs, the chamber dimension parallel with the particle trajectories can be so small that higher energy α particles lose less energy in the chamber than lower energy α particles, thereby producing less ionization than the lower energy α particles; hence, the appearance of the counting plateaus is reversed in order from that normally expected. The case becomes clear by considering the Bragg ionization curves depicted in Fig. 10.13. In this case the highest energy α particles deposit the least amount of energy before striking the chamber wall. As a result, it is the lowest

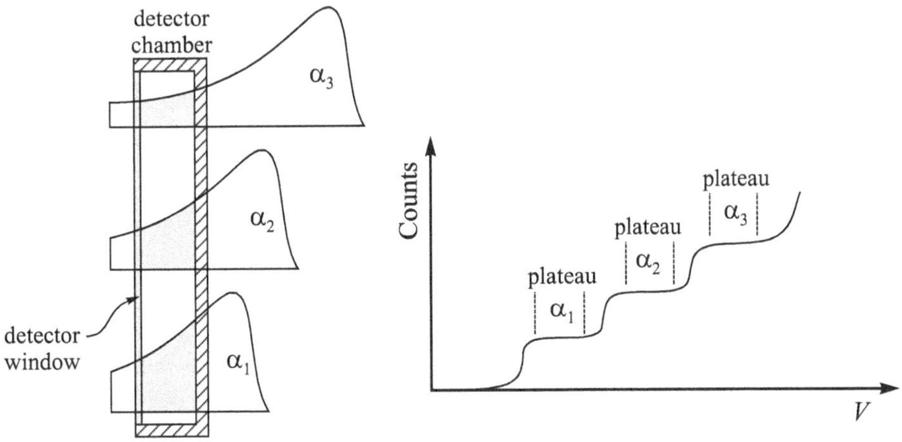

Figure 10.13. Depiction of Bragg ionization curves for multiple α particle energies as they intersect a narrow detector chamber and the resulting counting curve. The shaded regions in the Bragg ionization curves depict the integrated energy deposited in the chamber.

energy α particle that produces a counting plateau at the lowest operating voltage. Example radioactive sources that can exhibit this property with pancake proportional counters include ^{228}Th and ^{226}Ra. Note that particles entering the detector at oblique angles deposit more energy in the detector; however, the flat portion of the counting plateau appears when nearly all particles of a particular energy produce pulses above the LLD. Under appropriate detector design and operating conditions, it is possible that the different alpha particles have overlapping counting plateaus, making distinction among them difficult.

10.5.4 Fluctuations of the Gas Multiplication Process

Because the proportional counter maintains a proportional relation between absorbed energy and the input pulse $V_{in}(t)$, it is possible to operate proportional counters as energy spectrometers. Consider the case of a monoenergetic radiation source, in which the emitted radiations deposit their total energy into the detector upon each interaction. The average number of secondary electrons m produced through avalanching by a single initial electron is given by a PDF $p_{aval}(m)$ with a mean $\langle m \rangle$ and a variance $\sigma^2(m)$. Various models for $p_{aval}(m)$ are presented later in this section.

For a *single event* that produces n initial electron-ion pairs, there is an average multiplication factor $M(n)$ for the *total* avalanche shower, that is given by

$$M(n) \simeq \sum_{i=1}^{n} m_i, \tag{10.62}$$

where the $\{m_i\}$ are random variables sampled from $p_{aval}(m)$ representing independent electron avalanches. The average multiplication per initial electron is

$$\overline{m} = \frac{1}{n}\sum_{i=1}^{n} m_i = \frac{1}{n} M(n), \tag{10.63}$$

whose variance is

$$\sigma^2(\overline{m}) = \frac{1}{n^2}\sum_{i=1}^{n} \sigma^2(m_i) = \frac{\sigma^2(m)}{n}. \tag{10.64}$$

The expected or mean value of $M(n)$, a sum of n independent random variables (the $\{m_i\}$), is

$$\langle M\rangle = \langle \sum_{i=1}^{n} m_i \rangle = \sum_{i=1}^{n} \langle m_i\rangle = n\langle m\rangle, \tag{10.65}$$

and the variance is [Dunn and Shultis 2012]

$$\sigma^2(M) = \sum_{i=1}^{n} \sigma^2 m_i + 2\sum_{i=1}^{n}\sum_{\substack{j=1\\ j<i}}^{n} \mathrm{covar}(m_i, m_j). \tag{10.66}$$

If the $\{m_i\}$ are independent and identically distributed,[7] then the covariances vanish leaving

$$\sigma^2(M) = \sum_{i=1}^{n} \sigma^2 m_i = \sum_{i=1}^{n} \sigma^2(m) = n\sigma^2(m). \tag{10.67}$$

For a given source, detector, geometry, etc., the probability that a radiation event produced n electron-ion pairs in the chamber gas is denoted by $p_{in}(n)$ with a mean $\langle n\rangle$ and variance $\sigma^2(n)$. This PDF is very complex because it involves modeling the electron-ion pairs produced along the trajectory of the incident radiation particle. As a result no explicit expression for $p_{in}(n)$ is usually available, but, conceptually, it exists.

A pulse height spectrum is a histogram depicting the frequency of observing the total charge Q liberated in a detector. For each monoenergetic radiation particle interacting within the detector, there are n_j electron-ion pairs initially liberated, and the shower produced has an average multiplication factor $\overline{m}_j$. As information is accumulated, i.e., more interaction events are observed, an average number of initial electron-ion pairs $\overline{n}$ becomes evident as does an average multiplication factor $\overline{M}$. The average charge collected over multiple monoenergetic events is, therefore, $\overline{Q} = q\overline{nM}$ where

$$\overline{nM} \simeq \langle nM\rangle = \sum_{i=1}^{\infty} n_i M(n_i) p_{in}(n_i). \tag{10.68}$$

The variance $\sigma^2(\overline{Q})$ defines the theoretical limit on energy resolution for a proportional counter. Two sources of variance are identified above, namely, the variance $\sigma^2(\overline{n})$ in the initial number of electron-ion pairs produced and the variance $\sigma^2(\overline{m})$ in the multiplication factor for each avalanche triggered by an initial electron. Because $\overline{n}$ and $\overline{m}$ are independent variables, simple error propagation can be used to give

$$\frac{\sigma^2(\overline{Q})}{\overline{Q}^2} = \left(\frac{\sigma(\overline{n})}{\overline{n}}\right)^2 + \left(\frac{\sigma(\overline{m})}{\overline{m}}\right)^2. \tag{10.69}$$

Substitution of Eq. (10.64) into Eq. (10.69) yields

$$\frac{\sigma^2(\overline{Q})}{\overline{Q}^2} = \left(\frac{\sigma(\overline{n})}{\overline{n}}\right)^2 + \frac{1}{n}\left(\frac{\sigma(m)}{\overline{m}}\right)^2. \tag{10.70}$$

For a purely stochastic Poisson process, the expected variance for n is $\sigma^2(\overline{n}) = n$, where $\overline{n} = E/w$, E is the energy deposited in the detector and w is the average ionization energy. Therefore, the expected standard

[7] This assumption is not valid if the positive ions create significant space-charge effects.

error for a Poisson distribution is $\sigma(\overline{n}) = \sqrt{E/w}$. However, Fano [1947] points out that the ionization process does not strictly follow a Poisson probability density function.

The process by which charged particles lose energy by impacting gas atoms to produce electron-ion pairs is complex. One can measure the distance in the gas that an alpha particle travels (Δx) and the associated energy loss ($-\Delta E_\alpha$). After numerous trials, fluctuations are observed in ΔE_α about some average value $\overline{\Delta E_\alpha}$. Now consider the inverse case, in which the change in energy ΔE_α is constant and the associated fluctuation in Δx is recorded. Such a problem is similar to the alpha particle straggling issue described in Chapter 4.

Suppose it is determined that the average energy lost over distance x_0 is E_0. If instead, the alpha particle loses energy E over distance x_0, then it is expected that the average energy loss E_0 is achieved within an additional travel distance Δx, such that

$$\Delta x = \frac{x_0(E_0 - E)}{E_0}. \tag{10.71}$$

Note that Δx may be positive or negative. For any amount of energy lost, there is an average number of electron-ion pairs w produced for that energy loss, thereby producing an average number of ions n_0 for a defined energy loss E_0. If, instead, the alpha particle produces n ionizations along Δx while losing the energy E, then it is expected that the difference in ionization is realized within,

$$\Delta n = \frac{(E_0 - E)}{w}. \tag{10.72}$$

Adding n to both sides of this result and using $E_0/w = n_0$ one obtains

$$n + \Delta n - n_0 = n - \frac{E}{w}. \tag{10.73}$$

Consider a charged particle that has a total of I ionization impacts with the host gas. For the ith impact, n_i is the number of ionizations produced and u_i is the energy lost. The result of each impact is independent of other impacts; hence, averaged over many histories it can be assumed that $\overline{n}_i = \overline{n}$ for all n_i, and $\overline{u}_i = \overline{u}$ for all u_i. It is therefore shown, for distance x_0, $I = E_0/\overline{u} = n_0/\overline{n}$, and the variance is [Fano 1947]

$$\sigma^2_{n_0} = \langle (n - E/w)^2 \rangle = I \langle (n_i - u_i/w)^2 \rangle = F n_0, \tag{10.74}$$

where F is

$$F = \frac{\langle (n_i - u_i/w)^2 \rangle}{\overline{n}}, \tag{10.75}$$

Table 10.9. Fano factors for several detector gases.

Gas	F	Citation
He	0.20	Alkhazov 1969
Ne	0.13	Grosswendt 1984
Ar	0.17	Alkhazov 1967
Kr	0.19	Krajcar-Bronić 1998
Xe	0.18–0.32	Dias 1997
CO_2	0.32	Alkhazov 1967
CH_4	0.26	Alkhazov 1967
C_2H_2	0.27	Alkhazov 1967

and is traditionally called the *Fano factor*. The Fano factor accounts for the deviation from statistical fluctuations expected with a Poisson distribution. Hence, the Fano factor can be defined as

$$F = \frac{\text{observed variance}}{\text{expected variance from Poisson statistics}}. \tag{10.76}$$

Equation (10.70) then can be rewritten as,

$$\frac{\sigma^2(\overline{Q})}{\overline{Q}^2} = \frac{F}{\overline{n}} + \frac{1}{\overline{n}}\left(\frac{\sigma(m)}{\overline{m}}\right)^2. \tag{10.77}$$

Experimental measurements and theoretical calculations of Fano factors for various gases and gas mixtures have been reported in the literature with varying agreement [for example, Alkhazov et al. 1967, Alkhazov and Vorobe'ev 1969; Grosswendt 1984; Dias et al. 1997; Pansky et al.1997; Krajcar-Bronić 1998]. Dias et al. [1997] and Pansky et al. [1997] report that the Fano factor is dependent upon the initial energy of the primary ionizing particle, particularly important near photon absorption edges. The Fano factors for several gases are given in Table 10.9.

The Avalanche Distribution Function $p_{aval}(m)$

The theory of electron multiplication seeks to find the probability distribution function $p_{aval}(m)$ that gives the probability that an avalanche initiated by a single electron produces m electrons. This distribution problem has been studied by Snyder [1947], Frisch [1947, 1959], and Wijsman [1949]. All make several simplifying assumptions. Secondary electrons are produced only by impact ionization (one per impact) and production by photoionization is neglected. Recombination, electron attachment, and space charge effects are likewise neglected. Finally, the probability of an impact ionization is solely dependent upon the electron distance from surface of the anode.

In this section the approach used by Wijsman [1949] to find p_{aval} for a planar device is followed, In such a device, avalanching can occur anywhere within the chamber, i.e., avalanching could occur anywhere between the cathode ($x = 0$) and the anode ($x = d$). In Wijsman's approach it is also assumed that an impact collision creates only one additional electron.

Consider a coaxial ion chamber with the central wire as the anode. As electrons approach the inner anode wire, the avalanche region begins at the critical radius r_c from the coaxial center. When electrons enter the region defined by radius r_c, they attain sufficient kinetic energy to cause impact ionization. A local radial coordinate x is now introduced such that $x = 0$ at r_c and $x = 1$ at anode radius a, i.e.,

$$x = \frac{r_c - r}{r_c - a}, \quad a \leq r \leq r_c. \tag{10.78}$$

The probability an electron causes an ionization within dx about x is $\alpha_T(x)dx$, where α_T is the first Townsend coefficient. Wijsman sought $p(n, x)$ the probability that a single electron that started at $x = 0$ had grown to an avalanche of n electrons at a distance x. The probability no ionization occurs between $x = 0$ and x is

$$p(1, x) = \exp[-\tau(x)], \qquad \text{where} \qquad \tau(x) = \int_0^x \alpha_T(x')\, dx'. \tag{10.79}$$

To find $p(n, x)$, proceed as follows. The probability the avalanche contains $n - 1$ electrons at x' is $p(n - 1, x')$. The probability that one and only one of these electrons causes an ionization between x' and $x' + dx'$ is the probability that one of the $(n - 1)$ electrons causes an ionization multiplied by the probability that the other $(n - 2)$ electrons produce no ionization, i.e., $(n - 1)\alpha_T(x')dx'[1 - \alpha_T(x')dx']^{n-2}$. As $dx' \to 0$ this probability becomes,

$$(n - 1)\alpha_T(x')dx'. \tag{10.80}$$

Now the avalanche contains n electrons and the probability that none causes further ionizations between x' and x is

$$\exp[-n\{\tau(x) - \tau(x')\}]. \tag{10.81}$$

To find $p(n, x)$ multiply $p(n - 1, x')$ by the factors of Eqs. (10.80) and (10.81) and integrate over $(0, x)$ to obtain, for $n > 1$,

$$p(n, x) = \int_0^x p(n - 1, x')(n - 1)\alpha_T(x') \exp[-n\{\tau(x) - \tau(x')\}]dx'. \tag{10.82}$$

To solve this equation for $p(n,x)$, subject to Eq. (10.79), use a change of variables $y(x) = \exp[\tau(x)]$ from which it is found $y(0) = 1$ and $dy/y = d\tau$. Let $q(n,y) \equiv y^n p(n,y)$. Equation (10.82) transforms to

$$q(n,y) = (n-1)\int_1^y q(n-1,y')\,dy'. \tag{10.83}$$

Differentiation of this equation gives

$$\frac{dq(n,y)}{dy} = (n-1)q(n-1,y) \qquad \text{with} \qquad q(1,y) = 1. \tag{10.84}$$

Solve these equations successively by induction for $n = 2, 3, 4, \ldots$ to find

$$q(n,y) = (y-1)^{n-1}. \tag{10.85}$$

Finally, transforming y back to x, the solution of Eq. (10.82) is

$$p(n,x) = \exp[-n\tau(x)]\{\exp[\tau(x)] - 1\}^{n-1}. \tag{10.86}$$

In particular, when the avalanche reaches the anode, $x = 1$ and $n \rightarrow m$ so

$$p(m,1) = \exp[-m\tau(1)]\{\exp[\tau(1)] - 1\}^{m-1}, \tag{10.87}$$

or in terms of the original coordinate system

$$p(m,a) = \exp[-m\tau(a)]\{\exp[\tau(a)] - 1\}^{m-1}, \tag{10.88}$$

where the "optical thickness" of the avalanche region is

$$\tau(a) = \int_a^{r_c} \alpha_T(r')\,dr'. \tag{10.89}$$

This distribution is known as the Furry distribution.[8] It is easily shown

$$\overline{m} \equiv \langle m \rangle = \sum_{m=1}^{\infty} mp(m,a) = \exp[\tau(a)]. \tag{10.90}$$

Thus, $p_{aval}(m|\overline{m})$ is given by

$$\boxed{p_{aval}(m|\overline{m}) = \frac{1}{\overline{m}}\left(1 - \frac{1}{\overline{m}}\right)^{m-1}}\ . \tag{10.91}$$

[8] This distribution was derived earlier by Furry [1937] who was studying electron showers produced by very high energy electrons incident on lead. Furry found the number n of secondary electrons produced in a lead sheet of thickness t was given by

$$p(n|\lambda t) = e^{-\lambda t}(1 - e^{-\lambda t})^{n-1}, \quad n \geq 1,$$

where λ is an interaction coefficient with units of reciprocal length. The mean of this PDF is $\langle n \rangle = e^{\lambda t}$ so the PDF can be written as

$$p(n|t) = \frac{1}{\langle n \rangle}\left(1 - \frac{1}{\langle n \rangle}\right)^{n-1}.$$

It is interesting that two different problems yield the same result. Wijsman was unaware of this earlier work, but because of Furry's primacy, the distribution of Eq. (10.91) is named after him.

The second moment of this PDF is $\langle m^2 \rangle = 2\overline{m}^2 - \overline{m}$, so the relative variance is

$$\frac{\sigma^2(m)}{\overline{m}^2} = 1 - \frac{1}{\overline{m}}. \tag{10.92}$$

For large values of $\overline{m}$, which is usually the case, Eq. (10.92) reduces to [Lansiart and Morucci 1962],

$$\frac{\sigma^2(\overline{m})}{\overline{m}^2} = 1. \tag{10.93}$$

Equation (10.91) can be expanded as

$$p_{aval}(m|\overline{m}) = \frac{1}{\overline{m}}\left[1 - (m-1)\frac{1}{\overline{m}} + \frac{1}{2}(m^2 - 3m + 2)\left(\frac{1}{\overline{m}}\right)^2 - \mathcal{O}\left(\frac{1}{\overline{m}}\right)^3\right]. \tag{10.94}$$

In the limit of large $\overline{m}$

$$p_{aval}(m|\overline{m}) \simeq \frac{1}{\overline{m}}\left[1 - \frac{m}{\overline{m}} + \frac{1}{2}\left(\frac{m}{\overline{m}}\right)^2\right] \simeq \frac{1}{\overline{m}}\exp\left(-\frac{m}{\overline{m}}\right). \tag{10.95}$$

For values of $\overline{m} \gtrsim 50$, Eq. (10.95) approximates Eq. (10.91) quite well.[9] It should be noted that this exponential PDF gives large fluctuations in m about the mean. Genz [1973] also points out that, for low values of the reduced electric field ($\mathcal{E}/P$), the exponential approximation of the Furry model of Eq. (10.95) adequately describes the probability distribution of avalanche electrons.

Effects of High Electric Fields

However, as noted by Byrne [1962], the Furry model fails to predict the observed results reported for the distribution function of m under high electric field conditions [Curran et al. 1949]. The Furry model was initially developed for high-speed particle impacts and it was assumed that the electrons have equal probability of the causing impact ionization, within a differential distance of travel, regardless of their origin. Thus, the effects of high electric fields, which accelerate the electrons, were ignored. In a coaxial gas detector, as the electron shower progresses, the distance between ionizing collisions decreases because of the increasing electric field as electrons approach the anode. Further, as the applied electric field is increased, eventually the average distance between collisions in the highest field region is about the same as the minimum travel distance for impact ionization, thereby fluctuations about $\overline{m}$ are suppressed [Schlumbohm 1958a]. As a result, it is not surprising the Furry model does not hold for high values of $\mathcal{E}/P$.

Distribution functions $p_{aval}(m)$ from avalanche data for different values of $\mathcal{E}/P$ are given in the literature for selected gases [Curran 1949; Schlumbohm 1958a, 1958b; Cookson and Lewis 1966a, 1966b; Genz 1973]. From the literature, it becomes apparent that at high values of $\mathcal{E}/P$ the data distribution functions deviate from the exponential approximation of the Furry PDF but, instead, develop a maxima in the distribution as can be seen in Fig. 10.14. Several alternative models have been proposed to account for the high-field measured distributions of $p_{aval}(m)$. These models are discussed by Genz [1973], and are shown to match the measured distributions with varying degrees of success. One of the more popular models is based on the *Pólya* statistical model [Pólya 1930]. Arley [1943] introduced the use of the Pólya model for the analysis of an electron shower in lead, a similar problem considered by Furry.

[9] From Eq. (10.95) the relative variance is found to be,

$$\sigma^2(m) = \int_0^\infty (m - \overline{m})^2 \frac{1}{\overline{m}} \exp\left(-\frac{m}{\overline{m}}\right) dm = \overline{m}^2,$$

so that $\sigma^2(m)/\overline{m}^2 = 1$, in agreement with Eq. (10.93).

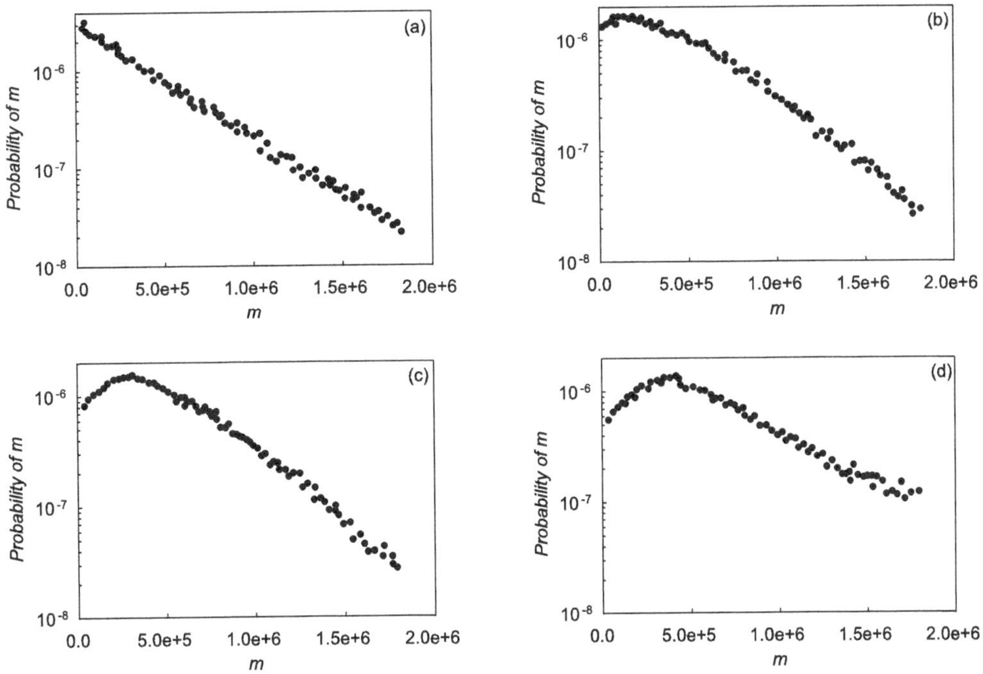

Figure 10.14. Measured multiplication factor distributions for methane gas at reduced electric fields of (a) 48, (b) 120, (c) 156, and (d) 218 V cm^{-1} $torr^{-1}$. Only under the lowest electric field reported (a) does the distribution follow Furry's prediction. The other distributions shown in (b)–(d) resemble a Pólya distribution. Data from [Cookson and Lewis 1966b].

Byrne [1962] relaxed many of the assumptions used by Wijsman and developed a more general expression for $p_{aval}(m)$. In the present notation, Byrne obtained

$$p_{aval}(m|\overline{m}) = \left(\frac{\overline{m}}{1+b\overline{m}}\right)^m \frac{(1)(1+b)...(1+(m-1)b)}{m!}(1+b\overline{m})^{-1/b}, \quad m \geq 1. \tag{10.96}$$

This PDF is the Pólya distribution, also known more commonly as the negative binomial distribution, in which $b \geq 0$ is a parameter.

The first and second moments of Eq. (10.96) are [Byrne 1962],

$$\langle m \rangle = \overline{m} \qquad \text{and} \qquad \langle m^2 \rangle = (1+b)\overline{m}^2 + \overline{m}. \tag{10.97}$$

Hence, $\sigma^2(m) = \langle m^2 \rangle - \langle m \rangle^2 = b\langle m \rangle^2 + \langle m \rangle$ and the relative variance is thus

$$\boxed{\frac{\sigma^2(\overline{m})}{(\overline{m})^2} = b + \frac{1}{\overline{m}}} \ . \tag{10.98}$$

Some measurements of the relative variance indicate that average values as low as 0.45 can be expected [Sipila 1976]. For large values of $\overline{m}$, the limiting case for Eq. (10.98) reduces to,

$$\frac{\sigma^2(\overline{m})}{(\overline{m})^2} = b, \tag{10.99}$$

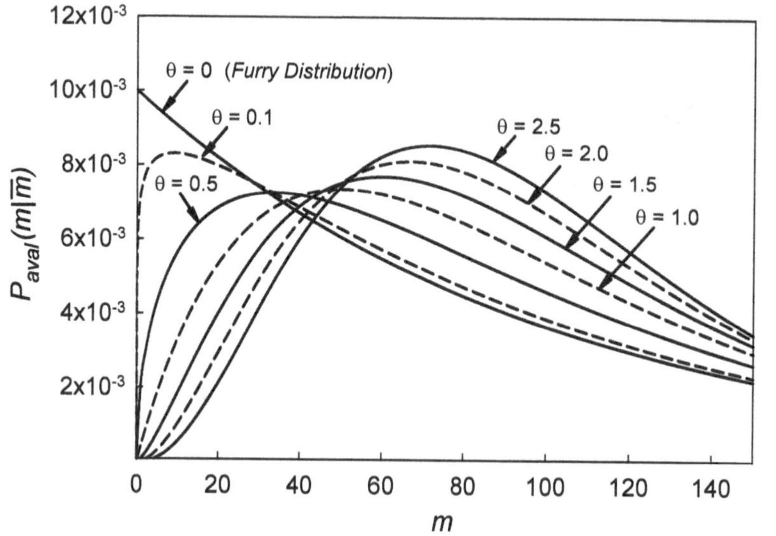

Figure 10.15. The Pólya distributions of (Eq. (10.102)) for $\overline{m} = 100$ and $\theta = 0, 0.1, 0.5, 1.0, 1.5, 2.0,$ and 2.5.

a result reported by Lansiart and Morucci [1962].

Cookson and Lewis [1966a] assumed that the first Townsend coefficient α_T can be represented as

$$\alpha_T = \alpha_{T(max)}\left(1 + \frac{\theta}{m}\right), \tag{10.100}$$

where $\alpha_{T(max)}$ is the limiting value of α_T for large m ($m \gtrsim 100$) and $\theta \geq 0$ is a parameter dependent upon the gas type and operating conditions. Byrne [1962] and Cookson and Lewis [1966b] demonstrated that the assumption of Eq. (10.100) leads to a Pólya distribution,

$$p_{aval}(m|\overline{m}) = \frac{\widetilde{b}}{\overline{m}}\frac{1}{(\widetilde{b}-1)!}\left(\frac{\widetilde{b}m}{\overline{m}}\right)^{\widetilde{b}-1}\exp\left(-\frac{\widetilde{b}m}{\overline{m}}\right), \tag{10.101}$$

where $\widetilde{b} = b^{-1} = (\theta + 1)$ and $\widetilde{b} \geq 1$. Byrne showed that as $\widetilde{b}$ increases above unity (or θ increases above zero), Eq. (10.101) has a maximum at increasing values of m, as shown in Fig. 10.15. The physical significance of θ (or $\widetilde{b}$) is debatable. Byrne [1962, 1969] suggests that it represents the fraction of electrons above the necessary gas ionization energy in the presence of large numbers of electrons. However, Cookson and Lewis [1966a] and Alkhazov [1970] have contested Byrne's explanation of θ (and $\widetilde{b}$). Limiting conditions indicate that for $\widetilde{b} \to 1$ (or $\theta \to 0$), Eq. (10.102) reduces to the Furry statistical model, whereas when $\widetilde{b} \to \infty$, Eq. (10.102) reduces to the Poisson statistical model.

With a value of $\widetilde{b} \simeq 3/2$, Eq. (10.101) agreed with the measured values of Curran et al. [1949]. This agreement supports the validity of using the Pólya model for the electron avalanche. Substitution of $\widetilde{b} = (1+\theta)$ into Eq. (10.101) yields,

$$\boxed{p_{aval}(m|\overline{m}) = \frac{(1+\theta)}{\Gamma(1+\theta)}\frac{1}{\overline{m}}\left(\frac{(1+\theta)m}{\overline{m}}\right)^{\theta}\exp\left(-\frac{(1+\theta)m}{\overline{m}}\right)} \,. \tag{10.102}$$

The total limiting variance for the avalanche electron population is found by substituting Eq. (10.92) or Eq. (10.98) into Eq. (10.77), thereby yielding

$$\frac{\sigma^2(\overline{Q})}{\overline{Q}^2} = \frac{F}{\overline{n}} + \frac{1}{\overline{n}}\left(1 - \frac{1}{\overline{m}}\right), \tag{10.103}$$

for low $\mathcal{E}/P$ values, and

$$\frac{\sigma^2(\overline{Q})}{\overline{Q}^2} = \frac{F}{\overline{n}} + \frac{1}{\overline{n}}\left(\frac{1}{\overline{b}} + \frac{1}{\overline{m}}\right), \tag{10.104}$$

for high $\mathcal{E}/P$ values.

Example 10.4: Determine the limiting resolution for 122-keV γ rays in a proportional counter backfilled with Ar for which $\overline{m} = 100$. Disregard any recombination effects.

Solution:

From Table 10.9, $F = 0.17$ for Ar, and from Table 9.1, $w = 26$ eV/iep pair for Ar. Assume a value of $\overline{b} \approx 0.45$. From Eq. (10.104) and use of $\overline{n} = E_\gamma/w$

$$\frac{\sigma^2(\overline{Q})}{\overline{Q}^2} = \frac{w}{E_\gamma}\left[F + \left(\frac{1}{\overline{b}} + \frac{1}{\overline{M}}\right)\right] = \frac{26 \text{ eV/eip pair}}{122000 \text{ eV}}\left[0.17 + \left(0.45 + \frac{1}{100}\right)\right] = 1.34 \times 10^{-4}.$$

The fractional energy resolution is defined as,

$$\frac{\text{FWHM}}{Q} = 2\sqrt{2\ln 2}\left(\frac{\sigma(\overline{Q})}{\overline{Q}}\right) = 0.0273. \tag{10.105}$$

This result predicts an energy resolution of 2.73%.

In practice, the energy resolution of a proportional counter of the magnitude predicted by Example 10.4 is unusually small because many other factors affect the energy resolution. Sources of resolution degradation include noisy electronics, varying gas pressure, gas impurities, avalanche saturation, anode wire eccentricity and end effects, and anode wire uniformity [Charles and Cooke 1968].

Sources of electronic noise include variations in operating voltages from the high voltage power supply, detector leakage current and amplifier electronic noise, all of which can be reduced to negligible levels for well-designed equipment. Gas pressure variations can be caused by temperature changes, or, for gas-flow systems, by variations in the gas flow, although such variations are usually small if the measurement period is relatively short. Gas impurities can alter ion speeds and increase electron attachment. However, if the gas detector system is thoroughly purged with a purified counting gas before use, the effect of gas impurities on energy resolution can be eliminated. The gas flushing should be conducted long enough to remove moisture contamination from the system interior surfaces.

Avalanche saturation causes the appearance of non-linear pulse heights, a result of operating the counter in Region IIIa, the transition region, as was discussed in Sec. 10.5.2. To reduce the shift in pulse heights, Spielberg [1967] reports that larger diameter wires produce less signal variance, and that tungsten wires show smaller shifts in pulse heights than do steel wires. Limiting counting rates below 2000 cps also helps minimize the variance [Charles and Cooke 1968]. Anode wire eccentricity can create axial non-uniformities in the electric field, thereby adding to the variance about $\overline{m}$, although this effect is generally small in well-built proportional counters [Rossi and Staub, 1949]. The uniformity of the wire has been shown also to affect the shift in pulse heights [Spielberg 1967], but thermal cleaning (heating to temperatures capable of evaporating a microscopic layer from the wire) improved performance. However, the benefit of thermally cleaning degraded over a period of a week, presumably from deposition of gas products upon the wires. Indeed, for proportional counters in a high radiation environment, the breakdown of counting gases, especially those with organic quenching gases, tends to coat the anode wires and leads to an increase in electrical resistance and wire diameter. As a result, the variance in the avalanching electric field increases the variance about $\overline{m}$. Despite these complications, energy resolutions below 8% FWHM have been reported for low-energy gamma rays [Charles and Cooke 1968, Sakurai and Ramsey 1992], and below 2% FWHM for neutron induced reaction products [Oed 1988].

10.6 Selected Proportional Counter Variations

Since their introduction by Geiger and Rutherford in 1908, gas-filled detectors have been used widely for many different radiation measurements, ranging from small pocket dosimeters to large chambers for high-energy physics measurements. Proportional counters are of particular interest, mainly due to their high signal gain while preserving energy deposition information. There are two basic types of proportional counters: gas flow detectors and sealed detectors. Proportional counters can detect charged particles, energetic photons, and neutrons. They are made in many sizes and shapes, too many to describe in this chapter, although many variations are described elsewhere [Rossi and Staub 1949; Staub 1953; Price 1964; Sauli 2014]. However, some common proportional counter geometries, along with a few uniquely interesting detectors, are described in the following sections.

10.6.1 Gas-Flow Proportional Counters

Gas flow detectors are operated with the detection gas being constantly replenished during operation. This constant gas replenishment allows gas molecules degraded from ionizing interactions to be removed and replaced with fresh gas molecules. Moreover, the gas supply can be shut off when not in operation. Gas flow detectors can operate indefinitely, provided that a gas source is available. Consequently, gas flow detectors are usually operated as laboratory instruments due to the inconvenient requirement of an external gas supply. This requirement makes portable instrumentation awkward.

2π and 4π Counters

The general construction of a 2π proportional counter is depicted in Fig. 10.16. The chamber is usually cylindrical, as shown in Fig. 10.17; but sometimes the chamber may be hemispherical. In either case, a radioactive sample placed in the chamber, usually through a rotating sample changer, is exposed to a detector solid angle of 2π steradians; hence the name "2π counter". An anode wire loop extends into the chamber through an insulating high-voltage electrical feedthrough. Typically, the anode wire is on the order of 25 to 75 microns in diameter, often composed of a gold-tungsten alloy. For safety reasons, positive high voltage is applied to the anode wire while casing of the chamber is kept at ground potential.

The usual gas used in a gas-flow counter is P-10, and flows into the chamber through an input port and exits through a gas metering device, such as a "bubbler" or gas flow meter. A common flow rate is "one bubble per second", which is about 50 to 60 sccm (standard cubic cm per minute). Because a detector remains unvented between uses, it is common practice to purge the detector with the chosen detection gas for several minutes before operation, usually at a higher flow rate than used during operation. For instance, the purge may use gas flows >300 sccm to remove any air from the chamber, but then is reduced to about 55 sccm during operation. The reduction in gas flow reduces the possibility of perturbing charge collection in the chamber, with the added advantage of conserving the detector gas.

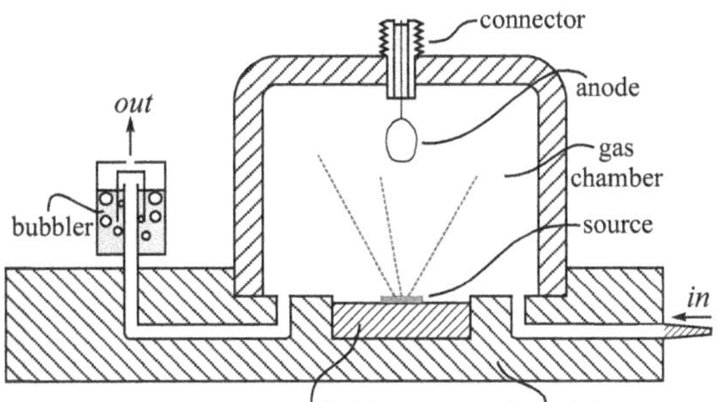

Figure 10.16. Diagram of a 2π gas-flow proportional counter.

Figure 10.17. A 2π gas-flow proportional counter, showing the detector base-plate, bubbler, 2π detector, and gas inlet.

A major feature of the "windowless" gas flow detector depicted in Fig. 10.16 and Fig. 10.17 is the absence of an attenuating medium between the radiation source and the detector, because the source is inside the detector. In the 2π detector approximately 50% of emissions enter the detector gas, although self-absorption and backscattering effects must still be taken into account. Alpha particles, with ranges of a few centimeters, deposit a large fraction of their energy in the chamber. Beta particles, usually having ranges up to tens of centimeters, usually deposit only a small portion of their energy in the chamber. Gamma rays usually interact in the walls of a proportional counter rather than with the gas. Interactions in the wall can eject energetic electrons into the gas chamber. The stopping power $-dE/dx$ is much higher for alpha particles than for beta particles (or electrons/positrons). As a result, for sources that emit both alpha and beta particles, a counting curve similar to that shown in Fig. 10.12 is expected, in which the alpha particle plateau(s) appears at a much lower voltage than the beta particle plateau. Consequently, beta particles and gamma rays are easily discriminated against by setting the high voltage to the center of the alpha particle counting plateau and adjusting the LLD above the pulse heights produced by beta particles and gamma rays. The windowless gas-flow detector also has the advantage of being able to detect low energy beta particles that would otherwise be attenuated by the intervening air and the detector window. This advantage is important for radionuclides that emit low energy beta particle such as ^{14}C and ^{3}H, provided that such sources are properly prepared so as to reduce self-absorption effects.[10]

The 4π windowless gas-flow detector is designed so the radioactive source is placed upon a thin platform between two 2π detectors [Martin and Green 1958]. If the combined thickness of the platform and radioactive sample is relatively thin compared to the ranges of the particles of interest, then almost all of emitted particles enter one of the detector halves. This configuration reduces uncertainty from backscattering, although self-absorption may still be an issue. These devices have been used to measure absolute beta particle activities. For radiation particles ejected simultaneously in relatively opposite directions, the two halves of a 4π counter can be operated in coincidence to record only those events (or operated in anti-coincidence to reject those events).

Thin Window Counters

Gas-flow detectors with thin windows are used when the radioactive source is outside of the detector, thereby restricting the detector solid angle to $<2\pi$. A common laboratory thin-window gas-flow detector configuration

[10]Low energy beta particles are more commonly measured with liquid scintillation detectors, covered in Chapter 13.

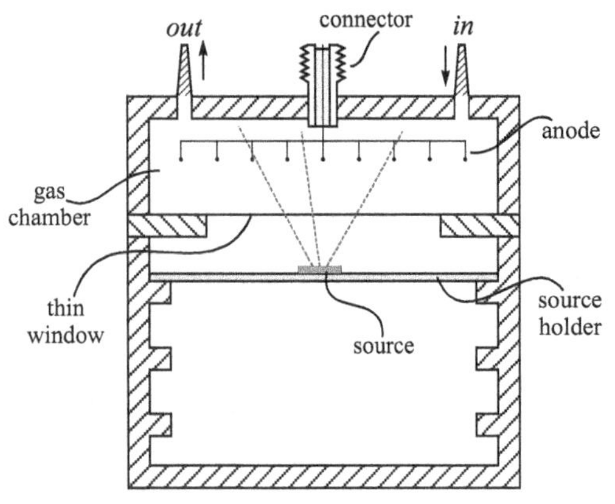

Figure 10.18. Diagram of a thin-window, gas-flow, proportional counter. A multi-wire anode is depicted in the figure; however, many thin-film proportional counters have a single wire loop for the anode.

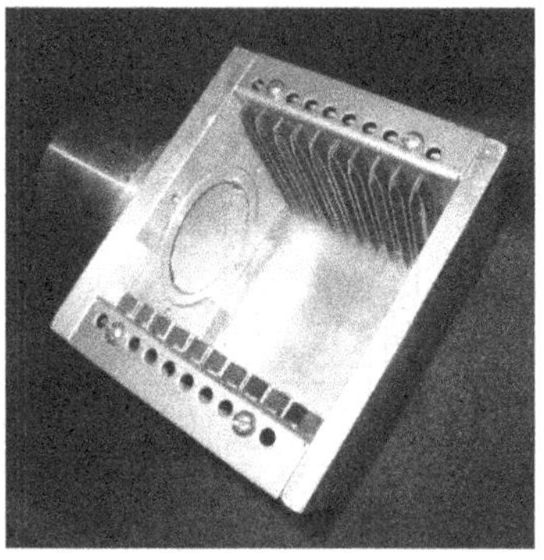

Figure 10.19. A thin-window, gas-flow, proportional counter, showing a metallized boPET window and sample holder slots.

is shown schematically in Fig. 10.18 and a photograph is given in Fig. 10.19. Commercial units commonly have windows fabricated from thin boPET material that has been aluminized to block interference from electromagnetic and visible light. The windows typically have mass thicknesses of about 0.4 to 2.0 mg/cm^2. To avoid damaging the window, the gas exit port must remain unobstructed.[11] Gas purge and operational flow rates are similar to those used for windowless gas-flow detectors.

Disadvantages of using a thin-window gas-flow proportional detector include decreased detector solid angle, parallax effects (see Sec. 7.5), and particle backscattering off the window. However, thin-window chambers are usually compact and relatively easy to load. Unlike windowless gas-flow detectors, the radioactive source distance for many thin-window detectors is variable, either incrementally with shelving slots, as shown in Fig. 10.18 and Fig. 10.19, or with the source mounted on a retractable threaded rod. Some commercial proportional counters have a head that can be mounted to a customized measurement system.

Thin-window gas-flow detectors, used as portable units at frisking stations, can have windows ranging from 25 to 100 cm^2 window area. They also have input ports for constant gas replenishment as shown in Fig. 10.20. Some commercial models have the gas input through the detector cable rather than through a separate gas supply line, thus simplifying cable/detector connections. The gas exits the detector through a pin-sized hole located at the top of the detector body. Thin-window gas-flow detectors are also configured as floor monitors, in which the detector height with respect to the floor is adjustable. The systems include a gas supply bottle, portable cart, readout ratemeter, and cable connections.

Specially designed thin-window, gas-flow, proportional counters are used for detection of low energy photons, such as characteristic x-rays [Goldstein et al. 1981], and are often used in measurement systems for wavelength dispersive x ray spectroscopy. The windows of these detectors can be made from boPET; however, higher transmission for low energies is obtained with lower density windows, such as polypropylene (1 to 6-microns thick) and Formvar® (polyvinyl formal), usually $<$ 1 micron thick. For example, a 6 micron thick boPET window absorbs 74% Al-K$_\alpha$ (1.487 keV), 85% Mg-K$_\alpha$ (1.254 keV), 92% Na-K$_\alpha$ (1.041 keV) and 99% F-K$_\alpha$ (0.677 keV) x rays (see Fig. 10.22). Overall, polypropylene of the same thickness (6 microns) absorbs approximately 60% fewer low energy photons than does boPET, consequently, it is routinely used for detection of characteristic x rays emitted by low Z elements. These low energy detectors typically have two windows, an entrance window usually fashioned from a low density organic material, and an exit window

[11] You can guess why one author of this book knows this.

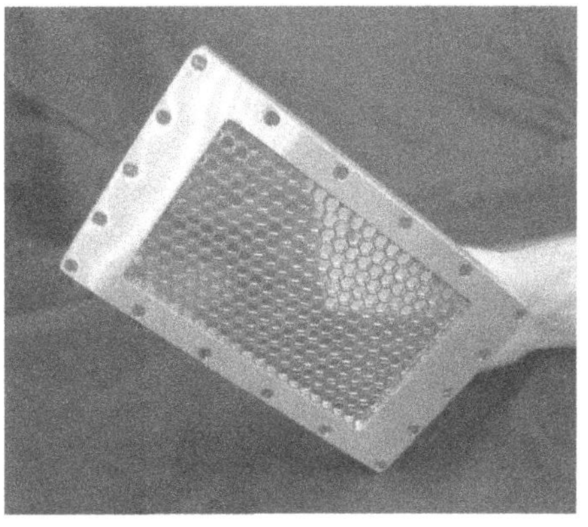

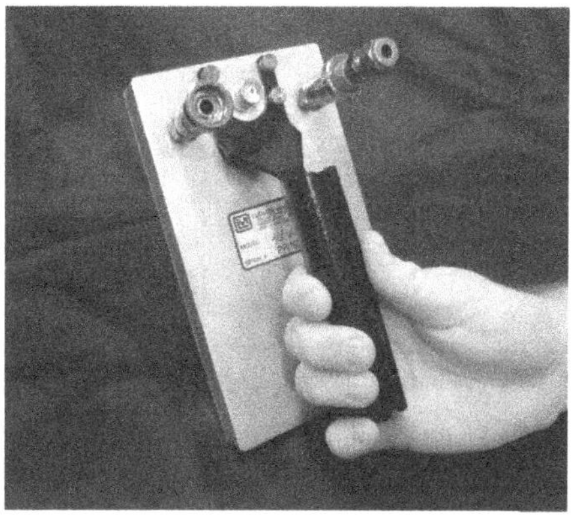

Figure 10.20. (left) Handheld thin-window gas-flow proportional counter (Ludlum 43-68) in which the thin aluminized boPET window and window guard are visible, and (right) handheld thin-window gas-flow proportional counter in which the gas and electronic connections are visible.

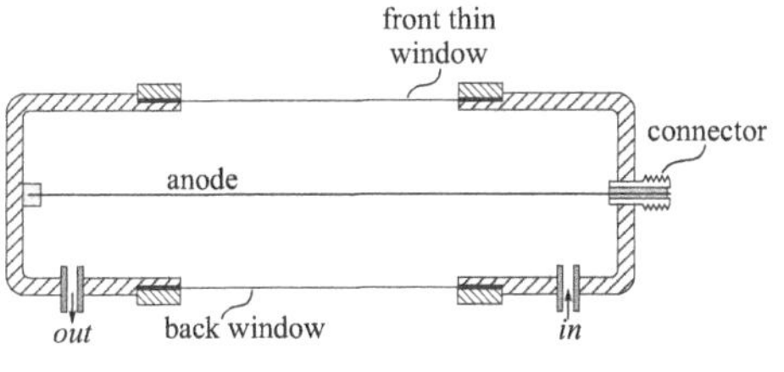

Figure 10.21. Diagram of a thin window proportional counter tube used for detection of low energy photons. After [Goldstein et al. 1981].

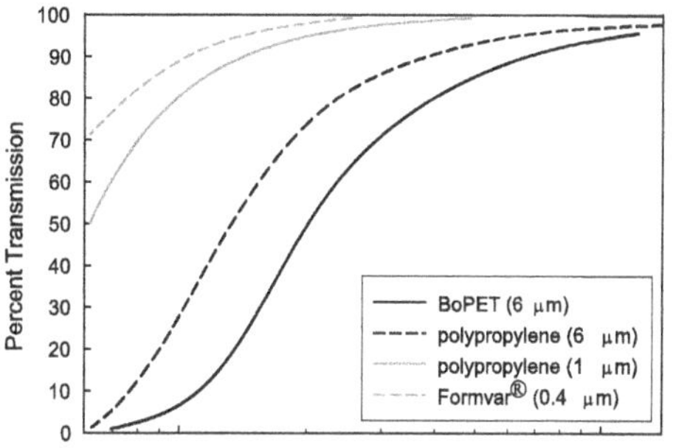

Figure 10.22. Photon transmission probability for several thin window materials. Data from [Potts 1987].

composed of 25 micron thick Be that helps to reduce the production of secondary radiation (ejected electrons) from photon interactions with the back wall (see Fig. 10.21).

10.6.2 Sealed Proportional Counters

Sealed proportional counters are available in a variety of geometries, although the most popular configurations are based on coaxial cylindrical designs. These proportional counters offer an economical method of radiation detection and come in sizes ranging from 1-cm diameter to several centimeters in diameter and with lengths of up to 1 to 2 meters. Unlike gas-flow detectors, the detector gas is not replenished, and consequently, the quenching gas can be depleted over time. Common commercial units are backfilled with H_2, Ne, Ar, Xe, Kr, P-10, ^{3}He, or $^{10}BF_3$.

Voltages applied to proportional counters are typically higher than those used for ion chambers; hence, it is necessary, as with ion chambers, to design the chamber to reduce electronic noise arising from leakage currents. As discussed in Sec. 9.5.1, a guard ring or guard tube is positioned around the anode wire, as depicted in Fig. 10.23. The voltage drop across the insulator between the guard tube and anode is effectively reduced, thereby lowering the leakage current as shown in Fig. 9.25. Leakage currents on the order of 10^{-16} amps can be achieved with the addition of a guard tube in the chamber [Borkowski 1947].

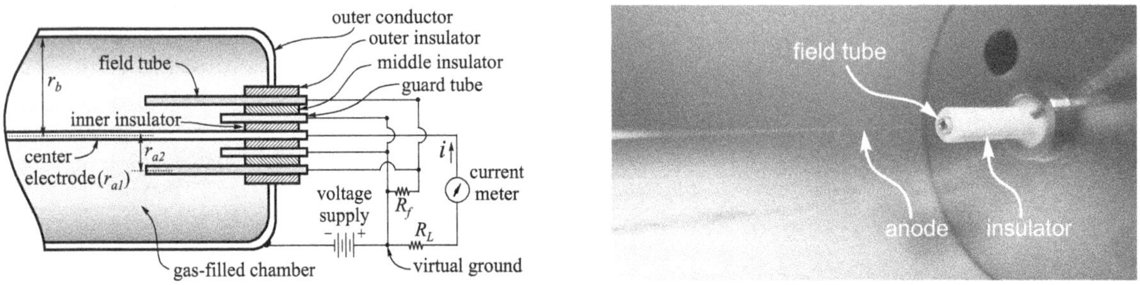

Figure 10.23. (left) Diagram of a sealed proportional counter tube with guard ring and field tube, and (right) inside chamber of a cylindrical steel proportional counter with a field tube.

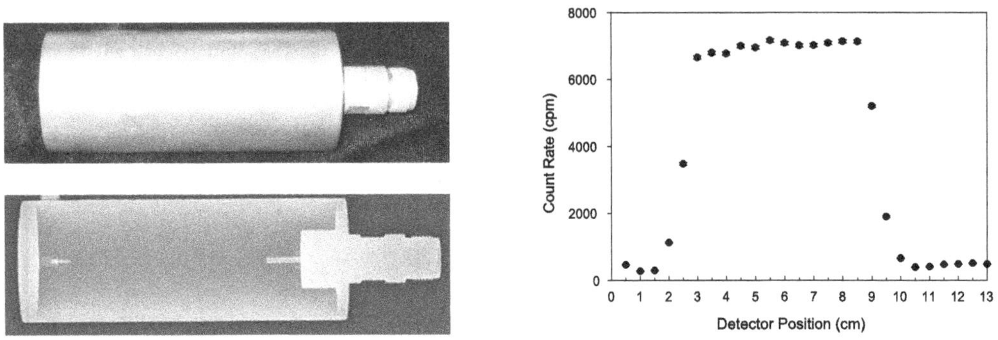

Figure 10.24. (upper left) Sealed ^{3}He-filled proportional counter, and (lower left) x-ray image of the sealed ^{3}He-filled proportional counter. The Teflon® insulator standoffs are barely visible in the x-ray image. (right) Detector response to a collimated thermal neutron beam as a function of position. The active region of the device is confined to the length between the ends of the standoff connectors, approximately 42% of the detector length.

The variation in the electric field around the anode near the insulating standoffs can have deleterious "end-effects" [Cockroft and Curran 1951]. Consequently, the sensitive region within a proportional counter is not well defined, as shown in Fig. 10.24, In this example, the sensitive region that limits the full development of the multiplication factor $\overline{M}$ amounts to only 42% of the detector length. However, the electric field end effects can be rendered negligible by fitting *field tubes* over and around the grounded guard tube (see Fig. 10.23). For a field tube of radius r_{a2}, the potential is adjusted such that the surface potential is the same as the radial potential in the gas at radial distance $r = r_{a2}$ near the detector center. The potential at the surface of the field tube and the resulting electric field are too small to produce an avalanche. If the field tube length is approximately the same as the detector radius r_b, then the electric field lines emanating from the anode wire between the ends of the field tubes are also radial with almost no distortion [Cockroft and Curran 1951].

Other sealed detector geometries include "pancake" detectors, side window detectors, end window detectors, quadrilateral detectors, spherical detectors, straw tube detectors, multi-wire detectors, microstrip detectors, and position sensitive detectors. A few of these detectors are described below.

10.6.3 Proportional Counters for Low Energy Gamma-Rays

As mentioned earlier, one important use of proportional counters is for detection of low energy x rays, especially for electron microprobe applications and wavelength dispersive spectroscopy. Such detectors can

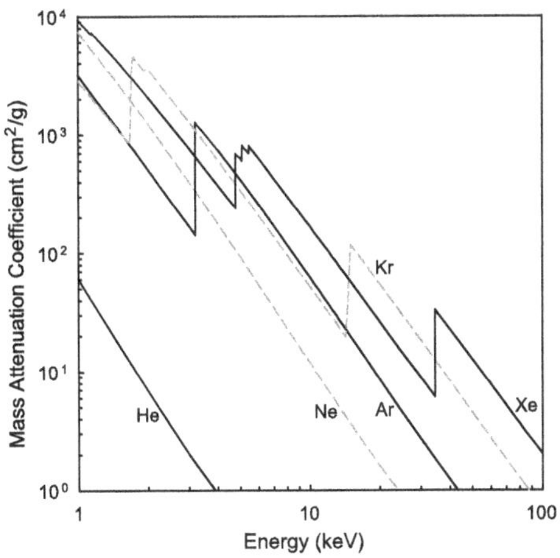

Figure 10.25. Gamma-ray total mass attenuation coefficients for common detector gases. Data from [NIST].

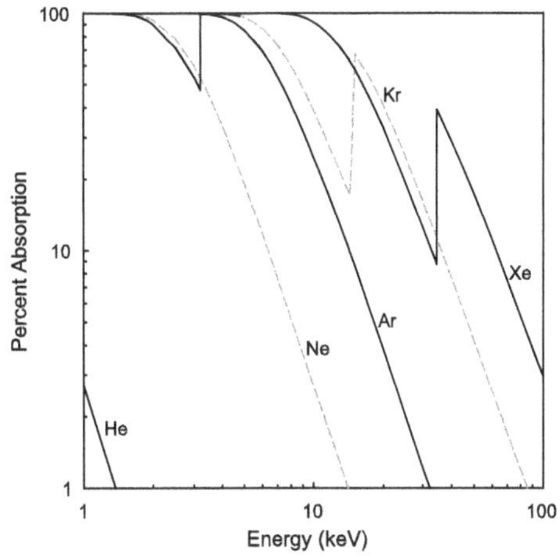

Figure 10.26. Gamma-ray percent absorption through a 2.54-cm diameter proportional counter at 1 atm pressure. Data from [NIST].

be fashioned as gas-flow detectors [Hendee et al. 1956; Dolby 1960; Riggs 1963] or sealed tubes [Lang 1951, 1956; Arndt et al. 1954]. To increase detection efficiency, the detector requires a thin entrance window that does not appreciably absorb the photons of interest (see Fig. 10.22). Additionally, the fill gas must be able to absorb the total energy of the photons, an ability that depends on atomic number of the gas, the gas pressure, and the detector width. Shown in Fig. 10.25 are the photon mass attenuation coefficients $\mu_m = \rho\mu$ for noble gases often used in proportional counters. An estimate of the expected intrinsic detection efficiency ϵ_I for a collimated photon beam intersecting the detector is

$$\epsilon_I = 1 - e^{-\mu_m \rho t} \tag{10.106}$$

where ρ is the gas density in g cm^{-3} and t is the proportional counter tube diameter. Results for several noble gases loaded into a 2.54-cm diameter tube are shown in Fig. 10.26. Careful selection of the fill gas can reject high energy photons while giving acceptable detection efficiency of characteristic x rays. For instance, Ar gas effectively detects characteristic x rays up to approximately 10 keV, while higher energy x rays (> 30 keV) generally pass through the device without detection. As mentioned in a prior section, a thin back window, usually constructed of Be or perhaps Al, can be used to reduce the probability that these higher energy x rays interact with the detector walls. Kr or Xe gas can be used to extend the detector sensitivity to higher energies.

Of special note are the gas K absorption edges that cause a reduction in efficiency at the absorption energy limit. For instance, the K edges for Ar, Kr, and Xe are 3.208 keV, 14.323 keV, and 34.579 keV, respectively. Directly below these energies for each respective gas, there is a sharp decrease in efficiency as can be seen from Figs. 10.25 and 10.26. Additionally, because of the characteristic x-ray transitions among different shells, energy can be lost from the detector. For instance, electrons excited from the K shell of Xe have K_α or K_β transitions of approximately 29.8 keV and 22.64 keV, respectively. If these Xe characteristic x rays escape the detector, the total energy deposited is reduced by the lost x-ray energy, which results in the formation of an *escape peak* in the pulse height spectrum.

10.6.4 Position Sensitive Proportional Counters

A method to turn a cylindrical gas filled detector into a position sensitive detector is described by Kuhlmann et al. [1966], McDicken [1967], and Borkowski and Kopp [1968]. The basic configuration takes advantage of

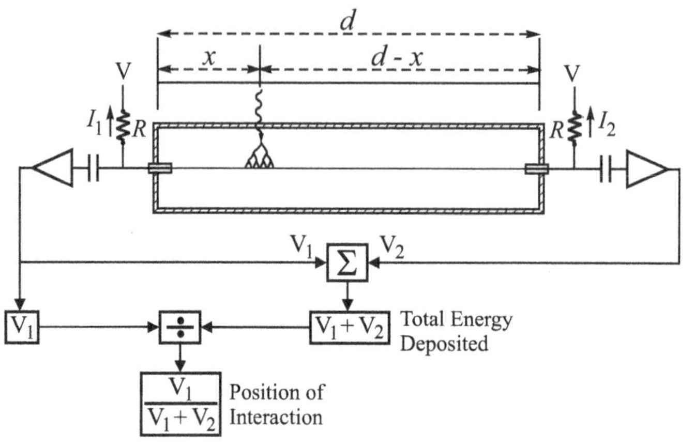

Figure 10.27. Schematic diagram of a position sensitive gas-filled proportional counter for the voltage division (charge division) method.

the distributed resistance of the central anode wire. An avalanche initiated by an ionizing particle causes current to flow through the anode and can be measured by preamplifiers at each end of the detector (as depicted in Fig. 10.27). The currents I_1 and I_2 flowing through the load resistors at each end of the detector are proportional to the input voltages of each preamplifier. The output voltages from each preamplifier, V_1 and V_2 are summed in a circuit to produce a signal that is proportional to the total ionization in the counter. The quotient of V_1 divided by V_1+V_2 produces a signal proportional to the location of the avalanche centroid with respect to the anode.[12] Although the configuration and concept are quite simple, the actual implementation can be somewhat complicated, as pointed out by Fischer [1977]. Regardless, for small localized avalanches, a spatial resolution FWHM below 1.6 mm is possible [Kuhlmann et al. 1966]. The position information has been shown to be remarkably linear with the method.

By measuring differences in the risetime of the signals from a detector, spatial resolution below 1 mm has been reported [Borkowski and Kopp 1968, 1970]. Charge collected from an anode of length d at some arbitrary position x causes a current to flow through the anode resistance over length x in one direction and over length $d-x$ in the other direction. The risetime of the input voltage into each preamplifier is a function of x and is independent of the total charge liberated in the detector. The input voltage is amplified and shaped with a double RC differentiator to obtain a bipolar pulse, and the risetime of the voltage is measured by the time of the crossover point of the resulting bipolar pulse. The difference between the risetimes of the pulses from each end of the detector is a function of the interaction location x [Borkowski and Kopp 1970].

10.6.5 Multiwire Proportional Counters

Multiwire proportional counters (MWPCs), as the name implies, are detectors with several anode wires, and sometimes several cathode wires, in the same chamber. Developed in 1968 by Charpak, position sensitive MWPCs use a criss-cross array of wires. Covering large surfaces with layers of classical cylindrical proportional tubes was regarded as impractical,[13] and the MWPC concept stems from this perceived limitation. The original MWPC design consisted of a parallel array of anode wires strung between two cathode planes,

[12]The technique has also been used with Geiger-Müller counters by performing the spatial measurement at a preset saturation threshold V_S where $V_S = V_1 + V_2$ at a set pulse height before the Geiger discharge is fully developed; therefore, the position x becomes proportional to V_1/V_S where V_S is approximately constant.

[13]This problem was greatly mitigated by the advent of the *straw tube* detector.

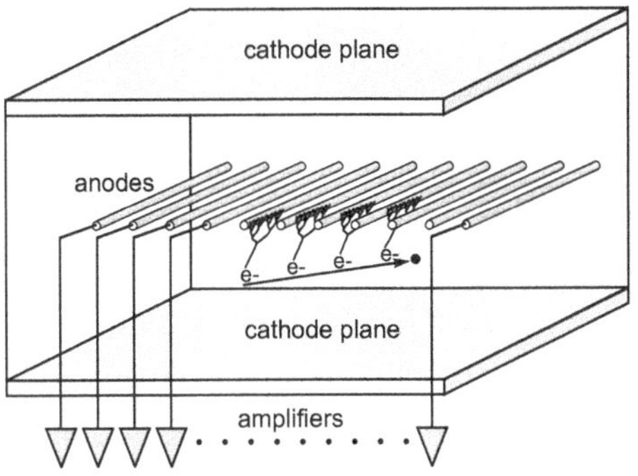

Figure 10.28. Cross section diagram of a MWPC.

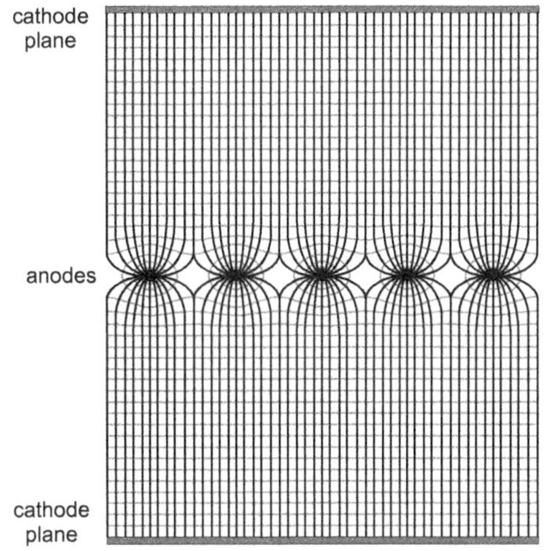

Figure 10.29. Electric field lines in a multiwire proportional counter (MWPC).

as shown in Fig. 10.28, with the parallel anode wires arranged only a few millimeters apart [Charpak 1970; Charpak and Sauli, 1979]. During operation, the detector electric field is perpendicular to the cathode planes, except in the close vicinity of the anodes as seen in Fig. 10.29. In the near field region of the anodes, the electric field lines must terminate perpendicularly to the anode wire surface, thereby producing a high electric field similar to a traditional cylindrical proportional counter. As a result, each wire operates as an individual proportional counter without a grounded barrier between the wires. Georges Charpak was awarded the 1992 Nobel Prize in Physics for his invention of the multiwire proportional chamber.

Particles entering the detectors produce a track of electron-ion pairs. It is possible with a MWPC to determine the (x, y, z) coordinates of track of the initial electron-ion pairs. Electrons drift towards the array of anodes, and when they enter the high field region, they initiate a Townsend avalanche. These electrons drift into the high field regions of the closest anode wires. The cloud of positive ions from the avalanche drifts towards the cathode planes, thereby inducing a fast rising current on the anodes from which they are departing. As discussed in Sec. 10.5.1, the short risetime of the pulse is a fraction to a few microseconds for most of the induced charge. The pulse is sensed upon those wires closest to the avalanche, thereby localizing in one dimension, x for instance, the particle track. The y dimension can be determined by connecting each wire in the same position sensing scheme as shown in Fig. 10.27. Although these anode wires are usually about 1 to 2 mm apart, the actual spatial resolution achieved is generally less than the wire spacing. Finally the third z coordinate of the original track location can be deduced from the time required for the electrons to drift to the anode wire array. A MWPC of this form is commonly referred to as a *drift chamber*, and is capable of determining the original track location with sub-millimeter spatial resolution.

Two-dimensional, position-sensitive MWPCs are an advancement on the original design. Typically there are two planar arrays of parallel cathode wires with the arrays positioned perpendicularly to each other [Sauli 1994]. One might consider one set of wires parallel to the x direction and the other set parallel to the y direction. In between the two cathode wire array planes is a parallel planar array of anode wires, which are typically arranged at a 45°-angle to the cathode wires, as shown in Fig. 10.30. As with the simple proportional counter, ionizing radiation produces primary electron-ion pairs in the detector gas. Electrons travel towards the nearest anode wires in the array, which then produce a Townsend avalanche of electron-ion pairs. The cloud of positive ions separates and travels towards the nearest cathode wires in the planes on

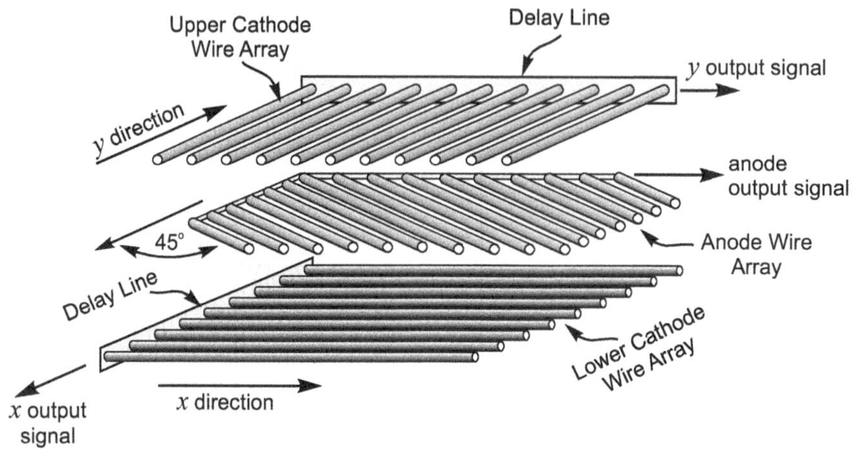

Figure 10.30. A multiwire gas filled proportional counter is composed of parallel layers of wire arrays. Shown is a system with three parallel wire arrays, in which the upper and lower arrays are cathode wires arranged orthogonally. The middle anode array is arranged at a 45°-angle to the cathode arrays.

both sides of the anodes. Hence, the position of the event is determined by which cathode wires deliver a signal on the x-y plane. Overall, the multiwire proportional counter can provide both energy information and position information of the ionizing event.

The multiwire concept has been used in a variety of detector designs, including open cell detector designs, gas-flow alpha particle detectors, and straw tubes. The interlacing of anodes and cathodes in a larger tube facilitate faster response times from these detectors. For planar devices, such as those shown in Figs. 10.18 and 10.20, the electric field is flattened throughout much of the active region, thereby providing a uniform response over most the detector area.

10.6.6 Microstrip Gas Chambers

A clever method to produce a microscopic electrode is to pattern the tiny anode upon an insulating substrate. The process is commonly used for microwave and millimeter wave transmission lines, in which a metallic strip is placed upon a dielectric substrate. The design has been used to produce compact proportional counters [Oed 1988].

The microstrip gas chamber (MSGC) is produced by photolithography means, in a similar fashion as used for semiconductor microelectronics. An insulating substrate, such as glass, precoated with a conductive metal, such as Cr, Cu or Au, is patterned with a photo sensitive chemical commonly referred to as a photoresist. A photomask with an image of the desired pattern is brought into close proximity of the substrate and the photoresist is exposed to ultraviolet light, usually with wavelengths of 360 to 420 nm. The transferred pattern is then developed on the substrate and the undesired regions of the conductive metal are etched away, leaving behind micropatterned microstrips. Afterwards, the photoresist is dissolved away, usually with an organic solvent.

This fabrication method can create microscopic metallic strips on the order of, or smaller than, common anode wires used in proportional counters and Geiger-Müller counters. The microstrips can be fabricated with great precision and have widths of about 10 ± 0.2 microns. Wider cathode strips with widths between 90 to 200 microns are also placed on the same substrate adjacent to the anodes, displaced by approximately 5 anode widths [Oed 1988; Workshop 1993]. As shown in Fig. 10.31, these MSPC plates are inserted into a gas chamber in which the entrance window also serves as the *drift cathode.*

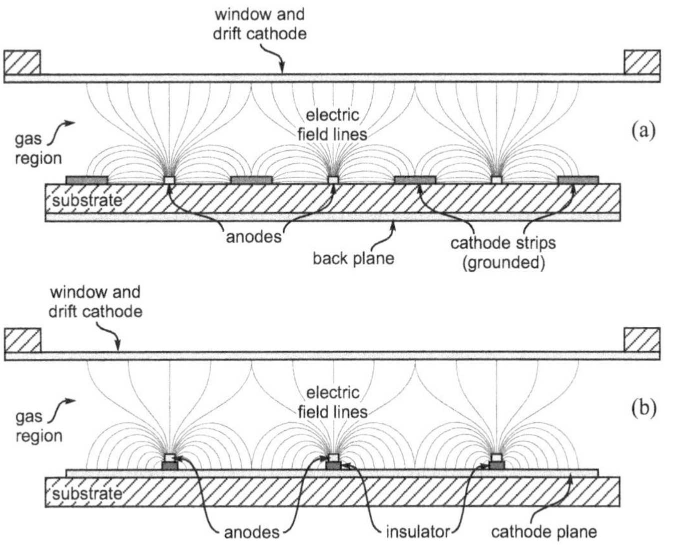

Figure 10.31. Cross section diagrams of (a) a gas-filled microstrip proportional counter and (b) a gas-filled micro-gap proportional counter. Typical dimensions for the anode width are 10 microns. The the MSPC, the cathode strip widths are between 90 and 200 microns with a repetitive pitch between 200 and 350 microns. The gas-filled region is on the order of 1 cm thick.

Radiation interactions in the chamber produce electron-ion pairs. The electrons drift towards the anodes, and when they come into the critical electric field region of the anode, a Townsend avalanche proceeds. The electrons are rapidly collected and the cloud of positive ions drift more slowly towards the adjacent cathode strips, thereby creating an induced signal that is input to the amplification electronics. Because of the very small anode dimensions, the applied potential required to produce an avalanching field is relatively low by comparison to traditional coaxial proportional counters. Further, the response times of the detectors is relatively short in the tens of nanoseconds range and are about 2 orders of magnitude faster than traditional coaxial proportional counters.

These detectors have demonstrated excellent energy resolution for thermal neutron reaction products when backfilled with a mixture of ^{3}He and an organic gas, and have a FWHM energy resolution of 1.43% [Oed 1988]. Further, energy resolution (FWHM) for 13.9-keV characteristic Np x rays from an ^{241}Am source was reported to be 18%, and 23% FWHM for 6.4-keV x rays from ^{57}Co.

Although positive ions drift towards the conductive cathode strips, some positive ions adhere to the insulating substrate. Those ions that come into contact with the insulating substrate tend to accumulate, creating a space charge field that, subsequently over time, reduces or negates the avalanching field. To eliminate space charge buildup, a conductive layer is applied to the back plane of the insulating substrate and biased at generally the same potential as the anodes. The back plane potential produces a repulsion field that emerges through the surface of the substrate not coated with conductive metal, thereby effectively protecting the insulating surface from buildup of space charge from the positive ions.

Another problem encountered in MSPCs is field sparking between the anodes and cathodes. Although the potential applied to the anodes is relatively low, the electric fields can be quite high and lead to electric arcing between the closely spaced electrodes. Such arcing can erode or damage the microstrips. One method to ameliorate the arcing is to apply an insulating polyimide spark guard, approximately 1 micron wide, over the edges of the cathode strips [Nagae et al. 1992].

An alternative design to reduce space charge buildup is the micro-gap gas chamber (MGC) [Angelini et al. 1993]. A pattern of insulating strips are applied on top of a substrate that remains completely coated with conductive material, as shown in lower part of Fig. 10.31. Because the substrate serves primarily as a mechanical foundation of the detector, the substrate no longer need be an insulator, and in fact may be a semiconductor coated with a conductive metal. The micropatterned insulating strips may be applied by common physical vapor deposition methods, such as evaporation or sputtering, or by chemical vapor deposition (CVD) methods, such as plasma enhanced CVD (PECVD). The thickness of the insulating strips is about two microns [Angelini et al. 1993]. Metal patterns serving as the anodes, slightly recessed by a few microns, are then micropatterned upon the insulating strips. The entire substrate surface of the resulting device is grounded, thereby preventing the accumulation of positive space charge.

The spectroscopic performance of a MGC is comparable to that of a MSPC, with reported FWHM energy resolution of 14.8% for 5.4-keV characteristic x rays from Cr. A major advantage of the MGC is its rapid response time. Because the positive ions travel such a short distance from the anode to cathode, the response times are on the order of a few nanoseconds comparable to those of semiconductor detectors [Angelini et al. 1993]. By comparison, common coaxial proportional counters have response times measured in microseconds, while the MSPC devices have response times approximately 10 times longer than those of MGC devices.

10.6.7 Straw Tubes

A variation of the sealed proportional counter is the *straw tube*, so named because the manufactured product resembles a drinking straw.[14] A common method used to fabricate straw tubes utilizes a cylindrical form, or mandrel, around which thin Mylar® strips are wound [Baringer et al. 1987; Marzec et al. 2002]. In some permutations, a 50-micron thick aluminized Mylar® strip is wound about a 7-mm diameter mandrel. Afterwards, a 35-micron thick clear Mylar strip is wound and glued over the first strip such that the outer strip covers the seam of the inner strip. The overlap for rotations is reportedly $\lesssim$2% of the area. After removal from the mandrel, insulating plugs are inserted into each end of the straw tube, through which a thin anode wire is threaded. By use of indexing tubules inserted into holes in each plug, the axial location of the wire is maintained within ±50 microns of center [Baringer et al. 1987].

According to Toki [1990], advantages of a straw tube chamber over multiwire proportional counters include:

1. They are inexpensive, relatively simple to construct, and robust.
2. The damage caused by anode wire breakage has a minimal effect, because it can be disconnected without affecting the response of neighboring anodes.
3. Minimal signal cross talk because the straw provides an electrostatic shield between anodes.
4. Electrostatic alignment distortions are minimal (if the anodes are carefully centered).

New methods of manufacture have been introduced, although the straw-tube name continues. For instance, Takuba et al. [2005] describes seamless straw-tube detectors fabricated by painting carbon-filled polyimide upon a mandrel followed by sintering. Afterwards the seamless tube is removed from the mandrel. Another reported method uses an ultra-sonic weld and a shaped extrusion method to produce thin copper tubes [Lacy 2012]. Regardless, the end product is a miniaturized tube with a diameter of 4 to 8 mm that serves as the cathode shell for a coaxial, gas-filled, proportional counter. Straw tubes can be manufactured in lengths ranging from a few centimeters to several meters. Straw tubes are often bundled together to form

[14] The Precision Paper Tube Company, still in business, is referenced as the manufacturer of aluminized Mylar® straw tubes in many journal articles during the late 1980s and early 1990s.

a larger tube or compartment to improve overall detection efficiency [Oh et al. 1991; Avery et al. 1993; Basile et al. 2004].

Thin-wall devices are reported to have been manufactured with two-sided wall thicknesses of less than 25 microns [Toki 1990]. These thin, straw-tube proportional counters are formed into drift chambers to track particles in high energy physics experiments [Baringer et al. 1987; Ash et al. 1987; Toki 1990]. Vertex tracking chambers often have hundreds of straw tube detectors arranged in concentric circles. The thin walls of the straw tubes minimize energy loss as particles pass through the detectors. Further, because so many detectors are required to build such an instrument, the simplicity of the detectors is economically attractive. Common gas mixtures reported for use in straw tube detectors include mixtures of Ar, CO_2, C_2H_6, and CH_4 [Avery et al. 1993]. Linear time-to-distance relationships have been observed for gas mixtures of CH_4:CO_2:Ar (4:3:93), CO_2:Ar (1:4), and ethane:Ar (1:1), each having drift speeds of about 40 to 80 microns/second. Experiments with C_2H_6/CO_2 (1:19) have yielded higher drift speeds but at the expense of non-linear time-to-distance responses.

Other straw tube designs are under investigation. Recently, multi-anode straw tubes have been introduced to improve performance of track detectors [Oh et al. 2011]. Straw tubes with B_4C-coated inner walls have also been explored for neutron detection [Lacy et al. 2011] and are covered in more detail in Chapter 17.

10.6.8 Gas Electron Multiplier

The gas electron multiplier (GEM) was introduced by Sauli in 1997, and is a unique method to produce high gain in a gas-filled detector [Sauli and Sharma 1999]. In general, a thin insulating film between 50 to 100 microns thick, typically a Kapton® film, is stretched on a frame and coated on both sides with a conductive metal (typically Cu). Afterwards, miniature holes approximately 100 microns in diameter are etched through the metal and insulating film as shown in Fig. 10.32. The GEM is placed in a gas-filled chamber between cathode and anode plates, usually closer to the anode. When a voltage is applied across the GEM, a high electric field is formed in each of the holes, high enough to cause ion pair avalanching (see Fig. 10.33). Typically, the GEM is held at a potential that is positive with respect to the cathode plate and negative to the anode plate.

When an ionizing particle enters the gas detector between the cathode and GEM, primary electron-ion pairs are created as in a typical ion chamber or proportional counter. The electrons drift towards the GEM and the positive ions drift towards the cathode. When electrons reach the GEM, they are funneled into the tiny holes where avalanching gas multiplication occurs. Hence, a much larger cloud of electrons emerges from the GEM with gains up to 200. These electrons continue to drift towards the anode where they induce an output signal. By using several GEMs in stages, much higher gains can be achieved, often exceeding 1000. Position sensitive GEMs can be manufactured by using segmented anodes. Various additional coatings on top of the actual GEM conductor coatings can make the GEM more sensitive to gamma rays and neutrons. For instance, a GEM coated with Gd, ^{10}B, or ^{6}LiF is sensitive to neutrons, while a GEM coated with CsI or NaI will have enhanced sensitivity to gamma rays.

10.6.9 Neutron-Sensitive Proportional Counters

Gas-filled proportional counters designed for neutron detection are briefly described in the present section. A detailed discussion of gas-filled neutron detectors is reserved for Chapter 17.

Detectors based on Neutron Reactive Gases

The most commonly used materials in proportional counters that can detect neutrons are the gases ^{3}He and $^{10}BF_3$, and the solid ^{10}B. Although neutron sensitive, neither $^{10}BF_3$ nor ^{3}He are ideal proportional gases, but they perform adequately well. Because the device operates in proportional mode, a low resolution spectrum associated with the reaction product energies of the $^{10}B(n,\alpha)^{7}Li$ reactions or the $^{3}He(n,p)^{3}H$ reactions can

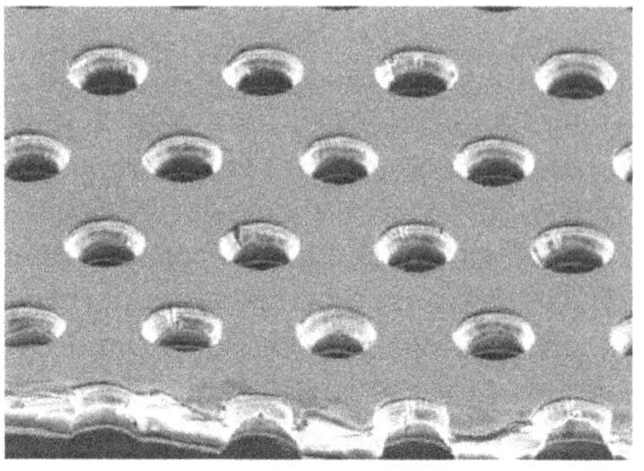

Figure 10.32. A scanning electron microscope photograph of a GEM section, showing the Kapton film, the Cu conductive surfaces, and the holes. The holes in the present photograph are approximately 75 microns in diameter. From Sauli and Sharma [1999]. Courtesy F. Sauli.

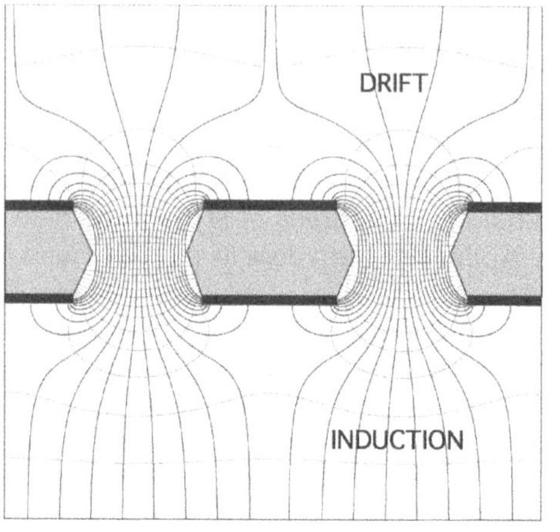

Figure 10.33. The electric fields around and in holes of a GEM detector. From Sauli [2016]. Courtesy F. Sauli.

be identified, depending on the gas used in the counter. The neutron detection efficiency can be increased by increasing the gas pressure of the counter so that there is more neutron absorber. Typical pressures for ^{3}He detectors range from 1 atm to 10 atm. Electron and ion speeds decrease inversely proportional to gas pressure and, consequently, increasing the gas pressure in the tube causes the dead time of the counter to increase. Gas-filled tubes come in a variety of sizes, ranging from only a few centimeterslong and one centimeter in diameter to several feet long and several inches in diameter.

Detectors based on Neutron Reactive Inserts

A traditional proportional gas may be used in the chamber, such as P-10, if instead of filling the chamber with a neutron reactive gas the walls are coated with ^{10}B. Unfortunately, the spectral features from such a device are harder to interpret due to interference from background gamma rays, and the total neutron detection efficiency is limited by the optimum ^{10}B absorber thickness, typically only 2 to 3 microns. The detectors are made more efficient by increasing the tube diameter, or by inserting additional ^{10}B-coated plates in the chamber. Neutron reactive inserts can also be introduced into the proportional counter, for example ^{6}Li blades. These detectors can be designed to preserve the reaction product spectra while delivering efficiencies competitive with detectors backfilled with ^{3}He or BF_3 [Nelson et al. 2012].

10.6.10 Selected Planar Proportional Counters

Finally there are several very specialized proportional counters that include detectors such as micro-mesh gaseous detectors (MicroMegas detectors) and resistive plate chambers (RPCs). The MicroMegas detector consists of a basic planar gas-filled chamber, on the order of 3 mm wide, that resembles the construction of a Frisch grid ion chamber [Giomataris et al. 1996]. The grid, or mesh, is located near the anode, usually within 100 microns. The voltage applied between the cathode and mesh drifts electrons towards and through the mesh ($\sim 10^3$ V cm^{-1}). The voltage applied to the region between the mesh and the anode produces an electric field above the critical field $\mathcal{E}_t$ required for avalanching ($> 10^5$ V cm^{-1}), and therefore, produces a Townsend avalanche. The resulting positive ions liberated in the avalanche are subsequently collected by the mesh electrode. Clever anode designs allow for MicroMegas detectors to be used as a position sensitive proportional counters [Derré and Giomataris 2001]. MicroMegas detectors are used as high-energy particle tracking detectors [Titov 2007].

The RPC is another planar style proportional counter dependent upon narrow gas-filled gaps. The basic construction has two parallel resistive plates, between which is a narrow gas-filled gap, which is usually a few hundred microns wide [Santonico and Cardarelli 1981; Cardarelli et al. 1988]. The resistivity of the materials is reportedly to be about 10^{10} to 10^{12} Ω-cm [Cardarelli et al. 1988]. A conductive electrode is applied to the outer surfaces of the insulators. A voltage applied across the detector will drop some voltage across the resistive materials; however, due to the high resistance of the gas-filled gap, most of the applied voltage appears across the gap, thereby producing an electric field above the critical electric field required for avalanching. Although there is signal reduction due to the capacitance of the resistive plates [Bacci et al. 1995], the increased signal from the avalanche is ample for detection. RPCs have reported timing resolution of around 100 ps [Santonico 2003]. Because the narrow gap absorbs little energy from the radiation particles of interest, these detectors generally consist of numerous RPCs in a stack. Patterned electrode designs and stacked configurations allow for the realization of relatively small position sensitive counters [Couceiri et al. 2007], with reported spatial resolution of 500 microns after the application of a reconstruction algorithm.

PROBLEMS

1. The electronic output from a cylindrical gas-filled chamber is primarily dominated by electron current when operated as an ion chamber. When operated as a proportional counter, the electronic output from the same detector becomes dominated by positive ion current. Explain why this change occurs.

2. A coaxial detector is backfilled with P-10 gas to 2 atm. The detector has anode wire with a radius of 25 microns and cathode radius of 1.5 cm. Determine r_c for an applied voltage of 1500 volts.

3. Use Stirling's approximation to prove that Eq. (10.102) becomes the Poisson distribution as $\theta \rightarrow \infty$.

4. Verify that the solution of Eq. (10.84) is given by Eq. (10.85).

5. Show that the second moment of Eq. (10.91) is $\langle m^2 \rangle = 2\overline{m}^2 - \overline{m}$.

6. Verify (a) that the Furry PDF of Eq. (10.88) is properly normalized, i.e., $\sum_{m=1}^{\infty} p(m|a) = 1$, (b) the first moment $\langle m \rangle = \exp[\tau(a)]$, (c) the second moment is $\langle m^2 \rangle = 2\langle m \rangle^2 - \langle m \rangle$, and (d) the relative variance is given by Eq. (10.92).

7. You have a coaxial detector with $r_a = 25$ microns and $r_b = 12.5mm$. The tube is pressurized to 1.5 atm with a mixture of 90% Xe/ 10% CH_4. What voltage must be applied to the anode to just barely produce a gas multiplication of $M = 20$?

8. A coaxial detector has $r_a = 25$ microns and $r_b = 15$ mm, and is backfilled with P-10 gas at 2 atm. The operating voltage is 1000 volts. A 5.5 MeV alpha particle interacts in the detector, producing an ionization trail parallel to the anode at a random distance r from the center. Approximate the time after the event ($t_{V_{max}/2}$) that 50% of the induced pulse is produced. How far did the ions travel to produce this pulse? You now change the gas pressure to 0.5 atm. What are the new values of ($t_{V_{max}/2}$) and the travel distance?

REFERENCES

AGRAWAL, P.C., B.D. RAMSEY, AND M.C. WEISSKOPF, "Study of Argon-Based Penning Gas Mixtures for use in Proportional Counters," *Nucl Instrum. Meth.*, **A277**, 557–564, (1989).

ALKHAZOV, G.D., A.P. KOMAR, AND A.A. VOROB'EV, "Ionization Fluctuations and Resolution of Ionization Chambers and Semiconductor Detectors," *Nucl. Instrum. Meth.*, **48**, 1–12, (1967).

ALKHAZOV, G.D. AND A.A. VOROB'EV, "Total Ionization and Ionization Fluctuations in Helium," *Phys. Lett.*, **29 A**, 25–26, (1969).

ALKHAZOV, G.D., "Statistics of Electron Avalanches and Ultimate Resolution of Proportional Counters," *Nucl. Instrum. Meth.*, **89**, 155–165, (1970).

ANGELINI, F., R. BELLAZZINI, A. BREZ, M.M. MASSAI, R. RAFFO, G. SPANDRE, M.A. SPEZZIGA, AND M. BOZZO, "The Micro-Gap Chamber," *Nucl. Instrum. Meth.*, **A335**, 69–77, (1993).

AOYAMA, T., "Generalized Gas Gain Formula for Proportional Counters," *Nucl. Instrum. Meth.*, **A234**, 125–131, (1985).

ARLEY, N., *On the Theory of Stochastic Processes and Their Application to the Theory of Cosmic Radiation*, New York: Wiley, 1943.

ARNDT, U.W., W.A. COATES, AND A.R. CRATHORN, "A Gas-Flow X-Ray Diffraction Counter," *Proc. Phys. Soc. London*, **B67**, 357–359, (1954).

ASH, W.W., ET AL., "Design, Construction, Prototype Tests and Performance of a Vertex Chamber for the MAC Detector," *Nucl. Instrum. Meth.*, **A261**, 399–419, (1987).

AVERY, R.E., ET AL., "Performance of 2mm Radius Straw Tube Drift Cells," *IEEE Trans. Nucl. Sci.*, **40**, 578–582, (1993).

BACCI, C., ET AL., "Test of a Resistive Plate Chamber Operating with Low Gas Amplification at High Intensity Beams," *Nucl. Instrum. Meth.*, **A352**, 552–556, (1995).

BARINGER, C. JUNG, H.O. OGREN, AND D.R. RUST, "A Drift Chamber Constructed of Aluminized Mylar Tubes," *Nucl. Instrum. Meth.*, **A254**, 542–548, (1987).

BASILE, E., ET AL., "A New Configuration for a Straw-Tube Microstrip Detector," *IEEE Nucl. Sci. Symp. Conf. Rec.*, Rome, Italy, Oct. 16–Oct. 22, 2004.

BORKOWSKI, C.J., "Ionization Chamber for Carbon-14 Measurements," U.S. Atomic Energy Commission Report MDDC-1099 (1947).

BORKOWSKI, C.J. AND M.K. KOPP, "New Type of Position-Sensitive Detectors of Ionizing Radiation Using Risetime Measurement," *Rev. Sci. Instrum.*, **39**, 1515–1521, (1968).

BORKOWSKI, C.J. AND M.K. KOPP, "Some Applications and Properties of One- and Two-Dimensional Position-Sensitive Proportional Counters," *IEEE Trans. Nucl. Sci.*, **NS-17**, 340–349, (1970).

BOWE, J.C., "Drift Velocity of Electrons in Nitrogen, Helium, Neon, Argon, Krypton, and Xenon," *Phys. Rev.*, **117**, 1411–1415, (1960).

BROWN, S.C., *Basic Data of Plasma Physics*, New York: MIT Press/Wiley, 1959.

BURKHALTER, P.G., J.D. BROWN, AND R.L. MYKLEBUST, "Pulse Amplitude Shifts in Gas Proportional X Ray Detectors," *Rev. Sci. Instrum.*, **37**, 1267–1268, (1966).

BYRNE, J., "III. Statistics of the Electron-Multiplication Process in Proportional Counters," *Proc. Roy. Soc. Edin.*, **66 A**, 33–41, (1962).

BYRNE, J., "Statistics of Electron Avalanches in the Proportional Counter," *Nucl. Instrum. Meth.*, **74**, 291–296, (1969).

CARDARELLI, R., R. SANTONICO, A. DI BAGIO, AND A. LUCCI, "Progress in Resistive Plate Counters," *Nucl. Instrum. Meth.*, **A263**, 20–25, (1988).

CHARPAK, G. "Evolution of the Automatic Spark Chamber," *Ann. Rev. Nucl. Sci.*, **20**, 195–254, (1970).

CHARLES, M.W., "Gas Gain Measurements in Proportional Counters," *J. Phys.*, **5**, 95–100, (1972).

CHARLES, M.W. AND B.A. COOKE, "Proportional Counter Resolution," *Nucl. Instrum. Meth.*, **61**, 31–36, (1968).

CHARPAK, G., AND F. SAULI, "Multiwire Proportional Chambers and Drift Chambers," *Nucl. Instrum. Meth.*, **162**, 405–428, (1979).

CHRISTOPHOROU, L.G., *Atomic and Molecular Radiation Physics*, New York: Wiley, 1971.

COCKROFT, A.L. AND S.C. CURRAN, "The Elimination of End Effects in Counters," *Rev. Sci. Instrum*, **22**, 37–42, (1951).

COOKSON, A.H. AND T.J. LEWIS, "Townsend First Ionization Coefficient for Methane and Nitrogen," *Brit. J. Appl. Phys.*, **17**, 891–903, (1966a).

COOKSON, A.H. AND T.J. LEWIS, "Variations in the Townsend First Ionization Coefficient for Gases," *Brit. J. Appl. Phys.*, **17**, 1473–1481, (1966b).

COTTRELL, T.L. AND I.C. WALKER, "Drift Velocities of Slow Electrons in Polyatomic Gases," Trans. Faraday Soc., **Vol. 61**, 1585–1593, (1965).

COUCEIRO, M., A. BLANCO, N.C. FERREIRA, R.F. MARQUES, P. FONTE, AND L. LOPES, "RPCPET: Status and Perspectives," *Nucl. Instrum. Meth.*, **A580**, 915–918, (2007).

CURRAN, S.C., H. FULBRIGHT, G. BISHOP, AND R. WILSON, "Particular Detection Methods," in *Beta- and Gamma-Ray Spectroscopy*, K. Siegbahn, Ed., Ch. VI, 165–200, 1955.

CURRAN, S.C., J. AANGUS, AND A.L. COCKROFT, "III. Investigation of Soft Radiations – II. The Beta Spectrum of Tritium," *Phil. Mag.* **40**, 53–60, (1949).

DERRÉ, J. AND I. GIOMATARIS, "Spatial Resolution and Rate Capability of MICROMEGAS Detector," *Nucl. Instrum. Meth.*, **A461**, 74–76, (2001).

DIAS, T.H.V.T., ET AL., "Full-Energy Absorption of X-Ray Energies Near the Xe L- and K-Photoionization Thresholds in Xenon Gas Detectors: Simulation and Experimental Results," *Appl. Phys. Lett.*, **82**, 2742–2753, (1997).

DIETHORN, W., *A Methane Proportional Counter System for Natural Radiocarbon Measurements*, U.S. Atomic Energy Commission Report NYO-6628 (1956).

DOLBY, R.M., "Absolute Intensity Measurements of the Carbon and Aluminum X-ray K-lines with a Proportional Counter," *Brit. J. Appl. Phys.*, **11**, 64–66, (1960).

DRUVESTEYN, M.J., AND F.M. PENNING, "The Mechanism of Electrical Discharges in Gases of Low Pressure," *Rev. Mod. Phys.*, **12**, 87–174, (1940).

DUNN, W.L. AND J.K. Shultis, *Exploring Monte Carlo Methods*, Amsterdam, The Netherlands: Academic Press, 2012.

FANO, U., "Ionization Yield of Radiations. II. The Fluctuations of the Number of Ions," *Phys. Rev.*, **72**, 26–29, (1947).

FERGASON, L., "Gain Shift of Proportional Counter with Counting Rate," *Rev. Sci. Instrum.*, **37**, 964–965, (1966).

FISCHER, B.E., "A Digital Processor for Position Sensitive Detectors," *Nucl. Instrum. Meth.*, **141**, 173–181, (1977).

FRISCH, O.R., "Statistics of Multiplicative Processes," unpublished lectures quoted by West [1954] and Curran [1955], 1947.

FRISCH, O.R., *Fluctuations in Chain Processes*, AECL Report 748, Chalk River, Ontario, 1959.

FURRY, W.H., "On Fluctuation Phenomena in the Passage of High Energy Electrons through Lead," *Phys. Rev.*, **52**, 569–581, (1937).

GENZ, H., "Single Electron Detection in Proportional Gas Counters," *Nucl. Instrum. Meth.*, **112**, 83–90, (1973).

GIOMATARIS, Y., PH. REBOURGEARD, J.P. ROBERT, AND G. CHARPAK, "MICROMEGAS: A High-Granularity Position-Sensitive Gaseous Detector for High Particle-Flux Environments," *Nucl. Instrum. Meth.*, **A376**, 29–35, (1996).

GOLDSTEIN, J.I., D.E. NEWBERRY, P. ECHLIN, D.C. JOY, C. FIORI, AND and E. LIFSHIN, *Scanning Electron Microscopy and X-ray Microanalysis*, New York: Plenum Press, 1981.

GOTT, R. AND M.W. CHARLES, "Attenuation of Proportional Counter Pulses by Equal Integrating and Differentiating Time Constants," *Nucl. Instrum. Meth.*, **72**, 157-160, (1969).

GROSSWENDT, B., "Statistical Fluctuations of the Ionisation Yield of Low-Energy Electrons in He, Ne and Ar," *J. Phys.*, **17 B**, 1391–1404, (1984).

HENDEE, C.F., S. FINE, AND W.B. BROWN, "Gas Flow Proportional Counter for Soft X-Ray Detection," *Rev. Sci. Instrum.*, **27**, 531–535, (1956).

HENDRICKS, R.W., "Space Charge Effects in Proportional Counters," *Rev. Sci. Instrum.*, **40**, 1216–1223, (1969).

HENDRICKS, R.W., "The Gas Amplification Factor in Xenon Filled Proportional Counters," *Nucl. Instrum. Meth.*, **102**, 309–312, (1972).

HRISTOV, L.G., "On the Dependence of the Gas-Amplification factor with Proportional Counter Voltage ", *Dokl. Bulg. Akad. Nauk./ Comptes Rendus de l'Academie des Sciences.*, **10**, 453–456, (1957).

JÄRVINEN, M.-L., AND H. SIPILÄ, "Improved Proportional Counters for Practical Applications," **NS-31**, 356–359, (1984).

KISER, R.W., "Characteristic Parameters of Gas-Tube Proportional Tubes," *Appl. Sci. Res.*, Sec. B, **8**, 183–200, (1960).

KORFF, S.A., *Electron and Nuclear Counters – Theory and Use*, New York: Van Nostrand Co., 1946.

KOWALSKI, T.Z., "Generalized Parametrisation of the Gas Gain in Proportional Counters," *Nucl. Instrum. Meth.*, **A243**, 501–504, (1986).

KRAJCAR-BRONIĆ, I., "W Values and Fano Factors for Electrons in Rare Gases and Rare Gas Mixtures," *Ionizing Radiation*, **24**, 101–125, (1998).

KUHLMANN, W.R., K.H. LAUTERJUNG, B. SCHIMMER, AND K. SISTEMICH, "Ortsempfindliche Zählrohre," *Nucl. Instrum. Meth.*, **40**, 118–120, (1966).

LACY, J.L., A. ATHANASIADES, L. SUN, C.S. MARTIN, t.D. LYONS, M.A. FOSS, H.B. HAYGOOD, "Boron-Coated Straws as a Replacement for ^{3}He-based Neutron Detectors," *Nucl. Instrum. Meth.*, **A652**, 359–363, (2011).

LACY, J.L., Public presentation during the *IEEE Nucl. Sci. Symp, Conf. Rec.*, Anaheim, California, Oct. 27- Nov. 3, 2012.

LANG, A.R., "The Proportional Counter in X-Ray Diffraction Work," *Nature*, **168**, 907–908, (1951).

LANG, A.R., "Some Notes on the Design and Performance of X-ray Proportional Counters," *J. Sci. Instrum.*, **33**, 96–102, (1956).

LANSIART, A. AND J.P. MORUCCI, "Amplification Gazeuse Dans un Compteur Proportionnel," *J. Phys. Radium*, **23**, Suppl. 6, 102A–104A, (1962).

MARTIN, T.C. AND O.E. GREEN, "Improved Four-Pi Proportional Gas Flow Counter," *Rev. Sci. Instrum.*, **29**, 1147–1148, (1958).

MARZEC, J., K. ZAREMBA, Z. PAWLOWSKI, AND B. KONARZEWSKI, "Transparency of the Straw Tube Cathode for the Electromagnetic Field," *IEEE Trans. Nucl. Sci.*, **49**, 548–552, (2002).

MCDICKEN, W.N., "A Position Sensitive Geiger Counter," *Nucl. Instrum. Meth.*, **54**, 157 (1967).

MORI, C., M. UNO, AND T. WATANABE, "Self-Induced Space Charge Effect on Gas Gain in Proportional Counters," *Nucl. Instrum. Meth.*, **196**, 49–52, (1982).

NAGAE, T., T. TANIMORI, T. KOBAYASHI, AND T. MIYAGI, "Development of Microstrip Gas Chambers with Multi-Chip Technology," *Nucl. Instrum. Meth.*, **A323**, 236 (1992).

NELSON, K.A., S.L. BELLINGER, B.W. MONTAG, J.L. NEIHART, T.A. RIEDEL, A.J. SCHMIDT, D.S. MCGREGOR, "Investigation of a Lithium Foil Multi-Wire Proportional Counter for Potential ^{3}He Replacement," *Nucl. Instrum. Meth.*, **A669**, 79–84, (2012).

OED, A., "Position-Sensitive Detector with Microstrip Anode for Electron Multiplication with Gases," *Nucl. Instrum. Meth.*, **A263**, 351–359 (1988).

OH, S.H., A.T. GOSHAW, AND W.J. ROBERTSON, "Construction and Performance of a 2.7 m Long Straw Drift Tube Prototype Chamber for the SSC," *Nucl. Instrum. Meth.*, **A309**, 368 (1991).

OH, S.H., W.L. EBENSTEIN, AND C.W. WANG, "Multi-Anode Wire Straw Tube Tracker," *Nucl. Instrum. Meth.*, **A640**, 160–163, (2011).

PANSKY, A., A. BRESKIN, AND, R. CHECHIK, "Fano factor and the Mean Energy per Ion Pair in Counting Gases, at Low X-Ray Energies," *Appl. Phys. Lett.*, **82**, 871–877, (1997).

PACK, J.L., AND A.V. PHELPS, "Drift Velocities of Slow Electrons in Helium, Neon, Argon, Hydrogen, and Nitrogen" *Phys. Rev.*, **121**, 798–806, (1961).

PENNING, F.M., "Über Ionisation Durch Metastabile Atome," *Die Naturwissenschaften*, **15**, 818, (1927).

PHELPS, A.V., J.L. PACK, AND L.S. FROST, "Drift Velocity of Electrons in Helium," *Phys. Rev.*, **117**, 470–474, (1960).

PÓLYA, G., "Sur Quelques Points de la Théorie des Probabilités," *Ann. l'Institut Henri Poincare*, **1**, 117–161, (1930).

POTTS, P.J., *A Handbook of Silicate Rock Analysis*, London: Blackie, 1987.

PRICE, W.J., *Nuclear Radiation Detection*, 2nd. Ed., New York: McGraw-Hill, 1964.

RAMSEY, B.D. AND P.C. AGRAWAL, "Xe-Based Penning Mixtures for Proportional Counters," *Nucl Instrum. Meth.*, **A278**, 576–582, (1989).

RICE-EVANS, P., *Spark, Streamer, Proportional and Drift Chambers*, London: The Richelieu Press, 1974.

RIGGS, JR., F.B., "New Design for a Gas Flow Proportional Counter," *Rev. Sci. Instrum.*, **34**, 392–395, (1963).

ROSE, H.E. AND S.A. KORFF, "An Investigation of the Properties of Proportional Counters. I," *Phys. Rev.*, **59**, 850–859, (1941).

ROSSI, B.B. AND H.H. STAUB, *Ionization Chambers and Counters Experimental Techniques*, New York: McGraw-Hill, 1949.

SAKURAI, H. AND B.D. RAMSEY, "Dependence of Energy Resolution on Anode Diameter in Xenon Proportional Counters," *Nucl. Instrum. Meth.*, **A313**, 155–160, (1992).

SANTONICO, R. AND R. CARDARELLI, "Development of Resistive Plate Chambers," *Nucl. Instrum. Meth.*, **187**, 377–380, (1981).

SANTONICO, R., "RPCs: the Challenges for the Next Two Years," *Nucl. Instrum. Meth.*, **A508**, 1–5, (2003).

SAULI, F., *Principles of Operation of Multiwire Proportional and Drift Chambers*, CERN 77-09, Geneva, 1977.

SAULI, F., "Potentials of Advanced Gas Detectors for Health Physics," CERN-PPE/94-195 (1994).

SAULI, F. AND A. SHARMA, "Micropattern Gaseous Detectors," *Ann. Rev. Nucl. Part. Sci.*, **49**, 341–388, (1999).

SAULI, F., *Gaseous Radiation Detectors*, Cambridge: Cambridge University Press, 2014.

SAULI, F., "The Gas Electron Multiplier (GEM): Operating Principles and Applications," *Nucl. Instrum. Meth.*, **A805**, 2–24, (2016).

SCHLUMBOHM, H., "Zur Statistik der Elektronenlawinen im Ebenen Feld. III," *Z. Physik*, **151**, 563 (1958a).

SCHLUMBOHM, H., "Zur Statistik der Elektronenlawinen im Proportionalzählrohr. IV," *Z. Physik*, **152**, 49 (1958b).

SIMPSON, JR., J.A., "Proportional Counter," *US Patent* 2532956 A, allowed May 2, 1950a.

SIMPSON, JR., J.A., "Air Proportional Counter," *US Patent* 2532956 A, allowed Dec. 5, 1950b.

SIPILÄ, H., "Energy Resolution of the Proportional Counter," *Nucl. Instrum. Meth.*, **133**, 251–252, (1976).

SNYDER, H.S., "Fluctuations for Proportional Counters.," *Phys. Rev.*, **72**, 181, (1947).

SPIELBERG, N., "Effect of Anode Material on Intensity Dependent Shifts in Proportional Counter Pulse Height Distributions," *Rev. Sci. Instrum.*, **38**, 291, (1967).

STAUB, H.H., "Detection Methods," Ch. 1 in *Experimental Nuclear Methods*, Vol. 1, E. SEGRÉ, Ed., New York: Wiley, 1953.

TAKUBO, Y., ET AL., "Development and Performance of Resistive Seamless Straw-Tube Gas Chambers," *Nucl. Instrum. Meth.*, **A551**, 271–284, (2005).

TITOV, M., "New Developments and Future Perspectives of Gaseous Detectors," *Nucl. Instrum. Meth.*, **A581**, 25–37, (2007).

TOKI, W., "Review of Straw Chambers," SLAC-PUB-5232, 1990.

TOWNSEND, J.S., *Electricity in Gases*, Oxford: Clarendon, 1915.

WARD, A.L., "Effect of Space Charge in Cold-Cathode Gas Discharge," *Phys. Rev.*, **112**, 1852–1857, (1958).

WEST, D., "Energy Measurements with Proportional Counters," *Progr. Nucl. Phys.*, O.R. FRISCH, Ed., **3**, 18–62, (1954).

WIJSMAN, R.A., "Breakdown Probability of a Low Pressure Gas Discharge," *Phys. Rev.*, **75**, 833–838, (1949).

WILKINSON, D.H., *Ionization Chambers and Counters*, Cambridge: University Press, 1950.

WILLIAMS, A. AND R.I. SARA, "Parameters Affecting the Resolution of a Proportional Counter," *Int. J. Appl. Radiation Isotopes*, **13**, 229–238, (1962).

WOLFF, R.S., "Measurement of the Gas Constants for Various Proportional-Counter Gas Mixtures," *Nucl. Instrum. Meth.*, **115**, 461–463, (1974).

Workshop on Progress in Gaseous Microstrip Proportional Counters, Conf. Rec., Grenoble, 1993.

ZASTAWNY, A., "Gas Amplification in a Proportional Counter with Carbon Dioxide," *J. Sci. Instrum*, **43**, 179-181, (1966).

ZASTAWNY, A., AND J. MIZERACZYK, "Gas Gain in a Proportional Counter Filled with Argon and Nitrogen," *Nukleonika*, **11**, 685–689, (1966).

ZASTAWNY, A., "On a New Formula for the Gas Amplification in a Proportional Counter," *J. Sci. Instrum*, **44**, 395, (1967).

Chapter 11

Gas-Filled Detectors: Geiger-Müller Counters

The sudden current through the gas due to the entrance of an α-particle in the testing vessel was thus increased...
Sir Ernest Rutherford and Hans Geiger

Although Hans Geiger originally created the gas-filled detector in 1908 (with Ernest Rutherford) [Rutherford and Geiger, 1908],[1] the "Geiger" counter in wide use today is based on an improved version that his first PhD student, Walther Müller, constructed in 1928 [Geiger and Müller 1928a, 1928b, 1929a, 1929b]. Hence, the proper name for the device is the *Geiger-Müller* counter. The original "Geiger" counter was sensitive to alpha particles, but not so much to other forms of ionizing radiation. Müller's improvements included the utilization of vacuum tube technology to create a device that was compact and portable and that was sensitive to alpha, beta, and gamma radiations. In 1947, Sidney Liebson further improved the device by including in the tube gas a small amount of a halogen as a quenching gas, an improvement that allowed the detector to operate at lower applied voltages and to last significantly longer [Liebson 1947a, 1947b].

11.1 Geiger Discharge

Geiger-Müller counters are operated in region IV of the pulse height curve for gas detectors as shown in Fig. 11.1 and depend upon gas multiplication for signal amplification, much like proportional counters do. However, a single important difference is that, at any specific applied voltage, *all* output pulses from a Geiger-Müller counter have the same amplitude *regardless of the energy or type of incident ionizing radiation.* Hence, Geiger-Müller counters cannot discern any difference among alpha, beta, or gamma radiations, nor can they distinguish among different energies of these radiations. To understand the reason for this behavior, it is important to understand the progression of the Geiger discharge.[2]

Geiger counters typically have a coaxial configuration, in which a thin anode wire is placed on the axis of a cylindrical tube that serves as the cathode. A high voltage is applied to the central anode wire, while the cathode is held at ground potential, as depicted in Fig. 11.3. When an ionizing particle enters a Geiger-Müller counter, the counting gas becomes locally ionized in which a small cloud of electron-ion pairs is formed as depicted in Fig. 11.2(1) and described in the previous chapter. Because a high voltage is applied to the anode and the tube is operated in region IV of the gas curve, the electrons drift rapidly to the anode while the ions slowly drift towards the cathode, as shown in Fig. 11.2(2). When the electrons enter the high electric field near the anode above the critical field strength $\mathcal{E}_t$ needed to produce

[1] Hans Geiger was nominated for the Nobel Prize in physics three times, in 1935 by Walther Nernst, in 1937 by Walther Bothe, and in 1955 by Boris Rajewsky, but never received it.

[2] A Geiger discharge consists of multiple Townsend avalanches.

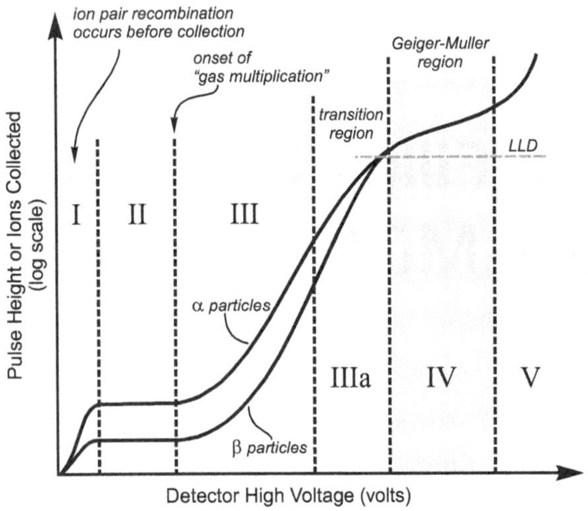

Figure 11.1. Geiger-Müller counters are operated in region IV of the pulse height curve for gas-filled detectors. The *LLD* is set to discriminate pulses produced in the proportional and lower regions.

avalanche ionization, they gain enough kinetic energy to produce more electron-ion pairs through impact ionization, and a large and dense cloud of electron-ion pairs is formed around the anode. Some of these energetic electrons lose energy, through bremsstrahlung, recombination, or de-excitation, processes that lead to the production of UV photons which, in turn, cause more ionization in either the cylinder gas or from interactions with the cathode wall. These newly freed electrons form new avalanches and yet more positive space charge builds up around the anode. This large accumulation of positive ions around the anode reduces the electric field near the anode as depicted in Figs. 11.2(3) and (4). There comes a point when the large accumulation of space charge around the anode is so high that the electric field is reduced below the critical field strength $\mathcal{E}_t$ needed to sustain avalanching. At this point the impact ionization ceases, as shown in Fig. 11.2(5). The positive ions then slowly drift towards the cathode where they are collected and produce the signal pulse for the detector. As they reach the cathode, the electric field near the anode recovers to full strength, and the detector is set to detect the next radiation interaction event, as depicted by Fig. 11.2(6).

Several features about the progression of the Geiger discharge should be noted. (a) The electric field in the detector increases with increasing applied voltage. (b) The discharge ceases when the electric field is reduced below $\mathcal{E}_t$ at the anode and, therefore, the required space charge to terminate the avalanching must also increase with applied operating voltage. (c) To prevent more electrons from being ejected when the ions strike the cathode, thereby preventing another avalanche, a form of quenching must be used. (d) The entire Geiger discharge process is slower than that of a proportional counter, mainly due to the longer time required to produce the dense cloud of positive ions around the anode. (e) Finally, the size of the output pulse is determined by how much space charge must accumulate to reduce the electric field below $\mathcal{E}_t$ and not by the initial energy absorbed or number of initial electron-ion pairs produced within the detector! Consequently, the pulse height for various energies of α-particles, β-particles, and γ-rays are all the same, within statistical variation, and the signal pulse height is predetermined solely by the applied operating voltage.

11.2 Basic Design

The basic design of a Geiger-Müller (G-M) counter is shown in Fig. 11.3 and consists of a coaxial chamber that serves as the detector cathode and the anode is threaded into the chamber through an insulator. The anode is small in diameter, yet robust enough to be rigidly suspended in the tube. Typically, wire diameters on the order of 100–200 microns can achieve the desired electric field. Thinner wires can be used if suspended by both ends. An insulating glass bead is attached to the anode wire, which helps stabilize the electric field at the end of the anode.

Older versions of G-M counters were constructed from glass envelopes, and a metal sleeve was inserted into the chamber to act as the cathode. The choice of cathode material affects the gamma-ray response of the chamber. High Z materials, such as Pb or Bi, have a fairly salient "over response" to gamma-ray energies below 500 keV (as discussed later in Section 11.5.3). Present day G-M counters are usually constructed from a stainless steel cathode tube without any such inserts, and rely on the stainless steel tube for gamma-ray interactions.

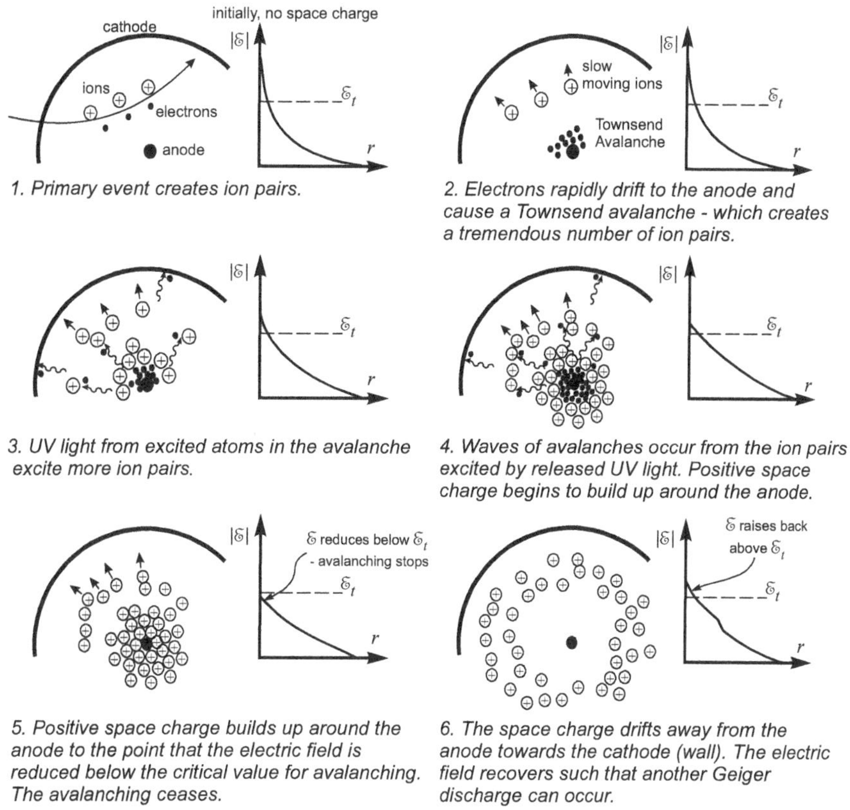

Figure 11.2. Geiger-Müller tube cross section depicting the progression of the Geiger discharge. The avalanching continues until the space charge accumulated around the anode wire decreases the electric field below the avalanche threshold, which then causes the progression to cease.

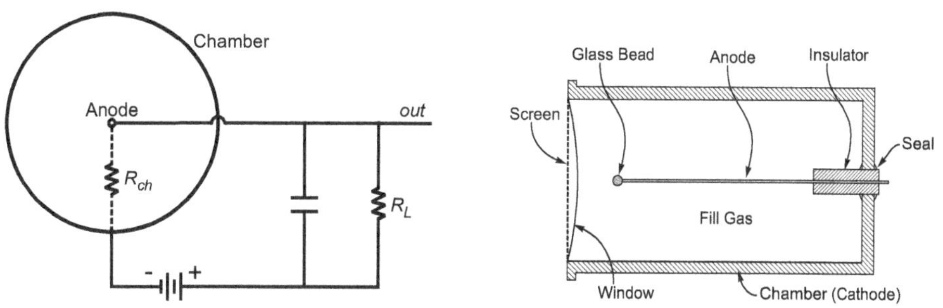

Figure 11.3. (left) simplified circuit for a G-M counter, and (right) the basic components of a G-M tube.

Because α and β particles cannot penetrate thick-walled G-M tubes, a thin entrance window is usually used to seal the end of the tube, thereby making a cantilevered anode wire essential. The window is usually made of either aluminized boPET or mica. The gas inside the G-M tube is under a mild vacuum (commonly between 0.1–0.2 atm) in order to reduce the detector dead time [recall Eq. (9.31)]. To prevent damage or contamination, many manufacturers of G-M tubes provide a protective screen over the entrance window.

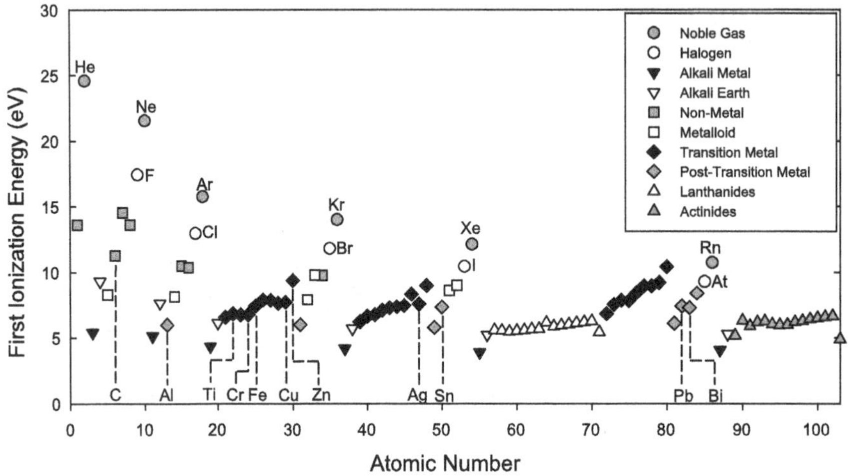

Figure 11.4. First ionization energies of the elements. Common gases and cathode materials that have been used in G-M counters are identified by their chemical symbol.

Care should be taken with G-M tubes and the ambient external pressure. At some altitudes, the external pressure can be lower than the internal tube pressure, thereby pressurizing the tube and causing a rupture in the window. Manufacturers usually document such restrictions on specification sheets.

11.3 Fill Gases

Although a variety of gases have been successfully used in G-M counters [Korff 1946; Sinclair 1956; Price 1963; Emery 1966], most often the host gas is one of the noble gases. From Fig. 10.10, neon allows the highest electron drift speeds, followed by those of He and Ar. Neon also has one of the highest ion drift speeds of the noble gases, second only to that of He [Munson and Tyndall 1941, Sharpe 1964, Barsuto et al. 2000].[3] Hence, Ne is usually the gas of choice, mainly because it helps reduce dead time by allowing the dense ion cloud to form more rapidly and to sweep out more rapidly the positive ions. However, Ne also has a higher first ionization coefficient (and higher average ionization energy) than Ar, although lower than that of He, as can be seen from Fig. 11.4. Thus, a higher specific ionization per unit length of electron travel is achieved in Ne than in He, while retaining a faster response than that of Ar. However, often a small amount of Ar is added to the gas to increase the specific ionization while still retaining a fast signal response.

There may be special applications where high atomic number noble gases are used in G-M counters, although this practice is uncommon. Under such conditions, it is possible for low energy gamma rays to interact directly with the gas. As can be seen from Fig. 10.26, high efficiencies can be achieved for gamma-ray energies below 30 keV. However, the electron velocities for Kr and Xe are significantly lower than those of Ne and Ar (see Fig. 10.10), as well as the ion mobilities [Munson and Tyndall 1941], thereby causing these G-M counters to have longer dead times.

Because of the large ionization clouds produced in G-M tubes, it is necessary to include a quenching gas along with the host detection gas inside sealed tubes. Organic polyatomic gases work well in G-M tubes, but these gases can be consumed after several months of continuous use as the molecules dissociate irreparably. Instead, it is a usual practice to use diatomic halogen gases to quench the Geiger discharge because these gases undergo a form of self-repair after dissociation.

[3]Recall from Fig. 10.10 that Ne has the highest electron drift speed of the noble gases.

11.3.1 Quenching

If a G-M tube is operated with a pure counting gas such as argon or neon, positive ions drifting to the cathode attract electrons to the surface, and upon contact, an electron from the wall neutralizes the positive ion. Ideally, the atom is neutralized and the collection process simply ends. However, the actual process is considerably more complicated.

When a gas ion comes into contact with the cathode wall, the absorption of an electron leaves the now neutralized atom in an excited state. This physical state occurs because of the differences in ionization energies between the gas atoms and the cathode atoms. Typically, a noble gas atom de-excites by releasing a photon with an energy comparable to the atom ionization energy. For example, the minimum ionization energy for Ar is 15.6 eV, corresponding to a photon wavelength of 78.47 nm and classified as belonging to the extreme ultraviolet (EUV) region of the electromagnetic spectrum. Almost all transition metals have minimum ionization energies ranging between 5–10 eV, as can be seen from Fig. 11.4. Consequently, if the EUV photon strikes the wall, a photoelectron may be ejected, thereby triggering another Townsend avalanche. Due to competing processes, the actual photo-efficiency for these UV photons is actually very small, on the order of 10^{-4} per photon [Korff 1946]. However, the number of electron-ion pairs produced for a single event and in the subsequent discharge usually produces many more than 10^4 electron-ion pairs per event so that there is an appreciable probability that a photoelectron is emitted from the cathode wall because of the absorption in the wall of a UV photon.

With sufficient excitation, it is also possible for atoms of the host gas to release, or excite, K-shell electrons. Further, it is possible that the host gas becomes multiply-ionized. By recombination and de-excitation, energetic x-ray photons can be emitted. For example, Ar photon emission energies can be over 3 keV ($K_\beta = 3.192$ keV; $K_{\alpha 1} = 2.957$ keV). Also, excited neutral gas atoms can emit one or more characteristic x rays. These characteristic photons have sufficient energy to cause direct ionization of the atoms of the host gas and, consequently, increase and prolong the avalanching process.

It is also possible for a gas ion, with sufficient kinetic energy, to directly cause the emission of an electron when it comes into contact with the cathode wall. The ejected electron then sets off another Geiger discharge.

In all these cases, there is a release of an electron that can cause another Geiger discharge, which is easily mistaken for another radiation interaction event, unless a mechanism is employed to prevent (quench) the release of another electron, or at least to render such secondary electrons incapable of starting another Geiger discharge. Geiger-Müller detectors are considered to be either "non-self-quenching" or "self-quenching" [Korff 1946]. Quenching in a non-self-quenched detector relies upon a combination of space charge accumulation and external resistance to terminate and quench the Geiger discharge. By contrast, self-quenching G-M detectors have a quenching gas added to the detection gas to prevent electron ejection by positive ions.

Non-Self-Quenching Detectors

Consider the circuit shown in Fig. 11.3, in which the potential across the detector is produced by a resistive divider, formed by the load resistor R_L and the chamber resistance R_{ch}. Because the chamber is filled with a neutral gas, $R_{ch} \gg R_L$. After a radiation particle enters the chamber and initiates a Geiger discharge, current from electron collection begins to flow through the anode wire, and also through the load resistor, thereby causing the voltage drop across R_L to increase at the expense of the voltage held across the chamber [Werner 1934a, 1934b]. As the Geiger discharge increases, so does the electron current and the space charge sheath surrounding the anode, until eventually the voltage held in the chamber falls below the threshold potential required to sustain a Geiger discharge. When the falling potential reaches a value such that the electric field at the anode surface falls below $\mathcal{E}_t$, the Geiger discharge terminates. Although this termination method was initially called "quenching resistance," Montgomery and Montgomery [1940] argue that a high load resistance actually prolongs the Geiger discharge, because the current leakage through R_L allows the anode wire to maintain its original potential for a longer period of time;[4] hence, it is mainly the space charge accumulation that stops the Geiger discharge and not the voltage reduction produced by the load resistance.

Typically, it is multiple cascades of electrons liberated from the cathode wall by UV photons that builds up the Geiger discharge. Montgomery and Montgomery [1940] report the observed formation of multiple discharges until the space charge sheath eventually terminates the progression. Regardless, if $R_L > 10^8\ \Omega$, then the RC time constant of the chamber and circuit is usually longer than the collection time of the ions, thereby allowing collection of the positive ions before the chamber resets. In other words, the detector recovery time is longer than the positive ion collection time. Hence, any electrons ejected from the cathode during the recovery time cannot cause another Geiger discharge. This method of quenching is seldom, if ever, used in modern Geiger-Müller counters, mainly because these detectors suffer from long resolving times ($\geq$ 10 ms), thereby limiting their use to low radiation fields.

Self-Quenching Detectors

The almost universal method used today for quenching G-M counters is to add a small amount of quenching gas to the main detection gas. Polyatomic organic gases were initially used with much success, as described earlier in Chapter 10 for proportional counters. One function of polyatomic organic gases in the mixture is to absorb the UV photons emitted by excited atoms of the primary gas. Without the quenching gas, these photons would produce additional ionization in the detector gas or cause photoemission of electrons from the cathode wall. Upon absorption of a UV photon, a polyatomic molecule, with relatively weak molecular bonds, generally dissociates before the absorbed energy can be emitted in the form of photoelectrons or photons. This dissociation causes a dramatic reduction in the number of UV photons that reach the cathode. Further, if the polyatomic gas has a lower ionization potential than the primary detector gas, electron exchange can take place such that the detection host gas ion is neutralized and the polyatomic molecule becomes ionized. When an ionized polyatomic gas molecule strikes the cathode, it becomes neutralized, and the excited state of the molecule causes dissociation rather than photon emission, thereby preventing electron emission from the cathode wall. The three major quenching processes are depicted in Fig. 11.5.

Neon is perhaps the most popular counting gas used in Geiger-Müller counters, followed by argon. Successful organic quenching gases include CH_4 (methane), C_2H_5OH (ethyl alcohol), and $C_3H_6O_2$ (ethyl formate), often added as 10% of the total gas concentration [Sinclair 1956]. Because G-M counters are typically closed tubes, the organic quenching gas inside can be exhausted over time, eventually leading to sporadic counts as the quenching gas is exhausted. Higher voltages require more space charge to stop the Geiger discharge and, consequently, reduce the lifespan of tubes using organic quench gases. Also, using these counters in relatively high radiation environments reduces the lifetime of the quench gas. It has been reported that the expected life of an organically quenched counter is $\simeq 10^{10}$ counts [Spatz 1943], in which approximately 10^9 molecules are destroyed per discharge [Price 1963].

The discovery that halogens also act as quenching gases for Geiger-Müller detectors led to their popular use [Liebson 1947a; Liebson 1947b; Present 1947; Liebson and Friedmann 1948]. Halogens, unlike organic gases, appear to undergo a healing recombination reaction back into diatomic molecules, thereby extending the lifetime of the quenching gas in the detector. Only low concentrations of halogens are needed to effect the quenching process, usually less than 1% of the total gas concentration [Sinclair 1956]. Reported gas mixtures include Ne or Ar as the primary counting gas with 0.1% of either Cl_2 or Br_2 gas added.

There are performance differences between organic and halogen quenched Geiger-Müller counters. For example, the required operating voltage of a halogen quenched detector is significantly lower than that of an organic quenched detector. Typical operating voltages for halogen quenched G-M tubes range from 250–700 volts, whereas the same tube quenched with an organic gas may require well over 1000 volts for operation.

[4] Korff [1946] suggested that a better name would be "recovery resistance".

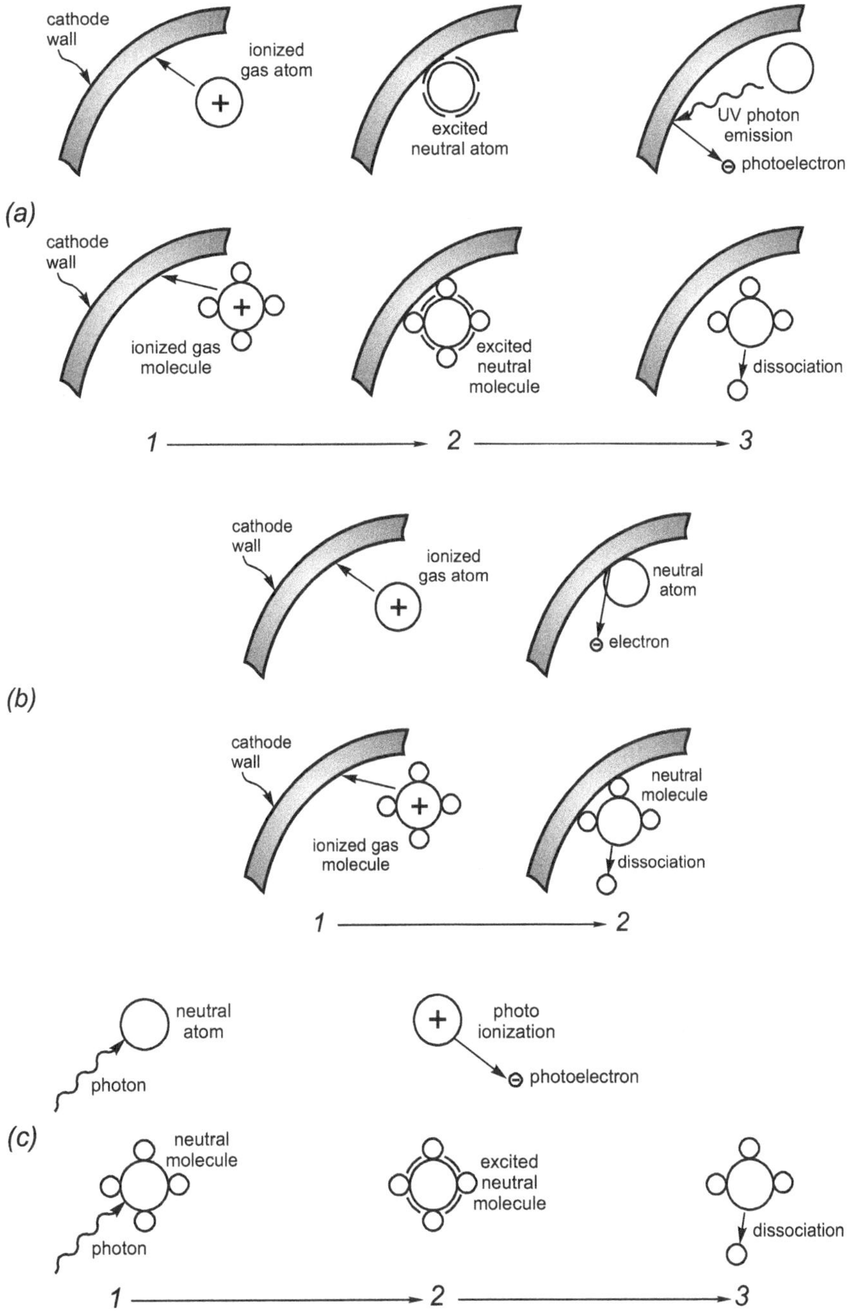

Figure 11.5. Quenching mechanisms of polyatomic gas molecules. In (a), an ionized gas atom comes into contact with the wall and neutralizes into an excited state. De-excitation produces an energetic photon that can eject photoelectrons from the cathode. However, excited gas molecules (organic or halogen), de-excite through dissociation and do not release additional electrons. Other processes of gas quenching through molecular dissociation are depicted in (b) and (c).

Also the plateau length of a halogen quenched G-M tube is usually shorter than one quenched with an organic gas, often having a plateau range no more than 150 volts. Further, the plateau slope is usually larger, typically on the order of a 10% increase over the plateau range.

Halogens are reactive oxidizers for many materials and this property can lead to corrosion over time of internal detector components. Although the concentration of halogen gas is small, caution is still necessary in selecting materials for the cathode and anode. Stainless steel as the anode and cathode has proven to work well with halogen quenched G-M counters and is frequently used in modern G-M counters. A popular choice of stainless steel for halogen quenched G-M counters is 446 SS (about 27% Cr, 73% Fe), a stainless steel with perhaps the best corrosion resistance of the Cr alloys.

Example 11.1: Consider a Geiger Müller tube pressurized to 0.2 atm filled with Ne and 10% CH_4. The tube has a diameter of 4 cm and a length of 15 cm. Estimate the number of counts that can be recorded before the quench gas is exhausted. Assume recombination of dissociated molecules is negligible.

Solution:

From the ideal gas law, the number of mols of gas in the tube is $n = PV/(RT)$ where the gas constant $R = 0.0821$ L atm K^{-1} mol^{-1} is

$$n = \frac{PV}{RT} = \frac{(0.2 \text{ atm})(\pi(2 \text{ cm})^2(15 \text{ cm}))}{(0.0821 \text{ L atm K}^{-1} \text{ mol}^{-1})(300 \text{ K})} = 1.53 \times 10^{-3} \text{ mol}.$$

The number of quench gas molecules N_{CH_4} initially in the tube is

$$N_{\text{CH}_4} = 0.1(nN_a) = 0.1(1.53 \times 10^{-3} \text{ mol})(6.022 \times 10^{23} \text{ molecules mol}^{-1})$$
$$= 9.21 \times 10^{19} \text{ CH}_4 \text{ gas molecules.}$$

where N_a is Avogadro's number.

If 10^9 quench molecules dissociate for each count, then the counter can have a maximum of 9.21×10^{10} counts before the quench gas is completely exhausted. In reality, the side effects of losing the quench gas molecules appear much earlier, manifested as an increase in counting plateau slope and the increasing appearance of sporadic counts.

11.4 Pulse Shape

Because the Geiger discharge produces a large signal voltage, Geiger-Müller counters can be operated without amplification circuitry, a clear advantage over other types of gas-filled detectors. The shape of the signal voltage pulse produced by a G-M counter is defined by the rise time and decay time of the detector.[5] The rise time constant is determined mainly by the load resistor and detector capacitance (see Fig. 11.3). There is a time delay between the initial ionizing event and the start of an avalanche, a delay that depends mainly on the electron drift speed and tube radius. The lighter high-mobility noble gases, such as He and Ne, have much shorter time delays than those in the heavier noble gases such as Kr and Xe. Hence, the observed signal pulse begins slightly after the radiation interaction, a delay of a few microseconds for low electron mobility gases.[6]

[5] The input voltage V_{in} is defined as the signal transferred from the detector to the initial amplification stage of a preamplifier, whereas the output voltage V_{out} is the output signal of the preamplifier. Because G-M counters are often operated without a preamplifier, the term "signal voltage" is adopted here for pulses produced by a G-M detector.

[6] For most applications, this fact is of little consequence. However, if one wishes to use a G-M counter for timing purposes, the delay time between the ionizing event and the pulse trigger must be taken into account.

The actual pulse from a G-M counter, during the progression of the Geiger discharge, develops from multiple avalanches. These avalanches eventually cause the space charge sheath to completely cover the anode wire. The electrons are rapidly collected, but contribute little to the signal pulse. It is instead the motion of positive ions towards the cathode that produces almost all of the signal. Recall from Chapter 10 that most of the induced signal from a coaxial detector is formed within the first few microseconds of ion drift. Hence, a load resistor is chosen to yield a time constant of a few microseconds.

A coupling capacitor between the load resistance and recording electronics effectively blocks DC current from the detector, while allowing signals to pass through. The signal continues to increase as the positive ions drift to the cathode; however, the slow motion of ions causes a slow rise in signal. The coupling capacitance is chosen to ensure that the decay time constant is sufficiently large to reduce signal attenuation, while still allowing reasonable count rates.

Consider the pulse shapes of Fig. 11.6. Large coupling capacitances increase the total pulse height, but also produce long decay times and, consequently, produce large dead times. Low coupling capacitances shorten the decay time and dead time, at the expense of a reduced pulse height. Although the signal is lowered by reducing dead time, it is of minor consequence, mainly because the ion density developed in a G-M counter is large to begin with so that a reduction in pulse height can be tolerated. Finally, because the space charge sheath needed to terminate the Geiger discharge is nearly the same (at a given operating voltage), then the signal pulses are also nearly the same, within statistical fluctuations.

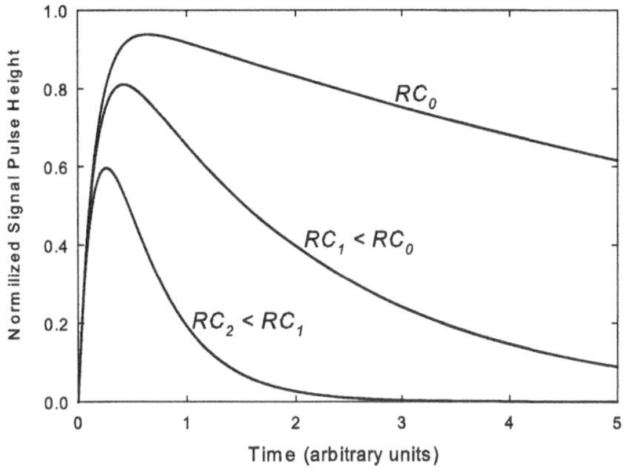

Figure 11.6. Depiction of Geiger-Müller tube signal pulse shapes for different differentiating RC time constants.

11.4.1 Dead, Resolving, and Recovery Times

Immediately following the Geiger discharge, the space charge sheath formed around the anode wire lowers the electric field such that avalanching ceases. Not until the electric field at the anode recovers above the critical avalanching threshold field $\mathcal{E}_t$ can another Geiger discharge occur. As the positive ions drift away from the anode, the electric field at the anode begins to increase, as depicted in Fig. 11.2. The dead time is the time interval between the initial pulse trigger and the electric field recovery to $\mathcal{E}_t$ and depends strongly upon the number of ions in the space charge sheath. In other words, the dead time is dependent upon the applied operating voltage. Dead times for Geiger-Müller counters can be on the order of 10 times longer than those of proportional counters of similar size, typically 80–300 microseconds, and represents the *minimum* time required between the arrival of successive incident radiation particles that allows their detection. The longer dead times associated with Geiger-Müller counters are a consequence of relatively long pulse formation process depicted in Fig. 11.2.

Although it is theoretically possible to register a count after the electric field has recovered to $\mathcal{E}_t$, Geiger-Müller counters are almost always operated with the *LLD* set to exclude electronic noise; hence, counts cannot be registered until the resulting pulse height surmounts the preset *LLD* threshold. The *resolving time* is the interval between the initial pulse trigger and when the pulse height reaches the *LLD* threshold. Clearly, the resolving time is a function of both the applied voltage and the preset *LLD* threshold; therefore, the resolving time is not an intrinsic feature of the detector, but instead changes with the operator's *LLD* setting. It is the resolving time that is usually reported as the detector dead time, although in the strictest

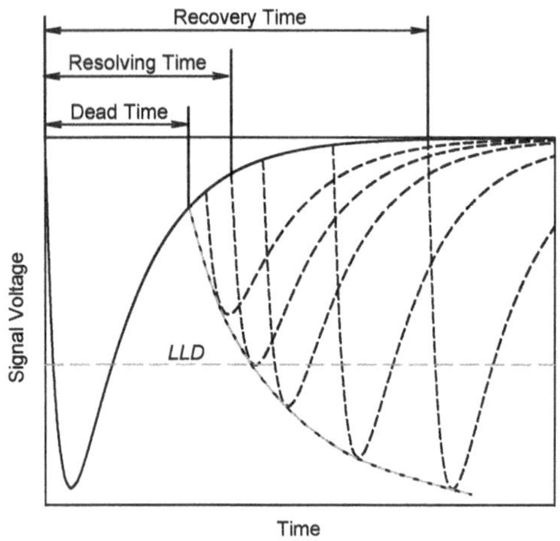

Figure 11.7. Illustration of dead, resolving, and recovery times with respect to signal pulses for a Geiger-Müller tube.

sense it is not the dead time, although it certainly has the same effect. Overall, count rates from a G-M counter are affected principally by the resolving time of the instrument.

The *recovery time* is defined as the interval between the initial pulse trigger and the time that another pulse signal can achieve the maximum amplitude possible. Intuitively, one might surmise that the recovery time is synonymous with the positive ion collection time; however, as explained previously for non-self-quenching detectors, the signal tail is also a function of the RC time constant of the detector. Hence, for a large load resistance R_L, the recovery time is a function of both the ion drift time and the detector time constant. Signal pulses formed during the time interval between the resolving time and the recovery time have lower amplitudes than those produced by a full Geiger discharge. Shown in Fig. 11.7 is an illustration of the relation between dead, resolving, and recovery times from a G-M counter.

11.5 Radiation Measurements

Although there are multiple variations and uses of Geiger-Müller counters, it is customary to determine first a proper operational voltage for the devices. This calibration process usually requires radioactive sources to determine the appropriate LLD setting and the boundaries of the counting plateau. G-M counters cannot distinguish among different types of incident radiation particles; however, the large signal pulses they produce make it relatively easy to detect radiation, be it alpha, beta, or gamma radiation.

11.5.1 Counting Plateau

As with proportional counters, proper operation of a G-M counter requires the appropriate voltage setting. At too low a voltage, the detector functions in the ion-chamber or proportional regions, and generally less efficiently than detectors designed for such operations. As previously discussed, the amplitude of signal pulses from a G-M counter operating in region IV of the gas pulse height curve is strongly dependent upon the space charge sheath required to terminate the Geiger discharge. Consequently, as voltage is increased in the Geiger-Müller region, the amount of space charge required to terminate the discharge also increases. If the operating voltage is further increased, eventually the detector is driven into a state of continuous discharge (region V).

An operator can determine a proper operating voltage as follows. First the detector is placed in a well-shielded or relatively radiation-free environment, and the high voltage is applied to the detector in increasing increments. In doing so, the operator establishes the discrimination level LLD necessary to eliminate electronic noise. Then the high voltage is reduced back to zero and a calibration source is placed in the vicinity of the detector. Signal input pulses produced by radiation particles from a detector operating in the ion-chamber or proportional regions are smaller than pulses produced in the G-M region and depend on the type of incident radiation. Also, pulses produced from mixed α and β sources have a large variation in signal amplitudes when the detector is operated in either the ion-chamber or proportional regions. As voltage is increased beyond the proportional region, the variation in pulse amplitudes reduces and the pulse amplitudes, although increasing with increasing voltage, converge to equal amplitudes in the G-M region.

If the signal pulse amplitudes produced by β and α particles are equal, within statistical fluctuations, then detector operation is at the beginning of the G-M region. The LLD is set at the signal voltage produced at the onset of the G-M region, as depicted in Fig. 11.1. Because Geiger discharges produce voltage pulses of equal amplitudes, at any set voltage, for any type of incident ionization radiation, the *LLD* is mostly ineffective at eliminating background radiations; however, the equality of the pulse amplitudes can be used to ensure that the detector is operating in the G-M region. With a source present, the operator can then track the count rate as a function of applied voltage, thereby producing a counting curve.

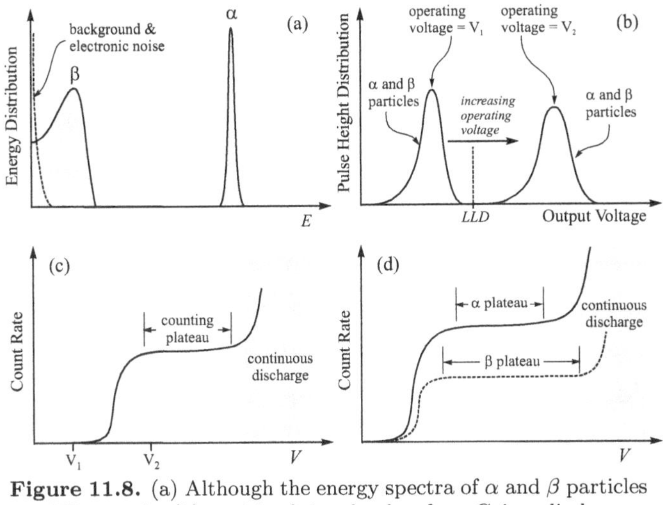

Figure 11.8. (a) Although the energy spectra of α and β particles are different, the (b) combined signal pulses from Geiger discharges, regardless of radiation energy or particle type, are nearly indistinguishable. At low operating voltages V, the resulting signal voltages fall below the *LLD*. As the operating voltage is increased, the signal voltages increase until most or all surpass the *LLD* threshold. (c) For a set *LLD* threshold, pulses below the *LLD* are not counted, whereas, for higher applied voltages that produce Geiger discharges above the *LLD* threshold, the counts are recorded. (d) Because the signal pulses are indistinguishable between α and β (and γ) particles, their counting plateaus (for separate and combined sources) appear at nearly the same applied high voltages.

Consider Fig. 11.8(a), where it is shown that alpha and beta particles have significantly different energy spectra. When these particles interact in a G-M counter, the pulses they produce are indistinguishable as shown in Fig. 11.8(b), thereby producing a continuous pulse height spectrum. With the *LLD* set to some arbitrary level, Geiger discharges that produce voltage pulses below the *LLD* threshold are not recorded. The combined spectrum of pulses from the α and β particles increases as the operating voltage increases. Further, the increase in signal amplitude allows some pulses to surmount the *LLD* threshold and causes a rapid increase in count rate as shown in Fig. 11.8(c). At higher voltages, all Geiger discharges exceed the *LLD* threshold and produce a counting plateau as shown in Fig. 11.8(c). At higher voltages, continuous discharging occurs and appears as a large increase in count rate. If the operator were to compare the counting plateaus for separate alpha and beta particle sources, both with the same *LLD* setting, then the counting plateaus would appear at approximately the same voltage, as depicted in Fig. 11.8(d) and shown in the measurements of Fig. 11.10. Typically, after the counting plateau is measured, the operator sets the operating voltage in the center of the plateau. Note that the plateau formed by the β-particle source is considerably flatter and wider than the plateau formed by the α-particle source, a feature not seen in a proportional counter.

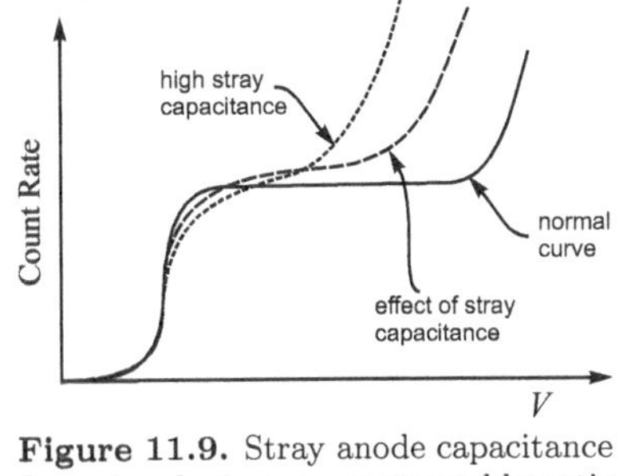

Figure 11.9. Stray anode capacitance from signal wires can cause problematic distortions in the counting curve.

The slope of the counting plateau is determined in the same manner described by Eq. (10.60) and Eq. (10.61). The slope of the curve is caused for two main reasons [Centronic 2014]. (1) The increased voltage causes the active volume of the detector to expand, thereby increasing the count rate. (2) Even with a quench gas present, there remains a small probability that residual positive ions or excited gas molecules may produce a spurious count, an effect that increases with voltage. The slope of the counting plateau can also be affected by stray capacitance associated with the detector anode (see Fig. 11.3). This anode stray capacitance, generally arising from signal wires connecting to the detector, has deleterious effects upon the detector performance [Centronic 2014], manifested as a shortened counting plateau and increased plateau slope (see Fig. 11.9). Other effects include an increase in dead time, shortened tube life, and spurious counts.

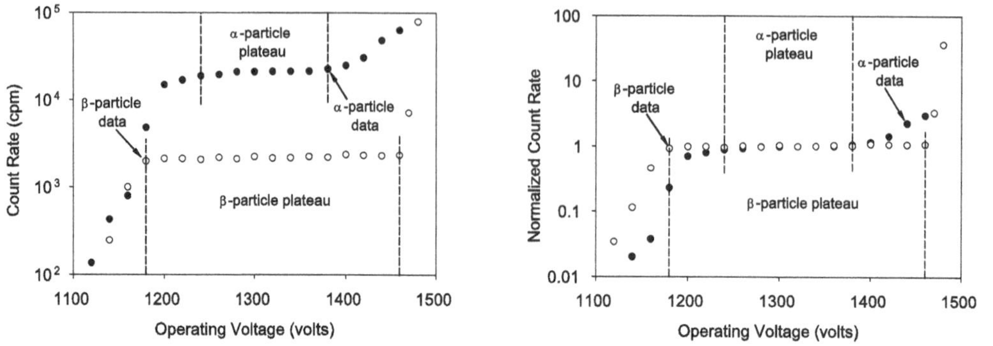

Figure 11.10. (left) A comparison of α- and β-particle counting plateaus from a single Geiger-Müller counter for different α- and β-particle sources with different activities, and (right) the count rate data normalized at 1320 volts. The LLD was kept constant for all measurements.

11.5.2 Alpha and Beta Particle Counting

Geiger-Müller counters have a thin central wire on the axis of a cylindrical tube, as depicted in Fig. 11.3. Most Geiger-Müller counters are completely sealed, and consequently, alpha and beta particles can be blocked by the detector wall from entering the sensitive volume of the detector. However, there are G-M tubes designed for α and β particle detection that have thin windows to allow charged particles to enter the sensitive gas-filled volume.

Although alpha particles typically have sufficient energy to pass through a window composed of boPET or mica, the transport of particles through air from the source to the detector window may result in so much energy loss they cannot pass through the window or, perhaps, not even reach the window. Consequently, the window portion of the detector should be positioned within a centimeter or so of the suspected source material so that alpha particles can enter the device.

It is also common for survey units to be enclosed and sealed in a plastic envelope or bag, mainly to prevent inadvertent contamination of the delicate entrance window. If the plastic bag becomes contaminated, it can be properly discarded and replaced with a new one. Unfortunately, such an envelope can prevent the entrance, and detection, of alpha particles and low-energy beta particles. Consequently, detection of alpha particles with a survey unit does run the risk of possible contamination of the detector window. Some manufacturers have a metal guard screen secured in front of the window to reduce the possible contamination or damage to the thin window. Plastic protection caps are also available that can be slipped over the entrance window, thereby protecting the thin window from damage or contamination while still permitting gamma-ray detection.

Energetic beta particles can pass through thin plastic and have manageable ranges in air. As a result, even enclosed in a plastic bag, a G-M counter can usually be used for counting beta particles. However, there are a few important radionuclides that emit low energy beta particles that present a detection challenge. These include ^{3}H, ^{14}C, and ^{45}Ca, which have, respectively, maximum emission energies of 18.6 keV, 154 keV, and 250 keV, low enough to be absorbed by any intervening materials such as air or the detector window, unless special precautions are exercised. For typical G-M counters with window mass thicknesses equivalent to 30 mg cm^{-2}, the transmission percentage is $\sim$ 0%, 0.04% and 3.0% for ^{3}H, ^{14}C, and ^{45}Ca, respectively.[7] There are commercial detectors with thin windows formed of mica or aluminized boPet, with

[7] Because it is difficult to measure low energy beta particles with a survey meter, liquid scintillation counting has become the preferred measurement choice for low energy beta particles.

mass thicknesses of 1–2 mg cm^{-2}, that can be used to detect beta particles from ^{14}C and ^{45}Ca, although tritium still remains a problem largely because of the short range of its beta particles in air. Higher energy beta particles can enter detectors with thicker windows; hence, common G-M counters can be used to detect beta particles with $E_{max} \gtrsim 250$ keV.

11.5.3 Gamma-Ray Detection

Gamma-ray detection with G-M counters is fairly straightforward. To reduce dead time, the gas in G-M tubes usually has pressures of less than 1 atm, and it is unlikely that a gamma ray interacts with the rarefied detection gas. Instead, x rays and gamma rays mostly interact with the walls of a G-M counter to produce an energetic electron, either through photoelectric, Compton scattering, or pair-production interactions, as shown in Fig. 11.11. If an energetic electron escapes into the detector gas, then it produces electron-ion pairs in the gas, which subsequently cause a Geiger discharge.

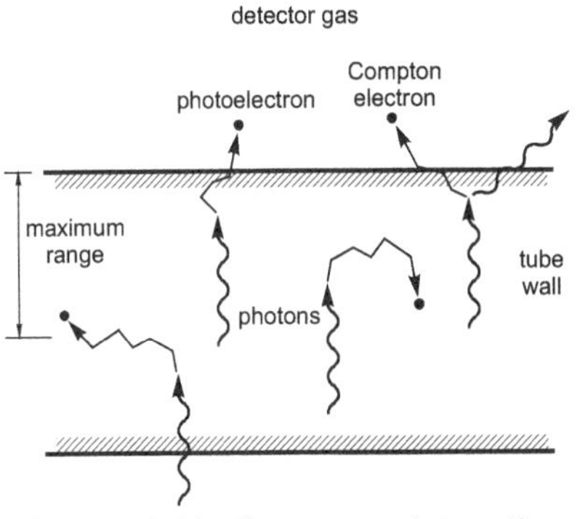

Figure 11.11. Gamma-ray interactions in the G-M tube wall produce energetic electrons that can enter the detector gas and create electron-ion pairs. Electrons liberated too far from the gas interface, or scattered away from the gas-wall interface, cannot enter the detector.

The response of Geiger-Müller tubes to gamma-rays is dependent upon the cathode material. From Fig. 11.12, it is evident that high Z elements have an "over-response" to low-energy gamma rays. The intrinsic efficiency of a G-M counter is defined as the number of primary electrons emitted into the counter gas divided by the number of photons incident on the detector. Hence, intrinsic efficiency includes gamma-ray absorption and electron escape probabilities. For low energy gamma rays, for which photoelectric absorption is dominant, $\sigma_{ph} \propto Z^4/E^3$ and large Z materials have higher absorption probabilities for gamma rays than do small Z materials. Consequently, the response of G-M counter to gamma rays is much higher for large Z materials than small Z materials in the low energy region and, thus, causes the observed over response (see Fig. 11.12).

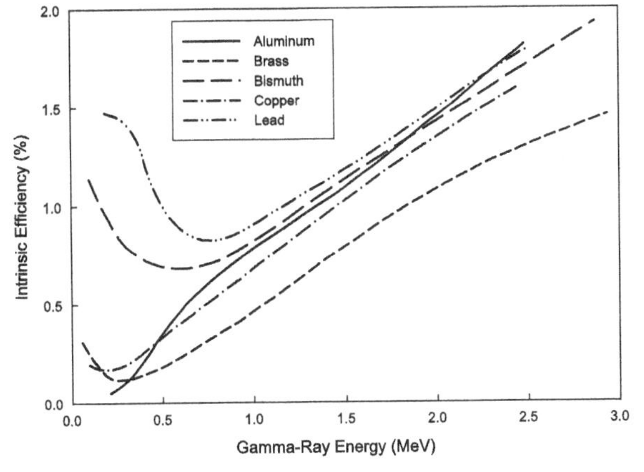

Figure 11.12. Gamma-ray detection efficiency in a Geiger-Müller counter as a function of energy and cathode material. Data from [Bradt et al. 1946, Sinclair 1956].

In the energy region largely dominated by Compton scattering, the gamma-ray interaction probability is approximately proportional to Z, whereas the electron escape probability is approximately proportional to Z^{-1}; hence the two probabilities more or less cancel. As a result, the G-M counter response to gamma-rays in the intermediate region between 500 keV and 2 MeV is relatively unaffected by the Z of the cathode material. From Fig. 11.12, it is seen that Al ($Z = 13$) and Bi ($Z = 83$) have nearly the same response for 1 MeV gamma rays. Although older G-M tube designs were lined with Bi or Pb, modern detectors generally have a Cr/Fe stainless steel alloy[8] cathode, and for enhanced gamma-ray response a Pt ($Z = 80$) cathode liner [Centronic 2014]. The Cr/Fe alloy can be processed to have a relatively high work function, thereby reducing the probability of secondary electron emission. Further, the Cr/Fe stain-

[8]446 stainless steel and is composed of 73% Fe and between 23% and 27% Cr.

less steel alloy is non-corrosive in the presence of the halogen quenching gas, while still having a reasonable response to gamma rays at approximately 1% detection efficiency.

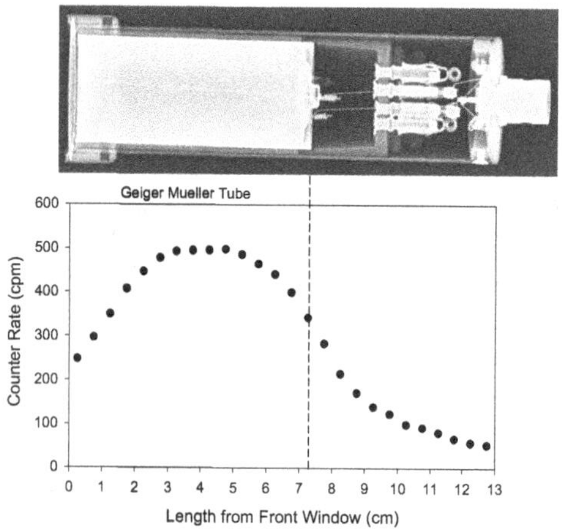

Figure 11.13. (top) X-ray photo of a Ludlum LND 723 G-M tube in its Al container. The detector gas occupies about 60% of the aluminum casing with the remainder filled with wire leads, connectors and resistive divider components. (bottom) The count rate response as a collimated ^{60}Co gamma-ray source moves along the container axis.

The response of G-M counters can be tailored by carefully selecting cathode materials inserted into the chamber. However, this practice is impractical for commercial sealed tubes, most of which are manufactured with stainless steel cathodes. Instead, instruments frequently come with an external shutter that the operator can open or close so as to change the G-M counter response to low- and high-energy gamma rays. Some commercial manufacturers offer detachable field flattening filters designed for special purposes, to reduce the over response to gamma rays at low energies.

The gamma-ray counting efficiency is also a function of the interaction location. An x-ray photograph of a Geiger-Müller tube is shown in Fig. 11.13. The tube, manufactured from stainless steel, rests inside an aluminum casing, connected to electronics through a "C"-type connector. The actual G-M tube takes up approximately 60% of the Al casing. Also shown in Fig. 11.13 is a count rate profile from a collimated ^{60}Co source. The count rate is highest near the center of the actual G-M tube, but drops off near the entrance window and also near the anode feed through. Counts observed from positions beyond the G-M tube are from Compton scattered gamma rays that interact with the Al casing and other materials surrounding the G-M tube.

11.6 Special G-M Counter Designs

Although G-M counters produce substantially similar output pulses regardless of the interacting radiation, special design features can be implemented to improve responses to specific forms of ionizing radiation. For instance, G-M tubes designed for alpha particle counting are outfitted with a thin mica window that allows easy penetration of alpha particles into the chamber. Other designs are optimized for beta particle penetration with either a thicker entrance window that blocks alpha particles or with thinned regions along the metal tube length. Some designs may include a shutter that can be optionally rotated into place to block beta particles while passing gamma rays (see Fig. 11.18). Such devices can be used to distinguish between beta-particle and gamma-ray induced counts.

Detectors may also be filled with different types of counting gas, including He, Ne, Ar, Kr, and Xe for the detection gas while CH_3 or various halogens (usually Cl_2 or Br_2) are used for the quenching gas. Electron mobility is higher in Ne gas, while gamma-ray interaction efficiency is higher in the heavier gases (Kr and Xe). Introduction of an organic gas can also improve electron speeds and, at the same time, serve as the quenching gas. Halogens, as already discussed, serve as a quenching gas while extending the tube life.

Detectors backfilled with Xe gas can be used for detecting low-energy gamma rays, and are often fabricated with thin beryllium or mica entrance windows. These devices are, therefore, designed to operate with the window-end pointing at the radiation source. To improve efficiency, these detectors have higher gas pressures (between 600 to 800 torr) than do conventional G-M tubes. Further, the length is longer than other G-M tubes to improve the gamma-ray absorption probability.

As already mentioned, the response to energetic gamma rays can be increased by lining the cathode with high Z material. Detectors designed for high energy gamma-ray detection generally do not have an entrance

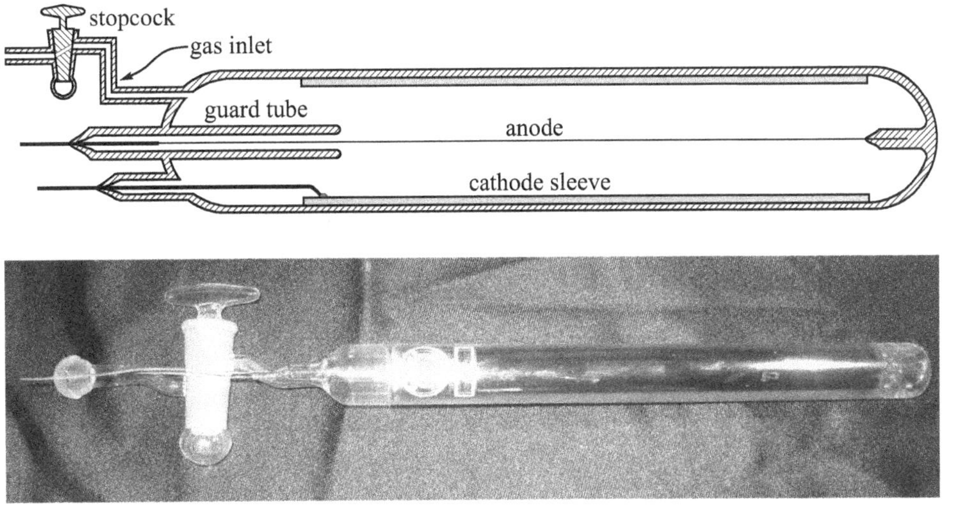

Figure 11.14. (top) Diagram of a refillable cylindrical glass GM counter and (bottom) photograph of a refillable cylindrical glass GM counter manufactured by N. Wood Counter Laboratory.

window and are usually designed for side irradiation to improve interaction efficiency. G-M counters can be made sensitive to slow neutrons by including a Cd liner, but this modification is traditionally not done because of intrinsic problems distinguishing between types of radiation.

11.7 Commercial G-M Counters

Commercial G-M counters of notable variety have been offered for decades. Many of the earliest G-M counters were manufactured as glass tubes, similar to vacuum tubes. These tubes could be acquired as refillable chambers, especially important if organic gases were used as the quenching gases. One such model is shown in Fig. 11.14. These refillable G-M tubes allow a variety of host gases and quenching gases to be used. Current manufacturers of G-M tubes supply numerous detector companies that install them in portable or laboratory instruments. Although many geometries are available, there are perhaps four main types, namely, end-window tubes, pancake detectors, thin wall beta/gamma sensitive tubes, and miniature tubes. The variety of Geiger-Müller tubes, and the equipment in which they are installed, is much too vast to cover in a single book chapter. More information on G-M tubes and equipment can be readily found at internet sites for many of the larger G-M counter manufacturers.[9] A few types of G-M counters are presented here as examples.

LND Inc. is one of the largest manufacturers of gas-filled Geiger-Müller counters and has provided G-M counters to commercial detector instrumentation companies for over 50 years, and a large variety of configurations with different detection gases are available. Their G-M counters are often filled with Ne gas under a mild vacuum, with a small amount of halogen added, usually Cl_2 or Br_2. The use of Ne as the gas, under slight vacuum, reduces the dead time of the detector by increasing the electron and ion mobilities and drift speeds (see Fig. 10.10). To reduce corrosion from the halogen quenching gas, the cathode and anode are constructed from 446 stainless steel. LND provides G-M counters to many detector manufacturers who package them in a variety of configurations, including laboratory detectors, handheld instruments, and portal

[9] For example Centronic, LND, Ludlum, Canberra, Thermo Fisher, and Saint Gobain have data sheets, schematics, and operating procedures available in varying detail.

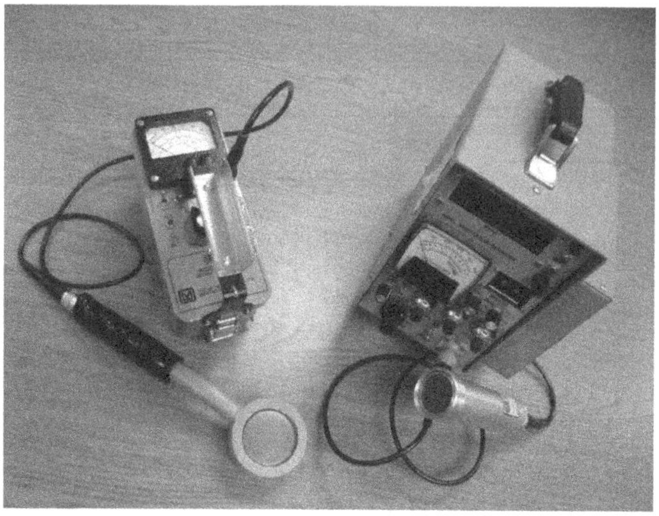

Figure 11.15. Shown are two commercial portable G-M counters manufactured by Ludlum Measurements, Inc. On the left is a 44-9 G-M pancake frisker attached to a model 14C G-M meter, and on the right is a 44-7 G-M tube attached to a model 2200 scaler rate meter.

monitors. An established radiation detector instrument manufacturer is Ludlum Measurements Inc., which offers several small and large instruments outfitted with G-M detectors. For instance, a popular instrument is the 44-9 pancake frisker hand held G-M counter used for general survey work. This instrument incorporates a LND 7311 pancake detector and can be attached to a variety of counters and rate meters as shown in Fig. 11.15.

Portable battery-operated G-M counters are relatively inexpensive and convenient for general survey work. The counters usually have a rate meter with either an analog scale or a digital counter. In either case, the device measures the rate that radiation interacts in the detector, but does not yield any information about the dose rate at a given location because G-M counters are insensitive to energy deposition.

G-M counters notoriously overestimate exposure and dose for low-energy gamma rays if the cathode is fabricated from moderately high atomic number materials. Most G-M counters are presently fabricated from stainless steel, composed primarily of Fe ($Z = 25$) and Cr ($Z = 24$). Although the atomic numbers of Fe and Cr are not considered overly large, the gamma-ray response is still exaggerated for energies below 100 keV. However, selective choice of filters placed over the sensitive portion of the G-M tube can flatten the response by reducing the counting efficiency at low energies, a method referred to as *energy compensation.*

Compensated Geiger-Müller tubes have a tubular collar as part of the detector, commonly an alloy of Sn and Pb, placed around them to flatten the response to gamma-rays at lower energies. Some filters are available as detachable units easily added to the sensitive window of a detector, as depicted in Fig. 11.16, while other detector probes have a rotatable filter as part of the probe unit. Shown in Fig. 11.17 are gamma-ray sensitivity response curves to a Ludlum standard 44-9 pancake G-M counter with and without the optional detachable filters [Ludlum 2014b]. The exposure filter is composed of Sn (1 mm) and plastic (.62 mm) inserts, while the dosimeter filter is composed of Al (0.4 mm), Sn (0.9 mm), and plastic (0.12mm) inserts. G-M counters designed primarily for gamma-ray and beta-particle counting may have a thinned portion of the tube that allows for beta particles to pass into the detector gas, but generally not as thin as windows needed for alpha-particle counting. Probes with rotating shields allow beta-particle counting

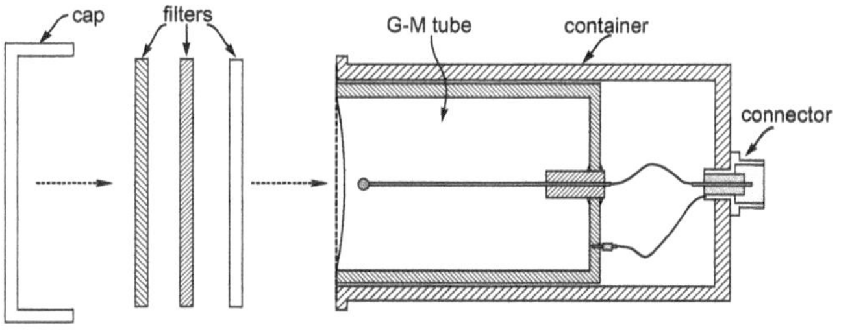

Figure 11.16. The gamma-ray response from a G-M counter can be altered by adding filters to the detector entrance window. With a proper selection of materials and orientation, the gamma-ray response of a G-M counter can be reasonably constant over a wide range of photon energies.

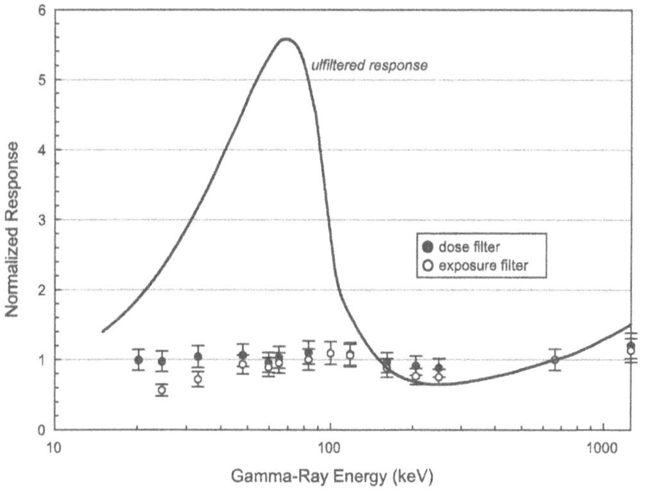

Figure 11.17. Gamma-ray responses from a commercial Ludlum 44-9 G-M pancake detector with and without filtration. The responses are normalized at 662 keV. The filters work to suppress the over-response to gamma rays at low energies. Data from [Ludlum Measurements, Inc. 2014a, 2014b].

when the shield is opened, and filtered gamma-ray counting when the shield is closed. Detectors of this sort are shown in Fig. 11.18, manufactured by Eberline[10] and Ludlum. The Ludlum 44-6 and 44-38 G-M probes come equipped with a Sn side shield that can be rotated in front of the thin G-M tube. The G-M tube listed as the detector in Ludlum 44-6 and 44-38 probes is a LND 725, which does not come equipped with a front window, but has a relatively thin stainless steel body designed to record counts from the side.[11] The LND 725 is backfilled with Ne with a halogen quenching gas, usually Cl_2 or Br_2. The capacitance of the tube is 3 pF, with a recommended load resistance of 1 MΩ, thereby yielding a rise time constant of 3 microseconds. When the Sn shield is rotated back, the G-M tube is exposed through open windows along the probe body (see Fig. 11.18), thereby allowing beta-particle counting for energies above 200 keV [Ludlum 2013a]. Shown in Fig. 11.19 is the gamma-ray response, with and without the shield, for a LND 725 G-M tube installed in a Ludlum 44-6 or Ludlum 44-38 handheld probe. The data were recorded with a few

[10]Now owned by Thermo Fisher.

[11]Some of the Ludlum 44-6 probes come equipped with a TGM N112 G-M tube with similar properties as the LND 725.

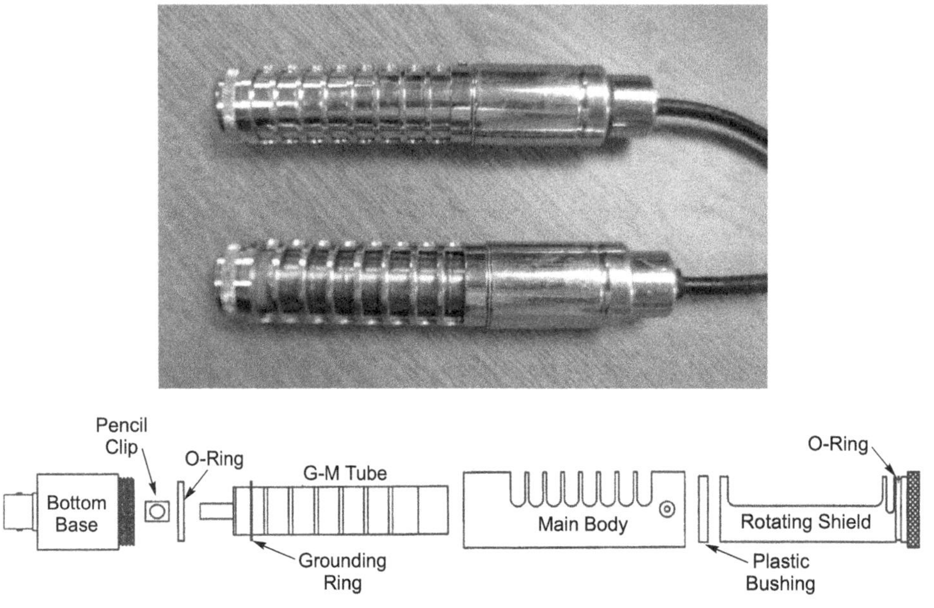

Figure 11.18. (top) Eberline HP177 G-M beta gamma hand probes with rotating shield, showing closed and open positions. The G-M tube was manufactured by Anton Electronics Laboratory. (bottom) Exploded view of a Ludlum 44-6 G-M beta gamma hand probe. From [Ludlum Measurements, Inc. 2013a].

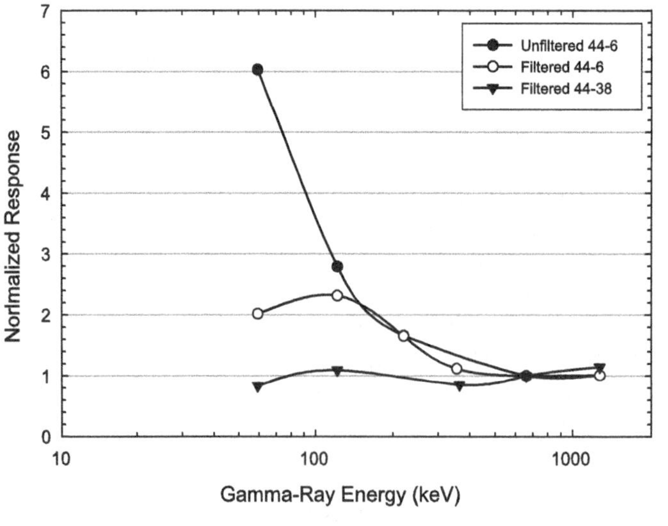

Figure 11.19. Gamma-ray response from an LND 735 G-M tube installed in commercial Ludlum 44-9 and 44-38 beta-gamma counters with and without filtration, normalized at 662 keV. The filters work to suppress the over-response to gamma rays at low energies. Data from [Ludlum Measurements 2013a, 2013b].

gamma-ray standard check sources, namely, ^{241}Am, ^{57}Co, ^{133}Ba, ^{137}Cs, and ^{60}Co. Because ^{133}Ba and ^{60}Co have multiple gamma-ray emissions, weighted averages for their emission energies were reported [Ludlum 2013a, 2013b].

Although G-M counters are used widely for portable handheld instrumentation, there are some larger instruments that have several G-M tubes operating as a single unit. For instance, portal and contamination monitors have multiple G-M counters installed in them. Hand and shoe monitors have multiple G-M counters installed in hand ports and underneath a foot screen for shoe monitoring. A single instrument may have over 20 pancake type G-M tubes, divided amongst the hand and shoe locations. Portal monitors have G-M tubes of varying geometries arranged about a frame that surveys the entire body.

PROBLEMS

1. Consider Fig. 11.10. Why is the β-particle counting plateau for a G-M counter wider and flatter than achieved with a proportional counter?

2. A G-M tube, which contains 0.12 atm of argon and 0.01 atm of alcohol, has a lifetime of 2×10^9 counts. If the end of the tube life occurs when all the alcohol molecules have been dissociated, estimate the number of alcohol molecules dissociated per discharge.

3. If the voltage applied to a G-M tube increases above that needed to achieve a full Geiger discharge, the pulse height continues to increase. Explain.

4. What would be the effect on the starting voltage of the Geiger plateau if (a) the pressure of the fill gas were doubled, or (b) the small concentration of the quench gas were doubled, or (c) the diameter of the anode wire were doubled?

5. An engineer decides to test a G-M counter with Ar as the host gas without the quenching gas. Estimate the number of electrons ejected from a cathode wall for a 5 MeV alpha particle fully absorbed in a G-M counter. Assume M = 1000 and the photoelectric efficiency of the wall is 10^{-4}.

6. Consider a typical G-M tube with dimensions of $D = 2.5$ cm and $L = 15$ cm. The tube is pressurized to 0.15 ATM with Ne gas along with 0.1% Cl_2 gas. If the quenching gas does not recombine, estimate the limiting number of counts possible with this G-M tube. If the actual number of counts observed is 10^{13}, what conclusions can you draw about the quenching gas?

7. While operating a G-M counter, the *LLD* is adjusted such that the resolving time is 220 μs. At what count rate do resolving time losses reach 5% for this detector? Plot the resolving time losses as a function of count rate from 0% up to 10% loss.

8. A G-M tube is used with a counting system that requires a full Geiger discharge in order to register a count. Is the dead time of such a system better described by an extendable or nonextendable model? Explain.

9. If a G-M counter internal wall is coated with ^{10}B, or is instead backfilled with ^{3}He or BF_3 gas, it can be made sensitive to neutrons; however, this is typically not done. Explain the disadvantage to using a G-M tube for neutron detection.

REFERENCES

Barsuto, E., J. de Urquijo, I. Alvarez, and C. Cisneros, "Mobility of He^+, Ne^+, Ar^+, N_2^+, O_2^+, and CO_2^+, in their Parent Gas," *Phys. Rev. E*, **61**, 3053–3057, (2000).

Bradt, H., P.C. Gugelot, O. Huber, H. Medicus, P. Preiswerk, and und P. Scherrer, "Empfindlichkeit von Zählrohren mit Blei-, Messing und Aluminiumkathode für γ-Strahlung im Energieintervall 0.1 MeV bis 3 MeV," *Helv. Phys. Acta*, **19**, 77–90, (1946).

Emery, E.W., "Geiger Müller and Proportional Counters," *Radiation Dosimetry II*, Ch. 10, F.H. Attix and W.C. Roesch, Eds., New York: Academic Press, 1966.

Geiger, H. and W. Müller, "Elektronenzhlrohr zur Messung Schwchster Aktivitten," *Die Naturwissenschaften*, **16**, 617–618, (1928a).

Geiger, H. and W. Müller, "Das Elektronenzhlrohr," *Phys. Zeitschrift*, **29**, 839–841, (1928b).

Geiger, H. and W. Müller, "Technische Bemerkungen zum Elektronenzhlrohr," *Phys. Zeitschrift*, **30**, 489–493, (1929a).

Geiger, H. and W. Müller, "Demonstration des Elektronenzhlrohrs," *Phys. Zeitschrift*, **30**, 523, (1929b).

Geiger-Müller Tubes, Croydon, UK: Centronic, Ltd., 2014.

Korff, S.A., *Electron and Nuclear Counters—Theory and Use*, New York: Van Nostrand Co., 1946.

Liebson, S.H., "Low Voltage Self-Quenching Counters," *Phys. Rev.*, **72**, 181–182, (1947a).

Liebson, S.H., "The Discharge Mechanism of Self-Quenching Geiger-Mueller Counters," *Phys. Rev.*, **72**, 602–608, (1947b).

Liebson, S.H. and H. Friedmann, "Self-Quenching Halogen-Filled Counters," *Rev. Sci. Instrum.*, **19**, 303–306, (1948).

Ludlum Measurements, Inc., *Ludlum Model 44-6 Beta-Gamma Detector*, Product Manual, 2013a.

Ludlum Measurements, Inc., *Ludlum Model 44-38 Beta-Gamma Detector*, Product Manual, 2013b.

Ludlum Measurements, Inc., *Ludlum Model 44-9 Alpha, Beta, Gamma Detector*, Product Manual, 2014a.

Ludlum Measurements, Inc., *Ludlum 44-9 Compensation Filter Design*, Internal Report 54992, 2014b.

Montgomery, C.G. and D.D. Montgomery, "The Discharge Mechanism of Geiger-Mueller Counters," *Phys. Rev.*, **57**, 1030–1040, (1940).

Munson, R.J. and A.M. Tyndall, "The Mobility of Positive Ions in Their Own Gas," *Proc. Royal Soc. London* **A177**, 187–191, (1941).

Present, R.D., "On Self-Quenching Halogen Counters," *Phys. Rev.*, **72**, 243–244, (1947).

Price, W.J., *Nuclear Radiation Detection*, 2nd. Ed., New York: McGraw-Hill, 1963.

Rutherford, E. and H. Geiger, "An Electrical Method of Counting the Number of a Particles from Radioactive Substances," *Proc. Royal Society of London*, Series A, **81**, 141–161, (1908).

Sharpe, J., *Nuclear Radiation Detectors*, London: Methuen & Co., 1964.

Sinclair, W.K., "Geiger-Mueller Counters and Proportional Counters," in *Radiation Dosimetry*, G.J. Hine and G.L. Brownwell, Eds., **Ch. 5**, 213–243, (1956).

Spatz, W.D.B., "The Factors Influencing the Plateau Characteristics of Self-Quenching Geiger-Mueller Counters," *Phys. Rev.*, **64**, 236–240, (1943).

Werner, S., "Die Entladungsforman im Zylindrischen Zählrohr," *Z. Phys.*, **90**, 384–402, (1934a).

Werner, S., "Die Entladungsforman im Zylindrischen Zählrohr II," *Z. Phys.*, **92**, 705–727, (1934b).

Chapter 12

Review of Solid State Physics

Nothing exists except atoms and empty space; everything else is opinion.
Democritus

12.1 Introduction

Before scintillation detectors are discussed in the next chapter and light collection devices are discussed in Chapter 14, it is necessary to have this introductory chapter on solid state physics. Further, the information in this chapter is also pertinent to the chapters on semiconductor radiation detectors (Chapters 15 and 16). Here many basic concepts are developed including the fundamental theory of solid state energy bands, effective mass, density of states, mobility, resistivity, impurity doping, and compensation methods. Note that those already familiar with basic solid state physics, including band theory and electronic properties of materials, may safely proceed on to Chapter 13 with, perhaps, little consequence.

12.2 Solid State Physics

Proficiency in the design and fabrication of scintillation and semiconductor detectors requires an understanding of basic solid state and semiconductor physics; and addressing this requirement is the purpose of the present chapter. Here the necessary background about crystalline solid state materials is presented. The physics reviewed here is needed for the chapters that follow.

12.2.1 Crystals and Periodic Lattices

A crystal is composed of atomic or molecular structures (the *basis*) arranged and repeated in a *lattice* arrangement [Kittel 1966]. The basis of the system is the grouping of atoms, which may be a single atom or may be a large molecular group of atoms. The lattice is the structure onto which the basis is projected and repeated. Together they form a crystal structure. Figure 12.1 depicts the basis and lattice of a crystal. When the basis is placed upon each of the lattice points, a crystal is formed. The lattice itself is a periodic structure that can be reduced to a single *unit cell* which, when repeated, reproduces the entire crystal. In other words, by simply repeating the unit cell over and over with each unit cell placed adjacent to the last, a crystal is formed. Figure 12.2 shows two possible unit cells for the lattice, both of which are valid. Unit cells do not have to be *primitive cells*, which are the smallest possible unit cell that, when repeated, forms the crystal. In fact, unit cells are not unique and can be formed in many geometric shapes, provided that the shapes, when stacked, again form the crystal.

The unit cell is related to the lattice in terms of *basis vectors*. Consider the two lattices shown in Fig. 12.3. Then $\mathbf{a}$ is a basis unit vector of length a, and $\mathbf{b}$ is a basis unit vector of length b. An equivalent

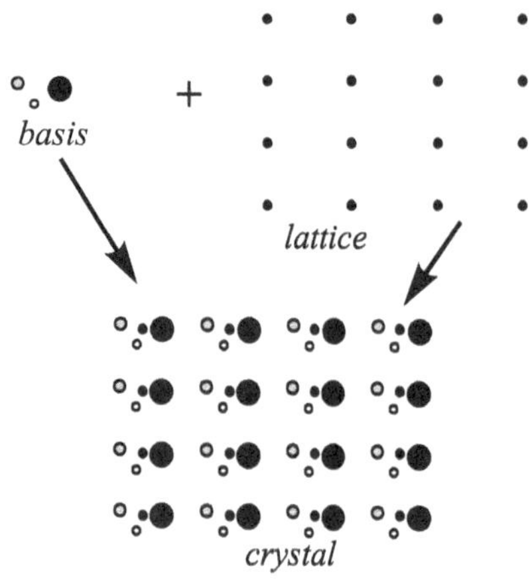

Figure 12.1. A crystal is formed by repeating a basis over the points of a lattice.

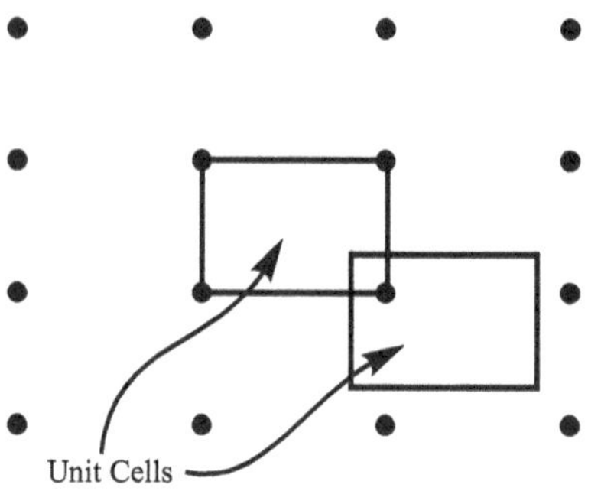

Figure 12.2. Two possible unit cells for the lattice shown.

point anywhere on the lattice can be found by translating the basis vectors by integers. In other words, equivalent points in a two-dimensional lattice can be represented by

$$\mathbf{r} = h\mathbf{a} + k\mathbf{b} \quad , \tag{12.1}$$

where h and k are positive or negative integers. As a result, the entire crystal can be reproduced by translating the unit cell by allowing the integers h and k to assume all possible values.

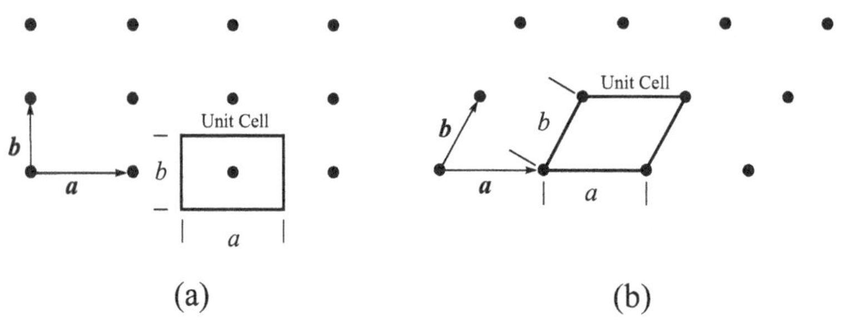

Figure 12.3. The unit cell can be defined by the unit basis vectors, **a** and **b** with magnitudes a and b, respectively. Shown are (a) rectangular and (b) oblique 2-dimensional lattices. Notice that the unit basis vectors are oriented in space according to the lattice alignment points, and is not necessarily a Cartesian system.

12.2.2 Bravais Lattice

It turns out that there are only five possible lattices for a 2-dimensional crystal. The five possibilities, referred to as the 2-dimensional Bravais lattices, named after the man who categorized them [Bravais 1850], are shown in Fig. 12.4. There are four crystal systems, namely the square, rectangular, oblique and hexagonal systems. The centered rectangular is part of the rectangular system, in which an additional lattice point is centered within the rectangle. Hence, the simple rectangle has one lattice point per unit cell (1/4 lattice point per corner), whereas the centered rectangle has two lattice points per unit cell. However, it should be

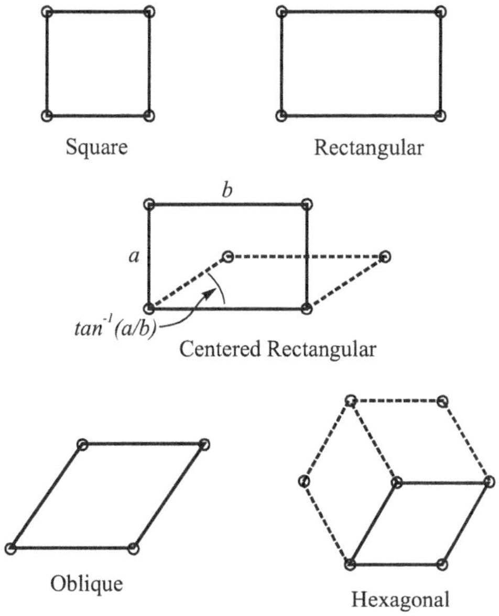

Figure 12.4. The 5 different 2-dimensional Bravais lattices.

pointed out that the centered rectangle can be represented as a special form of the oblique lattice, in which the angle formed between the basis vectors is represented by $\tan^{-1}(a/b)$ with a and b being the magnitude of the two different sides of the rectangle. Hence, the reduced oblique lattice has only one lattice point per unit cell and half the volume of the centered rectangular cell.

The concept can be expanded to 3 dimensions, in which a 3-dimensional crystal can be described by

$$\mathbf{r} = h\mathbf{a} + k\mathbf{b} + l\mathbf{c}, \tag{12.2}$$

where h, k and l are integers. As before, the entire crystal can be reproduced by translating a unit cell by consecutively translating the basis vectors alternately in the a, b and c directions, i.e., by allowing the integers h, k and l to assume all possible values. Although the number of possible unit cells is enormous, there are only 14 possible point lattices for a 3-dimensional system, which are shown in Fig. 12.5. Table 12.1 lists the conditions that define the 7 different crystal systems and the 14 different three-dimensional Bravais lattices. The angles between the basis vectors are denoted α, β and γ, in which

$$\mathbf{b}\bullet\mathbf{c} = |b||c| \cos\alpha \qquad \mathbf{a}\bullet\mathbf{c} = |a||c| \cos\beta \qquad \mathbf{a}\bullet\mathbf{b} = |a||b| \cos\gamma. \tag{12.3}$$

The cubic system has three unique lattice arrangements, the most fundamental unit being the simple cubic. By placing a lattice point in the center of the cubic unit cell structure, the body-centered cubic unit cell is formed. The third fundamental unit cell is formed by placing centered lattice points on the faces of the cube, thereby producing a face-centered cubic unit cell. The simple unit cell is denoted as P (for primitive), the body-centered as I (from the German *Innenzentrierte*), and the face-centered as F. Notice that some crystal systems have a single set of faces with a lattice point, denoted as C for base-centered. However, such an arrangement for the cubic system can be shown to produce a tetragonal P unit cell, and therefore does not constitute a unique cubic lattice arrangement. The tetragonal system has P and I unit cells, yet does not have unique F and C unit cells. It can be shown that a tetragonal F cell can be reoriented such that it is actually a tetragonal I unit cell. Further, the tetragonal C cell can be reoriented such that it is actually

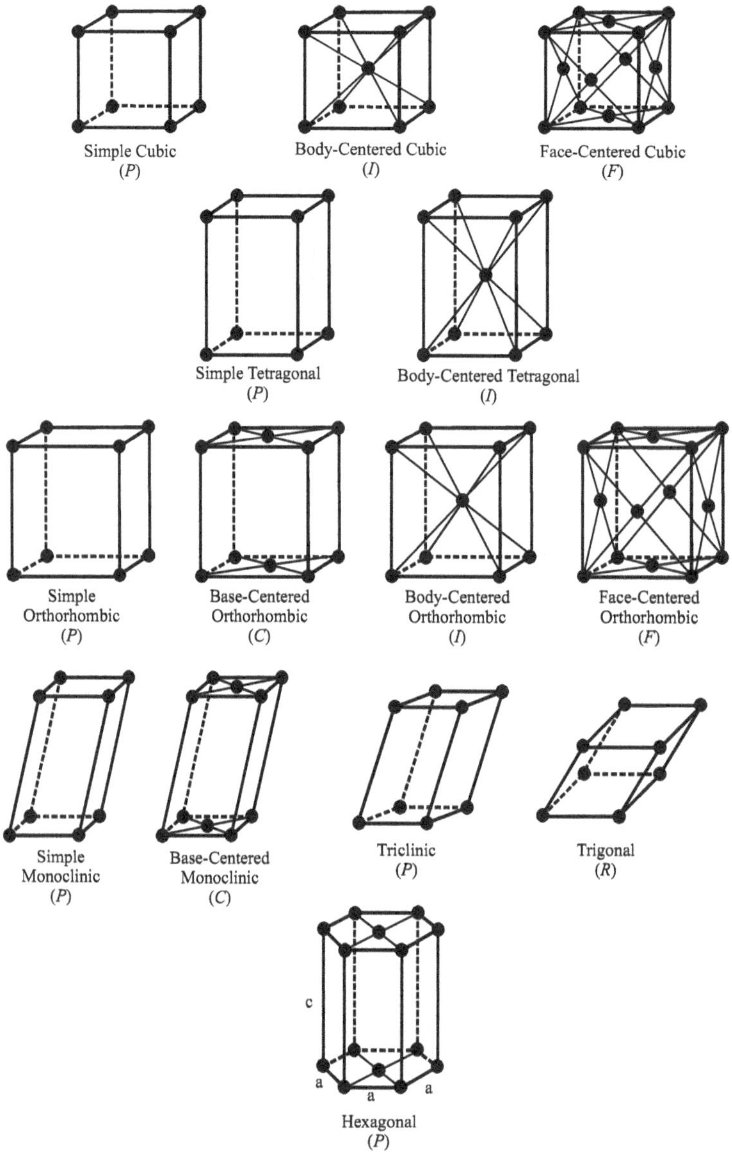

Figure 12.5. The 14 different 3-dimensional Bravais lattices.

a simple tetragonal P unit cell. The trigonal system has a single rhombohedral primitive unit cell, denoted as R. The hexagonal unit cell can actually be shown to be composed of three primitive rhombohedral cells, much like the 2-dimensional counterpart.

12.2.3 Miller Indices

Thus far it has been shown that a crystal is composed of a basis and a lattice. Further, there are only 14 possible Bravais lattices for 3-dimensional crystals. Shown in Fig. 12.6 is a coordinate system for a cubic

Table 12.1. Conditions for the 7 crystal systems and 14 Bravais lattices.

System	Dimensions and Angles	Lattices in System
Cubic	$a = b = c$ $\alpha = \beta = \gamma = 90^o$	P primitive I body centered F face centered
Tetragonal	$a = b \neq c$ $\alpha = \beta = \gamma = 90^o$	P primitive I body centered
Orthorhombic	$a \neq b \neq c$ $\alpha = \beta = \gamma = 90^o$	P primitive C base centered I body centered F face centered
Monoclinic	$a \neq b \neq c$ $\alpha = \gamma = 90^o \neq \beta$	P primitive C base centered
Triclinic	$a \neq b \neq c$ $\alpha \neq \beta \neq \gamma$	P primitive
Trigonal	$a = b = c$ $120^o > \alpha = \beta = \gamma \neq 90^o$	R rhombohedral primitive
Hexagonal	$a = b \neq c$ $\alpha = \beta = 90^o, \gamma = 120^o$	P primitive rhombohedral

three-dimensional lattice. Lattice planes can be defined by *Miller indices*[1]. To describe a single plane, the following procedure is used:

1. Set up a coordinate system along the edges of a unit cell.
2. Note where the plane of interest intercepts the axes and record those integer values. If a plane is parallel to the axis, then it is stated that it intercepts it at ∞. A minus sign is placed *over* the number for intercepts in the negative direction.
3. Invert the intercept values and convert to the smallest set of whole numbers.
4. Enclose the numbers in curved brackets (hkl).

For example, the plane shown in Fig. 12.6 has intercepts at $h = 3$, $k = 2$, and $l = 4$. Inversion gives $\frac{1}{3}$, $\frac{1}{2}$ and $\frac{1}{4}$. Multiplication by 12 reduces the set to the smallest whole numbers, hence the plane is identified as the (463) plane.

Finally, the set of equivalent planes by symmetry are designated by curly brackets, $\{hkl\}$. Note that in the cubic system that a set of planes applies to all permutations with the same indices. For instance, (100), (010), (001), $(\bar{1}00)$, $(0\bar{1}0)$ and $(00\bar{1})$ all belong to the $\{100\}$ set of planes. This is not necessarily the case for other crystal systems. For instance, in the tetragonal system, (100), (010), $(\bar{1}00)$ and $(0\bar{1}0)$ all belong to the $\{100\}$ set of planes, but the (001) and the $(00\bar{1})$ planes are dissimilar to the $\{100\}$ planes, and instead belong to the set of $\{001\}$ planes. Hence, extra care should be taken with non-cubic coordinate systems.

The *direction* vector describes a specific direction in the crystal lattice, and is found in the following manner:

[1]In 1829, the system of Miller indices was introduced by W. Whewell, the Chair of Mineralogy at Cambridge University. His successor was W.H. Miller who further developed the system, also receiving his name, and in 1839 published in his book "A Treatise on Crystallography" [Faria 1990].

1. Set the vector in the direction under investigation.
2. Compose a basis vector **r** by noting its projections along the coordinate axes.
3. Use a multiplier to convert to the smallest set of whole numbers.
4. Assign a direction vector, $\mathbf{r} = h\mathbf{a} + k\mathbf{b} + l\mathbf{c}$.
5. Enclose the numbers in square brackets, $[hkl]$.

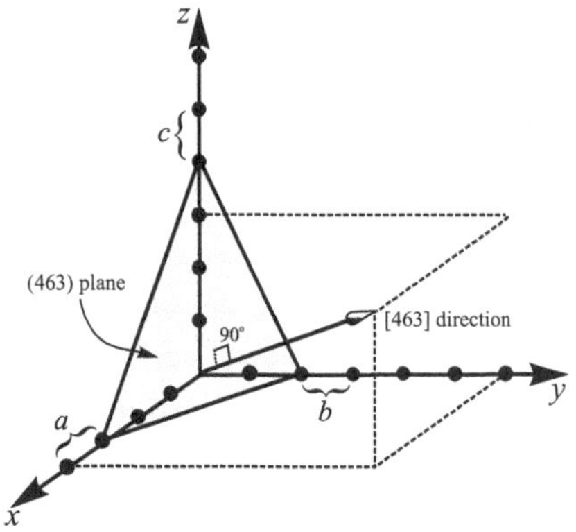

Figure 12.6. Miller indices are used to identify crystal planes and directions. In the cubic system, the crystal plane (hkl) has direction $[hkl]$ orthogonal to the plane.

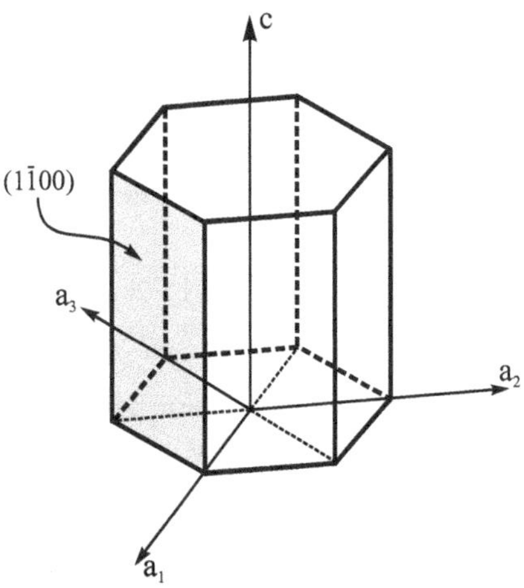

Figure 12.7. Coordinate system for hexagonal crystals, showing the $(1\bar{1}00)$ plane. The system is typically described with four index values.

Direction vectors perpendicular to lattice planes have identical Miller indices for the three cubic crystal lattices, a condition that is *not* necessarily true for any of the other 11 Bravais lattices. For example, Cartesian coordinates with axes x, y and z can be used for a cubic lattice. The direction shown in Fig. 12.6 has projections at $h = 4$, $k = 6$, and $l = 3$. The numbers are already reduced to the smallest of whole numbers, so that the direction is denoted as the [463] direction. As with planes, the minus sign is placed *over* the number for projections in the negative direction. The set of symmetrically equivalent directions are designated by triangular brackets, $\langle \text{hkl} \rangle$.

One of the more common lattices encountered in semiconductors, although definitely not the only one, is the cubic lattice. Si, Ge, GaAs, CdTe, CdZnTe, and InP are all cubic lattice materials, whereas HgI_2, GaSe, and PbI_2 are not. Common planes referenced when dealing with cubic crystals are shown in Fig. 12.8. There are occasions in which planes must be described that fall between lattice planes, as shown in Fig. 12.9. For instance, the (200) plane falls between two (100) planes, where the use of the numeral 2 indicates that it falls halfway between the planes. As mentioned above, cubic lattices have a unique feature that the $[hkl]$ directions are orthogonal to the (hkl) planes, as with the example in Fig. 12.6. This feature is a direct result of the high symmetry along the three axes. In the cubic system the spacing d between the atomic planes is

$$d = \frac{a}{\sqrt{h^2 + k^2 + l^2}}, \tag{12.4}$$

where a is the lattice constant. The cosine of the angle between two different crystal directions for the cubic system, h_1, k_1, l_1 and h_2, k_2, l_2, is

$$\cos\theta = \frac{h_1 h_2 + k_1 k_2 + l_1 l_2}{\sqrt{h_1^2 + k_1^2 + l_1^2}\sqrt{h_2^2 + k_2^2 + l_2^2}}. \tag{12.5}$$

Finally, not all systems are defined by only three Miller indices. The hexagonal lattice requires four Miller indices, namely $hkil$, in which h, k and i are integers along the a_1, a_2 and a_3 axes, and l is along the c axis, as shown in Fig. 12.7. Also shown in Fig. 12.7 is a plane that intercepts the a_1 axis at 1, the a_2 axis at -1, and is parallel to the a_3 and c axes. Inversion and reduction to the smallest integer set of whole numbers, the plane is found to be the $(1\bar{1}00)$ plane. Because a hexagonal unit cell is really three primitive rhombohedral cells, it is really only necessary to have three Miller indices for the hexagonal lattice; however, the established convention remains with four.

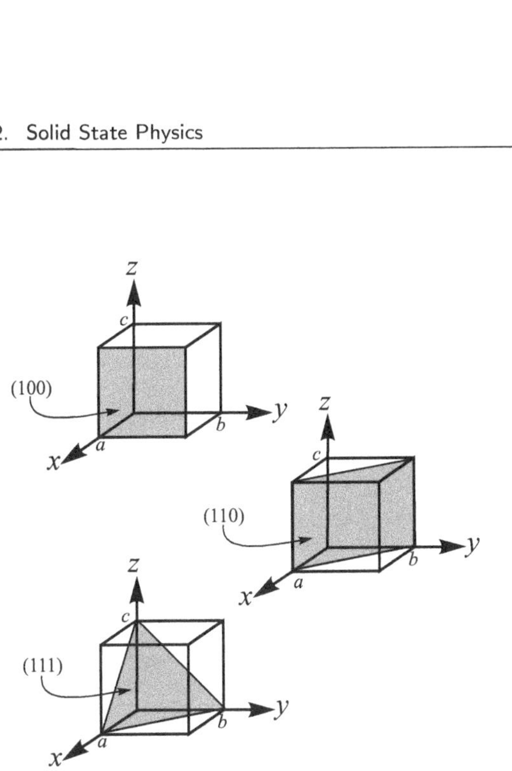

Figure 12.8. Common planes referred to and used for cubic crystals, showing the (100), (110) and (111) planes. The directions for the planes are perpendicular to the planes, namely [100], [110] and [111], respectively.

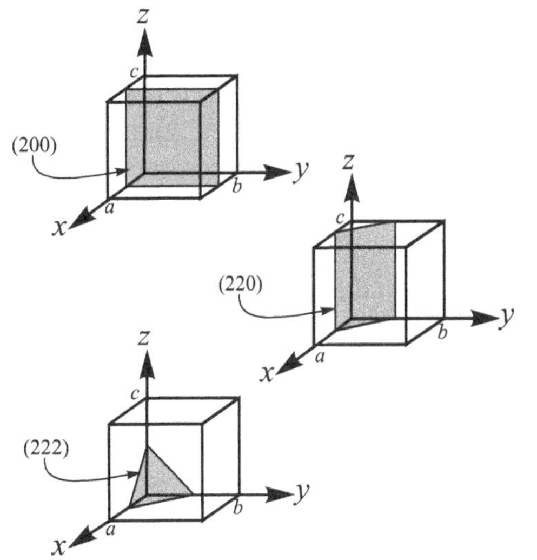

Figure 12.9. Planes that fall between lattice planes can be described without reducing the Miller indices to the lowest integer. Shown are three examples in which the (200) plane falls halfway between two (100) planes, the (220) plane falls halfway between two (110) planes and the (222) plane falls halfway between two (111) planes.

12.2.4 Reciprocal Lattice

Crystalline planes are described in what is usually referred to as *reciprocal lattice space*, a concept used throughout this text. From the real lattice space, which uses unit basis vectors **a**, **b**, and **c**, the fundamental reciprocal lattice vectors are [Kittel 1986]

$$\mathbf{a}^* = \frac{2\pi\mathbf{b}\times\mathbf{c}}{\mathbf{a}\bullet(\mathbf{b}\times\mathbf{c})}, \qquad \mathbf{b}^* = \frac{2\pi\mathbf{c}\times\mathbf{a}}{\mathbf{a}\bullet(\mathbf{b}\times\mathbf{c})}, \qquad \text{and} \qquad \mathbf{c}^* = \frac{2\pi\mathbf{a}\times\mathbf{b}}{\mathbf{a}\bullet(\mathbf{b}\times\mathbf{c})}. \tag{12.6}$$

It is easily seen that

$$\mathbf{a}^* \perp \mathbf{b} \text{ and } \mathbf{c}, \qquad \mathbf{b}^* \perp \mathbf{a} \text{ and } \mathbf{c}, \qquad \mathbf{c}^* \perp \mathbf{b} \text{ and } \mathbf{a}, \tag{12.7}$$

and that $\mathbf{a}^*\bullet\mathbf{a} = \mathbf{b}^*\bullet\mathbf{b} = \mathbf{c}^*\bullet\mathbf{c} = 2\pi$. Notice that the denominators in Eq. (12.6) are the volume of the space-lattice unit cell. Vectors in the real crystal lattice have dimensions of length; hence the reciprocal vectors have dimensions of inverse length. Further, the factor of 2π adjusts the reciprocal space to be numerically and dimensionally consistent with wave-vector space. Multiplication of each coordinate by $\hbar$ converts reciprocal space into momentum space (length^{-1}). From the conditions of Eq. (12.7), it is seen that

$$\mathbf{a}^*\bullet\mathbf{b} = \mathbf{a}^*\bullet\mathbf{c} = \mathbf{b}^*\bullet\mathbf{a} = \mathbf{b}^*\bullet\mathbf{c} = \mathbf{c}^*\bullet\mathbf{a} = \mathbf{c}^*\bullet\mathbf{b} = 0. \tag{12.8}$$

The reciprocal lattice vector is

$$\mathbf{G} = h\mathbf{a}^* + k\mathbf{b}^* + l\mathbf{c}^* \tag{12.9}$$

where the reciprocal lattice vector is perpendicular to the *hkl* plane in real space. A reciprocal lattice can be constructed in much the same way that a real space lattice is constructed, by simply translating each lattice point in **G** by having the integers h, k, and l assume all possible values. The distances between the points in the reciprocal lattice are inversely proportional to the distances between lattice points in real space.

Suppose one wished to find the fundamental reciprocal lattice vectors for a cubic lattice. The cubic lattice, as represented with Eq. (12.2), has all unit vectors of equal length, namely lattice constant a. For a simple cubic lattice with lattice constant a and the use of basic Cartesian coordinates, the basis vectors are

$$\mathbf{a} = a\hat{\mathbf{i}} \quad \mathbf{b} = a\hat{\mathbf{j}} \quad \text{and} \quad \mathbf{c} = a\hat{\mathbf{k}} \quad . \tag{12.10}$$

From Eqs. (12.6) and Eqs. (12.10), the fundamental vectors of the reciprocal lattice are found to be

$$\mathbf{a}^* = \frac{2\pi}{a}\hat{\mathbf{i}} \quad \mathbf{b}^* = \frac{2\pi}{a}\hat{\mathbf{j}} \quad \mathbf{c}^* = \frac{2\pi}{a}\hat{\mathbf{k}} \tag{12.11}$$

where $\hat{\mathbf{i}}$, $\hat{\mathbf{j}}$, and $\hat{\mathbf{k}}$ are unit vectors along the x, y, and z axes. The reciprocal space is widely used in solid state physics, and is generally referred to as *k-space*. The concept of reciprocal lattice space is used throughout the text.

12.2.5 Energy Band Gap

In free space, a single atom has discrete quantized energy states for the orbital electrons. Further, the Pauli exclusion principle states that no two electrons of an atom can have the same values for the four quantum numbers (n, l, m_l, m_s), where n is the principle number, which determines the electron's energy, l is the angular momentum quantum number, m_l is the magnetic quantum number, and m_s is the spin.

However, a solid material, such as a semiconductor crystal, is a matrix of atoms arranged in a lattice such that the various potentials of each of the atoms affect the surrounding atoms and their associated electrons. If two atoms, whose electrons initially have identical quantum numbers, are forced into close proximity then something must change so that Pauli's exclusion principle is not violated as the atoms' electrons begin to interact with each other. What happens is the appearance of new degenerate electron energy states. In other words, the original quantum energy levels split such that two states appear where there was only one before the atoms began interacting with each other.

Consider the case of a solid, in which typical atomic densities range from 10^{21} to 10^{23} atoms cm^{-3}. The total number of energy states must split to accommodate the increased electron density in solids. These degenerate energy states form bands of states in place of what were once individual states for a single atom. These bands may overlap, they may be relatively close to each other in energy with a small energy gap between them, or they may form with energy gaps between the bands (see Fig. 12.10). Electrons in the bands behave almost as though they are in an energy continuum, but it is actually a *quasi-continuum*, in which there is a density of available states that each of which can be occupied by an electron. The density of states is determined by the original total number of states of the individual atoms.

The electrical conductivity of a solid is determined by many parameters such as the charge carrier mobility, the density of free charge carriers available in a partially filled energy band, and the availability of unfilled energy states in the partially filled band. Figure 12.11 shows simplistic band diagrams for (a) insulators, (b) conductors, and (c) semiconductors. In Fig. 12.11(a) the *valence band*, which is active in chemical binding of electrons in compounds, is filled, and the next available higher energy band is devoid of electrons. In typical notation, the upper energy limit of the valence band is denoted E_V and the lower energy limit of the *conduction band* is denoted E_C. The energy difference between E_C and E_V is a forbidden energy region, referred to as the *energy band gap* and is denoted by E_g. No states exist between energies E_V and E_C, i.e., in the band gap. If the band-gap energy is sufficiently large, electrons cannot be thermally excited from the valence band into the conduction band. Such materials are said to be insulators. Conduction occurs only if there are empty states in the same energy band for charge carriers to occupy as they move through

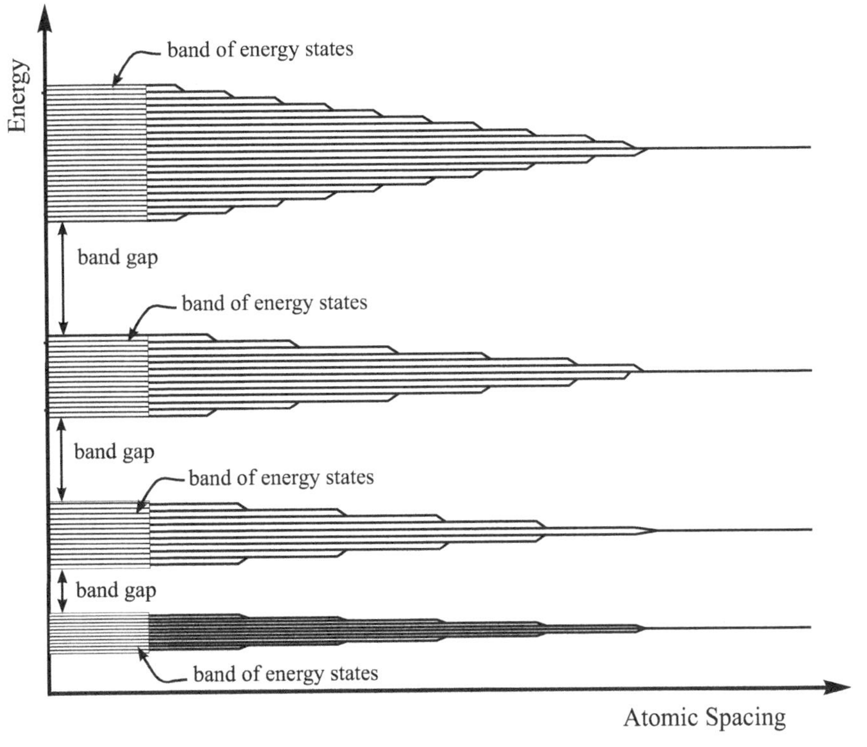

Figure 12.10. As atoms are brought closer together, their allowed energy states split into degenerate states. In a solid medium, the high atomic density causes these degenerate states to form quasi-continua referred to as energy bands.

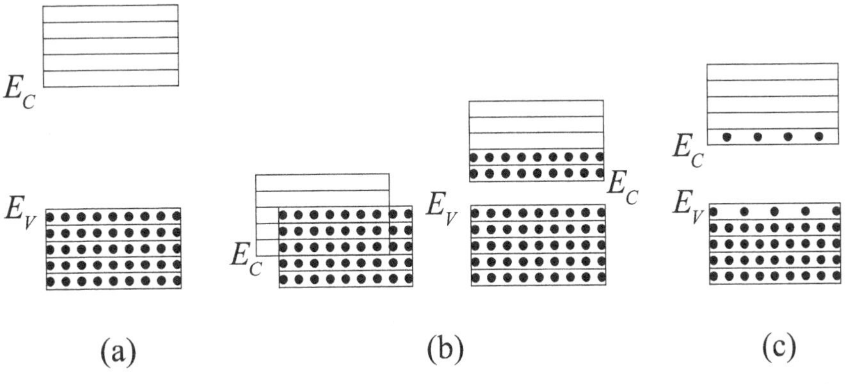

Figure 12.11. Depictions of simple band diagrams for (a) insulators, (b) conductors and (c) semiconductors. In (b) there are two depictions for conductors, one in which a filled valence band overlaps the conduction band, and the other in which the valence band is full with a partially filled conduction band.

the crystal. Because the valence band is completely full of charge carriers in Fig. 12.11, there are no empty states and hence no conduction. Further, the conduction band has empty states, but no charge carriers, hence again conduction does not take place.

In Fig. 12.11(b), there are two examples of conductors. In the first example, the valence band and the conduction band overlap such that electrons can easily move from the filled valence band into the partially filled conduction band without a change of energy. Hence, there are plenty of unfilled states with an overlapping reservoir of electrons (the valence band) that can move to the conduction band, thereby producing free conduction. In the other example, the valence band is filled, and the conduction band is partially filled with a high density of electrons, even at low temperature. Again the conditions exist for free conduction.

Table 12.2. Constants used for Eq. (12.12).

Semiconductor	E_{g0}	α	β
Si	1.170 eV	4.73×10^{-4}	636
Ge	0.7437 eV	4.774×10^{-4}	235
GaAs	1.519 eV	5.405×10^{-4}	204

In Fig. 12.11(c), there is a band gap, similar to that of the insulator example, except the band gap is relatively small. As a result, some electrons are thermally excited from the valence band into the conduction band where they can freely move through the crystal. However, the density of the electrons in the conduction band is determined largely by the band-gap energy and the temperature. At sufficiently low temperatures, the electrons all return to the valence band and the material behaves as an insulator. As the temperature is increased, more and more electrons cross the band gap into the conduction band, and the material conductivity increases. The band-gap energy is an inverse function of temperature, with empirical dependence,

$$E_g(T) \simeq E_{g0} - \frac{\alpha T^2}{T+\beta}, \tag{12.12}$$

where T is the absolute temperature, E_{g0} is the band gap at absolute zero, and α and β are experimentally determined constants. Table 12.2 shows constants reported for semiconductors Si, Ge, and GaAs [Sze, 1985].

Often the class of materials represented by Fig. 12.11(c) is separated into *semiconductors* and *semi-insulators*, the distinction roughly defined by the band-gap energy. Typically, band-gap energies up to about 1.4 eV constitute a class of materials usually designated as semiconductors, while band-gap energies from 1.4 eV to about 5 eV are considered semi-insulators. Band gaps exceeding 5 eV form the insulator class of materials. However, these ranges are not rigidly classified, and often semi-insulators and semiconductors are treated as the same as done in this chapter.

12.3 Quantum Mechanics

Quantum mechanics play a key role in understanding the behavior of electrons in semiconductor materials. An introduction to some quantum mechanics concepts and models is presented in Chapter 3. In this section two more important quantum mechanics models are presented.

12.3.1 Potential Barriers

Quantum mechanical tunneling is a phenomenon often observed and *used* for the design and development of semiconductor devices, including semiconductor radiation detectors. For instance, potential barriers form when a metal is attached to a semiconductor surface, which can cause the contact to have undesirably high resistance. Often, low-resistivity contacts, sometimes referred to as "ohmic contacts", are generally made by introducing impurity dopants on the surface along with the metal to such concentrations that barriers forming between the metal and the semiconductor are thin, typically only a few angstroms thick. As a result, electrons can tunnel through the potential barrier, and the effective contact resistance is reduced. By contrast, in classical mechanics an electron that encounters the potential barrier would simply be reflected.

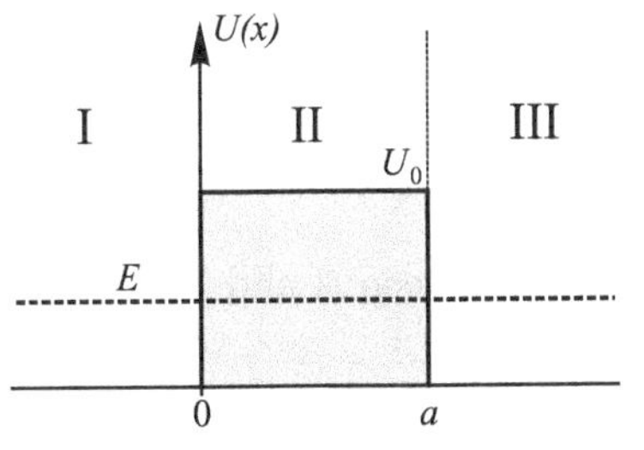

Figure 12.12. Square potential barrier in which the particle energy is less than the barrier height. In classical mechanics, such particles would be reflected by the barrier.

Depicted in Figure 12.12 is a potential barrier of height U_0 and width a. To find a solution for transmittance of electrons, with energy E less than U_0, through the potential barrier, three regions are defined: in region I ($x < 0$) $U(x) = 0$, in region II ($0 < x < a$) $U(x) = U_0$, and in region III ($x > a$) $U(x) = 0$.[2]

From Schrödinger's time independent wave equation, Eq. (3.33), namely,

$$\frac{d^2\psi}{dx^2} + \frac{2m_e}{\hbar^2}(E - U_0)\psi = 0, \tag{12.13}$$

the general solutions for the three regions can be written in the form

$$\psi_1(x) = Ae^{ikx} + Be^{-ikx},\ \psi_2(x) = Ce^{-\phi x} + De^{\phi x},\ \text{and}\ \psi_3(x) = Fe^{ikx}, \tag{12.14}$$

where

$$k^2 = \frac{2m_eE}{\hbar^2} \quad \text{and} \quad \phi^2 = \frac{2m_e}{\hbar^2}(U_0 - E). \tag{12.15}$$

The wave function in region I represents a wave traveling in the positive x direction and a reflected wave traveling in the $-x$ direction. In region II, reflections occur at the boundaries such that waves also travel in both directions. In region III, it can be assumed that waves travel only in the positive x direction after passing beyond the barrier. From the boundary condition at $x = 0$, namely, $\psi_1(0) = \psi_2(0)$, it is found that

$$A + B = C + D. \tag{12.16}$$

The boundary condition $\psi_2(a) = \psi_3(a)$ gives

$$Ce^{-\phi a} + De^{\phi a} = Fe^{ika}. \tag{12.17}$$

The boundary condition

$$\frac{d\psi_1}{dx}(0) = \frac{d\psi_2}{dx}(0)$$

yields

$$ik(A - B) = \phi(D - C), \tag{12.18}$$

and the boundary condition

$$\frac{d\psi_2}{dx}(a) = \frac{d\psi_3}{dx}(a)$$

gives

$$\phi(De^{\phi a} - Ce^{-\phi a}) = ikFe^{ika}. \tag{12.19}$$

Now seek a relation between A and F. Towards this end, divide Eq. (12.18) by ik and add the result to Eq. (12.16) to yield

$$2A = C\left(1 - \frac{\phi}{ik}\right) + D\left(1 + \frac{\phi}{ik}\right). \tag{12.20}$$

[2]Strictly speaking this problem should be treated with the time-dependent Schrödinger equation with the particle represented as a wave packet moving along the x axis. But it is also possible to solve this problem as a time-independent one, by imagining that, instead of a single particle, there is a continuous stream of particles moving towards the barrier, some of which are reflected and the rest transmitted through the barrier.

Now add and subtract Eq. (12.19) with Eq. (12.17) to obtain

$$2De^{\phi a} = Fe^{ika}\left(1 + \frac{ik}{\phi}\right) \quad \text{and} \quad 2Ce^{-\phi a} = Fe^{ika}\left(1 - \frac{ik}{\phi}\right). \tag{12.21}$$

Substitute Eqs. (12.21) into Eq. (12.20) to obtain the following relationship between A and F

$$4A = Fe^{ika}\left[\left(1 - \frac{\phi}{ik}\right)\left(1 - \frac{ik}{\phi}\right)e^{\phi a} + \left(1 + \frac{\phi}{ik}\right)\left(1 + \frac{ik}{\phi}\right)e^{-\phi a}\right]. \tag{12.22}$$

The above relationship is important because A defines the amplitude of the initial electron wave striking the barrier and F defines the portion of the wave that tunnels through the barrier. By finding the ratio of the fluxes defined by F and A, the transmission coefficient T can be determined. With a small amount of algebra, Eq. (12.22) becomes

$$4A = \frac{Fe^{ika}}{ik\phi}\left[(2ik\phi + k^2 - \phi^2)e^{\phi a} + (2ik\phi - k^2 + \phi^2)e^{-\phi a}\right], \tag{12.23}$$

which can be rewritten as

$$4A = \frac{Fe^{ika}}{ik\phi}\left[2ik\phi(e^{\phi a} + e^{-\phi a}) - (\phi^2 - k^2)(e^{\phi a} - e^{-\phi a})\right]. \tag{12.24}$$

Equation (12.24) can be written in terms of hyperbolic sines and cosines such that

$$2A = \frac{Fe^{ika}}{ik\phi}\left[2ik\phi\cosh(\phi a) - (\phi^2 - k^2)\sinh(\phi a)\right]. \tag{12.25}$$

The magnitude of A is easily found from Eq. (12.25) to be

$$|A|^2 = \frac{|F|^2}{4k^2\phi^2}\left[4k^2\phi^2\cosh^2(\phi a) - (\phi^2 - k^2)^2\sinh^2(\phi a)\right]. \tag{12.26}$$

Because $|\psi(x)\psi^*(x)|\,dx$ is the probability of finding a particle in dx about x, the probability of finding an electron in region I is $|Ae^{ikx}A^*e^{-ikx}|\,dx = |A|^2\,dx$, i.e., independent of position. Likewise the reflected wave in region I Be^{ikx} and the transmitted wave in region III Fe^{ikx} are also independent of position in their respective regions. Because the electrons in the incident beam must be conserved, $|A|^2 = |B|^2 + |F|^2$ or

$$1 = \frac{|B|^2}{|A|^2} + \frac{|F|^2}{|A|^2} \equiv R + T, \tag{12.27}$$

where R and T are the reflection and transmission probabilities or coefficients.

From Eq. (12.26)

$$T = \frac{|F|^2}{|A|^2} = \frac{4k^2\phi^2}{\left[4k^2\phi^2\cosh^2(\phi a) - (\phi^2 - k^2)^2\sinh^2(\phi a)\right]}. \tag{12.28}$$

With the identity that $\cosh^2(\phi a) - \sinh^2(\phi a) = 1$, this result reduces to

$$\frac{1}{T} = 1 + \frac{(k^2 + \phi^2)^2 \sinh^2(\phi a)}{4k^2\phi^2}. \tag{12.29}$$

Finally, the definitions of k and ϕ are substituted into Eq. (12.29) to yield

$$\frac{1}{T} = 1 + \frac{U_0^2}{4E(U_0 - E)} \sinh^2\left(a\sqrt{\frac{2m_e(U_0 - E)}{\hbar^2}}\right). \tag{12.30}$$

It should be obvious to the reader that the reflection probability is simply

$$R = 1 - T. \tag{12.31}$$

From Eq. (12.30), it can be observed that the transmission coefficient approaches unity as the electron energy approaches U_0.

Example 12.1:

For the potential barrier diagram of Fig. 12.12, suppose the following properties exist: $U_0 = 30$ eV, $a = 0.5$ Å, and $E = 5$ eV (for an applied voltage V_0 of 5 volts). Under such conditions, how many electrons strike the barrier per second to produce a tunneling current of 5 mA?

Solution:

Use the substitution $m_e = 511$ keV where $E/c^2 = m$, and from Eq. (12.30),

$$\frac{1}{T} = 1 + \frac{(30\,\text{eV})^2}{4(5\,\text{eV})(30\,\text{eV} - 5\,\text{eV})} \times$$
$$\sinh^2\left((0.5 \times 10^{-10}\,\text{m})\sqrt{\frac{2(5.11 \times 10^5\,\text{eV})(30\,\text{eV} - 5\,\text{eV})}{(2.998 \times 10^8\,\text{m s}^{-1})^2(6.582 \times 10^{-16}\,\text{eV s})^2}}\right)$$
$$= 1 + (1.8)\sinh^2\left[(0.5 \times 10^{-10}\,\text{m})(2.56 \times 10^{10}\,\text{m}^{-1})\right] = 5.965$$

It is found that $T = 0.1676$. At 5 mA, the number passing through the barrier per second is

$$n_b = \frac{5 \times 10^{-3}\,\text{A}}{1.6 \times 10^{-19}\,\text{C/electron}} = 3.13 \times 10^{16}\,\text{electrons s}^{-1}.$$

The number of electrons n_s striking the potential barrier per second is simply the number tunneling through the barrier per second divided by T and here is found to be

$$n_s = \frac{(3.13 \times 10^{16}\,\text{e}^-\,\text{s}^{-1})}{(0.1676)} = 1.865 \times 10^{17}\,\text{electrons s}^{-1}.$$

Note that potential barriers found in nature are typically not square functions, hence alternative methods to determine the transmission coefficient T are needed. One such method employs the Wentzel-Kramers-Brillouin (WKB) approximation, the details of which can be found in the literature [Gasiorowicz 1974].

12.3.2 Kronig-Penney Model

The electrons surrounding the atoms in a crystal can be thought of as confined in a series of identical potential wells. The functions describing the individual atomic potential wells and the electron confinement are generally quite complex. However, insight about the effect on electrons produced by a periodic series of potentials can be obtained with the idealized assumption that the potentials are a series of square well functions. The resulting quantum mechanical model is referred to as the Kronig-Penney model, and it is fundamental to understanding the behavior of electrons in a semiconductor [Kronig and Penney 1931].

Figure 12.13 shows a one-dimensional infinite series of potential rectangular wells of width a and separated by barriers of width b. Therefore the potential is a one-dimensional periodic function such that $U(n(a+b)) = U(x)$ where n is an integer. It is assumed that the crystal extends to $\pm\infty$; hence one need not be concerned about effects at the termination of the periodicity at the edge of the crystal. Although such a description of the potential an electron experiences in a crystal may strike the reader as unrealistic, it does provide a model for the development of important concepts fundamental to solid state physics, such as *crystal momentum* and carrier *effective mass*. These concepts are introduced in the next section.

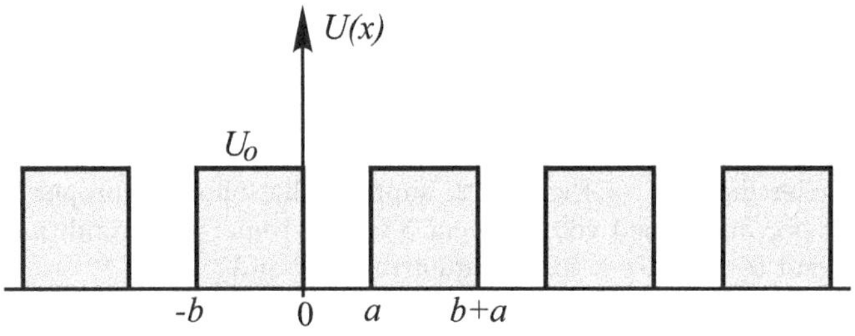

Figure 12.13. The Kronig-Penney model uses a series of rectangular wells to represent the periodicity of the atomic potentials in a crystalline solid. The potential is periodic, i.e., $U(x + a + b) = U(x)$.

It seems reasonable that the wave function describing an electron in the potential of Fig. 12.13 should also be periodic with the period of the crystal $T = a + b$. Indeed, in 1928 Felix Bloch showed that the energy wave function must have the form

$$\psi_k(x) = u_k(x)e^{ikx}, \tag{12.32}$$

where the amplitude $u_k(x)$ also is periodic such that $u_k(x) = u_k(x + T)$ and where k is the wavenumber defined as

$$k = \frac{2n\pi}{a+b}, \quad n = 0, 1, 2, \dots \quad . \tag{12.33}$$

The one-dimensional Schrödinger wave equation is used to describe an electron of mass m_e in the potential of Fig. 12.13

$$-\frac{\hbar^2}{2m_e}\frac{d^2\psi(x)}{dx^2} + U(x)\psi(x) = E\psi(x). \tag{12.34}$$

Here E is the energy eigenvalue (a positive quantity $< U_o$). The solution for $0 < x < a$ when $U(x) = 0$ (dropping the subscript k for the moment) is

$$\psi(x) = Ae^{igx} + Be^{-igx}, \tag{12.35}$$

in which A and B are arbitrary constants to be found later from the boundary conditions. This solution is the sum of two plane waves, one traveling to the right and the other to the left and has energy

$$E = \hbar^2 g^2/2m_e. \tag{12.36}$$

Within the potential barrier, $-b < x < 0$, the solution of Eq. (12.34) is

$$\psi(x) = Ce^{\gamma x} + De^{-\gamma x}, \tag{12.37}$$

where the positive real number $\gamma^2 = 2m_e(U_o - E)/\hbar^2$. Again C and D are constants to be found from boundary conditions.

Because the wave function must have the Bloch form of Eq. (12.32), the solution $\psi(x)$ for $x \in (a, a+b)$ must be related to the solution for $x \in (-b, 0)$ by

$$\psi(a < x < a+b) = \psi(-b < x < 0)e^{ik(a+b)}. \tag{12.38}$$

From this constraint the wavenumber k is determined and used as an index to the wave function.

The constants A, B, C, and D are chosen to make ψ and $d\psi/dx$ continuous at $x = 0$ and $x = a$. Thus at $x = 0$

$$A + B = C + D \tag{12.39}$$

$$ig(A - B) = \gamma(C - D) \tag{12.40}$$

At $x = a$ with the use of Eq. (12.38) to express $\psi(a)$ in terms of $\psi(-b)$, the boundary conditions require

$$Ae^{iga} + Be^{-iga} = (Ce^{-}\gamma b + De^{\gamma b})e^{ik(a+b)} \tag{12.41}$$

$$ig(Ae^{iga} - Be^{-iga} = \gamma(Ce^{-}\gamma b - De^{\gamma b})e^{ik(a+b)}. \tag{12.42}$$

For the four homogeneous equations for A, B, C, and D to have a non-trivial solution the coefficient matrix for the four constants must vanish. After much algebra, it is found that the determinant vanishes only if

$$\frac{\gamma^2 - g^2}{2\gamma g} \sinh \gamma b \sin ga + \cosh \gamma b \cos ga = \cos k(a+b). \tag{12.43}$$

Because the large potential around a nucleus is located very near the nucleus, i.e., $b \ll a$, the above result can be simplified by approximating the potential as a periodic delta function. Thus let $b \to 0$ and $U_o \to \infty$ such that $P \equiv \gamma^2 ab/2$ remains a finite positive quantity [Kittel 1986]. Thus, in this approximation $\gamma >> g$, $\gamma b << 1$, and $\sinh \gamma b \simeq \gamma b$. Then Eq. (12.43) reduces to

$$\frac{P}{ga} \sin ga + \cos ga = \cos ka. \tag{12.44}$$

Constraints to the solutions of Eq. (12.44) can be inferred from Fig. 12.14 for different values of P, in which the left-hand side of Eq. (12.44) is plotted versus ga. Because the right-hand side of Eq. (12.44) is bounded between ± 1, it is apparent that a solution can exist for the Kronig-Penney model only for those values of ga for which the left-hand side of Eq. (12.44) also is between $\pm$ 1. For all other regions of ga, there are no solutions, and *energy gaps* appear between the regions that have solutions. This situation is analogous to the expected effect of a periodic potential upon electrons in a solid, in which bands of allowed energy states form, and regions also appear in which states are not allowed.

From Eq. (12.44) an *allowed* value of ga on the left-hand side produces a corresponding value of ka from the right-hand side. Because $k \propto 1/\lambda$, where λ is the wavelength of the periodic wave function, then, from de Broglie's relation $p = h/\lambda$ between a particles momentum p and its wave length, it is seen that k is proportional to the electron's momentum. Likewise, from the definition of g in Eq. (12.36), $(ga)^2$ represents the allowed electron energies multiplied by a constant value, $2m_e a^2/\hbar^2$. Therefore, plotting $(ga)^2$ against ka

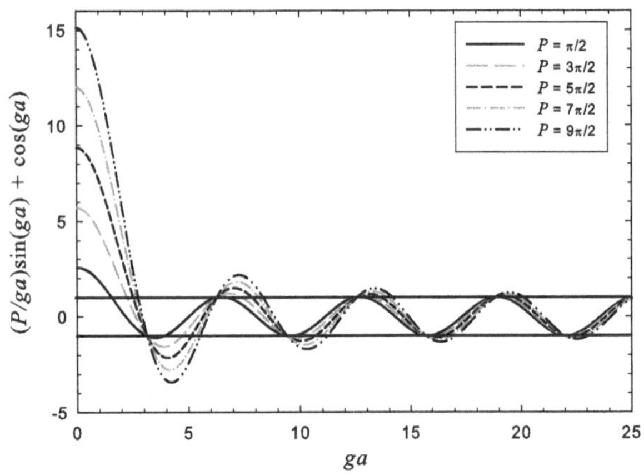

Figure 12.14. Solutions to the Kronig-Penney model for different values of P. Both axes are dimensionless.

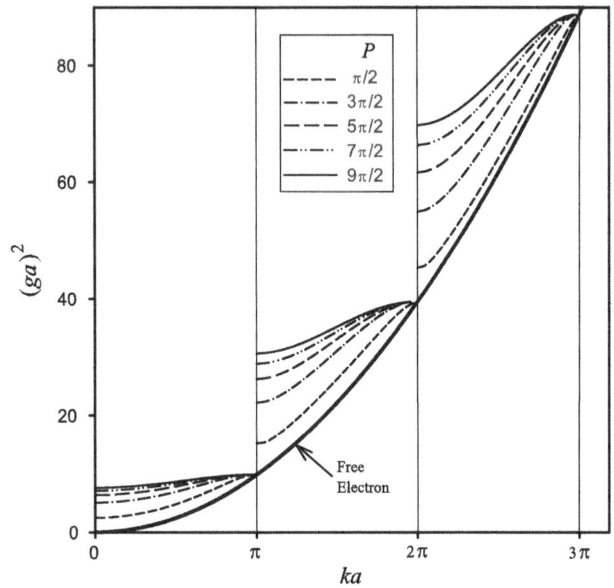

Figure 12.15. Depiction of energy bands from the Kronig-Penney model, showing the relationship between P, $(ga)^2$, and ka. Both axes are dimensionless with ka proportional to the electron's energy and $(ga)^2$ proportional to its momentum.

produces a dimensionless *energy-momentum diagram.* Figure 12.15 shows the results for various values of P for the first three regions of ka where solutions to Eq. (12.44) exist. From this figure, several important features can be seen. Values of ka equal to multiples of π bound momentum regions in which the allowed energies are smoothly continuous. Further, a discontinuity occurs at these boundaries in k-*space*, for which *gaps* between the allowed energy *bands* appear. Notice also that as P increases, the range of the allowed solutions of $(ga)^2$ (or energies) in the band regions decreases. Lastly, this diagram, usually referred to as an E-k diagram, shows that the allowed energies vary from the free electron model,[3] although the basic parabolic shape is still observed.

12.4 Semiconductor Physics

Thus far, the theory developed in the preceding sections applies in general to solid-state crystalline material. Yet, it is the subset of solid-state materials, usually identified as semiconductors, that is generally associated with solid-state radiation detectors. The result for the Kronig-Penney model can be used to understand many concepts in semiconductor physics.

12.4.1 Brillouin Zones

Because of the symmetry in ka of the one-dimensional Kronig-Penney model, Fig. 12.15 can be mirrored upon itself to produce the diagram of Fig. 12.16. The center zone is bounded by $-\pi$ and $+\pi$, and it has a continuum of allowed energy states. This first zone of allowed energies is referred to as the first Brillouin zone in one dimension. The next Brillouin zone is defined by the regions bound between -2π and $-\pi$ and

[3]For a free non-relativistic electron, it can be shown that $(ga)^2 = (ka)^2$.

also between $+\pi$ and $+2\pi$. The third Brillouin zone is bounded by -3π and -2π and also between $+2\pi$ and $+3\pi$, and so on. The total width in k-space is identical for the 1st, 2nd, and 3rd Brillouin zones and all higher Brillouin zones [Brillouin 1953]. At the boundaries of each Brillouin zone is a discontinuity at which an energy gap, or band gap, appears. Hence, Brillouin zones are defined as regions within which a continuum, or more correctly a quasi-continuum, of energy states appear and terminate at the k-space boundaries where the energy gaps appear.

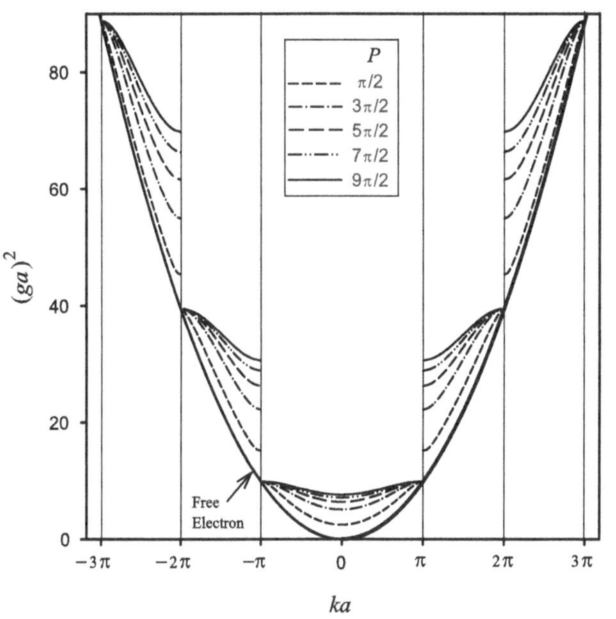

Figure 12.16. Energy band formation and Brillouin zones from the Kronig-Penney model for different values of P.

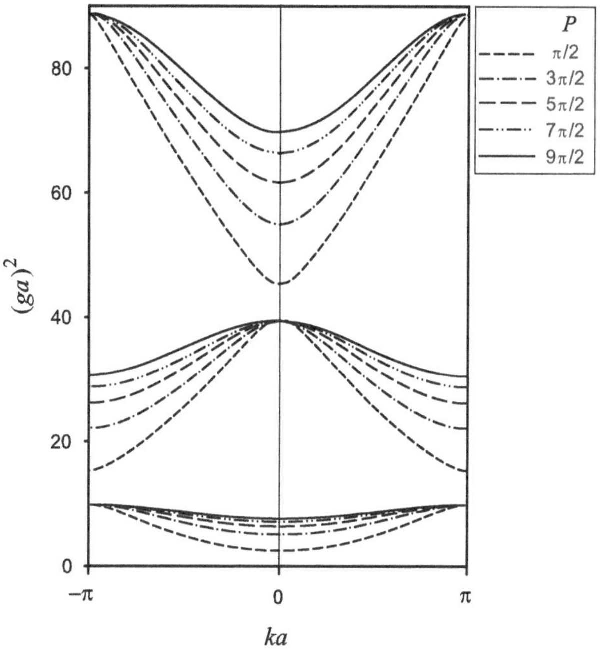

Figure 12.17. Reduced zone diagram, showing the allowed energy bands and forbidden bands for several values of P.

Because one-dimensional Brillouin zones (BZ) all have the same total width in k-space, it is usual to represent Fig. 12.2 in reduced form, as shown in Fig. 12.17. This type of plot is known as the reduced E-k diagram. The plot is created by simply folding each consecutive BZ backwards into the first BZ. The reduced zone diagram is used here to develop basic concepts about electron transport through the periodic potential of a semiconductor. In a three-dimensional solid state crystal, atomic electron orbitals form the energy bands and define the various BZs.

12.4.2 Effective Mass

Consider a wave packet of electrons influenced by an external force. To describe the motion of electrons under such conditions the "group velocity" v_g of the electron wave packet is often used. In general,

$$v_g = \frac{dx}{dt} = \frac{d\omega}{dk}, \tag{12.45}$$

where ω is the "center frequency" of the wave packet, and k is the wavenumber. The frequency is related to energy by $\omega = E/\hbar$ so that

$$v_g = \frac{d}{dk}\left(\frac{E}{\hbar}\right) = \frac{1}{\hbar}\frac{dE}{dk}. \tag{12.46}$$

Application of force (F) to the wave packet, such as occurs when an electric field is placed across the solid, produces a change in energy of the electrons

$$dE = Fdx = Fv_g dt, \tag{12.47}$$

which can be rearranged as

$$F = \frac{1}{v_g}\frac{dE}{dt} = \frac{1}{v_g}\frac{dE}{dk}\frac{dk}{dt} = \hbar\frac{dk}{dt} = \frac{d(\hbar k)}{dt}. \tag{12.48}$$

From Eq. (12.46),

$$\frac{dv_g}{dt} = \frac{1}{\hbar}\frac{d}{dt}\left[\frac{dE}{dk}\right] = \frac{1}{\hbar}\frac{d^2E}{dk^2}\frac{dk}{dt}. \tag{12.49}$$

Substitution of Eq. (12.48) for the force into the above result yields

$$\frac{dv_g}{dt} = \frac{1}{\hbar^2}\frac{d^2E}{dk^2}\frac{d(\hbar k)}{dt} = \left(\frac{1}{\hbar^2}\frac{d^2E}{dk^2}\right)F. \tag{12.50}$$

Finally, Newton's classical relationship $F = ma$ is used to define an "effective mass" m^* for the electrons as

$$F = m_e^*\frac{dv_g}{dt} = \left(\frac{1}{\hbar^2}\frac{d^2E}{dk^2}\right)^{-1}\frac{dv_g}{dt}, \tag{12.51}$$

which, upon rearrangement, gives the electron's effective mass as

$$m_e^* = \left(\frac{1}{\hbar^2}\frac{d^2E}{dk^2}\right)^{-1}. \tag{12.52}$$

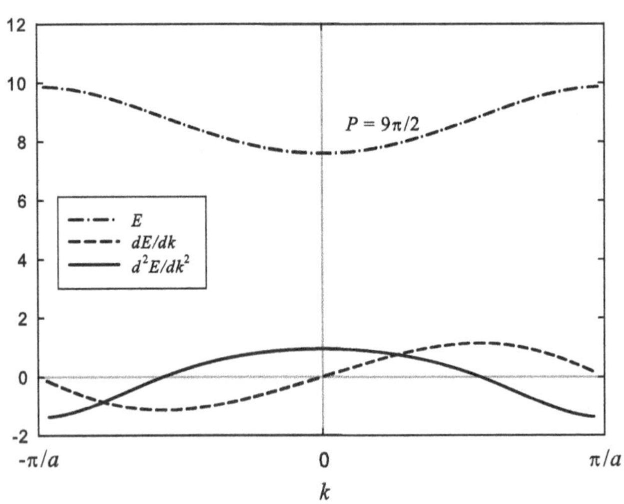

Figure 12.18. The expected values for E, dE/dk, and d^2E/dk^2 of the lowest energy band for the Kronig-Penney model with $P = 9\pi/2$.

This result, of course, does not imply the electron's physical mass has somehow changed, but rather that the periodic potential defined by E and k has changed the influence of an external force upon the electron from that expected upon a free electron. Often the effective mass for electrons in a semiconductor is quoted as a fraction of m_e, the rest mass of a free electron.

A few important concepts can be observed from Eq. (12.52). In order to do so, the result for the lowest energy band in the Kronig-Penney model is used with $P = 9\pi/2$. In Fig. 12.18, allowed energies E of the lowest band, dE/dk, and d^2E/dk^2 are shown as a function of k. Because the effective mass varies inversely with d^2E/dk^2, it is observed that the effective mass can be *negative* whenever the particle moves in the *opposite* direction to that expected for the applied electrical force [Pierret 1989]. This simply means that the particle behaves as though it has the opposite polarity, or in other words, the electron behaves as a positively charged electron. As a result, the electron moves in the opposite direction of the applied electrical force. It is also observed that the largest magnitude of d^2E/dk^2 occurs in k-space at the top and bottom of the energy band, so that the smallest effective mass occurs at the points of highest curvature with respect to k of the energy band. From Fig. 12.19, in which results for the lowest energy band for different values of P are shown, it is seen that the effective mass decreases as the energy band curvature increases.

The bands, as depicted in Fig. 12.20, are formed from *allowed* energy states, although the available energy states *may not be filled*. The lowest band formed is filled with tightly bound electrons which typically do not share in the chemical bonding process (except for the atoms with very low Z numbers). At higher energies, the band that actively takes part in chemical binding is referred to as the *valence band*. Above the valence band are various partially filled and empty energy bands. The energy gap between the valence band and the next band up, referred to as the *conduction band*, determines, to a large degree, the electronic properties of the solid. For instance, overlapping valence and conduction bands generally form in a conductive material, such as Cu. Those materials with small band-gap energies between the valence and conduction bands, typically no greater than 1.6 eV, generally form a class of materials regarded as *semiconductors*, such as Ge, Si, and GaAs. Larger band gaps, ranging from approximately 1.6 eV to 5 eV compose a class of materials often regarded as *semi-insulators*, which would include CdTe and SiC. Materials with band gaps above 5 eV behave as insulators at room temperature.

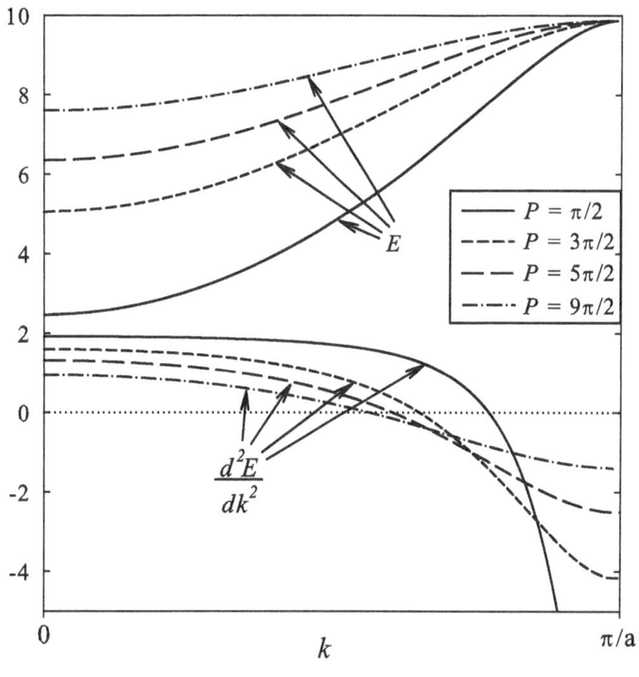

Figure 12.19. E and d^2E/dk^2 for the lowest energy band in the Kronig-Penney model for different values of P. Notice that the largest absolute values of d^2E/dk^2 occur in k space at the top and bottom of the energy bands.

Figure 12.20 depicts a material in which the tightly bound band is filled, the valence band is partially empty, the lower conduction band is partially filled, and the upper conduction band is completely empty. Those electrons populating the upper portions of the bands have negative effective masses, and those electrons populating the lower portions of the energy bands have positive effective masses. Under equilibrium conditions, the electrons in the bands are evenly distributed over k-space such that their net velocities add up to zero. This must be the case because no net current should flow from a simple solid block of semiconductor suspended in space.

Suppose electrical contacts are placed upon the semiconductor block and a voltage is applied. Completely filled bands do not contribute to conduction because there are no empty states in the band for electrons to move into. Conversely, completely empty bands do not contribute to conduction because there are no electrons available. However, for partially filled bands, the force offsets the equilibrium condition so that electrons are displaced in k-space, as depicted in Fig. 12.21. The importance of the effective-mass concept is that those electrons with positive effective masses flow in the direction of the applied electric field force, but those electrons with negative effective masses flow in the *opposite* direction of the applied force.

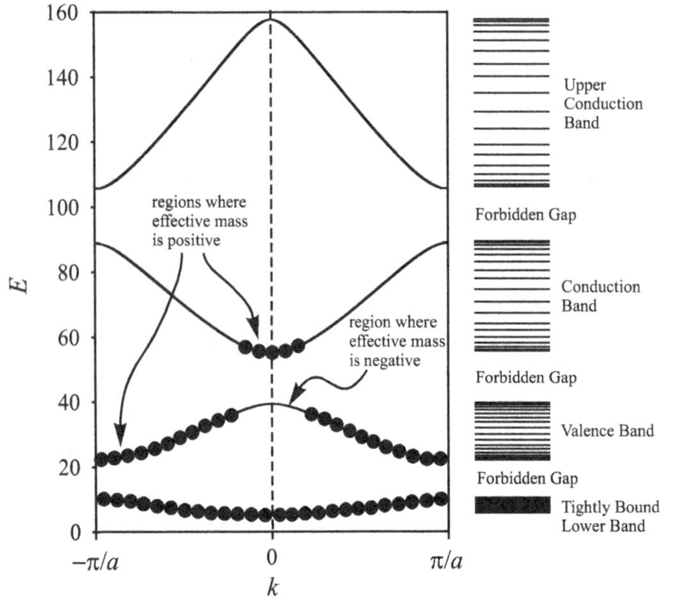

Figure 12.20. The reduced zone diagram depicts energy bands in one dimension. The bands may be completely filled, partially filled, or void of electrons.

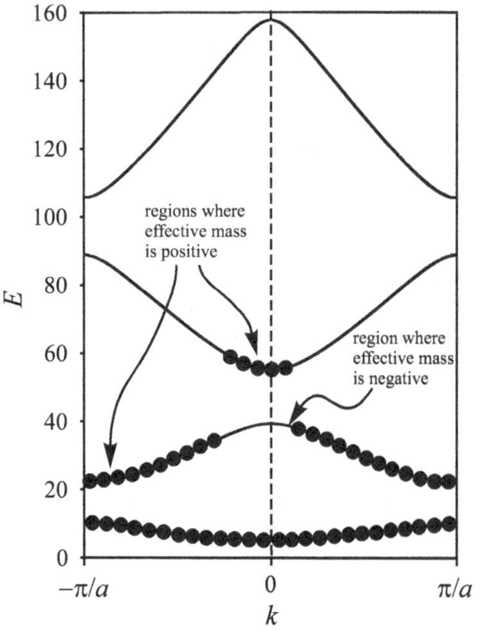

Figure 12.21. A force applied to the periodic structure, such as electric field force, displaces the electron in k-space, thereby perturbing the equilibrium condition.

To determine the current, the total motion of electrons in all of the bands must be taken into consideration [Pierret 1989]. Electric current from contributions in the lower conduction band can be written as

$$I_c = -\frac{q}{d} \sum_{\substack{i \\ \text{filled}}} v_i, \tag{12.53}$$

where the speed contributions from only those states in the conduction band filled with electrons are considered, and d is the width of the semiconductor block between the two contacts. Next, the current contributions from the valence band are found as

$$I_v = -\frac{q}{d} \sum_{\substack{i \\ \text{filled}}} v_i \quad , \tag{12.54}$$

where, again, the speed contributions from only those states in the valence band filled with electrons are used. The summation of currents for the valence band can be difficult because the electrons have both positive and negative effective mass. As a result, positive and negative velocities must also be taken into account. This situation can be greatly simplified by introducing a few approximations. First, note that electrons in the conduction band congregate at the bottom of the band, as occurs in semiconductors operating at ambient temperatures. It can also be assumed that all of the electrons have positive effective mass. To simplify further, it is apparent from Fig. 12.19 that the effective mass, inversely proportional to d^2E/dk^2, is fairly constant near the top and bottom of the energy bands. Hence, those electrons in the conduction band are assumed to have an average constant value for their effective masses. As a result, the evaluation of

Eq. (12.53) requires knowledge only of the number of filled states in the conduction band and the average effective mass at the bottom of the band.

Evaluation of the current from electrons in the valence band is slightly more complicated, but the same assumptions used for the conduction band electrons apply. Here it is assumed that the empty states are all at the top of the band where electrons would have negative effective mass. Again, it is assumed that the effective mass is fairly constant at the top of the band, and that an average effective mass can be used to describe the transport of electrons. Because there is an offset of filled states in k-space, where more positive effective mass states are filled than negative effective mass states, the net electron motion is from electrons moving in the direction of the electric force. That net flow of electrons is simply the difference between the total density of allowed valence band states and the number of valence band filled states. Thus,

$$I_v = \frac{q}{d}\sum_{\substack{i \\ \text{all}}} v_i - \frac{q}{d}\sum_{\substack{i \\ \text{filled}}} v_i = \frac{q}{d}\sum_{\substack{i \\ \text{empty}}} v_i \quad . \tag{12.55}$$

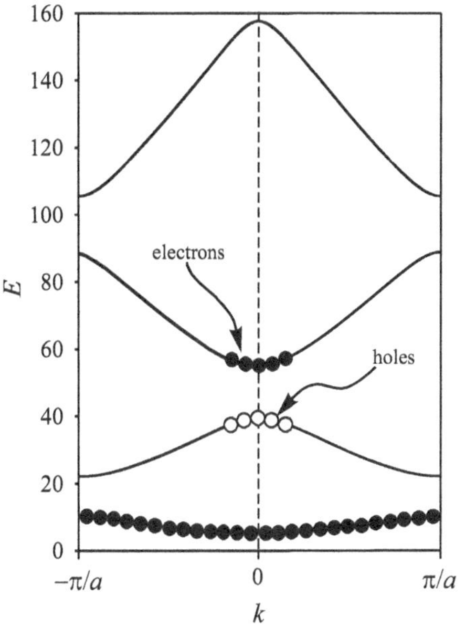

Figure 12.22. The valence band can be thought of as having "positive" charge carriers referred to as "holes" that occupy the empty states. By so doing, the motion of the valence band electrons can be replaced with a simple approximation in which only the motion of holes is considered.

which depends only on the number of empty states, all of which are assumed to be in the negative effective mass region of the valence band. In other words, the number of unfilled states determine the net number of positive effective mass electrons adding to the conduction mechanism. Because the missing electrons at the top of the valence band would have an average negative effective mass, the concept of a *hole* is introduced. A hole is an empty state that behaves like a positively charged particle in the valence band, and mimics the situation in which the valence band is partially filled with positively charged holes that travel in the opposite direction of the negatively charged electrons, as depicted in Fig. 12.22. As a result, the evaluation of Eq. (12.55) requires knowledge of only the number of empty states in the valence band and the average effective mass at the top of the band.

The conduction band of *cubic* semiconductors has three minima, although in some cases all three may not be prevalent. Second, the band gap refers to the energy difference between the lowest point of the conduction band and the highest point of the valence bands. In Fig. 12.23(a), the lowest point in the conduction band is in the [111] direction (the L valley), and is not directly above the highest point of the valence band in k-space. Off-centered band gaps are referred to as *indirect* band gaps. Semiconductors with the Γ valley being the lowest in the conduction band are referred to as having *direct* band gaps because the lowest electron energy is directly across from the highest energy of the valence bands. Third, it should be noted that *three* valence bands form at Γ. Recall that the effective mass in a band decreases as the curvature of the band increases, hence the effective masses for holes in the two upper valence bands are different. The band with narrow curvature is the *light hole* band, and the band with the wider curvature is the *heavy hole* band. The third and lower band is the *split-off band*, and is usually completely filled with electrons, hence has no holes. Typically, the split-off band's effect on free carrier densities is regarded as negligible and is largely ignored in calculations.

So far the discussion has been limited to the one-dimensional case, in which regions of continuous energy in reciprocal space, or k-space, were defined by boundaries of $2\pi/a$. A graphical

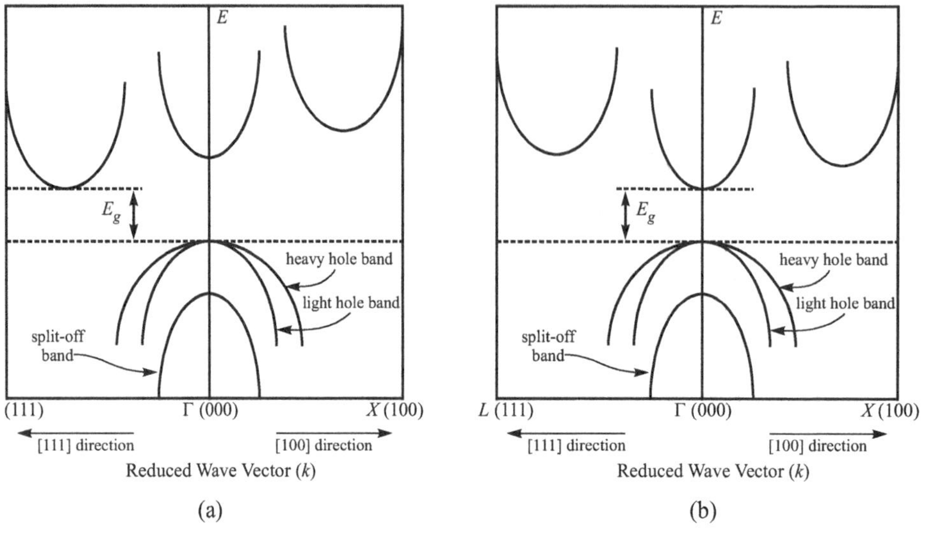

Figure 12.23. Cubic crystals typically have three conduction band minima and three valence band maxima. The conduction band minima are displaced in momentum space. Shown in (a) is an indirect band-gap semiconductor and in (b) a direct band gap semiconductor.

method to develop Brillouin zones from a 2-dimensional (2D) reciprocal lattice is performed as follows. Consider the square lattice depicted in Fig. 12.24. Choose a lattice point, depicted here as the white point in the center. To find the first Brillouin zone, draw straight lines from this point to the closest lattice points, indicated here by the numeral 1. These lines are now bisected by perpendicular lines, shown as dotted lines, and the zone confined in these boundaries is the first Brillouin zone, shown as the white area. These bisecting dotted lines define *Bragg planes*. The second Brillouin zone is found by drawing straight lines from the center to the set of next closest lattice points, indicated here with the numeral 2. These lines are now bisected with perpendicular dotted lines, and the regions confined by these new dotted lines and the previous dotted lines define the second Brillouin zone, shown as the gray area. This procedure is continued for higher order Brillouin zones.

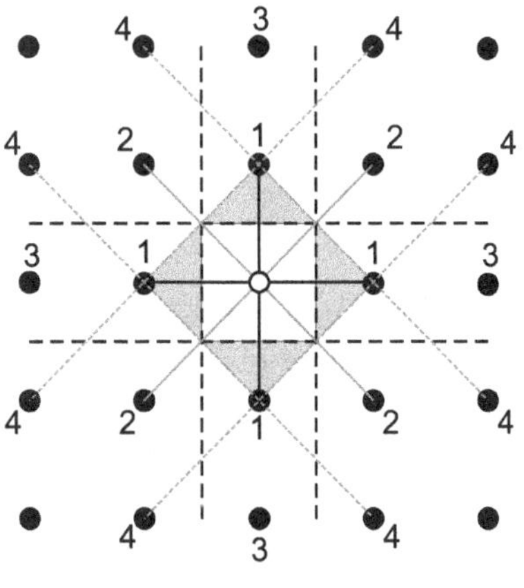

Figure 12.24. Method of developing Brillouin zones from a 2D reciprocal lattice, in this case a square lattice. The first two Brillouin zones are shown for the reciprocal lattice.

These Brillouin zones in two dimensions can also be determined from the reciprocal lattice vector [McKelvey 1966]

$$|\mathbf{G}|^2 + 2\mathbf{k}\cdot\mathbf{G} = 0, \tag{12.56}$$

where $\mathbf{G}$ is 2π multiplied by the reciprocal lattice vector. Hence, in two dimensions, for a square lattice with lattice constant (spacing) of a, the reciprocal lattice is also a square lattice with lattice spacing of $1/a$. Hence it is found that [Kittel 1986],

$$\mathbf{G} = \frac{2\pi}{a}(n_1\hat{\mathbf{x}} + n_2\hat{\mathbf{y}}), \tag{12.57}$$

where n_1 and n_2 are $\pm$ integer values and $\hat{\mathbf{x}}$ and $\hat{\mathbf{y}}$ are unit vectors in the k_x and k_y directions. From Eq. (12.56) and (12.57),

$$-\frac{\pi}{a}(n_1^2 + n_2^2) = k_x n_1 + k_y n_2, \tag{12.58}$$

which produces a family of straight lines in two-dimensional k-space where k_x and k_y are the axes, the boundaries at which Bragg reflection occurs. The intercepts are

$$\begin{aligned} k_x \text{ intercepts} &= -\frac{\pi}{a}\frac{(n_1^2+n_2^2)}{n_1}, \\ k_y \text{ intercepts} &= -\frac{\pi}{a}\frac{(n_1^2+n_2^2)}{n_2} \quad . \end{aligned}$$

Take the first case, in which $n_1 = 0, n_2 = \pm 1$ and $n_1 = \pm 1, n_2 = 0$. The resulting lines are defined as the boundaries at $k_x = \pm\frac{\pi}{a}, k_y = 0$ and $k_y = \pm\frac{\pi}{a}, k_x = 0$, thereby forming a square in two-dimensional k-space (see Fig. 12.25). This is the *first* Brillouin zone. The second Brillouin zone is found by setting $n_1 = \pm 1, n_2 = \pm 1$, producing the four lines described by $k_x \pm k_y = 2\pi/a$ and $k_x \pm k_y = -2\pi/a$. The reader should notice, just as with the one-dimensional Brillouin zones in which the total length in k-space was always the same, that the total area of the 1st BZ is the equal to the total area of the 2nd BZ. As with the 1-D BZ cases, the 2-D BZs can also be folded back into the first BZ, as shown in Fig. 12.26. The third Brillouin zone is found by setting $n_1 = 0, n_2 = \pm 2$ and $n_1 = \pm 2, n_2 = 0$, which forms the large square with boundaries of $\pm 2\pi/a = k_x, k_y$ shown in Fig. 12.25. Clearly the square formed has more area than the first two constructions, which cannot be. Fact is, there are more boundaries at which Bragg scattering can occur. Hence, to complete the construction of the third Brillouin zone, draw a vertical line from 0 to $k_y = 2\pi$. Rotate the line clockwise and note where the line crosses the next construction lines outside the 2nd BZ, which traces out the third zone. The same procedure is used to form each consecutive BZ. The 2D Brillouin zones form regions of continuous allowed energies, and there are gaps in the allowed energies at the boundaries, much like the result found with 1-D Brillouin zones.

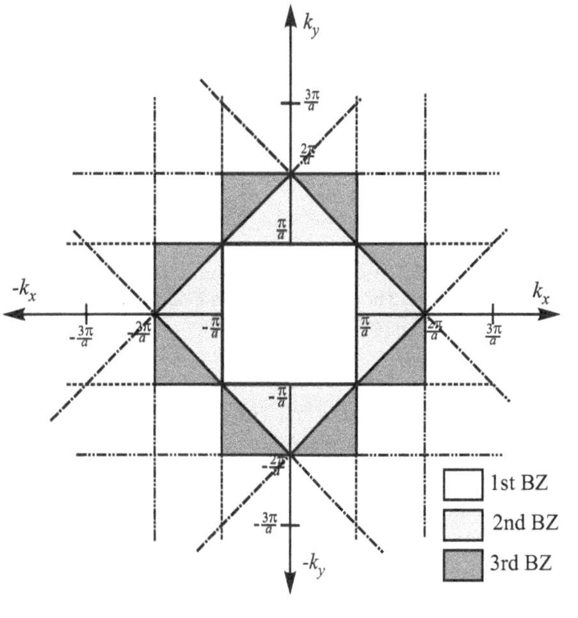

Figure 12.25. Formation of 2D Brillouin zones, showing the 1st, 2nd, and 3rd zones for a square lattice.

These findings can now be extrapolated into three dimensions. The 3D Brillouin zones are *volumes* in reciprocal (k) space, the boundaries of which satisfy the Bragg diffraction condition. As before, each Brillouin zone has the same k-space volume as the 1st Brillouin zone. Using the example for the 2D square, the 1st Brillouin zone for the simple cubic lattice is easily shown to be a cube. The Brillouin zones for the FCC and the BCC lattices are quite different, and are formed in k-space by finding $\mathbf{r}$ for the closest lattice points, and then finding the reciprocal lattice vectors so as to find $\mathbf{G}$. The first Brillouin zone of a BCC lattice is a rhombic dodecahedron, as can best be described, and the first Brillouin zone for the FCC lattice, which are generally of most concern, is a tetrakaidecahedron, which is a 14-faced polyhedron (see Fig. 12.27). The 14 faces consist of 6 planes with normal vectors belonging to the $\langle 100 \rangle$ directions, and 8 planes with normal vectors belonging to the $\langle 111 \rangle$ directions. The formation of the allowed energies and bands is different for

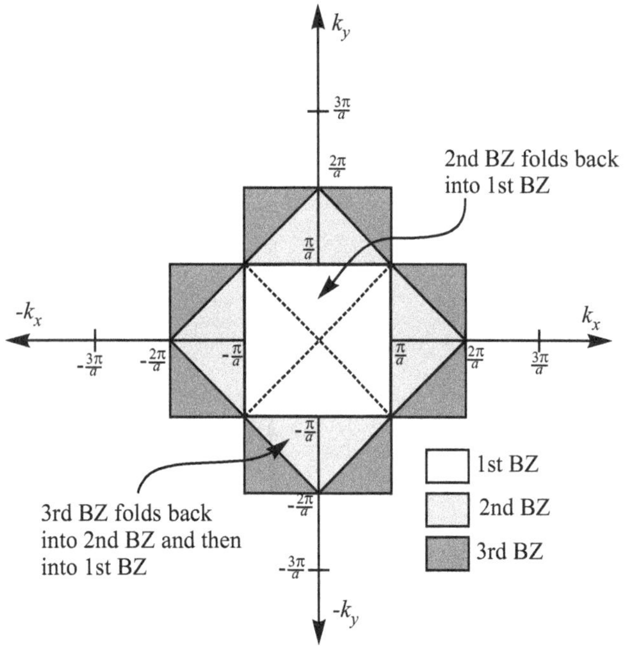

Figure 12.26. As with 1D Brillouin zones, the 2D Brillouin zones each have the same area in k-space, and all can be folded back into the 1st zone.

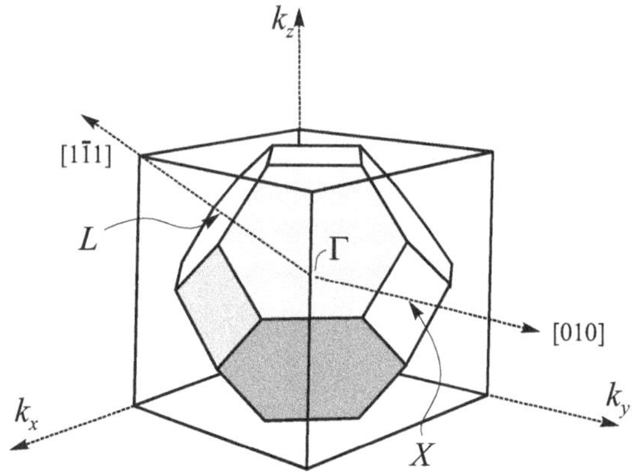

Figure 12.27. The 1st Brillouin zone of the FCC lattice.

the [100] direction than for the [111] direction, most obviously, and from results of the previous sections on the band formation with the Kronig-Penney model, because the distance in k-space to the BZ boundary in the [100] direction, termed X, is not the same as the distance in k space to the BZ boundary in the [111] direction, termed L. Typically, the band diagram for FCC crystals in k space is depicted with the bands formed in the [100] direction to the BZ boundary and the [111] direction to the BZ boundary.

Figure 12.28 depicts a typical FCC E-k diagram showing the energy bands folded back into the 1st Brillouin zone. Only the valence bands and the first conduction band are depicted in the diagram (although other bands may be included as well). The band gap depicted in Fig. 12.28 is indirect, displaced along the [111] direction.

Within the first BZ (see Fig. 12.27) for a FCC semiconductor are regions of constant energy. The regions can be depicted as constant energy *surfaces.* Refer back to Fig. 12.28 and notice again that there are three minima in the conduction band. Free electron carriers accumulate at the lowest energy in one of these bands such that they occupy energies that are quasi-continuous and bound within a region of energy defined by the band curvature. If the Γ valley is the lowest conduction band minima, then the band gap is direct, and electrons need only have an increase in energy to move from the valence bands up to the conduction band, and do not need to change momentum. However, if either the L or X valleys are the lowest conduction band minima, the band gap is indirect, and electrons must both increase in energy and have a change in momentum. The energy bands depicted in Fig. 12.28 are three-dimensional, so that the energy surface within which the electrons are confined is simply the band curvature rotated about the orientation axis, i.e., about [111] for the L valley, and about [100] for the X valley, and about the zone center for the Γ valley. In Chapter 16, modeled results are shown as reported from the literature for several semiconductor energy bands.

From the Γ-valley conduction band depicted in Fig. 12.28, consider a two-dimensional crystal, where the energy is shown as a function of wave vectors k_x and k_y projected along the $\langle 100 \rangle$ directions, as shown in

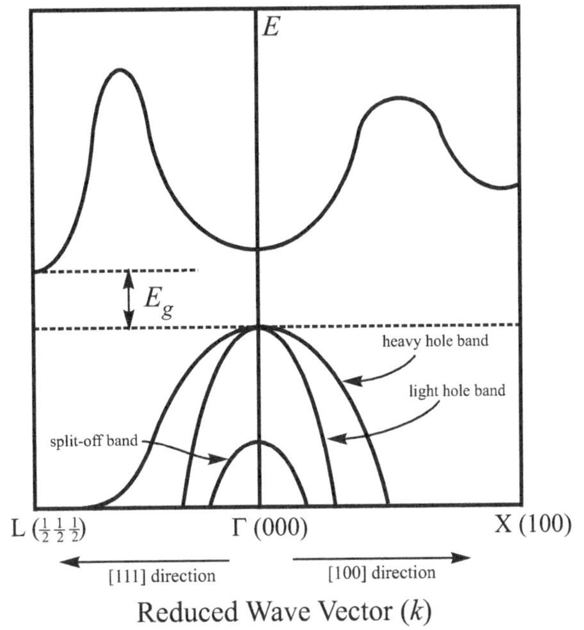

Figure 12.28. A typical E-k diagram for an FCC semiconductor crystal.

Fig. 12.29. Allowed energies are available provided that $E > E_C$. A constant energy, denoted as E_1 for this example, can be depicted as an ellipse around the paraboloid surface. Hence, an ellipse with constant energy around the periphery is described by

$$E_1 - E_C = Ak_x^2 + Bk_y^2, \tag{12.59}$$

where the intercept points along the axes can be identified by,

$$k_x = \pm\sqrt{\frac{E_1 - E_C}{A}},\ k_y = 0, \tag{12.60}$$

$$k_y = \pm\sqrt{\frac{E_1 - E_C}{B}},\ k_x = 0, \tag{12.61}$$

In the case of Fig. 12.29, the ellipse is centered about zone center, meaning that through symmetry the parabolic energy band is identical in the $\pm k_x$ and $\pm k_y$ directions. Consequently, $A = B$, and Eq. (12.59) reduces to the formula for a circle.

If the ellipse is displaced along an axis, as with an indirect band gap (e.g., the X valley in Fig. 12.28), depicted in Fig. 12.30(a) by k_{y1} in the $[0\bar{1}0]$ direction, the constant energy ellipse can now be described by

$$E_1 - E_C = Ak_x^2 + B(k_y + k_{y1})^2, \tag{12.62}$$

where the intercept points along the axes can be identified by,

$$k_x = \pm\sqrt{\frac{E_1 - E_C}{A}},\ k_y = 0, \tag{12.63}$$

$$k_y = \pm\sqrt{\frac{E_1 - E_C}{B}} - k_{y1},\ k_x = 0. \tag{12.64}$$

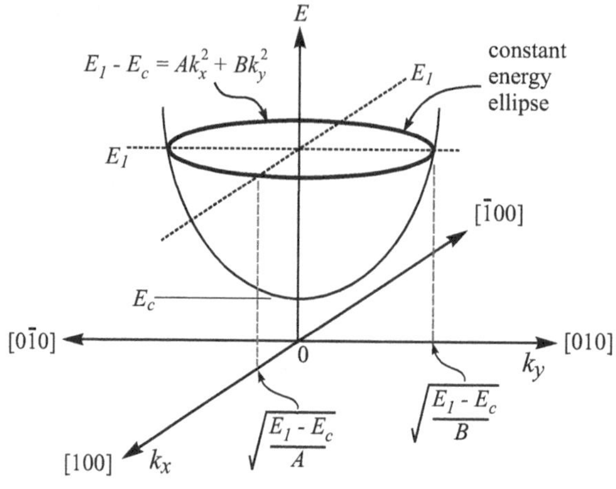

Figure 12.29. Depiction of a constant energy ellipse with energy $E_1 - E_C$ centered at $k_x = k_y = 0$.

However, because the energy band minimum is no longer at zone center, the symmetry between the k_x and k_y directions is lost. Hence, one can expect symmetry about zone center along the $\pm k_y$ directions, finding another energy band minimum at $+k_{y1}$. Also, the energy bands along $\pm k_x$ directions are symmetric about the location $-k_{y1}$ (and also $+k_{y1}$), but not equal to the k_y bands, as depicted in Fig. 12.30(b). Here, $A \neq B$, and the constant energy is depicted by an ellipse. There is a natural consequence to this discovery. Recall from Eq. (12.52) that the effective mass m_e^* is a strong function of the energy band curvature. Because the curvature at the energy minimum is different in the k_x and k_y directions, the electron effective masses are also different. Traditionally, the effective mass along the displaced axis (k_y in the example) is called the

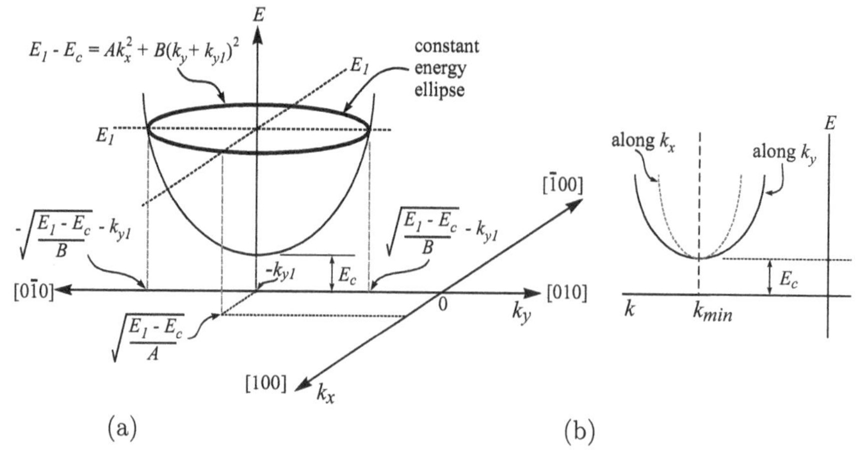

Figure 12.30. (a) Depiction of a constant energy ellipse with energy $E_1 - E_C$ centered at $k_x = 0, k_y = -k_{y1}$. (b) The curvature of the energy bands are different in the k_x and k_y directions.

longitudinal effective mass m_l^* while the effective mass in the perpendicular direction to the long axis (k_x in the example) is called the transverse effective mass m_t^*.

This same concept can now be expanded to three dimensions. Choosing E to lie within the energy ranges populated by the charge carriers in the conduction band, where $E_C \leq E$, the allowed k values form the surface in k-space. Because k is three-dimensional, and providing that the energy E is close to the conduction band edge E_C, the energy of the conduction band surfaces can be approximated by

$$E - E_C \simeq Ak_1^2 + Bk_2^2 + Ck_3^2, \tag{12.65}$$

where A, B and C are constants, and k_1, k_2, and k_3 are k-space coordinates measured from the *center* of an energy band minimum along the principal axis. Eq. (12.65) is the equation for an ellipsoid, meaning that the surface of the ellipsoid in all directions is a constant value.

With cubic crystals, the energy band maxima and minima are nearly parabolic, as was discussed earlier. Also, due to symmetry and expanding on the two-dimensional example, at least two constants in Eq. (12.65) are equal for cubic semiconductors. Hence, the constant energy surface for an indirect band gap, where the energy minimum is displaced from zone center, can be rewritten as

$$E - E_C = Ak_1^2 + B(k_2^2 + k_3^2) \tag{12.66}$$

where A is the axis of revolution.[4] Here, k_1 is the longitudinal axis while k_2 and k_3 are transverse axes. One can imagine spinning the ellipse of Fig. 12.30 about the k_y longitudinal axis to form such an ellipsoid.

For direct band-gap cubic semiconductors, all constants in Eq. (12.65) are equal so that

$$E - E_C = A(k_1^2 + k_2^2 + k_3^2), \tag{12.67}$$

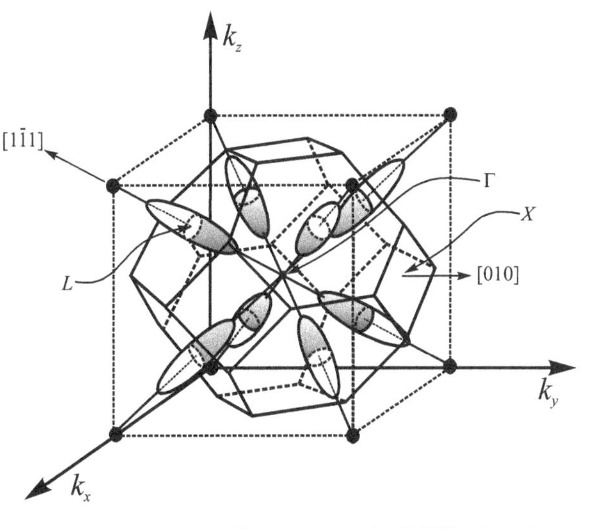

Figure 12.31. First Brillouin zone in the FCC lattice, showing ellipsoids of constant energy for the L valley.

which is the equation of a spherical surface. Hence, it is expected to have a single spherical constant energy surface for direct band-gap cubic semiconductors, six ellipsoidal constant energy surfaces for cubic semiconductors with the lowest band minima at X, and eight ellipsoidal constant energy surfaces for semiconductors with the lowest band minima at L. Figures (12.31), (12.32), and (12.33) depict the first Brillouin zone for the FCC lattice along with the constant energy ellipsoids for the lowest band minima at L, X, and Γ, respectively.

12.5 Charge Transport

To describe the transport of charges in a semiconductor, the concept of effective mass is revisited. The three-dimensional inverse effective mass tensor is described by

$$\frac{1}{\underline{\mathbf{m}}^*} = \begin{pmatrix} \mathbf{m}_{\mathbf{xx}}^{*\ -1} & \mathbf{m}_{\mathbf{xy}}^{*\ -1} & \mathbf{m}_{\mathbf{xz}}^{*\ -1} \\ \mathbf{m}_{\mathbf{yx}}^{*\ -1} & \mathbf{m}_{\mathbf{yy}}^{*\ -1} & \mathbf{m}_{\mathbf{yz}}^{*\ -1} \\ \mathbf{m}_{\mathbf{zx}}^{*\ -1} & \mathbf{m}_{\mathbf{zy}}^{*\ -1} & \mathbf{m}_{\mathbf{zz}}^{*\ -1} \end{pmatrix} \tag{12.68}$$

[4]This type of ellipsoid is called a "spheroid" or "ellipsoid of revolution".

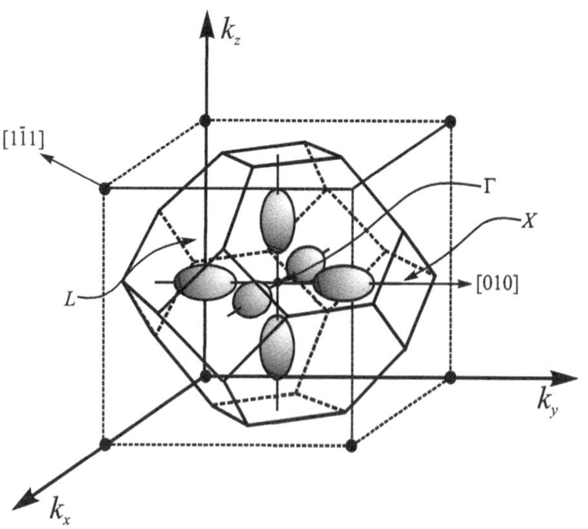

Figure 12.32. First Brillouin zone in the FCC lattice, showing ellipsoids of constant energy for the X valley.

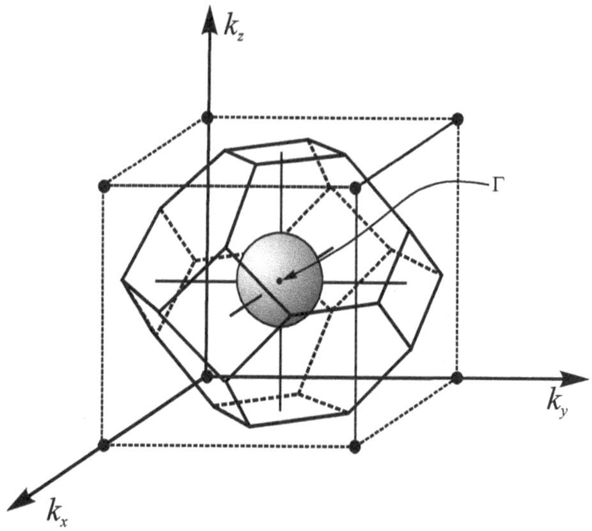

Figure 12.33. First Brillouin zone in the FCC lattice, showing a sphere of constant energy for the Γ valley.

and each inverse effective mass component is

$$\frac{1}{\mathbf{m}^*_{\mathbf{ij}}} = \frac{1}{\hbar^2}\frac{\partial^2 E}{\partial k_i \partial k_j} \quad i,\, j = x,\, y,\, z \quad . \tag{12.69}$$

From Eq. (12.66), the energies in the indirect conduction band valleys are differentiated according to Eq. (12.69) to find, in the case of cubic crystals, that the inverse effective mass tensor reduces to

$$\frac{1}{\underline{\mathbf{m}}^*} = \begin{pmatrix} {\mathbf{m}^*_{\mathbf{xx}}}^{-1} & 0 & 0 \\ 0 & {\mathbf{m}^*_{\mathbf{yy}}}^{-1} & 0 \\ 0 & 0 & {\mathbf{m}^*_{\mathbf{zz}}}^{-1} \end{pmatrix}, \tag{12.70}$$

in which the tensor components ${\mathbf{m}^*_{\mathbf{ij}}}^{-1}$ are non-zero only when $i = j$. The conduction mechanism in a semiconductor solid is affected by the total effective mass of the carriers, and can be defined by the relation

$$\mathbf{I} = \underline{\sigma} \bullet \mathbf{E} \tag{12.71}$$

or by

$$\mathbf{E} = \underline{\rho} \bullet \mathbf{I}, \tag{12.72}$$

where $\mathbf{I}$ is the current vector, $\mathbf{E}$ is the electric field vector, $\underline{\sigma}$ is the conductivity tensor and $\underline{\rho}$ is the resistivity tensor. The unit tensor is simply

$$\underline{\sigma} \bullet \underline{\rho} = \begin{pmatrix} \mathbf{1} & 0 & 0 \\ 0 & \mathbf{1} & 0 \\ 0 & 0 & \mathbf{1} \end{pmatrix}, \tag{12.73}$$

from which it is seen that $\underline{\sigma}^{-1} = \underline{\rho}$. The total conductivity is described by

$$\underline{\sigma} = \sum \underline{\sigma}^k \tag{12.74}$$

and

$$\underline{\sigma}^k = (n^k q^2 \tau^*) \begin{pmatrix} \mathbf{m}_{\mathbf{xx}}^{*\ -1} & \mathbf{m}_{\mathbf{xy}}^{*\ -1} & \mathbf{m}_{\mathbf{xz}}^{*\ -1} \\ \mathbf{m}_{\mathbf{yx}}^{*\ -1} & \mathbf{m}_{\mathbf{yy}}^{*\ -1} & \mathbf{m}_{\mathbf{yz}}^{*\ -1} \\ \mathbf{m}_{\mathbf{zx}}^{*\ -1} & \mathbf{m}_{\mathbf{zy}}^{*\ -1} & \mathbf{m}_{\mathbf{zz}}^{*\ -1} \end{pmatrix}, \tag{12.75}$$

where n^k is the number of free electrons in the equivalent ellipsoids in k-space, $\underline{\sigma}^k$ is the conductivity component for ellipsoids in k-space of direction k, q is the unit electronic charge and τ^* is the free carrier mean free drift time. For cubic crystals, the total effective mass always adds up to be the same, regardless of the direction of force applied to the semiconductor. For instance, a block of semiconducting material may have parallel electrodes placed at opposite ends of the block. Regardless of the planes upon which the electrodes are placed, or rather the direction by which an electric field force is applied across the crystal by an external voltage, the summed effective mass, and consequently the conductivity and charge carrier mobility, always sum to the same average value for *cubic lattice* semiconductors. Take for example an indirect band-gap semiconductor with the lowest conduction band minima in the [100] direction, or at X. The free carriers are confined in constant energy ellipsoidal surfaces as depicted in Fig. 12.32.

Suppose an electric field force is applied in the $[0\bar{1}0]$ direction, as shown in Fig. 12.34. The common convention has the electric field vector pointing in the negative direction. As a result, electrons in the ellipsoid surfaces are pushed towards the [010] direction, as depicted in Fig. 12.34. The curvature of the energy bands with respect to k are different in the transverse directions than in the longitudinal directions, hence the effective masses are also different. By invoking symmetry arguments, it can be seen from Fig. 12.34 that the force vector is directed longitudinally along two ellipsoids and transversely along four ellipsoids. The magnitudes of each effective mass tensor component is denoted as m_l^* for longitudinal contributions and m_t^* for transverse contributions.

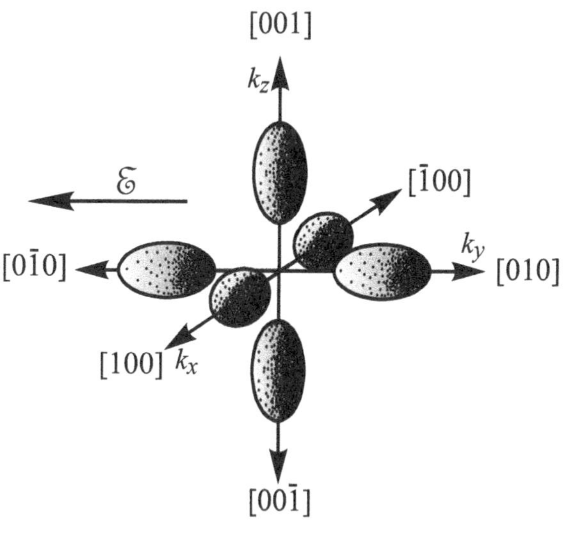

Figure 12.34. Constant energy ellipsoids for the X valley, depicting the effect of an electric field vector applied in the $[0\bar{1}0]$ direction. The small dots are for illustration purposes only, depicting that electrons are distributed preferentially according to the electric field vector.

For the conductivity component along the [100] direction, and by symmetry for the $[\bar{1}00]$ direction,

$$\underline{\sigma}^{[100]} = \underline{\sigma}^{[\bar{1}00]} = (n^{[100]} q^2 \tau^*) \begin{pmatrix} \mathbf{m}_\mathbf{l}^{*-1} & 0 & 0 \\ 0 & \mathbf{m}_\mathbf{t}^{*-1} & 0 \\ 0 & 0 & \mathbf{m}_\mathbf{t}^{*-1} \end{pmatrix}. \tag{12.76}$$

Similarly, for the remaining axial directions,

$$\underline{\sigma}^{[010]} = \underline{\sigma}^{[0\bar{1}0]} = (n^{[010]} q^2 \tau^*) \begin{pmatrix} \mathbf{m}_\mathbf{t}^{*-1} & 0 & 0 \\ 0 & \mathbf{m}_\mathbf{l}^{*-1} & 0 \\ 0 & 0 & \mathbf{m}_\mathbf{t}^{*-1} \end{pmatrix} \tag{12.77}$$

and

$$\underline{\sigma}^{[001]} = \underline{\sigma}^{[00\bar{1}]} = (n^{[001]} q^2 \tau^*) \begin{pmatrix} \mathbf{m}_\mathbf{t}^{*-1} & 0 & 0 \\ 0 & \mathbf{m}_\mathbf{t}^{*-1} & 0 \\ 0 & 0 & \mathbf{m}_\mathbf{l}^{*-1} \end{pmatrix}. \tag{12.78}$$

All six of the $\langle 100\rangle$ ellipsoids are equivalent energy minima, so that the total number of available conduction electrons can be represented by

$$\frac{n_0}{6} = n^{[100]} = n^{[\bar{1}00]} = n^{[010]} = n^{[0\bar{1}0]} = n^{[001]} = n^{[00\bar{1}]}$$

From Eq. (12.74), each conductivity direction tensor is added to find the summed conductivity tensor, which yields

$$\underline{\sigma} = \sum \underline{\sigma}^k = (n_0 q^2 \tau^*) \begin{pmatrix} \frac{1}{3}\left(\frac{2}{\mathbf{m}_t^*} + \frac{1}{\mathbf{m}_l^*}\right) & 0 & 0 \\ 0 & \frac{1}{3}\left(\frac{2}{\mathbf{m}_t^*} + \frac{1}{\mathbf{m}_l^*}\right) & 0 \\ 0 & 0 & \frac{1}{3}\left(\frac{2}{\mathbf{m}_t^*} + \frac{1}{\mathbf{m}_l^*}\right) \end{pmatrix}, \tag{12.79}$$

which can be rewritten as

$$\underline{\sigma} = \left(\frac{n_0 q^2 \tau^*}{3}\right)\left(\frac{2}{m_t^*} + \frac{1}{m_l^*}\right)\begin{pmatrix} \mathbf{1} & 0 & 0 \\ 0 & \mathbf{1} & 0 \\ 0 & 0 & \mathbf{1} \end{pmatrix} = \left(\frac{n_0 q^2 \tau^*}{m_c^*}\right)\underline{\mathbf{1}}, \tag{12.80}$$

where $\underline{\mathbf{1}}$ is the unit tensor and the *conductivity* effective mass is defined as the scalar value

$$\frac{1}{m_c^*} = \left(\frac{1}{3}\right)\left(\frac{2}{m_t^*} + \frac{1}{m_l^*}\right). \tag{12.81}$$

Despite the direction chosen for adding the effective mass tensor components, the result of Eqs. (12.80) and (12.81) for X valley semiconductors are always the same. Hence, regardless of the direction by which an electric field is applied to a cubic lattice semiconductor, the conductivity effective mass, m_c^*, is always the same, implying that mobility and conductivity can be treated as constant values for cubic semiconductors. A similar analysis for L valley semiconductors again produces a constant value for the effective mass, as does the trivial analysis needed for Γ valley semiconductors.

Example 12.2:

The constant energy ellipsoids of L valley cubic semiconductors, such as Ge, are shown in Fig. 12.31. Notice, however, they are aligned along the $\langle 111\rangle$ directions. Determine the conductivity tensor for the [111] direction ellipsoid.

Solution:

A new coordinate system is applied in which one ellipsoid is aligned along a major axis, z' for instance. In such a case, the new coordinate system can be rewritten in terms of the conventional Cartesian coordinate system. Each ellipsoid can then be rewritten in its appropriate coordinate representation.

To simplify matters a bit, note that there are eight ellipsoids in Fig. 12.31, but each ellipsoid terminates in the middle at the Brillouin zone boundary. Therefore, there are only four complete ellipsoids, which can be redrawn as shown in Fig. 12.35. Taking the ellipsoid labeled A, the z' direction is aligned along the longitudinal axis, which, using the diagram of Fig. 12.35, the unit vector can be written as

$$\mathbf{i}_z' = \frac{1}{\sqrt{3}}(\mathbf{i}_x + \mathbf{i}_y + \mathbf{i}_z). \tag{12.82}$$

The unit vectors for the x' and y' directions must be perpendicular to the z' direction and to each other. The plane they must lie in is depicted in Fig. 12.36. Notice that this plane must also intersect three planes, namely the zy, the xy, and the xz planes. Any of these intersections can be chosen for the second unit vector. For the present example, the $k_y k_z$-plane is chosen. Vectors in the $k_y k_z$ plane have no k_x component.

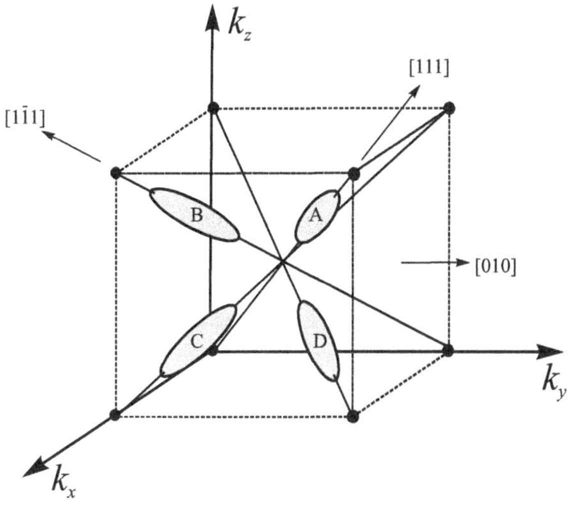

Figure 12.35. For the effective mass calculation, the eight half ellipsoids for the L valley can be rearranged as four complete ellipsoids as shown.

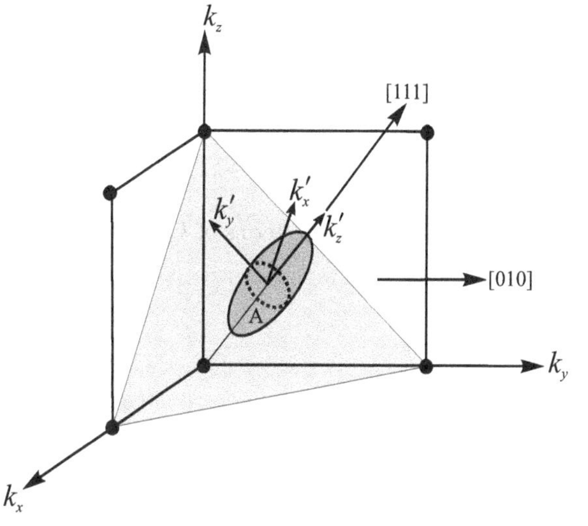

Figure 12.36. Coordinate systems used for the example, showing the relationship between k_x', k_y' and k_z' to k_x, k_y and k_z.

The dot product of the unit vector in this new vector and Eq. (12.82) must vanish, and is a suitable choice for $\mathbf{i}_y'$.

Hence, the y' unit vector is

$$\mathbf{i}_y' = \frac{1}{\sqrt{2}}(-\mathbf{i}_y + \mathbf{i}_z).$$

The x' direction unit vector is found from the cross product of $\mathbf{i}_y'$ and $\mathbf{i}_z'$,

$$\mathbf{i}_x' = \mathbf{i}_y' \times \mathbf{i}_z' = -\frac{2}{\sqrt{6}}\mathbf{i}_x + \frac{1}{\sqrt{6}}\mathbf{i}_y + \frac{1}{\sqrt{6}}\mathbf{i}_z = \frac{1}{\sqrt{6}}(-2\mathbf{i}_x + \mathbf{i}_y + \mathbf{i}_z).$$

The unit vectors are arranged in matrix form to redefine the x, y and z unit vectors for the [111] ellipsoid A,

$$\begin{pmatrix} -\frac{2}{\sqrt{6}} & \frac{1}{\sqrt{6}} & \frac{1}{\sqrt{6}} \\ 0 & -\frac{1}{\sqrt{2}} & \frac{1}{\sqrt{2}} \\ \frac{1}{\sqrt{3}} & \frac{1}{\sqrt{3}} & \frac{1}{\sqrt{3}} \end{pmatrix} \begin{pmatrix} \mathbf{i}_x \\ \mathbf{i}_y \\ \mathbf{i}_z \end{pmatrix} = \begin{pmatrix} \mathbf{i}_x' \\ \mathbf{i}_y' \\ \mathbf{i}_z' \end{pmatrix}$$

So the system of equations to determine the original Cartesian coordinates is represented by,

$$\begin{pmatrix} -\frac{2}{\sqrt{6}} & 0 & \frac{1}{\sqrt{3}} \\ \frac{1}{\sqrt{6}} & -\frac{1}{\sqrt{2}} & \frac{1}{\sqrt{3}} \\ \frac{1}{\sqrt{6}} & \frac{1}{\sqrt{2}} & \frac{1}{\sqrt{3}} \end{pmatrix} \begin{pmatrix} \mathbf{i}_x' \\ \mathbf{i}_y' \\ \mathbf{i}_z' \end{pmatrix} = \begin{pmatrix} \mathbf{i}_x \\ \mathbf{i}_y \\ \mathbf{i}_z \end{pmatrix}.$$

A method is now available to describe both coordinate systems in terms of the other. Consider an arbitrary vector $\mathbf{k}$ to be described in both coordinate systems, then

$$\mathbf{k} = \sum_j k_j \mathbf{i}_j \ , \ j = x, y, z = \sum_m k_m' \mathbf{i}_m' \ , \ m = x, y, z.$$

Hence,

$$\mathbf{k} = k_x\left(-\frac{2}{\sqrt{6}}\mathbf{i}'_x + \frac{1}{\sqrt{3}}\mathbf{i}'_z\right) + k_y\left(\frac{1}{\sqrt{6}}\mathbf{i}'_x - \frac{1}{\sqrt{2}}\mathbf{i}'_y + \frac{1}{\sqrt{3}}\mathbf{i}'_z\right)$$

$$+k_z\left(\frac{1}{\sqrt{6}}\mathbf{i}'_x + \frac{1}{\sqrt{2}}\mathbf{i}'_y + \frac{1}{\sqrt{3}}\mathbf{i}'_z\right) \quad .$$

Now $\mathbf{k}$ can be written in terms of k'_x, k'_y, k'_z by collecting the terms from the respective unit vectors,

$$k'_x = -\frac{2}{\sqrt{6}}k_x + \frac{1}{\sqrt{6}}k_y + \frac{1}{\sqrt{6}}k_z$$

$$k'_y = -\frac{1}{\sqrt{2}}k_y + \frac{1}{\sqrt{2}}k_z$$

$$k'_z = \frac{1}{\sqrt{3}}k_x + \frac{1}{\sqrt{3}}k_y + \frac{1}{\sqrt{3}}k_z$$

The constant energy surface of ellipsoid A is described by Eq. (12.66)

$$E - E_C = \hbar^2\left[\frac{(k'_z - k'_0)^2}{2m_l^*} + \frac{{k'_x}^2 + {k'_y}^2}{2m_t^*}\right].$$

Upon setting k'_0 to 0, and substitution of the terms for k'_x, k'_y, and k'_z one obtains

$$E - E_C = \frac{\hbar^2}{6m_l^*}\left(k_x^2 + k_y^2 + k_z^2 + 2k_xk_y + 2k_xk_z + 2k_yk_z\right)$$

$$+\frac{\hbar^2}{3m_t^*}\left(k_x^2 + k_y^2 + k_z^2 - 2k_xk_y - 2k_xk_z - 2k_yk_z\right)$$

The conductivity tensor for the [111] direction ellipsoid from Eq. (12.69) is,

$$\underline{\sigma}_{[111]} = \left(\frac{n_0q^2\tau^*}{4}\right)\begin{pmatrix} \frac{1}{3}\left(\frac{2}{\mathrm{m}_t^*} + \frac{1}{\mathrm{m}_l^*}\right) & -\frac{1}{3}\left(\frac{2}{\mathrm{m}_t^*} - \frac{1}{\mathrm{m}_l^*}\right) & -\frac{1}{3}\left(\frac{2}{\mathrm{m}_t^*} - \frac{1}{\mathrm{m}_l^*}\right) \\ -\frac{1}{3}\left(\frac{2}{\mathrm{m}_t^*} - \frac{1}{\mathrm{m}_l^*}\right) & \frac{1}{3}\left(\frac{2}{\mathrm{m}_t^*} + \frac{1}{\mathrm{m}_l^*}\right) & -\frac{1}{3}\left(\frac{2}{\mathrm{m}_t^*} - \frac{1}{\mathrm{m}_l^*}\right) \\ -\frac{1}{3}\left(\frac{2}{\mathrm{m}_t^*} - \frac{1}{\mathrm{m}_l^*}\right) & -\frac{1}{3}\left(\frac{2}{\mathrm{m}_t^*} - \frac{1}{\mathrm{m}_l^*}\right) & \frac{1}{3}\left(\frac{2}{\mathrm{m}_t^*} + \frac{1}{\mathrm{m}_l^*}\right) \end{pmatrix}$$

The remaining three ellipsoid conductivity tensors must now be calculated and summed together with this one to find the total conductivity tensor for the L valley. When performed, which is left as a problem exercise, the result is identical to the result for X valley semiconductors, demonstrating that the conductivity, and consequently mobility, in a cubic semiconductor is the same regardless of the direction of electrical force.

It should be noted that the result in the above example is not necessarily true for other semiconductor crystal systems and lattices. Non-cubic semiconductors, such as HgI_2 and PbI_2, can have quite different conductivities and charge carrier mobilities for different k-space directions.

Next the *density-of-states effective mass* is developed and should not be confused with the *conductivity effective mass.* Centering the indirect energy surface ellipsoids at zero, the relation between effective mass and energy is

$$E - E_C = \frac{\hbar^2}{2m_t^*}(k_1^2 + k_2^2) + \frac{\hbar^2}{2m_l^*}k_3^2. \qquad (12.83)$$

where the subscripts 1, 2 and 3 can be interchanged for the x, y and z axes. Substitution of

$$a_1 = \left[\frac{2m_l^*}{\hbar^2}(E - E_C)\right]^{1/2} \quad \text{and} \quad a_2 = \left[\frac{2m_t^*}{\hbar^2}(E - E_C)\right]^{1/2}$$

into Eq. (12.83), where $2a_1$ is the length of the major ellipsoid axis and $2a_2$ is the length of the minor ellipsoid axis, and division by $(E - E_C)$ yields

$$1 = \frac{k_1^2 + k_2^2}{a_2} + \frac{k_3^2}{a_1}.$$

The "volume" of the ellipsoid in k-space is

$$V = \frac{4}{3}\pi a_1 a_2^2 \quad .$$

Define m_e^* as the *isotropic effective mass*, which is an average effective mass within all of k-space, along with a spherical constant energy surface bounded by k as shown in Fig. 12.37. The volume of this sphere is

$$V = \frac{4}{3}\pi k_{eff}^3,$$

where

$$k_{eff} = \left[\frac{2m_e^*(E - E_C)}{\hbar^2}\right]^{1/2}.$$

Equating the ellipsoid volumes in k-space to the isotropic effective mass volume yields

$$N_{BZ}\frac{4}{3}\pi a_1 a_2^2 = \frac{4}{3}\pi k_{eff}^3$$

$$N_{BZ}\frac{4}{3}\pi\left[\frac{2m_l^*}{\hbar^2}(E - E_C)\right]^{1/2}\left[\frac{2m_t^*}{\hbar^2}(E - E_C)\right] = \frac{4}{3}\pi\left[\frac{2m_e^*(E - E_C)}{\hbar^2}\right]^{3/2}$$

$$N_{BZ}(m_l^* m_t^{*2})^{1/2} = m_e^{*3/2} \quad ,$$

where N_{BZ} is the number of constant energy ellipsoid surfaces lying within the first Brillouin zone. As a result, the isotropic effective mass for a cubic semiconductor is

$$m_e^* = N_{BZ}^{2/3}(m_l^* m_t^{*2})^{1/3}. \tag{12.84}$$

For the case of Si, there are six ellipsoid surfaces within the first BZ with $N_{BZ} = 6$, whereas for Ge only half of each ellipsoid within the first BZ (see Fig. 12.38), resulting in $N_{BZ} = 4$.
Their effective masses can be described by,

$$m_e^* \text{ (Si)} = 6^{2/3}(m_l^* m_t^{*2})^{1/3} \quad \text{and} \quad m_e^* \text{ (Ge)} = 4^{2/3}(m_l^* m_t^{*2})^{1/3}.$$

For direct band-gap semiconductors there is only one spherical constant energy surface, and $m_t^* = m_l^*$, hence,

$$m_e^* \text{ (GaAs)} = (m_l^* m_t^{*2})^{1/3} = m_e^*,$$

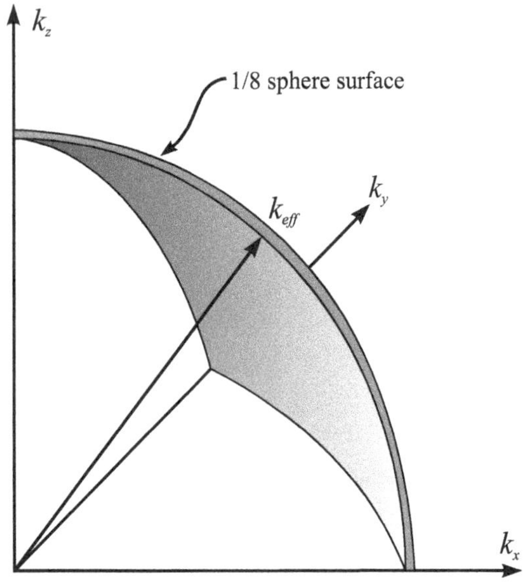

Figure 12.37. A sphere of constant energy in k-space with radius of k_{eff} is used to define m_e^*.

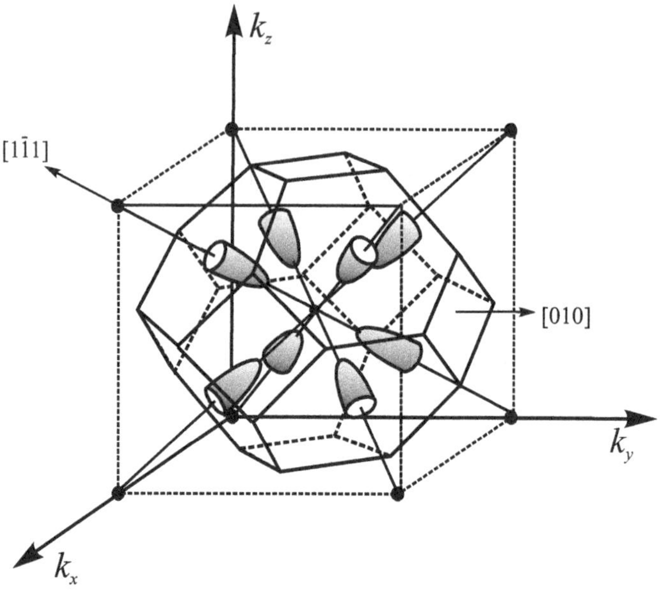

Figure 12.38. Constant energy ellipsoids for the L valley, depicting that only half of each ellipsoid lies within the first Brillouin zone.

where m_e^* is the radial effective mass in all directions, because m_l^* and m_t^* are essentially the same. Overall, a method of defining an isotropic effective mass is developed, which is commonly referred to as the *density-of-states* effective mass.

Next, the effective mass for the holes is addressed. Notice again in Fig. 12.28 that there are three valence band maxima, of which only two are considered to play an important part in hole conduction. Assuming that the upper portion of the heavy hole and light hole bands are parabolic and have spherical energy surfaces, it is usual to write an equivalent effective mass as

$$m_h^{*3/2} = m_{lh}^{*\,3/2} + m_{hh}^{*\,3/2} \tag{12.85}$$

or

$$m_h^* = (m_{lh}^{*\,3/2} + m_{hh}^{*\,3/2})^{2/3}. \tag{12.86}$$

Equation (12.86) is only an approximation, because (1) the bands are not actually parabolic at zone center, nor are the energy surfaces perfectly spherical, and (2) the split-off band may in fact contribute holes to the conduction process.

12.5.1 Charge Carrier Mobility

The motion of a charge carrier can be influenced by diffusion, scattering, trapping, magnetic fields and electric fields. The strength of this influence is characterized by the carrier mobility, described as [Sze 1981]

$$\mu_i = \frac{q\,\tau_{ci}}{m_i^*}, \qquad i = n \text{ or } p, \tag{12.87}$$

where τ_c is the average time between scattering collisions and m^* is the effective mass of the hole or electron.[5] Note that mobility increases with decreasing effective mass. The valence and conduction bands have different periodic potentials, and for this reason electron mobility in the conduction band is different than hole mobility in the valence band. Electron mobility is denoted μ_e and hole mobility is denoted μ_h.

The speed of a charge carrier can be approximated with

$$v_i = \mu_i \mathcal{E}, \tag{12.88}$$

where $\mathcal{E}$ is the electric field magnitude. Equation (12.88) is a good approximation provided that the electric field is below the *saturation* field strength (typically below about 2×10^3 V cm^{-1}). Above the saturation field the charge carrier velocities begin to asymptotically approach a limiting or saturation speed. Carrier velocities for indirect band-gap semiconductors are often described by the empirical formula [Sze 1981],

$$v \simeq v_s \left[1 + \left(\frac{\mathcal{E}_o}{\mathcal{E}}\right)^{\gamma}\right]^{-1/\gamma}, \tag{12.89}$$

where v_s is the saturation speed and E_o is an experimentally determined constant. Although Eq. (12.89) may be useful for some semiconductors, such as Si and Ge, it is altogether incorrect for semiconductors such as GaAs and InP in which more than one band gap may play an important function in the free carrier population. What is important to understand is that speeds for electrons and holes saturate, usually somewhere between 6×10^6 cm s^{-1} and 2×10^7 cm s^{-1}, depending on the type of semiconductor and dopant concentration. Consequently, the usefulness of Eq. (12.88) is limited.

12.5.2 Material Resistivity and Capacity

The ability of a semiconductor to conduct electrons is referred to as the material resistivity, with units of ohm centimeter. The resistivity of a semiconductor is

$$\rho = \frac{1}{q(\mu_e n + \mu_h p)}, \tag{12.90}$$

where q is the unit electronic charge, n and p and the free electron and hole densities, respectively, and μ_e and μ_h are the electron and hole mobilities, respectively. If $n \gg p$, then Eq. (12.90) reduces to

$$\rho \simeq \frac{1}{q\mu_e n}, \tag{12.91}$$

and if $p \gg n$, then Eq. (12.90) reduces to

$$\rho \simeq \frac{1}{q\mu_h p}. \tag{12.92}$$

The resistance of a right parallelepiped block of the semiconductor is

$$R = \rho \frac{d}{A}, \tag{12.93}$$

where d is the detector width or length, and A is the contact area. The parallel plate capacitance for a bar of semiconductor is found by

$$C = \frac{\epsilon_o \kappa A}{d}, \tag{12.94}$$

[5] Often in the literature the subscripts n and p, used to indicate *negative* and *positive* charge carriers, are replaced by e and h to refer, equivalently, to *electrons* and *holes*.

where ϵ_o is the permittivity of free space and κ is the semiconductor dielectric constant. The RC time constant of the material, commonly referred to as the *dielectric relaxation time*, is

$$RC = \frac{\rho d}{A}\frac{\epsilon_o \kappa A}{d} = \rho\epsilon_o\kappa. \tag{12.95}$$

This is the time needed by a semiconductor to return to electrical neutrality after injection or extraction of charge carriers.

12.5.3 Intrinsic Semiconductors

Semiconductors that are mostly void of impurities and defects that can alter the electrical characteristics of the pure material are referred to as *intrinsic semiconductors.* The term does not necessarily mean a perfect semiconductor, because materials are intrinsic if the *intrinsic properties* dominate over *extrinsic properties.* The conductivity in an intrinsic semiconductor is dominated by free charge carriers excited directly across the band gap rather than carriers excited from energy sites appearing in the band gap from crystal defects. Because the electrons excited into the conduction band come from the valence band, there must be an equal density of empty states in the valence band. To determine the conductivity of the semiconductor, the density of filled states in the conduction band and the density of empty states in the valence band must be determined.

Density of Energy States

To determine the density of energy states, revisit the problem of an electron trapped in a potential well, discussed in Section 12.3.4, but now with a slight change. The electron is now trapped in a *three*-dimensional potential box. The potential box is bounded in the x, y, and z directions by dimensions a, b and c. The three-dimensional Schrödinger wave equation that describes the possible states of the particle is [Nussbaum 1962],

$$-\frac{\hbar^2}{2m_e}\nabla^2\psi(x,y,z) + U(x,y,z)\psi(x,y,z) = E\psi(x,y,z), \tag{12.96}$$

with $0 < x < a$, $0 < y < b$, and $0 < z < c$.

The particle is trapped within the box, from which infinite energy is required to escape, i.e., the particle has zero potential energy so that $U(x,y,z) = 0$ within the box. With this simplification, Eq. (12.96) can be rewritten as

$$\nabla^2\psi(x,y,z) + k^2\psi(x,y,z) = 0, \tag{12.97}$$

where

$$k = \sqrt{\frac{2m_eE}{\hbar^2}} \qquad \text{and} \qquad E = \sqrt{\frac{\hbar^2k^2}{2m_e}}. \tag{12.98}$$

To solve Eq. (12.97) employ the method of separation of variables, in which the three-dimensional wave function is expressed as

$$\psi(x,y,z) = X(x)Y(y)Z(z). \tag{12.99}$$

Substitute for ψ in Eq. (12.97) and divide by XYZ to obtain

$$\frac{1}{X(x)}\frac{d^2X(x)}{dx^2} + \frac{1}{Y(y)}\frac{d^2Y(y)}{dy^2} + \frac{1}{Z(z)}\frac{d^2Z(z)}{dz^2} + k^2 = 0. \tag{12.100}$$

For Eq. (12.100) to be valid for a point (x,y,z) in the box, each of the first three terms must equal a constant, i.e.,

$$\frac{1}{X(x)}\frac{d^2X(x)}{dx^2} = -k_x^2, \quad 0 < x < a, \tag{12.101}$$

$$\frac{1}{Y(y)}\frac{d^2Y(y)}{dy^2} = -k_y^2, \quad 0 < y < b, \tag{12.102}$$

$$\frac{1}{Z(z)}\frac{d^2Z(z)}{dz^2} = -k_z^2, \quad 0 < z < c, \tag{12.103}$$

in which

$$k^2 = k_x^2 + k_y^2 + k_z^2. \tag{12.104}$$

The solution of Eq. (12.101), Eq. (12.102), and Eq. (12.103) is subject to the boundary conditions that $X(x)$, $Y(y)$, and $Z(z)$ must vanish on the surfaces of the potential box. For example, because $X(0) = X(a) = 0$ the only solutions of Eq. (12.101) are $X(x) = A\sin(k_x x)$, where $k_x = n_x\pi/a$, $n_x = \pm 1, \pm 2 \pm 3, \ldots$. Similar results are obtained for $Y(y)$ and $Z(z)$. In this manner, the solution of Eq. (12.97) is

$$\psi(x, y, z) = A\sin(k_x x)\sin(k_y y)\sin(k_z z) \tag{12.105}$$

with

$$k_x = \frac{n_x\pi}{a}, \qquad k_y = \frac{n_y\pi}{b}, \quad \text{and} \quad k_z = \frac{n_z\pi}{c}. \tag{12.106}$$

where n_x, n_y, and n_z are integers $\pm 1, \pm 2, \pm 3, \ldots$.

From this solution one can calculate the number of allowed electronic states, per unit volume defined by the dimensions a, b and c. Recall in reciprocal lattice space, the reciprocal-space unit-cell volume is defined by

$$\text{cell volume} = \left(\frac{\pi}{a}\right)\left(\frac{\pi}{b}\right)\left(\frac{\pi}{c}\right) = \frac{\pi^3}{abc} \tag{12.107}$$

and represents one state or solution. Hence, the number of solutions per reciprocal space volume is

$$\frac{\text{no. of solutions}}{\text{unit cell volume}} = \frac{1 \text{ state}}{\pi^3/abc} = \frac{abc}{\pi^3}. \tag{12.108}$$

Note that the number of solutions per reciprocal space volume does not equal the number of allowed states. From Eq. (12.98) and Eq. (12.101) the energy of an allowed state depends on (n_x^2, n_y^2, n_z^2) so the 8 possible solutions $\psi_{\pm n_x, \pm n_y, \pm n_z}(x, y, z)$ all have the same energy level. Hence, for each unit volume of reciprocal space, the number of solutions must be divided by the 8 possible combinations of $(\pm n_x, \pm n_y, \pm n_z)$ to get the density of allowed states. However, because the particles can have 2 values of spin, the allowed states must then be multiplied by 2. As a result, the density of allowed states within a reciprocal space unit volume becomes

$$\frac{\text{no. of allowed states}}{\text{unit cell volume}} = \frac{abc}{4\pi^3}. \tag{12.109}$$

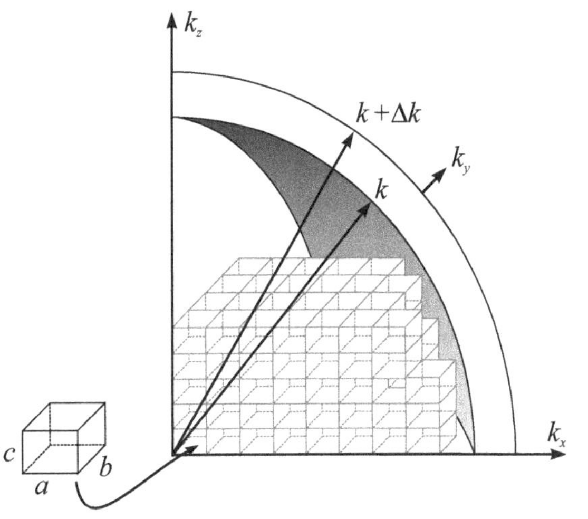

Figure 12.39. The number of allowed states can be modeled as quantum well boxes inside a spherical shell in k-space.

To determine the density of allowed states per unit volume in real space, consider a spherical shell in reciprocal space, as shown in Fig. 12.39, in which the axes are defined by k_x, k_y and k_z. For any particular value of $k = (k_x^2 + k_y^2 + k_z^2)^{1/2}$, the volume in dk about k is

$$\text{volume} = 4\pi k^2 dk. \tag{12.110}$$

Thus, the number of allowed states within this volume can be estimated as

$$\text{no. of allowed states} = (4\pi k^2 dk)\left(\frac{abc}{4\pi^3}\right) = \frac{(abc)k^2 dk}{\pi^2}. \tag{12.111}$$

From Eqs. (12.98),

$$dk = \frac{1}{\hbar}\sqrt{\frac{m_e}{2}}\frac{dE}{\sqrt{E}}, \qquad \text{where} \qquad dE = \frac{\hbar^2 k dk}{m_e}. \tag{12.112}$$

Substitution of these results into Eq. (12.111) gives the density of allowed states $N(E)dE$ with energies between E and $E + dE$, namely

$$N(E)dE = \frac{abc}{\pi^2}\frac{\sqrt{2}m_e^{3/2}}{\hbar^3}\sqrt{E}dE. \tag{12.113}$$

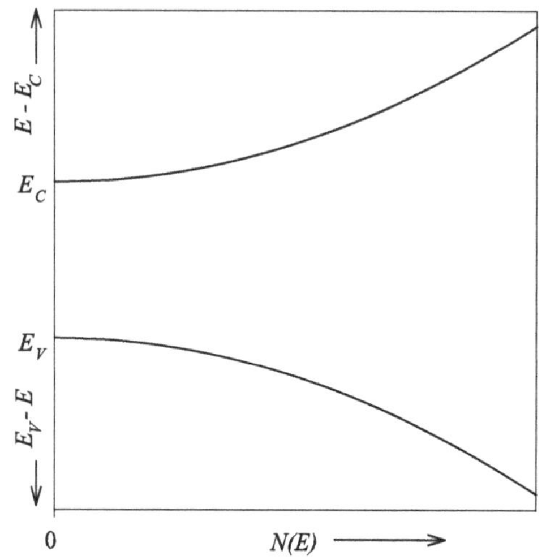

Figure 12.40. The density of states for the conduction and valence bands. In the band gap, there are no states, hence $N_C(E) = N_V(E) = 0$.

The density of states $N(E)$, per unit volume and per unit energy about E, is then obtained by dividing the number of states $N(E)dE$ by the unit volume per state in dE about E, i.e.,

$$N(E) = \frac{abc}{\pi^2}\frac{\sqrt{2}m_e^{3/2}}{\hbar^3}\sqrt{E}dE\left[\frac{1}{VdE}\right] = \frac{\sqrt{2E}m_e^{3/2}}{\pi^2\hbar^3}, \tag{12.114}$$

where $V = abc$.

Finally, because there are no allowed states permitted in the band gap, substitution of the effective masses for electrons and holes in Eq. (12.114), yields densities of allowed states for the conduction and valence bands, namely

$$N_C(E) = \frac{\sqrt{2(E - E_C)}m_e^{*3/2}}{\pi^2\hbar^3}, \qquad E \geq E_C, \tag{12.115}$$

and

$$N_V(E) = \frac{\sqrt{2(E_V - E)}m_h^{*3/2}}{\pi^2\hbar^3}, \qquad E \leq E_V, \tag{12.116}$$

where $N_C(E)$ is the density of states in the conduction band and $N_V(E)$ is the density of states in the valence band. Here m_e^* and m_h^* are the effective mass of an electron or hole, respectively. The variation of these densities with $E - E_C$ and $E_V - E$ is shown in Fig. 12.40.

Charge Carrier Concentration

Because electrons are fermions, they must obey the Pauli exclusion principle so each state at energy E can contain at most 1 electron with a given spin. Further, the probability an electron has energy E is governed by Fermi-Dirac statistics and states that the number $n(E)$ of electrons in quantum states in unit energy about E per unit volume is given by

$$n(E) = g(E)f_{\mathcal{FD}}(E), \tag{12.117}$$

where the degeneracy parameter $g(E)$ is the number of states in unit energy about E per unit volume available to an assembly of electrons and $f_{\mathcal{FD}}(E)$ is the Fermi function, named after its discoverer,

$$f_{\mathcal{FD}}(E) = \frac{1}{1 + \exp\left[\dfrac{E - E_F}{kT}\right]}. \tag{12.118}$$

Here k is Boltzmann's constant, T is the absolute temperature, and E_F is the Fermi energy. The Fermi energy is defined as the energy level, at 0 K temperature, for which all states below it are filled and all above it are empty. Equation (12.118) can be used to determine the density of electrons in the conduction band, i.e., for $E \geq E_C$. From Eq. (12.118), it is seen that the probability an electron crosses the band gap to the conduction band increases with increasing temperature. The Fermi-Dirac distribution is shown in Fig. 12.41.

The free electron density n in the conduction band is that of the *negatively* charged carriers, and the free hole density p in the valence band is that of the *positively* charged carriers. The empty states in the valence band are treated as positive charge carriers called "holes", a concept which greatly simplifies calculations. For a pure material, the density of electrons in the conduction band must equal the density of holes in the valence band, i.e., $n = p$. Such a situation is referred to as the *intrinsic* case.

The free electron density can be found by integrating the product of the conduction band density of states function and the Fermi-Dirac distribution function, namely

$$n = \int_{E_C}^{E_{top}} N_C(E) f_{\mathcal{FD}}(E) dE, \qquad (12.119)$$

where E_{top} is the top of the first conduction band. Almost all charge carriers in the conduction band are located near the bottom of the band with none of the states at the top filled; hence integration from E_C to E_{top} is equivalent to integration from E_C to ∞.

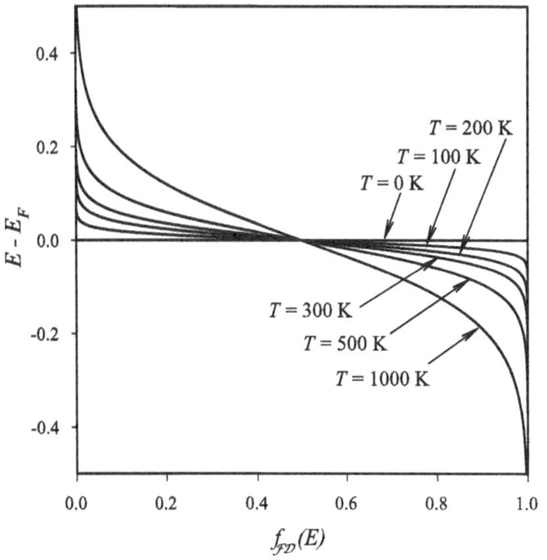

Figure 12.41. The Fermi-Dirac probability distribution function. At 0 K, the distribution indicates that all energy states above the Fermi energy are empty and all states below the Fermi energy level are filled. As the temperature is increased, the probability of states being filled above the Fermi energy increases.

Substitution of Eq. (12.115) and Eq. (12.118) into Eq. (12.119), with the upper integration limit replaced by ∞, gives

$$n = \frac{\sqrt{2} m_e^{*3/2}}{\pi^2 \hbar^2} \int_{E_C}^{\infty} \frac{\sqrt{E - E_C}}{1 + \exp[(E - E_F)/(kT)]} dE. \qquad (12.120)$$

To evaluate this integral, let $\xi = (E - E_C)/(kT)$, define $\eta_c \equiv (E_F - E_C)/(kT)$, recall that $\hbar = h/(2\pi)$, and recall $\Gamma(1 + \frac{1}{2}) = \sqrt{\pi}/2$ [Pierret 1989]. Eq. (12.120) then reduces to

$$\begin{aligned} n &= 2 \left[\frac{2\pi m_e^* kT}{h^2} \right]^{3/2} \frac{1}{\Gamma(1 + 1/2)} \int_0^{\infty} \frac{\xi^{1/2}}{1 + \exp[\xi - \eta_c]} d\xi, \\ &= N_C \frac{2}{\sqrt{\pi}} F_{1/2}(\eta_c) = N_C \mathcal{F}_{1/2}(\eta_c). \end{aligned} \qquad (12.121)$$

Here $N_C \equiv 2[2\pi m_e^* kT]/h^2]^{3/2}$ is energy-independent and referred to as the *effective* density of states in the conduction band. The quantity $F_{1/2}$ is the Fermi-Dirac integral of 1/2 order, while $\mathcal{F}_{1/2}$ is the Fermi-Dirac integral function of order 1/2.[6]

[6]This particular integral is one of the more important integrals encountered in semiconductor physics and belongs to a general set of Fermi-Dirac integral functions defined by [Blakemore 1985]

$$\mathcal{F}_j(t) = \frac{1}{\Gamma(1 + j)} \int_0^{\infty} \frac{\xi^j}{1 + \exp[\xi - t]} d\xi.$$

The density of empty states in the valence band must equal the density of filled states in the conduction band. Because $f_{\mathcal{FD}}(E)dE$ is proportional to the probability of filled states with energy in dE about E in the conduction band, then the probability of empty states in dE about E in the valence band must be proportional to $[1 - f_{\mathcal{FD}}(E)]dE$. Because the empty states are near the top of the valence band, the density p of empty states (or holes) in the valence band is

$$p = \int_{E_{bot}}^{E_V} N_V(E)[1 - f_{\mathcal{FD}}(E)]\, dE, \tag{12.122}$$

in which the integral is taken from the bottom of the valence band E_{bot} to the valence band edge at E_V. Because the empty states are near the top of the valence band with no unfilled states at the band bottom, the lower limit of integration may be replaced by $-\infty$. Substitution of Eq. (12.116) and Eq. (12.118) into Eq. (12.122), with the lower integration limit replaced by $-\infty$, gives

$$\begin{aligned} p &= \frac{m_p^{*3/2}\sqrt{2}}{\pi^2\hbar^3} \int_{-\infty}^{E_V} \frac{\sqrt{E_V - E}}{1 + \exp[(E - E_F)/(kT)]}\, dE \\ &= 2\left[\frac{2\pi m_h^* kT}{h^2}\right]^{3/2} \frac{2}{\sqrt{\pi}} F_{1/2}(\eta_v) = N_V \mathcal{F}_{1/2}(\eta_v), \end{aligned} \tag{12.123}$$

where $\eta_v = (E_V - E_F)(kT)$. The energy independent term N_V is the *effective* density of states in the valence band. It is important to note that the use of the Fermi-Dirac distribution function and the results of Eq. (12.121) and Eq. (12.123) are valid for all positions of the Fermi energy level, E_F.

Unfortunately, there are no closed-form analytical expressions for any Fermi function of positive order. Hence, closed form expressions for Eq. (12.121) and Eq. (12.123) are not available. Numerical integration may be used to evaluate $\mathcal{F}_{1/2}(\eta)$ or some approximate expression is used. There are several approximations for the Fermi-Dirac integral, some far more complex than others. One relatively simple approximation for the Fermi-Dirac integral of order $j = 1/2$ is [Pierret 1989]

$$\mathcal{F}_{1/2}(\eta) \simeq \exp(\eta) - 0.3\exp(2\eta) + 0.06\exp(3\eta) \tag{12.124}$$

with less than $\pm 0.75\%$ error for negative values of η and less than $\pm 1.5\%$ error for positive values of η no greater than 0.5.

The results of Eqs. (12.121) and (12.123) are quite general and apply to any semiconductor, doped or undoped, under the restriction that any added dopants or impurities do not appreciably change the band structure of the original semiconductor, i.e., the density of states functions of Eqs.(12.115) and (12.116) are not changed.

Approximations for the Charge Carrier Concentrations

A useful approximation for the non-degenerate case, in which the Fermi energy level is within the band gap and at least $3kT$ away from either the conduction or valence band edges, is to approximate the Fermi-Dirac distribution function $f_{\mathcal{FD}}(E)$ by the Boltzmann distribution $f_{\mathcal{B}}(E)$, i.e., if $E \geq E_C$ and $E_C - E_F \geq 3kT$

$$f_{\mathcal{FD}}(E) \equiv \frac{1}{1 + \exp\left[\dfrac{E - E_F}{kT}\right]} \simeq \exp\left[\frac{E_F - E}{kT}\right] \equiv f_{\mathcal{B}}(E). \tag{12.125}$$

With this approximation and Eq. (12.115), the concentration of free charge carriers (electrons) in the conduction band is found to be

$$n = \int_{E_C}^{\infty} N_C(E) f_{\mathcal{B}}(E)\, dE = \int_{E_C}^{\infty} \frac{m_e^{*3/2}\sqrt{2}}{\pi^2\hbar^3} \sqrt{E - E_C} \exp\left[\frac{E_F - E}{kT}\right] dE$$

$$= 2\left(\frac{2\pi m_e^* kT}{h^2}\right)^{3/2} \exp\left[\frac{E_F - E_C}{kT}\right],$$

which yields the following useful relation

$$\boxed{n = N_C \exp\left[\frac{E_F - E_C}{kT}\right].} \tag{12.126}$$

In a similar manner the concentration of free charge carriers (holes) in the valence band can be found approximately with Boltzmann's approximation. Here an approximation is needed for $1 - f_{\mathcal{FD}}(E)$ when $E \leq E_V$ and $E_F - E_V > 3kT$. In this case

$$1 - f_{\mathcal{FD}}(E) = \frac{1}{1 + \exp\left[\frac{E_F - E}{kT}\right]} \simeq \exp\left[\frac{E - E_F}{kT}\right]. \tag{12.127}$$

With this approximation and Eq. (12.116) the concentration p of holes in the valence band is

$$p = \int_{-\infty}^{E_V} N_V(E)[1 - f_{\mathcal{FD}}(E)]\, dE \simeq \int_{-\infty}^{E_V} \frac{m_h^{*3/2}\sqrt{2}}{\pi^2\hbar^3} \sqrt{E_V - E} \exp\left[\frac{E - E_F}{kT}\right] dE$$

$$= 2\left(\frac{2\pi m_h^* kT}{h^2}\right)^{3/2} \exp\left[\frac{E_V - E_F}{kT}\right],$$

from which the following useful relation is obtained

$$\boxed{p = N_V \exp\left[\frac{E_V - E_F}{kT}\right].} \tag{12.128}$$

Comparison of Eq. (12.126) to Eq. (12.121) and Eq. (12.128) to Eq. (12.123), shows that

$$\mathcal{F}_{1/2}(\eta) \simeq e^{\eta} \tag{12.129}$$

provided $E_C - E_F \geq 3kT$ (or $\eta_c \leq -3$) and $E_F - E_V \geq 3kT$ (or $\eta_v \leq 3$).

WARNING: Eq. (12.126) and Eq. (12.128) are only approximations and should not be used for degenerate semiconductors![7] If the material is intrinsic, i.e., the background impurities have negligible effect upon the semiconductor properties, then the following must be true for the free carrier concentrations,

$$n_i = n = p = N_C \exp\left[\frac{E_i - E_C}{kT}\right] = N_V \exp\left[\frac{E_V - E_i}{kT}\right], \tag{12.130}$$

[7] A degenerate semiconductor is a heavily doped semiconductor whose Fermi level may no longer be in the energy band gap.

where n_i is the *intrinsic* free carrier concentration and E_i is the *intrinsic* Fermi energy level.[8] From this result it is apparent that

$$np = n_i^2. \tag{12.131}$$

Equation (12.131) is the so-called *mass-action* law, and, because it was derived with the Boltzmann approximation, it is only valid for non-degenerate semiconductors. Equation (12.131) can be used to derive the intrinsic free carrier concentration of a semiconductor,

$$n_i^2 = np = N_C \exp\left[\frac{E_F - E_C}{kT}\right] N_V \exp\left[\frac{E_V - E_F}{kT}\right]$$
$$= N_C N_V \exp\left[\frac{E_V - E_C}{kT}\right] = N_C N_V \exp\left[\frac{-E_g}{kT}\right].$$

where E_g is the band-gap energy. Thus the intrinsic carrier concentration is

$$n_i = \sqrt{N_C N_V} \exp\left[\frac{-E_g}{2kT}\right]. \tag{12.132}$$

Notice that the free charge carrier concentration increases exponentially with absolute temperature. Notice also that the band gap of a semiconductor largely affects the free carrier concentration, in which large band-gap materials has higher electrical resistivity than small band-gap materials, a property that is important for semiconductor radiation detectors.

In Fig. 12.42 the Fermi-Dirac function is shown along with the density of states functions, $N_C(E)$ and $N_V(E)$, and the distribution of charge carriers in the conduction and valence bands. Notice in Fig. 12.42(a), that the Fermi-Dirac distribution is symmetric about the Fermi energy level, i.e., there are just as many filled states above E_F as there are empty states below E_F. Notice also that the density of states function increases with $\sqrt{E}$ from the band edges, as required by Eqs. (12.115) and (12.116), and that there are no states in the band gap. As a result, the product, $f_{\mathcal{FD}}(E)N_C(E)$, or $[1 - f_{\mathcal{FD}}(E)]N_V(E)$ is zero in the band gap. Further, as $N_C(E)$ or $N_V(E)$ increases with E, the probability of a state being filled decreases with increasing E. Hence, the product $f_{\mathcal{FD}}(E)N_C(E)$ is non-linear, as is the distribution of free charge carriers in the conduction band and the valence band, as shown in Fig. 12.42(c).

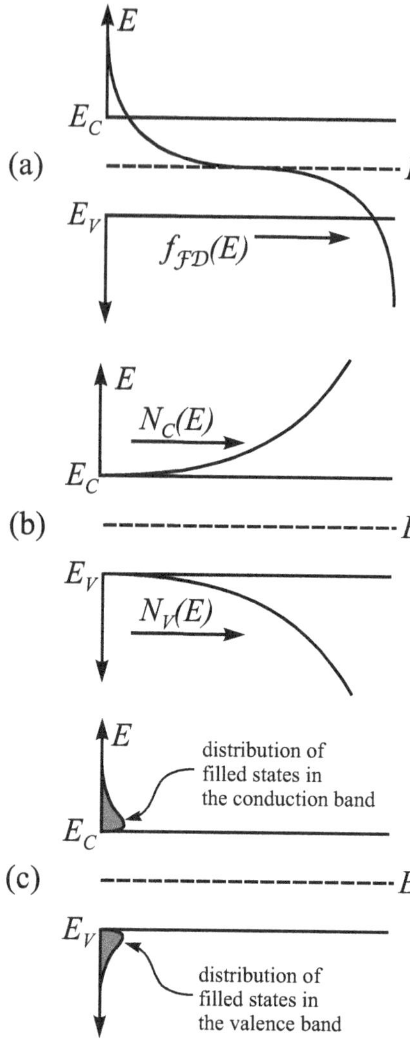

Figure 12.42. Shown are (a) the Fermi-Dirac function for an intrinsic semiconductor, (b) the density of states functions, and (c) the distribution of free charge carriers, electrons and holes, in the conduction and valence bands, respectively.

The total number of free carriers excited from the valence band to the conduction band per unit time in a semiconductor crystal depends upon the band gap energy, temperature, and the actual volume of the crystal.

[8]The intrinsic Fermi level E_i is a constant for a given semiconductor and equals the actual Fermi level E_F if impurities are negligible. However, as impurities or dopants are added the Fermi level changes although the intrinsic Fermi level remains constant.

Hence, one can quickly surmise that, given a limit on allowable leakage current, wide band-gap materials have lower leakage currents from thermal generation of charge carriers than narrow band-gap materials. Further, under ideal conditions, wide band-gap materials can be fashioned into larger devices than narrow band-gap materials, and they can still exhibit low leakage currents. Lastly, for narrow band-gap materials, the operating temperature of the material can be reduced so as to lower the leakage current, as is usually done with Ge and large Si devices.

12.5.4 Impurities and Extrinsic Semiconductors

One of the more useful properties of semiconductors is that their electrical conductivities can be altered with the simple addition of various impurities. Depending on the valence of the impurity atoms and the host semiconductor, the electrical conductivity can be dominated by hole motion or electron motion. Further, with proper concentrations, impurities can be added to an unintentionally contaminated (doped) semiconductor to make it behave much like an intrinsic semiconductor.

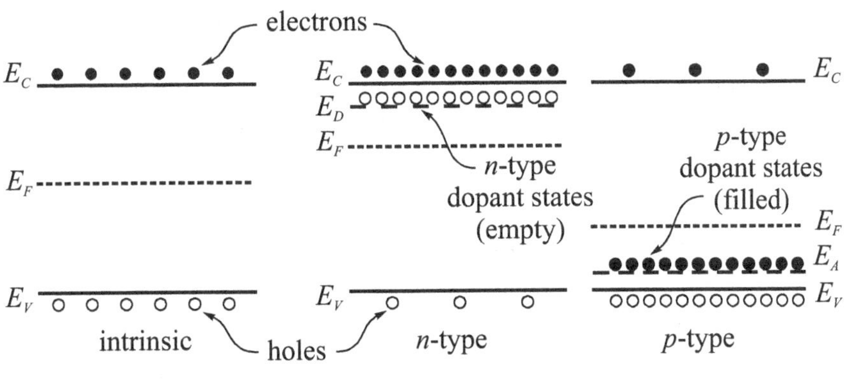

Figure 12.43. The simplified energy band structures for intrinsic, n-type, and p-type materials. The electron and hole carrier concentrations are equal for the intrinsic case. Materials doped n-type have an excess of electrons in the conduction band, and materials doped p-type have an excess of holes in the valence band.

Dopant Impurities

Dopant impurities are often added to a semiconductor to control its electrical properties. Dopants that add excess electrons to the chemical binding are called *donors*, because they need only a slight amount of thermal energy to transfer these excess electrons into the conduction band. Dopants that lack an electron to complete the valence bonding are called *acceptors*, because they need only a slight amount of energy to accept electrons from the valence band into their unfilled states. The simplified energy band structures for intrinsic, n-type, and p-type materials are depicted in Fig. 12.43. The concentration of donor atoms is denoted by N_D with energy level E_D, and the concentration of acceptor atoms is denoted by N_A with energy level E_A. If donors are added to the semiconductor, then the concentration of holes is reduced. The opposite occurs if acceptors are added to the semiconductor. Because dopants are generally used in such a small concentration that the energy band structure is unaffected, the general relationship between the free electron concentration and the free hole concentration is given by Eq. (12.131).

Background

In the preparation and the growing of crystals, there are many ways contaminant atoms can enter the crystal even after rigorous purification processes. Typically, a variety of processes are used to remove contaminants from the starting crystal material. These include vapor distillation, zone melting, and zone refinement. All of these methods depend upon an ability to diffuse impurity ions from one region of the starting material to another. Often the segregation coefficients of impurities is not conducive to such processes and the impurities are, therefore, hard to remove. Ge and Si can be purified to levels below one ppb with zone refinement and float zone refinement. However, inadequate handling techniques can reintroduce impurities from chemicals used for growth preparations, crucibles and ampoules used during the growth process, heating coils and liners in the growth furnaces, as well as human introduced impurities such as Na and K. These undesirable impurities are commonly referred to as *background impurities.*

Although the presence of impurities generally affects the electronic performance of a semiconductor material, the effect can vary widely. For instance, some impurities may introduce scattering centers that can affect the free carrier mobility, but may not be electrically active, i.e., they do not introduce free charge carriers to the conduction or valence bands. For example, boron impurities in GaAs, in high concentrations, can affect the mobility of the charge carriers, but they are electrically inactive. As a result, pyrolitic BN is often used as the crucible for GaAs bulk crystal growth. By contrast carbon, sources of which are present in practically all steps of the GaAs materials preparation and crystal growth, is an electrically active shallow acceptor and behaves as a p-type dopant. Hence, carbon affects mobility and electrical conductivity.

If the background impurities, primarily those that are electrically active, can be kept to levels below the intrinsic free carrier concentration of the semiconductor of interest, then the material still behaves intrinsically. However, extrinsic behavior is observed for impurity concentrations approaching or exceeding the intrinsic concentration. Often impurities of opposite electrical behavior are purposely added during the crystal growth process to counteract the effect of background impurities, a process referred to as *compensation.* Dopants may also be added during post-growth processes, such as implantation and diffusion, so as to intentionally *dope* a semiconductor to behave as n-type or p-type. Under such processes, select regions of a semiconductor sample may be doped n-type while other regions of the same sample are doped p-type. Selective doping allows for the formation of rectifying and ohmic junctions.

Extrinsic semiconductors are those materials whose electrical properties are dominated by impurity atoms and intrinsic defects. Typically, there is a temperature range in which extrinsic semiconductors demonstrate such behavior, beyond which the thermal excitation of charge carriers directly from the valence band causes the material to behave intrinsically. At the top of Fig. 12.44 are depictions of the free carrier concentrations for intrinsic material (n_i), lightly doped material (n or p), and highly doped material (n^+ or p^+). At low temperatures, the shallow donor and acceptor sites are not fully ionized, a region referred to as the *freeze-out* region.

As shown in the bottom of Fig. 12.44, as the temperature is increased, more of the impurities become ionized, until they are all ionized. In the extrinsic region, few electrons are excited from the valence band to the conduction band, hence the extrinsic carrier concentration from impurities far exceeds the free carrier concentration excited from the valence band. At this point the material behaves *extrinsically,* i.e., the conductivity remains largely unchanged and is dominated by the concentration of impurities present in the semiconductor. An important aspect to impurity doping is that it allows the device designer to control the semiconductor electrical conductivity over a wide temperature range, something that cannot be done with intrinsic materials in which the free carrier concentration changes exponentially with temperature.

As the temperature is increased further, thermal generation increases the density of free carriers excited from the valence band to the conduction band until they surpass the free carrier density excited from impurity sites, and the material behaves *intrinsically.* The temperature at which the crossover from extrinsic to intrinsic behavior is located increases as the impurity concentration increases. Hence, highly doped materials retain extrinsic behavior at higher temperatures than do lightly doped materials.

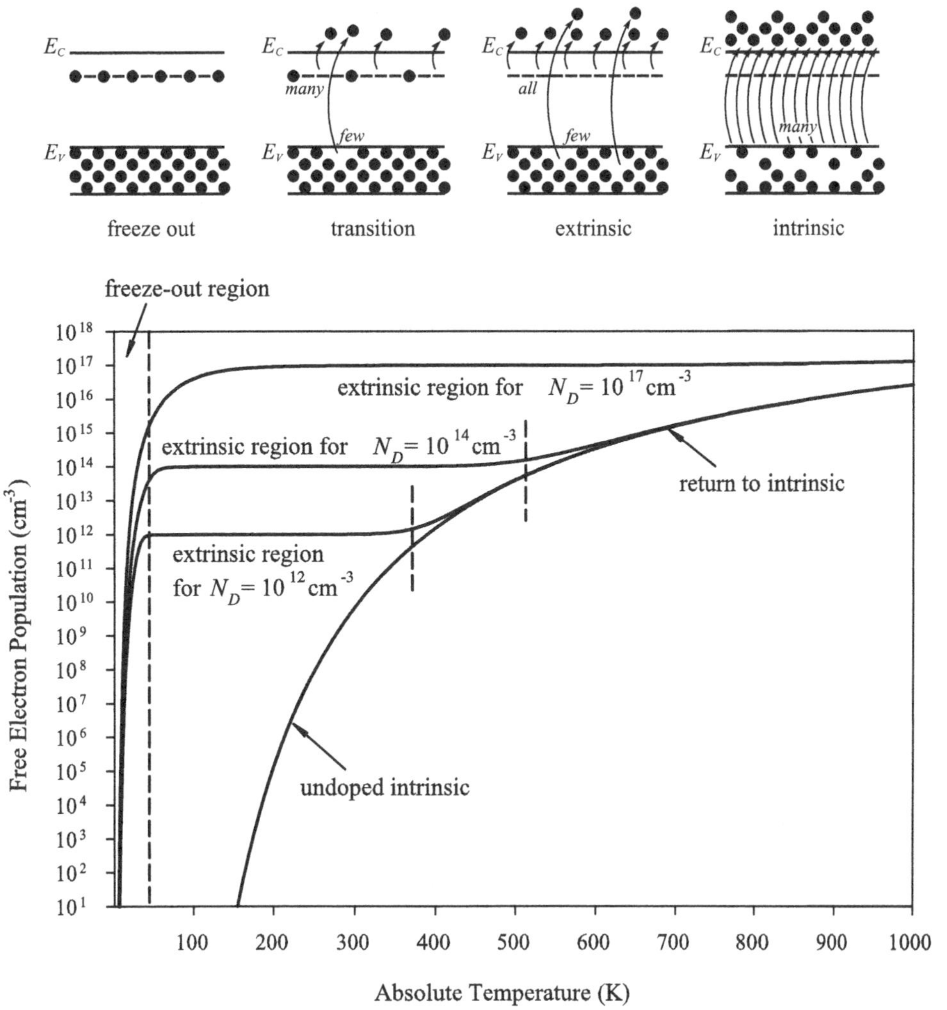

Figure 12.44. At low temperature, the impurities do not have enough thermal energy to become fully ionized, a temperature region known as *freeze-out*. As the temperature is increased, there is a transition in which the shallow impurities become ionized. At a high enough temperature, the thermal energy fully ionizes the shallow impurities, although the thermal energy is not enough to promote significant electron excitation from the valence to conduction bands. As a result, in the extrinsic region, the conductivity of the semiconductor is controlled, and limited, by the impurity concentration. At higher temperatures, the concentration of electrons excited from the valence band exceeds the impurity concentration, and the conductivity is controlled by the intrinsic free carrier concentration.

Degeneracy Factors

In the earlier use of the Fermi-Dirac distribution it was assumed that each available state could be occupied by only one electron. However, an electron occupying states in the band gap, whether donor or acceptor sites, can have either spin-up or spin-down, so that there are two different methods by which the state might be filled. As a result, the donor state is said to have a *degeneracy* g_D of 2. Typically, only one conduction band plays a dominant role in charge conduction, hence the Fermi-Dirac function is corrected for degeneracy, and the ionized donor concentration is

$$N_D^+ = N_D[1 - f_{\mathcal{FD}}(E_D)] = N_D - \frac{N_D}{1 + \dfrac{1}{g_D}\exp\left[\dfrac{E_D - E_F}{kT}\right]} = \frac{N_D}{1 + 2\exp\left[\dfrac{E_F - E_D}{kT}\right]}. \tag{12.133}$$

Recall that there are two active valence bands in cubic semiconductors; hence, there is a degeneracy of 2 for the spin states multiplied by a degeneracy of 2 for the overlapping heavy hole and light hole bands from which the electrons can fill the acceptor states. Typically, the hole degeneracy of a band-gap state, g_A, is assumed to be 4, thereby yielding

$$N_A^- = N_A f_{\mathcal{FD}}(E) = \frac{N_A}{1 + g_A \exp\left[\dfrac{E_A - E_F}{kT}\right]} = \frac{N_A}{1 + 4\exp\left[\dfrac{E_A - E_F}{kT}\right]}. \tag{12.134}$$

With these results one can now determine free carrier concentrations and conductivities for shallow dopant impurities.

Degeneracy values for deep levels in a band gap are far more difficult to determine, because impurity atoms can give rise to numerous energy states in a band gap. Typically, for deep donors, deep acceptors and traps, the ionized concentrations are written as

$$N_{DD}^+ = \frac{N_{DD}}{1 + g_{DD}\exp\left[\dfrac{E_F - E_{DD}}{kT}\right]}, \quad N_{AA}^- = \frac{N_{AA}}{1 + g_{AA}\exp\left[\dfrac{E_{AA} - E_F}{kT}\right]} \quad \text{and} \quad N_T^* = \frac{N_T}{1 + g_T \exp\left[\dfrac{E_F - E_T}{kT}\right]}.$$

where the subscripts DD and AA are used for deep donors and deep acceptors, respectively, and the subscript T is used for charge carrier traps. Depending on the lattice site into which an impurity atom falls, the impurity may act as a donor or an acceptor, complicating matters more. It is usual for a degeneracy of 1 to be assigned to deep donors, deep acceptors and traps; however, one should be aware that such an assumption could be seriously incorrect.

Shallow Dopant and Shallow Impurity Energy Levels

Atoms in a semiconductor crystal that are purposely introduced are commonly referred to as *dopant atoms*; whereas, those impurities that are inadvertently present as a consequence of the starting material or growth and handling processes are usually referred to as *background impurities*. In either case, the impurities can alter the electrical behavior of the host semiconductor, whether it is through changes in electrical conductivity, charge carrier mobility, or charge carrier mean free drift times. Ge, Si, and GaAs energy levels have been studied extensively, and energy levels associated with impurities and defects are shown in Tables 12.3, 12.4, and 12.5. Numerous other semiconductors used for radiation detectors have also been studied, yet the energy levels and associated impurities and defects are still under investigation.

A block of semiconductor material, whether it is intrinsic, n-type or p-type, must have the same number of negative and positive charges, because the material must have zero overall charge. This charge neutrality arises because any free electron with negative charge has an original host atom that became a positive ion, and for every free hole, there is an acceptor atom that became a negative ion. Additionally, every negative electron excited from the valence band to the conduction band leaves behind a positive hole in the valence band. This charge neutrality condition can be written, with the use of Poisson's equation, as

$$\nabla\mathcal{E} = \frac{\rho}{\kappa\epsilon_o} = \frac{q}{\kappa\epsilon_o}\left[p - n + N_D^+ - N_A^-\right] = 0, \tag{12.135}$$

Table 12.3. Energy levels (eV) reported as donors and acceptors for Ge (E_g = 0.66 eV). Donors are referenced from the conduction band edge ($E_C - E_D$) and acceptors are referenced from the valence band edge ($E_A - E_V$). Data are from [Milnes 1973] and [Sze 1981].

ID	n-type	p-type	ID	n-type	p-type
Al		0.01	In		0.011
Ag		0.13, 0.38, 0.57	Li	0.0093	
As	0.013		Mn		0.16, 0.37
Au	0.62	0.15, 0.46, 0.62	Ni		0.23, 0.36
B		0.01	O	0.04, 0.2	
Be		0.02, 0.06	P	0.012	
Cd		0.055, 0.16	Pt		0.04, 0.2, 0.54
Cr		0.07, 0.12	S	0.18	
Cu		0.04, 0.33, 0.4	Se	0.14, 0.28	
Co	0.36, 0.57	0.25	Sb	0.0096	
Fe		0.31, 0.39	Te	0.11, 0.3	
Ga		0.011	Tl		0.01
Hg		0.087, 0.23	Zn		0.035, 0.095

Table 12.4. Energy levels (eV) reported as donors and acceptors for Si (E_g = 1.12 eV). Donors are referenced from the conduction band edge ($E_C - E_D$) and acceptors are referenced from the valence band edge ($E_A - E_V$). Data from [Milnes 1973] and [Sze 1981].

ID	n-type	p-type	ID	n-type	p-type
Ag	0.79	0.76	Mn	0.43, 0.59	0.45
Al		0.067	Mo	0.33, 0.78, 0.82	
As	0.054		Na	0.77	
Au	0.83	0.58	Ni		0.23, 0.77
B		0.045	O	0.16, 0.51	0.41, 0.74
Ba	0.32	0.5	P	0.045	
Be		0.17, 0.42,	Pd		0.34
		0.87, 1.01	Pb	0.17	0.37
Bi	0.069		Pt	0.82	0.36, 0.87
C	0.25		S	0.26	0.48
Cd	0.45	0.55	Se	0.25, 0.4	
Co		0.35, 0.49, 0.59	Sb	0.039	
Cr	0.41		Si	0.63, 0.93	0.78
Cs	0.3	0.5	Sn	0.25	0.27
Cu		0.24, 0.4, 0.53	Sr	0.28, 0.62	
Fe	0.14, 0.51. 0.72		Ta	0.14, 0.43	
Ga		0.072	Te	0.14	
Ge	0.27, 0.62		Ti	0.21	
Hg	0.79, 0.87	0.76, 0.81	Tl		0.26
In		0.16	V	0.49	0.4
K	0.26, 0.77		W	0.22, 0.3, 0.37,	
Li	0.033			0.78, 0.81	
Mg	0.11, 0.25		Zn		0.26, 0.57

Table 12.5. Energies levels (eV) reported as donors and acceptors for GaAs ($E_g = 1.42$ eV). Donors are referenced from the conduction band edge ($E_C - E_D$) and acceptors are referenced from the valence band edge ($E_A - E_V$). Data are from [Milnes 1973], [Sze 1981], and [Pantelides 1992].

ID	n-type	p-type	ID	n-type	p-type
Ag		0.11	Mg		0.028
As_{Ga} (EL2)*	0.75, 1.0		Mn		0.095
Au		0.09	Ni		0.21
Be		0.028	O	0.4, 0.75**	
C		0.026	Pb		0.12
Cd		0.035	S	0.006	
Co		0.16	Se	0.006	
Cr		0.79	Si	0.006	0.03, 0.1, 0.22
Cu		0.023, 0.14, 0.19, 0.24, 0.44	Sn	0.006	0.17
			Te	0.03	
Fe		0.37, 0.52	V	0.22	
Ge	0.006	0.04, 0.07	Zn		0.031
Li		0.023, 0.05			

* EL2 means "energy level 2", associated with an intrinsic anti-site defect As_{Ga}.

** Has at times been confused with the EL2 level.

where n is the free electron concentration, p is the free hole concentration, N_D^+ is the ionized donor concentration and N_A^- is the ionized acceptor concentration. Substitution of Eq. (12.126), Eq. (12.128), Eq. (12.133), and Eq. (12.134) into Eq. (12.135) yields

$$N_V \exp\left[\frac{E_V - E_F}{kT}\right] - N_C \exp\left[\frac{E_F - E_C}{kT}\right] + \frac{N_D}{1 + 2\exp\left[\dfrac{E_F - E_D}{kT}\right]} - \frac{N_A}{1 + 4\exp\left[\dfrac{E_A - E_F}{kT}\right]} = 0. \quad (12.136)$$

For any specified temperature T, the concentrations, N_D^+, N_A^-, N_C, and N_V can be calculated. With these values p and n can be found as shown below. Finally, numerical iteration of Eq. (12.126) can be used to determine an exact value for the Fermi energy level E_F.

For the situation in which donors are the *majority* impurities, so that $N_D \gg N_A$, then with the use of Eq. (12.126),

$$n \simeq N_D^+ = \frac{N_D}{1 + g_D \exp\left[\dfrac{E_F - E_D}{kT}\right]} = \frac{N_D}{1 + g_D \left(\dfrac{n}{N_C}\right) \exp\left[\dfrac{E_C - E_D}{kT}\right]} = \frac{N_D}{1 + \dfrac{n}{N_{\xi d}}}, \quad (12.137)$$

where

$$N_{\xi d} = \left(\frac{N_C}{g_D}\right) \exp\left[-\frac{(E_C - E_D)}{kT}\right]. \quad (12.138)$$

Equating the first and last terms of Eq. (12.137) and with some algebraic rearrangement, one obtains

$$n^2 + nN_{\xi d} - N_{\xi d} N_D = 0, \quad (12.139)$$

whose solution for positive n is

$$n = -\frac{N_{\xi d}}{2} + \left[\frac{N_{\xi d}^2}{4} + N_{\xi d}N_D\right]^{1/2} = \frac{N_{\xi d}}{2}\left[\left(1 + \frac{4N_D}{N_{\xi d}}\right)^{1/2} - 1\right]. \tag{12.140}$$

In a similar fashion, it can be shown that when $N_A \gg N_D$,

$$p = \frac{N_{\xi a}}{2}\left[\left(1 + \frac{4N_A}{N_{\xi a}}\right)^{1/2} - 1\right], \quad \text{with} \quad N_{\xi a} = \left(\frac{N_V}{g_A}\right)\exp\left[-\frac{(E_A - E_V)}{kT}\right]. \tag{12.141}$$

Example 12.3: Suppose Si is doped with shallow donors at a concentration of $N_D = 10^{16}$ cm^{-3}. It is known that the donor site is 0.04 eV below the conduction band energy E_C. What is the electron free carrier concentration at 77 K and at 300 K?

Solution:
From Eq. (12.84) and the effective mass m_l^* and m_t^* values in Table 16.4 for Si at 4 K, the density of states effective mass at 77 K is estimated to be,

$$m_e^* \simeq 6^{2/3}(m_l^* m_t^{*2})^{1/3} = 6^{2/3}((0.9163)(0.1905)^2)^{1/3} = 1.062 m_e$$

where m_e is the mass of a free electron. Then from Eq. (12.120)

$$N_C = 2\left[\frac{2\pi m_e^* kT}{h^2}\right]^{3/2} = 2\left(\frac{2\pi(1.062 m_e k)(77\,\text{K})}{h^2}\right)^{3/2} = 3.57\times 10^{18}\ \text{cm}^3.$$

Substitution into Eq. (12.141) gives

$$N_{\xi d} = \frac{3.57\times 10^{18}\ \text{cm}^3}{2}\exp\left[\frac{-0.04\ \text{eV}}{(77\,\text{K})(8.617\times 10^{-5}\,\text{eV K}^{-1})}\right] = 4.3\times 10^{15}\ \text{cm}^3.$$

Finally, substitution of these values into Eq. (12.140) gives the free carrier concentration

$$n(77\ \text{K}) = \frac{4.3\times 10^{15}}{2}\left[\left(1 + \frac{4(10^{16}\ \text{cm}^{-3})}{4.3\times 10^{15}\ \text{cm}^3}\right)^{1/2} - 1\right] = 4.75\times 10^{15}\ \text{cm}^3.$$

From this result it is seen, at the cryogenic temperature of 77 K, 47.5% of the shallow donor atoms are ionized; hence, the material is still conductive and behaving extrinsically. If one wishes to design a detector with low free-carrier densities, as is needed for large-volume low-leakage current devices, it is often not enough to simply cryogenically cool the material; the semiconductor must also be purified so that only low concentrations of background impurities are present.

With the same calculations as above, the following results are obtained for $T = 300$ K: $m_e^* = 1.084 m_e$, $N_C = 2.83\times 10^{19}$ cm^{-3}, $N_{\xi d} = 3\times 10^{18}$ cm^{-3}, and $n(300\ \text{K}) = 9.97\times 10^{15}$ cm^{-3}. These values indicate that, at room temperature, almost all (99.7%) of the shallow donor states are ionized. Hence, at room temperature $N_D^+ \simeq N_D$.

The result in the above example can be used to simplify calculations for the free carrier concentrations. In the case that one dopant impurity concentration far exceeds the other type, it can be assumed that, at room temperature,

$$n \simeq N_D \quad \text{if} \quad N_D \gg N_A \qquad \text{or} \qquad p \simeq N_A \quad \text{if} \quad N_A \gg N_D. \tag{12.142}$$

For shallow dopants and impurities, the donor and acceptor states are almost completely ionized at room temperature. The free electron and hole concentrations, n and p, can be obtained from the condition of charge neutrality, i.e.,

$$p - n + N_D^+ - N_A^- = 0. \tag{12.143}$$

Then from the mass-action law of Eq. (12.131), $p = n_i^2/n$, and Eq. (12.143) can be rewritten as

$$n^2 - n(N_D^+ - N_A^-) - n_i^2 = 0. \tag{12.144}$$

This quadratic equation for n is readily solved for $n > 0$ as

$$n = \left(\frac{N_D^+ - N_A^-}{2}\right) + \left[\left(\frac{N_D^+ - N_A^-}{2}\right)^2 + n_i^2\right]^{1/2}. \tag{12.145}$$

In a similar fashion, the hole concentration is found to be

$$p = \left(\frac{N_A^- - N_D^+}{2}\right) + \left[\left(\frac{N_A^- - N_D^+}{2}\right)^2 + n_i^2\right]^{1/2}. \tag{12.146}$$

In the *extrinsic region*, where $N_D^+ \gg n_i$ then Eq. (12.145) reduces to

$$n = 2\left[\frac{N_D^+ - N_A^-}{2}\right] \simeq N_D^+ \simeq N_D \quad \text{when} \quad N_D \gg N_A. \tag{12.147}$$

Similarly, where $N_A^- \gg n_i$, Eq. (12.146) reduces to

$$p = 2\left[\frac{N_A^- - N_D^+}{2}\right] \simeq N_A^- \simeq N_A \quad \text{when} \quad N_A \gg N_D \quad . \tag{12.148}$$

In the *intrinsic region*, where $n_i \gg N_D^+$ or $n_i \gg N_A^-$, then Eq. (12.145) and Eq. (12.146) reduce to

$$n \simeq n_i \qquad \text{and} \qquad p \simeq n_i. \tag{12.149}$$

From the previous analysis, one can also determine the location of the Fermi level E_F. If $N_D = 0$ and $N_A = 0$ then $n = p = n_i$ and $E_F = E_i$. Then from Eqs. (12.126) and (12.128)

$$N_C \exp\left[\frac{E_F - E_C}{kT}\right] = N_V \exp\left[\frac{E_V - E_F}{kT}\right]. \tag{12.150}$$

Upon setting $E_F = E_i$ and taking the logarithm of both side of this equation, one obtains

$$\ln\left(\frac{N_C}{N_V}\right) = \frac{E_V + E_C - 2E_i}{kT}, \tag{12.151}$$

from which the intrinsic Fermi level is found to be

$$E_i = \left(\frac{E_V + E_C}{2}\right) + \frac{kT}{2}\ln\left(\frac{N_V}{N_C}\right). \tag{12.152}$$

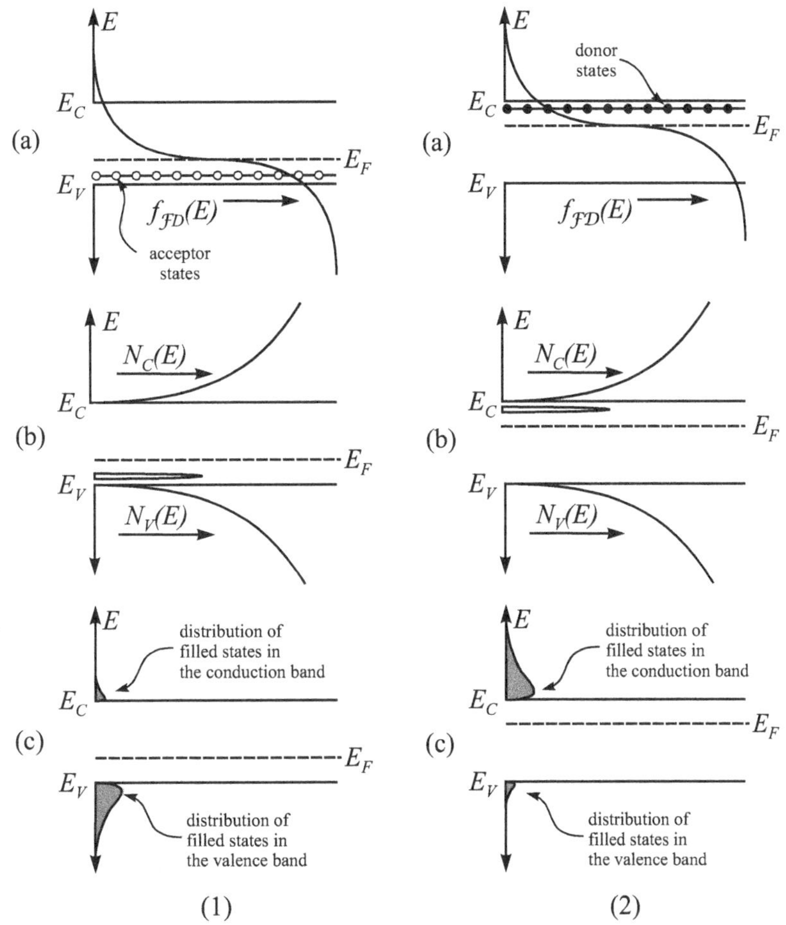

Figure 12.45. Shown are (a) the Fermi-Dirac distributions for extrinsic semiconductors, (b) the density of states functions, and (c) the distribution of free charge carriers, electrons and holes, in the conduction and valence bands, respectively. The distributions in column (1) are for p-type material and the distributions in column (2) are for n-type material.

From Eqs.(12.121) and (12.123), it is seen that $N_V/N_C = (m_h^*/m_e^*)^{3/2}$ so that

$$E_i = \left(\frac{E_V + E_C}{2}\right) + \frac{3kT}{4}\ln\left[\frac{m_h^*}{m_e^*}\right]. \tag{12.153}$$

In the *freeze-out region*, one can equate Eq. (12.126) and Eq. (12.140) and solve for E_F to obtain

$$E_F = E_C + kT\ln\left\{\frac{N_{\xi d}}{2N_C}\left[\left(1 + \frac{4N_D}{N_{\xi d}}\right)^{1/2} - 1\right]\right\}. \tag{12.154}$$

Similarly, from Eqs. (12.128) and (12.141), one finds

$$E_F = E_V - kT\ln\left\{\frac{N_{\xi a}}{2N_V}\left[\left(1 + \frac{4N_A}{N_{\xi a}}\right)^{1/2} - 1\right]\right\}. \tag{12.155}$$

Finally, in the *extrinsic region* one must first relate the intrinsic Fermi level E_i to the actual Fermi level. From Eq. (12.130) one has $N_C = n_i \exp[(E_C - E_i)/kT]$. Substitution of this result into Eq. (12.126) gives

$$n = n_i \exp\left[\frac{E_F - E_i}{kT}\right]. \tag{12.156}$$

Similarly from Eq. (12.130), it is seen that $N_V = n_i \exp[(E_i - E_V)/kT]$. Substitution into Eq. (12.128) then gives

$$p = n_i \exp\left[\frac{E_i - E_F}{kT}\right]. \tag{12.157}$$

The solution of these two equations for E_F in terms of n or p is

$$E_F = E_i + kT \ln(n/n_i) = E_i - kT \ln(p/n_i). \tag{12.158}$$

For the case $N_D \gg N_A$ and $N_D \gg n_i$, Eq. (12.158) reduces to

$$E_F = E_i + kT \ln(N_D/n_i), \tag{12.159}$$

and for the case $N_A \gg N_D$ and $N_A \gg n_i$

$$E_F = E_i - kT \ln(N_A/n_i). \tag{12.160}$$

The Fermi levels and charge carrier distributions for p-type and n-type materials are depicted in Fig. 12.45.

Deep Dopant, Impurity and Defect Energy Levels

Impurities and dopant atoms, as well as intrinsic defects such as antisites, vacancies, and interstitials, can introduce energy levels deep in the band gap of a semiconductor. Deep donors, acceptors, and traps are usually characterized as having an energy level greater than 0.1 eV from either band edge within the band gap. Unlike shallow energy levels, deep levels *should not* be assumed to be fully ionized at room temperature, and the Fermi-Dirac distribution function must be used to determine the fraction of deep states that are filled or empty.

Suppose a semiconductor exists in which there is only one deep level distributed throughout a material with a concentration of N_{DD} and a shallow acceptor level, N_A, of minimal concentration. The electron concentration is given by

$$n = \frac{N_C}{1 + \exp\left[\dfrac{E_C - E_F}{kT}\right]} \simeq N_C \exp\left[\frac{E_F - E_C}{kT}\right], \tag{12.161}$$

and the density of neutral deep donor sites is

$$N_{DD}^0 = \frac{N_{DD}}{1 + \dfrac{1}{g_{DD}} \exp\left[\dfrac{E_D - E_F}{kT}\right]}, \tag{12.162}$$

where N_{DD}^0 is the filled or *neutral* deep donor concentration. With the assumption that the shallow levels are fully ionized, and because the semiconductor total charge density must be neutral, the following balance relation holds

$$N_{DD} - N_{DD}^0 - N_A - n + p = 0. \tag{12.163}$$

The density of ionized deep donors is

$$N_{DD}^{+} = n + N_A - p = N_{DD} - N_{DD}^{0} = N_{DD}\left[1 - \frac{1}{1 + \frac{1}{g_{DD}}\exp\left[\frac{E_D - E_F}{kT}\right]}\right] = \frac{N_{DD}}{1 + g_{DD}\exp\left[\frac{E_F - E_D}{kT}\right]}. \tag{12.164}$$

To find the free electron concentration n, first consider the quantity $R \equiv nN_{DD}^{+}/(N_{DD} - N_{DD}^{+})$. Because the hole concentration is negligibly small, Eq. (12.163) gives

$$R = \frac{nN_{DD}^{+}}{N_{DD} - N_{DD}^{+}} = \frac{n(n + N_A)}{N_{DD} - n - N_A}. \tag{12.165}$$

From Eq. (12.161) and Eq. (12.164) one also finds

$$R = \frac{nN_{DD}^{+}}{N_{DD} - N_{DD}^{+}} = \frac{N_C}{g_{DD}}\exp\left[\frac{E_{DD} - E_C}{kT}\right]. \tag{12.166}$$

Equating these two expressions for R leads to a quadratic equation in n whose solution for positive n can be written as

$$n = 2(N_{DD} - N_A) \times \left[\left(1 + \frac{g_{DD}N_A}{N_C}\exp\left[\frac{E_C - E_{DD}}{kT}\right]\right) + \left\{\left(1 + \frac{g_{DD}N_A}{N_C}\exp\left[\frac{E_C - E_{DD}}{kT}\right]\right)^2 + \frac{4g_{DD}(N_{DD} - N_A)}{N_C}\exp\left[\frac{E_C - E_{DD}}{kT}\right]\right\}^{1/2}\right]^{-1} \tag{12.167}$$

Equation (12.167) can be simplified further by assuming N_A to be negligibly small such that $N_A \simeq 0$,

$$n \simeq 2N_{DD}\left[1 + \left(1 + \frac{4g_{DD}N_{DD}}{N_C}\exp\left[\frac{E_C - E_{DD}}{kT}\right]\right)^{1/2}\right]^{-1}. \tag{12.168}$$

Equation (12.168) provides a simple expression for the expected free electron concentration in the presence of a dominating deep donor level.

The opposite case is true if a deep acceptor is present with concentration N_{AA} and with a negligible number of shallow levels, either for N_D and N_A. Use of the same arguments used above to find n gives the hole concentration under the constraint that $N_{AA} \gg N_D$, namely,

$$p = 2(N_{AA} - N_D)\left[\left(1 + \frac{g_{AA}N_D}{N_V}\exp\left[\frac{E_{AA} - E_V}{kT}\right]\right) + \left\{\left(1 + \frac{g_{AA}N_D}{N_V}\exp\left[\frac{E_{AA} - E_V}{kT}\right]\right)^2 + \frac{4g_{AA}(N_{AA} - N_D)}{N_V}\exp\left[\frac{E_{AA} - E_V}{kT}\right]\right\}^{1/2}\right]^{-1} \tag{12.169}$$

Equation (12.169) can be simplified if it is assumed that N_A is negligibly small so that $N_A \simeq 0$,

$$p \simeq 2N_{AA}\left[1 + \left(1 + \frac{4g_A N_{AA}}{N_V}\exp\left[\frac{E_{AA} - E_V}{kT}\right]\right)^{1/2}\right]^{-1}. \tag{12.170}$$

Notice that the deep donor and acceptor impurities may not be completely ionized at room temperature (300 K); hence, the actual ionized concentrations must be calculated when determining the free carrier densities and material resistivities.

Deep Level Compensation

In the case of wide band-gap semiconductors, the free carrier concentration is typically low for the intrinsic material. Almost all wide band-gap semiconductors are compound materials with the exception of diamond which has a band gap of 5 eV. As a result, purification of wide band-gap semiconductors can be complicated. For example, Si and Ge are widely used semiconductors for radiation detectors. Both are materials composed of a single element and both can be prepared from a melt process in which a single crystal is grown by changing it directly from the liquid to solid state. Hence, both Ge and Si can be purified and processed as a single element. Zone refined Ge and Si has the impure regions removed after purification processes, remelted and grown into single crystals.

However, the case is not so simple for CdTe, which is a far more complex compound semiconductor. Both starting materials, Cd and Te, are typically purified separately. Afterwards, the purified materials must be reacted to form the compound CdTe, which unfortunately is never quite complete because there is always a small percentage of free Cd and free Te left in the molten material. The situation is further complicated by noting that segregation coefficients for some impurities in Cd may be negligible in Te. Hence, when the two are mixed, an impurity efficiently removed from one (Cd, for instance) may be introduced by the other element (Te in this example). Further, during growth, a Cd atom may fall into a Te location or vice versa, thus forming antisites that behave as electrically active impurities, a condition that does not occur for Si or Ge.

It is also important to be able to control the electrical properties of semiconductors that are to be used for radiation detectors. After purification, if the background doping levels remain too high, often impurities are added that act as *deep* impurities. Deep impurities have energy levels that fall deep in the band gap of a semiconductor, typically greater than 0.1 eV from the band edge. Often the most effective deep dopants have energy levels near the middle of the band gap.

Two-Level Model: In the case of two levels present in the semiconductor, for instance one shallow acceptor level and one deep donor level, a simplified compensation model can be developed for the free electron density in the conduction band [Lindquist and Ford 1982],

$$n \simeq \left(\frac{1}{x-1}\right) N_C \exp\frac{(E_A - E_C)}{kT}, \tag{12.171}$$

where $x = N_A/N_{DD}$.

Three-Level Model: Suppose one wants to compensate a material that has been unintentionally doped with p-type dopants from background contamination. In addition, suppose there is a residual concentration of n-type dopants as well, such that $N_A > N_D$. It is desired to add the correct amount of deep level donors to the semiconductor with the goal of making the material have a high resistivity. Hence, the goal is to force the Fermi energy level to the middle of the band gap, as shown in Fig. 12.46. For this example, the *three level model* is used in which $N_{DD} > N_A > N_D$ [Blanc and Weisberg 1960; Lindquist and Ford 1982]. Charge neutrality and Eq. (12.131) require

$$n = p + N_D^+ + N_{DD}^+ - N_A^- = \frac{n_i^2}{n} + N_D^+ + N_{DD}^+ - N_A^-. \tag{12.172}$$

At room temperature $N_D^+ \simeq N_D$ and $N_A^- \simeq N_A$. Hence, with the use of Eq. (12.162),

$$n = N_{DD}(1 - f(E_C - E_{DD})) - (N_A - N_D) + \frac{n_i^2}{n} = \frac{N_{DD}}{1 + g_{DD}\exp\left[\frac{E_F - E_{DD}}{kT}\right]} - (N_A - N_D) + \frac{n_i^2}{n}. \quad (12.173)$$

If it is assumed that $n \gg n_i$, the last term in the above result can be neglected. Then substitution of Eq. (12.126) into Eq. (12.173), with the last term set to zero, gives

$$(N_A - N_D) + n = \frac{N_C N_{DD}\exp\left[-\frac{E_C - E_{DD}}{kT}\right]}{n g_{DD} + N_C\exp\left[-\frac{E_C - E_{DD}}{kT}\right]}. \quad (12.174)$$

If the condition exists such that $N_A - N_D \gg n$, then

$$(N_A - N_D) \simeq \frac{N_C N_{DD}\exp\left[-\frac{E_C - E_{DD}}{kT}\right]}{n g_{DD} + N_C\exp\left[-\frac{E_C - E_{DD}}{kT}\right]}. \quad (12.175)$$

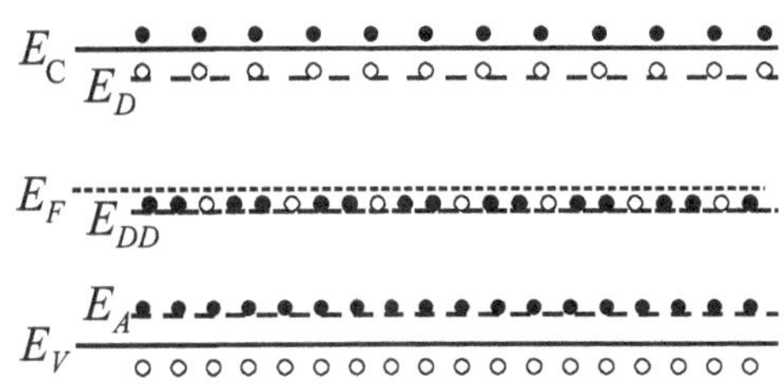

Figure 12.46. The basic three-level compensation model with two shallow levels and one deep level. Depicted is the case in which $N_{DD} > N_A > N_D$, in which the deep levels are only partially ionized. In such a case, the Fermi energy level is "*pinned*" near the middle of the band gap, thereby producing a relatively higher resistivity material.

The solution of this equation for ng_{DD} gives

$$n g_{DD} \simeq \frac{N_C(N_{DD} - 1)\exp\left[-\frac{E_C - E_{DD}}{kT}\right]}{(N_A - N_D)}. \quad (12.176)$$

Finally, if $N_{DD} \gg 1$, then the above result reduces to

$$n \simeq \frac{N_C N_{DD}\exp\left[-\frac{E_C - E_{DD}}{kT}\right]}{g_{DD}(N_A - N_D)}, \quad (12.177)$$

which gives an approximation for the expected free electron concentration in the conduction band for deep donor compensated material.

Using a similar argument, compensation with deep acceptor levels is achieved if $N_{AA} > N_D > N_A$. Through a similar, but opposite, system of equations, the resulting concentration of free holes is,

$$p \simeq \frac{N_V N_{AA}\exp\left[-\frac{E_{AA} - E_V}{kT}\right]}{g_{AA}(N_D - N_A)}. \quad (12.178)$$

Example 12.4:

The semiconductor GaAs is a moderately wide band-gap material (eV = 1.42 eV) that has been used for radiation detector development. Bulk grown material even in its purest form often has background impurities. Typically, GaAs is compensated by purposely incorporating excess As in the melt to give rise to the intrinsic anti-site defect, As_{Ga}, which produces a deep donor level in the material 0.75 eV below the conduction band edge (denoted EL2). Carbon, a shallow acceptor, appears as a background contaminant in concentrations near 10^{15} cm^{-3}, with shallow n-type materials appearing in concentration near 10^{14} cm^{-3}. What is the room temperature free electron concentration if the EL2 concentration is 10^{16} cm^{-3}?

Solution:

Substitution of given values into Eq. (12.177) with $g_{DD} = 2$ yields

$$n \simeq \frac{(4.7 \times 10^{17}\ \text{cm}^{-3})(10^{16}\ \text{cm}^{-3}) \exp\left[-\dfrac{0.75\ \text{eV}}{(300\ \text{K})(8.617 \times 10^{-5}\ \text{eV K}^{-1})}\right]}{2(10^{15}\ \text{cm}^{-3} - 10^{14}\ \text{cm}^{-3})} = 9.482 \times 10^{4}\ \text{cm}^{-3}.$$

With $\mu_e \simeq 8000$ cm^2 V^{-1} s^{-1}, this free electron concentration thus results in a resistivity from Eq. (12.91) of approximately $\rho = 8.24 \times 10^9$ ohm cm.

12.6 Summary

Solid state materials used for radiation detectors are generally composed of crystalline materials. A crystalline material is defined by a basis of atoms arranged upon a periodic lattice. There are 14 possible Bravais lattice systems. The periodic arrangement of atoms causes the appearance of periodic potentials. This potential periodicity causes bands of allowed states to form, producing quasi-continua of energy states in these bands. The density of allowed energy states in these bands is defined by the density of states function. Gaps between these bands are referred to as energy gaps. The energy band that plays the part of atomic bonding is the valence band, and the energy band that plays the part in electron conduction is the conduction band. The energy gap between the valence band and the conduction band is referred to as the band gap.

Energy bands have maxima and minima when defined by energy and crystal momentum in what are known E-k diagrams. Empty states in the valence band are treated as positive particles called “holes”. Because electrons need only lose energy to recombine with a hole in direct band gap semiconductors, the charge carriers have high recombination probabilities, thereby causing short charge carrier lifetimes. Electrons must change both momentum and energy to recombine with holes for semiconductors with indirect band gaps; hence, the charge carrier lifetimes are generally longer than observed for direct band-gap materials. Bands with sharp curvature d^2E/dk^2 cause the charge carrier effective mass for electrons or holes to decrease. Small effective mass increases the charge carrier mobility. The electrical behavior of semiconductors can be controlled by introducing dopant impurities. Deep dopants can be added into material to increase resistivity and reduce leakage currents in such a way that the material performs in a similar fashion as intrinsic material. The introduction of excess impurities or defect centers causes a reduction in charge carrier lifetime.

PROBLEMS

1. For the structure shown in Fig. 12.47, what are the Miller indices for the listed planes?

   (a) BSFT, (b) HMNP, (c) OTHO, (d) JVCW, (e) OKGR, (f) LTEL, (g) KSFO, (h) HTBP, (i) OTSO, (j) LFEL, (k) OKHP, (l) HFSP

2. For the structure shown in Fig. 12.47, what are the direction indices for the following directions?

   (a) OT, (b) LV, (c) SJ, (d) NK, (e) OJ, (f) LF, (g) SM, (h) NT

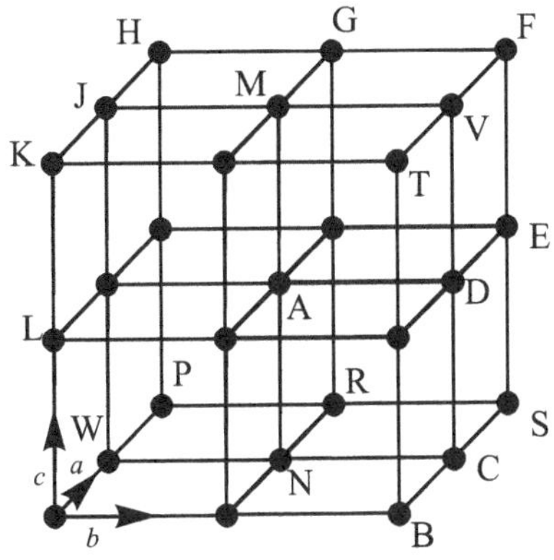

Figure 12.47. Lattice structure.

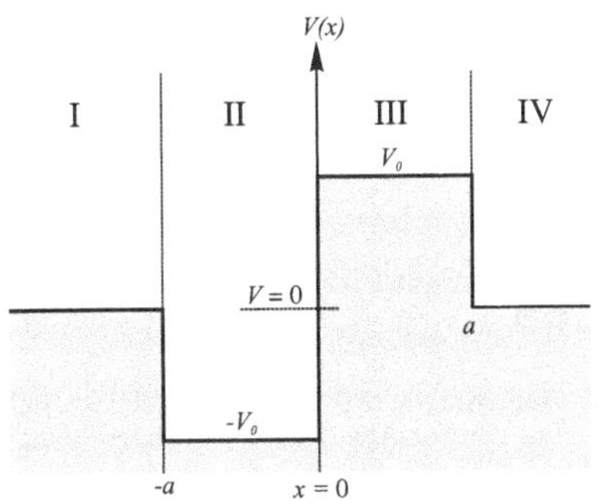

Figure 12.48. Potential well and barrier.

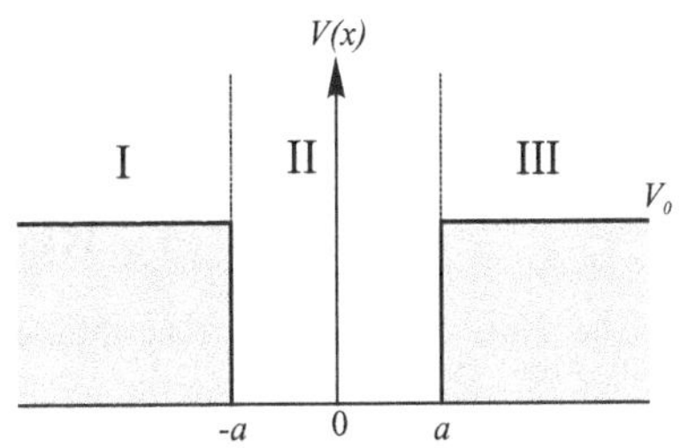

Figure 12.49. Potential well.

3. For a cubic unit cell, find the angles between the normals to the pairs of planes defined by the following Miller indices

(a) (100)(010), (b) (100)(111), (c) (100)(210), (d) (110)(210)

4. For the cubic lattice systems, determine the reciprocal lattice vectors $\mathbf{a}^*$, $\mathbf{b}^*$ and $\mathbf{c}^*$ for the FCC lattice and the BCC lattice.

5. For the potential shown in Fig. 12.48, solve for the reflection and transmission coefficients.

6. Quantum well cavities can be formed by growing consecutive layers of varying semiconductors that have different band-gap energies. For instance, a quantum well can be defined by epitaxially growing a layer of AlGaAs, with a band gap of 1.71 eV, followed by GaAs, with a band gap of 1.42 eV, followed by another AlGaAs layer. Typically, such films are only a few tens to hundreds of angstroms thick, and are grown on semiconductor buffer layers with protective contacts on top. Fig. 12.49 shows a square well where $a = 50$ Å, $V_0 = 0.3$ eV, and $m^* = 0.06\, m_0$. Find the following: (a) Find the allowed energies of the states in the well. (b) Find the wave functions and coefficient values. (c) Plot the wave functions for the allowed energies. (d) Determine the probabilities that an electron is located at the allowed energies.

7. Show that the solution to the Kronig-Penney model can be reduced to Eq. (3.58). Here is a hint to make matters simpler: Assume that the barrier is a delta function such that $\phi \gg k$ such that $d^2u/dx^2 \gg du/dx$ in the region, and simplify.

8. For a 2-dimensional square lattice, draw the first 4 Brillouin zones and clearly show their boundaries. Perform the same exercise for a 2-dimensional hexagonal lattice.

9. Determine the conductivity effective mass for cubic lattice L valley semiconductors by completing the example problem 12.2.

10. A silicon sample is doped with boron at a concentration of 5×10^{15} cm^{-3}. What is the expected resistivity of the material at a temperature of 194 K? What is it at a temperature of 300 K?

11. A silicon sample is doped with a concentration of 10^{16} cm^{-3} phosphorus atoms. At 300 K, what are the concentrations of free electrons and holes?

12. InP (E_g = 1.35 eV) is often doped with Fe ($E_C - E_{AA}$ = 0.66 eV) to increase resistivity. With $N_A = 3 \times 10^{14}$ cm^{-3}, $N_D = 2.5 \times 10^{15}$ cm^{-3}, and $N_{AA} = 1.5 \times 10^{16}$ cm^{-3}, what is the expected resistivity?

REFERENCES

BLAKEMORE, J.S., *Solid State Physics*, 2nd Ed., Cambridge: Cambridge University Press, 1985.

BLANC, J. AND L.R. WEISBERG, "Energy Level Model for High-Resistivity Gallium Arsenide," *Nature*, **192**, 155–156, (1960).

BLOCH, F., "Über die Quantenmechanik der Elektronen in Kristallgittern," *Z. Physik*, **52**, 555–600, (1928).

BRAVAIS, A., "Mémoire sur les Systèmes Formés par les Points Distribués Rguliérement sur un Plan ou Dans L'espace" *J. Ecole Polytech.*, **19**, 1-128, (1850).

BRILLOUIN, L., *Wave Propagation in Periodic Structures, 2nd. Ed.*, New York: Dover, 1953.

DE FARIA, J.L., Ed., *Historical Atlas of Crystallography*, Dordrecht: Kluwer Academic Publishers, 1990.

GASIOROWICZ, S., *Quantum Physics*, New York: Wiley, 1974.

KITTEL, C., *Introduction to Solid State Physics*, 6th Ed., New York: Wiley, 1986.

KITTEL, C., *Solid State Physics, A Short Course*, New York: Wiley, 1966.

KRONIG R. DE L. AND W.G. PENNEY, "Quantum Mechanics of Electrons in Crystal Lattices," *Proc. R. Soc. Lond. A*, **130**, 499–513, (1931).

LIMA DE FARIA, J., Ed., *Historical Atlas of Crystallography*, Dordrecht: Kluwer Academic Publishers, 1990.

LINDQUIST, P.F. AND W.M. FORD, "Semi-Insulating GaAs Substrates," Ch.1, *in GaAs FET Principles and Technology*, J. V. DILORENZO AND D.D. KHANDELWAL, Eds., DedHam: Artech House, 1982.

MCKELVEY, J.P., *Solid State and Semiconductor Physics*, New York: Academic Press, 1966.

MILLER, W.H., *A Treatise on Crystallography*, Cambridge: Pitt Press, 1839.

MILNES, A.G., *Deep Impurities in Semiconductors*, New York: Wiley, 1973.

NUSSBAUM, A., *Semiconductor Device Physics*, Englewood Cliffs: Prentice Hall, 1962.

PANTELIDES, S.T., *Deep Centers in Semiconductors*, 2nd Ed., Philadelphia: Gordon and Breach, 1992.

PIERRET, R.F., *Advanced Semiconductor Fundamentals*, Reading: Addison-Wesley, 1989.

SZE, S.M., *Physics of Semiconductor Devices*, 2nd Ed., New York: Wiley, 1981.

Chapter 13

Scintillation Detectors and Materials

In a few minutes there was no doubt about it. Rays were coming from the tube which had a luminescent effect upon the paper. I tried it successfully at greater and greater distances, even at two meters. It seemed at first a new kind of invisible light. It was clearly something new, something unrecorded.

Wilhelm C. Röntgen

13.1 Scintillation Detectors

Because Röntgen used a scintillating platinobarium cyanide plate to discover x rays in 1895, scintillation radiation detectors hold the distinction of being the first type of radiation detector ever invented. However, the emission of light was basically a qualitative measure of radiation interactions and not so much a quantitative measure. The fluorescence of the material indicated the presence of penetrating ionizing radiation, yet the actual intensity of radiation was not easily gauged.

Materials that scintillate are generally separated into two classes, namely inorganic and organic. The method by which either produces scintillation light is physically different, hence the distinction. Inorganic scintillators can be found as crystalline, polycrystalline, or microcrystalline materials. Organic scintillators come in many forms, including crystalline materials, plastics, liquids and even gases.

The scintillation principle is quite simple. Radiation interactions in a scintillator cause either the atomic or molecular structure in the scintillator to become excited such that electrons are increased in potential energy. These excited electrons then spontaneously de-excite with the decrease in potential energy emitted as radiant light energy. These light emissions can then be detected with light sensitive instrumentation.

Although simple in concept, there are certain attributes that a material must possess to be a useful scintillator. First, the material must be capable of absorbing the radiation of interest. Second, the subsequent energy released from the scintillator must be largely radiative, i.e., a large percentage of the absorbed energy is converted to photons. Third, the light emission must be spontaneous such that the scintillator *fluoresces* rather than *phosphoresces*.[1] Fourth, the scintillator must be transparent to its own scintillation light. Finally and fifth, the fluorescent light must be of a wavelength that can be detected with conventional

[1] Phosphorescence is a process in which light emission continues after the source of radiation is removed

light detecting systems, such as photomultiplier tubes or semiconductor photodiodes. There are many materials that scintillate, but only a select few have all of the necessary properties listed above. In fact, many scintillators were discovered and used in a limited sense for the first 50 years after Röntgen's discovery of x rays.

Many early discoveries depended on the observation of scintillation light. Wilhelm Röntgen used such light to discover x rays and Ernest Rutherford counted, through a microscope, flashes of light produced in a scintillating screen from alpha particle interactions in it.[2] Although scintillators were well known to be sensitive to radiation, their relatively low light yields made them difficult to use as practical detectors. Further, light detection devices needed to detect the scintillations were inadequate at the time. As a result, gas filled detectors dominated the detector industry prior to the 1940s.

An attempt to couple special Geiger-Müller counter tubes with scintillators is described by Krebs [1934, 1941] (also see [Daggs et al. 1952]), who used a thin window GM counter as a light detection device. A screen coated with zinc sulphide (ZnS) was the scintillator used by Krebs, and over the following years many other scintillators were also investigated by Krebs and other researchers [Mandeville and Scherb 1950, Krebs 1955]. However, GM tubes do not preserve energy deposition information, and these types of scintillation detectors served only as radiation counters.

In 1941, RCA introduced the first commercial *photo-multiplier tube* (PMT), a highly light-sensitive electron-amplification vacuum tube that could detect tiny amounts of visible light. In was the PMT that enabled scintillating materials to become practical radiation spectrometers. With improvements in stability and spectral response, eventually it was the PMT that became the light detector of choice for scintillators.

In 1944, Curran and Baker and later Broser and Kallman [1947a, 1947b] and Coltman and Marshall [1947] successfully used PMTs coupled to scintillators to measure alpha, beta, and gamma radiation. Broser and Kallman [1947b] and Deutsch [1948] used the organic scintillator naphthalene, because it was transparent to its own light emissions, to measure ionizing radiations.[3] In 1948 Robert Hofstadter discovered NaI:Tl, probably the most important scintillator in use over the last 65 years. The first efficient gamma-ray scintillation counters and spectrometers were demonstrated when a NaI:Tl crystal and a PMT were coupled together [Hofstadter 1948, 1949; McIntyre and Hofstadter, 1950; Hofstadter and McIntyre, 1950].

A typical scintillation spectrometer consists of a scintillating material hermetically sealed in an internally light-reflecting canister. Typical canisters are cylindrical, with an optically transparent window at one end of the cylinder and all remaining surfaces are Lambertian reflectors.[4] The optically transparent window is coupled to a light collection device, such as a PMT, with an optical compound. The optical compound helps match the indices of refraction between the scintillation canister and the light collection device so as to reduce reflective losses. The PMT provides a voltage output that is linear with respect to the light emitted from the scintillator. Hence, the voltage output "spectrum" is a linear indication of the radiation energy spectrum deposited in the detector. It is typical for commercial vendors to provide the scintillation canister and the PMT as one complete unit, although they can be acquired separately. Properties of commonly used inorganic scintillator materials are listed in Table 13.1.

13.2 Inorganic Scintillators

Inorganic scintillators depend primarily on the crystalline energy band structure of the material for the scintillation mechanism. In Fig. 13.1 an energy band diagram is shown for a typical inorganic scintillator.

[2]This fatiguing process led Hans Geiger and Rutherford to invent the Geiger gas-filled detector.

[3]Hofstadter notes in his 1975 review paper that "all nuclear physics laboratories smelled of mothballs" shortly after Broser and Kallman [1947b] published their work.

[4]A Lambertian surface, named after Johann Heinrich Lambert, has luminance independent of the angle of view. Typically these reflectors are white and *not* glossy or mirror-like.

Table 13.1. Widely used inorganic scintillator materials with some of their properties.

Scintillator	Density (g/cm^3)	Refractive Index (at λ_{max})	λ_{max} (nm)	Decay Time (ns)	Light Yield (photons per MeV)	Bi-Alkali PMT Response WRT NaI(Tl)	Reference
Alkali Metal Halides							
CsI - fast	4.51	1.95	315	16	2000	0.04–0.06	S
CsI - slow	4.51	1.95	450	1000	~200	ND	S
CsI:Na	4.51	1.84	420	460, 630, 1800	41000–49000	0.85	S, D
CsI:Tl	4.51	1.79	550	600, 3500	54000–61000	0.45	S, M
LiI:Eu	4.08	1.96	475	1400	15000	0.35	S2
NaI:Tl	3.67	1.82	415	230	43000	1.00	S
Alkali Earth Halides							
BaF_2 - fast	4.88	1.54	220	0.6–0.8	1800	0.03	S
BaF_2 - slow	4.88	1.5	310	630	10000	0.16	S
CaF_2:Eu	3.18	1.44	435	900	24000	0.50	S
CaI_2	3.96	1.8	410	550	86000	ND	H
CaI_2:Eu	3.96	1.8	470	790	86000	ND	H
$CaWO_4$	6.1	1.94	425	8000	~20000	0.3-0.5	Z
SrI_2:Eu	4.55	1.85	435	1200	115000	ND	C
Transition Metal Scintillators							
$CdWO_4$	7.90	2.30	470	14000	~1350	0.3–0.5	S
ZnS:Ag	4.09	2.36	450	110	50000	1.3	S
Post-Transition Metal Scintillators							
$Bi_4Ge_3O_{12}$	7.13	2.15	480	300	8200	0.13	S
Lanathanide Scintillators							
$CeBr_3$	5.2	2.09	371	17	68000	ND	R
Gd_2SiO_5	6.71	1.85	440	56, 400	9000	0.20	L
$LaCl_3$:Ce	3.86	2.05	350	28	49000	0.70–0.90	S
$LaBr_3$:Ce	5.29	2.05	380	16	63000	1.65	S
Elpasolites							
Cs_2LiYCl_6	3.31	1.81	305	6600	6535, 22420	ND	C2
Cs_2LiYCl_6:Ce	3.31	1.81	372, 400	600, 6000	9565, 18400	ND	C2
$Cs_2LiLaBr_6$:Ce	4.2	1.85	420	180, 1100	43000	1.15	S
$Cs_2LiLa(Br,Cl)_6$:Ce	4.08	1.9	420	120, 500, 1500	45000	~ 0.70	B2
Garnets, Perovskites, and Orthosilicates							
$LuAlO_3$:Ce	8.34	1.94	365	16.5, 74	11400	ND	M
$Lu_3Al_5O_{12}$:Pr	6.71	1.84	310	20-22	~18000	ND	N, D2
Lu_2SiO_5:Ce	7.4	1.82	420	47	25000	0.75	L
$Lu_{1.8}Y_{.2}SiO_5$:Ce	7.1	1.82	420	41	32000	0.75	S
$YAlO_3$:Ce	5.35	1.95	370	27	18000	0.45	B
$Y_3Al_5O_{12}$:Ce	4.55	1.82	550	88, 302	17000	0.50	S
Y_2SiO_5:Ce	4.45	1.8	420	56	24000	ND	L
Ceramic Scintillators							
Li glass:Ce	2.5	1.55	395	(n) 18, 57 (α) 16, 49 (β) 20, 58	~4650	~0.12	S1
Tb glass	3.0–3.52	1.5–1.62	550	3 ms–5 ms	13000–50000		P

B: Baryshevsky et al. 1991; B2: BNC 2020; C: Cherepy 2009; C2: Combes 1999; D: de Haas 2005; D2: Drozdowski 2008b; H: Hofstadter 1964a, 1964b; L: Leroq et al. 2005; M: Moszynski 1997a, 1997b; N: Nikl 2013; P: Pavan 1991; R: Shah 2005; R2: Shah 2017; S: Saint-Gobain 2016; S1: Scintacor 2016; S2: Syntfeld 2005; Z: Zdesenko 2005

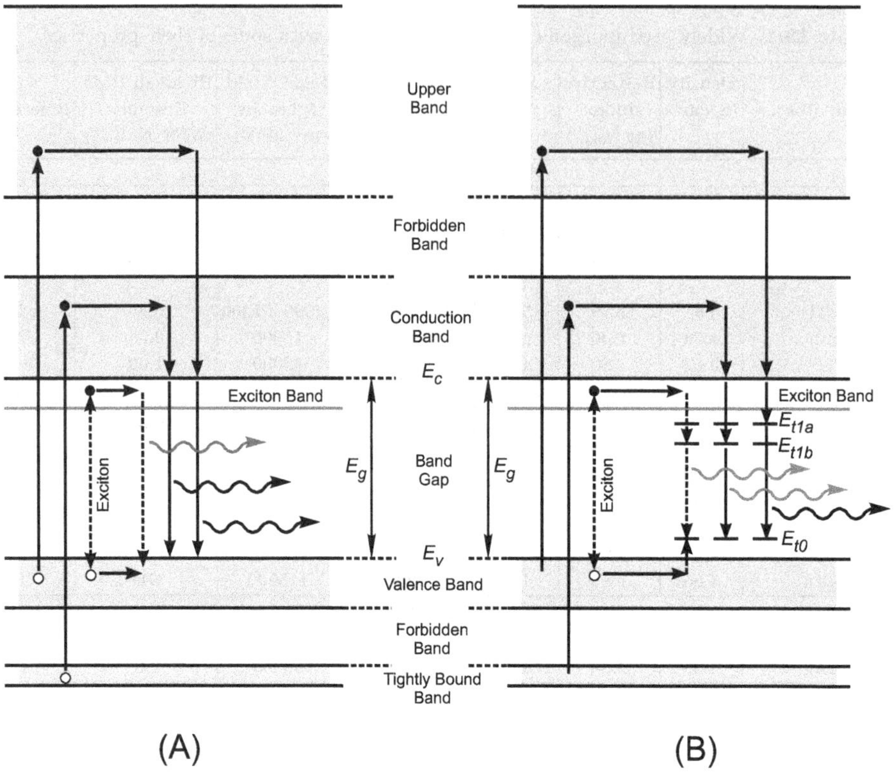

Figure 13.1. Shown are two basic methods by which an inorganic scintillator produces light, (A) is the intrinsic case and (B) is the extrinsic case.

A lower energy band, referred to as the valence energy band, has a reservoir of electrons. It is this band of electrons that participates in the binding of atoms. The next higher band is usually referred to as the conduction band, which for inorganic scintillators is usually devoid of electrons. Between the two bands is a forbidden region where electrons are not allowed to exist, typically referred to as the energy band gap.

If a radiation particle, such as a gamma ray or charged particle, interacts in the scintillation material, it can excite numerous electrons from the valence band and the tightly bound bands up into the conduction bands as shown in Fig. 13.1(A). These electrons rapidly lose energy and fall to the conduction band edge E_C. As they de-excite and drop back into the valence band, they can lose energy through light emissions. Unfortunately, because the radiated energy of the photons is equivalent to the band-gap energy, these same photons can be reabsorbed in the scintillator and again excite electrons into the conduction band. Hence, the scintillator usually is opaque to its own light emissions. There are exceptions in which intrinsic scintillators work well. For example, bismuth germanate (BGO) releases light through optical transitions of Bi^{+3} ions, which release light that is lower in energy than the band gap, hence is relatively transparent to its own light emissions.

However, if an *impurity* or *dopant* is added to the crystal, it can produce allowed states in the band gap, as depicted in Fig. 13.1(B). Such a scintillator is referred to as *activated.* In the best of cases, the impurity atoms are uniformly distributed throughout the scintillator. When electrons are excited by a radiation event, they migrate through the crystal and many subsequently drop into the excited state of the impurity atom. Upon de-excitation, a photon is produced equal in energy to the difference between the impurity atom's excited and ground states. Hence, it is unlikely to be reabsorbed by the scintillator material. Careful selection of the proper impurity *dopant* can allow the light emission wavelength to be tailored specifically to match the sensitivity of the light collection device.

13.2.1 Theory of Scintillation for Inorganic Scintillators

The scintillation process is described elsewhere as consisting of five distinctive stages [Rodyi 1997], namely

1. Radiation absorption and subsequent excitation of electron-hole pairs.
2. Relaxation of primary electron-hole pairs.
3. Thermalization of the low-energy of charge carriers.
4. Energy transfer to luminescent centers.
5. Photon emission from the luminescent centers.

It is difficult to observe the separate processes of stages 1 through 3 and stages 4 and 5. Instead, these processes are usually categorized as electron-hole production followed by photon emission. But, for the sake of completeness, all five stages are presented here.

Already presented in Chapter 4 are the basic processes by which radiation interacts in matter, and they are the same for scintillators. An ionizing particle excites multiple electron-hole pairs (see Chapter 12) by multiple ionizing processes including photoionization, elastic and inelastic scattering, Coulombic scattering, fission, and neutron transmutation. If the discussion is restricted to photon interactions, then the probability of exciting a K-shell electron is generally higher than exciting electrons from outer shells (L or M). The process by which the holes relax is either from radiative transitions, in which an outer shell electron falls into the empty state by releasing a photon (a characteristic x ray), or by a non-radiative transition (i.e., Auger electron emission). These subsequent emissions can be reabsorbed by outer shell electrons of nearby atoms, thereby producing more electron-hole pairs.

Meanwhile, the liberated energetic electrons travel through the crystal lattice and lose energy through Coulombic interactions and, thus, liberate more electron-hole pairs. The process continues until the resultant electrons and holes no longer have sufficient energy to produce more ionization. Overall, electrons and holes move to their respective band edges, electrons at the conduction band edge and holes at the valence band edge, within about 10^{-15} to 10^{-13} seconds.

Excited electrons still bound to their respective holes through Coulombic forces are also formed at this stage. This condition causes the electron-hole pair to have slightly lower energy than a free electron-hole pair. Thus, the pair exists at an energy below the band gap energy and can be represented as a temporary energy band perturbation traditionally called the exciton band, through which the electron-hole pair diffuse together as an electrically neutral quasiparticle called an *exciton*.

The intrinsic case (A) has no added activator dopants. Absorbed radiation energy excites electrons from the valence and tightly bound bands up into the higher energy conduction bands. These electrons quickly de-excite to the lowest conduction band edge E_C. As they drop to back to the valence band, they release light photons. Some intrinsic scintillators emit light through optical transitions from ionized elemental constituents, which can be of lower energy than the band gap. The extrinsic case (B) has activator dopants that produce energy levels in the band gap. Absorbed radiation energy excites electrons from the valence and tightly bound bands up into the higher conduction bands. These electrons quickly de-excite to the lowest conduction band edge E_C as in case (A). However, many drop into the upper activator site energy state E_{t1}. As they then drop to the activator ground state E_{t0}, they release light photons of lower energy than the scintillator band gap.

Emission Spectra

The emission spectrum is largely affected by the band-gap energy of the scintillator and the energy levels of the activator. Moreover, the emission spectrum is usually spread over a continuous range of energies

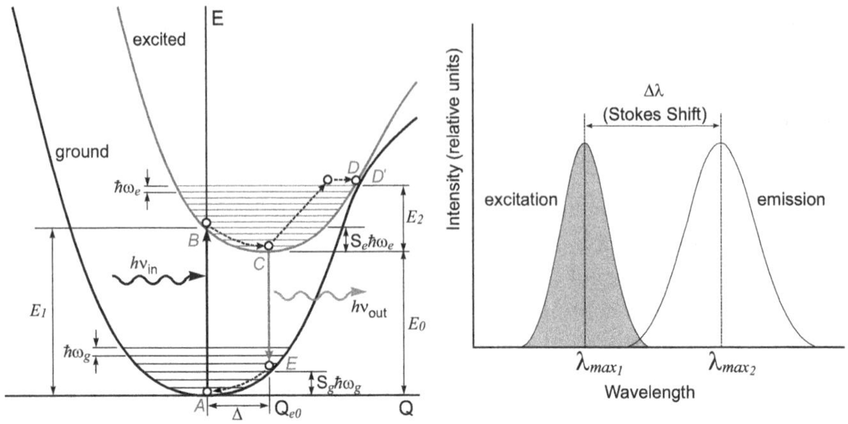

Figure 13.2. Configuration coordinate diagram for a luminescent center depicting luminescence with a Stokes shift and also a form of non-radiative de-excitation. The ordinate is energy E and the abscissa is the average distance Q between the luminescent center and the surrounding ions.

rather than concentrated at discrete energies. As shown in Fig. 13.2, an electron can be excited from the lowest energy state of a fluorescent center by a photon, it may capture a free electron, or it may capture an exciton to produce an excited electron at B. When produced by the absorption of a photon, this electron excitation is a vertical transition and, according to the Franck-Condon principle, occurs nearly instantaneously (femtoseconds). Because electrons are significantly less massive than the surrounding nuclei, the electron transition probabilities can be calculated with the assumption that the nuclear positions appear stationary during the transition. However, after the transition, thermal motion and the relocation of electron spatial positions cause vibrational states to appear, represented as horizontal lines in Fig. 13.2.

In inorganic scintillators, the photon assisted transitions are broadened, which, in turn, results in a widening of the luminescent bands [Rodnyi 1997]. If the luminescent center and its nearest ion neighbor can be modeled by Hooke's Law,[5] then the total energy of the ground state can be modeled as [Yamamoto 2007],

$$U_g = k_g \frac{Q^2}{2} \tag{13.1}$$

and the total energy of the excited state is,

$$U_e = k_e \frac{(Q - Q_{e0})^2}{2} + E_0, \tag{13.2}$$

where k_g and k_e are force constants of the chemical bonds, Q_{e0} is the interatomic distance when the ground state is at equilibrium, and E_0 is the total energy at $Q = Q_{e0}$. At location B, the total energy can be represented by

$$U_e = k_e \frac{(\Delta)^2}{2} + E_0. \tag{13.3}$$

where $\Delta = Q_A - Q_{e0} = -Q_{e0}$.

Elevation of an electron from the ground state to an excited state moves the electron into a different orbital. Consequently, spatial occupancy and the electron wavefunction changes with this transition from

[5]In classical physics, Hooke's Law describes the stored energy in a spring as a function of displacement, $U = kx^2/2$.

the ground state to the excited state. The electron also affects its neighboring atoms differently in its excited state, resulting in a change in the chemical binding (and force constant k), and the position of lowest energy changes from A to Q_{e0} (shown in Fig. 13.2). Further, the potential energy curve of the excited state is wider, thereby allowing the ground state to come close or overlap the excited state potential energy curve at point $D \to D'$.

An excited electron moves from point A to point B. This excited electron rapidly loses energy and moves to location C. It can now drop directly to the ground state at location E, thereby releasing a photon. The electron then loses energy, by non-radiative processes, to once again fall to location A. Ground state transitions are possible from the several vibrational energy states in the excited state. Consequently, there is usually a wide range of energies contributing to the emission spectrum, which appears as a continuous broad spectrum rather than discrete energy emissions.

The excitation energy ($A \to B$) is larger than the emission energy ($C \to E$). This change causes the resultant absorption spectrum of a luminescent center to be different than the emission spectrum because the absorption spectrum is centered about shorter wavelengths than is the centroid of the emission wavelengths. This shift in wavelength is known as the *Stokes shift*,

$$\Delta\lambda = \lambda_{max_{em}} - \lambda_{max_{ab}}, \tag{13.4}$$

where the subscripts em and ab indicate emission or absorption, respectively. If the Stokes shift is adequately large, photon emissions have too little energy to be reabsorbed by the scintillation crystal or other luminescent centers.

The vibrational states can be modeled as a harmonic oscillator bound by the energy wells depicted in Fig. 13.2, in which discrete energy states appear within the energy wells. The solutions give rise to vibrational energy states separated by the phonon energy $\hbar\omega$,[6] where ω is the vibrational mode angular frequency [Rodyni 1997], with even and odd wave functions for both of the wells,

$$E_{n_{g,e}} = (n + \frac{1}{2})\hbar\omega_{g,e}, \quad n = 0, 1, 2, \ldots, \tag{13.5}$$

where the g and e subscripts indicate either the ground or excited potential wells.

In Fig. 13.2, the case is shown in which the number of vibrational states involved with transition $A \to B \to C$, denoted by S_e, is different than the number of vibrational states involved with transition $C \to E \to A$, denoted by S_g. The energy and number of phonons involved with the relaxation from $B \to C$ are

$$S_e\hbar\omega_e = k_e\frac{(Q_{e0} - Q_A)^2}{2}. \tag{13.6}$$

In a similar fashion,

$$S_g\hbar\omega_g = k_g\frac{(Q_A - Q_{e0})^2}{2}. \tag{13.7}$$

Note that in Fig. 13.2, $Q_A = 0$. Suppose that the vibrational frequencies and force constants are the same for the ground and excited states, such that $k_g = k_e$ and $\omega_g = \omega_e$, then $S_e = S_g$ and the total energy difference between the transition $A \to B$ and the transition $C \to E$ is

$$\Delta E = \hbar\omega(S_g + S_e) = 2S\hbar\omega, \tag{13.8}$$

[6] Here $\hbar \equiv h/(2\pi)$ and is widely used to avoid repeated factors of 2π in quantum calculations.

where S is the number of phonons involved with the absorption or emission of photons, named the Huang-Rhys factor S [Huang and Rhys 1950]. The Stokes shift becomes,

$$\Delta\lambda = \frac{4\pi S\omega c\hbar^2}{E_o^2 - (2\hbar\omega)^2}. \tag{13.9}$$

At a given temperature T, the transition probability for phonon absorption is [Yamamoto 2007],

$$W_{ab}(\hbar\omega) = \frac{1}{\sqrt{2\pi}\sigma_{ab}} \exp\left[\frac{-(\hbar\omega - E_1)^2}{2\sigma_{ab}^2}\right], \tag{13.10}$$

where k is Boltzmann's constant,

$$E_1 = E_0 + \frac{k_e Q_{e0}^2}{2}, \qquad \text{and} \qquad [\sigma_{ab}(T)]^2 \approx 2S_e kT\frac{(\hbar\omega_e)^3}{(\hbar\omega_g)^2}. \tag{13.11}$$

Here the spectral absorption width is defined as the FWHM of the Gaussian distribution, $w_{ab} = 2.355\sigma_{ab}$.[7] For emission, the terms become [Ridley 1982],

$$W_{em}(\hbar\omega) = \frac{1}{\sqrt{2\pi}\sigma_{em}} \exp\left[\frac{-(\hbar\omega - E_3)^2}{2\sigma_{em}^2}\right], \tag{13.12}$$

where

$$E_3 = E_0 - \frac{k_g Q_{e0}^2}{2} = E_0 - S_g\hbar\omega_g \qquad \text{and} \qquad [\sigma_{em}(T)]^2 \approx 2S_g kT\frac{(\hbar\omega_g)^3}{(\hbar\omega_e)^2}. \tag{13.13}$$

If $\omega_e \approx \omega_g$, then the variance as a function of temperature is,

$$[\sigma_{em}(T)]^2 \approx 2S_g kT\hbar\omega, \tag{13.14}$$

which defines the spectral broadening of the emission spectrum, $w_{em} = 2.355\sigma_{em}$. From Eq. (13.14), it is observed that the variance of the emission spectrum increases with S_g, ω, and the absolute temperature T.

It is important that the choice of scintillator matches well to the choice of light sensing device. Perhaps the most common device employed to detect the relatively low light emissions from a scintillation is a photomultiplier tube, or PMT (covered in the following chapter). In Fig. 13.3, there are numerous emission spectra from common scintillator materials along with two comparison sensitivity responses for two types of PMTs. There is at least one characteristic that must be considered when coupling a PMT to a scintillator, mainly, that the spectra match well. For example, the emission spectrum from thallium doped NaI (or NaI:Tl) matches well to both the S-11 and the K-bialkali PMT responses. However, neither the responses from the S-11 or K-bialkali PMT match to the CsI:Tl emission spectrum.

Often the potential energy curves depicted in Fig. 13.2 either intersect or come into close proximity along the Q direction, represented as points D and D'. With thermal agitation, the electron can move to the ground state without releasing a photon. This process can be understood with the assistance of Fig. 13.2, in which electrons are excited from the ground state of a luminescent center to an excited state ($A \rightarrow B$). The electron quickly moves to the lowest potential energy of the excited state C. If the excited electron acquires thermal energy equivalent to E_2, it can be raised to point D and transfer to point D', thereby allowing the

[7] The spectral absorption width is defined here as the FWHM of the Gaussian distribution, or $2\sqrt{2\ln(2)}\sigma_{ab}$. However, the spectral width has been defined elsewhere as w where $W_{ab}(E_1)/q = W_{ab}(E_1 + w)$. Hence, under this alternative definition, $w_{ab} = \sqrt{2}\sigma_{ab}$ [Yamamoto 2007].

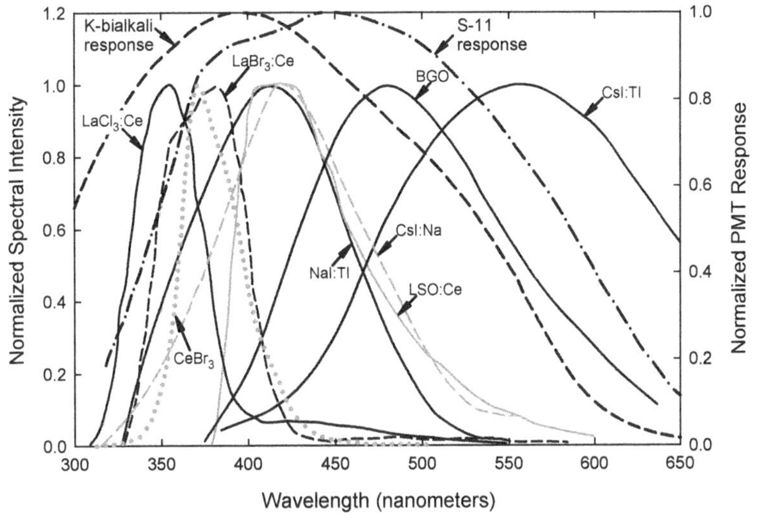

Figure 13.3. The emission spectra from several popular inorganic scintillators. Also shown is an overlaying sensitivity comparison to the normalized responses of two common photomultipliers (K-bialkali and S-11). Scintillation spectral response data from [Lindow et al. 1978, Shah et al. 2005; Valais et al. 2007, and Saint-Gobain 2016]. PMT response data from [Engstrom 1980].

electron to pass directly from the excited state into the ground state without releasing a photon. The excess energy is lost to thermal lattice vibrations (phonons). Ultimately, these electrons are lost to the luminescent process and do not contribute to the fluorescent light yield. The loss of these electrons, acting as information carriers, and the consequential light reduction is called *quenching.*

There are other competing processes that reduce the fraction of electrons contributing to the overall fluorescent scintillation spectrum. Electrons may fall into energy states that produce metastable states, whose transitions to the ground state are forbidden. With some additional energy, these electrons may move to higher allowed energy states and, thereby complete the radiative process. For instance, thermal energy from the ambient environment may be enough to cause these transitions. If these delayed transitions take a relatively long time, beyond several decay times for fluorescent photons, then the radiant decay is called *phosphorescence* or *afterglow*. Phosphorescence becomes a form of background noise in scintillation detectors. These afterglow light emissions can be detected, but the time delay for the emission decouples its correlation to the radiation event that caused the initial electron excitations. Note that some light-emitting materials use phosphorescent decay for benefit, such as the thermoluminescent detectors (TLDs) discussed in Chapter 19.

13.2.2 General Properties of Inorganic Scintillators

Price [1964] suggests that the charge measured at the output of the light collection device be represented by,

$$Q = Mqn_e, \tag{13.15}$$

where M is the multiplication factor of the light collection device (see Chapter 14), q is the electron charge, and

$$n_e = E_n F_n C_{np} T_p F_p S_m f F_c, \tag{13.16}$$

where E_n is the energy of the radiation quantum, F_n is the fraction of energy deposited in the scintillation medium, C_{np} is the conversion efficiency of energy to fluorescent light, T_p is the transparency of the scintillator to its own light emission spectrum, and F_p is the fraction of light photons that reach the light collection sensor. The terms $S_m f F_c$ pertain to the conversion efficiency of the light sensor, where S_m is the light to photoelectron conversion, f is a factor that corrects for spectral mismatch of the light output to the light sensor photoresponse, and F_c the photoelectron collection efficiency. Equation 13.15 can be rewritten as,

$$Q = q(MS_m f F_c)(E_n F_n C_{np})(T_p F_p) = qN_e L_p L_c. \tag{13.17}$$

The term N_e (charge production) is mostly addressed in Chapter 14, whereas the terms L_p (light production) and L_c (light collection) are addressed in this chapter.

Radiation Absorption Efficiency

The radiation absorption efficiency is a strong function of the atomic number and density of the material constituents. For gamma-ray detection, the three most important radiation interactions in a scintillator are photoelectric absorption, Compton scattering, and pair-production. For these three interactions, high Z values produce more efficient results than low Z values. There are some materials that have one moderately high Z constituent and one or more low Z constituents. Yet it is still the concentration of high Z elements that largely determines the absorption coefficient. For instance, the absorption coefficient for CsI (55/53) is only slightly higher than that for NaI (11/53) by about a factor of two or less.

Photoelectric absorption edges should also be considered when selecting a scintillator material for low energy gamma-ray detectors. Within some energy regions, a scintillator with a relatively large Z constituent may actually have lower absorption than another scintillator with lower Z constituents. For instance, the highest K edge for LSO (lutetium orthosilicate) is located at 63.3 keV below which its absorption coefficient drops from 66.87 cm^{-1} to 12.86 cm^{-1}. By comparison, the highest K edge for NaI is located at 33.16 keV. Consequently, the gamma-ray absorption efficiency is actually higher for NaI than LSO between 33.16 keV and 63.3 keV.

The exponential attenuation of uncollided gamma rays entering a scintillating crystal is given by Eq. (4.4), namely

$$I(x) = I_o e^{-\mu_t(E)x}, \tag{13.18}$$

where I_o is the initial gamma-ray intensity intersecting the material surface and x is the distance into the crystal material. Recall from Chapter 4 that the total interaction coefficient is given by

$$\mu_t(E) = \sum \mu_i(E) \approx \mu_{pe} + \mu_{cs} + \mu_{pp} \tag{13.19}$$

where the subscripts pe, cs, and pp refer to photoelectric, Compton scattering and pair-production, respectively. Note that other gamma-ray interactions are possible, as described in Chapter 4, but are less important to general gamma-ray detection. In the detector community, the ratio $I(x)/I_o$ is called the *interaction efficiency* and is the probability a photon interacts in some manner in the material while traveling a distance x in it.

Note the interaction efficiency describes the probability of an interaction, and *not* the amount of energy absorbed. A gamma ray absorbed through the photoelectric effect passes its energy (minus the electron binding energy) to a photoelectron. As the ionized atom de-excites, bound electrons fall into lower energy levels giving off x rays. These x rays may be reabsorbed in the medium from electrons in higher energy levels (L and M shells), or may in fact escape the material altogether. Gamma rays that interact through Compton scattering do not deposit their total energy. In fact, the fractional amount of energy deposited per scatter decreases as the gamma-ray energy increases. These concepts are described in detail in Chapter 4 and their consequences for radiation detection in Chapter 20. In summary, the absorption efficiency describes the probability of a photon interaction and *not* the probability that gamma rays are fully absorbed.

For other radiations, lower Z materials may be of more importance. For instance, the backscatter probability increases with Z for both α and β particles. Thus for a high Z material these radiations have a good chance of reflecting off the material rather than entering the detector and being detected. Hence, scintillators used for charged particle detectors are usually fabricated from low Z materials, such as organic scintillators. For fast neutrons, low Z materials facilitate higher energy loss per collision. Hence, organic scintillators, such as plastic scintillators, are usually used for fast neutron detectors. When doped with reactive materials with high absorption cross sections in the thermal region, these scintillators can be used for slow neutron detection. There are special scintillators with constituent materials that offer excellent thermal neutron absorption, such as lithium iodide (LiI) and gadolinium oxysulfide (Gd_2O_2S).[8]

Light Yield

Light yield is the number of fluorescent photons produced per energy unit absorbed in a scintillating medium. The *absolute* light yield is the total number of fluorescent photons released per unit absorbed energy, typically reported as photons per MeV. The absolute light yield is usually reported for a specific energy, such as 662 keV (^{137}Cs), 835 keV (^{54}Mn), or 1.836 MeV (^{88}Y), although in some cases a multitude of gamma-ray energies may have been used for the calibration (see for example [Holl et al. 1988]). For statistical reasons, a high absolute light yield is desirable for good gamma-ray energy resolution. In other words, the brighter the scintillator, the better the energy resolution. The absolute light yields of several common inorganic scintillators are listed in Table 13.1.

The *relative* light yield is the number of photons emitted per unit energy at different energies, reported as normalized at a specific energy. Recent publications have used energies near 450 keV as the normalization energy [Mengesha et al. 1998, Swiderski et al. 2009a, Payne et al. 2011]. The relative light yield is a measure of the fluorescent response linearity to absorbed energy. Ideally, the relative light yield should be constant; however, several physical factors can affect the light yield of scintillators, especially in the lower energy region below 500 keV [Jaffe et al. 2007].

Although the cause of non-linear light yield is not fully understood, progress has been made in recent years with predictive modeling. Discussions on different models that elucidate the non-linearity of light emission can be found in the literature [Payne et al. 2009, Payne et al. 2011, Moses et al. 2012]. An important factor affecting non-proportionality is the ionization density produced by an interacting charged particle (electron). From the Bethe-Bloche expression for electron energy loss in matter (see Chapter 4), the ionization density increases as the electron velocity (or energy) decreases. For most scintillators, there is a notable decrease in relative light yield in the low energy region, as shown in Fig. 13.4 and Fig. 13.5, which indicates that light yield for these scintillators decreases as ionization density increases. However, for several alkali-metal/halogen scintillators, there is a notable decrease in light yield also in the high energy region (see Fig. 13.4), an indication that the relative light yield is decreasing for low ionization densities. This ionization density is dependent upon the initial excitation energy of the ejected electron produced by Compton scattering, pair production, or the photoelectric effect. Further, the deexcitation processes, radiative or non-radiative, are also dependent on the incident particle energy. Over the past several decades, there have been numerous models proposed to predict or explain the relative light yield of inorganic scintillators, three of which are described here [Moses et al. 2012].

The "minimalist model" uses numerous simplifying assumptions to arrive at a manageable model. Foremost, it is assumed that luminous emissions are produced solely from exciton recombination at luminous defect centers. Further, it is assumed that free electrons and holes do not contribute to light emissions,

[8] Also known as GOS or Gadox.

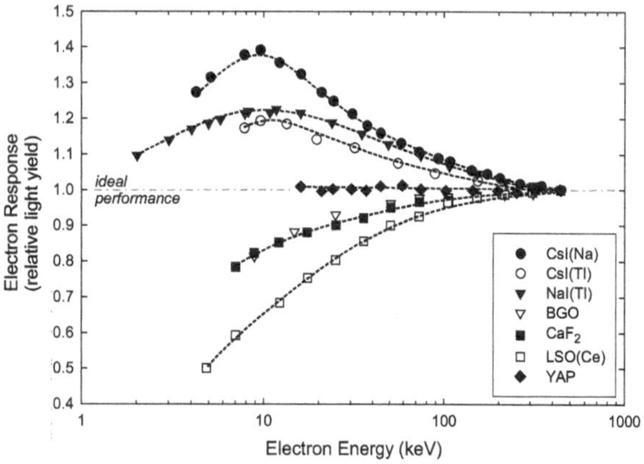

Figure 13.4. Relative light yield for several inorganic scintillating materials, all normalized at 440 keV. Data are from [Mengesha et al. 1998].

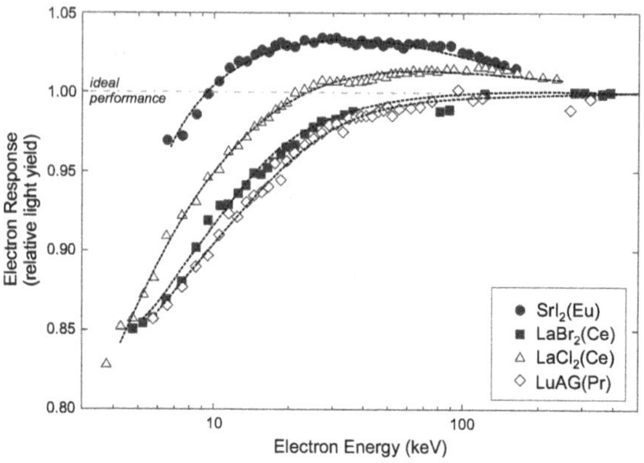

Figure 13.5. Relative light yield for several inorganic scintillating materials, normalized at 450 keV. Data are from [Swiderski et al. 2009a, Payne et al. 2011].

but instead recombine non-radiatively. It would, therefore, be the loss of excitons to competing processes that causes a reduction in relative light yield. The minimalist model seeks to explain this phenomenon as losses due to collisions between excitons in which one exciton pair is changed into an electron-hole pair that de-excites by non-radiative transitions but the other is ultimately preserved. The probability of this event increases with the square of the charge carrier density and, hence, becomes dominant for low energy photon interactions. Also assumed is that at higher energies, the electron and hole densities are less and are also spatially separated. Only those electrons and holes within the *Onsager* radius R_O can form mobile excitons.[9] Based on the non-linear behavior described by Birks [1964], the light yield thus becomes,

$$L = \frac{1 - \eta_{eh} \exp\left[-\dfrac{dE/dx}{(dE/dx)_{R_O}}\right]}{1 + kB\dfrac{dE}{dx}}, \tag{13.20}$$

where η_{eh} is the fraction of electrons and holes that do not form excitons, $(dE/dx)_O$ is the energy loss where the average spatial separation between electrons and holes is R_O, and kB is the quenching parameter described by Birks [1964]. Payne et al. [2011] successfully apply Eq. (13.20) to effectively model the response of numerous scintillator materials. Unfortunately, the minimalist model fails to predict relative light yield from first principles, and is more akin to an empirical data curve fitting method.

The "kinetic model" includes phenomena that are known to occur and affect scintillation yield, but are disregarded in the minimalist model. These effects include radiative emission from trap assisted electron-hole recombination and the reduction of charge carrier density as recombination progresses. Through either a rigorous system of mathematical rate equations, or through Monte-Carlo techniques, these kinetic models attempt to track changes in the electron-hole-exciton cloud as a function of time. The analytical techniques can be complicated, and attempt to include all processes that affect scintillation light yield. The stated advantage is that this modeling method provides much insight into the scintillation mechanism of a material. The disadvantage is the complexity involved with the model.

[9]The Onsager radius R_O is the physical distance between an electron and hole at which the electro-static energy is equal to the thermal energy, i.e., $q^2/(\epsilon R_O) = kT$, where ϵ is the dielectric constant and q is the unit electronic charge. At $R < R_O$, the electron (or hole) cannot escape from the Coulombic force field. [Payne et al. 2009].

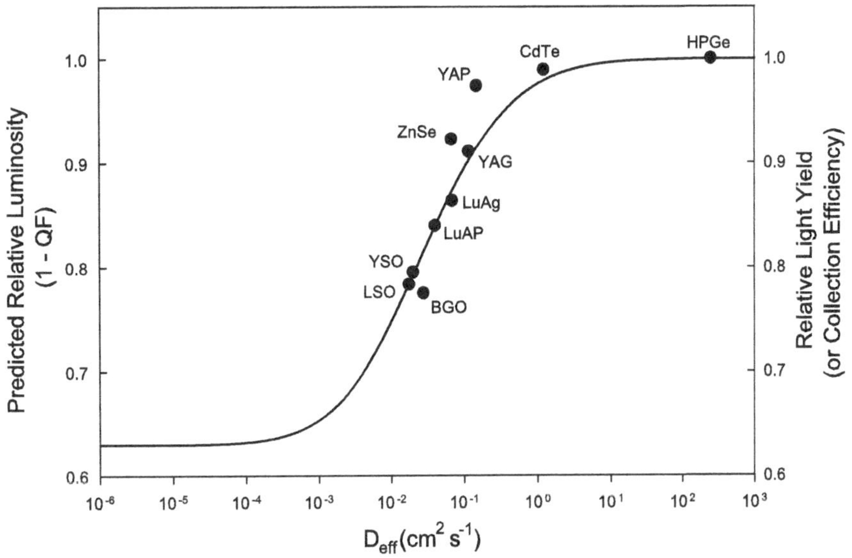

Figure 13.6. Measured light yield from several common scintillators compared to predicted results from a transport model. The responses are shown as functions of diffusion coefficients. The term (1 - QF) is the fraction of carriers that survive at least 10 ps after the charge cloud is formed. CdTe and HPGe are semiconductors, and the measured metric is the charge carrier collection efficiency. Data are from [Li et al. 2011].

The "diffusion model" includes the effect of electron and hole mobility, as well as ambipolar diffusion of excitons. Electrostatic forces and thermal diffusion affect the density of electrons and hole in a charge cloud. Recombination is a strong function of the product np and the trap density (covered in Chapter 15). As electrons and holes separate from diffusion, the probability of direct or trap assisted recombination decreases. Excitons are treated as neutral particles, affected only by thermal diffusion. The overall result can be used to modify the kinetic model. This carrier transport method has at least three major implications [Moses et al. 2012]. First, if one charge carrier type diffuses much faster than the other, for instance $\mu_e \gg \mu_h$, then these differing charge carriers become physically separated and, thus, significantly decreases recombination. Second, for materials with similar electron and hole diffusion coefficients, electrons and holes diffuse with similar spatial distributions, thereby producing a high light yield by efficient recombination. Third, charge carriers (and excitons) with high mobilities rapidly separate, thereby diminishing the carrier density and reducing light loss from quenching. Consequently, it is expected that scintillation materials with relatively high charge carrier mobilities have proportional light output, while the opposite is true for materials with low charge carrier mobilities. The diffusion model has been successfully used to predict luminosity from several common scintillators [Li et al. 2011] as shown in Fig. 13.6.

A scintillator is said to have a *linear response* if the light yield scales as

$$\langle N \rangle = kE + b \tag{13.21}$$

where $\langle N \rangle$ is the mean number of information carriers, E is the absorbed radiation energy, k is a proportionality constant, and b is a constant. For the case of scintillators, $\langle N \rangle$ is the average number of photons produced per event at energy E. In such a case, the scaling of a scintillator is predictable, but the average energy needed to produce a photon is a function of energy and is not constant ($w = E/\langle N \rangle \neq$ constant). If the constant b is zero, the scintillator is said to have a *proportional response*, in which case $w = E/\langle N \rangle =$ constant. Regardless of the value of b, the differential change in information carriers per differential change in energy is constant, or $d\langle N \rangle/dE = 1/k =$ constant for both linear and proportional responses.

Non-linear (or non-proportional) light yield response as a function of energy complicates energy calibration [Jaffe et al. 2007]. Ideally, the relative light yield (photons/MeV) should be constant, or at least proportional, for all energies. Consequently, pulse height spectral features do not appear in channels predicted with a linear formula. For instance, the relative light yield for NaI deviates from 1 (at 662 keV) to 1.25 (at 10 keV), whereas YAP has minimal deviation in relative light yield as shown in Fig. 13.4. With predictive modeling, a calibration can be developed to provide a reasonably accurate scaled response.

Example 13.1: You are using a NaI:Tl scintillator to measure gamma rays from different sources having energies of 59.5 keV and 662 keV. If system gain puts the 662 keV gamma ray at channel 1000, in what channel does the 59.5 keV gamma-ray center appear? Which channel should it appear in if the relative light yield was constant at all energies?

Solution:

Although the 59.5-keV gamma ray is lower in energy than the 662-keV gamma ray, it produces more light per unit energy. From Fig. 13.4, the relative light yield at 60 keV is about 1.15. To find the energy per channel:

$$E/Ch = 662 \text{ keV}/1000 = 0.662 \text{ keV/channel.}$$

The channel in which the 59.5-keV gamma ray appears, corrected for the increase in light yield from 1.00 to 1.15,

$$Ch = (1.15)(59.5 \text{ keV})/(0.662 \text{ keV/ch}) = \text{channel } 104.$$

rounded to the nearest channel. The 59.5-keV gamma ray should appear in channel 90 if the relative light yield was constant.

Large deviations in relative light yield as a function of energy also adversely affect the energy resolution by causing further statistical spread in the photon yield [Valentine et al. 1998]. Two factors identified by Valentine et al. that contribute to energy resolution broadening are (1) multiple interaction events resulting in one or more energetic primary electrons prior to photoelectric absorption and (2) the resulting electron cascade sequence that follows photoelectric absorption which produces Auger electrons and x-ray emissions. For instance, gamma rays of energy near 662 keV interact most often by Compton scattering in a medium and, hence, deposit only a fraction of their energy in the scintillator at a Compton scatter location. For each subsequent interaction, be it another Compton scatter event or photoelectric absorption, a different amount of energy is deposited in the scintillator until either the photon is absorbed through the photoelectric effect or the photon escapes the detector. Multiple events occur as the empty energy states created by Compton scattering of photoelectric absorption are filled, which includes a cascade of Auger electrons and x-ray emissions. These emissions can also be re-absorbed in the scintillation material. The overall light yield is the sum from all of these events, starting from the initial Compton scatter interaction, all of which produce an energetic electron of different initial energy. Because these energetic initial electrons are of different energies, they also produce a different relative light yield. The variance in light yield about the average light produced from this process must be added to the statistical variance in the average number of electron-hole pairs (and excitons) produced. For ideally proportional scintillators, this variance contribution to the light yield reduces to zero, although the fluctuations in the number of charges created from Poisson statistics remains.

Example 13.2: You are using a NaI:Tl scintillator to measure gamma rays from ^{137}Cs with gamma-ray emissions of 662 keV. Suppose that one gamma ray undergoes photoelectric absorption ejecting a photoelectron from the K shell of iodine. A second gamma ray produces a Compton electron of 150 keV, a second Compton electron of 200 keV, and is then captured by the photoelectric effect. Determine the difference in relative light yield and absolute light yield for these two events.

Solution:

The K-shell binding energy of iodine is 33.164 keV. Therefore, the first photoelectron is ejected with an energy of

$$662 \text{ keV} - 33.164 \text{ keV} = 628.84 \text{ keV}.$$

From Fig. 13.4, the relative light yield for the first gamma-ray absorption correlates to a relative light yield of 1.0.

The remaining energy is preserved by numerous cascades of electrons through atomic energy levels, but assume, to a first approximation, that a characteristic x ray from a $K\alpha_I$ emission is produced, which subsequently ejects a L_{II} electron. Note that a variety of transitions are possible, but the differences in these energy transitions are small. Hence,

$$28.61 \text{ keV(at } K\alpha_I) - 4.856 \text{ keV(binding energy at } L_{II}) = 23.75 \text{ keV}.$$

From Fig. 13.4, the relative light yield for the 23.75-keV Auger electron is about 1.2. The remaining energy (approximately 9.41 keV) is absorbed through various Auger electron transitions. Because two L electrons are missing, two allowed transitions from the M shell are chosen, namely $L\alpha_1$ and the $L\beta_1$. Auger electrons can be produced if these x rays are absorbed by the M shell. There are numerous possibilities, but we choose M_{II} or M_{III} electrons, then free electrons of energy 3.06 keV and 3.29 keV, respectively. From Fig. 13.4, these emissions would produce a relative light yield of approximately 1.15 each. The remaining 3.06 keV energy is regained by electrons relaxing into the M shell, producing free electrons with energies on the order of 0.50 keV to 0.88 keV. Regardless of the transitions, extrapolation from Fig. 13.4 gives a relative light yield near 1.0. If the relative light yield is tabulated in terms of fractions, then

$$\text{RLY} = \frac{628.64 \text{ keV}}{662 \text{ keV}}(1.0) + \frac{23.75 \text{ keV}}{662 \text{ keV}}(1.2) + \frac{3.06 \text{ keV} + 3.29 \text{ keV}}{662 \text{ keV}}(1.15) + \frac{3.06 \text{ keV}}{662 \text{ keV}}(1.0) = 1.0083.$$

The expected absolute light yield is,

$$43,000 \text{ photons/MeV} \left(\frac{1 \text{ MeV}}{1000 \text{ keV}}\right)(662 \text{ keV})(1.0083) = 28,702 \text{ photons}.$$

The same analysis can be done for the Compton scattered photons. The main difference is the ejection of a 150-keV Compton electron (relative light yield of 1.06), a 200-keV Compton electron (relative light yield of 1.03), and the subsequent absorption of the 312-keV Compton scattering photon, which produces a 278.8-keV photoelectron (relative light yield of 1.01). With a similar Auger electron cascade scheme, one arrives at the total relative light yield of 1.0345. The expected absolute light yield is,

$$43,000 \text{ photons/MeV} \left(\frac{1 \text{ MeV}}{1000 \text{ keV}}\right)(662 \text{ keV})(1.0345) = 29,448 \text{ photons}.$$

The total percent difference between the single photoelectric absorption and the Compton scattered example is

$$(100)\frac{1.0345 - 1.0083}{1.0083} = 2.6\%.$$

Decay Time and Pulse Formation

Suppose a radiation particle interacts in a material and excites N^* electron-hole pairs. Of these N^* electrons, an average fraction of electrons fall into luminescent centers, represented by $N = fN^*$. The rate at which these free electrons fall into luminescent centers can be represented as a decay rate,

$$\frac{dN(t)}{dt} = -\lambda_r N(t) = -\frac{N(t)}{\tau_r}, \tag{13.22}$$

where $N(t)$ is the population of free electrons that eventually fall into luminescent centers, λ_r is the probability, per unit differential time, that an electron falls into a luminescent center from its mobile state and τ_r is the mean decay time in which the subscript r designates 'rise'. The rate at which electrons, once in luminescent centers, decay and release a photon is given by

$$\frac{dN_c(t)}{dt} = -\lambda_d N_c(t) = -\frac{N_c(t)}{\tau_d}, \tag{13.23}$$

where λ_d is the probability, per unit differential time, that an electron falls into a luminescent center from its mobile state and τ_d is the mean decay time. Here the subscript d designates 'decay'. Hence, the rate of change in luminescent center population, immediately after a radiation interaction event, is

$$\frac{dN_c(t)}{dt} = \frac{N(t)}{\tau_r} - \frac{N_c(t)}{\tau_d}. \tag{13.24}$$

From the solution of this equation, the luminescent center population immediately after a radiation interaction is,

$$N_c(t) = N\frac{\tau_d}{\tau_r - \tau_d}\left(e^{-t/\tau_r} - e^{-t/\tau_d}\right). \tag{13.25}$$

The light emission rate, in photons per unit time, is expressed as a function of the decay time,

$$L(t) = \frac{N_c(t)}{\tau_d} = \frac{N}{\tau_r - \tau_d}\left(e^{-t/\tau_r} - e^{-t/\tau_d}\right). \tag{13.26}$$

Note that if the population 'rise time' is extremely short ($\leq$ a few ns), which it usually is, then Eq. (13.26) reduces to,

$$L(t) \simeq \frac{N}{\tau_d}\left(e^{-t/\tau_d}\right). \tag{13.27}$$

The solution of Eq. (13.26) is derived with the assumption that a single decay time can adequately represent the light emission rate of a scintillation material. The decay times listed in Table 13.1 are average values obtained from an experiment. Many luminescent states may play a part in the emission sequence. Often these decay times for separate energy levels are similar enough that an average decay time adequately defines the light emission behavior. However, there are many scintillators that have multiple decay times that are notably different. Some also have different emission wavelengths. As seen from Table 13.1, CsI:Na, CsI:Tl, and BaF_2 are examples of scintillators with multiple decay times. In fact, there is a significant difference in the decay times and the maximum wavelengths λ_{max} for BaF_2 emissions.

Differences in decay time constants can be used to distinguish between particles, energies, and in some circumstances the particle mass. One example is the phoswich[10] detector, where two different scintillators with different time constants are laid one upon the other, and read out by a single light detection device. By distinguishing between the times constants, the operator can discern in which detector the interactions occurred. Designs and uses of this configuration are discussed in Chapter 20.

[10] Phosphor sandwich.

Energy Resolution

The energy resolution of a radiation spectrometer is defined as the full width of the spectroscopic energy peak (usually ΔE) at half the maximum intensity (*FWHM*). The energy resolution from scintillators is traditionally reported in terms of percent of the most probable energy (the mode) in the pulse height spectrum,

$$R_{\%} = 100\left(\frac{\text{FWHM}}{E_{max}}\right). \tag{13.28}$$

The minimum value of the FWHM can be modeled, to a first approximation, with Poisson statistics, i.e.

$$\text{FWHM} = 2\sqrt{2\ln(2)\text{var}(E)}, \tag{13.29}$$

where $\text{var}(E)$ is the variance of the full energy peak in the pulse height spectrum. A FWHM is depicted in the example gamma-ray pulse height spectrum of Fig. 6.18, taken with a NaI:Tl detector. If the scintillator is coupled with a PMT light detector (Ch. 14), the light collected is converted to an electron signal. Hence, with the assumption that electronic noise from other electronic components is minimal, the variance in the number of electrons collected by the PMT is a measure of the variance of the pulse height spectrum, or $\sqrt{\text{var}(E)}/E_{max} \approx \sqrt{\text{var}(Q)}/Q_{max}$. The fractional variance in electrons collected by the PMT can be expressed as [Birks 1964]

$$v(Q) = \frac{var(Q)}{\bar{Q}^2} \approx \left[\frac{var(N)}{\bar{N}^2} - \frac{1}{\bar{N}}\right] + \frac{var(p)}{\bar{p}^2} + \frac{\bar{G}^2 + var(G)}{\bar{G}^2\bar{N}\bar{p}}, \tag{13.30}$$

where $\text{var}(N)$ is the variance in the number of scintillation photons, $\bar{N}$ is the average number of scintillation photons, $\text{var}(p)$ is the transfer variance, $\bar{p}$ is the average transfer efficiency, $\bar{G}$ is the PMT gain, and $\text{var}(G)$ is the variance in the PMT gain (an intrinsic characteristic of the PMT). The transfer efficiency, and associated variance, are factors that affect the efficiency of how well photons produce photoelectrons, which may include scintillation wavelength matching to the PMT response, coupling efficiency between the scintillator and the PMT window, internal reflection, index of refraction mismatch between the scintillator and the PMT window, photocathode nonuniformities, and photoelectron collection efficiency [Dorenbos et al. 1995].[11] For the case of scintillation light yield based on Poisson statistics, the variance and the average value are equal, i.e.,

$$\text{var}(N) = \bar{N}, \tag{13.31}$$

and the bracketed expression in Eq. (13.30) reduces to zero. However, there are other intrinsic factors that can affect the energy resolution. The main two factors are (1) non-proportional light yield, discussed previously, and (2) non-uniformities in the luminescent center (activators) distribution. The overall energy resolution measured from the scintillator detectors becomes,

$$R^2 = R_i^2 + R_p^2 + R_G^2 \tag{13.32}$$

where R_i is the intrinsic light yield resolution of the scintillation crystal, R_p is the transfer resolution, and R_G is the PMT limiting resolution. Although there are methods to independently measure the effects of R_i and R_p, these contributions to the energy resolution are often combined and defined as the detector resolution $R_d^2 = R_i^2 + R_p^2$. For an ideal detector perfectly coupled to a PMT, $R_d = 0$ and $\bar{p} = 1$; hence Eq. (13.30) reduces to,

$$v(Q) \approx \frac{1 + \text{var}(G)/\bar{G}^2}{\bar{N}}. \tag{13.33}$$

[11]These factors affecting PMT performance are explained in Chapter 14.

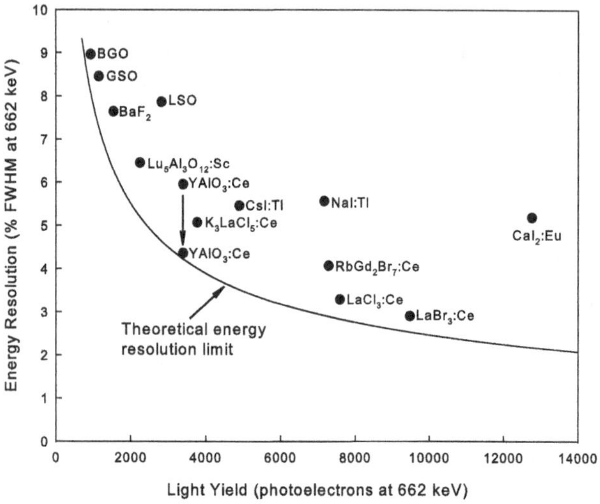

Figure 13.7. Measured energy resolution for several inorganic scintillating materials. Also shown is the theoretical limiting energy resolution. Data are from [Dorenbos 2002].

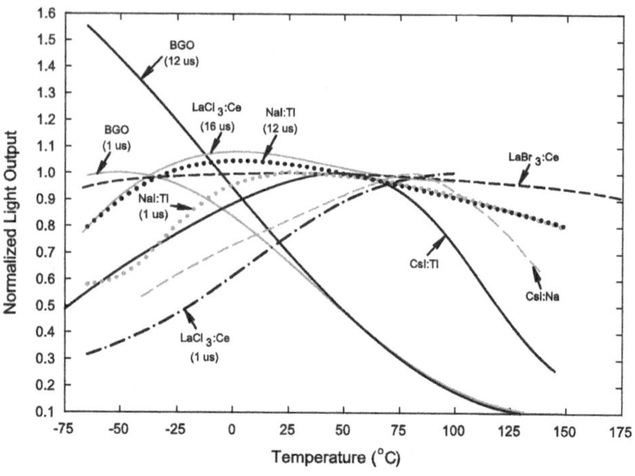

Figure 13.8. The light output, normalized at the maximum for $\tau = 1\mu s$, as a function of temperature for several common inorganic scintillators. The light output data for NaI:Tl, BGO, and $LaCl_3$ were normalized for data measured with a 1 μs shaping time, and comparison data is provided at longer shaping times. Data are from [Lindow et al. 1978, Saint-Gobain 2016].

Values for $\text{var}(G)/\bar{G}^2$ are usually between 0.1 to 0.2 [Dorenbos et al. 2002]. Hence, the ideal best energy resolution of a scintillator as a function of the number of photoelectrons produced becomes,

$$\%\ \text{FWHM} \approx 2\sqrt{2\ln(2)}\sqrt{\frac{1+0.1}{\bar{N}}} = 2.355\sqrt{\frac{1+0.1}{\bar{N}}}. \tag{13.34}$$

This ideal energy resolution is plotted as a solid line in Fig. 13.7, and is compared to measured values of many common inorganic scintillators. From Fig. 13.7, it is clear that almost all of the scintillators represented fall short of the predicted ideal energy resolution. Factors that broaden the energy resolution limit include non-proportional light yield, poor reflective properties of the crystal encapsulate, crystal transparency, crystal imperfections, and mismatch between the indices of refraction at the scintillator and light collection window interface. It is noteworthy that many of those scintillators shown to have large deviations from the predicted resolution also have significantly non-proportional light yield in the low energy region (< 200 keV), an indication of the importance of proportional light yield on scintillator performance. Energy resolution can sometimes be improved with better crystal synthesis and production methods. For instance, the energy resolution of the material $YAlO_3$:Ce (YAP) was reportedly reduced from 6% to 4.4% by improving the crystals (compare results of [Kapusta et al. 1999] to [Moszynski et al. 1998]).

Thermal Properties

From Fig. 13.2 and the explanation offered in the section on emission spectra, scintillation light yield at high temperatures (> 100°C) generally result in a reduction in relative light yield. One cause of the reduction is the increased numbers of excited electrons lost from direct transfer from the excited state to the ground state [Melcher 1989]. Another source of a decrease in light yield is the increase in $\sigma_e(T)$ with increasing temperature as can be seen from Eq. (13.14). This increase in $\sigma_e(T)$ spreads the light emission over more wavelengths and, consequently, reduces the maximum yield at λ_{max}. This reduction increases photon reabsorption from crystal defects as a result of increased wavelength mismatch between the light emissions and the light sensor (PMT, photodiode, etc.).

There is also a notable decrease in light yield for many scintillators at low temperatures, particularly at temperatures below room temperature. This decrease in relative light yield can be explained, at least in some circumstances, as a consequence of lengthening the primary decay constant. If the time constant of a light sensor is not lengthened to match the relative increase in τ, then an increased number of luminescent photons are not recorded. From Fig. 13.8, the light output of many common scintillators appears to decrease with lower temperature; however, changing the integration time of the light sensing system works to include previously lost photons, thereby increasing the relative light yield. Examples of such scintillators include BGO (1 μs $\rightarrow$ 12 μs), NaI:Tl (1 μs $\rightarrow$ 12 μs), and $LaCl_3$ (1 μs $\rightarrow$ 16 μs). Scintillators that have relatively flat thermal response to light yield over a broad range of energies include $LaBr_3$ and BaF_2 (220 nm emission only).

Material and Optical Properties

Important material properties for scintillators include fracture resistance, atomic number, mass density, radiation absorption length, and moisture resistance. The atomic number and material density affect the radiation absorption length. For gamma-ray detection, materials with at least one high Z component are best. Further, a high atomic density of the high Z constituent assists with photon absorption. For this fundamental reason, scintillators with high Z are sought for a variety of gamma-ray detection applications, including medical imaging, oil well logging, and gamma-ray spectroscopy. However, because of backscattering effects, low Z materials are best used for alpha-particle and beta-particle detection, and scintillators in which all constituents are low Z are best applied.

The fracture resistance is a measure of a scintillation crystal's ability to withstand shock. Many of the alkali-metal/halide and lanthanide/halide scintillators are soft and extremely fragile and crack under modest exposure to shock. For instance, NaI and CsI are both listed as '2' on the Moh hardness scale,[12] an indicator of resistance to fracture and plastic deformation. NaI is fragile and cracks easily, while CsI is pliable and can withstand more shock. Packaging improvements, which have internal shock resistant buffer regions, have allowed many of these delicate crystals to be used in rough environments such as oil well logging. Oxide based scintillators tend to be more resilient to shock than alkali halides, and they offer a solution for measurements conducted in rough environments. Some oxide based scintillators have high Moh hardness values, such as YAG (moh = 8.5) and YAP (moh = 8.6).

Many of the alkali-metal/halide and lanthanide/halide are chemically unstable in a humid environment, and require handling in special 'dry rooms' during the cutting and packaging process. Further, these *hygroscopic* crystals must be hermetically sealed in air tight containers to prevent decomposition. Otherwise they may absorb water moisture, which causes them to change color and/or decompose. The alkali-metal/halides NaI:Tl and LiI:Eu are very hygroscopic and are ruined if exposed to humidity. The same is true for the lanthanide/halides $LaBr_3$:Ce, $LaCl_3$:Ce, and $CeBr_3$. CsI:Na is also hygroscopic; however, pure CsI and CsI:Tl are only slightly hygroscopic. Oxide-based scintillators along with the alkali-earth/halides generally are not hygroscopic.

Two important optical properties are crystal clarity and the index of refraction, both at the scintillator peak luminescent wavelength. The scintillator materials should be transparent to the photons emitted by the luminescent centers; otherwise, they are reabsorbed and remain undetectable. The optical attenuation coefficient is defined as $\alpha = 1/d$, where d is the distance that a packet of photons with wavelength λ reaches an absorption loss of $1-e^{-1}$ (63.21%). A large Stokes shift assists with optical clarity, in that the luminescent photons are less likely to be absorbed in the crystal. The index of refraction is the ratio of light speed in vacuum to light speed in the medium ($n = c/v$). As presented in Chapter 14, this ratio changes with light

[12]This hardness scale, still used today, was initially developed in 1812 by Frederich Moh, who compared materials to ten commonly available minerals used as standards. The scale ranges from 1 to 10, with talc rated as '1' and diamond as '10'. It is neither linear nor logarithmic, but is ordinal and somewhat arbitrary in designation.

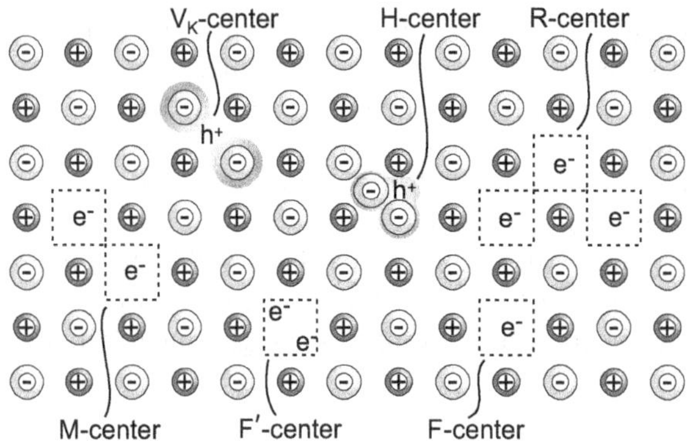

Figure 13.9. Depiction of several types of defects that can be formed in a crystal. F, F', M, and R centers involve vacancies, whereas V_K and H centers do not.

wavelength; hence the important value is the index of refraction at the most probable emission wavelength $\lambda_{\max}$. This index of refraction should be as close as possible to that of the light sensing instrument to obtain good coupling and light transmission. For typical glass windows on light sensing instruments, a scintillator index of refraction ≤ 1.5 is best.

Radiation Hardness

Radiation hardness is a relative measure of how well a substance resists changes in its important properties when exposed to large radiation doses. For scintillators, radiation hardness is usually used to describe the overall resistance to changes in its photon yield, pertaining either to luminescent intensity or emission spectra. Absorption of radiation causes changes in the crystalline structure, which can cause the introduction of defect centers called *color centers*. When a crystal takes on a visible hue, it has color centers in its makeup. These colors centers can absorb the scintillation light and re-emit light at a different wavelength.

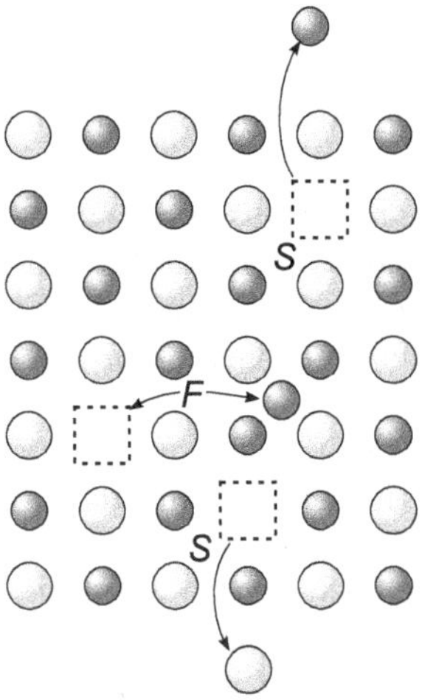

Figure 13.10. Depiction of a Frenkel pair and a Schottky defect pair (vacancies).

Although there are many types of color centers, perhaps one of the best understood is the F-*center*, so called from the German *Farbzentrum*, translated "color center". An F-center is formed by the displacement of a negative ion in the crystal structure, forming an anionic vacancy, which can trap an electron. The defect is formed by the coupling of a positive acting vacancy and the negative electron. This vacancy-electron combination produces energy levels in the band gap of the crystal, which can reabsorb luminescent photons, and through the Stokes shift, re-emit them at longer wavelengths, or lose the excited electron by phonon interactions. The net effect is to reduce the light yield. F-centers are native defects in alkali halide crystals, but can also be formed by x-ray, electron, or gamma-ray irradiation. In some cases, F-centers can be removed or reduced by annealing the crystal. There are many variants of these F-centers, all based on additional defects coupled to an F-center [Kittel 1956]. Included in the list of F-center variants are F'-centers, R-centers, and M-centers [Singh 2012], although the F-center is the most common. An F'-center is an anion vacancy with two trapped electrons [Kittel 1956], an M-center is a formation of two localized F-centers, and an R-center is a formation of three localized F-centers (see Fig. 13.9).

Trapped hole defects can also be formed, but do not involve a cation vacancy. Instead, a trapped hole is associated with either one or more negative ions [Ashcroft and Mermin 1976]. A V_k-center is formed when a hole is bound to two neighboring negative ions in the crystal, producing characteristics of a singly charged negative diatomic ion (A_2^-). The H-center is formed when a hole is bound to both an interstitial anion and neighboring anion in the lattice, forming a singly charged negative diatomic ion at the ion lattice location.

With sufficient energy transfer, lattice atoms can be removed from their respective sites into interstitial locations. If the removed ion is a cation, creating a positive ion interstitial in the crystal and leaving behind a negatively charged vacancy, a *Frenkel* defect is formed, consisting of the vacancy-interstitial pair (see Fig. 13.10). If a negative ion vacancy is formed along with a positive ion vacancy, the defect pair is a *Schottky* defect. By definition, a Schottky defect is not paired with interstitial ions, the Schottky defect pair maintaining electrical neutrality and crystal stoichiometry. Hence, Schottky defects are more likely formed near surfaces where the displaced ions can leave the crystal bulk.

The energy required to displace atoms in most solids is on the order of 25 eV. Hence, energy transfer in excess of 25 eV by radiation interactions can lead to the formation of both Frenkel and Schottky defects. Gamma-ray irradiation can form Frenkel (and Schottky) defects by transferring energy to electrons, through photoelectric, Compton scattering, or pair production interactions, which in turn transfers kinetic energy to lattice atoms.

A single energetic neutron can transfer enough energy to a recoil atom that can subsequently cause a cascade of displacements, forming a *void* in the crystal. The maximum energy transfer from a radiation particle to a lattice atom is given by Eq. (4.58), repeated here for convenience,

$$T_{\max} = \frac{\beta^2}{1-\beta^2}\left[\frac{2Mm_i^2c^2}{m_i^2 + M^2 + [2m_iM/\sqrt{1-\beta^2}]}\right].$$

where M is the target atom mass, m_i is the radiation particle mass, $\beta = v_i/c$ is the velocity ratio of m_i, and c is the speed of light. The magnitude of T_{max} must be greater than the displacement energy in order to produce these radiation-induced defects. For non-relativistic particles, Eq. (4.58) reduces to

$$T_{max} \approx 4E\frac{m_iM}{(m_i+M)^2}. \tag{13.35}$$

These radiation-induced traps can form color centers, as well as shallow and deep electron and hole traps.

Irradiation with hadrons[13] can cause activation and autodoping. Activated crystals possess intrinsic background radiation that is present during radiation measurements. Autodoping is a process in which constituent atoms are transmuted into different elements, thereby producing new activator sites (energy levels). It is possible that these newly formed energy levels reduce light yield by quenching the light emissions from luminescent centers.

The main effect of radiation induced defects, the most common of which are color centers, is to reduce the light emission from the scintillator. However, many of these defects heal over a relatively short time, thereby allowing the scintillation crystal to recover. Alkali halide scintillators tend to be more susceptible to hadron damage than photon (or electron) induced damage, while oxide-based scintillators tend to suffer more damage by photons than do alkali halides [Rodnyi 1997]. However, the extent of damage from radiation exposure depends on the type of crystal and the type of radiation. A study by Zhu [1998] on several common scintillators indicated that oxygen contamination in the alkali halides was the main source of radiation damage in these scintillators. Apparently the incorporation of hydroxyl groups (OH^-) into a scintillator can be broken down by radiation interactions in the crystals, after which the constituents migrate to become interstitials or substitutional atoms in the crystalline lattice [Zhu 1994]. Work performed by Zhu

[13] Hadrons are categorized as either baryons or mesons. Baryons have three quarks and include neutrons and protons. Mesons have one quark and one antiquark and include pions and kaons.

[1996] provides evidence to support this theory who showed the radiation hardness of CsI:Tl was improved by reducing the oxygen contamination. Zhu [1998] also concluded that radiation damage to oxide-based scintillators was due mainly to the formation of oxygen vacancies.

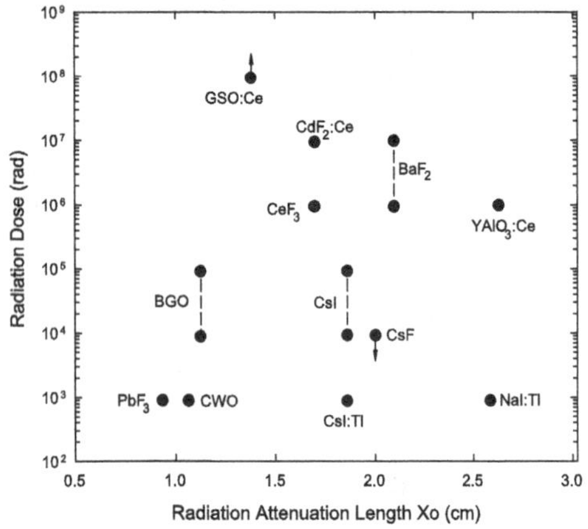

Figure 13.11. The radiation hardness of several scintillators versus unit radiation absorption length. Data are from Ishii and Kobayashi [1991].

Ishii and Kobayashi [1991] studied the effects of gamma radiation on scintillator performance. The study, conducted for applications in high energy physics, defined radiation hardness as that dose at which the light transmittance decreased by 2% per unit gamma-ray attenuation length (see Fig. 13.11). Although this measure may not apply to all scintillator applications, it does provide a relative measure of the radiation hardness between various common scintillator materials. From Fig. 13.11, of those scintillators studied, NaI:Tl, CsI:Tl, cadmium tungstate ($CdWO_4$), and PbF_2 had the lowest radiation resistance while gadolinium orthosilicate (GSO) had the highest tolerance. Zhu [1998] found that, depending on the dose rate, BGO and $PbWO_4$ can quickly (hours to weeks) recover from radiation damage at room temperature. Some of the more interesting properties regarding radiation hardness are described for a few specific scintillators in the following sections.

13.2.3 Properties of Several Common Inorganic Scintillators

Scintillators have been used for various radiation detection applications for over a century. Fluorescent screens and coatings were used in Crookes tubes, and $PtBa(CN)_4$ was the scintillating material that led to Röntgen's discovery of the x ray. ZnS:Ag was used for alpha particle counting by Rutherford, Geiger, and Crookes. In the early 1940s, organic scintillators were matched with efficient light collection devices to produce better scintillation detectors. However, it was the discovery of the bright inorganic scintillator NaI:Tl in 1948 by Hofstadter that allowed the practical application of scintillators to gamma-ray spectroscopy. Since then, the search has continued for better and brighter scintillators for higher energy resolution gamma-ray spectroscopy.

There have been some limited successes, which include those scintillators listed in Table 13.1. For instance, CsI:Na is similar in performance to NaI:Tl, but has a longer decay time. CsI:Tl has much higher light output than NaI:Tl, but the emission spectrum has a maximum at 560 nm, a wavelength that does not couple well to PMTs. However, CsI:Tl has been coupled to Si photodiode sensors quite successfully. Bismuth germanate (BGO) has lower light output, but is much denser and a better absorber of gamma rays. As a result, BGO has been used for medical imaging systems to reduce the overall radiation dose that a patient receives during the imaging procedure. In recent years, LSO:Ce has become an effective and useful alternative to BGO. LSO:Ce has higher light yield than BGO while also having comparable gamma-ray sensitivity. LiI:Eu is a scintillator that is primarily used for neutron detection because of the ^{6}Li content in the crystal. In recent years, $LaBr_3$:Ce, a relatively new scintillator with exceptional properties for gamma-ray spectroscopy, has become available. $LaBr_3$:Ce has a much higher light yield and a much shorter decay time constant than NaI:Tl. Further, it is composed of higher Z elements and, hence, is a better gamma-ray absorber than NaI:Tl. However, it is extremely hygroscopic and fragile and, thus, it is difficult to produce and handle. Although it has become commercially available, it is presently much more expensive than NaI:Tl because of production and fabrication problems.

Overall, there are numerous inorganic scintillators available for special radiation detection purposes. In fact, after more than 120 years of exploration for different scintillators, several hundreds have been identified. For example, tables of scintillators numbering in the hundreds, still incomplete, can be found in the literature

[Derenzo et al. 1991, 2016]. It is beyond the scope of the present text to be all inclusive of this extensive list of scintillators, and the authors limit the following descriptions to those scintillators that either have proven applications, historical significance, or show potential as scintillators for new applications.

Alkali-Metal Halide Scintillators

This family of scintillating salts composed of alkali metals and halogens is perhaps the most successful class of bright scintillators. Outlined here are some properties of a few notable scintillators in popular use, which are also listed in Table 13.1. A convenient and more extensive list of scintillator properties is provided by Derenzo et al. [2016].

Sodium Iodide (NaI, NaI:Tl) The most widely used inorganic scintillator is NaI:Tl. The notation ':Tl' indicates that the NaI crystal has been doped with the activator Tl. It is one of the brightest scintillators available, has acceptably good light yield proportionality, and can be grown in large sections. NaI:Tl was first reported by Hofstadter [1948] who prepared, initially, a powder sample of material with thallium added to the mixture that produced scintillation light when exposed to alpha particles. The sample was found to be hygroscopic, turning yellow within a few hours when exposed to room air. He then produced tiny NaI:Tl crystals (about 1 to 2 mm in size) in an evacuated ampoule. These crystals proved to be brighter than other scintillating materials known at the time.

NaI:Tl yields approximately 43,000 photons per MeV of energy absorbed in the crystal. Light emitted from NaI:Tl has a continuous wavelength spectrum, with a most probable emission at $\lambda_{max} = 415$ nm. The emission spectrum from NaI:Tl matches well to common K-bialkali and S-11 commercial photomultiplier tubes. NaI:Tl shows non-proportional light yield in the low energy region, below 200 keV, which consequently compromises spectroscopic performance. Regardless, NaI:Tl performs adequately well for a general use gamma-ray spectrometer. The decay time for NaI:Tl is 230 ns, which is relatively long compared to almost all organic scintillators and some inorganic scintillators. Although the decay time is manageable, it generally precludes NaI:Tl as a candidate for fast timing measurements.

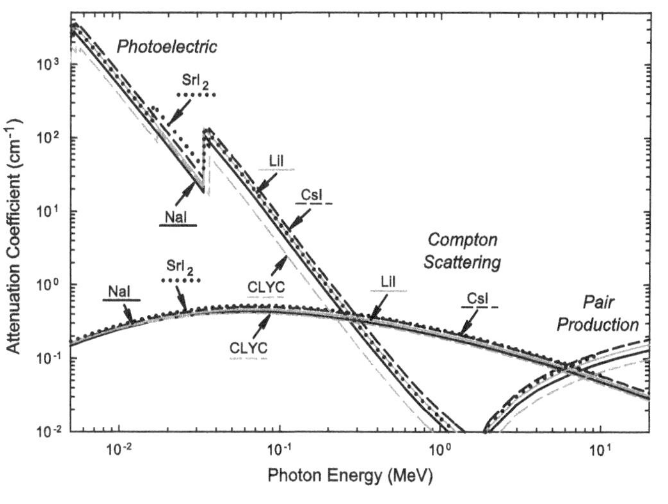

Figure 13.12. Gamma-ray linear attenuation coefficients for several common alkali-metal halide scintillators. Data are from [NIST XCOM].

For the most probable wavelength of 415 nm ($E = 2.99$ eV), a photon yield of 43,000/MeV represents only 12.8% of the energy absorbed. Other mechanisms for energy relaxation include phonons (heat) and phosphorescence. NaI:Tl has a slow 150 ms phosphorescent component, contributing about 9% of the photoelectron yield [Koički et al. 1973]. The emission rate is relatively low but usually does not interfere with low activity measurements, mainly because the low emission rate of phosphorescence produces low pulse height events easily rejected by a discriminator. This phosphorescent yield becomes important for high gamma-ray interaction rates during which it can become a steady source of background noise for radiation measurements and is often called *afterglow*.

Yet, it is the availability of large sizes and the relative linear response to gamma rays that makes NaI:Tl so important. Many different sizes are available, ranging in size from cylinders that are only 0.5 inches in diameter to almost a meter in diameter. Yet, the most preferred detector geometry remains the 3 × 3 inch right circular cylinder. It is the most characterized NaI:Tl detector size with extensive efficiency data in the literature. Further, it is the standard by which all other inorganic scintillators are measured.

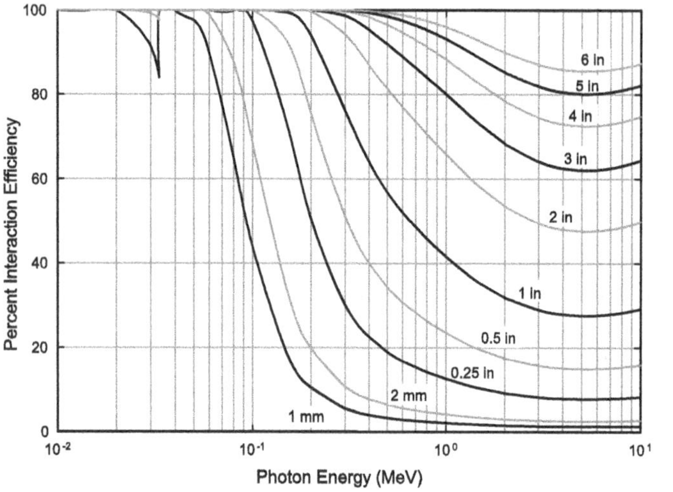

Figure 13.13. The calculated interaction efficiency for photons in NaI as a function of energy and detector depth. The probability of an initial interaction is shown and not the probability of subsequent or multiple interactions.

NaI:Tl is widely used to measure x rays and gamma rays because of its high efficiency for electromagnetic radiation, mainly due to the high interaction efficiency of iodine (see Fig. 13.12). The primary interaction efficiency as a function of gamma-ray energy and detector thickness is shown in Fig. 13.13.[14] X-ray detectors with a thin entrance window containing a very thin NaI:Tl detector are often used to measure the intensity and/or spectrum of low energy electromagnetic radiation. NaI:Tl detectors do not require cooling during operation and can be used in a great variety of applications. The bare NaI:Tl crystal is hygroscopic and fragile. However, when properly packaged in ruggedized containers, field applications are made possible and NaI:Tl detectors can operate over a long time period in warm and humid environments, resist a reasonable level of mechanical shock, and are resistant to radiation damage. For any application requiring a detector with a high gamma-ray efficiency and modest energy resolution, the NaI:Tl detector is clearly a good choice.

Pure NaI also produces scintillation light without the assistance of activator dopants. At room temperature, the light yield of pure NaI is low, having less than 2% energy conversion efficiency with a most probably emission wavelength of $\lambda_{\max}$ around 310 nm [Van Sciver 1956; Van Sciver and Bogart 1958]. Relatively recent measurements by Moszinzki et al. [2003b] with room temperature pure NaI indicated the release of 1,000 photoelectrons/MeV for a bialkali type PMT [Moszynski et al. 2003a]. The decay constant is substantially smaller for pure NaI than for NaI:Tl, being less than 20 ns at room temperature. At cryogenic temperatures, the light yield increases tremendously up to 25% energy conversion efficiency [Van Sciver and Bogart 1958], a yield of about 63,000 photons per MeV with a maximum light emission at approximately 77 K. Recent measurements put this yield for pure NaI closer to 44,000 ± 4,000 photons/MeV at 77 K [Moszynski et al. 2002]. Note that this trend of increased light yield with reducing temperature is observed for the entire family of pure alkali-metal/iodine scintillators [Boananami and Rossel, 1952]. The decay constant τ for pure NaI increases as the temperature is lowered, increasing from about 16 ns at room temperature up to about 95 ns at 4 K and about 63 ns at 77 K [Van Sciver and Bogart 1958].

The performance of pure NaI at low temperature (125 K), when coupled to a PMT, was reported as 18% FWHM for 122 keV gamma rays [Persyk et al. 1980]. As pointed out by Van Sciver and Bogart [1958], there are problems with operating scintillators coupled to PMTs at low temperature, which include efficient cooling of the PMT that is coupled to the scintillator, difficulties with coupling compounds that retain

[14]Figures such as Fig. 13.13 have been used for decades in textbooks and commercial literature and have been labeled as "absorption efficiency". However, this designation is misleading, because these dependencies are generated from only the total gamma-ray interaction coefficient μ by taking % "efficiency" as $100(1 - \exp[-\mu_t x])$. The calculation does not take into account Compton scatter losses or 511 keV losses from pair production, but instead represents only the probability of a primary interaction occurring. High energy gamma rays are predominantly forward scattered, leading to reduced energy attenuation and higher losses.

needed characteristics at cryogenic temperatures, and the fact the scintillators must be relatively small ($<$ 12.5 mm thick) to reduce photon self-absorption losses. At least one of these problems can be addressed by coupling pure NaI to a photodiode instead of a PMT, as reported by Moszynski et al. [2002; 2003a; 2003b]. This use of a photodiode gave $3.8 \pm 0.1\%$ FWHM for 662 keV gamma-rays when cooled to 100 K.

Cesium Iodide (CsI, CsI:Tl, CsI:Na) The absolute light yield of thallium doped CsI (CsI:Tl) has been measured to have a range between 54,000 photons/MeV to 61,000 photons/MeV by various sources [Saint Gobain 2016; de Haas et al. 2005; Mozynski et al. 1997b]. The refractive index is 1.79 and has a critical angle of 56.9° with a glass interface. The material is only slightly hygroscopic and is much more resilient than NaI:Tl. In fact, CsI:Tl can be machined (to some extent) and it has significantly fewer problems with shock and fracturing. Thin sheets of CsI:Tl are slightly malleable and can be formed over curved surfaces. With constituent elements Cs ($Z = 55$) and I ($Z = 53$), the gamma-ray attenuation coefficient is higher than that of NaI:Tl, as shown in Fig. 13.12. Hence, the gamma-ray intrinsic detection efficiency per unit volume is better for CsI:Tl than NaI:Tl. The primary interaction efficiency as a function of gamma-ray energy and detector thickness is shown in Fig. 13.14.

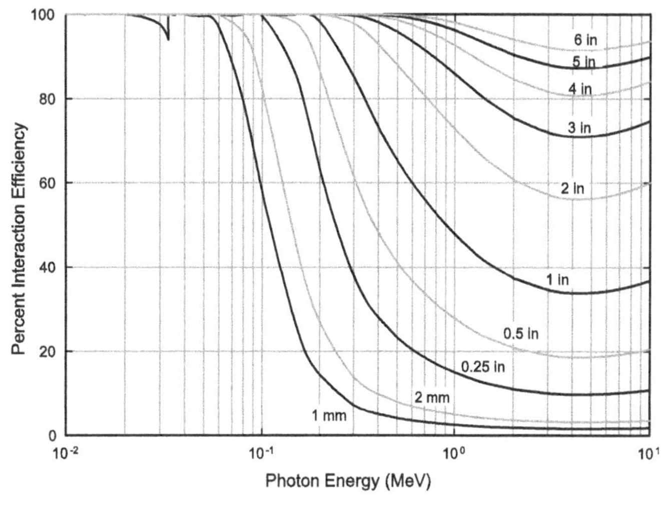

Figure 13.14. The calculated interaction efficiency for photons in CsI as a function of energy and detector depth. The probability of an initial interaction is shown and not the probability of subsequent or multiple interactions.

CsI:Tl has two decay components, 600 ns and 3.5 μs, with λ_{max} of 550 nm [Saint Gobain 2016]. Although CsI:Tl has a considerably higher light yield compared to most inorganic scintillators, the large mismatch of the luminescent spectrum with the response functions of common S-11 and K-bialkali PMTs, as can be seen from Fig. 13.3, compromises the spectroscopic performance. The spectroscopic performance can be improved by using long shaping times, thereby collecting more of the light from the luminescence. However, a longer collection time increases dead time problems and integrates more electronic noise. The maximum emission wavelength fortuitously matches well to the response function of Si photodiodes. Hence, reasonable spectroscopic performance can be achieved by coupling CsI:Tl crystals to either Si photodiodes or Si photomultipliers (or SiPMs; see Chapter 14). The maximum light output of CsI:Tl is observed at approximately 50°C, and reduces with lower or higher temperatures (see Fig. 13.8). Temperatures deviating only ±50°C from the optimum can cause a light yield reduction greater than 15%. Light loss to gamma radiation damage can become significant at doses approaching 10 Gray, with a measured light reduction of 15% [Saint Gobain 2016].

CsI:Tl can be grown by physical vapor deposition to form narrow vertical columns of the crystal [Jing et al. 1992]. Methods have been developed that create layer thicknesses up to 2 mm with column diameters as small as 5 μm [Nagarkar et al. 1998]. These columns perform as miniature optical fibers that propagate light, directionally, through the scintillator to the readout device. The CsI:Tl thin film can be patterned through optical lithography methods to place the scintillator on select locations [Jing et al. 1992]. When coupled to a Si diode or SiPM, the structure can be used to image a collimated beam of x rays. X-ray spatial resolution on the order of 75 microns was reported when the device was exposed to a collimated 25 μm wide $\leq$ 30 keV x-ray beam [Jing et al. 1992].

A variant of CsI has Na as the activation dopant. The absolute light yield of CsI:Na has been measured to range between 41,000 photons/MeV and 49,000 photons/MeV [Saint Gobain 2016; de Haas et al. 2005]. The refractive index is 1.84 with a critical angle of 54.6° to a glass interface. CsI:Na is much more hygroscopic than CsI:Tl, and care must be taken to prevent exposure to moisture. The primary decay time is also long, but less than CsI:Tl at 630 ns, with a λ_{max} of 420 nm. This luminescent spectrum from CsI:Na is a much better match to common K-bialkali and S-11 PMTs (see Fig. 13.3), with a reported photoelectron production about 85% of that observed with NaI:Tl [Saint Gobain 2016]. However, with the long decay time and the lower resolution performance, CsI:Na offers little advantage over NaI:Tl, except, perhaps, for its superior mechanical robustness. The maximum light output of CsI:Na is observed at approximately 80°C, and reduces with temperatures either lower or higher (see Fig. 13.8). The thermal dependence is greater than observed with CsI:Tl, and temperature deviation of ±50°C from the optimum can cause a light yield reduction greater than 20%.

Undoped CsI (or CsI(pure)) also produces scintillation light, but has very different properties than activated CsI:Tl or CsI:Na scintillators. At room temperature, it produces approximately 2,000 photons per MeV, with λ_{max} at 315 nm [Saint Gobain 2016]. The index of refraction for pure CsI is 1.95, which produces a critical angle of 50.3° at a glass interface. The primary decay time is much shorter than its doped counterparts, namely about 16 ns. Amsler et al. [2002] conducted experiments to measure the light yield and time response, and report the existence of three decay constants, all significantly shorter than doped CsI, namely 28 ± 2 ns, 6 ± 1 ns, and a third fast component reported to be 2 ns. Notably, the light yield of pure CsI increases significantly at low temperature, with a documented light yield of 50000 ± 5000 photons per MeV at 77 K [Amsler et al. 2002] and yielded an energy resolution of 8.3% FWHM for 511 keV photons when coupled to a PMT. Only a single decay constant was observed at low temperatures, which increased significantly up to about 1 μs at 85 K.

Both pure CsI and CsI:Tl recovered from radiation damage at room temperature, but at a slow rate of less than 0.5% per day. Thermal annealing of CsI or CsI:Tl does not assist with recovery, and appears to make matters worse if conducted at temperatures exceeding 300°C.

Lithium Iodide (LiI:Tl, LiI:Eu) Lithium iodide is an alkali-metal halide scintillator that is used as both a gamma-ray and neutron detector. LiI is hygroscopic and must not be exposed to humidity. Thus, LiI requires all handling to be performed in a dry environment. LiI has been used as a scintillator for several decades after being developed around the same time that NaI:Tl was introduced [Hofstadter 1948; Hofstadter at al. 1951]. These initial LiI crystals were doped with Tl as the activator [Hofstadter 1950]. Many of these crystals were yellowish in color, most likely a side effect from impurity contamination. It was also discovered that the activator dopant distribution in LiI:Tl was non-uniform, a side effect from crystal growth problems. The dopant source (LiTl) introduced into the crystal melt was not uniformly incorporated into the crystal lattice, a result which consequently compromised the spectroscopic performance [Bernstein and Schardt 1952]. Within a few years, alternative activator dopants were studied including In, Ag, Sn, Sm, Eu. Of these alternative activator dopants, Eu gave the best light output [Nicholson et al. 1955], and LiI:Eu remains the most studied and used of the LiI variants. With a mass of 4.08 g cm^{-3}, LiI:Eu has nearly the same gamma-ray interaction efficiency as NaI:Tl. LiI:Eu is a relatively slow scintillator with a decay time of 1.4 μs. The measured λ_{max} value is 475 nm with an absolute photon yield of approximately 1,500 per MeV, and has yielded 7.5% FWHM energy resolution for 662 keV gamma rays [Syntfeld et al. 2005]. LiI:Eu has a refraction index of 1.96, which gives a critical angle of 49.93° at an interface with common glass.

It is notable that in recent work LiI crystals doped with 0.5% Tl yielded $14,000 \pm 1,400$ photons per MeV [Khan et al. 2015]. Two decay constants were observed, one at 185 ns yielding 88% of the photons and a slower decay time at 1.089 μs yielding the remaining 12% of photons. The wavelength at the emission maxima (λ_{max}) was measured to be 470 nm. Energy resolution of LiI:Tl was reported to be 8.5% FWHM for 662 keV gamma rays.

The lithium constituent of LiI:Eu enables thermal-neutron detection through the ^{6}Li(n,^{3}H)^{4}He reaction. The natural abundance of ^{6}Li is 7.59%, yielding an atomic density of ^{6}Li in natural LiI:Eu of 1.39×10^{21} cm^{-3}. The resulting thermal macroscopic cross section for the ^{6}Li(n,^{3}H)^{4}He reaction is 1.31 cm^{-1} with an absorption length of 0.76 cm. If the Li is replaced with enriched ^{6}Li, the thermal-neutron absorption macroscopic cross section can be increased to 17.27 cm^{-1}, and the absorption length is reduced to 0.58 mm. Further discussion on LiI as a neutron detector is reserved for Chapter 17 on neutron detectors.

Some Other Alkali-Metal Halide Scintillators Once a scintillator has been developed from an alkali-metal/halogen salt, the obvious question to ask is, what else is there? Indeed, practically all binary combinations of alkali metals and halogens have been studied to determine if they have practical use as radiation detecting scintillators (except for those compounds based on francium or astatine). Some of these other combinations did show promise, but simply did not outperform those materials already described in this section. Combinations in which both constituent atoms are of low Z are less interesting for gamma-ray detectors and spectrometers, hence can be excluded based on their low interaction efficiency. Described here are a few of those other possibilities that have relatively high Z elements.

Potassium iodide (KI) was actually investigated by Hofstadter before NaI was, and led to the investigation of NaI:Tl powders as a scintillator [Hofstadter 1948]. Although relatively bright compared to other known scintillators at the time, KI:Tl does not have the high light yield produced by NaI:Tl and has a light yield of approximately only 9,700 photons/MeV [Holl et al. 1988]. It has two main decay branches at 240 ns and 2.5 μs [Robertson and Lynch 1961]. There also appears to be a change in decay constants when irradiated with alpha particles instead of gamma rays with decay times of 210 ns and 1.76 μs. Milton and Hofstadter [1949] report $\lambda_{max} = 400$ nm with a slightly narrower spread in the emission spectrum than observed with NaI:Tl. Similarly, Cook and Mahmoud [1954] report $\lambda_{max} = 425$ nm; hence, KI:Tl has a similar emission output as NaI:Tl.[15] The index of refraction for KI is 1.68 and has a critical angle of about 63° with a common glass interface. KI has a density of 3.12 g cm^{-3}, which is lower than that of NaI. Although K is a higher Z element than Na, it does little to improve the interaction efficiency of KI over NaI because iodine is the main photon absorber in both materials; hence, the gamma-ray interaction efficiencies are nearly the same. Along with the problems listed above, natural potassium has a radioactive isotope ^{40}K, which produces an intrinsic radiation background with K-based scintillators.

Cesium fluoride (CsF) is a fast inorganic scintillator initially studied by Van Sciver and Hofstadter [1952]. It has a mass density of 4.12 g cm^3, higher than that of NaI. As with many alkali-metal/halides, CsF is hygroscopic. CsF scintillates without the assistance of an activator dopant and has $\lambda_{max} = 390$ nm [Moszynski et al. 1983]. The absolute photon yield is approximately 1950/MeV at room temperature [Moszynski et al. 1983; Van Eijk et al. 1993]. CsF has an index of refraction of 1.48, below that of common glass, so that the critical angle of internal reflection is 90°. The measured decay time is 5 ± 1 ns, making it one of the fastest inorganic scintillators available. These studies revealed that the luminescent efficiency was much higher for gamma-ray interactions than for alpha particle interactions by approximately a factor of 4 $\pm$ 10%. It was also observed that the alpha particle pulses were produced by a much slower decay constant of about 200 ns. These multiple interesting properties make CsF an attractive scintillator for fast timing applications.

[15] KI:Tl has also been reported to have a broad emission spectrum at room temperature with a λ_{max} of about 301 nm [Edgarton and Teegarden 1963]. Other measurements indicated λ_{max} of about 410 nm [Smol'skaya et al. 1969] and is similar to that found by Milton and Hofstadter [1949]. It is quite possible that the samples studied by Edgarton and Teegarden [1963] were contaminated with additional impurities, yet this possibility is hard to ascertain from the references.

Cesium Bromide (CsBr) is a heavy element alkali-metal/halide with some interesting properties. It has a density of 4.44 g cm^{-3} with an refractive index of 1.7. It is also hygroscopic like many alkali-metal/halides. CsBr activated with Tl has been reported to have several decay times [Robertson and Lynch 1961; Robertson et al. 1961]. It was found that CsBr:Tl had two main decay times, which differ with the type of radiation particle. Notably, alpha particles (5.3 MeV) were observed to produce shorter decay times at 2.3±0.15 μs and 14.5±1.34 μs than electrons (from 662 keV gamma rays) at 2.57±0.13 μs and 24.23±1.5 μs [Robertson and Lynch 1961]. Pure CsBr also scintillates, although the luminescent yield is poor, reportedly 20 photons/MeV [Van Eijk et al. 1993]. The decay constant of pure CsBr is unusually short for an alkali-metal/halide at 0.07 ns with $\lambda_{max} = 250$ nm.

Rubidium iodide (RbI) has a density of 3.11 g cm^{-3} with a refractive index of 1.64. It has been reported to have two decay constants at 1 μs and 200 μs [Birks 1964]. RbI:Tl has a reported emission wavelength of 4.43 nm (at 80 K) [Babin et al. 1999; Zazubovich 2001], with the interesting property that light emission increases, and the appearance of a luminescent recombination peak of similar intensity appears at 318 nm when the crystal is placed under stress. Ultimately, the light yield is much lower than that observed for NaI:Tl. Pure RbI has a third decay constant reported to be 100 μs at low temperature (about 175 K) [Birks 1964]. Natural rubidium has a radioactive isotope ^{87}Rb, which produces an intrinsic radiation background with Rb-based scintillators.

Alkali-Earth Halide Scintillators

Alkali-earth halides generally consist of those scintillators possessing the form AB_2, where A is an alkali earth and B is a halide. These scintillators are usually hygroscopic and are often relatively soft. Many possess good properties for scintillator detectors, some with good properties for special applications.

Barium Fluoride (BaF_2) Barium fluoride is a commercially available scintillator that is slightly hygroscopic. It is relatively soft with a ranking of 3 on the Moh hardness scale. The mass density of BaF_2 is 4.88 g cm^{-3}, and the atomic numbers of the elemental constituents are 56 and 9. The primary interaction efficiency as a function of gamma-ray energy and detector thickness is shown in Fig. 13.15. It has two important luminescent emissions at 310 nm and 220 nm, both intrinsic emissions. The 310 nm luminescent emission has a decay constant τ of 630 ns, while the 220 luminescent emission has a much faster decay constant ranging between 600 to 800 ps. The refractive index of BaF_2 at 220 nm is 1.54 and at 310 nm it is 1.50, both practically the same as common glass windows on PMTs. Pure BaF_2 was initially studied by Faruhki and Swinehart in 1971, although there was no report of the short wavelength component at the time. The absorption of light in common soda-lime glass is much too high to allow the transmission of either of the two luminescent wavelengths. Although some borosilicate glasses can be used for the 310 nm emissions, they are inadequate for transmittance of the 220 nm emissions. Perhaps it is for this reason that the fast luminescent component remained unreported until 1983 [Laval et al. 1983]. Instead, a PMT with a fused quartz window is preferred if the 220 nm wavelength is to be used (see Chapter 14 on PMTs). Light yield for

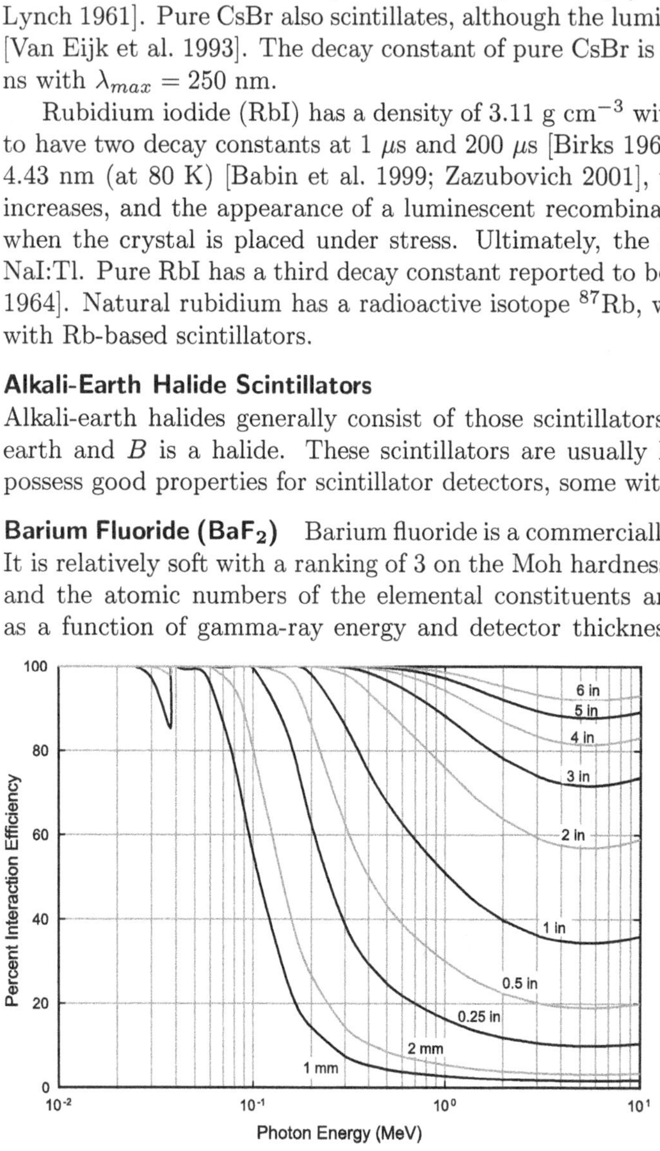

Figure 13.15. The calculated interaction efficiency for photons in BaF_2 as a function of energy and detector depth. The probability of an initial interaction is shown and not the probability of subsequent or multiple interactions.

the slow component (310 nm) is about 10,000 photons/MeV and for the fast component (220 nm) is about 1,800 photons/MeV.

As pointed out by Ishii and Kobayashi [1991], BaF_2 is mainly important for its fast decay time at the 220 nm emission, and radiation damage that reduces this transmission is of primary concern. Studies by Messner and Smakula [1960] and also Heath and Sacher [1966] predicted high radiation hardness for BaF_2. Measurements by Majewski and Anderson [1985] apparently confirm this prediction as fact. They irradiated BaF_2 with energetic protons, with exposures up to 1.3×10^7 rad, and found a loss of only 1%/cm light transmission. BaF_2 crystals have shown good resistance to gamma-ray irradiation up to 10^6 rad [Woody et al. 1989]. Another study showed that Pb contaminants in BaF_2 significantly reduced radiation hardness [Murashita et al. 1986]. Is was concluded by Ishii and Kobayashi [1991] that highly purified BaF_2 should yield excellent radiation hardened scintillators. BaF_2 did not recover from radiation damage at room temperature in the study performed by Zhu [1994], although there was indication of partial recovery at room temperature in the study performed by [Caffrey et al. 1986], or some damage recovery when exposed to sunlight [Murakami et al. 1991].

An interesting physical aspect to the light emissions from BaF_2 is that the 310 nm emission is greatly affected by temperature changes while the 220 nm emission is not. Studies performed by Schotanus et al. [1985] revealed that the 310 emission spectrum increases in intensity to approximately 3 times that of its room temperature output when the temperature is reduced to about 175 K, and practically disappears when the temperature is increased up to 358 K. However, the 220 nm emission is unaffected by these temperature changes.

Calcium Fluoride (CaF_2:Eu) Europium doped calcium fluoride, initially reported by Menefee et al. in 1966, is a commercially available scintillator that is not hygroscopic and is practically insoluble in water. It has a ranking of 4 on the Moh hardness scale. Because it is not hygroscopic and relatively robust, it can be used under rough circumstances which would be inappropriate for delicate hygroscopic scintillation crystals. The mass density of CaF_2:Eu is 3.18 g cm^{-3}, and the atomic numbers of the elemental constituents are 20 and 9, both relatively low for gamma-ray detection. With a photoelectric/Compton crossover near 70 keV, CaF_2:Eu is best applied for detection of only low energy gamma rays as can be seen from Fig. 13.16. However, CaF_2;Eu can be used as a beta particle detector because backscattering is reduced by the low Z constituents [Colmenares et al. 1974]. Its primary emission wavelength λ_{max} is 435 nm with a relatively long primary decay constant τ of 940 ns. The absolute light yield is reported to be 19,000 photons/MeV [Saint Gobain 2016], approximately 50% of the light yield of NaI:Tl. The refractive index at λ_{max} is 1.47, lower than common glass, indicating that internal reflection when coupled to the glass window of a PMT can be minimized to small values.

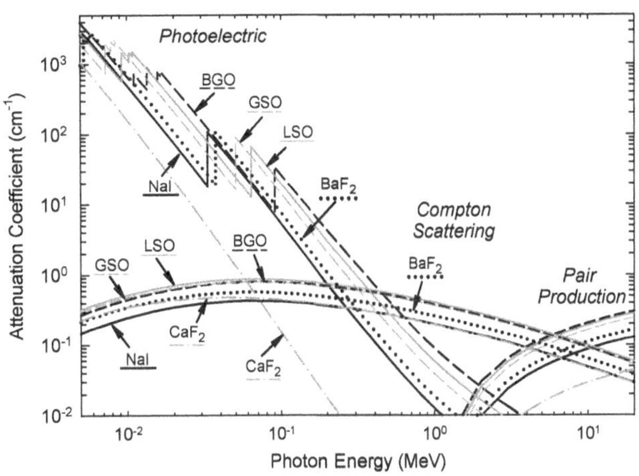

Figure 13.16. Gamma-ray linear attenuation coefficients for a few common scintillators as compared to NaI. Data are from [NIST XCOM].

Calcium Iodide (CaI_2, CaI_2:Tl, CaI_2:Eu) Calcium iodide (doped) is a hygroscopic rare-earth/halogen scintillator initially studied by Van Sciver and Hofstadter [1951] and Hofstadter et al. [1964a; 1964b; 1967]. Having a similar performance to that of NaI:Tl, this scintillator showed great promise. The pure form of

CaI_2, along with CaI_2 activated with either Tl or Eu, were reported as bright scintillators [Hofstadter et al. 1964b]. The mass density of CaI_2 is 3.96 g cm^{-3}. The initial work on CaI_2 was conducted with Tl as the activator [Van Sciver and Hofstadter 1951]. It was found that CaI_2:Tl had a measured decay constant of 1.1 μs and a light yield approximately the same as that for NaI:Tl. The most probable emission wavelength (λ_{max}) was 420 nm. Years later, pure and Eu-activated CaI_2 were studied by Hofstadter et al. [1964a; 1964b] and yielded better results than CaI_2:Tl. It was reported that both pure CaI_2 and CaI_2:Eu yielded up to 200% light yield as that for typical NaI:Tl crystals. The most probable emission wavelength for pure CaI_2 was 410 nm with a decay constant of 550 ns, and the most probable emission wavelength for CaI_2:Eu was 470 nm with a decay constant of 790 ns [Hofstadter et al. 1964b]. Energy conversion efficiency was calculated to be approximately 25%. In that same study, energy resolution of 5.19% FWHM was reported for 662 keV gamma-rays for a CaI_2:Eu sample. Recent work on CaI_2 indicates that the theoretically achievable light yield for CaI_2:Eu is 114,000 photons/MeV, while the measured value was 110,000 photons/MeV [Cherepy et al. 2009]. Hofstadter et al. [1964b] describe multiple problems with the production of CaI_2 crystals, mainly, the material is more hygroscopic than NaI:Tl and it is also quite easily fractured. CaI_2 has a rhombohedral crystalline lattice formed in a layered structure with weak bonds between the basal planes. Unfortunately, the crystal is prone to cracking and plastic deformation during both crystal growth and detector fabrication, and is most likely the reason why CaI_2 has not been extensively explored. Methods to improve crystal growth and material quality have been implemented [Boatner et al. 2015], although the scintillation performance of these newer high-quality crystals was significantly below that of the earlier results reported by Hofstadter.

Strontium Iodide (SrI_2, SrI_2:Eu) Strontium iodide doped with Eu is a hygroscopic rare-earth/halogen scintillator, initially studied by Hofstadter [1968], but it received little attention at the time. Recently this important scintillator has regained attention as a potential high-resolution scintillator. The measured light yield of 115,000 photons/MeV [Cherepy et al. 2009] is much higher than that of commercially available NaI:Tl and $LaBr_3$:Ce scintillators. The most probable emission λ_{max} for Eu-doped SrI_2 is 435 nm [Cherepy et al. 2009], and there is a reported broad emission peak that appears between 550 and 600 nm, depending on the scintillator temperature [Alekhin et al. 2011]. The density of SrI_2 (4.55 g cm^{-3}) combined with the relatively high Z constituent atoms (38/53) makes it an efficient gamma-ray absorber. Strontium iodide has a relatively long decay time of 1.2 μs, a common consequence associated with Eu activators. Unfortunately, the long decay constant makes SrI_2:Eu less attractive than NaI:Tl and $LaBr_3$:Ce for many radiation measurement applications, especially for those with high count rates. The relative light yield is good with a nominally proportional response down to 15 keV (see Fig. 13.5). It has a refractive index of 2.05 at λ_{max} and a critical angle with common glass of 47°. Energy resolution on the order of 2.7% has been reported for SrI_2:Eu crystals irradiated with 662 keV gamma rays [Cherepy et al. 2009; Cherepy et al. 2013]. The relative light yield increases at low temperature (80 K) and decreases significantly at temperatures above 300 K [Alekhin et al. 2011]. Alekhin et al. [2011] report spectral broadening of the emission spectrum from SrI_2:Eu as the temperature is increased from 100 K to 600 K. Also reported is an increase in decay constant τ with temperature going from 400 ns below 100 K to 7 μs at 600 K [Alekhin et al. 2011]. This unusual increase in τ with temperature is hypothesized to be a consequence of photon self-absorption of emissions from Eu luminescent centers. Pure SrI_2 also scintillates (λ_{max} is about 600 nm), but apparently lacks the property of an increasing decay constant with increasing temperature; rather there is a decrease in τ at higher temperatures, ranging from 1.5 μs at 78 K down to 450 ns at 300 K. Recent studies indicate that scintillations from pure SrI_2 are defect (or impurity) related, in that defect-free high purity crystals of undoped SrI_2 and SrI_2:Eu crystals lack this emission altogether [Kawai at al. 2016].

Transition Metal Scintillators

Yttrium Aluminum Perovskite ($YAlO_3$:Ce or YAP) Yttrium Aluminum Oxide (or YAP) is a transition element rare earth (Y) having the general chemical formula of ABO_3, making a mineral of the perovskite[16] family. Perhaps the most interesting property of $YAlO_3$:Ce (YAP:Ce) is its unusually good linearity in the low energy range, leading to energy resolution approaching the theoretical maximum. With an average light yield of about 18,000 photons/MeV, is has about 42% of the NaI:Tl light yield, yet the light yield linearity property of YAP:Ce also produces much better energy resolution. For example, a FWHM energy resolution of less than 4.4% has been reported for ^{137}Cs 662 keV gamma rays [Kapusta et al. 1999, Moszynski 2003]. Other advantages of YAP:Ce include a fast decay time of 27 ns, a relatively high mechanical strength and chemical stability, and thermal stability [Leroq et al. 2006]. The fast decay time enables fast timing applications, while the mechanical and chemical robustness reduces problems with detector fabrication (slicing and polishing). The thermal stability produces a minimal change in performance with temperature variations and allows its use at higher temperatures than other scintillators. However, disadvantages include relatively low Z numbers for the constituent elements (39,13,8), a high index of refraction (1.95), and a relatively low value for λ_{max} of 370 nm. Although the density of YAP is 5.35 g cm^{-3} (higher than that of NaI:Tl), the low Z elements yield a relatively low attenuation coefficient at moderate and high gamma-ray energies. The high index of refraction causes the critical angle for internal reflection at a glass interface to be high, approximately 50°. The low value of λ_{max} causes inefficient coupling to standard bialkali and multialkali PMTs.

Lutetium Aluminum Perovskite ($LuAlO_3$:Ce or LuAP) Lutetium Aluminum Oxide (LuAP), also a rare earth perovskite, has many similar properties to YAP. Perhaps the most important differences are LuAP's higher density of 8.34 g cm^{-3} and its higher Z constituent element Lu ($Z = 71$), both of which significantly increase its gamma-ray absorption coefficient. It certainly has one of the highest average gamma-ray attenuation coefficients of the known scintillators. It is also a fast decay scintillator, with two decay constants of 16.5 ns and 74 ns. However, the light yield of LuAP is only 11,400 photons/MeV with $\lambda_{max} = 365$ nm. Its index of refraction is 1.94; hence, the critical angle is also approximately 50° with most common glasses used for PMT windows. Consequently, these multiple disadvantages produce a measured energy resolution usually inferior to that observed with YAP:Ce, although there is at least one report indicating that relatively good energy resolution can be achieved, namely 6.8% FWHM at 662 keV for small crystals [Balcerzyk et al. 2005]. Although the light yield is only 26.5% of that of NaI:Tl, its energy resolution can be comparable, most likely a beneficial consequence of LuAP:Ce having a proportional relative light yield in the low energy region similar to that observed with YAP [Balcerzyk et al. 2005].

Yttrium Aluminum Garnet ($Y_3Al_5O_{12}$:Ce or YAG:Ce) Yttrium aluminum garnet[17] YAG:Ce is a non-hygroscopic relatively fast inorganic scintillator that has found use with scanning electron microscope (SEM) systems [Autrata et al. 1978; Bok and Schauer, 2014]. It has relatively high resistance to radiation damage, making it suitable for use under the expected high electron bombardment in a SEM. YAG has a density of 4.55 g cm^{-3}, and its highest Z component is yttrium (39) so that it has only modest gamma-ray absorption coefficients. YAG has a cubic crystal structure and rates 8.5 on the Moh hardness scale.

It is yellow in appearance, with a most probable wavelength of emission λ_{max} at 550 nm. Manufacturers recommend the use of an extended red PMT response, such as the S-20 PMT. However, the use of a silicon photodiode (Si PD) is a good alternative, mainly because the quantum efficiency is well matched between a Si PD and the emission spectrum from YAG:Ce. It has an index of refraction of 1.82 that produces a critical angle of 55.5° at a glass interface. Depending on the doping concentration and type of radiation,

---

[16]Discovered in the Ural Mountains in 1839 by Gustov Rose, this mineral is named after the Russian mineralogist Lev Perovski.
[17]The word "garnet" is derived from the Middle English word 'gernet', which means 'dark red'.

YAG:Ce has multiple decay times [Moszynski et al. 1994]. Moszynski et al. [1994] report decay times of 88 ns and 302 ns when irradiated with gamma rays, and decay times of 68 ns and 247 ns when irradiated with alpha particles. These differences in decay times for heavy charged particles and gamma rays have been used to discriminate radiation types through pulse shape discrimination [Ludziejewski et al. 1997]. The absolute light yield is approximately 15,000 photons/MeV if measured with a Si PD, but somewhat less at 1,270 photons/MeV when measured with a PMT [Ludziejewski et al. 1997]. Energy resolution of 11.1% was reported for 662 keV gamma rays when coupled to a PMT [Moszynski et al. 1994].

Lutetium Aluminum Garnet ($Lu_3Al_5O_{12}$:Ce or LuAG:Ce) Closely associated with YAG is LuAG, which is interesting for gamma-ray detection because of the higher Z component lutetium (71). A comprehensive review by Nikl et al. [2013] provides general information on its development. LuAG has a crystalline cubic structure, a density of 6.76 g cm^{-3}, and rates 8.5 on the Moh hardness scale. LuAG has been discovered to yield good performance with Ce [Chewpraditkul et al. 2009] and Pr activator dopants [Yanagida et al. 2011]. LuAG:Ce, also yellow in color, has a peak emission wavelength λ_{max} between 510 and 535 nm and a decay constant between 55 and 70 ns. The absolute light yield ranges between 18,000 and 26,000 photons/MeV and is relatively proportional above 100 keV [Nikl et al. 2013]. Gamma-ray performance is good, with a reported FWHM energy resolution between 5.5 and 7% at 662 keV [Nikl et al. 2013 and references therein]. LuAG:Pr has a reported peak emission wavelength λ_{max} of 310 nm [Świderski et al. 2009a] and a decay constant of 20 ns [Furukawa 2016]. A reported absolute light yield is 22,000 photons/MeV for commercial material [Furukawa 2016]. The light yield proportionality is quite excellent for energies above 100 keV [Świderski et al. 2009a], and most likely assists with the good energy resolution observed, with a best resolution reported as 4.2% at 662 keV for commercial LuAG:Pr [Furukawa 2016].

Zinc Sulfide (ZnS:Ag) Zinc sulfide is a relatively old scintillator, having been discovered by Théodore Sidot in 1866. It is a non-hygroscopic material with a density of 4.09 g cm^{-3}. Because of its low Z constituents, it is not an efficient gamma-ray detector, but is a relatively bright scintillator for heavy charged particles. ZnS was used as the scintillator for alpha particle counting by Rutherford and Geiger and is also the scintillating component in common spinthariscopes. There are two main crystalline forms of ZnS that can exist at room temperature, a cubic (zinc blend) form and a hexagonal (wurtzite) form, which unfortunately makes single crystal growth difficult and leads to polymorphism. Because of the difficulty in producing single crystals, ZnS is available either as a polycrystalline powder or as thin optical crystals. ZnS is also a wide band-gap semiconductor material with the hexagonal and cubic forms having band gaps of 3.91 eV and 3.54 eV, respectively.

When doped with Ag as an activator, ZnS:Ag has a most probable emission wavelength $\lambda_{max} = 450$ nm. Its refractive index is high at 2.36 that produces a critical angle of 39.5° when coupled with glass or acrylic.[18] ZnS:Ag has poor light response to gamma rays, but is relatively bright for heavy ion interactions. The light yield is about 54,000 per MeV for alpha particles, or approximately 126% of the light output of NaI:Tl and has a reported decay constant of 110 ns [Saint Gobain 2016]. ZnS:Ag in the form of polycrystalline powder is opaque to its own light emissions for mass thicknesses greater than 25 mg cm^{-2} (or 61 microns). Hence, most detectors utilizing ZnS:Ag are limited to thin films. ZnS:Ag is generally used for heavy ion detection. Two interesting configurations are the Lucas cell and the Hornyak detector.

A Lucas cell is a cylindrical cup configured as a gas-flow chamber, usually with dimensions on the order of 5 cm in diameter and 5 cm long. The gas-flow chamber has inlet and outlet ports, and the inside of the chamber is coated with a scintillator, usually ZnS:Ag [Lucas, 1957]. A Lucas cell is designed to accept a gas sample, filter out the radioactive particulates, and subsequently count radioactive decay products as they produce scintillation light when they strike the chamber walls. A PMT attached to the open top of

[18] Poly(methyl methacrylate) or PMMA, also known as plexiglass or lucite.

the chamber registers light pulses, thereby producing a count rate from the chamber. A Lucas cell is often used to measure radon gas concentrations [Abbady et al., 2004]. Radon, a radioactive gas, decays by α particle emission, which produces a chain of radiative daughter products. Inhalation of radon can result in radioactive daughter products becoming lodged on the surfaces of the respiratory system, and presents a health risk in regions of high radon accumulation. The standard for measuring radon gas can be found in the ANSI/AARST MS-PC 2015 report.

A Hornyak detector [Hornyak 1952], often called a Hornyak button, is a detector configuration used for fast-neutron detection while suppressing signals from slow neutrons and gamma rays. The original devices were composed of ZnS:Ag particles interspersed within an acrylic medium. Modern Hornyak buttons have interspersed concentric cylinders, similar to a "bull's eye" target, alternating between non-scintillating acrylic light waveguides and layers of ZnS:Ag. Fast neutrons interact in the plastic and eject protons into the ZnS:Ag. The energetic protons fluoresce the ZnS:Ag, and a fraction of the light propagates through the plastic to a light detection device, such as a PMT. Hornyak buttons are used for fast neutron detection in high gamma-ray environments in order to reduce the background emissions. More detail on this type of detector can be found in Chapter 18.

Post-Transition Metal Scintillators

Bismuth Germanate ($Bi_4Ge_3O_{12}$ or BGO) Bismuth germanate, commonly referred to as BGO, is a relatively rugged fracture resistant non-hygroscopic material, and it is rated as 5 on the Moh hardness scale. BGO has one high Z constituent (Bi = 83, Ge = 32, O = 8), along with a relatively high density of 7.13 g cm^{-3}, making it a good gamma-ray absorber. The primary interaction efficiency as a function of gamma-ray energy and detector thickness is shown in Fig. 13.17. Ishii and Kobayashi [1991] report that it is a relatively radiation hard material, capable of withstanding gamma-ray dose between 10^4 and 10^5 rad before showing significant light absorbance. Because of its high gamma-ray absorption efficiency, thereby reducing patient dose, it has found practical use in the medical imaging industry [Cho and Farukhi 1977]. The most probable emission wavelength λ_{max} is 480 nm with a decay time of 300 ns. The absolute light yield at room temperature is approximately 8,200 photons/MeV, with a PMT response of only 13% relative to NaI:Tl; hence the gamma-ray energy resolution is poor by comparison to NaI:Tl. BGO also has a lower luminosity fast emission, yielding about 700 photons/MeV with τ = 60 ns. With a relatively large refractive index of 2.15, it has a narrow critical angle of 44.2° at the glass interface of a PMT window, making light collection more difficult than with other inorganic compounds with lower indices of refraction. Overall, energy resolution of 15% FWHM is typical for 662 keV gamma rays [Nestor and Huang, 1975].

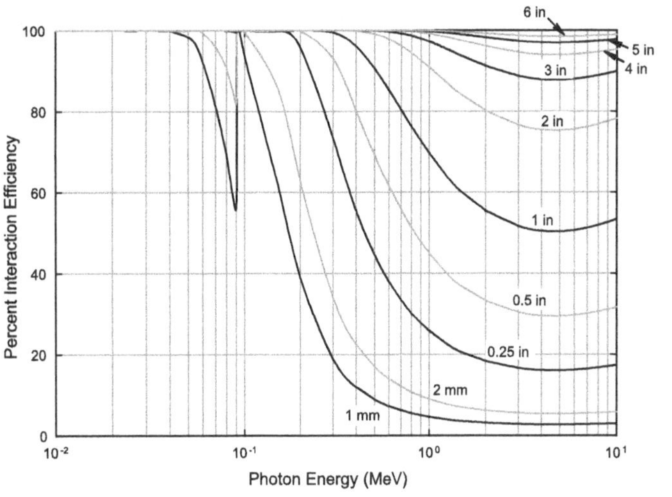

Figure 13.17. The calculated interaction efficiency for photons in BGO as a function of energy and detector depth. The probability of an initial interaction is shown and not the probability of subsequent or multiple interactions.

BGO scintillates without an activator, the luminescence caused by intrinsic $^3p_1 \rightarrow {}^1s_0$ transitions of the Bi^{3+} ions in the crystals [Weber and Monchamp 1973]. The Bi^{3+} ion actually has a small band gap (about 2 meV) between the excited states $^3p_1$ and $^3p_0$ and, thus, has two different transitions, namely the $^3p_1 \rightarrow {}^1s_0$ and the $^3p_0 \rightarrow {}^1s_0$ [Rodnyi 1997]. These two possible transitions help explain the appearance of two emission wavelengths and decay times. The Stokes shift is large enough such that BGO is transparent to its luminescent photons. BGO light yield is significantly dependent upon temperature and shows high thermal quenching above room temperature. Provided that an adequately long electronic shaping time is applied, the absolute light yield of BGO at 77 K can be double its room temperature yield (see Fig. 13.8). Ivanov et al. [1987] report an increase of relative quantum efficiency, compared to NaI:Tl, of 0.13 at room temperature, increasing to 0.44 at 80 K.

Lanthanide Scintillators

Lanthanide halides are those scintillators with the chemical makeup of AB_3, where A is an element in the lanthanide series and B is a halide. These scintillators are generally very hygroscopic and fragile, but many possess outstanding absolute light yield, good relative light yield proportionality, fast decay times, good gamma-ray absorption efficiency (Fig. 13.18), and ultimately superior energy resolution compared to NaI:Tl. Several are now commercially available, although mainly in sizes of 3 × 3 in right cylinders or smaller. For scintillators with the element La as a main constituent, a deficiency is the presence of ^{138}La, which is radioactive with a natural abundance of 0.09%. It emits low energy beta particles along with 1435.8 keV (BR = 0.7) and 788.7 keV (BR = 0.3) gamma rays. Although the gamma-ray emissions can be used as intrinsic calibration energies, they unfortunately also produce an intrinsic radiation background.

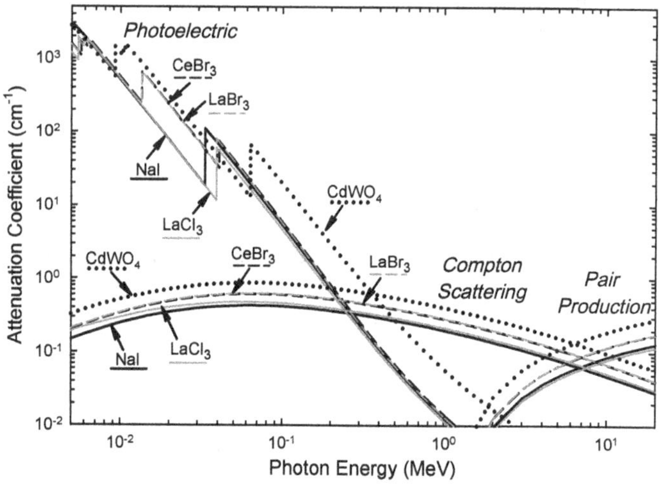

Figure 13.18. Gamma-ray linear attenuation coefficients for a few lanthanide scintillators as compared to NaI. Data are from [NIST XCOM].

Cerium Bromide ($CeBr_3$) Cerium bromide is a lanthanide halide scintillator that has received much attention in recent years. Although it was synthesized as early as 1899 [Muthman and Stüzel 1899], only recently was it recognized as a bright luminescent radiation detection material [Shah et al. 2005]. $CeBr_3$ has a density of 5.1 g cm^{-3}, and with constituent elements Ce ($Z = 58$) and Br ($Z = 35$) it has good gamma-ray absorption efficiency. $CeBr_3$ has a hexagonal crystal structure, is hygroscopic and fragile, easily cleaving along the slip planes [Doty et al. 2007]. The most probable light emission is at 390 nm with a decay constant of 19 ns. The reported light yield is approximately 68,000 photons per MeV [Shah et al. 2005], and the relative light yield is reasonably proportional above 100 keV, although less so than $LaBr_3$:Ce for energies below 100 keV [Khodyuk and Derenbos 2012, Quarati et al. 2013]. Luminescence from $CeBr_3$ arises from an excited state of the Ce^{3+}, and it does not require an activator dopant. It has a refractive index of 2.09 at λ_{max} with a critical angle of 45.9° at a common glass interface and with such a high refractive index, significant amounts of light can be lost from internal reflections. Regardless, energy resolution of 3.2% FWHM at 662 keV was reported for $CeBr_3$ doped with Ca^{2+} [Guss et al. 2014]. Methods have been implemented to address the fragility of $CeBr_3$, and other lanthanide halides, using aliovalent doping to inhibit cracking and had some promising success [Harrison et al. 2009]. Tests with gamma-ray irradiation indicate that $CeBr_3$ can withstand dose exceeding 100 kGy with only a minor decrease is energy resolution performance [Drozdowski et al. 2008b].

Lanthanum Chloride ($LaCl_3$:Ce) Cerium activated lanthanum chloride ($LaCl_3$:Ce) was reported by van Loef et al. in 2001 as a new bright scintillator that can perform in a fashion similar to that of NaI:Tl. The combination of a low mass density of 3.84 g cm^{-3} and only one substantial Z number atomic constituent La (57) yields $LaCl_3$:Ce gamma-ray detection efficiency similar to that of NaI:Tl. The material is very hygroscopic and brittle, and has a hexagonal crystal structure that cleaves easily along the {100} planes.[19] The most probable emission wavelength for $LaCl_3$:Ce λ_{max} is 350 nm, which unfortunately is not well matched to common PMTs and results in a loss in sensitivity to the light emissions. The refractive index is 1.9, which produces a critical angle of 52.1° with a glass interface. The absolute light yield is 49,000 photons/MeV, which is higher than NaI:Tl, but because of the spectral mismatch, its photoelectron response with a bi-alkali PMT ranges between 70% to 90% of that measured with NaI:Tl. Commonly reported energy resolution is near 4.0% FWHM at 662 keV for various crystal sizes. The good energy resolution observed from $LaCl_3$:Ce is most likely due to the relatively good proportionality of the crystal for energies above 30 keV. It is also a fast scintillator with a decay constant of $\tau = 28$ ns. The light yield from $LaCl_3$:Ce decreases with decreasing temperature with an integration time constant of 1 μs; however, if the integration time constant is increased to 16 μs, the light output response is fairly constant over a wide range of temperatures (see Fig. 13.8).

Lanthanum Bromide ($LaBr_3$:Ce) Cerium activated Lanthanum bromide ($LaBr_3$:Ce) was reported in 2002 by van Loef et al. as a new scintillator that can perform with better energy resolution than that of NaI:Tl. After a lull in the discovery of bright scintillators capable of competing with NaI:Tl performance, the discovery of Ce activated $LaBr_3$ helped reenvigorate the search for bright scintillators. $LaBr_3$:Ce has a mass density of 5.06 g cm^{-3}, and with relatively large atomic constituents of 57 (La) and 35 (Br), it has good gamma-ray absorption efficiency. The primary interaction efficiency as a function of gamma-ray energy and detector thickness is shown in Fig. 13.19. The material is very hygroscopic and brittle, and has a hexagonal crystal structure that cleaves easily along the {100} planes.[18] The probable emission wavelength for $LaBr_3$:Ce λ_{max} is 380 nm, which matches well-enough with common PMTs. The refractive index is 1.9 that produces a critical angle of 52.1° with a glass interface. The absolute light yield is 63,000 photons/MeV, and its photoelectron response with a bi-alkali PMT ranges between 130% to 165% of that measured for NaI:Tl. The energy resolution achieved is quite excellent for a scintillation, with reported values below 2.9% FWHM at 662 keV for various crystal sizes. The excellent energy resolution is a combined effect of the absolute light yield and the good relative light yield proportionality. It is also a fast scintillator with a decay constant of $\tau = 16$ ns, a time which requires that the light detection device be capable of following such a fast and bright light output without causing space charge buildup.[20]

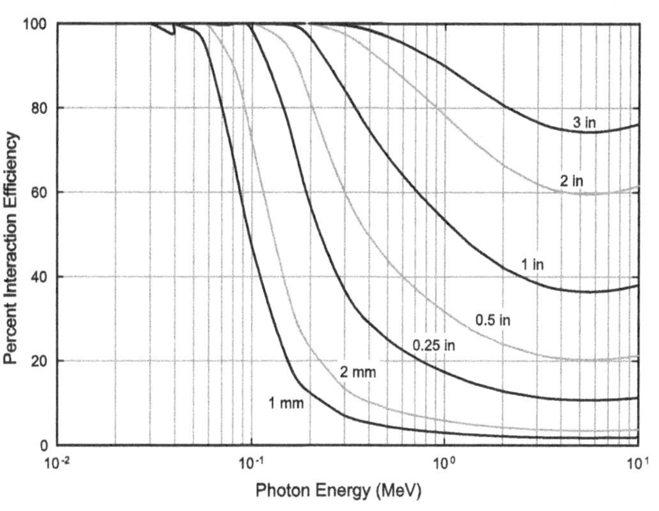

Figure 13.19. The calculated interaction efficiency for photons in $LaBr_3$ as a function of energy and detector depth. The probability of an initial interaction is shown and not the probability of subsequent or multiple interactions.

[19] or, more traditionally, the $\{10\bar{1}0\}$ planes.

[20] Charge buildup from dense electron packages in a PMT can alter the internal voltage and produce non-linear output spectra.

Alternative activator co-dopants have been explored with $LaBr_3$:Ce scintillators, including Li, Na, Mg, Ca, Sr, and Ba as co-dopants to Ce [Alekhin et al. 2013a]. Of these, usage of Sr as a co-dopant with Ce shows excellent improvement, reported as approaching 2% FWHM for 662 keV gamma rays from ^{137}Cs [Alekhin et al. 2013b]. This improvement in energy resolution is attributed to a reduction in scintillation light losses from competing radiationless recombination processes and also an observed increase in light yield. Consequently, the light response linearity is improved, even in the low energy region between 10 to 100 keV [Alekhin et al. 2013b]. A drawback to co-doping with Sr is typically longer, and multiple, decay times, including some phosphoresence with decay times in microseconds.

Rare Earth Orthosilicates

Gadolinium Oxyorthosilicate (Gd_2SiO_5:Ce or GSO:Ce) Gadolinium oxyorthosilicate is a lanthanide silicate scintillator discovered in 1982 by Takagi and Fukazawa. It has one element with a high Z constituent (Gd $Z = 64$, Si $Z = 14$, O $Z = 8$) and also a density of 6.71 g cm^{-3} and, thus, makes it a relatively efficient absorber of gamma rays. GSO has a monoclinic crystal structure and is not hygroscopic. Doped with the activator Ce, it has a fast decay component of 60 ns and a slow decay component of 600 ns. The light yield of the fast component is about 8,000 photons/MeV with about 20% light response to a common bi-alkali PMT compared to that of NaI:Tl, and the light yield of the slow component is about 1,000 photons/MeV [Rodnyi 1997]. The most probable decay component is centered at $\lambda_{max} = 430$ nm. This emission mechanism is believed to be mainly from self-trapped holes. GSO has an index of refraction of 1.85 which yields a critical angle of 54.2° at a glass interface and thus indicates that a large percentage of photons can be internally reflected. Energy resolution of less than 7.0% FWHM at 662 keV has been reported [Melcher et al. 1996, Kamae et al. 2002]. GSO has a natural radioactive background component from ^{152}Gd, which has a natural abundance of 0.2% and emits 2.14 MeV alpha particles.

The migration coefficient of Ce in GSO is near unity, meaning that the dopant is uniform throughout the crystal [Takagi and Fukazawa 1983]. A uniformity study performed by Ishibashi et al. [1998] indicated that the GSO doped with 0.5% mol Ce had less than 7% light variation over an ingot 280 mm long and 80 mm in diameter. With growth improvement, excellent results were achieved by Kurashige et al. [2004] with larger 100-mm diameter and 290 mm long crystals and produced only 0.42% standard deviation in light yield across the diameter and only 0.4% standard deviation along the length. Further, the decay constant was shown to be unaffected by small changes in Ce distribution and were found to have less than standard deviations of 1.7% longitudinally and 1.0% standard deviation radially.

Studies show that GSO is somewhat non-proportional over a wide energy range [Balcerzyk et al. 2000], but has better proportionality than either LSO or YSO. GSO is also a thermally stable scintillator, and had been used for oil well logging [Roscoe et al. 1992]. Due to its good gamma-ray detection efficiency and its superior energy resolution to that of BGO, it has also found some use in medical imaging PET instruments [Ficke et al. 1994].

The removal of impurities from the crystal improved light yield by a significant amount and increased the light yield by about 30% [Kamae et al. 2002]. Energy resolution for some of these samples was 6.7% FWHM at 662 keV. It was also found that co-doping with Zr improved light yield as well and increased the yield by 20% [Shimura et al. 2006], with best results having a Zr concentration of 200 ppm.

With the enormous thermal-neutron absorption cross section of natural Gd (49,000 b), GSO has been studied as a possible neutron detector by reliance upon either the prompt reactions ^{157}Gd(n,γ)^{158}Gd or ^{155}Gd(n,γ)^{156}Gd as the detection mechanism [Reeder 1994]. Conversion electrons and gamma-ray emissions can be detected after a neutron absorption, and the important characteristic emissions can be absorbed in only 75 μm of GSO. Unfortunately, GSO is also an efficient gamma-ray detector; hence, it is sensitive to background radiations that can be falsely interpreted as neutron induced events.

Lutetium Oxyorthosilicate (Lu_2SiO_5:Ce or LSO:Ce) Cerium activated lutetium oxyorthosilicate (LSO:Ce) is a relatively new scintillator discovered and developed by Melcher in 1990 [see Melcher 1990; Melcher 1991; Melcher and Schweitzer 1992; Daghighian et al. 1993]. The scintillator was originally developed for oil well logging measurements, but was found to have severely degraded energy resolution at the elevated temperatures typically encountered during oil well logging measurements. Further, the energy resolution was poor compared to common NaI:Tl detectors. However, it was measured to have 75% of the absolute light yield of NaI:Tl (about 31,000 photons/ MeV), which at the time was a tremendous breakthrough for inorganic scintillators [Daghighian et al. 1993]. LSO:Ce has a mass density of 7.4 g cm^{-3}, and with its large Z number constituent Lu (71), it has excellent gamma-ray absorption efficiency, comparable to that of BGO. The primary interaction efficiency as a function of gamma-ray energy and detector thickness is shown in Fig. 13.20. Daghighian et al. [1993] report two decay constants of 12 ns (30%) and 42 ns (70%), with a most probable emission wavelength λ_{max} of 420 nm. However, from various sources, the reported decay constant ranges between 37 ns to 47 ns [Rexon 2015; Ludziejewski et al. 1995]. It has relatively proportional light yield for gamma-ray energies greater than 200 keV, but not below 100 keV. The index of refraction is 1.82 which produces a critical angle of 55.5° at a glass interface. The crystal is robust, is not hygroscopic, and rates 5.8 on the Moh hardness scale. LSO:Ce has found good use as a replacement for BGO in PET scan systems [Daghighian et al. 1993]. It has 190% of the absolute light yield measured for common BGO crystals while having almost the same gamma-ray interaction efficiency. Energy resolution for 662 keV gamma rays as low as 10.3% FWHM have been measured, much better than that for BGO. Further, it is radiation hard, with a reported radiation limit of greater than 10^6 rads. A significant problem with LSO:Ce is that it is naturally radioactive because of ^{176}Lu (2.59% isotopic abundance) which emits beta particles and gamma rays. This problem is further exacerbated by unintentional doping with other radioactive impurities from the crystal growth source materials. However, for specific applications such as PET measurements, which uses coincidence counting, this natural background is of less concern. Another problem is that LSO:Ce has long phosphorescence, which has been measured to last several seconds [Szupryczynski et al. 2004]. Nassalski et al. [2007] generally concluded the long phosphorescence precludes LSO:Ce from application in CT systems because of the detector requirements.

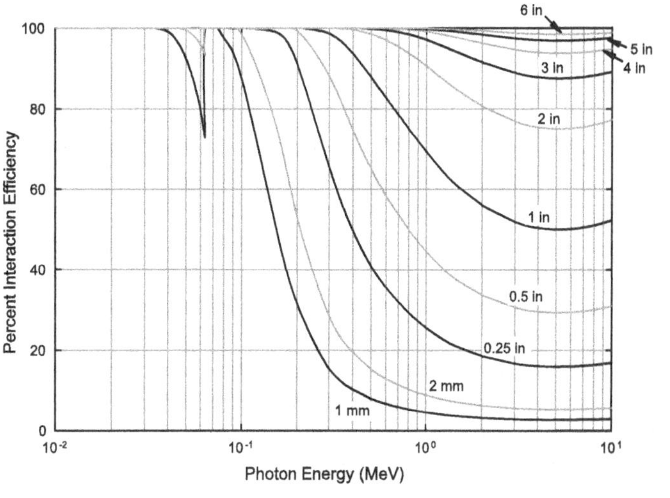

Figure 13.20. The calculated interaction efficiency for photons in LSO as a function of energy and detector depth. The probability of an initial interaction is shown and not the probability of subsequent or multiple interactions.

Yttrium Oxyorthosilicate (Y_2SiO_5 or YSO:Ce) Ce-doped yttrium orthosilicate (YSO:Ce) is a non-hygroscopic rare-earth oxyorthosilicate. The material has a density of 4.45 g cm^{-3}. YSO is disadvantaged as a gamma-ray spectrometer because of its relatively low Z number constituents (the largest is $Z = 39$ for Y) and moderate mass density. YSO:Ce has a most probable wavelength λ_{max} of 420 nm and relatively good energy proportionality above 100 keV [Cutler et al. 2009]. The decay constant τ is about 50 ns. Commercial YSO:Ce usually has an absolute light yield around 10,000 photons/MeV [Melcher 1996], although a higher light yield of about 24,000 photons/MeV has been reported for optimized crystals [Balcerzyk 2000, Cutler 2009]. Co-doping with Ce/Ca has been studied with reportedly small effect and produced an absolute light yield of 21,200 photons/MeV. The index of refraction is 1.8 to produce a critical angle of 56.4° at a glass interface. The reported energy resolution often is 9.0% to 9.4% FWHM for 662 keV gamma rays, although energy resolution as low as 7.4% FWHM has been reported [Dahlbom et al. 1997].

Lutetium Yttrium Oxyorthosilicate ($Lu_{1.8}Y_{0.2}SiO_5$:Ce or LYSO:Ce) Lutetium yttrium oxyorthosilicate (LYSO:Ce) has been grown in an attempt to include the main advantages of both LSO and YSO crystals [Cooke et al. 2000; Pepin et al. 2004; Chen et al. 2007]. LYSO:Ce is a commercially available scintillator with similar properties as LSO:Ce [Rexon 2015, Omega 2016]. In summary, the density of LYSO is 7.4 g cm^{-1} with a refractive index of 1.82. The decay constant has been measured to be between 40 and 44 ns with a most probable emission wavelength between 420 and 428 nm. It has similar radiation hardness (greater than 10^6 rad) as LSO, but it produces increased afterglow phosphorescence with increased radiation damage. The light yield is also the same as LSO, at approximately 31,000 photons/MeV. It is robust, non-hygroscopic with a Moh hardness of 5.8. Energy resolution is similar to that of LSO, near 10% FWHM for 662 keV [Chen et al. 2007]. Because of the natural abundance of ^{176}Lu, it also has an intrinsic radioactive background. Cooke et al. [2000] report lower temperature dependence on light yield than observed with LSO. Further, Cooke et al. [2000] note that the main advantages to LYSO over LSO are lower production costs, fewer inclusion defects in the crystals, and easier incorporation of the Ce activator dopant in the crystal lattice.

Elpasolites

Elpasolites[21] are a new class of scintillators of the form A_2BRX_6, where A and B are alkali metals, R is a rare earth, and X is a halogen. They have essentially a double perovskite structure, and many maintain a cubic structure over a broad range of formations [see Doty et al. 2012]. The possible combinations number in the thousands, and several have been identified as potentially good scintillators for gamma-ray spectroscopy and neutron detection [Gundiah et al. 2014; Wei et al. 2014; Doty et al. 2012]. Listed here are only a few combinations that show promise.

Cesium Lithium Yttrium Chloride (Cs_2LiYCl_6:Ce or CLYC:Ce) The relatively new cesium-based elpasolite scintillator Cs_2LiYCl_6:Ce, or CLYC:Ce, is a dual-use inorganic scintillator that can detect both gamma rays and neutrons. It has a density of 3.31 g cm^{-3}, with its highest Z component being Cs ($Z = 55$). However, the actual density of Cs atoms comes to 6.94×10^{21} cm^{-3}, approximately 66% of the atomic fraction for CsI, with an effective atomic number Z_{eff} of 44.5 [Lecoq et al. 2006]. CLYC:Ce is hygroscopic and, thus, must be encapsulated. The refractive index of CLYC is 1.81, which produces a critical angle at a common glass interface of 56°. CLYC:Ce has decay constants at 600 ns and 6 μs both of which are relatively long for most scintillator applications [Combes et al. 1999]. The luminescent emissions for Ce doped CLYC are centered at 372 nm and 400 nm, while undoped CLYC has a single emission centered at 305 nm [Combes et al. 1999]. With a 1 μs shaping time, between 7,000 photons/MeV (no Ce doping) to 10,200 photons/MeV (3% Ce doping) are observed. With a longer shaping time of 10 μs, between 22,000 photons/MeV (no Ce doping) to 10,800 photons/MeV (3% Ce doping) are measured. Despite the relatively low light yield and long decay constants, good energy resolution has been reported for CLYC:Ce, as low as 4.2% FWHM for 662 keV gamma rays [RMD 2016].

The lithium constituent of CLYC:Ce allows neutron detection through the ^{6}Li(n,^{3}H)^{4}He reaction. For natural Li as a constituent, the atomic density of ^{6}Li in CLYC is 2.6×10^{20} cm^{-3} to produce a macroscopic cross section for the ^{6}Li(n,^{3}H)^{4}He reaction of 0.248 cm^{-1}, or an absorption length of about 4 cm. If the Li is replaced with enriched ^{6}Li, the thermal-neutron absorption macroscopic cross section can be increased to 3.27 cm^{-1}, and the absorption length is reduced to 3 mm. Unfortunately, CLYC also has competing neutron absorbers, mainly Cl ($\sigma_{Cl} = 33.5$ b) and Cs ($\sigma_{Cs} = 28$ and 2.6 b), that ultimately limit its neutron detection

[21]The etymology comes from the location where these minerals were first found in El Paso County, Colorado, USA.

efficiency. With six halogen atoms per molecule, the neutrons lost to parasitic absorptions can be significant, limiting the maximum thermal neutron detection efficiency to approximately 22% for CLYC:Ce loaded with natural Li and approximately 78% neutron detection efficiency for CLYC:Ce loaded with enriched ^{6}Li.

There is also a core valence emission, with a decay constant of 2 ns with a yield of 660 photons/MeV (no Ce doping) to zero photons/MeV (3% Ce doping), produced by the radiative recombination of a valence electron with a core hole [Bessiere et al. 2005]. This core valence emission appears under gamma-ray irradiation, but not under alpha particle irradiation. Hence, detection of this emission can be used to distinguish between gamma-ray and heavy charged particle emissions such as those produced from neutron interactions.

Cesium Lithium Lanthanum Bromide ($Cs_2LiLaBr_6$:Ce or CLLB:Ce) Also new is the cesium-based elpasolite scintillator $Cs_2LiLaBr_6$:Ce, or CLLB:Ce, which also has dual use as a gamma-ray and neutron detector. It has a density of 4.2 g cm^{-3} and its highest Z components are Cs ($Z = 55$) and La ($Z = 57$). The density of Cs and La atoms comes to 8.51×10^{21} cm^{-3}, approximately 81% of the atomic density of Cs in CsI.

CLLB:Ce has multiple decay constants at 55 ns and at about 270 ns, and there is no reported core valence emission for CLLB. The luminescent emissions for Ce doped CLLB:Ce are centered at 390 nm and 420 nm [Shirwadkar et al. 2011]. With Ce doping, the luminescent yield is between 50,000 photons/MeV and 60,000 photons/MeV [Glodo et al. 2011; Shirwadkar et al. 2011]. This high light output, along with the fact that the luminescent yield is more proportional than that found with CLYC:Ce, indicates that CLLB:Ce is a promising new scintillator. Indeed, energy resolution of 3.0% to 2.9% FWHM for 662 keV gamma rays has been reported [Glodo et al. 2011; Shirwadkar et al. 2011].

As with CLYC:Ce, the lithium constituent of CLLB:Ce enables neutron detection through the ^{6}Li(n,^{3}H)^{4}He reaction. For natural Li as a constituent, the atomic density of ^{6}Li in CLLB is 2.16×10^{20} cm^{-3} to produce a macroscopic cross section for the ^{4}He(n,^{3}H)^{4}He reaction of 0.2027 cm^{-1}, or an absorption length of 4.93 cm. If the Li is replaced with enriched ^{6}Li, the thermal neutron absorption macroscopic cross section can be increased to 2.67 cm^{-1} and the absorption length is reduced to 3.75 mm.

Other Notable Elpasolites Other bright elpasolite scintillators that show promise include cerium activated Cs_2NaLaI_6:Ce, $Cs_2LiLaCl_6$:Ce, $Cs_2NaLaBr_6$:Ce, and Cs_2LiLaI_6:Ce. These scintillators are relatively proportional, provide good light yield, and have short decay constants on the order of 50 ns [Doty et al. 2012]. Kerisit et al. [2014] report that elpasolites with Br as the halogen rather than Cl should theoretically have higher light yields and quote theoretical light yields above 100,000 photons/MeV for $Cs_2LiLaBr_6$:Ce and Cs_2LiYBr_6:Ce. Gundiah et al. [2014] report a good light yield of about 46,000 photons/MeV for $Cs_2NaLaBr_6$:Ce with an energy resolution of 3.9% FWHM at 662 keV. Two emission maxima were observed at 387 nm and 415 nm, both adequately matched to a bi-alkali PMT. A recent introduction to the elpasolite family that shows great promise is $Cs_2LiLa(Br_{6-x}Cl_x)$ or CLLBC:Ce, which has good gamma-ray energy resolution ($\sim$ 3% FWHM at 662 keV) and retains the neutron detection characteristic of CLYC:Ce [Shirwadkar et al. 2012]. The substitution of Br for Cl works to reduce parasitic neutron losses, while the substitution of La for Y works to slightly increase parasitic neutron losses.

Rb_2LiYBr_6:Ce, and many others, show bright light output for neutron absorption [Birowosuto et al. 2008]. Rb_2LiYBr_6:Ce produces prompt reaction products from the ^{6}Li(n,^{3}H)^{4}He reaction, producing between 59,000 $\pm$ 5,400 to 83,000 $\pm$ 8,300 photons per neutron capture, depending on the Ce^{3+} doping concentration (between 0.1 and 0.5 %). The light yield for gamma rays varies between 16,500 $\pm$ 1,600 to 23,000 $\pm$ 2,300 photons/MeV for Ce^{3+} for concentrations between 0.1$ and 0.5%, respectively. It has at least three main decay components, reported as 71 ns (31%), 400 ns (53%), and 1400 ns (16%). Energy resolution for 662 keV gamma rays was reported as low as 4.6% FWHM [van Eijk et al. 2005]. The difference in light output between gamma-ray interactions, along with good energy resolution, allows pulse height discrimination between neutron and gamma-ray events.[22]

Ceramic and Glass Scintillators

Methods of producing fully transparent ceramic scintillators were initially developed by GE Corporation [Cusano et al. 1980; Dibianca et al. 1985] and ceramic casting offers an alternative method of scintillator production to that of growing a single crystal. Although each type of scintillator may have unique manufacturing processes particular to the material, there are several steps generally applicable to the manufacture of ceramic scintillators. The materials should be very pure (greater than 99.99%), near 100% phase purity, a total high specific surface area greater than 5 m^2g^{-1}, and a median particle size of 3 microns or less [Greskovich and Duclos 1997]. The general process involves the production and subsequent introduction of nanoparticles of scintillator material into a vessel for compaction. This unfired material is usually termed a "green" ceramic. The green ceramic is then mixed with the requisite scintillator host and fluorescent activator dopants. This mixing is one of the main advantages of the ceramic process because the desired activator concentration can be maintained in the ceramic far more easily than with the traditional crystal growth method. The compacted material is heat treated to temperatures of approximately 1,200°C in air, followed by vacuum sintering (at about 10^{-6} torr) at temperatures up to 2,000°C. Finally, hot pressing at pressures up to 29,000 psi in argon and at a temperature of 2,000°C is used to achieve a maximum density. The final product is a ceramic billet with grain sizes on the order of 10 microns.

There are multiple advantages to ceramic scintillators, namely they have a high radiation hardness compared to halide single crystals, they have relatively low production costs, they are hard and non-hygroscopic, they are machinable, they are environmentally robust, and they can be manufactured in large sizes that might be difficult to produce as single crystals. Among the earliest materials investigated as ceramic scintillators were BaFCl:Eu, LaOBr:Tb, CsI:Tl, $CaWO_4$, and $CdWO_4$ [Cusano et al. 1980]. Cubic crystal structure has been identified as a condition for successful production of high light yield ceramic scintillators [Debianca et al. 1985]. Non-cubic crystals have increased refraction at the grain boundaries that increases light self-absorption and ultimately decreases light yield. Unfortunately, the crystal grains for the materials listed in the patent by Cusano et al. [1980] are non-cubic, except for that of CsI:Tl which is hygroscopic. Bixbyite ceramic scintillator structures[23] were investigated by Debianca et al. [1985], and include Gd_2O_3, Y_2O_3, La_2O_3, Lu_2O_3. Neither Gd_2O_3 nor La_2O_3 are cubic in form, but are made cubic by incorporating about 50 mole percent of Y_2O_3 into the ceramic. Both Y_2O_3 and Lu_2O_3 are cubic structured. Lu_2O_3:Eu has been developed for x-ray imaging [Lempicki et al. 2002] and has reported properties of 9.4 g cm^{-3} mass density, a λ_{max} of 610 nm, a light yield of about 60% of that of CsI:Tl, and a long decay time constant τ of 1.3 ms. The production of $(Y,Gd)_2O_3$:Eu ceramic scintillators led to their incorporation in CT scanners [Duclos et al. 2003]. The reported properties of $(Y,Gd)_2O_3$:Eu are a λ_{max} of 611 nm, a decay constant τ of 960 μs, and an optical attenuation coefficient of 1 cm^{-1}.

Numerous ceramic scintillators have been investigated and much significant progress has been achieved in recent years. These ceramics include various garnets, perovskites, bixbyites, and pyrochlores[24] [Cherepy et al. 2008]. Two interesting ceramic scintillators recently developed are $Gd_{1.5}Y_{1.5}Ga_2Al_3O_{12}$ (GYGAG) and $Gd_{0.3}Lu_{1.6}Eu_{0.1}O_3$ (GLO) [Cherepy et al. 2014, 2015]. Cerium doped gadolinium yttrium gallium aluminum garnet (GYGAG), $(Gd,Y)_3(Ga,Al)_5O_{12}$ or GYGAG:Ce, have been under development for many years. The

[22]Note that the Q-value for the ^{6}Li(n,^{3}H)^{4}He reaction is 4.7 MeV, giving an upper light yield of about 17,300 photons/MeV as is similar to that for gamma-ray events. However, it is unlikely that gamma rays with energies approaching 4.78 MeV are fully absorbed, so discrimination is still possible.

[23]Named after Maynard Bixby, who discovered the ore in 1897.

[24]The etymology comes from the Greek $\pi\tilde{\upsilon}\rho$ (pyr = fire) and $\chi\lambda\omega\rho\acute{o}\varsigma$ (chloros = green) because pyrochlores often turn green with a common flame blowpipe analysis.

material is a robust transparent ceramic with mass density of 5.8 g cm^{-3} and an absolute light yield of approximately 50,000 photons/MeV. GYGAG:Ce has at least two decay times, a fast component at 250 ns and a slow component at 1.7 μs. This unfortunate slower component is thought to be a consequence of the energy transfer process in the scintillator [Cherepy et al. 2011]. Regardless, energy resolution of 4.5% FWHM at 662 keV has been reported for a GYGAG:Ce sample coupled to a PMT when operated with a 4 μs shaping time. Better energy resolution is achieved when coupled to a Si photodiode with reported values of 3% to 3.5% FWHM at 662 keV [Cherepy et al. 2015]. Cerium doped gadolinium lutetium europium oxide (GLO:Eu) is used for MeV-photon radiography. The material has a high mass density of 9.1 g cm^{-3} and an absolute light yield of approximately 55,000 photons/MeV [Cherepy et al. 2015].

Ce-activated glass scintillators were originally developed by Corning Glass Company, as evidenced from the patents listed in the literature [see Ginther and Schulman 1958]. Of these original scintillation glasses, it was Ce activated high silica glass that performed best [Ginther and Schulman 1958]. Shortly thereafter, Voitovetskii et al. [1960a, 1960b] reported different compositions of $Li_2O{\cdot}Si_2O$:Ce glass as neutron detecting scintillators. Ginther [1960] determined that a bright variation of a Li glass scintillator was a mixture of 26% MgO, 13% $LiO_{0.5}$, 10% $AlO_{1.5}$, 50% SiO_2, and 1% $CeO_{1.5}$, which yielded a light yield of 14% of that of NaI:Tl (nominally 5,000 to 6,000 photons/MeV) and has a most probable emission wavelength of about 390 nm. The Ginther and Schulman study revealed that contaminants in the glass can adversely affect brightness, especially for alkali earths Ca and Ba contaminants.

Glass scintillators can be machined or formed into various shapes that may not be achievable with common crystalline solids. Further, glass scintillators are relatively robust and shock resistant, and glass scintillators have a relatively constant light yield over a wide range of temperatures. These scintillators are used for a variety of applications, including neutron detection, oil well logging, and charged-particle detection in extreme environments. Modern commercial Ce-activated Li glass scintillators can be acquired with different Li activator concentrations and enrichments such as natural Li, 96% enriched ^{6}Li, and fully depleted of ^{6}Li. Glass scintillators with ^{6}Li can be used for combined neutron and gamma-ray detection, whereas glass scintillators depleted of ^{6}Li can be used in mixed radiation fields when neutron induced counts are undesirable. The resulting neutron attenuation efficiencies for various commercial Li glasses are shown in Fig. 13.21. The decay times generally range between 20 to 105 ns, and depend on the Li concentration and the type of radiation particles interacting in the scintillator. These scintillators have higher light yields for irradiation by electrons than by alpha particles, with an α/β ratio of about 0.23. Glass density is also a function of Li concentration, ranging from 2.42 g cm^{-3} for 24% Li up to 2.64 g cm^{-3} 7.5% concentration [Scintacor 2016]. The light yield of these glass scintillators is reported to be between 20% and 30% of that of anthracene (or 8.7% to 13% of NaI:Tl), depending on the type of glass and Li concentration. The index of refraction varies between 1.55 to 1.58 at the peak emission wavelength [see Table 13.2, also Applied Scintillation 2016].

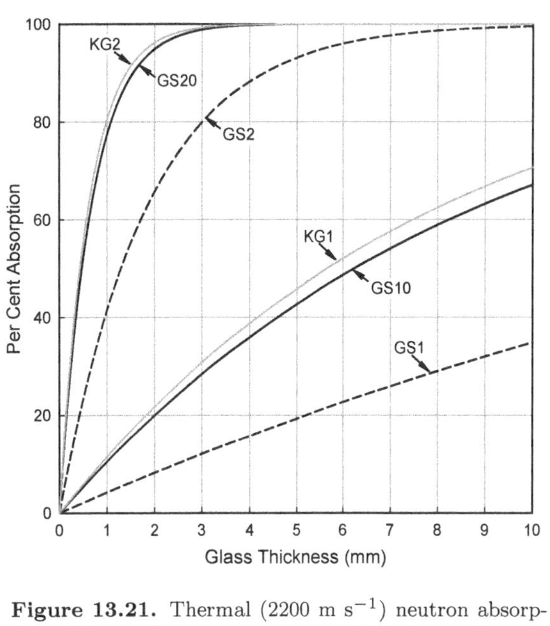

Figure 13.21. Thermal (2200 m s^{-1}) neutron absorption efficiency for various commercial Li glasses. Properties are listed in Table 13.2.

Other scintillating glasses have been explored and include boron-based scintillating glass and Li glasses that have been activated with either Tb or Eu. Boron-based glass scintillators have lower light yields than Li

Table 13.2. Properties of some commercial Ce activated ^{6}Li-glass (natural Li/ 95% enriched ^{6}Li) .

Glass Designation	GS1/GS2	GS10/GS20	KG1/KG2
Density (g cm^{-3})	2.64	2.5	2.42
Isotropic Ratio	2.4%	6.6%	7.5%
Refractive Index	1.58	1.55	1.57
Max Emission λ (nm)	395	395	395
Relative Light Yield*	22% - 34%	20% - 30%	20%
Decay Time (ns)	50 - 70	50 - 70	50 - 70
Available Thickness (mm)	0.1 - 10	0.1 - 10	0.1 - 10
Linear Attenuation Coefficient (cm^{-1})†	GS1: 0.428	GS10: 1.114	KG1: 1.225
	GS2: 5.35	GS20: 14.85	KG2: 16.34

*Relative to anthracene.
†For 2200 m s^{-1} neutrons.

glasses, and consequently have less practical application [Bollinger et al. 1962]. Glasses with Tb activators have higher light yields than Ce activated glasses, up to 50,000 photons/MeV [Pavan et al. 1991], and are reported to be 2 to 3 times more radiation hard than Ce activated glasses. However, the most probable wavelength of emission is around 550 nm with long decay times of about 3 to 5 ms. Glasses with Eu as the dopant were found to have a slightly higher light yield than that of Ce activated glass [Fujimoto et al. 2015], namely about 110% of that of a common GS20 Li glass scintillator. The most probable wavelength of emission is near 450 nm with a relatively long decay time of 1230 ns [Fujimoto et al. 2015].

13.3 Organic Scintillators

Because the energy bands and corresponding energy levels in inorganic solids appear as a consequence of a periodic potential, dissociation of the solid either through melting, decomposition, or solvation usually eliminates the luminescence. Organic scintillator luminescence originates from molecular π bonds and, thus, is fundamentally different from inorganic solids that emit light from luminescent centers such as defects and impurity dopants in a crystalline lattice. Organic scintillators depend primarily on the molecular structure of the material for the scintillation mechanism, and as a result, organic scintillators can emit photons as either solids, liquids, or gases.

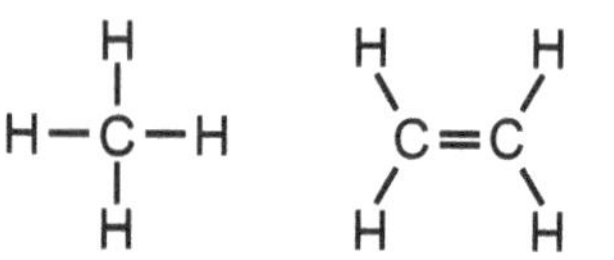

(a) (b)

Figure 13.22. Common depiction of σ and π bonds. In (a), methane is formed by four σ bonds between carbon and four hydrogen atoms. In (b), ethylene is formed with a σ bond and a π bond between the two carbon atoms, with the remaining four hydrogen atoms forming σ bonds with the carbon atoms.

Overlapping s and p orbitals can form a strong σ bond, for instance, as is the case of methane in which the four electrons in the carbon p orbitals combine with the s orbital electrons of four hydrogen atoms to produce CH_4. As described with fundamental organic chemistry, π states appear from the formation of π bonds, which occur when two p orbitals overlap in a side-by-side configuration. For instance, the simple molecule of ethylene (C_2H_4) is formed with the two carbon atoms sharing electrons in two p orbitals, the first producing a σ bond and the second forming a parallel π bond, while the remaining electrons form σ bonds with hydrogen atoms (Fig. 13.22).[25] Although a π bond is weaker than a σ bond, together they

[25]The graphic depiction in Fig. 13.22b is usually used in organic chemistry to depict a π bond. However, the bond should be more correctly thought of as a single strong electron bond between the carbon atoms with a weaker associated bonded electron traveling in the vicinity of the bond.

produce a stronger bond than either individually. The π bond is a common building block for many organic scintillation materials, such as found in benzene rings (Fig. 13.23).

In Fig. 13.24 an energy diagram is shown that is typical of an organic scintillator. An independent molecule of organic scintillation material can have an electron excited through the π states up from the ground state into an excited singlet state, of which there are many levels. Hence, electrons in the molecule can be excited through molecular states rather than atomic states.

There are many vibrational states associated with the ground states, typically denoted by S_{0x} in which x refers to one of the vibrational sub-states. There are also numerous excited singlet states as well as excited triplet states associated with the carbon π bonds. Electrons that gain energy rise to one of the excited vibrational states and generally fall rapidly to the lowest S_{10} state, which then de-excite through two possible channels. If the electrons de-excite directly from the S_{10} state to one of the S_{0x} states, the light emission is rapid and is referred to as scintillation fluorescence. Decay times for fluorescence are typically only a few nanoseconds, and fluorescent emission can be easily linked to individual radiation events. However, if the electrons de-excite by crossing to the triplet states T_{1x} and then fall to one of the S_{0x} states, the light emission is slow and is referred to as scintillation phosphorescence. This second light producing mechanism is undesirable because phosphorescent emission is slow and continues to produce afterglow for extended periods of time and, hence, cannot be directly linked to individual radiation events, especially in high-radiation fields. Regardless, the main point to notice is that organic scintillators depend on the organic structure, often a benzene ring structure, and do not need activator dopants for the scintillation mechanism. Hence, they also do not need to be crystalline or polycrystalline in structure. As a result, organic scintillators can be formed as solids, liquids, gases, and plastics. Some common organic scintillators are listed in Table 13.3.

Organic scintillators are composed mostly of hydrogen and carbon, both of which are poor absorbers of gamma rays. They are also notoriously non-linear in light output for heavy ion radiation. However, they are much more linear in response to electrons and beta particles, and the low atomic numbers for the constituents tend to make light ion backscattering almost negligible. Hence, organic scintillators are usually used for beta particle and electron detection.

Another use for hydrogen rich detector materials is fast neutron detection. As discussed in Chapter 4, and later Chapter 18, energetic neutrons lose more energy by scattering from low A materials than high A materials. Fast neutrons scatter off of the hydrogen and carbon in the organic scintillator, producing recoil hydrogen and carbon atoms that then slow down by causing ionization and excitation of other molecules in the organic scintillator. Because organic scintillators depend on molecular structure for light emission, other materials can be mixed in the scintillator without destroying the scintillation process. For instance ^{10}B, ^{6}LiF, or Gd can be mixed into an organic solution or a plastic to make them more sensitive to neutrons. Likewise, heavy metal particles, such as Pb, can be mixed into the organic or plastic to make them more sensitive to gamma rays. There is of course a limit to the amount of absorber material that can be added because the scintillator transparency reduces with increases in foreign material.

Overall, organic scintillators provide an excellent option when a larger less expensive detector is needed. Although not well suited for use in gamma-ray spectroscopy, they are quite useful as beta particle and fast neutron detectors. Moreover, because these detectors are composed of hydrogen, carbon, and oxygen with an average density of 1.032 g cm^{-3}, they make near "tissue equivalent" detectors, which is desired for dosimetry measurements.

13.3.1 Theory of Scintillation for Organic Scintillators

Radiation absorption within an organic scintillator produces energetic electrons, or recoil nuclei, that can continue to travel through the scintillator and excite more electrons. There are many possible vibrational and rotational energy levels and, consequently, the ground state (S_0) and the singlet and triplet states (S_1

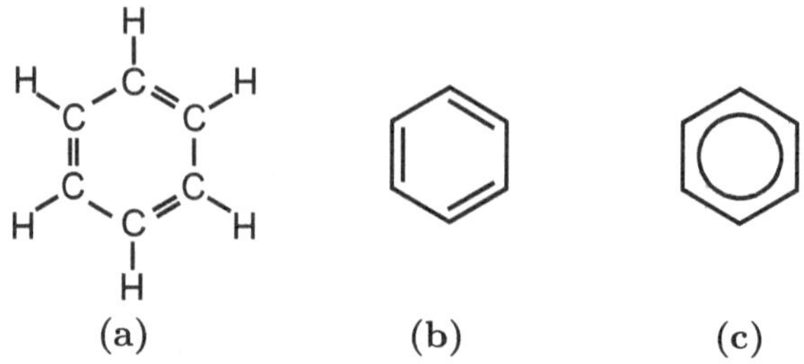

Figure 13.23. Three representations of a benzene ring. The shorthand version of (b) or the modern shorthand version (c) are now in more common usage. It is assumed that the "corner" carbon atoms of the benzene ring are attached to hydrogen atoms unless otherwise indicated.

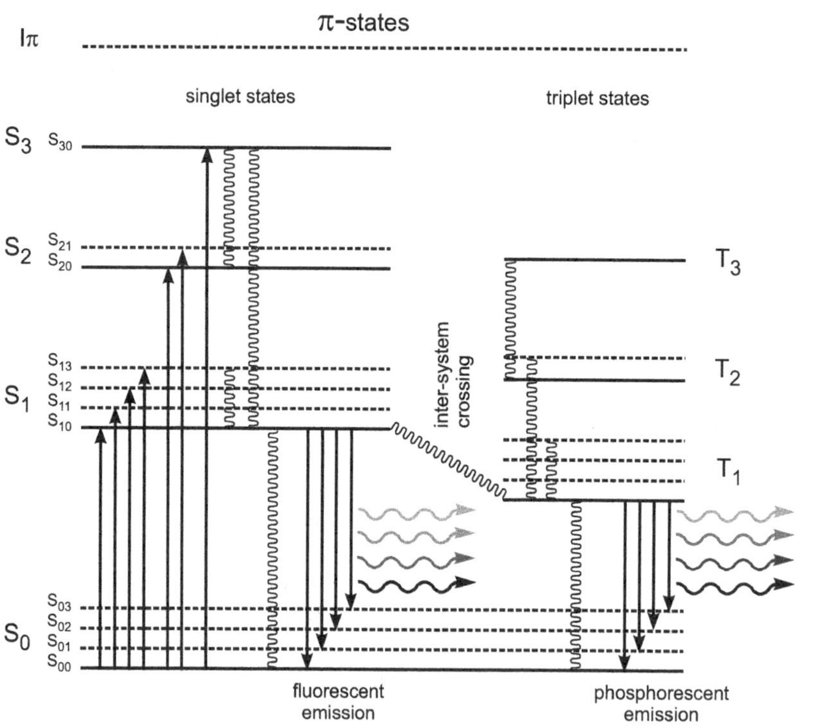

Figure 13.24. Jablonski diagram of the two basic methods by which an organic scintillator produces light. π electrons in the organic molecule are excited into upper vibrational states from a radiation event and rapidly de-excite to the lowest S_{10} state. Electrons that then de-excite directly to the S_{0x} states contribute to scintillation fluorescence. Those electrons that transfer to the triplet states fall to the T_{10} state, and gradually de-excite to the S_{0x} states, a process known as phosphorescence. Straight arrows are radiative transitions, while "squiggly" lines indicate non-radiative transitions. After Birks [1964].

Table 13.3. Common organic scintillator materials and their properties. The light yield is units of % that of anthracene, which, for comparison, is about 40% that of NaI:Tl.

Scintillator				Light Yield	λ_{max} (nm)	τ (ns)	Density (g/cm^3)	Refract. Index n	H/C Ratio	Loading	Softening Point (°C)	Notes
Crystalline												
Anthracene				100	447	30	1.28	1.595	0.714		215	β det. and spec.
(*E*)-Stilbene (or *trans*-Stilbene)				70	410	3.5	0.97	1.658	0.857		~125	α, β and n-spec.
p-Terphenyl (doped)				135	420	3.7	1.23	1.65	0.78		~213	doped, n-spec.
Pyrene				60[a]	477	90	1.27	1.85	0.625		~146	
p-Quaterphenyl				85[a]	438	8			0.75		~318	difficult to grow
Diphenylacetylene				40[a]	390	7	0.99	1.606	0.714		62.5	
Naphthalene				15	345	75	1.15	1.58	0.8		~79	low light yield
Liquid												
Legacy Standard[b]	Saint-Gobain	Eljen	Rexon									
NE 220	BC 220	EJ 351		65	425	3.8	1.036	1.44	1.65			
NE 213	BC 501A	EJ 301		78	425	3.2	0.87	1.505	1.212		26	γ, fast n
NE 224	BC 505	EJ 305		80	425	2.5	0.877	1.505	1.331		48	γ, fast n
		EJ 309		75	424	3.5	0.964	1.57	1.25		144	
NE 226	BC 509	EJ 313		20	425	3.1	1.61	1.377	0.0035	F	10	
NE 235H	BC 517H	EJ 321H		52	425	2	0.86	1.48	1.89		81	
NE 235L	BC 517L	EJ 321L		39	425	2	0.86	1.47	2.01		102	
	BC 517P			28	425	2.2	0.85		2.05		115	
	BC 517S			66	425	2	0.87		1.70		53	
NE 235C	BC 519	EJ 325		60	425	4	0.87	1.49	1.73		63	
NE 323	BC 521	EJ 331		60	425	4	0.89	1.5	1.31	Gd (1%)	44	n-spec, neutrino
	BC 523			65	425	3.7	0.916		1.74	B (5%)	-8	n-spec
NE 321A	BC 523A	EJ 339A		65	425	3.7	0.916	1.411	1.67	^{10}B (5%)	-8	n-spec
	BC 525	EJ 335		55	425	3.8	0.88	1.49	1,56	Gd (1%)	91	
	BC 531			59	425	3.5	0.87		1.63		93	
	BC 533			51	425	3	0.80		1.96		65	
NE 230	BC 537	EJ 315		61	425	2.8	0.954	1.498	0.99 D:C	^{2}H	-11	
NE 314A	BC 551			40	425	2.2	0.902		1.31	Pb (5%)	44	
	BC 553			34	425	3.8	0.951		1.47	Sn (10%)	42	

(cont.)

Table 13.3. *(cont.)* Common organic scintillator materials and their properties. The light yield is in units of % that of anthracene, which, for comparison, is about 40% that of NaI:Tl.

Scintillator				Light Yield	λ_{max} (nm)	τ (ns)	Density (g/cm^3)	Refract. Index n	H/C Ratio	Loading	Softening Point (°C)	Notes
Plastic												
Legacy Standard[b]	Saint-Gobain	Eljen	Rexon									
NE 102A	BC 400	EJ 212	RP 400	65	423	2.4	1.032	1.58	1.103		70	general purpose
		EJ 214		59	435	2	1.02	1.58	1.109		60	ultra-thin (250 nm)
NE 104	BC 404	EJ 204		68	408	1.8	1.032	1.58	1.107		70	large area
Pilot F	BC 408	EJ 200	RP 408	64	425	2.1	1.032	1.58	1.104		70	general purpose
NE 110	BC 412	EJ 208		60	434	3.3	1.032	1.58	1.104		70	
	BC 416		RP 200	38	434	4.0	1.032	1.58	1.110		70	
Pilot U	BC 418	EJ 228		67	391	1.4	1.032	1.58	1.100		70	
Pilot U2	BC 420	EJ 230		64	391	1.5	1.032	1.58	1.102		70	
NE 111A	BC 422	EJ 232	RP 422	55	370	1.6	1.032	1.58	1.102		70	fast timing
	BC 422Q			11	370	0.7	1.032	1.58	1.102	$C_{13}H_{10}O$ (0.5%)	70	
NE 103	BC 428	EJ 260		36	480	12.5	1.032	1.58	1.103		70	
	BC 430			45	580	16.8	1.032	1.58	1.108		70	
	BC 436			52	425	2.2	1.032	1.58	0.96 D:C	^{2}H (13.8%)	70	Thin disks
	BC 440	EJ 244	RP 440	60	434	3.3	1.032	1.58	1.104		99	
	BC 440M			60	434	3.3	1.039	1.58	1.104		100	
NE 115	BC 444	EJ 240	RP 444	41	428	285	1.032	1.58	1.109		70	phoswich detectors
	BC 448	EJ 248		60	425	2.1	1.032	1.58	1.104		99	
NE 142	BC 452	EJ 256	RP 452	32	424	2.1	1.080	1.58	1.134	Pb (5%)	60	
	BC 454	EJ 254		48	425	2.2	1.026	1.58	1.169	B (5%)	60	
NE 105	BC 470	EJ 252	RP 470	46	423	2.4	1.032	1.58	1.037		70	dosimetry
	BC 480			*[c]	425		1.032	1.58	1.100		70	
	BC 482A	EJ 280		*[d]	494	12.0	1.032	1.58	1.110		70	
		EJ 284			608	14.0	1.032	1.58	1.110		75	
NE 120	BC 490	EJ 290		55	425	2.3	1.032	1.58	1.103		70	
	BC 498			65	423	2.4	1.032	1.58	1.103		70	
		EJ 299-33A		56	420	γ13,35,270	1.08	1.58	1.106			n/γ PSD
		EJ 299-34				n 13,50,460						

[a] Adjusted to the S-11 response of a PMT.

[b] NE: Nuclear Enterprises, Inc., San Carlos, CA; Pilot: Pilot Chemical Company, Cambridge, MA.

[c] QE = 0.86.

[d] Ratio of Cerenkov light to scintillator light = 10:1.

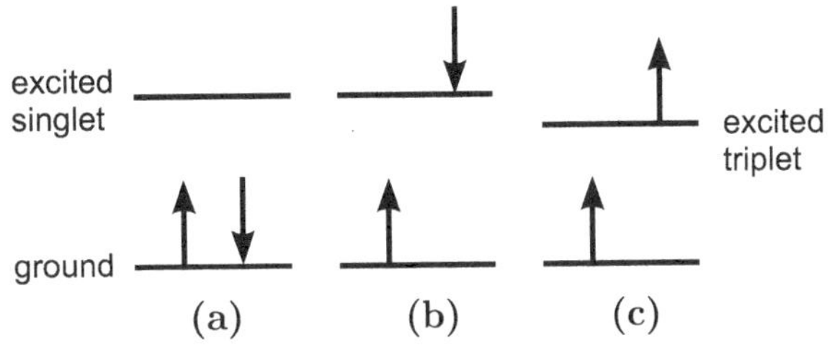

Figure 13.25. Shown are (a) an organic molecule singlet ground state, showing up and down spin of the electrons, (b) the singlet excited state, and (c) the triplet excited state. Note that the electron spins are alike in the triplet state, thereby through the Pauli exclusion principle, preventing the excited electron from dropping to the ground state.

and T_1) have many vibrational states. The energetic electrons are excited from the ground state into many of these possible vibrational states.

Electron spins are paired in singlet states, and the spin of an excited electron in a singlet state (S_{10} for instance) is still paired with the ground state electron. However, electrons in triplet states are no longer paired to electrons in the ground state, meaning they have identical spin and, therefore, cannot make the transition because of the Pauli exclusion principle (Fig. 13.25). Note that direct excitation to a triplet state from the ground state involves a forbidden spin transition and is, thus, improbable.

Electrons can be excited to upper singlet vibrational levels, which include the various S_1, S_2 and S_3 levels. The electrons in the higher S_1 levels quickly drop to the S_{10} state within about 10^{-12} seconds, while electrons in the higher S_2 and S_3 states rapidly relax to the S_1 through non-radiative processes. Electrons can also transfer to triplet states by $S_1 \rightarrow T_1$, a process referred to as non-radiative intersystem crossing, by which the spin of electron is reversed. The probability of intersystem crossing increases when the vibrational levels of the two excited states overlap, mainly because little or no energy must be gained or lost in the transition. Further, intersystem crossing apparently increases with molecules with high-Z atoms [Kearvell and Wilkinson 1968].

The relaxation process for $S_1 \rightarrow S_0$ is a spin-allowed transition and generally occurs on a times scale of a few nanoseconds. However, because $T_1 \rightarrow S_0$ is spin-forbidden, the transition takes significantly longer, on the order of μs to ms. This slow process of luminescence, thus, gives rise to phosphorescence.

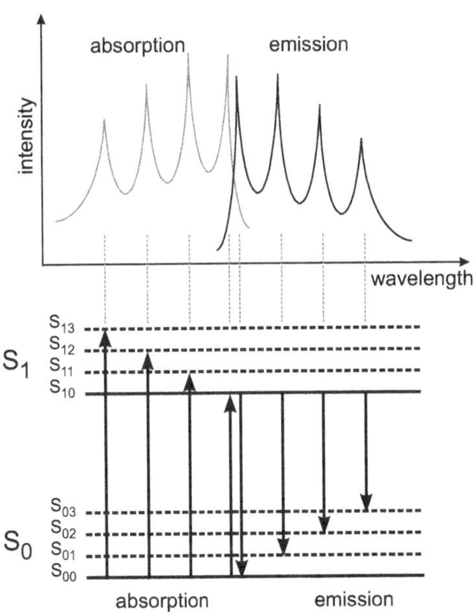

Figure 13.26. Depiction of absorption and emission spectra from an organic scintillator. Adapted from Adachi and Tsutsui [2007].

As with luminescent centers in inorganic scintillators, there is also a Stokes shift with organic scintillators and can be understood by reviewing Fig. 13.2. However, now the vibrational states are molecular for organic scintillators. Shown in Fig. 13.26 is a depiction of the origins of the absorption and emission spectra from an organic scintillator. Electrons excited from the ground state (S_0) up to an excited vibrational state (S_1 states) quickly deexcite to the S_{10} state through non-radiative processes. Because there are many possible excited vibrational states, the absorption efficiency is a function of the allowed transitions (see

Fig. 13.26). These excited electrons can radiatively deexcite to the many possible ground states, which release photons at wavelengths corresponding to the energy loss between the different states. The luminescent transition from $S_{10}\rightarrow S_{00}$ appears identical to the absorption transition $S_{00}\rightarrow S_{10}$, but is in fact of lower energy because of the Stokes shift.

Light Yield

Solid organic scintillators generally produce higher light yields than organic liquids or gases primarily because of the efficient transfer of energy between luminescent centers. Although organic scintillators should produce the same luminescent spectrum in either a solid or as independent molecules, significant changes can be observed because intermolecular complexes that can form between molecules in a solid and because of the differences in the energy transfer mechanism between molecules [Adachi and Tsutsui 2007]. Impurities in the organic crystal can also change the luminescent spectrum because efficient energy transfer allows trace impurities to become the primary luminescent center rather than the host molecule.

Studies performed by Bowen et al. [1949] with mixtures of organic crystals lead to the hypothesis that energy transfer between organic molecules was substantially due to exciton transfer. This transfer sequence would progress until the exciton was ultimately captured. Birks and Black [1951] showed that light output decreased in anthracene when electron irradiation was replaced by alpha particle irradiation. The explanation for this light reduction is that the damage caused by heavy ions created competing trapping centers, which could also capture excitons, and as a result caused luminescent quenching. From these observations, a semi-empirical dependence of light yield as a function of energy deposition was proposed by Birks [1951]. The number of excitons produced in the scintillator per unit path length of travel by the radiation particle can be described by $A(dE/dx)$ where A is a proportionality constant and dE/dx is the ion energy loss (or energy deposition) per unit pathlength of travel. Further, the local concentration of ionized molecules, excited molecules, and damaged molecules along the ion path, can be described by $B(dE/dx)$, where B is also a proportionality constant. If one defines k as the probability that an exciton is lost to a non-radiative transition, the specific fluorescence is[26]

$$\frac{dL}{dx} = \frac{A\ dE/dx}{1 + kB\ dE/dx}, \tag{13.36}$$

where the product kB is the quenching parameter. If the energy loss is small, or the corresponding damage is small, as observed with fast electrons and beta particles, then Eq. (13.36) reduces to

$$\frac{dL}{dx} \approx A\frac{dE}{dx}, \tag{13.37}$$

which shows a linear relationship with light yield and energy deposition. However, if the energy loss is large, and the corresponding damage is significant, such as that produced by protons, alpha particles and fission fragments,

$$\frac{dL}{dx} \approx \frac{A}{kB}, \tag{13.38}$$

which implies constant light output regardless of energy deposition. For most organic scintillators, fast electrons and beta particles with energies above 125 keV do in fact have a relatively linear light yield response. For slow electrons and beta particles with energies below 125 keV, dE/dx increases and produces

[26] Equation (13.36) is often referred to as "Birks' Equation" or "Birks' Law", although John Birks did not promote the name. Also, B is sometimes referred to as Birks' constant, although it appears in Eq. (13.36) mainly because the letter 'B' conveniently follows 'A' in the English alphabet, and not because it is the first letter in the name 'Birks'. Birks named the product kB the 'quenching parameter' [Birks 1964].

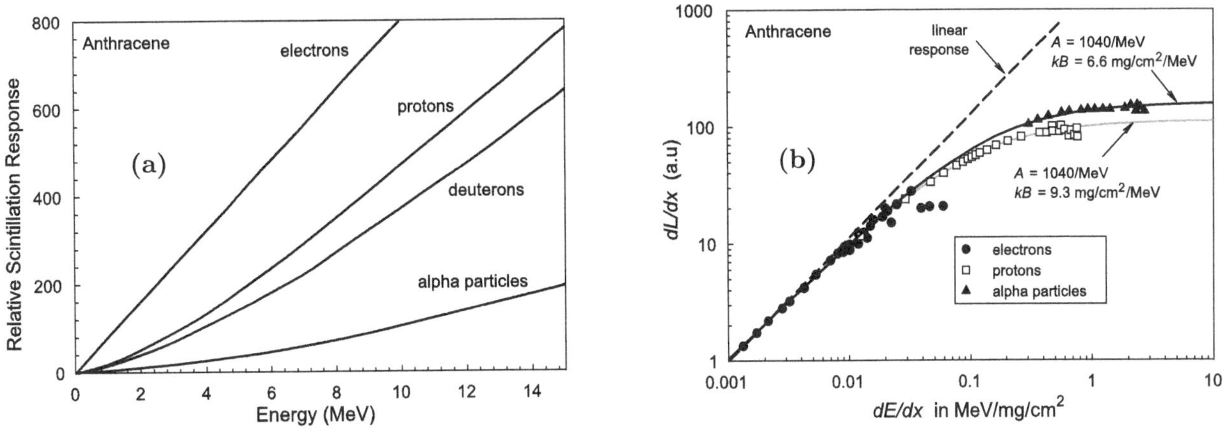

Figure 13.27. (a) Relative scintillation response for different energies of electrons, protons, deuterons, and alpha particles in anthracene. Data are from Birks [1951]. (b) Variation of specific fluorescence $\mathrm{d}L/dx$ with specific energy loss $\mathrm{d}E/dx$ for electrons, protons, and alpha particles in anthracene. Also shown are results from a linear dependence (Eq. (13.38)) and the dependence expressed by Eq. (13.36) with $kB = 6.6$ mg cm^{-2} MeV^{-1} and $kB = 9.3$ mg cm^{-2} MeV^{-1}. Data are from Brooks [1956].

a non-linear output in dL/dx. This effect is also apparent for protons, alpha particles and heavy ions. The severity of the effect in anthracene is shown in Fig. 13.27. In Fig. 13.27(a) there is shown a clear difference in total light yield as a function of particle type and energy. Shown in Fig. 13.27(b) the dramatic effect of Eq. (13.37) is apparent where it is seen that as dE/dx becomes large the value of dL/dx becomes constant. Note that specific energy loss described by the Bragg distribution is highest near the end of the particle range for protons, deuterons, and alpha particles. Hence, dL/dx tends towards a constant value as these heavy particles slow down.

A parameter sometimes quoted for organic scintillators is the α/β ratio, which is defined as the ratio of light produced by ^{210}Po (5.3 MeV) alpha particles to light produced by ^{137}Cs conversion/Auger electrons (624 keV),[27]

$$\frac{\alpha}{\beta} = \frac{(L/E)|_\alpha}{(L/E)|_{e^-}} \simeq \frac{(dL/dE)|_\alpha}{(dL/dE)|_{e^-}} = \frac{\frac{A}{kB}(dE/dx)|_\alpha}{A} = \frac{(dE/dx)|_\alpha}{kB}. \tag{13.39}$$

The quenching parameter can be approximated for a scintillator by measuring the α/β (or α/e^-) light ratio,

$$kB = \frac{(dL/dE)|_{e^-}}{(dL/dx)|_\alpha} = \frac{(dE/dx)|_\alpha}{(\alpha/e^-)}. \tag{13.40}$$

Quenching factors have been measured [Czirr 1964] and calculated for a few common organic and inorganic scintillators by Tretyak [2010] and Nyibule et al. [2014]. There has been work published aimed at refining the accuracy of Eq. (13.36) by adding more terms to the equation to mainly address the non-linear light output from heavy ions in organic scintillators [Blanc et al. 1964; Smith et al. 1968; Craun and Smith 1970]. Craun and Smith [1970] propose the addition of a third term,

$$\frac{dL}{dx} = \frac{A\ dE/dx}{1 + kB\ dE/dx + C\ (dE/dx)^2}. \tag{13.41}$$

[27] In reality, the 624-keV conversion electron is emitted by $^{137m}_{56}$Ba the daughter produced in the beta decay of ^{137}Cs. The same is true for the 662-keV gamma ray commonly attributed to ^{137}Cs. But the usage here is almost universally used.

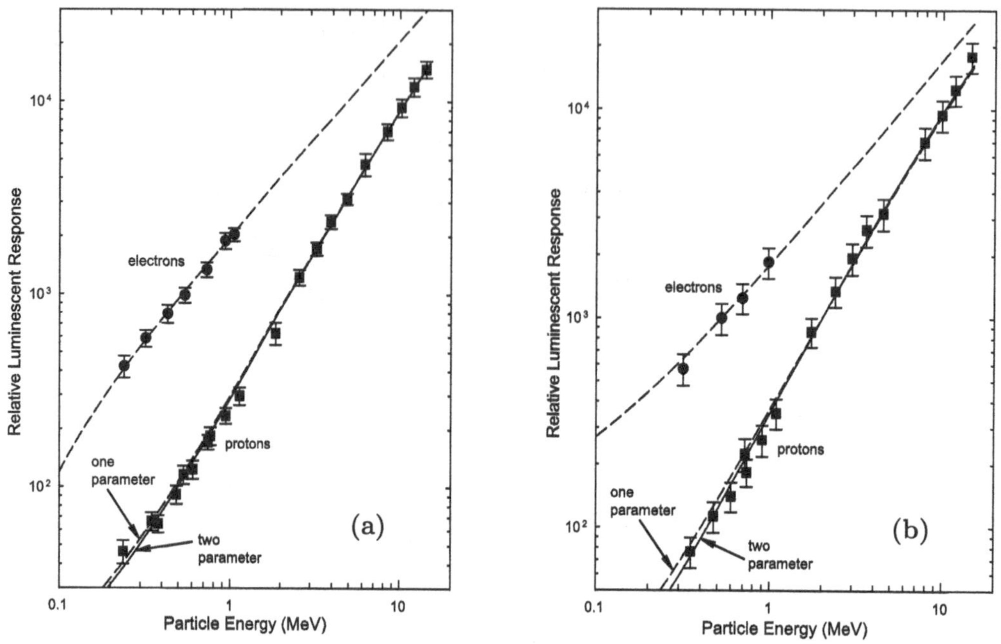

Figure 13.28. Semiempirical curve fits to luminescent yields from electrons and protons in (a) anthracene and (b) (*E*)-stilbene scintillators. Data are from Smith et al. (1968); curve fitting parameters are from Craun and Smith (1970).

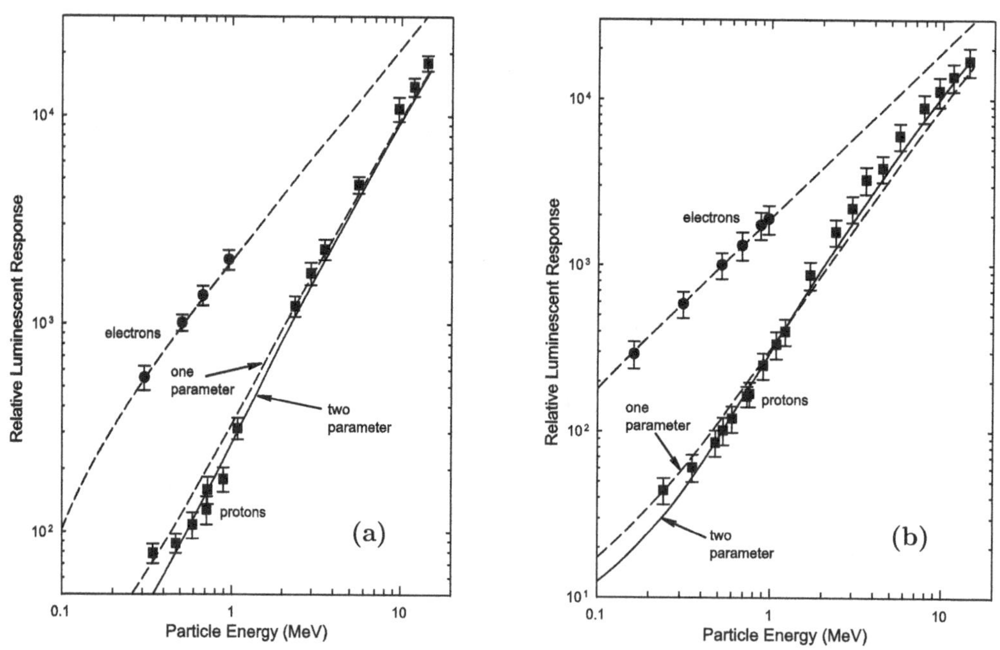

Figure 13.29. Semiempirical curve fits to luminescent yields from electrons and protons in (a) NE-102 (plastic) and (b) NE-213 (liquid) scintillators. Data are from Smith et al. (1968); curve fitting parameters are from Craun and Smith (1970).

Shown in Fig. 13.28 and Fig. 13.29 are comparisons of empirical results from Eq. (13.36) and Eq. (13.41) for organic crystals of anthracene and *trans*-stilbene, organic plastic NE-102, and organic liquid NE-213 luminescent light yields. Equation (13.41) with the additional parameter appears to fit the low energy data better than Eq. (13.36). Quenching parameters kB and C for a few organic scintillators are available in the literature [Craun and Smith 1970].

Decay Time

A generally accepted description of light emission uses the same theory of Eq. (13.26) (repeated here for convenience), in which there is a given rise time τ_r to populate the S_{10} π states and a given decay time τ_d to decay through the $S_{10} \rightarrow S_{00}$ process. The light emission as a function of time is thus given by

$$L(t) = \frac{N_c(t)}{\tau_d} = \frac{N}{\tau_r - \tau_d}\left(e^{-t/\tau_r} - e^{-t/\tau_d}\right).$$

The population rise time τ_r is usually only a fraction of a nanosecond, while τ_d is usually only a few nanoseconds. However, most organic scintillators have complex decay schemes that are difficult to model with a simple exponential equation. This difficulty is largely due to changes in decay time with the type of radiation particle and the fact that there often are multiple decay times involved.

By inspection, the two decay modes seem to appear, typically labeled *fast* and *slow* decay components. An example is seen from the emissions from *trans*-stilbene which are shown in Fig. 13.30. Owen simplified the luminescent decay by assigning only two decay constants to the light yield of organic scintillators, one fast decay time τ_f and one slow decay time τ_s [Owen 1959]. It is the fast component of an organic scintillator that is most often quoted in the literature. The short decay times were quite different for the organic crystals studied and listed in Table 13.3. Owen found that the long decay times were similar for anthracene, *trans*-stilbene, and quaterphenyl, ranging between 350 ns and 370 ns. These long decay times differed with the type of radiation absorbed, showing marked differences between neutron (n, p) events and gamma-ray events.

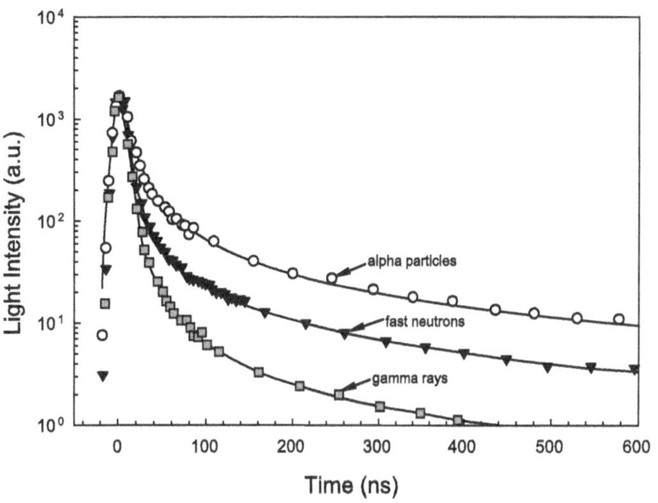

Figure 13.30. Luminescent decay from (E)-stilbene for alpha particles, neutrons, and gamma rays. Data are from Bollinger and Thomas [1961].

Another modification was suggested by Bengsten and Moszynski [1974], who proposed that the rise time of the pulse be represented by a Gaussian distribution $f(t)$ with $\acute{\sigma}_{\text{ET}}$ as the standard deviation. This additional modification takes into account energy transfer and wavelength shifting in the scintillator along with time spread in the PMT. Bengsten and Moszynski [1974] successfully apply this modification to specific plastic scintillators.

Pulse Shape Discrimination

Inspection of Fig. 13.30 reveals major differences among the light emission pulse shapes from alpha particles, neutrons, and gamma rays. This known phenomenon can thus be used to distinguish among different particle events in a mixed radiation field [Wright 1956; Owen 1958; Bollinger and Thomas 1961; Brooks and Jones 1974; Yanagida et al. 2015]. Pulse shape discrimination (PSD) can be used to match the decay time of the short component to the most probable interacting particle, mainly because the short decay time is strongly dependent on the type of incident particle. This comparison can be accomplished by plotting the ratio of light released by the slow component and the fast component (ordinate) against the light released by the fast component (abscissa), and thus produces a two-dimensional map that separates neutron induced events from gamma-ray events [Zaitseva et al. 2009; Yanagida et al. 2015].

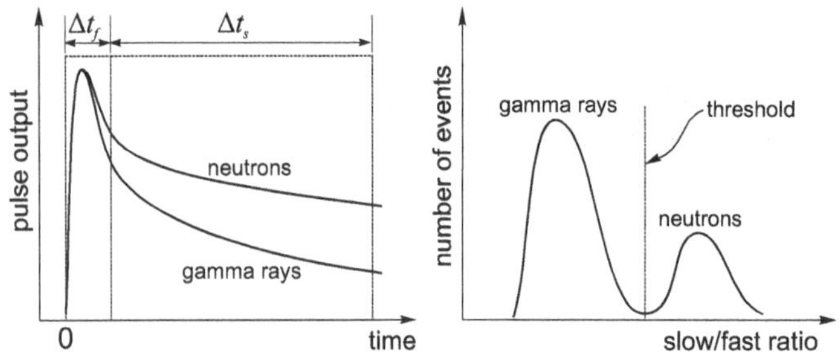

Figure 13.31. (left) Depiction of scintillation pulse output for gamma rays and neutrons. (right) Depiction of a histogram of events versus the slow/fast integrated signal ratio, which shows a clear difference between gamma-ray and neutron related events.

Another discrimination method is to find the ratio of integrated signals (the charge integration method) produced during separate time periods, Δt_f and Δt_s, that span the decay periods of the fast and slow components. A histogram of events (ordinate) is then plotted against this ratio as in Fig. 13.31(right). The method to produce a useful ratio of fast to slow signal components may vary, but ultimately the concept is the same. Several other methods have been explored to produce improved results for PSD [Back et al. 2008; Griffiths et al. 2018].

Often a clear spectral separation appears, especially between heavy ions and gamma rays so that events below a designated slow/fast threshold are attributed to gamma rays and events above the slow/fast ratio threshold are designated as neutron events. Once this cutoff threshold is determined, modern electronics can quickly process information in real time to discriminate between neutron and gamma-ray events. Because organic scintillators can be used as fast neutron detectors from (n, p) reactions, either method provides a means to discriminate between neutron induced events and gamma-ray induced events [Yanagida et al. 2015].

13.3.2 Organic Crystalline Scintillators

Although the first scintillators used for radiation detectors were inorganic (ZnS:Ag and $PtBa(CN)_4$), it was naphthalene, an aromatic hydrocarbon crystalline solid, coupled to a photomultiplier tube that was first used as a practical scintillation detector [Broser and Kallman 1947b; Hofstadter 1975]. Shortly thereafter, anthracene, an organic compound similar to naphthalene, was shown to be a relatively bright scintillator [Bell 1948]. Birks [1964] lists numerous organic crystalline solids as scintillators, a few of which are described here (see also Table 13.4).

Anthracene

Anthracene is one of the brightest organic scintillators available, and has a light yield of about 17,000 photons/MeV, or approximately 43.5% of the light yield of NaI:Tl (see Table 13.3). Anthracene is classified as an aromatic hydrocarbon. Just as inorganic compound scintillators are traditionally compared to NaI:Tl for relative performance, the relative brightness of most organic compounds is compared to that of anthracene whose brightness is taken as 100%. This relative benchmark comparison arises because anthracene is one of the first bright organic scintillators discovered. However, the relative light yield from anthracene is not only a function of particle type, but also the particle trajectory within the crystalline lattice [Tsukada and Kikuchi 1962; Tsukada et al. 1965; Oliver and Knoll 1968], a common trait for crystalline organic scintillators [Brooks and Jones 1974]. Birks [1964] considered this anisotropic light emission characteristic a fundamental drawback to having anthracene as the light yield scintillation standard for organic scintillators. The most probable wavelength of emission is approximately 447 nm, as depicted in the spectrum of Fig. 13.33. The decay time is approximately 30 ns and is fast compared to most inorganic scintillators. Anthracene can

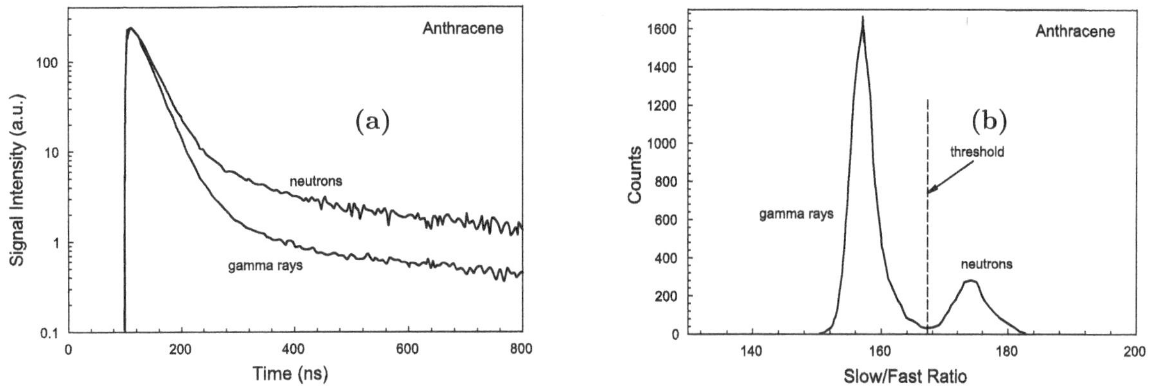

Figure 13.32. (a) Scintillation decay response in anthracene from gamma rays and neutrons. (b) Histogram of gamma-ray and neutron events versus the fast/slow integrated signal ratio for Δt_f = 20 to 40 ns and Δt_s = 40 to 150 ns. Data are from Yanagida et al. [2015].

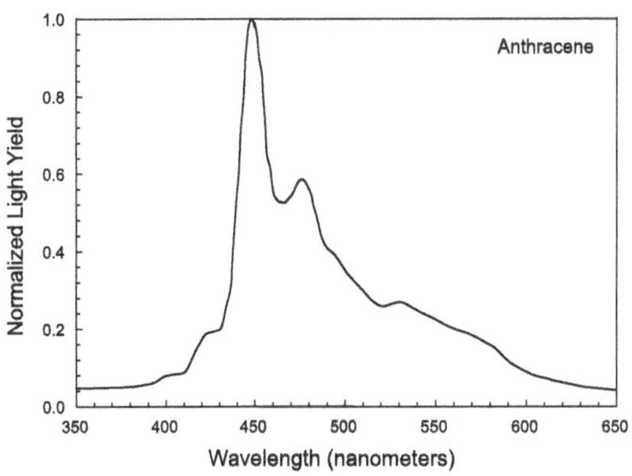

Figure 13.33. Emission spectrum from crystalline anthracene. Data are from Sangster and Irvine [1956].

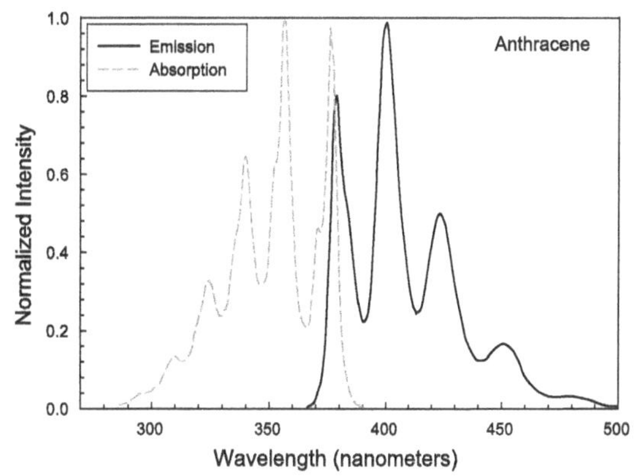

Figure 13.34. Absorption and emission spectra from anthracene dissolved in cyclohexane and fluoresced with 235.7 nm light. Data are from Berlman [1971].

be used for pulse shape discrimination between heavy ions, neutrons, and beta particles. Anthracene was shown in a recent study to be superior for neutron/gamma-ray pulse shape discrimination (PSD) over that for *trans*-stilbene and *p*-terphenyl and showed a clear distinction between neutron and gamma-ray events Fig. 13.32 [Yanagida et al. 2015].

The mass density of anthracene is 1.25 g cm^{-3} with a molecular weight of 178.23 g mol^{-1}. The refractive index of anthracene is 1.595, closely matching that of common borosilicate glass used for PMT windows. Anthracene crystals have a monoclinic lattice, and the chemical structure of anthracene ($C_{14}H_{10}$) is shown in Table 13.4. The melting point of anthracene is 215.76°C.

Anthracene can be used in crystalline form, but must be encapsulated to prevent the formation of an oxidation layer on the outer surface. This surface layer behaves as a quenching region, thereby reducing the absolute light yield and prolonging decay times. The decomposition of the surface through photo-oxidation is also promoted by exposure to light if not encapsulated. Anthracene is also fragile and can be easily fractured.

Table 13.4. Chemical formulas and structures of a few crystalline organic scintillators.

Fluor	Chemical Formula	Chemical Structure
Naphthalene	$C_{10}H_8$	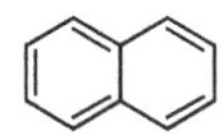
Anthracene	$C_{14}H_{10}$	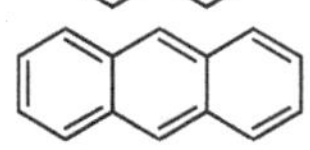
Diphenylacetylene	$C_{14}H_{10}$	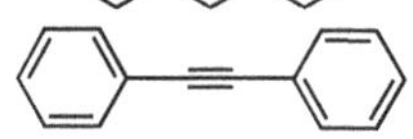
(*E*)-Stilbene	$C_{14}H_{12}$	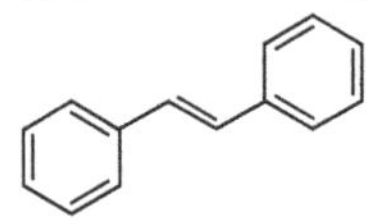
Pyrene	$C_{16}H_{10}$	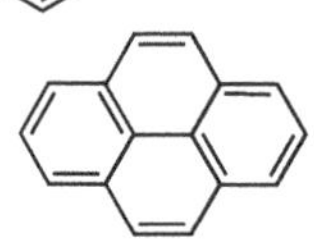
p-Terphenyl	$C_{18}H_{14}$	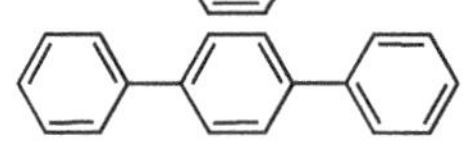
p-Quaterphenyl	$C_{24}H_{18}$	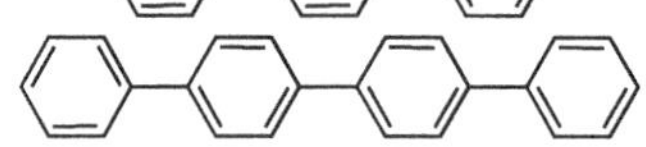

Anthracene can be dissolved in a solvent and used as a liquid scintillator, although the emission spectrum appears to shift (see Fig. 13.34). When dissolved in cyclohexane, the emission peak shifted to 401.6 nm with a decay time of 4.9 ns [Berlman 1971].

Pyrene

Pyrene is an aromatic hydrocarbon composed of four benzene rings with chemical formula $C_{16}H_{10}$ and monoclinic crystal structure (Table 13.4) [Camerman and Trotter 1965]. The relative light yield of pyrene is 60% that of anthracene and has a most probable emission wavelength of 477 nm [Birks 1964]. The emission spectrum is shown in Fig. 13.35. The decay time of pyrene is relatively long for an organic scintillator at 90 ns. The mass density of pyrene is 1.27 g cm^{-3} with a molecular weight of 202.25 g mol^{-1}. The refractive index of pyrene is 1.85, which produces a critical angle of 54° at a glass interface. The hydrogen to carbon (H/C) ratio is 0.625. Pyrene begins to soften at 146°C.

p-Quaterphenyl

One of the many *p*-oligophenylenes studied as scintillator solutes is *p*-quaterphenyl which has a molecular formula $C_{24}H_{18}$ (see Table 13.4). The *p*-oligophenylenes are aromatic hydrocarbon compounds composed of rings united in the para position by single σ bonds and their general formula is $C_{6n}H_{4n+2}$, where n is the number of rings and $n \geq 3$. It has a relatively high light yield of about 85% of that of anthracene [Birks 1964]. The maximum emission wavelength $\lambda_{\max}$ is 438 nm with a decay constant between 7 and 8 ns. The emission spectrum is shown in Fig. 13.36. The melting point is near 150°C. Pure crystals of *p*-quaterphenyl are difficult to produce and the material is usually available only in powder form and, thus, may have contributed to its limited use.

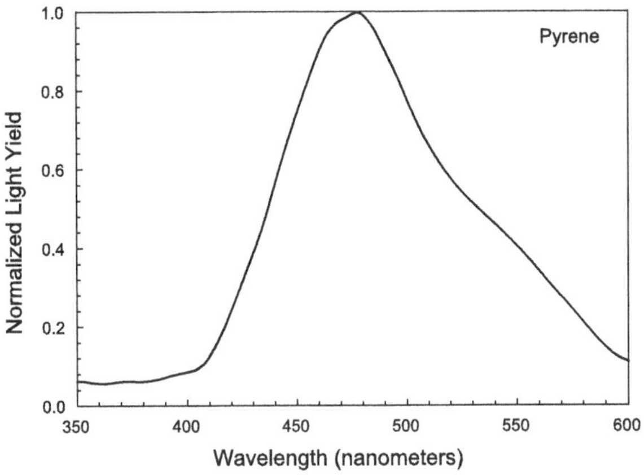

Figure 13.35. Emission spectrum from crystalline pyrene. Data are from Sangster and Irvine [1956].

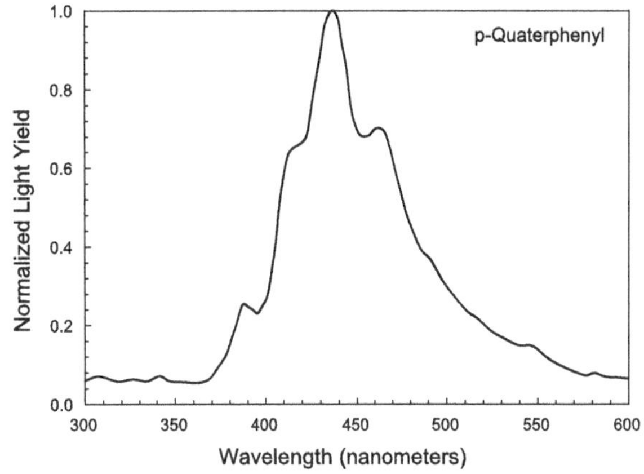

Figure 13.36. Emission spectrum from crystalline p-quaterphenyl. Data are from Sangster and Irvine [1956].

Trans-Stilbene

The organic *trans*-stilbene (or (*E*)-stilbene or diphenyl-ethylene) is a diphenylpolyene and is historically one of the more important crystalline scintillators. It has a mass density of 0.97 g cm^{-3}, a molecular density of 180.25 g mol^{-1}, and a chemical formula $C_{14}H_{12}$. The light yield is approximately 70% that of anthracene, which makes it a relatively bright organic crystal. The wavelength at maximum (λ_{max}) is near 400 nm with a fast decay time of 3.5 ns. The emission spectrum is shown in Fig. 13.37. The refractive index of *trans*-stilbene is 1.62 that produces a critical angle of 67.6° at a glass interface. As observed with many other organic crystalline scintillators, the light yield in *trans*-stilbene is directionally dependent upon the interacting particle trajectory [Hansen and Richter 2002; Cvachovec et al. 2002]. Stilbene is non-hygroscopic and relatively non-flammable (NFPA 704 rating of '1'), thus making it relatively easy to process.

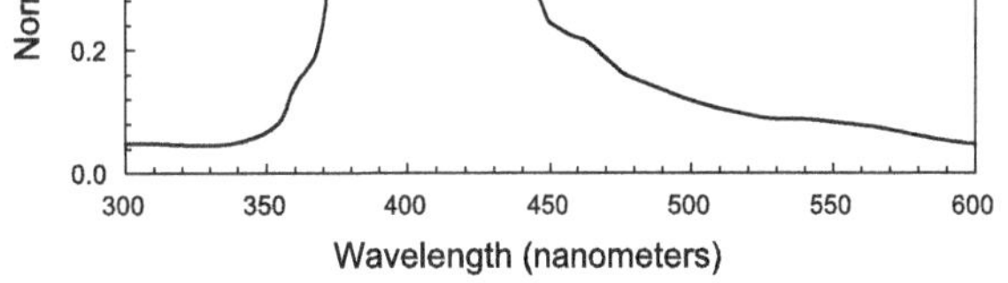

Figure 13.37. Emission spectrum from crystalline *trans*-stilbene. Data are from Sangster and Irvine [1956].

It has a melting point of 125°C and transparent crystals can be grown with the Bridgman technique. However, these melt-grown crystals are generally limited to small sizes. Recently, large faceted transparent crystals of *trans*-stilbene with 10-cm dimensions have been grown through solution methods [Zaitseva et al. 2011a; Carman et al. 2013; Zaitseva et al. 2015] and are now commercially available [Inrad 2016]. As a result there has been renewed interest in *trans*-stilbene as a fast neutron detector. These solution-grown crystals can be machined into varying shapes, and reportedly have similar performance as traditional Bridgman-grown crystals [Zaitseva et al. 2015]. Stilbene can be used for pulse shape discrimination between heavy ions, neutrons, and beta particles [Bollinger et al. 1961; Konobeevski et al. 2012; Yanagida et al. 2015]. The work of Harihar et al. [1977] indicates that the short decay time of *trans*-stilbene is mostly unaffected by the magnitude

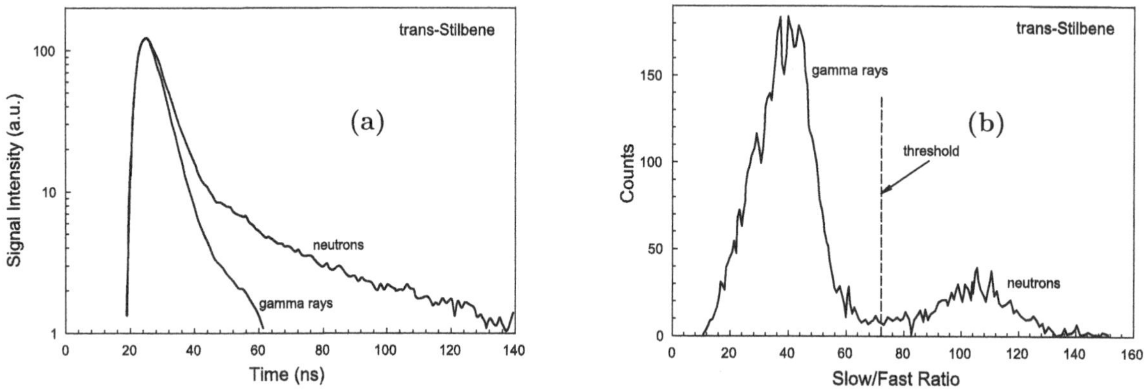

Figure 13.38. (a) Scintillation decay response in *trans*-stilbene from gamma rays and neutrons. (b) Histogram of gamma-ray and neutron events versus the fast/slow integrated signal ratio for $\Delta t_f = 20$ to 40 ns and $\Delta t_s = 40$ to 150 ns. Data are from Yanagida et al. [2015].

of the specific energy loss dE/dx while the long decay time increases with dE/dx. Zaitseva et al. [2011b] report on the use of *trans*-stilbene and mixed organic crystals of *trans*-stilbene for fast neutron detection using PSD methods. According to the results of Yanagida et al. [2015], shown in Fig. 13.38, *trans*-stilbene is a good scintillator for fast neutron PSD, yet not quite as good as anthracene and *p*-terphenyl. Overall, *trans*-stilbene is a good choice for fast neutron detection because it is relatively bright, can now be acquired in large sizes, is optically clear, fast, and couples well to common PMTs.

p-Terphenyl

Another *p*-oligophenylene studied as a scintillator solute is *p*-terphenyl, having molecular formula $C_{16}H_{14}$ (see Table 13.4). Pure crystals of *p*-terphenyl have a light yield of about 40% compared to that of anthracene [Birks 1964]. However, commercially produced large crystals of doped *p*-terphenyl are available with reported light yields that is 135% that of anthracene [Proteus 2016; Cryos-beta 2016]. The density of *p*-terphenyl is 1.23 g cm^{-3} with molecular weight of 230. The H/C ratio is 0.778 and the melting point is 213°C. This organic scintillator is relatively fast with decay times between 3.0 and 3.7 ns. The maximum emission wavelength $\lambda_{\max}$ is 420 nm, which matches well to common bi-alkali PMTs. The emission spectrum is shown in Fig. 13.39. The refractive index is 1.65 with a critical angle of about 65°C at the interface of common borosilicate glass. *p*-terphenyl can be used for pulse shape discrimination between heavy ions, neutrons, and beta particles [Yanagida et al. 2015]. The response of *p*-terphenyl to neutrons and gamma rays is shown in Fig. 13.40. A study with *p*-terphenyl in a composite scintillator in which the organic crystal was mixed into a polymer matrix, outperformed composite *trans*-stilbene

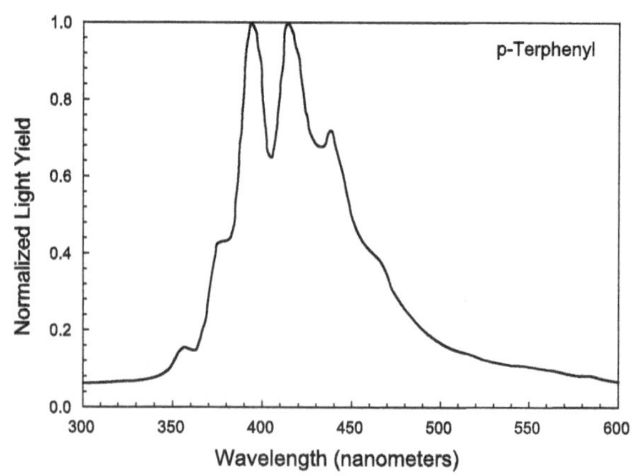

Figure 13.39. Emission spectrum from crystalline p-terphenyl. Data are from Sangster and Irvine [1956].

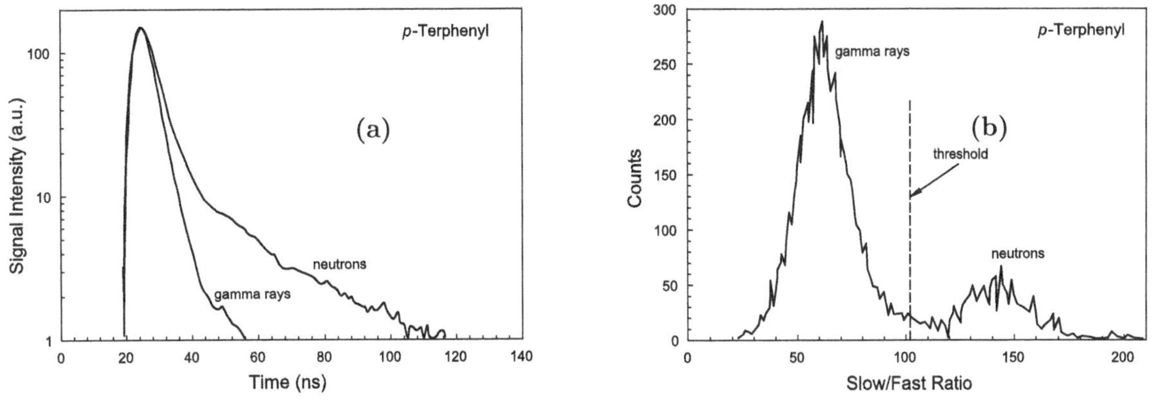

Figure 13.40. (a) Scintillation decay response in *p*-terphenyl from gamma rays and neutrons. (b) Histogram of gamma ray and neutron events versus the fast/slow integrated signal ratio for Δt_f = 20 to 30 ns and Δt_s = 30 to 150 ns. Data are from Yanagida et al. [2015].

mixtures for neutron/gamma-ray PSD [Iwanowska et al. 2011]. Further, *p*-terphenyl can be used as the solute in a liquid scintillation cocktail [Brooks 1956]. The absorption and emission spectrum of such a cocktail is shown in Fig. 13.41.

13.3.3 Liquid Scintillators

For many radiations the energies of the particles are too small to be measured by conventional means discussed up until this point. For instance, β-particle emissions from tritium (^{3}H) have a maximum energy of 18.59 keV, and the maximum energy of β-particles from ^{14}C is 156.47 keV. The energies emitted from these β-particle emitting sources are generally insufficient to penetrate the enclosure of most radiation detectors, including gas-filled detectors with thin BoPET windows. Sample preparation can also inhibit detection from energy self-absorption of the radiation particles within the sample. Although some of these radiations can be detected in a 2π or 4π gas-filled proportional counter, the detection method of choice is presently a technique referred to as liquid scintillation counting (LSC), can be used to measure α- and β-particle emissions, positrons, Auger electrons, internal conversion electrons, and Compton electrons.

The measurement method involves mixing a radioactive (typically aqueous) sample with the organic scintillation solution. The organic scintillation solution is commonly referred to as a "cocktail", and is composed of a solvent, a surfactant emulsifier and one or more fluors. The primary fluor is added to convert absorbed energy into fluorescent light. At times it is beneficial to mix a wavelength shifting scintillator, or secondary scintillator, into the solution to improve the scintillation response. Common fluors and solvents are listed in Tables 13.5 and 13.6, respectively.

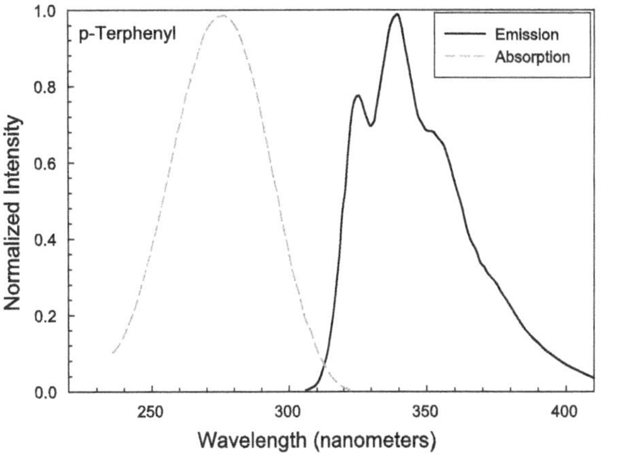

Figure 13.41. Absorption and emission spectra from p-terphenyl dissolved in cyclohexane and fluoresced with 303 nm light. Data are from Berlman [1971].

Table 13.5. Chemical Formulas and Structures of a Few Organic Liquid Scintillation Fluors (solutes).

Fluor	Type	Formula	Structure
p-Terphenyl	primary	$C_{18}H_{14}$	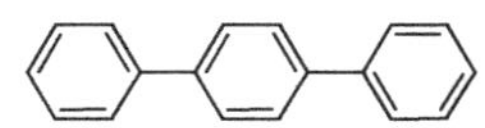
PPO	primary	$C_{15}H_{11}NO$	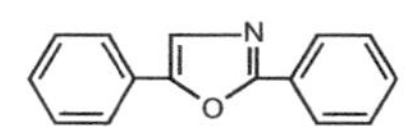
POPOP	wavelength shifter	$C_{24}H_{16}N_2O_2$	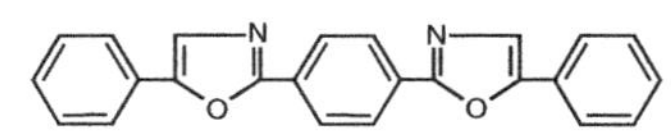
M_2-POPOP	wavelength shifter	$C_{26}H_{20}N_2O_2$	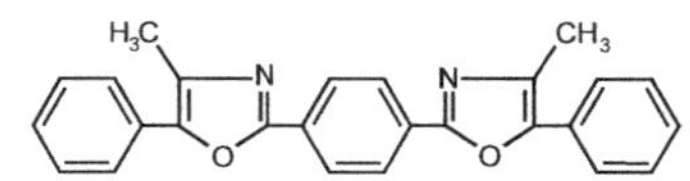
PBD	primary	$C_{20}H_{14}N_2O$	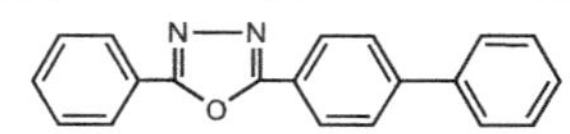
tert-Butly-PBD	primary	$C_{24}H_{22}N_2O$	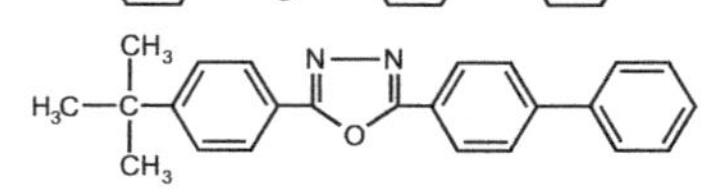
bis-MSB	wavelength shifter	$(CH_3C_6H_4CH{=}CH)_2C_6H_4$	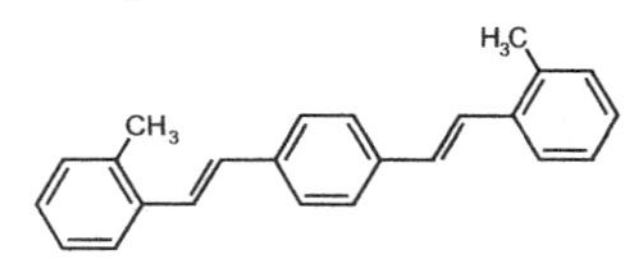

Table 13.6. Chemical formulas and structures of a few liquid organic scintillator solvents when mixed with 3 g/liter PPO.

Solvent	Chemical Formula	Chemical Structure	Relative Light Yield*
p-Xylene	C_8H_{10}		1.12
Pseudocumene	$C_6H_3(CH_3)_3$	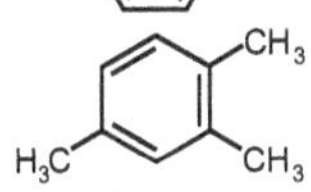	1.12
m-Xylene	C_8H_{10}	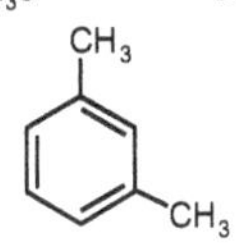	1.09
o-Xylene	C_8H_{10}	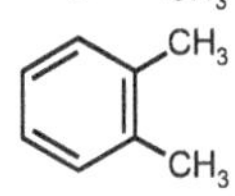	0.98
Xylene (mixed)	C_8H_{10}		1.07
Phenylcyclohexane	$C_{12}H_{16}$	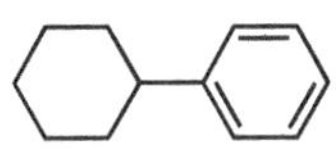	1.02
Toluene	C_7H_8	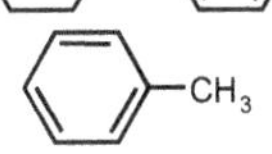	1.00
Ethylbenzene	C_8H_{10}	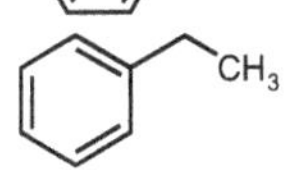	0.96
Triethylbenzene	$C_6H_3(C_2H_5)_3$	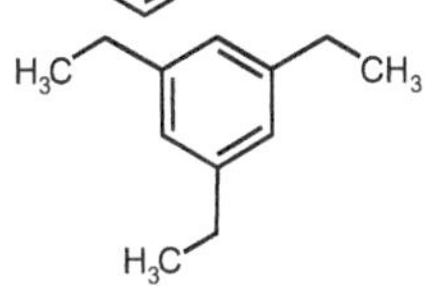	0.96
n-Butylbenzene	$C_{10}H_{14}$	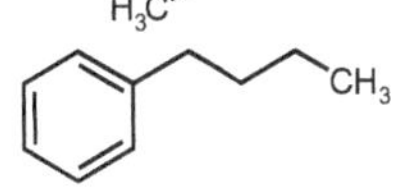	0.88
Benzene	C_6H_6		0.85
Anisole	C_7H_8O	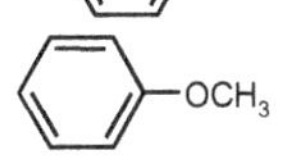	0.83
Mesitylene	$C_6H_3(CH_3)_3$	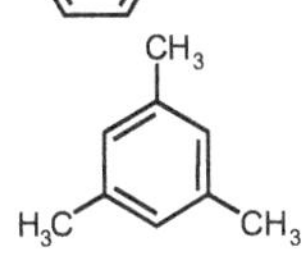	0.82
Cumene	$C_6H_5CH(CH_3)_2$	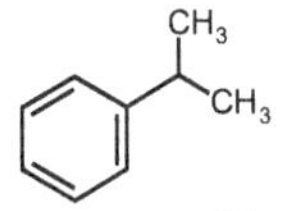	0.80
p-Cymene	$CH_3C_6H_4CH(CH_3)_2$	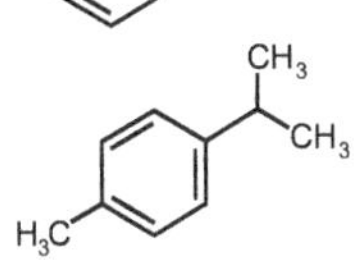	0.80

*As compared to toluene.

Liquid Scintillation Counting

The solvent serves as a suspension for the primary scintillator fluor; hence, the fluor must be soluble in the chosen solvent. The excitation energy of the solvent should be relatively low, thereby allowing efficient transfer of energy to the solvent from the radiation particles. Further, there must be efficient energy transfer from the solvent to the fluor in order to maintain efficient light output. Historically there are several solvents, termed "classical" solvents that have been used to great success and consist mostly of the aromatic organic benzene derivatives (see the list in Table 13.6). Aromatic organic molecules, such as toluene, *p*-xylene and *m*-xylene, have high densities of electrons, thereby producing efficient interaction between β-particles and molecular electrons so as to generate a relatively large light yield when mixed with a scintillation fluor. However, these same aromatic hydrocarbons have multiple health risks. Typically they are flammable with a low flash point ($<50°C$), pose an inhalation problem and can cause skin irritation with direct contact. Further, there is evidence that benzene and some associated aromatic hydrocarbons cause bone marrow failure and can lead to forms of cancer involving blood cells (such as leukemia), consequences which have led to stricter regulations regarding their use.

Until only a few decades ago, scintillation cocktails were usually mixed by the radiation practitioner. However, in modern times, premixed proprietary cocktail solutions are commercially available from vendors such as Beckman Coulter, Perkin-Elmer, Fisher Scientific, MP Biomedicals, National Diagnostics, and Research Products International. In recent times, most of these aromatic hydrocarbons, except, perhaps, pseudocumene (1,2,4-trimethylbenzene), have been replaced with safer solvents. However, pseudocumene continues to be used in a variety of commercially available "classical" cocktails. An alternative branch of solvents, named "safer" solvents, are in common use for LSC, considered safer because the safety aspects to these alternative solvents are safer than traditional classical aromatic hydrocarbons. For instance, a few alternative safer chemical solvents for LSC are benzyltoluene (BT), 1,2-diphenylethane (bibenzyl or DPE), 2,6-diisopropylnaphthalene (2,6-DIPN or DIN), phenyl xylylethane (PXE), and dodecylbenzene, a linear alkylbenzene (LAB) [Nemchenok et al., 2011], all of which have low vapor pressures with flash points above 140°C and are classified as safe for transport and storage.

There are many benefits to adding a wavelength shifting fluor to the cocktail [Schram 1963]. First, the emission wavelengths of many primary fluors have sufficient energy to be significantly reabsorbed in the cocktail solution. The wavelength shifter converts light to lower energies, thereby reducing the probability of reabsorption. Second, a wavelength shifter can be selected that emits photons that match well to the light collection device such as a PMT. For instance, a solution with *p*-terphenyl as the primary scintillator with λ_{max} near 340 nm may have a wavelength shifter added, such as POPOP with λ_{max} near 410 nm and this improves transparency with a better match to a common bi-alkali PMT. Third, the longer wavelength photons are less likely to be absorbed by the container structure, such as the vial walls that contain the cocktail. Fourth, the longer wavelength photons ($\lambda > 400$ nm) are more effectively reflected by TiO_2, a Lambertian reflector often used in scintillation counting. Typically, wavelength shifting fluors have lower solubility in the solvent than the primary fluor, hence the reason for their exclusion as a primary fluor. However, the energy transfer process between fluors requires only a small fraction of wavelength shifter, on the order of 1% concentration of the primary fluor. Energy lost during the energy transfer process between primary and wavelength shifting fluors is more than made up for by the improved luminescent transparency in the cocktail and efficient light coupling to the PMT (or other light sensitive device).

Details on liquid scintillation solutions and measurement methods can be found in several excellent books on the topic such as those by Ross, Noakes, and Spaulding [1991] and L'Annunziata [2012]. Here, only a brief overview of the art is presented.

Scintillator Vials LSC systems have conveyer systems that may consist of a belt with vial slots or racks capable of holding approximately 10 vials. Systems are designed for several vial sizes ranging from 4 ml to

25 ml capacity, although the most popular size seems to be 20 ml vials. The 20 ml size became the standard size limit mainly because of conventional photomultiplier tube diameters. A study conducted with 7 ml and 20 ml indicates that there is little difference in performance using either size of vial [Moore et al. 1977], although efficiency was better for low energy beta particles (^{3}H) for the 20 ml vials. It was determined that use of either should be filled more than halfway to reduce problems with quenching. Overall, the lower cost of operating with 7 ml vials seemed to be the main consideration. These vials have caps that include either a white Lambertian or a specular internal reflector.

These LSC vials are available fabricated from several different materials, including quartz, soda lime glass, borosilicate glass, polyethylene, polypropylene, nylon, and Teflon [Peng 1977]. Quartz has low background concentrations and is optically clear for UV light emissions, making it a preferred choice for vials. However, quartz vials are also relatively expensive, and are seldom used for routine samples. Borosilicate glass has a lower background, mainly from ^{40}K, than soda-lime glass and is optically clear, thereby making it a good choice. Further, glass is chemically inert to the organic compounds usually used in LSC. However, glass does adsorb a variety of other chemicals, including lipids, cations, anions, amino acids, polyethylene glycol, and many other compounds with hetero-atoms [Peng 1977]. Various additives and surface treatments can be used to reduce chemical adsorption in glass vials. Polyethylene vials do not appear to suffer from these adsorption effects. Glass can break, which has the risk of contaminating the workspace and counting system. Further, it is possible for radioactive particles to fluoresce the glass, which ultimately adds to the background light output. Finally, glass vials can be costly; therefore, many labs choose to clean used vials rather than dispose of them.

Polyethylene vials are more economical, but some of the solvents used in LSC can cause plastic vials to swell (mainly toluene and xylene) which can cause vials to become lodged in a system. Further, the continuous change in the cocktail solution because of diffusion of the solution through the vials causes a drift in the measured light output over time. To reduce risk of solvent loss and vial swelling, cocktails with volatile solvents contained in plastic vials are best measured within a day of preparation. Swelling is less of a problem with high density polyethylene (HDPE). Polypropylene vials suffer similar problems as polyethylene vials. Also, there are polypropylene vials available with internal anti-diffusion coatings that effectively reduce this problem. Polyethylene vials can have higher light transmission efficiency than glass for wavelengths between 360 and 400 nm, and they do not have the risk of breaking. Antistatic plastic vials are available if static electricity is a concern.

Nylon vials can soften with water-based or alcohol solutions, but are relatively resistant to toluene. Teflon vials are also inert to chemicals used for LSC, but are more expensive than nylon or polyethylene. Teflon vials also have a lower radiation background than glass, quartz, and polyethylene vials.

Radiation Measurement Laboratory LSC systems are configured in coincidence mode, such that a cocktail sample is inserted between two photomultiplier tubes (PMTs) gated to allow a 'count' only when both PMTs produce a signal above a discriminator setting within a defined time window Δt. The intensity of light, or electrical signal, is a relative measure of the energy deposited within the scintillator. Operating in coincidence mode reduces sporadic background counts from cosmic rays or other background radiations that may interact directly with either of the PMTs. To further reduce background, laboratory LSC systems have an abundance of lead shielding[28] surrounding the PMTs and internal source (if there is one). Laboratory units are designed to handle numerous samples in sequence. An operator can load hundreds of vials into racks or a conveyor, and each sample is systematically inserted between the PMTs, and the system automatically records data from each of the samples.

[28] One of the authors nearly crushed a perfectly good hand-cranked lift by raising a small Beckman laboratory LSC from the bottom floor of his lab to a second floor loft area.

The pulse height spectrum energy from a LSC is usually projected onto a logarithmic scale along the abscissa (channel number). This arrangement becomes necessary when comparing low and higher energy beta particle emissions that are significantly different such as those from ^{3}H and ^{14}C. In Fig. 13.42, pulse height spectra from three common beta particles sources are shown (^{3}H, ^{14}C, ^{32}P) with maximum energies of 18.3 keV, 156 keV, and 1.71 MeV, respectively. Each spectrum ranges from zero up to the end point energy, a consequence of the beta particle always sharing energy with an anti-neutrino decay product. Because the anti-neutrino does not produce scintillation light, only the energies from beta particles are recoded in the spectrum.

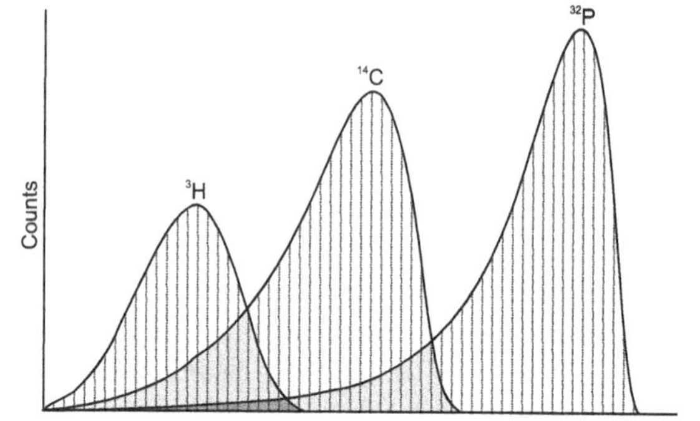

Figure 13.42. Depiction of output spectra from three beta particle sources. After [Beckman 1985].

Quench Correction Foreign matter and diluting fluids can cause light loss in a scintillation cocktail. If the amount of light produced in the scintillator falls below the discriminator setting, then the event is not recorded by the system. This light attenuation effect is usually categorized as "color" or "chemical" quenching. Color quenching is a result of chemicals in the solution that change its color and absorb scintillation light. For instance, yellow coloration produces strong color quenching, whereas blue coloration usually quenches very little. Chemical quenching occurs when chemicals added to the cocktail interfere with the energy transfer process and causes inefficient energy transfer to the scintillation phosphor. In either case, the result is a loss of light, which consequently changes the pulse height spectrum. For example, counts falling within ch2 in Fig. 13.43 may have 75% efficiency for a low quenched sample, but only 5% efficiency for a highly quenched sample of the same activity. All LSC samples are quenched and require some amount of quench correction to determine the actual sample activity. There are many quench correction techniques developed by LSC manufacturers, and are generally categorized as internal or external standard methods. Some of these methods are briefly described here.

Internal Standard Quench Correction The internal standard method typically requires that a known amount of radioactive sample be added to a scintillation cocktail. The sample is initially measured in the LSC system and afterwards a known amount of radionuclide, generally with high specific activity, is added to the cocktail followed by another measurement. The efficiency of the standard is determined by,

$$\epsilon_s = \frac{\text{Cps}(s+x) - \text{Cps}(x)}{\text{Dps}(s)}, \tag{13.42}$$

where $\text{Cps}(s+x)$ refers to the count rate from the mixed calibrated standard and sample, $\text{Cps}(x)$ refers to the count rate from the sample, and $\text{Dps}(s)$ refers to the known activity of the calibrated standard. The activity of the sample can then be found from,

$$\text{Dps}(x) = \frac{\text{Cps}(x)}{\epsilon_s}. \tag{13.43}$$

The internal standard method is best used for manual scintillation systems, such as field units, rather than automated systems. Although the internal standard method is considered accurate, it is time consuming and impractical for large numbers of samples. Note the method presumes that adding the known standard to the vial does not affect the overall quench. However, the addition of the standard to the vial changes the total liquid volume and can change the sample quenching, and, consequently, reduce the accuracy of the measurement. A disadvantage of the method is that the sample cannot be recounted, mainly because the activity is altered by the addition of the internal standard. Further, opening the vial, especially in a cool environment, my cause water condensation in the vial, and water is a strong quencher. Also, contamination from the transfer method (pipette, for instance) can add unintended quench matter into the cocktail. Addition of these quenches into the sample cocktail lead to exaggerated corrections to the calculated activity.

Internal standards for ^3H and ^{14}C are commercially available. Typically it is advised to use cocktail solutions for the radioassay with the composition as the quenched standards. Unquenched standards are best acquired from commercial vendors, mainly because they are usually flushed with argon to remove dissolved oxygen. Removal of the dissolved oxygen works to increase the scintillation efficiency. A variation of the internal standard method includes a unique cap on the vial that systematically pumps a known amount of calibrated quench fluid into the vial.

Channels Ratio Method LSC is most often used for beta emitting radionuclides; hence the pulse height spectrum extends from zero up to the maximum energy of the beta particles and the spectrum is very similar to the beta-particle energy distribution. If the pulse height spectrum is divided into two channels, then the ratio of counts in the channels can be used to determine the amount of quenching. An advantage of the method is that no additional standard need be added to the samples, and these samples can be recounted many times.

The LSC system usually has two independent channels, as shown in Fig. 13.43. Many ratio variations can be used, and do yield different results [Horrocks 1974]. For instance, the numerator may be ch 1, while the denominator is ch 2, or the numerator may be ch 2 while the denominator is the sum of ch1 and ch 2, and so on. The ratio choice should be selected so that the experimentally determined ratio does not produce a value close to zero.

The channels ratio quench correction method is performed by initially measuring a set of quench correction standards. These standards have a specific radionuclide of interest in the cocktails, ^3H or ^{14}C for instance, and each standard has a different level of quench agent added. At least one of the standards remains unquenched. With the unquenched standard, a measurement is conducted and the channels set at specific locations to split the spectrum into two portions. These channels remain unchanged for both the calibration and sample measurements. The remaining standards are measured, and the number of counts registered in both channels are recorded for each measurement.

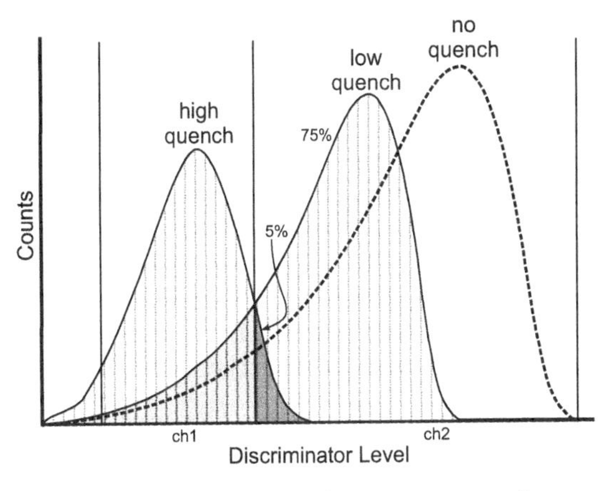

Figure 13.43. Depiction of output spectra from unquenched and quenched standards used for the channels ratio calibration. The number of counts in channel 1 increase with quenching, while the number of counts in channel 2 decrease with quenching.

The known activity of the unquenched standard is compared to the total counts summed from channels 1 and 2. This same exercise is repeated for all standards in the set. The channels ratio (for example, ch 1 divided by ch 2) is plotted against the counting efficiency for each measurement, thereby providing a characterization curve for the radionuclide samples. For each unknown sample, the ratio of ch1/ch2 is determined and compared to the calibration graph, which yields the efficiency. The corrected activity is then determined with Eq. (13.43).

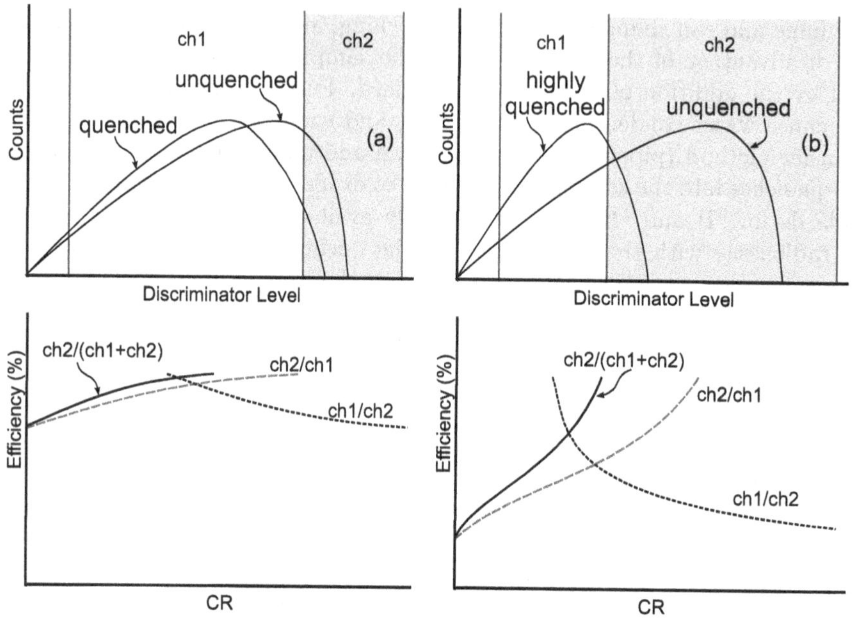

Figure 13.44. Counting efficiency as a function of the channels ratio for two different discriminator settings. In example (a) both channels are set relatively high, while a greater separation is between channel settings in example (b).

Proper selection of the channels is critical to this quench correction method. Consider the two examples shown in Fig. 13.44. The channel selection of Fig. 13.44(a) causes the counts in channel 2 to approach zero at relatively low quench. In other words, selection of the channel ratio method causes the ratio to either approach zero or infinity at a low amount of quench. Although it provides detail in the quench curve for low quenched samples, it has no information for highly quenched samples. For the channel selections of Fig. 13.44(b), the spectrum is more evenly distributed, thereby allowing the calibration of quench curves over a broader range of quenched solutions.

External Standard Method The basic external standard has a gamma-ray source of known activity, usually a ^{137}Cs source, as a component in the LSC system. This gamma-ray source (sometimes called a "pea") is withdrawn into a shielded region inside the LSC system to prevent unwanted background in a scintillation cocktail sample. With a sample in the system, the gamma-ray source is withdrawn from the shield and produces a Compton scattering continuum of energetic electrons in the scintillation cocktail, which adequately mimics the fluorescent yield of beta particles. The original method proposed by Higashimura et al. [1962] used the observed count rates and efficiencies from a set of known beta particle standards. For each quench standard, the counting efficiency is determined by,

$$\epsilon_{\rm i} = \frac{{\rm Cps}(S_i)}{{\rm Dps}(S_i)}, \tag{13.44}$$

where $\mathrm{Cps}(S_i)$ is the count rate for standard i and $\mathrm{Dps}(S_i)$ is the known activity of standard i. Each of these standards, immediately after (or before) the measurement, is exposed to the external standard gamma-ray source and the new count rate is recorded. This gamma ray source is placed in the same location with respect to the vial for each measurement. An efficiency calibration curve is developed using the net count rate response from the gamma-ray source,

$$\mathrm{NCps}(\epsilon_i) = \mathrm{NCps}(\gamma) = \mathrm{Cps}(S_i + \gamma) - \mathrm{Cps}(S_i), \tag{13.45}$$

where $\mathrm{Cps}(S_i + \gamma)$ is the observed count rate with both the gamma-ray source and the activity of the LSC standard. The LSC standard efficiency ϵ_i is plotted versus $\mathrm{NCps}(\gamma)$. For an unknown beta particle sample, the LSC sample efficiency is found by comparing the measured $\mathrm{NCps}(\gamma)$ against the calibration curve. Afterwards, Eq. (13.43) is used to determine the sample activity. Note that it is important that all measurements are conducted for the same amount of time, and that the lower level discriminator is also kept constant during a set of measurements. Calibration sets can be acquired for different beta particles sources, the most common being ^{3}H, ^{14}C, and ^{36}Cl to cover low, medium, and high energy beta particles, respectively. Calibration sets usually have ten quench vials to a set, typically with one "unquenched" standard.

External Standard Channels Ratio Method As the name implies, the external standard channels ratio (ESCR) method combines an external gamma-ray source measurement with the channels ratio quench correction method. In a system with ESCR, there are three wide channels. One channel spans the beta particle spectrum. The other two channels are set higher than the beta particle spectrum, and are used to establish a quench calibration curve with the external gamma-ray source. Hence, the efficiency of the beta particle standard is plotted versus the channels ratio of the gamma-ray source. During a measurement of a beta source of unknown activity, the channels ratio is determined by the gamma-ray source and the beta particle activity is corrected from the quench calibration curve. As with other techniques, two measurements per vial must be conducted so as to subtract the beta particle counts from the gamma-ray spectrum.

Although the ESCR method is convenient, making it popular, its accuracy is less than either the channels ratio or external standard methods [Peng 1977]. Sources of error include dissimilar spectra between the beta particles and the gamma rays, drift in pulse height (mainly with polyethylene vials), vial thickness differences between the set of standards and the samples, and solution volume differences.

Pulse Height Spectrum Shift (*H*-Number) Method A quench method developed by Horrocks [1964] uses the inflection point of the Compton spectrum for quench identification. Instead of comparing the total counts in the gamma-ray spectrum from the external source, the endpoint channels of the Compton spectra from a multichannel analyzer are compared. Consider the Compton spectra of Fig. 13.45. The endpoint energy, typically designated as the inflection point at the end of the Compton continuum, moves to lower channels in the pulse height spectrum as the quench increases. The maximum energy transferred from a single Compton scatter is 477 keV for ^{137}Cs gamma-rays. Hence, the LSC system is calibrated such that an unquenched sample exposed to ^{137}Cs has the inflection point channel aligned with 477 keV. Quenched samples cause the inflection point to appear in a lower channel.

A quench curve is derived in a similar fashion as the external standard method, in which a set of quench standards is used, except now it is the difference in the inflection point channel that is correlated to the efficiency instead of the total counts. This difference in inflection point is defined as the "H number" (after Horrocks who developed it), where

$$H\text{-number} = \mathrm{Ch(unquenched)} - \mathrm{Ch}_i\mathrm{(quenched)}. \tag{13.46}$$

The efficiency of each vial in the set of standards is determined from Eq. (13.44), and is plotted versus the corresponding H-number for each calibrated vial in the set. For an unknown beta particle sample, the LSC sample efficiency is found by comparing the measured H-number against the calibration curve. Afterwards, Eq. (13.43) is used to determine the sample activity. Note that the H-number method does not require that the beta spectrum be subtracted from the Compton spectrum because only the end point in Compton

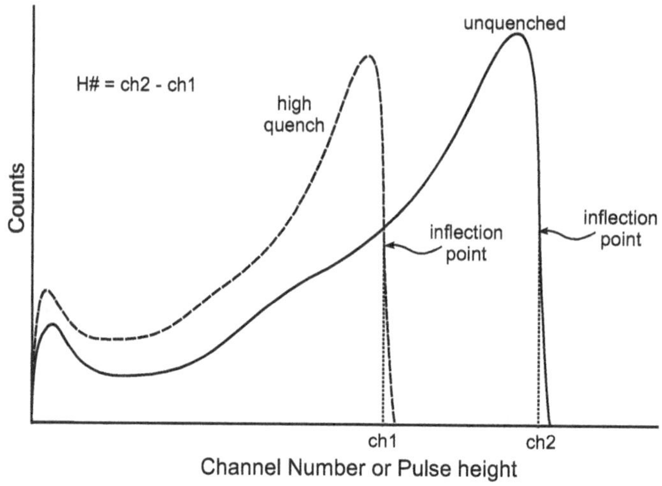

Figure 13.45. Depiction of different Compton spectra from a LSC system to determine the H-number.

continuum matters for the measurement.[29] Provided that sufficient counts are accumulated in the Compton continuum to produce meaningful data, the counting time can be varied between samples. The efficiency as a function of H-number changes tremendously for ^{3}H (18.3 keV) and ^{14}C (156 keV) beta particle sources. However, for high energy beta particles, such as for ^{32}P (E_{max} = 1710 keV) shown in Fig. 13.46, are less affected by quenching.

13.3.4 Plastic Scintillators

Initially introduced in 1950 by Schorr and Torney, plastic scintillators are in many ways similar to organic liquid scintillators in composition. Plastic scintillators contain both primary and secondary fluors in a solvent base that is subsequently polymerized to produce a plastic. This solvent may consist of a monomer that is polymerized by heating or by the addition of a solidifying catalyst. The first such reported plastic scintillators were produced by either introducing p-terphenyl into a hot solution of polystyrene $(C_8H_8)_n$, which was subsequently allowed to cool and solidify or, alternatively, by adding the catalyst 1% benzoil peroxide into a solution of styrene and fluors [Schorr and Torney 1950]. In like manner, methods of producing plastic scintillators from polyvinyl toluene [Koski 1951; Swank and Buck 1953; Buck and Swank 1953], 2,4-dimethylstyrene [Sandler et al. 1960]; and from polymethylmethacrylate (PMMA) [Kienzle et al. 1975; Klawonn et al. 1982; Salimgareeva and Kolesov 2005] were developed. Buck and Swank [1953] show that purity of starting materials has a profound effect upon luminosity, with clean and fresh chemicals producing higher light yields.

Contamination from air and the use of some catalysts that assist with polymerization degrade light yield. Removal of contaminants is performed by vacuum distillation of the starting materials. Fischer [1955] describes a method of passing nitrogen through the monomer solution before polymerization. If the plastic scintillator is not properly processed such that the monomer is completely polymerized, Pichat et al. [1953] showed that aged specimens have long-term oxidation that led to light self-absorption and reduced luminosity. It has also been reported that the process temperature to form plastic scintillators can significantly affect

[29] This convenience requires that the maximum beta particle energy is well below 477 keV. Otherwise, the beta particle counts still must be subtracted to ensure that the inflection point is properly identified.

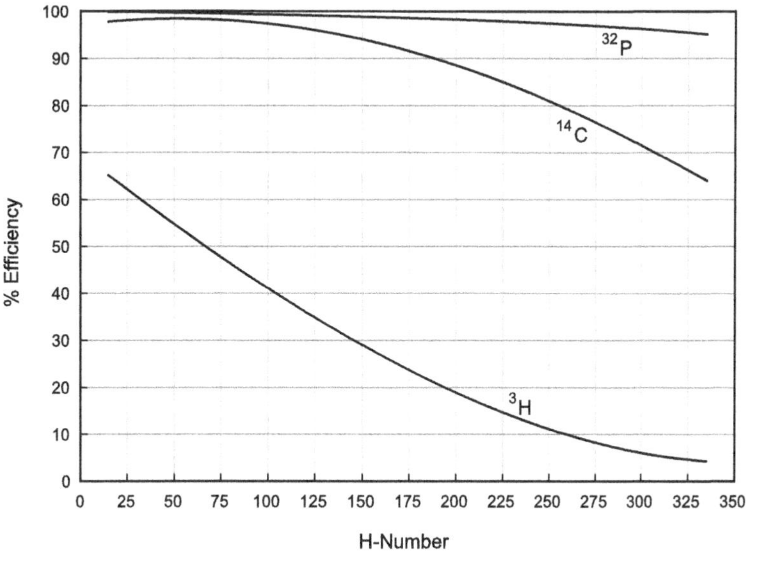

Figure 13.46. Calibration efficiency curves from a commercial LSC as a function of H-number for ^{3}H, ^{14}C, and ^{32}P beta particles. Data are from [Beckman 1985].

light yield [Basile 1957; Funt and Hetherington 1959]. Further, there is indication that light yield decreases for polymer molecular weights below about 100,000, becoming strongly dependent on molecular weight below 50,000 [Funt and Hetherington 1959].

Modern commercial plastic scintillators are based on similar chemistry, but the actual types and concentrations of primary and secondary fluors are kept proprietary. A casual glance at the varied types and specifications of plastic scintillators listed in Table 13.3 illustrates the variations in performance that can be achieved. Many of the legacy plastic scintillators originally developed by Pilot Chemical Co. and Nuclear Enterprises, Inc., are still being produced by contemporary vendors. Polyethylene naphthalate (poly(ethylene 2,6-naphthalate) or PEN) is a recently reported polymer capable of emitting scintillation photons without the assistance of fluor additives [Nakamura et al. 2011]. It has a light yield of about 10,500 photons/MeV with a λ_{max} of 425 nm. The density of PEN is 1.33 g cm^{-3} and has a refractive index of 1.65 that yields a critical angle of 65.4° at a glass interface. The simplicity of PEN can potentially offer a low cost alternative to traditional plastic scintillators.

Plastic scintillators have multiple advantages, namely they can be fabricated or machined into a variety of different shapes, they are relatively inexpensive, they are chemically stable and nonhygroscopic, and they are relatively fast (τ_f of about a few nanoseconds). The emission wavelengths are designed to couple well to conventional PMTs (λ_{max} near 425 nm). The refractive index for most commercial plastic scintillators is around 1.58, which yields a critical angle of 71.7° at a glass interface. Commercial plastic scintillators have a mass density of 1.038 g cm^{-3}. Many plastic scintillators also have a long decay component, similar to organic crystals, and can be used for PSD between neutron and gamma-ray events. The interaction efficiency of radiation particles can be increased by loading plastic scintillators with additional absorbers. For instance, neutron interaction rates can be increased by loading the monomer solution with ^{10}B, ^{6}LiF, or Gd. Gamma-ray interactions can be increased by loading the monomer solution with Pb or Sn. There is, of course, a limit to the amount of additional material that can be added, or otherwise the added material can decrease the luminosity of the plastic. Another notable use for plastic scintillators is that it can be used as a fast neutron detector with special plastics designed for PSD. Because they are composed of hydrogenous material, they can perform as fast neutron detectors through (n, p) reactions.

The multiple shapes and sizes of plastics include fibers and thin sheets. Scintillating optical fibers can be used to detect radiation and subsequently transport the luminescent signals relatively long distances to a light detector. The plastic fibers are typically composed of polystyrene with a refractive index of 1.58. In order to facilitate light propagation over long distances, a cladding material with lower refractive index must be applied to produce internal reflection. For instance, PMMA ($n = 1.49$) or fluorinated PMMA ($n = 1.42$) are often used as cladding materials. Plastics can also be manufactured as ultra-thin slices. These thin detectors are thin enough to allow the passage of relatively heavy charged particles, attenuating only a small portion of energy; hence can be operated as ΔE detectors for heavy ions such as protons and alpha particles. Thin film organics scintillators are available commercially, some with a thickness as low as 0.25 microns (see Table 13.3).

Recall that Birks' hypothesized non-linearity in light yield was due to damage by heavy ions, and indeed radiation damage can be of concern for plastics used over extended periods of time or in high dose environments. Over time, radiation damage, manifested as broken molecular bonds, cause a decrease in light yield. The effect may be directly due to damage of the base material, the primary fluor, or secondary fluors, and either the light emission is decreased or the transparency is decreased from internal absorption. Bross and Pla-Dalmau [1992] report that plastics operated in air (oxygen) are more affected by radiation damage than those operated in an inert atmosphere, and they hypothesize that oxygen reacts with radical species formed under irradiation as it diffuses into the plastic, and that these oxidation products contain carbonyl and hydroxyl groups which absorb at longer wavelengths and decrease transparency. Bross and Pla-Dalmau [1993] also report on radiation hard 3-hydroxyflavone (3HF) polystyrene plastic scintillators that had about 10% light loss after 10 Mrad of radiation exposure with ^{137}Cs. Studies performed by Vasil'chenko et al. [1996] on numerous (about 70) different polystyrene plastic scintillator formulations indicate that radiation hardness can depend upon the fluor type. Plastics that also contained plasticizers seemed to have higher radiation resistance, some withstanding little change after an exposure of 7 to 9 Mrad of gamma radiation from ^{137}Cs. Alternative polystyrene plastics were studied by Senchishin et al. [1995], who found that radiation hardness can be improved by the addition of diffusion enhancers (such as diphenyloxide). They also conclude that plastics prepared with polystyrene pellets rather than a monomer have superior radiation hardness.

13.4 Gaseous Scintillators

Fluorescence in air has been observed by humans living in northern and southern latitudes for millenia. However, the origin of the northern lights or aurora borealis or of the southern lights or aurora australis was unknown until the twentieth century. Protons and electrons emitted by the sun are usually deflected by earth's magnetic field. But at both poles the magnetic field is weaker and these charged particles can reach earth's atmosphere where they excite oxygen and nitrogen atoms and molecules. At high altitudes up to 200 miles the solar wind interacts with atomic oxygen producing 630.0 nm (red) light, but at lower altitudes the 630 nm emissions are suppressed and 557.7 nm (green) emissions dominate. Both the 557.7 and 630.0 nm emissions correspond to forbidden transitions of atomic oxygen and have decay constants of 0.7 s and 107 s, respectively. Thus, the red and green emissions are really phosphoresence. At even lower altitudes atomic oxygen is uncommon and the solar wind interacts with molecular nitrogen and excited molecular nitrogen to produce light in both the red and blue parts of the spectrum with 428 nm (blue) dominating. That air can be made to fluoresce by radiation particles is shown dramatically in Fig. 13.47. Monoenergetic deuterons emerging from the cyclotron travel in almost straight lines and have a well-defined range as can be seen from the abrupt ending of the fluorescent plume.

Figure 13.47. The white horizontal misty line is a beam of deuterons emerging from the 60-inch cyclotron at Argonne National Laboratory. The beam is visible because of excitation of air molecules by the energetic deuterons. From Kernan [1969].

13.4.1 Development of Gas Scintillator Counters

Discussed in Chapters 9 through 11 are three basic types of gas-filled detectors that operate on the concept of induced current from charge collection. Note that gas-filled detectors also release photons through deexcitation. Ordinarily these photons are purposely quenched to suppress the formation of additional electronic pulses. However, when the gas is used as a scintillator, this light quenching is undesirable.

The development of the first gas scintillation counter is generally credited to Grün and Schopper [1951]; however, the initial discovery of gas scintillation is not so well defined. It appears that the earliest mention of radiation induced luminosity from gas was reported by Huggins and Huggins [1903a; 1903b; 1905; 1906] who studied alpha particle induced (from radium bromide) luminosity from nitrogen, hydrogen, and air. They deduced that the familiar glow from the radium salt is actually caused by the fluorescing of the surrounding gas. Numerous studies were conducted on the emission spectra from various gases in the intervening years between the work of Huggins and Huggins and that of Grün and Schopper. These investigations are briefly described by Birks [1964]. Grün and Schopper [1951] used purified gases, H_2, Ar, N_2, O_2, and CO_2, which were irradiated with alpha particles, and the light emissions were monitored with a commercial PMT. They noted that impurities, including organic gases, caused a decrease in luminosity.

Gas scintillation detectors have some advantages over other types of detectors, especially when used to detect heavy ions. Gas scintillation detectors are relatively fast, usually having decay times on the order of a few nanoseconds. They also have a proportional response to energy deposition by heavy ions, an attribute generally not shared by other types of scintillators (organic and most inorganics). Multiple authors, outlined by Birks [1964], report a proportional response to charged particles over a wide range of stopping powers $-dE/dx$ regardless of the radiation type, and that the proportionality is superior to other scintillators (solid or liquid). Another important attribute is that gas scintillation detectors can be constructed in large sizes, which can be challenging to produce with solid-state scintillation and semiconductor detectors. Hence, for large, fast, and proportional detectors needed for the detection and energy spectroscopy of heavy ions, gas scintillation detectors may be a good choice.

13.4.2 Theory of Gas Scintillation Counters

The luminescent quantum efficiency, described by Stern and Volmer [1919], and modified by Birks [1964], can be expressed as follows. From the ideal gas law,

$$\frac{p_a V}{pV} = \frac{n_a RT}{nRT}, \tag{13.47}$$

where T is the absolute temperature, R is the gas proportionality constant, V is the volume, and p_a and n_a are the pressure and molar concentration, respectively, at STP. Rearrangement of terms, the molar concentration n of gas molecules for a closed container (V = constant) at constant temperature is

$$n = \frac{n_a p}{p_a}. \tag{13.48}$$

Stern and Volmer [1919] express the luminescent efficiency, without collisional quenching, as

$$\epsilon_1 = \frac{k_f}{k_f + k_i}, \tag{13.49}$$

where k_f is the rate constant for luminescent emissions and k_i is the internal quenching rate constant both with units of s^{-1}. The mean decay time is then

$$\tau_1 = \frac{1}{k_f + k_i}. \tag{13.50}$$

Birks [1964] adds the product nk_c to describe the quenching loss rate caused by collisions.[30] With collisional quenching, Eq. (13.49) is modified as

$$\epsilon' = \frac{k_f}{k_f + k_i + nk_c} = \frac{\epsilon_1}{1 + p/p'}. \tag{13.51}$$

where p' is defined as

$$p' = \frac{p_a(k_f + k_i)}{n_a k_c}. \tag{13.52}$$

Eq. (13.51) indicates that at low pressures, the intrinsic properties of fluorescence and quenching dominate the luminous efficiency, while at high pressures the collisional quenching becomes important and possibly dominant. A similar effect can be observed with the luminescent mean decay time

$$\tau' = \frac{1}{k_f + k_i + nk_c} = \frac{\tau_1}{1 + p/p'}. \tag{13.53}$$

With the above formulation for luminescence efficiency, the light yield for a gas can be computed as follows. Given a specific volume, the total number of ion pairs produced by a charged particle with energy E is

$$N_{ion} = \frac{E}{w}, \tag{13.54}$$

[30] The unit dimensions of k_c are not those of k_i and k_f, but must also include a factor to account for the gas density or pressure.

where w is the average energy required to produce an ion pair. The value of w is intrinsic to the gas atom (or molecule), and is not a function of the gas density. The number of ions produced for a partial energy loss ΔE is

$$N'_{ion} = \frac{\Delta E}{w}. \tag{13.55}$$

Hence, the scintillation yield per partial energy ΔE_i deposited is the product of the luminescent efficiency and the number of ion pairs produced, i.e.,

$$\frac{\Delta L}{\Delta E} \simeq A_i = \frac{\epsilon' \Delta E_i}{w}, \tag{13.56}$$

and the average light yield per unit energy is,

$$A = \frac{\epsilon_1}{1 + p/p'} \sum_i A_i = \frac{\epsilon_1 N_{ion}}{1 + p/p'}. \tag{13.57}$$

The light yield from a gas is linear with energy deposition as can be seen from Eq. (13.37). Because ΔE is proportional to p, the light yield for an energy deposition of ΔE is [Grün and Schopper 1954]

$$\Delta L = A \Delta E \propto \frac{p}{1 + p/p'}. \tag{13.58}$$

This result indicates that at low p the light yield decreases to small values (an intuitive result) and at high $p \gg p'$ the light yield levels off to a nearly constant value. This type of dependence was demonstrated by Grün and Schopper [1954] for a pure N_2 backfilled detector and an Ar + 8% N_2 backfilled detector.

13.4.3 Factors Affecting Performance

Grün and Schopper [1954] showed that the spectral response can be improved by the addition of a small amount of N_2 to an Ar backfilled detector, although they did not observe such a benefit when N_2 was added to a Xe backfilled detector. The reason for the improvement is a shift in wavelength from the short wavelength UV emission, characteristic of Ar (see Table 13.7), to the longer wavelength emission of N_2, thereby improving spectral matching to a PMT. The wavelength shift is most probably a consequence of ion collisions transferring energy from Ar to N_2 rather than UV photon absorption and reemission [Birks 1964]. Although there is apparent improvement in response with a slight amount of N_2 added to Ar, this result is an artifact of the measurement. In reality, the total number of photons emitted decreases because N_2 acts as a quenching gas; however, the total number of photons detected by the PMT light sensor used in the study increased. Also shown was that the presence of O_2 in small amounts works to reduce performance of Ar-filled gas scintillators.

Table 13.7. Properties of some gas scintillators at 0.974 atm irradiated with 4.7-MeV α particles.

Gas	w (eV/ion pair)	λ_m (nm)	Decay Time (ns)	Light Yield (photons per MeV)	Reference
He	42.3	390	$\leq$20	234	K1,B
Ar	26.4	250	$\leq$20	234	K1,B
Kr	24.1	318		447	K1,B
Xe	21.9	325		787	K1,B
N_2	35	390	$\leq$20	170	K1,B

B: Birks 1964; K1: Koch and Lesueur 1958.

Because the wavelengths emitted by almost all noble gases are in the UV range and do not match well to a PMT response (the exception is He), it is common to employ a wavelength shifter. These λ shifters are either coated upon the PMT window or upon the chamber walls. Several scintillating wavelength shifters that have been used including *trans*-stilbene, diphenylstilbene (DPS), *p*-terphenyl, *p*-quaterphenyl, POPOP, and tetraphenylbutadiene (TPB). Of these, Birks [1964] lists that DPS gives the best results when coupled to Ar, Kr, or Xe gas scintillators. There are precautions that should be implemented when using organic λ shifters. In particular, the outgassing of organic vapors contaminates the gas and causes a deleterious quenching effect that, consequently, lowers light output. Hence, the λ shifter should have a low vapor pressure to reduce outgassing, or the system can be cooled to also reduce the rate of outgassing. Alternatively, gas-flow chamber can be employed in which the scintillation gas is continuously replenished with fresh gas, or instead, the gas can be filtered and purified in a closed-loop system before re-entering the detector chamber. Finally, the light sensitivity can be increased by using PMTs with quartz windows through which the UV light has better transmittance than that of common soda-lime or borosilicate glass.

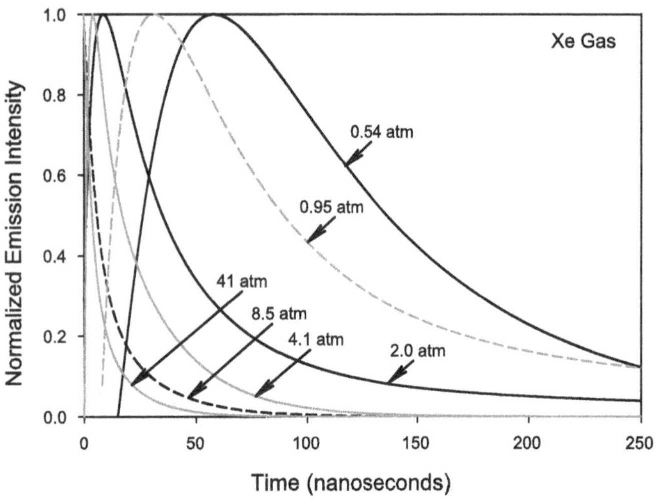

Figure 13.48. Normalized emission intensity for Xe gas as a function of gas pressure. [Parameters are from Koehler et al. 1974].

The emission efficiency and decay time are affected by the gas pressure, as can be seen from Eq. (13.51) and Eq. (13.53). Further, the purity of the gas also affects the emission spectrum, light yield, and decay time. A list of some gas scintillator values are provided in Table 13.7, although these values are limited to a specific case. The complex nature of energy transfer to luminescent levels in a gas scintillator proves difficult to model theoretically. However, Koehler et al. [1974] proposed the use of a summation of numerous exponential terms to model the scintillation decay response of Xe and Ar gases. They model the emitted light intensity by

$$I(t) = \sum_i A_i e^{-t/\tau_i}, \tag{13.59}$$

where A_i and τ_i are experimentally determined amplitudes and decay constants. The terms A_i and τ_i are listed in Koehler et al. 1974 for gas pressures ranging from 0.27 atm to 54.4 atm for Xe and ranging from 6.8 atm to 65 atm for Ar. The time responses for Xe gas at a few selected pressures, based on the work of Koehler et al. [1974], are plotted in Fig. 13.48.

13.4.4 Mixtures of Noble Gases

Gas mixtures of different noble gases have also been studied [See, for example Northrop and Gursky 1958b; Thiess and Miley 1974]. In general, the mixing of noble gases causes a decrease in light yield, with a mixture of He and Xe being the exception. Biondi and Brown [1949], Biondi and Holstein [1951], and Biondi [1952] propose that the dominating recombination process in a pure noble gas are (1) collisions between atomic ions and atoms that form stable molecular ions and (2) the subsequent capture of an electron by a molecular ion produces two atoms in excited states. These excited atoms can release scintillation photons. Apparently, the electron capture probability of the molecular ion is much greater than that of the atomic ion as suggested by the observation of significantly greater radiative recombination coefficients for the purified noble gases [Biondi 1952]. Recall from Chapter 10 (Table 10.6) that the ionization potentials of noble gases decrease with increasing atomic weight. In a gas mixture, the positive charge is transferred in collisions from the lighter gas molecule to the heavier molecule because the heavier gas always has the lower ionization potential. With the addition of small amounts of heavy gas to a lighter gas, the charge is rapidly passed to the heavy gas in a similar fashion as described for quenching gases in proportional counters. Because the concentration of heavy gas atoms is lower than in the pure condition, the probability of ion-atom collisions producing molecular ions is decreased, and consequently, the production of excited atoms is decreased. Consequently, the light yield typically decreases. The exceptions are a mixture of Xe-He, within a pressure ratio range of about 8%

to 100% Xe, and a mixture of Xe-Ne, within a pressure range of about 40% to 60% Xe. The optimum light yield was reported as about 10% Xe in He and yielded over 20% increase in light yield compared to that of pure Xe [Northrop and Gursky 1958b]. Shown in Fig. 13.49 are light yield curves for different gas mixture ratios of Xe fluoresced with alpha particles from ^{234}U. Other gas mixture combinations of He, Ne, Ar, and Kr produced lower light yield than the Xe mixtures.

13.4.5 Liquid and Solid Noble Elements

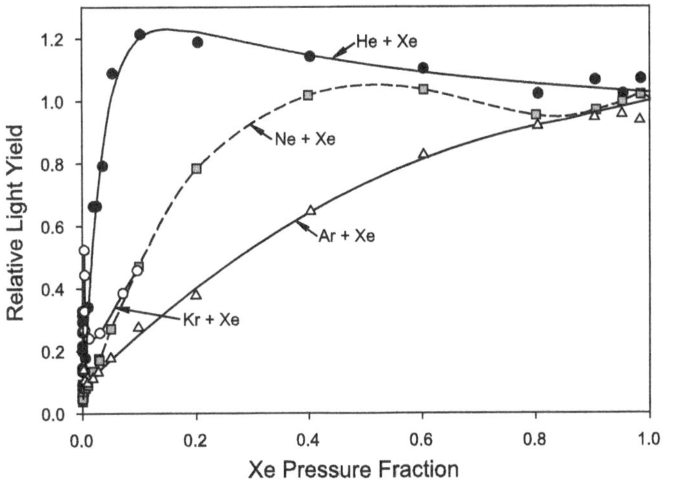

Figure 13.49. Light yield of noble gas mixtures to that of pure Xe gas. Shown are mixtures with the gas pressure fraction of Xe added to He, Ne, Ar, and Kr. Wavelength shifters were used in the work; diphenylstilbene for He-Xe, Ne-Xe, Ar-Xe, and *p*-quaterphenyl for Kr-Xe. Data are from [Northrop and Gursky 1958b].

The noble elements at high pressures have also been studied both in the liquid and solid state phases [see Aprile et al. 2006]. According to the observations of Simmons and Perkins [1961], the emission spectrum from liquid He is different than that from the gas phase and is in the far UV region (less than 160 nm) well below the transmission cutoff of quartz glass (about 180 nm). Similarly, Ar and Xe at high pressures (34 bar) were found to emit light centered at 126 nm and 169 nm, respectively [Koehler et al. 1974]. For such high energy UV photons, caution should be exercised to ensure that the photons are not reabsorbed in the detecting medium and can pass through the glass enclosure of the PMT. Consequently, soda-lime, borosilicate and even quartz windows should not be used for the detection of these emissions because they are effectively absorbed at these short wavelengths. However, the introduction of a wavelength shifter on the chamber walls and PMT window can effectively overcome this problem. The results reported by Northrop et al. [1958a] are listed in Table 13.8 for noble element scintillators when irradiated by alpha particles. Several cases are shown, including a comparison to NaI:Tl, which lists the use of a UV sensitive PMT (DuMont K1306) and a common bialkali PMT (DuMont 6292) with and without the assistance of a *p*-quaterphenyl wavelength shifter. Light yields from liquid and solid Xe were notably high, with liquid Xe having nearly the same light response as NaI:Tl and solid Xe having nearly twice the light response as NaI:Tl. Northrop et al. [1958a] note that the results of Table 13.8 were difficult to reproduce, but do reveal a trend in performance. More recent results by Doke et al. [1990] confirm Northrop's earlier work, reporting a light yield of about 4×10^4 photons/MeV for both liquid Ar and Ne. From the work of Kubota et al. [1978] and Hitachi and Takahashi [1983], fast and slow decay times have been observed for liquid Ar ($\tau_f = 6$ ns; $\tau_s = 1590$ ns) and liquid Xe ($\tau_f = 2.2$ ns; $\tau_s = 27$ ns). There is some indication that these decay times change with radiation type, but the differences reported by Hitachi and Takahashi [1983] were small. However, a relatively large difference in the light intensity ratio I_f/I_s was reported, with electrons yielding a small ratio (0.05), and increases for heavy charged particles for which $I_f/I_s = 0.45$ for alpha particles and $I_f/I_s = 1.6$ for fission fragments. Because of the low temperatures required to form liquid and solid noble elements, light coupling can be challenging. Although PMTs can be used, special models must be used that are capable of operating at cryogenic temperatures along with possessing enhanced sensitivity at UV wavelengths.

Table 13.8. Efficiencies of some noble element scintillators irradiated with α particles. Properties are normalized as compared to NaI:Tl without wavelength shifters.

PMT Type $\rightarrow$	DuMont K 1306				DuMont 6292			
PMT Converter $\rightarrow$	QP		None		QP			None
Reflector $\rightarrow$	Mo		Mo		Mo	Al	Mo	
Reflector Coat $\rightarrow$	none	QP	none	QP	none		QP	none
Xe								
Gas	0.34	0.25	0.74	0.66	0.5	0.55	0.49	
Liquid	0.43	0.37	0.92	1.29	0.78	0.52	0.75	
Solid	1.00		1.91		0.82	0.85	1.03	
Kr								
Liquid	0.43	0.43	0.80	1.19	0.86	0.69	0.70	
Solid	0.16		<0.04		0.31			
Ar								
Liquid	0.36		0.37	0.55	0.52		0.46	
Solid	0.35		<0.04		0.58			
NaI:Tl		0.59	0.97		0.90		0.87	1.00

Reproduced from Northrop et al. 1958 and Birks 1964.
QP $\equiv$ *p*-quaterphenyl.

13.4.6 Gas Proportional Scintillation Counters

It can be advantageous to apply voltage within the chamber of a gas scintillation detector. With an applied electric field, the liberated electrons can be drifted through the gas, much like as in an ion chamber. If the electric field is kept below the critical field required to produce avalanching, then collisions with bound electrons in the gas can be excited to higher energy levels, but do not produce impact ionization. As these excited electrons drop through lower states and eventually to the ground state, they lose energy sometimes by the emission of photons. Hence, this energy transfer produces a type of scintillation light gain. This operation is the basic principle behind a gas proportional scintillation counter (or gas PS counter) [Policarpo 1977].

An interesting advantage to the gas PS counter is its improved energy resolution over traditional proportional counters. Although the name given to the device is the "gas proportional scintillator counter," it is usually not operated in the proportional region of the gas pulse height curve. If operated in the proportional region, where avalanche multiplication occurs, the statistical fluctuation in electron production increases above that of the initial density of electron-ion pairs (see Chapter 10). Hence, the fluctuation in light production also increases thereby diminishing energy resolution. Policarpo et al. [1972] and Policarpo [1977] show that the gas PS counter can produce relatively good energy resolution for low-energy photons and for alpha particles. If a Gaussian model is assumed, the energy resolution of a detector can be approximated by

$$\mathrm{FWHM} = 2\sqrt{2\ln(2)}E(\sigma_{eh}^2 + \sigma_{pe}^2 + \sigma_g^2)^{1/2}, \tag{13.60}$$

where E is the radiation particle energy, σ_{eh}^2 is the fractional variance of ionization production, σ_{pe}^2 is the fractional variance of photo-electron production, and σ_g^2 is the fractional variance photomultiplier gain.

Equation (13.60) can be rewritten as,

$$\text{FWHM} = 2\sqrt{2\ln(2)}E\left[\left(\frac{\sqrt{Fn_0}}{n_0}\right)^2 + \left(\frac{\sqrt{n_{pe}}}{n_{pe}}\right)^2 + \left(\frac{\sigma_G}{G}\right)^2\right]^{1/2}, \tag{13.61}$$

$$= 2\sqrt{2\ln(2)}E\left[\frac{Fw}{E} + \frac{w}{EL\epsilon_c\epsilon_{pe}} + \frac{\sigma_G^2}{G^2}\right]^{1/2},$$

where the FWHM is in units of energy, F is the Fano factor, and w is the average ionization energy, L is the light yield at the operating voltage, ϵ_c is the light collection efficiency, ϵ_{pe} is the photoelectron emission efficiency, G is the PMT light collector gain, and σ_G^2 is the variance in the PMT gain. Typically, the largest statistical variance in gain is the fluctuation in the initial number of photoelectrons produced, accounted for in the second term of Eq. (13.62); hence the third term is usually small so that

$$\text{FWHM} \simeq 2\sqrt{2\ln(2)Ew}\left[\frac{FL\epsilon_c\epsilon_{pe} + 1}{L\epsilon_c\epsilon_{pe}}\right]^{1/2}. \tag{13.62}$$

As previously discussed, noble gas mixtures usually have reduced light yield. However, Alkhazov et al. [1967] show that optimized mixtures of noble gases can decrease the Fano factor. One can observe from Eq. (13.62) that a reduction in F improves the energy resolution, while a decrease in L works to degrade energy resolution; therefore, a proper optimization of the gas mixture should be developed for best results.

There are numerous configurations for the gas proportional scintillation counter [see Aprile et al. 2006 for examples], yet they almost all share a few attributes. The gas chamber has a drift volume designed with a fairly constant electric field which is adjacent to a high field scintillation region. Interactions in the constant field drift region produce electron-ion pairs. The electrons are drifted into the high field region where they gain enough energy to excite more electrons to upper energy levels of the noble gas, but not enough to produce more ionization (or a Townsend avalanche). The high field region is usually small by comparison to the drift region, and can be achieved with small diameter anodes in a similar fashion as in a proportional gas-filled counter. The scintillation light is usually measured with a PMT, although alternative structures with photodiodes have been explored. Other types of gas proportional counters may use microstrip or microgap detectors, or even gas electron multipliers (GEMs). In any event, the goal is to produce scintillations adjacent to a light sensing device with minimal reabsorption loss. As mentioned earlier, impurities can seriously quench the scintillation process, so gas proportional scintillation counters are usually operated as gas-flow detectors in which the high-purity gas is constantly replenished. Wavelength shifters applied to the PMT (or other light collector), and to the chamber walls can help enhance the luminescent response. Excellent energy resolution has been realized, on the order of 9.7% FWHM for 5.89 keV gamma rays [Palmer 1975] and 1.6% FWHM for 8.1 MeV alpha particles [Conde and Ferreira 1975].

PROBLEMS

1. Given that $S_e = S_g$ and $\omega_e = \omega_g$, derive the result for Eq. (13.9).

2. Determine the light conversion efficiency for NaI:Tl photoelectric absorption 898 keV gamma rays. Assume the maximum wavelength and light yield from Table 13.1 apply.

3. Prove Eq. (4.58) reduces to Eq. (13.35) for non-relativistic particles ($v_i \ll c$).

4. A beam of 1-MeV gamma rays is normally incident on a CsI crystal that is 30 mm thick. What is the probability that an incident gamma ray has one or more interactions in the crystal?

5. The dead time in a particular scintillator detector is 2 μs and in typical gas-filled detector is 250 μs. For a particular radionuclide sample, the gross counts obtained with the gas-filled detector suffer a 5% loss because of dead time. For the same source-detector geometry, what is the loss in the scintillator detector?

6. Given the task to efficiently detect gamma rays in the 30 to 60 keV range with a size restricted detector only 2 mm thick, which material would you use given the options of NaI:Tl, GSO, and LSO? Explain your selection choice.

7. Suppose you have a NaI:Tl detector with 7% FWHM energy resolution when operated with a 1 μs shaping time at 300K. What is the expected energy resolution if the temperature is increased to 325K?

8. Given a parallel beam of 662 keV gamma rays from a ^{137}Cs source, compare the efficiency for initial interactions in 1 cm thick slabs of NaI:Tl, $LaBr_3$:Ce, and LSO:Ce. What is it for 1.332 MeV gamma rays for these three scintillators?

9. The best choice of a scintillator material for a detector depends on many factors such as (a) cost, (b) availability, (c) response speed, (d) response linearity, (e) light output, (f) detector efficiency for high-energy photons, (g) radiation hardness, and (h) ability to discriminate between different types of radiation. Evaluate the attractiveness of a typical inorganic scintillator, such as NaI(Tl) and an organic scintillator, such as a plastic one, with regard to each of above performance factors.

10. Why are organic scintillators preferred for electron and beta-particle detection while inorganic scintillators are preferred for gamma-ray detection?

11. Explain the purpose of adding a trace amount of an activator element to an inorganic scintillator. Why is an activator not needed in organic scintillators?

12. A set of 10 vial standards, each having an activity of 0.1 μCi ^{14}C, has the following results for 10 second measurements in a LSC system:

ch1	ch2
420	26600
1000	23000
1600	20000
2000	17400
3500	13500
4200	11100
5000	8700
5900	6000
6800	3500
7800	1200

You then measure an unknown ^{14}C sample with 55120 in ch 1 and 106,000 counts in ch 2. What is the activity and standard error of the unknown sample?

REFERENCES

ABBADY, A., G.E. ADEL, AND R. MICHEL, "Indoor Radon measurement with the Lucas Cell Technique," *Appl. Rad. Isotopes*, **61**, 1469-1475, (2004).

ADACHI, C. AND T. TSATSUI, "Fundamentals of Luminescence," Ch. 1, Sec. 5, *in Fundamentals of Phosphors*, W.M. YEN, S. SHIONOYA AND H. YAMAMOTO, Eds., CRC Press, Boca Raton, 2007.

ALEKHIN, M.S., J.T.M. DE HAAS, K.W. KRÄMER, AND P. DORENBOS, "Scintillation Properties of and Self Absorption in $SrI_2:Eu^{2+}$," *IEEE Trans. Nucl. Sci.*, **NS-58**, 2519–2527, (2011).

ALEKHIN, M.S., D. BINER, K.W. KRÄMER, AND P. DORENBOS, "Improvement of $LaBr_3$:5%Ce Scintillation Properties by Li^+, Na^+, Mg^{2+}, Ca^{2+}, Sr^{2+}, and Ba^{2+} Co-doping," *J. Appl. Phys.*, **113**, 224904, 7 pages, (2013a).

ALEKHIN, M.S., J.T.M. DE HAAS, I.V. KHODYUK, K.W. KRÄMER, P.R. MENGE, V. OUSPENSKI, AND P. DORENBOS, "Improvement of γ-ray Energy Resolution of $LaBr_3:Ce^{3+}$ Scintillation Detectors by Sr^{2+} and Ca^{2+} Co-doping," *Appl. Phys. Lett.*, **102**, 161915, 4 pages,(2013b).

ALKHAZOV, G.D., A.P. KOMAR, AND A.A. VOROB'EV, "Ionization Fluctuations and Resolution of Ionization Chambers and Semiconductor Detectors," *Nucl. Instrum. Meth.* , **48**, 1–12, (1967).

AMSLER, C., ET AL., "Temperature Dependence of Pure Cs:I Scintillation Light Yield and Decay Time," *Nucl. Instrum. Meth.*, **A480**, 494–500, (2002).

ANSI/AARST MS-PC 2015, *Performance Specifications for Instrumentation Systems Designed to Measure Radon Gas in Air*, AARST Consortium on National Radon Standards, Fletcher, 2015.

APPLIED SCINTILLATION TECHNOLOGIES, "Li Doped Glass Scintillators," www.apace- science.com/ast/g_scint.htm, accessed 2016 (now renamed Scintacor).

APRILE, E., A.E. BOLOTNIKOV, A.I. BOLOZDYNYA, AND T. DOKE, *Noble Gas Detectors*, Mörlenbach: Wiley-VCH, 2006.

ASHCROFT, N.W. AND N.D. MERMIN, *Solid State Physics*, Philadelphia: Holt, Rinehart, Winston, 1976.

AUTRATA, R., P. SCHAUER, J. KVAPIL, AND J. KVAPIL, "A Single Crystal of YAG—New Fast Scintillator in SEM," *J. Phys. E: Sci. Instrum.*, **11**, 707–708, (1978).

BABIN, V., A. BEKESHEV, A. ELANGO, K. KALFER, A. MAAROOS, K. SHUNKEEV, E. VASIL'CHENKO, AND S. ZAZUBOVICH, "Effect of Uniaxial Stress on Luminescence of Undoped and Thallium-Doped KI and RbI Crystals," *J. Phys.: Cond. Matter*, **11**, 2303–2317, (1999).

BACK, H.O., ET AL., "Pulse-Shape Discrimination with the Counting Test Facility," *Nucl. Instrum. Meth.*, **A584**, 98–113, (2008).

BALCERZYK, M., M. MOSZYŃSKI, M. KAPUSTA, D. WOLSKI, J. PAWELKE, AND C.L. MELCHER, "YSO, LSO, GSO and LGSO. A Study of Energy Resolution and Nonproportionality," *IEEE Trans. Nucl. Sci.*, **NS-47**, 1319–1323, (2000).

BALCERZYK, M., M. MOSZYŃSKI, Z. GALAZKA, M. KAPUSTA, A. SYNTFELD, AND J-L. LEFAUCHEUR, "Perspectives for High Resolution and High Light Output LuAP:Ce Crystals," *IEEE Trans. Nucl. Sci.*, **NS-52**, 1823–1829, (2005).

BARYSHEVSKY, V.G., M.V. KORZHIK, V.I. MOROZ, ET AL., "$YAlO_3$:Ce-Fast-Acting Scintillators for Detection of Ionizing Radiation," *Nucl. Instrum. Meth.*, **B58**, 291–293, (1991).

BASILE, L.J., "Characteristics of Plastic Scintillators," *J. Chem. Phys.*, **27**, 801–806, (1957).

BECKMAN-COULTER LS 1801, 3801, 5801 Series Operating Manual, Beckman Instruments, Irvine, 1985.

BELL, P.R., "The Use of Anthracene as a Scintillation Counter," *Phys. Rev.*, **73**, 1405–1406, (1948).

BENGSTON, B. AND M. MOSZYŃSKI, "Energy-Transfer and Light-Collection Characteristics for Different Types of Plastic Scintillators," *Nucl. Instrum. Meth.*, **117**, 227–232, (1974).

BERLMAN, I.B., *Handbook of Fluoresence Spectra or Aromatic Molecules, 2nd Ed.*, New York: Academic Press, 1971.

BERNSTEIN, W. AND A.W. SCHARDT, "Activation of LiI Crystal Phosphors," *Phys. Rev.*, **52**. 919–920, (1952).

BESSIERE, A., P. DORENBOS, C.W.E. VAN EIJK, K. W. KRAMER, AND H. U. GÜDEL, "Luminescence and Scintillation Properties Of $Cs_2LiYCl_6:Ce^{3+}$ for Gamma and Neutron Detection." *Nucl. Instrum. Meth.*, **A537**, 242–246, (2005).

BIONDI, M.A. AND S.C. BROWN, "Measurement of Electron-Ion Recombination," *Phys. Rev.*, **76**, 1697-1700, (1949).

BIONDI, M.A. AND T. HOLSTEIN, "Concerning the Mechanism of Electron-Ion Recombination," *Phys. Rev.*, **82**, 962–963, (1951).

BIONDI, M.A., "Concerning the Mechanism of Electron-Ion Recombination. II.," *Phys. Rev.*, **83**, 1078–1080, (1952).

BIRKS, J.B. AND F.A. BLACK "Deterioration of Anthracene under α-Particle Irradiation," *Proc. Phys. Soc.*, **A 64**, 511–512, (1951).

BIRKS, J.B., "Scintillations from Organic Crystals: Specific Fluorescence and Relative Response to Different Radiation," *Proc. Phys. Soc.*, **A 64**, 874–877, (1951).

BIRKS, J.B., *The Theory and Practice of Scintillation Counting*, New York: MacMillan, 1964.

BIROWOSUTO, M.D., P. DORENBOS, J.R.M. DE HAAS, C.W.E. VAN EIJK, K.W. KRÄMER, AND H.U. GÜDEL, "Li-Based Thermal Neutron Scintillator Research: $Rb_2LiYBr_6:Ce^{3+}$ and Other Elpasolites," *IEEE Trans. Nucl. Sci.*, **NS-55**, 1152–1155, (2008).

BLANC, D., F. CAMBOU, AND Y.G. DE LAFOND, "Étude Cinétique de la Scintillation dans les Cristaux Organiques Purs," *J. de Physique*, **25**, 319–325, (1964).

BNC, BERKELEY NUCLEONICS CORPORATION, https://www.berkeleynucleonics.com/cllbc-scintillators, accessed March 31, 2020.

BOATNER, L.A., J.O. RAMEY, J.A. KOLOPUS, AND J.S. NEAL, "Divalent Europium Doped and Un-Doped Calcium Iodide Scintillators: Scintillator Characterization and Single Crystal Growth," *Nucl. Instrum. Meth.*, **A786**, 23–31, (2015).

BOK, J. AND P. SCHAUER, "Performance of SEM Scintillation Detector Evaluated by Modulation Transfer Function and Detective Quantum Efficiency Function," *Scanning*, **36**, 384–393, (2014).

BOLLINGER, L.M. AND G.E. THOMAS, "Measurement of the Time Dependence of Scintillation Intensity by a Delayed-Coincidence Method," *Rev. Sci. Instrum.*, **32**, 1044–1050, (1961).

BOLLINGER, L.M., G.E. THOMAS, AND R.J. GINTHER, "Neutron Detection with Glass Scintillators," *Nucl. Instrum. Meth.*, **17**, 97–116, (1962).

BONANOMI, J. AND J. ROSSEL, "Scintillations de Luminescence dans les Iodures D'Alcalins," *Helv. Phys. Acta.*, **25**, 725–752, (1952).

BOWEN, E.J., E. MIKIEWICZ, AND F.W. SMITH, "Resonance Transfer of Electronic Energy in Organic Crystals," *Proc. Phys. Soc.*, **A 62**, 26–31, (1949).

Brooks, F.D. and D.T.L. Jones, "Directional Anisotropy in Organic Scintillation Crystals," *Nucl. Instrum. Meth.*, **121**, 69–76, (1974).

Brooks, F.D., "Organic Scintillators," *Prog. Nuclear Phys.*, **5**, 252–313, (1956).

Broser, I. and H. Kallman, "Über die Anregung von Leuchtstoffen durch schnelle Korpuskularteilchen I," *Z. Natuf. A*, **2**, 439–440, (1947a).

Broser, I. and H. Kallman, "Über den Elementarprozeß der Lichtanregung in Leuchtstoffen durch α-Teilchen, schnelle Elektronen und γ-Quanten II," *Z. Natuf. A*, **2**, 642–652, (1947b).

Bross, A.D. and A. Pla-Dalmau, "Radiation Damage of Plastic Scintillators," *IEEE Trans. Nucl. Sci.*, **NS-39**, 1199–1204, (1992).

Bross, A.D. and A. Pla-Dalmau, "Radiation Effects in 3HF Scintillator," *Nucl. Instrum. Meth.*, **A327**, 337-345, (1993).

Buck, W.L. and R.K. Swank, "Efficient Plastic Scintillators," *Nucleonics*, **11**, 48–52, (1953).

Caffrey, A.J., R.L. Heath, P.D. Ritter, C.D. Van Siclen, D.F. Anderson, and S. Majewski, "Radiation Damage Studies on BaF_2 and BGO Scintillation Materials," *IEEE Trans. Nucl. Sci.*, **NS-33**, 230–234, (1986).

Camerman, A. and J. Trotter, "The Crystal and Molecular Structure of Pyrene," *Acta Cryst.*, **18**, 636–643, (1965).

Carmen, L., N. Zaitseva, H.P. Martinez, B. Rupert, I. Pawelczak, A. Glenn, H. Mulcahy, R. Leif, K. Lewis, and S. Payne, "The Effect of Material Purity on the Optical and Scintillation Properties of Solution-Grown Trans-Stilbene Crystals," *J. Cryst. Grow.*, **368**, 56–61, (2013).

Chen, J., L. Zhang, and R. Zhu, "Large Size LSO and LYSO Crystal Scintillators for Future High-Energy Physics and Nuclear Physics Experiments," *Nucl. Instrum. Meth.*, **A572**, 218-224, (2007).

Cherepy, N.J., J.D. Kuntz, J.J. Roberts, A.A. Hurst, B. Drury, R.D. Scanner, T.M. Tillotson, and S.A. Payne, *Transparent Ceramic Scintillator Fabrication, Properties and Applications*, LLNL-PROC-406543 Report, 2008.

Cherepy, N.J., S.A. Payne , S.J. Asztalos, G. Hull, J.D. Kuntz, T. Niedermayr, S. Pimputkar, J.J. Roberts, R.D. Sanner, T.M. Tillotson, E. van Loef, C.M. Wilson, K.S. Shah, U.N. Roy, R. Hawrami, A. Burger, L.A. Boatner, W.S. Choong, and W.W. Moses, "Scintillators With Potential to Supersede Lanthanum Bromide," *IEEE Trans. on Nucl. Sci.*, **56**, 873–880, (2009).

Cherepy, N.J., S.A. Payne, B.W. Sturm, S.P. O'Neal, Z.M. Seeley, O.B. Drury, L.K. Haselhorst, B.L. Rupert, R.D. Sanner, P.A. Thelin, and S.E. Fisher, "Performance of Europium-Doped Strontium Iodide, Transparent Ceramics and Bismuth-Loaded Polymer Scintillators," *Proc. SPIE*, **8142**, 81420W(1-8), (2011).

Cherepy, N.J., S.A. Payne, B. Sturm, O.B. Drury, S.P. O'Neal, P.A. Thelin, K.S. Shah, R. Hawrami, M. Momayezi, B. Hurst, A. Burger, B. Wiggins, P. Bhattacharya, L.A. Boatner, and J.O. Ramey, "Instrument Development and Gamma Spectroscopy With Strontium Iodide," *IEEE Trans. on Nucl. Sci.*, **60**, 955–958, (2013).

Cherepy, N.J., Z.M. Seeley, S.A. Payne, P.R. Beck, E.L. Swanberg, S. Hunter, L. Ahle, S.E. Fisher, C. Melcher, H. Wei, T. Stefanik, Y. Chung, and J. Kindem, *High Energy Resolution Transparent Ceramic Garnet Scintillators*, **LLNL-PROC-659775 Report**, 2014.

Cherepy, N.J., *Transparent Ceramic Scintillators for Gamma Spectroscopy and MeV Imaging*, **LLNL-PROC-676780 Report**, 2015.

Chewpraditkul, W., L. Swiderski, M. Moszynski, T. zczesnik, A. Syntfeld-Kazuch, C. Wanarak, and P. Limsuwan, "Scintillation Properties of LuAG:Ce, YAG:Ce and LYSO:Ce Crystals for Gamma-Ray Detection," *IEEE Trans. Nucl. Sci.*, **NS-56**, 3800–3805, (2009).

Cho, Z.H. and M.R. Farukhi, "Bismuth Germanate as a Potential Scintillation Detector in Positron Cameras," *J. Nucl. Med.*, **18**, 840–844, (1977).

Colmenares, C., E.G. Shapiro, P.E. Barry, and C.T. Prevo, "A Europium-Doped, Calcium-Fluoride Scintillator System for Low-Level Tritium Detection," *Nucl. Instrum. Meth.*, **114**, 277–289, (1974).

Coltman, J.W. and F. Marshall, "Photomultiplier Radiation Detector," *Nucleonics*, **1**, No. 3, 58–64, (1947).

Combes, C.M., P. Dorenbos, C.W.E. van Eijk, K.W. Krämer, and H.U. Güdel, "Optical and Scintillation Properties Pure and of Ce^{3+}-Doped Cs_2LiYCl_6 and Li_3YCl_6:Ce^{3+} Crystals," *J. Luminescense*, **82**, 299–305, (1999).

Conde, C.A.N. and M. Fátima A. Ferreira, "The Argon Uniform Field Gas Proportional Scintillation Counter: Improved Energy Resolution for α-Particles and its Behavior for X-Rays," *Nucl. Instrum. Meth.*, **124**, 307–308, (1975).

Cook, J.R. and K.A. Mahmoud, "Luminescence Characteristics of Some Scintllating Crystals," *Proc. Phys. Soc.* **67**, 817–824, (1954).

Cooke, D.W., K.J. McClellan, B.L. Bennett, J.M. Roper, M.T. Whittaker, R.E. Muenchausen, and R.C. Sze, "Crystal Growth and Optical Characterization of Cerium-Doped $Lu_{1.8}Y_{0}.2SiO_5$," *J. Appl. Phys.*, **88**, 7360–7362, (2000).

Craun, R.L. and D.L. Smith, "Analysis of Response Data for Several Organic Scintillators," *Nucl. Instrum. Meth.*, **80**, 239–244, (1970).

Cryos-beta, `www.cryos-beta.kharkov.ua/organic.php`, accessed May 16, 2016.

Curran, S.C. and W.R. Baker, "A Photoelectric Alpha Particle Detector," *U.S. Atomic Energy Comm. Rep. M.D.D.C. - 1296*, 1944.

Cusano, D.A., F.F. Holub, and S. Prochazka, "Ceramic-Like Scintillators," *US patent 4242221*, allowed Dec. 30, 1980.

Cutler, P.A., C. L. M elcher, M. A. Spurrier, P. Szupryczynski, and L.A. Eriksson, "Scintillation Non-Proportionality of Lutetium- and Yttrium-Based Silicates and Aluminates," *IEEE Trans. Nucl. Sci.*, **NS-56**, 915–919, (2009).

Cvachovec, F., J. Cvachovec, and P. Tajovsky, "Anisotropy of Light Output in Response of Stilbene Detectors," *Nucl. Instrum. Meth.*, **A476**, 200-202, (2002).

Czirr, J.B., "The α/β Ratio of Several Organic Scintillators," *Nucl. Instrum. Meth.*, **25**, 106-108, (1964).

Dabbs, V.M., W.H. Parr, and A.T. Krebs, "Tests on Scintillation GM Tubes," *Nucleonics*, **10**, 54–55, (1952).

Daghighian, F., P. Shenderov, K.S. Pentlow, M.C. Graham, B. Eshaghian, C.L. Melcher, and J.S. Schweitzer, "Evaluation of Cerium Doped Lutetium Oxyorthosilicate (LSO) Scintillation Crystal for PET," *IEEE Trans. Nucl. Sci.*, **NS-40**, 1045–1047, (1993).

Dahlbom, M., L.R. MacDonald, L. Eriksson, M. Paulus, M. Andreaco, I.E. Casey, and C. Moyers, "Performance of a YSO/LSO Phoswich Detector for use in a PET/SPECT System," *IEEE Trans. Nucl. Sci.*, **NS-44**, 1114–1119, (1997).

de Haas, J.T.M., P. Dorenbos, and C.W.E. van Eijk, "Measuring the Absolute Light Yield of Scintillators." *Nucl. Instrum. Meth.*, **A537**, 97–100, (2005).

Derenzo, S., M. Boswell, M. Weber, and K. Brennan, *Scintillation Properties*, scintillator.lbl.gov, accessed February 25, 2016.

Derenzo, S., W.W. Moses, J.L. Cahoon, T.A. Devol, and L. Boatner, "X-ray Fluorescense Measurements of 412 Inorganic Compounds," *Conf. Rec. IEEE Nuc. Sci. Symp.*, Sante Fe, NM, Nov. 2–9, 143–147, (1991).

Deutsch, M., "Naphthalene Counters for Beta and Gamma Rays," *Nucleonics*, **3**, No. 3, 58–59, (1948).

DiBianca, F.A., J-P. J. Georges, D.A. Cusano, and C.D. Greskovich, "Rare Earth Ceramic Scintillator," *US patent 4525626*, allowed June 25, 1985.

Doke, T., K. Masuda, and E. Shibamura, "Estimation of Absolute Photon Yields in Liquid Argon and Xenon for Relavistic (1 MeV) Electrons," *Nucl. Instrum. Meth.*, **A291**, 617–620, (1990).

Dorenbos, P., J.T.M. de Haas, and C.W.E. van Eijk, "Non-Proportionality in the Scintillation Response and the Energy Resolution Obtainable with Scintillation Crystals," *IEEE Trans. Nucl. Sci.*, **NS-42**, 2190–2202, (1995).

Dorenbos, P., "Light Output and Energy Resolution of Ce^{3+}-Doped Scintillators," *Nucl. Instrum. Meth.*, **A486**, 208–213, (2002).

Doty, F.P., D.S. McGregor, M. Harrison, K. Findley, and R. Polichar, "Structure and Properties of Lanthanide Halides," *Proc. SPIE*, **6707**, 670–705, (2007).

Doty, F.P., X. Zhou, P. Yang, and M.A. Rodriguez, "Elpasolite Scintillators," *Sandia Report*, **SAND2012-9951**, 2012.

Drozdowski, W., P. Dorenbos, J.T.M. de Haas, R. Drozdowski, A. Owens, K. Kamada, K. Tsutsumi, Y. Uusuki, T. Yanagida, and A. Yoshokawa, "Scintillation Properties of Praseodymium Activated $Lu_3Al_5O_{12}$ Single Crystals," *IEEE Trans. Nucl. Sci.*. **NS-55**, 2420–2424, (2008b).

Duclos, S.J., C.D. Greskovich, R.J. Lyons, J.S. Vartuli, D.M. Hoffman, R.J. Riedner, and M.L. Lynch, "Development of the HiLight^{TM} Scitillator for Computed Tomography Medical Imaging," *Nucl. Instrum. Meth.*, **A505**, 68–71, (2003).

Edgarton, R. and K. Teegarden, "Emission Spectra of KCL:Tl, KBr:Tl, and KI:Tl at 300, 80, and 12°K," *Phys. Rev.*, **129**, 169–174, (1963).

Engstrom, R.W. and unknown others, *Photomultiplier Handbook*, Lancaster: Burle Technologies, Inc., 1980.

Farukhi, M.R. and C.F. Swinehart, "Barium Fluoride as a Gamma Ray and Charged Particle Detector," *IEEE Trans. Nucl. Sci.*, **NS-18**, 200–204, (1971).

Ficke, D.C., M.M. Ter-Pogossian, R.S. Miyaoka, and T.K. Lewellen, "A GSO(Ce) Block Type Detector for High Count Rate PET Applications," *IEEE Trans. Nucl. Sci.*, Conf. Rec. IEEE Nuc. Sci. Symp., Norfolk, VA, Oct. 30–Nov. 5, 1859–1863, (1994).

Fischer, J., "Preparing Large-Area Plastic Scintillators," *Nucleonics*, **13**, 52–54, (1955).

Furukawa, "Pr:LuAG Scintillation Crystal," application note, Furukawa Denshi Co., LTD., see www.furukwawkk.co.jp/e_index.htm, accessed 2016.

Fujimoto, Y., T. Yanagida, M. Koshimizu, and K. Asai, "Photoluminescence and Scintillation Properties of SiO_2 Glass Activated with Eu^{2+}," *Sensors and Materials*, **27**, 263–268, (2015).

Funt, B.L. and A. Hetherington, "The Influence of Chain Length on the Luminescent Output of Plastic Scintillators," *Int. J. Appl. Rad. Isotopes*, **4**, 189–195, (1959).

Ginther, R.J. and J.H. Schulman, "Glass Scintillators," *IRE Trans. Nucl. Sci.*, **NS-5**, 92–95, (1958).

Ginther, R.J., "New Cerium Activated Scintillating Glasses," *IRE Trans. Nucl. Sci.*, **NS-7**, 28–31, (1960).

Glodo, J., E. van Loef, R. Hawrami, W.M. Higgins, A. Churilov, U. Shirwadkar, and K.S. Shah, "Selected Properties of Cs_2LiYCl_6, $Cs_2LiLaCl_6$, and $Cs_2LiLaBr_6$ Scintillators," *IEEE Trans. Nucl. Sci.*, **NS-58**, 333–338, (2011).

Greskovich, C. and S. Duclos, "Ceramic Scintillators," *Annu. Rev. Mater. Sci.*, **27**, 69-88, (1997).

Giffiths, J., S. Kleinegesse, D. Saunders, R. Taylor, and A. Vacheret, "Pulse Shape Discrimination and Exploration of Scintillation Signals Using Convolutional Neural Networks," *arXiv:1807.06853*, 13 pages, (2018).

Grün, A.E. and E. Schopper, "Über die Fluoreszenz von Gasen bei Anregung dürch α-Teilchen," *Z. Naturf.*, **6A**, 698–700, (1951).

Grün, A.E. and E. Schopper, "Die Untersuchung von Energieaustausch- und Löschprozessen in Gasen durch Anregung mit Schnellen Teilchen," *Z. Naturf.*, **9A**, 134–147, (1954).

Gundiah, G., K. Brennan, Z. Yan, E.C. Samulon, G. Wu, G.A. Bizarri, S.E. Derenzo, and E.D. Bourret-Courchesne, "Structure and Scintillation Properties of Ce^{3+}-Activated $Cs_2NaLaCl_6$, Cs_3LaCl_6, $Cs_2NaLaBr_6$, Cs_3LaBr_6, Cs_2NaLaI_6, and Cs_3LaI_6," *J. Luminescence*, **149**, 374–384, (2014).

Guss, P., M.E. Foster, B.M. Wong , F.P. Doty, K. Shah, M.R. Squillante, U. Shirwadkar, R. Hawrami, J. Tower, and D. Ying, "Results for Aliovalent Doping of $CeBr_3$ with Ca^{2+}," *J. Appl. Phys.*, **115**, 034908(1-10), (2014).

Hansen, W. and D. Richter, "Determination of Light Output Function and Angle Dependent Correction for a Stilbene Crystal Scitillation Neutron Spectrometer," *Nucl. Instrum. Meth.*, **A476**, 195–199, (2002).

Harihar, P., H. Chen, W.J. Stapor, and A.R. Knudson, "Scintillation Decays in a Trans-Stilbene Crystal," *Nucl. Instrum. Meth.*, **A336**, 176–178, (1993).

Harrison, M.J., C. Linnick, B. Montag, S. Brinton, M. McCreary, F.P. Doty, and D.S. McGregor, "Scintillation performance of Aliovalently Doped $CeBr_3$," *IEEE Trans. Nucl. Sci.*, **56**, 1661–1665, (2009).

Heath, D.F. and P.A. Sacher, "Effects of a Simulated High-Energy Space Environment on the Ultraviolet Transmittance of Optical Materials between 1050 Åand 3000 Å," *Appl. Optics*, **5**, 937–943, (1966).

Higashimura, T., O. Yamada, N. Nohara, and T. Shidei, "External Standard Method for the Determination of the Efficiency in Liquid Scintillation Counting," *Int. J. Appl. Radiat. Isot.*, **13**, 308–309, (1962).

Hitachi, A. and T. Takahashi, "Effect of Ionization Density on the Time Dependence of Luminescence from Liquid Argon and Xenon," *Phys. Rev. B*, **27**, 5279–5285, (1983).

Hofstadter, R., "Alkali Halide Scintillation Counters," *Phys. Rev.*, **74**, 100–101, (1948).

Hofstadter, R., "The Detection of Gamma-Rays with Thallium-Activated Sodium Iodide Crystals," *Phys. Rev.*, **75**, 796–810, (1949).

Hofstadter, R., "Properties of Scintillator Materials," *Nucleonics*, **6**, 70–73, (1950).

HOFSTADTER, R. AND J.A. MCINTYRE, "Gamma-Ray Spectroscopy with Crystals of NaI:Tl," *Nucleonics*, **7**, 32–37, (1950).

HOFSTADTER, R., J.A. MCINTYRE, H. RODERICK, AND H.I. WEST, "Detection of Slow Neutrons," *Phys. Rev.*, **82**, 749, (1951).

HOFSTADTER, R., E.W. ODELL, AND C.T. SCHMIDT, "CaI_2 and CaI_2(Eu) Scintillation Crystals," *Rev. Scientific Instrum.*, **35**, 246–247, (1964a).

HOFSTADTER, R., E.W. ODELL, AND C.T. SCHMIDT, "CaI_2 and CaI_2(Eu) Scintillation Crystals," *IEEE Trans. Nucl. Sci*, **NS-11**, 12–14, (1964b).

HOFSTADTER, R., "Europium Activated Calcium Iodide Scintillators," *US Patent 3,342,745*, allowed Sept. 19, 1967.

HOFSTADTER, R., "Europium Activated Strontium Iodide Scintillators," *US Patent 3,373,270*, allowed March 12, 1968.

HOFSTADTER, R., "Twenty Five Years of Scintillation Counting," *IEEE Trans. Nucl. Sci.*, **NS-22**, 13–25, (1975).

HOLL, I., E. LORENZ, AND G. MAGERAS, "A Measurement of the Light Yield of Common Inorganic Scintillators," *IEEE Trans. Nucl. Sci.*, **NS-35**, 105–109, (1988).

HORNYAK, W.F., "A Fast Neutron Detector," *Rev. Sci. Instrum.*, **23**, 264–267, (1952).

HORROCKS, D.L., *Applications of Liquid Scintillation Counting*, New York: Academic, 1974.

HORROCKS, D.L., "Measurement of Sample Quenching of Liquid Scintillator with X-ray and Gamma-ray Sources," *Nature*, **202**, 78–79, (1964).

HUANG, K. AND A. RHYS, "Theory of Light Absorption and Non-Radiative Transitions in *F*-Centres," *Proc. Roy. Soc. A*, **204**, 406–423, (1950).

HUGGINS, SIR W. AND LADY HUGGINS, "On the Spectrum of the Spontaneous Luminous Radiation of Radium at Ordinary Temperatures," *Proc. Roy. Soc.*, **A72**, 196–199, (1903a).

HUGGINS, SIR W. AND LADY HUGGINS, "Further Observations on the Spectrum of the Spontaneous Luminous Radiation of Radium at Ordinary Temperatures," *Proc. Roy. Soc.*, **A72**, 409–413, (1903b).

HUGGINS, SIR W. AND LADY HUGGINS, "On the Spectrum of the Spontaneous Luminous Radiation of Radium, Part III. — Radiation in Hydrogen," *Proc. Roy. Soc.*, **A76**, 488–492, (1905).

HUGGINS, SIR W. AND LADY HUGGINS, "On the Spectrum of the Spontaneous Luminous Radiation of Radium, Part IV. — Extension of the Glow," *Proc. Roy. Soc.*, **A77**, 130–131, (1906).

INRADOPTICS, "Production of Stilbene for Fast Neutron Detection," `www.inradoptics.com`, accessed Jan. 15, 2016.

ISHIBASHI, H., ET AL., "Scintillation Performance of Large Ce-doped Gd_2SiQ_5 (GSQ) Single Crystal," *IEEE Trans. Nucl. Sci.*, **NS-45**, 518–521, (1998).

ISHII, M. AND M. KOBAYASHI, "Single Crystals for Radiation Detectors," *Prog. in Crys. Growth and Char. of Mat.*, **23**, 245–311, (1991).

IVANOV, V.Y., A.V. KRUZHALOV, V.A. PUSTOVAROV, AND V.L. PETROV, "Electron Excitation and Luminescence in $Bi_4Ge_3O_{12}$ and $Bi_4Si_3O_{12}$ Crystals," **A261**, 150–152, (1987).

IWANOSKA, J., L. SWIDERSKI, M. MOSZYNSKI, T. SZCZESNIAK, P. SIBCZYNSKI, N.Z. GALUNOV, AND N.L. KARAVAEVA, "Neutron/Gamma Discrimination Properties of Composite Scintillation Detectors," *J. Instrum.*, **6**, P07007 (1–10), (2011).

JAFFE, J.E., D.V. JORDAN, AND A.J. PEURRUNG, "Energy Nonlinearity in Radiation Detection Materials: Causes and Consequences," *Nucl. Instrum. Meth.*, **A570**, 72–83, (2007).

JING, T., J. DREWERY, I. FUJIEDA, S.N. KAPLAN, A. MIRESHGHI, V. PERZ-MENDEZ, AND D. WILDERMUTH, "Enhanced Columnar Structure in CsI Layer by Substrate Patterning," *IEEE Trans. Nucl. Sci.*, **NS-39**, 1195–1198, (1992).

KAMAE, T., ET AL., "Improvement on the Light Yield of a High-Z Inorganic Scintillator GSO(Ce)," *Nucl. Instrum. Meth.*, **A490**, 456–464, (2002).

KAPUSTA, M., M. BALCERZIK, M. MOSZYNSKI, AND J. PAWELKE, "A High-Energy Resolution Observed from a YAP:Ce Scintillator," *Nucl. Instrum. Meth.*, **A421**, 610–613, (1999).

KAWAI, T., S. SAKURAGI, AND S. HASHIMOTO, "Luminescence Properties of Pure and Eu-doped SrI2 Crystals Purified by a Liquinert Process and Grown by Vertical Bridgman Method," *J. Luminescence*, **176**, 58–64, (2016).

KEARVELL, A. AND F. WILKINSON, "Fluorescence Quenching and External Spin-Orbit Coupling Effects," *Molecular Crystals*, **4**, 69–81, (1968).

KERISIT, S.N., F. GAO, Y.L. XIE, L.W. CAMPBELL D. WU, AND M.P. PRANGE, "Science Driven Candidate Search for New Scintillator Materials FY2014 Report," *Pacific Northwest Report*, PNNL-23752, 2014.

KERNAN, W.J., *Accelerators*, Understanding the Atom series, US Atomic Energy Comm., 1969.

KHAN, S., H.J. KIM, AND Y.D. KIM, "Scintillation Characterization of Thallium-Doped Lithium Iodide Crystals," *Nucl. Instrum. Meth.*, **A793**, 31–34, (2015).

KHODYUK, I.V. AND P. DORENBOS, "Trends and Patterns of Scintillator Nonproportionality," *IEEE Trans. Nucl. Sci.*, **NS-59**, 3320–3331, (2012).

KIENZLE, W., G. MATTHIAS, R. VANDERHAGEN, S. WEISZ, AND G. BURGUN, "Scintillator Developments and CERN," CERN NP Internal Report 75-12, Geneva, 1975.

KITTEL, C., *Introduction to Solid State Physics, 2nd Ed.*, New York: Wiley, 1956.

KLAWONN, K. KLEINKNECHT, AND D. POLLMANN, "A New Type of Acrylic Scintillator," *Nucl. Instrum. Meth.*, **195**, 483–489, (1982).

KOCH, L. AND R. LESUEUR, "Quelques Résultats Concernant les Scintillateurs Gazeux et Leurs états Condensés," *J. Phys. Radium*, **19**, 103–105, (1958).

KOEHLER, H.A., L.J. FERDERBER, D.L. REDHEAD, AND P.J. EBERT, "Vacuum-Ultraviolet Emission from High-Pressure Xenon and Argon Excited by High-Current Relativistic Electron Beams," *Phys. Rev. A*, **9**, 768–781, (1974).

KOIČKI, S., A. KOIČKI, AND A. AJDAČIĆ, "The Investigation of the 0.15 s Phosphorescence Component of NaI:Tl and its Application in Scintillation Counting," *Nucl. Instrum. Meth.*, **108**, 297–299, (1973).

KONOBEEVSKI, E.S., M.V. MORDOVSKOV, I.M. SHARAPOV, AND S.V. ZUYEV, "Comparison of Various Scintillation Detectors with n-γ Pulse Shape Discrimination," M. IVAN Ed., Proc. Part I and Part II, 4th Int. Conf. on Current Problems in Nuclear Physics and Atomic Energy (NPAE-Kyiv2012), In Vyshnevskyi, Ukraine, 632–636, 2012.

KOSKI, W.S., "Scintillations in Some Phosphor-Plastic Systems*," *Phys. Rev.*, **82**, 230–232, (1951).

KREBS, A.T., "Über die Messung Radioaktiver Niederschläge mit Hilfe von Spitzenzähler und Zählrohr," *Phys. Zeit.*, **35**, 796–800, (1934).

Krebs, A.T., "Ein Demonstrationsversuch zur Emanationsdiffusion," *Annalen der Physik*, **39**, 330–332, (1941).

Krebs, A.T., "Early History of the Scintillation Counter," *Science*, **122**, 17–18, (1955).

Kubota, S., M. Hasahiko, and J-Z. Raun(Gen), "Evidence for a Triplet State of the Self-Trapped Exciton State in Liquid Argon, Krypton and Xenon," *J. Phys. C: Solid State Physics*, **11**, 2645–2651, (1978).

Kurashige, K., et al., "Large GSO Single Crystals With a Diameter of 100 mm and Their Scintillation Performance," **NS-51**, 742–745, (2004).

L'Annunziata, M.F., Ed., *Handbook of Radioactivity Analysis*, San Diego: Academic Press, 2012.

Laval, M., M. Moszyński, R. Allemand, E. Cormoreche, P. Guinet, and R. Odru, "Barium Fluoride — Inorganic Scintillator for Subnanosecond Timing," *Nucl. Instrum. Meth.*, **206**, 169–176, (1983).

Lecoq, P., A. Annenkov, A. Gektin, M. Korzhik, and C. Pedrini, *Inorganic Scintillators for Detector Systems*, Berlin: Springer, 2006.

Lempicki, A., C. Brecher, P. Szupryczynski, H. Lingertat, V.V. Nagarkar, S.V. Tipnis, and S.R. Miller, "A New Lutetia-Based Ceramic Scintillator for X-Ray Imaging," *Nucl. Instrum. Meth.*, **A488**, 579–590, (2002).

Li, Q., J.Q. Grim, R.T. Williams, G.A. Bizarri, and W.W. Moses, "A Transport-Based Model of Material Trends in Nonproportionality of Scintillators," *J. Appl. Phys.*, **109**, 123716(1–17), (2011).

Lindow, J.T., E. Srader, M.R. Farukhi, R.L. Geiger, E.C. Stewart, R.W. Johnson, and R.J. Thompson, *Harshaw Scintillation Phosphors,* 3rd Ed., Solon: The Harshaw Chemical Company, 1978.

Lucas, H.F., "Improved Low-Level Alpha-Scintillation Counter for Radon," *Rev.of Sci. Instrum.*, **28**, 680-683, (1957).

Ludziejewski, T., K. Moszyńska, M. Moszyński, D. Wolski, W. Klamra, L.A. Norlin, E. Devitsin and V. Kozlov, "Advantages and Limitations of LSO Scintillator in Nuclear Physics Experiments," *IEEE Trans. Nucl. Sci.*, **NS-42**, 328–336, (1995).

Ludziejewski, T., M. Moszyński, D. Wolski, W. Klamra, and K. Moszyńska, "Investigation of Some Scintillation Properties of YAP:Ce Crystals," *Nucl. Instrum. Meth.*, **A398**, 287–294, (1997).

Majewski S. and D. Anderson, "Radiation Damage Test of Barium Fluoride Scintillator," *Nucl. Instrum. Meth.*, **A241**, 76–79, (1985).

Mandeville, C.E. and M.V. Scherb, "Photosensitive Geiger Counters: Their Applications," *Nucleonics*, **7**, 34–38, (1950).

McIntyre, J.A. and R. Hofstadter, "Measurement of Gamma-Ray Energies with One Crystal," *Phys. Rev.*, **78**, 617–619, (1950).

Melcher, C.L., "Scintillators for Well Logging Applications," *Nucl. Instrum. Meth.*, **B40/41**, 1214–1218, (1989).

Melcher, C.L., "Lutetium Orthosilicate Single Crystal Scintillator Detector," U.S. Patent 4958080, allowed Sept. 18, 1990.

Melcher, C.L., "Lutetium Orthosilicate Single Crystal Scintillator Detector," U.S. Patent 5025151, allowed Jun. 18, 1991.

Melcher, C.L. and J.S. Schweitzer, "Cerium-Doped Lutetium Oxyorthosilicate: A Fast, Efficient New Scintillator," *IEEE Trans. Nucl. Sci.*, **NS-39**, 502–505, (1992).

Melcher, C.L., J.S. Schweitzer, C.A. Peterson, R.A. Manente, and H. Suzuki, "Crystal Growth and Scintillation Properties of the Rare Earth Orthosilicates," in *Inorganic Scintillators and Their Applications*, Delft University Press (SCINT95), 309–315, (1996).

Menefee, J., C.F. Swinehart, and E.W. Oďell, "Calcium Fluoride as an X-ray and Charged Particle Detector," *IEEE Trans. Nucl. Sci.*, **13**, 720–724, (1966).

Mengesha, W., T.D. Taulbee, B.D. Rooney, and J.D. Valentine, "Light Yield Nonproportionality of CsI:Tl, CsI(Na) and YAP," *IEEE Trans. Nucl. Sci.*, **45**, 456–461, (1998).

Messner, D. and A. Smakula, "Color Center in Alkaline Earth Fluorides," *Phys. Rev.*, **120**, 1162–1166, (1960).

Milton, J.C.D. and R. Hofstadter, "Luminescent Spectra of Phosphors Useful in Scintillation Counters," *Phys. Rev.*, **75**, 1289–1299, (1949).

Moore, D.H., J.J. Marcec, A.R. Reich, and J.G. Ring, "The Performance of Liquid Scintillation Counters Using Large and Small Vials," *Liquid Scintillation Counting*, Vol. 5, Ch. 22, M.A. Crook and P. Johnson Eds., London: Heyden, pp. 182–190, (1977).

Moses, W.W., G.A. Bizarri, R.T. Williams, S.A. Payne, A.N. Vasil'ev, J. Singh, Q. Li, J.Q. Grim, and W-S Choong, "The Origins of Scintillator Non-Proportionality," *IEEE Trans. Nucl. Sci.*, **NS-59**, 2038–2044, (2012).

Moszyński, M., R. Allemand, M.L.R. Odru, and J. Vacher, "Recent Progress in Fast Timing with CsF Scintillators in Application to Time-of-Flight Positron Tomography in Medicine," *Nucl. Instrum. Meth.*, **205**, 239–249, (1983).

Moszyński, M., T. Ludziejewski, D. Wolski, W. Klamra, and L.O. Norlin, "Properties of the YAP:Ce Scintillator," *Nucl. Instrum. Meth.*, **A345**, 461–467, (1994).

Moszyński, M., D. Wolski, T. Ludziejewski, M. Kapusta, A. Lempicki, C. Brecher, D. Wisniewski, and A.J. Wojtowicz, "Properties of the New LuAP:Ce Scintillator," *Nucl. Instrum. Meth.*, **A385**, 123–131, (1997a).

Moszyński, M., M. Kapusta, M. Mayhugh, D. Wolski, and S.O. Flyckt, "Absolute Light Output of Scintillators," *IEEE Trans. Nucl. Sci.*, **NS-44**, 1052–1061, (1997b).

Moszyński, M., M. Kapusta, D. Wolski, W. Klamra, and C. Cederwall, "Properties of the YAP:Ce Scintillator," *Nucl. Instrum. Meth.*, **A404**, 157–165, (1998).

Moszyński, M., W. Czarnacki, M. Kapusta, M. Szawlowski, W. Klamra, and P. Schotanus, "Energy Resolution and Light Yield Non-Proportionality of Pure NaI Scintillator Studied with Large Area Avalanche Photodiodes at Liguid Nitrogen Temperatures," *Nucl. Instrum. Meth.*, **A486**, 13–17, (2002).

Moszyński, M., "Inorganic Scintillation Detectors in γ-Ray Spectrometry," *Nucl. Instrum. Meth.*, **A505**, 101–110, (2003).

Moszyński, M., W. Czarnacki, W. Klamra, M. Szawlowski, P. Schotanus, and M. Kapusta, "Intrinsic Energy Resolution of Pure NaI Studied with Large Area Avalanche Photodiodes at Liquid Nitrogen Temperatures," *Nucl. Instrum. Meth.*, **A505**, 63–67, (2003a).

Moszyński, M., M. Balcerzyk, W. Czarnacki, M. Kapusta, W. Klamra, P. Schotanus, A. Syntfeld, and M. Szawlowski, "Study of Pure NaI at Room and Liquid Nitrogen Temperatures," *IEEE Trans. Nucl. Sci.*, **NS-50**, 767–773, (2003b).

Murakami, A., H. Asakura, and H. Yoshinaka "Radiation Damage and Recovery in the Light Emittance Efficiency and the Attenuation Length of Barium Fluoride Scintillator," *Nucl. Instrum. Meth.*, **A301**, 435–444, (1991).

MURASHITA, M., H. SAITOH, K. TOBIMATSU, M. CHIBA, T. HIROSE, AND F. TAKASAKI, "Performance and Radiation Damage of a BaF_2 Calorimeter," *Nucl. Instrum. Meth.*, **A243**, 67–76, (1986).

MUTHMAN, W. AND STÜTZEL, L., "Eine Enfache Methode zer Darstellung der Schwefel-, Chlor- und Brom-Verbindungen der Ceritmetalle," *Berichte der Deutschen Chemischen Gesellschaft*, **32**, 3413–3419, (1899).

NAGARKAR, V.V., T.K. GUPTA, S.R. MILLER, Y. KLUGERMAN, M.R. SQUILLANTE, AND G. ENTINE, "Structured CsI:Tl Scintillators for X-ray Imaging Applications," *IEEE Trans. Nucl. Sci.*, **NS-45**, 492–496, (1998).

NAKAMURA, H., Y. SHIRAKAWA, S. TAKAHASHI, AND H. SHIMIZU, "Evidence of Deep-Blue Photon Emission at High Efficiency by Common Plastic," *EPL*, **95**, 22001(1–3), (2011).

NASSALSKI, A., M. KAPUSTA, T. BATSCH, D. WOLSKI, D. MÖCKEL, W. ENGHARDT, AND M. MOSZYŃSKI, "Comparative Study of Scintillators for PET/CT Detectors," *IEEE Trans. Nucl. Sci.*, **NS-54**, 3–10, (2007).

NEMCHENOK, I.B., V.I. BABIN, V.B. BRUDANIN, O.I. KOCHETOV, AND V.V. TIMKIN, "Liquid Scintillator Based on Linear Alkylbenzene," *Physics of Particles and Nuclei Lett.*, **8**, 129–135, (2011).

NESTOR, O.H. AND C.Y. HUANG, "Bismuth Germanate: A High-Z Gamma-Ray and Charged Particle Detector," *IEEE Trans. Nucl. Sci.*, **NS-22**, 68–71, (1975).

NICHOLSON, K.P. AND G.F. SNELLING, "Some Lithium Iodide Phosphors for Slow Neutron Detection," *Brit. J. Appl. Phys.*, **6**, 104–106, (1955).

NIKL, M., A. YOSHOKAWA, K. KAMADA, K. NEJEZCHLEB, C.R. STANEK, K.A. MARES, AND K. BLAZEK, "Development of LuAG-based scintillator Crystals A Review," *Prog. Crystal Growth Char. Mat.*, **59**, 47–72, (2013).

NIST XCOM, Natl. Inst. Sci. Tech., `https:// physics.nist.gov/PhysRefData/Xcom/html/xcom1.html`, accessed June 1, 2016.

NORTHROP, J.A., J.M. GURSKY, AND A.E. JOHNSRUD, "Further Work with Noble Element Scintillators," *IRE Trans. Nucl. Sci.*, **NS-5**, 81–87, (1958a).

NORTHROP, J.A. AND J.C. GURSKY, "Relative Scintillation Efficiencies of Noble Gas Mixtures," *Nucl. Instrum.*, **3**, 207–212, (1958b).

NOUJAIM, A.A., C. EDISS, AND L.I. WEIBE, *Liquid Scintillation Science and Technology*, New York: Academic, 1976.

NYIBULE, S., ET AL., "Birks' Scaling of the Particle Light Output Functions for the EJ 299-33 Plastic Scintillator," *Nucl. Instrum. Meth.*, **A 768**, 141–145, (2014).

OLIVER, D.B. AND G.F. KNOLL, "Anistropy of Scintillation Response of Anthracene to Neutron Generated Recoil Protons and Carbon Ions," *IEEE Trans. Nucl. Sci.*, **NS-15**, 122–126, (1968).

OMEGA PIEZO, "BGO, LYSO and GSO Crystal Scintillators," `www.omegapiezo.com/ crystal_scintillators.html`, accessed April 20, 2016.

OWEN, R.B., "The Decay Times of Organic Scintillators and Their Application to the Discrimination Between Particles of Differing Specific Ionization," *IRE Trans. Nucl. Sci.*, **NS-5**, 198–201, (1958).

OWEN, R.B., "The Variation of Phosphor Decay Time with Specific Ionization and its Applications," *Nuclear Electronics, 1958 Proc. Int. Symp. Nucl. Electronics*, **Vol. 1**, 27–35, (1959).

PALMER, H.E., "Recent Advances in Gas Scintillation Proportional Counters," *IEEE Trans. Nucl. Sci.*, **NS-22**, 100–103, (1975).

PAVAN, P., G. ZANELLA, R. ZANNONI, AND P. POLATO, "Radiation Damage and Annealing of Scintillating Glasses," *Nucl. Instrum. Meth.*, **B61**, 487–490, (1991).

PAYNE, S.A., ET AL., "Nonproportionality of Scintillator Detectors: Theory and Experiment," *IEEE Trans. Nucl. Sci.*, **NS-56**, 2506–2512, (2009).

PAYNE, S.A., ET AL., "Nonproportionality of Scintillator Detectors: Theory and Experiment. II," *IEEE Trans. Nucl. Sci.*, **NS-58**, 3392–3402, (2011).

PENG, C.T., *Sample Preparation in Liquid Scintillation Counting*, Arlington Heights: Amersham, 1977.

PEPIN, C.M., P. BÉRAD, A-L. PERROT, C. PÉPIN, D. HOUDE, R. LECOMTE, C.L. MELCHER, AND H. DAUTET, "Properties of LYSO and Recent LSO Scintillators for Phoswich PET Detectors," *IEEE Trans. Nucl. Sci.*, **NS-51**, 789–795, (2004).

PERSYK, D.E., M.A. SCHARDT, T.E. MOI, K.A. RITTER, AND G. MUEHLLEHNER, "Research on Pure Sodium Iodide as a Practical Scintillator," *IEEE Trans. Nucl. Sci.*, **NS-27**, 168–171, (1980).

PICHAT, L., P. PESTEIL, AND J. CLÉMENT, "Solides Fluorescents Non-Cristallins pour Mesures de Radioactvité," *J. Chim. Phys.*, **50**, 26–41, (1953).

POLICARPO, A.J.P.L., M.A.F. ALVES, M.C.M. DOS SANTOS, AND M.J.T. CARVALHO, "Improved Resolution for Low Energies with Gas Proportional Scintillation Counters," *Nucl. Instrum. Meth*, **102**, 337–348, (1972).

POLICARPO, A.J.P.L., "The Gas Proportional Scintillation Counter," *Space Sci. Instrum.*, **3**, 77–107, (1977).

PROTEUS, `www.space-science.com/proteus/organic.htm`, accessed May 16, 2016.

PRICE, W.J., *Nuclear Radiation Detection, 2nd Ed.*, New York: McGraw-Hill, 1964.

QUARATI, F.G.A., P. DORENBOS, J. VAN DER BIEZEN, A. OWENS, M. SELLE, L. PARTHIER, AND SCHOTANUS, "Scintillation and Detection Characteristics of High-Sensitivity $CeBr_3$ Gamma-Ray Spectrometers," *Nucl. Instrum. Meth.*, **A729**, 596–604, (2013).

RIDLEY, B.K., *Quantum Processes in Semiconductors*, Oxford: Clarendon Press, 1982.

REEDER, P.L., "Thin GSO Scintillator for Neutron Detection," *Nucl. Instrum. Meth.* **353**, 134–136, (1994).

REXON COMPONENTS, INC. "LYSO/LSO Technical Specifications," www.rexon.com, accessed Sept. 28, 2015.

ROBERTSON, J.C. AND J.G. LYNCH, "The Luminescent Decay of Various Crystals for Particles of Different Ionization Density," *Proc. Phys. Soc.*, **77**, 751–756, (1961).

ROBERTSON, J.C., J.G. LYNCH, AND W. JACK, "The Luminescence of CsBr(Tl) and CsI:Tl as a Function of Temperature," *Proc. Phys. Soc.*, **78**, 1188–1194, (1961).

RODNYI, P.A., *Physical Processes in Inorganic Scintillators*, Boca Raton: CRC Press, 1997.

ROSCOE, B., J.A. GRAU, R.A. MANENTE, C.L. MELCHER, C.A. PETERSON, J.S. SCHWEITZER, AND C. STOLLER, "Use of GSO for Inelastic Gamma-Ray Spectroscopy Measurements in the Borehole", *IEEE Trans. Nucl. Sci* **NS-39**, 1412–1416, (1992).

ROSS, H., J.E. NOAKES, AND J.D. SPAULDING, *Liquid Scintillation Counting and Organic Scintillators*, Chelsea: Lewis, 1991.

RMD, INC., "Gamma-Neutron Scintillation Detector," see also, `http://rmdinc.com/wp-content/uploads/2014/10/CLYC-Properties-PDF.pdf`, accessed Feb. 25, 2016.

SAINT-GOBAIN CRYSTALS, `https://www.crystals.saint-gobain.com/Scintillation_Materials.aspx` accessed Jan. 14, 2016.

SALIMGAREEVA, V.N. AND S.V. KOLESOV, "Plastic Scintillators Based on Polymethyl Methacrylate: A Review," *Instrum. Experimental Tech.*, **48**, 273–282, (2005).

SANDLER, S., S. LOSHAEK, E. BRODERICK, AND PH. BERSTEIN, "Dimethyl Styrene Yields More Efficient Scintillators," *Nucleonics*, **18**, 102–103, (1960).

SANGSTER AND IRVINE, "Study of Organic Scintillators," *J. Chem. Phys.*, **24**, 670–715, (1956).

SCHORR, M.G. AND F.L. TORNEY, "Solid Non-Crystalline Scintillation Phosphors," *Phys. Rev.*, **80**, 474, (1950).

SCHRAM, E., *Organic Scintillation Detectors*, Elsevier, Amsterdam, 1963.

SCHOTANUS, P., C.W.E. VAN EIJK, R.W. HOLLANDER, AND J. PIJPELINK, "Temperature Dependence of BaF_2 Scintillation Light Yield," *Nucl. Instrum. Meth.*, **A238**, 564–565, (1985).

SCHRAM, E., *Organic Scintillation Detectors*, Amsterdam: Elsevier, 1963.

SCINTACOR, "Glass Scintillators," www.scintacor.com, accessed 2016.

SENCHISHIN, V.G., ET AL., "A New Radiation Stable Plastic Scintillator," *Nucl. Instrum. Meth.*, **A364**, 253–257, (1995).

SHAH, K.S., J. GLODO, W. HIGGINS, E.V.D. VAN LOEF, W.W. MOSES, S.E. DERENZO, AND M.J. WEBER, "$CeBr_3$ Scintillators for Gamma-Ray Spectroscopy," *IEEE Trans. Nucl. Sci.*, **NS-52**, 3157–3159, (2005).

SHAH, K.S., J. GLODO, E.V.D. VAN LOEF, W. HIGGINS, R. HAWRAMI, U. SHIRWADKAR, AND S. MUKHOPADHAY, "New Detectors for Gamma and Neutron Studies," DOE SBIR Phase II, accessed Sept. 1, 2017.

SHIMURA, N., ET AL., "Zr Doped GSO:Ce Single Crystals and Their Scintillation Performance," *IEEE Trans. Nucl. Sci.*, **NS-53**, 2519–2522, (2006).

SHIRWADKAR, U., J. GLODO, E.V.D. VAN LOEF, R. HAWRAMI, S. MUKHOPADHYAY, A. CHURILOV, W.H. HIGGINS, AND K.S. SHAH, "Scintillation Properties of $Cs_2LiLaBr_6$ (CLLB) Crystals with Varying Ce^{3+} Concentration," *Nucl. Instrum. Meth.*, **A652**, 268–270, (2011).

SHIRWADKAR, U., R. HAWRAMI, J. GLODO, E.V.D. VAN LOEF, AND K.S. SHAH, "Novel Scintillation Material $Cs_2LiLaBr_{6-x}Cl_x$:Ce for Gamma-Ray and Neutron Spectroscopy," *Conf. Rec. IEEE Nuc. Sci. Symp.*, Anaheim, CA, Oct. 21–Nov. 3, paper N41-1, (2012).

SIDOT, T., "Sur les Propriétés de la Blende Hexagonale," *Compt. Rend.* **63**, 188-189, (1866).

SIMMONS, J.E. AND R.B. PERKINS, "Liquid Helium Scintillation Counter as a Neutron Polarimeter," *Rev. Sci. Instrum.*, **32**, 1173–1178, (1961).

SINGH, R.J., *Solid State Physics*, Kindersley, India: Dorling, 2012.

SMITH, D.L., R.G. POLK, AND T.G. MILLER, "Measurement of the Response of Several Organic Scintillators to Electrons, Protons and Deuterons," *Nucl. Instrum. Meth.*, **64**, 157–166, (1968).

SMOL'SKAYA, L.P., E.I. SHURALEVA, AND I.A. PARFIANOVICH, "Radiation of Paired Centers in the Phosphor KI(Tl)," *Soviet Physics Journal*, **12**, 1063–1965, (1969).

STERN, O. AND M. VOLMER, "Über die Abklingungszeit der Fluoreszenz," *Phys. Z.*, **20**, 183–188, (1919).

SWANK, R.K. AND W.L. BUCK, "The Scintillation Process in Plastic Solid Solutions," *Phys. Rev.*, **91**, 927–933, (1953).

ŚWIDERSKI, Ł., M. MOPSZYNSKI, A. NADDALSKI, A. SYNTFELD-KAŹUCH, T. SZCZEŚNIAK, K. KAMADA, K. TSUTSUMI, Y. USUKI, T. YANAGIDA, AND A. YOSHIKAWA, "Light Yield Non-Proportionality and Energy Resolution of Praseodymium Doped LuAG Scintillator," *IEEE Trans. Nucl. Sci.*, **NS-56**, 934–938, (2009a).

SYNTFELD, A., M. MOSZYNSKI, R. ARLT, M. BALCERZYK, M. KAPUSTA, M. MAJOROV, R. MARCINKOWSKI, P. SCHOTANUS, M. SWOBODA, AND D. WOLSKI, "^{6}LiI(Eu) in Neutron and Gamma-Ray Spectrometry — A Highly Sensitive Thermal Neutron Detector," *IEEE Trans. Nucl. Sci.*, **52**, 3151–3156, (2005).

SZUPEYCZYNSKI, P., C.L. MELCHER, M.A. SPURRIER, M.P. MASKARINEC, A.A. CAREY, A.J. WOJTOWICZ, W. DROZDOWSKI, D. WISNIEWSKI, AND R. NUTT, "Thermoluminescense and Scintillation Properties of Rare Earth Oxyorthosilicate Scintillators," *IEEE Trans. Nucl. Sci.*, **NS-51**, 1103–1110, (2004).

TAKAGI, K. AND T. FUKAZAWA, "Cerium-Activated Gd_2SiO_5 Single Crystal Scintillator," *Appl. Phys. Lett.*, **42**, 43–45, (1983).

THIESS, P.E. AND G.H. MILEY, "New Near-Infrared and Ultraviolet Gas- Proportional Scintillation Counters," *IEEE Trans. Nucl. Sci.*, **NS-21**, 125–145, (1974).

TRETYAK, V.I., "Semi-Empirical Calculation of Quenching Factors for Ions in Scintillators," *Astroparticle Physics*, **33**, 40–53, (2010).

TSUKADA AND S. KIKUCHI, "Directional Anisotropy in the Characteristics of the Organic-Crystal Scintillators," *Nucl. Instrum. Meth.*, **17**, 286–288, (1962).

TSUKADA, S. KIKUCHI, AND Y. MIYAGAWA, "Directional Anisotropy in the Characteristics of the Organic-Crystal Scintillators, II," *Nucl. Instrum. Meth.*, **37**, 69–76, (1965).

VALAIS, I., D. SRATOS, C. MICHAIL, A. KONSTANTINIDIS, I. KANDARAKIS, AND G.S. PANAYIOTAKIS, "Investigation of Luminescent Properties of LSO:Ce, LYSO:Ce and GSO:Ce Crystal Scintillators Under Low-Energy γ-Ray Excitation used in Nuclear Imaging," *Nucl. Instrum. Meth.*, **A581**, 99–102, (2007).

VALENTINE, J.D., B.D. ROONEY, AND J. LI, "The Light Yield Nonproportionality Component of Scintillator Energy Resolution," *IEEE Trans. Nucl. Sci.*, **NS-45**, 512–517, (1998).

VAN EIJK, C.W.E., J. ANDRIESSEN, P. DORENBOS, J. JANSONS, N.M. KHAIDUKOV, Z. RACHKO, AND J. VALBIS, "Experimental and Theoretical Studies of Cross Luminescence," *Heavy Scintillators for Scientific and Industrial Applications: Proc. of the 'Cristal 2000' Int. Workshop*, 161–166, (1993).

VAN EIJK, C.W.E., J.T.M. DE HAAS, P. DORENBOS, K. WRÄMER, AND H.U. GÜDEL, "Development of Elpasolite and Monoclinic Thermal Neutron Scintillators," *Conf. Rec. IEEE Nuc. Sci. Symp.*, Puerto Rico, Oct. 23–Oct. 29, 239–243, (2005).

VAN LOEF, E.V.D., P. DORENBOS, C.W.E. VAN EIJK, K. KRÄMER, AND H.U. GÜBEL, "Scintillation Properties of $LaCl_3$:Ce^{3+} Crystals: Fast, Efficient, and High-Resolution Scintillators," *IEEE Trans. Nucl. Sci.*, **NS-48**, 342–345, (2001).

VAN LOEF, E.V.D., P. DORENBOS, C.W.E. VAN EIJK, K. KRÄMER, AND H.U. GÜBEL, "Scintillation Properties of $LaBr_3$:Ce^{3+} Crystals: Fast, Efficient, and High-Energy-Resolution Scintillators," *Nucl. Instrum. Meth.*, **A486**, 254–258, (2002).

VAN SCIVER, W.J. AND R. HOFSTADTER, "Scintillations in Thallium-Activated CaI_2 and CsI*," *Phys. Rev.*, **84**, 1062–1063, (1951).

VAN SCIVER, W.J. AND R. HOFSTADTER, "Gamma-and Alpha-Produced Scintillations in Cesium Fluoride," *Phys. Rev.*, **87**, 522, (1952).

VAN SCIVER, W.J., "Alkali Halide Scintillators," *IRE Trans. Nucl. Sci.*, **NS-3**, 39–50, (1956).

VAN SCIVER, W.J. AND L. BOGART, "Fundamental Studies of Scintilllation Phenomenon in NaI," *IRE Trans. Nucl. Sci.*, **NS-5**, 90–92, (1958).

VASIL'CHENKO, V.G., V.G. LAPSHIN, A.I. PERESYPKIN, A.A. KONSTANTOMCHENKO, A.I. PYSHCHEV, V.M. SHERSHUKOV, B.V. SEMENOV, AND A.S. SOLOV'EV, "New Results on Radiation Damage Studies of Plastic Scintillators," *Nucl. Instrum. Meth.*, **A369**, 55–61, (1996).

VOITOVETSKII, V.K., N.S. TOLMACHEVA, AND M.I. A RSAEV, "Scintillating Glass for the Detection of Slow Neutrons," *Soviet J. Atomic Energy*, **6**, 203–207, (1960a).

VOITOVETSKII, V.K. AND N.S. TOLMACHEVA, "Lithium Silicate Scintillating Glasses for the Detection of Slow Neutrons," *Soviet J. Atomic Energy*, **6**, 335–336, (1960b).

WEBER, M.J. AND R.R. MONCHAMP, "Luminescence of $Bi_4Ge_3O_{12}$: Spectral and Decay Properties," *J. Appl. Phys.*, **44**, 5495–5499, (1973).

WEI, H., L. STRAND, M. ZHURAVLEVA, F. MENG, V. MARTIN, AND C.L. MELCHER, "Two New Cerium-Doped Mixed-Anion Elpasolite Scintillators: $Cs_2NaYBr_3I_3$ and $Cs_2NaLaBr_3I_3$," *Opt. Mater.*, **38** 154–160, (2014).

WOODY, C.L., P.W. LEVY, AND J.A. KLERSTEAD, "Slow Component Suppression and Radiation Damage in Doped BaF_2 Crystals," *IEEE Trans. Nucl. Sci.*, **NS-36**, 536–542, (1989).

WRIGHT, G.T., "Scintillation Decay Times of Organic Crystals," *Proc. Phys. Soc. B*, **69**, 358–372, (1956).

YAMAMOTO, H., "Fundamentals of Luminescence — Luminescence from a Localized Center," Ch. 1, Sec. 3, *Fundamentals of Phosphors*, W.M. YEN, S. SHIONOYA, AND H. YAMAMOTO, Eds., Boca Raton: CRC Press, 2007.

YANAGIDA, T., A. FUKABORI, Y. FUJIMOTO, A. IKEUSE, K. KAMADA, J. KATAOKA, Y. YOKOTA, A. YOSHIKAWA, AND V. CHANI, "Scintillation Properties of Transparent $Lu_3Al_5O_{12}$ Ceramics Doped with Different Concentrations of Pr^{3+}," *Phys. Stat. Sol.*, **C 8**, 140–143, (2011).

YANAGIDA, T., K. WANTANABE, AND Y. FUJIMOTO, "Comparative Study of Neutron and Gamma-Ray Pulse Shape Discrimination of Anthracene, Stilbene, and p-terphenyl," *Nucl. Instrum. Meth.*, **A784**, 111–114, (2015).

ZAITSEVA, N., J. NEWBY, S. HAMEL, L. CARMEN, M. FAUST, V. LORDI, N. CHEREPY, W. STOEFFL, AND S. PAYNE, "Neutron Detection with Single Crystal Organic Scintillators," *Proc. SPIE*, **7449**, paper 11, (2009).

ZAITSEVA, N., L. CARMEN, A. GLENN, J. NEWBY, M. FAUST, S. HAMEL, N. CHEREPY, AND S. PAYNE, "Application of Solution Techniques for Rapid Growth of Organic Crystals," *J. Cryst. Grow.*, **314**, 163–170, (2011a).

ZAITSEVA, N., A. GLENN, L. CARMEN, R. HATARIK, S. HAMEL, M. FAUST, BSCHABES, N. CHEREPY, AND S. PAYNE, "Pulse Shape Discrimination in Impure and Mixed Single-Crystal Organic Scintillators," *IEEE Trans. Nucl. Sci.*, **NS-58**, 3411–3420, (2011b).

ZAITSEVA, N., A. GLENN, L. CARMEN, H.P. MARTINEZ, R. HATARIK, H. KLAPPER, AND S. PAYNE, "Scintillation Properties of Solution-Grown *Trans*-Stilbene Single Crystals," *Nucl. Instrum. Meth.*, **A789**, 8–15, (2015).

ZAZUBOVICH, S., "Physics of Halide Scintillators," *Rad. Meas.*, **33**, 699–704, (2001).

ZDESENKO, Y.G., F.T. AVIGNONE, V.B. BRUDANIN, F.A. DANEVICH, S.S. NAGORNY, I.M. SOLSKY, V. TRETYAK, "Scintillation Properties and Radioactive Contamination of $CaWO_4$ Crystal Scintillators," *Nucl. Instrum. Meth.*, **A538**, 657–667, (2005).

ZHU, R., "On Quality Requirements for Barium Fluoride Crystals," *Nucl. Instrum. Meth.*, **A340**, 442–457, (1994).

ZHU, R., "CsI:Tl Radiation Damage and Quality Improvement," *Proc. of the VI Int. Conf. on Calorimetry in HEP*, A. ANTONELLI, ET AL., Eds., *Frascati Physics Series*, **Vol. VI**, pp. 589–598, (Frascati, June 8–14, 1996).

ZHU, R., "Radiation Damage in Scintillating Crystals," *Nucl. Instrum. Meth.*, **A413**, 297-311, (1998).

Chapter 14

Light Collection Devices

We have learnt through experience that when an electrical ray strikes the surface of an atom, an electron, and in some circumstances a second and even a third electron, can be detached.

Johannes Stark

Scintillators have been used as a method to observe radiation interactions from the time that Röntgen, in 1895, first observed x-ray fluorescence from a plate coated with barium-platinocyanide ($BaPt(CN)_4$). Hans Geiger meticulously counted alpha particle interactions under a microscope, indicated by flashes of light from a glass slide coated with zinc sulphide (ZnS), a fatiguing technique that inspired Geiger to invent his famous gas-filled detector alternative [Rutherford and Geiger 1908]. Crooke's spinthariscope was a more useful hand-held version of the scintillation/microscope setup used by Geiger. The first hand-held x-ray imaging fluoroscopes used scintillator screens to view various densities of attenuator for medical purposes. In fact, for over 50 years since the time Röntgen first discovered x rays, scintillators had been used for some basic radiation detection purposes. Yet, it was not until the development of a reliable quantitative method of detecting light emissions, other than the human eye, did scintillators become a standard used in research and industry for radiation detection and spectroscopy.

A series of discoveries and inventions led to the technologies used today for quantitatively recording radiation-induced emissions from scintillators. The most popular and widely used of these technologies is the photomultiplier tube (PMT), which is a vacuum tube that incorporates multiple amplification stages to produce large signal amplifications through secondary electron emission. Another important technology, also based on the same idea of the multiplication of secondary electrons, uses microcapillary tubes coated with electron emissive materials. Finally, a third photosensitive technology is based on excitation of conduction electrons in a semiconductor solid. An entire book could be written on the design, operation and application of these technologies to a wide variety of uses; however, that simply is not the goal here. Instead, these basic technologies, in their various forms, are described in this chapter as they are used in radiation detection and measurement.

14.1 Photomultiplier Tubes

In 1887, Heinrich Hertz [1887] made the observation that the voltage required to produce an electric arc between electrodes decreased if assisted by visible light illumination. He also observed that electrical arcing would also occur only for certain wavelengths of light, regardless of the intensity of light. These observed effects were later explained by Albert Einstein with what is now known as the photoelectric effect. It was in 1902 that Austin and Starke [Bruining 1954] published the observation that an electrical current, when directed onto a metallic surface, could produce a greater number of electrons emitted than those impinging upon it, a process often referred to as "secondary electron emission."

It was J.A. Slepian who first published, as U.S. patents, inventions that used the "Stark effect" to produce a current amplification device [Slepian 1923]. In the original description, a heated filament provided a thermionic stream of electrons, guided by magnetic and electric fields into an intermediate anode. Electrons striking this second electrode would then eject more electrons, thereby causing an increase in current, which was subsequently collected at another anode. Although the patent drawing is a simple device with only one electrode producing secondary electrons, Slepian explains that numerous intermediate electrodes can be inserted to increase the gain further, or in his words, "indefinitely." The Slepian device is the predecessor of modern day photomultiplier tubes (PMTs).[1]

During the 1930s, L.A. Kubetsky also designed several working prototypes of secondary emission amplifiers, all of which were based on Slepian's original concepts [Kubetsky 1937]. Kubetsky has been credited with demonstrating the first working electron amplification device, having built the first unit in 1930 [Lubsandorzhiev 2006], although these early "Kubetsky tubes" were more closely akin to microchannel plates than photomultiplier tubes. Various photomultiplier versions were later described by Kubetsky in his 1937 publication as having gains of greater than 10^3.

Also, in 1934, Krebs [1934] introduced a type of Geiger-Müller (GM) tube into which a photocathode was inserted so that the tube could respond to scintillation light. Photons striking the photocathode would eject electrons, which would subsequently produce a Geiger discharge. By 1941, Krebs had improved the device somewhat [Krebs 1941; Daggs et al. 1952], although these devices were largely dismissed due to limited spectral sensitivity. Further, they were mainly based on placing a scintillating ZnS screen near the tube to produce scintillation light which would then fall on the photocathode. However, such a radiation-to-light conversion device proved to have limited efficiency for gamma rays, although it worked well for alpha and beta particles [Daggs et al. 1952].

A chemical process introduced in 1948 by Scherb [1948] extended and improved the spectral response of the photo-sensitive GM tubes, which seemed to provide a promising method to produce a photo-sensitive detector for scintillators. Advancements also included the incorporation of scintillation materials as an integral part of the GM tube [Mandeville and Scherb 1950, Krebs 1955]. Photon-sensitive GM counters were commercially available during the early 1950s for scintillation counter measurements.[2] However, because GM counters do not preserve energy deposition information, these GM counters could be used only as radiation counters and could not function as spectrometers.

Meanwhile, work at the Radio Corporation of America (RCA) led to the demonstration of a single stage electron amplifier based on secondary electron emission that used electric fields to focus and guide electrons [Iams and Salzberg 1935]. The device had an average gain of approximately 8 and did not receive much interest. However, continued research on electron amplifiers eventually led to the demonstration of multistage photomultiplier tubes [Zworykin et al. 1936]. These early PMTs used magnetic and electric fields to accelerate and direct the electrons, a technique which proved difficult to control.

Some early attempts were made to wed charge amplification with vacuum tube technology [Zworykin and Rajchman 1939; Rajchman and Snyder 1940], but it was not until 1941 that the RCA Company released the first commercially successful multistage photomultiplier amplification tube that used secondary electron emission as the amplification mechanism [Janes and Glover 1941, Glover 1941]. The RCA Type 931, shown in

[1] There seems to be some contention regarding the inventor of the PMT, whether it is rightfully Kubetsky [1937] or the RCA group consisting of Zworykin, Morton, and Malter. Lubsandorzhiev [2006] writes in his short review paper that Zworykin visited Kubetsky's laboratory during 1934, implying that Zworykin was inspired by Kubetsky's work on electron amplifying tubes, thereby suggesting that Kubetsky is the originator. However, it was Slepian [1923] that suggested the use of multiple electrodes to increase the gain, as acknowledged by Kubetsky [1937]. Hence, by the authors' estimation, the fundamental operation of a PMT was actually first described by Slepian.

[2] The Nuclear Research Corp. of Philadelphia, Pennsylvania, for instance, commercialized these photon-sensitive GM tubes.

Fig. 14.1, was originally used as a signal amplifier for electronics, in particular it was used to jam radar during the second world war. Shortly thereafter, Curran and Baker [1944] reported the use of a PMT placed adjacent to a ZnS scintillator screen to detect alpha particles. This application is the first reported use of a PMT for radiation detection. A few years later, Marshall et al. [1947] demonstrated the use of a PMT wrapped in a fluorescent material to detect x rays.

Further, Marshall and Coltman [1947] and Coltman and Marshall [1947] demonstrated the coupling of various different scintillating screens to PMTs to detect alpha particles, beta particles, x rays, gamma rays, and neutrons. That same year, Broser and Kallman [1947a] also coupled scintillating screens to a PMT to detect alpha particles, and later that same year reported the use of the organic scintillator naphthalene with a PMT [Broser and Kallman 1947b]. Throughout the following few years, numerous groups reported on the advantage of coupling a light-collection of PMT amplifiers to radiation scintillators. Notably, Hofstadter [1948] discovered the inorganic scintillator NaI(Tl), which when coupled with a PMT produced the first practical solid-state gamma-ray spectrometer.

It is the introduction of the PMT that led to the practical realization of scintillation spectrometer detectors (see articles by Morton [1975a, 1975b]. Even today a PMT is the most popular light measuring device for most scintillating materials, mainly because of their high gain and low dark current.[3]

Figure 14.1. The RCA 931A was the first commercially available PMT.

14.1.1 Basic Design

The basic design of a photomultiplier includes (1) a transparent envelope, typically constructed of glass, (2) a photocathode, (3) a series of dynodes, and (4) an anode. Each component, described in the following sections, plays an important role in the efficiency, timing, and ultimate energy resolution of the spectrometer. The photomultiplier is a simple device to understand. The basic PMT, depicted in Fig. 14.2, has a photocathode that is located so as to absorb light emissions from a light source such as a scintillating material. When photons of light strike the coating on the photocathode, they emit electrons that can diffuse through the coating to the surface facing the vacuum of the tube. A fraction of these diffusing electrons then escape the surface and enter the vacuum tube. A voltage applied to the tube guides the liberated electrons to an adjacent electrode called a *dynode*. As an electron approaches the dynode, it gains speed and energy from the electric field formed by the applied voltage. Hence, when it strikes the dynode, it again causes more electrons to become liberated into the tube. These newly liberated electrons are then guided by the electric field to the next dynode where more electrons are liberated and so on. As a result, the total number of electrons released is a function of the number of dynodes in the PMT and the photoefficiency of the photocathode and the dynodes.

The total charge released in the PMT is

$$Q = q_e N_0 G^M \tag{14.1}$$

where q_e is the charge of an electron, N_0 is the initial number of photo electrons released at the photocathode, G is the number of electrons released by a dynode per incident electron (the gain), and M is the total number

[3]In 1941, RCA was the first company to introduce a commercial PMT into the marketplace. The RCA plant, located in Lancaster PA, was purchased by GE in 1986, whereupon in 1987 the former RCA management team purchased the PMT facility and technology, forming Burle Industries. Burle was later purchased by Photonis in 2005. Unfortunately, in 2009, Photonis discontinued the PMT line. ADIT Electron Tubes, a division of Ludlum Measurements Inc., purchased the production equipment and presently offers PMTs designed primarily for radiation detection.

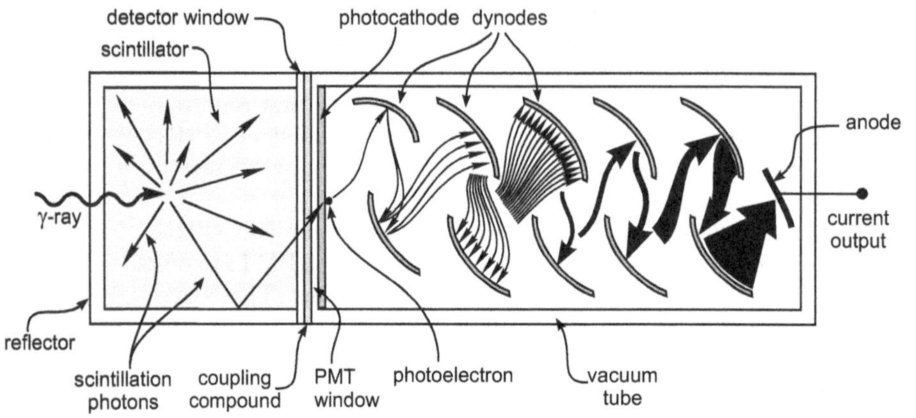

Figure 14.2. The basic mechanism of a photomultiplier tube (PMT). An absorbed γ ray causes the emission of numerous light photons which can strike the photocathode. A scintillation photon that strikes the photocathode creates a photoelectron. The photoelectron is accelerated and guided to the first dynode by an electric field, where it strikes the dynode and ejects more electrons. These electrons are accelerated to the next dynode and excite more electrons. The process continues through the dynode chain until the cascade of electrons is collected at the anode whose output current is used to produce a voltage pulse.

of dynodes in the PMT. For instance suppose that a PMT has 10 dynodes each operated with a gain of 4. An event that initially releases 1000 electrons (N_0) causes over 10^9 electrons to emerge from the PMT! Shown in Fig. 14.3 are the components of a NaI(Tl) detector (top) along with a complete detector (bottom). Also shown in Fig. 14.4 are the internal components of a NaI:Tl scintillation detector.

The photomultiplier tube is an extremely important tool in radiation detection because it is the device that allowed scintillation materials to become practical detectors. The PMT can take a small amount of light produced in a scintillator from a single radiation absorption and turn it into a large electrical signal. It is this electrical signal, typically converted to a voltage pulse, that is measured.

PMTs are stable and electronically quiet (low noise). Modern varieties have exceptional photocathode and dynode efficiencies, often referred to as the *quantum efficiency*, with gains that can exceed 30. However, there is an important design consideration. PMT materials used as photocathodes are generally fabricated from alkaline metals, which are most sensitive to light with wavelengths in the 350 to 450 nm range. Scintillators emitting light outside this range can still be used under some circumstances, although their effectiveness can be severely compromised.

14.1.2 Light Collection and Coupling

So far PMTs are treated simply as light collection devices, with a brief cautionary note about wavelength matching between the PMT and the scintillation light. In this section, this important aspect in the design of PMTs is discussed in much greater detail.[4]

[4]Throughout this chapter, "light" is characterized either by its wavelength $\lambda = c/\nu$ or by the energy of its photons $E = h\nu$. To convert between these two characterizations, a useful relation is

$$\lambda E = hc = 1240 \text{ nm eV}.$$

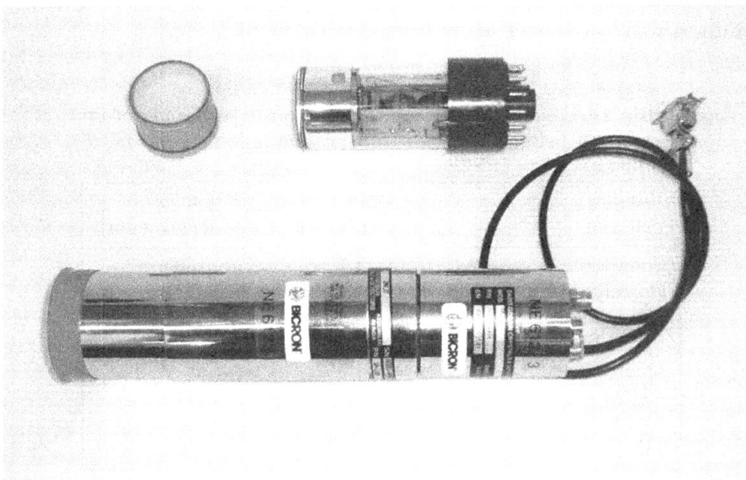

Figure 14.3. An encapsulated NaI(Tl) crystal in a reflective canister is shown in the upper left and a photomultiplier tube is shown in the upper right. The two components are coupled together to form a scintillation counter such as the one shown in the lower half. The NaI(Tl) scintillation spectrometer is composed of a 2×2 inch NaI(Tl) crystal coupled to a PMT, both of which are sealed in the container together. The combined voltage divider and preamplifier unit is attached to the back (right side) of the PMT.

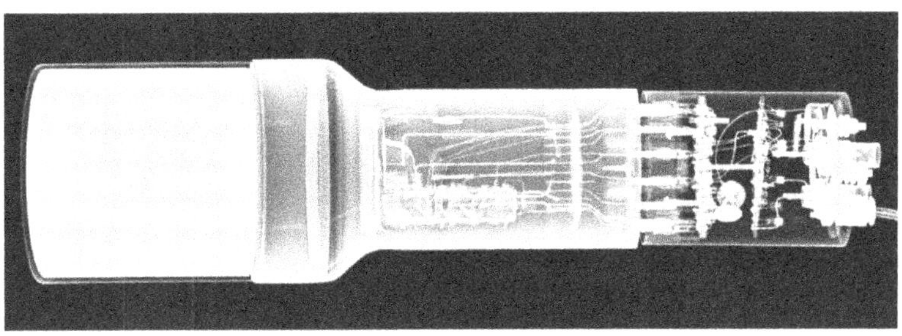

Figure 14.4. An x-ray photograph of a 3 in × 3 in NaI:Tl detector coupled to a linear focus PMT, resistive divider, and preamplifier.

Snell's Law

The *index of refraction* is defined as the ratio of light speed in vacuum to that in a particular medium,

$$n = \frac{c}{v}. \tag{14.2}$$

Snell's Law is a fundamental law describing the reflection and transmission of light at the interface between two adjacent optical surfaces. This law states, using the notation shown in Fig. 14.5(a),

$$\frac{\sin\theta_1}{\sin\theta_2} = \frac{n_2}{n_1} = \frac{v_1}{v_2}, \tag{14.3}$$

where n_i and v_i are the index of refraction and the speed of light in medium i, respectively. The index of refraction for some common window materials is given in Table 14.1.

Table 14.1. Index of refraction (at $\lambda = 410$ nm) and cut-off wavelength below which the transmission is less than 10% of the maximum.

Window material	Index of refraction	Cut-off wavelength (nm)
Soda Lime	1.536	300
Borosilicate	1.530	270
GE 214 (fused silica)	1.459	<160
Suprasil 310 (fused silica)	1.469	<155
Quartz	1.557	
Al_2O_3	1.785	145
MgF_2	1.383	115
LiF	1.398	105

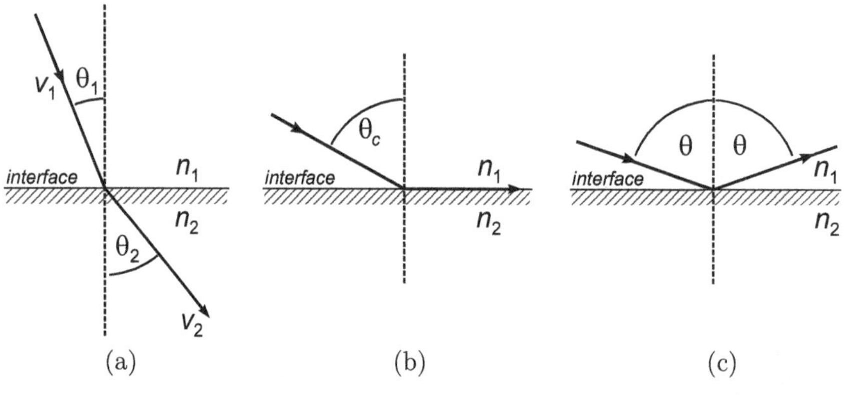

Figure 14.5. Snell's Law showing (a) refraction through two media, (b) total internal reflection at the critical angle, and (c) total internal reflection beyond the critical angle. In these figures it is assumed $n_1 \geq n_2$.

For the case $n_1 > n_2$, $\theta_2 > \theta_1$ as is shown in Fig. 14.5. There is a critical angle in medium 1, for which $\theta_2 = 90°$ beyond which no light is transmitted into medium 2 and all light is reflected back into medium 1. This critical angle for medium 1 is

$$\theta_{1c} = \sin^{-1}(n_2/n_1). \tag{14.4}$$

However, if $n_1 < n_2$, there is no critical angle for medium 1, i.e., light is transmitted for any incident angle.

Therefore, to reduce light losses from internal reflection at the scintillator/PMT interface, the index of refraction for the PMT window should closely match the index of refraction of the scintillation material. Because of a possible index mismatch between many common scintillation materials and PMT windows, a *coupling compound* is usually placed between the two materials. Coupling compounds are usually selected such that $n_{\text{coupling}} \simeq n_{\text{window}}$.

Example 14.1: Determine the critical angle for light propagating from common glass into air.

Solution:
The index of refraction for air is $n_2 = 1.0$, and glass has an index of refraction approximately equal to $n_1 = 1.5$. From Eq. (14.4),

$$\theta_c = \sin^{-1}(n_2/n_1) = \sin^{-1}(1.0/1.5) = 41.81^\circ.$$

Lambert's Cosine Law

It is usual for the scintillator to be encapsulated in a reflecting cannister in which some portion of the cannister has a transparent window for light to exit. For best results, the cannister window should have an index of refraction as close to those of the scintillation material and the entrance window to the PMT. Further, the relative amount of light (per unit energy deposited in the detector) that *enters* the PMT from the scintillator material should, ideally, be constant.

Consider a homogeneous half-space (a wall of the canister, for example) in which there is a constant light source emitting isotropically S photons cm^{-3} s^{-1}. As shown in Fig. 14.6, on the surface (the x-y plane) is a small area δA whose outward normal is the z-axis. The number δN of photons emitted per unit time from a differential volume dV a distance r from δA that pass through δA in a differential solid angle $d\Omega$ about $\mathbf{\Omega}$ is

$$\delta N = S\,dV\left(\frac{\delta A\cos\theta}{4\pi r^2}\right)\exp(-\mu r). \qquad (14.5)$$

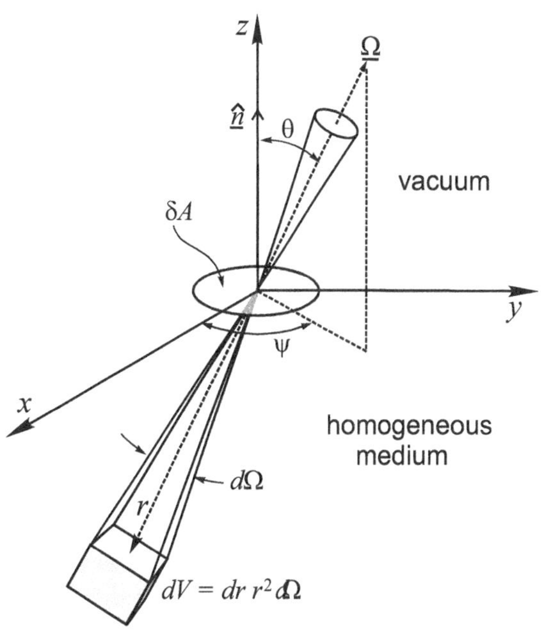

Figure 14.6. Coordinate system used to describe Lambert's cosine law.

Here $S\,dV$ is the number of photons emitted isotropically from dV per unit time. The term in parentheses is the probability photons emitted from dV are emitted towards δA, i.e., the fraction of the total solid angle subtended by δA at dV. The last term is simply the probability a photon travels to δA without interacting with the medium, which has an interaction coefficient μ.

The total flow through a unit area on the surface in direction $d\Omega$ about $\mathbf{\Omega}$ from all $dV = r^2 d\Omega dr$ is

$$J_n(\mathbf{\Omega})d\Omega = \int_0^\infty (\delta N/\delta A) r^2 d\Omega dr = \frac{S\cos\theta}{4\pi}d\Omega \int_0^\infty d^{-\mu r}dr = \frac{S\cos\theta}{4\pi\mu}d\Omega. \qquad (14.6)$$

From this result it is seen that the probability a photon emerges in a unit solid angle about a direction θ is proportional to $\cos\theta$, i.e., the intensity of light emitted from the surface is $I(\theta) = I(0)\cos\theta$ W cm^{-2} sr^{-1}. Any surface emitting light whose intensity is proportional to $\cos\theta$ is called a *Lambertian* surface.

Lambertian surfaces have an amazing property. As shown in Fig. 14.7, if one views an area A on such a surface the projected area perpendicular to θ is $A/\cos\theta$ cm^{-2}. The *sterance*[5] or light L observed by the eye or other light detector is

$$L = \frac{I(0)\cos\theta}{A\cos\theta} = \frac{I(0)}{A}. \tag{14.7}$$

This means the surface appears equally bright no matter from what angle the surface is viewed. For example, the surface of the sun is a Lambertian surface, so the sun's disk appears uniformly bright from the center to the edge.

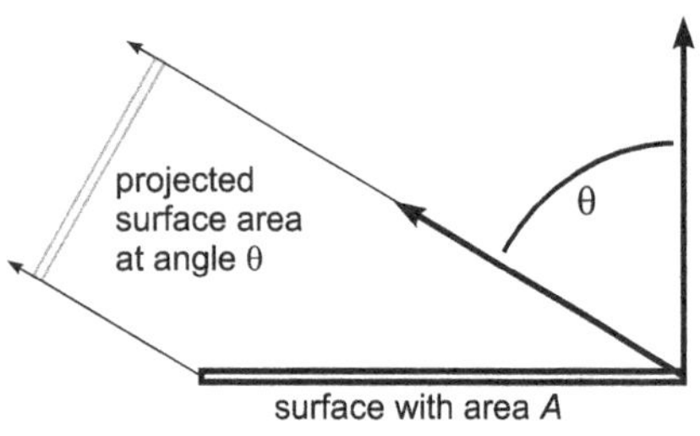

Figure 14.7. Depiction of projected area as a function of angle.

The important result of Eq. (14.7) indicates that the sterance from a Lambertian surface is constant, hence the reflected light from the cannister walls that reach the transparent interface between the scintillator and the PMT is almost constant regardless of where on the surface the reflected photons originated. Such is not true for a specular surface, at which the angle of light intersecting the transparent interface is a strong function of the initial location of the scintillation flash, thereby causing a large variance in the transmitted and internally reflected light. Hence, there is a larger variance in light transmitted into the PMT from a specular reflector compared to that from a Lambertian reflector.

Because energy resolution depends on the variance of the light entering the PMT, a Lambertian reflector wrapped around the scintillator produces better energy resolution than does a specular reflector. Common materials used to produce a near Lambertian surface include Teflon tape, Al_2O_3, TiO_2, and MgO powder. These materials produce a diffuse light reflection nearly independent of where the initial radiation interaction occurs in the scintillator. The product Spectralon®, a fluoropolymer, has nearly perfect Lambertian reflectance for wavelengths between 400 to 1500 nm.

Window Reflection and Transmission

After scintillation light transits the interface between the detector and the PMT, it must still pass through the transparent PMT window in order to reach the photocathode material. There are a variety of PMT window materials, each having a short wavelength cutoff beyond which photons with small wavelengths become strongly absorbed in the window (see Table 14.1). Many scintillators emit wavelengths near 40 nm (400 angstroms), which generally matches well to common glasses, including soda-lime and borosilicate glasses. However, there are many important scintillators that emit photons with wavelengths in the near UV region (300-400 nm) and are strongly absorbed by common glasses. The characteristic light response function of a scintillation detector can be severely compromised at short wavelengths simply by a poor choice of a PMT window.

Some of the light, with a wavelength λ, that impinges perpendicularly on an interface between two media undergoes reflection. The fraction of light reflected, or the *reflectance*, is [Hecht 1987]

$$R_{12}(\lambda) = \frac{[n_2(\lambda) - n_1(\lambda)]^2}{[n_2(\lambda) + n_1(\lambda)]^2}. \tag{14.8}$$

Here $n_1(\lambda)$ and $n_2(\lambda)$ are the refractive indices of the two media and both are functions of wavelength (see Fig. 14.8). Notice the reflectance is the same no matter from which medium the light is incident on the interface. The light not reflected is transmitted so the *transmittance* into the window is $T_{12} = 1 - R_{12}$.

[5]The sterance of a surface is the light-power emission rate per unit surface area per steradian and typically has units of W cm^{-2} sr^{-1}.

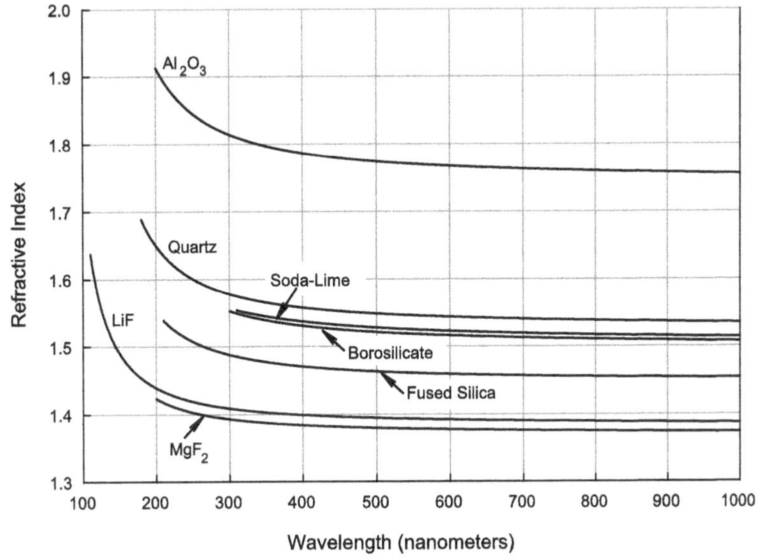

Figure 14.8. The index of refraction of a few PMT window materials as a function of wavelength. From [Polyanskiy, 2014].

As light passes from the scintillator (medium 1) through the window (medium 2) into the PMT (medium 3), it is reflected at the two surfaces of the window. If there is no absorption of light as it passes through the window, the minimum resulting *transmittance* T_{min} is

$$T_{min} = [1-R_{12}(\lambda)][1-R_{23}(\lambda)] = \left\{1 - \frac{[n_2(\lambda) - n_1(\lambda)]^2}{[n_2(\lambda) + n_1(\lambda)]^2}\right\}\left\{1 - \frac{[n_3(\lambda) - n_2(\lambda)]^2}{[n_3(\lambda) + n_2(\lambda)]^2}\right\}$$

$$= \frac{16n_1(\lambda)n_3(\lambda)n_2^2(\lambda)}{[n_2(\lambda) + n_1(\lambda)]^2[n_3(\lambda) + n_2(\lambda)]^2}, \tag{14.9}$$

where $n_3(\lambda)$ is the index of refraction of the medium light enters upon passing through the window (presumably a vacuum).

If one assumes medium 1 is air and medium 3 is a vacuum, then $n_1 \simeq n_3 = 1$, and Eqs. (14.8) and (14.9) reduce to

$$R(\lambda) = \frac{[n(\lambda) - 1]^2}{[n(\lambda) + 1]^2} \quad \text{and} \quad T_{min} \simeq \frac{16n^2(\lambda)}{[1 + n(\lambda)]^4}, \tag{14.10}$$

where n is the window refraction index.

Actually, slightly more light is transmitted to the PMT than described by Eq. (14.10). Light reflected at the PMT/window surface can be reflected at the other surface and some of this light is then transmitted into the PMT after making three crossings in the window. Similarly, light can be reflected many times inside the window and eventually be transmitted to the PMT. Hence the total transmittance is

$$T(\lambda) = [(1 - R(\lambda))^2][1 + R^2(\lambda) + R^4(\lambda) + \ldots] = \frac{(1 - R(\lambda))^2}{1 - R^2(\lambda)}. \tag{14.11}$$

With absorption losses, light transmission through the PMT window can be approximated by

$$\phi' = \tau(\lambda)\phi_0 e^{-\alpha(\lambda)x}, \tag{14.12}$$

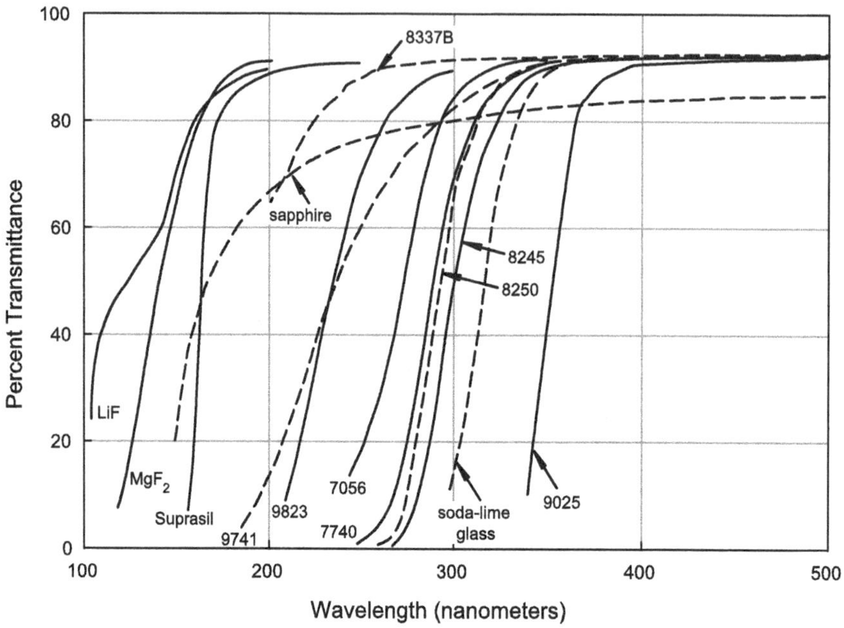

Figure 14.9. Transmission of light through various PMT windows, each 1 mm thick, as a function of wavelength. Corning borosilicate glasses 7740 (Pyrex), 7056, 9823, 9741 and Schott borosilicate glasses 8245 and 8250 have been usually used for PMT windows. Schott 8337B and Heraeus Suprasil (fused quartz) are often used for extended UV PMTs. Data are from Engstrom [1980], Donati [2000], and Schott [2014].

where ϕ' is the transmitted flux of light, ϕ_0 is the initial photon flux impinging perpendicular upon the surface, x is the glass thickness, and $\alpha(\lambda)$ is the wavelength-dependent attenuation coefficient. The reflectance $R(\lambda)$ is between 0.035 to 0.05 for most glasses for which the index of refraction $n \simeq 1.5$. The coefficient of reflective losses generally increases with increasing values of n.

For most scintillator detector applications, a window of borosilicate glass has a sufficiently low cutoff wavelength that adequately allows the transmission of scintillation photons. For instance, Pyrex (Corning 7740) efficiently transmits light with wavelengths above 350 nm (see Fig. 14.9). Common borosilicate glasses used for PMTs include Corning 7056, Corning 7740, Schott 8245, and Schott 8250. In high radiation fields, radiation damage to the glass can cause its clarity to change, slowly becoming brownish over time. PMTs with non-browning glass, such as Corning 9025, can be used in high radiation environments. The transmittance of a few materials used for PMT windows are shown in Fig. 14.9. In some applications, deeper UV emissions from a scintillator are of interest, but are attenuated by common borosilicate glasses. For deeper UV transmission, special PMT borosilicate windows with extended UV range are used, including Corning 9741, Corning 9823, and Schott 8337B glasses. Quartz and fused quartz (Suprasil) have excellent deep UV transmission properties and allow the passage of photons with wavelengths down to 160 nm. Further, the index of refraction is near 1.55, thereby producing minimal reflection losses. Because of their extended UV range, PMTs with fused quartz windows can be used as Čerenkov radiation detectors. Sapphire (Al_2O_3) windows also allow the transmittance of deep UV photons, but Al_2O_3 has a high index of refraction compared to that of glass and, thus, causes higher reflection losses, i.e., a higher $R(\lambda)$, especially at lower wavelengths.

For deep UV transmission, PMTs with either LiF or MgF_2 windows can be used. PMT windows manufactured from LiF windows transmit photons with wavelengths down to 105 nm, below which there is no transmittance. However, attenuation becomes significant below 120 nm, which has approximately 40% transmittance. LiF windows are fragile and expensive to produce; hence they are used only in special applications. LiF is also slightly hygroscopic, increasing so as the temperature increases, especially above 400°C.[6] Hence, precautions against moisture must be implemented. MgF_2 is not as fragile as LiF, and is perhaps a more popular solution for deep UV photon detection. This material allows transmission of UV photons with wavelengths down to 125 nm before significant attenuation losses occur. The cutoff wavelength for MgF_2 is about 115 nm. Finally, PMTs without windows must be used for the detection of UV light with wavelengths λ less than 105 nm.

Some PMT manufacturers emphasize the use of low background windows. For most applications, such a feature is probably unnecessary. However, for low background radiation measurements, contamination from ^{40}K, ^{238}U, or ^{232}Th may be an issue. Borosilicate glass often has considerable contamination from these radioactive elements, particularly from ^{40}K. Purified forms of borosilicate glass or fused silica (quartz), while more expensive, are offered as low background PMT windows.

14.1.3 Photocathode Materials and Design

The main function of a PMT is to convert photons into electrical charges. This conversion is accomplished by using the photoelectric effect, in which scintillation photons photoelectrically interact in a material coated onto a cathode in the PMT resulting in the ejection of a photoelectron.

Quantum Efficiency, η_{pc}

For PMT photocathodes, the *quantum efficiency* η_{pc} is defined as the number of electrons released from the photocathode per photon incident on the cathode,[7] and is usually defined at a specific photon wavelength. The quantum efficiency is affected by at least three different physical attributes of the absorber material, namely the absorption efficiency, the electron diffusion length, and the energy barrier.

Absorption efficiency can be affected by the glass enclosure, the type of photocathode, and, of course, the photocathode material. As just discussed in the section on window reflection and transmission, there are reflected and transmitted components of the light incident on the window. The size of these components generally depends on the wavelength of the light. Losses from reflection are generally near 8% when considering reflection on the inner and outer glass surfaces. Transmission is a function of wavelength, dropping to only 10% transmission at the cut-off wavelength. Hence, the η_{pc} can be severely reduced at photon wavelengths approaching the cut-off wavelength of the window. The absorption efficiency of the photocathode depends of the thickness of the material coating the photocathode. For semi-transparent (or transmission) photocathodes, a thin layer of photosensitive material is coated upon the inner portion of the glass window, usually only a few nanometers thick. An alternate design, less common for use with radiation detectors, is the internal opaque photocathode, which is not coupled to the PMT window, but rather is set apart inside the structure.

Photon absorption is restricted to those wavelengths with sufficient energy to elevate electrons into higher electronic energy levels. For instance, if a photon has sufficient energy to elevate an electron from the band of valence electrons into the band of conduction electrons, then the photon can be absorbed.

---

[6] Many manufacturers of LiF optics warn against using the material above 400°C due to being attacked by atmospheric moisture.

[7] The term *quantum efficiency* has several alternative definitions. It is also used to define the fraction η_λ of excited charge carriers that radiatively recombine to release photons, a definition generally used for photon emitting devices such as LEDs and laser diodes. For semiconductor photon detectors, quantum efficiency is also defined as the number of electron-hole pairs produced per unit photon energy at λ. In an attempt to remedy this confusion, the authors distinguish among the various definitions with an appropriate subscript.

Table 14.2. Properties of a few common photocathodes. Data are from Flyckt and Marmonier [2002] and Engstrom [1980].

Photocathode	Composition	Window	Emission threshold (nm)	λ_{max} (nm)	Radiant sensitivity (mA/W)	$\%\eta_{pc}$ at λ_{max}
S-1	AgOCs	BS, SL	1100	800	2.3	0.4
S-4	$SbCs_3$	BS, SL, FS, O	680	380	38	12
S-5	$SbCs_3$	UV, O	630	400	50	18
S-11	$SbCs_3$	BS, SL	700	440	80	22
S-13	$SbCs_3$	FS	700	440	80	22
S-20	$SbNa_2KCs$	BS, SL, FS	850	420	70	20
S-20R (ERMA)	$SbNa_2KCs$	BS, SL	900	550	35	8
S-24	$SbCs_3$	BS	640	380	50	16
bialkali	SbKCs	BS, SL, FS	630	400	90	28
bialkali (GEBA)	SbKCs	BS, SL	630	400	90	28
bialkali (HT)	SbNaCs	BS, SL	700	400	50	16
solar blind	CsTe	FS	340	235	20	10
NEA	GaAs:Cs-O	UV	860	800	80	12

BS: borosilicate glass
FS: fused silica
UV: Corning 9741 (UV transmitting)
GEBA: green extended bialkali
SL: soda lime glass
O: opaque (reflective) photocathode
ERMA: extended red multialkali
HT: high temperature, data for 130°C

Note that absorption of such a photon may still fail to give the electron sufficient energy to escape the conversion material. For most metals, electrons must absorb enough energy to surmount an energy barrier at the vacuum interface of the photocathode in order to enter the vacuum of the PMT. Semiconductors, by contrast, have relatively few electrons in the conduction band; hence electrons are generally excited from lower energy bands (such as the valence band), and must have sufficient energy to surmount both the energy gap between the bands plus energy to surmount the energy barrier at the vacuum interface. Long wavelength photons may fail to transfer the required energy to surmount the vacuum barrier and, thus, generally defines the long wavelength cut-off for many photocathodes.

Photoelectrons created very near the vacuum interface need diffuse only a short distance to the interface where, if they have an energy above the energy barrier, they escape into the PMT. However, electrons may diffuse in any random direction; hence, some electrons created very near the interface may diffuse in directions away from the vacuum interface. Also, electrons produced deeper in the photocathode material must diffuse greater distances to reach the vacuum interface, during which migration they lose energy from collisions with ambient electrons. If these electrons reach the vacuum interface, but no longer have sufficient energy to surmount the barrier, they de-excite to lower energy levels and can no longer escape into the vacuum and contribute to the PMT photocurrent. All of these processes tend to reduce the overall quantum efficiency of the photocathode. Listed in Table 14.2 are properties of a few common photocathode materials.

Work Function, ϕ_m

Photoemission is the process in which the absorption of photons in some material results in the ejection of energetic electrons from the material. Most PMTs use metals with low ionization energies as the photocathodes, although many PMTs also use special semiconducting materials.

To understand the photoemission process, consider the energy level diagrams shown in Fig. 14.10. In Fig. 14.10(a) an interface between a metal and vacuum is depicted based on the Sommerfeld model [Dekker 1957]. For an electron to escape from the metal into the vacuum, it must gain enough energy to reach the vacuum level, E_0. The energy difference between electrons at rest in the metal and electrons at rest in the

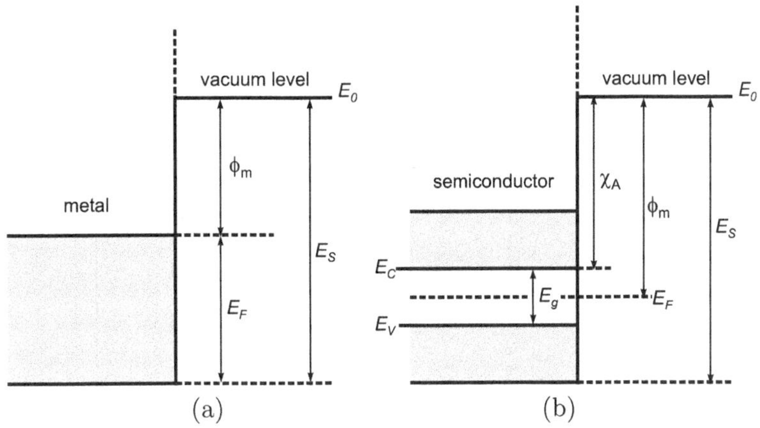

Figure 14.10. Simplified (a) band diagram of a metal based on the Sommerfeld model, and (b) a band diagram of a semiconductor/vacuum interface.

vacuum is denoted as E_S. The Fermi energy level E_F is defined as the energy below which all states are filled and above which all states are empty at absolute zero (0 K). The minimum energy required for electrons to move from the metal to the vacuum level is $E_S - E_F$, and is referred to as the metal *work function*, ϕ_m. The energy required to move an electron at rest in the metal to the vacuum level is referred to as the *outer work function* and equals E_S.

The probability that an electron occupies a state with energy E is given by the Fermi-Dirac distribution function of Eq. (12.118), repeated here for convenience,

$$f_{\mathcal{FD}}(E) = \frac{1}{1 + \exp\left[\dfrac{E - E_F}{kT}\right]},$$

where k is the Boltzmann constant, T is the absolute temperature, and E_F is the Fermi energy. At absolute zero, the Fermi-Dirac distribution function is a step function, i.e.,

$$f_{\mathcal{FD}}(E) = \begin{cases} 1 & E \leq E_F \\ 0 & E > E_F \end{cases} . \tag{14.13}$$

As the temperature increases, thermal energy allows some electrons to exceed the Fermi energy, provided that energy states are available.

With the method described in Sec. 12.5.3, the density of available states about differential energy dE in a metal is found to be (see Eq. (12.115))

$$N(E)dE = \frac{\sqrt{2}m_e^{3/2}}{\pi^2\hbar^2} = \frac{4\pi(2m_e)^{3/2}}{h^3}E^{1/2}dE, \tag{14.14}$$

or

$$N(E)dE = \frac{\sqrt{E}}{2\pi^2}\left(\frac{2m_e}{\hbar^2}\right)^{3/2} dE, \tag{14.15}$$

where h is the Plank constant and m_e is the electron rest mass. The product of $N(E)$ and $F(E)$ yields the filled state density as a function of energy and temperature, as shown in Fig. 14.11. The total number of

Table 14.3. Valence, Fermi energy, and work function for metals often used in PMTs.

Metal	Valence	Fermi Energy E_F (eV)	Work function ϕ_m (eV)
Li	1	4.72	2.9
Na	1	3.23	2.36
K	1	2.12	2.29
Rb	1	1.85	2.26
Cs	1	1.58	2.14
Cu	1	7.07	4.45
Ag	1	5.5	4.46
Ba	2	3.8	2.51
Al	3	11.7	4.20

filled states (electrons) per unit volume can be determined by integrating the product $N(E)F(E)$ over all possible energies,

$$n = \int_0^\infty N(E) f_{\mathcal{FD}}(E) dE = \frac{4\pi(2m_e)^{3/2}}{h^3} \int_0^\infty f_{\mathcal{FD}}(E) E^{1/2} dE. \tag{14.16}$$

At absolute zero, Eq. (14.16) simplifies to

$$n = \frac{4\pi(2m_e)^{3/2}}{h^3} \int_0^{E_F} \sqrt{E}\, dE = \frac{\pi}{3} \left(\frac{2m_e E_F}{h^2}\right)^{3/2} = 4.55 \times 10^{21}\ E_F^{3/2}\ \text{cm}^{-3}. \tag{14.17}$$

If the inner wall of the PMT tube is coated with an electron emissive material, electrons are more readily emitted by incident low-energy photons by using a material with a relatively low work function. Inspection of Fig. 14.12 reveals that the best candidates for photocathodes are the alkali metal elements, or mixtures thereof. Values of the Fermi energy and the work functions for these elements are given in Table 14.3. The overall spectral response to various photon wavelengths for the alkali metals is shown in Fig. 14.13. Some alkali earth elements (Group II in the Periodic Table) also have low work functions, and are sometimes used in PMTs as electron emissive materials.

Thermionic Emission

At a temperature of absolute zero, the Fermi-Dirac distribution indicates that all free electrons require an energy equal to the work function ϕ_m to surmount an energy barrier and escape into "vacuum" (see Fig. 14.11). As the temperature is increased, the probability that electrons are in states above the Fermi energy E_F increases. Such electrons require less energy than the work function to escape to the vacuum level. The process whereby thermal energy in the material leads to the liberation of electrons is referred to as *thermionic emission.*

The general expression for thermionic emission current density is described by the Richardson-Dushman equation

$$J = A(1 - \alpha_r)T^2 e^{-\phi_m/kT}, \tag{14.18}$$

where T is the absolute temperature, α_r is a reflection coefficient that quantifies the quantum mechanical reflection of the electron wave function at the barrier interface. The quantity $A(1-\alpha_r)$ is generally referred to as the "effective Richardson constant." It is this thermionic emission that contributes most to the thermal noise, or "dark current," of the PMT device. As described by Eq. (14.18), the dark current increases as ϕ_m decreases. However, at room temperature it turns out that the thermionic current is small and usually of

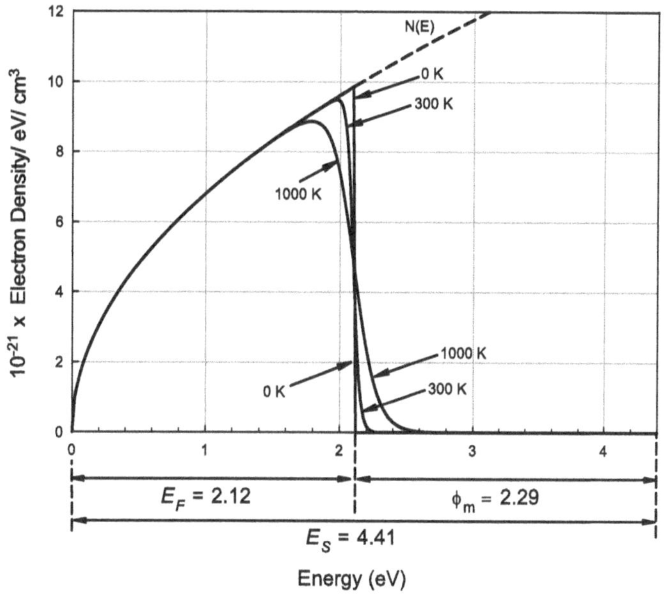

Figure 14.11. The density of filled states for potassium at absolute zero, 300 K, and 1000 K as a function of energy according to the free electron model.

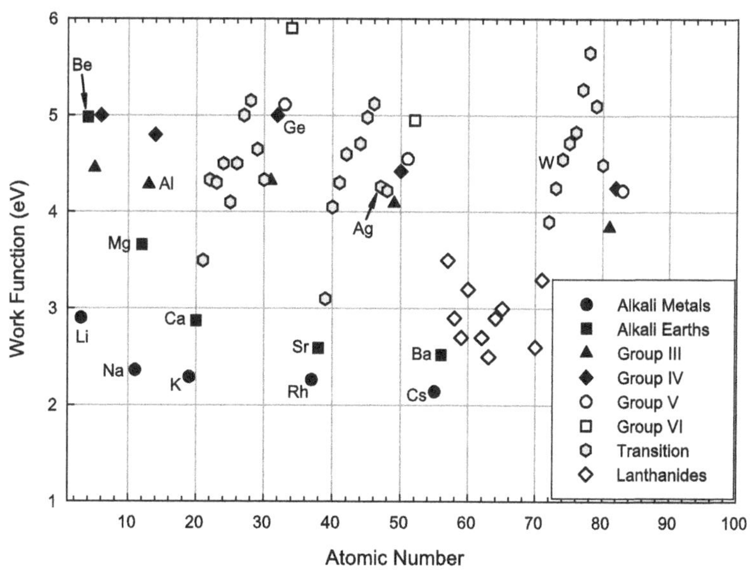

Figure 14.12. Work functions of most elements. Data are from Halas and Durakiewicz [1998].

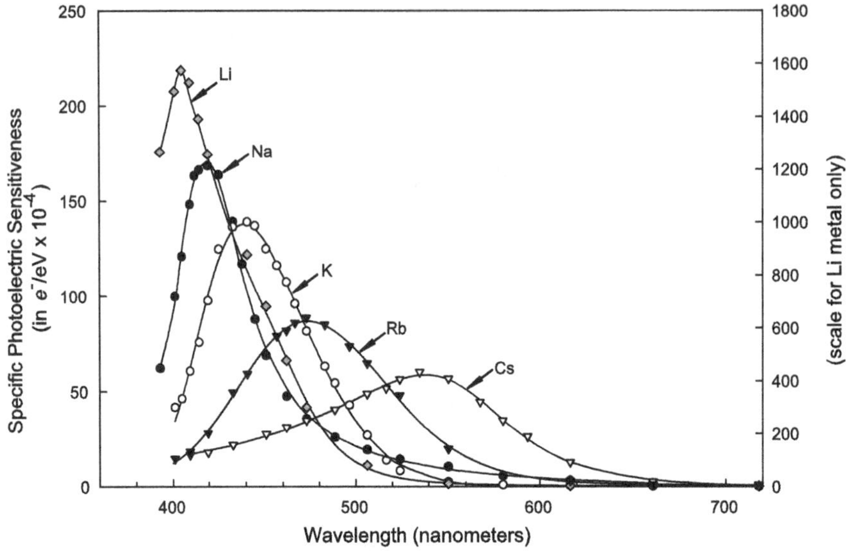

Figure 14.13. Measured relative spectral response for the alkali metals. Data are from Seiler [1920].

little consequence. With careful attention to the selection of the photocathode material, the production of ultra-low-noise PMTs is possible.

Photoemission

Photons interacting in a photocathode material can be absorbed by free electrons. If the photon energy is greater than the work function ($h\nu > \phi_m$), then electrons can be elevated to energies above the potential barrier, and are sometimes referred to as "hot electrons." If the electrons are excited from the Fermi energy E_F, then the resulting electron energy is simply

$$E_e = h\nu - \phi_m. \tag{14.19}$$

However, for electrons excited from deeper in the energy band, the electron energy is less, namely,

$$E_e = h\nu - (\phi_m + E'). \tag{14.20}$$

where E' is the additional energy required for the electron to reach the vacuum level. The probability that an electron is excited to the vacuum level from lower energy levels diminishes as E' increases. In order to actually escape the material, these excited electrons must first diffuse to the metal/vacuum interface and then still retain enough energy to surmount the barrier. Hence, it is less probable for electrons excited deep in the energy band to escape as they lose energy, through scattering from ambient electrons, as they diffuse to the boundary. Further, the deeper in the photocathode material that an electron interacts with a photon, the less likely it is to reach the surface with sufficient energy to escape, because of the longer diffusion path length it must travel and the increased number of energy loss interactions it undergoes as it diffuses. Hence, a combination of diffusion effects and energy losses produce a low energy tail in the photoelectron response distribution, as shown in Fig. 14.14.

The photocathode work function and the wavelength of photons impinging upon the photocathode generally determine the electron cutoff energy for a PMT. Photons that do not have adequate energy to excite

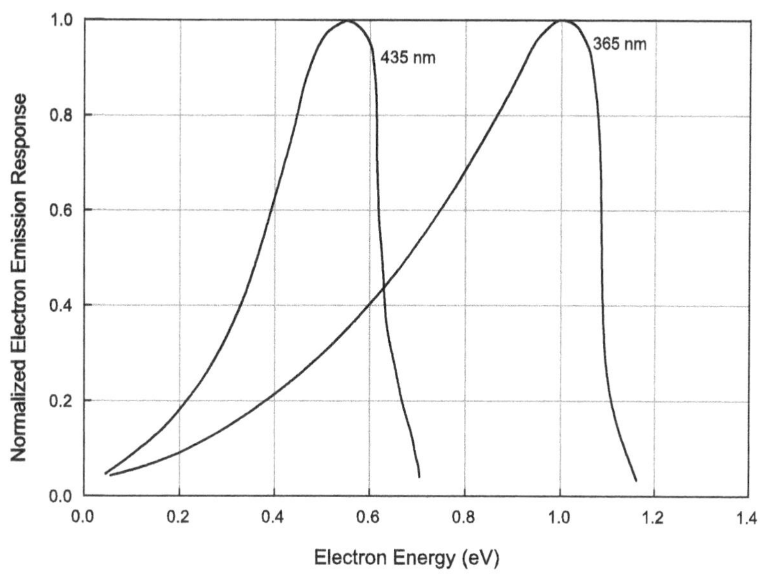

Figure 14.14. Normalized electron response for K for two light wavelengths, 435 nm and 365 nm. Data are from Brady [1934].

electrons above the potential barrier cannot produce photoexcition. Under ideal circumstances, such a cut-off should be represented by a step function. However, the high energy tail of the Fermi-Dirac distribution function indicates that some free electrons in the material occupy states that exceed the Fermi energy. Consequently, for photons of any given wavelength with $h\nu \geq \phi_m$, there are some electrons emitted with energies greater than $h\nu - \phi_m$, as is shown in Fig. 14.14.

Electron Affinity χ_A

Certain compounds of alkali metals form narrow gap semiconductors. These materials offer extended sensitivity for electron excitation, but, consequently, do not follow the Sommerfeld band model. Unlike a pure metal, electrons cannot freely flow between the valence and conduction bands, but must instead surmount an energy gap E_g, or the *band gap*, between the upper edge of the valence band and the lower edge of the conduction band. Shown in Fig. 14.10(b) is an energy band diagram for a semiconductor surface terminated at the vacuum level. The work function of the semiconductor is still defined as the difference between the vacuum energy level and the Fermi energy level.[8] However, the Fermi energy E_F is determined by the concentration and type of *dopants* introduced into the semiconductor. If the concentration of negative-type, or *n-type*, dopants (N_D) is greater that the concentration of positive-type, or *p-type*, dopants (N_A), then the Fermi energy level moves towards the conduction band edge energy E_C. Consequently, the work function decreases. If instead the concentration of *p*-type dopants is greater than the concentration of *n*-type dopants, then the Fermi energy level moves towards the edge energy of the valence band E_V, thereby causing the work function to increase.

Overall, both the Fermi energy level and the work function in a semiconductor are not constant, but instead are variable depending on the doping concentration. Consequently, the work function is difficult to use when calculating electron emission and escape probabilities. However, the electron affinity can be

[8]The physics of semiconductor devices is covered in Chapter 15. However, it is important to introduce some of these concepts here to further the understanding of a negative electron affinity photocathode.

Table 14.4. Properties of a few semiconductor photocathode materials.

Material	Band gap E_g (eV)	Electron affinity $\chi(E)$ (eV)	Threshold energy E_T (eV)	Threshold wavelength λ_T (nm)	Reference
Na_3Sb	1.1	2.1	3.2	388	Madelung et al. 1998
K_3Sb	1.1	1.5	2.6	477	Madelung et al. 1998
Rb_3Sb	1.0	1.2	2.2	564	Madelung et al. 1998
Cs_3Sb	1.6	0.45	2.05	605	Madelung et al. 1998
Na_2KSb	1.0	1.0	2.0	620	Madelung et al. 1998
Na_2KSb:Cs	1.0	0.55	1.55	800	Madelung et al. 1998
K_2CsSb	1.0	1.1	1.88	660	Sommer 1968
K_2CsSb:O	1.0	<1.10	1.59	780	Sommer 1968
Si	1.12	4.05	5.17	240	Sze 1981
Ge	0.66	4.0	4.66	266	Sze 1981
GaAs	1.42	4.07	5.49	225	Sze 1981
p^+-GaAs	1.42	−0.02	1.4	886	Donati 2000
Ga_xIn_{1-x}As:Cs	1.1	$\simeq 0$	1.1	1127	Donati 2000
p-Cs_2Te	3.2	0.2	3.4	365	Kong et al. 1995
p-InGaAsP/p-InP	<1.35/1.35	<0	0.886	1400	Niigaki et al. 1997
p-InGaAs/p-InP	<1.35/1.35	<0	0.729	1700	Hamamatsu 2007

treated as a constant value, defined as

$$\chi_A = E_0 - E_C, \tag{14.21}$$

where E_0 is the vacuum level energy. To calculate the energy barrier heights for semiconductors, it is the electron affinity that is typically used as a reference energy relative to the vacuum level. Hence, the photon energy generally required to excite an electron over the vacuum barrier is

$$h\nu > E_g + \chi_A = E_T, \tag{14.22}$$

where E_T is the threshold energy. Note that there is a considerable concentration of electrons in the conduction band that can be excited above the vacuum level, thereby allowing some electrons to be excited over the barrier with an energy of only χ_A. However, the concentration of these conduction band electrons is orders of magnitude below the concentration of electrons in the valence band, and, consequently, most of the excited charge comes from electrons elevated from the valence band.

Finally, it should be mentioned that not all photocathodes are constructed from alkali metals or metal compounds. Some photocathodes are constructed from more traditional semiconductors, such as *p*-type GaAs and *p*-type Cs_2Te. Listed in Table 14.4 are properties of a few semiconductors used as photocathodes, and shown in Fig. 14.15 are the absolute responsivities for many common photocathodes.[9]

Negative Electron Affinity $-\chi_A$

To construct some photocathodes, certain *doped* semiconductors are used. These semiconductors form no energy barrier at the vacuum interface! To understand the physics that cause this phenomenon, it is recommended that the reader study Fig. 14.16.

Most semiconductors have a number of unbounded states on the surface. These states fill with electrons up to a particular energy level, namely that of Fermi energy. In other words, the Fermi energy becomes

[9] The absolute responsivity, or radiant sensitivity, is defined as the photoelectric current generated by the photocathode divided by the number of incident photons, usually in units of mA/watt.

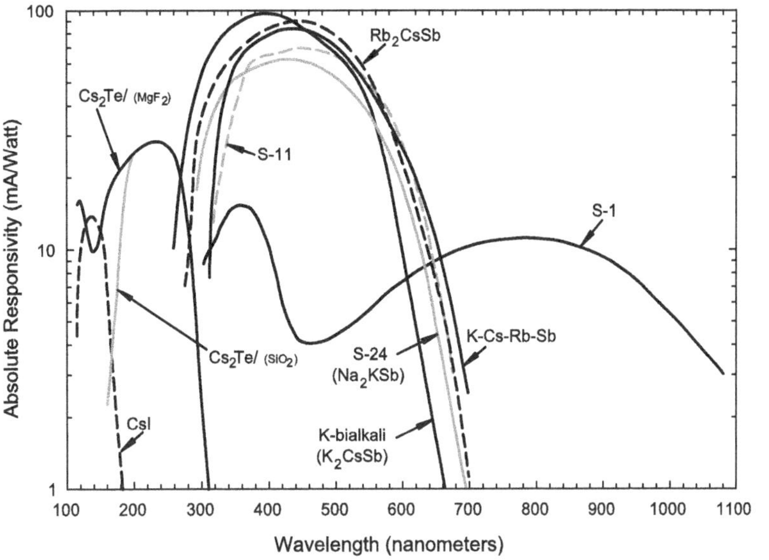

Figure 14.15. Response functions for several photocathode materials. The window material is soda-lime glass most of the responses. However, the S-24 response is for a 7056 borosilicate window, the CsI response is for a MgF_2 window, and the Cs_2Te responses are with either a MgF_2 or quartz window as marked. Data are from Engstrom [1980] and Hamamatsu [2007].

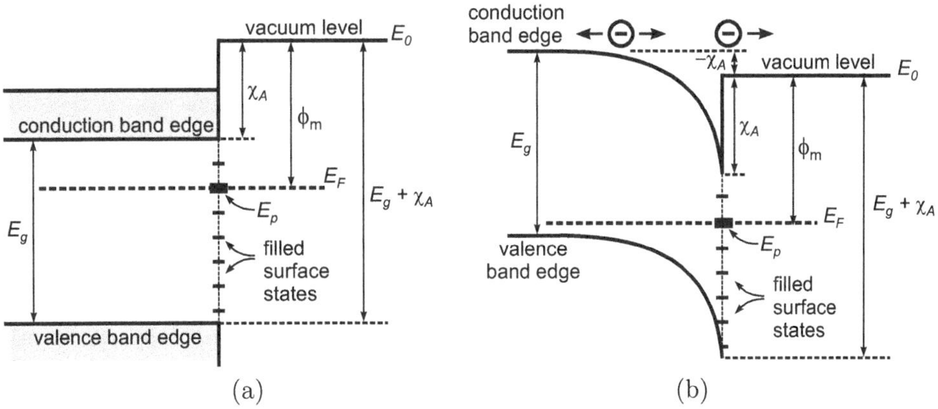

Figure 14.16. Semiconductor band diagrams for a surface pinned to a surface state energy E_p, showing (a) n-type material and (b) p^+-type material with negative electron affinities.

pinned at the energy E_p, below which states are filled and above which states are empty. Note, these energy states exist in the band gap *at the surface* as a consequence of unbound surface states, a condition that generally does not exist in the semiconductor material far from the interface.

The quantity N_D, used for n-type dopants, is an atomic density of impurity atoms that *donate* electrons to the conduction band, thereby increasing the density of negative charge carriers (electrons) in the conduction band. The quantity N_A, used for p-type dopants, is the atomic density of impurity atoms that *accept* electrons from the valence band, thereby removing electrons from the valence band and increasing the density of positive charge carriers (holes) in the valence band.

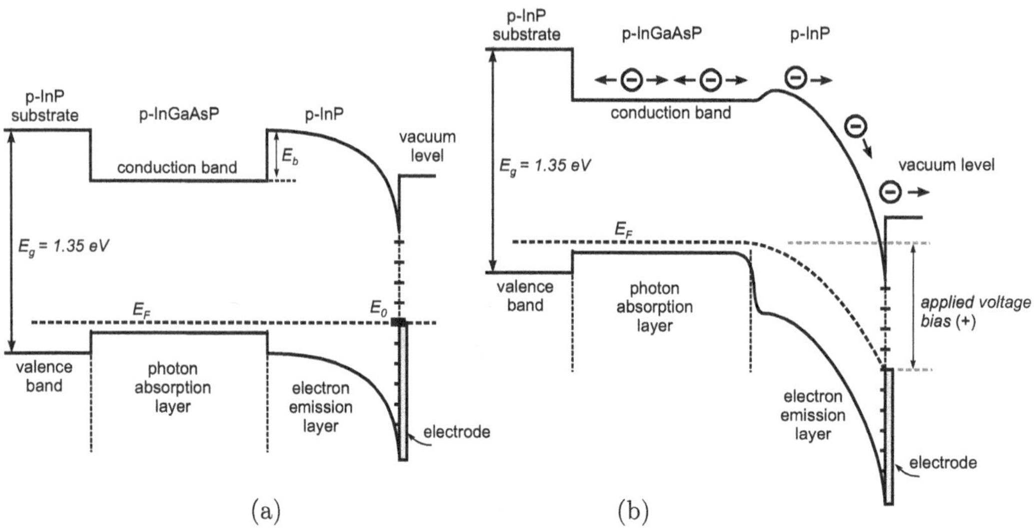

Figure 14.17. Semiconductor band diagrams for a field-assisted InP/InGaAsP/InP heterostructure photocathode (a) under zero bias and (b) under reverse bias operation.

By strongly doping a *p*-type semiconductor, the Fermi energy level in the *bulk* semiconductor is lowered close to E_V as shown in Fig. 14.16(b). At equilibrium, the Fermi energy must be constant across the interface and, hence, the energy bands bend accordingly to satisfy this condition such that the Fermi energy level in the bulk aligns with the pinned Fermi energy level at the surface. Consequently, the vacuum level E_0 at the surface moves below E_C in the bulk semiconductor, producing a *negative electron affinity* as calculated from Eq. (14.21). As a result, electrons can now move from the semiconductor bulk at energy E_C into the vacuum without encountering an energy barrier [Rougeot and Baud 1979].

Heterostructure negative electron affinity (NEA) photocathodes take advantage of semiconductor band-gap engineering to extend the IR response of PMTs while maintaining relatively high sensitivity. The structures are fabricated by growing an epitaxial *p*-type InP layer upon a *p*-type InGaAs or InGaAsP layer, as depicted in Fig. 14.17. The InP semiconductor layer has a much wider band-gap than the InGaAs (or InGaAsP) layer [Niigaki et al. 1997]. Because these epitaxial crystals are thin, typically with thicknesses of 0.2 to 0.5 μm, it is usual to grow them upon relatively thick InP substrates so they are forced to function as opaque (reflection) type photocathodes. Front windows with InP-based semi-transparent type heterostructure semiconductor photocathodes are also manufactured, but because of their poor absorption of photons above 1.35 eV, they suffer reduced sensitivity for photon wavelengths below 900 nm. The surface of the *p*-type InP layer forms a NEA boundary at the vacuum boundary, but there is still a barrier to surmount at the InGaAs/InP (or InGaAsP/InP) interface, as shown in Fig. 14.16(a). Application of a bias voltage to the photocathode lowers the conduction band barrier, thereby allowing electrons to more easily to pass into the vacuum, as shown in Fig. 14.16(b). These NEA photocathodes have sensitivity at photon wavelengths extending up to 1400 nm for the InGaAsP/InP type and up to 1700 nm for the InGaAs/InP type [Hamamatsu, 2007]. Advanced heterostructure photocathodes can operate up to 2300 nm [Niigaki et al. 2014]. Unfortunately, heterostructure semiconductor photocathodes also have a high dark current when used at room temperature; hence they must be cooled to temperatures between −60° C and −80° C during operation.

NEA photocathodes enable electrons to diffuse greater distances before dropping back to the conduction band edge and recombining with holes. This greater diffusion distance allows thicker photocathodes to

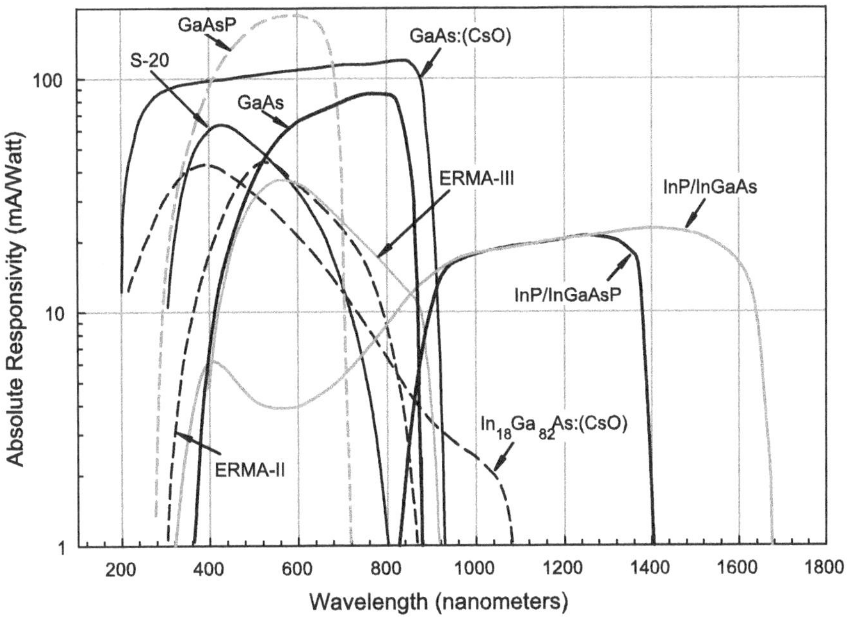

Figure 14.18. Response functions for several photocathode materials that are usually used for detection of longer wavelength light. Data are from Engstrom [1980] and Hamamatsu [2007].

be used while improving the electron emission. Hence, NEA photocathodes can have much higher photon response than common alkali metal photocathodes, especially in the long wavelength region, as can be seen from Fig. 14.18.

Semi-Transparent and Opaque Photocathodes

Photocathodes are either applied directly to the inner surface of the entrance window to the PMT enclosure or upon an internal secondary surface within the enclosure. Photocathodes applied to the inner surface of the transparent envelope are designed with thicknesses on the order of only a few electron diffusions lengths (tens of nanometers) and, therefore, are *semi-transparent.* Internal photocathodes are set back from the entrance window do not suffer the same problem and can have thicker material layers laid upon a substantial conductive backing, and are, therefore generally *opaque.* There are advantages and disadvantages to both of these designs.

Semi-transparent photocathodes are deposited upon the inside surface of the PMT entrance window, a geometry that enables the scintillation material to be placed adjacent to the photosensitive material. This usual arrangement serves to increase the solid angle between the light source and photosensitive area, thereby maximizing the detector signal. Such an arrangement is depicted in Fig. 14.2 and shown in Fig. 14.3. Because photoelectrons are ejected into vacuum from the side of the photocathode material opposite that upon which light is incident, this type of photocathode is sometimes called a *transmission* photocathode.

The thickness of a semi-transparent photocathode must be carefully controlled. If the photocathode is too thick, then photons can be absorbed beyond the critical escape depth for photoelectrons. Hence, any additional material serves only to decrease the electron signal. If the photocathode is too thin, then many of the incident photons pass through the photocathode without producing photoelectrons.

For opaque photocathodes, electrons are ejected from the same side as that upon which the light is incident. These photocathodes are often called *reflective photocathodes.* In these photocathodes the self-absorption of photoelectrons is no longer a limiting factor, and opaque photocathodes can be applied as thicker layers. However, because they are set back from the PMT entrance window, opaque photocathodes may suffer loss of light because of a decreased solid angle with respect to the light source.

Optical power transmission through the photocathode material is described by

$$P(x) = P(0)e^{-\alpha(\lambda)x}, \tag{14.23}$$

where $\alpha(\lambda)$ is the attenuation coefficient at a specific wavelength λ and x is the distance from the cathode surface along the direction of light travel. The inverse of $\alpha(\lambda)$ is the attenuation length in which a fraction $1-e^{-1} \simeq 63\%$ of the incident photons are absorbed. Optical power absorption is greatest at the vacuum interface for opaque photocathodes, whereas it is least at the vacuum interface for semi-transparent photocathodes. If one assumes that the number of photoelectrons is proportional to the profile of absorbed light energy, the distribution of electron-hole pairs created in the photocathode is

$$F_e(x) = k_e\alpha(\lambda)P(0)e^{-\alpha(\lambda)x}dx, \tag{14.24}$$

where k_e is a constant. Once a photon has been absorbed, and presumably produced an excited electron-hole pair, photoemission from the photocathode occurs only if (a) the photoelectron energy is greater than the threshold energy $E_T = E_g + \chi_A$, and (b) the photoelectron can diffuse to the vacuum boundary while retaining $E \geq E_T$.

It is possible for an electron to become excited from photon absorption but with an energy $h\nu < E_T$. However, the photoelectron does not have sufficient energy to surmount the energy barrier depicted in Fig. 14.10 and, consequently, cannot reach the vacuum. The threshold wavelength is defined as

$$\lambda_T = \frac{hc}{E_g + \chi_A} = \frac{1239.8}{E_T} \ \{\text{nm}\}, \tag{14.25}$$

where E_T is expressed in eV. The threshold wavelength for many photocathode materials are listed in Table 14.4.

The photoelectrons must also be able to diffuse to the vacuum boundary while maintaining sufficient energy to surmount the energy barrier. Although the threshold energy E_T is the theoretical minimum energy limit, photoelectrons actually require much more initial energy than E_T to escape the photocathode in order to compensate for energy losses experienced during their diffusion to the interface. Because the electrons diffuse in random directions, many photoelectrons diffuse away from the vacuum boundary. Also, as the photoelectrons diffuse through the photocathode material, they lose substantial energy with each scatter reaction. A scatter can also cause the electrons to radically change directions. Each scatter may result in the energy of an electron dropping below E_T or change direction so it diffuses away from the vacuum boundary, either case negating photoemission. Overall, for a wide spectral response, the initial energy E_{pe} of the photoelectron should be greater than the threshold energy, i.e.,

$$\frac{hc}{\lambda} = E_{pe} \gg E_T. \tag{14.26}$$

14.1.4 Dynode Materials

Energetic electrons, such as accelerated photoelectrons, impinging onto a surface can cause the emission of secondary electrons, provided that the energy transferred to secondary electrons is greater than the vacuum barrier. The secondary electron yield δ is defined as

$$\delta = \frac{N_s}{N_p}, \tag{14.27}$$

where N_s is the number of emitted secondary electrons and N_p is the initial number of primary electrons striking the material.

There are similarities between photoemission and secondary electron emission, the distinctive difference being that photoelectrons are a result of the photoelectric effect and secondary electrons are created through coulombic interactions between the incident primary electron and the ambient electrons of the material. Because a single primary electron can scatter from many secondary electrons in the material, there is a possibility that many electrons acquire enough energy to surpass the vacuum barrier. Some of the energetic secondary electrons may, in turn, interact with other electrons transferring sufficient energy to allow them to also escape. Thus, if the primary electron has sufficient initial energy, the gain of Eq. (14.27), is greater than unity.

Typically photoelectrons emerging from the photocathode surface do not have sufficient energy to produce secondary electron emission. However, this situation is easily remedied by applying a secondary electron emission material upon an electrode, referred to as a *dynode*, and applying an accelerating voltage to it. The photoelectron can then gain sufficient kinetic energy as it is attracted to the anode to promote secondary electron emission.

Secondary Electron Emission

As with photoelectron emission, a series of processes must take place for successful secondary electron emission. These processes include, (a) the primary electron acquiring an energy greater than E_T of the dynode material, (b) the excited secondary electrons then must diffuse in the direction of the vacuum barrier, and (c) the secondary electrons, upon reaching the vacuum barrier, must still retain an energy greater than E_T.

As the energy of the primary electron is increased, the maximum depth of secondary electron generation also increases. Consequently, secondary electrons generated deep in the dynode material must diffuse relatively long distances to reach the vacuum boundary compared to those secondary electrons generated near the surface. If the primary electron is normally incident on the dynode, the maximum depth a primary electron can reach in the dynode material is the continuous-slowing-down-approximation range R_{CSDA} discussed earlier in Sec. 4.6.3. To reach this maximum depth the electron trajectory must be almost a straight line, i.e., no large angle scatters. However, very few of the incident electrons produce such trajectories and reach such depths. Of more utility is the mean forward or projected range R_p defined as the expected or average maximum depth of penetration in the direction of the incident electron. However, the calculation of R_p from first principles requires detailed electron transport calculations that take into account quantum mechanical and electron binding effects in the ambient atoms. Alternatively, the CSDA range R_{CSDA} can be multiplied by the appropriate detour factor to give R_p. But tabulations of detour factors are hard to find and their calculation also requires extensive electron transport calculations. Because primary photons generated in a PMT seldom have energies above a few keV, it is better to base range calculations on experimentally determined ranges.

Young [1956] experimentally determined the primary electron penetration mass thickness in Al_2O_3 is described by

$$\rho R = 0.0115 E_o^{1.35}, \tag{14.28}$$

where E_o is in keV and ρR is in units of mg cm^{-2}. From Fig. 4.23 it is seen that the *mass-thickness* range for any given monoenergetic electron is nearly the same for a wide range of materials ranging from air to gold. Thus, the penetration range for all dynode materials can be estimated by

$$R \simeq \frac{1.15 \times 10^{-5}}{\rho} E_o^{1.35}, \tag{14.29}$$

where R is in cm, ρ is the dynode material density (g cm^{-3}) and E_o is in units of keV.

Example 14.2: Calculate the electron range as a function of energy for Al_2O_3, MgO, BeO, and GaP for accelerating voltages between 0 to 5 keV.

Solution:

The densities of Al_2O_3, MgO, BeO, and GaP are 3.99, 3.58, 3.01, and 4.14 g cm^{-3}, respectively. From Eq. (14.29), the penetration depth for Al_2O_3 is,

$$R \simeq \frac{1.15 \times 10^{-5}}{3.99 \text{ g cm}^{-3}} E_o^{1.35},$$

yielding the results depicted in Fig. 14.19. The ranges for the other materials are found in the same manner. These results indicate that, for accelerating voltages less than 1000 volts, the penetration depth R is limited to less than 50 nm for these materials.

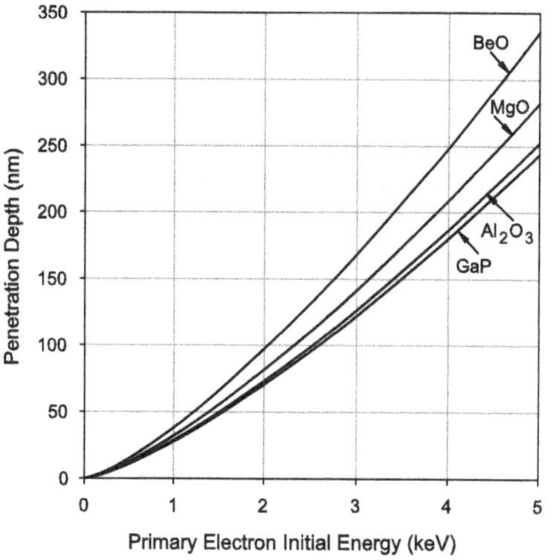

Figure 14.19. Calculated penetration depths for Al_2O_3, MgO, BeO, and GaP as a function of primary electron energy.

Models for Secondary Electron Emission

The secondary electron emission yield δ, defined as the ratio of emerging secondary electrons to impinging primary electrons, can be described by [Dekker 1958]

$$\delta = \int n(x, E_o) f_{esc}(x) dx. \tag{14.30}$$

Here $n(x, E_o)dx$ is the number of secondary electrons produced at depths in dx about x by a primary electron of initial energy E_o, and $f_{esc}(x)$ is the probability that a secondary electron produced at depth x reaches the vacuum interface and is emitted or escapes. The escape probability of secondary electrons, to a first approximation, may seem to be a simple solid angle problem defined by the diffusion length and excitation distance from the surface. However, the complex nature of electron scattering and transport causes the probability of emergence to be somewhat complicated [Simon and Williams, 1968].

Table 14.5. Maximum yields δ_{om} of secondary electrons and optimum initial energy E_{om} of the incident primary electron for some elements. From Dekker [1958].

Material	δ_m	E_{om} (eV)	Material	δ_m	E_{om} (eV)	Material	δ_m	E_{om} (eV)
Li	0.5	85	Be	0.5	200	B	1.2	150
Mg	0.95	300	Al	0.95	300	Si	1.1	250
K	0.7	200	Ti	0.9	280	Fe	1.3	400
Co	1.2	500	Ni	1.3	550	Cu	1.3	600
Ga	1.55	500	Ge	1.15	400	Rb	0.9	350
Zr	1.1	350	Nb	1.2	375	Mo	1.25	375
Pd	>1.3	>250	Ag	1.4	800	Cd	1.1	450
Sn	1.35	500	Sb	1.3	500	Cs	0.7	400
Ba	0.8	400	Ta	1.3	600	W	1.3	650
Pt	1.8	700	Au	1.4	800	Hg	1.3	600
Tl	1.7	650	Pb	1.1	500	Bi	1.15	550

For any element the yield δ of secondary electrons varies with the energy E_o of the incident primary electron. There is an optimum energy E_{om} that produces the maximum yield δ_m of secondary electrons for a particular element. Some optimum yields and energies are given in Table 14.5.

There are many models describing secondary electron emission yield. For example see Dekker [1958] and Adamson [1993]; however, only two of the more important models are described here, namely, the *power law* model and the *constant loss* model.

Power Law Model Baroody [1950] developed a theory for secondary electron yield based on the concepts that (1) the production of secondary electrons per unit pathlength of travel by the primary electron is proportional to the rate of energy loss $-dE/dx$, and (2) these secondary electrons are exponentially absorbed in the target material as they diffuse towards the surface. Hence, the secondary electron yield per incident primary electron is given by

$$\delta = -K \int_0^R \frac{dE}{dx} e^{-\alpha x} dx, \tag{14.31}$$

where K is a proportionality constant and α is the electron absorption coefficient. According to Whiddington's Law [Whiddington 1912], the energy loss per unit distance for the primary electron is inversely proportional to the energy at depth x from the surface, i.e.,

$$-\frac{dE(x)}{dx} = \frac{A}{E(x)}, \tag{14.32}$$

where A is a constant dependent on the dynode material. Rearrangement and integration yield

$$E(x) = \sqrt{E_o^2 - 2Ax}. \tag{14.33}$$

However, according to Lye and Dekker [1957], Whiddington's Law does not produce reliable results, and instead they proposed that Eq. (14.32) should be modified to

$$-\frac{dE(x)}{dx} = \frac{A}{E^n(x)}, \tag{14.34}$$

where again A is a constant dependent on the material. Rearrangement and integration then gives the energy of the primary electron at depth x as

$$E^{n+1}(x) = E_o^{n+1} - A(n+1)x. \tag{14.35}$$

Because $E^{n+1}(R) = 0$ setting $x = R$ in Eq. (14.35) gives the primary electron range as

$$R = \int_o^{E_o} \frac{E^n}{A} dE = \frac{E_o^{n+1}}{A(n+1)}. \tag{14.36}$$

Note that Eq. (14.36) with $n = 0.35$ agrees with the experimental result of Eq. (14.28).

With the assumption that the production of secondary electrons is proportional to $-dE/dx$, and that the function $f_{esc}(x)$ is proportional to $e^{-\alpha x}$, then Eq. (14.31) becomes

$$\delta = K \int_0^R -\frac{A}{E^n(x)} e^{-\alpha x} dx. \tag{14.37}$$

After much manipulation and change of variables, this equation can be written as

$$\delta = K \left[\frac{A(n+1)}{\alpha}\right]^{1/(n+1)} F_n(r), \tag{14.38}$$

where the function $F_n(r)$ is defined as

$$F_n(r) \equiv e^{-r^{(n+1)}} \int_0^r e^{y^{(n+1)}} dy, \tag{14.39}$$

and r is defined as

$$r^{n+1} = \alpha R = \frac{\alpha E_o^{n+1}}{A(n+1)}. \tag{14.40}$$

The maximum secondary electron yield δ_m is found by setting $d\delta/dE_o = 0$. The result is [Lye and Dekker 1957]

$$\delta_m = K \left[\frac{A(n+1)}{\alpha}\right]^{1/(n+1)} \frac{1}{(n+1)r_m^n}, \tag{14.41}$$

where r_m is the value of r at the maximum secondary electron yield. Lye and Dekker [1957], combined Eqs. (14.38) and (14.41) to obtain a normalized curve yield that is independent of the material properties, namely,

$$\frac{\delta}{\delta_m} = (n+1)r_m^n F_n(r) = \frac{F_n(r_m E_o/E_{om})}{F_n(r_m)}, \tag{14.42}$$

where E_{om} is the initial energy that maximizes the secondary electron yield, and $E_o/E_{om} = 1$ when $\delta/\delta_m = 1$.

Baroody [1950] evaluated Eq. (14.39) for $n = 1$ by using a series expansion to obtain

$$F_1(r) = e^{-r^2} \int_0^r e^{y^2} dy = \sum_{i=0}^{\infty} \frac{(-1)^i r^{2i+1}}{i!(2i+1)}, \tag{14.43}$$

although today with our PCs $F_n(r)$ is readily evaluated numerically. Baroody [1950] determined that $F_1(r)$ had a maximum value at $r_m = 0.92$, hence,

$$\frac{\delta}{\delta_m} = \frac{F_1(0.92E_o/E_{om})}{F_1(0.92)} = 1.85F_1(0.92E_o/E_{om}). \tag{14.44}$$

Baroody's treatment followed the general trend for secondary electron yield, but had large deviations from experimental data for $E_o/E_{om} > 1.1$, as can be seen from Fig. 14.20. Later, Lye and Dekker substituted $n = 0.35$ into Eq. (14.42), based on the findings of Young [1956], to yield a power law solution that agrees considerably better with experimental values than does Baroody's model. A comparison of these two power law models is shown in Fig. 14.20.

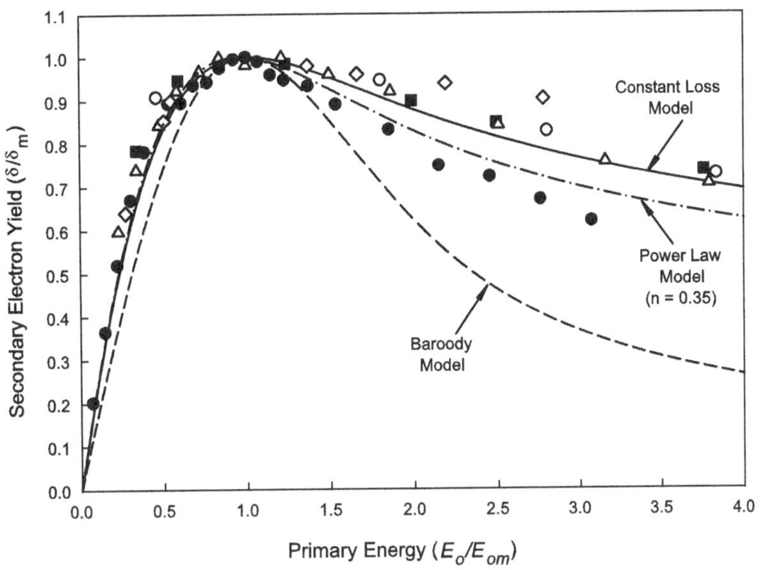

Figure 14.20. Comparison values of δ/δ_m for experimental data and theoretical models. Experimental data: △ Ge, ● MgO, ■ Mo, ◇ Pt, ○ Li; and are taken from from Kollath [1937], Baroody [1950], Whetten and Lamponsky [1957], Lye and Dekker [1957], Johnson and McKay [1954], and from Simon and Williams [1968].

Constant Loss Model The secondary production rate $n(x, E_o)$ in Eq. (14.37), according to Eq. (14.31), is given by

$$n(x, E_o) = -\frac{1}{w}\frac{dE}{dx}, \tag{14.45}$$

where w is the average absorbed energy needed to create an ion-electron pair, and the electron escape function $f_{esc}(x)$ in Eq. (14.37) is

$$f_{esc}(x) = B_1 B_2 e^{-\alpha x}, \tag{14.46}$$

where $1/\alpha$ is the mean-free-path length of secondary electrons or, equivalently, α is the probability, per unit differential path length of travel, a secondary electron is absorbed. B_1 is the probability that an electron diffuses in the direction of the vacuum boundary, and B_2 is the probability that upon reaching the vacuum boundary that the secondary electron still has sufficient energy to surmount the vacuum boundary, i.e., an energy above E_T. The term B_1 is assumed to have the value of approximately 0.5; however, when considering scattering and solid angle effects, B_2 should be a function of the depth at which the secondary electron is liberated. For simplicity, Simon and Williams [1968] treat B_2 as a function independent of x, thereby greatly reducing the complexity of calculating δ. For a secondary electron created at the surface $x = 0$, the product $B_1 B_2 = 0.5$. If it is assumed that energy loss of the primary electron is approximately constant over the primary electron range,[10] the change in energy with travel distance can be expressed as,

$$-\frac{dE}{dx} \simeq \frac{E_o}{R}. \tag{14.47}$$

[10]The assumption of a constant stopping power is valid for electrons with energies above 1 MeV, but not strictly true for lower energy electrons. In Sec. 4.6, a more rigorous treatment of electron ranges and stopping powers is presented.

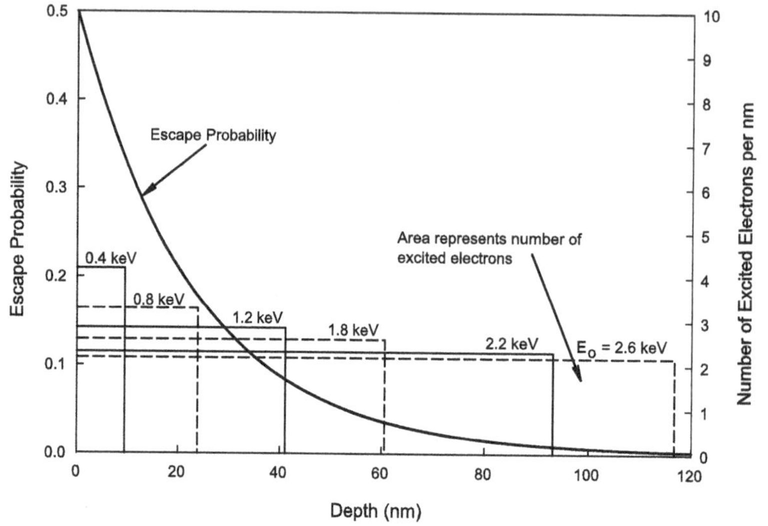

Figure 14.21. Escape probability and number of electrons liberated in MgO as a function of primary electron energy. After Simon and Williams [1968].

Substitution of Eqs. (14.45), (14.46), and (14.47) into Eq. (14.30) yields,

$$\delta = \int_0^R \frac{B_1 B_2}{w} \frac{E_o}{R} e^{-\alpha x} dx = \frac{B_1 B_2 E_o}{w \alpha R} \left(1 - e^{-\alpha R}\right). \tag{14.48}$$

Because it is assumed that there is a constant energy loss as the primary electron travels through the material, the total number of excited secondary electrons is E_o/w, and the number of electrons excited, per unit pathlength of the primary electron, is

$$-\frac{1}{w}\frac{dE}{dx} = \frac{E_o}{wR}. \tag{14.49}$$

In Fig. 14.21 $f_{esc}(x)$ and secondary electron production are shown for MgO.

From Eq. (14.36), the primary electron range can be written as $R = C_m E_o^{n+1}$ where C_m is a constant, so that Eq. (14.48) is rewritten as

$$\delta = K_m E_o \left[\frac{1 - e^{-\alpha C_m E_o^{n+1}}}{-\alpha C_m E_o^{n+1}}\right]. \tag{14.50}$$

where K_m is a combination of constants. There is some value E_o at which δ reaches a maximum value δ_m. This value can be found by setting $d\delta/dEo = 0$. Again the total gain of secondary electron emission, as a function of E_o, reaches a maximum value, beyond which δ begins to decrease.

To find a normalized dependence of δ on E_o, Lye and Dekker [1957] define a variable $z \equiv \alpha C E_o^{n+1}$ so, from Eq. (14.50) it is seen that δ is proportional to the function

$$f_n(z) \equiv \frac{1 - e^{(-z^{n+1})}}{z^n}. \tag{14.51}$$

Hence, the general normalized dependence of δ on E_o is given by

$$\frac{\delta}{\delta_m} = \frac{f_n(z_m E_o/E_{om})}{f_n(z_m)}. \tag{14.52}$$

Table 14.6. Secondary electron emission properties of some materials. From Dekker [1958].

Material	Density (g cm^{-3})	δ_m	E_{om} (keV)	$1/\alpha$ (nm)	w (eV)
Pt	21.45	1.8	0.7	2.0	80
Ge	5.35	1.15	0.4	3.5	71.5
MgO	3.58	24	1.2	23	10.25

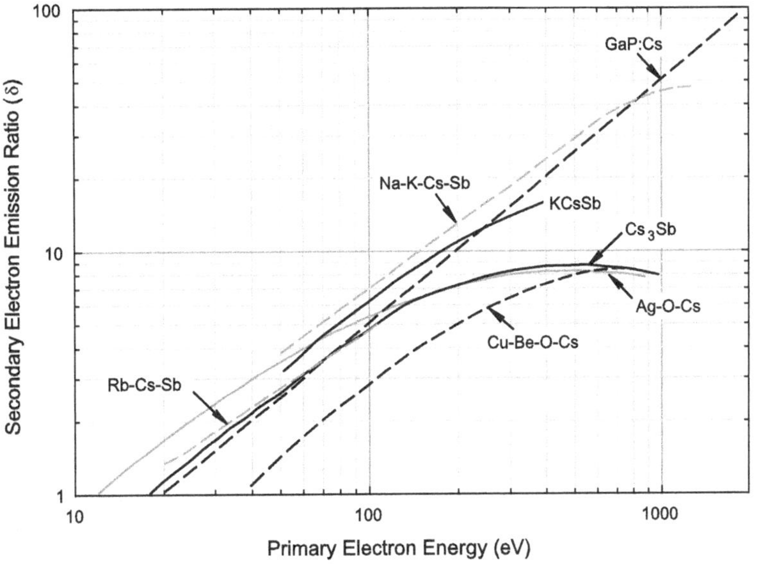

Figure 14.22. Secondary electron emission ratios for many materials used as dynodes in photomultiplier tubes. After [Engstrom 1980].

A plot of Eq. (14.52) with $n = 0.35$ is shown in Fig. 14.20. From the results in Fig. 14.20, it is seen that the constant loss model yields the best agreement with experiments, particularly when compared to the power law model and Baroody's analysis. Values of some model parameters for three materials are given in Table 14.6.

What important lessons are learned from the above analysis? Mainly, for any given dynode material, there is an optimum primary electron energy that yields the highest secondary electron yield. Further, the secondary electron emission yield of most materials can be described by a universal reduced yield curve, as shown in Fig. 14.20. The *yield ratio* is the number of secondary electrons emitted per incident electron. Experimentally measured yield ratios for many dynode materials are plotted in Fig. 14.22.

14.1.5 PMT Dynode Designs and Configurations

The shape and placement of the dynodes within the PMT determine the efficiency by which electrons are liberated and collected from each dynode stage. Calculated results can lead to efficient high-gain designs. It is interesting to note that, before the era of the modern electronic computers, such calculations were assisted with mechanical models, which included stretched rubber diaphragms and electrical resistor models [Engstrom 1980]. By arranging stretched rubber diaphragms between the "electrode" stage in the model,

Table 14.7. Properties of some common dynode designs. After Hamamatsu [2007].

Dynode design	Rise time response (ns)	Pulse linearity at 2% (mA)	Uniformity	Collection efficiency	Main characteristic
circular cage	0.9–3.0	1–10	poor	good	compact and high speed
box-and-grid	6–20	1–10	good	very good	high collection efficiency
venetian blind	6–18	10–40	good	poor	large diameter
linear focus	0.7–3	10–250	poor	good	high speed and high linearity
mesh	1.5–5.5	300–1000	good	poor	high magnetic immunity and high linearity
metal channel	0.65–1.5	30	good	good	compact and high speed
Micro-channel plate	0.1–0.3	700	good	poor	high speed

the height (potential energy) was adjusted to correspond to the electrical potential, and balls released upon the platform would follow the general path expected for electrons within the PMT. Hence, a simple dynode configuration could be designed with this tool. Electrically based models used a network of resistors in a grid pattern. By setting those resistors corresponding to a dynode plane in the model to a uniform potential, the electric potential between these planes could be directly measured between the electrodes.

Computer simulations of electron trajectories, voltage potentials, photocathode shape and focusing electrodes, dynode shapes and materials, are now routinely used for design of modern PMTs. Electrons emitted from the photocathode are focused onto a small area of the first dynode. Afterwards, secondary electrons emitted from the first dynode are directed to the next dynode, and so on until they reach the anode. The anode is specially designed to improve collection efficiency while suppressing space charge limitation effects. Most modern PMT designs have electron collection efficiencies η approaching 100%.

Designs for the Electron Optics

There are numerous photocathode and dynode geometries employed in PMTs, a few of which are described here. More details on these designs can be found elsewhere [Engstrom 1980; Flyckt and Marmonier 2002; Hamamatsu 2007]. A summary of some important properties of the different geometric designs, discussed below, is provided in Table 14.7.

Circular Cage Design The first successful commercial PMT was the 931A, developed by RCA and shown in Fig. 14.1, which used the circular cage design. This dynode design, shown in Fig. 14.23, is relatively compact and can be used for both the transmission or reflection type of PMTs. The original design had an opaque photocathode, depicted in Fig. 14.23(left), which was illuminated from the side. The 931A and 931B PMTs, both 9 stage circular cage dynode devices, are relatively small with diameters of 1.125 inches. Both provide high gains (931A $\simeq 5.5 \times 10^6$, 931B $\simeq 6 \times 10^6$), and can be used for scintillation samples that do not require coupling to a large photocathode. The envelope is composed of soda-lime glass. However, the photoresponses are different, mainly because the 931A has an Cs_3Sb photocathode and the 931B has a K_2CsSb photocathode. Typically, the 931B offers higher photocathode quantum efficiency, higher anode sensitivity, and lower dark current than the 931A. The compact sizes ensure a rapid time response or short

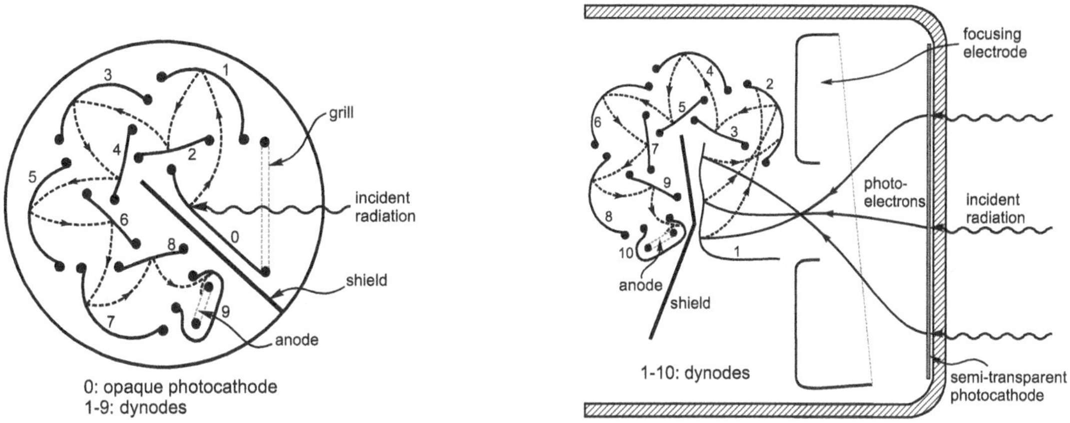

Figure 14.23. The circular cage dynode configuration showing (left) an opaque (reflection) photocathode arrangement and a (right) semi-transparent (transmission) photocathode arrangement. After [Engstrom 1980].

transit time of about 18 ns for the 931 tube series. The electron *transit time* is defined as the time interval between a photo-impulse impinging on the photocathode and the corresponding output pulse at the anode terminal to reach its peak amplitude. Advanced versions of the circular cage PMT have faster response times that range between 0.9 and 3.0 ns. A weakness in the design is the existence of a small electrostatic field at the apex of the grill and photocathode shown in Fig. 14.23(left). This field causes poor photoelectron collection efficiency in that region.

An alternative design of the circular cage dynode bank is shown in Fig. 14.23(right), in which the device has a much larger primary dynode and a semi-transparent photocathode. The flat semi-transparent photocathode allows a relatively large scintillation crystal to be directly coupled to the PMT. The design also has a focusing electrode, which, along with the large primary dynode, improves the photoelectron collection efficiency. Although this configuration of the PMT still has a rapid time response, the electron transit time is lengthened compared to the traditional circular cage dynode design because of the extra time required for electrons to traverse the distance between the photocathode and the primary dynode.

Venetian Blind Design The venetian blind design, see Fig. 14.24, offers excellent photoelectron collection and is ideally suited for "head-on" photomultiplier tubes because of its large photocathode diameter. The large photocathode permits good photon collection and is positioned relatively far from the first set of dynode blades. The photoelectrons are focused with steering electrodes onto the first set of dynode blades so as to achieve collection efficiency of nearly 100%, a feature necessary for good gamma-ray spectroscopy. Secondary electrons emitted from the dynode surfaces are steered into the next set of dynode vanes with accelerating grids. Consequently, some secondary electrons can be lost by collisions with these grids, thereby reducing overall secondary electron collection efficiency (ϵ_i) between dynode stages. Further, the potentials across the dynode paths are relatively low, thereby reducing the electron speeds and causing the transit time and response time of the PMT to lengthen. In summary, the venetian blind design enables PMTs to be coupled to relatively large scintillation crystals, offers high photoelectron collection efficiency, but has a relatively slow time response. Because of the slow response, this type of PMT now has somewhat limited use.

Box and Grid Design The box and grid design (Fig. 14.25), widely used in head-on photomultiplier tubes, has better photoelectron collection efficiency than most PMTs. This design incorporates a relatively large entrance window and photocathode. Accordingly, photomultiplier tubes using this dynode offer high detection efficiency and good uniformity. Similar to the venetian blind design, photoelectron collection is nearly

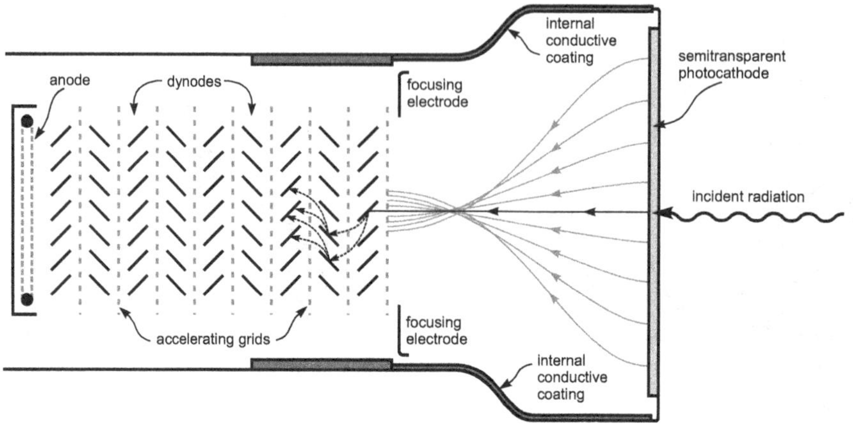

Figure 14.24. The venetian blind dynode configuration for a semi-transparent (transmission) photocathode. After [Engstrom 1980].

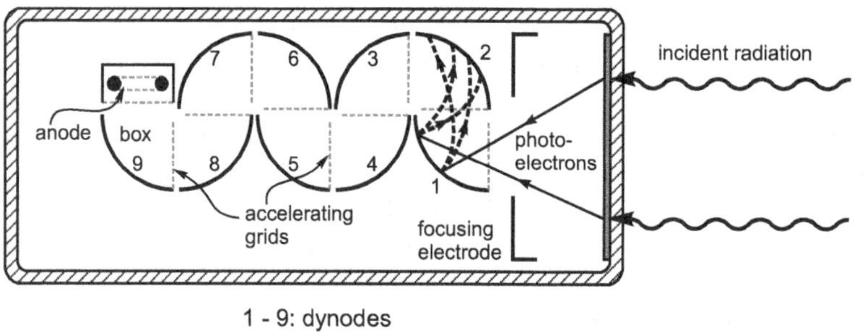

Figure 14.25. The box and grid dynode configuration for a semi-transparent (transmission) photocathode. After [Engstrom 1980].

100%, an attribute necessary for good gamma-ray spectroscopy. Each dynode has a grid electrode at the entrance, which assists with electron acceleration and steering. However, the weak electric fields between dynodes reduce electron speeds and, thus, produce a relatively slow time response. Regardless, this type of PMT still is used today as the light sensor for many scintillator/PMT detectors.

Linear Focus Design The linear focus, or in-line, dynode configuration is widely used in head-on photomultiplier tubes. Its attractive features include fast time response, good time resolution and excellent pulse linearity. A linear focus PMT attached to a NaI:Tl crystal is shown in Fig. 14.4. The linear focus dynode configuration is often employed with a curved photocathode as depicted in Fig. 14.26. The curved photocathode reduces the variance in electron pathlengths, as compared to a common flat photocathode. This smaller pathlength variance, in turn, reduces the variance in arrival time of electrons striking the first dynode. However, a curved photocathode makes it difficult to couple the PMT to scintillation crystals. As a consequence, linear focus dynode PMTs are also available with a flat photocathode surface. The dynodes aligned in the linear focus arrangement also reduces the variance in the arrival time of electrons as they eventually arrive at the anode. This compensation is accomplished with curved interposed dynodes designed to alternate between the longer and shorter electron pathlengths moving through weaker and stronger elec-

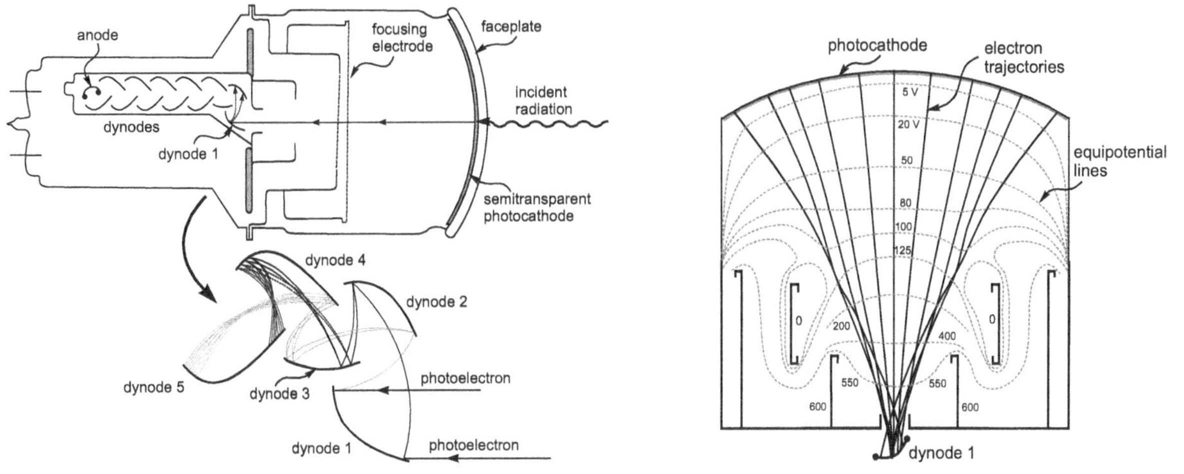

Figure 14.26. The linear focus dynode configuration showing (left) the dynode arrangement and (right) the focused electron trajectories from the curved photocathode. After [Engstrom 1980].

tric fields. Consequently, the total electron transit times for electrons passing through the dynode bank are nearly equal. An electron trajectory sequence is depicted in Fig. 14.26(bottom left). Overall the electron transit times and their variance are superior to most PMT designs.

Compact Dynode Designs A compact dynode configuration allows PMTs to be fabricated in relatively short containers. They typically have higher resistance to magnetic field interference, but they have lower gains. These compact PMTs are available as single channel PMTs or position sensitive PMTs. The dynodes most often used are described below.

Mesh Dynodes The mesh dynode is composed of interspersed parallel wire planes that form a system with the appearance of a wire screen; however, the wire planes are stacked, not interwoven. The wires may have a triangular cross section and are coated with secondary electron emission materials, as shown in Fig. 14.27(left). Each electrode layer is rotated 90° to the prior layer such that they are oriented in the x and y directions. Multiple stacked layers are used in close proximity to form a compact dynode array that can be used to provide x and y position information for imaging purposes. There are basically two types of design: a coarse mesh type and a fine mesh type. Both are excellent in output linearity and have good immunity to magnetic fields. Position sensing versions have either cross wire anodes or multianodes and can provide the location of photon interactions upon the photocathode. However, the devices generally have poor spatial resolution because of the spatial spread of the electron cascade, which is about 8 to 9 mm. Fine mesh types are developed primarily for photomultiplier tubes that are used in high magnetic fields. Some versions have magnetic immunity up to 1500 mT. The electric field and voltage drop between stages are low, hence the gain is typically lower than that of conventional PMTs. Gains on the order of 10^5 or more are achievable for systems with many stages (≥ 12).

Metal Foil Channel Metal foil channel dynodes are made with perforated foils coated with materials that produce secondary electrons. PMTs using these dynodes have properties similar to those of venetian blind PMTs, including the problem with electron collection. These dynodes can be used as position sensing PMTs with resolutions better than those of PMTs using mesh dynodes. With a resistive divider circuit, spatial resolution of approximately 4 to 5 mm can be achieved. The electric field and voltage drop between stages are low; hence, the gain is typically lower than conventional PMTs. Because the foils can be stacked in close proximity, metal foil channel PMTs can be fabricated in compact packages for special applications.

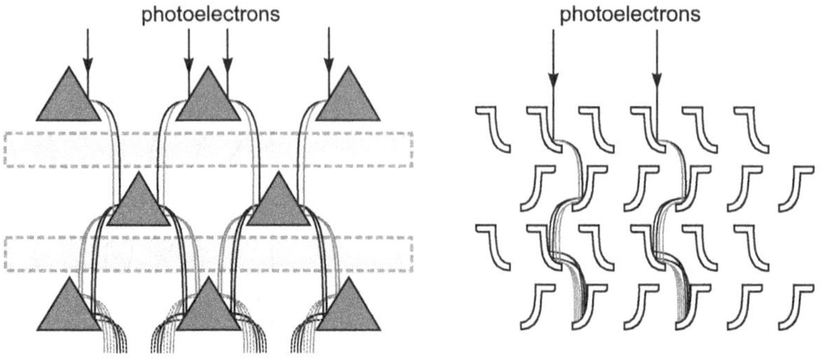

Figure 14.27. Compact photomultiplier dynode configurations, showing the basic scheme for (left) wire mesh dynodes and (right) metal foil channel dynodes.

Further, metal foil channel PMTs have higher magnetic immunity than most PMTs (up to fields of about 5 mT), except for mesh dynode PMTs. Gains on the order of 10^5 or more are achievable for systems with many stages (≥ 12).

Microchannel Plates Use of microchannel plates is another method for amplifying signals from a scintillator and are closely akin to the Kubetsky tube. Microchannel plates are bonded clusters of glass tubes, each usually having a diameter between about 15 to 50 μm, whose inside is coated with materials that emit secondary electrons as is shown in Fig. 14.28. The individual channels are referred to as "continuous channel electron multipliers" [Wiley and Hendee 1962]. A voltage is applied across the tube length which causes electrons to cascade down the tube. When an electron strikes the tube wall, more electrons are emitted, much as with dynodes in a PMT. Hence, a single electron can cause a cascade that can eventually produce more than 10^6 electrons that reach the other end of the tube. Typically, thousands of these continuous channel electron multipliers are bonded together to form a PMT.

At sufficiently high gains, the pulse can also become saturated, thereby compromising the pulse height distribution. In other words, regardless of the photon electron density striking the microchannel plate, all output pulses have nearly the same amplitude. A study conducted by Schmidt and Hendee [1966] indicates that space charge accumulation of electrons in the channel causes this saturation effect. Emission of energetic electrons can also cause sputtering as they strike the walls or anode, thereby liberating positive ions into the channel. These positive ions can travel in the opposite direction of the electrons and cause more electron emissions as they strike the walls, a process termed as "ion feedback." Consequently, the pulse rise time is extended by successive waves of secondary electrons, often over ten times the expected values. It was found that bending the continuous channel electron multipliers would remedy this feedback problem [Evans 1965], because electrons can drift towards the anode around the bend whereas ions cannot drift backwards before striking the wall with insufficient energy needed cause secondary electron emission. Modern microchannel plates used in PMTs are often arranged in a chevron configuration to eliminate ion feedback. Regardless, space charge effects can appear for gains of more than 10^7 and can still produce pulse height spectra saturation.

The microchannel plate can be substituted for the staged-dynode bank inside a vacuum tube to operate as a PMT [Chevalier et al. 1970]. Energetic photoelectrons entering can cause the ejection of secondary electrons, which cascade down the microchannels to liberate millions more electrons. The main advantage of a microchannel plate is its compact size. A microchannel plate only one-inch thick can produce a signal of similar strength as a conventional staged-dynode PMT. Microchannel plates have relatively fast rise times [Piétri 1975], typically between 100 and 300 ps, and a relatively small spread of transit times, usually less

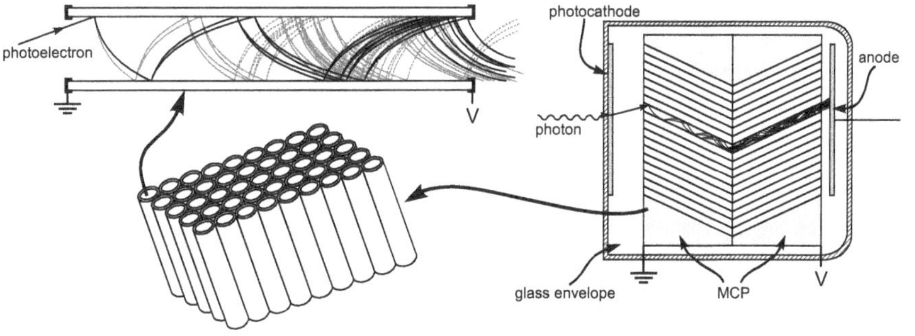

Figure 14.28. (left) Depiction of electron multiplication in a microchannel. (right) Depiction of a microchannel plate PMT.

than 100 ps. Further, microchannel plates have found use as image intensifiers, in which their compact and small channels provide excellent spatial resolution. The main problem with using microchannel plates is that the signal produced, per monoenergetic radiation event, is statistically much noisier than that produced by a PMT. Hence, spectroscopic energy resolution is usually worse than that produced by a multistage dynode PMT.

14.1.6 Gain

The *gain* G of a PMT is defined as the average total number of electrons collected at the anode N_a divided by the average total number N_{pc} of photoelectrons initially ejected from the photocathode, i.e.,

$$G = \frac{N_a}{N_{pc}} = \frac{q_e N_a}{q_e N_{pc}} = \frac{Q_a}{Q_{pc}}, \tag{14.53}$$

where q_e is the unit charge, Q_{pc} is the average charge emitted from the photocathode and Q_a is the average charge delivered to the PMT anode. The number of collected electrons N_a can be calculated from the dynode charge collection efficiency η and secondary emission factor δ for each dynode. The charge collection efficiency for the ith dynode located within the dynode chain is defined as

$$\eta_i = \frac{\text{no. electrons collected by dynode } i}{\text{no. electrons emitted from dynode } i-1}, \tag{14.54}$$

and the secondary emission factor for the same dynode is defined as

$$\delta_i = \frac{\text{no. electrons emitted by dynode } i}{\text{no. electrons collected by dynode } i}. \tag{14.55}$$

Suppose that η_1 is the charge collection efficiency of the first dynode for those electrons initially ejected from the photocathode N_{pc}, then the average number of electrons emitted by the first dynode N_1 is

$$N_1 = N_{pc}\eta_1\delta_1. \tag{14.56}$$

For a PMT with M number of dynodes, each with its own charge collection efficiency and secondary emission factor, the total number of electrons collected at the anode is

$$N_a = N_{pc}(\eta_1\delta_1)(\eta_2\delta_2)\ldots(\eta_{M-1}\delta_{M-1})(\eta_M\delta_M)\eta_a, \tag{14.57}$$

where η_a is the final charge collection efficiency of the anode. From Eq. (14.53), the average total gain is

$$G = \frac{N_a}{N_{pc}} = \eta_a \prod_{i=1}^{M} (\eta_i \delta_i). \tag{14.58}$$

The emission coefficient is strongly dependent upon the accelerating voltage, for it is the accelerating voltage that determines the energy of the electrons that strike the dynode. Further, the electron trajectories are also strongly dependent upon the applied voltage. In both cases, the overall dynode gain, $\eta_i \delta_i$, increases with interdynode voltage, and can be generally described by

$$G_i = \eta_i \delta_i = k_i V_i^{\alpha}, \tag{14.59}$$

where k_i is a proportionality constant and V_i is the voltage applied between dynode $i-1$ and i. Insertion of Eq. (14.59) into Eq. (14.58) gives the PMT gain as

$$G = \frac{Q_a}{Q_{pc}} = \eta_a \prod_{i=1}^{M} (k_i V_i^{\alpha}). \tag{14.60}$$

Values of α are typically reported between 0.6 and 0.8. Equation (14.60) is often simplified with the assumption that η_i and η_a are unity, i.e., 100% collection efficiency, and that δ_i and the interdynode voltages V_i are identical for all dynodes. With these assumptions Eq. (14.60) reduces to

$$G = \frac{N_a}{N_{pc}} = \delta^M = (k V_d^{\alpha})^M. \tag{14.61}$$

where V_d is the voltage applied to each dynode. Note that the assumptions used to derive Eq. (14.61) are not applicable to all PMTs. For instance, the interdynode voltage is often divided such that the highest possible η and optimum δ are achieved and this voltage is not always identical for each interdynode spacing.

Example 14.3: Given a 10-stage PMT where all $\eta_i = 0.97$, $\eta_a = 0.98$, $\delta_i = 3.5$ for the first 5 stages, and $\delta_i = 4.2$ for the last 5 stages, determine average gain for a single photoelectron.

Solution:

The gain is calculated from Eq. (14.60) as

$$G = \eta_a \prod_{i=1}^{M} (\eta_i \delta_i) = (\eta_a)(\eta_i)^{10} \left[\prod_{i=1}^{5} \delta_i\right] \left[\prod_{i=6}^{10} \delta_i\right] = (0.98)(0.97)^{10} (3.5)^5 (4.2)^5 = 4.96 \times 10^5.$$

14.1.7 Factors Affecting the Performance of a PMT

The performance of a PMT is affected by properties such as tube size, the glass enclosure, response speed, transit time spread (TTS), dark current, optimum wavelength, operating voltage, number of stages, and gain. These properties are usually reported by commercial manufacturers. However, there are other factors that affect the performance of PMTs such as dynode glow, space charge accumulation, and environmental effects. Summarized in the following subsections are descriptions of these other properties.

Commercial Specifications

Depending on the intended PMT usage, an operator must select the PMT that works best with the scintillator of choice. For instance, the sensitivity of the PMT must match well to the emission spectrum of the scintillator and the timing of the response speed must also match the decay constant of the scintillator. If the PMT has significant wavelength mismatch, then the overall gain and energy resolution suffer, whereas space charge build-up and spectral crowding occur for slow PMTs matched to high-speed scintillators. The reader is directed to Table 14.8 for a list of specifications for some PMTs used in radiation detection.

Dark Current and Noise

The dark current of a PMT is the current that flows in the cathode or anode circuits in the absence of light. The dark current flow of electrons is a source of random electromagnetic fields that add to electronic noise. The current that flows to the anode in total darkness has three main origins, namely (a) leakage current on the PMT surfaces, (b) intermittent emission of electrons, and (c) regenerative effects. The causes of these dark current contributions vary greatly with the operating conditions and also the prior storage conditions of the PMT. The dark current is dominated by leakage current under low voltage conditions, by thermionic emission at typical operating voltages, and by regenerative effects at higher voltages. These components are discussed in more detail below.

Ohmic Leakage Leakage current, sometimes referred to as "ohmic leakage," refers to current flow across the glass envelope and insulators when the PMT has voltage applied to it. Under dark conditions, it can be recognized by its nearly linear increase as the applied voltage is increased linearly. Although the leakage current is usually small, there may be residual conductive material remaining in the PMT from the photocathode and dynode manufacturing process. This residual material becomes an intrinsic property of the device. A controllable source of leakage current is surface contamination, which may be a result of water vapor, dirt, grease, or other foreign material that provides conduction paths. Thorough cleansing of the PMT with isopropyl alcohol (2-propanol) can eliminate such contaminants and reduces this portion of the leakage current. In humid environments, high voltage may cause arcing between electrodes. Precautions may be taken to insulate the base connectors with moisture resistant coatings. Most commercial scintillation detectors come as a single hermetic sealed unit, as a detector and PMT (see Fig. 14.3), thereby eliminating the arcing problem. Overall, the ohmic leakage current dominates the dark current at low voltages and also at low temperatures.

Thermionic Emission A thermally dependent source of sporadic electrons is thermionic emission of electrons from the surfaces of the photocathode and dynodes. The current density from thermionic emission, without gain, is described by Eq. (14.18). From Eq. (14.18), it is found that lowering the work function of photocathode materials increases thermionic emission. Hence, photocathode materials designed for low energy (extended red) photon sensitivity have a higher thermionic emission current. At low voltage, the thermionic emission current is negligible; however, once the voltage is sufficiently high to amplify the current through secondary electron emission, the thermionic current contribution becomes important. As the voltage is increased, thermionic emissions eventually dominate the dark current. Electrons emitted from the photocathode experience the entire gain described by Eq. (14.61), whereas electrons emitted from any one of the dynodes have the gain reduced by the number of dynodes skipped in the chain. Also, the photocathode surface is usually much larger than the dynode surfaces. As a result, most of the overall thermionic dark current comes from the photocathode emissions. Because the overall current increase is a function of the PMT gain, the thermionic current appears to also increase in proportion to the PMT gain. Thermionic emission from common photocathodes may range between 10^{-19} to 10^{-15} amps cm^{-2} over the maximum sensitivity range between 300 and 500 nm. For extended red photocathodes, such as the AgOCs S-1 response (see Table 14.2 and Fig. 14.15), thermionic emission can exceed 10^{-13} amp cm^{-2}. Finally, because thermionic

Table 14.8. Typical properties of some semitransparent PMTs recommended for scintillation detector spectroscopy. Data are from ADIT 2016; ET 2016; Hamamatsu 2016.

A	B	D	E	F	G	H	I	J	K	L	M	N	O	P	Q
A	L13D21H	KCs	13	B	1.49	~370	NR	0.8	800	1100	< 1.0	L-10	80	NR	NR
A	L13B03W	KCs	13	S	1.523	~385	25	1	NR	1100	< 1.0	L-10	80	NR	NR
A	B29B01W	KCs	28.6	S	1.523	~385	25	1	700	1500	0.05	B-11	80	15	NR
A	B29B10W	KCs	28.6	S	1.523	~385	25	1	700	1500	0.05	B-11	80	15	NR
A	B38D01W	KCs	38.1	S	1.523	~385	25	1	NR	1500	10	B-10	80	15	NR
A	B51D01W	KCs	50.8	S	1.523	~385	25	1	1100	1500	10	B-10	80	15	NR
A	B51D03W	KCs	50.8	S	1.523	~385	25	1	1100	1500	10	B-10	97	15	NR
A	B76D01W	KCs	76.2	S	1.523	~385	25	1	1100	1500	10	B-10	97	~20	NR
A	B89D01W	KCs	88.9	S	1.523	~385	25	1	1100	1500	10	B-10	97	~20	NR
A	B133D01W	KCs	133.4	S	1.523	~385	25	1	1100	1500	10	B-10	97	~20	NR
E	9106B	RbCs	29	B	1.49	380	28	0.01	500	1200	0.02	L-7	NR	4.5	26
E	9112B	RbCs	25	B	1.49	380	25	0.6	1000	1800	0.02	C-10	NR	1.8	20
E	9266B	KCs	51	B	1.49	380	30	0.6	850	1100	0.3	L-10	NR	4	40
E	9269B	RbCs	51	B	1.49	380	26	0.5	1100	1400	1.0	L-9	NR	7.5	42
E	9490B	RbCs	130	B	1.49	380	26	0.6	1100	1600	2.0	L-10	NR	13	60
E	9807B	KCs	51	B	1.49	370	30	7.0	1650	2300	3.0	L-12	NR	2	41
E	9813B	KCs	51	B	1.49	370	30	70.0	2100	2500	10.0	L-14	NR	2	46
E	9845KB	KCs	38	B	1.49	380	28	0.8	1000	1800	0.1	C-10	NR	5	35
E	9900B	RbCs	29	B	1.49	370	26	2.5	800	1100	0.3	B-11	NR	15	85
E	9903KB	RbCs	38	B	1.49	380	28	0.2	1000	1100	0.1	L-10	NR	3.5	30
E	9924B	RbCs	29	B	1.49	380	26	2	900	2000	0.2	B-11	NR	15	85
H	R329-02	B	51	B	NR	420	NR	1.1	1500	2700	6	L-12	85	2.6	48
H	R647	B	13	B	NR	420	NR	1.0	1000	1250	1.0	L-10	80	2.1	22
H	R877	B	127	B	NR	420	NR	0.42	1250	1500	10.0	B-10	88	20	115
H	R877-01	B	127	K	NR	420	NR	0.42	1250	1500	10.0	B-10	88	20	115
H	R878	B	51	B	NR	420	NR	1.0	1250	1500	5.0	B-10	90	7	70
H	R1306	B	51	B	NR	420	NR	0.27	1000	1500	2.0	B-8	95	7	60
H	R1307	B	76	B	NR	420	NR	0.27	1000	1500	2.0	B-8	95	8	64
H	R1450	B	19	B	NR	420	NR	1.7	1500	1800	3.0	L-10	88	3	19
H	R1635	B	10	B	NR	420	NR	1.0	1250	1500	1.0	L-8	80	0.8	9
H	R1924A	B	25	B	NR	420	NR	2.0	1000	1250	3	C/L-10	85	1.5	17
H	R2059	B	51	Q	NR	420	NR	20	2500	3000	50	L-12	85	1.3	28
H	R6091	B	76	B	NR	420	NR	5.0	1500	2500	10	L-12	85	2.6	48
H	R6234	B	H 55	B	NR	420	NR	0.27	1000	1500	2	B/L-8	95	1.3	52

A: Manufacturer* (A: ADIT, E: ET Enterprises, H: Hamamatsu).
Burle and Photonis no longer make PMTs. ADIT manufactures portions of their former line.
B: Model
D: Photocathode material (B: bialkali, M: multialkai, otherwise by response or chemical name)
E: Window size in mm (H: hexagonal, otherwise round)
F: Window glass (B: borosilicate, K: low ^{40}K borosilicate, Q; quartz, S: Schott B270)
G: Index of refraction
H: Wavelength of maximum response (nm)
I: Quantum efficiency at optimum wavelength (%)
J: Gain at recommended operating voltage (in units of 10^{-6}, i.e., $J \times 10^6$)
K: Recommended operating voltage (V)
L: Maximum operating voltage (V)
M: Typical anode dark current (nA)
N: Dynode type and number of stages (B: box & grid, C: circular cage, L: linear)
O: Cathode radiant sensitivity at optimum wavelength (mA/W)
P: Anode rise time (ns)
Q: Transit time (ns)
NR = not reported

emission is also a function of temperature, the thermionic effect can be reduced by cooling the PMT during operation.

Background Radiation Another contribution to dark current is due to background radiation from external and internal sources. When cosmic rays, primarily protons and alpha particles, interact with the Earth's atmosphere, they can produce a shower of secondary ionizing particles, composed mainly of neutrons, pions, positrons, and muons. Intersection of these particles with the PMT glass envelope can produce Čerenkov radiation, which can subsequently cause photoelectron emission from either the photocathode or the dynodes. Because significant energy is released from by Čerenkov radiation interactions, a pulse from such an event can be significant. Čerenkov radiation is emitted in a conical shape along the direction of the particle trajectory; hence dark current from Čerenkov radiation events is higher for a PMT pointing upwards rather than downwards.

Radioactive contaminants in the PMT glass envelope (mainly ^{40}K) can also cause electron emission from the photocathode and dynodes. Some commercial vendors offer PMTs with low concentrations of potassium in the glass envelope by using fused silica, for example, to reduce the intrinsic radiation background. Finally, radiation emitted by building walls and floors, again usually from ^{40}K and perhaps from other radioactive sources in the vicinity, can add to spurious counts and dark current. Usually direct interaction of background radiations with the PMT structure are rare and, thereby of little consequence. Background suppression becomes important for situations requiring low background counting, and perhaps for liquid scintillation counting in which the scintillation light output can be relatively small.

Exposure to Ambient Light Exposure of a PMT to normal lighting conditions while under operation can cause catastrophic damage to the PMT and, consequently, should always be avoided. However, exposure to typical lighting conditions, even while the PMT is not under bias, can cause an extended, but temporary, increase in anode dark current. This increase is a consequence of exciting the photocathode, glass envelope, and internal parts of the PMT, which may phosphoresce over an extended period of time. The intensity and length of the exposure dark current is a function of the exposure wavelength and exposure duration. Dark current produced by ambient light exposure decays somewhat exponentially with time. Exposure to short wavelength photons can cause exposure dark currents over two orders of magnitude higher than exposure to longer wavelength photons. Even a short burst from an intense source of UV photons can cause appreciable temporary dark current. PMTs stored in sunlight may need several days in complete darkness before the dark current is reduced to the usual unexposed condition. Overall, unused PMTs should always be stored in total darkness and never subjected to intense light, even while not in use.

Regenerative Effects Regenerative effects normally appear at potentials exceeding the recommended operating voltage of the PMT. These include field emission, afterpulsing, dynode glow, glass charging, and helium penetration. Operation with voltages in which regenerative effects appear can permanently damage a PMT.

Field emission occurs at irregularities and pinnacles formed on the photocathode and dynode surfaces. A strong electric field is formed at these points, thereby promoting the emission of electrons. The theory for this emission was first addressed by Fowler and Nordheim [1928]. Later this theory was modified to include the Schottky barrier effect and introduced two approximations to the Fowler-Nordheim model to make the emission model more numerically tractable [Spindt et al. 1976]. In this revised model, the current density J (A/cm^2) of emitted electrons is given by

$$J \simeq \frac{A\mathcal{E}^2}{\phi} \exp\left[-B\frac{\phi^{3/2}}{\mathcal{E}}v(y)\right], \tag{14.62}$$

where $A = 1.4 \times 10^{-6}$ and $B = 6.87 \times 10^7$ are constants, ϕ is the work function in eV, $\mathcal{E}$ is the electric field in V/cm, and y is the Schottky lowering of the work function barrier. Finally, the function $v(y)$ is approximated by $v(y) \simeq 0.95 - y^2$. Note that the applied electric field has a substantial effect on field emission, with a combined exponential and $\mathcal{E}^2$ increase. Field emitted electrons can fluoresce the surfaces inside the PMT, causing emissions of photons that may reach the photocathode. Field emission can also cause permanent damage and significantly shorten the life of a PMT.

Afterpulsing is the production of minor spurious pulses following a main event. There are two main sources of afterpulsing, luminous and ionic [Morton et al. 1967; Krall 1967]. Luminous afterpulses are produced by energetic electrons causing the anode or dynodes to fluoresce. Fluoresence photons can travel back to the photocathode and cause a subsequent event, usually lower in magnitude than the main event. Luminous afterpulses are delayed by the sum of the electron and fluorescent light transit times through the PMT, typically between 20 and 100 ns.

Ionic afterpulsing is caused by contamination gases in the tube becoming ionized and subsequently striking the photocathode. The sources of the contamination gases may be outgasing of components, desorption from the PMT walls, and residual gases remaining after the evacuation process. Energetic electrons drifting through the PMT can cause ionization of these trace gases, thereby producing positive ions. These positive ions, having much lower mobility than electrons, drift much more slowly than do electrons and, hence, strike the photocathode with a delay between 200 ns to over 1 μs. When these ions strike the photocathode, they can release numerous secondary electrons, which in turn produce another electronic avalanche and pulse. The magnitude of an ionic afterpulse is a function of the positive ion kinetic energy; hence the ionic afterpulse magnitude is strongly dependent upon the applied voltage. Troubling gas ions include N_2^+ and H_2^+, and gas sources may include N_2 from residual air and detachment of H_2 from water moisture. Under some conditions, CH_4^+ and He^+ may be present, also causing afterpulsing. Modern manufacturing procedures are set in place to reduce the concentration of these residual gases. For ordinary scintillation counting and spectroscopy, afterpulsing is usually of little consequence. However, for measurements requiring critical timing information, afterpulses can prove to be an annoyance. Techniques to suppress afterpulses are described in the literature [Lamaze et al. 1975; Mirzoyan et al. 1997].

If the PMT is operated or stored in an environment with high He concentrations, it is possible for the He to diffuse through the glass or connecting regions into the PMT enclosure. Because He is a non-reactive inert gas, He does not attach to the PMT structure, hence is not gettered. As He concentrations build, it may become a problem by producing afterpulsing when ionized. Relatively high concentrations of He^+ can cause electrical breakdown, and ultimately, render the PMT useless.

At high enough operating voltages, electron bombardment upon the dynodes can cause the emission of light, recognized as a blue glow, and occurs particularly in the latter stages of the dynode chain. Photons from this dynode glow can scatter back to the photocathode, thereby becoming an additional source for noisy current [Krall 1967]. This effect is largely suppressed by including shields in the PMT that work to block light from reaching the photocathode.

Acceleration of electrons, or ions, into the PMT glass envelope can produce the photoemission of secondary electrons, an effect referred to as *glass charging*. This particular problem occurs if the glass envelope has a positive potential that competes with that of the anode and can arise from external voltage sources. Installation of shielding material around the PMT reduces the problem of glass charging. The conductive shield should be grounded to the same potential as the cathode; otherwise, should the shield maintain a positive potential, there is a risk of permanently damaging the PMT [Engstrom 1980].

Timing

The timing characteristics reported for PMTs are the *transit time*, *rise time*, *fall time* (or decay time), and *statistical spread*. The transit time is the time between a delta-function light pulse striking the photocathode and when the output current from the PMT reaches peak amplitude [Frank 1977], as shown in Fig. 14.29.

The transit time has many components, including the diffusion time required for electrons excited in the photocathode material to reach the vacuum interface, the time for electrons to transverse the space between the photocathode to the first dynode, and the multiple transverse times between dynodes. Emission times for common alkali metal emissive materials is about 10 ps. However, the diffusion times of NEA materials, because of the longer electron mean free paths, may extend beyond 100 ps. Yet, these emission times have little consequence compared to collection times between dynodes. The longest transit time component is usually the time required for electrons to travel from the photocathode to the first dynode. This is especially true for configurations with large photocathode surfaces because large photocathodes must be placed further away from the dynodes to allow efficient focusing of electrons.

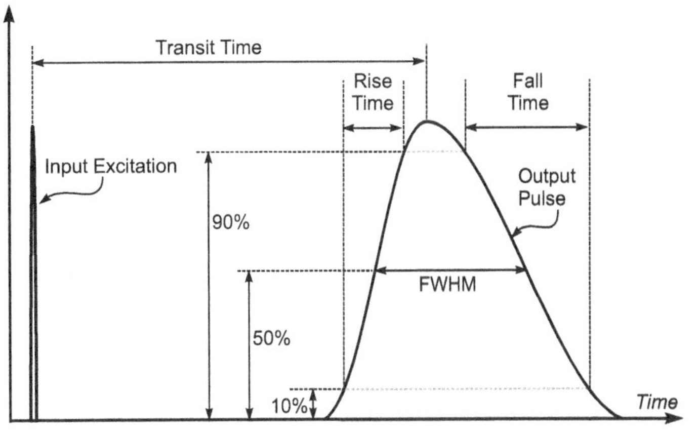

Figure 14.29. The different time constants used to characterize the output pulse from a PMT produced by a delta-function input.

The rise time of the PMT is mean time required for the output pulse to increase from 10% maximum to 90% maximum. It is the charge density and the charge collection time measured between the last dynode and the anode that determines the rise time, mainly because the anode output current is induced almost entirely from charges drifting between these last two electrodes (see Sec. 8.5 on weighting potentials). Rise times for common PMTs range between 1 and 20 ns. The PMT rise time is typically smaller than the fall time, and provides higher accuracy for timing measurements. The fall time of the PMT is the mean time required for the trailing edge of the output pulse to fall from 90% maximum to 10% maximum.

The statistical time spread is the full width at half maximum (FWHM) of the PMT output pulse. Contributions to statistical spread time include the time spread of the scintillation light source, time spread in electron trajectories from the photocathode to the first dynode, and time spread in the multiplication process as electrons move through the dynode chain. The light source time spread is a function of the decay time for the scintillation process, a physical process that cannot be altered for a given scintillation material. The time spread for electrons traveling between the photocathode and the first dynode can be minimized by including a curved inner photocathode, thereby providing nearly equidistant electron trajectories. The time spread from secondary electron emission between dynodes is complicated by the differences in electron emission speeds and directional trajectories. Overall, the timing resolution is determined mainly by the statistical spread in the secondary electron emissions.

The speed of an electron reaching a dynode is directly proportional to the applied electric field or voltage between the adjacent dynodes. Hence, the transit time is inversely proportional to voltage. Consequently, the time spread of electrons between dynodes is dependent on $1/V_i$ where V_i is the voltage between dynodes $i-1$ and i. If it is assumed that the speeds of electrons reaching a dynode have a Gaussian distribution, the standard deviation in speeds is proportional to $1/\sqrt{V_i}$. Hence, the broadening of the output pulse decreases with increasing voltage, and the timing characteristics of a PMT are usually maximized at the highest recommended operating voltage.

Fluctuations in the transit time between pulses, referred to as *jitter*, determines the accuracy limit for timing measurements. The variance in the transit time for a delta-function light input is described by

$$\sigma_{tt}^2 = \sigma_1^2 + \sigma_2^2, \tag{14.63}$$

where σ_1^2 is the variance for the cathode-to-first-dynode collection time, and σ_2^2 is the time variance of the combined collection times for the remaining interdynode spacings. The variance σ_1^2 has several contributions, mainly, absorption depth variance, emission location/trajectory variance, and chromatic variance. The variance in absorption depth affects the variance in the energies of the emitted electrons. Electrons that must diffuse far to the vacuum interface may have barely enough energy to surmount the energy barrier, whereas electrons excited at the vacuum interface may be emitted with the total residual energy $\phi - h\nu$. These electrons are emitted into the vacuum envelope in various directions, some directly towards the first dynode and others at varying angles away from the first dynode. Additionally, the emergence point on the photocathode may be anywhere on its surface. All together, the different diffusion times for the photoelectrons to reach the surface of the cathode and the different emergent locations and directions determine the variance in the time required for an electron leaving the photocathode surface to reach the first dynode. Wavelength variance is of little consequence for monochromatic light; however, scintillators emit light over a spectrum of energies, thereby adding to the variance in photoelectron energies.

Suppose a single electron is released from the photocathode. If the interdynode characteristics are assumed identical, then the variance in collection time between the dynodes as a function of transport between dynodes is [Gatti and Svelto 1966],

$$\sigma_d^2 = \frac{\sigma_{t1}^2}{\delta} + \frac{\sigma_{t2}^2}{\delta^2} + \frac{\sigma_{t3}^2}{\delta^3} + \ldots + \frac{\sigma_{tM}^2}{\delta^M}, \tag{14.64}$$

where σ_{ti}^2 is the variance in flight times from dynode $i-1$ to dynode i and the last term is the contribution to the flight time from the last dynode to the anode. Note that the last term is generally negligible, mainly because of the large value of δ^M and that charge induction only initiates as electrons travel towards the anode from the last dynode. If, in addition to the assumption that $\delta_i = \delta$ for all dynodes, one also assumes $\sigma_{ti} = \sigma_t$ for all dynodes, then for large values of M, Eq. (14.64) reduces to,

$$\sigma_d^2 \simeq \frac{\sigma_t^2}{\delta - 1}. \tag{14.65}$$

If the variance in gain is also included, then Gatti and Svelto [1966] show that the collection time variance for the dynode bank is,

$$\sigma_2^2 \simeq \frac{\sigma_t^2}{\delta - 1}\left(1 + \frac{1}{\delta - 1}\right) = \frac{\sigma_t^2 \delta}{(\delta - 1)^2}. \tag{14.66}$$

The total variance in the transit time, given by Eq. (14.63), is

$$\sigma_{tt}^2 = \sigma_1^2 + \frac{\sigma_t^2 \delta}{(\delta - 1)^2}. \tag{14.67}$$

Substitution of Eq. (14.59) into Eq. (14.67) yields,

$$\sigma_{tt}^2 = \sigma_1^2 + \frac{\sigma_t^2 k V^\alpha}{(kV^\alpha - 1)^2}, \tag{14.68}$$

which for large voltages reduces to

$$\sigma_{tt}^2 \simeq \sigma_1^2 + \frac{\sigma_t^2}{kV^\alpha}. \tag{14.69}$$

This result clearly shows that an increase in the operating voltage decreases the fluctuations in the transit time.

For typical scintillation spectroscopy with relatively slow inorganic scintillators, such as NaI:Tl, most PMTs have sufficiently fast responses that the time response of the photomultiplier is inconsequential compared to the decay time of the scintillator. However, with recent advances in producing scintillators with fast decay times, the transit and rise times of the PMT become important. Hence careful selection of a fast PMT is important for these short decay-time scintillators. In the case that a slow PMT is used with a relatively fast inorganic scintillator, such as $LaBr_3$, the high density of charge in the PMT can cause serious gain distortions and non-linear behavior, thereby distorting the resulting pulse height spectrum.

Photocathode Non-uniformities

The photocathode coating usually has non-uniformities across the surface that produce a variation in light response across the photocathode. For radiation spectroscopy, scintillation light striking different regions of the photocathode can produce a varied response, thereby degrading energy resolution. For PMTs with large photocathode surfaces, usually used for radiation detection applications, the light response decreases abruptly near the photocathode periphery. Scanning measurements across the surface of PMTs indicate that the variance is relatively small for well-processed photocathodes ($\pm$ 3%) [Engstrom 1980]. This variation in response is usually least at the optimum peak wavelength, but can be much larger and erratic as the photon wavelength is increased from the optimum wavelength.

There are still two principal reasons for the non-uniform photon response: (1) variations in the photocathode layer thickness produced by the physical vapor deposition used to form the coating, and (2) the design of the electron optics used to guide the photoelectrons to the first dynode. Light response uniformity is improved by producing a diffusive layer at the glass/photocathode interface, accomplished by sandblasting the inner surface of the glass before the photocathode material is applied. Light is scattered over a wider area, thereby decoupling the point of intersection from the overall photoresponse. Other methods to decrease the effects of photocathode non-uniformity include the installation of a permanent ring magnet or solenoid around the photocathode, thereby defocusing electron emissions around the periphery and nullifying their contributions. Also, placing a stopped-down aperture between the scintillator and photocathode reduces the sensitive photocathode region that contributes to the photocurrent and, thereby reduces effects from photocathode non-uniformities.

14.1.8 Ancillary Equipment

Support equipment needed to operate a PMT properly include the tube base, output electronics, and shielding. This ancillary equipment is briefly discussed in this section. Photomultiplier specification sheets typically provide a recommended socket and voltage divider configuration that is dependent on the PMT application. The socket serves as an electrical connector to the PMT dynode pins. Although separate sources can be used to provide voltage to each PMT dynode stage, such a configuration is somewhat impracticable. Instead, the applied high voltage is usually divided amongst the focusing electrodes, the anode, and the dynodes.

Sockets and Voltage Dividers

PMTs have various electrical pin configurations; hence, matching the PMT to the proper socket is, of course, mandatory. These sockets can be acquired with or without mounting flanges, depending on the intended application. For instance, the mounting flange allows the PMT to be installed on a stand-alone base, whereas sockets without the mounting flange must be connected directly to a circuit board. The preferred voltage divider can be connected directly to the socket and changed according to application.

Voltage Divider There are two basic types of voltage dividers, namely, resistive and active types. The resistive dividers consist of a series of resistors that divide the voltage between the electrical parts of the PMT. Of these, there are three main types of resistive divider: the uniform type, the progressive (or tapered) type, and the intermediate type. After the initial region between the photocathode and the first dynode,

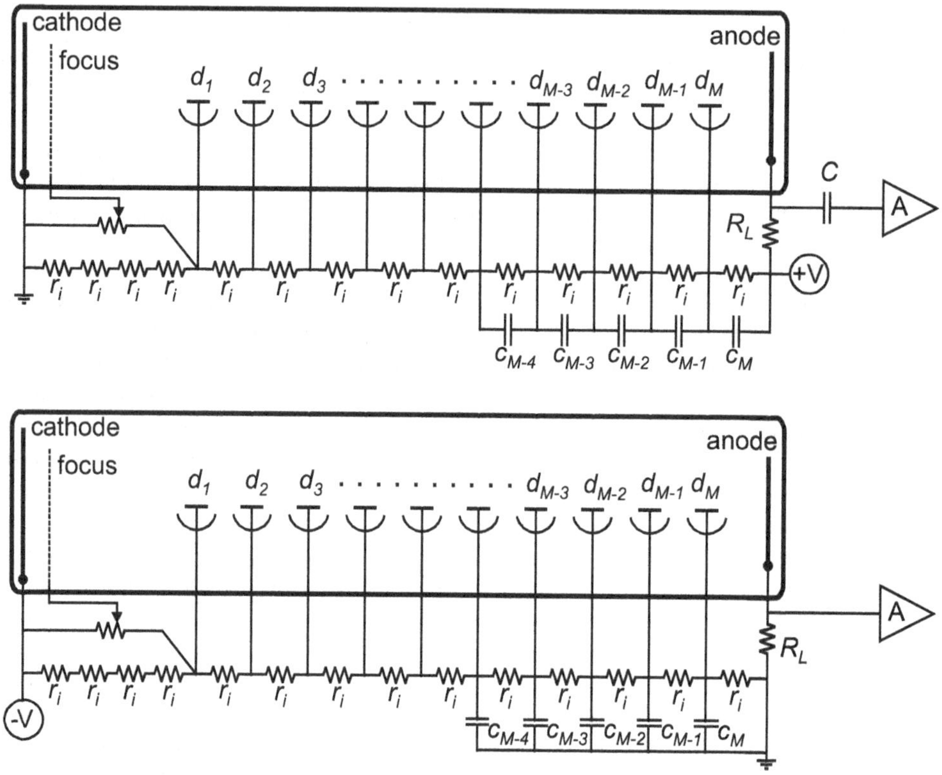

Figure 14.30. Voltage divider configurations with (top) positive high voltage applied to the anode, grounded cathode, and with series decoupling capacitors, and (bottom) negative high voltage applied to the cathode, grounded anode, and with parallel decoupling capacitors.

the uniform divider distributes the applied voltage equally over the dynodes stages. The uniform divider yields the highest gain of the three types of dividers, and is the most frequently used configuration for radiation measurements. The progressive divider increases the applied dynode voltage in the direction of the anode. The progressive divider helps overcome space charge effects and non-linear behavior for high current; however, the gain is typically lower than that of uniform dividers. Finally, the intermediate divider applies the same voltage to the first several dynodes stages, but then provides progressively larger voltages to the latter dynode stages. Intermediate dividers are used for fast response PMTs.

An active divider consists of a string of high impedance resistors connected in parallel with a string of field effect transistors [Kerns 1977; ET 2011]. These transistors help regulate the voltage between the dynodes to keep the potential drop constant. An active voltage divider ensures constant gain up to the maximum allowable current (about 100 μA). Power consumption is low, about 100 mW, so that portable PMTs can be realized. Active dividers often have current limiting circuitry to prevent damage to the PMT. An ideal active divider can sense the dynode voltages over the entire PMT dynamic range.

The applied PMT high voltage is often a positive bias applied to the anode with the photocathode region grounded. However, the PMT can work equally well with negative high voltage applied to the cathode with the anode grounded, as shown in Fig. 14.30. Specifications made by PMT manufacturers give the preferred bias arrangement along with recommended voltage dividers and sockets. The applied high voltage is often positive, thereby allowing the glass envelope and photocathode to remain at zero potential. One benefit of applying positive high voltage is the suppression of spurious pulses from glass electroluminescence. Further, it is typical that a scintillation crystal attached to the PMT be at zero potential, and the positive high voltage configuration practically eliminates spurious pulses from high-voltage leakage to nearby grounded objects. However, there are important cases when negative high voltage is preferred. With negative high voltage, the coupling capacitor at the output is no longer necessary because the anode is grounded. This divider then allows the resistive divider and PMT to be directly coupled to the output circuitry. The negative bias configuration is used for fast pulse applications in which the output is impedance matched to the transmission line, and for current mode operation with DC-coupling.

Decoupling Design of the resistive divider circuit depends upon the high voltage supply, the required voltage distribution, and the expected anode current (characterized by the FWHM of the output pulse). Variations in anode output current can be minimized by ensuring that $i_{leak}/i_A \geq 100$, where i_A is the anode output current and i_{leak} is the leakage current flowing through the resistive divider [Flyckt and Marmonier 2002]. The leakage current associated with the diode divider circuit can be estimated as

$$i_{leak} \simeq V_{bias} \Big/ \sum_{i=1}^{N} r_i \tag{14.70}$$

where the r_i are the resistances of the divider.

Example 14.4: Consider a 10-dynode PMT with a normal operating gain of 10^5 at 1000 volts potential and an anode output current FWHM of 10 ns. For a uniform resistive divider in which the voltage drop between the photocathode and the first dynode is five times the interdynode potentials, what is the minimum resistance required for each resistor in the divider?

Solution:

A single electron escaping the photocathode produces 10^5 photoelectrons. The anode current produced by a single electron is then,

$$i_A = \frac{qG}{t_A} = \frac{(10^5)(1.6 \times 10^{-19}\ \text{C/e}^-)}{10^{-8}\ \text{s}} = 1.6\ \mu\text{A}.$$

where G is the gain and t_A is the anode rise time. Conforming to the recommendation that $i_{leak}/i_A \geq 100$,

$$i_{leak} \geq 100\ i_A \approx 160\ \mu\text{A},$$

and the total resistance must be

$$\sum_i r_i = \frac{(1000\ \text{V})}{(160 \times 10^{-6}\ \text{A})} = 6.25\ \text{M}\Omega.$$

With 9 dynodes and the photocathode space, the voltage divisor is $9 + 5 = 14$, so that the interdynode resistance must be at least 446.43 kΩ.

Although high resistances can be employed, it is far more usual to use capacitors to decouple the dynodes from the resistive divider circuit. By so doing, the instantaneous anode current, or the currents between the dynode stages, can greatly exceed the average divider current. Decoupling may be accomplished by inserting capacitors either in parallel or series with the divider circuit resistors. Either configuration works; however, capacitors arranged in parallel require much higher voltage ratings. These decoupling capacitors function as charge storage to make up for the current drain from the voltage source as the electron cascade progresses through the PMT.

Suppose that the maximum allowable change in voltage at any dynode is ΔV and the expected anode current is

$$i_A = nq_eGt_A, \tag{14.71}$$

where n is the number of photoelectrons. Then the decoupling capacitance for the last dynode is

$$C_M = \frac{nq_eG}{\Delta V}. \tag{14.72}$$

where M is the number of dynodes. Consider the case in which all dynode gains are equal; then the required limiting capacitance for each dynode stage is,

$$C_i = \frac{nq_eG}{\delta_i \Delta V} = \frac{C_M}{\delta_i^{M-i}}, \tag{14.73}$$

where C_i and δ_i are the decoupling capacitance and gain from dynode number d_i, respectively. Typically, a safety margin equivalent to the stage gain is multiplied by the capacitance to yield,

$$C_i' = \delta\frac{C_M}{\delta_i^{N-i}} = \frac{C_M}{\delta_i^{N-i-1}}. \tag{14.74}$$

for all coupling capacitors except C_M. In most cases these decoupling capacitors are needed only if the required C_i is greater than the dynode stage capacitance (about 20 pF). The first several stages do not need decoupling capacitors, consequently only the latter stages of the divider circuit require these decoupling capacitors.

Preamplifier

Commercial scintillation/PMT units often come with a preamplifier, not to be confused with the voltage divider. In some circumstances, the output pulse may be large enough without further amplification, or the application may require current mode operation rather than pulse mode operation. However, for typical radiation detection measurements, it is important to amplify and shape the output pulse from the PMT. The preamplifier provides a small amount of gain, shape the output pulse (typically a tail pulse), and couple the PMT to the main amplification electronics for pulse mode operation.

Magnetic Shield

Photomultiplier tubes are affected by external magnetic and electrostatic fields because such fields can deflect the electrons from their designed trajectories. The end result is a loss of gain, and for radiation spectroscopy, deterioration in energy resolution from increased statistical spread of photoelectrons reaching the first dynode. Magnetic field effects generally become smaller as the high voltage increases, but usually not enough to eliminate the performance degradation of the PMT. In short, PMTs must be shielded against magnetic fields. Obviously, because the electrons within the tube are guided through successive potentials, external electric fields must also be shielded from the PMT. Most often the PMT is wrapped with "mu-metal" to terminate these fields and prevent them from interfering with the electron trajectories.[11] To prevent problems with glass charging and electroluminescence, the magnetic shield is kept at the same potential as the photocathode. Note that operation with the magnetic sleeve shield held at a high voltage (the situation when the cathode is operated at high voltage) presents a special safety issue, and precaution should be exercised to prevent hazardous shock. Preformed sleeves of mu-metal are usually available from the PMT manufacturers.

[11] The name "mu-metal" comes from the Greek letter μ, the usual symbol for magnetic permeability.

14.1.9 Environmental Effects

Finally, a word or two regarding environmental effects on PMTs. Most manufacturers specify maximum storage and operating temperatures, which according to Engstrom [1980] is usually near 75°C. Apparently, higher temperatures can cause a chemical reaction at the photocathode which may compromise the PMT performance. On the other hand, rapid cooling may fracture the glass envelope. Differences in the thermal expansion coefficient between the glass and PMT socket/base may also cause cracking, and it is recommended that the PMT temperature not go below −50°C. Cooling may also produce condensation that may cause short circuits between the dynodes. Coating with a moisture-resistant insulator can help mitigate this problem.

Photomultiplier tubes can withstand reasonable pressures, but it is advised that the external pressure be kept below 3 atm to avoid possible implosion.[12] Recall that gas diffusion into the PMT from a high pressure environment may be a problem, especially with He gas. Also photomultiplier tubes generate heat; hence operating them in vacuum can be problematic if a heat sink is not employed.

Photomultiplier tubes are relatively fragile, but can survive some level of shock force (up to about 10 g). Special robust PMTs are designed to withstand high shock (some up to 1500 g), and are used for space applications, military applications, and oil well logging. However, careful handling is always recommended. Damage to the internal structure can cause shorting or loss of gain and other performance characteristics. Obviously, excessive shock may also crack the glass envelope.

Of course, photomultiplier tubes are widely used in a radiation environment. High radiation environments, however, can cause an increase in background current by the production of fluorescence in the glass envelope, particularly in the face plate adjacent to the photocathode [Johnson 1973]. Permanent damage occurs from high radiation exposures that cause browning of the PMT faceplate. Soda lime and LiF faceplates are very susceptible to such radiation damage. Borosilicate glass has only slightly more radiation hardness. Quartz and sapphire windows show much less degradation than borosilicate or soda lime glasses. For shorter wavelength UV photons beyond the range of quartz or sapphire, MgF_2 windows are recommended for use in high radiation environments.

14.2 Semiconductor Photodetectors

Semiconductor photodetectors are based upon physics discussed in Chapters 12, 15, and 16, which include solid state physics, semiconductor device physics, and semiconductor device configurations. However, for completeness, the authors decided that these important photosensitive devices should be covered, to some extent, in the present chapter. Hence, a general explanation of these photodetectors and their operation is provided here, whereas detailed explanations on how these devices are designed and operated are offered in Chapters 15 and 16.

Attractive features of semiconductor photodetectors include their compact size, relatively low operating voltage, and fast response times. Semiconductor photodetectors, unlike PMTs, are also insensitive to magnetic fields, thereby allowing them to be used in environments with high magnetic fields. The overall quantum efficiency can be remarkably high for specific ranges of photon wavelengths as determined by the semiconductor's band-gap energy and the photon absorption efficiency. The main types of photon detectors used with scintillators are photodiodes, avalanche photodiodes, and semiconductor photomultipliers.

[12]This eventuality occurred on a devastating scale on Nov. 12, 2001 in the Super-Kamiokande neutrino detector located 1000 m below ground in an old zinc mine in Japan. The detector was a cylinder (34 m in diameter and 36 m high) filled with ultra-pure water and lined with 11,146 PMTs. On the fateful day, as the chamber was being refilled, a PMT imploded setting off a chain reaction that caused 6,600 of the $3,000 PMTs to also implode.

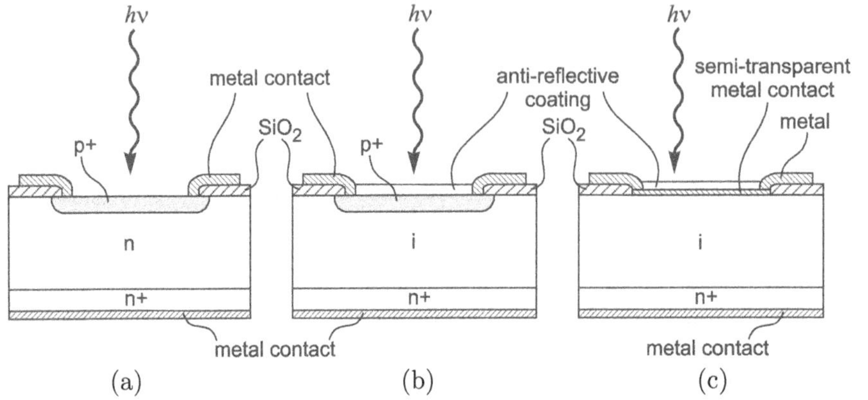

Figure 14.31. Semiconductor photodiodes of different designs, including (a) a common *pn* junction photodiode, (b) a *pin* photodiode and a (c) Schottky contact photodiode.

14.2.1 Photodiodes

Photodiodes are semiconductor devices formed into a *pn* or *pin* junction diode. The designation *p* refers to *p*-type impurity dopants that produce positive charge carriers, *n* refers to *n*-type impurity dopants that produce negative charge carriers, and *i* refers to the *intrinsic* material that is generally devoid of impurity dopants. Diodes formed from a simple *pn* junction offer an inexpensive method for detecting photons, yet have relatively thin active volumes. The reason for this thin volume, as explained in detail in Chapter 15, arises from buidup of space charge in the sensitive region that limits the electric field and, thus, narrows the photon active sensitive region or *depletion region.* The *n* and *p* doped regions are necessary for the device to function properly, yet this space charge appears as a consequence of the "impurity doping" in the semiconductor.

A configuration often used to increase the active region of a semiconductor photodiode is the *pin* junction diode. In general, the device has a *p*-type material adjacent to intrinsic material, with *n*-type material completing the diode opposite the intrinsic material. A simple *pin* device cross section is depicted in Fig. 14.31(b). The active region is increased by including an intrinsic, or low-doped (π or ν), region between the *n* and *p* region, thereby greatly reducing the space charge density and increasing the active region width.

A simple and common photodiode is based upon the Schottky effect (see Sec. 15.3), in which the application of an appropriate metal forms a rectifying blocking contact upon the semiconductor. Such a device functions similarly to a *pn* or *pin* junction photodiode, but generally has higher leakage current. The higher leakage current is fundamentally due to the physical process that suppresses the leakage current in reverse bias. The Schottky diode relies upon a barrier to reduce thermionic emission, whereas the *pn* junction relies upon the reduction in minority carriers available for leakage as explained in more detail in the next chapter.

When photons strike the semiconductor photodiode, usually fabricated from Si or GaAs based materials, electrons are excited into the conduction band of the material. A voltage bias across the diode causes the photoelectrons to drift across the device and induce charge in a similar fashion as that described for gas-filled ion chambers. The quantum efficiency of the semiconductor diode varies with the configuration and packaging of the diode. For instance, different commercial Si photodiodes have peak efficiencies at wavelengths ranging between 700 to 1000 nm. Regardless, they are typically more sensitive to longer wavelengths than commercial PMTs. For instance, CsI(Tl) scintillation emissions are better matched to Si photodiodes than to PMTs. Photodiodes operate with low voltage, are small, rugged and relatively inexpensive and, thus, they offer a compact method of sensing light emissions from scintillators.

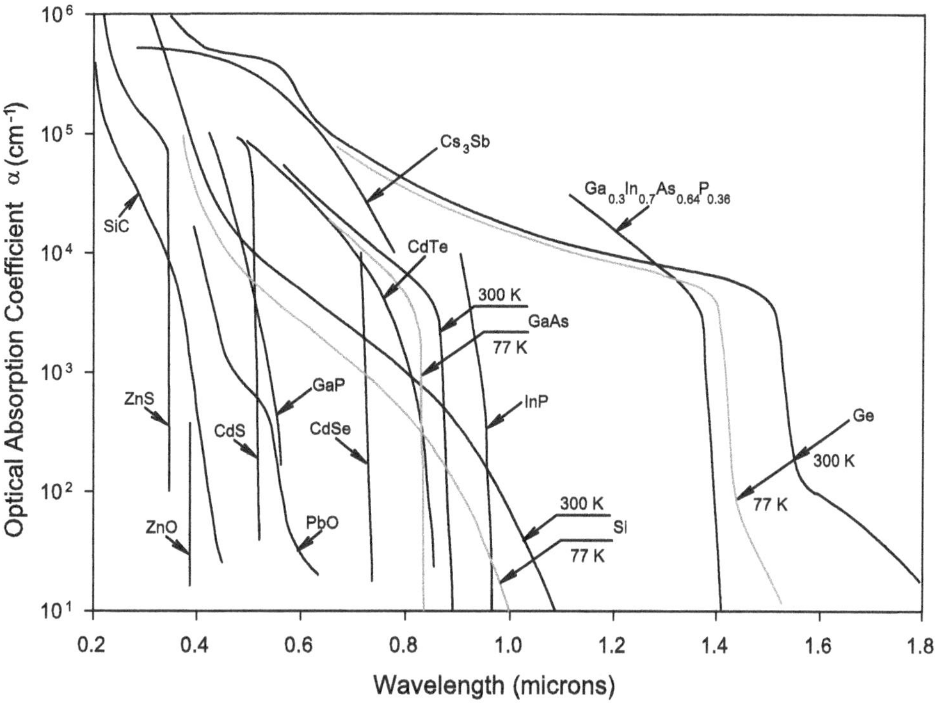

Figure 14.32. Optical absorption coefficients for various semiconductor photodetector materials. After [Melchior 1972; Sze 1981].

Performance Characteristics Quantum efficiency generally approaches a maximum as photon energies approach the band-gap energy. For photon energies below the band gap energy ($h\nu < E_g$), the material is generally transparent to the light, except for absorption by defect and impurity centers. For photon energies greater than the band gap ($h\nu > E_g$), absorption may occur with increasing likelihood as the photon energy increases. The increase in the absorption coefficient α occurs because electrons that are deeper in the valence band can be excited across the band gap into the conduction band as the energy of the absorbed photon increases. This property can be observed in Fig. 14.32 as the increase in absorption coefficient α as the light wavelength decreases.

For photons with wavelengths below the cutoff wavelength, determined by the semiconductor band-gap energy, optical absorption increases, as also shown in Fig. 14.32. However, the energy conversion efficiency[13] actually decreases as the photon energy increases, a consequence of excess energy being expended to excite a single electron-hole pair. For example, if an 800 nm photon (1.55 eV) is absorbed in GaAs ($E_g = 1.42$ eV), there is a likely chance that a single electron-hole pair is excited; hence $\eta_e = 0.916$. The same is true for a 700 nm photon (1.77 eV), but $\eta_e = 0.802$. Although the energy is higher for the second case, the photon has insufficient energy to excite two electron-hole pairs, causing η_e to decrease. Consequently, the energy conversion efficiency decreases even though the photon absorption efficiency increases.

The active region width of Si *pn* and *pin* photodiodes need only be about 1 to 4 microns to effectively absorb photons of wavelengths ranging between 450 nm to 650 nm [Melchior 1972]. For Ge photodiodes, these same active region widths correspond to photons of 950 nm to 1500 nm wavelengths. At shorter wavelengths,

[13]The energy conversion efficiency η_e is the band-gap energy divided by the energy expended to produce an electron-hole pair, i.e., $\eta_e = h\nu/E_g$.

absorption increases near the surface, consequently causing an increase in surface recombination and, thereby a decrease in the quantum efficiency. This behavior is apparent in Fig. 14.33 at the shorter wavelengths. Thus, the long wavelength cutoff is limited by the band-gape energy and the short wavelength cutoff is limited by high surface absorption and surface recombination. By contrast, Si Schottky contact photodiodes (also known as metal-semiconductor photodiodes), see Fig. 14.31(c), have thin metal contacts that allow photons to pass deeper into the active region of the photodiodes than do *pn* and *pin* diodes and, thereby extend the short wavelength range below 400 nm [Melchior 1972].

Both the energy conversion efficiency and the quantum efficiency determine the total charge excited by scintillation photons and, thereby affect the energy resolution of the detector. Silicon photodiodes can function at typical scintillation emissions with wavelengths near 400 nm, but excel in performance at longer wavelengths, generally in the 600 nm to 1000 nm range. Photodiodes generally do not have gain, and the charge signal is determined by the primary electron-hole pairs excited by the scintillation event. Consequently, the signals from photodiodes are small and need more amplification than signals from PMTs.

The quantum efficiency η_Q is defined as the average number of electron-hole pairs produced per scintillation photon. For light detection, a useful quantity used to determine the relative sensitivity of the diode to light is the *spectral responsivity* R, usually measured in amperes/watt [Donati 2000], and is defined as[14]

$$R = \frac{q_e \eta_Q \lambda}{hc} = \frac{q_e \eta_Q}{h\nu}. \tag{14.75}$$

From this result, R is seen to decrease with photon energy and to increase with quantum efficiency. Shown in Fig. 14.33 are spectral responses for several semiconductor photodiodes of interest for scintillation counting.

The details of semiconductor devices and operation are presented in Chapters 15 and 16. However, briefly stated, the number of charge carrier pairs produced is the average ionization energy w divided into the particle energy E, i.e., the number of electron-ion pairs is E/w. For high energy photons (x and gamma rays), most semiconductors have average ionization energies ranging between 2.5 and 5.5 eV/e-h pair. For low-energy photons emitted from scintillators, the average energy expended per electron-hole pair is a function of the absorption efficiency, the quantum efficiency, and the energy conversion efficiency. Consequently, without gain, the total number of charge pairs produced per scintillation event is considerably less than the number of charges produced by a PMT. Although the statistical spread in the number of electron-hole pairs produced is less than that predicted by Poisson statistics, the relatively low number of charge carriers per event compromises the overall energy resolution. Additionally, electronic noise from leakage and generation current can adversely affect the pulse height spectrum because of the relatively small signals produced by a common photodiode. This problem of poor energy resolution is particularly severe for low-energy scintillation photons.

Electronic Noise The electronic noise arising in common photodiodes is classified as either *series noise* or *parallel noise*. It is the fluctuation in these two current sources that produces a broad electromagnetic spectrum of frequencies in the noise, and, thus, it is usually referred to as *white noise*.

Parallel noise arises from dark current fluctuations within the diode, generally from thermionic leakage current and bulk generation current. Bulk generation current arises from charge carriers being thermally excited from the valence band into the conduction band, a process that increases with the active volume of the detector. Hence, bulk generation current is a function of both the contact area and the depth of the active region. Moreover, the dark current increases with applied voltage, mainly because the leakage current increases from non-ideal current saturation and the generation current increases as the active region increases with voltage. Parallel noise can be reduced by decreasing the leakage current of minority carriers

[14] This property is sometimes referred to as "spectral sensitivity" or "photosensitivity".

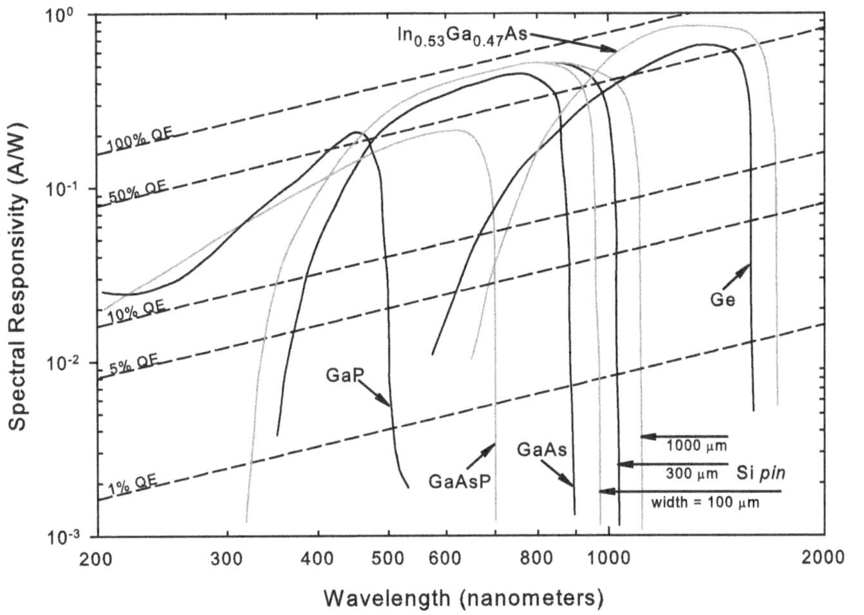

Figure 14.33. Spectral responsivity in amps/watt for various semiconductor photodetector materials of interest for scintillation counting. Quantum efficiency is also indicated. [after Donati 2000].

or by increasing the rectifying barrier height. Because the total leakage current increases as the total contact area increases, the practical size of photodiodes for scintillation spectroscopy is limited. For Si photodiodes, this area is usually restricted to approximately 1 cm^2.

Series noise arises from fluctuations in the series resistance and fluctuations in the current flowing into the input stages of the preamplifier circuitry, fluctuations that consequently produce short current bursts in the output of the preamplifier. This series noise is also referred to as *shot noise.* A large contribution to the series noise arises from the detector input capacitance C_{in} and is generally quantified as a function of C_{in}^2 [Iwanczyk and Patt 1985]. Hence, it is important to reduce the input capacitance as much as practically possible. This reduction can be accomplished by both reducing the detector contact area and increasing the active region, between the contacts. Note that both series and parallel noise can be decreased by selecting photodiodes with small contact areas; however, parallel noise decreases with the size of the active region whereas series noise increases with the size of the active region. Such counteracting trends in the two noise sources means there is an optimum operating voltage, which determines the size of the active region. Indeed, photodiodes are usually operated at active region depths well beyond those required to completely absorb the impinging scintillation light in order to optimize performance.

Photodiode Materials Silicon is of interest because of its good quantum efficiency in the extended red-wavelength range and because it couples well to common scintillators such as CsI(Tl) and even BGO, scintillators that typically are not optimized when coupled to PMTs. Although most photodiodes are manufactured from Si materials, there are many photodiodes manufactured from wide band-gap materials. These wide bandgap materials are almost all compound semiconductors, such as InP, GaAs, CdTe, GaP, CdS, and SiC, and generally have lower bulk generation current than that in Si. Further, because of their higher resistivity, wide band-gape semiconductors generally have lower leakage currents than Si diodes. Indeed, dark current can be reduced, in concept, by wider band-gap materials, but this higher resistivity is usually accomplished by the purposeful introduction of compensating energy levels into the material. The fluctuations in generation current can be significant, mainly a result of the shorter generation-recombination lifetimes characteristic of compensated compound semiconductors, and it is the fluctuation in dark current that contributes to the electronic noise. Wide band-gap experimental photodiodes have also been investigated, such as HgI_2. Indeed, a rather sizeable CsI(Tl) crystal (about 16 cm^3) coupled to a HgI_2 photodiode held the energy resolution record for many years for a scintillator/semiconductor device operated at room temperature, namely 4.98% FWHM at 662 keV [Markakis 1988]. Competing processes between direct gamma-ray detection by the photodiode and the CsI(Tl) crystal were evident, a consequence of using compound semiconductor photodiodes composed of high Z elements. Some other binary compound semiconductors and complex ternary and quaternary compound semiconductors, such as PbS, PbSe, InSb, InGaAs, InAsSb, and InGaAsP, have extended IR sensitivity, but typically also have higher dark current densities than those expected for wide band-gap materials.

Overall, photodiodes offer a relatively compact and inexpensive method to detect radiation induced light from scintillators. They usually require only a modest bias voltage, compared to that required by PMTs, and are generally unaffected by magnetic fields. Further, semiconductor photodiodes are robust and less prone to shock damage. However, semiconductor photodiodes have much smaller output signals, thereby compromising energy resolution. Parallel and series noise further reduce energy resolution. Finally, the leakage current from semiconductor photodiodes increases with temperature, thereby reducing their use in high temperature environments. Compact Peltier coolers can be used to help remedy the noise problem, yet this added complication with its own power requirement still restricts the use of photodiodes in many circumstances.

14.2.2 Drift Diodes

One method to increase the active area of a photodiode, while still maintaining relatively low capacitance, is to use a *silicon drift diode* (SDD). Details of the structure and operation of the drift diode are given in Sec. 16.4 on silicon detectors. However, the basic device has a relatively small anode with structured cathodes guiding electrons by way of consecutively divided potentials. The capacitance is dominated by the low anode capacitance. Further, modern designs incorporate the first amplifier stage directly onto the diode, thereby reducing the coupling capacitance. These detectors can have relatively larger areas compared to common photodiodes, often greater than 25 mm^2 and up to 1 cm^2 area. Modest cooling (about 0°C) can be implemented to ameliorate the bulk generation current caused by the larger diode volume.

Although the charges drift for a longer time than that in common photodiodes (typically a few microseconds) the rise time is surprisingly fast, mainly because it is mostly dependent upon the drift distance between the anode and the closest cathode in the structure. However, diffusion of the charge cloud causes differences in rise times, a consequence of photon interactions occurring randomly over the entire diode volume. Hence, the diode integration time should be set high enough to allow complete charge collection, regardless of interaction location.

Drift diodes have demonstrated excellent results when coupled to scintillators. For instance, a cylindrical CsI(Tl) crystal, (3-mm diameter and 5-mm thickness) produced 4.34% FWHM energy resolution for 662-keV gamma rays when coupled with Nuclear Enterprises' 586 optical grease to a SDD [Fiorini et al. 1998]. Later, Fiorini et al. [2006] produced 2.7% FWHM energy resolution for 662-keV gamma rays with a cylindrical $LaBr_3$(Ce) crystal (5-mm diameter and 5-mm thickness) when coupled to a 5 mm^2 SDD. Although the scintillators were comparatively small, the excellent reported performance rivals the best performance for systems incorporating PMTs. To increase the overall detector area, an array of hexagonal SDDs is also reported by Fiorini et al. [2007] as a method to produce a compact Anger camera [Anger 1958].

14.2.3 Avalanche Diodes

Semiconductor photodiodes based upon avalanche gain provide a method to combine the high quantum efficiency of semiconductors with large output signals. Avalanche photodiodes (APDs) are operated at high reverse bias voltages and are designed such that an electric field capable of promoting avalanche multiplication is produced within the device. Those properties, important for common photodiodes used for scintillation detection, are also required for APDs: namely, high quantum efficiency, efficient photon absorption, good response speed and high energy conversion efficiency. Thus, properties mentioned earlier for common photodiodes also apply to APDs. Quantum efficiencies ranging from 50% to 80%, depending on the wavelengths of the scintillation photons, are typical for commercial Si APDs. Additionally, gain and electronic noise properties are important for APD applications. Typical gains exceeding 100 can be achieved; however, APDs are characteristically noisier than common *pn* and *pin* junction photodiodes.

Avalanche Gain

Electron-hole pair production from energetic electrons and holes are usually described by ionization coefficients $\alpha(\mathcal{E})$ for electrons and $\beta(\mathcal{E})$ for holes and are the probabilities, per unit differential path-length of travel, that the charge carrier undergoes an ionizing collision. Typically, for most semiconductors, $\alpha(\mathcal{E}) > \beta(\mathcal{E})$, although there are exceptions. For instance, in Ge $2\alpha \simeq \beta$, yet for Si $\alpha > \beta$ and for GaAs $\alpha \gg \beta$ [Melchior 1972].

Consider a *pin* APD with a depletion region $x \in (0, W)$ in which a current of holes $I_h(x)$ flows from holes injected at $x = 0$, which increases from $x = 0$ to $x = W$ as a result of impact ionizations. Likewise a current of electrons $I_e(x)$ is formed that increase from $x = W$ to $x = 0$. Under steady-state conditions the total current $I = I_h(x) + I_e(x)$ is constant, as is the electric field so that α and β are constant throughout the depletion region. In this case the following balance equation holds:

$$dI_h(x)/q_e \text{ (increase in holes at } x) = (I_h/q_e)(\beta\, dx) + (I_e/q_e)(\alpha\, dx) \text{ (no. e-h pairs produced in } dx \text{ per s)},$$

or, because $I_e(x) = I - I_h(x)$,

$$\frac{dI_h(x)}{dx} - (\beta - \alpha)I_h(x) = \alpha I. \tag{14.76}$$

The solution of this differential equation, subject to the initial condition $I = I_h(W) = M_h I_h(0)$ or $I_h(0) = I/M_p$ where M_p is the multiplication factor for holes,

$$I_h(x) = I \exp((\beta - \alpha)x) \left\{ \frac{1}{M_h} + \frac{\alpha}{\beta - \alpha} [1 - \exp(-(\beta - \alpha))x] \right\}. \tag{14.77}$$

Sze [1981] considered a more general model in which α and β are functions of position in the depletion region and obtained

$$I_h(x) = I \left\{ \frac{1}{M_h} + \int_0^x \alpha(x') \exp\left[-\int_0^{x'} [\beta(x'') - \alpha(x'')]\, dx'' \right] dx' \right\} \Big/ \exp\left[-\int_0^x [\beta(x') - \alpha(x')]\, dx' \right], \tag{14.78}$$

which reduces to Eq. (14.77) when α and β are constants.

The hole multiplication factor M_h (or rather its inverse) is readily found from Eq. (14.77) by setting $x = W$ and setting $I_h(W) = I$ and performing some algebraic manipulations to obtain

$$1 - \frac{1}{M_h} = \frac{\beta}{\beta - \alpha}[1 - e^{-(\beta-\alpha)W}]. \tag{14.79}$$

Because the avalanche breakdown voltage is the voltage at which M_h approaches infinity, the breakdown condition is found from Eq. (14.79) by setting $1/M_h$ to zero. This condition can be written as

$$1 = \frac{\beta}{\beta - \alpha}[1 - e^{-(\beta-\alpha)W}]. \tag{14.80}$$

If the avalanche process were initiated by electrons injected at $x = W$, the same breakdown condition is obtained [Sze 1981]. This is to be expected since breakdown depends only on what is happening in the depletion region and not on which charge carriers (or primary current) initiate the avalanche. In the limit as $\beta \to \alpha$ the multiplication factor of Eq. (14.79), which is the APD gain M_{APD}, has the limiting value

$$M_{APD} = \frac{1}{1 - \alpha W}. \tag{14.81}$$

Note that as the product αW approaches unity, the gain progresses towards ∞, a generally undesirable result for the operation of common APDs. This situation is referred to as avalanche breakdown. However, there is a practical limitation to M_{APD}, in which series resistance and space charges limit the overall gain. An empirical formulation for gain in the pre-breakdown region was reported as [Miller 1955],

$$M_{\text{APD}} = \frac{1}{1 - (V_R/V_B)^n}. \tag{14.82}$$

where V_B is the breakdown voltage, V_R is the reverse bias voltage, and n depends upon the semiconductor material and device structure. As V_R approaches V_B, Melchior and Lynch [1966] developed a description for gain under high light illumination,

$$M_{\text{APD}} = \left. \frac{1}{1 - (V_R - IR/V_B)^n} \right|_{V_R \to V_B} \simeq \frac{V_B}{nIR}, \tag{14.83}$$

where IR is the voltage drop due to series resistance. If the gain is defined as the total amplified (avalanche enhanced) current I_{av} divided by the total unamplified primary current I_h, then Eq. (14.83) reduces to [Sze 1981] $M_{\text{APD}} = \sqrt{V_B/(nI_hR)}$. Thus the amplified photocurrent is $I_{av} = I_h M_{\text{APD}} = \sqrt{I_h V_B/(nR)}$.

Overall, the APD is operated as a linear amplification device, in which the total current, or integrated charge, is taken to be a linear function of the energy of light absorbed. However, because the quantum efficiency for scintillation light is less than unity (less than one electron-hole pair per photon), the APD is actually measuring the number of photons that strike the diode and not the energy absorbed in the scintillation crystal. As with PMTs, it is assumed that the total number of photons striking the APD is a linear function of energy absorbed in the scintillation detector. This assumption is mostly true despite issues with non-linear light output for many scintillators. Hence, the magnitude of the output current gives a measure of the initial energy absorbed in the scintillation crystal.

APD Designs

Various configurations for avalanche photodiodes have been explored, of which three are shown in Fig. 14.34. Most commercial APDs are fabricated from Si, although APDs made from commercial low-noise InGaAs and designed for photon wavelengths between 1.0 and 1.6 μm are also available. The design of Fig. 14.34(a) is perhaps one of the simplest, relying upon a high electric field at the n^+p junction to produce the avalanching electric field. The guard ring reduces leakage current and early breakdown at the pn junction corners. The $p^+ - \pi - p - n^+$ "hi-lo" structure of Fig. 14.34(b) has a similar structure to that of a common IMPATT diode [Sze 1981]. The $p^+\pi$ region provides a lower field drift region, in which electrons drift towards the high electric field region defined at the pn^+ junction. The low doped π region allows the active depletion region to extend nearly completely across the diode. When electrons reach the high field region, an avalanche occurs and produces the device gain. The beveled sidewalls of the APD design in Fig. 14.34(c) reduce the surface electric field [Huth et al. 1966], thereby allowing a higher electric field within the semiconductor bulk to promote avalanche gain. Note that the high field region, defined by the $p^+\nu$ junction, is near the thin semiconductor window. The etched thin window reduces photon absorption losses in the "dead" surface region. Overall, there are a variety of other APD designs that can be used as sensors for scintillation detectors, some of which are described elsewhere [Melchior 1972; Sze 1981].

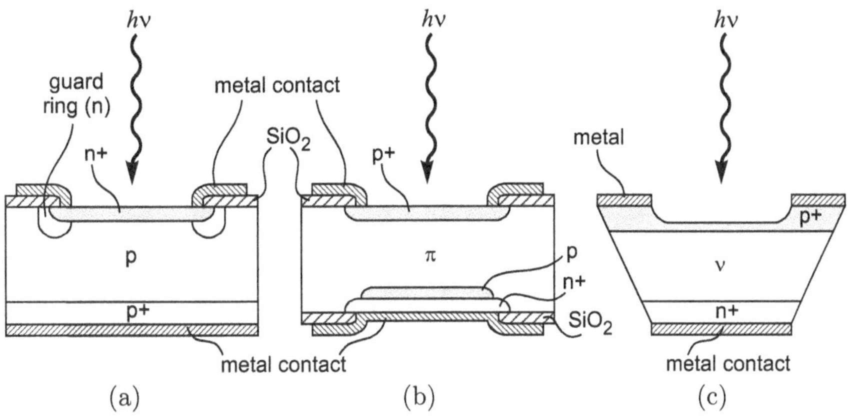

Figure 14.34. Avalanche photodiodes of three different designs: (a) a Si guard ring structure avalanche photodiode, (b) a $p^+ - \pi - p - n^+$ avalanche photodiode, and (c) a deep etched depletion avalanche photodiode with beveled sides.

Noise Factor

Ultimately, the desired feature of any photon detector is a proportional response to the total energy absorbed in the scintillation crystal attached to the detector. Hence, charge collected from an APD should be a linear measure of energy deposited in the scintillator, with an energy resolution defined by the variance of the output signal. Variations in the gain broaden the energy resolution of the scintillation light, and a non-linear output can make the energy calibration of the pulse height spectra difficult.

As developed in Sec. 10.5.4 for proportional counters, the avalanche process in an APD is statistical in nature. In other words, the amount of avalanching, and thus gain, for each electron is different, slight as it may be. Hence, the square of the average gain is not equal to the average gain squared but, rather the variance in the gain is

$$\sigma_M^2 = \langle M^2 \rangle - \langle M \rangle^2. \tag{14.84}$$

The "noise factor" is defined as

$$F = \frac{\langle M^2 \rangle}{\langle M \rangle^2}. \tag{14.85}$$

From Eqs. (14.84) and (14.85), the variance in the gain is

$$\sigma_M^2 = (F - 1)\langle M \rangle^2. \tag{14.86}$$

Note that if the gain variance is zero, then $F = 1$.

The mean square shot noise current for a reverse biased pn junction is defined as

$$\langle i^2 \rangle = 2q_e IB, \tag{14.87}$$

where B is the bandwidth in Hz. If a uniform avalanche multiplication region exists, the differential mean square shot noise $\langle di^2 \rangle$ for the differential current dI generated within element dx at x is multiplied by the mean square gain at x, $\langle M^2(x) \rangle$, to yield

$$d\phi(x) = \frac{\langle di^2 \rangle}{B} = 2q_e \, dI(x) \langle M(x)^2 \rangle, \tag{14.88}$$

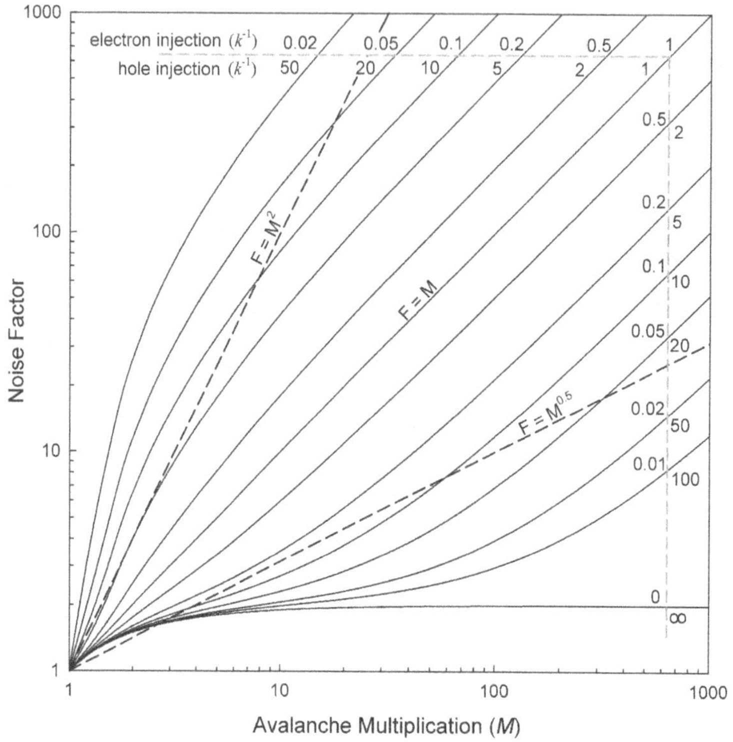

Figure 14.35. Noise factor as a function of gain M and the ratio α/β.

where ϕ is termed the "spectral density," and Eq. (14.87) represents a portion of the total mean square noise current. Integration of Eq. (14.87) yields the total noise contribution of the multiplication process. If $\beta = k\alpha$ (or $k^{-1} = \alpha/\beta$), where k is a scaling constant, and M is independent of x, McIntyre [1966] shows that the noise factors for electron and hole injection are

$$F_e = \frac{\phi}{2q_e I \langle M \rangle^2} = M\left[1 - (1-k)\left(\frac{M-1}{M}\right)^2\right] \quad \text{and} \quad F_h = \frac{\phi}{2q_e I \langle M \rangle^2} = M\left[1 + \frac{(1-k)}{k}\left(\frac{M-1}{M}\right)^2\right]. \tag{14.89}$$

The results from Eqs. 14.89 for several M and α/β are plotted in Fig. 14.35. In the case that $\alpha = \beta$, or $k = 1$, then both Eqs. 14.89 reduce to $F = \langle M \rangle$. For the case of electron injection, in which holes play no part in the avalanche process ($\beta = 0$), Eq. (14.89) reduces to,

$$F_e = \left(\frac{2M-1}{M}\right). \tag{14.90}$$

This result reduces to $F_e \simeq 2$ as $M \to \infty$. A similar result is achieved if holes play the major part in the avalanche current and electrons do not contribute to the avalanche.[15]

[15]Hence, $k^{-1} \to 0$, as shown in Fig. 14.35.

14.2.4 Semiconductor Photomultipliers

Junction breakdown of semiconductors can be used as another means of radiation counting and spectroscopy. For any *pn* junction operated with a reverse bias voltage, there is a voltage above which the junction "breaks down" and continues to conduct current across the semiconductor. This phenomenon is initiated by three different means, namely (1) thermal instabilities, (2) tunneling current, and (3) avalanche breakdown.

Here it is avalanche breakdown that is of interest. Avalanche breakdown occurs when Eq. (14.78) is satisfied. Should $\alpha = \beta$, then the breakdown condition is achieved when $\int_0^W \alpha(x)\,dx = 1$. The critical electric field required to sustain the avalanche condition is $\mathcal{E}_A = 2V_B/W$, where V_B is the breakdown voltage [Sze 1981].

The avalanche process is solely dependent upon what occurs within the depletion region W. Hence, an event that excites a charge carrier within the active depletion region may trigger an enormous, continuous avalanche from impact ionization with gains exceeding 10^6. This interesting mode of operation is usually not preferred for APD operation, the effects of which were studied by McIntyre [1961] and Haitz [1961].

However, there are particular applications in which the breakdown mode of operation is beneficial, mainly realized with the single-photon avalanche diode (SPAD), so-named because of its ability to detect low light levels, one photon at a time. The leading edge of an avalanche signal marks the arrival time of the interacting photon within the depletion region, and SPADs with 28 ps timing resolution at room temperature have been reported [Cova et al. 1989]. After the avalanche is triggered, the self-sustained current continues to flow unless quenched. Often a series resistance is included in the design that serves to quench the avalanche, much like an externally quenched GM counter. As a matter of fact, APDs designed and operated in this mode are often said to operate in "Geiger Mode" [Renker 2006]. As the avalanche current flows through the quenching resistance, it draws voltage from the source, thereby causing the voltage across the APD to drop below the breakdown voltage, causing the avalanche to cease. As a result, the output current and collected charge are nearly the same for each avalanche regardless of energy deposition in the device, statistical fluctuations notwithstanding. Consequently, if two or more photons interact in the device simultaneously, the total charge liberated from the quenched avalanche is roughly the same as if a single photon caused the avalanche, again much like a GM counter.

Silicon photomultipliers (SiPM)[16] use the breakdown condition with thousands of tiny silicon APDs arranged into a compact array, presently on the order of 6 mm × 6 mm area or less. The individual APDs in silicon photomultipliers (SiPM) have linear dimensions usually between 20 μm and 100 μm, a common size is 35 μm × 35 μm per APD pixel. Hence a single SiPM array may have more than 10^4 APD pixels. Each APD is electrically decoupled from adjacent APDs with polysilicon resistors fabricated on the same substrate. Typical operational bias voltage is 10% to 15% above the breakdown voltage. A scintillator may be fastened to the SiPM face, much like a conventional PMT, and light from radiation interactions within the scintillator triggers the APDs. Each APD provides information that is ultimately limited to recognizing the excitation of an electron-hole pair within the avalanche region. Because each pixel produces an avalanche of relatively the same magnitude per event, the total photon count can be determined by dividing the total measured charge by the avalanche gain. Hence, the device operates mainly as a photon counter, and the total measured current is proportional to the total number of photons. The enormous gain per pixel, when the pixel counts are added together, can produce more than 10^9 free electrons per event. Properties of commercially available SiPMs are given in Table 14.9.

General SiPM Characteristics

A particular advantage of SiPMs is the high quantum efficiency of Si for photons with wavelengths between 350 nm and 950 nm. The increased probability of absorption of scintillation photons, in the active volume of

[16] Also called multi-pixel photon counters (MPPC).

Table 14.9. Properties of some commercial silicon photomultipliers. The legend abbreviations are as follows: A manufacturer (H Hamamatsu, K Ketek, S SensL); B model; C SiPM chip size (x mm $\times$ x mm); D microcell pixel size (x μm $\times$ x μm); E number of microcells; F microcell fill factor (%); G gain at room temperature $\times 10^6$; H breakdown voltage, VBr (volts); I bias range above VBr (volts); J peak wavelength (nm); K maximum photon detection efficiency (%) at peak wavelength and max voltage; L dark count rate (kHz); M signal rise time (ns); N signal pulse width (FWHM in ns); O microcell recovery time (ns); P VBr temperature drift (mV/°C); and Q gain drift (%/°C). Non-reported table entries are indicated by NR.

A	B	C	D	E	F	G	H	I	J	K	L	M	N	O	P	Q
S	10010B	1	10	2880	28	0.2	24.5 ± 0.5	1-5	420	18	700	0.3	0.7	10	21.5	-0.8
S	10020B	1	20	1296	48	1	24.5 ± 0.5	1-5	420	31	700	0.3	0.7	50	21.5	-0.8
S	10035B	1	35	576	64	3	24.5 ± 0.5	1-5	420	41	700	0.3	0.7	180	21.5	-0.8
S	10050B	1	50	324	72	6	24.5 ± 0.5	1-5	420	47	800	0.3	0.7	350	21.5	-0.8
S	30020B	3	20	10998	48	1	24.5 ± 0.5	1-5	420	31	6600	0.6	1.3	100	21.5	-0.8
S	30035B	3	35	4774	64	3	24.5 ± 0.5	1-5	420	41	6700	0.6	1.3	180	21.5	-0.8
S	30050B	3	50	2668	72	6	24.5 ± 0.5	1-5	420	47	7500	0.6	1.3	350	21.5	-0.8
S	60035B	6	35	18980	64	3	24.5 ± 0.5	1-5	420	41	21500	1	3.2	210	21.5	-0.8
S	10010C	1	10	2880	28	2	24.65 ± 0.25	1-5	420	41	30	0.3	0.6	10	21.5	-0.8
S	10020C	1	20	1296	48	1	24.65 ± 0.25	1-5	420	41	30	0.3	0.6	90	21.5	-0.8
S	10035C	1	35	576	64	3	24.65 ± 0.25	1-5	420	41	30	0.3	0.6	180	21.5	-0.8
S	10050C	1	50	324	72	6	24.65 ± 0.25	1-5	420	41	30	0.3	0.6	350	21.5	-0.8
S	30020C	3	20	10998	48	1	24.65 ± 0.25	1-5	420	41	300	0.6	1.5	90	21.5	-0.8
S	30035C	3	35	4774	64	3	24.65 ± 0.25	1-5	420	41	300	0.6	1.5	180	21.5	-0.8
S	30050C	3	50	2668	72	6	24.65 ± 0.25	1-5	420	47	300	0.6	1.5	350	21.5	-0.8
S	60035C	6	35	18980	64	3	24.65 ± 0.25	1-5	420	41	1200	1	3.2	210	21.5	-0.8
H	S12571-010	1	NR	10000	33	0.135	65 ± 10	4.5	470	10	100	NR	0.3	NR	60	1.2
H	S12571-015	1	NR	4489	53	0.23	65 ± 10	4	460	25	100	NR	0.25	NR	60	1.5
H	S12571-025	1	NR	1600	65	0.515	65 ± 10	3.5	450	35	100	NR	0.25	NR	60	1.6
H	S12571-050	1	NR	400	62	1.25	65 ± 10	2.6	450	35	100	NR	0.25	NR	60	2.16
H	S12571-100	1	NR	100	78	2.8	65 ± 10	1.4	450	35	100	NR	0.3	NR	60	4.3
H	S12572-010	3	NR	90000	33	0.135	65 ± 10	4.5	470	10	1000	NR	0.3	NR	60	1.2
H	S12572-015	1	NR	40000	53	0.23	65 ± 10	4	460	25	1000	NR	0.25	NR	60	1.5
H	S12572-025	1	NR	14400	65	0.515	65 ± 10	3.5	450	35	1000	NR	0.25	NR	60	1.6
H	S12572-050	1	NR	3600	62	1.25	65 ± 10	2.6	450	35	1000	NR	0.25	NR	60	2.16
H	S12572-100	1	NR	900	78	2.8	65 ± 10	1.4	450	35	1000	NR	0.3	NR	60	4.3
H	S12576-050	1	NR	400	NR	1.25	62.9 ± 10	2.6	450	35	5	NR	0.25	NR	60	2.16
H	S12577-050	3	NR	3600	NR	1.25	62.9 ± 10	2.6	450	35	50	NR	0.25	NR	60	2.16
K	PM1150	1.2	50	576	70	1.7	25 ± 3	2.5-5	420	> 50	< 400	NR	NR	NR	NR	< 1
K	PM1150T	1.2	50	576	63	8	25 ± 3	2.5-5	420	> 50	< 400	NR	NR	NR	NR	< 1
K	PM3350	3	50	3600	70	8	25 ± 3	2.5-5	420	> 50	< 500	NR	NR	NR	NR	< 1
K	PM3350T	3	50	3600	63	6	25 ± 3	2.5-5	420	> 40	< 500	NR	NR	NR	NR	< 1
K	PM6660	6	60	10000	66	10	24	2.4-4.8	420	> 50	< 500	NR	NR	NR	NR	< 1

the APD, is a clear advantage of the SiPM when compared to the smaller absorption probabilities in PMT semitransparent photocathodes or PMT reflective photocathodes. Hence, the photon detection efficiency of a SiPM can be much higher than that of a common PMT. However, many commercial SiPM units usually have a peak sensitivity near 420 nm, much like that of conventional bialkali PMTs. SiPMs are compact and rugged, which allows them to be fabricated as compact instruments and dosimeters. This compactness is perhaps one of the most useful features of SiPMs compared to vacuum tube PMTs. Further, they need only modest power, usually requiring no more than a few tens of volts for operation, which is substantially less than that needed for a PMT. SiPMs are also insensitive to magnetic fields, thereby making them an attractive choice for applications in areas where high magnetic fields may arise. Typically, the total array of APDs in the SiPM has an extremely fast rise time, between 0.3 and 1 ns. However, the actual recovery time required for the SiPM is much longer and usually ranges from a few tens of ns up to 400 ns, depending on the pixel and array sizes. Hence, the rise times and count rates are comparable to conventional PMTs.

Presently the cost of a single SiPM array chip is modest, yet these devices range in area from only 1 mm $\times$ 1 mm up to 6 mm $\times$ 6 mm. Such small light detectors might be adequate for small projects and compact scintillation applications. However, to cover the same area as a 3-inch diameter bialkali PMT, typically used for scintillation spectroscopy, a minimum of 127 SiPM chips are needed, if the larger 6 mm $\times$ 6 mm chips are used. Commercially available SiPM arrays are available, presently with a maximum size of 6 mm $\times$ 6 mm chips arranged in an 8×8 array that has the equivalent area of a two-inch diameter PMT. At the time of writing, the cost of such an array is considerably higher than the PMT counterpart. However, commercial scintillation detector units have recently become available with a SiPM readout option.

Unlike a PMT photocathode, SiPMs have regions between the diodes that are insensitive to illumination. The ratio of the sensitive area to the total area is termed the "fill factor." Modern commercial units have fill factors ranging between about 30% to 70%. As fill factor increases, the possibility of cross talk between pixels also increases. Cross talk may manifest itself as photon emissions from de-exited electrons, and their subsequent absorption in adjacent pixel cells. Hence, a photon emitted in one pixel cell can trigger an avalanche in an adjacent cell. One method proposed to reduce this effect is the incorporation of etched trenches between the pixels. The trenches are then backfilled with an opaque material [Piemonte 2006]. However, if de-excitation photons are emitted from a pixel cell and towards the scintillator, it is also possible that they are subsequently reflected into a different pixel cell, thereby triggering an unwanted avalanche. Cross talk for commercial SiPMs ranges between $< 1\%$ to 35% and generally increases with the sizes of pixel cell and the array.

Another minor problem with APDs is the possibility of stray avalanches arising from the emission of charge carriers from trapping sites, an effect referred to as "afterpulsing". Electrons and holes flowing in the APD can become trapped in impurity and defect centers and are later emitted from the traps with a time constant, which strongly depends on the energy level of the trap. Whenever they are emitted, the detrapping of these charge carriers can trigger another avalanche, which may be included in the overall measurement of the initial scintillation event, thereby overestimating the light yield. Afterpulsing for commercial SiPMs ranges between 0.2% and 0.6%.

It is also possible that more than one photon strikes a pixel simultaneously (or at least within the pixel recovery time), so that only one of the photons is recorded. The compact pixel arrays are designed such that simultaneous interactions are relatively improbable, usually with negligible effect on performance. However, as the total light fluence from a scintillation event increases, and the decay time of the scintillator decreases, the probability of nearly simultaneous absorptions increases. Consequently, the measured current underestimates the measured light, producing non-linearity in the output spectrum. As the light intensity increases, the SiPM output becomes increasingly disproportionate to the number of scintillation photons produced.

A breakdown avalanche can also be triggered by thermally generated charge carriers and thus can cause problematic background count rates. The dark current, and count rate, from thermal excitation can be relatively large, ranging from 3×10^4 cps to 3×10^6 cps for commercial units and increases with pixel size and SiPM array size. This problem can be greatly mitigated by operating the device at reduced temperatures. Further, randomly triggered events ultimately produce an avalanche from a single pixel, which produces a substantially lower current than expected from a scintillation flash. Unless the device is being operated as a single photon detector for a low level light source, the use of a discriminator set at a reasonable threshold helps defeat this problem.

PROBLEMS

1. A NaI(Tl) crystal is coupled to a PMT having a Corning 7044 borosilicate window. The coupling compound Dow-Corning Q2-3067 has an index of refraction of 1.48. At the peak wavelength, determine the critical angle in the NaI(Tl) crystal for light passing through to the photocathode layer in the PMT.

2. Determine the value of $\tau(\lambda)$ for borosilicate glass and LiF at the wavelengths of 300 nm and 400 nm.

3. If there are no window absorption losses and the reflectance $R(\lambda)$ is small, show from Eqs. (14.10) and (14.11) that the total transmittance for a glass PMT window is described by

$$T(\lambda) = \frac{(1 - R(\lambda))^2}{1 - R^2(\lambda)} \simeq \frac{2n(\lambda)}{n(\lambda)^2 + 1}.$$

where one interface is surrounded by air and the other interface is at vacuum.

4. Determine the free electron concentration for Li, Na, K, Rb and Cs at absolute zero.

5. Determine the thermionic emission current density for a S-11 response PMT at a temperature of 32°C and compare it to that of a S-24 response PMT.

6. Determine the electron cutoff energy for the peak wavelength emission from a $LaBr_3$ detector coupled to a pure $CsSb_3$ photocathode. Repeat for a Na_2KSb photocathode.

7. Show that Eq. (14.37) can be cast into the form Eq. (14.38). Hint: Start with the substitution $y(x) \equiv E(x)$.

8. Given a 10-stage PMT with $\alpha = 0.72$ and a gain of 10^6 at 700 volts, find the value of k. What is the new gain if the voltage is increased to 800 volts?

9. The dark current from a PMT with an electron gain of 5×10^5 is measured to be 1.5 mA. What is the corresponding emission rate of electrons from the photodiode?

10. A $LaBr_3$:Ce detector is coupled to a PMT having a KCs photocathode in a borosilicate envelope. The detector is being used to detect gamma rays from a ^{137}Cs source. With 90% light impinging on the glass interface at all angles, determine the output charge from full energy absorption with a 10-stage PMT having the following properties: $\eta_i = 0.98$, $\eta_a = 0.98$, $\delta_i = 4$ for the first 5 stages, and $\delta_i = 3.8$ for the last five stages.

11. What are the advantages of using a microchannel PM tube compared to one of conventional design?

REFERENCES

ADAMSON, E.R.B., *Secondary Electron Emission Coefficient from Lexan: The Low Energy Crossover*, Doctoral Dissertation, Texas Tech University, Lubbock, 1993.

ADIT, http://www.aditpmt.com, accessed Jan. 15, 2016. (Now located at http://et-enterprises.com).

ANGER, H.O., "Scintillation Camera," *Rev. Sci. Instrum.*, **29** 27–33, (1958).

BAROODY, E.M., "A Theory of Secondary Electron Emission from Metals," *Phys. Rev.*, **78**, 780–787, (1950).

BRADY, J.J., "Energy Distribution of Photoelectrons as a Function of the Thickness of a Potassium Film," *Phys. Rev.*, **46**, 768–772, (1934).

BROSER, von I. AND H. KALLMANN, "Über die Anregung von Leuchtstoffen Durch Schnelle Korpuskularteilchen I," *Z. Naturforschung*, **2a**, 439-440, (1947a).

BROSER, von I. AND H. KALLMANN, "Über den Elementarprozess der Lichtanregung in Leuchtstoffen Durch Alpha-

Teilchen Schnelle Elektronen und Gamma-Quanten II," *Z. Naturforschung*, **2a**, 642-650, (1947b).

Bruining, H., *Physics and Applications of Secondary Electron Emission*, New York: McGraw-Hill, 1954.

Chevalier, P., J.P. Boutot, and G. Piétri, "A PM of New Design for High Speed Phsyics," *IEEE Trans. Nucl. Sci.*, **NS-17**, 75–78, (1970).

Coltman, J.W. and F-H. Marshall, "Photomultiplier Radiation Detector," *Nucleonics*, **1**, 58–64, (1947).

Cova, S., A. Lacaita, M. Ghioni, G. Ripamonti, and T.A. Louis, "20-ps Timing Resolution with Single-Photon Avalanche Diodes," *Rev. Sci. Instrum.*, **60**, 1104–1110, (1989).

Curran, S.C. and W.R. Baker, "A Photoelectric Alpha-Particle Detector", *U.S. Atomic Energy Commission Rpt. MDDC 1296*, November 17, 1944, declassified September 23, 1947. Later published as "Photoelectric Alpha-Particle Detector," *Rev. Sci. Instrum.*, **19**, 116, (1948).

Daggs, V.M., W.H. Parr, and A.T. Krebs, "Tests on Scintillation GM Tubes," *Nucleonics*, **10**, 54–55, (1952).

Dekker, A.J., *Solid State Physics*, Englewood Cliffs: Prentice Hall, 1957.

Dekker, A.J., "Secondary Electron Emission," in *Solid State Physics*, Vol. 6, 251–311, New York: Academic Press, 1958.

Donati, S., *Photodetectors*, Upper Saddle River: Prentice Hall, 2000.

Engstrom, R.W. and unknown others, *Photomultiplier Handbook*, Lancaster: Burle Technologies, Inc., 1980.

ET, *Photomultipliers*, `http://et-enterprises.com`, accessed Jan. 15, 2016.

ET, *Understanding Photomultipliers*, ET Enterprises, Ltd., 2011.

Evans, D.S., "Low Energy Charged-Particle Detection Using the Continuous-Channel Electron Multiplier," *Rev. Sci. Instrum.*, **36**, 375–382, (1965).

Fiorini, C., A. Longoni, F. Perotti, C. Labanti, P. Lechner, and L. Strüder, "First Experimental Results of a New Gamma-Ray Detector Based on a Silicon Drift Chamber Coupled to a Scintillator," *Nucl. Instr. Meth.*, **A409**, 395–399, (1998).

Fiorini, C., A. Gola, M. Zanchi, A. Longoni, P. Lechner, H. Soltau, and L. Strüder, "Gamma-Ray Spectroscopy with $LaBr_3$:Ce Scintillator Readout by a Silicon Drift Detector," *IEEE Trans. Nucl. Sci.*, **NS-53**, 2392–2397, (2006).

Fiorini, C., A. Gola, M. Zanchi, A. Longoni, H. Soltau, and L. Strüder, "Silicon Drift Photodetectors for Scintillation Readout in Medical Imaging," *Nucl. Instr. Meth.*, **A571**, 126–129, (2007).

Flyckt, S-O. and C. Marmonier, *Photomultiplier Tubes Principles and Applications*, Brive: Photonis, Inc., 2002.

Fowler, R.H. and L. Nordheim,"Electron Emission in Intense Electric Fields, *Proc. R. Soc. Lond. A*, **119**, 173–181, 1928.

Frank, J., *IEEE Standard Dictionary of Electrical and Electronic Terms*, New York: Wiley, 1977.

Gatti, E. and V. Svelto, "Review of Theories and Experiments of Resolving Time with Scintillation Counters," *Nucl. Instrum. Meth.*, **43**, 248–268, (1966).

Glover, A.M., "A Review of the Development of Sensitive Phototubes," *Proc. IRE*, **29**, 413–423, (1941).

Haitz, R.H., "Model of the Electrical Behavior of a Microplasma," *J. Appl. Phys.*, **35**, 1370–1376, (1961).

Halas, S. and T. Durakiewics, "Work Functions of Elements Expressed in Terms of the Fermi Energy and the Density of Free Electrons," *J. Phys., Cond. Matter*, **10**, 10815-10826, (1998).

Hamamatsu, *Photomultiplier Tubes: Basics and Applications*, Ed. 3a, Hamamatsu Photonics K.K., 2007.

Hamamatsu, *Photomultiplier Tubes (PMTs)*, `https://www.hamamatsu.com/us/en/product/optical-sensors/pmt/index.html`, accessed Jan. 15, 2016.

Hecht, E., *Optics, 2nd. Ed.*, Reading: Addison-Wesley, 1987.

Hertz, H., "Über den Einfluss des Ultravioletten Lichtes auf die Electrische Entladung," *A. Physik*, **267**, 983-1000, (1887).

Hofstadter, R., "Alkali Halide Scintillation Counters," *Phys. Rev.*, **74**, 100, (1948).

Huth, G.C., J.B. Trice, and R.A. Mckinney, "Internal Pulse Amplification in Silicon p-n Junction Radiation Detection Junctions ," *Rev. Sci. Instrum.*, **35**, 1220–1222, (1966).

Iams, H.E. and B. Salzberg, "The Secondary Emission Phototube," *Proc. IRE*, **23**, 55–64, (1935).

Iwanczyk, J.S. and B.E. Patt, "Electronic for X-Ray and Gamma-Ray Detectors," in *Semiconductors for Room Temperature Nuclear Detector Applications*, R.B. James and T.E. Schlesinger, Eds., 531–560, New York: Academic Press, 1985.

Janes, R.B. and A.M. Glover, "Recent Developments in Phototubes," *RCA Review*, **6**, 43–54, (1941).

Johnson, J.B. and K.G. McKay, "Secondary Electron Emission from Germanium," *Phys. Rev.*, **93**, 668–672, (1954).

Johnson, S.M., jr., "Radiation Effects on Multiplier Phototubes," *IEEE Trans. Nucl. Sci.*, **NS-20**, 313–124, (1973).

Kerns, C.R., "A High Rate Phototube Base," *IEEE Trans. Nucl. Sci.*, **NS-24**, 353–355, (1977).

Kollath, R., "Secondary Electron Emission from Solids," *Phys. Zeitschrift*, **38**, 202–224, (1937).

Kong, S.H., J. kinross-wright, D.C. Nguyen, and R.L. Sheffield, "Cesium Telluride Photocathodes," *J. Appl. Phys.*, **77**, 6031–6038, (1995).

Krall, H.R., "Extraneous Light Emission from Photomultipliers", *IEEE Trans. Nucl. Sci.*, **NS-14**, 455–459, (1967).

Krebs, A.T., "Über die Messung Radioaktiver Niederschläge mit Hilfe von Spitzenzähler und Zählrohr," *Phys. Zeit.*, **35**, 796–800, (1934).

Krebs, A.T., "Ein Demonstrationsversuch zur Emanationsdiffusion," *Annalen der Physik*, **39**, 330–332, (1941).

Krebs, A.T., "Early History of the Scintillation Counter," *Science*, **122**, 17–18, (1955).

Kubetsky, L.A., "Multiple Amplifier," *Proc. IRE*, **25**, 412–433, (1937).

Lamaze, G.P., J.K. Whittaker, R.A. Schrack, and O.A. Wasson, "After-Pulse Suppression for 8850 and 8854 Photomultipliers," *Nucl. Instrum. Meth.*, **123**, 403–404, (1975).

Lubsandorzhiev, B.K., "On the History of Photomultiplier Tube Invention," *Nucl. Instrum. Meth.*, **A567**, 236–238, (2006).

Lye, R.G. and A.J. Dekker, "Theory of Secondary Emission," *Phys. Rev.*, 977–981, (1957).

Madelung, O., U. Rössler, and M. Schulz, Eds., *Landolt-Börnstein - Group III Condensed Matter Numerical Data and Functional Relationships in Science and Technology*, Vol. 41C: Non-Tetrahedrally Bonded Elements and Binary Compounds I, Berlin: Springer-Verlag, 1998.

Mandeville, C.E. and M.V. Scherb, "Photosensitive Geiger Counters: Their Applications," *Nucleonics*, **7**, 34–38, (1950).

MARKAKIS, J.M., "Mercuric Iodide Photodetector-Cesium Iodide Scintillator Gamma Ray Spectrometers," *IEEE Trans. Nucl. Sci.*, **35**, 356–359, (1988).

MARSHALL, F.H. AND J.W. COLTMAN, "A Photomultiplier Radiation Detector," *Phys. Rev.*, **72**, 528, (1947).

MARSHALL, F.H., J.W. COLTMAN, AND L.P. HUNTER, "The Photomultiplier X-Ray Detector," *Rev. Sci. Instrum.*, **18**, 504–513, (1947).

MCINTYRE, R.J., "Theory of Microplasma Instability of Silicon," *J. Appl. Phys.*, **32**, 983–995, (1961).

MCINTYRE, R.J., "Multiplication Noise in Uniform Avalanche Diodes," *IEEE Trans. Elec. Dev.*, **ED-13**, 164–168, (1966).

MELCHIOR, H. AND W.T. LYNCH "Signal and Noise Response of High Speed Germanium Avalanche Photodiodes," *IEEE Trans. Elec. Dev.*, **ED-13**, 829–838, (1966).

MELCHIOR, H., "Demodulation and Photodetection Techniques," in *Laser Handbook*, F.T. AERCCHI, E.O. SCHULTZ-DUBOIS, Eds., **Vol. 1**, 725–835, Amsterdam: North-Holland, 1972.

MILLER, S.L., "Avalanche Breakdown in Germanium," *Phys. Rev.*, **99**, 1234–1241, (1955).

MIRZOYAN, R., E. LORENZ, D. PETRY, AND C. PPROSCH "On the Influence of Afterpulsing in PMTs on the Trigger Threshold of Multichannel Light Detectors in Self-Trigger Mode," *Nucl. Instrum. Meth.*, **A387**, 74–78, (1997).

MORTON, G.A., H.M. SMITH, AND R. WASSERMAN, "Afterpulses in Photomultipliers," *IEEE Trans. Nucl. Sci.*, **NS-14**, 443–448, (1967).

MORTON, G.A., "The Scintillation Counter Story, Part II," *IEEE Trans. Nucl. Sci.*, **NS-22**, 26–28, (1975a).[17]

MORTON, G.A., "The Scintillation Counter Story, Part III," *IEEE Trans. Nucl. Sci.*, **NS-22**, 29–30, (1975b).

NIIGAKI, M., T. HIROHATA, T. SUZUKI, H. KAN, AND T. HIRUMA, "Field-Assisted Photoemission from InP/InGaAsP Photocathode with p/n Junction," *Appl. Phys. Lett.*, **71**, 2493–2495, (1997).

NIIGAKI, M., T. HIROHATA, AND H. MIMURA, "Room Temperature Photoemission up to a Wavelength Threshold of 2.3 μm from n^+-$InAs_{0.4}P_{0.6}/p^-$-$InAs_{0.4}P_{0.6}/p^-$-$In_{0.7}Ga_{0.3}As$ Field-Assisted Photocathode," *Appl. Phys. Express*, **7**, 112201(1-4), (2014).

PIEMONTE, C., "Development of Silicon Photomultipliers at IRST," Research Techniques Seminar, Fermi National Accelerator Laboratory, Oct. 25, 2006; accessed at `http://www-ppd.fnal.gov/eppoffice-w/Research\_Techniques\_Seminar/` on Dec. 4, 2014.

PIÉTRI, G., "Toward Picoseond Resolution. Contribution of Microchannel Electron Multipliers to Vacuum Tube Design," *IEEE Trans. Nucl. Sci.*, **NS-22**, 2084–2092, (1975).

POLYANSKIY, M.N., Refractive Index Database, `http://refractive index.info`, accessed July 28, 2014.

RAJCHMAN, J.A. AND R.L. SNYDER, "An Electrostatically Focused Multiplier Phototube," *Electronics*, **13**, 20, (1940).

RENKER, D., "Geiger-Mode Avalanche Photodiodes, History, Properties and Problems," *Nucl. Instrum. Meth.*, **A567**, 48–56, (2006).

ROUGEOT, H. AND C. BAUD, "Negative Electron Affinity Photoemitters," *Advances in Electronics and Electron Physics*, **48**, 1–36, (1979).

RUTHERFORD, E. AND H. GEIGER, "An Electrical Method of Counting the Number of α-Particles from Radio-Active Substances," *Proc. Royal Soc. London, Series A*, **81**, 141–161, (1908).

SCHERB, M.V., "Photoelectric Effect in Self-Quenching Geiger-Mueller Counters," *Phys. Rev.*, **73**, 86–87, (1948).

SCHMIDT, K.C. AND C.F. HENDEE, "Continous Channel Electron Multiplier Operated in the Pulse Saturated Mode," *IEEE Trans. Nucl. Sci.*, **NS-13**, 100–111, (1966).

SCHOTT, *Technical Glasses, Physical and Technical Properties*, Schott North America, Inc., Duryea, PA, 2014; and Schott transmission data sheets for 8337B, 8245, and 8250 glasses.

SEILER, E.F., "Color-Sensitiveness of Photo-Electric Cells," *Astrophys. J.*, **52**, 129–153, (1920).

SIMON, R.E. AND B.F. WILLIAMS, "Secondary Electron Emission," *IEEE Trans. Nucl. Sci.*, **NS-15**, 167–170, (1968).

SLEPIAN, J., U.S. Patent 1,450,265, allowed April 3, 1923.

SOMMER, A.H., *Photoemissive Materials*, New York: Wiley, 1968.

SPINDT, C.A., I. BRODIE, L. HUMPHREY, AND E.R. WESTERBERG, "Physical Properties of Thin-Film Field Emission Cathodes with Molybdenum Cones," *J. Appl. Phys.*, **47**, 5248–5263, (1976).

SZE, S.M., *Physics of Semiconductor Devices, 2nd. Ed.*, New York: Wiley, 1981.

WHETTEN, N.R. AND A.B. LAPONSKY, "Secondary Electron Emission of Single Crystals of MgO," *J. Appl. Phys.*, **28**, 515–516, (1957).

WILEY, W.C. AND C.F. HENDEE, "Electron Multipliers Utilizing Continous Strip Surfaces," *IRE Trans. Nucl. Sci.*, **NS-9**, 103–106, (1962).

WHIDDINGTON, R., "The Transmission of Cathode Rays Through Matter," *Proc. Roy. Soc. Lond. (A)*, **86**, 360-370, (1912).

YOUNG, J.R., "Penetration of Electrons in Aluminum Oxide Films," *Phys. Rev.*, **103**, 292–293, (1956).

ZWORYKIN, V.K., G.A. MORTON, AND L. MALTER, "The Secondary-Emission Multiplier - A New Electronic Device," *Proc. IRE*, **24**, 351–375, (1936).

ZWORYKIN, V.K. AND J.A. RAJCHMAN, "The Electrostatic Electron Multiplier," *Proc. IRE*, **27**, 558–566, (1939).

[17] Interestingly, there is no "Part I" by G.A. Morton. In the beginning of the article, Morton mentions this paper as being added to the story written by R.A. Hofstadter which was published directly before this work in *IEEE Trans. Nucl. Sci.*, NS-22, 13–25, (1975a). Hence, this reference is "Part II" to the story begun by Hofstadter.

Chapter 15

Basics of Semiconductor Detector Devices

It has today occurred to me that an amplifier using semiconductors rather than vacuum is in principle possible.

William Shockley

15.1 Introduction

Semiconductors are much better for energy spectroscopy than are gas-filled or scintillation detectors because semiconductors have vastly superior energy resolution. This improved resolution is largely due to the better statistics caused by the larger numbers of signal carriers (charges) produced by a radiation interaction. On average, it takes only 3 to 5 eV to produce an electron-hole pair in a semiconductor. By comparison, it takes between 25 and 40 eV to produce an electron-ion pair in a gas-filled detector and between 100 eV and 1 keV to produce a single photoelectron ejection from the photocathode of a photomultiplier tube (PMT) in a scintillation/PMT detector, a consequence of light reflections and poor quantum efficiency at the photocathode. Hence, a semiconductor produces considerably more charge carriers from the primary ionization event and, thereby reduces the statistical fluctuations in the energy resolution.

Semiconductor radiation detectors have many different shapes, sizes, and configurations, yet there are some basic designs that, in some form or another, can be attributed to practically all semiconductor radiation detectors. These basic designs include *pn*-junction diodes, *pin*-junction diodes, Schottky diodes, resistive detectors, and photoconductors. The vast field of semiconductor device physics is much too large to describe in a single book chapter. As a result, only those concepts needed to understand basic detector operations and characteristics are offered here.

The general purpose of this text is to familiarize the reader with various aspects of radiation detectors so that the skills to design, fabricate, characterize, and deploy such devices are acquired. Proficiency in detector design and fabrication requires an understanding of basic solid state and semiconductor physics, hence the reason for the preceding and present chapters. In this chapter, the basic physics used to make semiconductors into radiation detectors is summarized. In this chapter much of the notation introduced in Chapter 12 is used extensively without any further explanation. Finally, values of the many physical constants used here and in other chapters are tabulated in Appendix A.

15.2 Charge Carrier Collection

Charge carriers in the conduction band spontaneously drop back into empty states in the valence band, while, at the same time, other electrons gain energy to cross the band gap. Overall, a somewhat constant density of electrons and holes remains in the conduction and valence bands, respectively. The time over which a charge carrier remains in either band is altered by impurity and defect states that appear in the band gap.

These traps and defects increase the probability of electrons or holes moving from either band edge to an intermediate state before completing the recombination process. If a charge carrier falls into a defect state, it is referred to as "trapped"; whereas, if the charge carrier's journey is completed, e.g., an electron falls completely back into a hole in the valence band, it is referred to as having "recombined". The average time over which an excited electron remains in the conduction band before being trapped or recombining is the *electron lifetime* τ_e. Similarly, the average time in which a hole remains in the valence band before being trapped or recombining is the *hole lifetime* τ_h.

15.2.1 Charge Carrier Generation, Recombination and Injection

The concentration of charge carriers in the conduction and valence bands of a semiconductor is temperature dependent. As the population of electrons and holes increases, the probability that electrons return to the valence band and recombine with holes also increases. The overall process reaches a thermal equilibrium, in which $np = n_i^2$.

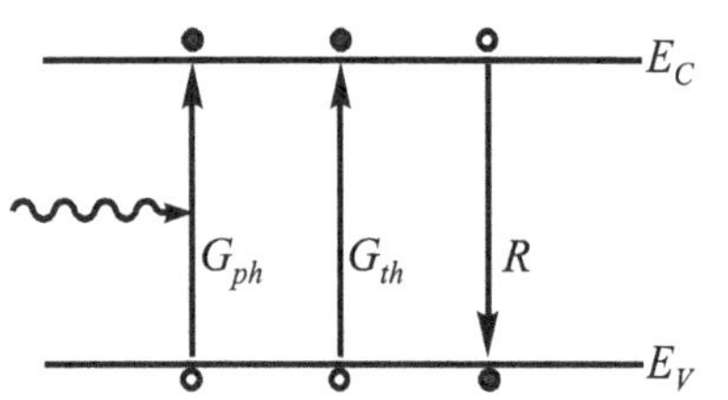

Figure 15.1. Thermal generation G_{th}, photo-generation G_{ph}, and recombination R of electron-hole pairs across the band gap of a semiconductor.

Other sources of energy deposition, such as the photoelectric effect, can excite charge carriers from the valence band into the conduction band (see Fig. 15.1). The introduction of these extra charge carriers is commonly referred to as charge carrier *injection*, and produces a non-equilibrium condition in the semiconductor. The continuous constant injection of charge may reach a steady-state, but the semiconductor cannot reach its equilibrium until the source of the charge injection is stopped. Semiconductor radiation detectors rely upon the charge injection process to excite electron-hole pairs, in which the deposition of radiation energy in the semiconductor produce excess electron and hole populations, Δn and Δp, thereby producing the condition $np > n_i^2$. These excess charge carriers can be measured to indicate the presence, and often energy, of the radiation particles interacting in the semiconductor.

15.2.2 Radiative Recombination

Consider a semiconductor material, initially at equilibrium, that has a package of charge Δn injected into the bulk material. Because these charges are excited from the valence band, an equal package of excess holes Δp appears in the valence band. Let the recombination rate with an excess of electrons be $r_n(n,p)$ and the equilibrium recombination rate be $r_n(n_o,p_o)$, in which the notation n_o and p_o are used for the equilibrium electron and hole concentrations. The change in electron recombination rate from the equilibrium case after charge injection is

$$R_n = r_n(n,p) - r_n(n_o,p_o). \tag{15.1}$$

Likewise, for hole recombination,

$$R_p = r_p(n,p) - r_p(n_o,p_o). \tag{15.2}$$

The electron lifetime is defined as the excess electron concentration divided by the change in recombination rate, i.e.,

$$\tau_n = \frac{n - n_o}{R_n} = \frac{\Delta n}{R_n} \tag{15.3}$$

and the hole lifetime is

$$\tau_p = \frac{p - p_o}{R_p} = \frac{\Delta p}{R_p}. \tag{15.4}$$

Radiative electron recombination refers to a direct transition from the conduction band to the valence band, during which a photon is released by the de-excitation of the electron. This process can occur only if a hole is available to complete the recombination process. Hence, the recombination rate is proportional to the product of electron and hole concentrations, i.e., $r_n = r_p = \beta np$, where β is an empirical, radiative-recombination, proportionality constant. From Eq. (15.1), the change in electron recombination after carrier injection becomes

$$R_n = R_p = \beta(np - n_i^2). \tag{15.5}$$

Because $\Delta n = \Delta p$, τ_n must equal τ_p. Substitution of Eq. (15.5) into Eq. (15.3) yields

$$\tau_n = \tau_p = \frac{n - n_o}{\beta(np - n_i^2)} = \frac{\Delta n}{\beta([n_o + \Delta n][p_o + \Delta n] - n_i^2)} = \frac{1}{\beta(n_o + p_o + \Delta n)}. \tag{15.6}$$

Here the equilibrium condition $n_o p_o = n_i^2$ has been invoked. This result shows that radiative recombination is inversely proportional, in all circumstances, to the total concentration of charge carriers.

15.2.3 Shockley-Read-Hall Recombination

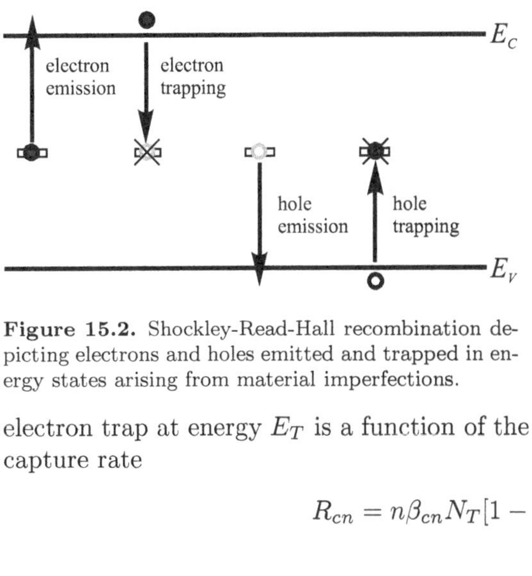

Figure 15.2. Shockley-Read-Hall recombination depicting electrons and holes emitted and trapped in energy states arising from material imperfections.

Material imperfections often introduce energy states or traps in the band gap that, in turn, can participate in the recombination process. Typically, this trap-assisted recombination, known as *Shockley-Read-Hall* (SRH) recombination, is more likely to occur than are direct radiative transitions (see arguments in Shockley and Read [1952], Hall [1952, 1959] and Varshni [1967]). The processes of SRH trapping are illustrated in Fig. 15.2.

Consider a concentration of imperfections in a semiconductor with trapping states at energy E_T. These trapping sites may arise from native defects, such as anti-sites, voids and interstitials, or may result from impurities in the material. The probability that an electron is captured by an electron trap at energy E_T is a function of the *unfilled* trap density N_T and the electron population n, with capture rate

$$R_{cn} = n\beta_{cn}N_T[1 - f_{\mathcal{FD}}(E_T)] = \frac{n\beta_{cn}N_T}{1 + \exp\left[\dfrac{E_F - E_T}{kT}\right]}, \tag{15.7}$$

where β_{cn} is a proportionality constant for electron capture. For those traps that are occupied by an electron, the rate at which electrons are emitted into the conduction band is

$$R_{en} = \beta_{en}N_T f_{\mathcal{FD}}(E_T) = \frac{\beta_{en}N_T}{1 + \exp\left[\dfrac{E_T - E_F}{kT}\right]}, \tag{15.8}$$

where β_{en} is a proportionality constant for electron emission.

At equilibrium, the rate at which electrons arrive in the conduction band must be equal to the rate at which they are captured from the conduction band, i.e.,

$$\beta_{en} = \beta_{cn}n_o\frac{[1 - f_{\mathcal{FD}}(E_T)]}{f_{\mathcal{FD}}(E_T)} = \beta_{cn}n_o\exp\left[\frac{E_T - E_F}{kT}\right]. \tag{15.9}$$

where n_o is the equilibrium electron concentration. For a non-degenerate semiconductor, n_o is described by Eq. (12.126), which when substituted into Eq. (15.9) gives

$$\beta_{en} = \beta_{cn} N_C \exp\left[\frac{E_T - E_C}{kT}\right]. \tag{15.10}$$

In the case that the trap energy E_T is equal to E_F

$$\beta_{en} = \beta_{cn} N_C \exp\left[\frac{E_F - E_C}{kT}\right] = \beta_{cn} n_1, \tag{15.11}$$

where n_1 is the electron concentration in the conduction band *if* $E_F = E_T$. The net rate R_n at which electrons are captured from the conduction band is the difference between Eq. (15.7) and Eq. (15.8), so that with Eq. (15.10) one obtains

$$R_n = R_{cn} - R_{en} = \beta_{cn} N_T \{n[1 - f_{\mathcal{FD}}(E_T)] - n_1 f_{\mathcal{FD}}(E_T)\}. \tag{15.12}$$

In the same manner used above for electrons, the rate at which holes are captured from the valence band into traps or, equivalently the rate at which traps are filled with electrons is

$$R_{cp} = p\beta_{cp} N_T f_{\mathcal{FD}}(E_T) = \frac{p\beta_{cp} N_T}{1 + \exp\left[\dfrac{E_T - E_F}{kT}\right]}, \tag{15.13}$$

where β_{cp} is the proportionality constant for hole capture. The rate at which holes are emitted from the trap as electrons move from the valence band into the traps is

$$R_{ep} = \beta_{ep} N_T [1 - f_{\mathcal{FD}}(E_T)] = \frac{\beta_{ep} N_T}{1 + \exp\left[\dfrac{E_F - E_T}{kT}\right]}, \tag{15.14}$$

where β_{ep} is the proportionality constant for hole emission. Making the emission and capture rates equal yields,

$$\beta_{ep} = \beta_{cp} p_o \frac{f_{\mathcal{FD}}(E_T)}{[1 - f_{\mathcal{FD}}(E_T)]} = \beta_{cp} p_o \exp\left[\frac{E_F - E_T}{kT}\right]. \tag{15.15}$$

At equilibrium, the hole capture and emission rates must be equal and the hole concentration in the conduction band is p_o. Substitution of Eq. (12.128) for p_o into Eq. (15.15), the emission rate for holes is found to be

$$\beta_{ep} = \beta_{cp} N_V \exp\left[\frac{E_V - E_T}{kT}\right]. \tag{15.16}$$

If the trap energy E_T is equal to E_F

$$\beta_{ep} = \beta_{cp} N_V \exp\left[\frac{E_V - E_F}{kT}\right] = \beta_{cp} p_1. \tag{15.17}$$

where p_1 is the hole concentration in the valence band at $E_F = E_T$. The net rate at which holes are captured from the valence band is the difference between Eq. (15.13) and Eq. (15.14), and with $\beta_{ep} = \beta_{cp} p_1$ from Eq. (15.17),

$$R_p = R_{cp} - R_{ep} = \beta_{cp} N_T \{p f_{\mathcal{FD}}(E_T) - p_1 [1 - f_{\mathcal{FD}}(E_T)]\}. \tag{15.18}$$

Unlike pure radiative recombination, with carrier injection, it is quite possible that $\Delta n \neq \Delta p$, mainly because the carrier capture coefficients for electrons and holes may be quite different. This issue becomes more important as the trap density in a semiconductor increases. For the following derivation, however, it is assumed that the trap density is relatively low so that, even in the non-equilibrium condition, $\Delta n \simeq \Delta p$. Hence, Eq. (15.12) and Eq. (15.18) are equal. Equating the right-hand sides of Eq. (15.12) and Eq. (15.18) and solving for $f_{\mathcal{FD}}(E_T)$ one obtains

$$f_{\mathcal{FD}}(E_T) = \frac{\beta_{cp}p_1 + \beta_{cn}n}{\beta_{cn}(n+n_1) + \beta_{cp}(p+p_1)}. \tag{15.19}$$

Substitute this result into either Eq. (15.12) or Eq. (15.18) to find

$$R_n = R_p = \frac{\beta_{cn}\beta_{cp}N_T(pn - n_i^2)}{\beta_{cn}(n+n_1) + \beta_{cp}(p+p_1)}, \tag{15.20}$$

in which it is recognized that $n_1 p_1 = n_i^2$. Division of the numerator and denominator of Eq. (15.20) by $\beta_{cn}\beta_{ce}N_T$ gives

$$R_n = R_p = \frac{pn - n_i^2}{\dfrac{1}{\beta_{cp}N_T}(n+n_1) + \dfrac{1}{\beta_{cn}N_T}(p+p_1)}. \tag{15.21}$$

Finally, define the trapping time for electrons as $\tau_n = 1/(\beta_{cn}N_T)$ and the trapping time for holes as $\tau_p = 1/(\beta_{cp}N_T)$, and substitute these definitions into Eq. (15.22) to obtain the general expression for SRH recombination rates, namely

$$\boxed{R_n = R_p = \frac{pn - n_i^2}{\tau_p(n+n_1) + \tau_n(p+p_1)}.} \tag{15.22}$$

Note that at equilibrium $np = n_i^2$ and Eq. (15.22) at the net recombination rate is zero.

15.2.4 Equations of Continuity

In a semiconductor, even in non-steady-state conditions, the hole generation rate must always equal the electron generation rate, i.e., $g_n = g_p$. Similarly, the hole and electron recombination rates must always be equal, i.e., $n/\tau_n = p/\tau_p$. The generation rates, carrier concentrations and lifetimes are, generally, functions of position $\mathbf{r}$ in the crystal and time.

In this section the continuity equations that determine the free electron concentration $n(\mathbf{r}, t)$ and the hole concentration $p(\mathbf{r}, t)$ are developed. Begin by first considering the free electrons in a small but arbitrary volume V of a semiconductor. In this volume, the following balance relation must apply:

$$\left\{\begin{array}{l}\text{the net}\\ \text{rate of}\\ \text{increase in}\\ V \text{ of free}\\ \text{electrons}\end{array}\right\} = \left\{\begin{array}{l}\text{generation}\\ \text{rate in } V\\ \text{of free}\\ \text{electrons}\end{array}\right\} - \left\{\begin{array}{l}\text{rate of recom-}\\ \text{bination of}\\ \text{electrons with}\\ \text{holes in } V\end{array}\right\} - \left\{\begin{array}{l}\text{rate at which}\\ \text{free electrons}\\ \text{flow out of } V\\ \text{through the}\\ \text{surrounding}\\ \text{surface } S\end{array}\right\}. \tag{15.23}$$

In symbols, this balance relation is written as

$$\int_V dV\,\frac{\partial n}{\partial t} = \int_V dV\,g_n - \int_V dV\,\frac{n}{\tau_n} - \int_S dS\,\mathbf{J}_n \bullet \hat{\mathbf{n}}, \tag{15.24}$$

where dS is a differential element of the surface with outward normal $\hat{\mathbf{n}}$ and $\mathbf{J}_n(\mathbf{r},t)$ is the current vector, i.e., the flow rate of free electrons across a unit area perpendicular to $\mathbf{J}_n$.

The last term in Eq. (15.24) can be written as a volume integral by use of Gauss's theorem, namely

$$\int_S dS\,\mathbf{J}_n\bullet\hat{\mathbf{n}} = \int_V dV\,\boldsymbol{\nabla}\bullet\mathbf{J}_n. \tag{15.25}$$

With this result, all the terms in Eq. (15.24) can be combined into a single integral, namely

$$\int_V dV\left\{\frac{\partial n}{\partial t} - g_n + \frac{n}{\tau_n} + \boldsymbol{\nabla}\bullet\mathbf{J}_n\right\} = 0. \tag{15.26}$$

But V is arbitrary, and the only integrand that integrates to zero for arbitrary integration limits is the zero integrand. Hence the balance condition is

$$\frac{\partial n}{\partial t} = -\boldsymbol{\nabla}\bullet\mathbf{J}_n + g_n - \frac{n}{\tau_n}. \tag{15.27}$$

A similar derivation can be made to derive a balance relation for the hole concentration $p(\mathbf{r},t)$. The result is

$$\frac{\partial p}{\partial t} = -\boldsymbol{\nabla}\bullet\mathbf{J}_p + g_p - \frac{p}{\tau_n}. \tag{15.28}$$

Equations (15.27) and (15.28) are often referred to as the *continuity* equations.

Before these equations can be used to find $n(\mathbf{r},t)$ and $p(\mathbf{r},t)$, the hole and electron current vectors must be written in terms of the concentrations. This can be done by writing $\mathbf{J}_n$ and $\mathbf{J}_p$ as a sum of a current from a diffusion process and a drift current caused by an electric field $\boldsymbol{\mathcal{E}}$.[1] In the diffusion process, charge carriers are assumed to flow away from regions of higher concentration according to Fick's law.[2] The currents of the free electrons and holes are written as

$$\mathbf{J}_n = -D_n\boldsymbol{\nabla}n - n\mu_n\boldsymbol{\mathcal{E}}, \tag{15.29}$$

$$\mathbf{J}_p = -D_p\boldsymbol{\nabla}p + p\mu_p\boldsymbol{\mathcal{E}}, \tag{15.30}$$

where μ_n is the electron mobility and μ_p is the hole mobility. The diffusion coefficients can be evaluated in terms of mobilities as [McKelvey 1966]

$$D_n = \frac{\mu_n kT}{q_e} \qquad \text{and} \qquad D_p = \frac{\mu_p kT}{q_e}, \tag{15.31}$$

which are often called the Einstein relationships.

Substitution of Eqs. (15.29) and (15.30) into Eqs. (15.27) and (15.28) gives

$$\frac{\partial n}{\partial t} = D_n\nabla^2 n + \mu_n\boldsymbol{\nabla}\bullet(n\boldsymbol{\mathcal{E}}) + g_n - \frac{n}{\tau_n}, \tag{15.32}$$

$$\frac{\partial p}{\partial t} = D_p\nabla^2 p - \mu_p\boldsymbol{\nabla}\bullet(p\boldsymbol{\mathcal{E}}) + g_p - \frac{p}{\tau_p}. \tag{15.33}$$

[1] It is assumed here that there are negligible magnetic fields.

[2] This is analogous to Fourier's law of heat conduction, in which heat energy flows away from regions at higher temperature into regions of lower temperature. Thus, free electrons flow, as a result of their random thermal motion, in the opposite direction of the electron density's greatest rate of increase, i.e., $\mathbf{J}_n = -D_n\boldsymbol{\nabla}n$, where D_n is the *diffusion coefficient.*

The drift terms in these continuity equations can be put into a more tractable form by use of the vector identity

$$\nabla \cdot (n\boldsymbol{\mathcal{E}}) = \boldsymbol{\mathcal{E}} \cdot \nabla n + n\nabla \cdot \boldsymbol{\mathcal{E}}. \quad (15.34)$$

Finally, it is convenient to express the generation rate (g_n or g_p) as the sum of the thermal generation rate (g_{on} or g_{op}) plus the rate at which the carriers are produced in excess of the thermal generation rate (g'_n or g'_p), i.e.,

$$g_n = g_{on} + g'_n \qquad \text{and} \qquad g_p = g_{op} + g'_p. \quad (15.35)$$

With these alterations, the continuity equations assume the form

$$\boxed{\begin{aligned} \frac{\partial n}{\partial t} &= D_n \nabla^2 n + \mu_n(\boldsymbol{\mathcal{E}} \cdot \nabla n + n\nabla \cdot \boldsymbol{\mathcal{E}}) + g'_n - \left(\frac{n}{\tau_n} - \frac{n_o}{\tau_n}\right) \\ \frac{\partial p}{\partial t} &= D_p \nabla^2 p - \mu_p(\boldsymbol{\mathcal{E}} \cdot \nabla p + p\nabla \cdot \boldsymbol{\mathcal{E}}) + g'_p - \left(\frac{p}{\tau_p} - \frac{p_o}{\tau_p}\right) \end{aligned}} \quad (15.36)$$

An Example of the Use of the Continuity Equation: Consider the case in which charge is injected, continuously at a constant rate, into the end of a semiconductor block without an electric field present, as depicted in Fig. 15.3. Eventually, the distribution of free electrons assumes a steady-state distribution along the length of the semiconductor. The one-dimensional, steady-state form of Eq. (15.36), with $\boldsymbol{\mathcal{E}} = 0$, gives the following equation for the right-hand portion of the semiconductor in which no charge is injected ($g'_n = 0$).

$$D_n \frac{d^2 n(x)}{dx^2} - \frac{n(x) - n_o}{\tau_n} = 0. \quad (15.37)$$

The general solution for $n(x) - n_o$ is

$$n(x) - n_o = \Delta n(x) = C_1 \exp\left[\frac{x}{\sqrt{D_n \tau_n}}\right] + C_2 \exp\left[\frac{-x}{\sqrt{D_n \tau_n}}\right], \quad (15.38)$$

where C_1 and C_2 are arbitrary constants and $\sqrt{D_n \tau_n}$ is the electron diffusion length L_n. Application of the boundary conditions that as $x \to \infty$ $n(x,t)$ must remain finite implies C_1 must be zero. So in the right-hand portion of the semiconductor the electron density must vary as

$$n(x) = n_o + C_2 \exp\left[\frac{-x}{\sqrt{D_n \tau_n}}\right], \quad (15.39)$$

where C_2 depends on $n(x)$ in the injection region. At the location $x = 0$, the injected charge is Δn, so that

$$n(x) = n_o + \Delta n \exp\left[\frac{-x}{\sqrt{D_n \tau_n}}\right], \quad (15.40)$$

from which the injected electron population is seen to exponentially decrease with x.

In a similar fashion, if holes are injected into the left-hand end of the semiconductor. The steady-state hole density $p(x)$ in the right portion of the semiconductor where there is no charge injection is the solution of

$$D_p \frac{d^2 p(x)}{dx^2} - \frac{p(x) - p_o}{\tau_p} = 0. \quad (15.41)$$

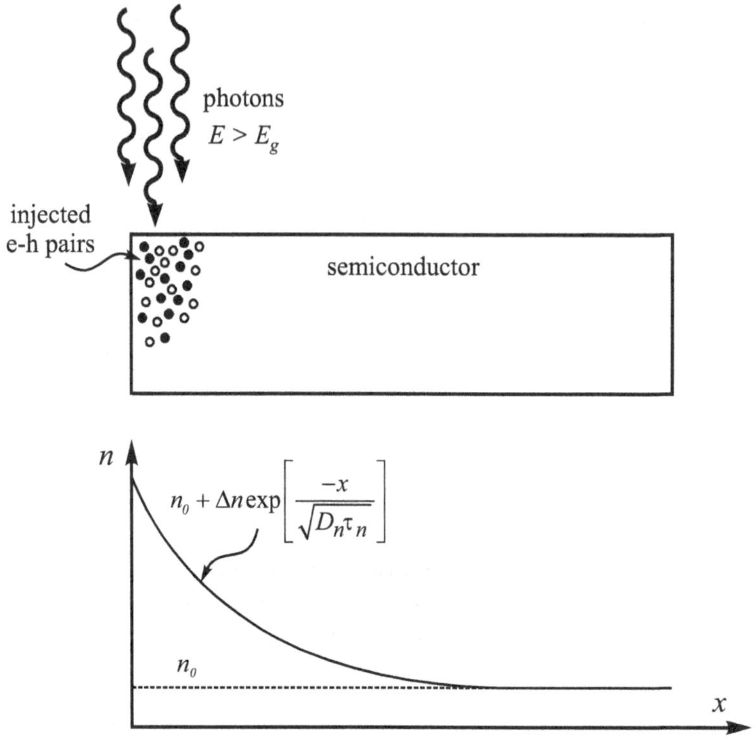

Figure 15.3. Electron-hole pairs can be injected through the photoelectric effect. Their population decreases through recombination as they diffuse through the semiconductor.

The general solution for $p(x) - p_o$ is

$$p - p_o = \Delta p(x) = C_3 \exp\left[\frac{x}{\sqrt{D_p \tau_p}}\right] + C_4 \exp\left[\frac{-x}{\sqrt{D_p \tau_p}}\right], \tag{15.42}$$

where again C_3 and C_4 are arbitrary constants and $\sqrt{D_p \tau_p}$ is the hole diffusion length L_p. Because p must remain finite as $x \to \infty$, C_3 must vanish so the solution to the present problem is

$$p(x) = p_o + C_4 \exp\left[\frac{-x}{\sqrt{D_p \tau_p}}\right]. \tag{15.43}$$

At the location $x = 0$, the injected charge is Δp so the above result becomes

$$p(x) = p_o + \Delta p \exp\left[\frac{-x}{\sqrt{D_n \tau_n}}\right], \tag{15.44}$$

from which the injected hole population is seen to exponentially decrease with increasing x.

15.3 Basic Semiconductor Detector Configurations

Semiconductor detectors can be fashioned into many different configurations, including junction diodes, Schottky barrier diodes, photoconductors, and photoresistors. To select the most appropriate configuration, one must consider the semiconductor material and the radiation detection application. Some semiconductor materials are composed of substances that are not easily converted into junction diodes, such as HgI_2 and PbI_2, whereas other semiconductors, such as Si and Ge, are easily fashioned into junction devices. Some materials, such as GaAs, can be fabricated easily into either photoconductors or Schottky barrier devices. Limited doping selection and other chemical constraints prevent some materials to be configured easily as either reverse biased diodes or photoconductors. The following section introduces the reader to basic semiconductor configurations.

15.3.1 The *pn* Junction

If two blocks of semiconductor material, one doped with concentration N_D donors and the other doped with concentration N_A acceptors, are brought into contact, then the *p*-type side of the junction boundary has an excess of free hole charge carriers and the *n*-type side of the junction has an excess of free electron charge carriers (see Fig. 15.4). The concentration gradient across the junction boundary causes holes to diffuse across the boundary into the *n*-type side, and electrons to diffuse to the *p*-type side (see Fig. 15.5). The free carriers leave behind the immobile host ions, which produce a region of *space charge* of opposite polarity, as depicted in Fig. 15.6. The result is the production of an internal electric field with an applied force in the opposite direction of the diffusion current. The presence of space charge distorts the band energy levels and causes the bands to bend across the junction boundary. The bands distort and bend until the diffusion current is equal to the electric-field current, thereby producing a state of equilibrium. In other words, the concentration gradient force that causes diffusion current matches the electric field force that causes the drift current in the opposite direction, such that a state of equilibrium is reached, as depicted in Fig. 15.6.

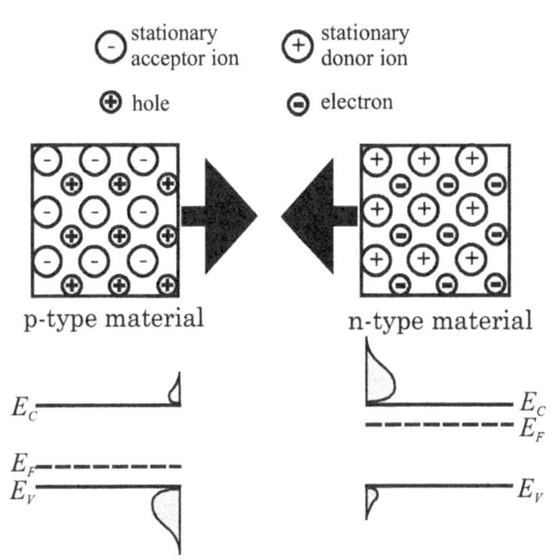

Figure 15.4. Depiction of two semiconductor materials and their energy band diagrams, one of which is *p*-type material and the other *n*-type material, suspended in vacuum before coming into contact.

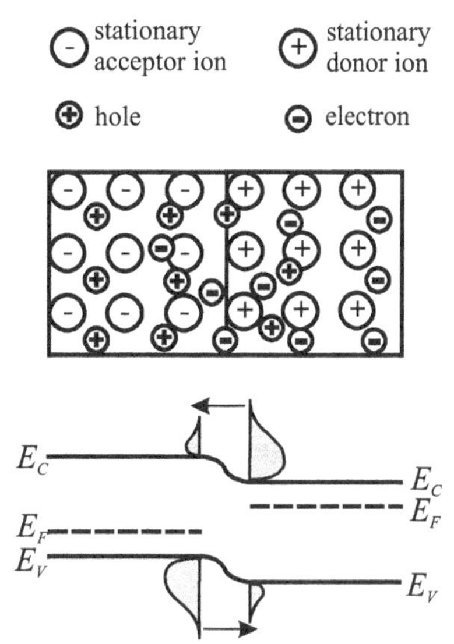

Figure 15.5. When the *p*-type and *n*-type materials come into contact, holes from the *p* side diffuse over to the *n* side, and electrons from the *n* side diffuse over to the *p* side. As they do, they leave the host ions behind, which behave as stationary space charge. Further, the Fermi energy level must be constant across the interface at equilibrium, hence the energy bands distort and bend as the system changes to the equilibrium condition.

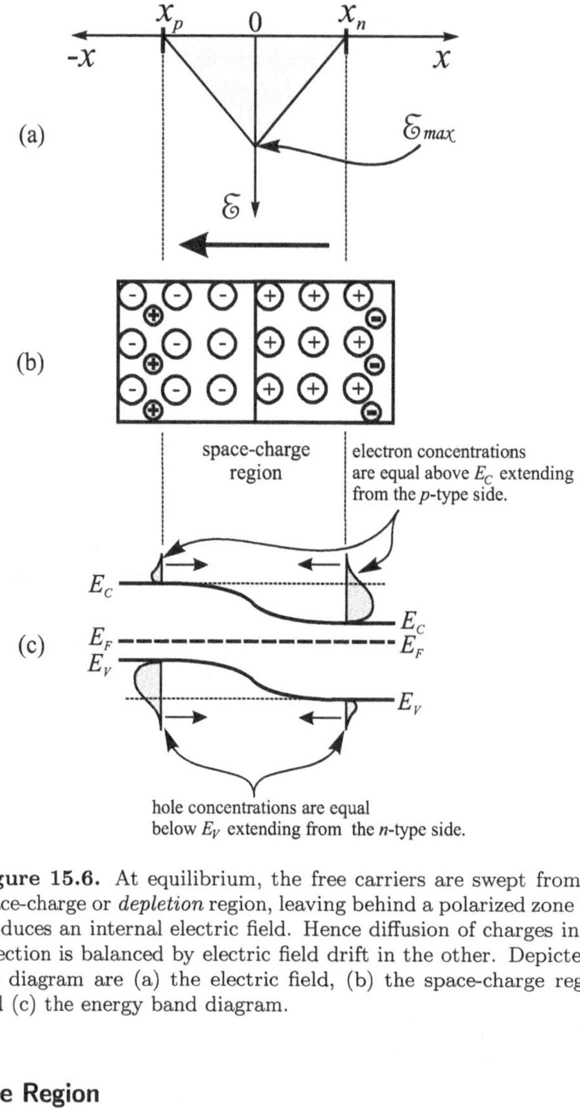

Figure 15.6. At equilibrium, the free carriers are swept from the space-charge or *depletion* region, leaving behind a polarized zone that produces an internal electric field. Hence diffusion of charges in one direction is balanced by electric field drift in the other. Depicted in the diagram are (a) the electric field, (b) the space-charge region, and (c) the energy band diagram.

Width of the Space-Charge Region

To determine the width of the space-charge region, begin with Poisson's equation and a few simplifying assumptions. The general Poisson equation is

$$\nabla \cdot \mathcal{D} = \nabla \cdot [\epsilon_0 \kappa \boldsymbol{\mathcal{E}}] = \nabla \cdot [\epsilon_0 \kappa \nabla V(\mathbf{r})] = \rho_c(\mathbf{r}), \tag{15.45}$$

$\mathcal{D}$ is the electrical displacement, ϵ_0 is the permittivity of free space, κ is the dielectric constant, $\boldsymbol{\mathcal{E}}$ is the electric field, V is the electric potential, and ρ_c is the volumetric charge density. For the one-dimensional case with a constant permittivity, Eq. (15.45) reduces to

$$\frac{\partial^2 \psi(x)}{\partial x^2} = -\frac{\rho(x)}{\epsilon_0 \kappa} = \frac{q_e}{\epsilon_s}(N_A^- - N_D^+ - p + n) \tag{15.46}$$

where q_e is the unit electron charge, $\psi(x)$ is the scalar voltage component in the x direction (the y and z voltage components are zero in the present example). Here n, p, N_D^+, and N_A^- are the concentrations of the free electrons, free holes, ionized donors, and ionized acceptors, respectively. Also the relation $\epsilon_s = \epsilon_0 \kappa$ has been used. To simplify the problem further, the following approximations are made.

1. Assume that the junction is abrupt at the interface between the p-type side and the n-type side and that all dopants in the materials are uniformly distributed.
2. Assume $N_D^+ \gg n$, p, or N_A for the n-type side and that $N_D^+ \simeq N_D$ at 300 K.
3. Assume $N_A^- \gg n$, p, or N_D for the p-type side and that $N_A^- \simeq N_A$ at 300 K.

Equation (15.46) then reduces to the following:

$$\frac{\partial^2 \psi(x)}{\partial x^2} \simeq \frac{-q_e N_D^+}{\epsilon_s}, \quad 0 \le x \le x_n, \qquad \text{and} \qquad \frac{\partial^2 \psi(x)}{\partial x^2} \simeq \frac{q_e N_A^-}{\epsilon_s}, \quad x_p \le x \le 0, \tag{15.47}$$

and outside the space-charge region $\partial^2\psi(x)/\partial x^2 = 0$, $x \notin (x_p, x_n)$.

The electric field on the n-type side of the junction is found by integration of Eq. (15.47) as

$$-\frac{\partial \psi(x)}{\partial x} = \mathcal{E}(x) = \int \frac{q_e N_D}{\epsilon_s}\, dx = \frac{q_e N_D}{\epsilon_s} x + C_1, \quad 0 \le x \le x_n, \tag{15.48}$$

and the electric field on the p-type side is

$$-\frac{\partial \psi(x)}{\partial x} = \mathcal{E}(x) = -\int \frac{q_e N_A}{\epsilon_s}\, dx = -\frac{q_e N_A}{\epsilon_s} x + C_2, \quad x_p \le x \le 0. \tag{15.49}$$

With the boundary conditions $\mathcal{E}(x_p) = \mathcal{E}(x_n) = 0$ the arbitrary constants C_1 and C_2 can be evaluated and the electric field is found to be

$$\mathcal{E}(x) = \begin{cases} \dfrac{q_e N_D}{\epsilon_s}(x - x_n), & 0 \le x \le x_n \\ -\dfrac{q_e N_A}{\epsilon_s}(x - x_p), & x_p \le x \le 0 \end{cases}. \tag{15.50}$$

Notice that the slope of the electric field magnitude is a linear function of the doping concentrations, N_D and N_A. The electric field must also be continuous across the junction, hence at $x = 0$ one has

$$-\frac{q_e N_D}{\epsilon_s} x_n = \frac{q_e N_A}{\epsilon_s} x_p \qquad \text{so that} \qquad N_D x_n = -N_A x_p. \tag{15.51}$$

These are important relations because they show that the relative width of the space charge region increases as the doping concentration decreases. For instance, if N_D is 10 times greater than N_A, then for any given applied voltage, x_p must be 10 times greater than x_n. Notice also that the largest magnitude of the electric field must be at $x = 0$ (the pn junction interface), i.e.,

$$\mathcal{E}_{max} = -\frac{q_e N_D}{\epsilon_s} x_n = \frac{q_e N_A}{\epsilon_s} x_p. \tag{15.52}$$

The change in voltage across the space-charge region can be found by integration of electric field,

$$\Delta V = -\int_{x_p}^{x_n} \mathcal{E}(x)\, dx = \int_{x_p}^{0} \frac{q_e N_A}{\epsilon_s}(x - x_p)\, dx - \int_0^{x_n} \frac{q_e N_D}{\epsilon_s}(x - x_n)\, dx$$

$$= \frac{q_e}{2\epsilon_s}\left(N_A x_p^2 + N_D x_n^2\right). \tag{15.53}$$

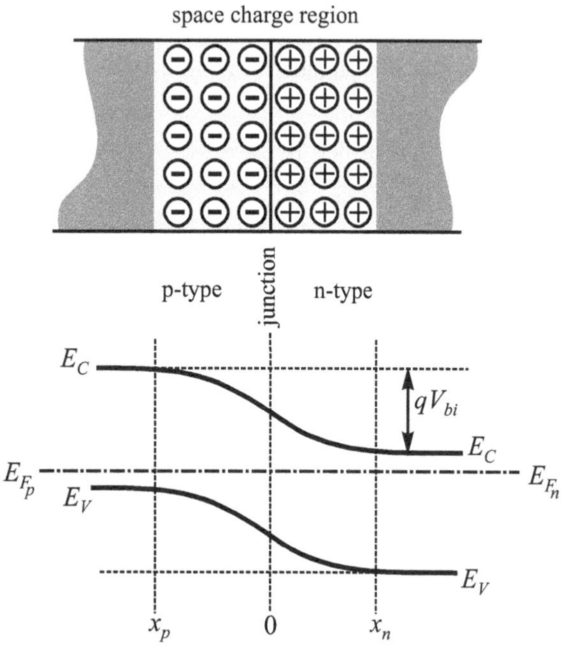

Figure 15.7. At equilibrium in a *pn* junction, the free charge carriers are swept from the space-charge or *depletion* region, leaving behind a polarized zone that produces an internal electric field. Diffusion of charges in one direction is balanced by electric field drift in the other. At equilibrium, the Fermi level is constant across the junction. At the *pn* junction, the energy bands bend and produce the built-in potential V_{bi}.

Hence, there is an electric field and a *built-in* voltage potential V_{bi} at the *pn* junction without the need of any externally applied voltage. Some semiconductor detectors operate well on this built-in potential alone. The built-in potential can be rewritten as

$$V_{bi} = \frac{q_e}{2\epsilon_s}\left(N_A x_p^2 + N_D x_n^2\right)$$
$$= -\frac{\mathcal{E}_{max}}{2}(x_n - x_p) = -\frac{\mathcal{E}_{max} W}{2}, \quad (15.54)$$

where $W = x_n - x_p$ is the total space-charge region width as shown in Fig. 15.7. Typically, *pn* junction diode detectors have a voltage applied in the reverse direction to increase the width of the space-charge or *active region.*

With the second result of Eq. (15.51), Eq. (15.53) can be rewritten as

$$V_{bi} = \frac{q_e}{2\epsilon_s}\left(\frac{N_D^2 x_n^2}{N_A} + N_D x_n^2\right),$$

which, upon solving for x_n, gives

$$x_n = \left\{\frac{2\epsilon_s V_{bi}}{q_e}\left(\frac{N_A}{N_D(N_D + N_A)}\right)\right\}^{1/2}. \quad (15.55)$$

In the same way, Eq. (15.53) can be rewritten as

$$V_{bi} = \frac{q_e}{2\epsilon_s}\left(N_A x_p^2 + \frac{N_A^2 x_p^2}{N_D}\right),$$

whose solution for x_p is

$$x_p = -\left\{\frac{2\epsilon_s V_{bi}}{q_e}\left(\frac{N_D}{N_A(N_D + N_A)}\right)\right\}^{1/2}. \quad (15.56)$$

Equation (15.55) and Eq. (15.56) can be combined to yield the active region width W, without a bias, as

$$W = x_n - x_p = \left\{\frac{2\epsilon_s V_{bi}}{q_e}\left(\frac{N_A}{N_D(N_D + N_A)}\right)\right\}^{1/2} + \left\{\frac{2\epsilon_s V_{bi}}{q_e}\left(\frac{N_D}{N_A(N_D + N_A)}\right)\right\}^{1/2},$$
$$= \left\{\frac{2\epsilon_s V_{bi}}{q_e(N_D + N_A)}\right\}^{1/2}\left\{\sqrt{\frac{N_A}{N_D}} + \sqrt{\frac{N_D}{N_A}}\right\} = \left\{\frac{2\epsilon_s V_{bi}}{q_e}\left(\frac{N_A + N_D}{N_A N_D}\right)\right\}^{1/2}. \quad (15.57)$$

When an external bias is applied, the active region width is

$$W = \left\{\frac{2\epsilon_s (V_{bi} - V)}{q_e}\left(\frac{N_A + N_D}{N_A N_D}\right)\right\}^{1/2}. \quad (15.58)$$

In the case that one side of the junction is doped much higher than the other side, by at least an order of magnitude, Eq. (15.58) can be approximated by

$$W \simeq \left\{ \frac{2\epsilon_s(V_{bi} - V)}{q_e N_b} \right\}^{1/2}, \tag{15.59}$$

where N_b is the doping concentration of the *lighter* doped side, commonly named the one-sided abrupt junction approximation.

One can also calculate V_{bi} as a function of the extrinsic and intrinsic Fermi energies. From Fig. 15.7, it is seen that $q_e V_{bi}$ equals the energy difference in the conduction band levels on the p-type side and on the n-type side. Thus, V_{bi} can be calculated as

$$q_e V_{bi} = (E_i - E_{Fp})\big|_{p-type} + (E_{Fn} - E_i)\big|_{n-type},$$

$$= kT \ln\left(\frac{p_p}{n_i}\right) + kT \ln\left(\frac{n_n}{n_i}\right) = kT \ln\left(\frac{n_n p_n}{{n_i}^2}\right).$$

At room temperature $p_p \simeq N_A$ and $n_n \simeq N_D$ so that

$$V_{bi} \simeq \frac{kT}{q_e} \ln\left(\frac{N_D N_A}{{n_i}^2}\right). \tag{15.60}$$

Example 15.1: Consider a Si diode structure with a p-type dopant of concentration $N_A = 10^{17}$ cm^{-3} diffused into the surface of lightly doped n-type material that has a background concentration of $N_D = 10^{14}$ cm^{-3}. What is the built-in potential and total active region width at room temperature (300 K)?

Solution:

For Si $\epsilon_s = 11.9\epsilon_o$, and $n_i(300\text{K}) = 1.45 \times 10^{10}$ cm^{-3}. Substitution into Eq. (15.60) gives

$$V_{bi} = \frac{kT}{q_e} \ln\left(\frac{N_D N_A}{{n_i}^2}\right) = \frac{(1.38 \times 10^{-23}\ \text{J/K})\,(300\ \text{K})}{1.6 \times 10^{-19}\ \text{C}} \ln\left(\frac{(10^{14}\ \text{cm}^{-3})\,(10^{17}\ \text{cm}^{-3})}{(1.45 \times 10^{10}\ \text{cm}^{-3})^2}\right) = 0.636 \text{ volts.}$$

From Eq. (15.58) the active region (or depletion region) width is

$$W = \left\{ \frac{2\epsilon_s(V_{bi} - V)}{q_e} \left(\frac{N_A + N_D}{N_A N_D}\right) \right\}^{1/2},$$

$$= \left\{ \left(\frac{(2)(11.9)(8.854 \times 10^{-14}\ \text{F/cm})(0.636\ \text{V} - 0\ \text{V})}{1.6 \times 10^{-19}\text{C}}\right) \left(\frac{10^{17}\text{cm}^{-3} + 10^{14}\text{cm}^{-3}}{(10^{14}\text{cm}^{-3})\,(10^{17}\text{cm}^{-3})}\right) \right\}^{1/2} = 2.895 \times 10^{-4} \text{ cm.}$$

Because the n-type side has a significantly lower doping concentration than the p-type side, one can also use the approximation of Eq. (15.59) with $N_b = N_D$ to give

$$W \simeq \left\{ \frac{2\epsilon_s(V_{bi} - V)}{q_e N_b} \right\}^{1/2} = \left\{ \frac{(2)(11.9)(8.854 \times 10^{-14}\ \text{F/cm})(0.636\ \text{V} - 0\ \text{V})}{(1.6 \times 10^{-19}\text{C})(10^{14}\ \text{cm}^{-3})} \right\}^{1/2} = 2.894 \times 10^{-4} \text{ cm.}$$

This result is essentially the same as the first solution.

At room temperature, the material resistivity ρ_s can be expressed as $\rho_s = 1/(q_e\mu_s N_s)$, where N_s and μ_s are the background dopant concentration and mobility for the lighter doped side of the junction, and Eq. (15.59) can be rewritten as

$$W \simeq \{2\epsilon_s\mu_s\rho_s(V_{bi} - V)\}^{1/2}. \tag{15.61}$$

The magnitude of the electric field across the device is

$$|\mathcal{E}(x)| \simeq \frac{q_e N_s}{\epsilon_s}(W - x), \quad 0 \le x \le W. \tag{15.62}$$

Device Capacitance

The capacitance of a parallel contact *pn* junction detector is given by

$$C_{det} = \frac{\epsilon_s A}{W}, \tag{15.63}$$

where A is the device active contact area. Substitution of Eq. (15.61) into Eq. (15.63) yields

$$C_{det} \simeq A\left[\frac{\epsilon_s}{2\mu_s\rho_s(V_{bi} - V)}\right]^{1/2}. \tag{15.64}$$

Detector capacitance affects the input voltage pulse height, with a large capacitance diminishing the input voltage from the detector that is measured by the amplification circuit such that

$$V_{in} \simeq \frac{Q}{C_{tot}}, \tag{15.65}$$

where C_{tot} is the total capacitance (detector and coupling) and Q is the total charge collected from the radiation detector after a radiation event. Hence, it is important to reduce detector capacitance and coupling capacitance between the detector and the shaping electronics. From Eq. (15.64), it is seen that increasing the reverse bias voltage decreases the detector capacitance, but at the expense of increasing the leakage current.

Fermi Level and Current at Equilibrium

The general equations describing current density for electrons and holes are given by Eqs. (15.29) and (15.30). The total conduction current density is $\mathbf{J}_c = \mathbf{J}_p + \mathbf{J}_n$. Under a condition of no applied external voltage and equilibrium conditions, these currents must all vanish. Thus, in one dimension,

$$J_p(x) = q_e\mu_p p(x)\,\mathcal{E}(x) - q_e D_p\frac{dp(x)}{dx} = 0, \tag{15.66}$$

in which μ_p is the hole mobility, p is the hole concentration, and

$$J_n(x) = q_e\mu_n n(x)\,\mathcal{E}(x) + q_e D_n\frac{dn(x)}{dx} = 0. \tag{15.67}$$

where μ_n is the electron mobility, n is the hole concentration. There is a relative change in the *intrinsic* Fermi level $E_i(x)$ location, with respect to the band edges, across a *pn* junction. From the definitions of $q_eV(x) = E(x)$ and D_n given by Eq. (15.31), Eqs. (15.66) and (15.67) can be written as

$$J_p(x) = \mu_p p(x)\,\frac{dE_i(x)}{dx} - kT\mu_p\frac{dp(x)}{dx} = 0, \tag{15.68}$$

and

$$J_n(x) = \mu_n n(x)\,\frac{dE_i(x)}{dx} + kT\mu_n \frac{dn(x)}{dx} = 0. \tag{15.69}$$

Eq. (12.130) can be divided by Eq. (12.128) to yield

$$\frac{p(x)}{n_i} = \frac{N_V \exp\left[\dfrac{E_V - E_F(x)}{kT}\right]}{N_V \exp\left[\dfrac{E_V - E_i(x)}{kT}\right]} = \exp\left[\frac{E_i(x) - E_F(x)}{kT}\right], \tag{15.70}$$

which is an expression for $p(x)$. The resulting derivative for dp/dx is then found to be

$$\begin{aligned}\frac{dp(x)}{dx} &= \frac{n_i}{kT}\exp\left[\frac{(E_i(x) - E_F(x))}{kT}\right]\left[\frac{dE_i(x)}{dx} - \frac{dE_F(x)}{dx}\right], \\ &= \left(\frac{p(x)}{kT}\right)\left[\frac{dE_i}{dx} - \frac{dE_F}{dx}\right].\end{aligned} \tag{15.71}$$

Substitution of Eq. (15.71) into Eq. (15.68) gives

$$J_p(x) = \mu_p p(x)\,\frac{dE_i(x)}{dx} - kT\mu_p\left(\frac{p(x)}{kT}\right)\left[\frac{dE_i}{dx} - \frac{dE_F}{dx}\right] = \mu_p p(x)\frac{dE_F}{dx} = 0. \tag{15.72}$$

A similar result can be derived for the n-type side, namely

$$J_n(x) = \mu_n n(x)\frac{dE_F}{dx} = 0. \tag{15.73}$$

so that the total current density is

$$J(x) = J_p(x) + J_n(x) = \mu_p p(x)\frac{dE_F}{dx} + \mu_n n(x)\frac{dE_F}{dx} = 0. \tag{15.74}$$

Recall that the Fermi level at equilibrium is continuous across a boundary. The interpretation of Eq. (15.74) is that the Fermi level at equilibrium is constant on the p and n sides of the junction, and must also be constant across the junction, as shown in Fig. 15.8(e). Hence, at equilibrium, the net current must be zero.

Current Under Forward Bias

The current density of a pn junction diode is described by [Sze 1981]

$$J = J_s\left\{\exp\left[\frac{q_e V}{\breve{n}kT}\right] - 1\right\}, \tag{15.75}$$

where $\breve{n}$ is an *ideality factor* to correct for non-ideal behavior, and saturation current density J_s is

$$J_s = \frac{q_e D_p p_{no}}{L_p} + \frac{q_e D_n n_{po}}{L_n}, \tag{15.76}$$

D_p is the diffusion coefficient for holes, D_n is the diffusion coefficient for electrons, p_{no} is the minority hole concentration in the n-type material, n_{po} is the minority hole concentration in the p-type material, and the diffusion lengths L_p and L_n are defined as

$$L_{p,n} = \sqrt{D_{p,n}\tau_{p,n}}. \tag{15.77}$$

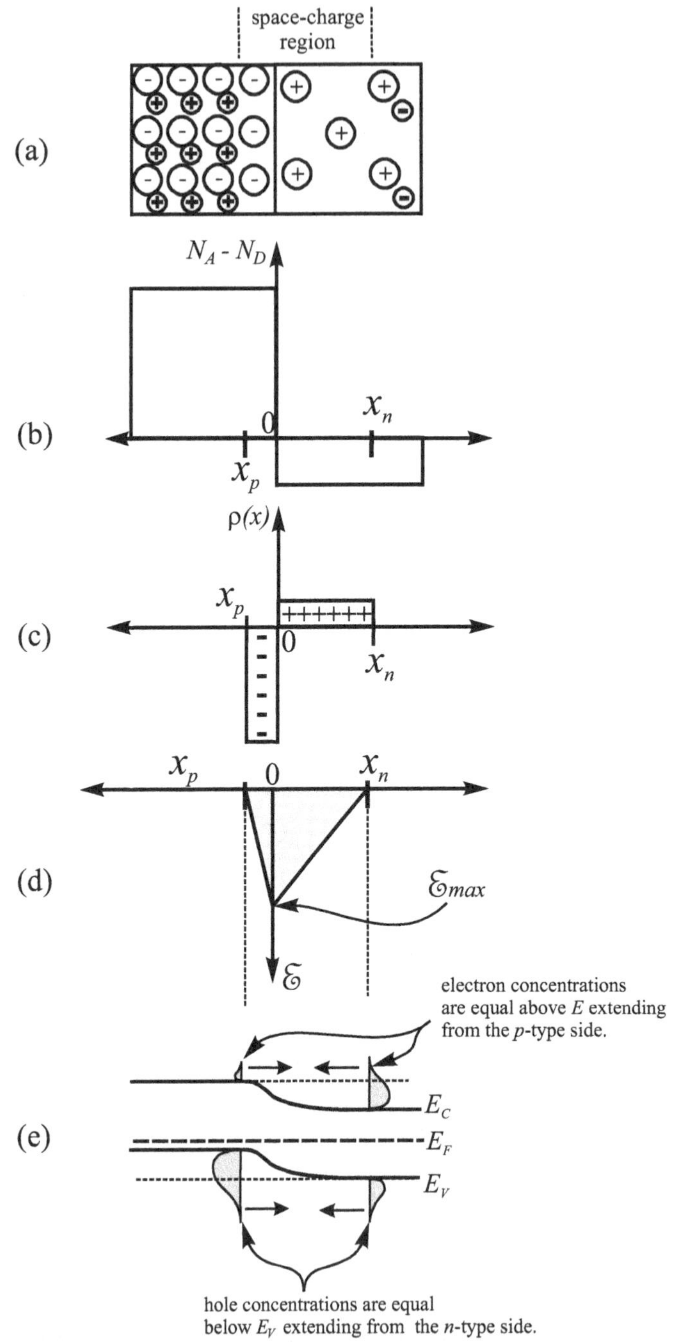

Figure 15.8. The depleted region extends furthest into the junction region of lowest doping. For the case in which the p-type region is of higher doping levels than the n-type region, shown are (a) the device space-charge region, (b) the dopant density, (c) the space-charge density (d) the electric field, and (e) the energy band diagram.

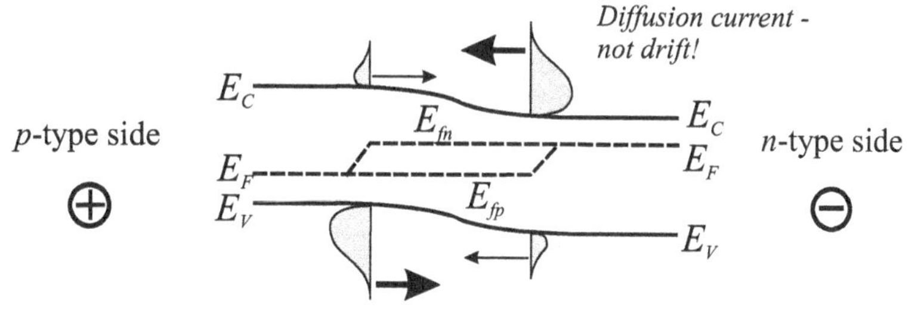

Figure 15.9. Diffusion current dominates in forward bias and, because of the non-linear distribution of free carriers in the conduction and valence bands, the observed current increases exponentially with increasing forward bias.

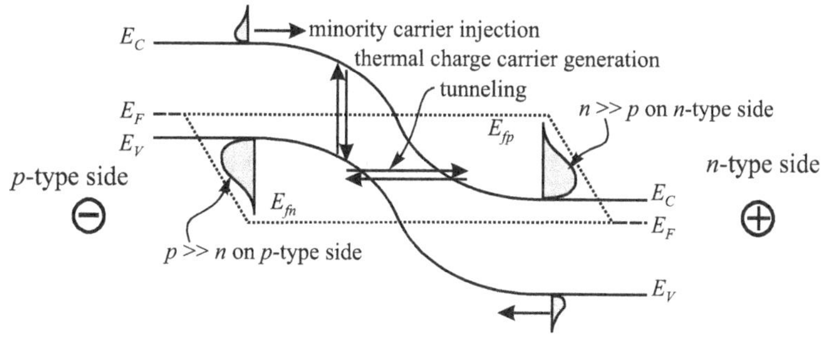

Figure 15.10. In reverse bias, the density of available carriers, dominated by the *minority* carrier concentration, determines the leakage current. At higher voltages, bulk generation and tunneling may increase the observed leakage currents.

Here $\tau_{p,n}$ is the corresponding charge carrier lifetime. With a forward bias (see Fig. 15.9), the electron distribution on the n-type side is raised in energy above the barrier on the p-type side, thereby allowing diffusion of charge carriers to flow into the p-type side. According to the integrand of Eq. (12.120), the electron distribution in the conduction band as a function of energy is non-linear, and the amount of current increases exponentially as more forward voltage is applied. From the integrand of Eq. (12.123), the same can be shown for holes diffusing from the p-type side over into the n-type side.

Also depicted in Fig. 15.9 are the quasi-Fermi levels E_{fn} for electrons and E_{fp} for holes.[3] A quasi-Fermi level describes the population of electrons and holes separately in the conduction band and valence band when no longer in equilibrium, such as the case with applied voltage (forward or reverse) or other cases of charge injection, such as photoexcitation.

Leakage Current Under Reverse Bias

There are three main sources of leakage current, as depicted in Fig. 15.10. Although the *majority* carriers on the p-type side are holes, according to Eq. (12.131) a small concentration of electrons is still present. These *minority* charge carriers (electrons) can diffuse into the depletion region, where they are swept across by the electric field and contribute to the leakage current. A similar case is true for holes diffusing from the n-type side into the depletion region. Leakage current can also occur from thermal generation of electrons directly across the band gap into the conduction band, producing electron and hole free carriers. Thermal generation of such charge carriers can be suppressed by cooling the detector while it is operating. Under high

[3] A quasi-Fermi level is also called an 'imref', the word 'fermi' spelled backwards [Hannay 1959].

voltage bias conditions, charge carriers can also tunnel directly across the band gap, again contributing to the leakage current. Variations about the leakage current can be a significant source of electronic noise (shot noise), which broadens the overall energy resolution of the detector. Basically, *pn* junctions are employed to minimize leakage currents in semiconductor detectors.

Example 15.2: Given a silicon *pn* junction with $N_A = 10^{15}$ cm^{-3} and $N_D = 10^{13}$ cm^{-3}, what is the minority carrier saturation leakage current density under reverse bias at 300 K? The electrons and hole mobilities are and 1320 cm^{2}V^{-1}s^{-1} and 470 cm^{2}V^{-1}s^{-1}, respectively (see Fig. 16.3), with $\tau_{n,p} = 10^{-3}$ s.

Solution:
From Eq. (15.31)

$$D = \frac{\mu k T}{q_e}$$

we have

$$D_p = \frac{(470 \text{ cm}^2 \text{ V}^{-1} \text{ s}^{-1})(1.38 \times 10^{-23} \text{ J K}^{-1})(300 \text{ K})}{1.6 \times 10^{-19}\text{C}} = 12.16 \text{ cm}^2 \text{ s}^{-1}.$$

and

$$D_n = \frac{(1320 \text{ cm}^2 \text{ V}^{-1} \text{ s}^{-1})(1.38 \times 10^{-23} \text{ J K}^{-1})(300 \text{ K})}{1.6 \times 10^{-19}\text{C}} = 34.16 \text{ cm}^2 \text{ s}^{-1}.$$

Because $np = n_i^2$, one has

$$J_s = q_e \left(\frac{D_p p_{n0}}{L_p} + \frac{D_n n_{p0}}{L_n} \right) = q_e n_i^2 \left(\sqrt{\frac{D_p}{n^2 \tau_p}} + \sqrt{\frac{D_n}{p^2 \tau_n}} \right),$$

which gives

$$J_s = (1.6 \times 10^{-19} \text{ C})(1.5 \times 10^{10} \text{ cm}^{-3})^2 \left(\sqrt{\frac{12.16 \text{ cm}^2 \text{ s}^{-1}}{(10^{13} \text{ cm}^{-3})^2 (10^{-3} \text{ s})}} + \sqrt{\frac{34.16 \text{ cm}^2 \text{ s}^{-1}}{(10^{15} \text{ cm}^{-3})^2 (10^{-3} \text{ s})}} \right),$$

$$= 4.04 \times 10^{-14} \text{ amperes}$$

Use of Reverse Bias

The active radiation-sensitive volume of the *pn*-junction detector is defined by the space-charge region (or depletion region), and the undepleted regions act as series resistances. Detectors based on *pn* junction diodes are typically operated under *reverse bias*, which is regarded as a negative voltage, but is actually a positive voltage applied to the *n*-type material with respect to the *p*-type material. Shown in Fig. 15.10 is the band diagram of a reverse biased *pn* junction. The large energy band barriers prevent electrons on the *n*-type side from diffusing over to the *p*-type side and prevent holes on the *p*-type side from diffusing into the *n*-type side. There remains the possibility of electrons on the *p*-type side diffusing to the *n*-type side, and holes on the *n*-type side diffusing into the *p*-type side, both of which produce current. However, from Eq. (12.131), the density of minority carrier holes and minority carrier electrons is small (depicted in Fig. 15.10), and consequently the leakage current is small. As a result, the *pn* junction suppresses leakage current that would be present if the semiconductor were operated as a resistor. This necessary condition reduces the electron current such that minute charge packets excited in the depletion region by radiation interactions can be measured.

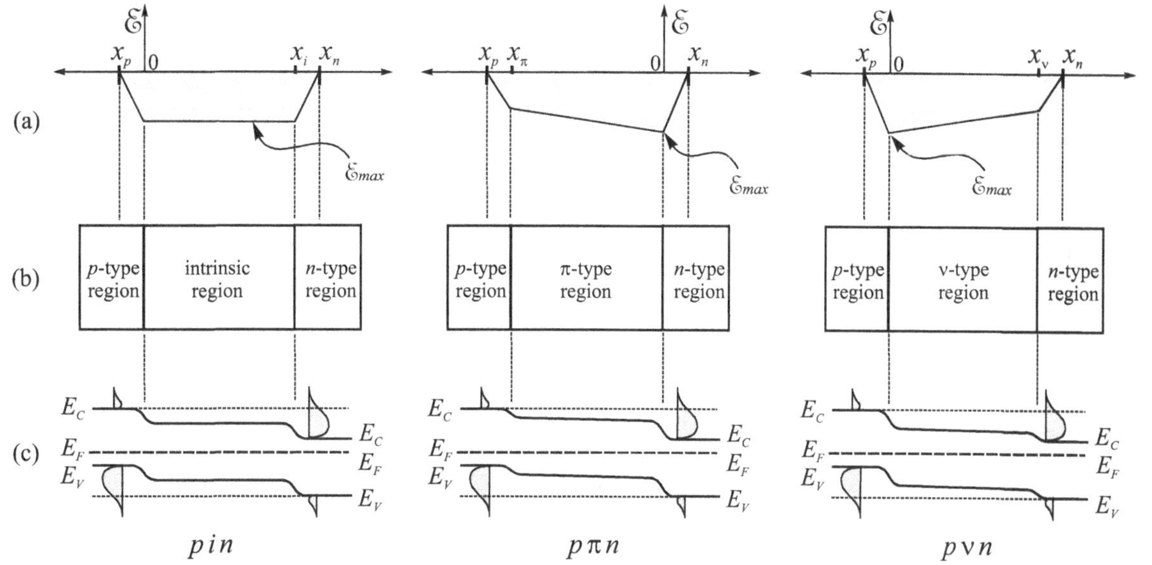

Figure 15.11. Electric field (a), material regiobs (b), and band diagrams (c) of *pin*, *pπn* and *pνn* devices. Note that the electric field across the intrinsic region of the *pin* diode is constant, but the lightly doped *pπn* and *pνn* have a definite slopes to their electric fields.

15.3.2 *pin* Junction Devices

Reverse biased diode detectors need a large depletion region to maximize radiation absorption. As found in Eq. (15.58), the depletion region width is proportional to the square root of the applied reverse voltage. This relationship indicates large voltages are required to produce large depletion regions in a common *pn* junction diode. The usual remedy to avoid high voltages is to construct a *pin* diode that has an intrinsic, or high purity region, between the *p*-type and *n*-type contacts. The *p* and *n* type contacts can be used as either blocking barriers or electrical *ohmic* contacts to the material. The application of *p*-type contacts to *p*-type material, or *n*-type contacts to *n*-type material, generally produce non-rectifying electrical contacts that follow Ohm's law.

A *pin* device may have a truly intrinsic region between the *p* and *n* type contacts, in which the electron and hole populations are identical, or the high purity region may be a lightly doped material. Lightly doped *p*-type material is commonly denoted π-type, and lightly doped *n*-type material is commonly denoted ν-type. Analysis of the diode construction should take into account the "punch-through" voltage, in which the depletion region extends completely across the high purity region, whether the material is i, ν, or π type. Semiconductor *pin* diode radiation detectors are commonly operated at biases above the punch-through voltage.

Intrinsic material either has dopant concentrations below the intrinsic concentration or has compensation dopants that cause the residual free carrier concentration to be below the intrinsic concentration. In either case, the residual space charge is practically zero ($\rho_c \simeq 0$), and from Eq. (15.46) the resulting electric field is constant across the *pin* diode under reverse bias. Many detectors are fashioned from high-purity semiconductor materials, yet such detectors may not behave intrinsically, because of a small concentration of residual dopants that is still above the intrinsic concentration. As a result, ρ_c is non-zero, even if very small, and the depletion region width is determined by V_{bi} and the applied voltage. If the lightly doped π or ν regions are relatively thin, the detector might become "fully depleted" without an applied voltage, as depicted in Fig. 15.11. However, complete depletion is usually not the case, and a reverse bias must then be applied to extend the depletion region and the electric field across the device, as determined from Eq. (15.58) and Eq. (15.59), respectively.

Example 15.3: Given a silicon $p\nu n$ diode with $N_A = 10^{15}$ cm^{-3}, $N_d = 10^{13}$ cm^{-3}, and $N_D = 10^{15}$ cm^{-3}, what is the depletion width for a 450 micron thick ν region with a bias of 200 volts reverse bias? With N_d reduced to 10^{12} cm^{-3}, what is the depletion width at 100 volts reverse bias?

Solution:

The one-sided abrupt junction approximation is used in the first case.

$$W \simeq \sqrt{\frac{2\kappa\epsilon_o V}{q_e N_d}} = \sqrt{\frac{(2)(11.9)(8.854 \times 10^{-14}\text{F cm}^{-1})(200 \text{ volts})}{(1.6 \times 10^{-19} \text{ C})(10^{13} \text{ cm}^{-3})}} = 162.3 \text{ microns.}$$

The ν region dopant concentration is now reduced to 10^{12} cm^{-3}, yielding,

$$W \simeq \sqrt{\frac{2\kappa\epsilon_o V}{q_e N_d}} = \sqrt{\frac{(2)(11.9)(8.854 \times 10^{-14}\text{F cm}^{-1})(200 \text{ volts})}{(1.6 \times 10^{-19} \text{ C})(10^{12} \text{ cm}^{-3})}} = 513.2 \text{ microns,}$$

which is wider than the ν region. Hence, in the second case the diode is fully depleted with W = 450 microns.

15.3.3 Metal-Semiconductor Contacts

Consider a situation in which a metal block and a semiconductor block are both suspended in free space, and that somehow they are spontaneously fused together. What then would happen? Entire books have been published on the physics of metal-semiconductor contacts. Such depth of coverage is not possible here; however, those major concepts, important to the design and fabrication of semiconductor detectors, are briefly presented.

Schottky Devices

Earlier it was shown that the Fermi energy level must be constant at equilibrium. Further, if the derivative of the Fermi level with respect to position were to change, it would imply the presence of current flow. However, it can be readily assumed that a free floating semiconductor substance onto which a metal is attached does not have current flowing through the structure. Hence, the Fermi level must remain constant. As a metal and semiconductor are brought together, the energy bands must deform to (1) ensure that the vacuum level is continuous and (2) ensure that the Fermi level is constant.

The application of a metal to a semiconductor surface acts to bend the semiconductor bands to form an energy barrier $q_e\phi_{bn}$, much like the pn junction diode, as shown in Fig. 15.12. The Fermi energies, defined by the metal and semiconductor work functions (ϕ_m and ϕ_s) must align when the two materials are brought into contact.

Shown in Fig. 15.12(a) are depictions of the energy bands for a metal and a semiconductor. The vacuum levels of both materials are continuous. The work functions of both materials are defined as the difference between the vacuum level and Fermi levels. Given a temperature and pressure, the work function of the metal, $q_e\phi_m$, may be regarded as a constant. However, the Fermi level of a semiconductor, $q_e\phi_s$, is a function of doping concentration, and, hence, is not considered a constant reference point. Instead, the electron affinity, $q_e\chi_s$, defined as the difference between the vacuum level and the conduction band edge, is generally used as a constant of reference. As the junction reaches equilibrium, the Fermi level on the metal

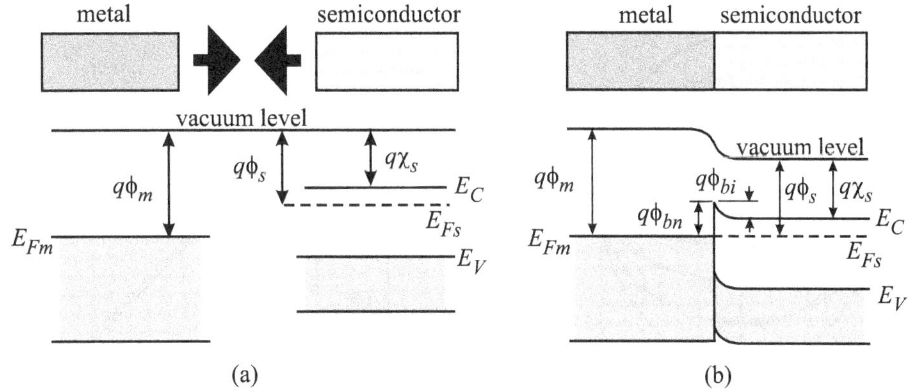

Figure 15.12. The band configuration depicted for a metal and semiconductor (a) before and (b) after contact, according to the Schottky-Mott model.

and semiconductor sides must align. As a result of the junction formation, a built-in potential $q_e\phi_{bi}$ forms and a potential barrier $q_e\phi_{bn}$ forms. The final barrier height is given by

$$q_e\phi_{bn} = q_e(\phi_m - \chi_s), \tag{15.78}$$

and the built in potential is

$$q_e\phi_{bi} = q_e(\phi_m - \phi_s). \tag{15.79}$$

This result is called the *simple* Schottky model. Under a reverse bias, this energy barrier serves to reduce leakage current. Typically the barrier height is lower than that for *pn* or *pin* junction diodes, hence detectors based on the Schottky barrier diode typically have higher leakage currents than *pn* or *pin* diodes.

Schottky Effect

Consider a single point charge in vacuum, an electron for instance, placed near the surface of a conductor at distance x, as shown in Fig. 15.13. The point charge repels similar charges in the metal, leaving a net positive charge on the surface, thereby giving the appearance of a positive charge of equal magnitude in the metal at the distance $-x$ from the metal surface. The forces exerted upon the point and image charges are equal and opposite, namely

$$F = \left.\frac{q_e^2}{4\pi\epsilon_s(2x)^2}\right|_{point} = \left.\frac{-q_e^2}{4\pi\epsilon_s(2x)^2}\right|_{image}. \tag{15.80}$$

The potential due to the *image force* is found as

$$V_1(x) = \int_{\infty}^{x} F\,dx = \frac{-q_e^2}{16\pi\epsilon_s x}, \tag{15.81}$$

and the electric field potential is $V_2(x) = -q_e xE$. Therefore, the total energy is

$$V_{tot} = V_1 + V_2 = \frac{-q_e^2}{16\pi\epsilon_s x} - q_e xE. \tag{15.82}$$

The situation is depicted in Fig. 15.14, in which the total potential of the image charge has a maximum value at $q_e\Delta\phi$ located at x_m from the metal/vacuum interface.

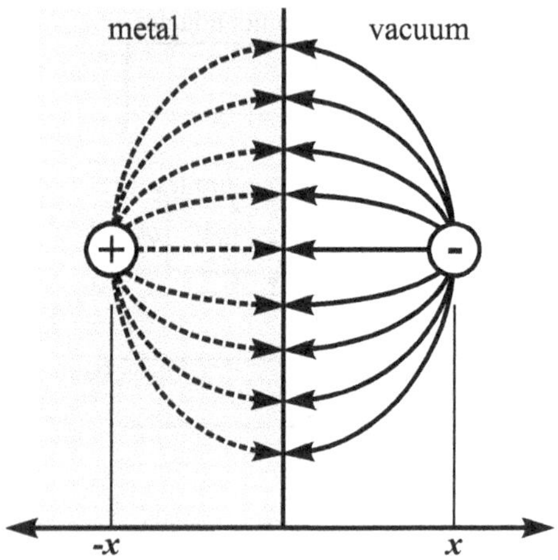

Figure 15.13. A single point charge in vacuum, such as an electron, near a conductive surface effectively produces an *image charge* of opposite magnitude to appear in the conductor.

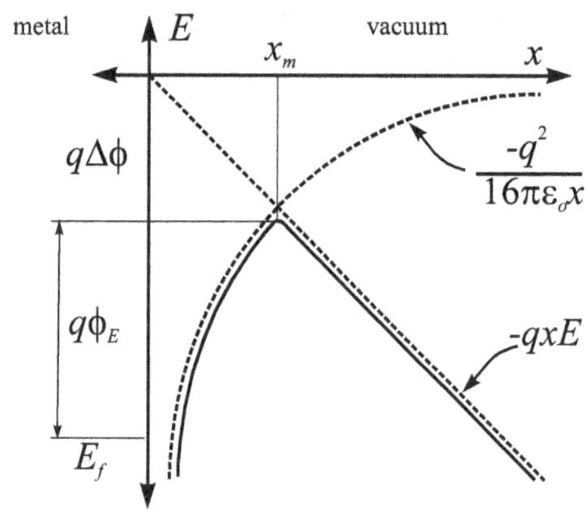

Figure 15.14. Schottky barrier lowering results from the combined effects of the image and electric field potentials.

The Schottky barrier effect thus lowers the Schottky barrier by an amount $q_e \Delta\phi$. This barrier lowering occurs at x_m, where the electric field has a maximum E_m. The distance x_m can be found by setting the derivative of Eq. (15.82) to zero, i.e.,

$$\frac{d}{dx}\left(\frac{-q_e^2}{16\pi\epsilon_s x} - q_e x E_m\right)\bigg|_{x_m} = \frac{q_e^2}{16\pi\epsilon_s x_m^2} - q_e E_m = 0. \tag{15.83}$$

Solution of Eq. (15.83) for x_m gives

$$x_m = \sqrt{\frac{q_e}{16\pi\epsilon_s E_m}}, \tag{15.84}$$

from which the maximum field strength is

$$E_m = \frac{q_e}{16\pi\epsilon_s x_m^2}. \tag{15.85}$$

The change in the Schottky barrier height is

$$\Delta\phi_{bi} = x_m E_m + \frac{q_e}{16\pi\epsilon_s x_m} = 2x_m E_m = 2E_m\sqrt{\frac{q_e}{16\pi\epsilon_s E_m}} = \sqrt{\frac{q_e E_m}{4\pi\epsilon_s}}, \tag{15.86}$$

which is also the potential of Schottky barrier lowering. What this result means is that the actual Schottky barrier for the diode is lower than expected from the simple Schottky model. The end result is a barrier height of

$$q_e\phi_{bn} = q_e(\phi_m - \chi_s - \Delta\phi_{bi}). \tag{15.87}$$

Bardeen Model: Surface State Effects

It should be noted that measured Schottky barrier heights typically do not match the theoretical values, even after considering the Schottky lowering effect. In fact, quite often the actual barrier height ϕ_{bn} is only modestly affected by ϕ_m, or in some cases the final barrier height ϕ_{bn} appears to be completely independent of the metal.

Because the surfaces of semiconductors have defects, interface states, and possible contaminants, the actual barrier height depends not only upon the choice of metal but also on how the surface is prepared. These surface states can effectively "pin" the detector barrier height, thereby predetermining the actual value of $q_e\phi_{bn}$ before the metal is applied. Bardeen offered an explanation to the difference by taking into account the effect of interface states on the Schottky barrier.

Assume that there is an unspecified density D_s (in units of $\mathrm{eV^{-1}\,cm^{-2}}$) of unbonded states upon the surface of a semiconductor material. These unbonded states can arise from a combination of open bonds from the semiconductor host atoms, referred to as Tamm-Shockley states, and also from contaminate atoms attached to the surface. The surface states are classified donor states if they are neutral when occupied, and these states are classified as acceptor states if they are neutral when empty. Fast surface states are close to the bulk material and reach equilibrium quickly with a rapid change in potential. Slow surface states are further from the bulk material and take a relatively longer time to reach equilibrium with a potential change.

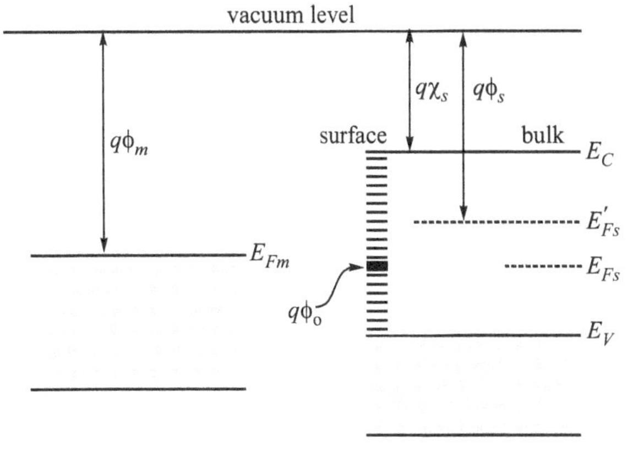

Figure 15.15. Band conditions for an n-type semiconductor with a high density of surface states, showing the neutral level $q_e\phi_o$. States below $q_e\phi_o$ are donors and states above $q_e\phi_o$ are acceptors.

Now assume that there is a thin interfacial insulator region δ between the metal and the semiconductor that is capable of supporting a potential. Also assume that the surface states are a property of the semiconductor alone and are independent of the metal. The states must be either filled or empty, and the filled surface states produce image charges of the opposite polarity in the semiconductor. In other words, for n-type material, a space-charge (depletion) region forms from the repulsion of electrons at the surface, and the bands bend to accommodate the potential formed at the surface. Overall, a specific density of these states is filled up to a neutral level $q_e\phi_o = E_o$, as depicted in Fig. 15.15. The states below the neutral level are filled donor states and hence are neutral, and the states above the neutral level are empty acceptors and are also neutral. If the Fermi level is raised to E'_f, then some of the surface acceptor states become filled and produce a net negative surface charge.

The surface charge Q_{ss} density, in units of $\mathrm{C\,cm^{-2}}$, is

$$Q_{ss} = -q_e D_s\left[E_g - q_e\Delta\phi_{bi} - q_e\phi_{bn} - q_e\phi_0\right], \tag{15.88}$$

in which ϕ_{bn} is the energy barrier height at the metal-semiconductor interface. According to Gauss' law, the semiconductor and metal cannot support an electric field, hence

$$\epsilon_i\mathcal{E}_i = Q_{ss} = -Q'_m, \tag{15.89}$$

where ϵ_i is the permittivity of the thin interface insulator, and Q_m is the mirror charge formed on the metal. The voltage across the thin insulator is

$$V_i = \delta\mathcal{E}_i = \frac{\delta Q_{ss}}{\epsilon_i}. \tag{15.90}$$

As depicted in Fig. 15.16, the metal work function is,

$$q_e\phi_m = q_e V_i + q_e\chi_s + q_e\phi_{bo}, \tag{15.91}$$

and therefore the barrier height is

$$q_e\phi_{bo} = q_e\phi_m - q_e\chi_s - q_e\frac{\delta Q_{ss}}{\epsilon_i}. \tag{15.92}$$

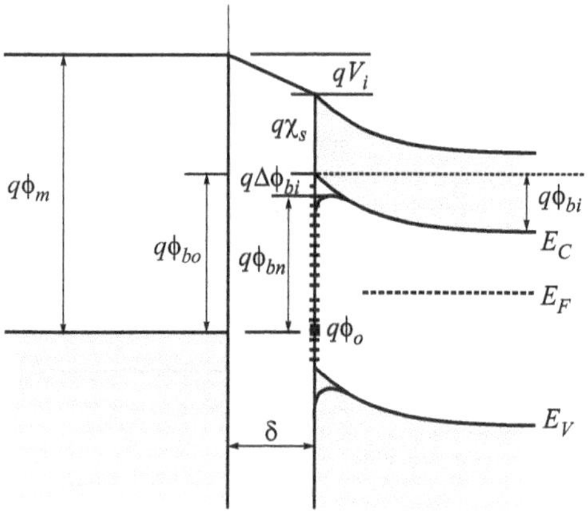

Figure 15.16. Band conditions for an n-type pinned Schottky barrier, showing the neutral level $q_e\phi_o$, below which states are filled and above which states are empty. The shaded regions indicate energy bands with allowed states.

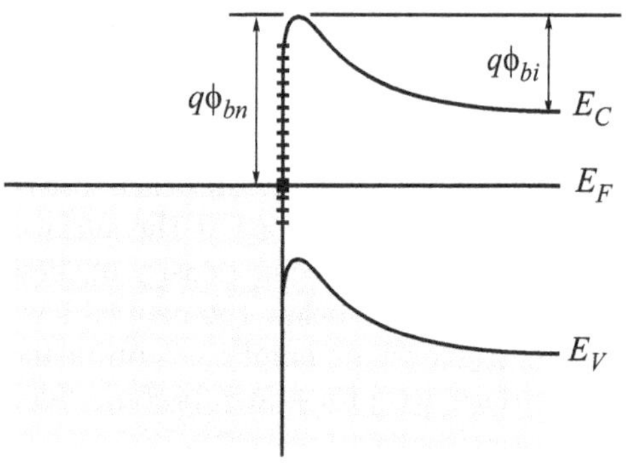

Figure 15.17. Band conditions for an n-type pinned Schottky barrier, showing the Fermi energy pinned at the neutral level $q_e\phi_o$.

Substitution for Q_{ss} from Eq. (15.88) gives

$$q_e\phi_{bo} = q_e\phi_m - q_e\chi_s + \frac{q_e\delta D_s\left[E_g - q_e\Delta\phi_{bi} - q_e\phi_{bn} - q_e\phi_0\right]}{\epsilon_i}. \tag{15.93}$$

Define $\zeta \equiv \epsilon_i/(\epsilon_i + q_e\delta_i D_s)$, and substitute it into Eq. (15.93) to obtain

$$\phi_{bo} = q_e\phi_{bn} + \Delta\phi_{bi} = \zeta(\phi_m - \chi_s) + (1-\zeta)\left(\frac{E_g}{q_e} - \phi_o\right). \tag{15.94}$$

As the width δ of the thin interface insulator decreases to zero, $\zeta \to 1$ and Eq. (15.94) reduces to

$$\phi_{bn} = \phi_m - \chi_s - \Delta\phi_{bi}, \tag{15.95}$$

which is the expected result of Eq. (15.87) found earlier. If the surface state density becomes large, then $\zeta \to 0$ and Eq. (15.94) becomes

$$\phi_{bn} = \frac{E_g}{q_e} - \phi_o - \Delta\phi_{bi}. \tag{15.96}$$

The barrier height is no longer a function of the Fermi energy, but is instead a function of the neutral surface state energy. Note that the neutral level may not be at the same energy as the bulk Fermi energy. The Fermi level in the semiconductor bulk must remain constant under equilibrium conditions, hence, if E_o is below E_{Fs}, the bands must bend up (as shown in Fig. 15.17) until the Fermi level is aligned with E_o. If the neutral level E_o is above E_{Fs}, then the energy bands must move downwards. In Fig. 15.18 the interface states and band conditions are shown for n-type and p-type Schottky barriers.

Depletion Region

The electric field and the depletion region width can be calculated from the one-dimensional Poisson's equation

$$\frac{d^2\psi}{dx^2} = -\frac{d\mathcal{E}(x)}{dx} = -\frac{\rho_c(x)}{\epsilon_s}, \tag{15.97}$$

where $\psi(x)$ is the voltage or potential in the device. Thus, Eq. (15.97) for n-type material becomes

$$\frac{d^2\psi(x)}{dx^2} = -\frac{d\mathcal{E}(x)}{dx} = -\frac{q_e N_D^+}{\epsilon_s}, \tag{15.98}$$

and for p-type material

$$\frac{d^2\psi(x)}{dx^2} = -\frac{d\mathcal{E}(x)}{dx} = \frac{q_e N_A^-}{\epsilon_s}. \tag{15.99}$$

At room temperature, it is reasonable to assume that the shallow dopants are fully ionized, i.e., $N_A^- \simeq N_A$ and $N_D^+ \simeq N_D$. For an n-type Schottky contact, the electric field is

$$\mathcal{E}(x) = \int \frac{d\mathcal{E}(x)}{dx}\,dx = \int \frac{q_e N_D}{\epsilon_s}dx = \frac{q_e N_D}{\epsilon_s}x + C. \tag{15.100}$$

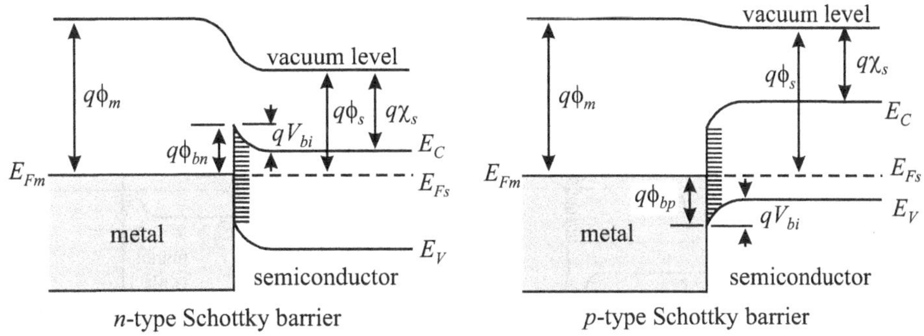

Figure 15.18. Interface states and band conditions for n-type and p-type "pinned" Schottky barriers.

With the boundary condition $\mathcal{E}(x_n) = 0$, the integration constant C in the above result is found to be $C = -(q_e N_D x_n)/\epsilon_s$, and the electric field is thus

$$\mathcal{E}(x) = \frac{q_e N_D}{\epsilon_s}(x - x_n). \tag{15.101}$$

The built-in potential is then evaluated as

$$V_{bi} = -\int_0^{x_n} \mathcal{E}(x)dx = -\int_0^{x_n} \frac{q_e N_D}{\epsilon_s}(x - x_n)dx = \frac{q_e N_D x_n^2}{2\epsilon_s}. \tag{15.102}$$

Rearrangement of terms and inclusion of a reverse bias voltage yield the width of the depletion region as

$$W = \left\{\frac{2\epsilon_s(V_{bi} - V)}{q_e N_D}\right\}^{1/2}, \tag{15.103}$$

a result similar to that for a one-sided abrupt pn junction.

For p-type Schottky junctions, it can be shown in a similar fashion

$$\mathcal{E}(x) = \frac{q_e N_A}{\epsilon_s}(x_p - x), \tag{15.104}$$

and

$$W = \left\{\frac{2\epsilon_s(V_{bi} - V)}{q_e N_A}\right\}^{1/2}. \tag{15.105}$$

Just as Eq. (15.59) can be rewritten as Eq. (15.61), Eqs. (15.103) and (15.105) can be rewritten as

$$W \simeq \{2\epsilon_s \mu_s \rho_s (V_{bi} - V)\}^{1/2}. \tag{15.106}$$

The value of V_{bi} for common Schottky diodes is 0.3 volts. Note that Schottky contacts can be formed on n-type or p-type semiconductors, as depicted in Fig. 15.19. Besides being simple to construct, Schottky barrier detectors have a thin entrance region at the contact, unlike most pn and pin diodes, and energy attenuation in this "dead region" is kept to a minimum. As a result, Schottky barrier detectors (sometimes called "surface barrier detectors") are useful as charged-particle spectrometers.

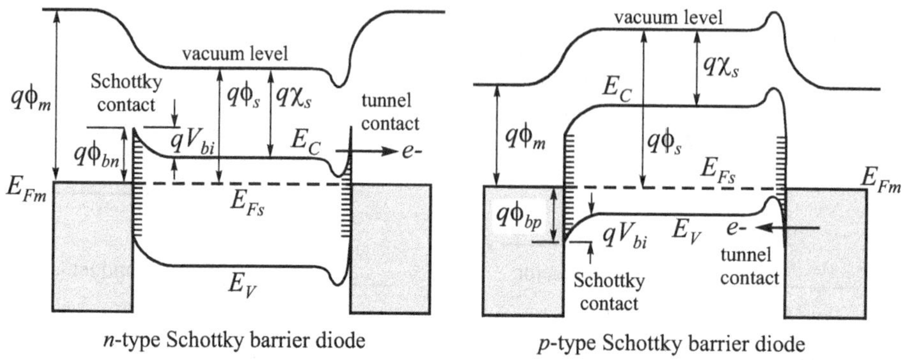

Figure 15.19. Interface states alter the interface potentials and effectively "pin" the Schottky barrier. Shown are n-type and p-type Schottky barriers, along with n-type and p-type tunneling ohmic contacts.

Current Under Forward Bias

The current density of a Schottky barrier rectifying junction can be written as [Rhoderick and Williams 1988]

$$J = J_s\left(\exp\left[\frac{q_e V}{kT}\right] - 1\right), \tag{15.107}$$

and

$$J_s = A^* T^2 \exp\left[-\frac{q_e \phi_{bn}}{kT}\right]. \tag{15.108}$$

and A^* is the effective Richardson constant. Under forward bias, the majority carrier concentration is elevated in energy above the semiconductor barrier ϕ_{bi}, thereby allowing a non-linear supply of electrons to diffuse across the junction.

In the case that the Schottky barrier height is affected by a bias voltage, an additional change must be included

$$\phi'_{bn} = \phi_m - \chi_s - \Delta\phi_{bi} + \beta V = \phi_{bn} + \beta V, \tag{15.109}$$

where the coefficient β is positive for forward bias [Rhoderick and Williams 1988]. Substituting ϕ'_{bn} for ϕ_{bn} into Eq. (15.107) yields

$$J = J_s \exp\left[\frac{-\beta q_e V}{kT}\right]\left(\exp\left[\frac{q_e V}{kT}\right] - 1\right), \tag{15.110}$$

The *ideality factor* is defined as

$$\breve{n} = \frac{1}{1-\beta}, \tag{15.111}$$

which, upon substitution into Eq. (15.110), yields

$$J = J_s \exp\left[\frac{q_e V}{\breve{n} kT}\right]\left(1 - \exp\left[\frac{-q_e V}{kT}\right]\right). \tag{15.112}$$

For large values of V ($> 3kT/q_e$), Eq. (15.112) can be approximated by

$$J \approx J_s \exp\left[\frac{q_e V}{\breve{n} kT}\right]. \tag{15.113}$$

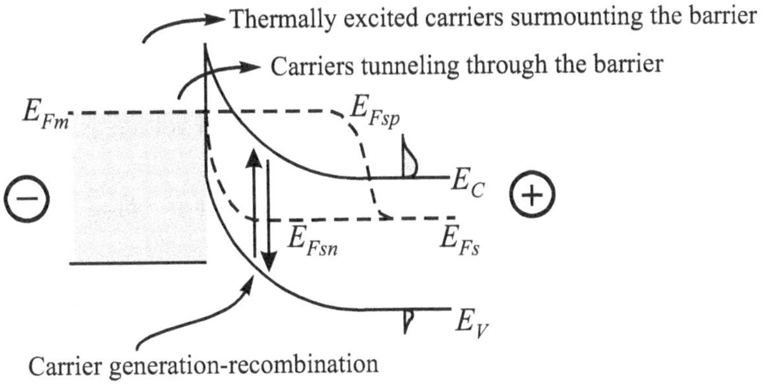

Figure 15.20. Energy band diagram for an *n*-type Schottky contact under reverse bias.

Reverse Leakage Current Under Reverse Bias

Sources of leakage current are shown in Fig. 15.20. Unlike the *pn* junction device, there is no supply of "minority" carriers at the junction boundary. Instead, there is a reservoir of free electrons from the metal, which must surmount the energy barrier before they can enter the semiconductor. Hence, the reverse bias for a Schottky barrier is a function of the barrier height ϕ_{bn}. Under reverse bias, these charge carriers (electrons) can be elevated by thermionic emission over barrier and contribute to the leakage current. A similar situation occurs for holes with a *p*-type Schottky contact. Leakage current can also occur from thermal generation of electrons directly across the band gap into the conduction band, producing electron and hole free carriers. Similar to the *pn* junction device, thermal generation can be suppressed by cooling the device while under operation. Finally, under high voltage bias conditions, charge carriers can tunnel quantum mechanically directly through the energy barrier, thereby also contributing to the leakage current.

Junction Breakdown

In Sec. 14.2.3, the concept of diode breakdown was briefly discussed as an introduction to SPAD and SiPM light detection devices. The phenomenon of breakdown can also be observed in semiconductor diodes such as *pn*, *pin*, and Schottky devices. Breakdown can be recognized as a tremendous increase in leakage current at bias voltages past a specific threshold. Two mechanisms can describe this junction breakdown, namely, *tunneling* and *avalanche* breakdown.

Tunneling Breakdown Under high reverse bias conditions, it is possible to cause such band bending across a *pn* junction that electrons can tunnel directly from the valence band into the conduction band. Such a condition is shown in Fig. 15.10 as a potential source of leakage current. In some circumstances, *pn* junction diodes are fabricated with relatively high dopant concentrations with the intentional purpose of causing this kind of breakdown. Such devices are called *Zener diodes*,[4] and are commonly used for overvoltage circuit protection. Typically, electric fields exceeding 10^6 V cm^{-1} are needed to achieve the band bending required for Zener breakdown in specially designed highly doped *pn* junctions [Grove 1967]. Radiation detectors are usually fabricated from high purity materials, and Zener breakdown is usually not reached before other breakdown mechanisms are reached. Other breakdown mechanisms are usually of more concern.

Avalanche Breakdown The topic of avalanche breakdown was discussed earlier in Sec. 14.2.3. Indeed, some light detection devices were seen to rely upon this effect for gain. However, avalanche breakdown can become a source of unwanted noise and can also cause irreparable damage to semiconductor detectors. The

[4]Named after Clarence M. Zener who discovered the effect.

breakdown occurs when electrons gain enough kinetic energy from an applied electric field such that some of their energy can be transferred to the lattice and excite more electrons into the conduction band.

Consider a uniformly doped semiconductor upon which is formed an abrupt one-sided rectifying junction. From Eqs. (15.50) and (15.52), the general expression for the applied voltage is

$$V = \int_0^W \mathcal{E}(x)dx = \int_0^W \frac{q_e N_b W}{\kappa \epsilon_o}\left(\frac{x}{W} - 1\right)dx = -\mathcal{E}_{max}\frac{W}{2}, \tag{15.114}$$

where N_b is the doping concentration of the lower doped side and W is the depletion width. Suppose that $\mathcal{E}_{max}$ is the critical field required for avalanche breakdown. Then the magnitude of the breakdown electric field is

$$|\mathcal{E}_{crit}| = \frac{2V_B}{W} = \frac{2(V_{bi} - V_{rev})}{W}, \tag{15.115}$$

where V_B is the breakdown voltage and V_{rev} is the applied reverse voltage. Substitution of the one-sided depletion approximation into Eq. (15.115) yields

$$V_B = V_{bi} - V_{rev} = \frac{\mathcal{E}_{crit}^2 \kappa \epsilon_o}{2 q_e N_b}. \tag{15.116}$$

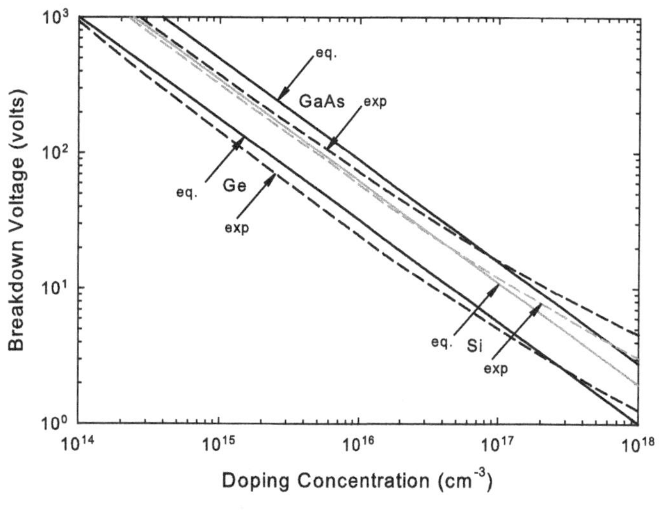

Figure 15.21. Avalanche breakdown voltage as a function of doping concentration for an abrupt one-sided *pn* junction diode. A comparison is made between the data of Sze and Gibbons [1966a] and the empirical formula of Eq. (15.116).

The breakdown voltage of Si, Ge, and GaAs as a function of background doping concentration was studied by Sze and Gibbons [1966a] and the experimental measurement results shown in Fig. 15.21. Sze and Gibbons [1966a] offer the following empirical relation to estimate the breakdown voltage

$$V_B \simeq 60 \left(\frac{E_g}{1.1}\right)^{3/2} \left(\frac{N_b}{10^{16}}\right)^{-3/4} \quad \text{in volts}, \tag{15.117}$$

where N_b is the background impurity concentration and E_g is the band-gap energy in electron volts. A comparison of Eq. (15.117) with the measured data is also shown in Fig. 15.21. It appears that Eq. (15.117) yields acceptable results for background doping concentrations below 10^{17} cm^{-3}, but is less accurate at higher doping levels. It is shown in Fig. 15.21 and also Eq. (15.117) that the breakdown voltage increases with band-gap energy. Intuitively, this observation makes sense, because wider band-gap materials require more energy, on average, to produce an electron-hole pair than do narrow band-gap materials.

Another breakdown mechanism is a lowering of the breakdown voltage from rounded junction interfaces. When dopants are introduced into substrates, the contact pattern is often defined by photolithography methods. Doped patterns of specific size and shape are defined by windows etched through diffusion masks, commonly a layer of SiO_2 or other high-temperature insulator. Dopants are introduced by various methods, including diffusion and implantation (for details, see Sze [VLSI 1985 and others]). These dopants are heated to high temperatures to drive them into the substrate, and in doing so, they diffuse from high to low concentrations. Consequently, at the pattern edges, it is usually the case that the junction interface becomes rounded in a semi-cylindrical volume, with radius r_j, as the dopants diffuse laterally from the point of introduction. At the pattern corners, the combined edges form a hemispherical interface. Recall from Sec. 8.7 and Sec. 8.8 that the electric field is higher at the smaller terminal than the larger terminal for cylindrical and hemispherical geometries. Hence, the electric field formed around the pattern edges is higher than expected with a simple planar *pn* junction. The overall effect is to lower the breakdown voltage of the *pn* junction [Armstrong 1957]. Reported results by Sze and Gibbons [1966b] on Si diodes with different junction curvatures are shown in Fig. 15.22.

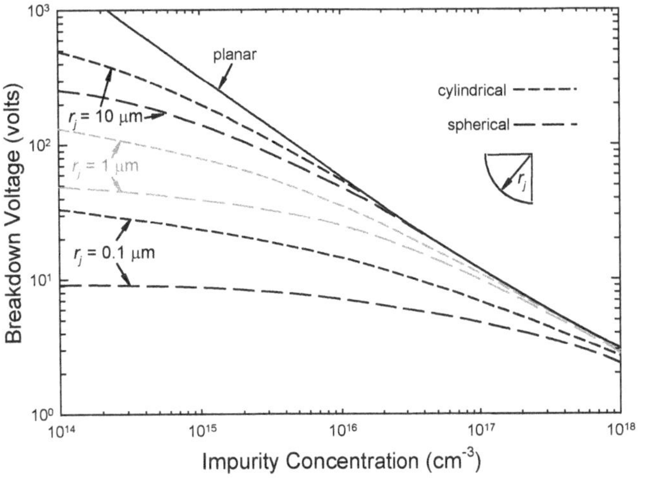

Figure 15.22. Breakdown voltage versus impurity concentration for one-sided abrupt junction Si diodes with different cylindrical and spherical curvatures. Data from Sze and Gibbons [1966b].

Ghandhi [1977] offers an empirical relation to determine the breakdown voltages V_{JB} of curved junctions in Si abrupt one-sided *pn* junctions. For cylindrical junctions

$$\frac{V_{JB}}{V_B} = 0.5(\eta^2 + 2\eta^{6/7})\ln(1 + 2\eta^{-8/7}) - \eta^{6/7}, \quad (15.118)$$

and for spherical junctions

$$\frac{V_{JB}}{V_B} = \eta^2 + 2.14\eta^{6/7} - (\eta^3 + 3\eta^{13/7})^{2/3}, \quad (15.119)$$

where V_B is the breakdown voltage for the dopant concentration and

$$\eta = \frac{r_j}{W} \quad (15.120)$$

where W is the depletion width of the material at V_B and r_j is the radius of curvature. The results from Eqs. (15.118) and (15.119) are shown in Fig. 15.23.

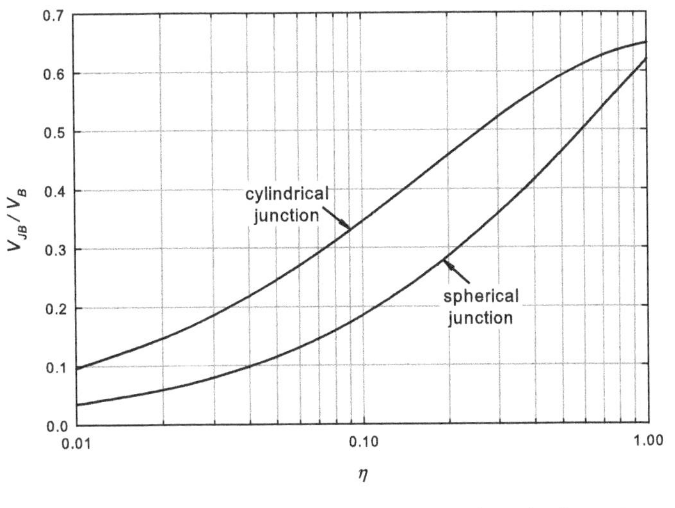

Figure 15.23. Normalized junction breakdown (V_{JB}/V_B) as a function of $\eta = r_j/W$ for cylindrical and hemispherical junctions in Si. After Ghandhi [1977].

Example 15.4: An abrupt p-type junction is diffused into a silicon substrate having an n-type background concentration of 10^{15} cm^{-3}. The edges are cylindrical with a radius of curvature equal to 1 micron. What is the junction breakdown voltage?

Solution:

First find the breakdown voltage from Eq. (15.117).

$$V_B \simeq 60\left(\frac{E_g}{1.1}\right)^{3/2}\left(\frac{N_B}{10^{16}}\right)^{-3/4} = 60\left(\frac{1.12\text{ eV}}{1.1}\right)^{3/2}\left(\frac{10^{15}}{10^{16}}\right)^{-3/4} = 346.65\text{V}.$$

Next find the depletion width at V_B.

$$W = \sqrt{\frac{2\kappa\epsilon_o V_B}{q_e N_b}} = \sqrt{\frac{2(11.9)(8.854\times10^{-14}\text{ F/cm})(346.65\text{ V})}{(1.6\times10^{-19}\text{ C})(10^{15}\text{ cm}^{-3})}} = 0.00214\text{ cm}.$$

Determine the value of η from Eq. (15.118),

$$\frac{V_{JB}}{V_B} = 0.5(\eta^2 + 2\eta^{6/7})\ln(1+2\eta^{-8/7}) - \eta^{6/7}$$

$$= 0.5\left[\left(\frac{1\ \mu\text{m}}{21.4\ \mu\text{m}}\right)^2 + 2\left(\frac{1\ \mu\text{m}}{21.4\ \mu\text{m}}\right)^{6/7}\right]\ln\left[1+2\left(\frac{1\ \mu\text{m}}{21.4\ \mu\text{m}}\right)^{-8/7}\right] - \left(\frac{1\ \mu\text{m}}{21.4\ \mu\text{m}}\right)^{6/7} = 0.2369.$$

The junction breakdown voltage is thus determined to be $V_{JB} = 0.2369(346.65\text{ volts}) = 82.12$ volts.

Punch Through Breakdown Semiconductor radiation detectors are usually fabricated from relatively pure materials with low background impurity concentrations. Consequently, it is possible to fully deplete a pn, $p\nu n$, or $p\pi n$ diode, a condition called *punch through*. As the voltage is increased, the maximum electric field also increases at the pn junction. At high enough voltages, the electric field reaches the critical field required for avalanching and reach a condition named *punch through breakdown*.

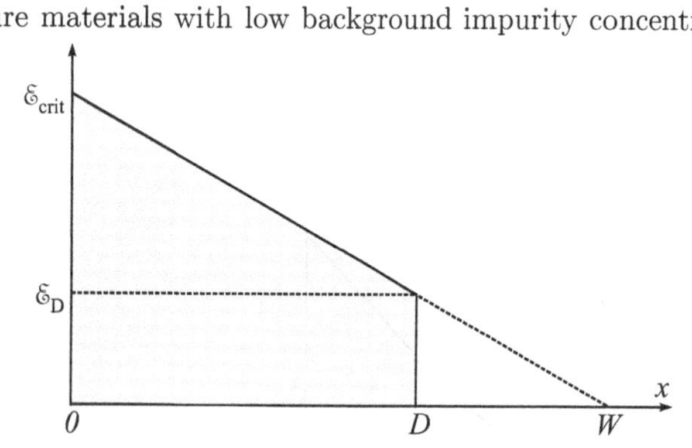

Figure 15.24. The breakdown electric field is a function of the background doping and space charge. The integral of the electric field as a function of x yields the resulting breakdown voltage, which is also a function of the substrate width.

Consider Fig. 15.24, in which an ideal electric field is depicted for an abrupt one-sided junction for a uniformly doped substrate. For a relatively thick substrate, a depletion width W is reached at the required breakdown electric field $\mathcal{E}_{crit}$, where W can be found with Eq. (15.115). The breakdown voltage V_B is described by Eq. (15.116) and can be estimated by Eq. (15.117); hence $\mathcal{E}_{crit}$ can be determined.

If instead the width of the substrate is less than W, labeled D in Fig. 15.24, the integral of the electric field yields the applied voltage, depicted by the shaded area of Fig. 15.24,

$$V = \int_0^D \mathcal{E}(x)dx.$$

Recall that the solution for the electric field in a uniformly doped *pn* junction yields a simple linear equation, shown in Eq. (15.50) with slope $(q_e N_b)/(\kappa\epsilon_o)$. Using simple geometric analysis with Fig. 15.24, the punch through breakdown voltage V_{BP} is

$$V_{BP} = V_B \frac{(\mathcal{E}_{crit} + \mathcal{E}_D)D}{2} \frac{2}{\mathcal{E}_{crit} W} = V_B \frac{D}{W}\left[2 - \frac{D}{W}\right]. \quad (15.121)$$

Insertion of Eq. (15.59) and the relation $V_{bi} - V = V_B$ yields

$$V_{BP} = V_B D \sqrt{\frac{q_e N_b}{2\kappa\epsilon_o V_B}} \left[2 - D\sqrt{\frac{q_e N_b}{2\kappa\epsilon_o V_B}}\right]. \quad (15.122)$$

Shown in Fig. 15.25 are the dependences of V_{BP} on doping concentration for several values of D for silicon diodes.

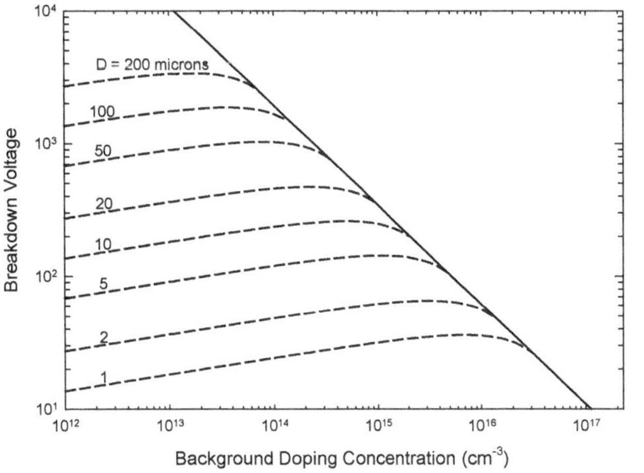

Figure 15.25. Breakdown voltage for Si $p\nu n$ and $p\pi n$ diodes as a function of background doping concentration and substrate thickness determined from Eq. (15.117) and Eq. (15.122).

Example 15.5: Given a 100-micron thick silicon $p\nu n$ diode with $N_A = 10^{15}$ cm^{-3}, $N_d = 10^{14}$ cm^{-3}, and $N_D = 10^{15}$ cm^{-3}, what is the punch through breakdown voltage?

Solution:

First find V_B by estimating the value with Eq. (15.117) or reading from the graph in Fig. 15.21.

$$V_B \simeq 60\left(\frac{E_g}{1.1}\right)^{3/2}\left(\frac{N_B}{10^{16}}\right)^{-3/4} = 60\left(\frac{1.12\text{ eV}}{1.1}\right)^{3/2}\left(\frac{10^{14}}{10^{16}}\right)^{-3/4} = 1949\text{V}.$$

At a voltage bias of 1949 volts, the expected depletion width is,

$$W = \sqrt{\frac{2\kappa\epsilon_o V_B}{q_e N_b}} = \sqrt{\frac{2(11.9)(8.854\times10^{-14}\text{ F/cm})(1949\text{ V})}{(1.6\times10^{-19}\text{ C})(10^{14}\text{ cm}^{-3})}} = 0.016\text{ cm}.$$

Consequently, at V_B, the diode is fully depleted and punched through. The punch through breakdown voltage is found with Eq. (15.121),

$$V_{BP} = V_B \frac{D}{W}\left[2 - \frac{D}{W}\right],$$

$$= (1949\text{ V})\frac{0.01\text{ cm}}{0.016\text{ cm}}\left[2 - \frac{0.01\text{ cm}}{0.016\text{ cm}}\right],$$

$$= 1674.9\text{ V}.$$

The punch through breakdown voltage V_{BP} is 85.94% of the breakdown voltage V_B.

15.3.4 The MOS Structure

Another common semiconductor device structure is the metal-oxide-semiconductor (MOS) capacitor. Radiation detectors usually do not have these structures, but there are some special radiation detection devices that operate with MOS capacitors. For that reason, this design and the operating physics are briefly discussed here. A MOS structure has three basic components, the semiconductor, the insulating oxide layer and a metallic gate upon the oxide as shown in Fig. 15.26. The MOS field effect transistor, or MOS-FET, also has an injecting source contact and a receiving drain contact.

An ideal MOS capacitor can be described as a structure in which both the metal work function ϕ_m and the semiconductor work function ϕ_s are equal, i.e.,

$$q_e\phi_m - q_e\phi_s = q_e\phi_m - q_e\chi_s - \frac{E_g}{2} - |q_e\psi_B| = 0, \quad \text{for } p\text{-type}, \tag{15.123}$$

and

$$q_e\phi_m - q_e\phi_s = q_e\phi_m - q_e\chi_s - \frac{E_g}{2} + |q_e\psi_B| = 0, \quad \text{for } n\text{-type}. \tag{15.124}$$

The MOS band structures for p-type and n-type substrates are depicted in Fig. 15.27. Hence, at zero voltage, the band structures have constant energy so they are flat. Further, the insulating oxide region of an ideal MOS device has no space charge imperfections and so there is no trapped charge at the oxide-semiconductor interface. Finally, the oxide acts as a perfect insulating barrier and no leakage current penetrates the oxide during operation. These conditions define an *ideal* MOS capacitor.

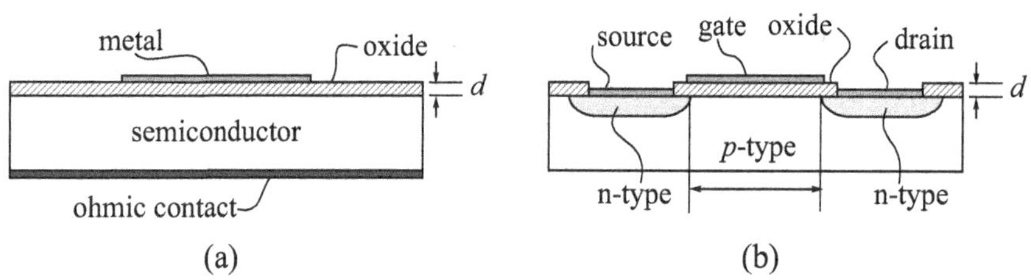

Figure 15.26. Cross section depiction of the main structural components of a (a) MOS capacitor and a (b) MOS field effect transistor.

The band structure of an ideal p-substrate MOS capacitor under bias is depicted in Fig. 15.28. Note that, because the oxide is a "perfect" insulator, no current can flow across the barrier, hence the Fermi energy level remains flat even under bias. If a net negative bias is applied to the gate metal, then the energy bands bend upwards, as shown in Fig. 15.28(a), and a potential well forms at the interface. The concentration of the majority carrier holes is

$$p_p = n_i \exp\left[\frac{E_i - E_{F_s}}{kT}\right], \tag{15.125}$$

and at the oxide-semiconductor interface $E_i - E_F$ becomes relatively large. Consequently, holes gather at the interface, a condition termed *accumulation.* A similar case occurs on a n-type substrate under positive bias in which the majority carriers are electrons which accumulate at the interface so that

$$n_n = n_i \exp\left[\frac{E_{F_s} - E_i}{kT}\right]. \tag{15.126}$$

Likewise the application of a small positive bias on the gate of a p-type MOS capacitor causes the energy bands to bend downwards, ultimately reducing the number of holes at the interface, thereby causing

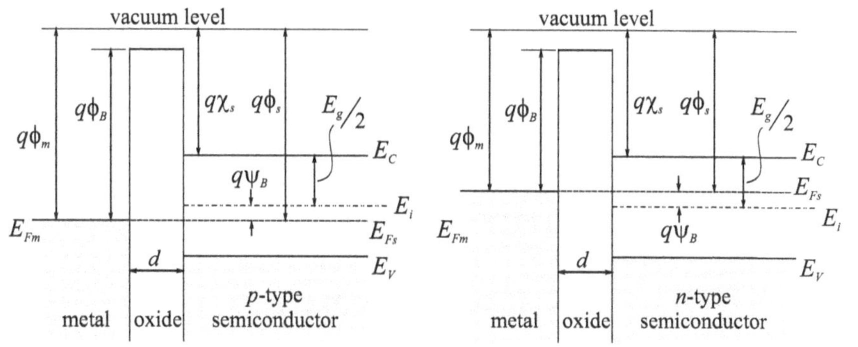

Figure 15.27. Energy band diagrams of *ideal* p-type and n-type MOS structures.

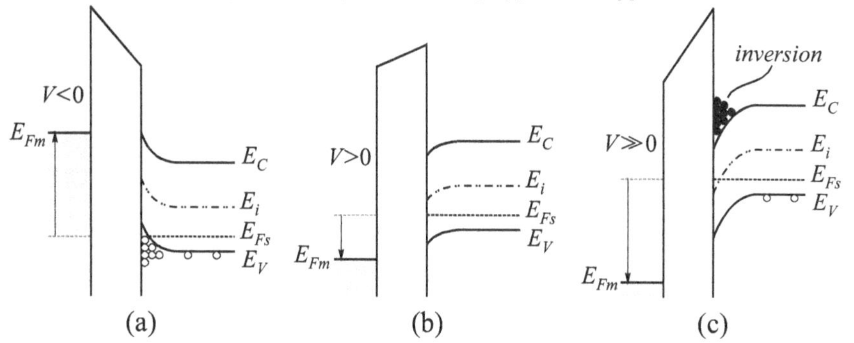

Figure 15.28. Energy band diagrams for an ideal p-type MOS structure under bias. (a) The band structure with the gate at negative voltage, termed *accumulation*. (b) The band structure with the gate at positive voltage, termed *depletion*. (c) The band structure with the gate at a large positive voltage, termed *inversion*.

a depletion region to form at the oxide-semiconductor interface. The depletion depth W is a function of the voltage and also the background doping concentration of the semiconductor substrate, where the space charge per unit area is approximated by

$$Q_{sc} = -q_e N_A W, \tag{15.127}$$

for a p-type substrate. A similar situation occurs if a negative voltage is applied to an n-type substrate so that

$$Q_{sc} = q_e N_D W. \tag{15.128}$$

If a radiation event produces electron-hole pairs in the semiconductor, the electrons accumulate in the potential well at the interface. With a readout contact adjacent to the oxide layer, this charge can be removed and measured, an ability that underlies the operation of the charge-coupled device discussed in the next chapter.

As the positive voltage is increased further, the depletion region continues to increase until the intrinsic Fermi energy E_i is bent down at the interface to the same value of the Fermi energy E_{F_s} in the semiconductor. Increasing the voltage even more causes E_i to cross over E_{F_s} and the Fermi energy approaches the conduction band. The minority electron concentration increases at the interface as

$$n_p = n_i \exp\left[\frac{E_{F_s} - E_i}{kT}\right], \tag{15.129}$$

and causes the p-type material to act as n-type material, an operating condition termed *inversion.* As the bias is increased further, the conduction band E_C eventually comes close to E_{F_s}, and the electron population near the interface increases rapidly. Hence, the charge density confined to a narrow inversion region of width x_i directly under the oxide becomes dominated by the electron population, and not the semiconductor space charge. Consequently, after the inversion layer is formed, additional voltage does not increase the depletion width, but instead tends to increase the electron density that opposes the additional applied voltage. In other words, with strong inversion, the depletion region no longer increases in width with higher voltage. With strong inversion, the total charge per unit area becomes

$$Q_s = -q_e n_p x_i - q_e N_A W_{max}. \tag{15.130}$$

The opposite situation occurs for an n-type substrate, namely

$$Q_s = q_e p_n x_i + q_e N_D W_{max}. \tag{15.131}$$

Consider the diagram of Fig. 15.26(b) under inversion. A electron rich population appears under the oxide connecting the source and drain. Hence, charge can flow across the device while under the inversion condition. Reduction of the voltage on the gate causes the device to return to the depletion condition, effectively stopping the charge flow. This behavior is fundamentally how the MOS-FET operates as a current switch.[5]

The maximum depletion width determines the potential at which the concentration of electrons at the surface is equal to the space charge density of the depleted volume. Here the electrostatic potential ψ is defined as the difference between the intrinsic Fermi energy in the semiconductor bulk (beyond the depletion region) to that near the surface region where the bands are bent so that

$$q_e\psi = E_{i_{bulk}} - E_{i_{bent}}. \tag{15.132}$$

Hence, ψ is zero in the semiconductor bulk. The surface potential is defined as

$$q_e\psi_s = E_{i_{bulk}} - E_{i_{sur}}. \tag{15.133}$$

The surface electron concentration can then be represented as

$$n_{sur} = n_i \exp\left[\frac{q_e(\psi_s - \psi_B)}{kT}\right], \tag{15.134}$$

and the surface hole population as

$$p_{sur} = n_i \exp\left[\frac{q_e(\psi_B - \psi_s)}{kT}\right], \tag{15.135}$$

in which $\psi_B = E_i - E_{F_s}$ in the bulk (see Fig. 15.27). The opposite conditions exist for the n-type substrate.

The potential distribution under the oxide can be calculated from Poisson's equation as [Sze 1981]

$$\psi(x) = \psi_s\left[1 - \frac{x}{W}\right]^2, \tag{15.136}$$

[5]Because the inversion layer behaves as an n-type channel, this type of device is called NMOS, after the type of inversion channel and not the substrate dopant. Similarly, a MOS device formed on an n-type substrate produces a hole-filled p-type inversion channel, and is thus called PMOS. Transistor circuits with complementary pairs of NMOS and PMOS transistors are called CMOS.

where the surface potential is

$$\psi_s = \frac{q_e N_A W^2}{2\kappa_s \epsilon_0}. \tag{15.137}$$

The surface potential at which strong inversion appears is generally accepted as approximately equal to $2\psi_B$ or

$$\psi_{inv} \simeq 2\psi_B = \frac{2kT}{q_e} \ln\left[\frac{N_A}{n_i}\right], \tag{15.138}$$

for p-type substrates, and

$$\psi_{inv} \simeq 2\psi_B = \frac{2kT}{q_e} \ln\left[\frac{N_D}{n_i}\right], \tag{15.139}$$

for n-type substrates. Substitution of Eq. (15.138) into Eq. (15.136) yields

$$W_{max} = \left[\frac{4\kappa_s \epsilon_0 kT}{q_e^2 N_A} \ln\left(\frac{N_A}{n_i}\right)\right]^{1/2}, \tag{15.140}$$

for p-type substrates where, from Eq. (15.127),

$$Q_{sc} = -q_e N_A W = -\left[4N_A \kappa_s \epsilon_0 kT \ln\left(\frac{N_A}{n_i}\right)\right]^{1/2} = -\left[4q_e N_A \kappa_s \epsilon_0 \psi_B\right]^{1/2}. \tag{15.141}$$

Similarly, for n-type substrates

$$W_{max} = \left[\frac{4\kappa_s \epsilon_0 kT}{q_e^2 N_D} \ln\left(\frac{N_D}{n_i}\right)\right]^{1/2}, \tag{15.142}$$

and

$$Q_{sc} = q_e N_D W = \left[4N_A \kappa_s \epsilon_0 kT \ln\left(\frac{N_D}{n_i}\right)\right]^{1/2} = \left[4q_e N_D \kappa_s \epsilon_0 \psi_B\right]^{1/2}. \tag{15.143}$$

Another important measure is the threshold voltage at deep inversion. Note that the potential applied to the gate drops partially through the oxide layer. Hence, the total voltage required to reach deep inversion includes both the voltage across the oxide V_o and the depletion region ψ_s, i.e., $V_G = V_o + \psi_s$. The capacitance per unit area across the oxide is defined as

$$C_o = \frac{\kappa_o \epsilon_0}{d}, \tag{15.144}$$

where d is the oxide thickness. Hence, the threshold voltage V_T becomes

$$V_T = \frac{-Q_{sc}}{C_o} + 2\psi_B = \left[\frac{4d^2 q_e N_A \kappa_s \psi_B}{\kappa_o^2 \epsilon_0}\right]^{1/2} + 2\psi_B. \tag{15.145}$$

or for n-type substrates

$$V_T = -\left[\frac{4d^2 q_e N_D \kappa_s \psi_B}{\kappa_o^2 \epsilon_0}\right]^{1/2} - 2\psi_B. \tag{15.146}$$

There are a few important features that can be gleaned from this simple ideal example. First, the maximum depletion region possible for a MOS structure is approximately proportional to the inverse of $\sqrt{N_{A,D}}$. Second, the threshold voltage for deep inversion increases with substrate background dopant concentration and oxide thickness. Third, the capacitance per unit area increases as the oxide thickness d is decreased.

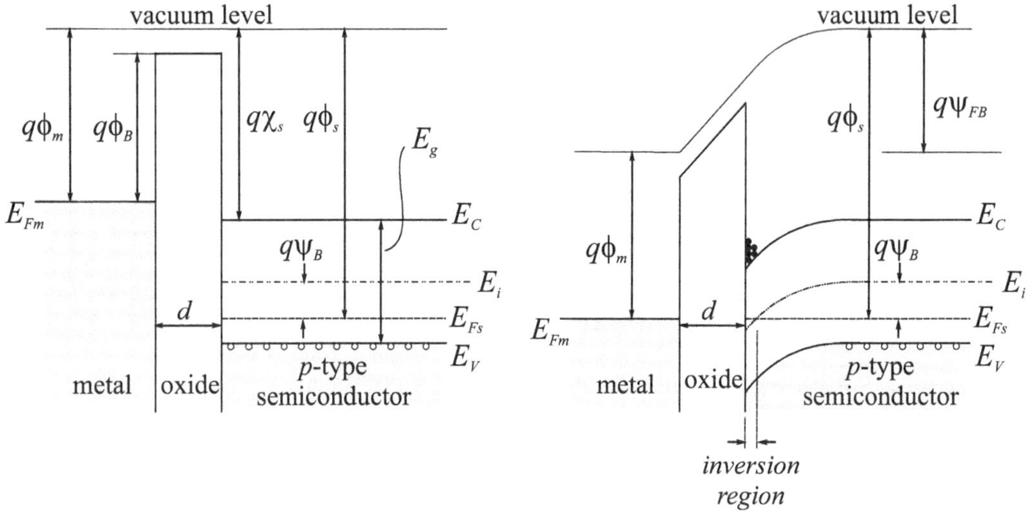

Figure 15.29. Energy band diagram for a more realistic p-type MOS structure. (left) The structure before equilibrium and (right) after equilibrium with a constant Fermi energy level.

However, there are additional properties of a common MOS device that are important to understand in order to use a MOS device in radiation detection applications. First, seldom are the work functions of both the metal and semiconductor equal. Second, impurities and defects in the oxide and at the oxide-semiconductor interface can produce space charge, which, in turn, affects the operating potentials of the device. Consider the diagram of Fig. 15.29 for a p-type substrate MOS structure. The work function of the metal is somewhat less than the semiconductor. As the materials are brought into contact, the bands bend to produce a flat Fermi energy, as shown in Fig. 15.29(right). Hence, to flatten the bands, a negative voltage must be applied to straighten the bands. This voltage is termed the *flat-band voltage* V_{FB}. The flat-band voltage must adjust for the difference in the work function $(\phi_m - \phi_s)$ and the oxide space charge (Q_{sc}), i.e.,

$$V_{FB} = q_e(\phi_m - \phi_s) + \frac{Q_{sc}}{C_o}. \tag{15.147}$$

The operating gate voltage must be adjusted to account for the flat-band voltage as

$$V'_G = V_G - V_{FB}, \tag{15.148}$$

and the threshold for the onset of deep inversion becomes

$$V'_T = V_T - V_{FB}. \tag{15.149}$$

Hence, it is possible that the device is in an inversion condition at zero voltage. It should be noted that the operating potential V_G changes as Q_{sc} changes, which is a major concern for MOS devices operated in a radiation environment. This space charge can arise from many sources, including interface trapped charge, fixed oxide charges, mobile ions, and trapped charges. Interface charge is a function of the semiconductor surface type and preparation. The ⟨100⟩ Si surface generally has the lowest density of surface states and is preferred over a ⟨111⟩ Si surface for MOS devices [Sze 1981]. Contamination of the oxide can produce a strong electric field which moves any mobile charges. Contamination from alkali metals, such as Na^+ ions, is often a source of mobile charges. Fixed charge, usually positive, is caused by oxide defects near the oxide-semiconductor surface and cannot be eliminated once formed. Finally, trapped charge is produced as a result of radiation damage, usually from charged particles and x-ray irradiation. As the trapped charge increases, the voltage required to operate the device increases, ultimately causing catastrophic failure. Dopant type inversion can also occur in the semiconductor, especially under neutron irradiation, where transmutation activation can change the semiconductor work function over time.

15.3.5 Ohmic Contacts

Efficient charge carrier extraction from a semiconductor detector requires ohmic-like contacts; hence, it can be important that the metal/semiconductor interface *not* form a rectifying potential barrier. For instance, metal contacts to the n and p regions of pn and pin detectors are constructed so as to be non-rectifying or *ohmic*, i.e., they have a current-voltage relationship that satisfies Ohm's Law. For "ideal" ohmic contacts, a careful choice of metals upon a selected semiconductor can produce a non-rectifying junction. Unfortunately, because of interface state pinning, a potential barrier is formed when almost any metal is applied to the surface of semiconductors. To remedy this problem, degenerate doping is applied with the metal and diffused, typically through thermal treatment, into the semiconductor. The process causes the Schottky barrier to become extremely thin such that electrons can tunnel directly through the barrier, thereby producing a contact with ohmic behavior.

Schottky diodes are typically constructed with one (or more) Schottky contact used as a rectifying barrier to reduce leakage current and one (or more) opposing ohmic contact to allow more efficient carrier extraction, thereby reducing electronic noise (see depiction in Fig. 15.19).

Ideal Ohmic Contacts

Consider the case in which the work function of the metal is smaller than the work function of an n-type semiconductor, as is shown in the left-hand part of Fig. 15.30. In order for the Fermi energies in the metal and semiconductor to align, the semiconductor energy bands bend downward, as shown in Fig. 15.31. For p-type material, the metal work function must be larger than the semiconductor work function, in which case the bands bend upwards, as is depicted in the right-hand portion of Fig. 15.31.

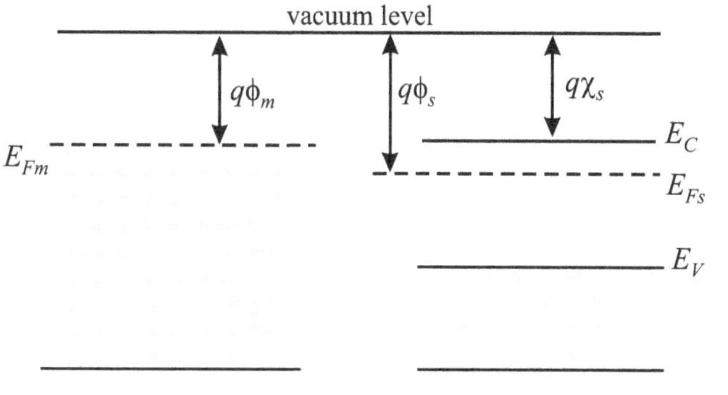

Figure 15.30. Energy band conditions for an ideal n-type ohmic contact.

For the n-type ohmic contact, electrons in the conduction band no longer encounter an energy barrier when drifting into the metal contact. Similarly, the holes no longer encounter and energy barrier at the metal-semiconductor interface. In both cases, the contact and resulting device have a linear current-voltage (IV) relationship that follows Ohm's law.

Tunneling Ohmic Contacts

As mentioned before, semiconductor surfaces typically have defects and impurity contamination. These imperfections may arise from the cutting and polishing processes that are usually incorporated during detector fabrication. Although subsequent liberal surface etching can remove much of the damage, sufficient surface damage usually remains that "pins" the energy barrier at the metal/semiconductor interface. Hence, the above ideal model of the ohmic contacts rarely applies.

Ohmic contacts can be made, however, by introducing degenerate doping at the semiconductor surface before or during the metal contact application, with degenerate n-type dopant levels introduced to n-type material and degenerate p-type dopant levels introduced to p-type material. The dopant is diffused into the surface, causing the bands to bend strongly at the interface. Although an energy barrier still appears, the

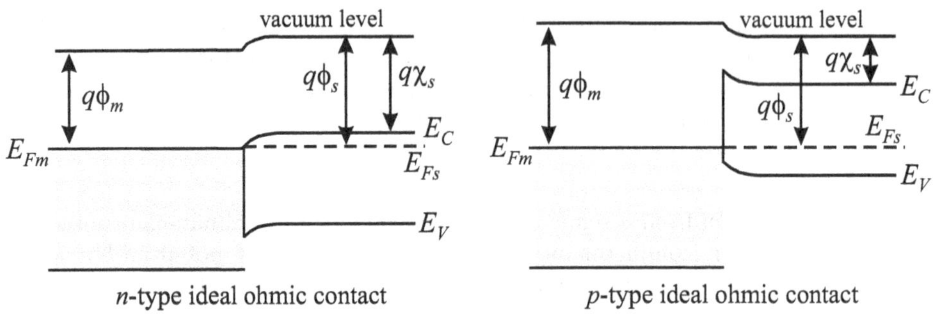

Figure 15.31. Energy band diagrams for ideal n-type and p-type ohmic contacts.

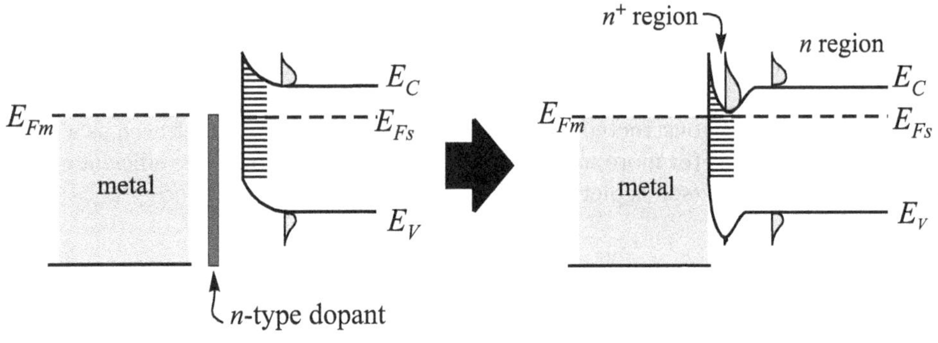

Figure 15.32. Interface states and band conditions for a tunneling n-type ohmic contact.

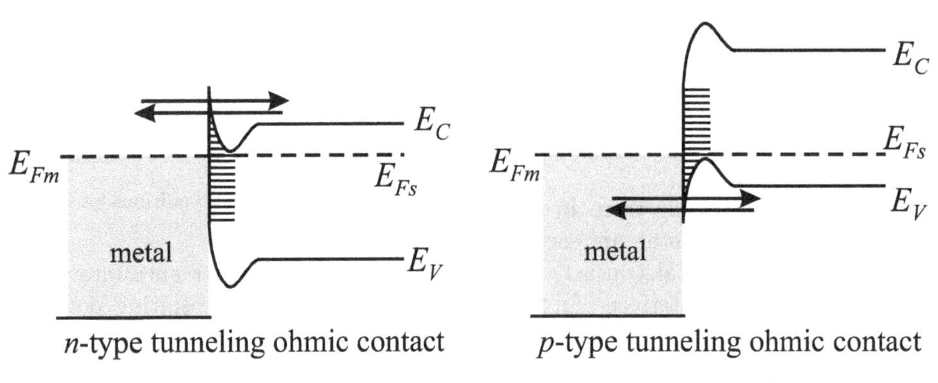

Figure 15.33. Energy band diagrams for n-type and p-type tunneling ohmic contacts.

actual barrier width is relatively small such that the charge carriers can tunnel directly through the barrier. The result is a linear IV relationship that follows Ohm's law. The interface states and energy bands for tunneling ohmic contacts are shown in Fig. 15.32 and Fig. 15.33, respectively.

The degenerate doping may be applied by traditional methods, such as ion implantation or thermal diffusion. Often a metal eutectic, partially composed of an element that is a dopant in the semiconductor, is applied to the surface and annealed so that the dopants become active in the semiconductor surface. Examples include AuGeNi annealed into n-type GaAs and AuZn annealed into p-type GaAs. Here Ge is an

n-type dopant and Zn is a p-type dopant. Often a chemical reaction may produce the desired effects, such as occurs in the application of $AuCl_3$ to p-type CdTe or CdZnTe. The $AuCl_3$ reacts with the Cd, displacing it, and deposits Au in the material, which acts as a p-type dopant.

15.3.6 Series Resistance and Space Charge Effects

Wide band-gap semiconductors are of interest because many can be operated at room temperature without the complications of cryogenic or mechanical cooling. Unfortunately, all of these room temperature semiconductors are compound materials,[6] and consequently, these semiconductors typically have low electrical resistivity from residual background impurities that are difficult to remove, or possess high concentrations of native defects that act as charge carrier donors or acceptors. The resistivity of these interesting materials can be significantly increased by clever additions of deep energy levels. These energy levels can be added through the purposeful introduction of deep dopant impurities or native defect states. The unintentional introduction of impurities and trapping centers and the purposeful introduction of compensating centers, work together to reduce the free charge carrier concentration, and thereby increase the resistivity of the semiconductor (as demonstrated in Chapter 12). These traps and compensating centers can produce space charge in the semiconductor when filled (or emptied, depending on type). The combined effect of series resistance and space charge alters the forward bias current characteristic in a semiconductor device.

However, the traps can cause issues with the internal electric field within the device, under either forward or reverse bias. Under forward bias, the high series resistance of the detector causes a reduction in the observed current, namely

$$J = J_s \left\{ \exp\left[\frac{q_e(V - JRA)}{kT}\right] - 1 \right\}, \tag{15.150}$$

where J_s is the saturation current density, V is the applied voltage, J is the current density, R is the diode forward resistance, and A is the diode effective area. As the resistance R in Eq. (15.150) is increased, the amount of voltage at the rectifying junction decreases, which causes a decrease in the observed current (Fig. 15.34). The result is a gradually rising forward current characteristic, often modeled with a corrective "ideality factor" $\breve{n}$ as

$$J \approx J_s \left\{ \exp\left[\frac{q_e V}{\breve{n}kT}\right] - 1 \right\}, \tag{15.151}$$

and the subsequent results are shown in Fig. 15.35. Rhoderick and Williams [1988] point out that the oft-used Eq. (15.151) in the literature is an incorrect form, and that the more correct form should actually be Eq. (15.112), mainly because of changes in the rectifying barrier physics. However, additional problems from series resistance and space charge limitations render that argument less effective, mainly because forward bias currents are controlled by other mechanisms than just the contact barrier height. Regardless, Eq. (15.151) is an approximation of the observed case.

The general case for electron current flow in an ideal insulating medium is approached with the current density Eq. (15.67). Consider the case in which electrical contacts are applied to an insulating slab, across which is applied a voltage V_0. For general conditions, drift current is substantially higher than the diffusion current. Hence, Eq. (15.67) reduces to

$$J \simeq q_e \mu_n n(x) \mathcal{E}(x). \tag{15.152}$$

If $n(x)$ is uniform, then from Poisson's equation

$$\int \mathcal{E}(x) d\mathcal{E} = \int \frac{J}{\mu_n \epsilon_s} dx, \tag{15.153}$$

[6]Diamond is an exception, but is inadequate as a gamma-ray detector because of the low Z number.

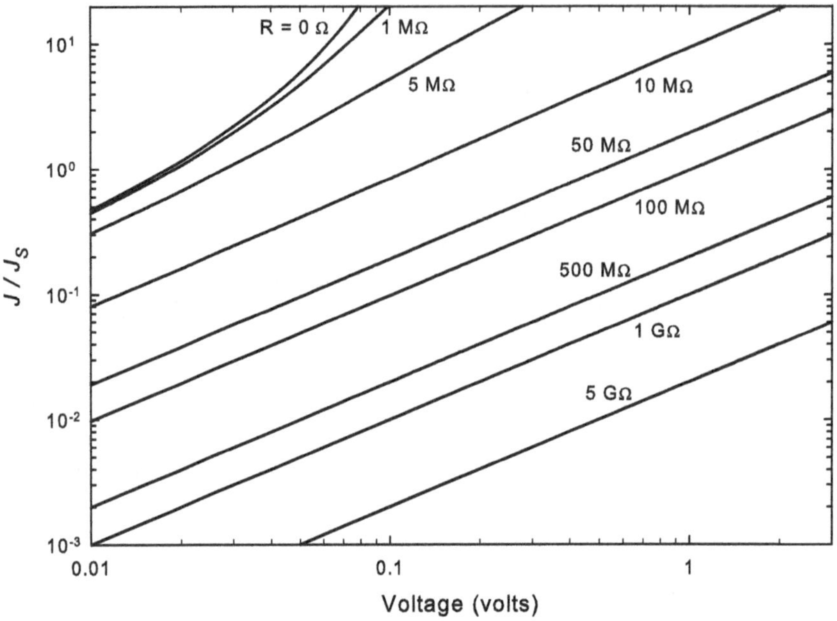

Figure 15.34. The ratio J/J_s plotted as a function of series resistance for the case $J_s = 1$ nA cm^{-2}.

or

$$\mathcal{E}(x) = \left[\frac{2Jx}{\mu_n \epsilon_s} + C\right]^{1/2}, \tag{15.154}$$

where C is an integration constant. For the simplified case, it is assumed that the electric field is zero at the cathode ($x = 0$) [Wright 1959], hence $C = 0$. The voltage drop across the insulating slab is

$$V = \int_0^W \mathcal{E}(x)dx = \int_0^W \left[\frac{2Jx}{\mu_n \epsilon_s}\right]^{1/2} dx = \frac{2}{3}\left[\frac{2J}{\mu_n \epsilon_s}\right]^{1/2} W^{3/2}, \tag{15.155}$$

where W is the slab thickness. Substitution of $i = JA$ and rearrangement of the terms in Eq. (15.155) yields the current

$$i = \frac{9}{8}\frac{\mu_n \epsilon_s A V^2}{W^3}, \tag{15.156}$$

where A is the effective contact area. Eq. (15.156) is known as Child's Law for solids, as developed by Mott and Gurney [1948].[7]

The forward current characteristic can also be affected by space charge buildup from the injected current. As the injected current is increased by the applied bias, the traps in the semiconductor material become filled at a rate high enough to produce a net space charge at steady state. More traps are filled as the forward voltage is increased, a case studied in detail by Rose [1963], Tredgold [1966], Lampert [1956], and Lampert and Mark [1970]. Consider that case in which a semiconductor has a trap density N_T. The voltage at which all traps are filled is similar to the one-sided depletion approximation, i.e., the trap-filled-limit

[7] Eq. (15.156) is also known as the "trap-free square law" or the "Mott-Gurney square law".

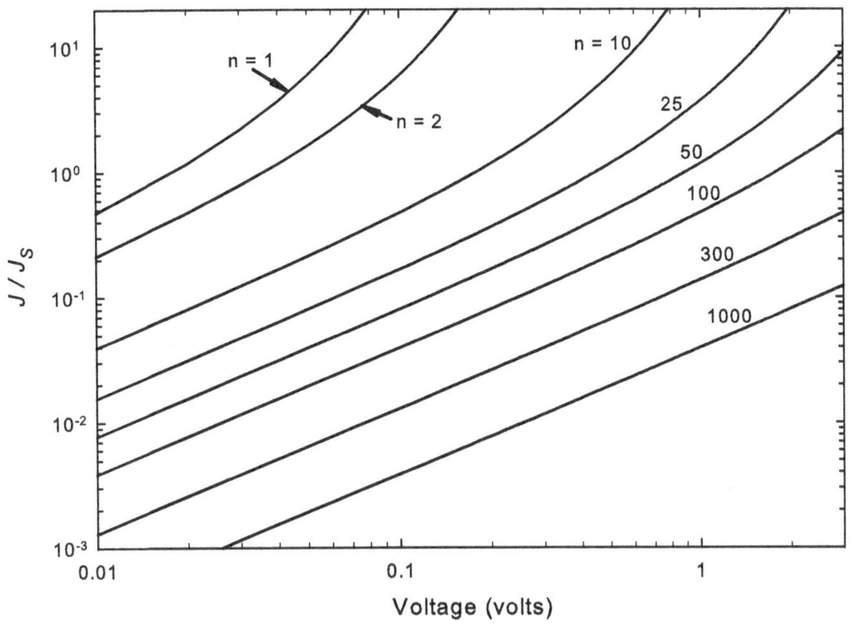

Figure 15.35. The ratio J/J_s plotted as a function of the ideality factor n.

voltage (V_{TFL}) is [Wright 1959]

$$V_{TFL} = \frac{q_e N_T W^2}{2\epsilon_s}. \tag{15.157}$$

Below V_{TFL}, the current is reduced by the trapping effect, which can be shown to be an alteration of Eq. (15.156) [Lampert and Mark 1970]

$$i = \frac{9}{8}\frac{\mu_n \epsilon_s \theta A V^2}{W^3} \tag{15.158}$$

where θ is the fraction of injected charge that is untrapped and remains available for current flow. Eq. (15.158) is referred to as the modified Child's Law. Three regions can usually be observed for space charge limited solids, namely, the ohmic region at very low voltages, the modified Child's Law region (Eq. (15.158)), and the Child's Law region (Eq. (15.156)). At low voltages, the current appears to have an ohmic response, i.e. $i \propto V$. As the voltage is increased, traps begin to fill and produce a space charge region, which affects the electric field and increases current flow ($i \propto V^2$). At V_{TFL} the traps are filled and an abrupt increase appears in the current characteristic from Eq. (15.158) to Eq. (15.156), above which is the full Child's Law region (see Fig. 15.36). This rectifying behavior is due to the overall trap density of the semiconductor (or semi-insulator) rather than a contact barrier or junction chemical potential.

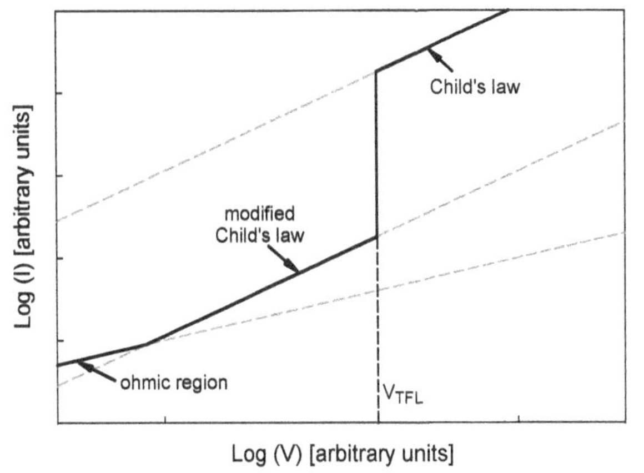

Figure 15.36. Rectifying behavior of space charge limited conduction through a solid.

Note that one of the assumptions with Eq. (15.156) and Eq. (15.158) is that the charge carrier speed is proportional to $\mu\mathcal{E}$, thus yielding one of the voltage terms. At high electric fields the charge carrier speeds saturate and it becomes possible that $i \propto V$ rather than V^2.

15.3.7 Resistive and Photoconductive Devices

There are detectors that operate without rectifying contacts, namely resistive and photoconductive detectors. Resistive detectors have such high resistivity that rectifying barriers are not needed to reduce the leakage current, and photoconductive devices use efficient charge injection to produce photoconductive gain.

Resistive Devices

Semiconductor detectors fabricated from wide band-gap materials (generally > 1.6 eV) have material resistivities high enough to reduce leakage currents to low levels, and, as a result, do not require rectifying contacts to suppress leakage currents. The devices typically have ohmic contacts as electrodes to prevent rectification and the formation of space-charge regions (which limits the active region volume). Radiation interactions in the detectors create electron-hole pairs that are swept out of the detectors by an applied electric field. The high resistivity of the device ensures that the leakage current is significantly lower than the current produced by radiation interactions in the device. For example, the current from a common 5-mm (width) $\times$ 10-mm $\times$ 10-mm CdZnTe detector, with a 1.62-eV band gap and electron mobility of 1350 $cm^2V^{-1}s^{-1}$, has a resistivity of 10^{11} Ω cm, so that the detector has a resistance of 5×10^{10} Ω. At a bias of 500 volts, the electric field is 10^3 V cm^{-1} with a leakage current of 10^{-8} amps. The electron population is found from Eq. (15.152), from which the steady-state charge population is found to be 4.6×10^4 cm^{-3}. Given the volume, the steady-state electron charge population in motion for the detector is 2.3×10^4 cm^{-3}. The average ionization energy of CdZnTe is 5 eV; hence, a 662-keV gamma ray produces approximately 1.32×10^5 electron-hole pairs, thereby increasing the total electron population in the device, momentarily, by 5.74 times. Should the resistivity decrease, the leakage current would add more to the output, whereas an increase in resistivity would decrease the influence of leakage current. Notably, energy resolution degrades from *fluctuations* in the leakage current, which generally increases as the leakage current increases.

Photoconductive Devices

There is a unique class of semiconductor detectors known as photoconductors. To understand how these devices work, consider a semiconductor to which ohmic contacts are affixed that prevent the creation of a blocking barrier and a space-charge region. Such a device is in essence an ohmic resistor. From Eq. (12.90) it is seen that the resistivity of a semiconducting material is inversely proportional to the free carrier concentration. Further, the free carrier concentrations are affected by thermal changes and impurity atoms.

Free carriers are also produced in the semiconductor by gamma-ray interactions. For example, fast electrons are produced as photoelectrons, Compton recoil electrons and, for photons of sufficient energy, as pair production electrons. These fast electrons create many electron-hole pairs as they slow down, producing a small localized charge cloud. Thus, for a single radiation event in a conventional diode detector, the local conductivity is changed spontaneously; however, the small charge cloud is surrounded by higher resistivity material on all sides. Further, the charge cloud is dissipated rapidly by the applied voltage.

Suppose, however, the semiconductor block is saturated with a *radiation pulse*, consisting of a huge number of radiation quanta, such that electron-hole pairs are evenly distributed throughout the crystal bulk. Now, the conductivity of the entire semiconductor block increases because of the large and uniform injection of free carriers throughout the semiconductor crystal. Thus, with a constant applied voltage, the current flowing through the device must increase. After the pulse, the current continues to flow, if the ohmic contacts are well-fabricated, because every electron exiting the device at the anode is replaced by another electron injected at the cathode. This *photocurrent* continues to flow, decaying away as a function of the free charge carriers lifetimes.

A photoconductor consists of a semiconductor material upon which ohmic contacts have been applied. For pure (intrinsic) semiconductors, the conductivity can be described by

$$\sigma = \rho^{-1} = q_e(n\mu_n + p\mu_p), \tag{15.159}$$

where q_e is the unit electronic charge, n is the negative electron population, p is the positive "hole" population, and μ is the mobility of either electrons or holes. Photons interacting with the semiconductor excite electrons into the conduction band and, thus, serve to increase n, p, and the conductivity. The resistance of the material is described by

$$R = \frac{\rho L}{A} = \frac{L}{\sigma A}, \tag{15.160}$$

where L is the effective device length and A is the effective cross sectional area of the device. Hence, the observed current is

$$I = \frac{\sigma AV}{L} = \sigma A\mathcal{E} = q_e A\mathcal{E}(n\mu_n + p\mu_p), \tag{15.161}$$

where V is the applied voltage and $\mathcal{E}$ is the electric field. Note that the photocurrent is a direct function of the charge carrier densities n and p. Suppose that Δn the increase in the electron charge carriers produced by an impulse of light. Hence, $n(t) = n_o + \Delta n \exp(-t/\tau)$, where n_0 is the electron charge carrier population without light, i.e., the dark current. These charge carriers recombine or become trapped, so that charge carriers have a mean lifetime of τ. Under steady-state irradiation, free charge carriers that recombine are regenerated at the same rate, hence the generation rate per unit volume becomes $G = n/\tau$ or

$$\frac{n}{\tau} = \frac{1}{AL}\frac{q_e\eta_Q P_o}{h\nu}, \tag{15.162}$$

where P_o is the optical power density and η_Q is the quantum efficiency.[8] The electron current contribution is found by substituting Eq. (15.161) into Eq. (15.160) to obtain

$$I = I_n + I_p = \frac{(\mu_n\tau_n + \mu_p\tau_p)\mathcal{E}}{L}\frac{q_e\eta_Q P_o}{h\nu}. \tag{15.163}$$

The primary photocurrent is described by

$$I_{prim} = \frac{q_e\eta_Q P_o}{h\nu}, \tag{15.164}$$

which is the contribution to the current from the initial photoelectric excitation. However, because the resistivity changes with irradiation, charge carriers are injected into one contact as they exit the opposite contact, as required by Ohm's law. Hence, the photocurrent gain is [Bube 1960]

$$M_{\text{PC}} = \frac{I}{I_{prim}} = \frac{(\mu_n\tau_n + \mu_p\tau_p)\mathcal{E}}{L} = \frac{(v_n\tau_n + v_p\tau_p)}{L} = \frac{\tau_n}{t_n} + \frac{\tau_p}{t_p}, \tag{15.165}$$

where t_n and t_p are the transit times across the device for electrons and holes, respectively. For a single flash of light, from a scintillator for instance, the current is described by

$$I = q_e A\mathcal{E}(\mu_n(n_o + \Delta n e^{-t/\tau_n}) + \mu_p(p_o + \Delta p e^{-t/\tau_p})). \tag{15.166}$$

[8]Here η_Q is defined as the number of electron-hole pairs excited per photon.

The current is largely affected by the excited charge carrier populations Δn and Δp, which eventually return to the dark current density of n_0 and p_0. If the photocurrent is primarily from band to band transitions with $n_0 = p_0$, then Eq. (15.166) reduces to

$$I = q_e A\mathcal{E}(n_o + \Delta n e^{-t/\tau_n})(\mu_n + \mu_p). \tag{15.167}$$

The gain for such a device is dominated by the charge carrier lifetimes. Hence, high speed devices require short lifetimes, whereas high gain devices require long lifetimes. Further, the fluctuation in current can be problematic and leads to poor energy resolution. Consequently, photoconductors are usually not the light sensing device of choice for scintillation detector spectroscopy.

For a planar device, Eq. (15.167) can be rewritten as

$$I(t) = \frac{q_e V A}{L}(n_o + \Delta n e^{-t/\tau_n})(\mu_n + \mu_p). \tag{15.168}$$

This result is important. First, the current decays away as a function of the free charge carrier lifetime, so that the current can continue to flow even after the primary charge carriers produced in the semiconductor reach the electrodes. Second, the duration of the detector current pulse can be tailored by changing the lifetime of the free charge carriers. High speed photoconductive radiation detectors can be manufactured by purposely adding lifetime shortening dopants, or by shortening the lifetimes with intentional radiation damage. These detectors are ideal for fast timing measurements of large radiation bursts.

15.3.8 Photon Drag Detectors

A special type of photoconductor is the *photon drag detector*, which operates by sensing momentum changes of free charge carriers. These special photoconductive detectors are used primarily to measure high intensity beams of infrared radiation, such as laser emissions. Photon drag detectors consist of a semiconductor bar, generally cylindrical, around which ring electrodes have been fashioned near the ends [Gibson and Kimmitt 1980]. The semiconductor is selected such that the band-gap energy is greater than the photon energy under investigation, thereby allowing a large fraction of photons to pass directly through the semiconductor without being absorbed. For instance, initial photon drag experiments were conducted with p-type Ge ($E_g \simeq 0.7$ eV) for $\lambda = 10.6$ μm (0.117 eV) light [Grinberg 1970; Danishevskiĭ et al. 1970; Gibson et al. 1970].

When photons and electrons interact, the electrons can gain energy and momentum. Consider a beam of photons having a cross section A impinging lengthwise upon a semiconductor cylinder of length L_0. These electrons, having gained some momentum, acquire an ordered motion in the direction of the light propagation and a momentum transfer rate of

$$M = \frac{\alpha W \mathrm{n} e^{-\alpha L}}{Ac}, \tag{15.169}$$

where c is the light speed, W is power in watts, n is the index of refraction, α is the optical absorption coefficient in cm^{-1}, L is the location along the semiconductor length. The motion of charge carriers produces an opposing electric field

$$q_e\mathcal{E} = \frac{\alpha W \mathrm{n} e^{-\alpha L}}{n_c A c}, \tag{15.170}$$

where n_c is the majority charge carrier density. If the majority carriers are holes, then the electric field polarity is reversed. Hence, momentum is transferred to holes in the valence band. Integration of Eq. (15.170) over the device length gives the resulting voltage per unit power illuminating the detector. After reflection at both ends of the device are included [Gibson et al. 1970], the voltage per unit power is

$$\frac{V}{W} = \frac{-\rho\mu \mathrm{n}(1-R)(1-e^{-\alpha L_0})}{Ac(1+Re^{-\alpha L_0})}, \tag{15.171}$$

where R is the semiconductor reflection coefficient (two facets), ρ is the semiconductor resistivity, μ is the majority charge carrier mobility, and L_0 is the length of the sample. For heavily doped samples, the product αL_0 is relatively large, and Eq. (15.171) reduces to

$$\frac{V}{W} \approx \frac{-\rho\mu\mathrm{n}(1-R)}{Ac}, \tag{15.172}$$

Hence, the photon drag detector operates as an infrared detector in which radiation passes through a doped semiconductor crystal, creating a measurable voltage drop associated with the photon beam intensity. Overall the effect is small and photon drag detectors are generally only useful for the detection of high fluxes of sub-band-gap radiation, such as from an infrared laser source. Numerous semiconductors have been investigated for these detectors, including Ge, Si, Te, GaAs, GaP, an InAs, as described in the literature [Gibson and Kimmitt 1980]. Photon drag detectors can be acquired through several commercial manufacturers of light sensors, the most common one composed of p-type Ge.

15.4 Measurements of Semiconductor Detector Properties

Synthesis and development of semiconductor materials and semiconductor detectors requires characterization of certain properties to better understand expected performance of a detector. For instance, the energy resolution from a semiconductor material is strongly dependent upon the product of the charge carrier mobility and the charge carrier mean free drift time, commonly referred to as the $\mu\tau$ product. Analysis of semiconductor detectors is often accomplished through measurements of fundamental properties, which include measurements of current, capacitance, the $\mu\tau$ product, electrical contact resistance, and charge carrier trapping. There are many excellent books devoted to this topic (see for examples [Look 1989; and Schroder 1990]). Only the more fundamental characterization methods are briefly described here.

15.4.1 IV Measurements

The current-voltage characteristic, or IV characteristic, is a fundamental property that yields leakage currents, normally under reverse bias. The quality of a blocking contact is generally revealed by the overall leakage current density that is observed. The measurement is typically performed with a probe station, manual or automated, that applies a voltage across a junction device and measures that resulting current. Common commercial systems allow reverse (negative) or forward (positive) voltage to be applied across a junction, automated with incrementally increasing steps. Many systems have built-in safety circuits that allow the establishment of a "compliance" limit to the measured current, beyond which the voltage is turned off.

The measurement is simple and the logarithm of the *absolute value* of the current is typically plotted against the applied voltage on a linear voltage scale showing both the forward and reverse currents. Recall the thermionic current for a pn junction is (see Eq. (15.75) and also Eq. (15.112))

$$I = \left| I_s \left(\exp\left[\frac{q_e V}{\breve{n} kT}\right] - 1 \right) \right| . \tag{15.173}$$

The resulting plot, shown in Fig. 15.37, is useful in (a) revealing the reverse leakage current and diode quality, and (b) issues with series resistance and space-charge limitations for the forward bias characteristic.

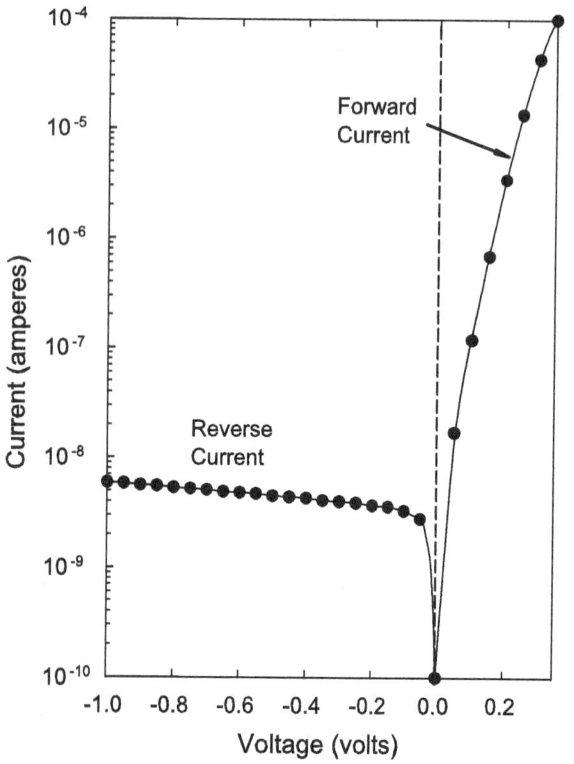

Figure 15.37. Conventional IV method of plotting the iv characteristic of a diode. The device is a 1 cm^2 Schottky barrier Si detector.

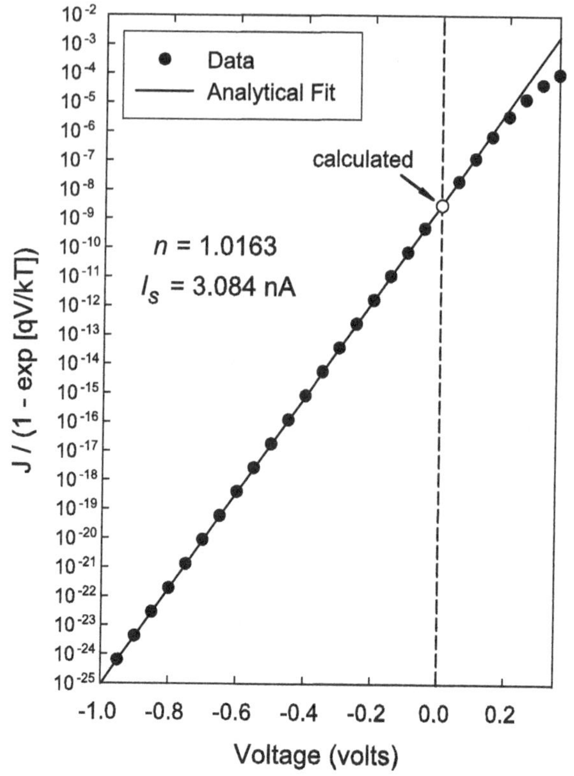

Figure 15.38. The IV plotting method proposed by Missous and Rhoderick [1986]. The plot yields $n = 1.0183$ and $i_s = 3.084$ nA.

Missous and Rhoderick [1986] propose another method, based on Eq. (15.112), that yields additional information about n. In this method a plot of the quantity

$$\ln\left(\frac{I}{1-\exp\left[\dfrac{-q_eV}{kT}\right]}\right), \tag{15.174}$$

versus the applied voltage V gives a linear plot of the IV characteristic at all values of V, including those values of $V < 3kT/q_e$, as shown in Fig. 15.38. From Eq. (15.112), the following relationship is obtained

$$\left[\ln(I_s) + \frac{q_eV}{\breve{n}kT}\right] \propto V. \tag{15.175}$$

The slope of the resulting graph yields the value of n, but primarily from the forward current. Also possible is a determination of I_s (or J_s) at the $V = 0$ intercept. From the relationship of Eq. (15.108), these measurements can yield information on the rectifying barrier height. Note that deviation from linearity can still occur from space charge limited current under forward bias, and charge carrier trapping and recombination can also cause distortions for reverse bias currents.

15.4.2 CV Measurements

The capacitance-voltage (or CV) characteristic can be used to determine a semiconductor doping concentration or the depletion width as a function of voltage for a junction device. The original CV measurement method was developed by Hilibrand and Gold [1960] for a parallel plate design. Recall the relationships

$$C = -\frac{dQ}{dV}, \tag{15.176}$$

in which the differential change in charge is negative from electron contributions, and

$$dQ = -q_e A|N_D(W) - N_A(W)|dW, \tag{15.177}$$

where q_e is the unit charge, A is the device area, $|N_D(W) - N_A(W)|$ is the net doping concentration as a function of depletion width W, and W is the depletion width. From these two results

$$C = -\frac{dQ}{dV} = q_e A|N_D(W) - N_A(W)|\frac{dW}{dV} = q_e A N_b \frac{dW}{dV}. \tag{15.178}$$

The general assumption for a planar device is that the capacitance is a function of the depletion layer width, i.e.,

$$C = \frac{\epsilon_s A}{W}. \tag{15.179}$$

Substitution of Eq. (15.59) or Eq. (15.103) into Eq. (15.179) yields,

$$C^2 = \frac{q_e \epsilon_s N_b A^2}{2(V_{bi} - V)}. \tag{15.180}$$

This result can be rearranged to give

$$V = -\frac{1}{C^2}\frac{q_e \epsilon_s N_b A^2}{2} + V_{bi}. \tag{15.181}$$

The result of Eq. (15.181) is rather important. A plot of $1/C^2$ versus the applied voltage V yields the semiconductor doping concentration and the built-in potential V_{bi} (see Fig. 15.39). Further, knowledge of A and C determines the active region width W from Eq. (15.179). Muller and Kamins [1986] point out that the value determined for N_b is generally accurate, although the value determined for V_{bi} can be have significant error.[9] The slope of $1/C^2$ can be determined by simple differentiation, namely

$$\frac{d}{dV}\left[\frac{1}{C^2}\right] = \frac{d}{dV}\left[\frac{2(V_{bi} - V)}{q_e \epsilon_s N_b A^2}\right] = \frac{-2}{q_e \epsilon_s N_b A^2}. \tag{15.182}$$

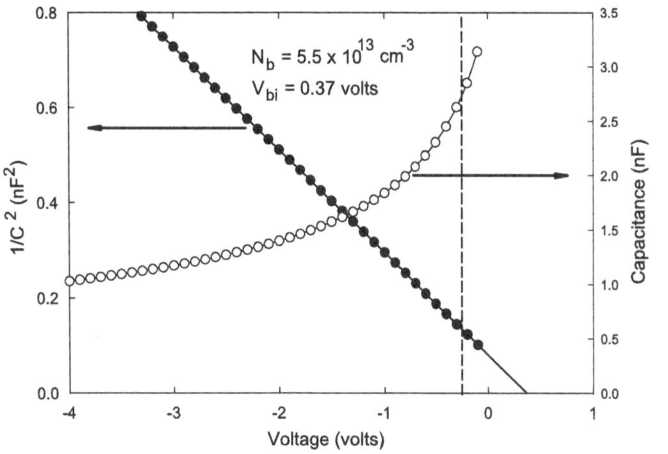

Figure 15.39. The two types of capacitance plots most often used in semiconductor radiation detector analysis. Shown is the case for a Si Schottky diode 1 cm^2 area, $N_b = 5.5 \times 10^{13}$ cm^{-3}, and $V_{bi} = 0.37$ volts.

[9] A small change in the slope causes a large change in the intercept value V_{bi}.

A common method of determining full depletion for a junction diode is to observe the voltage at which the CV characteristic curve stops changing and reaches a constant value.

The capacitance can be measured with a CV system consisting of an applied simultaneous DC voltage and a smaller AC voltage through a capacitance bridge, and yields the general relationship for current

$$I = \frac{V}{Z} = \frac{V}{(j\omega C)^{-1}} = j\omega VC, \tag{15.183}$$

where j is $\sqrt{-1}$, ω is the modulation frequency, Z is the device impedance, V is the applied voltage, and C is the capacitance. Typically, the modulation frequencies range between 10 kHz to 10 MHz. These measurements work well for the smaller band-gap semiconductors such as Ge and Si. However, problems can arise with the larger band-gap semiconductors having band energies above 1.2 eV.

For wide band-gap semiconductors, the series resistance in the substrate region can become problematic and yield incorrect results. The simple assumption of Eq. (15.183) is that the impedance of the diode is dominated by the depletion region capacitance and the undepleted substrate region capacitance is essentially shorted by the semiconductor conductivity. However, there is a parallel resistance with the depletion layer capacitance, which is also true for the undepleted substrate. Hence, there is a series resistance R that can confuse the capacitance measurement

$$I = \frac{V}{R + (j\omega C)^{-1}} = V\left[\frac{R\omega^2 C^2}{1 + (\omega RC)^2} + \frac{j\omega C}{1 + (\omega RC)^2}\right]. \tag{15.184}$$

The true capacitance can be determined from the measured capacitance C_m by [Look 1989]

$$C_m = \frac{C}{1 + (\omega RC)^2}. \tag{15.185}$$

The actual doping concentration, corrected for series resistance, is [Wiley 1975]

$$|N_D(W) - N_A(W)| = -\frac{C_m^3}{q_e \epsilon_s A^2 \sin^4\phi [dC_m/dV - 2C_m(d\phi/dV)\cot\phi]}, \tag{15.186}$$

where ϕ is the phase angle obtained from $\tan\phi = (\omega RC)^{-1}$. If the product $\omega RC \gtrsim 0.2$, then the measured doping profile is distorted [Look 1989].

If the substrate is a semi-insulator, the result can become unreliable. Consider the small signal model of a detector in (Fig. 15.40), simplified by showing a depletion region, a substrate region, and the electrical contacts. If it is assumed that the electrical contacts are well made, then the resistances short out the contact capacitance, rendering their contributions minor. However, that is not the case for the substrate region, where R is significant and cannot be ignored.

Hence, the simple small signal circuit has an impedance

$$Z = \left[\frac{R_D}{1 + j\omega R_D C_D} + \frac{R_S}{1 + j\omega R_S C_S}\right], \tag{15.187}$$

where the D and S subscripts indicate the depletion region and substrate regions, respectively. For high modulation frequencies ω and high substrate resistivities, the impedance takes the form [McGregor and Kammeraad 1995]

$$Z \approx \frac{\rho(W_D + W_S)}{A(j\omega\rho\epsilon_s + 1)}, \tag{15.188}$$

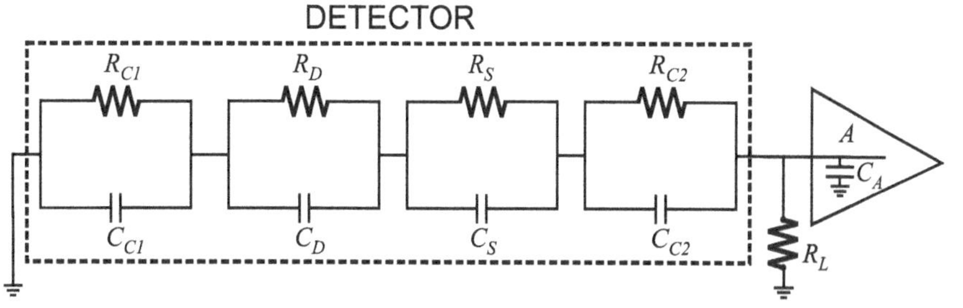

Figure 15.40. Small signal equivalent circuit model of a semiconductor detector. [Adapted from Walter 1960].

where ρ is the substrate resistivity, W_D is the depletion layer width, and W_S is the substrate layer width. The current is approximated by

$$I = V\left[\frac{1}{(R_D + R_S)} + j\omega C_T\right], \tag{15.189}$$

where C_T is the series combination of C_D and C_S, representing the total capacitance between the electrical contacts. This important result shows that (1) sufficiently high frequencies and (2) sufficiently high semiconductor resistivities reduce the measured capacitance to C_T. In either case, the observed capacitance does not change with voltage, erroneously misleading the measurement to indicate that the device active region[10] extends completely across the device, when in fact it does not [McGregor and Kammeraad 1995].

15.4.3 Measurement of Contact Resistance

Contact resistance is another fundamental property used as a quality metric for semiconductor detectors. The resistance of a blocking contact gives a measure of the rectification quality, while the resistance of an ohmic contact is used as a measure of conductance quality. The higher the blocking contact resistance, above the natural resistivity of the semiconductor, the more the leakage current is reduced. The lower the resistance of an ohmic contact, the smaller is the voltage drop at the contact so that there is better charge carrier collection and lower shot noise. IV measurements do provide a good first measurement of the rectifying behavior of a semiconductor diode detector, but do not disclose the contact resistance R_C. There are many methods to measure this contact resistance (see Look [1989] and; Schroder [1990]), a few of which are described here.

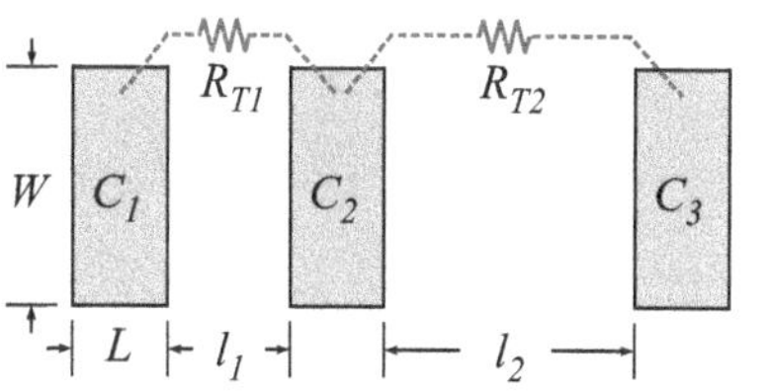

Figure 15.41. The transmission line model pattern, with ohmic contacts of width W, length L, and gap distances of l_1 and l_2. The contacts, labeled C_1, C_2, and C_3, are presumed identical.

The *transmission line model* (TLM), based on the pattern of Fig. 15.41, is a simple three-terminal design with which the contact resistance can be estimated. The total resistance from one contact to an adjacent contact is

$$R_{Ti} = \frac{r_s l_i}{W + 2R_c}, \tag{15.190}$$

where r_s is the sheet resistance,[11] W is the contact width, R_c is the contact resistance, and l_i is the distance between the contact edges. The measurement is made by placing a voltage across contacts C_1 and C_2 to

[10] The substitute use of 'active region' is much more accurate than the term 'depletion region'. The active region describes the region in a detector where the electric field is high enough to drift the charges, whereas the depletion region describes the region where the semiconductor is devoid of charge carriers. Because semi-insulating semiconductors, such as GaAs and CdTe, are already devoid of charge carriers, the description 'depletion region' has less meaning than 'active region'.

[11] The units for sheet resistance are $\Omega/\square$, meaning that any square of semiconductor surface yields the very same resistance. For instance, the sheet resistance of a 1 cm $\times$ 1 cm of semiconductor square surface is the same as a 2 cm $\times$ 2 cm square surface.

yield a current of I_1, followed by a second measurement across C_2 and C_3 to yield I_2. The total resistance for either measurement is simply $R_{Ti} = V/I_i$. The two equations formed from this pattern are solved to yield the contact resistance

$$R_c = \frac{R_{T1}l_2 - R_{T2}l_1}{2(l_2 - l_1)}. \tag{15.191}$$

The transmission line model is simple in application, but does have accuracy problems. Errors in the measured distances l_i can lead to measurement inaccuracy. Also, from Eq. (15.191), the difference between two large numbers, combined with small contact resistances, can lead to an inaccurate determination of R_c.

The *transfer length model* (unfortunately also termed TLM) is a popular method used to test the specific contact resistance of an ohmic contact. The method consists of placing consecutive planar contacts upon a semiconductor surface and performing a series of current measurements. Consider the pattern shown in Fig. 15.42 that has a series of identical contacts, labeled C_1 through C_5, each with a width W and length L, placed apart at increasing spacings. A resistance measurement is made by placing a known voltage across two contacts and measuring the resulting current. The total resistance is

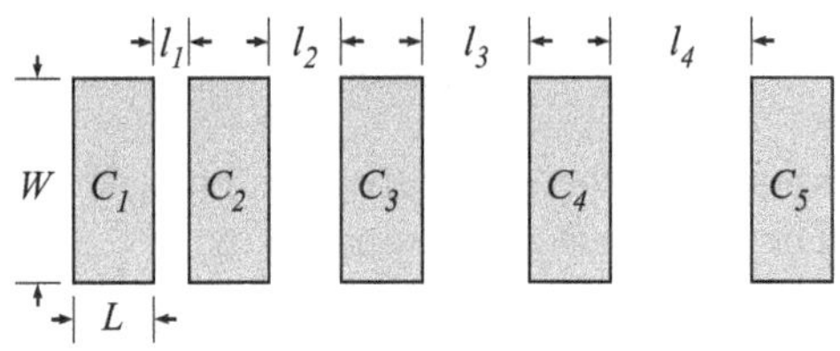

Figure 15.42. A common TLM pattern, with ohmic contacts of width W, length L, and gap distances of l_1, l_2, l_3, and l_4. The diagram depicts gap distances of increasing integer values, i.e., $l_4 = 4l_1$.

$$R_T = 2R_c + r_s \frac{l_i}{W}, \tag{15.192}$$

where R_c is the contact resistance of a single contact, r_s is the sheet resistance *between* the contacts, and l_i is the distance between the contacts. This measurement is repeated for the four pairs of contacts shown in Fig. 15.42. The total resistance R_T (where $R_T = V/I$) can be plotted against the intercontact distances l_i, and noting that Eq. (15.192) is a linear equation, the y intercept yields the value of $2R_c$ as shown in Fig. 15.43. The x intercept is a function of the transfer length, defined as

$$2L_T = \frac{2WR_c}{r_s}. \tag{15.193}$$

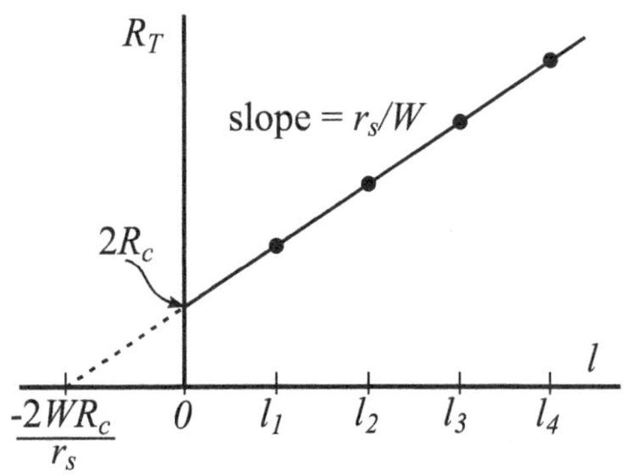

Figure 15.43. The measured resistance as a function of the distance l_i between the ohmic contacts, showing the method used to find R_c.

Three important values are derived from this transfer length method, namely, R_c, r_s, and the sheet resistance $R_s = r_s/W$.

This method is one of the most commonly used techniques to measure the quality of an ohmic contact, but does have a few potential problems. First, this method is based on the assumption that current flows uniformly from the contacts when, in reality, the current actually flows between the closest edges [Look 1989]. This non-uniform current is termed *current crowding*. To correct for current crowding, Look [1989] shows that the contact resistance is better described by

$$R_c = \frac{\sqrt{R_s \rho_c}}{W} \coth(kL), \tag{15.194}$$

where ρ_c is the specific contact resistivity under the metal-semiconductor interface, R_s is the sheet resistance *under* the contact, L is the contact length, and $k^2 = R_s/\rho_c$. If $kL \gtrsim 2$, then ρ_c is well approximated by

$$\rho_c \simeq \frac{W^2 R_c^2}{R_s}. \tag{15.195}$$

Note that it is assumed here that r_s and R_s are interchangeable, an assumption that may not necessarily be true. During contact processing and development, it is possible to alter the material sheet resistance such that $r_s \neq R_s$. The inclusion of effects from the *end resistance* can compensate for errors introduced when r_s and R_s are different. This compensation is made by either placing a guard ring around the TLM structure, or (originally) by placing an extra contact C_e to the left of C_1 (in Fig. 15.42) at distance L [Reeves and Harrison 1982]. This extra end resistance can be found by measuring the voltage at C_e while applying a bias between the C_1 and C_2. Hence, the end resistance is [Look 1989]

$$R_e = \frac{V(C_e) - V(C_1)}{I} = \frac{R_c}{\cosh(kL)} = \frac{R_c}{\cosh(L\sqrt{R_s/\rho_c})}, \tag{15.196}$$

where I is the measured current. Here the actual value of R_s is found from the ratio

$$\frac{R_c}{R_e} = \cosh(kL) = \cosh(L\sqrt{R_s/\rho_c}). \tag{15.197}$$

Reeves and Harrison [1982] show that the difference between R_s and r_s can be significant, with examples of GaAs having a r_s/R_s ratio of 19.55 and Si having r_s/R_s ratio of 4.88. Issues with end resistance have been addressed with a circular transmission line model proposed by Reeves [1980], in which a circular "bull's eye" pattern takes the place of the TLM pattern of Fig. 15.41. The method employs Bessel function solutions and requires careful design considerations for the method to converge on a solution.

There is another difficulty encountered with many of the high-resistivity semiconductors, such as SI-GaAs, CdTe, CdZnTe, and HgI_2. The substrate resistance is such a dominant factor that determining the contact resistance can be statistically challenging. One can observe from Eq. (15.192) that if the total resistance measured R_T is large, then the error associated with a small change in the slope r_s/W produces a large variance in the transfer length. Hence, the error associated with the measurement of r_s becomes large. To reduce this error, often a contact scheme is tested on a conductive semiconductor substrate and then applied to a semi-insulating substrate of the same type of material. However, the actual substrate resistance can make a significant contribution to the contact resistance and lead to an incorrect result.

The chemical compatibility of electrical contacts can also be a major issue, especially with the softer high-Z semiconductors such as HgI_2 and TlBr. For instance, there are only a few conductive materials that can be used on HgI_2 that do not destructively react with the compound, namely, carbon, Pd, and Pt. Gold rapidly diffuses in Si and, thus, produces a deep acceptor. Aluminum with gold produces a purple corrosion (the "purple plague" - $AuAl_2$) on GaAs. Hence, some amount of caution should be exercised when deciding upon the contact material.

Finally, there are some semi-conductors used for radiation detectors with band gaps exceeding 2 eV. Hence, they are better classified as semi-insulators. Examples include HgI_2, PbI_2, TlBr, and SiC. Contact formation to these materials need not be rectifying, mainly because the intrinsic resistivity is sufficient to reduce the leakage current to manageable levels. Hence, additional resistance from electrical contacts is less of a problem, although poor contact formation may cause capacitance issues that create multiple energy peaks in a pulse height spectrum.

15.4.4 Measurement of Resistivity

A fundamental property of a semiconductor is its resistivity, i.e.,

$$\rho = \frac{1}{q_e(\mu_e n + \mu_h p)}. \tag{15.198}$$

Perhaps the simplest method of measuring resistivity is to place ohmic contacts on opposite ends of a semiconductor bar of length L, followed by a measurement of the current (I) resulting from voltage (V)

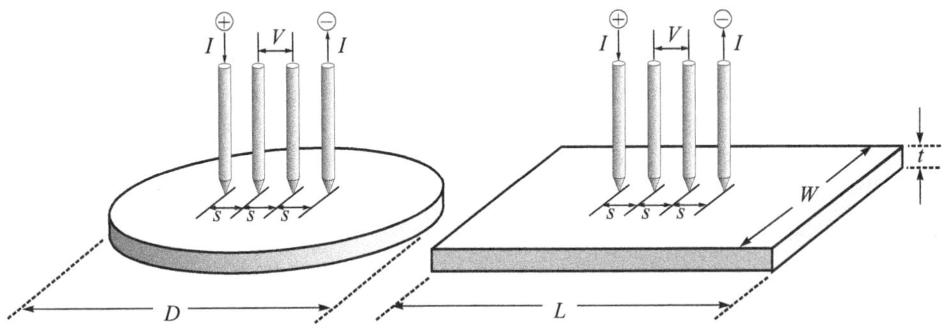

Figure 15.44. Geometries for four-point probe measurements on circular and rectangular samples.

applied across the bar. If the area of each end is A, then the current density is

$$\frac{I}{A} = J_n + J_p = \frac{V}{L}(q_e n \mu_n + q_e \mu_h p), \tag{15.199}$$

or

$$\rho = \frac{VA}{IL}. \tag{15.200}$$

Usually a semiconductor is dominated by one of the charge carriers, previously named the majority carrier. Hence, the resistivity is usually due to n-type or p-type charge carrier conduction.

A common method of measuring resistivity is with a four-point probe. The system has four equally spaced point probes, arranged in a straight line, that are brought into contact with a semiconductor surface (see Fig. 15.44). The spacing between each probe is denoted by s. A current is supplied to the outermost probes with a constant current source and the resulting voltage is measured between the two inner most probes. For a thin semiconductor sample in which the thickness under investigation is much smaller than the lateral dimensions of length L and width W, the sheet resistance is given by [Sze 1981]

$$r_s = \frac{\pi}{\ln(2)} \frac{V}{I} K = 4.5324 \frac{V}{I} K. \tag{15.201}$$

where K is a geometric correction factor. The volume resistivity is determined from

$$\rho = r_s t. \tag{15.202}$$

Schroder [1990] points out that geometric correction factors have been calculated by many researchers using a variety of techniques. However, all reduce the correction factor to three important components, $K = f_1 f_2 f_3$. The correction factor f_1 corrects for differences with the sample thickness, f_2 corrects for the lateral sample dimensions, and f_3 corrects for the location of the probes with respect to the sample edges. If the sample width W is much larger than the probe spacing ($W \gg s$), then Eq. (15.202) reduces to

$$\rho = 4.5324 \frac{Vt}{I}, \tag{15.203}$$

where t is the sample thickness.

For thinner samples of smaller widths W, the proper corrections must be applied. The factor f_1 addresses corrections for finite thickness t, and is given by

$$f_{1a} = \frac{\ln(2)}{\ln\left(\dfrac{\sinh(t/s)}{\sinh(t/2s)}\right)}, \tag{15.204}$$

if the bottom underlayer is an insulator and by

$$f_{1b} = \frac{\ln(2)}{\ln\left(\dfrac{\cosh(t/s)}{\cosh(t/2s)}\right)}, \tag{15.205}$$

if the bottom underlayer is a conductor. The limiting values are found for $t \ll s$ as

$$f_{1a} \approx 1 \qquad \text{and} \qquad \rho \approx 4.5324 \frac{Vt}{I} f_2, \tag{15.206}$$

$$f_{1b} \approx \frac{8}{3}\frac{s^2}{t^2}\ln(2) \qquad \text{and} \qquad \rho \approx \frac{8\pi}{3}\frac{s^2}{t}\frac{V}{I} f_2, \tag{15.207}$$

and for $t \gg s$ as

$$f_{1a} = f_{1b} \approx 2\ln(2)\frac{s}{t} \qquad \text{and} \qquad \rho \approx 2\pi s \frac{V}{I} f_2. \tag{15.208}$$

The factor f_2 addresses corrections for a finite width W. Smits [1958] calculated and published a table of these values as $\pi f_2/\ln(2)$ for rectangular and circular samples. The correction f_2 is plotted in Fig. 15.45 for values D/s and W/s between 1 and 40. As W/s becomes large, f_2 approaches unity. The third correction factor has four possible expressions, depending on whether the probes are near a conducting or non-conducting boundary [Valdes 1954], and also depending upon the orientation of the probes with that boundary. Plotted in Fig. 15.46 are corrections factors for f_3 for both conducting and non-conducting boundaries, according to the probe orientation and distance-to-spacing ratio d/s. Note that in all cases f_3 approaches unity provided that the probe is at least $3s$ distance away from the sample edge.

15.4.5 Measurement of Charge Carrier Mobility

Recall that the mobility of an extrinsic semiconductor is

$$\mu_e = \frac{1}{|q_e|\rho_e n} \quad \text{for } n\text{-type material,} \tag{15.209}$$

$$\mu_h = \frac{1}{|q_e|\rho_h p} \quad \text{for } p\text{-type material.} \tag{15.210}$$

This result shows that the charge carrier mobility can be found by measuring the carrier concentration and resistivity. The carrier concentration is commonly measured using the Hall effect, named after E.H. Hall [1879] who discovered the effect. Hall discovered that a magnetic field applied to a material perpendicular to a current flowing through the same material induces a voltage perpendicular to both the magnetic field and current. Putley [1968] devotes a textbook to the study of this effect. Consider the block of semiconductor material depicted in Fig. 15.47. Electrical contacts are applied to the ends, with additional electrical contacts applied to central portions on surfaces of the width. A voltage is applied across the longitude of the block (depicted in the x direction) and is labeled V_s. A magnetic field is applied perpendicular to the applied voltage as shown (depicted in the z direction). Connection across the semiconductor width measures the induced Hall voltage V_H.

The force on a charged particle is described by

$$\mathbf{F} = q_e(\mathcal{E} + \mathbf{v} \times \mathbf{B}), \tag{15.211}$$

where $\mathbf{v}$ is the charge carrier velocity vector and $\mathbf{B}$ is the magnetic field vector. A current I in the sample of Fig. 15.47 flows in the positive x direction. For p-type material this current is given by

$$I = \frac{V_s}{R_s} = \frac{WtV_s}{\rho L} = |q_e| pWtv_h, \tag{15.212}$$

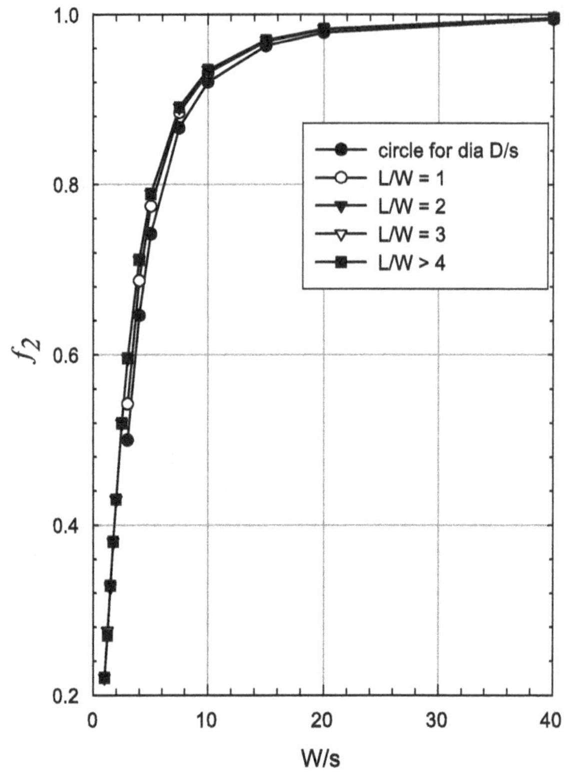

Figure 15.45. The correction factor f_2 as a function of W/s for many rectangular shapes L/W. Also shown is the correction f_2 for a circular sample as a function of the diameter D/s. Data from Smits [1958].

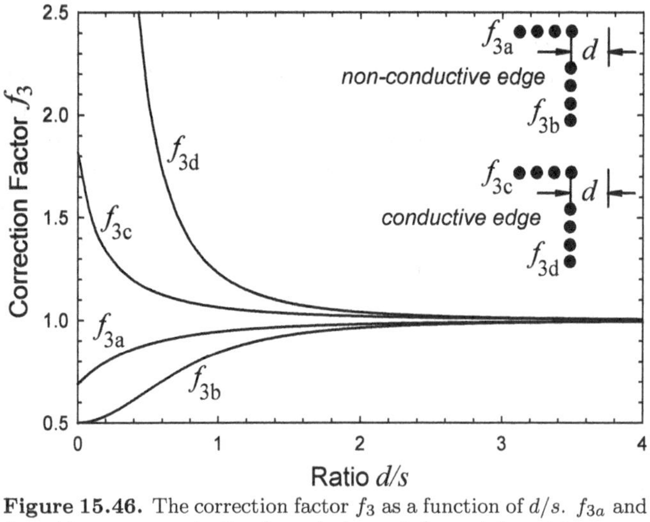

Figure 15.46. The correction factor f_3 as a function of d/s. f_{3a} and f_{3b} address non-conductive boundaries and f_{3c} and f_{3d} address conductive boundaries. The probes are placed distance d either parallel or perpendicular to the boundary edge.

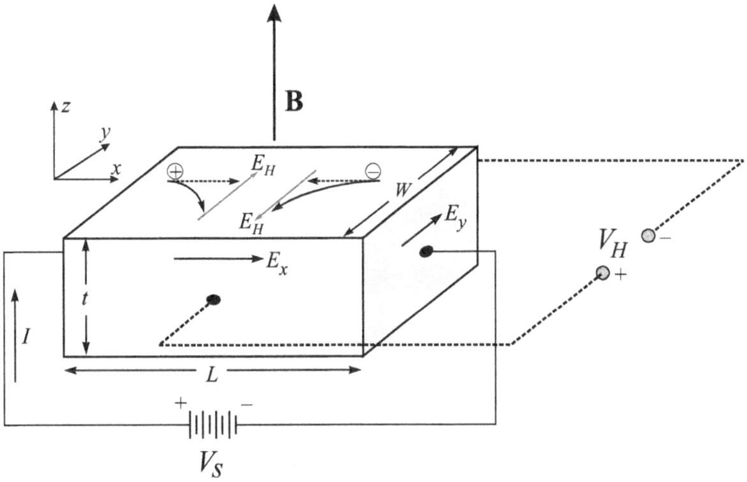

Figure 15.47. Configuration of a Hall effect measurement.

where R_s is the resistance in the longitudinal direction, t is the sample thickness, W is the sample width, and L is the sample length. Holes accumulate near the negative y surface and produce an electric field. But a current cannot flow out of the device. The net result is the y forces must be equal and opposite; hence, the force F_y in the y direction must be zero. Then from Eq. (15.211)

$$q_e \mathcal{E}_y = q_e v_{hx} B_z \qquad \text{or} \qquad \mathcal{E}_y = v_{hx} B_z, \tag{15.213}$$

where $\mathcal{E}_y$ is the *Hall field*, and the resulting induced voltage $V_H = \mathcal{E}_y W$ is the *Hall voltage*. Substitution of the hole drift speed gives

$$V_H = \mathcal{E}_y W = \frac{B_z I}{q_e p t W} W = R_H \frac{I B_z}{t}, \tag{15.214}$$

where R_H is the Hall coefficient and $R_H = 1/(q_e p)$.

In a similar manner, the Hall coefficient for n-type material is,

$$R_H = -\frac{1}{q_e n}. \tag{15.215}$$

Therefore, from the measured current and known magnetic field, the hole and electron concentrations are found as

$$p = \frac{1}{q_e R_H} \qquad \text{and} \qquad n = -\frac{1}{q_e R_H}. \tag{15.216}$$

Notice that the resulting positive or negative sign yields the type of majority charge carrier (n or p). The Hall mobility is determined from

$$\mu_H = \frac{|R_H|}{\rho}, \tag{15.217}$$

a useful result, but is not the carrier drift mobility. The actual drift mobility for holes is found by substituting the resistivity into Eqs. (15.209) and (15.210), namely

$$\mu_h = \frac{\mu_H}{q_e p |R_H|}, \tag{15.218}$$

and for electrons

$$\mu_e = \frac{\mu_H}{q_e n |R_H|}. \tag{15.219}$$

When both holes and electrons are present, the Hall coefficient is represented by [Schroder 1990]

$$R_H = \frac{\left[\left(p - n\left(\frac{\mu_e}{\mu_h}\right)^2\right) + (\mu_e B)^2(p-n)\right]}{q_e\left[\left(p + \frac{n\mu_e}{\mu_h}\right)^2 + (\mu_e B)^2(p-n)^2\right]}. \tag{15.220}$$

Unfortunately, if $p \approx n$, the result of Eq. (15.220) can be difficult to interpret.

Overall the Hall measurement can be used to determine the majority charge carrier drift mobility and the type of majority carrier (n or p). Although the geometry of Fig. 15.47 is perhaps the simplest, it can have practical difficulties when dealing with thin samples such as epitaxial crystals. Further, errors associated with the measurement can render R_H inaccurate, as outlined in Look [1989]. To address these issues, a variety of alternative contact geometries have been developed for such cases, as outlined in Look [1989] and Schroder [1990].

15.4.6 Measurement of the $\mu\tau$ Product

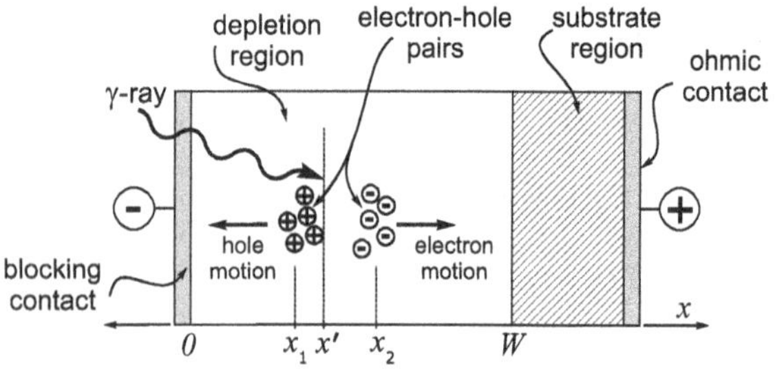

Figure 15.48. Electron-hole pairs are excited from a radiation interaction.

A useful measure of expected performance from a semiconductor is the mobility-lifetime (or $\mu\tau$) product. The speed of charge carriers can be estimated from the linear relation $v = \mu\mathcal{E}$ provided that the electric field is below the onset of scatter limited saturation speeds. The lifetime of a charge carrier is generally defined as the recombination time, which can be much longer than the mean free drift time. The mean free drift time is the average time period that a charge carrier conducts before being removed from the process, either by trapping or recombination. Consequently, it is the mean free drift time that defines the average time between charge carrier excitation and its removal from conduction, usually through traps. Hence, it is customary to replace the lifetime τ with the mean free drift time τ^*. For a planar device, the carrier extraction factor can be described by

$$\varrho = \frac{v\tau^*}{W} = \frac{\mu\mathcal{E}\tau^*}{W} = \frac{\mu\tau^* V}{W^2}. \tag{15.221}$$

In the next section a model is developed for the induced charge in a planar device (see Fig. 15.48) when trapping is present. The final result Eq. (15.240) is

$$Q(x') = -q_e N_o\left[\varrho_e\left(1 - \exp\left[\frac{x' - W}{\varrho_e W}\right]\right) + \varrho_h\left(1 - \exp\left[\frac{-x'}{\varrho_h W}\right]\right)\right], \tag{15.222}$$

where x' is the location of the ionizing event. Substitution of Eq. (15.221) into this result gives

$$-\frac{Q}{q_e N_o} = \frac{V}{W^2}\left\{\mu_e\tau_e^*\left(1 - \exp\left[\frac{(x' - W)W}{\mu_e\tau_e^* V}\right]\right) + \mu_h\tau_h^*\left(1 - \exp\left[\frac{-x'W}{\mu_h\tau_h^* V}\right]\right)\right\}. \tag{15.223}$$

From Eq. (15.223) it is observed that the $\mu\tau^*$ product yields a measure of the induced charge which has an effect on the detector energy resolution. The usual method of conducting a $\mu\tau^*$ measurement is to irradiate one of the contacts on a planar detector with an alpha-particle source. The reason for this arrangement is because the range of alpha-particles is short, confining the excitation of charge carriers in the semiconductor to a region adjacent to the electrical contact, i.e., $x' \simeq 0$. If a planar detector is irradiated from the cathode, Eq. (15.223) reduces to

$$-\frac{Q}{q_e N_o} = \frac{V}{W^2}\left\{\mu_e\tau_e^*\left(1-\exp\left[\frac{-W^2}{\mu_e\tau_e^* V}\right]\right)\right\}. \tag{15.224}$$

The $\mu\tau^*$ measurement is conducted by observing the relative pulse heights from the detectors at increasing intervals of voltage. The data is plotted and an interpolative method is used to fit the data to Eq. (15.224). For high-resistivity materials, W in Eq. (15.224) adequately describes the boundaries of the detector. However, for detectors requiring depletion to produce an active region, either Eq. (15.59) or Eq. (15.103) is substituted for W^2 to produce

$$-\frac{Q}{q_e N_o} = \frac{q_e N_b}{2\epsilon_s(V_{bi}-V)}V\left\{\mu_e\tau_e^*\left(1-\exp\left[\frac{-2\epsilon_s(V_{bi}-V)}{\mu_e\tau_e^* V q_e N_b}\right]\right)\right\}, \tag{15.225}$$

or

$$-\frac{Q}{q_e N_o} \approx \frac{q_e N_b}{2\epsilon_s}\left\{\mu_e\tau_e^*\left(1-\exp\left[\frac{-2\epsilon_s}{\mu_e\tau_e^* q_e N_b}\right]\right)\right\}, \tag{15.226}$$

which has the form

$$\frac{Q}{Q_o} \approx C_1\left\{\mu_e\tau_e^*\left(1-\exp\left[\frac{-1}{\mu_e\tau_e^* C_1}\right]\right)\right\}, \tag{15.227}$$

where the constant $C_1 = q_e N_b/(2\epsilon_s)$.

One difficulty with using Eq. (15.226) and Eq. (15.227) is that the value of Q/Q_0 appears constant until the full depletion condition is met, at which point W becomes constant and Eq. (15.224) adequately describes the measurement.[12] The measurement method based on Eq. (15.224) is commonly used to determine the $\mu\tau$ product, which can then be used in Eq. (15.223) to develop a Q-map for the detector. Measurements of electron $\mu\tau^*$ products are typically easier to conduct than hole $\mu\tau^*$ products, mainly because (1) the mobility of electrons is usually larger than for holes, and (2) τ^* is usually larger for electrons than holes.

If the hole $\mu\tau^*$ is so small that pulse heights are drastically reduced for anode irradiation, a method developed by Baciak and He [2003a] may be of practical use. This method involves the accumulation of pulse height spectra from both cathode and anode irradiation. The centroid of each distribution is chosen to represent $\langle Q_C\rangle$ (cathode irradiation) and $\langle Q_A\rangle$ (anode irradiation). The ratio of charge collected from the anode $\langle Q_A\rangle$ and cathode $\langle Q_C\rangle$

$$\frac{\langle Q_A\rangle}{\langle Q_C\rangle} = \frac{\mu_h\tau_h^*\left(1-\exp\left[\dfrac{-W^2}{\mu_h\tau_h^* V}\right]\right)}{\mu_e\tau_e^*\left(1-\exp\left[\dfrac{-W^2}{\mu_e\tau_e^* V}\right]\right)}, \tag{15.228}$$

can be used to extract information about $\mu_h\tau_h^*$. This method requires prior knowledge of $\mu_e\tau_h$ so that the denominator of Eq. (15.228) is known. Hence, the only unknown is $\mu_h\tau_h$, which is found from Eq. (15.228)

[12] The value of τ^* often does change with applied voltage, although determining τ^* requires knowledge of N_b and μ, and a new solution for C_1 for each voltage increment.

with iterative methods. There are limitations to the method, mainly because the pulse height distribution from holes can occupy the lower energy channels and, thereby can be severely contaminated by electronic noise in the system. Hole pulses lost in the noise and contamination in the pulse height spectrum from noise can introduce a significant uncertainty in $\langle Q_A \rangle$.

15.5 Charge Induction

A depiction of a planar semiconductor radiation detector is shown in Fig. 15.48. The detector is separated into a depletion (or active) region and a substrate region. The substrate region does not contribute to charge induction but rather acts as a series resistance in the circuit. Hence, charge motion in the depletion region produces the induced charge under investigation.[13] Upon the excitation of electron-holes pairs in the depletion region, these charges are drifted to their respective contacts, electrons towards the anode and holes towards the cathode, to produce an induced current. This drifting occurs because of the applied electric field produced by operating voltage applied across the detector. The charge induction produced by a planar radiation detector was developed in Chapter 8 (see Eq. (8.44)) and is represented here as

$$\Delta Q = q_e N_0 \left(\frac{x_2 - x_1}{W} \right), \tag{15.229}$$

where W is the detector width. As also derived in Chapter 8, full charge collection reduces Eq. (15.229) to $Q = q_e N_0$. Hence, the total charge excited is preserved on the output signal provided that the charge carriers are collected during the electronic integration time of the detector and circuitry. The other geometrical cases derived in Chapter 8 also apply to semiconductor detectors, in which the depletion or active region is the radiation sensitive region.

15.5.1 Charge Induction With Trapping

The ideal cases of charge carrier collection, induced current, induced charge, and the weighting potential were developed in Chapter 8. However, semiconductors are typically manufactured from crystalline solids or polycrystalline solids and typically have intrinsic and extrinsic defects. These defects may arise from vacancies, interstitials, impurities, and complex combinations of defects. Defects may be electrically active, or may be neutral and serve mainly as scattering centers. Regardless, the presence of crystalline defects affects the transport of charge carriers, and may serve to trap charge carriers, assist with recombination, or delay the conduction process so as to effectively remove them from adding to the induced charge of its corresponding signal.

Four distinguishable trapping and recombination conditions may exist in a semiconductor radiation detector [Dearnaley and Northrop 1966]. These are as follows.

1. Complete carrier extraction in which both electrons and holes are removed from the active region of the device. Hence, there are no trapping effects.

2. Short-term trapping in which carriers are briefly trapped, but undergo detrapping and are extracted from the detectors within the shaping time of the pulse. The output pulse may be elongated, and may suffer from some ballistic deficit, but the charge carriers are collected.

3. Partial recombination of the carriers in which some charge carriers are permanently lost.

[13] There are special cases in which there is no actual "depletion region," but instead only an "active region" defined by the electric field strength. Under such a condition, the entire semiconductor length can take part in producing the induced charge. These special cases arise in compensated wide band-gap materials.

4. Long-term trapping in which charge carriers that are trapped are not detrapping during the signal shaping time. Hence, these trapped charges are also effectively lost. Moreover, detrapping over time can produce a residual background current.

Consider the design of a common planar semiconductor detector in which electrical contacts are placed on opposite sides of a planar semiconductor slab. This detector is then irradiated from the side so that the probability of radiation interactions is constant at all points between the contacts. If a radiation event occurs at one of the electrodes, the cathode for instance, the pulse is entirely conditional on electron motion. If the event occurs at the anode, the pulse becomes entirely conditional on hole motion. At other event locations within the detector, both carriers contribute to the output signal to a greater or lesser degree. If a carrier collection time is longer than the trapping time, then the effective mobility changes as

$$\mu' = \mu \frac{\tau_t}{\tau_t + \tau_d}, \tag{15.230}$$

where μ is the charge carrier mobility without trapping effects, τ_t is the mean trapping time and τ_d is the mean detrapping time (or average time spent in the trap). Similarly, the charge carrier extraction time t' is reduced to

$$t' = t \frac{\tau_t}{\tau_t + \tau_d}, \tag{15.231}$$

where t is the expected collection time in the absence of trapping. If the situation exists in which one charge carrier follows condition 1 (no trapping), while the other charge carrier follows condition 2 (short-term trapping and detrapping), the pulse shape is deformed depending on the interaction location of the radiation absorption.

Building on the results of Sec. 8.6 for a planar detector, the charge induction when trapping occurs can be derived by first considering the contributions from electrons and holes, namely

$$\Delta Q = N_o \left(-q_e \left[\frac{x_2 - x_0}{W} \right] \Big|_{e^-} + q_e \left[\frac{x_1 - x_0}{W} \right] \Big|_{h^+} \right), \quad x_1 < x_0 < x_2, \tag{15.232}$$

$$= \frac{N_o}{W} \left(-q_e \Delta x_e - (+q_e) \Delta x_h \right), \tag{15.233}$$

where N_o is the number of electron-hole pairs created by a radiation event and x_0 is the radiation interaction location. The corresponding current is

$$I = \frac{dQ}{dt} = \frac{-q_e N_o}{W} \left(\frac{dx_e}{dt} + \frac{dx_h}{dt} \right), \tag{15.234}$$

in which the signs indicate the unit charge (positive or negative) and direction. Equation (15.234) is rewritten, after charge carrier speeds and trapping terms are inserted, to yield

$$I = \frac{dQ}{dt} = \frac{-q_e N_o}{W} \left(v_e e^{-t/\tau_e^*} + v_h e^{-t/\tau_h^*} \right), \tag{15.235}$$

where τ_e^* and τ_h^* are the mean free drift times for the electrons and holes, respectively, before the charge carriers are permanently removed from conduction. Integration of Eq. (15.235) yields the induced charge

$$Q = \frac{-q_e N_o}{W} \left(\int_0^{t_e} v_e e^{-t/\tau_e^*} dt + \int_0^{t_h} v_h e^{-t/\tau_h^*} dt \right) = \frac{-q_e N_o}{W} \left[v_e \tau_e^* \left(1 - e^{-t_e/\tau_e^*} \right) + v_h \tau_h^* \left(1 - e^{-t_h/\tau_h^*} \right) \right], \tag{15.236}$$

where the terms t_e and t_h are the time intervals required to collect all electrons and holes, respectively, from their point of origin. The speed is the distance $x_{e,h}$ that a charge carrier travels over time period $t_{e,h}$, so that Eq. (15.236) can be rewritten as

$$Q = \frac{-q_e N_o}{W}\left[\frac{x_e \tau_e^*}{t_e}\left(1 - e^{-t_e/\tau_e^*}\right) + \frac{x_h \tau_h^*}{t_h}\left(1 - e^{-t_h/\tau_h^*}\right)\right]. \tag{15.237}$$

The *carrier extraction factor* is defined as

$$\varrho \equiv \frac{v\tau^*}{W} = \frac{x\tau^*}{tW} = \frac{\lambda^*}{W}, \tag{15.238}$$

where λ^* is the mean free drift length. Equation (15.237) can be rewritten as

$$Q = -q_e N_o\left[\varrho_e\left(1 - \exp\left[\frac{-x_e}{\varrho_e W}\right]\right) + \varrho_h\left(1 - \exp\left[\frac{-x_h}{\varrho_h W}\right]\right)\right]. \tag{15.239}$$

Suppose the detector orientation shown in Fig. 15.48 is chosen for a coordinate system, such that electrons drift in the positive x direction and holes drift in the negative x direction. For any location x' within the detector, the drift distance for electrons is $W - x'$ and the drift distance for holes is simply x'. In terms of x' Eq. (15.239) becomes

$$Q(x') = -q_e N_o\left[\varrho_e\left(1 - \exp\left[\frac{x' - W}{\varrho_e W}\right]\right) + \varrho_h\left(1 - \exp\left[\frac{-x'}{\varrho_h W}\right]\right)\right]. \tag{15.240}$$

Radiation events occurring at the contacts, at either $x = 0$ or $x = W$, reduce Eq. (15.240) to

$$Q = -q_e N_o \varrho_e\left(1 - \exp\left[\frac{-1}{\varrho_e}\right]\right), \quad \text{for electron dominated signals,} \tag{15.241}$$

$$Q = -q_e N_o \varrho_h\left(1 - \exp\left[\frac{-1}{\varrho_h}\right]\right), \quad \text{for hole dominated signals.} \tag{15.242}$$

The probability distribution of induced charge is found by multiplying Eq. (15.240) by the interaction probability distribution function $P(x')$ where $P(x')dx'$ is the probability an interaction creating change Q_0 occurs in dx' about x'. The evaluation of $P(x')$ generally requires the use of numerical transport theory if all the various ways photons interact are to be considered. Further, the interaction PDF also generally depends on the amount of energy deposited as well as the interaction location. The only situation in which an analytic expression for the interaction probability can be obtained is one in which only the photoelectric effect is present. This situation is discussed further later in this section.

A special case is one in which $P(x') = 1$ for all x', i.e., radiation events occur uniformly throughout the detector volume. Such a case may be observed for highly penetrating radiation that irradiates the device from a side between the contacts. In this case

$$P(x')\frac{Q(x')}{Q_0} \rightarrow \frac{Q(x')}{Q_0}.$$

The pulse height distribution with $P(x') = 1$ and with both carrier extraction factors equal is shown in Fig. 15.49. This Q-map shows that if both charge carrier extraction factors are large (≥ 50), the variance in the pulse height spectrum is small. This difference is detectable by comparing the largest pulse height (at $x = 0.5$) to the smallest pulse height (at $x = 0$ and $x = 1$). The Q-map can be interpreted as follows:

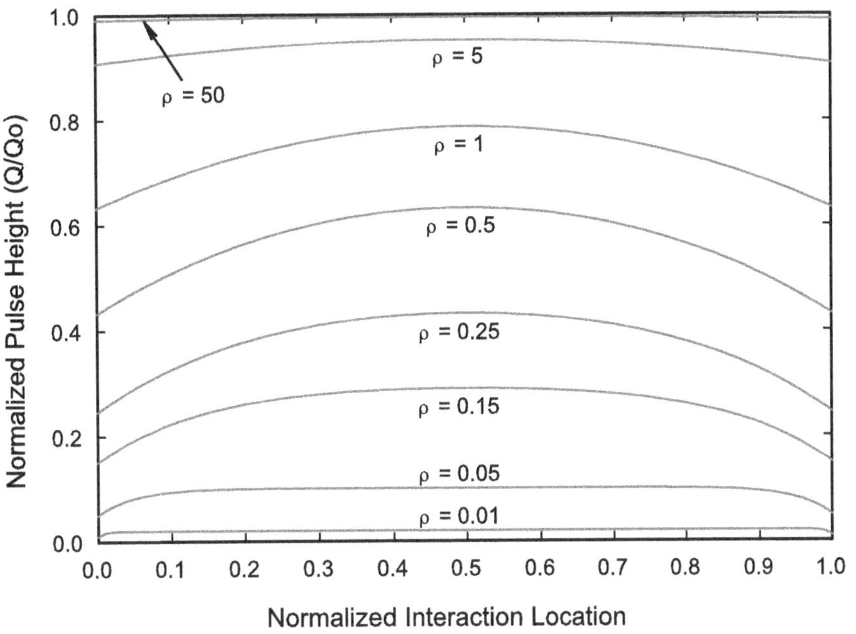

Figure 15.49. Q-map for different cases in which $\varrho_e = \varrho_h$. For large values of ϱ_e and ϱ_h, the variance in normalized pulse height is minimal. As ϱ for both charge carriers decreases, the variance broadens.

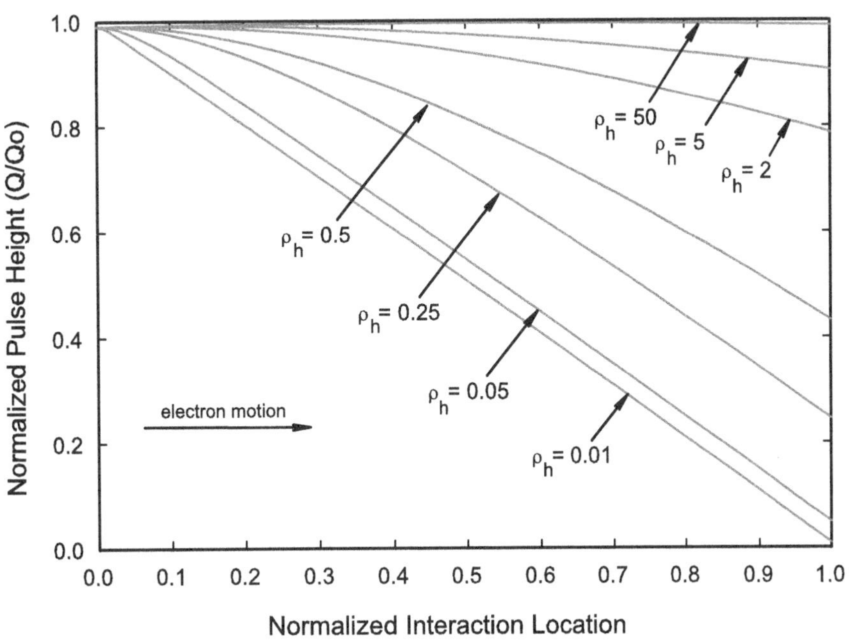

Figure 15.50. Q-map for different cases in which ϱ_e is large (= 50) and ϱ_h changes. As ϱ_h is decreased, the variance broadens.

a relatively horizontal region in a curve represents the formation of an energy peak in the pulse spectrum, whereas sloped regions represent the formation of a tail in the pulse height spectrum. The curve in Fig. 15.49 for $\varrho_e = \varrho_h = 50$ is horizontal and indicates that all pulses fall into an energy peak, thereby rendering good energy resolution. Further, the magnitude of the pulse is nearly 100% of the maximum possible and indicates that all of the charge carriers are collected. However, the curve for $\varrho_e = \varrho_h = 0.5$ shows a short horizontal region between $x = 0.3$ to $x = 0.7$, an indication that approximately 40% of pulses appear in an energy peak, while the remaining 60% of pulses appear in a tail. The largest pulse height appears at approximately 63% of the expected value, which indicates severe losses of the charge carriers. Further the tail extends from 63% of the expected pulse height down to 43% of the expected pulse height. Note that a considerable horizontal region appears for low values of $\varrho < 0.15$. However, the expected pulse height is small for these conditions and indicates that system electronic noise ultimately competes with the radiation signals, thereby compromising the energy resolution of the detector.

The energy resolution is much worse if $\varrho_e \neq \varrho_h$, as is usually the case. For example, shown in Fig. 15.50 are several cases in which ϱ_e is constant at 50 while ϱ_h is changed. Notably, the signal variance is worse in all cases than those shown in Fig. 15.49 (except for $\varrho_e = \varrho_h = 50$). For instance, the comparative case with $\varrho_e = 50$ and $\varrho_h = 0.5$ indicates the appearance of a small energy peak from events in the region between $x = 0$ and $x = 0.15$. Only 15% of interactions accumulate in an energy peak. The remaining signals accumulate in a tail region extending from approximately 100% of the expected pulse height down to 43% of the expected pulse height. An example of this result is shown in Fig. 15.51, from which it is clear that the full energy peak is not Gaussian and most of the pulse height spectrum appears in an elongated tail.

The solution of the variance for Eq. (15.240) can be shown to be [Day et al. 1967; Knoll and McGregor 1993]

$$\begin{aligned}\sigma_Q^2 = q_e^2 N_o^2 \Big[& \varrho_e^2 + 2\varrho_e^3(e^{-1/\varrho_e} - 1) + \frac{\varrho_e^3}{2}(1 - e^{-2/\varrho_e}) + \varrho_h^2 + 2\varrho_h^3(e^{-1/\varrho_h} - 1) + \\ & \frac{\varrho_h^3}{2}(1 - e^{-2/\varrho_h}) 2\varrho_e\varrho_h + 2\varrho_e^2\varrho_h(e^{-1/\varrho_e} - 1) + 2\varrho_e\varrho_h^2(e^{-1/\varrho_h} - 1) \\ & +2\frac{(\varrho_e\varrho_h)^2}{\varrho_e - \varrho_h}(e^{-1/\varrho_e} - e^{-1/\varrho_h}) \Big] - \langle Q\rangle^2,\end{aligned} \tag{15.243}$$

where

$$\langle Q\rangle = q_e N_0 \left[\varrho_e + \varrho_e^2(e^{-1/\varrho_e} - 1) + \varrho_h + \varrho_h^2(e^{-1/\varrho_h} - 1) \right]. \tag{15.244}$$

The percent fractional standard deviation can be found from

$$\sigma_q = \frac{100\sigma_Q}{\langle Q\rangle}, \tag{15.245}$$

and is plotted as a function of electron and hole extraction factors in Fig. 15.52. Note that the values in Fig. 15.52 do not depict the energy resolution, typically defined as the FWHM, where the usual definition is derived from a Gaussian peak, namely

$$\mathrm{FWHM} = 2\sqrt{2\ln 2}\sigma = 2.355\sigma. \tag{15.246}$$

For moderate to severe charge carrier trapping, the peaks in pulse height spectra are usually not Gaussian, but instead are skewed to the lower energies with a salient tail region and, thus, the use of Eq. (15.246) is inaccurate. For example, the pulse height distribution for full energy absorption (photoelectric effect), without Gaussian spreading, is shown in Fig. 15.51 for the case in which $\varrho_e = 50$ and $\varrho_h = 0.5$. Most of the

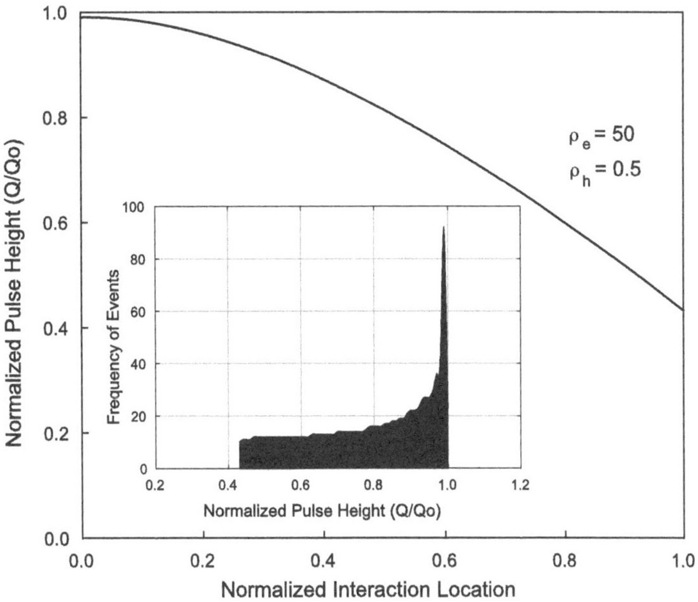

Figure 15.51. A comparison of the Q-map for $\varrho_e = 50, \varrho_h = 0.5$ and the resulting pulse height spectrum with 1000 solution bins divided over 100 channels.

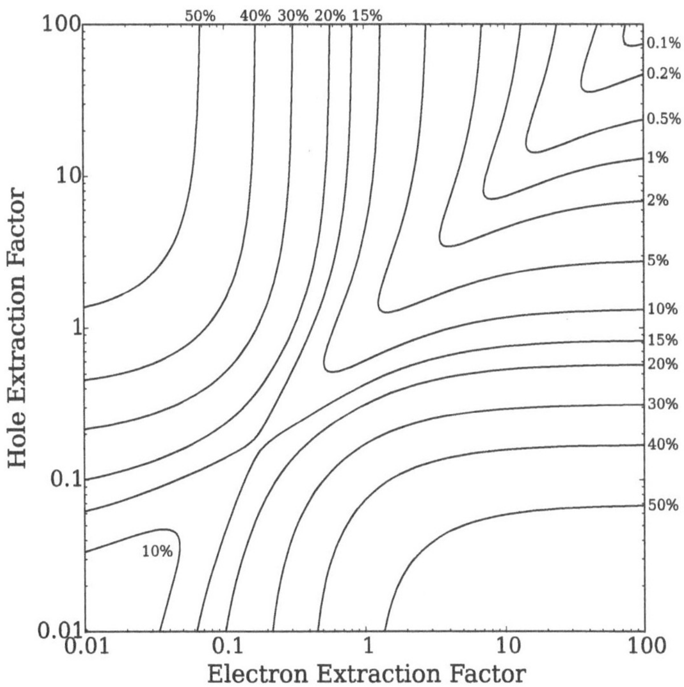

Figure 15.52. The expected percent standard deviation in the observed induced charge signal as function of ϱ_e and ϱ_h. From [Knoll and McGregor 1993].

distribution appears in a long tail region, with approximately 19% of the events appearing in the full energy peak. Overall, Eq. (15.240), Eq. (15.243), and Eq. (15.244) serve to illustrate the effect of trapping on the energy resolution of a planar radiation detector.

Radiation detectors are often irradiated with one side exposed to the source. For instance, the condition for Eq. (15.240), Eq. (15.243), and Eq. (15.244) is uniform irradiation between the detector contacts, as with irradiating a semiconductor block from the side between the contacts. Often a detector is irradiated towards one of the electrical contacts, in which exponential absorption from one contact to the other render a non-uniform gamma-ray interaction distribution. The resulting pulse height distribution is weighted by the probability the gamma ray interacts within the detector, thereby changing the result of Eq. (15.240) and Eq. (15.243), a condition especially true for low-energy gamma rays. If it is assumed that a gamma-ray deposits all energy for each event in a defined small region (i.e., the photoelectric effect dominates), then the pulse height distribution for a planar detector can be corrected by multiplying Eq. (15.240) and the gamma-ray interaction probability density function, namely

$$P(x')\frac{Q(x')}{Q_0} = \frac{Q(x')}{Q_0}\frac{\mu_p \exp^{-\mu_p x'}}{1-\exp^{-\mu_p W}}, \tag{15.247}$$

where μ_p is the photoelectric interaction coefficient which, for low energy photons, is much greater than μ_C and μ_{pp} for scattering and pair production, respectively. However, the predominant method of gamma-ray interactions for moderate to high-energy gamma rays is Compton scattering, which further confuses the evaluation of Q, $\langle Q \rangle$, and σ_Q^2. Geometries other than planar can be used to improve the energy resolution by increasing the charge induction influence of one charge carrier over the other, as already described in Secs. 8.7 and 8.8. Corrections must also be made to the expected value of $Q(x)$ to account for differences in the induced charge and differences in the gamma-ray absorption probability. Such a calculation is best accomplished by using transport theory methods to determine location and energy deposited, coupled to the weighting potential to determine the effect on induced charge, and multiplying the results by Eq. (15.240). These concepts are further explored in a latter chapter on radiation spectroscopy.

15.5.2 Energy Resolution Improvement Methods and Designs

Because of the severe hole trapping and the degradation in energy resolution, multiple methods have been developed to improve the spectroscopic performance of CdZnTe detectors [Zhang et al. 2013]. Electronic corrections such as rise time rejection and rise time correction were introduced to mitigate hole trapping problems. The usual difficulties accompanied these methods. For instance, rise time rejection helps to eliminate undesirable pulses, thereby decreasing the effective volume of the detector and consequently reducing the efficiency. Rise time correction has also been successfully used, a method that artificially adds "charge" to the signal as a function of the initial slope of the current and induced charge. More clever methods used device weighting fields and geometric effects to counter charge loss from hole trapping. These techniques used for Si and Ge detectors altogether eliminated the need for complicated electronics and required only common preamplifier and amplifier circuits. Finally, there are hybrid detectors that used both clever weighting potential corrections and rise time corrections to achieve improved energy resolution and detection efficiency.

Pulse Shape Discrimination and Pulse Rise Time Compensation

Electronic corrections have been used to improve the gamma-ray energy resolution of many different semiconductor detectors, most of them fabricated as planar-style detectors. Jones and Woollam [1975] are credited with first introducing the idea of improving the results of poor charge collection and trapping effects by discarding pulses with rise times greater than 100 ns. In so doing, pulses dominated by hole drift were eliminated, leaving only those pulses that were mainly electron dominated. This technique of discarding undesirable pulses is referred to as *pulse shape discrimination* (PSD), and can be used to improve the spectroscopic performance of detectors in which one type of charge carrier undergoes problematic trapping [Holzer 1983]. The use of PSD has been reported for CdTe detectors [Richter and Siffert 1992 (shown in Fig. 15.53); Bargholtz et al. 1999], CdZnTe detectors [Cardoso et al. 2003] and HgI_2 detectors [Holzer 1983]. Unfortunately, PSD also reduces the effective volume of the detectors, thereby also reducing the detection efficiency.

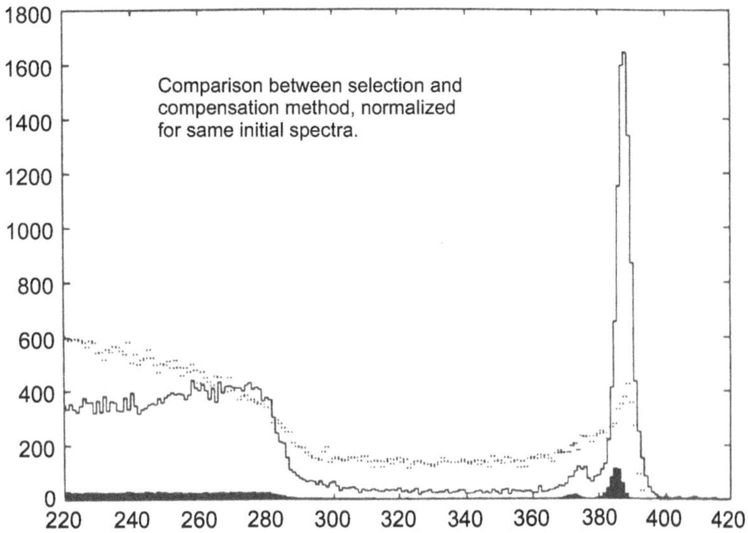

Figure 15.53. A comparison of results from a planar 5 mm × 5 mm × 2 mm CdTe detector. The comparison shows the result for ^{137}Cs without correction (light dots), with PSD (blackened spectra), and PRC (black outline). From Richter and Siffert [1992]; copyright Elsevier (1992), reproduced with permission.

Another electronic method to improve energy resolution was introduced by Matsushita et al. [1981; 1982], and is generally referred to as *pulse rise-time compensation* (PRC).[14] Matsushita et al. applied the technique to correct for charge losses in neutron damaged Ge or Ge(Li) detectors. The technique can involve (1) the recognition that the initial slope of the pulse is an indicator of the number of charge pairs liberated by the radiation event, and (2) the resultant pulse height is a manageable function of the trapping charge loss. Various methods of accomplishing PRC have been reported, including results by Richter and Siffert [1992; 1993], Keele et al. [1996], and Auricchio et al. [2005]. Although the method was originally applied to planar detectors, it was found that PRC could be applied to coaxial and trapezoidal geometries, although the corrections were complicated by the non-uniform electric fields Matsushita et al. [1981; 1982]. Richter and Siffert [1992] found that the full energy peak shifts linearly with pulse height, thereby allowing an electronic correction by artificially adding value to a smaller pulse height. The PRC technique gives improved spectroscopic results while retaining relatively good counting efficiency. Combining PSD and PRC, Ivanov et al. [1995] report best energy resolution of 1% FWHM for 662-keV gamma rays with a 2 mm × 5 mm × 5mm detector at room temperature without a decrease in full energy peak efficiency. Although both the PSD and PRC methods give improved performance, Richter et al. [1993] point out that differences in detector

[14] Matsushita et al. [1982] originally named the method the "pulse-height correction" (PHC) technique.

performance, including problems with space charge limited currents, can diminish the effective use of these correction methods.

Coaxial and Hemispherical Two-Terminal Devices

In Secs. 8.7 and 8.8 it was seen that the weighting function can be made strongly non-linear by adjusting the relative sizes of the cathode and anode contacts. The fundamental geometries evaluated were planar, coaxial and spherical. The problem of using planar designs in the presence of highly trapped charge carriers was evaluated in Sec. 15.5.1 and the consequences shown in prior sections of this chapter. Lund et al. [1996] attempted to reduce the effect of hole trapping in CdZnTe detectors by producing coaxial devices; however, the performance was only marginally improved over that of a planar detectors. Unlike the coaxial geometry commonly used for HPGe detectors, in which charge carrier trapping for both holes and electrons is negligible, the non-uniform electric field in a coaxial CdZnTe detector produces non-constant charge carrier speeds and, consequently, changes in the mean free drift times compromising the performance [McGregor and Rojeski 1999]. There are relatively few reported results for coaxial compound semiconductor detectors, mainly because of the difficulty with producing coaxial detectors from soft and brittle semiconductors. A partial coaxial geometry, literally a wedge segment from a coaxial geometry, was reported by McGregor et al. [1999] with improved results.

Hemispherical detectors have also been investigated [Malm et al. 1975; Zanio 1977; Alekseeva et al. 1985]. The detectors are semiconductor hemispheres with the rounded surface acting as the cathode with a small dot anode positioned at the hemisphere focus. The geometric weighting of the device causes most pulses to originate near the cathode, thereby most pulses are electron dominated and consequently the energy resolution is improved. The field dependent speed and mean free drift times must also be considered when predicting the spectroscopic performance. Khusainov [1992] describes the performance of CdTe hemispherical detectors, slightly chilled with a Peltier cooler, and obtained an energy resolution of 10 keV FWHM at 662 keV. Back in 1992, these CdTe hemispherical detectors were commercially available and in use by the IAEA for control of SNM. Sensitive volumes reported by Khusainov were, at maximum, only 32 mm^3, while Richter and Sifferet [1992] report energy resolution averaging about 5% FWHM at 662 keV and show results from a 65 mm^3 device. Bale and Szeles [2006] report the results from a commercial "quasi-hemispherical" CdZnTe detector, fabricated by covering five of the six faces of a semiconductor right cuboid with an electrode that serves as the cathode, and then placing a small dot anode in the center of the sixth face. A right cuboid shape is much easier to fabricate than a hemisphere, and the spectroscopic performance is practically the same, reported to be 20 keV FWHM (3%) for 662 keV gamma rays. The $\mu_e\tau_e$ product limits the practical radius of the device (see Fig. 15.54), a limitation which is also true for a coaxial detector. While a coaxial detector volume can be increased without affecting the radius by simply lengthening the device, the same is not true for a hemisphere. Consequently, the useful volume of a hemispherical device is limited by a radial dimension that has efficient electron collection.

There is a fundamental problem with almost all two terminal devices designed to enhance the induced current of one charge carrier over the other; mainly, the weighting potential is the normalized distribution of the actual operating potential. Although the presence of space charge complicates the analysis by further altering the voltage and electric field distribution in a detector, for simplicity consider the case without space charge. Detectors with strongly non-linear weighting potentials also have strongly non-linear operating potentials, thereby also producing large regions of low electric field. Consequently, the charge carrier speeds of both electrons and holes are also low in the low electric field region and reduces the $\mu\tau$ products. Detectors with three or more terminals can be used to counter this problem by decoupling the weighting potential from the operating potential.

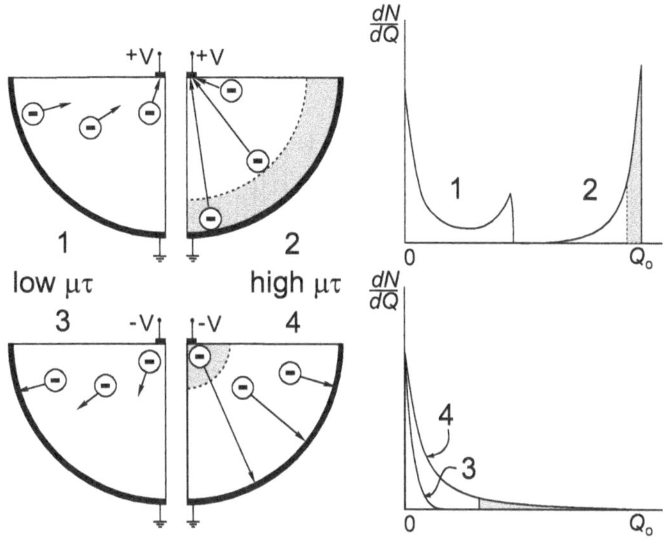

Figure 15.54. Expected pulse height spectra from a hemispherical detector with severe hole trapping for four cases: (1) positive focus with low $\mu_e\tau_e$, (2) positive focus with high $\mu_e\tau_e$, (3) negative focus with low $\mu_e\tau_e$, and (4) negative focus with high $\mu_e\tau_e$. After Malm et al. [1975].

Weighting Field Optimization

Another method to improve spectroscopic resolution employs clever electrode designs to force the appearance of strongly non-linear weighting potentials [Zhang et al. 2013]. These electrode designs are used to enhance the importance of charge induction of one charge carrier type (usually electrons) while reducing or negating the influence of the other charge carrier (usually holes). Quite often these detectors are referred to as single carrier detectors. There are many such designs and a few important types are briefly described here.

A common method used to solve the detector weighting potential is to employ Laplace's equation

$$\nabla^2\phi = 0, \tag{15.248}$$

where ϕ is the electrical potential. As explained in Section 8.6, space charge does not affect the weighting potential and, hence, Laplace's equation is used instead of Poisson's equation. Equation (15.248) is solved with the boundary conditions that the output conductive contact is normalized to one volt and *all* other contacts are grounded at 0 volts. This computational method is described and used by Shockley [1938].[15]

Radiation spectrometers with optimized weighting fields to improve performance are often designed with the methods introduced in Chapter 8. Although simple designs using planar, coaxial, and spherical geometries are easily calculated with analytical methods, complex geometries and electrode patterns require numerical methods to analyze the weighting field and weighting potential. There are available many finite-difference software programs designed to analyze the electric fields and voltage potentials within a dielectric material.

Co-Planar Grid Luke [1994, 1995] introduced a contact geometry referred to as the *coplanar grid detector*. The method mimics the Frisch grid ion chamber. The detector is generally planar in form, with a planar cathode contact on the lower surface and a series of parallel contact strips on the upper surface. It has

[15]The method is derived from Green's theorem, although over the years it has been named Ramo's theorem or the Shockley-Ramo theorem. The core solution is derived from either Green's theorem or the reciprocal Green's theorem.

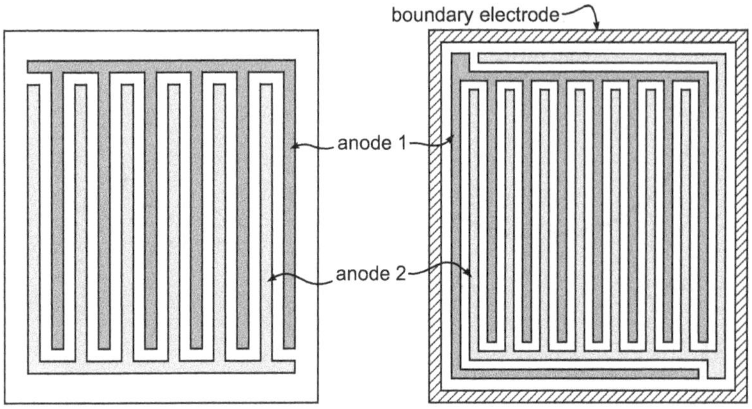

Figure 15.55. Simple coplanar grid design (on left) and a balanced coplanar grid design (on right).

no actual drift region as with the Frisch grid ion chamber, but instead has a virtual drift region defined by the strip widths and pitch as shown in Fig. 15.55. In the original work, alternating contact strips were connected to two different preamplifiers, labeled in the present case as high and low preamplifiers. The high contacts operate as collecting anodes and the low contacts act as secondary non-collecting anodes with a lower positive voltage than the collecting anodes. Both contacts are biased positively with respect to the planar back cathode.

The signal from the low preamplifier is subtracted from the signal of the high preamplifier, thereby producing a virtual Frisch grid effect. As electron charge carriers move toward the anodes, they induce current to flow on all anodes, both collecting and non-collecting. At distances relatively far from the anodes, these induced currents are practically identical, hence, the net output result is zero. The same is true for hole charge carriers moving in the opposite direction. However, as charge carriers reach the strip electrodes, the higher bias applied through the high preamplifier causes electrons to turn into the collecting electrodes, producing a positive current flow on the high preamplifier and negative current flow on the low preamplifier. As electrons are completely collected, the net subtracted result between both outputs is the appearance of full charge collection on the combined output. Because holes move the opposite direction towards the planar back electrode (cathode), their net contribution is practically zero, thereby negating the deleterious effect of trapping. The biasing scheme and resultant induced signal are depicted in Fig. 15.56, and the resulting weighting potential is shown in Fig. 15.57. Although the improvement was dramatic and an important innovation for room temperature radiation spectroscopy, effects from electron trapping and weighting field imbalances between the anode strips caused broadening of energy peaks. The resulting full energy peaks typically had low and high energy tails, producing a flared appearance to the base of the energy peak, and thus limited the improvement in energy resolution.

He et al. [1998] and Sturm et al. [2004] corrected the weighting field imbalance by optimizing the coplanar strip design, which included two balancing strips around the coplanar strips with an additional boundary electrode around the detector perimeter as shown in Fig. 15.55. In doing so, He et al. [1998] dramatically improved the performance of the coplanar grid detector. The flared base of the full energy peak was practically eliminated, and with charge compensation through electronic methods, energy resolution below 2% FWHM at 662 keV was usually realized. This balanced configuration, in various forms, is now a standard design feature for commercial coplanar grid CdZnTe detectors.

Small Pixel Detectors Detectors with pixels have been fabricated from a variety of semiconductors, including Ge, Si, GaAs, and CdZnTe, for the purposed of two-dimensional imaging of radiation. Ultra-small pixels on CdZnTe detectors were initially introduced by Doty et al. [1992]; however, it was Barrett et al. [1995] who identified what is now commonly referred to as the *small pixel effect* by analyzing the results from pixelated CdZnTe detectors. The basic design is essentially another two-terminal device, but there is instead an array of small pixel anodes on one surface and a single planar cathode contact on an opposing surface as shown in Fig. 15.58. Barrett et al. [1995] correctly point out that the weighting potential is strongly non-linear for each pixel, giving stronger influence to charge carriers moving near the anode pixels. As charge carriers, electrons for instance, approach any particular pixel, the induced charge accumulates almost entirely on a single pixel. Hence, connecting separate outputs to each pixel allows one to determine the charge location while benefiting from improved charge collection. For pixel size $\epsilon \times \epsilon$ and detector thickness L, the total induced charge for electron-hole pairs excited at location $z = 0.2L$ near a pixel is plotted as a function of ratio ϵ/L in Fig. 15.59. From this figure it becomes clear that the single carrier effect becomes stronger as the ϵ/L ratio diminishes.

If the same *operating* potential is applied to all pixels, the overall effect is to have a near constant electric field across the detector thickness, thereby negating the non-uniform electric field effects observed with cylindrical, trapezoidal, and hemispherical detectors. The simple pixel design, however, suffers from charge sharing and charge loss problems. Charge sharing is caused by electrons produced by a single initial event are collected by two or more electrodes, and causes pulse height reduction on all electrodes involved. Charge loss is caused by trapping of slow moving electrons in the low field region between pixels, ultimately reducing the pulse height.

The small pixel effect can be enhanced by adding a third electrode around the pixels (see Fig. 15.58), which performs as a non-collecting steering electrode much like that in the co-planar grid [He 1995]. Under such a condition, the pixels can be biased positive with respect to the steering grid, while the steering grid

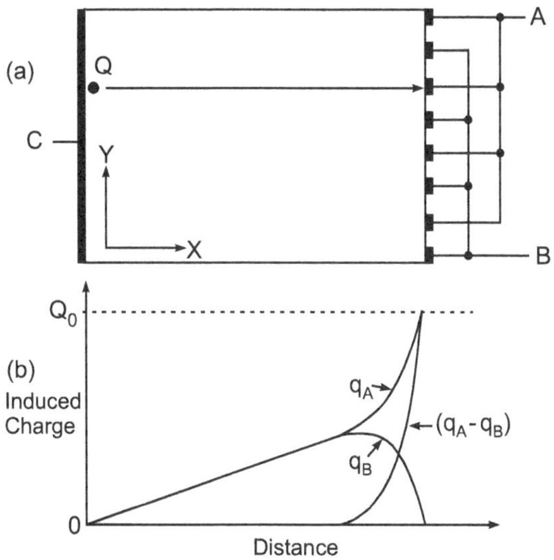

Figure 15.56. Shown are (a) the basic structure of a coplanar grid detector, (b) induced charge on the electrodes A (q_A) and electrodes B (q_B) and the resultant difference signal. After Luke [1995].

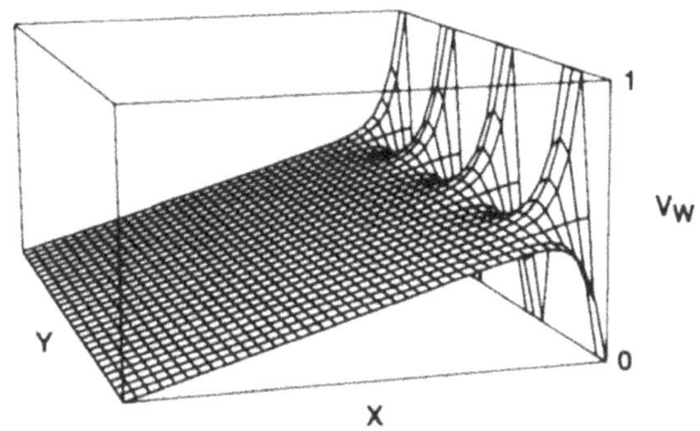

Figure 15.57. Weighting potential distribution for the set of collecting grid electrodes in the coplanar grid configuration. Reproduced from P.N. Luke, *Appl. Phys. Lett.*, **65**, 2884, (1994), with the permission of AIP publishing.

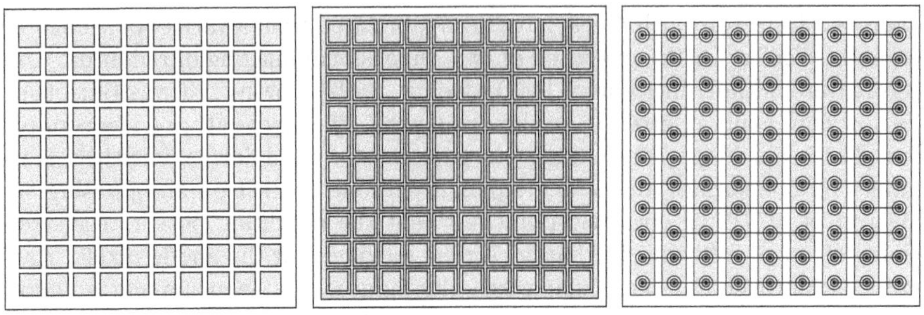

Figure 15.58. Simple pixel detector design (left), a pixel detector design with a steering grid (middle), and a hybrid pixel-strip design (right).

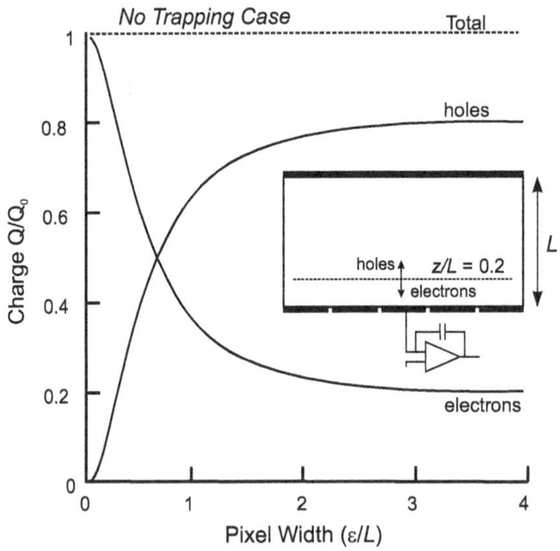

Figure 15.59. Total collected charge for holes and electrons excited at depth $Z = 0.2L$ as a function of pixel size. After Barrett et al. [1995].

is biased positive with respect to the cathode. Usually the pixels are just a few volts higher in positive bias than the grid, enough to efficiently collect electrons while minimizing leakage current between the anodes and grid. Hence, the steering grid functions as a non-collecting electrode which (1) assists with electrically steering electrons into a collecting pixel and (2) reduces the leakage current impact on the anode outputs [He et al. 1999]. Energy resolution was further enhanced by applying electronic correction methods to the pixel/grid structure. Because a lateral electric field is applied between the grid and pixels, the steering grid helps reduce the problem with charge losses observed in simple grid designs.

Another pixelated electrode configuration, a hybrid between a pixel detector and a strip detector, was studied by Mayer et al. [1999] and Montemont et al. [2007]. Each strip electrode has an embedded row of pixels, in which the strip acts as the non-collecting electrode. The design attempts to produce a simple scheme for two-dimensional radiation imaging, thereby greatly reducing the number of electronics readout points from a basic pixel detector, while retaining the advantage of the small pixel effect. The strips each have a separate readout and operate as steering grids, while the rows of pixels are interconnected, each row having a separate readout (see Fig. 15.58). Readout information from the x-y grid yields position information, while the small pixel effect improves gamma-ray energy resolution.

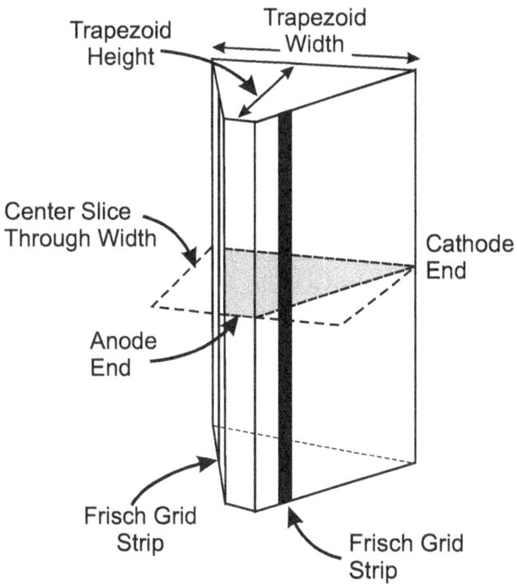

Figure 15.60. A 3-D depiction of a trapezoid frustum semiconductor Frisch grid detector. Also shown is a plane sliced through the middle of the device. From McGregor et al. [2001].

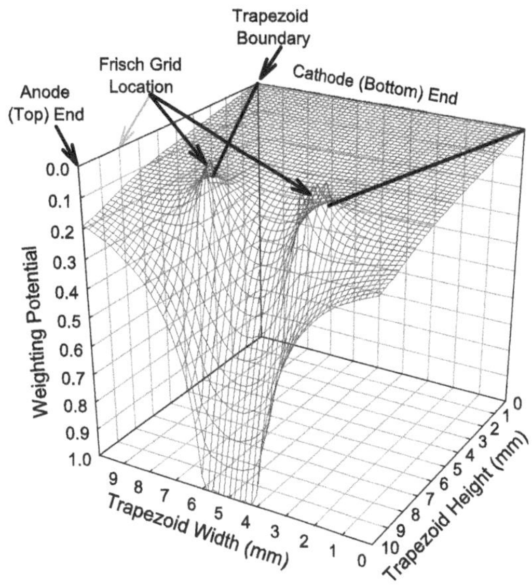

Figure 15.61. The weighting potential of a trapezoid frustum geometrically weighted CdZnTe Frisch grid detector, sliced through the center of the detector as shown to the left. After McGregor et al. [1999a].

Frisch Style Constructions McGregor et al. [1997; 1998] introduced the parallel strip and geometrically weighted parallel strip CdZnTe detector designs in 1997. The simplistic design required only one preamplifier output, and was designed to mimic the operation of a Frisch grid ion chamber. The detector consisted of a CdZnTe bar, nominally 2-3 mm thick, with surface areas of 5 mm × 5 mm or 10 mm × 10 mm. Parallel edges of the planar device had electrical contacts applied. On opposite surfaces, centering approximately 1 to 2 mm from one edge contact, gate contacts were applied through physical vapor deposition. One edge contact was biased negatively, the parallel gate contacts were grounded, and the output anode edge contact was biased positively through a commercial preamplifier. Room temperature energy resolution of 41 keV FWHM for 662-keV gamma rays was reported for a 1 cm × 1 cm × 2 mm device. The device worked to demonstrate the concept, and was later improved by using a trapezoidal frustum shape. The small end of the frustrum operated as the output anode, the large bottom of the frustum operated as the cathode, that the gate electrodes were centered at 1.5 mm from the anode (see Fig. 15.60 and Fig. 15.61). As explained in the literature [McGregor et al. 1999a; McGregor and Rojeski 1999b, 2001], more gamma-ray interactions occur in the drift region near the base, thereby causing the majority of pulses to be electron dominated. The multiple-electrode design provides the benefits of both a non-linear weighting potential and also, through appropriate application of voltages to the anode and grid, a nearly uniform electric field through the detector. Room temperature energy resolution of 17.76 keV FWHM for 662-keV gamma rays was reported for a 1 cm^3 device. Initially there were some issues with leakage current between the grid and anode. Theory shows that the grid need not be in direct contact with the semiconductor; hence, surface leakage current was practically eliminated by applying a thin Al_2O_3 layer atop the semiconductor sides, followed by the application of the grid strips. Although the trapezoid Frisch grid detectors worked well,[16] the detector cutting, shaping, and polishing proved challenging and difficult to fabricate.

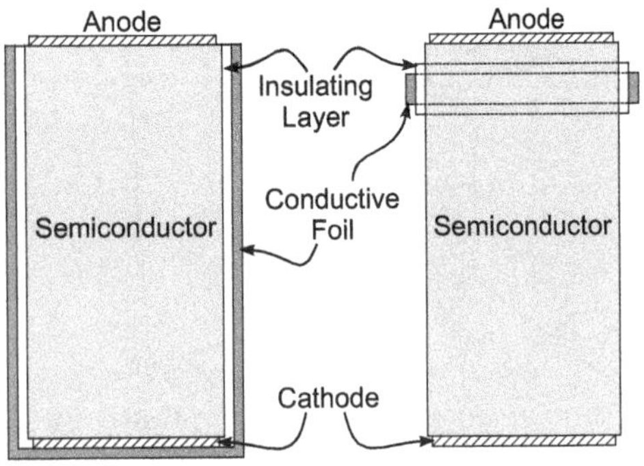

Figure 15.62. Cross section depictions of a (left) Frisch collar and (right) Frisch ring semiconductor detector.

McGregor and Rojeski [2001] (and McGregor [2004]) introduced another simple device, commonly referred to as either a *Frisch collar detector* or *Frisch ring detector*, depending on the length of the gate contact (see Fig. 15.62). The detector is designed as an elongated semiconductor bar with contacts at each end. The bar is wrapped in an insulating coating followed by a conductive wrapper or sleeve. The insulating layer eliminates leakage current between the Frisch ring/Frisch collar and the collection electrode. Much like the strip grid designs, the Frisch ring design produces both a non-linear weighting potential and, with the proper application of potentials to the grid ring and anode, a nearly uniform electric field throughout the detector (see Fig. 15.63). However, the Frisch collar design is simpler to manufacture than the trapezoid or Frisch ring designs, and generally produces better energy resolution, despite being a two-terminal device. The conductive wrapper or sleeve is grounded and the remaining anode contact is positively biased.

The first such detector was reported by McNeil et al. [2004] whose detector had dimensions of 3 mm × 3 mm (area) × 6 mm (length). Electroless gold contacts were applied to the 3 mm × 3 mm ends and Teflon tape was wrapped around the bar as the insulator. Afterwards, thin Cu foil was used as the conductive wrapper. Using a commercial charge sensitive preamplifier, the detector delivered room temperature energy resolution of 15.3 keV FWHM (2.31%) for 662 keV gamma rays. Kargar et al. [2006; 2009] and Harrison et al. [2009] worked to optimize the design and performance of the Frisch collar detector, eventually achieving room temperature energy resolution of 5.9 keV FWHM (0.89%) for 662 keV gamma rays [Ugorowski et al. 2011]. It was discovered that performance improved as the dielectric constant of the insulating layer decreased, and Teflon tape, with a dielectric constant of only 2.1, was an obvious choice and yielded superior results over alternative insulators, including Al_2O_3 and SiO_2. The Frisch collar design has also been successfully applied to HgI_2 detectors [Ariesanti et al. 2010, 2011; Kargar et al. 2011a, 2011b], and a 2.1 mm × 2.1 mm × 4.1 mm device yielded room temperature energy resolution of 11.9 keV FWHM (1.8%) for 662 keV gamma rays. A considerable advantage of the Frisch collar design is the simple construction from a common planar bar of semiconductor material and low cost materials, thereby eliminating the need for electrode patterning altogether. These detectors can be arranged in an array for imaging purposes [McGregor 2004; Cui et al. 2007, 2008; Ugorowski et al. 2011]. The robustness of the Frisch collar design has been verified by other research groups [Kim et al. 2015; Jeong et al. 2015; Bolotnikov et al. 2012].

Drift Style Detectors The semiconductor drift chamber (SDC), originally designed for large area Si detectors [Gatti and Rehak 1984; Rehak et al. 1985, 1986], has been adapted to compound semiconductors as a means of countering the effects of charge carrier trapping. The original intent of the SDC was to increase the active area of a semiconductor detector while reducing capacitive noise. The basic design uses a series of microstrip electrodes, either in a linear array or a series of concentric circles (bull's eye) on opposing sides of a semiconductor block [Rehak et al. 1985; Bertuccio et al. 1992]. At least one of the electrodes serves as an anode, while the remaining electrodes serve as guiding electrodes. These guiding electrodes are biased

[16] At the time, the trapezoid parallel strip detectors set the room temperature resolution record for CdZnTe, but was surpassed by other technologies within only a few years.

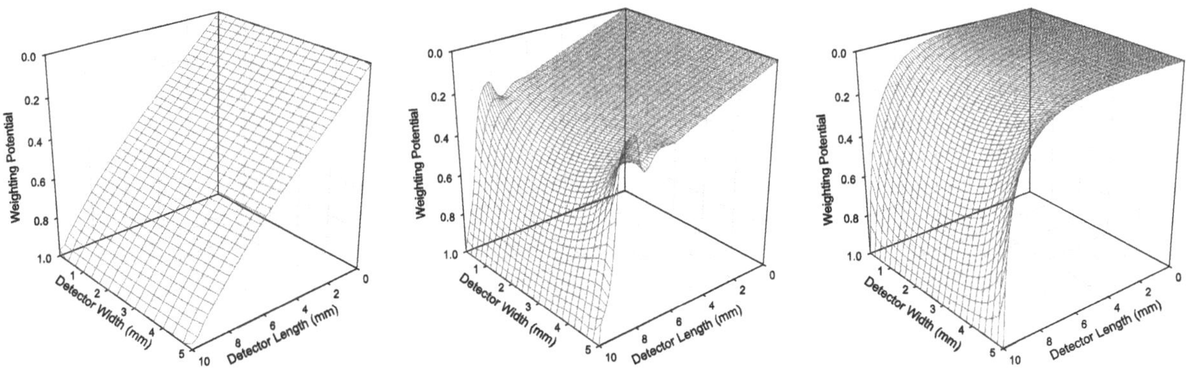

Figure 15.63. Comparison of weighting potentials from a (left) planar device, (middle) Frisch ring device, and (right) Frisch collar device. Each detector is modeled as a 5 mm × 5 mm × 10 mm CdZnTe detector. The insulating material is 100-micron thick Teflon.

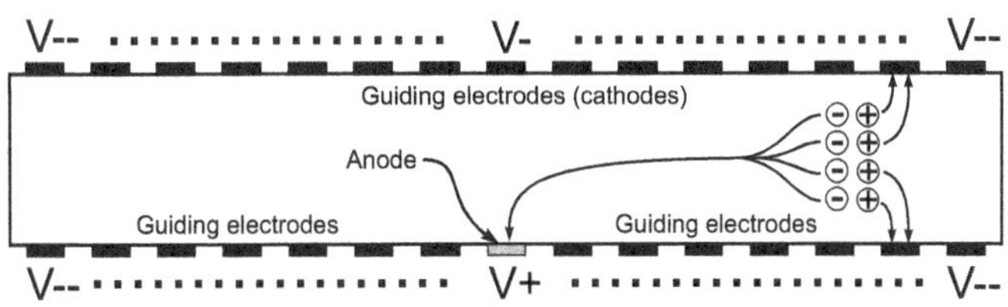

Figure 15.64. Side view depicting the operation of a semiconductor drift chamber.

so as to drift electrons to the anode, typically a bias scheme with sequentially increasing positive electrodes towards the anode or sequentially decreasing negative electrodes toward the anode (see Fig. 15.64).

Immediately one can observe that the electrode geometry has much in common with the coplanar, pixel, and Frisch style detectors, in which most of the induced charge accumulates on the anode as electrons are transported in the region nearest the anode [Patt et al. 1996]. Consequently, the SDC can be operated as a single carrier device for materials with hole trapping problems. Patt et al. [1996], van Pamelen and Budtz-Jørgensen [1998], and Kuvvetli et al. [2001, 2005] describe HgI_2 and CdZnTe SDCs with linear array strips while Owens et al. [2007] reports on a CdZnTe SDC with a concentric ring pattern. These detector configurations show improved performance over simple planar detectors. Van Pamelen reports 6.89 keV FWHM (1.1%) for 662-keV gamma rays on a small (1.5 mm thick) CdZnTe detector and Patt et al. [1996] report 6.62 keV FWHM (1%) for 662-keV gamma rays on a small 6 mm × 6 mm area × 2 mm thick HgI_2 detector. An alternative drift style detector was reported by Butler [1997], composed of a single pixel surrounded by a drift/focusing electrode. A planar cathode was applied upon the opposing surface. The detector was operated by applying equal negative voltage to the focusing electrode and cathode with respect to the anode[17] [Apotovsky et al. 1997]. A best energy resolution of 3.67 keV (0.72%) at 511 keV was reported for a 3 mm × 3 mm × 3 mm cube of CdZnTe.

Hybrid Improvements

Resolution improvement has also been realized by combining multiple effects, particularly, PRC methods with single carrier detector designs. The use of PRC methods helps to offset energy degradation from electron

[17]Named "Spectrum Plus" detectors, these devices were commercially available for a short time during the late 1990s.

trapping. He et al. [1998] introduced hybrid solutions with balanced co-planar grid detector geometries to achieve energy resolutions on the order of 1% FWHM at 662 keV. The combination of the small pixel effect with PRC methods has also been explored by He et al. [1999] who reported a high energy resolution of less than 1.5%. An added benefit of the hybrid pixelated detector is the ability to determine the interaction location within the bulk of the semiconductor detector as a function of x, y and z. Consequently, using Compton scattering deconvolution methods, the actual direction of a gamma-ray source (or sources) can be determined in a relatively short time [Wahl et al. 2015]. These hybrid methods have been successfully applied to different compound semiconductors, including CdZnTe [Wahl et al. 2015], HgI_2 [Baciak and He 2003a, 2003b], and TlBr [Hitomi et al. 2008].

PROBLEMS

1. Determine the following:
   (a) kT at 300 K
   (b) kT/q_e at 300 K
   (c) The band-gap energy of Si and Ge at 300 K

2. Calculate the electron and hole concentrations, the material resistivity, and the Fermi energy level for Si doped with $p = 5 \times 10^{15}$ cm^{-3} with background dopants $n = 3 \times 10^{15}$ cm^{-3}.

3. An Si pn junction diode has $N_A = 8 \times 10^{17}$ cm^{-3} and $N_D = 5 \times 10^{15}$ cm^{-3}, with $\tau_p = 10^{-7}$ s and $\tau_n = 10^{-6}$ s, with the device area of $A = 25$ mm^{-2}. (a) Determine the theoretical saturation current at 300 K. (b) Calculate the forward and reverse currents at ± 0.7 V.

4. A Si pn junction diode has $N_A = 10^{17}$ cm^{-3} and $N_D = 10^{15}$ cm^{-3}. Calculate the value of V_{bi} for T = 100 K, 200 K, 250 K, 300 K, 350 K, 400 K, and 500 K. Determine both W and $\mathcal{E}_{max}$ at 300 K.

5. It is the goal in making a semiconductor detector to create the largest possible depletion width. Explain how this goal is best realized by using the purest semiconductor material available.

6. Given a 600-micron thick Si semiconductor $p\nu n$ diode with $p = 10^{16}$ cm^{-3}, $\nu = 10^{12}$ cm^{-3}, and $n = 10^{16}$ cm^{-3}, find the depletion width when reverse bias of 100 volts is applied. What is the depletion width if the reverse bias is increased 300 volts?

7. You have a Si $p\nu n$ junction device that is 150 microns wide with $\nu = 10^{13}$ cm^{-3}. Determine the punch through voltage. Determine the breakdown voltage.

8. Describe quantitatively how the magnitude of the reverse bias applied to a p-n junction affects (a) the device capacitance, (b) the width of the depletion layer, and (c) the maximum electric field.

9. Given a 1.2-micron sample of CdS with a shallow trap density of 10^{15} cm^{-3}, what is the expected value of V_{TFL}?

10. A young engineer under your direction comes up with the idea of stacking two identical semiconductor detectors in series and applying the bias across both detectors. What is the expected output and the consequence of this device circuit?

11. An n-type Si photoconductor bar (dimensions $W_1 = 0.1$ mm, $W_2 = 0.01$ mm, $L = 2$ mm) is uniformly illuminated for a period of time $\gg \tau_n$, where $n = 10^{14}$ cm^{-3}, $\tau = 1$ μs, and $T = 300$ K. If the initial illumination produces a generation carrier density of $G(0) = G_0 = 10^{16}$ cm^{-3}, and turned off at $t = 0$, what is the time dependent value of Δn? At 5 volts bias, what is the measured current as a function of time?

12. Explain how and why the energy resolution of a surface barrier detector changes as the surface area of the detector is increased.

13. Calculate and plot the Q-map for the case in which the ratio of ϱ_e/ϱ_h is constant at 10, with ϱ_e values of 50, 25, 0.5, 2.5, 0.5 and 0.05. Comment on the expected pulse height distributions. How would you design a planar detector to ensure good energy resolution?

14. For an ideal p-type MOS structure ($N_A = 5 \times 10^{15}$ cm^{-3}) with an oxide thickness of 30 nm, determine the threshold voltage V_T for strong inversion. Note that $\kappa = 3.9$ for SiO_2 and 11.9 for Si.

15. Derive the result of Eq. (15.244).

16. Derive the result of Eq. (15.243).

17. Starting with the material resistivity ρ very high, such as with a semi-insulating material, derive the result of Eq. (15.189). Explain the consequence of this finding.

18. You have introduced an n-type dopant into a Si surface, and the backplane has a conductive layer. The doping depth is 1.5 microns and the feature size is 1 cm L $\times$ 0.5 cm W. Using a four-point probe with spacing of 1.5 mm between the probes, you measure 5 μA at 3 volts. What is the resistivity of the doped region?

19. Derive Eq. (15.247).

20. Given a CdZnTe hemispherical detector with anode $r_1 = 1$ mm and cathode $r_2 = 6$ mm, calculate the weighting potential of the device. What fraction of the induced charge is from electrons if a gamma-ray is completely absorbed at $r = 2$ mm, 3 mm, 4 mm, and 5 mm? Suppose that trapping is present with $\tau_e = 10^{-6}$ s, if $\tau_e \gg \tau_h$, explain why these outcomes are important.

REFERENCES

Alekseeva, L.A., P.G. Dorogov, V.I. Ivanov, and A.K. Khusainov, "Cadmium Telluride Gamma Radiation Detectors with a High Energy Resolution," *Pribory i Tehnika Eksperimenta*, **No. 1**, 54–58, (1985).

Apotovsky, B., C.L. Lingren, A. Oganesyan, B. Pi, J.F. Butler, F.P. Doty, R.L. Conwell, and S.J. Friesenhahn, "Semiconductor Radiation Detector with Enhanced Charge Collection," US Patent 5677539; allowed Oct. 14, 1997.

Ariesanti, E., A. Kargar, and D.S. McGregor, "Fabrication and Spectroscopy Results of Mercuric Iodide Frisch Collar Detectors," *Nucl. Instrum. Meth.*, **A624**, 656–661, (2010).

Ariesanti, E., A. Kargar, and D.S. McGregor, "Mercuric Iodide Crystal Growth and Frisch Collar Detector Fabrication," *Nucl. Tech.*, **175**, 124–130, (2011).

Armstrong, H.L., "A Theory of Voltage Breakdown of Cylindrical *P-N* Junctions, with Applications," *IRE Trans. Electron Devices*, **9**, 15–16, (1957).

Auricchio, N., L. Amati, A. Basili, E. Caroli, A. Donati, T. Franceschini, F. Frontera, G. Landini, A. Roggio, F. Schiavone, J.B. Stephen, and G. Ventura, "Twin Shaping Filter Techniques to Compensate the Signals From CZT/CdTe Detectors," *IEEE Trans. Nucl. Sci.*, **NS-52**, 1982–1988, (2005).

BACIAK, J.E. AND Z. HE, "Spectroscopy on Thick HgI_2 Detectors: A Comparison Between Planar and Pixelated Electrodes," *IEEE Trans. Nucl. Sci.*, **NS-50**, 1220–1224, (2003a).

BACIAK, J.E. AND Z. HE, "Comparison of 5 and 10 mm Thick HgI_2 Pixelated γ-Ray Spectrometers," *Nucl. Instrum. Meth.*, **A 505**, 191–194, (2003b).

BALE, D.S. AND C. SZELES, "Design of High Performance CdZnTe Quasi-Hemispherical Gamma-Ray CAPture^{TM} Plus Detectors," *Proc. SPIE*, **6319**, 1–11, (2006).

BARGHOLTZ, CHR., E. FUMERO, AND L. MÅRTENSSON, "Model-Based Pulse Shape Correction for CdTe Detectors," *Nucl. Instrum. Meth.*, **A434**, 399–411, (1999).

BARRETT, H.H., J.D. ESKIN, AND H.B. BARBER, "Charge Transport in Arrays of Semiconductor Gamma-Ray Detectors," *Phys. Rev. Lett.*, **75**, 156–159, (1995).

BERTUCCIO, G., M. SAMPIETRO, AND A. FAZZI, "High Resolution X-Ray Spectroscopy with Silicon Drift Detectors and Integrated Electronics," *Nucl. Instrum. Meth.*, **A322**, 538–542, (1992).

BOLOTNIKOV, A.E., J. BUTCHER, G.S. CARMARDA, Y. CUI, ET AL., "Array of Virtual Frisch-Grid CZT Detectors with Common Cathode Readout for Correcting Charge Signals and Rejection of Incomplete Charge-Collection Events," *IEEE Trans. Nucl. Sci.*, **NS-59**, 1544–1551, (2012).

BUBE, R.H., *Photoconductivity of Solids*, New York: Wiley, 1960.

BUTLER, J.F., "Novel Electrode Design for Single-Carrier Charge Collection in Semiconductor Nuclear Radiation Detectors," *Nucl. Instrum. Meth.*, **A 396**, 427–430, (1997).

CARDOSO, J.M., J.B. SIMÕES, T. MENEZES, AND C.M.B.A. CORREIA, "CdZnTe Spectra Improvement Through Digital Pulse Amplitude Correction Using the Linear Sliding Method," *Nucl. Instrum. Meth.*, **A505**, 334–337, (2003).

CUI, Y-H., A. BOLOTNIKOV, G. CARMADA, G.A. CARINI, G.D. GERONIMO, P. O'CONNER, R.B. JAMES, A. KARGAR, M.J. HARRISON, AND D.S. MCGREGOR, "Readout System for Arrays of Frisch-Ring CdZnTe Detectors," *IEEE Trans. Nucl. Sci.*, **NS-54**, 849–853, (2007).

CUI, Y-G., A. BOLOTNIKOV, G. CARMADA, A. HOSSAIN, R.B. JAMES, G.D. GERONIMO, J. FRIED, P. O'CONNER, A. KARGAR, M.J. HARRISON, AND D.S. MCGREGOR, "Hand-Held Gamma-Ray Spectrometer Based on High-Efficiency Frisch-Ring CdZnTe Detectors," *IEEE Trans. Nucl. Sci.*, **NS-55**, 2765–2769, (2008).

DANISHEVSKIĬ, A.M., A.A. KASTAĹSKIĬ, S.M. RYVKIN, AND I.D. YAROSHETSKIĬ, "Dragging of Free Carriers by Photons in Direct Interband Transitions," *Sov. Phys. JEPT*, **31**, 292–294, (1970).

DAY, R.B., G. DEARNALEY, AND J.M. PALMS, "Noise, Trapping and Energy Resolution in Semiconductor Gamma-Ray Spectrometers," *IEEE Trans. Nucl. Sci.*, **NS-14**, 487-491, (1967).

DEARNALEY, G. AND D.C. NORTHROP, *Semiconductor Counters for Nuclear Radiations*, 2nd Ed., New York: Wiley, 1966.

DOTY, F.P., J.F. BUTLER, J.F. SCHETZINA, AND K.A. BOWERS, "Properties of CdZnTe Crystals Grown by a High Pressure Bridgman Method," *J. Vac. Sci. Tech.*, **B 10**, 1418–1422, (1992).

GATTI, E. AND P. REHAK, "Semiconductor Drift Chamber - An Application of a Novel Charge-Transport Scheme," *Nucl. Instrum. Meth.*, **225**, 608–614, (1984).

GATTI, E., P. REHAK AND J.T. WALTON, "Semiconductor Drift Chamber - First Results and Optimum Processing of Signals," *Nucl. Instrum. Meth.*, **226**, 129–141, (1984).

GHANDHI, S.K., *Semiconductor Power Devices*, New York: Wiley, 1977.

GIBSON, A.F., M.F. KIMMITT, AND A.C. WALKER, "Photon Drag in Germanium," *Appl. Phys. Lett.*, **17**, 75–77, (1970).

GIBSON, A.F. AND M.F. KIMMITT, "Photon Drag Detection," in *Infrared and Millimeter Waves*, K.J. BUTTON, Ed., **Vol. 3**, *Submillimeter Techniques*, Chap. 4, 181–217, New York: Academic Press, 1980.

GRINBERG, A.A., "Theory of the Photoelectric and Photomagnetic Effects Produced by Light Pressure," *Soviet Phys. JETP*, **31**, 531–534, (1970).

GROVE, A.S., *Physics and Technology of Semiconductor Devices*, New York: Wiley, 1967.

HALL, E.H., "On a New Action of the Magnet on Electric Circuits," *Amer. J. Math.*, **2**, 287–292, (1879).

HALL, R.N., "Electron-Hole Recombination in Germanium," *Phys. Rev.*, **87**, 387, (1952).

HALL, R.N., "Recombination Processes in Semiconductors," *Proc. IEEE*, **106B**, *Suppl. 17*, 923–931, (1959).[18]

HANNAY, N.B., *Semiconductors*, New York: Reinhold, 1959.

HE, Z., "Potential Distribution within Semiconductor Detectors using Coplanar Electrodes," *Nucl. Instrum. Meth.*, **A 365**, 572–575, (1995).

HE, Z., G.F. KNOLL, D.K. WEHE, AND Y.F. DU, "Coplanar Grid Patterns and Their Effect on Energy Resolution of CdZnTe Detectors," *Nucl. Instrum. Meth.*, **A 411**, 107–113, 1998.

HE, Z., W. LI, G.F. KNOLL, D.K. WEHE, J. BERRY, AND C.M. STAHLE, "3-D Position Sensitive CdZnTe Gamma-Ray Spectrometers," *Nucl. Instrum. Meth.*, **A422**, 173–178, (1999).

HILIBRAND, J. AND R.D. GOLD, "Determination of the Impurity Distribution in Junction Diodes from Capacitance-Voltage Measurements," *RCA Rev.*, **21**, 245-252, (1960).

HITOMI, K., T. ONODERA, T. SHOJI, AND Z. HE, "Investigation of Pixelated TlBr Gamma-Ray Spectrometers with the Depth-Sensing Technique," *Nucl. Instrum. Meth.*, **A591**, 276–278, (2008).

HOLZER, A., "Rise Time Study of Mercuric Iodide Detectors," *Nucl. Instrum. Meth.*, **213**, 83–88, (1983).

IVANOV, V.I., V.A. GARBUSIN, P.G. DOROGOV, A.E. LOUTCHANSKI, AND V.V. KONDRASHOV, "Spectrometric Characteristic Improvement of CdTe Detectors," *IEEE Trans. Nucl. Sci.*, **NS-42**, 258–262, (1995).

JEONG, M., W.J. JO, H.S. KIM, AND J.H. HA, "Characteristics of a Frisch Collar Grid CdZnTe Radiation Detector Grown by Low-Pressure Bridgman Method," *J. Korean Phys. Soc.*, **66**, 1669-1674, (2015).

JONES, L.T. AND P.B. WOOLLAM, "Resolution Improvement in CdTe Detectors Using Pulse-Shape Discrimination," *Nucl. Instrum. Meth.*, **124**, 591–595, (1975).

KARGAR, A., A.J. JONES, W.J. MCNEIL, M.J. HARRISON, AND D.S. MCGREGOR, "CdZnTe Frisch Collar Detectors for γ-Ray Spectroscopy," *Nucl. Instrum. Meth.*, **A558**, 497–503, (2006).

KARGAR, A., A.C. BROOKS, M.J. HARRISON, K.T. KOHMAN, R.B. LOWELL, R.C. KEYES, H. CHEN, G. BINDLEY, AND D.S. MCGREGOR, "The Effect of Dielectric Thickness on Spectral Performance of CdZnTe Frisch Collar Gamma Ray Spectrometers," *IEEE Trans. Nucl. Sci.*, **NS-56**, 824–831, (2009).

[18] In some library systems, this reference can be a challenge to locate.

KARGAR, A., E. ARIESANTI, S. JAMES, AND D.S. MCGREGOR, "Charge Collection Efficiency Characterization of a HgI_2 Frisch Collar Spectrometer with Collimated High Energy Gamma Rays," *Nucl. Instrum. Meth.*, **A652**, 186–192, (2011a).

KARGAR, A., E. ARIESANTI, AND D.S. MCGREGOR, "A Comparison Between Spectroscopic Performance of HgI_2 and CdZnTe Frisch Collar Detectors," *Nucl. Tech.*, **175**, 131–137, (2011b).

KEELE, B.D., R.S. ADDLEMAN, AND G.L. TROYER, "A Method to Improve Spectral Resolution in Planar Semiconductor Gamma-Ray Detectors," *IEEE Trans. Nucl. Sci.*, **NS-43**, 1365–1368, (1996).

KHUSAINOV, A.KH., "CdTe Detectors with Thermoelectric Cooling," *Nucl. Instrum. Meth.*, **A322**, 335–340, (1992).

KIM, K.H., C. PARK, P. KIM, S. CHO, J. LEE, T.K. HONG, A. HOSSAIN, A.E. BOLOTNIKOV, AND R.B. JAMES, "Spectroscopic Properties of Large-Volume Virtual Frisch-Grid CdZnTe Detectors," *J. Korean Phys. Soc.*, **66**, 1761–1765, (2015).

KNOLL, G.F. AND D.S. MCGREGOR, "Fundamentals of Semiconductor Detectors for Ionizing Radiation," *Proc. MRS*, **302**, 3-17, (1993).

KUVVETLI, I., C. BUDTZ-JØRGENSEN, L. GERWARD, AND C.M. STAHLE, "Response of CZT Drift-Strip Detector to X- and Gamma Rays," *Rad. Phys. Chem.*, **61**, 457–460, (2001).

KUVVETLI, I. AND C. BUDTZ-JØRGENSEN, "Pixelated CdZnTe Drift Detectors," *IEEE Trans. Nucl. Sci.*, **NS-52**, 1975–1981, (2005).

LAMPERT, M.A., "Simplified Theory of Space-Charge-Limited Currents in an Insulator with Traps," *Phys. Rev.*, **103**, 1648-1656, (1956).

LAMPERT, M.A. AND P. MARK, *Current Injection in Solids*, New York: Academic Press, 1970.

LOOK, D.C., *Electrical Characterization of GaAs Materials and Devices*, New York: Wiley, 1989.

LUKE, P.N., "Single-Polarity Charge Sensing in Ionization Detectors Using Coplanar Electrodes," *Appl. Phys. Lett.*, **65**, 2884–2886, (1994).

LUKE, P.N., "Unipolar Charge Sensing with Coplanar Electrodes-Application to Semiconductor Detectors," *IEEE Trans. Nucl. Sci.*, **NS-42**, 207–213, (1995).

LUND, J.C., R.W. OLSEN, R.B. JAMES, J.M. VAN SCYOC, E.E. EISSLER, MM. BLAKELEY, J.B. GLICK, AND C.J. JOHNSON, "Performance of a Coaxial Geometry $Cd_{1-x}Zn_xTe$ Detector," *Nucl. Instrum. Meth.*, **A377**, 479–483, (1996).

MALM, H.L., C. CANALI, J.W. MAYER, M-A. NICOLET, K.R. ZANIO, AND W. AKUTAGAWA, "Gamma-Ray Spectroscopy with Single-Carrier Collection in High-Resistivity Semiconductors," *Appl. Phys. Lett.*, **26**, 344–346, (1975).

MATSUSHITA, N., WM.C. MCHARRIS, R.B. FIRESTONE, J. KASAGI, AND W.H. KELLY, "On improving Ge Detector Energy Resolution and Peak-to-Compton Ratios by Pulse-Shape Discrimination," *Nucl. Instrum. Meth.*, **179**, 119–124, (1981).

MATSUSHITA, N., J. KASAGI, AND WM.C. MCHARRIS, "The Pulse-Height Correction Technique for Improving γ-Ray Spectra from Coaxial Ge Detectors," *Nucl. Instrum. Meth.*, **201**, 433–438, (1982).

MAYER, M., L.A. HAMEL, O. TOUSIGNANT, J.R. MACRI, J.M. RYAN, M.L. MCCONNELL, V.T. JORDANOV, J.F. BUTLER, AND C.L. LINGREN, "Signal Formation in a CdZnTe Imaging Detector with Coplanar Pixel and Control Electrode," *Nucl. Instrum. Meth.*, **A 422**, 190–194, (1999).

MCGREGOR, D.S. AND J.E. KAMMERAAD, "Gallium Arsenide Radiation Detectors and Spectrometers", **Chapter 10**, in *Semiconductors for Room Temperature Nuclear Detector Applications*, T.E. SCHESLINGER, R.B. JAMES, Vol. Eds., **Vol. 43** of *Semiconductors and Semimetals*, R.K. WILLARDSON, A.C. BEER, E.R. WEBER, Eds., San Diego: Academic Press, pp. 383–442, 1995.

MCGREGOR, D.S. AND H. HERMON, "Room-Temperature Compound Semiconductor Radiation Detectors, *Nucl. Instrum. Meth.*, **A 395**, 101–124, (1997).

MCGREGOR, D.S., Z. HE, H.A. SEIFERT, R.A. ROJESKI, AND D.K. WEHE, "CdZnTe Semiconductor Parallel Strip Frisch Grid Radiation Detectors, *IEEE Trans. Nucl. Sci.*, **NS-45**, 443-449, (1998).

MCGREGOR, D.S., R.A. ROJESKI, Z. HE, D.K. WEHE, M. DRIVER, AND M. BLAKELY, "Geometrically Weighted Semiconductor Frisch Grid Radiation Spectrometers," *Nucl. Instrum. Meth.*, **A422**, 164–168, (1999a).

MCGREGOR, D.S. AND R. ROJESKI, "Performance of Geometrically Weighted Semiconductor Frisch Grid Radiation Spectrometers, *IEEE Trans. Nucl. Sci.*, **NS-46**, 250-259, (1999b).

MCGREGOR, D.S. AND R. ROJESKI, "High Resolution Ionization Detectors and Array of Such Detectors," US Patent 6175120; allowed Jan. 16, 2001.

MCGREGOR, D.S., "Collimated Radiation Detector Assembly, Array of Collimated Radiation Detectors and Collimated Radiation Detector Module," US Patent 6781132; allowed August 24, 2004.

MCKELVEY, J.P., *Solid State and Semiconductor Physics*, New York: Academic Press, 1966.

MCNEIL, W.J., D.S. MCGREGOR, A.E. BOLOTNIKOV, G.W. WRIGHT, AND R.B. JAMES, "Single-Charge-Carrier-Type Sensing with an Insulating Frisch Ring CdZnTe Semiconductor Radiation Detector," *Appl. Phys. Lett.*, **84**, 1988–1990, (2004).

MISSOUS, M. AND E.H. RHODERICK, "New Way of Plotting Current/Voltage Characteristics of Schottky Diodes," *Elec. Lett.*, **22**, 477–478, (1986).

MONTÉMONT, G., M-C. GENTET, O. MONNET, J. RUSTIQUE, AND L. VERGER, "Simulation and Design of Orthogonal Capacitive Strip CdZnTe Detectors," *IEEE Trans. Nucl. Sci.*, **NS-54**, 854–859, (2007).

MOTT, N.F. AND R.W. GURNEY, *Electronic Processes in Ionic Crystals*, Oxford: Oxford Univ. Press, 1948.

MULLER, R.S. AND T.I. KAMINS, *Device Electronic for Integrated Circuits, 2nd Ed.*, New York: Wiley, 1986.

OWENS, A., R. DEN HARTOG, F. QUARATI, V. GOSTILO, V. KONDRATJEV, A. LOUPILOV, A.G. KOZOREZOV, J.K. WIGMORE, A. WEBB, AND E. WELTER, "Hard X-Ray Response of a CdZnTe Ring-Drift Detector," *J. Appl. Phys.*, **102**, paper 054505, (2007).

PATT, B.E., J.E.. IWANCZYK, G. VILKELIS, AND Y.J. WANG, "New Gamma-Ray Structures for Electron Only Charge Carrier Collection Utilizing High-Z Compound Semiconductors," *Nucl. Instrum. Meth.*, **A380**, 276–281, (1996).

PUTLEY, E.H., *The Hall Effect and Semi-Conductor Physics*, New York: Dover, 1968.

REEVES, G.K., "Specific Contact Resistance Using a Circular Transmission Line Model," *Solid-State Elec.*, **23**, 487-490, (1980).

REEVES, G.K. AND H.B. HARRISON, "Obtaining the Specific Contact Resistance from Transmission Line Model Measurements," *IEEE Elec. Dev. Lett.*, **3**, 111-113, (1982).

REHAK, P., J. WALTON, E. GATTI, A. LONGONI, M. SANPIETRO J. KEMMER, P. HOLL, R. KLANNER, G. LUTZ, A. WYLIE, AND H. BECKER, "Progress in Semiconductor Drift Detectors," *Nucl. Instrum. Meth.*, **A248**, 367–378, (1986).

REHAK, P., E. GATTI, A. LONGONI, J. KEMMER, P. HOLL, R. KLANNER, G. LUTZ, AND A. WYLIE, "Semiconductor Drift Chambers for Position and Energy Measurements," *Nucl. Instrum. Meth.*, **A235**, 224–234, (1985).

RHODERICK, E.H. AND R.H. WILLIAMS, *Metal-Semiconductor Contacts, 2nd Ed.*, Oxford: Clarendon Press, 1988.

RICHTER, M. AND P. SIFFERT, "High Resolution Gamma Ray Spectroscopy with CdTe Detector Systems," *Nucl. Instrum. Meth.*, **A322**, 529–537, (1992).

RICHTER, M., M. HAGE-ALI, Z.T. KUZNICKI, AND P. SIFFERT, "Pulse Processing for Planar Cadmium Telluride Detectors," *Proc. MRS, Semiconductors for Room-Temperature Radiation Detector Applications*, **302**, 195–204, (1993).

ROSE, A., *Concepts in Photoconductivity and Allied Problems*, New York: Interscience Pub., 1963.

SCHRODER, D.K., *Semiconductor Material and Device Characterization*, New York: Wiley, 1990.

SHOCKLEY, W., "Currents to Conductors Induced by a Moving Point Charge," *J. Appl. Phys.*, **9**, 635-636, (1938).

SHOCKLEY, W. AND W.T. READ "Statistics of the Recombinations of Holes and Electrons," *Phys. Rev.*, **87**, 835–842, (1952).

SMITS, F.M., "Measurement of Sheet Resistivities with the Four-Point Probe," *Bell Syst. Tech. J.*, **37**, 711–718, (1958).

STURM, B.W., Z. HE, E. R.HODES, T.H. ZURBUCHEN, AND P.L. KOEHN, "Coplanar Grid CdZnTe Detectors for Space Science Applications," *Proc. SPIE*, **5540**, 14–21, (2004).

SZE, S.M., *Physics of Semiconductor Devices*, 2nd Ed., New York: Wiley, 1981.

SZE, S.M. AND G. GIBBONS, "Avalanche Breakdown Voltages of Abrupt and Linearly Graded *p-n* Junctions in Ge, Si, GaAs, and GaP," *Appl. Phys. Lett.*, **8**, 111–113, (1966a).

SZE, S.M. AND G. GIBBONS, "Effect of Junction Curvature on Breakdown Voltages in Semiconductors," *Solid State Elec.*, **9**, 831–845, (1966b).

TREDGOLD, R.H., *Space Charge Conduction in Solids*, Amsterdam: Elsevier, 1966.

UGOROWSKI, P.B., A,C, BROOKS, R.D. TAYLOR, T.J. SOBERING, AND D.S. MCGREGOR, "Preliminary Results of KSU Frisch-Collar Array," *Conf. Rec. IEEE Nucl. Sci. Symp.*, Valencia, Spain, Oct. 23–29, 4674–4676, (2011).

VALDES, L.B., "Resistivity Measurements on Germanium for Transistors," *Proc. IRE*, **42**, 420–427, (1954).

VAN PAMELEN, M.A.J. AND C. BUDTZ-JØRGENSEN, "CdZnTe Drift Detector with Correction for Hole Trapping," *Nucl. Instrum. Meth.*, **A411**, 197–200, (1998).

VARSHNI, Y.P., "Band-to-Band Recombination in Groups IV, VI, and III-V Semiconductors (II)," *Phys. Stat. Sol.*, **20**, 9–36, (1967).

WAHL, C.G., W.R. KAYE, W. WANG, F. ZHANG, J.M. JAWORSKI, A. KING, Y.A. BOUCHER, AND Z. HE, "The Polaris-H Imaging Spectrometer," *Nucl. Instrum. Meth.*, **A784**, 377–381, (2015).

WALTER, F.J., "Discussion of Pulse Rise Times and Equivalent Circuits," *Semiconductor Nuclear Particle Detectors*, Proc. of an Informal Conference, J.W.T. DABBS AND F.J. WALTER, Eds., Ashville, N.C., Sept. 28–30, 237–246, (1960).

WILEY, J.D., "Mobility of Holes in III-V Compounds," in *Semiconductors and Semimetals*, **Vol. 10**, R.K. WILLARDSON AND E.C. BEER, New York: Academic, pp. 91–174, 1985.

WRIGHT, G.T., "Some Properties and Applications of Space-Charge-Limited Currents in Insulating Crystals," Paper No. 2928 E, *Inst. Electrical Engineers*, 915–922, (1959).

ZANIO. K., "Use of Various Device Geometries to Improve the Performance of CdTe Detectors," *Rev. Phys. Appl.*, **12**, 343–347, (1977).

ZHANG, Q., C. ZHANG, Y. LU, K. YANG, AND Q. REN, "Progress in the Development of CdZnTe Unipolar Detectors for Different Anode Geometries and Data Corrections," *Sensors*, **13**, 2447–2474, (2013).

Chapter 16

Semiconductor Detectors

I have lately met with an extraordinary case ... which is in direct contrast with the influence of heat upon metallic bodies ... On applying a lamp ... the conducting power rose rapidly with the heat ... On removing the lamp and allowing the heat to fall, the effects were reversed.

Michael Faraday

16.1 Introduction

Semiconducting behavior was first discovered by Faraday in 1833 when he observed that current flow through AgS increased with temperature,[1] the opposite effect observed with conductive metals. We now know this effect to be a consequence of increased free charge carrier densities in the conduction and valence bands as temperature is increased. Decades later, Braun [1874] observed rectifying behavior from a point contact on a PbS crystal, in which current would flow in only one direction, but not the other, when positive voltage was applied, alternately, to the sample. This fundamental semiconductor physical property would be used nearly a century later as a building block for semiconductor diode and transistor designs.

Soon after the development of the gas-filled detector by Hans Geiger and Ernest Rutherford, some interest was placed in discovering a solid-state material for radiation detection. The concept, correctly, was to increase the radiation detection efficiency by substituting the gas medium with a solid-state medium. Perhaps the earliest recording of such an effort was a study performed by Röntgen and Joffé [1913] on various solids exposed to high levels of x radiation, followed by another study in 1921. Additional work by Schiller [1926] and Jaffé [1932] revealed conduction through dielectric crystals upon the application of high voltages and high irradiation fields. Yet, none of these works resulted in the practical application of solid state conduction counters.

The first semiconductor-based radiation detector was reported by Stetter in 1941, in which several materials were exposed to α-particles, β-particles, and protons. As with earlier works, most of those materials were insulators (alkali halides, sulphur, cerrusite, fluorite) resulting in no observed effect; however, Stetter also reported the observation of electrical current from a diamond sample exposed to ionizing radiation. Several years later, Van Heerden [1945; 1950a; 1950b] published his work on AgCl crystals, "The Crystal Counter",

[1]Published account is in his book *Experimental Researches in Electricity* [1839].

Table 16.1. Chronological initial exploration dates as radiation detectors for selected semiconductors.

Semiconductor	Common Abbreviation	Year	Reference
Diamond	diamond (C)	1941	Stetter
		1947	Wooldridge, Ahearn & Burton
		1948	Curtiss and Brown; Jentschke
Silver Chloride	AgCl	1945	Van Heerden
		1947	Hofstadter, Milton, Ridgway
Cadmium Sulphide	CdS	1947	Frerichs
		1948	Kallman and Warminsky
Thallium Bromide	TlBr	1947	Hofstadter
Thallium Iodide	TlI	1947	Hofstadter
Zinc Sulfide	ZnS	1948	Ahearn
Argon (solid)	Ar	1948	Hutchinson
Silver Bromide	AgBr	1949	Yamakawa
Mercury Sulfide	HgS	1949	Ahearn
Sulphur†	S	1949	Georgesco
Germanium	Ge	1949	McKay
Sodium Iodide	NaI	1950	Witt
Silicon	Si	1958	Davis
Gallium Arsenide	GaAs	1960	Harding
Cadmium Telluride	CdTe	1967	Autagawa, Zanio and Mayer
Mercuric Iodide	HgI_2	1971	Willig
Lead Iodide	PbI_2	1971	Roth and Willig
Cadmium $Zinc_{1-x}Telluride_x$	CZT	1992	Doty et al.

†Stetter [1941] also studied sulphur, but with negative results.

widely accepted as the first reported semiconductor detector.[2] Upon the publication of Van Heerden's work, numerous semiconductor materials were subsequently investigated, and the investigators along with their introductory years are tabulated in Table 16.1.

Up until 1960, semiconductor detectors, then called "crystal counters", were considered mostly a novelty for radiation detection and were not commercially produced. The main reason, as expressed by Price [1958, 1964], was that semiconductors were of "very little use because of the preponderance of advantages of the scintillation detectors".[3] A significant disadvantage of semiconductors at the time was the high level of background impurities and crystal defects. Under reverse bias, the impurities could become ionized, as explained in Sec. 15.3.5, and trapping of electrons or holes in the defect states would reduce the induced current. In both cases, space charge would be created which tends to counter the applied voltage and reduce both the electric field and the depletion region width. However, in 1960 while working for the General Electric Co. (GE), Erik M. Pell [1960a, 1960b] developed a method of introducing a compensating Li ion into Si that countered the effect of *p*-type dopants. The method included the application of Li material to a heated bulk Si sample and driving Li ions deep into the semiconductor with a strong electric field, a method known as "Li drifting". The technique was successfully applied to Si [Pell 1960b; Elliot 1961; Blankenship and Borkowski 1962] and Ge crystals [Freck and Wakefield 1962; Tavendale 1963, 1966; Goulding and Hansen 1964; Goulding 1965, 1966], and allowed relatively thick depletion layers to be formed. It was this fundamental improvement

[2]Although Stetter reported his observation years earlier, the work went largely unnoticed. Consequently, Van Heerden has often been quoted as the originator of the first semiconductor detector. Jeschke attempts to correct the record in his 1948 letter to the editor.

[3]Price retracted this opinion in his 1964 book.

that allowed semiconductors to become important radiation detectors, and Ge(Li) and Si(Li) detectors[4] came into common use as high resolution gamma-ray and x-ray detectors.

A drawback to all Ge(Li) detectors was that they required constant cryogenic cooling. If the crystal were to rise in temperature, the Li ions would diffuse away from their compensation locations and the crystal would be ruined. Ge crystals could be re-drifted once, maybe twice, but almost always resulted in lower energy resolution performance. Si(Li) detectors did not suffer the same Li ion diffusion consequence at room temperature, but both Ge(Li) and Si(Li) detectors perform best when operated at or near the temperature of liquid nitrogen (LN2).

A solution to the nuisance of cryogenic cooling for Ge(Li) was explored by two research teams, one team led by Robert Hall at GE and the other team led by William Hansen at the Lawrence Berkeley Laboratory (LBL) [Haller 1996]. The concept was to apply the zone refinement purification technique developed by William Pfann [1966], thereby purifying Ge to such a level that it was practically devoid of impurities and, hence, no longer in need of Li compensation. Both teams succeeded in the mission and produced virtually intrinsic high-purity Ge crystals, or HPGe crystals. Detectors fabricated from this material no longer required constant cooling, but still required cryogenic cooling during operation in order to reduce leakage currents. By the early 1970s, zone refined HPGe, high purity Si, and Si(Li) detectors became standard instruments for radiation spectroscopy.

Yet, the required cryogenic cooling during operation remains inconvenient. There are solutions, mainly, low vibration refrigerators or small handheld dewars can be attached to HPGe detectors. Wide band-gap semiconductors, such as GaAs, CdTe, HgI_2, and CdZnTe (or CZT), that can operate at room temperature (RT) without cryogenic cooling have been investigated as alternative semiconductors for gamma-ray spectroscopy. Perhaps to date the most successful of these RT candidates is CZT, mainly because of the introduction of "single carrier" detector designs by Eskin et al. [1995], Luke [1995], McGregor and Rojeski [1998, 1999], and McGregor [2004]. Included in the present chapter are descriptions of properties for semiconductors important for radiation detection. Descriptions of semiconductor detectors that are either commercially available, or at least show great promise for various radiation measurement applications, are also described in the present chapter.

16.2 General Semiconductor Properties

Desired properties of a semiconductor radiation detector vary with its application. For instance, materials with low atomic numbers are best used for detection of β particles and electrons while materials with high atomic numbers are best for detection of x rays and gamma rays. High energy resolution generally requires a low ionization energy, best accomplished by choosing semiconductors with relatively narrow band gaps. A small ionization energy causes increased numbers of excited charge carriers, thereby improving statistics and enhancing spectroscopic energy resolution. However, semiconductor detectors fashioned from narrow band-gap materials generally require cryogenic cooling during operation. A wide band-gap energy (> 1.5 eV) and high resistivity allow room temperature operation that otherwise would require cryogenic cooling to reduce electronic noise, but at the price of reduced spectroscopic energy resolution. Other properties sought for gamma-ray detectors include high atomic density, long charge carrier lifetime, high resistivity, and high electron and hole mobilities. High atomic density and high Z components increase the γ-ray interaction probability (see Fig. 16.1). Long charge carrier lifetimes and high charge carrier mobilities increase the charge collection efficiency and produce better spectroscopic results. Unfortunately, no existing semiconductor actually has all of these characteristics; hence the investigator should select a semiconductor

[4] Pronounced "jelly" and "silly" detectors.

Table 16.2. Common semiconductors used as radiation detectors and their properties at 300K.

Semiconductor	Atomic Number (Z)	Density (g cm^{-3})	Structure	Fano Factor	Ionization Energy (eV/e-h pair)	Dielectric Constant (ϵ_s/ϵ_0)
C (diamond)	6	3.5	FCC/Dia.	0.38	13	5.7
Si	14	2.33	FCC/Dia	0.115	3.61	11.9
Ge	32	5.33	FCC/Dia	0.13	2.98	16
CdSe	48/34	5.8	Hex/W		5.5	10.6
CdTe	48/52	5.86	FCC/ZB	0.11	4.43	10.36
$Cd_{0.85}Zn_{0.15}Te$ (CZT)	48/30/52	6.0	FCC/ZB	0.089	5.0	10.63
GaAs	31/33	5.32	FCC/ZB	0.1	4.2	13.1
GaSe	31/34	5.03	Hex		4.5	8
HgI_2	80/53	6.4	Tetragonal	0.19	4.42	8.8
InP	49/15	4.79	FCC/ZB		4.2	12.5
PbI_2	82/53	6.2	Hex	0.19	4.9	4.3
SiC(3C)	14/6	3.21	3C/ZB		5.5	9.72
SiC (4H)	"	"	Hex/4H/W		7.8	9.66
SiC (6H)	"	"	Hex/6H/W	0.09	6.9	9.66
TlBr	81/35	7.56	BCC		6.5	29.8

Table 16.3. Common semiconductors used as radiation detectors and their properties at 300K.

Semiconductor	Band Gap (eV)	Intrinsic Resistivity (Ω-cm)	Electron Mobility (cm^2/V·s)	Hole Mobility (cm^2/V·s)	Electron Lifetime (sec.)	Hole Lifetime (sec.)
C (diamond)	5.47 (I)	$> 10^{12}$	2400	2100	$< 10^{-9}$	$< 10^{-9}$
Si	1.12 (I)	$> 5 \times 10^4$	1500	450	$> 10^{-3}$	$> 10^{-3}$
Ge (77 K)	0.72 (I)	$> 10^{18}$	36000	42000	$> 10^{-3}$	$> 10^{-3}$
Ge (300 K)	0.68 (I)	≈ 50	3900	1900	$> 10^{-3}$	$> 10^{-3}$
CdSe	1.73 (D)	10^8	720	75	10^{-6}	10^{-6}
CdTe	1.52 (D)	10^9	1050	100	3×10^{-6}	2×10^{-6}
CZT	1.60 (D)	10^{11}	1350	120	10^{-6}	5×10^{-8}
GaAs	1.42 (D)	$\approx 10^8$	8500	400	$10^{-9} - 10^{-8}$	$10^{-9} - 10^{-8}$
GaSe	2.1 (D)		60	215	2×10^{-6}	5×10^{-6}
HgI_2	2.13 (D)	10^{13}	100	4	$> 10^{-6}$	$> 10^{-6}$
InP	1.35 (D)	10^7	4600	150	1.5×10^{-9}	$< \times 10^{-7}$
PbI_2	2.3 (D)	10^{12}	8	2	10^{-6}	
SiC (3C)	2.36 (I)	150	800	320		
SiC (4H)	3.23 (I)	$> 10^{15}$	900	120	4×10^{-7}	6.7×10^{-7}
SiC (6H)	3.05 (I)		400	90		
TlBr	2.68 (I)	10^{12}	6-25	1.5-6.2	$\sim 2.5 \times 10^{-6}$	$\sim 4.3 \times 10^{-6}$

detector best suited for the desired application. The basic properties of several semiconductors used for semiconductor detectors are listed in Tables 16.2, 16.3, and 16.4.

16.2.1 Atomic Numbers and Mass Density

The choice of detector material depends on the type of radiation to be detected. As discussed in Sec. 7.2.1 and Sec. 7.2.2, the backscatter coefficient for charged particles (both electrons and alpha particles) increases with the atomic number of the absorber. Likewise the absorption coefficient for photons (see Sec. 4.4) also increases with the atomic number of the absorber. Hence, it is generally better to fabricate charged particle detectors from low Z materials, such as Si. By contrast, high Z materials are preferred for gamma-ray

Table 16.4. Common semiconductors used as radiation detectors and their properties at 300K.

Semiconductor	N_C (cm^{-3})	N_V (cm^{-3})	m_t^*	m_l^*	m_o^*	m_{lh}^*	m_{hh}^*
C (Diamond)	$\sim 10^{20}$	$\sim 10^{19}$	0.36	1.4	–	0.3	1.1
Si (4 K)	4.22×10^{16}	1.75×10^{16}	0.1905	0.9163	–	0.153	0.537
Si (300 K)	2.8×10^{19}	1.04×10^{19}	0.19	0.98	–	0.16	0.49
Ge (4 K)	1.58×10^{16}	8.24×10^{15}	0.08152	1.588	–	0.0429	0.347
Ge (300 K)	1.04×10^{19}	6×10^{18}	0.082	1.64	–	0.044	0.28
CdSe	2×10^{18}		–	–	0.13	0.45	–
CdTe	7.5×10^{17}	1.8×10^{19}	–	–	0.11	0.13	~0.8
CZT							
GaAs	4.7×10^{17}	7×10^{18}	–	–	0.067	0.45	0.82
GaSe							
HgI_2							
InP	5.7×10^{17}	1.1×10^{19}	–	–	0.077	0.089	0.6
PbI_2							
SiC (3C)	1.5×10^{19}	1.9×10^{19}	0.677	0.247		0.6	1.5
SiC (4H)	1.8×10^{19}	2.1×10^{19}	0.33	0.42		0.66	1.75
SiC (6H)	8.8×10^{19}	2.2×10^{19}	2.0	0.48		0.66	1.85
TlBr							

detection. There are special materials that react to neutrons, which can be adapted if neutron detection is the goal, or should be avoided if neutron discrimination is desired. One such semiconductor material is CdTe (and CdZnTe), which offers good gamma-ray detection, but also reacts to low energy neutrons to produce prompt gamma-rays. Special attention to material selection should be applied to low-energy gamma-ray and x-ray detection, for which the K absorption edge may become important. For example, as seen in Fig. 16.1, the total photon absorption coefficient below the two K edge energies of CdTe drops below that of Ge; hence the lower Z material actually performs better than the higher Z material at energies below 26.727 keV. Si is generally selected over Ge for characteristic x-ray analysis because its K edge is at a lower energy (1.840 keV) than Ge (11.115 keV); hence, the transition K edge has less effect on the photon absorption efficiency.

The mass density is of obvious importance, because increased mass density generally indicates higher electron density. Further, photoelectric absorption is a strong function of Z; hence, the interaction efficiency is strongly influenced by the atomic density of the highest Z component. Selection of materials with at least one high Z component is important for gamma-ray detection. For instance, InP has one component with a relatively high atomic number $Z_{In} = 49$, and one relatively low atomic number $Z_P = 15$. Overall, the gamma-ray absorption characteristic is predominately influenced by the density of indium because the phosphorus component makes little contribution. Atomic number and mass density for several semiconductors are listed in Table 16.2.

16.2.2 Band Gap

The energy band gap of a semiconductor E_g, introduced in Sec. 12.2.5, is

$$E_g = kT \ln \left[\frac{N_C N_V}{n_i^2} \right] \tag{16.1}$$

or, more traditionally, is described by

$$n_i = \sqrt{N_C N_V}\, e^{-E_g/2kT}, \tag{16.2}$$

where n_i is the intrinsic charge carrier concentration, N_C is the effective density of state in the conduction band, N_V is the effective density of states in the valence band, k is Boltzmann's constant, and T is the

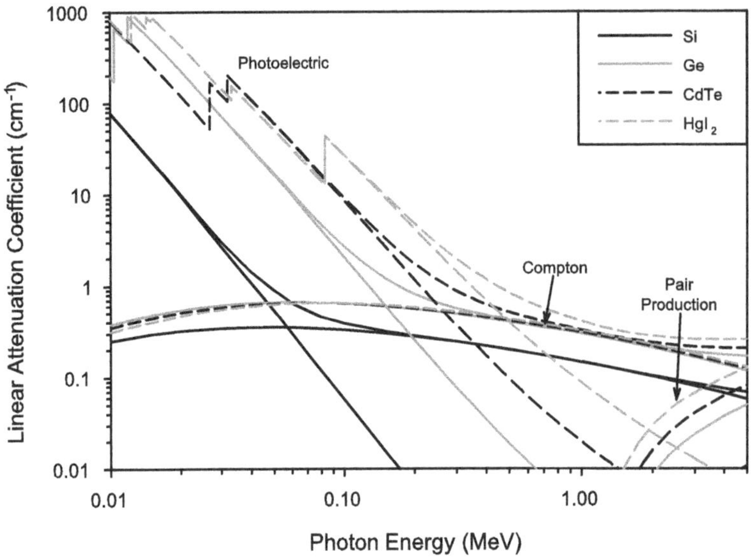

Figure 16.1. The linear attenuation coefficients for several representative semiconductors.

absolute temperature. The relationship between n_i and E_g is of major importance because a small increase in the band-gap energy has a significant effect upon the free charge carrier concentration. Hence, to reduce leakage current, it is desirable to use semiconductors with relatively large band-gap energies. Note that lowering the temperature also reduces the value of n_i, a method often used to suppress leakage current for semiconductors that have relatively small values or E_g.

For most semiconductor radiation detectors, indirect band gaps are preferred over direct band gaps mainly because the charge carrier lifetimes are theoretically much longer [Shockley and Read 1952; Hall 1952, 1959; Varshni 1967]. The reason for the difference is that electrons in the conduction band of an indirect band-gap semiconductor must change in both momentum and energy to recombine with a hole, whereas electrons in the conduction band of a direct band-gap semiconductor need only to lose energy to complete the recombination process. Consequently, indirect band-gap semiconductors generally have higher carrier extraction factor values than direct band-gap semiconductors for similar detector dimensions. Listed in Table 16.3 are band-gap energies and types (indirect or direct) for several semiconductors used in radiation detection. Note that the defect and impurity density of a semiconductor, if significant, can have a greater effect on the overall mean free lifetime rather than the type of band gap.

16.2.3 Ionization Energy

A property strongly influenced by the energy band gap is the average ionization energy w, which limits the minimum spectral energy resolution. For materials and detectors that suffer little to no charge carrier trapping losses, the expected variance in charge pair excitation should follow Gaussian statistics so that

$$\sigma_{N_0}^2 = \frac{FE_\gamma}{w} = FN_0, \tag{16.3}$$

where F is the semiconductor Fano factor and E_γ is the energy of the radiation quantum. The resulting pulse height spectrum, representing full energy peaks of radiation quanta, should also have a Gaussian appearance.

Hence, the full width at half maximum (FWHM) of these Gaussian peaks is, in terms of charge pairs N_0,

$$\Gamma = \text{FWHM}|_{N_0} = 2\left[2\ln(2)\right]^{1/2}\sigma, \tag{16.4}$$

which is transformed into units of energy by multiplying by $w = E_\gamma/N_0$ as

$$\Gamma = \text{FWHM}|_{eV} = 2w\left[2\ln(2)\,\frac{FE\gamma}{w}\right]^{1/2} \simeq 2.355\left[wFE_\gamma\right]^{1/2}. \tag{16.5}$$

This result can be written in terms of percent energy by dividing Eq. (16.5) by E_γ

$$\Gamma = \text{FWHM}|_{\%} = 100\frac{2\left[2\ln(2)\;wFE_\gamma\right]^{1/2}}{E_\gamma} \simeq 235.5\left[\frac{wF}{E_\gamma}\right]^{1/2}. \tag{16.6}$$

It becomes clear from Eqs. (16.5) and (16.6) that w significantly affects Γ and a small value of w is desirable.

The dependence of w upon the band-gap energy was studied by Klein [1968], and it was determined that energy was expended by three mechanisms, namely (1) excitation across the band-gap energy, (2) residual kinetic energy of excited charges, and (3) optical phonon losses. It is obvious that the minimum energy required to produce an electron-hole pair must be E_g. Yet, the energy released from Coulombic interactions of primary excited electrons, from either charged particle or gamma-ray interactions, can excite energetic electrons from deep in the valence band or lower tightly bound energy bands. Further, the energy absorbed may be more than required to transition the band gap into the conduction band. This excess energy is manifested as kinetic energy of the charge carriers. If this excess kinetic energy is less than E_g, then it becomes expended in the material and does not contribute to additional electron-hole pair creation.

The average ionization energy can be represented by

$$w = E_g + \langle W_L \rangle, \tag{16.7}$$

where $\langle W_L \rangle$ is the average energy expended for purposes other than electron-hole production. Klein [1968] breaks $\langle W_L \rangle$ into two fundamental components

$$\langle W_L \rangle = \langle E_R \rangle + \langle E_K \rangle, \tag{16.8}$$

where $\langle E_R \rangle$ represents optical phonon losses and $\langle E_K \rangle$ represents thermalization losses. Optical phonons are out of phase movements of the crystalline lattice and can be produced by energetic charged particles having sufficient energy to excite electron-hole pairs. Their average energy is

$$\langle E_R \rangle = r(\hbar\omega_R),\; 0.5\text{ eV} \leq r(\hbar\omega_R) \leq 1.0\text{ eV}, \tag{16.9}$$

where r represents the average number of phonons per electron-hole pair created and ω_R is the angular frequency of optical phonon oscillations. Hence, $\hbar\omega_R$ represents the energy per phonon (Raman quanta).

From experimental data of several semiconductors,[5] Klein [1968] estimates a threshold for secondary ionization as

$$E_I \simeq \frac{3}{2}E_g \tag{16.10}$$

and the average residual kinetic energy of excited charge pairs as

$$\langle E_K \rangle \simeq 2\mathcal{L}E_I = 3\mathcal{L}E_g, \tag{16.11}$$

[5]Klein [1968] lists Ge, Si, GaAs, InP as the sources of data.

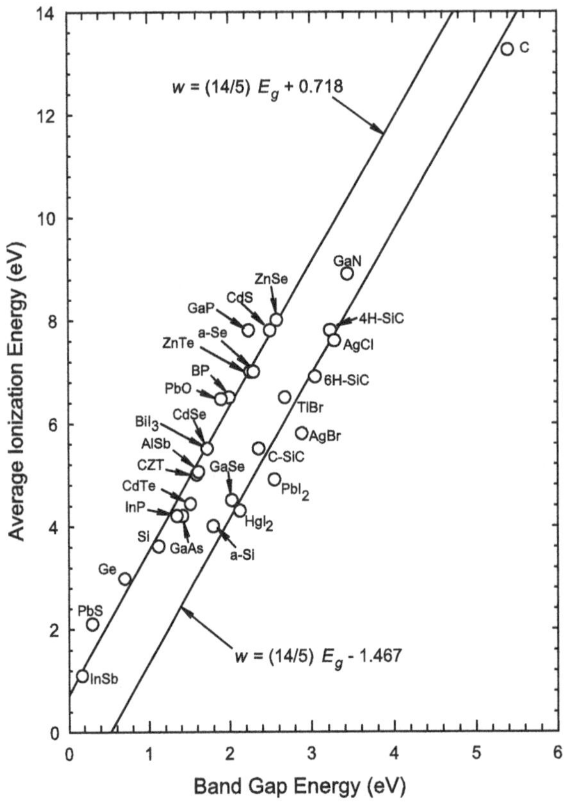

Figure 16.2. The average ionization energy (w) of several semiconductors as a function of the band-gap energy (E_g).

where $\mathcal{L}$ is a dimensionless scaling constant.[6] Klein [1968] assigns a value of $\mathcal{L} = 3/5$, yielding $\langle E_K \rangle = (9/5)E_g$, slightly below the minimum energy required to excite two electrons across the band gap. Klein [1968] thereby produces the following formulation for the average ionization energy:

$$w = \frac{14}{5}E_g + r(\hbar\omega_R). \tag{16.12}$$

Lund et al. [1988] and Shah et al. [1989] note that there appear to be two salient linear dependencies, both following Eq. (16.12) as shown in Fig. 16.2, but having different ordinate intercepts.[7] Using literature values of w, a linear least squares method yields an ordinate intercept of $r(\hbar\omega_R) = 0.718$ for the upper (traditional) dependence. The ordinate intercept for the lower second dependence is found to be $r = -1.467$. The use of $r(\hbar\omega_R) = 0.718$ is within the limits of Klein's theory and therefore has physical significance. However, as pointed out by Owens and Peacock [2004], the use of $r = -1.467$ for the lower dependence does not follow the theory of Klein [1968] and is used strictly as a curve fitting parameter.

[6]This dimensionless scaling constant refers to the average wasted residual kinetic energy described in Van Roosbroeck's [1965] paper on "crazy carpentry".

[7]This discovery seems to have been missed by Alig and Bloom [1975], although it now seems obvious from their data.

16.2.4 Mobility

As discussed in Sec. 12.5, the charge carrier mobility is a strong function of the curvature of the energy bands. Rapid charge carrier extraction reduces trapping and dead time problems and is fundamentally dependent upon the charge carrier drift mobilities. Recall also from Sec. 12.5.1 that the charge carrier mobility is inversely proportional to the charge carrier effective mass, and the effective mass is related to the E-k diagram band curvature by

$$m^* \propto \left[\frac{d^2E}{dk^2}\right]^{-1}. \tag{16.13}$$

The curvature at the minima of the conduction band for most semiconductors is larger than the curvature at the valence band maxima. Consequently, electron mobilities are almost always much larger than the hole mobilities in a semiconductor. A few methods of measuring the charge carrier mobility of both electrons and holes are described in the Sec. 15.4.5.

Also discussed in Sec. 12.5, the mobility tensor sums to a constant value for cubic crystals regardless of the electric field force vector direction. However, this convenient condition does not necessarily hold true for non-cubic lattice systems, and mobilities can be quite different depending on the electric field direction. The mobility symmetry makes cubic crystals generally more desirable for radiation detectors than non-cubic crystals.

Mobility decreases as the probability of charge carriers encountering motion impeding collisions increases. Sources of motion impeding collisions include phonon (lattice) scattering (μ_l), ionized impurity scattering (μ_i), neutral impurity scattering (μ_n), crystal defect scattering (μ_D), ionized deep donor and deep acceptor scattering (μ_{di}), carrier-carrier scattering (μ_{cc}), and piezoelectric scattering (μ_p). Phonon scattering refers to collision events caused from lattice vibrations of thermally excited lattice atoms. Ionized impurity scattering refers to collisions from coulombic interactions between ionized impurities and charge carriers. Neutral defect scattering becomes important at low temperatures and in polycrystalline materials. Piezoelectric scattering is important in piezoelectric materials (for instance, GaAs and CdTe), in which displacement of atoms from their lattice sites produces an internal electric field. Carrier-carrier scattering, in which charge carriers scatter off each other, is important for high carrier densities, which is usually not the case for most radiation detector applications. The resulting mobility is

$$\mu = \left[\frac{1}{\mu_l} + \frac{1}{\mu_i} + \frac{1}{\mu_n} + \frac{1}{\mu_D} + \frac{1}{\mu_{di}} + \frac{1}{\mu_{cc}} + \frac{1}{\mu_p} + \ldots\right]^{-1}. \tag{16.14}$$

Of the many possible scatter mechanisms, the most influential on mobility are phonon scattering and ionized impurity scattering. The mobility dependence of phonon scattering (μ_l) is described by [Bardeen and Shockley 1950]

$$\mu_l = \frac{\sqrt{8\pi} q_e \hbar^4 C_{11}}{3E_{ds}(m^*)^{5/2}(kT)^{3/2}}, \tag{16.15}$$

where C_{11} is the average longitudinal elastic constant of the semiconductor and E_{ds} is the band-edge displacement per unit dilation of the lattice from vibrational motion, and m^* is the charge carrier effective mass. Eq. (16.15) shows that phonon (lattice) scattering decreases mobility with increasing temperature and effective mass. The mobility dependence of ionized impurity scattering is [Conwell and Weisskopf 1950]

$$\mu_i = \frac{64\sqrt{\pi}\epsilon_s^2(2kT)^{3/2}}{N_i q_e^3 (m^*)^{1/2}} \left\{ \ln\left[1 + \left(\frac{12\pi\epsilon_s kT}{q_e^2 (N_i)^{1/3}}\right)^2\right]\right\}^{-1}, \tag{16.16}$$

where N_i is the ionized impurity concentration. From this result it is clear that increasing the impurity concentration reduces mobility, typically becoming significant at ionized impurity densities greater than

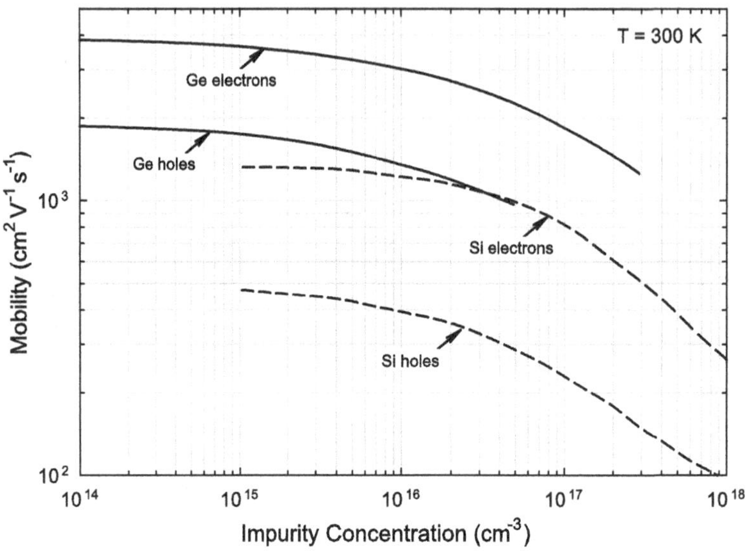

Figure 16.3. The mobility of Si and Ge as a function of impurity concentration at 300 K. Data are from Prince [1953] and Jacoboni et al. [1977].

10^{15} cm^{-3} as can be seen in Fig. 16.3. Notice also from Eq. (16.16) that the mobility from ionized impurities decreases as T decreases. Hence, μ_l decreases with increased T while μ_i increases with increased T. Hence, the overall dependence of mobility on temperature can be approximated by

$$\mu \approx \left[\frac{1}{\mu_l} + \frac{1}{\mu_i}\right]^{-1}. \tag{16.17}$$

The general expression for charge carrier group velocity at relatively low electric fields is

$$v_g\Big|_{\text{low } \mathcal{E}} = \mu\mathcal{E}, \tag{16.18}$$

where the mobility μ is an intrinsic property of the semiconductor. Hence, it becomes clear that large mobilities are desirable for rapid charge carrier extraction from the detector. At high electric fields, the charge carrier group velocity for many indirect band-gap semiconductors can be modeled as [Sze 1981]

$$v_g\Big|_{\text{high } \mathcal{E}} = v_s\left[\frac{1}{1 + (\mathcal{E}_o/\mathcal{E})^\beta}\right]^{1/\beta}, \tag{16.19}$$

where v_s is the saturation velocity at high electric fields and both $\mathcal{E}_o$ and β are experimentally determined constants.

The simple empirical model of Eq. (16.19) does not take into account the crystal orientation, which does in fact affect the charge carrier velocity (see Figs. 16.4 and 16.5). Indeed, under equilibrium conditions the total effective mass tensor is constant regardless of electric field direction; however, at high electric fields the charge carrier distribution in the transverse and longitudinal conduction band energy ellipsoids does change, depending on the force vector of the electric field (Jacoboni et al. [1977]). Consequently, the effective mass tensor also changes, and the charge carrier velocity characteristics are slightly different for electric fields applied in different crystal directions.

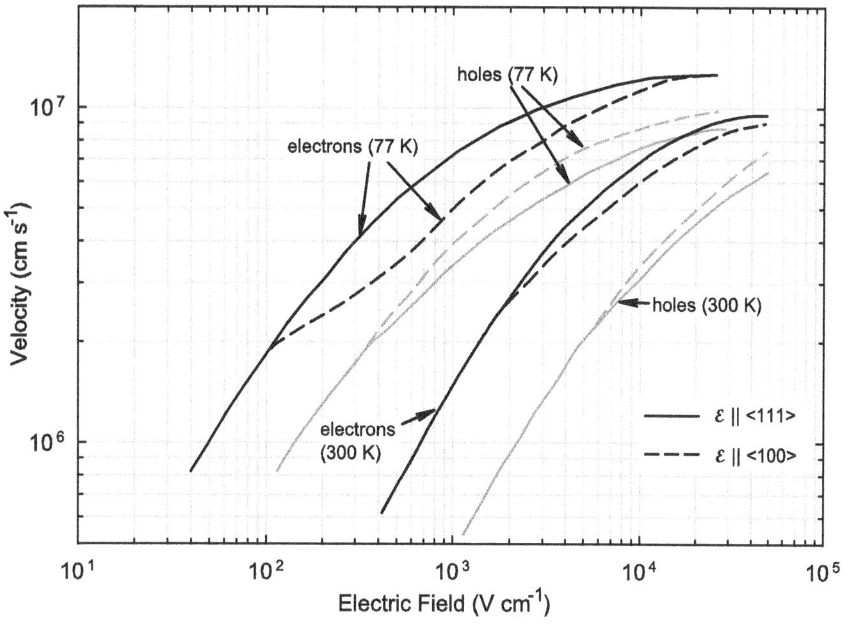

Figure 16.4. The charge carrier velocities of Si as a function electric field and temperature (77 K and 300 K). Data are from Jacoboni et al. [1977].

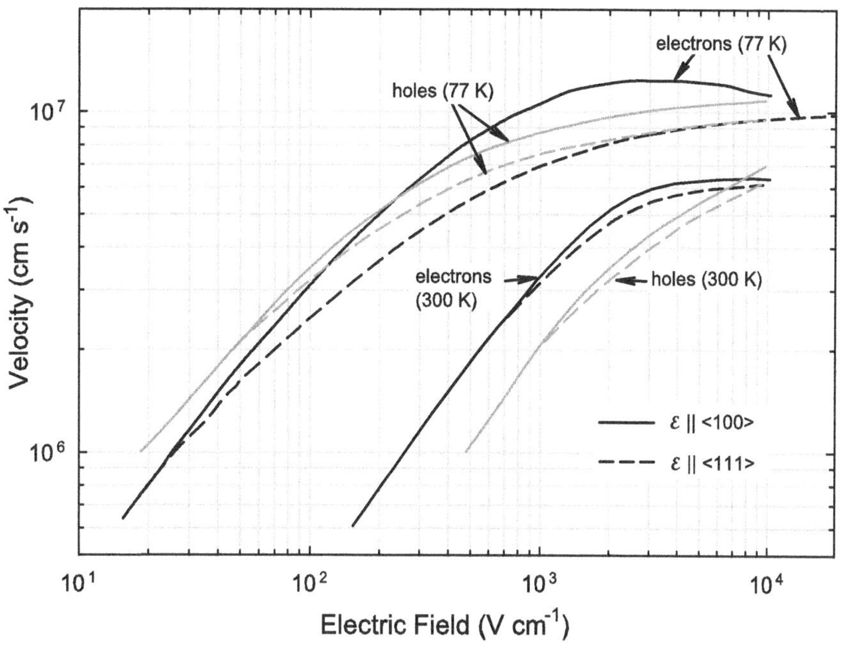

Figure 16.5. The charge carrier velocities of Ge as a function electric field and temperature (77 K and 300 K). Data are from Reggiani [1978] and Jacoboni et al. [1981].

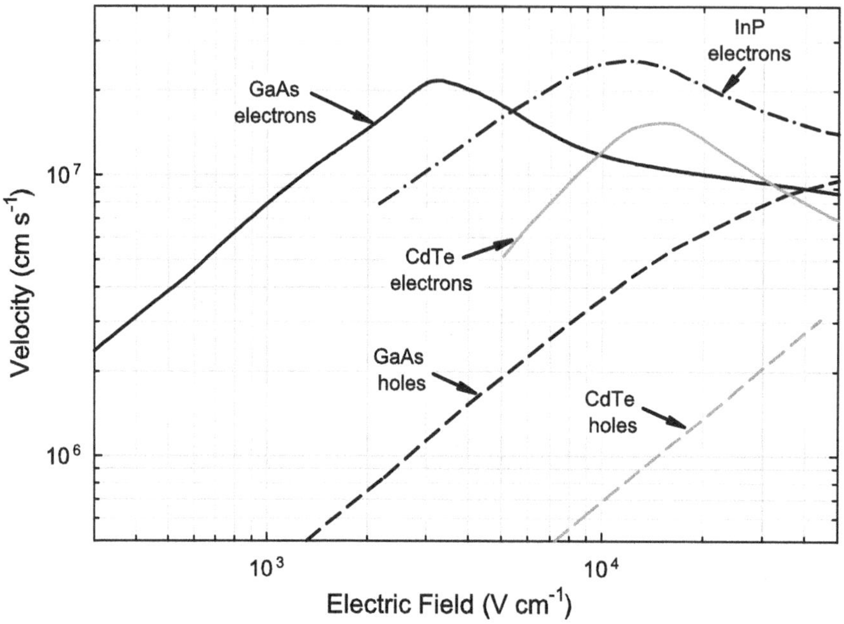

Figure 16.6. The charge carrier velocities of GaAs, InP, and CdTe as a function electric field at 300 K. Data are from Ruch and Kino [1967], Dalal et al. [1971], Canali et al. [1971a, 1971b], Smith et al. [1980], Blakemore [1982], and González Sánchez et al. [1992].

There are a few semiconductors in which more than one conduction band minima can play a part in electron transport. For instance, GaAs, InP, and CdTe are direct band-gap semiconductors, yet also have indirect band-gaps that can accept electrons at high enough electric fields, and these indirect band gaps have larger electron effective masses than their corresponding direct band gaps. Consequently, the high electric field velocity for direct band-gap semiconductors, in which electron transfer to upper indirect bands is possible, is much more complicated than given by Eq. (16.19) (see, for example, Pozela and Reklaitis [1980]), and must take into account the change in effective mass for the upper valley minima. Hence, at low voltages, the velocity is defined by the direct band-gap minimum, at medium voltages the velocity is defined by a sharing of mobilities between the direct and indirect band-gap minima, and at high electric fields the mobility is determined by one or more indirect band-gap minima. The sharing of electron mobility between conduction band minima produces a region of *negative differential resistance* [Sze 1981], in which the electron velocity and, consequently, the electrical current decrease with increased applied voltage (see Fig. 16.6). A side effect of the negative differential resistance is to produce an instability, the Gunn effect, that can produce oscillations at high voltage [Gunn 1963].

16.2.5 Resistivity

High resistivity for a material is important to reduce thermally generated current and other forms of leakage current. For pure materials, the resistivity is mainly a function of band-gap energy. Examples include high-purity Ge and float-zone refined Si. However, with a few exceptions, the purification of most compound semiconductors is difficult. Consequently, high resistivity compound semiconductors are usually produced through compensation methods (as outlined in Sec. 12.5.4). The natural resistance of a semiconductor block can be described by

$$R = \frac{\rho L}{A} = \frac{L}{q_e A(n\mu_n + p\mu_p)}, \tag{16.20}$$

where L is the length of the block, A is the area of each electrical contact at the ends, n is the free electron concentration, p is the free hole concentration, μ_n is the electron mobility, and μ_p is the hole mobility. An impulse current should be significantly higher than the background leakage current.

The type of detector contacts required is strongly influenced by the material resistivity. If the resistivity is $\gtrsim 5 \times 10^{10}$ Ω-cm, a resistive device, or photoconductor, can be fashioned. If the resistivity is below 5×10^{10} Ω-cm, it is usually best to construct a junction diode, as described in Sec. 15.3.1 (and in Sections 15.3.2 and 15.3.3), to suppress electronic noise from leakage currents. There are methods to increase the resistivity of a semiconductor with compensation methods, but there are physical consequences that can arise from the operation of junction diodes fabricated from compensated high-resistivity materials and are addressed later in this chapter.

16.2.6 Mean Free Drift Time

The recombination time τ_r is the average lifetime of an electron before it recombines with a hole, whereas the mean free drift time τ_m is the average time a charge carrier is in motion before it encounters a physical impediment that stops its motion. In a pure defect-free semiconductor, the recombination time is essentially the same as the mean free drift time. However, they are not the same in most semiconductors, mainly because recombination is usually assisted by transitions through band-gap defect states.

The lifetime of a charge carrier is largely a function of the band-gap energy and defect density. Hall [1959] and Varshni [1967] show that the recombination time for indirect band-gap semiconductors can be significantly longer than that observed for direct band-gap semiconductors. The fundamental reason for the difference is electrons must lose energy and change momentum to transcend the band gap to recombine with a hole; however, electrons in a direct band gap need only to lose energy to recombine with a hole. The recombination times predicted by either Hall [1959] or Varshni [1967] are relatively long by comparison to observed values. This difference is explained by the theory of Shockley and Read [1952] and Hall [1952], which attributes recombination mainly to the trap density in a semiconductor.

Ultimately, it is the mean free drift time that affects the performance of a semiconductor detector, and it is τ_m that is usually measured to determine the often quoted $\mu\tau$ product. Recall from Sec. 15.5 that charge carrier collection is directly related to the $\mu\tau$ product, in which the carrier extraction factor for a planar detector is given by Eq. (15.221), namely,

$$\varrho = \frac{\mu\tau_m \mathcal{E}(x)}{W}, \tag{16.21}$$

where W is the detector width. Sufficiently good energy resolution is observed if both carrier extraction factors are greater than 50. The value of ϱ for holes and electrons can be increased by reducing W or increasing $\mathcal{E}(x)$. However, τ_m is an intrinsic property of the material, increased only by improving the crystal quality. Semiconductor materials with relatively large mean free drift times are generally sought for semiconductor spectrometers.

16.2.7 Linearity

The process by which information carriers are produced in semiconductors differs from scintillators and gas detectors. In Sec. 12.3 it was shown how the electron energy states in solids form a quasi-continuum of energy bands and energy gaps. As discussed in Chapter 4, charged particles interact in a medium primarily through Coulombic interactions while photons may interact by several different processes, usually the photoelectric effect, Compton scattering, or pair production (if $E_\gamma > 1.022$ MeV). These initial electrons excited from

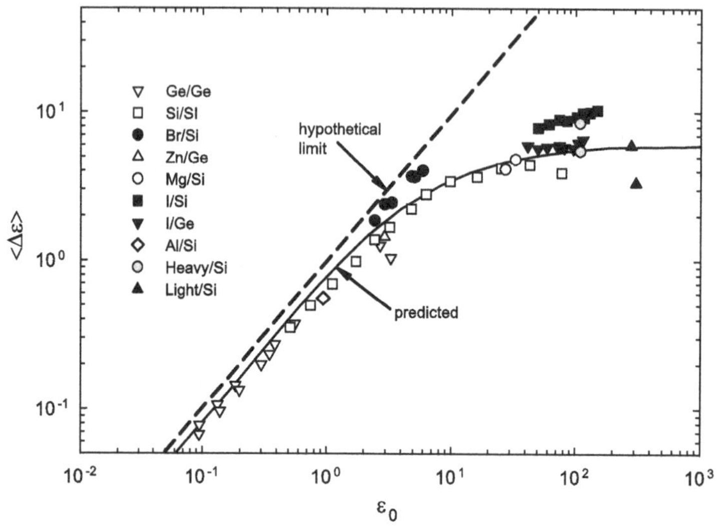

Figure 16.7. Shown are dimensionless measures of the pulse height defect ($\langle \Delta\epsilon \rangle$) vs. dimensionless energy (ϵ_0) of heavy ions absorbed in Si and Ge. The dotted line is hypothetical limit of the expected charge production, where $\langle \Delta\epsilon \rangle = \epsilon_0$. Data are from Haines and Whitehead [1966].

lower bands are either removed from the nucleus (ionization) transferred to the upper bands (excitation), and electrons with sufficient kinetic energy can ionize more free electrons (delta rays) that further excite electrons. This process is common to both semiconductors and scintillators.

Scintillators rely on electrons transitioning through energy states to produce photons subsequently detected by a light sensitive device. However, in a semiconductor, the free electrons are measured directly by the current they produce under an applied voltage, much like a gas-filled detector. For mature semiconductors with relatively long charge carrier mean free pathlengths, and large $\mu\tau$ products, the measured current or charge is a linear function of energy deposition. Such is not the case for scintillators for which phosphor saturation (organic scintillators), exciton losses (inorganics), or other competing mechanisms lead to non-linear responses with respect to energy and particle type. Overall, the average number of electron-hole pairs produced per unit energy in a semiconductor is remarkably linear over a wide energy range.

However, there are some situations that can produce a non-linear output from a semiconductor. For instance, the active thickness of the detector must be capable of absorbing the total energy deposited by an ionizing particle. If the ionizing particle range extends beyond the active region range, then some energy is lost and the resulting signal is diminished. The probability of this form of non-linearity increases with particle energy. For low-energy charged particles, the dead layer surrounding a detector absorbs energy, resulting in energy loss and some amount of energy non-linearity in the output. Charge carrier trapping can also cause the appearance of a non-linear spectrum. Although Si and Ge detectors tend not to suffer much from trapping effects, compound semiconductors almost always do. Hence, it is common to irradiate the cathode of compound semiconductors in order to emphasize electron charge induction over hole charge induction. Still, the average charge *measured* per unit energy deposited reduces with increasing particle or photon energy, typically degrading the energy resolution. For heavy ions, the dense charge cloud produced can cause appreciable recombination, which can also contribute to non-linear output. Recombination problems can be reduced by increasing the detector operating voltage.

Also, heavy ions can lose energy by direct nuclear collisions, which do not produce electron-hole pairs [Lindhard and Nielsen 1962; Haines and Whitehead 1966]. Consequently, there can be appreciable non-linearity in the number of electron-hole pairs produced from heavy low-energy charged particles. The average dimensionless energy losses $\Delta\epsilon$ as a function of dimensionless energy ϵ_0 for a few heavy ions interacting in either Si or Ge are shown in Fig. 16.7. The greatest fraction of energy loss occurs at ϵ_0 values less than 10, corresponding to approximately 6 MeV for I ions, 3 MeV for Br ions, and 400 keV for Si ions interacting in Si [Haines and Whitehead 1966]. The dashed line indicates the situation in which all energy is lost to nuclear non-ionizing recoil events. Below $\epsilon_0 = 5$ nearly half of the ion energy is lost to recoils, and the situation asymptotically approaches the hypothetical limit with decreasing ϵ_0.

16.3 Semiconductor Detector Applications

Briefly described are common radiation detection applications that employ the use, at times, for semiconductor radiation detectors. In most cases, these applications exploit the low ionization energy of semiconductors to produce radiation spectrometers, and semiconductors are used to detect charged particles, energetic photons, and neutrons. Some systems use the semiconductor properties to make electronic devices and arrays into compact low-power detectors. Some details on specific types of semiconductor detectors are provided as later sections.

16.3.1 Charged Particle Detectors

Semiconductor charged particle spectrometers offer high energy resolution for energetic ions. Typically, the devices are operated in vacuum, along with the source, to eliminate energy losses from a charged particle as its passes from the source to the detector. The detectors are typically designed with thin contacts and/or thin *pn* junctions in order to reduce particle energy loss in the non-sensitive (or dead) region of the contact. Because low Z elements have less problems with ion backscattering, Si is typically the material choice for particle detectors. These detectors can be used for a variety of charged particle identification and characterization purposes, including high resolution spectroscopy of α particles, β particles, protons, and heavy ions, continuous air monitoring, and particle telescopes.

Spectroscopy requires that the full energy of the particle be deposited in the active region of the detector, hence charged particle spectrometers are constructed in such a manner that the depleted region extends beyond the range of the particle under investigation. For most alpha particle spectroscopy applications, a depletion width of 100 microns is adequate. Proton ranges are much longer than those of alpha particles of the same energies, and special detectors that can achieve wider depletion widths may be required. Electrons generally have longer ranges in semiconductors than heavy ions, but also have higher backscatter coefficients.

There are some particle detectors designed for the detection of minimum ionizing particles (MIPs). Note that by definition MIPs can be β particles, protons or any other ionizing charged particles, provided that their kinetic energy is at least twice their rest mass energy. For electrons and β particles, they are classified as MIPs if their kinetic energy exceeds 1.022 MeV. For protons, they must exceed 1.88 GeV energy to be classified as MIPs.

β particles are emitted from a radioactive source with a continuous energy spectrum (see, for example, Fig. 3.20) and, thus, do not have discrete energy peaks in the pulse height spectrum. Further backscattering and self-shielding can often distort the measured spectrum. Because MIPs tend to deposit an even distribution of energy with distance (dE/dx is constant), detectors for MIPs are designed to detect a fraction of the particle energy (ΔE) rather than the total energy. Hence, they are commonly designed to absorb enough energy to identify a MIP, but usually not all of the MIP energy.

16.3.2 Gamma-Ray and X-Ray Detectors

Semiconductors excel above gas-filled and scintillators as energy spectrometers. This nearly universal property is a consequence of the lower ionization energy that allows the excitation of a greater number of charge carriers (electron-hole pairs) per unit energy than in gas-filled or scintillation detectors. By pure Gaussian statistics, the fractional standard deviation is smaller for semiconductors than other detectors. Further, with few exceptions, the Fano factor is smaller for semiconductors than gases and scintillators. Recall from Sec. 6.8.1 that the FWHM of a Gaussian distribution is

$$\text{FWHM} = 2\sqrt{2\ln(2)\sigma^2} \simeq 2.355\sigma, \tag{16.22}$$

and conversion of this result to the number of charge pairs is done by inserting the Fano factor correction so that the energy resolution is

$$\text{FWHM} = 2\sqrt{\frac{2\ln(2)E_\gamma F}{w}}. \tag{16.23}$$

This number is easily converted to units of energy by multiplying Eq. (16.23) by the semiconductor average ionization energy (w), so that

$$\text{FWHM(energy)} = 2\sqrt{2\ln(2)E_\gamma wF} \simeq 2.355\sqrt{E_\gamma wF}. \tag{16.24}$$

Indeed, the standard for reporting energy resolution for semiconductors is in terms of energy, commonly in units of keV. This standard is different for scintillators, in which the FWHM is usually reported as a percent of the gamma-ray energy. The main reason for the change from the traditional standard used for scintillators is the exceptional energy resolution of semiconductors, commonly sub-percent for Si and Ge semiconductors.[8] Note the energy resolution of a semiconductor detector is expected to increase with the square root of the gamma-ray energy, a dependence that holds relatively well for Si and Ge detectors. However, trapping effects cause the full energy peak to appear non-Gaussian for most compound semiconductors, and Eq. (16.24) is usually not a good predictor of the FWHM.

Semiconductor gamma-ray detectors are usually smaller than their scintillator counterparts, and it is common for a semiconductor detector spectral response to have a relatively high Compton continuum in the spectrum. This problem has been addressed by manufacturing large semiconductor detectors (there are rather sizeable Ge detectors available in the commercial market for high cost), or by producing detectors from large Z materials to improve gamma-ray absorption. However, the improved energy resolution typical of semiconductor detectors is usually enough to permit easy identification of gamma-ray full energy peaks and any additional spectral features (such as escape peaks).

The compactness of semiconductors accompanied by the generally high energy resolution makes them attractive radiation detectors. Further, they do not have a secondary detection device, such as a photomultiplier tube or a SiPM array required to properly operate a scintillation detector. Overall, scintillators are often used when modest energy resolution is acceptable and semiconductors are used when high energy resolution is most important. Note also that semiconductor devices can be fashioned into arrays and electronic readout that can identify the location of where the interaction occured in the detector volume, an advantage not easily duplicated with scintillation detectors. These advantages allow the production of directionally sensitive gamma-ray detectors, either as physical arrays composed of many individual detectors or as pixelated devices with sophisticated electronics that localize events in a semiconductor block. A variety of semiconductor gamma-ray detectors, commercial and experimental, fabricated from different materials are described throughout the remainder of this chapter.

---

[8]In recent times, energy resolution for compound semiconductor detectors, such as CdZnTe and HgI_2, is often reported in terms of percent. However, this departure from an established system is not strictly correct, and is an unfortunate consequence of the deleterious effect that crystal imperfections and charge carrier trapping have on energy resolution, causing many compound semiconductors to perform with similar energy resolution as the best scintillators.

16.3.3 Neutron Detectors

Semiconductor radiation detectors used as neutron detectors are typically configured as *pn*-junction or Schottky-junction diodes coated with a neutron reactive material. The basic construction of such a detector is shown in Fig. 16.8, in which a Schottky or *pn* junction diode detector has a coating of neutron reactive material applied to the surface. Typically the devices have either ^{10}B or ^{6}LiF as the active coating. The absorption cross sections for both ^{10}B and ^{6}Li follow a $1/v$ dependence. The ^{10}B(n,α)^{7}Li neutron reaction yields two possible de-excitation branches from the excited ^{11}B compound nucleus, namely

$$^1_0\text{n}+^{10}_{5}\text{B} \longrightarrow \begin{cases} ^4_2\text{He}\,(1.4721\text{ MeV}) + ^7_3\text{Li}^*\,(0.8398\text{ MeV}) & [93.7\%] \\ ^4_2\text{He}\,(1.7762\text{ MeV}) + ^7_3\text{Li}\,(1.0133\text{ MeV}) & [6.3\%], \end{cases}$$

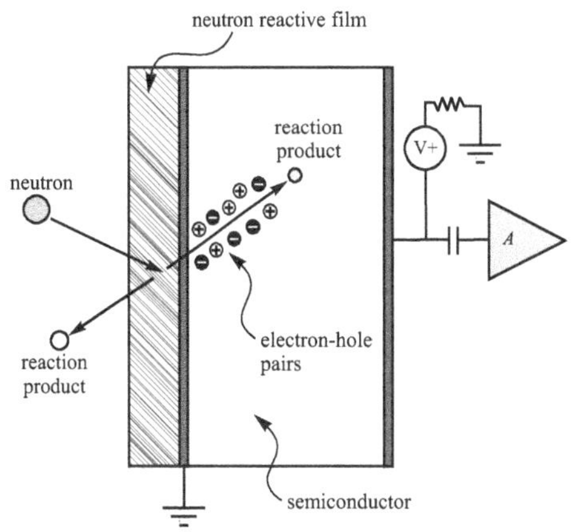

Figure 16.8. Cross section of a coated semiconductor neutron detector.

where the Li ion in the 94% branch is ejected in an excited state, which deexcites through the emission of a 480-keV gamma ray. Fully enriched ^{10}B has a microscopic absorption cross section for thermal neutrons (2200 m/s) of 3840 barns. With a mass density of 2.15 g cm^{-3}, the solid structure of ^{10}B has a macroscopic thermal absorption cross section of 500 cm^{-1}.

The ^{6}Li(n,t)^{4}He neutron reaction yields a single product branch,

$$^1_0\text{n} + ^6_3\text{Li} \longrightarrow ^3_1\text{H}\,(2.7276\text{ MeV}) + ^4_2\text{He}\,(2.0553\text{ MeV}).$$

The reaction products from the ^{6}Li(n,t)^{4}He reaction are more energetic than those of the ^{10}B(n,α)^{7}Li reaction and, hence, are much easier to detect and discriminate from background radiations. ^{6}Li has a relatively large microscopic thermal neutron absorption cross section of 940 b, although it is less than that of ^{10}B. Unfortunately, Li is a chemically reactive metal, and, therefore, it is the stable compound ^{6}LiF, with a 2200-m/s macroscopic cross section of 57.51 cm^{-1}, that is usually used as the reactive coating.

For thermal neutrons, the charged particle reaction products are ejected in opposite directions, meaning that only one reaction product can actually enter the semiconductor detector. Further, the reaction products lose energy as they pass through the neutron absorber to the semiconductor detector, thereby limiting the effective absorber thickness. Detectors of this type are limited to less than 5% thermal neutron detection efficiency. These devices are generally not commercially available as independent units; rather they are sold as an active component inside some electronic dosimeter modules. Advanced versions of coated semiconductor detectors, having etched features backfilled with neutron reactive materials, have reported efficiencies greater than 65%. A detailed discussion on these advanced coated-semiconductor neutron detectors is in Chapter 17.

There have been attempts to use semiconductors in which at least one type of constituent atom is neutron reactive. Semiconductors with neutron reactive isotopes that promptly emit charged-particle reaction products include B-based semiconductors (such as BC_4, BN, BP, and BAs) and Li-based semiconductors (such as LiZnP, LiZnAs, and $LiInSe_2$). Semiconductors with neutron reactive isotopes that promptly emit gamma-ray reaction products include Cd-based semiconductors (such as CdTe and CdZnTe) and Hg-based semiconductors (such as HgI_2). The concept is simple; neutrons interact with the reaction isotopes that promptly emit ionizing radiation back into the semiconductor. The ionizing radiation produces electron-hole pairs in the semiconductor, and these charges are drifted with an electric field to produce an induced current or induced voltage on an electronic output. To date, these types of neutron detectors have been

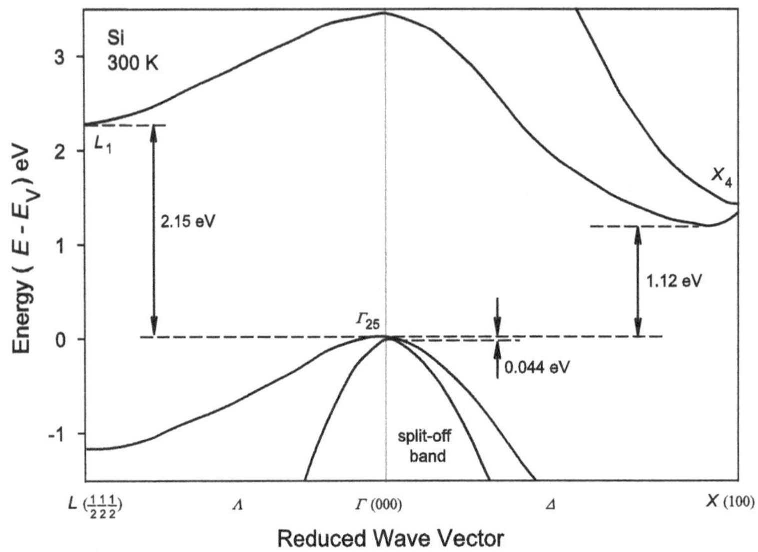

Figure 16.9. The theoretical energy bands of Si at 300 K calculated with a pseudopotential method. Data are from Chelikowsky and Cohen [1976].

largely unsuccessful. Regardless, a section in Chapter 17 is reserved for discussion on these types of neutron detectors.

16.4 Detectors Based on Group IV Materials

The family of semiconductor materials based on group IV materials include Si, Ge, diamond, and SiC. Both Si and Ge detector processing are mature technologies, having been developed as semiconductors for electron devices for over 50 years. By far, most semiconductor radiation detectors are fabricated from either high-purity Si or Ge, although there is a modest market for other semiconductor detector materials. Diamond is also used as a radiation detector material, but its low value of Z diminishes its effectiveness as a gamma-ray detector. Hence, diamond is mainly explored for used as a charged particle detector. SiC is also a group IV material, but it is also a compound semiconductor, which can suffer from intrinsic defects (antisites, for instance) typically not a problem with Si, Ge, or C (diamond) crystals. The processes by which diamond and SiC are produced increase the material costs, hence these materials are used only in cases in which alternative semiconductors are inadequate. For instance, both diamond and SiC can be used in higher temperature environments than Si or Ge, and both diamond and SiC tend to have higher radiation hardness than either Si or Ge. Basic detectors fashioned from these semiconductor materials are described in following subsections.

16.4.1 Detectors Based on Silicon

Silicon is by far the most studied of the semiconductor materials for electronic devices, and it is popular for its many beneficial physical and electronic properties. As can be seen from Fig. 16.9, the energy of the band gap for Si is 1.12 eV, which is just wide enough for room temperature operation of small volume devices. Further, the band gap is indirect, and charge carrier mean free drift times are relatively long for highly purified material. The density of Si is only 2.33 g cm^{-3} and it has an atomic number of 14, low for most gamma-ray detector applications, but still useful as an x-ray and low-energy gamma-ray detector. The average ionization energy of Si is 3.6 eV per e-h pair with 0.115 as the Fano factor. The backscatter coefficient of Si is relatively small, making it a good choice for charged particle spectrometers. It has a native oxide (SiO_2) that is robust, stable, and electrically and thermally resistant. The production of thicker SiO_2 layers is straightforward, easily grown through 'dry' and 'wet' processes at high temperature (> 950°C) in a tube furnace. The crystal has a cubic FCC lattice structure.[9] Recalling from Sec. 12.5, the average charge carrier mobility is practically constant regardless of charge carrier direction. It is easily etched by wet and dry processes, which include anisotropic and isotropic methods. Doping is straightforward, and can be performed with thermal diffusion and ion implantation. Because of its high melting point of 1414°C, activation of these dopants is also straightforward by many methods, including rapid thermal annealing and furnace annealing. Si is notably a material that can be doped through neutron activation, mainly because

$$^{30}\text{Si} + n \rightarrow {}^{31}\text{Si}\ (t_{1/2} = 2.62\ \text{hrs}) \rightarrow {}^{31}\text{P} \tag{16.25}$$

reactions lead to the production of stable phosphorus, an n-type dopant in Si. Uniformity is ensured by the low thermal neutron absorption cross section of only 0.171 barns. Si is a relatively tough material, with a Mohs hardness of 7, and it has high tensile strength. Si crystalline material can be acquired as large diameter ingots and sizeable prepolished wafers. Electronic grade Si wafers can be acquired as 12-inch diameter wafers or even greater for special cases. High purity float-zone-refined Si wafers of 6-inch diameter can also be acquired commercially.

Charged Particle Detectors Based on Si

Charged particle detectors are designed with thin contact regions to minimize energy loss in the electrical contact. Such energy loss can increase the variance on the measured energy, while reducing the total integrated energy deposited within the detector active region. For semiconductors with small band gaps, such as Si and Ge, thin rectifying contacts are fabricated as either Schottky barriers and shallow pn junctions. The Schottky barrier design is perhaps one of the easiest semiconductor detectors to fabricate, requiring a clean semiconductor surface upon which are applied select metals to form a Schottky barrier. These detectors are commonly referred to as *surface barrier detectors.*[10] Shallow pn junctions, prepared by diffusion or ion implantation, require more process steps, but tend to be more robust with lower leakage currents than SSB detectors.

The most commonly used semiconductor for charged particle detection is Si. Si possesses many physical attributes that make it attractive as the select detector material, including low Z to reduce backscatter, a strong and easy to grow high resistivity oxide layer, a relatively strong crystal structure with high tensile strength, and chemical stability that enables a relatively straightforward fabrication of detectors. Further, Si has an indirect band gap of 1.12 eV that produces long charge carrier lifetimes and allows room temperature operation for detectors of conservative size.

Silicon surface barrier (SSB) detectors were amongst the earliest versions of practical semiconductor radiation detectors, introduced in 1958 by Davis as alpha particle detectors. The simplistic construction requires only a pristine etched Si surface upon which a gold layer is evaporated, forming a rectifying Schottky barrier.

An etch commonly used for SSB detectors is CP4A,[11] because it produces a pristine mirror finish on a Si surface [Dearnaley and Northrop 1966]. High purity n-type or p-type Si is etched, mounted and epoxied

[9] Often referred to as a 'diamond lattice', this identification is not one of the Bravais lattices. A diamond structure has two identical atoms, a part of a tetrahedron, on each FCC lattice point.

[10] The origin of "surface barrier" as a detector name is not exactly known by the authors; however, the earliest reference discovered that uses this name is the 1955 Ph.D. dissertation by Simon.

[11] CP4A is a 3:5:3 mixture of HF(48%):HNO_3(73%):CH_3COOH(100%) and must be kept in a plastic container (not glass). The formula rapidly etches Si, and must be cooled (such as in an ice bath) to control the etch rate. This etch is excellent for producing home-made SSB detectors, and leaves a pristine mirror finish if the samples are agitated during the etch process.

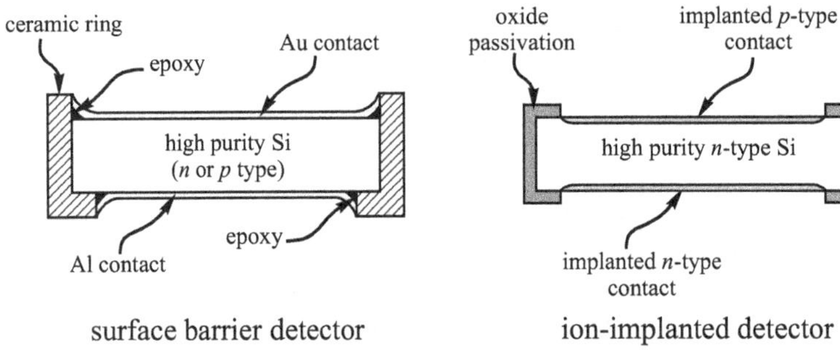

Figure 16.10. General configurations for Si surface barrier detectors and implanted junction detectors.

into a ceramic ring. The samples are bonded into insulating collars with an amine-free epoxy and desiccated, followed by evaporating a metal layer ranging from 80-200 nm thick on opposite surfaces, commonly Au on one side and Al on the other. The thin contact region minimizes the amount of energy lost by particles that enter the device, a necessary precaution to preserve high energy resolution.

At high voltages, it is common to have electrical breakdown at the perimeter of the diode, where the metal crosses from the epoxy bead to the semiconductor surface (see Fig. 16.10). Introduction of an iodine-doped amine epoxy around the perimeter atop the amine-free epoxy bead on the Au side (before evaporation) can produce a reverse biased surface diode, which mitigates electrical breakdown [Gibbs 1988]. Addition of a guard ring helps with the reduction of leakage current noise [Dearnaley and Northrop 1966]. Environmental conditions can affect the performance of SSB detectors, mainly the ambient temperature and humidity. Modernized process methods help mitigate these problems, and include improved surface preparation methods and the addition of a perimeter oxide around Si sample [Fretwurst et al. 1990].

These delicate surface barrier detectors can be easily damaged by improper handling and are often difficult to clean. Detectors can be obtained in a variety of sizes, ranging from a few mm to 50 mm diameter. These detectors are usually light sensitive and must be operated in darkness, although commercial companies do offer versions that can operate in ambient light, at the expense of energy resolution. Depletion depths range from approximately 100 microns up to, for special cases, 5 mm. A simple nomogram representing the depletion width as a function of voltage and impurity concentration can be produced by linearizing Eq. (15.61) [Allcock et al. 1963; Blankenship and Borkowski 1960],

$$\log_{10}(W) = \frac{1}{2}\log_{10}\left[\left(\frac{2\epsilon_s}{q_e}\right)\left(\frac{V_{bi}-V}{N_b}\right)\right]$$

or

$$\log_{10}(W) - \frac{1}{2}\log_{10}(2\epsilon_s\mu) = \frac{1}{2}\left[\log_{10}(\rho) + \log_{10}(V_{bi}-V)\right], \tag{16.26}$$

where N_b has been replaced with $1/q_e\mu\rho$ and the applied voltage V is negative (reverse bias). With the properties for Si listed in Table 16.4, such a nomogram was developed for a silicon surface barrier detector, shown in Fig. 16.11, in which the depletion width is a function of material resistivity (n-type and p-type).

Implanted junction detectors rely upon an abrupt junction pn diode for rectification (see Fig. 16.10). These devices are often fabricated from high purity n-type Si. An oxide is grown on the devices for passivation, followed by etching windows back to the Si surface. Shallow p-type and n-type dopants are implanted on opposite sides of the Si surface and thermally activated. An implanted junction of dopants, usually boron into n-type Si, initially produces a Gaussian distribution displaced from the semiconductor surface by the average ion range. These implants must then be electrically activated, usually through an annealing process,

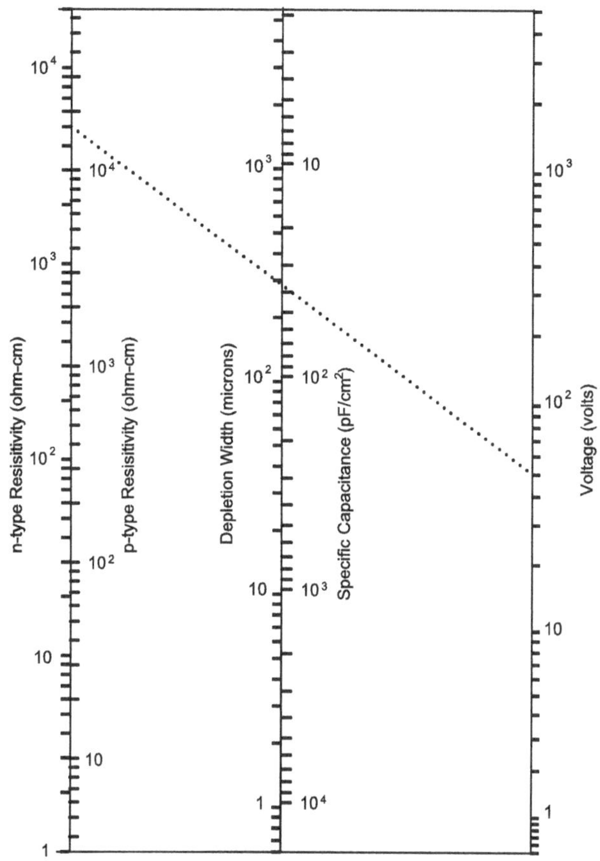

Figure 16.11. The depletion region as a function of material resistivity and reverse voltage for Si particle detectors. A straight line will yield material resistivity (n or p type), the depletion depth and applied voltage, according to Eqs. (15.61) and (15.64). The example shown with the dotted line is for 5 kΩ n-type material at 50 volts reverse bias, which yields W = 282 microns.

which serves to also redistribute the dopants. It is typical to model the final doping distribution as an error function (see Sec. 6.8.2). This implantation process produces a dead layer junction with a thicknesses of about 50 nm that serves as the conductive contact so that there is no need for a metal overlayer. Because there is no thin metalization layer over the detector, implanted junction detectors are more robust and easier to clean than common SSB detectors. Implanted junction detectors can be used for the same basic detection functions that SSB detectors are used. SSB and implanted junction detectors can be acquired in numerous shapes, sizes and configurations, making them a versatile choice for particle detection and spectroscopy. Further, the detectors can be acquired as multi-element arrays for position sensing and timing purposes. The adaptation of very large scale integration (VLSI) processing technology to Si detectors allows detector arrays to be fabricated in a vast number of detector designs, including custom devices contracted to commercial vendors.

Both SSB and shallow junction pn diodes operate practically the same. Both rely on a reverse biased rectifying junction to reduce leakage current, and the active region is a function of applied reverse voltage and impurity concentration, as described in Sec. 15.3.1. In both designs, the rectifying contact is insensitive

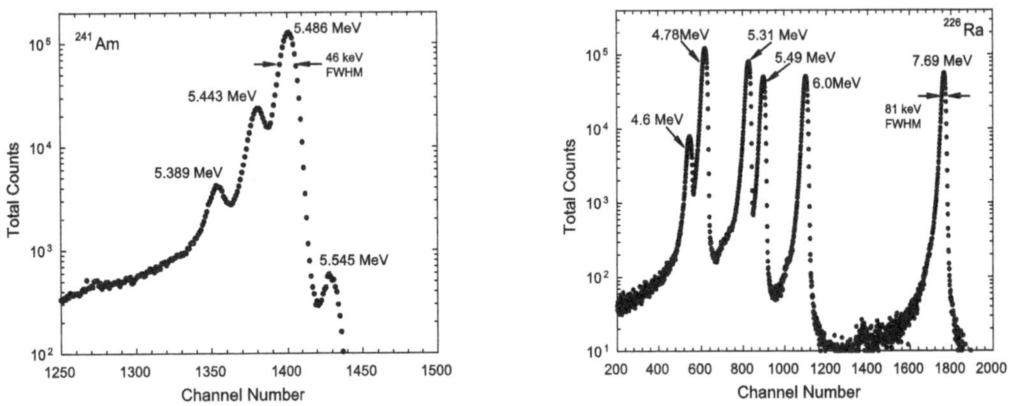

Figure 16.12. Room temperature alpha-particle differential pulse height spectra of spectroscopic grade ^{241}Am and ^{226}Ra sources taken with a 24-mm diameter implanted-junction Si detector operated in vacuum.

to ionizing radiation, considered to be a "dead region". Modern commercial implanted junction detectors have slightly less dead region thickness, an indication, in principle, that they can achieve higher energy resolution than an SSB detector. However, there are many factors that affect the overall energy resolution and include detector area, reverse bias voltage, background impurities, leakage current, thickness uniformity, and crystal orientation. Example alpha-particle spectra from a common commercial implanted-junction Si detector are shown in Fig. 16.12.

The area of a charged particle detector, whether an SSB detector or ion implanted detector, directly affects the capacitance as well as the depletion depth. Capacitance increases linearly with detector area and decreases with depletion width. Consequently, high-purity float zone refined Si is required to achieve large depletion depths at manageable operating voltages, as indicated by the nomogram of Fig. 16.11. The reverse bias leakage current generally increases with applied voltage. Recall from Sec. 15.3 that the leakage current is smaller for a *pn* junction than that of a Schottky barrier, the former being a function of the minority carrier injection condition and the latter being a function of thermionic emission over a potential barrier. Hence, it is expected that for a given detector area the implanted junction detector has a lower leakage current than the SSB detector.

A type of Si particle detector, typically fashioned as an SSB detector, is a transmission device that allows particles to transport all the way through the detector and out the opposite side. These ΔE detectors measure only a fraction of the energy of a charged particle, and because the total energy deposited is a function of the detector mass encountered, flatness and uniformity are important. The resolution of the energy peak from a transmission device is largely dependent upon the total planar flatness of the detector. For instance, a surface deviation of only 1% causes a similar deviation in energy resolution. Transmission detectors can be acquired from commercial vendors with thicknesses less than 20 microns to over one or two millimeters, and are designed to operate under full depletion conditions. Transmission detectors are used for particle identification, time-of-flight measurements, backscatter experiments, and particle telescopes.

Example 16.1:

Assume that you have an SSB transmission detector with average thickness of 15.5 microns, and that it has a thickness deviation from 14 to 15 microns across its thickness. The alpha particle source under investigation is ^{241}Am, measured in vacuum. Assume that the average energy loss for an alpha particle through 15.5 microns of Si is 15.5 eV/Å. How does this thickness deviation affect the energy resolution of that transmission peak?

Solution:

The total energy deposited in the detector is,

$$E \simeq 15.5\ \mu\text{m}\left(\frac{15.5\ \text{eV}}{\text{Å}}\right)\left(\frac{1\ \text{Å}}{10^{-4}\ \mu\text{m}}\right) = 2.403\ \text{MeV}.$$

The total energy spread from the thickness variation is

$$\Delta E \simeq 1\ \mu\text{m}\left(\frac{15.5\ \text{eV}}{\text{Å}}\right)\left(\frac{1\ \text{Å}}{10^{-4}\ \mu\text{m}}\right) = 155\ \text{keV},$$

which is a 6.45% difference.

The crystal orientation is also important. The lattice of a semiconductor forms channels that ions can travel through without suffering ionic collisions, an effect named *channeling*. For cubic semiconductors, channels form as hexagons in the {110} directions, rhombuses (squares) in the {100} directions, and squares in the {111} directions. Ions that enter into the surface below a critical angle with respect to normal incidence, usually 6° or less, scatter down these channels so they penetrate deeply into the semiconductor. To prevent a wide doping distribution depth from implantation, Si wafers are usually oriented beyond this critical angle, and often 7° to 11° from normal is adequate. Otherwise, the implanted dopant can travel much further into the semiconductor, thus causing a much thicker dead layer. The same is true for particle detection. Consequently, SSB and implanted-junction detectors are commonly made with the flat surface polished at an off angle to conventional [100], [111], and [110] directions. Performance data for implanted and SSB Si detectors is given in Table 16.5.

Position Sensitive Particle Detectors Position sensitive detectors are devices that have the capability of simultaneously sensing the absorbed energy and the location of the ionizing event. Semiconductor detectors

Table 16.5. Typical alpha-particle energy resolution of a few representative commercially available semiconductor detectors. Many other sizes are offered by the different vendors.

detector	area (mm^2)	energy (keV)	FWHM (keV)	comments	source*
implanted Si diode	100	5486 5486	13 12	$W = 100\ \mu$m $W = 500\ \mu$m	C,O
implanted Si diode	450	5486 5486	17–21 15–19	$W = 100\ \mu$m $W = 500\ \mu$m	C,O
implanted Si diode	900	5486 5486	27–33 22–28	$W = 100\ \mu$m $W = 500\ \mu$m	C,O
p-type SSB	50	5486 5486	15–17 15–17	$W = 100\ \mu$m $W = 500\ \mu$m	O
p-type SSB	150	5486 5486	16–19 16–18	$W = 100\ \mu$m $W = 500\ \mu$m	O
p-type SSB	900	5486 5486	30–40 30–53	$W = 100\ \mu$m $W = 500\ \mu$m	O

*C = Canberra, O = Ortec.

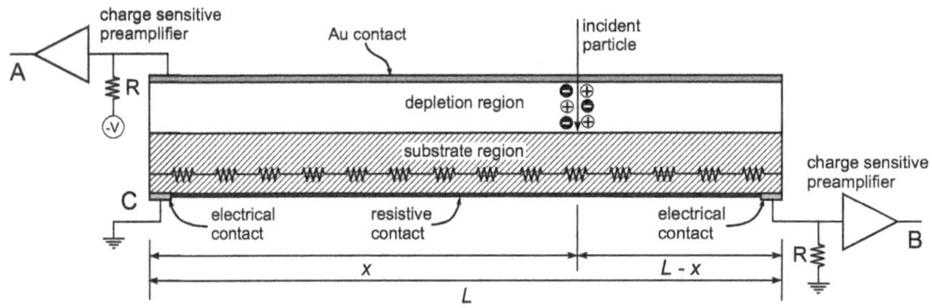

Figure 16.13. The resistive divider position sensitive detector operation principle.

are often used as position sensitive devices because of compactness from VLSI process methods, low operating voltages, and relatively high stopping power as charged-particle detectors. Double-sided cross strip detectors can have spatial resolutions as low as 25 μm, and pixellated detectors can have spatial resolution as small as 0.4 mm. Large arrays of position sensitive Si detectors can be used in collider facilities, x-ray scattering, and Compton cameras. Drift diode configurations, a variant design that drifts electronic charge carriers laterally along the detector to a small collection contact, offer low capacitances with large sensitive areas. These detectors are available in many forms, with some of the more popular configurations described here.

Resistive Divider Detectors Resistive divider detectors operate on the same principle as their gas-filled detector counterparts, described in Sec. 10.6.4. The detector consists of a rectifying contact in the form of a long strip, typically a few millimeters wide and a few centimeters long [Clegg et al. 1966; Bock et al. 1966]. The rectifying contact may be a simple Schottky contact, or can be fashioned as a *pn* junction. A conductive electrical contact, usually gold, is applied to the rectifying junction, and the detector is operated under partial depletion. The back contact has two conductive contacts, one at each end, and a thin relatively resistive contact (such as a layer of Bi, Cr, Ni, or Al) is applied between these two contacts [Bock et al. 1966; Bertolini and Coche 1968]. Because there are essentially three conductive contacts (see Fig. 16.13), it inspired the original name of these devices as 'nuclear triode detectors'. The literature suggests that unannealed implanted surfaces perform better as resistive contacts than evaporated metal layers [Laegsgaard et al. 1968; Elad and Sareen 1974], because the implantation process has superior uniformity and the lattice damage produced by the implantation process increases the surface sheet resistance.

Although in depth analyses of the device operation are available in the literature [Kalbitzer and Melzer 1967; Alberi and Radeka 1976], the general operation of the detector can be simplified as follows. The detector is operated with at least two charge sensitive preamplifiers. A reverse bias is placed on the rectifying contact, which produces a depletion region. One of the bottom conductive contacts is grounded and the conductive contact at the other end of the resistive region is connected through a second preamplifier. Ionizing events that occur at position x produce a total induced charge on the upper preamplifier at position A in Fig. 16.13 proportional to the total absorbed energy (E) in the depletion region. However, the conductive contacts at points B and C are affected by the distributed resistance of the lower contact. Consequently, the current flowing to points B and C is divided by the fractional resistances represented for lengths L-x and x. From Ohms law

$$E_B = E_A \frac{x}{L} = \frac{E_A R_C}{R_C + R_B}, \tag{16.27}$$

where L is the length of the bottom resistive contact, and R_C and R_B represent the resistances of lengths x and L-x, respectively. To find the linear position of the initial ionizing event, one needs to divide only the output from location B by the output of the upper preamplifier at location A, i.e., $x = LE_B/E_A$.

As simple as the concept is, these position sensitive detectors can provide a spatial resolution of about 30 μm FWHM at room temperature, reducing to 10 μm at 77 K [Elad and Sareen 1974]. Further, this type

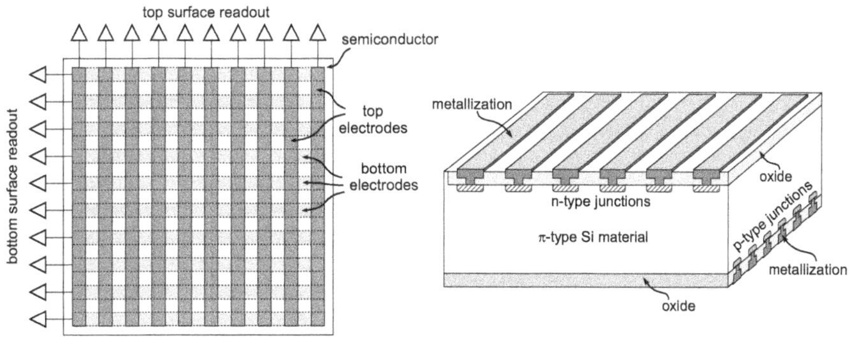

Figure 16.14. Basic structure of a cross-strip position sensitive particle detector. (left) Top view of the readout pattern and (right) isometric view of the basic construction.

of detector can be fashioned as an x-y position sensitive detector by aligning numerous parallel strips, for which the resistive divider identifies the x location, and the strip number identifies the y position. Spatial resolution of approximately 0.2 mm FWHM was reported for one such device [Lamport et al. 1976].

Double-Sided Cross-Strip Detectors Introduced originally by Hofker et al. in 1966, the dual-sided cross-strip detector has orthogonal conductive strips applied to opposite sides of a semiconductor substrate. From the top view depiction of Fig. 16.14, it is easy to understand the original name given by Hofker et al. [1966] as 'checker board counters'. These detectors are designed to operate under full depletion, with one side of the device having rectifying contact strips and the other side having orthogonal ohmic contact strips. The readout scheme becomes obvious by simple inspection. Ionizing events that occur within the detector produce electron-hole pairs. Electrons are swept to the anode strip in closest vicinity and the holes are swept to the closet cathode strip. Consequently, an ionizing particle produces an induced current pulse through the leads of one of the front strips and of one of the back strips. The event location is identified on the x-y grid because the opposing signal strips overlap. The scheme requires at least one preamplifier for each strip on the front and back. It its simplest form, a cross-strip detector can be manufactured by applying conductive contacts to opposite sides of various semiconducting of semi-insulating substrates, such as Si, Ge, GaAs, CdTe, CdZnTe, and HgI_2. For most applications, Si is a preferred choice.

A method to reduce the number of preamplifiers employs a network of external resistors, arranged as a series of resistive dividers, to identify the activated strip [Gerber 1976]. The top array of strips has two preamplifiers, one each attached to outermost strips, connected consecutively to each inner strip through a resistor division circuit. The difference or ratio of signals from each end is used to determine which of the strips was activated by an event. The sum of the signals preserves the energy information. Employing the same scheme on the backside orthogonal strip array, the two perpendicular strips closest to the ionizing event can be identified. Gerber et al. [1977] manufactured these detectors as position sensitive gamma-ray detectors, and, hence, they use Ge wafers instead of Si in order to increase the interaction probability. To prevent cross conduction and isolate the strips from each other, the strips have trenches etched between them.

Particle detectors employing the same external resistive divider scheme as Gerber et al. [1977] have been fabricated with common Schottky barrier designs by applying Au strips and Al strips on opposite sides of a ν-type Si wafer [Lamport et al. 1976, 1981; Heijne et al. 1980]. Position resolution of 150 μm was achieved for energy deposits of 60 MeV from ^{60}Fe ions. To reduce leakage currents between adjacent strips, it is common to use SiO_2 insulation between the strips. This oxide is grown through high-temperature methods, and photolithography is used to pattern and define the locations of the microstrips. The oxide is etched back

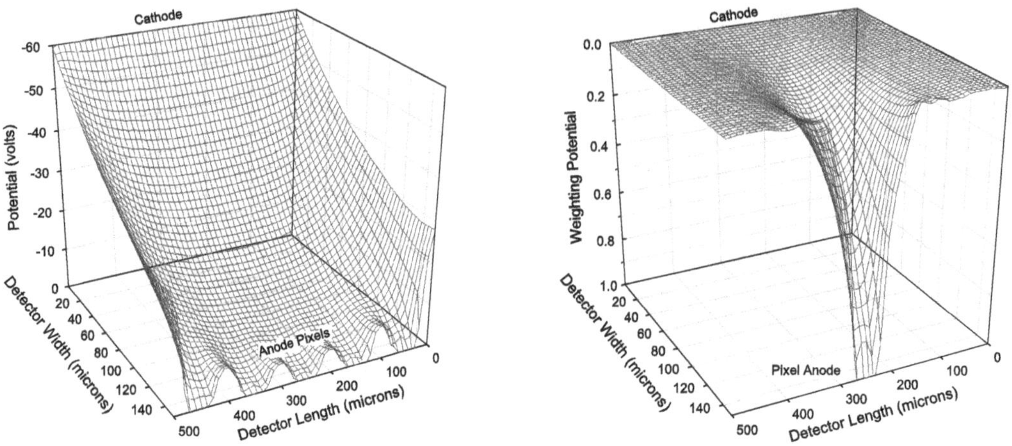

Figure 16.15. (left) For a space charge density of $N_D^+ = 3.33 \times 10^{11}$ cm^{-3}, the calculated voltage distribution for a section of a 150 μm wide pixellated diode with 50 μm $\times$ 50 μm pixels. (right) The resulting weighting potential for a single pixel of the same device.

to the bare Si to which the strips are to be applied. Implantation or traditional dopant diffusion is used to apply the p and n contacts, over which metal is applied, producing the end result shown in Fig. 16.14. However, Lutz [1999] points out that the oxide may be a cause of eventual device failure unless critical design features are not in place. Silicon dioxide can have residual impurity ions (such as Na^+) or may incur radiation damage that produces positive space charge. This space charge attracts electrons at the oxide-silicon interface and can form a conduction path between the strips, consequently, shorting adjacent strips. Proposed solutions to the problem include p-type implantation between the strips and producing negatively biased MOS structure atop the oxide strips [Lutz 1999].

Pixellated Detectors Spatial resolution can be improved by simply placing an array of contacts on a semiconductor surface and connecting each contact to a separate output. These contacts are referred to as pixels (or sometimes pads) and this detector structure has several advantages, which include increased energy resolution from the asymmetric weighting potential, reduced capacitance of each pixel, reduced leakage current flowing to each pixel, and relatively fast response times.

Shown in Fig. 16.15 are the calculated operating and weighting potentials for a $p\nu n$ pixel detector fabricated on high-resistivity Si ($N_D^+ = 3.3 \times 10^{11}$). The back plane of the design is a p-type contact fabricated on a ν-type substrate, with n-type anode ohmic contacts. The voltage distribution is nearly the same as a planar detector, indicating that charge carriers are collected as expected from a common planar device. However, the weighting potential for each pixel is highly non-linear with most of the induced current appearing on the output as free charges drift in the near-anode region. This effect is commonly referred to as "the small pixel effect," and was initially explored by Barrett et al. [1995] (see also Eskin et al. [1995]). If the pixel size is smaller than the final charge carrier cloud size, largely affected by the initial ionizing event and the charge carrier diffusion thereafter, then more than one pixel is expected to sense the arrival of electrons. This problem is denoted as "pixel-sharing" and leads to confusion about where the initial event occurred in the detector. This effect is reduced by increasing the pixel size, but at the expense of spatial resolution.

The capacitance of a pixel strongly affects the electronic noise and ultimately the potential developed by the induced charge. A single square pixel has an area of n^2, where n represents a side dimension, so that a microstrip of length L has an area of $n \times L$. The expected increase in capacitance for the microstrip is L/n; hence, the output voltage for the microstrip decreases by a similar amount for the same induced charge. The lower leakage current produced by a single pixel also reduces the electronic noise sensed by the preamplifier circuit. Mouthuy [2005] reports on comparisons of various pixel detectors for high energy physics applications with the claim that some pixel devices can have 15× lower electronic noise than microstrip counterparts. Consequently, because of the lower signal-to-noise ratio, thinner detectors can be used while still discerning events from noise, even with the reduction in energy deposition. Finally, because of the asymmetric charge induction from the small pixel effect, most of the charge induction appears on the output within 10 ns, thereby improving timing resolution of events.

However, there are complications with high-density pixel detectors. The pixel detector design requires as many readouts as detector contacts, a requirement that can be very demanding on the readout electronics. As pixel resolution increases, the readout density also must increase and can be cumbersome for traditional wire bonding methods. Instead, it is common to bond the pixel detector to an application specific integrated circuit (ASIC) designed as a miniaturized preamplifier readout for a high-density array of detectors. The detector contacts are connected to the readout electronics by a "flip-chip" method, originally developed by General Electric [1963]. This method permits the face of the pixel device to be aligned directly atop a readout microchip. The contact pads of the microchip have solder patterned on each contact. When heated, the solder forms into a ball from its own surface tension, and is afterwards allowed to cool. Hence, the ASIC chip has an array of soft solder balls exactly placed at the locations of the detector pixels and the readout contacts. The detector is pressed onto the soft solder, commonly an indium alloy, and heated to solder the two parts together [Broennimann et al. 2006]. Indium solder is commonly used because the different alloys have low melting points, ranging from 7°C to 310°C. Under some conditions, the mechanical compression of the lower temperature alloys is adequate to form a bond, thereby removing the heating requirement [Lozano et al. 2001]. Indium is also malleable and less prone to fracture under stress or from thermal expansion mismatch.

Drift Diode Detectors Large area planar detectors can have significant capacitance that reduces pulse height and increases detector electronic noise. Consequently, surface barrier detectors and the like are usually limited to modest diameters, with most commercial units having less than 40 mm diameter, although there are some special-application implanted-junction detectors with up to 60 mm diameter active regions. This problem of increased capacitance is largely mitigated with the drift diode design, a clever device that has a small anode with numerous guiding electrodes, invented by Gatti and Rehak in 1984 (see also Gatti et al. 1985). There are several variants on the design, the most common are the linear drift diode and the concentric ring drift diode [Lutz 1999].

The linear drift diode design concept is shown is Fig. 16.16, having a high purity substrate with multiple reverse biased *pn* junction strips aligned parallel on the front and back surfaces. One or more collection electrodes are established on one side, usually at one end of the strip array or near the center of the device between several microstrips. Because electrons have higher mobility in Si, it is usual to fabricate the device from ν-type material with p-type cathode strips on the surfaces. The collection electrode(s) is an n-type ohmic contact and is the location of the output electronics.

The device is operated under reverse bias, with gradually increasing positive voltage applied to each drift field electrode from the outermost strips towards the collection anode. This potential distribution can be accomplished by employing a resistive divider along the distribution of microstrips. At full depletion, from both sides of the substrate, the residual space charge forms a potential valley near the center of the volume as shown on the left of Fig. 16.17. It is interesting to note that the potential valley does not form without the benefit of some space charge (N_D^+ or N_A^-) being present. Ionizing events that occur in the device produce electron-hole pairs. Electrons drift to the bottom of this potential valley and are guided to the anode. The resultant weighting potential of the configuration causes most of the induced charge to be sensed as the electrons transit past the last drift electrodes and to the anode (see Fig. 16.17 (right)).

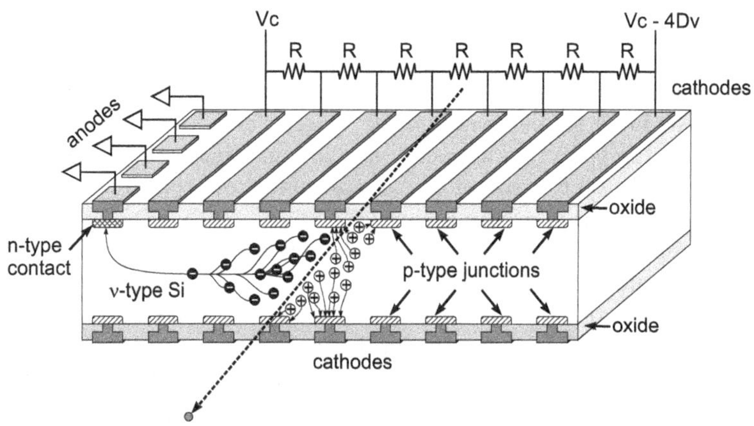

Figure 16.16. Structure of a linear Si drift diode detector. After excitation, electrons are drifted through a central potential channel and collected at an anode. Holes are collected at nearby cathodes. The output signal is heavily weighted on the electron induced charge as they near the anodes. After Gramegna et al. [1995].

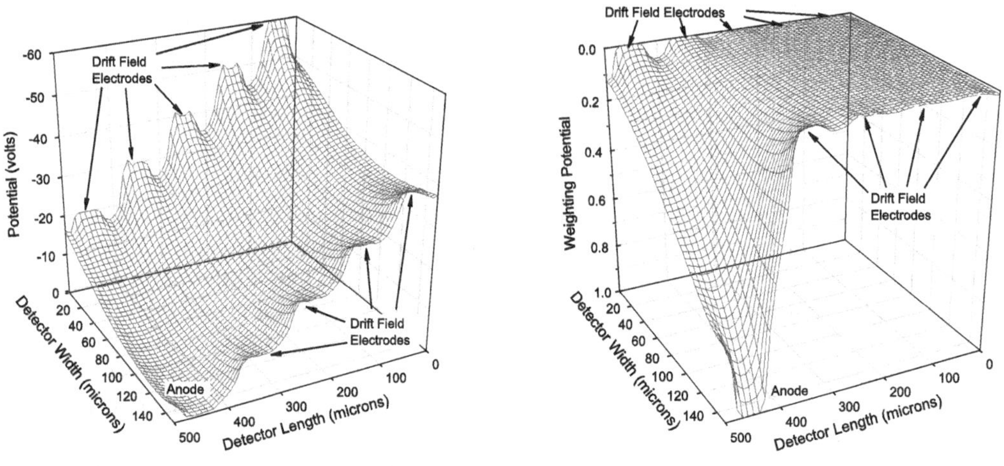

Figure 16.17. (left) For a space charge density of $N_D^+ = 10^{12}$ cm^{-3}, the calculated voltage distribution for a 150 μm wide linear Si drift diode near the anode region. (right) The resulting weighting potential for the same device.

A drift detector design employing ν-type Si material with a series of p^+ contact concentric rings is described by Gatti et al. [1985] and Lechner et al. [1993; 1996]. The device can be larger than the linear design but still have the advantage of low capacitance. Lechner et al. [1993; 1996] include a first stage field effect transistor (FET) input on the detector, which further reduces capacitance and electronic noise. The detector anode is near the center, but actually surrounds the output FET. The design has another difference from the basic linear design first proposed by Gatti et al. [1984a, 1984b], in that the drift field electrodes are only on one side of the device. The backside of the detector is a simple p^+ planar contact, used to promote full depletion of the device. Performance for a 300 μm thick concentric drift diode of 3.5 mm^2 area yielded room-temperature (300 K) FWHM energy resolution of 220 eV for 5.9 keV photon emissions from ^{55}Fe. When cooled to 150 K with a thermoelectric cooler (Peltier cooler), the energy resolution reduced to

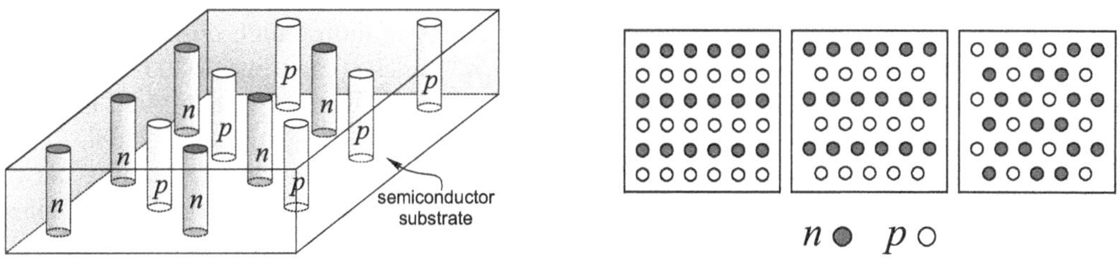

Figure 16.18. (left) Isometric view of the basic 3D architecture, showing interspersed n and p columnar dopant columns in a semiconductor substrate. (right) Top view of a few proposed 3D patterns.

139 eV at FWHM. Lechner et al. [1996] also describe an array of drift diodes on a single substrate that can be read out individually for position sensing information.

There are several advantages to the drift diode design. First, the device design (linear, concentric ring, or variant) allows the realization of a relatively large area device with significantly reduced capacitance compared to a planar device of the same area. Second, these detectors can be fabricated by common VLSI processing techniques, thereby enabling mass production. Finally, drift diode detectors can be used as direct conversion devices for x rays or can be used as light sensitive photodetectors for scintillation detectors, especially for scintillators with longer wavelength emissions which are usually mismatched for bi-alkali photomultiplier tubes. Avset et al. [1991] note that non-uniformities with material resistivities and charge carrier diffusion broadening can adversely affect drift times and position information for two-dimensional readout schemes. They further conclude that best results are obtained with drift fields that exceed 100 V cm^{-1}.

Silicon Three-Dimensional (3D) Detectors Radiation detectors at the Large Hadron Collider (LHC) receive over 10^{15} $\text{n}_{[1\text{MeV}]_{eq}}\,\text{cm}^{-2}$,[12] an indication that these detectors must survive this excessive dose. With upgrades to a high-luminosity LHC, this fluence is expected to exceed 10^{16} $\text{n}_{[1\text{MeV}]_{eq}}\,\text{cm}^{-2}$ [Bates et al. 2011]. Introduced in 1997 by Parker et al., three-dimensional detectors offer an alternative architecture for radiation hardened particle physics detectors than do microstrip or pixel designs. These detectors are constructed as vertically oriented *pin* diodes penetrating deep into a semiconductor substrate (see Fig. 16.18).

There are two salient advantages to the 3D design. First, the short distance between electrodes tremendously reduces the collection time, thereby countering the expected reduction in τ^* and λ^* for electrons and holes after extreme neutron damage. The recoil damage to the crystalline structure increases with fast neutron fluence, and consequently, the charge carrier trap density also increases. This increase in trap density causes a reduction of the electron and hole mean free drift times. Unfortunately, it has been documented that irradiation of 10^{16} $\text{n}_{[1\text{MeV}]_{eq}}\,\text{cm}^{-2}$ for a Si detector reduces the electron and hole mean free drift times τ^* to 0.24 ns and 0.16 ns, respectively [Bates et al. 2011]. These low values of τ^* correspond to mean free drift lengths λ^* of only 24 μm for electrons and 16 μm for holes. Recall from Sec. 15.4.6 the advantage of increasing the carrier extraction factors $\varrho = v\tau/W$ for electrons and holes leads to improved charge collection. The value of $\varrho_{e,h}$ can be increased by reducing the drift distance W between the detector contacts, accomplished with the 3D design by producing adjacent penetrating contacts only tens of microns apart. The resulting induced charge output ΔQ does not suffer as badly for the 3D design as compared to the signal produced for charges drifting across the substrate width of a planar detector. Second, the capacitance of the small electrodes is lower than a planar detector of the same total mass (thickness). Hence, the electronic

[12]The compound unit "$\text{n}_{[1\text{MeV}]_{eq}}\,\text{cm}^{-2}$" describes the equivalent dose damage per square centimeter fluence from a radiation source as expected if the detector were irradiated with monoenergetic neutrons of 1 MeV energy.

noise level should be reduced compared to a planar design, thereby enabling the observation of small pulses. Together these advantages yield particle detectors capable of withstanding high doses of neutron damage.

The first designs had alternating offset rows of p^+ and n^+ holes etched into a 300 to 600 micron thick Si substrate. These holes had diameters of about 15 to 30 microns [Kenney et al. 1999]; later designs had smaller holes down to 10 microns in diameter [Da Viá et al. 2008; Pellegrini et al. 2008]. The holes are etched deep into the substrate, sometimes completely through, with depths usually ranging between 250 to 300 microns. The pitch between adjacent holes is varied, but can be as low as 30 microns [Pellegrini et al. 2008].

The original 3D detectors were fabricated on one surface of a Si semiconductor substrate and are often referred to as "full 3D detectors" [Bates et al. 2011]. Da Viá et al. [2008] comment that p-type wafers are probably a better choice for the detector substrate and apparently are not susceptible to type-inversion.[13] The penetrating holes were etched with inductively coupled plasma reactive ion etching (ICP-RIE) methods [Kenney et al. 1999]. To provide additional mechanical strength during the fabrication process, these first devices were bonded to a Si support wafer. An array of holes was etched and doped p^+, followed by a backfill of polysilicon to produce a relatively flat surface for the next process steps. A second array of holes was then patterned between the first array, etched, and doped n^+. This second array of holes was also backfilled with polysilicon. The final steps included patterning conductive leads to the separate arrays, dicing the detectors from the wafer, and removing the support substrate. To mitigate dead regions around the detector perimeter, electrodes were later patterned around the edges of the detectors [Kenney et al. 2001, 2006]. Da Viá et al. [2008] report that full 3D detectors retained 38% charge collection efficiency after irradiation of 6.8×10^{15} $n_{[1\text{MeV}]_{eq}}$ cm^{-2}, equivalent to approximately 10 years of operation within the Super-LHC.

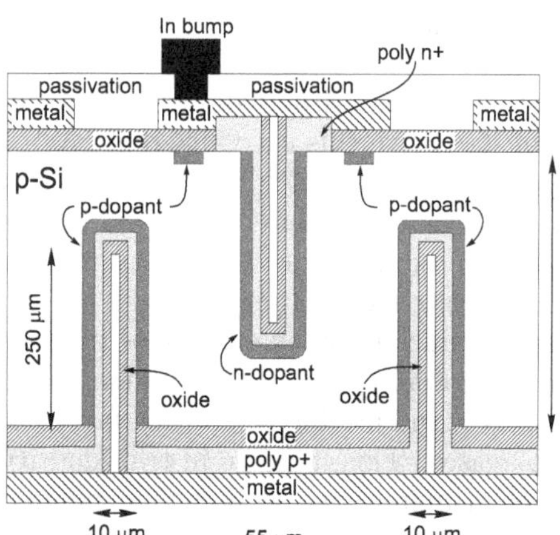

Figure 16.19. Structure of a p-type Si double-sided 3D detector (not to scale). After Pelligrini et al. [2008] and Bates et al. [2011].

The double-sided 3D architecture was introduced to simplify the fabrication process while reducing cross contamination of the p and n doped columns [Pelligrini et al. 2008; Zoboli et al. 2008]. The holes are etched from both sides of a semiconductor substrate, with one side doped p^+ and the other side doped n^+. The holes are arranged in an interspersed pattern such that the p^+ and n^+ columns are adjacent. The holes are etched blind such that they do not penetrate completely through the wafer (see Fig. 16.19). Bates et al. [2011] reports on Si 3D detectors with dimensions shown in Fig. 16.19, and for which the detectors continued to deliver 47% charge collection after irradiation equivalent damage of 10^{16} $n_{1\text{MeVeq}}\text{cm}^{-2}$ at 150 volt bias. At higher voltage, charge multiplication was observed from the device, indicating impact ionization gain. Additional experiments conducted with double-sided 3D detectors fabricated from both p-type and n-type substrates continued to deliver up to 70% charge collection efficiency at 2×10^{16} $n_{[1\text{MeV}]_{eq}}$ cm^{-2} [Köhler et al. 2011]. The n-type detectors did not show type-inversion. Because of the straightforward fabrication process and the high radiation hardness, up to 25% of the insertable B-Layer (IBL) Si detector ring has been populated with double-sided 3D detectors during a recent upgrade [CERN 2010; Lange 2015].

Charged Coupled Devices (CCD) The charged coupled device, or CCD, is an imaging device used to register the intensity of energy deposited as a function of position. When color filters are added to the

[13]The reversal of a rectifying junction that occurs from radiation damage turning n-type silicon into p-type silicon.

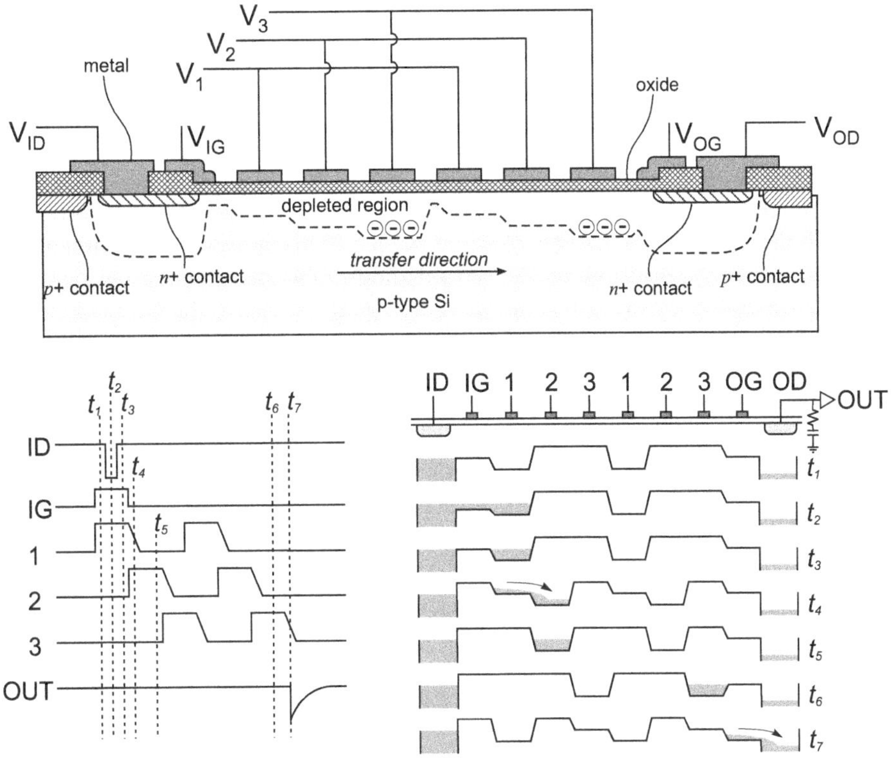

Figure 16.20. Depiction of a MOS n-channel CCD and its operation. (top) A cross section depiction of a surface channel CCD showing only six transfer registers. (left) The clock waveforms and output signal. After Kim [1979].

system, the CCD is used for optical photography and video imaging. The history of the CCD extends back to 1970, when the inventors Boyle and Smith first described its function and operation. The first report on an operating device was also published in 1970 [see Amelio et al. 1970]. This first device was designed as a series of metal pads placed atop a SiO_2 layer grown on a n-type Si substrate to produce a series of MOS capacitors in line on a substrate. Although these first-generation devices had multiple problems with charge sharing and incomplete charge transfer, modern optical devices are still based on improved MOS designs. Damerell et al. [1981, 1987, 1990] and Bailey et al. [1983] report the use of MOS CCDs for radiation detection applications. Advantages of the CCD concept include a reduced number of readout electronics for the large number of pixels, small pixel sizes for good spatial resolution, and relatively rapid response [Stefanov 2006]. Further, CCDs can be tiled to produce a much larger detector array.

The general operation of the CCD is often described with a simple MOS structure as an example, such as the 3-phase surface-channel MOS CCD depicted in Fig. 16.20. The device of Fig. 16.20 has a series of MOS diodes fabricated on a continuous surface of SiO_2. At the ends of the MOS diodes are input and output structures that consist of an input diode and gate (ID and IG) and is used to inject charge and an output diode and gate (OD and OG) that is used to collect charge [Sze 1981]. Because the charge to be measured is excited by radiation events in the depletion region, charge injection is not a necessary requirement for radiation detection CCDs but is used here to provide an example of the charge transport. Every third gate contact, or transfer register, is biased at the same potential, labeled V_1, V_2, and V_3. The transfer registers can be fabricated as strips, and the orthogonal crossing of each register strip along the transverse direction constitutes a pixel. A digital clock is used to time and apply voltages to the gates in a sequential order. The voltages applied at the input and output diodes are biased high enough to produce deep depletion regions underneath the input and output gates. At time t_1, the voltage applied to V_1 is higher than V_2 and V_3 and produces deeper depletion wells at all positions 1. At t_2, the voltage is adjusted to reduce the barrier between ID and V_1, which allows charge to flow into the first well. At t_3, V_{ID} is increased to close the gate, while excess charge is removed through the ID contact. A t_4, V_1 is lowered and V_2 is increased, causing the depletion wells at 1 to decrease and depletion wells at 2 to increase. Consequently, charge flows from positions 1 into positions 2, a process named *charge transfer*. This process is continued, as depicted for t_5 through t_7, until the charge is transferred to the output diode. The process is often described as analogous to a 'bucket brigade'. A similar shift register, oriented orthogonal to the CCD gate transfer registers, can be used to read the output from each row of transfer registers. There are many variations of MOS CCDs, and Fig. 16.20 depicts only one of many [Kim 2009; Sze 1981; Schroder 1990; Lutz 1999]. Each entire readout of the CCD array is called a 'frame' and the pixels are often termed 'gates'. Each row of registers is read out such that the synchronized clocking electronics can identify the pixel origin when the charge appears on the output.

Recall the description of MOS devices given in Sec. 15.3.4, where again the terms for flat band voltage V_{FB} and gate voltage V_G are used here. The surface potential ψ_s under a CCD gate is defined as

$$V_G - V_{FB} = V_i + \psi_s = \frac{q_e N_A W^2}{C_o} + \psi_s, \tag{16.28}$$

where W is the depletion layer width, N_A is the semiconductor doping concentration, C_o is the oxide layer capacitance in F cm^{-2}, V_i is the potential drop across the insulating oxide layer, and

$$\psi_s = \frac{q_e N_A W^2}{2\kappa_s \epsilon_0}, \tag{16.29}$$

where κ_s is the dielectric constant for the semiconductor. Rearrangement and substitution of the definition of W into Eq. (16.28) gives

$$V_G - V_{FB} = \psi_s + \frac{1}{C_o}\left[2\kappa_s \epsilon_0 q_e N_A \psi_s\right]^{1/2}. \tag{16.30}$$

As charge is accumulated in the potential well under the gate, the gate voltage is altered to become

$$V_G - V_{FB} = \frac{Q_s}{C_o} + \frac{\left[2\kappa_s \epsilon_0 q_e N_A \psi_s\right]^{1/2}}{C_o} + \psi_s. \tag{16.31}$$

Solving this equation for ψ_s gives

$$\psi_s = V_G - V_{FB} + V_0 + \frac{Q_s}{C_o} - \left[2\left(V_G - V_{FB} + \frac{Q_s}{C_o}\right)V_0 + V_0^2\right]^{1/2}, \tag{16.32}$$

where Q_s is the stored charge per unit surface area and

$$V_0 = \frac{q_e N_A \kappa_s \epsilon_0}{C_o^2}. \tag{16.33}$$

The maximum charge that can be stored in a single pixel is described by [Schroder 1984] as

$$Q_{max} = -C_o\left(V_G - V_{FB} - 2\psi_B - V_B\right), \tag{16.34}$$

where

$$\psi_B = \frac{kT}{q_e} \ln\left(\frac{N_A}{n_i}\right) \quad \text{and} \quad V_B = \left[\frac{4q_e \kappa_s \epsilon_0 N_A \psi_B}{C_o^2}\right]^{1/2}. \tag{16.35}$$

Under the assumption that both V_B and ψ_B are small by comparison to the applied effective operating voltage $V_G - V_{FB}$, Eq. (16.34) becomes

$$Q_{max} \approx -C_o \left(V_G - V_{FB}\right). \tag{16.36}$$

The gate oxide is typically about 100 to 200 nm thick and the difference in gate voltage is usually 5 to 10 V. Pixel dimensions for scientific CCDs can range from 50 microns down to 1 micron, and CCDs can have over 2 megapixels in a single device. Commercial CCDs for x-ray imaging applications typically have 20 × 20 micron pixels. If the charge accumulated in a potential well exceeds the capacity, then mobile charges can diffuse into adjacent pixel wells, causing a blurring or a washed out line to appear in the image.[14] A typical charge limit for a MOS CCD potential well is about 5×10^6 electrons [Schroder 1990].

Example 16.2: A common surface channel MOS CCD has pixel sizes of 20 × 20 microns. The oxide thickness is 0.2 microns and the effective operating voltage is 10 volts. Determine the maximum charge that can be stored under the gate.

Solution:

The capacitance per unit area C_o is

$$C_o = \frac{\kappa_o \epsilon_0}{W_o} = \frac{(3.9)(8.854 \times 10^{-14} \text{ F cm}^{-1}}{2 \times 10^{-5} \text{ cm}} = 1.73 \times 10^{-8} \text{F cm}^{-2}$$

The maximum charge capacity is

$$Q_{max} = -C_o \left(V_G - V_T\right) = (1.73 \times 10^{-8} \text{F cm}^{-2})(10 \text{ V})(2 \times 10^{-3} \text{ cm})^2 = 6.9 \times 10^{-13} \text{ C}.$$

This charge corresponds to

$$\text{number of electrons} = \frac{6.9 \times 10^{-13} \text{ C}}{1.6 \times 10^{-19} \text{ C/e}^-} = 4.3 \times 10^6 \text{ electrons}.$$

The surface n-channel device has some problems with efficient charge transfer from one gate register to an adjacent register, mainly because the lowest part of the potential well is adjacent to the oxide layer. To improve transport efficiency, the buried n-channel CCD, or BCCD, was introduced, in which an n-type tub is deep diffused into the p-type bulk, upon which the oxide is grown [Walden et al. 1972]. The BCCD architecture moves the lower point of the potential well away from the surface and into the semiconductor bulk [Schroder 1990].

MOS-based devices can have many problems when used as radiation detectors. First, the oxide can be destroyed by ionizing radiation by producing stationary trapped charge, ultimately changing the operating voltage of individual pixels and causing failure [Killiany et al. 1974, 1975, 1980]. Further, common small pixel CCDs do not have the capacity to store charge for high-energy particles, such as alpha particles, causing charge overflow into adjacent pixels. Finally, the pixel sizes are usually small with depletion region depths

[14] You may have seen this effect when pointing your CCD camera at a candle or other bright light.

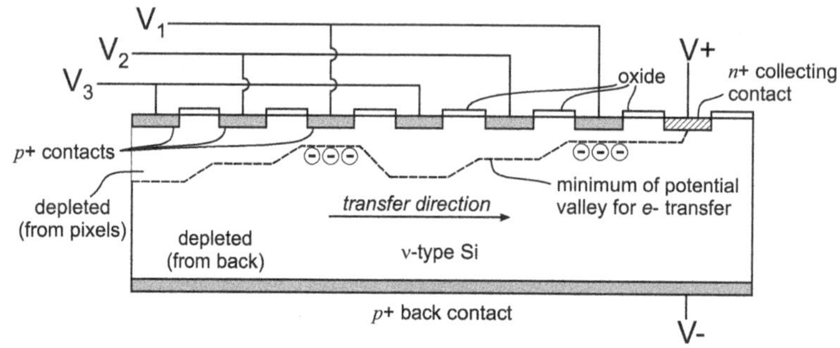

Figure 16.21. Fully depleted *pn* junction type CCD, depicting a cross section portion. After Strüder et al. [1987a, 1990].

of about 1 to 3 μm [Lutz 1999], and energetic particles deposit only a small portion of energy in a single pixel. In some cases, a particle can transit across many pixels.

The *pn*-junction design, proposed by Gatti and Rehak [1984a, 1984b], has received interest as an alternative for radiation imaging applications. A cross section of this device is depicted in Fig. 16.21. The *pn*-junction CCD has three operating voltages synchronized by a clocking circuit [Strüder et al. 1987b, 1990]. The device is reverse biased from an *n*-type contact against a *p*-type back contact and the multiple *p*-type pixel contacts, resulting in a fully depleted ν-type Si substrate. A pixel is defined by the orthogonal crossing of a *p*-type strip transfer register with that of a *n*-type channel guide (see Strüder et al. 1987b). The three operating voltages (V_1, V_2, V_3) are altered to move the potential minimum along the length of register pixel contacts towards the anode. As these voltages are changed, the lowest potential is moved along the contacts towards the anode, ultimately forcing charges to sequentially move from the region under one contact to an adjacent contact. The design is improved by adding channel guides in the direction of motion towards the anode and are composed of an implanted *n*-type region perpendicular to the *p*-type transfer pixels [Strüder et al. 1987a]. These channel guides prevent punchthrough currents between the *p*-type contacts. They also provide a route for the moving charges and reduce lateral diffusion of charges to pixel locations at the sides.

There are additional advantages of the fully depleted *pn*-junction CCD. Because the entire volume is depleted, the *pn*-junction CCD has a large active region, leading to enhanced efficiency over the MOS type. Further, a relatively thin backside *p*-type contact allows the device to be operated with the backside under illumination, thereby producing a much more uniform response of the device area. Because the volume is depleted from top and bottom, the potential minimum is displaced from the surface, unlike that in a traditional MOS CCD, and the charges can travel much faster through the device to the anode. The *pn* junction is more radiation hard than the MOS version; hence the *pn*-junction CCD is more suited for operation in a radiation environment.

There are two common methods by which CCDs function for radiation imaging: (1) as optical detectors for scintillating materials and (2) as direct conversion detectors. Scintillators can be attached to a CCD and the light response spatially measured by the CCD. Methods include the direct attachment of scintillating material to the CCD, although this method puts the CCD in the radiation field. Resolution can be enhanced by having the CCD displaced from a large scintillating screen with the image focused and projected onto the CCD, often with mirrors or a light waveguide [Gruner et al. 2002]. Thick scintillators increase efficiency, but also cause a reduction in spatial resolution. Phosphor-coated screens preserve the spatial resolution, but often have reduced efficiency for high-energy particles. However, for low energy particles and x rays, phosphor screens can provide a suitable detection medium. In a study comparing phosphor screens for x-ray computed tomography, having either lens-coupled CCDs and fiber-coupled CCDs, the lens method provided higher spatial resolution [Uesugi et al. 2011]. However, the light transmission of the fiber-coupled system was four times more efficient than the lens-coupled device. Consequently, the higher transmission efficiency enables a fourfold reduction in the exposure time. Further, samples prone to deformation are less affected by the reduced exposure time. Finally, the radiation dose to patients is lower because of the shorter exposure time.

The direct imaging method relies on ionizing radiation producing electron-hole pairs in the depleted sensitive region of the CCD. The obvious problem with this approach is that the CCD is placed in the radiation field, and defect formation can eventually cause device failure. This failure is especially problematic when irradiated with heavy ions and reaction products emitted from neutron conversion films. Regardless, the direct conversion method has been used for detection of MIPs in Vertex detectors for high energy physics applications [Damerell 1998, 2003, 2005; Turala 2005; Stefanov 2006; Ishikawa 2015]. Direct conversion CCDs can be commercially acquired for x-ray imaging applications [Hammamatsu 2016]. Windowless CCDs are front irradiated and are designed to detect x-ray energies between 0.5 keV and 10 keV. The back irradiated versions are designed with a thinned surface layer to detect lower energy x rays and also have thicker depletion regions for higher x-ray energies.

For low energy photons, CCDs can be used as photon counters and direct conversion spectrometers. However, charge sharing between pixels can degrade energy resolution and confuse the photon counting, especially for surface channel and BCCD MOS devices in which a large portion of the semiconductor is not depleted. Photon interactions can cause ionization in the undepleted portion, and through charge diffusion, electrons can be shared with many pixels [Lumb and Nousek 1992]. To reduce this problem, CCDs operated as photon counters should have no more than approximately 0.15 photons per pixel per frame to stay within 1% count error. Photon spectroscopy is even more demanding, reportedly requiring about 0.018 photons per pixel per frame [Lumb and Nousek 1992]. The chance of additional interactions during charge transfer process increases with increasing radiation fields, and may pose a problem if radiation spectroscopy is the desired outcome [Lumb and Nousek 1992], mainly because the total background events per frame increases. The clock sampling rate can be increased, but unfortunately this change also causes the signal to noise ratio to decrease. To reduce background charges during the readout process, Lumb and Nousek [1992] also suggest the implementation of a shutter during readout.

As with all semiconductors, leakage current has a generation-recombination component. Hence, CCDs operated at room temperature thermally generate electron-hole pairs during operation and during charge transfer. Usually room temperature operation does not pose a problem. However, for low light or low radiation environments, thermal noise can cause image degradation. This problem is remedied by cooling the CCD during operation between 173 K to 223 K.

Si(Li) X-ray Detectors

The low Z number and low volume density of electrons causes the γ-ray absorption coefficient to be small for Si. Further, the energy at which the photoelectric absorption and the Compton scattering are equal is relatively low at only 60 keV. The Compton scattering coefficient is much lower than most semiconductors used for radiation detection (see Fig. 16.1). Hence, Si is a poor choice for high energy γ-ray spectroscopy. However, its K absorption edge appears at 1.838 keV, meaning that the absorption edge discontinuity does not adversely affect x-ray absorption at higher energies, nor does the appearance of x-ray escape peaks cause significant issues in spectra. By comparison, the K absorption edge for Ge is 11.103 keV. Because higher energy γ rays have less chance of interacting in Si, this property serves to reduce background effects. For these reasons, Si does have importance as an x-ray spectrometer for applications such as x-ray fluorescence, x-ray microanalysis, particle induced x-ray emission (PIXE), x-ray absorption spectroscopy (XAS), x-ray diffraction, and Mössbauer spectroscopy at energies generally below 50 keV. Energy resolution is the full width at half the maximum (FWHM) of a spectral full energy peak. Silicon detectors deliver excellent energy resolution, and the FWHM, reported in units of energy, is

$$\mathrm{FWHM_{eh}} = 2\sqrt{2\ln(2) w E_\gamma F}, \tag{16.37}$$

where w is the average energy to produce an electron-hole pair, E_γ is the photon energy, and F is the Fano factor (typically 0.12). The Fano factor is a correction factor that accounts for the typically higher energy resolution than predicted from pure Gaussian statistics. With the inclusion of noise sources, the energy resolution becomes

$$\mathrm{FWHM} = \left[(\mathrm{FWHM_{noise}})^2 + (\mathrm{FWHM_{eh}})^2\right]^{1/2}. \tag{16.38}$$

where the $\mathrm{FWHM_{noise}}$ contribution includes electronic noise and fluctuations in the leakage current.

Highly purified Si can be fashioned into a type of pn, $p\pi n$, or $p\nu n$ diode; yet, even with zone-refined high-purity material, Eq. (15.61) (or Fig. 16.11) show these devices are limited to depleted regions less than 2-mm wide, an unsatisfactory thickness for efficient x-ray absorption. The problem is remedied by compensating the remaining impurities in Si with the Li-drifting technique. Pell [1960b] introduced the process of Li drifting in Si, a method to electrically drive Li^+ ions deep into a semiconductor, in which Li behaves as an n-type dopant. If the Si material is π-type, these Li^+ ions exactly compensate the negative space charge effects of acceptor impurities and, thus, the Si material behaves as if it were intrinsic-like. Although other Group I alkali metals such as Na and K also act as donor impurities in Si, it is the small atomic radius and high interstitial drift mobility of Li that makes the drift process possible [Pell 1960b]. Unlike common impurity dopants such as B, P, and As that become electrically active by displacing Si atoms in the crystalline lattice, Li ions become interstitial donors lodged between lattice sites with an ionization energy of 0.03 eV [Mann et al. [1962].

Li drifting is fundamentally a three-step process, starting with producing a pn junction diode, into which Li is introduced as the n-type dopant onto a p-type slab. A purified single crystal of p-type or π-type Si sample is sliced and etched to appropriate dimensions. Afterwards, Li is applied to the surface of the Si slab. There are many processes to apply Li to the Si surface, including physical vapor deposition [Blankenship and Borkowski 1962; Mann et al. 1962; Ristenen et al. 1967] and application of a suspension of Li in oil [Ammerlaan and Mulder 1963; Mayer 1962]. Painting the Li-oil suspension on the sample seems straightforward; however, if done improperly, the layer can be uneven and form cracks, consequently producing a uneven Si surface once the diffusion process is finished. The Si sample is then heated to approximately 400°C for several hours. This Li predeposition process produces a pn junction at the surfaces in contact with the Li dopant. The Li distribution after this predeposition step is

$$N_D(x) = N_0 \operatorname{erfc}\left[\frac{x}{\sqrt{4Dt_p}}\right], \tag{16.39}$$

where N_0 is the initial donor concentration, D is the diffusion coefficient, and t_p is the process time. Pell [1960a] measured the diffusion of Li in Si at temperatures between 25°C and 125°C. Combined with data acquired elsewhere [Fuller and Severiens 1954], Pell [1960a] estimates the diffusion coefficient dependence as[15]

$$D(T) = (25 \pm 2) \times 10^{-4} \exp\left[\frac{-q_e(0.655 \pm 0.01)}{kT}\right] \mathrm{cm^2/s}, \tag{16.40}$$

[15] Based on the data from Fuller and Severiens [1954], the Li^+ ion mobility as a function of temperature can be estimated with the result of a least squares fit

$$D(T) = 28.91 \times 10^{-4} \exp\left[\frac{-7733.45}{T}\right] \mathrm{cm^2/s}.$$

Although this result differs from that reported by Fuller and Severiens [1954], it is similar to that found by Pell [1960a].

where T is the absolute temperature in Kelvin. After the Li deposition step, Pell [1960b] approximates the Li distribution by expanding Eq. (16.39) as

$$N_D(x) \simeq \frac{2N_0\sqrt{Dt_p}}{x\sqrt{\pi}} \exp\left[\frac{-x^2}{4Dt_p}\right]. \tag{16.41}$$

Because $N_A = N_D$ at location c in Fig. 16.22

$$N_A \simeq \frac{2N_0\sqrt{Dt_p}}{c\sqrt{\pi}} \exp\left[\frac{-c^2}{4Dt_p}\right], \tag{16.42}$$

so that

$$N_0 \simeq N_A \frac{c}{2}\sqrt{\frac{\pi}{Dt_p}} \exp\left[\frac{c^2}{4Dt_p}\right]. \tag{16.43}$$

Upon substitution of Eq. (16.43) into Eq. (16.41) one finds

$$N_D(x) \simeq N_A \frac{c}{x} \exp\left[\frac{c^2 - x^2}{4Dt_p}\right]. \tag{16.44}$$

Afterwards, the temperature is lowered, usually between 100-200°C, and a reverse bias is applied to the diode [Ammerlaan and Mulder 1963; Miller et al. 1963]. This primary drift serves to force the Li$^+$ ions deep into the Si material and enlarge the compensated thickness. Miller et al. [1963] point out that the thermally generated charge carrier density (and reverse leakage current) increase with drift temperature. At too high a drift temperature, the thermally generated electron and hole concentrations can exceed the donor space charge concentration. Consequently, the compensation process can become non-uniform and incomplete. Goulding and Hansen [1964] point out the importance of temperature and current control over the process to maintain good compensation uniformity. Periodically during the drifting process, it is possible to stop the drift and measure the reverse current and capacitance to yield a measure of the compensation region width [Ammerlaan and Mulder 1963; Dearnaley and Lewis 1964], mainly because capacitance is inversely proportional to the width while the generation current is proportional to the width.

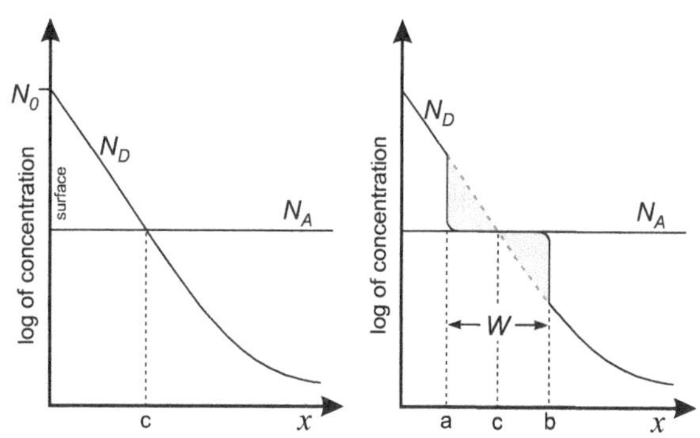

Figure 16.22. The impurity concentrations for Li (N_D) and the acceptors (N_A) (left) directly after the Li diffusion process and (right) after the ion drift process. W represents the compensated region.

For the general analysis of the Li$^+$ ion drift, it is assumed that the distribution of acceptor impurities N_A is uniform and constant. The Li$^+$ ion mobility in Si can be described by [Blankenship and Borkowski 1962]

$$\mu_{\mathrm{Li}}(T) = 2.78 \times 10^{-2}\left(\frac{1000}{T}\right) \exp\left[-7.5\frac{1000}{T}\right], \tag{16.45}$$

where T is the absolute temperature in Kelvin. Assuming that the drift force $\mu_{\mathrm{Li}}\mathcal{E}N_D$ is much greater than the diffusion force $D\nabla N_D$, the initial Li$^+$ concentration at locations $x < c$ decreases and the Li$^+$ concentration at locations $x > c$ increases. As the compensated region expands, the Li$^+$ concentration at $x < a$ continues to decrease while the Li$^+$ concentration at $x > b$ continues to increase. Note that the Li$^+$ concentration does not reduce below N_A for $x < c$ and does not exceed N_A for $x > c$. If it could, then the space charge polarity would reverse in either region, causing Li^+ to drift towards the negative space charge until compensation was complete. The number of Li^+ ions per cm^2, the ion current density, moving across the junction boundary in time t is $\mathcal{E}\mu_{Li}N_A t$ and it is assumed the electric field is primarily in the compensated region where $N_D \simeq N_A$. Overall, the total number of Li^+ ions drifted is represented by the gray areas in Fig. 16.22. Near $t = 0$ at the beginning of the drift process, when $W \ll 2Dt_p/c$, Pell [1960b] approximates the amount of Li drifted over time t as

$$\int_0^t \mu_{\text{Li}}\mathcal{E}N_A dt = \int_a^c N_D(x)\,dx - (c-a)N_A = (b-c)N_A - \int_c^b N_D(x)\,dx. \tag{16.46}$$

Insertion of Eq. (16.39) into Eq. (16.46) and use of

$$\int \text{erfc}(az)\,dz = z\,\text{erfc}(az) - \frac{1}{a\sqrt{\pi}}\exp(-a^2z^2), \tag{16.47}$$

which is obtained by integration by parts, yields

$$\int_0^t \mu_{\text{Li}}\mathcal{E}N_A dt = \int_a^c N_D(x)\,dx - (c-a)N_A = -(c-a)N_A + cN_0\text{erfc}\left[\frac{c}{\sqrt{4Dt_p}}\right]$$
$$-aN_0\text{erfc}\left[\frac{a}{\sqrt{4Dt_p}}\right] + \frac{N_0}{\pi}\sqrt{4Dt_p}\left[\exp\left(\frac{-a^2}{4Dt_p}\right) - \exp\left(\frac{-c^2}{4Dt_p}\right)\right]. \tag{16.48}$$

A similar expression can be obtained in terms of a and c using the right-hand side of Eq. (16.46). A Taylor expansion of the erfc and exponential functions in Eq. (16.48) about c gives

$$\int \mu\mathcal{E}dt \simeq \frac{(c-a)^2}{2L} + \frac{(c-a)^3}{6L^2}, \tag{16.49}$$

where $L = 2Dt_p/c$. Pell [1960b] assumed that the compensated width $W = (b-a) \ll L$ if t_p is very small. If the slope of the Li concentration is assumed linear at point c in Fig. 16.22, then at $t_p \approx 0$ the distance $c - a \approx b - c$, such that the first expansion term of Eq. (16.49) yields

$$\int \mu_{\text{Li}}\mathcal{E}dt \simeq \frac{W^2}{8L}. \tag{16.50}$$

As the drift process continues, it can be assumed that $(b-c) \gg (c-a)$ and $W \simeq (b-c)$, so that [Goulding 1966]

$$dW \simeq \mu_{\text{Li}}\mathcal{E}dt. \tag{16.51}$$

For a planar device where $\mathcal{E} = V/W$, the solution for the compensated detector width is

$$W = \sqrt{2\mu_{\text{Li}}Vt}. \tag{16.52}$$

By using both Eq. (16.45) and Eq. (16.52), Blankenship and Borowski [1962] develop a plot predicting the compensated width of a planar Si(Li) detector as a function of temperature, time, mobility, and voltage. The resulting devices usually have active regions with thicknesses between 3 and 5 mm; however, thicker widths are possible [Ristinen et al. 1967; Bertolini and Coche 1968]. Sharma and Divatia [1986] report a table of drift time and temperature that indicates a 5-mm compensated region takes 280 kilovolt hours at 110°C while raising the temperature to 130°C reduces the drifting duration to 115 kilovolt hours. Adequate detail of a few drift procedures are provided in the literature [Blankenship and Borkowski 1962; Goulding 1965, 1966; Sharma and Divatia 1986].

The performance of Si(Li) detectors can be affected by detector volume and the initial distribution of background impurities. Although the analysis performed above is a common treatment for the Li drift process, there are many non-ideal situations that affect the outcome. Gibbons and Iredale [1967] and Lauber [1969] analyze numerous alternative solutions that include nonuniform distribution of acceptors, the affect of free charge carries, generation and recombination current, and the effect of trapping and detrapping on the drift process. Gibbons and Iredale [1967] also analyze the expected Li drift result for a detector with coaxial geometry. Contaminants in the Si can also cause problems with the process, further evidence that it is best to start with relatively pure Si material. Pell [1961] and Ammerlaan and Mulder [1963] report that oxygen contamination in the Si can adversely affect the drift process and drastically lower the Li^+ ion mobility from the formation of LiO^+-complex.

A second "clean up" drift at a lower temperature is usually performed after the primary drift process [Lauber 1969], although this drift procedure is still conducted at a temperature higher than the detector operating temperature. The second drift is conducted such that the thermally generated free charge carrier density is much lower, thereby lowering the total space charge concentration. Sharma and Divatia [1986] describe a "clean up" process in which the temperature is slowly reduced under bias as the drifting process continues. Overall, the second drift process produces a much more uniform and complete compensation as the Li^+ ions redistribute.

Unfortunately, even with the second drift, there are usually some residual uncompensated centers, resulting in a net space charge density [Gibbons and Ireland 1967]. These residual uncompensated defects alter the electric field, pulse shape, and performance of the detector [Moszyński et al. 1968; Moszyński and Przyborski 1968; Moroz and Moszyński 1969]. As explained in Sec. 15.3, space charge limits the depletion thickness, and from Sec. 16.2.4, ionized impurities reduce the charge carrier mobilities. If the residual space charge is uniform, then the electric field drops linearly, being lowest near the n-type contact. However, if the residual space charge varies over the detector width, the electric field can have a much more complicated distribution. The research of Moszińiski and Przyborski [1968] indicates that the actual electric field distribution is parabolic in shape, a consequence of non-uniform compensation of the background impurities, which appears to have a linearly varying distribution of space charge. The pulse shapes are dependent on the radiation interaction location, and ultimately the operating voltage must be increased to sustain good performance.

Llacer [1964, 1966] determined that the surface conditions and the junction geometry affect the performance of a Si(Li) detector, noting that most of the surface current originates at the i-p junction where the electric field is highest. Further, it was observed that an inversion layer can form on the periphery of the Li-drifted region, where a n-type layer forms, thereby forming a junction around the device including atop the p-type region. The consequence is low-breakdown voltage and relatively high surface leakage current, both contributing to electronic noise. By changing the detector geometry, it was found that the surface breakdown voltage could be increased and the leakage current decreased [Goulding 1965; Llacer 1966]. The added advantage of a higher operating voltage is faster charge carrier collection times. Two popular Si(Li) detector geometries used to increase surface breakdown voltage are depicted in Fig. 16.23, both pin-type devices (discussed in Chap. 15). The inverted-T design includes a deeply etched trench surrounding the n-type contact, while the etched planar design isolates the n-type contact away from the bulk p-type region. Malm and Dinger [1976] and Goulding [1977] point out that the inverted-T design has regions with only partial collection of charge carriers, mainly near the boundary of the etched trench where charges can be collected at the trench. This charge becomes lost and consequently degrades the signal. An isolated guard ring around the n-type contact isolates the detector sensitive region from the trench and improves overall performance [Goulding 1977; Hau et al. 2003].

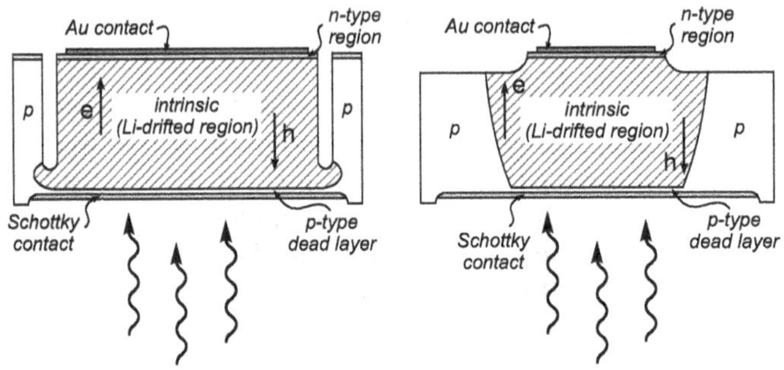

Figure 16.23. Popular configurations for Si(Li) detectors, showing (left) the inverted-T and (right) the etched planar.

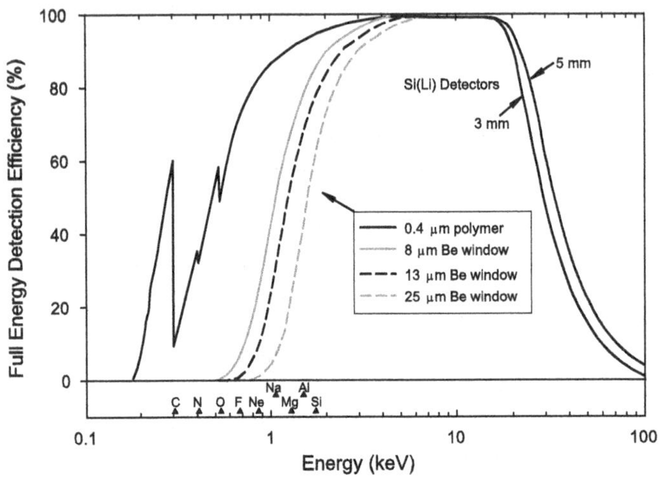

Figure 16.24. The absorption efficiency of a Si(Li) detector as a function of x-ray energy, depletion thickness, and entrance window. Data from Canberra Industries, Inc. 2003.

Unlike the lithium-drifted version of Ge detectors, or Ge(Li) detectors, the Li^+ ions remain locked into position at room temperature. Although Si(Li) detectors can be operated at room temperature, they perform best when cooled to low temperatures. Various Si(Li) detectors are available coupled to either liquid nitrogen (LN2) dewars or Peltier coolers.

The detectors are encapsulated in a protective container with a thin entrance window, typically constructed from Be. The entrance window of the detector affects the low energy sensitivity limit. Shown in Fig. 16.24 is the detection efficiency for various Si(Li) detectors with different thicknesses and different entrance windows. Note that thicker detectors increase the efficiency for higher energy x rays, and an appropriate choice of material for the entrance window can increase the detector efficiency for low-energy x rays.

Si(Li) detectors are best used for spectroscopy of low-energy gamma and x rays. When cooled to cryogenic temperatures (77 K), the resolution performance can be excellent. The energy resolution of a Si(Li) detector can be predicted from Eq. (16.38), as shown in Fig. 16.25. An example x-ray spectrum from a chilled Si(Li) detector is shown in Fig. 16.26.

Si(Li) detectors can be commercially acquired in a variety of segmented patterns, including strips, triangular, and square patterns. The detectors consist of *pin* diode structures individually fabricated into a

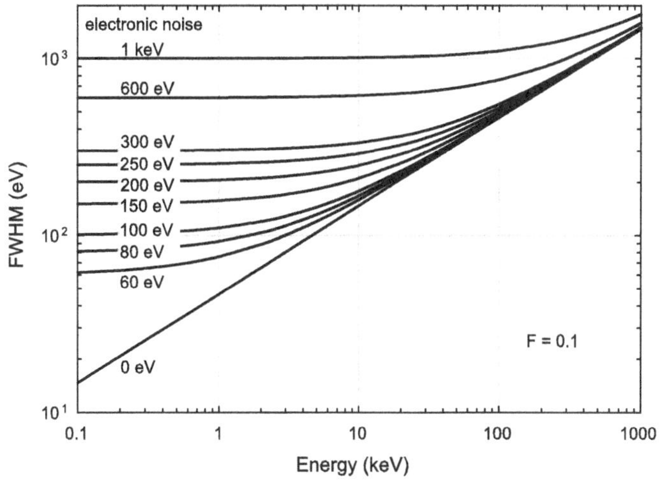

Figure 16.25. The theoretical energy resolution of a Si(Li) detector, which includes contributions from electronic noise. A Fano factor of 0.1 was used for the calculation. After Canberra Industries, Inc. 2003.

single Si substrate, thereby reducing "dead zones" between neighboring detectors. These detectors offer high x-ray energy resolution and spatial interaction information. Further, clever designs can actually improve count rate efficiency for ion probe instrumentation, such as PIXE, by surrounding the target region with multiple detectors. Used in conjunction with other γ-ray detectors, the segmented Si(Li) detectors can be used for Compton scatter γ-ray cameras. Properties of commercially available Si(Li) detectors are presented in Table 16.6.

Table 16.6. Typical gamma-ray energy resolution of a few representative commercially available silicon semiconductor detectors. Many other sizes are offered by the different vendors.

detector	area (mm^2)	energy (keV)	FWHM (keV)	comments	source*
Si(Li)	12.5	5.9	.155–.175	LN2 cooled	C,O
Si(Li)	20	5.9 59.54	.180 .450	Peltier cooled	B
Si(Li)	28–30	5.9	.165–.180	LN2 cooled	C,O
Si(Li)	80	5.9	.175–.190	LN2 cooled	C,O
Si(Li)	200	5.9	.220	LN2 cooled	C,O
Si *pin*	13 25	5.9 5.9	.18–.22 .127–.230	Peltier cooled	A

*A = AmpTek, B = Baltic Scientific, C = Canberra, O = Ortec.

Si(Li) β-Particle Detectors

At energies above 500 keV, the linear energy transfer (collisional) for fast electrons between 500 keV and 10 MeV is nearly constant (see Fig. 4.20). In Si the linear energy transfer ranges between 1.6 MeV cm^2g^{-1} ($\sim$ 3.8 MeV cm^{-1}) at 500 keV to 1.92 cm^2g^{-1} ($\sim$ 4.5 MeV cm^{-1}) at 100 MeV. Unfortunately, excessive reverse voltage is required to deplete surface barrier and *pn* junction Si diodes to thicknesses capable of recording

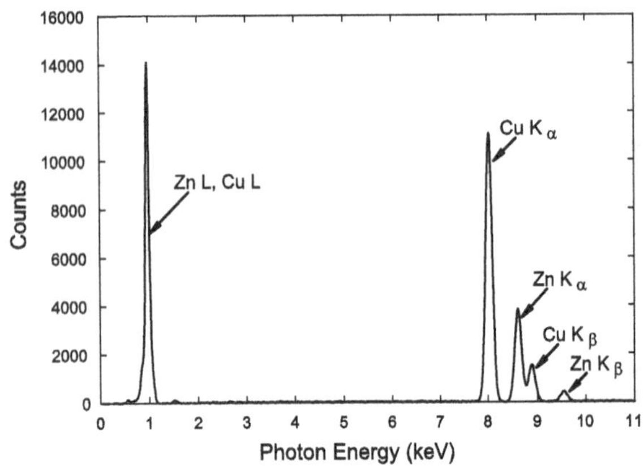

Figure 16.26. X-ray spectrum of a brass sample taken with a Si(Li) detector.

Table 16.7. Maximum particle energy all of which is deposited in Si(Li) within a thickness x. Particles are normally incident on the detector and the thickness x is the $R_{\text{CSDA}}(E_{\max})$.

Thickness (mm)	Electrons Max Energy† (keV)	Protons Max Energy† (MeV)	α particles Max Energy† (MeV)
0.1	116	3.19	12.5
0.2	177	4.75	19.0
0.5	321	8.13	32.5
0.8	444	10.7	42.6
1	522	12.1	48.4
2	889	18.0	71.8
5	1964	30.2	121*
8	3085	39.4	157*
10	3861	44.6	178*

†Determined with the STAR series of codes [Berger 1992].
*Determined with TRIM 2013. See Zeigler et al. [2013].

meaningful energy deposition exceeding a few hundred keV. For instance, a surface barrier detector with $N_b \simeq 10^{12}$ cm^{-3} biased at -100 volts has a depletion region 250 microns wide, yet is capable of absorbing the full energy of only about a 200 keV or lower energy electron (see also Table 16.7).[16] This problem of incomplete energy deposition is mitigated, somewhat, by using Si(Li) detectors, and indeed special planar Si(Li) detectors are designed specifically for the purpose of energetic particle detection.

Detectors with active region depths between 4 mm to 8 mm are commercially available and are capable of absorbing full electron energies ranging from 1.6 MeV to 3 MeV, respectively. Berger et al. [1969b] report lower electron energy limits for some Si detectors, ranging between 20 to 40 keV, depending on the thickness of dead layer which is the detector's front contact. Here the 20 keV limit is for a surface barrier detector and the 40 keV limit is for a Si(Li) detector. Should the detector width be less than the electron range, then the electrons can pass completely through the detector and deposit only a fraction of their energy [Berger et al. 1969a]. Consequently, a low energy tail, a consequence of partial energy deposition, forms in the pulse height spectrum accompanied by a diminished full energy peak (as discussed further in Sec. 20.9). Backscatter losses and bremsstrahlung energy losses can also add to the low energy tail [Frommhold et al. 1991]. Berger et al. [1969a] published tables of response functions that offer a method to determine the energy deposition as a function of detector thickness. Modern Monte Carlo programs can also be used to generate the same data.

Positrons deposit energy in the detector in much the same way as do electrons. However, when they come to rest, through annihilation, two 511-keV photons are released in opposite directions. If either photon undergoes Compton scattering in the device, then the total positron energy plus the added energy left by the scattered photon appear summed in the pulse height spectrum. Consequently, a high energy tail can appear with positron spectroscopy (discussed and shown later in Sec. 20.9). Because Si has a low Z number, the spectrum from a Si(Li) detector suffers less from Compton scattering effects than do detectors composed of higher Z materials.

[16]The maximum energy listed in Table 16.7 is the energy of the particle whose CSDA range is x. So heavy charged particles with $E < E_{\max}$, that mostly travel in straight lines as they slow, are completely stopped within a distance x. By contrast, the tortuous paths traveled by electrons allow some electrons with $E > E_{\max}$ to still deposit all their initial energy within a thickness x.

The electron backscatter coefficient for Si is relatively small, being approximately 0.14 at 100 keV (see Fig. 7.7). As discussed in Sec. 7.2.2, the backscatter coefficients in Fig. 7.7 are for the special case in which the particle enters perpendicular to the surface. However, backscattering becomes worse as the entrance angle decreases from 90°; hence the backscattering coefficients of Fig. 7.7 depict overly optimistic results. Regardless, Si is one of the best semiconductor materials for electron spectroscopy.[17] To improve the measured electron energy spectrum the source can be collimated so electrons are normally incident on the detector. In this manner, improved spectral features can be obtained without excessive backscattering adding to the low energy tail.

There are special precautionary measures that must be taken for charged particle spectroscopy. First, to eliminate energy loss from the environment, the particle source and detector should be in an evacuated vessel. Further, Si(Li) detectors can be operated at room temperature, but better energy resolution is obtained if the detector is cooled to some extent. Commercial spectrometers often come with an integrated method of cooling the detector. However, when chilling the detector, precaution should be taken to prevent condensation forming on the detector itself and, thereby attenuating the incident charged particles. The prevention of condensation is typically achieved by operating the detector in vacuum. Also, some commercial units may have a thin window protecting the entrance contact, but such windows add to the overall dead layer thickness. For spectroscopy of low-energy electrons, it is best to enclose both source and detector, without a window, in the evacuated chamber. In fact, some commercial vendors offer a Si(Li) detector with a mount that can be directly attached to a user's vacuum chamber. For special cases requiring transmission measurements, where the goal is to measure ΔE for a particle passing through the detector, transmission devices, sometimes called *sidelookers*, with thin contacts on front and back are available. Aligning several transmission Si(Li) detectors can be used to measure electron energies that have ranges beyond that of a single detector.

16.4.2 Detectors Based on Ge

Germanium (Ge) was first reported as a semiconductor detector by McKay in 1949. This initial Ge detector was manufactured from a point-contact Ge rectifier, and was reported as a particle counter.[18] Ge is a mature and commercially available semiconductor manufactured in many radiation detector configurations. With a density of 5.33 g cm^{-3} and atomic number of 32, it has modest gamma-ray absorption efficiency, but certainly superior to Si detectors. Ge has a narrow band-gap energy of 0.66 eV at room temperature, increasing to 0.72 eV at 77 K. Although a narrow band gap has the disadvantage of increasing the thermal leakage current, it has the advantage of decreasing the average ionization to only 2.98 eV/ e-h pair. Ge has a cubic FCC lattice formed as a diamond structure. Ge also has an indirect band gap, with a band minimum at L in the [111] direction (see Fig. 16.27), which helps to increase charge carrier lifetimes, often greater than 10^{-3} s for highly pure material. The room temperature charge carrier mobilities are μ_e of 3900 and μ_h of 1900 cm^2 V^{-1}s^{-1}, which are relatively high. However, Ge detectors are usually operated at 77 K, and the mobilities are much higher at μ_e of 39000 and μ_h of 42000 cm^2 V^{-1}s^{-1}. A Fano factor below 0.13 is commonly reported.

Germanium (Ge) Radiation Detectors

Presently, the most popular high resolution γ-ray spectrometers are constructed from high purity Ge (HPGe). The material is purified through zone refinement, resulting in a nearly intrinsic material. Although numerous detector configurations exist, including special order devices, a standard unit is a coaxial $p\pi n$ or $p\nu n$ design

[17] Although both SiC and diamond may have lower backscatter coefficients, these detectors presently do not have the required active region width for high energy electron spectroscopy.

[18] Although Si is the most commonly used semiconductor for electronic components, at one time Ge was widely used for diodes and transistors. Ge transistors are still preferred for some specialized electronics, such as electric guitar pedals.

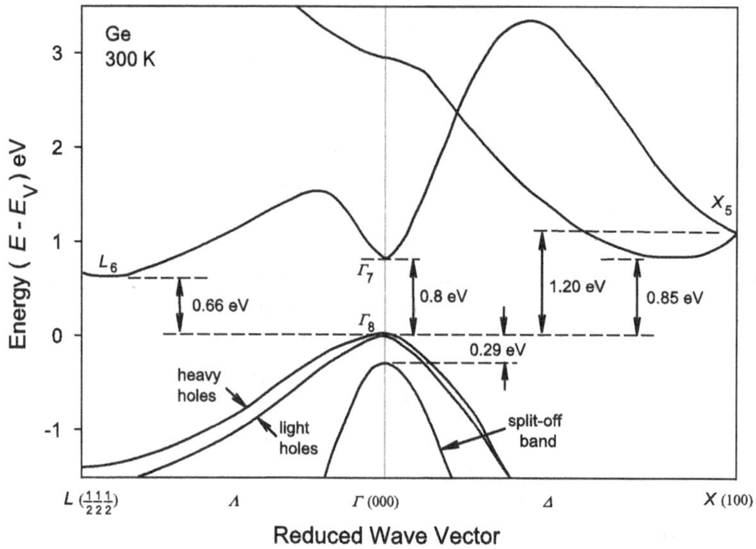

Figure 16.27. The theoretical energy bands of Ge at 300 K calculated with a pseudopotential method. Data are from Chelikowsky and Cohen [1976].

with the rectifying junction on the outer surface (see Fig. 16.36). The coaxial design permits large detectors to be fabricated while minimizing the detector capacitance. With the rectifying surface on the outer surface, rather than on the inner surface, the active volume is increased and detection efficiency for low energy γ-rays is improved.

One difficulty with HPGe detector operation is the need to chill the device during operation. Because of the small band-gap energy (0.66 eV at 300 K), the intrinsic carrier concentration of electrons and holes is much too high at room temperature and significant leakage current is produced when operated at high voltage, which can damage the detector. For this reason, HPGe detectors are typically attached to a dewar and cooled with LN2, or they are attached to a low noise refrigerator system. To ensure that damage does not occur from excessive leakage current should the LN2 be exhausted, most modern systems have a safety shut-off that disconnects the high voltage if the HPGe detector increases to a preset temperature. Portable survey detectors and laboratory units are available with either LN2 dewars or electrically cooled refrigerators. Hybrid LN2/electrical cryostats have become available, in which the main cooler is electrical, backed up with LN2 cooling in case of a power outage.

The usual standard for quality comparisons of HPGe detectors is to quote the energy resolution for 1.33 MeV γ rays from ^{60}Co. The expected energy resolution can be approximated by Eq. (16.37), in which the Fano factor is approximately 0.08. Efficiency, for historical reasons, is quoted most often as a comparison to a 3 in $\times$ 3 in (7.62 cm $\times$ 7.62 cm) right cylindrical NaI:Tl detector with the source placed 25 cm from the face of either detector. For instance, a relative 30% HPGe detector has 30% of the efficiency expected from a 3 in $\times$ 3 in NaI:Tl detector at 1.33 MeV. Although useful as an approximation of detector performance, due to differences in detector geometries and mounting apparatuses, such sweeping generalizations can be erroneous for accurate measurements. It is best to characterize the detector efficiency and resolution by using a method described by ANSI/IEEE 325-1996. A calibrated National Institute of Standards ^{60}Co check source is placed 25 cm from the front of the detector face. The number of counts appearing in the full energy peak for the 1.332 MeV γ ray are divided by the number of emissions over that same time interval, which yields the absolute efficiency. The relative efficiency is found by dividing the absolute efficiency by 1.2×10^{-3}, which is the standard efficiency for a 3 in × 3 in NaI:Tl detector under the same irradiation conditions.

Li Drifted Ge [Ge(Li)] Detectors

Although Ge was first reported as a radiation detector in 1949 [McKay 1949], it was over ten years later that it became a serious contender as a radiation detector (spectrometer) [Freck and Wakefield 1962]. Impurity concentrations were still too high in processed Ge, even for zone-refined materials, in which the background impurities predetermined the depletion width of diodes and consequently also limited the overall detector volume. Even at low temperature, a significant density of background impurities would produce high leakage current. Processed Ge ingots were predominantly p-type from the background impurities. In 1960, Pell introduced the process of Li drifting, a method to electrically drift Li^+ ions through a semiconductor. Li behaves as an n-type dopant in Ge, and counterbalances the p-type contaminants [Tavendale and Ewan 1963; Tavendale 1966]. Further, Li^+ is a small ion and has high mobility in Ge crystals, which allows Li^+ ions to be driven deep into a Ge semiconductor block. Li may be applied to one end of a detector with a grease, oil, or with physical vapor deposition [Goulding 1965; IAEA 1966; Brownridge 1972]. A high voltage of approximately 500 volts is applied to the crystal by metallic clamps such that the positive Li^+ ions drift into the crystal. The Ge sample is heated to approximately 30-60°C, usually from the ohmic current, which can be controlled with thermoelectric coolers. The Li^+ ions drift through the lattice as interstitial ions, individually coming to rest as they neutralize the negative electric field formed by the impurity acceptors, such that $N_D^+ = N_A^-$ [Pell 1960a]. Concurrently, unmatched Li^+ ions continue to drift through the lattice while the voltage is applied. The Li^+ drifting proceeds provided that the drift velocity of the Li^+ ions is much greater than the diffusion force, or $\mu_{\mathrm{Li}^+}\mathcal{E}N_D \gg D\nabla N_D$ [Pell 1960a]. Hence, Li^+ drifting produces a region of intrinsic behaving material. After the drift process is complete, the crystal is etched and mounted in a cryostat and evacuated down to at least 10 millitorr. The crystal is then cooled to 77 K, and this temperature must be maintained *at all times* whether in operation or not. The Li drifting process enabled sizeable Ge detectors to be realized, and Ge detectors became a standard for high-resolution gamma-ray spectroscopy. If the Ge(Li)[19] detector warms up, the Li diffuses away from the compensation sites and the crystal is ruined, no longer performing well as a spectrometer. Because of the problem of Ge(Li) becoming destroyed if allowed to become warm, which unfortunately happened too often, Ge(Li) detectors have been almost entirely replaced by high-purity Ge (HPGe) detectors.[20]

High-Purity Ge Production

If a Ge(Li) detector loses cooling, the Li becomes redistributed through diffusion and an expensive detector is ruined as a result. This Li redistribution was a serious problem. Although such a damaged crystal could be redrifted once, maybe twice, the redrifted detector inevitably had poorer energy resolution. Hence, around 1970 an active program began to purify Ge to such levels that intrinsic material was achieved solely through extreme purification. Two research groups sought to accomplish this purification task, one led by Robert N. Hall [1971] at the General Electric Company and another led by William L. Hansen [1971] at Lawrence Berkeley Laboratory [Haller 1996].

The process adopted to purify Ge was originally introduced by Pfann [1952; 1966], referred to as zone-refinement.[21] In zone refining, a series of molten zones are passed in one direction through an ingot. Depending on the segregation coefficient of the various impurities, they travel with, or opposite, the direction of the molten passes. The segregation coefficient is defined as the solute concentration in the *freezing solid*

[19]Pronounced "jelly".

[20]The Li-drifted Si(Li) detector, pronounced "silly" detector, is still in production because the Li ion is immobile at room temperature and, unlike Ge(Li) detectors, is not ruined if it accidently loses cooling.

[21]Sometimes referred to as zone melting or zone leveling, Pfann [1966] clearly distinguishes the differences between the methods.

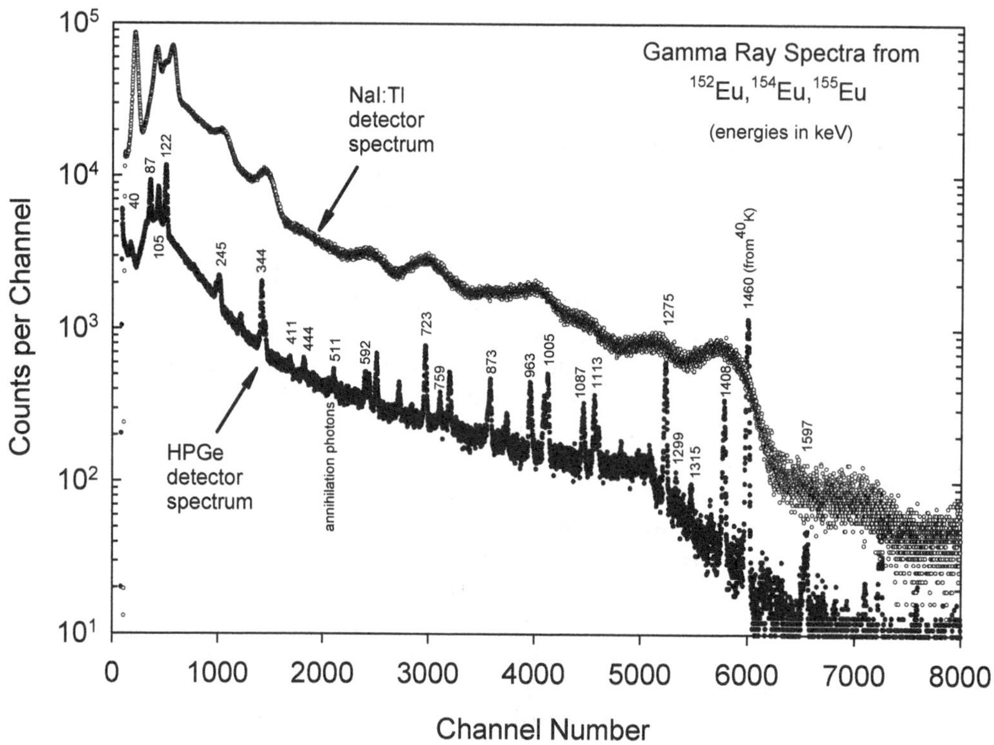

Figure 16.28. A comparison of HPGe detector performance to that of a 3 × 3 NaI:Tl scintillation detector. The gamma-ray source is a mixture of ^{152}Eu, ^{154}Eu, and ^{155}Eu.

to that of the concentration in the *molten liquid.* The segregation coefficient is greater than unity if the impurity raises the melting point of the material, and less than unity if it lowers the melting point. After numerous passes of the molten zones, the impurities collect at either end of the ingot, leaving a relatively pure substance in the middle.

The impure end regions are sliced from the ingot, and the remaining purified material is used to grow a single crystal of Ge (commonly through Czochralski methods). Impurity levels below 2×10^{10} cm^{-3} are common for zone-refined Ge. Although the purified material is nearly devoid of impurities, the finished product is usually slightly p-type (denoted π-type) or slightly n-type (denoted ν-type). The single crystals of high-purity Ge (HPGe) are then sliced into sections and ground to a desired shape, typically a cylinder. The grinding and slicing damage to the crystal is subsequently etched away and p and n junctions are applied to the sample to produce p-π-n or p-ν-n diodes.

The gamma-ray absorption efficiency for Ge (Z = 32) is much less than that for the iodine (Z = 53) in NaI:Tl. Due to the higher atomic number and generally larger size, NaI:Tl detectors often have higher detection efficiency for high-energy gamma rays than do HPGe detectors, but much poorer energy resolution, as shown in Fig. 16.28. When first introduced, Ge detector efficiency was commonly compared to that of a 3 inch diameter × 3 inch long (3 × 3) right circular cylinder of NaI:Tl for 1332-keV gamma rays from ^{60}Co. Even today, efficiency of a Ge detector is quoted in terms of a 3 × 3 NaI:Tl detector (Fairstein et al. 1996). For instance, an HPGe detector denoted as 60% relative efficiency has 60% of the efficiency that a 3 × 3 NaI:Tl detector would have for 1332-keV gamma rays from ^{60}Co. HPGe detectors are much more expensive than NaI:Tl detectors, hence are best used when gamma-ray energy resolution is most important for measurements. If efficiency is of greatest concern, it is often wiser to use a NaI:Tl detector. Comparison spectra between a 20% HPGe detector and a 3 × 3 NaI:Tl detector are shown in Fig. 16.28. Although very expensive, modern manufacturers do produce larger HPGe detectors with 200% relative efficiency.

Various Designs

HPGe detectors are manufactured in various shapes, although most conform to either a planar or coaxial design. Small detectors are commonly manufactured as planar detectors. Relatively large HPGe detectors are manufactured as coaxial devices mainly to keep detector capacitance low. Small HPGe detectors usually have better energy resolution than larger devices, and the larger detectors have better γ-ray detection efficiency.

Planar HPGe planar detectors are typically fabricated by applying a doped region upon a highly pure slab of Ge so that the abrupt junction approximation is valid. The solutions for the electric field and the width of the depletion region were derived in Sec. 15.3. The electric field was determined to be

$$\mathcal{E}(x) = -\frac{|q_e|N_b}{\kappa\epsilon_0}(W - x),\ 0 \leq x \leq W, \tag{16.53}$$

where W is the depletion region width, N_b is the net background doping concentration, κ is the semiconductor dielectric constant, and ϵ_0 is the permittivity of free space. The depletion width, as a function of applied voltage, is

$$W \simeq \left[\frac{2\kappa\epsilon_0(V_{bi} - V)}{|q_e|N_b}\right]^{1/2}, \tag{16.54}$$

where V_{bi} is the junction built-in potential. The full depletion voltage is obtained by setting $W = W_D$,

$$V_D = V_{bi} - \frac{|q_e|N_b W_D^2}{2\kappa\epsilon_0}. \tag{16.55}$$

Remember that V_D is a reverse (negative) voltage by convention. If the applied reverse voltage is higher than the required depletion voltage, then the electric field becomes

$$\mathcal{E}(x) = -\frac{|q_e|N_b}{\kappa\epsilon_0}(W_D - x) - \frac{|V - V_D|}{W_D},\ V > V_D. \tag{16.56}$$

Coaxial The electric field in a coaxial detector can be found by solving Poisson's equation for a cylinder with space charge

$$\nabla^2 V = \frac{1}{r}\frac{\partial}{\partial r}\left(r\frac{\partial V}{\partial r}\right) + \frac{1}{r^2}\frac{\partial^2 V}{\partial\phi^2} + \frac{\partial^2 V}{\partial z^2} = \frac{-q_e(N_A^- + N_D^+ + n + p)}{\kappa\epsilon_0}, \tag{16.57}$$

where κ is the semiconductor dielectric constant, ϵ_0 is the permittivity of free space. For the following analysis, assume that the detector is a straight coaxial cylinder, thereby eliminating the angular and longitudinal dependences of ϕ and z. Under operation, it is assumed that the device is fully depleted of charges, leaving behind a net space charge, denoted here as ρ_c. Eq. (16.57) reduces to

$$\frac{1}{r}\frac{d}{dr}\left(r\frac{dV(r)}{dr}\right) = \frac{d^2V(r)}{dr^2} + \frac{1}{r}\frac{dV(r)}{dr} = -\frac{\rho_c}{\kappa\epsilon_0}, \tag{16.58}$$

The rectifying contact is usually fabricated on the outer surface of a coaxial detector, although it could also be formed on the inner surface. The choice of placing the pn junction on the outer surface allows the depletion region to increase from the largest geometric volume, easily understood by approximating

$$\Delta v \simeq 2l\pi r\Delta r \tag{16.59}$$

where Δv is an incremental change in volume, l is the coaxial length, Δr is a incremental increase in radius, and r is a radial distance located between r_1 and r_2. Here r_1 is the coaxial inner radius and r_2 is the outer radius. The volume incrementally increases the most where r is the largest, *i.e.*, at r_2.

HPGe coaxial detectors are usually operated with the high voltage applied to the outer contact, with the inner contact held at ground potential. For gas-filled coaxial detectors, the high voltage is almost always applied to the central wire with the outer cylinder wall held at ground potential. However, because the actual Ge crystal is contained in a vacuum sealed protective housing, the high potential is not accessible to the user. Hence, the high voltage is applied to the outer contact at r_2, with the electronic readout connected to the contact at r_1. This bias configuration reduces electronic noise introduced into the input FET (at r_1).

To solve Eq. (16.58) first use the change of variable $r = e^y$, from which

$$\frac{dV}{dr} = \frac{dV}{dy}\frac{dy}{dr} = \frac{dV}{dy}e^{-y}, \tag{16.60}$$

and

$$\frac{d^2V}{dr^2} = e^{-2y}\left(\frac{d^2V}{dy^2} - \frac{dV}{dy}\right). \tag{16.61}$$

Substitution of Eq. (16.60) and Eq. (16.61) into Eq. (16.58) then yields the equation

$$\frac{d^2V}{dy^2} = -\frac{\rho_c}{\kappa\epsilon_0}e^{2y}, \tag{16.62}$$

whose general solution is

$$V(y) = \frac{-\rho_c e^{-2y}}{4\kappa\epsilon_0} + C_1 y + C_2. \tag{16.63}$$

Substitution of $r = e^y$ back into Eq. (16.63) yields

$$V(r) = \frac{-\rho_c r^2}{4\kappa\epsilon_0} + C_1 \ln(r) + C_2, \tag{16.64}$$

where C_1 and C_2 are arbitrary constants.

The values of C_1 and C_2 can be found in terms of the boundary voltages $V(r_1)$ and $V(r_2)$. In terms of $V_0 = V(r_2) - V(r_1)$, C_1 is found to be

$$C_1 = \frac{V_0 - \dfrac{\rho_c}{\kappa\epsilon_0}\left[\dfrac{r_2^2 - r_1^2}{4}\right]}{\ln\left[\dfrac{r_2}{r_1}\right]}, \tag{16.65}$$

and C_2 is

$$C_2 = \frac{\rho_c r_1^2}{4\kappa\epsilon_0} - \frac{V_0 - \dfrac{\rho_c}{\kappa\epsilon_0}\left[\dfrac{r_2^2 - r_1^2}{4}\right]}{\ln\left[\dfrac{r_2}{r_1}\right]}\ln(r_1). \tag{16.66}$$

Hence, at full depletion of the HPGe volume, the voltage distribution in the HPGe crystal is

$$V(r) = \frac{-\rho_c r^2}{4\kappa\epsilon_0} + \frac{V_0 - \dfrac{\rho_c}{\kappa\epsilon_0}\left[\dfrac{r_2^2 - r_1^2}{4}\right]}{\ln\left[\dfrac{r_2}{r_1}\right]}\ln\left[\frac{r}{r_1}\right] \tag{16.67}$$

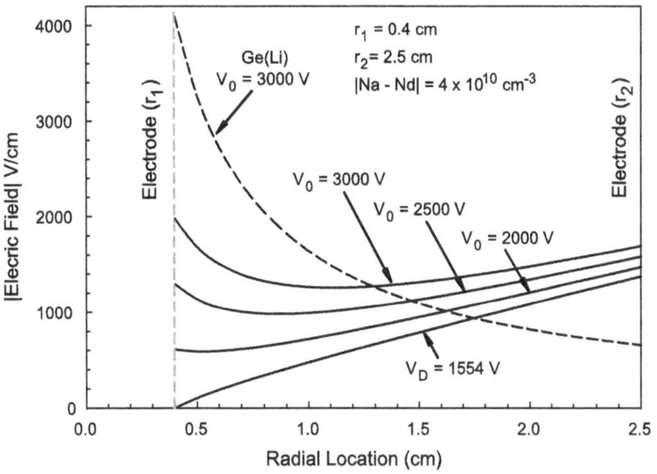

Figure 16.29. Calculated electric field profiles for a HPGe detector with $r_1 = 0.4$ cm and $r_2 = 2.5$ cm. The background impurity concentration is $|N_A - N_D| = 10^{10}$ cm^{-3}. The calculated minimum full depletion voltage is 1554 volts. Also shown is a comparison to the calculated electric field within a Ge(Li) detector of the same size.

and the corresponding electric field is

$$\mathcal{E}(r) = -\frac{dV}{dr} = \frac{\dfrac{\rho_c}{\kappa\epsilon_0}\left[\dfrac{r_2^2 - r_1^2}{4}\right] - V_0}{r \ln\left[\dfrac{r_2}{r_1}\right]} - \frac{\rho_c r}{2\kappa\epsilon_0}. \tag{16.68}$$

If the electric field just barely reaches the contact at r_1, then the device is said to be fully depleted. Hence, with the condition $\mathcal{E}(r_1) = 0$, the voltage required for full depletion is,

$$V_D = \frac{\rho_c}{\kappa\epsilon_0}\left(\left[\frac{r_2^2 - r_1^2}{4}\right] - \frac{r_1^2}{2}\ln\left[\frac{r_2}{r_1}\right]\right). \tag{16.69}$$

An example of the electric field expected in an HPGe detector of average dimensions is shown in Fig. 16.29 compared to that expected from a Ge(Li) detector of the same size and dimensions. Note that the weighting potential for both the HPGe and Ge(Li) detectors is the normalized field distribution of the Ge(Li) detector.

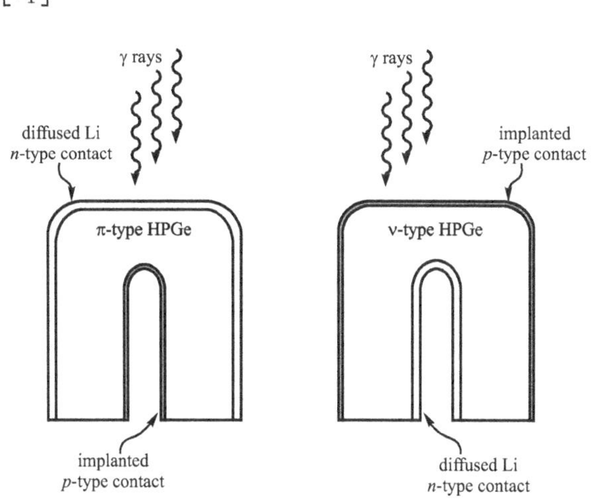

Figure 16.30. Typical configurations for π-type and ν-type coaxial HPGe detectors.

Coaxial detectors are usually fabricated as blind right cylinders, in which the coaxial center hole is not drilled completely through the cylinder. The blind section of the detector is used as the entrance window, a geometry that serves to increase the gamma-ray absorption efficiency. However, for a right circular cylinder, the electric field strength around the detector edge perimeter is weak and causes reduced charge carrier velocity, usually observed as longer pulse rise times.

To remedy the problem, blind coaxial detectors are bulletized (end is rounded) to even out and increase the electric field strength around the edge (see Fig. 16.30).

The capacitance of a straight coaxial HPGe detector can be determined from the depletion depth. For a simple coaxial capacitor, the difference in potential from the inner conductor to the outer conductor can be found from

$$V = \int_{r_1'}^{r_2} \frac{V_0}{r} dr = V_0 \ln\left(\frac{r_2}{r_1'}\right), \tag{16.70}$$

where the undepleted region may be too conductive to behave as a dielectric; hence, r_1' is the depletion region boundary. The total charge at the depletion boundary is

$$Q = \int_0^L dz \int_0^{2\pi} r_1' d\phi \frac{\kappa\epsilon_0 V_0}{r_1'} = 2\pi L \kappa\epsilon_0 V_0, \tag{16.71}$$

where L is the cylinder length. Capacitance is defined as $C = Q/V_0$, hence, at full depletion, it is obvious that the capacitance per unit length becomes,

$$\frac{C}{L} = \frac{2\pi\kappa\epsilon_0}{\ln(r_2/r_1)}. \tag{16.72}$$

Pulse Shape

Determining the pulse shape for planar detectors is relatively straightforward. At 77 K, the hole mobility is higher than the electron mobility, with $\mu_e = 36000$ cm^2 V^{-1} s^{-1} and $\mu_h = 42000$ cm^2 V^{-1} s^{-1}, giving a μ_e/μ_h ratio of 0.875. Consequently, it takes 1.143 times longer to collect electrons over the same distance as holes. It is expected that holes are collected faster than electrons, linearly, with electric field. From Eq. (8.44) the charge induction, without trapping, for a planar detector is

$$\frac{dQ}{dt} = \frac{q_e}{W}[v_e + v_h] = \frac{q_e\mathcal{E}}{W}[\mu_e + \mu_h] \tag{16.73}$$

while both types of charge carriers are in motion. Otherwise, as one charge carrier is collected, and the other is in motion, only one charge carrier type contributes to the induced current. If only electrons are in motion,

$$\frac{dQ}{dt} = \frac{q_e\mu_e\mathcal{E}}{W}, \tag{16.74}$$

and if only holes are in motion,

$$\frac{dQ}{dt} = \frac{q_e\mu_h\mathcal{E}}{W}. \tag{16.75}$$

Normalized to the total current

$$\left(\frac{W}{q_e\mathcal{E}(\mu_e + \mu_h}\right)\frac{dQ}{dt} = \frac{\mu_e}{\mu_e + \mu_h}\Bigg|_0^{t_e'} + \frac{\mu_h}{\mu_e + \mu_h}\Bigg|_0^{t_h'}, \tag{16.76}$$

where t_e' is the electron collection time and t_h' is the collection time for holes. Substitution of the 77 K mobilities for Ge, the normalized pulse increases as

$$\left(\frac{W}{q_e\mathcal{E}(\mu_e + \mu_h}\right)\frac{dQ}{dt} = 0.4615\Bigg|_0^{t_e'} + 0.5385\Bigg|_0^{t_h'}. \tag{16.77}$$

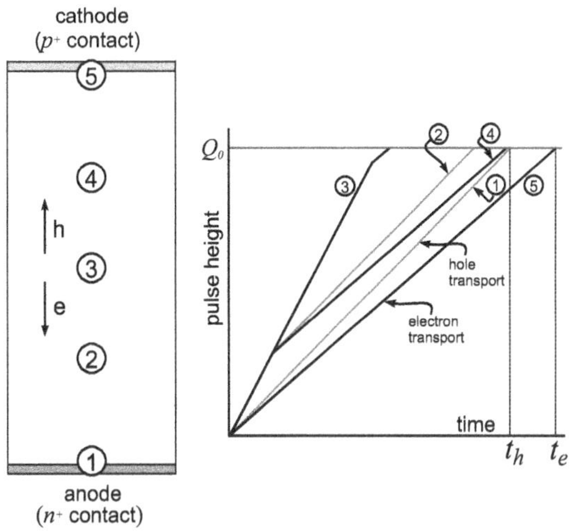

Figure 16.31. The expected position-dependent pulse shapes from a planar HPGe detector operated at 77 K.

Ballistic deficit from the *RC* time constant notwithstanding, the position dependent pulse shape is depicted in Fig. 16.31. If the event occurs at the anode, then the entire pulse is dependent upon hole collection, extending to the maximum hole collection time of t_h. Similarly, if the event occurs at the cathode, then the entire pulse is dependent upon electron collection, extending to the maximum electron collection time of t_e.

The pulse shape of a coaxial HPGe detector is much more complicated, a consequence of both the non-uniform electric field and the non-uniform weighting potential. If the electric field through the device is high enough to reach velocities near saturation, the pulse shapes can be predicted primarily from the weighting potential, assuming no trapping complications. The normalized induced charge for a coaxial detector is given by Eq. (8.80), namely

$$\frac{Q(t)}{Q_0} = \left[\ln\left(\frac{r_2}{r_1}\right)\right]^{-1} \ln\left(\frac{r_2'}{r_1'}\right). \tag{16.78}$$

where r_1' and r_2' are the free charge carrier locations after moving from the initial starting location at r_0. By substitution of $r' - r_0 = vt$ into Eq. (16.78), the pulse shape of a $p^+\pi n^+$ detector, where electrons are collected at the outer contact, is described by

$$\frac{Q(t)}{Q_0} = \left[\ln\left(\frac{r_2}{r_1}\right)\right]^{-1} \left[\ln\left(1 + \frac{v_e t}{r_0}\right)\Big|_0^{t_e'} - \ln\left(1 - \frac{v_h t}{r_0}\right)\Big|_0^{t_h'}\right], \tag{16.79}$$

where t_e' and t_h' are the electron and hole collection times, respectively. An example of the expected pulse shapes, as calculated from Eq. (16.79), is shown in Fig. 16.32 for a $p^+\pi n^+$ HPGe detector operated at saturation velocity (10^7 cm s^{-1}). For a $p^+\nu n^+$ HPGe detector, the roles of electrons and holes are switched in Eq. (16.79).

Cryogenics and Cooling

HPGe detectors no longer need to be constantly chilled to low temperature as did the older Ge(Li) detectors. However, they still must be chilled to cryogenic temperatures during operation to suppress thermal leakage current. Otherwise, the energy resolution is poor and the detector may suffer severe damage from excessive thermal leakage current. Modern HPGe detector units have a thermal sensor that turns off the detector voltage if the detector temperature rises too high, which can happen if the coolant is allowed to expire or

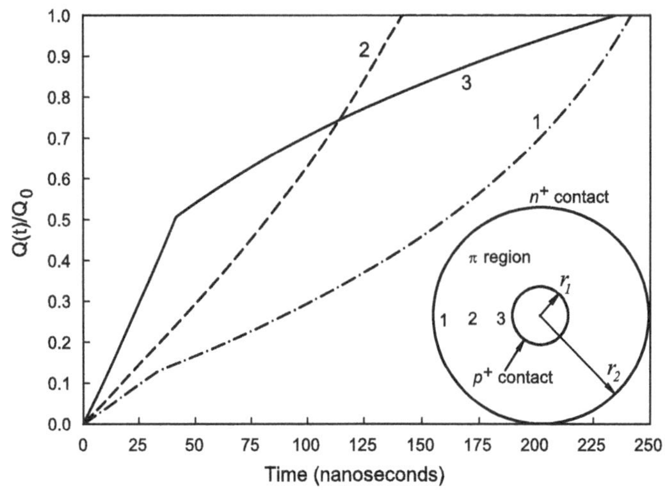

Figure 16.32. The expected position-dependent pulse shapes from a coaxial HPGe detector operated at 77 K, determined from Eq. (16.79). The calculation was performed for a $p^+\pi n^+$ detector with $r_1 = 5.85$ mm and $r_2 = 33.5$ mm. The start locations are identified as $r_0 = 30$ mm, 20 mm, and 10 mm for locations 1, 2, and 3, respectively.

prematurely boils off from vacuum failure in the dewar/stem assembly. This thermal sensor precaution is added to prevent severe damage to the preamplifier circuit. Cooling is achieved by two common methods. The first is the incorporation of a cryogenic dewar filled with liquid nitrogen (LN2). The second is a mechanical refrigerator unit operating on a modified Stirling cycle.

The HPGe detector crystal is mounted inside a mounting cup with standoffs to hold it in place. Usually the mounting cup is electrically attached as an outer contact to the crystal and a conductive pin is used as the inner contact within the hollow penetration. The mounted crystal is placed inside a housing, hermetically sealed, and evacuated. Older designs had the entire stem and housing evacuated; newer designs use a modular approach, in which only the detector compartment is evacuated. In the newer designs, an insulator plug partitions the electronics from the evacuated detector chamber. In either case, the detector crystal is shielded by the mounting cup and the housing, which unfortunately reduce efficiency for low-energy gamma rays. It is common to include a package of molecular sieve in the evacuated chamber to absorb residual gas molecules when the chamber temperature is lowered. The mounted detector is attached to a cold stem that is inserted into an LN2-filled dewar. The dewar is lined with "superinsulation", made of multiple thin sheets of aluminized Mylar®, to reduce LN2 losses and lengthen the time between periodic refills. The cold finger is usually made of copper, and is axially located inside a vacuum-tight jacket. This jacket is also partially filled with molecular sieve, used to getter residual gas molecules at low temperature. The preamplifier electronics are stationed close to the detector to reduce line capacitance. In many models, the preamplifer is situated adjacent to the detector house. In modern modular models, the entire preamplifier is situated next to the detector within the housing (see Figs. 16.33 and 16.34).

The actual detector is much smaller in size than the housing, and often not exactly a perfect cylinder. From experience, some detectors may have a small section sliced from the crystal (to remove a twin or bad region). The actual spectroscopic performance is within the vendors specification, as is the quoted efficiency. However, casual solid angle calculations without taking into account possible geometry differences produces measurement error. The accepted method of calibrating a HPGe detector is documented in IEEE 325-1996 [Fairstein et al. 1996], which mostly negates problems with detector shape and size.

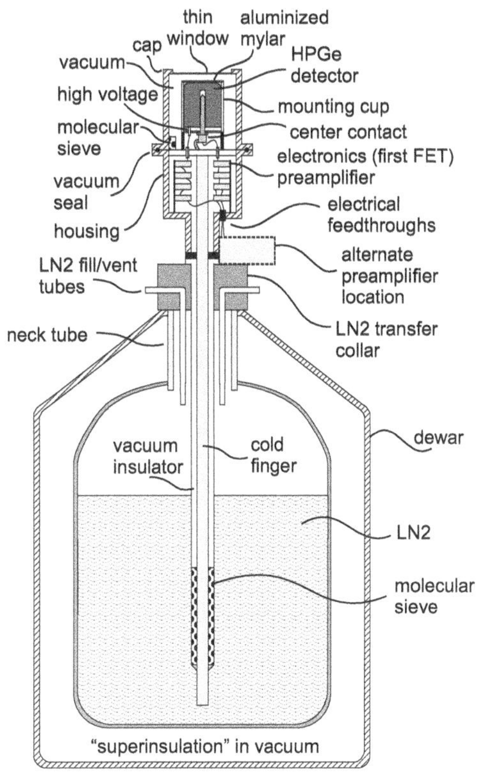

Figure 16.33. A common configuration for an HPGe detector is the vertical dipstick.

LN2 loss of about 15 to 45 ml per hour is expected, although the loss rate can change with environment. Common 30 liter laboratory dewars are refilled at least every two weeks to avoid accidents or other problems. Measurements should not be conducted during refills to avoid microphonics noise that could ruin the results. If the dewar becomes damaged, such that the insulation is compromised, unusually fast boil-off may occur. A cross section diagram of a common HPGe dewar system in the vertical dipstick configuration is shown in Fig. 16.33. Alternative configurations include the horizontal dipstick ("side looker"), downlooking device with the detector pointing down positioned under the dewar, the sidelooking detector attached near the bottom of the dewar, and the "J" type detector with detector attached near the dewar bottom, but the detector and housing positioned to point upwards. There are smaller dewars for portable units, usually rated for approximately 24 hours of use. Some modern dewar systems use a combined system between a refrigerator and conventional LN2 cooling. These units have a refrigerator that recycles the LN2 boil-off back into the LN2 dewar. In the case of a power outage, the LN2 keeps the system cool for up to a week. Because of the recycling of the LN2 in the chamber, these detectors can go without refilling for over a year.

In the last decade, major improvement with mechanical coolers have allowed modern HPGe detectors to operate without LN2, a convenience for locations where LN2 is hard to come by. Originally, these mechanical coolers were used to bring the HPGe detector down to operating temperature ($\sim$ 77 K), and then switched off during a measurement, a procedure requirement because vibrations would cause excessive microphonic noise that would compromise the spectrum. However, vast improvements in refrigerator design now allow these systems to continue operation during a spectroscopic measurement. Some laboratory systems have a large compressor with an umbilical cooler attachment that couples to a modular HPGe detector. Portable

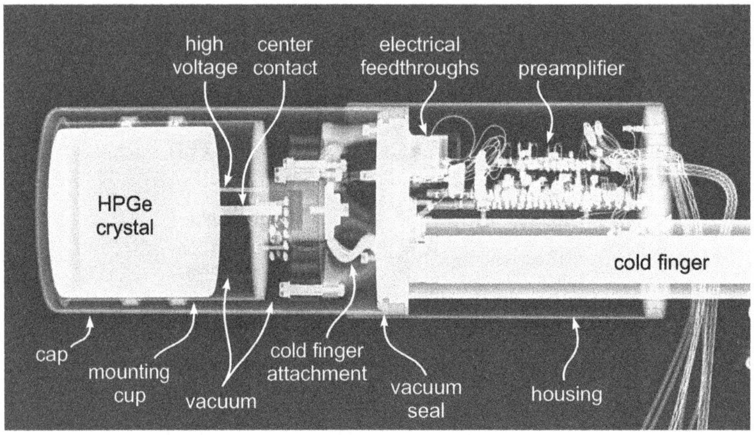

Figure 16.34. X-ray image of an HPGe detector mounting and electronics configuration for a side-looking detector.

units have either modified Stirling cycle refrigerators or pulse tube refrigerators. In either case, the modern units have negligible vibrations and can operate without LN2.

Detection Efficiency, Absorption Losses, and Dead Layers

The response functions for ν-type and π-type HPGe detectors are quite different at low energies. High purity π-type detectors are fabricated with Li, an n-type dopant, as the rectifying contact diffused at a depth of approximately 700 μm thick around the outer surface. A much thinner implanted junction of p-type dopant (typically boron), approximately 300 nm deep, is formed as the ohmic contact on the inner surface. Consequently, the relatively thick "dead" layer formed by the outer contact significantly reduces the detector sensitivity to low energy photons (typically below 40 keV). From Eqs. (16.78) and (16.79), the reverse bias configuration and geometry of $p\pi n$ HPGe detectors cause the average output pulse to be dominated by hole motion. These $p\pi n$ HPGe detectors typically have slightly better energy resolution at high γ-ray energies than $p\nu n$ HPGe detectors.

High purity ν-type detectors are fabricated with p-type dopants implanted and activated at a depth of approximately 300 nm for the outer rectifying contact. A much thicker diffused Li junction up to 700 μm thick is fabricated as the ohmic contact. As a result, low energy γ rays and x rays encounter less "dead" layer in the outer contact, thereby increasing the efficiency for these low energy photons. To take further advantage of the thin surface contact, these ν-type detectors are typically packaged in a can that has a thin Be window, thereby minimizing γ-ray and x-ray attenuation through the detector container. An additional advantage with ν-type HPGe detectors is their increased radiation hardness to neutron radiation. Neutron damage tends to form hole trapping sites, hence the electron dominated pulses from ν-type HPGe detectors are somewhat less affected.

In order to reduce absorption losses from low-energy gamma rays, the front window of the detector cap is fabricated from thin low-density material, commonly aluminum approximately 1.5 mm thick. Aluminum windows are used for efficient transmission of energies above 30 keV. However, detectors with special composite carbon windows or beryllium windows can be purchased that allow transmission for low-energy gamma rays and x rays down to 10 keV and 3 keV, respectively. Depending on the vendor and the design, the detector mounting cup may have an open end design, or instead may have a thin aluminized Mylar® window, apparently in place to reduce infrared absorption. The HPGe crystal is kept in vacuum, which insulates the cryogenically cooled crystal from the external environment. Otherwise, ice crystals can form on

Table 16.8. Typical gamma-ray energy resolution of a few representative commercially available HPGe semiconductor detectors. Many other sizes are offered by the different vendors.

HPGe detector	relative eff. (%)	energy (keV)	FWHM (keV)	comments	source*
p-type coaxial	20	122	.715–.975	LN2 cooled	B,C,O,P
	50		.9–1.2	LN2 cooled	B,C,P
	100		1.2–1.4	LN2 cooled	B,C,O,P
p-type coaxial	20	1332	1.8–2.0	LN2 cooled	B,C,O,P
	50		1.9–2.1	LN2 cooled	B,C,P
	100		2.0–2.3	LN2 cooled	B,C,O,P
n-type coaxial	20	122	.69–1.0	LN2 cooled	C,P
	50		.86–1.2	LN2 cooled	C,P
	70		1.1–1.3	LN2 cooled	C,P
n-type coaxial	20	1332	1.8–2.0	LN2 cooled	C,O,P
	50		2.1–2.3	LN2 cooled	C,O,P
	70		2.3–2.5	LN2 cooled	C,P

*B = Baltic Scientific, C = Canberra, O = Ortec, P = PGT (now defunct).

the cap, and in the worst case, inside the cap on the actual detector. The vacuum is also necessary to reduce the accumulation of contaminants on the Ge crystal surface. Over time, an accumulation of contaminants adhering to the Ge surface can increase the detector leakage current, a value that must be maintained below 20 pA during operation. Because the detector windows are thinned at the front entrance, these detectors are designed for irradiation from the front. Note that gamma rays may enter from the sides with relatively good efficiency for moderately high energy gamma rays, yet the user should be aware that detection efficiency changes between front irradiation and alternative orientations.

Detectors to be used primarily for low energy measurements need not have a large block of germanium to accomplish the task, nor is the coaxial configuration required. Instead, the planar design is used, with effective diameters available between 6 mm to 70 mm and ranging between 6 mm to 20 mm thick. The planar detectors are $p\nu n$ devices, with an outer *p*-type implanted contact to reduce the dead layer. The edges are not bulletized, but instead maintain 90° edges, which improves the efficiency of gamma-ray measurements for sources placed close to the detector face. These detectors have better overall absorption efficiency than a Si(Li) detector because of the higher Z (32 opposed to 14); however, they suffer a noticeable decrease in absorption efficiency at 11.1 keV from the Ge K-edge, a fact that should be taken into consideration when deciding on a type of detector for low energy gamma-ray measurements.

As previously stated, the standard reporting procedure for efficiency calibration is outlined in IEEE Standard 325-1996, and efficiency is reported for 1.33 MeV gamma rays from ^{60}Co as a comparative ratio to that of a 3 in × 3 in right circular cylinder NaI:Tl detector [see Fairstein et al. 1996]. HPGe detectors can be obtained commercially with efficiencies ranging from 10% up to 200% [Sangsingkeow et al. 2003], with expense significantly increasing with efficiency. Energy resolutions for some commercially available HPGe detectors are listed in Table 16.8.

Efficiency response examples for a few HPGe variations are shown in Fig. 16.35. Notice in Fig. 16.35 the dip in efficiency at the Ge K absorption edge (11.1 keV). Also note the efficiency reduction below 100 keV for the *p*-type HPGe detector, which becomes only an issue for the *n*-type devices represented in Fig. 16.35

at energies below 10 keV. The drop in efficiency is due to a combination of photon absorption in the detector contact dead region and the container holding the detector. Detectors specifically designed for low energy γ-ray spectroscopy typically have thin Be windows that do not appreciably attenuate γ rays entering the device. Overall, the decision regarding which HPGe detector is best for an application requires some knowledge of the preferred energy resolution, necessary detection efficiency, and the photon energy range of interest.

Radiation Damage

Radiation interactions in Ge can cause displacement damage. Under common use, charged particles typically cannot penetrate the detector housing, and gamma-ray damage is minimal. It is damage from fast neutrons that can cause degradation in detector performance by producing Frenkel defects that give rise to charge carrier trapping. Consequently, HPGe detectors operated in radiation fields with appreciable fast neutron exposure suffers, over time, from performance degradation, manifested as a trapping tail (Sec. 15.5.1) and broadening of the energy resolution.

These damage induced defect sites preferentially trap holes over electrons. Consequently, coaxial ν-type detectors are less susceptible to radiation damage effects than coaxial π-type detectors. This difference can be understood by inspecting the geometric weighting and the weighting potential of the detectors. Recall from Eq. (16.59) that the outer detector volume has a higher probability of gamma-ray interactions than inner portions, obvious from volumetric arguments. A π-type detector has the rectifying barrier on the outside perimeter, and holes are drifted in the direction of $r_2 \rightarrow r_1$. With a linear dependence of r, the geometrical dependence indicates that more hole dominated pulses appear than electron dominated pulses.

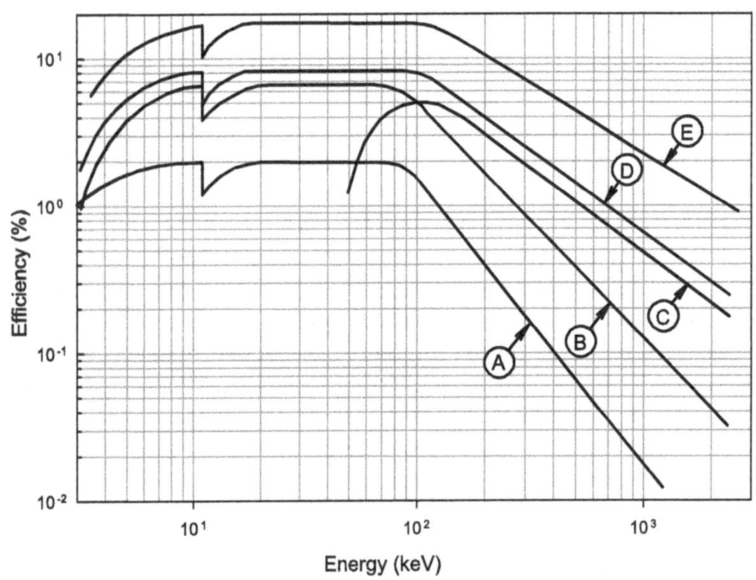

Figure 16.35. The absolute detection efficiency for several HPGe detector configurations with a 2.5 cm source to end cap spacing, showing a (A) 200 $mm^2 \times 10$ mm thick low energy ν-type nominally planar HPGe detector, (B) 10 $cm^2 \times 15$ mm thick low energy ν-type nominally planar HPGe detector, (C) coaxial π-type HPGe detector with 10% relative efficiency, (D) coaxial thin window ν-type 15% relative efficiency HPGe detector, and a (E) broad energy range π-type 5000 $mm^2 \times 30$ mm thick nominally planar HPGe detector. After Canberra, Incorporated [2016].

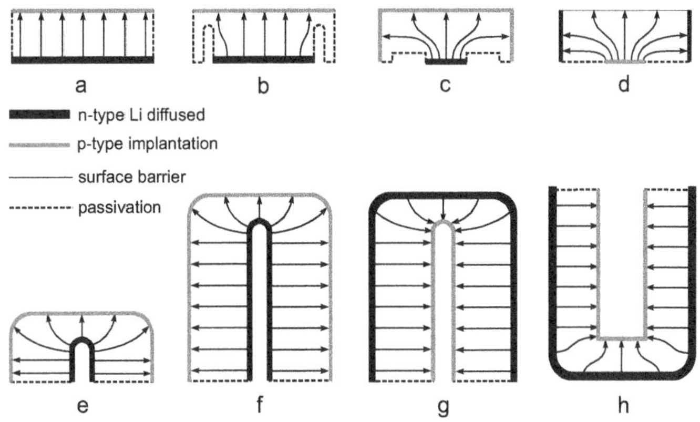

Figure 16.36. Different commercially available HPGe detector structures with electric field lines depicted: (a) basic planar detector, (b) grooved planar, (c) low capacitance planar, (d) surface barrier planar, (e) truncated $p\nu n$ blind coaxial, (f) $p\nu n$ blind coaxial, (g) $n\pi p$ blind coaxial, (h) $p\pi n$ well configuration. After Darken and Cox [1995], Canberra [2016] and Ortec [2016].

From the coaxial design, the induced current is highest as charge carriers drift in the vicinity of r_1 (see Sec. 8.7). Under severe hole trapping, holes originating from the outer regions near r_2 may not reach r_1, and consequently, the pulse output is compromised and lower than expected, producing low energy tails for energy peaks in pulse height spectra.

The opposite case is true for ν-type detectors, for which the majority of pulses are electron dominated. Because the neutron induced defects are primarily hole traps, electron transport from r_2 to r_1 is less affected by charge carrier losses. Holes drifting from r_1 towards r_2 may suffer from trapping, but most of their induced charge is produced in the region nearest r_1. Hence, holes can still contribute a significant amount to the induced current before they are lost to traps.

Both devices can be heated and annealed to remove neutron radiation damage. Detectors manufactured from ν-type Ge can be annealed at approximately 100°C for one day to effectively remove defects. However, π-type HPGe detectors must be annealed at slightly higher temperature ($\sim$ 120°C) for up to a week to effectively remove neutron induced defects. Unfortunately, because Li is used as the n-type dopant on the outer surface of π-type HPGe detectors, the high diffusivity of Li allows it to drive deeper into the detector and increase the outer surface dead region. Consequently, the low energy sensitivity of π-type detectors reduces with the annealing procedure. However, the implanted p-type contact on the outer surface of ν-type detectors is not affected by diffusion at the relatively low annealing temperatures, and consequently does not suffer a reduction in low energy efficiency. In short, ν-type HPGe detectors are better suited for operation in a fast neutron environment than are π-type HPGe detectors.

Detector Structures

There are several types of HPGe detector structures available for specific uses. These structures include detectors optimized for low energy gamma-ray measurements, specifically designed with thin dead regions and thin entrance windows. Other detectors are optimized for best energy resolution, increased radiation hardness, or high detection efficiency. Some of the more popular structures are depicted in Fig. 16.36 and discussed in the following sections.

Although detection efficiency is mostly determined by the detector volume, the energy resolution is a function of the statistical fluctuations in the production of electron-hole pairs, charge carrier losses during

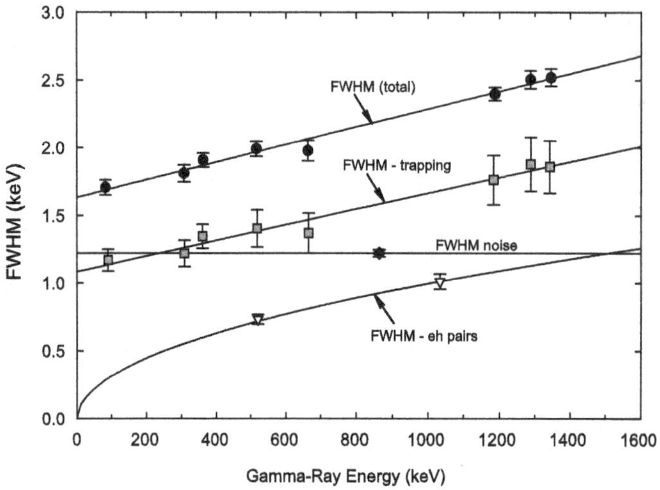

Figure 16.37. Example of energy resolution dependence on gamma-ray energy, electronic noise, and trapping. The curves for the total and the trapping FWHMs were found with a linear least squares fit and the curve for the eh pairs FWHM was modeled with $F = 0.06$ in Eq. (16.83). Data are from Owens [1985].

transport, and signal fluctuations from electronic noise, so that

$$\text{FWHM} = \left[(\text{FWHM}_{\text{noise}})^2 + (\text{FWHM}_{\text{trap}})^2 + (\text{FWHM}_{\text{eh}})^2\right]^{1/2}. \tag{16.80}$$

The contribution from statistical fluctuations in excited charge carriers is presumed Gaussian in nature, the variance in number of charge carriers produced is

$$\sigma_{eh}^2 = \frac{FE_\gamma}{w} = F\bar{N}, \tag{16.81}$$

where E_γ is the gamma-ray energy, w is the average excitation energy, F is the Fano factor correction constant, and $\bar{N}$ is the average number of charge carriers produced. The fractional standard deviation is

$$\sigma_{eh(f)} = \sqrt{\frac{FE_\gamma}{w}}\frac{1}{\bar{N}} = \sqrt{\frac{wF}{E_\gamma}}, \tag{16.82}$$

and by multiplying this fraction by the original gamma-ray energy, the contribution of the excited charges to the energy resolution is found as

$$\text{FWHM}_{\text{eh}} = 2\sqrt{2\ln(2)}\sigma_{eh(f)}E_\gamma = 2.355\sqrt{wFE_\gamma}. \tag{16.83}$$

Contributions from electronic noise can generally be measured with a pulser added into the circuit under normal detector operation, usually through a connection port provided on the preamplifier. The value of $\text{FWHM}_{\text{noise}}$ can be measured from the resulting pulse height spectrum of the pulser input. Sources include the detector and line capacitance, possible ground loops, and impedance mismatch of connectors.

As seen in Sec. 15.5.1, the pulse height spectrum is skewed towards a lower energy tail, an effect that is important for high trapping losses. Hence, the FWHM is not strictly Gaussian in nature.[22] Further, the contribution from charge carrier trapping changes with detector voltage and size. Larger detectors suffer more trapping, and consequently a higher variance contribution. Increasing the voltage can reduce the effect of trapping, but with limited results. Ultimately, charge carriers reach a saturation velocity (see Fig. 16.5), and increasing the voltage actually has an adverse effect if electronic noise increases. Further, at high enough voltages the diode can reach a voltage breakdown condition. Manufacturers usually have a suggested optimum voltage that produces the best overall energy resolution.

Overall, the total energy resolution FWHM of HPGe detectors, in units of energy, increases with gamma-ray energy. This dependence becomes obvious from Eq. (16.83) for the FWHM in electron-hole pairs produced, but should also be intuitive from the trapping consequences described in Sec. 15.5.1. As the gamma-ray energy increases, the photon absorption becomes more uniform, which decreases the energy resolution in the presence of trapping. Note also from Fig. 16.37 that electronic noise has more effect on resolution for low-energy gamma rays than with high-energy gamma rays.

Planar Detectors The physical attributes of planar semiconductor detectors have been presented in Chapter 15 and in prior sections of this chapter. Of particular note is that planar structures are much easier to fabricate than coaxial detectors. However, the linear weighting potential of a planar detector is more susceptible to resolution degradation from charge carrier trapping. Fortunately, HPGe detectors have low trap densities, long charge carrier lifetimes, and high charge carrier mobilities at low temperature. It is found that smaller planar HPGe detectors yield better energy resolution, with many quoted as having a resolution of less than 200 eV FWHM at 10 keV. HPGe detectors designed for low energy gamma-ray spectroscopy do not need large active volumes, and hence, do not need to be configured in a coaxial structure. Planar HPGe detectors have special use as low energy gamma-ray spectrometers; however, because of their reduced mass, they are considerably less efficient than coaxial detectors at high gamma-ray energies. Shown in Fig. 16.36 are three common planar structures, the simple planar, the grooved planar and the low capacitance planar structures. The grooved structure (Fig. 16.36(b)) suppresses the surface leakage current and thus reduces electronic noise. The small electrical contact design shown in Fig. 16.36(c) takes advantage of a reduced detector capacitance, non-linear weighting potential, and high energy-resolution typical of a small detector to improve performance at low energies. All of the shown planar configurations have a shallow p-type implanted region on ν-type material which serves as the rectifying contact and entrance surface. Coupled with a low Z thin cap window, these detectors can be used for gamma-ray and x-ray energies down to 3 keV. With low Z polymer window materials [Moxtek 2017] as replacements for beryllium windows, it is possible to detect photons with energies below 1 keV. A relatively new configuration fabricated from π-type material has a Schottky barrier front contact, a Li diffused n-type contact around the periphery, and a small p-type contact at the collection electrode. This detector type can have outstanding efficiency and energy resolution performance at energies ranging from 3 keV up into the higher energy range generally relegated to coaxial devices (see Figs. 16.36 and 16.38).

Coaxial Detectors By far the most popular geometry in current use, the coaxial detector design offers outstanding energy resolution and gamma-ray detection efficiency. Capacitance is reduced from common planar detectors, as already mentioned. Coaxial detectors are fabricated from either ν-type or π-type materials, both having advantages. Coaxial $p\pi n$ detectors generally have better energy resolution performance than coaxial $p\nu n$ detectors of similar size, but the thick dead region formed by the Li-diffused junction limits

[22]Owens [1985] argues that HPGe coaxial detectors have nearly Gaussian features even in the presence of trapping provided that the applied electric field is high. This observation is perhaps true, but takes advantage of the coaxial geometry with the non-linear weighting potential, for which the charge carriers suffering higher trapping are collected on the outer electrode.

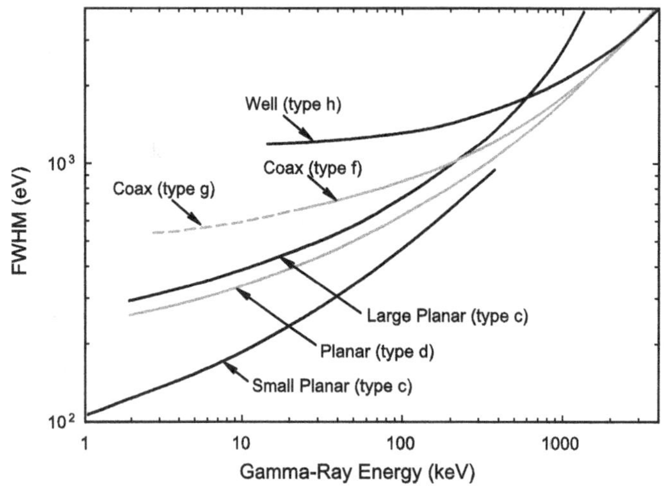

Figure 16.38. Energy resolution at FWHM as a function of gamma-ray energy for several commercial HPGe detector types identified in Fig. 16.36. Data are from Canberra [2016].

detection to gamma-rays above 40 keV. Coaxial $p\nu n$ detectors can be used for lower energy gamma-ray detection, down to 3 keV, and they are much more radiation hard to neutrons than their $n\pi p$ counterparts. The configuration is harder to manufacture than a common planar detector. Straight coaxial detectors have a hole, usually 8 mm to 10 mm diameter, drilled and etched straight through a right circular cylinder. Blind coaxial detectors have the hole drilled nearly through, but with approximately 5 mm of mass left at the detector end. Recall that blind coaxial detectors have the end bulletized to help produce a uniform electric field in the device. The detectors are designed for front side irradiation, and can be acquired in a number of different dewar/detector configurations, the most popular are the dipstick and the top side-looker. Shallow coaxial detectors (Fig. 16.36 (e)) have lower capacitance from the smaller size and are generally thicker than most planar designs. These shallow coaxial detectors are fabricated as $p\nu n$ devices, allowing their use for low-energy gamma-ray spectroscopy with better energy resolution than their larger coaxial counterparts. Unfortunately, because of their smaller size, truncated coaxial detectors have reduced efficiency at higher gamma-ray energies.

Well Detectors The well detector, shown in Fig. 16.36 (g), is a device that nominally produces a 4π counting geometry, in which the radiation source is placed inside the well structure. To achieve highest performance at low energies, the well hole is doped with p-type implants in a π-type blind coaxial device. The outer surface is doped with diffused Li to produce a n-type contact. For this configuration, it must be fully depleted to take advantage of the thin p-type implanted ohmic junction. The construct of the well cap is thin aluminum, usually thin enough to allow detection down to 10 keV. Because of the geometry, well detectors suffer compromised energy resolution compared to other geometries, but can still achieve FWHM near 3 keV for 1332-keV gamma rays. There have been some commercial well hole units in the past that were true coaxial detectors, with the hole drilled completely through the detector. The configuration of the blind hole well requires a longer lead to the first amplifier stage which is manifested in microphonic noise [Gilmore 2008]. The lead on a through-well structure can be shorter, thereby reducing the electronic noise problem. Although the loss of mass at the bottom of the well may reduce efficiency, this problem can be mitigated by making the coaxial detector longer.

Special HPGe Detectors HPGe detectors can be fashioned into numerous different geometries for application-specific measurements. For instance, special instruments for high energy physics [Eberth and Simpson 2008], Compton scatter studies [Choong et al. 2008], gamma-ray tracking [Wieland et al. 2003], and gamma-ray imaging [Niedermayr et al. 2005] have been built for these various applications. In some cases, an array of common coaxial HPGe detectors may be adequate, although the geometry of the array may require innovative electronics arrangements. These arrays improve efficiency and reduce Compton scattered photon losses while yielding information of the gamma-ray trajectories. Eberth and Simpson [2008] provide a good review of several detector array projects and results for assemblies of commercial HPGe detectors.

The maturity of germanium crystals enables the production of segmented arrays of detectors that come in many forms. The most basic are large crystal volumes snuggly fit into a single package and interconnected to form a single device. For instance, the CLOVER device introduced by Canberra Industries [Canberra 2012] has four detectors that when connected provide over 130% efficiency at 1.33 MeV by using a coincidence counting method to sum the full energies of Compton scattered and pair production events. Other segmented detectors have been fashioned by defined sections on either planar and coaxial detectors through photolithography methods [Sandsingkeow et al. 2003], or by directly cutting grooves with a wire saw [King et al. 2008]. In either case, the end result is an array of rectifying contacts individually isolated on the outer surface regions. Cylindrical detectors are divided into zones, and individual preamplifiers attached to these zones are used to sense independent signals originating from these zones. By doing so, Compton scatter imaging and gamma-ray tracking can be performed. Commercial manufacturers offer segmented detectors in either planar or coaxial geometries, some with over 30 segments to a single coaxial device. Segmented detectors can also be arranged in complex arrays to increase the total volume of a detection system.

16.4.3 Diamond Detectors

Another single element group IV semiconductor used occasionally as a radiation detector is carbon in the form of diamond. Diamond was the first semiconductor investigated, successfully, as a radiation detector [Stetter 1941], although it went largely unnoticed at the time. Diamond forms as an FCC lattice,[23] has an indirect band gap of 5.47 eV, and requires on average 13.25 eV to produce an electron-hole pair. Overall, the high ionization energy results in fewer charge pairs produced per unit energy than either Si or Ge. Consequently, there is a reduction in relative pulse height for diamond detectors compared to other semiconductors, yet the average ionization energy is still approximately two to three times less than most gases used in proportional counters. With a Z of only 6, it is the lightest of the group IV semiconductors. The dielectric constant of diamond is relatively low at 5.7 and the index of refraction is 2.424 at a wavelength 546 nm, reducing to 2.418 at 500 nm. The atomic density is 1.763×10^{23} cm^{-3} and the mass density is 3.52 g cm^{-3}.

Diamond has a thermal conductivity coefficient of 25 W cm^{-1} K^{-1}, which is nearly 17 times greater than silicon. The room-temperature electron and hole mobilities for natural diamond reach 2800 and 2000 cm^{2}V^{-1}s^{-1}, respectively. Shown in Figure 16.39 are variations of charge carrier mobilities with temperature and impurity concentration. Near 300 K, these mobilities are only slightly altered with changing temperature. From the literature, highly doped diamond suffers reduction in mobility at low temperature, not an intuitive result [Nebel 2003].

The room-temperature charge carrier saturation velocities are higher than most semiconductors, reaching velocities greater than 1.6×10^7 cm s^{-1} for electrons and greater than 10^7 cm s^{-1} for holes (see Fig. 16.40). Because of its low Z, it is impractical as a gamma-ray spectrometer; however, it has found use on occasion as a particle detector, especially for high-temperature applications. Having a Mohs hardness of 10, it is also more radiation resistant than all other semiconductors. Diamond has a displacement energy ranging between

[23] It is often classified as a diamond lattice, but this is actually incorrect terminology. Instead it is the FCC Bravais lattice with two atoms arranged at each lattice point.

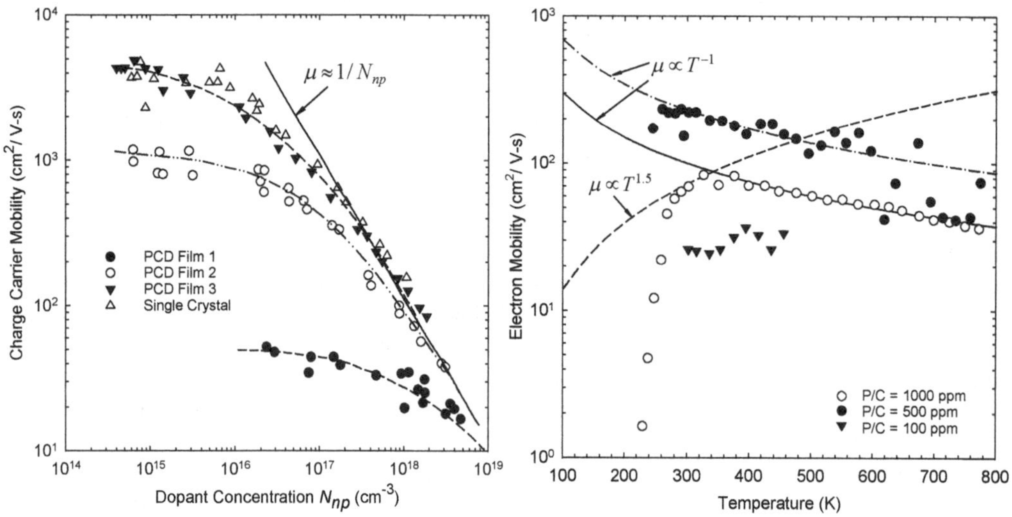

Figure 16.39. The charge carrier mobility in diamond as a function of (left) impurity concentration and (right) temperature. Data are from Nebel [2003].

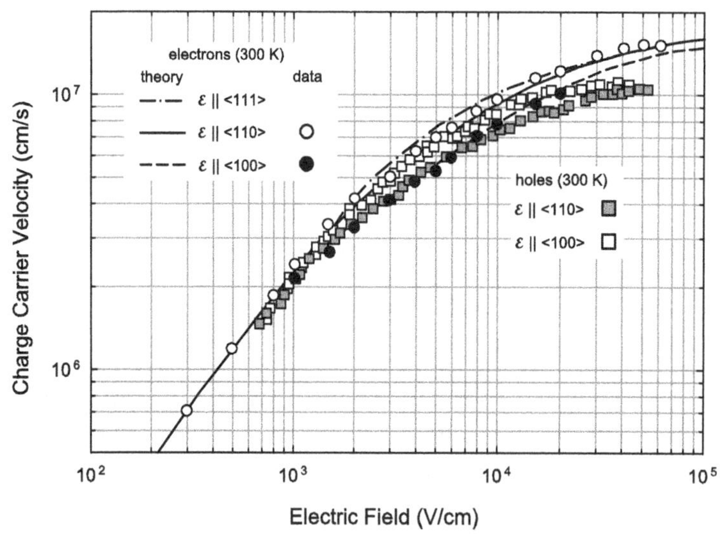

Figure 16.40. The charge carrier velocity in pure natural diamond as a function of electric field and crystal orientation. Data are from Nava et al. [1980] and Reggiani et al. [1981].

35 and 48 eV per atom depending on the irradiation direction in the crystal [Bourgoin and Massarani 1976; Koike et al. 1992], higher than common semiconductors used for radiation detection. The displacement energies for most semiconductors generally range between 6 to 25 eV per atom [for examples, see Loferski and Rappaport 1959; Larin 1968; Bryant and Cox 1968]. This radiation resistance should not be interpreted as being radiation proof, because diamond also accumulates displacement damage from charged-particle and neutron interactions. Consequently diamond detectors also suffer eventual catastrophic failure in a harsh

radiation environment. However, the onset of failure from charged particle damage is usually at a much higher fluence, approximately an order of magnitude greater than observed with either Si or Ge.

Diamond is synthetically produced by chemical vapor deposition (CVD) in a microwave chamber, generally limited to a thickness of 1 mm or less [Hemley et al. 2005]. The process requires a reaction between hydrogen and methane gases at low pressure and at a relatively high temperature, typically above 700°C [Schwartz et al. 2004; Wolfer et al. 2009]. The concentration of methane is much lower than the hydrogen, usually only 1-2% of the total gas mixture [Liang et al. 2009]. The organic gas molecules decompose in the microwave field and deposit the carbon in layers on a prepared substrate, after which a post-growth anneal at higher temperatures (> 2000°C) eliminates graphite structures that may have formed [Liang et al. 2009]. Growth rates increase with temperature, and apparently growth rates also increase with the introduction of nitrogen [Yan et al. 2002]. Including a small amount of oxygen in the CVD plasma also increases the growth rate of diamond, improves crystallinity, and suppresses the formation of graphite [Tapper 2000]. CVD-grown diamond reportedly has higher mobilities than natural diamond, reaching up to 4500 $cm^2V^{-1}s^{-1}$ for electrons and 3800 $cm^2V^{-1}s^{-1}$ for holes [Wort and Balmer 2008]. Effective doping of diamond is difficult, but can be performed. The ionization energies of traditional dopants in group IV materials are relatively large. For instance, the ionization energy of boron (p-type dopant) is 0.37 eV and the ionization energy of arsenic and phosphorus (n-type dopants) are 0.41 eV and 0.59 eV, respectively. Regardless, *pn* junction diodes have been successfully produced with diamond. Except for impure materials, a rectifying junction is probably unnecessary for diamond detectors because the intrinsic resistivity is greater than 10^{13} Ω cm.

The construction of diamond detectors is simple, usually consisting of a parallel plate design with metal contacts on opposing surfaces. Alternative designs have a pair of interdigitated electrodes on a single surface [Mainwood 2000]. One of the major advantages of diamond is its relatively higher radiation hardness to other semiconductors. Because the band gap is wide and also indirect, the probability of band-to-band radiative transitions is low. Instead, performance is strongly dependent on the trap density, and Shockley-Read-Hall recombination dominates. The radiation hardness of diamond is usually measured by the degradation in charge collection efficiency, defined by a comparative decrease in the combined charge carrier drift lengths,

$$\lambda_e + \lambda_h = (\mu_e \tau_e^* + \mu_h \tau_h^*)\mathcal{E} = v_e \tau_e^* + v_h \tau_h^*. \tag{16.84}$$

In other words, as defects are introduced through radiation absorption events, the trap density increases and the mean free drift time (τ^*) decreases. The overall induced charge loss is also a function of detector thickness. Hence, it is expected that as the thickness of a diamond detector is increased, the relative radiation hardness decreases. Damage tends to become severe for charged particle fluences exceeding 10^{15} cm^{-2} [Mainwood 2000], an order of magnitude higher than observed for Si [Adam et al. 2000] and bulk GaAs [Gersch 2002].

16.5 Compound Semiconductor Detectors

Compound semiconductors have been under investigation since 1945 when Van Heerden first explored the use of AgCl as a radiation counter. Unfortunately, compound semiconductors have been substantially less successful than group IV semiconductors. One main reason for the lower success is because of increased number of possible defects that can form with compound semiconductors, commonly arising from antisite defects and non-stoichiometric effects, neither of which is a problem with common group IV semiconductors such as Si and Ge. Added to the complications with purification, these multiple defects can increase charge carrier trapping and affect the electric field profile. However, there are a few compound semiconductors that are useful for special applications, a few of which are described in the following sections.

Compound semiconductor detectors have become more important in recent years, with commercial units now available. Typically these detectors are somewhat smaller than Si- and Ge-based detectors, mainly due to material imperfections. Regardless, a few materials, namely CdTe, CdZnTe, and HgI_2, have desirable properties for room-temperature operated devices, an advantage not shared by Si(Li) or HPGe detectors. The reason for this advantageous property is that, at 300 K, their larger band gap energies reduce their intrinsic carrier concentrations and substantially increase their resistivities. Further, CdTe, CdZnTe, and HgI_2 all have relatively high Z atomic constituents, hence have larger gamma-ray absorption coefficients over those of Si and Ge. Still, because of their typical smaller size, energy resolution for these compound semiconductor detectors are usually reported relative to 662-keV gamma rays of ^{137}Cs instead of 1.33 MeV. Because of the importance of these properties, lengthy reviews have been periodically published that are devoted entirely to the status of compound semiconductors as radiation detectors, notably Hofstadter [1949a; 1949b], Chynoweth [1952], Mayer [1966], Sakai [1982], Cuzin [1987], Squillante and Shah [1995], McGregor and Hermon [1997], and Owens and Peacock [2004]. Further, a few technical books have devoted much material to discussions on compound semiconductor radiation detectors, such as Bertolini and Coche [1968], Schlesinger and James [1995], and Owens [2012]. In the present text, the authors attempt to condense that vast amount of information on compound semiconductor radiation detectors to a manageable level, incorporating some of the more important developments and findings.

The total charge collected is usually affected by crystalline imperfections that serve as *trapping* sites, which are energy states that remove free charge carriers from the conduction and valence bands. Charge is induced as long as these charge carriers are in motion; hence, their removal diminishes the output voltage. Although the actual trapping process is complicated, it is typical to describe the relative charge collection efficiency as a simplified function of trapping. The induced charge for planar detectors is given by Eq. (15.240), which is written here as

$$\frac{Q(x)}{Q_0} = \varrho_e \left(1 - \exp\left[\frac{x-W}{\varrho_e W}\right]\right) + \varrho_h \left(1 - \exp\left[\frac{-x}{\varrho_h W}\right]\right), \tag{16.85}$$

where W is the detector active region width, Q_0 is the initial excited charge magnitude, x is the event location in the detector, and

$$\varrho_{e,h} = \left(\frac{\tau_{e,h} v_{e,h}}{W}\right) = \left(\frac{\mu_{e,h}\tau_{e,h} V}{W^2}\right) , \tag{16.86}$$

where τ is the charge carrier lifetime, v is the charge carrier velocity, and V is the applied operating voltage. Note, that the relative charge collection is dependent upon the interaction location x, and for low values of ϱ, the energy resolution can be poor. Good energy resolution is achieved if $\varrho > 50$ for both electrons and holes, for which Q/Q_0 has little deviation over the detector width W. In some cases, even if the value of ϱ is low for one charge carrier (holes, for instance), energy resolution can often be preserved by irradiating the cathode side of the detector provided that the gamma-ray energy is relatively low, below the energy at which the cross sections for Compton scattering and photoelectric effect are equal. The reason for the preserved energy resolution is because electron-hole pairs are predominantly produced near the cathode, thereby most of the induced charge is from electron motion.[24] For more penetrating gamma rays, the effect diminishes as electron-hole pairs are produced more uniformly over the detector bulk.

The value of ϱ can be increased by decreasing the detector size (W), increasing carrier lifetimes (τ) through material improvement, or increasing the applied voltage V. Due to practical voltage limitations and the fundamental difficulty with improving materials, most compound semiconductor detectors are manufactured with small active widths to improve detector energy resolution, and, hence, the devices are relatively small. The $\mu\tau$ values for electrons and holes are often quoted measures of quality for compound semiconductors used as γ-ray spectrometers.

Several other methods have been employed to improve the performance of compound semiconductor radiation spectrometers. Electronic correction methods aimed at addressing charge loss have been used,

[24] If $\varrho_h \gg \varrho_e$, seldom the case, then one should obviously irradiate the anode instead of the cathode.

which include pulse-height *rejection* and pulse-height *correction* methods. Advanced detector designs that manipulate the weighting potential to effectively turn the detectors into "single carrier devices" have also been used to improve energy resolution. In more complex systems, a combination of weighting potential manipulation combined with pulse height correction have been used to produce excellent results. Examples of these methods are described in the following sections.

16.5.1 SiC Detectors

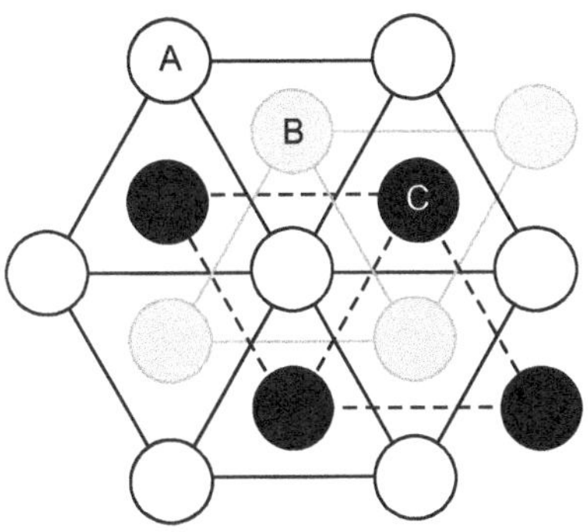

Figure 16.41. In a close pack lattice, layers can stack in positions A, B, or C. The sequence by which atoms stack upon these locations determines the polymorph.

SiC is a group IV compound semiconductor that has long been investigated as a potential radiation detector. SiC has a density of 3.21 g cm^{-3} with atomic numbers 14 and 6. SiC can form in over 200 different polymorphs, and the first identified polytype was a rhombohedral lattice (15R) [Ott 1925; Thibault 1944a, 1944b]. The three most common commercial forms of SiC are 4H, 6H, and 3C. The number indicates the repetitive layering sequence and the letter indicates the Bravais lattice, H for hexagonal, R for rhombohedral, and C for cubic.[25] The structural unit of SiC is a tetrahedron with four carbon atoms about a Si atom, although the basis can be considered the Si-C pair. Each of these pairs is located at a lattice point. A layer of the basis type can be formed in a two-dimensional hexagonal close pack lattice, such as a rack of billiard balls (see Fig. 16.41), which can serve for reference as the basal plane labeled A. A second layer can be laid atop the depressions of the first layer, of which there are two possibilities, labeled B and C. Essentially all of the SiC polytypes can be assembled by arranging these three possible layering structures in a proper sequence.

From Fig. 16.41, a 2H SiC polytype has stacking sequence ABAB where AB is the repetitive sequence. In similar fashion, the 4H polytype has sequence ABCB, the 6H polytype has sequence ABCACB, and the 3C polytype has sequence ABC (zinc blende) [Iwami 2001]. Although the mass densities are the same for these polytypes, the band-gap energies are not: the band gap for 3C is 2.36 eV, for 6H is 3.05 eV, and for 4H is 3.23 eV. SiC is also extremely hard with a Mohs hardness of 9. At 9.66, SiC has a slightly lower dielectric constant than most semiconductors. The intrinsic resistivity of 4H SiC is greater than 10^{15} Ω cm. SiC has a high melting point (decomposition point) of 2830°C (3100 K), a temperature that precludes traditional melt growth methods of production. Instead, SiC bulk crystals are grown by vapor phase transport (VPT), chemical vapor deposition (CVD), or liquid phase epitaxy (LPE) [Dmitriev and Spencer 1998; Sudarshan et al. 2005]. Vapor epitaxial methods are used to grow high quality SiC layers, usually on a bulk SiC substrate. With any of these methods, the production costs are relatively high, thereby making SiC wafers expensive by comparison to most other semiconductors.

Because SiC has such low Z numbers, it is relatively uninteresting as a gamma-ray detector. With the wider band gap, 4H SiC has an average ionization energy of approximately 7.8 eV per electron-hole pair, much larger than the average ionization energy of Si at 3.6 eV per electron hole pair. Further, the high temperature properties and the chemical resistance of SiC make *pn* junction formation a challenge, requiring much higher temperatures (greater than 2000°C) than that needed for common Si diode fabrication [Saxena and Steckl

[25] Ramsdell [1947] suggested this classification method rather than the previous arbitrary method of designating SiC types in sequential order of discovery. Instead, the number represents the number of layers per unit cell and the letter designates the Bravais lattice; H for hexagonal and R for rhombohedral. Thibault [1944a] named the cubic polytype β-SiC and named all other polytypes α-SiC with Roman numeral designations. Because there is only one cubic form, Ramsdell kept the identification β-SiC, yet even this was eventually changed to C for cubic. Hence, Type I became 15R, Type II became 6H, Type III became 4H, Type IV became 3C, etc.

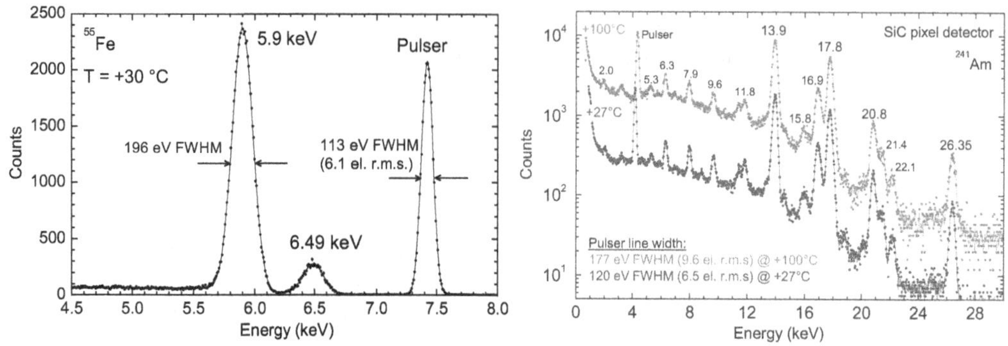

Figure 16.42. (left) Spectrum of ^{55}Fe acquired at 100°C with a SiC pixel detector. An 8 μs triangular pulse shaping was used and the electronic noise was 8.9 electrons r.m.s. (right) Comparison spectra of ^{241}Am acquired at 27°C and 100°C with a SiC pixel detector. From Bertuccio et al. [2011]; copyright Elsevier (2011), reproduced with permission.

1998]. Junction devices can be formed with CVD epitaxial growth of doped layers or with ion implantation of dopant species. Because *pn* junction formation introduces fabrication complications, Schottky contacts are a common method of producing rectifying junctions on SiC substrates. Ohmic contacts generally require a doped sublayer of material, produced by CVD or implantation, before the ohmic metal is applied. The hardness of SiC also presents a challenge for chemical and plasma-based etching. Surfaces that terminate with either C or Si atoms have different chemical properties.[26]

Regardless, there are several attributes to SiC that are of interest for radiation detection. First, because of the high melting point and the wide band gap, SiC detectors can be operated at high temperatures (greater than 700°C) with negligible leakage current [Babcock et al. 1963]. Second, it is relatively radiation hard by comparison to most other semiconductors [Babcock 1965; Barry et al. 1991; Kinoshita et al. 2005; Ruddy et al. 2007; Lefévre et al. 2009]. Hence, a thrust for SiC detector research is for high temperature radiation environments, such as those found in the vicinity of a nuclear reactor [Dulloo et al. 1999]. Third, SiC has a high thermal conductivity coefficient 4.9 W cm^{-1} K^{-1} at room temperature. Fourth, SiC has a saturation velocity of 2×10^7 cm s^{-1}, nearly twice that of Si. Finally, SiC has a high breakdown electric field, 2.2×10^6 V cm^{-1} for 4H material and 2.5×10^6 V cm^{-1} for 6H material so that high voltages can be applied to the detectors. Of the commercially popular SiC polytypes, 4H has emerged as the favored material for radiation detectors mainly because it has higher charge carrier mobilities than 6H and much higher resistivity than 3C. SiC wafers up to 150 mm diameter can be acquired commercially.[27] Because epitaxial SiC is generally of higher quality than bulk SiC, it is common that a high-quality epitaxial layer, grown upon a bulk grown SiC substrate, serves as the detection volume.

SiC has been studied as a neutron detector since the early 1960s [Brussels 1960; Weisman 1960, 1962; Belgium 1962; Babcock et al. 1963]. The primary reason for interest in SiC is because it has unique properties that permit their operation at high temperatures in a harsh radiation environment [Canepa et al. 1964; Seshadri et al. 1999; Nava et al. 2008; Ha 2009]. Detectors for slow neutron detection are usually fashioned as *pn* or Schottky junction diodes coated with a converter material [Dulloo et al. 2003; Manfredotti et al. 2005; Ha et al. 2009]. A type of SiC fast neutron detector relies on fast $^{12}C(n,n')^{12}C$ and $^{28}Si(n,n')^{28}Si$ elastic and inelastic collisions, along with numerous possible threshold reactions, as the detection mechanisms [Ruddy et al. 2006a; Ha et al. 2011]. Ruddy et al. [2006a] show spectral features for several fast reactions in a 4H SiC *pin* structure, and also demonstrate different pulse height spectra from three different fast neutron

---
[26] This trait is common to compound semiconductors.
[27] Cree, Inc.

sources, ^{252}Cf, AmBe, and a D-T generator. An alternative fast neutron detection method is to cover the SiC diode with a hydrogenous material such as HDPE and use the (n,p) recoil reactions as the detection method [Flammang et al. 2007]. A more detailed discussion about these neutron detectors is reserved for Chapters 17 and 18. Ruddy et al. [2006b] report on the performance of 4H SiC material as an alpha particle detector. These detectors were fabricated from a 100 micron thick high purity SiC epitaxial layer grown upon a 300 micron thick n-type SiC substrate. Ruddy et al. [2006b] report room temperature energy resolution of 41.5 keV FWHM at 3.18 MeV and 55.4 keV at 8.38 MeV. SiC has also been studied for low-energy x-ray detectors [Bertuccio et al. 2004a, 2004b, 2011; Moscatelli 2007; Nava et al. 2008]. Although these detectors are not suitable for high energy gamma-ray detectors, they performed well for photon energies below 60 keV. Bertuccio et al. [2004a, 2004b, 2011] report on photon detectors fabricated from SiC epitaxial layers, nominally 70 microns thick, grown upon SiC substrates. Operated at 373 K, energy resolution of 233 eV FWHM has been reported for 5.9-keV gamma rays for a pixel detector with a 200 micron diameter Schottky junction (Fig. 16.42) [Bertuccio et al. 2011].

16.5.2 Detectors Based on Group III-V Materials

Group III-V semiconductors are produced from elements in group III and group V of the periodic table. These materials include binary compounds such as GaAs, GaP, InP, and AlSb, ternary compounds such as InGaAs and InGaP, and quaternary compounds such as InGaAsP. For ionizing radiation detector applications, the binary compounds have been investigated the most, although some ternary and quaternary materials have been explored, primarily as contact materials. Both GaAs and InP are relatively mature semiconductors that are in common use within the VLSI industry. Both of these important semiconductors have direct band gaps and show negative differential resistance at high electric fields. Because of their direct band gaps, these two semiconductors have been used most commonly for photonic emission devices such as LEDs and laser diodes. The relatively wide band-gap energies of many III-V semiconductors have motivated researchers to investigate their potential use as room temperature semiconductor radiation detectors.

GaAs

Gallium arsenide (GaAs) was first reported as a semiconductor detector by Harding et al. in 1960. It is a relatively mature and commercially available semiconductor with some properties that make it attractive as a radiation detector [McGregor and Kammeraad 1995]. With a density of 5.32 g cm^{-3} and atomic numbers 31 and 33, it has similar gamma-ray absorption properties as Ge. It has a band gap of 1.42 eV and, thus, allows room temperature operation for moderate size detectors. GaAs has a direct band gap, which is beneficial for photonic emission devices, but also decreases the charge carrier lifetime. Electrons confined in the direct Γ valley have a high mobility, theoretically up to 8500 cm^2 V^{-1} s^{-1} and, thus, generate high electron velocities. The indirect L valley of GaAs is 1.71 eV, as shown in Fig. 16.43, which produces negative differential resistance as electrons transfer from the Γ valley to the L valley for electric fields exceeding 3×10^3 V cm^{-1} (see Fig. 16.6).

The intrinsic resistivity of GaAs is approximately 10^8 Ω cm, but GaAs detectors still must have blocking contacts to reduce leakage current to manageable levels. Detectors are most often produced with Schottky blocking contacts, generally composed of varied thicknesses and combinations of Au, Ti, and Pt [Williams 1990]. Almost all radiation detectors reported were fabricated with $\langle 100 \rangle$ GaAs substrates. Consequently, the $\langle 100 \rangle$ surface states typically "pin" the Fermi energy level near mid-gap, thereby largely decoupling the work function of the metal from the resulting Schottky barrier height. Regardless, there is some experimental evidence that modest alterations in the energy barrier can be achieved with appropriate surface treatments and choice of metal contact [Williams 1990]. The alternative to Schottky barrier blocking contacts is the application of p and n junctions. However, these junctions are not easily achieved with GaAs.

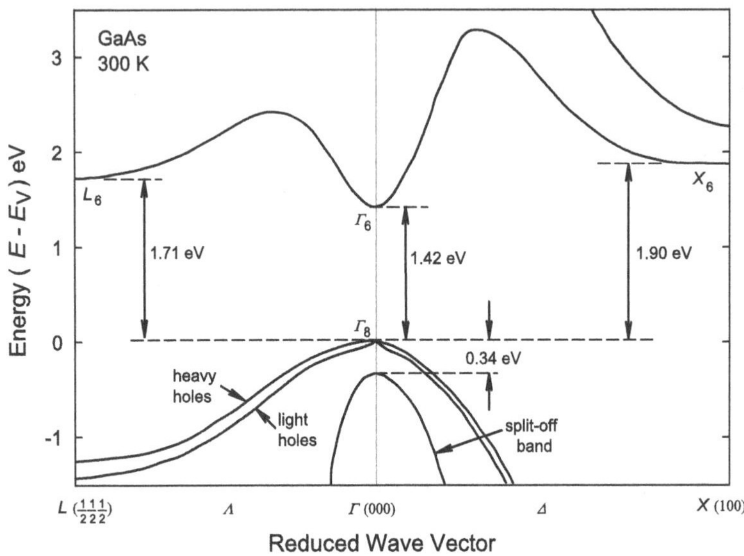

Figure 16.43. The theoretical energy bands of GaAs at 300 K calculated with a pseudopotential method. After Blakemore [1982].

Purification through traditional methods, such as zone refinement or zone leveling, are rendered difficult because arsenic outstreams from GaAs when heated above 400°C for prolonged periods of time, increasing in rate with temperature, thereby ruining the material.[28] Consequently, impurity purification is performed on starting materials (Ga and As) before crystal growth, with purity levels exceeding 7N being commercially available.[29] However, high resistivity is usually achieved by deep level compensation rather than intrinsic properties through purity. Deep level compensation is commonly achieved in bulk GaAs with the antisite As_{Ga} with As located on a Ga lattice location. This antisite defect produces a double deep donor with energies, measured at 0.75 eV and 1.0 eV below the conduction band edge, label EL2 [Martin et al. 1980; Thomas et al. 1984; Bourgoin et al. 1988; Martin and Makram-Ebeid 1992; Baraff 1992].

Historically, GaAs was the first compound semiconductor to demonstrate high energy resolution at room temperature (295 K) for gamma rays [Eberhardt et al. 1970, 1971]. Those detectors were fabricated with GaAs crystals grown by liquid phase epitaxy (LPE), a crystal growth method that intrinsically purifies materials through thermal migration of impurities. Unfortunately, the detectors were also limited to thicknesses of approximately 130 microns and, thus, severely limiting the gamma-ray detection efficiency. A typical pulse height spectrum from such a detector at room temperature is shown in Fig. 16.44. The LPE method of crystal growth is relatively expensive and impractical as a production method for large volume detectors. Epitaxially grown GaAs detectors have demonstrated excellent energy resolution [Owens et al. 1999], some with thicknesses up to 200 microns [Alexiev and Butcher 1992]. Attempts to manufacture thicker epitaxial GaAs have been explored (up to 600 microns), but the authors state that quality was compromised as the thickness increased [Hesse et al. 1972]. Unfortunately, thin devices combined with the relatively low atomic numbers of GaAs offer little advantage over alternative higher Z compound semiconductors, such as CdZnTe and HgI_2.

[28] If performed within a sealed ampoule backfilled with As gas, GaAs can be annealed at elevated temperatures. The cooling method can greatly alter the electrical resistivity of the material [Lagowski et al. 1986].

[29] In the usual jargon for purity, an 'N' represents a '9' in purity. Hence, 5N (or five nines) pure is 99.999% pure.

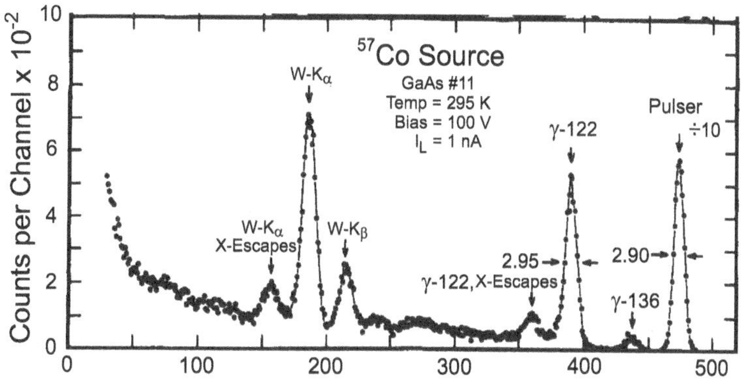

Figure 16.44. Room temperature differential pulse height spectrum from a GaAs Schottky diode fabricated from a 75 micron LPE layer. Reproduced from Eberhardt et al., *Appl. Phys. Lett.*, **17**, 427, (1970), with the permission of AIP Publishing.

GaAs produced by bulk crystal growth methods was also studied during the 1980s and 1990s, also with limited success [McGregor and Kammeraad 1995]. Motivation for studying GaAs as a radiation detector included application as room-temperature operated gamma-ray spectrometers and as a radiation-hard particle detectors for high-energy colliders [Beaumont et al. 1992, 1993, 1994]. High resistivity GaAs is generally compensated with either Cr [Cronin and Haisty 1964; Lin and Bube 1976] or with the native EL2 defect [Martin et al. 1980; Kaminska 1987; Bourgoin et al. 1988]. In either case, the charge carrier lifetimes are relatively short, on the order of nanoseconds, thereby drastically reducing the useful size of these detectors. For a bulk GaAs device 130 microns thick, an acceptable energy resolution of 8.83 keV FWHM was achieved for 122 keV gamma rays [McGregor and Hermon 1997]. Energy spectra from a bulk GaAs spectrometer is shown in Fig. 16.45. A modest amount of cooling below room temperature produces a dramatic improvement in performance, as demonstrated by Bertuccio et al. [1997], namely an energy resolution for 60-keV gamma rays measured with 100-micron-thick bulk-grown GaAs detectors reduced from 15.5 keV FWHM at 295 K to 2.4 keV FWHM at 243 K. Yet, the small size also rendered these devices useful only for low energy gamma rays. The problem with compensated bulk GaAs is that the field dependent capture cross section of EL2 [Panousis et al. 1969; Prints and Bobylev 1981; Prinz and Rechkunov 1983] alters the electric field within the detectors and causes the electric field to increase with a non-traditional voltage dependence [McGregor et al. 1994a].

The deep level EL2 is an intrinsic antisite defect As_{Ga} that naturally produces a double deep donor with energy levels approximately 0.75 eV and 1.0 eV below the conduction band edge E_C. EL2 levels are neutral when filled with an electron and positively charged when empty. This antisite defect is purposely produced in concentrations of about 10^{16} cm^{-3} by introducing additional As into the GaAs melt source [Holmes et al. 1982]. A careful balance of the EL2 concentration with that of the residual carbon acceptor concentration (usually near 10^{15} cm^{-3}), and possibly other acceptor impurities, produces semi-insulating (SI) material with resistivities of about 10^8 Ω cm. This relatively high resistivity is still inadequate to produce resistive devices, and instead SI GaAs detectors are fabricated as rectifying diodes operated under reverse bias. Under equilibrium conditions, only a fraction of the EL2 deep donors are ionized and the Fermi level is pinned near midgap, as explained by the three-level model of Sec. 12.5.4. However, as the detector is reverse biased, the EL2 deep donor becomes fully ionized, and produces excess positive space charge near the reverse biased junction as shown in Fig. 16.46. Consequently, the non-uniform ionization distribution promotes the appearance of a two-zone electric field distribution [McGregor and Kammeraad 1995].

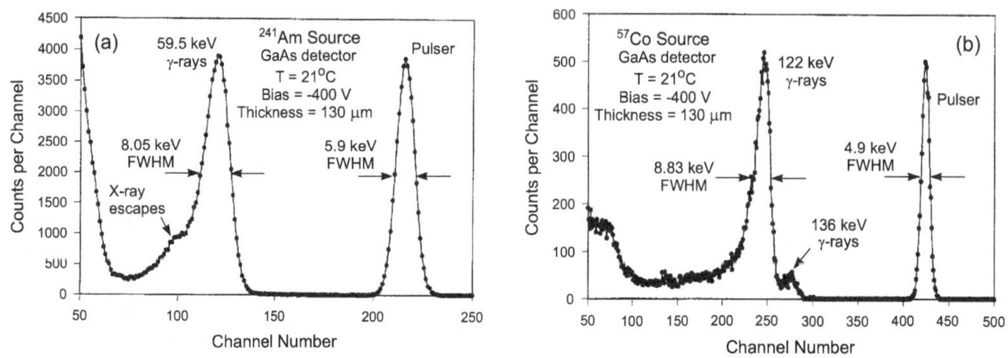

Figure 16.45. Room temperature differential pulse height spectra from a 130 μm thick, 0.5 mm^2 bulk SI GaAs detector. From McGregor and Hermon [1997]; copyright Elsevier (1997), reproduced with permission.

As the bias is increased further, the capture cross section of the EL2 increases [Panousis et al. 1969; Prints and Bobylev 1981; Prinz and Rechkunov 1983], and increases the likelihood that electrons are recaptured, thereby effectively neutralizing a fraction of the deep donors [McGregor et al. 1992, 1994a, 1994b]. McGregor et al. [1994b] propose that the unexpected voltage dependence is a consequence of dynamic trap neutralization of deep centers from electric field enhanced electron capture. Consequently, a quasi-stable constant electric field appears which rapidly decreases to low values at the end of the active region. This balance between trapping and emission at these EL2 sites is a probable reason for low frequency oscillations observed in GaAs biased at high voltages [Holanyak and Bevacqua 1963; Barraud 1963; Sacks and Milnes 1970; Kaminska et al. 1982]. This hypothesis is based on an arbitrary "filling function" applied in the model [McGregor 1994a, 1994b], yet measurements by Berwick et al. [1993] seem to confirm the hypothesis. McGregor et al. [1992, 1994a] and Berwick et al. [1993] found that the electric field active region increased nearly linear with applied voltage, at approximately 1 micron per volt, leaving a considerably wide low-field substrate region with poor charge collection. Unfortunately, with such a peculiar voltage dependence, it takes 1000 volts to produce a 1-mm-thick active region, a voltage that is impractical for most gamma-ray detector applications. Kubicki et al. [1994] attempt to explain the unexpected electric field and depletion characteristic by incorporating leakage currents in their model instead of including the field enhanced capture of electrons. Although the model has merit in explaining the low-field characteristic, it fails to reproduce the dependencies measured by McGregor et al. [1992, 1994b] and Berwick et al. [1993]. Similar results to that of McGregor et al. [1994b] were produced by Cola et al. [1994, 1997], who incorporated the model of McGregor et al. [1994b] and portions of the model of Kubicki et al. [1994]. Overall, Cola et al. [1997] appear to provide additional evidence to field enhanced electron capture by EL2 as the cause of the unusual electric field and depletion characteristic of undoped SI GaAs.

Despite these problems, undoped SI GaAs was found to be useful as a high-speed radiation detector because of the short charge carrier lifetimes [Wagner et al. 1986; Wang et al. 1986, 1989; Friant et al. 1989]. Detectors of this type were used to measure the radiation wave from nuclear weapons tests. Although these detectors were capable of following high-speed radiation pulses, response times were still tens of nanoseconds in duration. The high-speed response was improved by irradiating GaAs materials in a nuclear reactor, which produced a high density of electron and hole traps than appear in bulk-grown GaAs from the damage, thereby reducing the response times to 30 picoseconds [McGregor and Kammeraad 1995].

GaAs pixelated detectors have found use for radiation imaging of low-energy gamma rays and x rays [Manolopoulos et al. 1998; Owens et al. 2001]. The advantages of using GaAs for x-ray imaging over Si

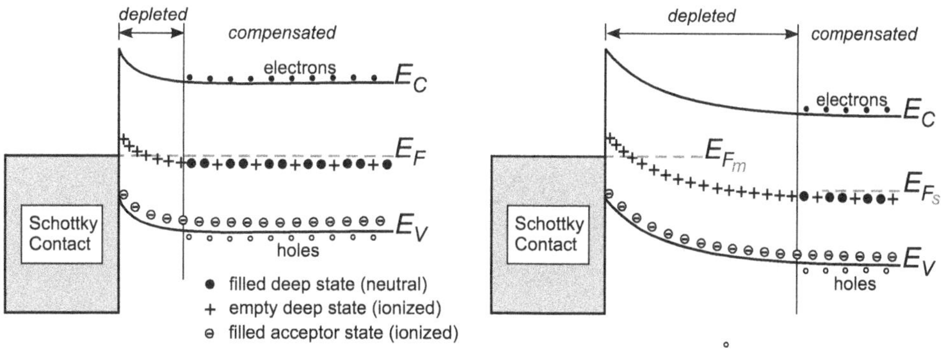

Figure 16.46. Band diagrams of a Schottky contact diode with shallow acceptors and compensating deep donors. (left) With no applied bias, a space charge region appears adjacent to the Schottky contact. (right) With reverse bias applied, the space charge region extends further into the semiconductor as more deep donors become completely ionized. After McGregor et al. [1994a].

include the higher density and Z of GaAs, the higher intrinsic resistivity and lower leakage current, the high Γ valley electron mobility (greater than 8000 cm^2 V^{-1} s^{-1}), and pixel isolation is achieved with the semi-insulating substrate rather than a field oxide. There are other semiconductors with higher Z than GaAs, such as CdTe, CdZnTe, and HgI_2, yet the maturity of GaAs and the resultant high quality crystal structure enables common VLSI processing of the pixelated structures.

InP

Indium phosphide (InP) is a compound III-V semiconductor that has received modest attention, sporadically, as a radiation detector. Perhaps one of the more attractive attributes of InP is its relatively high electron charge carrier mobility of 4500 cm^2 V^{-1} s^{-1}, with holes having a mobility of 150 cm^2 V^{-1} s^{-1}. The saturation velocity is higher than that of GaAs, reaching over 2.5×10^7 cm s^{-1} at an electric field of 1.2×10^4 V cm^{-1}. The average ionization energy is approximately 4.2 eV per e-h pair. The gamma-ray absorption is dominated by the higher Z indium component with atomic number 49 (where phosphorus is only 15), and the density of InP is 4.79 g cm^{-3}. With a direct band gap of approximately 1.35 eV (Fig. 16.47), the intrinsic resistivity is approximately 10^7 Ω cm, just high enough to operate small devices at room temperature. However, this high resistivity is usually achieved by compensation doping with Fe impurities. Notice from Fig. 16.6 that InP also possesses negative differential resistance for electrons at electric fields above 1.2×10^4 V cm^{-1} and, thus, makes InP attractive for Gunn effect devices.

InP is usually grown by either the liquid encapsulated Czochralski (LEC) method or the vertical gradient freeze (VGF) method. Undoped material is usually n-type, although resistivity control can be attained by purposely doping with the n-type dopants S or Sn. InP can be acquired as p-type material if doped with Zn. High resistivity material is produced by adding a deep accepter compensating dopant to the melt. For detector purposes, Fe-doped semi-insulating (SI) InP has been investigated the most.

SI InP doped with Fe was explored as a high speed radiation photoconductor to measure short bursts of radiation. Fe acts as a deep acceptor site and compensates residual shallow donor impurities to produce a semi-insulating material as described in Sec. 12.5.4. These detectors routinely had response times between 100 ps to 200 ps, determined to be largely dependent upon the bulk recombination of electrons and holes [Hammond et al. 1981, 1983, 1984]. The response times were shown to be dependent upon the Fe concentration and had an obvious reduction in response time as the deep acceptor Fe concentration was increased [Wagner et al. 1986]. Wagner et al. [1986] also found that neutron irradiation damage further improved the high speed response of InP:Fe photoconductors. There is a limit to the neutron damage that can be

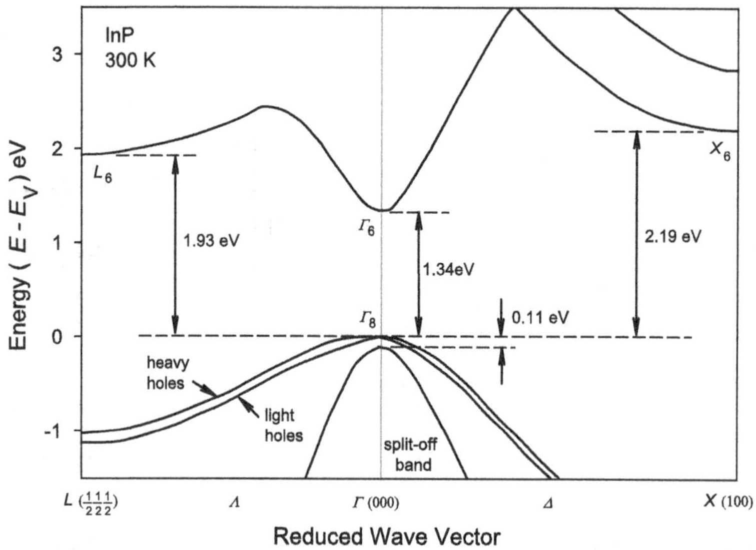

Figure 16.47. The theoretical energy bands of InP at 300 K calculated with a projector-augmented-wave (PAW) method. Data are from Y-S. Kim et al. [2009] and Ioffe [2016].

sustained, mainly because the reaction ^{115}In(n,γ)^{116}In transmutes into ^{116}Sn (14.1 s half-life) through β-particle emission, an n-type dopant for InP and, thus, ultimately reduces the material resistivity over time. Although the transient response of the detectors was shown to closely follow a short radiation impulse, they failed to follow the input power of longer duration pulse trains, in particular, square-pulse inputs from a laser, and instead showed a superlinear response function with elongated tails [Kania et al. 1986; Wagner et al. 1986]. Neutron irradiation damage failed to show improvement for this particular issue. Wagner et al. [1986] surmise that the superlinear behavior and observed failure to follow long pulse excitations was due to an imbalance of electron and hole trapping. Electrons are trapped more efficiently, leading to the slower moving holes carrying the majority of current. Consequently, steady state output is reached only when the electron and hole recombination and generation rates become balanced. This same effect diminishes as the radiation intensity increases. Although these InP:Fe detectors had excellent performance for transient responses, neutron damaged semi-insulating (SI) GaAs demonstrated faster response times without the superlinear problem for longer pulses. Consequently, investigation of high-speed InP radiation detectors diminished in favor of SI GaAs detectors.

InP was also investigated as a possible room temperature semiconductor detector for ionizing radiation and neutrinos [Lund et al. 1988; Olschner et al. 1989; Suzuki et al. 1989]. These initial investigations demonstrated that InP:Fe was sensitive to ionizing radiation, although the spectral performance was poor. The neutrino detection application is generally impractical because of the miniscule solar neutrino cross section [Raghaven 1976] and would require a rather sizeable InP detector as pointed out elsewhere [McGregor and Hermon 1997].[30] Regardless, this concept continues to be of interest [Fukuda et al. 2010]. As for detection of ionizing radiation, trapping and the non-optimal resistivity of SI InP leads to poor performance at room temperature. The charge carrier losses from trapping and the noise from excessive leakage current [El-Abassi et al. 2001] diminish the signal and cause resolution degradation. Owens et al. [2002a, 2002b]

[30] Raghaven [2001] states that a detector with a "modest 4 ton mass of In" would be required, amounting to approximately 1 cubic meter of InP.

reported improved results for small InP:Fe detectors (3 mm × 3 mm × 0.18 mm thick) used as x-ray detectors. By cooling the detectors to 213 K to reduce thermal leakage current, Owens et al. [2002a] achieved 2.5 keV FWHM for 5.9 keV-gamma rays (^{55}Fe) and 9.2 keV FWHM for 59.5-keV gamma rays (^{241}Am). By reducing the detector temperatures even lower to 103 K, Owens et al. [2003b] observed 911 eV FWHM for 5.9-keV gamma rays and 2.5 keV FWHM for 59.5-keV gamma rays.

From the work of Wagner et al. [1986], it was already observed that Fe compensation reduces the electron mean free drift times. From Section 12.5.4, it was shown that the concentration of deep acceptors must be greater than the concentration of shallow donors to produce semi-insulating material. Consequently, there is a lower concentration limit of Fe that must be introduced to maintain high resistivity and reasonably low leakage current. Gorodynskyy et al. [2005] and Yatskiv et al. [2009] proposed that Fe doping could be replaced with alternative co-dopants, such as Ti and Mn or Ti and Zn, and still maintain higher mobilities and improved charge carrier collection. This hypothesis seems to have merit, in that material produced by Gorodynskyy et al. [2005] was measured to have over 90% charge collection efficiency (CCE) for 200-micron-thick devices operated at 250 K with corresponding energy resolution of 3.4% FWHM for 5.48 MeV alpha particles. However, both CCE and energy resolution were compromised at room temperature (300 K), yielding a CCE of 52% and FWHM of 17% for 5.48 MeV alpha particles. Zdansky et al. [2006] report on results for small detectors fabricated from Ta-doped InP irradiated with α-particles, which produced energy resolution of about 6% FWHM at 5.48 MeV. Yatskiv et al. [2009] experimented with InP co-doped with Ti and Zn instead of Fe to achieve high resistivity material with superior CCE. By designing detectors with guard rings and cooling the detectors to 230 K to reduce leakage current noise, CCE up to 99.9% were achieved with a best energy resolution of 0.9% FWHM for 5.48 MeV α particles for 250-micron-thick detectors. The improved α-particle spectral results indicate material improvement, but remain an impractical application for InP, mainly because alpha particle backscattering is significantly less for Si detectors than InP detectors. Because of the relatively poor room temperature performance by comparison to alternative compound semiconductors, InP radiation detectors continue to be mainly experimental.

16.5.3 Detectors Based on Group II-VI Materials

Group II-VI semiconductors are produced from elements in group II and group VI of the periodic table. These materials include binary compounds such as CdTe and CdZnTe, perhaps the two most studied II-VI compounds for radiation detection. The relatively wide band-gap energies of CdTe and CdZnTe allow room temperature operation of these radiation detectors.

CdTe

Cadmium Telluride (CdTe) is a wide band-gap semiconductor of interest as a room-temperature operated gamma-ray spectrometer. CdTe has a cubic zincblende crystal structure. CdTe has a direct band-gap energy of about 1.52 eV, with an indirect L valley of 2.73 eV as shown in Fig. 16.48. A saturation velocity of approximately 1.5×10^7 cm s^{-1} is reached at an electric field of 1.5×10^4 V cm^{-1}, above which CdTe can also exhibit negative differential resistance much like GaAs and InP (see Fig. 16.6). The average ionization energy is 4.43 eV per e-h pair with a 0.11 Fano factor. The dielectric constant for CdTe is 10.36 [Strauss 1977]. The wide band gap produces an intrinsic resistivity of 10^9 Ω cm; however, because of the presence of contaminant impurities, high resistivities greater than 10^9 Ω cm are usually achieved through impurity and defect compensation. CdTe is relatively soft, rating 54 on the Knoop hardness scale (approximately 2.2 on Mohs scale). The elemental constituents have atomic numbers 48 and 52 and the density of CdTe is 5.86 g cm^{-3}. Because of the relatively high Z numbers, photoelectric absorption dominates up to approximately 260 keV. Charge carrier mobilities are 1050 cm^2 V^{-1} s^{-1} for electrons and 100 cm^2 V^{-1} s^{-1} for holes. Electron and hole mean free drift times are material dependent, but often quoted near 3×10^{-6} s and 2×10^{-6} s, respectively.

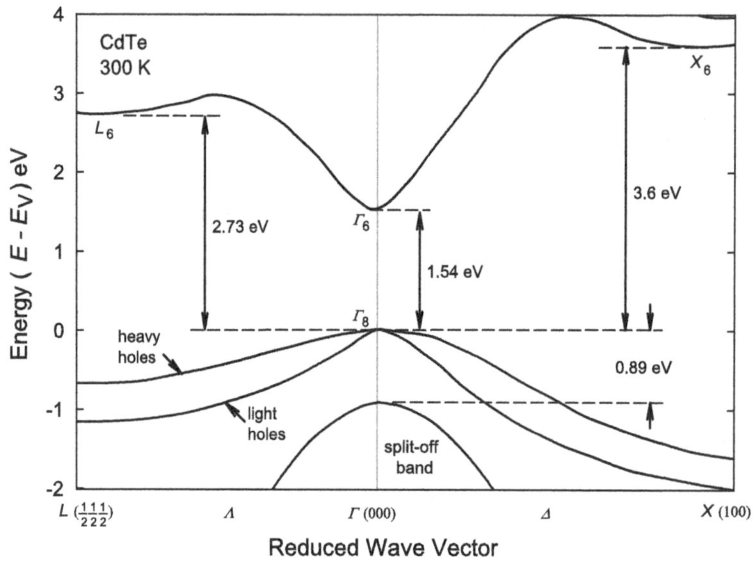

Figure 16.48. The theoretical energy bands of CdTe at 300 K calculated with a pseudopotential method. Data are from Chelikowsky and Cohen [1976].

CdTe has been investigated as a room-temperature gamma-ray detector since the mid-1960s [Autagawa et al. 1967; Mayer 1967; Vavilov et al. 1970; Zanio et al. 1970], and various detailed review articles and books cover CdTe material and detector performance [Whited and Schieber 1979; Kröger and de Nobel 1955; de Nobel 1959; Hage-Ali et al. 1995a, 1995b, 1995c; McGregor and Hermon 1997; Owens and Peacock 2004; Owens 2012]. Because of material imperfections, which include a combination of impurities and native defects, these devices continue to be manufactured as small detectors. Although the band gap is high enough for room temperature operation, background impurity contamination causes the leakage currents to be too high. Dopant compensation, typically with Cl, is used to create high resistivity material. Still the detectors must be manufactured as *pn* junction or Schottky junction diodes to reduce leakage current to manageable levels. As a result, the detector volumes are usually no more than a few millimeters thick. Commercial units are available as small gamma-ray spectrometers. Typically, the best energy resolution is achieved with the assistance of small electronic Peltier coolers.

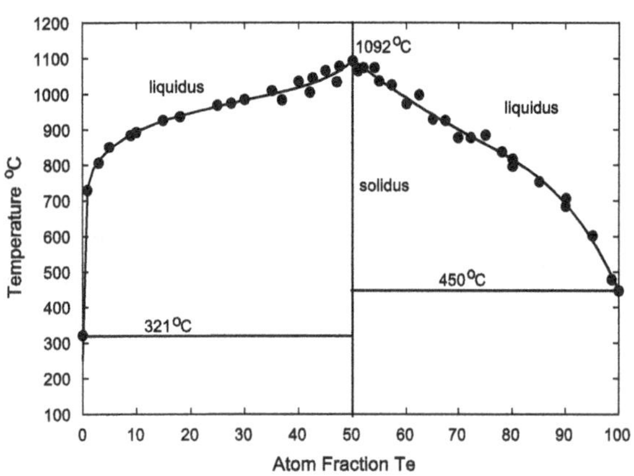

Figure 16.49. Temperature versus composition (T-x) diagram for the Cd-Te system. After Zanio [1978] and references thereof.

The melting point of CdTe is 1092°C and the crystal can be grown through melt-growth processes such as Bridgman and Stockbarger [Zanio 1978; Hage-Ali and Siffert 1995a; Lindstrom et al. 2016]. However, Te precipitates are a chemical consequence of CdTe growth, resultant from the retrograde solubility of tellurium [Zanio 1978]. The CdTe phase diagram shows that the melt temperature decreases dramatically off stochiometry, especially salient for Te-rich solutions (Fig. 16.49). Essentially, a higher concentration of Te can be suspended in a mixture of CdTe liquid and solid, which as the material solidifies into a CdTe crystal, the excess Te forms precipitates in the solid. To reduce the effects of traditional high-temperature melt growth, the traveling heater method (THM) is commonly used for radiation detector materials. This method has slowly moving heating coils that travel upwards along a vertical sealed ampoule of Te-rich CdTe stock materials. The Te melts at a lower temperature (450°C) and acts as a solvent for the CdTe stock material. Consequently, the CdTe stock dissolves at the hotter leading edge of the melt solution and crystallizes at the cooler trailing edge, much like liquid phase epitaxy. The advantages include a lower process temperature (typically below 900°C), lower Te precipitate concentrations, and lower defect and impurity concentrations. However, THM is a much slower crystal growth method than alternatives, such as high-pressure Bridgman or Stockbarger growth. CdTe grown by THM is usually doped with Cl, by adding $CdCl_2$ to the stock material, which compensates native defects and impurities to produce material with high resistivities of about 10^9 Ω cm.

Ohmic contacts are usually formed with an electroless process, in which noble metal salt solutions of $AuCl_3$, $AgNO_3$ or $PrCl_4$ are applied to a polished and etched surface.[31] The noble metal salt solution etches the CdTe surface and displaces Cd from the surface region while redepositing layers of both the noble metal and Cl [Hage-Ali and Siffert 1995b]. The Cl incorporation can extend beyond 0.5 microns into the surface [McGregor and Hermon 1997]. Calculations show that ohmic contacts are inadequate to reduce leakage currents below the induction current produced by radiation events except for small detectors. Unfortunately, although the resistivity is relatively high, it is still too low to operate CdTe as resistive devices, and instead must either be fabricated with rectifying blocking contacts or small enough such that leakage current is minimal. To reduce leakage current while allowing high operating voltages, detectors have been produced as Schottky junction devices and ion implanted *pn* and *pin* devices [Cornet et al. 1970, 1972], indium diffusion (on p-type material) [Squillante et al. 1989], and LPE growth [Shin et al. 1985].

Although the incorporation of a blocking contact permits higher bias voltages and generally reduced leakage current, the performance of these spectrometers varied. The rectifying junction causes an imbalance in the compensation such that deep level ionization produces space charge that ultimately limits the depletion region depth [Khusainov 1992]. Depletion studies on SI CdTe indicate the presence of non-uniform electric fields within reverse biased diodes [Malm at al. 1974; Siffert et al. 1978a, 1978b; Hage-Ali et al. 1993]. In some cases, the electric field did not extend completely across the device until appreciable voltage was applied, indicating the build-up of a space charge region much like the situation depicted in Fig. 16.46. Polarization is often observed with Cl compensated CdTe, manifested as a gradual change in gamma-ray sensitivity and relative pulse height [Siffert 1978a]. Low resistivity CdTe detectors usually suffer less polarization than high-resistivity detectors, and the problem of polarization can be reduced by using In-doped or low resistivity CdTe [Whited and Schieber 1979]. Illumination with IR light [Bell et al. 1974] and alternative conductive contacts have apparently also helped reduce polarization [Bell and Wald 1972; Puschert et al. 1976]. Takahashi et al. [2002] report on a 2 mm × 2 mm × 1 mm thick CdTe Schottky diode with Pt contacts. In that study, the CdTe detector had minimal polarization and yielded RT energy resolution of approximately 2.3 keV FWHM for 59.5-keV gamma rays, decreasing to 810 eV FWHM at 248 K.

CdTe detectors generally suffer from poor hole collection. Consequently, CdTe detectors are usually small in size, yet the small size also serves to reduce generation current. Regardless, the literature is replete with examples of excellent room-temperature gamma-ray energy resolution for small devices, usually about 1 to 2 mm thick (see references in [McGregor and Hermon 1997]). Methods to compensate the poor hole collection problem include single-carrier geometries and electronic correction methods. Hemispherical detector designs (see Sec. 12.5.2) were explored as room-temperature detectors with good results [Zanio 1977; Siffert 1994; Alt et al. 1994; Hage-Ali et al. 1995c], resulting in energy resolution of 26 keV FWHM for 662-keV gamma rays

[31] Electroless plating is an auto-catalytic chemical reaction process applied to a surface and does not require the application of a voltage. The chemical reaction displaces surface ions and causes the deposition of the metal component onto the surface. De Nobel [1959] provides a description of metal plating processes used on CdTe.

Table 16.9. Typical gamma-ray energy resolutions of a few representative commercial CdTe and CdZnTe semiconductor detectors. Many other sizes are offered by the different vendors.

detector	area (mm^2)	energy (keV)	FWHM (keV)	comments	source*
CdTe	9	122	≤1.2	Peltier cooled	A
Schottky	25	122	≤1.5	Peltier cooled	
CdZnTe	≈ 100	122	≤6.1	room temp	K
hemisphere	≈ 100	662	≤20	room temp	B,K
CdZnTe	100	662	13.2–26.4	room temp	K
coplanar	225	662	16.5–26.4	room temp	

*A = AmpTek, B = Baltic Scientific, K = Kromek.

with a 65 mm^3 active volume. Pulse height discrimination (PHD) methods [Jones and Woollam 1975], in which short pulse heights are eliminated from spectra, help to reduce spectral distortion from hole trapping. These electronic methods have yielded 7.5 keV FWHM at RT for 662-keV gamma rays for a 5 mm × 5 mm × 2 mm thick CdTe detector [Richter and Siffert 1992; Richter et al. 1993]. Unfortunately, PHD is a method that electronically eliminates a sizeable volume of the detector, ultimately reducing the detection efficiency. The method of pulse height correction (PHC) (charge loss compensation) improves energy resolution while preserving detection efficiency (see Sec. 15.5.2), where the slope of the pulse rise with respect to time is used to project the expected true pulse height [Richter and Siffert 1993; Richter et al. 1993, 1993; Eisen and Horovitz 1994]. Energy resolution of 6.1 keV FWHM and 6.4 keV FWHM for 122-keV and 356-keV gamma rays, respectively, were achieved by Eisen and Horowitz [1994] with this method. The energy resolution of commercially available detectors are presented in Table 16.9.

With the relatively good energy resolution of CdTe detectors, they have been demonstrated as detectors for slow neutrons [Vradii et al. 1977]. The strong neutron absorption cross section of Cd below 0.5 eV (σ_{th} of ^{113}Cd = 20,000 b) produces prompt gamma-ray emissions, which can be subsequently detected by the CdTe detector. Unfortunately, the gamma-ray capture efficiency is low, and the actual neutron detection efficiency is relatively poor. This type of neutron detector is presented with more detail in Chapter 17.

CdTe detectors are commercially available and have been used for various unique applications for which small room-temperature gamma-ray spectrometers are best applied. These applications include medical instruments, heat shield ablation sensors, nuclear power station activity monitors, nuclear fuel probes, capillary electrophoresis detectors, narcotics and explosive scanners, portable dosimeters, ammunition cartridge control monitors, and various x-ray and gamma-ray imaging devices [see references in McGregor and Hermon 1997]. Small CdTe detectors (5 mm × 5 mm × 1 mm thick), coupled to Peltier coolers having excellent energy resolution, of about 850 eV at 122 keV, are commercially available [Amptek 2017]. The operating voltage is nominally 500 volts for these thin devices, thereby requiring cooling below 263 K to reduce the leakage current to manageable levels.

$Cd_{1-x}Zn_xTe$

The introduction of Zn in the growth process of CdTe, nominally between 2% to 15%, has led to the production of CdZnTe detectors.[32] Overall, CdZnTe detectors have the same advantages as CdTe detectors with several added benefits. By adding a small amount of ZnTe to the melt, many important semiconductor

[32] This particular semiconductor is denoted CZT (common), (Cd,Zn)Te (less common), $Cd_{1-x}Zn_xTe$, or CdZnTe. By adhering to traditional elemental symbols, and because of the many different elemental combinations in use for substrates and detectors, the authors choose to use CdZnTe.

properties are drastically improved. The band gap of CdZnTe increases with Zn concentration, with band-gap energy of $Cd_{1-x}Zn_xTe$ at 300 K is well approximated by [Olego et al. 1985]

$$E_g(x) \text{ eV} = (1.51 \pm 0.005) + (0.606 \pm 0.01)x + (0.139 \pm 0.01)x^2 \text{ eV}. \qquad (16.87)$$

The band gap ranges from 1.52 to 1.64 eV for Zn contents ranging from $x = 0.02$ to $x = 0.2$; hence the detectors can operate at room temperature without serious leakage current concerns. With an increased band-gap energy, the intrinsic free carrier concentration diminishes, thereby increasing the resistivity while reducing detector leakage current. Butler et al. [1992] report that adding $x = 0.2$ amount of Zn changes the resistivity of CdTe from 3×10^9 Ω cm to 2.5×10^{11} Ω cm for $Cd_{0.8}Zn_{0.2}Te$. The dielectric constant is approximately 10.6 for CdZnTe, although this property is also a function of the Zn concentration. The addition of Zn increases the hardness of CdZnTe over CdTe and decreases the dislocation density [Anand 2013]. The etch pit density (EPD) is a measure of the dislocation defect density. Butler et al. [1993] show a marked reduction in the EPD as the Zn content was increased, with the densities of 1.5×10^5 cm^{-2}, 10^4 cm^{-2}, and 5×10^3 cm^{-2} for Zn contents of 0, 0.04, and 0.2, respectively. Further, CdZnTe detectors do not exhibit the polarization phenomenon at low irradiation levels, which are often observed with CdTe detectors. However, Wang et al. [2013] report the observation of polarization from CdZnTe detectors under high irradiation conditions. An additional benefit of incorporating Zn is that CdZnTe devices, because of the increased band gap and resistivity, can be operated at higher temperatures than CdTe devices, and they can also resolve lower energy photon energies (x rays and gamma rays) traditionally difficult to observe with CdTe detectors. CdZnTe detectors are presently used in handheld gamma-ray spectrometers and for smaller medical imaging apparatuses.

The gamma-ray absorption efficiency of CdZnTe is similar to that of CdTe, with a slight decrease in absorption efficiency with increased concentrations of Zn ($Z = 30$). The ionization energy is approximately 5 eV per e-h pair, although this number changes as a function of band-gap energy. The Fano factor of CdZnTe has been measured to be approximately 0.11 [Bale et al. 1999] at 233 K, with at least one research group [Redus et al. 1997] reporting a lower Fano factor of 0.089 also at 233 K.

Many of the same growth problems with CdTe are also experienced with CdZnTe, including the retrograde Te solubility. Further, the Zn component has a segregation coefficient larger than unity (about 1.3 at 1100°C) and can behave as a migrating impurity during growth, much like a zone-refinement pass [James et al. 1995]. In fact, CdZnTe behaves more like an alloy of CdTe and ZnTe [Owens 2012]. Consequently, there is a Zn concentration gradient through the crystal ingot that causes a change in the band gap over the length of the crystal ingot. This gradient appears to be more salient for relatively short ingots over that of long ingots. Usually, about 2 to 3 cm of material at both ends of the ingot are removed and discarded, and the remainder is used for detector fabrication. The Zn concentration gradient adversely affects the ionization energy by increasing the variance in the number of charge carriers excited per unit energy. Consequently, detectors with a relatively large variance in Zn concentration experience a broadening in spectroscopic energy resolution. Further, detector shot noise increases with the variance in leakage current, also a consequence of non-uniform Zn concentration.

CdZnTe substrates were originally developed as low defect substrates with less lattice mismatch for epitaxial growth of HgCdTe [Bell and Sen 1985; Sen and Stannard 1993], a shallow band-gap semiconductor used for IR detectors. Variations of Bridgman growth were initially used to grow bulk CdZnTe crystals, as outlined in James et al. [1995]. However, it was the high pressure method introduced by Raiskin and Butler [1988] that yielded the material used for the first reported CdZnTe radiation detectors [Butler et al. 1992; Doty et al. 1992a, 1992b]. For many years, high-pressure (HP) Bridgman growth, pressured with Ar to about 100 atm, was used to produce high-resistivity CdZnTe radiation detectors. CdZnTe material was grown on the Te rich side of the phase diagram, and as the crystal cooled and solidified in the growth process, Te precipitates would decorate the grain boundaries in the ingot. Single crystal pieces with low precipitate and

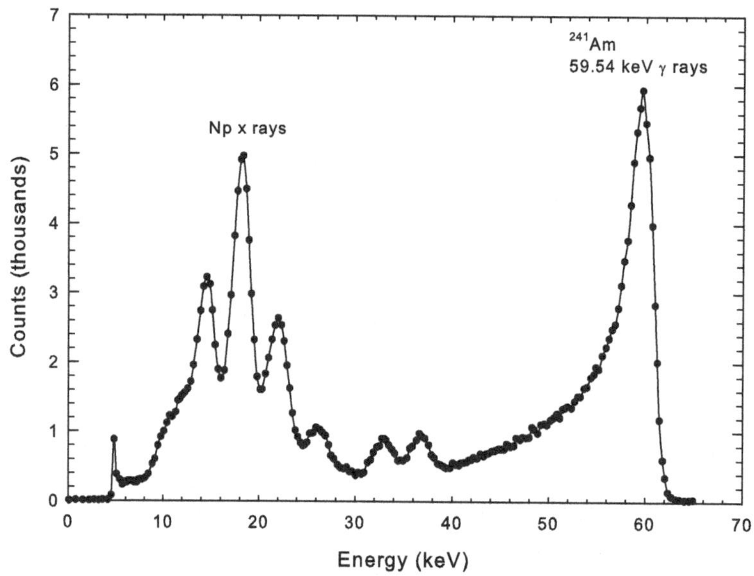

Figure 16.50. Typical spectrum of 59.54-keV gamma rays from ^{241}Am taken with a planar CdZnTe detector. The detector dimensions were 5 mm × 5 mm × 1.2 mm with a bias of 300 volts. Data from Butler et al. [1993].

defect concentrations were "mined" from the CdZnTe ingots, and these pieces were fabricated into detectors. Unfortunately, large defect-free pieces were rare, thereby limiting the useful detector sizes. Due to problems with twinning and the formation of Te precipitates, the useful detector yield with HP Bridgman growth was poor. Much like CdTe, the traveling heater method is now the preferred growth method for CdZnTe. CdZnTe can be grown with the THM at lower temperatures (less than 800°C). This low temperature growth reduces the defect concentration, reduces the precipitate concentration, and also reduces Zn segregation. Although the growth rate is intrinsically limited by the THM method [Peterson et al. 2016], the resultant product is generally of higher quality than that of HB Bridgman grown material [Roy 2013]. Post-growth annealing helps to reduce further the size and concentration of Te precipitates.

Large high-resistivity single crystal CdZnTe ingots are now produced commercially for radiation detectors [Chen et al. 2007, 2008]. Because of the high resistivity, usually greater than 10^{10} Ω cm, CdZnTe detectors are manufactured with ohmic contacts instead of blocking contacts and operated as resistive detectors. Electrical contacts are usually fabricated with electroless $AuCl_3$, although sputtered Pt contacts have shown good results.

Planar detectors fabricated from CdZnTe typically have long pulse height spectrum tails (Fig. 16.50), a consequence of severe hole trapping as described in Section 15.5.1. The severity of the effect increases with the detector width and gamma-ray energy. Because electron transport is superior to hole transport, CdZnTe detectors are invariably designed for cathode irradiation, thereby taking some advantage of the increased gamma-ray absorption nearest the cathode, although the effect diminishes with increasing gamma-ray energy, as shown in Fig. 16.51.

Most planar CdZnTe detectors are less than 3 mm thick so as to increase the values of ϱ_e and ϱ_h. Yet the poor hole transport properties still produce relatively poor energy resolution for penetrating high-energy gamma rays. Owens et al. [2002c] show that energy resolution for low-energy gamma rays can be improved by reducing the detector operating temperature, optimized near 253 K for 59.5 keV gamma rays (see Fig. 16.52).

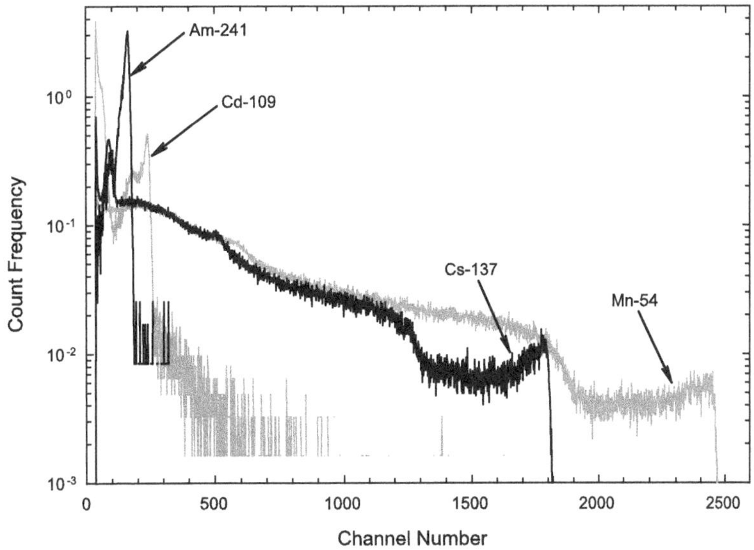

Figure 16.51. Comparison spectra taken with a 1 cm × 1 cm area × 2 mm thick CdZnTe detector irradiated on the cathode side with several monoenergetic gamma-ray sources. Charge carrier trapping effects increase with gamma-ray energy, causing the tailing effect to become more important.

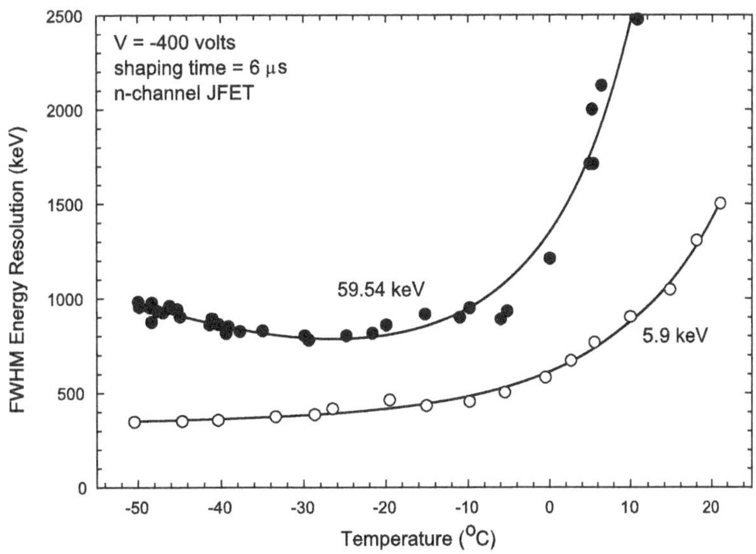

Figure 16.52. The energy resolution temperature dependence of a CdZnTe detector for 5.9-keV and 59.54-keV gamma rays. The detector was 2.5 mm thick by 3.2 mm^2 area. Data are from Owens et al. [2002c].

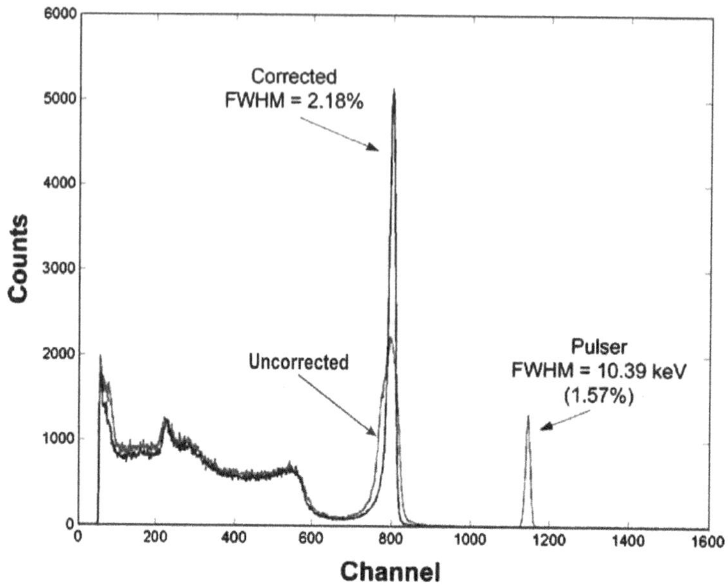

Figure 16.53. Energy resolution of a 1.5 cm × 1.5 cm × 1 cm thick CdZnTe coplanar grid detector, showing the results with and without electronic depth correction. From He and Sturm [2005]; copyright Elsevier (2005), reproduced with permission.

Several methods have been introduced to improve energy resolution and efficiency [see review by Zhang et al. 2013], a few of which are described here. CdZnTe detectors have adequate electron transport properties, but poor hole transport properties due to the combined effects of short hole mean free drift times and low hole mobility. As a result, conventional planar detectors seldom produce useful energy resolution for moderate to high energy gamma rays (greater than 300 keV). Instead, clever geometric detector shapes and electrode contact shapes are used to modify the weighting potential and electronic signal so that electrons dominate signal formation rather than holes. Energy resolution below 7 keV FWHM for 662-keV gamma rays can be achieved at room temperature for these *single carrier* detector designs. Further, electronic correction methods are often applied to improve the spectral performance of CdZnTe.

Luke [1994, 1995] introduced the co-planar grid (CPG) CdZnTe detector, described in Sec. 15.5.2, as a method to remedy poor hole transport. These initial detectors demonstrated superior energy resolution at 662 keV of 24.5 keV FWHM.[33] He et al. [1997] used position sensing to improve the energy resolution with charge compensation methods. He et al. [1998] also demonstrated that an optimized and balanced grid, including a boundary electrode, on a CPG device yielded superior results, and the use of a balanced grid is now routine for CPG CdZnTe detectors. He and Sturm [2005] combined the balanced coplanar design with rise time compensation to achieve room temperature energy resolution of 14.4 keV FWHM at 662 keV for a 15 mm × 15 mm × 10 mm thick device as shown in Fig. 16.53. Owens et al. [2006] also reported similar results, but without electronic correction methods, reporting room temperature energy resolution of about 12 keV FWHM at 662 keV.

[33]Because CdZnTe detector volumes are relatively small by comparison to most HPGe detectors, the 662 keV gamma-ray emission from ^{137}Cs is generally used to report energy resolution rather than the traditional 1332-keV gamma-ray emission from ^{60}Co.

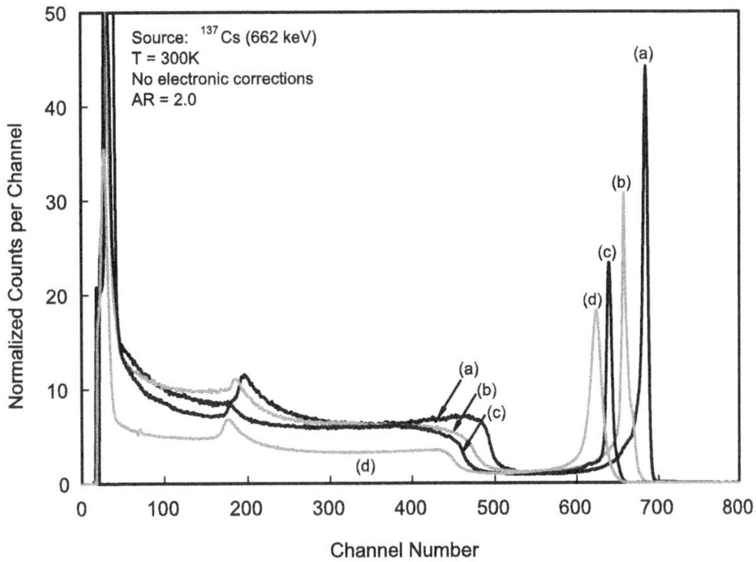

Figure 16.54. Spectroscopic results from ^{137}Cs for several Frisch collar CdZnTe detectors, each having a 2:1 aspect ration. The sizes and FWHM resolutions are (a) 0.9% for a 4.7 mm × 4.7 mm × 9.5 mm device, (b) 1.1% for a 6.5 mm × 6.5 mm × 13 mm device, (c) 1.2% for a 7.8 mm × 7.8 mm × 15.6 mm device, and (d) 2.4% for a 11 mm × 11 mm × 22 mm device. From Dunn and McGregor [2012].

Quasi-hemispherical CdZnTe detectors were introduced by Butler [Apotovsky et al. 1997; Butler 1997], and were commercially available for a short time. These detectors were three terminal devices formed on a parallelepiped CdZnTe block. The detector had a small anode on one surface, and planar cathode on the opposing surface, and a "control" electrode surrounding the anode [Lingren et al. 1998]. These detectors performed well, but were usually small. Energy resolution of 3.68 keV FWHM at 511 keV was reported for a 3 mm × 3 mm × 3 mm device [Butler 1997].[34] Parnham et al. [2000] report the development of a quasi-hemispherical CdZnTe fabricated on a parallelepiped block, on which a top surface has an electrical contact and the bottom half (approximately) of the remaining five sides of the devices were coated with an electrical contact. The design worked to improve the energy resolution above that expected from a planar device, and yielded room temperature (23°C) energy resolution of 10.4 keV at 662 keV for a 10 mm × 10 mm × 5 mm thick device [Bale and Szeles 2006; Szeles et al. 2006].[35] Related to the quasi-hemispherical CdZnTe detector is the ring-drift design [Owens et al. 2007; Alruhalli et al. 2015]. These devices have additional concentric rings around a central anode instead of the single control electrode described by Butler [1997]. These additional electrode rings enable improved steering of electrons to the anode and help to reduce low electric field regions present in the Butler [1997] design. Owens et al. [2007] report 4.85 keV FWHM at 662 keV for a 5 mm × 5 mm × 1 mm thick ring-drift CdZnTe detector.

Various quasi-Frisch grid designs were reported by McGregor, including the parallel strip design [McGregor et al. 1998a; McGregor and Rojeski 1998b], the geometrically weighted design [McGregor et al. 1999a; McGregor et al. 1999b], the Frisch ring design [McGregor and Rojeski 2001; McGregor 2004], and the Frisch collar design [McGregor and Rojeski 2001; Montemont et al. 2001; McGregor 2004; McNeil et al. 2004]. All of these detector designs use an external electrode to alter the weighting potential to enhance the contribu-

[34]These devices were commercially available for a short while from Digirad.

[35]Named CAPture^{TM} Plus detectors, these devices were commercially available for a short time from eV Products.

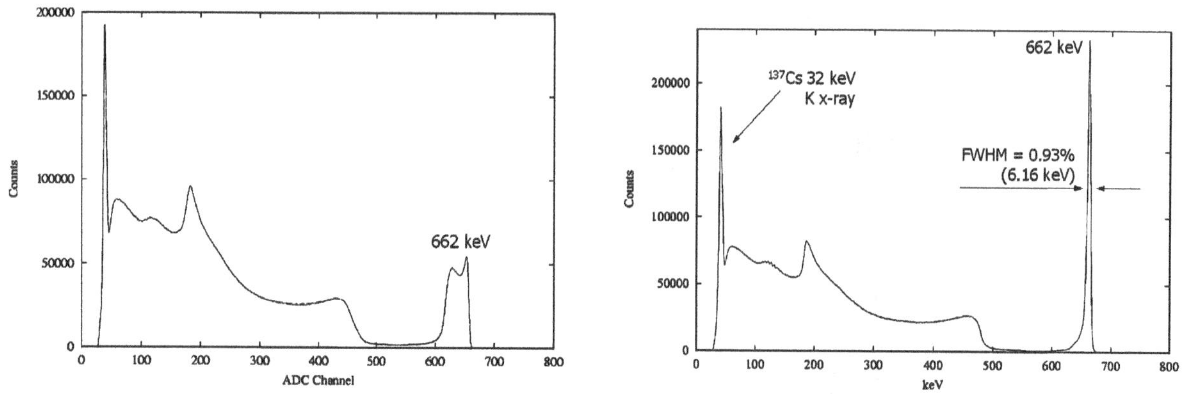

Figure 16.55. Room temperature spectrum of ^{137}Cs acquired with a CdZnTe pixelated detector (left) without depth correction and (right) with depth correction. All pixels are summed for both spectra. Bias conditions were -2200 volts on the cathode and -65 volts on the grid isolating the pixels. ©[2005] IEEE. Reprinted, with permission, from Zhang et al., *IEEE Trans. Nucl. Sci.*, **52**, 2009–2016, [2005].

tion of electron transport to the spectral output. The geometrically weighted Frisch Grid detector combines orientation, geometrical shape (either a trapezoid or pyramid frustum), the small pixel effect, and the charge induction isolation of the Frisch grid to produce enhanced energy resolution. These detectors were capable of delivering room temperature energy resolution of 17.2 keV FWHM at 662 keV for a 1 cm^3 device without electronic corrections [McGregor et al. 1999b]. However, these designs proved difficult to fabricate and led to simpler permutations, denoted either as Frisch ring or Frisch collar detectors [Kargar et al. 2011a, 2011b; Harrison 2009] and the basic distinction between the two is described in Sec. 15.5.2. Without electronic corrections, the Frisch collar design has delivered room temperature energy resolution of 5.9 keV FWHM at 662 keV for a 4.7 mm × 4.7 mm × 9.5 mm device, with comparable energy resolution for larger geometries (see Fig. 16.54).

CdZnTe has been investigated for medical imaging instrumentation, for which a block of CdZnTe is patterned with an array of small electrical contacts called pixels. Doty et al. [1994] report on the first monolithic CdZnTe pixelated device designed for medical imaging purposes. These CdZnTe pixel arrays were fabricated by traditional VLSI methods. This work led to the discovery of the "small pixel effect" [Barrett et al. 1995], described in Sec. 15.5.2, and is commonly used to enhance the energy resolution for gamma-ray spectrometers. CdZnTe pixel devices were successfully used for gamma-ray imaging (examples include Polichar et al. [1994, 1996], and Barber et al. [1997]). This technology has also proved important for gamma-ray spectroscopy [He et al. 1999, 2000; Zhang et al 2005]. He et al. [2000] combined CdZnTe pixel detectors with electronic energy-resolution correction methods to enhance the performance of pixelated detector arrays (see example in Fig. 16.55). Room temperature energy resolution of 6.16 keV FWHM at 662 keV was achieved with large (1.5 cm × 1.5 cm × 1 cm thick) CdZnTe pixelated detectors [Zhang et al. 2005]. With Compton scatter backprojection methods, these devices can be used to determine the gamma-ray energy and the general source direction by producing a near 360° image of radiation source locations and include the perceived intensity [Boucher et al. 2012; Wahl et al. 2015]. After many years of development, these devices are now commercially available.[36]

[36] H3D, Inc.

16.5.4 Detectors Based on Halide Compounds

The halogen semiconductors are produced from materials with at least one constituent atom from group VII of the periodic table. These materials include compounds such as HgI_2, PbI_2, and TlBr. Select materials have wide band-gap energies and can be operated at room and elevated temperatures.

Mercuric Iodide (HgI_2)

Attractive for its large Z components, mercuric iodide (HgI_2) has long been studied as a gamma-ray spectrometer with varying degrees of success [Willig 1971, 1972; Malm 1972; Swierkowsky et al. 1973; Llacer et al. 1974; Ponpon et al. 1975]. HgI_2 has atomic numbers of 80 and 53 with a volume density of 6.4 g cm^{-3}. The α-phase (red) of HgI_2 has measured charge carrier mobilities of 100 cm^2 $V^{-1}s^{-1}$ for electrons and 4 cm^2 $V^{-1}s^{-1}$ for holes [Ponpon et al. 1975]. Thus a substantial voltage is required to ensure acceptably high charge carrier collection. HgI_2 has a measured average ionization energy of 4.42 eV per e-h pair with a Fano factor of 0.19 [Ricker et al. 1982]. Because of the large atomic number of mercury (80), the photoelectric effect is predominant for gamma-ray energies below 400 keV (see Fig. 16.1). As a result, HgI_2 detectors can be relatively thinner than other common semiconductors and still have comparable efficiencies. For example, the absorption efficiency for 140-keV gamma rays in a 2-mm-thick HgI_2 detector is similar to 4.2-mm-thick CdTe or CdZnTe detectors and 1.9-cm-thick Ge or GaAs detectors [McGregor and Hermon 1997]. The α-phase of HgI_2 has a band-gap energy of 2.13 eV and has a resistivity of about 10^{13} Ω cm.

The melting point of HgI_2 is 259°C, but it also has a relatively low vapor pressure and can be sublimed at lower temperatures if processed in a vacuum. HgI_2 has numerous polymorphs that can form during crystal growth, most of which are metastable tetragonal structures [Hostettler et al. 2001, 2003, 2005]. The two most important polymorphs are the preferred α-phase tetragonal crystals (red) formed and stable at temperatures below 130°C and the β-phase monoclinic crystals (yellow) that form above 130°C. Growing β-phase HgI_2 crystals and subsequently cooling them below 130°C causes a destructive change in the crystals because they can form any number of metastable crystals or polycrystalline α-phase crystals [Newkirk 1956; Hostettler et al. 2003]. Moreover, with a melting point of 259°C, which is higher than the phase change temperature, a suitable method to grow detector-quality HgI_2 that maintains a growth region below 130°C is required.

Solution growth methods were first used to grow large HgI_2 crystals for radiation detectors [Willig 1971], and optically clear crystals have been produced with saturated solutions ranging in temperature between 25°C and 40°C [Mellet and Friant 1989; Nicolau and Joly 1980; Friant et al. 1980]. Unfortunately, detector performance of these solution grown crystals is compromised by impurity contamination introduced from the solvent suspension. Because of the improved crystalline quality and the reduction in impurity contamination, HgI_2 vapor growth methods are preferred [McGregor and Hermon 1997]. There are two main methods of vapor growth, referred to here as the vertical method, initially introduced by Scholz [1974], and the horizontal method, initially introduced by Schieber et al. [1976].

The vertical method has a relatively large ampoule, typically ranging in diameter between 10 cm to 15 cm and 15 cm high. The bottom of the ampoule has a raised pedestal approximately 3 cm in diameter that sits atop a temperature controlled cooler region, usually a cold finger or an air jet. The ampoule is either rotated within a heating coil [Scholz 1974; Schieber et al. 1976] or has a heating framework surrounding a stationary ampoule [Hermon et al. 1992]. There are three main elements to control the thermal profile consisting of a controlled cool region under the pedestal, a base heating element around the pedestal, and an axial heating element. The temperature around and under the ampoule is controlled such that a thermal gradient nucleates a seed crystal on the pedestal. The process takes several weeks to produce a single crystal between 250 g to 1000 g.

Schieber et al. [1976] describe an alternate method to growth HgI_2 in an evacuated horizontal ampoule of 5 to 10 cm diameter and 40 cm long. After loading with purified HgI_2 materials, ampoules are inserted into a two-zone furnace with a crystal formation zone held near 103°C and other source zone varied between

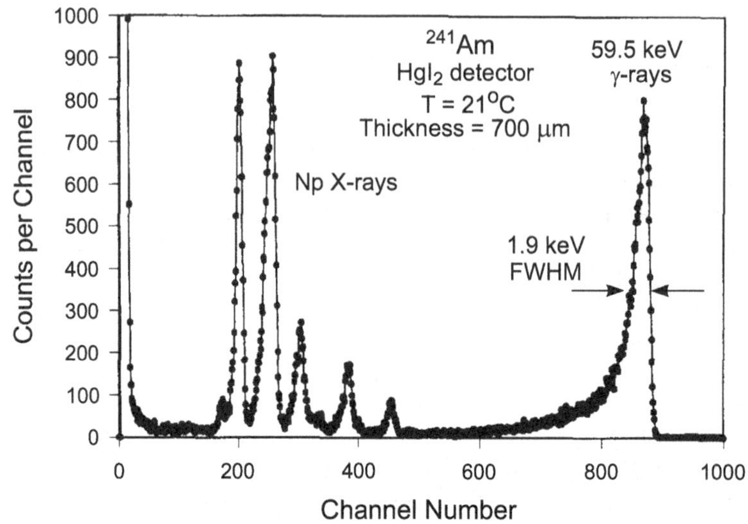

Figure 16.56. Room temperature differential pulse height spectrum from a 700-micron-thick HgI_2 detector. From McGregor and Hermon [1997]; copyright Elsevier (1997), reproduced with permission.

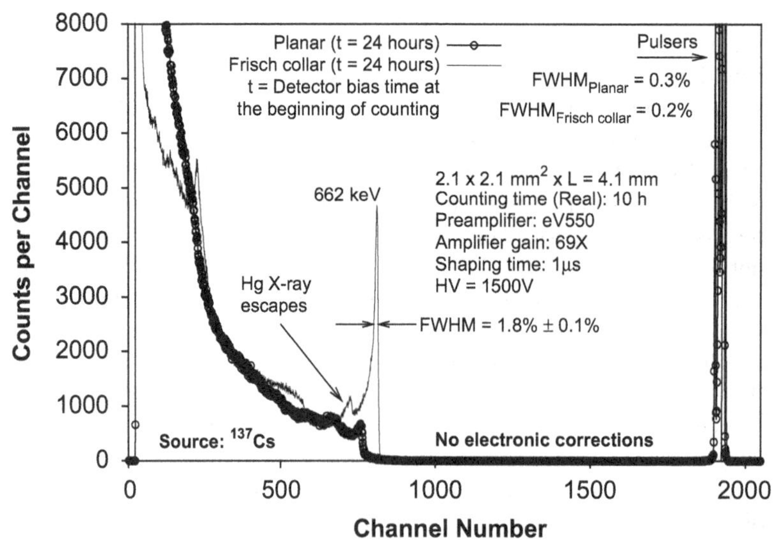

Figure 16.57. Pulse height spectra taken with a 2.1 × 2.1 × 4.1 mm^3 HgI_2 device with and without a Frisch collar. A 1.8% FWHM energy resolution was measured with 662-keV gamma rays from ^{137}Cs when using the Frisch collar. From Ariesanti et al. [2010]; copyright Elsevier (2010), reproduced with permission.

97°C to 110°C. The source zone temperature is raised to nucleate a crystal, or is lowered to etch away poorly formed samples. When one or more good crystals are nucleated, temperature in the source zone is no longer varied and is held at a temperature above 103°C to continue the growth of the HgI_2 singles crystals. The process dynamics prevent the growth of crystals larger than approximately 30 g without twinning. Faile et al. [1980] (also Faile [1981]) simplified the horizontal method by introducing purified HgI_2 into an evacuated growth ampoule approximately 2 to 4 cm diameter and 40 cm long along with low density polyethylene. This ampoule was inserted into a two-zone furnace with the temperature held at 120°C in the source zone and 100°C in the nucleation zone. Several "chance" crystals nucleated and grew near the middle of the ampoule. The size of these crystals ranged from about a few mm^3 to a cm^3 in volume and many appeared as single crystals with better physical and electronic performance than those produced without the addition of polyethylene.

Faile proposed several hypotheses about how polyethylene aided the growth of HgI_2 crystals that included impurity capturing, ampoule surface priming, free radicals or hydrogen reacting with crystal defects, and polymers alloying with HgI_2, to name a few [Faile et al. 1980]. However, none of these hypotheses could explain the observed results. McGregor and Ariesanti [2013] found that the function of low molecular weight polyethylene is to be a source of polar organic ketone molecules, impurities present in low molecular weight HDPE from the manufacturing process.[37] These aromatic ketones attach to the HgI_2 surfaces by weak Van der Waals bonds and inhibit growth on the {110} faces. When tested with purified ketones instead of polyethylene, such as 3-hexadecanone or 14-heptacosanone, improved results were observed and near perfect spectroscopic-grade tetragonal prismatic crystals were produced [Ariesanti and McGregor 2017]. Still the resulting crystals are only a few mm^3 to 1 cm^3 in volume.

HgI_2 detectors have fewer commercial applications than CdTe or CdZnTe detectors, mainly because of process and fabrication issues. Slicing and sectioning of a layered crystal like HgI_2 is difficult, often leading to delamination and mechanical damage. To reduce damage, sectioning is often performed with a chemical string saw, usually soaked with KI solution. A KI solution etch is usually used to remove surface damage after the crystal is reduced to the proper size and shape. HgI_2 is also highly reactive and only a few materials are known to be compatible for electrical contacts, those being carbon (Aquadag®), Pd, and Pt. HgI_2 detectors are usually encapsulated with parylene to prevent decomposition over time. Also, the material is known to *polarize* and is manifested as a gradual change in spectral features over time.

HgI_2 detectors are mostly fabricated as planar detectors, but because of the relatively low hole mobility, significant hole trapping can occur for thick detectors. Such an effect is shown by the lower energy tail in the spectrum of Fig. 16.56. Alternative detector designs with weighting potentials emphasizing electron charge induction have improved energy resolution, with some success realized with pixelated imaging detectors [Patt et al. 1995, 1996, 1997; Zhu et al. 2007]. The Frisch collar design has also been used with good results (see Fig. 16.57), evidently showing no signs of polarization [Ariesanti et al. 2010, 2011]. During the 1980s, HgI_2 was also explored as the photodetector for scintillators having relatively longer wavelengths [Markakis 1988a, 1988b]. Unfortunately, gamma-ray events were also directly detected in the HgI_2 photodetector, which added to the background. This effect was reduced by limiting the thickness of HgI_2 photodetectors to only 2 mm. One such device, when coupled to a 2.54-cm diameter × 2.54-cm-thick block of CsI:Tl, yielded room-temperature energy resolution of 5% FWHM at 662 keV [Markakis 1988b]. A modestly successful application of HgI_2 was as x-ray spectrometers for analysis of lead paint [Wang et al. 1993], and was offered

[37] Crystal improvement is observed only with low molecular weight polyethylene. Addition of high-purity spectroscopic grade polyethylene or HDPE does not produce any observed change in the crystal quality.

commercially for many years.[38] These small detectors proved efficient for low energy gamma rays and x rays and had excellent room temperature energy resolution (see Fig. 16.56, for example). Individual HgI_2 detectors have been offered as commercial units from time to time, but presently are relatively difficult to acquire.

Thallium Bromide (TlBr)

Thallium bromide (TlBr) is a wide band-gap semiconductor that is also of interest as a room-temperature gamma-ray spectrometer. TlBr is a yellowish-green transparent crystal with a 2.68 eV indirect band-gap energy. The crystal forms as a body centered cubic[39] with a density of 7.56 g cm^{-1}. The high Z components ($Z = 81/35$) make it especially attractive for gamma-ray detection. TlBr has an average ionization energy of 6.5 eV per e-h pair, relatively large by comparison to other semiconductors used for radiation detection. With $\epsilon_s/\epsilon_0 = 29.8$, the dielectric constant is also high by comparison to most semiconductors. Consequently, the high dielectric constant increases detector capacitance, and combined with the higher ionization energy reduces the output voltage signal by comparison to traditional semiconductors (such as Si or Ge) of similar dimensions. With such a wide band gap, the expected intrinsic resistivity is about 10^{12} Ω cm, although, in practice, TlBr usually has resistivities between 10^{10} and 10^{11} Ω cm. The electron mobility is only 6 cm^2 V^{-1} s^{-1}. Mean free drift times vary with crystal quality, and electron $\mu\tau$ products as high as 6.5×10^{-3} cm^2 V^{-1} s^{-1} along with the hole $\mu\tau$ products of 4×10^{-4} cm^2 V^{-1} s^{-1} have been reported for highly purified material [Shorohov et al. 2009b; Churilov et al. 2010; Kim et al. 2011; Dönmez et al. 2012] and indicates that high quality TlBr has relatively long charge carrier mean free drift times. Although these $\mu\tau$ values allow relatively efficient charge carrier collection, the low charge carrier mobilities require longer electronic shaping times to reduce the ballistic deficit. TlBr has a melting point of 460°C, making it attractive for traditional melt growth. Several melt growth techniques have been successfully used to produce quality TlBr ingots [for example, Gostilo et al. 2003; Kozlov et al. 2004; Hitomi et al. 2007b; Zhou et al. 2009; Churilov et al. 2010]. Detector fabrication requires careful handling, mainly because TlBr is a soft crystal with a Knoop hardness of 11.9 (approximately 1.38 on the Mohs scale) and is easily damaged.

TlBr was one of the first semiconductors explored as a radiation detector, having been investigated by Hofstadter in 1947 [see also Hofstadter 1949a, 1949b, 1950]. Hofstadter reported that the detectors made during these early investigations were sensitive to radiation, but failed to produce acceptable spectroscopic results. Many years passed after these initial investigations, but TlBr was revisited by Ijaz-Ur-Rahman and Hofstadter starting in 1984 [also Ijaz-Ur-Rahman et al. 1987]. Although these investigations yielded better results than before, they still produced poor energy resolution and also showed polarization when operated at room temperature. From the time of these early studies, much effort has been expended to purify TlBr materials and study the polarization effect.

Several reports indicate that cooling of TlBr detectors seems to improve performance. Ijaz-Ur-Rahman and Hofstadter [1984] note that the polarization effect seemed to become negligible when the detectors were operated at a temperature of 253 K. At even lower temperatures (183 K), alpha particle energy resolution was reported to improve, but still remained relatively poor and comparable to "a crude CsI(Tl) scintillation counter," although gamma-ray contributions to the spectra were non-existent. Shortly thereafter, Shah et al. [1989, 1990] reported improved results from zone-refined material with the appearance of gamma-ray spectra, especially at a reduced temperature (213 K). Elshazly et al. [2010] report that the mean free drift

[38] Offered by TN Technologies KayRay Sensall, Inc., or TN/KSI, for many years, these XRF analyzers used small platelet grown HgI_2 detectors. The TN/KSI name was later changed to Thermo MeasureTech, a subdivision of Thermo Fisher, who no longer produces this device.

[39] By placing Tl-Br pairs at each lattice site, this crystal qualifies as a simple cubic structure. The BCC structure comes from one constituent atom centered between eight of the other constituent atoms, where the eight other constituents populate the corners of a cube. Each unit cell has two atoms, the usual number for BCC. The Pearson notation for the crystal is cP2.

times increase as the temperature is lowered, especially below 160 K, and hypothesize that the notable change may be due to a low temperature phase change.[40]

Purification of the TlBr material can be conducted directly with the compound, and has been shown to make a significant difference in detector performance. Hitomi et al. [2007b] show that zone refinement improves the carrier transport properties and has little effect on the resistivity. Energy resolution was shown to improve for 410-micron-thick planar detectors as the number of zone refinement passes increased. For example, energy resolution improved from 33% FWHM to 11% FWHM for 59.5-keV gamma rays when the refinement passes were increased from 1 to 300. Gostilo et al. [2003] studied the effect of the growth method on performance, mainly as a comparison between Stockbarger and traveling molten zone (TMZ) methods, and concluded that the results were comparable. However, hole transport was observed to improve with the TMZ method while electron transport improved with the Stockbarger method.

TlBr detectors are most commonly constructed as planar detectors. The general process includes slicing wafers from an ingot with a precision wire saw followed by mechanical polishing and surface etching. Examples include Owens et al. [2003a]; Vaitkus et al. [2004]; Hitomi et al. [2009]; and Churilov et al. [2010]. Diamond wire saws and SiC slurry wire saws have both been used for the slicing process. Polishing with Al_2O_3 slurries are the most common, although diamond pastes have also been used. The mechanical polish is often followed by an etch in 5% bromine/methanol solution. The reported final detector thicknesses are usually between 200- to 400-microns thick in most cases. Contact patterns are applied through a shadow mask, and have been fabricated from a variety of metals, including Au, Cr, Tl, Pt, and colloidal carbon, as well as layered structures such as Cr/Au and Tl/Al. Thin Pd wires are often bonded to the conductive contacts with colloidal carbon solutions. Detectors fabricated with a pixel array were processed in the same manner as above, with the exception that the pixel side of the device was patterned with a pixelated shadow mask, while the other was coated as with a common planar detector [Owens et al. 2003c; Hitomi et al. 2007].

Planar detector designs are susceptible to resolution degradation from charge carrier losses, primarily hole trapping, yet have yielded in some cases an energy resolution at room temperature of about 2.2% FWHM at 662 keV for thin devices (nominally 2 mm) [Shorohov et al. 2009a]. Improved performance is achieved with alternative designs using electron-dominated weighting potentials and in some cases electronic corrections. By using the Frisch collar design of McGregor [2004], energy resolution for planar detectors was improved by Kim et al. [2012], and achieved a room temperature energy resolution of 2.4% FWHM at 662 keV for 8.4-mm long TlBr detectors. In that same work, using the depth correction method of He et al. [1996], a 13-mm-thick array, having 25 pixels each 2 mm $\times$ 2 mm, achieved 2% FWHM energy resolution for 662-keV gamma rays at a temperature of 255 K. Room temperature energy resolution of 2% FWHM at 662 keV was achieved with a combination of depth correction and pulse height rejection for a pixel array 1.8-mm thick [Hitomi et al. 2007a]. In recent times (as of this writing), most TlBr detectors are fabricated as pixelated arrays to take advantage of the small pixel effect. Usually one or more pixels outperform the other pixels, an indication of remaining problems with material or contact uniformity. Regardless, energy resolution of about 1.5% FWHM was reported for some pixels on these TlBr detectors [Kim et al. 2009; Churilov et al. 2010].

TlBr has a serious problem with destructive polarization when operated over extended periods of time. As noted originally by Ijaz-Ur-Rahman and Hofstadter [1984], the polarization issue was rendered negligible when detectors were operated at temperatures below 253 K. Studies by Vaitkus et al. [2005] on Au contacted TlBr detectors provide evidence that polarization is caused by the drifting of Tl^+ ions through the TlBr

[40]Notably, TlBr grown as thin films reportedly has a phase change at low temperatures, from the usual body-centered cubic lattice (such as CsCl) at room temperature to either an orthorhombic or NaCl cubic phase [Schulz 1951; Blackman and Khan 1961]. Because these earlier measurements were conducted on thin TlBr films, and that thicker films appear to lack this property, Smith [2013] concludes that a bulk phase change is unlikely. Regardless, there appears to be no additional evidence that a phase change is responsible for the improved performance.

crystal. Datta and Motakef [2015] observed migration and ionic conduction from Tl^+ ions and Br^- ions through TlBr detectors fabricated with thin Au contacts. Using optical transmission, Datta and Motakef [2015] recorded the progression of defect formation across a detector over several hundred hours and found that the defects nucleate at the cathode surface and propagate to the anode. From calculations on the formation of native defects in TlBr, Du [2010] produced three important observations. First, the native defects have low formation energies and do not produce electron traps and, thus, can explain the relatively large $\mu\tau$ products. Second, the resistivity of the material is not caused by traditional deep level compensation, but instead by paired defects that do not induce the formation of deep traps. These defects are nearly stoichiometric TlBr and are mainly Schottky defects that pin the Fermi level near midgap. Third, the material has low diffusion barriers to several native defects, thereby enabling ionic conduction. It is this ionic conduction that can cause the appearance of polarization as these native defects drift through the material, as proposed earlier by Vaitkus et al. [2004, 2005]. Along with these polarization mechanisms generally observed with high-quality TlBr, Du [2010] recommends that temperature reduction can reduce this ionic conduction (already observed by Ijaz-Ur-Rahman and Hofstadter [1984]), and the use of Tl electrodes to neutralize the diffusing (drifting) defects (as observed by Hitomi et al. [2008, 2009]). Kozorezov et al. [2010] report the observation of polarization of TlBr with x rays of energies greater than 50 keV, manifested by a gradual degradation of the detector charge collection efficiency. Consequently, spectroscopic features were also observed to degrade over time. Kozerezov et al. [2010] conclude that the radiation induced polarization is caused by hole trapping and the resultant space charge buildup. Hitomi et al. [2008] studied various metal layered contacts with combinations of Au and Tl. Evidently, long-term operation has been successful with Tl contacts, for which Hitomi et al. [2009] report 600 hours of room-temperature operation without observing the polarization effect. Datta and Motakef [2015] also report improved results with Pt contacts, although measurements still indicated degradation of the Pt contact with time [Motakef and Datta 2015; Datta and Motakef 2015]. TlBr detectors presently remain mainly experimental devices, although some commercial vendors are offering detectors on a custom order basis.

Lead Iodide (PbI_2)

The properties of lead iodide (PbI_2) indicate that it is a promising material for use as a room-temperature radiation detector. It has periodically been investigated as a potential room-temperature operated detector since the initial investigations by Roth and Willig [1971]. PbI_2 has a hexagonal close-pack (HCP) crystal structure and a mass density of 6.16 g cm^{-3}. The dielectric constant reportedly ranges between 4.3 and 5.2 [Glasser et al. 1967]. The band-gap energy is reportedly between 2.3 eV and 2.55 eV. The relatively large band-gap energy leads to low leakage currents even at elevated temperatures. The reported charge carrier mobilities are 8 cm^2 V^{-1} s^{-1} for electrons and 2 cm^2 V^{-1} s^{-1} for holes [Manfredotti et al. 1977]. Unfortunately, these low mobilities, coupled with short mean free drift times, produce small $\mu\tau$ products, the largest measured at $\mu_e\tau_e = 1 \times 10^{-5}$ cm^2 V^{-1} and $\mu_h\tau_h = 3 \times 10^{-7}$ cm^2 V^{-1} [Lund et al. 1992]. Deich and Roth [1996] report similar $\mu\tau$ values of $\mu_e\tau_e = 0.8 \times 10^{-6}$ cm^2 V^{-1} and $\mu_h\tau_h = 6.4 \times 10^{-7}$ cm^2 V^{-1}.

Unlike HgI_2, PbI_2 does not suffer from a destructive phase change at lower temperatures below the melting temperature of 402°C. Hence, PbI_2 can be grown directly from the melt. The high mass density coupled with the high atomic numbers ($Z_I = 53$, $Z_{Pb} = 82$) ensure a relatively high gamma-ray interaction efficiency, similar to that of HgI_2. Initial results from research on PbI_2 detectors indicated that the mean free drift times of the charge carriers are short, and spectra obtained with this material have satellite energy peaks for alpha particles. However, zone refinement purification of the PbI_2 starting material reportedly improved radiation spectrometer performance [Lund et al. 1988, 1992; Oliveira et al. 2002]. Detectors have been fabricated by various methods, and samples are generally cleaved with a fine blade or sliced with a string saw. Detector contacts have been fabricated from Au, Pd, and colloidal graphite (Aquadag®) and all have yielded similar results. The measured mean free drift times are 1.0 μs for electrons and 0.3 μs for holes. A reported room-temperature energy resolution for a 107-μm-thick device is 712 eV FWHM for 5.9-keV gamma rays and 1.83 keV FWHM for 59.5-keV gamma rays [Deich and Roth 1996]. PbI_2 detectors and materials remain experimental and are not commercially available.

16.6 Additional Semiconductors of Interest

There are a multitude of additional compound semiconductors that have been investigated as radiation detectors, many listed in review articles describing special applications and results [Hofstadter 1949a, 1949b; Chynoweth 1952; Mayer 1966; Swierkowski 1976, Yee et al. 1976, Armantrout et al. 1977, Sakai 1982; Cuzin 1987; McGregor and Hermon 1997; and Owens and Peacock 2004]. Unfortunately, space limitations preclude any lengthy discussion of all these potential alternative semiconductors that could be used for radiation detectors. Here only three are briefly reviewed. The reader is referred to the referenced works for more information about other semiconductor materials.

$Cd_{1-x}Mn_xTe$

Related to CdZnTe is the ternary compound CdMnTe, a magneto-optical material with a large Verdet constant.[41] This material is used for Faraday rotation devices, LEDs, spintronics, electro-magnetic interference free devices, field tunable phase shifters, small coupled solar cells, phased array radar, and harsh environment sensors [International 2017]. The band-gap dependence on elemental composition of $Cd_{1-x}Mn_xTe$ was studied by numerous research groups [Gaj et al. 1978; Bottka et al. 1981; Abreu et al. 1981; Diouri et al. 1982; Lemasson et al. 1983; Lee et al. 1984; El Amrani et al. 1983; Bücker et al. 1985], all of which reported a direct band gap with a linear energy dependence on Mn concentration between $0.0 \leq x \leq 0.7$. The band-gap energy of $Cd_{1-x}Mn_xTe$ at 300 K is generally described by [from data of Lee and Ramdas 1984]

$$E_g(x) \text{ eV} = 1.524 + 1.328x \text{ eV}, \ 0.0 \leq x \leq 0.7. \tag{16.88}$$

The band gap can be expanded to 1.6 eV, the same as $Cd_{0.85}Zn_{0.15}Te$, by using $x = 0.057$, or $Cd_{0.943}Mn_{0.057}Te$, thereby permitting room-temperature operation. The theoretical resistivity of such material should be high enough so that injecting contacts can be used rather than blocking contacts as are needed for CdTe:Cl.

Non-uniformities of Mn concentration causes variances in the band-gap energy and, consequently, also increases the variance in the average ionization energy; hence uniformity is important. The segregation coefficient (k) of Mn in a solution of CdTe is almost 1.0 so that Mn does not segregate during the growth process. For example, Triboulet et al. [1990] report a Mn concentration gradient of only 0.48 to 0.52, when x is 0.5 over a crystal length of 8 cm. The potential uniformity of CdMnTe is a clear benefit over CdZnTe, for which the segregation coefficient (k) of Zn in a CdTe solution is reportedly 1.35 [Tanaka et al. 1989], so that it is difficult to grow homogeneous CdZnTe crystals.

Triboulet et al. [1990] report that $Cd_{1-x}Mn_xTe$ forms as a cubic zincblende crystal for x less than 0.75. However, the crystal structure of MnTe is hexagonal, and consequently there is an increased probability of having mixed crystalline structures between cubic zincblende and hexagonal as the Mn concentration is increased in $Cd_{1-x}Mn_xTe$, especially for x greater than 0.75 for which mixed phases have been shown to appear [Triboulet and Didier 1981]. Further, it is possible to produce hexagonal inclusions within an otherwise cubic lattice system [Mycielski et al. 2005]. Also Triboulet and Didier [1981] report the appearance of cubic $MnTe_2$ precipitates within Bridgman grown crystals, and Triboulet et al. [1990] report the appearance of twinning for Mn concentrations exceeding 15%. However, they also report that both problems can be reduced by subsequently regrowing the ingot by the traveling heater method. From Eq. (16.88) small amounts

[41] The Verdet constant describes the strength of the Faraday effect, a phenomenon in which the plane of light polarization is rotated linearly proportional to the magnetic field in the direction of light propagation.

of Mn greatly affect band-gap changes, reportedly 13 meV [Lee and Ramdas 1984] per atomic percent of Mn,[42] an indication that relatively low concentrations of Mn are required to produce high resistivity material.

Similar to CdZnTe, the formation of Cd vacancies (V_{Cd}) produces intrinsic acceptor centers, usually rendering as-grown material p-type with a resistivity of about 10^3 Ω cm. The resistivity can be increased by annealing the ingot in Cd vapor, thereby reducing the concentration of V_{Cd}. The resistivity can also be increased by including one or more compensating deep donor dopants into the crystal, including V [Burger et al. 1999] and In [Kim et al. 2009]. Owens [2012] notes an additional problem with CdMnTe production is the relatively high reactivity of Mn, which tends to bond with oxides on the growth ampoule surface, as reported by Triboulet and Didier [1981].

The first reported use of CdMnTe for radiation detectors was by Burger et al. [1999], with x equal to either 0.15 or 0.45, and with vanadium added as a compensation dopant. All materials studied reached resistivities exceeding 5×10^{10} Ω cm. Spectroscopic results for detectors fabricated from $Cd_{0.55}Mn_{0.45}Te$ were relatively poor, but the work did show that CdMnTe has promise as a radiation detector material. Since the time of that initial work, there has been some, but limited, pursuit of CdMnTe as a radiation detector [Parkin et al. 2007; Zhang et al. 2007; Babalola et al. 2009; Kim at al. 2009; Rafiei et al. 2012, 2013].

Planar detectors fabricated from $Cd_{1-x}Mn_xTe$ by Kim at al. [2009], with measured values of x between 0.074 and 0.078, had resistivities greater than 1.5×10^{10} Ω cm and $\mu_e\tau_e$ of 10^{-3} cm^2 V^{-1}. However, spectroscopic measurements from low energy gamma rays delivered non-uniform results, with the best being approximately 1.8 keV FWHM (3%) for 59.54-keV gamma rays. By employing the Frisch collar design [see McGregor and Rojeski 2001; McGregor 2004], Kim et al. [2015] produced a 6 mm $\times$ 6 mm $\times$ 12 mm detector from CdMnTe that yielded a room temperature energy resolution of 13.9 keV FWHM at 662 keV, the best results reported as of this writing. CdMnTe indeed shows promise as an alternative material for room-temperature operated semiconductor gamma-ray spectrometers, but material improvements are still required before commercial detectors are expected to appear.

CdSe

Cadmium selenide (CdSe) has been studied as a possible room-temperature (RT) gamma-ray and x-ray detector periodically over the last several decades [for example Burger et al. 1983; Roth 1989; Chen et al. 1998]. CdSe has atomic numbers of 48/34 with a mass density of 5.8 g cm^{-3}. The crystalline structure of CdSe is wurtzite, a type of hexagonal crystal, and the band-gap energy of CdSe is 1.73 eV. Canali et al. [1972] measured the electron and hole mobilities of CdSe, reporting $\mu_e = 720$ cm^2 V^{-1} s^{-1} and $\mu_h = 75$ cm^2 V^{-1} s^{-1} (the direction was not specified).

CdSe was first identified as a potential RT operated radiation detector by Prince and Polishuk [1967], although it was also recognized that impurity reduction and *pn* junction formation were major obstacles. A theoretical study explored the use of compound semiconductors, including CdSe, as room temperature radiation detectors [Armantrout et al. 1977]. However, Armantrout et al. were not optimistic about CdSe and proposed that AlSb,[43] InP, ZnTe, and CdTe would be better detector candidates.

Burger et al. [1983] were first to report results for CdSe nuclear detectors. These crystals were grown with a vertical unseeded vapor growth technique (VUVG). Typically, this crystal growth method produces n-type material with a low resistivity ranging between 1 and 10 Ω cm [Roth 1989] and is a consequence of Se depletion in the crystals. These samples of 250-micron thickness were heat treated at 900°C in a Se environment, resulting in a resistivity increase up to 10^{12} Ω cm, high enough to reduce leakage current for radiation detectors. The samples were polished and etched with bromine based solutions, and subsequently detectors were fabricated by applying Au contacts (non-symmetric) to both sides of the planar samples.

[42] By comparison, the band gap of $Cd_{1-x}Zn_xTe$ increases at 6.7 meV per atomic percent of Zn.

[43] Although studied often, AlSb has not developed into a commercial semiconductor for radiation detection, mainly because of the multiple problems with the material, including the rapid decomposition of AlSb in air.

The energy resolution of the samples when irradiated with 59.54-keV gamma rays from ^{241}Am was poor (or nonexistent). Roth [1989] notes that Se treatment can produce compositional and resistivity inhomogeneities, thereby adversely affecting charge transport properties and probably contributed to the poor resolution performance. Notably, the samples showed temporal stability and no signs of polarization.

Burger and Roth [1984a] and Roth and Burger [1986] improved the performance of CdSe detectors by growing crystals in a Se solution with the temperature gradient solution zoning (TGSZ) method, a technique originally developed by Steininger [1968]. The TGSZ growth method resulted in crystals with resistivities between 10^7 and 10^8 Ω cm. Samples ranging from 200 μm to 1.0 mm thick were prepared by cleaving or slicing the crystals along the $[1\bar{2}11]$ planes and subsequently etching in a bromine-methanol solution, slowly diluted with methanol to prevent the formation of a $CdBr_2$ layer on the surface [Roth and Burger 1986]. Contacts were formed with a carbon Aquadag® colloidal solution. Results reported by Roth et al. [1987] and Roth [1989] gave $\mu_e\tau_e = 1.2 \times 10^{-5}$ cm^2 V^{-1} and $\mu_h\tau_h \approx 7 \times 10^{-7}$ cm^2 V^{-1} for $\mu\tau$ products.[44] With these various changes and improvements (crystal growth, surface preparation, electrical contacts), improved performance was realized, yielding 1.4 keV FWHM at 5.9 keV (^{55}Fe), 4 keV FWHM at 27.4 keV (^{125}I), and 8.5 keV FWHM at 59.54 keV (^{241}Am).

Chen et al. [1998] reported on CdSe detectors fabricated from commercially produced CdSe crystals with resistivity of 10^{10} Ω cm. The detectors were nominally 4 mm $\times$ 4 mm area. The process used to fabricate the devices seemed to return to the earliest detector fabrication methods and employed mechanical polishing with alumina abrasive followed by a 2-minute etch in a 5% bromine-methanol solution. Final sample thicknesses ranged from 150 μm to 1.2 mm, and Au contacts were applied to opposite detector sides by physical vapor deposition. The detectors were exposed to 59.54-keV gamma rays, and produced an energy resolution of 18 keV FWHM for 150 μm thick detectors, 25 keV FWHM for 300 μm thick detectors, and no discernible resolution for the 1.2-mm-thick detectors.

Ternary compounds such as CdZnSe [Burger and Roth 1984b; Burger et al. 1985] and CdTeSe [Fiederle et al. 1994; Kim et al. 2008] have been explored as alternative semiconductors for radiation detection. Recall that CdSe is a hexagonal crystal, it is interesting to note that the crystal structures of both ZnSe and CdTe are cubic (zincblende). Hence, the varied concentrations of Se and Zn, or Se and Te, can cause not only band-gap changes, but also crystal structural changes. Burger et al. [1985] report that the leakage current from a $Cd_{0.7}Zn_{0.3}Se$ sample, a wurtzite structure, was approximately an order of magnitude lower than that of a CdSe detector of comparable size. However, the spectroscopic results were similar to CdSe reported by Roth [1989], having 1.8 keV FWHM at 5.9 keV (^{55}Fe) and 4 keV FWHM at 27.4 keV (^{125}I). Fiederle et al. [1994] reported on detectors fabricated from $CdTe_{0.9}Se_{0.1}$, a cubic zincblende crystal. Devices 1.34 mm thick yielded 6 keV FWHM for 59.54-keV gamma rays. Kim et al. [2008] also describe the performance of *p*-type Cl compensated CdTeSe, which had reported resistivity of 4.5×10^9 Ω cm. The measured charge carrier mobilities were $\mu_e = 59$ cm^2 V^{-1} s^{-1} and $\mu_h = 33$ cm^2 V^{-1} s^{-1}, yet the reported $\mu\tau$ products were high, reported to be $\mu_e\tau_e = 6.55 \times 10^{-2}$ cm^2 V^{-1} and $\mu_h\tau_h = 8.12 \times 10^{-2}$ cm^2 V^{-1}. However, despite these seemingly beneficial properties, the detectors had high leakage current and performed marginally with an energy resolution of 18 keV FWHM for 59.54-keV gamma rays.

There has been limited investigation of this wide band-gap semiconductor in recent years, perhaps due in part to the success of other compound semiconductors such as CdZnTe. Regardless, CdSe has enjoyed renewed success as a nanomaterial for quantum dot fabrication. Apparently accurate control over the quantum dot size is achievable, thereby allowing the band gap to be tailored with great precision. Potential alternative uses include highly efficient lasers and light emitting diodes, quantum computing, and photovoltaic

[44] Chen et al. [1998] report much different values from various sources, listing $\mu_e\tau_e = 6.3 \times 10^{-5}$ cm^2 V^{-1} and $\mu_h\tau_h = 7.5 \times 10^{-5}$ cm^2 V^{-1}.

applications through multiple exciton generation, an improvement over conventional Si solar panels [Colvin et al. 1994; Robel et al. 2006; Somers et al. 2007].

GaSe

Gallium selenide (GaSe) was investigated as a possible room temperature radiation detector several decades ago [for example, Manfredotti et al. 1974a; Mancini et al. 1976; Sakai et al. 1988; Nakatani et al. 1989], although in recent years interest appears to have subsided. GaSe has atomic numbers of 31/34 with a mass density of 4.55 g cm^{-3}. The crystal is hexagonal with covalently bonded Se-Ga-Ga-Se layers held together by van der Waals bonds. GaSe has a direct band gap of 2.03 eV, wide enough for room-temperature operation. The motivation of pursuing GaSe as a radiation detector included the search for materials capable of operating at elevated temperatures, low cost fabrication, and minimal leakage current, all of which GaSe seemed to possess [Manfredotti et al. 1974a; Castellano 1986].

Manfredotti et al. [1974a] communicate two methods used to produce the GaSe crystals, namely Bridgman growth [Cardetta et al. 1972a] and a chemical transport method [Cardetta et al. 1972b]. The Bridgman growth produced sizeable crystals of about 1 cm$^2\times$ 4 cm length with resistivities ranging between 10^3 to 10^6 Ω cm [Manfredotti et al. 1974a, 1974b]. The reported charge carrier mobilities were $\mu_e = 60$ cm^2 V^{-1} s^{-1} and $\mu_h = 215$ cm^2 V^{-1} s^{-1} measured along the c axis. The $\mu\tau$ products were about $\mu_e\tau_e = 1.4 \times 10^{-6}$ cm^2 V^{-1} and $\mu_h\tau_h = 3.65 \times 10^{-6}$ cm^2 V^{-1}, thereby indicating mean free drift times of $\tau_e = 2 \times 10^{-6}$ s and $\tau_h = 5\times10^{-6}$ s. Current-voltage tests produced space charge limited characteristics as described by Lampert and Mark [1970]. Samples grown by chemical transport were smaller in size, the largest being about 50 mm^2 area and 150 μm thick and had substantially higher resistivities between 10^8 to 10^9 Ω cm. However, the reported transport properties were inferior to the Bridgman samples, with $\overline{\mu_e\tau_e} = 1.3\times10^{-7}$ cm^2 V^{-1} yielding $\tau_e \approx 2 \times 10^{-9}$ (results for holes were not reported). Detectors were fabricated by evaporating 7 mm^2 area Au contacts onto opposite sides of planar samples ranging between 50 to 120 μm thick. Measurements with these detectors yielded a poor energy resolution of 1.13 MeV FWHM for 5.486 MeV alpha particles, along with the observation of satellite peaks, an indication of inhomogeneous electrical properties [Manfredotti et al. 1974a]. By collimating the alpha particles to a 0.8 mm^2 area on the detector the energy resolution was improved to 370 keV FWHM.

Sakai et al. [1988] and Nakatani et al. [1989] report on Bridgman grown GaSe material with measured charge carrier mobilities of $\mu_e = 70$ cm^2 V^{-1} s^{-1} and $\mu_h = 45$ cm^2 V^{-1} s^{-1}. The average ionization energy was found to be between 4.0 and 4.5 eV per e-h pair, and falls on the alternate dependence of the Klein model (see Fig. 16.2), like many other layered semiconductors (HgI_2, PbI_2). The best RT energy resolution reported by Sakai et al. [1988] and Nakatani et al. [1989] was approximately 250 keV FWHM for 5.486 MeV alpha particles. Yamazaki et al. [1993] experimented with GaSe doping of various elements, primarily Si, Ge, and Sn, a doping which decreased the leakage currents. Spectroscopic measurements with these doped samples yielded varied energy resolution, ranging from 220 keV (4%) FWHM to 850 keV FWHM for 5.486 MeV alpha particles. Yamakazi et al. [1993] also report that the best of the undoped GaSe detectors they fabricated yielded an energy resolution of 252 keV (4.6%) FWHM for 5.846 MeV alpha particles. Most of the undoped GaSe samples generally produced leakage current too high to yield spectroscopic results.

GaSe has also been explored for x-ray and low energy gamma-ray detection [Castellano 1986; Arutyunyan et al. 1989; Mandal et al. 2007], and also as muon beam monitors [Manfredotti et al. 1975; Mancini et al. 1976]. Castellano [1986] and Arutyunyan et al. [1989] studied GaSe as an x-ray beam dose monitors, and reported primarily on the photocurrent produced by x-ray beams. Castellano [1986] promoted the concept of using GaSe for x-ray dosimetry because of its apparent thermal stability, low dark current, and simplistic detector construction. Mandal et al. [2007] report on the spectroscopic performance of In-doped Bridgman grown GaSe exposed to 59.54-keV gamma-rays, yielding 2.4 keV (4%) FWHM.

Problems that exacerbate the progress of GaSe as a radiation detector include the fragility of the crystals, relatively low charge carrier mobilities, and the resultant relatively low $\mu\tau$ products. The crystals are soft (Mohs hardness of 2) and usually separate or cleave along the van der Waals bonds, thereby making slicing and detector fabrication challenging. The addition of In as a dopant appears to have a positive effect on strengthening the crystal structure [Voevodin et al. 2004]. Improvements with purification and crystal growth may increase charge carrier mean free drift times and result in higher $\mu\tau$ products; however, the direct band gap indicates relatively short charge carrier lifetimes still persist. The charge carrier mobilities are intrinsic properties, thereby meaningful improvements are not expected.

The layered structure of GaSe produces highly anisotropic electrical properties that are beneficial for non-linear optics [Brebner 1964]. Non-linear GaSe optical materials are used in the frequency conversion of laser light, which involves shifting the wavelength of monochromatic laser light to alternate wavelengths [Kübler et al. 2005], and has been exploited recently for THz generation with an extremely large bandwidth of up to 41 THz [Huber et al. 2000; Liu et al. 2004]. GaSe is one of the most widely employed crystals for non-linear optical generation and detection of THz radiation. Consequently small GaSe crystals are commercially available, but primarily with common sizes about 1 $cm^2 \times$ 200 μm to 1 mm thick.

16.7 Summary

Semiconductor materials are attractive as radiation detectors for two principle reasons. First, because of their low average ionization energy w, semiconductors produce a large number of signal charge carriers per unit energy, thereby decreasing statistical fluctuations beyond that of gas-filled and scintillation detectors and, as a consequence, produce a much better energy resolution. Second, semiconductor materials have energy band structures that allow their electrical properties to be altered through the addition of impurities. These materials can be manipulated to have a majority of negative (n-type) electrical charge carriers, or electrons, or positive (p-type) charge carriers, denoted as "holes". Adjacent n-type and p-type materials can be manipulated to form detectors with rectifying contacts, in which both leakage current and electrical noise are reduced.

Moreover, semiconductor detectors can be fashioned into various types of detectors especially designed for x-ray detection, γ-ray detection, or charged-particle detection. Detector performance is optimized by semiconductor choice and device design. Charged-particle detectors are generally designed with low Z material, such as Si, to reduce backscattering. Higher Z materials, because of their improved absorption efficiencies, are generally used for γ-ray detectors. Low energy x-ray and γ-ray detectors are often fabricated from Li-drifted Si (Si(Li) detectors), although the most commonly used semiconductor for photon detection is high-purity Ge (HPGe detectors). Both Si(Li) and HPGe detectors must be cooled to low temperature for best operation. Wide band-gap semiconductors, such as CdZnTe, can be used as room-temperature operated γ-ray spectrometers.

Finally, semiconductor materials can be fashioned into arrays to yield spatial interaction information. These arrays can be arranged from the tiling of numerous individual detectors, or they can be fabricated as pixels upon a single semiconductor substrate. Commercial vendors offer numerous varieties of semiconductor detectors, which include particle, x-ray, and γ-ray detectors, in the form of individual devices or as arrays.

PROBLEMS

1. A 7-mm slab of π-type Si with $N_A = 10^{14}$ cm^{-3} has a layer of Li evaporated upon it. The temperature is raised to 400°C for 3 hours. Afterwards, the temperature is lowered to 120°C, and 500 volts are applied for a period of 100 hours. What is the resulting compensated width?

2. A surface-channel CCD device is to be used to measure alpha particles from ^{148}Gd. The background concentration is $N_A = 10^{12}$ cm^{-3}, the effective operating bias is 10 volts, and the oxide thickness is 0.15

microns. If the pixel sizes are 5 micron × 5 micron area, what is Q_{max}? Will this device operate for the intended use?

3. A Si(Li) detector has a linearly varying residual positive space charge. Derive a general expression for the electric field distribution in a planar detector under such a condition.

4. Determine the electron and hole velocities for a 5 mm × 5 mm × 1 cm thick CdZnTe detector operated at 1000 volts.

5. How would a change in the Fano factor in a germanium detector affect the energy resolution of the detector?

6. Plot your estimate of the energy resolution of a HPGe detector as a function of the incident photon energy between 100 keV and 10 MeV. Assume that there is complete charge collection and that there is negligible noise.

7. How thick must the depletion region in a Si(Li) be to record the full energy of a 4-MeV alpha particle?

8. You have a Si surface barrier detector with the following characteristics: n-type Si, $n = 2 \times 10^{14}$ n-type dopant impurities cm^{-3}, and $V_{bi} = 0.3$ volts. You also have an ^{241}Am alpha particle source. The detector is operated in vacuum. (a) What does the differential pulse height spectrum appear as with no voltage on the detector? (b)What does the differential pulse height spectrum appear as with 25 volts applied? (c) How much voltage is required to extend the detector depletion region just beyond the alpha particle range?

9. Given a Si(Li) detector with a 3.5-mm-thick drifted region, determine the energy that the detector can measure from an energetic electron of 20 MeV if the electron passes through the detector.

10. You measure spectra from a detector fabricated from CdTe that suffers much charge carrier trapping. A few measurements reveal the following: electron mean free drift times are 0.5 microseconds, electron mobility = 1000 cm^2 V^{-1} s^{-1}, hole mean free drift times are 0.2 microseconds, electron mobility = 80 cm^2 V^{-1} s^{-1}, and saturation velocity for both carriers is 10^7 cm s^{-1}.

The operating voltage is kept at 100 volts. (a) With highly penetrating gamma rays (high energy), what is the maximum detector thickness that the detector can be to keep the measured pulses below 20% deviation? (b) 10% deviation? (c) 1% deviation?

11. What is the best theoretical energy resolution that you can achieve with Ge for 122-keV gamma rays in Ge? Give answers in energy and percent. Repeat for 1.17 MeV gamma rays.

12. You are given pure Ge material with the properties: $N_C = 10^{19}$ cm^{-3} and $N_V = 6 \times 10^{18}$ cm^{-3}. (a) What is the value of n_i at T = 77K? (b) What is the value of n_i at T = 300K? (c) Explain why germanium detectors are operated at cryogenic temperatures.

13. Why are escape peaks generally more prominent in spectra from HPGe detectors to those observed with NaI:Tl detectors of equal volume?

14. What properties of a semiconductor material are of most concern to you when considering using the material as a radiation detector? Consider both technical and practical aspects of using the material.

REFERENCES

Abreu, R.A., W. Giriat, and M.P. Vecchi, "Temperature Dependence of the Absorption Edge in $Cd_{1-x}Mn_xTe$," *Phys. Lett. A*, **85**, 399–401, (1981).

Adam, W. et al., "Pulse Height Distribution and Radiation Tolerance of CVD Diamond Detectors," *Nucl. Instrum. Meth.*, **A447**, 244–250, (2000)

Ahearn, A.J., "Conductivity Pulses Induced in Single Crystals of Zinc Sulfide by Alpha-Particle Bombardment," *Phys. Rev.*, **73**, 524, (1948).

Ahearn, A.J., "Search for Crystals That Exhibit Conduction Pulses Under Alpha-Particle Bombardment," *Phys. Rev.*, **75**, 1966–1967, (1949).

Alberi, J.L. and V. Radeka, "Position Sensing by Charge Division," *IEEE Trans. Nucl. Sci.*, **NS-23**, 251–258, (1976).

Alexiev, D. and K.S.A. Butcher, "High Purity Liquid Phase Epitaxial Gallium Arsenide Nuclear Radiation Detector," *Nucl. Instrum. Meth.*, **A 317**, 111–115, (1992).

Alig, R.C. and S. Bloom, "Electron-Hole-Pair Creation Energies in Semiconductors," *Phys. Rev. Lett.*, **35**, 1522–1525, (1975).

Allcock, H.J., J.R. Jones, and J.G.L. Michel, *The Nomogram, 5th Ed.*, London: Pitman, 1963.

Alruhaili, A., P.J. Sellin, A. Lohstroh, V. Boothman, P. Veeramani, M.C. Veale, K.J.S. Sawhney, and V. Kachkanov, "Development of a CZT Drift Ring Detector for X and γ Ray Spectroscopy," *JINST*, **10**, paper P04005, 13 pages, (2015).

Alt, R., P. Ambrosi, J. Böhm, G. Hilgers, M. Jordan, and K-H. Ritzenhoff, "The Generation of the Response Matrix of Hemispherical CdTe Detectors," *Nucl. Instrum. Meth.*, **A353**, 71–75, (1994).

Amelio, G.F., M.F. Tompsett, and G.E. Smith, "Experimental Verification of the Charge Coupled Device Concept," *Bell. Syst. Tech. J.*, **49**, 593–600, (1970).

Ammerlaan, C.A.J. and K. Mulder, "The Preparation of Lithium-Drifted Semiconductor Nuclear Particle Detectors," *Nucl. Instrum. Meth.*, **21**, 97–100, (1963).

AMPTEK, http://amptek.com/products/xr-100cdte-x-ray-and-gamma-ray-detector/, accessed July, 2017.

Anand, K.G., "Defects in Cadmium Zinc Telluride (CdZnTe) - A Review," *Int. J. Engin. Management Sci.*, **4**, 113–120, (2013).

Apotovsky, B., C.L. Lingren, A. Oganesyan, B. Pi, J.F. Butler, F.P. Doty, R.L. Conwell, and S.J. Friesenhahn, "Semconductor Radiation Detector with Enhanced Charge Collection," US Patent 5677539; allowed Oct. 14, 1997.

Ariesanti, E., A. Kargar, and D.S. McGregor, "Fabrication and Spectroscopy Results of Mercuric Iodide Frisch Collar Detectors," *Nucl. Instrum. Meth.*, **A624**, 656–661, (2010).

Ariesanti, E., A. Kargar, and D.S. McGregor, "Mercuric Iodide Crystal Growth and Frisch Collar Detector Fabrication," *Nucl. Tech.*, **175**, 124–130, (2011).

Ariesanti, E. and D.S. McGregor, "Process of Forming a Crystal Having a Particular Shape and the Crystal Formed by the Process," US Patent 9797065, allowed Oct. 24, 2017.

Armantrout, G.A., S.P. Swierkowski, J.W. Sheroman, and J.H. Y—SC EE, "What Can be Expected From High-Z Semiconductor Detectors?" *IEEE Trans. Nucl. Sci.*, **NS-24**, 121–125, (1977).

Arutyunyan, V.M., M.L. Dimaksyan, V.L. Élbakyan, and G.E. Grigoryan, "X-Ray Sensitivity of Gallium Monoselenide," *Sov. Phys. Semicond.*, **23**, 315–316, (1989).

Autagawa, W., K. Zanio, and J.W. Mayer, "CdTe as a Gamma Detector," *Nucl. Instrum. Meth.*, **55**, 383–385, (1967).

Avset, B.S., L. Evensen, G. Hall, S. Roe, and, R. Wheadon, "Practical Limitations to the Performance of Silicon Drift Chambers Imposed by Material Quality," *Nucl. Instrum. Meth.*, **A310**, 203–207, (1991).

Babalola, O.S., A.E. Bolotnikov, M. Groza, A. Hossain, S. Egarievwe, R.B. James, and A. Burger, "Study of Te Inclusions in CdMnTe Crystals for Nuclear Detector Applications," *J. Crys. Growth*, **311**, 3702–3707, (2009).

Babcock, R.V. and H.C. Chang, "SiC Neutron Detectors for High-Temperature Operation," *Neutron Dosimetry: Proc. Symp. Neutron Detection, Dosimetry,and Standardization*, Vienna, IAEA, **Vol. 1**, pp. 613–622, 1963.

Babcock, R.V., "Radiation Damage in SiC," *IEEE Trans. Nucl. Sci.*, **NS-12**, 43–47, (1965).

Bailey, R., C.J.S. Damerell, R.L. English, A.R. Gillman, A.L. Lintern, S.J. Watts, and F.J. Wickens, "First Measurement of Efficiency and Precision of CCD Detectors for High Energy Physics," *Nucl. Instrum. Meth.*, **213**, 201–215, (1983).

Bale, G., A. Holland, P. Seller, and B. Lowe, "Cooled CdZnTe Detectors for X-Ray Astronomy," *Nucl. Instrum. Meth.*, **A436**, 150–154, (1999).

Bale, D.S. and C. Szeles, "Design of High Performance CdZnTe Quasi-Hemispherical Gamma-Ray Capture^{TM} Plus Detectors," *Proc. SPIE*, **6319**, 1–11, (2006).

Baraff, G.A., "The Mid-Gap Donor Level EL2 in Gallium Arsenide: Recent Developments," Ch. 6, in *Deep Centers in Semiconductors*, 2nd Ed., S.T. Pantelides, Ed., Philadelphia: Gordan and Breach, pp. 547–589, 1992.

Barber, H.B., H.H. Barrett, F.L. Augustine, W.J. Hamliton, B.A. Apotovsky, E.L. Dereniak, F.P. Doty, J.D. Eskin, J.P. Garcia, D.G. Marks, K.J. Matherson, J.M. Woolfenden, and E.T. Young, "Development of a 64 × 64 CdZnTe Array and Associated Readout Integrated Circuit for Use in Nuclear Medicine," *J. Elec. Mater.*, **26**, 765–772, (1997).

Barber, H.D., "Effective Mass and Intrinsic Concentration of Silicon," *Solid-State Electronics*, **10**, 1039–1051, (1967).

Bardeen, J. and W. Shockley, "Deformation Potentials and Mobilities in Non-Polar Crystals," **80**, 72–80, (1950) .

Barraud, A., "Le Comportement de l'Arséniure de Gallium de Haute Résistivité Soumis á dea Champs Électriques Intenses," *Comptes Rendus*, **265**, 3632–3635, (1963).

Barrett, H.H., J.D. Eskin, and H.B. Barber, "Charge Transport in Arrays of Semiconductor Gamma-Ray Detectors," *Phys. Rev. Lett.*, **75**, 156–159, (1995).

Barry, A.L., B. Lehmann, D. Fritsch, and D. Bräunig, "Energy Dependence of Electron Damage and Displacement Threshold Energy in 6H Silicon Carbide," *IEEE Trans. Nucl. Sci.*, **NS-38**, 1111–1115, (1991).

Bates, R.L., C. Parkes, B. Rakotomiaramanana, C. Fleta, G. Pellegrini, M. Lozano, J.P. Balbuena, U. Parzefall, M. Koehler, M. Breindl, and X. Blot, "Charge Collection Studies and Electrical Measurements of Heavily Irradiated 3D Double-Sided Sensors and Cmoparison to Planar Strip Detectors," *IEEE Trans. Nucl. Sci.*, **NS-58**, 3370–3383, (2011).

Beaumont, S.P., et al., "First Results from GaAs Double-Sided Detectors," *Nucl. Instrum. Meth.*, **A 321**, 414–517, (1992).

Beaumont, S.P., et al., "GaAs Solid State Detectors for Physics at the LHC," *IEEE Trans. Nucl. Sci.*, **NS-40**, 1225–1230, (1993).

Beaumont, S.P., et al., "Gallium Arsenide Microstrip Detectors for Charged Particles," *Nucl. Instrum. Meth.*, **A 348**, 172–179, (1994).

Belgium, (conducted in Mol, Belgium) "Solid State Neutron Detector," Centr d'Etude de l'Energi Nucléaire, work performed under the United States-Euratom Joint Research and Development Program. This is a series of quarterly reports from 1962–1964.

Bell, R.O. and F.V. Wald, "Recent Advances in the Preparation of CdTe for Nuclear Detectors," *IEEE Trans. Nucl. Sci.*, **NS-19**, 334–338, (1972).

Bell, R.O., G. Entine, and H.B. Serreze, "Time-Dependent Polarization of CdTe Gamma-Ray Detectors," *Nucl. Instrum. Meth.*, **117**, 267–271, (1974).

Bell, S.L. and S. Sen, "Crystal Growth of $Cd_{1-x}Zn_xTe$ and its use as a Superior Substrate for LPE Growth of $Hg_{0.8}Cd_{0.2}Te$," *J. Vac. Sci. Tech.* **A 3**, 112–115, (1985).

Berger, M.J., S.M. Seltzer, S.E. Chappell, J.C. Humphreys, and J.W. Motz, "Response of Silicon Detectors to Monoenergetic Electrons with Energies Between 0.15 and 5.0 MeV," *Nucl. Instrum. Meth.*, **69**, 181–193, (1969a).

Berger, M.J., S.M. Seltzer, S.E. Chappell, J.C. Humphreys, and J.W. Motz, "Table of Response Functions for Silicon Electron Detectors," *Tech. Note 489*, U.S. Nat. Bureau of Standards, (1969b).

Berger, M.J., *ESTAR, PSTAR, and ASTAR: Computer Programs for Calculating Stopping Power and Range Tables for Electrons, Protons, and Helium Ions*, Report NISTIR 4999, National Institute of Standards and Technology, Gaithersburg, MD, (1992). [Distributed as Peripheral Shielding Routine PSR-330 by Radiation Shielding Information Center, Oak Ridge National Laboratory, Oak Ridge, TN.]

Bertolini, G. and A. Coche, *Semiconductor Detectors*, New York: Wiley, 1968.

Bertuccio, G., M. Sampietro, and A. Fazzi, "High Resolution X-Ray Spectroscopy with Silicon Drift Detectors and Integrated Electronics," *Nucl. Instrum. Meth.*, **A 322**, 538–542, (1992).

Bertuccio, G., A. Pullia, C. Canali, F. Nava, and C. Lanzieri, "Schottky Junctions on Semi-Insulating LEC Gallium Arsenide for X- and γ-Ray Spectrometers Operated at and Below Room Temperature," *IEEE Trans. Nucl. Sci.*, **NS-44**, 117–124, (1997).

Bertuccio, G., R. Casiraghi, A. Cetronio, C. Lanzieri, and F. Nava, "A New Generation of X-Ray Detectors Based on Silicon Carbide," *Nucl. Instrum. Meth.*, **A518**, 433–435, (2004a).

Bertuccio, G., R. Casiraghi, A. Cetronio, C. Lanzieri, and F. Nava, "Silicon Carbide for High Resolution X-Ray Detectors Oeprating up to 100°C," *Nucl. Instrum. Meth.*, **A522**, 413–419, (2004b).

Bertuccio, G., S. Caccia, D. Puglisi, and D. Macera, "Advances in Silicon Carbide X-Ray Detectors," *Nucl. Instrum. Meth.*, **A652**, 193–196, (2011).

Berwick, K., M.R. Brozel, C.M. Buttar, M. Cowperthrwaite, and Y. Hou, "Imaging of High Field Regions in SI GaAs Particle Detectors," *Inst. Phys. Conf. Ser., Defect Recognition and Image Processing in Semiconductors and Devices*, No. 135, 305–310, (1993).

Blackman, M. and I.H. Khan, "The Polymorphism of Thallium and Other Halides at Low Temperatures," *Proc. Phys. Soc.*, **77**, 471–475, (1961).

Blakemore, J.S., "Semiconducting and Other Major Properties of Gallium Arsenide," *J. Appl. Phys.*, **53**, R123–R181, (1982).

Blankenship, J.L. and C.J. Borkowski, "Silicon Surface-Barrier Nuclear Particle Spectrometer," *IRE Trans. Nucl. Sci.*, **NS-7**, 190–195, (1960).

Blankenship, J.L. and C.J. Borkowski, "Improved Techniques for Making P^+-I-N^+ Diode Detectors," *IRE Trans. Nucl. Sci.*, **NS-9**, 181–189, (1962).

Bock, R., H.H. Duhm, W. Melzer, F. Pühlhofer, and B. Stadler, "An Array of Position Sensitive Surface Barrier Detectors for Use in a Broad Range Magnetic Spectrograph," *Nucl. Instrum. Meth.*, **41**, 190–194, (1966).

Bottka, N., J. Stankiewicz, and W. Giriat, "Electroreflectance Studies in $Cd_{1-x}Mn_xTe$ Solid Solutions," *J. Appl. Phys.*, **52**, 4189–4193, (1981).

Boucher, Y., J. Jaworski, W. Kaye, F. Zhang, and Z. He, "Results from Testing of 145 3D Position-Sensitive, Pixelated CdZnTe Detectors," *IEEE Trans. Nucl. Sci.*, **59**, 3332–3338, (2012).

Bourgoin, J.C. and B. Massarani, "Threshold Energy for Atomic Displacement in Diamond," *Phys. Rev.*, **B14**, 3690–3694, (1976).

Bourgoin, J.C., H.J. von Bardeleben, and D. Stiévenard, "Native Defects in Gallium Arsenide," *J. Appl. Phys.*, **64**, R65–R91 (1988).

Boyle, W.S. and Smith, "Charge Coupled Semiconductor Devices," *Bell Syst. Tech. J.*, **49**, 587–593, (1970).

Braun, F., "Uber die Stromleitung durch Schwefelmetalic," *Ann. Phys. Chem.*, **153**, 556–563 (1874).

Brebner, J.I., "The Optical Absorption Edge in Layer Structures," *Solid State Comm.*, **2**, xii, (1964).

Broennimann, Ch., F. Glauss, S. Heising, M. Horisberger, R. Horisberger, H.C. Kästli, J. Lehmann, T. Rohe, and, S. Streuli, "Deveopment of an Indium Bump Bond Process for Silicon Pixel Detectors at PSI," *Nucl. Instrum. Meth.*, **A565**, 303–308, (2006).

Brownridge, I.C., *Lithium-Drifted Germanium Detectors, Their Fabrication and Use. An Annotated Bibliography.*, New York: IFI/Plenum, 1972.

Brussels, (conducted in Brussels) "Solid State Neutron Detector," Centre d'Etude de l'Energi Nucléaire, work performed under the United States-Euratom Joint Research and Development Program. This is a series of quarterly reports from 1960–1961.

Bryant, F.J. and A.F. Cox, "Atomic Displacement Energies for Binary Semiconductors," *J. Phys. C*, **1**, 1734–1736, (1968).

Bückner, R., H-E. Gumlich, and M. Krause, "The Influence of the Temperature and the Composition on the Reflectivity of $Cd_{1-x}Mn_xTe$ Within the Spectral Range of $1.5 \leq E \leq 4$ eV," *J. Phys. C: Solid State Phys.*, **18**, 661–667, (1985).

Burger, A., I. Shilo and M. Schieber, "Cadmium Selenide: A Promising Novel Room Temperature Radiation Detector," *IEEE Trans. Nucl. Sci.*, **NS-30**, 368–370, (1983).

Burger, A. and M. Roth, "Growth of Medium Electrical Resistivity CdSe Single Crystals by the Temperature Gradient Solution Zoning Technique," *J. Crys. Growth*, **67**, 507–512, (1984a).

Burger, A. and M. Roth, "Temperature Gradient Solution Zoning Growth and Characterization of $Zn_xCd_{1-x}Se$ Single Crystals," *J. Crys. Growth*, **70**, 386–392, (1984b).

Burger, A., M. Roth, and M. Schieber, "The Ternary $Cd_{0.7}Zn_{0.3}Se$ Compound, A Novel Room Temperature X-Ray Detector," *IEEE Trans. Nucl. Sci.*, **NS-32**, 556–558, (1985).

Burger, A., K. Chattopadhyay, H. Chen, J.O. Ndap, X, Ma, S. Trivedi, S.W. Kutcher, R. Chen, and R.D. Rosemeier, "Crystal Growth, Fabrication and Evaluation of Cadmium Manganese Telluride Gamma Ray Detectors," **198/199**, 872–876, (1999).

Butler, J.F., C.L. Lingren, and F.P. Doty, "$Cd_{1-x}Zn_xTe$ Gamma Ray Detectors," *IEEE Trans. Nucl. Sci.*, **39**, 605–609, (1992).

Butler, J.F., F.P. Doty, B. Apotovsky, J. Lajzerowicz, and L. Verger, "Gamma- and X-Ray Detectors Manufactured from $Cd_{1-x}Zn_xTe$ Grown by a High Pressure Bridgman Method," *Mater. Sci. Eng.*, **B16**, 291–295, (1993).

Butler, J.F., "Novel Electrode Design for Single-Carrier Charge Collection in Semiconductor Nuclear Radiation Detectors," *Nucl. Instrum. Meth.*, **A 396**, 427–430, (1997).

Canali, C., M. Martini, G. Ottaviani, and K.R. Zanio, "Transport Properties of CdTe," *Phys. Rev. B*, **4**, 422–431, (1971a).

Canali, C., G. Ottaviani, M. Martini, and K.R. Zanio, "Hole Drift Velocity in Semi-Insulating CdTe," *Appl. Phys. Lett.*, **19**, 51–53, (1971b).

Canali, C., F. Nava, G. Ottaviani, and C. Paorici, "Hole and Electron Drift Velocity in CdSe at Room Temperature," *Solid State Comm.*, **11**, 105–107, (1972).

Canberra, Silicon (Li) Detector Systems, Canberra Industries, Inc., Document CSP0157, 2003.

Canberra, Clover Detectors, Canberra Industries, Inc., Document C39840, 2012.

Canberra, Germanium Detectors, Mirion Technologies, Inc., Document C39606, 2016.

Canepa, P.C., P. Malinaric, R.B. Campbell, and J. Ostroski, "High Temperature Nuclear Particle Detector," *IEEE Trans. Nucl. Sci.*, **NS-11**, 262–270, (1964)

Cardetta, V.L., A.M. Mancini, and A. Rizzo, "Melt Growth of Single Crystal Ingots of GaSe by Bridgman-Stockbarger's Method," *J. Crys. Growth*, **16**, 183–185, (1972a).

Cardetta, V.L., A.M. Mancini, and A. Rizzo, "Growth and Habit of GaSe Crystals Obtained from Vapour by Various Methods," *J. Crys. Growth*, **17**, 155–161, (1972b).

Castellano, A., "GaSe Detectors for X-Ray Beams," *Appl. Phys. Lett.*, **48**, 298–299, (1986).

Cern, *Atlas Insertable B-Layer Technical Design Report*, CERN-LHCC-2010-013, ATLAS TDR 19, (2010).

Chelikowsky, J.R. and M.L. Cohen, "Nonlocal Pseudopotential Calculations for the Structure of Eleven Diamond and Zinc-Blende Semiconductors," *Phys. Rev. B*, **14**, 556–582, (1976).

Chen, H., M. Hayes, X. Ma, Y.-F. Chen, S.J. Egarievwe, J.O. Ndap, K. Chattopadhyay, A. Burger, and J. Leist, "Physical Properties and Evaluation of Spectrometer Grade CdSe Single Crystal," *Proc. SPIE*, **3446**, 17–28, (1998).

Chen, H., S.A. Awadella, J. Mackenzie, R. Redden, G. Bindley, A.E. Bolotnikov, G.S. Carmarda, G. Carini, and R.B. James, "Characterization of Traveling Heater Method (THM) Grown $Cd_{0.9}Zn_{0.1}Te$ Crystals," *IEEE Trans. Nucl. Sci.*, **54**, 811–816, (2007).

Chen, H., S.A. Awadella, F. Harris, P. Lu, R. Redden, G. Bindley, A. Copete, J. Hong, J. Grindlay, M. Amman, J.S. Lee, P. Luke, I. Kuvvetli, and C. Budtz-Jorgensen, "Spectral Response of THM Grown CdZnTe Crystals," *IEEE Trans. Nucl. Sci.*, **55**, 1567–1572, (2008).

Choong, W-S., K.M. Vetter, W.W. Moses, G. Hull, S.A. Payne, N.J. Cherepy, and J.D. Valentine, "Design of a Facility for Measuring Scintillator Non-Proportionality," *IEEE Trans. Nucl. Sci.*, **NS-55**, 1753–1758, (2008).

Churilov, A.V., G. Ciampi, H. Kim, W.M. Higgins, L.J. Cirignano, F. Olschner, V. Biteman, M. Minchello, and K.S. Shah, "TlBr and $TlBr_xI_{1-x}$ Crystals for γ-Ray Detectors," *J. Crys. Growth*, **312**, 1221–1227, (2010).

Chynoweth, A.G., "Conductivity Crystal Counters," *Am. J. Phys.*, **20**, 218–226, (1952).

Clegg, T.B., A.C.L. Barnard, and J.B. Swint, "A Nuclear Reaction Polarimeter Using Position-Sensitive Particle Detectors," *Nucl. Instrum. Meth.*, **40**, 45–53, (1966).

Cola, A., L. Vasanelli, L. Reggiani, A. Cavallini, and F. Nava, "Microscopic Modelling of Semi-Insulating GaAs Detectors," *Nucl. Instrum. Meth.*, **A395**, 98–100, (1994).

Cola, A., F. Quaranta, L. Vasanelli, C. Canali, A. Cavallini, F. Nava, and M.E. Fantacci, "A Study of the Trap Influence on the Performance of Semi-Insulating GaAs Detectors," *Nucl. Instrum. Meth.*, **A395**, 349–354, (1997).

Colvin, V.L., M.C. Schlamp, and A.P. Allvisatos, "Light-Emitting Diodes made from Cadmium Selenide Nanocrystals and a Semiconducting Polymer," *Science*, **370**, 354-357, (1994).

Conwell, E. and V.F. Weisskopf, "Theory of Impurity Scattering in Semiconductors," *Phys. Rev.*, **77**, 388–390, (1950).

Cornet, A., P. Siffert, A. Coche, and R. Triboulet, "Cadmium Telluride Surface Barrier Detectors," *Appl. Phys. Lett.*, **17**, 358–361, (1970).

Cornet, A., J. Hage-Ali, J.J. Grob, R. Stuck, and P. Siffert, "Ion Implanted Contacts on Cadmium Telluride Detectors," *IEEE Trans. Nucl. Sci.*, **NS-19**, 358–361, (1972).

Cronin, G.R. and R.W. Haisty, "The Preparation of Semi-Insulating Gallium Arsenide by Chromium Doping," *J. Electrochem. Soc.*, **111**, 874–877, (1964).

Cuzin, M., "Some New Developments in the Field of High Atomic Number Materials," *Nucl. Instrum. Meth.*, **A 253**, 407–417, (1987).

Da Viá, C., J. Hasi, C. Kenney, V. Linhart, S. Parker, T. Slavicek, S.J. Watts, P. Bem, T. Horazdovsky, and S. Pospisil, "Radiation Hardness Properties of Full-3D Active Edge Silicon Sensors," *Nucl. Instrum. Meth.*, **A 587**, 243–249, (2008).

Dalal, V.L., A.B. Dreeben, and A. Triano, "Temperature Dependence of Hole Velocity in p-GaAs," *J. Appl. Phys.* **42**, 2864–2867, (1971).

Damerell, C.J.S., F.J.M. Farley, A.R. Gillman, and F.J. Wickens, "Charge-Coupled Devices for Particle Detection with High Spatial Resolution," *Nucl. Instrum. Meth.*, **185**, 33–42, (1981).

Damerell, C.J.S., R.L. English, A.R. Gillman, A.L. Lintern, and F.J. Wickens, "CCDs for Vertex Detection in High Energy Physics," *Nucl. Instrum. Meth.*, **A253**, 478–481, (1987).

Damerell, C.J.S., R.L. English, A.R. Gillman, A.L. Lintern, D. Phillips, D. Su, F.J. Wickens, G. Agnew, P. Clarke, S. Hedges, A. McKemey, and S.J. Watts, "A CCD-Based Vertex Detector for SLD," *Nucl. Instrum. Meth.*, **A288**, 236–239, (1990).

DAMERELL, C.J.S., " Charge-Coupled Devices as Particle Tracking Detectors," *Rev. Sci. Instrum.*, **69**, 1549–1573, (1998).

DAMERELL, C.J.S., "A CCD-Based Vertex Detector for the Future $e^+ e^-$ Linear Collider," *Nucl. Instrum. Meth.*, **A512**, 289–298, (2003).

DAMERELL, C.J.S., "CCD-Based Vertex Detectors," *Nucl. Instrum. Meth.*, **A541**, 178–188, (2005).

DARKEN, L.S. AND C.E. COX, "High-Purity Germanium Detectors," in *Semiconductors for Room Temperature Nuclear Detector Applications*," **Chap. 2**, T.E. SCHLESINGER AND R.B. JAMES, Vol. Ed., part of the *Semiconductors and Semimetals* Series, R.K. WILLARDSON, A.C. BEER AND E.R. WEBER, Eds., **Vol. 43**, San Diego; Academic, pp. 23-83, 1995.

DATTA, A. AND S. MOTAKEF, "Cathode Degradation in Thallium Bromide Devices," *IEEE Trans. Nucl. Sci.*, **NS-62**, 1244–1250, (2015).

DAVIS, W.D., "Silicon Crystal Counters," *J. Appl. Phys.*, **29**, 231–232, (1958).

DE NOBEL, D., "Phase Equilibria and Semiconducting Properties of Cadmium Telluride," *Philips Res. Rep.*, **14**, 361–399; 430–492, (1959).

DEARNALEY, G. AND J.C. LEWIS, "A Lithium-Drifted Silicon-Barrier Detector for Nuclear Radiaions," *Nucl. Instrum. Meth.*, **25**, 237–243, (1964).

DEARNALEY, G. AND D.C. NORTHROP, *Semiconductor Counters for Nuclear Radiations*, 2nd Ed., New York: Wiley, 1966.

DEICH, V. AND M. ROTH, "Improved Performance Lead Iodide Nuclear Radiation Detectors," *Nucl. Instrum. Meth.*, **A 380**, 169–172, (1996).

DMITRIEV, V.A. AND M.G. SPENCER, "SiC Fabrication Technology: Growth and Doping," *SiC Materials and Devices*, **Chap. 3**, Y.S. PARK, Vol. Ed., in the *Semiconductors and Semimetals* Series, R.K. WILLARDSON, A.C. BEER AND E.R. WEBER, Eds., **Vol. 32**, San Diego: Academic Press, pp. 21–76, 1998.

DIOURI, J., J.P. LASCARAY, AND R. TRIBOULET, "Study of the Absorption Edge Pinning of $Cd_{1-x}Mn_xTe$ by Transmission and Piezotransmission Measurements," *Solid State Comm.*, **42**, 231–232, (1982).

DÖNMEZ, B., Z. HE, H. KIM, L.J. CIRIGNANO, AND K. SHAH, "Collection of Holes in Thick TlBr Detectors at Low Temperatures," *Nucl. Instrum. Meth.*, **A689**, 7–11, (2012).

DOTY, F.P., J.F. BUTLER, J.F. SCHETZINA, AND K.A. BOWERS, "Properties of CdZnTe Crystals Grown by a High Pressure Bridgman Method," *J. Vac. Sci. Tech.*, **B 10**, 1418–1422, (1992a).

DOTY, F.P., C.L. LINGREN, AND J.F. BUTLER, "$Cd_{1-x}Zn_xTe$ Gamma-Ray Detectors," *IEEE Trans. Nucl. Sci.*, **NS-39**, 605–609, (1992b).

DOTY, F.P., H.B. BARBER, F.L. AUGUSTINE, J.F. BUTLER, B.A. APOTOVSKY, E.T. YOUNG, AND W. HAMILTON, "Pixellated CdZnTe Detector Arrays," *Nucl. Instrum. Meth.*, **A353**, 356–360, (1994).

DU, M-H., "First-Principles Study of Native Defects in TlBr: Carrier Trapping, Compensation, and Polarization Phenomenon," *J. Appl. Phys.*, **108**, paper 053506, (2010).

DULLOO, A.R., F.H. RUDDY, J.G. SEIDEL, C. DAVISON, T. FLINCHBAUGH, AND T. DAUBENSPECK, "Simultaneous Measurement of Neutron and Gamma-Ray Radiation Levels from a TRIGA Reactor Core using Silicon Carbide Semiconductor Detectors," *IEEE Trans. Nucl. Sci.*, **NS-46**, 275–279, (1999).

DULLOO, A.R., F.H. RUDDY, J.G. SEIDEL, J.M. ADAMS, J.S. NICO, AND D.M. GILLIAM, "The Thermal Neutron Response of Miniature Silicon Carbide Semiconductor Detectors," *IEEE Trans. Nucl. Sci.*, **A498**, 415–423, (2003).

DUNN, W.L. AND D.S. MCGREGOR, "Gamma-Ray Detectors," Chapter 17, in *Applications of Particle Detectors in Particle Physics, Medicine, Astrophysics and Related Fields*, I. BUVAT AND C. GRUPEN, Eds., Berlin: Springer, 411–452, 2012.

EBERHARDT, J.E., R.D. RYAN, AND A.J. TAVENDALE, "High-Resolution Nuclear Radiation Detectors from Epitaxial n-GaAs," *Appl. Phys. Lett.*, **17**, 427–429, (1970).

EBERHARDT, J.E., R.D. RYAN, AND A.J. TAVENDALE, "Evaluation of Epitaxial n-GaAs for Nuclear Radiation Detectors," *Nucl. Instrum. Meth.*, **94**, 463–476, (1971).

EBERTH, J. AND J. SIMPSON, "From Ge(Li) Detectors to Gamma-Ray Tracking Arrays 50 Years of Gamma Spectroscopy with Germanium Detectors," *Prog. Part. Nucl. Phys.*, **60**, 283–337, (2008).

EISEN, Y. AND Y. HOROVITZ, "Correction of Incomplate Charge Collection in CdTe Detectors," *Nucl. Instrum. Meth.*, **A353**, 60–66, (1994).

EL-ABASSI, H., S. RATH, AND P.J. SELLIN, "Characterisation of Vertical Gradient Freeze Semi-Insulating InP for use as a Nuclear Radiation Detector," *Nucl. Instrum. Meth.*, **466**, 47–51, (2001).

EL AMRANI, M., J.P. LASCARAY, AND J. DIOURI, "Fundamental Absorption of $Cd_{1-x}Mn_xTe$ Crystals," *Solid State Comm.*, **45**, 351–353, (1983).

ELAD, E. AND R. SAREEN, "Low Temperature Characteristics of Ion Implanted Silicon Position Sensitive Detectors," *IEEE Trans. Nucl. Sci.*, **NS-21**, 75-84, (1974).

ELLIOT, J.H., "Thick Junction Radiation Detectors Made by Ion Drift," *Nucl. Instrum. Meth.*, **12**, 60–66, (1961).

ELSHAZLY, E., G. TEPPER, AND A. BURGER, "Charge Trapping in Detector Grade Thallium Bromide and Cadmium Zinc Telluride: Measurement and Theory," *Nucl. Instrtum. Meth.*, **A620**, 279–284, (2010).

ESKIN, J.D., H.H. BARRETT, H.B. BARBER, AND J.M. WOOLFENDEN, "The Effect of Pixel Geometry on Spatial and Spectral Resolution in a CdZnTe Imaging Array," *Conf. Rec. IEEE Nucl. Sci. Symp.*, San Francisco, Oct. 21-28, 544–548, (1995).

FAILE, S.P., A. DABROWSKI, G.C. HUTH, AND J.S. IWANCZYK, "Mercuric Iodide (HgI_2) Platelets for X-Ray Spectroscopy Produced by Polymer Controlled Growth," *J. Crys. Growth*, **50**, 752–756, (1980).

FAILE, S.P., "Vapor Growth of Mercury Iodide for use as High Energy Detectors," US Patent 4282057, Allowed Aug. 4, 1981.

FAIRSTEIN, E., S. WAGNER, ET AL., *IEEE Standard Test Procedures for Germanium Gamma-Ray Detectors*, **IEEE Std 325-1996**, NIDC, 1996.

FARADAY, M., *Experimental Researches in Electricity, Vol. 1*, London: Taylor, 1839.

FIEDERLE, M., D. EBLING, C. EICHE D.M. HOFMANN, M. SALK, W. STADLER, K. BENZ, AND B.K. MEYER, "Comparison of CdTe, $Cd_{0.9}Zn_{0.1}Te$ and $CdTe_{0.9}Se_{0.1}$ Crystals: Application for γ- and X-Ray Detectors," *J. Crys. Growth*, **138**, 529–533, (1994).

FLAMMANG, R.W., J.G. SEIDEL, AND F.H. RUDDY, "Fast Neutron Detection with Silicon Carbide Semiconductor Radiation Detectors," *Nucl. Instrum. Meth.*, **A579**, 177–179, (2007).

FRECK, D.V. AND J. WAKEFIELD, "Gamma-Ray Spectrum Obtained with a Lithium-Drifted *p-i-n* Junction in Germanium," *Nature*, **193**, 669, (1962).

FRERICHS, R., "The Photo-Conductivity of "Incomplete Phosphors," *Phys. Rev.*, **72**, 594–601, (1947).

FRETWURST, E., H. HERDAN, G. LINDSTRÖM, U. PEIN, M. ROLLIWAGEN, H. SCHATZ, P. THOMSEN, AND R. WUNSTORF, "Silicon Detector Developments for Calorimetry: Technology and Radiation Damage," *Nucl. Instrum. Meth.*, **A 288**, 1–12, (1990).

FRIANT, A., J. MELLET, C. SALIOU, AND T.M. BRAHIM, "HgI_2 Detectors for Solution Grown Crystals," *IEEE Trans. Nucl. Sci.*, **NS-27**, 281–285, (1980).

FRIANT, A., C. SALIOU, R. GALLI, AND S. BARDAY, "Picosecond GaAs X- and Gamma-Ray Photodetectors," *Nucl. Instrum. Meth.*, **A 283**, 318–322, (1989).

FROMMHOLD, Th., W. ARNOLD, H. FRIEDRICHS, R. GÖBELS, R.D. HEIL, U. KNEISSL, U. SEEMAN, F. STEIPER, AND C. KOZHUHAROV, "Response Functions of Si Detectors to Monoenergetic Electrons and Positrons in the Energy Range 0.8–3.5 MeV," *Nucl. Instrum. Meth.*, **A310**, 657-664, (1991).

FUKUDA, Y., T. IZAWA, Y. KOSHIO, S. MORIYAMA, T. NAMBA, AND M. SHIOZAWA, "InP Solid State Detector for Measurement of Low Energy Solar Neutrinos," *Nucl. Instrum. Meth.*, **A623**, 460–462, (2010).

FULLER, C.S. AND J.C. SEVERIENS, "Mobility of Impurity Ions in Germanium and Silicon," *Phys. Rev.*, **96**, 21–24, (1954).

GAJ, J.A., J. GINTER, AND R.R. GALAZKA, "Exchange Interaction of Manganese $3d^5$ States with Band Electrons in $Cd_{1-x}Mn_xTe$," *Phys. Stat. Sol. (b)*, **89**, 655–662, (1978).

GATTI, E. AND P. REHAK, "Semiconductor Drift Chamber - An Application of Novel Charge Transport Scheme," *Nucl. Instrum. Meth.*, **225**, 608–614, (1984a).

GATTI, E., P. REHAK, AND J.T. WALTON, "Silicon Drift Chambers - First Results and Optimum Processing of Signals," *Nucl. Instrum. Meth.*, **226**, 129–141, (1984b).

GATTI, E., P. REHAK, A. LONGONI, J.KEMMER, P. HOLI, R. KLANNER, G. LUTZ, A. WYLIE, F. GOULDING, P.N. LUKE, N.W. MADDEN, AND J. WALTON, "Semiconductor Drift Chambers," *IEEE Trans. Nucl. Sci.*, **NS-32**, 1204–1208, (1985).

GENERAL ELECTRIC, "Flip-Chip Interconnections," *Aviation Week & Space Technology*, **79**, 96–97, (1963).

GEORGESCU, M., "Détection des Particules Ionisantes, par les Cristaux de Soufre," *Compt. Rend.*, **228**, 383–385, (1949).

GERBER, M.S. AND D.W. MILLER "The Noise Analysis and Optimum Filtering Techniques for a Two-Dimensional Position Sensitive Orthogonal Strip Gamma Ray Detector Employing Resistive Charge Division," *Nucl. Instrum. Meth.*, **138**, 445–456, (1976).

GERBER, M.S., D.W. MILLER, P.A. SCHLOSSER, J.W. STEIDLEY, AND A.H. DEUTCHMAN, "Position Sensitive Gamma Ray Detectors Using Resistive Charge Division Readout," *IEEE Trans. Nucl. Sci.*, **NS-24**, 182–187, (1977).

GERSCH, H.K., D.S. MCGREGOR, AND P.A. SIMPSON, "A Study of the Effect of Incremental Gamma-Ray Doses and Incremental Neutron Fluences Upon the Performance of Self-Biased ^{10}B-Coated High-Purity Epitaxial GaAs Thermal Neutron Detectors," *Nucl. Instrum. Meth.*, **A489**, 85–98, (2002).

GIBBONS, P.E. AND P. IRELAND, "On The Accuracy of Acceptor Compensation by Lithium Drift Method," *Nucl. Instrum. Meth.*, **53**, 1–6, (1967).

GIBBS, A., Los Alamos National Laboratory, 1988, *private communication.*

GILMORE, G.R., *Practical Gamma-Ray Spectrometry, 2nd Ed.*, Chichester: Wiley, 2008.

GLASSER, L., P.G. HALL, AND D.D. LIEBENBERG, "Dielectric Properties of Lead Iodide," *J. Chem. Soc. A*, **0**, 295–296, (1967).

GOLDBERG Yu., M.E. LEVINSHTEIN, AND S.L. RUMYANTSEV, "Silicon Carbide (SiC)," Ch. 5, in *Properties of Advanced Semiconductor Materials GaN, AlN, SiC, BN, SiC, SiGe*, M.E. LEVINSHTEIN, S.L. RUMYANTSEV and M.S. SHUR, Eds. New York: Wiley, pp. 93-148, 2001.

GONZÁLEZ SÁNCHEZ, T., J.E. VELÁZQUEZ PÉREZ, P.M. GUTIÉRREZ CONDE, AND D. PARDO COLLANTES, "Electron Transport in InP Under High Electric Field Conditions," *Semicond. Sci. Technol.*, **7**, 31–36, (1992).

GORODYNSKYY, V., K. ZDANSKY, L. PEKAREK, V. MALINA, AND S. VACKOVA, "Ti and Mn Co-doped Semi-insulating InP Particle Detectors Operating at Room Temperature," *Nucl. Instrum. Meth.*, **A555**, 288–293, (2005).

GOSTILO, V., A. OWENS, M. BAVDAZ, I. LISJUTIN, A. PEACOCK, H. SIPILA, AND S. ZATOLOKA, "A Comparison of the X-Ray Performance of TlBr Crystals Grown by the Bridgman-Stockbarger and Travelling Molten Zone Methods," *Nucl. Instrum. Meth.*, **A509**, 47–51, (2003).

GOULDING, F.S. AND W.L. HANSEN, "An Automatic Lithium Drifting Apparatus for Silicon and Germanium Detectors," *IEEE Trans. Nucl. Sci.*, **NS-11**, 286–290, (1964).

GOULDING, F.S., *Semiconductor Detectors for Nuclear Spectroscopy*, UCR:-16231, July 1965, unpublished notes prepared for lectures at Herceg-Novi Yugoslavia, Aug. 1965.

GOULDING, F.S., "Semiconductor Detectors for Nuclear Spectroscopy, 1," *Nucl. Instrum. Meth.*, **43**, 1–54, (1966).

GOULDING, F.S., "Some Aspects of Detectors and Electronics for X-Ray Fluorescence Analysis," *Nucl. Instrum. Meth.*, **142**, 213–223, (1977).

GRAMEGNA, G., F. CORSI, D. DE VENUTO, C. MARZOCCA, A. VACCHI, V. MANZARI, F. NAVACH, S. BEOLÉ, G. CASSE, P. GIUBELLINO, L. RICCATI, AND P. BURGER, "Designing a Linear Silicon Drift Detector," *IEEE Trans. Nucl. Sci.*, **NS-42**, 1497–1504, (1995).

GRUNER, S.M., M.W. TATE, AND E.F. EIKENBERRY, "Charge-Coupled Device Area X-Ray Detectors," *Rev. Sci. Instrum.*, **73**, 2815–2842, (2002).

GUNN, J.B., "Microwave Oscillations of Current in III-V Semiconductors", *Solid-State Commun.*, **1**, 88–91, (1963).

HA, J.H., S.M. KANG, S.H. PARK, H.S. KIM, N.H. LEE, AND T-Y. SONG, "A Self-Biased Neutron Detector Based on an SiC Semiconductor for a Harsh Environment," *Appl. Rad. Isotopes*, **67**, 1204–1207, (2009).

HA, J.H., S.M. KANG, H.S. KIM, S-H. PARK, N.H. LEE, T-Y. SONG, AND J.H. LEE, "4H-SiC PIN-Type Semiconductor Detector for Fast Neutron Detection," *Prog. Nucl. Sci. and Tech.*, **1**, 237–239, (2011).

HAGE-ALI, M. AND P. SIFFERT, "Recent Developments in High Resistivity Detector-Grade CdTe," *Proc. MRS*, **302**, 479–486, (1993).

HAGE-ALI, M. AND P. SIFFERT, "Growth Methods of CdTe Nuclear Detection Materials," in *Semiconductors for Room Temperature Nuclear Detector Applications*," **Chap. 6**, T.E. SCHLESINGER AND R.B. JAMES, Vol. Ed., part of the *Semiconductors and Semimetals* Series, R.K. WILLARDSON, A.C. BEER AND E.R. WEBER, Eds., **Vol. 43**, San Diego: Academic Press, pp. 219–257, 1995a.

HAGE-ALI, M. AND P. SIFFERT, "Characterization of CdTe Nuclear Detector Materials," in *Semiconductors for Room Temperature Nuclear Detector Applications*," **Chap. 7**, T.E. SCHLESINGER AND R.B. JAMES, Vol. Ed., part of the *Semiconductors and Semimetals* Series, R.K. WILLARDSON, A.C. BEER AND E.R. WEBER, Eds., **Vol. 43**, San Diego: Academic Press, pp. 259–289, 1995b.

HAGE-ALI, M. AND P. SIFFERT, "CdTe Nuclear Detectors and Applications," in *Semiconductors for Room Temperature Nuclear Detector Applications*, **Chap. 8**, T.E. SCHLESINGER AND R.B. JAMES, Vol. Ed., part of the *Semiconductors and Semimetals* Series, R.K. WILLARDSON, A.C. BEER AND E.R. WEBER, Eds., **Vol. 43**, San Diego: Academic Press, pp. 291–334, 1995c.

HAINES, E.L. AND A.B. WHITEHEAD, "Pulse Height Defect and Energy Dispersion in Semiconductor Detectors," *Rev. Sci. Instrum.*, **37**, 190–194, (1966).

HALL, R.N., "Electron-Hole Recombination in Germanium," *Phys. Rev.*, **87**, 387, (1952).

HALL, R.N., "Recombination Processes in Semiconductors," *Proc. IEE*, **106B**, **Suppl. 17** 923–931, (1959).[45]

HALL, R.N. AND T.J. SOLTYS "High Purity Germanium for Detector Fabrication," *IEEE Trans. Nucl. Sci.*, **NS-18**, 160–165, (1971).

HALLER, E.E., "Ultra-pure Germanium: From Applied to Basic Research or an Old Semiconductor Offering New Opportunities," in *Fundamentals of Semiconductors*, by P.Y. YU AND M. CARDONA, Berlin: Springer, pp. 539-541, 1996.

HAMAMATSU, *X-Ray Direct CCD Camera C8000-30D*, Cat. No. SCAS0013E07, Hamamatsu, (2016).

HAMMOND, R.B., N.G. PAULTER, A.E. IVERSON, AND R.C. SMITH, "Sub-100 ps Bulk-Recombination-Limited InP:Fe Photoconductive Detector," *IEEE Int. Electron Devices Meet. Tech. Dig*, 157–160, (1981).

HAMMOND, R.B., N.G. PAULTER, R.S. WAGNER, AND T.E. SPRINGER, "Excitation and Fe Concentration Dependences in the Impulse Photoconductance of InP:Fe," *Appl. Phys. Dev.*, **ED-30**, 412–415, (1983).

HAMMOND, R.B., N.G. PAULTER, R.S. WAGNER, T.E. SPRINGER, AND M.D.J. MACROBERTS, "InP:Fe Photoconductors as Photodetectors," *IEEE Trans. Elec. Dev.*, **44**, 620–622, (1984).

HANSEN, W.L., "High-Purity Germanium Crystal Growing," *Nucl. Instrum. Meth.*, **94**, 377–380, (1971).

HARDING, W.R., C. HILSUM, M.E. MONCASTER, D.C. NORTHROP, AND O. SIMPSON, "Gallium Arsenide for γ-Ray Spectroscopy," *Nature*, **187**, 405, (1960).

HARRISON, M.J., A. KARGER, A.C. BROOKS, K.H. CARPENTER, AND D.S. MCGREGOR, "Improved Techniques for the Fabrication of Frisch Collar CdZnTe Gamma Ray Spectrometers," *IEEE Trans. Nucl. Sci.*, **56**, 1671–1676, (2009).

HAU, I.D., C. TINDALL, AND P.N. LUKE, "New Contact Development for Si(Li) Orthogonal-Strip Detectors," *Nucl. Instrum. Meth.*, **A505**, 148–154, (2003).

HE, Z., G.F. KNOLL, D.K. WEHE, R. ROJESKI, C.H. MASTRANGELO, M. HAMMIG. C. BARRETT, AND A. URITANI, "1-D Position Sensitive Single Carrier Semiconductor Detectors" *Nucl. Instrum. Meth.*, **A380**, 228–231, (1996).

HE, Z., G.F. KNOLL, D.K. WEHE, AND J. MIYAMOTO, "Position Sensitive Single Carrier Semiconductor Detectors" *Nucl. Instrum. Meth.*, **A388**, 180-185, (1997).

HE, Z., G.F. KNOLL, D.K. WEHE, AND Y.F. DU, "Coplanar Grid Patterns and Their Effect on Energy Resolution of CdZnTe Detectors," *Nucl. Instrum. Meth.*, **A411**, 107–113, (1998).

HE, Z., W. LI, G.F. KNOLL, D.K. WEHE, J. BERRY, AND C.M. STAHLE, "3-D Position Sensitive CdZnTe Gamma-Ray Spectrometers," *Nucl. Instrum. Meth.*, **A422**, 173–178, (1999).

HE, Z., W. LI, G.F. KNOLL, D.K. WEHE, AND Y.F. DU, "Effects of Charge Sharing in 3-D Position Sensitive CdZnTe Gamma-Ray Spectrometers," *Nucl. Instrum. Meth.*, **A439**, 619–624, (2000).

HE, Z. AND B.W. STURM, "Characteristics of Depth-Sensing Coplanar Grid CdZnTe Detectors," *Nucl. Instrum. Meth.*, **A554**, 291–299, (2005).

HEIJNE, E.H.M., L. HUBBLING, B.D. HYAMS, P. JARRON, P. LAZEYRAS, F. PIUZ, J.C. VERMEULEN, AND A. WYLIE, "A Silicon Surface Barrier Microstrip Detector Designed for High Energy Physics," *Nucl. Instr. Meth.*, **179**, 331–343, (1980).

HEMLEY, R.J., Y-C. CHEN, AND C-S YAN, "Growing Diamond Crystals by Chemical Vapor Deposition," *Elements*, **1**, 105–108, (2005).

HERMON, H., M. ROTH, AND M. SCHIEBER, "Improved Technique for HgI_2 Growth," *Nucl. Instrum. Meth.*, **A322**, 432–434, (1992).

HESSE, K., W. GRAMANN, AND D. HÖPPNER, "Room-Temperature GaAs Gamma-Detectors," *Nucl. Instrum. Meth.*, **101**, 39–42, (1972).

HITOMI, K., T. ONODERA, T. SHOJI, AND Z. HE, "Pixellated TlBr Detectors with the Depth Sensing Technique," *Nucl. Instrum. Meth.*, **A578**, 235–238, (2007a).

HITOMI, K., T. ONODERA, AND T. SHOJI, "Influence of Zone Purification Process on TlBr Crystals for Radiation Detector Fabrication," *Nucl. Instrum. Meth.*, **A579**, 153–156, (2007b).

HITOMI, K., T. SHOJI, AND Y. NIIZEKI, "A Method of Suppressing Polarization in TlBr Detectors," *Nucl. Instrum. Meth.*, **A585**, 102–104, (2008).

HITOMI, K., Y. KIKUCHI, T. SHOJI, AND K. ISHII, "Improvement of Energy Resolutions in TlBr Detectors," *Nucl. Instrum. Meth.*, **A607**, 112–115, (2009).

HOFKER, W.K., D.P. OOSTOEK, A.M.E. HOEBERECHTS, R. VAN DANTZIG, K. MULDER, J.E.J. O—SC BERSKI, L.A. CH. KOERTS, J.H. DIEPERINK, E. KOK, AND R.F. RUMPHORST, "The Checker Board Counter: A Semiconductor dE/dx Detector with Position Indication," *IEEE Trans. Nucl. Sci.*, **NS-13**, 208–213, (1966).

HOFSTADTER, R., J.C.D. MILTON, AND S.L. RIDGWAY "Behavior of Silver Chloride Crystal Counters," *Phys. Rev.*, **72**, 977–978, (1947).

HOFSTADTER, R., "Thallium Halide Crystal Counter," *Phys. Rev.*, **72**, 1120–1121, (1947).

HOFSTADTER, R., "Crystal Counters - I," *Nucleonics*, **4**, No. 4, 2–28, (1949a).

HOFSTADTER, R., "Crystal Counters - II," *Nucleonics*, **4**, No. 5, 29–43, (1949b).

HOFSTADTER, R., "Crystal Counters," *Proc. IRE*, **38**, 726–740, (1950).

HOLONYAK JR., N AND S.F. BEVACQUA, "Oscillations in Semiconductors Due to Deep Levels," *Appl. Phys. Lett.*, **2**, 71–73, (1963).

[45] In some library systems, this reference can be a challenge to locate.

HOLMES, D.E., R.T. CHEN, K.R. ELLIOT, AND C.G. KIRKPATRICK, "Stoichiometry-Controlled Compensation in Liquid Encapsulated Czochralski GaAs," *Appl. Phys. Lett.*, **40**, 46–48, (1982).

HOSTETTLER, M., H. BIRKEDAL, AND D. SCHWARZENBACH "Polymorphs and Structures of Mercuric Iodide," *Chimia*, **55**, 541-545, (2001).

HOSTETTLER, M., H. BIRKEDAL, AND D. SCHWARZENBACH "The Yellow Polymorphs of Mercuric Iodide (HgI_2)," *Helvetica Chimica Acta* **86**, 1410-1422, (2003).

HOSTETTLER, M., AND D. SCHWARZENBACH "Phase diagrams and structures of HgX_2 (X = I, Br, Cl, F)," *Comptes Rendus Chimie*, **8**, 147-156, (2005).

HUBER, R., A. BRODSCHELM, F. TAUSER, AND A. LEITENSTORFER, "Generation and Field-Resolved Detection of Femtosecond Electromagnetic Pulses Tunable up to 41 THz," *Appl. Phys. Lett.*, **76**, 3191–3193, (2000).

HUTCHINSON, G.W., "Ionization in Liquid and Solid Argon," *Nature*, **162**, 610–611, (1948).

IAEA, *Lithium-Drifted Germanium Detectors*, Proc. of a Panel, IAEA, Vienna, June 6-10, 1966.

IJAZ-UR-RAHMAN AND R. HOFSTADTER, "Thallium Halide Radiation Detectors," *Phys. Rev. B*, **29**, 3500–3507, (1984).

IJAZ-UR-RAHMAN, W.A. FISHER, R. HOFSTADTER, AND S. JING, "Behavior of Thallium Bromide Conduction Detectors," *Nucl. Instrum. Meth.*, **A261**, 427–439, (1987).

INTERNATIONAL CRYSTAL LABORATORIES, "Cadmium Manganese Telluride (CdMnTe) Magneto Optical Crystals," *Data Sheet*, `http://www.internationalcrystal.net/optics\_05.htm`, accessed Jan. 9, 2017.

IOFFE PHYSICO-TECHNICAL INSTITUTE, "New Semiconductor Materials, Characteristics and Properties," http://www.ioffe.ru/SVA/NSM/, accessed Dec. 7, 2016.

ISHIKAWA, A., et al., "Neutron Irradition Test of the Fine Pixel CCD Vertex Detector for the ILC," *IEEE Nucl. Sci. Symp. Conf. Rec.*, San Diego, CA, Oct. 31 - Nov.7, 2015.

IWAMI, M., "Silicon Carbide: Fundamentals," *Nucl. Instrum. Meth.*, **A466**, 406–411, (2001).

JACOBONI, C., C. CANALI, G. OTTAVIANI AND A.A. QUARANTA, "A Review of Some Charge Transport Properties of Silicon," *Solid-State Elec.*, **20**, 77–89, (1977).

JACOBONI, C., F. NAVA, C. CANALI, AND G. OTTAVIANI, "Electron Drift Velocity and Diffusivity in Germanium," *Phy. Rev. B*, **24**, 1014–1026, (1981).

JAFFÉ, G., "Über den Einfluß von α-Strahlen auf den Elektrizitätsdurchgang durch Kristalle," *Phys. Z.*, **33**, 393–399, (1932).

JAMES, R.B., T.E. SCHLESINGER, J. LUND, AND M. SCHIEBER, "$Cd_{1-x}Zn_xTe$ Spectrometers for Gamma and X-Ray Applications," in *Semiconductors for Room Temperature Nuclear Detector Applications*," **Chap. 9**, T.E. SCHLESINGER AND R.B. JAMES, Vol. Ed., part of the *Semiconductors and Semimetals* Series, R.K. WILLARDSON, A.C. BEER AND E.R. WEBER, Eds., **Vol. 43**, San Diego: Academic Press, pp. 336–382, 1995.

JENTSCHKE, W., "The Crystal Counter," *Phys. Rev.*, **73**, 77-78, (1948).

JONES, L.T. AND P.B. WOOLLAM, "Resolution Improvement in CdTe Detectors Using Pulse-Shape Discrimination," *Nucl. Instrum. Meth.*, **124**, 591–595, (1975).

KALBITZER, S. AND W. MELZER, "On the Charge Dividing Mechanism in Position Sensitive Detectors," *Nucl. Instrum. Meth.*, **56**, 301–304, (1967).

KALLMAN, V.H. AND R. WARPINSKY, "Ober den Versfärkungseffekt der Elektrischen Leitfihigkeit von Cadmiumsulfid-Kristallen bei Bestrahlung mit α-Teilchen, Elektronen und γ-Quanten," *Ann. Physik*, **439**, 69–84, (1948).

KAMINSKA, M., J.M. PARSEY, J. LAGOWSKI, AND H.C. GATOS, "Current Oscillations in Semi-Insulating GaAs Associated with Field-Enhanced Capture of Electrons by the Major Deep Donor EL2," *Appl. Phys. Lett.*, **41**, 989–991, (1982).

KAMINSKA, M., "EL2 Defect in GaAs," *Phys. Scripta*, **T19**, 551–557, (1987).

KANIA, D.R., A.E. IVERSON, D.L. SMITH, R.S. WAGNER, R.B. HAMMOND, AND K.A. STETLER, "Flat Response Detectors for the Vacuum Ultraviolet and Soft X-Ray Region: InP:Fe Photoconductors," *J. Appl. Phys.*, **60**, 2596–2599, (1986).

KARGAR, A., E. ARIESANTI, S. JAMES, AND D.S. MCGGREGOR, "Charge Collection Efficiency Characterization of a HgI_2 Frisch Collar Spectrometer with Collimated High Energy Gamma Rays," *Nucl. Instrum. Meth.*, **A652**, 186–192, (2011a).

KARGAR, A., E. ARIESANTI, AND D.S. MCGGREGOR, "A Comparison Between Spectroscopic Performance of HgI_2 and CdZnTe Frisch Collar Detectors," *Nucl. Tech.*, **175**, 131–137, (2011b).

KENNEY, C., S. PARKER, J. SEGAL, AND C. STORMENT, "Silicon Detectors with 3-D Electrode Arrays: Fabrication and Initial Test Results," *IEEE Trans. Nucl. Sci.*, **NS-46**, 1224–1236, (1999).

KENNEY, C.J., S. PARKER, AND E. WALCKIERS, "Results from 3-D Silicon Sensors with Wall Electrodes: Near-Cell-Edge Sensitivity Measurements as a Preview of Active-Edge Sensors," *IEEE Trans. Nucl. Sci.*, **NS-48**, 2405–2410, (2001).

KENNEY, C.J., J. SEGAL, E. WESTBROOK, S. PARKER, J. HASI, C. DA VIA, S. WATTS, AND J. MORSE, "Active Edge Planar Radiation Sensors," *Nucl. Instrum. Meth.*, **A 565**, 272–277, (2006).

KHUSAINOV, A.KH., "CdTe Detectors with Thermoelectric Cooling," *Nucl Instrum. Meth.*, **A322**, 335–340, (1992).

KILLIANY, J.M., W.D. BAKER, N.S. SAKS, AND D.F. BARBE, "Effects of Ionizing Radiation on Charge-Coupled Device Operations," *IEEE Trans. Nucl. Sci.*, **NS-21**, 193–200, (1974).

KILLIANY, J.M., N.S. SAKS, AND W.D. BAKER, "Effects of Ionizing Radiation on a 256-Stage Linear CCD Imager," *IEEE Trans. Nucl. Sci.*, **NS-22**, 2634–2638, (1975).

KILLIANY, J.M., "Radiation Effects in Silicon Charge-Coupled Devices," *Topics in Appl. Phys.*, **38**, 147–176, (1980).

KIM, C.K., "The Physics of Charge-Coupled Devices," in *Charge-Coupled Devices and Systems*, M.J. HOWES AND D.V. MORGAN, Eds., New York: Wiley, 2009.

KIM, H., L. CIRIGNANO, A. CHURILOV, G. CIAMPI, W. HIGGINS, F. OLSCHNER, AND K. SHAH, "Developing Larger TlBr Detectors - Detector Performance," *IEEE Trans. Nucl. Sci.*, **56**, 819–823, (2009).

KIM, H., A. CHURILOV, G. CIAMPI, L. CIRIGNANO, W. HIGGINS, S. KIM, P. O'DPIGHERTY, F. OLSCHNER, K. SHAH, AND K. SHAH, "Continued Development of Thallium Bromide and Related Compounds for Gamma-Ray Spectrometers," *Nucl. Instrum. Meth.*, **A629**, 192–196, (2011).

KIM, H., A. KARGAR, L. CIRIGNANO, A. CHURILOV, G. CIAMPI, W. HIGGINS, F. OLSCHNER, AND K . SHAH, "Recent Progress in Thallium Bromide Gamma-Ray Spectrometer Development," *IEEE Trans. Nucl. Sci.*, **59**, 243–248, (2012).

KIM, K-H., J-K. HONG, AND S-U. KIM, "Electrical Properties of Semi-Insulating $CdTe_{0.9}$ $Se_{0.1}$:Cl Crystal and its Surface Preparation," *J. Crys. Growth*, **310**, 91–95, (2008).

KIM, K-H., S-H. CHO, J-H. SUH, J. HONG, AND S-U. KIM, "Gamma-Ray Response of Semi-Insulating CdMnTe Crystals," *IEEE Trans. Nucl. Sci.*, **56**, 858–862, (2009).

KIM, K-H., C. PARK, P. KIM, S. CHO, J. LEE, T.K.. HONG, A. HOSSAIN, A.E. BOLOTNIKOV, AND R.B. JAMES, "Spectroscopic Properties of Large-Volume Virtual Frisch-Grid CdMnTe Detectors," *J. Korean Phys. Soc.*, **66**, 1761–1765, (2015).

KIM, Y-S., K. HUMMER, AND G. KRESSE, "Accurate Band Structures and Effective Masses for InP, InAs, and InSb Using Hybrid Functionals," *Phys. Rev.*, **B 80**, 035203-1–035203-19, (2009).

KING III, G.S., F.T. AVIGNONE III, C.E. COX, T.W. HOSSBACK, W. JENNINGS, AND J.H. REEVES, "Fabrication and Performance Tests of a Segmented p-type HPGe Detector," *Nucl. Instrum. Meth.*, **A595**, 599–604, (2008).

KINOSHITA, A., M. IWAMI, K-I. KOBAYASHI, I. NAKANO, R. TANAKA, T. KAMIYA, A. OHI, T. OHSHIMA, AND Y. FUKUSHIMA, "Radiation Effect on pn-SiC Diode as a Detector," *Nucl. Instrum. Meth.*, **A541**, 213–220, (2005).

KLEIN, C.A., "Bandgap Dependence and Related Features of Radiation Ionization Energies in Semiconductors," *J. Appl. Phys.*, **39**, 2029–2038, (1968).

KÖHLER, R. BATES, C. FLETA, K. JAKOBS, M. LOZANO, C. PARKES, U. PARZEFALL, G. PELLEGRINI, AND J. PREISS, *Nucl. Instrum. Meth.*, **A659**, 272–281, (2011).

KOIKE, J., D.M. PARKIN, AND T.E. MITCHELL, "Displacment Threshold Energy for Type IIa Diamond," *Apl. Phys. Lett.*, **60**, 145–1452, (1992).

KOZLOV, V., M. LESKELÄ, T. PROHASKA, G. SCHULTHEIS, G. STINGEDER, AND H. SIPILA, "TlBr Crystal Growth, Purification and Characterisation," *Nucl. Instrum. Meth.*, **A531**, 165–173, (2004).

KOZOREZOV, A., V. GOSTILO, A. OWENS, F. QUARATI, M. SHOROHOV, M.A. WEBB, AND J.K. WIGMORE, "Polarization Effects in Thallium Bromide X-Ray Detectors," *J. Appl. Phys.*, **108**, paper 064507, (2010).

KRÖGER, F.A. AND D. DE NOBEL, "Preparation and Electrical Properties of CdTe Single Crystals," *J. Electronics*, **1**, 190–202, (1955).

KUBICKI, T., K. LÜBELSMEYER, J. ORTMANNS, D. PANDOULAS, O. SYBEN, M. TOPOROWSKY, AND W.J. XIAO, "Calculation of the Electric Field in GaAs Particle Detectors," *Nucl. Instrum. Meth.*, **A345**, 468–473, (1994).

KÜBLER, C., R. HUBER, S. TÜBEL, AND A. LEITENSTORFER, "Ultrabroadband Detection of Multi-THz Field Transients with GaSe Electro-Optic Sensors," *Springer Series in Chemical Physic*, **79**, 753–755, (2005).

LAEGSGAARD, E., F.W. MARTIN, AND W.M. GIBSON, "Position-Sensitive Semiconductor Particle Detectors Fabricated by Ion Implantation," **NS-15**, 239–245, (1968).

LAGOWSKI, J., H.C. GATOS, C.H. KANG, M. SKOWRONSKI, K.Y. KO, AND D.G. LIN, "Inverted Thermal Conversion - GaAs, a New Alternative Material for Integrated Circuits," *Appl. Phys. Lett.*, **49**, 892–894, (1986).

LAMPERT, M.A. AND P. MARK, *Current Injection in Solids*, New York: Academic Press, 1970.

LAMPORT, J.E., G.M. MASON, M.A. PERKINS, AND A.J. TUZZOLINO, "A Large Area Circular Position Sensitive Si Detector," *Nucl. Instrum. Meth.*, **134**, 71–76, (1976).

LAMPORT, J.E., M.A. PERKINS, A.J. TUZZOLINO AND R. ZAMOW, "Thin, Circular, Two-Dimensional Position Sensitive Si Detectors for Trajectory Determination," *Nucl. Instr. Meth.*, **179**, 105–112, (1981).

LANGE, J., "Recent Progress on 3D Silicon Detectors," *Proc. 24th Int. Workshop on Vertex Detectors*, June 1 -5, Santa Fe, USA, 2015.

LARIN, F., *Radiation Effects in Semiconductor Devices*, New York: Wiley, 1968.

LAUBER, A., "On the Theory of Compensation in Lithium Drifted Semiconductor Detectors," *Nucl. Instrum. Meth.*, **75**, 297–308, (1969).

LECHNER, P., ET AL., "New DEPMOS Applications," *Nucl. Instrum. Meth.*, **A326**, 284–289, (1993).

LECHNER, P., ET AL., "Silicon Drift Detectors for High Resolution Room Temperature X-Ray Spectroscopy," *Nucl. Instrum. Meth.*, **A377**, 346–351, (1996).

LEE, Y.R. AND A.K. RAMDAS, "A Piezomodulation Study of the Absorption Edge and Mn^{++} Internal Transition in $Cd_{1-x}Mn_xTe$, A Prototype of Diluted Magnetic Semiconductors," *Sol. State Comm.*, **51**, 861–863, (1984).

LEFÉVRE, J., J-M. COSTANTINI, S. ESNOUF, AND G. PETITE, "Silicon Threshold Displacement Energy Determined by Photoluminescence in Electron-Irradiated Cubic Silicon Carbide," *J. Appl. Phys.*, **195**, paper 023520, 6 pages, (2009).

LEMASSON, P., B-L. WU, R. TRIBOULET, AND J. GAUTRON, "The Semiconductor-Electrolyte Interface: Optical Transitions in CdMnTe Alloys," *Solid. State Comm.*, **47**, 669–671, (1983).

LIANG, Q., C-S. YAN, Y. MENG, J. LAI, S. KRASNICKI, H-K. MAO, AND R.J. HEMLEY, "Recent Advances in High-Growth Rate Single-Crystal CVD Diamond," *Diamond and Related Materials*, **18**, 698–703, (2009).

LIN, A.L. AND R.H. BUBE, "Photoelectric Properties of High-Resistivity GaAs:Cr," *J. Appl. Phys.*, **47**, 1859–1867, (1976).

LINDHARD, J. AND V. NIELSEN, "Nuclear Collisions and Ionisation Fluctuations in Charged Particle Detectors," *Phys. Lett.*, **2**, 209–211, (1962).

LINDSTRÖM, A., S. MIRBT, B. SANYAL, AND M. KLINTENBERG, "High Resistivity in Undoped CdTe: Carrier Compensation of Te Antisites and Cd Vacancies," *J. Phys. D: Appl. Phys.*, **49**, paper 035101, 12 pages, (2016).

LINGREN, C.L., B. APOTOVSKY, J.F. BUTLER, R.L. CONWELL, F.P. DOTY, S.J. FRIESENHAHN, A. OGANESYAN, B. PI, AND S. ZHAO, "Cadmium-Zinc-Telluride, Multiple Electrode Detectors Achieve Good Energy Resolution with High Sensitivity at Room Temperature," *IEEE Trans. Nucl. Sci.*, **45**, 433–437, (1998).

LIU, K., J. XU, AND X-C. ZHANG, "GaSe Crystals for Broadband Terahertz Wave Detection," *Appl. Phys. Lett.*, **85**, 863–865, (2004).

LLACER, J., "Study of Surface Effects in Thick Lithium Drifted Silicon Radiation Detectors," *IEEE Trans. Nucl. Sci.*, **NS-11**, 221–231, (1964).

LLACER, J., "Geometric Control of Surface Leakage Current and Noise in Lithium Drifted Silicon Radiation Detectors," *IEEE Trans. Nucl. Sci.*, **NS-13**, 93–103, (1966).

LLACER, J.L., M.K. WATT , M. SCHIEBER, R. CARLSTEN, AND D. SCHNEPPLE, "Prelinminary Studies of Charge Carrier Transport in Mercuric Iodide Radiation Detectors," *IEEE Trans. Nucl. Sci.*, **NS-21**, 305–314, (1974).

LOFERSKI, J.J. AND P. RAPPAPORT, "Displacement Thresholds in Semiconductors," *J. Appl. Phys.*, **30**, 1296–1299, (1959).

LOZANO, M., E. CABRUJA, A. COLLADO, J. SANTANDER AND M. ULLÁN, "Bump Bonding of Pixel Systems," *Nucl. Instrum. Meth.*, **A473**, 95–101, (2001).

LUKE, P.N., "Single-Polarity Charge Sensing in Ionizing Detectors Using Coplanar Electrodes," *Appl. Phys. Lett.*, **65**, 2884–2886, (1994).

LUKE, P.N., "Unipolar Charge Sensing with Coplanar Electrodes-Application to Semiconductor Detectors," *IEEE Trans. Nucl. Sci.*, **NS-42**, 207–213, (1995).

LUMB, D.H. AND J.A. NOUSEK, "Energy and Time Response of CCD X-Ray Detectors," *IEEE Trans. Nucl. Sci.*, **39**, 1379–1383, (1992).

LUND, J.C., F. OLSCHNER, F. SINCLAIR, AND M.R. SQUILLANTE, "Indium Phophide Particle Detectors for Low Energy Solar Neutrino Spectroscopy," *Nucl. Instrum. Meth.*, **A272**, 885–888, (1988a).

LUND, J.C., K.S. SHAH, M.R. SQUILLANTE, AND F. SINCLAIR, "Lead Iodide Nuclear Spectrometers," *IEEE Trans. Nucl. Sci.*, **NS-35**, 89–92, (1988b).

LUND, J.C., K.S. SHAH, F. OLSCHNER, J. ZHANG, L MOY, S. MEDRICK, AND M.R. SQUILLANTE, "Recent Progress in Lead Iodide X-Ray Spectrometer Development," *Nucl Instrum. Meth.*, **A322**, 464-466, (1992).

LUTZ, G., *Semiconductor Radiation Detectors*, Berlin: Springer, 1999.

MAINWOOD, A., "Recent Developments of Diamond Detectors for Particles and UV Radiation," *Semicond. Sci. Technol.*, **15**, R55-R63, (2000).

MALM, H.L., "A Mercuric Iodide Gamma-Ray Spectrometer," *IEEE Trans. Nucl. Sci.*, **NS-19**, 263–265, (1972).

MALM, H.L. AND M. MARTINI, "Polarization Phenomena in CdTe Nuclear Radiation Detectors," *IEEE Trans. Nucl. Sci.*, **NS-21**, 322–330, (1974).

MALM, H.L. AND R.J. DINGER, "Charge Collection in Surface Channels on High-Purity Ge Detectors," *IEEE Trans. Nucl. Sci.*, **NS-23**, 76–80, (1976).

MANCINI, A.M., C. MANFREDOTTI, R. MURRI, A. RIZZO, A. QUIRINI, AND L. VASANELLI, "The Use of GaSe Semiconductor Detectors for Monitoring High Energy Muon Beams," *IEEE Trans. Nucl. Sci.*, **NS-23**, 189–197, (1976).

MANDAL, K.C., M. CHOI, S.H. KANG, R.D. RAUH, J. WEI, H. ZHANG, L. ZHENG, Y. CUI, M. GROZA, AND A. BURGER, "GaSe and GaTe Anisotropic Layered Semiconductors for Radiation Detectors," *Proc. SPIE*, **6706**, 67060E1 - 67060E10, (2007).

MANFREDOTTI, C., R. MURRI, AND L. VASANELLI, "GaSe as Nuclear Particle Detector," *Nucl. Instrum. Meth.*, **115**, 349–353, (1974a).

MANFREDOTTI, C., R. MURRI, A. RIZZO, S. GALASSINI, AND L. RUGGIERO, "Deep Hole Traps in p-Type GaSe Single Crystals," *Phys. Rev.*, **B 10**, 3387–3393, (1974b).

MANFREDOTTI, C., R. MURRI, A. QUIRINI, AND L. VASANELLI, "A Particular Application of GaSe Semiconductor Detectors in the Neutrino Experiment at CERN," *Nucl. Instrum. Meth.*, **131**, 457–462, (1975).

MANFREDOTTI, C., R. MURRI, A. QUIRINI, AND S. VASANELLI, "PbI_2 as Nuclear Particle Detector," *IEEE Trans. Nucl. Sci.*, **NS-24**, 126–128, (1977).

MANFREDOTTI, C., A. LO GIUDICE, F. FASOLO, E. VITTONE, C. PAOLINI, F. FIZZOTTI, A. ZANINI, G. WAGNER, AND C. LANZIERI, "SiC Detectors for Neutron Monitoring," *Nucl. Instrum. Meth.*, **A552**, 131–137, (2005).

MANN, H.M., J.W. HASLETT AND F.J. JANAREK, "Lithium-Drifted p-i-n Junction Detectors," *IRE Trans. Nucl. Sci.*, **NS-9**, 43–54, (1962).

MANOLOPOULOS, S., R. BATES, M. CAMPBELL, C. DA VIA, E. HEIJNE, M. HEUKEN, J. JURGENSEN, J. LUDWIG, D. MARDER, K. MATHIESON, V. O'SHEA, C. RAINE, M. ROGALLA, AND K.M. SMITH, "Developments in GaAs Pixel Detectors for X-Ray Imaging," *IEEE Trans. Nucl. Sci.*, **45**, 394–400, (1998).

MARKAKIS, J.M., "High Resolution Scintillation Spectroscopy with HgI_2 as the Photodetector," *Nucl. Instrum. Meth.*, **A263**, 499–503, (1988a).

MARKAKIS, J.M., "Mercuric Iodide Photodetector-Cesium Iodide Scintillator Gamma Ray Spectrometers," *IEEE Trans. Nucl. Sci.*, **35**, 356–359, (1988b).

MARTIN, G.M., J.P. FARGES, G. JACOB, J.P. HALLAIS, AND G. POIBLAUD, "Compensation Mechanisms in GaAs," *J. Appl. Phys.*, **51**, 2840–2852, (1980).

MARTIN, G.M. AND S. MAKRAM-EBEID, The Mid-Gap Donor Level EL2 in Gallium Arsenide, Ch. 6, in *Deep Centers in Semiconductors*, 2nd Ed., S.T.. PANTELIDES, Ed., Philadelphia: Gordan and Breach, pp. 457–545, 1992.

MAYER, J.W., "Characteristics of *p-i-n* Junctions Produced by Ion-Drift Techniques in Silicon," *J. Appl. Phys.*, **33**, 2894–2902, (1962).

MAYER, J.W., "Semiconductor Detectors for Nuclear Spectroscopy, II," *Nucl. Instrum. Meth.*, **43**, 55–64, (1966).

MAYER, J.W., "Evaluation of CdTe by Nuclear Particle Measurements," *J. Appl. Phys.*, **38**, 296–301, (1967).

MCGREGOR, D.S, G.F. KNOLL, Y. EISEN, AND R. BRAKE, "Development of Bulk GaAs Room Temperature Radiation Detectors," *IEEE Trans. Nucl. Sci.*, **NS-39**, 1226–1236, (1992).

MCGREGOR, D.S, R.A. ROJESKI, G.F. KNOLL, F.L. TERRY, JR., J. EAST, AND Y. EISEN, "Present Status of Undoped Semi-Insulating LEC Bulk GaAs as a Radiation Spectrometer," *Nucl. Instrum. Meth.*, **A343**, 527-538, (1994a).

MCGREGOR, D.S, R.A. ROJESKI, G.F. KNOLL, F.L. TERRY, JR., J. EAST, AND Y. EISEN, "Evidence for Field Enhanced Electron Capture by EL2 Centers in Semi-Insulating GaAs and the Effect on GaAs Radiation Detectors," *J. Appl. Phys.*, **75**, 7910-7915, (1994b); errata *J. Appl. Phys.*, **77**, 1331, (1995).

MCGREGOR, D.S. AND J.E. KAMMERAAD, "Gallium Arsenide Radiation Detectors and Spectrometers, in *Semiconductors for Room Temperature Nuclear Detector Applications*," **Chap. 10**, T.E. SCHLESINGER AND R.B. JAMES, Vol. Ed., part of the *Semiconductors and Semimetals* Series, R.K. WILLARDSON, A.C. BEER AND E.R. WEBER, Eds., **Vol. 43**, San Diego: Academic Press, pp. 383–442, 1995.

MCGREGOR, D.S. AND H. HERMON, "Room-Temperature Compound Semiconductor Radiation Detectors," *Nucl. Instrum. Meth.*, **A395**, 101–124, (1997).

MCGREGOR, D.S., Z. HE, H.A. SEIFERT, D.K. WEHE, AND R.A. ROJESKI, "Single Charge Carrier Type Sensing with a Parallel Strip Pseudo-Frisch-Grid CdZnTe Semiconductor Radiation Detector," *Appl. Phys. Lett.*, **72**, 792–794, (1998a).

MCGREGOR, D.S., Z. HE, H.A. SEIFERT, R.A. ROJESKI, AND D.K. WEHE, "CdZnTe Semiconductor Parallel Strip Frisch Grid Radiation Detectors," *IEEE Trans. Nucl. Sci.*, **NS-45**, 443-449, (1998b).

MCGREGOR, D.S., R.A. ROJESKI, Z. HE, D.K. WEHE, M. DRIVER, AND M. BLAKELEY, "Geometrically Weighted Semiconductor Frisch Grid Radiation Spectrometers," *Nucl. Instrum. Meth.*, **A422**, 164–168, (1999a).

McGregor, D.S. and R.A. Rojeski, "Performance of Geometrically Weighted Semiconductor Frisch Grid Radiation Spectrometers," *IEEE Trans. Nucl. Sci.*, **NS-46**, 250-259, (1999b).

McGregor, D.S. and R. Rojeski, "High Resolution Ionization Detectors and Array of Such Detectors," US Patent 6175120; allowed Jan. 16, 2001.

McGregor, D.S., "Collimated Radiation Detector Assembly, Array of Collimated Radiation Detectors and Collimated Radiation Detector Module," US Patent 6781132; allowed August 24, 2004.

McGregor, D.S. and E. Ariesanti, "Platelet Growth of Mercuric Iodide," *J. Crys. Growth*, **379**, 7–15, (2013).

McKay, K.G., "A. Germanium Counter," *Phys. Rev.*, **76**, 1537, (1949).

McNeil, W.J., D.S. McGregor, A.E. Bolotnikov, G.W. Wright, and R.B. James, "Single Charge Carrier Type Sensing with an Insulated Frisch Ring CdZnTe Semiconductor Radiation Detector," *Appl. Phys. Lett.*, **84**, 1988–1990, (2004).

Mellet, J. and A. Friant, "I(t), I(V) and Surface Effect Studies of Vapor Grown and Solution Grown HgI_2 Detectors," *Nucl. Instrum. Meth.*, **283**, 199-207, (1989).

Miller, G.L., B.D. Pate, and S. Wagner, "Production of Thick Semiconductor Radiation Detectors by Lithium Drifting," *IEEE Trans. Nucl. Sci.*, **NS-10**, 220–229, (1963).

Montémont, G., M. Arques, and J. Rustique, "A Capacitive Frisch Grid Structure for CdZnTe Detectors," *IEEE Trans. Nucl. Sci.*, **48**, 278–281, (2001).

Moroz, Z. and M. Moszynski, "Process of Charge Collection in Silicon Drifted Detectors with Unhomogeneous Distribution of the Electric Field," *Nucl. Instrum. Meth.*, **68**, 261–267, (1969).

Moscatelli, F., "Silicon Carbide for UV, Alpha, Beta, and X-Ray Detectors," *Nucl. Instrum. Meth.*, **A583**, 157–161, (2007).

Moszynski, M., W. Kurcewicz, and W. Przyborski, "Time Spreads of Pulses Generated by Beta-Particles in Silicon Drifted Detectors," *Nucl. Instrum. Meth.*, **61**, 173–180, (1968).

Moszynski, M. and W. Przyborski, "The Shape of Pulses Generated by Alpha-Particles in Silicon Drifted Detectors," *Nucl. Instrum. Meth.*, **64**, 244–250, (1968).

Motakef, S. and A. Datta, "Room Temperature Performance of Thallium Bromide Detectors," *Conf. Proc. IEEE Nucl Sci. Symp.*, San Diego, CA, Oct. 31–Nov. 7, 2015.

Mouthuy, T., "Silicon Pixel Detector Research and Development," *Nucl. Instrum. Meth.*, **A368**, 213–216, (2005).

Moxtek, Ultra-thin AP5 X-ray Windows Datasheet, Document WIN-DATA-1006, Rev B, 2017.

Mycielski, A., S. Burger, M. Sowinska, M. Groza, A. Szadkowski, P. Wojnar, B. Witkowska, W. Kaliszek, and P. Siffert, "Is the (Cd,Mn)Te Crystal a Prospective Material fo X-Ray and γ-Ray Detector?," *Phys. Stat. Sol. (c)*, **2**, 1578–1585, (2005).

Nakatani, H., E. Sakai, C. Tatsuyama, and F. Takeda, "GaSe Nuclear Particle Detectors," *Nucl. Instrum. Meth.*, **A 283**, 303–309, (1989).

Nava, F., C. Canali, C. Jacoboni, and L. Reggianni, "Electron Effective Masses and Lattice Scattering in Natural Diamond," *Solid State Commun.*, **33**, 475-477, (1980).

Nava, F., G. Bertuccio, A. Cavallini, and E. Vittone, "Silicon Carbide and its use as a Radiation Detector Material," *Meas. Sci. Technol.*, **19**, paper 102001, 25 pages, (2008).

Nebel, C.E., "Electronic Properties of CVD Diamond," *Semicond. Sci. Tech.*, **18**, S1–S11, (2003).

Newkirk, J.B., "A Crystallographic Study of Mercuric Iodide," *Acta Metallurgica*, **4**, 316-330, (1956).

Nicolau, I.F. and J.P. Joly, "Solution Growth of Sparingly Soluble Single Crystals from Soluble Complexes III. Growth of α-HgI_2 Single Crystals from Dimethylsulfoxide Complexes," *J. Crys. Growth*, **48**, 61–73, (1980).

Niedermayr, T., K. Vetter, L. Milhailescu, G.J. Schmid, D. Beckedahl, J. Blair and J.E. Kammeraad, "Gamma-ray Imaging with a Coaxial HPGe Detector," *Nucl. Instrum. Meth.*, **A553**, 501–511, (2005).

Olego, D.J., J.P. Faurie, S. Sivananthan, and P.M. Raccah, "Optoelectronic Properties of $Cd_{1-x}Zn_xTe$ Films Grown by Molecular Beam Epitaxy on GaAs Substrates," *Appl. Phys. Lett.*, **47**, 1172–1174, (1985).

Oliveira, I.B., F.E. Costa, M.J. Armelin, L.P. Cardoso, and M.M. Hamada, "Purification and Growth of PbI_2 Crystals: Dependence of the Radiation Response on the PbI_2 Crystal Purity," *IEEE Trans. Nucl. Sci.*, **49**, 1968–1973, (2002).

Olschner, F., J.C. Lund, M.R. Squillante, and D.L. Kelly, "Indium Phosphide Particle Detectors," *IEEE Trans. Nucl. Sci.*, **NS-36**, 210–212, (1989).

Ortec, *PROFILE HPGe Photon Detector Product Configuration Guide*, Document 012517, Ametek, 2016.

Ott, H., "Das Gitter des Karborunds (SiC) II," *Zeits. Krist.* **62**, 201–217, (1925).

Owens, A., "Spectral Degradation Effects in an 86 cm^3 Ge(HP) Detector," *Nucl. Instrum. Meth.*, **A238**, 473–478, (1985).

Owens, A., M. Bavdaz, S. Kraft, A. Peacock, S. Nenonen, H. Anderson, M.A. Gagliardi, T. Gagliardi, F. Scholze, and G. Ulm, "X-Ray Response of Epitaxial GaAs," *J. Appl. Phys.*, **85**, 7522–7527, (1999).

Owens, A., M. Bavdaz, A. Peacock, A. Poelart, H. Anderson, S. Nenonen, H. Sipila, L. Tröger, and G. Bertuccio, "High Resolution X-Ray Spectroscopy with GaAs Arrays," *J. Appl. Phys.*, **90**, 5376–5381, (2001).

Owens, A., M. Bavdaz, V. Gostilo, D. Gryaznov, A. Loupilov, A. Peacock, and H. Sipila, "The X-ray Response of InP," *Nucl. Instrum. Meth.*, **A487**, 435–440, (2002a).

Owens, A., A. Peacock, M. Bavdaz, G. Brammertz, F. Dubecky, V. Gostilo, D. Gryaznov, N. Haack, M. Krumrey, and A. Loupilov, "The X-ray Response of InP: Part B, Synchotron Radiation Measurements," *Nucl. Instrum. Meth.*, **A491**, 444-451, (2002b).

Owens, A., M. Bazdaz, H. Andersson, T. Gagliardi, M. Krumrey, S. Nenonen, A. Peacock, I. Taylor, and L. Tröger, "The X-Ray Response of CdZnTe," *Nucl. Instrum. Meth.*, **A484**, 242–250, (2002c).

Owens, A., M. Bavdaz, G. Brammertz, V. Gostilo, H. Graafsma, A. Kozorezov, M. Krumrey, I. Lisjutin, A. Peacock, A. Puig, H. Sipila, and S. Zatoloka, "The X-ray Response of TlBr," *Nucl. Instrum. Meth.*, **A497**, 370–380, (2003a).

Owens, A., A. Peacock, and M. Bavdaz, "Progress in Compound Semiconductors," *Proc. SPIE*, **4851**, 1059–1070, (2003b).

Owens, A., M. Bavdaz, G. Brammertz, V. Gostilo, N. Haack, A. Kozorezov, I. Lisjutin, A. Peacock, and S. Zatoloka, "Hard X-Ray Spectroscopy Using a Small-Format TlBr Array," *Nucl. Instrum. Meth.*, **A497**, 359–369, (2003c).

Owens, A. and A. Peacock, "Compound Semiconductor Radiation Detectors," *Nucl. Instrum. Meth.*, **A531**, 18-37, (2004).

OWENS, A., T. BUSLAPS, V.. GOSTILO, H. GRAAFSMA, R. HIJMERING, A. KOZOREZOV, A. LOUPILOV, D. LUMB, AND E. WELTER, "Hard X- and γ-Ray Measurements With a Large Volume Coplanar Grid CdZnTe Detector," *Nucl. Instrum. Meth.*, **A563**, 242–248, (2006).

OWENS, A., R. DEN HARTOG, F. QUARATI, V. GOSTILO, V. KONDRATJEV, A. LOUPILOV, A.G. KOZOREZOV, J.K. WIGMORE, A. WEBB, AND E. WELTER, "Hard X-Ray Response of a CdZnTe Ring-Drift Detector," *J. Appl. Phys.*, **102**, paper 054505, 9 pages, (2007).

OWENS, A., *Compound Semiconductor Radiation Detectors*, Boca Raton: CRC Press, 2012.

PANOUSIS, P.T., R.H. KRAMBECK AND W.C. JOHNSON, "Field-Enhanced Emission and Constant-Field Domains in Semiconductors with Deep Traps," *Appl. Phys. Lett.*, **15**, 79–81, (1969).

PARKER, S., C.J. KENNEY, AND J. SEGAL, "3D - A Proposed New Architecture for Solid-State Radiation Detectors," *Nucl. Instrum. Meth.*, **A 395**, 328–343, (1997).

PARKIN, J., P.J. SELLIN, A.W. DAVIES, A. LOHSTROH, M.E. ÖZSAN, AND P. SELLER, "α Particle Response of Undoped CdMnTe," *Nucl. Instrum. Meth.*, **A573**, 220–223, (2007).

PARNHAM, K., J.B. GLICK, C. SZELES, AND K.G. LYNN, "Performance Improvement of CdZnTe Detectors Using Modified Two-Terminal Electrode Geometry," *J. Cryst. Growth*, **214-215**, 1152-1154, (2000).

PATT, B.E., J.E. IWANCZYK, M.P. TORNAI, C.S. LEVIN, AND E.J. HOFFMAN, "Development of a Mercuric Iodide Detector Array for Medical Imaging Applications," *Nucl. Instrum. Meth.*, **A366**, 173–182, (1995).

PATT, B.E., J.E.. IWANCZYK, G. VILKELIS, AND Y.J. WANG, "New Gamma-Ray Structures for Electron Only Charge Carrier Collection Utilizing High-Z Compound Semiconductors," *Nucl. Instrum. Meth.*, **A 380**, 276–281, (1996).

PATT, B.E., M.P. TORNAI, J.E. IWANCZYK, C.S. LEVIN, AND E.J. HOFFMAN, "Development of an Intraoperative Gamma Camera Based on a 256-Pixel Mercuric Iodide Detector Array," *IEEE Trans. Nucl. Sci.*, **NS-44**, 1242–1248, (1997).

PELL, E.M., "Diffusion Rate of Li in Si at Low Temperatures*," *Phys. Rev.*, **119**, 1222–1225, (1960a).

PELL, E.M., "Ion Drift in an *n-p* Junction," *J. Appl. Phys.*, **31**, 291–302, (1960b).

PELL, E.M., "Study of Li-O Interaction in Si by Ion Drift," *J. Appl. Phys.*, **32**, 1048–1051, (1961).

PELLIGRINI, G., M. LOZANO, M. ULLÁN, R. BATES, C. FLETA, AND D. PENNICARD, "First Double-Sided 3-D Detectors Fabricated at CNM-IMB," *Nucl. Instrum. Meth.*, **A 592**, 38–43, (2008).

PETERSON, J.H., A. YECKEL, AND J.J. DERBY, "A Fundamental Limitation on Growth Rates in the Traveling Heater Method," *J. Crystal Growth*, **452**, 12–16, (2016).

PFANN, W.G.,"Principles of Zone-Melting," *Trans. AIME*, **194**, 747–754, (1952).

PFANN, W.G., *Zone Melting*, 2nd Ed., New York: Wiley, 1966.

POLICHAR, R., R. SCHIRATO, AND J. REED, "Development of CdZnTe Energy Selective Arrays for Industrial and Medical Radiation Imaging," *Nucl. Instrum. Meth.*, **A353**, 349–355, (1994).

POLICHAR, R., R. SCHIRATO, D. SHREVE, J. REED, AND A. POLICHAR, "Energy Selective Imaging Systems Based on Monolithic CdZnTe Arrays Operated in a High Speed Counting Mode," *Nucl. Instrum. Meth.*, **A380**, 323–330, (1996).

PONPON, J.P., R. STUCK, P. SIFFERT, B. MEYER, AND C. SCHWAB, "Properties of Vapour Phase Grown Mercuric Iodide Single Crystal Detectors," *IEEE Trans. Nucl. Sci.*, **NS-22**, 182–191, (1975).

POZELA, J. AND A. REKLAITIS, "Electron Transport Properties in GaAs at High Electric Fields," *Solid-State Elec.*, **23**, 927–933, (1980).

PRICE, W.J., *Nuclear Radiation Detection*, New York: McGraw-Hill, 1958.

PRICE, W.J., *Nuclear Radiation Detection*, 2nd Ed., New York: McGraw-Hill, 1964.

PRINCE, M.B., "Drift Mobilities in Germanium. I. Germanium," *Phys. Rev.*, **92**, 681–687, (1953).

PRINCE, M.B. AND P. POLISHUK, "Survey of Materials for Radiation Detection at Elevated Temperatures," *IEEE Trans. Nucl. Sci.*, **NS-14**, 537–543, (1967).

PRINTS [sic] (PRINZ), V. YA., AND B.A. BOBYLEV, "Influence of a Strong Electric Field on the Electron Capture by Nonradiative GaAs Centers," *Sov. Phys. Semi.*, **14**, 1097–1098, (1980).

PRINZ, V. YA. AND S.N. RECHKUNOV, "Influence of a Strong Electric Field on the Carrier Capture by Nonradiative Deep-Level Centers in GaAs," *Phys. Stat. Sol. (b)*, **118**, 159–166, (1983).

PUSCHERT, W. AND H. SCHOLZ, *Appl Phys. Lett.*, "γ-Ray Spectra Detected with HgI_2 at Room Temperature," **28**, 357–359, (1976).

RAFIEI, R., D. BOARDMAN, A. SARBUTT, D.A. PROKOPOVICH, K. KIM, M.I. REINHARD, A.E. BOLOTNIKOV, AND R.B. JAMES, "Investigation of the Charge Collection Efficiency of CdMnTe Radiation Detectors," *IEEE Trans. Nucl. Sci.*, **NS-59**, 634–641, (2012).

RAFIEI, R., M.I. REINHARD, K. KIM, D.A. PROKOPOVICH, D. BOARDMAN, A. SARBUTT, G.C. WATT, A.E. BOLOTNIKOV, L.J. BIGNELL, AND R.B. JAMES, "High-Purity CdMnTe Radiation Detectors: A High-Resolution Spectroscopic Evaluation," *IEEE Trans. Nucl. Sci.*, **NS-60**, 1450–1456, (2013).

RAGHAVEN, R.S., "Inverse β Decay of $^{115}In \rightarrow {}^{115}Sn^*$: A New Possibility for Detecting Solar Neutrinos from the Proton-Proton Reaction," *Phys. Rev. Lett.*, **37**, 259–262, (1976).

RAGHAVEN, R.S., "pp-Solar Neutrino Spectroscopy: Return of the Indium Detector," *Bell Laboratory Technical Memo*, **10009622-010606-19TM**, (2001).

RAISKIN, E. AND J.F. BUTLER, "CdTe Low Level Gamma Detectors Based on a New Crystal Growth Method," *IEEE Trans. Nucl. Sci.*, **35**, 81–84, (1988).

RAMSDELL, L.S., "Studies on Silicon Carbide," *Am. Mineral.*, **32**, 64–82, (1947).

REDUS, R., J. PANTAZIS, A. HUBER, V. JSC ORDANOV, J. BUTLER, AND B. APOTOVSKY, "Fano Factor Determination for CZT," *Proc. MRS*, **487**, 101–107, (1997).

REGGIANI, L., "Effect of Impurity Scattering on the Hole Drift Velocity in Germanium," *Phys. Rev.*, **17**, 2800–2802, (1978).

REGGIANI, L., S. BOSI, C. CANALI, F. NAVA, AND S.F. KOZLOV, "Hole-Drift Velocity in Natural Diamond," *Phys. Rev. B*, **23**, 3050–3057, (1981).

RICHTER, M. AND P. SIFFERT, "High Resolution Gamma Ray Spectroscopy with CdTe Detector Systems," *Nucl. Instrum. Meth.*, **A322**, 529–537, (1992).

RICHTER, M., M. HAGE-ALI, Z.T. KUZNICKI, AND P. SIFFERT, "Pulse Processing for Planar Cadmium Telluride Detectors," *Proc. MRS, Semiconductors for Room-Temperature Radiation Detector Applications*, **302**, 195–204, (1993).

RICKER, G.R., J.V. VALLERGA, A.J. DABROWSKI, J.S. IWANCZYK, AND G. ENTINE, "New Measurement of the Fano Factor of Mercuric Iodide," *Rev. Sci. Instrum.*, **53**, 700–701, (1982).

RISTINEN, R.A., D.A. LIND, AND J.L. HPOLMAN, "The Manufacture of Thick Lithium-Drifted Silicon Detectors," *Nucl. Instrum. Meth.*, **56**, 55–60, (1967).

ROBEL, I., V. SUBRAMANIAN, M. KUNO, AND P.V. KAMAT, "Quantum Dot Solar Cells. Harvesting Light Energy with CdSe Nanocrystals Molecularly Linked to Mesoscopic TiO_2 Films," *J. Am. Chem. Soc.*, **128**, 2385–2393, (2006).

RÖNTGEN, W.C. AND A. JOFFÉ, "Über die Elektrizitätsleitung in einigen Kristallen und über den Einfluß der Bestrahlung darauf," *Ann. der Phys.*, **41**, 449–498, (1913).

RÖNTGEN, W.C. AND A. JOFFÉ, "Über die Elektrizitätsleitung in einigen Kristallen und über den Einfluß einer Bestrahlung darauf," *Ann. der Phys.*, **64**, 1–195, (1921).

ROTH, S. AND W.R. WILLIG, "Lead Iodide Nuclear Particle Detectors," *Appl. Phys. Lett.*, **18**, 328–330, (1971).

ROTH, M. AND A. BURGER, "Improved Spectrometer Performance of Cadmium Selenide Room Temperature Gamma-Ray Detector," *IEEE Trans. Nucl. Sci.*, **NS-33**, 407–410, (1986).

ROTH, M., A. BURGER, J. NISSENBAUM, AND M. SCHIEBER, "Transient Charge Technique Investigation of HgI_2 and CdSe Nuclear Detectors," *IEEE Trans. Nucl. Sci.*, **NS-34**, 465–469, (1987).

ROTH, M., "Advantages and Limitations of Cadmium Selenide Room Temperature Gamma Ray Detectors," *Nucl. Instrum. Meth.*, **A283**, 291–298, (1989).

ROY, U.N., A. BURGER, AND R.B. JAMES, "Growth of CdZnTe Crystals by the Traveling Heater Method," *J. Crystal Growth*, **379**, 57–62, (2013).

RUCH, J.G. AND G.S. KINO, "Measurement of the Velocity-Field Characteristic of Gallium Arsenide," *Appl. Phys. Lett.*, **10**, 40–42, (1967).

RUDDY, F.H., A.R. DULLOO, J.G. SEIDEL, M.K. DAS, S-H. RYU, AND A.K. AGARWAL, "The Fast Neutron Response of 4H Silicon Carbide Semiconductor Radiation Detectors," *IEEE Trans. Nucl. Sci.*, **NS-53**, 1666–1670, (2006a).

RUDDY, F.H., J.G. SEIDEL, H. CHEN, A.R. DULLOO, AND S-H. RYU, "High-Resolution Alpha-Particle Spectrometry using 4H Silicon Carbide Semiconductor Detectors," *IEEE Trans. Nucl. Sci.*, **NS-53**, 1713-1718, (2006b).

RUDDY, F.H. AND J.G. SEIDEL, "The Effects of Intense Gamma-Irradiation on the Alpha-Particle Response of Silicon Carbide Semiconductor Radiation Detectors," *Nucl. Instrum. Meth.*, **B263**, 163–168, (2007).

SACKS, H.K. AND A.G. MILNES, "Low Frequency Oscillations in Semi-Insulating Gallium Arsenide," *Int. J. Electronics*, **28**, 565–583, (1970).

SAKAI, E., "Present Status of Room Temperature Semiconductor Detectors," *Nucl. Instrum. Meth.*, **196**, 121–130, (1982).

SAKAI, E., H. NAKATANI, C. TATSUYAMA, AND F. TAKEDA, "Average Energy Needed to Produce an Electron-Hole Pair in GaSe Nuclear Particle Detectors," *IEEE Trans. Nucl. Sci.*, **NS-35**, 85–88, (1988).

SANGSINGKEOW, P., K.D. BERRY, E.J. DUMAS, T.W. RAUDORF, AND T.A. UUNDERWOOD, "Advances in Germanium Detector Technology," *Nucl. Instrum. Meth.*, **A505**, 183–186, (2003).

SAXENA, V. AND A.J. STECKL, "Building Blocks for SiC Devices: Ohmic Contacts, Schottky Contacts, and p-n Junctions," in *SiC Materials and Devices*, Y.S. PARK, Vol. Ed., part of the *Semiconductors and Semimetals* Series, R.K. WILLARDSONAND E.R. WEBER, Eds., **Vol. 32**, San Diego: Academic Press, pp. 77–160, 1998.

SCHIEBER, M., W.F. SCHNEPPLE, AND L. VAN DEN BERG, "Vapor Growth of HgI_2 by Periodic Source or Crystal Temperature Oscillation," *J. Crys. Growth*, **33**, 125–135, (1976).

SCHILLER, H., "Untersuchungen über die Elektrizitätsleitung in festen Dielektriken bei hohen Feldstärken," *Ann. der Phys.*, **81**, 32–90, (1926).

SCHLESINGER, T.E., AND R.B. JAMES, Vol. Eds., *Semiconductors for Room Temperature Nuclear Detector Applications*, in *Semiconductors and Semimetals*, **43**, R.K. WILLARDSON, A.C. BEER AND E.R. WEBER, Treatise Eds., San Diego: Academic Press, 1995.

SCHOLZ, H., "On Crystallization by Temperature-Gradient Reversal," *Acta Electron.*, **17**, 69–73, (1974).

SCHRODER, D.K., "Schottky-Barrier Gate Charge-Coupled Devices," Chap. 8, in *Metal-Semiconductor Schottky Barrier Junctions and Their Applications*, B.L. SHARMA Ed., New York: Plenum, pp. 293–333, 1984.

SCHRODER, D.K., *Advanced MOS Devices*, Reading: Addison-Wesley, 1990.

SCHULZ, L.G., "Polymorphism of Cesium and Thallium Halides," *Acta Cryst.*, **4**, 487–489, (1951).

SCHWARTZ, S., C. ROTTMAIR, J. HIRMKE, S. ROSIWAL, AND R.F. SINGER, "CVD-Diamond Single-Crystal Growth," *J. Crys. Growth*, **217**, 425–434, (2004).

SEN, S. AND J. STANNARD, "Improved Quality of Bulk II-VI Substrates for HgCdTe and HgZnTe Epitaxy," *Proc. MRS*, **302**, 391–401, (1993).

SESHADRI, S., A.R. DULLOO, F.H. RUDDY, J.G. SEIDEL, AND L.B. ROWLAND, "Demonstration of an SiC Neutron Detector for High-Radiation Environments," *IEEE Trans. Nucl. Sci.*, **NS-46**, 567–571, (1999).

SHAH, K.S., J.C. LUND, F. OLSCHNER, L. MOY, AND M.R. SQUILLANTE, "Thallium Bromide Radiation Detectors," *IEEE Trans. Nucl. Sci.*, **NS-36**, 199–202, (1989).

SHAH, K.S., F. OLSCHNER, L. MOY, J.C. LUND, AND M.R. SQUILLANTE, "Characterization of Thallium Bromide Nuclear Detectors," *Nucl. Instrum. Meth.*, **A299**, 57–59, (1990).

SHARMA, R.P. AND A.S. DIVATIA, "Fabrication of Lithium-Drifted Silicon Detectors by Constant Temperature Method," *Pramana - J. Phys.*, **26**, 191–203, (1986).

SHIN, S.H., G.T. NIIZAWA, J.G. PASKO, G.L. BOSTRUP, F.J. RYAN, AND M. KHOSHNEVISAN, "P-I-N CdTe Gamma Ray Detectors by Liquid Phase Epitaxy (LPE)," *IEEE Trans. Nucl. Sci.*, **NS-32**, 487–491, (1985).

SHOCKLEY, W. AND W.T. READ JR., "Statistics of the Recombination of Holes and Electrons," *Phys. Rev.*, **87**, 835–842, (1952).

SHOROHOV, M., V. GOSTILO, V. IVANOV, AND A. OWENS, "Probe Prototypes Based on TlBr Detectors," *Nucl. Instrum. Meth.*, **A607**, 116–119, (2009a).

SHOROHOV, M., M. KOUZNETSOV, I. LISITSKIY, V. IVANOV, V. GOSTILO, AND A. OWENS, "Recent Results in TlBr Detecor Crystals Performance," *IEEE Trans. Nucl. Sci.*, **56**, 1855–1858, (2009b).

SIFFERT, P., "Part 1. CdTe: Current Possibilities and Limitations of Cadmium Telluride Detectors," *Nucl. Instrum. Meth.*, **150**, 1–12, (1978a).

SIFFERT, P., B. RABIN, H.Y. TABATABAI, AND R. STUCK, "Extension of the Depletion Region Layer in Cadmium Telluride Nuclear Radiation Detectors," *Nucl. Instrum. Meth.*, **150**, 31–37, (1978b).

SIFFERT, P., "Cadmium Telluride and Related Materials as X-Ray and Gamma-Ray Detectors: A Review of Recent Progress," *Proc. SPIE*, **2305**, 98–109, (1994).

SIMON, E., *Area Germanium Photovoltaic Cells*, Ph.D. Dissertation, West Lafayette: Purdue University, 1955, *unpublished*.

SMITH, P.M., M. INOUE, AND J. FREY, "Electron Velocity in Si and GaAs at Very High Electric Fields," *Appl. Phys. Lett.*, **37**, 797–708, (1980).

SMITH, H.M., *Characterization of Thallium Bromide (TlBr) for Room Temperature Radiation Detectors*, PhD Dissertation, Berkeley: UC Berkeley, 2013.

SOMERS, R.C., M.C. BAWENDI, AND D.G. NOCERA, "CdSe Nanocrystal Based Chem-/Bio-Sensors," *Chem. Soc. Rev.*, **36**, 579–591, (2007).

SQUILLANTE, M.R., G. ENTINE, E. FREDERICK, L. CIRIGNANO, AND T. HAZLETT, "Development of Two New M-π-n CdTe Sensors," *Nucl. Instrum. Meth.*, **A283**, 323–329, (1989).

SQUILLANTE, M.R. AND K.S. SHAH, "Other Materials: Status and Prospects," in *Semiconductors for Room Temperature Nuclear Detector Applications*," **Chap. 12**, T.E. SCHLESINGER AND R.B. JAMES, Vol. Ed., part of the *Semiconductors and Semimetals* Series, R.K. WILLARDSON, A.C. BEER AND E.R. WEBER, Eds., **Vol. 43**, San Diego: Academic Press, pp. 465–492, 1995.

STEFANOV, K.D., "CCD-Based Vertex Detecor for ILC," *Nucl. Instrum. Meth.*, **A569**, 48–52, (2006).

STEININGER, J, "Growth of CdSe Single Crystals by Temperature Gradient Solution Zoning in Excess Se," *Mater. Res. Bull.*, **3**, 595–598, (1968).

STETTER, G., "Durch Korpuskularstrahlen in Kristallen Hervorgerufene Elektronenleitung," *Verh. d. D. Phys. Ges.*, **22**, 13-14, (1941).

STRAUSS, A.J., "The Physical Properties of Cadmium Telluride," *Revue de Physique Appliquee*, **12**, 167–184, (1977).

STRÜDER, L., P. HOLL, AND G. LUTZ, "Device Modeling of Fully Depletable CCDs," *Nucl. Instrum. Meth.*, **A253**, 386–392, (1987a).

STRÜDER, L., P. HOLL, G. LUTZ, AND J. KEMMER, "Development of Fully Depletable CCDs for High Energy Physics Applications," *Nucl. Instrum. Meth.*, **A257**, 594–602, (1987b).

STRÜDER, L., ET AL., "The MPI/AIT X-Ray Imager (MAXI) - High Speed pn CCDs for X-Ray Detection," *Nucl. Instrum. Meth.*, **A288**, 227–235, (1990).

SUDARSHAN, T.S., D. CHEREDNICHENKO AND R. YAKIMOVA, "Growth of Silicon Carbide", in *Bulk Crystal Growth, Chap. 15*, P. CAPPER, Ed., Chichester, UK: Wiley, pp. 433-450, 2005.

SUZUKI, Y., Y. FUKUDA, AND Y. NAGASHIMA, "An Indium Phosphide Solid State Detector: A Possible Low Energy Gamma and Neutrino Detector," *Nucl. Instrum. Meth.*, **275**, 142–148, (1989).

SWIERKOWSKI, S.P., G.A. ARMANTROUT, AND R. WICHNER, "High-Resolution HgI_2 Spectrometers," *Appl. Phys. Lett.*, **23**, 281–282, (1973).

SWIERKOWSKI, S., "A Comprehensive Model for Predicting Semiconductor Detector Performance," *IEEE Trans. Nucl. Sci.*, **NS-23**, 131–137, (1976).

SZE, S.M., *Physics of Semiconductor Devices*, 2nd Ed., New York: Wiley, 1981.

SZELES, C., D. BALE, J. GROSHOLZ, JR., G.L. SMITH, M. BLOSTEIN, AND J. EGER, "Fabrication of High Performance CdZnTe Quasi-Hemispherical Gamma-Ray CAPture^{TM} Plus Detectors," *Proc. SPIE*, **6319**, 631909-1–6310909-9, (2006).

TAKAHASHI, T., T. MITANI, Y. KOBAYASHI, M. KOUDA, G. SATO, S. WATANABE, K. NAKAZAWA, Y. OKADA, M. FUNAKI, R. OHNO, AND K. MORI, "High-Resolution Schottky CdTe Diode Detector," *IEEE Trans. Nucl. Sci.*, **NS-49**, 1297–1303, (2002).

TANAKA, A., Y. MASA, S. SETO, AND T. KAWASAKI, "Zinc and Selenium Co-Doped CdTe Substrates Lattice Matched to HgCdTe," *J. Crys. Growth*, **94**, 166–170, (1989).

TAPPER, R.J., "Diamond Detectors in Particle Physics," *Rep. Prog. Phys.*, **63**, 1273–1316, (2000).

TAVENDALE, A.J. AND G.T. EWAN, "A High Resolution Lithium-Drift Germanium Gamma-Ray Spectrometer," *Nucl. Instrum. Meth.*, **25**, 185–187, (1963).

TAVENDALE, A.J., "Coaxial Lithium-Drifted Germanium Diodes for Gamma-Ray Spectroscopy. Fabrication and Characteristics," *Lithium Drifted Germanium Detectors*, Proc. IAEA, Vienna, 4–28, June 6–10, 1966.

THIBAULT, N.W., "Morphological and Structural Crystallography and Optical Properties of Silicon Carbide (SiC), Part I" *Am. Mineral.*, **29**, 249–278, (1944a).

THIBAULT, N.W., "Morphological and Structural Crystallography and Optical Properties of Silicon Carbide (SiC), Part II" *Am. Mineral.*, **29**, 327–362, (1944b).

THOMAS, R.N., H.M. HOBGOOD, G.W. ELDRIDGE, D.L. BARRETT, T.T. BRAGGINS, L.B. TA, AND S.K. WANG, "High-Purity LEC Growth and Direct Implantation of GaAs for Monolithic Microwave Circuits," Ch. 1, in *Semi-Insulating GaAs*, of the *Semiconductors and Semimetals* book series, R.K. WILLARDSON AND A.C. BEER, Eds., **20**, Orlando: Academic Press, pp. 1–87, 1984.

TRIBOULET, R. AND G. DIDIER, "Growth and Characterization of $Cd_{1-x}Mn_xTe$ and MnTe Crystals; Contribution to the CdTe-MnTe Pseudo-Binary Phase Diagram Determination," *J. Crys. Growth*, **52**, 614–618, (1981).

TRIBOULET, R., A. HEURTEL, AND J. RIOUX, "Twin-Free (Cd,Mn)Te Substrates," *J. Crys. Growth*, **101**, 131–134, (1990).

TURALA, M., "Silicon Tracking Detectors - Historical Review," *Nucl. Instrum. Meth.*, **A541**, 1–14, (2005).

UESUGI, K., M. HOSHINO, AND N. YAGI, "Comparison of Lens- and Fiber-Coupled CCD Detectors for X-Ray Computed Tomography," *J. Synchrotron Rad.*, **18**, 217–223, (2011).

VAITKUS, J., V. GOSTILO, R. JASINSKAITE, A. MEKYS, A. OWENS, S. ZATOLOKA, AND A. ZINDULIS, "Investigation of Degradation of Electrical and Photoelectrical Properties in TlBr Crystals," *Nucl. Instrum. Meth.*, **A531**, 192–196, (2004).

VAITKUS, J., J. BANYS, V. GOSTILO, S. ZATOLOKA, A. MEKYS, J. STORASTA, AND A. ZINDULIS, "Influence of Electronic and Ionic Processes on Electrical Properties of TlBr Crystals," *Nucl. Instrum. Meth.*, **A546**, 188–191, (2005).

VAN HEERDEN, P.J., *The Crystal Counter, A New Instrument in Nuclear Physics*, Amsterdam: Noorde-Hollandsche Uitgevers-Mij, 1945.

VAN HEERDEN, P.J., "The Crystal Counter, A New Apparatus in Nuclear Physics for the Investigation of β- and γ-rays, Part I," *Physica*, **16**, 505–516, (1950a).

VAN HEERDEN, P.J., "The Crystal Counter, A New Apparatus in Nuclear Physics for the Investigation of β- and γ-rays, Part II," *Physica*, **16**, 517–527, (1950b).

VAN ROOSBROECK, W., "Theory of the Yield and Fano Factor of Electron-Hole Pairs Generated in Semiconductors by High-Energy Particles," *Phys. Rev.*, **139**, A1702–A1716, (1965).

VARSHNI, Y.P., "Band-to-Band Recombination in Groups IV, VI, and II-V Semiconductors (I)," *Phys. Stat. Sol.*, **19**, 459–514, (1967).

VAVILOV, V.S., R.KH., VAGAPOV, V.A. CHAPNIN, AND M.V. CHUKICHEV,"Radiation Detectors Using Semi-Insulating Cadmium Telluride," *Sov. Atomic Energy*, **28**, 641–642, (1970). (English translation).

VOEVODIN, V.G., O.V. VOEVODINA, S.A. BEREZNAYA, Z.V. KOROTCHENKO, A.N. MOROZOV, S.YU. SARKISOV, N.C. FERNELIUS, AND J.T. GOLDSTEIN, "Large Singel Crystals of Gallium Selenide: Growing, Doping by In and Characterization," *Optic. Mater.*, **26**, 495–499, (2004).

VRADII, A.G., M.I. KRAPIVIN, L.V. MASLOVA, O.A. MATVEEV, A.KH. KHUSAINOV, AND V.K. SHASHURIN, "Possibilities of Recording Thermal Neutrons with Cadmium Telluride Detectors," *Sov. Atomic Energy*, **42**, 64–66, (1977). (English translation).

WAGNER, R.S., J.M. BRADLEY, AND R.B. HAMMOND, "Picosecond Photoconductors as Radiation Detectors," *IEEE Trans. Nucl. Sci.*, **33**, 250–253, (1986).

WAHL, C.G., W.R. KAYE, W. WANG, F. ZHANG, J.M. JAWORSKI, A. KING, Y.A. BOUCHER, AND Z. HE, "The Polaris-H Imaging Spectrometer," **A784**, 377–381, (2015).

WALDEN, R.H. KRAMBECK, R.J. STRAIN, J. MCKENNA, N.L. SCHRYER, AND G.E. SMITH, "The Buried Channel Charge-Coupled Device," *Bell Syst. Tech. J.*, **51**, 1635–1640, (1974).

WANG, C.L., J.D. ECKELS, W.V. MORGAN, M.D. POCHA, D.R. SLAUGHTER, B.A. DAVIS, D.R. KANIA, AND R.S. WAGNER, "Photoconductor X/γ-Ray Detectors and X-Ray Bolometers," *Rev. Sci. Instrum.*, **57**, 2182–2183, (1986).

WANG, C.L., M.D. POCHA, J.D. MORSE, M.S. SINGH, AND B.A. DAVIS, "Neutron Treated, Ultrafast, Photoconductor Detectors," *Appl. Phys. Lett.*, **54**, 1451–1453, (1989).

WANG, Y.J., J.S. IWANCZYK, AND W.R. GRAHAM, "Evaluation of HgI_2 Detectors for Lead Detection in Paint," *IEEE Trans. Nucl. Sci.*, **NS-40**, 846–850, (1993).

WANG, X., S. XIAO, M. LI, L. ZHANG, Y. CAO, AND Y. CHEN, "Further Process of Polarization Within a Pixellated CdZnTe Detector Undeer Intense X-Ray Irradiation," *Nucl. Instrum. Meth.*, **A700**, 75–80, (2013).

WEISMAN, J., "Development of Miniature SiC Neutron Detectors," Westinghouse Electric Corp., Atomic Power Dept., Pittburgh, Reports WCAP-1874, WCAP-1929, WCAP-1989, over years 1960–1962.

WEISMAN, J., "Miniature Neutron Detector Development," Westinghouse Electric Corp., Atomic Power Dept., Pittburgh, Reports WCAP-1874, WCAP-2053, WCAP-2117, WCAP-2138, WCAP-2175, WCAP-2274, WCAP-2346, WCAP-2382, WCAP-2403, WCAP-2562, WCAP-2596, over years 1962–1964.

WHITED, R.C. AND M.M. SCHIEBER, "Cadmium Telluride and Mercuric Iodide Gamma Radiation Detectors," *Nucl. Instrum. Meth.*, **162**, 113–123, (1979).

WIELAND, O., ET AL., "Gamma-ray Tracking with Segmented HPGe Detectors," *Brazilian J. Phys.*, **33**, 206–210, (2003).

WILLIAMS, R., *Modern GaAs Processing Techniques*, Boston: Artech House, 1990.

WILLIG, W.R., "Mercury Iodide as a Gamma Spectrometer," *Nucl. Instrum. Meth.*, **96**, 615–616, (1971).

WILLIG, W.R., "Large Bandgap Mercury and Lead Compounds for Nuclear Particle Detection," *Nucl. Instrum. Meth.*, **101**, 23–24, (1972).

WOLFER, M., J. BIENER, B.S. EL-DASHER, M.M. BIENER, A.V. HAMZA, A. KRIELE, AND C. WILD, "Crystallographic Anisotropy of Growth and Etch Rates of CVD Diamond," *Diamond and Related Materials*, **18**, 713–717, (2009).

WOOLDRIDGE, D.E., A.J. AHEARN, AND J.A., BURTON "Conductivity Pulses Induced in Diamond by Alpha-Particles," *Phys. Rev.*, **71**, 913, (1947).

WORT, C.J.H. AND R.S. BALMER, "Diamond as an Electronic Material," *Materials Today*, **11**, 22-28, (2008).

YAMAKAWA, K.A., "Silver Bromide Crystal Counters," *Phys. Rev.*, **82**, 522–526, (1951).

YAMAZAKI, T., H. NAKATANI, AND N. IKEDA, "Characteristics of Impurity-Doped GaSe Radiation Detectors," *Jpn. J. Appl. Phys.*, **32**, 1857-1858, (1993).

YAN, C-S., Y.K. VOHRA, H-K. MAO, AND R.J. HEMLEY, "Very High Growth Rate Chemical Vapor Deposition of Single-Crystal Diamond," *PNAS*, **99**, 12523–12525, (2002).

YATSKIV, R., K. ZDANSKY, AND L. PEKAREK, "Room-Temperature Particle Detectors with Guard Rings Based on Semi-Insulating InP Co-doped with Ti and Zn," *Nucl. Instrum. Meth.*, **A598**, 759–763, (2009).

YEE, J.H., J.W. SHEROMAN, AND G.A. AARMANTROUT, "Theoretical Band Structure Analysis on Possible High-Z Detector Materials," *IEEE Trans. Nucl. Sci.*, **NS-23**, 117–123, (1976).

ZANIO. K., J. NEELAND, AND H. MONTANO, "Performance of CdTe as a Gamma Spectrometer and Detector," *IEEE Trans. Nucl. Sci.*, **NS-17**, 287–295, (1970).

ZANIO. K., "Use of Various Device Geometries to Improve the Performance of CdTe Detectors," *Rev. Phys. Appl.*, **12**, 343–347, (1977).

ZANIO. K., *Cadmium Telluride*, Ch. 1, in *Semiconductors and Semimetals*, **13**, New York: Academic Press, 1978.

ZDANSKY, K., V. GORODYNSKYY, L. PEKAREK, AND H. KOZAK, "Evaluation of Semiinsulating Annealed InP:Ta for Radiation Detectors," *IEEE Trans. Nucl. Sci.*, **NS-53**, 3956–3961, (2006).

ZEIGLER, J.F., J.P. BIERSACK, AND M.D. ZEIGLER, *SRIM: The Stopping and Range of Ions in Matter*, available through the web at `www.srim.org`, 2013.

ZHANG, F., Z. HE, G.F. KNOLL, D.K. WEHE, AND J.E. BERRY, "3-D Position Sensitive CdZnTe Spectrometer Performance Using Third Generation VAS/TAT Readout Electronics," *IEEE Trans. Nucl. Sci.*, **52**, 2009–2016, (2005).

ZHANG, J., W. JIE, T. WANG, D. ZENG, AND B. YANG, "Growth and Characterization of In Doped $Cd_{0.8}Mn_{0.2}Te$ Single Crystal," *J. Crys. Growth*, **306**, 33–38, (2007).

ZHANG, Q., C. ZHANG, Y. LU, K. YANG, AND Q. REN, "Progress in the Development of CdZnTe Unipolar Detectors for Different Anode Geometries and Data Corrections," *Sensors*, **13**, 2447-2474, (2013).

ZHOU, D., L. QUAN, X. CHEN, S. YU, Z. ZHENG, AND S. GONG, "Preparation and Characterization of Thallium Bromide Single Crystal for Room Temperature Radiaiton Detectors Use," *J. Crys. Growth*, **311**, 2524–2529, (2009).

ZHU, Y., W. KAYE, Z. HE, AND F. ZHANG, "Stability and Characteristics of 3D HgI_2 Detectors at Different Cathode Bias," *Conf. Rec. IEEE Nuc. Sci. Symp.*, Honolulu, Hawaii, paper N24-302, Oct. 27–Nov. 3, 2007.

ZOBOLI, M. BOSCARDIN, L. BOSISIO, G-F. DALLA BETTA, C. PIEMONTE, S. RONCHIN, AND N. ZORZI, "Double-Sided, Double-Type-Column 3-D Detectors: Design, Fabrication, and Technology Evaluation," *IEEE Trans. Nucl. Sci.*, **NS-55**, 2775–2784, (2008).

Chapter 17

Slow Neutron Detectors

When a neutron enters a nucleus, the effects are about as catastrophic as if the moon struck the earth. The nucleus is violently shaken up by the blow, especially if the collision results in the capture of the neutron. A large increase in energy occurs and must be dissipated, and this may happen in a variety of ways, all of them interesting.

I.I. Rabi

Neutron detection is challenging primarily because neutrons, like photons, do not have an electrical charge, i.e., they are classified as indirectly ionizing radiation. Unlike α and β radiations, neutrons must first interact in a medium to produce directly ionizing reaction products, or at least some other observable effect. Neutron detectors usually rely upon an interaction with the atomic nuclei within the detector medium and most often it is an absorption or scattering interaction. Slow neutron detectors usually incorporate neutron reactive materials with relatively large neutron absorption cross sections in the low energy region below about 1 keV, whereas fast neutron detectors quite often have relatively large scattering cross sections for neutron energies above 500 keV. Because the differences between the detection methods are considerable, this chapter is devoted to *slow* neutron[1] detectors while the next chapter discusses *fast* neutron detectors.

Slow neutron detectors usually rely upon reactive neutron absorptions for the detection mechanism. Neutrons are absorbed in the target material and the resulting compound nucleus emits ionizing radiations. Most of the neutron absorbing materials used for the neutron detection emit heavy charged particles, although some of the less frequently used materials emit conversion electrons, beta particles, or gamma rays. The three most popular reactive materials for slow neutron detection are ^{3}He, ^{10}B, and ^{6}Li while alternative reactive materials include Gd, U, Hg, and Cd.

17.1 Cross Sections in the $1/v$ Region

An important characteristic of the ^{3}He, ^{10}B, and ^{6}Li isotopes is that they are all so-called $1/v$ absorbers. Radiative capture of low-energy neutrons can be modeled with the Breit-Wigner one-level formula [Lamarsh 1966]

$$\sigma_\gamma(E_c) = \pi\gamma_1^2 g\sqrt{\frac{E_1}{E_c}}\frac{\Gamma_n\Gamma\gamma}{(E_c - E_1)^2 + \Gamma^2/4}, \tag{17.1}$$

[1] In most instances the totality of slow neutrons are called thermal neutrons, i.e., neutrons with speeds that are in thermal equilibrium with the thermal motion of the ambient medium through which they diffuse (see Problem 17-1). An exception would be a beam of monoenergetic slow neutrons produced by a neutron diffractometer.

where E_1 is the energy of the lowest absorption resonance peak, E_c is the center-of-mass energy between the neutron and target nucleus, γ_1 is the reduced de Broglie wavelength for a neutron at energy E_1, g is a statistical factor that is a function of the spin of the target nucleus and the resulting compound nucleus, and Γ_n, Γ_γ, and $\Gamma = \Gamma_n + \Gamma_\gamma$ are, respectively, the width for neutron emission, the width for the (γ, n) reaction, and the total width of the resonance.[2]

The statistical factor g gives the probability that a particular compound state is formed and is given by

$$g = \frac{1}{2}\left(\frac{2J+1}{2I+1}\right), \quad I \neq 0 \tag{17.2}$$

or $g = 1$ if $I = 0$, where I and J are, respectively, the spins of the target nucleus and the compound nucleus. The Breit-Wigner formula is derived for the center-of-mass system so that for an incident neutron of speed v the value of γ_1 is given by

$$\gamma_1 = \frac{\hbar}{\mu v}, \tag{17.3}$$

in which μ is the reduced mass of the nucleus-neutron system, i.e.,

$$\mu = \frac{Am_n}{A+1}, \tag{17.4}$$

where m_n is the neutron mass and Am_n is the mass of the target nucleus. For heavy nuclei $A \gg m_n$ and $\mu \simeq m_n$ so these center-of-mass nuances are not important; but they are for light target nuclei.

Because E_1, γ_1, Γ_γ, Γ_n, and Γ are constants in Eq. (17.1), the radiative capture cross section produces the observed $1/v_c$ behavior at low neutron energies ($E_c \ll E_1$), namely

$$\sigma_\gamma \approx K_1\sqrt{\frac{E_1}{E_c}} = \frac{K_2}{v_c}, \tag{17.5}$$

where K_1 and K_2 are proportionality constants.

Similarly, the cross section for neutron capture resulting in the release of charged particle reaction products can be expressed as [Lamarsh, 1966]

$$\sigma_{(a,b)} = \sqrt{\frac{E_b}{E_a}}H(E_a) \approx \frac{K_3}{v_a}, \tag{17.6}$$

for the reaction $X(a,b)Y$, where E_a and E_b are the kinetic energies of the particles in the center-of-mass system and $H(E_a)$ is a correction factor for non-$1/v$ behavior for moderately large $E_a \ll E_1$, and K_3 is a proportionality constant. Neutrons at low energy ($E_a \ll E_1$) have little effect on the reaction Q value, so the Q value of the reaction is nearly equal to E_b. In such cases, $H(E_a) \approx 1$.

Consider the laboratory system in which neutrons approach target nuclei with relative velocities

$$\mathbf{v_r} = \mathbf{v} - \mathbf{V}, \tag{17.7}$$

where $\mathbf{v}$ is the neutron velocity vector and $\mathbf{V}$ is the target nuclei velocity vector in the laboratory system. As observed from the target nuclei, the incident neutrons are a differential beam of intensity

$$dI = n(\mathbf{v})v_r\, d\mathbf{v}, \tag{17.8}$$

[2]The "width" is a measure of the probability the compound nucleus decays in a specific manner; it is equal to $\hbar$ times the decay constant for the particular process.

where n($\mathbf{v}$) is the neutron density distribution and $v_r = |\mathbf{v}_r|$. The neutrons interact with the nuclei at a reaction density rate

$$dF = n(\mathbf{v})N(\mathbf{V})\sigma(v_r)v_r \, d\mathbf{v} \, d\mathbf{V} \ (\text{cm}^{-3} \ \text{s}^{-1}), \tag{17.9}$$

where $N(\mathbf{V})$ is the target nuclei velocity density distribution and $\sigma(v_r)$ is the neutron cross section at the relative speed v_r. The total interaction rate density is then

$$F = \iint n(\mathbf{v})N(\mathbf{V})\sigma(v_r)v_r \, d\mathbf{v} \, d\mathbf{V}. \tag{17.10}$$

If the target nuclei exhibit a $1/v$ absorption behavior, then from Eq. (17.5) and Eq. (17.6), any arbitrary absorption cross section $\sigma_a(v_{ro})$ for relative speed v_{ro} is related to the cross section at v_r by

$$\sigma_a(v_r) = \frac{\sigma_a(v_{ro})v_{ro}}{v_r}. \tag{17.11}$$

Substitution of Eq. (17.11) into Eq. (17.10) yields

$$\begin{aligned} F_a &= \iint n(\mathbf{v})N(\mathbf{V})\sigma_a(v_r)v_r \, d\mathbf{v} \, d\mathbf{V} \\ &= \iint n(\mathbf{v})N(\mathbf{V})\frac{\sigma_a(v_{ro})v_{ro}}{v_r}v_r \, d\mathbf{v} \, d\mathbf{V} \\ &= N\sigma_a(v_{ro})nv_{ro} = \Sigma_a(E_o)nv_o, \end{aligned} \tag{17.12}$$

where $\Sigma_a(E_o)$ is the macroscopic reaction cross section at an arbitrary reference speed v_o (most often chosen as 2200 m/s), n is the total neutron density, and N is the total target nuclei density. The result of Eq. (17.12) is very important and shows that the reaction rate of neutrons with a $1/v$ material is *independent of the neutron or target nuclei velocity distributions.* Hence, a measurement using a reaction rate to determine detection efficiency is independent of the neutron energy distribution for $1/v$ absorbers. Note that the result of Eq. (17.12) is restricted to the $1/v$ region of the neutron cross section, generally applicable to only slow neutrons. In the case that the absorber nuclei are non-$1/v$, a temperature dependent Westcott correction factor $g_a(T)$ is required to correct for non-$1/v$ interaction behavior [Lamarsh 1966]

$$F = g_a(T)\Sigma_a(E_0)nv_0, \tag{17.13}$$

where T is the neutron temperature of the assumed Maxwellian spectrum of interacting thermal neutrons.

17.2 Slow Neutron Reactions Used for Neutron Detection

As previously noted, the three most popular isotopes for slow neutron detection are ^{3}He, ^{10}B, and ^{6}Li, all having $1/v$ dependence in the slow neutron region (see Fig. 17.1). In all three cases, reactions with slow neutrons cause the reaction products to be ejected in opposite directions (180°). While ^{3}He is used for gas-filled detectors, both ^{10}B and ^{6}Li are used in a variety of detector configurations, including gas-filled, scintillator, and semiconductor devices.

There are also numerous large Z absorbers used for neutron detection that do not exhibit $1/v_c$ behavior because of resonances appearing in the epithermal and thermal-neutron range. Unfortunately, the interaction rate for such absorbers is dependent on the neutron and target nuclei velocities. Consequently, more information must be known about the neutron flux and energy distribution to accurately calibrate instruments utilizing these absorbers than expressed by the simple expression of Eq. (17.12). Nevertheless, many are useful for neutron detection, with some of the more important ones that have very large thermal-neutron absorption cross sections being ^{113}Cd, ^{157}Gd, ^{199}Hg, and ^{235}U.

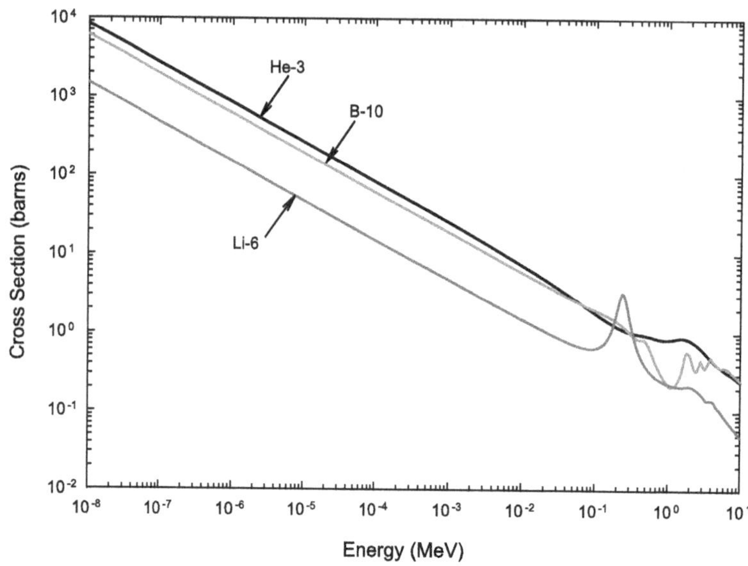

Figure 17.1. Neutron absorption cross sections in barns for common $1/v$ neutron detector materials. The energy is the incident neutron energy in the center-of-mass coordinate system. Data are from [ENDFPLOT, 2015].

17.2.1 The ^{3}He Reaction

There are two natural isotopes of helium, namely ^{3}He and ^{4}He with natural abundances of 0.000137% and 99.999863%, respectively. The isotope ^{3}He has a relatively large total thermal-neutron cross section of 5320 barns at 2200 m/s,[3] of which the ^{3}He(n,p)^{3}H reaction constitutes 5316 barns. By contrast the thermal-neutron absorption cross section for ^{4}He is essentially zero. The microscopic thermal-neutron absorption cross-section for ^{3}He decreases with increasing neutron energy, with a dependence proportional to the inverse of the neutron speed ($1/v$) up to about 0.1 MeV, as shown in Fig. 17.1.

For an incident neutron with negligible kinetic energy, the ^{3}He(n,p)^{3}H reaction has the following energetics:

$$^3\text{He} + {}^1_0\text{n} \rightarrow {}^3\text{H (0.191 MeV)} + {}^1\text{H (0.573 MeV)} \quad \text{with} \quad Q = 0.764 \text{ MeV}. \tag{17.14}$$

The reaction products, while emitted isotropically, are emitted opposite to each other in order to preserve the almost zero incident momentum of the neutron.

The density of natural He gas is 0.1786 g L^{-1} at standard temperature and pressure (STP),[4] with an atom density of 2.686 × 10^{19} cm^{-3}. Hence, the macroscopic absorption cross section (Σ_a) for pure ^{3}He at STP is 0.1492 cm^{-1}. Most neutron detectors based on ^{3}He are pressurized to increase the value of Σ_a. A common pressure for ^{3}He gas-filled neutron detectors is 2.8 atm; however, detectors with pressures ranging from 1 atm up to 20 atm are commercially available. Because of the pressurization, various national aircraft regulations may limit the transport of these detectors.

[3]Thermal neutrons generally have a wide range of energies that are often well-described by a Maxwellian flux distribution, which has a most probable energy of 0.0253 eV (corresponding to a speed of 2200 m/s) for a neutron temperature of 293.5 K. In this chapter, the thermal-neutron cross section refers to the 2200 m/s cross section whose values are taken from the 2017 ENDF/B-VIII library.

[4]STP is 273.15 K (0 °C) and 100 kPa (0.9869 atm).

For low energy neutrons, the neutron contributes very little to the overall reaction product energies; hence the final energy shared by the reaction products is essentially the reaction Q value. As the neutron kinetic energy increases, the reaction products absorb and share this additional energy, yet for energies in the slow to epithermal region, this additional energy is usually masked by the Q value energy; hence, spectroscopic energy information is lost. The same is true for the neutron momentum, which contributes little, thereby causing almost all momentum to be shared between the two reaction products. Hence, for slow neutrons, reaction product net momentum is essentially zero. If one neglects the energy and momentum of the neutron, conservation of energy and momentum yields,

$$E(^3\text{H}) + E(^1\text{H}) = Q = 0.764 \text{ MeV}, \tag{17.15}$$

and

$$M(^3\text{H})v(^3\text{H}) = M(^1\text{H})v(^1\text{H}), \tag{17.16}$$

or

$$\sqrt{2E(^3\text{H})M(^3\text{H})} = \sqrt{2E(^1\text{H})M(^1\text{H})}. \tag{17.17}$$

From Eq. (17.15) and Eq. (17.17), the reaction product energies shown in Eq. (17.14) are readily obtained.

Neutron detectors based on ^{3}He are generally considered a standard for neutron measurements and are widely used in various nuclear industries. However, the natural abundance of ^{3}He is low. The problem is further exacerbated by the dwindling supply of ^{3}H gas,[5] which decays by β-particle emission into ^{3}He. Liquid ^{3}He also has unique superfluid properties for ultra-low-temperature physics, thereby creating another demand for this scarce resource. Consequently, ^{3}He gas and detectors are relatively expensive, costing thousands of US dollars per liter at STP.

17.2.2 The ^{10}B Neutron Reaction

There are two natural isotopes of boron, namely ^{10}B and ^{11}B with natural abundances of 19.9% and 80.1%, respectively. The isotope ^{10}B has a relatively large thermal-neutron (2200 m/s) cross section of 3847 barns, of which the ^{10}B(n,α)^{7}Li cross section constitutes 3844 barns. The microscopic thermal-neutron absorption cross section for ^{10}B decreases with increasing neutron energy, with a dependence proportional to the inverse of the neutron speed ($1/v$) over much of the slow energy range, as shown in Fig. 17.1. The density of natural B is 2.35 g cm^{-3}, and the density of enriched ^{10}B is 2.16 g cm^{-3}. The resulting macroscopic thermal-neutron cross section (Σ_a) for natural B material is 100 cm^{-1} and 497 cm^{-1} for fully enriched ^{10}B material.

The ^{10}B(n,α)^{7}Li reaction leads to the following reaction energetics when the incident neutron has negligible energy:

$$^{10}\text{B} + {}^1_0\text{n} \rightarrow \begin{cases} ^7\text{Li}^* \text{ (0.840 MeV)} + \alpha \text{ (1.470 MeV)} & Q = 2.310 \text{ MeV} \quad 94\% \\ ^7\text{Li (1.015 MeV)} + \alpha \text{ (1.777 MeV)} & Q = 2.792 \text{ MeV} \quad 6\% \end{cases}. \tag{17.18}$$

The reaction products are emitted isotropically but in opposite directions when thermal neutrons (with negligible momentum) are absorbed by ^{10}B. After absorption, 94% of the reactions leave the ^{7}Li ion in its first excited state, which rapidly deexcites to the ground state in about 10^{-13} s by releasing a 480-keV gamma ray. For a variety of neutron detectors, especially gas-filled neutron detectors, the 480-keV gamma ray usually escapes and does not contribute to the detection output. The remaining 6% of the reactions

[5]Much of the world's ^{3}H came from the tritium generated for nuclear weapons production, which today has been significantly reduced by international treaties.

produce a ^{7}Li ion in its ground state. Conservation of energy and momentum then requires

$$E(^7\text{Li}^*) + E(^4\text{He}) = Q = 2.310 \text{ MeV}, \tag{17.19}$$

$$E(^7\text{Li}) + E(^4\text{He}) = Q = 2.792 \text{ MeV}, \tag{17.20}$$

and

$$\sqrt{2E(^7\text{Li}^*)M(^7\text{Li}^*)} = \sqrt{2E(^4\text{He})M(^4\text{He})}, \tag{17.21}$$

$$\sqrt{2E(^7\text{Li})M(^7\text{Li})} = \sqrt{2E(^4\text{He})M(^4\text{He})}. \tag{17.22}$$

Simultaneous analysis of Eq. (17.19) with Eq. (17.21), and Eq. (17.20) with Eq. (17.22) yields the reaction product energies shown in Eq. (17.18).

17.2.3 The ^{6}Li Neutron Reaction

There are two natural isotopes of lithium, namely ^{6}Li and ^{7}Li, with natural abundances of 7.59% and 92.41%, respectively. The isotope ^{6}Li has a relatively large thermal-neutron cross section of 938.8 barns with the ^{6}Li(n,t)^{4}He reaction contributing 938.0 barns. The nuclide ^{7}Li has a relatively small thermal-neutron absorption cross section of approximately 45.5 mb. The microscopic thermal-neutron absorption cross-section of ^{6}Li decreases with increasing neutron energy, with a dependence proportional to the inverse of the neutron speed ($1/v$) over much of the energy range, with a notable resonance near 240 keV as shown in Fig. 17.1. The density of natural Li is 0.534 g cm^{-3}, and the density of enriched ^{6}Li is 0.463 g cm^{-3}. With an atomic density of 4.65 × 10^{22} atoms cm^{-3}, the resulting macroscopic thermal neutron cross section (Σ_a) for natural Li material is 3.300 cm^{-1} and 43.48 cm^{-1} for enriched ^{6}Li material. The ^{6}Li(n,t)^{4}He reaction leads to the following products:

$$^6\text{Li} + {}^1_0\text{n} \rightarrow {}^4\text{He (2.05 MeV)} + {}^3\text{H (2.73 MeV)} \qquad Q = 4.78 \text{ MeV}. \tag{17.23}$$

Again they are emitted isotropically but in opposite directions if the incident neutron energy is sufficiently small. Conservation of energy and momentum, for negligible incident neutron momentum, yields

$$E(^4\text{He}) + E(^3\text{H}) = Q = 4.78 \text{ MeV}, \tag{17.24}$$

and

$$\sqrt{2E(^4\text{He})M(^4\text{He})} = \sqrt{2E(^3\text{H})M(^3\text{H})}. \tag{17.25}$$

Simultaneous solutions of Eq. (17.24) with Eq. (17.25) yields the reaction product energies shown in Eq. (17.23).

Because of its higher absorption cross section, the ^{10}B(n,α)^{7}Li reaction leads to a generally higher reaction probability than the ^{6}Li(n,t)^{4}He reaction for neutron energies below 100 keV. However, the higher energy reaction products emitted from the ^{6}Li(n,t)^{4}He reaction lead to greater ease of detection than do the particles emitted from the ^{10}B(n,α)^{7}Li reaction.

Although there are neutron detectors based on enriched ^{6}Li metal, careful handling is required because Li metal is highly reactive in the presence of water moisture or other oxidizers. A popular and relatively non-reactive Li compound used for neutron detection is ^{6}LiF. The molecular density of ^{6}LiF is 6.12 × 10^{22} molecules cm^{-3} which also equals the atomic density of the ^{6}Li atoms.[6] The mass density of ^{6}LiF is 2.54 g cm^{-3}, which produces a macroscopic thermal-neutron (n,α) cross section of 57.41 cm^{-1}.

[6] It is interesting to observe that the atomic density of Li atoms in LiF is actually higher than in pure Li metal.

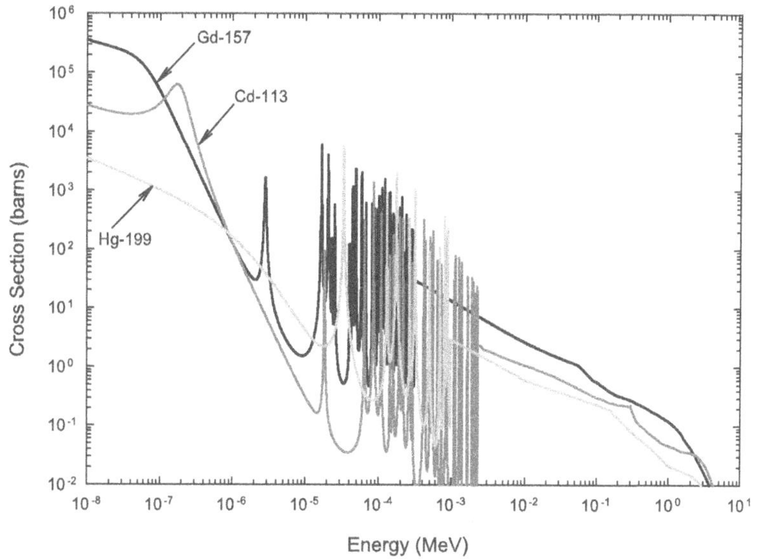

Figure 17.2. Neutron absorption cross sections in barns for common non-$1/v$ neutron detector materials, showing the radiative capture cross sections for ^{113}Cd, ^{157}Gd, and ^{199}Hg. For such heavy nuclide targets the center-of-mass neutron energy almost equals the neutron energy in the laboratory system. Data are from [ENDFPLOT, 2015].

17.2.4 The ^{155}Gd and ^{157}Gd Neutron Reactions

The ^{157}Gd(n,γ)^{158}Gd reaction leads to the emission of low energy internal conversion electrons, most of which are emitted at energies below 220 keV, along with an assortment of gamma rays. The attractiveness of using ^{157}Gd is because of its large (n,γ), thermal-neutron cross-section of 252.9 kb. ^{157}Gd has low-energy absorption resonance at 2.82 eV and, consequently, its microscopic neutron absorption cross section does not have a $(1/v)$ dependence in the sub eV range as is evident from Fig. 17.2. Another neutron absorbing Gd isotope that cannot be ignored is ^{155}Gd, which has a thermal-neutron (n,γ) cross section of 60.74 kb. Together these two isotopes of Gd are perhaps the largest of the solid-state thermal-neutron absorbers.[7] The density of natural Gd is 7.901 g cm^{-3}, and the natural abundances of ^{155}Gd and ^{157}Gd are 14.80% and 15.65%, respectively.

Table 17.1. Conversion electron energies and emission fractions for neutron interactions in natural Gd. The conversion electrons are produced from both ^{155}Gd(n,γ)^{156}Gd and ^{157}Gd(n,γ)^{158}Gd reactions. Data are from [Schulte et al., 1994].

Energy (keV)	Electrons/ neutron (%)	Energy (keV)	Electrons/ neutron (%)
29	9.82	149	0.84
39	4.19	173	1.46
71	26.80	180	0.31
78	6.17	191	0.30
81	4.97	198	0.06
88	1.16	228	0.40
131	3.41	246	0.02

Although some of the other existing Gd isotopes have significant thermal-neutron (n,γ) cross sections, they do not contribute to the neutron detection capability of this element. For example, ^{152}Gd has a 735.1 b 2200-m/s (n,γ) cross section and an abundance of 0.020% while ^{154}Gd has a 85.2 b cross section and a

[7]The radioactive gas fission product ^{135}Xe has a thermal-neutron (n,γ) cross section of 2.6 Mb, the largest of any isotope.

2.18% abundance. The three other stable isotopes ^{156}Gd, ^{158}Gd and ^{160}Gd have negligible cross sections of 1.80 b, 2.03 b, and 1.41 b, respectively.

Given the natural abundances of the isotopes in natural Gd, the natural Gd macroscopic thermal-neutron absorption cross is 1467 cm^{-1}. This macroscopic absorption cross section could be increased by using enriched ^{157}Gd; however, in practice it is impractical due to the high cost of isotope separation.

Neutron absorptions in Gd produce an abundance of spontaneous internal conversion electrons and gamma-rays from the excited ^{156}Gd and ^{158}Gd nuclei. Although the gamma-ray emissions can be used to indicate neutron interactions, it is usually the conversion electrons that are used in the detection process. Shown in Table 17.1 is a list of conversion electron energies and relative emissions (%) per captured neutron, totaling only 59.91% of the product yield [Aoyama et al. 1992]. The most salient emission of conversion electrons is that from the excited state of ^{158}Gd nucleus and produces an electron with an energy of 71 keV. There are also two conversion electron emissions near 80 keV totaling 11.14% of the product yield. The total fraction of conversion electrons above 85 keV amounts to only 7.96% of the total emissions. Hence, the lower level discriminator must be set below 71 keV equivalent in order to detect most of the conversion electrons. These relatively low energy conversion electrons (and accompanying gamma-ray emissions) can be difficult to distinguish from background gamma rays in a high radiation field.

17.2.5 The ^{113}Cd Neutron Reaction

Natural cadmium has eight existing isotopes, of which radioactive ^{113}Cd ($T_{1/2} = 8.04 \times 10^{15}$ y) strongly absorbs thermal neutrons, and has a thermal-neutron (n,γ) cross section of 19.96 kb. The density of natural Cd is 8.65 g cm^{-3}, and the isotope ^{113}Cd has a natural isotopic abundance of 12.23%. The natural Cd macroscopic thermal-neutron absorption cross section is 113.1 cm^{-1}. The cross section as a function of neutron energy is shown in Fig. 17.2. This macroscopic absorption cross section could be greatly increased by using Cd enriched in ^{113}Cd, but high isotopic separation costs preclude this approach.

The ^{113}Cd(n,γ)^{113}Cd reaction produces numerous gamma-ray emissions, ranging from relatively low energies up to and beyond 9 MeV. However, there are two salient prompt gamma-ray emissions of interest at 558.6 keV and 651.3 keV [Lone et al. 1981; Tuli 1997], with branching ratios of 72.73% and 13.9%, respectively. These gamma rays can be discerned with a relatively high-resolution gamma-ray spectrometer. Unfortunately, a gamma-ray spectrometer capable of detecting the ^{113}Cd(n,γ)^{113}Cd emissions also detects other background gamma rays, a signal which can potentially confound interpretation of a measurement. Also, unlike neutron reactive detector materials that emit charged-particle reaction products, neutron detectors that rely on the absorption of prompt gamma-ray emissions suffer a higher signal loss because of reaction products escaping the detector and, thus, the detection efficiency is reduced. Further, at the prompt emission energies of 558.6 keV and 651.3 keV, the most probable gamma-ray interaction in detectors is Compton scattering, which produces a continuum of energies that can be difficult to distinguish from background gamma rays that undergo Compton scattering in the detector and surroundings.

17.2.6 The ^{199}Hg Neutron Reaction

The element Hg is also an interesting material for neutron detection. It has one strongly neutron reactive isotope, ^{199}Hg, with a thermal-neutron (n,γ) cross section of 2150 b (see Fig. 17.2).[8] The density of natural Hg is 13.534 g cm^{-3}, with a ^{199}Hg natural abundance of 16.94%, and has a natural Hg macroscopic thermal-neutron (n,γ) cross section of 14.80 cm^{-1}. As with ^{157}Gd and ^{113}Cd, enriched ^{199}Hg is rare and not an economically practicable option. Hg is a toxic liquid at room temperature, which causes additional handling

[8] The naturally occurring isotope ^{196}Hg has an even larger thermal-neutron (n,γ) cross section of 3079 b, but its small natural abundance of 0.15% has negligible effect on the ability of Hg to absorb thermal neutrons. The five other naturally occurring isotopes of Hg have thermal cross sections of, at most, a few barns.

problems.[9] Detectors that use ^{199}Hg as the neutron reactive material are usually a room temperature solid compound or alloy of Hg.

Similar to the ^{113}Cd(n,γ)^{113}Cd reaction, there are numerous gamma-ray emissions from the ^{199}Hg(n,γ)^{200}Hg reaction; however, there is one prominent gamma-ray emission of interest at 368.1 keV [Lone et al. 1981; Tuli 1997] with a branching ratio of 81.35%. When Hg is coupled with a high resolution gamma-ray spectrometer, the 368.1-keV energy peak can be used as a measure of neutron interactions. Note that this prompt gamma-ray emission energy is low enough to have a reasonably high probability of being fully absorbed in the detector, which works to improve the visibility of a full energy peak. However, this spectral feature can still overlap with background gamma-ray spectra, thereby making measurements difficult to analyze.

17.2.7 Fission Reactions

Fissile materials are also used as neutron detection materials, primarily in gas-filled proportional counters named "fission chambers". Fissile materials, such as ^{235}U and (rarely) ^{239}Pu, are also used for slow neutron detectors (see Fig. 17.3). The fission reaction releases approximately 200 MeV of energy, approximately 150 MeV of which is transferred to the heavy charged particle reaction products. Unlike other reactions described in this chapter, these reaction products vary, with over 100 possible reaction products emitted after a fission reaction. However, the reaction products generally fall into a low energy branch and a high energy branch, centered near 60 MeV and 90 MeV, respectively.

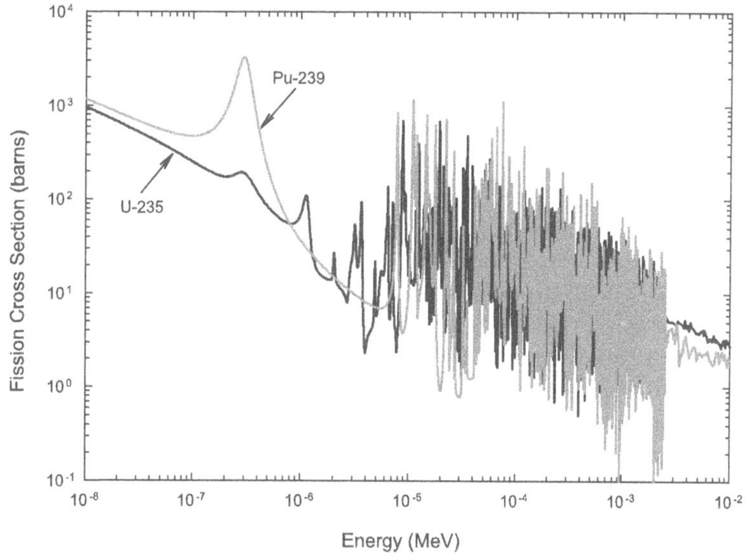

Figure 17.3. Neutron fission cross sections in barns for materials used in slow neutron fission chambers, ^{235}U and ^{239}Pu [data are from ENDFPLOT, 2015].

The 2200 m/s thermal-neutron fission cross section for ^{235}U is 585.0 barns, although the fission plus capture thermal-neutron absorption cross section is higher at 683.7 barns. The nuclide ^{235}U is also radioactive with a halflife of 7.038×10^8 years, and decays by alpha-particle and gamma-ray emission. The gamma-ray emissions have energies of 143.76 keV, 185.72 keV, and 204 keV with branching ratios of 11%, 55%, and 5%,

[9]Many organic compounds of Hg are extremely toxic.

respectively. Principal alpha particles are emitted with energies of 4.366 MeV, 4.398 MeV, and 4.58 MeV with branching ratios of 18%, 57%, and 8%, respectively. Consequently, detectors fashioned with ^{235}U as the neutron reactive converter always have an intrinsic radiation background. However, the signal produced by the fission reaction is enormous by comparison to signals produced by the background radiations; hence these background signals can usually be rejected by a discriminator circuit when operated in pulse mode, and they add very little to the output current if the detector is operated in current mode. The main advantage of using fissile materials for neutron detection is the large output signal per event that can stand out in the midst of a high radiation background.

Slow neutron detectors based on fission reactions are usually not fabricated with highly enriched ^{235}U, but instead with natural U or low enriched ^{235}U. Hence, a large constituent of the material is ^{238}U, which is fissionable but not fissile. ^{238}U has a long half life of 4.468×10^9 years, and decays mainly by alpha-particle emission with energies of 4.15 MeV (25%) and 4.20 MeV (75%). Slow neutron capture in ^{238}U, with a thermal-neutron capture cross section of 2.684 barns, can transmute into ^{239}U ($T_{1/2} = 23.47$ minutes) which upon beta particle decay creates ^{239}Np ($T_{1/2} = 2.345$ days) and upon another beta particle decay, eventually into fissile ^{239}Pu. Hence, a detector with a proper mixture of ^{235}U and ^{238}U can be used to prolong the useful detector life through neutron transmutation of ^{238}U into ^{239}Pu.

The thermal-neutron fission cross section for ^{239}Pu is 747.4 barns, although the total thermal-neutron absorption cross section is slightly higher at 1017.5 barns due to a thermal-neutron capture cross section of 270.1 barns. A cross resonance at 0.295 eV yields a fission cross section of 3258 b (see Fig. 17.3). ^{239}Pu is radioactive with a halflife of 2.411×10^4 years, and decays mainly by alpha-particle decay (5.105 MeV at 11% and 5.156 MeV at 88%), although there are also several gamma-ray and conversion electron emissions with low branching ratios. Hence, an intrinsic radiation background component gradually increases with detector use. As with ^{235}U, the fission products produce much larger signals than the background alpha particle radiations, and can be easily discerned with a discriminator circuit.

17.3 Gas-Filled Slow Neutron Detectors

Gas-filled slow neutron detectors can be generally classified as one of two types: those that use neutron reactive gases and those that use conventional proportional gases with a solid neutron reactive material inside the detector. Grosshoeg [1979] provides a review describing the origins of these detectors. There are two main types of neutron detectors that are backfilled with either of two neutron reactive gases, namely ^{3}He gas or $^{10}BF_3$ gas. There are several variants for gas-filled detectors having internal components composed of solid neutron reactive materials, including detectors based on ^{10}B, ^{6}Li, or ^{235}U materials. In all of these detectors, slow neutrons react with the neutron absorbing material and induce currents from the charged-particle reaction products with energies much greater than the kinetic energy of the slow neutrons. These large reaction Q-values mask the small kinetic energies of slow neutrons and, hence, do not preserve the original information of the neutron energy. Consequently, they cannot be used directly as neutron spectrometers, but rather they serve as neutron counters. However, spectral features of the reaction products do play an important role in neutron detection, mainly, the pulse height spectra allow the user to set the discriminator at an acceptable level to remove background signals while optimizing the neutron response.

17.3.1 Detectors with Neutron Absorbing Fill Gases

The class of detectors containing neutron reactive gases all have similar energy deposition features, a consequence of *wall* and *end* effects. Gas pressure and detector volume strongly affect the limiting neutron detection efficiency in these detectors. Increased pressure generally increases the probability of absorption, although it may also affect signal proportionality and response speed. Further, these detectors are also

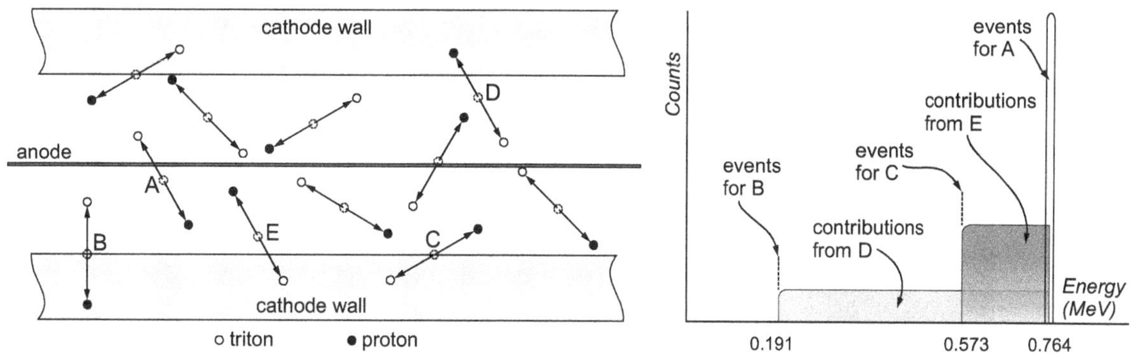

Figure 17.4. (left) Trajectories of reaction products that cause the appearance of the wall effect. (right) Contributions to the pulse height spectrum for the various trajectories. See the text for an explanation of the consequences.

relatively radiation hard and temperature resistant so they are able to operate at elevated temperatures and higher neutron fluxes than other neutron detectors.

^{3}He Gas-Filled Detectors

^{3}He detectors were originally introduced in 1952 [Batchelor 1952; see also Batchelor et al. 1956] and are presently the most popular commercially available neutron detectors. These detectors are operated in the proportional region of the gas pulse-height curve (see Fig. 10.1). Because ^{3}He is a rare and expensive gas, these detectors are not available as gas-flow proportional counters. Instead, they are generally designed as hermetically sealed coaxial detectors. The detection method is straightforward. Neutrons enter the gas volume and are absorbed by the ^{3}He to produce a compound nucleus, which spontaneously decays by emitting two energetic charged-particle reaction products. The two reaction products pass through the gas and produce electron-ion pairs. The electrons rapidly drift towards the anode and create a Townsend avalanche. The drifting motion towards the cathode of the positive ion cloud from the avalanche induces current to flow which is transformed into a voltage pulse by the attached amplification circuitry.

Wall Effect Neutron interactions can occur randomly within the ^{3}He detector, and in some of these cases, the interaction occurs near the chamber cathode wall. If a reaction product strikes the wall, its residual kinetic energy is deposited in the wall rather than the gas and does not contribute to gas ionization. Hence, the resulting amplitude of the output pulse is reduced below that of that for a full energy deposition event. Consequently, there are certain spectral features that are intrinsically observed from a ^{3}He gas-filled detector, primarily a consequence of this *wall effect.*

Shown in Fig. 17.4 are various possible reaction product trajectories in a ^{3}He detector that form the main wall effect features. Trajectory A represents cases in which a neutron is absorbed in the gas and both reaction products deposit all of their initial kinetic energy in the gas volume. Reactions like type A form a full energy peak in the pulse height spectrum. Trajectories B and C represent cases in which the neutron is absorbed next to the cathode wall and only one of the reaction products enters the gas, while the other reaction product deposits all of its energy in the wall. Cases B and C form the limiting *minimum* energy cases for each reaction product type and produce a stairstep feature in the pulse height spectrum. Because each reaction product has a different energy, there are two such stairstep features that appear at the triton energy (0.573 MeV) and at the proton energy (0.191 MeV). Trajectories D and E represent those cases in which the neutron is absorbed near the cathode wall and all of one reaction product energy is absorbed in the gas, while only part of the other reaction product energy is absorbed in the gas before striking the wall. Cases D and E fill in the wide range of possibilities between the limiting cases of partial energy deposition B and C and that of full energy deposition A. Hence, cases for D and E produce a continuum of energies in the

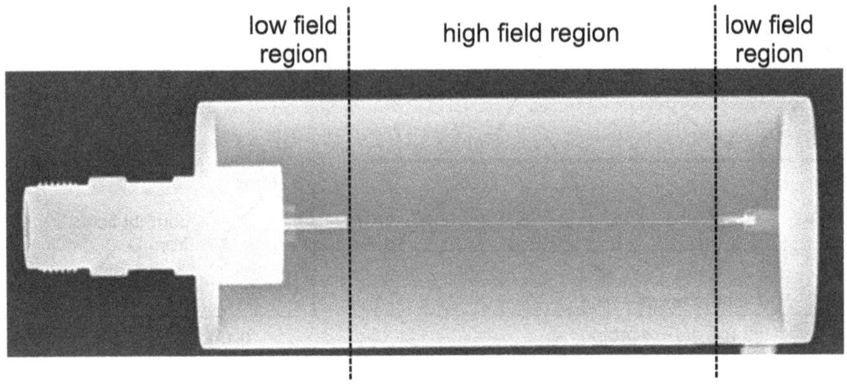

Figure 17.5. X-ray image of a sealed ^{3}He gas-filled neutron detector. The detector gas chamber is 13 cm long; however, only 8 cm of that length are in the high field region.

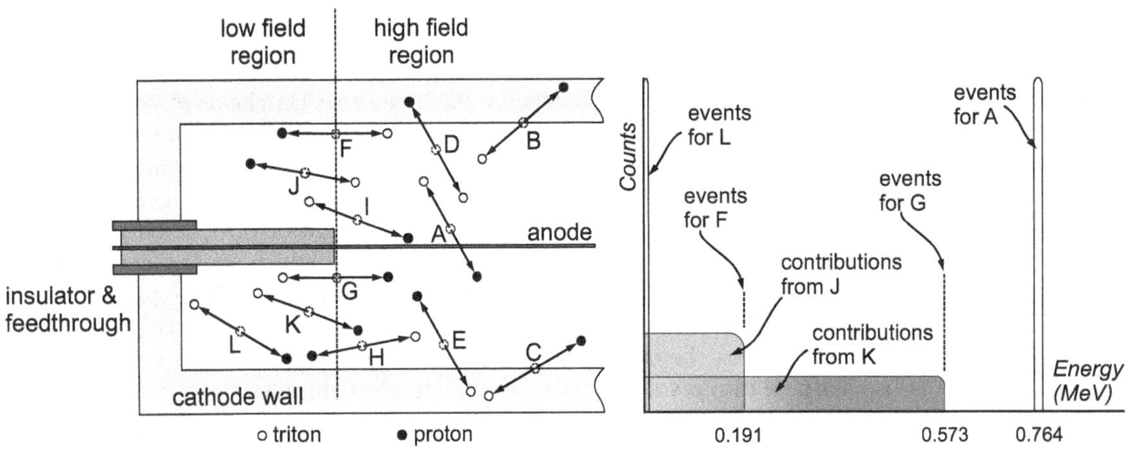

Figure 17.6. (left) Reactions produce trajectories that cause the appearance of both the end effect and the wall effect. (right) Additional features that appear in the pulse height spectrum from the end effect. See the text for an explanation of the consequences.

pulse height spectrum between the limiting minimum energy and the full energy peak. The wall effect forms a "valley" between the radiation background (and electronic noise) and the neutron reaction pulse height spectrum and is a natural location to set the lower level discriminator (LLD).

End Effect The wall effect appears in a spectrum regardless of the geometry for a ^{3}He detector if any portion of it is placed in a neutron field. However, if the detector is entirely immersed in a neutron field, another effect, which is often called the *end effect*, can also be observed. Many gas-filled proportional counters have long insulating stand-offs or field tubes at the ends designed to reduce leakage current, arcing, and spurious counts (see Fig. 17.5). These structures reduce the electric field below the critical field strength necessary for impact avalanche ionization. Hence, these low field regions operate more like an ion chamber instead of a proportional counter. Ionization produced in the low field region appears in the lower energy spectral channels and usually does not contribute to the output signal.

Included in Fig. 17.6 are additional reaction product trajectories that produce the end effect. Trajectories H and I are cases in which a neutron reaction occurs in the high field region near the low field region, in which one particle deposits some energy in the low field region and the other particle deposits all energy in the high field region. Hence, the low field region forms a virtual wall effect, essentially having the same effect as cases D and E. Similarly, the trajectories of F and G have the same effect as cases B and C. Additional effects, however, are from trajectories J and K, along which a neutron is absorbed in the low field region and only one particle emerges into the high field region. The case is analogous to the self-attenuation effect observed with neutron reactive coatings. Variations of cases J and K produce a continuum of energies in the pulse height spectrum ranging from zero up to each of the reaction product energies (either 0.191 MeV or 0.573 MeV), thereby forming a reverse stair step spectrum from that of the wall effect (see Fig. 17.6). Finally events described by trajectory L are essentially lost. The added counts appear in the pulse height spectrum in the low energy valley (or gap) region between the background counts and the wall effect spectrum. Further, the end effect adds pulses to the lower portion of the wall effect spectrum, consequently enhancing that part of the spectrum. The main spectral features expected from a combination of the wall and end effects are depicted in Fig. 17.7.

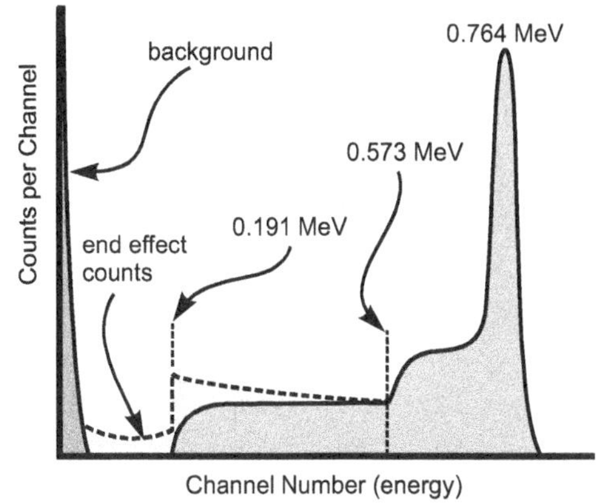

Figure 17.7. Depiction of the main features expected in a pulse height energy spectrum from a ^{3}He gas-filled neutron detector. End effect counts are depicted in light gray.

The end effect is less likely to appear in detector configurations in which a beam of neutrons is incident perpendicularly to the cylindrical side of the detector. However, for detectors operated in a radiation field in which neutrons interact more or less uniformly over the detector volume, the end effect is likely be observed. Shown in Fig. 17.8 are two spectra for different irradiation conditions. The gray spectrum is the result of a collimated thermal-neutron beam intersecting perpendicular to the central region of the detector of Fig. 17.5. In this spectrum, features of the main wall effect are discernible, but end effect features are absent. The white spectrum of Fig. 17.8 is a summed composite in which the collimated thermal-neutron beam was stepped along the detector length at 5 mm intervals from the end to end, which shows the combined spectral features from both the end and wall effects.

Both the wall effect and end effect features increase with surface to volume ratio of a detector. Hence, both of these effects become less prominent for large volume ^{3}He detectors. However, the spectral energy resolution of larger detectors is usually lower than smaller counterparts.

$^{10}BF_3$ Gas-Filled Detectors

Another gas of interest for neutron detection is BF_3, introduced in 1939 [Korff 1939; Korff and Danforth 1939; see review by Wilpert 2012]. Neutrons interact with the ^{10}B and cause the spontaneous ejection of the reaction products of Eq. (17.18). These reaction products ionize the host gas, $^{10}BF_3$, in the same manner as reaction products emitted in a ^{3}He detector. To increase interaction efficiency, modern BF_3 detectors usually have an enrichment of about 96% in $^{10}BF_3$.

Much like ^{3}He gas-filled detectors, $^{10}BF_3$ gas-filled proportional counters also show wall and end effects in the pulse height spectra. Because there are two reaction product branches, there is the possibility of observing four wall effects features, as shown in Fig. 17.9. However, the energy resolution of a $^{10}BF_3$ usually is not adequate to clearly define the 6% branching ratio wall effect features at 1.05 MeV and 1.78 MeV (see Fig. 17.10). The resulting pulse height spectrum is formed from the reaction products and is not defined by the energy of the absorbed neutrons. However, good spectral energy resolution assists with defining an appropriate location for the LLD, thereby allowing pulse height discrimination between neutron and

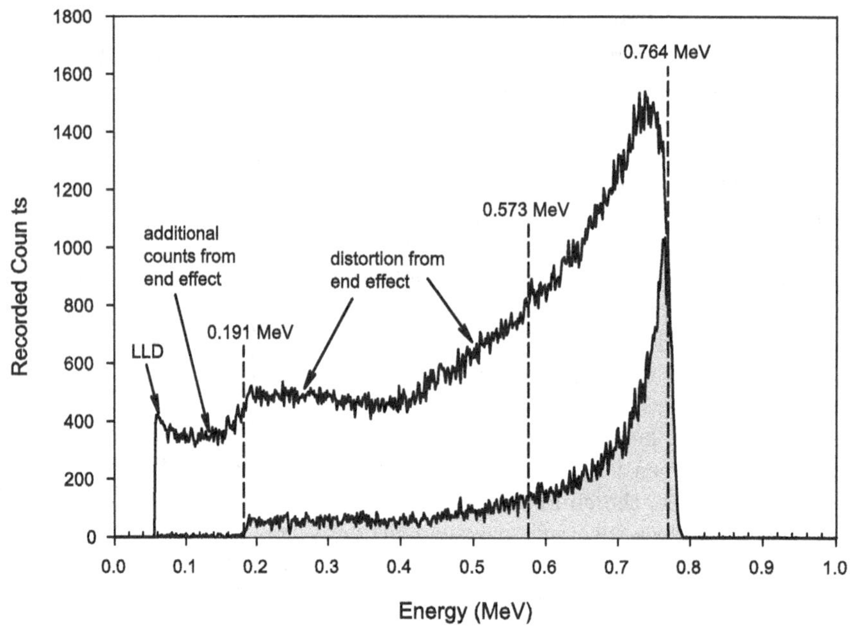

Figure 17.8. Comparison of pulse height spectra from a ^{3}He gas-filled neutron detector irradiated with a thermal-neutron beam directed perpendicular through the center (gray) and a composite spectrum from neutron irradiations at incremental positions along the length (white). The detector was a 2 inch diameter, 6 inch long ^{3}He detector pressurized to 4 atm, which had insulating standoffs >2 cm long at each end. Both wall and end effects are apparent in the composite spectrum.

gamma-ray events. For most applications, the LLD is set in the valley between the wall effect features and the background counts (see Fig. 17.9).

The pressure of the BF_3 gas affects the observed energy resolution of the detector, and ultimately works to limit the maximum neutron detection efficiency. A study by Fowler [1963] clearly showed spectral deterioration as BF_3 gas pressure was increased from 0.131 atm (or 13.33 kPa) up to 0.789 atm (80 kPa), in which the 94% full energy peak FWHM broadened from 8% up to 56%. Although the detector continued to register counts at the higher pressures, the energy resolution was poor, which under some circumstances remains important for neutron detection. Described by Fowler [1963], the best performance was observed for backfill pressures near 0.26 atm (26 kPA). Unfortunately, at these low pressures, the advantage of using the high absorption of $^{10}BF_3$ gas is compromised, yielding typically lower efficiencies than observed with common ^{3}He gas-filled detectors.

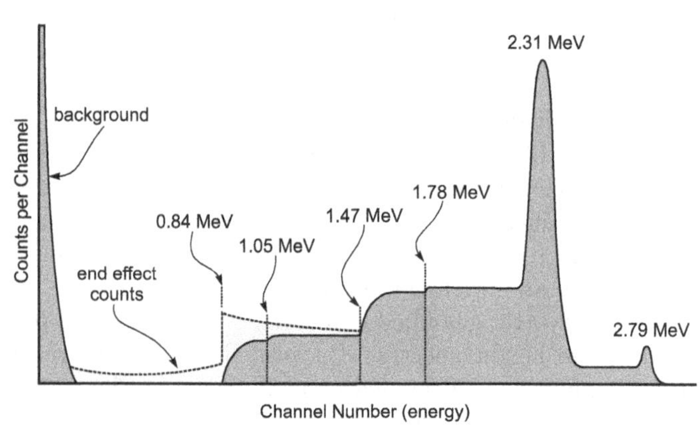

Figure 17.9. Depiction of the main features expected in a pulse height spectrum from a $^{10}BF_3$ gas-filled neutron detector. End effect counts are depicted in light gray.

The efficiency of a BF_3 gas-filled detector can be estimated by calculating the total neutron absorption in the ^{10}B gas region within the active volume of the detector [Sampson and Vincent 1971],

$$\epsilon(E) = \prod_{i=1}^{M} e^{-x_i \Sigma_t^i(E)} \left[\frac{\Sigma_a^{^{10}\mathrm{B}}(E)}{\Sigma_t(E)}\right](1 - e^{-L\Sigma_t(E)}), \tag{17.26}$$

where $\Sigma_t^i(E)$ and x_i denote the macroscopic total cross sections and thicknesses of the ith ($i = 1 \ldots M$) non-sensitive regions, L is the length of the sensitive BF_3 gas volume, and where

$$\Sigma_t = \Sigma_a^{^{10}\mathrm{B}} + \Sigma_s^{^{10}\mathrm{B}} + \Sigma_t^{^{11}\mathrm{B}} + \Sigma_t^{\mathrm{F}} + \Sigma_t^{\mathrm{Ar}} \tag{17.27}$$

with t, s, and a denoting the total, scattering, and absorption cross sections of the homogeneous gas components. Note in the approximation of Eq. (17.26), all scattered neutrons are neglected and are assumed not to produce any subsequent counts.

BF_3 detectors can be acquired with relatively thin entrance windows at one end, which allows the long axis of the detector to be pointed towards the neutron source. Such devices have increased absorption in the sensitive volume and improved detection efficiency. However, many BF_3 detectors are designed with thick end caps, along with relatively large insulating standoffs for the central wire, thereby producing significant dead regions at the ends. Consequently, pointing such a device towards the neutron source may reduce the count rate and observed detection efficiency. Hence, when modeling a detector for a particular neutron detection application, one must be aware of the orientation of the detector with respect to the neutron source. Most commercial manufacturers do not quote efficiency for BF_3 gas-filled detectors in percent efficiency, but rather as cps/nv (counts per second per unit thermal-neutron flux), and are evaluated with the entire detector irradiated with neutrons incident from all directions, a condition in which Eq. (17.26) is inaccurate and not applicable.

Example 17.1: Consider a 2200-m/s neutron beam of 10-mm diameter intersecting a $^{10}BF_3$ gas-filled detector with the following specifications: 25-mm inner diameter, 2 mm thick Al cylindrical casing, 0.8 atm pressure, 250 mm total length with anode wire insulating standoffs each 20 mm long at the ends. Assume an enrichment of 96% ^{10}B with a fill pressure of 0.8 atm. The density of BF_3 is 2.76 mg cm^{-3} at 1 atm and the molar mass is 67.82 g mol^{-1}. What is the thermal-neutron intrinsic efficiency if the neutron beam intersects perpendicular to the BF_3 detector midsection? Afterwards, if the detector is pointed endwise directly at the neutron beam, what is the thermal-neutron intrinsic detection efficiency?

Solution:
First, determine the macroscopic absorption cross section for the ^{10}B atoms,

$$\Sigma_a^{^{10}\mathrm{B}} = \frac{0.96 P\sigma\rho N_a}{A} = \frac{0.96(0.8 \text{ atm})(3844 \text{ b})(0.00276 \text{ g cm}^{-3})(6.022 \times 10^{23} \text{ mol}^{-1})}{67.82 \text{ g mol}^{-1}}$$
$$= (1.88 \times 10^{19}\ {}^{10}\mathrm{B}\text{ cm}^{-3})(3844 \times 10^{-24} \text{ cm}^2) = 0.0723 \text{ cm}^{-1}$$

The total 2200-m/s cross section of Al is 1.68 b so the macroscopic absorption cross section of Al is,

$$\Sigma_t^{\mathrm{Al}} = \frac{\sigma\rho N_a}{A} = \frac{(1.68 \times 10^{-24} \text{ cm}^2)(2.7 \text{ g cm}^{-3})(6.022 \times 10^{23} \text{ mol}^{-1})}{26.98 \text{ g mol}^{-1}} = 0.101 \text{ cm}^{-1}.$$

Here it is assumed that attenuation from F and ^{11}B in the gas is negligible, i.e., $\Sigma_t \approx \Sigma_a^{^{10}\mathrm{B}}$, and all neutrons absorbed by the ^{10}B in the gas produce a detectable count. Because the beam diameter is small, the incident neutrons all pass through 0.2 cm of Al to enter the tube. Then from Eq. (17.26)

$$\epsilon = [e^{-x\Sigma_t^{\rm Al}}][1 - e^{-D\Sigma_a^{^{10}\rm B}}] = [e^{-(0.2\ \rm cm)(0.101\ \rm cm^{-1})}][1 - e^{-(2.5\ \rm cm)(0.0723\ \rm cm^{-1})}] = 0.162.$$

Hence, the maximum intrinsic detection efficiency is 16.2% for a crosswise thermal-neutron beam. Absorption or scattering with the detector tube and F in the gas, along with unfavorable reaction product trajectories, causes the efficiency to be lower than this ideal case.

Turning the detector lengthwise into the beam, the neutron sensitive region is only 210 mm long with 20 mm dead regions at each end,

$$\epsilon = [e^{-x_{\rm Al}\Sigma_t^{\rm Al}}][e^{-x_{^{10}\rm B}\Sigma_a^{^{10}\rm B}}][1 - e^{-L\Sigma_a^{^{10}\rm B}}]$$

$$= [e^{-(0.2\ \rm cm)(0.101\ \rm cm^{-1})}][e^{-(2\ \rm cm)\ (0.0723\ \rm cm^{-1})}][1 - e^{-(21\ \rm cm)(0.0723\ \rm cm^{-1})}] = 0.662.$$

yielding 66.2% intrinsic thermal-neutron detection efficiency.

Although ^{10}B is a $1/v$ neutron absorber with a relatively high thermal neutron absorption cross, the absorption efficiency for epithermal neutrons is quite low. The efficiency for these higher energy neutrons can be increased by increasing the gas pressure, but pressurizing the gas above about 1 atm unfortunately compromises detector performance and energy resolution; hence, it is usually avoided. Consequently, most BF_3 detectors are limited mainly to slow neutron detection.

Argon gas is sometimes added to BF_3 gas-filled detectors, mainly because the Ar (a common proportional gas) helps preserve the spectral energy resolution [Fowler, 1963; Padalakshmi and Shaikh 2008]. Most commercial detectors have significantly more Ar gas in the detector tube than actual $^{10}BF_3$ gas and have pressure ratios greater than about 40:1 Ar:BF_3. The addition of Ar reduces the avalanche voltage, decreases the gas multiplication and shifts the counting plateau to lower voltages. For instance, as detailed by Padalakshmi and Shaikh [2008], the addition of 100 torr Ar to a vessel filled with 10 torr BF_3 effectively reduced the average ionization energy from 35.5 eV for pure BF_3 gas down to 26.8 eV for the mixtures of BF_3 and Ar mixture. Hence, equivalent avalanche gains appear at much lower operating voltages for the BF_3 and Ar mixture at 110 torr than for the pure BF_3 gas at 10 torr, despite the total gas pressure being increased by more than a factor of 10. Further, the onset of the high voltage counting plateau was observed to decrease for the BF_3 and Ar mixtures below that observed for the pure BF_3 gas samples. Padalakshmi and Shaikh [2008] note that excellent resolution and stable operating performance was observed for mixtures between 600–800 torr Ar and 100 torr BF_3. Fowler [1963] notes that the increased pressure of a BF_3 and Ar mixture suppresses the wall effect while retaining good energy resolution.

An increase in the pressure of the BF_3 within the gas chambers increases the probability of neutron absorption (and detection efficiency), but also causes the maximum gain of the detector to decrease as a function of voltage. For instance, Fowler [1963] reports an avalanche gain of 100 achieved with 1000 volts bias for 10 torr BF_3, but to achieve the same gain at 600 torr BF_3 required a bias of 6000 volts. Further, electron attachment and columnar recombination become problematic at higher pressures and ultimately cause the formation of smaller electronic signals. Although BF_3 and Ar mixtures can be adjusted to achieve optimal results at a given total pressure, the general trend indicates that at a set applied voltage the avalanche gain decreases with BF_3 partial pressures above 10 torr, and increasing the BF_3 pressure also causes degradation in spectral energy resolution. Typical pressures average about 600 torr for commercial BF_3 detectors, but detectors can be acquired with pressures ranging from 100 torr up through 1300 torr.

As with most proportional counters, BF_3 detectors have dead times on the order of tens of microseconds, which ultimately limits the practical pulse mode count rate to less than about 2×10^4

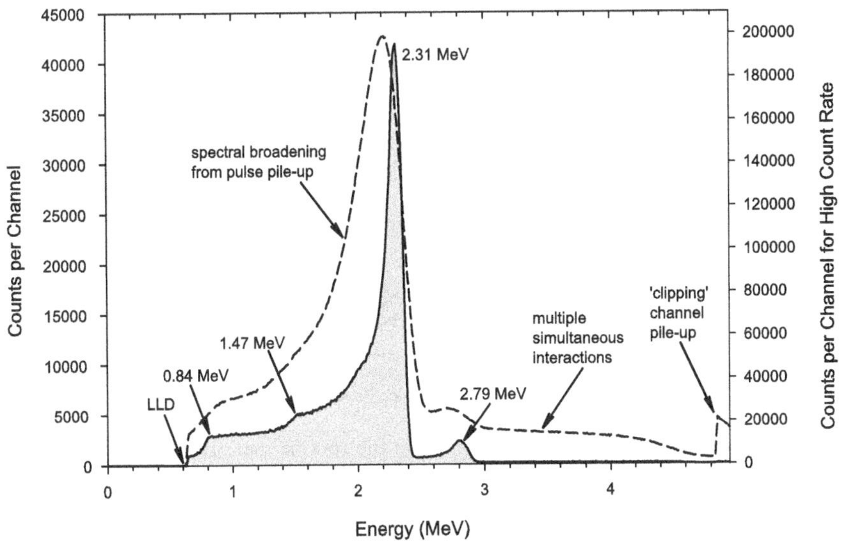

Figure 17.10. Pulse height spectrum from a $^{10}BF_3$ gas-filled neutron detector at low (shaded region) and high (dotted line) neutron interaction rates. The detector was irradiated with a collimated thermal-neutron beam intersecting the mid-section; hence, end effects are not apparent. The wall effect features at 1.05 MeV and 1.78 MeV are not easily discernible.

cps before dead time issues become troublesome. At low count rates with low probability of pulse pile-up, the spectral features remain recognizable. However, as the count rate increases, the spectral features become smeared because of multiple coincident events distorting the pulse height spectrum, as shown in Fig. 17.10. Another feature of BF_3 detectors is that the gas becomes exhausted over an extended operation time. Qian et al. [1998] report performance degradation due to gas depletion appearing for total neutron interactions in the detector gas between 10^{10} and 10^{11}. Depending on the size of a BF_3 detector, typical operating voltages may range from 500 volts for small diameter detectors (about 12 mm) up to 3000 volts for some large diameter detectors (about 50 mm). Also, usually quoted operating temperatures range from -50 °C to 100 °C. Operating the detectors at much higher temperatures than room temperature degrades energy resolution, and in some cases, at elevated temperatures, appears to cause irreversible damage.

BF_3 gas is designated as a level 3 health hazard by the National Fire Protection Association (NFPA 704). This classification means that short exposure to the gas can cause serious temporary or residual injury. Many organizations have abandoned the use of $^{10}BF_3$ primarily because of the health risks. Because of the toxic nature of BF_3 gas, the government has imposed special restrictions on the transport of these devices; however, because $^{10}BF_3$ detectors are an important type of neutron detector, there is not a complete ban on the shipping of certain types of these detectors. The International Air Transport Association (IATA) Special Provision A190 permits the aircraft shipment of BF_3 detectors containing more than 1 gram up to 12.8 grams of BF_3 and filled to pressures no more than 105 kPa absolute at 20°C. It also applies to radiation detection systems containing up to 51.2 grams of BF_3. The U.S. Department of Transportation (DOT) special provision 238 basically reiterates the same provisions. Overall, these detectors must be packed in a sealed plastic bag or plastic liner along with an absorbent capable of absorbing all of the gas contained in the detectors should the gas detector leak during transportation. Detector manufacturers can obtain special transport permits for detectors with higher BF_3 mass and gas pressures than allowed by IATA SP A190 or DOT SP 238.[10]

17.3.2 Detectors with Neutron Reactive Coatings and Layers

The class of neutron detectors that incorporate reactive materials as coatings on the walls, or reactive inserts inside the gas container, are often called coated neutron detectors. This term actually applies to variations

[10] IATA SP A190 and DOT SP 238 were still effective at the time of writing this book.

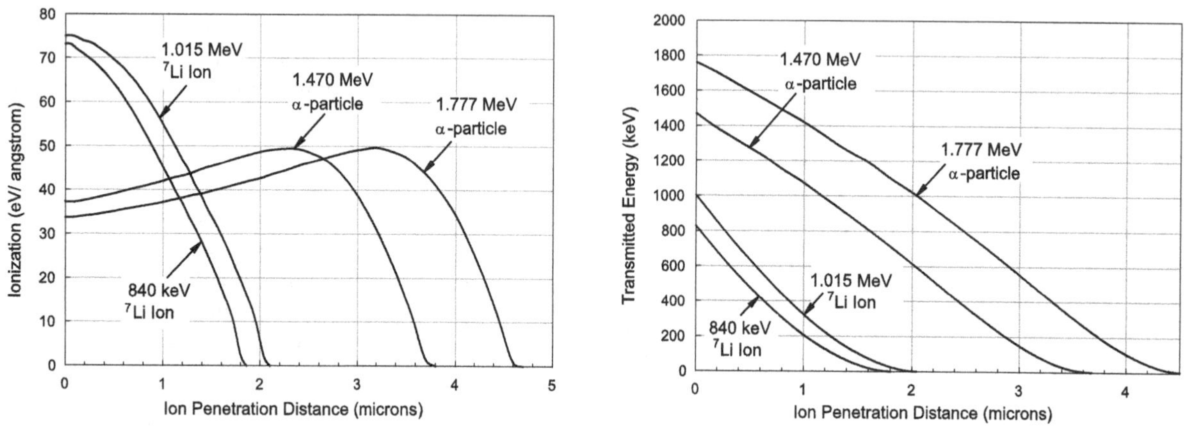

Figure 17.11. (left) The Bragg ionization curve for pure ^{10}B material, showing the specific ionization of ^{10}B(n,α)^{7}Li reaction products as they pass through the material. (right) The residual energy retained by ^{10}B(n,α)^{7}Li reaction products as they pass through pure ^{10}B material [from McGregor et al. 2003].

of coated gas-filled, scintillator, and semiconductor neutron detectors, and the basic analysis applies to all, with minor differences.

It is important to note that the charged particles have discrete ranges within a material, and they lose energy as a function of the specific stopping power of the material through which they pass. Charged particle energy loss is not linear with distance, and the energy loss for high-energy heavy charged particles can generally be described by Eq. (4.147). For non-relativistic particles this equation reduces to

$$-\frac{1}{N}\frac{dE}{dx} = \frac{4\pi z^2 e^4}{4\pi\epsilon_o^2 m_e v^2} Z \ln\left(\frac{m_e v^2}{\overline{I}}\right), \tag{17.28}$$

in which ze is the charge of the heavy ion, $v(x)$ the speed of the charged particle after traveling a distance x, N is the atomic density, m_e is the rest mass of an electron, Z is the material atomic density, and $\overline{I}$ is an experimentally determined mean excitation energy describing the average excitation potential of a representative atom in the material.

Although the average range is treated as constant, the actual range fluctuates from particle to particle because of deviations in the number of Coulombic scatters per unit distance and deviations in energy loss per collision. The fluctuation may exceed 5% of the total range, and this fluctuation is referred to as the straggle in the particle range. Recall from Fig. 4.26, R_p is the mean or average range and $R_e > R_p$ is the extrapolated range. The slope of the fraction reaching R_p is $1/(\alpha\sqrt{\pi})$ so that the two ranges are related by

$$R_p = R_e - \frac{\sqrt{\pi}}{2}\alpha. \tag{17.29}$$

A Bragg curve is a plot of the ionization versus penetration distance, averaged over a large number of individual identical source ions. Let $I(x)$ be the Bragg curve ordinate, i.e., the average ionization per unit differential path length at a distance x from the source. Then $I(x)$, accounting for straggling, is [Evans 1955]

$$I(x) = \int_x^\infty \frac{i(r-x)}{\alpha\sqrt{\pi}} \exp[(R_p - r)/\alpha]^2 \, dr, \tag{17.30}$$

where $i(r-x)$ is the specific ionization (LET) along the path of an individual particle at a distance $(r-x)$ from the *end* of its path, and x is the distance an individual particle is from the source. An interesting

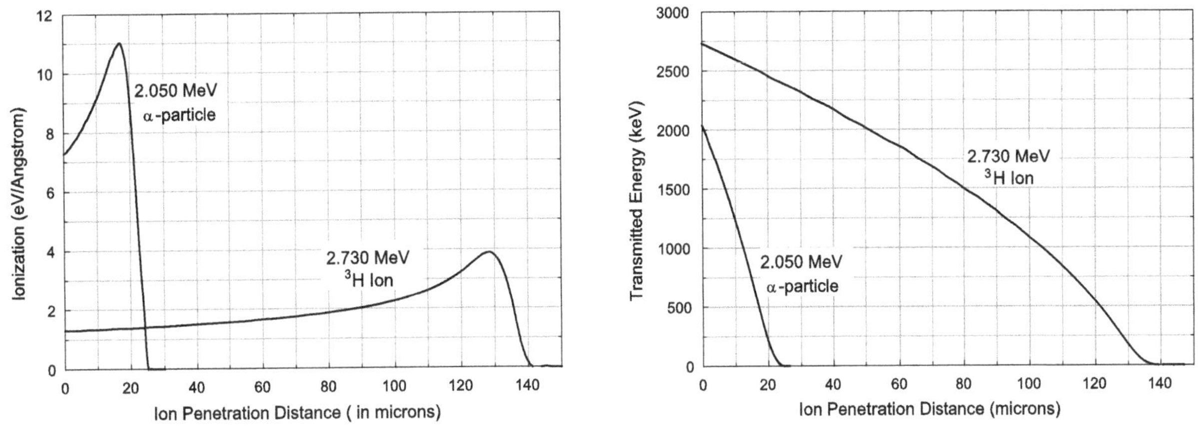

Figure 17.12. (left) The Bragg ionization curve for pure ^{6}Li metal, showing the specific ionization of ^{6}Li(n,t)^{4}He reaction products as they pass through the metal. (right) The residual energy retained by ^{6}Li(n,t)^{4}He reaction products as they pass through pure ^{6}Li metal [from McGregor et al. 2003].

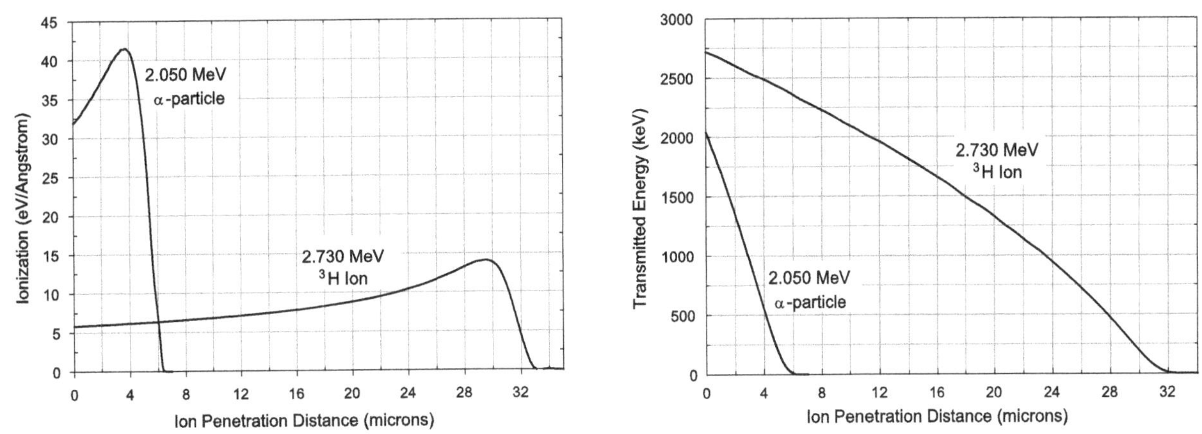

Figure 17.13. (left) The Bragg ionization curve for pure ^{6}LiF material, showing the specific ionization of ^{6}Li(n,t)^{4}He reaction products as they pass through the material. (right) The residual energy retained by ^{6}Li(n,t)^{4}He reaction products as they pass through pure ^{6}LiF material [from McGregor et al. 2003].

characteristic of the Bragg ionization curve for alpha (α) particles is the increase in columnar ionization per unit volume as the ion energy decreases. Hence, the highest density of electron-hole pairs is excited near the end of the range for α-particles.

In the present treatment the average effective range, denoted by $\overline{L}$, is the distance through which an ion may travel within the neutron reactive film before its energy decreases below the set minimum detectable threshold, typically defined by the electronic lower-level-discriminator (LLD) setting. The effective range $\overline{L}$ does not take into account additional energy losses from contact "dead regions". $\overline{L}_{SR}$ and $\overline{L}_{LR}$ denote the average effective ranges for the short-range and long-range reaction products, respectively. The graphs of Fig. 17.11, Fig. 17.12, and Fig. 17.13 can be used to determine the reaction product effective ranges as a function of LLD for pure enriched ^{10}B, ^{6}Li, and ^{6}LiF, respectively.

Neutrons may interact anywhere within the reactive film. From Fig. 17.11, Fig. 17.12, and Fig. 17.13, it is apparent that the reaction products also lose energy as they move through the reactive film, thus limiting the energy that can be transferred to the detector gas (or semiconductor material). The finite specific energy loss in the reactive film limits the usable film thickness that can be deposited on the walls of the gas chamber or over the semiconductor device. The voltage signal measured is directly proportional to the number of ion pairs created within the detector gas or semiconductor. Reaction products that deposit most or all of their energy in the active volume (the gas or semiconductor where ionization is recorded) produce much larger voltage signals than those reaction products that lose most of their energy in the surface films and contacts.

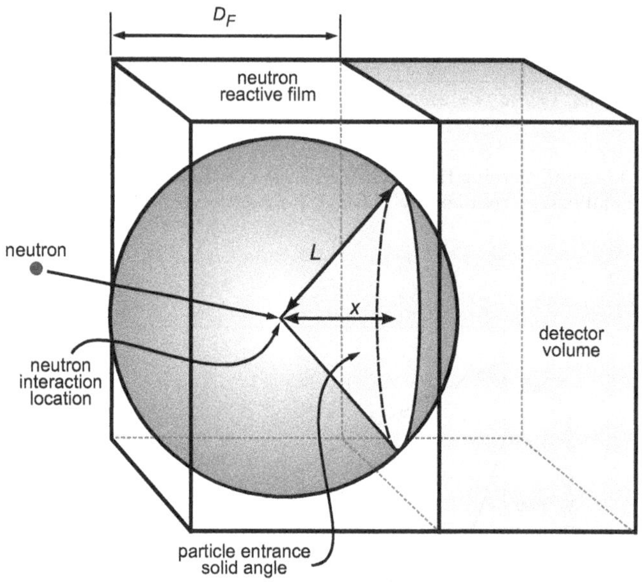

Figure 17.14. The solid angle formed by a reaction product range that subtends the detector at the interface determines the neutron detection probability.

The neutron flux transported through the film a distance x, ignoring neutron scattering, is

$$I(x) = I_o \exp[-x\sigma_a N_a] = I_0 \exp[-x\Sigma_a]. \tag{17.31}$$

where I_0 is the initial neutron flux, N_a is the atomic density of the neutron reactive isotope in the film, σ_a is the microscopic thermal-neutron absorption cross-section of the film, and Σ_a is the film macroscopic thermal-neutron absorption cross section. The fraction of neutrons absorbed in the film while traveling a distance dx about x is,

$$p(x)dx = \Sigma_a \exp[-x\Sigma_a]dx. \tag{17.32}$$

Besides the attenuation of neutrons as they travel in the reactive film, the reaction products are also attenuated as they traverse the reactive film. This ion self-attenuation depends on the ion emission angle with respect to the normal of the film. Once a neutron is absorbed and the reaction products are emitted, the probability that a reaction product enters the detector is determined by the solid angle subtending the surface within the average effective range $\overline{L}$ of the reaction particle. From Fig. 17.14, a neutron interaction taking place at distance x from the semiconductor surface has a probability of entering the active volume equal to the fractional solid angle that subtends the detector, i.e.,

$$p_{\rm p}(x) = \frac{\Omega(x)}{4\pi} = \frac{2\pi}{4\pi}\left(1 - \frac{x}{\overline{L}}\right) = 0.5\left(1 - \frac{x}{\overline{L}}\right), \quad x \le \overline{L}, \tag{17.33}$$

where the subscript "p" refers to the reaction ion of interest. Because the reactions of interest in the present work release two different reaction products per event, the total probability of detecting a neutron reaction consists of adding the detection probabilities of both reaction products. Note that at some interaction locations only one reaction product may be able to reach the active volume of the detector.

Efficiency Analysis for Detectors with ^{10}B, ^{6}Li, or ^{6}LiF Coatings

First consider a simple case. Suppose a layer of neutron reactive material is placed within gas-filled detector upon an inside surface, as depicted in Fig. 17.15. The thermal-neutron reactive materials ^{10}B and ^{6}Li emit these reaction products in opposite directions; hence, only one of the particles can enter the detector gas. The neutrons may be absorbed in the neutron reactive coating on the entry side of the detector (cases A, C,

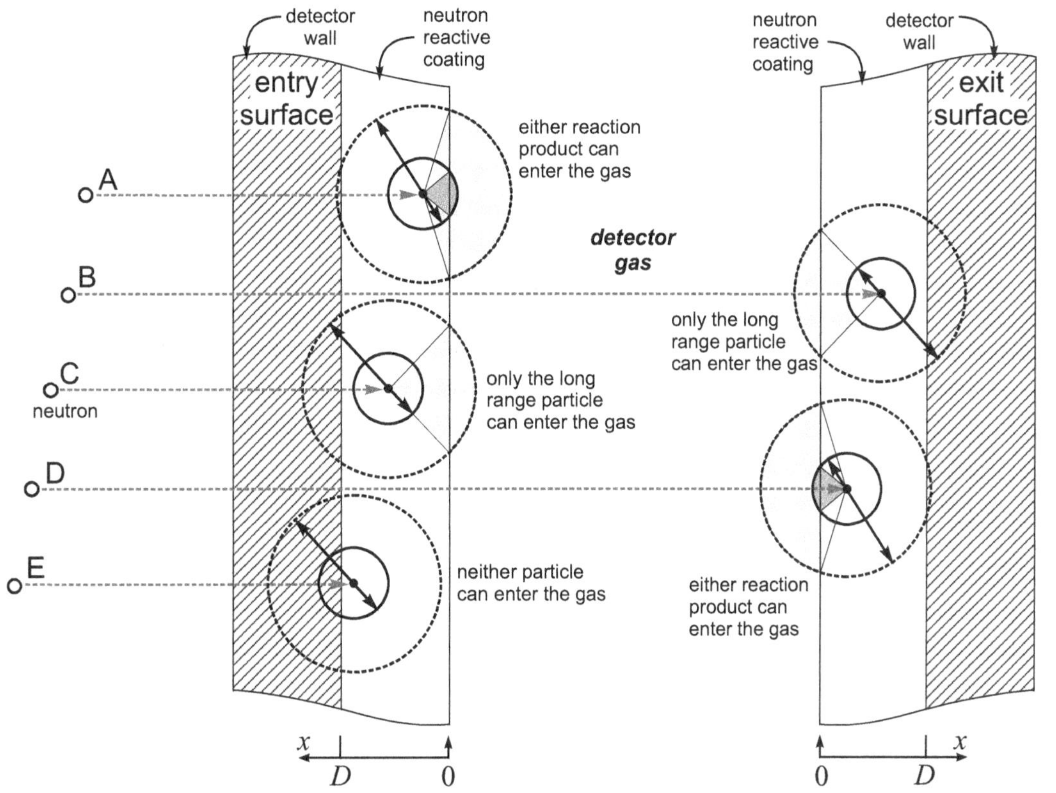

Figure 17.15. Escape of reaction products from a neutron reactive film coating inside the gas-filled region of a detector.

and E in Fig. 17.15), or may pass through and become absorbed in the neutron reactive coating on the exit side of the detector (cases B and D in Fig. 17.15), or may pass through the detector without being absorbed.

Consider cases in which neutrons interact with the coating on the entrance side in the gas-filled detector. Because the reaction products are emitted in opposite directions for thermal neutron absorptions, only one reaction product per event can enter the detector active region. Each reaction product can be analyzed separately and the results are subsequently summed to determine the total detection efficiency. Situation A in Fig. 17.15 depicts the case in which either reaction product can enter the gas region of the detector, and situation C depicts the case in which only the long range reaction product can reach the detector gas. Situation E depicts that case in which neither reaction product can reach the active volume of the detector. Note that the calculation must account for any neutron interactions in the detector wall, because those neutrons are lost before entering the detector and, consequently, reduce the overall detection efficiency.[11]

For the coordinate orientation for the entry surface shown in Fig. 17.15, Eq. (17.32) and Eq. (17.33) are multiplied and integrated over the coating thickness D to find the efficiency for each reaction product. Here the neutrons are assumed to be normally incident on the wall. Also the wall/coating interface is placed at

[11] Actually, some wall scattered neutrons do reach the detector gas and may produce a count. Thus, the analysis in this section, which treats a scatter as an absorption, represents a lower bound on the detector efficiency.

$x = 0$ so that $y = D - x$. If the overall coating thickness D is less than the reaction product effective range $\overline{L}$, the sensitivity contribution (or probability of entering the detector gas or semiconductor) of that reaction product is,

$$\begin{aligned} S_P(D) &= \frac{F_p}{4\pi}\prod_{i=1}^{M} e^{-\Sigma_t^i T_i}\int_0^D 2\pi\Sigma_a e^{-\Sigma_a(D-x)}\left(1-\frac{x}{\overline{L}}\right)dx, \\ &= 0.5F_p\prod_{i=1}^{M} e^{-\Sigma_t^i T_i}\left[\left(1+\frac{1}{\Sigma_a\overline{L}}\right)(1-e^{-\Sigma_a D})-\frac{D}{\overline{L}}\right], \quad D \le \overline{L}, \end{aligned} \tag{17.34}$$

where F_p is the branching ratio of the reaction product, Σ_t^i and T_i denote the different material total thermal macroscopic cross sections and thicknesses of those materials through which neutrons must pass before encountering the reactive coating, M is the number of these materials encountered, D is the thickness of the reactive coating, Σ_a is the macroscopic thermal-neutron absorption cross section of the reactive coating, and $\overline{L}$ is the effective range of the reaction product.

If the overall coating thickness D is greater than the reaction product effective range $\overline{L}$, the self-absorption neutron losses within the entry coating layer must be included. The sensitivity contribution of that reaction product is, for $D > \overline{L}$,

$$\begin{aligned} S_P(D) &= \frac{F_p e^{-\Sigma_a(D-\overline{L})}}{4\pi}\prod_{i=1}^{M} e^{-\Sigma_t^i T_i}\int_0^{\overline{L}} 2\pi\Sigma_a e^{-\Sigma_a(D-x)}\left(1-\frac{x}{\overline{L}}\right)dx \\ &= 0.5F_p e^{-\Sigma_a(D-\overline{L})}\prod_{i=1}^{M} e^{-\Sigma_t^i T_i}\left[\left(1+\frac{1}{\Sigma_a\overline{L}}\right)(1-e^{-\Sigma_a\overline{L}})-1\right]. \end{aligned} \tag{17.35}$$

For a cylindrical tube detector, neutrons that do not interact in the coating as they enter the detector may transit the detector gas and interact in the coating on the opposite surface, as shown by cases B and D in Fig. 17.15. These interactions with neutrons interacting with the exit surface of the detector can produce additional detectable events that must be added to the events recorded for interactions with the coating on the entry surface. The analysis is simplified if the axis is reversed, as shown in Fig. 17.15 for the exit surface. Eq. (17.32) and Eq. (17.33) are multiplied and integrated over the coating thickness D to find the efficiency for each reaction product. If the coating thickness D is less than the reaction product effective range $\overline{L}$, the sensitivity contribution of that reaction product is, for $D \le \overline{L}$,

$$\begin{aligned} S_P(D) &= \frac{F_p e^{-\Sigma_a D}}{4\pi}\prod_{i=1}^{M} e^{-\Sigma_t^i T_i}\int_0^D 2\pi\Sigma_a e^{-x\Sigma_a}\left(1-\frac{x}{\overline{L}}\right)dx, \\ &= 0.5F_p e^{-\Sigma_a D}\prod_{i=1}^{M} e^{-\Sigma_t^i T_i}\left[\left(1-\frac{1}{\Sigma_a\overline{L}}\right)(1-e^{-\Sigma_a D})+\frac{D}{\overline{L}}e^{-\Sigma_a D}\right], \end{aligned} \tag{17.36}$$

where losses in the entrance coating are included. If the coating thickness D is greater than the reaction product effective range $\overline{L}$, the sensitivity contribution of that reaction product (also accounting for reaction product losses in the entrance surface coating) is

$$\begin{aligned} S_P(D) &= \frac{F_p e^{-\Sigma_a D}}{4\pi}\prod_{i=1}^{M} e^{-\Sigma_t^i T_i}\int_0^{\overline{L}} 2\pi\Sigma_a e^{-x\Sigma_a}\left(1-\frac{x}{\overline{L}}\right)dx, \\ &= 0.5F_p e^{-\Sigma_a D}\prod_{i=1}^{M} e^{-\Sigma_t^i T_i}\left[\left(1-\frac{1}{\Sigma_a\overline{L}}\right)(1-e^{-\Sigma_a\overline{L}})+e^{-\Sigma_a\overline{L}}\right]. \end{aligned} \tag{17.37}$$

Usually a coated gas-filled detector is backfilled with a gas that is relatively unreactive with neutrons; hence it is usually neglected in sensitivity calculations. However, if the gas is an absorber of neutrons, then those losses imposed by the gas must also be taken in account. After the sensitivity components are calculated for all reaction products and branching ratios, they are summed to determine the intrinsic neutron detection efficiency, i.e.,

$$\epsilon_{int} = \sum_{p=1}^{M} S_p(D)\bigg|_{entry} + \sum_{p=1}^{M} S_p(D)\bigg|_{exit} . \tag{17.38}$$

Example 17.2: Consider a two-inch inner diameter gas-filled detector backfilled at 1 atm with P-10 counting gas. The detector is fabricated from 2-mm-thick Al tube. The walls are coated with $D = 10$ microns of $\gamma = 94\%$ enriched ^{6}LiF. The detector is best operated with the LLD set to an equivalent of 250 keV. For a 1-cm diameter beam of 2200 m s^{-1} neutrons intersecting perpendicularly through a cross section of the detector, what is the expected neutron detection efficiency ϵ_{int}?

Solution:

The mass density of common LiF is 2.635 g cm^{-3} and the atomic weight is $A = 25.939$. From these data the macroscopic cross section of the ^{6}LiF material is found as

$$\Sigma_a = \sigma_a N = \frac{\sigma_a \gamma \rho N_a}{A}$$
$$= \frac{(938 \times 10^{-24}\ \text{cm}^2)(0.94)(2.635\ \text{g cm}^{-3})(6.022 \times 10^{23}\ \text{mol}^{-1})}{25.939\ \text{g mol}^{-1}} = 53.9\ \text{cm}^{-1}.$$

The residual energy graph (at an LLD of 250 keV) for pure enriched ^{6}Li, shown in Fig. 17.13, is used here to estimate the values of $\overline{L}$, with $\overline{L}_\alpha \approx 4.8$ microns and $\overline{L}_t \approx 29.5$ microns even though the density used in the figure is slightly different than in this problem. Because $\overline{L}_t > D$, Eqs. (17.34) and (17.36) are used to determine the triton contribution to the efficiency. The 2200-m/s total microscopic cross sections for Al and Ar are 1.68 b and 1.32 b, respectively; hence, it is assumed that 2 mm of Al and 5 cm of Ar at 1 atm have little effect on neutron losses and can be neglected. Further, the branching ratio for the reaction products is 100%; hence, $F_p = 1$. Equations (17.34) and (17.36) reduce to

$$S_P(D)\bigg|_{entry} = 0.5\left[\left(1 + \frac{1}{\Sigma_a \overline{L}_t}\right)(1 - e^{-\Sigma_a D}) - \frac{D}{\overline{L}_t}\right],$$
$$= 0.5\left[\left(1 + \frac{1}{(53.9\ \text{cm}^{-1})(29.5 \times 10^{-4}\ \text{cm})}\right)\left(1 - e^{-(53.9\ \text{cm}^{-1})(10^{-3}\ \text{cm})}\right)\right.$$
$$\left. - \frac{10^{-3}\ \text{cm}}{29.5 \times 10^{-4}\ \text{cm}}\right] = 0.0218.$$

and

$$S_P(D)\bigg|_{exit} = 0.5e^{-\Sigma_a D}\left[\left(1 - \frac{1}{\Sigma_a \overline{L}_t}\right)(1 - e^{-\Sigma_a D}) + \frac{D}{\overline{L}_t}e^{-\Sigma_a D}\right] = 0.0207.$$

Because $\overline{L}_\alpha < D$, Eqs. (17.35) and (17.37) are used to determine the alpha particle contribution to efficiency, which along with the previous assumptions, yield

$$S_P(D)\bigg|_{entry} = 0.5e^{-\Sigma_a(D - \overline{L}_\alpha)}\left[\left(1 + \frac{1}{\Sigma_a \overline{L}_\alpha}\right)(1 - e^{-\Sigma_a \overline{L}_\alpha}) - 1\right] = 0.0062,$$

and

$$S_P(D)\bigg|_{exit} = 0.5e^{-\Sigma_a D}\left[\left(1 - \frac{1}{\Sigma_a \overline{L}_\alpha}\right)(1 - e^{-\Sigma_a \overline{L}_\alpha}) + e^{-\Sigma_a \overline{L}_\alpha}\right] = 0.0061.$$

These results are summed to determine the intrinsic efficiency,

$$\epsilon_{int} = \sum_{p=1}^{M} S_p(D)\bigg|_{entry} + \sum_{p=1}^{M} S_p(D)\bigg|_{exit} = 0.0218 + 0.0062 + 0.0207 + 0.0061$$
$$= 0.0548 \text{ or } 5.5\%.$$

Self-Attenuation Effect

Because the reaction products are released in opposite directions, only one of these particles can enter the gas region, as explained previously. Consequently, the pulse height spectrum extends from zero up to the highest energy of any single reaction product. The fact that energy can be lost by a reaction ion before depositing the remainder of its energy in the gas is known as the "self-attenuation effect".

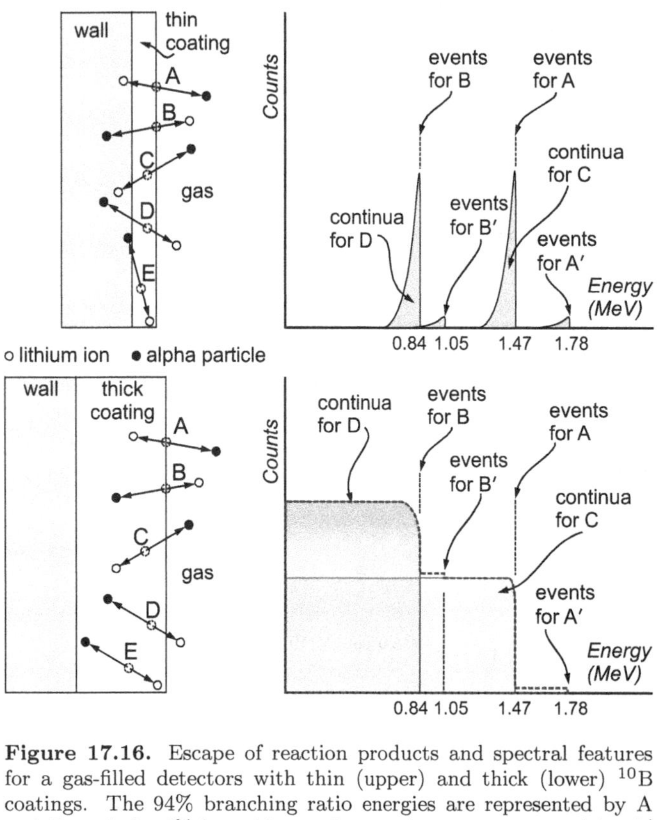

Figure 17.16. Escape of reaction products and spectral features for a gas-filled detectors with thin (upper) and thick (lower) ^{10}B coatings. The 94% branching ratio energies are represented by A and B, and the 6% branching ratio energies are represented by A′ and B′.

Consider the case of a gas-filled detector with a ^{10}B coating on the inner cathode surface, as depicted in Fig. 17.16. Neutron absorptions that occur at the gas/coating interface, depicted as cases A and B, allow either the alpha particle or ^{7}Li ion to deposit all their energy in the gas. Note that in either case none of the energy from the other reaction product is absorbed in the gas, but instead is completely lost in the coating and detector wall. Under no circumstance is the full Q-value of the reaction recorded for this detector configuration. For all interactions that occur deeper in the ^{10}B coating, depicted by cases C and D, energy self-attenuation lowers the amount of energy registered for those particles that emerge into the detector gas. Consequently, these events appear at lower energies in the pulse height spectrum and range from zero up to the full energy of the four possible reaction product energies. Case E depicts those events in which neither particle emerges into the detector gas; hence the events are not recorded.

For thin ^{10}B coatings, the probability of case E occurring is relatively low, and energy self-attenuation is minimized, thereby allowing spectral features of the energies of the different reaction products to appear in the pulse height spectrum, as depicted in Fig. 17.16. For thick ^{10}B coatings, these spectral features are lost, and instead a "stair-step" spectrum is produced. The stair-step feature is a consequence of the combined effects of solid angle, interaction depth, and energy self-attenuation. The resulting probability of depositing energy between zero and the maximum is approximately constant for each reaction product, thereby producing a rectangular shaped pulse height spectrum for each reaction product, that when added together forms the familiar stair-step spectrum. The calculated results for different converter thicknesses of a ^{10}B-lined detector are shown in Fig. 17.17.

^{10}B-Lined Tubes

Detectors coated with ^{10}B, generally called boron-lined detectors, and (along with lithium-lined detectors) are perhaps the first type of gas-filled slow neutron detectors introduced to the scientific community [Dunning et al. 1935a; Mitchell 1936].[12] They are commercially available with casings usually constructed from either aluminum or stainless steel. The coatings, in units of mass thickness, range from 0.025 mg cm^{-2} up to 1 mg cm^{-2} or in thicknesses of 0.1 micron up to 4.33 microns. Coatings thicker than 4.6 microns are greater than the range of the most penetrating reaction product (see Fig. 17.11) and do not increase the detection efficiency but serve only to reduce the neutron flux. The thicker coatings, often quoted with mass thicknesses between 0.5 and 1.0 mg cm^{-2}, produce pulse height spectra with the self-attenuation stair-step feature shown in Figs. 17.16 and 17.17. Efficiency is usually quoted in units of total count rate from the detector per unit flux (cps/nv), with the detector completely immersed in a neutron field. If used in a neutron beam, the orientations described in Example 17.1 and method of Example 17.2 should be used to estimate the expected efficiency.

Boron-lined tubes are available in various sizes, with diameters usually ranging from 13 mm up to 77 mm. Because neutrons interact in the coating attached to the walls, detectors are designed for irradiation through the side wall; hence, there is dead space near the tube ends from the insulating standoffs and/or field tubes.

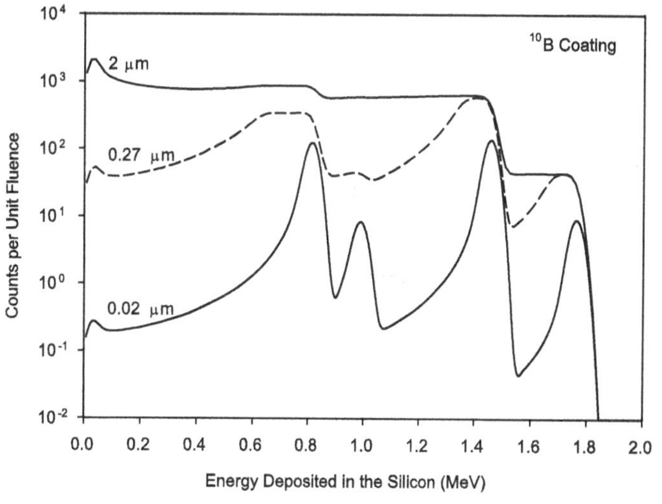

Figure 17.17. Pulse height spectra for a boron-lined detector for three boron layer thicknesses. Although the total number of registered counts increases with boron thickness, the self-attenuation effect is also enhanced and degrades the reaction product spectrum.

Boron-lined gas-filled detectors can use practically any counting gas, an advantage over detectors based on ^{3}He or $^{10}BF_3$ gases. Boron-lined tubes designed as ion chambers can be commercially acquired with hydrogen as the backfill gas, while boron-lined proportional counters are usually backfilled with traditional proportional gases such as argon. Boron-lined gas-filled detectors compensated for gamma-ray background can be acquired with a variety of fill gases, including hydrogen, nitrogen, and xenon. Boron-lined tubes can be used under relatively higher temperature conditions (up to about 200°C) than is possible with $^{10}BF_3$ gas-filled detectors.

Unfortunately, the self-attenuation effect limits boron-lined gas-filled neutron detectors to approximately 9% maximum intrinsic thermal-neutron detection efficiency, much lower than that measured for ^{3}He or $^{10}BF_3$ gas-filled detectors. Neutron measurements often require the LLD be set to reduce gamma-ray background counts. Unfortunately, any non-zero setting of the LLD also eliminates neutron counts, effectively reducing the detection efficiency below the maximum possible. This problem is a consequence of the reaction product self-attenuation, which produces no "valley" feature in the pulse height spectrum in the low energy region (unlike ^{3}He or $^{10}BF_3$ gas-filled detectors). Detectors with axially aligned ^{10}B-coated fins, inserts, or honeycomb structures inside a boron-lined tube have been investigated as a method to increase the efficiency [McGregor et al. 2013a; Nelson et al. 2012, 2014; Edwards et al. 2018]. The total neutron detector efficiency increases with the added boron-coated surface area in the gas chamber.

[12] Curie and Joilet [1933] and Bonner [1933] had already shown that ionization by neutron recoil collisions can be detected.

Compensated Ion Chambers

A form of ion chamber widely used for nuclear reactor controls is the *compensated ion chamber*. Ion chambers, when operated in *current mode*, can be used in high radiation environments. If a gas-filled neutron detector is placed near a nuclear reactor, it responds to both neutrons and gamma rays. Yet, current mode operation does not permit pulse height discrimination between neutron and gamma-ray interactions as does pulse mode operation. The compensated ion chamber design is used to distinguish between the two types of radiations.

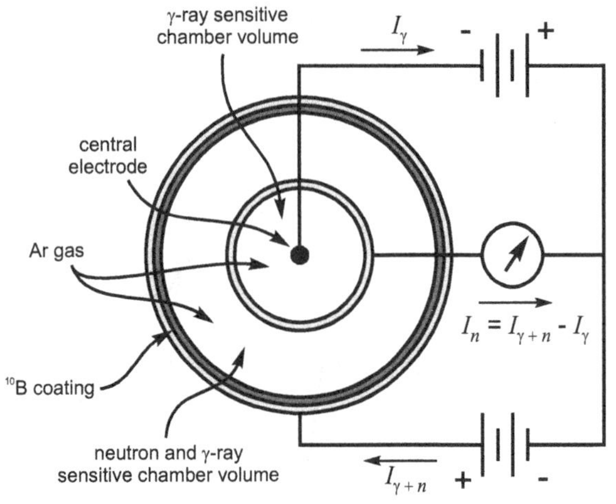

Figure 17.18. Cross section diagram of concentric compensated ion chamber. The configuration allows both chambers to experience the same radiation field. Differences between the two chambers can be properly calibrated by adjusting the operating voltages.

Typically the chamber has three concentric electrodes, one coated with a neutron sensitive material such as ^{235}U or a compound containing ^{10}B. As seen from Fig. 17.18, the ^{10}B (or ^{235}U) coated chamber is generally referred to as the "working" chamber and the uncoated chamber is referred to as the "compensating" chamber. When exposed to a combined gamma-ray and neutron source, the voltage potential for the working chamber causes current to flow that deflects the meter in one direction. The voltage potential in the compensating chamber, sensitive only to gamma rays, causes current to flow in the opposite direction. The voltage potentials on the chambers can be adjusted so that the two gamma-ray induced currents are equal. As a result, in the compensating chamber the two gamma-ray currents cancel each other so the net signal is solely from the neutron induced current. Compensated ion chambers are widely used in nuclear reactors because of their ability to respond to neutron fields that vary up to ten orders of magnitude, i.e., these detectors have a very large "dynamic range".

$^{10}B_4C$-Coated Straw Tubes

Miniaturized version of the boron-lined neutron detector, dubbed "boron straw tubes", have been introduced in recent years [Lacy et al. 2009, 2013]. These detectors have similar properties to those of the straw tubes described in Sec. 10.6.7. Boron straw tubes are fabricated from either Al or Cu metal foil that has been coated with BC_4, enriched to approximately 96% ^{10}B [Lacy et al. 2013]. The BC_4 material is applied with physical vapor deposition (PVD) to produce a durable layer that can withstand shock and relatively high temperatures. Afterwards, the metal foil is rolled and welded to form the envelope. Each straw tube is designed as a coaxial detector, with a central anode wire and the metal foil envelope acting as the cathode. These detectors range in diameter from 2 mm to 15 mm, and can be over 1 meter long, depending on the design.

A single detector has the same efficiency limitations as any other boron-coated gas-filled neutron detector. However, because of their relatively low production cost, compact size, and durability, large groups of boron straw tubes are usually bundled together in a larger vessel, thereby increasing the overall neutron detection efficiency. Recent designs deviate from the conventional cylindrical tube, adapting instead a "star-shaped" straw tube. The star shape design increases the overall neutron reactive surface while allowing the straws to pack together in an interlocking manner [Lacy et al. 2013]. Such a configuration increases the amount of ^{10}B-coated surfaces encountered by an intersecting neutron beam. Approximately 31 straw tubes, each nominally 4.43 mm diameter, can be packaging inside a 1.15 in (29.2 mm) diameter Al tube and operated as a single detector. Lacy et al. [2016] have also adopted the axial fin design [McGregor et al. 2013a] inside straw tubes to improve thermal neutron detection efficiency.

A disadvantage of such small geometries is that the reaction product ranges are typically much greater than the straw-tube cross dimensions. Consequently, much of a reaction product energy is lost when it encounters an opposing surface; hence only a portion of the reaction energy is actually deposited in the straw tube, thereby producing a smaller output signal. Energy deposition can be increased by increasing the pressure inside the straw tube. One advantage of the small size is the reduced ionization in the detectors from background gamma rays so that the background radiation interference is reduced. Charged particles are naturally shielded from entering the vessel, and electrons excited through gamma-ray and x-ray interactions inside a straw tube can not physically deposit much energy (low $-dE/dx$) before encountering a surface. Hence, the discriminator can be set lower than that of a conventional boron-coated detector and, thereby allows more neutron-induced counts to be tallied.

The overall efficiency is a function of the BC_4 thickness, enrichment, gas pressure, LLD setting, and the number of surfaces encountered. Shown in Fig. 17.19 are calculated efficiencies, using a combination of Eqs. (17.34), (17.35), (17.36), and (17.37) for each ^{10}B surface encountered in a stack, illustrating the efficiencies that are achievable by stacking multiple boron-coated straw tubes. For optimal conditions at an LLD setting of 73 keV equivalent with a neutron beam intersecting perpendicular to the vessel, a 29.2 mm diameter vessel with 31 round straw tubes can achieve an efficiency exceeding 35%, while 31 star shaped close packed straw tubes can achieve efficiencies exceeding 50% [Lacy et al. 2013]. These numbers vary with LLD setting. Presently, boron coated straw tubes are being incorporated into field instruments, including portal monitors, backpack monitors, and vessels having traditional form factors as those of conventional ^{3}He detectors [Lacy et al. 2013].

Gas-Filled Detectors with ^{6}Li Metal

A relatively new technology that shows great promise is a neutron proportional counter loaded with ^{6}Li metal foil sheets, made possible by film advancements by the Li battery industry. These ^{6}Li foil sheets can either be attached to the container walls or, more effectively, several ^{6}Li foil inserts can be suspended inside the detector chamber. A system of analytical equations can be used to determine the theoretical efficiencies of stacked detector variations [McGregor et al. 2003] or, instead, Monte-Carlo methods can be used to determine the expected efficiencies [Nelson et al. 2012].

Relatively large neutron detectors can be constructed by laminating the walls of a chamber with thin ^{6}Li foils, between each foil a multiwire proportional counter is constructed (see Fig. 17.20). A single detector box typically has two foils, one on each side, and each additional detector increment adds two more foils. This type of Li foil detector has similar properties to the boron lined gas-filled detector or any of the coated detector variants, in which only one of the reaction products from the $^{6}Li(n,t)^{4}He$ reaction can escape into the detection gas. The efficiency can be calculated using the method introduced by Eqs. (17.34) to (17.37), while accounting for neutron absorption losses through each detector box. The resulting calculated efficiencies are shown in Fig. 17.21. Although foils less than 20 micron in thickness can reach thermal-neutron detection efficiencies exceeding 50%, this value is presently impractical due to limitations with manufacturing the thin Li foils. Presently, the quoted lower limit is 55 microns,[13] and yields a saturated maximum intrinsic thermal neutron detection efficiency of 47%. Because there are additional neutron losses from the box materials, it is expected that the actual efficiency is somewhat lower.

Large neutron detectors can also be constructed by suspending framed sheets of ^{6}Li foils inside a gas-filled chamber interspersed between multiwire anodes as shown on the right-hand side of Fig. 17.22 [McGregor et al. 2013a, 2017]. Unlike coated detectors, the thin ^{6}Li foil may allow both reaction products to reach

[13]Source: Rockwood Lithium.

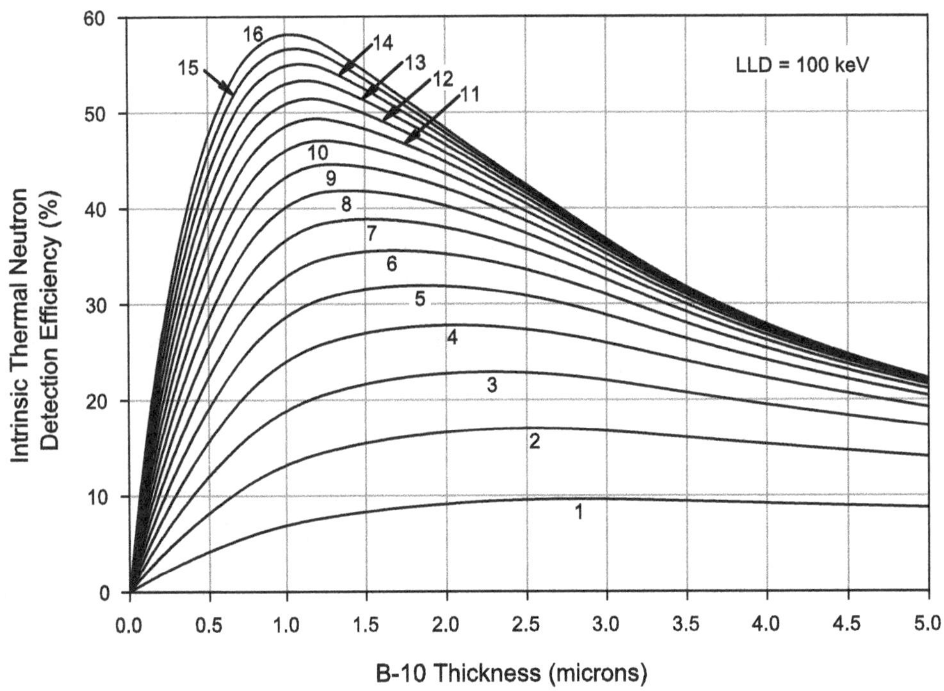

Figure 17.19. Calculated efficiencies for a stack of ^{10}B-lined tubes irradiated with a thermal-neutron beam normal to the axis. The efficiency is shown as a function of the ^{10}B film thickness and the number of tubes intersected by the thermal-neutron beam.

the detection gas, as depicted in Fig. 17.22. Situation A in Fig. 17.22 depicts the case in which either reaction ion can enter the gas region of the detector with a small possibility of both ions entering the gas on opposite sides. Situation B depicts the case in which only the long range reaction product can reach the detector gas from either side of the foil. Situation C depicts that case in which either one of the reaction ions can reach the detector gas but not both. The thermal-neutron detection efficiency for this type of detector can be calculated with the system of equations listed in the literature [McGregor et al. 2003], an analytical method that can be somewhat tedious. Alternatively, widely used Monte-Carlo codes, such as MCNP-6 or GEANT-4, can also be used to calculate the thermal-neutron detection efficiency [Nelson et al. 2012]. The resulting calculated efficiencies are shown in Fig. 17.23. For the minimum ^{6}Li foil thickness of 55 microns, a detector with 10 foils can deliver 72% thermal-neutron detection efficiency. The efficiency reported for ^{6}Li 5-foil detectors exceeds 56%. It is interesting to note that a ^{6}Li-foil detector with a single suspended foil (maximum 24% efficiency) outperforms a two foil detector of the laminated design (maximum 20% efficiency).

The main advantage of the laminated ^{6}Li design is that each detector box in a stack requires only one bank of multiwire anodes. However, the laminated design cannot achieve the high efficiency of the suspended foil design. The main advantage of the suspended foil design is that it can achieve efficiencies similar to those of ^{3}He gas-filled detectors. In either case, the detector is backfilled with a proportional counter gas rather than a rare or caustic reactive gas (which may also not exhibit ideal signal proportionality). A variety of detector shapes and sizes are described in the literature [Nelson et al. 2012, 2014, 2015].

Although Li foils react and decompose in the presence of water moisture, fabrication problems can be mitigated by constructing the detectors within a "dry room", in which water moisture can be kept below

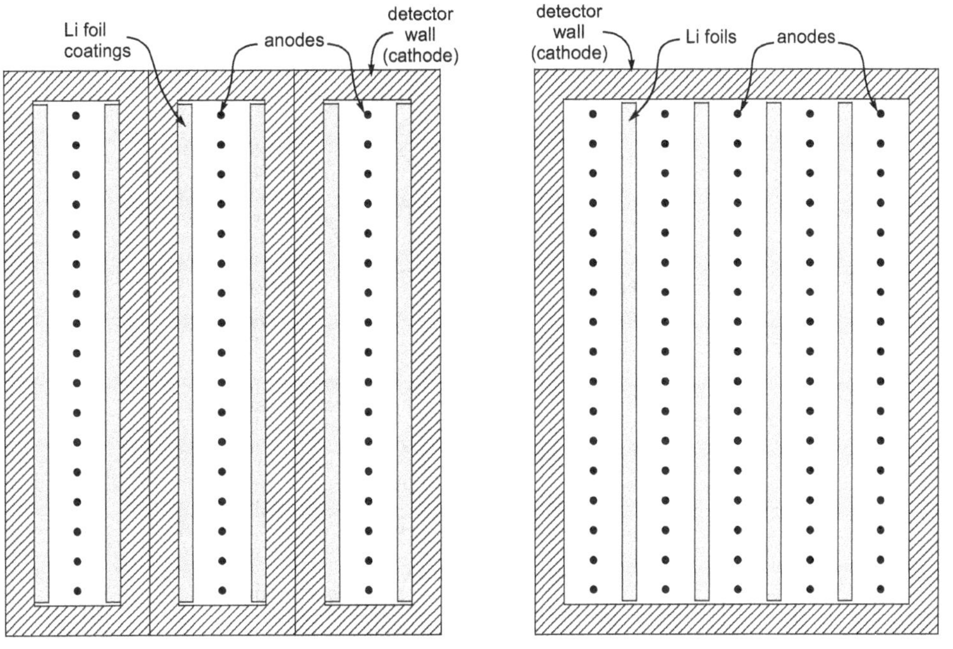

Figure 17.20. For large area Li-foil detectors, there are two basic design constructions. Depicted are (left) a laminated construction with six ^{6}Li foils and (right) a suspended construction with four ^{6}Li foil inserts in the gas container.

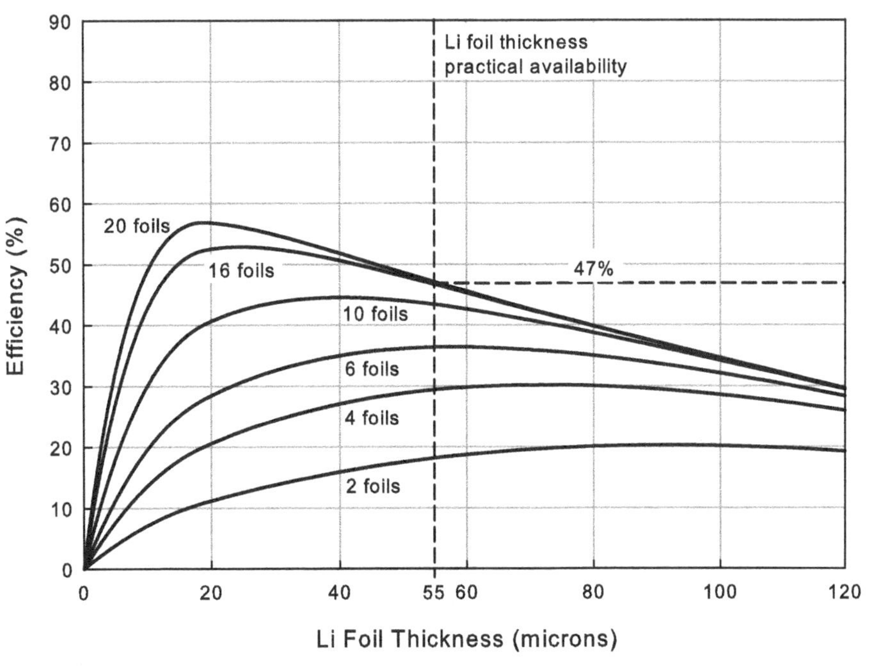

Figure 17.21. Calculated thermal-neutron detection efficiencies for ^{6}Li foils attached to the walls of stacked gas-filled chambers.

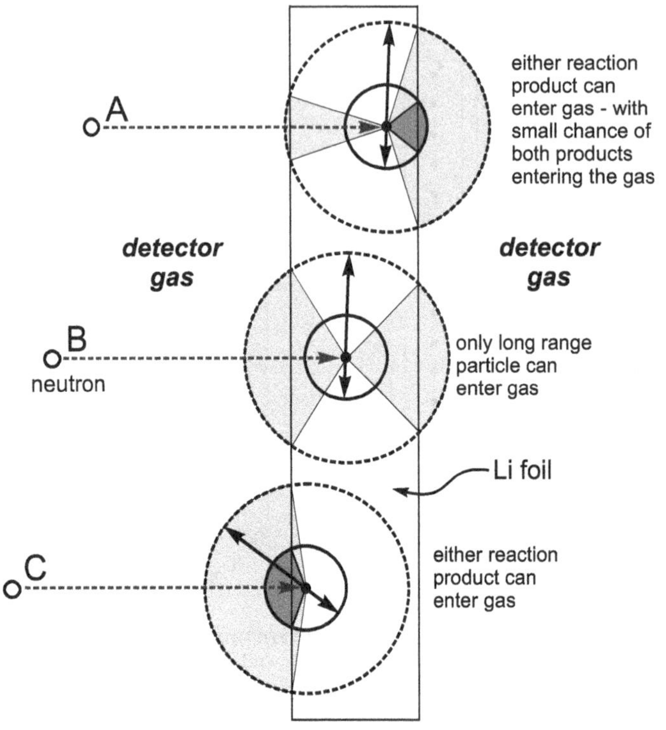

Figure 17.22. Escape of reaction products from a thin suspended foil of ^{6}Li inside a gas-filled detector.

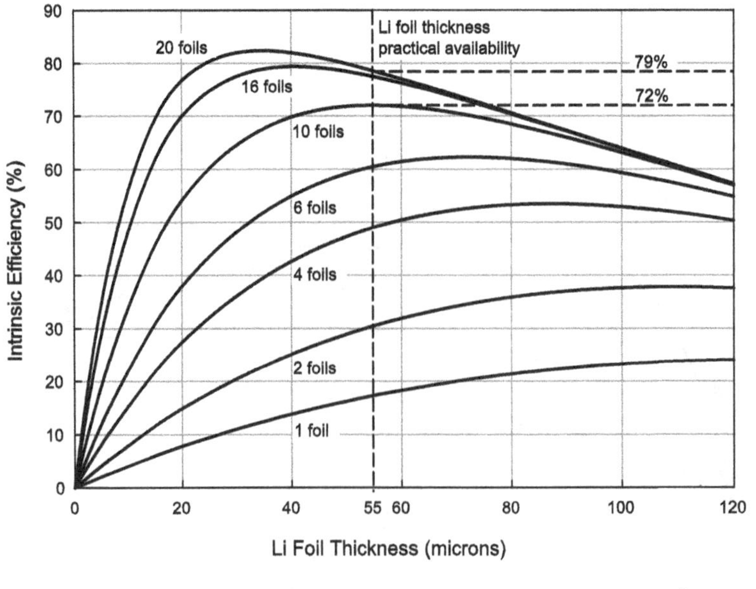

Figure 17.23. Calculated thermal-neutron detection efficiencies for ^{6}Li foil inserts suspended in a gas-filled chamber.

10 ppb. The detectors operate with proportional gases, such as argon. However, it is inadvisable to use quenching gases that contain oxygen, such as CO_2, because these gases cause the Li foils to decompose over time. Instead, saturated hydrocarbons such as the linear alkanes methane, ethane, and propane can be used as quenching gases. To prevent foil decomposition, all gases introduced into the chamber should have ultra-high purity, generally with less than 1 ppm of either O_2 or water moisture contamination.

Fission Chambers

A special type of coated gas-filled radiation detector is a fission chamber. This type of detector has fissile materials (such as ^{233}U, ^{235}U, ^{239}Pu) or fissionable materials (such as ^{237}Np, ^{232}Th, ^{238}U) as coatings in which a fission reaction emits two highly energetic fission products. The energy eventually released per ^{235}U fissioned is approximately 207 MeV, with about 168 MeV of that energy as the kinetic energy of the two fission fragments. The remaining 39 MeV is released in the form of gamma rays, fast neutrons, beta particles, and neutrinos, most of which escape detection. Usually two fission fragments share 168 MeV as kinetic energy, although there is a slight chance of triplet fission, that produces two relatively large fission products and a much smaller fission product (such as a triton, alpha particle, Li ion, or B ion). The two fission fragments are released with different kinetic energies and masses, with average energies of 68.1 MeV for the heavy fragment and 99.2 MeV for the light fragment. The variety of possible fission fragments numbers in the hundreds with atomic masses ranging from 70 to 170. The fission product mass and energy distributions for fissile materials (^{233}U, ^{235}U, ^{239}Pu) have similar distributions, although there are slight differences. Threshold detectors can be used for fast fission reactions, which include ^{237}Np, ^{238}U, and ^{232}Th to measure only the fast flux while discriminating epithermal and slow neutrons (see Fig. 17.24).

The large kinetic energy of the fission provides an attractive converter for neutron detectors, mainly because the energy from either fission product (fragment) can be measured without the need for avalanche multiplication. As a result, most fission chambers are designed as ion chambers. The ranges of fission products in 1 atm of Ar are between 1.5 cm and 2.9 cm, depending on their mass and energy; hence many fission chambers are designed with dimensions of a few centimeters. In a fission event, the fission fragments are ejected in approximately opposite directions, although the other energetic fission emissions, such as prompt fission neutrons, cause a slight change in the opposite fission product trajectories. Consequently, much like other coated neutron detectors, only one of the fission products is likely to cause ionization of the chamber gas and be measured.

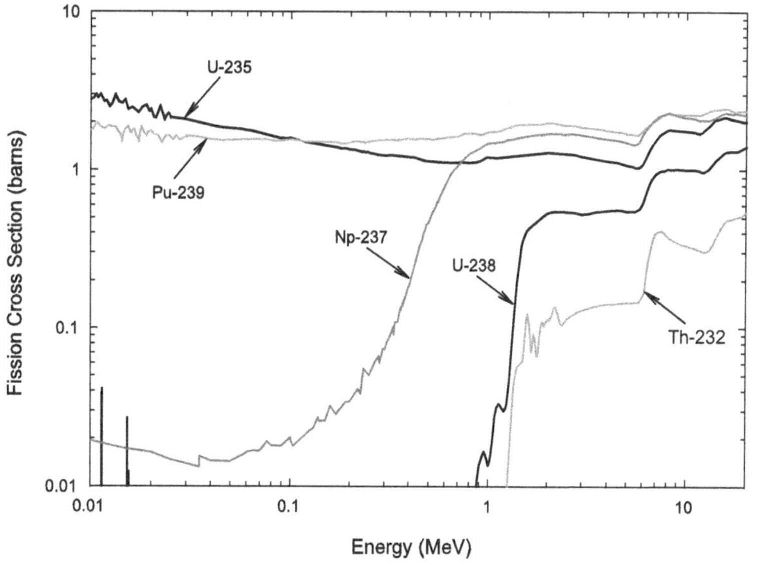

Figure 17.24. Fast neutron fission cross sections for several materials used for fission chambers. The isotopes ^{238}U, ^{237}Np, and ^{232}Th are used as threshold converters, in which neutrons below an energy threshold are unlikely to interact in the detector. Data are from [ENDFPLOT 2015]. (The slow neutron cross sections for ^{235}U and ^{239}Pu are shown in Fig. 17.3).

Although the number of electron-ion pairs created in the gas depends on the amount of energy deposited by the fission fragments, it is also a function of the effective Z of the fission products. The average number of ion pairs created, per unit energy deposited, is higher for an alpha particle than for fission fragments, assuming full energy deposition and identical operation conditions [Rao et al. 1990]. This outcome is a consequence of heavy ions losing energy to competing processes, such as direct nuclear collisions, that do not produce electron-ion pairs (discussed in Sec. 16.2.7). Hence, the actual pulse height produced by a fission fragment does not scale linearly with its initial energy, unlike the more linear relation for alpha particles. Also, the average charge on a fission fragment reduces rapidly as it passes through a medium as the fission fragment captures ambient electrons. Consequently, the specific ionization of a fission fragment is *highest* upon entering a medium and decreases as the fission fragment loses energy. This changing charge of the ion is quite different than that experienced by alpha particles and protons. Overall, for full energy deposition, the total amount of energy converted to electron-ion pairs for fission fragments is considerably higher than observed by alpha particles, beta particles, or gamma rays.

Theoretical models, developed by Kahn et al. [1965], predict the pulse height spectra produced by fission fragments emitted within a 2π solid angle from a thin film coating of UO_2 on a surface within a gas detector. Selected predictions are compared to measured pulse height spectra in Fig. 17.25. For relatively thin films, the two fission fragment branches are clearly discernible. For thicker films, the energy distribution becomes skewed towards the lower energies, a consequence of energy self-absorption for fission products that lose energy as they transit the UO_2 film before emerging into the detector gas. Because the lightest and most energetic fission products cannot reach the detector gas if born at a distance greater than about 8.0 microns from the coating surface, there is no practical reason to apply films any thicker. Hence, the thermal-neutron detection efficiency for a $^{235}UO_2$-coated detector is limited to less than about 0.5%. Fission chambers with U metal instead of UO_2 are also manufactured, and have a higher density of U per unit volume. However, the fission product ranges in U are also shorter (less than about 6 μm) than in UO_2; consequently, the resultant thermal-neutron detection efficiencies are also less than about 0.5%.

One method to increase the thermal-neutron detection efficiency of fission chambers is to include in the chamber multiple plates each of which is coated with fissile material [Lamphere 1960]. These plates are separated by small gaps usually of about 2 to 5 mm [Bogdzel et al. 1982; Wender et al. 1993]. The chambers are usually backfilled with Ar, often at elevated pressures to improve energy absorption of the fission products in the small gaps. The plates are biased with opposing voltages to collect electron-ion pairs liberated in the gas by the fission fragments.

Fission chambers deployed in a nuclear reactor suffer from burnup just as does the nuclear fuel. Over time the amount of fissile material becomes depleted and the detector response decreases. The initial sensitivity of a fission chamber can be increased by using enriched ^{235}U, but the signal rapidly diminishes with neutron fluence and decreases to approximately half the initial response after an exposure of 10^{21} thermal neutrons cm^{-2} [Jackson 1967]. Consequently, without periodic recalibration, the signal output no longer accurately measures the reactor power. This problem of decreasing sensitivity can be mitigated by including fissile and fertile materials in the fission chamber, thereby producing a type of breeder detector known as a *regenerative* fission chamber. Some regenerative chambers include ^{235}U and ^{234}U [Jackson 1967], ^{239}Pu and ^{238}U [Böck and Balcar 1975], ^{235}U and ^{238}U and natural U with ^{232}Th [Reichenberger et al. 2014]. With appropriate mixing ratios of these isotopes, the signal variation can be constrained to less than 5% for neutron fluences extending, in some cases, beyond 10^{22} cm^{-2}, which correspond to fluences accumulated over several years in the core of a power reactor.

The build-up of fission products inside the chamber increasingly makes the chamber radioactive. Consequently, a background signal from the radioactive emissions of the fission products increases with time, and remains even after the reactor is shut down. This effect is often called the *memory effect* [Roux 1966]. The amount of fission product build-up and memory current is a function of the fissile material mass, neutron flux, and irradiation time. Consequently, the residual memory effect, as described by normalized currents, differs among fission chamber designs. To understand the general effect of fission product build-up, Roux [1966] describes the memory effect as a normalized current which varies with the irradiation time in a nuclear

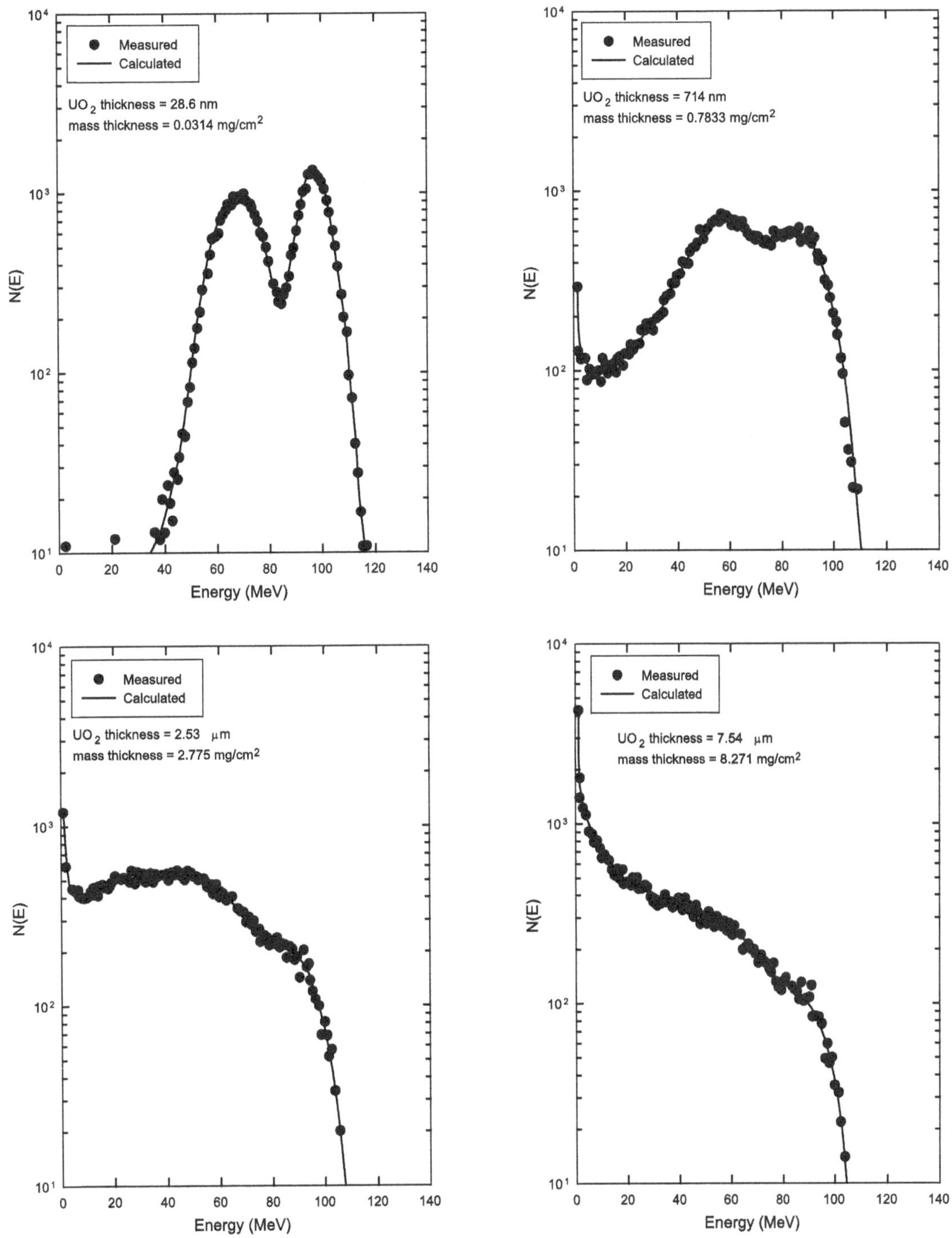

Figure 17.25. Calculated and experimentally measured pulse height spectra of fission fragments from UO_2 film thicknesses of 28.6 nm, 714 nm, 2.53 μm, and 7.54 μm. Data are from [Kahn et al. 1965].

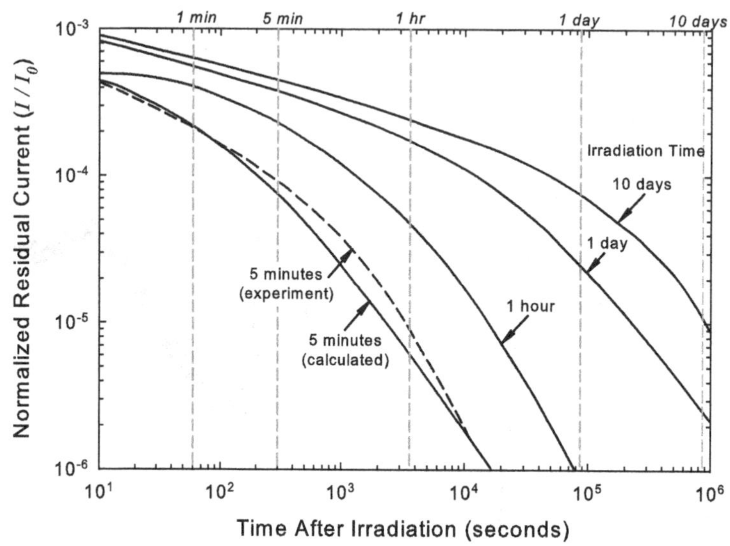

Figure 17.26. Calculated and experimental fission-chamber normalized residual currents from fission products as a function of time after irradiation. Shown are normalized currents for varied irradiation times. Data are from [Roux 1966].

reactor and the subsequent time after irradiation (see Fig. 17.26). A few features can be observed from the Roux results following irradiation: first within a few seconds, the memory current drops by a factor of 10^3, and second it can take many days for the memory current to reduce by a factor of 10^5.

Fission chambers can be operated in pulse-mode, current mode, and Campbelling mode. Because of the multiple operational modes, fission chambers are often used in "wide range" channels in nuclear instrumentation systems. For low flux situations, pulse mode operation offers the advantage that each interaction can be discriminated from background. The disadvantage is that operation is restricted to lower count rates to reduce dead time problems. Pulse mode is used mainly for start-up because it can be used to distinguish between neutrons and gamma rays. At elevated count rates, fission chambers can be operated in current mode, which has the advantage of no longer suffering dead time problems. The measured time-averaged current is proportional to the average power density at the detector location. However, signals from various radiations are no longer distinguishable. The use of Campbelling mode (or mean square voltage mode) does assist with gamma-ray and background discrimination [Campbell and Francis 1946; Vermeeren et al. 2011]. The output signal is still a form of current mode, in which the output is a combination of signals from steady-state and a time varying components. The Campbelling circuit is used to block the steady-state component while squaring the time varying component. The resulting signal is proportional to the square of the charge that is created by each incident particle of radiation, thus enhancing the difference between types of radiation [Geslot et al. 2011]. Hence, the neutron induced portion of the signal is enhanced, while the gamma-ray component is reduced. Thus, Campbell mode operation is useful for neutron measurements in a high gamma-ray background.

Commercial units designed as beam port monitors may have a "pancake" design so that only a small fraction of the neutrons passing through a sensitive volume interact and are detected. Typical diameters of such detectors are about 70 to 115 mm. The purpose of the design is to provide an in-situ neutron measurement of the beam intensity while minimally perturbing the beam.

Cylindrical detectors are fashioned generally in a coaxial-type design, but one in which the inner anode diameter is relatively large, often having nearly the same diameter as the outer cathode. Hence, these detectors are not designed to produce an avalanching field as occurs in common proportional counters. The gap between the anode wall and the cathode wall is typically only a few mm. Consequently, the device performs more as a parallel plate detector rather than a coaxial gas-filled detector (as described in Chapters 10 and 11). Designs may have the cathode or anode coated with the fissile material. In either case, fission fragments are ejected into the small gas volume between the cathode and anode walls and deposit in the gas only a fraction of the total fission product energy. Elevating the gas pressure can increase the energy deposition, although most commercial units are pressurized at only 1 atm. The fill gas is often a combination of Ar and N_2. The argon is more effective at absorbing energy from the fission fragments and serves as the detection gas (see Table 9.1), while the N_2 has a higher electron mobility so as to produce faster pulse rise times (see Fig. 10.10). As shown in Table 9.1, the ionization energy for N_2 is slightly less than that of argon so that charge can be transferred from ionized argon to the nitrogen. The small gas-filled gap limits energy deposition from fission fragments and background radiations. However, because the energy deposited in this gap by fission fragments is significantly larger than that deposited by beta particles, gamma rays, and alpha particles, pulse height discrimination can be used to distinguish between neutron induced events and background.

Sub-miniature fission chambers (SMFC) have been under development for in-core nuclear reactor monitoring for many years [Poujade and Lebrun 1999; Blandin et al. 2003; Lamirand et al. 2014]. These detectors are manufactured with a traditional coaxial design but with the distinction that the outer diameter is approximately 1.5 mm. The anode is also thicker in diameter than that required for proportional counters and is coated with a enriched fissile material such as 97% ^{235}U [Blandin et al. 2003]. The sensitive lengths of such SMFCs are approximately 1 cm. The detectors are backfilled with Ar gas pressurized between 1 to 1.5 atm. Sub-miniature detectors are designed to operate in high neutron fluxes in the core of a nuclear reactor with some designs now under investigation for transient measurements. Commercial variations of the sub-miniature design are available.

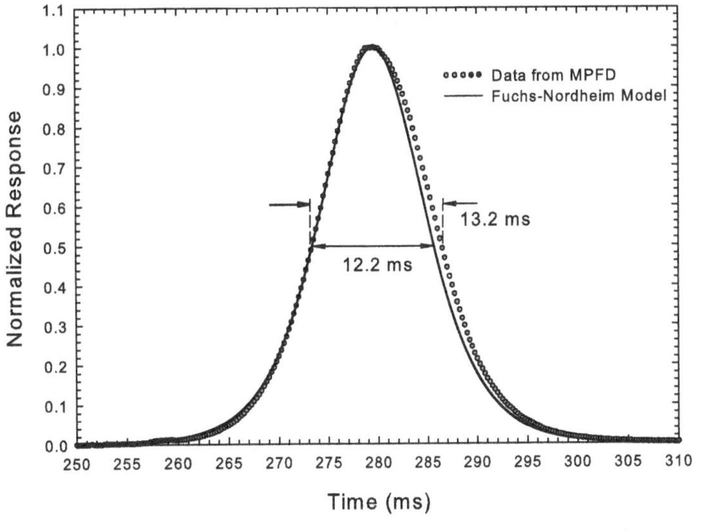

Figure 17.27. Pulse shape comparison of the Fuchs-Nordheim model prediction and measured results from an MPFD for a \$2.77 reactivity insertion. The experiment was conducted in a TRIGA nuclear reactor with thermal peak flux of 9.28×10^{15} cm^{-2} s^{-1}. Data are from [Nichols et al. 2018].

Another type of miniaturized fission chamber is the micro-pocket fission detector (MPFD). These tiny devices are significantly smaller than normal fission chambers and have dimensions of approximately 1 mm with gas volumes of 0.5 mm^3 or less. Although the small volumes are incapable of absorbing the total energy from any of the fission products, the amount of energy absorbed is about 3 to 7 MeV for all fission products, regardless of the branch [McGregor et al. 2005]. The small volume ensures that background gamma-rays and beta particles deposit less than 1 keV, too small to be detected. Alpha particle emission from the reactive coating can deposit more energy, but also is small, typically only 60 keV or less. Hence, simple pulse height discrimination can be used to separate neutron-induced events from the radiation background.

MPFDs are intentionally designed to be inefficient because they are intended as in-core instrumentation for nuclear reactors. Because of their small size, they can be inserted into tiny test ports without causing flux perturbations. Dead time is also less of an issue, mainly because the electron-ion drift times are much smaller for these detectors than traditional fission chambers. These detectors have been operated in pulse mode for neutron fluxes ranging up to 10^{13} n cm^{-2} s^{-1} [McGregor et al. 2005].

Variants of these miniaturized fission chambers are under investigation for different reactor cores and applications [Reichenberger et al. 2016, 2017]. One variant eliminates the need to plate fissile material on an electrode and uses a parallel electrode wire geometry to collect the charges from the miniature gas pocket. With reactor excursion experiments, current mode operation shows that these detectors can track reactor excursions up to a thermal flux of at least 9.28×10^{15} n cm^{-2} s^{-1} as demonstrated in a TRIGA nuclear reactor [Nichols et al. 2018]. When compared to the Fuchs-Nordheim model [Fuchs 1946; Nordheim 1946], the experimental results match well to the model (see Fig. 17.27), showing experimentally a FWHM of 13.2 ms compared to the predicted value of 12.2 ms.

17.4 Scintillator Slow Neutron Detectors

Scintillators used for slow neutron detectors can be categorized into two groups, (1) those composed of a single compound that both fluoresces and also interacts strongly with thermal neutrons and (2) those composed of a mixture of two compounds, one of which fluoresces and the other of which contains neutron reactive material. For instance, LiI and CLYC are in group 1, whereas boron loaded plastic and liquid scintillators are in group 2. Some examples are briefly discussed here. Greater detail about these scintillators is provided in Chapter 13 and also by Van Eijk et al. [2004] and Birowosuto [2007].

17.4.1 Neutron Reactive Scintillators

LiI Scintillator

LiI:Eu crystal, described in Chapter 13, is a highly hygroscopic alkali-metal halide scintillator, recognized by Hofstadter et al. [1951] for its unique properties suitable for thermal and slow neutron detection. It has a density of 4.076 g cm^{-3} and a relative molecular mass of 133.89. The natural abundance of ^{6}Li is 7.59% with a 2200-m/s (n,t) cross section of 938.0 b.[14] It then follows that the macroscopic (n,t) cross section for 2200-m/s neutrons is 1.31 cm^{-1}; hence over 70% of thermal neutrons are absorbed in a crystal 1 cm thick. The attenuation length can be decreased by growing LiI enriched with ^{6}Li. The most probable emission wavelength λ_{max} is approximately 475 nm. As with many Eu activated scintillators, LiI is a relatively slow scintillator with a decay constant of 1.4 μs so that its response rate is somewhat limited. The use of Tl as the activator introduces fluorescence with short decay times [Khan et al. 2015], although a relatively long fluorescence component persists. Further, the light yield decreases compared to LiI:Eu with a most probable emission wavelength λ_{max} remaining near 475 nm for activator concentrations between 0.02% and 0.5% Tl.

Not only is LiI a good neutron absorber, the high iodine content also makes it a good gamma-ray absorber. Consequently, LiI neutron detectors can also have a relatively high sensitivity to background radiation counts. Because the ^{6}Li(n,t)^{4}He reaction releases 4.73 MeV, often simple pulse height analysis is enough to distinguish between gamma-ray and neutron events. However, pulse height discrimination may not be an adequate discrimination technique in the presence of gamma-ray energies of approximately 4 MeV or greater, or in intense gamma-ray environments that induce pulse pile-up [Nicholson and Snelling 1955].

[14] This cross section was often called the *absorption* cross section, a term no longer in wide usage because it includes fission plus 16 reactions in which the neutron disappears such as (n,γ) and 15 reactions that produce positively charged reaction product(s) such as (n,p), (n,t), (n,2α), and (n,t2α). However, here the total 2200-m/s is 938.8 b, an indication that the (n,t) reaction is essentially the only reaction.

LiI:Eu has been reinvestigated in recent years as a compact neutron imager. LiI grown as a microcolumnar scintillator has been developed with the impetus of using the natural fiber optic structure for light segregation [Nagarkar et al. 2001]. Enriched to 96% in ^{6}LiI, crystals were grown on fiber optic substrates to promote light propagation to a light sensing device, in this case, a charged coupled device (CCD). The LiI films were 1.2 mm thick with 30 μm diameter microcolumns. Measurements indicated that a spatial resolution of 110 μm is achievable.

LiI:Eu scintillators attached to either a Si PIN diode [Pausch and Stein 2008] or a SiPM [Foster and Ramsden 2008] have been investigated as compact neutron detectors. The advantage of such a structure is the lower operating voltages required and the smaller overall volume. Pausch and Stein [2008] report results from a 3-mm-thick LiI:Eu detector exposed to a mixed gamma-ray/neutron field, in which a moderated AmBe source and ^{137}Cs and ^{60}Co sources were used. It was found that counts from gamma-rays interacting in the LiI scintillator or directly interacting in the semiconductor diode can be significant and, consequently, lead to misinterpretation of gamma-ray counts as neutron counts. Pausch and Stein [2008] report a pulse processing method that can discriminate between gamma-ray and neutron events. Foster and Ramsden [2008] report the appearance of artifacts in the pulse height spectrum when the LiI/SiPM combination was exposed to a mixed gamma-ray/neutron field, in which a modestly moderated ^{252}Cf source and a ^{137}Cs source were used. It is less likely that the artifacts are from direct gamma-ray interactions in the SiPM, mainly because the depletion region in a SiPM is relatively small (smaller than in a PIN diode). Instead, it is hypothesized that the artifacts result from neutron scattering from Li nuclei [Foster and Ramsden 2008]. Also observed was a strong dependence of the neutron induced pulse height spectrum and the operating temperature. Foster and Ramsden [2008] argue that the thermal dependence is of little consequence, mainly because the pulse height spectrum is only used for a LLD setting and not for spectroscopy, and that bias adjustments to the SiPM can help compensate thermal changes in the pulse height spectrum.

Elpasolite Scintillators

Cesium lithium yttrium chloride (Cs_2LiYCl_6 or CLYC) and other elpasolites containing Li can be used for slow neutron detection [van Eijk et al. 2005]. Lithium-based elpasolites are sensitive to both neutrons and gamma rays, and have been used to detect both. For low count rate measurements, the relatively good energy resolution of CLYC:Ce and also cesium lithium yttrium bromide (Cs_2LiYBr_6 or CLYB) allows pulse height discrimination (PHD) between neutron and gamma-ray counts, just as with LiI:Eu detectors. The α/β ratio for CLYC:Ce is 0.66, an indication that pulse shape discrimination (PSD) could be used to discriminate between gamma rays and neutrons. CLYC:Ce also possesses core valence luminescence (CVL) that produces only a fast component from gamma-ray events. The same is not true for neutron events. Hence, PSD can be performed by simply distinguishing between events with and without CVL. Note that the long decay times characteristic of elpasolites can be a problem in high mixed radiation fields for either PHD or PSD, mainly because pulse pile-up contaminates the neutron-induced spectrum. The elpasolite $Cs_2LiLaBr_6$ (or CLLB) has better energy resolution than CLYC:Ce [Glodo et al. 2011], but apparently lacks core valence luminescence [Bessiere et al. 2005]. However, neutron PSD can be achieved by comparing the decay time of the decay components (as described in Chapter 13), mainly because the slow components differ with the radiation type, where the induced gamma-ray tail has a longer duration than the neutron induced tail.

CLYC:Ce can be obtained commercially and can be coupled to either a traditional PMT or a SiPM. Spectral performance is good with approximately 4.5% FWHM for 662-keV gamma rays. It performs adequately well for gamma-ray spectroscopy and neutron detection. However, for high energy resolution gamma-ray spectroscopy, there are other scintillators that can outperform CLYC:Ce (such as $LaBr_3$). The same is true for compact neutron detectors. Hence, the main novelty of CLYC:Ce is the ability to detect efficiently both gamma rays and neutrons with a single device.

A significant disadvantage to CLYC:Ce is that both ^{133}Cs (100% natural abundance) and ^{35}Cl (75.78% natural abundance) are moderate neutron absorbers that transmute into radioisotopes that emit beta particles. Further, their atomic ratios per molecule are much higher than Li. For natural constituents, approximately 78% of neutron absorptions in CLYC:Ce do not produce a spontaneous neutron reaction, but rather increase the background from emissions of beta particles and gamma rays. Consequently, the intrinsic thermal-neutron detection efficiency of CLYC:Ce is limited to approximately 22%. If natural Li is replaced with enriched ^{6}Li, the thermal-neutron detection efficiency has an upper limit of 80%. To remedy this problem, other elpasolites without Cs and Cl have been explored and a successful candidate was found to be Rb_2LiYBr_6 (or RLYB:Ce) [van Eijk et al. 2005]. Overall, the light output and performance of RLYB:Ce is similar to CLYC:Ce although, at the time of this writing, it is not commercially available. Another introduction to the elpasolite family with promising results is $Cs_2LiLa(Br_{6-x}Cl_x)$ or CLLBC:Ce, which has good gamma-ray energy resolution along with the neutron detection characteristic of CLYC:Ce [Shirwadkar et al. 2012]. The substitution of Br for Cl works to reduce parasitic neutron losses, while the substitution of La for Y works to slightly increase parasitic neutron losses. CLLB:Ce has become commercially available with reported energy resolution for 662 keV gamma rays of approximately 4% FWHM.[15] The popularity of CLLB:Ce is likely to increase in the coming years because of its dual application to neutron and gamma-ray detection.

LGBO Scintillator

Many other scintillators with reactive components have been investigated. For instance, $Li_6Gd(BO_3)_3$:Ce scintillators with varying concentrations of Gd, B, or Li enrichments have been studied [Czirr et al. 1999]. $Li_6Gd(BO_3)_3$:Ce can be produced as a powder or as a transparent crystal [Czirr et al. 1999]. Three of the main constituents react with neutrons, namely, ^{6}Li(n,t)^{4}He, ^{10}B(n,α)^{7}Li, and ^{155,157}Gd(n,γ)^{156,158}Gd. The neutron absorption response of this interesting scintillator can be changed by adjusting the ratios of the neutron absorbing nuclides to their non-reactive isotopes. When using materials of natural abundance, the Gd component dominates neutron absorption. Consequently, the reaction products are mainly gamma-rays and low energy conversion electrons, and the Li and B components contribute little to the response. Recapture of the gamma rays is inefficient and the conversion electrons have short ranges, a property which unfortunately limits the response. Van Eijk [2001] reports that natural $Li_6Gd(BO_3)_3$:Ce produces a low-end average light yield of 14,000 photons/neutron. Note that this average is a compilation of three uniquely different emissions from the four possible reactions. According to Czirr et al. [1999], light emission ratios for Gd, B, and Li per neutron are approximately 2, 11, and 52, respectively. In these results the representative conversion electron energy of 85 keV has been substituted for the Gd emissions (see Schulte et al. [1994]). Hence, the pulse height from ^{6}Li(n,t)^{4}He is significantly higher than those from other reactions.

If the Gd is depleted of ^{155}Gd and ^{157}Gd, then the boron component can dominate the absorption. If the ^{10}B is then also omitted, the Li component dominates neutron absorption even further. Czirr et al. [1999] conclude that a good compromise is to produce $^6Li_6{}^{[nat-(155,157)]}Gd(^{11}BO_3)_3$:Ce so as to take advantage of the high energy reaction products from ^{6}Li and the resultant light yield of 56,900 photons/neutron. The reaction product ranges, and subsequent light attenuation lengths, are still problematic in a manner similar to that for coated gas-filled detectors. In an attempt to overcome this problem, Allier [2001a, 2001b] suggest Si diode well-structures backfilled with $Li_6Gd(BO_3)_3$:Ce powder. In summary, Van Eijk [2001] notes that advantages include the low index of refraction of 1.66 and the high neutron absorption efficiency. Disadvantages include the longer decay times that range from 200 ns to 800 ns and the high cost of Gd depleted of ^{155}Gd and ^{157}Gd.

[15] Through Saint-Gobain Crystals.

GSO Scintillator

Gadolinium orthosilicate (Gd_2SiO_5:Ce or GSO) is another gadolinium-based scintillator that has been studied as a possible neutron detector [Reeder 1994a, 1994b; Uozumi et al. 1997]. This scintillator has a fast decay component (56 ns) that is responsible for 85-90% of the light yields and a slow component (600 ns) that is responsible for the remaining 10-15% of the light yield [Melcher et al. 1990]. The maximum wavelength is 430 nm and matches well to bialkali PMTs. However, the light yield is relatively low at only 12,500 photons/MeV. Reeder [1994a] notes that the pulse height spectrum peak is approximately 77 keV equivalent and matches well to the Gd conversion electron spectrum [Schulte et al. 1994]. Because GSO is sensitive to both gamma-rays and neutrons, it can be used to detect both. Reeder [1994a] lists advantages of GSO as simplistic because it has dual detection capabilities as well as compact and efficient because a small crystal readily absorbs neutrons. Disadvantages include competition between low-energy background gamma rays and the Gd conversion electron emissions as well as contamination by characteristic x-rays from the shielding (especially from Pb).

Because conversion electrons, Compton electrons, and photoelectrons produced by neutrons and gamma rays have fundamentally similar properties, discriminating between neutron and gamma-ray events with pulse shape discrimination does not appear to be an option [Uozumi et al. 1997]. Further, the complex nature of the combined spectra makes simple pulse height discrimination difficult. Reeder [1994b] attempts to address this problem by coupling thin GSO crystals to PMTs, so thin that the probability of gamma-ray interactions is significantly reduced below that of thermal-neutron absorptions, all the while retaining the ability to capture the conversion electron reaction product emissions. To demonstrate the concept, Reeder [1994b] compared the response of a 60 μm thick GSO crystal and a 1 cm thick GSO crystal to gamma-ray and neutron emissions from a moderated PuBe source, as well as gamma-ray emissions from a ^{137}Cs source. Although the work qualitatively demonstrates the concept, the sample areas (and solid angles) were different, thereby making quantitative conclusions difficult. Regardless, it is clear that thick crystals of GSO neccessarily have problems with background gamma rays interfering with the neutron-induced pulse height spectrum.

$LiBaF_3$ Scintillator

The scintillator $LiBaF_3$ has also been studied as a neutron detector [Knitel et al. 1996; Combes et al. 1998]. $LiBaF_3$ has multiple emission wavelengths with different decay times, ranging from 1 ns up to 13 μs depending on the doping type and concentration [van Eijk 2001]. The light yields and decay times are significantly different for gamma-ray and neutron interactions. A bright fast component is produced by gamma-ray interactions (about 1,200 photons/MeV with a mean lifetime τ of about 0.8 ns), which is absent for neutron interactions [Knitel et al. 1996]. A method that rejects these high amplitude fast pulses is used to discern between neutron and gamma-ray events [Knitel et al. 1996; Combes et al. 1998]. Van Eijk [2001] argues that the fast emission produced by gamma-ray interactions is due to core-valence luminescence, which is accompanied by the delayed luminescence from self-trapped holes. The fast core-valence luminescence is altogether missing for neutron interactions. Combes et al. [1998] report that the use of radiation damaged $LiBaF_3$ crystals have improved discrimination between gamma-ray and neutron events. Along with co-doping with Ce and Rb, the damaged material produced a fast decay time for the self-trapped hole luminescence, thereby reducing issues with pulse pile-up while improving the signal to noise ratio [Combes et al. 1998].

17.4.2 Scintillators Loaded with Neutron Reactive Materials

Organic Scintillators

Organic scintillators loaded with either ^{10}B, ^{6}Li, or natural Gd are commercially available as neutron detectors. These materials are usually applied to fast neutron measurements, but can also be used for thermal or slow neutron detection. The concept is simple; neutrons are absorbed in the absorber, either ^{10}B or ^{6}Li, which release energetic reaction products that subsequently cause fluorescence in the scintillator. These detectors can be economically manufactured in large sizes and formed into multiple shapes. The light emission can be significant, mainly because of the high Q-values from the ^{10}B(n,α)^{7}Li and ^{6}Li(n,t)^{4}He reactions.

The amount of boron that can be loaded into an organic plastic scintillator (polyvinyltoluene) is limited by the reduction in scintillator transparency. Loading is usually no more than 5% by weight for commercial boron-loaded organic scintillators, and these scintillators have lower light yields than their unloaded counterparts by approximately 20% [Drake et al. 1986] (see also Table 13.2). Commercial plastic scintillators, such as EJ-254 and BC-454, are loaded with natural boron; hence, the actual ^{10}B loading amounts to approximately 1%. Higher boron loadings are available by special request. Although the Q-value of the boron reaction is high, in Chapter 13 it was mentioned that light yield in organic scintillators is lower for heavy ions than for electrons. The work of Verbinski et al. [1968] can be used to find an approximate light yield for the ^{10}B(n,α)^{7}Li reactions products compared to energetic electrons. For instance, the scintillation signals from ^{10}B(n,α)^{7}Li reactions are approximately equivalent to that of 76-keV electrons [Drake et al. 1986; Eljen 2016a]. Consequently, gamma-ray background can interfere with neutron measurements. For fast neutrons, the initial (n,p) reaction in the plastic produces a prompt light response, followed by a time delay of approximately 2.7 μs before the light response from a ^{10}B(n,α)^{7}Li reaction appears. By measuring both light emission pulses from the proton recoil and the subsequent ^{10}B(n,α)^{7}Li reaction products, fast neutron interactions can be readily identified [Drake et al. 1986].

Liquid Scintillators

Liquid scintillators loaded with neutron reactive absorbers are also commercially available, and, in many ways, are more attractive as slow neutron detectors than are plastic scintillators. For liquid scintillators loaded with boron, pulse shape discrimination methods can be used to distinguish between gamma-ray and neutron events [Chou and Horng 1993]. Commercial products are available loaded with either natural or 90% enriched ^{10}B. Loading is typically 5% by weight, which yields 1% ^{10}B loading by weight for natural boron and 4.6% ^{10}B loading by weight for fully enriched boron. A variety of loadings can be obtained by special request. For electron and gamma-ray events, the light yield is approximately 65% that of anthracene, which is similar to that of other organic scintillators. The pulse height from a ^{10}B(n,α)^{7}Li event is approximately equivalent to a 90-keV electron pulse [Eljen 2016b]. For fast neutrons, the average capture time is a function of the ^{10}B loading, ranging from 0.3 μs for 5% enriched ^{10}B loading to 1.4 μs for 5% natural boron loading. Also, there is on average a 2.7 μs delay between the initial (n,p) reaction emission and the ^{10}B(n,α)^{7}Li emission, a delay which permits neutron identification with delayed time gating methods.

Liquid scintillators loaded with natural Gd rely on prompt gamma rays and conversion electron emissions from the ^{157}Gd(n,γ)^{158}Gd and ^{155}Gd(n,γ)^{156}Gd reactions. These scintillators are used for neutron spectroscopy and neutrino detection. Gd loading for commercial products is typically 0.5% by weight, but special Gd concentrations are available.

Because liquid scintillators are often employed in large containers, the solvents are chosen to have high flash points to reduce the risk of combustion.[16] Also, liquid scintillators can be contaminated with water moisture, a condition that compromises performance. Hence, the user must take care to keep exposure to air at a minimum. Typically these liquid scintillators are packaged under an inert environment (gas).

Ceramic Neutron Sensitive Scintillators

Li Glass Scintillators Cerium activated silicate glasses were introduced in the late 1950s for radiation detection [Ginther and Schulman 1958]. Although boron loaded glasses were also investigated for neutron detection [Bollinger et al. 1959], it is the Li-loaded class of silicate glasses [Voitovetskii et al. 1960a, Voitovetskii and

[16] The flash point of a liquid is the temperature at which the liquid has sufficient vapor pressure in equilibrium with the liquid to just form a flammable atmosphere.

Table 17.2. Properties of Ce activated ^{6}Li-Glass (nat. Li/ 95% enriched ^{6}Li/ 99.99% enriched ^{7}Li).

Property	GS1/GS2/GS3	GS10/GS20/GS30	KG1/KG2/KG3
Total Li by weight	2.4%	6.6%	7.5%
Density g cm^{-3}	2.66	2.5	2.42
Refractive index	1.58	1.55	1.57
Melting point (K)	1475	1475	1475
Max emission λ (nm)	395	395	395
Relative Light Yield*	22%–34%	20%– 30%	20%
Decay times, neutron (ns)**	21, 69, 115	18, 57, 98	18, 62, 93
Decay times, alpha (ns)**	20, 65, 106	16, 49, 78	15, 45, 56
Decay times, beta (ns)**	21, 62, 114	20, 58, 105	17, 51, 96
Available thickness (mm)	0.1–10	0.1–10	0.1–10
Resolution of peak for moderated PoBe neutrons	13%–22%	15%–28%	20%–30%
Linear attenuation coefficient for thermal neutrons (cm^{-1})	0.393/4.95/0.0002	1.016/12.79/0.0006	1.118/14.07/0.0007

*Relative to anthracene.

**Fast component, slow component, 90%–10%, respectively. Data for GS2, GS20, and KG2 glasses only.

Tolmacheva 1960b; Ginther 1960; Bollinger et al. 1962] that have been researched and widely commercialized for neutron detection. The scintillation mechanisms and properties of glass scintillators were briefly covered in Sec. 13.2.3, with more details on the properties of Li-loaded Ce-activated silicate glasses detailed in the literature [Spowart 1976, 1977; Fairley and Spowart 1978].

Li-loaded glass scintillators, activated with Ce, are commercially available in a variety of concentrations and enrichments, as summarized Table 17.2.[17] The absorption characteristics of such glasses are shown in Fig. 17.28. These glass scintillators can be obtained in various sizes, thicknesses, and ^{6}Li doping concentrations. Perhaps the most common of the Li glass materials used by commercial instrumentation companies is GS20 with 6.6% doping of 95% enriched ^{6}Li. GS20 has a maximum emission wavelength of 395 nm with between 20% and 30% of the light yield compared to that of anthracene. The light yield for GS20 glass ranges between 3,270 photons/MeV to 5,550 photons/MeV, depending upon the glass thickness and scintillation yield standard. For full energy absorption within the scintillator, the total light yield per interaction ranges between 15,450 and 26,200 photons/MeV centered about a wavelength of 395 nm. There are other Li glasses available, with a general labeling convention in which '1' designates natural lithium, '2' designates 95% enriched ^{6}Li, and '3' designates 99.99% enriched with ^{7}Li, which is insensitive to neutrons.

The decay constants are a function of radiation type and the Li loading, and generally have fast and slow decay components [Fairley and Spowart 1978]. Yet, from Table 17.2, these decay times vary only small amounts between radition types. There are varied Li loadings, with commercial codes [Spowart 1976, 1977] listed as 2.4% (GS1,GS2,GS3), 6.6% (GS10,GS20,GS30) and 7.5% (KG1,KG2,KG3) by weight. The ranges of the ^{6}Li(n,t)^{4}He reaction products in Li glass, tritons and α particles, are approximately 36.3 microns and 6.3 microns, respectively. Absorptions near the glass surface can lead to reaction products escaping the scintillator and decreasing the light output, an effect known as the "edge effect" [Yamaguchi 1989]. This situation can be remedied by fastening sheets of ^{7}Li-glass, such as GS3, GS30, or KG3, that is depleted of ^{6}Li

[17] The traditional designations for the different Li glass scintillators come from Levy West Laboratories, which became Applied Scintillation Technologies and is now named Scintacor. The 'GS' designation is, presumably, named for Ginther and Schulman.

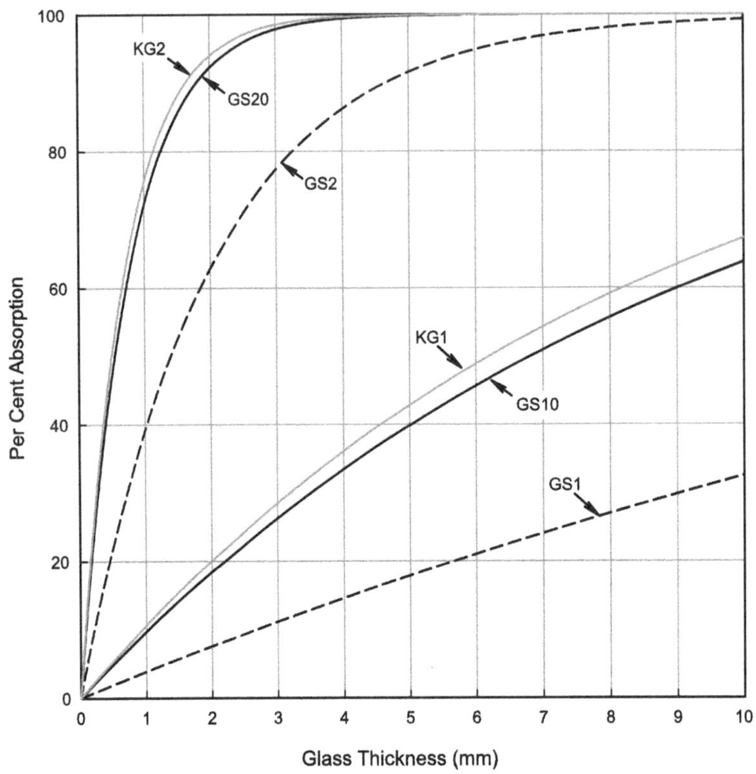

Figure 17.28. Absorption characteristics of 2200 m s^{-1} neutrons in various Ce activated Li glasses.

to 99.99% on both sides of the neutron sensitive Li glass. Reaction products escaping the neutron sensitive region fluoresce the ^{7}Li-glass and, thus, allow emitted photons to be recovered. Lithium glasses provide a convenient way to produce large area sheets of a neutron scintillator and are used to detect both thermal and fast neutrons. They can be used under high pressure and at high temperatures because of their high melting point of 1475 K. Spowart [1977] reports a luminescent thermal dependence that reaches a maximum near 450 K, reducing with either higher or lower temperatures.

Gadolinium Oxysulfide Ceramics Gadolinium oxysulfide (Gd_2O_2S, GOS or Gadox) is used for thin scintillator screens. It is available as a powder, but can be hot pressed into a translucent ceramic [Greskovich and Duclos 1997]. Co-dopants include Pr, Ce, and F, and have a central emission wavelength of 580 nm and emit 40,000 photons/MeV [Lecoq et al. 2006]. The emission decay time is relatively long at 2.1 μs. Gadox is primarily used for digital radiography imaging panels for x-ray excitations. However, its elemental Gd constituent produces a high neutron absorption efficiency; hence, it has been evaluated as a neutron scintillator. Because Gd is a good absorber of photons and neutrons, and the resulting reaction products are nearly indistinguishable (Compton electrons, photoelectrons, conversion electrons), there are difficulties in distinguishing between neutron and background gamma-ray events.

However, because Gd is such a strong thermal-neutron absorber ($\sigma_{\rm Gd}^{\rm nat}(\rm n,\gamma) = 48,610$ b at 2200 m/s), only a thin layer is necessary for adequate performance. For instance, the density of Gadox is 7.34 g cm^{-3} with a corresponding 2200-m/s macrocroscopic cross section of 1,360 cm^{-1}. These values indicate that a layer only 10 μm thick absorbs over 85% of intersecting thermal neutrons. Lindsay et al. [1986] report the use of a thin layer of Gadox on a LIXI (Light Intensifier X-ray Image) scope, in which the background from gamma-ray interactions is suppressed for real-time neutron radiography. Kardilov et al. [2011] report methods to improve performance of Gadox films for neutron imaging by imbedding the scintillator in sodium silicate (or "water glass") and applying an Al reflective coating. Greskovich and Duclos [1997] report that the hexagonal crystal structure of Gadox limits light collection, a consequence of optical anisotropy, residual porosity, and foreign phases from densification additives. They also report that Gadox is susceptible to x-ray damage, more so than many other ceramic scintillators.

Zinc Sulfide Scintillator Zinc sulfide activated with silver is a bright scintillator (see Chapter 13) with a light yield that is approximately 130% of that of NaI:Tl. This material is often mixed with neutron reactive materials to produce a thermal-neutron detector. For instance, LiF or B_2O_3 mixed with ZnS:Ag make effective scintillating films. ZnS:Ag is available mainly in powder form, and it is also opaque to its own light emissions, a property that limits the working thickness of the material to no more than 25 mg cm^{-2} or about 60 microns. The most probable light emission wavelength is centered at 450 nm and matches adequately well to most PMTs. The index of refraction is 2.36 at 450 nm that can become a significant source of internal reflection if coupled directly to glass or PMMA ($\theta_c = 39.5\,^\circ$). ZnS:Ag is often (incorrectly) stated to be insensitive to gamma rays and beta particles. Actually, ZnS:Ag is sensitive to both gamma rays and electrons; however, the amount of energy deposited in a thin film of only 60 microns or less by an energetic electron is relatively small and easily distinguished from heavy ion reaction products from either ^{6}Li(n,t)^{4}He or ^{10}B(n,α)^{7}Li by pulse height discrimination.

17.5 Semiconductor Slow Neutron Detectors

Semiconductor slow-neutron detectors can be classified as either coated detectors or bulk detectors. Coated detectors are electron devices, usually Schottky or junction diodes, upon which a neutron reactive coating has been applied. Bulk detectors are semiconductors with one or more neutron reactive constituent isotopes. There are variations of both types [Caruso 2010], and a few of the more important structures are described below.

Coated Semiconductor Neutron Detectors

Coated semiconductor neutron detectors are generally fashioned as semiconductor diodes with a neutron reactive coating applied to the rectifying contact.[18] The first such detector was a small Si *pn* junction diode, only a few square milimeters in area, with a ^{10}B film vapor deposited upon the rectifying junction [Babcock et al. 1959]. Although studied for decades, these coated semiconductor neutron detectors were viewed as novel devices for dosimeter packages, primarily because of their inherently low efficiencies. However, recently, with the introduction of microstructured semiconductor neutron detectors (MSNDs), they are receiving renewed attention because of their high neutron detection efficiencies. Overall, there are two main types of coated semiconductor neutron detectors, namely planar diode detectors and MSNDs.

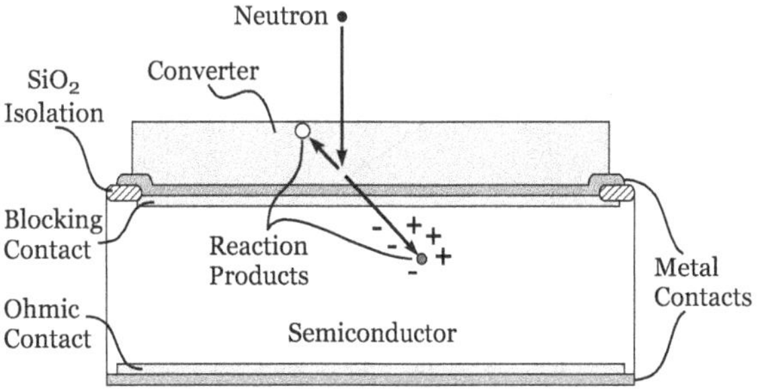

Figure 17.29. The basic construction of a coated planar semiconductor diode neutron detector.

[18] In some of the original literature on these devices, they are called "foil detectors" because a thin foil of the neutron reactive material was positioned adjacent to the semiconductor diode detector. However, almost all detectors of this sort now have the reactive material applied directly to the rectifying junction of the semiconductor diode; hence, "foil detector" is no longer a suitable description.

Planar Diodes Thin film neutron detectors consist of semiconductor diodes, preferably with relatively thin contact layers, upon which a layer (or layers) of neutron reactive material has been deposited. The basic concept has been used by a variety of research groups (see the review by Caruso [2010]). As shown in Fig. 17.29, neutrons absorbed in the neutron reactive film release charged-particle reaction products in opposite directions, one of which may enter the semiconductor diode detector. Charged particles entering the detector lose their energy through Coulombic scattering, thereby creating a high density cloud of columnar ionization in the form of electron-hole pairs. The semiconductor diode detector is voltage biased to separate the electron-hole pairs and drift the charges to their respective contacts. The mobile charges each induce an image charge on the contacts as they move through the device, and the induced charge is integrated and measured by an external preamplifier and accompanying electronics. The basic structure of a planar coated semiconductor neutron detector is depicted in Fig. 17.29 that shows neutron reactive material fastened to the electrical contact adjacent to the diode rectifying contact. Advantages of these detectors include a compact size and low operating voltages and, thus, make possible the production of compact neutron dosimeters [Aoyama et al. 1992; Sasaki et al. 1998; Ndoye et al. 1999] and compact fieldable instrumentation [Schulte and Kesselman 1999]. It is also possible to integrate multiple detectors into an array upon a single substrate to create compact linear neutron imaging arrays [Campbell et al. 1997; Schelten et al. 1997; Petrillo et al. 1999; McGregor 2002b] or 2-dimensional neutron imaging arrays [Mirashghi et al. 1992, 1994; McGregor et al. 1996b; Unruh et al. 2009].

Semiconductor diodes coated with a neutron reactive material have been studied as neutron detectors for well over fifty years and date to the original work by Babcock et al. [1959]. Most of the devices reported in the literature either have a coating of ^{10}B [Rose 1967] or ^{6}LiF [Pospíšil et al. 1993], although Gd [Feigl and Rauch 1968] and pure Li metal coatings [McGregor et al. 2003] have also been explored. Si is the semiconductor substrate most often used as the substrate, mainly because it is relatively inexpensive and has a small thermal-neutron interaction cross section of $\sigma_{(n,\gamma)} = 0.17$ b. However, gallium arsenide (GaAs) [McGregor et al. 2000] and chemical vapor deposited (CVD) diamond [Foulon et al. 1998] have also been explored as semiconductor substrates for coated neutron detectors. Simple semiconductor diodes coated

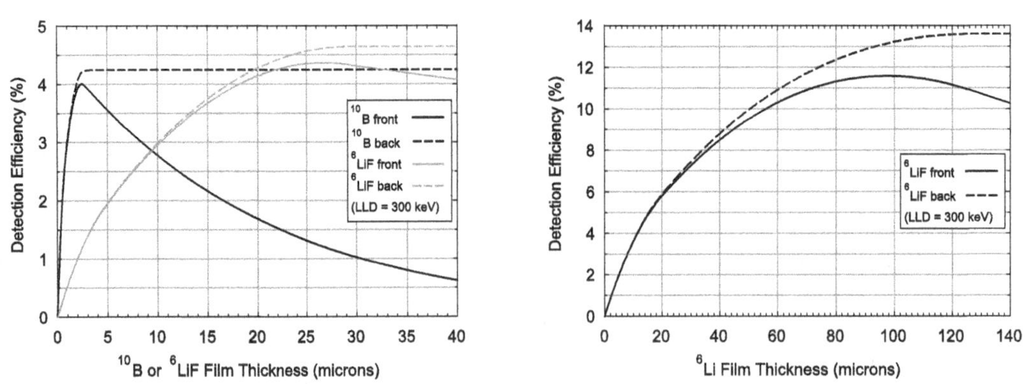

Figure 17.30. (left) Thermal neutron (2200 m s^{-1}) detection efficiency vs. film thickness for front and back irradiation of a Si detector coated with ^{10}B or ^{6}LiF. (right) Thermal neutron (2200 m s^{-1}) detection efficiency vs. film thickness for front and back irradiation of a Si detector coated with ^{6}Li metal. [from McGregor et al. 2003].

with either ^{10}B or ^{6}LiF are limited to approximately 5% thermal-neutron detection efficiency [McGregor et al. 2003], and consequently have received limited application as neutron detectors. Detectors can be stacked in various configurations to improve efficiency [McGregor et al. 2003], similar to stacking gas-filled neutron detectors. The same equations, Eqs. (17.34) to (17.38), used to calculate efficiencies for coated gas-filled

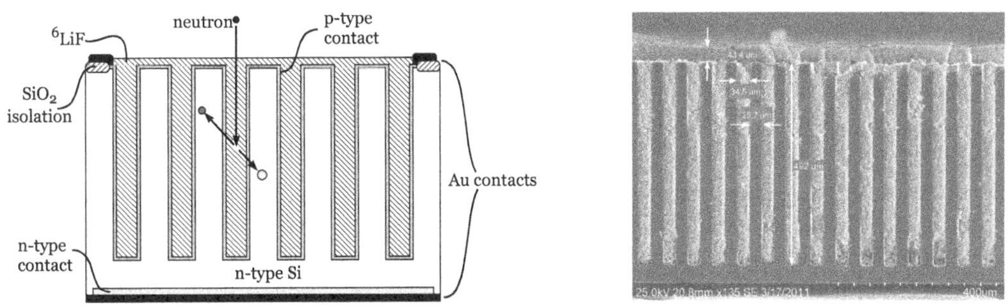

Figure 17.31. (left) The basic structure of an MSND. (right) An MSND with 492 μm deep trenches, each 25.7 μm wide, backfilled with ^{6}LiF nanoparticles. The Si fins are 34.5 μm wide. [from McGregor et al. 2013b; McGregor et al. 2015].

detectors also apply to coated-semiconductor detectors, the results of which are shown in Fig. 17.30. The efficiency can be improved by using pure ^{6}Li metal, in which the thermal neutron detection efficiency can be increased above 13% [McGregor et al. 2003]. However, pure Li metal is reactive and requires moisture proof encapsulation in an inert environment. Semiconductor detectors coated with natural Gd have also been explored [Feigl and Rauch 1968; Schulte et al. 1994; Schulte and Kesselman 1999]. The appeal of using Gd is because of its large thermal-neutron (n,γ) cross section of 48,610 b, but the resulting low-energy conversion-electron and gamma-ray emissions can be difficult to distinguish from background gamma rays in a high radiation field. A compensated detector design can help discern between background gamma rays and neutron signals [McGregor et al. 2002b].

Microstructured Diodes Muminov and Tsvang [1987] suggested that channels etched in a semiconductor substrate backfilled with neutron reactive material might increase the neutron detection efficiency. The first such device was reported by McGregor et al. [2002a], and consisted of an array of holes etched into a GaAs substrate backfilled with enriched ^{10}B nanoparticles. Although these initial detectors exhibited modest efficiencies, they indicated the technology of etching microstructures into the semiconductor held promise. Since that time, these Microstructured Semiconductor Neutron Detectors (MSNDs) have matured into a highly efficient and compact type of neutron detector.

The basic MSND has microcavity features extending into a semiconductor surface. Within and around the microstructures is a rectifying contact [McGregor et al. 2013b]. Many of the first generation devices had simple Schottky barrier rectifying contacts; however, almost all recent generation devices have p-type contacts diffused into high-resistivity n-type Si. Although the first such detectors were backfilled with enriched ^{10}B, the relatively higher energy reaction-products from the ^{6}Li(n,t)^{4}He reaction produces improved detection efficiency with better gamma-ray discrimination [Uher et al. 2007; Shultis and McGregor 2009].

Hence, modern MSNDs are mostly constructed with nanoparticles of enriched ^{6}LiF backfilled into the Si microstructures. The detector configuration and operation are depicted in Fig. 17.31 (left). Neutrons interact in the ^{6}LiF material, which has a macroscopic thermal-neutron (n,α) cross section of 57.4 cm^{-1}. Absorption of a neutron causes the compound nucleus ^{7}Li* to decay into an alpha particle with kinetic energy of 2.06 MeV and a triton with kinetic energy of 2.73 MeV. For thermal neutrons, these reaction products are emitted in opposite directions. With narrow channels, the probability of one or more reaction products emerging from the ^{6}LiF absorber into the adjacent semiconductor can be relatively high. Those reaction products that do enter the semiconductor produce a dense cloud of electron-hole pairs. An applied reverse bias across the detector junction drifts the released electrons and holes to the electrical contacts, which induces a current to flow, and ultimately produces a voltage pulse in the attached preamplifier circuitry. Deep channels, on the order of 400–500 microns, increase the neutron absorption probability. Narrow channels, on the order of 20 microns wide, improve the probability of capturing and measuring a reaction product in the adjacent semiconductor. The width of the semiconductor region between channels is also generally close to the channel width of about 20 microns.

These microstructured detectors are manufactured with VLSI methods, methods which permit mass production and, thereby lower the cost of the detectors. Si is the preferred substrate, mainly because it has a low neutron interaction cross section and it is inexpensive. Further, the (110) orientation of Si can be easily and rapidly anisotropically etched to form deep narrow trenches on the order of 450 microns (Fig. 17.31 (right)). The *pn* junction contacts are formed with traditional dopant diffusion in high-temperature furnaces, and metallization is applied with physical vapor deposition. The backfilling of the neutron reactive material is economically and rapidly performed by using a centrifuge to push ^{6}LiF nanoparticles in solution into the channels. These detectors are fabricated on the wafer scale, yielding up to fifty 1-cm^2 detectors on a single 100-mm diameter wafer. MSNDs also require little to no applied voltage, mainly because of the deep depletion region in the high purity Si.

In recent years, MSNDs have been fabricated as dual sided devices, with p-type regions on both sides of a high purity n-type substrate [Fronk et al. 2015]. Neutrons that stream through the Si regions on one side of the device encounter the neutron absorber on the other side of the device. The manufacture processing for dual-sided MSNDs is actually easier than the singled-sided design, mainly because there is only one diffusion step [Fronk et al. 2015]. Ochs et al. [2019] explain that the optimum packing fraction of ^{6}LiF in the trenches is approximately 65% rather than solid, a consequence of the combined effects of neutron absorption lengths and reaction product ranges. These new detectors have reached over 69% thermal-neutron detection efficiency ϵ_{tn}, and theoretical models suggest that values of ϵ_{tn} up to 80% are possible [Ochs et al. 2019]. Both single and dual-sided MSNDs are commercially available, routinely fabricated with $\epsilon_{tn} > 30\%$ and $\epsilon_{tn} > 60\%$, respectively.[19] The detectors are mass produced in compact packages as either 1-cm^2 or 4-cm^2 detectors.

MSNDs have several advantages as alternative neutron detectors. They are rugged and compact, and require less than 2 volts to operate. Under certain conditions, no voltage is required to operate the detector (although it is still required for the coupling electronics). MSNDs are relatively gamma-ray insensitive, with discrimination ratios n/γ greater than 10^6. Because MSNDs are mass produced, they can be purchased inexpensively compared to ^{3}He detectors. They can be arranged in arrays with several different form factors [McGregor et al. 2005], or can be used as single detectors for neutron dosimetry and remote neutron monitoring. The detectors can also be arranged in arrays with form factors of common ^{3}He gas-filled detectors. A device loaded with single-sided MSNDs, having similar dimensions of a 4-atm ^{3}He gas-filled detector, matched the performance of the much more expensive detector [McGregor et al. 2015]. Work with an array of dual-sided MSNDs yielded greater efficiency than a compact 6-atm ^{3}He gas-filled detector of the same size [Ochs et al. 2017], and calculations show that MSNDs can also give comparable performance to small 10-atm ^{3}He gas-filled detectors. Large arrays of MSNDs arranged in blocks of HDPE have been demonstrated as a system to locate and identify fast neutrons sources [Hoshor et al. 2015]. From the average detection location in the 3D array of detectors, an algorithm determines the most probable type of source producing the neutron field as well as the direction of the source.

There are reports of MSNDs with boron backfilling instead of LiF [Nikolic et al. 2008; Huang et al. 2014]. The single advantage to using boron backfilling is the higher neutron absorption cross section, thereby reducing the mean absorption length of 180 microns for ^{6}LiF to only 20 microns for pure ^{10}B. Hence, the microstructures can be 9 times shallower for the boron-filled device than the ^{6}LiF-filled device and produce similar absorption efficiency, provided that the backfilling densities are similar. However, because the reaction product energies and ranges are much less for ^{10}B than ^{6}LiF, there are disadvantages to the use of boron

[19] Through Radiation Detection Technologies, Inc.

[Shultis and McGregor 2009]. These disadvantages include relatively high energy loss from dead layers at the microstructure boundaries, significant count losses from the wall effect, and the requirement of much smaller features to etch and subsequently backfill (about 2 microns for ^{10}B opposed to about 25 microns for ^{6}LiF). Nikolic et al. [2008] have produced MSNDs as"pillar" detectors while Huang et al. [2014] report Si detectors with holes backfilled with boron. Wu et al. [2017] have adopted the single-sided straight trench structure for boron-filled MSNDs, reporting 32% thermal neutron detection efficiency for small devices.

17.5.1 Bulk Semiconductor Neutron Detectors

Semiconductor detectors, in which one of more constituent atoms are neutron reactive, are called *bulk semiconductor neutron detectors.*[20] The obvious advantages of a solid state version of a gas-filled ion chamber is that they should have high-efficiency and a compact form. Bulk solid-state neutron detectors can be divided into two basic categories: those that rely on the detection of charged particle reaction products and those that rely on prompt capture gamma rays. In general, this type of neutron detector is difficult to make reliably and at the time of writing this book is not commercially available. However, for the sake of completeness, a brief discussion of this class of detectors is included.

The bulk materials that rely upon charged particle emissions are based on boron and lithium containing semiconductors. In the search for bulk semiconductor neutron detectors, the boron-based materials, such as BP, BAs, BN, and B_4C, have been investigated more than other potential materials. See, for example, Ananthanarayanan et al. [1974], Kumashiro et al. [1987], Emin and Aselage [2005], Caruso et al. [2006], McGregor et al. [2008], and Doan et al. [2015]. Boron-based semiconductors in cubic form are difficult to grow as bulk crystals, mainly because they require high temperatures and high pressure for synthesis. BP and BAs can decompose into undesirable crystal structures (cubic to icosahedral form) unless synthesized under high pressure. B_4C also forms icosahedral units in a rhombohedral crystal structure [Clark and Hoard 1943], an undesirable transformation because the icosahedral structure has relatively poor charge collection properties [Domnich et al. 2011] which make these icosahedral forms unsuitable for neutron detection.

BN can be formed as either simple hexagonal, cubic (zincblende) or wurtzite crystals, depending on the growth temperature, and it is usually grown by thin film methods. It is the simple hexagonal form of BN that has been most studied as a neutron detector. Thin film chemical vapor deposition methods are usually employed to produce BP, BAs, BN, or B_4C. These boron-based films are often grown upon n-type Si substrates, which can form a *pn* junction with the Si and, therefore, produce a simple coated Si diode as described at the beginning of this section. Consequently, the neutron response from the device can be easily mistaken as a bulk response when it is actually a coated diode response. To date, there is sparse evidence of boron-based semiconductors producing intrinsic neutron signals.[21]

Li-containing semiconductors, categorized as Nowotny-Juza compounds, have also been investigated as bulk neutron detectors. The Nowotny-Juza compound LiZnAs has been demonstrated as a neutron detector [Montag et al. 2016]; however, the material is difficult and expensive to synthesize, and only small semiconductor crystals have been reported. Traditional semiconductor materials with neutron reactive dopants have been investigated, namely, Si(Li) detectors. Neutrons interact with the lithium dopant in the material and produce energetic reaction products. However, the dopant concentration is relatively low in Li drifted Si detectors (or other doped semiconductors), typically less than 10^{19} cm^{-3}. For a degenerate concentration of Li on the order of 10^{19} cm^{-3}, a 5-cm-thick block of natural Si(Li) would have less than 1% thermal-neutron

[20] The expression "solid-form" semiconductor neutron detectors has been used in the past by the authors, but it seems awkward and has thus been abandoned.

[21] The spectral output from many of these boron-based semiconductors grown on Si is often later identified correctly as the spectrum from a boron-coated diode as in Fig. 17.17 [McGregor and Shultis 2004; 2005]. Films grown on insulating substrates also may produce signals from reaction products ionizing the surrounding air.

detection efficiency, while a 5-cm-thick block of a Si(^{6}Li) detector would have only 4.6% thermal-neutron detection efficiency.

Another material currently being studied as a bulk semiconductor neutron detector is lithium indium diselenide ($LiInSe_2$) [Tupitsyn et al. 2012]. Originally studied for its optical properties [Isaenko et al. 2002], $LiInSe_2$ has garnered interest for its semiconductor properties. With a bandgap of 2.85 eV, the material can operate as a room temperature detector. $LiInSe_2$ detectors rely on the ^{6}Li(n,t)^{4}He reaction as the neutron detection mechanism. Energy from these heavy ion reaction products is easily absorbed provided the crystal is reasonably thicker than the combined ion ranges (that are about 80 μm). Lukosi et al. [2016, 2017] demonstrated that $LiInSe_2$ could be used as a neutron detector and a neutron imager. Although there appear to still be problems with producing uniform $LiInSe_2$ crystals, this relatively new neutron detector material shows promise. The relatively high Z value of In suggests that gamma-ray background could be a potential problem; however, pulse height discrimination can help distinguish between gamma rays and neutrons because of the high 4.67-MeV Q-value of the ^{6}Li(n,t)^{4}He reaction.

The mass density of $LiInSe_2$ is 4.25 g cm^{-3} and the atom density is 9.15×10^{21} cm^{-3}. The resulting 2200-m/s macroscopic cross section for the ^{6}Li(n,t)^{4}He reaction is 0.653 cm^{-1}. If 95% enriched ^{6}Li is substituted, the 2200-m/s macroscopic absorption cross section for the ^{6}Li(n,t)^{4}He reaction increases to 8.18 cm^{-1}. The indium in the compound, and to a lesser extent the selenium, combine to parasitically absorb neutrons without producing a useful signal. Consequently, the total 2200-m/s macroscopic absorption cross sections for natural and enriched forms of $LiInSe_2$ are 2.64 cm^{-1} and 10.17 cm^{-1}, respectively. The combined absorption cross sections of the various elements yield a limiting 2200-m/s neutron detection efficiency of 24.6% from the ^{6}Li(n,t)^{4}He reaction. Also, the activation of ^{115}In leads to the gradual buildup of a background beta-particle and gamma-ray signal, which decays with a half-life of 54.2 minutes after irradiation. If 95% enriched ^{6}Li is substituted, the limiting thermal neutron detection efficiency from the ^{6}Li(n,t)^{4}He reaction increases to 80.4%.

Prompt gamma-ray emitting semiconductors, such as CdTe [Vradii et al. 1977], CdZnTe [McGregor et al. 1996a], and HgI_2 [Beyerle and Hull 1987; Bell et al. 2004], have been successfully used as neutron detectors. Both natural Cd and Hg have relatively large thermal-neutron (n,γ) cross sections of 2444 b and 370 b for 2200 m s^{-1} neutrons, respectively. These detectors rely upon the prompt gamma-ray emissions from the ^{113}Cd(n,γ)^{114}Cd reaction (producing 558.6-keV and 651.3-keV gamma rays) and the ^{199}Hg(n,γ)^{200}Hg reaction (producing 368.1-keV and 661.1-keV gamma rays). Because the emissions can be reabsorbed at distances relatively far from the neutron interaction, the spatial information of the original neutron interaction is lost. These semiconductor materials and detectors are best used as gamma-ray spectrometers and, hence, are intrinsically sensitive to the gamma-ray background. The effective neutron detection efficiency is compromised because of the relatively small Compton ratio. In other words, a large fraction of events add to the Compton continuum rather than to the full energy peak, thus, making discrimination between neutrons and background gamma rays difficult. However, with adequate energy resolution, pulse height discrimination can be used to distinguish the prompt gamma-ray emissions from neutron interactions.

CdTe and CdZnTe have mass densities of 5.86 and about 5.8 g cm^{-3} and 2200-m/s macroscopic absorption cross sections of 37.2 cm^{-1} and 34.8 cm^{-1}, respectively. Hence, over 95% of thermal neutrons are absorbed within a 1-mm thickness of either material with most thermal neutrons being absorbed near the detector surface. For detectors of typical thicknesses between 2 and 10 mm, nearly half of the gamma rays are emitted in directions away from the detector bulk. This geometric problem, coupled with the low gamma-ray attenuation coefficients for 558.6-keV and 651.3-keV gamma rays (see Fig. 16.1), mostly from Compton scattering, yields a relatively low neutron detection efficiency. For instance, McGregor et al. [1996a] report a thermal-neutron detection efficiency of only 3.7±1.9% for a 3-mm-thick CdZnTe detector, although the device was capable of absorbing over 99% of the intersecting thermal neutrons. Using Cd starting material slightly depleted of ^{113}Cd tends to distribute neutron absorption more evenly throughout thick crystals; however, this approach can be costly.

The problem of non-uniform neutron absorption is less pronounced for HgI_2, which has a density of 6.4 g cm^{-3} and a 2200-m/s macroscopic cross section of 3.17 cm^{-1}. Because the neutron absorption is more evenly distributed through the material, it takes approximately 1 cm of material to absorb 95% of thermal neutrons. Further, the prompt gamma-ray emission energy of 368.1 keV has a higher linear attenuation coefficient, with the photoelectric coefficient being slightly higher than the Compton scattering coefficient. Hence, the prompt gamma rays resulting from ^{199}Hg(n,γ)^{200}Hg reactions have a higher probability of reabsorption compared to that for the prompt gamma rays from CdTe and CdZnTe devices. Unfortunately, material problems with HgI_2 have not yet yielded results that support its widespread use for neutron detection. Further, because of the high Z components, interference from background gamma rays becomes more of a problem so that good energy resolution of the full energy peak is important.

17.6 Neutron Diffraction

The concept of wave-particle duality was introduced in Secs. 3.2.5 and 3.2.6, and Bragg diffraction was briefly mentioned in Sec. 12.4. Described here is a method to use the wave nature of neutrons with Bragg diffraction to measure the energy of slow neutrons. Recall from Sec. 3.2.6 that the de Broglie wavelength is

$$\lambda = \frac{h}{p} = \frac{h}{m_n v}. \tag{17.39}$$

where h is Planck's constant, m_n is the neutron rest mass, and v is the neutron speed. Given a neutron energy, then Eq. (17.39) can be rewritten as

$$\lambda = \frac{h}{\sqrt{2m_n E}}. \tag{17.40}$$

At low energies, relativistic effects can be ignored, hence the rest mass of a neutron can be used to find the de Broglie wavelength

$$\lambda_n = \frac{0.286}{\sqrt{E}} \text{ Å} \tag{17.41}$$

where λ_n is in angstroms and E is expressed in electron volts (eV).

The Bragg condition for diffraction is related to wavelength by

$$n\lambda = 2d \sin\theta, \tag{17.42}$$

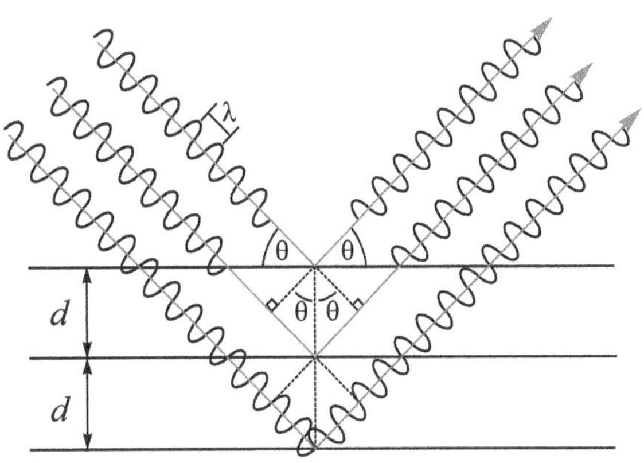

Figure 17.32. Geometry for Bragg diffraction of particles, showing the case for $n = 4$.

where n is an integer, d is the spacing between parallel atomic planes in an ordered crystal, and θ is the angle at which radiation intersects the crystal from the direction parallel to the planes, as depicted in Fig. 17.32. Elsasser [1936] predicted that neutrons could be diffracted from crystal planes, shortly thereafter proven by experiments conducted by Halban and Preiswerk [1936] and Mitchell and Powers [1936]. Substitution of Eq. (17.41) into Eq. (17.42) yields

$$\sin\theta = \frac{0.286n}{2d\sqrt{E}}, \tag{17.43}$$

where d is in angstroms and E is in eV. Consider the fundamental diffraction condition with $n = 1$. Here it is shown that given a distance d between crystal planes there is an lower limit to the neutron energy that can be observed. Also, given a practical working angle θ, an upper energy limit can also be determined.

Example 17.3: Given a lower diffraction angle limit of 1°, what are the neutron diffraction energy limits for $n = 1$ of a Si crystal oriented for diffraction off the {111} planes.

Solution:

For silicon, $a = 5.43$ Å and the distances between planes are found from Eq. (12.4),

$$d = \frac{a}{\sqrt{h^2 + k^2 + l^2}} = \frac{5.43 \text{ Å}}{\sqrt{3}} = 3.135 \text{ Å}$$

where (hkl) refer to the Miller indices of the crystal. The limits must lie between sin(1°) and 1, hence

$$\sin(1^\circ) \leq \frac{0.286}{2d\sqrt{E}} \leq 1.$$

Rearranging terms yields

$$6.83 \text{ eV} \geq E \geq 0.00208 \text{ eV}.$$

Although reflection and diffraction seem similar, in that the angle of incidence is equal to the angle of emission, they are fundamentally different. First, all atoms in the crystal along the path of the incident beam participate in diffraction, whereas reflection is more or less a surface phenomenon (e.g., reflection of light). Diffraction occurs only if the Bragg condition of Eq. (17.42) is met, while reflection can occur at practically any angle. Reflection is a far more efficient process than diffraction. For instance, practically all photons are reflected from a specular surface, whereas a diffracted beam comprises only a few percent of the incident beam. Neutron diffraction can be used in either reflection or transmission modes [Bacon and Lowde 1948], as depicted in Fig. 17.33. In either case, the effect of diffracted neutrons leaving the beam is called *extinction*. In other words, the strength of the neutron beam diminishes as neutrons are diffracted out, thereby causing the diffracted beam to weaken as the diffraction depth increases.

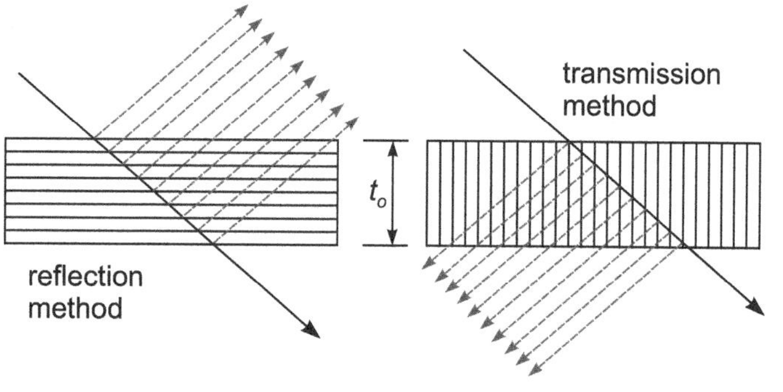

Figure 17.33. Distinction between reflection and transmission cases of diffraction by a parallel crystal slab of thickness t_o. After Bacon [1975].

The method of neutron diffraction can be used to isolate specific energies of neutrons, most effectively for slow neutrons. Such an arrangement is shown in Fig. 17.34. Consider the case in which moderated neutrons from a source are collimated onto a perfect crystal. If the origin of the neutrons is considered a point centered at the collimator entrance, then neutrons

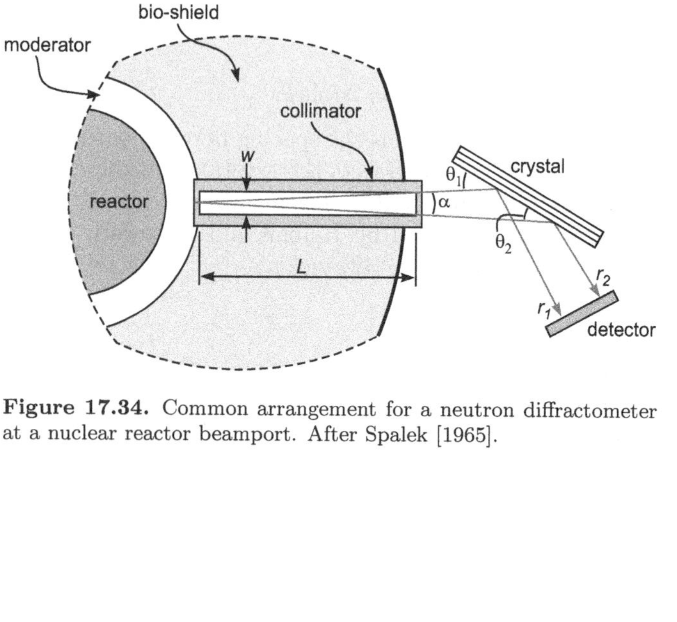

Figure 17.34. Common arrangement for a neutron diffractometer at a nuclear reactor beamport. After Spalek [1965].

may be diffracted at limits defined by rays r_1 and r_2. Consequently, only those neutrons with wavelength (energy) defined by θ_1 in Eq. (17.42) diffract along ray r_1, while only those neutrons with wavelength (energy) defined by θ_2 in Eq. (17.42) will diffract along ray r_2. Obviously, neutrons confined within these limits will diffract only if the Bragg condition is also met. Hence, there is a spread of neutron wavelengths that can diffract from the crystal with this configuration

$$n\,d\lambda = 2d\cos\theta d\theta. \tag{17.44}$$

The angle α defined by the collimator is small, hence

$$d\theta = \alpha \simeq \frac{w}{L} \tag{17.45}$$

where w and L are the collimator diameter and length, respectively. Note that neutrons may enter the collimator at any location along the entrance orifice, which broadens the total spread in angle to 2α. Substitution of these results into Eq. (17.44) yields

$$n\,d\lambda = \frac{4wd}{L}\cos\theta. \tag{17.46}$$

The *resolving power* η is a measure of the diffractometer ability to distinguish between neutron wavelengths, generally defined as

$$\eta = \frac{\lambda}{d\lambda} = \frac{L\tan\theta}{2w}. \tag{17.47}$$

In reality, crystals have numerous imperfections in the structure. In fact, a so-called perfect crystal without point defects will still have structural misorientations of crystal segments on the order of 500 nm long, referred to as the *crystal mosaic.* The magnitude of the angular misorientation of these mosaic blocks, called the *mosaic spread,* can range from a few minutes of arc up to a half a degree. Consequently, neutrons can actually penetrate deeper into the crystal and diffract more efficiently over a range of mosaic angles.[22] This increase in diffracted neutrons is generally proportional to the square of the crystal structure factor ($\boldsymbol{F}^2_{hkl}$) [Bacon 1966].

17.6.1 The Structure Factor for Crystals $\boldsymbol{F}_{hkl}$

The crystal structure factor $\boldsymbol{F}_{hkl}$ is a mathematical method of describing the diffraction of waves in a crystalline solid, which describes the amplitude and phase of the diffracted wave

$$\boldsymbol{F}_{hkl} = \sum_{j=1}^{N} b_j e^{[2\pi i(hx_j+ky_j+lz_j)]}, \tag{17.48}$$

where the sum is over all atoms (N) in a unit cell, x_j, y_j, z_j are the position coordinates of the jth atom, and b_j is the neutron scattering length of the jth atom. From Euler's formula

$$\boldsymbol{F}_{hkl} = \sum_{j=1}^{N} b_j \cos\left[2\pi\left(hx_j + ky_j + lz_j\right)\right] + i\sum_{j=1}^{N} b_j \sin\left[2\pi\left(hx_j + ky_j + lz_j\right)\right]. \tag{17.49}$$

The notation for the structure factor is often abbreviated as simply $\boldsymbol{F}$.

[22] In fact, diffraction crystals are purposely compressed under pressure to produce an optimum mosaic spread in order to increase the strength of the diffracted neutron beam.

Recall from Chapter 12 that the terms *hkl* refer to the Miller indices. Hence, the structure factor is the result of all waves scattered in the [*hkl*] direction of reflection by the N atoms contained in the unit cell. The structure factor for neutrons is described in terms of length with units on the order of 10^{-12} cm. When considering a *small* volume of the crystal, dV, the total integrated diffraction under the reflection orientation from this segment is

$$QdV = \frac{\lambda^3 M^2}{\sin(2\theta_B)} \boldsymbol{F}^2_{hkl}, \tag{17.50}$$

where M is the reciprocal of the unit cell volume, θ_B is the Bragg scattering angle, and λ is the de Broglie neutron wavelength.

As neutrons proceed through a *large* crystal at the Bragg angle, the diffracted amplitude is reduced by both neutron absorption and neutron diffraction. Thus, even after allowing for the effect of absorption, the deeper parts of the crystal contribute less than the upper parts, because neutrons corresponding to the Bragg condition of Eq. (17.42) are being removed before reaching lower crystal planes. Consequently, Eq. (17.50) fails to describe the integrated diffraction for thick crystals, the phenomenon referred to earlier as extinction [Bacon and Lowde 1948].

Example 17.4: Determine the structure factor of an FCC lattice.

Solution:

The unit cell has four complete atoms, with one located at the origin and three located at the adjacent faces of the unit cubic cell. Hence, the coordinates for four adjacent atoms in the unit cell can be written as

$$x_1, y_1, z_1 = (0,0,0); \; x_2, y_2, z_2 = \left(\frac{1}{2}, \frac{1}{2}, 0\right);$$

$$x_3, y_3, z_3 = \left(0, \frac{1}{2}, \frac{1}{2}\right); \; x_4, y_4, z_4 = \left(\frac{1}{2}, 0, \frac{1}{2}\right)$$

For integer values of *hkl* and from Eq. (17.49) the structure factor becomes,

$$\boldsymbol{F}_{hkl} = b\left(1 + \cos[\pi(h+k)] + \cos[\pi(k+l)] + \cos[\pi(h+l)]\right).$$

Consequently, if *hkl* are *all* even or are *all* odd, then $\boldsymbol{F}_{hkl} = 4b$. However, if *hkl* is mixed odd and even, then $\boldsymbol{F}_{hkl} = 0$.

17.6.2 Angular Response to a Maxwellian Neutron Distribution

Consider the distribution of slow neutrons in a nuclear reactor, generally described by a Maxwellian distribution

$$dn = K_1 v^2 e^{-v^2/v_o^2}\, dv, \tag{17.51}$$

where dn is the number of neutrons within differential speed dv about v per unit volume, and v_o is the most probable neutron velocity. The flux of neutrons of some speed v is simply the product vdn so intensity or flux of neutrons traveling down the beam port is

$$dn_b = v\, dn = K_1 v^3 e^{-v^2/v_o^2}\, dv. \tag{17.52}$$

These neutrons strike the diffraction crystal, as depicted in Fig. 17.34. Neutron diffraction off the crystal varies inversely with the neutron energy E or v^2. Consequently, the diffracted beam (reflection condition) spectrum from the neutron beam becomes

$$dn_d = \frac{v\,dn}{v^2} = K_1 v e^{-v^2/v_o^2}\,dv. \tag{17.53}$$

According to Eq. (17.53), as the neutron velocity increases, the number of diffracted neutrons decreases with v. Given any Bragg angle θ_B, the neutron velocities must satisfy Eq. (17.42) to undergo diffraction. Neutrons that reach the neutron detector for select Bragg angles are of constant velocity. Hence, to map a select neutron energy range, the crystal can be rotated in the neutron beam to change θ_B. The detector position must also be changed by $2\theta_B$ to match the Bragg angle.

From Eq. (17.39) and Eq. (17.44), one can derive

$$-\frac{nh}{mv^2}dv = 2d\cos\theta\,d\theta. \tag{17.54}$$

Considering only the fundamental diffraction ($n = 1$), Eq. (17.54) can be rearranged to find

$$dv = -\frac{2m_n v^2 d}{h}\cos\theta\,d\theta. \tag{17.55}$$

Substitution of Eq. (17.55) into Eq. (17.53) yields

$$\frac{dn_d}{d\theta} = -K_1\frac{2m_n v^3 d}{h}e^{-v^2/v_o^2}\cos\theta. \tag{17.56}$$

The collimator geometry defines $d\theta$ and can be regarding as constant. Consequently, the detector response C, or the count rate, is described by

$$C = K_2 v^3 e^{-v^2/v_o^2}\cos\theta, \tag{17.57}$$

and when divided by $\cos\theta$ yields

$$\frac{C}{\cos\theta} = K_2 v^3 e^{-v^2/v_0^2}, \tag{17.58}$$

the same relative distribution as the slow neutron distribution in the beamport (Eq. (17.52)). Hence, this outcome yields the thermal neutron *flux* in the nuclear reactor, while dividing by v yields the slow neutron *distribution* in the nuclear reactor.

17.6.3 Measurements with Diffracted Neutron Beams

Diffracted neutron beams can be used to identify neutron energies provided that the Bragg condition under use is well calibrated. Such uses apply to polyenergetic neutron sources, such as from a beamport from a nuclear reactor. Another use of diffracted neutron beams is to study the crystallinity of crystals, where the crystal under study is rotated through the beam about the Bragg angle. The resolution of the response gives a measure of the crystal perfection, the response often referred to as a *rocking curve*. Another common application is as a nearly pure monoenergetic neutron beam for measurements and experiments. Although contamination may be present from higher orders of diffracted neutrons ($n > 1$) and scattered background neutrons, the combined effects of Eq. (17.50) and Eq. (17.53) diminish this effect. This contamination can be reduced further with a second diffraction crystal between the primary crystal and the detector. This second crystal must be set at the same Bragg condition as the first crystal, which produces a somewhat stringent alignment requirement between the two crystals, a tedious process, and comes with the expense of a lower overall neutron flux at the work location. From Eq. (17.47), the limiting energy resolution dE/E from the beamport is,

$$\left|\frac{dE}{E}\right| = \frac{2d\lambda}{\lambda} = \frac{4w}{L\tan\theta}. \tag{17.59}$$

With simple substitutions and use of trigonometric identities,

$$\left|\frac{dE}{E}\right| = \frac{4wd\sqrt{E}\cos\theta}{nLk} = \frac{4wd\sqrt{E-(k^2n^2/4d^2)}}{nLk}, \tag{17.60}$$

where d is the interplane distance and $k = 0.286$ from Eq. (17.43). If the Bragg angle is well known from the crystal diffractometer, then the neutron energy is also known, and accurate measurements of neutron cross sections can be performed for various intervening materials (between the crystal and detector), limited by the energy resolution of the system.

17.7 Calibration of Slow Neutron Detectors

There are numerous methods used by credible organizations to calibrate and determine the neutron detection efficiency of thermal neutron detectors. These methods can differ significantly with varying degrees of measurement error. The reason for the differing calibration methods stems primarily from the difficulty in securing a calibrated standardized neutron source. For thermal-neutron detection efficiency, a diffracted neutron beam from a nuclear reactor is possibly the best calibration source. However, access to a nuclear research reactor is often inconvenient, and an organization may instead resort to a moderated ^{252}Cf or AmBe source. Fortunately, calibration of slow-neutron detectors can be a relatively straightforward process for $1/v$ absorbers [McGregor and Shultis 2011]. A few of these calibration methods are described here, although they do not represent the many methods used by various organizations. The different measurement protocols, although reproducible, are often not easily related to each other.

According to National Institute of Standards and Technology (NIST), there is no reliable method to produce an omnidirectional thermal neutron source to test a detector in an isotropic flux [McGregor and Shultis 2011]. Monte Carlo codes such as MCNP can be used to understand the experimental conditions and optimize a measurement, but it is inadequate to predict the detection efficiency, mainly because of the uncertainty associated with the code (about 5%) and uncertainty with the neutron measurement (also about 5%) together produce a result that has an unacceptable error for NIST-traceable detector calibrations.

17.7.1 Method of the NIST

The NIST uses a combination of a calibrated neutron current and calibrated neutron detectors to determine the unknown efficiency of a neutron detector [McGregor and Shultis 2011]. The calibration consists of using a ^{252}Cf neutron source placed in a heavy-water moderator pool that approximates the thermal-neutron environment inside a light-water reactor. However, the preferred method for testing a thermal-neutron detector is with a single-crystal diffracted beam, whose spatial profile is determined from activation analysis of a Dy foil, or is measured with a neutron-imaging detector so that the neutron beam profile is documented within the workspace. The neutron flux is then measured with a calibrated fission chamber. Hence, the basic method is to (1) profile the relative thermal-neutron beam intensity as a function of position and then (2) measure the neutron current (or fluence) with a calibrated standard fission chamber. The count rate of the unknown detector is then directly compared to the measured neutron current. Note that fission chambers, typically fabricated with uranium as the neutron reactive material, deviate slightly from the ideal $1/v$ behavior. Hence, knowledge of the diffraction angle, and resulting neutron energy, is important for making thermal corrections to the NIST calibration method.

17.7.2 Method of Reuter Stokes

A calibration method used by some commercial manufacturers of neutron detectors is to archive a so-called "gold standard" for each neutron detector product. For instance, for one such manufacturer, a large box formed by walls of 6-inch-thick high-density polyethylene (HDPE) is used as a moderator. The box has a large open cavity in the center, and a neutron source (^{252}Cf) is inserted into the box, centered near one of the walls. Gold activation foils (see next section) are then used to measure the thermal-neutron flux at various locations in the box after the foil measurements are corrected for neutron self-absorption. In this manner the neutron flux is mapped spatially inside the HDPE box. A detector is then placed inside the box, at an angle, near a wall opposite that of the location of the ^{252}Cf source. A measurement is conducted, correcting for spatial differences in the neutron flux. Afterwards, a NIST traceable detector is placed in the box at the prior location of the test detector, and another identical measurement is conducted.

This measurement method yields relative efficiencies as compared to a NIST-traceable standard. Hence, the method relies upon a fast neutron source moderated in a box that produces a thermal-neutron flux incident on the detector from nearly 4π directions. Due to the nature of the source-detector configuration, the neutron flux intersecting the detector is not uniform in direction or space. The Au foil calibration allows approximate corrections for non-uniform distribution of thermal neutrons. This method is adequate for in-house calibrations, but is not easily used for calibrations at other facilities. Finally, the method is probably adequate for large neutron detectors, but has larger measurement uncertainty for small neutron detectors.

17.7.3 Method of ORNL

A method introduced at Oak Ridge National Laboratory (ORNL) also uses a ^{252}Cf source. The neutron source is placed within a large block of HDPE capable of moderating and minimizing the leakage of neutrons. A blind hole beamport is drilled into the HDPE block that is oriented towards the source, but still leaves enough moderator between the drilled beamport hole and the ^{252}Cf to produce a thermalized neutron beam. A borated shutter, composed of perpendicular jaws, is attached to the moderator block around the beamport opening and is used to reduce the beam cross sectional area. The thermal-neutron flux at the testing location is measured with a calibrated neutron detector. For additional verification, MCNP is used to calculate the expected neutron flux and profile at the testing location and is quoted as having approximately 10% error [McGregor and Shultis 2011]. The observed count rate, per unit area, from the detector being studied is then divided by the measured thermal-neutron flux to determine the intrinsic neutron detection efficiency.

17.7.4 Method of Sampson and Vincent

The method of Sampson and Vincent, introduced primarily as a calibration system for $^{10}BF_3$ detectors, is used to determine the efficiency for a narrow beam geometry of diffracted neutrons [Sampson and Vincent 1971]. Here the BF_3 detector is placed end-on facing into the neutron beam from the diffractometer. The monoenergetic neutron flux from the diffracted beam is calibrated by activation of a NIST-traceable Au foil. The Au foil is placed between the detector and the emerging neutron beam. The beam is set at an arbitrary diffraction angle θ from the monodirectional neutrons incident on the diffractometer. The energy and beam angle are described by the Bragg diffraction condition

$$n\lambda = n\frac{h}{\sqrt{2E_n m_n}} = 2d\sin\theta, \qquad n = 1, 2, 3, \ldots, \tag{17.61}$$

where λ is the neutron wavelength, d is the atomic interplane distance, m_n is the neutron mass, h is Plank's constant, and E_n is the neutron kinetic energy. For a given direction θ, the diffracted beam has neutrons of energy E_1 (for $n = 1$) and also for neutrons of higher energy, the "harmonics" ($n > 1$).

The gold foil is used to measure the neutron flux ϕ incident on the end of the $^{10}BF_3$ detector. If the foil is totally immersed in the beam, the saturation activity of the Au foil A_∞^{Au} is

$$\phi = \frac{A_\infty^{Au}}{N_d \sigma_a^{Au}(E)}, \tag{17.62}$$

where N_d is the total number of Au atoms in the foil, and $\sigma_a^{Au}(E)$ is the Au microscopic absorption neutron cross section at energy E. For negligible absorption in the foil, the detector count rate when exposed to the beam is

$$R = \epsilon_{in}(E)\phi\mathcal{A}, \tag{17.63}$$

where $\mathcal{A}$ is the cross sectional area of the neutron beam incident on the detector and $\epsilon_{in}(E)$ is the intrinsic efficiency of the detector. Combine these last two equations to obtain

$$\epsilon_{in}(E) = \frac{RN_d\sigma_a^{Au}(E)}{A_\infty^{Au} S_b}. \tag{17.64}$$

In this simplistic overview, the details of correcting the gold foil activity for self-absorption, fast neutron flux components, and counter efficiency used to measure the foil activity have been omitted. Sources of uncertainty with the Sampson and Vincent method include flux contamination from higher order diffraction harmonics (mainly from the second-order contamination), uncertainty in the flux calibration due to a sizeable neutron resonance in the Au foil at 4.9 eV, uncertainty in the measurement of the neutron beam area (S_b), and finally uncertainly in the dead region at the end of the detector that reduces the neutron flux before encountering the high-field region of the detector. Sampson and Vincent report a correction for second order effects, and also report a theoretical efficiency that takes into account the end window absorption effects and neutron scattering.

17.7.5 Method of McGregor and Shultis

A method developed by McGregor and Shultis [2011] provides an in-house calibration standard and works especially well for neutron detectors based on $1/v$ absorbers without introducing multiple sources of error. This method requires a moderated neutron beam without contamination from fast neutrons. Isolation of the thermal neutrons is accomplished by diffracting, from a graphite crystal, a neutron beam from a nuclear reactor. The graphite crystal structure destructively interferes with diffraction of even harmonics, thereby eliminating problems with the contamination from second harmonic diffraction. Two $1/v$ neutron detectors are required for the calibration, along with a duplicate, but empty shell, of at least one of the two detectors.

The test detector to be used as the calibration standard is placed in the neutron beam, as shown in Fig. 17.35(b), such that neutrons pass through the container walls and neutron reactive gas. A reference detector is placed in the neutron beam beyond the calibration detector, also shown in Fig. 17.35(b). The reference detector measures a count rate of

$$R_{out} = R_{in}e^{-\Sigma_w t_w}e^{-\Sigma_g t_g}\exp^{-\Sigma_w t_w} = R_{in}e^{-2\Sigma_w t_w}e^{-\Sigma_g t_g}, \tag{17.65}$$

where R_{in} is the neutron count rate from the reference detector before neutrons are attenuated by the test detector and R_{out} is the neutron count rate measured by the reference detector after attenuation through the test detector, Σ_w is the macroscopic cross section of the empty detector shell material, Σ_g is the macroscopic cross section for the detection material (gas) in the test detector, t_w is the shell material thickness, and t_g is the thickness of the neutron reactive material within the test detector. The transmission factor for the test detector is

$$T_d \equiv \frac{R_{out}}{R_{in}} = e^{-2\Sigma_w t_w}e^{-\Sigma_g t_g}. \tag{17.66}$$

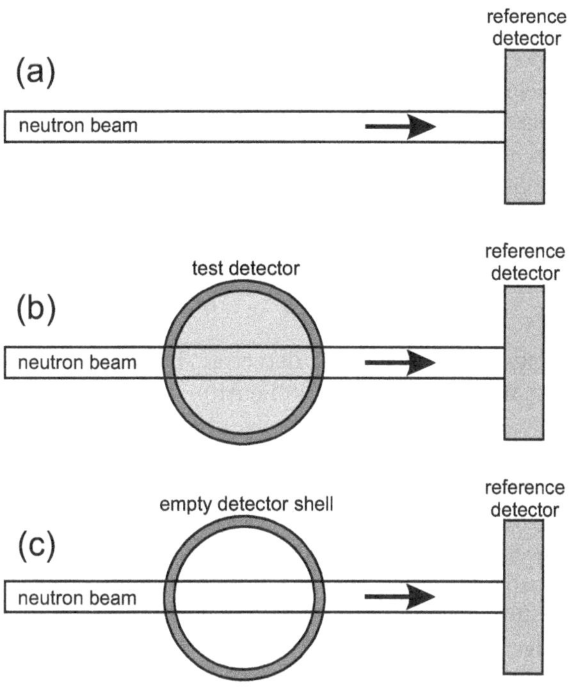

Figure 17.35. Method to produce a calibrated standard detector. After the calibration, the test detector can be used as a calibration standard for other $1/v$ neutron detectors. For the method to function properly, the reference detector must also be a $1/v$ detector [McGregor and Shultis 2011] .

Transmission through the empty shell detector is,

$$T_{2w} = \frac{R_w}{R_{in}} = e^{-2\Sigma_w t_w}, \tag{17.67}$$

where R_w is the count rate from the beam attenuated by the empty detector shell. The transmission factor for neutrons passing through the test detector $1/v$ absorber is,

$$T_g = \frac{R_{out}}{R_w} = e^{-\Sigma_g t_g}. \tag{17.68}$$

The fraction of the neutron beam attenuated by the absorber in the test detector reveals the fraction of interactions in the test detector, but does not yield the detector intrinsic efficiency. The beam attenuated by the neutron absorber was already attenuated by one wall of the detector shell. Hence, the fraction of the neutron beam transmitted through one wall of the detector shell is,

$$T_{1w} = \sqrt{\frac{R_w}{R_{in}}} = [e^{-2\Sigma_w t_w}]^{1/2}. \tag{17.69}$$

The intrinsic slow-neutron detection efficiency is,

$$\epsilon_{in} = \frac{\sigma_{ao}}{\sigma_{to}} e^{-\Sigma_w t_w}(1 - e^{-\Sigma_g t_g}), \tag{17.70}$$

where σ_{ao} is the absorption cross section leading to detectable events in the test detector and σ_{to} is the total cross section at an arbitrary energy E_o. The total cross section σ_{to} is the total probability that a neutron interacts with a target nuclei, including capture, fission, elastic scattering, and inelastic scattering. Because inelastic and elastic scattering usually do not follow a $1/v$ dependence, the ratio of σ_{ao}/σ_{to} deviates from $1/v$ behavior at moderately high neutron energies, and proper corrections must be made to account for the differences. However, at low energies (generally below 10 eV), $\sigma_{ao} \approx \sigma_{to}$ is a good approximation. Therefore, substitution of Eq. (17.68) and Eq. (17.69) into Eq. (17.70) yields,

$$\epsilon_{in} = \sqrt{\frac{R_w}{R_{in}}}\left(1 - \frac{R_{out}}{R_w}\right), \tag{17.71}$$

which is the intrinsic detection efficiency for the test detector. The test detector can be used thereafter as a calibration standard, provided that the totality of the neutron beam intersects within the perimeter of the detector active region.

From the discussion in Section 17.1, the count rates represented by R_{in}, R_{out}, and R_w are independent of neutron energy for materials with $1/v$ cross sections. Because of the σ_{ao}/σ_{to} ratio, corrections may apply for $1/v$ absorbers if the neutrons in the beam have energies exceeding approximately 10 eV. Hence neutron moderation is required for the measurements in order to avoid such corrections. Finally, if the detector under test is composed of non-$1/v$ absorbers for the neutron reactive material, such as, for example, ^{157}Gd, ^{113}Cd, or ^{235}U, then it is necessary to apply the appropriate non-$1/v$ corrections to the calculated efficiency, and generally requires some knowledge of the neutron energies.

17.8 Neutron Detection by Foil Activation

In the early days before modern neutron detectors, measurement of a neutron field was often done by irradiating small samples (usually in the form of foils) composed of elements (or isotopes) whose cross sections varied differently with the incident neutron energy. The radionuclides produced in the sample by neutron absorptions decay with the emission of radiation that can be measured. The rate at which radiation is emitted by an irradiated sample after it was removed from the neutron field depended on the flux used to irradiate the sample, the irradiation time, and the composition of the detecting material.

From the activity produced in the foils and the use of clever unfolding algorithms, the magnitude and energy dependence of the neutron flux could be estimated. The foil activation method is also discussed in Sec. 18.5.7 for fast neutron measurements. Although once widely used, the use of the foil activation method is not as prevalent today as it once was, primarily as a result of better and newer neutron detectors. However, one foil activation method is still in wide use and is the basis of several thermal neutron flux standards. In this method a gold foil is activated and from its resulting activity, both the thermal and fast neutron fluxes can be inferred. This one-foil method is widely accepted and, with appropriate correction factors, can give excellent results.[23] Other foils that have been used to measure the thermal neutron flux with the Cd covered/bare foil technique described in this section are listed in Table 17.3.

If the flux $\phi(E)$ is constant throughout the volume V_d of a bare gold foil, then the rate of production R_b of the radionuclide produced by the absorption of a neutron in gold is

$$R_b = N_d \int_0^\infty \sigma_a(E)\phi(E)\,dE = V_d \int_0^\infty \Sigma_a(E)\phi(E)\,dE, \tag{17.72}$$

where N_d is the number of gold atoms in the foil, $\sigma_a(E)$ the microscopic absorption cross section of gold, $\phi(E)$ the energy-dependent flux density, and E the neutron energy.

[23] One could say, in more ways than one, this foil technique is the "gold standard" for measuring thermal neutrons.

Table 17.3. Some activation foils used for thermal neutron measurements. Only the most abundant gamma rays are listed. Data are extracted mostly from ENDF/B [2017] and LM-KAPL [2002].

Element	Isotope	Nat. abundance	2200-m/s (n,γ) cross section (b)	Trans. isotope	γ-rays (keV) [yield%]	Half-Life
Ag	^{107}Ag	51.84	45	^{108}Ag	633.0 [1.76]	2.382 m
	^{109}Ag	48.16	4.4	^{110m}Ag	657.8 [95.61]	249.8 d
					677.6 [10.70]	
					763.9 [22.60]	
					884.7 [75.0]	
					937/5 [35.0]	
					1384.3 [25.1]	
					1475.8 [4.08]	
					1505.0 [13.33]	
					1562.3 [1.22]	
Au	^{197}Au	100	98.65	^{198}Au	411.8 [95.62]	2.694 d
Co	^{59}Co	100	16.9	^{60m}Co	58.60 [2.060]	10.47 m
			20.2	^{60}Co	1173.2 [99.85]	5.271 y
					1332.5 [99.98]	
Cu	^{63}Cu	69.17	4.503	^{64}Cu	345.8 [0.475]	12.70 h
	^{65}Cu	30.83	2.169	^{66}Cu	1039.2 [9.23]	5.120 m
Dy	^{164}Dy	28.18	$\simeq$1700	^{165m}Dy	515.5 [1.53]	1.257 m
					0.5155 [0.015]	
			$\simeq$ 950	^{165}Dy	94.7 [3.80]	2.334 h
In	^{113}In	4.29	56 ^{114m}In	190.3 [15/56]	49.51 d	
					558.4 [4.4]	
					725.2 [4.4]	
			$\simeq$2	^{114}In	1299.8 [0.139]	71.9 s
	^{115}In	95.71	$\simeq$160	^{116m1}In	4.16.9 [27.2]	54.29 m
					818.7 [12.13]	
					1097.3 [58.5]	
					1293.6 [84.8]	
					1507.6 [9.92]	
					1752.5 [2.36]	
					2112.3 [15.09]	
			42	^{116}In	1293.4 [1.30]	14.10 s
Mn	^{55}Mn	100	13.28	^{56}Mn	846.8 [98.85]	2.5789 h
					1810.7 [26.9]	
					2113.1 [14.2]	

In Fig. 17.36 the thermal energy dependence of the neutron flux and of the total cross sections for gold and cadmium is shown. Several features should be noted. Below the so-called thermal cutoff $E_{TC} \simeq 0.2$ eV the neutron flux has a Maxwellian distribution $\phi_M(E)$. As can be seen from the cadmium cross section, a cadmium covered gold foil is activated primarily by neutrons with energies above the *cadmium cutoff energy* $E_{CC} \simeq 0.4$ eV. Further below E_{CC} gold is seen to be almost a perfect "$1/v$" absorber.

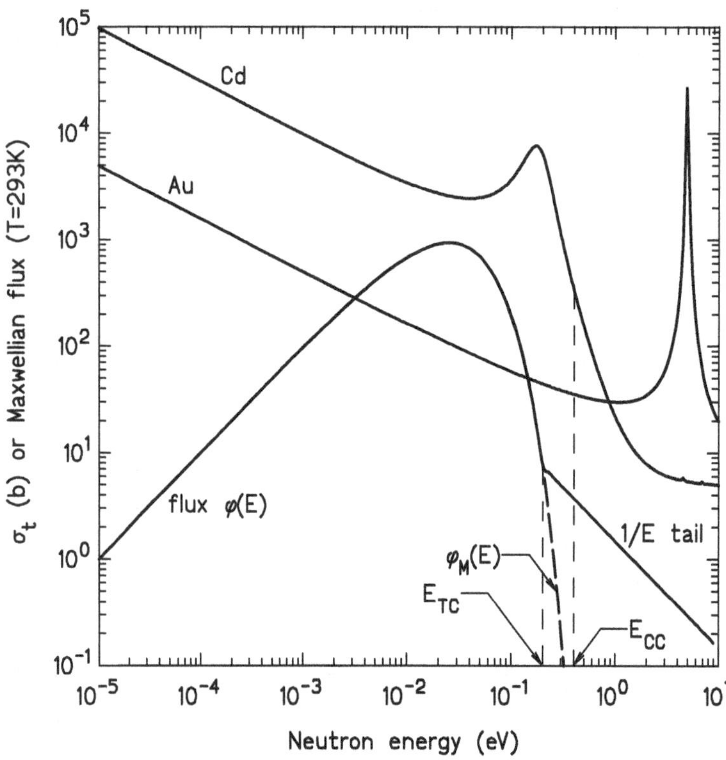

Figure 17.36. The energy dependence of the thermal flux $\phi_M(E)$ for neutrons in thermal equilibrium at room temperature (T = 293 K), the idealized epithermal $1/E$ slowing-down flux above the thermal cutoff E_{TC}, and the total (absorption) cross sections for gold and cadmium.

The *total thermal flux density*, or *thermal flux* for short, is defined as

$$\phi_t \equiv \int_0^{E_{TC}} \phi(E)\, dE, \tag{17.73}$$

and the *total fast flux* as

$$\phi_f \equiv \int_{E_{TC}}^{E_{\max}} \phi(E)\, dE, \tag{17.74}$$

where $E_{\max}$ is the maximum neutron energy. Below the *thermal cutoff* E_{TC} the energy distribution of the flux can often be well approximated by the Maxwellian distribution (see Problem 17-1)

$$\phi(E) \simeq \phi_M(E,T) = \phi_t \frac{E}{E_T^2} e^{-E/E_T}, \tag{17.75}$$

where $E_T \equiv kT$ (= 0.0253 eV for a neutron temperature of T = 293.5 K) is the most probable neutron energy for the thermal flux density.

For $E > E_{TC}$, the neutron flux in and from a reactor varies as $1/E$ or $\phi(E) = \phi_o/E$, characteristic of a neutron slowing down spectrum. The total fast flux is thus

$$\phi_f \equiv \int_{E_{TC}}^{E_{\max}} \phi(E)\, dE = \int_{E_{TC}}^{E_{\max}} \frac{\phi_o}{E} dE = \phi_o \ln\left(\frac{E_{\max}}{E_{TC}}\right). \tag{17.76}$$

For $E_{TC} \simeq 0.2$ eV and $E_{\max} \simeq 20$ MeV, $\phi_f \simeq 18\phi_o$. Below E_{TC} gold is a $1/v$ absorber, i.e., $\sigma_a(E) = \sigma_a(E_o)\sqrt{E_o/E}$ where $\sigma_a(E_o)$ is the gold absorption cross section at an arbitrary thermal reference energy usually taken as $E_o = E_{T=293.5\text{K}} = 0.0253$ eV.

The activation rate for a bare gold foil may then be written as

$$R_b = N_d \left\{ \int_0^{E_{TC}} \frac{\sigma_a(E_o)\sqrt{E_o}}{\sqrt{E}} \phi_M(E,T)\, dE + \int_{E_{TC}}^{E_{\max}} \sigma_a(E) \frac{\phi_o}{E}\, dE \right\} \tag{17.77}$$

or

$$R_b = N_d \overline{\sigma}_a \phi_t + N_d \int_{E_{TC}}^{E_{\max}} \sigma_a(E) \frac{\phi_o}{E}\, dE \tag{17.78}$$

where the thermal-averaged absorption cross section for gold is

$$\overline{\sigma}_a \equiv \frac{\int_0^{E_{TC}} \sigma_a(E_o)\sqrt{\frac{E_o}{E}}\phi_M(E,T)\, dE}{\int_0^{E_{TC}} \phi_M(E,T)\, dE} \simeq$$

$$\frac{\int_0^{\infty} \sigma_a(E_o)\sqrt{\frac{E_o}{E}}\phi_M(E,T)\, dE}{\int_0^{\infty} \phi_M(E,T)\, dE} = \frac{\sqrt{\pi}}{2}\sqrt{\frac{T_o}{T}}\sigma_a(E_o). \tag{17.79}$$

If it is assumed $\phi(E) = \phi_o/E$ for $E > E_{TC}$ then the integral in Eq. (17.78) can be evaluated as

$$\int_{E_{TC}}^{E_{\max}} \sigma_a(E) \frac{dE}{E} = \phi_o \int_{E_{TC}}^{E_{CC}} \sigma_a(E_o)\sqrt{\frac{E_o}{E}}\frac{1}{E}\, dE + \phi_o \int_{E_{CC}}^{E_{\max}} \frac{\sigma_a(E)}{E}\, dE$$

$$= \sigma_a(E_o)\phi_o\sqrt{4E_o}\left(\frac{1}{\sqrt{E_{TC}}} - \frac{1}{\sqrt{E_{CC}}}\right) + \phi_o \int_{E_{CC}}^{E_{\max}} \frac{\sigma_a(E)}{E}\, dE.$$

$$= \overline{\sigma}_a\phi_o\sqrt{\frac{16E_T}{\pi}}\left(\frac{1}{\sqrt{E_{TC}}} - \frac{1}{\sqrt{E_{CC}}}\right) + \phi_o \int_{E_{CC}}^{E_{\max}} \frac{\sigma_a(E)}{E}\, dE. \tag{17.80}$$

Substitution of this result into Eq. (17.78) gives the activation rate of a bare foil as

$$R_b = N_d \left\{ \overline{\sigma}_a \phi_t + \phi_o \overline{\sigma}_a \sqrt{\frac{16E_T}{\pi}} \left(\frac{1}{\sqrt{E_{TC}}} - \frac{1}{\sqrt{E_{CC}}} \right) + \phi_o \int_{E_{CC}}^{E_{\max}} \frac{\sigma_a(E)}{E}\, dE \right\}. \tag{17.81}$$

To measure the thermal flux ϕ_t, one must differentiate between the activation produced by thermal neutrons and by fast neutrons. One way to do this experimentally is to make two measurements, one with a bare foil and one with a detector foil covered with cadmium of sufficient thickness to absorb all thermal neutrons below the cadmium cutoff energy E_{CC} from reaching the gold foil. However, neutrons with energies above E_{CC} reach the gold foil and activate it at a rate

$$R_c = N_d \phi_o \int_{E_{CC}}^{E_{\max}} \sigma_a(E) \frac{dE}{E} \equiv N_d \phi_o I_d. \tag{17.82}$$

The integral in this expression is called the *resonance integral* I_d of a detector foil element. Values of the resonance integral for many elements have been measured and tabulated [ANL 5800]. For example, the resonance integral for gold is 1558 b. From the rate of activation of the cadmium-covered foil, the value of ϕ_o can be obtained. However, it must be recognized that the detector foil absorption cross section generally has many resonances and there can be significant self-shielding in the foil at these resonance energies. Thus the assumption that the flux is constant throughout the foil may be a poor assumption for the fast-neutron activation of the foil. Correction for this and other non-ideal effects are considered later.

The difference between the activation rate of the bare and cadmium-covered foils is thus

$$R_b - R_c = N_d \overline{\sigma}_a \left\{ \phi_t + \phi_o \sqrt{\frac{16E_T}{\pi}} \left(\frac{1}{\sqrt{E_{TC}}} - \frac{1}{\sqrt{E_{CC}}} \right) \right\}. \tag{17.83}$$

For $E_T = 0.03$ eV, $E_{TC} = 0.2$ eV, and $E_{CC} = 0.4$ eV, the coefficient of ϕ_o is about 0.25. The value of ϕ_o is often sufficiently small that the second term on the right-hand side of Eq. (17.83) can be neglected. However, one should, in principle, distinguish between the *thermal flux* contribution and the *subcadmium flux* contribution from epithermal neutrons, even if, in practice, the thermal flux is assumed to be proportional to $R_b - R_c$.

17.8.1 Cadmium Ratio

In flux measurements, one often encounters the term *cadmium ratio* (CR). This ratio is simply the specific activation rates of bare and cadmium-covered foil detectors, i.e.,

$$\text{CR} \equiv \frac{\widehat{R_b}}{\widehat{R_c}} \tag{17.84}$$

where the specific activation rate $\widehat{R} \equiv R/m$ with the foil mass denoted by m. From the results of Eqs. (17.81) and (17.82), the CR is

$$\text{CR} = 1 + \frac{\overline{\sigma}_a}{I_d} \left\{ \frac{\phi_t}{\phi_o} + \sqrt{\frac{16E_T}{\pi}} \left(\frac{1}{\sqrt{E_{TC}}} - \frac{1}{\sqrt{E_{CC}}} \right) \right\} \simeq 1 + \frac{\phi_t}{\phi_o} \frac{\overline{\sigma}_a}{I_d}. \tag{17.85}$$

where the last simplification is obtained by ignoring the usually very small subcadmium epithermal activation. From this result, it is evident that the larger the measured value of the CR, the larger is the proportion of thermal neutrons in the total neutron population reaching the detecting foils.

17.8.2 Measuring Activation Rates

Thus far, only the rates of production of radioactive foil atoms have been considered. In this section production rates are related to measurable quantities. After irradiation, it is the activity of the foil that is actually measured, which usually persists some time after the irradiation procedure. Thus it is necessary to relate the foil activity A to the production rate R of radionuclides in the foil during the irradiation process.

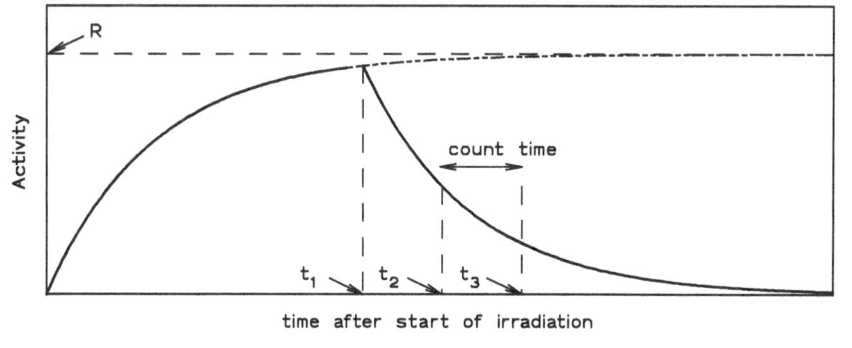

Figure 17.37. Activity of a foil that is irradiated by a constant flux between times 0 and t_1. At time t_1 the foil is removed from the neutron beam. The foil is subsequently counted between times t_2 and t_3.

The buildup of radionuclides in a foil during irradiation is described by

$$\frac{dN(t)}{dt} = R - \lambda N(t), \tag{17.86}$$

where $N(t)$ is the number of radionuclides (with decay constant λ) in the foil at time t after the commencement of irradiation. Burnup of the parent and daughter nuclides can usually be neglected. If the foil was "clean" at the time ($t = 0$) irradiation began, then the number of radionuclide daughters at a later time t is

$$N(t) = \frac{R}{\lambda}\left(1 - e^{-\lambda t}\right), \tag{17.87}$$

and the foil activity at time t is

$$A(t) \equiv \lambda N(t) = R\left(1 - e^{-\lambda t}\right). \tag{17.88}$$

If the foil remains in the neutron flux for a time much greater than the half-life of the daughter radionuclide, the activity reaches its saturation value $A_\infty = R$, the sought quantity. After removal of the foil at time t_1, the activity exponentially decreases as

$$A(t) = A(t_1)e^{-\lambda(t-t_1)} = R\left(1 - e^{-\lambda t_1}\right)e^{-\lambda(t-t_1)}, \quad t > t_1. \tag{17.89}$$

By counting the radiation emitted by the radioactive foil we can determine R. Suppose, the radionuclide emits a gamma ray of energy E_0 with a frequency f per decay. A gamma-ray spectrometer with known efficiency $\epsilon(E_0)$ at gamma-ray energy E_0 is used to measure the activity of E_0 emitted from the foil between times t_2 and t_3 (from Fig. 17.37). The total observed counts in the photopeak is

$$C = f\epsilon \int_{t_2}^{t_3} A(t)\,dt + BG = f\epsilon \int_{t_2}^{t_3} A(t_1)e^{-\lambda(t-t_1)}\,dt + BG$$

$$= \frac{f\epsilon A(t_1)e^{\lambda t_1}}{\lambda}\left(e^{-\lambda t_2} - e^{-\lambda t_3}\right) + BG$$

$$= \frac{f\epsilon R}{\lambda}\left(e^{\lambda t_1} - 1\right)\left(e^{-\lambda t_2} - e^{-\lambda t_3}\right) + BG \tag{17.90}$$

where BG is the background count that occurred between t_2 and t_3. From this result the production rate R (equal to the saturation activity A_∞) is found to be

$$R = A_\infty = \frac{\lambda(C - BG)}{\epsilon f(e^{\lambda t_1} - 1)(e^{-\lambda t_2} - e^{-\lambda t_3})}. \tag{17.91}$$

To obtain R from the above expression, it is necessary to first obtain the counter efficiency ϵ for the counting geometry and radiation emitted by the foil's radionuclides.

Foils used in a neutron flux measurement usually have different masses. To correct for the different masses, the *specific* foil activity is calculated, i.e., the activity per unit mass of the foil. From Eq. (17.91) the specific activity is

$$\widehat{R} \equiv \frac{R}{m} = \frac{\lambda(C - BG)}{m\epsilon f(e^{\lambda t_1} - 1)(e^{-\lambda t_2} - e^{-\lambda t_3})}. \tag{17.92}$$

Finally, the confidence limits on the variance of $\widehat{R}$ are obtained from the variances of the foil and background counts, and the uncertainties in the times t_1, t_2, and t_3.

Natural gold consists of only the stable ^{197}Au isotope. Upon absorption of neutrons, radioactive ^{198}Au is produced. This isotope has a half-life of 2.696 d and decays by β^- emission. In the decay process a 0.4118-MeV gamma ray is emitted with a frequency of 95.51%. Other emitted gamma and x-ray photons have frequencies of less than 2%. Activity measurements of the gold foils are performed with a gamma-ray spectrometer (usually a high-resolution HPGe detector) by measuring the counts in the 411-keV full energy peak. The absolute spectrometer efficiency ϵ at this energy and the sample position must be obtained by using calibration sources.

17.8.3 Flux Correction Factors

The theory presented so far has neglected some minor effects which require corrections to obtain accurate flux values. Specifically, the following facts have been neglected:

1. Gold is not exactly a $1/v$ absorber at the high end of the Maxwellian spectrum (see Fig. 17.36).
2. Some neutrons with $E < E_{CC}$ pass through the cadmium cover and activate the covered gold foil.
3. The cadmium cover also stops some neutrons with $E > E_{CC}$, so fewer fast neutrons activate a covered gold foil compared to a bare foil.
4. If the detector foils are measuring the flux in a medium through which neutrons are diffusing, the presence of the foil with its relatively high absorption cross section causes the flux around the foil to be depressed. For neutrons streaming through a void or in air, no such flux depression occurs.

5. At neutron energies for which the gold foil has a large cross section, the outer layer of the foil shields the inner foil volume, i.e., the activation per unit volume is not uniform throughout the foil. This is particularly important for the fast neutron activation of gold foils.

A detailed discussion of these complicating factors is beyond the scope of this book. Here only a brief description of how these effects can be investigated is presented. Greater detail for the theory and methods used to correct experimental data for these complicating factors are given in the references.

17.8.4 Correction for Non-$1/v$ Absorption

To account for the first effect, a corrected thermal averaged cross section is used. The thermal averaged cross section of Eq. (17.79) is more correctly defined as

$$\overline{\sigma}_a \equiv \frac{\displaystyle\int_0^\infty \sigma_a(E)\phi_M(E,T)\,dE}{\displaystyle\int_0^\infty \phi_M(E,T)\,dE} = \frac{\sqrt{\pi}}{2} g_a(T)\sqrt{\frac{T_o}{T}}\sigma_a(E_o), \tag{17.93}$$

where $g_a(T)$ is the Westcott non-$1/v$ correction factor (see Eq. (17.13)) obtained by evaluating the above integrals numerically. For a perfect $1/v$ absorber $g_a = 1$, and for a gold foil in a Maxwellian neutron energy distribution at room temperature (T = 293 K), g_a(293 K) = 1.005 [ANL 5800]. However, as the neutron temperature T increases the Maxwellian "hardens" as it shifts towards the gold absorption resonance at E = 4.90 eV (see Fig. 17.36) and $g_a(T)$ increases as gold becomes more non-$1/v$. However, near room temperature one can set $g_a(T) \simeq 1$ for gold without introducing significant error.[24]

17.8.5 Correction for Cadmium Filter Effects

Ideally, a high-pass neutron filter would absorb all neutrons below a sharp cutoff energy and be transparent to neutrons with energies above the cutoff. A cadmium cover with a thickness of between 0.5 to 1 mm comes close to this ideal, absorbing almost all neutrons below about 0.4 eV and only a few above about 0.6 eV.

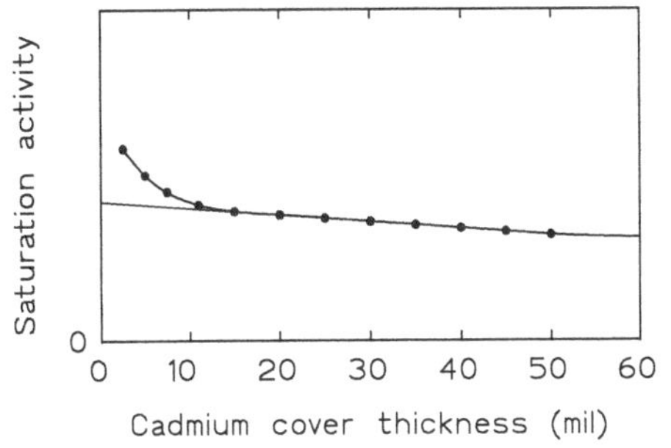

Figure 17.38. Foil saturation activity for different thicknesses of cadmium covering the foil.

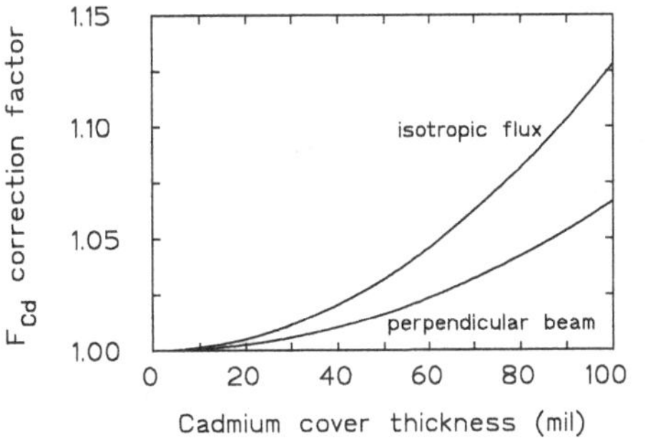

Figure 17.39. Epithermal cadmium correction factor F_{Cd} for covers of different thicknesses.

The effect of a cadmium filter covering a gold foil can be investigated by irradiating detector foils with different thicknesses of the cadmium cover. A typical plot of such results is shown in Fig. 17.38. The actual

[24]Data for $g_a(T)$ and the resonance integral needed for other foils used in activation analyses can be found in ANL [2003], Mughabghab [2003], and on-line at `wwwnde.jaea.go.jp/j40/j40.htm`.

shape of the curve depends on several factors such as the detector foil material, foil size, the medium in which the measurements are made, and the cadmium ratio. From Fig. 17.38, the rapid rise below approximately 20 mil ($\sim$ 500 microns) of cadmium indicates that a significant number of subcadmium neutrons reach the foil. More than 20 mils ($\sim$ 500 microns) of cadmium seems to reduce the subcadmium transmission enough so that the curve has only a small slope (caused by the cadmium absorption of epi-cadmium neutrons). For cadmium thicknesses greater than about 20 mils, one can correct for the absorption of epi-cadmium neutrons by extrapolating to zero cadmium thickness (the straight line in Fig. 17.38).

If a cadmium cover of thickness 500 microns or more is used, negligible thermal neutrons reach the gold foil. However, this cover absorbs some epithermal neutrons. To correct for this, a cadmium filter correction factor F_{Cd} is defined as

$$F_{Cd} \equiv \frac{R_{epi}}{R_c}, \tag{17.94}$$

where R_{epi} is activation rate in a bare foil by epithermal (above thermal) neutrons and R_c is the measured activation rate in a cadmium covered foil by epithermal neutrons. For a gold foil most of the epithermal activation is caused by neutrons absorbed in the gold resonance at $E_r = 4.91$ eV. If α represents the probability a 4.91-eV neutron passes through a cadmium cover of thickness t_c

$$F_{Cd} \equiv \frac{R_{epi}}{R_c} = \frac{R_{epi}}{\alpha R_{epi}} = \frac{1}{\alpha}. \tag{17.95}$$

For a beam of 4.91-eV neutrons normally incident on the cadmium cover, $\alpha = \exp[-\Sigma_{Cd}(E_r)t_c] \equiv \exp[-\tau_c]$. Then for a normally incident beam of neutrons

$$F_{Cd} = \exp[\tau_c]. \tag{17.96}$$

For a covered foil placed in an isotropic flux, it can be shown [Profio 1976], $\alpha = 2E_3(\tau_c)$ so that

$$F_{Cd} = \frac{1}{2E_3(\tau_c)}, \tag{17.97}$$

where $E_3(x)$ is the exponential integral function of order 3. These factors for a gold foil are shown in Fig. 17.39. Thus, to correct for epithermal-neutron absorption in the cadmium covered foil, we should use $R_{epi} = F_{Cd}R_c$ in place of R_c in Eqs. (17.81) to (17.83) which are used to calculate the fast and thermal flux densities.

17.8.6 Correction for Flux Perturbation and Self-Shielding

The complicating effects of foil self-shielding and the depression of the neutron field around the foil can be investigated by irradiating foils with the same area but different thicknesses. A typical result is shown in Fig. 17.40. The shape of such a plot depends on the area of the foil, the foil material, and the medium in which the measurements are made. The effect of flux depression and self-shielding can be eliminated by extrapolating the experimental results to zero foil thickness. Previous investigators have analyzed these complicating factors both analytically and experimentally, and several semiempirical correction factors have been developed [ANL 5800].

When an absorbing foil is used to measure the flux (fast or thermal) at some point in a moderator, the foil causes a depression in the flux around the foil. Moreover, the flux is depressed in the center of the foil because of self-shielding. A typical flux-profile is shown in Fig. 17.41. The average flux $\overline{\phi}$ is found through the through measuring the foil activity. However, it is the unperturbed flux ϕ_∞ that is sought. These two

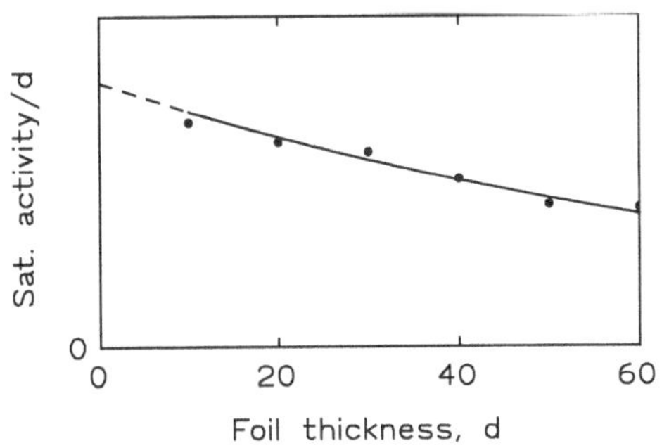

Figure 17.40. Saturation activity per unit volume for different thicknesses of foil.

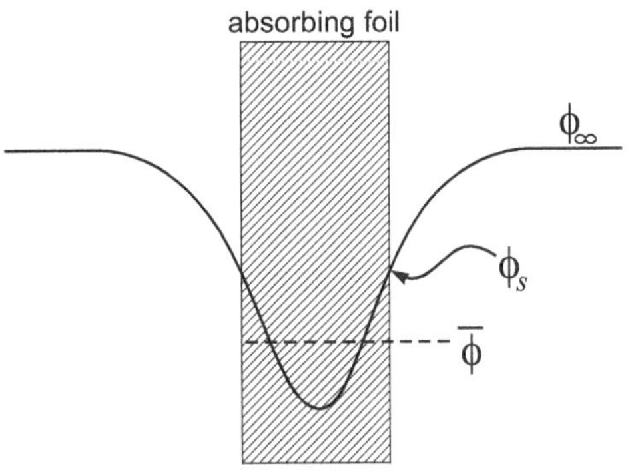

Figure 17.41. The depression of the flux and the self-shielding effect produced by an absorbing foil in a moderator.

flux densities are related by

$$\overline{\phi} = \left(\frac{\overline{\phi}}{\phi_s}\right)\left(\frac{\phi_s}{\phi_\infty}\right)\phi_\infty \qquad \text{or} \qquad \phi_\infty = \frac{1}{F_1F_2}\overline{\phi}, \tag{17.98}$$

where the *self-shielding factor* $F_1 \equiv (\overline{\phi}/\phi_s)$ and the *flux-depression factor* $F_2 \equiv (\phi_s/\phi_\infty)$. In general, these correction factors depend on the angular distribution of the flux density, foil size and thickness, and on the scattering and absorption cross sections of the surrounding material. Two extreme cases of angular distribution of the flux are considered below.

Perpendicular Beam in Air

In this geometry, the foil is placed in a non-diffusing medium and irradiated by a perpendicular beam of neutrons. The flux of the beam is not affected by the foil so that $F_2 = 1$ and $\phi_s = \phi_\infty$.

Thermal Flux-Density Self-Shielding: For a foil with $\Sigma_a \gg \Sigma_s$ whose radius is much greater than its thickness d, and which is irradiated by a perpendicular beam of thermal neutrons with a Maxwellian energy distribution, the position dependent neutron flux is,

$$\phi(x) = \phi_s e^{-\overline{\Sigma}_a x}, \tag{17.99}$$

where x is distance into the foil. For a thin foil, there is negligible spectrum hardening of the Maxwellian as the neutrons pass through the foil so that $\overline{\Sigma}_a = (\sqrt{\pi}/2)\Sigma_{ao}$. Then the average flux in the foil to the incident flux is

$$F_1 \equiv \left(\frac{\overline{\phi}}{\phi_s}\right) = \left(\frac{\overline{\phi}}{\phi_\infty}\right) = \frac{1}{d}\int_0^d e^{-\overline{\Sigma}_a x}dx = \frac{1}{\tau}[1-e^{-\tau}], \tag{17.100}$$

where $\tau \equiv \overline{\Sigma}_a d$.

Epithermal Flux-Density Self-Shielding: In the epithermal region, the absorption cross section is divided into two portions (see Eq. (17.80)): a $1/v$ component between the thermal and cadmium cutoff energies, and the component above E_{CC} whose absorption is dominated by absorption resonances. The self-shielding in the $1/v$ region is negligible compared to that in the resonance region. In gold most of the fast absorption occurs in the large resonance at 4.91 eV with a peak value of $\sigma_{ar}^{\max} \simeq 30,600$ b. In the case of an isolated resonance without significant Doppler broadening, the fast self-shielding factor for a thin foil without appreciable scattering can be approximated by [Profio 1976]

$$F_1 = e^{-\tau_r/2}[I_0(\tau_r/2) + I_1(\tau_r/2)], \tag{17.101}$$

where τ_r is the thickness of the foil in mean-free-path lengths evaluated at the peak resonance energy, i.e., $\tau_r = N_d \sigma_{ar}^{max} d$, and I_0 and I_1 are the modified Bessel functions of the first kind of the zeroth and first order, respectively.

Nearly Isotropic Flux Density

In this geometry, the foil is placed in a diffusing medium where the neutron field has an angular distribution that is well described by just the P_0 and P_1 Legendre components, i.e., is nearly an isotropic field. In such a field the volume averaged neutron absorption in the foil is nearly independent of the foil orientation.

Thermal Flux-Density Self-Shielding: The self-shielding correction factor for a nearly isotropic thermal flux is given by [ANL 5800]

$$F_1 = \frac{1}{\tau}\left[\frac{1}{2} - E_3(\tau)\right], \tag{17.102}$$

where E_3 is the exponential integral function of order three.

Resonance Self-Shielding: For isotropic neutron incidence on the foil, the self-shielding correction factor for resonances is [Profio 1976]

$$F_1 = \frac{\tau_r}{2}\int_{\tau_r/2}^{\infty} y^2 e^{-y}[I_0(y) + I_1(y)]\, dy. \tag{17.103}$$

Other expressions for F_1 are [ANL 5800]

$$F_1 = \frac{4}{3\sqrt{\pi}}\frac{1}{\sqrt{\tau_r}} \qquad \text{“thick foils”} \tag{17.104}$$

and

$$F_1 = 1 + \frac{\tau_r \ln \tau_r}{4} - 0.3274\tau_r \qquad \text{“thin foils”}. \tag{17.105}$$

Thermal-Neutron Flux-Density Depression: The thermal-neutron flux can be severely depressed in the vicinity of a strong absorber and thermal neutrons absorbed in the foil can no longer pass through the foil and subsequently be reflected back to the foil and increase the foil activity. By contrast, fast neutrons passing through a foil at resonance energies are degraded below the resonance energies in the moderator before they are reflected back to the foil. Thus, for fast neutrons, the flux depression factor correction is close to unity, i.e., $F_2^{\text{fast}} = 1$.

For thermal neutrons, the flux-depression correction factor can be expressed as [ANL 5800]

$$F_2^{\text{th}} = \left\{1 + \left[\frac{1}{2} - E_3(\tau)\right] g(R,\gamma)\right\}^{-1}, \tag{17.106}$$

where $\gamma \equiv \overline{\Sigma_s}/\overline{\Sigma_t} \simeq 1$ for the moderator. Several expressions for the g-factor have been derived [ANL 5800]. One expression for $g(R,\gamma)$ is [Profio 1976]

$$g(R,\gamma) = \left\{\frac{3}{2}\left(\frac{L}{\lambda}\right) S\left(\frac{2R}{L}\right) - K\left[\left(\frac{2R}{L}\right), \gamma\right]\right\} r_g. \tag{17.107}$$

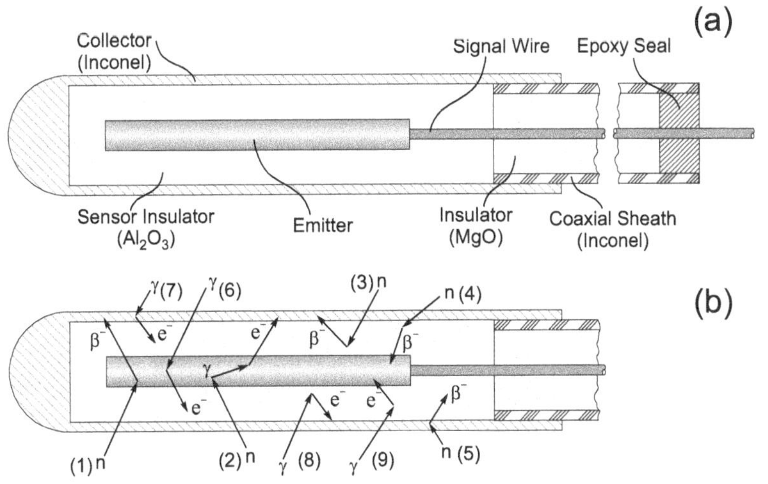

Figure 17.42. (a) The basic components of an SPND; (b) Various radiation interactions that can occur in an SPND.

Here S is the Skyrme function,

$$S(x) = 1 - \frac{4}{x}\int_0^1 e^{-xt}\sqrt{1-t^2}\,dt, \tag{17.108}$$

K is a small transport correction that can be approximated by

$$K \simeq \begin{cases} \dfrac{1}{15}\dfrac{2R}{\lambda_s} & 2R \leq \lambda_s \\ 0.15 & 2R > \lambda_s \end{cases}, \tag{17.109}$$

and r_g is a factor depending on L/λ and varies between 0 and 0.9 [Ritchie and Eldridge 1960]. The moderator parameters L, λ, and λ_s are the diffusion length, total mean-free-path length, and the scattering mean-free-path length for thermal neutrons, respectively, in the moderator.

17.9 Self-Powered Neutron Detectors

The self-powered neutron detector (SPND) was originally introduced by J.W. Hilborn [1964].[25] The name "self-powered" is appropriate because these detectors operate without the need for an applied bias. Indeed, they provide a current that is proportional to the neutron absorption rate, a current that can be easily read with a femtoammeter. They are designed as in-core instrumentation for nuclear reactors, usually for operation under steady-state conditions. SPNDs operate on the same principle as neutron activation analysis, except the radioactive decay is rapid and the resulting activation is recorded in real time.

The construction of a basic SPND is shown in Fig. 17.42(a). The components include a neutron absorbing emitter, an electrically conductive collector, and an insulating medium between them. Because these detectors are inserted directly into the core of a nuclear reactor, the materials of choice must be radiation hard in order to survive the total radiation fluence over their operating lifetime. The insulator is typically composed of either aluminum oxide or magnesium oxide, both of which have relatively small neutron interaction cross

[25] In some older texts and publications, they are referred to as Hilborn detectors.

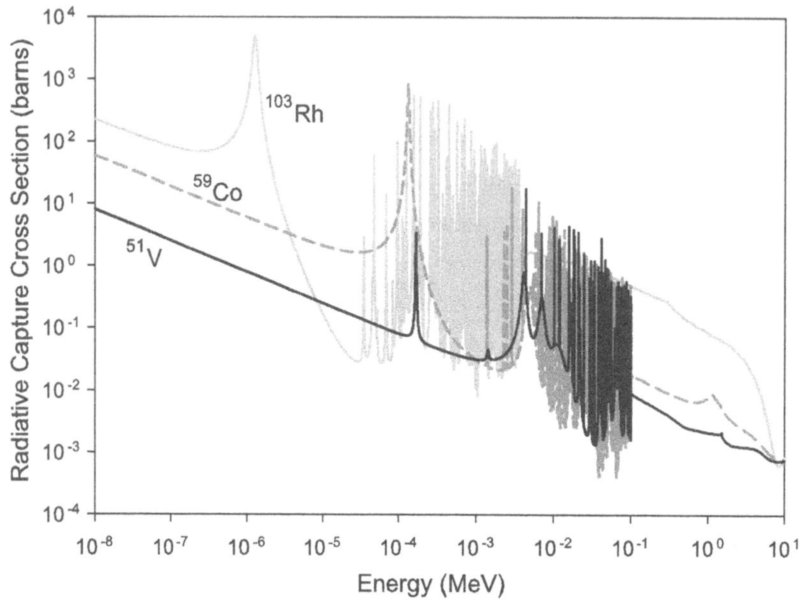

Figure 17.43. Radiative capture cross sections for ^{103}Rh, ^{51}V, and ^{59}Co. Data are from [ENDFPLOT, 2015].

sections. Both insulators have high resistivity at room temperature and at the usual operating temperature of a nuclear reactor core, where the resistivity varies from 10^{12} Ω-cm at room temperature to 5×10^{8} Ω-cm at 300 °C. The resistivity of the insulating material will change over time with radiation damage, yet these changes are comparable for both Al_2O_3 and MgO for high purity materials [Angelone et al. 2014]. The neutron cross section is lower for MgO than Al_2O_3 and is typically used for the signal cable insulator. Both Al_2O_3 and MgO are used for the emitter insulator [Stevens 1973].

The conductive metal sheath is usually fabricated from Inconel, a relatively oxidation and corrosion resistant austenite metal alloy of Ni, Cr, and Fe. Different Inconel alloys may include other elements to improve specific properties. Inconel 600 is a common collector choice because of its resistance to corrosion by high purity water, excellent oxidation resistance, high tensile strength, and resistance to stress cracking at high temperatures. The emitter is a neutron activation material that produces daughter products with relatively short decay times. These emitter materials must have thermal-neutron (n,γ) cross sections large enough to be useful for activation, yet small enough so that burnup does not cause large changes in the activation rate over time. The emitter sensitivity should be selective for neutrons or gamma rays. Further, the preferred decay products are beta particles or conversion electrons, although gamma-ray emissions can be used. Typical choices include rhodium, vanadium, and cobalt as emitter materials (whose cross sections are shown in Fig. 17.43), although several other candidate materials have been studied [Hilborn 1964; Ramírez and David 1970; Goldstein and Todt 1979; Amu and Petitcolas 1989; Imel and Hart 1996]. SPNDs are available with short emitter active lengths of only 3 to 4 cm up to long active lengths greater than 20 cm.

For activated beta particle emitters, such as ^{103}Rh and ^{51}V, the emission of beta particles from the emitter creates an electron current, i.e., the electrons that leave the emitter induce charge to flow through the emitter lead and thus produce a current that can be monitored by a femtoammeter. For activated gamma-ray emitters, the gamma rays can interact in the emitter and produce energetic electrons (photoelectric, Compton), which also leave the emitter and produce an induced current. These beta particles, which are electrons, may strike the collector; however, it is not necessary that these particles reach the collector to produce a current. The mere action of the electrons leaving the emitter produces the induced current. Recall from Sec. 8.5 that mobile charge moving between electrodes induces an image charge to appear on the electrodes. Throughout much of this text, charges were influenced into motion by an electric field. However, in the case of an SPND the electron velocity is produced by the kinetic energy imparted through the decay process. Because beta particles are emitted as a continuous spectrum, with each decay sharing energy with the recoil nucleus and an antineutrino, some beta particles have nearly zero energy while others retain the maximum allowable energy. Consequently, some beta particles will not emerge from the emitter. Regardless, the amount of current produced is a function of the emitter activity and, hence, the measured current is a function of the neutron interaction rate in the emitter.

An analysis on the operation of SPNDs by Warren and Shah [1974] takes into consideration the influence of space charge. Electrons do not freely conduct across the insulator between the emitter and collector. Consequently electrons that come to rest in the insulator build up space charge over time, which ultimately can produce an electric field. Already proven in Sec. 8.6.1 is the fact that space charge does not change the induced current of an electron as it passes through a medium. However, the negative space charge can work to produce a repulsive force that slows down the beta particle. Warren and Shah [1974] calculate that a toroidal electric field maxima will appear in the insulator at some distance $(d_\rho - d_l)/2$ from the emitter surface, where d_l and d_p are the diameter of the emitter and the electric field maxima, respectively. Hence, the amount of current induced by the beta particles and electrons can be reduced if the linear pathlength is reduced. Also, electrons that are coulombically repelled backwards will produce a net induced current representative of the net distance traveled between the emitter and collector. Electrons repelled back to the emitter surface will, consequently, produce a net induced current of zero. A similar situation can be expected for electrons emerging from the collector into the insulator.

Given a thermal-neutron, free-field, flux ϕ_∞, the average thermal flux $\overline{\phi}$ in the emitter is given by Eq. (17.98) or $\overline{\phi} = F_1 F_2 \phi_\infty$. For this simplistic analysis, the nuances of thermal neutron absorption and non $1/v$ cross section behavior discussed in Section 17.1 are ignored. The rate, per unit volume, of production of the radioactive transmutation product from the emitter material is given by

$$R_1(t) = N_e \sigma_e \overline{\phi} \exp[-\sigma_e \overline{\phi} t], \tag{17.110}$$

where N_e is the atomic density of emitter atoms and σ_e is the microscopic *activation* cross section of the emitter material. Here it is assumed there is negligible burnup of the emitter atoms, i.e., N_e remains constant. The atomic density $N_1(t)$ of the radioactive transmutation product accounting for both radioactive decay and transmutation is described by

$$\frac{dN_1(t)}{dt} = R_1(t) - \sigma_1 \overline{\phi} N_1(t) - \lambda_1 N_1(t), \tag{17.111}$$

where σ_1 is the thermal absorption cross section of the radioactive transmutation product and λ_1 is its radioactive decay constant.

Substitution of Eq. (17.110) into Eq. (17.111) and use of the initial condition $N_1(0) = 0$ yields the following result for the activity of the daughter

$$\lambda_1 N_1(t) = \frac{\lambda_1 N_e \sigma_e \overline{\phi}}{\lambda_1 + (\sigma_1 - \sigma_e)\overline{\phi}} \left(\exp[-\sigma_e \overline{\phi} t] - \exp[-(\lambda_1 + \overline{\phi}\sigma_1)t] \right). \tag{17.112}$$

The amount of current produced is also a function of the escape probability of beta particles and electrons from the emitter. A correction constant K is often used to account for this escape probability, which is an intrinsic function of the emitter geometry and the beta particle emission spectrum. The current is then determined by multiplying the above activity by the emitter volume V_e and by the unit charge constant, i.e.,

$$I(t) = \frac{q\bar{n}KV_e\lambda_1 N_e\sigma_e\overline{\phi}}{\lambda_1 + (\sigma_1 - \sigma_e)\overline{\phi}} \left(\exp[-\sigma_e\overline{\phi}t] - \exp[-(\lambda_1 + \sigma_1\overline{\phi})t]\right), \tag{17.113}$$

where $\bar{n}$ is the average number of radiation quanta emitted per decay of N_1. Typically the emitter material is relatively thin, on the order of 0.5–2 mm in diameter so as to reduce the neutron self-shielding. If emitters with low absorption cross sections for both the initial and transmuted materials are used, then both the neutron self-shielding and flux depression are negligible. Hence, Eq. (17.113) reduces to

$$I(t) \simeq q\bar{n}KV_e N_e\sigma_e\phi_\infty \left(1 - e^{-\lambda_1 t}\right). \tag{17.114}$$

The time response of an SPND is strongly dependent upon λ_1, with a large λ_1 producing a rapid response and a small λ_1 yielding a delayed response. Properties of several emitters are listed in Table 17.4. As t increases to relatively long operating times ($t \to \infty$), the saturation current is reached and Eq. (17.114) becomes

$$I_{sat}(t) \simeq q\bar{n}KV_e N_e\sigma_e\phi_\infty. \tag{17.115}$$

For practical applications, $0.984I_{sat}$ is reached at six half lives of the activated material. Hence, after a step change in nuclear power, it takes 4 min 14 sec for a Rh emitter and 22 min and 36 sec for a vanadium emitter to reach 98.4% of the new steady-state condition.

Shown in Fig. 17.42(b) are several possible reactions that can occur in an SPND [Moreira and Lescano 2013]. Events (1) through (5) are from neutron captures and events (6) through (9) are from direct gamma-ray interactions. For events (1), (2), and (6), the loss of an electron or beta particle from the emitter causes a net positive charge on the emitter, which causes electrons to flow towards the emitter from the signal cable, thereby producing a positive current on the electrometer. For events (5) and (7), the loss of electrons or beta particles from the collector produces a net positive charge on the collector, which causes negative current to flow in the circuit. Events (3) and (8) add negative charge to the collector, which causes positive current to flow in the signal cable, while events (4) and (9) produce negative charge in the emitter and cause negative current flow in the signal cable. Ultimately, all of these events are possible, and the overall measured signal is the net induced current response. If the emitter is designed such that most neutron interactions occur in it while reducing gamma-ray interactions in the surrounding insulator and collector, then most of the measured signal is from neutron absorptions in the emitter, and the neutron flux can be determined from Eq. (17.115).

Neutron activation and gamma-ray interactions in the signal cable may cause background current, which would overestimate the actual neutron flux at the emitter. To remedy this situation, many SPNDs are available with twin leads in which both extend the distance of the entire cable, but only one is connected to the emitter. The second lead produces a compensating signal, which can be subtracted from the primary emitter lead signal in a similar fashion as in a compensated ion chamber.

Rhodium SPNDs are the most popular version in commercial use, mainly because of their higher sensitivity to neutrons than other emitter materials (see Table 17.4). Increased sensitivity, consequently, also means higher burnup at a rate of $\approx$ 3.5% per month in a thermal-neutron flux of 10^{14} cm^{-2} s^{-1}. Rhodium emitters have a relatively short half-life of 42.3 seconds for approximately 92.4% of the emissions, but also have a longer delayed component for the remaining 7.6% of emissions. These delayed emissions are a consequence of neutron captures producing an excited state of ^{104m}Rh with a decay to ground state half-life of 261.6 seconds [Lederer and Shirley 1978].[26] After this decay to ground, the ^{104}Rh decays by beta emission, with

[26] This half-life of 4.4 minutes is sometimes incorrectly presented as beta emission decay, which it is not. It is the half-life of the metastable state ^{104m}Rh to decay to the ground state of ^{104}Rh, which then subsequently decays by beta particle emission

Table 17.4. Properties of several SPND emitter materials with 508-μm diameters. The type of reaction refers to how current is induced in the emitter, i.e., $[n,\beta^-]$ indicates the daughter nuclide undergoes beta decay and $[n,\gamma]$ indicates capture gamma rays produce recoil electrons from scatters in the SPND. The monthly burnup is for the SPND exposed to a flux of 10^{14} cm^{-2} s^{-1}. Data are from Hilborn [1964]; Böck [1976]; Goldstein and Todt [1979].

Nuclide	Isotopic Abund.	2200-m/s σ (b)	daughter $T_{1/2}$ (s)	Reaction	max E_β (MeV)	Sensitivity (A s/cm^3)	Mon. % burnup
^{103}Rh	100%	134 $\rightarrow$ ^{104}Rh	42.3	$[n, \beta^-]$	2.44	1.2×10^{-21}	3.9
^{103}Rh	100%	11 $\rightarrow$ ^{104m}Rh	261.6	none			
^{51}V	99.75%	4.92	226	$[n,\beta^-]$	2.47	7.7×10^{-23}	0.12
^{59}Co	100%	37.2	prompt	$[n,\gamma]$		1.7×10^{-23}	1.0
Nat. Pt		10	prompt	$[n,\gamma]$		1.3×10^{-22}	0.25
^{195}Pt	33.83%	27.5	prompt	$[n,\gamma]$			
^{107}Ag	51.35%	27.6	143	$[n,\beta^-]$	1.56		0.9
^{109}Ag	48.65%	90.3	24.6	$[n,\beta^-]$	2.87		2.3
^{27}Al	100%	0.233	135	$[n,\beta^-]$	2.87		0.006

$T_{1/2} = 42$ seconds, to stable ^{104}Pd. The expected activity, and therefore current, is depicted in Fig. 17.44 for a step increase in reactor power at $t = 0$ and a subsequent step decrease to zero power at $t = 30$ min. Prompt increases in reactor power cannot be monitored with Rh-SPNDs because of these delayed signals. Further, a step decrease in reactor power as shown in Fig. 17.44 serves to illustrate the issue; however this scenario is somewhat artificial because reactor power cannot be reduced in a step-like manner but decreases at its fastest with an 80 s period because of delayed neutrons. In short, Rh-SPNDs, like most SPNDs, are good for steady-state power monitoring, but not for measuring power transients.

Vanadium SPNDs are also commercially available. With a half-life of 226 seconds, V-SPNDs have a slower response than Rh-SPNDs. V-SPNDs are also less sensitive to neutrons and produce a much lower output current. Although the lower sensitivity produces lower signal strengths, it serves to reduce burnup and extend the SPND life to over 30 times that of a Rh emitter of the same size. For longevity, V-based SPNDs have become popular in power reactors, mainly because they can be operated over a long period of time with minimal burnup. As with Rh-SPNDs, V-SPNDs are good for steady-state power measurements, but are not useful for monitoring power transients.

Emitters fashioned from Co perform as prompt flux monitoring devices. The prompt reaction ^{59}Co(n,γ)^{60}Co produces energetic gamma rays between 58 keV and 7.49 MeV, with predominant emissions at 230 keV, 277 keV, 447 keV, 556 keV [see Lone et al. 1981; Tuli 1997]. The activation gamma rays can be reabsorbed or scattered in the emitter material, both processes producing energetic electrons, although these reabsorptions are relatively inefficient (1–2%) in such a small device [Goldstein and Todt 1979]. If the electron exits the emitter, the net positive condition of the emitter causes charge to flow in the signal wire, thereby producing current. The main advantage of Co-SPNDs is the prompt response that allows them to follow better reactor transients and step changes in power.

However, the gamma rays can also interact in the sheath material and the insulator material. If energetic electrons exit the sheath from gamma-ray interactions, then the result opposes the desired emitter emissions. The same is true if a Compton or photo-electron enters the emitter from reactions in the insulator. Another problem with SPNDs that rely upon (n,γ) reactions is the large gamma-ray background of a nuclear reactor.

with a half life of 42 seconds. The two half lives are not additive. The beta particle activity is determined by simultaneously solving the differential equations of the decay scheme.

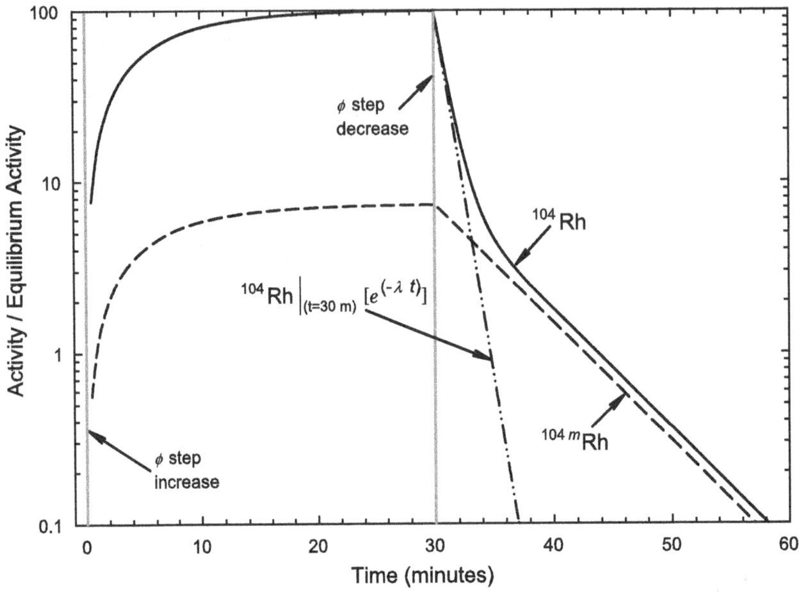

Figure 17.44. Percent ratio of activity from ^{104}Rh and ^{104m}Rh to that of the activity of ^{104}Rh at steady-state equilibrium. The left side of the graph depicts the increase to saturation for an *immediate* step increase in power from zero, and the right side of the graph depicts the activity for an *immediate* step decrease to zero. See text for details regarding issues with this simple model.

These background gamma rays can also interact in the emitter, insulator, or collector, adding either to the emitter or collector currents. Hence, Co-SPNDs are designed such that the net current signal is produced from emitter absorptions, which may be only 10–20% of the total interaction rate [Goldstein and Todt 1979]. Consequently, the inefficiency of capturing the prompt gamma-rays along with the competing interaction processes in the SPND cause the mass sensitivity to be much lower for Co-based SPNDs than for Rh-based SPNDs. Over time the transmutation product ^{60}Co builds up, gradually producing a radioactive product that may require special handling when the SPND is removed from the reactor core. Also, ^{60}Co is a beta particle emitter with $T_{1/2} = 5.27$ years. The result is an increasing background signal with neutron fluence. Some modern systems have compensating electronics for this gradual increase in background current while shielded emitter designs have been proposed to limit the number of ^{60}Co beta particles emerging from the emitter [Goldstein and Todt 1979].

Platinum has also been investigated as an emitter material [Shields 1973; Lynch et al. 1977], although the name SPND is misapplied because it is not strictly a neutron detector.[27] However, for completeness the use of Pt as an emitter is discussed here. Instead of neutron activation, Pt emitters rely directly upon gamma-ray interactions. Further, it is actually desirable that it is insensitive to neutrons because the measurement relies on gamma-ray interactions. Platinum is a candidate because of it high atomic number (78), relatively low average neutron absorption cross section ($\sim$ 10 b), and high density (21.45 g cm^{-3}). Gamma rays interact in the emitter by either photoelectric or Compton events, which eject energetic electrons from the emitter. As with the Co-SPND, gamma rays also interact in the insulator and the collector, giving rise to opposing

[27] Shields named them "self-powered flux detectors".

current to the emitter current. Also as with the Co-SPND, the measured current is the net response from the summed emitter and collector currents. A Pt-SPND has a prompt response to power step changes in a reactor and reportedly a higher sensitivity than Co [Goldstein and Todt 1979]. However, it is assumed with Pt-SPNDs that the gamma-ray flux scales proportionally with the neutron flux, which may not be exact, especially at the reactor core boundaries.

There are additional drawbacks to using Pt as an emitter material; mainly, it does in fact respond to both gamma-rays and neutron interactions [Lynch et al. 1977]. Neutron capture by ^{195}Pt, with a thermal-neutron capture cross section of $\sigma_\gamma = 29$b, produces prompt high-energy gamma-rays of 5.9 MeV and 6.1 MeV [Tuli 1997]. The thermal neutron capture cross section for ^{196}Pt is relatively small at $\sigma_\gamma = 0.4$b; however, the resonance integral capture cross section is $\sigma_{I,\gamma} = 5$ b, producing predominantly 333 keV and 356-keV gamma rays [Lone et al. 1981]. Finally, neutron capture by ^{198}Pt, with $\sigma_\gamma = 3.7$b and $\sigma_{I,\gamma} = 56$b, produces a undesirable beta particle emission response ($T_{1/2} = 30.8$ min). Lynch et al. [1977] report that as much as 55% of the total signal can result from neutron interactions for some emitter geometries, but also provide a method to discern between neutron and gamma-ray induced signals.

17.10 Time-of-Flight Methods

Another method used to measure the energy of a neutron involves the use of timing circuitry (covered in Chapter 22) with which the speed of a neutron is recorded. Suppose that a finite package of neutrons is released from a source, where the starting time gate is linked in coincidence with a neutron detector some distance from the source. Then the detector can be connected to record pulse height events as a function of arrival time. Usually this time spectrum is recorded with a multichannel analyzer where the channels are increments in time. For low energy neutrons, relativistic effects are negligible. The arrival time is a direct measure of the speed, or energy,

$$t = \frac{d}{v} = d\sqrt{\frac{m_n}{2E}}, \tag{17.116}$$

where t is the time duration between pulse initiation and detection, v is the neutron velocity, d is the distance between the neutron source and detector, m_n is the neutron mass, and E is the neutron energy. Given any velocity v or energy E, it becomes clear from Eq. (17.116) that the time spread increases with distance d.

To produce a simulated pulse from a neutron source, a device named a *chopper* is generally used.[28] The first such devices consisted of a large spinning disk with open spaces for neutrons to pass through, while the remaining portions of the disk were composed of a neutron absorber [Amaldi et al. 1935, Frisch and Sorenson 1935, Dunning et al. 1935b, Fink 1936]. The device described by Frisch and Sorenson [1935] was a 50-cm diameter wooden wheel that was 2 cm thick, and relied on neutron collisions with protons in the wood to attenuate slow neutrons. The upgraded design by Dunning et al. [1935b] used an aluminum wheel with portions blocked by Cd. In the Dunning et al. [1935b] design, and also the Fink [1936] design, a second stationary Cd collimator, adjacent to the neutron detector, was used downstream from the rotating disk to accept only those neutrons that emerged from chopper opening.

An improved design, shown in Fig. 17.45 and conceived by Fermi et al. [1947], has a series of parallel slots arranged in a rotating cylinder (or drum). These collimator slots need not be voids, but are usually thin plates of Al clamped between layers of Cd. Notably, the Cd works well for low-energy neutrons with energies below approximately 0.5 eV, but because neutron absorption reduces tremendously above 0.5 eV, it is less effective for higher energies of neutrons. The chopper is rotated at high speeds, where the slots align with the source every half-rotation of the chopper. When the slots are aligned with the neutron stream, a burst of neutrons from the source passes through the chopper and heads towards the detector. The start time is

[28] These devices were originally named *velocity selectors*, but by the mid-1950s the name *chopper* had been largely adopted.

measured by producing a gate signal when the chopper is aligned with the detector, often performed with a magnetic or optical actuator. If it is assumed that the uncertainty in d is relatively small, the uncertainty in the energy calculation can be approximated with Eq. (17.116)

$$\Delta E = \left[\left(\frac{dE}{dt}\right)^2 (\Delta t)^2\right]^{1/2} = 2E\frac{\Delta t}{t} = \sqrt{\frac{8}{m}}\frac{\Delta t}{d}E^{3/2}, \tag{17.117}$$

where Δt is the uncertainty in the flight time.

As shown in Section 17.7.5, the interaction rate in a $1/v$ absorbing detector is independent of neutron energy for slow neutrons with energies in the $1/v$ region. Hence, the number of counts recorded as a function of time yields a measure of the energy distribution of slow neutrons emerging from the chopper. To reduce neutron scattering effects, a beam collimator that is either evacuated or backfilled with ^{4}He or oxygen should be positioned between the chopper and the detector (see Fig. 17.46).[29]

Seidl et al. [1954] describe a fast neutron chopper with eight pairs of collimated channels on a relatively large wheel (30-inch diameter). The device was designed for neutron energies ranging between 10 eV up to 10 keV, where the stopping material is phenolic laminate. The chopper was designed to operate at about 12000 rpm with a temporal emission distribution that appears approximately as an isosceles triangle with FWHM of 0.7 μs. The distance between the chopper and a bank of $^{10}BF_3$ detectors was 20 meters, between which was a large plastic helium-filled "balloon".

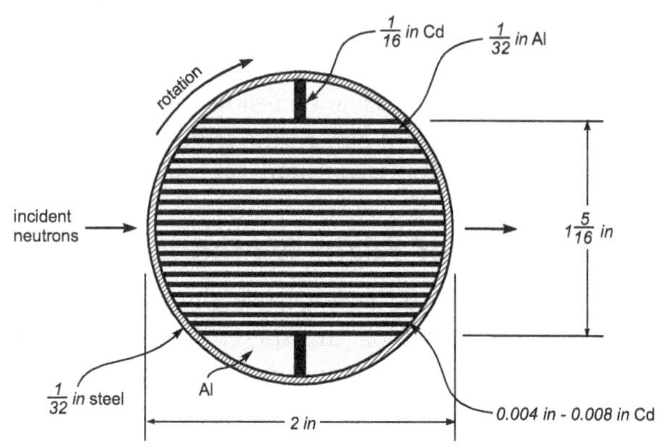

Figure 17.45. Cross section view of the chopper design from Fermi et al. [1947].

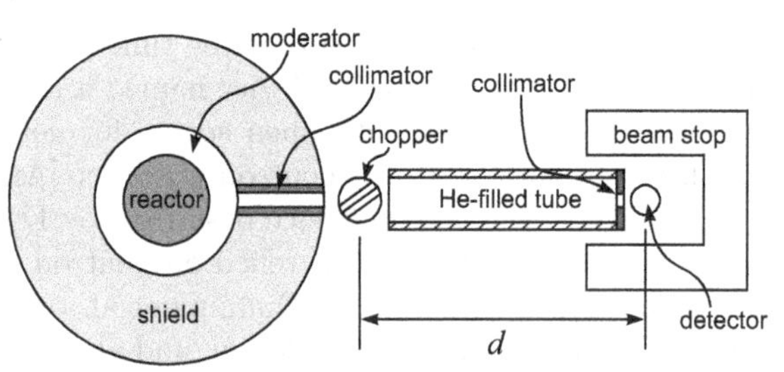

Figure 17.46. One method to arrange a TOF neutron energy measurement from a nuclear reactor.

[29]One of the earliest applications of neutron beam choppers was to measure the neutron lifetime. An alternative method is to place neutrons in a cryogenic bottle, whose walls are perfect neutron reflectors, and measure how many remain after a period of time. At first both methods gave slightly different results, a difference in the mean lifetime of 9-s that persists to this day and is well outside experimental errors. One explanation could be a neutron decay mode that does not produce a proton, in which case the beam method would overestimate the neutron lifetime.

PROBLEMS

1. For an infinite homogeneous medium with weak sources and absorption, the flux of thermal neutrons in thermal equilibrium with the thermal motion of the ambient atoms is given by the Maxwellian distribution

$$\phi_M(E) = \frac{2\pi n}{(\pi kT)^{3/2}} \sqrt{\frac{2}{m_n}} E e^{-E/kT}.$$

Here n is the total neutron density, k is Boltzmann's constant, T is the neutron temperature (K), and m_n is the neutron mass. Show (a) the most probable energy (the energy at which $\phi_M(E)$ has a maximum is $E_T \equiv kT$) and (b) the average energy is $2E_T = 2kT$.

2. Other forms of the Maxwellian distribution include
(a) *neutron number density energy distribution*

$$n_M(E) = \frac{2\pi n}{(\pi kT)^{3/2}} \sqrt{E} e^{-E/(kT)}$$

(b) *neutron flux velocity distribution*

$$\phi_M(v) = 4\pi n \left(\frac{m_n}{2\pi kT}\right)^{3/2} v^3 e^{-m_n v^2/(2kT)}$$

(c) *neutron number density velocity distribution*

$$n_M(v) = 4\pi n \left(\frac{m_n}{2\pi kT}\right)^{3/2} v^2 e^{-m_n v^2/(2kT)}$$

Sketch these three distributions and that of Problem 1. Find the average and most probable speed or energy for each.

3. Determine the thermal-neutron absorption length for a ^{3}He gas-filled detector pressurized to 4 atm. If the detector has a diameter of 1 inch (2.54 cm), what is the maximum attenuation for a beam of neutrons intersecting perpendicular the tube cross section?

4. You have a $^{10}BF_3$ detector built from a 25-mm diameter (I.D.) Al tube, backfilled with 0.2 atm of $^{10}BF_3$ and 2 atm of Ar gas. If a 0.5 cm diameter thermal-neutron (2200 m s^{-1}) beam intersects the detector mid-section, what is the expected intrinsic thermal-neutron detection efficiency?

5. Given a thermal-neutron beam (0.025 eV) intersecting a 10-mm-thick sample of CLYC:Ce, determine the intrinsic thermal-neutron detection efficiency of the CLYC:Ce sample. If the natural Li is replaced with 96% enriched ^{6}Li, what is the new thermal-neutron detection efficiency?

6. A monoenergetic beam of neutrons is normally incident on a thin sheet of material. Confining your attenuation to a single atom in the material, how long must you wait, on the average, before a neutron interacts with the atom? The beam intensity is 10^{10} neutrons cm^{-2} s^{-1} and the total microscopic cross section for the atoms is 10 barns.

7. Determine the intrinsic (2200 m s^{-1}) thermal-neutron detection efficiency for a neutron beam normally incident on the front surface of a Si diode coated with 1.5 microns of 96% enriched ^{10}B. In the coating, the range of the ^{4}He ion is 2.3 microns and that for the Li ion is 0.8 microns. Assume absorption in the contact layer is negligible.

8. Sketch the expected pulse height spectrum from a thermal-neutron beam intersecting the midpoint of a P-10 gas-filled detector with inner walls coated with (first) 1 micron 90% enriched ^{10}B over which a 10-micron layer of 95% enriched ^{6}LiF layer has been applied.

9. Explain the possibilities and limitations of operating ^{3}He and $^{10}BF_3$ detectors as ion chambers, Geiger-Müller counters, or proportional counters. What practical consideration dictates using the latter (proportional) operation?

10. The sensitivity or efficiency of a BF_3 tube or a boron lined tube decreases with use because of the burnup of the ^{10}B. If such a tube is exposed to a constant thermal flux ϕ, show that the tube sensitivity decreases in time as $\exp[-\sigma_a \phi t]$, where σ_a is the absorption cross section for ^{10}B.

11. A young engineer decided that it was time for change with present day neutron detectors and designed a $^{10}BF_3$ proportional tube with ^{6}LiF-coated walls. Sketch the expected pulse height spectrum if the LiF is very thin. Sketch the expected pulse height spectrum if the LiF is as thick as the maximum range of the highest energy reaction product from the ^{6}Li(n,t)^{4}He reaction. Discuss advantages and disadvantages of the design.

12. Why can't the counting efficiency of a boron-lined proportional counter be increased indefinitely by simply increasing the coating thickness?

13. The government of the rogue nation Haides seeks to develop an atomic bomb. But enriching its abundant natural uranium reserves to extract ^{235}U is beyond the technical and financial abilities of the nation. However, Haides' chief nuclear scientist, Prof. Lou Cipher, thinks there is another way to acquire sufficient ^{235}U. Because fission chambers contain 93% enriched U, they could provide an easy, commercially available, alternative to obtaining the needed ^{235}U. One such chamber with an active cylindrical region of length 35 cm and diameter of 18 cm has a 93% enriched UO_2 coating of thickness 0.015 μm. Prof. Cipher has learned from his spies that a reflected sphere of 15 kg of 93% enriched ^{235}U is needed for a nuclear weapon. How many fission chambers must Haides acquire?

14. In what situations is a ^{3}He tube preferred over a BF_3 tube for counting thermal neutrons?

15. What is the expected activity of an 1 cm^2 Au foil 500 microns thick irradiated in a thermal neutron beam of 10^6 cm^{-2} s^{-1} for 30 minutes?

16. Although the gold foil activation technique is designed for thermal neutron measurements in which the neutrons have a Maxwellian energy distribution, it can also be used in other situations involving slow neutrons. For example, a neutron diffractometer is sometimes used to extract neutrons of (almost) a single energy from the neutrons with thermal Maxwellian distribution leaving a beam port of a nuclear reactor. This monodirectional monoenergetic beam can then be used to calibrate a slow neutron detector as described in Sec. 17.6.5. Modify the analysis of Sec. 17.7 to obtain an expression for the activation of a gold foil in such a beam to determine the intensity (flux) of the monoenergetic beam.

17. Show that Eq. (4.146) for a relativistic heavy charged particle reduces to Eq. (17.28) for a non-relativistic heavy ion.

18. Derive Eq. (17.112).

19. Consider an SPND with ^{103}Rh as the emitter with an atom density of N_o. Let the subscripts "1" and "2" refer to ^{104m}Rh and ^{104}Rh, respectively. The atom density, decay constant, and activation cross

section are denoted by $N_i(t)$, λ_i, and σ_i, respectively, $i = 1, 2$. A new SPND is inserted into a reactor operated at a constant flux of ϕ_o. Show that the buildup to equilibrium of the two daughters is given by

$$N_1(t) = \frac{N_o\sigma_1\phi_o}{\lambda_1}(1 - e^{-\lambda_1 t}),$$

and

$$N_2(t) = -\frac{N_o\phi_o(\sigma_1 + \sigma_2)}{\lambda_2}[1 - e^{-\lambda_1 t}].$$

20. Consider the ^{103}Rh SPND of the previous problem. Show that the transients from equilibrium as $\phi_o \to 0$ at $t = 0$ are given by

$$N_1(t) = \frac{N_o\phi_o\sigma_1}{\lambda_1}e^{-\lambda_1},$$

and

$$N_2(t) = N_o\phi_o\left\{\frac{\sigma_1 + \sigma_2}{\lambda_2}e^{-\lambda_2} + \frac{\sigma_1}{\lambda_2 - \lambda_1}\left(e^{-\lambda_1 t} - e^{-\lambda_2 t}\right)\right\}$$

21. Given a Rh SPND operating continuously in a BWR for 6 months at a thermal-neutron flux of 10^{14} cm^{-2} s^{-1}, what is the expected percent decrease in observed current?

22. A beam chopper is used to select neutrons with speeds of 300 m/s. After traveling down a 30-m beam line, 3.40% fewer neutrons reached the detector than left the chopper. Estimate the mean lifetime of neutrons.

REFERENCES

Agu, M.N. and H. Petitcolas, "Self-Powered Neutron Detectors for Measuring Epithermal Neutron Flux," *Reactor Dosimetry: Methods, Applications, and Standardization*, ASTM STP 1001, H. Farrar IV and E.P. Lippincott, Eds., Philadelphia: American Society for Testing and Materials, pp. 720–725, 1989.

Allier, C.P., *Micromachined Si-Well Scintillator Pixel Detectors*, The Netherlands: Delft University Press, 2001a.

Allier, C.P., R.W. Hollandrer, C.W.E. van Eijk, P.M. Sarro, M. de Boer, J.B. Czirr, J.P. Chaminade, and C. Fouassier, "Thin Photodiodes for a Neutron Scintillator Silicon-Well Detector," *IEEE Trans. Nucl. Sci.*, **NS-48**, 1154–1157, (2001b).

Amaldi, E., O. d'Agostino, E. Fermi, B. Pontecorvo, F. Rasetti, and E. Segré, "Radioattivitá Provocata da Bombardamento di Neutroni - VIII," *La Ricerca Scientifica*, **6**, 123–125, (1935).

Ananthanarayanan, K.P., P.J. Gielisse, and A. Choudry, "Boron Compounds for Thermal Neutron Detection," *Nucl. Instrum. Meth.*, **118**, 45–48, (1974).

Angelone, A. K.lix, M. Pillon, P. Batistoni, U. Fischer, and A. Santagata, "Development of Self-Powered Neutron Detectors for Neutron Flux Monitoring in HCLL and HCPB ITER-TBM," *Fusion Eng. Design*, **89**, 2194–2198, (2014).

ANL, "Reactor Physics Constants," 2nd Ed., Argonne National Laboratory, Report 5800, USAEC, 1963.

Aoyama, T., Y. Oka, K. Honda, and C. Mori, "A Neutron Detector Using Silicon PIN Photodiodes for Personal Neutron Dosimetry," *Nucl. Instrum. Meth.*, **A314**, 590–594, (1992).

Babcock, R.V., R.E. Davis, and S.L. Ruby, "Coated Semiconductor is Tiny Neutron Detector", *Nucleonics*, **17**, 116–122, (1959).

Bacon, G.E. and R.D. Lowde, "Secondary Extinction and Neutron Crystallography," *Acta Cryst.*, **1**, 303–314, (1948).

Bacon, G.E., *X-ray and Neutron Diffraction*, Oxford: Pergamon, 1966.

Bacon, G.E., *Neutron Diffraction*, 3rd Ed., Oxford: Clarendon, 1975.

Batchelor, R., "Neutron Energy Measurements with a Helium-3 Filled Proportional Counter," *Proc. Phys. Soc. A*, **65**, 674–675, (1952).

Batchelor, R., R. Aves, and T.H.R. Skyrme, "Helium-3 Filled Proportional Counter for Neutron Spectroscopy," *REv. Sci. Instrum.*, **26**, 1037–1047, (1956).

Bell, Z.W., K.R. Pohl, and L. van den Berg, "Neutron Detection with Mercuric Iodide," *IEEE Trans. Nucl. Sci.*, **51**, 1163–1165, (2004).

Bessiere, A., P. Dorenbos, C.W.E. van Eijk, K.W. Kramer, and H.U. Güdel, "Luminescence and Scintillation Properties Of Cs_2LiYCl_6:Ce^{3+} for Gamma and Neutron Detection," *Nucl. Instrum. Meth.*, **A537**, 242–246, (2005).

BEYERLE, A.G. AND K.L. HULL, "Neutron Detection with Mercuric Iodide Detectors," *Nucl. Instrum. Meth.*, **A256**, 377-380, (1987).

BIROWOSUTO, M.D., *Novel γ-ray and Thermal Neutron Scintillators: Search for High-Light-Yield and Fast-Response Materials*, Amsterdam: IOS Press, 2007.

BLANDIN, CH., S. BREAUD, L. VERMEEREN, AND M. WEBER, "Development of New Sub-Miniature Fission Chambers; Modeling and Experimental Tests," *Prog. Nucl. Energy*, **43**, 349–355, (2003).

BÖCK, H. AND E. BALCAR, "Long-Time Behaviour of Regenerating In-Core Neutron Detectors with ^{238}U-^{239}Pu Electrodes During Power Cycling," *Nucl. Instrum. Meth.*, **124**, 563–571, (1975).

BÖCK, H., "Miniature Detectors for Reactor Incore Neutron Flux Monitoring," *Atomic Ener. Rev.*, **14**, 87–132, (1976).

BOGDZEL, A.A., A. DUKA-ZÓLYOMI, J. KLIMAN, V. PRESPERIN, S.P. AVDEEV, V.D. KUZNETSOV, AND Z. DLOUHÝ, "Fast Ionization Chamber for Fission Fragment Detection with a Large Amount of U-235," *Nucl. Instrum. Meth.*, **200**, 407–410, (1982).

BOLLINGER, L.M., G.E. THOMAS, AND R.J. GINTHER, "Glass Scintillators for Neutron Detection," *Rev. Sci. Instrum.*, **30**, 1135–1136, (1959).

BOLLINGER, L.M., G.E. THOMAS, AND R.J. GINTHER, "Neutron Detection with Glass Scintillators," *Nucl. Instrum. Meth.*, **17**, 97–116, (1962).

BONNER, T.W., "Ionization of Gases by Neutrons," *Phys. Rev.*, **43**, 871–874, (1933).

CAMPBELL, N.R. AND V.J. FRANCIS, "A Theory of Valve and Circuit Noise," *J. Inst. Elec. Eng.*, **93**, 190, (1946).

CAMPBELL, M., E.H.M. HEIJNE, J. KUBAŠTA, P. MIDDELKAMP, S. POSPÍŠIL, P. ŠÍCHO, W. SNOEYS, AND V. VRBA, "Silicon Pixel Devices as a Slow Neutron Precise Position Detector," *Nucl. Instrum. Meth.*, **A395**, 457–458, (1997).

CARUSO, A.N., P.A. DOWBEN, S. BALKIR, N. SCHEMM, K. OSBERG, R.W. FAIRCHILD, O.B. FLORES, S. BALAZ, A.D. HARKEN, B.W. ROBERTSON, AND J.I. BRAND, "The All Boron Carbide Diode Neutron Detector: Comparison and Theory," *Mat. Sci. and Eng.: B*, **135**, 129–133, (2006).

CARUSO, A.N., "The Physics of Solid-State Neutron Detector Materials and Geometries," *J. Phys.: Condens. Matter*, **22**, article 443201, 32 pages, (2010).

CHOU, H.P. AND C.Y. HORNG, "Integral Test of a Boron-10 Loaded Liquid Scintillator for Neutron Detection," *Nucl. Instrum. Meth.*, **A328**, 522–525, (1993).

CLARK, H.K. AND J.L. HOARD, "The Crystal Structure of Boron Carbide," *J. Am. Chem. Soc.*, **65**, 2115–2119, (1943).

COMBES, C.M., P. DORENBOS, R.W. HOLLANDER, AND C.W.E. VAN EIJK, "A Thermal Neutron Scintillator with n/γ Discrimination $LiBaF_3$:Ce,Rb," *J. Nucl. Instrum. Meth.*, **A416**, 364–370, (1998).

CURIE, I. AND F. JOLIOT, "Preuves Expérimentales de L'Existence du Neutron," *J. Phys. Radium*, **4**, 21–33, (1933).

CZIRR, J.B., G.M. MACGILLIVRAY, R.R. MACGILLIVRAY, AND P.J. SEDDON. "Performance and Characteristics of a New Scintillator," *Nucl. Instrum. Meth.*, **A424**, 15–19, (1999).

DOAN, T.C., S. MAJETY, S. GRENADIER, J. LI, J.Y. LIN, AND H.X. JIANG, "Hexagonal Boron Nitride Thin Film Thermal Neutron Detectors with High Energy Resolution of the Reaction Products," *Nucl. Instrum. Meth.*, **A783**, 121–127, (2015).

DOMNICH V., S. REYNAUD, R.A. HABER, AND M. CHHOWALLA, "Boron Carbide: Structure, Properties, and Stability Under Stress," *J. Am. Ceram. Soc.*, **94**, 3605–3628, (2011).

DRAKE, D.M., W.C. FELDMAN, AND C. HURLBUT, "New Electronically Black Neutron Detectors," *Nucl. Instrum. Meth.*, **247**, 576–582, (1986).

DUNNING, J.R., G.B. PEGRAM, G.A. FINK, AND D.P. MITCHELL, "Interaction of Neutrons with Matter," *Phys. Rev.*, **48**, 265–280, (1935a).

DUNNING, J.R., G.B. PEGRAM, G.A. FINK, D.P. MITCHELL, AND E. SEGRÉ, "Velocity of Slow Neutrons by Mechanical Velocity Selector," *Phys. Rev.*, **48**, 704, (1935b).

EDWARDS, N.S., K.A. NELSON, N.J. HINSON, C.N. TINER, M. REICHENBERGER, R. FRONK, AND D.S. MCGREGOR, "Neutron Sensitivity of $^{10}B_4C$-Coated Aluminum Honeycomb using a Single-Anode Wire, P-10 Continuous-Gas-Flow Proportional Counter," *Nucl. Instrum. Meth.*, **A898**, 85–89, (2018).

ELJEN TECHNOLOGY, "Boron Loaded EJ454," **http://www.eljentechnology.com/index.php/products/plastic-scintillators/ej-254**, accessed 2016a.

ELJEN TECHNOLOGY, "Boron Loaded EJ-339, EJ-339A, EJ-339A2," **http://www.eljentechnology.com/index.php/products/liquid-scintillators/ej-339**, accessed 2016b.

ELSASSER, W.M., "Sur la Diffraction des Neutrons Lents par les Substances Cristallines," *Compte Rendus, Academy of Science of Paris*, **202**, 1029–1030, (1936).

EMIN, D. AND T.L. ASELAGE, "A Proposed Boron-Carbide-Based Solid-State Neutron Detector," *J. Appl. Phys.*, **97**, (013529)1–3, (2005).

ENDF/B, National Nuclear Data Center, Brookhaven National Laboratory, `www.nndc.bnl. gov/exfor/endf00.jsp`, accessed 2017.

ENDFPLOT-2, Korea Atomic Energy Research Institute, `/http://atom.kaeri.re.kr /endf/plot.shtmlS`, accessed 2015.

EVANS, R.D., *The Atomic Nucleus*, New York: McGraw-Hill, 1955.

FAIRLEY, E.J. AND A.R. SPOWART, "Neutron Scintillating Glasses: Part III," *Nucl. Instrum. Meth.*, **150**, 159–163, (1978).

FEIGL, B. AND H. RAUCH, "Der Gd-Neutronenzähler," *Nucl. Instrum. Meth.*, **61**, 349–356, (1968).

FERMI, E., J. MARSHALL, AND L. MARSHALL, "A Thermal Neutron Velocity Selector and Its Application to the Measurement of the Cross Section of Boron," *Phys. Rev.*, **72**, 193–196, (1947).

FINK, G.A., "The Production and Absorption of Thermal Energy Neutrons," *Phys. Rev.*, **50**, 738–747, (1936).

FOSTER, M. AND D. RAMSDEN, "A Compact Neutron Detector Based on the use of a SiPM Detector," *Conf. Rec. Nucl. Sci. Symp.*, Dresden, Germany, paper N26-3, Oct. 19–25, 1882–1886, (2008).

FOULON, F., P. BERGONZO, A. BRAMBILLA, C. JANY, B. GUIZARD, AND R.D. MARSHALL, "Neutron Detectors Made From Chemically Vapor Deposited Semiconductors," *Proc. MRS*, **487**, 591–596, (1998).

FOWLER, I.L., "Very Large Boron Trifluoride Proportional Counters," *Rev. Sci. Instrum.*, **34**, 731–739, (1963).

FRISCH, O.R., AND E.T. SORENSON, "Velocity of 'Slow Neutrons'," **136**, 258, (1935).

FRONK, R.G., S.L. BELLINGER, L.C. HENSON, T.R. OCHS, C.T. SMITH, J.K. SHULTIS, AND D.S. MCGREGOR, "Dual-Sided Microstructured Semiconductor Neutron Detectors," *Nucl. Instrum. Meth.*, **A804**, 201–206, (2015).

FUCHS, K., *Efficiency for Very Slow Assembly*, LA-596, Los Alamos: Los Alamos National Laboratory, 1946.

GESLOT, B., T.C. UNRUH, P. FILLIATRE, C. JAMMES, J. DISALVO, S. BRÉAUD, AND J.F. VILLARD, "Method to Calibrate Fission Chambers in Campbelling Mode," *Animma Int. Conf.*, Ghent, Belgium, June 6-9, 4 pages, (2011).

GINTHER, R.J. AND J.H. SCHULMAN, "Glass Scintillators," *IRE Trans. Nucl. Sci.*, **NS-5**, 92–95, (1958).

GINTHER, R.J., "New Cerium Activated Scintillating Glasses," *IRE Trans. Nucl. Sci.*, **NS-7**, 28–31, (1960).

GLODO, J., E. VAN LOEF, R. HAWRAMI, W.M. HIGGINS, A. CHURILOV, U. SHIRWADKAR, AND KS. SHAH, "Selected Properties of Cs_2LiYCl_6, $Cs_2LiLaCl_6$, and $Cs_2LiLaBr_6$ Scintillators," *IEEE Trans. Nucl. Sci.*, **NS-58**, 333–338, (2011).

GOLDSTEIN, N.P. AND W.H. TODT "A Survey of Self-Powered Detectors—Present and Future," *IEEE Trans. Nucl. Sci.*, **NS-26**, 916–923, (1979).

GRESKOVICH, C. AND S. DUCLOS, "Ceramic Scintillators," *Ann. Rev. Mater. Sci.*, **27**, 69–88, (1997).

GROSSHOEG, G., "Neutron Ionization Chambers," *Nucl. Instrum. Meth.*, **162**, 125–160, (1979).

VON HALBAN, H., JR. AND P. PREISWERK, "Preuve Expérimentale de la Diffraction des Neutrons," *Compte Rendus, Academy of Science of Paris*, **203**, 73, (1936).

HILBORN, J.W., "Self-Powered Neutron Detectors for Reactor Flux Monitoring," *Nucleonics*, **22**, 69–74, (1964).

HOFSTADTER, R., J.A. MCINTYRE, H. RODERICK, AND H.I. WEST, "Detection of Slow Neutrons," *Phys. Rev.*, **82**, 749, (1951).

HOSHOR, C.B., T.M. OAKES, E.R. MYERS, B.J. ROGERS, J.E. CURRIE, S.M. YOUNG, J.A. CROW, P.R. SCOTT, W.H. MILLER, S.L. BELLINGER, T.J. SOBERING, R.G. FRONK, J.K. SHHULTIS, D.S. MCGREGOR, AND A.N. CARUSO, "A Portable and Wide Energy Range Semiconductor-Based Neutron Spectrometer," *Nucl. Instrum. Meth.*, **803**, 68–81, (2015).

HUANG, K-C., R. DAHAL, J. J-Q. LU, A. WELTZ, Y. DANON, AND I.B. BHAT, "Scalable Large-Area Solid-State Neutron Detector with Continuous p-n Junction and Extremely Low Leakage Current," *Nucl. Instrum. Meth.*, **763**, 260-265, (2014).

IMEL, G.R. AND P.R. HART, "The Performance of Hafnium and Gadolinium Self Powered Neutron Detectors in the TREAT Reactor," *Nucl. Instrum. Meth.*, **B111**, 325–336, (1996).

ISAENKO, L., A. YELISSEYEV, S. LOBANOV, V. PETROV, F. ROTERMUND, G. SLEKYS, AND J-J. ZONDY, "$LiInSe_2$: A Biaxial Ternary Chalcogenide Crystal for Nonlinear Optical Applications in the Midinfrared," *J. Appl. Phys.*, **91**, 9475–9480, (2002).

JACKSON, C.N., *Reactor In-Core Regenerative Neutron Detectors Interim Development Report*, AEC Research and Development Report BNWL-430, Richland, WA: Pacific Northwest Laboratory, 1967.

KAHN, S., R. HARMAN, AND N. FORGUE, "Energy Distributions of Fission Fragments from Uranium Dioxide Films," *Nucl. Sci. Eng.*, **23**, 8–20, (1965).

KARDJILOV, N., M. DAWSON, A. HILGER, I. MANKE, M. STROBL, D. PENUMADU, F.H. KIM, F. GARCIA-MORENO, AND J. BANHART, "A Highly Adaptive Detector System for High Resolution Neutron Imaging," *Nucl. Instrum. Meth.*, **A651**, 95–99, (2011).

KHAN, S., H.J. KIM, AND Y.D. KIM, "Scintillation Characterization of Thallium-Doped Lithium Iodide Crystals," *Nucl. Instrum. Meth.*, **A793**, 31–34, (2015).

KNITEL, M.J., P. DORENBOS, J.T.M. DE HAAS, AND C.W.E. VAN EIJK, "$LiBaF_3$, a Thermal Neutron Scintillator with Optimal n-γ Discrimination," *Nucl. Instrum. Meth.*, **A374**, 197–201, (1996).

KORFF, S.A., "Fast Neutron Measurements with Recoil Counters," *Phys. Rev.*, **56**, 1241–1242, (1939).

KORFF, S.A. AND W.E. DANFORTH, "Neutron Measurements with Boron Trifluoride Detectors," *Phys. Rev.*, **55**, 980, (1939).

KUMASHIRO, Y., Y. OKADA, S. MISAWA, AND T. KOSHIRO, "The Preparation of ^{10}BP Single Crystals," *Proc. Tenth International Conference Chemical Vapor Deposition*, **8788**, 813–818, (1987).

LACY, J.L., A. ATHANASIADES, L. SUN, C.S. MARTIN, AND G.J. VAZQUEZ-FLORES, "Boron Coated Straw Detectors as a Replacement for ^3He," *Conf. Rec. Nucl. Sci. Symp.*, Orlando, Florida, paper N06-4, Oct. 25–31, 119–125, (2009).

LACY, J.L., A. ATHANASIADES, C.S. MARTIN, L. SUN, AND G.J. VAZQUEZ-FLORES, "Evolution of Neutron Straw Detector Applications in Homeland Security," *IEEE Trans Nucl. Sci.*, **60**, 1140–1146, (2013).

LACY, J.L., ET AL., "Pie-Shaped Boron-Coated Straws With Enhanced Neutron Sensitivity," *Conf. Rec. IEEE Nucl. Sci. Symp.*, Strasbourg, France, Oct. 29 - Nov. 5, (2016).

LAMARSH, J.R., *Nuclear Reactor Theory*, Reading: Addison-Wesley, 1966.

LAMIRAND, V., B. GESLOT, J. WAGEMANS, L. BORMS, E. MALAMBU, P. CASOLI, X. JACQUET, G. ROUSSEAU, G. GRÉGOIRE, P. SAUVECANE, D. GARNIER, S. BRÉAUD, F. MELLIER, J. DISALVO, C. DESTOUCHES, AND P. BLAISE, "Miniature Fission Chambers Calibration in Pulse Mode: Interlaboratory Comparison at the SCK•CEN BR1 and CEA CALIBAN Reactors," *IEEE Trans. Nucl. Sci.*, **61**, 2306–2311, (2014).

LAMPHERE, R.W., "Fission Detectors," in *Fast Neutron Physics, Part I*, Ch. IIIE, J.B. MARION AND J.L. FOWLER, Eds., New York: Wiley, pp. 449–506, 1960.

LECOQ, P., A. ANNENKOV, A. GEKTIN, M. KORZHIK, AND C. PEDRINI, *Inorganic Scintillators for Detector Systems*, Berlin: Springer, 2006.

LEDERER, C.M. AND V.S. SHIRLEY, Eds., *Table of Isotopes, 7th Ed.*, New York: Wiley, 1978.

LINDSAY, J.T., J.D. JONES, C.W. KAUFFMAN, AND B. VAN PELT, "Real Time Neutron Radiography Using a LIXI Neutron Imaging Device," *Nucl. Instrum. Meth.*, **242**, 525–530, (1986).

LM-KAPL (LOCKHEED MARTIN-KAPL), *Nuclides and Isotopes: Chart of the Nuclides*, 16th ed., prepared by E.M. BAUM, H.D. KNOX, AND T.R. MILLER, Schenectady, NY: KAPL Inc., 2002.

LONE, M.A., R.A. LEVITT AND D.A. HARRISON, "Prompt Gamma Rays from Thermal-Neutron Capture," *Atomic Data Nucl. Data Tables*, **26**, 511–559, (1981).

LUKOSI, E., O. CHVALA, AND A.C. STOWE, "Response Functions of Semiconducting Lithium Indium Diselenide," *Nucl. Instrum. Meth.*, **A822**, 9–14, (2016).

LUKOSI, E., E.H. HERRERA, D.S. HAMM, A. BURGER, AND A.C. STOWE, "Neutron Imaging with Lithium Indium Diselenide: Surface Properties, Spatial Resolution, and Computed Tomography," *Nucl. Instrum. Meth.*, **A872**, 181–186, (2017).

LYNCH, G.F., R.B. SHIELDS, AND P.G. COULTER, "Characterization of Platinum Self-Powered Detectors," *IEEE Trans. Nucl. Sci.*, **NS-24**, 692–695, (1977).

McGregor, D.S., J.T. Lindsay, and R.W. Olsen, "Thermal Neutron Detection with Cadmium$_{1-x}$Zinc$_x$Telluride Semiconductor Detectors," *Nucl. Instrum. Meth.*, **A381**, 498–501, (1996a).

McGregor, D.S., J.T. Lindsay, C.C. Brannon, and R.W. Olsen, "Semi-insulating Bulk GaAs Thermal Neutron Imaging Arrays," *IEEE Trans. Nucl. Sci.*, **NS-43**, 1357–1364, (1996b).

McGregor, D.S., S.M. Vernon, H.K. Gersch, S.M. Markham, S.J. Wojtczuk, and D.K. Wehe, "Self-Biased Boron-10 Coated High-Purity Epitaxial GaAs Thermal Neutron Detectors," *IEEE Trans. Nucl. Sci.*, **NS-47**, 1364–1370, (2000).

McGregor, D.S., R.T. Klann, H.K. Gersch, E. Ariesanti, J.D. Sanders, and, B. Van der Elzen, "New Surface Morpohology for Low Stress Thin-Film-Coated Thermal Neutron Detectors," *IEEE Trans. Nucl. Sci.*, **NS-49**, 1999–2004, (2002a).

McGregor, D.S., R.T. Klann, J.D. Sanders, J.T. Lindsay, K.J. Linden, H.K. Gersch, P.M. De Lurgio, C.L. Fink, and E. Ariesanti, "Recent Results From Thin-Film-Coated Semiconductor Neutron Detectors," *Proc. SPIE*, **4784**, 164–182, (2002b).

McGregor, D.S., M.D. Hammig, Y-H Yang, H.K. Gersch, and R.T. Klann, "Design Considerations for Thin Film Coated Semiconductor Thermal Neutron Detectors I: Basics Regarding Alpha Particle Emitting Neutron Reactive Films," *Nucl. Instrum. Meth.*, **A500**, 272–308, (2003).

McGregor, D.S. and J.K. Shultis, "Spectral Identification of Thin-Film-Coated and Solid-Form Semiconductor Neutron Detectors," *Nucl. Instrum. Meth.*, **A517**, 180–188, (2004).

McGregor, D.S. and J.K. Shultis, "Response to the "Comment" by S. Hallbeck et al.," *Nucl. Instrum. Meth.*, **A536**, 232-234, (2005).

McGregor, D.S., M.F. Ohmes, R.E. Ortiz, A.S.M.S. Ahmed, and J.K. Shultis, "Micro-Pocket Fission Detectors (MPFD) for In-Core Neutron Flux Monitoring," *Nucl. Instrum. and Meth.*, **A554**, 494–499, (2005).

McGregor, D.S., T. Unruh, and W.J. Mc Neil, "Thermal Neutron Detection with Pyrolytic Boron Nitride," *Nucl. Instrum. Meth.*, **A591**, 530–533, (2008).

McGregor, D.S. and J.K. Shultis, "Reporting Detection Efficiency for Semiconductor Neutron Detector: A Need for a Standard", *Nucl. Instrum. Meth.*, **A632**, 167–174, (2011).

McGregor, D.S., S.L. Bellinger, W.J. McNeil, K.A. Nelson, and M.F. Ohmes, "Gas-Filled Neutron Detectors Having Improved Detection Efficiency," US Patent 8519350; allowed Aug. 27, 2013a.

McGregor, D.S., S.L. Bellinger, and J.K. Shultis, "Present Status of Microstructured Semiconductor Neutron Detectors," *J. Crys. Growth*, **379**, 99–110, (2013b).

McGregor, D.S., S.L. Bellinger, R.G. Fronk, L. Henson, D. Huddleston, T. Ochs, J.K. Shultis, T.J. Sobering, and R.D. Taylor, "Development of Compact High Efficiency Microstructured Semiconductor Neutron Detectors," *Rad. Phys. Chemistry*, **116**, 32–37, (2015).

McGregor, D.S., S.L. Bellinger, and K.A. Nelson, "Gas-Filled Detectors and Imaging System and Array of Such Detectors," US Patent 9817138; allowed Nov. 14, 2017.

Melcher, C.L., J.S. Schweitzer, "Scintillation Properties of GSO," *IEEE Trans. Nucl. Sci.*, **NS-37**, 161–164, (1990).

Mireshghi, A., G. Cho, J. Drewery, T. Jing, S.N. Kaplan, V. Perez-Mendez, and D. Wildermuth, "Amorphous Silicon Position Sensitive Neutron Detector," *IEEE Trans. Nucl. Sci.*, **NS-39**, 635–640, (1992).

Mireshghi, A., G. Cho, J. Drewery, W.S. Hong, T. Jing, H. Lee, S.N. Kaplan, and V. Perez-Mendez, "High Efficiency Neutron Sensitive Amorphous Silicon Pixel Detectors," *IEEE Trans. Nucl. Sci.*, **NS-41**, 915–921, (1994).

Mitchell, D.P., "The Absorption of Neutrons Detected by Boron and Lithium," *Phys. Rev.*, **49**, 453–458, (1936).

Mitchell, D.P. and P.N. Powers, "Bragg Reflection of Slow Neutrons," *Phys. Rev.*, **50**, 486–487, (1936).

Montag, B.W., M.A. Reichenberger, N. Edwards, P.B. Ugorowski, M. Sunder, J. Weeks, and D.S. McGregor, "Device Fabrication, Characterization, and Thermal Neutron Detection Response of LiZnP and LiZnAs Semiconducting Devices," *Nucl. Instrum. Meth.*, **A836**, 30–36, (2016).

Montag, B.W., S.L. Bellinger, N.S. Edwards, J. Lage, K.A. Nelson, L.C. Henson, and D.S. McGregor, "Recent Progress in the Commercialization of the Li Foil Multi-wire Proportional Counter Neutron Detectors," *Rad. Chem. Phys.*, **155**, 158–163, (2019).

Moreira, O. and H. Lescano, "Analysis of Vandium Self Powered Neutron Detector's Signal," *An. Nucl. Energy*, **58**, 90–94, (2013).

Mughabghab S.F., "Thermal Neutron Capture Cross Sections, Resonance Integral and G-Factors," Report INDC(NDS)-440, International Nuclear Data Comm., IAEA, Austria, 2003.

Muminov, R.A. and L.D. Tsvang, "High Efficiency Semiconductor Thermal-Neutron Detectors," *Sov. At. Energy*, **62**, 316–319, (1987).

Nagarkar, V.V., S.V. Tipnis, V. Gaysinskiy, Y. Klugerman, M.R. Squillante, and G. Entine, "Structured LiI Scintillator for Thermal Neutron Imaging," *IEEE Trans. Nucl. Sci.*, **NS-48**, 2330–2334, (2001).

Ndoye, A., F. Cosset, B. Barelaud, A. Celerier, J.L. Decossas, and J.C. Vareille, "Neutron Radiation of an Electronic Sensor Using Coincidence Method for an Active Dosimeter," *Nucl. Instrum. Meth.*, **A423**, 414–420, (1999).

Nelson, K.A., S.L. Bellinger, B.W. Montag, J.L. Neihart, T.A. Riedel, A.J. Schmidt, and D.S. McGregor, "Investigation of a Lithium Foil Multi-Wire Proportional Counter for Potential ^{3}He Replacement," *Nucl. Instrum. Meth.*, **A669**, 79–94, (2012).

Nelson, K.A., J.A. Geuther, J.L. Neihart,T.A. Riedel, R.A. Rojeski ,P.B. Ugorowski,and D.S. McGregor, "A Novel Method for Detecting Neutrons Using Low Density High Porosity Aerogel and Saturated Foam," *Nucl. Instrum. Meth.*, **A686**, 100-105, (2012).

Nelson, K.A., M.R. Kusner, B.W. Montag, M.R. Mayhugh, A.J. Schmidt, C.D. Wayant, J.K. Shultis, SC AND D.S. McGregor, "Characterization of a Mid-Sized Li Foil Multi-Wire Proportional Counter Neutron Detector," *Nucl. Instrum. Meth.*, **A762**, 119–124, (2014).

Nelson, K.A., N.S. Edwards, N.J. Hinson, C.D. Wayant, and D.S. McGregor, "A Suspended Boron Foil Multi-Wire Proportional Counter Neutron Detector," *Nucl. Instrum. Meth.*, **A767**, 14-18, (2014).

Nelson, K.A., N.S. Edwards, M.R. Kusner, M.R. Mayhugh, B.W. Montag, A.J. Schmidt, C.D. Wayant, and D.S. McGregor, "A Modular Large-Area Li Foil Multi-Wire Proportional Counter Neutron Detector," *Rad. Phys. Chem.*, **116**, 165–169, (2015).

Nicholson, K.P. and G.F. Snelling, "Some Lithium Iodide Phosphors for Slow Neutron Detection," *Brit. J. Appl. Phys.*, **6**, 104–106, (1955).

Nichols, D., M. Reichenberger, S. Stevenson, C. Hilger, T. Swope, K. Kellogg, J. Hewitt, J. Roberts, and D.S. McGregor, "TRIGA Pulse Tracking Utilizing a Multi-Node Micro-Pocket Fission Detector," *Trans. Amer. Nucl. Soc.*, **119**, 328–331, (2018).

Nikolić, R.J., A.M. Conway, C.E. Reinhardt, R.T. Graff, T.F. Wang, N. Deo, and C.L. Cheung, "6:1 Aspect Ratio Silicon Pillar Based Thermal Neutron Detector Filled with ^{10}B," *Appl. Phys. Lett.*, **93**, 133502, 3 pages, (2008).

Nordheim, L.W., *Pile Kinetics*, **MDDC-35**, Oak Ridge: Oak Ridge National Laboratory, 1946.

Ochs, T.R., S.L. Bellinger, R.G. Fronk, L.C. Henson, D.E. Huddleston, Z.I. Lyric, J.K. Shultis, C.T. Smith, T.J. Sobering, and D.S. McGregor, "Present Status of the Microstructured Semiconductor Neutron Detector-Based Direct Helium-3 Replacement," *IEEE Trans. Nucl. Sci.*, **64**, 1846–1850, (2017).

Ochs, T.R., S.L. Bellinger, R.G. Fronk, L.C. Henson, R.M. Hutchins, and D.S. McGregor, "Improved Manufacturing and Performance of the Dual-Sided Microstructured Semiconductor Neutron Detector (DS-MSND)," *Nucl. Instrum. Meth.*, **A**, (2019), in press.

Padalakshmi, M. and A.H. Shaikh, "Effect of Adding Ar Gas on the Pulse Height Distribution of BF_3-Filled Neutron Detectors," *Pramana J. Phys.* **71**, 1177–1181, (2008).

Pausch, F. and J. Stein, "Application of ^{6}Li(Eu) Scintillators With Photodiode Readout for Neutron Counting in Mixed Gamma-Neutron Fields," *IEEE Trans. Nucl. Sci.*, **NS-55**, 1413–1419, (2008).

Petrillo, C., F. Sacchetti, G. Maehlum, and M. Mancinelli, "Space Resolution of a Si/Gd Microstrip as Linear Position-Sensitive Detector for Thermal Neutrons," *Nucl. Instrum. Meth.*, **A424**, 523–532, (1999).

Pospíšil, S., B. Sopko, E. Havránková, Z. Janout, J. Koníček, I. Mácha, and J. Pavlů, "Si Diode as a Small Detector of Slow Neutrons," *Rad. Prot. Dosimetry*, **46**, 115–118, (1993).

Poujade, O. and A. Lebraun, "Modeling of the Saturation Current of a Fission Chamber Taking Into Account the Distortion of Electric Field Due to Space Charge Effects," *Nucl. Instrum. Meth.*, **A433**, 673–682, (1999).

Profio, A.E., *Experimental Reactor Physics*, New York: Wiley, 1976.

Qian, T., P. Tonner, N. Keller, and W.J.L. Buyers, "A Method for Comparing Degradation of Boron Trifluoride and Helium Detectors in Neutrons and Gamma Fields," *IEEE Trans. Nucl. Sci.*, **45**, 636–642, (1998).

Ramírez, G. and L. David, "A Study of Self-Powered Detectors for Mixed Neutron-Gamma Radiation Fields," *Nucl. Instrum. Meth.*, **85**, 279–283, (1970).

Rao, M.N., D.C. Biswas, and R.K. Choudhury, "Charge Determination of Fission Fragments Via Energy Loss and X-Ray Measurements," *Nucl. Instrum. Meth.*, **B51**, 102–106, (1990).

Reeder, P.L., "Neutron Detection Using GSO Scintillator," *Nucl. Instrum. Meth.*, **A340**, 371–378, (1994a).

Reeder, P.L., "Thin GSO Scintillator for Neutron Detection," *Nucl. Instrum. Meth.*, **A353**, 134–136, (1994b).

Reichenberger, M.A., P.B. Ugorowski, J.A. Roberts, and D.S. McGregor, "First-Order Numerical Optimization of Fission-Chamber Coatings Using Natural Uranium and Thorium," *IEEE Nucl. Sci. Symp.*, Seattle, WA, Nov. 8–15, paper N25-6, 4 pages, (2014).

Reichenberger, M.A., T.C. Unruh, P.B. Ugorowski, T. Ito, J.A. Roberts, S.R. Stevenson, D.M. Nichols, and D.S. McGregor, "Micro-Pocket Fission Detectors (MPFDs) for In-Core Neutron Detection," *An. Nucl. Energy*, **87**, 318–323, (2016).

Reichenberger, M.A., D.M. Nichols, S.R. Stevenson, T.M. Swope, C.W. Hilger, R.G. Fronk, J.A. Geuther, and D.S. McGregor, "Fabrication and Testing of a 5-Node Micro-Pocket Fission Detector Array for Real-Time, Spatial, Iron-Wire Port Neutron-Flux Monitoring," *An. Nucl. Energy*, **110**, 995–1001, (2017).

Ritchie, R.H. and H.B. Eldridge, "Thermal Neutron Flux Depression by Absorbing Foils," *Nucl. Sci. Eng.*, **8**, 300–311, (1960).

Rose, A., "Sputtered Boron Films on Silicon Surface Barrier Detectors," *Nucl. Instrum. Meth.*, **52**, 166–170, (1967).

Roux, D.P., *Parallel-Plate Multisection Ionization Chambers for High-Performance Reactors*, **ORNL-3929**, Oak Ridge: Oak Ridge Nat. Lab., 1966.

Sampson, T.E. and D.H. Vincent, "An Absolute Measurement of the Efficiency of a BF_3 Proportional Counter," *Nucl. Instrum. Meth.*, **95**, 563–569, (1971).

Sasaki, M., T. Nakamura, N. Tsujimura, O. Ueda, and T. Suzuki, "Development and Characterization of Real-Time Personal Neutron Dosemeter with Two Silicon Detectors," *Nucl. Instrum. Meth.*, **A418**, 465–475, (1998).

Schelten, J., R. Reinartz, R. Engels, M. Balzhäuser, J. Lauter, W. Schäfer, and K.D. Müller, "Position-Sensitive Neutron Detection with ^{6}LiF Layers on Silicon Semiconductor," *Nucl. Instrum. Meth.*, **A389**, 447–453, (1997).

Schulte, R.L., F.Swanson, and M. Kesselman, "The Use of Large Area Silicon Sensors for Thermal Neutron Detection," *Nucl. Instrum. Meth.* **A353**, 123–127, (1994).

Schulte, R.L. and M. Kesselman, "Development of a Portable Directional Thermal Neutron Detection System for Nuclear Monitoring," *Nucl. Instrum. Meth.* **A422**, 852–857, (1999).

Scintacor, "Glass Scintillators," www.scintacor .com, accessed 2016.

Seidl, F.G.P., D.J. Hughes, H. Palevsky, J.S. Levin, W.Y. Kato, and N.G. Sjöstrand, " "Fast Chopper" Time-of-Flight Measurement of Neutron Responses," *Phys. Rev.*, **95**, 476–499, (1954).

Shields, R.B., "A Platinum In-Core Flux Detector," *IEEE Trans. Nucl. Sci.*, **NS-20**, 603–608, (1973).

Shirwadkar, U., R. Hawrami, J. Glodo, E.V.D. van Loef, and K.S. Shah, "Novel Scintillation Material $Cs_2LiLaBr_{6-x}Cl_x$:Ce for Gamma-Ray and Neutron Spectroscopy," *Conf. Rec. IEEE Nuc. Sci. Symp.*, paper N41-1, Anaheim, CA, Oct. 21–Nov. 3, (2012).

Shultis, J.K. and D.S. McGregor, "Design and Performance Considerations for Perforated Semiconductor Thermal-Neutron Detectors," *Nucl. Instrum. Meth.*, **A606**, 608–636, (2009).

Spalek, J., "Neutron Diffraction," in *A Manual of Reactor Laboratory Experiment*, ANL-6990, by D.A. Daavatila, W.F. Gilbert, L.L. Lawyer, K.C. Ruzich, J. Spalek, and W.J. Sturm, Argonne: Argonne Nat. Lab., 1965.

Spowart, A.R., "Neutron Scintillating Glasses: Part I," *Nucl. Instrum. Meth.*, **135**, 441–453, (1976).

Spowart, A.R., "Neutron Scintillating Glasses: Part II," *Nucl. Instrum. Meth.*, **140**, 18–28, (1977).

STEVENS, H.H., "Neutron Sensors - In-Core," in *Nuclear Power Reactor Instrumentation Systems Handbook*, Ch. 3, HARRER, J.M. AND J.G. BECKERLEY, Eds., TID-25952-PI, Vol. 1, pp. 42–56, 1973.

TULI, J.K., *Thermal-Neutron-Capture Gamma Rays II*, BNL-NCS- 64434 Report, Vol. 2, Upton, NY: Brookhaven National Laboratory, 1997.

TUPITSYN, E., P. BHATTACHARYA, E. ROWE, L. MATEI, M. GROZA, B. WIGGINS, A. BURGER, AND A. STOWE, "Single Crystal of $LiInSe_2$ Semiconductor for Neutron Detector," *Appl. Phys. Lett.*, **101**, paper 202101, 3 pages, (2002).

UHER, J., C. FRÖJDH, J. JAKŮBEK, C. KENNEY, Z. KOHOUT, V. LINHART, S. PARKER, S. PETERSSON, S. POSPÍŠIL, AND G. THUNGSTRÖM, "Characterization of 3D Thermal Neutron Semiconductor Detectors," *Nucl. Instrum. Meth.*, **A576**, 32–37, (2007).

UNRUH, T.C., S.L. BELLINGER, D.E. HUDDLESTON, W.J. MCNEIL, E. PATTERSON, T.J. SOBERING, AND D.S. MCGREGOR, "Design and Operation of a 2-D Thin-Film Semiconductor Neutron Detector Array for Use as a Beamport Monitor," *Nucl. Instrum. Meth.*, **A604**, 150–153, (2009).

UOZUMI, Y., K. ANAMI, A. NOHTOMI, T. SAKAE, AND M. MATOBA, "Absolute Neutron Sensitivity of a GSO(Ce) Scintillation Detector," *J. Nucl. Sci. and Tech.*, **34**, 80–82, (1997).

VAN EIJK, C.W.E., "Inorganic-Scintillator Development," *Nucl. Instrum. Meth.*, **A460**, 1–14, (2001).

VAN EIJK, C.W.E., A. BESSIÉRE, AND P. DORENBOS, "Inorganic Thermal Neutron Scintillators," *Nucl. Instrum. Meth.*, **A529**, 260–267, (2004).

VAN EIJK, C.W.E., J.T.M. DE HAAS, P. DORENBOS, K.W. KRAMER, AND H.U. GÜDEL, "Development of Elpasolite and Monoclinic Thermal Neutron Scintillators," *Conf. Rec. Nucl. Sci. Symp.*, paper N12-3, pp. 239–243, Puerto Rico, Oct. 23–29, (2005).

VERBINSKI, V.V., W.R. BURRUS, T.A. LOVE, W. ZOBEL, N.W. HILL, AND R. TEXTOR, "Calibration of an Organic Scintillator for Neutron Spectrometry," *Nucl. Instrum. Meth.*, **65**, 8–25, (1968).

VERMEEREN, L., M. WEBER, L. ORIOL, S. BRÉAUD, P. FILLIATRE, B. GESLOT, C. JAMMES, S. NORMAND, AND B. LESCOP, "Experimental Verification of the Fission Chamber Gamma Signal Suppression by the Campbelling Mode," *IEEE Trans. Nucl. Sci.*, **58**, 362–369, (2011).

VOITOVETSKII, V.K., N.S. TOLMACHEVA, AND M.I. A RSAEV, "Scintillating Glass for the Detection of Slow Neutrons," *Soviet J. Atomic Energy*, **6**, 203–207, (1960a).

VOITOVETSKII, V.K. AND N.S. TOLMACHEVA, "Lithium Silicate Scintillating Glasses for the Detection of Slow Neutrons," *Soviet J. Atomic Energy*, **6**, 335–336, (1960b).

VRADII, A.G., M.I. KRAPIVIN, L.V. MASLOVA, O.A. MATVEEV, A. KH. KHUSAINOV, AND V.K. SHASHURIN, "Possibilities of Recording Thermal Neutrons with Cadmium Telluride Detectors," *Sov. Atomic Energy*, **42**, 64–66, (1977).

WARREN, H.D. AND N.H. SHAH, "Neutron and Gamma-Ray Effects on Self-Powered In-Core Radiation Detectors," *Nucl. Sci. Eng.*, **54**, 395–415, (1974).

WENDER, S.A., S. BALESTRINI, A. BROWN, R.C. HAIGHT, C.M. LAYMON, T.M. LEE, P.W. LISOWSKI, W. MCCORKLE, R.O. NEISON, W. PARKER, AND N.W. HILL, "A Fission Ionization Detector for Neutron Flux Measurements at a Spallation Source," *Nucl. Instrum. Meth.*, **A336**, 226–231, (1993).

WILPERT, T., "Boron Trifluoride Detectors," *Neutron News*, **23**, 14–19, (2012).

WU, J-W, A. WELTZ, M. KOIRALA, J.J-Q. LU, R. DAHAL, Y. DANON, AND I.B. BHAT, "Boron-10 Nanoparticles Filled Silicon Trenches for Thermal Neutron Detection Application," *Appl. Phys. Lett.*, **110**, paper 192105, 5 pages, (2017).

YAMAGUCHI, S., "Edge Effect of Thin Lithium-Glass Scintillators," *Nucl. Instrum. Meth.*, **A274**, 573–578, (1989).

Chapter 18

Fast Neutron Detectors

These results, and others I have obtained in the course of the work, are very difficult to explain on the assumption that the radiation from beryllium is a quantum radiation, if energy and momentum are to be conserved in the collisions. The difficulties disappear, however, if it be assumed that the radiation consists of particles of mass 1 and charge 0, or neutrons.

Sir James Chadwick

18.1 Detection Mechanisms

Because interaction cross sections for neutrons in the epithermal to fast region are generally relatively small (except in narrowly defined resonance regions) compared to thermal-neutron cross sections, detectors designed for fast neutrons must rely on alternative approaches to those traditionally used for thermal-neutron detectors that rely on a select few isotopes with large reaction cross sections. An additional difficulty is encountered in the detector design if the goal is to distinguish between thermal and fast neutrons.

Fast neutron detection can be divided into three basic methods. First, fast neutrons can be detected by moderating them into the thermal or epithermal energy region for which a traditional thermal-neutron detector can be used as the sensor. Second, fast neutrons, upon scattering, can introduce recoil atoms into a detecting medium, which is either attached to the detector or is part of the detector material. Third, a fast neutron may undergo in the detector material a nuclear reaction that can, by some means, be distinguished from those produced by thermal neutrons.

18.1.1 Neutron Moderation and Scattering

A common method of detecting fast neutrons is through neutron moderation. This technique places low atomic number materials close to a traditional thermal-neutron detector. Neutrons scatter in the moderator material and lose energy with each scatter, until, ultimately, they lose enough energy such that absorption in the thermal-neutron detector becomes likely.

The kinematics of inelastic neutron scattering are discussed in Sec. 4.3.4. In this chapter only elastic scattering ($Q = \Delta = 0$ and $\gamma = 1/A$) is considered and the results of Chapter 4 are considerably simplified. In particular, the interrelationships among the cosine of the scattering angles are repeated here for convenience.[1] From Eq. (4.47)

$$\omega_s = \frac{1 + A\omega_c}{\sqrt{1 + 2A\omega_c + A^2}}, \qquad -1 \le \omega_c \le 1. \tag{18.1}$$

From Eq. (4.48)

$$\omega_c = -(1 - \omega_s^2)/A + \omega_s\sqrt{1 - (1 - \omega_s^2)/A^2}, \quad -1 \le \omega_s \le 1. \tag{18.2}$$

[1]Here the subscript c and s refer to the center-of-mass and laboratory systems, respectively. The scattering angle is θ_i and its cosine is $\omega_i = \cos\theta_i$, $i = c, s$.

For the special case of $A = 1$ in which $\omega_s \geq 0$ this result reduces to $\omega_c = -1 + 2\omega_s^2$, which implies $\theta_c = 2\theta_s$. The cosine of the angle of the recoil nucleus for elastic scattering is given by Eq. (4.55) as

$$\omega_r = \sqrt{\frac{1-\omega_c}{2}} = \sin(\theta_c/2), \quad 0 \leq \omega_r \leq 1, \tag{18.3}$$

and, from Eq. (4.54), the recoil energy of the nucleus is

$$\frac{T}{E} = \frac{2A}{(1+A)^2}(1-\omega_c) = \frac{4A}{(1+A)^2}\omega_r^2. \tag{18.4}$$

This result indicates that the maximum energy transfer occurs when the laboratory scattering angle of the recoil nucleus θ_r is 0, or when $\theta_s = \theta_c = \pi$.

For a given incident neutron energy E, any one of ω_s, ω_c, ω_r, E', or T is independent, i.e., specification of one determines values of the other parameters. Because scattering is azimuthally symmetric about the direction of the incident neutron in polycrystalline materials, the singly differential scattering cross sections in terms of the different scattering parameters are related by

$$\frac{d\sigma_s(E,\omega_c)}{d\Omega_c}2\pi d\omega_c = \frac{d\sigma_s(E,\omega_s)}{d\Omega_s}2\pi d\omega_s = \frac{d\sigma_s(E,E')}{dE'}dE' = -\frac{d\sigma_s(E,T)}{dT}dT. \tag{18.5}$$

Notice the minus sign in the last term, an indication that as T decreases, ω_c, ω_s, and E' all increase.

Example 18.1: Determine the maximum fractional energy transfer for neutron scattering off H and ^{4}He.

Solution:
From Eq. (18.4) for hydrogen, in which $A = Z = 1$,

$$\frac{T}{E} = \left[\frac{4A\cos^2\theta_r}{(1+A)^2}\right] = \left[\frac{4(1)\cos^2(0)}{(1+(1))^2}\right] = 1.$$

This result shows that the total energy E of the neutron can be transferred to the proton in a single collision, provided $\theta_s = \pi$. For elastic scattering from ^{4}He,

$$\frac{T}{E} = \left[\frac{4A\cos^2\theta_r}{(1+A)^2}\right] = \left[\frac{4(4)\cos^2(0)}{(1+(4))^2}\right] = 0.64,$$

which indicates that a maximum of 64% of the neutron energy can be transferred to a ^{4}He nucleus in a single scatter.

18.1.2 Multiscattered Neutrons

Although a single scatter with hydrogen can transfer a fast neutron to the thermal region, generally several scatters are required. On average, the number n of scatters required to degrade a neutron with initial energy E_o to a lower energy E is [Lamarsh 1966]

$$n = \frac{1}{\xi}\ln\frac{E_o}{E}, \tag{18.6}$$

Table 18.1. Elastic slowing down parameters. Data are from [Lamarsh 1966].

Nucleus	Mass No.	ξ
^{1}H	1	1.000
H_2O		0.920
^{2}H (or D)	2	0.725
D_2O		0.509
Be	9	0.209
C	12	0.158
O	16	0.120

where ξ is[2]

$$\xi = 1 - \frac{(A-1)^2}{2A} \ln\left(\frac{A+1}{A-1}\right). \tag{18.7}$$

For example, using the data of Table 18.1, to thermalize a neutron from 2 MeV to 1 eV requires 14.51, 15.77, 28.5, and 91.83 scatters in H, H_2O, D_2O, and C, respectively.

18.1.3 Absorption

Epithermal and fast neutrons can also be absorbed in materials, although the probability of this is usually lower than that observed for slow neutrons in most materials. However, fast neutrons must pass through the absorption resonance energies of materials as they slow down. The rate at which these slowing neutrons are absorbed is defined by the *resonance integral* which, for a $1/E$ slowing down spectrum is defined as

$$I = \int_{E_{CC}}^{E_{\max}} \sigma_a(E)\frac{dE}{E}. \tag{18.8}$$

where $E_{\max}$ is the maximum neutron energy or energy at which absorption resonances can be resolved. Thus the fast neutron absorption rate in a foil, for example, is given by Eq. (17.82).

There are several select isotopes that absorb neutrons only at energies above a specific threshold energy. Discussion of these threshold reactions is deferred to later in this chapter where foil activation methods are discussed.

18.2 Detectors Based on Moderation

A simple method used to detect fast neutrons is to provide a means to moderate the neutrons and slow them to thermal energies so that a traditional thermal-neutron detector can be used, such as a ^{3}He gas-filled detector. The methods may include simply detecting the presence of neutron or may include complicated unfolding algorithms used to determine the neutron energy spectrum. Some of these moderation detection methods are described in the following sections.

18.2.1 Bonner Spheres

A small thermal-neutron detector is placed in the center of a sphere of high density polyethylene (HDPE). Fast neutrons entering the HDPE sphere lose energy through scattering and, thus, increase the probability of interacting in the thermal-neutron detector. Usually the thermal-neutron detector is a ^{6}LiI:Eu scintillator, although $^{10}BF_3$ and ^{3}He gas-filled detectors have been used [Rotondi and Geiger 1968; Goldhagen 2011], as well as activation foils [Thomas et al. 2002a].

The spheres can be obtained commercially in a set of different diameters, often referred to as a *Bonner sphere set*, named after one of the pioneering authors of the method [Bramblett et al. 1960; Bonner 1961]. A set of commercial Bonner spheres usually has between 6 to 10 different sizes so that several different measurements of the neutron field can be made including one with the bare detector. The main components to a Bonner sphere detector are shown in Fig. 18.1, along with a depiction of a few fast neutron trajectories. Also shown in Fig. 18.1 is a photograph of a commercial Bonner sphere set.

For a small sphere, there is little moderation and it is unlikely that a high energy neutron loses enough energy to be efficiently detected by the neutron detector should it enter that sphere. As the size of the

---

[2] A neutron of energy E has a *lethargy* $u(E) \equiv \ln(E_{\max}/E)$ where $E_{\max}$ is the maximum energy of a neutron in a system. As E decreases, u increases, i.e., the neutron moves more lethargically, and hence the name *lethargy*. Here ξ is the average gain in lethargy per scatter averaged over all scattering angles.

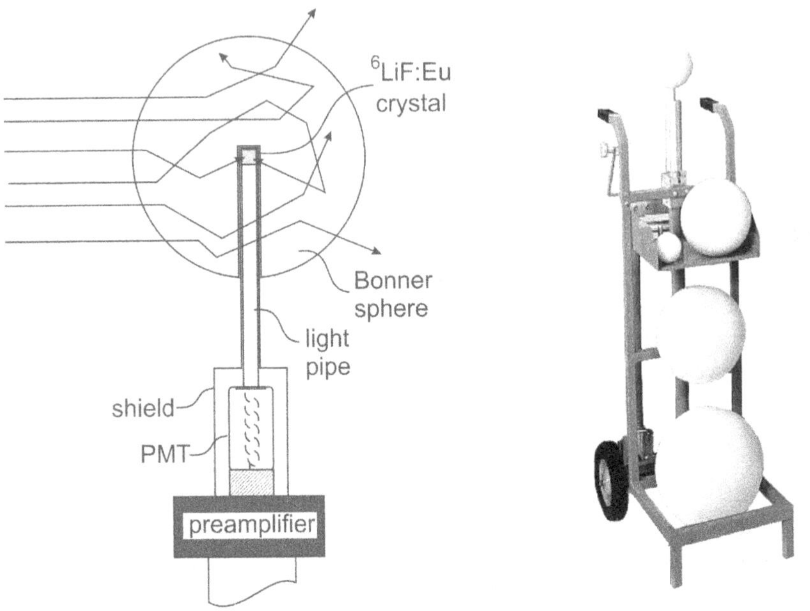

Figure 18.1. (left) Depiction of the main Bonner sphere components. Also shown are several possible neutron trajectories along which some neutrons scatter out of the Bonner sphere, some neutrons are absorbed by the Bonner sphere, and some neutrons are detected by the ^{6}LiF:Eu scintillator. (right) The Ludlum model 42-5 is a commercial Bonner sphere set and rack with 6 different Bonner sphere sizes. Courtesy of Ludlum Measurements.

sphere is increased, the moderation becomes more pronounced for fast neutrons, but overly so for low energy neutrons. Hence, epithermal neutrons may lose too much energy before reaching the detector and be absorbed in the moderator. As the size of the sphere increases further, moderated neutrons with sufficient range to reach the detector may miss it entirely and either be absorbed in the moderator or even escape from the Bonner sphere. Overall, the detector response is a complex function of how the neutrons enter a sphere, the detector absorbing medium (such as ^{6}Li or ^{10}B), the detector absorber mass thickness, the Bonner sphere size, and the distance in air (or other medium) between the neutron source and the detector. Thomas and Alevra [2002b] discuss many of these points in a review on the operation of Bonner spheres for neutron spectroscopy [see also the results of Burgett et al. 2009].

Slow neutrons interact readily with the thermal-neutron detector and, therefore, need no moderation. For each sphere under test, neutrons interacting in the spheres undergo scattering reactions that lead to some moderation. Thermal neutrons may be completely absorbed in even the smallest moderator sphere, usually 2 inches (5 cm) in diameter, and thus go undetected. However, slow neutrons in the low eV range have a relatively high probability of being detected because this sphere is optimized for 1 eV neutrons as can be seen from Fig. 18.2. The next size up is optimized for a slightly higher neutron energy, and so on for the remaining spheres. Most sets have a 12-inch (30.5 cm) diameter sphere as the largest size, although some sets have spheres as large as 18 inches (45.7 cm).

The response function for the ith sphere, $R_i(E)$, such as those shown in Fig. 18.2, gives the average number of counts per incident neutron of energy E. The response function depends on the size and type of thermal-neutron detector at the center of each sphere and how the sphere is illuminated by neutrons. For example, a 1-in diameter incident neutron beam has a different response function than that for broad beam illumination. In such cases the response functions depend on what portion of a sphere is illuminated

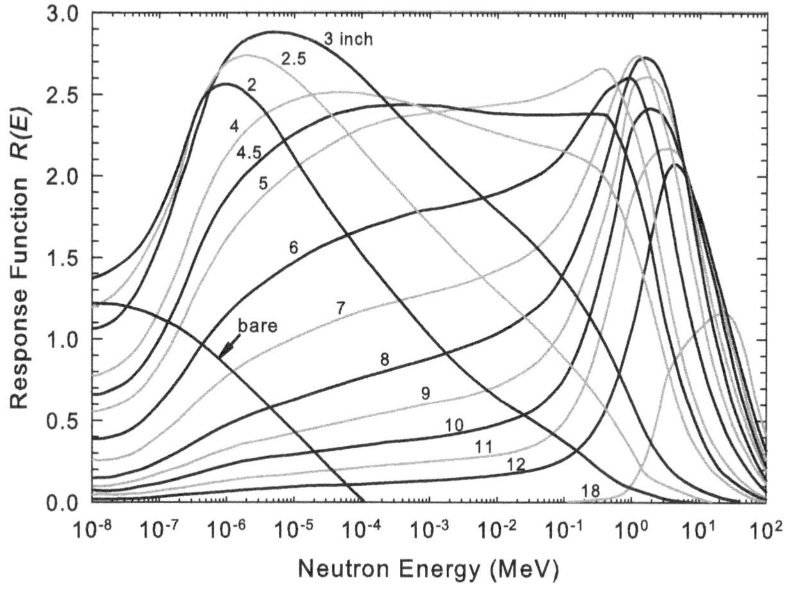

Figure 18.2. Calculated response functions for different sized Bonner spheres over a thermal-neutron detector (^{6}LiI scintillator). Data extracted from Jacobs and van den Bosch [1980].

and the intersecting location on the sphere (centered, off-centered, etc.). Likewise, response functions for a point neutron source near the sphere differs from that for the source placed far from the sphere. Most often reported response functions are for spheres uniformly illuminated by a plane parallel beam of monoenergetic neutrons. In this case $R_i(E)$ gives the average count rate per unit intensity of the beam, i.e., counts per neutron cm^{-2}.

For a particular neutron energy, the probability that the detector records a count can be determined with Monte Carlo techniques [Braga and Dias 2002; Bedogni et al. 2007]. The response functions calculated by Monte Carlo simulations have associated statistical errors and are also obtained for only discrete incident neutron energies. These results are then subjected to some smoothing procedure to make each $R_i(E)$ a smooth continuous function of E.

The neutron field incident on a Bonner sphere is characterized by its energy-dependent intensity or flux $\phi(E)$ (cm^{-2} MeV^{-1} s^{-1}), which here is assumed to be monodirectional and constant in time over a sphere. The detector response function $R_i(E)$ is the count rate obtained with the ith sphere per unit incident beam intensity of energy E. Hence the count rate observed the ith of N spheres is given by the Fredholm equation

$$c_i = \int_{E_{\min}}^{E_{\max}} R_i(E)\phi(E)\,dE, \qquad i = 1, \ldots, N, \tag{18.9}$$

where $E_{\max}$ and $E_{\min}$ are the maximum and minimum neutron energies in the neutron spectrum. Given the N measured c_i and the response functions $R_i(E)$, the unfolding problem is to solve these equations for the neutron energy spectrum $\phi(E)$.

Neutron spectra are usually plotted versus the logarithm of energy. Thus to accurately fit the data, the unfolding problem is better handled in terms of the logarithm of energy. Equivalently, the neutron lethargy

u can be used. The lethargy is defined as

$$u \equiv \ln\left(\frac{E_{max}}{E}\right), \tag{18.10}$$

whose derivative with respect to energy is simply

$$\frac{du}{dE} = -\frac{1}{E}. \tag{18.11}$$

The maximum lethargy u_{max} corresponds to the minimum energy, and Eq. (18.9) can be written as

$$c_i = \int_0^{u_{max}} du\, R(u)\, \phi(u), \qquad i = 1, \ldots, N, \tag{18.12}$$

where $R(u) = R(E(u))$ and $\phi(u) = E(u)\phi(E(u))$.

Typically, the Bonner sphere response functions are determined by using Monte Carlo methods, which with cubic spline or some other method for smoothing result in a set of curves such as those shown in Fig. 18.2. These response function calculations can be verified by using known fast neutron sources (see Tables 5.6 and 5.7 for examples of neutron sources). The next step in unfolding an unknown spectrum is to convert the Fredholm equations, Eq. (18.9) or Eq. (18.12), into linear algebraic equations by using some numerical quadrature method. The simplest such discretization is to assume $\phi(E)$ is constant over each energy interval formed by a contiguous grid of energies (or lethargies) E_j and E_{j+1}, $j = 1, \ldots, M$. Thus, Eq. (18.9) or Eq. (18.12) are transformed to

$$c_i = \sum_{j=1}^{M} R_{i,j}\phi_j, \qquad i = 1, \ldots, N. \tag{18.13}$$

where M is the total number of assigned flux groups. In matrix form these equations can be written compactly as

$$\mathbf{c} = \mathbf{R}\bullet\boldsymbol{\phi} \tag{18.14}$$

where $\mathbf{c}$ and $\boldsymbol{\phi}$ has elements c_i and ϕ_j and the response matrix $\mathbf{R}$ has elements R_{ij}.

Many unfolding methods and algorithms have been developed since the introduction of the Bonner sphere method by Bramblett et al. in 1960. Some of these unfolding techniques take various types of neutron source spectra into consideration, and can identify the most probable source [Matzke 2002; Goldhagen 2011]. However, the method is time consuming and relatively cumbersome because between 6 or more measurements, depending on the set, must be made with long enough measurement times to ensure reasonable counting statistics. Further, during the measurement process, the user assumes that the neutron flux environment is unchanging. Finally, although the result is a neutron spectrum, it is usually of relatively low resolution. Hence, Monte Carlo codes and archived data are often used as a comparison to the measured results to determine the most probable neutron source (or spectrum) [Thomas and Alevra 2002b].

If the number of Bonner spheres N is greater than or equal to the number of energy groups M, then Eqs. (18.9) can be solved outright. The resolution of the solution increases with the number of Bonner spheres (and energy groups). However, if the number of energy groups exceeds the number of Bonner sphere measurements, then one must deal with an underdetermined set of equations, which, generally, have an infinity of solutions, most of which are physically unrealistic such as negative or complex fluences. Two basic approaches for solving Eqs. (18.13) have evolved over time [Ryan 1998]. The oldest is one based on some iteration scheme wherein some initial spectrum guess is modified to give better agreement with Eqs. (18.9). The second and more recent approach is to add constraints (such as positivity) to the possible solutions and obtain as many equations as there are unknowns so as to *regularize* the unfolding problem. The bases of both of these approaches are outlined below.

The Iterative Method of Doroshenko

An iterative algorithm developed by Doroshenko et al. [1977] is used in many commercial unfolding codes for the Bonner sphere system and is the basis of comparison for the Backus-Gilbert method in this study. The Doroshenko method utilizes a simple recursion relationship to infer the approximate solution from the measured data. The algorithm requires an initial guess of the neutron spectrum $\mathbf{\Phi}^0$, which produces estimated count data given by $c_i^0 = \mathbf{r}_i \bullet \boldsymbol{\phi}^0$, where $\mathbf{r}_i$ is ith row vector of $\mathbf{R}$ for moderator i. Successive estimates of the flux $\boldsymbol{\phi}^{k+1}$ are found by iterating the relation

$$\boldsymbol{\phi}^{k+1} = \boldsymbol{\phi}^k \left[\sum_{i=1}^{N} \frac{\mathbf{r}_i}{c_i^k} \Big/ \sum_{i=1}^{N} \frac{\mathbf{r}_i}{c_i} \right], \qquad k = 0, \ldots, k_{max}, \tag{18.15}$$

where k_{max} is the maximum number of iterations and $c_i^k = \mathbf{r}_i \bullet \boldsymbol{\phi}^k$. Essentially, this method attempts to bring the estimated count data c_i^k into agreement with the measured count data c_i by adjusting the estimated neutron spectrum $\boldsymbol{\phi}^k$.

A measure of error ϵ in the iteration is the root mean squared difference between the estimated and measured count rates

$$\epsilon^2 = \frac{1}{N} \sum_{i=1}^{N} \left[\frac{(c_i^k - c_i)}{c_i} \right]^2. \tag{18.16}$$

Although this outcome does not actually reflect the error between the true and measured solutions, ϵ does reflect the convergence rate of the method. In most problems ϵ decreases rapidly and approaches a fixed value, making ϵ a poor cutoff criterion for iteration. Therefore, it is more useful to observe the change, for the k-th iteration, $\delta\epsilon^k = \epsilon^k - \epsilon^{k-1}$, which approaches a uniform value more slowly than ϵ.

Provided that the response functions $\mathbf{r}_i$ and the estimate $\boldsymbol{\phi}^0$ are both positive, positivity of solutions is guaranteed by Eq. (18.15). Good numerical stability is demonstrated for solutions that are similar to the initial estimate. Most importantly, the method gives places more importance on regions where R_i is large, or where each detector combination is the most sensitive. Unfortunately, the method is limited in its ability to handle unknown neutron spectra, where no estimate of $\boldsymbol{\phi}^0$ is possible. Because $\boldsymbol{\phi}^0$ is modified by Eq. (18.15) so as to agree with the measured count rates, it has a strong influence on the final solution $\boldsymbol{\phi}^{k_{max}}$. Secondly, the cutoff criterion k_{max} is an arbitrary parameter that also depends on $\boldsymbol{\phi}^0$. For small values of k_{max} the estimated solution closely resembles $\boldsymbol{\phi}^0$. For large values, numerical instabilities yield non-realistic results.

The Regularization Method

The neutron spectrum $\phi(u)$ can be recovered by minimization principles, using numerical methods as outlined in Press et al. [1992]. To utilize *a priori* information in calculating a solution, one might choose two positive functionals $\mathcal{A}[\boldsymbol{\phi}] > 0$ and $\mathcal{B}[\boldsymbol{\phi}] > 0$, such that $\boldsymbol{\phi}$ can be determined by minimization of $\mathcal{A}[\boldsymbol{\phi}]$ or $\mathcal{B}[\boldsymbol{\phi}]$. $\mathcal{A}$ can be chosen to estimate the accuracy of the solution, and $\mathcal{B}$ to estimate some fixed solution property, such as smoothness.

One obvious functional to minimize that quantifies how good a possible solution $\widehat{\phi}$ is the χ^2 statistic

$$\chi^2 = \sum_{i=1}^{N} \left[c_i - \sum_{j=1}^{M} R_{ij} \widehat{\phi}_j \right]^2, \tag{18.17}$$

where the variances are neglected in this brief discussion. In many regularization schemes $\mathcal{A}$ is taken as χ^2. For the Bonner sphere problem (and many other inverse problems) $M \gg N$ (more unknowns than equations) there are an infinity of solutions some of which make χ^2 very small, if not zero. However, such solutions almost always oscillate widely and otherwise are physically unrealistic (negative or complex values of $\widehat{\phi}$). To obtain realistic solutions, some additional information or constraints on ϕ (contained in the functional $\mathcal{B}$) are specified.

Thus, the goal is to minimize $\mathcal{A}[\phi]$ subject to the constraint that $\mathcal{B}[\phi] = b$. The method of Lagrange multipliers allows this type of minimization problem to be solved by introducing a multiplier μ. The Lagrange functional to be minimized is given by

$$\mathcal{L} = \mathcal{A}[\phi] + \mu_1(\mathcal{B}[\phi] - b), \tag{18.18}$$

The minimum of Eq. (18.18) is found by setting its first derivative with respect to ϕ equal to zero,

$$\frac{\partial}{\partial \phi}\{\mathcal{A}[\phi] + \mu_1(\mathcal{B}[\phi] - b)\} = \frac{\partial}{\partial \phi}(\mathcal{A}[\phi] + \mu_1\mathcal{B}[\phi]) = 0. \tag{18.19}$$

Alternately, $\mathcal{B}[\phi]$ can be minimized subject to the constraint that $\mathcal{A}[\phi] = a$ using Lagrange multiplier μ_2.

$$\frac{\partial}{\partial \phi}\{\mathcal{B}[\phi] + \mu_2(\mathcal{A}[\phi] - a)\} = \frac{\partial}{\partial \phi}(\mathcal{B}[\phi] + \mu_2\mathcal{A}[\phi]) = 0. \tag{18.20}$$

In either case, the inverse problem is changed to

$$\text{minimize} \quad \mathcal{A} + \mu\mathcal{B}. \tag{18.21}$$

As $\mu = \mu_1 = 1/\mu_2$ varies between 0 and ∞, the solution $\phi(\mu)$ varies along a trade-off curve between minimizing $\mathcal{A}$ (small μ) and minimizing $\mathcal{B}$ (large μ). Thus any solution along this trade-off curve can be interpreted as (1) minimizing $\mathcal{A}$ subject to constraint $\mathcal{B} = b$, (2) minimizing $\mathcal{B}$ subject to constraint $\mathcal{A} = a$, or (3) minimizing the sum $\mathcal{A} + \mu\mathcal{B}$.

A benefit of this minimization scheme is that a unique solution ϕ can be obtained even where $\mathcal{A}[\phi]$ is degenerate. The general theory and several methods for regularizing underdetermined problems are given by Press et al. [1992]. Application of this method to unfolding Bonner sphere data is given by Ryan [1998] who used the Backus-Gilbert approach in which $\mathcal{A}$ and $\mathcal{B}$ are chosen so as to maximize the stability of the solution $\widehat{\phi}$. Reginatto et al. [2002] used the maximum entropy method in which $\mathcal{B}$ is taken as the negative of the entropy, or negentropy, and the inverse problem becomes

$$\text{minimize} \quad \chi^2(\widehat{\phi}) + \mu \sum_{j=1}^{M} \widehat{\phi}_j \ln(\widehat{\phi}_j). \tag{18.22}$$

18.2.2 REM Counters

The REM counter is a special Bonner sphere that is 10 inches (25.4 cm) in diameter, and is a common device used to measure neutron dose equivalent H. The detector used with many commercial units is a ^{3}He detector embedded in the ball, the entire detector/sphere combination attached to a handheld survey unit. The reason for the chosen 10-inch diameter is that the detector response is close to that for an average human dose equivalent [Hankins 1962; Bramblett et al. 1960].

Hankins [1962, 1965] compares the calculated response of a 10-inch Bonner sphere with a central ^{6}LiF:Eu detector with the inverse radiation protection guide (RPG) response.[3] This comparison is shown in Fig. 18.3.

[3]This quantity is more frequently called the fluence-to-dose conversion factor or dose response function. It is the dose rate per unit neutron flux in units of dose per neutron.

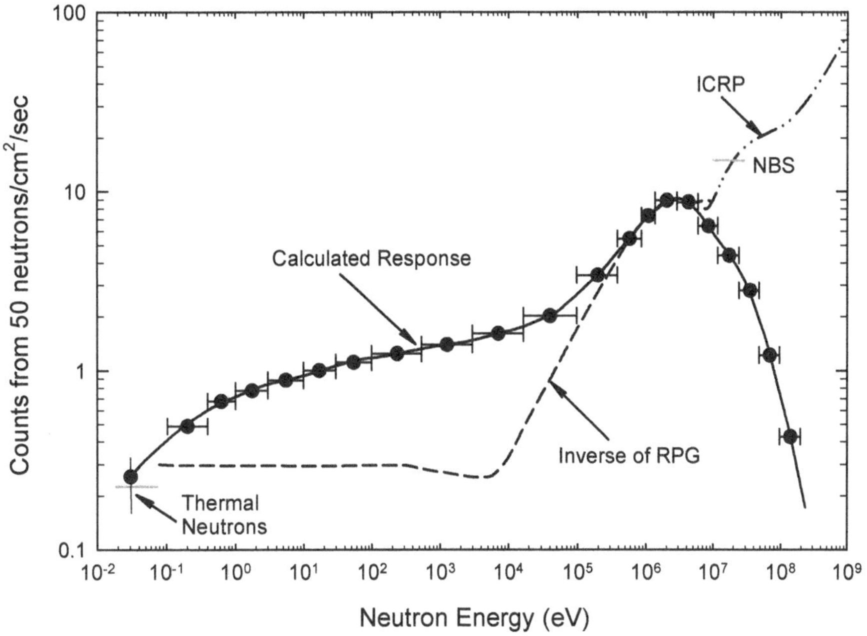

Figure 18.3. Calculated responses from a 10-inch Bonner sphere rem counter compared to (the then) recommended inverse RPG curve. Data are from Hankins [1965]. Also shown are results from the International Commission on Radiological Protection (ICRP) and the National Bureau of Standards (NBS) which is now the National Institute of Standards and Technology (NIST).

At energies below 200 keV, the instrument overestimates the neutron dose by as much as a factor of 5 for some energies. At energies above 7 MeV, the instrument severely underestimates the neutron dose. For these reasons, Hankins [1965] recommends that neutron dosimetry with the basic 10-inch Bonner sphere be restricted to the 200 keV to 7 MeV range. However, modifications to the rem counter design permits the use of smaller diameter spheres, usually ranging between 7 inches and 9 inches, to obtain results that adequately match the inverse RPG curve for energies between thermal and 7 MeV.

A cylindrical neutron dosimeter instrument based on the same concept as the Bonner sphere was introduced by Andersson and Braun [1964]. The detector is designed with $^{10}BF_3$ gas-filled detector with a diameter of about 30 mm and an active length of 60 mm placed inside cylinders of polyethylene. There is an inner annulus of polyethylene with a 0.63 inch (16 mm) wall thickness encapsulated by a perforated 5-mm-thick blanket of boron-loaded polyethylene. Surrounding the entire configuration is another annulus of polyethylene with a 2.56 inch (65 mm) wall thickness and length of 9.6 inches (244 mm). Lastly, a 43-mm-thick plug of polyethylene is placed inside the outer annulus in front of the $^{10}BF_3$ detector. The detector design has several basic features in common with the first long counter designs (discussed in the next section).

Although the response to neutrons above 7 MeV is similar to that determined by Hankins [1965] for the 10-inch Bonner sphere, the energy region below 200 keV gives a better fit to the RPG curve. However, this response improvement comes at the expense of a heavier instrument, reportedly 8.5 kg. These types of detectors are commercially available, which still use the basic cylindrical design and gas-filled neutron detector.[4] In some cases, the $^{10}BF_3$ detector has been replaced with a ^{3}He detector. Modern designs have

[4]The Canberra SNOOPY and the Thermo-Fisher Wendi-2 (see Fig. 18.4) are based on the Andersson-Braun design.

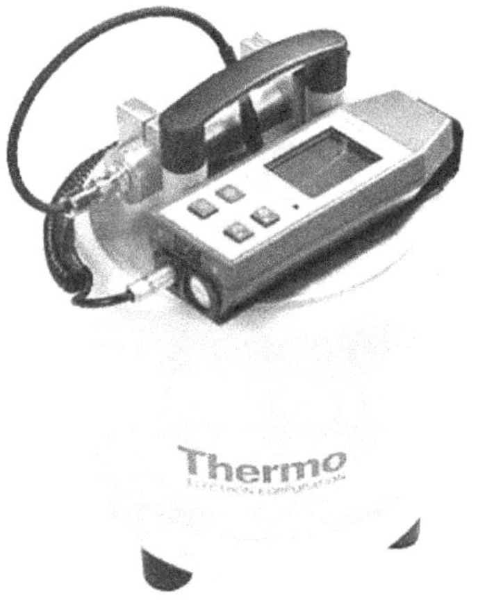

Figure 18.4. The Thermo Fisher model FHT 762 Wendi-2 Wide-Energy neutron detector is a commercial unit based on the Andersson-Braun design. Reproduced with permission from Thermo Fisher Scientific.

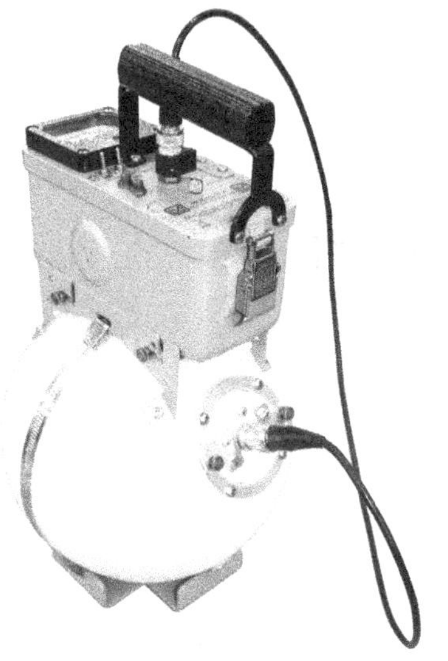

Figure 18.5. The Ludlum model 12-4 is a commercial 9-inch spherical rem counter based on the Leake design. Reproduced with permission from Ludlum Measurements, Inc.

replaced the boron loaded polyethylene with a boron doped synthetic rubber layer. The introduction of heavy elements sheets into the Andersson-Braun design have extended the useful energy range of the instrument. Mares et al. [2002] report a variation that includes a lead layer encapsulating a perforated boron doped rubber layer, both surrounding the polyethylene encapsulated $^{10}BF_3$ detector, a change that extends the neutron detection range beyond 1 GeV. This increase in response is due to (n,xn) fast reactions in the lead.

Leake [1966] modified the Andersson-Braun design to improve the detector response with respect to the RPG curve while reducing the weight of the instrument. The first such modification was styled much like the Hankins [1962] device, but included concentric spheres around the $^{10}BF_3$ detector. A smaller Cd-covered polyethylene sphere encapsulated the detector, while a larger 8.2 inch diameter polyethylene concentric ball surrounded the device. The Cd covered ball suppressed the over-response of neutrons in the low energy region (below 0.5 eV) while reducing the overall weight by eliminating boron-loaded polyethylene sections. Overall, the device response is similar to the inverse RPG curve and weighs only 5 kg.

Another improvement reported by Leake [1968] was to replace the $^{10}BF_3$ detector with a spherical ^3He detector. The change reduced background gamma-ray counts while improving the efficiency, and the entire device weighed approximately 6.6 kg. This *Leake detector*, as it became known, was for many years a standard commercially available instrument.[5] Because of changes made by the International Commission on Radiological Protection (ICRP) in their fluence-to-dose conversion factors, minor alterations in the Leake design have been implemented over the years [Leake 1999; 2004] to remain in compliance. These commercial units are usually designed to follow the inverse RPG curve from thermal neutron energies up to 7 MeV. They also detect neutrons up to 10 MeV, although the results are not dose equivalent. Additional modifications have been explored to increase the detection range of neutron energies. For instance, Hsu et al. [1994] report the responses of a lead shell surrounding the detector, within various polyethylene ball sizes, yielding up to

[5]For instance, Ludlum Measurements, Inc., produces a wide variety of Leake-style handheld units (see Fig. 18.5).

a 5 times increase in sensitivity to 800 MeV neutrons. The use of Cu, Pb, and W to extend the neutron energy range was also explored by Burgett et al. [2009].

18.2.3 Long Counter

Early measurements of the number of neutrons emitted from fast sources were performed by Amaldi and Fermi [1936] and Amaldi et al. [1937] by immersing a neutron source in a water bath and measuring the thermal-neutron signal at various locations in the bath. The technique was later adopted by O'Neal [1946] to measure the neutron energy spectra of different photo-neutron sources. The radiation detector used in the former (Amaldi et al. 1937) case was an ion chamber, while activation foils were used in the latter (O'Neal) case. Although the methods proved useful, the inconvenience of the water bath and the time consuming approaches led Hanson and McKibben [1947] to seek a practical and portable solution, called the long counter.

The first long counter was based on the concept introduced by Amaldi and Fermi [1946], but replaced the water bath with paraffin and the detector with a $^{10}BF_3$ gas-filled proportional counter. The original device had a 1-inch diameter $^{10}BF_3$ detector, with a sensitive length of 8 inches, centrally inserted into a 12-inch long cylinder of paraffin. Slow neutrons could enter the detector front (one end of the $^{10}BF_3$ tube) or the adjacent paraffin. The escape probability for slow neutrons from the paraffin is high near the front of the device but decreases for higher energy neutrons that penetrate further into the paraffin cylinder. Overall, the probability of a neutron being moderated and absorbed somewhere along the length of the $^{10}BF_3$ detector is nearly independent of the neutron energy so a flat energy response is achieved. Through experimentation, it was determined that an 8-inch-diameter paraffin cylinder produced the most uniform counting response from a wide spectrum of neutron energies.

The design was modified to exclude neutrons scattered in air surrounding the device by including a shield surrounding the inner portion. The shield consisted of a B_2O_3 blanket and another thick encapsulating blanket of paraffin, as shown in Fig. 18.6. Holes were drilled into the inner paraffin cylinder to assist with the detection of the lower energy fast and epithermal neutrons. Further, the $^{10}BF_3$ detector active length was 10.5 inches. The basic design of the shielded long counter, as developed by Hanson and McKibben [1947], is still used today, although there have been modifications and improvements over the years [Allen 1960; De Pangher and Nichols 1966; East and Walton 1969].

McTaggart's device [see Allen 1960] implemented a few design modifications to optimize the long counter response. The holes were drilled deeper to a length of 6 inches. McTaggart's experiments demonstrated that the flatness of the long counter response would deteriorate if the detector was pushed from its flush position further into the paraffin moderator (see Fig. 18.7). Hence, the $^{10}BF_3$ detector was inserted such that the front end was flush with the paraffin moderator. Also, a 1.5-inch-long boron sleeve was placed over the $^{10}BF_3$ detector with the front edge 3 inches from the paraffin front face. Finally, the inner moderator length was extended to 14 inches, while the outer moderator length was extended to 20 inches. These changes produced a flat response within 5% for neutrons with energies of 25 keV to 5 MeV. The optimized shielded long counter of McTaggart has become known as the *standard long counter.*

Other changes, such as those described by De Pangher and Nichols [1966] rendered additional improvements. A reported problem with the standard long counter was the response reproducibility between different counters [Tagziria and Thomas 2000]. De Pangher and Nichols [1966] sought to remedy this problem with many changes to the standard long counter design aimed at regularizing the components. The paraffin was replaced with high density polyethylene (HDPE), the B_2O_3 glass thermal neutron shield was replaced with boron-loaded polyethylene, and the outermost housing was fabricated from aluminum. The 1-inch diameter holes were replaced with an annular trench and thicker Cd shields were positioned on both ends of the device. According to Marshall [1970], the De Pangher design, also called the *precision long counter*, rendered reproducible results superior to the standard long counter. Harvey et al. [1976] report that the precision

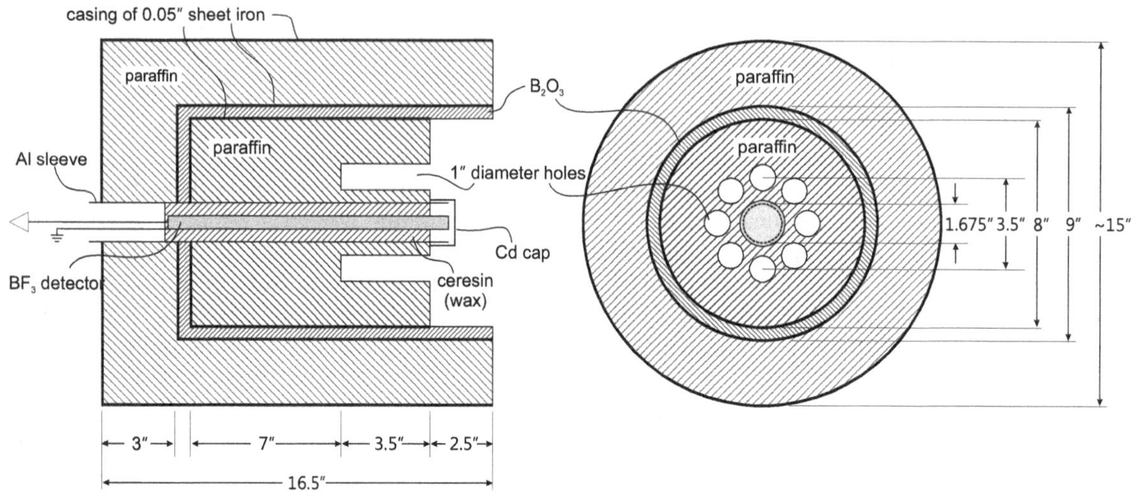

Figure 18.6. The original shielded long counter design as described by Hanson and McKibben [1947]. Depicted are (left) a cross section and (right) front view.

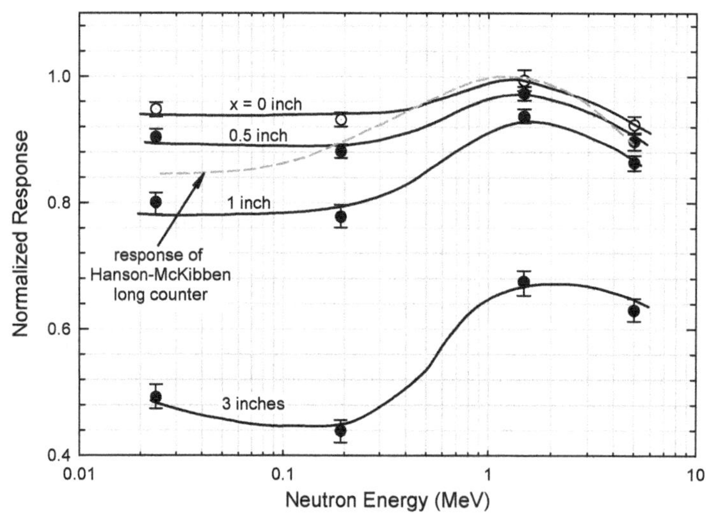

Figure 18.7. The response of McTaggart's long counter as a function of the axial location of the $^{10}BF_3$ detector end (as x) from the paraffin face. Also shown is a comparison between the optimized McTaggart detector ($x = 0$ inch) and the original Hanson-McKibben detector. Data are combined from Allen [1960].

long counter response was relatively independent of neutron energy. Various researchers have calibrated both the standard long counter and the precision long counter [Slaughter and Rueppel 1977; Hunt and Mercer 1978; Tagziria and Thomas 2000], and found that both long counter designs perform well for neutrons with energies up to about 10 MeV.

Another design called the *modified long counter* and reported by East and Walton [1969] replaced the $^{10}BF_3$ detector with five ^3He gas-filled proportional counters in an attempt to increase the overall neutron detection efficiency. This detector has the usual cylindrical shape, but has incorporated many changes to the moderator. The inner moderator is HDPE with a diameter of 9 inches, while the outer moderator is boron-loaded paraffin. Instead of 8 holes drilled into the inner moderator, there are 12 holes, each 1-inch in diameter and 3.5 inches deep in the HDPE. Although the hole pattern is symmetric, they are no longer on a single radial axis, but rather staggered around a radius of approximately 3.5 inches. Another torus of HDPE, approximately 0.75 inch thick, covers most of the hole openings on the device. The resulting device did show improved detection efficiency with a near constant value of $11.5 \pm 0.5\%$ for neutron energies between 25 keV and 4 MeV. However, because the Q value of the ^{3}He(n,p)^{3}H is much lower than the Q value of the ^{10}B(n,α)^{7}Li reaction, the authors report that background gamma-ray interference is worse than with a standard long counter, and recommend the standard long counter for neutron measurements in a high gamma-ray field.

18.2.4 Directional Neutron Spectrometer

First introduced by Oakes et al. [2010], and Cooper et al. [2011; 2012], the directional neutron spectrometer is a relatively new device that uses the fundamental concept of the Bonner sphere. However, instead of having to change several spheres, one for each measurement, the unit has numerous slices of moderators arranged as a cylinder. Between each moderator slice is an array of compact thermal-neutron detectors, usually of the MSND variety (see Sec. 17.5). A single spectral measurement is performed with all detectors in place, and does not require the removal or replacement of moderators [Oakes et al. 2013]. The output data is processed and template-matched to the most probable neutron source from a library containing hundreds of environment conditions and neutron source spectra. A depiction of the first prototype detector arrangement is shown in Fig. 18.8.

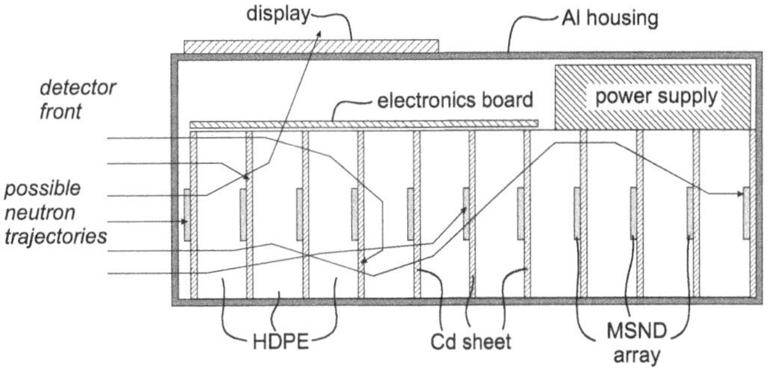

Figure 18.8. The basic arrangement of a sectioned neutron spectrometer from Cooper et al. [2011, 2012].

The initial sectioned neutron spectrometers had up to 20 HDPE layers with a fourplex array of dual-stacked MSNDs between each layer, and a bare MSND array at the front of the stack [Cooper et al. 2011]. In later experimental versions, the small fourplex array was increased to as many as 108 detectors per layer with 30 moderator layers to give a total of 3240 detector channels [Hoshor et al. 2015]. With a mass of nearly 50 pounds (22.7 kg), this large device pushed the limits of portability. Commercial models are considerably lower in weight and have approximately 16 MSNDs per layer. The total number of detectors varies with the number of moderator slices, usually between 7 to 12 slices per unit. The response of an eight-layer device to an AmBe and ^{252}Cf source is shown in Fig. 18.9. Commercial handheld devices range in weight from approximately 8 pounds (3.62 kg) to 20 pounds (9.1 kg). The detector can rapidly identify the neutron spectrum from numerous fast neutron sources under a variety of environmental circumstances [Hoshor et al. 2015; 2017].

This neutron spectrometer also doubles as an imaging device, capable of locating the general direction(s) of the fast neutron source(s). Although the direction finding algorithm can be complicated, it fundamentally operates by tallying counts from the MSNDs in the total array and identifying an imaginary vector drawn from the lowest average count rate per device to the highest average count rate per device. The vector is used to point towards the most probable source location and is usually accurate to within a few degrees.

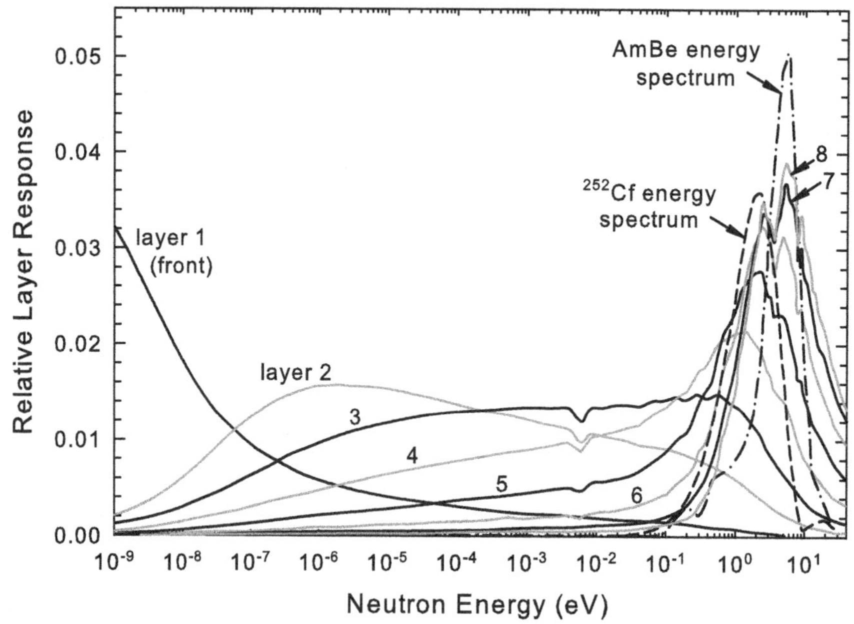

Figure 18.9. Calculated relative response functions for an eight-layer sectioned neutron spectrometer. Also shown for comparison are energy spectra from ^{252}Cf and AmBe neutron sources. Data from Hoshor et al. [2015].

18.2.5 Other Moderated Detectors

There are other types of detectors that take advantage of the moderation detection method. In fact, practically any thermal-neutron detector can be inserted into a moderator material, usually HDPE, to improve overall sensitivity to fast neutrons. The gas-based neutron detectors described in Chapter 17 are commonly inserted into HDPE blocks to improve fast neutron responses. Along with the gas-filled devices already described in this chapter, other examples include moderated ^3He detectors reported by Kamboj et al. [1978], and moderated $^{10}BF_3$ detectors reported by Barrett et al. [1969] and Sekharan et al. [1976]. The system reported by Kamboj et al. [1978] has several ^3He detectors in a layered block of polyethylene with the goal of producing another neutron source calibration device similar to the long counter. They report a linear neutron energy response between 2.1 and 4.5 MeV. The system described by Barret et al. [1969] has numerous BF_3 detectors embedded in a large single block of paraffin all positioned radially about a neutron source. The method is based on the analysis by Thies [1963] and provides a relatively flat response for neutron energies ranging from 30 keV to 5 MeV. A similar 4π system reported by Sekharan et al. [1976] was designed to measure the cross sections for various neutron producing reactions. The performance of moderated ^6Li-foil detectors were described by Nelson et al. [2014; 2015]. High densities of MSNDs have been inserted into moderators with reportedly similar performance as high-pressure ^3He gas-filled detectors [Ochs et al. 2017].

18.3 Detectors Based on Recoil Scattering

Detectors based on recoil scattering rely on the recoil atoms transferring energy to a detection device. For gas detectors, the recoil ion generates electron-ion pairs that can be measured as the output. Scintillators generally rely either on scattered recoil ions that are a part of the scintillator to fluoresce the scintillator, such as an organic scintillator, or rely on recoil products entering and fluorescing an adjacent scintillator,

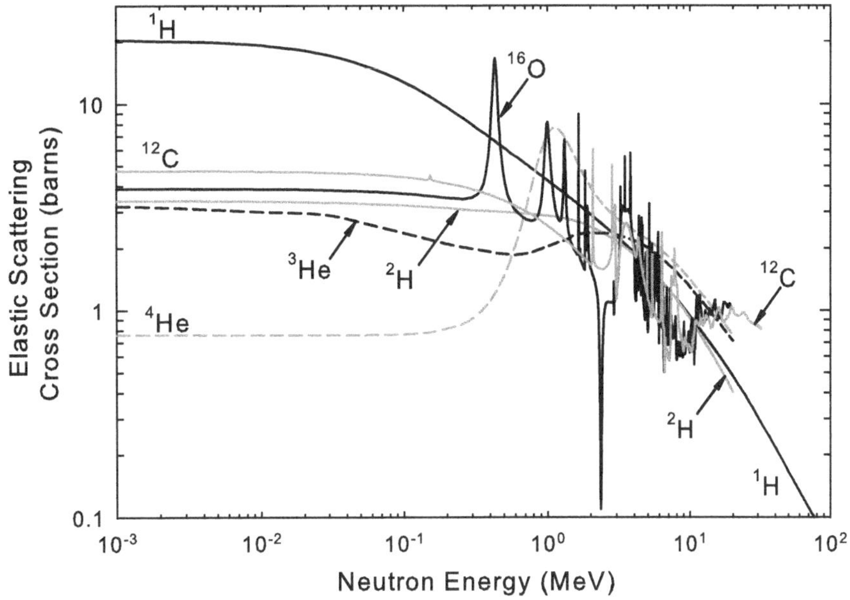

Figure 18.10. Elastic scattering cross sections for many materials used for recoil detectors. Data are from [ENDFPLOT 2017].

such as a ZnS:Eu scintillator. Semiconductors can also be used, whose constituent elements in the material recoil and produce electron-hole pairs. From Eq. (18.4), it becomes clear that more energy is transferred per scatter from low A recoil materials than from high A materials. Consequently, most detectors based on recoil ion scatter are manufactured from the lighter elements. The neutron elastic scattering cross sections for several isotopes used for recoil detectors are shown in Fig. 18.10.

Differential Scattering Cross Sections

The differential scattering cross section quantifies how likely a neutron is to scatter into some new direction from its initial direction. With this information the energy and angular distribution of a recoil nucleus can be determined. This topic is explored in this section.

Of course all scattering cross section measurements are performed in the laboratory coordinate system but, primarily for historical reasons, scattering information was often converted to the center of mass system, fitted to a sum of Legendre polynomials, and the resulting expansion coefficients were then archived in the early cross section libraries. To use such data, the results, upon extraction, had to be reconverted to the laboratory system. This cumbersome approach was used because generally much less scattering data needed to be saved. For low energy neutron scattering in which a compound nucleus is formed, which is at rest in the center of mass system, the subsequent neutron decay occurs isotropically because information about the incident neutron direction is "forgotten" during the relatively long lifetime of the compound nucleus. Thus, the differential scattering cross section would appear almost constant in the center of mass system and only one Legendre coefficient had to be saved.

Today, computer memory and data storage is much less expensive and data from scattering in the laboratory system can readily be archived. The many cross section libraries available from nuclear institutions around the world, now generally have a mixture of scattering data in the two coordinate systems and one must be careful in the use of retrieved scattering data as to which coordinate system was used for the data.

The conversion of the differential scattering cross section from the laboratory system into the center of mass system is found from Eq. (18.5) as

$$\frac{d\sigma_s(E,\omega_c)}{d\Omega_c} = \frac{d\sigma_c(E,\omega_s)}{d\Omega_s}\frac{d\omega_s}{d\omega_c}, \tag{18.23}$$

where $\omega_s = \cos(\theta_s)$ and $\omega_c = \cos(\theta_c)$. Differentiation of Eq. (18.1) with respect to ω_c and substitution of the result into Eq. (18.23) gives

$$\frac{d\sigma_c(E,\omega_c)}{d\Omega_c} = \frac{d\sigma_s(E,\omega_s)}{d\Omega_s}\frac{A^2[A+\cos(\theta_c)]}{[1+2A\cos(\theta_c)+A^2]^{3/2}}. \tag{18.24}$$

Energy Distribution of the Recoil Nucleus

The probability that a neutron of initial energy E transfers energy into dT about T to the scattering nucleus is

$$P(E,T)\,dT = \frac{1}{\sigma_s(E)}\frac{d\sigma_s(E,T)}{dT}\,dT, \tag{18.25}$$

which, from Eq. (18.5), can be written as

$$P(E,T)\,dT = \frac{1}{\sigma_s(E)}\frac{d\sigma_s(E,\omega_c)}{d\Omega_c}\left(-\frac{2\pi d\omega_c}{dT}\right)dT. \tag{18.26}$$

Differentiation of Eq. (18.4) with respect to ω_c gives

$$\frac{d\omega_c}{dT} = -\frac{(1+A)^2}{2AE} = -\frac{2}{(1-\alpha)E}, \tag{18.27}$$

where $\alpha = (A-1)^2/(A+1)^2$. Substitution of this result into Eq. (18.26) gives the energy distribution of the recoil nucleus

$$P(E,T) = \frac{1}{\sigma_s(E)}\frac{d\sigma_s(E,\omega_c)}{d\Omega_c}\frac{4\pi}{(1-\alpha)E}, \qquad 0 \le T \le (1-\alpha)E. \tag{18.28}$$

Thus the ideal pulse height spectrum from a single scatter is proportional to the differential scattering cross section in the center of mass coordinate system.

Example 18.2: In some cases scattering in the center of mass system is isotropic, i.e., $d\sigma_s(E,\omega_c(E,T))/d\Omega_c$ is simply $\sigma_s(E)/(4\pi)$. For this case, derive an expression for $d\sigma_s(E,T)/dT$.

Solution:

From Eq. (18.2) one finds

$$\frac{d\omega_c}{dT} = \frac{-2}{(1-\alpha)E}$$

so that, from Eq. (18.5),

$$\frac{d\sigma_s(E,T)}{dT} = -\frac{d\sigma_s(E,\omega_c)}{d\Omega_c}\frac{4\pi}{(1-\alpha)E},$$

which upon substitution of $d\sigma_s(E,\omega_c(E,T))/d\Omega_c \simeq \sigma_s(E)/(4\pi)$, gives the desired result

$$\frac{d\sigma_s(E,T)}{dT} \simeq \frac{\sigma_s(E)}{(1-\alpha)E}.$$

18.3.1 Gas Detectors Based on Recoil Scattering

The most popular backfill gases are low mass gases, usually ^1H, ^3He, ^4He, and CH_4. The reaction cross sections for these gases are shown in Fig. 18.11. Except for ^3He detectors, detectors filled with these gases all respond in similar manners. All gas-based recoil detectors function with a gas composed, at least in part, of a low Z material. The usual candidates include pure hydrogen, CH_4, and helium, although other hydrogenous gases can work. Energetic neutrons scatter from the detector gas and transfer some kinetic energy to the recoiling atom, as described by Eq. (18.4).[6] Typically these detectors are designed as a type of proportional counter, although gridded ion chambers can also work. Gas recoil detectors can in principle detect relatively low energy scatters, yet the practical lower energy threshold is near 20 keV [Ferguson 1960]. The detector cavity should be of adequate dimensions beyond the expected range of the recoiling ions in order to reduce wall effect distortions in the pulse height spectrum. The ion ranges can be reduced by increasing the gas pressure; however, as reported in Chapter 10, increasing the gas pressure reduced the charge carrier velocity and can reduce the response proportionality. At operating pressures up to 20 atm, these detectors can respond to neutron energies up to 20 MeV.

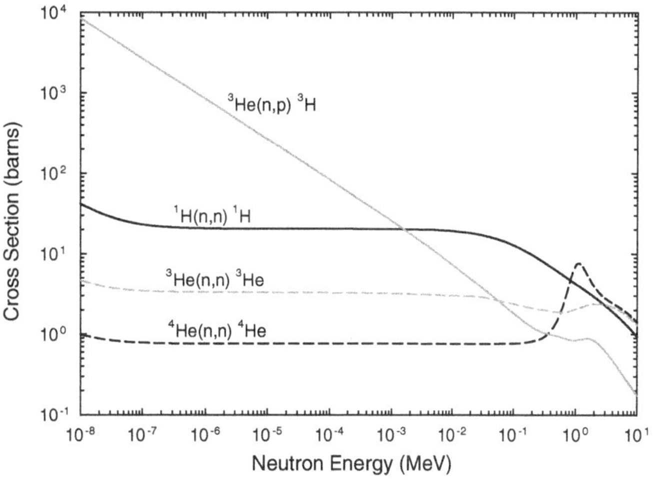

Figure 18.11. Neutron reaction cross sections for ^1H, ^3He, and ^4He. Data are from [ENDF VIII 2018].

The possibility of carbon scattering is high in a CH_4 chamber, and the differential scatter cross section for carbon is not at all uniform. Also, the maximum energy transfer for carbon is only 28%, significantly less than proton recoil contributions. Further, the pulse height spectrum of the recoil energy becomes even more complicated when the proton recoil wall effect is included. Detectors based on the use of ^4He are a commercial alternative. These detectors do not have the complications of multiple nuclides, but the differential scattering cross section is not isotropic. Hence, the pulse height spectrum is not a simple step function, although the endpoint energies for monoenergetic neutrons should be near 64% of the initial neutron energy (see Example 18.1).

Typical commercial units are pressurized between 0.526 atm (400 torr) up to 10 atm (7600 torr). Advantages of the high pressure tubes include increased detection efficiency and extended energy range. However, the operating voltage must also be increased, usually to several thousand volts. Further, the gamma-ray background is higher for the pressured recoil detectors. Detectors with lower pressures respond faster to neutrons, require much lower operating voltages (often less than 1000 volts), and suffer less background contamination. However, the recoil particle ranges are longer such that the recoil range increases approximately inversely proportional to the gas pressure. Because of the importance of hydrogen as a moderator and a source of recoil ions, these detectors are considered first.

[6] The recoil atom may be an ion having lost one or more shake-off electrons. As the speed of the recoil atom increases above the speed of K-shell electrons, it becomes increasingly likely the recoil atom has shed one or more electrons.

Proton Recoil Detectors with Hydrogen

Neutron scattering from hydrogen is nearly isotropic in the center-of-mass coordinate system. Hence, the differential scattering cross section $d\sigma_s^H(E,\omega_c)/d\Omega_c$ for hydrogen can be well approximately as $\sigma_s^H(E)/(4\pi)$ for neutron energies well beyond 20 MeV. From Eq. (18.28) with $\alpha = 0$ for H, the ideal recoil energy distribution $P(E,T)$ is a constant or step function ranging from $T = 0$ to $T = E$, the incident neutron energy.

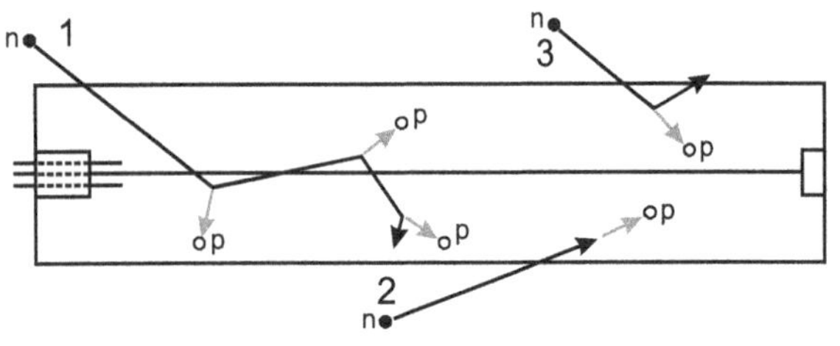

Figure 18.12. Three possible ways in which neutrons can interact in a $^1H_2$ gas detector.

To preserve energy information, these detectors are generally fashioned as cylindrical proportional counters. The interaction efficiency for a given neutron energy can be estimated by a simple exponential approximation,

$$\epsilon(E) = 1 - \exp[-N_H\sigma_s^H(E)], \qquad (18.29)$$

where N_H is the atomic density of hydrogen in the detector, σ_s^H is the elastic scatter cross section of hydrogen, and w is the detector effective width. Three possible neutron trajectories are depicted in Fig. 18.12. For trajectory 1, multiple scatters cause the neutron to pass nearly all of its energy to hydrogen in the detector. For trajectory 2, the neutron escapes the detector after leaving partial energy behind. Finally, in trajectory 3, the neutron imparts all of its energy to a proton in a single scatter. Because the range of energetic protons in hydrogen gas is relatively long, the recoil protons often strike the wall or enter the dead regions of the counter (see end effect in Sec. 17.3.1) so as to produce a type of wall effect. These effects distort the pulse height spectrum from the expected step function. Although hydrogen seems to be an obvious choice for a recoil detector gas, it is usually not used in spectroscopic applications because of the wall-effect distortions.

Several types of gas-filled recoil detectors use high purity hydrogen gas, which would be an obvious choice considering the previously reviewed kinematics for scattering from ^{1}H. Shown in Fig. 18.13 are the differential elastic scattering cross sections in the laboratory system for different incident neutron energies on ^{1}H. The results of combining the differential scattering cross section for hydrogen in Fig. 18.13 at a few select energies with the conversion of Eq. (18.24) are shown in Fig. 18.14. Note that when $A = 1$ that Eq. (18.2) reduces simply to $\theta_c = 2\theta_s$. It is seen that the differential elastic scattering cross section for hydrogen is remarkably constant (isotropic) at all scatter angles in the center of mass system, i.e., $d\sigma_s(E_c,\omega_c)/d\Omega_c = \sigma_s(E)/4\pi$.

With A equal 1 and isotropic center-of-mass scattering, Eq. (18.28) reduces to $P(E,T) = 1/E$, $0 \le T \le E$ so the expected pulse height spectrum from hydrogen scattering is simply a rectangular step function terminating at $T = E$, the energy of the incident neutron. There are a few notable points when dealing with ^{1}H as the scattering medium. First, the neutron scatter angle in the laboratory system is restricted between angles $\pi/2 \le \theta_s \le 0$, i.e., there is no neutron backscatter. Second, the neutron can transfer its entire energy to the recoil proton. Third, the probability of scattering through any angle between $\pi \le \theta_c \le 0$ in the center-of-mass system is practically constant.

Although it would seem that ^{1}H is the ideal gas for a recoil spectrometer, there are some problems with the use of ^{1}H as the recoil gas. The range of the recoil proton can be significant in hydrogen gas, approximately 163 ± 2.13 mm for a 2 MeV proton in 2 atm of $^1H_2$ gas. Even if the detector is oriented longitudinally to the fast neutron source, there is still a significant probability that the recoil proton collides with the chamber wall before expending all of its energy, thus producing a wall effect (see Section 17.3.1). The wall effect can be reduced by increasing the gas pressure, but this remedy comes at the expense of longer response time and increased gamma-ray background. Recall from Chapter 10 that gamma-rays primarily interact in the detector shell and not directly in the gas. Hence, the energy deposited by photoelectrons or Compton electrons entering the gas increases approximately linearly with gas pressure. Consequently, the LLD must be increased as the pressure is increased and, hence, reduces the epithermal neutron response

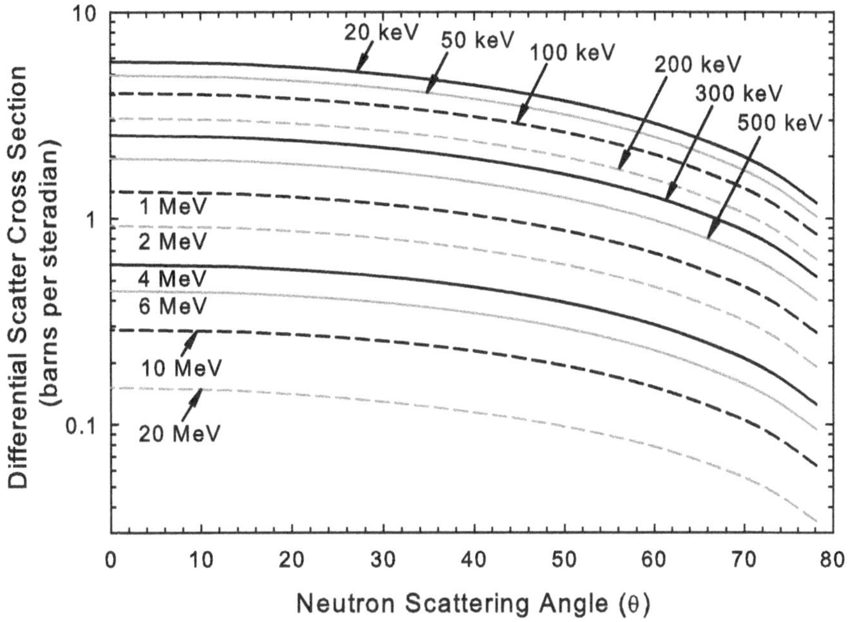

Figure 18.13. The laboratory system ^{1}H differential scattering cross sections in barns per steradian ($d\sigma/d\Omega$) for several incident neutron energies. Data are from [ENDF VIII 2018].

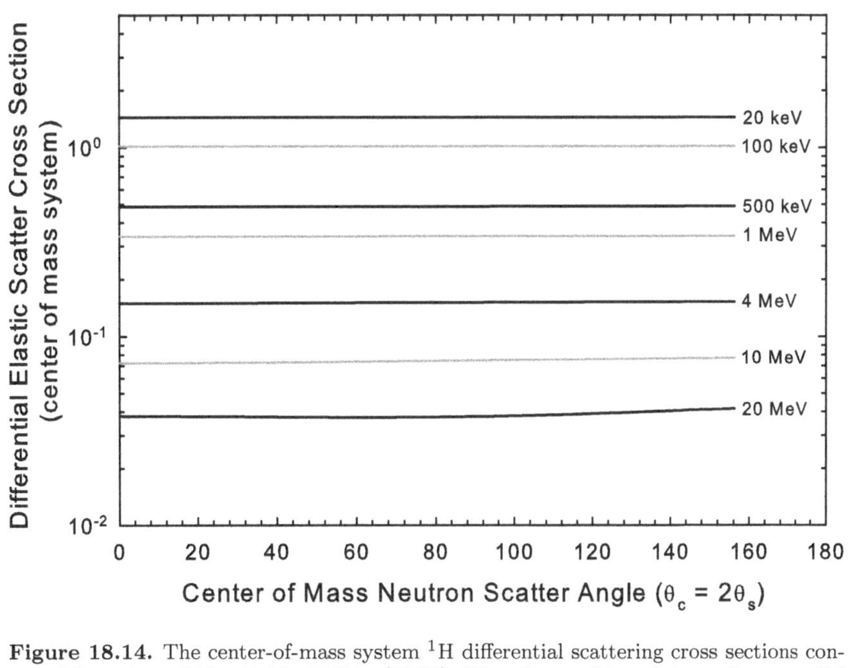

Figure 18.14. The center-of-mass system ^{1}H differential scattering cross sections converted from data of Fig. 18.13 with Eq. (18.24). That the resulting plots are lines parallel to the abscissa is confirmation that scattering is isotropic in the center-of-mass system.

of the detector. Further, the total interaction cross section is relatively constant over a large energy range from 0.1 eV up to approximately 10 keV (see Fig. 18.10). Beyond 10 keV, the elastic scattering cross section decreases, dropping from 18 barns at 20 keV eventually to 0.48 barn at 20 MeV. To counter this reduction in the microscopic cross section at fast neutron energies, the pressure in the detector container may be increased; however, the interaction efficiency for gamma-rays increases as well. Many of these effects are discussed in detail in the literature [Snidow and Warren 1967; Gold and Bennett 1968; Benjamin et al. 1968; Weise et al. 1991].

Proton Recoil Detectors with Non-Hydrogen Gases

There are other gases of interest for proton recoil detectors, mainly CH_4 and ^{4}He. Note that ^{3}He gas can also be used, but because it also has a considerable neutron absorption reaction cross section, is covered in a latter section. To overcome the range problem, in part, methane can be used instead of pure hydrogen [Fink et al. 1980], or at least a partial pressure of methane can be added to the detector chamber [Bennett 1962]. Consequently, the recoil proton range is reduced and allows the detector to respond to higher energy neutrons. For instance, the range of a 2-MeV recoil proton in 2 atm of methane is 39.7 $\pm$ .6 mm, less than a quarter of that in 2 atm of $^1H_2$. Benjamin et al. [1968] describe a spherical proportional counter backfilled with P-10 gas that has reduced wall and end effects. However, neutron scattering from carbon, which transfers at most only 28% of its energy per scatter, distorts the pulse height spectrum [Bennett 1962]. Although the carbon may leave some residual energy in the gas that appears in the low energy region of the spectrum, it is the energy reduction of the scattered neutron that distorts the high energy cutoff. The scattered neutron may subsequently interact with a hydrogen atom, and, because of its reduced energy, produces recoil protons also of reduced energy, thereby shifting the spectrum to lower energies.

The use of ^{4}He as a proportional gas in neutron spectrometers was suggested many decades ago, perhaps initially by Baldinger et al. [1938]. The use of ^{4}He gas is attractive because the range of an energetic ^{4}He ion is significantly lower to that of a proton of equal energy. The effect is enhanced because a ^{4}He recoil ion can carry at most only 64% of the initial fast neutron energy, and the proportional gas (^{4}He) has a higher stopping power than H_2 gas. From Eq. (18.4), the maximum scatter energy of a ^{4}He recoil ion from a 2 MeV neutron is found to be 1.28 MeV, and the range of this recoil ion in 2 atm of He is 17.5 $\pm$ 0.37 mm, over 9 times less than in 2 atm of $^1H_2$ gas. Further, the total microscopic elastic scatter cross section for ^{4}He surpasses that of ^{1}H at approximately 800 keV. Hence, the interaction efficiency is higher in a ^{4}He proportional counter than in a $^1H_2$ counter with an equal atomic density. For neutron energies up to 15 MeV, it is necessary to increase the stopping power of recoil protons by adding another noble gas. Because Xe and Kr are more susceptible to gas impurities, and there is a natural radioactive isotope ^{85}Kr that can add to the background, a common choice is to use a large percentage of Ar. Birch [1988] reports on a 40% Ar/60% ^{4}He mixture for the measurement of 15 MeV neutrons.

Unlike ^{1}H, the differential cross section for ^{4}He is not uniform at all center-of-mass scattering angles.[7] Consequently, the pulse height spectrum becomes somewhat complicated. The lab system and center-of-mass system differential cross sections for ^{4}He are shown in Figs. 18.15 and 18.16, respectively. The results of using the data of Fig. 18.16 combined with Eqs. (18.4) and (18.28) are shown in Figs. 18.17 and 18.18, which show the idealized features expected from ^{4}He proportional gas detector. The pulse height spectral distortions imposed by variations in charge carrier collections times, wall effect, and detector dimensions have been modeled by various research groups [for example Atwater 1972; McDaniel and Hungerford 1979; Weyrauch et al. 1998]. Atwater [1972] provides a model that includes the wall effect, gas pressure, and Gaussian smearing to produce a more realistic recoil spectrum for a ^{4}He detector and is shown in Figs. 18.19 and 18.20. The similarity of the results of Figs. 18.17 and 18.18 to those of Figs. 18.19 and 18.20 are apparent.

[7]In fact, ^{1}H is the only nuclide that possesses this property in the MeV energy range.

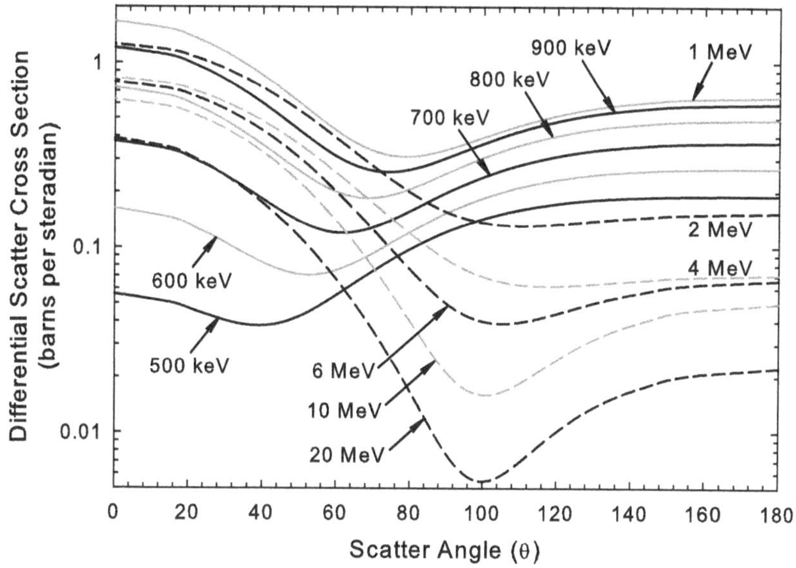

Figure 18.15. The laboratory system ^{4}He differential scattering cross sections in barns per steradian ($d\sigma/d\Omega$) for several energies. Data are from [ENDF VIII 2018].

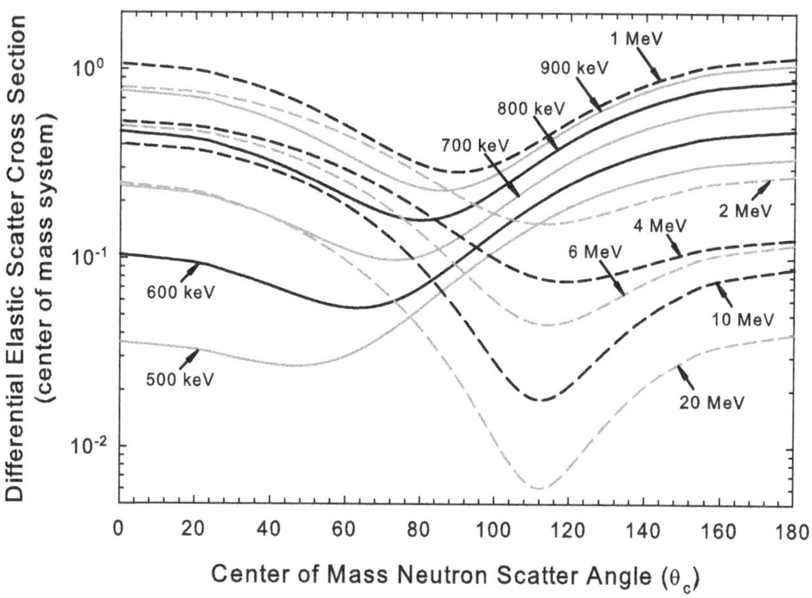

Figure 18.16. The center-of-mass system ^{4}He differential scattering cross sections in barns per steradian ($d\sigma/d\Omega$) for several energies. The graphs were generated from the data of Fig. 18.15 and Eq. (18.24).

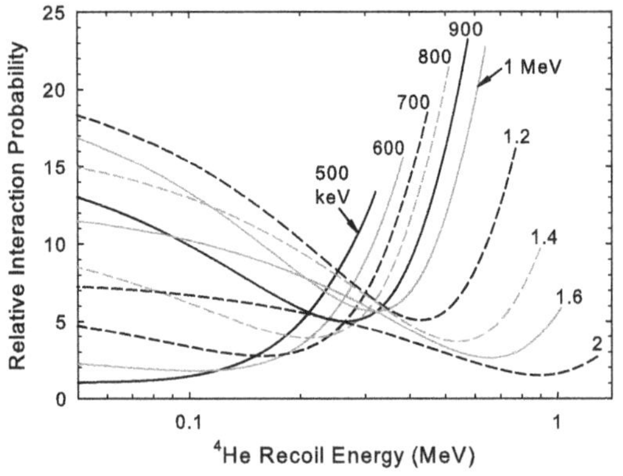

Figure 18.17. The ideal pulse height distribution for recoil ions scattered from fast neutrons. The ordinate is $P(E,T)$ and is plotted as a function of the ^{4}He recoil energy. Curves are labeled with initial neutron energy.

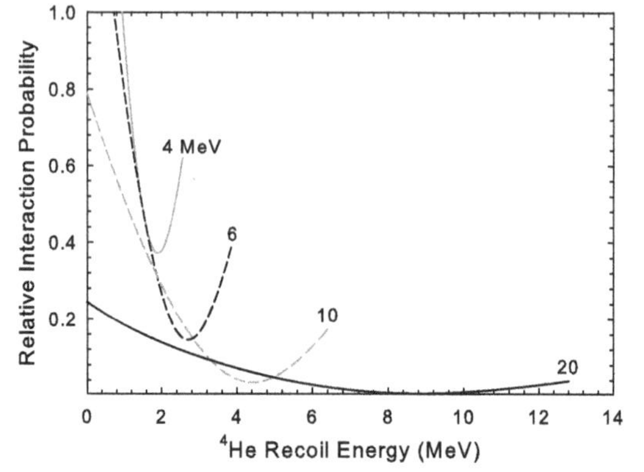

Figure 18.18. The ideal pulse height distribution for recoil ions scattered from fast neutrons. The ordinate is $P(E,T)$ and is plotted as a function of the ^{4}He recoil energy. Curves are labeled with initial neutron energy.

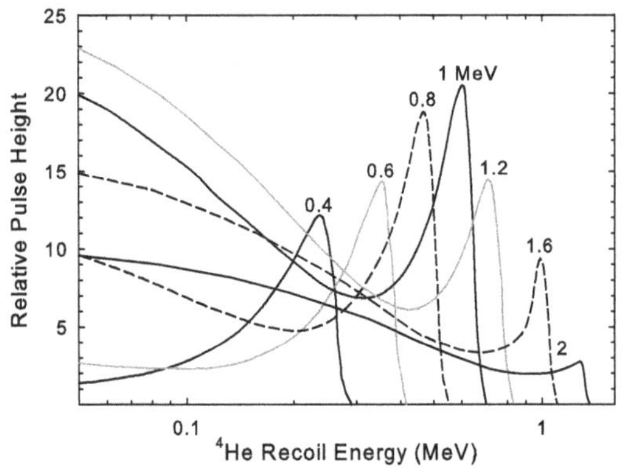

Figure 18.19. Calculated differential pulse height spectrum in 8 atm ^{4}He from monoenergetic neutrons. Data from Atwater [1972].

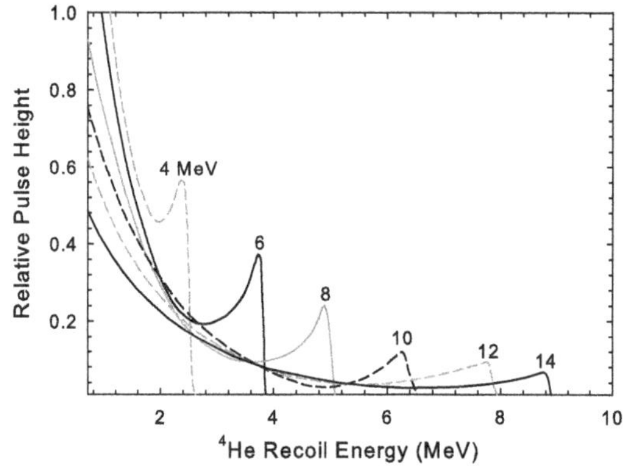

Figure 18.20. Calculated differential pulse height spectrum in 8 atm ^{4}He from monoenergetic neutrons. Data from Atwater [1972].

Birch [1988] reports that the proton background from (n,p) reactions in the steel wall of the chamber become considerable for neutron energies above 10 MeV. To mitigate this problem a 0.5-mm-thick lead liner is inserted into the gas-chamber. Weyrauch et al. [1998] further develop a model for commercial ^{4}He proportional counters and give a variety of comparisons of theoretical to measured results. It is interesting to note that, although ^{4}He gas-filled recoil detectors work adequately as neutron spectrometers, the problems with charge collection are greatly mitigated if the device is instead used as a gas scintillation detector [Jebali et al. 2015].

18.3.2 Unfolding the Recoil Energy Spectrum

If a fast neutron detector is to be used for neutron spectroscopy to determine the energy dependent flux $\phi(E)$ of the neutrons incident on the device, the signals recorded by the instrument must be processed to obtain an estimate of $\phi(E)$. None of the fast neutron detectors described in this chapter directly give $\phi(E)$ and some unfolding of the measured data must be made to obtain the energy information. Briefly mentioned here is how unfolding a measured recoil energy spectrum can be approached. The measured recoil energy T is related to the incident neutron flux $\phi(E)$ by a Fredholm integral equation

$$T = \int_0^{E_{\max}} R(T,E)\phi(E)\,dE, \qquad 0 \le T \le T_{\max}, \tag{18.30}$$

where $E_{\max}$ is the maximum neutron energy, $T_{\max} = (1-\alpha)E_{\max}$ is the maximum recoil energy, and $R(T,E)$ is the spectrometer response function which gives the probability of observing a recoil energy T per unit intensity of an incident monoenergetic (E) neutron beam. The response function, as a first approximation, is given by Eq. (18.28). But an accurate determination of $R(T,E)$ requires Monte Carlo simulations to account for multiple scatters, wall and end effects, statistical fluctuations, non-linearities in the scintillators, and non-ideal behavior of the electronics.

To obtain $\phi(E)$ from Eq. (18.30), the integral equation (like that for a Bonner sphere) is first converted to a set of linear algebraic equations. This conversion can be performed by several methods. The simplest method is to form a contiguous energy grid $E_1 = 0, E_2, \ldots E_i, E_{i+1}, \ldots, E_{I+1} = E_{\max}$ with $\Delta_i \equiv E_{i+1} - E_i$. Similarly, the recoil energy range $(0, T_{\max})$ is split into J contiguous energy intervals Δ_j. Typically, the Δ_j are equi-width and are the interval widths of the MCA channels used to record the recoil energies.

Equation (18.30) can be written as

$$T = \sum_{i=1}^{I} \int_{E_i}^{E_{i+1}} R(E,T)\phi(E)\,dE. \tag{18.31}$$

Integration over the jth recoil energy interval gives

$$\int_{T_j}^{T_{j+1}} T\,dT \equiv T_j = \sum_{i=1}^{I} \int_{E_i}^{E_{i+1}} dE \int_{T_j}^{T_{j+1}} dT R(E,T)\phi(E) \qquad j = 1, \ldots, J, \tag{18.32}$$

where T_j is the measured count rate in channel j of the MCA spectrum of the recoil energy spectrum. If the flux $\phi(E) \simeq \phi_i$ is constant in each neutron energy interval Δ_i, then Eq. (18.32) becomes

$$T_j = \sum_{i=1}^{I} R_{ji}\phi_i, \qquad j = 1, \ldots, J, \tag{18.33}$$

where the elements of the response matrix are

$$R_{ji} = \int_{E_i}^{E_{i+1}} dE \int_{T_j}^{T_{j+1}} dT R(E,T). \tag{18.34}$$

Hence, J linear algebraic equations for the I unknowns of the ϕ_i are obtained. If the recoil atoms and incident neutrons use the same energy grid, then $I = J$ and the solution to Eq. (18.34) can be obtained by standard means. Birch [1988] describes two unfolding methods to determine the incident neutron spectrum from the ^{4}He recoil spectrum, namely matrix inversion and iterative unfolding. He reports better resolution for the iterative unfolding method, especially after differentiating the pulse height spectrum and the response functions.

18.3.3 Scintillators Used in Recoil Neutron Scattering

The basic physics regarding scintillation detectors are described in Chapters 12 and 13 and are not repeated here, except perhaps for some clarification. Both inorganic and organic scintillators are used for fast neutron detectors. The inorganic scintillators can be used as adjacent devices to a recoil medium, such as a hydrogenous material, and which fluoresces from the interactions of recoil ions. Organic scintillators are usually deployed as a combined recoil-fluorescence detector, in which proton (or carbon) recoils from fast neutrons in the organic compound subsequently fluoresce the scintillator. Already discussed in Chapter 13, organic compounds suffer non-linear light yield as the differential energy deposition increases. A few of the detectors and configurations that primarily use recoil reactions as the detection mechanism are briefly described in the following sections.

Response of Organic Scintillators

Organic scintillators, discussed in Section 13.3, are compositions of hydrogen, carbon, and some with other elements such as nitrogen. Proton recoil scattering is nearly isotropic in the center-of-mass system, and the resulting pulse height spectrum for monoenergetic incident neutrons resembles closely a simple step function that extends from zero energy up to the incident neutron energy E. Further, from Eq. (18.4), a fast neutron may impart all of its energy to a proton recoil in a single scatter, but at most only 28% of its energy to a carbon atom with a single scatter. Consequently, the response function from an organic scintillator deviates from the ideal step function. In fact, there are multiple factors that transform the observed pulse height spectrum, including the characteristic non-linear response of organic scintillators, statistical fluctuations, multiple scattering, edge effects, and directional asymmetry [Swartz and Owen 1960].

The Ideal Case There are numerous organic scintillators that contain only hydrogen and carbon atoms, such as the crystalline organics anthracene and stilbene, numerous plastics, and many liquids. Notably, many popular organic fluors contain nitrogen. However, for simplicity, the discussion is limited here to organic scintillators that contain only H and C.

With the definition used by Swartz and Owen [1960], efficiency is the ratio of the number of recoil protons to incident neutrons. The molecular density N_{hc} of the organic scintillator is

$$N_{hc} = \frac{\rho N_a}{A}, \tag{18.35}$$

where ρ is the mass density, and N_a and A have the usual designations of Avogadro's number and the substance's molecular weight. The atomic densities of hydrogen and carbon can be expressed, respectively, as

$$N_H = f_H N_{hc} = m_H \frac{\rho N_a}{A^2} \qquad \text{and} \qquad N_H = f_C N_{hc} = 12 m_C \frac{\rho N_a}{A^2}, \tag{18.36}$$

where f_H and f_C are, respectively, the mass fractions of hydrogen and carbon per molecule. Here m_H and m_C are, respectively, the masses of hydrogen and carbon per molecule, which, to a good approximation, are the number of hydrogen and carbon atoms per molecule. Multiplication by the microscopic scattering cross section gives the macroscopic scattering cross sections,

$$\Sigma_H = \sigma_H N_H \qquad \text{and} \qquad \Sigma_C = \sigma_C N_C. \tag{18.37}$$

The expected efficiency for proton recoil interactions involving a single scatter is,

$$\epsilon_{1h} = \frac{N_1(E_0,\ell)}{N_0} = \frac{\Sigma_H}{(\Sigma_H + \Sigma_C)}\left[1 - \exp\left[-(\Sigma_H + \Sigma_C)\ell\right]\right], \tag{18.38}$$

where $N_1(E_0,\ell)$ is the number of recoil protons produced by a single scatter of a neutron with initial energy E_0 in a scintillator of length ℓ and N_0 is the fluence of neutrons impinging on the scintillator. Equation (18.38) indicates that, provided that the crystal is of adequate length, a sufficiently high efficiency can be expected. In practice, large volumes of organic detector are impractical, mainly because the absorption efficiencies of gamma rays and charged particles also increase and ultimately interfere with the recoil signal. The use of various discrimination methods can reduce background interference. These methods include pulse height analysis, pulse shape analysis, coincidence (or anticoincidence) counting, capture gating with loaded scintillators, and the selection of clever scintillator physical dimensions to reduce energy deposition from energetic electrons. The ideal pulse height spectrum can be defined by a step function,

$$N(T_p) = \begin{cases} N_1(E_0,\ell)/E_0 & T_p \leq E_0 \\ 0 & T_p > E_0 \end{cases} . \tag{18.39}$$

Unlike gas detectors which measure the energy distribution of the recoil atoms directly, scintillators produce a light yield distribution from the energy deposited by the recoil atoms. It is much more difficult to extract the energy distribution of the incident neutrons from the pulse height spectrum of the light output than from the pulse height distribution of the energy from recoil atoms as produced by a gas detector. Further, there are various distortion effects that alter the ideal differential pulse height spectrum of light. Some of these effects include non-proportional light yield, statistical fluctuations, carbon scattering, multiple proton scatters, recoil protons escaping the detector, and in some cases direction asymmetry in light yield. The combined effects cause the distortion from the ideal case Fig. 18.21(a) to that shown in Fig. 18.21(f). A brief discussion of these effects follows.

Scintillator Non-linearity The non-linearity of the light output L of an organic scintillator is given by Eq. (13.36), which with slight changes in notation, can be written as[8]

$$\frac{dL}{dx} = \frac{\widehat{A}\, dT/dx}{1 + k\widehat{B}\, dT/dx}, \tag{18.40}$$

where k is the probability that an exciton is lost to a non-radiative transition, $k\widehat{B}$ is the quenching parameter, $\widehat{A}$ is a proportionality constant, and L is the luminous output.

The variable P (same energy units as L) for pulse height is introduced as $P = L/\widehat{A}$, such that Eq. (18.40) becomes

$$\frac{dP}{dT} = \frac{1}{1 + k\widehat{B}\, dT/dx}. \tag{18.41}$$

If $k\widehat{B}$ is known, then numerical integration of Eq. (18.41) yields values of P. In the ideal case, $k\widehat{B}$ is zero and $P = T$; hence P would also be a rectangular step function for proton recoils. Also, for energetic electrons, dT/dx is small and $P = T$ once again. However, for recoil protons, scintillation quenching can have a significant impact on the luminescent yield and the quenching parameter cannot be ignored. The number of recoil protons with energy between T_p and $T_p + dT_p$ is defined as $N(T_p)dT_p$, where T_p is the initial energy of the recoil proton. The pulse height spectrum of the light output is written as $N(P)dP$ so that

$$N(P) = N(T_p)\frac{dT_p}{dP}. \tag{18.42}$$

Typically a solution to Eq. (18.42) is obtained through numerical methods and unfolding algorithms [Slaughter and Strout 1982]. However, in some cases, empirical formulas fit to measured data can simplify the evaluation of Eq. (18.42). For instance, the light yield L or P can be modeled as a power function of the

[8]Here the empirical constants are denoted by $\widehat{A}$ and $\widehat{B}$ to avoid confusion with the symbols without the accent. Also the charged particle energy is denoted by T instead of E to avoid confusion with the neutron energy.

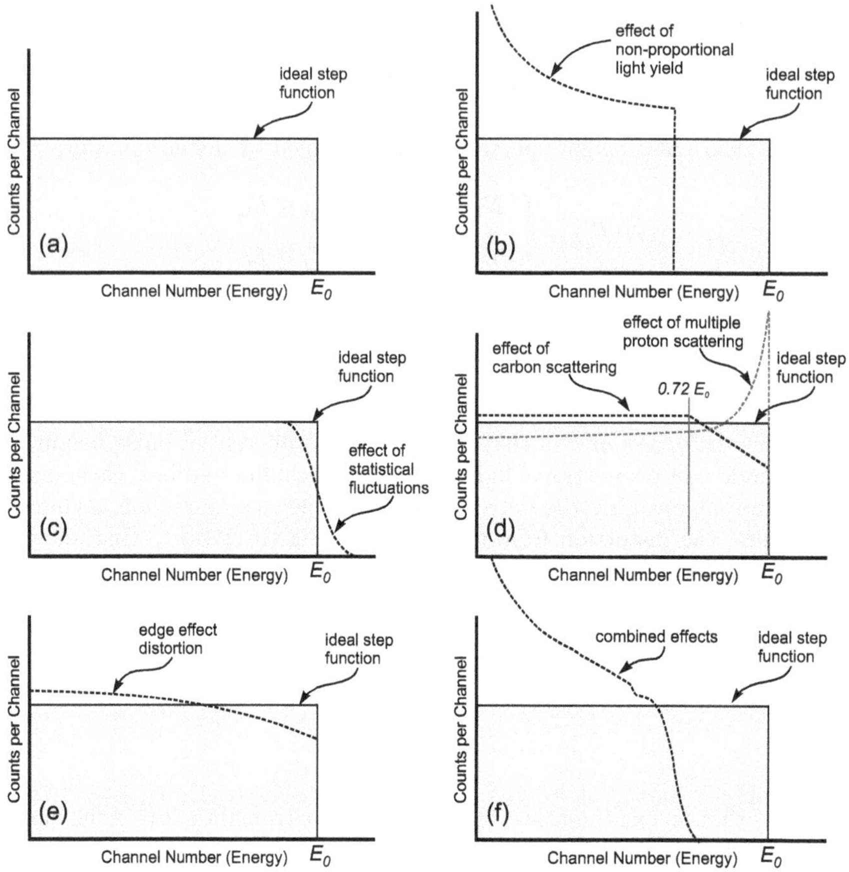

Figure 18.21. Representative pulse-height distortions from neutron scattering in an organic scintillator. (a) The ideal case with no distortions; (b) non-linear light yield; (c) statistical fluctuations; (d) carbon scattering and multiple proton scattering; (e) edge effects; and (f) the resulting spectrum from the combined effects.

recoil energy, usually a form of $T_p^{1.5}$. Based on the data given by Craun and Smith [1970] for stilbene, this approach yields

$$P(T_p) = \left[0.0204 + \frac{5.397}{T_p^{1.5}}\right]^{-1} \simeq \frac{T_p^{1.5}}{5.397}, \tag{18.43}$$

where the approximation ignores the small offset of 0.0204, and P and T_p are both in units of MeV. Given a value of $k\widehat{B} \approx 0.01$ mg cm^{-2} MeV^{-1} and the data of Craun and Smith [1970], the values of dT_p/dP for stilbene are approximated by,

$$\frac{dT_p}{dP} \approx \frac{-1.781 + 64.237\sqrt{T_p} + 6.77T_p}{1 - 5.239\sqrt{T_p} + 26.157E_p - 1.72T_p^{1.5}}, \quad 0.005 < T_p < 15 \text{ MeV}. \tag{18.44}$$

The distortion in the pulse height spectrum from the idealized proton recoil spectrum, which is normalized to unit height, is shown in Fig. 18.22. It is notable that the general shape of the pulse height spectrum due solely to scintillator non-linearity is determined analytically with the use of empirical fits. The normalized luminous output, from Eq. (18.43), is described by

$$P(T_p) \approx c_1 T_p^{1.5}, \qquad (18.45)$$

where c_1 is a constant of proportionality. Then from Eq. (18.42)

$$N(P) = \frac{N(T_p)}{dP/dT_p}. \qquad (18.46)$$

Because $N(T_p)$ is constant for $T_p < E_0$

$$N(P) = \frac{c_2}{1.5\sqrt{T_p}}. \qquad (18.47)$$

From Eq. (18.45) it is seen that $\sqrt{T_p}$ is proportional to $P^{1/3}$ and substitution of this result into Eq. (18.47) yields the expected shape of the pulse height distribution, namely

$$N(P) = \frac{c_3}{P^{1/3}}, \qquad (18.48)$$

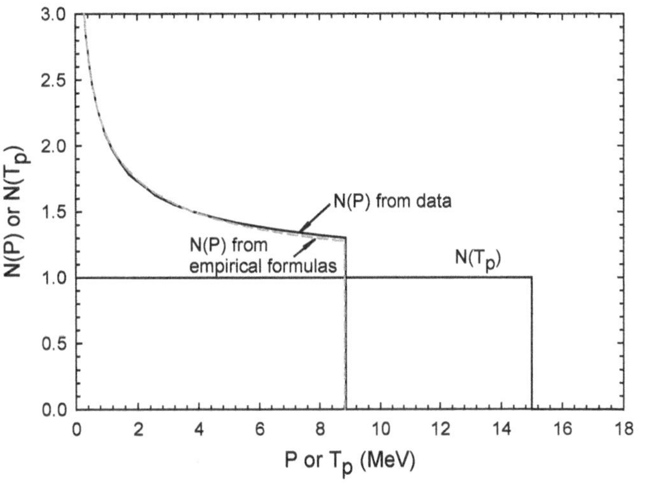

Figure 18.22. Distortion of the ideal pulse height spectrum for stilbene due to non-linear light yield. Shown is the expected pulse height spectrum from 15-MeV neutrons.

where c_3 is a constant of proportionality. This dependence of $N(P)$ on P is clearly seen in Fig. 18.22.

Statistical Fluctuations The effect of statistical fluctuations on detector energy resolution was covered in prior chapters, mainly in Sections 6.9 and 6.10 on counting statistics and error propagation and in Section 10.5.4 for proportional counters. In general, the energy resolution of a detector is limited by the fluctuation in charge carriers. For a scintillator, there are multiple sources of fluctuations, such as fluctuations in the number of excited free electrons and excitons produced, the fraction of charges that decay by fluorescence, the number of photons arriving at the light detection device, the charge carriers produced at the light detector (photoelectrons, for instance), and the gain. These fluctuations produce a variation in the light output near $T = E_0$ such that $dN(P)/dP$ assumes a Gaussian shape with mean $\langle P \rangle \simeq E_0$ and a variance σ_P^2. This effect softens the high energy edge by producing a sigmoid transition instead of the step function change, as shown in Fig. 18.21(c).

Multiple Scattering Protons entering the detector may undergo scattering with carbon atoms. From Eq. (18.4), at most only 28% of the initial proton energy can be transferred to a carbon atom with a single scatter so that the scattered neutron has a residual energy ranging from $0.72E_0$ to E_0. The recoil carbon atom also has kinetic energy, but the luminescence is low and these events are considered to add negligible fluorescence [Swartz and Owen 1960], and hence do not contribute (appreciably) to the pulse height spectrum.

The efficiency of neutrons scattering once by carbon can be expressed by,

$$\epsilon_{1c} = \frac{N_C(E_0,\ell)}{N_0} = \frac{\Sigma_C}{(\Sigma_H + \Sigma_C)}\left[1 - \exp\left[-(\Sigma_H + \Sigma_C)\ell\right]\right]. \qquad (18.49)$$

Suppose that the energy of the carbon scattered neutrons can be defined by an average neutron energy $\overline{E}_1$ and that these neutrons must pass through some average amount of material $\overline{\ell}_1$ of scintillator. If the organic scintillator is sufficiently thick to allow at least two neutron scatters, the efficiency of proton scattering first from carbon followed by scattering from hydrogen is

$$\epsilon_{2(ch)} = \frac{N_2(E_0,\ell,\overline{\ell}_1)}{N_0} = \frac{N_C(E_0,\ell)}{N_0}\frac{\Sigma_H(\overline{E}_1)}{[\Sigma_H(\overline{E}_1) + \Sigma_C(\overline{E}_1)]}\left[1 - \exp\left\{-[\Sigma_H(\overline{E}_1) + \Sigma_C(\overline{E}_1)][\overline{\ell}_1]\right\}\right]. \qquad (18.50)$$

The total proton scattering efficiency becomes

$$\epsilon_t = \epsilon_{1h} + \epsilon_{2(ch)}. \quad (18.51)$$

Swartz and Owen [1960] note that the result of Eq. (18.51) is often nearly the same as if no carbon scattering occurred, i.e., $\Sigma_C = 0$. The effect on the pulse height spectrum is shown in Fig. 18.21(d). Although some carbon scattered neutrons may exit the detector, this problem can be greatly mitigated by using detectors that are relatively wide. Hence, carbon scattering tends to reduce the energy of some neutrons, but not necessarily the number of proton recoils. Instead, the spectrum is shifted such that the number of counts is slightly increased in the lower energy region, while the number of counts is slightly decreased in the energy range between $0.72E_0$ to E_0. Overall the total number of counts recorded in the pulse height spectrum would change little.

There is also the possibility of multiple proton scatters, in which two proton recoils from a single neutron occur rapidly in succession and are recorded as a single event. Such double (n,p) scattering emphasizes the higher energy region of the pulse height spectrum [Segel et al. 1954], an opposite effect to that of carbon scattering. A depiction of this effect on the spectrum is shown in Fig. 18.21(d). Note that if the detector were sufficiently thick, the total neutron energy would be deposited in the detector for each event. Because of non-linear light yield, there would be a variance about the average luminescence per neutron, even from a monoenergetic neutron source. Overall, these two different scattering effects tend to partially offset each other in small detectors. In either case, the knowledge of the differential scattering cross section combined with the knowledge of the scintillator specific light yield can be used to model the pulse height spectrum with Monte-Carlo methods.

Edge Effects Edge effects occur as a consequence of losing recoil protons from the detector periphery. This effect is enhanced for detectors with dimensions less than the average range of the recoil protons. A count is still recorded for the recoil proton, but the pulse height is diminished by the amount of energy not deposited in the detector. Intuitively, one can understand that energy loss from protons escaping the detector increases counts in the lower energy channels while reducing events in the higher channels, as depicted in Fig. 18.21(e).

Threshold Reactions At sufficiently high neutron energies, there are threshold reactions that can be of concern [Mösner at al. 1966; Grin et al. 1969], mainly the ^{12}C$(n,\alpha)^9$Be and the ^{12}C$(n,n)3\alpha$ reactions with Q-values of 5.70 MeV and 7.27 MeV, respectively. Both of these reactions release energetic heavy ions, and consequently add to the total neutron detection efficiency [Hermsdrof et al. 1973]. However, because of non-proportional light yield, especially for ions heavier than protons (see Chapter 13), the added luminescence further distorts the differential pulse height spectrum. From the result of Problem 18.11, the threshold energy for these reactions are 6.176 MeV and 7.881 MeV, respectively, as shown in Fig. 18.23.

Directional Asymmetry Thus far, it has been assumed that the light yield is relatively constant for all proton trajectories. However, both anthracene and stilbene show directional dependence on light emission, reportedly as much as 15% variation [Swartz and Owen 1960]. However, directional asymmetry is generally not a problem with plastic and liquid scintillators. In the case of varied light yield with proton direction, the overall result is an increased variance in the pulse height spectrum, further smearing and distorting the idealized spectrum and is shown in Fig. 18.21(f).

Dependence on the Lower Level Discriminator (LLD) From Fig. 18.21(a), it becomes evident that the pulse height spectrum ranges from channel 0 (no energy transfer) up to the total fast neutron energy E_0. If the LLD is set to zero, then electronic noise and background radiation contaminate the spectrum.[9] Hence, it is traditional to set the LLD to discriminate neutron events from other events such as gamma ray

[9]The condition with LLD = 0 is at times referred to as the zero bias condition, an unfortunate choice, mainly because this condition is also used more accurately to describe radiation detectors that operate with no voltage applied.

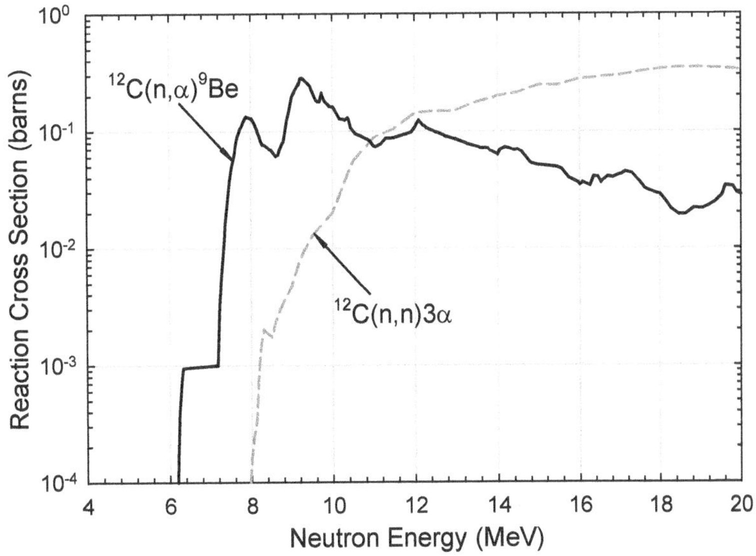

Figure 18.23. Fast neutron reaction cross sections for ^{12}C. Data are from [ENDF VIII 2018].

interactions; however, such discrimination artificially reduces the overall detection efficiency by eliminating low energy (n,p) recoil events. This reduction in efficiency is non-linear, being zero for proton recoils below the LLD, and also decreases as the fast neutron energy E_0 is increased.

Suppose the LLD is set at 500 keV. Under such a condition, only a few select events with $\theta_c \simeq 180°$ allow 500 keV neutrons to be detected. The number of possible detectable scatter angles increases as E_0 increases and produces higher efficiency for neutron energies greater than E_0, in this example, 500 keV. The overall scatter cross section for ^1H decreases with energy (see Fig. 18.14), which decreases the detection efficiency even though the detectable scatter angle θ_c increases. Combined, these effects produce a maximum in the efficiency curve as a function of neutron energy and LLD setting, as depicted in Fig. 18.24.

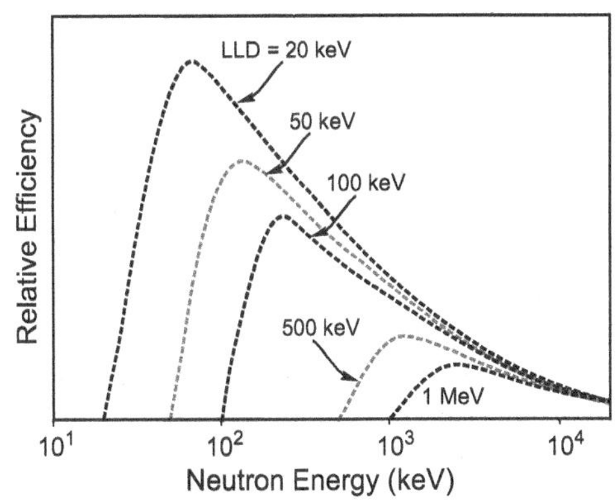

Figure 18.24. Efficiency is a function of the LLD setting, neutron energy, and elastic scattering cross section.

Multiple Neutron Energies For multienergetic neutron sources, the varied emission energies will work to further smear the features in the output spectrum. For instance, the varied energies from a PuBe source range from hundreds of keV up to 11.5 MeV (see Fig. 5.5), with all energies being susceptible to the response deviations described in the previous sections on ion recoils. Hence, the spectral features for proton recoils from a multienergetic source will be a convolution of the various energies, their emission probabilities, and combined distortion effects for each of these emission energies.

Pulse Shape Discrimination Detectors

Pulse shape discrimination (PSD) is used as a method to distinguish between different ionizing events in a scintillation detector, already outlined in Sec. 13.3.1 under organic scintillators. The basic method takes advantage of the different pulse output pulse shapes produced by a scintillator from heavy ions, electrons, and

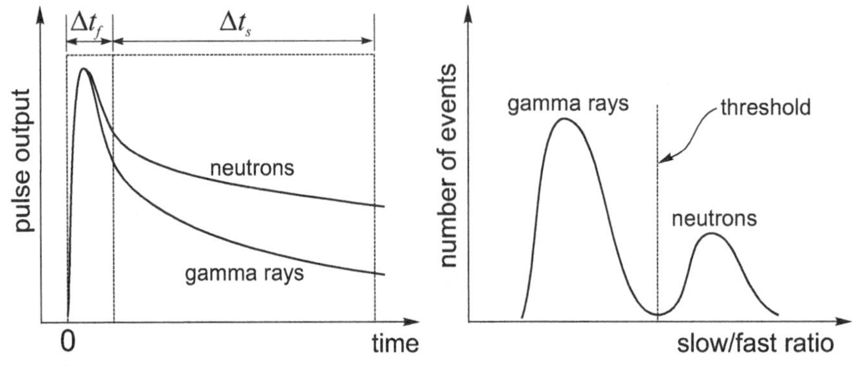

Figure 18.25. Depiction of two different pulse shapes for a scintillator. Divided into two regions, the ratios of the integrated output produce a PSD spectrum that distinguishes between gamma-ray and heavy ion events.

gamma-rays. Gamma-ray pulses tend to be shorter with a faster decay time than those observed for heavy ion events. Consequently, by time gating the pulse shape into two separate regions a distinct separation between heavy ion and gamma-ray events becomes apparent. Plastic scintillators with high hydrogen-to-carbon ratios can be used as fast neutron detectors from proton recoil events. For many scintillators, both organic and inorganic, the pulse shapes for fast protons are much different than those for gamma-ray reactions.

Consider the depiction of scintillator output pulse shapes in Fig. 18.25(left). A time window width Δt_f is set for the fast decay component while a second time window width Δt_s is set for the slow decay component. The integrated output over Δt_s divided by the integrated output over Δt_f yields unique signatures characteristic of gamma-rays or neutrons. The number of these pulse height ratios as a function of the ratio value often reveals a clear distinction between gamma-ray and neutron induced pulses, as depicted in Fig. 18.25(right). Several examples of pulse shape discrimination spectra from organic scintillators are shown in Chapter 13.3.

Although many organic scintillators can distinguish between neutron and gamma-ray events with PSD, there are a select few that perform better than most and are so identified in Sec. 13.3.2. Also, some inorganic scintillators show discrimination between pulse heights for various forms of ionizing radiation. Because most inorganic materials are more efficient at absorbing gamma rays than the organic scintillators, they are generally not used for such purposes. However, there are some exceptions such as the use of ZnS:Ag adjacent to a plastic waveguide.

Shown in Figs. 18.26 and 18.27 are proton recoil spectra from a ^{252}Cf source and an AmBe source taken with a plastic scintillator. The *pulse shapes* were separated into gamma-ray and neutron events with the pulse shape analysis technique, and then plotted as a function of *pulse height* digitally sorted into a sequence of channels. Both ^{252}Cf and AmBe neutron sources emit a broad spectrum of energies as described in Section 5.4. Consequently, the pulse height spectra in Figs. 18.26 and 18.27 also show continua, blurred by the multiple effects described in Fig. 18.21, a similar result to that reported by Cester et al. [2014].

Proton Recoil Telescope

The various problems with the degradation of the energy resolution in a plastic scintillator can be reduced if the energies of recoil protons are accurately measured. Further, if the scatter angle of the recoil proton is known accurately, the initial neutron energy can be determined from Eq. (18.4) with $A = 1$. A device to accomplish such measurements was introduced during the 1940s and is described by Johnson [1960]. The basic device, named a *proton recoil telescope*, consists of a thin hydrogenous material acting as a proton radiator with high resolution detectors positioned, in some manner, behind the radiator. This configuration

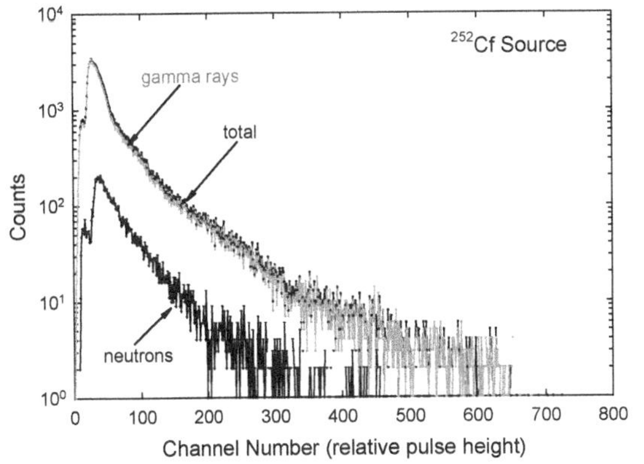

Figure 18.26. Proton recoil spectrum from ^{252}Cf taken with a plastic scintillator Elgin EJ-299-33A. Pulse shape discrimination was used to separate gamma-ray and neutron events. Courtesy Priyarshini Ghosh and Taylor Ochs (Kansas State).

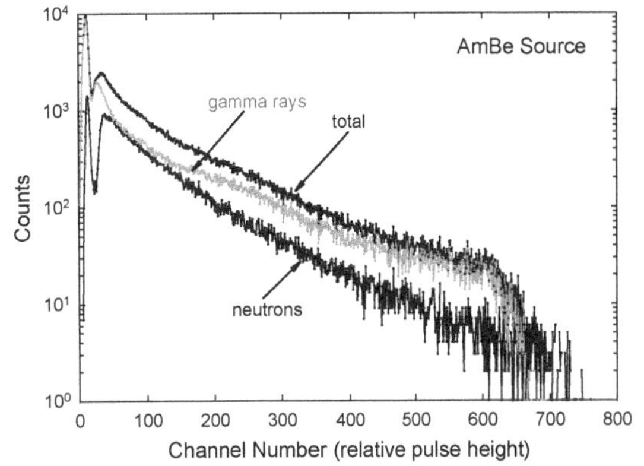

Figure 18.27. Proton recoil spectrum from an AmBe source taken with a plastic scintillator Elgin EJ-299-33A. Pulse shape discrimination was used to separate gamma-ray and neutron events. Courtesy Priyarshini Ghosh and Taylor Ochs (Kansas State).

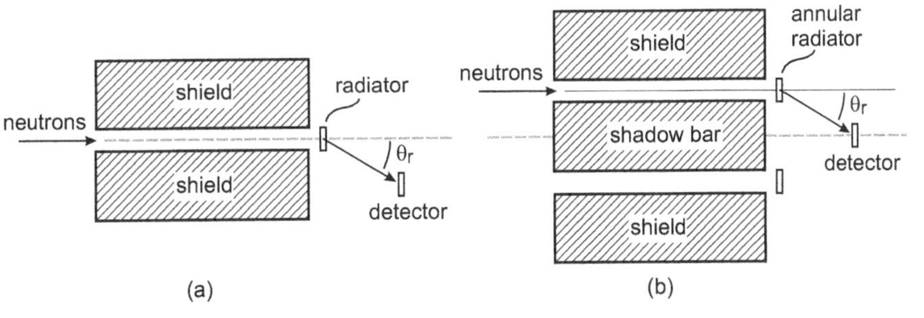

Figure 18.28. Proton recoil telescope spectrometers; (a) with out-of-line geometry and (b) with annular geometry. After on Hawkes et al. [2002].

mostly mitigates the multiple problems encountered with organic scintillator spectrometers as described in the previous sections. The purpose of the proton recoil telescope is to accurately measure the energy of impinging fast neutrons. However, this goal also requires that the radiator be relatively thin in order to diminish energy self-absorption and to preserve the energy of the recoiling proton. Consequently the efficiency of the device is quite low, often below 10^{-4}.

There are several designs proposed for proton recoil telescopes, each intended for a specific use. In some cases, the proton recoil telescope is used as a neutron spectrometer and the proton energy and the scatter angle must be known accurately. Collimators can be used to ensure that only specific proton angular trajectories are allowed to reach the particle detector [Johnson 1960]. Collimators may also be used to define the fast neutron direction with respect to the radiator and the proton detectors located at known angles to the radiator. For instance, Hawke et al. [2002] report on two such configurations, depicted in Fig. 18.28, to operate as proton recoil spectrometers. The collimator is used to align the neutrons to the radiator and the detector is offset from the neutron beam at a known angle. This out-of-line configuration reduces direct neutron interactions in the proton detector. The annular configuration (Fig. 18.28(b)) has the added advantage of a larger source solid angle, thereby increasing the overall detection efficiency.

To remove background interference, a thin ΔE detector can be placed in front of the proton detector and operated in coincidence [Ryves 1976; Cazzaniga et al. 2015; Marini et al. 2017]. The thin ΔT detector samples a small portion of energy from the proton while the main detector absorbs the remaining energy T'. For detectors responding linearly to energy deposition, the initial neutron energy E is obtained from Eq. (18.4) as

$$E = (\Delta T + T') \cos^2 \theta_r. \tag{18.52}$$

The product of $(\Delta T)(T')$ gives a measure of the particle mass [Parkinson and Bodandsky 1965; Chaminade et al. 1967; Homeyer 1967], and can be used to assure that the impinging particle is a proton and not another charged particle type. Donzella et al. [2010] describe a proton recoil telescope with a multilayer segmented radiator, two position sensitive ΔT Si detectors, followed by a CsI:Tl detector, all operated in coincidence. The segmented radiator doubles as a position sensitive organic scintillator, giving the originating location of the (n,p) reaction. The reported energy resolution of the fast neutrons ranged from 20% FWHM at "a few" MeV to 2% FWHM at approximately 160 MeV, and the device had an average efficiency of about 3×10^{-5}. An advanced version, with multiple position sensitive ΔT CMOS detectors aligned with a silicon surface barrier detector is reported by Combe et al. [2018]. This device has an in-line configuration, and the proton recoil angles are tracked between the three CMOS pixel detectors before capturing the residual energy in the silicon surface barrier detector.

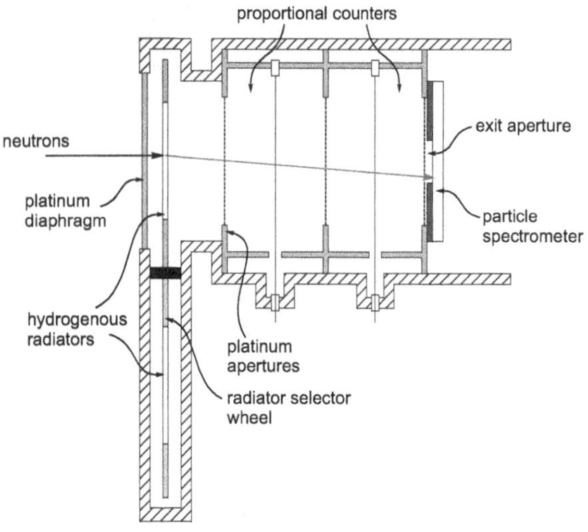

Figure 18.29. A proton recoil telescope. The radiator and the exit aperture define the acceptance solid angle of the proton recoils. Based on Bame et al. [1957].

For monoenergetic neutron sources, the efficiency of the spectrometer may be the most important factor, especially for measurements of cross sections. Early designs, still in use, have a dual chamber proportional counter arranged between the proton radiator and a spectroscopic detector [Johnson and Trail 1956; Bame et al. 1957; Ryves 1976; Lövestam 2006], as shown in Fig. 18.29. The geometry of these devices is an in-line configuration. These proton recoil telescopes have several different radiators on a selector wheel that allows the user to easily change the targets, each radiator target usually having different mass thicknesses of the same substance. The reason for these different targets is to account for changes in (n,p) cross section as the fast neutron energy changes [Bame et al. 1957]. The particle spectrometer can be a scintillator [Johnson and Trail 1956; Bame et al. 1957] or semiconductor detector [Ryves 1976; Lövestam 2006]. The proportional counters and the neutron spectrometer are operated in triple coincidence to reduce background.

Capture Gated Neutron Spectrometer

The capture gated concept is implemented with a loaded scintillator [Drake e al. 1986], most commonly an organic scintillator loaded with ^{10}B, or with loaded scintillator immersed in an organic scintillator [Czirr and Jensen 1989]. From the previous section describing the pulse height spectrum from elastic scattering in an organic scintillator, fast neutrons produce a continuous spectrum of energies with distortions arising from many effects. The interaction time for elastic scattering, including multiple scatters, is relatively short, typically tens to hundreds of nanoseconds. If the neutron loses nearly all of its kinetic energy from elastic scattering in the scintillator, the thermalized neutron continues to diffuse until it is absorbed by either a hydrogen atom (332 mb) or (less probable) with a carbon atom (3.53 mb). The $^1H(n,\gamma)^2H$ reaction releases a 2.22 MeV gamma ray, but this gamma ray usually escapes the organic scintillator and goes undetected. Should the neutron react with carbon, there are three possible gamma-ray emissions (1.26 MeV, 3.68 MeV, 4.95 MeV) also of high energy and most likely to escape the scintillator. If instead a small fraction of ^{10}B atoms are distributed within the organic scintillator, for instance 1% loading, the chance of absorption in the boron is over 100 times greater than in the hydrogen. Upon absorption in the boron, the $^{10}B(n,\alpha)^7Li$ reaction releases reaction product energy of either 2.31 MeV (94%) or 2.79 MeV (6%). In either case, these reaction products also produce scintillation light from the detector.

Capture gated neutron spectroscopy is possible because the time required to thermalize the neutron is much less than the time it takes for the thermalized neutron to diffuse to a ^{10}B location. Hence, there are two releases of scintillation photons, first from the elastic scattering and second from the $^{10}B(n,\alpha)^7Li$ reaction, separated by some time period Δt, usually between 10 and 20 μs. Because it is unlikely that an epithermal or fast neutron interacts in the ^{10}B, these paired scintillations occur predominantly when the initial fast neutron has lost all of its kinetic energy. Hence, the magnitude of the first scintillation release is indicative of the initial neutron energy and the second scintillation release indicates that the neutron did indeed lose all (or most) of its kinetic energy. Because the Q-value of the $^{10}B(n,\alpha)^7Li$ reaction is recognizable, it is a simple matter of calibrating the detection system to recognize within a given Δt the two separate pulses, that the latter emission pulse has the distinctive energy signature from the ^{10}B reaction. Hence, the energy of the second pulse can also be discriminated with an energy window to ensure that it is truly from a $^{10}B(n,\alpha)^7Li$ reaction. If the second pulse does not appear within some set coincidence time Δt, or if the relative energy indicates that it is not from the ^{10}B reaction, the former pulse is rejected. The method reduces the overall fast neutron counting efficiency of the scintillation detector, although the quality of the energy spectrum is improved.

Ideally, the scintillation light produced by fast neutrons that have been thermalized in an organic scintillator would be indicative of the total energy deposited, regardless of the number of scatters required for thermalization. In reality, there is an average number of scatters needed to thermalize a neutron and also a variance about that average. For mono-energetic neutrons, this physical result means that the scatter angles and recoil proton energies of the scattering events are different, although the total energy loss for fully absorbed neutrons must still be the same for all incident neutrons. As discussed in Sec. 13.3.1, organic scintillators respond non-linearly to heavy charged particles, especially as the specific energy loss dT/dx increases. The light yield per scatter is a function of the recoil proton energy and also the energy loss per unit distance dE/dx, thereby adding to the variance in light yield. Consequently, the energy resolution of the neutron spectrum is much broader than predicted by simple Gaussian statistics based on charge carrier production (see Fig. 18.30). Attempts to mitigate the problem of non-linear light yield have been tried by using segmented detectors to compartmentalize the light emission, thereby allowing light yield corrections to the energy deposited [Abdurashitov et al. 2002; Bowden et al. 2009].

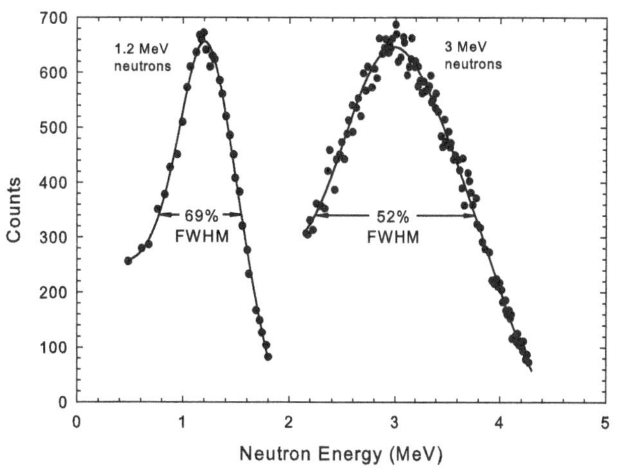

Figure 18.30. Capture gated spectra from monoenergetic fast neutrons. The system consisted of a plastic scintillator annulus around a ^3He gas-filled detector. Data are from Drake et al. [1986].

Although boron-loaded plastics are a popular choice for capture-gated neutron spectroscopy [Drake et al. 1986; Feldmann et al. 1991; Kamykowski 1992; Holm et al. 2014], many other configurations have been reported with good results. Boron-loaded liquid scintillators have been reported with the added function of pulse shape discrimination to eliminate gamma-ray background [Jastaniah and Sellin 2004; Flaska and Pozzi 2009]. One of the earliest implementations of capture-gated neutron spectroscopy used a plastic scintillator annulus around a ^{3}He gas-filled detector [Drake et al. 1986]. Lithium can also be added to plastic or liquid scintillators [Czirr et al. 2002; Fisher et al. 2011 ; Wilhelm et al. 2017], the advantage is a higher Q value (4.73 MeV), a single decay branch, and no gamma-ray emissions from an excited state. A disadvantage of using LiF is the lower neutron absorption cross section (microscopic and macroscopic) [McGregor et al. 2003]. Pulse shape discrimination methods have also been used with a Li-glass/plastic scintillator layered structured, also operated as a capture gated detector [Czirr and Jensen 1994]. Nattress et al. [2016] describe the use of Li-loaded glass rods arrayed in a matrix inside a plastic scintillator.

Hornyak Buttons

ZnS:Ag has a unique property of having relatively low light yield for electron interactions while being reasonably bright for heavy ion interactions. This particular property was put to use by Hornyak in 1952 (see discussion on Zinc Sulfide in Section 13.2.2). The original *Hornyak buttons* were fabricated by mixing a fine powder of ZnS:Ag with PMMA[10] molding powder and curing the samples in shapes as small plugs [Hornyak 1952]. That fact that PMMA is a non-scintillating organic helps to reduce background light from gamma-ray events. However, in high radiation fields of mixed neutrons and gamma rays, the production of Čerenkov emissions from the PMMA are possible ($n = 1.49$) and, thus, increase the background. Regardless, Hornyak [1952] reported good efficiency for fast neutrons while still being relatively insensitive to gamma-ray background.

At mass thicknesses of 25 mg cm^{-2} or greater, ZnS:Ag becomes opaque to its own light emissions. This mass thickness is equivalent to 61 microns of pure ZnS:Ag (4.09 g cm^{-3}). However, ZnS:Ag is usually deployed as a powder mixed with a binder, which reduces the volume density of ZnS:Ag, but the opaqueness problem remains. Because of self-absorption light losses, modern Hornyak button designs deviate from the original design. Instead of mixing the ZnS:Ag into the PMMA, modern Hornyak buttons are usually arranged in a bullseye pattern, with concentric alternating rings of PMMA and ZnS:Ag. Protons scatter from the PMMA rings into the ZnS:Ag and produce light. The light subsequently emerges from the ZnS:Ag back into the PMMA and is guided to a light sensor, usually a PMT.

Gamma-ray interactions in PMMA can produce Čerenkov radiation emissions. In low radiation environments, simple pulse height discrimination is sufficient to effectively distinguish between neutron and gamma-ray events. However, in mixed-field high-radiation environments, the production of significant Čerenkov radiation can require that the lower level discriminator be set at much higher channels, a setting that consequently also removes neutron events and lowers neutron detection efficiency. It has been demonstrated that PSD can improve the detection efficiency of Hornyak-style devices [Johnson et al. 2016], more than doubling the efficiency than with pulse height discrimination alone.

PRESCILA Detector

The PRESCILA (Proton REcoil SCIntillator – Los Alamos) detector is another recoil detector based primarily on proton recoils entering a ZnS:Ag scintillator layer (see Fig. 18.31). The PRESCILA detector was developed at Los Alamos National Laboratory with the goal of producing a new REM meter of much lighter weight than conventional polyethylene REM balls [Olsher and Seagraves 2003; Olsher et al. 2004]. The basic device has a large central cube of Lucite that serves both as a light guide and also as a moderator to thermalize neutrons. The device has concentric ring style Hornyak buttons on four sides (EJ-410P in Fig. 18.32),

[10]Polymethyl methacrylate with brand names such as Lucite, Plexiglas, and Acrylite.

each recessed into borated polyethylene plates, used for fast neutron counts. On top, the detector has a thermal-neutron detector made from a combination of ^{6}LiF and ZnS:Ag. The Lucite block is attached to a PMT, and the PRESCILA wand output can be connected to a handheld survey meter (Fig. 18.31).

The response of the PRESCILA detector is reasonably flat over a wide range of energies. The relative response per unit dose equivalent to both H*(10) and NCRP-38, in which a uniform result of 1.0 for all energies is the desired outcome, are shown in Figs. 18.33 and 18.34. The response to neutrons is overestimated for energies below 100 keV while the response to fast neutrons is underestimated between the energies of 100 keV and 2 MeV [Olsher et al. 2004]. The overall response was designed to provide a relatively uniform dose equivalent. Comparatively, the response from the Hankins rem ball and the Andersson-Braun cylinder seem to be flatter over these same energy regions. At energies between 2 MeV to 20 MeV, the response from the PRESCILA becomes relatively flat, while the responses from the Hankins rem ball and the Andersson-Braun cylinder both, almost identically, decrease as energy is increased above 7 MeV. Finally, the PRESCILA detector is considerably smaller and weighs only 4.5 lbs (2 kg), which is much less than the rem ball or cylinder. For handheld field instruments, size and weight are important considerations.

Figure 18.31. A commercial proton recoil dosimeter. The black cubic object is a PRESCILA detector (see description in text).

Gas-Filled Scintillation Detectors

Recall from Sec. 13.4 that several gases were discovered to scintillate, often in the UV region.[11] Of these gases, He is the best candidate for use in a gas recoil scintillation detector, mainly because of its lower atomic number compared to those of the other noble gases so that more energy is transferred to a recoil atom from a neutron scattering event. For ^{3}He gas, there is also the possibility of producing energetic ions from the (n,p) reaction that can also cause the gas to fluoresce. The use of ^{4}He gas as a recoil scintillator, rather than as a proportional counter, mitigates problems associated with charge collection and avalanche production [Chandra et al. 2012]. Further, the response speed is faster than in a proportional counter and the scintillation mechanism allows the use of pulse shape discrimination methods. The kinematics of (n,α) scattering are the same as described for proportional counters; hence the maximum energy transfer per neutron scatter is only 64%. The recoil ^{4}He ions fluoresce the gas, and photomultiplier tubes (PMT) [Jebali et al. 2015] or compact silicon photomultipliers (SiPM) can be used to measure the light yield [Arktis 2017]. A pressurized tube filled with ^{4}He gas is attached to a light collection device, and because charge carrier collection is not required and charge mobility is not an issue, these chambers can be pressurized, in some cases up to 150 atm [Kelley et al. 2015]. The high pressure reduces the recoil ^{4}He ion range and, consequently, also the wall effect so that the overall efficiency for the neutron interactions is increased.

18.4 Semiconductor Fast Neutron Detectors

The use of a hydrogenous radiator, such a polyethylene, attached to a semiconductor diode has been explored as a compact fast neutron detector and as a beam port monitor [Klann and McGregor 2000, 2002; Klann et al. 2001, 2002; De Lurgio et al. 2003]. These detectors rely on the recoil proton entering a Schottky or *pn* junction diode. The energy recorded is the proton recoil energy minus the energy self-absorption losses as the proton traverses the radiator. The maximum range of the proton limits the useful radiator thickness and, consequently, also determines the efficiency as a function of neutron energy. The maximum range of 14-MeV

[11] The only noble gas that appears not to effectively scintillate is neon [Northrop and Gursky 1958], although liquid neon does have use as a scintillator [Nikkel et al. 2008].

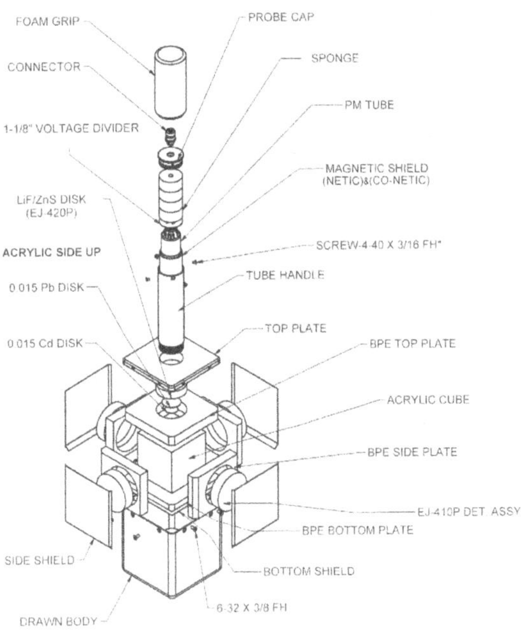

Figure 18.32. Exploded view of a PRESCILA Detector. Courtesy Ludlum Measurements, Inc.

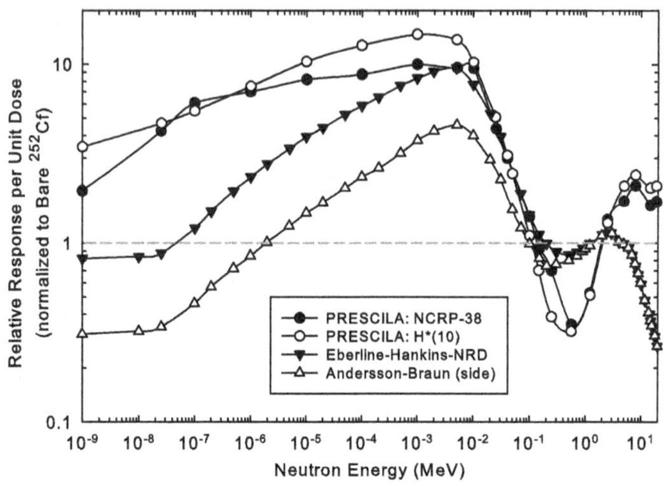

Figure 18.33. The energy response functions per unit dose for the PRESCILA detector compared to a Hankins rem ball and an Andersson-Braun cylinder. Data are from Olsher et al. [2004].

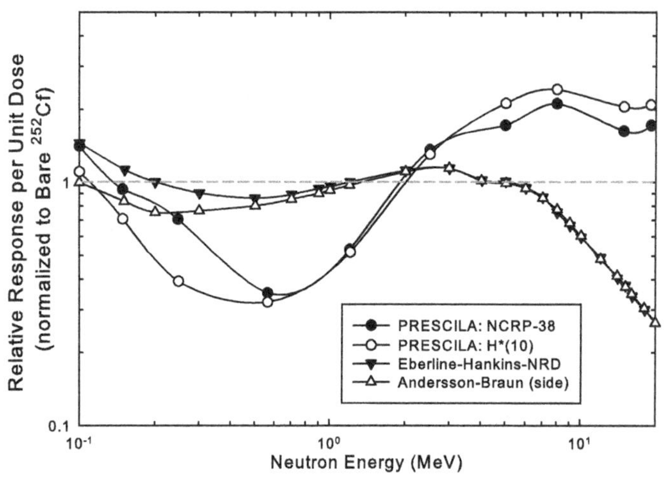

Figure 18.34. Detail of the fast neutron energy response functions per unit dose for the PRESCILA detector compared to a Hankins rem ball and an Andersson-Braun cylinder. Data are from Olsher et al. [2004].

protons in high density polyethylene is approximately 2.2 mm, with lower energies having smaller ranges. The neutron interaction efficiency is combined with the recoil energy and solid angle, the escape probability of the recoil proton, and the LLD setting to determine the efficiency as a function of energy [Klann and McGregor 2000]. Models indicate that 0.11% detection efficiency is possible for 14-MeV neutrons with a 2.2 mm HDPE attached to a semiconductor diode. In general, the fast neutron detection efficiency is relatively low for these detectors, but their compactness and low power requirement are advantageous under special conditions [De Lurgio et al. 2003].

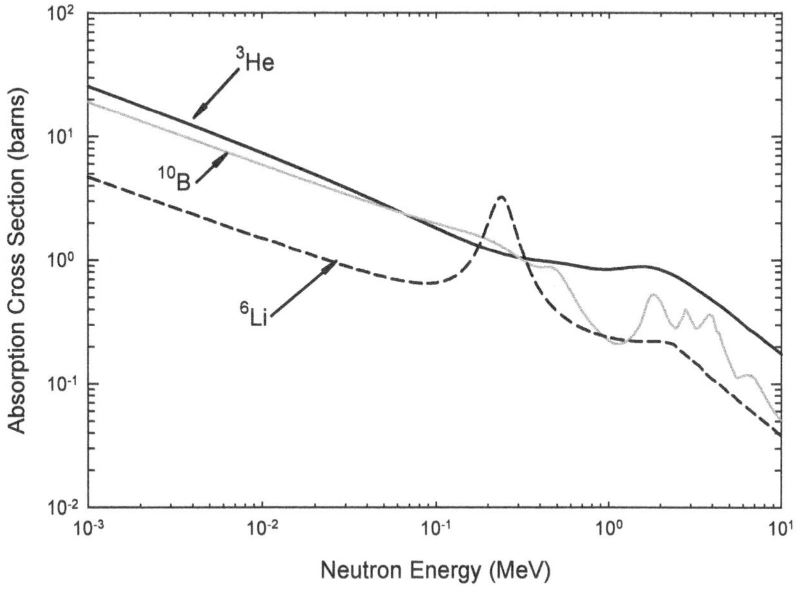

Figure 18.35. Absorption cross sections for ^{3}He, ^{6}Li, and ^{10}B in the fast neutron region. Data are from [ENDFPLOT 2017].

There are several direct interactions that can also induce pulses from the semiconductor detector. For instance, Si is a frequent choice for a particle detector, mainly because of its relatively low absorption coefficient for gamma rays. However, fast neutrons can interact in Si and release ionizing particles. For instance, the (n,p) and (n,α) reactions can produce energetic particles directly in the semiconductor, ions which do not suffer self-absorption. Although produced by neutrons, these events are considered part of the background as are possible gamma-ray interactions [Dearnaley and Northrop 1966]. One method used to reduce background is a compensated diode in which one section of the diode is covered with a hydrogenous radiator while the remaining area, of equal size, remains bare. The hydrogenous side records both proton recoils and background, while the bare side records background, and, thus, can be subtracted.

18.5 Detectors Based on Absorption Reactions

As described in Chapter 17, there are many thermal-neutron detectors based on absorption reactions. However, there are only a few isotopes with high thermal-neutron absorption cross sections that produce spontaneous ionizing reaction products. These same materials can be used for epithermal and fast neutron detection, although the absorption cross sections are reduced. The two most popular absorption reactions for fast neutron detection are the ^{3}He(n,p)^{3}H and the ^{6}Li(n,t)^{4}He reactions, whose cross section is shown in Fig. 18.35.

18.5.1 ^{3}He Detectors

^{3}He proportional counters are a popular choice for thermal-neutron detection; however, they are also used for fast-neutron detection with combined effects from recoil scattering, fast-neutron absorption, and thermal-neutron absorption. A few possible event trajectories are depicted in Fig. 18.36. As with thermal-neutron detection, the fast neutrons, with energy E, may become moderated outside of the detector, and then absorbed in the detector with negligible kinetic energy transferred to the reaction products. Such a trajectory, depicted in track 4, results in the deposition of the ^{3}He(n,p)^{3}H Q-value of 0.764 MeV in the gas. Alternatively, the neutron may be fully moderated in the detector (unlikely) and then be absorbed in the gas, as depicted in track 1, so that a total energy of $E + 0.764$ MeV is absorbed in the gas. Although the absorption cross section for ^{3}He is small for fast neutrons, there is also a small possibility that the neutron becomes fully absorbed with the reaction products producing an energy deposition of $E_n + 0.764$. Such an event is depicted as track 2 in Fig. 18.12. Finally, a recoil scatter may occur and the scattered neutron leaves the system, as shown by track 3, so that a continuous distribution of energies up to $0.75E$ is deposited, the maximum recoil energy as given by Eq. (18.4) (see Example 18.3). The idealized shape of this continuum component is given by Eq. (18.28). When all of these possible outcomes are combined, the resulting pulse height spectrum of the energy deposited in the gas has fast effects superimposed onto the traditional thermal-neutron response as shown in Fig. 18.37.

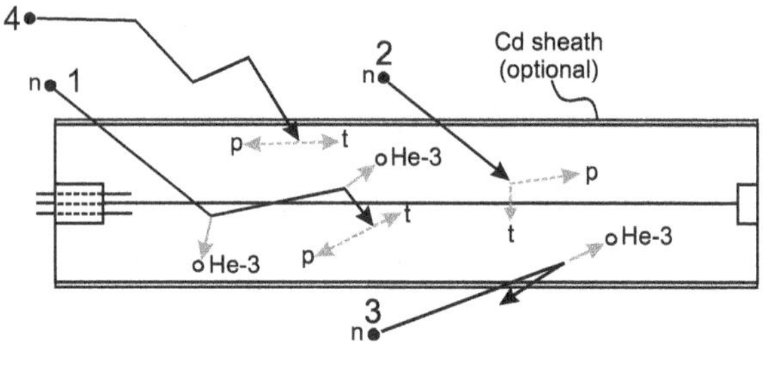

Figure 18.36. Neutron interaction possibilities for a ^{3}He gas detector.

Example 18.3: Determine the energies of the spectral features produced by monoenergetic 6 MeV neutrons interacting in a ^{3}He detector.

Solution:

1. There is a thermal-neutron peak at 0.764 MeV from slow neutrons interacting directly with the ^{3}He by the ^{3}He(n,p)^{3}H reaction.
2. The elastic scatter reaction results in an endpoint energy, given by Eq. (18.4), of

$$T = \frac{4A}{(A+1)^2}E = \frac{(4)(3)}{4^2}(6)\text{ MeV} = 0.75(6\text{ MeV}) = 4.5\text{ MeV}.$$

3. The full energy peak is at an energy equal to the total kinetic energy added to the Q value, namely, 6.764 MeV.

Typically ^{3}He gas-filled neutron detectors are designed as cylindrical proportional counters [Shalev and Cuttler 1973], mainly because of the many advantages already described in Chapter 10. Recall that the electronic pulse is formed primarily from ion motion, and that after the Townsend avalanche has progressed, over 50% of the signal is generated as the ion cloud travels a distance of only a few microns from the anode wire. Consequently, a spectrum can be formed, albeit a relatively low resolution one. To improve the energy absorption efficiency of the recoil He ions, aimed at decreasing the wall effect, a heavy gas with greater stopping power can be added. Candidates include Ar, Kr, and Xe, each having a lower ionization potential than He. However, all of these gases also have lower electron and ion mobilities than He. The use of long shaping times several microseconds long are encouraged for best results [Shalev and Cuttler 1973]. The addition of CO_2 or CH_4 can help increase the average charge carrier velocities [Shalev and Cuttler 1973; Owen et al. 1981]. Regardless, the pulse height spectrum becomes more complicated from recoils with these added fill gases. There have been reports on the rejection of elastic scatter events through pulse height discrimination [Sayres and Coppola 1964].

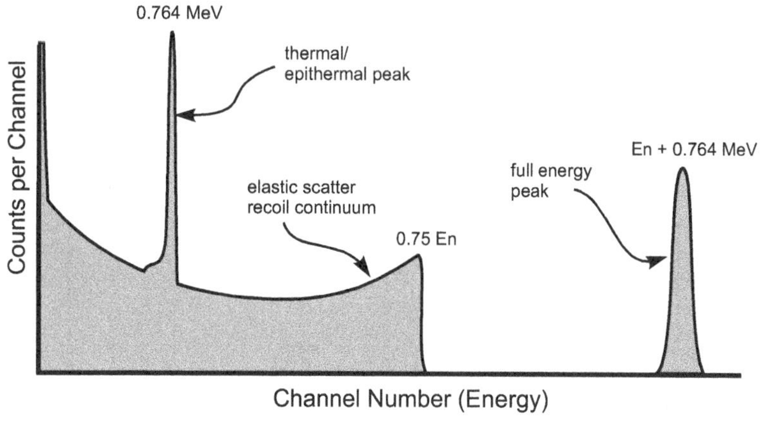

Figure 18.37. The main spectral features resulting from fast neutrons interacting in a ^{3}He detector.

Other designs include the gridded ion chamber, configured as a cylindrical design with a concentric grid[12] surrounding the central anode wire [Franz et al. 1977; Sailor and Prussin 1980; Owen et al. 1981]. Recall from Sec. 9.5.5 that gridded ion chambers rely almost entirely on electron motion and can have better energy resolution than proportional counters. The lower electric fields used to operate gridded ion chambers reduce the electron velocities, and, consequently, increases dead time. The counting gas is commonly a mixture of different gases, including ^{3}He, Ar, and CH_4 [Owen et al. 1981). The results of Franz et al. [1977] and Owen et al. [1981] demonstrate the relatively good energy resolution that is achievable. An example neutron spectrum from a gridded ion chamber is shown in Fig. 18.38.

Although the recoil spectrum is complicated by the addition of other gases than ^{3}He, Owen et al. [1981] discuss unfolding methods that can be used to produce a neutron energy pulse height spectrum. Franz et al. [1977] list many difficulties encountered with these detectors, including distortions from the wall effect, proton recoil contributions (from the CH_4 in the gas mixture), interaction location dependent rise times, gamma-ray background, and microphonics. The thermal-neutron signal can be suppressed by inserting the ^{3}He detector into filter composed of cadmium and boron [Franz et al. 1977]. Such methods diminish the elastic scatter and thermal-neutron portion of the spectrum while preserving the full energy peak of the fast neutrons $(Q + E)$. With the techniques outlined in Franz et al. [1977], reasonably high energy resolution of complex fast neutron spectra can be achieved.

18.5.2 ^{6}LiI:Eu Scintillators

Scintillators can also be used as direct conversion detectors if at least one element in the scintillator compound produces spontaneous reaction products upon the absorption of a fast neutron. A drawback is that the most commonly used neutron absorption solid state materials, ^{6}Li and ^{10}B, have very reduced interaction cross sections in the fast neutron region.

Discussed in Chapters 13 and 17, and also previously in this chapter for moderated devices, the scintillator ^{6}LiI:Eu has also been applied directly for fast neutron measurements. The concept is similar to the use of ^{3}He as a fast neutron detector. Fast neutrons that directly interact ^{6}Li(n,t)^{4}He yield reaction products with a total Q-value of 4.78 MeV. The kinetic energy of the reaction is added to Q-value to produce a full energy

[12]The reference Shalev and Cuttler [1973] is the often quoted source of the cylindrical gridded ^{3}He chamber design. However, there is no mention of a grid in this paper, Frisch is not referenced, and the authors clearly refer to the device as a proportional counter and not a gridded ion chamber.

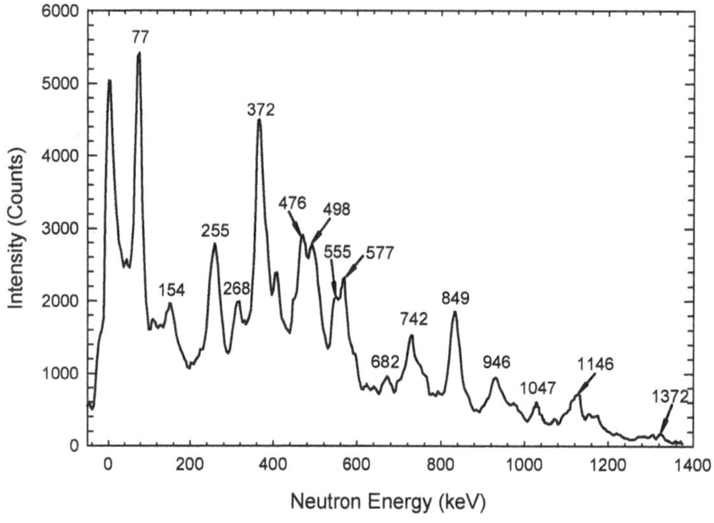

Figure 18.38. Delayed neutron spectrum from ^{137}I measured with a ^{3}He gas-filled gridded ion chamber. Data are extracted from [Franz et al. 1977].

of the reaction at $E+Q$. Murray [1958] and Johnson et al. [1969] describe competing fast neutron reactions that can complicate the spectrum. The spectrum is further confused by the intrinsic non-linear response of ^{6}LiI:Eu to fast neutrons at room temperature. Consequently, the energy resolution for fast neutrons in ^{6}LiI:Eu is relatively poor. Murray [1958] and Johnson et al. [1969] found that cooling the ^{6}LiI:Eu crystal to cryogenic temperatures, near that of liquid nitrogen, greatly improved both the linearity and resolution of the detector. Example pulse height spectra of fast neutrons (5.3 MeV) at different temperatures are shown in Fig. 18.39.

18.5.3 ^{6}Li Sandwich

Thermal-neutron detection with ^{6}Li foils or ^{6}LiF layers between detection media has been described thoroughly in Sec. 17.3.2. This "sandwich" structure, possibly first explored by Love and Murray [1961] is depicted in Fig. 18.40. It can also be used for fast neutron detection, although there are some notable differences in performance. Analysis of the detection efficiency for thermal neutrons is simplified by assuming that the reaction particle trajectories are emitted in opposite directions. However, for neutron energies above thermal (0.0259 eV), their kinetic energy must be added to the Q-value of the ^{6}Li(n,t)^{4}He reaction (see Problem 13). For slow neutrons, this added energy has little effect on the reaction product trajectories, so that the simplifying approximation that the reaction products are generally emitted in opposite directions can still be used. However, for epithermal and fast neutrons the emission angle between reaction products is less than 180°. The emission angle decreases as E increases and the reaction products are emitted increasingly in the general direction of the original incident neutron.

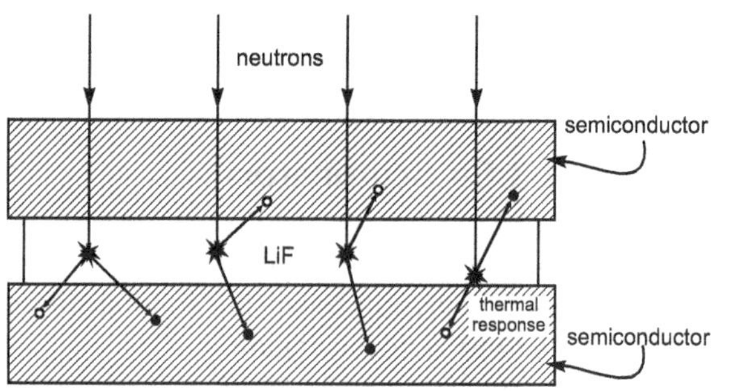

Figure 18.40. The LiF sandwich depicting possible reaction product trajectories for fast neutron interactions.

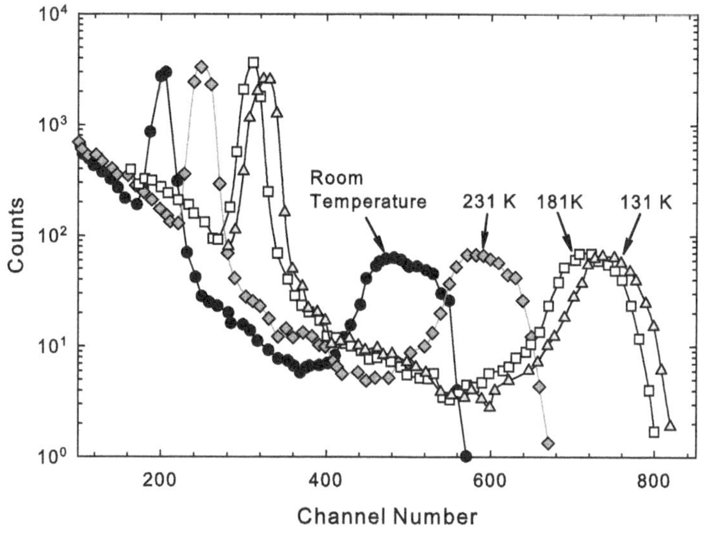

Figure 18.39. ^{6}LiI:Eu scintillator pulse height spectra at different temperatures for 5.3 MeV neutrons. The energy resolution of the full energy peak improves as the temperature is lowered, with FWHM of 18% (RT), 15% (231 K), 12% (181 K), and 10% (131 K). Data are extracted from [Murray 1958].

Background can be significantly discriminated from thermal-neutron detection by operating the two detectors in coincidence, but this operation comes at the expense of detection efficiency [McGregor et al. 2003]. Unfortunately, it can be an inefficient detection method for fast neutrons, mainly because the chance of coincidence decreases as the neutron energy increases. Because of energy self-absorption as the reaction products pass through the LiF foil, the total energy deposited in the detector has a broad spectrum instead of a single energy peak. This energy loss effect can be reduced by using LiF foils that are much thinner than the combined reaction product ranges, although this approach also reduces the efficiency [Dearnaley and Northrop 1966], which is further diminished by the reduced absorption cross section at high neutron energies. Detailed responses and efficiencies for a ^{6}Li sandwich structure are described by Rydin [1968].

Bishop [1968] and Silk [1968] report on results from a ^{6}LiF sandwich. Most of the calibrations were conducted with thermal neutrons, although both works report results from fast neutrons. In both cases, a thin LiF radiator produced higher resolution results, and produced rough neutron spectra from nuclear reactors. Rickard [1973] describes methods to improve the fast neutron energy resolution, using either a summing technique or isolating and measuring only the triton emission. Unfolding methods have also been employed to improve resolution [Seghour and Sens 1999]. The drawback to this particular use of semiconductors in a reactor environment is the rapid destruction of the semiconductor detector [Gersch et al. 2002].

Although gamma-ray interactions in low Z semiconductors (Si, SiC) are generally small, neutron events may not be easily discriminated from other possible radiations and interactions. Sources of background can also arise from direct interactions in the semiconductor. For instance, direct (n,p) reactions in Si produce energetic reaction products. Because these protons are released directly in the semiconductor, they do not suffer self-absorption effects. Consequently, the protons can produce pulses similar in magnitude to those produced by the ^{6}Li(n,t)^{3}H reaction [Rydin 1968]. This damage effect in silicon has been proposed as a possible fast-neutron dosimeter [Kramer 1966].

18.5.4 The Grey Detector

Closely related to the Hankins rem meter (or Bonner spheres) is the so-called "grey" detector [Pönitz 1968]. Recall that the Hankins rem meter is operated by moderating fast neutrons in a hydrogenous ball of material that, upon thermalization, could migrate to an internal thermal-neutron detector. The grey detector operates in a similar fashion, except that the neutron detector is removed. Instead, the thermalized neutron may be absorbed in the hydrogenous material, promptly emitting a 2.2-MeV gamma ray. A NaI:Tl detector placed adjacent to the moderator is used to detect these gamma rays, thereby yielding a measure of the neutron interactions. Pönitz [1968] assumes that (1) the moderating ball is of adequate size to moderate and absorb the neutrons with negligible leakage and (2) gamma rays escape the moderating material with no absorption. These two assumptions are valid for neutron energies ranging from thermal to 2 MeV for a moderator sphere (water or paraffin) between 20 and 25 cm diameter [Pönitz 1968]. The efficiency response for neutrons with energies from thermal to 2 MeV neutrons is relatively flat. However, for neutrons with higher energies, knowledge of the neutron leakage becomes increasingly important in order to determine the detector efficiency [Poenitz 1969].

18.5.5 Cryogenic Detectors

A low temperature solution to neutron spectroscopy was investigated by de Marcillac et al. [1993] and Richardson et al. [1998]. These devices are called both *bolometers* or *microcalorimeters* interchangeably, although there is a difference in function between the two (see Chapter 19). The difference between the two devices is that a bolometer measures the amount of radiation incident upon the device (similar to a current mode device), whereas a calorimeter measures the energy of individual interactions in the device (similar to a pulse mode spectrometer) [Kraus 1996]. In a bolometer the radiation absorber is attached to a thermally sensitive electronic component, typically a thermistor.[13] Both of these cryogenic detectors measure the thermal energy deposited in a low temperature (milli-Kelvin or mK) radiation absorber. The increase in heat in the absorber is converted into an electronic pulse, yielding a measure of the heat change and, thus, the energy deposited.

Bolometers for radiation detection are cryogenically cooled through a heat sink that returns the absorber and thermistor to the pre-irradiation equilibrium condition. A voltage is applied to the thermistor to produce a steady-state current. When radiation deposits energy in the absorber, the thermistor temperature increases, causing a change in current flow. This change can be measured as a current or voltage pulse, either being indicative of energy absorbed.

A microcalorimeter can be fashioned by attaching an absorber to a superconducting film. Below a specific transition temperature, the resistance of the superconductor is zero. If the temperature is set slightly below that temperature, then heat added to the system moves the superconductor above the transition temperature so it no longer has zero resistance. Consequently, current flow through the superconductor material now produces a voltage, which gives a measure of the energy deposited in the absorber. This type of calorimeter is called a transition edge superconducting (TES) spectrometer.

Bolometers and microcalorimeters have been demonstrated as functional neutron spectrometers [de Marcillac et al. 1993; Silver et al. 2002]. These devices work, usually, by attaching a material that incorporates either ^{6}Li or ^{10}B to a thermally sensitive electronic device. For either coating, the neutron reaction results in energy deposition of the Q-value plus the neutron kinetic energy. The work of de Marcillac et al. [1993] describes the use of a 2-g sample of ^{6}LiF operated below 100 mK and describes the difference in response for different ionizing particles. Measured results indicated that the (n,α) reaction produced counts in a lower energy channel than anticipated from calibrations with monoenergetic alpha particles.

[13] A type of resistor in which the resistance is a function of the temperature.

Neutron sensitive bolometers or microcalorimeters use energetic neutron reactions to deposit energy in the absorber, most commonly the ^{6}Li(n,t)^{4}He and the ^{10}B(n,α)^{7}Li reactions. While the work by de Marcillac et al. [1993] focused on the use of LiF as the absorber, Richardson et al. [1998] investigated the use ^{6}LiPb as an absorber for a neutron bolometer. In both studies, thermal neutrons were the radiation of interest. Energy resolution as low as 16 keV FWHM was observed for a 2-g sample of LiF operated at 10 K [de Marcillac et al. 1993; Silver et al. 2002] also report on an optimization study with enriched ^{6}LiF as a thermal-neutron detector, the largest sample being 1 cm^{3}. The best results were obtained from the smaller sample, $4 \times 4 \times 2$ mm^{3} and produced an energy resolution of 39 keV FWHM with pulses having a 80 μs rise time and 3.1 ms decay time.

Niedermayr et al. [2004] investigated the ^{10}B(n,α)^{7}Li reaction with a TiB_2 calorimeter, and also the ^{6}Li(n,t)^{4}He reaction with ^{6}LiF bolometers [Niedermayr et al. 2007; Hau et al. 2006]. Although the purpose of exploring TiB_2 as a TES microcalorimeter was to produce a fast neutron spectrometer, Niedermayr et al. [2004] describe results only for thermal-neutron reactions and report 10.5 keV FWHM for the excited state (2.31 MeV) and 5.5 keV FWHM for the ground state (2.792 MeV).

Fast neutron spectroscopy was explored by Hau et al. [2006] with ^{6}LiF samples, in which fast neutrons of energy E deposit total energy of $E_{total} = E + Q$ in the absorber. Energy resolution of 55 keV FWHM was reported for thermal neutrons. Spectra were also reported for a ^{252}Cf source. There was a clear spectral separation between neutron absorption events and the other events resulting from either elastic scattering or gamma-ray interactions. A single elastic recoil from ^{6}Li can, at most, deposit only $0.49E$. The residual energy can be recovered if the scattered neutron is subsequently absorbed. Otherwise, a substantial gap appears between the elastic scatter spectrum and the absorption spectrum. A similar situation occurs for gamma-rays, mainly because they most likely interact in the LiF crystal through Compton scattering, thereby also depositing only a fraction of their energy. Also, Hau et al. [2006] note that certain features in the absorption pulse height spectrum reflect the energy dependence of the ^{6}Li(n,t)^{4}He energy-dependent microscopic absorption cross section. Niedermayr et al. [2007] expand on the ^{6}Li fast neutron detection by describing a simple unfolding method to recover the actual energy distribution from the neutron source, which can be quite different from the elastic scattering or absorption spectrum.

Merlo et al. [2015] describe a TES microcalorimeter made of a Nb strip coated with 450 nm of natural boron. The difficulty with this structure is similar to that of a thin neutron absorbing detector [McGregor and Shultis 2004], namely, the absorber is too thin to (efficiently) absorb the total Q-value energy of the (n,α) reaction or that of a fast neutron. Consequently, the device may operate as a neutron counter, but not as a neutron spectrometer. Shishido et al. [2017] expand on this same concept by applying ^{10}B to a serpentine Nb superconducting electrode. Instead of using the TES method, Shishido et al. [2017] use another superconducting phenomenon, in which energy deposited in the superconductor causes a change in the magnetic inductance by a decrease in the Cooper pair density.[14] This device, called a microwave kinetic inductance detector (MKID), allows the position of interacting neutrons to be determined in one dimension along the serpentine pattern, reportedly with 1.3 mm spatial resolution.

18.5.6 Foil Activation Methods

One of the earliest methods for determining the energy dependence of the neutron field was the placement of different foils in the neutron field, and from the activity produced in the irradiated foils, the energy spectrum of the neutrons could be inferred. One of the earliest publications on the subject [McElroy et al. 1967] used a hand analysis to analyze the foil data. Because each foil material has cross sections with different energy dependence, the activation of each foil changes as the neutron spectrum changes. In Sec. 17.8 a detailed description of how a gold foil, bare and cadmium covered, can be used to determine characteristics of a

[14]Cooper pairs are the fundamental conduction mechanism in superconductors. See Cooper [1956] and also Ogg [1946].

typical neutron spectrum from a thermal reactor. In this section it is seen that the use of multiple foils can be used to determine a neutron energy spectrum in a way that is very similar to that used by the Bonner spheres. In fact, unfolding Bonner sphere measurements is exactly the same as unfolding foil activities and the same unfolding algorithms have been used for both.

The term foil as used here can be in various forms: powder, wires, or actual metallic foils although they are "small" to reduce self-absorption effects and to give good spatial resolution. For fluxes that have high energy neutrons, materials that have threshold cross sections are attractive. Typically ten or so different foils are used and activities are measured when the foils are irradiated, both when bare and when covered by some slow neutron filter material (typically, cadmium or gold). There appears to be no best set of foils to use in all situations. Some widely used foil reactions are listed in Table 18.2.

18.5.7 The Foil Inversion Problem

With the method discussed in Sec. 17.8.2, the saturation activity can be determined by counting the radiation emitted by a particular radioisotope produced by the neutron irradiation of a foil. Usually a gamma ray of a well-defined energy is counted, although beta-particle counting is also sometimes used. Division by the number of activation target atoms that are in the foil gives the specific saturated activity A_i per target atom at infinite dilution (so the burnup of the activation product can be ignored) for the ith activation reaction. Further, this specific activity is corrected for self-shielding and flux depression. This activity is given by the Fredholm integral equation

$$A_i = \int_0^\infty R_i(E)\phi(E)\,dE \qquad i = 1,\ldots,N. \tag{18.53}$$

Here $\phi(E)$ is the energy dependent flux at the point of interest and the response function is

$$R_i(E) = \sigma_i(E)\exp[-N_c x_c \sigma_c(E)], \tag{18.54}$$

where $\sigma_i(E)$ is the microscopic activation cross section for the ith reaction. The exponential term accounts for a cover material, such as cadmium, placed around the foil to suppress activation by thermal or epithermal neutrons. The cover has an atom density of N_c, thickness x_c, and absorption cross section $\sigma_c(E)$. For a bare foil $x_c = 0$ and the exponential term in Eq. (18.54) is absent.

Estimation of the Energy-Dependent Flux

In principle, any method used to solve the Bonner sphere problem of Eqs. (18.9) can be used to invert the foil activation problem. A general overview is provided by Cross and Ing [1987]. Here the basis of the SAND-II code [McElroy et al. 1969] is reviewed because it was the first code upon which many subsequent codes with minor algorithmic changes and additional features, such as enhanced error analysis, are based. To convert Eqs. (18.53) a contiguous energy grid E_j, $j = 1,\ldots,M$ is defined such that $\phi(E) = 0$ if $E < E_1$ or $E > E_M$. The grid spacing is $\Delta_j = E_{j+1} - E_j$. In SAND-II, 620 energy intervals are used covering the range 10^{-10} to 18 MeV (45 intervals per decade up to 1 MeV and 170 intervals between 1 and 18 MeV).

Equations (18.53) can be written as

$$A_i = \sum_{j=1}^{M} \int_{E_j}^{E_{j+1}} R_i(E)\phi(E)\,dE \equiv \sum_{j=1}^{M} A_{ij} \qquad i = 1,\ldots,N. \tag{18.55}$$

The solution procedure begins by assuming an initial flux guess ϕ_j^0, $E_j \leq E_{j+1}$ and then this flux is refined iteratively ϕ_j^k, $k = 1, 2, \ldots, K$ by making the *calculated* partial activities A_{ij}^k, which depend on ϕ_j^k, to agree with the *measured* activities A_i to within some specified tolerance. To calculate the activity, it is first assumed

Table 18.2. Some activation foils used for thermal- and fast-neutron measurements. Most foils are irradiated with Cd, Au, or ^{10}B covers to suppress contributions from thermal and epithermal neutrons.

Element	Reaction	Nat. abundance	Threshold energy (MeV)	γ-rays (MeV) [yield %]	Half-Life
Al	^{27}Al(n,p)^{27}Mg	100	1.896	0.8437 [71.8] 1.0144 [28.0]	9.45 min
	^{27}Al(n,α)^{24}Na	100	3.250	1.369 [100] 2.754 [99.9]	14.95 h
Au	^{197}Au(n,2n)^{196}Au	100	8.114	0.3557 [23.07]	6.167 d
	^{197}Au(n,γ)^{198}Au	100	0	0.41180 [95.51]	2.6952 d
Cu	^{63}Cu(n,2n)^{62}Cu	69.1	11.038	0.511 [195.6]	9.74 min
	^{65}Cu(n,2n)^{64}Cu	30.83	10.064	0.511 [35.7]	12.701 h
	^{63}Cu(n,α)^{60}Co	69.17	0	1.173 [99.86] 1.332 [99.98]	5.271 y
	^{63}Cu(n,γ)^{64}Cu	69.17	0	0.511 [35.7]	12.701 h
Co	^{59}Co(n,γ)^{60}Co	69.1	0	1173 [99.86] 1332 [99.98]	5.271 y
Fe	^{56}Fe(n,p)^{56}Mn	91.7	2.966	0.8469 [99.87]	2.578 h
	^{54}Fe(n,p)^{54}Mn	5.85	0	0.8348 [99.97]	312.1 d
In	^{115}In(n,n$'$)^{115m}In	95.71	0.336	0.3363 [46.71]	4.485 h
Mg	^{24}Mg(n,p)^{24}Na	78.99	4.932	1.368 [99.99]	14.95 h
Mn	^{55}Mn(n,γ)^{56}Mn	100	0	0.8468 [98.87]	2.578 h
Ni	^{58}Ni(n,2n)^{57}Ni	68.08	12.478	1.378 [77.9]	36.1 h
	^{58}Ni(n,p)^{58}Co	68.08	0	0.8107 [99.44]	70.88 d
Th	^{232}Th(n,f)^{140}La	99.27	–	1.5962 [95.49]	1.678 d
Ti	^{47}Ti(n,p)^{47}Sc	7.44	0	0.1594 [68.0]	3.349 d
	^{46}Ti(n,p)^{46}Sc	8.25	1.619	0.8896 [99.98] 1.211 [99.98]	83.81 d
	^{48}Ti(n,p)^{48}Sc	73.72	3.274	0.9835 [100.0] 1.0375 [97.5] 1.3121 [100.0]	43.7 h
U	^{238}U(n,f)^{140}La	99.27	–	1.5962 [95.5]	1.678 d
	^{238}U(n,f)^{140}Ba	99.27	0	0.5373 [23.5]	12.75 d
Zr	^{90}Zr(n,2n)^{89}Zr	51.45	12.102	0.9091 [99.04]	3.27 d

the cross sections are constant over an energy interval, i.e.,

$$\overline{\sigma}_{ij} \simeq \frac{1}{\Delta_j} \int_{E_j}^{E_{j+1}} \sigma_i(E)\, dE. \tag{18.56}$$

Likewise the cover cross section is a constant $\overline{\sigma}_c$ over each energy interval. Thus for the k iteration

$$A_{ij}^k = \overline{\sigma}_{ij} \exp[-N_c x_c \overline{\sigma}_{cj}]\phi_j^k \Delta_j = \overline{R}_{ij}\phi_j^k \Delta_j, \tag{18.57}$$

from which the calculated activity A_i for the ith reaction is found from Eq. (18.55).

To update the flux profile, two parameters are used: an activity weighting term $W_{ij} = A_{ij}^k/A_j^k$ and a so-called activity-weighted correction term

$$C_j^k = \frac{\sum_{i=1}^N W_{ij}^k \ln(A_i/A_i^k)}{\sum_{i=1}^N W_{ij}^k}. \tag{18.58}$$

The next estimate of the flux is then computed as

$$\phi_j^{k+1} = \phi_j^k \exp(C_j^k). \tag{18.59}$$

This iterative procedure is continued until the standard deviations of the ratios of measured to calculated activities is less than some specified value. There is also a foil discard test such that if the measured activity of a particular reaction deviates more than some specified number of standard deviations from the calculated value that foil is removed and another solution without that foil is obtained.

The background and methodology of the SAND-II code has been extensively studied (see references cited by McElroy et al. [1969]). There are many other unfolding methods and codes available, some of which are reviewed by Moghari [1979].

18.6 Summary

In this chapter many different types of fast neutron detectors have been briefly described. The detection of fast neutrons is accomplished by using one of the following three techniques. (1) The fast neutrons are first moderated to lower energies and those that become thermalized are detected using thermal neutron detectors. Bonner spheres and rem balls are examples of fast neutron detectors based on the moderation principle. (2) Charged recoil atoms produced by fast neutron scattering can be measured in charged particle detectors. This technique is the basis for many types of recoil spectrometers. (3) Finally, fast neutrons can often produce nuclear reactions not possible with thermal neutrons and products of these fast neutron interactions can often be measured and the fast neutron flux can be inferred. Foil activation methods typify the use of this method to distinguish neutrons with a wide range of energies.

In almost all of these detectors indirect information about the energy dependence of the neutron field is obtained and, hence, they can also be used as neutron energy spectrometers and not just detectors. However, their use as energy spectrometers is not as straightforward as, say, with gamma-ray spectrometers where full energy peaks in the output pulse height spectrum immediately give the energies of incident monoenergetic gamma rays. The output spectrum from neutron spectrometers, by contrast, is not easy to interpret for two main reasons. First, most neutron fields have a continuum of neutron energies without any superimposed monoenergetic peaks. Second, the output pulse height spectrum from a fast neutron spectrometer often bears little resemblance to the neutron spectrum. Consequently, the output must be post-processed or unfolded to recover energy information about the incident neutrons. In fact, the output from any radiation spectrometer needs to be unfolded to recover *all* details of the energy dependence of the incident radiation. Although a few brief discussions are given in this chapter about the unfolding or deconvolution of the output from fast neutron devices, the general topic of unfolding is worthy of several chapters, or even a book, to be treated thoroughly.

PROBLEMS

1. For $A = 1$, show that Eq. (18.2) implies that $\theta_c = 2\theta_s$.

2. Determine the thermal neutron absorption length for a ^{3}He gas-filled detector pressurized to 4 atm. If the detector is 1 inch in diameter (2.54 cm), what is the maximum attenuation for a beam of neutrons intersecting perpendicularly the tube cross section?

3. You have a $^{10}BF_3$ detector built from a 25 mm diameter (I.D.) Al tube, backfilled with 0.2 atm of $^{10}BF_3$ and 2 atm of Ar gas. If a 0.5 cm diameter thermal neutron (2200 m s^{-1}) beam intersects through the detector mid-section, what is the expected intrinsic thermal neutron detection efficiency?

4. Given a a thermal neutron beam (0.025 eV) intersecting a 10-mm-thick

5. sample of CLYC:Ce, determine the intrinsic thermal neutron detection efficiency of the CLYC:Ce sample. If the natural Li is replaced with 96% enriched ^{6}Li, what is the new thermal-neutron detection efficiency?

6. Plot ω_c versus the corresponding ω_s for $A = 1$, 2, 4, 10, 50, 100, and 200. What does this graph reveal about the center of mass and laboratory coordinate systems as A becomes large?

7. The speed of the K-shell electrons in ^{4}He, as given by the Bohr model [Shultis and Faw 2016] as $v_K = e^2/(\epsilon_o h)$. Compare this speed to that of a 4-MeV recoil ^{4}He atom. At this recoil speed, the shake-off effect has removed almost all of the electrons from the helium, so that a 4-MeV ^{4}He beam in gas has about an average charge of $+1.99e$ per atom [Evans 1955].

8. Derive Eq. (18.24) from Eq. (18.23).

9. Derive Eq. (18.27).

10. Consider a general binary nuclear reaction X(x,y)Y and use the conservation of energy and momentum constraints, as used in Chapter 4, to show

$$\sqrt{E_y} = \sqrt{\frac{m_x m_y E_x}{(m_y + m_Y)^2}}\,\omega_y \pm \sqrt{\frac{m_x m_y E_x}{(m_y + m_Y)^2}\omega_y^2 + \left[\frac{m_Y - m_x}{(m_y + m_Y)}E_x + \frac{m_Y Q}{(m_y + m_Y)}\right]},$$

where the Q-value is

$$Q = (m_x + m_X - m_y - m_Y)c^2.$$

11. From the previous problem for exoergic reactions $Q < 0$, show the kinematic threshold for this reaction is

$$E_x^{th} = -\frac{m_y + m_Y}{m_y + m_Y - m_x}Q \simeq \left(1 + \frac{m_x}{m_X}\right)Q.$$

12. Find the Q-values and threshold energies for the reactions (a) $^{12}C(n, \alpha)^9Be$ and (b) $^{12}C(n, n)3\alpha$. The threshold energies can be calculated from Eq. (4.46) and from the formula developed in the previous problem.

13. From the result of Problem 18-9 show that for exoergic reactions ($Q > 0$) as $E_x \to 0$

$$E_y = \frac{m_Y}{m_y + m_Y} Q \quad \text{and} \quad E_Y = \frac{m_y}{m_y + m_Y} Q,$$

so that

$$Q = E_y + E_Y.$$

14. The energetics of the ^{6}Li(n,t)^{4}He reaction are given in Sec. 17.2.3 for a thermal neutron. If the same reaction is caused by a 1-MeV neutron, what are the kinetic energies of the reaction products? As the energy of the neutron increases, how does the angle between the reaction products change?

REFERENCES

ABDURASHITOV, J.N., V.N. GAVRIN, A.V. KALIKHOV, V.L. MATUSHKO, A.A. SHIKHIN, V.E. YANTS, O.S. ZABORSKAIA, J.M. AADAMS, J.S. NICO, AND A.K. THOMPSON, "A High Resolution, Low Background Fast Neutron Spectrometer," *Nucl. Instrum. Meth.*, **A476**, 318–321, (2002).

ALLEN, W.D., "Flat Response Counters," in *Fast Neutron Physics, Part I*, Chap. III.A, J.B. MARION AND J.L. FOWLER, Eds., New York: Interscience, 1960.

AMALDI, E. AND E. FERMI, "On the Absorption and the Diffusion of Slow Neutrons," *Phys. Rev.*, **50**, 899–928, (1936).

AMALDI, E., L.R. HAFSTAD, AND M.A. TUVE, "Neutron Yields from Artificial Sources," *Phys. Rev.*, **51**, 896–912, (1937).

ANDERSSON, I.O. AND J. BRAUN, *A Neutron REM Counter*, Report **AE-132**, Sweden: Aktiebolaget Atomenergi, 1964.

ARKTIS, "S670e: Combined Fast and Thermal Neutron Detector," Zurich, Switzerland: Arktis Radiation Detectors Ltd., 2017.

ATWATER, H.F., "Monte Carlo Calculation of Recoil Spectra in ^{4}He Proportional Counters," *Nucl. Instrum. Meth.*, **100**, 453–457, (1972).

BALDINGER, E., P. HUBER, AND H. STAUB, "Ausbeute und Energiebestimmung der Neutronen aus der Kernreaktion," *Helv. Phys. Acta*, **11**, 245–268, (1938).

BAME, S.J., Jr., E. HADDAD, J.E. PERRY, AND R.K. SMITH "Absolute Determination of Monoenergetic Neutron Flux in the Energy Range 1 to 30 Mev," *Rev. Sci. Instrum.*, **28**, 997–1006, (1957).

BARRETT, R.F., J.R. BIRKELUND, AND H.H. THIES, "Investigation of the Flatness of Response of a Multi BF_3-Counter Neutron Detection System," *Nucl. Instrum. Meth.*, 245–268, **68**, 277–283, (1969).

BEDOGNI, R., C. DOMINGO, A. ESPOSITO, AND F. FERNÁNDEZ, "FRUIT: An Operational Tool for Multisphere Neutron Spectrometry in Workplaces," *Nucl. Instrum. Meth.*, **A 580**, 1301–1309, (2007).

BENJAMIN, P.W., C.D. KEMSHALL, AND J. REDFERN, "High Resolution Spherical Proportional Counter," *Nucl. Instrum. Meth.*, **59**, 77–85, (1968).

BENNETT, E.F., "Proportional Counter Proton-Recoil Spectrometer with Gamma Discrimination," *Rev. Sci. Instrum.*, **33**, 1153–1160, (1962).

BIRCH, R., *An Alpha-Recoil Proportional Counter to Measure Neutron energy Spectra Between 2 MeV and 15 MeV*, AERE Report R-13002, Oxfordshire: Harwell Laboratory, 1988.

BISHOP, G.B., "A ^{6}Li Sandwich Semiconductor Detector for Fast Neutron Spectrum Measurement," *Nucl. Instrum. Meth.*, **62**, 247–252, (1968).

BONNER, T.W., "Measurements of Neutron Spectra from Fission," *Nucl. Phys.*, **23**, 116–121, (1961).

BOWDEN, N.S., P. MARLEAU, J.T. STEELE, S. MROWKA, G. AIGELSDINGER, AND W. MENGESHA, "Improved Fast Neutron Spectroscopy via Detector Segmentation," *Nucl. Instrum. Meth.*, **A609**, 32–37, (2009).

BRAGA, C.C. AND M.S. DIAS, "Application of Neural Networks for Unfolding Neutron Spectra Measured by Means of Bonner Spheres," *Nucl. Instrum. Meth.*, **A476**, 252–255, (2002).

BRAMBLETT, R.L., R.I. EWING, AND T.W. BONNER, "A New Type of Neutron Spectrometer," *Nucl. Instrum. Meth.*, **9**, 1–12, (1960).

BURGETT, E.A, N.E. HERTEL, AND R.M. HOWELL, "Energy Response and Angular Dependence of a Bonner Sphere Extension," *IEEE Trans. Nucl. Sci.*, **56**, 1325–1328, (2009).

CAZZANIGA, C., M. REBAI, M. TARDOCCHI, G. ROCI, M. NOCENTE, S. ANSELL, C.D. FROST, AND G. GORINI, "A Telescope Proton Recoil Spectrometer for Fast Neutron Beam-Lines," *Prog. Theor. Exp. Phys.*, 073H01, 12 pages (2015) .

CESTER, D., M. LUNARDON, G. NEBBIA, L. STEVANATO, G. VIESTI, S. PETRUCCI, AND C. TINTORI, "Pulse Shape Discrimination with Fast Digitizers," *Nucl. Instrum. Meth.*, **A748**, 33–38, (2014).

CHAMINADE, R., J.C. FAIVRE, AND J. PAIN, "A Simple Particle Identifier," *Nucl. Instrum. Meth.*, **49**, 217–219, (1967).

CHANDRA, R., G. DAVATZ, H. FRIEDERICH, U. GENDOTTI, AND D. MURER, "Fast Neutron Detection with Pressurized ^{4}He Scintillation Detectors," *JINST*, **7**, C03035, 12 pages (2012).

COMBE R., N. ARBOR, Z. BITAR, S. HIGUERET, AND D. HUSSON, "Conception of a New Recoil Proton Telescope for Real-Time Neutron Spectrometryin Proton-Therapy," EPJ Web of Conf., **170**, 09001, 6 pages (2018).

COOPER, B.W, S.L. BELLINGER, A. CARUSO, R.G. FRONK, W.H. MILLER, T.M. OAKES, J.K. SHULTIS, T.J. SOBERING, AND D.S. MCGGREGOR, "Neutron Energy Spectrum with Microstructured

Semiconductor Neutron Detectors," *Conf. Rec. IEEE Nucl. Sci. Symp.*, Valencia, Spain, Oct. 23-29, paper R16-6, 4783–4786, (2011).

COOPER, B.W, S.L. BELLINGER, S.R. BOLDING, A. CARUSO, R.G. FRONK, W.H. MILLER, T.M. OAKES, J.K. SHULTIS, T.J. SOBERING, R. TAYLOR, D. HUDDLESTON, AND D.S. MCGGREGOR, "Portable Neutron Energy Spectrometer Utilizing Microstructured Semiconductor Neutron Detectors (MSND)," *Conf. Rec. IEEE Nucl. Sci. Symp.*, paper R12-7, pp. 4262–4265, Anaheim, CA, Oct. 27 – Nov. 3, (2012).

COOPER, L.N., "Bound Electron Pairs in a Degenerate Fermi Gas," *Phys. Rev.*, **104**, 1189–1190, (1956).

CRAUN, R.L. AND D.L. SMITH, "Analysis of Response Data for Several Organic Scintillators," *Nucl. Instrum. Meth.*, **80**, 239–244, (1970).

CROSS, W.G. AND H. ING, "Neutron Spectroscopy," Ch.2 in *The Dosimetry of Ionizing Radiation*, K.R. KASE, B.E. BJÄRNGARD AND F.H. ATTIX, Eds. Academic Press, Orlando, FL, 1987.

CZIRR, J.B. AND G.L. JENSEN, "A Neutron Coincidence Spectrometer," *Nucl. Instrum. Meth.*, **A284**, 365–369, (1989).

CZIRR, J.B. AND G.L. JENSEN, "A Compact Neutron Coincidence Spectrometer, its Measured Response Functions and Potential Applications" *Nucl. Instrum. Meth.*, **A349**, 532–539, (1994).

CZIRR, J.B., D.B. MERRILL, D. BUEHLER, T.K. MCKNIGHT, J.L. CARROLL, T. ABBOTT, AND E. WILCOX, "Capture-Gated Neutron Spectroscopy," *Nucl. Instrum. Meth.*, **A476**, 309–312, (2002).

DEARNALEY, G. AND D.C. NORTHROP, *Semiconductor Counters for Nuclear Radiations*, 2nd Ed., New York: Wiley, 1966.

DE LURGIO, P.M., R.T. KLANN, C.L. FINK, D.S. MCGREGOR, P. THIYAGARAJAN, AND I. NADAY, "A Neutron Detector to Monitor the Intensity of Transmitted Neutrons for Small-Angle Neutron Scattering Instruments," *Nucl. Instrum. and Meth.*, **A505**, 46–49, (2003).

DE MARCILLAC, N. CORON, J. LEBLANC, C. BOBIN, I. BERKES, M. DE JESUS, J.P. HADJOUT, L. GONZALEZ-MESTRES, AND J.W. ZHOU, "Characterization of a 2 g LiF Bolometer," *Nucl. Instrum. Meth.*, **A337**, 95–100, (1993).

DE PANGHER, J. AND L.L. NICHOLS, *A Precision Long Counter for Measuring Fast Neutron Flux Density*, Report **BNWL-260**, 1966.

DONZELLA, A., M. BARBUI, F. BOCCI, G. BONOMI, M. CINAUSERO, D. FARBRIS, A. FONTANA, E. GIROLETTI, M. LUNARDON, S. MORETTO, G. NEBBIA, M.M. NECCHI, S. PESENTE, G. PRETE, V. RIZZI, G. VIESTI, AND A. ZENONI, "A Proton Recoil Telescope for Neutron Spectroscopy" *Nucl. Instrum. Meth.*, **A613**, 58–64, (2010).

DOROSHENKO, J., S.N. KRAITOR, T.V. KUZNETSOVA, K.K.KUSHNEREVA, AND E.S. LEONOV, "New Methods for Measuring Neutron Spectra with Energy from 0.4 eV to 10 MeV," *Nucl. Tech.*, **33**, 296–304, (1977).

DRAKE, D.M., W.C. FELDMAN, C. HURLBUT, "New Electronically Black Neutron Detectors," *Nucl. Instrum. Meth.*, **A247**, 576–582, (1986).

EAST, L.V. AND R.V. WALTON, "Polyethylene Moderated ^{3}He Neutron Detectors," *Nucl. Instrum. Meth.*, **72**, 161–166, (1969).

ENDFPLOT-2, Korea Atomic Energy Research Institute, `/http://atom.kaeri.re.kr/endfplot.shtmlS`, accessed 2017.

EVANS, R.D., *The Atomic Nucleus*, New York: McGraw-Hill, 1955; republished by Melborne, FL: Krieger Publishing Co., 1982.

FELDMAN, W.C., G.F. AUCHAMPAUGH, R.C. BYRD, "A Novel Fast-Neutron Detector for Space Applications," *Nucl. Instrum. Meth.*, **A306**, 350–365, (1991).

FERGUSON, A.T.G., "Gas Recoil Counters," in *Fast Neutron Physics, Part I*, Chap. II.A, J.B. MARION AND J.L. FOWLER, Eds., Interscience, New York, 1960.

FINK, C.L., J.H. EICHHOLZ, D.R. BURROWS, AND A. DEVOLPI, "Proton-Recoil Proportional Counter Tests at TREAT," *IEEE Trans. Nucl. Sci.*, **NS-27** 833–837, (1980).

FISHER, B.M., J.N. ABDURASHITOV, K.J. COAKLEY, V.N. GAVRIN, D.M. GILLIAN, J.S. NICO, A.A. SHIKHIN, A.K. THOMPSON, D.F. VECCHIA, AND V.E. YANTS, "Fast Neutron Detection with ^{6}Li-Loaded Liquid Scintillator," *Nucl. Instrum. Meth.*, **A646**, 126–134, (2011).

FLASKA, M., AND S.A. POZZI, "Digital Pulse Shape Analysis for the Capture-Gated Liquid Scintillator BC-523A," *Nucl. Instrum. Meth.*, **A599**, 221-225, (2009).

FRANZ, H., W. RUDOLPH, H. OHM, K.-L. KRATZ, G. HERRMANN, F.M. NUH, D.R. SLAUGHTER, AND S.G. PRUSSIN, "Delayed-Neutron Spectroscopy with ^{3}He Spectrometers," *Nucl. Instrum. Meth.*, **144**, 253–261, (1977).

GERSCH, H.K., D.S. MCGREGOR, AND P.A. SIMPSON, "A Study of the Effect of Incremental Gamma-Ray Doses and Incremental Neutron Fluences Upon the Performance of Self-Biased ^{10}B-Coated High-Purity Epitaxial GaAs Thermal Neutron Detectors," *Nucl. Instrum. Meth.*, **A489**, 85–98, (2002).

GOLD, R., AND E.F. BENNETT, "Effects of Finite Size in 4π-Recoil Proportional Counters," *Nucl. Instrum. Meth.*, **63**, 285–299, (1968).

GOLDHAGEN, P., "An Extended-Range Multisphere Neutron Spectrometer with High Sensitivity and Improved Resolution," *Nucl. Tech.*, **175**, 81–88, (2011).

GRIN, G.A., B. VAUCHER, AND J.C. ALDER, "^{12}C(n,n), (n,n'), and (n,n')3 Reactions at 14.1 MeV," *Helv. Phys. Acta*, **42**, 990–1002, (1969).

HANKINS, D.A., *A Neutron Monitoring Instrument Having a Response Approximately Proportional to the Dose Rate from Thermalto 7.0 MeV*, LA-2717, Los Alamos, NM: Los Alamos Scientific Laboratory, 1962.

HANKINS, D.A., *Determination of the Neutron Contribution to the REM Dose*, LA-DC-7323A, Los Alamos, NM: Los Alamos Scientific Laboratory, 1965.

HANSON, A.O., AND J.L. MCKIBBEN, "A Neutron Detector Having Uniform Sensitivity from 10 keV to 3 MeV," *Phys. Rev.*, **72**, 673–677, (1947).

HARVEY, J.R., A. LAVENDER, AND I.M.G. THOMPSON, "The Calibration of a Number of Neutron Sensitive Instruments with a Novel Source Emitting Neutrons Around 0.5 keV," *Health Phys.*, **31**, 363–369, (1976).

HAU, I.D., T. NIEDERMAYR, O.B. DRURY, A. BURGER, Z.W. BELL, AND S. FRIEDRICH, "Neutron Absorption Spectroscopy for Identification of Light Elements in Actinides," *Nucl. Instrum. Meth.*, **A559**, 745–747, (2006).

HAWKE, N.P., D.S. HAWKESA, D.S. BONDA, S. CROFTA, O.N. JARVIS, AND A.C. SHERWOODA, "The Design of a Proton Recoil Telescope for 14MeV Neutron Spectrometry," *Nucl. Instrum. Meth.*, **A476**, 506–510, (2002).

HERMSDORF, D., K. PASIEKA, AND, D. SEELIGER, "EFFI - Ein Monte-Carlo-Progamm zur Berechnung der Nachweiseffektivität enies Neutronendetektors mit Organischem Szintillator," *Nucl. Instrum. Meth.*, **107**, 259–268, (1973).

HOLM, P., K. PERÄJÄRVI, S. RISTKARI, T. SIISKONEN, AND H. TOIVNONEN, "A Capture-Gated Neutron Spectrometer for Characterization of Neutron Sources and Their Shields," *Nucl. Instrum. Meth.*, **A751**, 48–54, (2014).

HOMEYER, H., "A Note on $\Delta E \cdot E$ Particle Discrimination," *Nucl. Instrum. Meth.*, **56**, 355–356, (1967).

HORNYAK, "A Fast Neutron Detector," *Rev. Sci. Instrum.*, **23**, 264–267, (1952).

HOSHOR, C.B., T.M. OAKES, E.R. MYERS, B.J. ROGERS, J.E. CURRIE, S.M. YOUNG, J.A. CROW, P.R. SCOTT, W.H. MILLER, S.L. BELLINGER, T.J. SOBERING, R.G. FRONK, J.K. SHULTIS, D.S. MCGREGOR, AND A.N. CARUSO, "A Portable and Wide Energy Range Semiconductor-Based Neutron Spectrometer," *Nucl. Instrum. Meth.*, **A803**, 68–81, (2015).

HOSHOR, C.B., E.R. MYERS, T. OAKES, S.M. YOUNG, J. CURRIE, P.R. SCOTT, W.H. MILLER, S.L. BELLINGER, D.S. MCGREGOR, AND, A.N. CARUSO, "Real-time Neutron Source Localization and Identification with a Handheld, Volumetrically-Sensitive, Moderating-Type Neutron Spectrometer," *Nucl. Instrum. Meth.*, **A866**, 252–264, (2017).

HSU, H.H., K.R. ALVAR, AND D.G. VASILIK, "A New Bonner-Sphere Set for High Energy Neutron Measurements: Monte Carlo Simulation," **41**, 938–940, (1994).

HUNT, J.B., AND R.A. MERCER, "The Absolute Calibration of a Long Counter by the Associated Activity Technique," *Nucl. Instrum. Meth.*, **156**, 451–457, (1978).

JACOBS, G.J.H., AND R.L.P VAN DEN BOSCH, "Calibration Measurements with the Multi-Sphere Method and Neutron Spectrum Analysis using the SAND-II Program," *Nucl. Instrum. Meth.*, **175**, pp. 483–489, (1980).

JASTANIAH, S.D., AND P.J. SELLIN, "Digital Techniques for n/γ Pulse Shape Discrimination and Capture-Gated Neutron Spectroscopy Using Liquid Scintillators," *Nucl. Instrum. Meth.*, **A517**, 202–210, (2004).

JEBALI, R., J. SCHERZINGER, J.R.M. ANNAND, R. CHANDRA, G. DAVATZ, K.G. FISSUM, H. FRIEDERICH, U. GENDOTTI, R. HALL-WILTON, E. HÅKANSSON, K. KANAKI, M. LUNDIN, D. MURER, B. NILSON, A. ROSBORG, AND H. SVENSSON, "A First Comparison of the Responses of a ^{4}He-Based Fast-Neutron Detector and a NE-213 Liquid-Scintillator Reference Detector," *Nucl. Instrum. Meth.*, **A794**, 102–108, (2015).

JOHNSON, C.H., "Recoil Telescope Detectors," in *Fast Neutron Physics, Part I*, Chap. II.C, J.B. MARION AND J.L. FOWLER, Eds., New York: Interscience, 1960.

JOHNSON, C.H. AND C.C. TRAIL, "Proton-Recoil Neutron Spectrometer," *Rev. Sci. Instrum.*, **27**, 468–474, (1956).

JOHNSON, D.R., J.H. THORNGATE, AND P.T. PERDUE, "A Sensitive Spectrometer for Fast Neutrons Using ^{6}Li(Eu)," *Nucl. Instrum. Meth.*, **75**, 61–65, (1969).

JOHNSON, J.T., S.J. THOMPSON, S.M. WATSON, AND D.L. CHICHESTER, "Optimization of a Fast Neutron Scintillator for Real-Time Pulse Shape Discrimination in the Transient Reactor Test Facility (TREAT) Hodoscope," *IEEE Nucl. Sci. Symp. Conf. Rec.*, 4 pages, Strasbourg, France, Oct. 29 - Nov. 6, (2016).

KAMBOJ, B.K., M.G. SHAHANI, U.V. PHADNIS, AND D. SHARMA, "A Fast Neutron Detector with Linear Energy Response," *Nucl. Instrum. Meth.*, **148**, 57–60, (1978).

KAMYKOWSKI, E.A., "Comparison of Calculated and Measured Spectral Response and Intrinsic Efficiency for a Boron-Loaded Plastic Neutron Detector," *Nucl. Instrum. Meth.*, **A317**, 559–566, (1992).

KELLY, R.P., D. MURER, H. RAY, AND K.A. JORDAN, "Analysis of the Scintillation Mechanism in a Pressurized ^{4}He Fast Neutron Detector Using Pulse Shape Fitting," *AIP Advances*, **5**, 037144, (2015).

KLANN, R.T. AND D.S. MCGREGOR, "Development of Coated Gallium Arsenide Neutron Detectors," Conf. Rec. of ICONE-8, 8th International Conference on Nuclear Engineering, paper 8110, 6 pages, Baltimore, MD, USA, April 2–6, (2000).

KLANN, R.T., C.L. FINK, D.S. MCGREGOR, AND H.K. GERSCH "Development of Semi-Conductor Detectors for Fast Neutron Radiography," Conf. Rec. of the 16th Int. Conf. on Appl. of Accelerators in Res. and Industry, pp. 1118–1121, Nov., (2001).

KLANN, R.T., C.L. FINK, D.S. MCGREGOR, AND H.K. GERSCH "Development of Coated Gallium-Arsenide for Neutron Detection Applications," Conf. Rec. of the 12th Biennial RPSD Topical Meeting, Am. Nucl. Soc., 9 pages, 2002.

KLANN, R.T., AND D.S. MCGREGOR, "Coated Semiconductor Devices for Neutron Detection," US patent 6479826; allowed Nov. 12, 2002.

KRAMER, G., "The Semiconductor Fast-Neutron Dosimeter—Its Characteristics and Applications," *IEEE Trans. Nucl. Sci.*, **NS-13**, 104–110, (1966).

KRAUS, H., "Superconductive Bolometers and Calorimeters," *Supercond. Sci. Technol.*, **9**, 827–842, (1996).

LAMARSH, J.R., *Introduction to Nuclear Reactor Theory*, Reading, MA: Addison-Wesley, 1966.

LEAKE, J.W., "A Spherical Dose Equivalent Neutron Detector," *Nucl. Instrum. Meth.*, **45**, 151–156, (1966).

LEAKE, J.W., "An Improved Spherical Dose Equivalent Neutron Detector," *Nucl. Instrum. Meth.*, **63**, 329–332, (1968).

LEAKE, J.W., "The Effect of ICRP (74) on the Response of Neutron Monitors," *Nucl. Instrum. Meth.*, **A 421**, 365–367, (1999).

LEAKE, J.W., T. LOWE, AND R.S. MASON, "Improvements to the Leake Neutron Detector I," *Nucl. Instrum. Meth.*, **A519**, 636–646, (2004).

LOVE, T.A. AND R.B. MURRAY, "The use of Surface-Barrier Diodes for Fast-Neutron Spectroscopy," *IRE Trans. Nucl. Sci.*, **8**, 91–97, (1961).

LÖVESTAM, G., "The Recoil Proton Telescope in Non-Coincidence Mode for Neutron Fluence Measurements," *Nucl. Instrum. Meth.*, **A 566**, 609–614, (2006).

MARES, V., A.V. SANNIKOV, AND H. SCHRAUBE, "Response Functions of the Andersson-Braun and Extended Range REM Counters for Neutron Energies from Thermal to 10 GeV," *Nucl. Instrum. Meth.*, **A 476**, 341–346, (2002).

MARSHALL, T.O., "Some Tests on the Consistency of the Performance of Six Precision Long Counters Intended as Secondary Standards for the Measurement of Fast Neutron Flux Densities," *Health Phys.*, **18**, 427–429, (1970).

MARINI, P., L. MATHIEUA, L. ACOSTAB, M. AICHEA, S. CZAJKOWSKIA, B. JURADOA, AND I. TSEKHANOVICHA, "Systematic Investigation of Background Sources in Neutron Flux Measurements with a Proton-Recoil Silicon Detector," *Nucl. Instrum. Meth.*, **A 841**, 56–64, (2017).

MATZKE, M., "Propagation of Uncertainties in Unfolding Procedures," *Nucl. Instrum. Meth.*, **A 476**, 230–241, (2002).

MCDANIEL, P.J. AND H.E. HUNGERFORD, "An Alpha-Recoil Spectrometer," *Nucl. Instrum. Meth.*, **165**, 509–516, (1979).

MCELROY, N.M., S. BERG, AND G. GIGAS, "Neutron-Flux Determination by Foil Activation," *Nucl. Sci. Eng.*, **27**, 533–541, (1967).

McElroy, N.M., S. Berg, T.B. Ccrockett, and R.J. Tuttle, "Measurement of Neutron Fluc Spectra by a Multiple Foil Activation Iterative Method and Comparison with Reactor Physics Calculations and Spectrometer Measurements," *Nucl. Sci. Eng.*, **36**, 15–37, (1969).

McGregor, D.S., M.D. Hammig, Y-H Yang, H.K. Gersch, and R.T. Klann, "Design Considerations for Thin Film Coated Semiconductor Thermal Neutron Detectors I: Basics Regarding Alpha Particle Emitting Neutron Reactive Films," *Nucl. Instrum. Meth.*, **A500**, 272–308, (2003).

McGregor, D.S. and J.K. Shultis, "Spectral Identification of Thin-Film-Coated and Solid-Form Semiconductor Neutron Detectors," *Nucl. Instrum. Meth.*, **A517**, 180–188, (2004).

Merlo, V., M. Salvato, M. Cirillo, M. Lucci, I Ottaviani, A. Scherillo, G. Celentano, and A. Pietropaolo, "Hybrid Superconducting Neutron Detectors," *Appl. Phys. Lett.*, **106**, 113502, 4 pages, (2015).

Moghari M., "Fast Neutron Spectrum Measurement with Threshold Detectors," Ph.D. Dissertation, Iowa State University, 1979.

Mösner, J., G. Schmidt, and J. Schintlmeister, "Unterschung der Reaktion ^{12}C(n,n′) 3α," *Nucl. Phys.*, **75**, 113–127, (1966).

Murray, R.B., "Use of Li^{6}I(Eu) as A Scintillation Detector and Spectrometer for Fast Neutrons," *Nucl. Instrum. Meth.*, **2**, 237–248, (1958).

Nattress, J., M. Mayer, A. Foster, A.B. Meddeb, C. Trivelpiece, Z. Ounaies, and I. Jovenovic, "Capture-Gated Spectroscopic Measurements of Monoenergetic Neutrons with a Composite Scintillation Detector," *IEEE Trans. Nucl. Sci.*, **63**, 1227-1235, (2016).

Nelson, K.A., M.R. Kusner, B.W. Montag, M.R. Mayhugh, A.J. Schmidt, C.D. Wayant, J.K. Shultis, and D.S. McGregor, "Characterization of a Mid-Sized Li Foil Multi-Wire Proportional Counter Neutron Detector," *Nucl. Instrum. and Meth.*, **A762**, 119–124, (2014).

Nelson, K.A., N.S. Edwards, M.R. Kusner, M.R. Mayhugh, B.W. Montag, A.J. Schmidt, C.D. Wayant, and D.S. McGregor, "A Modular Large-Area Li Foil Multi-Wire Proportional Counter Neutron Detector," *Rad. Phys. Chem.*, **116**, 165-169, (2015).

Niedermayr, T., I.D. Hau, T. Miyazaki, S. Terracol, A. Burger, V.E. Lamberti, Z.W. Bell, J.L. Vujic, S.E. Labov, and S. Friedrich, "Microcalorimeter Design for Fast-Neutron Spectroscopy," *Nucl. Instrum. Meth.*, **A520**, 70–72, (2004).

Niedermayr, T., I.D. Hau, A. Burger, U.N. Roy, Z.W. Bell, and S. Friedrich, "Unfolding of Cryogenic Neutron Spectra," *Nucl. Instrum. Meth.*, **A579**, 165–168, (2007).

Nikkel, J.A., R. Hasty, W.H. Lippincott, and D.N. McKinsey, "Scintillation of Liquid Neon from Electronic and Nuclear Recoils," *Astroparticle Phys.*, **29**, 161–168, (2008).

Northrup, J.A. and J.C. Gursky, "Relative Scintillation Efficiencies of Noble Gas Mixtures," *Nucl. Instrum.*, **3**, 207–212, (1958).

Oakes, T.M., W.H. Miller, S. Karki, P.R. Scott, A.N. Caruso, S.L. Bellinger, D.S. McGregor, J.K. Shultis, and T.J. Sobering, "A Li-6 Diode Based, Moderating Detector for Neutron Detection," *Trans. Amer. Nucl. Soc.*, **105**, 337-338, (2010).

Oakes, T.M., S.L. Bellinger, W.H. Miller, E.R. Myers, R.G. Fronk, B.W Cooper, T.J. Sobering, P.R. Scott, P. Ugorowski, D.S. McGregor, J.K. Shultis, and, A.N. Caruso, "An Accurate and Portable Solid State Neutron Rem Meter," *Nucl. Instrum. Meth.*, **A719**, 6–12, (2013).

Ochs, T.R., S.L. Bellinger, R.G. Fronk, L.C. Henson, D.E. Huddleston, Z.I. Lyric, J.K. Shultis, C.T. Smith, T.J. Sobering, and D.S. McGregor, "Present Status of the Microstructured Semiconductor Neutron Detector (MSND)-Based Direct Helium-3 Replacement (HeRep)," *IEEE Trans. Nucl. Sci.*, **64**, 1846–1850, (2017).

Ogg, R.A., "Bose-Einstein Condensation of Trapped Electron Pairs. Phase Separation and Superconductivity of Metal-Ammonia Solutions," *Phys. Rev.*, **69**, 243–244, (1946). See also erratum in *Phys. Rev.*, **69**, 544, (1946).

Olsher, R.H. and D.T. Seagraves, "Proton Recoil Scintillator Neutron Rem Meter," US Patent 6529573, allowed March 4, 2003.

Olsher, R.H., D.T. Seagraves, S.L. Eisele, C.W. Bjork, W.A. Martinez, L.L. Romero, M.W. Mallett, M.A. Duran, and C.R. Hurlbut, "PRESCILA: A New, Lightweight Neutron REM Meter," *Health Phys.*, **86**, 603–612, (2004).

O'Neal, R.D., "The Slowing Down of Low Energy Neutrons in Water. II. Determination of Photo-Neutron Energies," *Phys. Rev.*, **70**, 1–4, (1946).

Owen, J.G., D.R Weaver, and J. Walker, "The Calibration of a ^{3}He Spectrometer and its use to Measure the Neutron Spectrum from an Am/Li Source," *Nucl. Instrum. Meth.*, **188**, 579–593, (1981).

Parkinson, E.R. and D. Bodansky, "A dE/dx-E-T Particle Identification System," *Nucl. Instrum. Meth.*, **35**, 347–349, (1965).

Pönitz, W.P., "A "Grey" Neutron Detector for the Intermediate Energy Region," *Nucl. Instrum. Meth.*, **58**, 39–44, (1968).

Poenitz, W.P., "Experimental Determination of the Efficiency of the Grey Neutron Detector," *Nucl. Instrum. Meth.*, **72**, 120–122, (1969).

Press, W.H., B.P. Flannery, S.A. Teukolsky, and W.T. Vetterling, *Numerical Recipes*, Cambridge, UK: Cambridge University Press, 1986.

Reginatto, M., P. Goldhagen, and S. Neumann,"Spectrum Unfolding, Sensitivity Analysis and Propagation of Uncertainties with the Maximum Entropy Deconvolution Code MAXED," *Nucl. Instrum. Meth.*, **A476**, 242–246, (2002).

Richardson, J.M., W.M. Snow, Z. Chowdhuri, and G.L. Greene, "Accurate Determination of Thermal Neutron Flux via Cryogenic Calorimetry," *IEEE Trans. Nucl. Sci.*, **45**, 550–555, (1998).

Rickard, I.C., "Neutron Spectroscopy in the Energy Range 10-500 keV Using the ^{6}Li Sandwich Detector," *Nucl. Instrum. Meth.*, **113**, 169–174, (1973).

Rotondi, E. and K.W. Geiger, "A Neutron Dosemeter with Spherical Moderator Containing Absorbers and a BF_3 Counter," *Proc. First Int. Congress Rad. Prot.*, W.S. Snyder, H.H. Abee, I.K. Burton, R. Maushart, A. Benco, F. Duhamel, and B.M. Wheatley, Eds., Amsterdam: Elsevier, pp. 141–145, 1968.

Ryan, B.C., *Analysis Methods for Bonner Sphere Spectrometry*, M.S. Thesis, Kansas State University, Manhattan, KS, 1998.

Rydin, R.A., "Neutron Detection and Spectroscopy," in *Semiconductor Detectors*, Ch. 4.5, G. Bertonlini and A. Coche, Eds., New York: Wiley, 1968.

Ryves, T.B., "A Proton Recoil Telescope for 12–20 MeV Neutrons," *Nucl. Instrum. Meth.*, **135**, 455–458, (1976).

Sailor, W.C. and S.G. Prussin, "Calculation of the Energy Dependent Efficiency of Gridded ^{3}He Fast Neutron Ionization Chambers," *Nucl Instrum. Meth.*, **173**, 511–515, (1980).

Sayres, A. and M. Coppola, "^{3}He Neutron Spectrometer Using Pulse Risetime Discrimination," *Rev. Sci. Instrum.*, **35**, 431–437, (1964).

Segel, R.E., C.D. Swartz, and G.E. Owen, "Detection of 2.5-MeV Neutrons with a Single-Crystal Scintillation Counter," *Rev. Sci. Instrum.*, **25**, 140–144, (1954).

Seghour, A. and J.C. Sens, "Neutron Spectrometry at Strasbourg University Reactor," *Nucl. Instrum. Meth.*, **A420**, 243–248, (1999).

Sekharan, K.K., H. Laumer, B.D. Kern, and F. Gabbard, "A Neutron Detector for Measurement of Total Neutron Production Cross Sections," *Nucl. Instrum. Meth.*, **133**, 253–257, (1976).

Shalev, S. and J.M. Cuttler, "The Energy Distribution of Delayed Fission Neutrons," *Nucl. Sci. Engin.*, **51**, 52–66, (1973).

Shishido, H., H. Yamaguchi, Y. Miki, S. Miyajima, K. Oikawa, M. Harada, M. Hidaka, T. Oku, M. Arai, A. Fujimaki, and T. Ishida, "Neutron Detection Using the Superconducting Nb-Based Current-Biased Kinetic Inductance Detector," *Supercond. Sci. Technol.*, **30**, 094003, 5 pages (2017).

Shultis, J.K. and R.E. Faw, *Fundamentals of Nuclear Science and Engineering*, 3rd Ed., Boca Raton, FL: CRC Press, 2017.

Silk, M.G., "The Determination of Fast-Neutron Spectra in Thermal Reactors by Means of a High-Resolution Semiconductor Spectrometer," *Nucl. Instrum. Meth.*, **66**, 93–101, (1968).

Silver, C.S., J. Beeman, L. Piccirillo, P.T. Timbie, and J.W. Zhou, "Optimization of a ^{6}LiF Bolometric Neutron Detector," *Nucl. Instrum. Meth.*, **A485**, 615–623, (2002).

Slaughter, D.R. and D.W. Rueppel, "Calibration of a Depangher Long Counter from 2 keV to 19 MeV," *Nucl. Instrum. Meth.*, **145**, 315–320, (1977).

Slaughter, D.R. and R. Strout II, "Flyspec: A Simple Method of Unfolding Neutron Energy Spectra Measured with NE213 and Stilbene Spectrometers," *Nucl. Instrum. Meth.*, **198**, 349–355, (1982).

Snidow, N.L. and H.D. Warren, "Wall Effect Corrections in Proportional Counter Spectrometers," *Nucl. Instrum. Meth.*, **51**, 109–116, (1967).

Swartz, C.D. and G.E. Owen, "Recoil Detection in Scintillators," in *Fast Neutron Physics, Part I*, Chap. II.B, J.B. Marion and J.L. Fowler, Eds., New York: Interscience, 1960.

Tagziria, H. and D.J. Thomas, "Calibration and Monte Carlo Modelling of Neutron Long Counters," *Nucl. Instrum. Meth.*, **A452**, 470–483, (2000).

Thies, H.H., "Calculation of Flat-Response Fast-Neutron Detection Systems Employing Finite Size BF_3-Counters," *Nucl. Instrum. Meth.*, **25**, 89–92, (1963).

Thomas, D.J., A.G. Bardell, and E.M. Macaulay, "Characterisation of a Gold Foil-Based Bonner Sphere Set and Measurements of Neutron Spectra at a Medical Accelerator," *Nucl. Instrum. Meth.*, **A 476**, 31–35, (2002).

Thomas, D.J. and A.V. Alevra, "Bonner Sphere Spectrometers - a Critical Review," *Nucl. Instrum. Meth.*, **A 476**, 12–20, (2002).

Weise, K., M. Weyrauch, and K. Knauf, "Neutron Response of a Spherical Proton Recoil Proportional Counter," *Nucl. Instrum. Meth.*, **A309**, 287–293, (1991).

Weyrauch, M., A. Casnati, P. Schillebeeckx, and M. Clapham "Use of ^{4}He-Filled Proportional Counters as Neutron Spectrometers," *Nucl. Instrum. Meth.*, **A403**, 442–454, (1998).

Wilhelm, K., J. Nattress, and I. Jovanovic, "Development and Operation of a ^{6}LiF:ZnA(Ag) - Scintillating Plastic Capture-Gated Detector," *Nucl. Instrum. Meth.*, **A842**, 54–61, (2017).

Chapter 19

Luminescent, Film, and Cryogenic Detectors

The crystals were swallowed by the patients, recovered one or two days later, and the accumulated dosage in roentgens was measured by matching the thermoluminescence intensity with that produced in the crystals by a known roentgen dosage.

Farrington Daniels

19.1 Luminescent Dosimeters

A luminescent dosimeter is a type of phosphorescent detector that can store energy information for extended periods of time. This passive readout detector has the property of storing electronic charges in deep traps that are later depopulated with an intentional and controlled sampling method. Hence, a luminescent detector is a type of integrating dosimeter. Typically, the device can be reused after the reading of the accumulated dose is complete. Luminescent dosimeters are popular for radiation dosimetry because of their wide sensitivity range, compact size, linear response to radiation, relative energy independence, and their reusability. However, because a readout system is required to obtain doses from these dosimeters, they cannot provide immediate dose information. Further, once read, the accumulated dose is effectively erased and, consequently, there is no permanent record.

19.1.1 Thermoluminescent Dosimeters

Thermoluminescence (TL), as the name implies, is the release of photons from a substance when heat is applied. The time when thermoluminescence was discovered is unknown, although the phenomenon was recorded in alchemist texts dating back at least to 1602 [Arnold 1991]. The first scientific report on thermoluminescence was presented by Robert Boyle in 1663 to the Royal Society of London [Becker 1973], where he described glimmering light from diamond samples when heated. TL was first used as early as 1895 by Weidemann and Schmidt to detect ionizing radiation. Hence, the thermoluminescent phenomenon has been known for quite a while.

The physical aspects of thermoluminescence were studied by many researchers after the pioneering work of Weidemann and Schmidt, and include, for example, work by Randall and Wilkins [1945], Boyd [1949], and numerous references provided in Becker [1973] and Horowitz [1984a]. However, it was not until 1954 that Farrington Daniels, after studying the radiation effects of thermoluminescent minerals [Daniels et al. 1949; 1953; Daniels and Saunders 1950; see also Boyd 1949; Saunders et al. 1953], suggested the use of thermoluminescent materials, particularly LiF, as dosimeters [Morehead and Daniels 1957; Cameron 1991]. Several additional thermoluminescent materials have since been studied as dosimeters as well, including CaF_2, $CaSO_4$, and $Li_2B_4O_7$ [Kossel et al. 1954; Ginther and Kirk 1957; Hecklesberg and Daniels 1957; Schulman et al. 1967].

General Operation

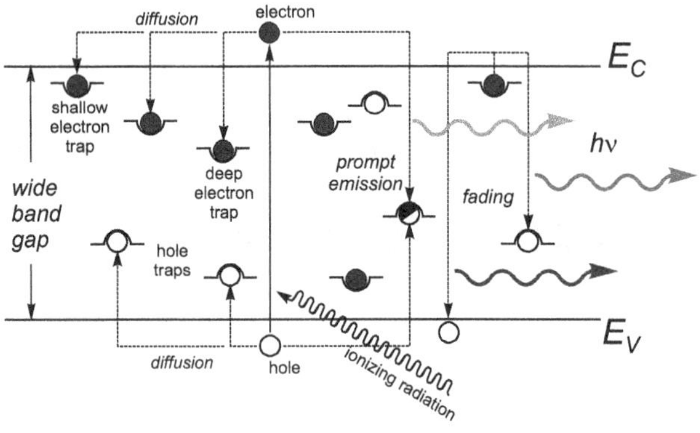

Figure 19.1. Band diagram depicting electronic trapping and emission during the irradiation and storage time.

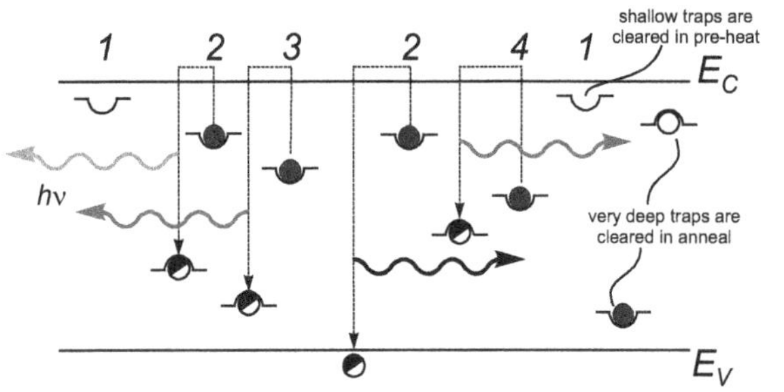

Figure 19.2. Band diagram depicting detrapping and emission during the heating time. The shallow traps (1) are emptied during the preheat phase. After preheat, the PMT counter is activated and the total photon yield is integrated as the traps release charges in order of trap energy (2, 3, and 4). These deeper traps are used to provide a reading of the accumulated dose. After the read time, the counter is deactivated and the TLD temperature is increased to release charges from the deepest traps.

Thermoluminecesnce involves two main processes: (1) radiation absorption with subsequent electron trapping, and (2) electron and hole recombination with light emission. During the absorption process, TL materials absorb energy from ionizing radiation and excite electron-hole pairs. These electrons and holes diffuse through the crystal, much as they do in a scintillator, and during their migration they can recombine or fall into electronic traps located in the band gap (see Fig. 19.1). However, unlike scintillators that rapidly release photons via *fluorescence*, thermoluminescent materials undergo a delayed release of photons and should more correctly be referred to as *phosphorescence.* The phosphorescence emission time is a function of the trap energy. Shallow traps tend to release charge carriers in a relatively short period of time while deep traps can retain charge carriers for long periods of time, in some cases for many decades. These storage traps can be designed into the material by adding select amounts of specific impurities.

If the trap is unstable, or shallow, recombination may occur at room temperature, an effect often termed as "fading". However, the stable deeper traps retain charge carriers provided that they do not absorb enough energy to excite the electrons or holes back into the conduction or valence bands, respectively. When energy is deposited in a TL material, trapped electrons and holes may acquire sufficient energy to be dislodged and allow them to return to the conduction or valence bands, respectively, and then rapidly recombine at a luminescent center. As these electrons and holes recombine, the excess energy is released as UV or visible light as shown in Fig. 19.2. By recording the number of photons released with a light sensor (such as a photomultiplier tube), the number of trapped charges can be measured. Ultimately, the total number of photons detected gives a measure of the sample radiation dose.

The governing differential equations that predict the dynamic behavior of a TLD are generally coupled and non-linear. Only for simple models, and even then with restrictive assumptions, are analytic solutions possible. Nevertheless, much can be learned from these simple models. Below a few simple TLD dynamic models are summarized.

Modeling the Thermoluminescence Intensity

Randall and Wilkins [1945] proposed a model for luminescent output from a thermoluminescent dosimeter (TLD).[1] In their model all of the electron traps are at a single energy E_t below the conduction band. The

[1] See also Garlick and Wilkins [1945].

probability per unit time of releasing an electron from a trap back to the conduction band is

$$p = \overline{\omega} \exp\left[\frac{-E_t}{kT}\right], \qquad E_V < E_t < E_C, \tag{19.1}$$

where $\overline{\omega}$ is a *frequency factor* associated with the trap, and k and T have their usual meanings. Randall and Wilkins [1945] explain that the trapped electron is like a particle in a potential box and that the frequency factor $\overline{\omega}$ is the product of the frequency with which the electron strikes the sides of the box times the Mott and Gurney [1948] reflection coefficient. Alternatively, $\overline{\omega}$ can be thought of as the frequency with which a trapped electron interacts with the lattice phonons times a transition probability. The frequency factor $\overline{\omega}$ can never be greater than the lattice vibrational frequency and is usually several orders of magnitude less than this limiting frequency, i.e., $\overline{\omega} \ll 10^{13}$ s^{-1} [Mott and Gurney 1948; Randall and Wilkins 1945].

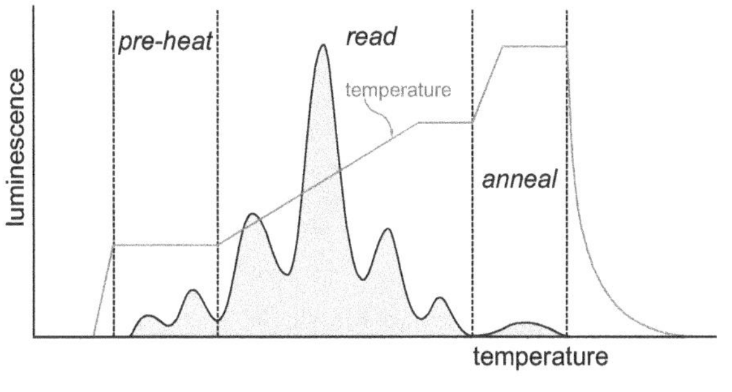

Figure 19.3. Depiction of a temperature-dependent glow curve from a TLD with multiple trap energies, showing the three regions of operation for the TLD reader.

If $N_{te}(t)$ is the density of electrons located in traps at time t, then the average number of trap releases per unit time per unit volume is $N_{te}(t)p$ or

$$-\frac{dN_{te}(t)}{dt} = N_{te}(t)\overline{\omega} \exp\left[\frac{-E_t}{kT}\right]. \tag{19.2}$$

If the emitted electrons are not retrapped, then the intensity of the thermoluminescent glow is proportional to the electron release rate, i.e.,

$$I(t) = -C\frac{dN_{te}(t)}{dt} = CN_{te}(t)\overline{\omega} \exp\left[\frac{-E_t}{kT}\right], \tag{19.3}$$

where C is a constant describing the luminescent efficiency and volume of the TLD. If the TLD is heated at a uniform rate

$$R_T = \frac{dT}{dt}, \tag{19.4}$$

the independent variable t can be replaced by T through $dt = (1/R_T)\,dT$ so that Eq. (19.3) can be written as

$$\frac{dN_{te}(T)}{N_{te}(T)} = -\frac{\overline{\omega}}{R_T} \exp\left[\frac{-E_t}{kT}\right] dT. \tag{19.5}$$

If $N_{te}(T_0)$ is the density of electrons in traps at time t_0 and temperature T_0, then Eq. (19.5) can be integrated to find

$$\ln\left[\frac{N_{te}(T)}{N_{te}(T_0)}\right] = -\frac{\overline{\omega}}{R_T}\int_{T_0}^{T} \exp\left[\frac{-E_t}{kT'}\right] dT'. \tag{19.6}$$

Substitution of the result of Eq. (19.6) into Eq. (19.3) yields

$$\begin{aligned} I(T) &= -C\frac{dN_{te}(t)}{dt} = -C\frac{dN_{te}(T)}{dT}\frac{dT}{dt} = -C\frac{dN_{te}(T)}{dT}R_T \\ &= N_{te}(T_0)C\,\overline{\omega} \exp\left[\frac{-E_t}{kT}\right] \exp\left[-\frac{\overline{\omega}}{R_T}\int_{T_0}^{T} \exp\left[\frac{-E_t}{kT'}\right] dT'\right]. \end{aligned} \tag{19.7}$$

Equation (19.7) describes the emission of a temperature-dependent *glow curve* of a TLD. Figure 19.3 shows a typical glow curve for a TLD with multiple trap energies and Fig. 19.4 shows results for the above model of a

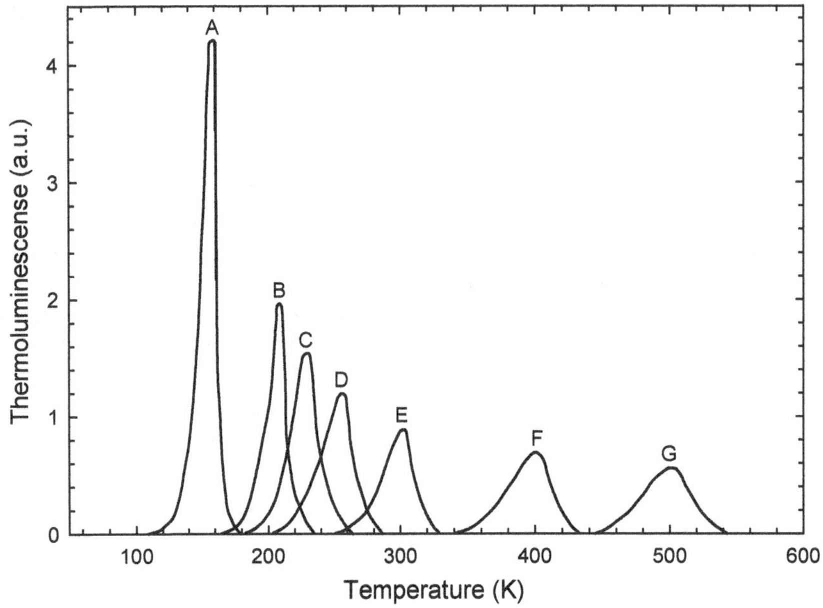

Figure 19.4. Theoretical glow curves versus temperature for phosphors with a single trap level, assuming no retrapping of electrons, after irradiation at $T = 90$ K with $R_T = 2.5$ °C/sec. Parameters for the various cases are:

(A)	$E_t = 0.3$ eV, $\overline{\omega} = 10^9$ s^{-1};	(B)	$E_t = 0.4$ eV, $\overline{\omega} = 10^9$ s^{-1}
(C)	$E_t = 0.4$ eV, $\overline{\omega} = 10^8$ s^{-1};	(D)	$E_t = 0.4$ eV, $\overline{\omega} = 10^7$ s^{-1}
(E)	$E_t = 0.6$ eV, $\overline{\omega} = 10^9$ s^{-1};	(F)	$E_t = 0.8$ eV, $\overline{\omega} = 10^9$ s^{-1}
(G)	$E_t = 0.4$ eV, $\overline{\omega} = 10^6$ s^{-1}		

After Garlick and Gibson [1948].

TLD with a single trap energy. The maximum intensity $I_{\max}$ for a glow peak can be found by differentiation of Eq. (19.7) with respect to T and setting the result to zero. In this manner, the temperature T_m at peak luminosity is found from the relationship

$$\frac{E_t}{kT_m^2} = \frac{\overline{\omega}}{R_T} \exp\left[\frac{-E_t}{kT_m}\right]. \tag{19.8}$$

There are several properties that can be gleaned from Eq. (19.7) and Eq. (19.8) [Garlick 1949]. First, the temperature of the glow peak maximum increases with E_t for constant values of $\overline{\omega}$ and R_T. Second, as the heating rate R_T is increased or the value of $\overline{\omega}$ is decreased, the glow peak shifts to higher temperatures for constant values of E_t. For any constant value of filled trap states $N_{te}(T_0)$, the total luminosity, the area of the glow curve, should remain constant, although the shape changes with $\overline{\omega}$, E_{te}, and R_T. Finally, at low temperatures ($T \ll T_m$), the initial portion of the glow curve should follow the prediction of Eq. (19.3).

Schayes and Brooke [1963] derived an expression for the maximum luminosity that includes the effects of total dose and heating rate,

$$I_m \simeq \frac{k}{E_t} R_T N_{te}(t_0) \left[\frac{e^{-1+2u^*-6(u^*)^2}}{(u^*)^2}\right], \tag{19.9}$$

where the dimensionless temperature at the glow curve maximum is $u^* = (kT_m)/E_t$. From Eq. (19.9), it would seem that the maximum output intensity would remain constant provided that the radiation exposure is adjusted to be proportional to $1/R_T$. In other words, if N_{te} is linearly adjusted to exactly counter the effect of R_T, the glow peak maximum output is constant. These predictions were tested with experiments performed by Gorbics et al. [1969], in which glow curves were measured for CaF_2:Mn, natural fluorite, LiF (TLD-100), $CaSO_4$:Mn, $Li_2B_4O_7$:Mn, and lithium aluminosilicate (LAS) glass. The glow curves were obtained at different heating rates for different gamma-ray exposures. The important results indicate that thermal quenching occurs for rapid heating rates, consequently reducing the total light yield and maximum light intensity. Shown in Fig. 19.5 and Fig. 19.6 are temperature and time dependent glow curves, respectively, for CaF_2:Mn (TLD-400) at several different heating rates. The TLDs were exposed to gamma rays from ^{60}Co for a time proportional to $1/R_T$. The results should yield identical peak maxima but, as can be seen from the results of Fig. 19.5 and Fig. 19.6, do not. The deviation from Eq. (19.9) is a consequence of TLD *thermal quenching* [Gorbics et al. 1969].

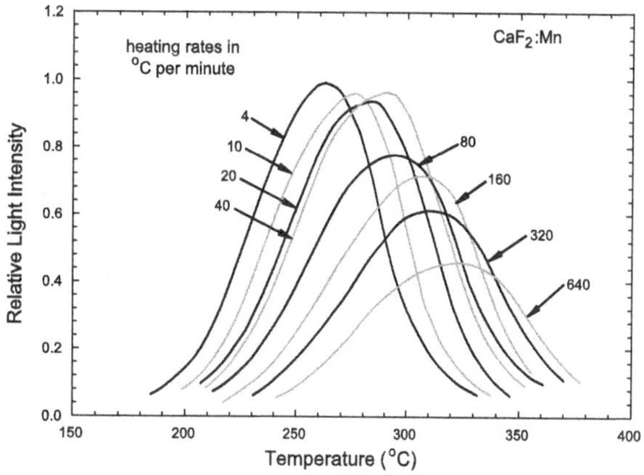

Figure 19.5. Glow curves versus temperature for CaF_2:Mn for eight different heating rates. For each case the gamma-ray exposure was adjusted to be proportional to $1/R_T$, an adjustment which should lead to a nearly constant glow curve maximum given by Eq. (19.8). Data are from Gorbics et al. [1969].

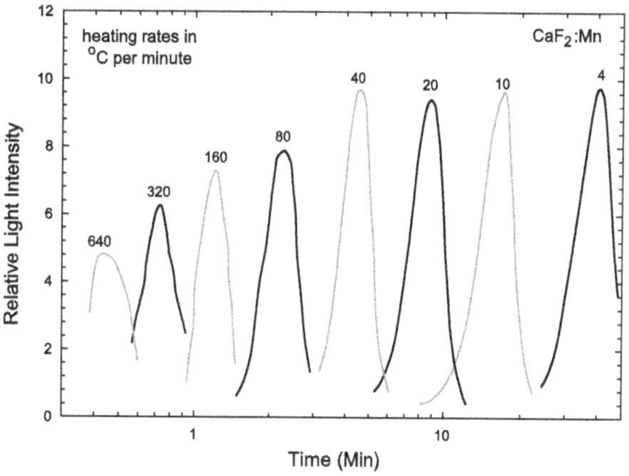

Figure 19.6. Glow curves versus time for CaF_2:Mn for eight different heating rates. For each case the gamma-ray exposure was proportional to $1/R_T$. Data are from Gorbics et al. [1969].

Attix [1986] points out a few other properties that can be seen from the results of Fig. 19.5 and Fig. 19.6. First, the length of time required to reach the glow peak maximum is nearly inversely proportional to the heating rate. Second, for heating rates below 40°C per min, the glow curve areas, as a function of temperature, are nearly identical. Yet, the dose for these glow curves was decreased by a factor of 10 as the heating rate was increased from 4°C per min to 40°C per min. At heating rates greater than 40°C per min, the glow curve area is observed to decrease. Third, the glow curve area as a function of time is observed to be nearly proportional to the dose for heating rates less than 40°C per min. Fourth, the glow curve peak temperature increases with heating rate. Finally, for CaF_2:Mn, the temperature for the onset of thermal quenching was approximately 290°C. Overall, for dosimetry purposes, it was concluded by Attix that the total light emission over time is a superior dose metric, compared to the light emission maximum, because there is a smaller error associated with fluctuations in the heating rate.

For most TLDs there are numerous levels of electron and hole traps that are mostly designed into the TLD by purposely adding dopants that act as traps and luminescent centers. The photons emitted from the multiple traps can be added to produce the total luminescent glow. Hence, the glow curve is described by [Leverenz 1950],

$$I(T) = \sum_{i=1}^{M} N_{te}^i \overline{\omega}^i C^i \exp\left[-\frac{E_t^i}{kT}\right] \exp\left[-\frac{\overline{\omega}^i}{R_T}\int_{T_0}^{T} \exp\left[-\frac{E_t^i}{kT'}\right] dT'\right], \tag{19.10}$$

where the superscript i identifies the trap and properties and M is the number of different trap levels. Luminescent centers that are close in energy may not be resolved and, thus, cause the appearance of asymmetric and wider glow peaks.

For the derivation of Eq. (19.7), Randall and Wilkins [1945] assumed that no electron retrapping occurred once excited from traps. Garlick [1949] considered a simplistic case for retrapping, for which detrapped electrons have equal probability of falling into either luminescent centers or empty trapping centers. For simplicity, assume that the material has one type of trapping center distributed uniformly with density N_t. Further, it is assumed that there are no empty luminescent centers[2] in the unexcited phosphor. The density of filled trapping centers is N_{te} and the density of empty trapping centers is N_{t0}; hence

$$N_{t0} = N_t - N_{te}. \tag{19.11}$$

If the material is an insulator, and most TLD materials are, the density of electrons in the conduction band should be considerably less than the density of trapped electrons N_{te}. Heating causes centers to depopulate so they become a source of electrons for the traps. Hence, the density of trapped electrons becomes the same as the density of empty luminescent centers ($N_{te} = N_{\ell 0}$).[3]

For a density of electron-filled traps of N_{te}, the probability that a detrapped electron falls into a luminescent center, which has a density of $N_{\ell t}$, is

$$p = \frac{N_{te}}{N_{t0} + N_{\ell 0}} = \frac{N_{te}}{(N_t - N_{te}) + N_{te}} = \frac{N_{te}}{N_t}. \tag{19.12}$$

Equation (19.12) indicates that when few traps are populated, then there is a low probability that detrapped electrons recombine at luminescent centers. Further, if the traps are completely populated, i.e., $N_{te} = N_t$, a detrapped electron has no option but to recombine at a luminescent center. From this interpretation, Eq. (19.2) is modified to

$$-\frac{dN_{te}(t)}{dt} = \frac{N_{te}(t)^2}{N_t}\overline{\omega}\exp\left[\frac{-E_t}{kT(t)}\right]. \tag{19.13}$$

Note that it is assumed in Eq. (19.12) and Eq. (19.13) that the capture probabilities of the trap and luminous centers are equal, but this is an assumption that may not be true.

To solve the above separable differential equation, first change the independent variable to T with the use of Eq. (19.4). The solution is

$$N_{te}(T) = \left\{\frac{1}{N_{te}(T_0)} + \frac{1}{N_t}\int_{T_0}^{T} \frac{\overline{\omega}}{R_T}\exp\left[-\frac{E_t}{kT'}\right] dT'\right\}^{-1}. \tag{19.14}$$

From this result the thermoluminescent intensity I as a function of temperature, which is increased at a uniform rate [Garlick and Gibson 1948], is

$$I(T) = -C\frac{dN_{te}}{dt} = -C\frac{dN_{te}}{dT}\frac{dT}{dt} = \frac{C\dfrac{N_{te}(T_0)^2}{N_t}\overline{\omega}\exp\left[\dfrac{-E_t}{kT}\right]}{\left[1 + \dfrac{N_{te}(T_0)}{N_t}\displaystyle\int_{T_0}^{T}\frac{\overline{\omega}}{R_T}\exp\left[\frac{-E_t}{kT'}\right] dT'\right]^2}. \tag{19.15}$$

[2] An imperfection in the crystal responsible for luminescent emissions. In the present case, these imperfections are hole traps. When occupied by an electron, they are not capable of producing light from recombination. If the electron is excited from the center, then it is effectively occupied by a hole. An electron recombining with a hole at the luminescent center releases a photon.

[3] Garlick and Gibson [1948] appear to use the condition that the only significant source of trapped electrons is from depopulated luminescent centers.

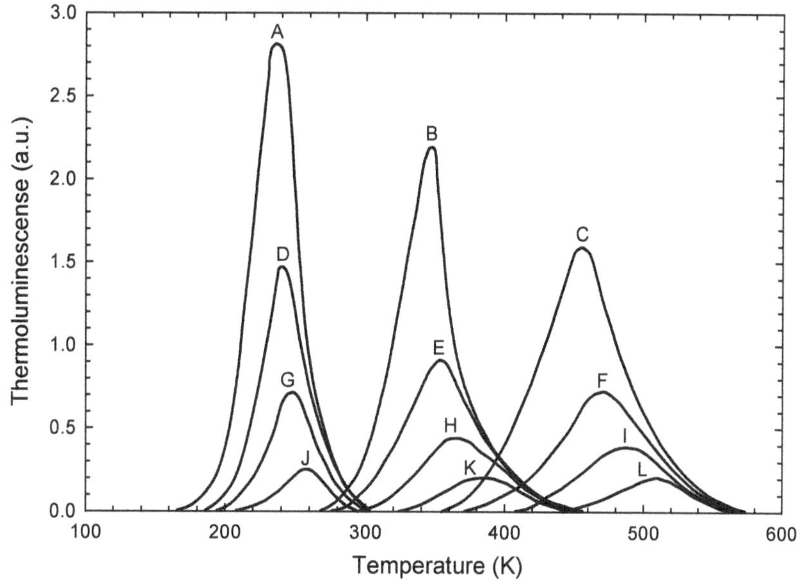

Figure 19.7. Theoretical glow curves versus temperature for phosphors with a single trap level, including the effects of electron retrapping, after irradiation at $T = 90$ K with $R_T = 2.5$ °C/sec and $\overline{\omega} = 10^9$ s^{-1}. Parameters for the various curves are as follows:

(A)	$E_t = 0.4$ eV, $N_{te}(0) = N_T$	(B)	$E_t = 0.6$ eV, $N_{te}(0) = N_T$
(C)	$E_t = 0.8$ eV, $N_{te}(0) = N_T$	(D)	$E_t = 0.4$ eV, $N_{te}(0) = 0.5N_T$
(E)	$E_t = 0.6$ eV, $N_{te}(0) = 0.5N_T$	(F)	$E_t = 0.8$ eV, $N_{te}(0) = 0.5N_T$
(G)	$E_t = 0.4$ eV, $N_{te}(0) = 0.25N_T$	(H)	$E_t = 0.6$ eV, $N_{te}(0) = 0.25N_T$
(I)	$E_t = 0.8$ eV, $N_{te}(0) = 0.25N_T$	(J)	$E_t = 0.4$ eV, $N_{te}(0) = 0.1N_T$
(K)	$E_t = 0.6$ eV, $N_{te}(0) = 0.1N_T$	(L)	$E_t = 0.8$ eV, $N_{te}(0) = 0.1N_T$

After Garlick and Gibson [1948].

Example results of Eq. (19.15) are shown in Fig. 19.7 and a few characteristics are observed. First, for constant values of $N_{te}(T_0)$, $\overline{\omega}$, and R_T, the temperature of the luminescent peak maximum increases nearly proportional with the trap energy. Second, for constant values of $N_{te}(T_0)$ and N_t, the temperature of the luminescent peak maximum increases with R_T or decreases with $\overline{\omega}$. Third, the area under each curve is proportional to the number of initial trapped electrons, or $N_{te}(T_0)/N_T$. Further, the shape of the glow curve and position of the maximum is also a function of $N_{te}(T_0)/N_T$. Finally, at low temperatures, the integral in the denominator portion of Eq. (19.15) is much smaller than unity, and the beginning of the glow curve should follow

$$I(T) \simeq C\,[N_{te}(T_0)]^2\,\frac{\overline{\omega}}{N_t}\exp\left[\frac{-E_t}{kT}\right], \tag{19.16}$$

a prediction that differs slightly from the result of Eq. (19.3).

A much more rigorous treatment on retrapping is offered by Halperin and Braner [1960], which leads to [Horowitz 1984a],

$$I(T) \simeq C\,[N_{te}(T_0)]^2\,\frac{\overline{\omega}A_m}{N_tA_n}\exp\left[\frac{-E_t}{kT}\right], \tag{19.17}$$

where A_n is the probability of falling, into a trap while A_m is the probability of recombining at a luminescent center. Note that Eq. (19.17) is the same result as Eq. (19.16) for $A_m = A_n$, as originally assumed in the model of Garlick and Gibson [1948].

These models, as earlier stated, serve to provide a basic understanding of the thermoluminescent process. Yet, as pointed out by Bos [2001], the actual thermoluminescent process can be complex, meaning that the absorption and decay processes can deviate from these ideal cases. For instance, the luminescent centers may be complex defect structures that are difficult to identify or characterize. In some cases, the defects may not be stable when heated. Overall, to help mitigate the impact of these physical issues, Bos [2001] recommends that the steps taken to measure the radiation dose be carefully controlled, including the irradiation process, read-out, and annealing.

TLD Systems A TLD response is measured with a TLD reader. There are many different models, but all have a few features in common (see Fig. 19.8). A TLD reader has a drawer unit that may have a small filament pan that a TLD chip may be placed upon, or it may have a cradle that the inner part of a TLD badge may be inserted. The drawer is shut in a light-tight enclosure with a backfill gas flowing into the chamber, usually N_2 or Ar.

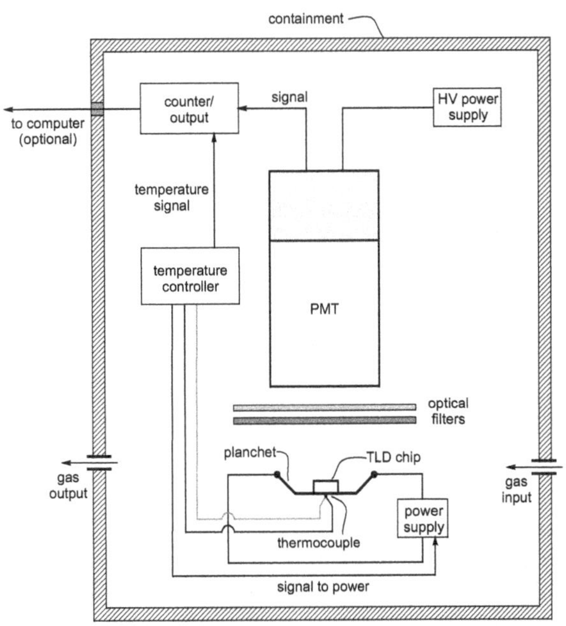

Figure 19.8. The basic components of a TLD reader.

The backfill gas is necessary to prevent background luminescent from TLD surface contamination, such a oxidation, during the heating process. After the chamber is purged of air, usually about a minute, the TLD reader is activated to operate in the regions depicted in Fig. 19.3. The TLD is heated, usually by a current flowing through the resistive filament pan or through radiant heating of the pan. A thermocouple is used to measure the pan temperature and provide feedback in order to control the filament temperature during the heating process. A photomultiplier tube (PMT) is used to record the light output from the TLD. Optical filters are placed between the TLD and the PMT to reduce contamination from spurious light emissions during the readout process. The current from the PMT is a measure of the light emission and, thus, is also a relative measure of the dose received by the TLD. This current is recorded and displayed as a glow curve on a monitor. The TLD reader itself may have a built-in display or, more frequently, is connected to a computer with a display monitor. The system is programmable with a variable stage input so that the user can select the temperature range, temperature ramp rate, temperature hold times, and temperatures between which the electronic output is measured.

Low energy traps are unreliable for dosimetry because they release electrons at room temperature, an effect referred to as *trap leakage*. Consequently, the light emission from shallow traps is a strong function of TLD storage time after the TLD irradiation. To avoid measuring the emissions from these shallow traps, current from the PMT is not recorded until a preset threshold temperature is reached that is sufficient to empty these traps. The TLD reader increases the temperature of the sample up to the lower threshold temperature and then begins recording the current, or accumulated charge, from the PMT output. This initial heating phase is called the preheat step. A typical lower threshold temperature is approximately 150°C, but varies with the TLD material selected. Above the low temperature threshold, the current

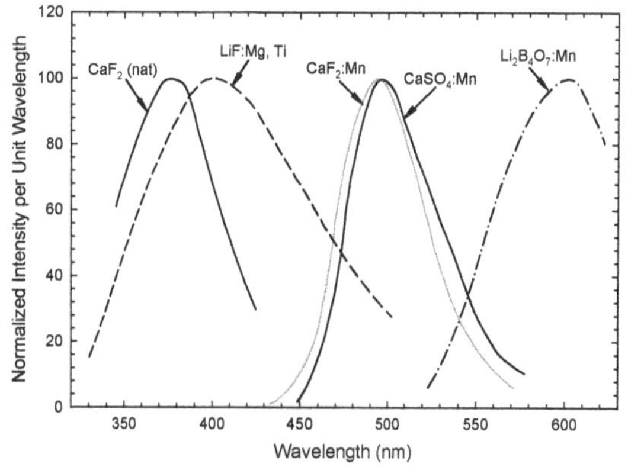

Figure 19.9. Emission spectra from several TLD phosphors normalized to a peak intensity of 100 per unit wavelength. Data from Fowler and Attix [1966] and Gorbics [1966].

and total charge from the PMT is recorded and, in some systems, the glow curve is displayed. Because the PMT is an amplification device, the total charge measured is an amplification of the total number of initial electrons producing thermoluminescence. Under constant operating conditions, it is expected that the measured charge is proportional to the actual TLD dose so the dose can be determined by multiplication of the measured charge by a scaling constant. At a high temperature threshold or limit, the TLD reader stops recording the output but continues to heat the TLD in order to clear the deep traps. A typical higher threshold temperature is approximately 300°C, enough to clear the important dosimetry traps for most TLD materials, although this limit may be set higher for certain TLDs that have important glow curvs features above 300°C. Although deeper traps could be useful because of their long-term electron storage properties, spurious infrared emissions from the heated TLD and surroundings can leak into the PMT and produce an exaggerated dose. For reliability reasons, deep traps are usually not included in the dose measurement. After the anneal step, the heating is terminated and the TLD is allowed to cool. It is important that the fill gas continues to flow as the TLD is cooled to avoid oxidation and possible TLD damage. A typical operating temperature range may go up to 400°C to clear the deep traps. Properties of some of the most widely used TLDs are given in Table 19.1.

From the results of Gorbics et al. [1969], it is clear that precise control of the heating rate is important for reliable and repeatable TLD dosimetry results. An established calibration method should be used to ensure that the measured charge corresponds to a known dose standard. Further, the temperature measurement of the pan must be rapid and accurate so that the reader measures current during the proper temperature limits. Although this precaution may seem insignificant, the pan temperature may differ from the actual TLD temperature (an insulator) or that of the thermocouple. This precaution is especially important for rapid heating rates. It was seen in Sec. 14.1.6 that the voltage has a strong influence on the gain of a PMT, so the TLD output should be compared periodically with a calibrated TLD phosphor to ensure that gain is constant and with insignificant variance. The TLD reader enclosure is sealed when the pan drawer is closed, and should be kept in such manner when not in operation to avoid contamination of the optical surfaces in the system. Further, to avoid possible reader damage, the system gas should be used to purge the enclosure of air before operating the heater.

Emission Spectra As with scintillators, it is preferable to use TLD phosphors with emission spectra that match well with the response functions of widely available photo-multiplier tubes. Shown in Fig. 19.9 are the emission spectra for some common TLD phosphors. It is notable that LiF:Mg,Ti (TLD-100) matches well to common bi-alkali PMTs. CaF_2:Mn, $CaSO_4$:Mn, and $Li_2B_4O_7$ on the other hand do not match well to common PMTs. However, their peak wavelengths near 500 nm and 600 nm present interesting possibilities for TLD readout systems based on Si diode technology, including silicon photomultipliers (SiPMs). The drawback is that long wavelength readout systems also have increased sensitivity to long wavelength photons from infrared emissions, a source of background from the heating process.

Sensitivity Sensitivity is determined by the amount of light released per unit energy absorbed in the TLD. The lower limit of detectable light emission is a function of the TLD reader and, thus, also determines the lower limit of detectable ionizing dose. The photoelectric process dominates at low energy, especially with

Table 19.1. Common thermoluminescent materials used for radiation dosimetry.

Designation	Material	Atomic Number (Z)	Density (g cm^{-3})	Emission Spectrum Range (nm)	Temp. Main Peak (°C)	Useful Dose Range (R)	Sensitivity to that of TLD-100	Energy Response*	Max Temp (°C)	Fading at 20°C
TLD-100	nat. LiF:Mg,Ti	3/9	2.635	350–600	195	10^{-3} – 10^{5}	1	1.35	400	5%/yr
TLD-100H, GR-200	nat. LiF:Mg,Cu,P	3/9	2.635	380–400		10^{-5} – 1.4×10^{3}	30	1.35	240	5%/yr
TLD-200	CaF_2:Dy	20/9	3.18	483–577	180	10^{-5} – 10^{6}	30	13		10%/1 day 16%/2 wks
TLD-300	CaF_2:Tm	20/9	3.18					13		
TLD-400	CaF_2:Mn	20/9	3.18	440–600	260	10^{-3} – 3×10^{5}	10	13		10%/16 hrs 15%/2 wks
TLD-500	Al_2O_3:C	13/8	3.987				72	3.5	400	
TLD-600	95.6% ^{6}LiF:Mg,Ti	3/9	2.545	350–600	195		1	1.35	400	
TLD-600H	95.6% ^{6}LiF:Mg,Cu,P	3/9	2.545	400				1.35	240	
TLD-700	^{7}LiF:Mg,Ti	3/9	2.653	350–600	195		1	1.35	400	
TLD-700H	^{7}LiF:Mg,Cu,P	3/9	2.653	400			12	1.35	240	
TLD-800	$Li_2B_4O_7$:Mn	3/5/8	2.4	530–630	200	10^{-3} – 10^{6}	0.15	0.8		
TLD-900	$CaSO_4$:Dy	20/16/8	~2.96**	480–570	220	10^{-6} – 10^{5}	20	12		2%/1 mon 8%/6 mon
	$CaSO_4$:Mn	20/16/8	~2.96**	450–600	110	10^{-6} – 10^{4}	70	12		50%/day
	BeO	4/8	3.01	150–325	200,240,290†					

*Sensitivity at 30-keV photons relative to sensitivity at 1-MeV photons (^{60}Co).

**Affected by water content.

†Depends on preparation method.

TLDs made of relatively high Z materials. Shown in Fig. 19.10 are calculated sensitivity responses to photon energy for several different TLD materials. Note that TLDs manufactured from relatively high Z materials, in particular CaF_2, $CaSO_4$, $CaCO_3$, SiO_2, and Al_2O_3, have a much higher response to photons below 100 keV.

Stability and Fading Low energy traps are unreliable for dosimetry because they release electrons at room temperature, an effect known as *trap leakage* or *fading* with minor distinction. Trap leakage is short-term electron loss to recombination from relatively shallow traps and usually occurs in hours to days after irradiation. Fading refers to a longer term loss of electrons from deep traps over an extended period of time. Both cases are primarily due to thermal vibration of the crystal lattice from the ambient temperature causing the ejection and recombination of electrons. Consequently, the TLD reader is usually programmed to clear shallow traps before the system begins recording the electronic signal from the PMT.

Recall the probability of electron emission from a trap, described by Eq. (19.1), has a strong temperature dependence. The half-life of the trap is defined as the time for the filled population to emit half of the electrons

$$t_{1/2} = \frac{\ln(2)}{p} = \frac{\ln(2)}{\overline{\omega}} \exp\left[\frac{E_t}{kT}\right], \qquad (19.18)$$

where p is the emission probability and $\overline{\omega}$ is the frequency factor. Energy levels shallower than approximately 1 eV fade rapidly, while energy levels exceeding 1.5 eV are generally long lived, even with large frequency factors [Shani 1991].

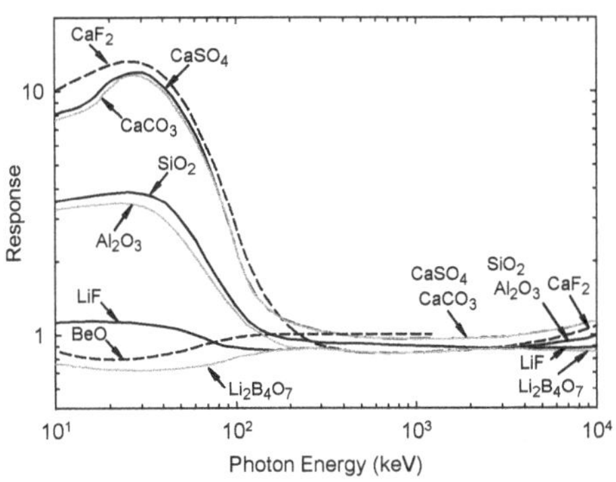

Figure 19.10. Theoretical sensitivity of TL phosphors as a function of photon energy, calculated as a function of ratio of the energy deposited in the phosphor to the energy deposited in tissue. Data combined from Cluchet and Joffre [1965] and Scarpa [1970].

Methods to reduce the effect of fading include a pre-irradiation anneal to clear the traps and reduce their relative contribution to the glow curve. After irradiation, a pre-heat step is routinely performed to empty these traps before the PMT is used to record the thermoluminescence. For most TLDs, the pre-heat step is performed for temperatures below 150°C, although this temperature threshold may vary with the TLD material. Storage of the TLDs in darkness at cool temperature can reduce fading. However, because shallow traps are still unreliable as a dosimetry record, mainly due to fading during the exposure period, the pre-heat step is still recommended.

Superlinearity and Sensitization At low radiation doses, the response of TLD phosphors is nearly linear. However, at high doses the response of a TLD deviates from linearity and becomes supralinear as shown in Fig. 19.11. The threshold for supralinear behavior is a function of the phosphor types, occurring at approximately 10 röntgens (R) for $Li_2B_4O_7$:Mn and as high as 10^4 R for CaF_2:Mn. At excessive doses the thermoluminescence decreases with dose.

If one simply regards the retrapping model of Garlick and Gibson [1948], the luminescence would be expected to increase as $N_{te}(T_0) \rightarrow N_t$, because the electrons are more likely to fall into luminescent centers than traps that are already filled with electrons, a prediction that can be observed from Fig. 19.7. Hence, the luminescent efficiency should increase with dose. At some degree of irradiation, the traps become saturated and no additional electrons can be stored in the traps. Consequently, the TLD material reaches a luminosity limit. Additional irradiation can cause damage to the material, thereby introducing more defect centers that act as traps that compete with the luminescent centers and consequently reduce the luminosity. The supralinear effect for several TLD materials is shown in Fig. 19.12.

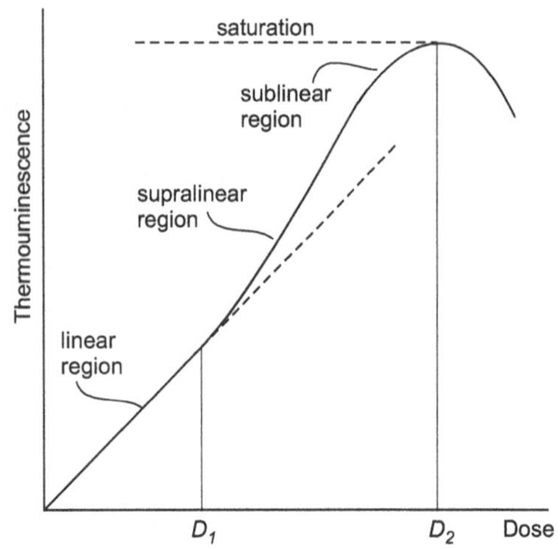

Figure 19.11. Representation of thermoluminescent growth curve depicting the linear, supralinear, sublinear, and saturation regions.

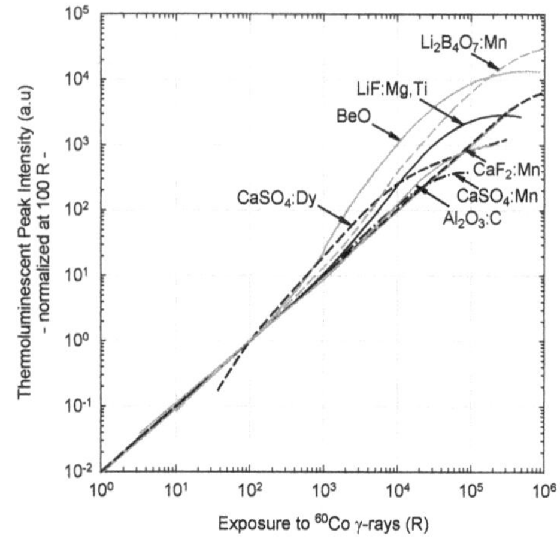

Figure 19.12. Thermoluminescent growth curves for several common TLD materials. The graphs were all normalized at 100 R. Data are from Attix [1986]; Lakshmanan et al. [1981]; Marrone and Attix [1964]; Portal et al. [1980]; Sunta et al. [1971]; Tochilin et al. [1969]; and Wilson and Cameron [1968].

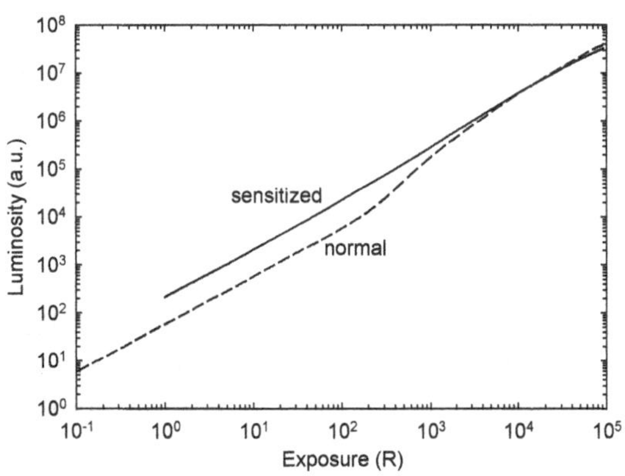

Figure 19.13. TL as a function of exposure for $Li_2B_4O_7$:Mn before and after radiation-induced sensitization from exposure to ^{60}Co gamma rays. Data are from West [1967].

Sensitization is the process whereby a TLD becomes more sensitive to radiation (S/S_o) after exposure to a large radiation fluence. Here S is the sensitivity after the process and S_o is the response for unsensitized samples. The production of additional traps from radiation damage makes a TLD produce more light as a function of radiation exposure, as shown for $Li_2B_4O_7$:Mn in Fig. 19.13. S/S_o values of 5 can be achieved with TLD-100 [Horowitz 1984b]. The sensitization process is different for each type of TLD material. For instance, Cameron et al. [1968] describe a process in which LiF is irradiated to about 10^5 R and subsequently annealed at 300°C for an hour. Sensitization can lead to increased linearity over a wider range than that observed with an unsensitized sample, and TLDs may be purposely sensitized to take advantage of this property. However, the deep traps produced during the process may become a source of low background phosphorescence, a nuisance for low-level dosimetry measurements [Horowitz 1984b]. Mayhugh and Fullerton [1975] describe a method to reduce interference effects of deep levels on low exposure measurements, in which TLD-100 samples are illuminated with UV radiation while being annealed at 300°C. Such samples retain high sensitivity while remaining useful in the mR range.

Radiation Damage The response of TLDs declines at higher radiation exposures than used for sensitization, usually greater than 10^5 R [Cameron et al. 1968; Shani 1991]. Multiple models have been proposed for the high radiation behavior observed for TLDs that involve the formation of competing trap centers in the materials [Horowitz 1984b]. If a significant density of shallow traps are created from radiation damage, they effectively accumulate electrons that might otherwise lodge in the useful phosphorescent centers. These shallow traps are neglected during the measurement process, and consequently the overall signal decreases below the predicted linear output. If deep traps are introduced by radiation damage, they also do not contribute to the glow curve signal because they require higher temperatures to depopulate. In either case, the competing traps emit electrons at temperatures outside of the read region depicted in Fig. 19.3. This radiation damage may be permanent and cannot be removed by annealing. Further, at high doses, electrons filling the usual phosphorescent centers can cause a saturation effect, in which the probability of filling these useful centers diminishes as fewer become available. At high radiation doses, TLDs become discolored, an indicator of significant radiation damage. With increasing radiation exposure, TLDs change in appearance from translucent white, to yellow, brown, dark brown, and eventually to black. The formation of these color centers causes re-absorption of luminescent photons, resulting in a decrease in light output. Overall, the useful linear range of most TLDs is from tens of mR up to approximately 10^3 R, and the useful sensitized range is from hundreds of mR up to approximately 10^5 R. The thermoluminescent intensity as a function of dose for several TLD materials are compared in Fig. 19.12.

Calibration Calibration procedures are based on comparisons between known TLD radiation exposures and the response of unknown TLD exposures. The procedure requires a number of stringent controls, including sample size and mass, calibration standard location with respect to the radiation source, and temperature and time control of the glow curves. Further, it is important that the actual TLD reader be maintained by periodically verifying the correct response to a known TLD exposure (dose).

Calibration generally consists of irradiating a TLD with a known dose and calibrating the TLD reader to correlate the glow curve output with that known dose. Suppose that the sample dosimeter is deployed and irradiated in a known environment with a known dose of D_k and produces a glow curve response of R_k. Then a simple calibration factor becomes

$$f_D = \frac{D_k}{R_k}. \tag{19.19}$$

Given an unknown exposure of ionizing radiation to a dosimeter under the same conditions, the unknown dose D_u is a function of the glow curve response R_u, and is determined from

$$D_u = f_D R_u. \tag{19.20}$$

If the same TLD chip is used for the calibration and then, shortly after being cleared, is used for dosimetry measurements, then the correlation seems straightforward. However, complexities arise from the non-linear response of TLDs to gamma-ray energies as shown in Fig. 19.10. The identified dose may be incorrect unless the same radiation type and energy are used for both the calibration and the measurement. Further, because TLDs are often used to determine dose, a correction must be made to relate the TLD response to that which would occur within tissue (or bone). The outcome becomes even more complicated because the response of a TLD changes with accumulated dose, storage time, thermal history, and even the environment. Consequently the value of f_D must be measured periodically.

If the individual TLDs are also calibrated, this process may account for variances in the TLD output. However, complications can arise if a general average response for a type of TLD is used for calibration to characterize additional TLDs of supposedly the same type. Clearly such an assumption can promote error, mainly because of variances in the responses between individual TLD chips.

To correct the response for different radiations and energies, the TLD can be exposed to various sources and the independent values of f_D can be recorded. However, this cumbersome method is usually replaced by calibrating the TLD with a single reference source. For instance, the source in the calibration table of Fig. 19.14 is ^{137}Cs, while another common source is ^{60}Co. This reference source can be used to determine a reference dose D_r and the resulting reference TLD reader response R_r. This result can be directly compared

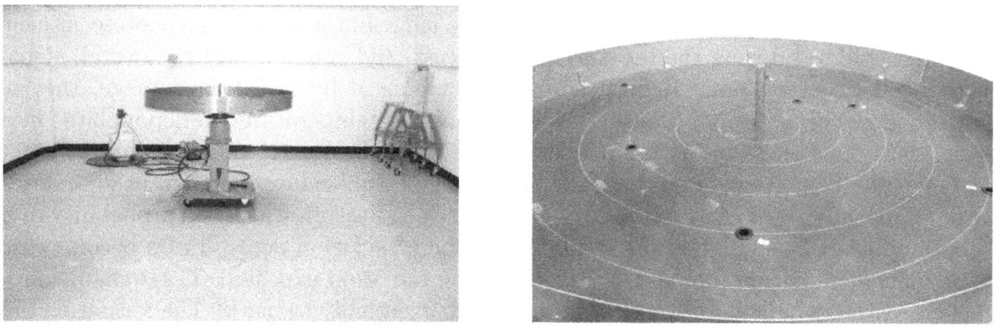

Figure 19.14. A dosimetry calibration table in a low scatter environment. (left) A ^{137}Cs source is stored in the shield below the table. (right) The source is raised through the port stem when in use. Concentric rings increasing by 10 cm radial increments mark placement locations.

to the known dose at the working location, where the badge is deployed, to find a relative scaling factor η_r. The calibration factor for the reference source is

$$f_r = D_r/R_r. \tag{19.21}$$

The relative response is defined as

$$\eta_r = f_D/f_r, \tag{19.22}$$

which gives the user a method to calibrate TLDs from the reference source. For high energy electrons and gamma rays, above 300 keV, the response is almost constant for most TLD materials, hence η_r is also essentially constant. For lower energies, this simply is not true. Further, at high doses, the TLD may become sensitized, thereby also changing the response. Consequently, the reference source can be used only for a limited energy range and dose.

A calibration system typically consists of a table, preferably made from low Z material such as plastic or wood to minimize photon scattering, that contains a port for a gamma-ray source. The gamma-ray source is stored in a radiation shield below the table, and is mechanically raised up through the port to expose the calibration TLDs placed at various distances from the port. Although ^{60}Co is a source frequently used for calibration, ^{137}Cs and ^{226}Ra are also widely used. Source strengths of a few mCi are often used to reduce statistical uncertainty and exposure time. The table is usually circular with concentric distance markings surrounding the source port. These tables typically are between 2 to 4 meters in diameter. If the source-to-TLD distance is greater than 2 m, air scattered photons may become significant and can skew the calibration from the ideal, point-source, uncollided flux model usually assumed in the calibration process. Likewise placement of the TLDs too close to the source may introduce complexities arising from the finite source size and uncertainties in the solid angle of irradiation. Hence, TLDs are usually placed a distance of 10 cm to 2 m away from the source port. In Fig. 19.14 an example of a TLD calibration table is shown.

The calibration system is placed in an empty room of such size that the detectors are always much closer to the source than the floor, ceiling or walls. This precaution minimizes the TLD absorption of scattered gamma rays. The TLDs should be placed at known locations from the source, horizontally and vertically, usually on premarked lines on the table as shown in Fig. 19.14(right). The gamma-ray exposure for each TLD can then be calculated from the known source activity based on an uncollided photon flux model to produce a calibration curve.

The TLD reader can also suffer from slight changes in light sensitivity, also causing error in the dose measurement. Lindskoug and Lundberg [1984] recommend testing the TLD reader as often as each day (presumably for daily usage of the instrument) with a constant light source. They also suggest the use of a TLD dosimeter as that light source, irradiated at a constant known dose beforehand, used each day and stored in a refrigerator between uses. This new calibration factor f_L is multiplied by Eq. (19.20) to correct for TLD reader response changes. Overall, the measured dose can be determined with

$$D_u = \eta_r f_L f_r R_u. \tag{19.23}$$

Some TLDs, such as LiF, have a density close to that of tissue, which explains, in part, the popularity of LiF. In other words, the light yield from the glow curve must be calibrated to the dose in tissue. A simple correction method is to assume that the dose of the TLD is related to tissue dose by ratio of the mass collisional stopping powers. Energy deposition is also a function of radiation type. Shani [1991] presents several formulations describing the relation between dose and TLD response for different radiations, including electrons, heavy charged particles, gamma rays, and neutrons. Many are sensitive to both neutrons and gamma rays, while others have almost no sensitivity to neutrons; hence, by strategically pairing different TLDs, separate gamma-ray and neutron doses can be determined.

TLD Forms

Thermoluminescent materials are available in many physical forms such as powder, compressed ribbon, dispersed in Teflon, pellets, plates, and single crystals. Many TLD materials are initially grown as an ingot with the activator dopant included. Unfortunately, the impurity concentration is distributed through the crystal according to its migration coefficient and, consequently, the activator concentration varies throughout the crystal. If TLDs are sliced directly from such an ingot, they usually have a wide variation in radiation response. To avoid this variability, the ingot is often ground into powder, usually with an average grain size of about 200 microns, and the powder is then thoroughly mixed to produce a uniform activator concentration [Mayhugh 2000]. This thermoluminescent powder can be deployed in a gelatin capsule, a glass capillary, or plastic tubing. For plastic tubing or gelatin capsules, the irradiated TLD powder is poured into a planchet pan and read. Powder encapsulated in potassium-free glass capillaries can be read in the capsule. The more convenient and commercial TLD forms have the powder hot-pressed into solid ribbons, rods, or sliced into small chips and disks. These forms are easy to handle and can be placed directly on the heater planchet for readout.

Often the sensitivity or stability of some TLD materials is improved when embedded in an inert material such as Teflon. The TLD powder is mixed with the inert material and then compressed into pellets or chips with thicknesses of about 100 to 300 microns. These chips can be inserted into containers that are inserted directly into the TLD reader. Finally, under special conditions, single crystal TLDs cut directly from a single crystal ingot are sometimes used. However, because of the wide variance in doping distribution in the bulk crystal, and variance in response among different chips, this form is unpopular for generally dosimetry.

Summary of Advantages and Disadvantages

The advantages of thermoluminescent dosimeters as a dose detector are many. TLDs are relatively inexpensive and can be acquired in many different forms. By properly using established annealing methods, they can be reused multiple times. Further, with periodic calibration, the variance in dose measurement can be kept small. Most TLDs are made from relatively low Z materials and have a low mass density; hence, they can be modeled as nearly tissue equivalent dosimeters. Their small size is convenient for dosimetry because they can be modeled as point detectors. The dose measurement range is large with some TLDs being capable of measuring radiation doses below a millirem while also being able to measure doses above 10^4 rem. Finally, modern TLD readers can be connected to personal computers to provide straightforward readout and data interpretation.

However, TLDs also have some disadvantages. First, once a TLD is read, the dosimetry information is erased from the dosimeter and, thus, is quite different from photographic dosimetry methods. All TLDs fade with time, some more rapidly than others. Consequently, the luminescence is a function of storage time and storage environment. Further, the luminescence for many TLDs is a function of the irradiation rate as well as the total dose, and some TLDs show strong supralinearity. The output luminescence and glow peak maxima for many TLDs are functions of the heating rate. TLDs are sensitive to UV light and proper storage and annealing before deployment is necessary. The emission spectrum for some TLDs is mismatched to the response function of many photomultiplier tubes and thus decreases the measured luminosity. TLDs can have different sensitivities, even when made from similar materials. Thus each individual TLD must be calibrated. Finally, TLD sensitivity can change after receiving large radiation doses, an effect termed as sensitization.

Common Materials for Thermoluminescent Dosimetry

Numerous materials have been investigated as thermoluminescent materials, including many common minerals. In fact, there are so many that it is impractical to discuss all of them here, and authors of books devoted to this topic seldom wander down that path. Instead, discussed here are a few common materials successfully used as TLDs, many of which are commercially available and widely used.

Lithium Fluoride

Lithium fluoride (LiF) is the most studied of the TLD materials and also is the most popular TLD in use today. The relatively low density of LiF is closer to tissue than most other TLD materials, an important property for dosimeters. LiF is stable at low temperatures and is mostly chemically inert. Manufacturers of LiF optics warn that the material is attacked by atmospheric moisture at temperatures above 400°C.[4]

LiF TLDs are available with many different ^{7}Li enrichments and activators, the most popular being the TLD-100, formed from natural LiF doped with Mg and Ti. Enriched and highly sensitive LiF TLDs are also available for special applications. Those most used are listed in Table 19.1.

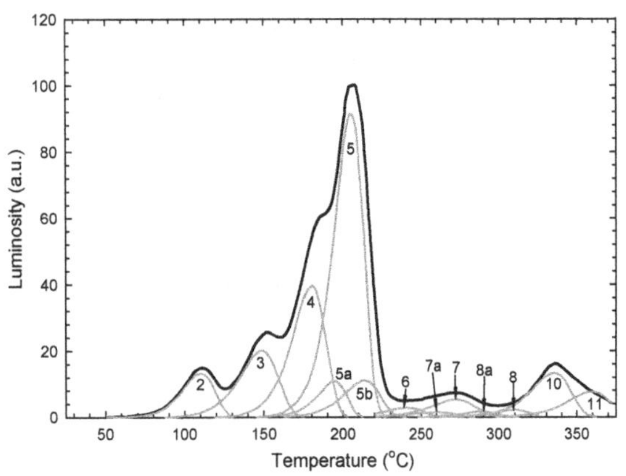

Figure 19.15. The normalized glow curve and its various composite components for LiF TLD-100 exposed to 1 Gy gamma irradiation. The luminosity is in arbitrary or relative units (a.u.). Composite data acquired and adjusted from Livingstone et al. [2009] and Horowitz and Moscovitch [2013].

Although LiF can be grown as a single crystal, this form is generally not used for TLDs. Instead, dopants are added to the purified LiF starting material and single crystals are grown from a melt process, typically with Czochralski or Stockbarger methods. The unfavorable segregation coefficient of Mg causes it to migrate during growth (as with other activators during zone melt purification) and, consequently, a non-uniform doping distribution is produced. As mentioned above, this non-uniform activator distribution can be ameliorated by pulverizing the crystal, mixing the resulting powder, and hot pressing or sintering the mixed powder into rods and ribbons. The end product is a translucent material with a relatively uniform distribution of the activator dopants. A normalized composite representation of a LiF:Mg,Ti glow curve with the separate trap luminosity components is shown in Fig. 19.15. The emissions denoted 2 and 3 in this figure are usually excluded from a measurement, typically set to begin when the temperature ramp reaches 150°C and extends up to approximately 225°C. Hence, the main luminescent contributions

[4]Apparently, LiF in the presence of water moisture at elevated temperatures hydrolizes and forms HF, which consequently attacks glass.

come from peaks 4, 5, 5a, and 5b. The sensitivity is linear in the range between 1 mR and 1 kR, and is supralinear between 1 and 50 kR.

To clear the deeper traps, some care is required for proper annealing of TLD-100 dosimeters to maintain reliable reuse. The annealing is usually performed at 400°C for 1 hour, followed by a slow cooling step and a longer term (~ 24 hours) anneal at 80°C. However, to prevent damage to LiF material embedded in a temperature sensitive binder, these TLDs can also be annealed at lower temperatures ranging between 280° to 300°C [Becker 1973].

The higher temperature annealing is important to empty traps 6 through 11, thereby reducing background phosphorescence. The IR contamination of the spectrum from the heater is the reason these higher temperature traps are not used in the dose measurement. Although fading is relatively slow, different low fade rates arising from different annealing procedures have been reported. When the annealing method is rigorously controlled, the fading can be limited to approximately 5 to 10% per year [McKeever 1985].

TLD-100 dosimeters are sensitive to charged particles, gamma rays, and neutrons. Gamma rays excite primary electrons in the material through the photoelectric effect, Compton scattering, or pair production. These energetic electrons excite secondary electrons in the material through Coulombic scattering. It is unlikely that energetic electrons deposit all of their energy in a small TLD chip before reaching the detector boundaries; hence, usually only a fraction of the incident energy is measured. The same is true for energetic beta particles. However, dosimetry is based on the amount of ionization produced in the body, primarily body tissue, and *cavity theory* attempts to correct for these energy losses.[5]

Heavy charged particles produce electron-hole pairs directly through Coulombic scattering. Because natural Li has a natural abundance of approximately 7.6% ^{6}Li, neutrons are also detected through the ^{6}Li(n,t)^{4}He reaction, in which the triton and helium reaction products directly ionize the material. The ranges of alpha particles and heavy ions are relatively short in LiF, usually only tens of microns, and most of the energy of the reaction products can be absorbed in the dosimeter. The glow curve from electrons and heavy ions is significantly different because a glow curve peak appears at 275°C from alpha particles [McKeever 1985]. Hence, the glow curve peak 7 (Fig. 19.15) is enhanced by high *linear energy transfer* (LET) radiation [Budd et al. 1979]. In some cases, the enhanced response of peak 7 has been used to extend the sensitivity of TLD-100 dosimeters out to 10^7 R [Jones and Martin 1968].

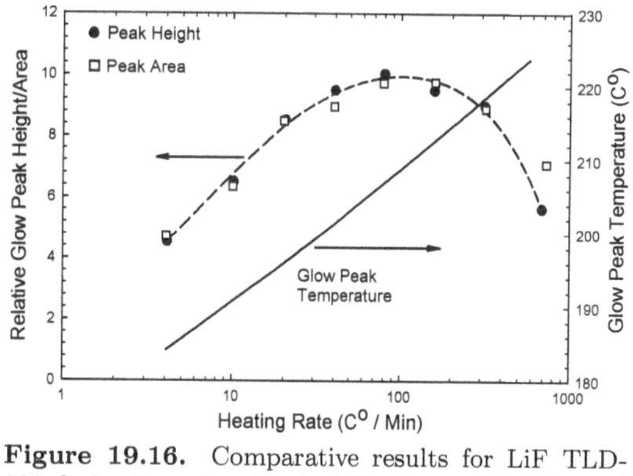

Figure 19.16. Comparative results for LiF TLD-100 dosimeters, showing deviation in peak height and area as a function of heating rate. The gamma-ray exposure was adjusted to be proportional to $1/R_T$ for the peak height, while the exposure was normalized for the peak area data. Also shown is the change in peak temperature with heating rate. Data are from Gorbics et al. [1969].

The glow peak properties of LiF:Mn,Ti changes with heating rate, and thus, this TLD requires a calibrated and constant measurement procedure. Both the glow curve peak height (peak 5) and the glow curve area have an output maxima near a heating rate of 80°C per minute (see Fig. 19.16). Further, the thermal location of the glow peak increases with heating rate, changing from 185°C at a rate of 4°C per minute up to 225°C at a rate of 700°C per minute. If operated at the luminosity maximum, the glow peak appears near 205°C.

The neutron response of LiF can be enhanced by replacing the natural Li with approximately 95.6% enriched ^{6}Li, and TLDs of this type are labeled TLD-600. The emission of heavy charged particles with high

[5]Part of cavity theory is the assumption that an electron leaving the dosimeter is replaced by an electron leaving the surrounding environment and into the dosimeter. This condition is known as *charged particle equilibrium*.

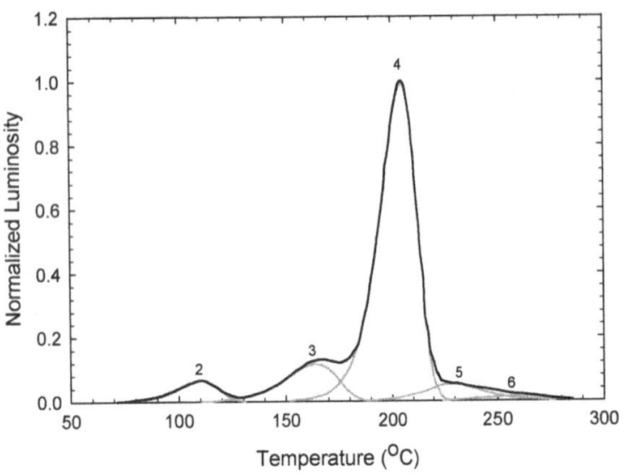

Figure 19.17. The normalized glow curve with its various components for LiF TLD-100H exposed to 5 Gy gamma irradiation. Data were acquired and adjusted from those of Triolo et al. [2006].

LET produces a glow curve with a significant increase in peak 7 with respect to peak 5 for neutron reactions. With proper calibration, the ratio of peak 7 to peak 5 can give a measure of the neutron dose [McKeever 1985]. LiF depleted of ^{6}Li, having 99.99% ^{7}Li, are relatively insensitive to neutrons and are labeled TLD-700. Often TLD-600 dosimeters are paired with TLD-700 dosimeters in a single dosimeter badge. By comparing the signals from TLD-600 and TLD-700 pairs, a separate measure of the neutron and gamma-ray dose can be determined.

The sensitivity of LiF is enhanced by changing the activation phosphors from a combination of Mg and Ti to a combination of Mg, Cu, and P, and natural LiF with this activator combination are labeled TLD-100H (also, GR-200). The glow curve is different than observed with TLD-100 with the main peak appearing near 200°C [Triolo et al. 2006], as shown in Fig. 19.17. Overall, these dosimeters can usefully measure lower doses than TLD-100, ranging from 10^{-5} R to about 10^3 R. However, they do not have the extended upper range observed with TLD-100. Further, the maximum thermal range is limited to no more than 240°C or the TLD is permanently damaged. Unfortunately, this restriction also means that TLD-100H dosimeters cannot be annealed to clear the deep traps. Eventually, after many exposures, the background emissions from these deep traps can begin to interfere with low dose measurements.

A relatively new LiF TLD with Mg, Cu, Si activators is under investigation [Kim at al. 2010; Horowitz and Moscovich 2013]. The main peak appears near 250°C, higher than TLD-100, but still low enough to minimize IR contamination. These LiF TLDs have similar sensitivity as the TLD-100H, but the deep traps can be annealed without damaging the dosimeter. Kim at al. [2010] report that after a multi-step anneal process, the residual deep trap signal is reduced to only 0.05%.

Calcium Fluoride

Calcium fluoride (CaF_2), or fluorite, is a mineral with attractive properties for thermoluminescent dosimetry. There are four prevalent CaF_2 TLD types, namely natural fluorite, Dy-doped (TLD-200), Tm-doped (TLD-300), and Mn-doped (TLD-400). A characteristic shared by all CaF_2 TLDs is the elevated gamma-ray energy absorption at energies below 100 keV (see Fig. 19.10), a consequence of the generally higher mass density of CaF_2. To reduce the effect, filters can be employed to flatten the gamma-response at these lower energies. Reviews of important CaF_2 TLD properties can be found in the literature [Becker 1973; Harvey et al. 2010; Bidyasar 2014].

CaF_2:MBLE Natural CaF_2, sometimes called MBLE CaF_2, was studied extensively and was a commercial TLD product for many years.[6] However, CaF_2:natural is far from an ideal TLD material because material obtained from different sources can have significantly different glow curves as shown in Fig. 19.18. Peaks in the glow curves, labeled in Roman numerals by Schayes et al. [1967], appear between the temperatures of 70°C and 600°C. The useful glow peaks for dosimetry are identified as III and III′, and appear with different luminosity for the samples of Fig. 19.18. Cameron et al. [1968] report that the integrated light yield from peaks III and III′ comprise approximately 75% of the total light yield. Although there are some smaller glow

[6] Studied and developed by Manufacture Belge des Lampes Électriques (or MBLE) in Belgium, these TLDs are no longer manufactured.

peaks at higher temperatures, they are generally not used for dosimetry because temperatures above 450°C can permanently damage CaF_2 TLDs.

The linearity of CaF_2 is quite good, and these TLDs show good linearity for gamma-ray exposures up to 10^4 R and exhibit only slight supralinearity up to 10^5 R. CaF_2 becomes sensitized at high gamma-ray exposures between 10^4 and 4×10^4 R, with a maximum gain in sensitivity ranging between 1.6 to 4 if annealed at 450°C for 30 minutes [Cameron et al. 1968; Sunta and Kathuria 1971]. From Fig. 19.19, the main glow peak does shift in temperature with increasing heating rate; however, the total peak area and the total peak height remain remarkably constant for a wide range of heating rates.

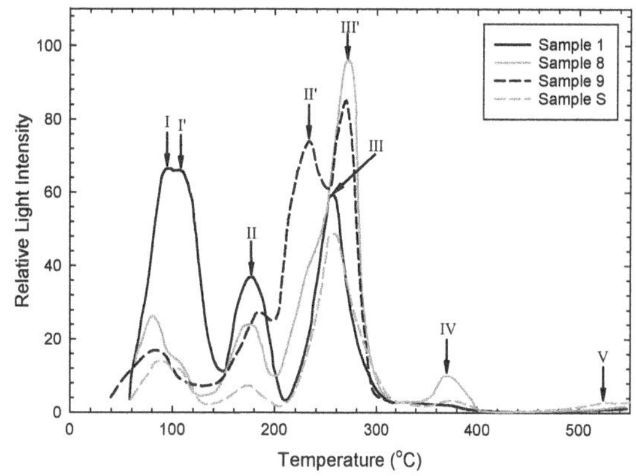

Figure 19.18. Glow curves from several samples of natural CaF_2. Data were adopted and adjusted from those of Schayes et al. [1967].

Natural CaF_2 is strongly sensitive to UV light, which primarily affects the output of glow peak III′. It is believed that UV irradiation assists with the movement of electrons from deep traps to shallow traps, producing additional luminescence in glow peak III. Brooke and Schayes [1967] describe a method of UV sensitization, or thermoluminescent build-up, that allows rereading of TLDs numerous times from a single radiation exposure, provided that the initial exposure exceeds approximately 5 R. Overall, the overexposure of CaF_2:natural causes a *bleaching* of the TLD and the signal changes from that of competing processes, including thermoluminescent build-up, charge transfer, and fading. TLDs unexposed to ionizing radiation tend to increase in luminescence, while the radiation induced signals tend to decrease with UV exposure [Becker 1973].

CaF_2:Dy (TLD-200) Developed by Harshaw for commercial use,[7] TLD-200 is a synthetic Dy-doped form of CaF_2. It is highly sensitive, much more so than TLD-100, and thus can be used for low-dose measurements. The glow curve is relatively complicated, as shown in Fig. 19.20, with two salient composite peaks, one glow peak appearing between 80°C and 210°C and another high-temperature composite glow peak appearing between 250°C and 450°C [Yazici et al. 2002; Bidyasagar et al. 2014]. Each of these composite glow peaks is formed from several trap levels, and their intensity changes with radiation type, dose, and preheat method.

TLD-200 is not as linear as CaF_2:MBLE or CaF_2:Mn, and begins showing peak height supralinearity near a dose of 5×10^3 rad, extending up to approximately 5×10^4 rad before becoming sublinear [Hasan and Charalambous 1983]. The integrated luminescence (all peaks between 50°C and 370°C) shows supralinearity starting as low as 800 rad. Saturation occurs between an exposure of 10^5 R and 10^6 R [Becker 1973]. Becker [1973] reports that linearity can be improved by pre-irradiation and annealing TLD-200 at 600°C for two hours.

Fading can be significant, reportedly up to 25% in a month if post-irradiation treatments are not applied [Binder and Cameron 1969]. If illuminated during storage, for instance in a diffuse room light, fading can increase up to 30% in a single day. The main cause of fading is from the low temperature glow peak components. If eliminated by post-irradiation annealing, nominally 80°C for 10 minutes, the fading can be reduced to 13% per month [Binder and Cameron 1969], but at the expense of 20%-30% reduction in sensitivity.

CaF_2:Tm (TLD-300) Thulium-doped CaF_2 was introduced as a dosimeter material in 1977 [Lucas and Kapsar 1977] and put into commercial use as TLD-300 by Harshaw. The glow curve of TLD-300 (see Fig. 19.21) displays two main composite peaks in the spectrum, centered near 175°C and 260°C, generally

[7]Now produced by Thermo-Fisher Scientific. See Thermo-Fisher [2016].

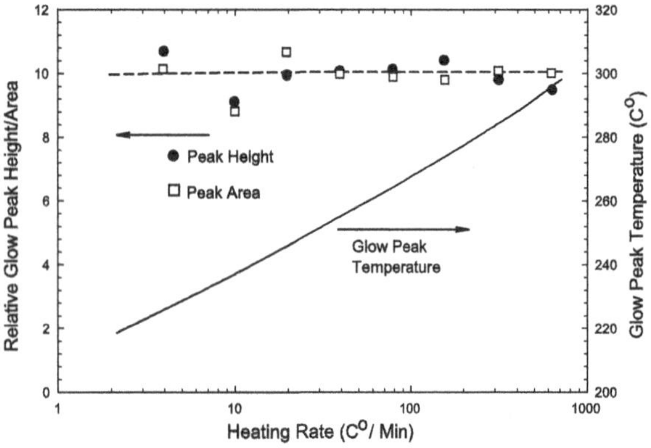

Figure 19.19. Comparative results for natural CaF_2, showing deviation in peak height and area as a function of heating rate. The gamma-ray exposure was adjusted to be proportional to $1/R_T$ for the peak height, while the exposure was normalized for the peak area data. Also shown is the change in peak temperature with heating rate. Data are from Gorbics et al. [1969].

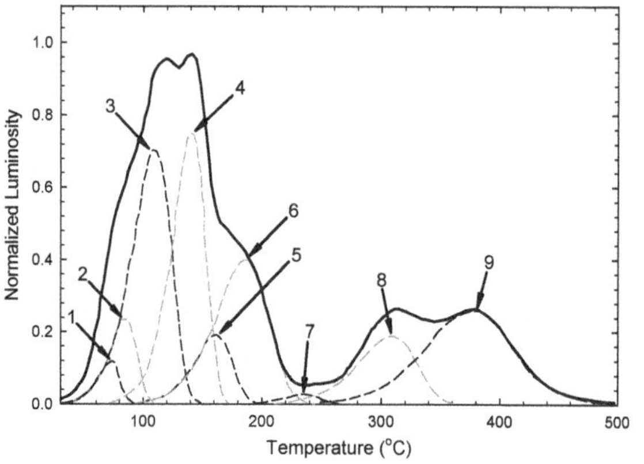

Figure 19.20. Normalized glow curves for typical TLD-200 exposure measurements after gamma-ray irradiation. Data adopted and adjusted from those of Yazici et al. [2002] and Massillon-JL et al. [2008].

labeled as peak 3 and peak 5, although the observed peak 5 is really a combination of peaks 5 and 6 [Massilon-JL et al. 2008]. A third glow peak is present if readout occurs within 24 hours of exposure and is centered near 100°C; however, this glow peak usually fades after 24 hours and usually does not appear in the glow curve. The output characteristics of TLD-300 are relatively stable and appear not to be affected by the heating rate [Kafadar et al. 2013].

The luminosity of the two main peaks is strongly dependent on the linear energy transfer and the ratio of peak 3 to peak 5 decreases with increasing LET. This feature can be used in dose interpretation in mixed radiation fields. Furretta and Tuyn [1985] show that the ratio of peak 3 to peak 5 increases from 1.22 for ^{3}H beta-particle emissions up to 3.15 for ^{90}Sr beta-particle emissions. Because the ratio also shows a strong dependence with heavy ions, Massilon-JL et al. [2008] suggest its use as an indicator of ion beam quality.

The recommended anneal process for TLD-300 is one hour at 400°C followed by a rapid quench cooling in air to room temperature at a rate of 75°C per minute [Massillon-JL et al. 2008]. Fading is reduced to approximately 5% over a 4-week period for both peak 3 and peak 5, while the contribution from peak 6 appears unaffected [Kafadar et al. 2013].

CaF_2:Mn (TLD-400) Mn-doped CaF_2 was developed for thermoluminescent dosimetry at the Naval Research Laboratory in the late 1950s and early 1960s [Ginther and Kirk 1957, Schulman et al. 1960]. The general method of producing TLD-400 material is by first chemically precipitating CaF_2 and MnF_2 into solution, following by filtering and drying the residue in an inert atmosphere. The sintered material is then ground to particle sizes of about 100 microns and thoroughly mixed. Afterwards, the material is formed into various shapes and sizes and sintered.

TLD-400 has a simple glow curve consisting of just one main glow peak appearing between 260 to 320 °C, depending on the heating rate as shown in Fig. 19.22. The relatively higher glow peak temperature indicates emission from a deep trap, yet CaF_2:Mn has high fading rate of up to 10% per month. TLD-400 has a higher intrinsic sensitivity than TLD-100, and can be used to measure radiation exposures below 1 mR. The response of TLD-400 is linear over a wide range, up to 3×10^5 R before sublinearity is observed (See Fig. 19.12). Further, there appears to be no dependence on the type of ionizing radiation or LET as observed with LiF-based TLDs. The dose rate dependence is relatively constant up to a heating rate of

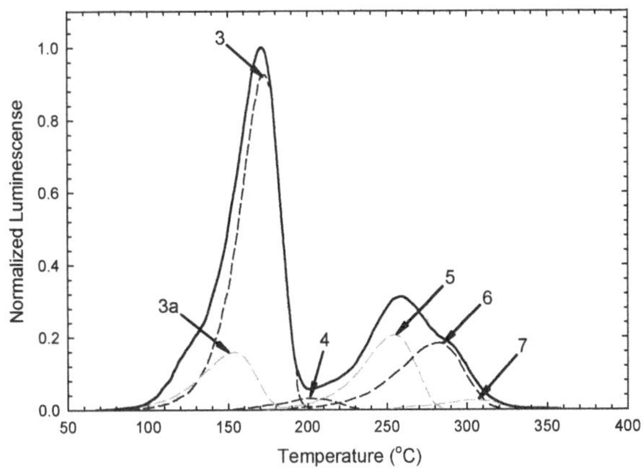

Figure 19.21. Normalized glow curves for typical TLD-300 exposure measurements after gamma-ray irradiation. Data adopted and adjusted from those of Yazici et al. [2002] and Massillon-JL et al. [2008].

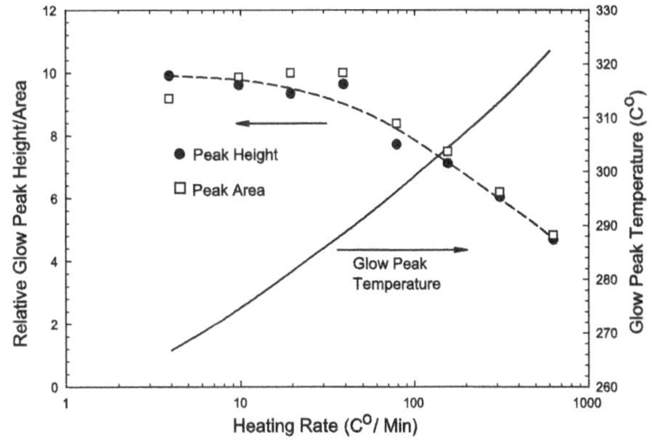

Figure 19.22. Comparative results for CaF_2:Mn dosimeters, showing deviation in peak height and area as a function of heating rate. The gamma-ray exposure was adjusted to be proportional to $1/R_T$ for the peak height, while the exposure was normalized for the peak area data. Also shown is the change in peak temperature with heating rate. Data are from Gorbics et al. [1969].

approximately 30°C per minute before both the relative glow curve peak height and integrated area begin to decline. As with all CaF_2 TLDs, TLD-400 is overly sensitive to gamma-ray energies below 100 keV. Although filters can be used to soften the response to lower energy gamma rays and x rays, the same filters can also over-attenuate soft x rays, thereby producing a lower measured dose.

Calcium Sulphate

Calcium sulphate ($CaSO_4$) is a relatively high Z material that has been studied for use as a thermoluminescent dosimeter. Mn-doped $CaSO_4$ has a low temperature glow curve peak (see Fig. 19.23), which fades rapidly at room temperature. Fading is reportedly to be as high as 40% within one day, and up to 85% after three days [Fowler and Attix 1966]. Consequently, $CaSO_4$:Mn is not useful as a dosimeter except, perhaps, for short-term laboratory experiments in which the TLD is read soon after irradiation. Regardless of this drawback, $CaSO_4$:Mn is very sensitive to ionizing radiation and capable of measuring exposures in the μR range [Fowler and Attix 1966]. The main reason for this perceived higher sensitivity is actually a fortuitous benefit of the low-temperature glow curve, in which the background IR emissions from the heater and TLD are small by comparison to the thermoluminescence. Hence, the signal-to-noise ratio is high so that measurement of low doses is possible. The emission spectrum from $CaSO_4$:Mn peaks near 500 nm, which is somewhat long in wavelength to be a good match to most bi-alkali PMTs. All forms of $CaSO_4$ are overly sensitive to gamma rays below 100 keV, much like CaF_2. Although the glow peak maximum shifts from approximately 80°C to over 120°C for heating rates between 10 C°/min up to 500 C°/min, the peak height and peak area remain remarkably constant for heating rates below 150 C°/min (see Fig. 19.24).

Because of the high fading, other phosphors for $CaSO_4$ were studied. For instance, Sm was tested as an activator with encouraging results [Cameron et al. 1968], but the main glow peaks appear at 100°C and 400°C, with an emission spectrum centered at 600 nm, too high for efficient coupling to conventional PMTs. Of the activators explored, both Dy-doped and Tm-doped have proven to be relatively successful. The main glow peak for both $CaSO_4$:Dy (also called TLD-900) and $CaSO_4$:Tm appears at approximately 225°C. Although the glow peak appears simple, Souza et al. [1993] report at least ten peaks in the TLD-900 glow curve, although only seven actually comprise the main glow peak. Of these deconvolved peaks Harvey et al. [2010] note that only two or three are important for doses below 10 R.

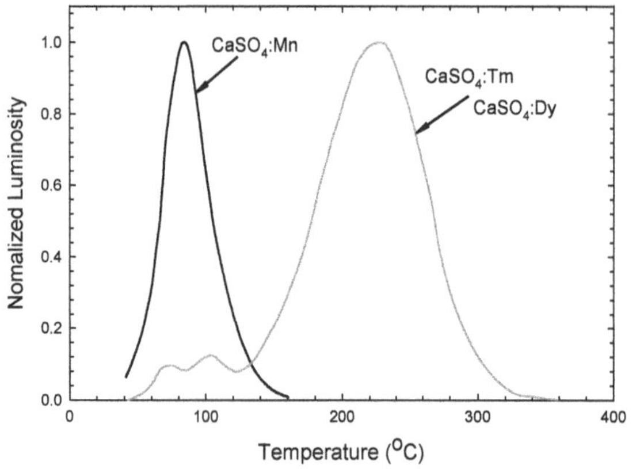

Figure 19.23. Normalized glow curves for typical $CaSO_4$:Mn, $CaSO_4$:Tm, and $CaSO_4$:Dy (TLD-900) for exposure measurements after gamma-ray irradiation. Data adopted and adjusted from those of Yamashita et al. [1971] and Ekdal et al. [2014].

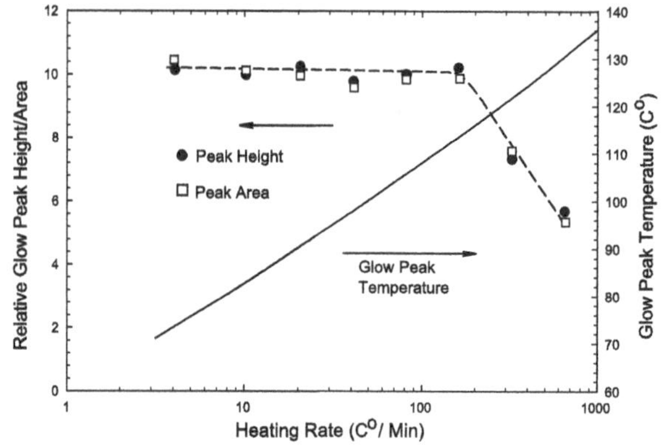

Figure 19.24. Comparative results for $CaSO_4$:Mn dosimeters, showing deviation in peak height and area as a function of the heating rate. The gamma-ray exposure was adjusted to be proportional to $1/R_T$ for the peak height, while the exposure was normalized for the peak area data. Also shown is the change in peak temperature with heating rate. Data are from Gorbics et al. [1969].

The main luminescent emissions from TLD-900 appear near 480 nm and 570 nm. By comparison, the luminescent emission maximum from $CaSO_4$:Tm is near 450 nm, a better match to traditional alkali metal PMTs, although $CaSO_4$:Tm is less sensitive to ionizing radiation than TLD-900. The luminescent response for TLD-900 is linear up to approximately 3×10^3 R. Supralinearity appears between 3×10^3 R and 10^4 R, beyond which sublinear behavior appears [Becker 1973]. As with many TLDs, saturation is reached at 10^5 R.

TLD-900 is significantly more sensitive than TLD-100 by about a factor of 30, and low gamma-ray doses below a mrad have been measured. Fading is minor, reportedly as low as 2% in the first month and 6% over a five-month period when stored at room temperature [Becker 1973]. Similar results were reported by Guelev et al. [1994], who observed approximately 5% fading over a six-month period. However, Harvey et al. [2010] report a much higher degree of fading near 30% in one month, attributed to low temperature glow peaks apparently not considered in the Guelev et al. [1994] study. The annealing procedure is straightforward, requiring no more than a 15-minute soak at 400°C. Yamashita et al. [1971] report that TLD-900 is significantly sensitive to thermal neutrons while $CaSO_4$:Tm is not and conclude that the use of $CaSO_4$:Tm is preferred in a mixed radiation field because of the high response that TLD-900 has to thermal neutrons.

Lithium Borate ($Li_2B_4O_7$:Mn)

Lithium borate (LBO) is mainly attractive as a thermoluminescent dosimeter because of it low Z components and low mass density. As a result, the response is nearly flat from 10 keV to 10 MeV, as shown in Fig. 19.10. Its low density is closer to bone and soft tissue than most other TLD materials, which helps to simplify dose conversions. $Li_2B_4O_7$ is water soluble and hygroscopic, unless mixed and processed with 0.25% SiO_2 by weight, which seems to stabilize it [Becker 1973].

Mn is the best known activator for $Li_2B_4O_7$ and is often named TLD-800 for short. The emission spectrum maximum for TLD-800 is reportedly at 605 nm [Gorbics 1966], a relatively long wavelength for most PMTs. This maximum can be shifted to shorter wavelengths by doping with Cu (or Ag). $Li_2B_4O_7$:Mn has emission maximum near 352 nm [Tiwari et al. 2010]. TLD-800 has a linear response to gamma rays up

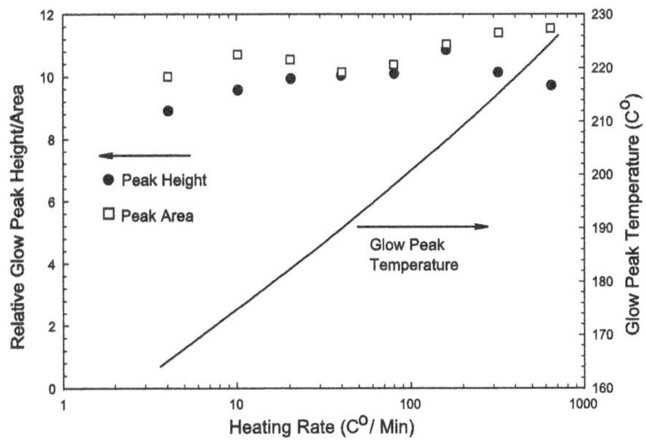

Figure 19.25. Comparative results for $Li_2B_4O_7$:Mn dosimeters, showing deviation in peak height and area as a function of heating rate. The gamma-ray exposure was adjusted to be proportional to $1/R_T$ for the peak height, while the exposure was normalized for the peak area data. Also shown is the change in peak temperature with heating rate. Data are from Gorbics et al. [1969].

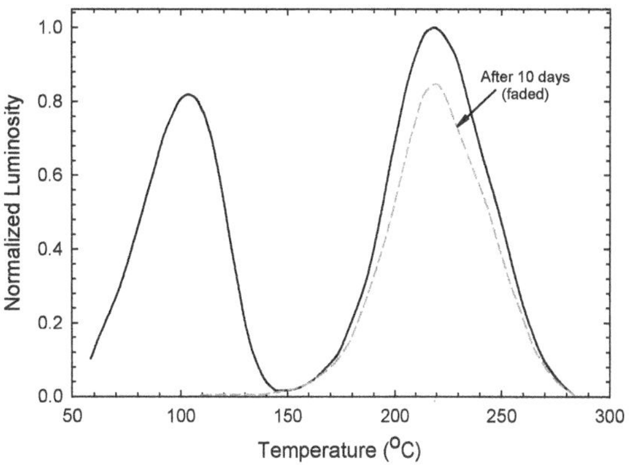

Figure 19.26. Normalized glow curves for typical $Li_2B_4O_7$:Mn exposure measurements after gamma-ray irradiation. Data adopted and adjusted from those of Yamashita et al. [1971] and Ekdal et al. [2014].

to 150 R, becomes supralinear between 150 R and 10^5 R, eventually saturating at $\times 10^5$ R. TLD-800 has lower sensitivity to gamma rays than LiF, generally ten times less sensitivity than TLD-100. West et al. [1967] report that TLD-800 sensitizes when irradiated with 10^5 R gamma rays and annealed for 15 minutes at 300°C (see Fig. 19.13), achieving approximately 4 times the sensitivity for gamma-ray doses below 200 R. Becker [1973] reports that annealing is simple, requiring only a 15-minute soak at 300°C.

Gorbics et al. [1969] report that the peak height and peak area are relatively constant over a wide range of heating rates; however, this result is markedly different from that reported by Kafadar et al. [2014], who report large changes in peak area for TLD-800 when heated at rates above 120°C/min. Both Gorbics et al. [1969] and Kafadar et al. [2014] report glow peak temperature changes from below 170°C to nearly 220°C over heating rates ranging between 10°C/min to 500°C/min as shown in Fig. 19.25. Two glow peaks are prevalent shortly after irradiation for both TLD-800 and $Li_2B_4O_7$:Cu. TLD-800 has one glow peak appearing at approximately 100°C and the other at approximately 225°C (see Fig. 19.26). $Li_2B_4O_7$:Cu has a lower temperature glow peak near 120°C, and also has a higher temperature peak reported as 205°C by Takenaga et al. [1980], but reported as much higher at approximately 240°C by Aydin et al. [2013]. For all cases the lower temperature glow peak fades rapidly, usually disappearing within a day [Schulman et al. 1967; Aydin et al. 2013]. The higher temperature peak for TLD-800 fades by approximately 8% over a 30-day period [Cameron et al. 1968], but there are differing fading results reported for $Li_2B_4O_7$:Cu. Wall [1982] reports rapid fading for $Li_2B_4O_7$ that was doped with Cu or co-doped with Cu and Ag. However, Aydin et al. [2013] reports fading is approximately the same as TLD-800, fading about 10% over a 40-day period. Most likely the preparation method is responsible for these differences, as demonstrated by Christenson [1967] who increased the dosimetric response and range of TLD-800 by changing the preparation method.

Because of the constituent elements B and Li, the thermal neutron response of TLD-800 is relatively high [Gambarini 2003], higher than TLD-100, but less than TLD-600 and TLD-900. Sunta et al. [1972] report that the fast neutron response (from recoil protons) to gamma-ray response ratio was nearly 3 while TLD-100 and TLD-900 response ratios were much less at 1.24 and 1.6, respectively.

Beryllium Oxide (BeO)

Another low density thermoluminescent material is BeO, generally regarded as a "tissue equivalent" material. The often quoted effective Z (or Z_{eff}) is 7.12.[8] BeO is toxic in powdered form, but not as a ceramic, and comes in three main forms, namely hot pressed (HP), slip cast (SC), and nuclear quality (NQ) [Scarpa 1970]. These three forms have different densities, with SC at 2.2 g cm^{-3} and the other forms approaching 2.8 g cm^{-3}. It is relatively chemically inert and resistant to mechanical shock.

The glow curve spectra and the peak temperatures were observed to shift according to TLD manufacturing process [Scarpa 1970]. Three different emission spectra were observed by Scarpa [1970] with peaks centered differently at 210°C, 240°C, and 290°C for SC, NQ, and HP BeO, respectively (see Fig. 19.27). The response to gamma rays is fairly linear from 10 mR up to approximately 100 R, beyond which the output becomes supralinear (see Fig. 19.12). Although HP BeO showed up to 20% fading in three days, the other forms demonstrated approximately 4% fading within 24 hours with no significant fading thereafter.

Measurements conducted by Crase and Gammage [1975] indicate that various forms of BeO all produce spectra in the UV region, generally between 200 nm to 400 nm with a peak near 265 nm. Consequently, most commercial PMTs are insensitive to much of the spectrum, and special PMTs (or TL readers) with deep UV sensitivity must be used (see Sec. 14.1.2 for more information on PMT windows).

BeO is more sensitive to gamma rays than $Li_2B_4O_7$, but much less so than LiF by about a factor of 3 [Shani 1991]. The sensitivity of SC BeO is practically linear for gamma-ray energies ranging from less than 20 keV to over 1 MeV, while the other forms are slightly higher in response for energies below 500 keV [Scarpa 1970]. Although linear at low doses, competition from non-radiation induced emissions from pyroelectric incandescensce contribute to background, making it difficult for accurate low-dose measurements. Exposure to light can depopulate BeO traps, inducing fading. Consequently, BeO TLDs should be handled carefully and not exposed to light during exposure or readout.

Aluminum Oxide (Al_2O_3)

Aluminum oxide is the chemical name for a mineral often regarded as a precious gem. If it has a small amount of Cr impurity (Cr_2O_3) it is called ruby. Other gem quality samples are called sapphires, associated with a wide range of colors according to the residual impurities. Otherwise, the common name for Al_2O_3 is corundum. Aluminum oxide is one of the hardest known substances, rated as 9 on the Mohs hardness scale. Aluminum oxide is mostly chemically inert and it is water insoluble. It has high electrical resistance and its melting point is extremely high at 2072°C.

Aluminum oxide was first studied for its thermoluminescent properties by Rieke and Daniels [1957], and later as a dosimeter by McDougall and Rudin [1970]. Samples with Cr contamination (ruby) produced a glow curve peak between 350°C and 400°C with a maximum emission wavelength near 650 and 700 nm. Unfortunately, at such high temperatures infrared emissions contaminate the luminescent emissions, and the emission wavelength is well beyond the sensitivity range of most photomultiplier tubes.

Pure Al_2O_3 has a thermoluminescent glow peak wavelength near 410 nm. Portal et al. [1980] report several glow peaks for Al_2O_3 with 0.7% Na and 4% Ti, appearing at approximately 140°C, 265°C, 450°C, and 630°C. See Fig. 19.28 for example glow curves. The lowest temperature glow peak fades quickly, while the higher two glow peaks have IR contamination during readout. It is the peak at 265° that has importance

---

[8]For the purpose of gamma-ray attenuation through a compound material, a fictitious Z_{eff} is often used. This number is often determined by the empirical relation [Murty 1965]

$$(Z_{\text{eff}})^{2.94} = \sum_{i=1}^{N} f_i Z_i^{2.94},$$

where f_i is the fraction of electrons associated with element i having atomic number Z_i. Although often used, the reader is cautioned that this is an empirical expression, and the exponent of 2.94 may not be accurate. In fact, Murty [1965] states in his publication that measurements indicate that 3.1 instead of 2.94 for the exponent yields improved results.

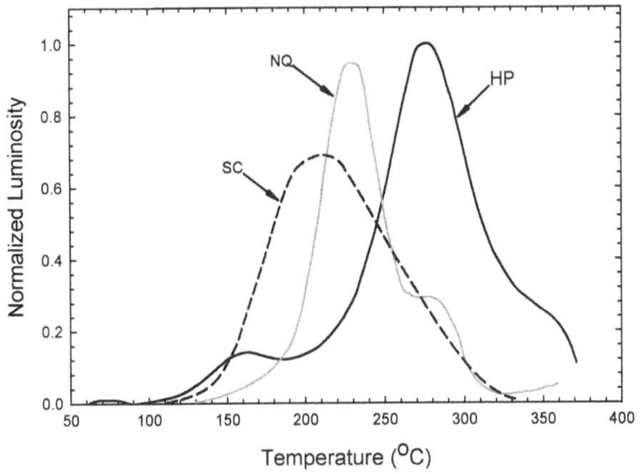

Figure 19.27. Normalized glow curves for typical BeO exposure measurements after gamma-ray irradiation. Data are adopted and adjusted from Scarpa [1970].

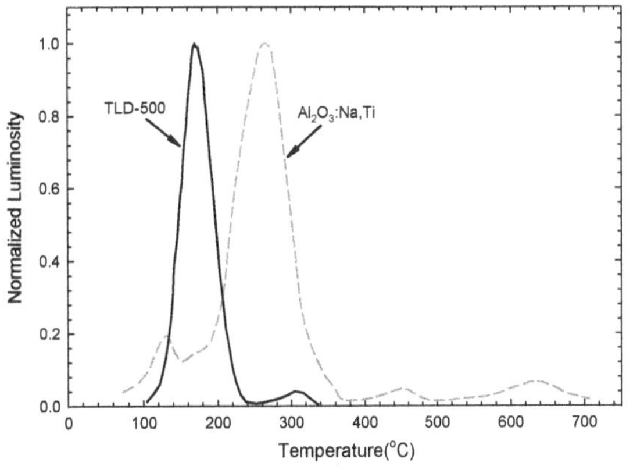

Figure 19.28. Normalized glow curves for typical Al_2O_3 measurements after gamma-ray irradiation. Data are adopted and adjusted from those of Portal et al. [1980] and Kortov and Ustyantsev [2013].

in dosimetry. The effective Z is 10.2, and below 200 keV Al_2O_3 has higher sensitivity than TLD-100, but less so than CaF_2 (Fig. 19.10).

When doped with carbon, Al_2O_3 has relatively good thermal stability, and is often denoted as TLD-500. The peak emission wavelength is 420 nm, well matched to the response function of common PMTs. Further, the thermal fading is approximately 3% per year and the dosimetric sensitivity can be as low as 5 μR. Kortov and Ustyantsev [2013] report multiple different glow curves dependent on the dose level. For instance, a single salient glow peak appears near 185°C at a dose of 800 mR from ^{60}Co, shifting downward to 165°C at a dose of 10^6 R. Several high temperature glow peaks appear for doses above 3×10^6 R, the main one appearing near 425°C. Although Al_2O_3:C functions as a thermoluminescent material, Akselrod et al. [1998] note that thermal quenching changes luminosity with heating rate. Consequently, because optically stimulated luminescence (OSL) demonstrates increased sensitivity and also the lack of thermal quenching, TLD-500 has instead gained popularity as an OSL dosimeter.

19.1.2 Optically Stimulated Luminescent Dosimeters

Closely related to thermally stimulated luminescence is optically stimulated luminescence (OSL) [McKeever 2001], with the main difference being in the method used to induce photon emissions. Both an OSLD and a TLD use a ceramic material in which intrinsic or activated traps store charge produced from radiation absorption events. However, instead of using thermal methods to release electrons from traps and into luminescent centers, energetic photons of sufficient energy are used to accomplish this task. Because the OSLD is not heated, problems with thermal destruction of the dosimeter are nearly eliminated. Further, select portions of the dosimeter can be sampled with a focused UV laser, a feature not possible with the uniform heating used in thermoluminescent dosimetry. Hence, an OSLD may have only a portion of the sample area illuminated, leaving other areas of the dosimeter unaltered for later measurements if required.

TLDs produce a glow curve that corresponds to light emissions from increasing trap energies as a function of increasing temperature. As temperature is increased, definitive features in the glow curve appear in the output spectrum. However, OSL dosimeters are illuminated with a light source that can affect shallow and deep traps simultaneously. This illumination is usually in the form of focused laser light. The output response, i.e., the emitted light, depends on several characteristics of the input illumination such as the wavelength, light intensity or power, and transient nature of the illumination.

Because of the similarities between TL and OSL dosimetry, Fig. 19.1 adequately depicts the initial process of radiation absorption. However, the stimulated emission process is quite different, as illustrated in Fig. 19.29. Instead of heat, energetic photons of energy greater than or equal to the trap energies are used to depopulate the dosimeter traps. The ejected electrons can then fall into luminescent centers accompanied by photon emissions, or the electrons may instead fall back into traps. However, if the electrons fall back into traps, they can subsequently be re-emitted by the optical stimulation, and can still contribute to the output signal if they fall into a luminescent center. To obtain a useful output signal, it is important that the stimulation wavelength be longer than the emission wavelengths, thereby facilitating filtering and reducing induced emissions unrelated to the radiation dose. For instance, the most commonly used commercial OSL dosimeter is Al_2O_3:C, which emits a broad wavelength spectrum centered at 420 nm (violet-blue). However, the stimulation wavelength is larger, optimized at 525 nm (green light). In this manner the stimulated luminescence can be separated from the interrogation light.

Modeling the Stimulated Emission

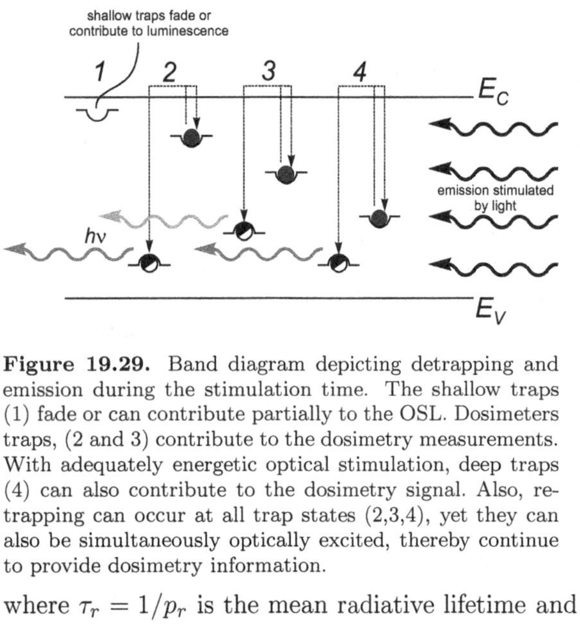

Figure 19.29. Band diagram depicting detrapping and emission during the stimulation time. The shallow traps (1) fade or can contribute partially to the OSL. Dosimeters traps, (2 and 3) contribute to the dosimetry measurements. With adequately energetic optical stimulation, deep traps (4) can also contribute to the dosimetry signal. Also, re-trapping can occur at all trap states (2,3,4), yet they can also be simultaneously optically excited, thereby continue to provide dosimetry information.

Electrons in an excited state can go to a lower energy state with or without photon emission with probabilities p_r and p_{nr}, respectively. Recall from the configuration coordinate diagram of Fig. 13.2, there is a probability p_{nr} that an electron in an excited state can move directly to the ground state at the crossover $D \to D'$ without any photon emission. This probability is described by

$$p_{nr} = p_0 e^{-\Delta E/kT}, \tag{19.24}$$

where ΔE is the activation energy required to move an electron from the bottom of the conduction band C to reach the crossing point at DD' in Fig. 13.2 and p_0 is a scaling constant referred to as the frequency factor. The radiative transition probability p_r is unaffected by temperature.

The optical luminescent efficiency is defined as

$$\eta = \frac{p_r}{p_r + p_{nr}} = \frac{\tau_{nr}}{\tau_r + \tau_{nr}}, \tag{19.25}$$

where $\tau_r = 1/p_r$ is the mean radiative lifetime and $\tau_{nr} = 1/p_{nr}$ is the mean non-radiative lifetime. Substitution of Eq. (19.24) into this definition gives

$$\eta = \frac{p_r}{p_r + p_0 e^{-\Delta E/kT}} = \frac{1}{1 + ce^{-\Delta E/kT}}, \tag{19.26}$$

where $c = p_0/p_r = p_0\tau_r$, a constant.

From Eq. (19.26), it is readily apparent that as the temperature of the OSL material increases, the luminescent efficiency should decrease. The loss of luminescence with temperature is referred to as *thermal quenching*. The total lifetime of excited charges is also a function of the radiative and non-radiative transition probabilities, namely

$$\tau(T) = \frac{\tau_r \tau_{nr}}{\tau_r + \tau_{nr}} = \frac{\tau_r}{1 + \tau_r p_0 e^{-\Delta E/kT}}. \tag{19.27}$$

From this result it is seen that the total lifetime also decreases as the temperature increases. Akselrod et al. [1998] show that a plot of τ versus temperature can yield both the values of the frequency factor p_0 and the activation energy ΔE.

The use of OSL as a dosimetry technique was first explored by Huntley et al. [1985] as a method to date geological samples. The test samples were natural quartz irradiated with 514.5 nm light. The technique was developed over several years for geology, and years later was recognized as a potential technique that can be used for radiation dosimetry [Bøtter-Jensen and McKeever 1996]. Numerous materials have been investigated for use as OSL dosimeters, including many traditional TLD materials. The most successful, and most popular, is carbon activated alumina, Al_2O_3:C.

For constant illumination, the rate of stimulation from a trap, with units s^{-1}, for photons of energy E_{ph} is given by

$$R_{te} = \sigma_t \phi_{ph} = \frac{1}{\tau_d}, \tag{19.28}$$

where τ_d is the decay constant, σ_t is the photoionization cross section, and ϕ_{ph} is the photon flux of energy E_{ph} [Bøtter-Jensen et al. 2003]. For simplicity, suppose all electron traps are at the same energy level and all luminescence centers, likewise, are at a single lower energy level. Because there is no charge on the material (the *charge neutrality* condition), the concentration n of electrons in the conduction band and the concentration N_{te} of electrons in traps must equal the concentration p of holes in the valence band and the concentration N_{rh} of holes stored in recombination centers. After the OSLD has been exposed to radiation, the free electrons and holes either fall into traps or recombine so that $n = p = 0$. Here it is assumed that there is a wide band gap with negligible thermal generation of free charges and that the stimulation photons have insufficient energy to excite electrons from the valence band to the conduction band.

However, upon illumination this equilibrium balance is altered. The rate of gain of electrons in the conduction band equals the rate at which electrons leave their traps plus the rate at which electrons fall from the conduction band into recombination centers, i.e.,

$$\frac{dn(t)}{dt} = -\frac{dN_{te}(t)}{dt} + \frac{dN_{rh}(t)}{dt}. \tag{19.29}$$

The rate of increase in the density of trapped electrons equals the rate per unit volume that electrons are excited out of the traps (negative change) plus the rate per unit volume that free electrons fall into the traps, namely

$$\frac{dN_{te}(t)}{dt} = -R_{te}\left[\frac{N_{te}(t)}{N_t}\right]N_t + p_{ec}\left[\frac{N_t - N_{te}(t)}{N_t}\right]n(t)N_t, \tag{19.30}$$

where p_{ec} is the free electron capture probability per unit time, N_t is the trap concentration and $(N_t - N_{te})/N_t$ is the fraction of traps that are empty and able to receive an electron. The rate of decrease in the hole concentration at recombination centers is the rate at which electrons from the conduction band fall into open recombination centers, i.e.,

$$\frac{dN_{rh}(t)}{dt} = -p_{er}\frac{N_{rh}(t)}{N_r}N_r n(t), \tag{19.31}$$

where p_{er} is the probability, per unit time, of an electron in the conduction band recombining with a hole at a recombination center, which has a concentration of N_r, and N_{rh}/N_r is the fraction of recombination centers which contain a hole and are able to receive an electron from the conduction band.

To obtain an expression for the intensity of the stimulated luminescence beginning at $t = 0$, two simplifying assumptions are made. First, it is assumed that the probability of exciting an electron out of a trap and falling into a recombination center is much higher than that of a free electron returning to a trap center. Hence Eq. (19.30) is simplified to

$$\frac{dN_{te}(t)}{dt} = -\sigma_t \phi_{ph} N_{te}(t), \tag{19.32}$$

whose solution is

$$N_{te}(t) = N_{te}^0 \exp[-\sigma_t \Phi_{ph} t], \tag{19.33}$$

where N_{te}^0 is the trap-filled population at $t = 0$. The second assumption is that the change in recombination center population equals the change in trap population, i.e., $dn/dt \approx 0$. Thus, from Eq. (19.29) the stimulated luminescence can be described by

$$I = -C\frac{dN_{rh}(t)}{dt} = -C\frac{dN_{te}(t)}{dt}. \tag{19.34}$$

where C is a constant to account for the size of the interrogation volume and the luminescence efficiency. For a constant optical stimulation for $t \geq 0$ with a constant photon flux of energy E_{ph}, the luminescent intensity is found by substitution of Eqs. (19.32) and (19.33) into Eq. (19.34) to give

$$I = CN_{te}^0 \sigma_t \phi_{ph} \exp[-\sigma_t \phi_{ph} t], \qquad t > 0, \tag{19.35}$$

where N_{te}^0 is the trap-filled population at $t = 0$. Constant power illumination is referred to as CW-OSL (continuous wave OSL) in which the OSL dosimeter is illuminated with a selected bandwidth of wavelengths at constant power for a specific time. Electrons depopulate the storage traps, fall into recombination centers, and produce luminescence. Under CW operation, Eq. (19.35) indicates that the expected luminescent output is a rapidly decaying exponential function of time. This behavior is expected because light emissions decrease with time as the traps are depopulated. These trap emissions are concurrent. Spectral emissions develop simultaneously according to the trap emission probability.

The integrated luminescence I_{lum} over time period t_1 is

$$I_{lum} = \int_0^{t_1} CN_{te}^0 \sigma_t \phi_{ph} e^{-\sigma_t \phi_{ph} t}\, dt = -CN_{te}^0 e^{-\sigma_t \phi_{ph} t}\Big|_0^{t_1} = CN_{te}^0 \left[1 - e^{-\sigma_t \phi_{ph} t_1}\right]. \tag{19.36}$$

This result shows that the luminescent yield is proportional to the initial filled trap density N_{te}^0. If the initial filled trap density is a linear function of radiation dose, then this integrated luminescence can be used to measure the dose [Yukihara and McKeever 2008].

Multiple Traps for Electrons and Holes Because an OSL dosimeter usually has multiple trap energies as well as several luminescent center energies, the output spectrum is, therefore, not a simple exponential function of time. Instead it is superposition of emissions from multiple trap energies and luminescence of different energies. Other factors that contribute to the spectral output are competing trapping centers, thermally induced emission (usually from shallow traps), retrapping, and photo-excitation of electrons from deep to shallower traps. In some circumstances, the combined output spectrum may be best modeled as an inverse power function rather than as a simple exponential dependence [Yukihara et al. 2004].

The effect of shallow traps in an OSL dosimeter is to produce a concentration of competing traps that initially reduces the luminescence. These shallow traps fade rapidly after irradiation and a large fraction of the centers depopulate before optical stimulation. Consequently, these empty traps compete with luminescent centers for electrons during the optical stimulation step, causing a reduction in the emitted luminosity. As the intensity measurement continues, a steady-state condition is reached in which trapping at shallow traps is matched by detrapping and results in an increase in luminosity. As the measurement continues further, the luminosity characteristic eventually decreases with time, similar to the ideal case without shallow traps.

The effect of deep traps is similar to that observed with TLDs and causes an apparent increase in dose sensitivity, or sensitization. Upon irradiation, a new and pristine OSL dosimeter has energetic electrons excited across the band gap. Many of these electrons diffuse through the material and drop into traps, including shallow, medium, and deep traps. In the present case, the medium level traps that participate

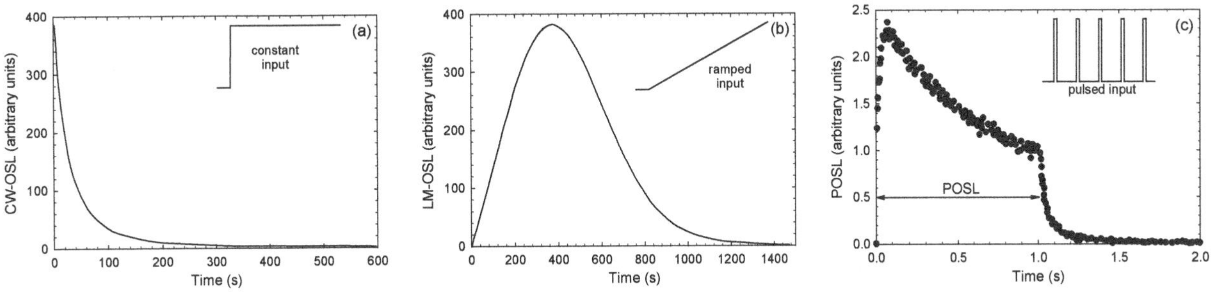

Figure 19.30. Luminescent output from Al_2O_3:C OSLDs for three types of laser stimulation, (a) CW-OSL, (b) LM-OSL, and (c) POSL. The OSLDs in (a) and (c) are Al_2O_3:C (Luxel^{TM}) and in (b) it is a Al_2O_3:C TLD-500. After Bøtter-Jensen et al. [2003].

in the dosimetric signal are referred to as dosimetry traps. During optical stimulation, a large percentage of shallow and dosimetry traps are emptied, but the deep traps are not. As these deep traps are filled, and remain filled, they are eliminated as competing traps. Consequently, subsequent irradiation of an OSL produces a higher percentage of filled dosimetry traps because there are fewer competing deep traps. The end result is higher luminosity per unit dose during optical stimulation.

The excitation wavelength is selected to correspond to sub-bandgap energies. The purpose for this choice is simple; the stimulation process is responsible for exciting more electrons into dosimetry traps. However, it is still possible for photons during the optical stimulation process to excite electrons across competitive traps. For instance, electrons in deep traps may be transferred to dosimetry traps.

OSL Systems

OSL readers have at least two main components, namely, a light source to stimulate the dosimeter and a photosensitive receptor to detect the stimulated luminescence. A challenge for these components is the vast difference in luminescent intensity between the light stimulation source and the luminescent emissions, a difference that can span over 14 orders of magnitude ranging between 10^{16} photons cm^{-2} s^{-1} for stimulation and 10^2 photons cm^{-2} s^{-1} for low dose luminescence [Yukihara and McKeever 2011]. OSL systems are designed to reduce photo-ionization and photo-transference of electrons. A long-pass filter[9] is placed between the light source (usually a laser, photodiode, or lamp) and the OSL dosimeter. Another filter is also placed between the OSL dosimeter and the light collection device (typically a PMT), usually a bandpass filter which transmits photons with wavelengths between 290 nm to 370 nm [Yukihara and McKeever 2008]. The bandpass filter reduces scattered light from the light source from entering the PMT, while passing the prominent wavelengths from optically stimulated luminescence. Additional components may include optical components, such as mirrors and lenses, to guide and focus the light, both stimulated and luminescent. There are at least three types of stimulation systems in use for OSL, listed as continuous wave (CW-OSL), linear modulation (LM-OSL), and pulsed (POSL). These systems are briefly described here but more detailed descriptions can be found in the literature [Bøtter-Jensen et al. 2003; Yukihara and McKeever 2011].

CW-OSL The CW system focuses light of constant power onto an OSL dosimeter in a step input, as depicted in Fig. 19.30(a). CW-OSL remains the most popular method to sample an OSL dosimeter, mainly because of its simplicity and reliability. The output response is described by Eq. (19.35) in which the expected output signal is largely modeled as an exponentially decreasing function. The output signal is mainly a function of the initial trap filled density N_{te}^0, the stimulation intensity ϕ_{ph}, and the time duration of the stimulation t. For a fixed illumination time period $t_1 = \Delta t$ with constant power ϕ_{ph}, the integrated

[9]These filters are used to filter short wavelengths that may excite electrons across a bandgap or produce prominent photo-transference while at the same time passing longer wavelengths that still can excite electrons out of dosimetry traps.

intensity produces a luminescent output proportional to the initial trap-filled population. A sample taken over a short time period, 1 second for instance, is sufficient to gain the required information on dose. After such a short sampling time, many of the trap states may remain filled. The OSL can then be stored and re-read if so required, taking into account the time period of the prior optical stimulation.

Notably, integration of Eq. (19.35) for the total luminescence I_{tot}, i.e., $\Delta t \to \infty$, is solely a function of the initial number of filled traps N_{te}^0, namely

$$I_{tot} = C \int_0^{\infty} N_{te}^0 \sigma_t \phi_{ph} e^{-\sigma_t \phi_{ph} t} \, dt = C N_{te}(t_0). \tag{19.37}$$

Consequently, the intensity of the stimulation light ϕ_{ph} does not affect the total luminescence or resulting measured dose. Hence, the light stability and intensity of the stimulation source have, theoretically, no effect on the measured output or dose provided that the volume of the OSL dosimeter is cleared of all trapped electrons by this long-term stimulation. However, the initial luminescence $I(t = 0^+)$ emitted by the stimulated OSL dosimeters is in fact a function of the intensity ϕ_{ph} of the light source, as shown in Eq. (19.35), and the total light output for a finite Δt is, from Eq. (19.36), also a function of ϕ_{ph}. Hence, if the user reads only part of the OSL dosimeter traps during a measurement, it is important that the light source be stable during the measurement time to obtain reliable results.

The simple model of Eq. (19.36) does not take into account the possibility of different trap energies participating in the luminescent output. Further, it does not consider retrapping of charges, effectively prolonging the stimulated luminescence. Curve fitting indicates that a power function may in fact describe better the luminescent behavior of a real OSLD than does a simple exponential.

POSL Although filters are used to reduce contamination from scattered light from the source, these same filters can also absorb some of the optically stimulated light. If there is significant overlap between the stimulation spectrum and the OSL spectrum, then part of the OSL signal is sacrificed to eliminate scattered light from the laser source. Because the intensity of the stimulation light is orders of magnitude greater than the OSL output, the *stimulation* light leakage that arrives at the light sensor may still have an intensity comparable to that of the optically *stimulated* luminescent light reaching the sensor.

Pulsed mode systems were developed as a readout system that can reduce scattered light from contaminating the OSL signal. The technique involves the use of a pulsed laser to stimulate luminescence from the OSL dosimeter, while measuring the luminescence between pulses [McKeever et al. 1996]. Such a system still has filters to reduce the detection of scattering photons. However, between pulses, the counting system is gated such that current from the PMT is measured only while the laser is off. During the pulse, the electrons in traps are elevated to the conduction band and can drop into luminescent centers. When the light is switched off, these free electrons continue to fall back into luminescent centers and traps, hence continue to produce luminescence. If the concentration of electrons in excited luminescent centers, before decay and photon emission, is N_{re}^*, the process can be modeled as

$$\frac{dN_{re}^*(t)}{dt} = -\frac{N_{re}^*(t)}{\tau_d} - \frac{dN_{te}(t)}{dt}, \tag{19.38}$$

where the change in the excited luminescent center population is equal to decay of the luminescent centers with time constant τ_d plus the decrease of trapped electrons into luminescent centers. Substitution of Eq. (19.32) and Eq. (19.33) into Eq. (19.38) gives

$$\frac{dN_{re}^*(t)}{dt} + \frac{N_{re}^*(t)}{\tau_d} = R_{te} N_{te}(t) = N_{te}(0) R_{te} \exp[-R_{te} t], \tag{19.39}$$

where the stimulation rate $R_{te} = \sigma_t \phi_{ph}$ has units s^{-1}. The solution of this differential equation is

$$N_{re}^*(t) = A\exp[-t/\tau_d] + \frac{N_{te}(0)R_{te}\tau_d}{1 - R_{te}\tau_d}\exp[-R_{te}t], \tag{19.40}$$

where A is a constant defined by an initial condition.

Now apply this result to the POSL chain of laser pulses shown in Fig. 19.31. Suppose that at the beginning of a pulse at time t_b, the concentration $N_{re}^*(t_b)$ of luminescence centers is known. Then from Eq. (19.40) the constant A is found to be

$$A = e^{t_b/\tau_d}\left[N_{re}^*(t_b) - \frac{N_{te}(0)R_{te}\tau_d}{1 - R_{te}\tau_d}\exp[-R_{te}t_b]\right]. \tag{19.41}$$

Substitution of A into Eq. (19.40) gives for $t \ge t_b$

$$N_{re}^*(t) = N_{re}^*(t_b)e^{-(t-t_b)/\tau_d} + \frac{N_{te}(t_b)R_{te}\tau_d}{1 - R_{te}\tau_d}\left\{e^{-R_{te}(t-t_b)} - e^{-(t-t_b)/\tau_d}\right\}, \tag{19.42}$$

where $N_{te}(t_b) = N_{te}(0)\exp[-R_{te}t_b]$. The decay of the excited luminescent center, and the emitted luminescent intensity, during a laser pulse, is $I(t) = CN_{te}^*(t)/\tau_d$ or

$$I(t)/C = \frac{N_{te}^*(t_b)}{\tau_d}e^{(t_b-t)/\tau_d} + \frac{N_{te}(t_b)R_{te}\tau_d}{\tau_d(1 - R_{te}\tau_d)}(e^{R_{te}(t_b-t)} - e^{(t_b-t)/\tau_d}), \quad t_b \le t \le t_e$$
$$\text{where } t_b = (m-1)(t_p + \Delta t), \quad m = 1, 2, \ldots, M. \tag{19.43}$$

Here Δt is the time duration between pulses and t_p is the optical pulse width. Between pulses, there is no optical stimulation ($R_{te} \to 0$), and the emitted luminescent intensity is

$$I(t)/C = \frac{N_{te}^*(t_e)}{\tau_d}e^{(t_e-t)/\tau_d}, \quad \text{where } t_e = t_b + t_p = (m-1)\Delta t + mt_p. \tag{19.44}$$

During the stimulation time $t_p = t_e - t_b$, it is usual to close the electronic gate on the PMT signal, thereby avoiding the inclusion of scattered light. Consequently only the light emitted during the time period between pulses Δt is measured. Examples of the emitted luminescence are shown in Fig. 19.31.

Consider the light emitted during Δt to that over a single period $(t_p + \Delta t)$. For a constant measurement/period ratio, the ratio of measured light to the total luminescence increases with frequency. If the stimulation time is relatively long, followed by decay, a large fraction of light is lost during stimulation, with the luminescence decaying to low values. In the example of Fig. 19.31(left), the amount of light emitted after a stimulation pulse is approximately 40% for a frequency of 6.67 Hz. Because the optical signal is not measured during stimulation, this result means that 60% of the signal is lost. By increasing the frequency, the ratio of lost luminescence decreases. From the example of Fig. 19.31(b), with all parameters the same except that the frequency is increased a hundredfold to 667 Hz, the percentage of measured light increases to approximately 80%. Although Fig. 19.31 serves to demonstrate the general trend for POSL, the duty cycle[10] in the example is only 33.3%. In practice, the duty cycle is often shorter and on the order of 10%.

The ratio of lost luminescence decreases as the decay constant τ_d increases. To improve the measurements, it is best to select a stimulation pulse width t_p equal to or less than the luminescence decay constant τ_d.

[10]The fraction of a period that the stimulation source is active.

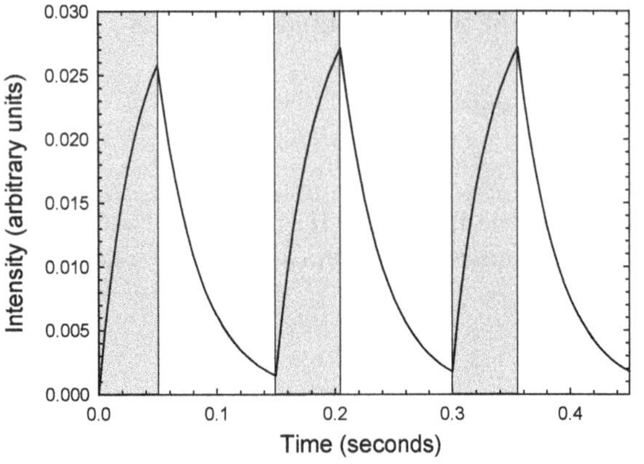

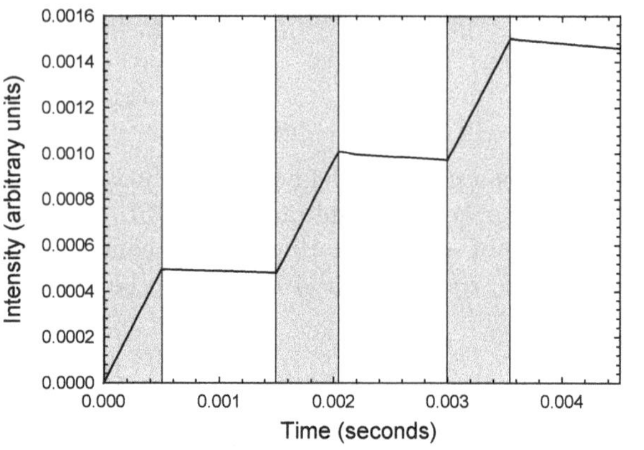

Figure 19.31. POSL luminescence signal with the stimulation pulse width equal to 1/3 of the total period. The gray region indicates the time t_p when the laser pulse is activated and the output signal is switched off, and the clear area represents the time period Δt when the luminescence is measured. On the left, the period is 150 ms (6.67 Hz) with $t_p = 50$ ms and on the right the period is 1.5 ms (667 Hz) with $t_p = 500$ μs. All other parameters are the same.

For a given value of τ_d, the pulse frequency and pulse width t_p can be adjusted to optimize the light yield while reducing contamination from scattered light. The general luminescent trend for a POSL measurement is shown in Fig. 19.30(c) where it is seen that the luminescence initially increases, and begins to decrease as the filled traps depopulate. At the end of the measurement, when the laser is switched off after a series of pulses, the remaining excited luminescent centers decay away.

LM-OSL Linear modulation is an optical stimulation method in which the optical intensity is linearly increased over time so that the stimulation rate (s^{-1}) of removing a trapped electron is

$$R_{te}(t) = \sigma_t \phi_{max} \frac{t}{T}, \tag{19.45}$$

where ϕ_{max} is the maximum optical power and T is the stimulation time period which begins at $t = 0$. The luminescent decay time constant is now a function of time

$$\tau_d(t) = \frac{1}{R_{te}} = \frac{T}{\sigma_t \phi_{max}}. \tag{19.46}$$

The trapped electron concentration is given by

$$\frac{dN_{te}(t)}{dt} = -R_{te}(t) N_{te}(t), \tag{19.47}$$

whose solution is

$$N_{te}(t) = N_{te}(0) \exp\left[-\frac{\sigma_t \phi_{max} t^2}{2T}\right]. \tag{19.48}$$

The optical luminescence produced is thus

$$I(t) = -C\frac{dN_{te}(t)}{dt} = CN_{te}(0)\sigma_t \phi_{max} \frac{t}{T} \exp\left[\frac{-\sigma_t \phi_{max} t^2}{2T}\right]. \tag{19.49}$$

The intensity maxima is easily found to be,

$$t_{max} = \sqrt{\frac{T}{\sigma_t \phi_{max}}}. \tag{19.50}$$

Substitution of Eq. (19.50) into Eq. (19.49) yields the maximum luminescence

$$I_{max} = N_{te}(0)\sqrt{\frac{\sigma_t \phi_{max}}{T}} \exp\left[-\frac{1}{2}\right]. \tag{19.51}$$

The general luminescent shape is shown in Fig. 19.30(b), where it is seen that the luminescence increases from zero to a maxima and then decreases as the traps depopulate. It is these maxima features that Bular [1996] argues as advantageous, especially in the presence of multiple traps. Unlike the CW-OSL method in which multiple traps blend together to produce a single decay curve, each trap level shows a maxima with the LM-OSL method, thereby making identification of the individual traps possible. Nonetheless, for the same sample material, the total area under the luminescence curves are the same for CW-OSL and LM-OSL. Although numerous groups have experimented with LM-OSL (see references in Yukihara and McKeever [2011]), it has not yet received widespread acceptance. Complications arise from non-linear light increase associated with the stimulation source, spectral shift of the stimulated luminescence as the power is increased, and changes in background light leakage measured by the PMT as the light source power is increased.

OSL Materials

Materials for optical stimulation luminescence include a variety of synthesized and natural minerals. Many of the synthesized materials can be used for radiation measurements, mainly because the storage traps and luminescent centers can be controlled during synthesis. Of these synthetic materials, Al_2O_3:C is the most popular for dosimetry. Many natural minerals that exhibit OSL characteristics can be used to determine post-factum doses and are used for geological dating. A select few of these materials are described here.

Aluminum Oxide (Al_2O_3:C) Carbon activated aluminum oxide (Al_2O_3:C) was originally proposed as a TLD, yet has gained popularity as an OSL dosimeter [McKeever and Akselrod 1999; Akselrod and McKeever 1999]. Al_2O_3 has a high sensitivity to low radiation doses, and is now in wide-spread use as a commercial dosimeter, having replaced TLD materials for many applications. Carbon doped alumina has a main emission line near 420 nm with a decay constant of 35 ms, and a lesser fast emission at 335 nm with a time constant less than 7 ns. Both of these emissions are produced by stimulation of the dosimeter traps with green light, optimized at 525 nm. Reported sensitivity ranges from 1 mrem (10 μSv) up to 10^4 rem (100 Sv); however, the uncertainly for low doses can be greatly affected by the environmental background radiation [Gilvin and Perks 2010]. When stimulated with CW-OSL, the maximum luminescence occurs with a stimulation wavelength near 500 nm as shown in Fig. 19.32. Shown in Fig. 19.33 is a decay curve for Al_2O_3:C after a dose of 60 mGy from a ^{90}Sr source. Room-temperature thermal fading is reportedly not appreciable up to 85 days after irradiation. Bøtter-Jensen et al. [1997] report fading is approximately only 1.8% after 27 days and, subsequently, becomes relatively stable [Schembri and Heijmen 2007].

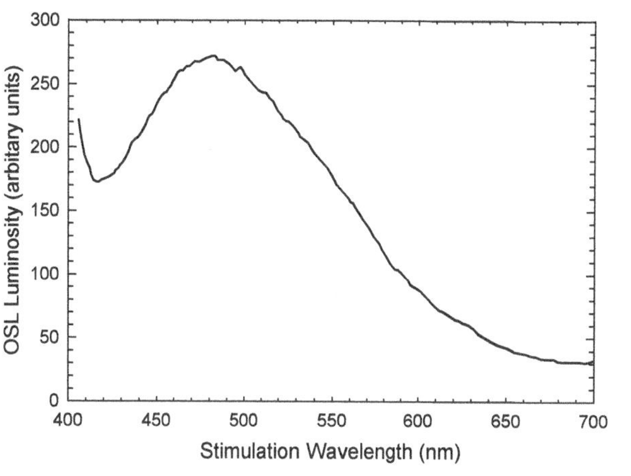

Figure 19.32. OSL luminosity versus the stimulation wavelength for Al_2O_3:C. Data are from Bøtter-Jensen et al. [1997].

Although the fast UV emission at 335 nm increases the total optical stimulated luminescence, it is the longer wavelength emissions with maxima near 420 nm that are usually used for dosimetry. The 420 nm emission is attributed to F-center luminescence [Yukihara and McKeever 2008]. Optical excitation from the dosimeter traps produces free electrons that recombine with F^+-centers. The inclusion of carbon in the growth process enhances the formation of these F^+-centers by replacing a fraction of the trivalent Al^{3+} ions with C^{2+} ions [McKeever et al. 1999]. The capture of an electron produces an excited F-center, and these excited F-centers de-excite by releasing a photon. Dosimetric linearity of up 5×10^3 rad (50 Gy) has been reported for the F-center emission from Al_2O_3:C OSL dosimeters [Yukihara and McKeever 2006]. Without proper filtration, the contribution to the total signal by the UV emission can be significant, approximately 20% of the F-center signal for low doses and increasing to nearly the same as the F-center emission at high doses. Bandpass filters designed to pass F-center emission wavelengths can be used to reduce the influence of the UV emission.

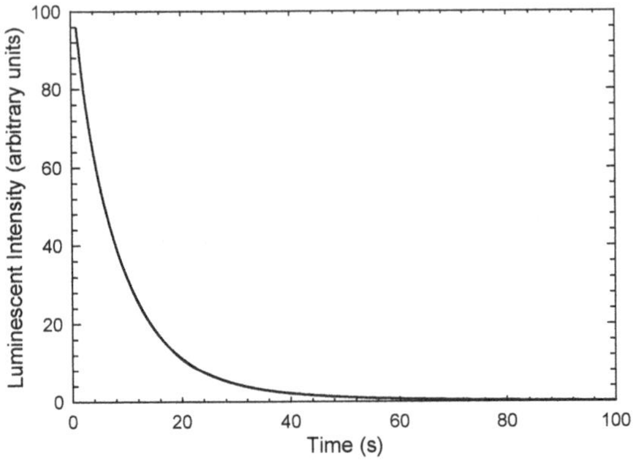

Figure 19.33. An OSL decay curve for Al_2O_3:C after irradiation with 60 mGy from a ^{90}Sr beta particle source. The OSL decay curve is for constant power stimulation (16 mW cm^{-2}) with wavelength band of 420 to 550 nm. Data are from Bøtter-Jensen et al. [1997].

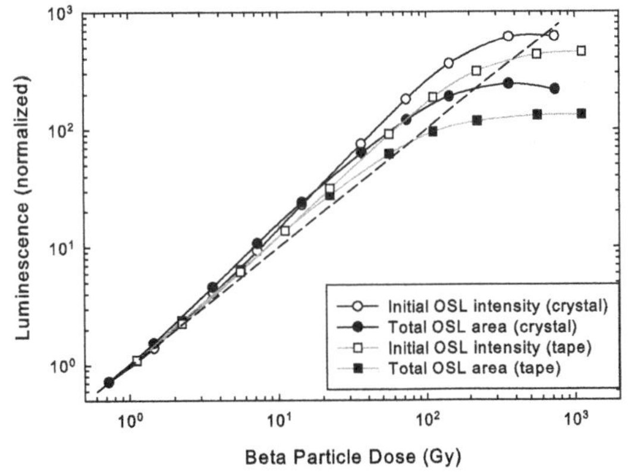

Figure 19.34. OSL beta particle dose response from a Al_2O_3:C single crystal and from a Luxel Al_2O_3:C tape. Initial OSL response is for the first 3 seconds of optical stimulation. Total OSL area is for the integrated response over 300 seconds of optical stimulation. Data are from Yukihara et al. [2004].

The CW-OSL responses (Fig. 19.34) reveal sensitization (superlinearity) with both the initial intensity (measured within the first 3 seconds of stimulation) and the total luminescence (integrated over 300 seconds of stimulation). Experiments that remove the UV component indicate that these emission centers are responsible for the supralinearity observed for doses above 10^3 rad (10 Gy). Because of the short time constant, the UV component can be eliminated from the measurement by using POSL. During the stimulation pulse, most of the UV component is released and does not contaminate the measurements between pulses. Sublinearity is observed for doses above 10^4 rad (100 Gy) for both UV and F-center emissions.

The dose response can be affected by the dose history as deep traps become filled over time and are not emptied by annealing. An advantage of single crystal Al_2O_3:C dosimeters over OSLD formed from powder on tape is that crystals can be annealed at high temperatures to remove electrons from deep traps, up to 900°C [Bøtter-Jensen et al. 1997]. This annealing procedure ensures that background and sensitization from deep traps does not build up over time. However, much like TLDs, the uniformity of the carbon doping determines the variability of response, and large variations can be present within and between batches. Hence, crystalline Al_2O_3:C OSL dosimeters can have a large variance in luminescence among samples.

Just as with TLDs, commercial Al_2O_3:C OSL material can be pulverized after crystal growth and mixed with multiple batches to improve uniformity amongst samples. The pulverized materials are bound with plastic to produce a length of tape. This tape is then sliced into sections for commercial use as OSL dosimeters. These OSL dosimeters cannot be annealed at high temperatures because the plastic binder is destroyed. Although optical bleaching can be used to empty most of the dosimeter traps, it is not adequate to empty the deepest traps. Consequently, the background luminescence increases with integrated dose.

Drawbacks to the use of Al_2O_3:C OSL dosimeters include the high effective Z number and the presence of shallow traps that makes the material sensitive to light. The effective Z number is 11.3, and the low energy response to gamma rays can be quite high, comparable to SiO_2 (see Fig. 19.10), so that Al_2O_3:C OSL dosimeters overrespond to x rays and low-energy gamma rays. The shallow traps can affect the total light output, but generally fade rapidly at room temperature. Edmund and Andersen [2007] report that shallow traps in Al_2O_3:C can affect the luminescent yield. If the irradiation temperature cannot be maintained at a constant temperature, they recommend a correction factor between -0.2% to 0.6% per °C over a range from 10°C to 50°C.

Beryllium Oxide Beryllium oxide (BeO), or beryllia, is a thermoluminescent material and has been used as an OSL dosimeter. The luminescent spectrum reported by Bular and Göksu [1998] has a broad spectrum, ranging from 420 to 550 nm, with a maximum peak at 435 nm. Sommer et al. [2008] report the use of 455 nm light to stimulate these emissions and a Schott DUG11X optical filter to make luminescence measurements. BeO OSL dosimeters are linear over a wide range, extending from 100 μrad (1 μGy) [Sommer et al. 2007] up to 1000 rad (10 Gy) [Bular and Göksu 1998]. Sub-linearity appears at approximately 2000 rad (20 Gy), and the onset of saturation is evident at approximately 10 krad (100 Gy) [Sommer et al. 2008]. The low effective Z of 7.14 makes these OSLDs almost "tissue equivalent" and they have a relatively flat response for low- to high-energy gamma rays (see Fig. 19.10) with only a slight under-response to low-energy gamma rays.

Short-term fading from shallow traps amounts to approximately 6% loss within the first few hours of irradiation. The fading becomes almost negligible thereafter and is less than 1% over a 6-month period [Sommer et al. 2007]. There is a temperature dependence on the decay constant τ_d of the main luminescent centers, ranging from 27 μs at room temperature down to 800 ns at 140°C [Yukihara and McKeever 2011]. This decrease in decay time constant is largely due to non-radiative transitions. Also, the luminescent efficiency decreases with temperature and, characteristically, higher temperatures decrease the total luminescent yield [Akselrod et al. 1998]. Bular and Göksu [1998] and Sommer et al. [2008] investigated the use of Thermalox®, a commercial form of BeO,[11] for dosimetry with promising results.

Other Synthetic Materials There are numerous other synthetic materials that have been investigated for OSL dosimetry. However, none are as promising as Al_2O_3:C and BeO, mainly because of the high fading from shallow traps exhibited by these other materials. These alternatives include MgO:Tb^{3+} [Bos et al. 2006], $MgSi_2O_4$:Tb, Al_2O_3:Cr,Mg,Fe, $MgAl_2O_4$ (cubic spinal crystals), Brazilian fluorite (CaF_2), and various barium aluminoborate glasses ($20Al_2O_3 \cdot 50B_2O_3 \cdot 30BaO$) [Yoshimura and Yakihara 2006]. Potassium compounds, such as KBr:Eu, KCl:Eu, KBr:Cu^+, and KBr:Cu^+, have also been investigated as OSL materials [Nanto et al. 1993; Douguchi et al. 1999; Bandyopadhyay et al. 1999] and they all have relatively short decay times of about 1 μs. All have a luminescent emission near 420 nm and the optimum stimulation wavelength for KBr:Eu is 620 nm and for KCl:Eu is 560 nm. Problems associated with these potassium-based materials include substantial fading and the high effective Z of 18.1 and 31.5 for KCl and KBr, respectively, and thus they overcompensate for doses from low-energy gamma rays.

Kearfott et al. [2015] studied several known thermoluminescent materials for application in OSL dosimetry, namely LiF:Mg,Tl, $Li_2B_4O_7$:Cu, $CaSO_4$:Tm, and CaF_2:Mn. Of these materials, only $CaSO_4$:Tm produced appreciable optically stimulated luminescence; however, its luminescent yield was only 10% of the yield from commercial Al_2O_3:C OSL dosimeters.

Natural Materials Luminescent spectra from several minerals can be used for dating and dosimetry characterization [Krbetschek et al. 1997]. Both thermoluminescence and optically stimulated luminescence have been used to characterize these minerals, mainly to obtain information on recombination site energies,

[11]Developed by Brush Wellman, Inc., for the thermal and electrical properties of BeO, Thermalox® is now produced by the renamed Materion Corporation. However, there is no indication of its use for OSL dosimetry in the Materion literature.

possible information on trap energies, and also for use in retrospective dosimetry. Possibly the two most investigated minerals for OSL dosimetry applications are quartz and feldspar, originally studied by Huntley et al. [1985]. However, difficulties with using natural minerals for dosimetry arise from compositional differences, activator differences, and various inclusions which usually contaminate natural crystals. Consequently, luminescent spectra are variable and depend on the origin of the mineral. Further, OSL emissions are a function of the thermal history of the samples, with samples exposed to high temperature having much lower luminosity [Bøtter-Jensen and McKeever 1996].

For quartz samples taken from archaeological fired clay, Bluszcz and Bøtter-Jensen [1995] determined that gamma-ray doses of less than 100 mrad (1 mGy) can be measured. Optical stimulation from long wavelength light (541 nm or 647 nm) produces a single emission band from quartz near 365 nm [Huntley et al. 1991]. The luminescent lifetime τ of this 365 nm emission is approximately 35 μs. Although Huntley et al. [1985] first studied feldspar with 514 nm light stimulation, Hütt et al. [1988] discovered that IR stimulation produced superior results. The optically stimulated emissions from feldspar are much more complex, with multiple emissions reported by various researchers [Krbetschek et al. 1997], mainly at 280 nm, between 320 to 340 nm, 400 nm, between 560 to 570 nm, and in some cases a broad emission band between 330 nm and 620 nm, with results depending on the choice of OSL filters [Clarke and Rendel 1997; Huntley et al. 1985; Krbetscheck and Rieser 1995]. Anomalous fading is another problem with feldspars in which trapped charges are lost by mechanisms other than thermal quenching.

Luminescent Detector Summary

High heating temperatures can lead to non-ideal behavior for some TLDs. IR radiation from the heat source can cause interference and alter the measured TL spectrum. Also, because several TLD materials may suffer irreversible damage to the TLDs at high temperatures, a thermal heating limit is often imposed to prevent such damage. Consequently, deep traps that require temperatures above this thermal limit cannot be efficiently depopulated. Further, buildup within these deep traps over time causes sensitization of the TLD and this buildup can also become a source of background emissions during readout.

There are several advantages of OSL over TL. Commercial OSL dosimeters can also suffer damage if heated to too high a temperature because of damage to the binding material. However, because the read-out process does not require a heating step, OSL dosimeters generally are unaffected by this encumbrance. With sufficient energy of the stimulation photons, many of these deep traps not only can be depopulated, but can also be used in the dosimetry measurement. This increased density of useful trapping centers improves the radiation sensitivity of OSL dosimeters.

The luminescence from OSL dosimeters is very faint by comparison to the stimulation light source, hence filters must be used to reduce background light. POSL can be also be used to reduce or almost eliminate this background light, provided that the decay time of the luminescent centers is much longer than the optical pulse and the stimulation light source duty cycle is relatively small. As with a TLD, once the OSLD is read, the dose information is lost. However, an OSLD does not necessarily need to be completely cleared during the measurement process. A spot on thc OSLD can be read with a focused laser, leaving the remaining unilluminated regions for later measurements. Also, an OSL dosimeter may be illuminated for a time that is small compared to the mean decay time of the trap and can, thus, leave a high density of filled dosimetry traps for a later measurement.

19.2 Photographic Film

Somehow it seems fitting that near the end of the long journey through this book, which began with the discovery of x rays from exposed photographic plates, the use of photographic film as a radiation detector is finally described with some respectable amount of detail. Radiation exposure of photosensitive emulsion was first observed in 1842 by Ludwig Moser who reported that when certain bodies were brought into contact with a silver iodide plate in darkness, an image would still form. He concluded that some materials must be "self-luminous". There is evidence that natural radiation had also been observed from uranium salts by Claude Félix Abel Niépce de Saint-Victor, in which a uranyl nitrate sample had been brought into close proximity of a photo sensitive silver bromide paper sheet [Niépce de Saint-Victor 1857, 1859]. The Niépce de Saint-Victor experiments were reported to the French Academy of Sciences in 1858, and apparently Antoine Henri Becquerel was aware of the work, primarily through work of his father, Edmond Becquerel, who mentioned the findings of de Saint-Victor in his book *La Lumiére: Ses Causes et Ses Effets.* Regardless, A. Henri Becquerel was apparently the first to explain and understand the significance of the discovery, having observed images formed from uranium salts on photographic plates in 1896.

19.2.1 Basics of Photographic Film

Photographic film consists of a photosensitive emulsion spread upon a base material. This base is either made of glass or a cellulose material such as cellulose acetate.[12] Photographic emulsions consist of a binder and a photosensitive compound, usually a distribution of microscopic silver halide crystals suspended in gelatin. Silver bromide (AgBr) is the photosensitive compound usually used for nuclear emulsions. The process for producing photographic film has some interesting aspects. A photographic emulsion can be produced by adding the two reacting solutions

$$\text{KBr (Aq)} + \text{AgNO}_3\text{ (Aq)} \rightarrow \text{AgBr (s)} + \text{KNO}_3\text{ (Aq)}$$

to a gelatin. The mixture is heated between 50°C to 70°C to cause *Ostwald ripening* of the mixture [Becker 1966]. It is also usual to add ammonia (NH_3) to the solution, which produces various amine complexes and forms an ammoniacal emulsion. During this ripening process, the size of AgBr crystals, referred to as grains, increases by growing between 100 to 1000 times in size. Larger grain sizes are usually associated with *faster* emulsions. The *speed* of the film can be increased by adding a small amount of mustard oil to the solution,[13] which forms particles of silver sulfide (Ag_2S) on the AgBr grains. Afterwards, the emulsion is shredded and washed to remove excess KNO_3. During a reheat process, in which the product is melted, more gelatin is added to the emulsion along with dyes and stabilizers. The process of "after-ripening" increases the sensitivity and affects the contrast, fogging, and storage properties of the emulsion. The finished emulsion is then applied to the base. The thickness of the finished emulsion varies with the intended use and ranges from 5 microns up to 600 microns. The AgBr content in common photographic emulsions is approximately 30%, but is increased to up to 80% for nuclear emulsions.

The silver halide grains in the emulsion are believed to be in the ionized state, even though they are in crystalline form. Exposure to ionizing radiation can cause the ejection of an electron from the halide ion and the electron subsequently is trapped at an impurity location. It is believed that the silver sulfide specks in the emulsion may be such locations. A positive silver ion can migrate to a trapped electron site and combine with the electron, becoming neutral silver. A single silver atom is unstable, and eventually decomposes back to a silver ion unless more silver atoms accumulate together. The neutral silver atom may also acquire an extra electron, thereby becoming a negative ion. If a second positive silver ion migrates to the negative silver ion, they can combine to form a two-atom cluster. This process continues as long as there are free electrons. Stability requires accumulation of at least three to four silver atoms. The sensitized silver grains that acquired neutral silver clusters produce a *latent image* in the emulsion, and it is this latent image that can then be developed to form a darkened image.

[12]Nitrocellulose is considered a much better base for film emulsions. Is it produced by soaking cotton in nitric acid. Unfortunately, it is also extremely flammable, almost explosively flammable. For this reason, cellulose acetate, a much slower burning base formed from soaking cotton in acetic acid, is a preferred choice.

[13]This process was discovered when film speeds were observed to increase for gelatin bases made from the collagen of cows that had eaten mustard plants.

The development process consists of several chemical treatments after the latent image is formed. In the case of nuclear emulsions used for dosimetry, the latent image may be accumulated over an extended period, weeks for instance. Photographic films are generally sensitive to visible light and must be developed in total darkness. The photographic film can be slipped into a light-tight development container for easier handling. Chemical developers are used to change positive silver ions in the emulsion into stable neutral silver atoms. This conversion is done more efficiently for silver grains already having silver atom clusters. The developer chemical has *reducing* agents attached to a benzene ring. Here the act of reducing refers to lowering the positive charge of an ion, in this case changing positive silver ions to neutral silver atoms.[14] The reduction potential of a chemical indicates the relative ability to reduce the photographic film. It is also usual to blend two or more reduction agents together to increase the activity [Becker 1966], and two popular reduction agents that are used together are metol and hydroquinone. Because the unexposed silver grains can also become developed, the developer temperature and development times must be strictly controlled to avoid over- or under-developing the film.

Most developers have several components, consisting of a reducing agent, an accelerator, a preservative, a restrainer, and a solvent. Because the reducing agent can be slow, an accelerator usually consisting of an alkali metal compound is added to the developer. Although KOH and NaOH work well as accelerators, they can also cause softening of the emulsion [Larmore 1965]. Effective accelerators include Na_2CO_3 and K_2CO_3, but release CO_2 gas that can cause film blistering if the development temperature is above 27°C. Less effective, but adequate accelerators that do not soften the gelatin and do not release gases include $Na_2B_4O_7$ (borax) and $NaBO_3$ (sodium metaborate). Because the developer is designed to chemically reduce the silver halide, it also has a strong affinity for oxygen, and can become ineffective over time. Preservatives are added to the developer to mitigate oxidation and prolong the developer lifetime. Sodium sulfite (Na_2SO_4) is one of the most commonly used preservatives. Although Na_2SO_4 reacts with oxygen, it does not adversely affect the development process, and it also prevents oxidation products (such as negative bromine ions) from accumulating on the film. These halogen ions decrease the pH of the developer. Further, the excess alkali metal ions added to accelerate the emulsion development can also *fog* the film, an effect that appears as a general darkening of the emulsion. Hence, a restrainer is also added to balance the pH during development, which, interestingly, turns out to be more bromine ions (often KBr). An overall chemical balance is achieved by adding excess alkali metal ions to the developer, which is compensated by the excess bromine ions. The addition of the small number of bromine ions released during development is insufficient to upset the chemical balance. Finally, a solvent is used to dilute the developer concentrate and, in most cases, is simply water.

To quickly stop the film development, the film is soaked in a *stop bath* solution, usually an acetic acid solution, that neutralizes the development process. Although the stop bath step is not strictly necessary, and thorough rinsing with water can be substituted, the development process continues, even if it is at a lower rate, unless a stop bath is used. Once the film is developed, the remaining undeveloped AgBr must be removed and is accomplished by *fixing* the emulsion. Otherwise, the film continues to darken when exposed to light. The fixer step is performed by soaking the film in a solution of sodium thiosulfate ($Na_2S_2O_3$), which dissolves the remaining AgBr crystals. The film is then thoroughly washed in water and dried. The darkness of the image that is formed gives a measure of the ionizing dose absorbed by the film and can be measured with a *densitometer*.

19.2.2 Photographic Film Characteristics

There are several important film properties that determine the adequacy of a film to perform well as a nuclear emulsion. Among these properties are the contrast, film speed, and the linearity of the film with exposure.

[14] Contrarily, *oxidation* is a chemical process that causes an atom or ion to lose an electron to become more positive.

These properties are discussed below through a description of the general characteristics of photographic film.

The Characteristic Curve

The passage of ionizing radiation through photographic film generally deposits only a small amount of energy in the emulsion. For visible light, the film may actually absorb all energy of the incident radiation. However, for x rays, gamma rays, and beta particles, generally only a small portion of the radiation energy is deposited. The absorbed energy produces a proportional amount of ionization in the emulsion. Hence, film provides a measure of the total ionization produced in the film over a period of time, and so film should be thought of as a type of time-integrating radiation detector.

After exposure, the opacity of a film emulsion is measured to determine the absorbed ionization. The opacity is measured by the amount of visible light transmitted through the film. The film density D is the ratio of incident and transmitted light, or I_0/I_t. Traditionally for most radiation absorption applications, this relationship is determined by an exponential expression, but this is not so for photographic film. Instead, the density is determined from the transmitted intensity on a base 10 logarithmic scale. Hence, the density of the image on film is generally expressed as

$$D = \log_{10}\left[\frac{I_0}{I_t}\right]. \tag{19.52}$$

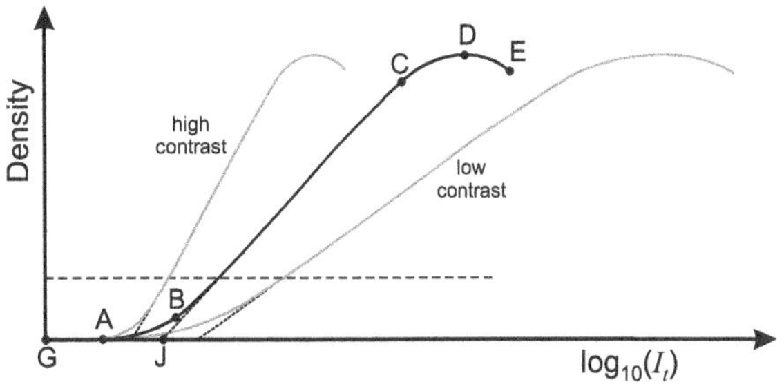

Figure 19.35. The characteristic, or H-D, curves for high, medium, and low contrast films depicting typical features. The dashed line indicates a location on the H-D curves all having the same density for different exposure intensities.

The density of unexposed film is 0, while film with a density capable of blocking 90% of light would have a density of 1. To achieve a density of 2, then 99% of light must be blocked, increasing to 99.9% for a density of 3, and so on. The response of film to ionizing radiation can be documented with the *characteristic curve* that is a plot of the film density versus $\log_{10}$ of the transmitted light intensity as depicted in Fig. 19.35. The characteristic curve was first reported and employed by Hurter and Driffield [1890] as a film characterization method.[15] The region between A and B in Fig. 19.35 is the *toe* of the curve, while the region between C and D is the *knee* of the curve. The region between B and C is a nearly linear function of density versus $\log_{10}(I_t)$ and is the most useful portion of the film characteristic. In the region between D and E, film overexposure causes the density to actually decrease because a latent image reverses from a negative image to a positive image, an effect referred to as a *solarized* image. The consequence for dosimetry is obvious, namely in this region a higher dose diminishes the density and results in an underestimate of the dose.

Contrast

The contrast of the film, denoted γ, is the slope of the characteristic curve region between points B and C, i.e.,

$$\gamma = \frac{D_1 - D_2}{\log_{10}(I_{t1}) - \log_{10}(I_{t2})}, \tag{19.53}$$

where D_1 and D_2 are any two different densities on the curve. High contrast films have $\gamma > 1$ while low contrast film has $\gamma < 1$. The implication is that high contrast film exhibits more sensitivity to small exposure differences than low contrast film. However, the dynamic range of exposure, referred to as the film *latitude*, is greater for low contrast film, that is, it takes more absorbed energy to push the density into the knee

[15]Sometimes the curve is referred to as the H-D curve after Hurter and Driffield.

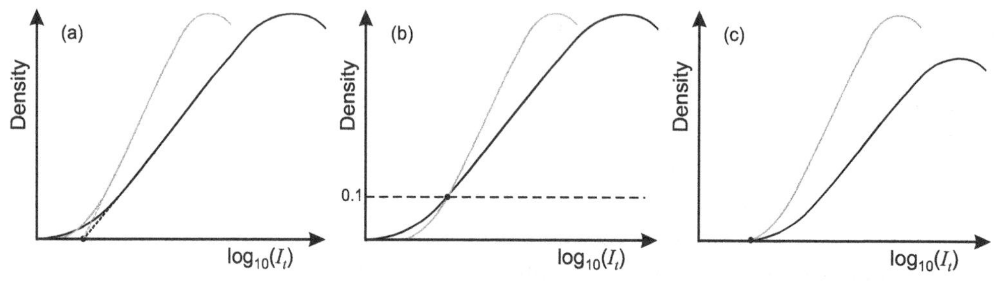

Figure 19.36. Ambiguity with early speed designations, showing H-D curves with (a) the same inertia but different latitudes, (b) the same intersection at D = 0.1, but with different latitudes, and (c) the same Scheiner speeds, but with different latitudes.

region. Although film has a characteristic latitude for the recommended development process, the contrast is also a function of developer type and development time. Hence, there is some degree of control over the film contrast through changes in the development process. Low contrast films often have a wide distribution of grain sizes with the large grains sensitive to low levels of light while the small grains require a greater exposure before accumulating enough stable Ag atoms to produce a developable image.

The Speed Index

The amount of absorbed energy that is required to produce an image is largely dependent on the speed of a film. Unfortunately, there are numerous speed indices for films. Three historically popular speed indices are depicted in Fig. 19.36. One system referenced the *inertia* of a film, which in Fig. 19.35 is the distance along the $\log_{10}(I_t)$ axis from point G (the origin) to point J where the linear portion of the characteristic curve, if projected, would intersect the abscissa. This speed rating presumes that a smaller inertia indicates a larger γ, but this is not necessarily true.

The Scheinergrade (Sch.) system was introduced by the German astronomer Julius Scheiner in about 1894. The Scheiner number is based on the minimum exposure threshold required to produce a latent image (point A in Fig. 19.35), but even this system for a speed index can produce ambiguous results, mainly because this index is based on an unused portion of the characteristic curve. The Scheiner number is reported in degrees between 1° (Sch.) to 20° (Sch.). A 19° increment produced a 100-fold increase in the sensitivity while a 3° increment almost doubles the sensitivity ($\sqrt[19]{100} = 2.06914... \simeq 2$).

The DIN (Deutsches Institut für Normung) speed index system, which replaced the Schneiner system in 1934, refers to the exposure required to produce an image density of 0.1 on the film. Similar to the Scheiner number, the DIN number is expressed in degrees, but the increment to 100-fold increase in sensitivity is 20° instead of 19°. Further, the system is based on a $\log_{10}$ scale so that the exposure increase is

$$S = 10^{\mathrm{DIN}/10}. \qquad (19.54)$$

A DIN increase of 3° would indicate a doubling in speed ($10^{3/10} = 1.9953... \simeq 2$). For instance, a film with DIN 27 would be double in speed to film of DIN 24. Again, this system does not necessarily produce an accurate depiction of γ. This system, originally developed for black and white film, was superseded in the 1980s by the ISO system.

Edward Weston and his father founded the Weston Instrument Corporation which made one of the first photographic light meters. One of their employees, William Goodwin, who designed the light meter also invented the Weston film speed index which was based on the exposure required to make the image density equal to γ. This system was introduced in the early 1930s. About the same time, General Electric introduced its own film speed rating system called the GE film values. However, both the Weston and GE systems were replaced by the ASA system in the 1960s.

Based on early work by Kodak and inspired by the Weston and GE film speed systems the American Standards Association[16] produced the ASA system in the 1950s. The ASA system follows the rule of reciprocity in which the amount of energy needed to produce a specific density on the film on the linear portion of the characteristic curve was inversely proportional to the film speed. Hence, film with rating 400 ASA would be twice as fast as 200 ASA and require half the exposure to produce equal density on the characteristic curve.

The International Organization for Standardization combined the qualities of the DIN and ASA system to produce the ISO number for speed, which uses the linear ASA number and the DIN number in degrees, denoted as ASA/DIN. For instance, a DIN number of 21 is equal to ASA 100 and ISO 100/21°. This system with several revisions is the current system used by most film manufacturers.

Example 19.1: Given a film package marked DIN 24, what is the equivalent speed in ASA and ISO?

Solution:
Noting that DIN 24 is 3° greater than DIN 21, we know that the film is doubled in speed. Hence

$$\text{ASA} = 10^{(24-21)/10} \times 100\,\text{ASA} \simeq 200\,\text{ASA}.$$

Also, the ISO designation would then be 200/24°.

Fast films usually have large grain sizes in the emulsion, while slow films have substantially smaller grain sizes. The advantage to the fast film is that they are useful for low exposure situations, but have the disadvantage of compromised resolution (graininess). The slower films require more exposure to produce a latent image, but have much higher image resolution.

Non-Linear Effects

Photographic film fails in low light to record an image whose density is linear with accumulated exposure, an effect known as *reciprocity failure.* A classic example of this effect is encountered in astrophotography in which celestial images take several minutes to hours for proper exposure. If the time between the reduction of individual Ag^+ ions is long compared to the dissociation time of a single Ag atom, then a latent image takes longer to form than predicted by reciprocity. Further, during the exposure time, the photographic emulsions are exposed to oxygen and possibly moisture in the air. The gelatin is hygroscopic and can absorb water moisture, thereby incorporating contaminants in the film emulsion. These contaminants act as oxidizing agents and compete for the isolated Ag atoms, thus causing them to decompose to Ag^+ before stabilizing with Ag clusters. Consequently, absorbed energy information is lost, and the law of reciprocity fails. In some cases, reciprocity failure in a photographic emulsion can be so severe that no additional amount of low-level light exposure increases the density of the latent image. In such situations an equilibrium condition is established in which the number of Ag^+ ions capturing electrons is equal to the number of Ag atoms losing an electron.[17]

Because the radiation environment is usually low for most radiation workers, the period between ionizing interactions is typically long compared to an astrophotograph exposure. However, Becker [1966] and Herz [1969] note that a single energetic electron interacting in the film most likely produces enough localized

---

[16] Now the American National Standards Institute or ANSI.

[17] Complicated processes were developed to reduce reciprocity failure with astronomical emulsions, including reducing the photographic emulsion temperature with dry ice or LN2, and drying the film in a pure hydrogen environment. The former process used an instrument named a *cold camera* and the process was named *hypersensitization.*

excited electrons to effectively produce Ag clusters. Hence, the reciprocity failure effect is less pronounced for this type of exposure than for exposures produced by low intensity and low energy (optical) photons in which one photon statistically produces a single silver ion.

As a consequence, low-intensity reciprocity failure is of little concern with most photographic films used for dosimetry. For high-energy gamma rays, the energy passed to photoelectrons or Compton electrons, as well as secondary ionization electrons, leads to the asymptotic minimum ionizing power of 380 eV per micron. Chassende-Baroz [1961] demonstrated that such a low ionization power can cause reciprocity failure for slow photographic emulsions.

Another form of reciprocity failure is caused by excessive exposure that propels the film density past the knee region of the characteristic curve, producing a solarized image of reduced density. Ehrlich and McLaughlin [1956] and Ehrlich [1956] show that the effect can be severe, and is a function of irradiation rate and dose. At high enough doses, x-ray films were shown to return nearly to zero density, as though the samples were barely exposed at all. The tested films reached the knee of the H-D curve at exposures between 10 R up to 200 R, depending on the irradiation rate (R s^{-1}).

Another source of dosimetric uncertainty is related to the variation in time of how the film is exposed to radiation, or the so-called intermittency. For low to medium exposures, the image density of the photographic film falls in the linear portion of the H-D curve. If the film is exposed to radiation intermittently over time, the density of the latent image increases additively with total exposure. In fact, it is this property that makes photographic film an important integrating dosimeter. However, for high total exposures generally exceeding 10 R, intermittent exposure can cause reciprocity failure and solarizes the film. Results from Ehrlich [1956] indicate that intermittent exposures can actually cause solarization at lower exposures than continuous exposures, even though the irradiation rates are identical during the exposure periods.

19.2.3 Film Dosimetry Badges

Personnel dosimetry film badges once were widely used, although TLDs and OSLDs have now replaced film in many applications. Regardless, photographic film emulsions can be used to measure dose from x rays and beta particles. If a neutron reactive foil is placed adjacent to the film, it can also be used for neutron detection. The film is contained inside a film holder or film badge which serves as a light proof, vapor proof envelope that protects the film from light, chemicals, and water moisture which would otherwise alter the response or fog the film.

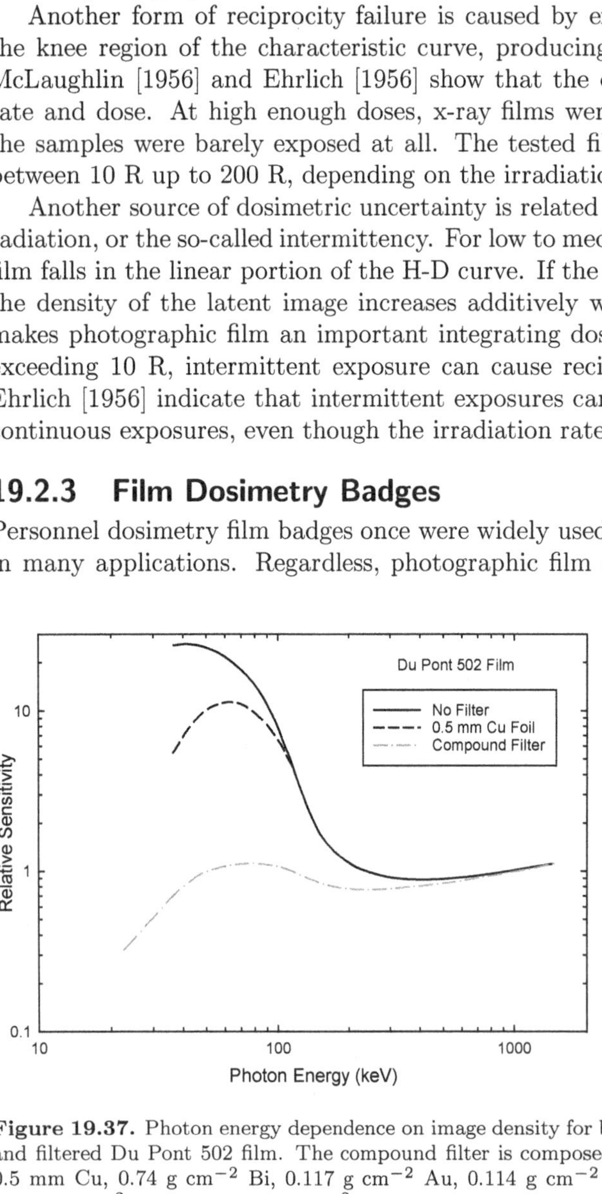

Figure 19.37. Photon energy dependence on image density for bare and filtered Du Pont 502 film. The compound filter is composed of 0.5 mm Cu, 0.74 g cm^{-2} Bi, 0.117 g cm^{-2} Au, 0.114 g cm^{-2} Ta, 0.148 g cm^{-2} Er, and 0.148 g cm^{-2} Gd. Data are extracted and adopted from Storm and Shlaer [1965].

Dosimetry film has two emulsion layers, one on each side of a backing layer. One layer is a fast large grain emulsion, while the other is a much slower small grain emulsion. Hence, the large grain emulsion is sensitive to low radiation exposures, while the smaller grain emulsion is sensitive to high radiation exposures. Should the fast emulsion be overexposed to radiation, it can be removed and the slower emulsion be used to interpret the dose.

Filters can be used to distinguish among different radiations. For instance, a shield can be used to block beta particles while allowing gamma rays to interact in the film. Also, filters can be used to flatten the energy response of gamma rays by attenuating a portion of low energy photons interacting in the film (see Fig. 19.37). With different attenuators over one or more films and with knowledge of the film energy response with each attenuator, a crude energy response to the total photon fluence can be calculated.

Photographic film has the advantage that it can be preserved as a permanent record of dose, with accurate response down to 100 mrem. However, disadvantages include the added processing with chemical development, data interpretation (densitometer readings), and the fact that doses less than 20 mrem are difficult to analyze. Low doses may be susceptible to reciprocity failure, a form of fading. Also, photographic film can be fogged if left in relatively high heat for too long.

The position of a film badge on the human body is important. Usually the badge should be worn on the torso, often clipped to a shirt pocket, a belt near the wasteline, or clipped to a lanyard around the neck. For persons working in a radiation area with their hands, such as a beam port or with laboratory sources, special rings loaded with photographic film can be used to measure dose to fingers and hands.

19.3 Track Detectors

Track detectors belong to a class of sensors that can provide visual evidence of radiation interactions. For instance, the cloud chamber that was introduced in Sec. 9.5.8 is a type of gaseous detector that produces visible ion tracks made by charged particles. In fact, discussion of the cloud chamber could have been deferred to this chapter, but because it operates with a cooled saturated vapor and often has a voltage applied to counter gravitation effects, the authors chose to include it in the chapter on ion chambers. There are numerous other types of track detectors, many described here, all providing visual evidence of the passage of radiation through the detector.

19.3.1 Nuclear Track Emulsions

For a photographic film to be used as radiation dosimeters, it is assumed that the film is uniformly exposed to a radiation field so as to produce a uniform density over the developed film. However, it is also possible to detect the passage of individual radiation particles from the tracks they make in the emulsion. Nuclear track emulsions are a special type of film with a higher density of nanoparticle silver halide (AgBr) crystal grains. The silver halide density can be up to 82% by weight with grain sizes of about 300 nm. To increase the probability that incident radiation particles or reaction products produced by interactions in the emulsion are completely captured in the film, these emulsions may be up to 600 microns thick. Individual tracks are produced by high LET particles, and the developed tracks are tediously visually counted through a microscope. Yet, much information can be extracted from the tracks, including trajectory, energy, and LET. Examples of alpha particle tracks in film are shown in Fig. 19.38 for film that has been permeated with fine thorium and uranium dust.

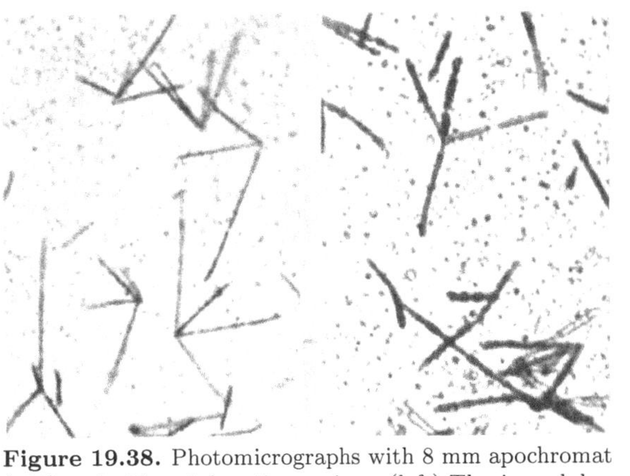

Figure 19.38. Photomicrographs with 8 mm apochromat (x20) and x10 periplanatic eyepiece. (left) Thorium alpha-particle stars; (right) similar tracks after uranium intensification. Reproduced from Powell et al., *J. Sci. Instrum.*, **23**, 102–106, (1946). Copyright IOP Publishing. Reproduced with permission. All rights reserved.

Several charged-particle ranges as a function of energy and mass are shown in Fig. 19.39 for a specific nuclear emulsion (Ilford C2). Except at low energies below 1 MeV, particle ranges are almost proportional to the initial energy of the particle. Note also that 1-MeV particles have ranges of 2 to 15 microns.

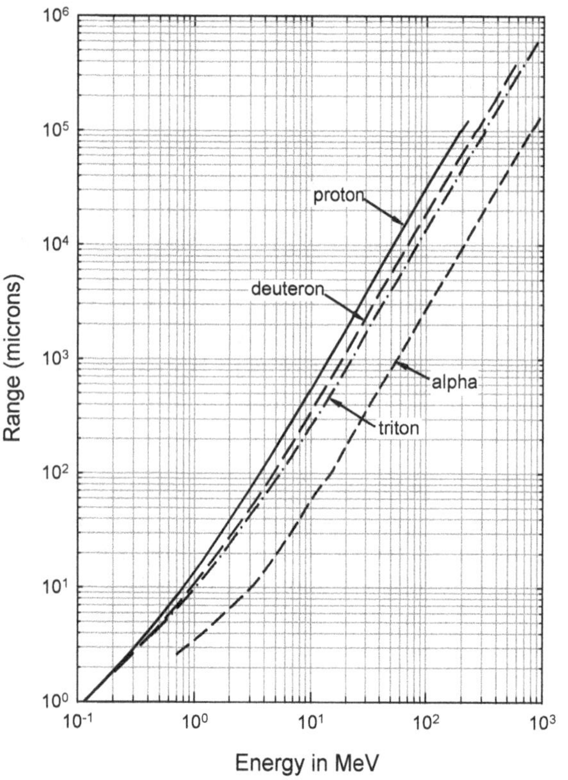

Figure 19.39. The range-energy curves for protons, deuterons, tritons, and alpha-particles in Ilford C2 photographic emulsion. Data are extracted and adopted from Vigeron [1953].

At higher energies, generally above 20 MeV, emulsions exceeding 600 microns are needed, a thickness that can be realized by stacking several film emulsions together [De Serio et al. 2005]. Such an emulsion stack allows the entire track of a particle to be followed. Also, because the intrinsic interaction efficiency can be low for a single emulsion layer, use of a stack of emulsion layers, therefore, increases the detection efficiency.

Each film layer is called a *pellicle*, and it records the track of a particle as it traverses the emulsion of the pellicle. The stack layers are pressed together to form a 3-dimensional block of emulsion and may consist of several hundred pellicles. Because of non-uniformities across the emulsion surface, the pellicles must be pressed together with a uniform pressure, recommended as 44 lbs per square inch [Barkas 1963]. At room temperature, with the relative humidity kept below 60%, the pellicles should not stick together. Each pellicle is indexed so that, after the pellicle stack is developed, the track image sequence can be followed through the different levels.

To develop a stack, the pellicles are separated and all are developed simultaneously with or without a backing. The simultaneous developing ensures that each pellicle experiences the same development conditions. If developed without backing, each pellicle is suspended in a frame protected by a wire mesh on each side. Development of a pellicle without a backing produces a more uniform result throughout the emulsion than if a backing were used, and the developing time is shortened because development proceeds from both sides of the emulsion. Further, unequal emulsion swelling, a consequence of attaching pellicles to a glass backing before development, is greatly diminished. Also, the ionization tracks can be viewed from both sides of a pellicle without backing. However, pellicles without backings seldom return to their original dimensions after fixing and drying and this can distort the actual length and geometry of the tracks. Free pellicles must be handled with great care, and it is possible that they adhere to each other or other surfaces during the processing. Finally, a free developed pellicle usually does not lay flat, and so may still require adhesion to a stiff backing.

If the emulsion is developed after attachment to a glass plate backing, the film receives mechanical support and can be developed in special processing racks. However, during the preparation process that requires washing of the pellicles, water can be trapped and distort the image. Further, unequal shrinkage can alter the image. An indexing grid printed on the pellicle can be used to correct distortions. Also, it is usual for the emulsion thickness to shrink after the development and drying process. This shrinkage changes the measured track length from that in the original emulsion. However, irradiation with x rays or other ionizing particles, shadow shielded to enter a select location at a 45° angle, can be used to determine the thickness change [Barkas 1963]. The shrinkage factor can be determined from

$$S = \frac{lD}{Ld}, \tag{19.55}$$

where the variables l, L, d and D are defined in Fig. 19.40.

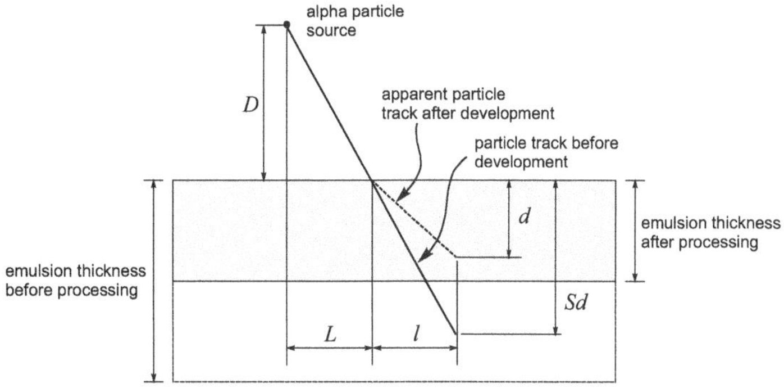

Figure 19.40. Determination of shrinkage for nuclear emulsions. After Becker [1966].

Nuclear track emulsions can be used to study nuclear reactions because a reaction in the emulsion preserves the energy deposition as a function of position. Measurements of these tracks can be conducted by visual inspection under a microscope, but automated processes relieve this tedious method. The track angle into the emulsion is important to correct for the foreshortened projected length. Emulsions are used to measure the track length of a particle, the angle and curvature of a particle, scattering angles, star formation, and various other reaction product results. An example of a nuclear reaction captured on film is shown in Fig. 19.41. When exposure is conducted in a magnetic field, the curvature of the track yields charge and mass. Star formation is an indicator of a reaction, with the resulting reaction products emitted in various directions. The star formation products may be simultaneous, or the result of decay chains. Neutrons can be measured by way of recoil protons from the water in the emulsion. Also, reactive materials can be incorporated in the emulsion to enhance specific reactions of interest (^{10}B, for instance). Photographic track film was used to discover a number of subatomic particles, including the π^+ and π^- mesons (pions),[18] the K^+ and K^- mesons (kaons), the Σ^+ baryon, and the Λ^0 baryon.

19.3.2 Track Etch Detectors

In 1958, Young reported on a technique of etching out the damaged tracks left behind by heavy ions penetrating a solid material. The substrate under investigation was a LiF crystal bombarded by fission fragments. After etching in a solution of hydrogen fluoride (HF) and water, pits appear where the fission fragments damaged the crystal. A few years later, Price and Walker [1962] reported on a similar experiment with mica that was irradiated by recoil ions from spallation reactions, also etched in hydrofluoric acid. Shortly thereafter, Fleischer and Price [1963] used the method to investigate damage tracks in polymers, which they noted could be applied to radiation dosimetry. This technique is recognized as an important alternative detection method for ionizing radiation dose, and is referred to as *track etching* [Fleischer et al. 1965a].

Fleischer et al. [1972] point out that the technique of particle track counting is simple, the result produces a straightforward permanent record (a hole), and there is a natural sharp discrimination between particles that leave tracks and those that do not. Track etch measurements are useful in radiation dosimetry, as well as other applications, including dating of geological samples and archaeological artifacts, quantitative chemical analysis for fissionable nuclei or nuclei susceptible to (n, α) reactions, uranium concentrations in water and other substances, the measurement of diffusion coefficients in metals, the production of uniform pores in filters, and many other applications as listed elsewhere [Fleischer et al. 1972; Becker 1973; Fleischer et al. 1975].

Track Formation

Energetic ionizing particles that interact in a substance can produce a dense track of material damage, an effect that can be observed in organic and inorganic materials. There are at least four models used to describe

[18]Cecil Frank Powell discovered the pion by photographic means in 1946, and was awarded the Nobel Prize in Physics in 1950 for the development of the photographic method to study nuclear processes and the discovery of the pion.

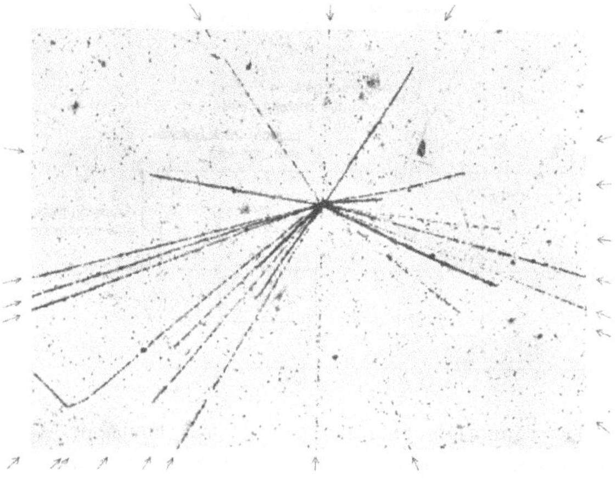

Figure 19.41. Mosaic of photo-micrographs of the 'explosive' disintegration of the nucleus of a silver atom by a cosmic-ray particle of great energy. The tracks of twenty-five fragments can be distinguished under the microscope, including those of protons, α-particles, and heavier nuclei. Many of the tracks 'dip' steeply and end in one of the surfaces of the emulsion. Reproduced from Powell and Occhialini, *Nuclear Physics in Photographs*, Oxford, Clarendon, 1947; by permission of Oxford University Press.

the observed tracks in inorganic solids, namely the *ion-explosion spike model*, the *atomic displacement model*, the *thermal spike model*, and the *total energy loss model*. The latter three models may contribute some to the damage formation, but Fleischer et al. [1975] argue that they are ultimately improbable explanations of the observed results.

In the displacement model track formation is due to elastic collisions with the material lattice, which causes a column of displacements along the ion path. This model predicts that damage paths should appear more prevalently toward the end of the particle range and that these paths should appear equally for nearly all materials. However, the theory fails to explain the observed results for most ion tracks in materials. Fleischer et al. [1975] point out that there are some special cases where ion displacement does participate as an important mechanism for track formation, mainly for heavy low-energy particles and heavy recoil fragments from alpha particle decay.

The thermal spike model is based on the assumption that intense ionization and excitation of electrons along the ion trajectory causes rapid heating of the material. This heat is initially confined to a small cylindrical core around the ion path, which is then rapidly dissipated and cooled by conduction to the surrounding material lattice. It is proposed that the action of rapid heating and cooling produces distortions and disorder in the material lattice [Seitz 1949; Bullough and Gilman 1966]. However, the model fails to explain observed results, and Fleischer et al. [1975] discard this model as an explanation of track formation.

The total energy loss model is based on the stopping power equations of Bethe and Bloch (from Chapter 4). This model assumes that the ion tracks form above a critical stopping power $S(E) = -dE/dx$, which depends on the radiation particle, its energy, and the target material [Fleischer et al. 1964]. Below this critical dE/dx, the tracks become distorted and eventually disappear at sufficiently low particle energies. However, Fleischer et al. [1965] also discard this model in favor of the ion-explosion model, mainly because the total energy loss model fails to predict observed results.

The ion-explosion spike model maintains that a heavy ion of sufficient energy causes the dense removal of electrons from host nuclei by Coulombic attraction. Because of the large mass differences, the electrons are easily removed and ionized, while the heavy ion trajectory remains unchanged along a straight path. The remaining dense concentration of positive charge left behind causes repulsion along a cylindrical core, thereby producing lattice distortions usually on the order of several atomic spacings [Fleischer et al. 1965b; Fleischer et al. 1975; Nordlund et al. 1998]. Ultimately, the end result is the production of a strained lattice with vacancies and interstitials produced along the ion path. These weakened regions etch in caustic chemicals at a faster rate than undamaged regions, and consequently produce a residual visible track. There are various other theories for track formation [Fleischer et al. 1975], but it is the ion-explosion model that remains the favored explanation. Price and Fleischer [1971] describe a track formation metric J that depends on the particle's atomic number Z and its relative speed $\beta = v/c$ where c is the speed of light in vacuum, namely

$$J = \frac{\alpha Z_e^2 \left[\ln\left(\beta^2/(1-\beta^2)\right) + K - \beta^2 - \delta(\beta)\right]}{\beta^2}, \tag{19.56}$$

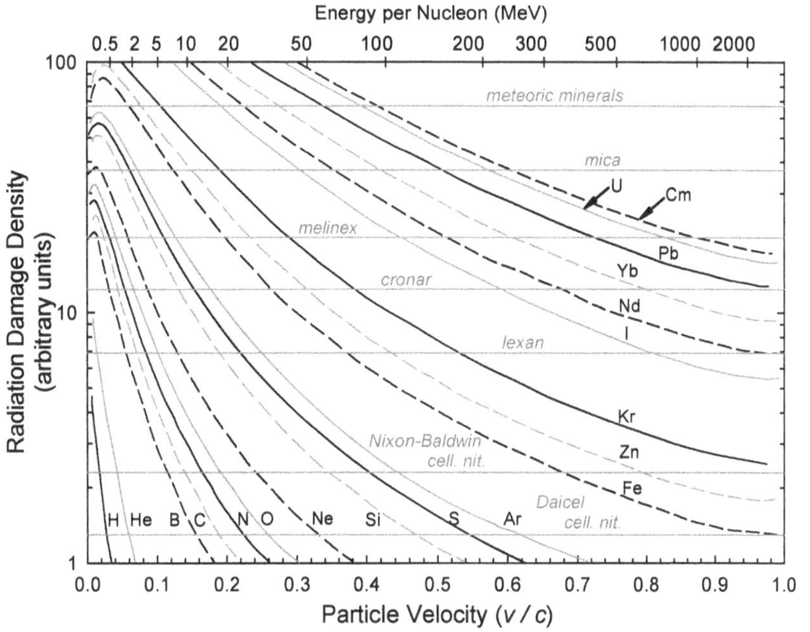

Figure 19.42. The relative density of radiation damage as a function of velocity for various nuclei. Horizontal lines indicate the approximate threshold required for several materials before a track forms. After Price and Fleischer [1971].

where K and α are constants, Z_e is the effective charge of the projectile ion, and $\delta(\beta)$ accounts for relativistic corrections. For $\beta \lesssim 0.8$, the term $\beta(\delta)$ is zero. The constant K is chosen experimentally with the boundary condition that the etch rate depends solely on J (within experimental error). Shown in Fig. 19.42 are a set of curves, developed from Eq. (19.56) with $K = 16$. Also shown are experimentally established ionization thresholds for a variety of materials, showing the dependence on ion mass. Price and Fleischer [1971] note the actual value of K changes with the type of track material with K being higher for organic materials such as Lexan than for glass and minerals such as flint glass. Just as fission fragments capture ambient electrons as they slow (see Sec. 4.5.6), the effective charge Z_e of a heavy ion changes, decreasing as the ion loses energy and captures ambient electrons, an effect that is included in Fig. 19.42.

Visible tracks in organic polymers are apparently related to delta-ray electrons in the specific channel region of the ion trajectory (see Fig. 19.43). The local dose in the ion track channel can exceed several Mrad, caused mainly by these delta-ray electrons [Becker 1973]. Further, measurements indicate that the bulk etch rate of plastics can increase with gamma-ray and electron irradiation [Goland and Mateosian 1973]. Hence, the formation of tracks in organic polymers is mainly attributed to a radiochemical damage mechanism. The etchable tracks are formed by radiolytic scission of polymer chains into short fragments. The resulting lower molecular weight fragments are far easier to dissolve than the surrounding undamaged material. Consequently, etched pits appear as these localized regions etch faster than the surrounding undamaged plastic.

Track Etching

The damage region from a heavy ion is approximated as a cylinder with a diameter of approximately 5 nm (50 Å). The highest damaged region is at the center of the track with the damage becoming less with increasing

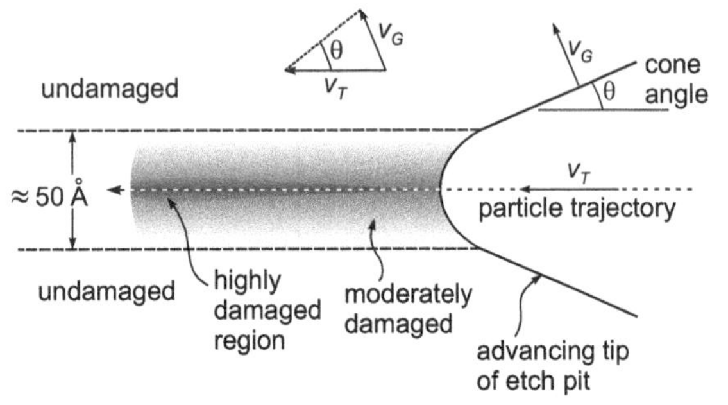

Figure 19.43. Ion damage track and etching on a submicroscopic scale. After Henke and Benton [1971].

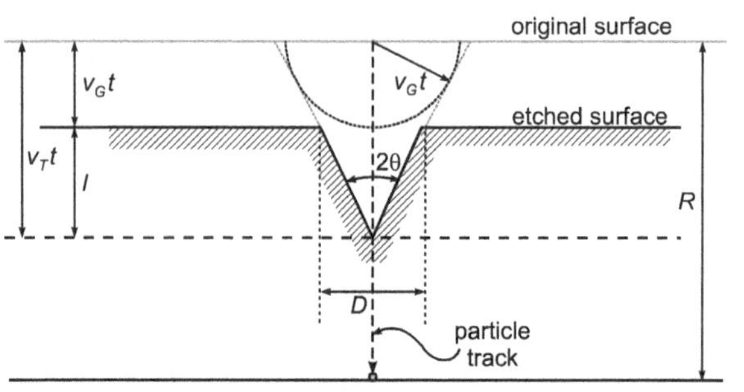

Figure 19.44. Track geometry with constant V_T and V_G for a vertical incidence. After Price and Fleischer [1971].

radial distance from the track axis to the track edge, beyond which the substrate remains undamaged (see Fig. 19.43). The substrate in which the tracks occur and which subsequently are made visible by etching is often referred to as a *track etch foil* or simply foil.

The etch rate in the track center is denoted as v_T while the etch rate of the undamaged substrate is denoted as v_G. Because the damaged region diminishes radially from the track core, the etch rate in the track region varies between these two extremes from the track axis to the edge of the track. However, for the sake of simplicity, consider the case in which the etching rate within the track is constant at v_T. With the notation shown in Fig. 19.44, the depth of the substrate etched away, within time t, from the original substrate surface is $v_G t$ while the depth etched along the ion track is $v_T t$. For an ion trajectory normal to the track substrate, the net depth l of the etched track is

$$l = v_T t - v_G t, \tag{19.57}$$

in which some of the original ion track is etched away during the chemical etching process. The diameter of the track at the widest point, after etching, is

$$D = 2v_G t \left[\frac{v_T - v_G}{v_T + v_G}\right]^{1/2}. \tag{19.58}$$

Note that both Eq. (19.43) and Eq. (19.44) vanish if $v_T = v_G$.

The ratio of the two etch rates is

$$\frac{v_T}{v_G} = \frac{2}{D}\left[\left(\frac{D}{2}\right)^2 + l^2\right]^{1/2} = \csc\theta. \tag{19.59}$$

Note also, by simple trigonometric manipulations [Price and Fleischer 1971]

$$v_G t = \frac{D}{2l}\left[\frac{D}{2} + \left[\left(\frac{D}{2}\right)^2 + l^2\right]^{1/2}\right] = \frac{D}{2}\left[\tan\theta + \sec\theta\right], \tag{19.60}$$

and

$$v_T t = \left[\left(\frac{D}{2}\right)^2 + l^2\right]^{1/2}\left[\frac{D}{2l} + \frac{1}{l}\left[\left(\frac{D}{2}\right)^2 + l^2\right]^{1/2}\right] = \frac{D}{2}\csc\theta\left[\tan\theta + \sec\theta\right]. \tag{19.61}$$

If the etch is allowed to progress for too long, the track broadens and eventually becomes undiscernible from background pits. This is especially true if the depth $v_G t$ becomes greater than the original ion range R.

For track orientations at angles less than 90° to the target surface, the conic section that intersects the surface forms an ellipse instead of a circle (Fig. 19.45). The actual track position becomes displaced from the center, more so as angle ϕ decreases. Also, as ϕ decreases towards the angle θ, the major axis of the ellipse increases towards infinity. With the notation of Fig. 19.45, the geometric relations are

$$\phi = \frac{1}{2}\left[\arctan\left(\frac{r}{a+p}\right) + \arctan\left(\frac{r}{p}\right)\right], \quad (19.62)$$

and

$$\theta = \frac{1}{2}\left[-\arctan\left(\frac{r}{a+p}\right) + \arctan\left(\frac{r}{p}\right)\right]. \quad (19.63)$$

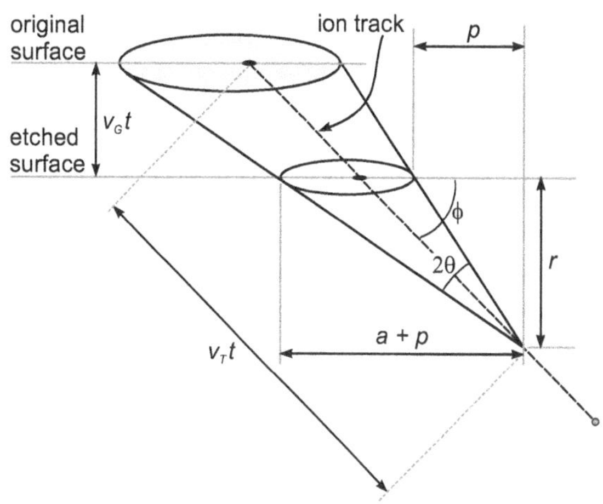

Figure 19.45. Track geometry with constant V_T and V_G for angled incidence at ϕ. After Henke and Benton [1971] and Price and Fleischer [1971].

Note that the vertical track etch depth becomes less than the substrate etch depth if

$$v_G > v_T \sin\phi \quad (19.64)$$

so that a visible etch pit does not form. Hence, no etch pit forms for angles less than the critical angle, defined by

$$\phi_c = \arcsin\left(\frac{v_G}{v_T}\right). \quad (19.65)$$

Thus to be useful as a track etch material, the track etch speed v_T should be much greater than the etch speed of the substrate v_G. As v_T becomes greater than v_G, the etch pit contrast improves while also the critical angle ϕ_c decreases. Goland and Meteosian [1973] showed for organic polymers that the ratio of etch rates v_G/v_T decreases as the total sample gamma-ray dose increases, ultimately becoming unity at a high enough dose. Under such a condition, where $v_G/v_T = 1$, no etch pits form; however, the threshold to enhance bulk etchability is relatively high, quoted to be above 10 Mrad [Attix 1986].

There are a multitude of etch formulas for the various track etch materials, several of which are listed in Table 19.2. The etch rate can be altered by changing the solvent concentration and also varies significantly with the foil type. Although the general trend is an increase in etch rate with increasing solvent concentration, not all foils exhibit this dependence. For instance, the etch rate for cellulose nitrate does not increase for concentrations above 6N NaOH solutions and remains relatively constant for higher concentrations [Blanford et al. 1970]. Other polymers exhibited increasing etch rates with solvent concentration, but with varying concentration dependences. Not surprisingly, periodic agitation or stirring of the etchant solution also increases the etch rate [Becker 1973], most likely a benefit from the physical removal of solute which, otherwise, would obstruct the fresh polymer surface.

Besides the etchant concentration, other environmental factors can have significant influence on the etch rate and foil sensitivity. For instance, increasing the etchant solution temperature can significantly increase the etch rate, and the user should be careful to terminate the process well before the tracks disappear. There are experimental results that showed a dramatic increase in sensitivity, i.e., number and size of pits, for some organic foils when irradiated in dry oxygen as opposed to irradiation in a vacuum or nitrogen gas

Table 19.2. Etching conditions for several track etch materials. From Becker [1973], Fleischer et al. [1975], and associated references.

Detector Material	Etchant	Time	Temperature (°C)
	Minerals		
lepidolite mica	15% HF	20 sec	50°
	48% HF	3-70 sec	23°
muscovite mica	20% HF	2 h	23°
	20% HF	12 m	52°
	15% HF	20 m	50°
	48% HF	10-40 m	23°
phlogopite mica	20% HF	5 m	23°
	15% HF	1 m	50°
	48% HF	1-5 m	23°
	Glasses		
Quartz	48% HF	24 h	23°
Fused Silica	48% HF	1 m	23°
Borate Glass	H_2O	1 m	23°
Obsidian	48% HF	30 s	23°
Phosphate Glass	48% HF	5-20 m	23°
Soda Lime	48% HF	5 s	23°
	5% HF	2 m	23°
	Organics		
Cellulose Acetate	28% KOH	30 m	60°
Cellulose Acetate Butyrate	6.25N NaOH	12 m	70°
Cellulose Nitrate	6.25N NaOH	2-4 h	23°
Cellulose Propionate	28% KOH	100 m	60°
Cellulose Triacetate	28% KOH	60 m	60°
Cormophenol	NaOH	1 h	40°
Poly-Allyl-Diglycol Carbonate (CR-39)	7M NaOH†	24 h	70°
Polycarbonate	6.25N NaOH	20 m	50°
Polyethylene	10 g $K_2Cr_2O_7$		
	+35 ml 30% H_2SO_4	30 m	85°
Ionomeric Polyethylene	10 g $K_2Cr_2O_7$		
	+35 ml 30% H_2SO_4	1 h	85°
Polyethylene Terephthalate	6.25N NaOH	10 m	70°
(PET, Melinex®, Cronar®)			
Polyimide	25%(aq) $KMnO_4$	1.5 h	100°
Polymethylmethacrylate	sat. $KMnO_4$	50 m	85°
Polyoxymethalene	5% $KMnO_4$	10 h	60°
Polyphenoxide	25%(aq) $KMnO_4$	4 m	100°
Polystyrene	sat. $KMnO_4$	2.5 h	85°
Polyvinylacetochloride	25%(aq) $KMnO_4$	30 m	100°
Polyvinylchloride	sat. $KMnO_4$	2.5 h	85°
	25%(aq) $KMnO_4$	2 h	55°
Polyvinyl Toluene	(sat aq) $KMnO_4$	30 m	100°
Silicone-Polycarbonate Co-polymer	6.25N NaOH	20 m	50°
Siloxane-Cellulose Co-polymer	8N NaOH + Benax	3 h	85°

†Recommended by Kodaira et al. [2016].

Table 19.3. Relative sensitivity of various track etch materials. From Fleischer et al. [1975].

Detector Material	Atomic Composition	Least Ionizing Observable Ion
	Minerals	
Hypersthene	$Mg_{1.5}Fe_{0.5}Si_2O_6$	100 MeV ^{56}Fe
Diopside	$CaMg(SiO_3)_2$	170 MeV ^{56}Fe
Augite	$CaMg_3Fe_3Al_2Si_4O_{19}$	170 MeV ^{56}Fe
Oligoclase	$Na_4CaAl_6Si_{14}O_{40}$	4 MeV ^{28}Si
Bytownite	$NaCa_4Al_9Si_{11}O_{40}$	4 MeV ^{28}Si
Orthoclase	$KAlSi_3O_8$	100 MeV ^{40}Ar
Muscovite Mica	$KAl_3Si_3O_{10}(OH)_2$	100 MeV ^{40}Ar
	Glasses	
Flint Glass	$18SiO_2$:$4PbO$:$1.5Na_2O$:K_2O	2-4 MeV ^{20}Ne
Quartz	SiO_2	100 MeV ^{40}Ar
Silica Glass	SiO_2	16 MeV ^{40}Ar
Soda Lime Glass	$23SiO_2$:$5Na_2O$:$5CaO$:Al_2O_3	20 MeV ^{20}Ne
	Organics	
Amber	$C_2H_3O_2$	Fission Frag.
Bisphenol A-polycarbonate (Lexan™, Makrofol®)	$C_{16}H_{14}O_3$	0.3 MeV ^4He
Cellulose Nitrate (Daicell®)	$C_6H_8O_9N_2$	0.55 MeV ^1H
Poly-Allyl-Diglycol Carbonate (CR-39)	$C_{12}H_{18}O_7$	13 MeV $^1H^{\dagger}$
Polyethylene	CH_2	Fission Frag.
Ionermeric Polyethylene (Surlyn®)		36 MeV ^{16}O
Polyimide	$C_{11}H_4O_4N_2$	36 MeV ^{16}O
Polyoxymethylene (Delrin®)	CH_2O	28 MeV ^{11}B
Polypropylene	CH_2	1 MeV ^4He
Polymethylmethacralate (PMMA, Plexiglas®, Lucite®)	$C_5H_8O_2$	3 MeV ^4He
Polyvinylacetochloride	$C_6H_9O_2Cl$	42 MeV ^{32}S
Polyvinylchloride + Polyvinyledene Chloride	$C_2H_3Cl + C_2H_2Cl_2$	42 MeV ^{32}S

†Lounis et al. [2001].

environment [Becker 1968]. Becker [1968] also found that exposure to water before irradiation increases the etching speed of the solvent and the number of visible tracks, and is especially pronounced if the track etch foil is first soaked in a H_2O_2 solution before irradiation. Heinzelmann and Haschke [1971] also observed an increase in sensitivity when the track etch foils are soaked in ethanol/ammonia before the etching process.

Another method shown to increase sensitivity is to expose the track etch foil to specific wavelengths of UV light in the presence of oxygen before conducting the etching process [Henke et al. 1970]. This improvement in sensitivity was found to increase with shorter wavelengths of the sensitizing light. The etch rate also increases in the presence of an electric field [Tommasino 1970; Tommasino et al. 1981; 1982], with methods and results reported for various track etch foils [Somogyi 1977; Tommasino 1981; Sohrabi 1981]. Other factors that can affect the etch rate are bake hardening and foil aging, both methods causing a reduction in size and number of visible etch tracks [Becker 1973]. A probable cause for the reduction in etch pit density is thermal annealing of track damage, which causes tracks made by low LET radiation to disappear before those made by high LET radiation [Attix 1986].

Track Etching for Neutron Dosimetry

The high threshold gamma-ray dose required for track formation makes track etch foils impractical for gamma-ray dosimetry. Further, heavy charged particles have relatively short ranges in air. Combined with energy attenuation through the dosimeter package, track etch detectors are a poor choice for heavy charged-particle dosimeters. Although beta particles may penetrate the encapsulate, it takes high doses before notable changes occur in the track etch foil [Becker 1973]. However, track etch detectors have utility in fast and thermal neutron detection, mainly through absorption and scatter reactions. The relative insensitivity to other radiation interactions is a natural discrimination benefit. Neutron-induced track dosimetry is performed by (a) first using energy conversion radiator foils to produce energetic neutron-reaction ions that subsequently enter the track etch foil, or (b) producing neutron-reaction ions from elastic scattering in the track etch foil itself.

There are a few clear advantages of track etch foils for neutron dosimetry. First, as already stated, they naturally discriminate background radiations. Second, after etching, they provide a relatively permanent dose record, and are largely unaffected by the environment if kept at room temperature (no fading). The development process is straightforward and can be automated, including the readout, which can also yield highly sensitive measurements. Aside from the obvious visual method, many readout methods have been developed, including automated optical, radiation, and spark transmission methods [Becker 1973; Fleischer et al. 1975]. For track etch foils coupled to radiator foils, the dose range is adjustable according to the density of the fissile element. Finally, the sensitivity can be functionally reduced by separating the radiator foil from the track foil.

Many materials have trace amounts of uranium, which can produce fission fragments from fast or slow neutrons. Fissions from slow neutrons depend on the presence of ^{235}U, which has a much higher fission cross section than does ^{238}U but has a significantly lower natural abundance. Hence, the track density depends on the atomic density and the absorption cross sections of the fissionable nuclides [Becker 2003]. It is by this means that many of the natural minerals are used for geological age and historical dose calculations, and many useful minerals are listed in Tables 19.2 and 19.3.

Personnel neutron dosimetry can be performed by outfitting users with a track etch badge. The badge has track etch foils coupled to radiator foils, often consisting of two samples, one with ^{232}Th for fast neutrons and the other with ^{235}U for slow neutrons. Other candidate converter foils include ^{237}Np, ^{239}Pu, and ^{32}S, all having different cross sections and threshold energies. The $1/v$ absorbers ^{10}B and ^{6}Li can also be used as converter foils. By combining different foils with separate track etch dosimeters, a crude estimation of neutron dose as a function of neutron energy can be determined.

Track Etch Materials

Cellulose nitrate was once a favorite for nuclear track etch dosimeters, mainly because it is relatively sensitive to low LET ions (see Table 19.3). Direct fast neutron interactions in cellulose nitrate produce energetic H, C, N, and O recoil ions. The recoil protons do not produce etchable tracks (unless the energy is below 550 keV), yet tracks from the other recoil ions can be resolved. Recall from Sec. 18.3 that fast neutrons transfer only a small fraction of their energy to the recoiling heavy ion, the fraction increasing as the neutron scattering angle increases. Consequently, the energy of these recoil ions is much smaller than that of the initial neutron energy and the track lengths the ions travel are relatively small. However, the visible size of the tracks can be enhanced by electrochemical etch methods (mentioned above). Recall from Sec. 19.2 (footnote 12) that cellulose nitrate (nitrocellulose) is explosively flammable, and has been largely replaced with cellulose acetate as a track etch material. Listed in Table 19.3 are numerous other track etch materials that have utility in dosimetry.

Despite the success of many organic foils, one that stands out as a most useful track etch material for neutron dosimetry is the polymer poly-allyl-diglycol carbonate ($C_{12}H_{18}O_7$ or PADC), or by the more common name, CR-39.[19] It has a relatively low-energy threshold and is capable of forming etch pits from 13 MeV protons [Lounis et al. 2001], a capability significantly better than that of cellulose nitrate and cellulose acetate. Attix [1986] points out that a direct correlation between radiation dose and the track density from proton recoils is not actually possible, although superficially it may seem so. The main reason for the discrepancy, given by Attix [1986], is that protons of high enough energy do not produce etchable tracks until their energy is reduced below a threshold.[20] Consequently, the track does not correctly represent the total amount of energy dissipated in the foil. There is also a low energy threshold, below which ions are undetectable. Although the specific ionization may be high, the small residual track is not visible when etched [Cross 1986]. When etched by electrochemical assistance at 25 kV cm^{-1}, the practical lower energy limit is approximately 50 keV for protons in CR-39.

19.3.3 Spark Chambers

A moving charged particle produces a column of electron-ion pairs in a gas medium. In the earlier discussion on gas filled proportional and Geiger-Müller counters, it was seen that electrons in a high electric field were accelerated by the field and could produce even more ionization and excited atoms that ended in a Townsend avalanche. When the excited atoms deexcite, they can give off visible light which can, in turn, produce further ionization. This light-producing phenomenon can also be used to create a different type of particle detector. Such an application was first reported by Greinacher [1934] who described the detection of alpha-particles with a point electrode near a planar surface. Years later, Chang and Rosenblum [1945] devised a wire spark counter that accomplished the same task, while eliminating many of the problems associated with Greinacher's original design [Payne 1949; Eichholz 1952]. Their device had a single wire stretched parallel to a metal plate. In 1949 Keuffel devised a counter with parallel metal plates (either Mo or Cu) spaced 2.5 to 3 mm apart. The surface areas for these parallel plate detectors ranged up to 35 cm^2 in order to create a large electrically active region.[21] These types of devices are conventionally named *spark counters*. Although early versions of spark counters operated in air, pure noble gases are preferred with a mixture of He and Ne frequently being used [Rice-Evans 1974].

Wire grids or plates stacked parallel to each other can be used to record the tracks of high energy particles over a very large volume. Such a structure is called a *spark chamber*. One example is shown in Fig. 19.46. The wire grid design allows charged particles to pass through the gas while encountering far less solid material than in the parallel plate design. These devices had multiple stacked planes of wires or plates so as to produce a large detection volume [Henning 1957]. Early reviews of these chambers were published by Roberts [1961] and Higenbotham [1965]. A neutron sensitive parallel plate spark chamber (no wires) which used corrugated metal sheets for the plates was reported by Eicholz [1966; 1969].

Suppose that the high electric field is produced between two parallel plates, with a gas between them, and an ionizing particle can pass relatively perpendicular to the inner surfaces. In a parallel plate configuration, the electric field is relatively constant for an applied voltage, and liberated electrons can produce avalanches along the entire particle path. As the avalanche grows, the build-up of space charge shields the inner-most charge pairs from the externally applied potential, while the outer most electrons and ions are rapidly drifted towards their respective electrodes. Hence, the outermost electrons gain energy and produce more ionization. Rice-Evans [1974] notes that a single avalanche must produce at least 10^8 free electrons as a condition for an arc to form. As the charges drift in the gap, these avalanches can combine to produce a relatively dense

[19] Columbia Resin number 39.

[20] Attix [1986] puts this lower energy limit at 4 MeV; however, the work of Lounis et al. [2001] indicates that it can be as high as 13 MeV in CR-39.

[21] Roberts [1961] notes that Keuffel [1949] was the first to report a parallel plate spark counter, but this work was largely unnoticed. Not until several others published similar works did it become recognized as important.

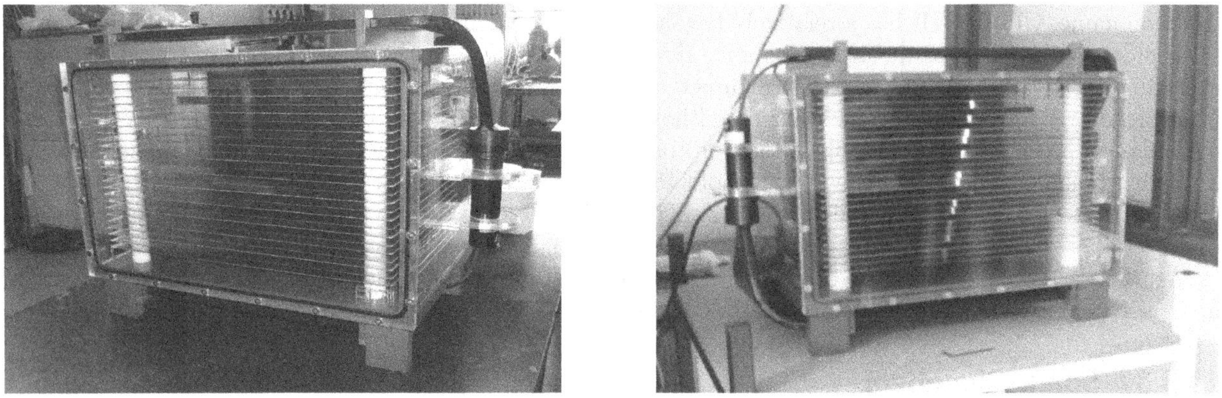

Figure 19.46. (left) A parallel plate spark chamber. (right) A cosmic ray track in a spark chamber. These spark chambers were manufactured and demonstrated by the Physics Department at the University of Coimbra, Portugal.

columnar track of electron-ion pairs creating a low resistance track between the plates so that electrons and ions can move freely between the plates much like a lightning strike. The excited ions produce light upon returning to their ground state, light which is observed as an arc or spark discharge. These stages in the development of a spark are illustrated in Fig. 19.47. The voltage across the gap is usually too low to cause coronal discharge, but high enough to produce an arc if an ionizing particle crosses the gap.

It is notable that the basic spark counter structure is similar to a simple Geiger-Müller counter. However, the data recording method is different and much faster for the spark counter than the Geiger-Müller counter. Recall that the rise time and collection time of a GM Counter is relatively slow, usually having resolving times between tens to hundreds of microseconds. The process that produces a spark, however, is significantly faster and does not rely on charge collection. In fact, spark counter responses can be on the order of 10 ns [Roberts 1961].

Spark chambers can be used to image the passage of cosmic rays and highly penetrating ionizing particles a shown in Fig. 19.46(right). In order to reduce sporadic arcing and possible damage from coronal discharge, they are typically not operated under constant high voltage. Instead a gate detector triggers a coincidence circuit that applies additional voltage to the spark chamber stack, high enough to produce avalanching. The gate detector can be a pair of scintillators over the top and bottom of the device (depicted in Fig. 19.48). After the spark occurs, the chamber still has mobile ions between the gaps. These ions can be rapidly removed with a "clearing field" that is a voltage applied across the spark chamber after the spark event. Usually the clearing field voltage is applied in the opposite direction to the high voltage and lasts no more than a few milliseconds. Images of the trajectory of the high energy particle as it travels through the detector, producing sparks between the plates, can be recorded with a photograph. The device, for instance, can be operated in darkness and a time-extended photograph can record multiple events. Individual events can also be recorded by acoustic methods, video, and magnetic sampling methods as explained by Rice-Evans [1974].

19.3.4 Bubble Chambers

Another track detector, the bubble chamber, is closely related to cloud chambers (see Sec. 9.5.8), and these devices were once important instruments for tracking particles. A reduction of pressure or an increase in temperature of a liquid can drive the liquid into the gaseous state (see Fig. 19.49). However, under some circumstances, a liquid can be slowly increased in temperature across the liquid/vapor phase boundary (from point 1 to 3 in Fig. 19.49), and still exist as a liquid in a metastable state. The same is true if the pressure is reduced (point 2 to 3 in Fig. 19.49), again producing a superheated liquid in a metastable state.

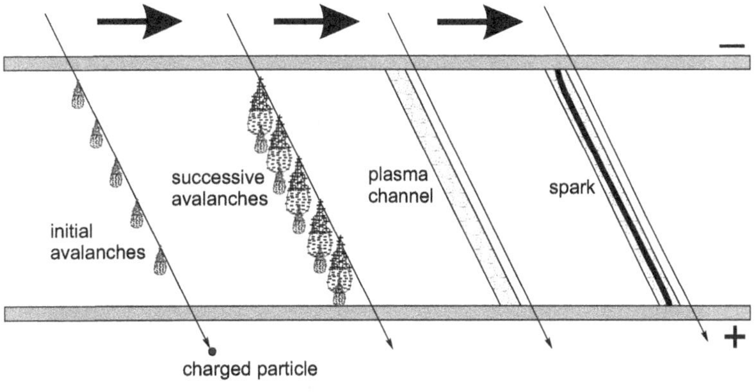

Figure 19.47. Formation of a spark channel at four different times.

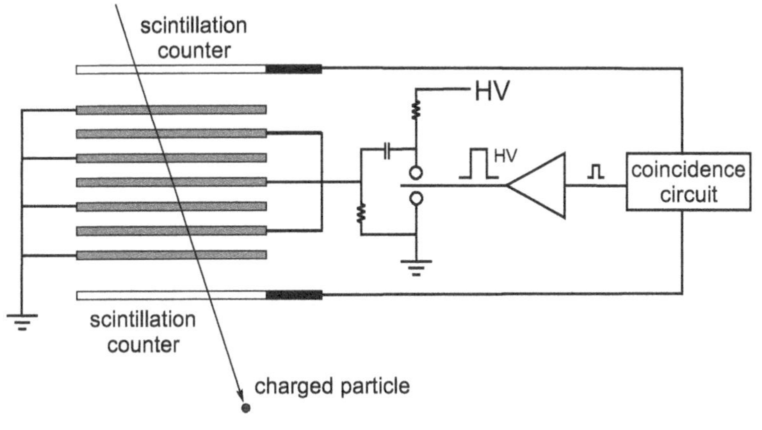

Figure 19.48. Depiction of basic spark chamber components. After Rice-Evans [1974].

Once in a metastable superheated state, a disruption can cause instantaneous boiling as the liquid converts to vapor. Examples of disruptive stimuli include small bubbles in the liquid, contaminants, and ionization produced by radiation interactions.

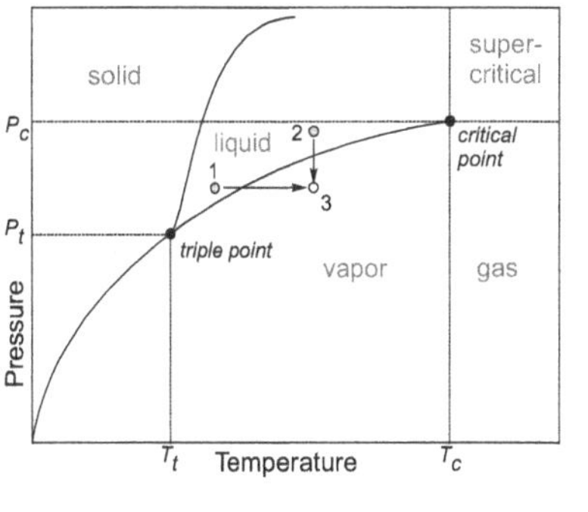

Figure 19.49. A P-T phase diagram, showing the triple and critical points.

Like the cloud chamber, a bubble chamber has an expansion chamber whose pressure can be changed by an attached piston. Unlike a cloud chamber containing supersaturated vapor, the bubble chamber is filled with a liquid brought up to a temperature just below vaporization (boiling point). If the piston is rapidly withdrawn, the liquid becomes spontaneously superheated, and charged particles interacting in the liquid produce ionization events that erupt into small vapor pockets, or bubbles [Glaser 1952].[22] These bubbles continue to grow in size as the piston is further withdrawn. Cameras are used to photograph the bubble tracks, much as is done with the cloud chamber. Incorporating multiple cameras around a transparent chamber, or a chamber with view ports, can be used to produce a three-dimensional image of the bubble

[22]Donald Glasser received the 1960 Nobel Prize in Physics for the invention of the bubble chamber.

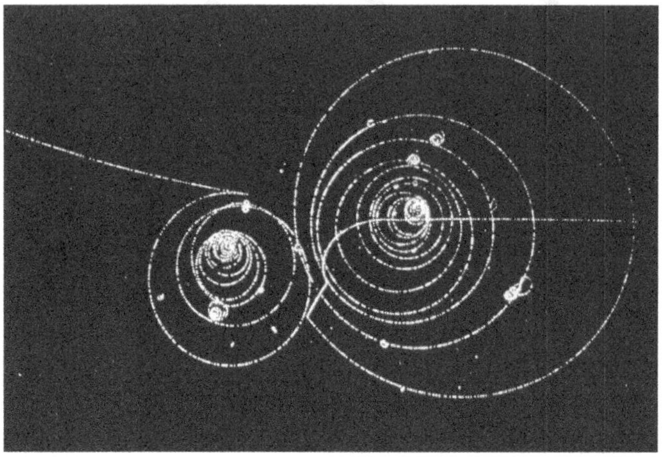

Figure 19.50. This image is taken from one of CERN's bubble chambers and shows the decay of a positive kaon in flight. The decay products of this kaon can be seen spiraling in the magnetic field of the chamber. Photo courtesy of CERN [1973].

tracks in the chamber. A magnetic field can also be applied to the chamber to cause charged particles to spin about the magnetic field described by $\mathbf{F} = q\mathbf{v} \times \mathbf{B}$ (see Fig. 19.50). Hence, the force on the particle changes with velocity, and the curvature of the bubble paths can be used to determine the particle velocity and momentum. Liquid hydrogen is probably the most used medium in a bubble chamber. Other liquids used include liquid deuterium, a mixture of liquid neon and hydrogen, liquid propane, and bromotrifluoromethane ($CBrF_3$). Laboratory bubble chambers may be only a few liters in size, while bubble chambers used for high energy physics can be very large. For instance, the Gargamelle bubble chamber was designed to hold approximately 12 cubic meters of liquid, while the volume of the Big European Bubble Chamber (BEBC) contains almost 35 cubic meters. While bubble chambers at one time played an important part in high energy physics, they have been largely replaced by multiwire gas-filled chambers.

19.3.5 Superheated Drop Detectors

Another type of bubble chamber is the superheated drop detector (SDD), shown in Fig. 19.51. First introduced by Apfel [1979], the device consists of a closed vessel backfilled with polymerized gel saturated with liquid droplets, usually a type of halocarbon with a boiling point well below room temperature. For instance, an elastic polymer or gelatin may be saturated with liquified R-12 refrigerant (dichlorodifluoromethane, CCl_2F_2) droplets. The saturation process is usually performed under high pressure [Roy et al. 1987].

These superheated droplets are on the order of 25 microns in diameter. At room temperature, the halocarbon liquid droplets would ordinarily convert to the vapor phase. However, the pressure on the gelatinous polymer provides sufficient surface force around the droplets to keep them in the liquid state at a temperature above their ordinary boiling point. By physically separating and suspending the superheated droplets in the gel, the phase change of a single drop from superheated liquid to vapor does not cause additional eruptions. Consequently, the other droplets stay in the superheated state.

Neutron interactions in the fluid produce recoil ions, and these ions can track near or through a superheated drop. Recoil ions can produce a trail of vapor embryos inside a drop, which are small vapor bubbles

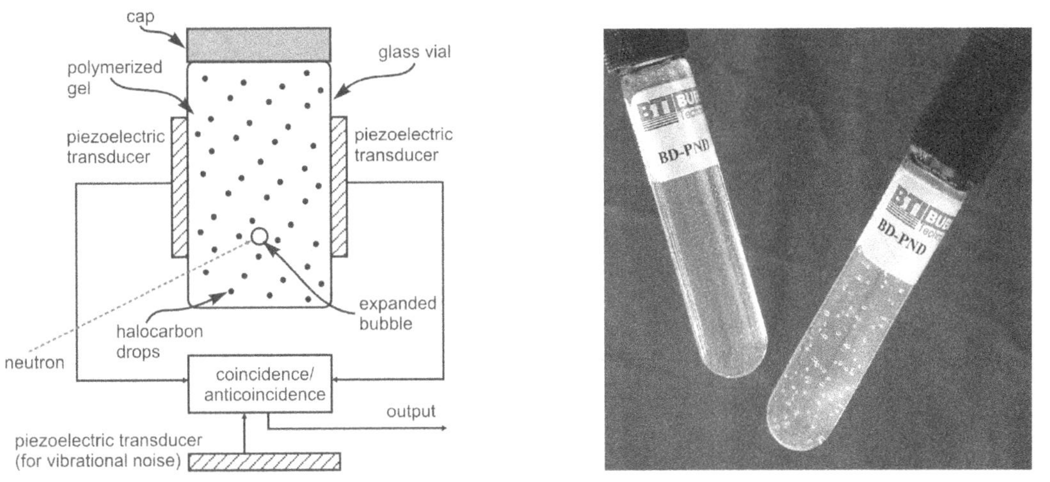

Figure 19.51. (left) Components of a superheated drop detector. (right) Before and after neutron irradiation of a superheated drop detector. Photo courtesy of Bubble Technology Industries, Inc., [2018].

on the order of 100 nm diameter. Consider a vapor embryo at a specific temperature T and pressure P. The critical energy required for bubble nucleation is [Roy et al. 1987; Seth et al. 2013]

$$E_c = \frac{16\pi\sigma^3(T)}{3(P_v(T) - P_0)^2}, \tag{19.66}$$

where $\sigma(T)$ is the surface tension[23] of the droplet at temperature T, $P_v(T)$ is the vapor pressure of the liquid at T, and P_0 is the pressure of the emulsion suspension (and the droplet fluid). If the radius of the droplet exceeds a certain critical radius r_c, then the droplet expands until it is entirely converted to vapor. If instead the radius is less than r_c, then it shrinks back into the superheated liquid drop. This critical radius is given by [Seitz 1958; Roy et al. 1987]

$$r_c = \frac{2\sigma(T)}{P_v(T) - P_0}. \tag{19.67}$$

Through this expansion process, particle interactions within the volume, by both charged and neutral particles, cause the spontaneous state change of liquid to vapor, forming suspended vapor bubbles on the order of 1 mm diameter. These bubbles appear at the particle interaction locations, and the density of bubbles yields a measure of the particle interaction density, imparted dose, and, in some cases, the particle flux. Cooling or repressurizing the emulsion in the vial can, in some designs, return the gas bubbles back into small droplets in the superheated liquid state and so render the device usable for subsequent measurements.

Unlike the bubble chamber, these detectors are portable and do not require moving parts (such as a piston). Superheated drop detectors are extremely sensitive to epithermal and fast neutrons, but virtually insensitive to gamma rays. If chlorinated halocarbons are used, superheated drop detectors can also be sensitive to thermal neutrons through the ^{35}Cl(n,p)^{35}S reaction [d'Errico 2001]. Readout of the devices can be conducted optically or acoustically.

The optical measurement mode consists of counting the bubbles formed in the chamber, either manually or with an optical scanner. Another measurement method is to incorporate a graduated expansion tube as

[23]Here σ should not be confused with the same symbol used throughout this book for a microscopic cross section.

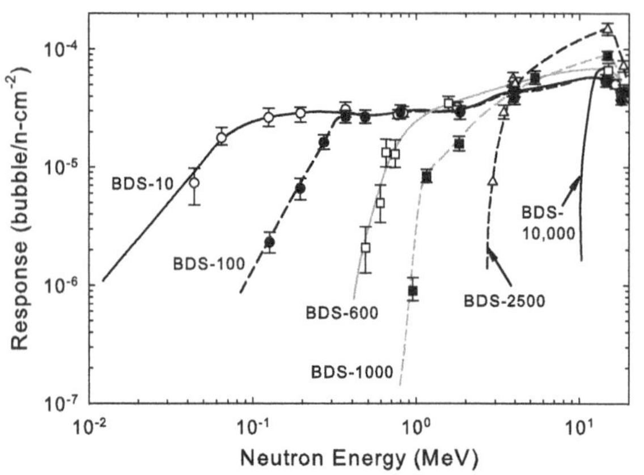

Figure 19.52. Response of a superheated drop detector threshold spectrometer set (Bubble Detector Spectrometer or BDS) from Bubble Technology Industries. Data are from Ing [2001].

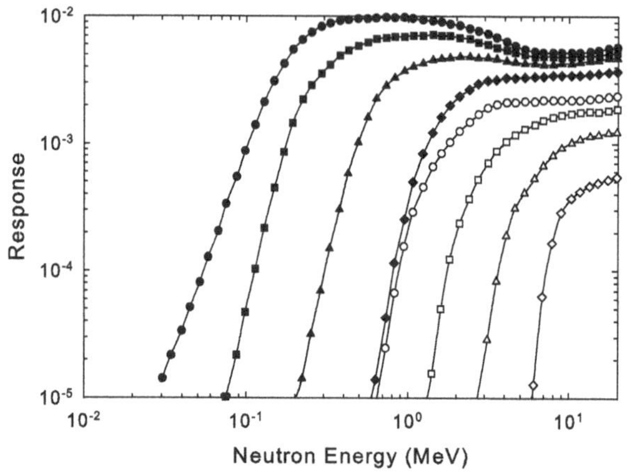

Figure 19.53. Thermal response dependence for two types of threshold superheated drop detectors. Black data are for SDD-1000 and white data are for SDD-6000. Here ○ = 40°C, □ = 35°C, △ = 30°C, and ◇ = 25°C. Data are from Apfel and d'Errico [2002].

part of, or extending from, the vial, much like a thermometer [Apfel 1992]. As bubbles are produced, the emulsion expands and moves along the metered tube, yielding a rapid measure of the number of bubbles or the neutron dose. Because the bubbles in the vial are measured after irradiation, these SDDs are considered to be *passive* counters that accumulate a radiation response over their exposure time.

If the SDDs are used acoustically, the noise produced by an expanding bubble is measured by transducers arranged around the chamber during radiation exposure. These transducers are connected in coincidence mode to discriminate out background acoustical noise. An additional acoustical detector can be added to operate in anticoincidence to discriminate out vibrational noise from electronics and the readout system. In some cases, the transducers can be connected at opposite ends of the vile, and the arrival time of the sound wave at each transducer can be used to localize the 1-dimensional position of the erupted bubble to within approximately 1 mm [Lim and Wang 1995; 1996]. By adding more microphones, including one inside the vial, a 2-dimensional localization method can produce a spatial resolution as small as 1.21 mm^2 [Felizardo et al. 2009]. SDDs that report counts in real time are considered *active* counters.

The threshold for reactions is a function of halocarbon type, emulsion pressure, and the operating temperature. As the temperature is increased, the bubble density for a given neutron exposure may change significantly [d'Errico 2001]. The neutron energy threshold for bubble formation decreases with increasing temperature [Apfel and d'Errico 2002], and increases with increasing pressure [Rezaeian et al. 2015]. The cause may be understood with Eq. (19.67) which shows that an increase in temperature causes the vapor pressure in the embryo bubble to increase and so results in a reduction in r_c. For example, d'Errico [2001] reports that the value of r_c decreases from 326 nm at 18°C down to 77 nm at 40°C for R-114 (ClF_2CCF_2Cl) refrigerant. Consequently, a larger fraction of embryos erupt as the emulsion temperature increases and r_c decreases. To counter the effect caused by pressure changes in the nucleated embryos, a small amount of volatile solvent can be added into the vial above the polymerized gel so as to increase the pressure P_0 on the gel as the temperature rises. This addition helps to counteract the increase in $P_v(T)$ and a reduction in the threshold for r_c [Apfel 1992]. Although not perfect, Apfel [1992] reports that this method decreases the bubble formation variation from 0.05 °C^{-1} to 0.01 °C^{-1} for the temperature range of 15°C to 35°C. In another application, d'Errico et al. [1996] added cooling strips to the vial to keep the temperature constant during the measurement, thereby eliminating variation in neutron sensitivity with temperature.

The strong dependence on halocarbon type, pressure, and temperature can be used to produce a type of threshold neutron spectrometer based on a set of different SDDs [Bonin et al. 1993; Ing 2001]. Such a spectrometer is very similar in concept to the foil activation and Bonner sphere methods, discussed in Ch. 18, for determining the energy spectrum of neutrons. The SDD method uses a set of (usually) six different SDDs with an unfolding algorithm to provide a relatively rough estimate of important aspects of a neutron energy profile. Data from a set of superheated drop detectors, named a "bubble detector spectrometer" (BDS), are shown in Fig. 19.52.

A similar active, acoustically operated, form of superheated drop detector was also described by d'Errico et al. [1995] and called the bubble interactive neutron spectrometer (BINS). The SDD response to neutron energy is shown in Fig. 19.53 for two different superheated drop detectors.[24] The change in neutron energy required to trigger an event is large, decreasing by an order of magnitude for both detectors for a temperature increase from 25°C to 40°C. This version of a neutron spectrometer has two temperature controlled SDDs that complement each other over the range of neutron energies to be investigated. This range can be adjusted for each SDD by changing the SDD temperature. Notice that the threshold energy for SDD-1000 at 25°C is the same as that of SDD-6000 at 40°C (see Fig. 19.53) and, thus, provides a smooth transition of sensitivity over a wide range of neutron energies. Spectral unfolding with the BINS detector can theoretically yield on the order of 10% FWHM energy resolution for monoenergetic fast neutrons between 250 keV to 5 MeV [d'Errico 2002].

Superheated drop detectors, or bubble detectors, are used for dosimetry measurements, and match well to the ICRP 60 recommendations [Portal and Dietze 1992]. Commercial dosimeters and threshold spectrometer sets are available, along with automatic reading systems.[25] The bubbles can be counted and subsequently recompressed into superheated droplets so that the device can be reused for future measurements. There is a practical limit on the total neutron fluence that can be measured and is reached when the radiation dose is so large that bubbles become so numerous that they begin to coalesce. If the bubble density becomes too high, then it becomes difficult to accurately count the number of bubbles in the vial. For a commercial unit, this limit corresponds to approximately 300 bubbles. Also, after the initial eruption, the bubbles continue to slowly grow is size. At some point, it is possible that the bubble grows large enough to damage the elastic emulsion surrounding the bubble by exceeding the elastic limit. Consequently, the vapor bubble cannot be recompressed into a superheated droplet. Generally, it is best that the SDD be read and recompressed on a short time schedule. Ing [2001] recommends resetting on a daily basis.

19.4 Cryogenic Detectors

As discussed throughout this book, there are many detector characteristics to be considered in the design of an "ideal" energy spectrometer system. Factors such as good intrinsic efficiency and short dead times are important; but probably the paramount factor is the energy resolution. HPGe systems have largely replaced NaI:Tl systems as energy spectrometers because of the significantly higher resolution of the former. To obtain high resolution, a detector must have a small excitation energy to produce a signal carrier so that the statistical noise is reduced. However, thermal excitation of these signal carriers must be reduced to minimize electronic noise. This latter requirement means the detector can be operated only at low temperatures. Sometimes liquid nitrogen temperatures are sufficient as, for example, with HPGe detectors. To achieve even better resolution, several types of detectors have been designed to operate at temperatures of about 0.1 K. Such detectors are termed *cryogenic* detectors.

---

[24]These two SDDs originally commercialized by Apfel Enterprises, Inc., are no longer available. The superheated halocarbons in SDD-1000 and SDD-6000 are C-318 and R-114, respectively [d'Errico et al. 1995], and neither of these superheated halocarbons is sensitive to gamma rays or thermal neutrons.

[25]Bubble Technology Industries, Inc., manufactures these detectors and systems.

Cryogenic detectors can be separated into two broad categories: (1) *thermal* detectors that measure temperature changes caused by radiation interactions in the detector, and (2) *athermal* detectors that measure charge (or some other signal carrier) produced by the radiation [Friedrich 2006]. Examples of each of these two types of cryogenic detectors are discussed below. A much more comprehensive presentation by experts in the field is provided by Enss [2005].

19.4.1 Methods of Cooling

There are several methods used to achieve the low temperatures required for the different types of cryogenic detectors. In some cases, detectors such as HPGe need only be cooled to temperatures near 77K, achievable with a liquid nitrogen (LN2) bath, a relatively simple and inexpensive technique by contemporary standards. Yet, a key requirement for many cryogenic detectors is a means to cool the detector to operating temperatures below 0.1 K. Typically, cooling to such low temperatures is performed in two or more steps. For lower temperatures near 4 K, liquid He can be used as the cooling bath which is often surrounded by a LN2 jacket. Liquid He is more expensive than LN2, yet laboratory compressors are available that can provide this liquid. Although many He-based refrigerators are designed for laboratory use, portable precooling systems have become available. Finally a second cooling step is used to take the detectors from about 4 K to mK temperatures. Several such cooling techniques are summarized below.

Precooling Pulse-Tube Refrigerators

Over the past 20 years refined versions of the pulsed-tube refrigerator, introduced in 1954 by Gifford and Longsworth [1964], offer an attractive method for the initial precooling of cryogenic detectors. The pulse-tube refrigerator is a He closed-cycle, vibrationless, "dry" device with no moving parts. The lack of vibrations is critical for most cryogenic detectors. The original pulsed-tube device of Gifford and Longsworth cooled only to 174 K. In 1984 Mukulin added an orifice to provide resistance to the He flow and, thereby reached a temperature of 105 K. Since then, various design refinements have continued to lower the temperature of the cold end of the device to about 4 K, just below the boiling point of He (4.2 K). With a two-stage double-inlet pulse tube cooler, Wang et al. [1997] achieved a temperature of 2.23 K. Then in the early 2000s an even lower temperature of 1.73 K was achieved by replacing the He gas, which is mostly ^{4}He, by ^{3}He. Today commercial pulse-tube refrigerators are available from several vendors.[26] Hence, replacement of traditional cryogenic liquids by a modern mechanical cryocooler for precooling makes cryogenic detectors truly portable.

^{3}He Refrigerators

Many cryogenic detectors require temperatures below a few tenths of a degree kelvin, and various refrigerators have been developed to produce such low temperatures [Lounasmaa 1974]. A ^{3}He refrigerator operates by evaporative cooling of ^{4}He to produce a 1 K reservoir or *pot* of ^{3}He. The liquid ^{3}He is further evaporatively cooled to reduce the pot temperature to about 0.2 K. Due to the rarity and high cost of ^{3}He, it is recycled in a closed loop within the system. Single-stage ^{3}He refrigerators go down to about 0.3 K and two-stage ^{3}He refrigerators can go down to about 0.2 K.

Adiabatic Demagnetization Refrigerators

A relatively compact system, by comparison to a dilution refrigerator, is an adiabatic demagnetization refrigerator (ADR). Adiabatic demagnetization is a single cycle method of cooling, unlike a dilution refrigerator that operates continuously. Adiabatic demagnetization is also the oldest known method of reducing temperatures below 1 K [Lounasmaa 1974] and was originally introduced by Debye [1926] and Giauque [1927]. The magnetocaloric effect operates by first heating a paramagnetic material with magnetism, a process in which the entropy is lowered in a magnetic field. The heat is removed by a liquid He bath through a contact

[26]see `starcryo.com`, `entropy-cryogenics.com`, `bluefors.com`, `oxint.com`, and `leiden-crygenics.com`

switch. After opening the switch, the magnetic field is slowly lowered, adiabatically, to keep the entropy of the coolant paramagnet constant, thereby causing the temperature to decrease. The device can produce low temperatures on the order of 1 mK. Seminal work for development of the ADR was made by Hagmann and Richards [1994] whose device sustained temperatures of 100 mK for 100 hours. These refrigerators may have more than one type of paramagnet. Friedrich et al. [2001] describes such a refrigerator with two paramagnet stages. The first stage is cooled to about 1 K with a gadolinium gallium garnet ($Gd_3Ga_5O_{12}$ or GGG), and a second stage is cooled to approximately 0.1 K with a paramagnetic salt $Fe(NH_4)(SO_4)\cdot 12H_2O$ or FAA. The entire system has LN2 and liquid He cooling tanks. ADRs with CPA (chromium potassium alum) can go down to about 12 mK and those with CMN (cerium magnesium nitrate) can theoretically reach 2 mK. However, few cryogenic detector researchers use ADRs. The exception is those in NASA who need them because ADRs work in zero-gravity environments.

Dilution Refrigerators

The ^{3}He/^{4}He dilution refrigerator can produce temperatures down to 2 mK. It has no moving parts in the low temperature region of the refrigerator, relying solely on the phase change of ^{3}He and ^{4}He. For He mixtures containing more that 6% ^{3}He, a phase separation occurs at 870 mK. The "denser" phase is composed almost entirely (100%) of ^{3}He floating atop a "dilute" phase composed approximately of 94% ^{4}He and 6% ^{3}He. Cooling is achieved by forcing the pure ^{3}He to move across the phase boundary to the lower dilute phase, much like evaporation in which the denser phase of ^{3}He is the liquid and the dilute phase is vapor. This continued action is achieved by circulating the ^{3}He with a room temperature pump. The ^{3}He is recirculated, going through an LN2 precool step, a liquid helium 4 K precool, finally through several heat exchangers before arriving at the phase separation and mixing chamber, the coldest part of the refrigerator. The process of moving ^{3}He through the dilute phase is endothermic, and consequently, heat is removed from the system. Dilution refrigerators are effective in producing temperatures below 1 K, but are relatively bulky devices.

19.4.2 Cryogenic Microcalorimeters

There are a few basic configurations of cryogenic detectors, generally described as thermal/equilibrium detectors and athermal/nonequilibrium detectors [Friedrich 2006]. Thermal/equilibrium cryogenic detectors can be described as either bolometers or microcalorimeters, with the distinction that a bolometer measures the amount of radiation incident upon the device (similar to a current mode device) and a calorimeter measures the energy of individual interactions in the device (similar to a pulse mode spectrometer). The origin of the bolometer dates to Langley, who in 1880 invented the device to measure small quantities of infrared radiation [Langley 1880]. Much later, Andrews et al. [1942] introduced a superconducting version of the bolometer fabricated from tantalum wire operated at its superconducting transition temperature (3.22 K to 3.23 K).

Cryogenic detectors, bolometers and microcalorimeters, measure thermal energy deposited in a low temperature (mK) absorber. The increase in heat in the absorber is converted into a signal and is, thus, a measure of the energy deposited by incident radiation. Low temperature operation is required to reduce thermal noise that would otherwise mask the small signal. There are a variety of thermal detectors, some of which are reviewed in this section.

The simplest cryogenic detector, and possibly the oldest, is a radiation absorber attached to a thermometer. The absorber and thermometer together are weakly attached to a low temperature reservoir bath or heat sink. These detectors are called, interchangeably, *bolometers*, *calorimeters*, *thermal detectors*, and other variants. However, there is a distinct difference in the way that bolometers and microcalorimeters operate. In either case, a cryogenic thermal detector has an absorber attached to a thermally sensitive electronic component, usually a thermistor.[27] Some designs are monolithic so that temperature sensor is the radiation absorber itself.

Consider the simple cryogenic detector depicted in Fig. 19.54. The absorber is heated by incident radiation interacting in it at the rate $P_{in}(t)$ (J s^{-1}). The total heat capacity of the absorber is denoted by c_a (J K^{-1}) and the total heat conductance between the absorber and heat sink by g (J K^{-1} s^{-1}). A small constant current I flows through the thermistor of resistance R and creates a voltage V across it. The current in the thermistor generates a small thermal power of $P_o = IV = I^2/R$, which at equilibrium, causes the absorber/thermistor to have a temperature T_o. Incident radiation interacting in the absorber raises the absorber/thermistor temperature and, thus, the resistance of the thermistor changes. This change produces a change in the voltage V that is measured and used to infer the temperature change of the absorber/thermistor from its equilibrium value as well as to record the occurrence of a radiation interactions in the absorber.

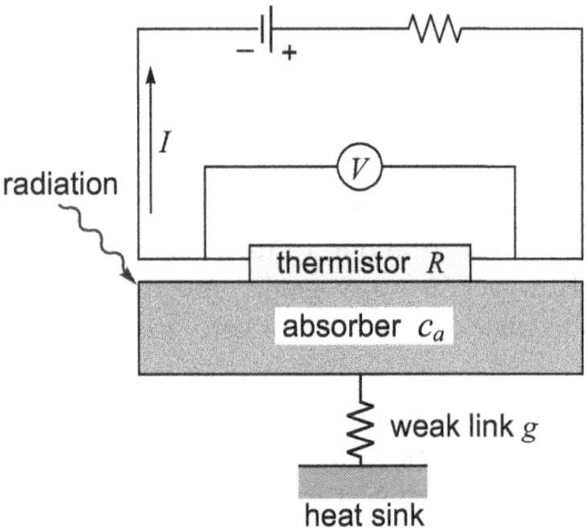

Figure 19.54. A cryogenic detector has an absorber with heat capacity c_a coupled to a thermistor, both weakly attached with conductance g to a low-temperature heat sink.

The behavior of the detector depends on its design. Consider a detector at thermal equilibrium with $P_{in} = 0$ for $t < 0$ with the absorber/thermistor at temperature T_o. At $t = 0$ radiation begins to heat the absorber, producing a change in temperature $\Delta T(t) = T(t) - T_o$ that is measured by the thermistor circuitry. Conservation of energy requires

$$\frac{\Delta T(t)}{dt} = \frac{1}{c_a}\left[P_{in}(t) - g\Delta T(t)\right]. \tag{19.68}$$

The general solution of this first order differential equation with constant coefficients subject to the initial condition $\Delta T(0) = 0$ is

$$\Delta T(t) = \frac{\exp[-t/\tau_F]}{c_a}\int_0^t \exp[t'/\tau_F]P_{in}(t')\,dt', \qquad t > 0, \tag{19.69}$$

where the time constant $\tau_F = c_a/g$.

If a quantum of radiation deposits energy E in the absorber at $t = 0^+$, i.e., $P_{in}(t) = E\delta(t - 0^+)$, the output temperature pulse is found from Eq. (19.69) as

$$\Delta T(t) = \frac{E}{c_a}\exp(-t/\tau_F). \tag{19.70}$$

Typically these devices are used for measurement of low energy infrared and x-ray photons which deposit the photon energy E in the absorber in a single interaction, although other particles can be detected. Thus the output from the detector is a pulse with an amplitude proportional to the energy deposited in the absorber by a radiation particle. The decay tail of the pulse can be made small by making τ_F very small. If the average time τ_γ between radiation interactions in the absorber is large so that $\tau_F/\tau_\gamma \ll 1$ then the cryogenic detector produces a train of output pulses whose amplitudes are proportional to the energy deposited by each incident photon from which the energy spectrum of the incident radiation can be formed. Thus the device can be used as an energy spectrometer, i.e., a microcalorimeter.

[27] A type of resistor in which the resistance is a function of the temperature.

However, if $\tau_\gamma \ll \tau_F$, the output pulses pile up upon the tails of earlier pulses and the device behaves more like a detector operated in current mode. For steady incident illumination with a flux of particles ϕ which, on average, deposit an energy E per interaction in the absorber, the incident power $P_{in} = \phi\mu V_a E$, where μ is the interaction coefficient and V_a is the volume of the absorber. If the illumination of the absorber begins at $t = 0$, then, from Eq. (19.69)

$$\Delta T(t) = \frac{\phi\mu V_a E}{g}\left[1 - \exp(-t/\tau_F)\right]. \tag{19.71}$$

Thus, for times greater than a few τ_F, the output is a signal that is constant and proportional to the incident flux ϕ. Thus in this case the cryogenic device is seen to act as a bolometer. In this simple analysis it is assumed that the time for deposited energy to be manifested as heat is negligible compared to τ_F or τ_γ.

Consequently, by changing the readout circuit of a thermal detector, the device can be operated as either a bolometer or calorimeter. For radiation energy spectroscopy, it is the microcalorimeter that is of most interest. There are many types of microcalorimeters, yet these devices are separated into two classes, thermal/equilibrium detectors and athermal/non-equilibrium detectors. Disadvantages of these detectors include their small volume, which places a practical limit on measurable photon energies, their relatively slow responses (which limit the measurable gamma-ray interaction rates), and their need for low operating temperatures. Friedrich [2006] describes these various microcalorimeters in a review article, summarized here in the following subsections.

Energy Resolution Thermal or equilibrium cryogenic detectors consist of a radiation absorber with heat capacity c_a attached to a thermometer (usually a thermistor). The absorber and thermometer are *weakly* coupled to a cold bath by a conductance g. Photon absorption increases the absorber temperature according to Eq. (19.70) in proportion to the absorbed energy E. The resulting temperature change is measured by the thermometer before the absorber relaxes to the bath temperature. However, thermal fluctuations of $4gkT^2$ limit the energy resolution. These resolution-limiting thermal fluctuations are a result of the random exchange of phonons across the weak thermal link.

For a fixed absolute T, the total energy of the absorber is c_aT and the average energy of the phonons is kT. The ratio of these terms gives the average number of phonons, $\overline{N}_p = c_a/k = \sigma^2(\overline{N}_p)$ if a Gaussian distribution is assumed. The variance in the phonon energy is $\sigma^2(E) = \sigma^2(kT\overline{N}_p) = k^2T^2c_a/k$. The energy resolution limit, measured by the FWHM of the phonon energy distribution, is [Kerson 1987]

$$\Delta E_{\text{FWHM}} = 2\xi\sqrt{2\ln 2}\,\sigma(E) = 2.355\xi\sqrt{kT^2c_a}. \tag{19.72}$$

Here ξ is a dimensionless parameter that depends on the temperature dependence of c_a [Moseley et al. 1984] which otherwise would be unity. However, the lattice heat capacity is a function of the absorber volume with a T^3 dependence [McClintock et al. 1984], i.e.,

$$c_a = \left(\frac{12\pi^4 N V_a k}{5}\right)\left(\frac{T}{\theta_D}\right)^3 \text{ J K}^{-1}\text{ mol}^{-1}, \tag{19.73}$$

where N is the atom density and V_a is the volume of the absorber. Here

$$\theta_D = \frac{\hbar}{k}\left(6\pi^2 v^3 N\right), \tag{19.74}$$

referred to as the Debye temperature, in which v is the mean thermal atomic speed.

From Eqs. (19.72) and 19.73, it is seen that the energy resolution improves both as the temperature decreases and as the absorber volume decreases (to make c_a smaller). However, a smaller absorber volume means fewer radiation quanta interact in the absorber so the efficiency of the detector decreases. This decrease in efficiency then leads to the use of arrays of small absorbers in an effort to offset the decrease in efficiency of each absorber. The need for high energy resolution means *all* microcalorimeters must operate at temperatures as low as possible and with absorbers that are as small as possible, but with sufficient size to produce an acceptable detection efficiency.

Early microcalorimeters used semiconductor thermistors as the temperature sensor. An x-ray absorption causes the resistance in the thermistor to increase which produces a change in voltage across the thermistor for current biased devices. These measured voltage changes can be used to infer the ΔT change in the detector as a result of a radiation interaction in it. Although effective, as shown by a measured energy resolution of less than 8 eV for 5.9 keV gamma rays from ^{55}Fe, the resolution is limited by the heat capacity of the absorbers.

Microcalorimeters are operated at such low temperatures, near 0.1 K, and with small absorber volumes that energy resolution of a few eV FWHM is possible only for low energy x rays. The maximum achievable energy resolution is also a function of Johnson noise in the thermistor, amplifier noise, and photon background [McCammon 2005a]. Moseley et al. [1984] show that a practical limit is approximately twice the predicted value of Eq. (19.72). The thermal coupling g between the microcalorimeter is weak at low temperatures, and consequently these detectors are relatively slow. The thermal pulse decay time $\tau_F = c_a/g$ is generally on the order of 1 ms.

Semiconductor Microcalorimeters

Semiconductor thermistors based on doped Si or Ge are used for high resolution x-ray spectroscopy, and are probably the oldest type of cryogenic microcalorimeter [Low 1961; Moseley et al. 1984]. These thermistors require a relatively large change in resistance with temperature (dR/dT). The absorber can be formed from a high Z material to improve both the photon absorption efficiency and heat capacity c_a; however, the material may experience a decrease in thermal conductance between the absorber and the thermistor, g_a. Consequently, the added noise degrades the energy resolution by [Bühler et al. 1994],

$$\Delta E_{\text{FWHM}} = 2.355\xi\sqrt{4kT^2c_a}\left[\frac{g}{g_a} + \left(\frac{g}{g_a}\right)^2\right]^{1/4} . \tag{19.75}$$

As discussed Sec. 12.5.3, the intrinsic charge carrier concentration of a semiconductor increases exponentially with temperature. At low temperatures the number of free charge carriers diminishes to zero, effectively causing the semiconductor to become an insulator so that it no longer is useful as a thermistor. The conductivity can be increased by doping the semiconductor with shallow donors or acceptors so the semiconductor retains some conductivity even at lower temperatures. However, at the very low temperatures required for high resolution microcalorimeters, even shallow dopants enter the "freeze-out" region and thermal ionization becomes negligible. The work of Anderson [1958] and Mott and Twose [1961] describes the conduction mechanism of doped semiconductors, including their low temperature behavior. There is a critical doping concentration below which a semiconductor at 0 K is an insulator but for $T > 0$ the semiconductor retains conductivity [McCammon 2005b]. If the semiconductor is doped slightly below the limit, the conduction mechanism is governed by phonon-assisted tunneling between impurity sites called "variable range hopping" (*vrh*) [Shklovskii and Efros 1984].

For *vrh*, the electrical resistance can be described by,

$$R(T) = R_0 \exp\left[\frac{T_o}{T}\right]^p , \tag{19.76}$$

where p is approximately 1/2 [Efros and Shklovskii 1975] with R_0 and T_0 fitting parameters. Equation (19.76) indicates that $\ln R(T) \propto T^{-1/2}$ and is valid over the very low temperature operating range of the detector.

Note that Eq. (19.76) indicates that R_0 is the resistance at very high temperatures; but Eq. (19.76) is an approximation valid over only a limited low temperature range and fails at high temperatures. The variation of resistance with temperature is characterized by the temperature coefficient defined as

$$\alpha = \frac{d\ln R}{d\ln T} = \frac{T}{R}\frac{dR}{dT}. \tag{19.77}$$

For $R(T)$ given by Eq. (19.76)

$$\alpha = -p\left(\frac{T_0}{T}\right)^p. \tag{19.78}$$

The utility of the temperature coefficient is that, for small temperature changes ΔT, the associated change in resistance ΔR can be estimated from Eq. (19.77) as

$$\Delta R = \alpha\frac{R}{T}\Delta T. \tag{19.79}$$

The temperature dependence of R for Si and Ge, at first, did not appear to follow the model of Eq. (19.76). The experimental data varied between samples, at least at the time when measurements were first made. However, neutron transmutation doped (NTD) semiconductors on Si and Ge [Haller et al. 1996] do, in fact, follow the prediction of Eq. (19.76) [McCammon 2005b]. This improvement is probably a consequence of the superior dopant uniformity of NTD along with the reduction of interstitial defects.[28] Uniform doping with deep ion implantation also yields good results [Friedrich 2006].

For high energy resolution, the absorber heat capacity must be kept low, on the order of 10 pJ K^{-1} [Friedrich 2006]. Bryant and Keesom [1961] found that this requirement for Ge limited the absorber volume to about 0.001 mm^3. With a typical thermal conductance g of 10 nW K^{-1} between the absorber and heat sink, the decay time $\tau_F = g/c_a$ is approximately 1 ms [Bühler et al. 1994]. The absorber and thermistor volumes are kept small to sustain high energy resolution and good performance.

The relatively high specific heat of doped semiconductors, even at low temperature, reduces the sensitivity of the detector. Kelly et al. [1993] describe a *monolithic* semiconductor device in which only a small portion of the semiconductor absorber is doped. Hence, the absorber is the undoped region, and the thermistor is the smaller doped region, all a part of a single substrate. The *composite* microcalorimeter has a separate low c_a absorber and thermistor. The combined pair can be optimized to produce a larger device while reducing τ_F. To increase the overall volume, these small individual detectors can be fabricated into arrays with individual absorbers atop each thermistor [Stahle et al. 2004].

Superconducting Transition Edge Microcalorimeters

The electrical resistance of a superconducting material is zero below a transition temperature T_c and exhibits a characteristic increase in resistance as the temperature is increased above the transition temperature as shown in Fig. 19.55. The transition temperature changes in a magnetic field, decreasing as the B field is increased [Rose-Innes and Rhoderick 1978]. Above a critical magnetic field H_o for a Type I superconductor, superconduction is not possible, even at 0 K. A Type II superconductor has two phases defined by two limiting B fields. Below T_C and the boundary defined by H_{c1}, the material superconducts as observed with a Type I superconductor. However, between magnetic field boundaries defined by H_{c1} and H_{c2}, the material

[28] NTD transforms Ge or Si into dopant atoms through activation, typically n-type dopants, whereas diffusion and implantation doping replace Ge or Si by forcing them from their lattice locations by the impurities introduced during either the diffusion drive-in or implantation process.

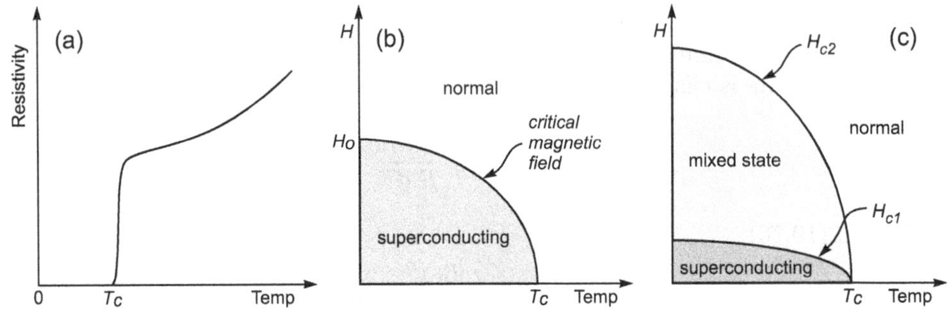

Figure 19.55. (a) Superconductivity abruptly occurs below a transition temperature T_c. (b) Type I superconductivity in which the transition temperature T_c is also a function of the applied magnetic field, with a critical field H_o above which superconductivity is not possible. (c) Type II superconductor with H_{c1} defining the boundary of the mixed state and H_{c2} defining the critical field above which superconductivity is not possible.

transitions into the mixed state, in which both superconductivity and normal conductivity coexist. Consequently, for magnetic fields bounded by H_{c1}, Type II superconductors may have two transition temperatures defined by T_{c1} (superconducting) and the other T_{c2} (mixed state) above which superconductivity vanishes.[29]

If a superconductor is set slightly below T_c, then heat added to the system moves the superconductor above the transition temperature so it no longer has zero resistance. Consequently, current flow through the material (no longer superconducting) now produces a voltage, which gives a measure of the energy deposited in the absorber. This type of calorimeter is called a transition edge superconducting (TES) spectrometer. In normal operation, the device is chilled well below the transition edge, and heated ohmically by applying a constant voltage bias to the superconducting film that is in contact with the absorber. The bias is adjusted such that the temperature of the device is maintained slightly below the transition edge. The absorption of an x ray causes the superconducting film to become normal conducting, thereby increasing the resistance and decreasing the current.

Because the TES film in contact with the absorber increases in resistance with temperature (just opposite of that for a doped semiconductor calorimeter), the TES detector can use electrothermal feedback (ETF). With ETF, the power $P = V^2/R$ decreases as the resistance increases. Consequently, with lower power, the return to the cryogenic bath temperature is quicker and the energy resolution is improved [Irwin 1995],

$$\Delta E_{\text{FWHM}} = 2.355 \left[\frac{4kT^2C}{\alpha} \sqrt{\frac{n}{2}} \right]^{1/2}, \tag{19.80}$$

where n is a parameter characterizing the temperature dependence of the thermal conductance as given by [Irwin 1995]

$$g = \frac{dP}{dT} = nKT^{n-1}. \tag{19.81}$$

Here K is a material and geometry related parameter. When an interaction event heats the film, the return to equilibrium is described by

$$c_a \frac{d\Delta T}{dt} = -\frac{P_0 \alpha}{T} \Delta T - g\Delta T, \tag{19.82}$$

[29]The normal state is confined to threads of *normal cores* of conductivity in a triangular pattern with each core surrounded by superconducting material [McClintock et al. 1984].

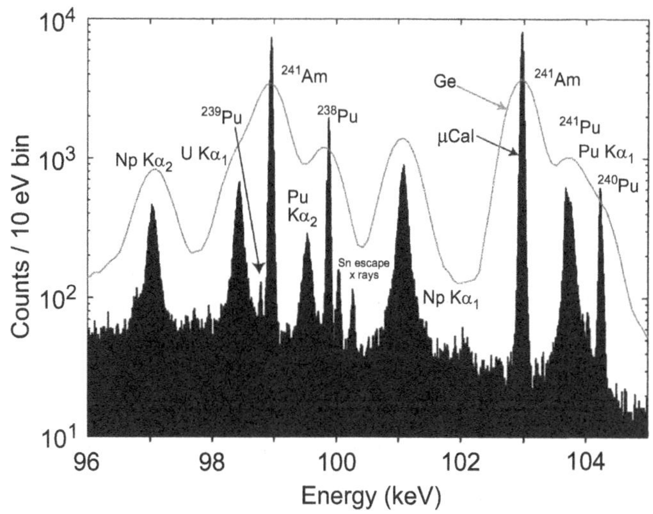

Figure 19.56. Pu spectrum from a microcalorimeter array using data from 11 of 13 active pixels. The combined array resolution is approximately 45 eV. At this resolution, the broad x-ray peaks can be readily distinguished from gamma-ray peaks. The solid curve is a spectrum taken with a conventional HPGe detector. ©[2009] IEEE. Reprinted, with permission, from Bacrania et al., *IEEE Trans. Nucl. Sci.*, **56**, 2299–2302, [2009].

where P_0 is the equilibrium Joule power. When the superconducting film is well below the transition temperature, $P_0 = KT^n = gT/n$, and the pulse time constant is [Irwin et al. 1998],

$$\tau_F = \frac{c_a}{g}\frac{n}{n + \alpha\varphi_b}. \tag{19.83}$$

Here φ_b is a measure of how far the detector is biased above the bath temperature T_b and is defined as

$$\varphi_b \equiv 1 - \left(\frac{T_b}{T_e}\right)^n, \tag{19.84}$$

where T_e is the equilibrium temperature of the TES calorimeter. Equation (19.83) indicates that high values of α improve recovery time.

The current is measured through induction with a superconducting quantum interference device (SQUID) current amplifier. This device is used to reduce amplifier noise because a SQUID does not have voltage noise. TES calorimeters must have low resistance to ensure rapid and homogeneous thermalization of the absorber and superconducting film. Typically the choice of absorber depends greatly upon the photon energy of interest. An energy resolution of 2.37 ± 0.17 eV has been achieved for 5.9-keV gamma rays from Mn K_α lines using Mo-Au TES thermistors [Iyomoto et al. 2008]. Higher energy gamma rays, yet generally below 100 keV, produce good results with a Sn absorber bonded to superconducting bilayer of Mo and Cu with a transition temperature of 100 mK [Doriese et al. 2007]. The best reported energy resolution for a single device was 25 eV for 103-keV gamma rays. Arrays of microcalorimeters can be used to maintain a fast response time while increasing detection efficiency [Doriese et al. 2007; Iyomoto et al. 2008; Bacrania et al. 2009]. Shown in Fig. 19.56 are comparison x-ray spectra for a typical HPGe semiconductor detector and a TES microcalorimeter array.

Magnetic Microcalorimeters

Another type of low-temperature microcalorimeter measures the change in magnetism as small amounts of heat are added to the system. These devices have a radiation absorber coupled to a paramagnetic sensor, together weakly attached to a low-temperature thermal reservoir [Fleischmann 2005]. Thermal energy absorbed by the system causes the magnetism to decrease, a decrease which can be measured by an inductively coupled SQUID.

Originally introduced by W. Seidel in 1986,[30] the technology was subsequently investigated for its potential as an ultra-high resolution x-ray spectrometer [Bühler and Umlauf 1988; Bühler et al. 1993; Fausch et al. 1993; Bühler et al. 1994]. In these original works, energy resolution of 320 eV was achieved for 5-MeV alpha particles with the compound cerium magnesium nitrate or CMN [see Fleischmann 2005]. However, the response times were slow with a 40-ms rise times and up to 10-s thermalization times. This slow response, which is a common problem with paramagnetic dielectric materials, is unsuitable for most x-ray spectroscopy applications. Bandler et al. [1993] suggested that the response problem can be overcome by embedding magnetic ions into a conductive metallic absorber, particularly gold doped with erbium. Fleischmann et al. [2000] adopted this concept and have extensively studied Au:Er metallic magnetic calorimeters (MMC).

The magnetization change from a paramagnetic sensor is described by [Fleischmann et al. 2005],

$$\delta M = \frac{\partial M}{\partial T}\delta T = \frac{\partial M}{\partial T}\frac{\delta E}{c_a} \tag{19.85}$$

clearly showing that a small value of c_a is required to produce a relatively large change in temperature and, consequently, a large change in magnetization. The energy resolution of the detector is limited by [Fleischmann et al. 2003]

$$\Delta E_{\rm FWHM} = 2.355\sqrt{4kT^2c_a}\left[\frac{4\tau_0}{\tau_1}\right]^{1/4}. \tag{19.86}$$

where $\tau_0 = c_a/g_a$, approximately equal to 1 μs, is the average time for deposited energy in the absorber to change the spin in the paramagnetic sensor. The time constant $\tau_1 = c_a/g$ describes the time needed for the (weak) thermal coupling to transfer heat to the low-temperature heat sink. It is notable that MMCs do not have a bias current, because they are inductively coupled to a SQUID amplifier and thus have no bias voltage, and therefore do not suffer from resistive heating or associated noise [Enss et al. 2004]. For a fixed value of c_a, MMCs can have better energy resolution than other types of calorimeters, or, instead, have the same energy resolution but with a larger absorber so that they have improved efficiency. Although these detectors are relatively slow, they are theoretically capable of less than 1 eV energy resolution. Fleischmann et al. [2004] reported energy resolution of 3.4 eV for 6.5-keV x rays with a Au:Er MMC, while Gastaldo et al. [2013] reported 2-eV FWHM for 6-keV x rays.

19.4.3 Athermal Cryogenic Charge Detectors

Another class of cryogenic detectors does not measure thermal effects produced by radiation but instead is based on measuring the number or the energy of quasiparticles produced by radiation in a superconductor. Before describing these athermal detectors, some more properties of superconductivity must be understood.

Properties and Mechanism of Superconductivity There is a distinctive difference between a perfect conductor and a superconductor in the presence of a magnetic field. If a perfect conductor is cooled to low temperatures to obtain zero resistance, and is then introduced into a magnetic field, according to Lenz's Law the magnetic field is canceled inside the conductor and effectively dispelled. A similar condition is observed for superconductors, where the magnetic field is expelled from the superconductor. However, if a perfect conductor is cooled while already in a magnetic field, the magnetic field continues to permeate the conductor, and remains trapped in the perfect conductor once the magnetic field is removed, causing the sample to become magnetized. However, if a superconductor is cooled below the transition temperature in the presence

[30] Fleischmann et al. [2005] reference the unpublished thesis of W. Seidel as the introduction of this concept.

of a magnetic field, it expels the magnetic field upon being reduced below T_c. This phenomenon associated with superconductors is called the Meissner effect. Upon removal from the magnetic field, a superconductor does not trap the magnetic field.

Superconductivity results from the pairing of electrons, called Cooper pairs, created at low temperature by phonon-electron interactions.[31] An electron can cause lattice distortions by attracting nearby positive atomic nuclei, causing them to move slightly closer to the electron, their movement producing a phonon, and ultimately appearing as a location with positive charge. This apparent positive charge attracts another electron, even at relatively long distances on the order of several hundred nanometers,[32] and overcomes the usual repulsion between electrons. These two electrons have opposite spin and momentum and temporarily form a pair, physically producing a composite boson particle with a lower potential energy than two single electrons. In fact, the sea of Cooper pairs are represented by a single wave function and follow Bose-Einstein statistics. They form a state of lowest energy called the condensed state with an energy gap ϵ_{gT} between this condensed state and the lowest energy of two unpaired electrons. This energy gap becomes nearly constant for temperatures below approximately $0.6T_c$.

Cooper pairs do not stay paired, but repair with other electrons, such that the superconductivity condition is preserved. The energy of the pair is on the order of 10^{-3} eV, and can be broken by sufficient thermal energy. Consequently it takes low temperatures to produce substantial numbers of Cooper pairs. If a Cooper pair is scattered, the total momentum of the pair is conserved and their motion along the electric field lines is unimpeded. To destroy a Cooper pair, an energy of at least $2\epsilon_{gT}$ is required, which is improbable at temperatures well below T_c. However, the value of ϵ_{gT} decreases at the temperature is increased, ultimately becoming zero at T_c.

Superconducting Tunnel Junction Detector

As just described, ϵ_{gT} is the small band gap formed between the Cooper-pair state and the normal state. The limiting energy gap at 0 K is denoted ϵ_{g0} and is on the order of 1 meV. A semiconductor, such as silicon, has a band-gap energy on the order of 1 eV, and so the superconducting band gap is 1000 times smaller. Consequently, the number of electrons changing from bound Cooper pairs into electrons can be vastly greater per unit absorbed energy. Much like a semiconductor, there is an average energy required to produce a *quasiparticle*[33] from a broken Cooper pair.

Although the minimum energy must be at least $2\epsilon_{gT}$ to make two quasiparticles, the average energy w_q needed to produce a single quasiparticle is slightly more than ϵ_{gT}. Initial estimates put the upper limit of w_q at $2\epsilon_{gT}$ per quasiparticle [Kurakado and Mazaki 1981a, 1981b]. Later estimates, produced by Monte Carlo calculations, put the average value of w_q closer to $1.7\epsilon_{gT}$ [Kurakado 1982]. Consequently, the average number $\langle N \rangle$ of quasiparticles produced per unit energy in a superconductor is extremely high, an effect which reduces the statistical fluctuations in the measured signal carriers.[34] Ultimately, because approximately 1000 times more quasiparticle charge carriers are produced than in a semiconductor, the energy resolution can be approximately $\sqrt{1000} \simeq 32$ times better.

Generally the voltage applied to a superconductor causes a current that renders undetectable the small current changes produced by radiation events. However, the quasiparticles can be measured with a superconducting tunnel junction (STJ) device. The basic construction of an STJ is a thin layer of insulating material

[31] The concept is credited to Leon N. Cooper, who described the idea in 1956. However, it was actually Richard A. Ogg, Jr., who first proposed the concept in 1946, ten years earlier than Cooper, but Ogg's work was apparently disregarded at the time.

[32] Yes, this distance is small, but typical lattice spacings are a fraction of a nanometer, so this distance is relatively large by comparison.

[33] A particle whose fundamental behavior is altered by its physical environment. For instance, electrons or holes in semiconductors are quasiparticles treated classically as particles traveling in vacuum by applying an effective mass.

[34] Here $\langle N \rangle = E_\gamma / w_q$.

sandwiched between two superconducting materials, in which the insulating barrier impedes current flow. The insulating barrier is made extremely thin such that electrons and holes produced from decoupled Cooper pairs can tunnel through the barrier. With a potential applied to the device, the tunneling current becomes a function of the quasiparticle population crossing the barrier [Rando et al. 1992].

To improve detection efficiency and reduce diffusion of quasiparticles away from the barrier, a multilayered superconducting STJ can be fabricated [Booth 1987; Frank et al. 1998]. The insulating barrier is sandwiched between two superconductors each with a band gap of ϵ_{gT1}. On the outside of these superconductors, a much thicker second superconductor with $\epsilon_{gT2} > \epsilon_{gT1}$ is applied on both sides. The structure forms quantum wells adjacent to the insulating layer, and thereby assists with confinement of quasiparticles at the barrier. At least one type of device has Al_2O_3 as the barrier ($\sim$ 2 nm thick), Al as the lower ϵ_{gT} superconductor ($\sim$ 50 nm) overcoated with Nb as the higher ϵ_{gT} superconductor ($\sim$ 265 nm) [Frank et al. 1998]. Other STJ detector configurations used Ta instead of Nb [Kraft et al. 1999]. Quasiparticles excited across the larger energy gap fall into the well by phonon emissions. If a phonon emission is of adequate energy, more quasiparticles can be excited across the lower energy gap [Mears et al. 1993]. Under a voltage bias, the quasiparticles tunnel through the barrier and can recombine as Cooper pairs on the other side (see Fig. 19.57).

Booth [1987] describes a quasiparticle multiplier device that increases the signal by fabricating successive layers of quantum well tunneling barriers. This layered structure enables multiple tunneling events to occur within some recombination time t_r [Mears et al. 1993]. The effect increases the overall measured signal, but also broadens the statistical spread in energy resolution. With corrections to the expected energy resolution including the Fano factor F (≈ 0.2) and statistical fluctuations in the number of tunneling quasiparticles $(1 + 1/\langle N \rangle)$, the overall energy resolution is described by,

$$\Delta E_{\mathrm{FWHM}} = 2.355\sqrt{w_q E \left(F + 1 + \frac{1}{\langle N \rangle} \right)}. \quad (19.87)$$

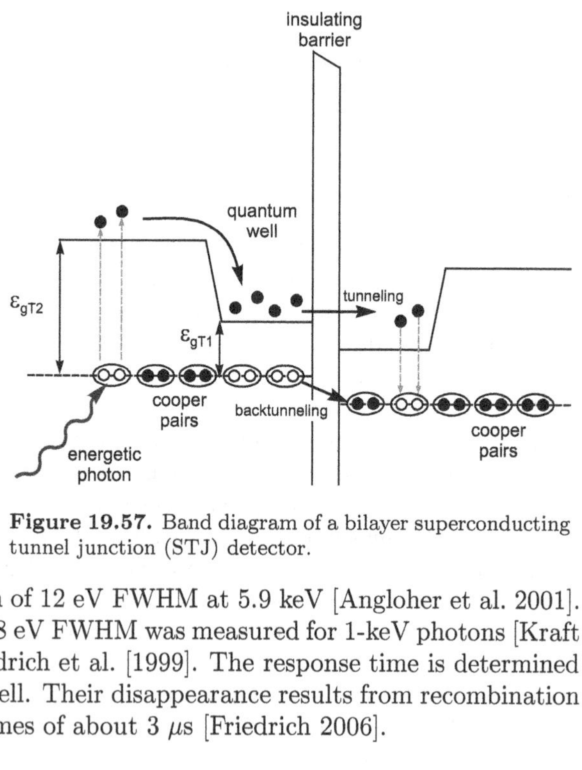

Figure 19.57. Band diagram of a bilayer superconducting tunnel junction (STJ) detector.

STJ detectors have been reported with energy resolution of 12 eV FWHM at 5.9 keV [Angloher et al. 2001]. Also, for lower energy photons, an energy resolution of 5.8 eV FWHM was measured for 1-keV photons [Kraft et al. 1999], and 1.9 eV FWHM for 70-eV photons [Friedrich et al. [1999]. The response time is determined by the time that quasiparticles remain in the quantum well. Their disappearance results from recombination or diffusion from the region which results in response times of about 3 μs [Friedrich 2006].

Microwave Kinetic Inductance Detectors

In the past ten years a new type of athermal detector has attracted considerable attenuation, particularly in the astrophysical community, because these detectors can be easily fabricated as large imaging arrays with high quantum efficiency for photons with wavelengths in the mid infrared (5 μm) to UV photons (0.1 μm). These detectors are known as Kinetic Inductance Detectors (KIDs) and operate by measuring the change in kinetic inductance caused by absorption of photons in a thin strip of superconducting material. The change in the kinetic inductance caused by the liberation of quasiparticles in the semiconductor by the incident photon is measured as the change in the resonance frequency of a microwave resonator and, hence these detectors are known better as Microwave Kinetic Inductance Detectors (MKIDs).

The seminal paper that began the flurry of interest in these detectors is that by Day et al. [2003], upon which most of the following discussion is based. Recall that in a superconductor a DC supercurrent, carried by Cooper pairs, flows without resistance. A Cooper pair, which is bound together by electron-phonon interactions, has a binding energy $w_q < 1.7\epsilon_{gT_c}$. However, superconductors have a non-zero impedance for AC currents. An electric field applied near the surface of a superconductor accelerates the Cooper pairs and, thereby stores energy in the superconductor in the form of kinetic energy. Because the supercurrent is non-dissipative, the stored energy may be extracted by reversing the electric field. Similarly, energy may be stored in the magnetic field inside the superconductor, which penetrates only a short distance, $\lambda \simeq 50$ nm, from the surface. Consequently, a superconductor has a surface inductance $L_s = \mu_o \lambda$ caused by the reactive energy flow between the superconductor and the electromagnetic field. The surface impedance $Z_s = R_s + j\omega L_s$[35] includes a surface resistance R_s to account for AC losses at angular frequency ω caused by the small fraction of electrons that are not in Cooper pairs, i.e., the quasiparticles. For temperatures $T \ll T_c$ we find that $R_s \ll \omega L_s$, thereby having negligible losses in the resonance circuit [Baselmans 2011].

Photons with energy ($h\nu > 2w_q$) may dissociate one or more Cooper pairs. The absorption of a high-energy photon creates $N_{qp} < \eta h\nu / w_q$ quasiparticles where $\eta < 0.57$ is the efficiency with which the photon energy is converted to quasiparticles. The quasiparticles recombine into Cooper pairs on timescales of $\tau_{qp} = 10^{-3}$ to 10^{-6} s during which they diffuse over a distance $\ell \simeq \sqrt{D\tau_{qp}}$ where D is the diffusion constant for quasiparticles in the superconductor material. Similarly, the absorption of a steady stream of low-energy (millimeter/submillimeter) photons would raise the steady-state quasiparticle density by an amount δn_{qp} above its thermal equilibrium value. MKIDs make use of the dependence of the surface impedance Z_s on the quasiparticle density. Although the changes δZ_s are quite small, very sensitive measurements may be made using a resonant circuit. Changes in L_s and R_s affect the frequency and width of the resonance, respectively. These changes, in turn, alter the amplitude and phase of a microwave signal transmitted through the measurement circuit.

The resonant circuit is often a parallel LC circuit which is capacitively coupled to a coaxial through line. The effect of the surface inductance L_s is to increase the total inductance L, while the effect of the surface resistance R_s is to make the inductor slightly lossy (adding a series resistance). On resonance, the LC circuit loads the through line, producing a dip in its transmission. The quasiparticles produced by the photons increase both L_s and R_s. These changes move the resonance to lower frequency (due to L_s), and make the dip broader and shallower (due to R_s). Both of these effects contribute to changing the amplitude and phase of a microwave probe signal transmitted though the circuit.

The choice of a parallel LC circuit coupled to a through line has high transmission at frequencies away from the resonance and, thus, is very well suited for frequency-domain multiplexing because multiple resonators operating at slightly different frequencies can all be coupled to the same through line. These detectors can count individual photons with no false counts and determine the energy and arrival time of each photon with good quantum efficiency.

Other Athermal Detectors

Superconducting Nanowire Single-Photon Detector A superconducting nanowire single-photon detector (SNSPD) uses an array of superconducting wires cooled to well below T_c and biased with a DC current just below the superconducting critical current. Typically these wires are niobium nitride 5 nm thick and 100 nm wide patterned on a substrate. If a photon is absorbed in a wire dissociating Cooper paired electrons, a small region is created in which the critical current is momentarily reduced below the bias current to form a small non-superconducting region. A detectable voltage pulse of about 1 ns duration indicates the photon

[35] $j = \sqrt{-1}$ is used to avoid confusion with the multiple uses of i.

absorption event [Natarajan et al. 2012]. Advantages of this detector include its high speed (up to 2 GHz count rate) and low dark current. However, SNSPDs provide little information about the photon energy.

Roton Detectors The elementary collective excitations in superfluid ^{4}He are *phonons* and *rotons*.[36] A particle interacting in this fluid can produce rotons. These rotons can be detected bolometrically or by the evaporation of helium atoms [Bandler et al. 1992]. Because ^{4}He is very pure, rotons travel ballistically and are stable and thus a large volume of fluid can be used.

Quasiparticles in Superfluid ^{3}He Below 0.001 K ^{3}He behaves as a *superfluid* in which pairs of atoms are bound as quasiparticles similar to Cooper pairs, ultimately having zero viscosity and flows without kinetic energy losses. The paired atoms have an exceptionally small energy gap of about 100 neV. Thus a detector similar to a STJ could be fabricated [Bradley et al. 1995]. The advantage of such a detector is that a huge number (10^9) of unpaired normal ^{3}He atoms are produced per particle interaction. But this detector is not very practical because it is very difficult to detect the normal ^{3}He atoms and to produce and maintain much superfluid at such low temperatures.

Capabilities of the Two Classes of Cryogenic Detectors

The principal advantage of charge detectors over microcalorimeters is that the former is much faster [Frank et al. 1998]. STJs and MKIDs can produce several thousand counts per second per pixel. By contrast, microcalorimeters are limited to a few tens of counts per second. However, charge detectors, unlike microcalorimeters, are limited to low energy photons because superconductors cannot sustain electric fields to sweep out the charges and charges diffuse little before they recombine. Microcalorimeters, while somewhat more robust in measuring the energy of the radiation quanta, cannot absorb all the energy of very energetic particles, such as a 10-MeV gamma ray, in their typically small absorbers.

19.5 Wavelength-Dispersive Spectroscopy

Ultra-high resolution can be achieved for low energy gamma and x rays with wavelength dispersive spectroscopy (WDS), which can yield x-ray peak resolution better than that of semiconductor or cryogenic detectors. The method utilizes Bragg scattering, in which the Bragg condition for constructive interference must be satisfied, i.e,

$$n\lambda = 2d \sin\theta, \tag{19.88}$$

where n is an integer, d is the spacing between crystalline planes, λ is the wavelength of the photon under inspection, and θ is the angle at which radiation intersects the crystal from the parallel condition as shown in Fig. 19.58.

Because it is difficult to make a portable diffraction system, this instrument is generally attached to an electron microprobe or scanning electron microscope (SEM) [Goldstein et al. 1981]. These instruments accelerate a focused energetic electron beam onto a sample to produce induced radiations such as secondary electrons, backscattered electrons (from the beam), and various x-rays as shown in Fig. 19.59. Different radiation detectors are used with these types of instruments, including gas-filled, scintillation, and semiconductor detectors. For instance, secondary electrons are generally detected by a scintillation detector[37] shrouded by a Faraday cage, all inside the SEM or microprobe chamber. Backscattered electrons have lower energy, and because of their wider spatial dispersement, an annular semiconductor detector around the SEM pole piece is used to detect these electrons. X-ray emissions are often detected with a cryogenically-cooled

[36] Rotons are excitations or quasiparticles with a very short wavelength of about 3Å, which is about the same as the spacing between the helium atoms in the liquid. At temperatures of 1 K and above, it is the rotons which make the main contribution to the specific heat and the thermal energy.

[37] Named a Everhart-Thornley detector after the inventors.

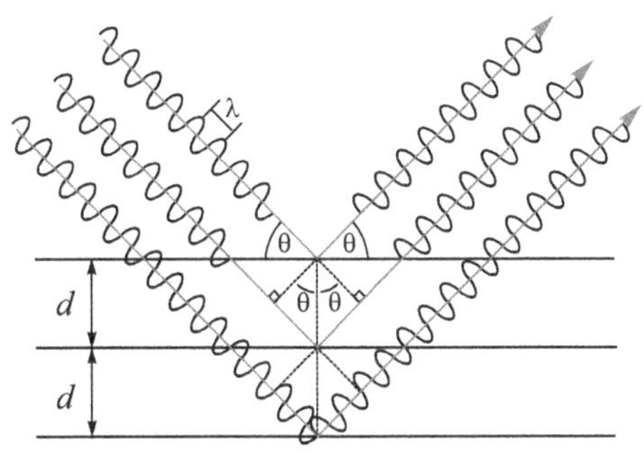

Figure 19.58. Geometry for Bragg diffraction of photons.

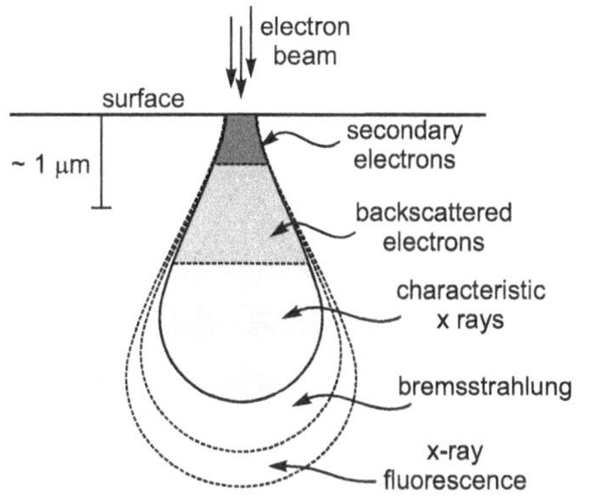

Figure 19.59. Radiations emitted from different parts of the interaction volume probed with an electron microprobe.

Si(Li) detector (see Sec. 16.4.1), which yields excellent spectroscopic energy resolution.[38] However, in some cases, higher x-ray energy resolution is required than that provided by a chilled Si(Li) detector and in such cases WDS can produce the higher energy resolution.

Shown in Fig. 19.60 is a common arrangement for the WDS tool, which allows a sample to be inspected while it is irradiated with an electron beam which produces characteristic x rays from the sample. These x rays intersect a slightly bent diffraction crystal. Those x rays satisfying the Bragg condition diffract into a detector and can be recorded, whereas other x rays are absorbed in the crystal, scatter randomly, or pass through the crystal. Because only the number of counts at a given diffraction angle need be recorded, the detector need not be a high-resolution spectrometer, hence a gas-filled proportional counter is usually used as the x-ray detector. During operation, the crystal and detector are rotated through a Rowland circle, which permits the Bragg condition to be maintained as the arrangement rotates through a continuum of wavelengths.

As a result, the x ray detector records the number of counts as a function of wavelength. The stringent requirement for the Bragg condition results in ultra-high resolution, which can be plotted as a function of photon energy. The important advantage of WDS is the superior identification ability it provides to the user. Comparison spectra between a Si(Li) spectrometer, a microcalorimeter, and a WDS spectrometer are shown in Fig. 19.61. Unfortunately, the system can be used only for photons with sufficiently low energy that Bragg diffraction can occur. Several commercial systems have a rotating rack of different diffraction crystals that extend the sensitive range. WDS systems are laboratory-based, hence are not generally considered portable.

19.6 Čerenkov (Cherenkov) Detectors

A charged particle passing through a dielectric material can cause the molecules along its path of travel to form dipoles, forming a segment of polarization. These molecules promptly return to the relaxed condition by the spontaneous emission of light, forming wavelets of light to propagate through the dielectric. If the particle is passing through the material at velocities below that of light in the same media, these successively released wavelets destructively interfere with each other, resulting in the suppression of luminescence. However, if the particle is traveling faster than the speed of light in the medium, these wavelets add constructively to produce a wavefront of light (depicted in Fig. 19.62, left). This pulse of light is known as Čerenkov

[38]This method is often called EDS, or energy dispersive spectroscopy, but is the same as energy pulse height analysis.

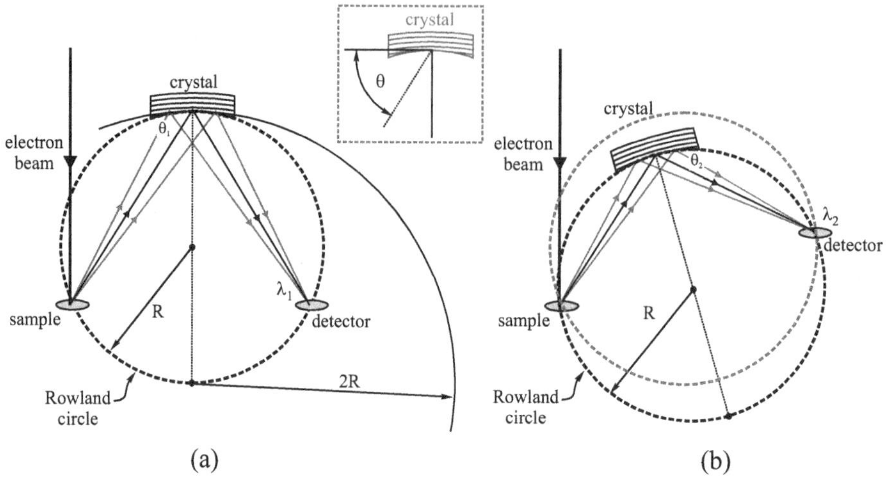

Figure 19.60. A typical WDS diffraction arrangement is aligned on a Rowland circle. The sample location remains stationary. The diffraction crystal is bent with a radius twice that of the Rowland circle radius R, and it is typically ground with radius R. The Bragg condition is maintained for various values of λ by moving both the crystal and detector, with the sample remaining stationary, such that all points remain on the Rowland circle.

(Cherenkov) radiation, after the Russian physicist who first observed the phenomenon in 1934 by noticing blue light emissions from water irradiated with gamma rays [Čerenkov 1934; see also Jelley 1955].[39]

The speed of light c' in a dielectric medium is determined by the index of refraction n of the medium as

$$c' = \frac{c}{n}, \tag{19.89}$$

where c is the speed of light in a vacuum. If the particle speed v is greater than c', then conditions arise that produce Čerenkov radiation.[40] Consider the diagram of Fig. 19.62(right) in which the distance traveled by a particle from point **A** to point **C** is represented by vt. If the speed of the particle $v > c'$, i.e., $v/c \equiv \beta > 1/n$ then the dipoles along its path emit light as the particle slows. The time t for the light wave to travel from point **A** to point **B** is the same as the time needed for the particle to travel from point **A** to point **C**, namely

$$vt \cos\theta = \frac{c}{n}t. \tag{19.90}$$

Rearrangement gives the following condition for Čerenkov radiation emissions

$$\cos\theta = \frac{1}{\beta n}. \tag{19.91}$$

There are some important observations to make from this result. First, the particle speed v greatly affects the light emission angle θ because Čerenkov radiation is emitted anisotropically, i.e., this light is emitted perpendicularly to the surface of a conical shell subtended by an angle 2θ along the particle path. Second, if βn falls to unity, then θ goes to zero and there is no emission angle or light emission. As the particle loses

[39] Pavel Čerenkov, Il'ja Frank, and Igor Tamm were awarded the 1958 Nobel Prize in Physics for the discovery and interpretation of this phenomenon.

[40] Oliver Heaviside had already predicted Čerenkov radiation in two papers published in 1888 and 1889.

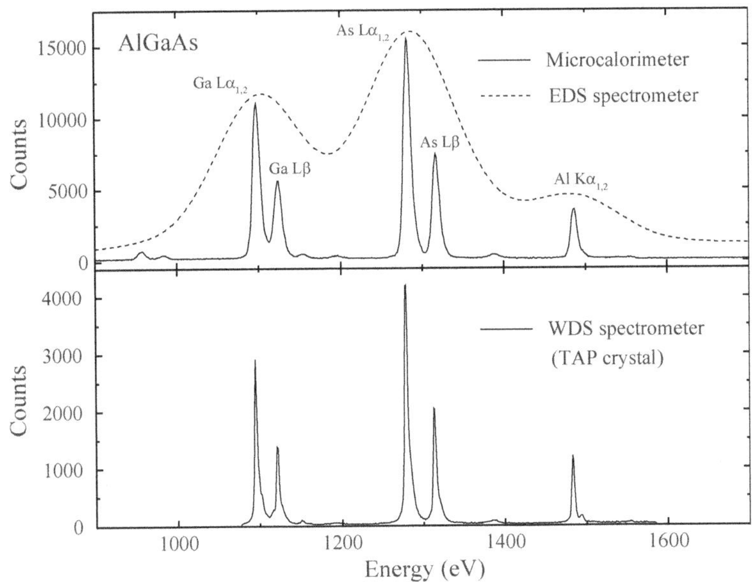

Figure 19.61. Shown are (top) a comparison of EDS spectra from a Si(Li) detector and a microcalorimeter detector and (bottom) an additional comparison to a WDS detector. Reproduced from Wollman et al., *J. Microscopy*, **188**, 196–223, (1997), with the permission of Wiley Publishing.

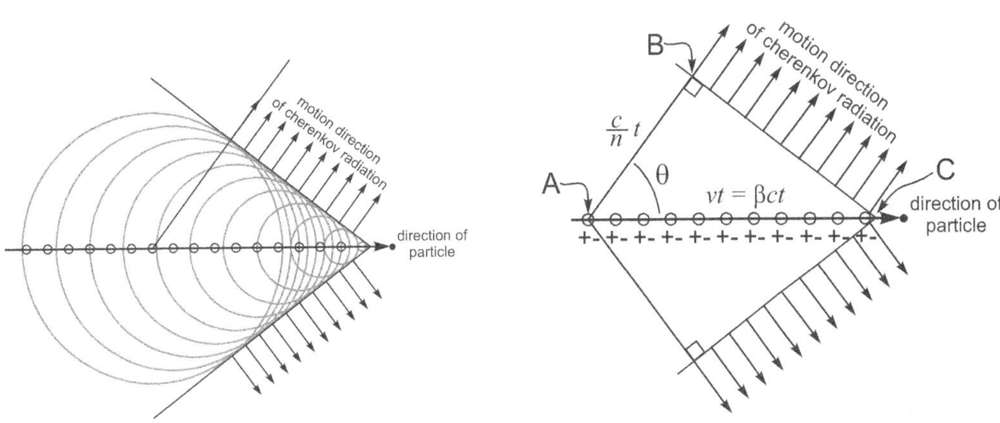

Figure 19.62. (left) A particle moving through a medium faster than the velocity of light produces polar-aligned waves. (right) The angle of light emission is defined by the particle speed and the index of refraction.

energy through light emissions, the conical angle diminishes and eventually becomes zero. The emission anisotropy was photographically documented by Čerenkov by irradiating water and benzene with gamma rays [Čerenkov 1937]. In that experiment, a vial filled with either water or benzene was surrounded by an angled toroidal mirror to reflect the light up into a camera. The resulting images are shown in Fig. 19.63. The general outline of the cone is apparent as two fuzzy bright spots in the photographs. Also apparent is an increase in cone angle with an increase in the index of refraction as predicted by Eq. (19.91).

Frank and Tamm [1937] used Maxwell's equations to obtain the following expression for the number of Čerenkov photons emitted, per unit differential path length of travel for a particle with charge number z,

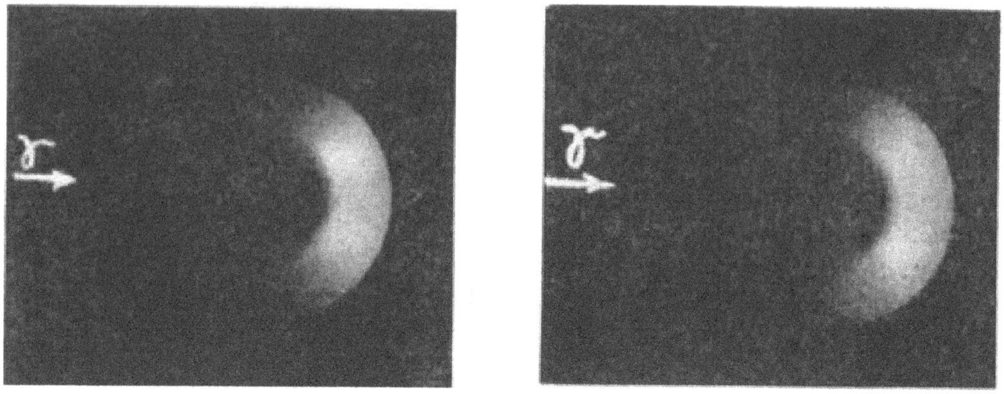

Figure 19.63. Čerenkov radiation from (left) water ($n = 1.337$) and (right) benzene ($n = 1.501$). The semiannular feature is an artifact of the angled toroidal mirror wrapped around the dielectric samples. However, the two fuzzy bright spots in both photographs do show the perimeter of the emission cone. Reproduced from Čerenkov, *Phys. Rev.*, **52**, 378–379, (1937), with the permission of APS Publishing.

per unit wavelength,

$$\frac{d^2N}{dx\,d\lambda} = 2\pi\alpha z^2\left(1 - \frac{1}{\beta^2 n^2}\right)\frac{1}{\lambda^2} = \frac{2\pi\alpha z^2}{137\lambda^2}\left(1 - \cos^2\theta\right) = \frac{2\pi\alpha z^2}{137\lambda^2}\sin^2\theta, \tag{19.92}$$

where λ is the emission wavelength and $\alpha \equiv q^2/(2\epsilon_0 hc) = 1/137.03567$ is the mysterious dimensionless fine-structure constant.[41] From Eq. (19.92), it becomes clear that the light yield increases as λ decreases so that the emission spectrum is greatest in the violet-blue end of the visible spectrum. However, ultraviolet light is readily absorbed in many dielectrics and, thus, reduces that portion of the spectrum. Consequently, Čerenkov radiation usually appears blue in color, although many other visible light wavelengths are emitted but with lower intensities. In the derivation of this expression for Čerenkov radiation, Frank and Tamm assumed that the index of refraction n is constant, i.e., independent of the light wavelength. However, for wavelengths in the x- and gamma-ray portion of the spectrum, n decreases to below unity and Čerenkov emission cannot occur.

The number of Čerenkov photons emitted between λ_1 and $\lambda_2 < \lambda_1$, per unit path length of travel of the charged particle, is

$$\frac{dN}{dx} = \int_{\lambda_2}^{\lambda_1} \frac{d^2N}{dx\,d\lambda}\,d\lambda = \frac{2\pi z^2}{137}\left[\frac{1}{\lambda_2} - \frac{1}{\lambda_1}\right]\sin^2\theta. \tag{19.93}$$

Because the most probable wavelengths are also in the sensitive region of most photomultiplier tubes, Čerenkov radiation can contribute to background for some scintillation detectors, especially organic scintillators. Also, because common glass ($n = 1.5$) can produce Čerenkov radiation, the glass enclosure of a PMT can produce background Čerenkov photons and it is especially problematic if the photocathode or first few dynode stages are illuminated by these photons.

The total threshold particle energy E_{th} for Čerenkov emission occurs when $\beta = 1/n$ and is found to be

$$E_{th} = \frac{m_0c^2}{\sqrt{1-\beta^2}} = m_0c^2\sqrt{\frac{n^2}{n^2-1}}, \tag{19.94}$$

[41] No one knows why $1/\alpha = 137.03567$ or how to calculate it. But if it were to change even slightly, our universe would be very different. For example, if it changed by even 0.4% fusion in stars could not produce carbon and carbon-based life would not be possible.

which includes rest mass energy equivalent m_0c^2. The kinetic energy T_{th} required to first produce Čerenkov radiation is found by subtracting the rest mass energy of the particle from the above total energy to give

$$T_{th} = m_0c^2 \left[\sqrt{\frac{n^2}{n^2 - 1}} - 1 \right]. \tag{19.95}$$

Compton scattering is the most probable interaction for gamma rays in low Z dielectrics, and the gamma-ray energy E_γ to produce a maximum energy $T_{\max}$ of the recoil electron can be found from Eq. (4.74) with $\theta_s = \pi$ as

$$E_\gamma = \frac{T_{\max} + \sqrt{T_{\max}^2 + 2T_{\max}m_oc^2}}{2}. \tag{19.96}$$

Substituting T_{th} of Eq. (19.95) for $T_{\max}$ into Eq. (19.96) then gives an expression for the minimum energy of gamma rays to produce recoil electrons that in turn produce Čerenkov radiation in a medium with an index of refraction n. The threshold energies for electrons and gamma rays are plotted in Fig. 19.64.

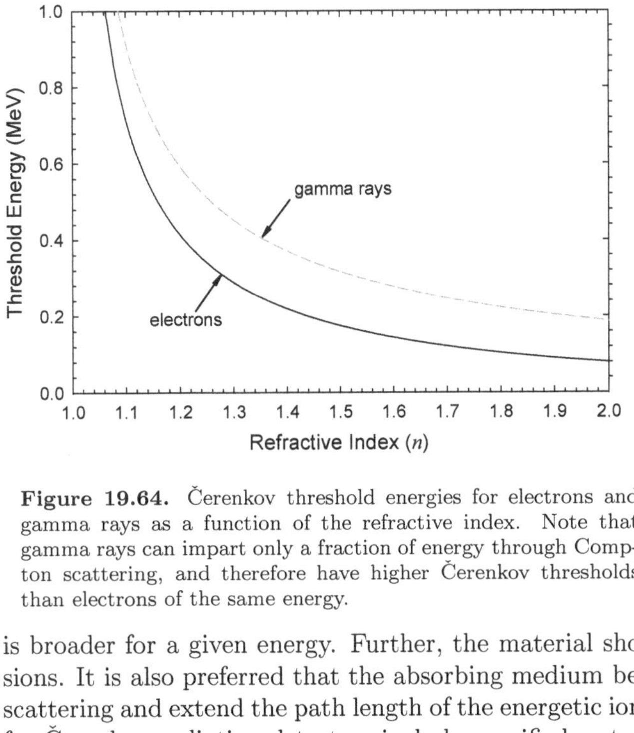

Figure 19.64. Čerenkov threshold energies for electrons and gamma rays as a function of the refractive index. Note that gamma rays can impart only a fraction of energy through Compton scattering, and therefore have higher Čerenkov thresholds than electrons of the same energy.

Čerenkov detectors are used for a number of different applications for the detection of highly energetic particles, which include gamma rays (producing energetic electrons), electrons, β particles, π mesons, neutrinos, and cosmic rays. They usually consist of a transparent interaction medium, such as water or heavy water (D_2O), plastic, or other organic material. Attached to the absorbing medium are light detectors, usually photomultiplier tubes (PMTs). Materials chosen for Čerenkov detectors should not produce scintillation luminescence or phosphorescence. For high n materials, the energy threshold is lower and the Čerenkov cone angle is broader for a given energy. Further, the material should be optically transparent to the Čerenkov emissions. It is also preferred that the absorbing medium be composed of low Z lightweight materials to reduce scattering and extend the path length of the energetic ionizing particles. A short list of materials investigated for Čerenkov radiation detectors includes purified water, heavy water, lucite, silica aerogel, benzene, glass, and various gases (some listed in Table 19.4).

Čerenkov radiation detectors have a fast time response, typically measured in picoseconds. Although the fast response may have discrimination advantages from other forms of luminescence, this fast timing property ultimately is limited by the PMT response. The Čerenkov light flashes are relatively dim compared to those from scintillators consisting of several hundred visible light photons per centimete of particle travel in the medium. Distinguishing this dim light from background light can be a problem.

The light wave direction and cone angle can be used to determine both particle energy and speed. The light ring formed by the cone can be projected upon a surface, a method used in Ring Imaging Čerenkov Counters (RICH) [Seguinot and Ypsilantis 1994; Ypsilantis and Seguinot 1994]. A RICH counter has an interaction medium adjacent to a reflecting mirror that projects Čerenkov light rings upon a light imaging sensor. These measurements can also be conducted by arranging multiple PMTs around (or in) the absorber

Table 19.4. Čerenkov detector properties for pure materials.

Material	Refractive Index (n) $\lambda = 589.29$ nm	Threshold Energy (MeV) (electrons)	Threshold Energy (MeV) (gamma rays)
Air (STP)	1.000277	21.2	21.46
Air (0°C)	1.000293	20.6	20.86
Ar (0°C)	1.00028	21.09	21.34
H_2 (0°C)	1.000132	30.94	31.2
He (0°C)	1.000036	59.71	59.97
Silica Aerogel	1.0026–1.26	6.589–0.329	6.835–0.498
Liquid He	1.25	0.34	0.511
Water	1.33	0.264	0.423
Water (Ice)	1.31	0.28	0.442
Heavy Water (D_2O)	1.328	0.266	0.425
Water + 10% Glucose Sol.	1.3477	0.251	0.408
Water + 20% Glucose Sol.	1.3635	0.241	0.396
Water + 60% Glucose Sol.	1.4394	0.199	0.3465
Ethanol	1.361	0.242	0.398
Benzene	1.501	0.174	0.315
PMMA (Lucite)	1.49	0.178	0.320
Pyrex	1.47	0.186	0.330
Quartz	1.5	0.175	0.316
Flint Glass	1.6	0.144	0.276
Diamond	2.417	0.050	0.141

and tracking the PMTs that respond to the Čerenkov cone. The Sudbury Neutrino Observatory (powered off on Nov. 28, 2006) had a 6-m diameter spherical acrylic tank filled with 1102 tons of heavy water, surrounded by 9600 PMTs. The Super-Kamiokande Neutrino Detection Experiment, or Super-K, has 50,000 tons of purified water with over 11,100 PMTs surrounding the tank.[42]

There is an enormous Čerenkov detector array in Antarctica underneath the Amundsen-Scott Pole Station, formerly the Antarctic Muon and Neutrino Detector Array (AMANDA), but now officially a part of the IceCube Neutrino Observatory array. It consists of digital optical modules (DOMs), each containing one photomultiplier tube (PMT), sunk in the Antarctic ice cap at depths between 1500 to 1900 meters. AMANDA was originally composed of 677 optical modules mounted on 19 separate strings spread out in a circle roughly 200 meters in diameter. The newer IceCube array has 60 DOMs per string, with 86 strings deployed ranging in depths from 1450 m to 2450 m. There are 5160 DOMs in operation within the IceCube Neutrino Observatory. Each string was placed in the ice by first drilling holes with a hot water hose, lowering the cable string and attached optical modules into the hole, and then the water was allowed to refreeze around the string.

AMANDA detects neutrinos that pass through the Earth from the northern hemisphere and then react just as they are leaving upwards through the Antarctic ice. A neutrino collides with nuclei of oxygen or hydrogen atoms contained in the surrounding water ice, producing a muon and a hadronic shower. The optical modules detect the Čerenkov radiation produced by these particles, and by analysis of the timing and location of when and where photons strike the PMTs, the direction of the original neutrino can be determined with a spatial resolution of approximately 2 degrees.

[42]On November 12, 2001, the Super-Kamiokande suffered a severe setback when one of the PMTs imploded and caused a shock wave in the water that subsequently led to a chain reaction and implosion of over 7000 of the PMTs. The accident occurred when the water tank was being refilled after routine maintenance. The detector has been rebuilt and is back in operation.

The IceTop detector array, which is a surface array above IceCube arranged to detect neutrino air showers, is composed of an additional 320 optical sensors. The main goal of the IceCube experiment is to detect neutrinos in the energy range from 10^{11} eV to about 10^{21} eV.

An additional array called the Deep Core Low-Energy Extension, part of the IceCube Neutrino Observatory, is located in the overall IceCube array at depths between 1760 m and 2450 m. The array location was chosen for the clarity of the ice at those depths. The Deep Core array is targeted at measuring neutrinos with energy below 100 GeV and is focused on identifying and characterizing extra-solar sources of neutrinos. Compared to other large underground neutrino detectors consisting of large water tanks and hundreds of PMTs, the IceCube is capable of looking at higher energy neutrinos because it is not limited in volume. However, accuracy is compromised because the PMTs are spaced further apart and the conditions of the Antarctic ice are not controllable. Traditional neutrino observatories, such as the former Sudbury Neutrino Observatory or the Super-Kamiokande Neutrino Observatory, were designed to provide greater detail about neutrinos emanating from the Sun or generated in the Earth's atmosphere. However, higher neutrino energies in the spectrum should be dominated by neutrinos from sources external to our solar system.

PROBLEMS

1. Prove the result of Eq. (19.8) from Eq. (19.7).

2. Derive the result of Eq. (19.15) beginning from Eq. (19.13).

3. Both scintillators and TLDs emit photons from radiation absorption, and the phenomenon is used to measure radiation. Explain the basic difference between these two types of detectors. Why are electron traps preferred in TLDs but not in scintillators?

4. Consider an extension to the simple model of single-trap level TLD discussed in Sec. 19.1.1. Assume the traps have a continuous distribution of energies $N_{te}(E_t)$. Derive an expression for the glow curve from such a TLD analogous to that of Eq. (19.10) for multiple discrete trap levels.

5. An OSL has decay constant $\tau = 6$ ms and $\upsilon = 1\ \text{s}^{-1}$. Given a duty cycle of 10%, determine the fraction of measured light from an OSL measurement if (a) $\tau = 30$ ms and (b) if $\tau = 3$ ms. Assume $P_r^*(0) = 0$.

6. Discuss the many advantages and disadvantages of a TLD versus an OSLD.

7. Why is the logarithm of the *transmitted* intensity through photographic film used instead of simply the intensity to characterize the response of film to light?

8. What is the critical angle for a track etch film where $v_T = 2v_G$?

9. Quartz is perpendicularly irradiated with 5.5 MeV alpha particles with average range of 25 microns. The track is then etched in a dilute solution of hydrofluoric acid. Given an etch rate of 50 nanometers per minute for a general undamaged rate, with the damaged rate being 3 times the undamaged rate, what is the diameter of the track etch after 30 minutes? What is the track length?

10. Derive Eq. (19.78) from Eq. (19.76).

11. Consider the simplified cryogenic calorimeter discussed in Sec. 19.4.2. Calculate and plot the output response produced by a series of impulse inputs, i.e., $P_{in}(t) = \sum_{i=1}^{N} E\delta(t - t_i)$ where $t_i = (i-1)\tau_F/2$.

12. What is the minimum electron energy required to produce Čerenkov radiation in flint glass? Compare this energy to the minimum required energy to produce Čerenkov radiation in pure water.

13. Based on Eq. (19.92), show that the number of Čerenkov photons emitted per unit path length of charged-particle travel and per unit frequency is independent of frequency and is given by

$$\frac{d^2N}{dx\,d\nu} = 2\pi\alpha z^2 \left(1 - \frac{1}{\beta^2 n^2}\right).$$

Show that integration of this result from frequency ν_1 to $\nu_2 > \nu_1$ (corresponding to wavelengths λ_1 and λ_2) to obtain the number of photons emitted between these two frequencies is the same as that given by Eq. (19.93).

14. Equation (19.92) suggests that as $\lambda \to \infty$ an infinite number of Čerenkov photons which have an infinite amount of energy are emitted. Clearly this cannot be the case. What is the explanation for this false inference?

15. Derive Eq. (19.96) from Eq. (4.74).

16. Gamma rays of various energies irradiate a translucent material and Čerenkov radiation is observed when the gamma rays have an energy greater than 1.6 MeV. What is the index of refraction for the material?

REFERENCES

AKSELROD, M.S. AND S.W.S. MCKEEVER, "A Radiation Dosimetry Method Using Pulsed Optically Stimulated Luminescence," **81**, 167–176 (1999).

AKSELROD, M.S., N.A. LARSEN, V. WHITLEY, AND S.W.S. MCKEEVER, "Thermal Quenching of F-Center Luminescence in Al_2O_3:C," *J. Appl. Phys.*, **84**, 3364–3373, (1998).

ANDERSON, P.W., "Absence of Diffusion in Certain Random Lattices," *Phys. Rev.*, **109**, 1492–1505, (1958).

ANDREWS, D.H., W.F. BRUCKSCH, JR., W.T. ZIEGLER, AND E.R. BLANCHARD, "Attenuated Superconductors I. For Measuring Infra-Red Radiation," *Rev. Sci. Instrum.*, **13**, 281–292, (1942).

ANGLOHER, G., P. HETTI, M. HUBER, J. JOCHUM, F.V. FEILITZSCH, AND R.L. MÖßBAUER, "Energy Resolution of 12 eV at 5.9 keV from Al-Superconducting Tunnel Junction Detectors," *J. Appl. Phys.*, **89**, 1425–1429, (2001).

APFEL, R.E., "The Superheated Drop Detector," *Nucl. Instrum. Meth.*, **162**, 603–608, (1979).

APFEL, R.E., "Characterisation of New Passive Superheated Drop (Bubble) Dosemeters," *Rad. Prot. Dosim.*, **44**, 343–346, (1992).

APFEL, R.E. AND F. D' ERRICO, "A Neutron Spectrometer Based on Temperature Variations in Superheated Drop Compositions," *Nucl. Instrum. Meth.*, **A476**, 298–303, (2002).

ARNOLD W.A., "Experiments," *Photosythesis Research*, **27**, 73–82, (1991).

ATTIX, F.H., *Introduction to Radiological Physics and Radiation Dosimetry*, New York: Wiley, 1986.

AYDIN, T., H. DEMIRTAS, AND S. AYDIN, "TL/OSL Studies of $Li_2B_4O_7$:Cu Dosimetric Phosphors," *Radiat. Meas.*, **58**, 24–32, (2013).

BACRANIA, M.K., A.S. HOOVER, P.J. KARPIUS, M.W. RABIN, C.R. RUDY, ET AL., "Large-Area Microcalorimeter Detectors for Ultra-High-Resolution X-Ray and Gamma-Ray Spectroscopy," *IEEE Trans. Nucl. Sci.*, **56**, 2299–2302, (2009).

BANDLER, S.R., R.E. LANOU, H J. MARIS, T. MORE, F.S. PORTER, G.M. SEIDEL, AND R.H. TORIL, "Particle Detection by Evaporation from Superfluid Helium," *Phys. Rev. Lett.*, **68**, 2429–2432, (1992).

BANDLER, S.R., C. ENSS, R.E. LANOU, H.J. MARIS, T. MORE, F.S. PORTER, AND G.M. SEIDEL, "Metallic Magnetic Bolometers for Particle Detection," *J. Low Temp. Phys.*, **93**, 709–714, (1993).

BANDYOPADHYAY, P.K., G.W. RUSSELL, AND K. CHAKRABARTI, "Optically Stimulated Luminescence in KCl:Cu X-Irradiated at Room Temperature," *Radiat. Meas.*, **30**, 51–57, (1999).

BARKAS, W.H., *Nuclear Research Emulsions, I. Techniques and Theory*, New York: Academic Press, 1963.

BASELMANS, J., "Kinetic Inductance Detector," *J. Low Temp. Phys.*, **167**, 292–304, (2012).

BECKER, K., *Photographic Film Dosimetry*, London: Focal Press, 1966.

BECKER, K., "The Effect of Oxygen and Humidity on Charged Particle Registration in Organic Foils," *Radiat. Res.*, **36**, 107–118, (1968).

BECKER, K., *Solid State Dosimetry*, Cleveland: CRC Press, 1973.

BECQUEREL, E., *La Lumiére: Ses Causes et Ses Effets*, Tome 2: Effets de la Lumire, Paris: F. Didot, 1868.

BIDYASAGAR, M., Th.B. SINGH, A.G. BARUA, AND R.K. GARTIA, "Trap Spectroscopy and Thermoluminescense of CaF_2 Based TLDs," *Indian J. Pure Appl. Phys.*, **52**, 13–18, 2014.

BINDER, W. AND J.R. CAMERON, "Dosimetric Properties of CaF_2:Dy," *Health Phys.*, **17**, 613–618, 1969.

BLANFORD, G.E., JR., R.M. WALKER, AND J.P. WEFEL, "Track Etch Parameters for Plastics," *Radiation Effects*, **3**, 267–270, (1970).

BLUSZCZ, A. AND L. BØTTER-JENSON, "Dosimetric Properties of Natural Quartz Grains Extracted from Fired Materials," *Rad. Measurements*, **24**, 465–468, 1995.

BONIN, H.W., G.R. DESNOYERS, AND T. COUSINS, "Fast Neutron Dosimetry and Spectroscopy Using Bubble Detectors", *Rad. Prot. Dosim.* **46**, 265–271, (1993).

BOOTH, N.E., "Quasiparticle Trapping and the Quasiparticle Multiplier," *Appl. Phys. Lett.*, **50**, 293–295, (1987).

BOS, A.J.J.,"High Sensitivity Thermalluminescence Dosimetry," *Nucl. Instrum. Meth.*, **B184**, 3–28, (2001).

BOS, A.J.J., M. PROKIĆ, AND J.C. BROUWER, "Optically and Thermally Stimulated Luminescence Characteristics of MgO:Tb^{3+}," *Radiat. Prot. Dosi.*, **119**, 130–133, (2006).

BRADLEY, D.I., Yu.M. BUNKOV, D.J. COUSINS, M.P. ENRICO, S.N. FISHER, M.R. FOLLOWS, A.M. GUÉNAULT, W.M. HAYES, G.R. PICKETT, AND T. SLOAN, "Potential Dark Matter Detector? The Detection of Low Energy Neutrons by Superfluid ^{3}He," *Phys. Rev. Lett.*, **75**, 1887-1890, (1995).

BØTTER-JENSEN, L. AND S.W.S. MCKEEVER, "Optically Stimulated Luminescence Dosimetry Using Natural and Synthetic Materials," *Rad. Prot. Dosim.*, **65**, 273–280, (1996).

BØTTER-JENSEN, L., N. AGERSNAP, B.G. MARKEY, AND S.W.S. MCKEEVER, "Al_2O_3:C as a Sensitive OSL Dosemeter for Rapid Assessment of Environmental Photon Dose Rates," *Rad. Meas.*, **27**, 295–298, (1997).

BØTTER-JENSEN, L., S.W.S. MCKEEVER, AND A.G. WINKEL, *Optically Stimulated Luminescence Dosimetry*, Amsterdam: Elsevier, 2003.

BOYD, C.A., "A Kinetic Study of Thermoluminescense of Lithium Fluoride," *J. Chem. Phys.*, **17**, 1221–1226, (1949).

BROOKE, C. AND R. SCHAYES, "Recent Development in Thermoluminescent Dosimetry; Extensions in the Range of Applications," *Proc. Int. Symp. Solid-State and Chem. Radiat. Dosimetry*, IAEA, Vienna, 31–40, 1967.

BRYANT, C.A. AND P.H. KEESOM, "Low-Temperature Specific Heat of Germanium," *Phys. Rev.*, **124**, 698–700, (1961).

BUDD, T., M. MARSHALL, L.H.J. PEAPLE, AND J.S. DOUGLAS, "The Low- and High-Temperature Response of Lithium Fluoride Dosemeters to X-Ray," *Phys. Med. Biol.*, **24**, 71–80, (1979).

BÜHLER, M. AND E. UMLAUF, "A Magnetic Bolometer for Single-Particle Detection," *Europhys. Lett.*, **5**, 297–301, (1988).

BÜHLER, M., T. FAUSCH, AND E. UMLAUF, "Measurement of Spin-Lattice Relaxation Times at Millikelvin Temperatures," *Europhys. Lett.*, **23**, 529–534, (1993).

BÜHLER, M., E. UMLAUF, AND J.C. MATHER, "Noise of a a Bolometer with Vanishing Self-Heating," *Nucl. Instrum. Meth.*, **A346**, 225–229, (1994).

BULAR, E., "An Alternative Technique for Optically Stimulated Luminescence (OSL) Experiment," *Radiat. Meas.*, **26**, 701–709, (1996).

BULAR AND GÖKSU, "OSL from BEO Ceramics: New Observations from an Old Material," *Rad. Measurements*, **29**, 639–650, (1998).

BULLOUGH, R. AND J.J. GILMAN, "Elastic Explosions in Solids Caused by Radiation," *J. Appl. Phys.*, **37**, 2283–2287, (1966).

CAMERON, J.R., N. SUNTHARALINGAM, AND G.N. KENNEY, *Thermoluminescent Dosimetry*, Madison: U. Wisconsin Press, 1968.

CAMERON, J., "Radiation Dosimetry," *Environ. Health Perspectives*, **91**, 45–48, (1991).

ČERENKOV, P.A., "Visible Light in Clear Liquids Under γ-Radiation," *Compt. Rend. de l'Acad. des Sci. IURSS*, **3**, 451–454, (1934).

ČERENKOV, P.A., "Visible Radiation Produced by Electrons Moving in a Medium with Velocities Exceeding that of Light," *Phys. Rev.*, **52**, 378–379, (1937).

CERN, photograph CERN-EX-23296-1, 1973, (accessed 2018).

CHANG, W.Y. AND S. ROSENBLUM, "A Simple Counting System for Alpha-Ray Spectra and the Energy Distribution of Po Alpha-Particles," *Phys. Rev.*, **67**, 222–227, (1945).

CHASSENDE-BAROZ, N., "Utilisation des Émulsions Photographiques pour las Measure de Forte Doses - Influence du Débit de la Source de Rayonnement," *Selected Topics in Radiation Dosimetry*, IAEA, Vienna, 285–288, (1961).

CHRISTENSON, P., "Manganese-Activated Lithum Borate as a Thermoluminescent Dosimetry Material," *RISÖ Report 161*, Danish AEC Res. Est. Risö, Denmark: Roskilde, 1967.

CLARKE, M.L. AND H.M. RENDELL, "Infra-red Stimulated Luminescence Spectra of Alkali Feldspars," *Radiat. Meas.*, **27**, 221–236, (1997).

CLUCHET, J. AND H. JOFFRE, "Application of Thermoluminescence Dosimetry in Health Physics," *Luminescence Dosimetry, Proc. Int. Conf.*, AEC Symp. Ser., **Vol. 8**, CONF-650637, US-AEC, 1965.

COOPER, L.N., "Bound Electron Pairs in a Degenerate Fermi Gas," *Phys. Rev.*, **104**, 1189–1190, (1956).

CRASE, K.W. AND R.B. GAMMAGE, "Improvements in the Use of Ceramic BeO in TLD," *Health Phys.*, **29**, 739–746, (1975).

CROSS, W.G., "Characteristics of Track Detectors for Personnel Neutron Dosimetry," *Int. J. of Rad. Appl. Instrum. Part D. Nucl. Tracks and Rad. Meas.*, **12**, 533–542, (1986).

DANIELS, F., C.A. BOYD, AND D.F. SAUNDERS, "Thermoluminescence as a Research Tool," *Science*, **109**, 440, (1949).

DANIELS, F. AND D.F. SAUNDERS, "The Thermoluminescence of Rocks," *Science*, **111**, 462, (1950).

DANIELS, F., C.A. BOYD, AND D.F. SAUNDERS, "Thermoluminescence as a Research Tool," *Science*, **117**, 343–349, (1953).

DAY, P.K., H.G. LEDUC, B.A. MAZIN, A. VAYONAKIS, AND J. ZMUIDZINAS, "A Broadband Superconducting Detector Suitable for Use in Large Arrays," *Nature*, **425**, 817–821, (2003).

DEBYE, P., "Einige Bemerkungen zur Magnetisierung bei tiefer Temperatur," *Ann. Phys*, **386**, 1154–1160, (1926).

D'ERRICO, F., W.G. ABERTS, G. CURZIO, S. GULDBAKKE, H. KLUGE, AND M. MATZKE, "Active Neutron Spectrometry with Superheated Drop (Bubble) Detectors," *Rad. Prot. Dosim.*, **61**, 159–162, (1995).

D'ERRICO, F., W.G. ABERTS, E. DIETZ, G. GUALDRINI, J. KURKDJIAN, P. NOCCIONI, AND B.R.L. SIEBERT, "Active Neutron Spectrometry with Superheated Drop (Bubble) Detectors," *Rad. Prot. Dosim.*, **65**, 397–400, (1996).

D'ERRICO, F., "Radiation Dosimetry and Spectrometry with Superheated Emulsions," *Nucl. Instrum. Meth.*, **B184**, 229–254, (2001).

DE SERIO, M., M. IEVA, M.T. MUCIACCIA, S. SIMONE, M. COZZI, ET AL., "High Precision Measurements with Nuclear Emulsions using Fast Automated Microscopes," *Nucl. Instrum. Meth.*, **A 554**, 247–254, (2005).

DORIESE, W.B., J.N. ULLOM, J.A. BEALL, W.D. DUNCAN, L. FERRIEARA, ET AL., "14-Pixel, Multiplexed Array of Gamma-Ray Microcalorimeters with 47 eV Energy Resolution at 103 keV," *Appl. Phys. Lett.*, **90**, paper 193508, 3 pages, (2007).

DOUGUCHI, Y., H. NANTO, T. SATO, A. IMAI, S. NASU, E. KUSANO, SC AND A. KINBARA, "Optically Stimulated Luminescence in Eu-Doped KBr Phosphor Ceramics," *Radiat. Prot. Dosim.*, **84**, 143–148, (1999).

EDMUND, J.M. AND C.E. ANDERSEN, "Temperature Dependence of the AL_2O_3 Response in Medical Luminescence Dosimetry," *Radiat. Meas.*, **42**, 177–189, (2007).

EFROS, A.L. AND B.I. SHKLOVSKII, "Coulomb Gap and Low Temperature Conductivity of Disordered Systems," *J. Phys. C: Solid State Phys.*, **8**, L49–L51, (1975).

EHRLICH, M. AND W.L. MCLAUGHLIN, "Reciprocity Law for X-Rays Part. I. Validity for High Intensity Exposures in the Negative Region," *J. Opt. Soc. Am.*, **46**, 797–800, (1956).

EHRLICH, M., "Part II. Validity for High Intensity Exposures in the Negative Region," *J. Opt. Soc. Am.*, **46**, 801–804, (1956).

EICHHOLZ, G.G., "The Rosenblum Spark Counter," *Nucleonics*, **10**, 46–49, (1952).

EICHHOLZ, G.G. AND W.M. TEMPLETON, "Neutron Detection by Corrugated-Plate Spark Counters," *IEEE Tran. Nucl. Sci.*, **NS-13**, 725–728, (1966).

EICHHOLZ, G.G., *Spark Counter Neutron Detector for High Temperature Applications*, Final Report for contract AT-(40-1)-3563, June, 1969.

ENSS, C., A. FLEISCHMANN, S.R. BANDLER, T.R. SSTEVENSON, AND, G.M. SEIDEL, "Design Considerations for Arrays of MMCs for X-Ray Astronomy," *Nucl. Instrum. Meth.*, **A 520**, 407–410, (2004).

ENSS, C., Ed., *Cryogenic Particle Detectors*, Vol. 99 in *Topics in Applied Physics*, New York: Springer, 2005.

EKDAL, E, T. KARAH, A. KELEMAN, M. IGNATOVYCH, V. HOLOVEY AND C. HARMANSAH, "Thermoluminescense Characteristics of $Li_2B_4O_7$, Single Crystal Dosimeters Doped with Mn," *Rad. Phys. Chem.*, **96**, 201–204, (2014).

FAUSCH, M. BÜHLER, AND E. UMLAUF, "Signal Rise Time of the Magnetic Bolometer," *J. Low Temp. Phys.*, **93**, 703–708, (1993).

FELIZARDO, M., R.C. MARTINS, T.A. GIRARD, A.R. RAMOS, AND J.G. MARQUES, "Acoustic Spatial Localization of Events in Superheated Droplet Detectors," *Nucl. Instrum. Meth.*, **A599**, 93–99, (2009).

FLEISCHER, R.L. AND P.B. PRICE, "Tracks of Charged Particles in High Polymers," *Science*, **140**, 1221–1222, (1963).

FLEISCHER, R.L., P.B. PRICE, R.M. WALKER, AND E.L. HUBBARD, "Track Registration in Various Solid State Nuclear Track Detectors," *Phys. Rev.*, **133A**, 1443–1449, (1964).

FLEISCHER, R.L., P.B. PRICE, AND R.M. WALKER, "Tracks of Charged Particles in Solids," *Science*, **149**, 383–393, (1965a).

FLEISCHER, R.L., P.B. PRICE, AND R.M. WALKER, "Ion Explosion Spike Mechanism for Formation of Charged-Particle Tracks In Solids," *J. Appl. Phys.*, **36**, 3645–3652, (1965b).

FLEISCHER, R.L., H.W. ALTER, S.C. FURMAN, P.B. PRICE, AND R.M. WALKER, "Particle Track Etching," *Science*, **178**, 255–263, (1972).

FLEISCHER, R.L., P.B. PRICE, AND R.M. WALKER, *Nuclear Tracks in Solids*, Berkeley: Univ. California Press, 1975.

FLEISCHMANN, A., J. SCHÖNEFELD, J. SOLLNER, C. ENSS, J.S. ADAMS, S.R. BANDLER, Y.H. KIM AND G.M. SSC EIDEL, "Low Temperature Properties of Erbium in Gold," *J. Low Temp. Phys.*, **118**, 7–21, (2000).

FLEISCHMANN, A., T. DANIYAROV, H. ROTZINGER, M. LINCK, C. ENSS AND G.M. SEIDEL, "Magnetic Calorimeters for High Resolution X-Ray Spectroscopy," *Rev. Sci. Instrum.*, **74**, 3947–3954, (2003).

FLEISCHMANN, A., M. LINCK, T. DANIYAROV, H. ROTZINGER, C. ENSS, AND G.M. SEIDEL, "Metallic Magnetic Calorimeters (MMC): Detectors for High-Resolution X-Ray Spectroscopy," *Nucl. Instrum. Meth.*, **A 520**, 27–31, (2004).

FLEISCHMANN, A., C. ENSS, AND G.M. SEIDEL, "Metallic Magnetic Calorimeters," *in Cryogenic Particle Detection, Topics in Appl. Phys.*, **99**, 151–217, (2005).

FOWLER, J.F. AND F.H. ATTIX, "Solid State Integrating Dosimeters," *Radiation Dosimetry, 2nd Ed.* **Ch. 13**, Vol. II, F.H. ATTIX AND W.C. ROESCH, Eds., New York: Academic Press, pp. 2410–290, 1966.

FRANK, I. AND IG. TAMM, "Coherent Visible Radiation of Fast Electrons Passing Through Matter," *Compt. Rend. de l'Acad. des Sci. URSS*, **14**, (3), 109–144, (1937).

FRANK, M., L.J. HILLER, J.B. LE GRAND, C.A. MEARS, S.E. LABOV, M.A. LINDEMAN, H. NETEL, D. CHOW, AND A.T. BARFKNECHT, "Energy Resolution and High Count Performance of Superconducting Tunnel Junction X-Ray Spectrometers," *Rev. Sci. Instrum.*, **69**, 25–31, (1998).

FRIEDRICH, S., J.B. LE GRAND, L.J. HILLER, J. KIPP, M. FRANK, AND S. LABOV, "High-Resolution Tunnel Junction Extreme Ultraviolet Detectors Limited b Quasiparticle Counting Statistics," *IEEE Trans. Appl. Superconduct.*, **9**, 3330–3333, (1999).

FRIEDRICH, S., T. NIEDERMAYR, O. DRURY, M.F. CUNNINGHAM, M.L. VAN DEN BERG, J.N. ULLOM, A. LOSHAK, T. FUNK, S.P. CRAMER, J.D. BATTEUX, E. SEE, M. FRANK, AND S.E. LABOV, "A Superconducting Detector Endstation for High-Resolution Energy-Dispersive SR-XRF," *Nucl. Instrum. Meth.*, **A 467–468**, 1117–1120, (2001).

FRIEDRICH, S., "Cryogenic X-Ray Detectors for Synchrotron Science," *J. Synchrotron Rad.*, **13**, 159–171, (2006).

FURRETTA, C. AND J.W.N. TUYN, "A Study of the Peak Ratios in CaF_2:Tm (TLD-300) as a Function of Beta Energy," *Rad. Prot. Dosim.*, **11**, 263–266, (1985).

GAMBARINI, G., "Thermoluminescent (TL) Dosimetry of Slow-Neutron Fields at Radiotherapy Dose Level," *6 Int. Conf. 16 National Congress on Solid State Dosimetry*, Mexico City, Mexico, Sept. 10-12, 7 pages, (2003).

GARLICK, G.F.J. AND M.H.F. WILKINS, "Short Period Phosphorescence and Electron Traps," *Proc. Royal Soc. London*, **184**, 408–433, (1945).

GARLICK, G.F.J. AND A.F. GIBSON, "Luminescence Processes in Zinc Sulphide Phosphors," *Nature*, **148**, 704–705, (1946).

GARLICK, G.F.J. AND A.F. GIBSON, "The Electron Trap Mechanism of Luminescence in Sulphide and Silicate Phosphors," *Proc. Phys. Soc.*, **60**, 574–590, (1948).

GARLICK, G.F.J., *Luminescent Materials*, London: Oxford, 1949.

GASTALDO, L., P.C.-O. RANITZSCH, F. VON SEGGERN, J.-P. PORST, S. SCHÄFER, C. PIES, S. KEMPF, T. WOLF, A. FLEISHMANN, C. ENSS, A. HERLERT, AND K. JOHNSTON, "Characterization of Low Temperature Metallic Calorimeters Having Gold Absorbers with Implanted ^{163}Ho Ions," *Nucl. Instrum. Meth.*, **A 711**, 150–159, (2013).

GIAUQUE, W.F., "A Thermodynamic Treatment of Certain Magnetic Effects. A Proposed Method of Producing Temperatures Considerably Below 1° Absolute," *J. Am. Chem. Soc.*, **49**, 1864–1870, (1927).

GIFFORD, W.E. AND R.C. LONGSWORTH, "Pulse-Tube Refrigeration," *J. Eng. Ind.*, **86(3)**, 264-268, (1964).

GILVIN, P. AND C.A. PERKS, "Practical Low Dose Limits for Passive Personal Dosemeters and the Implications for Uncertainties Close to the Limit of Detection," European Conf. on Individual

Monitoring of Ionizing Radiation, Athens, Greece, March 8-12, (2010).

Ginther, R.J. and R.D. Kirk, "The Thermoluminescence of CaF_2:Mn," **104**, 365–369, (1957).

Glaser, D.A., "Some Effects of Ionizing Radiation on the Formation of Bubbles in Liquids," *Phys. Rev.*, **87**, 665, (1952).

Goland, A.N. and E. der Mateosian, "Effect of Energy Electron Irradiation Upon Charged-Particle Track Registration in Polycarbonate Films," *Nucl. Instrum. Meth.*, **106**, 295–299, (1973).

Goldstein, J.I., D.E. Newbury, P. Echlin, D.C. Joy, C. Fiori, and E. Lifshin, *Scanning Electron Microscopy and X-Ray Microanalysis*, New York: Plenum Press, 1981.

Gorbics, S.G., *The Emission Spectra of Various Thermoluminescent Materials*, NRL Report 6408, Washington, D.C.: Naval Research Laboratory, 1966.

Gorbics, S.G., A.E. Nash, and F.H. Attix, "Thermal Quenching of Luminescence in Six Thermoluminescent Dosimetry PhosphorsII: Quenching of Thermoluminescence," *Appl. Rad. Isotopes*, **20**, 843–852, (1969).

Greinacher, H., "Über Einen Hydraulischen Zähler für Elementarstrahlen," *Helv. Phys. Acta*, **7**, 360–367, (1934).

Guelev, M.G., I.T. Mischev, B. Burgkhardt, and E. Piesch, "A Two-element $CaSO_4$:Dy Dosemeter for Environmental Monitoring," *Rad. Prot. Dosim.*, **51**, 35–40, (1994).

C. Hagmann, C. and P.L. Richards, "Two-Stage Magnetic Refrigerator for Astronomical Applications with Reservoir Temperatures above 4 K," *Cryogenics*, **34(3)**, 221-226, (1994).

Haller, E.E., K.M. Itoh, and J.W. Beeman, "Neutron Transmution Doped (NTD) Germanium Thermistors for Sub-MM Bolometer Applications," *Proc. 30th ESLAB Symp., Submillimetre and Far-Infrared Space Instrumentation*, Noordwijk, The Netherlands, Sept. 24–26, (1996).

Halperin, A. and A.A. Braner, "Evaluation of Thermal Activation Energies from Glow Curves," *Phys. Rev.*, **117**, 408–415, (1960).

Harvey, J.A., N. P. Haverland, and K.J. Kearfott, "Characterization of the Glow-Peak Fading Properties of Six Common Thermoluminescent Materials," *Appl. Radia. Isotopes*, **68**, 1988–2000, (2010).

Hasan, F. and S. Charalambous, "The Thermoluminescense Behaviour of CaF_2-Dy (TLD-200) for Low up to High Doses," *J. Phys. C: Solid State Phys.*, **16**, 5921–5928, (1983).

Heinzelmann, M. and W. Haschke, "Geeignete Verfahren zur Empfindlichkeitserhöhung bei der Nichtfotografischen Kernspurregistrierung," *Report Jül-881-ST*, KFA, Jülich, Germany, (1971).

Heaviside, O., "The Electromagnetic Effects of a Moving Charge," *The Electrician*, Part II, **Nov. 23**, 83-84, (1888).

Heaviside, O., "On the Electromagnetic Effects due to the Motion of Electrification through a Dielectric," *Philos. Mag.*, **27**, 324–339, (1889).

Hecklesberg, L.F. and F. Daniels, "Thermoluminescence of Fourteen Alkali Halide Crystals," *J. Phys. Chem.*, **61**, 414–418, (1957).

Henke, R.P., E.V. Benton and H.H. Heckman, "Sensitivity Enhancements of Plastic Nuclear Track Detectors Through Photo-Oxidation," *Radiation Effects*, **3**, 43–49, (1970).

Henke, R.P. and E.V. Benton, "On Geometry of Tracks in Dielectric Nuclear Track Detectors," *Nucl. Instrum. Meth.*, **97**, 483–489, (1971).

Henning, P.G., "Die Ortsbestimmung Geladener Teilchen mit Hilfe von Funkenzählern," *Atomkernenergie*, **2**, 81–88, (1957).

Herz, R.H., *The Photographic Actions of Ionizing Radiations*, New York: Wiley, 1969.

Higenbotham, W.A., "Wire Spark Chambers," *IEEE Trans. Nucl. Sci.*, **NS-12**, 199–202, (1965).

Horowitz, Y.S., *Thermoluminescence and Thermoluminescent Dosimetry*, Vol. I, Boca Raton: CRC Press, 1984a.

Horowitz, Y.S., *Thermoluminescence and Thermoluminescent Dosimetry*, Vol. II, Boca Raton: CRC Press, 1984b.

Horowitz, Y.S. and J. Moscovitch, "Highlights and Pitfalls of 20 years of Application of Computerized Glow Curve Analysis to Thermoluminescence Research and Dosimetry," *Rad. Prot. Dosim.*, **153**, 1–22, (2013).

Huntley, D.J., D.I. Godfrey-Smith, and M.L.W. Thewalt, "Optical Dating of Sediments," *Nature*, **313**, 105–107, (1985).

Huntley, D.J., D.I. Godfrey-Smith, and E.H. Haskell, "Light-Induced Emission Spectra from Some Quartz and Feldspars," *Nucl. Tracks Radiat. Meas.*, **18**, 127–131, (1991).

Hurter, F. and V.C. Driffield, "Photochemical Investigations and a New Method of Determination of the Sensitiveness of Photographic Plates," *J. Soc. Chem. Ind.*, **9**, 455–469, (1890).

Hütt, G., I. Jaek, and J. Tchonka, "Optical Dating: K-Feldspars Optical Response Stimulation Spectra," *Quat. Sci. Rev.*, **7**, 381–385, (1988).

Ing, H., "Neutron Measurements Using Bubble Detectors – Terrestrial and Space," *Rad. Meas.*, **33**, 275–286, (2001).

Iyomoto, N., S.R. Bandler, R.P. Brekovsky, A.-D. Brown, J.A. Chervenak, F.M. Finkbeiner, R.L. Kelley, C.A. Kilbourne, F.S. Porter, J.E. Sadleir, S.J. Smith, and F. Figueroa-Feliciano, "Close-Packed Arrays of Transition-Edge X-Ray Microcalorimeters with High Spectral Resolution at 5.9 keV," *Appl. Phys. Lett.*, **92**, paper 013508, 3 pages (2008).

Irwin, K.D., "An Application of Electrothermal Feedback for High Resolution Cryogenic Particle Detection," *Appl. Phys. Lett.*, **66**, 1998-2000, (1995).

Irwin, K.D., G.C. Hilton, D.A. Wollman, and J.M. Martinis, "Thermal-Response Tie of Superconducting Transition-Edge Microcalorimeter," *J. Appl. Phys.*, **83**, 3978–3985, (1998).

Jelley, J.V., "Čerenkov Radiation and its Applications," *Brit. J. Appl. Phys.*, **6**, 227–232, (1955).

Jones, J.L., and J.A. Martin, "Use of LiF (TLD 700) For Doses Greater than 0.1 Mrad," *Health Phys.*, **14**, 521–522, (1968); (in the "Notes" section).

Kafadar, V.E., M. Bedir, A.N. Yazici and T. Günal, "The Analysis of Main Dosimeteric Glow Peaks in CaF_2:Tm (TLD-300)," *Chin. Phys. Lett.*, **30** paper 057802, 5 pages (2013).

Kafadar, V.E., R.G. Yildirim, H. Zebari and D. Zebari, "Investigation of Thermoluminescence Characteristics of $Li_2B_4O_7$:Mn (TLD-800)," *Thermochimica Acta*, **575**, 300-304, (2014).

Kearfott, K.J., W.G. West and M. Rafique, "The Optically Stimulated Luminescence (OSL) Properties of LiF:Mg,Tl, $Li_2B_4O_7$:Cu, $CaSO_4$:Tm, and CaF_2:Mn Thermoluminescent (TL) Materials," Appl. Rad. Isotopes, **99**, 155-161, (2015).

Kelley, R.L., S.H. Moseley, C.K. Stahle, A.E. Szymkowiak, M. Juda, D. McCammon and J. Zhang, "Development of Microcalorimeters for High Resolution X-Ray Spectroscopy," *J. Low Temp. Phys.*, **93**, 225–230, (1993).

Kerson, H., *Statisical Mechanics*, 2nd Ed., New York: Wiley, 1987.

KEUFFEL, J.W., "Parallel-Plate Counters," *Rev. Sci. Instrum.*, **20**, 202–208, (1949).

KIM, J.L., J.I. LEE, A.S. PRADHAN, I. CHANG, B.H. KIM, K.S. CHUNG, AND H.S. CHOE, "Optimization of Preparation Procedure of LiF:Mg,Cu,Si TLD for Improving the Reusability," *Rad. Meas.*, **45**, pp. 583–585, (2010).

KODAIRA, S., K. MORISHIGE, H. K—SC AWASHIMA, H. KITAMURA, M. KURANO, N. HASEBE, Y. KOGUCHI, W. SHINOZAKI, AND K. OGURA, "A Performance Test of a New High-Surface-Quality and High-Sensitivity CR-39 Plastic Nuclear Track Detector – TechnoTrak," *Nucl. Instrum. Meth.*, **B383**, 129–135, (2016).

KORTOV, V. AND Y. USTYANTSEV, "Advantages and Challenges of High-Dose Thermoluminescent Detectors," *Rad. Meas.*, **56**, 299-302, (2013).

KOSSEL, W., U. MAYER, AND H.C. WOLF, "Simultan-Dosimetrie von Strahlungsfeldern im lebenden Objekt," *Die Naturwissenschaften*, **41**, 209, (1954).

KRAFT, S., P. VERHOEVE, A. PEACOCK, N. RANDO, D.J. GOLDIE, R. HART, D. GLOWACKA, F. SCHOLZE, AND G. ULM, "On the Factors Governing the Energy Resolution of Ta-Based Superconducting Tunnel Junctions," *J. Appl. Phys.*, **86**, 7189–7191, (1999).

KRBETSCHEK, M.R. AND U. RIESER, "Luminescence Spectra of Alkalifeldspars and Plagioclases," *Radiat. Meas.*, **24**, 473–477, (1995).

KRBETSCHEK, M.R., J. GÖTZE, A. DIETRICH, AND T. TRAUTMANN, "Spectral Information from Minerals Relevant for Luminescence Dating," *Radiat. Meas.*, **27**, 695–748, (1997).

KURAKADO, M. AND H. MAZAKI, "Possibility of a High-Resolution Nuclear Radiation Detector using a Superconductor, I. Mean Energy Loss per Excess Quasiparticle," *Nucl. Instrum. Meth.*, **185**, 141–147, (1981a).

KURAKADO, M. AND H. MAZAKI, "Possibility of a High-Resolution Nuclear Radiation Detector using a Superconductor, II. Tunnel Junction," *Nucl. Instrum. Meth.*, **185**, 149–155, (1981b).

KURAKADO, M., "Possibility of a High-Resolution Detectors using Superconducting Tunnel Junctions," *Nucl. Instrum. Meth.*, **196**, 275–277, (1982).

LAKSHMANAN, A.R., R.C. BHATT, AND S.J. SUPE "Mechanism of Nonlinearity in the Response Characteristics of Thermoluminescent Dosimeters," *J. Phys. D: Appl. Phys.*, **14**, 1683–1706, (1981).

LANGLEY, S.P., "The Bolometer," *Proc. Amer. Metrological Soc.*, **2**, 184-190, (1880).

LARMORE, L., *Introduction to Photographic Principles*, New York: Dover, 1965.

LEVERENZ, H.W., *Luminescence of Solids*, New York: Wiley, 1950.

LIM, W. AND C.K. WANG, "One-Dimensional Position-Sensitive Superheated-Liquid-Droplet In-Phantom Neutron Dosimeter," *RevŚci. Instrum.*, **66**, 5442–5449, (1995).

LIM, W. AND C.K. WANG, "Position-Sensitive Superheated-Liquid-Droplet Neutron Dosimeter," *Nucl. Instrum. Meth.*, **A370**, 568–572, (1996).

LINDSKOUG, B.A., AND L-M. LUNDBERG, "Clinical Applications of Thermoluminescent Dosimetry," Ch. 2, in *Thermoluminescence and Thermoluminescent Dosimetry*, Y.S. HOROWITZ, Ed., Vol. III, Boca Raton: CRC Press, 1984.

LIVINGSTONE, Y.S. HOROWITZ, L. OSTER, H. DATZ, M. LERCH, A. ROSENFELD, AND A. HOROWITZ, "Experimental Investigation of the 100 keV X-Ray Dose Response of the High-Temperature Thermoluminescence in LiF:Mg,Tl (TLD-100): Theoretical Interpretation Using the Unified Interaction Model," *Rad. Prot. Dosimetry*, **138**, 320–333, (2009).

LOUNASMAA, O.V., *Experimental Principles and Methods Below 1 K*, New York: Academic Press, 1974.

LOUNIS, Z., S. DJEFFAL, K. MORSLI, AND M. ALLAB, "Track Etch Parameters in CR-39 Detctors for Proton and Alpha Particles of Different Energies," *Nucl. Instrum. Meth.*, **B179**, 543–550, (2001).

LOW, F.J., "Low-Temperature Germanium Bolometer," *J. Optical Soc. Amer.*, **31**, 1300–1304, (1961).

LUCAS, A.C. AND B.M. KAPSAR, *Proc. 5th Int. Conf. on Luminescence Dosimetry*, A. SHARMAN, Ed., Justus Liebig Univ., Sao Paulo, Brazil, 131–139, (1977).

MARRONE, M.J., AND F.H. ATTIX, "Damage Effects in CaF_2:Mn and LiF Thermoluminescent Dosimeters," *Health Phys.*, **10**, 431–436, (1964).

MASSILLON-JL, G., I. GAMBOA-DEBUEN, A.E. BUENFIL, M.A. MONROY-RODRÍGUEZ, AND M.E.BRANDAN, "CaF_2:Tm (TLD-300) Thermoluminescent Response and Glow Curve induced by γ-Rays and Ions," *Nucl. Instrum. Meth.*, **B266**, 772–780, (2008).

MAYHUGH, M.R., AND G.D. FULLERTON, "Thermoluminescence in LiF, Sensitization Useful at Low Exposures," *Health Phys.*, **28**, 279–299, (1975).

MAYHUGH, M.R., *private communication*, 2000.

MCCAMMON, D., "Thermal Equilibrium Calorimeters," in *Cryogenic Particle Detection*, C. ENSS, Ed., Berlin: Springer, pp. 14–34, 2005a.

MCCAMMON, D., "Semiconductor Thermistor," in *Cryogenic Particle Detection*, C. ENSS, Ed., Berlin: Springer, pp. 35–62, 2005b.

MCCLINTOCK, P.V.E., D.J. MEREDITH, AND J.K. WIGMORE, *Matter at Low Temperatures*, Glasgow: Blackie, 1984.

MCDOUGALL, R.S. AND S. RUDIN, "Thermoluminescent Dosimetry of Aluminum Oxide," *Health Phys.*, **19**, 281–283, (1970).

MCKEEVER, S.W.S., *Thermoluminescence of Solids*, Cambridge: Cambridge Univ. Press, 1985.

MCKEEVER, S.W.S., J.S. AKSELROD, AND B.G. MARKEY, "Pulsed Optically Stimulated Luminescence Dosimetry Using α-Al_2O_3:C," *Radiat. Prot. Dosim.*, **65**, 267-272, (1996).

MCKEEVER, S.W.S. AND J.S. AKSELROD, "Radiation Dosimetry using Pulsed Optically Stimulated Luminescence of α-Al_2O_3:C," *Radiat. Prot. Dosim.*, **84**, 317-320, (1999).

MCKEEVER, S.W.S., J.S. AKSELROD, L.E. COLYOTT, N.A. LARSEN, J.C. POLF, AND V. WHITELEY, "Characterisation of Al_2O_3 for use in Thermally and Optically Stimulated Luminescence Dosimetry," *Radiat. Prot. Dosim.*, **84**, 163–168, (1999).

MCKEEVER, S.W.S., "Optically Stimulated Luminescence Dosimetry," *Nucl. Instrum. Meth.*, **B184**, 29–54, (2001).

MEARS, C.A., S.E. LABOV, AND A.T. BARFKNECHT, "Energy-Resolving Superconducting X-Ray Detectors with Charge Amplification Due to Multiple Quasiparticle Tunneling," *Appl. Phys. Lett.*, **63**, 2961–2963, (1993).

MIKULIN, E.I., A.A. TARASOV, AND M.P. SHKREBYONOCK, "Low Temperature Expansion Pulse Tubes," *Ad. Cryo. Eng.*, **29**, 629–637, (1984).

MOREHEAD, F.F., AND F. DANIELS, "Thermoluminescence and Coloration of Lithium Fluoride Produced by Alpha Particles, Electrons, Gamma Rays, and Neutrons," *J. Chem. Phys.*, **27**, 1318–1324, (1957).

MOSELEY, S.H., J.C. MATHER AND D. MCCAMMON, "Thermal Detectors as X-Ray Spectrometers," *J. Appl. Phys.*, **56**, 1257–1262, (1984).

MOSER, L., "Über den Process des Sehens und die Wirkung des Lichts auf alle Körper," *Ann. Phys. Chem.*, **6**, 177–234, (1842).

MOTT, N.F. AND R.W. GURNEY, *Electronic Processes in Ionic Crystals*, 2nd Ed., London: Oxford University Press, 1948.

MOTT, N.F. AND W.D. TWOSE, "The Theory of Impurity Conduction," *Advan. Phys.*, **10**, 107–163, (1961).

MURTY, R.C. "Effective Atomic Numbers of Heterogeneous Materials," *Nature*, **207**, 398–399, (1965).

NANTO, H., K. MURAYAMA, T. USADA, S. TANIGUCHI, AND N. TAKEUCHI, "Optically Stimulated Luminescence in KCl:Eu Single Crystals," *Radiat. Prot. Dosim.*, **47**, 281–284, (1993).

NATARAJAN, C.M., M.G. TANNER, AND R.H. HADFIELD, "Superconducting Nanowire Single-Photon Detectors: Physics and Applications," *Supercond. Sci. Technol.*, **25**, 063001, 16 pages, (2012).

NIÈPCE DE SAINT-VICTOR, C.F.A., "Memoire sur une Nouvelle Action de la Lumière," *Compte Rendus*, **45**, 811–815, (1857).

NIÈPCE DE SAINT-VICTOR, C.F.A., "Influence de la Lumiere dans les Action Moleculaires," *Compte Rendus*, **47**, 1002–1006, (1859).

NORDLUND, K., M. GHALY, R.S. AVERBACK, M. CATURLA, T. DIAZ DE LA RUBIA, AND J. TARUS, "Defect Production in Collision Cascades in Elemental Semiconductors and FCC Metals," *Phys. Rev. B*, **57**, 7556–7570, (1999).

OGG, JR., R.A., "Bose-Einstein Condensation of Trapped Electron Pairs. Phase Separation and Superconductivity of Metal-Ammonia Solutions," *Phys. Rev.*, **69**, 243–244, (1946).

PAYNE, R.M., "The Rosenblum Spark Counter: A New Counter for the Detection of Fast Ionizing Particles," *J. Sci. Instrum.*, **26**, 321–324, (1949).

PORTAL, G., S. LORRAIN, AND G. VALLADAS, "Very Deep Traps in Al_2O_3 and $CaSO_4$:Dy," *Nucl. Instrum. Meth.*, **175**, 12–14, (1980).

PORTAL, G. AND G. DIETZE, "Implication of New ICROP and ICRU Recommendations for Neutron Dosimetry," *Rad. Prot. Dosim.*, **44**, 165–170, (1992).

POWELL, C.F., G.P.S. OCCHIALINI, D.L. LIVESEY, AND L.V. CHILTON, "A New Photographic Emulsion for the Detection of Fast Charged Particles," *J. Sci. Instrum.*, **23**, 102–106, (1946).

POWELL, C.F. AND G.P.S. OCCHIALINI, *Nuclear Physics in Photographs*, Oxford: Clarendon, 1947.

PRICE, P.B. AND R.L. FLEISCHER, "Identification of Energetic Heavy Nuclei with Solid Dielectric Track Detectors: Applications to Astrophysical and Planetary Studies," *Annu. Rev. Nucl. Sci.*, **21**, 295–334, (1971).

PRICE, P.B. AND R.M. WALKER, "Electron Microscope Observation of Etched Tracks from Spallation Recoils in Mica," *Phy. Rev. Lett.*, **8**, 217–219, (1962).

RANDALL, J.T. AND M.H.F. WILKINS, "Phosphorescence and Electron Traps I. The Study of Trap Distributions," *Proc. Royal Soc. London*, **184**, 366–389, (1945).

RANDO, N., A. PEACOCK, A. VAN DORDRECHT, C. FODEN, R. EENGELHARDT, B.G. TAYLOR, P. GARE, J. LUMLEY, AND C. PEREIRA, "The Properties of Niobium Superconducting Tunneling Junctions as X-Ray Detectors," *Nucl. Instrum. Meth.*, **A313**, 173–195, (1992).

REZAEIAN, P., G. RAISALI, A. AKHAVAN, H. GHODS, AND B. HAJIZADEH, "Development of a New Pressure Dependent Threshold Superheated Drop Detector for Neutrons," *Nucl. Instrum. Meth.*, **A776**, 50–56, (2015).

RICE-EVANS, P., *Spark, Streamer, Proportional and Drift Chambers*, London: Richelieu Press, 1974.

RIEKE, J.K. AND F. DANIELS, "Thermoluminescence Studies of Aluminum Oxide," *J. Phys. Chem.*, **61**, 629–633, (1957).

ROBERTS, A., "Development of the Spark Chamber: A Review," *Rev. Sci. Instrum.*, *32*, 482–485, (1961).

ROSE-INNES, A.C. AND, E.H. RHODERICK, *Introduction to Superconductivity*, 2nd Ed., Oxford: Pergamon, 1978.

ROY, S.C., R.E. APFEL, AND Y.-C. LO, "Superheated Drop Detector: A Potential Tool in Neutron Research," *Nucl. Instrum. Meth.*, **A255**, 199–206, (1987).

SAUNDERS, D.F., F.F. MOREHEAD, AND F. DANIELS, "A Convenient Source of Gamma Radiation," *J. Am. Chem. Soc.*, **117**, 343–349, (1953).

SCARPA, G., "The Dosimetric Use of Beryllium Oxide as a Thermoluminescent Material: A Preliminary Study," *Phys. Med. Bio.*, **15**, 667–672, (1970).

SCHAYES, R., AND C. BROOKE, "Thermoluminescent Dosimetry," Part III, *Rev. M.B.L.E.*, **6**, 5–29, (1963).

SCHAYES, R., C. BROOKE, I. KOZLOWITZ, AND M. LHEUREUX, "Thermoluminescent Properties of Natural Calcium Fluoride" *AEC Symp. Ser.*, CONF-650637, 138–157, (1967).

SCHEMBRI, V. AND B.J.M. HEIJMEN, "Optically Stimulated Luminescence (OSL) of Carbon-Doped Aluminum Oxide (Al_2O_3:C) for Film Dosimetry in Radiotherapy," *Med. Phys.*, **34**, 2113–2118, (2007).

SCHULMAN, J.H., F.H. ATTIX, E.J. WEST, AND R.J. GINTHER, "New Thermoluminescent Dosimeter," *Rev. Sci. Instrum.*, **31**, 1263–1269, (1960).

SCHULMAN, J.H., R.D. KIRK, AND E.J. WEST, "Use of Lithium Borate for Thermoluminescent Dosimeters," *Luminescent Dosimetry*, Proc. Int. Conf., AEC Symp. Ser., **Vol. 8**, Stanford, (1967).

SEGUINOT, J. AND T. YPSILANTIS, "A Historical Survey of Ring Imaging Cherenkov Counters," *Nucl. Instrum. Meth.*, **A343**, 1–29, (1994).

SEITZ, F., "The Disordering of Solids by the Action of Fast Massive Particles," *Disc. Faraday Soc.*, **5**, 271–282, (1949).

SEITZ, F., "On the Theory of the Bubble Chamber," *Phys. Fluids*, **1**, 2–13, (1958).

SETH, S., M. DAS, S. BHATTACHARYA, P. BHATTACHARJEE, AND S. SAHA, "The Nucleation Parameter for Heavy-Ion Induced Bubble Nucleation in Superheated Emulsion Detector," *J. Instrumentation*, **8**, paper p05001, 15 pages, (2013).

SHANI, G., *Radiation Dosimetry Instrumentation and Methods*, Boca Raton: CRC Press, 1991.

SHKLOVSKII, B.I. AND A.L. EFROS, *Electronic Properties of Doped Semiconductor*, Berlin: Springer, 1984.

SOHRABI, M., "Electrochemical Etching of Fast-Neutron-Induced Recoil Tracks: The Effects of Field Strength and Frequency," *Nucl. Tracks*, **4**, 131–140, (1981).

SOMOGYI, G., "A Study of the Basic Properties of Electrochemical Track Etching," *Rad. Effects*, **34**, 51–56, (1977).

SOMMER, M., R. FREUDENBERG AND J. HENNIGER, "New Aspects of a BeO-Based Optically Stimulated Luminescence Dosimeter," *Radiat. Meas.*, **42**, 617–620, (2007).

SOMMER, M., A. JAHN AND J. HENNIGER, "Beryllium Oxide as Optically Stimulated Luminescence Dosimeter," *Radiat. Meas.*, **43**, 353–356, (2008).

SOUZA, J.H., L.A.R. DA ROSA, C.L.P. MAURICIO, "On the Thermoluminescence Glow Curve of $CaSO_4$:Dy," *Rad. Prot. Dosim.*, **47**, 103–106, (1993).

STAHLE, C.K., C.A. ALLEN, K.R. BOYCE, R.P. BREKOSKY, G.V. BROWN, ET AL., "The Next-Generation Microcalorimeter Array of XRS on Astro-E2," *Nucl. Instrum. Meth.*, **A 520**, 466–468, (2004).

STORM, E. AND S. SHLAER, "Development of Energy-Independent Film Badges with Multi-Element Filters," *Health Phys.*, **11**, 1127–1144, (1965).

SUNTA, C.M., V.M. BAPAT, AND, S.P. KATHURIA, "Effects of Deep Traps on Supralinearity, Sensitization and Optical Thermoluminescence in LiF TLD," *Proc. Third Int. Conf. Luminescence Dosimetry*, 146–155, (1971).

SUNTA, C.M. AND, S.P. KATHURIA, "Effects of Residual Thermoluminescence in Fluorite and Lithium Fluorite Thermoluminescent Dosimeters," *Proc. IAEA Symp. Adv. Phys. Biol. Radiat. Detect.*, 125–130, (1971).

SUNTA, C.M., K.S.V. NAMBI, AND, V.N. BAPAT, "Fast Neutron Response of Thermoluminescent Detectors Using Proton Radiator Technique," *Proc. Symp. Neutron Monitor. Radiation Protection, IAEA*, Vienna, (1972).

TAKENAGA, M., O. YAMAMOTO, AND T. YAMASHITA, "Preparation and Characteristics of $Li_2B_4O_7$:Cu Phosphor," *Nucl. Instrum. Meth.*, **175**, 77–78, (1980).

THERMO-FISHER SCIENTIFIC, www.thermofisher.com/order/catalog/product/SN010106, accessed Oct. 27, 2017. See also Thermo Scientific Harshaw TLD Materials and Dosimeters, BR-RMSI-TLD-0616, (2016).

TIWARI, B., N.S. RAWAT, D.G. DESAI, S.G. SINGH, M. TYAGI, P. RATNA, S.C. GADKARI AND M.S. KULKARNI, "Thermoluminescence Studies on Cu-Doped $Li_2B_4O_7$ Single Crystals," *J. Luminescence*, **130**, 2076–2083, (2010).

TOCHILIN, E., N. GOLDSTEIN, AND W.G. MILLER, "Beryllium Oxide as a Thermoluminescent Dosimeter," *Health Phys.*, **16**, 1–7, (1969).

TOMMASINO, L., "Electrochemical Etching of Damaged Track Detectors by H.V. Pulse and Sinusoidal Waveforms," *Proc. 7th Int. Colloq. Corpuscular Photography and Solid Detectors*, CNEN Report RT//PROT, **71**, Barcelona, Spain, (1970).

TOMMASINO, L., G. ZAPPAROLI AND R.V. GRIFFITH, "Electrochemical Etching," *Nuclear Tracks*, **4**, 191–196, (1981).

TOMMASINO, L., G. ZAPPAROLI AND F. CAIAZZO, "A New Method for Electrochemical Etching: I. Results with DC Voltage," *Solid State Nuclear Track Detector, Proc. 11th Int. Conf.Bristol, UK, 1981*, P.H. FOWLER, V.M. CLAPHAM, Eds., 211–214, (1982).

TRIOLO, A., M. BRAI, A. BARTOLOTTA, AND M. MARRALE, "Glow Curve Analysis of TLD-100H Irradiated with Radiation of Different LET: Comparison Between Two Theoretical Method," *Nucl. Instrum. Meth.*, **A560**, 413–417, (2006).

VIGNERON, L., 'Calcul Général de la Relation Parcours-Énergie des Particules dans les Émulsions ou un Milieu Ralentisseur Quelconque. Application Numeérique à L'Émulsion Ilford C2," *J. Phys. Radium*, **14**, 145–159, (1953).

WALL, B.F., C.M.H. DRISCOLL, J.C. STRONG, AND E.S. FISHER, "The Suitibility of Different Preparations of Thermoluminescent Lithium Borate for Medical Dosimetry, *Phys. Med. Bio.*, **27**, 1023–1034, (1982).

WANG, C., G. THUMMES, AND C. HEIDEN, "A Two-Stage Pulse Tube Cooler Operating below 4 K," *Cryogenics*, **37(3)**, 159–164, (1997).

WEST, E.J., A.E. NASH, F.H. ATTIX, R.D. KIRK, AND J.H. SCHULMAN, *Thermoluminescent Dosimeters of Lithium Borate*, Prog. Rep. No. TID-24523, June, NRL, 1967. Republished in CAMERON, J.R., N. SUNTHARALINGAM AND G.N. KENNEY, *Thermoluminescent Dosimetry*, Madison: U. Wisconsin Press, 1968.

WIEDEMANN, E. AND G.C. SCHMIDT, "Ueber Luminescenz," *Ann. Phys.*, **54**, 604–625, (1895).

WILSON C.R. AND J.R. CAMERON, "Dosimetric Properties of $Li_2B_4O_7$:Mn (Harshaw)," Proc. Sec. Int. Luminescence Dosimetry, CONF-680920, 161–178, (1968).

WOLLMAN, D.A., K.D. IRWIN, G.C. HILTON, L.L. DULCIE, D.E. NEWBERRY, AND J.M. MARTINIS, "High-Resolution, Energy-Dispersive Microcalorimeter Spectrometer for X-Ray Microanalysis," *J. Microscopy*, **188**, 196–223, (1997).

YAMASHITA, T., N. NADA, H. ONISHI, AND S. KITAMURA, "Calcium Sulfate Activated by Thulium or Dysprosium for Thermoluminescence Dosimetry," *Health Phys.*, **21**, 295–300, (1971).

YAZICI, A.M., R. CHEN, S. SOLAK, AND Z. YEGINGIL, "The Analysis of Thermoluminescent Glow Peaks of CaF_2:Dy (TLD-200) After β-Irradiation," *J. Phys. D: Appl. Phys.*, **35**, 2526–2535, (2002).

YOSHIMURA, E.M. AND E.G. YUKIHARA, "Optically Stimulated Luminescence: Searching for New Dosimetric Materials," *Nucl. Instrum. Meth.*, **B 250**, 337–341, (2006).

YOUNG, D.A., "Etching of Radiation Damage in Lithium Fluoride," *Nature*, **182**, 375–377, (1958).

YPSILANTIS, T. AND J. SEGUINOT, "Theory of Ring Imaging Cherenkov Counters," *Nucl. Instrum. Meth.*, **A343**, 30–51, (1994).

YUKIHARA, E.G. AND S.W.S. MCKEEVER, "Spectroscopy and Optically Stimulated Luminescence of Al_2O_3 using Time-Resolved Measurements," *J. Appl. Phys.*, **100**, paper 083512, 9 pages, (2006).

YUKIHARA, E.G. AND S.W.S. MCKEEVER, "Spectroscopy and Optically Stimulated Luminescence of Al_2O_3 Using Time-Resolved Measurements," *J. Appl. Phys.*, **100**, paper 083512, 9 pages, (2006).

YUKIHARA, E.G., V.H. WHITLEY, S.W.S. MCKEEVER, A.E. AKSELROD, AND M.S. AKSELROD, "Effect of High-Dose Irradiation on the Optically Stimulated Luminescence of Al_2O_3:C," *Rad. Measurements*, **38**, 317–330, (2004).

YUKIHARA, E.G. AND S.W.S. MCKEEVER, "Optically Stimulated Luminescence (OSL) Dosimetry in Medicine," *Phys. Med. Biol.*, **53**, R351–R378, (2008).

YUKIHARA, E.G. AND S.W.S. MCKEEVER, *Optically Stimulated Luminescence Fundamentals and Applications*, New York: Wiley, 2011.

Chapter 20

Radiation Measurements and Spectroscopy

Experimental confirmation of a prediction is merely a measurement.
An experiment disproving a prediction is a discovery.

Enrico Fermi

In this book many detectors have been described that respond not only to the presence of radiation but also have a response that depends upon the energy of the radiation. Unfortunately, the energy response of such spectrometers is generally not directly proportional to the energy of the incident radiation. Consequently, the output from such detectors must first be analyzed or unfolded to obtain information about the true nature of the radiation energy emitted by a source.

By far the most frequently used energy spectrometers are those used to obtain information about x-ray and gamma-ray sources. In this chapter, methods of analyzing gamma-ray spectra obtained from detectors capable of energy discrimination are discussed. However, the methods used for gamma-ray spectroscopy are somewhat general and can also be applied to spectroscopy of other types of radiation. Thus, also included in this chapter are some basic aspects of charged-particle spectroscopy. Not included here are energy spectroscopic methods for neutrons, a topic discussed earlier in Chapter 18 on neutron detectors.

The general objective of spectroscopy is to obtain, at a minimum, the qualitative identification of the source (e.g., source energies or the radionuclides present). However, most spectroscopy applications also seek quantitative information such as the strengths of the sources or the concentrations of the radionuclides in a source. Several different methods for qualitative and quantitative analyses are summarized here and illustrative examples are provided.

20.1 Introduction

Gamma rays and x rays are photons of electromagnetic radiation that are capable of producing ionization through their interactions with ambient atoms. Technically, x rays differ from gamma rays in their source of origin but, for practical purposes, this nuance is irrelevant and photon spectra can be analyzed by the same methods for both x rays or gamma rays. As a practical matter, photons of energy less than about 10 keV are difficult to detect because they are easily absorbed by the detector housing or source material itself. The concepts that are discussed here can be applied, in principle, to spectra from photons of energy less than 10 keV and also to spectra generated by other particles, such as electrons. For instance x-ray photoelectron spectroscopy (XPS) and electron scattering for chemical analysis (ESCA) lead to electron spectra that can be analyzed by the methods discussed here for photons. Thus, it is understood that reference to gamma-ray spectroscopy is an oversimplification and many of the methods discussed here can be applied to spectra generated by x rays, gamma rays, or other types of radiation.

Gamma-ray spectroscopy is a general area of study within which spectra are analyzed in order to determine qualitative and, if possible, quantitative information about a sample under investigation. Spectra generally refer to collections of data for which the independent variable is the channel number (or a related quantity such as momentum, energy, or wavelength) and the dependent variable is a detector response that depends on the independent variable. Spectra are generated in many radiation measurement techniques, such as energy-dispersive x-ray fluorescence (EDXRF), neutron activation analysis (NAA), prompt-capture gamma-ray neutron activation analysis (PGNAA), electron photoelectron spectroscopy (XPS), and general counting of unknown radioactive sources. Often, spectroscopy is directed at the quantitative objective of identifying concentrations of specific elements or isotopes (henceforth the generic term "nuclides" is used) that are present in samples, but it also can be employed in a qualitative manner to identify whether specific gamma-emitting nuclides, such as those in special nuclear materials, are present in samples.

In general, a sample to be analyzed emits photons whose energies are characteristic of the nuclides present in the sample. The emitted photons may arise from excitation from an external energy source or the sample may emit these photons naturally. A careful spectroscopic investigation generally seeks to determine either the energies emitted and their intensities or the nuclides present and their concentrations. In previous chapters on radiation detectors, mainly Chapters 13, 14, and 16–18, examples of radiation measurements with specific detector materials and configurations have been presented, and the reader is directed to those chapters for particular detector applications applied to radiation measurements and spectroscopy. For example, the performance of HPGe detectors is described in Sec. 16.4.2. However, there are measurement and spectroscopy methods that can be generally applied to most spectrometers. In this chapter, attention is given to photon spectra that are generated from samples interrogated by any of a number of means, active or passive, to determine information about the constituents of the sample.

20.2 Basic Concepts

Various photon detectors can provide responses that are proportional to the energy deposited in the detector. These include proportional counters, scintillation detectors, and semiconductor detectors. Regardless of the detector, a voltage pulse is created whose amplitude h, generally called the *pulse height*, is a function of the energy deposited E_d in the detector, i.e.,

$$h = f(E_d), \tag{20.1}$$

where f is some function. For many detectors, f is linear and

$$h = \alpha + \beta E_d, \tag{20.2}$$

where α and β are constants unique to a detector. However, scintillation detectors, in particular, can exhibit non-linearities, especially at low photon energies (see Sec. 13.2.2), and this non-linearity should be taken into account. (Linearity of a detector is often expressed in terms of the pulse height per unit energy as a function of the deposition energy, a quantity which is constant for a linear detector.) In any event, a good spectroscopist should know f, the functional relationship between pulse height and deposited energy, for any spectroscopic detector used.

In gamma-spectroscopy, the pulse height h is measured and the deposited energy E_d is obtained by inversion of Eq. (20.1), namely

$$E_d = f^{-1}(h), \tag{20.3}$$

or, if the spectrometer is linear, then

$$E_d = \frac{h - \alpha}{\beta}. \tag{20.4}$$

Moreover, the pulse heights produced by repeated deposition of energy E_d in a detector are not exactly the same; rather, they are distributed about a mean value $h_0 = f(E_d)$ by some kernel $g(h|E_d)$ such that $g(h|E_d)dh$ is the probability an event that deposits energy E_d in the detector results in a pulse with an amplitude in dh about h. In gamma-ray spectroscopy, this *spreading* kernel has a shape similar to a Gaussian shape although, in practice, it is usually skewed slightly towards lower amplitudes as a consequence of different collection times of charge carriers produced in different regions of the detector and the different amount of recombination and trapping they experience as they are collected. Shallow angle Compton scattering also contributes slightly to this asymmetry. An example of this amplitude kernel for a HPGe detector is shown in Fig. 20.1. In effect, the detector systems that produce the voltage pulses operate on the energy deposited E_d with an energy kernel $z(E_d, E')$ that transforms E_d into a range of "apparent" energies E', which are centered on E_d. The resolution of a spectrometer is determined by the degree to which pulse amplitudes are spread out around h_0 or around E_d and is quantified by the full width at half maximum (FWHM) either as a percent or in energy units.

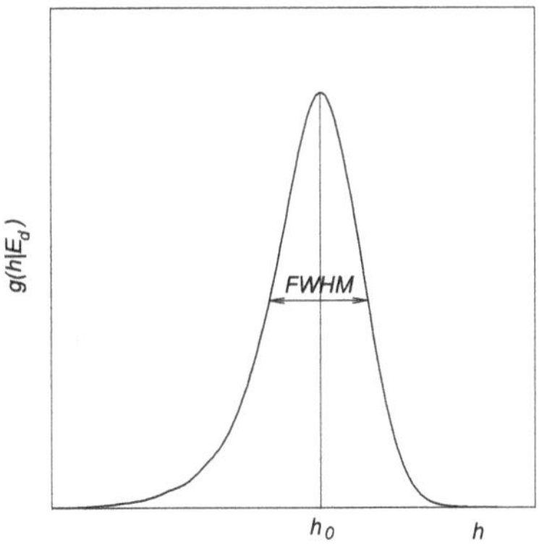

Figure 20.1. A pulse-height spreading kernel for a HPGe detector as measured for the 1173-keV ^{60}Co gamma ray [IEEE-325 1996]. Note the peak is not quite symmetric about the centroid $h_{\max}$.

Often the amplitude spreading kernel is approximated by the Gaussian probability distribution

$$g(h|h_0, \sigma) = \frac{1}{\sqrt{2\pi}\sigma} \exp[-(h - h_0)^2/(2\sigma^2)], \tag{20.5}$$

where h_0 is the mean value or centroid and σ is the standard deviation. The peak value of the Gaussian PDF occurs at the centroid and is given by $g_{max} = 1/(\sqrt{2\pi}\sigma)$. As discussed in Sec. 6.8 the FWHM for a Gaussian is FWHM $= 2\sqrt{2\ln 2}\sigma \simeq 2.355\sigma$.

It is worth noting that the energy resolution of semiconductor detectors is, in general, significantly better than that for scintillation detectors or proportional counters. McGregor [2008] gives a good comparative description of the resolution of various detector types. Each type of detector has its advantages and disadvantages. The chapters in this book on the various detector types provide useful information on the characteristics of each type of detector.

The pulse heights are scaled to be within a finite interval $[h_{min}, h_{max}]$. Typically, these limits are $h_{min} = 0$ and $h_{max} = 10$ V. In any case, the variables E_d and h are continuous variables. Whatever the pulse height limits are, the response of a detector is typically binned into discrete "channels." If n denotes an individual discrete (integral) channel number and N is the total number of channels, then the channel width is

$$\Delta = \frac{h_{max} - h_{min}}{N}.$$

The channel numbers are related to the detected magnitudes of the voltage pulses (pulse heights) by the relations

$$\begin{aligned} n &= 1, \text{ if } h_{min} < h \le h_{min} + \Delta \\ &= 2, \text{ if } h_{min} + \Delta < h \le h_{min} + 2\Delta \\ &\vdots \\ &= N, \text{ if } h_{min} + (N-1)\Delta < h \le h_{min} + N\Delta. \end{aligned} \tag{20.6}$$

Thus, the measured continuous pulse heights are converted into discrete channels and each pulse registers a count in one and only one channel. NIM multichannel analyzer systems confine voltage signals or pulse heights to be within 0–10 volts. Hence, a binary system may be subdivided with 2^N voltage bins over the 10-volt range. For $N = 10$, there are 1024 bins, or *channels*, over the 10 volt range having, in this case, 9.8×10^{-3} volts per channel. Typical spectroscopic systems now often operate with $N = 13$ where $2^{13} = 8,192$ channels, with some having $N = 14$, or 16,384 channels.

Although there are a finite number of discrete channels, corresponding to the pulse-height intervals specified in Eq. (20.6), the peak centroid can occur at any continuous value of h. Thus, it is customary to specify a linear relationship between the continuous values of h and a continuous channel number $\tilde{n}$, given by

$$\tilde{n} = \frac{h - h_{min}}{\Delta}, \tag{20.7}$$

which varies continuously between 0 and N. Henceforth, the term channel number is used to mean either the discrete integer channel number n or the continuous channel number $\tilde{n}$; the context can be used to infer the intent.[1]

20.3 Detector Response Models

Consider a source that emits photons at a rate with some energy distribution $s(E)$ and a spectrometer that detects the photons and produces a pulse-height spectrum. The basic spectroscopic relationship in its continuous form can be written [Dunn and McGregor 2012],

$$r(h) = \frac{dR(h)}{dh} = \int_0^\infty s(E)\mathcal{R}(h|E)dE + b(h), \quad h \in [h_{min}, h_{max}], \tag{20.8}$$

where

- $r(h)$ is the detector response such that $r(h)dh$ is the expected number of counts within dh about h per unit time.
- $s(E)$ is the source strength, in photons per unit time, such that $s(E)dE$ is the number of source photons emitted within dE about E.
- $\mathcal{R}(h|E)$ is the detector response kernel such that $\mathcal{R}(h|E)dh$ gives the probability that a particle of energy E interacting in the detector produces a pulse whose height is within dh about h. Here $\mathcal{R}(h|E) = p(E, E_d)g(h|E_d)$ where $p(E, E_d)dE_d$ is the probability that is a source gamma ray of energy E interacts in the detector and deposits an energy in dE_d about E_d in the detector.
- $b(h)$ is the background count rate such that $b(h)dh$ is the expected number of counts within dh about h, per unit time, that are due to background radiation.

Note that $R(h) = \int_0^h r(h')dh'$ is the cumulative number of counts per unit time due to pulses whose heights are less than h and that $R(h_2) - R(h_1) = \int_{h_1}^{h_2} r(h)dh$ is the total number of counts, per unit time, whose pulse heights are between h_1 and h_2. Note also that $\mathcal{R}(h|E)dh$ accounts for the probability of transport of source photons to and within the detector, deposition of energy E_d in the detector, and spreading of the deposited energy into an apparent energy E' that produces a pulse whose pulse height is within dh about h.

[1]In this book n is also used as a dependent variable that represents a net count rate $n = c - b$ where c and b are the source and background count rates, respectively. Likewise N refers sometimes to the total net counts, the number of spectral channels or the number density of some particle. Again the context should make clear the meaning of n and N.

In general, a source can emit photons at J discrete energies and also over a continuum of energies. For such sources

$$s(E) = \sum_{j=1}^{J} s_j \delta(E - E_j) + s_c(E), \tag{20.9}$$

where s_j is the emission rate of photons of energy E_j from the source, $\delta(E - E_j)$ is the Dirac delta function, and $s_c(E)$ is the source emission rate of the photons with continuum of energies such that $s_c(E)dE$ is the expected number of photons emitted within dE about E per unit time. For a detector that sorts counts into discrete channels, one can substitute Eq. (20.9) into Eq. (20.8), integrate over each channel width, and write the discrete form of the pulse height spectrum in the form

$$c(n) = \sum_{j=1}^{J} s_j \mathcal{R}_n(E_j) + \int_0^\infty s_c(E)\mathcal{R}_n(E)dE + b_n, \quad n = 1, 2, \ldots, N, \tag{20.10}$$

where $c(n)$ is the number of counts recorded in channel n (per unit time), b_n is the expected number of counts recorded in channel n (per unit time) from background radiation, and $\mathcal{R}_n(E)$ is the *detector response function* that is the probability that a particle of initial energy E that is emitted by the source produces a count within the nth channel. Here

$$\mathcal{R}_n(E) = \int_{(n-1)\Delta}^{n\Delta} \mathcal{R}(h|E)\, dh.$$

In general, spectra are accumulated over a counting time T, in order to increase the number of recorded counts and, hence, improve the statistical precision of the measurements. In this chapter, it is assumed that the source strength $S(E)$ is constant in time. If the photon source is from the decay of a radionuclide, then the measurement time T must be much less than the half-life of the radionuclide; otherwise, corrections must be made to correct for the decrease in photon emission rate as was done in the discussion of foil activation in Sec. 17.7.2.

The total counts obtained over counting time T per channel can be obtained by integrating Eq. (20.10) over the counting time. The continuum source term can be due to photons emitted from the source over a continuum of energies and/or to photons emitted at discrete energies from the source that scatter into the detector from material around the detector. Because the primary objective of spectroscopy is to find the E_j and s_j, for $j = 1, 2, \ldots, J$, it is customary to combine the continuum and background terms into a generalized background. Doing so and integrating over the counting time T, one obtains

$$\boxed{C(n) = \sum_{j=1}^{J} S_j \mathcal{R}_n(E_j) + B(n), \quad n = 1, 2, \ldots, N,} \tag{20.11}$$

where $C(n) \equiv C_n = c(n)T$ is the detector response in channel n, $S_j = s_j T$, and

$$B(n) = \left[\int_0^\infty s_c(E)\mathcal{R}_n(E)dE + b(n)\right] T. \tag{20.12}$$

A plot of $C(n)$ versus n is called a *pulse height spectrum.* Summation over any range of channels gives the total counts within those channels.

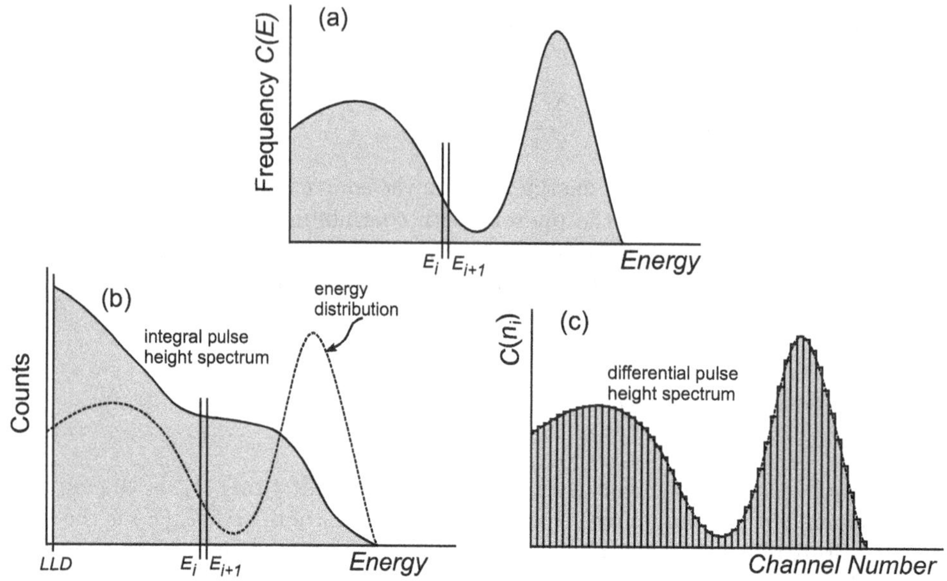

Figure 20.2. Shown in (a) is the continuous energy distribution recorded with a radiation spectrometer. If a counter is connected to the detector, shown in (b) is the resulting count rate as a function of the lower level discriminator setting, called the integral pulse height spectrum, superimposed on the radiation energy distribution. Shown in (c) is the resulting discrete differential pulse height spectrum from (b).

Equation (20.11) describes an inverse problem, in which the specific E_j and $S_j, j = 1, 2, \ldots, J$ are sought given $J \leq N$ measured pulse-height responses $C(n)$. The generalized background B_n is typically not known explicitly and, thus, further complicates the inversion process. There are different methods that can be used to approach this inverse problem and the number M of channels, $M \leq N$, used depends on the method chosen. Because most spectrometers have thousands of channels, most spectroscopy inverse problems are over-determined in the sense that there are considerably more responses available than unknowns.

It is noted that a variant of this inverse problem often is posed in terms of the $k = 1, 2, \ldots, K$ nuclides present in a sample, each of which can emit photons at one or more discrete energies. Rather than look for the J discrete energies and their intensities, one looks for the K nuclides and their concentrations. The form of this inverse problem is similar to the form of Eq. (20.11) and is defined later more precisely by Eq. (20.95).

20.4 Gamma-Ray Spectroscopy

Suppose that the energy deposition in the detector is that shown in Fig. 20.2. A simple counter records events in the detector that produced a signal exceeding the lower level discriminator (LLD). In the hypothetical case with no electronic noise or background, *all* events interacting in the detector within measurement time T are recorded if the LLD is set to zero. As the LLD is increased, those lower energy events in Fig. 20.2(a) are excluded from the count rate, and the total number of recorded counts within time T decreases. This function can be plotted as shown in Fig. 20.2(b), where the total integrated counts above the energy equivalent LLD setting are plotted as a function of LLD setting. This plot is called the *integral pulse height spectrum.* There are notable features in the integral pulse height curve that can be interpreted as follows. Flatter features indicate energy regions where few events appear, usually from valleys in the energy spectrum. Steeper features indicate energy regions where many counts are located and produce a large change in counts as a function of LLD, often caused by energy maxima or peaks in the energy spectrum.

Although an experienced spectroscopist might be able to interpret the data of an integral pulse height spectrum, it is usually the derivative of this spectrum that is used in spectroscopy, mainly because interpretation is more straightforward. Suppose the energy spectrum of Fig. 20.2(b) is divided into n number of channels, each channel having width of ΔE, then the total number of channels describing the spectrum is

$$n = (E_{max} - E_0)/\Delta E, \tag{20.13}$$

where E_{max} is the highest energy recorded as a function of channel number and E_0 is an experimentally determined *zero offset*. Ideally, the value of E_0 is zero, but in practice usually is not. Suppose that each energy bin is defined by the energies between two adjacent boundaries, i.e., by $E_{i+1} - E_i = \Delta E$. Then the number of counts within each boundary, or channel n, is described by

$$C_i = \int_0^\infty C(E)dE - \int_0^{E_i} C(E)dE - \int_{E_{i+1}}^\infty C(E)dE = \int_{E_i}^{E_{i+1}} C(E)dE. \tag{20.14}$$

This result is the same as that found by taking the difference between the counts at E_i and E_{i+1} in Fig. 20.2(b). If this change in the recorded counts from the integral pulse height spectrum is plotted as a function of the channel number (or energy), then a discrete *differential pulse height spectrum* is produced,

$$C(n_i) = \frac{-(C_{E_{i+1}} - C_{E_i})}{E_{i+1} - E_i} = \frac{-\Delta\text{Counts}}{\Delta E}, \tag{20.15}$$

as depicted in Fig. 20.2(c).

Figure 20.2(c) is a histogram called the *differential pulse height spectrum*, which mimics the energy distribution seen by the detector, i.e., that of Fig. 20.2(a). For a good spectrometer, the measured energy distribution is very similar to the actual distribution of energy deposited in the detector. However, for detectors that have non-linear effects, recombination, or charge carrier trapping problems, the energy deposited in the detector and the output signal are not necessarily proportional. Although a spectrum such as that depicted in Fig. 20.2(c) can be developed with a single channel analyzer (Sec. 2.5.6) by sequentially moving the energy window ΔE from zero up to 10 volts, it is a multichannel analyzer that is generally used to display the energy spectrum (Sec. 2.5.13).

20.4.1 Gamma-Ray and X-Ray Spectral Features

Already discussed in Chapter 4 are the various ways that photons interact in a material and, for the present application, in radiation detectors. The three main mechanisms are the photoelectric interaction, Compton scattering, and pair production. Although Raleigh (coherent) scattering can be significant at low energies, the photoelectric effect dominates (in the same energy region), usually by more than an order of magnitude. Consequently, Rayleigh scattering is usually ignored in practical gamma-ray spectroscopy. Likewise, binding effects become apparent only at low energies and are usually ignored.

Photoelectric Effect Features

Photons of relatively low energy (less than a few hundreds of keV) interact with the ambient medium mostly through photoelectric absorption in which some of the photon energy is transferred to a photoelectron to give it an energy

$$T_{pe} = E_\gamma - E_b, \tag{20.16}$$

where E_b is the binding energy of the electron which depends on its electron shell of origin. The liberated photoelectron moves through the detector medium causing more ionization through Coulombic interactions. Ultimately, an average number of electrons per unit energy is liberated (or excited) in the detector medium.

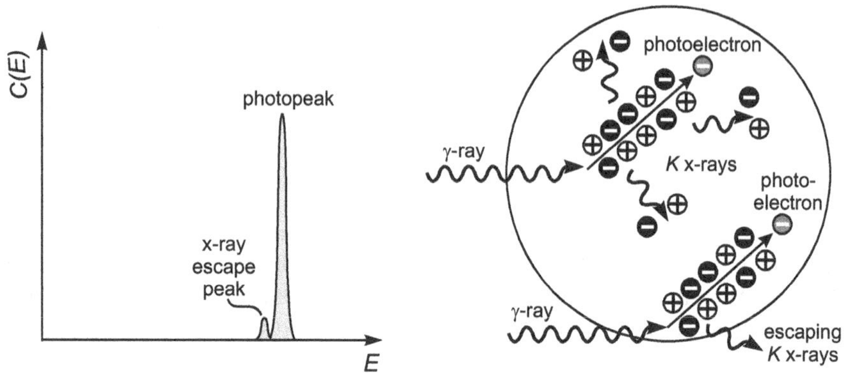

Figure 20.3. A photoelectron absorbs the full gamma-ray energy, minus the binding energy ($h\nu - E_b$). This photoelectron produces free charges in the detector. Through electron deexcitation, the binding energy is recovered by the emission and reabsorption of characteristic x rays. Some x rays emitted near surfaces can escape the detector. The resulting common spectral features are shown.

For scintillators, the important quantity is the average number of excited electrons that produce fluorescent light. For gas detectors the important quantity is the average energy required to produce an electron-ion pair. Similar to gas detectors, in a semiconductor it is the average energy required to produce an electron-hole pair that is important. Ultimately, the number of information carriers is a function of the average energy w required to produce the carriers and the absorbed photon energy. The effect is observed as a *photopeak* in the differential pulse height spectrum (depicted in Fig. 20.3).

The binding energy E_b is recovered as electrons fall into lower energy levels. For instance, an ejected K shell electron leaves an empty state, which can be filled by an electron falling from the L shell into the K shell, producing a characteristic K_α x ray. The energy may also be recovered by the emission of an Auger electron. If the x ray is absorbed by a lower energy bound electron (L or M), it produces more ionization, and the energy is recovered. The same is true for M shell electrons falling into K or L states, producing either K_β or L_α x rays, respectively. If generated near the detector surface, it is possible that these characteristic x rays escape the detector, consequently causing the total deposited energy to be reduced by the characteristic x ray energy. The photopeak under such a circumstance forms at the initial photon energy less the x-ray energy, thereby creating an x ray escape peak as depicted in Fig. 20.3. This escape peak results primarily from K shell x rays escaping and not from the loss of L shell x rays because L shell x rays have much lower energies and are less likely to escape and, if they do escape, the lost energy usually blends with the main photopeak and is less noticeable. These x ray escape peaks are a function of the detector material, and appear mainly when (1) the detector is small so that a large fraction of the characteristic x rays can escape and (2) when the detector is made of high Z material so that the characteristic x rays have larger energies.

Compton Scattering Features

At higher photon energies, ranging between tens of keV up to several MeV, depending on the material, the Compton scattering effect becomes dominant compared to the photoelectric effect. The energy of a Compton scattered electron is

$$T_{cs} = \frac{E_\gamma^2(1-\cos\theta_s)}{m_ec^2 + E_\gamma(1-\cos\theta_s)}, \tag{20.17}$$

where θ is the photon scatter angle and $m_ec^2 = 511.0$ keV is the rest-mass energy equivalent of an electron. The scattered gamma rays have a continuum of energies from zero up to the maximum possible energy

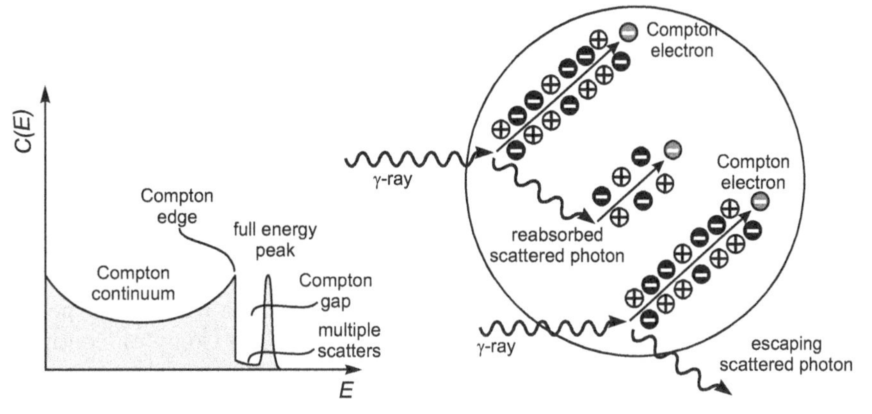

Figure 20.4. Compton scattered electrons produce free charges in the detector up to the maximum allowable energy transfer. If the Compton scattered photon is recaptured and full energy deposited, the energy adds to the full energy peak. If the scattered photon escapes the detector, the energy is added to the Compton continuum.

transfer to a Compton electron (for a single scatter at $\theta_s = \pi$), namely

$$T_{cs} = \frac{2E_\gamma^2}{m_e c^2 + 2E_\gamma}. \tag{20.18}$$

A simple example is the case in which there is a single Compton scatter and the scattered gamma ray escapes the detector. Under such a condition, the energy lost from the detector is given by Eq. (20.17). Consequently, the pulse height spectrum is a continuum of energies from zero up to the energy described by Eq. (20.18), termed the *Compton continuum* and is depicted in Fig. 20.4. The high energy limit of the Compton continuum is called the *Compton edge.* If a considerable fraction of photons are Compton scattered multiple times, ultimately terminating with photoelectric absorption, then the total initial photon energy is represented by the energy peak in the pulse height spectrum. Because more than one type of interaction contributed to the energy absorption, the proper term for this peak is the *full energy peak.* A gap appears between the full energy peak and the Compton edge, termed the *Compton gap.* Multiple Compton scatters that still result in some energy escaping the detector produce a small continuum in the Compton gap. Ultimately, the Compton continuum is more prominent in small detectors than large detectors, mainly because more Compton scattered photons escape the smaller detector.

Compton scattering of source photons in the surroundings and shielding of a detector can result in some of these scattered photons reaching the detector and being absorbed in it. Theoretically, this backscatter spectrum can have energies described by

$$E_{bs} = \frac{E_\gamma m_e c^2}{m_e c^2 + E_\gamma(1 - \cos\theta)}, \tag{20.19}$$

where θ is the scattering angle needed for the scattered photon to reach the detector. The minimum energy of these scattered photons that reach the detector occurs when $\theta = \pi$ and is

$$E_{bs} = \frac{E_\gamma m_e c^2}{m_e c^2 + 2E_\gamma}. \tag{20.20}$$

Recall from Sec. 4.4.2 the discussion on scatter angle probability and photon energy, where forward scattering becomes more probable with increasing gamma-ray energy. Hence, the differential backscatter cross section actually decreases with higher gamma-ray energies (see Fig. 4.6). However, from Fig. 4.7, photons that scatter with angle greater than about 100° emerge with nearly the same energies.

In practice, for geometrical reasons, it is unlikely that forward scattering in the surroundings or shielding results in a significant number of scattered photons entering the detector. In fact, it is the larger scattering angles that are most likely to return to the detector, so that the backscattering spectrum typically increases at low energies in the pulse height spectrum. The effect is enhanced because low energy photons are more readily absorbed in a detector than are high energy photons. Further, from Fig. 4.7, the energies of backscattered photons are similar. A wide peak appears at the lower limit, called the *backscatter peak*, as depicted in Fig. 20.5. This backscatter peak is actually a continuum with a somewhat discernible maxima. Multiple scatters can cause the appearance of photon energies below the backscatter peak, but in practical spectroscopy, these energy additions are difficult to distinguish from the Compton continuum.

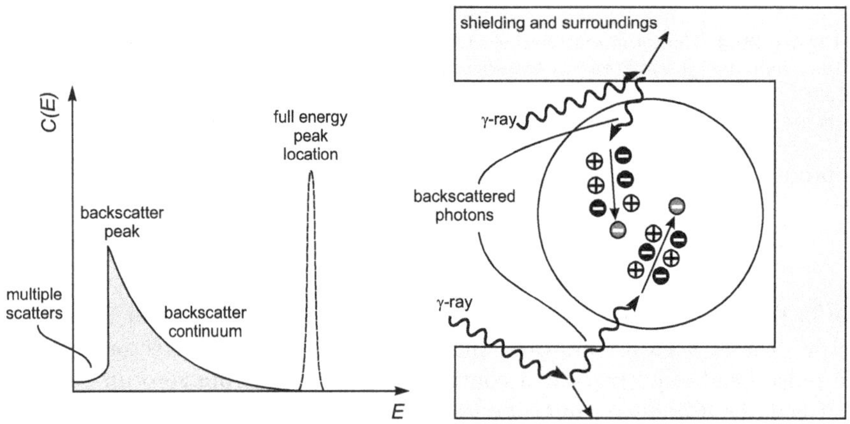

Figure 20.5. Gamma rays that are Compton scattered in the materials surrounding the detector may deposit energy in the detector. Consequently a backscatter spectrum appears.

Pair Production Features

If the gamma-ray energy is greater than 1.022 MeV, then pair production is possible. When this photon interaction occurs in a detector, 1.022 MeV of the photon energy is converted into the masses of the electron-positron pair and the remaining energy shared as kinetic energy between the two particles. The particles produce ionization in the detector just like photoelectrons and Compton electrons. After the electron loses almost all its initial energy, it is absorbed in the material and returns to an allowed state. However, when the positron comes to rest, it combines with an electron and annihilates, producing two 511-keV photons emitted in opposite directions to preserve the zero linear momentum condition. If both of the annihilation photons are reabsorbed in the detector, the initial gamma-ray energy is represented in the full energy peak. If multiple events result in one 511-keV annihilation photon escaping the detector, then an energy peak forms in the pulse height spectrum at $E_\gamma - 511$ keV, named a *single escape peak.* If multiple events result in both 511-keV annihilation photons escaping the detector, then an energy peak forms in the pulse height spectrum at $E_\gamma - 1.022$ MeV, named a *double escape peak.* Single and double escape peaks can be identified by (1) noticing that no Compton edge forms and (2) noting their energy location with respect to the full energy peak. Escape peaks are more prominent in small detectors than large detectors, mainly because a larger fraction of annihilation photons escape the smaller detector.

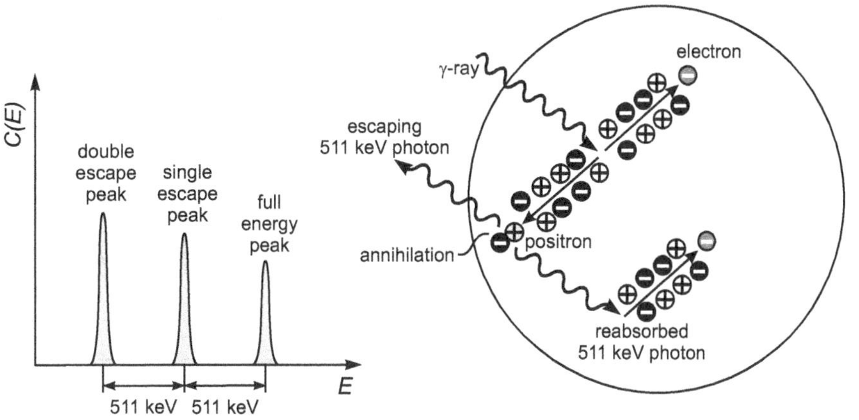

Figure 20.6. Gamma rays that undergo pair production produce energetic electron-positron pairs. These particles produce free charge in the detector. When the positron loses nearly all kinetic energy, it annihilates with an electron and emits two 511-keV photons in opposite directions. Spectral features form accordingly if these annihilation photons are reabsorbed, one escapes the detector, or if both escape the detector.

Pair Production and Fluorescence in the Surroundings

Pair production can also can occur in the shielding and surroundings, causing the emission of 511 keV annihilation photons that can enter the detector and add to the pulse height spectrum (Fig. 20.7). Because of this possibility of contamination, 511 keV annihilation photons are typically not used for efficiency calibration, but can still be used for energy calibration.

The shielding and surrounding material, when irradiated by energetic photons, can also fluoresce by the emission of characteristic x rays. These x rays appear in the pulse height spectrum at low energies, usually below 90 keV. Because lead is a common shielding material, several energy lines ranging from 72.8 keV to 87.3 keV may appear in the low energy portion of the pulse height spectrum. Unless low energy spectroscopy is required, usually these additional lines are of little consequence. However, directional shielding can be applied to reduce the appearance of these lines, usually realized by a layer of copper placed between the detector and lead shielding (as described further in Chapter 21).

Finally, bremsstrahlung fluorescence is generally present with gamma-ray detection. Many gamma-ray emitting radioisotopes also emit beta particles, which in turn emit bremsstrahlung as they slow down in a target material (Sec. 5.3.2). From Eq. (5.6), the energy intensity of bremsstrahlung increases with target Z and the electron energy. A low Z beta-particle filter between the radionuclide source and the detector can soften the bremsstrahlung spectrum. Also, photon interactions in the surroundings produce energetic electrons, and these energetic electrons radiate bremsstrahlung as they slow down. Sample emission distributions in lead are shown in Fig. 5.2. Although much of bremsstrahlung is self-absorbed in the shielding, some of it can still interact in the detector. Because bremsstrahlung is generally featureless, this form of fluorescent contamination increases the background but does not produce energy peaks. Regardless, the addition of bremsstrahlung does increase the total counts in the pulse height spectrum, generally in the lower energy channels, which can cause an overestimation of the gamma-ray flux.

Summary

To summarize, consider a radiation source that emits monoenergetic photons of energy E_o. A source photon that enters the detector may experience any of several outcomes, including the following:

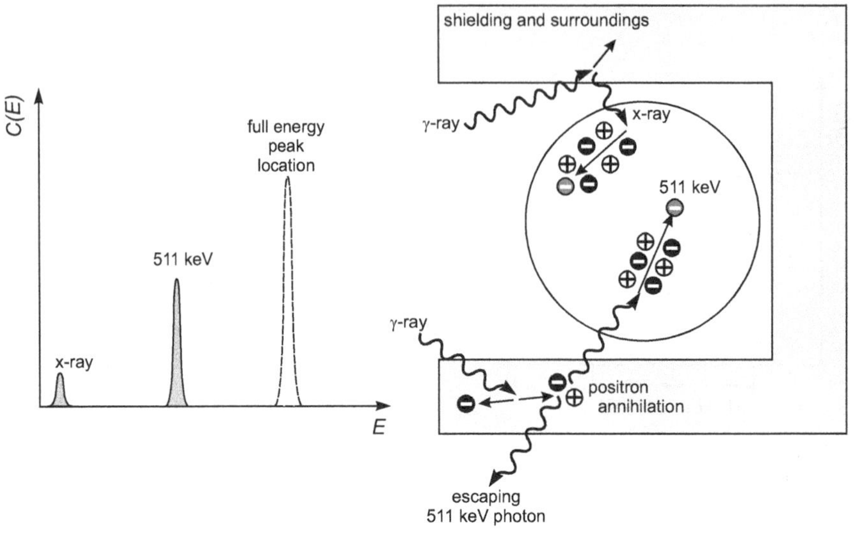

Figure 20.7. Gamma rays can also produce pair production in the surroundings. Consequently, 511 keV annihilation photons can enter the detector from outside sources. Energetic photons can also fluoresce the surroundings and the detector container to produce x-rays that can enter the detector.

1. It may be completely absorbed by photoelectric absorption, in which case the deposited energy is the photon energy, i.e., $E_d = E_o$.

2. It might undergo a sequence of one or more scatters within the detector and then be absorbed within the detector by photoelectric absorption, which again leads to $E_d = E_o$.

3. It might scatter one or more times in the detector and then escape with an energy E_r, which leads to a deposited energy of $E_d = E_o - E_r$.

4. If $E_o > 1.022$ MeV, it might undergo pair production in the detector, producing an electron-positron pair that usually leads to the production of two 0.511-MeV photons by positron annihilation.[2] If both annihilation photons deposit all of their energy in the detector, then $E_d = E_o$. If one of the annihilation photons escapes and the other is absorbed, the energy deposited is $E_d = E_o - 0.511$ MeV. If both annihilation photons escape the detector, the deposited energy is $E_d = E_o - 1.022$ MeV. If either or both annihilation photons scatter within the detector and then escape, an intermediate energy within the range $E_o - 1.022 < E_d < E_o$ is deposited.

5. It might not interact at all in the detector and so $E_d = 0$.

Note that outcomes 1 and 2 both contribute to a "full-energy" peak. Voltage pulses result whose pulse heights are distributed from zero up to a maximum determined by the energy E_d and the energy resolution of the detector. Thus a monoenergetic source produces a pulse-height spectrum that is distributed over the many channels. The features shown in Fig. 20.3 through Fig. 20.6 can be combined to produce the expected features that appear in a gamma-ray pulse height spectrum. Depicted in Fig. 20.8 are differential pulse height spectra for monoenergetic gamma rays with energies below and above 1.022 MeV.

[2]Positron annihilation occasionally takes place at higher than thermal energies. Consequently, annihilation photons of higher energies are (infrequently) produced.

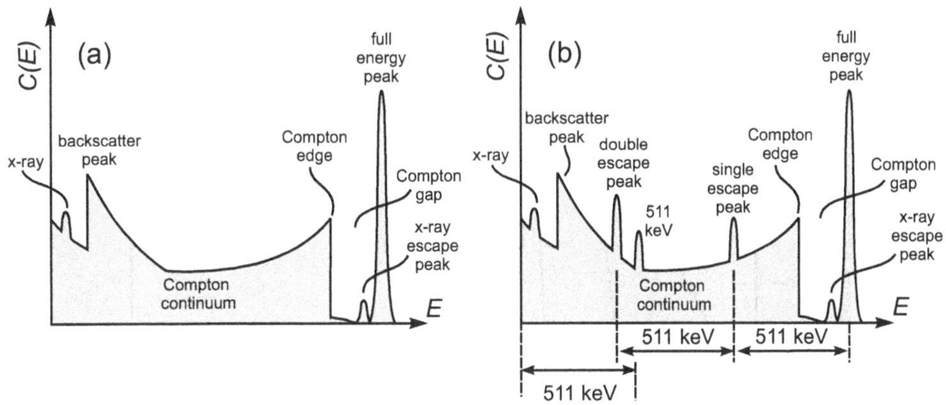

Figure 20.8. Composite pulse height spectra for monoenergetic gamma rays formed from the features shown in Fig. 20.3 - Fig. 20.6, with (a) $E_\gamma < 1.022$ MeV, and (b) with $E_\gamma > 1.022$ MeV.

An example pulse height spectrum obtained from a scintillation spectrometer exposed to the photons from a ^{22}Na source, which emits 1.28-MeV photons and 0.511-MeV annihilation photons, is shown in Fig. 20.9. The features in the spectrum, from right to left, include a full-energy peak centered about the channel corresponding to $E_1 = 1.28$ MeV, a Compton edge and a continuum extending from about channel 1250 downward, a full-energy peak due to 0.511-MeV annihilation photons, a Compton edge and Compton continuum for the annihilation photons, and a backscatter peak. The maximum amount of energy that a photon can give up in a Compton scatter occurs in a "backscatter" through an angle of π radians, and so the Compton edge occurs over those channels that represent Compton scatters in the detector through angles near π. Photons that are emitted by the source and backscatter (within either the source or the material behind the source) have energies that are near the energy given by Compton scatter through π radians. The backscatter peak in the spectrum of Fig. 20.9 is caused by source gamma rays that scatter, in or near the sample, through angles close to π and then deposit full energy in the detector. This peak typically has a FWHM that is larger than the FWHM of a full-energy peak for monoenergetic photons. The larger FWHM occurs because source photons can backscatter in or near the source over a range of angles near π radians and thus these scattered photons that reach the detector do not all have the same energy.

Summation Peaks

Other features that may appear in the pulse height spectrum are *sum peaks*, which are artifacts from two or more photons interacting nearly simultaneously (i.e., within the resolving time) in the detector. Sum peaks appear to be valid energies because they have many of the aforementioned features including Compton edges and Compton gaps. Unknown energy peaks appearing in a pulse height spectrum can be checked to determine if combinations of known energy emissions from the source sum to the unknown energy.

Given a source with multiple independent gamma-ray energy emissions E_i, each having intrinsic full energy peak absorption efficiency of ϵ_{pi}, the probability of N coincident photons leading to a full energy absorption is represented by

$$P_N = KW(0°)\prod_i^N \Omega_{fi}\epsilon_{pi}B_i \tag{20.21}$$

where Ω_{fi} is the fractional solid angle for the source energy, B_i is the branching ratio (or frequency) of the emission, and K is a geometric correction for parallax and detector shape, and $W(0°)$ is a factor included to account for the angular correlation between the coincident gamma rays [Heath 1964; Rose 1953]. Gamma-ray sources with multiple emissions per decay, such as ^{60}Co and ^{133}Ba, can be expected to develop sum peaks

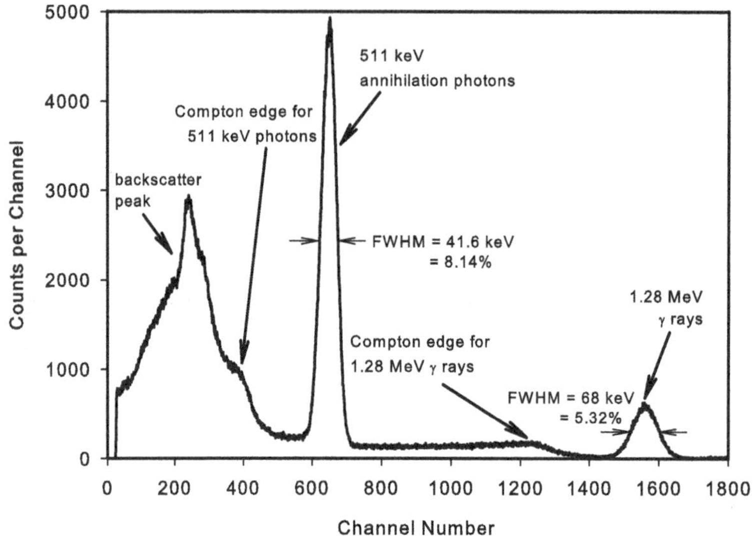

Figure 20.9. The pulse height spectrum obtained by a NaI:Tl scintillation detector exposed to a ^{22}Na source. The source emits photons at 0.511 MeV, as a result of positron annihilations, and at 1.28 MeV, emitted as the product ^{22}Na transitions from the excited state to the ground state [McGregor 2008].

in the pulse height spectrum. Events of this type, in which two or more photons are released from a single decay and are coincident in the detector, are often called *true coincidences.* The expected number of counts in the sum peak are determined by multiplying the total source emission At by P_N, where A is the source activity and t is the counting time.

For geometries with a single point source and detector, the value of Ω_{fi} is the same for all E_i. Consider those cases in which there are only two simultaneous gamma-ray emissions such as occurs, for instance, with ^{60}Co. Then for a point source and single detector Eq. (20.21) becomes

$$P_2 = KW(0°)\Omega_f^2\epsilon_1\epsilon_2B_1B_2. \tag{20.22}$$

The parallax correction factor K approaches unity as the source-to-detector becomes large. Also, it is often assumed that the emission directions of the different photons are uncorrelated so that $W(0°) = 1$.[3] Under these conditions the true coincidence sum rate for two simultaneous emissions reduces to the often used expression

$$R_{\text{true}} = A\Omega_f^2\epsilon_1\epsilon_2B_1B_2. \tag{20.23}$$

From Eqs. (20.21) and (20.23) it is clear that decreasing the fractional solid angle Ω_f reduces the formation of sum peaks more effectively than reducing counts in the real energy peaks because coincidences are a function of Ω_f^N while true energy peaks are proportional only to Ω_f.

Sum peaks can also appear for high count rates, when two gamma rays from independent radioactive decays are coincident within the processing time of the detection system. Recall from the discussion on dead time coincidences in Sec. 7.3.2 that a random sequence of incoming pulses with average interaction rate $\overline{r}$. The average number of interactions within time t is simply $\overline{r}t$. The Poisson distribution of Eq. (7.26) describes the probability of multiple interactions in the detector within some time period t from a single

[3]This assumption of course does not pertain to positron annihilation photons, which are emitted in opposite directions.

radiation source, namely,

$$\mathcal{P}(n|\mu) = \frac{(\overline{r}t)^n}{n!} e^{-\overline{r}t}, \tag{20.24}$$

where n is the number of interactions. If t_r represents the pulse resolving time of the detector system, and should these emissions enter a single detector within the pulse processing time, then they can produce a coincidence sum peak. These events are sometimes referred to as *accidental coincidences*, mainly because of the stochastic nature of radioactive decay allows such events to randomly occur by accident from unrelated radioactive disintegrations. For a monoenergetic source, where the efficiencies, branching ratios, and solid angles are the same, from Eq. (7.26) the probability that *only one* gamma ray interacts within the resolving time of the same detector t_r reduces to

$$P = \mathcal{P}(1|\mu) = \overline{r}t_r e^{-\overline{r}t_r}. \tag{20.25}$$

For small $\overline{r}t \ll 1$, Eq. (20.25) reduces to

$$P = \overline{r}t_r e^{-\overline{r}t_r} \simeq \overline{r}t_r(1 - \overline{r}t_r) \simeq \overline{r}t_r = \mu, \tag{20.26}$$

which is also the average number of photons interacting in the detector within time t_r. Suppose that an initial gamma ray event occurs in the detector at $t = 0$, from gamma rays interacting at an average rate of $\overline{r}$. The rate of accidental coincidences, therefore, is the product of the average interaction rate $\overline{r}$ and the average number of photons interacting within time t_r

$$R_{acc} = (\overline{r})(\mu) = \overline{r}^2 t_r. \tag{20.27}$$

Equation 20.27 is the widely used expression for accidental coincidences from a monoenergetic source [Vincent 1973].

There is a (smaller) possibility that two or more photons interact within the resolving time, which would also reduce counts in the ideal spectrum without coincident effects. The probability that at least one event occurs in time t_r is given by

$$P = 1 - \mathcal{P}(0|\mu) = 1 - e^{-\overline{r}t_r}, \tag{20.28}$$

which yields a total accidental coincidence rate of

$$R_{acc} = \overline{r}(1 - e^{-\overline{r}t_r}). \tag{20.29}$$

By adjusting (reducing) either $\overline{r}$ or t_r, the rate of accidental coincidences can be reduced.

Shown in Fig. 20.10 is a spectrum of ^{60}Co emissions taken with an HPGe detector. The features include a true coincidence sum peak from the two gamma-ray emissions along with two chance coincidence sum peaks of the two emissions. Note that the counts in the 1173.2 keV + 1332.5 keV sum peak may be from both true and chance coincidences, which are difficult to discern without some added analysis (e.g., with measurements taken at different distances). Also shown in Fig. 20.10 are other features discussed in the previous sections, including two Compton continua along with the coincidence sum Compton continua, single and double escape peaks, 511-keV annihilation photons from the surroundings, Pb x rays, background from ^{40}K, and a backscatter peak.

A consequence of summation peaks is their ultimate reduction in the counts represented in the actual full energy peaks. Consider the sum peaks of Fig. 20.10. For each count in the sum peak formed from 1173.2 keV + 1332.5 keV gamma rays, one gamma ray each is removed from their corresponding real full energy peaks. Also, the accidental sum peaks remove two gamma rays each from their corresponding full energy peaks. One might correct the problem by simply adding the counts tallied in the sum peaks to the appropriate full

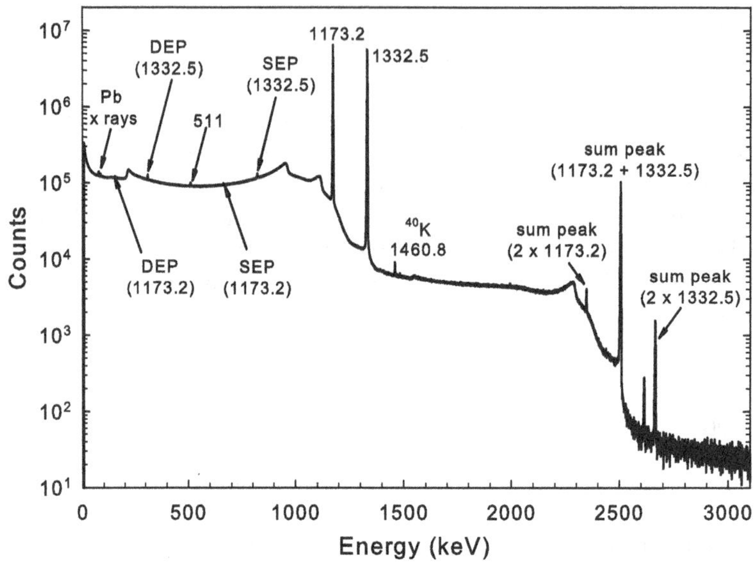

Figure 20.10. A gamma-ray spectrum of ^{60}Co taken with an HPGe detector. Included in the features are three sum peaks from true and chance coincident absorptions of the 1173 and 1332 keV gamma rays. Courtesy of Nathaniel Edwards, KSU.

energy peak, yet this alone does not account for all coincident losses. The coincident absorptions where at least one photon is Compton scattered produces a coincidence Compton spectrum below the sum peak(s). Any coincidence of one photon with full energy deposition of a second photon alters the number of counts in full energy peaks. For instance, a full energy absorption of one energy (say, E_1) coincident with a Compton scatter from another photon (E_2) still removes a count from the full energy peak of E_1. Unfortunately, these events fall into the coincidence Compton spectrum, and the original photon energies are difficult to determine. Consequently, proper correction to the full energy peaks becomes difficult.

The total counts observed C_{oi} in the full energy peak for E_i is related to the total counts C_i that should appear in the peak by

$$C_{oi} = C_i - (\text{real coincident counts} + \text{accidental coincident counts}), \tag{20.30}$$

where the full energy of E_i is deposited in the detector for each coincidence. For conceptual simplicity, consider a source with only two gamma ray emissions per decay, such as a ^{60}Co source. The observed counts in the full energy peak for the gamma rays of energy E_1 are represented by

$$C_{o1} \simeq At\left[\Omega_{f1}\epsilon_{p1}B_1(1 - K\Omega_{f2}\epsilon_{I2}B_2 - Att_r(\Omega_{f1}\epsilon_{I1}B_1) - Att_r(\Omega_{f2}\epsilon_{I2}B_2))\right], \tag{20.31}$$

where ϵ_{I2} is the total intrinsic counting efficiency of the second photon of energy E_2. For a single point source and detector, the solid angle is the same, and the counts for E_1 become

$$C_{o1} \simeq At\Omega_f\epsilon_{p1}B_1\left[(1 - \Omega_f(K\epsilon_{I2}B_2 + Att_r(\epsilon_{I1}B_1 + \epsilon_{I2}B_2))\right]. \tag{20.32}$$

A similar expression can be found for the full energy peak of E_2. For decay schemes with more than two gamma-ray emissions per decay, then the multiple coincident possibilities must be considered. These expressions can become quite complicated as shown by Vincent [1975] and Smith [1978]. Byun et al. [2004] describe a Monte Carlo method to obtain coincidence correction factors for multiple coincidences and detectors.

20.4.2 Spectral Response Function

A spectrum such as that in Fig. 20.9, normalized to unit concentration, is called the *spectral response function* for a given radionuclide. Each nuclide has a characteristic spectral response function which is different for each spectrometer for a specified source-detector geometry. Let the subscript k denote a specific nuclide. Then the detector response function u_{nk} is the expected response (number of counts) of the spectrometer in channel n per unit concentration of nuclide k.

Spectral response functions also can be associated with monoenergetic photons. If a source of monoenergetic photons of energy E_j, in some specific source-detector geometry, irradiates a detector then the detector response function u_{nj} gives the expected response in channel n per source particle of energy E_j emitted from the source. In either case, it is not the full energy peaks that are of only interest; the entire spectrum contains information about the source. Ways to exploit this dependence are considered later.

20.4.3 Qualitative Analysis

For some purposes, it is necessary to identify only whether or not a sample emits photons at certain discrete energies E_j. This may be the case, for instance, if one wants to know if a sample contains a particular radionuclide. Alternatively, a procedure such as EDXRF, NAA, or PGNAA can be used to excite characteristic photons from a sample under investigation. If the element of interest is present, the characteristic photons emitted from the sample should create full-energy peaks in a pulse-height spectrum collected from the sample.

Photons of energy E_j that are emitted by the source produce full-energy pulses whose magnitudes are distributed about a mean value of

$$h_j = f^{-1}(E_j). \tag{20.33}$$

A pulse of magnitude h_j produces a count in discrete channel n_j if

$$h_{min} + (n_j - 1)\Delta < h_j \leq h_{min} + n_j\Delta.$$

For purposes of energy determination, it is useful to consider the continuous, non-integer, channel number n_j corresponding to the pulse-height h_j of the full-energy peak. Then the gamma-ray energy can be estimated by determining the continuous or fractional channel number $\tilde{n}_j$ centroid of each peak.

It is then straightforward to obtain the corresponding h_j from

$$h_j = h_{min} + \tilde{n}_j\Delta. \tag{20.34}$$

For the usual case, where $h_{min} = 0$ and $h_{max} = 10$ V, this reduces to

$$h_j = \frac{10}{N}\tilde{n}_j. \tag{20.35}$$

The expected source energy is then easily obtained from

$$E_j = f^{-1}(h_j). \tag{20.36}$$

For a linear detector whose response is given by Eq. (20.2), then Eq. (20.36) is simply

$$E_j = \frac{h_j - \alpha}{\beta}. \tag{20.37}$$

If a nuclide emits several characteristic-energy photons, then one can gain confidence in the conclusion that the element is present if peaks occur at all energies for which the photon abundance and the detection efficiency would lead one to suspect that a peak should occur.

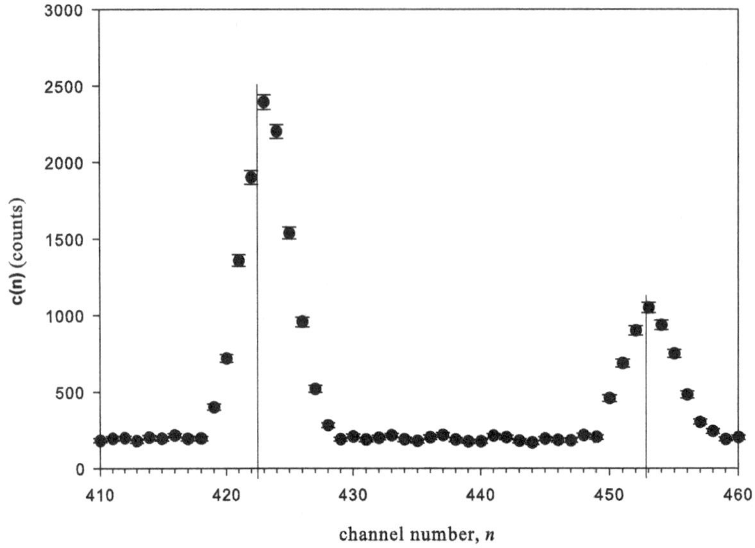

Figure 20.11. An example of visual inspection to yield the channel numbers corresponding to the centroids of two peaks in a spectrum. The vertical lines are used to connect the apparent peaks to their centroid fractional channel numbers.

The simplest way to estimate the channel corresponding to the peak is by inspection of the plotted spectrum. This technique often is adequate. One merely estimates the continuous channel number n_j that corresponds to the apparent centroid of the full-energy peak due to photons of energy E_j. For instance, in the spectrum shown in Fig. 20.11, one can obtain estimates of the fractional channel numbers of the centroids of the two peaks shown.[4] It should be apparent that this procedure is subjective (different researchers may estimate slightly different centroids) and, thus, has limited accuracy. Nevertheless this procedure may suffice for some applications.

When estimation of the centroid by inspection is deemed insufficient, other methods must be employed. The wavelet transform has been used in various spectroscopic applications, including nuclear magnetic resonance [Barache et al. 1997] and Raman spectroscopy [Xu et al. 1994]. However, this approach is not commonly employed in gamma-ray spectroscopy and, thus, is not further considered here. Rather, it is common that the centroids of the peaks present are identified as part of a fitting process that can determine quantitative information about both the characteristic energies E_j and their relative abundances. Alternatively, such methods may seek to identify particular radionuclides and their concentrations in a sample. Such methods are discussed in the next sections.

20.4.4 Quantitative Analysis

The model of Eq. (20.11) identifies a general inverse problem in which many channels contain information about the source distribution. One typically seeks to determine not only the characteristic energies E_j emitted by the source but also the individual source strengths s_j from the measured responses. Alternatively, one may use the spectral responses to identify the nuclide k and its concentration ξ_k in a sample. Quanti-

[4]In this example, the analyst is inexperienced because the maximum of a symmetric Gaussian distribution must lie between the midpoint of the channel with the most counts and the channel boundary adjacent to the boundary with the second most counts. Thus, both vertical lines should be moved slightly to the right.

tative analysis refers to the determination of quantities such as s_j and ξ_k. There are several approaches to quantitative analysis in spectroscopy, including the following:

1. Area under isolated peaks
2. Model fitting
3. Spectrum stripping
4. Library least-squares
5. Symbolic Monte Carlo

Summaries of how these methods are typically implemented are given in the following sections. In general, one seeks to obtain both the $E_j, j = 1, 2, \ldots, J$, and the net areas under each of the J peaks. For spectra that are linearly related to the source strengths, the net area A_j under the jth full-energy peak is related to S_j, the number of gamma rays emitted by the source during the counting time T, by

$$S_j = \frac{A_j}{\eta_j},$$

where η_j is the detector efficiency, presumed known, at energy E_j,

$$A_j = \sum_{n_1}^{n_2} [C(n) - B(n)], \tag{20.38}$$

and n_1 and n_2 are channel numbers over which the peak is spread. Similarly, in linear systems, the concentration ξ_k of nuclide k is directly related to the net peak area.

A gamma-ray spectrum, generally, consists of a series of peaks atop a continuum produced by sources emitting a continuum of energies and Compton plateaus from scattered photons. Ideally such a spectrum can be described by a sum of K Gaussian peaks plus a continuum as

$$y(n) = \sum_{k=1}^{K} g_k(n, \mu_k, \sigma_k) + B(n), \tag{20.39}$$

where

$$g_k(\tilde{n}, \mu_k, \sigma_k) = \frac{A_k}{\sqrt{2\pi}\sigma_k} \exp[-(\tilde{n} - \mu_k)^2/(2\sigma_k^2)], = B_k \exp[-((\tilde{n} - \mu_k)/\tau_k)^2]. \tag{20.40}$$

Here $A_k = \sqrt{\pi}\tau_k B_k$ is the number of counts in peak k and $\sigma_k = \tau_k/\sqrt{2}$ is the standard deviation of the peak. Usually, the continuum is represented by a (piecewise) polynomial of $\tilde{n}$ of low order. The principal purpose of quantitative analysis of the spectrum is to determine values of A_k and μ_k from which radioisotope identification and concentrations can be determined. Also in calibrated MCA spectrometer systems, the continuous channel number $\tilde{n}$ is replaced by the photon energy E and the channel number n by the energy E_n at the channel midpoint.

20.4.5 Area Under an Isolated Peak

For isolated peaks, i.e., those that do not overlap with other peaks, a simple approach can be used to estimate the net peak area. Generally, the peak is superimposed on a generalized background, as shown, for an ideal case, in Fig. 20.12. To obtain the net area A, one identifies the channel numbers, n_1 and n_2, at which the peak disappears into the background. Then, the net peak area is estimated as

$$A = \sum_{n=n_1}^{n_2} \left[C(n) - (n_2 - n_1)\frac{C(n_1) + C(n_2)}{2} \right]. \tag{20.41}$$

The net area is the total number of counts between channels n_1 and n_2 minus the area under an (assumed) linear background between $C(n_1)$ and $C(n_2)$. For instance, for the spectrum shown in Fig. 20.12, one might choose $n_1 = 833$ and $n_2 = 861$. The total area between these channels is the sum of the counts in each channel and the background is the area under the straight line connecting $C(833)$ and $C(861)$; thus, the net area A is the area beneath the peak and above the background line in the figure.

This simple procedure cannot be applied to overlapping peaks and gives only approximate net areas because the background may not be linear under the peak and identifying the channels n_1 and n_2 is subjective, especially when the standard deviations of the responses, $\sigma(C(n_1))$ and $\sigma(C(n_2))$, are large relative to the values $C(n_1)$ and $C(n_2)$.

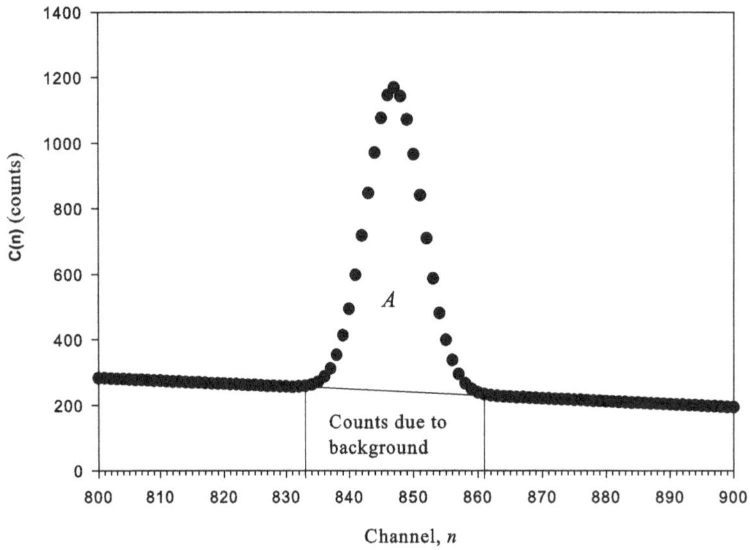

Figure 20.12. The net area A under the peak but above background.

20.4.6 Linear Least Squares Method for a Straight Line

Before methods to identify energy peaks and their corresponding energies, a review of the least squares method to fit models to data is appropriate. The least squares method presented here and in subsequent sections draws heavily on the discussion by Price et al. [2002].

Begin by considering the problem of fitting N data points (x_i, y_i) to a straight-line model

$$y(x) \equiv y(x|a, b) = a + bx. \tag{20.42}$$

This problem is often called *linear regression* by social scientists who invented the phrase a long time ago. Here it is assumed that the standard deviation σ_i for the measurements y_i are known and that the values of

x_i are known exactly. To quantify how well this linear model agrees with the data, the chi-square statistic is often used, which in this case is

$$\chi^2(a,b) = \sum_{i=1}^{N} \left(\frac{y_i - a - bx_i}{\sigma_i} \right)^2, \tag{20.43}$$

where σ_i is the standard deviation of y_i and is assumed to be known. If the σ_i are not known, then simply set $\sigma_i = 1$. If the measurement errors are normally distributed, as in counting data, then minimizing χ^2 with respect to a and b gives the maximum likelihood estimators of a and b [Press et al. 1992]. Even if the errors are not normally distributed the following approach is still useful.

At the minimum of Eq. (20.43) its derivatives with respect to a and b vanish, i.e.,

$$\begin{aligned} 0 &= \frac{\partial \chi^2}{\partial a} = -2 \sum_{i=1}^{N} \frac{y_i - a - bx_i}{\sigma_i^2} \\ 0 &= \frac{\partial \chi^2}{\partial b} = -2 \sum_{i=1}^{N} \frac{x_i(y_i - a - bx_i)}{\sigma_i^2} \end{aligned} \tag{20.44}$$

The equations can be rewritten more conveniently by first defining the following summations:

$$S \equiv \sum_{i=1}^{N} \frac{1}{\sigma_i^2}, \quad S_x \equiv \sum_{i=1}^{N} \frac{x_i}{\sigma_i^2}, \quad S_y \equiv \sum_{i=1}^{N} \frac{y_i}{\sigma_i^2}, \quad S_{xx} \equiv \sum_{i=1}^{N} \frac{x_i^2}{\sigma_i^2}, \text{ and } S_{xy} \equiv \sum_{i=1}^{N} \frac{x_i y_i}{\sigma_i^2}. \tag{20.45}$$

With these definitions, Eqs. (20.44) become

$$aS + bS_x = S_y \quad \text{and} \quad aS_x + bS_{xx} = S_{xy}. \tag{20.46}$$

The solution of these two equations for a and b is

$$a = \frac{S_{xx}S_y - S_x S_{xy}}{\Delta} \quad \text{and} \quad b = \frac{SS_{xy} - S_x S_y}{\Delta}, \tag{20.47}$$

where $\Delta \equiv SS_{xx} - (S_x)^2$.

This is not the end of the problem. The uncertainties in a and b must still be estimated because the uncertainties in the measurements makes the above estimates of a and b also somewhat uncertain. Recall from Sec. 6.9 that the variance in some function f of the y_i is, for independent y_i,

$$\sigma_f^2 = \sum_{i=1}^{N} \sigma_i^2 \left(\frac{\partial f}{\partial y_i} \right)^2. \tag{20.48}$$

For the straight line the derivatives of a and b with respect to the y_i are found from Eqs. (20.47)

$$\frac{\partial a}{\partial y_i} = \frac{S_{xx} - S_x x_i}{\sigma_i^2 \Delta} \quad \text{and} \quad \frac{\partial b}{\partial y_i} = \frac{Sx_i - S_x}{\sigma_i^2 \Delta}. \tag{20.49}$$

Sum over the points as in Eq. (20.48) to obtain

$$\sigma_a^2 = S_{xx}/\Delta \quad \text{and} \quad \sigma_b^2 = S/\Delta. \tag{20.50}$$

Finally, the covariance of a and b is $\text{covar}(a, b) = -S_x/\Delta$ from which the coefficient of correlation between the uncertainty in a and b is

$$\rho_{ab} = \frac{-S_x}{\sqrt{SS_x x}}, \tag{20.51}$$

which is a number between -1 and 1. A positive value of r_{ab} indicates that the errors in a and b are likely to have the same sign, while a negative value indicates the errors are anti-correlated, i.e., the errors are likely to have opposite signs.

Numerical Considerations

The formulas of Eqs. (20.47) are notoriously susceptible to roundoff error [Press et al. 1992]. To avoid roundoff problems, define

$$t_i = \frac{1}{\sigma_i}\left(x_i - \frac{S_x}{S}\right), \quad i = 1, 2, ..., N \qquad \text{and} \qquad S_{tt} \equiv \sum_{i=1}^{N} t_i^2. \tag{20.52}$$

Then, as can be verified by direct substitution into Eqs. (20.47)–(20.51),

$$a = \frac{1}{S_{tt}}\sum_{i=1}^{N}\frac{t_i y_i}{\sigma_i} \qquad \text{and} \qquad b = \frac{S_y - S_x b}{S}, \tag{20.53}$$

$$\sigma_a^2 = \frac{1}{S}\left(1 + \frac{S_x^2}{SS_{tt}}\right) \qquad \text{and} \qquad \sigma_b^2 = \frac{1}{S_{tt}}, \tag{20.54}$$

$$\text{cov}(a, b) = -\frac{S_x}{SS_{tt}} \qquad \text{and} \qquad r_{ab} = \text{cov}(a, b)\sigma_a\sigma_b. \tag{20.55}$$

Application to Power and Exponential Functions

The linear least squares curve fit method can be used on power and exponential functions with appropriate substitutions. For example, the function $y(x) = \alpha x^\beta$ can be curve fit by substituting $Y = \ln(y)$, $X = \ln(x)$, $a = \ln(\alpha)$, and $b = \beta$. Likewise an exponential function $y(x) = \alpha e^{\beta x}$ can be fit to a straight line with substitutions $Y = \ln(y)$, $X = x$, $a = \ln(\alpha)$ and $b = \beta$. These substitutions can then be used to fit the linear model

$$a + bX = Y \tag{20.56}$$

to the observed data $y(x_i)$ and x_i, $i = 1, \ldots, N$. The use of such a transformation to make a non-linear model into a linear one is illustrated in Example 20.1.

Application to Unweighted Least Squares Fitting

Sometimes the uncertainties σ_i of the data points are unknown (such as in Example 20.1). In this case, an unweighted straight line can be fit to the data by setting all the $\sigma_i = 1$. The values of a and b obtained from Eq. (20.50) are then used to evaluate $\chi^2(a, b)$ from Eq. (20.43). The variances in the fit parameters are then estimated by multiplying the values obtained from Eq. (20.54) (with $\sigma_i = 1$) by $\sqrt{\chi^2/(N-2)}$ [Price et al. 1992].

Example 20.1: An experiment is conducted with a set of lead absorbers of different thicknesses with a ^{137}Cs gamma-ray source which emits a 661.7-keV gamma ray. A spectrometer is used to measure the number of counts per measurement acquired in the gamma-ray full energy peak as a function of absorber thickness. Each measurement has the same counting time. Determine the linear and mass attenuation coefficients for Pb for the gamma-ray energy emitted from ^{137}Cs.

Solution:
The data are best described by an exponential function,

$$y = \alpha e^{-\beta x} = \alpha e^{-(\beta/\rho)(\rho x)},$$

where β is the linear attenuation coefficient in units of cm^{-1} and β/ρ is the mass attenuation coefficient in units of $cm^2\ g^{-1}$. The absorbers were calibrated according to mass thickness ρx with units of g cm^{-2}. The measured data are listed the following table.

Data for attenuation of ^{137}Cs gamma rays through lead.

Mass Thickness (g cm^{-2}) $x_i = X_i$	X_i^2	Counts in Peak y_i	$Y_i = \ln y_i$	X_iY_i
0	0	10136	9.22385	0
0.9253	0.85618	8990	9.10387	8.4238
1.8213	3.31713	8290	9.02281	16.4332
2.6507	7.02621	7683	8.94677	23.7152
4.4936	20.19244	6184	8.72972	39.2279
7.1408	50.99102	4727	8.46105	60.4187
$S_x = 17.032$	$S_{xx} = 82.3830$		$S_y = 53.4881$	$S_{xy} = 148.219$

With $S = N = 6$ and all $\sigma_i = 1$, the set of linear fitting equations Eqs. (20.47) becomes

$$6a + 17.0322b = 53.4881 \qquad \text{and} \qquad 17.0322a + 82.3830b = 148.219,$$

which are easily solved to give

$$a = 9.2127 \qquad \text{and} \qquad b = -0.10560$$

With these values, the $\chi^2(a, b)$ of Eq. (20.43), with all $\sigma_i = 1$, is found to be $\chi^2 = 8.516 \times 10^{-4}$. The standard deviations of a and b are then estimated (as explained above) by multiplying the values obtained from Eqs. (20.50), with all $\sigma_i = 1$, by $\sqrt{\chi^2/(N-2)}$. The results are

$$\sigma_a = 0.00927 \qquad \text{and} \qquad \sigma_b = 0.00257.$$

Converting back to an exponential $\alpha = e^a = 10024$ and $\beta = b/\rho = -0.1056 \pm 0.0026$ so the best expected fit to the data is

$$y = \alpha e^{-\beta x} = 10,024(\text{cts}) \exp[-0.1056(\text{cm}^2\ \text{g}^{-1})\rho x(\text{g cm}^{-2})].$$

Hence, the mass attenuation coefficient at 661.7 keV is 0.1056 ± 0.00257 $cm^2 g^{-1}$. Multiplication by the mass density of lead of $\rho = 11.34$ g cm^{-3}, the linear attenuation coefficient at 661.7 keV is estimated to be 1.198 ± 0.029 cm^{-1}.

20.4.7 General Linear Least-Squares Model Fitting

Here is considered the problem of fitting a linear combination of M *basis functions* $X_k(x)$ to a set of data points (x_i, y_i), i.e.,

$$y(x|a_1, ..., a_M) \equiv y(x|\mathbf{a}) = \sum_{m=1}^{M} a_m X_m(x), \tag{20.57}$$

where $X_m(x)$ can be highly non-linear in x. The term "linear" refers to the linear dependence of the model on the parameters a_m. This problem is a more general problem of fitting a straight line to a set of data in which $X_1(x) = 1$ and $X_2(x) = x$. As before, one seeks values of the parameters a_m that minimize the merit function

$$\chi^2 = \sum_{i=1}^{N} \left[\frac{y_i - \sum_{m=1}^{M} a_m X_m(x_i)}{\sigma_i} \right]^2. \tag{20.58}$$

First define an $N \times M$ *design* matrix $\mathbf{A}$ with elements

$$A_{ij} = \frac{X_j(x_i)}{\sigma_i}. \tag{20.59}$$

In general $N \geq M$ because there must be at least as many data points as there are model parameters. Also define a vector $\mathbf{b}$ with components

$$b_i = \frac{y_i}{\sigma_i}, \quad i = 1, \ldots, N \tag{20.60}$$

and a vector $\mathbf{a}$ whose components are the parameters a_m, $m = 1, \ldots, M$.

The minimum of χ^2 occurs when $\partial\chi^2/\partial a_m = 0$ or when

$$0 = \sum_{i=1}^{N} \frac{1}{\sigma_i^2} \left[y_i - \sum_{j=1}^{M} a_j X_j(x_i) \right] X_m(x_i), \quad m = 1, \ldots, M. \tag{20.61}$$

This is a set of M linear algebraic equations in the M unknown a_m. These so-called *normal equations* can be written compactly as

$$\sum_{j=1}^{M} \alpha_{mj} a_j = \beta_m, \quad m = 1, \ldots, M, \tag{20.62}$$

where

$$\alpha_{mj} = \sum_{i=1}^{N} \frac{X_j(x_i) X_m(x_i)}{\sigma_i^2} \quad \text{or equivalently} \quad \boldsymbol{\alpha} = \mathbf{A}^T \cdot \mathbf{A} \tag{20.63}$$

and

$$\beta_m = \sum_{i=1}^{N} \frac{y_i X_m(x_i)}{\sigma_i^2} \quad \text{or equivalently} \quad \boldsymbol{\beta} = \mathbf{A}^T \cdot \mathbf{b}. \tag{20.64}$$

Equations (20.62) can be solved by any standard linear equation solver such as Gauss-Jordan or Cholesky decomposition techniques. However, these normal equations are susceptible to roundoff errors and often these simple solution methods fail and more sophisticated methods such as the *singular value decomposition* technique should be used. Formally, the solution of Eq. (20.62) can be written as

$$a_j = \sum_{m=1}^{M} [\boldsymbol{\alpha}^{-1}]_{jm} \beta_m = \sum_{m=1}^{M} C_{jm} \left[\sum_{i=1}^{N} \frac{y_i X_m(x_i)}{\sigma_i^2} \right], \tag{20.65}$$

where $\mathbf{C} = \boldsymbol{\alpha}^{-1}$.

However, finding the model parameters is not the end of the data fitting problem. The standard errors for the estimated parameters must also be estimated. With the usual formula for the propagation of errors (see Eq. (6.75)) the variance of a fit parameter is estimated as

$$\sigma^2(a_j) = \sum_{i=1}^{N} \sigma_i^2 \left(\frac{\partial a_j}{\partial y_i} \right)^2. \tag{20.66}$$

Because C_{jm} is independent of y_i differentiation of Eq. (20.65) with respect to y_i gives

$$\frac{\partial a_j}{\partial y_i} = \sum_{m=1}^{M} C_{jm} X_m(x_i)/\sigma_i^2 \tag{20.67}$$

so that

$$\left(\frac{\partial a_j}{\partial y_i}\right)^2 = \frac{1}{\sigma_i^2} \sum_{m=1}^{M} \sum_{k=1}^{M} C_{jm} C_{jk} \left[\sum_{i=1}^{N} X_m(x_i) X_k(x_i)\right]. \tag{20.68}$$

Substitution of Eq. (20.68) into Eq. (20.66) produces

$$\sigma^2(a_j) = \sum_{m=1}^{M} C_{jm} \sum_{k=1}^{M} C_{jk} \left[\sum_{i=1}^{N} \frac{X_m(x_i) X_k(x_i)}{\sigma_i^2}\right] = C_{jj}. \tag{20.69}$$

The last simplification in this result arises because the term in square brackets is α_{km} (see Eq. (20.63)) and, because $\boldsymbol{\alpha}^{-1} = \mathbf{C}$, one has

$$\sum_{k=1}^{M} C_{jk} \left[\sum_{i=1}^{N} \frac{X_m(x_i) X_k(x_i)}{\sigma_i^2}\right] = \sum_{k=1}^{M} C_{jk} C_{km}^{-1} = [\mathbf{C} \cdot \mathbf{C}^{-1}]_{jm} = \delta_{jm}. \tag{20.70}$$

Hence the diagonal elements of the $\mathbf{C}$ matrix are the variances of the estimated parameters. Not surprisingly, the off-diagonal elements of $\mathbf{C}$ are the covariances covar(a_j, a_m).

Computer programs and subroutines for performing linear least squares fits are provided by Press et al. [1992], Bevington [1969] and Moré et al. [1980]. An application of this fitting procedure for an isolated peak is given in the following example.

Example 20.2:

Consider a full-energy peak produced in a HPGe spectrometer from 350-keV gamma rays emitted by the naturally occurring ^{214}Pb radionuclide. The measured data in the spectrum around this peak are listed in Table 20.1. Fit the data to a Gaussian distribution and determine the area of the peak.

Table 20.1. A portion of a spectrum around the 351.93-keV ^{214}Pb full-energy peak in the background spectrum shown in Fig. 21.7.

i	E_i (keV)	y_i cnts	i	E_i (keV)	y_i cnts	i	E_i (keV)	y_i cnts	i	E_i (keV)	y_i cnts
1	348.3	2626	9	350.4	2651	17	352.6	4927	25	354.8	2303
2	348.5	2574	10	350.7	2969	18	352.9	3561	26	355.1	2457
3	348.8	2594	11	351.0	3669	19	353.2	2888	27	355.3	2388
4	349.1	2588	12	351.3	4952	20	353.4	2461	28	355.6	2433
5	349.4	2558	13	351.5	6388	21	353.7	2417	29	355.9	2310
6	349.6	2579	14	351.8	7701	22	354.0	2399	30	356.2	2477
7	349.9	2658	15	352.1	7931	23	354.3	2550	31	356.4	2385
8	350.2	2650	16	352.4	6735	24	354.5	2405			

Solution:

From Fig. 20.13 the data appears to consist of a single Gaussian on top of an almost linear background. Thus, the general model of Eq. (20.39) reduces to

$$y(E, E_o, \sigma) = \frac{A}{\sqrt{2\pi}\sigma} \exp[-(E - E_o)^2/(2\sigma^2)] + a_1 + a_2 E. \tag{20.71}$$

It is this model that is to be fitted to the data in Table 20.1. However, to use the general linear least-squares fitting method described in this section, the peak energy E_o and its standard deviation σ must be known *a priori* so that the fitting function depends only linearly on A, a_1 and a_2.

Because the radioisotope is given as ^{214}Pb, then the energy of the decay gamma ray shown in Table 20.1 is $E_o = 351.9$ keV (see Appendix C), which, for this example is rounded to 520 keV. The variances of the counts is taken as the number of counts, i.e., $\sigma_i^2 = y_i$. From past experience with the spectrometer used to obtain the data of Table 20.1, the value of $\sigma \simeq 0.47$ keV. Thus, it is decided to look at two cases, one in which $\sigma = 0.45$ keV and another in which $\sigma = 0.50$ keV. The results of the fits for the two choices of σ are shown in Fig. 20.13. Clearly, $\sigma = 0.50$ kev is a better choice.

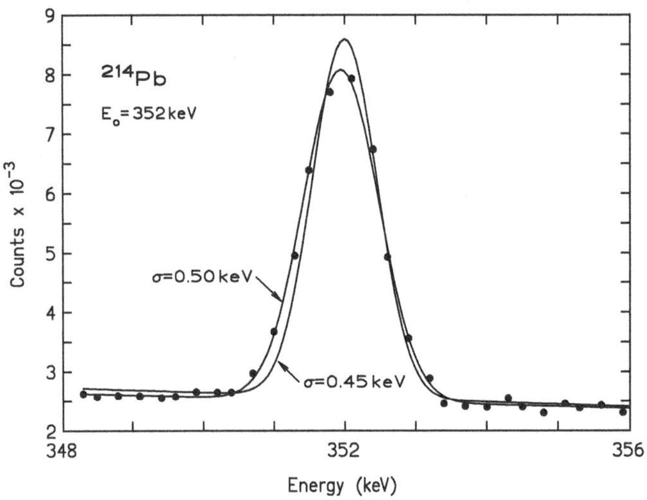

Figure 20.13. Two linear least-squares fit of Eq. (20.71) to the data of Table 20.1 for two assumed values of σ.

To determine the number of counts in the peak, the method described in Sec. 20.4.5 could be used. By inspection $n_1 = 9$ and $n_2 = 20$. Then the area is

$$\begin{aligned} \text{area}_1 &= \sum_{n=9}^{20} c(n) - \text{ trapezoid area} \\ &= 56,833 - \frac{1}{2}[c(9) + c(20)][11] = 28,717. \end{aligned}$$

From the linear least squares fit (with $E_o = 352$ keV and $\sigma = 0.50$) it is found that $A = 7161.79 \pm 62.19$ keV, $a_1 = 15819 \pm 1316$, and $a_2 = -37.75 \pm 3.73$ keV^{-1}. The energy width per channel is

$$\Delta E = \frac{E(31) - E(1)}{31 - 1} = 0.2700 \text{ keV/ch.}$$

Thus, the least-squares area of the Gaussian peak is estimated as

$$\text{area}_2 = [A \pm \sigma(A)]/\Delta E = \frac{7161.79 \pm 62.188 \text{ keV}}{0.2700 \text{ keV}} = 26,525 \pm 230.$$

20.4.8 Non-Linear Least-Squares Model Fitting

Often the model $y(x|\mathbf{a})$ to be fitted to the data y_i, $i = 1, ..., N$ depends both linearly on some of the parameters in $\mathbf{a}$ and non-linearly on the remaining M parameters. As before, in the least squares method, values of the parameters are determined by choosing parameter values that minimize the merit function

$$\chi^2(\mathbf{a}) = \sum_{i=1}^{N} \left[\frac{y_i - y(x_i|\mathbf{a})}{\sigma_i} \right]^2. \qquad (20.72)$$

The normal equations (similar to Eq. (20.61)) are no longer linear in the parameters $\mathbf{a}$ and so cannot be solved directly. Rather, iterative techniques must be used to find the minimum of $\chi^2(\mathbf{a})$.

Equation (20.72) describes a hypersurface in the M-dimensional hyperspace whose axes are the M parameters in $\mathbf{a}$. One must incrementally traverse this hypersurface in some methodical fashion to find its minimum. Such searches range from brute-force grid searches, in which one moves incrementally along each axis to find a local minimum before searching along another axis, to more sophisticated searches, in which each incremental step is taken in a direction opposite to the gradient of $\chi^2(\mathbf{a})$. However, the hypersurface often has many local minima in which an incremental search can become trapped and, thus miss the sought-for global minimum. Moreover, these searches can often venture into regions of the hyperspace with physically unrealistic parameter values, such as those producing negative Gaussian distributions. Such a problem is frequently encountered if a search is begun far from the global minimum. Consequently, it is often necessary to do a constrained search to prevent the search path from entering unrealistic regions of the hyperspace.

In the analysis of gamma-ray spectroscopic data (x_i, y_i), x_i is either the channel number n or the channel midpoint energy E_n and y_i is the measured number of counts in a channel. Typically, data in a gamma-ray spectrum, or more often, a portion of the spectrum are fit to a sum of Gaussian peaks plus a polynomial to represent the continuum upon which the peaks sit. For this case, the fitting model is

$$y(x|\mathbf{a}) = \sum_{k=1}^{K} B_k \exp\left[-\left(\frac{x - E_k}{\tau_k} \right)^2 \right] + a_0 + a_1 x + a_2 x^2 + \ldots + a_n x^n. \qquad (20.73)$$

The model parameters B_1, E_1, $\tau_1, ..., B_K$, E_K, τ_K, a_0, $a1, ..., a_n$ are the components of $\mathbf{a}$. However, other models are easily treated with the non-linear least squares technique described below.

Non-Linear Least-Squares Search

There are several ways to search the M dimensional hyperspace to find $\mathbf{a}_{\rm min}$ that minimizes $\chi^2(\mathbf{a})$ of Eq. (20.72). Here the Leverberg-Marquardt (LM) method is presented and follows closely that presented by Press et al. [1992]. Near the minimum of $\chi^2(\mathbf{a})$ the hypersurface should be well approximated by a hyperparaboloid of the form

$$\chi^2(\mathbf{a}) \simeq \gamma - \mathbf{d} \cdot \mathbf{a} + \frac{1}{2} \mathbf{a} \cdot \mathbf{D} \cdot \mathbf{a}, \qquad (20.74)$$

where γ is a constant, $\mathbf{d}$ is an M component vector, and $\mathbf{D}$ is an $M \times M$ matrix. Take the gradient of this expression and rearrange to obtain

$$\mathbf{a} = \mathbf{D}^{-1} \cdot \mathbf{d} - \mathbf{D}^{-1} \cdot \boldsymbol{\nabla} \chi^2(\mathbf{a}). \qquad (20.75)$$

At the minimum $\boldsymbol{\nabla}\chi^2(\mathbf{a}_{\rm min}) = 0$, so $\mathbf{a}_{\rm min} = \mathbf{D}^{-1} \cdot \mathbf{d}$. Substitute this result into Eq. (20.75) to see that $\mathbf{a}_{\rm min}$ can be reached from the current position in hyperspace $\mathbf{a}_{\rm cur}$ in a single step as

$$\mathbf{a}_{\rm min} = \mathbf{a}_{\rm cur} + \mathbf{D}^{-1} \cdot \boldsymbol{\nabla}\chi^2(\mathbf{a}_{\rm cur}). \qquad (20.76)$$

However, if Eq. (20.74) is a poor approximation, then the best that can be done is to step down the gradient, i.e.,

$$\mathbf{a}_{\text{next}} = \mathbf{a}_{\text{cur}} - \text{constant} \times \boldsymbol{\nabla}\chi^2(\mathbf{a}_{\text{cur}}) \tag{20.77}$$

where the constant is sufficiently small to avoid overshooting the minimum. To use either Eq. (20.76) or Eq. (20.77) of the above stepping procedures the gradient of $\chi^2(\mathbf{a})$ must be known for any $\mathbf{a}$ and to use Eq. (20.76) one must also know $\mathbf{D}$, the so-called *Hessian* matrix, whose components are the second derivatives of $\chi^2(\mathbf{a})$.

The gradient of $\chi^2(\mathbf{a})$ with respect to the parameters $\mathbf{a}$, which equals zero at the minimum of $\chi^2(\mathbf{a})$, is

$$\frac{\partial \chi^2}{\partial a_m} = -2\sum_{i=1}^{N} \frac{[y_i - y(x_i|\mathbf{a})]}{\sigma_i^2}\frac{\partial y(x_i|\mathbf{a})}{\partial a_m}. \tag{20.78}$$

Then differentiate this result to obtain the second derivatives, namely

$$\frac{\partial^2 \chi^2}{\partial a_m \partial a_l} = 2\sum_{i=1}^{N} \frac{1}{\sigma_i^2}\left[\frac{\partial y(x_i|\mathbf{a})}{\partial a_m}\frac{\partial y(x_i|\mathbf{a})}{\partial a_l} - [y_i - y(x_i|\mathbf{a})]\frac{\partial^2 y(x_i|\mathbf{a})}{\partial a_l \partial a_m}\right]. \tag{20.79}$$

Now define a vector $\boldsymbol{\beta}$ and matrix $\boldsymbol{\alpha}$ as[5]

$$\beta_m \equiv -\frac{1}{2}\frac{\partial \chi^2}{\partial a_m} \qquad \text{and} \qquad \alpha_{ml} \equiv \frac{1}{2}\frac{\partial^2 \chi^2}{\partial a_m \partial a_l}. \tag{20.80}$$

Let $\delta\mathbf{a} = \mathbf{a}_{\text{cur}} - \mathbf{a}_{\text{next}}$, which is the next step in the search through hyperspace to find the minimum of $\chi^2(\mathbf{a})$. If the iterative search is based on Eq. (20.76) with $\mathbf{a}_{\text{min}}$ replaced by $\mathbf{a}_{\text{next}}$ then manipulation of Eq. (20.76) gives

$$\mathbf{D}\,\delta\mathbf{a} = \boldsymbol{\nabla}\chi^2(\mathbf{a}) \qquad \text{or} \qquad \boldsymbol{\alpha}\,\delta\mathbf{a} = \boldsymbol{\beta}, \tag{20.81}$$

which can be rewritten as a set of linear algebraic equations

$$\sum_{l=1}^{M} \alpha_{ml}\delta a_l = \beta_m. \tag{20.82}$$

If, on the other hand, one wishes to use the steepest descent formula of Eq. (20.77) then

$$\delta a_l = \text{constant} \times \beta_l. \tag{20.83}$$

The Hessian matrix of Eq. (20.79), in general, contains the second derivatives $\partial^2 y(x_i|\mathbf{a})/\partial a_l \partial a_m$. These terms can be ignored if they are zero (as in a linear model) or if they are small compared to the other terms. Because the second derivatives are multiplied by $[y_i - y(x_i|\mathbf{a})]$ a quantity that, for a good fitting model, is the random measurement error. These errors can have either sign and should tend to cancel out when averaged over all i. Further, the second derivatives can destabilize the iterative search for the minimum if the spectrum is contaminated by outlier points. For these reasons, the second derivative terms in Eq. (20.79) are almost always neglected. Additionally, their neglect makes the search somewhat simpler and, although altering the matrix $\boldsymbol{\alpha}$ changes the path taken in hyperspace to reach the minimum of $\chi^2(\mathbf{a})$, the values of the parameters at the minimum are unaltered. Henceforth the elements of the $\boldsymbol{\alpha}$ are taken as

$$\alpha_{ml} = \sum_{i=1}^{N} \frac{1}{\sigma_i^2}\left[\frac{\partial y(x_i|\mathbf{a})}{\partial \alpha_m}\frac{\partial y(x_i|\mathbf{a})}{\partial \alpha_l}\right]. \tag{20.84}$$

[5]The $\boldsymbol{\alpha}$, which equals one-half of the Hessian matrix, is often called the *curvature* matrix.

The Levenberg-Marquardt Search

The Levenberg-Marquardt (LM) method is widely used for non-linear least-squares analyses. It varies smoothly between the inverse-Hessian method of Eq. (20.82) when close to the minimum and the steepest descent method of Eq. (20.83) when far from the minimum. The LM method is based on two key insights made by Levenberg and Marquardt.

The first insight concerns the constant in Eq. (20.83). It is not dimensionless as is $\chi^2(\mathbf{a})$; but because β_m has dimensions of those of $1/a_m$, which may be different for each m. Thus the constant of proportionality between β_m and δa_m must have dimensions of a_m^2. In the matrix $\boldsymbol{\alpha}$ the only element with this dimension is $1/\alpha_{mm}$, which defines the scale of the constant. However, this scale may produce too large a step size. To reduce the step size, a factor $\lambda \gg 1$ is introduced so that Eq. (20.83) can be written as

$$\delta a_l = \frac{1}{\lambda \alpha_{ll}} \beta_l \qquad \text{or} \qquad \lambda \alpha_{ll} \delta a_l = \beta_l \tag{20.85}$$

Clearly α_{ll} must be positive as is guaranteed by its definition in Eq. (20.84).

The second important key observation made by Marquardt is that Eqs. (20.85) and (20.82) can be combined by defining a new matrix $\boldsymbol{\alpha}'$ as

$$\alpha'_{jj} \equiv \alpha_{jj}(1+\lambda) \qquad \text{and} \qquad \alpha'_{jm} \equiv \alpha_{jm}, \; j \neq k, \tag{20.86}$$

so Eqs. (20.85) and (20.82) can be replaced by the single equation

$$\sum_{l=1}^{N} \alpha'_{ml} \delta a_l = \beta_m. \tag{20.87}$$

Notice that when λ is very large, $\boldsymbol{\alpha}'$ becomes diagonally dominant and Eq. (20.87) becomes Eq. (20.85). But as λ becomes very small, Eq. (20.87) converges to Eq. (20.82).

Given an initial guess for the parameters $\mathbf{a}$, the algorithm for the LM search for the minimum of $\chi^2(\mathbf{a})$ proceeds as follows:

1. Compute $\chi^2(\mathbf{a})$.
2. Pick a modest value for λ, say $\lambda = 0.001$.
3. Solve Eqs. (20.87) for $\delta\mathbf{a}$ and evaluate $\chi^2(\mathbf{a} + \delta\mathbf{a})$.
4. If $\chi^2(\mathbf{a} + \delta\mathbf{a}) \geq \chi^2(\mathbf{a})$, increase λ by a factor of 10 or other substantial amount. Then go to step 3.
5. If $\chi^2(\mathbf{a} + \delta\mathbf{a}) < \chi^2(\mathbf{a})$, decrease λ by a factor of 10 and go back to step 3.

Continue this incremental walk through hyperspace until $\chi^2(\mathbf{a})$ decreases by a small amount such as 0.001 or some small fractional amount such as 0.001. Once an acceptable minimum has been found, the covariance matrix

$$\mathbf{C} = \boldsymbol{\alpha}^{-1} \tag{20.88}$$

is computed (after setting $\lambda = 1$). The diagonal elements of this matrix give estimates of the variances of the corresponding parameters.

In the examples below, the LM method is applied to the case of an isolated peak considered in Example 20.2 and to a case in which two peaks overlap. In both cases, the calculations use the LM computer subroutines given by Press et al. [1992].

Isolated Peaks

To assure that the entire isolated peak is completely fit, it is suggested that the portion of the spectrum extend to at least $\pm 3\sigma$ above and below the centroid channel. These limits are chosen because the area under a Gaussian over the range $\mu \pm 3\sigma$ is 0.997 of the total area under the Gaussian and thus the parameter A or B in Eqs. (20.39) and (20.40) is a very good approximation to the total area under the Gaussian.

For the portion of the spectrum containing an isolated peak, the appropriate fitting model $y(x|\mathbf{a})$ is that of Eq. (20.39) with $K = 1$ and a polynomial of low order n to describe the background upon which the peak sits. The merit function to be minimized in this case is

$$\chi^2(\mathbf{a}) = \sum_{i=1}^{N} \left[\frac{y_i - y(x_i|\mathbf{a})}{\sigma_i} \right]^2 = \sum_{i=1}^{N} \frac{1}{\sigma_i^2} \left[y_i - \left\{ B \exp\left[- \left(\frac{x_i - \mu}{\tau} \right)^2 \right] + \sum_{m=0}^{n} a_m x_i^m \right\} \right]^2, \tag{20.89}$$

where the components of the parameter vector $\mathbf{a}$ are B, μ, τ, $a_0, ..., a_n$. To use the LM method, the derivatives of $y(x|\mathbf{a})$ with respect to each parameter are needed. Here one has

$$\frac{dy(x|\mathbf{a})}{dB} = f(x, \mu, \tau), \qquad \frac{dy(x|\mathbf{a})}{d\mu} = Bf(x, \mu, \tau)\frac{2(x-\mu)}{\tau^2},$$
$$\frac{dy(x|\mathbf{a})}{d\tau} = f(x, \mu, \tau)\frac{2(x-\mu)^2}{\tau^3}, \quad \text{and} \quad \frac{dy(x|\mathbf{a})}{da_j} = x^j, \quad j = 0, 1, ..., n. \tag{20.90}$$

To illustrate this analysis approach, the data of Table 20.1 for a ^{214}Pb peak are fit to a Gaussian plus a linear function ($n = 1$) in Example 20.3.

Example 20.3:

For the portion of a spectrum containing an isolated peak, as given by the data in Table 20.1, the fitting model $y(x|\mathbf{a})$ is given by Eq. (20.73) with $K = 1$ and a linear background function so $n = 1$ in Eq. (20.73). This is the same model used in Example 20.2, but now estimate the centroid of the peak and its standard deviation by including them in the fitting parameters.

Solution:

The merit function to be minimized is that of Eq. (20.89) with $n = 1$. The initial guess of the parameter values was

$$B = 3000 \text{ MeV}, \quad \mu = 0.350 \text{ MeV}, \quad \tau = 0.002 \text{ MeV}, \quad a_0 = 1000, \quad a_1 = -200 \text{ MeV}^{-1}.$$

After 23 LM steps through 5-dimensional parameter hyperspace, the $\chi^2(\mathbf{a})$ was reduced from an initial value of 16,653 to 72.55, the minimum of $\chi^2(\mathbf{a})$. The best fit parameters at the $\chi^2(\mathbf{a})$ minimum were

$$B = 5583.6 \pm 58.4, \quad \mu = 351.960 \pm 0.0051 \text{ keV}, \quad \tau = 0.74146 \pm 0.00744 \text{ keV},$$
$$a_0 = 14062 \pm 1331 \quad \text{and} \quad a_1 = -32823 \pm 3776 \text{ MeV}^{-1}$$

The normalization of the Gaussian is $A = \sqrt{\pi}\tau B = 7338.0 \pm 70.64$ keV counts/channel, which, upon dividing by the energy width per channel of 0.2700 keV/channel, gives the total counts in the peak as

$$A_3 = 27,178 \pm 262.$$

This result is midway between the two estimates $A_1 = 28,717$ and $A_2 = 26,525 \pm 230$ obtained in Example 20.2. Finally, the standard deviation of the peak is $\sigma = \tau/\sqrt{2} = 0.52429 \pm 0.00526$ keV. The resulting model fit and its two components are shown in Fig. 20.14. Of note is the estimated gamma-ray energy of 351.960 keV for this peak. This value compares very favorably to the accepted NUDAT value of 351.9321

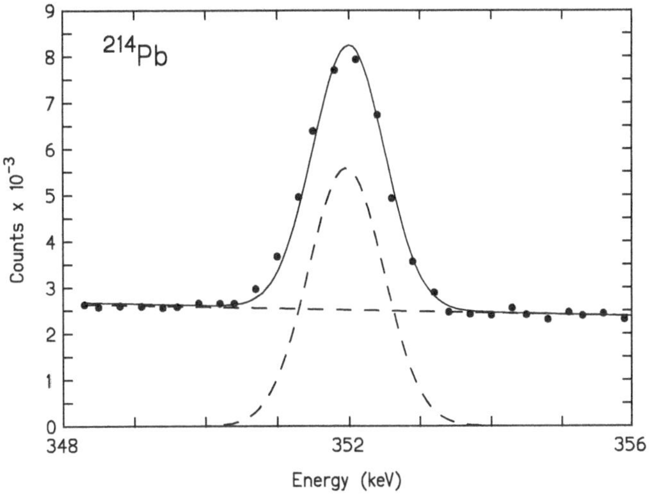

Figure 20.14. The LM fit to the data of Table 20.1. Also shown by dashed lines are the Gaussian peak component and the linear background.

(18) keV. The small difference is easily accounted for by inevitable small errors in the energy calibration of the channels of the spectrometer system.

Overlapping Peaks

Sometimes two or more peaks overlap. In situations where a peak is asymmetric or has a FWHM larger than expected, one can try to fit multiple peaks to the data locally. In such cases, one might try a fitting model of the form (see Eq. (20.39))

$$y(x|\mathbf{a}) = \sum_{j=1}^{K} B_j g_j(x, \mu_j, \tau_j) + a_0 + a_1 x + a_2 x^2, \quad \text{where } g(x, \mu_j, \tau_j) = \exp\left[-\left(\frac{x-\mu_j}{\tau_j}\right)^2\right] \qquad (20.91)$$

and K is the number of peaks one suspects might be overlapping.

Values of the model parameters $\mathbf{a} = (B_1, \mu_1, \tau_1, \ldots, B_K, \mu_K, \tau_K, a_0, a_1, a_2)$ are found as those values that minimize the merit function

$$\chi^2(\mathbf{a}) = \sum_{i=1}^{N} \left[\frac{y_i - y(x_i|\mathbf{a})}{\sigma_i}\right]^2 = \sum_{i=1}^{N} \frac{1}{\sigma_i^2}\left[y_i - \left\{\sum_{k=1}^{K} B_k \exp\left[-\left(\frac{x_i-\mu_k}{\tau_k}\right)^2\right] + \sum_{j=0}^{2} a_j x_i^j\right\}\right]^2. \qquad (20.92)$$

As a practical matter, you might try $K = 2$ first and see if you obtain reasonable results. If not, try larger values of K. Of course this model introduces new non-linear model parameters for each additional peak. The fitting of multiple overlapping peaks, with asymmetric peak models, to XPS spectra is considered in detail by Dunn and Dunn [1982].

Generally, the y_i are either gross counts C_i in channel i or, for background-subtracted spectra, net counts N_i in channel i. (Note that background subtraction removes only part of the generalized background, the B_n of Eq. (20.12).) If the $C(i)$ are gross counts, then one presumes that Poisson statistics apply and $\sigma^2(y_i)) \equiv \sigma_i^2 = C(i)$. If the spectrum is background-subtracted, then $\sigma^2(y_i) = C(i) + B_i$. If the $C(i)$ result from some other process, then one should use the appropriate variances in Eq. (20.92). For the non-linear least-squares approach followed here, values of the model parameters **a** are chosen that minimize the merit function $\chi^2(\mathbf{a})$. To use the LM minimization method, partial derivatives of $\chi^2(\mathbf{a})$ with respect to each parameter are needed and for the above model are given by Eqs. (20.90).

An example of fitting two overlapping peaks is given in Example 20.4. In this example, and in many non-linear least-squares fitting analyses, the incremental path through the $\chi^2(\mathbf{a})$ hypersurface may often lead to physically unrealistic values of some of the parameters. This is particularly true for overlapping peaks. Unless one starts the minimizing search with very good guesses for the μ_k for each peak, often a broad positive Gaussian ($B_1 > 0$) with a negative Gaussian ($B_2 < 0$) results. Such is the case for Example 20.4. One way to avoid these unrealistic parameter values is to start the search very near the minimum of $\chi^2(\mathbf{a})$, which is seldom known *a priori*, or to perform a *constrained* search along the hypersurface. In such a constrained search, the current values of the parameters are examined after each step and, if outside some preset range, an offending parameter is reset to the nearest range limit.

Example 20.4: Fit the model of Eq. (20.91) with $K = 2$ to the count data given in the table below.

Table 20.2. A portion of a spectrum giving channels and gross counts. The data points (x_i, y_i) to be fit are $x_i = i$ the channel number and $y_i = c(i)$ the counts in channel i.

i	$C(i)$	i	$C(i)$	i	$C(i)$	i	$C(i)$
117	2210	127	4756	137	45568	147	3827
118	2253	128	6176	138	36698	148	3416
119	2333	129	9761	139	23773	149	3156
120	2487	130	17016	140	14669	150	2896
121	2763	131	26462	141	10054	151	2713
122	2869	132	30846	142	7666	152	2448
123	2984	133	28392	143	6284	153	2335
124	3312	134	26822	144	5387	154	2119
125	3629	135	32667	145	4792	155	1957
126	4077	136	42618	146	4224	156	1754

Solution:
The LM method was used to find values of the best fit parameters. The derivatives needed for this method are

$$\frac{dy(x|\mathbf{a})}{dB_i} = f(x, \mu_i, \tau_i), \quad i = 1,2 \qquad \frac{dy(x|\mathbf{a})}{d\mu_i} = Bf(x, \mu_i, \tau_i)\frac{2(x-\mu_i)}{\tau_i^2}, \quad i = 1,2$$

$$\frac{dy(x|\mathbf{a})}{d\tau_i} = f(x, \mu_i, \tau_i)\frac{2(x-\mu_i)^2}{\tau_i^3}, \quad i = 1,2 \qquad \frac{dy(x|\mathbf{a})}{da_j} = x^j, \quad j = 0,1,2. \tag{20.93}$$

The search for the minimum of $+\chi^2(\mathbf{a})$ began with the following dimensionless starting values.

$$\mu_1 = 130. \quad \sigma_1 = 2.200 \quad B_1 = 22,000. \qquad \mu_1 = 140. \quad \sigma_2 = 2.800 \quad B_2 = 38,000.$$

$$a_0 = -177000. \quad a_2 = 2680 \quad a_2 = -9.8400$$

After 15 steps, the initial value of $\chi^2(\mathbf{a})$ was reduced from 80,978. to 2034.92. Further iterations did not change $\chi^2(\mathbf{a})$. Values of the best fit parameters are

$$\mu_1 = 131.664 \pm 0.015 \qquad \sigma_1 = 2.1298 \pm 0.0157 \qquad B_1 = 23,403.4 \pm 137.7$$

$$\mu_2 = 136.734 \pm 0.0120 \qquad \sigma_2 = 2.8700 \pm 0.0138 \qquad B_2 = 38538.5 \pm 129.0$$
$$a_0 = -166321. \pm 1970. \qquad a_1 = 2517.03 \pm 29.22 \qquad a_2 = -9.2409 \pm 0.10771$$

The areas of the two Gaussians are

$$A_1 = \sqrt{\pi}\tau_1 B_1 = 88348.1 \pm 552.78 \qquad A_2 = \sqrt{\pi}\tau_2 B_2 = 196039. \pm 761.10$$

and the standard deviations of the peak are

$$\sigma_1 = \tau_1/\sqrt{2} = 1.5060 \pm 0.01108 \qquad \sigma_1 = \tau_1/\sqrt{2} = 2.0295 \pm 0.009724.$$

The fit function is shown in Fig. 20.15.

It is seen that the model fits the data quite well. One would expect that a good model would be within the error intervals for about 68% of the data points and this seems to be the case here. The uncertainty in each of the model parameters is small, relative to the parameter value, which also indicates a good fit. It is noted that the values of $i_1 = 117$ and $i_2 = 156$ are well beyond the 3σ range for the σ_1 and σ_2 obtained. Thus, the A_1 and A_2 values should be good estimates of the net areas under the two peaks.

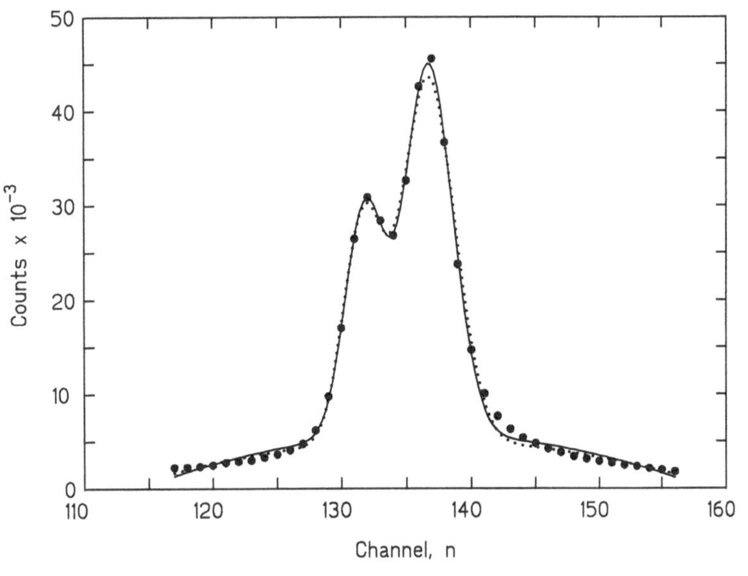

Figure 20.15. A fit of two overlapping peaks with a quadratic background. The solid line is a weighted least-squares fit with $\sigma_i = \sqrt{c(i)}$ and the dotted line is an unweighted least-squares fit with $\sigma_i = 1$.

20.4.9 Spectrum Stripping

If response functions can be collected or generated for all the sources (or radionuclides) that are expected to be present in an unknown sample, and if the dependence of the response model is linearly related to the source strengths or radionuclide concentrations, then the method of spectrum stripping can be applied. This procedure is as follows:

- Collect a spectrum from the unknown sample.

- Identify the highest-energy peak in the spectrum.
- Subtract the response function for that energy peak, weighted by a constant, such that the peak is effectively removed.
- Proceed down the spectrum while subtracting other weighted response functions until all peaks are removed.

If the residuals are randomly distributed about zero or about some smooth background, then the specific response functions stripped from the spectrum for the unknown identify the radionuclides present and the weighting constants estimate the source strength or concentration of each radionuclide.

20.4.10 Library Least-Squares

Because a detector produces a spectrum, even for a monoenergetic input one can try to utilize the entire spectral response, or at least a significant part of it, rather than just the response values near each peak or set of overlapping peaks. The library least-squares approach, originally introduced by Marshall and Zumberge [1989], asks the following question. Why focus on only the peaks since other parts of the spectrum also contain information related to the abundances of the radionuclides that produce the peaks? One approach for using all of the information in a spectrum is the library least-squares (LLS) method, as implemented, for instance, by Gardner and Sood [2004] and Gardner and Xu [2009]. The LLS method is based on a library that contains detector response functions for all radionuclides that might possibly be present in the sample whose spectrum is to be analyzed in order to determine the abundances of specific radionuclides. Then some fitting technique, such as least-squares or weighted least-squares, is used to fit the library spectra to the spectrum from a sample whose radionuclide abundances are sought.

This approach was not possible many years ago because one could not experimentally measure good spectra from all candidate radionuclides or a sufficient number of monoenergetic gamma rays. However, Monte Carlo modeling has become sufficiently robust that detector response functions can be calculated for almost any radionuclide. When Monte Carlo is used to generate detector response functions, the method often is referred to as Monte Carlo library least-squares or MCLLS. In this method, one calculates monoenergetic detector response functions $\mathcal{R}_n(E)$ for a wide range of discrete energies that can be emitted by sources of interest. Examples of such MC calculated detector response functions are shown in Fig. 20.16 and were calculated by special MC software and empirical resolution functions [Gardner and Sood 2004]. The response function $\mathcal{R}_n(E)$ is the probability that a photon of energy E emitted by the source produces a count in channel n. This response function, as before, includes the probability a source photon reaches the detector, a probability which depends on the specific source-detector geometry but it, generally, does not include photons scattered into the detector by material around the detector, i.e., so-called *roomshine.*

To determine the concentrations of various radioisotopes present in a sample, response functions, per unit activity concentration, for each possible radioisotope likely to be in the sample must first be constructed. A library of such radioisotopic response functions $\mathcal{R}_n^k$ can be constructed from the monoenergetic response functions $\mathcal{R}_n(E)$, for a given source-detector geometry, as

$$\mathcal{R}_n^k = \sum_{i=1}^{I} f_i^k \mathcal{R}_n(E_i^k)\hat{\xi}^k, \tag{20.94}$$

where f_i^k is the frequency a photon of energy E_i^k, $i = 1, ..., I$ is emitted per decay of radionuclide k and $\hat{\xi}^k$ is the decay rate of a unit activity concentration of the radionuclide.

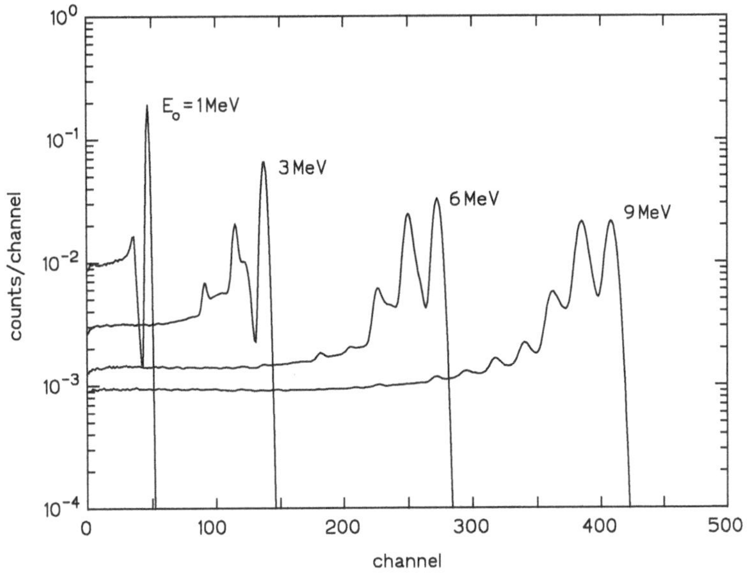

Figure 20.16. Examples of library response functions $\mathcal{R}_n(E)$ for a 6 in × 6 in cylindrical NaI detector for monoenergetic photons normally and uniformly incident on the circular end of the crystal. No contribution from scattering in material around the detector is included as indicated by the absence of backscatter and annihilation peaks although several annihilation escape peaks from the crystal are evident. These response functions are a small part of the library for a 512-channel spectrum over the photon range 0–11.38 MeV. Courtesy of Robin Gardner, NCSU.

Then the expected number of counts $y(n)$ recorded in channel n in a measurement time T produced by a sample of K possible radioisotopes is

$$y(n|\xi) = T \sum_{k=1}^{K} \mathcal{R}_n^k \xi_k, \qquad n = 1, ..., N, \tag{20.95}$$

where ξ_k are the radioisotopic activity concentrations being sought. Again it is assumed all the activity concentrations remain constant over the measurement time T. If $C(n)$ are the observed number of counts in channel n from the sample, then the vector $\boldsymbol{\xi}$, whose components are the concentrations ξ_k, can be determined as those values that minimize the merit function

$$\chi^2(\boldsymbol{\xi}) = \sum_{n=1}^{N} W_n [c(n) - y(n, |\boldsymbol{\xi})]^2, \tag{20.96}$$

where W_n is a weight factor, often taken as $W_n = 1/\sigma^2(C(n)) = C(n)$. Because the response functions do not include background and roomshine, the counts $C(n)$ must first be corrected for these contributions. If a fitted ξ_k value is negative or near zero for one or more k, then the corresponding radionuclides might not be present in the sample. In this case, it is a good practice to remove the detector response functions for these radionuclides and repeat the analysis to see if a good fit is obtained.

The process just described assumes the response model is linear in the radionuclide concentrations and, thus, can be referred to as the *linear LLS* approach. This approach is very similar to the previous linear weighted least-squares process used to fit Gaussian distributions to spectral peaks. But here more features of the spectrum other than just the full-energy peaks are used. There are instances, however, in which the model is not linear in the radionuclide concentrations. Such cases arise in prompt gamma-ray neutron activation analysis (PGNAA) and energy dispersive x-ray fluorescence spectroscopy (EDXRF).

Non-Linear Spectra

Generally, the samples used in a neutron activation analysis (NAA) are small and their mass can be accurately measured and, because one is usually seeking concentrations of trace elements, a NAA analysis is well approximated as a linear process and application of linear LS technique is appropriate. However, for bulk samples, a prompt-gamma neutron activation analysis (PGNAA) is non-linear, primarily for the following reasons:

- Sample mass, which often is not known, affects the flux density and the macroscopic capture cross section of the sample.
- The composition of the sample, which is unknown in advance, affects the spectrum. In particular, moisture content strongly affects the thermal-neutron flux density, which is what gives rise to the prompt gamma rays. Also, neutron absorbers affect the thermal flux density.

Thus, non-linear models are needed for a PGNAA. Such models are iterative in nature and require the need to calculate spectral responses. Usually, Monte Carlo methods are used for such calculations. The general Monte Carlo Library Least-Squares (MCLLS) approach in the non-linear case proceeds as follows.

1. Assume values for the concentrations and use Monte Carlo to generate a complete spectrum for a sample of this assumed composition.
2. Keep track of the individual spectral responses for each element within the Monte Carlo code, so as to provide library spectra u_{nk} for each radionuclide.
3. Use linear LLS to estimate the radionuclide concentrations $\xi_k,\ k = 1, 2, \ldots, K$ from the sample spectrum.
4. If the calculated $\xi_k,\ k = 1, 2, \ldots, K$ match the assumed composition closely enough, you are done. If not, pick a new composition, based on the calculated concentrations, and repeat the process.
5. Iterate until you converge to the actual composition, to within a desired tolerance.

Results of this general procedure are given, for instance, by Gardner and Xu [2009].

20.4.11 Symbolic Monte Carlo

In X-ray fluorescence, the responses are due to the *elements* present, but each element is composed of radionuclides and the convention was introduced earlier to refer only to radionuclides. Hence in the discussion below, elemental concentrations are called nuclide concentrations. Non-linear matrix effects lead to absorption and enhancement in EDXRF. For instance, the characteristic x rays of nuclide a can be absorbed by elements with lower atomic numbers, which reduces the signal from nuclide a and enhances the signals for the lower atomic number nuclides. This effect means that the models in EDXRF also are not linear in the nuclide concentrations. Another implementation of Monte Carlo has been used in the EDXRF case. The method, originally called Inverse Monte Carlo (IMC) [Dunn 1981], was applied to EDXRF by Yacout and Dunn [1987] for primary and secondary x rays. Mickael [1991] extended the work to include tertiary fluorescence.

The term IMC has been used for other purposes, e.g. to solve inverse problems by iterative Monte Carlo simulations in which the unknown parameters are varied until simulated and measured results agree sufficiently. The acronym IMC also has been used for "implicit Monte Carlo" [Gentile 2001]. Thus, Dunn and Shultis [2009] recently proposed renaming the version of IMC that is non-iterative in the Monte Carlo simulations *symbolic Monte Carlo* (SMC) because the method proceeds by using symbols in the Monte Carlo scores for the unknown parameters.

SMC is a specialized technique in which the inverse problem of estimating the k and ξ_k is solved by a system of algebraic equations generated by a single Monte Carlo simulation. For purposes of illustration, a ternary system (one that contains three elemental nuclides) is considered. In essence, SMC creates models, with symbols for the unknown concentrations ξ_k, for the areas under all of the various x-ray peaks (e.g., K_α and K_β) in a single simulation. The models depend on the detector efficiency as a function of energy, the nuclide concentrations, the primary, secondary, and tertiary fluorescence produced in the sample, and the background. For a ternary system, three equations result. The equations are rather complex (see Yacout and Dunn [1987] and Mickael [1991]) but can be developed using only a single Monte Carlo simulation. The advantage of this approach is that there is no need to iteratively run full Monte Carlo simulations as the assumed concentrations are varied. The disadvantage is that development of the model is involved and the algebraic equations are quite complex. Nevertheless, the method has been shown to work well in x-ray spectroscopy and can, in principle, be applied to other spectroscopic applications, such as PGNAA, in which the responses are non-linear functions of the concentrations.

20.5 Radiation Spectroscopy Measurements

The purpose of radiation spectroscopy is to identify energetic emissions from radioactive materials. Such emissions are reported with several metrics, mainly the confidence in the identified energy or the *energy resolution* and the source activity (a function of the detector efficiency). The choice of detector is determined by the application and required spectroscopic performance. Detectors needed for field applications with sufficient spectroscopic performance to identify common isotopes may be best performed with lower resolution scintillation detectors. Measurements requiring high resolution spectroscopy generally require more expensive semiconductor spectrometers. No matter the type of spectrometer, it must undergo an energy calibration before any measurements can be made.

20.5.1 Channel Calibration

Calibration of the spectrometer channels in energy units is relatively simple, particularly if a linear energy/channel response is assumed. Typically two particle energies are chosen with relatively wide separation. For instance, ^{60}Co and ^{57}Co might be chosen. The peak channel $\tilde{n}_H$ for the higher energy E_H and the peak channel $\tilde{n}_L$ for the lower energy E_L are observed and a simple linear fit between these two channels provides a channel energy calibration, i.e.,

$$E(\tilde{n}) = \tilde{n}\left(\frac{E_H - E_L}{\tilde{n}_H - \tilde{n}_L}\right) + E_0, \tag{20.97}$$

where E_0 is the energy offset at channel $\tilde{n} = 0.0$, which is found from

$$E_0 = E_H - \tilde{n}_H\left(\frac{E_H - E_L}{\tilde{n}_H - \tilde{n}_L}\right). \tag{20.98}$$

For simplicity, the continuous $\tilde{n}$ is often taken as the channel number n which equals $\tilde{n}$ at the midpoint of a channel. The term in parentheses in the above equations is the energy width per channel ΔE and is an important spectrometer parameter because it defines the minimum energy resolution possible with the spectrometer.

Semiconductor materials usually show good signal linearity with energy deposition, but there are many scintillators that exhibit a non-linear output with energy, especially at lower energies (see Sec. 13.2.2). Example scintillators include NaI:Tl and CsI:Tl in the energy region below 500 keV. For such non-linear detectors, several energies in the non-linear region should be used to fit a polynomial function to the channel energy calibration.

20.5.2 Quality Metrics

Several quality metrics have been established, some official and others less so, that depend on the system and type of detector. These metrics are used to give the user an idea of the expected detector performance under different measurement conditions. Metrics include measurements of detection efficiency, energy resolution, channel width corrections, figure of merit, noise resolution, energy rate limit, and peak-to-Compton ratio. Many of these metrics are described in the IEEE Std 325-1996 document for HPGe detectors, outlined in this section.

Detection Efficiency

There are multiple methods for defining and measuring the efficiency of a gamma-ray spectrometer. The efficiencies of most interest and utility are the absolute efficiency, the intrinsic efficiency, and the escape peak efficiency. In Sec. 7.1 a general description of detector efficiencies was given. Here efficiencies for a full energy peak from a gamma-ray spectrometer are defined.

Total Intrinsic Detection Efficiency The total intrinsic detection efficiency is defined by

$$\epsilon_{\mathrm{I}} = \sum_{i}^{N} \frac{A_{\mathrm{sp}}}{A\Omega_f t B_i} \tag{20.99}$$

where A_{sp} are the recorded counts from the entire detector spectrum, t is the live time of the counts, A is the source activity, Ω_f is the fractional solid angle subtended by the detector, and B_i are the branching ratios of the radiation emissions. For instance, ^{60}Co emits two gamma rays per decay ($B = 2$), while only 85% of decays from ^{137}Cs result in the emission of a 661.7 keV gamma ray ($B = 0.85$). This metric ϵ_{I} yields some information about the overall detector counting efficiency, but does not give information on the energy resolution performance. Further, this particular metric is subject to changes with the LLD setting, and background contamination can skew the results.

Intrinsic Peak Efficiency The intrinsic full energy peak efficiency, described in Sec. 7.1 and repeated here for convenience, is defined by

$$\epsilon_{\mathrm{peak}} = \frac{A_p}{A\Omega_f B t} \tag{20.100}$$

where A_p are the recorded counts from the detector in the full energy peak, t is the live time of the counts, A is the source activity, Ω_f is the fractional solid angle subtended by the detector, and B is the branching ratio of the emission under investigation. The intrinsic peak efficiency is similar to the intrinsic detection efficiency, except the A_p pertains only to those counts located in the background subtracted full energy peak. It is notable that the intrinsic peak efficiency may be difficult to determine accurately for several detector types. For instance, the size and shape of an HPGe detector may be difficult to assess because the actual crystal is much smaller than the apparent cryogenic packaging (see Fig. 16.34). A similar situation exists with scintillation counters whose crystal is hermetically packed in a light-tight reflecting cannister. The practical outcome of the intrinsic peak efficiency is a function of the packing cannister, the absorbing

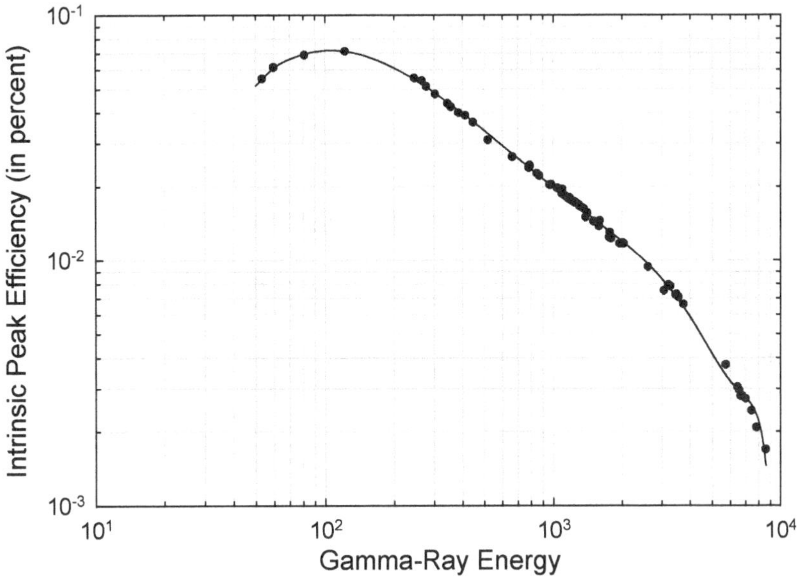

Figure 20.17. Measured intrinsic peak efficiency for a 25% relative efficient n-type HPGe detector. The sources used for the calibration were ^{241}Am, ^{133}Ba, ^{152}Eu, ^{137}Cs, ^{60}Co, ^{188}Ta, ^{56}Co, ^{49}Ti, and ^{36}Cl. Data acquired from Kis et al. [1998].

material (NaI:Tl, HPGe, CdTe, etc.), the type of electrical contacts, the detector size, and the detector shape. Energy absorption in the detector encapsulation and contact dead layer affect the low energy efficiency, while the atomic number and volume of the detector determine the high energy efficiency (see Fig. 16.35). An example of the efficiency variation is shown in Fig. 20.17 for a 25% relative efficient, bulletized coaxial n-type HPGe detector [Kis et al. 1998].

A variety of gamma-ray sources can be used to measure the intrinsic peak efficiency over the energy region of interest. Afterwards, a curve fit can be used as a predictive measure of efficiencies at other energies within the measured span [Kis et al. 1998]. Although there are several curve fitting functions offered in the literature, many examples listed in Kis et al. [1998], modern commercial curve fitting computer programs are available that can provide high fit r^2 values (> 0.999).

Escape Peak Efficiency The escape-peak efficiency, mainly the single and double escape peaks from pair production, is a measure of the detector's ability to effectively recapture 511-keV annihilation photons. Larger escape peaks are indicative of larger losses. Although there are multiple definitions of this metric in the literature, a generally accepted definition for the intrinsic escape peak efficiency is that described by Cline [1968] and Nafee [2011],

$$\epsilon_{es} = \frac{A_{\rm ep}}{A\Omega_f Bt}, \tag{20.101}$$

where $A_{\rm ep}$ pertains to the number of counts in either the single escape peak or the double escape peak. To determine the number of counts in either escape peak, the subtraction method of Eq. (20.41) is employed. Equation 20.101 can be rewritten as

$$\epsilon_{es} = \epsilon_{\rm peak}\frac{A_{\rm ep}}{A_p}, \tag{20.102}$$

where the escape peak under investigation is for the corresponding full energy peak.

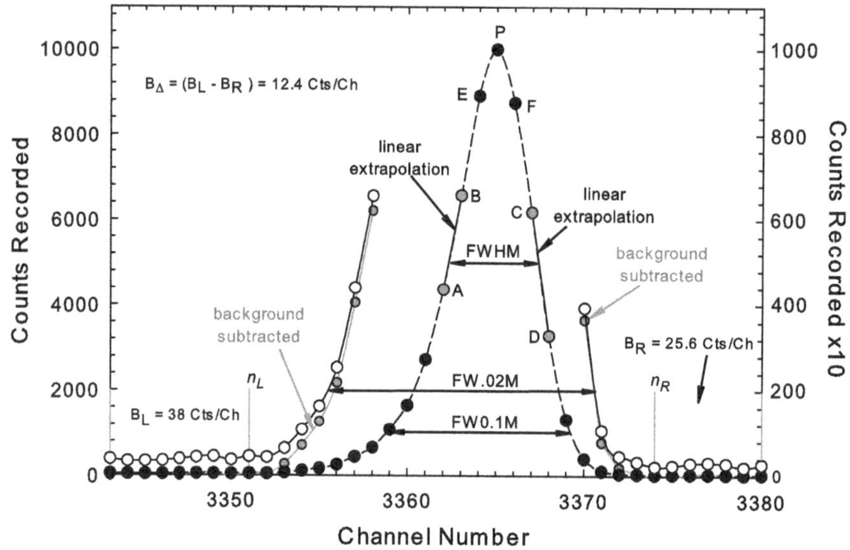

Figure 20.18. Spectral calibration and features for the ^{60}Co 1332.5-keV full energy peak taken with a HPGe detector. The left axis is for the full energy peak and the right axis is for the tails and the background subtracted data on either side of the peak. The open circles are measured counts $C(n)$ and the solid circles are counts corrected for background, i.e., $N(n) = C(n) - B(n)$. Adopted from IEEE 325-1996.

It is notable that escape peaks can be used to identify the initial gamma-ray energies. Escape peaks do not have a Compton gap or Compton edge, and the apparent lack of these features can assist with their identification. Although gamma rays equal to or greater than 1.022 MeV can be absorbed through pair production, escape peaks usually do not become apparent for gamma-ray energies below approximately 1.5 MeV.

Energy Resolution

A method to determine the energy resolution of a gamma-ray spectrometer is prescribed by the IEEE Std 325-1996 and was developed for a HPGe detector and a ^{60}Co check source. The method, which appears to be used by few if any researchers, makes no use of statistical uncertainties in the spectral count data (unlike the weighted least squares fitting methods used earlier). Rather it depends on a very high number of counts in the full energy peak so statistical uncertainties are of little consequence. The one unique feature of this standard, however, is that the method makes no Gaussian symmetry assumption about the shape of the full energy peak and it is capable of treating asymmetric peaks that are wider at energies below the peak energy than above—a feature observed in many spectrometers. In essence, the method relies solely on manipulations of the observed channel counts $C(n) \equiv C_n$ to estimate the FWHM.

The IEEE Std 325-1996 recommends a 8192 (2^{13}) channel multichannel analyzer with the 1332.5-keV ^{60}Co full energy-peak scaled to appear in channel 4032. The reason for the location is to ensure that at least 5 channels are included within the FWHM of the energy peak. If the detector under test has worse resolution than normally expected from a HPGe detector, a lower channel can be used provided that the 5 channel criteria is maintained (see Fig. 20.18). At least 10,000 counts must appear in peak maximum channel, and the spectrum and live time are to be recorded. If a pulse generator is used to measure electronic noise, the pulser peak is to be set between the ^{60}Co energy peaks (1173.2 keV and 1332.5 keV), and as close as possible to the 1332.5-keV full energy peak without interfering with the recorded counts. The method does allow for other gamma-ray sources, but the requirement that at least 5 channels be included in the FWHM remains.

Begin by identifying the first channel n_L to the left of the peak which shows no contribution from the peak and consists solely of background events. Likewise identify n_R the first channel above the peak that shows no peak contribution. Then average the 6-8 channels below n_L and above n_R to find the average background B_L immediately below the peak and B_R above the peak. Use a linear equation between (n_L, B_L) and (n_R, B_R) to represent the background under the full peak, i.e.,

$$B(n) = B_L - (n - n_L)\frac{B_L - B_R}{n_R - n_L}, \quad n_L \leq n \leq n_R. \tag{20.103}$$

The net peak counts corrected for background are then $N(n) = C(n) - B(n)$.

As shown in Fig. 20.18 the channels containing the seven highest counts are identified with the letters A, B, C, D, E, F, and P and the counts in these channels are used to characterize the peak. The peak may not be exactly at the centroid of channel n_P, which occurs only if $N_E = N_F$, and may be slightly shifted to $\tilde{n}_{\rm max}$ where the peak has the true maximum $N_{\rm max}$. To determine the true maximum fit, the three highest counts to the parabola

$$N(\tilde{n}) = \alpha\tilde{n}^2 + \beta\tilde{n} + \gamma. \tag{20.104}$$

The maximum count $N_{\rm max}$ occurs when $dN(\tilde{n})/d\tilde{n} = 0$ or at $\tilde{n}_{\rm max} = -\beta/(2\alpha)$. The maximum is found to occur at

$$\tilde{n}_{\rm max} = n_P + \frac{(N_E - N_F)/2}{N_E + N_F - 2N_P}, \tag{20.105}$$

where $N_E \equiv N(n_E)$ and so on for the other peak channels. The maximum number of counts is

$$N_{\rm max} = N_P - \frac{(N_E - N_F)^2}{8(N_E + N_F - 2N_P)}. \tag{20.106}$$

The FWHM maximum can be measured by producing linear extrapolations for the data points that straddle the FWHM location at a height $N_{\rm max}/2$. This procedure is performed after the background has been subtracted. For example from Fig. 20.18, these extrapolations would be applied to data points at positions A and B and also at positions C and D. Note $n_B - n_A = n_C - n_D = 1$. Thus, the location $\tilde{n}_l$ between A and B of the left end of the horizontal FWHM line is found to be

$$\tilde{n}_l = n_A + \frac{N_{\rm max}/2 - N_A}{N_B - N_A}. \tag{20.107}$$

Similarly the right end $\tilde{n}_r$ between C and D is

$$\tilde{n}_r = n_C - \frac{N_{\rm max}/2 - N_C}{N_C - N_D}. \tag{20.108}$$

The difference between these two limits then gives the FWHM in channels, namely

$$\text{FWHM}_{\rm ch} = \tilde{n}_r - \tilde{n}_l = (n_C - n_A) + \frac{N_C - N_{\rm max}/2}{N_C - N_D} - \frac{N_{\rm max}/2 - N_A}{N_B - N_A}. \tag{20.109}$$

The energy resolution in energy units and as a percent are

$$\text{FWHM}_{\rm energy} = \Delta E \ \text{FWHM}_{\rm ch} \quad \text{and} \quad \text{FWHM}_{\%} = 100\frac{\Delta E}{E_\gamma}\text{FWHM}_{\rm ch}, \tag{20.110}$$

where ΔE is the channel width in energy units (typically keV) and E_γ is the energy of the gamma ray.

The same method can be used to find the full width at $(1/m)C_{\max}$. Shown in Fig. 20.18 are the FW.1M ($m = 10$) and FW.02M ($m = 50$). These metrics are used to quantify the asymmetry of the full energy peak with a *skewness factor* defined as

$$S_R(m) = \frac{\text{FW}(1/m)\text{M}}{\text{FWHM}}\sqrt{\frac{\ln 2}{\ln m}}. \tag{20.111}$$

For a symmetric Gaussian peak $S_R(m) = 1$ for all m but for asymmetric peaks the skewness factors are greater than unity. That full-energy peaks in a HPGe detector are asymmetric arises from the longer collection times for charges created in low electric field regions of the detector compared to that for charges created in normal regions. This longer collection time produces some signal attenuation when passing through an amplifier's linear pulse shaping network.[6] Also Compton scattering decreases the signal in some pulses. The overall result is an increase in the FW$(1/m)$M compared to the FWHM.

Example 20.5: Determine the corrected peak channel and counts for the peak shown in Fig. 20.18. The peak is located between channels $n_L = 3351$ and $n_R = 3374$ and the observed three highest gross counts are: $G_P = 10,000$ at channel 3365 and $G_E = 8910$ and $G_F = 8748$.

Solution:
First find the background corrections. From Eq. (20.103)

$$B(n) = B_L - (n - n_R)\frac{B_L - B_L}{n_R - n_L} = 38 - (n - 3351)\frac{38 - 25.6}{3374 - 3351} = 38 - (n - 3351)(0.5391),$$

from which $B_E = 30.99$, $B_P = 30.45$, and $B_F = 29.91$. Then $N(n) = G(n) - B(n)$ so that

$$N_E = 8879.01 \qquad N_P = 9969.55 \qquad N_F = 8718.09.$$

From Eq. (20.105) the continuous channel for the maximum counts is

$$\tilde{n}_{\max} = n_p + \frac{(N_E - N_F)/2}{N_E + N_F - 2N_P} = 3365 + \frac{80.46}{-2342} = 3364.96.$$

Finally from Eq. (20.106) the true maximum number of counts is

$$N_{\max} = N_P - \frac{(N_E - N_F)^2}{8(N_E + N_F - 2N_P)} = 9969.55 - \frac{(-160.92)^2}{-2342.00} = 9980.61.$$

In this case the peak correction is very small because N_E is very close to N_F. In fact if $N_E = N_F$ there would be zero correction to N_P. The largest correction would occur when $N_P = N_E$ or $N_P = N_F$.

Peak-to-Compton Ratio

Another metric specified for HPGe radiation spectrometers is the peak-to-Compton ratio (PCR). As described in the ANSI/IEEE standard 325-1996, the measurement of the PCR is performed with the 1332.5 keV gamma-ray energy from ^{60}Co. The PCR is defined as

$$\text{PCR} = \frac{C_{\max}}{\overline{N}}, \tag{20.112}$$

[6] Gated-integrating pulse processing is less prone to this attenuation.

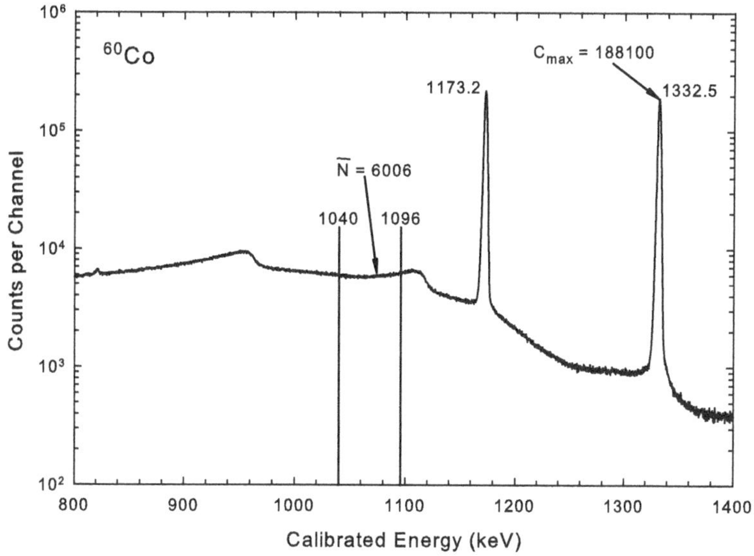

Figure 20.19. The peak-to-Compton ratio (PCR) is calculated by dividing $C_{\max}$ by $\overline{N}$. In this example for a 20% η_{rel} HPGe coaxial detector the measured PCR is 31.3. Courtesy of Nathaniel Edwards, KSU.

where $C_{\max} = C_P$ is the number of counts in the peak channel for the 1332.5-keV gamma ray and $\overline{N}$ is the average number of counts per channel between the channels represented by 1040 keV and 1096 keV. The energy range between 1040 keV and 1096 keV appears in the Compton gap of the 1173.2-keV gamma ray of ^{60}Co; therefore, is almost entirely associated with the Compton continuum of the 1332.5-keV peak (see Fig. 20.19). The environmental background B should be subtracted from the measurements before calculating the PCR. The PCR metric is analogous to a sort of signal-to-noise ratio because the PCR is a measure of the spectrometer's ability to discern lower energy and lower count rate gamma-ray peaks in the presence of higher energy gamma-rays and their corresponding spectral features.

As the energy resolution of a HPGe detector improves, the number of counts in the 1332.5-keV peak channel increases. Also, as the size of the HPGe detector increases, so do counts in the 1332.5-keV peak channel because fewer scattered gamma rays can escape the detector. Hence, improved energy resolution and improved efficiency both tend to increase the PCR. PCR values can range from 30:1 for smaller HPGe detectors up to over 90:1 for relatively large detectors [Gilmore 2008].

It is notable that prior versions of the IEEE ANSI standard 325 also included a PCR for ^{137}Cs in which the maximum counts for the peak channel at 661.7 keV is divided by the average counts per channel in those channels between 358 keV to 382 keV. Although the ^{137}Cs based PCR is generally not used to characterize HPGe detectors, it still has utility as a benchmark for comparing PCR values for much smaller semiconductor spectrometers, such as those with CdZnTe or HgI_2 detectors whose peak efficiency at 1332.5 keV can be small compared to that of a HPGe detector. However, this particular definition does not appear in the most recent version ANSI 325-1996.

Peak-to-Valley Ratio

Another, less official, metric is the peak-to-valley ratio (PVR), which is measured with a ^{137}Cs source. This metric is defined as the number of counts in the peak channel divided by the number of counts in the middle of the Compton gap at 569 keV. This metric is seldom used for HPGe detectors; rather, it is more often used for compound semiconductor detectors and (rarely) used for experimental scintillation detectors. This metric provides a measure of the detector efficiency, energy resolution, and for semiconductors, a measure of the charge carrier trapping effects.

Peak-to-Total Ratio

The peak-to-total ratio (PTR) is the ratio of the total counts in a full energy peak to the number of counts in the entire spectrum. The metric is useful with monoenergetic gamma-ray sources, such as ^{137}Cs and ^{54}Mn, but is not well defined or interpreted for polyenergetic gamma-ray sources. The peak-to-total ratio is proportional to the intrinsic full energy peak efficiency $\epsilon_{\rm peak}$ because

$$\epsilon_{\rm peak} = \epsilon_{\rm I}\left(\frac{\text{peak counts}}{\text{total counts}}\right) = \epsilon_{\rm I}\text{PTR}, \tag{20.113}$$

where $\epsilon_{\rm I}$ is the total intrinsic counting efficiency. By carefully selecting monoenergetic gamma-ray sources, the detector intrinsic peak efficiency can be calibrating over a broad energy range [Heath et al. 1967]. Unfortunately, there are a limited few practical monoenergetic gamma-ray sources to conduct such a measurement.

Another method to accomplish this same task is to use two detectors in coincidence, with the detector under investigation receiving the scattered gamma ray. The method entails capturing a Compton scattered gamma ray that escapes from a high energy resolution primary (first) detector. The residual energy deposited in the second detector is determined by subtracting the energy deposited in the first detector from the known emission energy. Collimation and pulse height discrimination can be used to select and allow only scattered gamma rays of a specific energy, thereby producing a pulse height spectrum, at the energy under investigation, from the second detector.

20.5.3 Detection and Spectroscopy with Scintillators

The properties and advantages of scintillators are described in Chapter 13. Added here are some specific details on spectroscopic measurements and performance pertinent to scintillators. The quality metrics used to assess the photon spectroscopy performance of scintillation detectors include the energy resolution, the peak-to-total ratio (PTR), the intrinsic peak efficiency ($\epsilon_{\rm I}$), and the escape peak efficiency ($\epsilon_{\rm es}$) [Heath 1964]. Usually performance comparisons between different scintillation materials and detectors are conducted with 661.7-keV gamma rays from ^{137}Cs.

Recall from Chapter 4 that the photon interactions important to gamma-ray counting and spectroscopy are photoelectric, Compton scattering, and with sufficient energy, pair production. The combined intrinsic gamma-ray interaction efficiencies that any one of these interactions occurs are plotted in Chapter 13, Figs. 13.13, 13.14, 13.15, 13.17, 13.19, and 13.20 as a function of photon energy for several scintillation materials. An often used method to predict gamma-ray detection performance is to couple the solid angle calculations (from Chapter 7) with the interaction efficiencies, a method repeated in many textbooks on radiation detection and measurements. These complicated calculations often include parallax effects as well. Although useful in determining the probability of an interaction occurring, and perhaps assisting with the overall determination of counting efficiency, these methods do not yield information on the energy deposited nor the expected spectral features. Monte Carlo techniques can instead be used to calculate the photon energy deposition and the resulting spectroscopic features and respective intensities with increased accuracy above that of previously used graphical and empirical methods.

Detector Geometries

Commercial scintillators can be acquired in a variety of geometries and configurations, ranging from individual crystals to integrated packages with light collection device, voltage divider, and preamplifier. Many of the alkali metal and halogen based scintillators, such as NaI:Tl, LiI:Eu, CLYC:Ce, SrI_2:Eu, $LaBr_3$:Ce and $CeBr_3$ are highly hygroscopic. Hence, commercial material is usually obtained already packed in a hermetically sealed reflective cannister that can be attached to a light collection device of choice. Large devices are usually configured as right circular cylinders, while smaller devices can be obtained as either cylinders or cuboids. The cylindrical shape is convenient for coupling to common circular PMTs, while the small cuboid shapes are convenient for coupling to SiPMs. Proper selection of a scintillation crystal requires that the response function of the chosen light collector match the scintillator emission spectrum (see Sec. 13.2.2 and Fig. 13.3). Also, to avoid space charge distortion in a PMT, the response speed of the PMT should be fast enough to follow the light emissions. This requirement is especially important for bright and fast scintillators such as $LaBr_3$:Ce, $LaCl_3$:Ce, and $CeBr_3$. If a proper response speed is not selected, current crowding and space charge buildup result in a non-linear output (see discussions in Sections 13.2.3, 14.1.7, and 14.1.8). For high-energy gamma rays interacting in bright scintillators, similar problems occur with SiPMs in which the probability of coincident light photons interact in the SPADs (see Sec. 14.2.4).

Although commercial vendors may provide just the scintillator for custom work, it is far more common that scintillators are obtained already coupled to the light collection device. Hence, the total package is usually already optimized with the proper coupling compound, shielding, glass envelope, spectral response and electronic output. Commercial vendors quote the optimum operating conditions, including recommended voltage and count rate limitations. Specification sheets usually list the performance metrics mentioned above.

NaI:Tl, discovered by Hofstadter in 1948, remains the most commonly used scintillator for general gamma-radiation spectroscopy applications. The reasons for this widespread usage include its relatively low cost, the fact that large crystals can be produced, and the gamma-ray interaction efficiency is adequately large (because of the iodine component). The general application of NaI:Tl to radiation spectroscopy is reviewed by Smith and Kearfott [2018]. It is the 3 in × 3 in right circular cylinder of NaI:Tl that is quoted most often in performance comparisons to other spectrometers (scintillators and semiconductors), although a wide variety of NaI:TL sizes are available. Some of these alternative shapes include large plates for Anger cameras, pixelated arrays, well detectors, long bars, cuboids, spheroids, annuli for Compton suppression, and curved crystals.

Right Circular Cylinder Detectors The right circular cylinder is the most common geometry for scintillation detectors, mainly because the flat circular face conveniently couples well to cylindrical PMTs. The nomenclature for detector size is the diameter × length. For example, a 2 in × 3 in size describes a 2 in diameter by 3 in long detector. Common commercial sizes are 1 in × 1 in, 2 in × 2 in, 3 in × 3 in, and 5 in × 5 in, although some companies can provide custom sizes for an added charge. The common use of $m \times n$ dimensions where $m = n$ is with purpose, mainly, the light is distributed more uniformly over the photocathode of the PMT. This uniformity produces the best energy resolution, and deviations from this ratio have a poorer resolution. Solid angle corrections for cylindrical geometry are described in Sec. 7.4, and spectroscopic performance for NaI:Tl detectors are well documented by several authors [see for example Heath 1964; Crouthamel 1970].

Well and Annular Detectors Some scintillation detectors, called *well* detectors, are available as a cylinder with a blind hole or well drilled in them. These detectors are designed to perform, nearly, as a 4π detector, where the sample under investigation is placed at the bottom of the well. The probability that an isotropically emitted gamma ray enters the crystal is

$$P \approx .5\left[1 + \frac{d}{\sqrt{d^2 + a^2}}\right], \tag{20.114}$$

where d is the well depth and a is the well radius. However, because of different attenuation lengths through the crystal, especially at corners, the probability of interacting in the detector changes with the gamma-ray trajectory. Further, as explained previously, the interaction efficiency (actually the interaction probability) does not describe the efficiency of total energy deposition. Hence, well detectors may increase the total counting efficiency, but the intrinsic peak efficiency may suffer from added Compton losses at the extended boundary, especially as the gamma-ray energy increases.

If, instead, a smaller detector (small scintillator or semiconductor) is located inside the well, the well detector can be used as a Compton suppression coincidence detector (reviewed later in this chapter). Annular detectors are also used for Compton suppression, where the scintillation detector is a larger ring that surrounds a relatively large detector such as an HPGe detector. The annular detector may be a single annulus, or may be composed of annular segments that form the ring.

Position Sensitive Detectors A long bar of scintillator with a light detector at each end can be used as a position sensitive gamma-ray detector. Light radiates from the ionizing path of the particle, be it a photoelectron, Compton electron, or charges produced by pair production. This light generally decreases exponentially in magnitude as it propagates through the crystal. Consider the configuration of Fig. 20.20 where an ionizing event produces light originating at position x along a scintillator bar of length L. For isotropic generation of scintillation photons and with the same notation used in Sec. 13.2.2, PMT A produces an induced charge signal

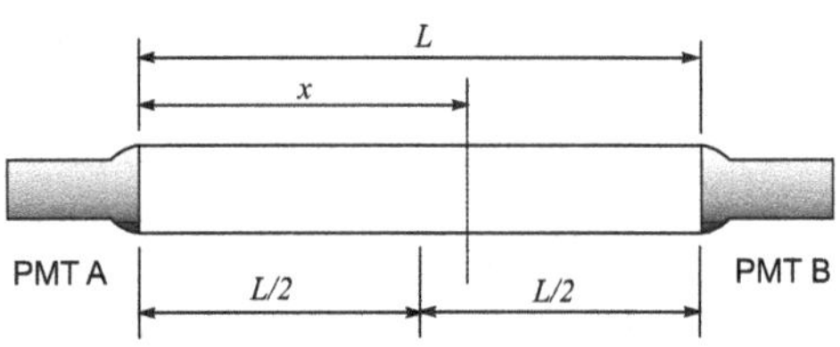

Figure 20.20. Depiction of a scintillator bar detector used for position sensitive measurements.

$$Q_A = f_A q_e M N_e E_n F_n C_{np} U_A e^{-\alpha x} \simeq K_A E_n e^{-\alpha x}, \tag{20.115}$$

where E_n is the ionizing particle energy, q_e is the unit charge, F_n is the fraction of energy deposited in the detector, C_{np} is the average light conversion efficiency (light yield), N_e is the charge released from the photocathode, M is the PMT gain, f_A is the fraction of photons that propagate generally in the direction of PMT A, and U_A accounts for nonuniformities in light collection from geometrical effects, and α is the attenuation coefficient. The induced charge from PMT B is, therefore,

$$Q_B = f_B q_e M N_e E_n F_n C_{np} U_B e^{-\alpha(L-x)} \equiv K_B E_n e^{-\alpha(L-x)} \tag{20.116}$$

The ratio of the PMT signals is

$$\frac{S_A}{S_B} \equiv \frac{Q_A}{Q_B} = \frac{K_A E_n e^{-\alpha x}}{K_B E_n e^{-\alpha(L-x)}} = C_1 e^{-2\alpha x + L} \tag{20.117}$$

By adjusting the gain M of either PMT, the value of K_A can be made equal to K_B, hence C_1 becomes equal to unity. Carter et al. [1982] define $x' = x - (L/2)$ so that Eq. (20.117) yields

$$\frac{S_A}{S_B} = e^{-2\alpha x'}. \tag{20.118}$$

Hence, taking the natural log of Eq. (20.118) yields a relationship for the interaction position

$$x' = \frac{1}{2\alpha} \ln\left(\frac{S_B}{S_A}\right) \tag{20.119}$$

The corresponding standard error is

$$\sigma_{x'} = \frac{e^{\alpha L/4}}{2\alpha\sqrt{E_n}} \left(\frac{e^{\alpha x'}}{K_A} + \frac{e^{-\alpha x'}}{K_B}\right)^{1/2}. \tag{20.120}$$

If $K_A = K_B = K$, then Eq. (20.120) becomes

$$\sigma_{x'} = \frac{e^{\alpha L/4}}{2\alpha}\left(\frac{e^{\alpha x'} + e^{-\alpha x'}}{E_n K}\right)^{1/2}. \tag{20.121}$$

Substituting Eq. (20.119) into Eq. (20.118) yields the interaction position

$$x = \frac{1}{2\alpha}\ln\left(\frac{S_B}{S_A}\right) + \frac{L}{2}. \tag{20.122}$$

and

$$\sigma_x = \frac{e^{\alpha L/4}}{2\alpha}\left(\frac{e^{\alpha(x-L/2)} + e^{-\alpha(x-L/2)}}{E_n K}\right)^{1/2}. \tag{20.123}$$

Finally, to find the initial gamma-ray energy E_n multiply Eqs. (20.115) and (20.116) to obtain

$$Q_A Q_B = K_A K_B E_n^2 e^{-\alpha L} = K^2 E_n^2 e^{-\alpha L}. \tag{20.124}$$

Then solving this result for E_n gives

$$E_n = \sqrt{\frac{Q_A Q_B}{K^2}}e^{\alpha L/2} = C_2\sqrt{S_A S_B}e^{\alpha L/2}, \tag{20.125}$$

where the constant C_2 accounts for signal conversion from the induced charges. From the results of Eqs. (20.119) and (20.124), it is shown that the linear position and energy of an event can be measured with the relatively simple arrangement of Fig. 20.20. For a NaI:Tl bar with dimensions of 5 cm × 5 cm × 50 cm long, Carter et al. [1982] report a position resolution for 661.7-keV gamma rays of 10 mm ± 10% at the center and approximately 12 mm ± 12% at the ends.

Pixelated scintillation detectors are banks of scintillator rectangular prisms fastened together with thin Lambertian layers called "separators". These separators range in thickness from 0.5 mm down to 0.04 mm depending on type, and differ in relative reflectivity, with powders of TiO_2 and MgO_2 being amongst the highest in both reflectivity and thickness. The light emissions from these segmented crystals are isolated; hence, by attaching the array to a position sensitive PMT or SiPM, the xy spatial location of an event can be determined [Saint Gobain, 2019b]. To increase light collection, a Lambertian reflector is applied over the array surface opposing the light detector. There are several types of commercial arrays of prismatic crystals, including CsI:Tl, BGO, LYSO, and $CdWO_4$. The prism sizes are constrained by material limitations, but are generally on the order of 0.3 to 0.8 mm long with side dimensions on the order of 0.3 to 1 mm.

These pixelated arrays are designed for radiation imaging such as computed tomography or other radiation imaging techniques. In principle, the same method described for linear position sensing can be used to determine the z spatial component, making the device a volumetric position sensitive device. However, because of the small prism sizes, Compton escapes into adjacent crystals are relatively probable. To ameliorate this effect, the output can be summed to assist with energy identification and to increase the overall intrinsic peak efficiency. The energy resolution, however, is decreased because of the variation in energy lost in the insensitive separator layers.

An alternative approach has been to grow columnar structures of CsI:Tl, each crystal column being on the order of a micron in diameter and typically are less than 600 microns long, but apparently special units up to 1.5 mm can be obtained [Bugby et al. 2016]. The columns perform as optical light guides that direct light to the column ends. Due to the small thickness, this imaging scintillator is best used for x-ray energies.

The columnar crystal structure is attached to a photon imaging device, which delivers the xy spatial position. A reflective coating can be used to help collect light that transports, initially, in the opposite direction of the position sensitive light detector. Plates on the order of 440 mm $\times$ 440 mm are commercially available [Hamamatsu 2016].

Ruggedized Units Many scintillators are extremely fragile, and mishandling can lead to catastrophic fractures induced by mechanical shock that can permanently damage the crystal. However, there are certain ruggedized designs that resist this problem. For instance, polycrystalline NaI:Tl compressed under high temperature conditions can eliminate differences in the index of refraction at the grain boundaries. Hence, light transports through the device in the same fashion as a single crystal. However, fractures induced in a single crystal grain are confined to that grain and do not propagate across the device. These ruggedized units have been developed for the oil well logging industry to meet the stringent demands on instrumentation for in situ measurements, including heat resistant seals and electronics.[7]

Low Background and Low Energy Detectors For low-level radiation measurements, special units composed of scintillators and supporting electronics with relatively low background levels are commercially available. Much of the low-level background contamination comes directly from ^{40}K in the PMT glass envelope. The problem is greatly mitigated by using purified quartz as the PMT envelope, or at least as the main coupling window. Also, the metal housing is made with purified low background stainless steel or copper. The scintillator itself may be contaminated with a background source.[8] Hence, careful selection of purified scintillators without radioactive constituents is preferred.

Low energy gamma- and x-ray applications require special crystal thicknesses that are adequately efficient only at low energies, with higher energy photons either passing through without interaction or depositing only a fraction of their energy. Also, the thickness and material of the entrance window must be altered to allow the efficient passage of low-energy photons. Common choices are beryllium and aluminum, with the former having an energy range between 3 to 100 keV with the latter having an energy range between 10 to 200 keV. Note that NaI:Tl is often selected for low-energy low-background scintillation detectors. The relative light yield at low energy increases (see Fig. 13.4), improving the crystal performance at energies below approximately 100 keV. The phoswich detector, presented later in this chapter, is another detector configuration specifically designed for low-energy gamma-ray measurements. Anti-coincidence methods can reduce the number of Compton scattering events, thereby enhancing the appearance of the gamma-ray full energy peaks.

Energy Resolution

The energy resolution describes the spectroscopic performance of a scintillation. For scintillators this particular metric is traditionally reported as a percent FWHM of the peak energy

$$\mathrm{FWHM}_{\%} = 100\frac{\Delta E}{E_\gamma}, \tag{20.126}$$

where ΔE is the FWHM in units of energy. It is common to quote energy resolution for scintillation detectors at 661.7 keV for gamma rays from ^{137}Cs. Typical energy resolution ranges from approximately 3% FWHM at 661.7 keV for some of the high-performance scintillators ($LaBr_3$:Ce, $CeBr_3$, CLLB:Ce), while traditional scintillators such as NaI:Tl have energy resolution on the order of 7% FWHM at 661.7 keV, or approximately 47 keV FWHM. The energy resolution affects the ability to distinguish neighboring gamma-ray energies. Hence, gamma-ray peaks closer than 47 keV apart significantly overlap.

[7] For instance, the Polyscin® NaI:Tl detector developed by Saint Gobain Crystals is designed to withstand shock and vibration for temperatures between 55°C and 205°C.

[8] Such as ^{138}La and ^{227}Ac in $LaCl_3$ and $LaBr_3$.

The energy resolution of a scintillator is a function of multiple processes, which include the light conversion efficiency, the relative light yield linearity, light collection efficiency, and conversion gain and linearity of the light detector. Given an average photon yield per energy deposited, the percent energy resolution can be generally described by Eq. 13.36, rewritten here as

$$\mathrm{FWHM}_{\%} \approx 235.5\sqrt{\frac{1.1}{\overline{N}}}, \tag{20.127}$$

where $\overline{N}$ is the average number of photons released for the energy under investigation. Assume that the value of $\overline{N}$ is a linear function of the gamma-ray energy, where

$$\overline{N} = C_{np}E_\gamma \tag{20.128}$$

and C_{np} is average light yield per MeV. Then

$$R = \mathrm{FWHM}_{\%} \approx 235.5\sqrt{\frac{1.1}{C_{np}E_\gamma}} = C_1 E_\gamma^{-1/2}, \tag{20.129}$$

where C_1 is a constant. Beattie and Bryne [1972] show that Eq. (20.129) can be described as a linear equation

$$\ln(R) = \ln(C_1) - \frac{1}{2}\ln(E_\gamma) \tag{20.130}$$

as plotted in Fig. 20.21 for several gamma ray energies recorded with a NaI:Tl detector.

As a general rule, bright scintillators are sought because the increased photon yield should increase the value of $\overline{N}$, and from Eq. (20.127) should also improve the energy resolution. However recall from Sec. 13.2.2 that linearity of the relative light yield also has a significant effect. For example, from Table 13.1, NaI:Tl has a light yield of 43000/MeV, while $LaBr_3$:Ce has a light yield of 63000/MeV. Hence, it would be expected that $LaBr_3$:Ce would have approximately 1.21 times improved energy resolution over similar sized NaI:Tl detectors. However, energy resolution also depends on spectral matching to the light collection device and the linearity of the relative light yield. From Fig. 13.3, NaI:Tl matches better to common PMTs than does $LaBr_3$:Ce, yet comparing Figs. 13.4 and 13.5, $LaBr_3$:Ce has a superior linear response at low energies. Consequently, the energy resolution of $LaBr_3$:Ce is far superior to that of NaI:Tl than might be surmised from Eq. (20.127), as can be observed by the spectral comparisons in Fig. 20.22. Overall, energy resolution generally improves with scintillator brightness and relative light yield linearity.

Peak-to-Total Ratio

Shown in Fig. 20.23 are PTR values for several NaI:Tl detector sizes, source distances, and gamma-ray energies. Intuitively, the PTR decreases with increasing gamma-ray energy and increases with detector size [Leutz et al. 1966], being best for a collimated gamma-ray beam striking the detector. From Fig. 20.23, it is also notable that the PTR is nearly the same for similar detector sizes regardless of source distance and collimation. The PTR increases, generally, with the effective Z and mass density of the scintillation material. Shown in Fig. 20.24 are calculated PTR values as a function of energy for several different commercial scintillator materials. Clearly the materials with higher Z components have the highest peak-to-total ratios, while the lightest material shown (CaF_2) has the lowest.

Intrinsic peak efficiency

Because the PTR is the ratio of $\epsilon_{\mathrm{peak}}/\epsilon_{\mathrm{I}}$ (see Eq. (20.113)), it is also noted that the full energy peak efficiency increases with PTR. When counting efficiency is most important, for instance in medical imaging applications, then often materials with high Z, and ultimately a high PTR, are attractive choices. Materials that

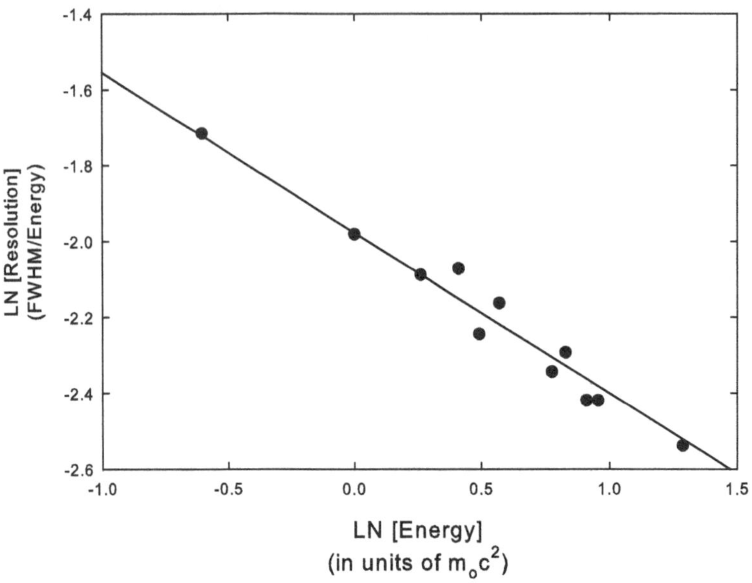

Figure 20.21. Experimental values of energy resolution for a NaI:Tl detector compared to a least squares fit. The gamma ray energies have been converted into units of electron rest mass (511 keV). Data from Beattie and Byrne [1972].

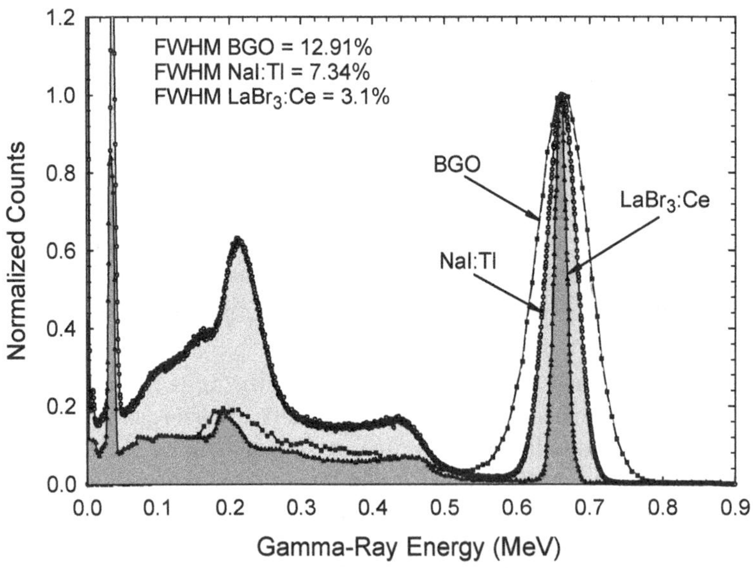

Figure 20.22. Comparison of normalized spectral performance for a 2×2 NaI:Tl detector, a 2×2 $LaBr_3$:Ce detector, and a BGO detector of similar size for 662 keV gamma rays from ^{137}Cs. BGO data from Vincke et al. [2002] and $LaBr_3$:Ce data from Quarati et al. [2007]. NaI:Tl data courtesy of Taylor Ochs, KSU.

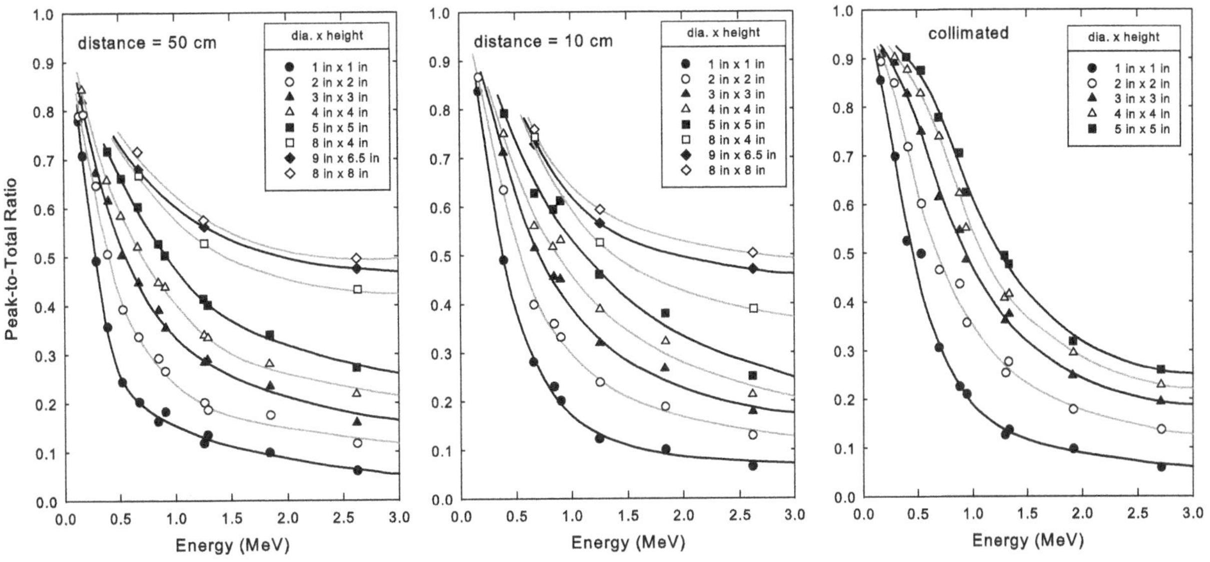

Figure 20.23. Peak-to-total ratios for NaI:Tl detector of different sizes for different gamma ray energies, showing from left to right the results of a point source located 50 cm from the face center, 10 cm from the face center, and a collimated beam 0.7 cm diameter. Data from Leutz et al. [1966].

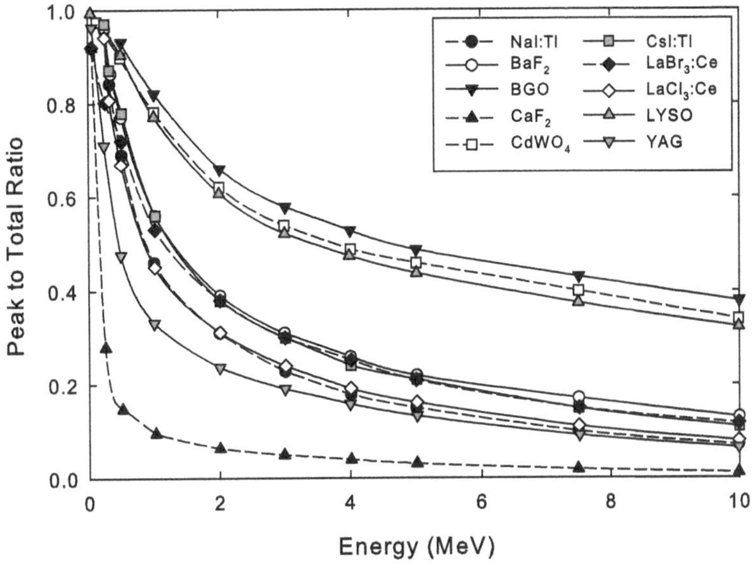

Figure 20.24. Monte Carlo calculated comparisons of peak-to-total ratios for a collimated beam of gamma rays for several commercial scintillators. The data is for 3 in × 3 in right circular cylinders. Data from Saint Gobain [2016].

generally have high Z elements typically have better intrinsic peak efficiencies than those composed of low Z materials. For instance, the plot in Fig. 20.24 indicates that BGO should have a relatively higher intrinsic peak efficiency than $LaBr_3$:Ce, which should have higher peak efficiency than NaI:Tl. This correlation can be observed from the normalized spectra shown in Fig. 20.22, where clearly BGO has the highest ϵ_{peak} value while NaI:Tl has the lowest ϵ_{peak} value of the three scintillators in the example. Note also that the Compton continuum has relatively more counts for the NaI:Tl detector than either the BGO or $LaBr_3$:Ce detector, an indication that the NaI:Tl detector has more scattered photon losses. Overall, the plots like Fig. 20.24 can also yield some indication of the expected intrinsic peak efficiencies.

Summary

The primary disadvantages of scintillator spectrometers are (1) that a photon detector such as a PMT must be incorporated into the device and (2) that the energy resolution is inferior to that of a semiconductor spectrometer. Historically, scintillation crystals were coupled to PMTs so that the spectrometers were fragile and bulky. A special μ-metal is also needed to shield the PMT from magnetic fields. Modest success was realized with the coupling of scintillators to semiconductor photodiodes, yet the energy resolution was inferior to that obtainable with a PMT. The introduction of the silicon photomultiplier (SiPM) has greatly reduced the problem of fragility of PMTs and their bulkiness while nearly preserving the energy resolution achieved with PMTs. Scintillator spectrometers of many varieties coupled to SiPMs are now commercially available as rugged and compact devices unaffected by magnetic fields and provide nearly the same energy resolution as their PMT-coupled counterparts. Regardless of these new benefits, the spectroscopic performance of scintillators remains inferior to semiconductor spectrometers, especially to that of high-purity Ge and Si spectrometers (see Fig. 20.25).

20.5.4 Spectroscopy with Semiconductors

The physics and performance of semiconductor detectors are described in Chapters 15 and 16, including sections on device designs, efficiencies, and energy resolution. A few more details are offered here that are pertinent to radiation spectroscopy with semiconductor detectors. Otherwise, the general discussions on spectroscopy found elsewhere in this chapter also apply to semiconductor detectors.

IEEE Standard for HPGe Peak Efficiency The detection efficiency of a HPGe detector is defined in ANSI 325-1996, and is actually an absolute efficiency under a specific geometry. A calibrated point ^{60}Co source is placed exactly 25 cm from the center of the cryogenic end cap face of a HPGe detector. The source activity is calculated and the total number of counts appearing in the 1332.5 keV full energy peak is divided by the total number of 1332.5-keV gamma rays emitted by the source during the measurement. The ANSI 325-1996 document suggests that a pulser set at an energy channel 5% higher than 1332.5 keV ($\approx$ 1400 keV) can be used for timing calibration.[9]

A background measurement is conducted without the ^{60}Co source present. After subtracting the background spectrum (as described earlier in Sec. 20.5.2), the *absolute* efficiency of the HPGe detector is then quoted as

$$\eta_a = 100\frac{A_p}{N_s}\%, \tag{20.131}$$

where A_p is the integrated number of counts (area of) in the 1332.5-keV background-corrected full energy peak and N_s is the total number of 1332.5-keV gamma rays emitted by the ^{60}Co source during the measurement live time. It is assumed any source or air absorption is negligible.

[9]The concrete of radiation counting laboratories may be contaminated with ^{40}K, which emits a 1460.8 keV gamma ray. Although the energy 1400 keV falls in the Compton gap of the ^{40}K spectrum, the possible interference from ^{40}K can be eliminated altogether by setting the pulser at 15% above 1332.5 keV (1530 keV).

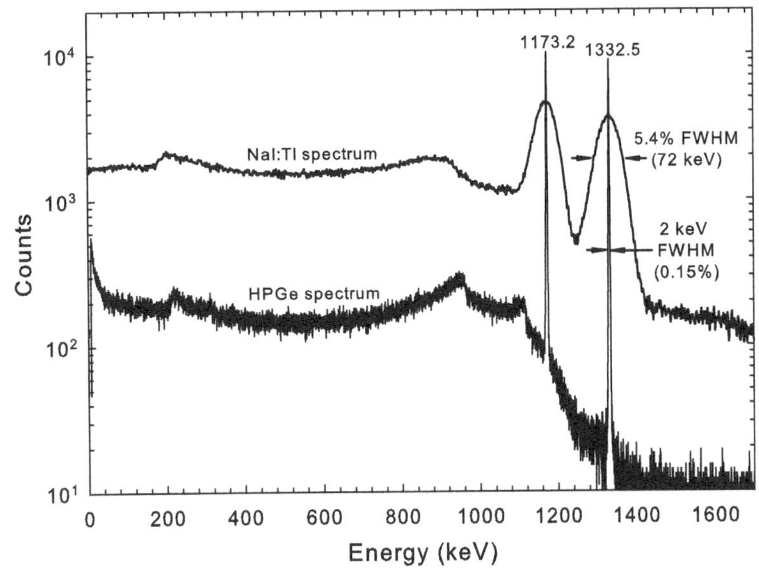

Figure 20.25. Comparison gamma-ray spectra of ^{60}Co taken with a NaI:Tl detector and an HPGe detector.

Traditionally, the efficiency of a HPGe detector is quoted as a relative comparison to the expected efficiency of a 3 in × 3 in (or 7.62 cm × 7.62 cm) NaI:Tl detector.[10] Such an efficiency is referred to as the *relative* efficiency. The expected number of counts per unit activity in the 1332.5-keV full energy peak of such a NaI:Tl detector, when set back 25 cm from the source, is well established as 0.0012 cps per Bq. The measurement is conducted, as before, with a calibrated ^{60}Co source placed exactly 25 cm from the center of the cryogenic end cap face of a HPGe detector. After first subtracting the background spectrum, the relative efficiency is found as

$$\eta_{rel} = 100 \frac{A_p}{0.0012 \times A_{^{60}\mathrm{Co}}} \%, \tag{20.132}$$

where $A_{^{60}\mathrm{Co}}$ is the activity of the source at the time of the measurement. It is this relative efficiency that is usually quoted for coaxial HPGe detectors. However, because other (non-coaxial) HPGe detector configurations may not be efficient at 1332.5 keV, the relative efficiency is usually not used for non-coaxial designs. Usually the manufacturer makes clear the standard being used to report detector efficiency.

Reporting Energy Resolution The energy resolution dependence for semiconductors was derived in Sec. 16.3.2, repeated here for convenience,

$$\mathrm{FWHM}_{\mathrm{energy}} = 2\sqrt{2\ln(2) w F E_n} \approx 2.355\sqrt{w F E_n}, \tag{16.24}$$

where w is the average ionization energy, E_n is the radiation particle energy, and F is the Fano factor. For semiconductor detectors the energy resolution is traditionally reported in units of keV for both gamma-ray and charged particle detectors. The reason for this standard is that the energy resolution is generally below 1% FWHM for high-performance semiconductors, and reporting the FWHM in terms of percent generally is less descriptive than reporting the actual energy. Regardless, there has been a departure from this historic

[10]This historic standard dates from the early introduction of semiconductor detectors, when it was common to directly compare efficiency to what had become the standard for gamma-ray spectroscopy, the 3 × 3 NaI:Tl detector.

standard for some of the lower resolution compound semiconductors, and often the energy resolution FWHM is reported in terms of percent for compound semiconductors.

HPGe detector energy resolution has historically been measured at 1332.5 keV (one of the gamma rays emitted by ^{60}Co) and is also the IEEE standard. However, because compound semiconductors are typically smaller and have less efficiency at 1332.5 keV, the energy resolution of these devices is commonly reported at 661.7 keV (a gamma ray emitted by ^{137}Cs). Si(Li) detectors are relatively inefficient at energies below 60 keV, and are generally reserved for use at lower energies. Typical gamma-ray energy resolution is on the order of approximately 2 keV FWHM at 1332.5 keV for some of the high-performance HPGe detectors, and around 200 eV FWHM at 5.9 keV for Si(Li) x-ray detectors. The more mature compound semiconductors (CdTe, HgI_2, CdZnTe) typically have between 6 to 20 keV FWHM at 661.7 keV.

20.6 Factors Affecting Energy Resolution

The energy resolution achievable from a spectrometer is largely determined by the average energy required to produce a fundamental signal carrier, statistical fluctuations in signal carrier production, non-linearity in signal carrier production, and electronic noise. Energy resolution is also compromised by non-linear light yield in scintillators. In semiconductors, charge carrier recombination and trapping produces non-linear charge collection and affects the resolution. Electronic noise also adds to the variance in the pulse height distribution. Common factors that affect the energy resolution of radiation spectrometers are discussed below, with the emphasis on scintillation and semiconductor spectrometers. Also discussed are the primary sources of electronic noise, which also affects the resolution.

Variation in Charge Collection

The production of information carriers in a spectrometer is usually modeled as a Gaussian process. For instance, the average number $\overline{N}_{eh}$ of electron-hole pairs produced by a gamma-ray in an HPGE semiconductor detector is the gamma-ray energy absorbed E_γ divided by the average ionization energy w. In Sec. 16.4.2, a Fano factor F was used to correct for differences from Gaussian predications, i.e., $\sigma^2(\overline{N}_{eh}) = F\overline{N}_{eh}$. For semiconductors, the Fano factor is usually on the order of 0.1, but varies between materials as can be seen from Table 16.2. Hence, the energy resolution in terms of energy is defined by

$$\text{FWHM}_{\text{energy}} = 2\sqrt{2\ln(2)}\sigma_{eh}(f)E_\gamma = 2.355\sqrt{wFE_\gamma}. \tag{20.133}$$

It is apparent the FWHM increases as the square root of the gamma-ray energy. Traditionally, the energy resolution of a semiconductor spectrometer is reported in units of keV. However, for some applications, the percent energy resolution is of interest, namely

$$\text{FWHM}_{\%} = 235.5\sqrt{\frac{wF}{E_\gamma}}\ \%. \tag{20.134}$$

Equations (20.133) and (20.134) are relatively good predictors for well-behaved semiconductors that do not suffer from severe recombination or trapping. However, large losses from charge carrier trapping cause the full energy peak to distort, as shown by examples in Sec. 15.5.1, with the development of a low-energy tail on the low-energy side of the full energy peak. If trapping is severe, the tail can cause such a large distortion that a full energy peak is difficult or impossible to discern.

Scintillator Non-linearities

Another source of variance in the number of signal carriers is due to a non-linear light yield response to deposited gamma-ray energy in many scintillators as shown in Figs. 13.4 and 13.5. Although the average photon yield is often estimated by the product E_γ and the average light yield per unit energy (photons/MeV), such an estimate is based on a linear response. For many scintillators, the non-linearity in light yield can be severe for gamma-ray energy deposition below 500 keV. Hence, the manner by which a photon interacts in the scintillator affects the total light yield. For example, a 1.332-MeV gamma ray completely absorbed by the photoelectric effect produces less light in NaI:Tl than if the photon scattered twice, depositing approximately 400 keV for each scatter before finally being absorbed by the photoelectric effect. Even though the same amount of energy is absorbed in both cases, the amount of light produced is different and, thus, increases the variance in the output signal. Consequently, scintillators with a linear response, but perhaps lower light yields, can still perform well. An example is YAP:Ce, which is a linear scintillator with about 40% of the light yield produced by NaI:Tl. Although not as bright, the energy resolution of YAP:Ce has been reported at 4.4% FWHM at 661.7 keV, much lower than NaI:Tl (typically greater than 6% FWHM at 661.7 keV). The output from the light collection device, a PMT for example, also contributes to the total variance as quantified by Eq. (13.30). Traditionally, the energy resolution at FWHM is reported in terms of percent for scintillators.

Electronic Noise

Electronic noise arises from fluctuations in the speed and number of electrons and is present in both pulse mode and current mode detector systems. The fluctuations may be from thermal effects, leakage currents, or generation-recombination effects and may originate in the actual detectors and/or in the system electronics. Regardless of the source, electronic noise determines the minimum FWHM energy resolution that is possible for a given detector and its associated electronics. Noise from current fluctuations is commonly named *parallel noise*, while noise from voltage fluctuations is often called *series noise*. Thermionic leakage current from a reverse biased diode produces *shot noise*, while resistors produce *Johnson noise*. The appearance of so-called periodic $1/f$ noise, where f is the noise frequency, is a consequence of non-random emissions, such as carrier generation and recombination, or trapping and detrapping, in the detector or system electronics. The goal in designing a detector system is to reduce the electronic noise as much as possible, so as to improve the signal-to-noise ratio η (SNR), defined as the ratio of the detector mean signal pulse voltage V_{avg} to the square root of the mean-square noise voltage (the variance in the noise V_n^2), i.e.,

$$\eta = \frac{V_{avg}}{\sqrt{V_n^2}}. \tag{20.135}$$

As discussed in Chapter 22, if the noise is modeled as a Gaussian distribution, then the contribution to the detector and system energy resolution can be approximated by

$$\frac{V_{\mathrm{FWHM}_{\mathrm{noise}}}}{V_{avg}} = 2.355\frac{V_n}{V_{avg}}. \tag{20.136}$$

Ultimately, this contribution to the variance must be added, by error propagation, to the other resolution limiting variances to predict the overall detector resolution

$$\begin{aligned}\mathrm{FWHM} = \big[&(\mathrm{FWHM}_{\mathrm{noise}})^2 + (\mathrm{FWHM}_{\mathrm{trap}})^2 \\ &+ (\mathrm{FWHM}_{\mathrm{eh}})^2 + (\mathrm{FWHM}_{\mathrm{non-linear}})^2 + \ldots\big]^{1/2}.\end{aligned} \tag{20.137}$$

The electronic noise component $\mathrm{FWHM}_{\mathrm{noise}}$ can be measured by inserting an electronic pulser into the system. The pulser delivers a constant voltage pulse into the system, usually through the preamplifier, and the FWHM of the pulse can be directly measured. The square of this measurement can be subtracted from the square of Eq. (20.137) to determine the limiting resolution of the actual detector.

20.7 Experimental Design

Proper experimental design can make the difference between an accurate and a questionable measurement. Previously in Chapter 6 there was some discussion about the minimal detectable activity (MDA) and the critical detection limit (CDL). These measurements are used when the radiation activity of interest is similar to that of the background. There are also situations in which discernment between similar energies is important, an issue partly covered in Sec. 20.4.8 of this chapter. Rather than just start by making measurements, an experimental outcome can be generally much improved by designing a procedure or protocol for the experiment before the taking measurements. Some experimental design considerations are described here.

20.7.1 Optimization of Measurement Time

Sometimes the time allocated t_A to an experimenter to use a particular system that is in high demand is limited, such as with a low-level deep-underground counting facility. In such cases it is important to optimize the counting times for the background and the source or foreground measurements so as to minimize the uncertainty in the background corrected source measurements.

A Single Source Measurement

In this case the allowed time t_A must be divided between the time t_G for the foreground or source measurement and the time t_B for the background measurement. Typically, the background count rate b is subtracted from the count rate g observed for the source measurement to determine the net counting rate n, and the two independent measurements are used to calculate the experimental error. Minimizing this error can be achieved by optimizing the foreground measurement time t_G and the background measurement time t_B, subject to the constraint $t_A = t_G + t_B$. The radiation counting rate g, which represents contributions from both the radioactive or foreground source and the background, is

$$g = \frac{G}{t_G}, \tag{20.138}$$

where G is the total number of counts recorded during time t_G. The background counting rate is

$$b = \frac{B}{t_B}, \tag{20.139}$$

where B is the total number of background counts recorded during time t_B. The net counting rate is

$$n = g - b = \frac{G}{t_G} - \frac{B}{t_B}, \tag{20.140}$$

The estimated net counting error, as derived in Sec. 6.9.2, is

$$\sigma(n) = \left[\frac{G}{t_G^2} + \frac{B}{t_B^2}\right]^{1/2} \tag{20.141}$$

or

$$\sigma^2(n) = \frac{g}{t_G} + \frac{b}{t_B}. \tag{20.142}$$

To optimize this result with the total allowed time fixed at $t_A = t_G + t_B$ Eq. (20.142) is differentiated as

$$\frac{d\sigma^2(n)}{dt_G} = \frac{d}{dt_G}\left\{\frac{g}{t_G} + \frac{b}{t_B}\right\} = -\frac{g}{t_G^2} + \frac{b}{(t_A - t_G)^2} \tag{20.143}$$

and this result is set to zero to find t_G/t_B to produce the minimum σ_n^2. The result is

$$\frac{g}{t_G^2} = \frac{b}{(t_A - t_G)^2} = \frac{b}{t_B^2} \tag{20.144}$$

or

$$\frac{t_G}{t_B} = \sqrt{\frac{g}{b}}. \tag{20.145}$$

This result may seem counter-intuitive because it implies that the higher the source count rate is to the background count rate, the smaller is the fraction of the available time that should be given to determine b. But if $g \gg b$, then the background has a very minor influence on the net count rate and hence need not be determined very accurately. However, there is one obvious problem with this optimization procedure. To use it, the experimenter must already have knowledge of the expected background count rate b and the expected radiation response rate g. Such *a priori* knowledge must be obtained from past experience or initial measurements of small duration.

20.7.2 Discernment Between Two Outcomes

Determination of the appropriate number of counts in a measurement is straightforward for a measurement with little or no background and no overlapping energy peaks. However, for low-level counting and for identification of isotopes emitting gamma rays of almost equal energies, it is necessary to design the experiment to discern the correct result among multiple possibilities.

Suppose there is a measurement background with rate b recorded by your detector and you have a sample which yields a count rate g with a source. This result can be for any source that deposits sufficient energy to provide a count in the measurement. Hence, g is the sum of the background count rate b and the source count rate or net rate n. The problem is that the measurement of g includes both background and source detector events. Typically, it is sufficient to simply subtract the background rate from g to obtain n, but at times the rate of the measurements may be extremely low. Hence, some understanding of the minimum number of total counts required is necessary.

If the background measurement time is t_B and source measurement time is t_G, how many counts must be recorded to provide confidence in the measurement? For a Gaussian model, the average number of background counts is

$$\overline{B} = bt_B \pm \sqrt{bt_B}, \tag{20.146}$$

and the average number of recorded source counts is

$$\overline{G} = gt_G \pm \sqrt{gt_G}. \tag{20.147}$$

If gt_G is relatively large compared to bt_B, then it is easy to distinguish between the two measurements. However, if gt_G is similar to bt_B, it is difficult to determine the actual source activity. This problem is addressed in Sec. 6.10.1, in which the minimum detectable activity (MDA) and critical detection limit (CDL) were examined and practical results were determined. Here the discernment between two possible outcomes is considered.

Consider a situation in which identification of a collection of events is a function of a theoretical average. The tallied events may be a collection of counts or a collection of pulse heights. If the outcome of two such averages are similar, how does one identify the proper outcome? For instance, if two gamma-ray energies from different isotopes are similar, how does one identify the proper isotope? Detectors with high energy resolution reduce this uncertainty, a consequence of the Gaussian spread about the peak channel, but do not totally eliminate the uncertainty. A similar situation exists in radiation counting where two predictions from

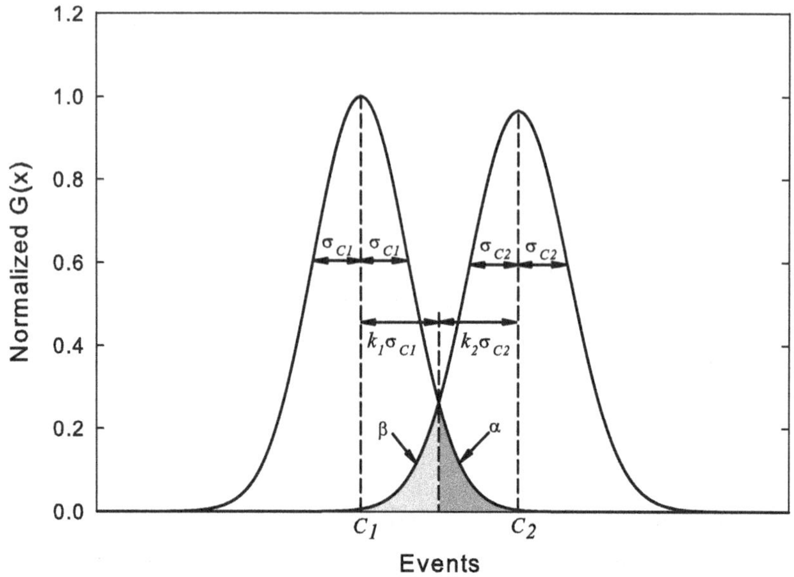

Figure 20.26. The situation in which two similar possible experimental outcomes exist, either $C_1 = c_1 t$ or $C_2 = c_2 t$.

different phenomena (e.g., counting rate or efficiency) may be similar. Hence, a measurement is conducted that results in a collection of counts within a time period, the outcome of which can be easily misinterpreted.

Assume a count rate c_1 for one outcome and a count rate of c_2 for a second but similar outcome. The total number of events is either $c_1 t$ or $c_2 t$ where t is the observation time. In either case, the expected outcome has experimental uncertainty such that the number of events observed for $c_i t$ is

$$C_i = r_i t \pm \sqrt{r_i t}, \qquad i = 1, 2. \tag{20.148}$$

What then is the probability of an incorrect conclusion? The situation is best understood with the help of Fig. 20.26. If the outcome is identified as the average count rate c_1, what is the probability that it is actually c_2, and if c_2 is chosen, what is the probability that it is actually c_1? The uncertainty in this situation can be decreased by first deciding on an acceptable error $k\sigma$, and then conducting a measurement to ensure that adequate counts are acquired. Note that the average count rate does not change for either situation, but the uncertainty for either situation decreases as more counts are accumulated.

The probability distributions for C_1 and C_2 are assumed to be Gaussians, and a portion of the two probability distributions overlap, represented by the shaded area in Fig. 20.26. If a condition is set that the outcome is identified as C_1 provided that the measured result is below $C_1 + k_1\sigma(C_1)$, where k_1 is the number of standard deviations from the mean of C_1, then the shaded area β represents the probability that the outcome is actually C_2. If instead the outcome is identified as C_2 if the measured result is above $C_1 + k_1\sigma(C_1)$, then the shaded area α represents the probability that the outcome is actually C_1. The difference between the averages of the two possible outcomes is represented by

$$\Delta C = C_2 - C_1 = [k_1\sigma(C_2) + k_2\sigma(C_1)], \tag{20.149}$$

where k_2 is the number of standard deviations from the mean of C_2. Because the two distributions are relatively close, it can be assumed that at the crossover point $k_1 \simeq k_2$ so that

$$\Delta C = C_2 - C_1 \simeq k(\sigma(C_1) + \sigma(C_2)) = k(\sqrt{C_1} + \sqrt{C_2}). \tag{20.150}$$

From this result

$$C_1 = C_2 - k(\sqrt{C_1} + \sqrt{C_2}). \tag{20.151}$$

The measurement time can now be adjusted to increase the value of k by increasing the C_i, which are related by

$$k = \frac{C_2 - C_1}{\sqrt{C_1} + \sqrt{C_2}}. \tag{20.152}$$

The theoretical outcomes C_1 and C_2 are calculated from the measurement time t. For example, if $k = 1.65$, then the shaded area α or β represents 5% of either Gaussian distribution and indicates the possibility of a 5% error for either outcome at the point $C_1 + k\sigma_{C1}$. To decrease this error to only 1%, then the calculated outcomes must change such that $k = 2.33$, which can be accomplished by increasing the measurement time t.

Example 20.6: An experimental neutron detector is under development that uses a semiconducting compound with boron as a constituent in the form of B_4C. These films are usually grown on well-established semiconductors such as Si. Because B_4C is difficult to produce, the layers are very thin. However, boron is a p-type dopant in Si, and instead of a thin B_4C semiconductor detector, the device may actually be a boron-coated pn-junction Si diode. For thin films, it is difficult to discern between the two possible outcomes. Find an expression to determine the measurement time required to fulfill the desired confidence for any k given a film thickness x and monodirectional thermal-neutron beam of intensity ϕ_n normally incident on the film.

Solution:

The counts obtained in measurement time t from a bulk device should be

$$C_2 = \phi_n At\,(1 - \exp[-x\Sigma_F])\,,$$

where ϕ_n is the neutron flux intersecting the detector, A is the area, Σ_F is the macroscopic cross section of the film material, and x is the thickness. The counts from a coated device can be described by the ratio of counts recorded from thin-film device to a that of a bulk device, namely

$$C_1 = \phi_n At\frac{\epsilon_{TF}}{\epsilon_{BK}}\,(1 - \exp[-x\Sigma_F])\,,$$

where ϵ_{TF} is the expected efficiency from the thin-film device and ϵ_{BK} is the efficiency of the bulk device. The difference between the two possible outcomes is

$$C_2 - C_1 = \phi_n At\left(1 - e^{-x\Sigma_F}\right)\left(1 - \frac{\epsilon_{TF}}{\epsilon_{BK}}\right) = k\,(\sigma_{TF} + \sigma_{BK})\,.$$

Square the expressions,

$$\left[\phi_n At\left(1 - e^{-x\Sigma_F}\right)\left(1 - \frac{\epsilon_{TF}}{\epsilon_{BK}}\right)\right]^2 = k^2\,(\sigma_{TF} + \sigma_{BK})^2\,.$$

From this result the measurement time is found to be

$$t = \frac{k^2}{\phi_n A}\frac{\left(\left[\frac{\epsilon_{TF}}{\epsilon_{BK}}\left(1 - e^{-x\Sigma_F}\right)\right]^{1/2} + \left[\left(1 - e^{-x\Sigma_F}\right)\right]^{1/2}\right)^2}{\left[\left(1 - e^{-x\Sigma_F}\right)\left(1 - \frac{\epsilon_{TF}}{\epsilon_{BK}}\right)\right]^2}.$$

20.7.3 Coincidence and Anti-Coincidence Measurements

Detectors can be connected to interact together so as to record or reject events from radiation interactions. For instance, two detectors may be connected in such manner that both must receive a radiation interaction and produce a pulse within a certain time period Δt in order to record a count. Interaction events that do not comply with this restriction are rejected. Such measurements are named *coincidence counting*. Further, with spectrometers, an energy window can be used for one or more detectors to reduce the recording of down-scattered or background gamma-ray coincidences. The opposite case can also be employed so that, if two or more detectors record radiation interactions within a time Δt, all detectors reject the events. This mode of operation is named *anti-coincidence counting*. The operation of such systems is summarized in Chapter 21.

Compton Suppression

Compton suppression is an anti-coincidence method used to enhance the full-energy peaks in a gamma-ray pulse height spectrum. In this method the gamma-ray spectrometer is surrounded by a second detector (or a ring of detectors). For example, a HPGe detector is often used as the spectrometer, with a scintillator detector (or detectors), formed into a ring around the HPGe detector, operated as a coincidence detector. Because the total coincidence energy is of less concern than the recording of a full-energy event in the HPGe detector, low-resolution scintillators with relatively high gamma-ray absorption efficiency are preferred (e.g., BGO and NaI:Tl). With this configuration an incident photon that Compton scatters in the HPGe detector and then escapes has a reasonable chance of subsequently interacting in the surrounding scintillation detector(s). Both interactions occur at almost the same time. The partial photon energy deposited by the Compton scattered photon in the HPGe detector can thus be suppressed by an anti-coincidence gate that accepts only pulses in the HPGe detector that are anti-coincident with pulses in the scintillator(s). Thus, events in the HPGe detector are recorded only if no simultaneous event occurs in the surrounding scintillator detector(s). The resulting HPGe spectrum has a much reduced Compton continuum at all energies, thereby enhancing the full-energy peak in the spectrum. Dramatic improvements in the pulse height spectrum can be achieved [Molnar 2004; Bender et al. [2015]. Example spectra recorded with the technique are shown for $LaBr_3$:Ce in Fig. 20.27 and HPGe in Fig. 20.28, both having an annular NaI:Tl guard detector for the coincidence gate [Bender et al. [2015].

The Compton suppression method can be used to improve significantly prompt gamma-ray analyses. Because prompt gamma rays tend to be of high energy (most are $1 < E < 12$ MeV) there normally is a large Compton component to the spectra, especially for relatively small semiconductor detectors. Suppression of the Compton continuum with the anti-coincidence method, thus, produces a pulse height spectrum in which full-energy peaks are more easily identified and quantified.

Positron Emission and Pair Production

Recall that when a positron comes nearly to rest, it combines with an electron to produce two 511-keV annihilation photons. These 511-keV photons are emitted in opposite directions in order to preserve the zero total linear momentum.[11] If two detectors are placed on opposite sides of a source emitting positrons isotropically (such as ^{22}Na), the two detectors can be connected such that a count is tallied only when both detectors record a count within a small time interval Δt. The probability of recording a count is determined by the geometry of the counting arrangement. Suppose that two cylindrical detectors with radii a_1 and a_2 are connected in coincidence and are placed opposite and at distances d_1 and d_2 from a positron source with the cylindrical axes pointed at the source. The probability of recording a count is proportional to the

[11] Very rarely does the positron annihilate with an ambient electron before it has completely slowed. In this case the annihilation gamma rays are not emitted in opposite directions.

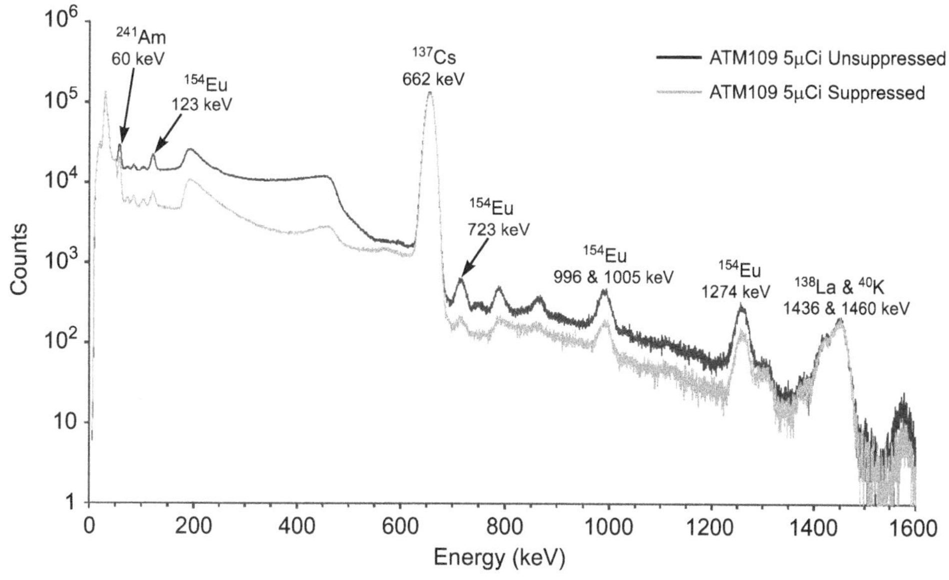

Figure 20.27. Unsuppressed and suppressed measured spectra from a 5 μCi ATM 109 spent fuel sample using a $LaBr_3$:Ce-based Compton suppression system. From Bender et al. [2015]; courtesy Kenan Ünlü, Penn State.

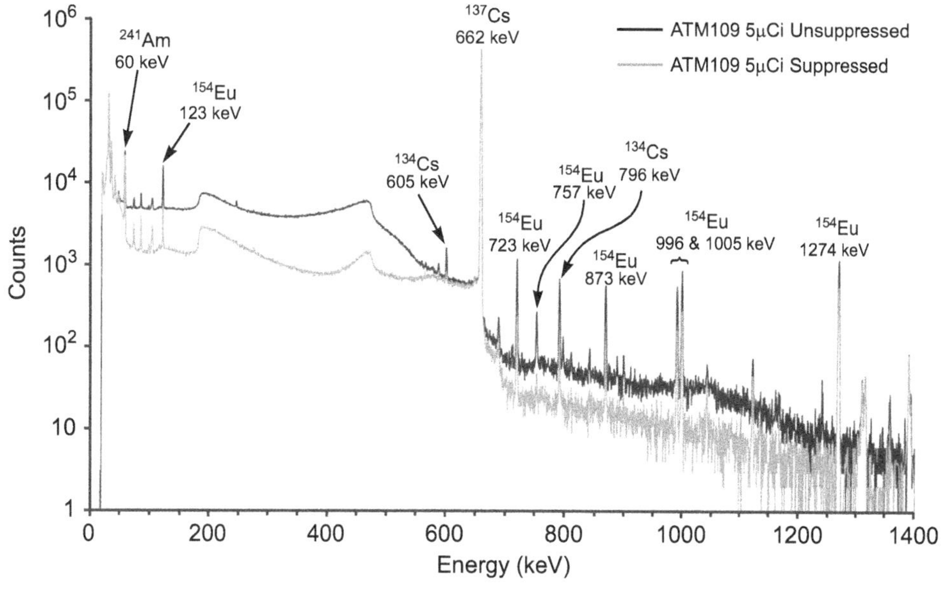

Figure 20.28. Unsuppressed and suppressed measured spectra from a 5 μCi ATM 109 spent fuel sample using an HPGe-based Compton suppression system. From Bender et al. [2015]; courtesy Kenan Ünlü, Penn State.

smallest fractional solid angle[12] Ω_f for either detector, namely

$$P_{\text{record}} \simeq \epsilon_1\epsilon_2 \min(\Omega_{f1}, \Omega_{f2}) = \epsilon_1\epsilon_2 \frac{1}{2} \min\left(\left[1 - \frac{d_1}{\sqrt{d_1^2 + a_1^2}}\right], \left[1 - \frac{d_2}{\sqrt{d_2^2 + a_2^2}}\right]\right), \tag{20.153}$$

where ϵ_i $i = 1, 2$ is the detector gamma-ray interaction efficiency at 511 keV for detector i. Here it is assumed that if one photon enters the detector with the smaller solid angle, then the other photon traveling in the exact opposite direction must enter the opposite detector. However, any intervening material reduces this ideal recording probability. The recorded pulse height spectrum consists almost entirely of a spectrum produced by the 511-keV photons. The method almost entirely eliminates background, although the possibility of accidental coincidence from background and other source emissions remains. Also, for example, with a ^{22}Na source, the 511-keV annihilation spectrum would appear, but the 1.28-MeV gamma-ray spectrum would be almost entirely absent, with perhaps a few counts from accidental coincidences.

Another radiation source that spontaneously emits two photons per decay is ^{60}Co; however, these gamma-rays are randomly emitted in all directions. Using the same arrangement for coincidence counting with identical detectors, the probability of a coincidence is now

$$P_{\text{record}} \simeq \epsilon_1\Omega_1\epsilon_2\Omega_2, \tag{20.154}$$

mainly because the two photons are not necessarily emitted in opposite directions and the solid angles of both detectors must be taken into account. Hence, for radiation sources that emit more than one gamma ray (e.g. ^{57}Co, ^{133}Ba), coincidence counting requires accounting for the solid angle of the two (or more) radiation detectors as well as the branching ratios of the emissions. Shown in Fig. 20.10 are several sum peaks from a ^{60}Co source recorded with an HPGe spectrometer.

Positron Emission Tomography

An example of coincidence counting, which is widely used for medical imaging applications, is positron annihilation tomography (PET). The system generally has a ring of individual detectors connected in coincidence as depicted in Fig. 20.29. The patient is administered a positron emitting radiopharmaceutical, usually one of Na^{18}F, ^{82}RbCl, $^{13}NH_3$, or ^{18}FDG (fluorodeoxyglucose), which accumulates in the organ or tissue under investigation. For instance, Na^{18}F accumulates in bone tissue while ^{18}FDG accumulates rapidly in brain tumors. The inoculated patient is passed through the detector ring, incrementally, in lengths more or less equivalent to the detector ring width. Annihilation photons emitted from the patient may interact in two detectors arranged in the ring about the patient, and through coincidence circuitry and ray tracing algorithms, the two detectors responding within Δt to 511-keV annihilation photons can be used to identify a straight line through the patient between the two detectors. An energy window can be set about 511 keV to help reject scattered photons and accidental coincidences. Although BGO has been used in PET scanning systems, mainly because of its high gamma-ray absorption from Bi, the superior energy resolution of LSO, also having high gamma-ray absorption from Lu, is now used as a replacement in most modern PET scanning systems. A backprojection algorithm is used to produce a three-dimensional image of the patient, particularly the location where the positron emitters reside. An example of a PET scan image is shown in Fig. 20.30.

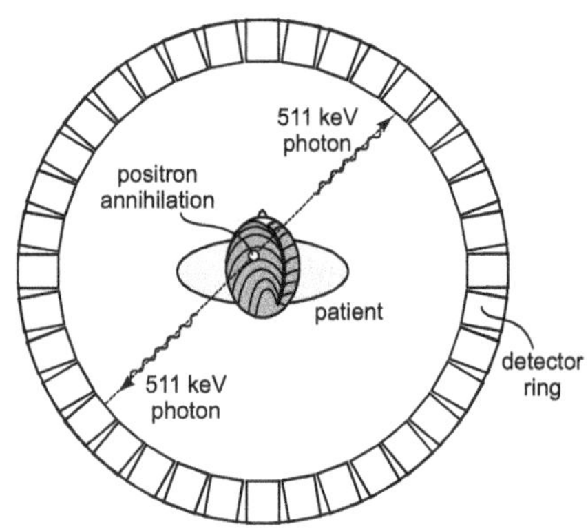

Figure 20.29. The basic layout for positron emission tomography.

[12]The *fractional* solid angle is the solid angle subtended by a detector at the point source divided by 4π sr.

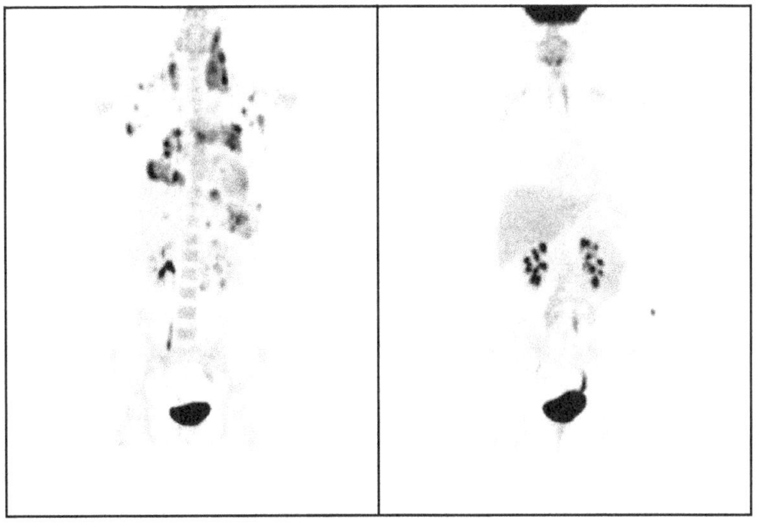

Figure 20.30. PET images of a lymphoma patient. Left panel: prior to chemotherapy. Right panel: two months post-chemotherapy. Illustration courtesy of Dr. Frederic Fahey, Children's Hospital, Boston, MA.

Phoswich Detector

The phoswich detector is a special scintillation spectrometer designed to measure low radiation levels in the presence of high-energy ambient background radiation. The name comes from "phosphor sandwich", where two scintillation detectors are layered one atop the other [Wilkinson 1952]. The crystals are arranged as shown in Fig. 20.31, where light from both crystals is measured by a single photomultiplier tube. The signals from the different crystals are discerned by measuring the different light decay times.

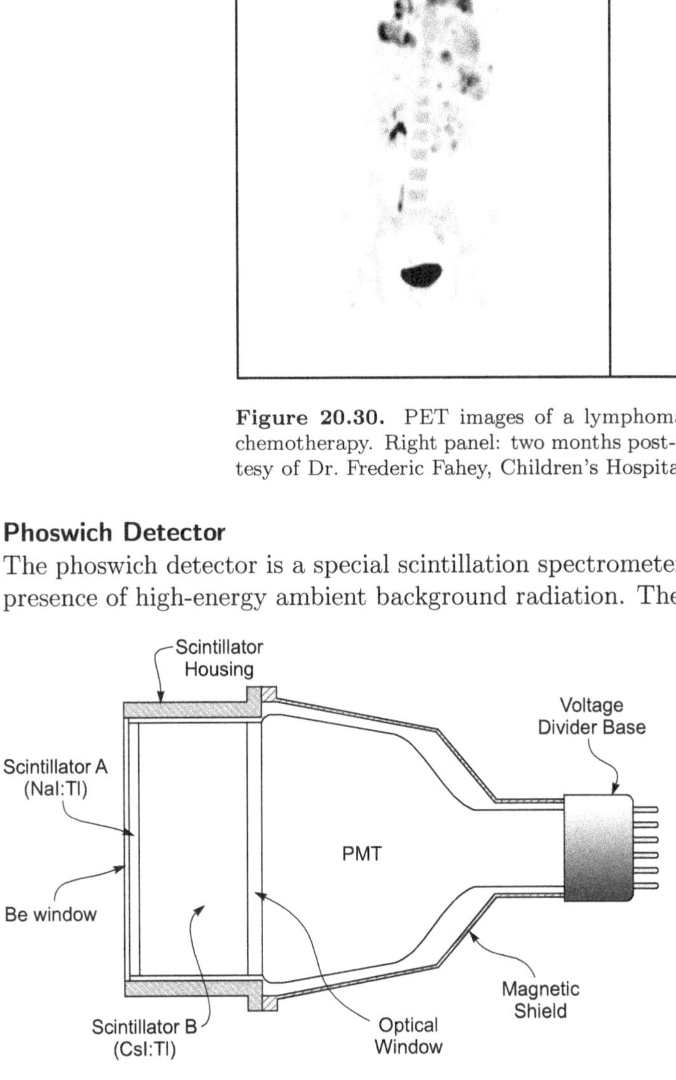

Figure 20.31. The basic components of a phoswich detector. In this example, the front detector is a thin crystal of NaI:Tl, while the back detector is a thicker slab of CsI:Tl. After Lindow et al. [1978].

The phoswich detector has a thin Be window that allows the passage of low-energy photons. These low-energy photons can be absorbed in the thin front detector, labeled A, which is designed to have sufficient thickness to absorb the radiation energies of interest, commonly on the order of 1.5 to 3 mm thick. Higher energy photons generally pass through the front detector and can be absorbed in the much thicker back detector, labeled B, commonly on the order of 50 mm thick. Pulse shape discrimination is used to determine the pulses from different scintillators, which includes photons that Compton scatter from scintillator B into scintillator A. Hence, pulses recognized as coming from scintillator A can be accepted while those from scintillator B are rejected, thereby producing a spectrum primarily from only low energy photons. Coincident events from both scintillators can be rejected outright. Operated in anti-coincidence mode, background events are substantially reduced. Consequently, the resulting spectrum has substantially reduced background so that low energy photon events can be seen. One interesting feature of this design is that the signals are developed from a single PMT output. Coincident events are determined from pulses of different rise times arriving within a preset time window. Detectors for low-energy gamma rays commonly have combinations of either a thin NaI:Tl crystal upon a thicker CsI:Tl crystal, or a thin CaF_2:Eu crystal upon a thicker NaI:Tl crystal. Other combinations have been explored, which for example include the use of BGO, GSO, and BaF_2 in various combinations [Costa et al. 1986; Kamae et al. 1993; Futami et al. 1993]. The use of YSO with LSO and LuYAP with LSO has also been explored for positron emission tomography applications [Dahlbom et al. 1997; Eriksson et al. 2013].

Although phoswich detectors are often designed for low-energy gamma-ray and x-ray detection, they have also found use with charged particle detection [de Celis et al. 2007]. For instance, the combining of ZnS:Ag as a thin front detector with an organic (plastic) scintillator as a back detector, alpha particle events can be separated from beta particle events [Usuda 1992]. Recall from Chapter 13 that ZnS:Ag is used for heavy particle detection, but is inefficient as a gamma-ray or electron detector. Hence, alpha particles can lose all energy in the thin ZnS:Ag front detector (usually only tens of microns thick), while the back detector can absorb the full electron energy. The small amount of energy deposited by electrons in the front detector is negligible. Hence, the particle counts can be separated by either pulse shape or pulse height discrimination.

Note that the same method can be used to discriminate gamma rays and charged particles from neutrons, where neutrons slip through the first detector (low cross section) and scatter in the second (organic scintillator) [Watanabe et al. 2008]. If the first detector has adequate gamma-ray and charged-particle interaction efficiency, coincident events can be rejected while recognizing neutron scatters in the second detector.

Special designs with fast (front) and slow (back) plastic scintillators are used for particle energy spectroscopy and identification [Wilkinson 1952; Bodansky and Eccles 1957; Saint-Gobain 2019a]. Particles passing through the first detector (fast response) yield information about the stopping power, or dE/dx, while summing the total integrated light from both detectors yields the particle energy.

Anger Camera

The Anger camera, or gamma-ray camera, was introduced by Hal Anger [Anger 1958]. A cross section of the basic components of the device is depicted in Fig. 20.32. The device can be used with single photon emission computed tomography (SPECT) or PET. The system consists of an array of photomultiplier tubes populated upon a thin sheet of NaI:Tl scintillator. The scintillator is usually on the order of 6 mm to 12 mm thick. The patient is administered a radiopharmaceutical, commonly containing ^{99m}Tc which emits 140-keV gamma rays, and placed adjacent to the Anger camera. Other radiopharmaceuticals used for SPECT include ^{123}I (159 keV) and ^{111}In (mainly 23 keV, 171.3 keV, and 245.4 keV, although there are other emissions). Gamma-ray photons emerging from the patient can be absorbed in the relatively thin NaI:Tl scintillator, spontaneously producing scintillation photons emitted isotropically. To reduce the measurement of scattered gamma rays, which can blur the image and produce an erroneous correlation to the emission point, a collimator is aligned in front of the NaI:Tl scintillator. The collimator also im-

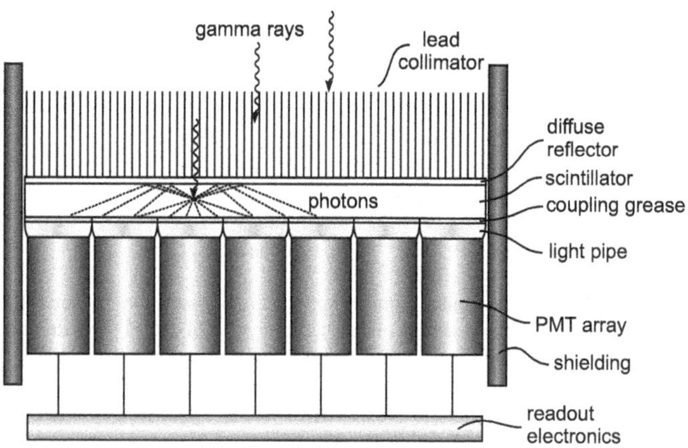

Figure 20.32. The basic components of a gamma-ray camera (Anger camera).

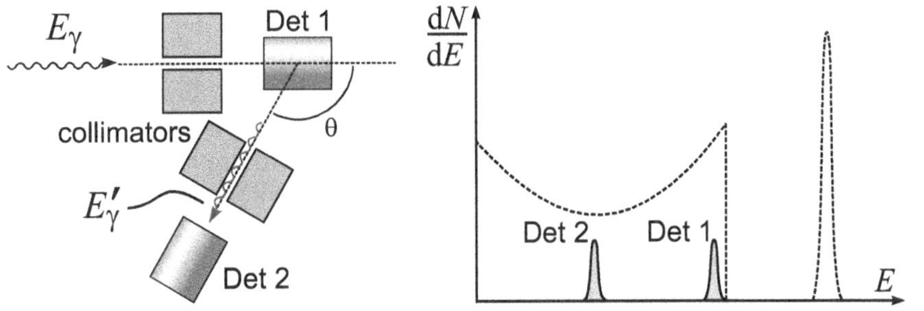

Figure 20.33. Geometry for a Compton spectrometer, in which detector 1 and detector 2 are operated in coincidence. The dotted line spectrum represents the energy spectrum absorbed in detector 1, while the shaded spectra represent the coincident energy spectra *recorded* by either detector 1 or detector 2 (as indicated).

proves the spatial resolution, but consequently also greatly reduces the detection efficiency. Multiple PMTs are usually involved with measuring the light and the closest PMT to the gamma-ray event produces the largest voltage signal. An algorithm weights the different PMT signals in the x and y directions and ensures the signals arrive within some Δt coincidence time interval to be included in the measurement. This algorithm was originally a clever resistor circuit capable of identifying the interaction location within 5 mm. Advanced electronics and algorithms in modern Anger cameras can reproduce the interaction location with spatial FWHM of 3 mm [Cherry et al. 2003).

Compton Spectrometer

A Compton spectrometer is a coincidence mode device used to extract information from the combination of two or more detectors. The basic geometry of the method is depicted in Fig. 20.33. Two radiation detectors are connected in coincidence, with one acting as the main detector (detector 1, for example) and the second detector acting as the coincidence witness. A collimator is used for detector 1 and also (preferably) detector 2, to narrow the directions of both the initial and scattered gamma rays. Sometimes a narrower witness detector is used instead of the collimator for detector 2. The geometric arrangement is important, because detector 2 is set at a specific angle θ with respect to the collimated gamma-ray trajectory.

The Compton spectrometer is a method, albeit inefficient, originally intended to produce a gamma-ray spectrum from counters instead of spectrometers. Obviously, the energy resolution depends on the uncertainty in the scattering angle. Hence, better collimation of both the initial gamma-ray and the Compton scattered gamma-ray decreases this uncertainty, but it also reduces the efficiency and, consequently, requires an increased measurement time. Gamma rays absorbed in the first detector by the photoelectric effect do not produce a coincidence event with the witness detector and are excluded. The method can be used to map out the Compton continuum, the Compton edge energy yielding the initial gamma-ray energy. Yet, there is still the possibility that multiple Compton scatters occur in detector 1, with the final scattered photon emerging and still producing a coincidence in detector 2, giving a false indication of the original gamma-ray energy. This problem can be reduced by using smaller detectors or replacing the counters with detectors capable of spectroscopy.

Compton Spectrometer for Light Yield Measurements In Figs. 13.4 and 13.5, it is seen that many scintillators have non-linear light yields for electron energies below about 500 keV, i.e., the emitted light is not proportional to the electron energy deposited in the scintillator. One method to measure the light yield as a function of energy deposition is to measure directly the light yield with multiple gamma-ray sources, preferable monoenergetic sources. However, this method relies, in practice, on only a few sources and, thus, leaves wide energy gaps in the measurements. A type of Compton spectrometer can be used to measure the light yield of scintillators over a much larger range of deposition energies than possible with common monoenergetic gamma-ray sources [Valentine and Rooney 1994]. This device is a regular Compton spectrometer, shown in Fig. 20.33, with the change that the first detector is now made from the scintillation material being studied and the second detector is a gamma-ray spectrometer. A known monoenergetic gamma-ray source, ^{137}Cs for instance, is used to irradiate the test scintillator (detector 1). Scattered gamma-ray events in detector 2 are tallied in coincidence with events in detector 1. The energy of the scattered gamma rays is known from the collimated scattering angle. At the same time the light yield from the test detector is measured with a photomultiplier tube. With this technique the light yield for many different deposition energies, determined by the scattered photon energies $(E_\gamma - E'_\gamma)$, can be measured using the same monoenergetic source by simply changing the coincidence scattering angle.

Also designed for light yield measurements, a relatively complicated Compton coincidence spectrometer system was reported, named the SLYNCI for Scintillator Light Yield Non-proportionality Characterization Instrument, with five HPGe detectors connected in coincidence with the test scintillator [Choong et al. 2008]. This device still requires source collimation between the source and the test scintillator, although no collimators are used between the test scintillator and the multiple HPGe coincidence detectors. The SLYNCI system was designed to reduce the measurement time required to characterize the light yield from a test scintillator.

The design of the Compton coincidence light yield system was significantly simplified by Ugorowski et al. [2008] by placing a small gamma-ray check source adjacent to the test scintillator and using a high-resolution HPGe detector for the coincidence witness detector. Because there are no collimators in this design, the HPGe detector is placed near the test scintillator, thereby increasing the solid angle of the witness detector. These three simple design changes remove the need for any collimators, decrease the uncertainty in the measurement, and reduce the measurement time. This simpler arrangement also negates the need to know the Compton scattering angle because the HPGe detector provides the residual energy, from which the energy deposited in the test scintillator can be inferred.

Compton Camera System The Compton imaging system also depends on the properties of photon scattering to locate gamma-ray sources. The basic device has a position sensitive front detector following by another position sensitive back detector [Schönfelder et al. 1973]. The device is designed so that some gamma rays scattered in the front detector are subsequently absorbed in the back detector. The two detectors are operated in a coincidence mode to ensure that scattering and absorption events are correlated. Because this system does not require a collimator, the overall gamma-ray detection efficiency is improved compared to gamma-ray cameras that require collimators to exclude background. In this device the location of where an incident gamma ray interacts in the first detector and the location of where the scattered photon is absorbed in the second detector are measured, within the spatial resolution limits of the front and back detectors, and a ray can be traced between the two interaction locations.

If the incident gamma-ray energy is known, the energy deposited in the first detector and the interaction locations in both detectors can then be used to extract information about the location of the gamma-ray source. From the Compton scattering condition of Eq. (4.32), the energy deposited in the first detector is

$$E_e = E_\gamma - E'_\gamma = E_\gamma \left(\frac{E_\gamma(1-\cos\theta)}{511 \text{ keV} + E_\gamma(1-\cos\theta)} \right). \tag{20.155}$$

The direction of the scattered gamma ray is traced between the two interaction positions in the two detectors. Hence, E_γ and E_e are both known, leaving θ as the only unknown. This angle θ can be reconstructed as a cone with the surface forming all possible directions as depicted in Fig. 20.34(a) [Everett et al. 1977]. A second gamma-ray scatter produces a new angle, depicted as ϕ in Fig. 20.34(b), and a new cone is reconstructed, overlapping the first cone in two locations. After several additional cones from more gamma

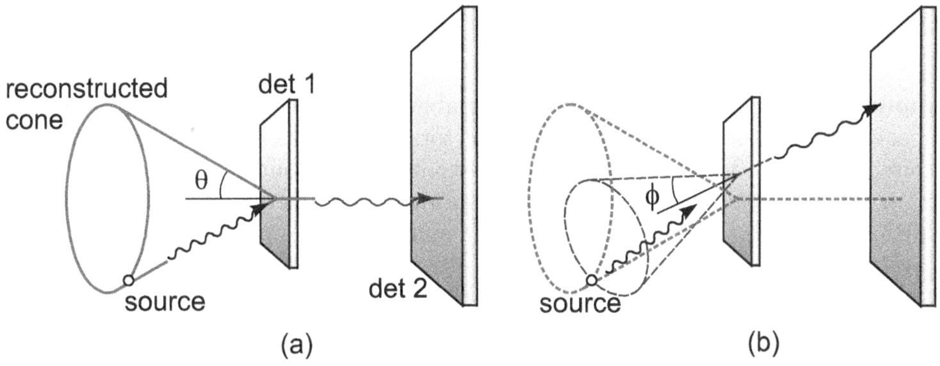

Figure 20.34. The Compton camera concept; detector 1 is the position sensitive scatter spectrometer; detector 2 is the position sensitive absorber detector. (a) A reconstructed cone with angle θ from a Compton scatter. (b) Another reconstructed cone with angle ϕ. The cone perimeters overlap at the source location.

rays are reconstructed, a specific location where all of the cones overlap becomes apparent and so the source location is identified. The method becomes more complicated if multiple sources are in the field of view. However, the reconstructed cones can still image the location of these sources because the images increase at the multiple crossover points.

If the energies of the gamma rays emitted by the source are not known *a priori*, then, with E_γ unknown, the scattering angle θ cannot be calculated from Eq. (20.155). However, this problem can be overcome by using a position-sensitive high-resolution gamma-ray spectrometer as the second detector. The sum of energies deposited in the first and second detectors can then be used to determine E_γ, preferably with the scattered gamma ray completely absorbed in the second detector by the photoelectric effect. The weakness of this approach is that the Compton scattered gamma ray may not be completely absorbed in the second detector but instead be scattered out of it. This incomplete energy absorption leads to an erroneous estimation of E_γ. The probability of fully absorbing the scattered gamma ray is improved if the second detector is made from high Z materials. Also, the device thickness can be increased with the consequence that multiple scatters may occur in the second detector, thereby confusing the interaction location. Another remedy is to use multiple stacked position sensitive spectrometers as scatter planes to back trace the gamma-ray path, ultimately producing a reconstructed cone that intercepts the source location(s) [Dogan et al. 1990]. Also, an anti-coincidence detector surrounding or near the Compton camera can eliminate a significant number of events when the scattered gamma rays escape the system [Dogan et al. 1990].

Compton Spectrometer and Imager The Compton spectrometer and imager (COSI) is a Compton camera instrument designed for the study of astronomical gamma-ray bursts [Lowell et al. 2016, 2017]. It has a 2 × 2 × 3 position-sensitive array of double-sided HPGe strip detectors [Amman et al. 2007] operated in coincidence as a Compton camera with an additional CsI:Tl scintillator shield surrounding the array that operates in anti-coincidence to reduce background [Kiernans et al. 2016]. The CsI:Tl shield rejects escaping Compton scattered gamma rays that interact in the shield so they are not used in the estimation of the incident gamma-ray energy. The shield also passively and actively blocks albedo radiation and vetoes high-energy charged particles. The position resolution of the HPGe strip detectors is reported to be 2 mm in the x and y directions and approximately 0.2 mm in the z direction. Designed to detect gamma rays between 200 keV to 5 MeV, the FWHM energy resolution of the COSI is 2.7 keV at 661.7 keV. The detector is kept cool with a mechanical Stirling cycle cryocooler. The device is deployed on a high pressure balloon, complete with solar panels to maintain power. When deployed, the COSI field of view is approximately 25% of the sky. In at least one launch, the COSI stayed aloft and collected gamma-ray data for 46 days [Kiernans et al. 2016].

Portable Compton Scatter Cameras There are portable variants of the Compton camera used mainly as gamma-ray imagers and source identifiers. These detectors are used to rapidly locate gamma-ray sources in a region and often need only a few dozen interactions to locate the sources. A relatively complex Compton scatter camera is the position sensing CdZnTe gamma-ray spectrometer, named the Polaris spectrometer [Wahl et al. 2015; H3D 2018]. This device has various numbers of position sensitive CdZnTe spectrometers arranged in layers of arrays. Each CdZnTe detector can sense the interaction position within the detector volume using a depth sensing technique based on a pixelated electrode configuration.

A gamma-ray camera with dual capability as a Compton scatter camera and a pin-hole camera is reported by Hull et al. [PHDS 2018]. The detector, named GeGI for germanium gamma-ray imager, has a single slab of pixelated HPGe attached to a Stirling cycle mechanical cryocooler. The 9-cm diameter and 1-cm thick spectrometer is a position sensitive strip detector, with strip pitches of 5 mm. Instead of several detectors operating as scatter and absorption spectrometers, the GeGI operates by position tracking the signals produced within separate pixels formed by the x and y positions of the strips. The z direction is inferred from the charge collection signal by using the three dimensional sensing system developed by He and Zhang [2008].

A scintillation Compton camera alternative is a device built with GAGG:Ce scintillators [Kishimoto et al. 2014; Kataoka et al. 2018]. The GAGG:Ce Compton camera is based on the original concept proposed by Schönfelder et al. [1973], in which two arrays of position sensitive scintillators are aligned as the scatter and absorption spectrometers. The three-dimensional sensing device is constructed with individual GAGG:Ce scintillators, with nominal dimensions of 2 mm $\times$ 2 mm $\times$ 4 mm, arranged in 11 $\times$ 11 pixel arrays. The scintillators are separated by reflectors and have multi-pixel photon counters (MPPC, alternatively named SiPMs) attached to both sides. Both the scattering plane and the absorbing plane have four scintillator array blocks each separated by 12 mm. The x and y interaction positions are determined by the location of the affected pixels while the z component is determined by pulse height ratios measured with the dual MPPCs. Although the energy resolution ($\sim$ 8.6% FWHM at 661.7 keV) is not as good as the CdZnTe and HPGe counterparts, the device is relatively lightweight ($\sim$ 1.9 kg) and provides an angular resolution of about 8° to 9° FWHM.

20.8 Gamma-Ray Spectroscopy—Summary

Gamma-ray spectroscopy seeks to determine, first, the gamma-ray energies emitted by radionuclides (which emit gamma rays of distinct energies) that are present within a sample. This goal can be called *qualitative analysis.* However, generally, one also seeks either the source strength of the particular gamma rays or the concentration of the radionuclide emitting the gamma rays. This more demanding task often is referred to as *quantitative analysis.*

The objectives of gamma-ray spectroscopy are realized by analyzing the pulse height spectra. In this chapter, pulse-height spectra in which the number of counts are specified by discrete channel number are considered. This approach is used because channel number is directly proportional to pulse height and, thus, to the photon energy deposition. Consequently, quantitative information can then be obtained from the pulse height spectrum whether or not the spectrometer is linear. Spectroscopic analysis techniques begin by attempting to determine the continuous channel number that corresponds to the centroids of the full energy peaks in the spectrum that are of interest. The MCLLS approach focuses on the entire spectrum and seeks to determine the parameters of models that best fit all or major portions of the spectra. The symbolic Monte Carlo approach holds promise for spectroscopic applications in which the model is non-linear in terms of the radionuclide concentrations of samples.

Depending upon the need, there are several devices that can be used for gamma-ray spectroscopy. Efficiency with adequate energy resolution can be provided with large volume scintillators, whereas high energy resolution can be achieved with semiconductor detectors. Both scintillator and semiconductor detectors can be acquired as portable units with good detection efficiency for gamma rays. For ultra-high energy resolution, microcalorimeters or WDS diffractometers offer excellent performance (as described in Chapter 19). Yet, microcalorimeters and WDS spectrometers are generally restricted to laboratory-based instrumentation for low energy gamma rays and x rays. Note that the spectrometers discussed in the present chapter represent only a select sample of variations that are commercially available. More information can be found in earlier chapters dedicated to semiconductor and scintillation detectors.

20.9 Charged-Particle Spectroscopy

With charged-particle spectroscopy it is possible to identify some radioisotopes, provided that certain experimental precautions are exercised. As charged particles travel in a medium, they lose energy, mainly through Coulombic interactions. The amount of kinetic energy lost depends on the material stopping power, the particle energy, and the particle mass and charge. Regardless, energy is lost, even if the particle is emerging from its own source material. If some small and variable energy is lost before the particle enters the active region of the detector where it is subsequently stopped, the measured deposited energy is not the initial energy of the particle. Consequently, the uncertainty in the energy of a particle is increased and the energy resolution of the spectrometer decreases. To reduce energy losses while improving energy resolution, charged-particle measurements of relatively small samples are often conducted in a vacuum. Sample preparation is also important to ensure the mass thickness of a laboratory sample is kept relatively low so as to reduce energy self-absorption losses. It is also important to reduce the mass thickness of any material between the source and detector.

20.9.1 Electrons, Positrons, and Beta Particles

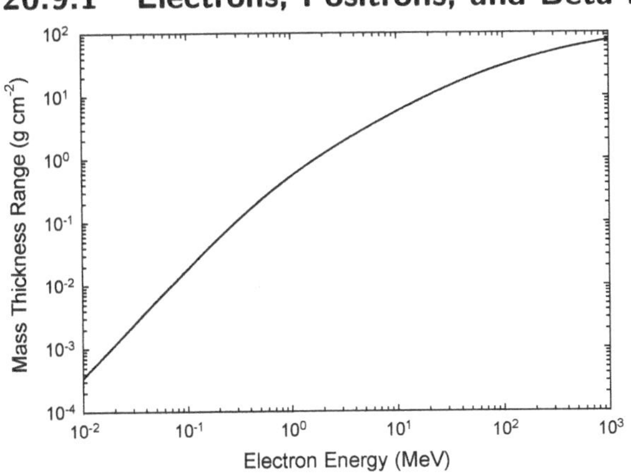

Figure 20.35. Average electron range in Si as a function of energy. Data are from ESTAR [NIST 2018].

Electron, conversion electron, positron, and beta particle spectroscopy can all be conducted with organic and liquid scintillation counters; however, the preferred method to achieve high-energy resolution is to use low Z semiconductor detectors, mainly Si detectors. The reason for low Z materials becomes evident from Fig. 7.7 in which it is seen that the backscattering of electrons and beta particles from a material increases with atomic number. The electron backscatter coefficient ranges between 0.14 at 50 keV down to 0.06 at 2 MeV, reducing to 0.0125 at 10 MeV. Any variety of planar Si detector can detect electrons, yet because the range of electrons can be significant (see Fig. 4.23 and Fig. 20.35), the type of detector is important. For instance, the mass thickness range for 100 keV and 1 MeV electrons is 0.018 g cm^{-2} and 0.54 g cm^{-2}, corresponding to ranges in Si of 77 μm and 2.36 mm, respectively. In the former case, a silicon surface barrier or implanted junction detector can be depleted, with reasonable voltage, to absorb all energy. However, the latter case would require excessive voltage to achieve a 2.31-mm-thick depletion region, and it would be better to instead use a Si(Li) detector. Note that the trajectory of electrons in a solid undergoes significant scattering as shown in Fig. 4.22. Hence, it is still possible to absorb the total electron energy in the detector even if the active thickness is somewhat less than the average linear range.

To preserve the electron energy distribution of the source, the measurement should be conducted in a vacuum. Such a measurement is straightforward for SSB and *pn* junction detectors, but Si(Li) detectors are often cooled during operation and, hence, require encapsulation to prevent contamination. Consequently, the use of a Si(Li) detector for low-energy electron spectroscopy requires a thin, low Z, window (such as Be) and the ability to produce a vacuum between the detector window and the source. Ahmad and Wagner [1974] describe a Si(Li) detector operated in vacuum with apparently no window between the source and detector face, although there was a 2-micron-thick contact layer.[13]

The pulse height energy deposition of monoenergetic electrons in Si was studied by Berger et al. [1969a] who compared Monte Carlo calculations to measured energy spectra. A few of their comparisons are shown in Fig. 20.36. Three detector types were used (of six different thicknesses) for the measurements, including silicon surface barrier (SSB) detectors, a diffused junction detector, and Si(Li) detectors. In Fig. 20.36 (left column), different detectors were used to measure 250-keV electrons with an average range of 343 microns in Si. Although the 105-micron-thick detector was less than the electron range, some scattered electrons did deposit their full energy in the detector. The full energy peak also appears in the two thicker detectors. In Fig. 20.36 (right column), the detectors were used to measure 1-MeV electrons (range of 2.36 mm). These electrons pass through the SSB detector, depositing only a fraction of their energy. The thicker diffused junction detector does show a full energy peak, but of low efficiently. The thicker 3 mm Si(Li) detector showed a significant full energy peak. Detailed response functions and range tables for electrons in Si detectors can be found in the literature [Berger et al. 1969b].

A low-energy tail, a consequence of partial energy deposition in the detector, usually appears, regardless of detector thickness. For relatively thin detectors, electrons may pass directly through the device and leave only a fraction of energy, as shown in Figs. 20.36(a, left and right). Backscatter losses can also cause partial energy deposition, as shown in Figs. 20.36(b,c left and right). Self-absorption energy loss as electrons or beta particles emerge from a sample, or energy attenuation between the sample and detector active volume, are other sources of energy loss that can contribute to the low-energy tail. The angle that electrons enter the detector can also cause a difference in backscattering, increasing as the angle increases from that of normal incidence. Also, the escape of characteristic and bremsstrahlung x rays may contribute to the energy loss tail. For positron detection, there is a small chance that 511-keV annihilation photons emitted from the eventual recombination of an electron and positron may simultaneously deposit some (or all) energy in the detector, thereby producing a high-energy tail above the full-energy positron peak. Measured response functions for monoenergetic electrons and positrons were reported by Frommhold et al. [1991] and compared to Monte Carlo simulations. The basic components of a monoenergetic electron or position spectrum are identified in Fig. 20.37(left). Unlike an electron spectrum, a positron spectrum has a high-energy tail above the full energy positron peak as a result of the Compton scatter of 511-keV annihilation photons.

Additional distortions of the true electron energy distribution arise from detector and electronic effects. If $T(E)$ is the number of electrons, per unit energy, incident on the detector, the measured energy deposition $M(E')$, per unit energy, in the detector by these electrons is

$$M(E') = \int_0^\infty \mathcal{R}(E, E')T(E)\,dE, \tag{20.156}$$

[13]The authors refer to the Au contact as a window, but this is traditionally improper nomenclature. Such contacts are usually referred to as "dead layers."

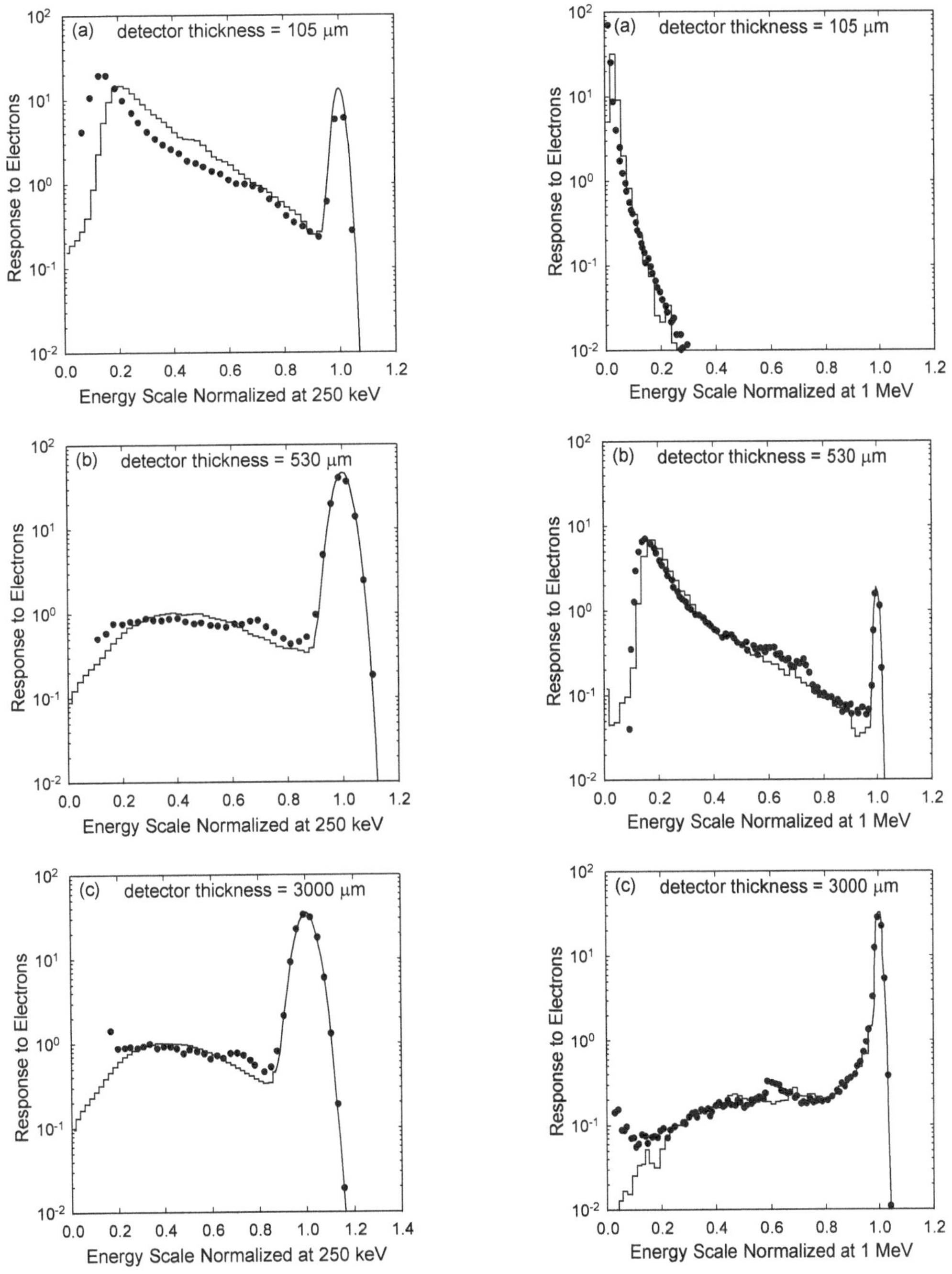

Figure 20.36. Comparison of Monte Carlo calculated and measured pulse height spectra for (left column) 250-keV electrons and (right column) 1-MeV electrons for different Si detector thicknesses: (a) 105-μm SSB detector; (b) 530-μm diffused-junction detector; (c) 3-mm Si(Li) detector. The points are experimental and the lines are calculated results. Data are from Berger et al. [1969a].

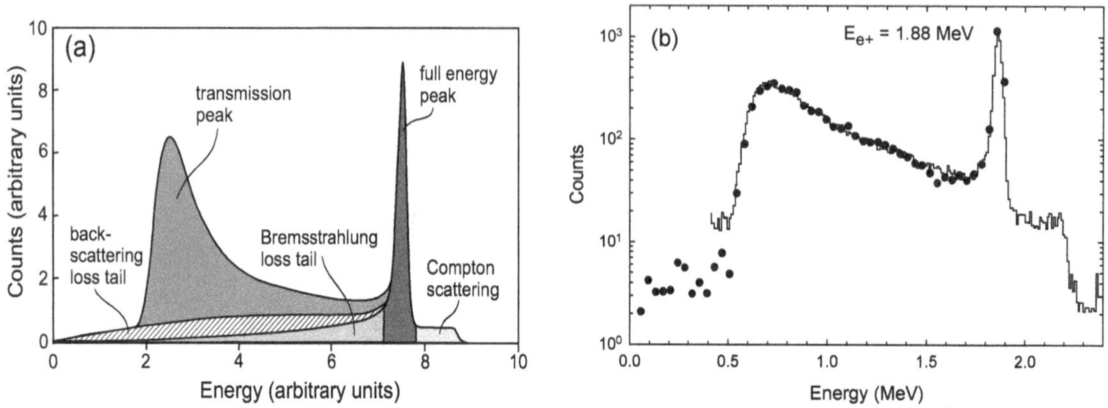

Figure 20.37. (a) Response function components of a monoenergetic electron or positron spectrum (after Frommhold et al. [1991]). The Compton scattering distribution from 511-keV annihilation photons is absent from an electron spectrum. (b) A measured positron spectrum compared to a Monte Carlo calculation. Data are from Frommhold et al. [1991].

where $\mathcal{R}(E, E')$ is the detector response function normalized such that

$$\int_0^\infty \mathcal{R}(E, E')dE' = 1. \tag{20.157}$$

Tsoulfanidis et al. [1969] developed the following response function to describe the energy deposition E' by monoenergetic electrons of energy E in a plastic scintillator:

$$\mathcal{R}(E, E') = \frac{b}{2E'}\text{erfc}\left[\frac{E - E'}{\sigma\sqrt{2}}\right] + \frac{1-b}{\sigma\sqrt{2\pi}}\left[\exp\frac{-(E-E')^2}{2\sigma^2}\right] \tag{20.158}$$

where the complimentary error function is

$$\text{erfc}(z) = 1 - \frac{2}{\sqrt{\pi}}\int_0^z \exp(-t^2)dt, \tag{20.159}$$

σ is the standard deviation of the modeled Gaussian function and b is the fraction of electrons appearing in the tail region,

$$b^{-1} = 1 + \sqrt{2\pi}\left(\frac{g}{t} - 0.5\right)\frac{\sigma}{E'}, \tag{20.160}$$

where g is the height of the full energy peak and t is the height of the backscattering tail region. The energy response function for beta particles with their continuous energy distribution is certainly more complex, but can be addressed by dividing the beta source spectrum into energy groups. The matrix of response functions can then be used to unfold energy deposition spectral data. For instance, beta particle spectra have been produced from pulse height energy deposition spectra acquired with scintillators and Si(Li) detectors with the use of unfolding methods [Tsoulfanidis et al. 1969; Dakubu and Gilroy 1978].

20.9.2 Alpha Particles

The best detectors for alpha particle spectroscopy are low-Z semiconductor detectors such as SSB and implanted junction Si detectors. The range of most alpha particle emissions from calibration check sources is less than 60 microns in Si, which is a depletion region thickness that can be produced in SSB and implanted junction detectors with a reasonable voltage. With very pure Si, the width of the depletion region may be sufficiently wide from just the junction potential (V_{bi}). However, it is recommended that reverse voltage be applied to improve charge collection and reduce detector capacitance.

Because each alpha particle emission is monoenergetic, although some sources have multiple alpha particle emission energies, the general expectation is the formation of a Gaussian full-energy peak for each alpha particle energy. However, the full-energy deposition peak usually is not perfectly Gaussian for several reasons. First, the variance is actually less than that predicted with Poisson or Gaussian statistics because the FWHM with units of energy must be corrected with a Fano factor F as

$$\text{FHWM} = 2\sqrt{2\ln(2)E_\alpha wF} = 2.355\sqrt{E_\alpha wF}, \tag{20.161}$$

where w is the average ionization energy and E_α is the alpha particle energy. The Fano factor for Si is generally quoted to be about 0.115 (see Table 16.2).

Second, a low-energy tail inevitably appears in the energy-deposition spectrum because some of the alpha particle's initial energy is lost as a result of self-absorption as the particle emerges from the source material and also as the particle passes through the detector dead layer. The energy loss from the source self-absorption and detector dead layer also decrease the energy resolution. Another contributor to the tail arises from backscattering of the alpha particle which varies with the incident direction of the particle on the detector although this backscatter effect is usually minor [Bertolini and Coche 1968]. Backscattering and attenuation effects were addressed in Sec. 7.2.1. As is observed in electron and positron energy-deposition tails, escaping x-ray emissions from excited electron states produced by the passage of alpha particles in the detector material may also contribute to counts in the spectral tail. Alpha particles can also backscatter from the source backing. Consequently, the count rate recorded by an alpha particle detector may need several corrections to account for the many contributors of the low energy tail.

Third, the value of mean ionization energy w is usually somewhat greater for alpha particles than that for gamma rays. The reason for this difference is that alpha particles have sufficient energy and mass to cause nuclear recoil events in Si, events which do not produce electron-hole pairs. Consequently, there is an additional source of variance in the pulse height spectrum from the distribution of Si recoils, along with a reduction in the total number of charge carriers produced. Bertolini and Coche [1968] estimate this contribution to the energy resolution is about 6 keV for 5-MeV alpha particles. Fourth, although Si is a mature semiconductor, some part of the low-energy tail may be a consequence of incomplete charge collection [Steinbauer et al. 1994].

Finally, at certain orientations, ions can enter the semiconductor crystal and pass between adjacent atomic planes, an effect called *channeling*. The consequences of channeling include extended ion ranges and reduced linear energy transfer (LET or stopping power) [Gibson 1965]. Channeled alpha particles can have significantly longer ranges than those shown in Fig. 4.23. Hence, channeling alpha particles may pass well beyond the depletion region, thereby depositing only a fraction of their energy in the active region of the detector [Dearnaley 1964] and producing counts only in the low-energy tail region. The LET is lower for channeled alpha particles, and thin transmission detectors designed for dE/dx measurements produce incorrect results [Madden and Gibson 1964]. Channeling also has an effect on detector fabrication in which ion implantation is used to introduce dopant impurities to form a *pn* junction. The actual junction depth increases beyond the predicted depth, producing a much thicker surface dead layer. Detectors are usually oriented at a slight angle from normal towards the source to prevent channeling.[14] There is a critical angle θ_c from a major crystal plane beyond which channeling is negligible [Seidel 1983] and, for heavy ions, is

[14]Semiconductor wafers are usually oriented between 7° to 11° from normal to the ion trajectories.

calculated as [Morgan 1973]

$$\theta_c = \left(\frac{2Z_1Z_2q_e^2}{Ed}\right)^{1/2} \quad \text{radians,} \tag{20.162}$$

where Z_1 is the ion, Z_2 is the target material, q_e is the fundamental electronic charge, E is the particle energy, and d is the interplane spacing along the channeling direction. The transition between high and low energy ions is generally defined by

$$E \simeq \frac{2dZ_1Z_2q_e^2}{a^2}, \tag{20.163}$$

where a is the Thomas-Fermi screening radius in angstroms

$$a = 0.8853\, a_0 \left(Z_1^{2/3} + Z_2^{2/3}\right)^{-1/2}, \tag{20.164}$$

where a_0 is the Bohr radius of 0.529 Å (5.29 nm). For low-energy ions, the critical angle is [Middleman and Hochberg 1993]

$$\theta_{cL} = \left(\frac{\sqrt{3}a}{\sqrt{2}d}\theta_c\right)^{1/2} \quad \text{radians.} \tag{20.165}$$

It turns out for Si, the transition energy for alpha particles defined by Eq. (20.163) is 104.4 keV. Alpha particles are invariantly emitted at higher energies than 100 keV, hence the use of Eq. (20.162) is appropriate for alpha particle spectroscopy with Si detectors.

The response signal from protons and alpha particles is linear for semiconductors in the absence of heavy charge carrier trapping. Deviations from linearity may be a consequence of energy loss in the contact dead layer, recombination of electron-hole pairs (especially near the end of the particle range), trapping, and nuclear recoils. Energy calibration with zero offset can be accomplished with the assistance of spectrometer grade check sources (see Sec. 20.5.1). There are some alpha particle sources of special interest for calibration purposes, including ^{241}Am, ^{148}Gd, ^{226}Ra, and ^{228}Th. Perhaps one of most used calibration sources is ^{241}Am, mainly because of its long half-life (432.7 years), average emission energies, and absence of interfering radiations, save for a 59.5-keV gamma-ray emission and some Np x-ray emissions at low energies. Although it appears to be monoenergetic, it actually has five energy emissions, all near 5.5 MeV [Chanda and Deal 1970]. When operated in vacuum with a well-prepared source, surface barrier and implanted junction Si detectors with excellent energy resolution can resolve all five emissions. However, usually only four alpha-particle emissions are discernible (see Fig. 16.12).

The ^{148}Gd source is of special interest for two main reasons: it is truly monoenergetic with a (relatively) low-energy alpha-particle emission of only 3.18 MeV and it has a comfortably long half-life of 70.9 years. ^{148}Gd has at times been difficult to acquire, mainly because it is a spallation byproduct from accelerator targets that are not always available.[15] Two sources with relatively high-energy alpha-particle emissions are ^{226}Ra and ^{228}Th. Both of these sources emit multiple alpha-particle energies from daughter products. ^{226}Ra has a half-life of 1599 years, with its highest energy emission being from daughter ^{214}Po of 7.69 MeV (see Fig. 16.12). However, a daughter product is also ^{222}Rn, which emits a 5.49-MeV alpha particle. Because ^{222}Rn is a gas, this daughter product can diffuse from the source and contaminate the detector, causing the appearance of a background peak. Fortunately, the half-life of ^{222}Rn is only 3.82 days and almost all daughters in the chain also have short half-lives; hence the background peak disappears after a few weeks. Regardless, for measurements requiring a low background environment, the operator may wish to avoid using

[15] It took over five years of waiting for the authors to get a spectroscopic grade ^{148}Gd source.

this source with a dedicated low-background detector. ^{228}Th has a relatively short half-life of only 1.91 years. ^{228}Th, with daughter products, is also a multi-energetic alpha-particle emitter. The highest emission energy is 8.78 MeV from daughter ^{212}Po, amongst one of the highest alpha-particle energies available as a check source.

20.9.3 Heavy Ions

Heavy ions, such as fission fragments, generally do not produce a linear spectrometer response. Instead, there is a marked signal loss with heavy ions, i.e., they produce a significantly smaller signal than that predicted for lighter ions of the same energy [Schmitt et al. 1965]. This difference in the expected and the observed results is called the *pulse height defect* (PHD) [Miller 1961]. The specific energy loss is highest for heavy charged particles at the entrance point in the detector, quite the opposite of alpha particles and protons which have their highest energy loss at the end of their paths as shown in Fig. 4.28. Consequently, energy loss in the dead layer of the detector is higher for heavy charged particles than that observed for alpha particles and protons. As is well documented, the pulse height defect can be sizeable for heavy ions [Bertolini and Coche 1968] and is reported to be between 10±3 MeV to 20±3 MeV [Miller 1961; Finch 1973].

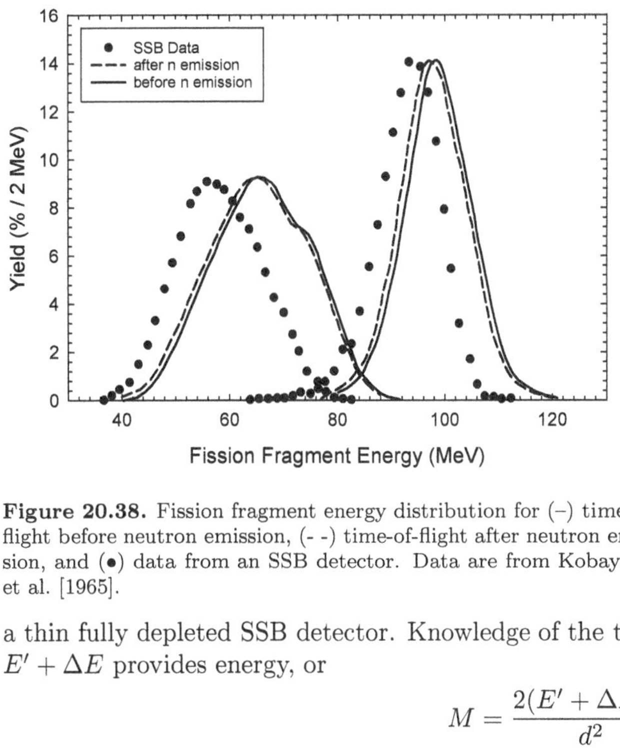

Figure 20.38. Fission fragment energy distribution for (–) time-of-flight before neutron emission, (- -) time-of-flight after neutron emission, and (•) data from an SSB detector. Data are from Kobayashi et al. [1965].

A measurement method used to reduce the effect of PHD is the time-of-flight (TOF) technique, briefly mentioned in Sec. 17.10 for thermal neutron detectors. However, a thin *zero time detector* is used to discern the timing start condition instead of a "chopper", while a second detector is used to define the transit time and (with limited resolution) particle energy. A particle passing through the zero time detector triggers the start time, while the second detector determines the stop time or t and residual energy E'. The zero time detector may be active or passive. In the passive case, the ion passes through a thin window (e.g., polymer or carbon film) that causes the emission of electrons [Fraser and Milton 1958; Milton and Fraser [1962]; DiIorio and Wehring 1977], and these electrons are then deflected into a detector that provides the fission fragment start time. A simple active zero time detector is a thin ΔE device, such as a thin gas-filled detector, a thin plastic scintillator, or a thin fully depleted SSB detector. Knowledge of the travel distance provides the velocity, while the sum of $E' + \Delta E$ provides energy, or

$$M = \frac{2(E' + \Delta E)t^2}{d^2} = \frac{2Et^2}{d^2}, \qquad (20.166)$$

where d is the distance between the detectors and M is the particle mass. If ΔE is negligible, then $E \simeq E'$. Time resolution is usually on the order of 250 ps or less, and the uncertainty in timing can be decreased by increasing t (i.e., the transit distance) [Butler et al. 1970; Zeidmen et al. 1974]. Hence, timing resolution can be designed to contribute less than 0.1% to the overall measurement uncertainty. Consequently, it is the energy resolution that strongly affects the mass resolution. Because fission fragments are emitted with various charge states, the energy resolution of the stop detector is limited. For example, an SSB detector

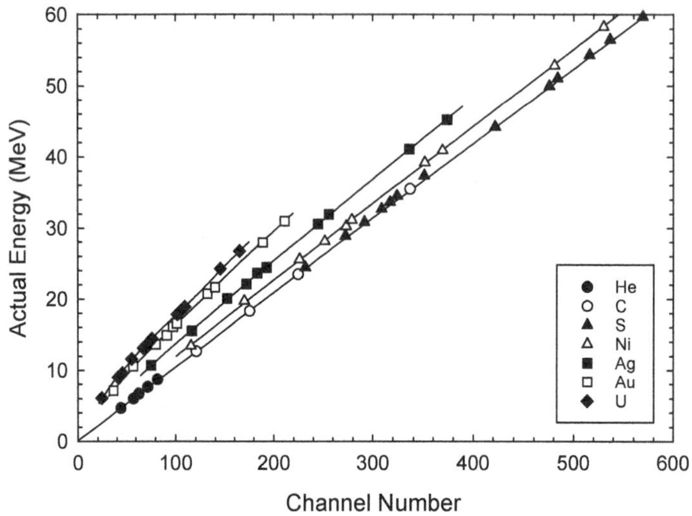

Figure 20.39. Actual ion energy versus pulse height channel number for a few heavy ions measured with an SSB detector. Data are from Wilkins et al. [1971].

is limited to 2%–4% FWHM for fission fragments. DiIorio and Wehring [1977] use an electrostatic analyzer between the zero time detector and an SSB stop detector to sort different kinetic-energy-to-ionic-charge ratios, and, thereby improve the particle mass resolution to less than 0.5 amu.[16] A comparison of the energy measured for fission fragments using time-of-flight methods and an SSB detector is shown in Fig. 20.38. The solid curve depicts the energy distribution of fission fragments before prompt neutron emission, the dashed curve depicts the energy distribution of fission fragments after prompt neutron emission, and the data points were collected by the Si detector (after prompt neutron emission). The PHD is clearly shown by the difference with the SSB detector response and the time-of-flight responses.

However, the high energy loss from source self-absorption and dead layer losses (nominally only about 2 MeV) are not the only loss mechanisms. Dearnaley and Northrop [1966] emphasize that non-ionizing nuclear recoils (in Si) are a major source of energy loss that contributes to the PHD. There are even more mechanisms that contribute to the PHD. These include electron and hole recombination in the dense charge cloud produced by the heavy charged particles. Also charge carriers are lost by being trapped in crystal lattice defects, an effect which increases as the semiconductor is exposed to increasing numbers of heavy particles. Finally, increased recombination losses also occur as the density of recombination centers increases with crystal damage (see Sect. 15.2.3). Thus, in general, one must correct for the PHD in heavy ion spectroscopy. The measured pulse height P and the energy E of the heavy charged particle is often related by

$$E = Sn + \delta, \tag{20.168}$$

[16]The active zero time detector provides another method to determine the mass-energy relationship, albeit generally less accurate [Sachs et al. 1966]

$$E\frac{dE}{dx} = (E' + \Delta E)\Delta E = k_1 M Z^2 \ln\left[\frac{E}{k_2 M}\right] \simeq KMZ^2. \tag{20.167}$$

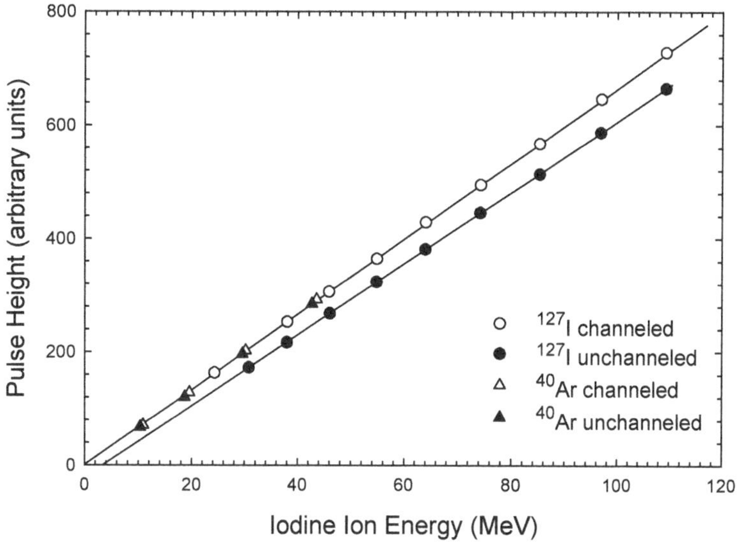

Figure 20.40. Pulse height dependence for channeled and non-channeled ^{40}Ar and ^{127}I ions along the [110] axis. Data are from Moak et al. [1966].

where S is a scaling constant and δ is the pulse height defect. For a common multichannel analyzer system, S may be the equivalent energy per channel and n would be the channel number. Shown in Fig. 20.39 is a comparison between the actual ion energies of various heavy ions and the associated pulse heights indicated by channel number on a multichannel analyzer [Wilkins et al. 1971] and from which it is apparent that the PHD increases with the Z number of the heavy ion.

It is notable that in at least one study the PHD seems to disappear with channeled ions [Moak et al. 1966]. The comparison was drawn between ions of two different masses, ^{40}Ar and ^{127}I. Much like sulphur in Fig. 20.39, the PHD for Ar ions (with an atomic mass close to that of S) was small with little difference between ions with or without channeling. However, there was clearly the appearance of a PHD for the unchanneled iodine ions, which have a slightly larger atomic mass than that of silver, and the PHD for iodine was similar to that of silver in Fig. 20.39. However, the PHD was similar to that of Ar when channeled, as shown in Fig. 20.40. The reason for the difference is attributed to a lower stopping power $-dE/dx$ of the channeled ions and the reduction or absence of nuclear collisions. Further, the energy resolution reportedly improved by a factor of three for the channeled ions. The experiment performed by Moak et al. [1966] seems to confirm that it is nuclear collisions that are responsible for most of the PHD observed with heavy ions, although the lower $-dE/dx$ also reduces recombination losses. The observed energy resolution improvement is probably because the variance term for non-ionizing energy losses becomes negligible for the channeled ions. Attempts to correlate measured PHD to empirical formulas are in the literature (for example, Schmitt et al. [1965]; Wilkins et al. [1971]). One general correlation offered by Schmitt et al. [1965] is based on the assumption that two linear functions can be used to describe both the scaling constant and the PHD in Eq. (20.168), namely

$$E = (a + a'M)P + (b + b'M), \tag{20.169}$$

where $M \simeq A$ is the mass (in u) of the ion and a, a', b, and b' are experimentally determined constants particular to the detector. The constants can be found by using a spectroscopic grade fission source, such as ^{252}Cf. If P_L denotes the 3/4-maximum in the pulse height for the lower mass (higher energy) peak ($P_L = 0.75N_L$ in Fig. 20.41) and P_H denotes the 3/4-maximum in the higher mass (lower energy) peak

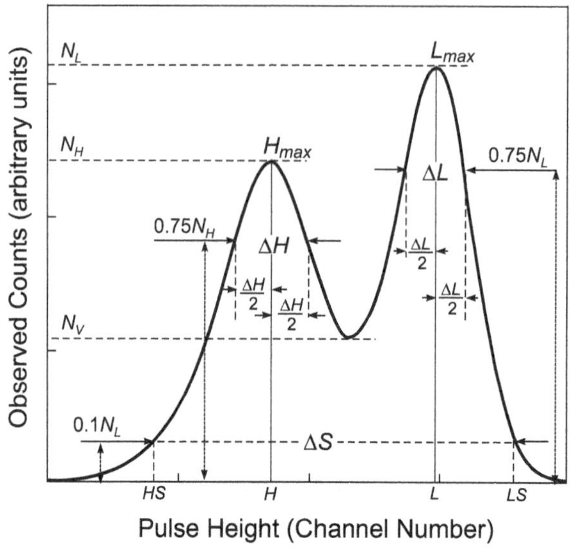

Figure 20.41. Identifying parameters for a ^{252}Cf pulse height spectrum. After Schmitt and Pleasonton [1966].

Table 20.3. Summary of SSB detector spectral metrics for Fig. 20.41.

Parameter	Reasonable Limit	^{252}Cf Expected Value	^{235}U Expected Value
N_L/N_V	> 2.85	~ 2.9	~ 19
N_H/N_V	~ 2.2	~ 2.2	~ 12.5
N_L/N_H	–	~ 1.3	1.49-1.55
$\Delta L/(L-H)$	< 0.38	~ 0.36	0.22-0.23
$\Delta H/(L-H)$	$\lesssim 0.45$	$\lesssim 0.44$	0.35-0.36
$(H-HS)/(L-H)$	< 0.70	$\lesssim 0.69$	0.38-0.39
$(LS-L)/(L-H)$	$\lesssim 0.49$	$\lesssim 0.48$	0.27-0.28
$(LS-HS)/(L-H)$	$\lesssim 2.18$	~ 2.17	~ 1.66

($P_H = 0.75N_H$ in Fig. 20.41) , the constants are found to be [Weissenberger et al. 1986]

$$a = \frac{a_0}{P_L - P_H}, \quad a' = \frac{a'_0}{P_L - P_H}, \quad b = b_0 - aP_L \quad \text{and} \quad b' = b'_0 - a'P_L. \tag{20.170}$$

For a ^{252}Cf source,[17]

$$a_0 = 24.300, \quad a'_0 = 0.0283, \quad b_0 = 90.397, \quad \text{and} \quad b'_0 = 0.1150, \tag{20.171}$$

all in units of MeV.

[17]These constants are slightly different from the original work by Schmitt et al. [1965]. Weissenberger et al. [1986] state the prior values systematically overestimated the kinetic energies by 1-2%.

For a ^{252}Cf source measured with an SSB detector, certain parameters can be used to provide quantitative metrics for the particle detector [Schmitt and Pleasonton 1966] that are shown in Fig. 20.41. The vertical axis, with labels N, refer to the number of observed counts at relative pulse heights (channel numbers). These metrics include peak ratios, such as N_L/N_V, N_H/N_V, and N_L/N_H. Of these, Schmitt and Pleasonton [1966] indicate that most important is the peak-to-valley ratio N_L/N_V with a reasonable lower limit of 2.85. With N_L/N_V, the FWHM resolution for a 70-MeV ion (Br, for instance) should be approximately 1 MeV, while that for a corresponding iodine ion is 1.4 MeV. Listed in Table 20.3 are other performance metrics.

PROBLEMS

1. Consider the design of an experiment to obtain the data in Example 20.1. The total interaction coefficient μ of the absorbing foils contains a scattering component which, instead of absorbing photons traveling from the point source to the detector, scatter some of the source photons. Some of these scattered photons still reach the detector and are counted as uncollided or unattenuated photons. Design an experiment, i.e., the size and placement of the foils with respect to the source and detector, that minimizes the scattered component of photons that reach the detector.

2. Show that for small $\overline{r}t_r$ Eq. (20.29) is approximately the same as Eq. (20.27).

3. Derive Eqs. (20.90).

4. Consider the count data in Table 20.4. (a) Write a program to find the best fit parameters of this isolated peak. With the fit model of Eq. (20.71), use the linear least-squares method as in Example 20.2 and by trial and error find the peak centroid and standard deviation to obtain the best fit. (b) How many counts are there under the peak?

Table 20.4. A portion of a spectrum around the 911.2-keV ^{228}Ac full-energy peak in the background spectrum shown in Fig. 21.7.

E_i (keV)	y_i cnts	E_i (keV)	y_i cnts	E_i (keV)	y_i cnts	E_i (keV)	y_i cnts
0.9066	589	0.9088	601	0.9109	2349	0.9131	537
0.9069	625	0.9090	692	0.9112	1985	0.9134	568
0.9071	596	0.9093	878	0.9115	1467	0.9137	591
0.9074	582	0.9096	1039	0.9118	1148	0.9139	597
0.9077	590	0.9099	1511	0.9120	831	0.9142	586
0.9079	640	0.9101	1867	0.9123	721	0.9145	581
0.9082	624	0.9104	2287	0.9126	585	0.9148	606
0.9085	615	0.9107	2430	0.9128	590	0.9150	594

5. Use the non-linear least-squares method to fit simultaneously all the parameters of Eq. (20.71) to the data in the above problem.

6. Derive Eq. (20.155).

7. Show that fitting Eq. (20.104) to N_E, N_F and N_P the parameters in the quadratic fit are

$$\alpha = (1/2)[N_E + N_F - 2N_P], \qquad \beta = (1/2)(N_F - N_E) - n_P(N_F + N_E - 2N_P)$$

and

$$\gamma = N_P + n_P \frac{N_E - N_F}{2} + n_P^2 \frac{N_E + N_F - 2N_P}{2}.$$

Then derive the expressions for the channel number and the maximum peak count given by Eqs. (20.105) and (20.106).

8. Show that the skewness factors defined in Eq. (20.111) are unity for a Gaussian peak.

9. Determine the corrected peak channel and counts for the peak shown in Fig. 20.18, but shifted slightly so that now $C_P = 10,000$ at channel 3365 and $C_E = 9910$ and $C_F = 7748$. The peak is still located between channels $N_L = 3351$ and $n_R = 3374$.

10. In Tables 20.5 and 20.6 are the channels and counts observed for the ^{214}Bi 2204.1-keV and ^{40}K 1460.8-keV full-energy peaks, respectively. Estimate the resolution in keV and % of the HPGe detector used to obtain these data.

Table 20.5. A portion of a spectrum around the 2204.1-keV ^{214}Bi full-energy peak in the background spectrum shown in Fig. 21.7.

n	y_i cnts	n	y_i cnts	n	y_i cnts	n	y_i cnts
8073	71	8081	175	8089	682	8097	95
8074	82	8082	225	8090	609	8098	79
8075	74	8083	280	8091	543	8099	92
8076	81	8084	441	8092	390	8100	74
8077	88	8085	542	8093	332	8101	68
8078	70	8086	621	8094	207	8102	71
8079	102	8087	687	8095	157	8103	75
8080	155	8088	735	8096	138	8104	65

Table 20.6. A portion of a spectrum around the 1460.8-keV ^{40}K full energy peak in the background spectrum show in Fig. 21.7.

n	y_i cnts	n	y_i cnts	n	y_i cnts	n	y_i cnts
5343	362	5351	3871	5359	36336	5367	450
5344	360	5352	7064	5360	29415	5368	288
5345	351	5353	12202	5361	21395	5369	226
5346	381	5354	19092	5362	13901	5370	189
5347	509	5355	26669	5363	8260	5371	177
5348	672	5356	34041	5364	4260	5372	141
5349	1159	5357	38791	5365	1992	5373	175
5350	1996	5358	39769	5366	968	5374	157

11. Explain why the resolution of a gamma-ray spectrometer that uses a scintillator is usually expressed as a percentage, while that of a semiconductor spectrometer is usually given in energy units.

12. Using the detector configuration of Fig. 20.20, derive the expression for Eq. (20.121).

13. With the result of Eq. (20.145) and the constraint that $T_A = t_G + t_B$, derive an expression for the optimum source count time t_G in terms of the available time t_A.

14. You have a total of 5 minutes each to conduct measurements of numerous radioactive activation samples, including the background measurement and the gross measurement. Given that on average $g \simeq 3500$

cm and the average background is $b \simeq 100$ cpm, find the optimized times for the gross and background measurements.

15. You have a 300-nm-thick B_4C detector claimed to be a semiconductor neutron detector. The film was grown with natural boron on n-type Si and has area 2 mm × 2 mm. Given a thermal neutron (2200 m s^{-1}) flux of 10^4 cm^{-2} s^{-1}, how long should a measurement be to discern with 98% confidence that it really is a B_4C semiconductor detector and not a thin-film Si pn-junction detector? Hint: the thin-film neutron detector expressions in Chapter 17 might be helpful.

16. Given the two different cylindrical NaI:Tl detector configurations in Fig. 20.42, find the detection efficiency of coincidence events if the source is ^{22}Na for both configurations. Repeat for ^{60}Co.

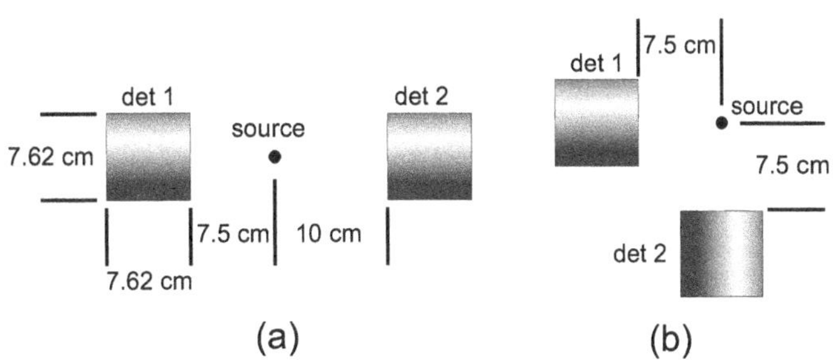

Figure 20.42. Coincidence counting geometry for two NaI:Tl detectors.

17. Determine the critical angle required to prevent alpha particle channeling in a Si SSB detector with surface direction ⟨100⟩.

REFERENCES

Ahmad, I. and F. Wagner, "A Simple Cooled Si(Li) Electron Spectrometer," *Nucl. Instrum. Meth.*, **116**, 465-469, (1974).

Amman, M., P.N. Luke, and S.E. Boggs, "Amorphous-Semiconductor-Contact Germanium-Based Detectors for Gamma-Ray Imaging and Spectroscopy," *Nucl. Instrum. Meth.*, **A579**, 886–890, (2007).

Anger, H.O., "Scintillation Camera," *Rev. Sci. Instrum.*, **29**, 27–33, (1958).

Barache, D., J-P. Antoine, and J-M. Dereppe, "The Continuous Wavelet Transform, an Analysis Tool for NMR Spectroscopy," *J. Magnetic Res.*, **128**, 1–11, (1997).

Bender, S., B. Heidrich, and K. Ünlü, "Compton Suppressed $LaBr_3$ Detection System for use in Nondestructive Spent Fuel Assay," **A784**, 474–481, (2015).

Berger, M.J., S.M. Seltzer, S.E. Chappell, J.C. Humphreys, and J.W. Motz, "Response of Silicon Detectors to Monoenergetic Electrons with Energies Between 0.15 and 5.0 MeV," *Nucl. Instrum. Meth.*, **69**, 181–193, (1969a).

Berger, M.J., S.M. Seltzer, S.E. Chappell, J.C. Humphreys, and J.W. Motz, "Table of Response Functions for Silicon Electron Detectors," *Tech. Note 489*, U.S. Nat. Bureau of Standards, (1969b).

Bertolini, G. and A. Coche, *Semiconductor Detectors*, New York: Wiley, 1968.

Bevington, P.R., *Data Reduction and Error Analysis for the Physical Sciences*, New York: McGraw-Hill, Inc., 1969.

Bodansky, D. and S.F. Eccles, "Application of a Phoswich to Particle Identification," *Rev. Sci. Instrum.*, **28**, 464–465, (1957).

Bugby, S.L., L.K. Jambi, and J.E. Lees, "A Comparison of CsI:Tl and GOS in a Scintillator-CCD Detector for Nuclear Medicine Imaging," *JINST*, **11**, p09009, 12 pages, (2016).

Butler, G.W., A.M. Poskanzer, and D.A. Landis, "Identification of Nuclear Fragments by a Combined Time-of-Flight, ΔE-E Technique," *Nucl. Instrum. Meth.*, **89**, 189–198, (1970).

Byun, S.H., W.V. Prestwich, K. Chin, F.E. McNneill, and D.R. Chettle, "Efficiency Calibration and Coincidence Summing Correction for a 4π NaI(Tl) Detector Array," *Nucl. Instrum. Meth.*, **A535**, 674–685, (2004).

CARTER, J.N., P. CHARALAMBOUS, A.J. DEAN, J.B STEPHEN, R.C. BUTLER, G. DI COCCO, E. MORELLI, G. SPADA, A. SPIZZICHINO, L. BARBARESCHI, G. BOELLA, F. PEROTTI, G. VILLA, M. BADIALI, C. LA PADULA, F. POLCARO, AND P. UBERTINI, "A Position Sensitive Detector for a Gamma-Ray Imaging Telescope," *Nucl. Instrum. Meth.*, **196**, 477–482, (1982).

CHANDA, R.N. AND R.A. DEAL, *Catalogue of Semiconductor Alpha-Particle Spectra*, IN-1261, 1970.

CHERRY, S.R., J.A. SORENSON, AND M.E. PHELPS, *Physics in Nuclear Medicine*, 3rd Ed., London: Elsevier Health Sciences, 2003.

CHOONG, W.-S., K.M. VETTER, W.W. MOSES, G. HULL, S.A. PAYNE, N.J. CHEREPY, AND J.D. VALENTINE, "Design of a Facility for Measuring Scintillator Non-Proportionality," *IEEE Trans. Nucl. Sci.*, **55**, 1753–1758, (2008).

CLINE, J.E., "Studies of Detection Efficiencies and Operating Characteristics of Ge(Li) Detectors," *IEEE Trans. Nucl. Sci.*, **NS-15**, 198–213, (1968).

COSTA, E., E. MASSARO, AND L. PIRO, "A BGO-CsI(Tl) Phoswich: A New Detector for X- and γ-Ray Astronomy," *Nucl. Instrum. Meth.*, **A243**, 572–577, (1986).

CROUTHAMEL, C.E., *Applied Gamma-Ray Spectrometry*, revised by F. ADAMS AND R. DAMS, in *International Series of Monographs in Analytical Chemistry*, Vol. 41, Oxford: Pergamon, 1970.

DAHLBOM, M., L.R. MACDONALD, L. ERIKSSON, M. PAULUS, M. ANDREACO, M.E. CASEY, AND C. MOYERS, "Performance of a YSO/LSO Phoswich Detector for use in a PET/SPECT System.," *IEEE Trans. Nucl. Sci.*, **44**, 1114–1119, (1997).

DAKUBU, T. AND W.B. GILBOY, "An Unfolding Procedure for Slowly Varying Pulse Height Spectra Produced by β Particles in Si(Li) Detectors," *Nucl. Instrum. Meth.*, **150**, 479–482, (1978).

DEARNALEY, G., "The Channeling of Ions Through Silicon Detectors," *IEEE Trans. Nucl. Sci.*, **NS-11**, 249–253, (1964).

DEARNALEY, G. AND D.C. NORTHROP, *Semiconductor Counters for Nuclear Radiations*, 2nd Ed., Aylesbury: Barnes and Noble (Watson & Viney), 1966.

DE CELIS, B., R. DE LA FUENTE, A. WILLIART, AND DE CELIS ALONSO, "Coincidence Measurements in $\alpha/\beta/\gamma$ Spectroscopy with Phoswich Detectors Using Digital Pulse Shape Discrimination Analysis," *Nucl. Instrum. Meth.*, **A580**, 206–209, (2007).

DIIORIO, G. AND B.W. WEHRING, "HIAWATHA, A Fission-Fragment Recoil Mass Spectrometer," *Nucl. Instrum. Meth.*, **147**, 487–499, (1977).

DOGAN, N., D.K. WEHE, AND G.F. KNOLL, "Multiple Compton Scattering Gamma Ray Imaging Camera," *Nucl. Instrum. Meth.*, **A299**, 501–506, (1990).

DUNN, W.L., "Inverse Monte Carlo Analysis," *J. Comput. Phys.*, **41**, 154–166, (1981).

DUNN, W.L. AND T.S. DUNN, "An Asymmetric Model for XPS Analysis," *Surf. Interface Anal.*, **4**, 77–88, (1982).

DUNN, W.L. AND J.K. SHULTIS, "Monte Carlo Analysis for Design and Analysis of Radiation Detectors," *Radiat. Phys. Chem.*, **78**, 852–858, (2009).

DUNN, W.L. AND D.S. MCGREGOR, "Gamma-Ray Detectors," Vol. 1, Ch. 17 in *Applications of Particle Detectors in Particle Physics, Medicine, Astrophysics and Related Fields*, I. BUVET AND C. GRUPEN Eds., Berlin: Springer, 411–452, 2012.

ERIKSSON, L., M. CONTI, C.L. MELCHER, M. ZHURAVLEVA, M. ERIKSSON, AND H. ROTHFUSS, "LuYAP/LSO Phoswich Detectors for High Resolution Positron Emission Tomography," *IEEE Trans. Nucl. Sci.*, **60**, 194–196, (2013).

EVERETT, D.B., J.S. FLEMING, R.W. TODD, AND J.M. NIGHTINGALE, "Gamma-Radiation Imaging System Based on the Compton Effect," *Proc. IEEE*, **124**, 995–1000, (1977).

FINCH, E.C., "An Analysis of the Causes of the Pulse Height Defect and its Mass Dependence for Heavy-Ion Silicon Detectors," *Nucl. Instrum. Meth.*, **113**, 41–49, (1973).

FRASER, J.S. AND J.C.D. MILTON, "Automatic Recording Apparatus for Fission-Fragment Time-of-Flight Measurements," *Nucl. Instrum.*, **2**, 275–281, (1958).

FROMMHOLD, Th., W. ARNOLD, H. FRIEDRICHS, R. GÖBELS, R.D. HEIL, U. KNEISSL, U. SEEMAN, F. STEIPER, AND C. KOZHUHAROV, "Response Functions of Si Detectors to Monoenergetic Electrons and Positrons in the Energy Range 0.8–3.5 MeV," *Nucl. Instrum. Meth.*, **A310**, 657-664, (1991).

FUTAMI, Y., T. MIZOTA, Y.H. PU, Y. HONJO, K. YUASA-NAKAGAWA, H. TOYOKAWA, S.M. LEE, K. FURUTAKA, T. MURAKAMI, J. KASAGI, K. YOSHIDA, AND T. NAKAGAWA, "Performance of a Phoswich Detector Composed of Thin Plastic and Thick BaF_2 Scintillators," **A326**, 513–516, (1993).

GARDNER, R.P. AND A. SOOD, "A Monte Carlo Simulation Approach for Generating NaI Detector Response Functions (DRFs) that Accounts for Nonlinearity and Variable Flat Continua," *Nucl. Instrum. Meth. in Phys. Res.*, **B213**, 87–99, (2004).

GARDNER, R.P. AND L. XU, "Status of the Monte Carlo Library Least-Squares (MCLLS) Approach for Non-Linear Radiation Analyzer Problems," *Radiat. Phys. Chem.*, **78**, 843–851, (2009).

GENTILE, N.A., "Implicit Monte Carlo Diffusion–An Acceleration Method for Monte Carlo Time-Dependent Radiative Transfer Simulations," *J. Comput. Phys*, **172**, 543–571, (2001).

GIBSON, W.M., "The Role of Particle Channeling in Detector Systems," *IEEE Trans. Nucl. Sci.*, **NS-13**, 162–175, (1965).

GILMORE, G., *Practical Gamma-Ray Spectrometry*, 2nd Ed., New York: Wiley, 2008.

HAMAMATSU, *X-Ray Scintillator*, Technical Note TXPR1024E02, 2016.

H3D, `h3dgamma.com`, accessed June 2018.

HE, Z. AND F. ZHANG, "'Three-Dimensional, Position-Sensitive Radiation Detection," US-7411197, allowed Aug. 12, 2008.

HEATH, R.L., *Scintillation Spectrometry Gamma-Ray Spectrum Catalogue*, 2nd Ed., AEC Report IDO-16880, Vols. 1 and 2, 1964.

HEATH, R.L., R.G. HELMER, L.A. SCHMITTROTH AND G.A. CAZIER, "Method for Generating Single Gamma-Ray Shapes for Analysis of Spectra," *Nucl. Instrum. Meth.*, **47**, 281–304, (1967).

IEEE/ANSI, "IEEE Standard Test Procedures for Germanium Gamma-Ray Detectors," Standard 325-1996, 1996.

KAMAE, T., S. GUNJI, M. HIRAUAMA, S. MIYAZAKI, T. NAGATO, A. NAKAO, Y. SEKIMO- TO, K. SUZUKI, T. TAKAHASHI, T. TAMURA, M. TANAKA, N. YAMAOKA, T. YAMAGAMI, M. NOMACHI, AND N. MURAKAMI, "Well-Type Phoswich Counter for Low-Flux X-Ray /γ-Ray Detection," *IEEE Trans. Nucl. Sci.*, **40**, 204–207, (1993).

KATAOKA, J. AND 25 CO-AUTHORS, "Ultracompact Compton Camera for Innovative Gamma-Ray Imaging," *Nucl. Instrum. Meth.*, **A912**, 1–5, (2018).

KIERNANS, C.A., S.E. BOGGS, J.-L. CHIU, A. LOWELL, C. SLEATOR, J.A. TOMSICK, M. AMMAN, H.-K. CHANG, C.-H. TSENG, C.-Y. YANG, C.-H. LIN, P. JEAN, AND P. VON BALLMOOS, "The 2016 Super Pressure Balloon flight of the Compton Spectrometer and Imager," *Proc. Science, 11th INTEGRAL Conf. Gamma-Ray Astrophysics in Multi-Wavelength Perspective*, Oct. 10-14, Amsterdam, (2016).

KIS, Z., B. FAZEKAS, J. ÖSTÖR, Z. RÉVAY, T. BELGYA, G.L. MOLNÁR, AND L. KOLTAY, "Comparison of Efficiency Functions for Ge Gamma-Ray Detectors in a Wide Energy Range," *Nucl. Instrum. Meth.*, **A418**, 374–386, (1998).

KISHIMOTO, A., J. KATAOKA, T. NISHIYAMA, T. FUJITA, K. TAKEUCHI, H. OKOCHI, H. OGATA, H. KUROSHIMA, S. OHSUKA, S. NAKAMURA, M. HIRAYANAGI, S. ADACHI, T. UCHIYAMA, AND H. SUZUKI, "Performance and Field Tests of a Handheld Compton Camera using 3-D Position-Sensitive Scintillators Coupled to Multi-Pixel Photon Counter Arrays," *J. Instrum.*, **9**, P11025, (2014).

KOBAYASHI, H., A. NAKAMOTO, AND M. HOSOE, "Energy Dependence of Pulse Height Defect with Fission Fragments in a Silicon Surface Barrier Detector," *Nucl. Instrum. Meth.*, **34**, 222–224, (1965).

LEUTZ, H., G. SCHULZ, AND L. VAN GELDEREN, "Peak/Total-Ratios for NaI(Tl) Crystals," *Nucl. Instrum. Meth.* **40**, 257–260, (1966).

LINDOW, J.T., E. SHRADER, M.R. FARUKHI, R.G. GEIGER, E.C. STEWART, R.W. JOHN- SON, AND R.J. THOMPSON, *Harshaw Scintillation Phosphors*, 3rd. Ed., Solon, Ohio: Harshaw Chemical Company, 1978.

LOWELL, A., S. BOGGS, J.L. CHIU, C. KIERANS, S. MCBRIDE, C.H. TSENG, A. ZOGLAUER, M. AMMAN, H.K. CHANG, P. JEAN, C.H. LIN, C. SLEATOR, J. TOMSICK, P. VON BALLMOOS, AND C.Y. YANG, "Positional Calibrations of the Germanium Double Sided Strip Detectors for the Compton Spectrometer and Imager," *Proc. SPIE*, **9915**, paper 99152H, 10 pages, (2016).

LOWELL, A.W., S.E. BOGGS, C.L. CHIU, C.A. KIERANS, C. SLEATOR, J.A. TOMSICK, A.C. ZOGLAUER, H.-K. CHANG, C.-H. TSENG, C.-Y. YANG, P. JEAN, P. VON BALL- MOOS, C.-H. LIN, AND M. AMMAN, "Polarimetric Analysis of the Long Duration Gamma-Ray Burst GRB 160530A with the Balloon Borne Compton Spectrometer and Imager," *Astrophysical Journal*, **848**, paper 119, 10 pages, (2017).

MADDEN, T.C. AND W.M. GIBSON, "Uniform and Stable dE/dx P-n Junction Particle Detectors," *IEEE Trans. Nucl. Sci.*, **NS-11**, 254–261, (1964).

MARSHALL III, J.H. AND J.F. ZUMBERGE, "On-line Measurements of Bulk Coal Using Prompt Gamma Neutron Activation Analysis," *Nucl. Geophys.*, **3**, 445–459, (1989).

MCGREGOR, D.S., "Detection and Measurement of Radiation," Ch. 8, in *Fundamentals of Nuclear Science and Engineering*, 2nd Ed., by J.K. SHULTIS AND R.E. FAW, New York: CRC Press, 2008.

MICKAEL, M.W., "A Complete Inverse Monte Carlo Model for Energy-Dispersive X-Ray Fluorescence Analysis," *Nucl. Instrum. Meth.*, **A301**, 523–542, (1991).

MIDDLEMAN, S. AND A.K. HOCHBERG, *Process Engineering Analysis in Semiconductor Device Fabrication*, New York: McGraw-Hill, 1993.

MILLER, G.L., "Diffused Junction Detectors," *Semiconductor Nuclear Particle Detectors*, N.A.S. Pub. 871, 19–27, (1961).

MILTON, J.C.D. AND J.S. FRASER, "Time-of-Flight Fission Studies on U^{233}, U^{235}, and Pu^{239}," *Can. J. Phys.*, **40**, 1626–1663, (1962).

MOAK, C.D., J.W.T. DABBS, AND W.W. WALKER, "Response of Silicon Fission Detectors to Channeled ^{127}I and ^{40}Ar," *Rev. Sci. Instrum.*, **37**, 1131–1134, (1966).

MOLNAR, G.L., *Handbook of Prompt Gamma Activation Analysis with Neutron Beams*, Boston: Kluwer Academic Publishers, 2004.

MORGAN, D.V., *Channeling: Theory, Observation and Applications*, New York: Wiley, 1973.

MORÉ, J.J., B.S. GARBOW, AND K.E. HILLSTROM, *User's Guide for MINPACK-1*, Report ANL-80-74, Argonne National Laboratory, 1980.

NAFEE, S.S., "A Mathematical Approach to Determine Escape Peak Efficiencies of High-Purity Germanium Cylindrical Detectors for Prompt Gamma Neutron Activation Analysis," *Nucl. Tech.*, **175**, 162–167, (2011).

NIST, ESTAR, `https://physics.nist.gov/PhysRefData/Star/Text/ESTAR.html`, accessed 2018.

NUDAT, *Nuclear Structures and Decay Data*, v. 2.7, National Nuclear Data Center, Brook- haven National Laboratory. Available at `http://www.nndc.bnl.gov/`; accessed July 2017.

PHDS, `phdsco.com`, accessed June 2018.

PRESS, W.H., S.A. TEUKOLSKY, W.T. VETTERLING, AND B.P. FLANNERY, *Numerical Recipes in FORTRAN 77*, 2nd Ed., New York: Cambridge University Press, 1992.

QUARATI, F., A.J.J. BOS, S. BRANDENBURG, C. DATHY, P. DORENBOS, S. KRAFT, R.W. OSTENDORF, V. OUSPENSKI, AND A. OWENS, "X-Ray and Gamma-Ray Response of a 2" × 2" $LaBr_{+}3$:Ce Scintillator Detector," *Nucl. Instrum. Meth.*, **A574**, 115–120, (2007).

ROSE, M.E., "The Analysis of Angular Correlation and Angular Distribution Data," *Phys. Rev.*, **91**, 610–615, (1953).

SACHS, M.W., C. CHASMAN, AND D.A. BROMLEY, "Heavy Ion Reaction Product Identification by Measurement of dE/dx and E," *Nucl. Instrum. Meth.*, **41**, 213–225, (1966).

SAINT-GOBAIN, *Efficiency Calculations for Selected Scintillators*, Saint-Gobain Crystals, Tech Note 08-16, (2016).

SAINT-GOBAIN, *Phoswich Detectors For High Energy Backgrounds*, Saint-Gobain Crystals, Tech Note 02-19, (2019a).

SAINT-GOBAIN, *Scintillation Arrays*, Saint-Gobain Crystals, Tech Note 05-19, (2019b).

SCHMITT, H.W., W.E. KIKER, AND C.W. WCILLIAMS, "Precision Measurements of Correlated Energies and Velocities of ^{252}Cf Fission Fragments," *Phys. Rev.*, **137**, B837–B847, (1965).

SCHMITT, H.W. AND F. PLEASONTON, "Evaluation of Semiconductor Detectors for Fission Fragment Energy Measurements," *Nucl. Instrum. Meth.*, **40**, 204–208, (1966).

SCHÖNFELDER, V., A. HIRNER, AND K. SCHNEIDER, "A Telescope for Soft Gamma Ray Astronomy," *Nucl. Instrum. Meth.*, **107**, 385–394, (1973).

SEIDEL, T.E., "Ion Implantation," in *VLSI Technology*, S.M. Sze, Ed., Ch. 6, New York: McGraw-Hill, 1983.

SMITH, D., "Improved Correction Formulae for Coincidence Counting," *Nucl. Instrum. Meth.*, **152**, 505-519, (1978).

SMITH, T. AND K.J. KEARFOTT, "Practical Considerations for Gamma Ray Spectroscopy with NaI(Tl): A Tutorial," *Health Physics*, **114**, 94–106, (2018).

STEINBAUER, E., G. BORTELS, P. BAUER, J.P. BIERSACK, P. BURGER, AND I. AHMAD, "A Survey of the Physical Processes which Determine the Response Function of Silicon Detectors to Alpha Particles," *Nucl. Instrum. Meth.*, **A339**, 102–108, (1994).

TSOULFANIDIS, N., B.W. WERHRING, AND M.E. WYMAN, "The Use of an Analytical Response Function for Unfolding Beta Spectra," *Nucl. Instrum. Meth.*, **73**, 98–102, (1969).

UGOROWSKI, P., M.J. HARRISON, C. LINNICK, S. BRINTON, AND D.S. MCGREGOR, "A Compton-Scattering Coincidence System for Light Yield Measurements on Aliovalently-Doped $CeBr_3$," *Proc. IEEE Nucl. Sci. Symp.*, Dresden, Germany, Oct. 19-25, 1246–1250, (2008).

USUDA, S., "Development of ZnS(Ag)/NE102A and ZnS(Ag)/Stilbene Phoswich Detectors for Simultaneous α and $\beta(\gamma)$ Counting," *J. Nucl. Sci. Tech.*, **29**, 927–929, (1992).

VALENTINE, J.D. AND B.D. ROONEY, "Design of a Compton Spectrometer Experiment for Studying Scintillator Non-Linearity and Intrinsic Energy Resolution," *Nucl. Instrum. Meth.*, **A353**, 37–40, (1994).

VINCENT, C.H., *Random Pulse Trains: Their Measurement and Statistical Properties*, London: Peter Peregrinus, 1973.

VINCENT, C.H., "Estimation of Multiple Coincidences," *Nucl. Instrum. Meth.*, **127**, 421–427, (1975).

VINCKE, H., E. GSCHWENDTNER, C.W. FABJAN, AND T. OTTO, "Response of a BGO Detector to Photon and Neutron Sources: Simulation and Measurements," *Nucl. Instrum. Meth.*, **A484**, 102–110, (2002).

WAHL, C.G., W.R. KAYE, W. WANG, F. ZHANG, J.M. JAWORSKI, A. KING, Y.A. BOU- CHER, AND Z. HE, "The Polaris-H Imaging Spectrometer," *Nucl. Instrum. Meth.*, **A784**, 377–381, (2015).

WATANABE, T., H. ARAKAWA, T. KAJIMOTO, Y. IWAMOTO, D. SATOH, S. KUNIEDA, S. NODA, N. SHIGYO, K. ISHIBASHI, T. NAKAMURA, AND R.C. HAIGHT, "Performace of a Phoswich Detector Composed of an Inner NaI(Tl) Crystal and Surrounding NE102A Plastic Scintillator for Neutron Spectrometry," *Nucl. Instrum. Meth.*, **A587**, 20–28, (2008).

WEISSENBERGER, E., P. GELTENBORT, A. OED, F. GÖNNENWEIN, AND H. FAUST, "Energy Calibration of Surface Barrier Detectors for Fission Fragments," *Nucl. Instrum. Meth.*, **A248**, 506–515, (1986).

WILKINS, B.D., M.J. FLUSS, S.B. KAUFMAN, C.E. GROSS, AND E.P. STEINBERG, "Pulse-Height Defects for Heavy Ions in Silicon Surface-Barrier Detector," *Nucl. Instrum. Meth.*, **92**, 381–391, (1971).

WILKINSON, D.H., "The Phoswich - A Multiple Phosphor," *Rev. Sci. Instrum.*, **23**, 414–417, (1952).

XU, Y., J.B. WEAVER, D.M. HEALY, JR., AND J. LU, "Wavelet Transform Domain Filters: A Spatially Selective Noise Filtration Technique," *IEEE Trans. Image Proc.*, **3**, 747–758, (1994).

YACOUT, A.M. AND W.L. DUNN, "Application of the Inverse Monte Carlo Method to Energy-Dispersive X-Ray Fluorescence," *Adv. X-Ray Anal.*, **30**, 113–120, (1987).

ZEIDMAN, B., W. HENNING, AND D.G. KOVAR, "A High-Resolution Mass-Identification System for Heavy Ions," **118**, 361–365, (1974).

Chapter 21

Mitigation of Background in Gamma-Ray Spectroscopy

Eventually, we reach ... the utmost limits of our [instruments]. There, we measure shadows, and we search among ghostly errors of measurement for landmarks that are scarcely more substantial.

Edwin Powell Hubble

Everything in the universe is constantly being irradiated by different types of directly or indirectly ionizing radiation. On Earth we and our scientific instruments are bathed in radiation from many different sources. In particular, radiation detectors respond in varying degrees to the different types of background radiation incident on them (or even produced by them). Thus, any measurement of the radiation from a particular source with any of the many detectors discussed in this book also contains responses from the ever present background radiation field. The separation of the foreground response from the source under study from the background response is a key element in radiation measurements. This separation generally becomes increasingly difficult as the foreground component decreases in the combined responses and as the stochastic nature of the detector response becomes more pronounced.

To increase the strength of the foreground component and to reduce stochastic fluctuations in the detector measurement, several approaches can be used. Generally placing a detector close to a compact radiation source that is being analyzed increases the response rate by way of the inverse square law while keeping the background response rate relatively unchanged. Increasing the measurement time also generally reduces statistical uncertainties in the output response provided that the radiation source does not vary appreciably during the measurement time. The use of a detector with a larger sensitive volume, while increasing the foreground response rate, also increases the background response rate; however, because of the larger number of recorded events, statistical uncertainties are reduced. Often shielding structures are used to reduce the amount of background radiation that reaches the detector. The use of detector shielding is especially important to reduce the background in *low-level* counting applications involving weak sources. Finally, in some cases multiple detectors with coincident and anti-coincidence circuitry can be used to reject signals produced by background radiation.

To understand how background radiation affects the response of a detector, it is important to know how the detector responds to different types of radiation (well covered in the preceding chapters), the sources and strengths of background radiation, and how different materials and techniques can be used to shield a detector from background radiation. Background radiation and methods to reduce its impact are the subject of this chapter.

21.1 Sources of Background Radiation

There are many sources of background radiation which can affect the response of a detector. These include (1) cosmic rays incident on the earth's atmosphere, (2) secondary particles produced by cosmic rays interacting in the atmosphere, (3) naturally occurring radionuclides in materials near the detector, (4) naturally occurring radionuclides in the detector itself, (5) radioactivity in the earth that produces *terrestrial radiation* and in materials far from the detector, (6) radioactivity in the air surrounding the detector, and (7) man-made radiation sources from particle accelerators, x-ray machines, and radionuclides from weapons tests or created for medical or industrial use.

21.1.1 Cosmic Radiation

The earth is bombarded continuously by radiation originating from our sun, from sources within our galaxy, and from sources far beyond our galaxy. The radiation as it reaches the earth's atmosphere consists of 99% of high-energy atomic nuclei and 1% electrons. Cosmic rays also contain some photons but with energies far below those of the atomic nuclei. The highest observed photon energy is about 10^{14} eV, while the highest energy for a fermionic cosmic ray is an astounding 3×10^{20} eV. The origins of the extra galactic cosmic rays is still somewhat mysterious, but recent satellite observations suggest the supernova explosions and active galactic centers produce some of this radiation.

Hydrogen nuclei (protons) constitute the major material component (90%) of cosmic rays, along with alpha particles (9%) and some heavier atoms, decreasing in importance with increasing atomic number. Cascades of nuclear interactions in the atmosphere, initiated by primary cosmic rays, give rise to many types of secondary particles such as neutrons, kaons and pions. The kaons almost all decay to muons before reaching the surface of the earth.[1] The fluxes (in units of $\text{cm}^{-2}\ \text{s}^{-1}\ \text{MeV}^{-1}$) of the various cosmic rays at the earth's surface are shown in Fig. 21.1. The primary and secondary cosmic rays also interact in the soil (or water) and produce yet more photons and neutrons. The neutron flux is affected mostly by the water content or the soil as is shown in Fig. 21.2. Below 1 MeV the secondary atmospheric produced neutrons begin to slow down as they are moderated by the soil and air producing a characteristic $1/E$ slowing down spectrum that at thermal energies assumes Maxwellian shape. The data for Fig. 21.1 and Fig. 21.2 were generated by the freely downloadable program EXPACS [2018] which is based on the work of Sato et al. [2006, 2008]. With this program the cosmic ray components can be determined at any altitude, location, and at any time in the solar cycle.

As can be seen in Fig. 21.1, the intensity of cosmic rays reaching the surface of the earth is anti-correlated with the solar activity, i.e., when the sun is most active in its 22-year cycle the cosmic ray intensity is the lowest and, when the sun is least active, the cosmic ray intensity is at its highest. This counterintuitive anti-correlation is a result of changes in the solar magnetic field. When the sun spot activity is at its greatest, the solar magnetic field is at its strongest and deflects more cosmic rays away from the inner solar system. Likewise, when the sun is least active, its magnetic field is weakest and allows more cosmic rays to reach earth. Although this effect appears small on the logarithmic scale of Fig. 21.1, the intensity swing is about 50% for neutrons and 30% for photons at an energy of 1 MeV.

[1]Muons μ^+ and antimuons μ^- are radioactive elementary particles with a rest mass energy equivalent of 105.7 MeV and a mean lifetime of 2.197 μs. They decay as

$$\mu^- \rightarrow e^- + \overline{\nu}_e + \nu_\mu \qquad \text{or} \qquad \mu^+ \rightarrow e^+ + \nu_e + \overline{\nu}_\mu .$$

They interact with electrons, but lose little energy per interaction, and produce little bremsstrahlung. Consequently, they are very penetrating. For example, a 1-MeV muon has a CSDA range of 470 cm in water, while a 1000-MeV muon has a 2426-m range [Bevelacqua 2008].

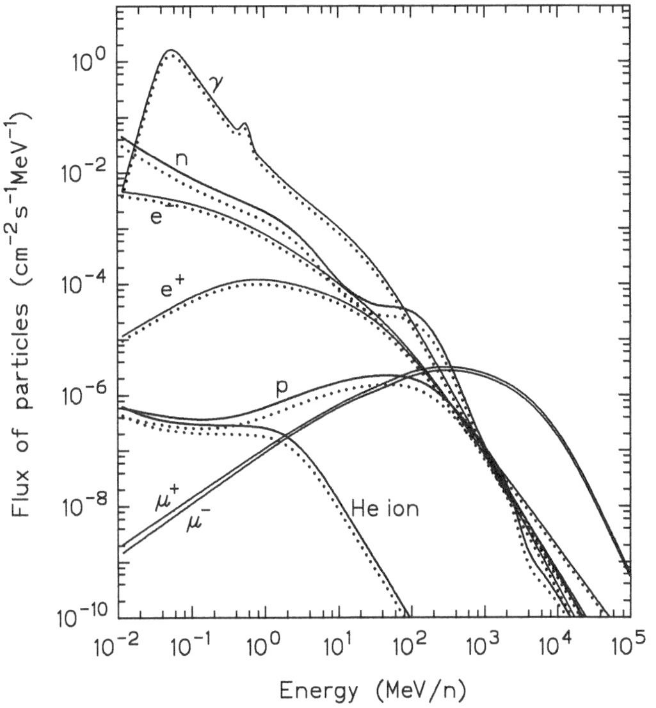

Figure 21.1. Ground-level fluxes of the various components of cosmic rays at the minimum (solid lines) and at the maximum of the solar cycle (dotted lines). The ground is at average elevation of the U.S. (2493 ft). Data generated by EXPACS [2018].

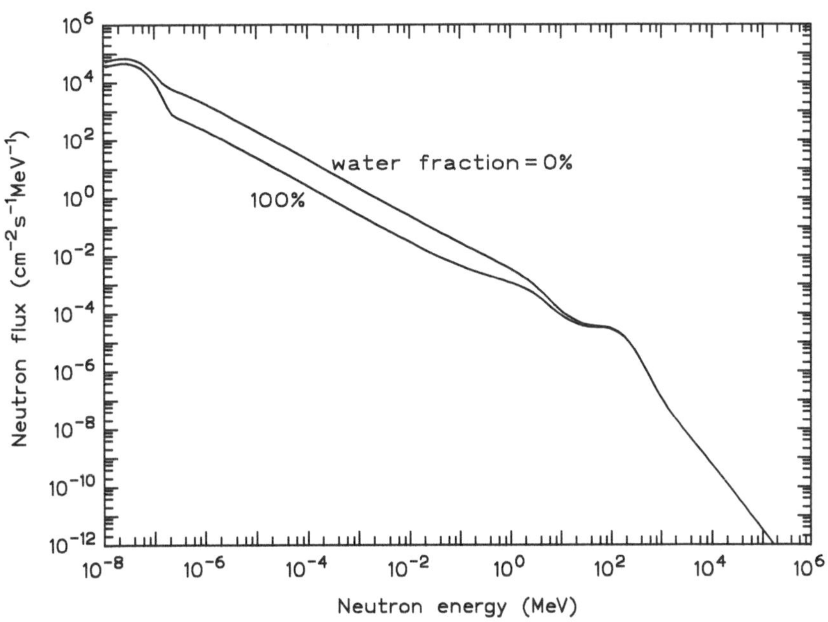

Figure 21.2. The neutron flux produced by cosmic radiation at ground-level for two extreme volume fractions of water in the ground. Data generated by EXPACS [2018].

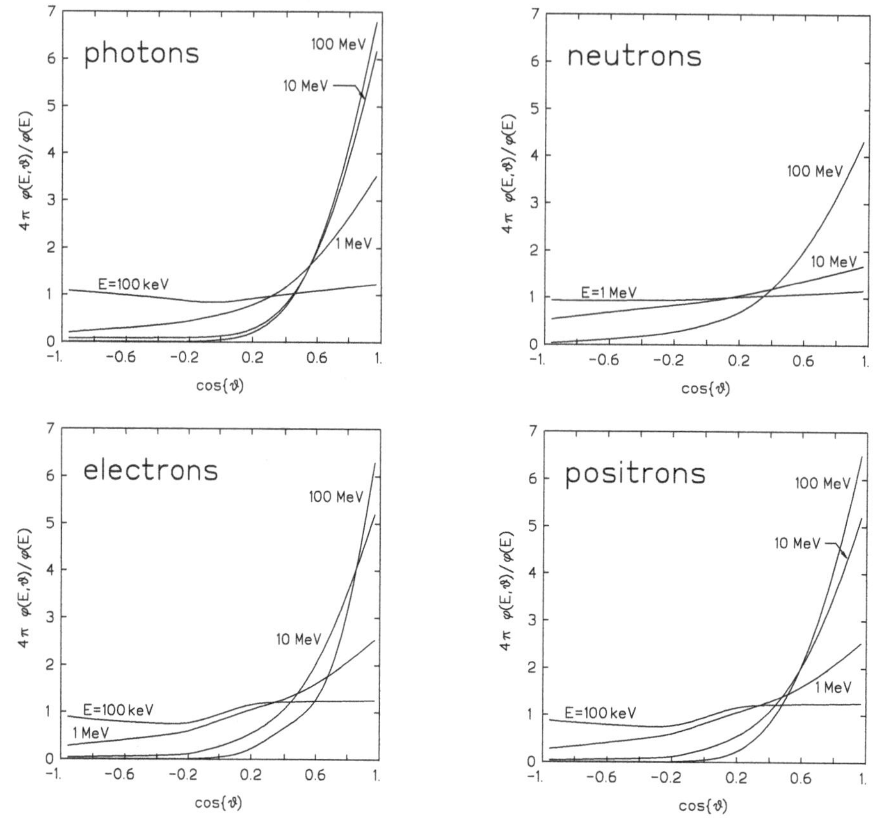

Figure 21.3. Anisotropy of gamma rays, neutrons, electrons and positrons produced by cosmic radiation at ground-level. Plotted is the ratio $R = 4\pi\phi(E,\theta)/\phi(E)$ versus cosine of the zenith angle. For an isotropic flux $R = 1$. The direction $\cos(\theta) = 1$ is the vertically downward direction. The 4π integrated flux is shown in Fig. 21.1. Data were obtained from EXPACS [2018].

The angular distribution of cosmic radiation that reaches the earth's surface depends on the radiation particle since the earth's surface can affect, especially at lower energies, the reflection of the radiation into the atmosphere. For example, muons are negligibly reflected and their angular flux $\phi(E,\theta) \equiv d^2\phi(E,\theta)/\{dE\,d\Omega(\theta)\}$ cm^{-2} s^{-1} MeV^{-1} sr^{-1} is essentially zero in the upward directions, i.e., for nadir angles $\theta > \pi/2$. By contrast, neutrons at energies below about 0.1 MeV are almost isotropically distributed as can be seen in Fig. 21.3. In the figure the ratio $R \equiv 4\pi\phi(E)/\phi(E,\cos(\theta))$ is plotted versus the cosine of the nadir angle. Here $\cos\theta = 1$ refers to the directly downward direction. If the angular flux is isotropic $R = 1$. As can be seen from Fig. 21.3 the angular flux becomes increasingly anisotropic in the downward direction.

Both as a useful measure and as a single cosmic-ray metric, the effective dose equivalent is widely used to describe the intensity of cosmic rays. At earth's surface, cosmic radiation dose rates are largely due to muons and electrons. The intensity and angular distribution of galactic radiation reaching the earth is affected by the earth's magnetic field and perturbed by magnetic disturbances generated by solar flare activity. Consequently, at any given location, cosmic ray doses may vary in time by a factor of 3. At any given time, cosmic ray dose rates at sea level may vary with geomagnetic latitude by as much as a factor of 8, being greatest at the pole and least at the equator. The global average cosmic ray dose equivalent rate at sea level is about 240 μSv per year for the directly ionizing component and 20 μSv per year for the neutron component [UN 1988]. Cosmic ray dose rates also increase with altitude. At geomagnetic latitude 55° N, for example, the absorbed dose rate in tissue approximately doubles with each 2.75 km (9000 ft) increase in altitude, up to 10 km (33,000 ft). The neutron component of the dose equivalent rate increases more rapidly with altitude than does the directly ionizing component and dominates at altitudes above about 6 km [UN 1988].

Solar cosmic rays associated with flares are mainly hydrogen and helium nuclei. While of too low an energy to contribute to radiation doses at the surface of the earth, solar-flare radiation, which fluctuates cyclically with an 11-year period (from to minimum to minimum sun spot activity), perturbs earth's magnetic field and thereby modulates galactic cosmic-ray dose rates with the same period. As the sun becomes more active, the solar wind from the sun increases and deflects more of the cosmic radiation from reaching earth. This decrease in cosmic ray activity at ground level with increasing solar activity is seen in Fig. 21.1 where the dotted lines are ground-level fluxes at maximum solar activity which are seen to be slightly lower than the solid lines at the solar minimum. Thus, maxima in solar flare activity lead to minima in dose rates. Solar flare radiation, in comparison to galactic cosmic rays, is of little significance as a hazard in aircraft flight or low orbital space travel. On the other hand, solar-flare radiation presents a considerable hazard to personnel and equipment in space travel outside the earth's magnetic field.

Released continuously from the sun, as an extension of the corona, is the solar wind, a plasma of low-energy protons and, presumably, very low-energy electrons. The solar wind does not present a radiation hazard, even in interplanetary space travel. However, it does affect the interplanetary magnetic field and the shape of the geomagnetically trapped radiation belts. These radiation belts are thought to be supplied by captured solar-wind particles and by decay into protons and electrons of neutrons created by interactions of galactic cosmic rays in the atmosphere. The trapped radiation can present a significant hazard to personnel and equipment in space missions.

Galactic Cosmic Rays

Cumulative energy spectra for the various components of the galactic cosmic radiation incident on the earth's atmosphere are illustrated in Fig. 21.4. These spectra may be approximated as

$$\phi(E > E_o) \simeq \frac{A}{(M_o c^2 + E_o)^n}, \tag{21.1}$$

in which $\phi(E > E_o)$ is the flux density of particles with energies greater than E_o, $M_o c^2$ is the rest-mass energy equivalent of the particles, A and n are parameters which, generally, depend on the state of the solar cycle.

As a result of nuclear spallation reactions with constituents of the atmosphere, secondary neutrons, protons, and pions, mainly, are produced. Subsequent pion decay results in electrons, photons, neutrons, and muons. Muon decay, in turn, leads to secondary electrons, as do Coulombic scattering interactions of charged particles in the atmosphere.

Except for short-term influences of solar activity, galactic cosmic radiation has been constant in intensity for at least several thousand years. The influence of solar activity is cyclical. The principal variation is on an 11-year cycle, but there are very small diurnal variations and evidence of a 22-year cycle as well as 27-day quasi-periodic variations [Haffner 1967].

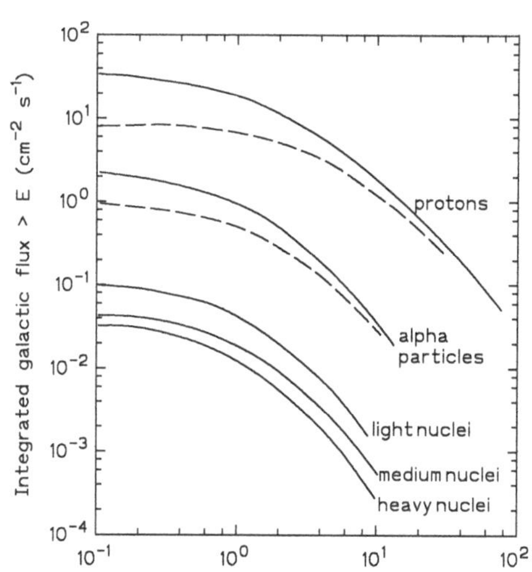

Figure 21.4. Integral energy spectra (i.e., the flux of particles with energies $> E$) for various components of the galactic cosmic radiation. Light nuclei, $3 \leq Z \leq 5$, medium nuclei $6 \leq Z \leq 9$, and heavy nuclei $Z \geq 10$. Solid lines are for the solar minimum and dashed lines are for the solar maximum. Data from Haffner [1967].

The geomagnetic field of the earth is responsible for limiting the number of cosmic rays which can reach the atmosphere. For particles of atomic number Z vertically incident at geomagnetic latitude λ, the minimum momentum for a cosmic ray to reach the atmosphere is proportional to $Z \cos^4 \lambda$ [Haffner 1967]. This dependence accounts for a strong effect of latitude on cosmic-ray dose rates.

Solar Flare Particulate Radiation

Solar flare particles are mostly protons and alpha particles, predominantly the former. Electrons are thought to be emitted as well, but with energies less than those of protons by a factor equal to the ratio of the rest masses. These flares can disrupt earth's ionosphere which affects long-range radio communications. They have little effect at the earth's surface and produce little change in the background experienced by radiation detectors. However, solar flares can be accompanied by coronal mass ejections (CME) which can cause large geomagnetic storms that, in turn, can disable satellites and shut down electrical grids for hours. For space missions outside the atmosphere, solar flares must be considered in order to protect both astronauts and their equipment.

Table 21.1. Measured fluences (cm^{-2}) outside earth's atmosphere produced by the solar flare that occurred July 10, 1959. Source: Haffner [1967].

protons		alpha particles	
$E > 10$ MeV	4.5×10^9	$E > 40$ MeV	1.6×10^8
$E > 30$ MeV	1.0×10^9	$E > 120$ MeV	2.4×10^7
$E > 100$ MeV	1.4×10^8	$E > 400$ MeV	5.0×10^5

Energy spectra are highly variable, as are temporal variations of intensity. Haffner [1967] describes records of solar flares during the 1956-1961 interval which encompasses the 1959 maximum in activity. The large event of 10 July, 1959, for example, resulted in the fluences (cm^{-2}) outside the earth's atmosphere given in Table 21.1. A typical course of events for a flare is as follows. Gamma and x-ray emission takes place over about 4 hours as is evidenced by radio interference. The first significant quantities of protons reach the earth after about 15 hours and peak proton intensity occurs at about 40 hours after the solar eruption [NCRP 1989].

21.1.2 Natural Occurring Radioactivity

Cosmogenic Radionuclides

Cosmic-ray interactions with constituents of the atmosphere, sea, or earth, but mostly with the atmosphere, lead directly to radioactive products. Capture of secondary neutrons produced in primary interactions of cosmic rays leads to the formation of many more radionuclides. However, the thermal-neutron flux at sea level is only about 8 cm^{-2} h^{-1} [Morgan 1967] and the total neutron flux density is only about 30 cm^{-2} h^{-1} [UN 1988]. Thus, neutron capture in the earth's crust or the sea is of little importance in comparison with capture in the atmosphere, except for production of long-lived ^{36}Cl [NCRP 1975]. Of the nuclides produced in the atmosphere, only ^{3}H, ^{7}Be, ^{14}C, and ^{22}Na contribute appreciably to human radiation exposure. The distribution of these cosmogenic radionuclides is shown in Table 21.2. Radionuclides borne by the 10^7 kg of "space dust" reaching the earth annually have a total specific activity of less than 450 pCi kg^{-1} and result in only extremely low atmospheric concentrations [NCRP 1975].

Table 21.2. Global distribution of cosmogenic radionuclides produced in the atmosphere.

	^{3}H	^{7}Be	^{14}C	^{22}Na
Global Inventory (PBq)	1300	37	8500	0.4
Distribution (%)				
stratosphere	6.8	60	0.3	25
troposphere	0.4	11	1.6	1.7
land surface/biosphere	27	8	4	21
mixed ocean layer	35	20	2.2	44
deep ocean	30	0.2	92	8
ocean sediments			0.4	

Source: UN [1982].

Over the past century, combustion of fossil fuels and the emission of CO_2 not containing ^{14}C has diluted the cosmogenic content of ^{14}C in the environment. Moreover, since World War II, artificial introduction of

^{3}H, ^{14}C, and other radionuclides,[2] into the environment by human activity has been significant, especially as a result of atmospheric nuclear-weapons tests. Consequently, these nuclides no longer exist in natural equilibria in the environment.

Tritium is produced in the atmosphere mainly from the ^{14}N(n,t)^{12}C and the ^{16}O(n,t)^{14}N reactions. It has a half-life of 12.3 years and, upon decay, releases one beta particle with maximum energy 18.6 keV (average energy 5.7 keV). Tritium exists in nature almost exclusively as HTO but may be partially incorporated into organic compounds such as those in wood, a common construction material. In continental surface waters the ratio of ^{3}H to stable hydrogen is 3.3×10^{-18} (corresponding to 0.00039 Bq cm^{-3}). Because ^{3}H emits only a low-energy beta particle, it is generally of little concern in most low-level counting applications. Only if a liquid scintillator is used to look for tritium leaks from nuclear facilities that also produce tritium does the incorporation of cosmogenic tritium in the scintillator become a concern.

The nuclide ^{14}C is produced mainly from the ^{14}N(n,p)^{14}C reaction. It exists in the atmosphere as CO_2, but the main reservoir is the ocean. It has a half-life of 5730 years and decays by beta particle emission, each decay resulting in an electron of maximum energy 157 keV (average energy 49.5 keV). The natural atomic ratio of ^{14}C to stable carbon is 1.2×10^{-12} (corresponding to 0.226 Bq ^{14}C per gram of carbon) [NCRP 1975]. Because ^{14}C emits only beta particles, it contributes negligibly to the photon background save low-energy bremsstrahlung photons below 157 keV. Only in low-energy beta counting measurements is it of concern.

The ^{7}Be radionuclide, with a half-life of 53.4 days, is also produced by cosmic ray interactions with nitrogen and oxygen in the atmosphere. It decays by electron capture, 10.4 percent of the captures resulting in the emission of a 478-keV gamma ray. Environmental concentrations in temperate regions are about 3000 Bq m^{-3} in surface air and 700 Bq m^{-3} in rainwater [UN 1982]. A background gamma-ray spectrum often shows a very small 478-keV full energy peak.

The radionuclide ^{22}Na is produced in spallation interactions of atmospheric argon with high-energy cosmic-ray secondary neutrons. It has a half-life of 2.602 y, decaying by positron emission (90 percent) and electron capture (10 percent). The positron has a maximum energy of 546 keV (average energy 216 keV). Essentially all decays are accompanied by emission of a 1.275-MeV gamma ray from the excited ^{22}Ne daughter. Although the 1.275-MeV gamma ray may be observed in a background spectrum taken by a very sensitive gamma-ray spectrometer, the 0.511-MeV positron annihilation photon which usually is present in a background spectrum is attributable mostly to cosmic-ray positrons.

Activation by Cosmic Ray Neutrons

The neutrons produced by spallation reactions in the atmosphere and ground by cosmic rays can lead to the buildup of radionuclides in materials stored at ground level. Although of little radiological concern, these radionuclides can contribute to the background spectrum in underground ultra-low-level facilities. For example, pre-World War II steel, i.e., steel without any fission products, was tried as a shield against the gamma rays emitted from lead shielding but found to be too contaminated by ^{54}Mn ($T_{1/2} = 312.1$ d) that was produced by cosmic ray neutrons interacting with ^{54}Fe when the old steel was stored for years at ground level [Brodzinski et al. 1985]. Thus, steel should not be used in such ultra-low-level facilities unless it has been stored underground for many years. However, for less sensitive facilities, activation of steel is usually of little concern.

One other example is the use of copper, which is used in spectrometer facilities for both shielding and detector electronics. Electrolytic copper, while expensive, generally has very low impurity levels from the

[2] One such nuclide is ^{137}Cs ($T_{1/2} = 30.0$ y) which, upon decay, emits a 661.7-keV gamma ray, which is found in most background gamma-ray spectra. The presence of this radionuclide in a sample measured in a low-level counting chamber indicates the sample dates from the post 1950 era and is used, for example, to determine if a purportedly old wine (or its bottle) is truly made of pre-1950 material.

naturally occurring radionuclides. However, storage at ground level allows the buildup of ^{60}Co from energetic cosmic ray neutrons through the ^{63}Cu(n,α)^{60}Co reaction. So copper for use in ultra-low-level counting facilities must be moved underground as soon as possible after its purification.

Primordial Radionuclides

Of the many radioactive species present when the earth formed 4 billion years ago, some 17 extremely long-lived radionuclides still exist as singly occurring or isolated radionuclides, i.e., as radionuclides not belonging to a decay chain. These terrigenous (primordial) radionuclides are listed in Table 5.1. Of these primordial radionuclides, only ^{40}K contributes significantly to the gamma-ray background.

The radionuclide ^{40}K is a major contributor to human exposure from natural radiation. Present in an isotopic abundance of 0.0118 percent, it has a half-life of 1.227×10^9 y, decaying both by electron capture (11 percent) and beta-particle emission (89 percent). The beta particle has a maximum energy of 1312 keV (average energy 509 keV). Electron capture results in emission of a 1461-keV gamma ray in 10.7 percent of decays. Very low energy Auger electrons and x rays are also released from electron capture (Kocher 1981). The average elemental concentrations of potassium in Reference Man[3] is 2 percent. Annual doses in Reference Man are 140 μGy to bone surface, 170 μGy on average to soft tissue, and 270 μGy to red marrow (UN 1982). ^{40}K also contributes in a major way to external exposure. The average specific activity of the nuclide in soil, 12 pCi g^{-1} (0.44 Bq g^{-1}), results in an annual whole-body dose equivalent of 120 μSv (12 mrem) [UN 1982]. However, actual specific activities can vary considerably depending on the type of soil as is shown later in Table 21.5.

Decay Series of Primordial Origin

Each naturally occurring radioactive nuclide with $Z > 83$ is a member of one of three long decay chains, or radioactive series, stretching through the upper part of the Chart of the Nuclides. These radionuclides decay by α or β^- emission and they have the property that the number of nucleons (mass number) A for each member of a given decay series can be expressed as $4n + i$, where n is an integer and i is a constant (0, 2, or 3) for each series. The three naturally occurring series are named the thorium ($4n$), uranium ($4n + 2$), and actinium ($4n + 3$) series, named after the radionuclide at, or near, the head of the series.

Two of these primordial decay series, identified by the long-lived parents ^{238}U and ^{232}Th contribute appreciably to the natural gamma-ray background. Another series headed by the primordial ^{235}U radionuclide contributes very little to the radiation background and makes a negligible contribution to the radiation background.[4]

The three naturally occurring series are shown schematically in Figs. 21.5 and 21.6, and the principal radiations emitted by members of the two important series are given in Tables 21.3 and 21.4. Within each series are subseries headed by a radionuclide with a half-life much greater than those of its daughters. Although all the members of a series are not likely to be in radioactive secular equilibrium in nature because of chemical or physical reasons, members of a subseries are more likely to be so. The subseries headed by the gases ^{220}Rn and ^{222}Rn are of special importance and are treated separately in a subsequent section.

The radiations emitted by members of the primordial decay chains and listed in Tables 21.3 and 21.4 are but a small fraction of all the radiations emitted. Most have frequencies less than the 0.5% cutoff used in

---

[3]The Reference Man is a hypothetical person with the anatomical and physiological characteristics of an average individual and is widely used in calculations that assess internal doses to various organs from various radiation sources. Specifically, he is 20-30 years of age, weighs 70 kg, is 170 cm in height, and lives in a climate with an average temperature of 10° to 20°C [ICRP 1975].

[4]There is no naturally occurring series represented by $4n+1$. This series was recreated after $^{241}_{94}$Pu was made in nuclear reactors. This series does not occur naturally since the half-life of the longest lived member of the series, $^{237}_{93}$Np, is only 2.14×10^6 y, much shorter than the lifetime of the earth. Hence, any members of this series that were in the original material of the solar system have long since decayed away.

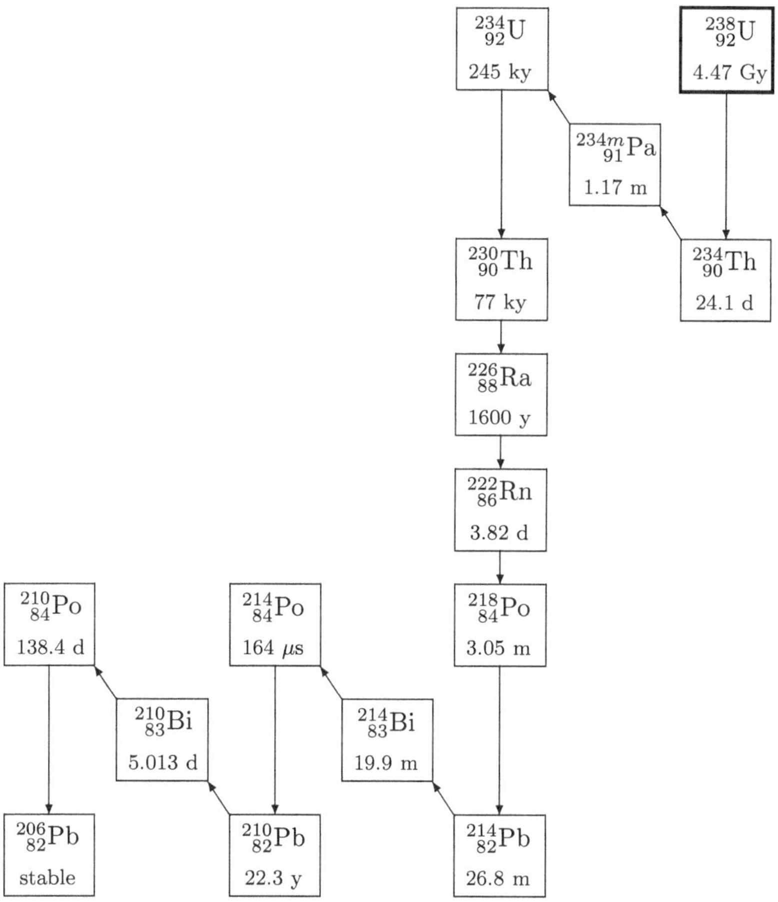

Figure 21.5. The $^{238}_{92}$U $(4n+2)$ natural decay series. Alpha decay is depicted by downward arrows and β^- decay by arrows upward and to the left. Not shown are (1) the isomeric transition to $^{234}_{91}$Pa (0.16%) followed by beta decay to ^{234}U, (2) beta decay of ^{218}Po to ^{218}At (0.020%) followed by alpha decay to ^{214}Bi, (3) alpha decay of ^{214}Bi to ^{210}Tl (0.0210%) followed by beta decay to ^{210}Pb, and (4) alpha decay of ^{210}Bi to ^{206}Tl (0.000132%) followed by beta decay to ^{206}Pb. From Shultis and Faw [2017].

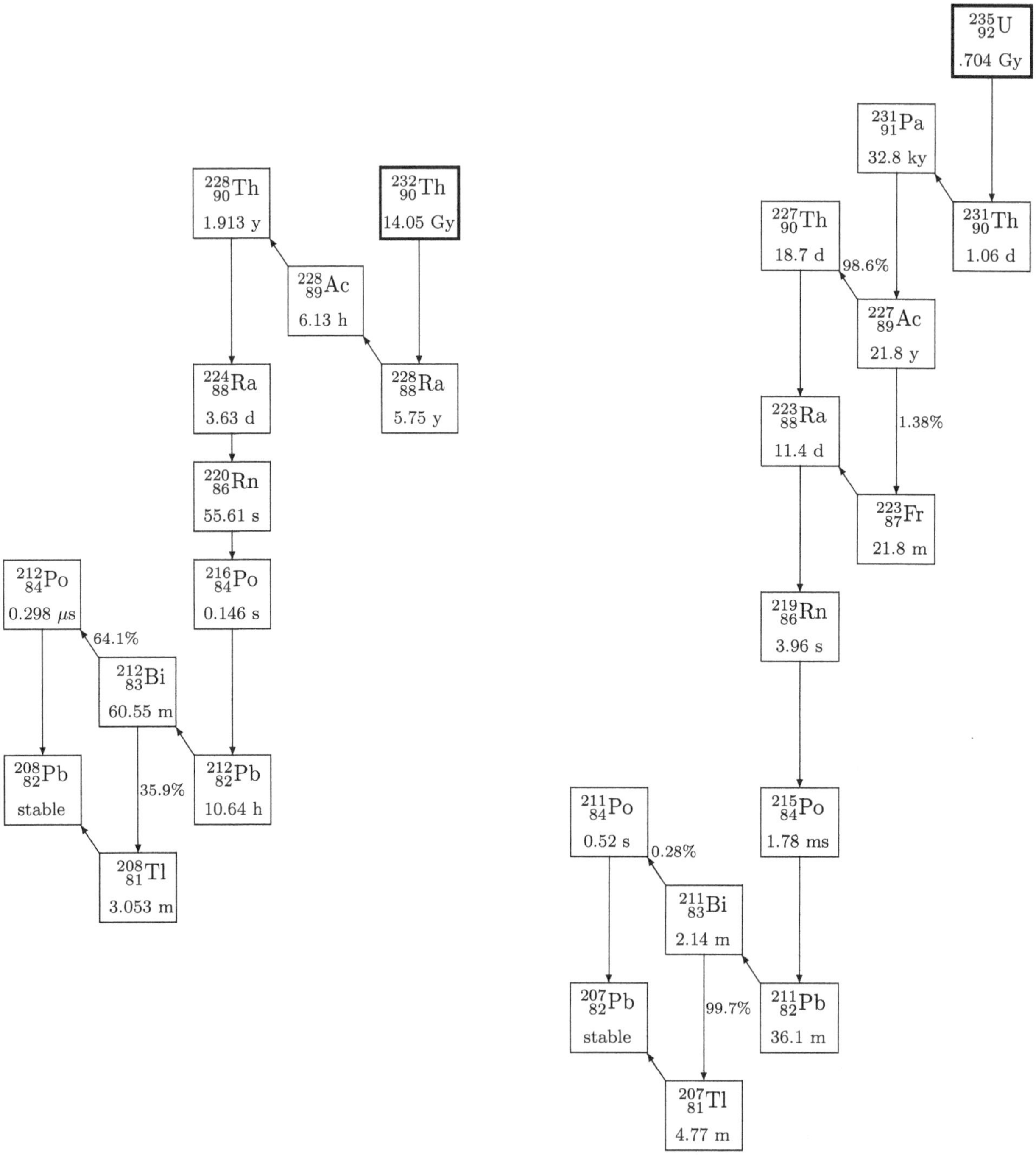

Figure 21.6. The $^{232}_{90}$Th ($4n$) natural decay series (left) and the $^{235}_{92}$U ($4n+3$) natural decay series (right). Alpha decay is depicted by downward arrows and β^- decay by arrows upward and to the left. Not shown in the $^{235}_{92}$U series on the right are two very minor side chains: (1) alpha decay of ^{223}Fr (0.004%) to ^{85}At which beta decays to ^{223}Ra (3%) or alpha decays to ^{215}Bi which beta decays to ^{215}Po, and (2) the beta decay of ^{215}At ($< 0.001\%$) followed by alpha decay to ^{211}Bi. From Shultis and Faw [2017].

Table 21.3. Principal radiations from nuclides in the ^{232}Th decay series. Not listed are x rays, conversion and Auger electrons, and beta or gamma rays with frequencies less than 0.5%.

Nuclide	Alpha particles E_α (keV)	Alpha particles freq. (%)	Beta particles E_{max} (keV)	Beta particles E_{av} (keV)	Beta particles freq. (%)	Gamma rays E_γ (keV)	Gamma rays freq. (%)
^{232}Th	3811.1	0.069				12.3	7.1
	3947.1	21.7					
	4012.3	78.2					
^{228}Ra		39.5	7.24	100			
^{228}Ac			410	117.50	1.76	99.51	1.26
			446	129.00	2.43	129.07	2.42
			451	130.82	1.12	153.98	0.72
			488	142.79	4.19	209.25	3.89
			491	143.73	3.0	214.85	0.76
			496	145.32	1.15	270.25	3.46
			603	181.0	7.6	328.00	2.95
			684	208.9	0.60	338.32	11.27
			702	215.3	1.2	409.46	1.92
			907	288.8	0.67	463.00	4.40
			966	310.2	3.11	562.50	0.87
			981	315.7	5.8	674.75	2.10
			1011	327.0	5.90	726.86	0.62
			1111	364.7	3.11	755.32	1.00
			1165	385.0	29.9	772.29	1.49
			1806	609.3	0.59	794.95	4.25
			1738	609.7	11.65	830.49	0.54
			2076	747.0	7	835.71	1.61
						904.20	0.77
						911.20	25.8
						964.77	4.99
						968.97	15.8
						1247.08	0.50
						1459.14	0.83
						1588.20	3.22
						1630.63	1.51
^{228}Th	5340.3	26.0				84.37	1.19
	5423.2	72.7					
^{224}Ra	5448.6	5.06				240.0	4.10
	5685.4	94.92					
^{220}Rn	6288.1	99.90					
^{216}Po	6779	99.89					
^{212}Pb			154.6	41.1	5.08	115.2	0.60
			331.3	93.5	83.1	238.6	43.6
			569.9	171.7	11.9	300.1	3.30
^{212}Bi	6010	5.0				39.86	1.06
	6300	26.0					
	6340	35.0					
^{212}Po	8784.9	100.					
^{208}Tl			1038.1	342.9	3.18	252.6	0.78
			1079.2	358.7	0.63	277.4	6.8
			1290.6	441.5	24.2	510.8	22.6
			1523.9	535.4	22.2	583.2	85.0
			1801.3	649.5	49.1	763.1	1.79
						860.6	12.5
						2614.5	99.7

Source: NuDat [2017].

Table 21.4. Principal radiations from nuclides in the ^{238}U decay series. Not listed are x rays, conversion and Auger electrons, and beta or gamma rays with frequencies less than 0.5%.

Nuclide	Alpha particles E_α (keV)	Alpha particles freq. (%)	Beta particles E_{max} (keV)	Beta particles E_{av} (keV)	Beta particles freq. (%)	Gamma rays E_γ (keV)	Gamma rays freq. (%)
^{238}U	4147	23					
	4196	77					
^{234}Th			86	22.3	1.5	63.29	3.7
			106	27.7	6.4	92.38	2.13
			107	27.8	14.0	92.80	2.10
			199	53.9	78.0		
^{234m}Pa			1224	405.6	1.00	1001.0	0.842
			1459	496.0	0.95		
			2269	820.5	97.57		
^{234}U	4722.4	28.42					
	4774.6	71.38					
^{230}Th	4620.5	23.40					
	4687.0	76.3					
^{226}Ra	4601	6.61				186.2	3.64
	4784.3	93.84					
^{222}Rn	5489.5	99.92					
^{218}Po	6002.4	99.98					
^{214}Pb			180	48.3	2.75	53.23	1.08
			485	143.1	1.04	241.99	7.25
			667	205.5	45.9	258.86	0.531
			724	225.6	40.2	295.22	18.42
			1019	334.9	11.0	351.93	35.60
						785.96	1.06
						839.07	0.583
^{214}Bi			788	248.2	1.244	609.32	45.49
			822	260.9	2.78	665.45	1.531
			977	318.2	0.557	768.36	4.89
			1066	352.1	5.60	806.18	1.264
			1077	356.5	0.855	934.06	3.107
			1151	385.1	4.345	1120.29	14.92
			1253	424.6	2.450	1155.21	1.633
			1259	427.1	1.431	1238.12	5.834
			1275	433.5	1.177	1280.98	1.434
			1380	474.9	1.588	1377.67	3.988
			1423	492.0	8.14	1385.31	0.793
			1505	525.3	16.96	1401.51	1.330
			1540	539.4	17.57	1407.99	2.394
			1609	567.2	0.61	1509.21	2.130
			1727	615.4	3.12	1583.20	0.705
			1855	668.1	0.89	1661.27	1.047
			1892	683.7	7.35	1729.60	2.878
			2661	1007.5	0.58	1764.49	15.30
			3270	1268.8	19.10	1847.43	2.025
						2118.51	1.160
						2204.06	4.924
						2447.70	1.548
^{214}Po	7686.8	99.989					
^{210}Pb			17.0	4.16	84	46.5	4.05
			63.5	16.16	16		
^{210}Bi			1162.2	389.0	100.0	265.6	51.0
						304.6	28.0
						329.6	0.66
						344.3	0.71
						368.9	0.66
^{210}Po	5304.3	100.0					

Source: NuDat [2017].

these tables. For example ^{228}Ac emits a known 54 different beta particles and an astounding 246 different energy gamma rays! The background energy spectrum, thus, is almost a continuum of different energy particles emitted by the decay chain radionuclides, of which only a few occur sufficiently frequently that their contribution to the gamma-ray spectrum is identifiable.

Radionuclide Concentrations in Different Materials

Primordial radionuclides and their radioactive daughters are present in all natural materials. Such radioactive material is classified as NORM (Naturally Occurring Radioactive Material). The concentrations of these radionuclides is often altered by extraction, manufacturing, and processing of the natural materials. These materials with altered radionuclide concentrations are sometimes classified as TENORM (Technologically Enhanced Normal Radioactive Material). For regulatory purposes some man-made radioactivity such as fallout from nuclear weapons tests, is also considered part of the natural background.

Table 21.5. Typical naturally occurring radionuclide concentrations in U.S. rocks and soils.

Rock/soil	Average specific activity (Bq kg^{-1})			absorbed dose rate in air (nGy h^{-1})
	^{40}K	^{238}U	^{232}Th	
Igneous Rock				
acidic	100	60	80	120
intermediate	70	23	32	62
mafic	240	11	11	23
ultrabasic	150	0.4	25	23
Sedimentary Rock				
limestone	90	2.8	7	20
carbonate	–	2.7	8	17
sandstone	370	20	8	32
shale	700	45	45	79
Soil Type				
serozem	650	30	50	74
gray-brown	700	28	40	69
chestnut	550	25	37	60
chernozem	400	20	36	51
gray forest	370	18	27	41
sodpodzolic	300	15	22	34
podzolic	150	9	12	19
boggy	90	6	6	11
world average	370	25	25	46
Coal				
United States	50	20	22	

Source: UN [1977].

Typical natural radionuclide activity concentrations in several soils and rocks are listed in Table 21.5. Also listed are dose rates in air 1 m above the nuclide-bearing material, calculated on the basis of radioactive equilibrium throughout the entire series. These dose rates are seen to be quite comparable to the background dose rate of 88 nGy/h to which the average person in the U.S. is exposed from external terrestrial radionuclides.

21.1.3 Airborne Radioactivity

Sources of radiation are also present in the air. Dust from soil or construction activities can lead to airborne naturally occurring radionuclides. However, a constant source of airborne radioactivity is from radon and its daughters. Decay of radon and its daughters in the atmosphere leads first to individual unattached ions or neutral atoms. The ions or atoms may become attached to aerosol particles, the attachment rate depending in a complex manner on the size distribution of the particles. Radioactive decay of an attached ion or atom, because of recoil, usually results in an unattached daughter ion or atom. Either attached or unattached species may be deposited (plate out) on surfaces, especially in indoor spaces, the rate depending on the surface-to-volume ratio of the space. Because of plate out, radon daughter products in the atmosphere are not likely to be in equilibrium with the parent.

^{222}Rn and its daughters ordinarily present a greater radiological hazard than ^{220}Rn (thoron) and its daughters, largely because the much shorter half-life of ^{220}Rn makes decay more likely prior to release into the atmosphere. Relatively little is known about the rates of release, diffusion, inhalation, plate-out, attachment, etc. for ^{220}Rn and its daughters, and most studies have emphasized the more important ^{222}Rn series shown below.

$$^{222}_{86}\text{Rn} \xrightarrow[3.82\text{ d}]{\alpha} {}^{218}_{84}\text{Po} \xrightarrow[3.05\text{ m}]{\alpha} {}^{214}_{82}\text{Pb} \xrightarrow[26.8\text{ m}]{\beta^-} {}^{214}_{83}\text{Bi} \xrightarrow[19.9\text{ m}]{\beta^-} {}^{214}_{84}\text{Po} \xrightarrow[164\mu\text{s}]{\alpha} {}^{210}_{82}\text{Pb (22.2 y)}. \tag{21.2}$$

Airborne decay products beyond ^{214}Po are of little consequence since ^{210}Pb has such a long half-life that it becomes bound to surfaces before it decays.

The alpha particles emitted by these airborne radionuclides are of little concern in most radiation measurements except in the case of alpha-particle spectrometers for which they contribute to the energy spectrum. The alpha particles from radon daughters are also of radiological concern because they can become lodged in the pulmonary-bronchial system and produce large doses to the epithelial lung tissues [Faw and Shultis 1999]. Of greater concern from a radiation measurement perspective are the gamma rays emitted by ^{214}Pb and ^{214}Bi (see Table 21.4). These gamma rays are prominent in most gamma-ray background spectra.

Outdoor Radon Activity

The noble gas radon diffuses into the atmosphere from rocks, soils, and building materials containing progenitor radionuclides. The exhalation rate of ^{222}Rn from rocks and soils is highly variable, ranging from 0.2 to 70 mBq m^{-2} s^{-1}. An area-weighted average for continental areas, exclusive of Antarctica and Greenland, is 16 mBq m^{-2} s^{-1} [UN 1988]. Exhalation from the surface of the sea is only about 1% of that from land areas. Rain, snow, and freezing decrease exhalation rates so that they are generally lower in winter than in summer. Barometric pressure and wind speed also affect exhalation rates, decreasing pressure or increasing wind speed causing the rate to increase.

Radon and daughter products are dispersed in the atmosphere by turbulent diffusion and convection. Extreme conditions of atmospheric stability can lead to ground-level concentrations differing by as much as a factor of 100. Associated with conditions of atmospheric stability are marked diurnal variations in ground-level concentrations, with minima at noon and maxima at midnight being associated respectively with greater and lesser atmospheric instability. Concentrations decrease with altitude, with those of ^{220}Rn decreasing more rapidly because of the shorter half-life of the parent. In the absence of precipitation, parent-daughter equilibrium is approached at elevations exceeding about 100 m. However, rainfall may remove daughter products from the atmosphere, causing absorbed dose rates in air at ground level to be as much as twice normal.

Mean annual concentrations of ^{222}Rn above continental areas range from 1 to 10 Bq m^{-3}, with 5 Bq m^{-3} and an equilibrium factor of 0.8 being typical (UN 1988). Typical mean annual concentrations over the ocean and over island areas are 0.1 Bq m^{-3}. Concentrations of ^{220}Rn daughters are typically 10% of those of ^{222}Rn daughters. Anomalously high levels often exist near coal-fired and geothermal power stations and near uranium-mine tailings.

Indoor Radon Activity

Most spectroscopy is performed inside some structure or in rooms that are located at or below ground level where concentrations of radon and its daughters can build up to levels several times those outdoors. Because of the short lifetimes of the daughters, the members of the radon decay chain ideally would be in secular equilibrium, i.e., the activity concentration of the daughters C_i would equal that of ^{222}Ra C_o. However, because of plate-out and settling, members of the decay chain are seldom in equilibrium. An *equilibrium factor* f_i, defined as $f_i = C_i/C_o$, typically has values around 0.5.

Indoor radon sources include exhalation from soil, building materials, water, ventilation air, and natural gas if unvented (as used in cooking). Exhalation from soil can be a significant contribution if a building has cracks or other penetrations in the basement structure or if the building has unpaved and unventilated crawl spaces. Water usage in buildings can present a significant source of Rn if concentrations in the water exceed about 10 kBq m^{-3}. In the United States, concentrations of ^{222}Rn in natural gas average 0.7 kBq m^{-3} but in some states the average is as high as 2 kBq m^{-3}. Radon precursor concentrations in building materials are highly variable. Greater concentrations occur in phosphogypsum and in concrete based on fly ash or alum shale. Sealing concrete surfaces with materials such as epoxy-resin paints greatly reduces radon exhalation. Indoor airborne radon and daughter concentrations depend not only on exhalation of the parent gas from surfaces, but also on intake and loss of radionuclides through ventilation and plate-out of daughters on interior surfaces. The rate of plate-out depends on the unattached fraction which is influenced greatly by humidity and by the presence of aerosols (smoke, dust, etc.).

21.1.4 Modern Radiation Sources

In our modern technological world, new sources of ionizing radiation have been introduced. Radioisotopes made by humans are used in many applications in medicine and industry. Likewise machines that produce penetrating radiation are now widely used. For example, electron accelerators are widely used in cancer therapy and in x-ray machines for inspecting welds and forgings. Radiation measurement facilities, such as border inspection portals are often subjected to industrial sources of x rays. Although sources of high-energy photon radiation are usually well shielded horizontally by thick building walls, they often have limited shielding provided by a roof and, hence, radiation escaping upwards can scatter in the atmosphere and contribute to the background radiation at locations far removed from the radiation facility. This source of background is termed *skyshine* and requires special techniques to assess its severity [Shultis and Faw 2000].

21.2 Mitigation of the Radiation Background

The design goal for a radiation measurement system is to reduce the background to sufficiently low levels so that meaningful measurements of a sample can be made and analyzed. For the case of samples with extremely small activities, the background must be made very low, sometimes to a few counts per day or per week. Fortunately, most measurement requirements are not as severe. Ideally, the background signals should be comparable or less than those produced by the sample being analyzed.

In this section the background for gamma-ray and x-ray measurements are primarily considered. To appreciate the background we all live in, a typical gamma-ray energy spectrum is shown in Fig. 21.7 and the most often seen contributors to the various full energy peaks seen in such a spectrum are listed in Table 21.6. The background spectrum of Fig. 21.7 has an overall (sum over all energy channels) of about 35 cps, which is higher than one might like, but is unavoidable in this measurement laboratory with little protection from cosmic rays and which also is enclosed by thick concrete walls, floor and ceiling, all containing significant ^{40}K.

Table 21.6. Full energy gamma-ray peaks observed in a Ge detector background spectrum. The most intense are indicated by boldface type.

Origin	Energy (keV)	Origin	Energy (keV)	Origin	Energy (keV)
Bi $K_{\alpha 2}$	74.82	^{125}Sb	427.9	^{208}Tl	860.6
Bi $K_{\alpha 1}$	77.11	^{228}Ac	463.0	^{228}Ac	**911.2**
^{228}Th	84.40	^{7}Be	477.5	^{214}Bi	934.1
Bi $K_{\beta 1}$	87.35	^{208}Tl	510.7	^{228}Ac	964.8
Bi $K_{\beta 2}$	89.78	annilation	**511.0**	^{228}Ac	**969.0**
^{234}Th	92.58	^{106}Ru	511.8	^{214}Bi	**1120.3**
^{228}Ac	129.1	^{208}Tl	**583.2**	^{214}Bi	1155.2
^{144}Ce	133.5	^{214}Bi	**609.3**	^{214}Bi	**1238.1**
^{228}Ac	153.9	^{106}Ru	621.8	^{214}Bi	1281.0
^{235}U	185.7	^{137}Cs	**661.7**	^{214}Bi	**1377.7**
^{226}Ra	186.2	^{214}Bi	665.4	^{214}Bi	1401.5
^{228}Ac	209.3	^{212}Bi	727.3	^{214}Bi	1408.0
^{212}Pb	**238.6**	^{228}Ac	755.3	^{40}K	**1460.8**
^{224}Ra	240.9	^{95}Zr	756.8	^{228}Ac	1495.9
^{214}Pb	242.0	^{95}Nb	765.8	^{228}Ac	1588.2
^{228}Ac	270.2	^{214}Bi	768.4	^{214}Bi	1729.6
^{228}Ac	328.0	^{228}Ac	795.0	^{214}Bi	**1764.5**
^{228}Ac	338.3	^{214}Bi	806.2	^{214}Bi	1847.4
^{214}Pb	**351.9**	^{228}Ac	835.7	^{214}Bi	2204.1
^{228}Ac	409.5	^{228}Ac	840.4	^{208}Tl	**2614.5**

Source: Gilmore [2008] and Zvara et al. [1994].

Gamma-ray counting facilities can be classified into three broad categories. First *ultra-low-level* facilities are characterized by extremely low background count rates, often less than a few counts per day. Such facilities require heroic effort and expense: detector, electronics, and support material, all constructed from radiopure materials, are placed deep underground at depths up to 10,000 m (water-equivalent). There are several tens of these facilities currently in operation most intended to investigate various rare phenomena such as proton decay, neutrinoless double beta decay, magnetic monopoles, dark matter, neutrino physics, etc. A few are also used to measure very low levels of radioactivity.

In the second category are *very low-level* facilities, which use, in addition to passive lead and copper shielding around the spectrometer, electronic coincidence shielding to suppress the Compton plateau and the cosmic ray background. Many are placed several hundred to a few thousand meters (water-equivalent) underground. Background count rates in such facilities can thus be reduced to a few cps. The third category of counting facilities are at or near ground level and simply rely on passive shielding around the spectrometer and isolation from man-made radiation sources. These relatively inexpensive facilities typically have background count rates of several tens of counts per second.

Almost all the peaks seen in Fig. 21.7 are attributable to ^{40}K and members of the ^{238}U and ^{232}Th decay chains.[5] Although not seen in Fig. 21.7, the radionuclide ^{9}Be ($T_{1/2}$ = 53.3 d), produced by cosmic rays, can also sometimes be seen. The fission product ^{137}Cs, produced by atomic bomb tests, can be seen. Other background spectra sometimes also show the fission products ^{95}Zr, ^{95}Nb, ^{106}Ru and ^{144}Ce.

[5]These radionuclides are sometimes referred to simply as the KUT nuclides.

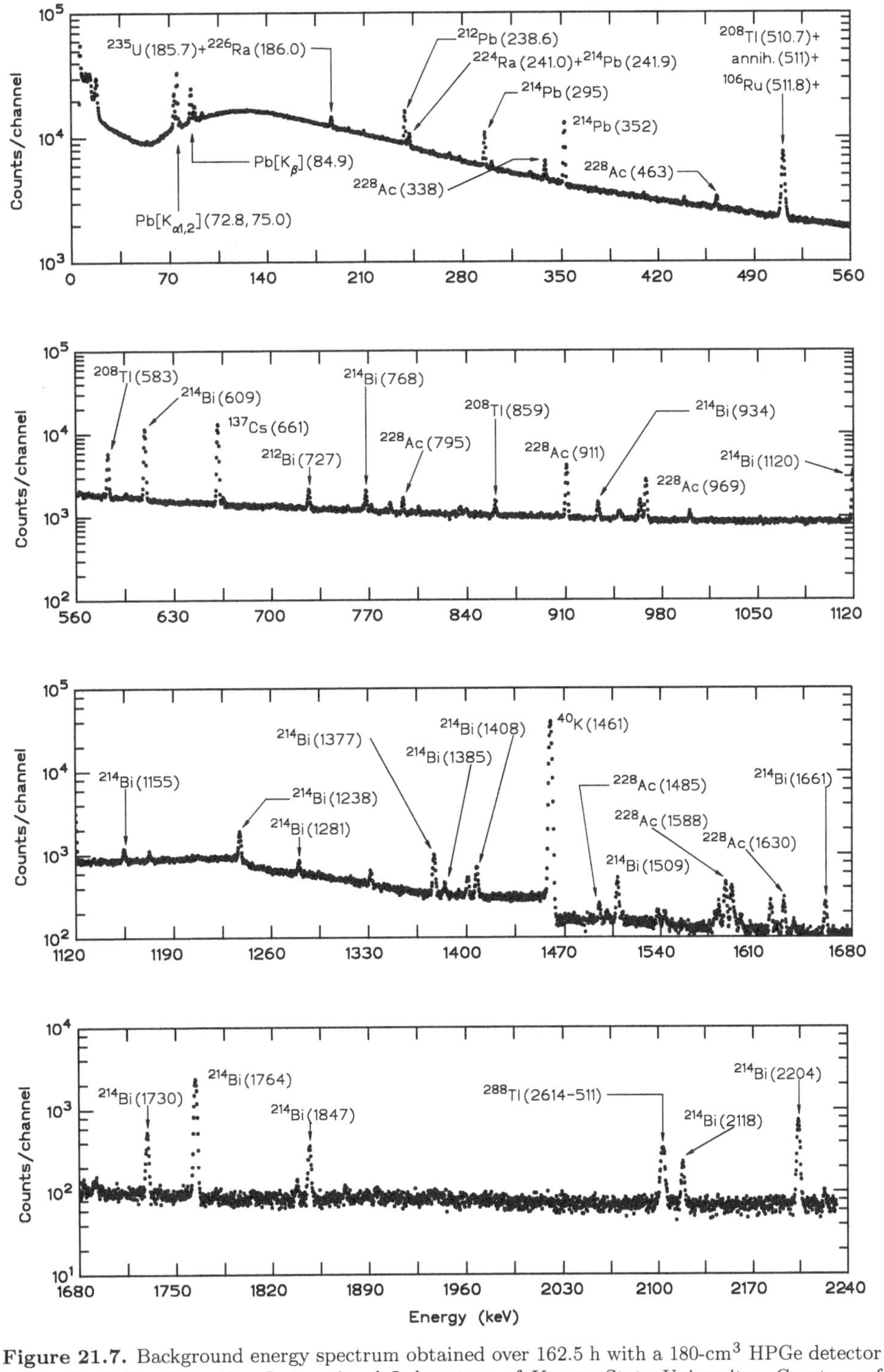

Figure 21.7. Background energy spectrum obtained over 162.5 h with a 180-cm^3 HPGe detector in the Radiation Detection Instructional Laboratory of Kansas State University. Courtesy of Nathaniel Edwards, KSU.

Before discussing ways to minimize the effect of background on the analysis of a sample, it must be recognized that the choice of detector type depends on the specific application. If radionuclide identification is the purpose, then a high-resolution spectrometer should be selected. However, if detection of contamination is of interest, then a high-efficiency detector, generally with poorer resolution, should be chosen. In either case, the larger the active volume the better the statistics for a given count time. Nevertheless, while increasing the detector size increases the measured foreground counts, it also proportionally increases the number of background counts. Thus, increasing a detector's size does little to improve the foreground-to-background ratio, although it is usually easier to separate the background from a foreground spectrum with a high-resolution detector such as HPGe compared to a NaI:Tl spectrometer.

There are several ways to improve the foreground-to-background ratio so that the background-corrected spectrum depends less on the background spectrum. These methods are discussed below.

21.2.1 Sample Placement

To increase the foreground count rate from a sample, the sample should be placed as close to the source as possible. Unless the sample activity is weak, source placement too close to the source may increase the deadtime of the detection system beyond acceptable limits. Also the position of a sample with respect to the detector is further limited if activities of the radionuclides in the sample are sought (as distinct from just the identification of the radionuclides through the energies of their gamma rays). For absolute counting it is important that the efficiency of the spectrometer as a function of gamma-ray energy is known. This energy-dependent efficiency is determined by "point" calibration sources placed at well-defined positions, typically tens of centermeters on the axis from the detector. Thus samples must be made physically small and placed at a calibration position to appear as a point source. Sometimes the distance between source and detector may be very difficult or impossible to change, such as measurement systems designed to study solar neutrinos and cosmic rays.

If background radiation is weak compared to the sample activity, the source (and detector) should also be placed as far as possible from walls and other objects from which gamma rays emitted by the source could be reflected into the detector. The minimization of this reflection contribution when the detector must be shielded from background radiation is considered in more detail later. Also if a sample is physically large compared to the detector volume, then corrections must be made to account for photon absorption by the sample itself, thereby preventing sample radiation from reaching the detector.

21.2.2 Minimize Radioactivity in the Detector System

Because all materials contain radionuclides to a lesser or greater degree, it is important in the construction of a low-level detection system to use materials with very low radionuclide (RN) concentrations for both the material that detects the energy of the incident radiation as well as the detector assembly itself.

In Table 21.7 are listed representative specific radionuclide activities in materials commonly used in many spectrometer systems. Most of the KUT radioisotopes appear as impurities and their concentrations often vary by factors of ten or more depending on how the materials were prepared. Materials with low levels of radionuclide contaminants are essential to reduce the background in a spectrometer. This is particularly true for components that are very close to the detector and include the detector material itself, the casing of the detector, radiation shields near the detector, the air surrounding the detector, and various system components such as preamplifiers, photomultipliers, and cryostat endcaps. Selection of materials for low-level spectrometer systems is a lengthy but vital aspect of the design of a spectrometer system with an inherent low background. Guidance can be found from published accounts of the construction of systems in existing facilities such those described by Brodzinski et al. [1985] and Leonard et al. [2008] who also provide specific radionuclide activities in over 200 materials considered for the EXO-200 facility.

Table 21.7. Typical naturally occurring radionuclide concentrations in various materials used in detector systems.

Material	Average specific activity (Bq kg^{-1})		
	^{40}K	^{238}U or ^{214}Bi	^{232}Th or ^{208}Tl
aluminum	< 0.4–15	0.07	0.7
		< 0.07–35†	0.1–4‡
aluminum (6061 Harshaw)	< 0.83	0.7	7.0
aluminum (1100 Harshaw)	< 1.0	< 0.28	4.0
aluminum (1100 ALCOA)	< 1.8	< 1.8	1.3
aluminum (3003 Harshaw)	9.3	< 0.43	1.7
apiezon grease	5	8	8
berylium	15	120†	< 0.2‡
cement, Portland	75	22	4
copper	0.2	0.06	0.04
copper, sheet	< 1.7	< 0.013–0.05†	< 0.005‡
copper (101)	< 0.008	< 0.0008†	< 0.0005‡
cooper(OFHC)	0.2	0.015–0.17†	< 0.005†
epoxy	< 20–120	14–900†	0.8–70†
flint glass	0.1	0.03	0.03
germanium	0.2	0.02	0.02
glass, flint	0.1	0.03	0.03
glass, pyrex	8	0.6	0.8
glass, quartz	35	0.35–17†	0.01–1‡
iron	0.1	0.01	0.01
indium	< 0.35	0.05†	0.02‡
lead	< 0.02	< 0.0007†	< 0.0003‡
magnesium rod	1.7	< 0.67	1.0
MgO	0.4	0.3	0.2
molecular sieve	18–150	6–50	8–75
neoprene	6	< 0.2	< 0.14
plastic tubing	< 15	< 0.07†	< 0.07‡
polyvinyl resin	3	2	2
printed circuit board	70	70†	35‡
rubber, sponge	< 7–35	1.3–200†	1–35‡
silica, fused	< 20	< 0.2†	< 0.35‡
silicone, foam	< 35	1†	0.35‡
sodium iodide	< 0.5	< 0.07†	< 0.05‡
solder	< 0.15	< 0.015†	< 0.005‡
steel, stainless	< 3.5	< 0.1†	< 0.035‡
steel, stainless (304)	< 0.33	< 0.12	< 0.10
steel, stainless (304-L)	1.7	< 0.083	< 0.083
steel, pre-WWII	< 0.15	< 0.015†	< 0.01‡
teflon	< 0.35	< 0.002–0.1†	< 0.005‡
titanium	0.02	0.06	0.004

$\dagger$ ^{214}Bi $\ddagger$ ^{208}Tl.
Sources: Zvara et al. [1994], Brodzinski et al. [1985] and Camp et al. [1974].

In gas-filled detectors, the working gas may contain ^{222}Rn and ^{220}Rn and their daughters. This contribution to the background can be reduced by purifying the gas or by just letting the gas age for several half-lives of ^{222}Rn ($T_{1/2} = 3.38$ d). The material of inorganic detectors such as NaI:Tl and CsI:Tl contribute to the background mainly through ^{40}K and to a lesser extent through U and Th impurities, although with modern manufacturing methods these contributors to the background have been largely eliminated. Germanium detectors require a very high degree or purity to produce crystals suitable for radiation spectral measurements and so background from RN impurities is quite small. Nevertheless, manufacturers of ultra-low-level spectrometers can, for an additional cost, provide Ge detectors with extremely low concentrations of impurities. Typically such highly pure crystals have an inherent activity about ten times lower than that of conventional ones [Zvara et al. 1994].

Bismuth germanate (BGO) crystals can have an appreciable background from the impurity ^{207}Bi ($T_{1/2}$ = 38 y), which is thought to be produced by cosmic ray protons in (p,n) reactions with ^{206}Pb in the same ore from which the Bi was extracted. BGO produced from lead-free ores does not have this background component. Some of the newer inorganic scintillators that are presently attractive are compounds containing lanthanum or lutetium (see Sec. 13.2). However, these elements contain naturally occurring radioisotopes and, thus, have an inherent built-in background gamma-ray source. To eliminate these radionuclides would require isotopic separation, an expensive process not routinely available. Consequently, these scintillators are not well suited for low-level spectrometric applications.

Certain materials such as pyrex glass have KUT radionuclides as normal constituents and, thus, can be a significant background source. Most materials contain these radioisotopes as impurities and so they can be reduced by various purification processes. Electrolytic copper and magnesium generally have low levels of radioactivity and are good choices for the construction of low-background counters. Stainless steel also generally has low-background activity. However, aluminum, which is widely used structurally for many experimental systems because of its reasonable cost and ease of cutting and machining, can have relatively high levels of radioactivity from uranium and/or radium contaminants. Brass also generally has low specific radioactivities (if its lead content is low). Indeed, one facility replaced stainless steel screws with brass ones to reduce ever so slightly the background [Brodzinski et al. 1985]. Electric solder and circuit boards often have high levels of radioactive impurities and, thus, electrical components in a spectrometer system must be evaluated as a source of background radiation.

Scintillation spectrometers rely on photomultiplier tubes (PMTs), but the glass enclosures of these tubes and the tube base or socket are often a significant source of radiation background. Manufacturers of PMTs offer, at a higher price, versions that have significantly smaller radioactive content. Likewise, tube bases are also available that have much less activity that normal sockets. Detector housings are usually made of aluminum, magnesium or beryllium for scintillator materials and of stainless steel for gas tube detectors. These housing materials must be chosen to have the lowest radioactivity possible. Germanium detectors must be cooled cryogenically for proper operation, and so another possible source of background is the radioactivity contained in the endcap of the cryostat which is usually positioned adjacent to the Ge crystal. Material for such endcaps must contain very low levels of radioactivity.

Although not widely used in the construction of radiation spectrometers, materials such as plastics and carbon fibers, which do not derive from ores, contain extremely low levels of radioactivity. Almost all radioactivity in these materials comes from the cosmogenic radionuclides of ^{3}H and ^{14}C, which emit only beta particles and, thus, do not contribute to the background in a gamma-ray spectrometer (save through low-energy bremsstrahlung). Indeed, sample holders and calibration source encapsulations are usually made from plastic to avoid the introduction of background gamma rays.

21.2.3 Passive Shielding of Radiation Spectrometers

To reduce the background response inherent in any radiation measurement, it is routine to use some sort of radiation shield to minimize the amount of background radiation that reaches a detector. Shields are usually composed physically of matter, although shields of a non-corporeal nature could be used. In fact one of the most important shields for us on earth are magnetic fields produced by the sun and earth that deflect charged cosmic radiation from our detectors on the earth's surface. In principle, one could use strong electro magnetic fields around a detector to shield the detector from incident charged particles.

Another extreme shield is to use the earth itself as a shield by moving the detector deep underground in old mine shafts, for example, to prevent almost all cosmic rays from reaching the detector. This is done in large research facilities that seek to study neutrinos or seek the half-life for proton decay. Such experiments have extremely weak foreground signals (some as low as counts per week or even months) and even the smallest background would swamp the foreground response.

However, for most radiation measurement situations, far less heroic shielding effort is needed to obtain statistically useful results. In the next two sections the use of common shielding materials and their arrangement to reduce background is discussed.

21.2.4 Shielding Against Gamma Rays

To reduce the background in a gamma-ray spectrometer, passive methods employing photon shields are almost always used. Because photons are exponentially attenuated as they pass through a shield, unlike charged particles which have a finite range, no photon shield can stop all photons from reaching the spectrometer. However, proper shield design can be quite effective in reducing the spectrometer background.

Lead

Because of its large interaction coefficient (macroscopic cross section) for gamma and x rays, lead ($>$ 99.9% pure) is routinely used as a shielding material against these radiations. However, lead is a poor construction material and is not suited for high temperature applications (melting point 327.4°C). To increase its strength, lead is often alloyed with antimony. Lead shields are relatively easy to fabricate because lead may be cast, extruded, rolled, machined and welded. Casting requires special care because lead increases in mass density from the liquid at 10.678 g cm^{-3} to the solid at 11.34 g cm^{-3} [Kirshenbaum et al. 1961], leading to a volume decrease of 5.83% upon solidification. Also, the fumes from lead are toxic and safety precautions should be exercised. Lead is resistant to chemical corrosion, but it is a relatively expensive material (currently about $2.30 per kg).

Lead often is cast into rectangular bricks (typically, 2$\times$ 4$\times$ 6 inches) which are then used to create simple gamma-ray shields by stacking the bricks. Streaming through the cracks between bricks can be minimized by staggering the brick layout or shaping the bricks so there are no straight line streaming paths through the shield. Alternatively, lead may be cast into cylindrical annuli to produce 2π shielding without corners. When casting lead, care must be taken to avoid voids in the cast. Sometimes lead shot is used to fill containers to produce a lead shield with slightly lower density than solid lead, but with more flexibility in the shape of the shield. Epoxies and plastic putty-like compounds containing high concentrations of lead are commercially available which can be molded and shaped to fill shield gaps or shield penetrations through which radiation could stream.

Lead and uranium are usually found together in ore and, although the uranium and most of its decay daughters are easily removed upon smelting and purification, the primordial Pb decay daughters are retained in the lead. Of these ^{210}Pb ($T_{1/2}$ = 22.3 y) is of concern for shielding for ultra-low-level gamma-ray spectrometers because it emits a 46 keV gamma ray followed by 205 and 304 keV gamma rays from the decay of its daughter ^{210}Bi. After several hundred years this component of background radiation from a lead shield disappears through decay. Lead has been extracted and used by humans for several thousand years. This ancient lead is ideal for constructing shields with low radioactivities for ultra-low-level counting spectrometers. Although quite expensive, sources of old lead are still sometimes available. Lead, first obtained as a byproduct of silver extraction from ore, has been used by humans since about 4000 BC. Romans used it as liners for aquaducts, baths, and almost all other plumbing applications. Later in history, the building of medieval churches consumed much lead as roof liners, masonry joints and even as mortar.[6]

[6] One source of old lead used for the Italian Cryogenic Underground Observatory for Rare Events (CUORE) was from a Roman ship that sank between 80 and 50 BC off the coast of Sardinia. From this shipwreck over 1000 lead ingots, each 33 kg in

Water

Large water-filled tanks offer an inexpensive and flexible way to create gamma-ray shields. On a *per unit mass basis*, water is almost as effective in reducing the energy of photons through Compton scattering, as is lead.[7] Furthermore, water is far less expensive and can be more easily purified to remove essentially all radioactive impurities save tritium which is of little concern for gamma-ray spectroscopy. Only the radioactive impurities in the tank materials are of concern. However, lead has a larger photoelectric cross-section and is a superior material for absorbing low-energy photons. Thus, rather than using a monolithic shield, a layered approach can be used with a water shield used to reduce the high-energy background gamma rays followed by a thinner lead shield near the spectrometer system to absorb the low-energy photons.

To reduce the background, the spectrometer could be lowered into water, thereby creating a much larger water shield. At depths of about 10 meters, the background for surface terrestrial radionuclides and all cosmic rays (except for muons) are eliminated. Only the radionuclides dissolved in the water and incorporated in the spectrometer produce a background. This approach has been taken to an extreme with the IceCube Neutrino Observatory at the Amundsen-Scott South Pole Station in Antarctica where thousands of sensors are spread over 1 km^3 at depths between 1450 and 2450 meters in the Antarctic ice cap. Here the water, in the form of ice, serves both as the neutrino detection medium as well as a shield against cosmic rays.

Steel

Steel (or iron) is sometimes used to fabricate gamma-ray shields although thicker iron shields compared to lead shields are needed to produce the same reduction in transmitted radiation. For example, the tenth thickness[8] for 1-MeV photons in iron is almost 5 cm (compared to 1.2 cm in lead and 14 cm in water). Often a shield has a steel outer layer followed by a thinner inner lead layer.

Steel generally has low concentrations of radioactive impurities and is often a less expensive alternative to a large lead shield. As mentioned earlier in this chapter, energetic cosmic ray neutrons can lead to the buildup of radioactive ^{54}Mn in iron, so steel produced to provide shielding for low-level spectrometers should be moved underground as soon after production as possible. Also because post WW II steel contains fission products from open-air nuclear tests in the 1950s, old pre-WW II steel, typically from old ships, is often sought. But unless stored properly, cosmic ray neutrons may have contaminated this old steel. Also steel produced from scrap iron may also contain unacceptable radioactive impurities as a consequence of radionuclide sources being inadvertently included with the scrap iron.

Tungsten

Tungsten, because of its large $Z = 74$ number and its high density of 19.3 g cm^{-3} is an excellent material for gamma-ray shielding. But pure metal is hard to machine or to cast. To form various shapes of tungsten components, powder technology is often used to produce a sintered product with almost the same density as the metal. More frequently, tungsten is alloyed with nickel, iron, and copper to produce a metal with a density of about 90% of tungsten that is far easier to machine. But because of its cost, machinability, and toxic properties, tungsten is used in limited quantities and usually only for shields adjacent to a source. Its use in low-level spectroscopy is to produce collimated photon beams in a small space.

---

weight, were recovered. Italy's National Institute of Nuclear Physics financially supported the archaeologists in their recovery efforts and, in return, received 120 of the most damaged ingots from which a 3-cm-thick low-activity shield was constructed for the underground Laboratory at Gran Sasso.

[7]The mass interaction coefficient μ/ρ for 1-MeV photons is 0.07066 and 0.06803 cm^2 g^{-1} for water and lead, respectively.

[8]The tenth thickness is the thickness of a material required to reduce the uncollided radiation to 10% of its incident value, and is given by

$$x_{1/10} = \ln 10/\mu(E),$$

where μ is the total interaction coefficient for the material for photons of energy E.

Table 21.8. Typical compositions of representative concrete after curing.

Element	Partial density (g/cm^3)						
	Ordinary (NBS 03)	Ordinary (NBS 04)	Magnetite[a]	Barytes[b]	Magnetite and steel	Limonite and steel[c]	Serpentine[d]
H	0.020	0.013	0.011	0.012	0.011	0.031	0.035
O	1.139	1.171	1.168	1.043	0.638	0.708	1.126
Si	0.342	0.743	0.091	0.035	0.073	0.067	0.460
Ca	0.582	0.194	0.251	0.168	0.258	0.261	0.150
C	0.118						0.002
Na		0.040					0.009
Mg	0.057	0.006	0.033	0.004	0.017	0.007	0.297
Al	0.085	0.107	0.083	0.014	0.048	0.029	0.042
S	0.007	0.003	0.005	0.361			
K	0.004	0.045		0.159		0.004	0.009
Fe	0.026	0.029	1.676		3.512	3.421	0.068
Ti			0.192		0.074		
Cr			0.006				0.002
Mn			0.007				
V			0.011		0.003	0.004	
Ba				1.551			
Ni	0.026						
P	0.007						
Density (g/cm^3):	2.41	2.35	3.53	3.35	4.64	4.54	2.1

[a] Magnetite ($FeO{\cdot}Fe_2O_3$) as aggregate.
[b] Barytes, a $BaSO_4$ ore, as aggregate.
[c] Limonite, a hydrated Fe_2O_3 ore, plus steel punchings, as aggregate.
[d] Serpentine ($3MgO{\cdot}2SiO_2{\cdot}2H_2O$) as aggregate; a concrete usable at high temperatures with minimal water loss.
Source: From Shultis and Faw [2000].

Concrete

Concrete is the most widely used shielding material for both gamma rays and neutrons because it is relatively inexpensive and, while still a liquid slurry, can be poured into forms to create shields of all shapes and sizes. It can also be cast into concrete blocks which can later be stacked to form radiation shields. Concrete is prepared from a mixture, by weight, of about 13% cement, 7% water (including water in the aggregate), and 80% aggregate. There are many different types of concretes each with distinctive properties as determined by the choice of aggregate [see Shultis and Faw 2000]. For example, to enhance its photon-attenuation properties, scrap iron or iron ore may be incorporated into the sand-and-gravel aggregate. Examples of elemental compositions of concrete frequently encountered in radiation shielding applications are given in Table 21.8.

However, concrete often has appreciable concentrations of the KUT radionuclides. In any spectrometer facility with concrete walls, floors, and ceilings, the 1.46-MeV gamma ray from ^{40}K is almost always observed in a gamma-ray spectrum. From Table 21.8 it is seen that barytes concrete, which is the most effective concrete for attenuating gamma rays, has the highest K concentration. To reduce the ^{40}K presence, special (and expensive) low potassium sand and gravel can be used. However, as seen in Table 21.7, common cements also have high ^{40}K activities. To ameliorate the radionuclide impurities in concrete shields, steel, or lead liners can be placed on the inner surface of a concrete shield.

Mercury and Molybdenum

Like tungsten, mercury also has low intrinsic radioactivity and, because of its high $Z = 80$ number and density $\rho = 13.53$ g cm^{-3}, is an excellent material for attenuating gamma rays. However, also like tungsten, it is relatively expensive and because it is a liquid it must be placed in containers, which may themselves be another source of gamma rays. Mercury is also toxic and thus requires special care in handling. Nevertheless, shields made from mercury have been used in ultra-low-level counting facilities.

Molybdenum is another material with good gamma-ray shielding properties ($Z = 42$ and $\rho = 10.2$ g cm^{-3}). It generally has low intrinsic radioactivity, but like tungsten and mercury is relatively expensive.

21.2.5 Shielding Against Neutrons

Although most gamma-ray spectrometers are inherently insensitive to neutrons, background neutrons are still of concern in the design of low-level counting facilities. Neutrons produced by cosmic rays have supra-MeV energies, and as they slow through scattering interactions they produce inelastic scattered gamma rays. Further these neutrons are eventually absorbed creating highly excited nuclei which decay by the emission of energetic capture gamma rays. The capture gamma rays are often very energetic, some with energies of 8 to 10 MeV. Also neutrons, particularly after they have slowed, can interact with elements in the detector itself to produce radionuclides, such as ^{24}Na in a NaI:Tl detector whose decay produces gamma rays, or produce reaction products, such as those in the ^{6}Li(n,t)^{4}He reaction in Si(Li) or Ge(Li) detectors. Thus, one must use neutron shielding outside the gamma-ray shielding to minimize the gamma-ray background in the spectrometer system.

Because neutrons are absorbed mainly when they have near thermal energies, a neutron shield for fast neutrons first attempts to slow the neutrons through scattering interactions (or *moderate* them) and then absorb them with materials such as cadmium, boron, or lithium, that have large thermal-neutron cross sections. To moderate neutrons rapidly, materials such as water or polyethylene with a high hydrogen content are normally used because the lighter the scattering nucleus, the more energy a neutron loses on a scatter. However, for cosmic ray neutrons with energies above 10 MeV, the cross section for scattering from hydrogen decreases rapidly and becomes ineffective at slowing the neutrons. Consequently,concrete used for neutron shielding against fast neutrons usually contains high Z materials, such as iron, which slow fast neutrons through inelastic scattering (and also produce unwanted inelastic scattering gamma rays). Once the neutrons are slowed below 1 MeV, hydrogen can then quickly moderate the neutrons towards thermal energies where they are absorbed by impurities or by materials with large neutron absorption cross section that are added to the moderating material.

From ease of fabrication, fast neutron shields are usually composed of laminates of iron and high density polyethylene (HDP). As the neutrons approach thermal energies, ^{6}Li can be incorporated into the HDP to absorb the slow neutrons without the emission of any capture gamma rays. Often a high Z layer is added to the inner surface of the neutron shield to attenuate capture-gamma, inelastically scattered, and bremsstrahlung photons. For instance, cadmium or boron is widely used to absorb thermal neutrons, but such absorptions produce gamma rays that must be shielded from the detector.

21.2.6 Minimize Radioactivity in Air around Spectrometer

Gaseous radon and its airborne daughters are always present in the air in any low-level counting facility. Indeed without adequate ventilation these radionuclides can build up, thereby producing air concentrations many times those in ambient outside air. This is especially a problem for underground facilities. To avoid this activity in the air in a spectrometer counting cavity two main approaches have been used. One is to seal the cavity after the sample has been inserted and then wait several days before counting to allow these

Table 21.9. Typical naturally occurring radionuclide concentrations in various building and structural materials.

Material	Average specific activity (Bq kg^{-1})		
	^{40}K	^{226}Ra	^{232}Th
asbestos	200	10	15
brick, clay (red)	670	50	50
brick, snad-lime	330	10	10
cement	200	20	15
ceramics	500	30	40
clinker	600	60	70
concrete	400	40	30
concrete, aerated	430	60	40
granite	1000	80	80
gravel	150	10	10
limestone	60	10	5
marble	150	10	5
natural gypsum	80	10	10
pearlite	1200	50	80
plaster	250	15	45
phosphogypsum	60	390	20
sand	200	10	10

Source: Eisenbud [1987], UN [1977], and Zvara et al. [1994].

airborne radionuclides to decay. However, this approach severely limits the rate at which samples can be analyzed because the insertion of a new sample requires a multiday wait before analysis.

The usual approach to minimize airborne radioactivity is to purge the counting cavity with purified gas, such as argon, which has little inherent radioactivity. Some facilities whose spectrometer requires cryogenic cooling with liquid nitrogen use the boil-off nitrogen as a purge gas.

Radionuclides from the soil or radon decay can also be attached to dust particles, which also contribute to the air activity. Thus, it is important to effectively filter the ventilation air to the counting room.

21.2.7 Use Construction materials with Low Radioactivity

Typical concentrations of ^{40}K, ^{226}Ra, and ^{232}Th in common building materials are given in Table 21.9. However, these activity concentrations can vary by several hundred percent depending on the source of the material. Typical radionuclide activities in materials used in construction of radiation detection systems are listed in Table 21.7. Again these activities can vary substantially depending on the manufacturer, but to minimize the effect of background on radiation spectrometers, it is imperative that close attention be given to the choice of materials used.

21.2.8 Counting Enclosures

Almost all low-level counting spectrometers and the radioactive sample to be analyzed are placed in some small enclosure or cavity whose walls act as the primary passive shields against background radiation in the laboratory walls and air. These cavities typically have a rectangular parallelepiped shape because the walls are most conveniently constructed from shielding bricks or plates of various materials. Some cavities are cylindrical in shape with walls made from lead and other materials cast into cylindrical annuli shapes, but these are more expensive and difficult to fabricate than parallelepipeds. The shield material used for these cavities should, of course, contain as little radioactivity as possible. The size of the counting cavity is usually quite small so as to minimize the amount of shielding required.

As background radiation incident on the walls of the counting cavity is attenuated, it also creates a buildup of secondary radiation in the form of Compton scattered photons, bremsstrahlung, and x rays. Even if the shield material is radiopure, low-energy photons produced in the shield walls can still enter the counting cavity. Although lead is a excellent shield material, photoelectric interactions in it from background photons produce significant low-energy x rays ($K_\alpha = 75$ keV, $K_\beta = 84.9$ keV, and $L_\alpha = 10.5$ keV). To absorb these x rays, lower Z materials, such as copper and aluminum, are often used as inner shields. One example of a passive shield described by Zvara et al. [1994] is that at the Comenius University in Czechoslovakia has a mass of 18,000 kg with outer dimensions of 80 × 90 × 172 cm. It is composed, starting from the outside, with 1 cm of Fe, 10 cm of Pb, 5 cm of Cu, 8 cm of polyethylene with boric acid, 1 mm Cu, and 1 cm of Plexiglas. To stop the low-energy, non-muonic cosmic rays, the top of the enclosure has a 12-cm Fe slab shield. Modern materials with no special selection were used.

As an illustration of the use of a Cu inner liner to reduce Pb x rays, the energy dependent flux of ^{60}Co photons leaving a 10-cm bare and 0.5-cm Cu clad lead shield is shown in Fig. 21.8. The flux shown is that entering a NaI:Tl crystal on the cold side of the shield. No Gaussian smearing of the spectrum is used to account for the actual resolution of the detector. Notice the Cu liner totally eliminates all the Pb x rays as well and also suppresses some of the transmitted Compton lead scattered photons, except near the Pb x ray region where the Cu slightly enhances the photon flux. With modern radiation transport codes, calculations such as those in Fig. 21.8 can be used to design shields for counting cavities and can eliminate much of the trial and error and intuition used a few decades ago to design passive shielding for low-level counting facilities.

In measuring the spectrum emitted by a sample with small amounts of radioactivity, radiation from the sample that reflects from the inside of the passive shielding is usually not of too much concern. However, such radiation incident on the inner portions of a lead shield will produce x rays that will be seen by the spectrometer. Again, lining the inner cavity with lower Z materials can ameliorate this effect. An example of the reflected photon flux from ^{137}Cs gamma rays emitted by a sample is shown in Fig. 21.9. Here it is seen that a 0.5-cm-thick Cu liner almost totally eliminates the lead x rays but at the price of significantly enhancing the low-energy part of the reflected spectrum. Such calculations with modern radiation transport codes should be used to customize the inner cavity shielding for a particular type of radiation source that is to be analyzed.

21.2.9 Laboratory Location

Terrestrial and cosmic ray background are greatest at the earth's surface. Consequently, construction of a low-level spectrometer facility at ground level is seldom done. Even a few tens of meters below the surface, the background radiation is greatly reduced, although the cosmic-ray muonic component is affected little. Thus, in a multistoried building, the low-level spectrometer system should be built in the lowest basement room possible.

Cosmic ray background can be almost completely eliminated by building the counting system deep underground or underwater. To this end, there are several tens of low-level counting systems around the world that have been built in abandoned mines, some of which are close to 10,000 m (water equivalent) deep. At such depths all the muons are stopped, leaving only neutrinos, which are the particles being studied in many of these facilities. Of course, the walls of underground caverns or rooms still contain primordial radionuclides so that spectrometer shielding is still required. Ideally, surrounding wall material should have as small as possible radionuclide content. For this reason, abandoned salt mines are very attractive locations. Shown in Fig. 21.10 are the background spectra for four different low-level facilities, two at ground level and two deep underground away from cosmic ray muons. The superiority of an underground location is obvious.

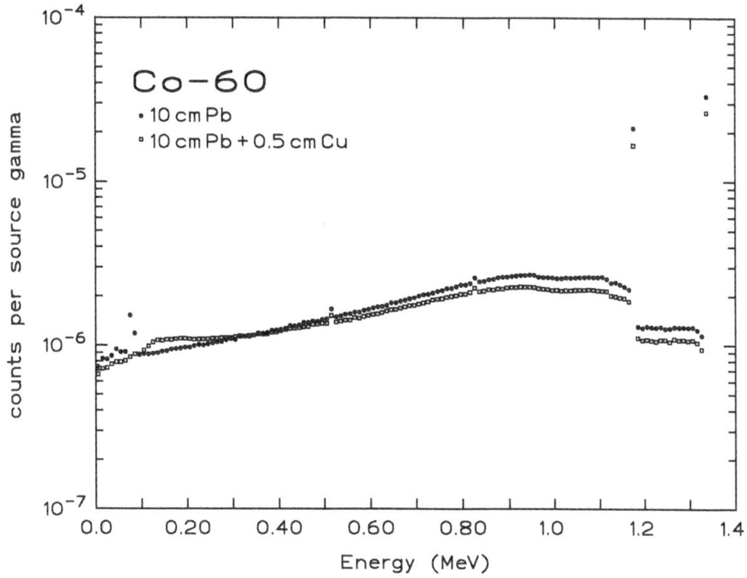

Figure 21.8. ^{60}Co gamma rays transmitted through 10 cm of lead and from the same amount of lead plus 0.5 cm of copper on the exit surface as calculated by MCNP6. Gamma rays are normally incident on the outer surface.

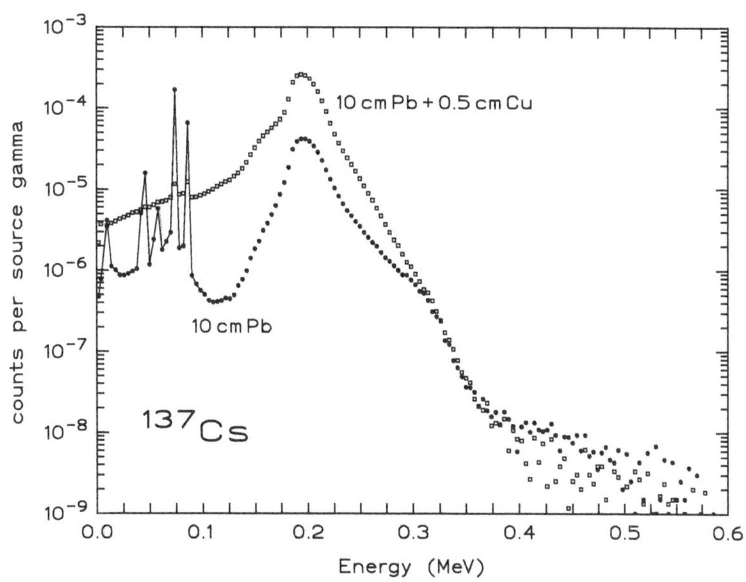

Figure 21.9. ^{137}Cs 0.6617-MeV gamma rays reflected from 10 cm of lead and from the same amount of lead plus 0.5 cm of copper on the reflecting surface as calculated by MCNP6. The gamma rays are normally incident on the inner shield surface.

21.2.10 Other Considerations

The measurement of background spectra from low-activity sources almost always requires long count times, sometimes of multiday or, in extreme cases, multimonth duration. For such long count times, there are a

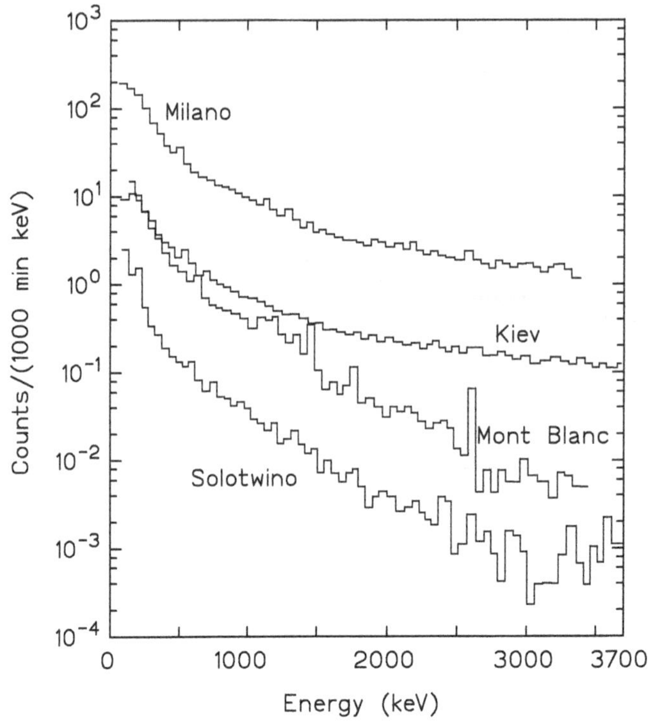

Figure 21.10. Comparison of Ge:Li background spectra in above ground (Milano and Kiev) and underground (Mont Blanc and Solotwino) laboratories. Data are from [Zvara et al. 1994].

myriad of small factors that can affect the spectrometer system and confound the subsequent analysis of the measured spectra. It is essential that stability of the system be maintained during long data collection times. It is always a good practice to interrupt a long accumulation time of a spectrum and perform periodic calibration verifications and recalibrations if necessary.

The response of the electronics is often susceptible to small changes in the ambient temperature. For example, Rizzacacca [1986] reports that a temperature change of 2°C over a 24-hour period can shift the spectrum from a Ge:Li detector by about one FWHM (2 keV at 1.33 MeV). Today the use of electronic spectrum stabilizers reduces such shifts substantially. Also underground facilities are less susceptible to temperature changes because the huge thermal inertia of the surrounding rock and soil minimizes any temperature fluctuations.

Also during long count times, the background radiation field can fluctuate. This is of particular concern for the airborne radionuclides of radon and its daughters. Following rain storms or thawing cycles, the emission of radon from soil and rocks may increase substantially and, depending on how readily airborne contaminants can reach the gas surrounding the spectrometer detector, the count rate can be affected. Likewise, activity such as housekeeping activities can alter the concentration of radionuclides in the air.

It stands to reason that any low-level counting facility be located as far as possible from known sources of radiation such as industrial x-ray sources or other sources used for industrial or research purposes. It is not a good idea to store even weak radioactive "check" sources or radioactive samples near the spectrometer. The authors are amazed at how often they have found radiation sources in desk drawers or on shelves in or near a room used for spectroscopic measurements. It should also be remembered that people are radiation sources (notably from ^{40}K) and they also act as radiation reflectors and shields. Thus, their presence in and around a spectrometer as it collects data alters the background. Such human activity near a spectrometer can perturb the measured spectrum depending on the effectiveness of the passive shielding surrounding the spectrometer.

Finally, it should be mentioned that contamination of surfaces near and on the spectrometer can occur over time, especially if many unsealed calibration sources and/or samples have been processed. Periodic wipes and checks for such contamination should be made and appropriate remediation taken if such contamination is found. Sometimes sensitive detector windows are protected by plastic film that can be replaced if contamination occurs.

21.3 Self-Absorption of Photons

So far in the discussion about reducing the background in photon spectrometers, the emphasis has been on reducing the radiation reaching the spectrometer detector and radioactive content of various components of the low-level counting facility. However, a radioactive impurity in some facility component does not necessarily lead to a background count. The radiations from a radioactive decay in some material may be absorbed in the material before they can escape and deposit their energy elsewhere. This *self-absorption* effect can be important in assessing the effectiveness of thick shields and walls of the spectrometer facility.

21.3.1 Infinite Slab

Consider a homogeneous slab shield of thickness T which contains a uniform volumetric source of a radionuclide which emits a gamma ray of energy E_o per decay. The number of photons escaping the slab without interaction is, by symmetry, the same for either surface. The probability the gamma ray escapes through one surface is [Case et al. 1953]

$$P_1^o(T) = \frac{1}{2\mu T}\left[\frac{1}{2} - E_3(\mu T)\right], \tag{21.3}$$

where μ is the total interaction coefficient for photons of energy E_o and $E_3(x)$ is the exponential integral function of order three.[9]

The one-sided escape probability of Eq. (21.3) can be approximated to within 1% by

$$\frac{1}{P_1^o(T)} = \begin{cases} 2.077 + 2.951\mu T + 0.1348(\mu T)^2, & 0.003 < \mu T < 3.5 \\ 4\mu T, & \mu T \geq 3.5 \end{cases} . \tag{21.4}$$

The uncollided escape probability for the slab, i.e., through either face is $P_0 = 2P_1^o$. The uncollided escape probability is independent of the slab material and depends only on the thickness of the slab in units of mean free path lengths $\mu(E_o)T$.

Also leaving a slab surface are secondary photons produced by the decay gamma ray interacting in the slab to produce fluorescence, bremsstrahlung, Compton scattered photons, and annihilation photons. These photons have a distribution of energies. The total number of secondary photons escaping through one surface, per source gamma ray, is denoted by P_1^s. No analytic expression is known for the energy distribution or total number of these secondary photons. However, with modern transport codes, it is a fairly simple matter to

[9]The exponential integral function of order n is defined as

$$E_n(a) = \int_1^\infty x^{-n} e^{-ax} dx, \; n = 0, 1, 2, \ldots .$$

make such calculations. In Fig. 21.11 results for a concrete slab are shown. Of particular note is the relative insensitivity of P_1^s to the energy of source gamma rays. Unlike P_1^o, P_1^s depends not only on the mean free path thickness of the slab, but it also depends on the slab material because the interaction coefficients for the various reactions that produce secondary photons vary with the material.

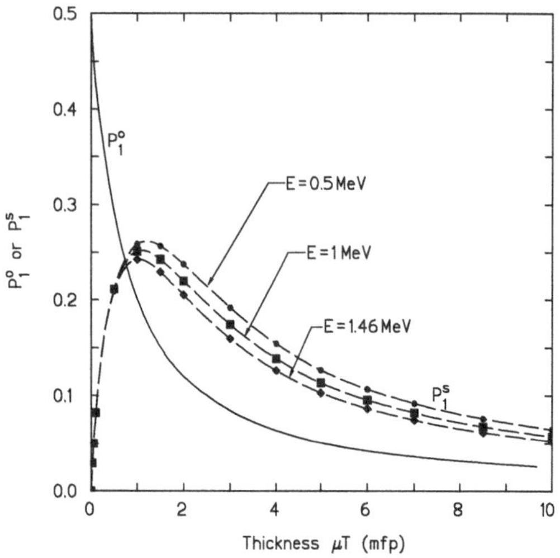

Figure 21.11. One-sided escape probabilities from a concrete slab for a monoenergetic, isotropic, uniformly distributed, gamma-ray source. Concrete is ANSI/ANS-6.6.3 standard concrete with a density of 2.3 g cm^{-3}. The solid line is calculated from Eq. (21.3) and the data points for secondary photons were calculated by MCNP6.

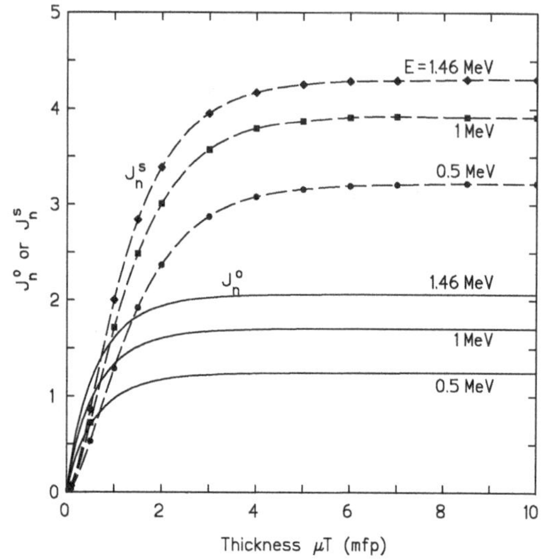

Figure 21.12. The current or flow rate of photons through a unit area of the surface of a concrete shield containing a unit strength volumetric source ($S_v = 1$ cm^{-3}s^{-1}) of monoenergetic photons of energy E_o. The solid lines for J_n^o are from Eq. (21.5) and the data points are values obtained with MCNP6.

For a radioactive contaminant that emits gamma rays of energy E_o and that is uniformly distributed in the slab, the volumetric gamma-ray source strength is $S_v = Af$ cm^{-3} s^{-1}, where A is the activity concentration Bq cm^{-3} and f is the gamma-ray frequency, i.e., the probability a gamma ray of energy E_o is emitted per decay. Then the total flow of uncollided photons crossing a unit area on the surface of the slab is the photon current leaving the slab and is given by $J_n^o = S_v T P_1^o(T)$, where $S_v T$ is the total number of gamma rays of energy E_o emitted per unit time in a 1 cm^2 prism of length T. If gamma rays leave this prism while migrating in the slab they are compensated, by symmetry, by gamma rays that enter the prism. Thus, the uncollided current leaving the slab is from Eq. (21.3)

$$J_n^o \equiv \int_{\mathbf{n}\bullet\mathbf{\Omega}>0} \mathbf{n}\bullet\mathbf{\Omega}\phi(E_o, \mathbf{\Omega})\, d\mathbf{\Omega} = \frac{S_v}{2\mu}\left[\frac{1}{2} - E_3(\mu T)\right], \tag{21.5}$$

where $\mathbf{n}$ is the outward normal to the slab surface. Likewise the flow through a unit area on the slab's surface is $J_n^s = S_v T P_1^s$. In Fig. 21.12 these surface flows of escaping photons for a unit strength source strength, i.e., $S_v = 1$. Notice for shields greater than a few mean free paths thick, these surface flow rates become almost independent of the shield thickness.

The uncollided flux at a point exterior to the slab is [Shultis and Faw 2000]

$$\phi^o = \frac{S_v}{2\mu}\left[1 - E_2(\mu T)\right]. \tag{21.6}$$

This result assumes that the photons, once outside the slab, no longer interact. Notice the uncollided flux is not dependent of the distance to the slab surface.

The above analysis is for an infinite slab. However, all slab shields are finite in all dimensions. But if the point of interest on the surface of the slab is a few mean free paths away from the slab edges, the above results serve as a good (conservative) approximation.

Example 21.1: A concrete wall 30-cm-thick contains ^{40}K with a specific activity of 250 Bq kg^{-3}. The total interaction coefficient for the 1.46-MeV gamma ray emitted by ^{40}K is 0.1212 cm^{-1} and the concrete has a mass density of 2.3 g cm^{-3}. Estimate the number of uncollided and secondary photons that escape per unit time through a unit area on the concrete wall.

Solution:

The radionuclide ^{40}K emits a 1.46-MeV gamma ray with a frequency of 0.107 per decay. Thus, the volumetric source strength is

$$S_v = 250 \text{ Bq/kg} \times 0.001 \text{ kg/g} \times 2.3 \text{ g/cm}^3 \times 0.107 = 0.0615 \text{ cm}^{-3} \text{ s}^{-1}.$$

The wall thickness in mean free path lengths is μT=(0.1212 cm^{-1})(30 cm) = 3.636. The one-sided escape probability from a concrete slab is, from Fig. 21.11, about $P_1^o = 0.06$ or from Eq. (21.3) is found to be $P_1^o = 0.068176$. The approximation of Eq. (21.4) gives $P_1^o = 0.06855$. Thus, the number of uncollided photons escaping is

$$J_n^o = TS_vP_1^o = (30 \text{ cm})(0.0615 \text{ cm}^{-3} \text{ s}^{-1})(0.068176) = 0.126 \text{ cm}^{-2} \text{ s}^{-1}.$$

Alternatively, from Fig. 21.12 J_n^o for S_v =1 is about 2.0. Multiplication by $S_v = 0.0615$ gives $J_n^o = 0.123$ cm^{-2} s^{-1}.

The flow of secondary photons for $S_v = 1$ is found from Fig. 21.12 to be about 4.1 cm^{-2} s^{-1}. Multiplication by $S_v = 0.0615$ gives $J_n^s = 0.252$ cm^{-2} s^{-1}.

21.3.2 Infinite Cylinder

Analytic expressions are known for the uncollided escape probability for other simple geometric shapes such as spheres, hemispheres, oblate spheroids, and cylinders [Case et al. 1953]. Of interest here are impurity radionuclides in a cylindrical crystal. For a infinite cylinder of radius R, the uncollided escape probability is

$$P_0(R) = \frac{2\mu R}{3}\left\{2\left[\mu a K_1(\mu R)I_1(\mu R) + K_0(\mu R)I_0(\mu R) - 1\right] + \frac{K_1(\mu R)I_1(\mu R)}{\mu R} - K_0(\mu R)I_1(\mu R) + K_1(\mu R)I_0(\mu R)\right\}, \tag{21.7}$$

where I_n and K_n are the modified Bessel functions of the first and second kind and of order n. For radii small compared to a mean free path length $1/\mu$, Eq. (21.7) may be approximated by

$$P_0(a) \simeq 1 - \frac{4\mu R}{3} + \frac{1}{2}(\mu R)^2\left\{\ln\left(\frac{2}{\mu R}\right) + \left(\frac{5}{4} - \gamma\right)\right\}, \tag{21.8}$$

where γ = Euler's constant = 0.577216.... For a much larger than one mean free path length

$$P_0(a) \simeq \frac{1}{2(\mu a)} - \frac{3}{32(\mu a)^3}. \tag{21.9}$$

For a cylindrical crystal of height $2H$ and radius R that is uniformly contaminated by a radionuclide emitting photons of energy E_o, the uncollided flux on the axis at the center of the crystal is given by [Shultis and Faw 2000]

$$\phi^o = \frac{S_v}{\mu} G_2(\mu H, \mu R), \tag{21.10}$$

where the function G_2 is defined as

$$G_2(a,b) \equiv \frac{1}{\pi} \int_0^a [E_1(x) - E_1(\sqrt{x^2+b^2})]\, dx, \tag{21.11}$$

and which must be evaluated numerically. This function is shown in Fig. 21.13.

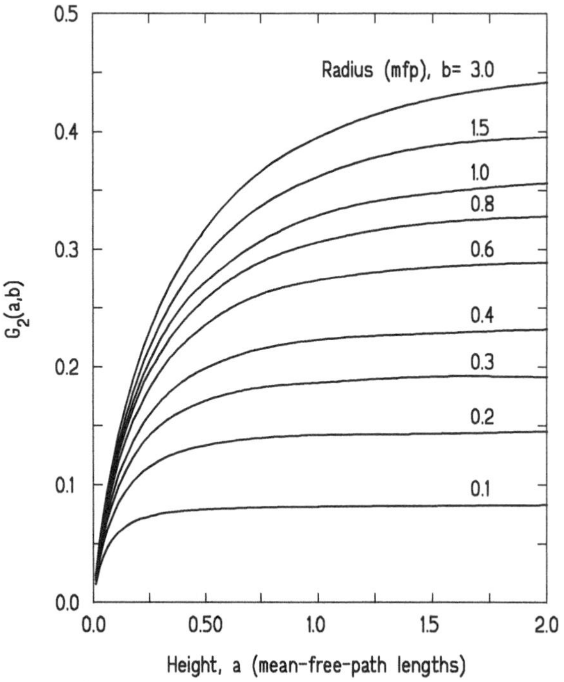

Figure 21.13. The function $G_2(a, b)$ used for the calculation of the uncollided flux at the center of a cylinder of radius $b = \mu R$ and height $a = 2\mu H$ in mean free pathlengths.

21.4 Electronic Methods for Background Reduction

To substantially reduce cosmic ray background generally requires massive amounts of "passive" shielding with radiopure materials and/or the placement of the spectrometer deep underground where the overlying earth provides the needed massive passive shield. An alternative is to use "active" shielding techniques in which multiple detectors and associated electronics are used to reduce the recorded background. Coincidence and anti-coincidence methods were introduced in Section 20.7.3, with examples of the effectiveness of the technique (see Figs. 20.27 and 20.28).

21.4.1 Anti-coincident Background Reduction

Active (anti-coincidence) shielding is a powerful method for reducing the detector background. This method uses several detectors placed near the spectrometer detector to reject a spectrometer detector pulse that coincides within a microsecond or so of a pulse originating in one or more of the surrounding detectors. Such coincident pulses arise from Compton scattered gamma rays from the passive shield or spectrometer detector. Likewise cosmic rays may produce nearly simultaneous pulses in both the spectrometer and a surrounding detector. Thus, not only is the background reduced, but the Compton continuum in the spectrometer is also reduced.

An active anti-coincidence shield can be devised using GM-tubes, proportional counters, plastic or liquid scintillators, NaI:Tl or BGO crystals. If the anti-coincident system is placed inside a passive shield enclosure, then NaI:Tl detector(s) are attractive because they generally have smaller space requirements and have a high detection efficiency. The random direction of gamma rays from the environment calls for a uniform anti-coincidence shield around the spectrometer detector. Cosmic ray muons are quite anisotropic, but do not significantly affect the required shape of the active shield. This outcome is mainly due to the fact that a muon path length of only a fraction of a centimeter, is generally sufficient to produce a usable pulse in the guard detector(s) for the anti-coincidence circuit to reject a spectrometer event.

The Compton scattering process in the detector produces a continuum spectrum that raises the lower detection limit for gamma rays within this continuum spectrum. The scattered photons that escape the spectrometer detector can be detected with a sufficiently large NaI:Tl or BGO crystal surrounding the detector. NaI:Tl has a better energy resolution than that of BGO, but a lower density. As a result NaI:Tl is about 2.5 times less efficient than BGO (averaged over several common gamma-ray energies) at detecting the Compton scattered photon. Because BGO has only about 10–20% the light output of NaI:Tl, it has a higher threshold that consequently causes a lower Compton suppression factor at low gamma-ray energies. If an HPGe spectral detector is placed above the center of a large anti-coincidence NaI:Tl crystal (30×30 cm), a Compton suppression factor of about 10–12 can be achieved for gamma-ray energies below 1.5 MeV [Zvara et al. 1994].

21.4.2 Coincident Counting

Many radionuclides emit two or more radiation quanta upon decay. With the use of coincidence counting techniques using multiple detectors, the background count rate in the spectrometer detector can be sharply reduced. Only if the spectrometer and another detector both detect an event at the same time is the event recorded. Although background events are thus most rejected, some events which should be recorded are rejected because only the spectrometer detector experienced the event. Thus, the efficiency of coincident counting usually reduces the efficiency of the spectrometer for the radiation of interest while at the same time much of the background is eliminated. There are several types of coincidence schemes as briefly mentioned below.

Double Coincidence Spectrometer

This type of spectrometer is used to measure the activity of a radionuclide that emits two or more nearly simultaneous gamma rays. ^{60}C for example is one such radionuclide which upon beta decay to ^{60}Ni, 99+% of the time emit both a 1.173-MeV and 1.333-MeV gamma ray within 0.7 ps of each other. Usually two NaI:Tl crystals, a NaI:Tl and HPGe detector or two HPGe detectors, are used. The HPGe-NaI:Tl spectrometer, which combines the good energy resolution of the HPGe crystal with the high efficiency of the NaI:Tl detector, is preferred for low-level gamma-ray activity.

Triple Coincidence Spectrometer

This type of spectrometer is used as a pair spectrometer for the simultaneous measurement of the annihilation and gamma photons of radionuclides that decay by positron emission. Three NaI:Tl detectors or two

NaI:Tl detectors with one HPGe detector are combined for this type of spectrometer. Sometimes a large NaI:Tl crystal split into two optically isolated regions is used, which can also be used as an anti-Compton spectrometer. A triple coincidence scintillation spectrometer with active and passive shields has been used for the measurement of ^{22}Na, ^{26}Al, and ^{60}Co activities in meteorite samples [Vartanov and Samojlov 1975]. With this spectrometer, the sensitivity for ^{22}Na was greatly increased in the presence of higher ^{26}Al and ^{60}Co activities.

Beta-Gamma Coincidence Spectrometer

If a sample has a radioisotope that emits two different types of radiation that can be measured simultaneously, then coincident techniques between the detectors for each type of radiation can be used to substantially reduce the background signal. For example, many radionuclides decay by beta particle emission whose excited daughter nucleus decays almost immediately by emitting one or more prompt gamma rays that are unique to the parent [Dewaraja et al. 1994]. The sample is placed adjacent to a beta counter and a gamma-ray spectrometer detector. By recording only those events which coincidentally produce signals in each detector, the background is reduced by eliminating events for which only one detector responds.

PROBLEMS

1. The muons μ^- and μ^+ are produced by cosmic ray interactions in earth's atmosphere miles above the surface. These particles are radioactive with a half-life of 1.523 μs and travel at speeds of $\beta = 0.9997c$. Thus half of the muons decay after traveling $(1.523 \times 10^{-6}\text{s})(3 \times 10^8\text{m/s}) \simeq 450$ m. How then do muons reach the earth's surface and contribute to our background dose?

2. At what speed must a 58-g tennis ball travel to have a kinetic energy equal to that of the most energetic cosmic rays, protons with energies of about 10^{20} eV. How high would the tennis ball jump if 10^{20} eV of energy were converted into gravitational potential energy?

3. The energy-dependent neutron flux produced by cosmic rays at ground level which has a water fraction of 25% is shown in the table below. Estimate the total thermal flux $\phi_T = \int_0^{.1eV} \phi(E)\, dE$ in units of cm^2 h^{-1}.

E MeV	$\phi(E)$ cm^{-2} s^{-1} MeV^{-1}	E MeV	$\phi(E)$ cm^{-2} s^{-1} MeV^{-1}
1.13E-08	5.22E+04	4.50E-08	5.41E+04
1.42E-08	5.83E+04	5.66E-08	4.30E+04
1.79E-08	6.32E+04	7.13E-08	3.05E+04
2.25E-08	6.61E+04	8.97E-08	1.89E+04
2.84E-08	6.59E+04	1.13E-07	1.00E+04
3 .57E-08	6.19E+04		

4. From the data in Fig. 21.1, how many muons per minute with energies between 100 and 1000 MeV enter a 4×5 inch NaI:Tl detector near the ground at an elevation of 2493 ft?

5. Potassium and sodium are both Group I elements and thus have similar chemical properties. As a consequence, K is a natural impurity found in NaI:Tl crystals. Estimate the concentration of K in ppm (by mass) to produce two cps in a NaI:Tl detector.

6. (a) What is the activity from the decay of ^{238}U in a 1-kg sample of uranium? (b) what is the total activity in the sample? State any assumptions made.

7. Consider a two-storied house 20 × 20 m in size with walls 8 m high and a basement 3 m deep. The basement floor and walls are of concrete 30-cm thick, and the outside walls are brick 1-cm thick. Plaster 1-cm-thick lines along all walls and the ceilings. Estimate the activity in Bq of ^{40}K, ^{226}Ra, and ^{232}Th in the structural material of the house.

8. You are responsible for determining the activity of material to be used to create a very low-level gamma spectrometer facility. You have access to both a HPGe spectrometer and a large NaI:Tl spectrometer. Explain how you would go about determining the specific gamma-ray energies and emission rates and the radionuclides impurities of the construction and shield materials being delivered to the facility.

9. Show that for very thin slabs ($\mu T \to 0$), Eq. (21.3) reduces to $P_1^o = 0.5$. Explain why this result should be expected.

10. Plot the probability that a monoenergetic photon produced by a radionuclide uniformly distributed in an iron slab escapes uncollided through one side of the shield as a function of the photon energy from 0.1 to 10 MeV. Consider iron slabs of thicknesses 0.5, 1 and 2 mfp.

11. Explain in words why most of the variation in J_n^o and J_n^s in Fig. 21.12 occurs in the first few mean free paths of the slab's thickness.

12. Twenty percent of the 1.46-MeV gamma rays produced by the radioactive decay of ^{40}K uniformly distributed in a concrete slab are found to escape uncollided (half from one side and half from the other). Estimate the thickness of the concrete slab.

13. Uncollided ^{40}K gamma rays from a thick concrete wall are measured to be 3.2 per second per cm^2 of the wall surface. Estimate the activity concentration in the concrete in units of Bq/kg.

14. The MCNP6 code used to generate data for Fig. 21.11 and Fig. 21.12 is shown below.

```
Esc. prob. from infin. slab, w/ uniform isotropic src.
c  For standard concrete 1 mfp at 0.5  MeV = 4.98833 cm
c ***  cell cards ***
  1  1   -2.32       10 -20 -99     imp:p=1      $ slab
  2  0              -10 -99         imp:p=0      $ left void
  3  0               20 -99         imp:p=0      $ right void
  50 0               99             imp:p=0      $ graveyard

c *** surface cards ***
  10  px  0.0    $slab left face
  20  px 4.98833  $slab right face
  99  so  100000. $graveyard sphere

c *** block three -- problem definition
NPS  1000000
MODE P
PHYS:P 10.0  0 0
c
c ----- Src uniform 0 to 1 mfp
SDEF   X=d1 Y=0 Z=0  ERG=0.500  PAR=2
SI1  0.00001 4.98833
SP1  -21 0
c
c ----- Detectors: F1 current (uncollided and collided)
F1:p   10 20 T
FT1  INC        $ separate by no. scat.
```

```
FU1  0 10000 T  $ 0 scat; <=10000 scat; total
c
c     CONCRETE: ANSI/ANS-6.4.3;  rho = 2.32 g/cm^3  mass frac.
m1     1001   -.0056   8016   -.4983   11023   -.0171
      12000   -.0024   13027   -.0456  14000   -.3158
      16000   -.0012   19000   -.0192  20000   -.0826
      26000   -.0122
```

Modify this input deck to calculate the escaping gamma flux of 0.5-MeV photons by using the f2 tally. Note that you must use an FM card to multiply the tally by slab surface area to obtain flux per unit volumetric source. Check your results by comparing the uncollided flux to that of Eq. (21.6).

REFERENCES

BEVELACQUA, J.J., *Health Physics in the 21st Century*, Weinheim: Wiley-VCH, 2008.

BRODZINSKI, R.L., D.P. BROWN, J.C. EVANS, W.K. HENSLEY, J.H. REEVES, N.A. WOG- MAN, F.T. AVIGNONE III, AND H.S. MILEY, "An Ultralow Background Germanium Gamma-Ray Spectrometer," *Nucl. Instr. Meth.*, **A239**, 207–213, (1985).

CAMP, D.C., C. GATROUSIS, AND L.A. MAYNARD, "Low Background Ge(Li) Detector Systems for Radioenvironmental Studies," *Nucl. Intru. Meth.*, **117**, 189-211, (1974).

CASE, K.M., F. DE HOFFMANN, AND G. PLACZEK, *Introduction to the Theory of Neutron Diffusion*, Vol. I, Los Alamos, NM: Los Alamos Sci. Lab., 1953.

DEWARAJA, Y., Z. HE, R.F. FLEMING, D.K. WEHE, S.V. GURU, J.C. FERREIRA, AND R.H. FLEMING, "A Position Sensitive β-γ Coincidence Technique for Multiplexed Gamma Spectrometry of Many Small Samples," *Nucl. Instrum. Meth.*, **A353**, 588–592, (1994).

EIESENBUD, M., *Environmental Radioactivity*, 3rd Ed., Orlando, Fl: Academic Press, 1987.

EXPACS ver. 4.0; T. SATO, "Analytical Model for Estimating the Zenith Angle Dependence of Terrestrial Cosmic Ray Fluxes," PLOS ONE, 11(8): e0160390. From `https://phits.jaea.go.jp/expacs/`, accessed June 2018.

FAW, R.E. AND J.K. SHULTIS, *Radiological Assesssment: Sources and Exposures*, Am. Nucl. Soc., LaGrange Park, IL, 1999.

GILMORE, G.R., *Practical Gamma-Ray Spectroscopy*, 2nd Ed., New York: Wiley, 2008.

HAFFNER, J.W., *Radiation Shielding in Space*, New York: Academic Press, 1967.

ICRP, *Report of the Task Group on Reference Man*, International Commission on Radiological Protection, Report 23, Oxford: Pergamon Press, 1975.

KIRSHENBAUM, A.D., J.A. CAHILL, AND A.V. GROSSE, "The Density of Liquid Lead from the Melting," *J. Inorganic & Nucl. Chem.*, **22**, 33–38, (1961).

KOCHER D.C., *Radioactive Decay Data Tables*, Report DOE/TIC-11026, Washington, DC: US Dept. of Energy, 1981.

LEONARD, D.S. AND 56 COAUTHORS,"Systematic Study of Trace Radioactive Impurities in Candidate Construction Materials for EXO-200," *Nucl. Instru. Meth.*, **A591**, 490–504, (2008).

MCNP6, *MCNP6 User's Manual*, LA-CP-13-00634, Los Alamos National Laboratory, 2013.

MORGAN, K.Z., "History of Damage and Protection from Ionizing Radiation," in *Principles of Radiation Protection*," K.Z. MORGAN AND J.E. TURNER, Eds., New York: Wiley, 1967.

NCRP, *Natural Background Radiation in the United States*, Report 45, National Council on Radiation Protection and Measurements, Washington, DC, 1975.

NCRP, *Guidance on Radiation Received in Space Activities*, Report 98, Washington, DC: National Council on Radiation Protection and Measurements, 1989.

NUDAT, *Nuclear Structures and Decay Data*, v. 2.7, National Nuclear Data Center, Brookhaven National Laboratory. Available at `http://www.nndc.bnl.gov/`; accessed July 2017.

RIZZACASSA, P., "Nuclear Instruments and Methods in Physics Research Section B: Beam Interactions with Materials and Atoms," *Nucl. Instru. Meth.*, **B17**, 519–523, (1986).

SATO T., H. YASUDA, K. NIITA, A. ENDO, AND L. SIHVER, "Development of PARMA: PHITS Based Analytical Radiation Model in the Atmosphere," *Radiat. Res.*, **170**, 244-259 (2008).

SATO T. AND K. NIITA, "Analytical Functions to Predict Cosmic-Ray Neutron Spectra in the Atmosphere," *Radiat. Res.*, **166**, 544-555 (2006).

SHULTIS, J.K. AND R.E. FAW, *Radiation Shielding*, Lagrange Park, IL: Am. Nucl. Soc., 2000.

SHULTIS, J.K. AND R.E. FAW, *Fundamentals of Nuclear Science and Engineering*, 3rd Ed., Boca Raton, FL: CRC Press, 2017.

UN, *Report of the United Nations Scientific Committee on the Effects of Atomic Radiation*, New York, 1977.

UN, *Report of the United Nations Scientific Committee on the Effects of Atomic Radiation*, New York, 1982.

UN, *Report of the United Nations Scientific Committee on the Effects of Atomic Radiation*, New York, 1988.

VARTANOV, N.A. AND P.S. SAMOJLOV, *Applied Scintillation Gamma-Spectroscopy*, (in Russian), Moscow: Atomizdat, 1975.

ZVARA, P., P. OVINEC, AND L. YKORA, "Determination of Very Low Levels of Radioactivity," *Pure & Appl. Chem.*, **68**, No. 12, 2537–2586, (1994).

Chapter 22

Nuclear Electronics

Despite many great achievements in the science of radio and electronics, what we know today is far less than what we have still to learn.

David Sarnoff

The bulk of this textbook has been dedicated to the description of radiation detectors and their functional operation. However, it would seem incomplete not to discuss, at least cursorily, some of the common electronics used to operate these detectors. Because the field of radiation detector electronics is vast, a complete and thorough examination of the art is simply not possible in this book, and the authors refer the reader to other respected works on the topic, including Gillespie [1953], Chiang [1969], Kowalski [1970], Nicholson [1974], and Spieler [2005]. What follow are brief introductory descriptions of the electronics often used with radiation detectors and how these electronic circuits operate.

In Sec. 2.4 a brief historical review was presented about the reason for and development of modular units for nuclear radiation detection. From this early attempt at standardizing detector electronics, Nuclear Instrument Modules (NIM) were developed. These modules are typically inserted into a universal power supply unit called a Nuclear Instrument Module BIN (or NIM bin). This system has a set of standards, described in U.S. DOE Report ER/DOE-0457T, which specifies a 10-volt processing limit. Pulses greater than 10 volts are rejected or "clipped" upon entering any of the NIM units.

The NIM system is not the only electronic system available. Perhaps the second most popular modular system is Computer Automated Measurement and Control (CAMAC) system. Analogous to the NIM bin, CAMAC instruments are powered by a CAMAC crate. CAMAC is popular for circumstances requiring complicated timing and measurement controls, such as with high-energy collider experiments. Moreover, CAMAC components are not restricted to only complicated radiation measurements but can be used in other electronic applications. But because of the typically higher cost of CAMAC components compared to NIM components, they are usually used only when NIM electronics are less effective. Besides the NIM and CAMAC systems there are stand-alone units, such as those found in survey instruments, which have multiple integrated electronic components, all combined within a single unit. In NIM and CAMAC systems, these different components are separate components. Regardless of the system, the basic electronics described in this chapter are universally applied within the various electronic systems.

22.1 Mathematical Transforms

Before electronic networks and components are discussed, it is first necessary to briefly discuss the traditional methods used to mathematically analyze a network of passive elements. Consider a circuit with an input signal $f_{in}(t)$, be it a voltage or current, to which the resulting output $f_{out}(t)$ is sought. Generally the relation between the input and output is described by a set of time-dependent ordinary differential equations. Rather than solve these equations explicitly, they are often first transformed by either a Fourier or Laplace integral transform. Under such a transform or mapping, the system of differential equations is converted to a set of coupled linear algebraic equations. In essence a calculus problem is transformed into an algebraic problem, which is much easier to solve than are the differential equations. Once the transformed output is obtained by algebraic means, it can then be mapped back to the time domain by an inverse integral transform to give $f_{out}(t)$. The two most common transforms used to describe linear circuits are the Fourier and the Laplace transforms.

22.1.1 The Fourier Transform

The Fourier transform is a generalization of a Fourier series for functions $f(t)$ that have no periodicity. If $f(t)$ is absolutely integrable, i.e., $\int_{-\infty}^{\infty} |f(t)|\, dt < \infty$ then the Fourier transform $F(\omega)$ of $f(t)$ is given by

$$F(\omega) \equiv \mathcal{F}[f(t)] = \int_{-\infty}^{\infty} e^{-j\omega t} f(t)\, dt \tag{22.1}$$

and the inverse transform by

$$f(t) \equiv \mathcal{F}^{-1}[F(\omega)] = \frac{1}{2\pi} \int_{-\infty}^{\infty} e^{j\omega t} F(\omega)\, d\omega, \tag{22.2}$$

where $j \equiv \sqrt{-1}$ and ω is the angular frequency.[1] Here $\mathcal{F}[f(t)]$ transforms $f(t)$ from the time domain to the frequency domain and $\mathcal{F}^{-1}[F(\omega)]$ transforms $F(\omega)$ back from the frequency domain to the time domain.

22.1.2 The Laplace Transform

Often $f(t)$ is not absolutely integrable (e.g., $f(t)$ is any polynomial in t or $f(t)$ is a step function $H(t)$). However, the modified function $e^{-\sigma t} f(t) H(t)$, with σ real and positive, may be integrable, i.e.,

$$\int_0^{\infty} dt\, e^{-\sigma t} |f(t)| < \infty.$$

Then instead of taking the Fourier transform of $f(t)$ (which does not exist), we take the Fourier transform of

$$g(t) = \begin{cases} e^{-\sigma t} f(t), & \text{if } t \geq 0 \\ 0, & \text{if } t < 0 \end{cases} \tag{22.3}$$

The Fourier transform of this function and its inverse are

$$\mathcal{F}[g(t)] \equiv G(\omega) = \int_0^{\infty} e^{-j\omega t} e^{-\sigma t} f(t)\, dt, \tag{22.4}$$

and

$$\mathcal{F}^{-1}[G(\omega)] \equiv g(t) = \frac{1}{2\pi} \int_{-\infty}^{\infty} e^{j\omega t} G(\omega)\, d\omega. \tag{22.5}$$

This pair of transforms can be simplified by introducing a complex variable $s = \sigma + j\omega$. Clearly, $\omega = (s - \sigma)/j$ and $d\omega = -j\, ds$. Hence for $t > 0$

$$g(t) \equiv f(t) e^{-\sigma t} = \frac{1}{2\pi} \int_{-\infty}^{\infty} e^{j\omega t} G(\omega)\, d\omega$$

$$= \frac{1}{2\pi} \int_{-\infty}^{\infty} e^{j\omega t} \int_0^{\infty} f(t')\, e^{-\sigma t'} e^{-j\omega t'}\, dt'\, d\omega = \frac{-j}{2\pi} \int_{\sigma - j\infty}^{\sigma + j\infty} ds\, e^{st} e^{-\sigma t} \int_0^{\infty} f(t')\, e^{-st'}\, dt'\, ds$$

[1] Although the usual mathematical symbol is $i = \sqrt{-1}$, the letter i is also commonly used in electrical engineering calculations for current. To avoid confusion, it is common to use j for $\sqrt{-1}$ in circuit theory.

Table 22.1. Some commonly encountered Laplace transforms.

$F(s)$	$f(t)$	$F(s)$	$f(t)$
$\frac{1}{s}$	$H(t)$ unit step fn	$\frac{s-a}{(s-a)^2+b^2}$	$e^{at}\cos bt$
$\frac{1}{s^n}(\ n=1,2,...)$	$\frac{t^{n-1}}{(n-1)!}$	$\frac{s^2-a^2}{(s^2+a^2)^2}$	$t\cos at$
$\frac{\Gamma(k)}{s^k}\ (k>0)$	t^{k-1}	$\frac{1}{s(s^2+a^2)}$	$\frac{1}{a^2}(1-\cos at)$
$\frac{1}{s-a}$	e^{at}	$\frac{1}{s^2(s^2+a^2)}$	$\frac{1}{a^3}(at-\sin at)$
$\frac{1}{(s-a)(s-b)}$	$\frac{1}{a-b}(e^{at}-e^{bt})$	$\frac{1}{(s^2+a^2)^2}$	$\frac{1}{2a^3}(\sin at - at\cos at)$
$\frac{s}{(s-a)(s-b)}$	$\frac{1}{a-b}(ae^{at}-be^{bt})$	$\frac{s}{(s^2+a^2)^2}$	$\frac{1}{2a}\sin at$
$\frac{1}{s^2+a^2}$	$\frac{1}{a}\sin at$	$\frac{s^2}{(s^2+a^2)^2}$	$\frac{1}{2a}(\sin at + at\cos at)$
$\frac{s}{s^2+a^2}$	$\cos at$	$\frac{1}{(s-a)^2+b^2)^2}$	$\frac{1}{b}e^{a}t\sin bt$
$\frac{1}{s^2-a^2}$	$\frac{1}{a}\sinh at$	$\frac{s-a}{(s-a)^2+b^2)^2}$	$e^{at}\cos bt$
$\frac{s}{s^2-a^2}$	$\cosh at$	$\frac{s}{s^4+4a^4}$	$\frac{1}{2a^2}\sin at\sinh at$

or

$$f(t) = \frac{1}{2\pi j}\int_{\sigma-j\infty}^{\sigma+j\infty} e^{st}\int_0^{\infty} f(t')\,e^{-st'}\,dt'\,ds. \tag{22.6}$$

From this identity, one identifies the transform pair, known as the *Laplace Transform*,

$$f(t) = \frac{1}{2\pi j}\int_{\sigma-j\infty}^{\sigma+j\infty} e^{st}F(s)\,ds \equiv \mathcal{L}^{-1}[F(s)] \tag{22.7}$$

$$F(s) = \int_0^{\infty} f(t)e^{-st}\,dt \equiv \mathcal{L}[f(t)]. \tag{22.8}$$

Here σ is a sufficiently large and positive constant chosen so that $\int_0^\infty e^{-\sigma t}|f(t)|\,dt < \infty$. The Laplace transform is particularly useful for functions $f(t)$ that are zero for negative t.

Unlike that for Fourier transforms, the inversion of the Laplace transforms from Eq. (22.7) is not straightforward because it involves contour integration in the complex s-plane. An alternative is to prepare a 'dictionary' of Laplace transforms of common functions and, when faced with the need to invert a transform, simply look up in the dictionary the parent function that has the transform of interest. An example of such a listing is given in Table 22.1. Perhaps the most complete listing of Fourier and Laplace transforms and their inverses is that of Bateman [1954].

Example 22.1: Find the Fourier and Laplace transforms for the function

$$f(t) = \begin{cases} 0, & \text{if } t < 0 \\ e^{-at}, & \text{if } t \geq 0 \end{cases}$$

Solution:
The Fourier transform is

$$G(\omega) = \mathcal{F}[f(t)] = \int_0^\infty e^{-\alpha t} e^{-j\omega t} dt = \frac{-1}{\alpha + j\omega} e^{-(\alpha + j\omega)t}\Big|_0^\infty = \frac{1}{\alpha + j\omega}.$$

The Laplace transform is

$$F(s) = \mathcal{L}[f(t)] = \int_0^\infty e^{-\alpha t} e^{-st} dt = \frac{-1}{\alpha + s} e^{-(\alpha + s)t}\Big|_0^\infty = \frac{1}{\alpha + s}$$

In this case the results are very similar for the Fourier and Laplace transforms. The general naive, and incorrect, impression may be that $j\omega$ in the Fourier transform can be switched for s in the Laplace transform, and vice versa. However, this is actually not always true, and in fact, the Fourier transform is a smaller subset of the Laplace transform with the term $\sigma = 0$.

22.1.3 Properties of the Laplace Transform

The Laplace transform has many useful properties that are used throughout this chapter. These properties are summarized below. The Fourier transform, likewise, has many similar properties.

Integration:

$$\mathcal{L}\left[\int_0^t dt'\, f(t')\right] = \frac{1}{s}\,\mathcal{L}[f(t)] = \frac{F(s)}{s}. \tag{22.9}$$

Example: Given $f(t) = a + \int_0^t dt'\, f(t')$, find the solution $f(t)$. Take the Laplace transform to obtain

$$F(s) = \frac{a}{s} + \frac{F(s)}{s} \qquad \text{or} \qquad F(s) = \frac{a}{s-1},$$

from which $f(t) = a\, e^t$.

Differentiation:

$$\mathcal{L}\left[\frac{df(t)}{dt}\right] = s\,\mathcal{L}[f(t)] - f(0) = s\,F(s) - f(0). \tag{22.10}$$

Example: Given $df(t)/dt - af(t) = c$ with $f(0) = 0$, find $f(t)$. Taking the Laplace transform gives

$$sF(s) - f(0) - aF(s) = \frac{c}{s} \qquad \text{or} \qquad F(s) = \frac{c}{s(s-a)} = -\frac{c/a}{s} + \frac{c/a}{s-a}$$

Inversion of this result gives the solution $f(t) = (c/a)[-1 + e^{at}]$.

Higher Order Derivatives:

$$\mathcal{L}\left[\frac{d^n f(t)}{dt^n}\right] = s^n F(s) - \sum_{m=1}^{n} s^{n-m} f^{(m)}(0^-), \tag{22.11}$$

where $f^{(n)}(0^-) \equiv d^n f(t)/dt^n|_{t=0^-}$.

Convolution:

$$\mathcal{L}^{-1}[F(s)G(s)] = \int_0^t d\tau\, f(t-\tau)g(\tau) = \int_0^t d\tau\, f(\tau)g(t-\tau) \tag{22.12}$$

Example: Given $F(s) = \omega/[s(s^2+\omega^2)]$, find $f(t)$. If one lets $G(s) = 1/s$ and $F(s) = \omega/(s^2+\omega^2)$, then $g(t) = 1$ and $f(t) = \sin \omega t$. From this theorem, one then has

$$f(t) = \int_0^t d\tau\, g(t-\tau)f(\tau) = \int_0^t d\tau\, \sin \omega\tau = \frac{1}{\omega}(1 - \cos \omega t).$$

Shift Formula:

$$\mathcal{L}^{-1}[e^{-as}F(s)] = \begin{cases} f(t-a), & t \geq a \\ 0, & t < a \end{cases} \tag{22.13}$$

Example: Given $F(s) = e^{-2s}/(s+1)$, what is $f(t)$? Recall that if $G(s) = 1/(s+1)$ then $g(t) = e^{-t}$. By this theorem, one then has

$$f(t) = \begin{cases} e^{-(t-2)}, & t \geq 2 \\ 0, & t < 2 \end{cases}$$

Translation Formula:

$$\mathcal{L}[e^{-at}f(t)] = F(s+a) \tag{22.14}$$

Example: Given $F(s) = s/[(s+a)^2 + \omega^2]$, find $f(t)$. One can rewrite $F(s)$ as

$$F(s) = \frac{(s+a)-a}{(s+a)^2+\omega^2} \equiv G(s+a)$$

where

$$G(s) = \frac{s}{s^2+\omega^2} - \frac{a}{s^2+\omega^2} \quad \text{for which} \quad g(t) = \cos \omega t - \frac{a}{\omega} \sin \omega t$$

Then from this theorem, $f(t) = e^{-at}(\cos \omega t - \frac{a}{\omega} \sin \omega t)$.

Initial Value Theorem:

$$\lim_{|s|\to\infty} [sF(s)] = \lim_{t\to 0} f(t) \tag{22.15}$$

Example: If $F(s) = s/(s^2+\omega^2)$, what is $f(0)$? From tables one knows $f(t) = \cos \omega t$ so that $f(0) = 1$. We can get this same result from the theorem, namely,

$$f(0) = \lim_{|s|\to\infty} \left[s \frac{s}{s^2+\omega^2} \right] = 1.$$

Final Value Theorem:

$$\lim_{|s|\to 0} [sF(s)] = \lim_{t\to\infty} f(t) \tag{22.16}$$

Example: If $F(s) = 1/[s(s+2)]$, what is $\lim_{t\to\infty} f(t)$? From tables one finds $f(t) = (1 - e^{-2t})/2$ so that $f(\infty) = 1/2$. From this theorem, the same result is obtained, namely

$$f(\infty) = \lim_{|s|\to 0} \left[s \frac{1}{s(s+2)} \right] = \frac{1}{2}.$$

22.1.4 The Transfer Function

The dynamic response (output) $y(t)$ of a *linear* network of components to an input $x(t)$ is generally described by a set of linear coupled differential equations. By taking the Fourier or Laplace transforms of these dynamic equations, they are transformed into a set of algebraic equations involving the transforms $X(s)$ and $Y(s)$ (or $X(\omega)$ and $Y(\omega)$).

The transfer function describes the ratio of two signals, usually the signal output of a circuit divided by a corresponding input signal. Denoting $x(t)$ as the time dependent input signal and $y(t)$ as the time dependent output signal, then the Fourier transfer function is

$$T(\omega) = \frac{Y(\omega)}{X(\omega)} = \frac{\mathcal{F}(y(t))}{\mathcal{F}(x(t))}, \tag{22.17}$$

and the Laplace transfer function is

$$H(s) \equiv \frac{Y(s)}{X(s)} = \frac{\mathcal{L}(y(t))}{\mathcal{L}(x(t))}. \tag{22.18}$$

For the Laplace transfer function, it is usual to assume the initial values $x(0)$ and $y(0)$ are both zero. Values that cause the numerator of Eq. (22.18) to become zero are called *zeros*, and values that cause the denominator to become zero are *poles*. The poles of the transfer function play a key role in the behavior of $y(t)$ when $Y(s)$ is transformed back into the time domain. This behavior and the importance of poles are discussed later in the chapter. A hint of the importance of poles can be seen from the entries in Table 22.1 where the poles in $F(s)$ determine the time and scaling constants in $f(t)$.

Often a dynamic system (here a circuit) is considered that is in steady-state for $t < 0$ and then for $t \geq 0$ some time-dependent input $x(t)$ (or deviation from the steady-state input) is applied to the system so as to produce an output $y(t)$ (or deviation from its steady-state output). The output $y(t)$ could be obtained by solving the differential equations describing the dynamic behavior of the system. However, the same solution can be obtained from the transfer function of Eq. (22.18) as

$$y(t) = \mathcal{L}^{-1}[X(s)H(s)]. \tag{22.19}$$

In particular, if $x(t) = \delta(t)$ the output $h(t)$ produced by this unit impulse input is $h(t) = \mathcal{L}^{-1}H(s)$ since $X(s) = \mathcal{L}[\delta(t)] = 1$. Once the impulse response is known, the output $y(t)$ for any input $x(t)$ can be obtained with the convolution property of Laplace transforms as

$$y(t) = \mathcal{L}^{-1}[H(s)Y(s)] = \int_0^\infty h(\tau)x(t-\tau)\,d\tau. \tag{22.20}$$

The frequency response of electric circuits is important in characterizing the circuit. This response can also be obtained directly from the transfer function. If the input for $t \geq 0$ is $x(t) = x_o \sin\omega t$ the output can be shown to be[2]

$$y(t) = x_o|H(j\omega)|\sin(\omega t + \phi) + \text{ decaying transient terms.} \tag{22.21}$$

Thus, a sinusoidal input produces, asymptotically, a sinusoidal output with a phase shift ϕ. The *gain* $G(\omega)$ of the circuit is the ratio of the amplitude of the asymptotic sinusoidal output to that of the input, i.e.,

$$G(\omega) = |H(j\omega)|, \tag{22.22}$$

and the *phase shift* $\phi(\omega)$ is

$$\phi(\omega) = \arg H(j\omega) = \tan^{-1}\frac{\text{Im}H(j\omega)}{\text{Re}H(j\omega)}. \tag{22.23}$$

[2]If $H(s)$ has a pole at $s = 0$, then a constant shift also appears in the right-hand side of Eq. (22.21).

Example 22.2: What are the zeros and poles of the transfer function?

$$H(s) = \frac{s(s+3)}{(s-2)(4+s)^2}.$$

Solution:
This transfer function has simple zeros at $s = 0$ and $s = -3$. It also has a pole of order 1 at $s = 2$ and a pole of order 2 at $s = -4$.

22.2 Pulse Shaping

The production of induced charge in a radiation detector was described earlier in Chapter 8 where it was seen that the amount of charge created is linearly proportional to the energy deposited by a radiation event in the detector. Signals from a radiation detector usually develop rapidly and are of short duration, mainly because the charge induction continues only while the free charge carriers are in motion inside the detector medium. The goal of a detection system is to measure accurately the charge produced in a detector and to generate an output signal that corresponds linearly to the liberated charge. The method used to measure this charge involves pulse shaping, wherein the input current from the detector is changed to an output voltage pulse with a maximum that is proportional to the initial charge liberated in the detector.

Imperfections with the pulse processing electronics often introduce artifacts into the output stream of pulses as a result of *pulse pile-up*, *ballistic deficit*, and *electronic noise*. Radiation induced pulses arriving at the input of the processing electronics are stochastic, meaning that they arrive at random times. Consider the pulse train shown in Fig. 22.1 generated from monoenergetic radiation interactions. The ideal situation would be to have the processed pulses all have identical heights for identical energies. However, in the case depicted by Fig. 22.1(a), voltage pulses arrive before the electronic tail of previous pulses have relaxed to zero, effectively causing them to appear larger than they really are. In the case depicted by Fig. 22.1(b), the pulse tails have long undershoots, i.e., they extend to negative values before the electronics reset. Consequently, some pulses appear to be smaller than they actually are. Both of these cases are caused by pulse pile-up in which radiation events occur faster than the electronics can respond. Both cases can cause serious deleterious effects on radiation spectroscopy by widening the pulse height spectrum.

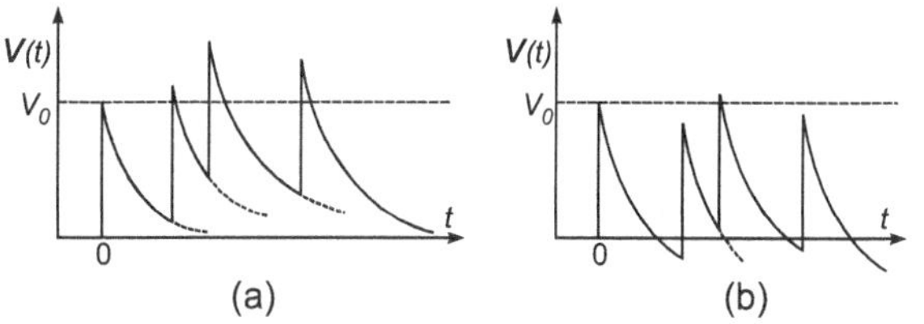

Figure 22.1. The effects of pulse pile-up for (a) tail pulses with relatively long tails and (b) tail pulses with long undershoot. The pulse height without distortions is v_0.

The term "ballistic" deficit refers to the situation in which the output pulses have a lower output amplitude than expected because the processing electronics prematurely ends the signal formation. Unfortunately, this premature pulse termination is often necessary to increase the pulse processing speed in order to accommodate high rates of radiation interactions. Under ideal conditions, the ballistic deficit would be proportional to the amount of induced charge and have little effect on spectroscopic performance. However, gas-filled and semiconductor detectors have induced currents that vary with the trajectory of an ionizing particle and with the interaction location. This effect, thus, broadens the energy resolution of the spectrometer.

Electronic noise produces a variation among ideally identical radiation-induced signals and, thus, causes broadening in the energy resolution of a spectrometer. The average noise voltage should be zero with some standard deviation about zero, or about the root mean square (RMS) voltage. Input pulses have the randomized noise added, positively or negatively, to the signal so that the energy resolution is broadened. Noise is most problematic in the early stages of amplification. Sources of noise and mitigation methods are discussed in a latter section of this chapter.

22.2.1 Circuit Element Impedances

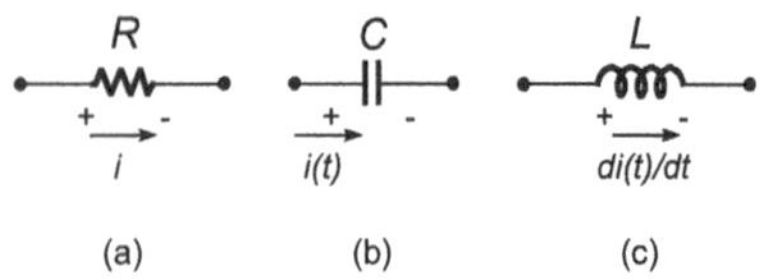

Figure 22.2. Symbols for passive circuit elements, (a) resistor, (b) capacitor, (c) inductor.

There are three main passive elements used in electronic circuits, namely *resistors*, *capacitors*, and *inductors*. The resistor is a passive element that, as the name implies, resists the flow of electrical current in accordance with Ohm's law

$$v(t) = i(t)R = R\frac{dq(t)}{dt}, \tag{22.24}$$

where $v(t)$ is the voltage, $i(t)$ is the current, $q(t)$ is the charge, and R is the resistance with units of ohms (Ω).[3] The symbol for a resistor is a zig-zag line as depicted in Fig. 22.2(a).

A capacitor is a passive device in which two conductors are separated by an insulator (see Sec. 8.3) capable of storing charge when a voltage is applied across the terminals. *Capacitance* C is the ratio of stored charge $q(t)$ to the resulting voltage $v(t)$ (charge per volts) and has units of farads (F), or ampere-second per volt. Thus

$$v(t) = \frac{q(t)}{C}. \tag{22.25}$$

If the stored charge changes with time, then the resulting current is

$$i(t) = \frac{dq(t)}{dt} = C\frac{dv(t)}{dt}. \tag{22.26}$$

An inductor produces a magnetic field when a current flows through it. Further, if the current flowing into the inductor changes with time, an induced voltage is produced that is proportional to the rate of change, namely,

$$v(t) = L\frac{di(t)}{dt} = L\frac{d^2q(t)}{dt^2}, \tag{22.27}$$

where the inductance L is the ratio of change between the induced voltage and the change in current and has units of henries (H), or volt-second per ampere.

Phasors and Impedances

If a circuit contains capacitors and/or inductors, equations for currents and voltages involve time derivatives. So, even for the simplest circuits, transient responses are generally described by ordinary differential equations. Only for circuits composed solely of resistors are the voltage and current transients given by simple algebraic equations. However, there is one important case where the differential equations can be replaced by algebraic equations, namely when the voltages and currents are sinusoidal functions of time.[4] To obtain such a simplification, it is necessary to use the complex plane and introduce complex quantities such as *phasors* and *impedances*.

Sinusoidal voltages and currents can always be written as

$$v(t) = v_o \sin(\omega t + \phi_v) \qquad \text{and} \qquad i(t) = i_o \sin(\omega t + \phi_i), \tag{22.28}$$

[3]Here lowercase v, q, and i are used to denote generally time-dependent quantities, whereas their uppercase counterparts, used for example in Chapter 8, generally denote steady-state values.

[4]The same simplification is obtained with the more complex Laplace transform technique, which is applicable to any type of transient. But the special case of sinusoidal transients is a well-established fixture of circuit analysis.

where v_o and i_o are amplitudes, ϕ_v and ϕ_i are the phase shifts, and ω is the angular frequency. Any complex number z can be written in terms of its real and imaginary parts as $\Re\{z\} + j\Im\{z\}$. Then with Euler's theorem $e^{j\omega t} = \cos\omega t + j\sin\omega t$, the real parts of $v(t)$ and $i(t)$ can be written as

$$v(t) = \Re\{v_o e^{j(\omega t+\phi_v)}\} = \Re\{\mathbf{v}e^{j\omega t}\} \quad \text{and} \quad i(t) = \Re\{i_o e^{j(\omega t+\phi_i)}\} = \Re\{\mathbf{i}e^{j\omega t}\},$$

where the complex numbers $\mathbf{v} \equiv v_o e^{j\phi_v}$ and $\mathbf{i} \equiv i_o e^{j\phi_i}$ are often called *phasors*. Note that the phasors are independent of time. Then the derivatives of the voltage and current can be written as

$$\frac{dv(t)}{dt} = \Re\{j\omega\mathbf{v}e^{j\omega t}\} \qquad \text{and} \qquad \frac{di(t)}{dt} = \Re\{j\omega\mathbf{i}e^{j\omega t}\}. \tag{22.29}$$

Hence, in the complex plane, differentiation with respect to t is the same as multiplication by $j\omega$. In this manner, the differential equations for the analysis of a circuit are converted into algebraic equations. The only difference is that currents and voltages are now phasors and the resistances, capacitances and inductances are replaced by their complex *impedances* defined as the phasor ratio $\mathbf{v}/\mathbf{i}$ with units of *ohms* (Ω).

As an example of an impedance, consider a sinusoidal voltage placed across a capacitor. The complex voltage and current are $\mathbf{v}e^{j\omega t}$ and $\mathbf{i}e^{j\omega t}$, respectively. Then from Eq. (22.26)

$$\mathbf{i}e^{j\omega t} = C\frac{d}{dt}\left(\mathbf{v}e^{j\omega t}\right) \qquad \text{or} \qquad \mathbf{i} = j\omega C\mathbf{v}.$$

The capacitor impedance is thus

$$Z_C = \frac{\mathbf{v}}{\mathbf{i}} = \frac{1}{j\omega C}. \tag{22.30}$$

Equation (22.30) is worth a closer look. At low frequencies, when $\omega \approx 0$, Eq. (22.30) becomes extremely large, i.e., the impedance becomes infinite or equivalent to an open circuit. At high frequencies Eq. (22.30) approaches zero, and the capacitor behaves as a short circuit. The impedances of resistors and inductors can be found in a similar fashion to give,[5]

$$\frac{\mathbf{v}}{\mathbf{i}} = Z_R = R, \;\; \text{(resistor)}, \qquad \frac{\mathbf{v}}{\mathbf{i}} = Z_C = \frac{1}{j\omega C} \;\; \text{(capacitor)}, \qquad \frac{\mathbf{v}}{\mathbf{i}} = Z_L = j\omega L \;\; \text{(inductor)}. \tag{22.31}$$

Note that the impedances of capacitors and inductors change with signal frequency, and lumped circuits must include the complex algebraic manipulations of these terms. Further, opposite of the capacitor, the inductor has high impedance at high frequency and behaves as a short circuit at low frequency. The resistor has no frequency dependence.

Combining Circuit Elements

The impedances of these elements of resistance, capacitance, and inductance generally follow the rules of series and parallel resistive circuits, that is, impedances in series

$$Z_{eq} = Z_1 + Z_2 + Z_3 + \ldots + Z_N, \tag{22.32}$$

and impedances in parallel

$$\frac{1}{Z_{eq}} = \frac{1}{Z_1} + \frac{1}{Z_2} + \frac{1}{Z_3} + \ldots + \frac{1}{Z_N}, \tag{22.33}$$

[5]The Laplace transform representations are $Z_C = 1/(sC)$ and $Z_L = sL$, which are also their transfer functions.

where the complex terms remain appropriately separated. The real portion of the impedance is called resistance R (or active component), while the imaginary part is called the reactance X. The magnitude of a complex impedance is

$$|Z| = |R + jX| = \sqrt{(R + jX)(R - jX)} = \sqrt{R^2 + X^2}. \tag{22.34}$$

The phase angle of a complex circuit is

$$\theta = \arctan\left(\frac{X}{R}\right), \tag{22.35}$$

denoted $\angle\theta$ relative to 0°, that is, positive for leading angle and negative for lagging angle with respect to the voltage versus current. Consequently, a purely inductive circuit has voltage phase angle of 90° leading the current, while a purely capacitive circuit has voltage phase angle of −90° lagging the current.

Example 22.3: Find the impedance of the following circuit:

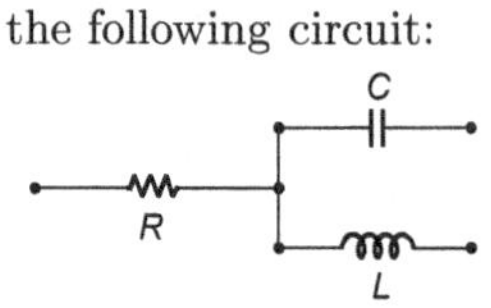

Solution:
First, the parallel combination of the impedance is

$$Z_{\|} = \left[\frac{1}{Z_1} + \frac{1}{Z_2}\right]^{-1} = \frac{Z_1 Z_2}{Z_1 + Z_2} = \frac{(j\omega L)\left(\dfrac{1}{j\omega C}\right)}{j\omega L + \dfrac{1}{j\omega C}} = \frac{j\omega L}{1 - \omega^2 LC}$$

Now add the series resistance to find

$$Z_{eq} = R + \frac{j\omega L}{1 - \omega^2 LC} = \frac{R(1 - \omega^2 LC) + j\omega L}{1 - \omega^2 LC}.$$

The magnitude is

$$Z_{eq} = \sqrt{R^2 + \left(\frac{\omega L}{1 - \omega^2 LC}\right)^2},$$

with phase angle

$$\theta = \frac{\omega L}{R(1 - \omega^2 LC)},$$

which, at low frequency ($\omega^2 LC \ll 1$), becomes

$$\theta \approx \frac{\omega L}{R},$$

and at high frequency ($\omega^2 LC \gg 1$)

$$\theta \approx \frac{-1}{\omega RC}.$$

What then does this mean? The phase angle is zero for DC voltage applied to the circuit. At low frequencies, the phase angle is positive, and at high frequencies the phase angle is negative. A resonant frequency appears at $\omega_o^2 = 1/LC$, meaning that the magnitude of the impedance effectively becomes infinite, behaving as a notch (or band-stop) filter at the resonant frequency.

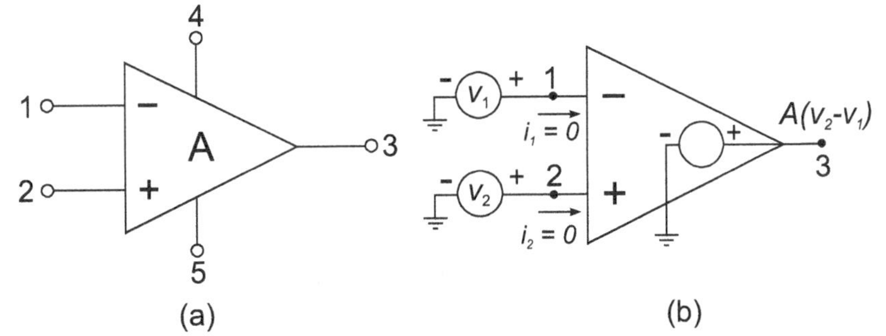

Figure 22.3. Usual depiction of an operational amplifier showing the basic inputs 1 and 2 and output terminal 3. Power is provided at terminals 4 and/or 5.

22.2.2 Operational Amplifier

A basic building block of a preamplifier is often the operational amplifier (or "op amp"), usually depicted as a right pointing triangle with two inputs and one output. This simplistic figure takes the place of a complex network of transistors, resistors, and capacitors that generally make up the actual op amp [Irvine 1981]. For instance, the common 741 op amp has three amplification stages and 20 or so transistors.

The operational amplifier has two input terminals, identified as 1 and 2 in Fig. 22.3(a), and one output terminal, denoted as terminal 3. The devices require power to operate, usually applied at terminal 4 (positive) and 5 (negative). A common operating voltage requirement is ± 9 volts, although modern amplifiers may operate on only a few volts. The *open-loop gain* is denoted by A, which is typically very large usually between 10^5 to 10^7 for constant input voltages. An op amp may also have additional terminals designated for special purposes, such as frequency compensation and offset nulling.

The main function of an op amp is to sense the difference between the voltage inputs at terminals 1 (v_1) and 2 (v_2), each referenced to ground, and supply amplification $A(v_2 - v_1)$ at terminal 3. For low frequency inputs, the open-loop gain A has a large value. But at some breakpoint frequency the gain begins to decrease with frequency at -20 db/decade, and where the open-loop gain crosses 0 db is defined as the open-loop bandwidth. The *closed-loop bandwidth* of an op amp is the frequency range over which the gain remains above -3 db of its value at $\omega = 0$. The frequency dependence of the open-loop gain, as discussed in the next sections, is important as it is a major contributor to the non-ideal performance of an op amp circuit. Detailed analysis for op amp performance can be found in a number of references, including Carter and Brown [2016], Carter and Mancini [2017], Jung [2005], and Zumbahlen [2008].

Ideally, an op amp should have an infinite input impedance, thereby eliminating input current. Further, the output impedance at terminal 3 should be zero, such that there is no voltage loss from the ideal gain $A(v_2 - v_1)$. The equivalent circuit is shown in Fig. 22.3(b). Terminal 1 is called the inverting terminal, and terminal 2 is the non-inverting terminal.

Consider the situation of Fig. 22.3(b). If the output at terminal 3 is in fact $A(v_2 - v_1)$, then the following applies for the output voltage at terminal 3

$$\frac{v_3}{A} = (v_2 - v_1). \tag{22.36}$$

Because A is large, and ideally treated as infinite, then it is found that

$$\frac{v_3}{A} = (v_2 - v_1) \simeq 0. \tag{22.37}$$

Consequently, if A is very large, and approaching infinity as in the ideal case, then $v_1 \simeq v_2$, a condition emulating a short circuit between terminals 1 and 2. For this reason, there is a *virtual short circuit* between terminals 1 and 2. Also, if terminal 2 is connected to ground, then terminal 1 is effectively also connected to ground through the virtual short circuit, and is referred to as a *virtual ground.*

22.2.3 The General Case for Feedback Transfer Function

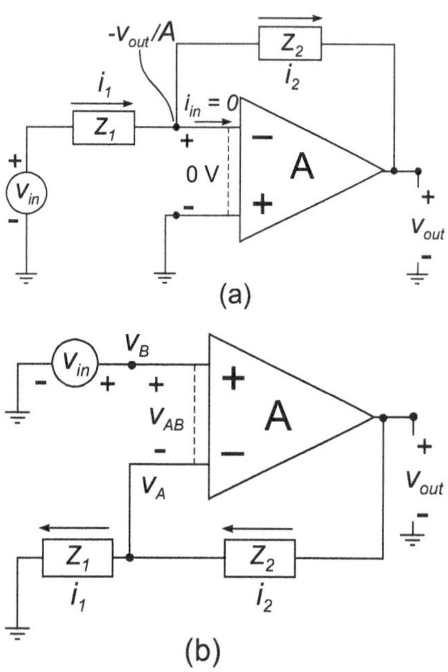

Figure 22.4. The (a) inverting configuration and the (b) non-inverting configuration with generalized impedances.

Consider the op amp configuration of Fig. 22.4(a). There is an input voltage applied across an impedance into the inverting input. Also, another impedance is placed across the inverting input terminal and the output terminal. With the positive terminal grounded, the voltage at the inverting terminal must be $v_{in} = -v_{out}/A$.

The current i_1 is defined as

$$i_1 = \frac{(-v_{out}/A) - v_{in}}{Z_1} = -\frac{1}{Z_1}\left(v_{in} + \frac{v_{out}}{A}\right). \tag{22.38}$$

Because of the high input impedance at the inverting input, the current across the feedback impedance Z_2 is practically the same as that across the impedance Z_1 ($i_2 \approx i_1$). The output voltage v_{out} must be the sum of the voltages at the inverting input terminal and across Z_2

$$v_{out} = -\frac{v_{out}}{A} - \frac{1}{Z_1}\left(v_{in} + \frac{v_{out}}{A}\right) Z_2 \tag{22.39}$$

Rearranging terms yields

$$\frac{v_{out}}{v_{in}} = -\frac{Z_2}{Z_1}\left[\frac{1}{1 + (1 + Z_2/Z_1)/A}\right]. \tag{22.40}$$

Recalling that the value of A is large, Eq. (22.40) reduces to

$$\frac{v_{out}}{v_{in}} \approx -\frac{Z_2}{Z_1}. \tag{22.41}$$

Hence, the transfer function for Fig. 22.4(a) is

$$H(s) = -\frac{Z_2(s)}{Z_1(s)}. \tag{22.42}$$

A similar analysis can be used to show that the non-inverting case of Fig. 22.4(b) results in the following:

$$\frac{v_{out}}{v_{in}} \approx 1 + \frac{Z_2}{Z_1}, \quad \text{with} \quad H(s) = \frac{Z_1(s) + Z_2(s)}{Z_1(s)}. \tag{22.43}$$

22.2.4 Passive Low-Pass Filter

A simple low-pass filter is depicted in Fig. 22.5, in which a time dependent input voltage $v_{in}(t)$ is placed across the resistor and the capacitor. As a result the voltage $v_{out}(t)$ is formed across the capacitor and is measured as a function of time. From the figure the voltage across the resistor must equal the difference between the input voltage and the voltage across the capacitor, i.e.,

$$v_{in}(t) - v_{out}(t) = Ri_{in}(t). \tag{22.44}$$

The current flowing to the capacitor is

$$i_{in}(t) = \frac{dQ(t)}{dt} = C\frac{dv_{out}(t)}{dt}, \tag{22.45}$$

where $Q(t) = Cv_{out}(t)$ is the charge accumulated on the capacitor as a function of time. Substitution of Eq. (22.45) into Eq. (22.44) and rearrangement gives

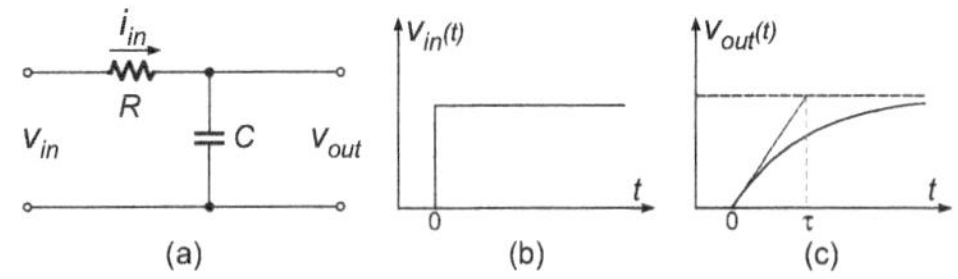

Figure 22.5. The (a) basic low-pass circuit and result from a (b) step function input $v_{in}(t)$ with (c) output response $v_{out}(t)$.

$$\frac{dv_{out}(t)}{dt} + \frac{1}{\tau}v_{out}(t) = \frac{1}{\tau}v_{in}(t), \tag{22.46}$$

where $\tau = RC$ is the time constant. The most general solution of this linear, first-order, ordinary differential equation is

$$v_{out}(t) = e^{-t/\tau}\left[v_{in}(0^-) + \int_0^t e^{t'/\tau}\frac{v_{in}(t')}{\tau}\right] dt', \tag{22.47}$$

where $v_{in}(0^-)$ is the initial voltage condition. Often it is assumed that $v_{in}(0^-) = 0$ for many solutions.

Step Input

For $V_{in}(0^-) = 0$, or no initial charge, and for the step input shown in Fig. 22.5(b) with $v_{in}(t) = v_{in}^o$ for $t > 0$, this general solution reduces to

$$v_{out}(t) = v_{in}^o\left(1 - e^{-t/\tau}\right), \quad t \geq 0. \tag{22.48}$$

Consequently, the output signal is an increasing exponential function that approaches the input voltage v_{in}^o. Note that the rate of increase is a function of the time constant, which can be altered by changing either C, R, or both. The smaller the time constant, the faster the rise to v_{in}. Depicted in Fig. 22.5(c) is the response to a voltage step input, where the tangent of the initial signal rise intersects the maximum possible output value at $t = \tau$. Stated differently, the difference between $v_{out}(\infty)$ and $v_{out}(\tau)$ equals $v_{out}(\infty)/e \simeq 0.368 v_{out}(\infty)$.

Oscillatory Input

To determine the frequency characteristics of this device, first take the Laplace transform of Eq. (22.46) and rearrange to obtain the transfer function

$$H(s) = \frac{1}{1 + sRC} = \frac{1}{1 + s\tau}. \tag{22.49}$$

The voltage gain, defined here by Eq. (22.22), is

$$G(\omega) = |H(j\omega)| = \left|\frac{1}{1 + j\omega\tau}\right| = \left[\left(\frac{1}{1 + j\omega RC}\right)\left(\frac{1}{1 - j\omega RC}\right)\right]^{1/2} = \frac{1}{\sqrt{1 + (\omega RC)^2}} = \frac{1}{\sqrt{1 + \omega^2\tau^2}}, \tag{22.50}$$

which at low frequencies is $G(\omega) \simeq 1$, i.e., the amplitude of the output sine wave equals that of the input sine wave. But at high frequencies $G(\omega) \simeq 1/\omega\tau$ and the output amplitude decreases as $1/\omega$. Because low frequencies of the input pass, without change, to the output while high frequencies are attenuated, the circuit of Fig. 22.5 is called a *low-pass filter*. The cut-off frequency is given by

$$f_c = \frac{1}{2\pi RC} = \frac{1}{2\pi\tau} \quad \text{or} \quad 2\pi f_c = \omega_o = \frac{1}{RC} = \frac{1}{\tau}. \tag{22.51}$$

The voltage gain is often expressed in decibels[6] as

$$G_{db}(\omega) = 20\log_{10}|H(j\omega)|. \tag{22.52}$$

The phase shift is

$$\phi(\omega) = \tan^{-1}\left(\frac{\mathrm{Im}H(j\omega)}{\mathrm{Re}H(j\omega)}\right) = -\tan^{-1}(\omega\tau). \tag{22.53}$$

At the frequency $\omega = 1/\tau$, then $\phi = -45°$.

Example 22.4: Given $R = 1$ MΩ and $C = 10$ pF, what is the cutoff frequency of the low-pass filter?

Solution:

$$f_c = \frac{1}{2\pi CR} = \frac{1}{2\pi(10^6\ \Omega)(10^{-11}\ \text{F})} = 15.916\ \text{kHz}.$$

So the cutoff frequency is 15.916 kHz. To test this result $w_c = 2\pi f_c$ so

$$|G_{db}(\omega)| = 20\log_{10}\left[\frac{1}{\sqrt{1+(\omega_o RC)^2}}\right] = 20\log_{10}\left[\frac{1}{\sqrt{2}}\right] = -3\ \text{dB},$$

Hence, at the cutoff frequency, the output magnitude has dropped by 3 dB.

22.2.5 Active Low-Pass Filter

Two simple *active* low-pass filters (inverting configurations) are depicted in Fig. 22.6. In the case depicted by Fig. 22.6(a)

$$i_1(t) = \frac{v_{in}(t)}{R_1}. \tag{22.54}$$

If the initial condition has the voltage across the feedback capacitor C_2 of V_0 then

$$v_{out}(t) = V_0 - \frac{1}{C_2}\int_0^t i_1(t)\,dt = V_0 - \frac{1}{R_1 C_2}\int_0^t v_{in}(t)\,dt. \tag{22.55}$$

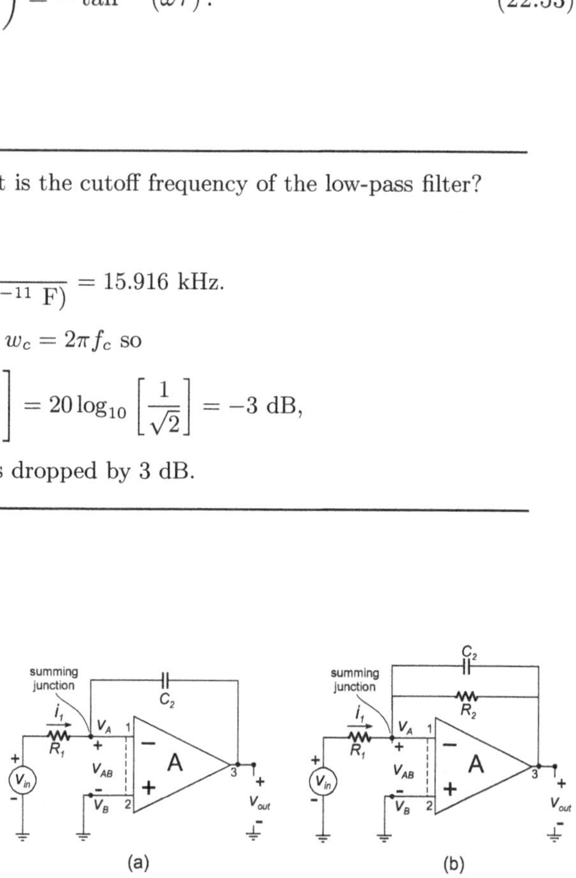

Figure 22.6. Simple low-pass amplification circuits depicting the (a) ideal integrator and the (b) practical integrator.

[6]The decibel is actually a logarithmic measure of the ratio of power output to input. The power gain is

$$G\Big|_{\text{power}} = \frac{P_{out}}{P_{in}} = \frac{v_{out}^2/Z_{out}}{v_{in}^2/Z_{in}}$$

where $P = v^2/Z$. Expressed in decibels,

$$G_{db}(\omega)\Big|_{\text{power}} = 10\log_{10}\left(\frac{v_{out}^2}{v_{in}^2}\right) - 10\log_{10}\left(\frac{Z_{out}}{Z_{in}}\right).$$

Note that if $Z_{in} = Z_{out}$, then power gain and voltage gain are equivalent when expressed in decibels

$$G_{db}(\omega)\Big|_{\text{power}} = 10\log_{10}\left(|H(j\omega)|^2\right) = 20\log_{10}\left(|H(j\omega)|\right) = G_{db}(\omega)\Big|_{\text{voltage}}.$$

Hence, the output is the integral of the input and, thus, the reason this circuit is called an *integrator*. Also, R_1C_2 is the integration time constant. The negative output is a consequence of the inverting op amp configuration. If the input is taken as $v_{in}(t) - V_0$ so as to give a zero initial condition, the closed loop transfer function is

$$H(s) = -\frac{Z_2(s)}{Z_1(s)} = -\frac{\frac{1}{sC_2}}{R_1} = \frac{-1}{sR_1C_2}, \tag{22.56}$$

which has gain

$$G(\omega) = |H(j\omega)| = \frac{1}{\omega R_1 C_2}. \tag{22.57}$$

A positive voltage input produces a negative output, while a negative input produces a positive output. The output voltage remains constant if there is a zero value input. Suppose a series of positive square pulses is applied to the input. With the ideal integrator, every positive pulse produces a progressively increasing negative voltage on the output. This sequential increase occurs because the feedback capacitor C_2 does not discharge, but instead continues to accumulate charge for every pulse. However, this simple integrating circuit has several drawbacks. Eventually the integrator circuit saturates, and the circuit no longer responds to input voltages until the capacitor is discharged. The saturation is a result of a small difference in voltage across the input terminals due to slight transistor mismatching in the amplifier circuit. Over time, this small offset voltage continues to accumulate on the capacitor until, eventually, it changes the baseline V_0 enough to distort the output signal or to saturate the preamplifier. Another difficulty arises because the bandwidth of an ideal integrator is small so that it can be used over only a small input range of frequencies.

These problems can be remedied by placing a feedback resistor R_2 in parallel with C_2, as shown in Fig. 22.6(b), a design traditionally labeled the *practical integrator*. The transfer function for this arrangement is

$$H(s) = \frac{Z_2(s)}{Z_1(s)} = \frac{\left[\frac{1}{R_2} + sC_2\right]^{-1}}{R_1} = \left(\frac{R_2}{R_1}\right)\frac{1}{(1 + sR_2C_2)} \tag{22.58}$$

with gain

$$|G(\omega)| = \left(\frac{R_2}{R_1}\right)\left[\frac{1}{(1 + (\omega R_2 C_2)^2}\right]^{1/2}. \tag{22.59}$$

At low frequency, the gain is R_2/R_1, and the angular cutoff frequency is

$$\omega_o = 2\pi f_c = \frac{1}{R_2C_2}. \tag{22.60}$$

Also with this integrator configuration, capacitor C_2 now discharges with time constant $\tau = R_2C_2$.

22.2.6 Passive High-Pass Filter

A simple passive high-pass filter is depicted in Fig. 22.7, in which a time-dependent input voltage $v_{in}(t)$ is placed across the capacitor and the resistor so that a voltage $v_{out}(t)$ is formed across the resistor. The difference between the input and output voltages is the voltage Q/C developed across the capacitor C, i.e.,

$$v_{in}(t) - v_{out}(t) = \frac{Q(t)}{C}.$$

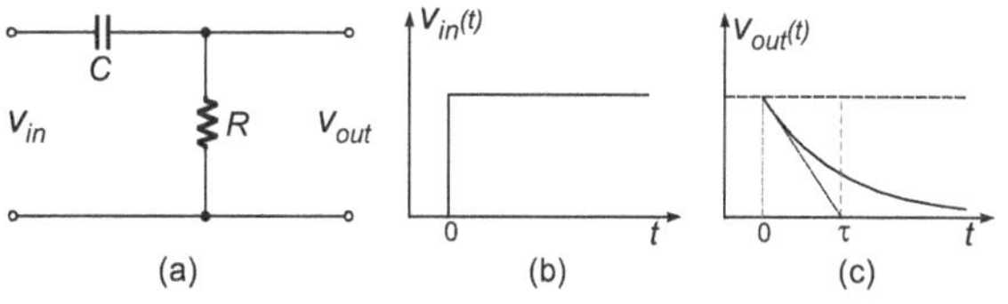

Figure 22.7. The (a) basic high-pass circuit and result from a (b) step function input $v_{in}(t)$ with (c) output response $v_{out}(t)$.

Differentiation with respect to time gives

$$\frac{dv_{in}}{dt} - \frac{dv_{out}(t)}{dt} = \frac{1}{C}\frac{dQ(t)}{dt}. \tag{22.61}$$

But $i(t) = dQ/dt = v_{out}(t)/R$ and, with $\tau = RC$, Eq. (22.61) becomes

$$\frac{dv_{in}(t)}{dt} = \frac{dv_{out}(t)}{dt} + \frac{v_{out}(t)}{\tau}. \tag{22.62}$$

Take the Laplace transform of Eq. (22.62) to obtain the transfer function

$$H(s) = \frac{sRC}{1+sRC} = \frac{s\tau}{1+s\tau}. \tag{22.63}$$

The gain of this circuit is then

$$|G(\omega)| = \left|\frac{v_{out}(t)}{v_{in}(t)}\right| = \left[\left(\frac{j\omega RC}{1+j\omega RC}\right)\left(\frac{-j\omega RC}{1-j\omega RC}\right)\right]^{1/2} = \sqrt{\frac{\omega^2R^2C^2}{1+\omega^2R^2C^2}}. \tag{22.64}$$

From Eq. (22.64), it is seen that at high frequencies $G(\omega) \simeq 1$, i.e., $v_{out} \approx v_{in}$. By contrast, at low frequencies $\omega \ll 1$ $G(\omega) \simeq \omega$ so v_{out} decreases as ω decreases. Thus, the name *high-pass* filter for this device. The low-frequency cutoff is also

$$2\pi f_c = \omega_o = \frac{1}{RC},$$

where $|G(\omega_o)| = -3$ dB.

Step Input to Passive High-Pass Filter

Given a positive step input of height $v_{in}(t=0)$, as depicted in Fig. 22.7(b), there is an instantaneous change in v_{in} at $t = 0$. Otherwise, the input voltage remains constant at v^o_{in}. Thus Eq. (22.62) becomes

$$\frac{dv_{in}(t)}{dt} = 0 = \frac{dv_{out}(t)}{dt} + \frac{v_{out}(t)}{\tau}, \tag{22.65}$$

which, upon integration, gives

$$v_{out}(t) = v^o_{in}e^{-t/\tau}, \qquad t > 0. \tag{22.66}$$

Thus, a step input produces a tail pulse with a decay constant $\tau = RC$.

Rectangular Pulse Input to a Passive High-Pass Filter

If a square wave pulse input $v_{in}(t)$ with width t_p and height v^o_{in} is applied to the passive high-pass filter at $t = 0$, the initial response is the same as the step input. However, at the end of the pulse, the input voltage returns to zero. However, the voltage across the capacitor in the circuit cannot immediately return to zero because the discharge current is limited by the resistor. Consequently, the output must decrease by the pulse input v^o_{in}, or,

$$v_{out}(t_p) = v^o_{in}e^{-t_p/\tau} - v^o_{in}, \tag{22.67}$$

which shows that the output becomes *negative*. Beyond t_p, the pulse exponentially decays back to zero as

$$v_{out}(t) = v^o_{in}\left(e^{-t_p/\tau} - 1\right)e^{-(t-t_p)/\tau}, \qquad t \ge t_p. \tag{22.68}$$

To explore the consequences of this result, consider the cases where $\tau \gg t_p$ and $\tau \ll t_p$. If $\tau \gg t_p$, then the output pulse has decayed very little at time t_p so that the negative undershoot is very small and has little effect on the output. If instead $\tau \ll t_p$, then much decay occurs well before t_p is reached so that the positive output voltage spike and the negative undershoot has the same magnitude as v_{in}^o and appears as a negative spike. Another less obvious result can be found by integrating and comparing both Eqs. (22.66) and (22.68) within their respective operating times, namely

$$\int_0^{t_p} v_{in}^o e^{-t/\tau} = -\tau v_{in}^o e^{-t/\tau}\Big|_0^{t_p} = \tau v_{in}^o \left(1 - e^{-t_p/\tau}\right)$$

and

$$\int_{t_p}^{\infty} v_{in}^o \left(e^{-t_p/\tau} - 1\right) e^{-(t-t_p)/\tau} = v_{in}^o \left(1 - e^{t_p/\tau}\right)\left(-\tau e^{-t/\tau}\right)\Big|_{t_p}^{\infty} = -\tau v_{in}^o \left(1 - e^{-t_p/\tau}\right). \tag{22.69}$$

Hence, the area under the output pulse before t_p and then after t_p are equal (with opposites signs). Consequently, the time it takes to return to zero can be unacceptably long if the circuit has large τ.

Series of Rectangular Pulses to a Passive High-Pass Filter

The natural extension of the analysis for the input of a single rectangular pulse is to consider a series of rectangular pulses. Suppose that the input voltage is a rectangular pulse of duration t_1 and constant voltage v_1 followed by a lower constant voltage v_2 of duration t_2. The total period of a cycle is, therefore, $T = t_1 + t_2$ with a total voltage swing of $V = v_1 - (-v_2) = v_1 + v_2$. This voltage swing can be displaced by a DC voltage shift v_{DC} as shown in Fig. 22.8(a). In other words, the actual high-input voltage is $v_1 + V_{DC}$ and the actual low-input voltage is $V_{DC} - v_2$.

Because the output voltage $v_{out}(t)$ is unaffected by V_{DC} (a high-pass filter does not pass a constant voltage), for the following analyses one can assume any value of V_{DC}. If one chooses $V_{DC} = -v_2$ then Eq. (22.61) for $v_{out}(t)$ can be written as

$$\frac{dv_{out}(t)}{dt} + \frac{1}{\tau} v_{out}(t) = \frac{dv_{in}(t)}{dt}, \tag{22.70}$$

where the input wave form for cycle n is

$$v_{in}(t) = \begin{cases} v_1 - v_2 \equiv v_{in}^o & (n-1)T \le t < (n-1)T + t_1 \\ v_2 - v_2 = 0 & (n-1)T + t_1 \le t < nT \end{cases}. \tag{22.71}$$

The solution of Eq. (22.70) for any $v_{in}(t)$ is

$$v_{out}(t) = e^{-t/\tau}\left[v_{out}(t_o^-) + \int_{t_o^-}^{t} e^{t'/\tau} \frac{dv_{in}(t')}{dt'}\, dt'\right], \qquad t > t_o, \tag{22.72}$$

where t_o^- is the time just before the start of transient calculation.

The output voltage during any cycle has two parts: (1) that during the rectangular pulse when $v_i(t) = v_{in}^o$ and (2) that during the time when $v_{in}(t) = 0$. For the first portion t_o is the time the pulse begins, and for the second portion t_o is the time the pulse ends. The transients for both portions can be calculated from Eq. (22.72) using the appropriate value of $v_{in}(t_o^-)$. Because $v_{in}(t)$ consists of two step functions, dv_{in}/dt consists of two delta functions since the derivative of a step function is a delta function. In particular, for the nth cycle

$$\frac{dv_{in}(t)}{dt} = v_{in}^o \delta(t - [n-1]T)) - v_{in}^o \delta(t - [n-1]T - t_1). \tag{22.73}$$

The output voltage is now calculated cycle by cycle.

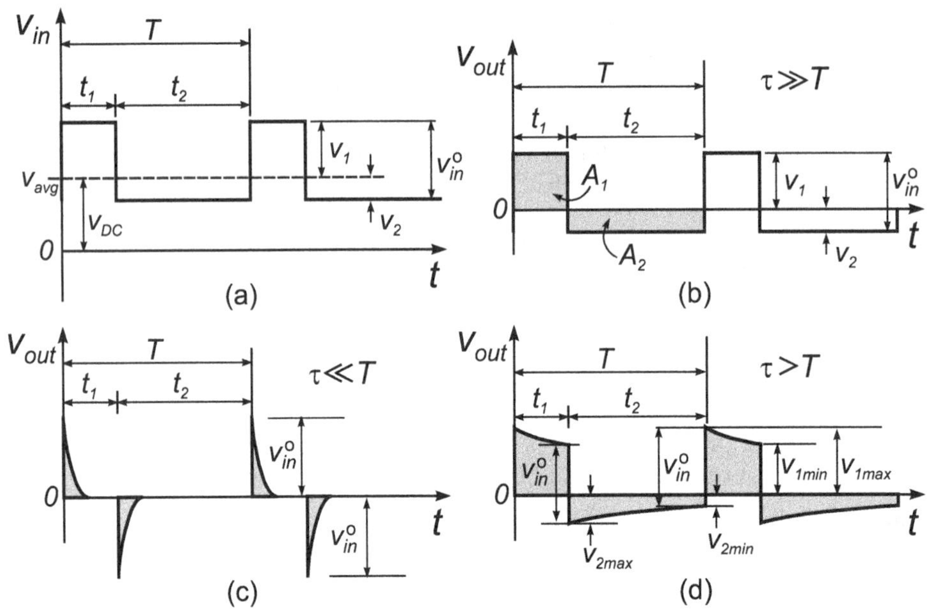

Figure 22.8. (a) Square wave input with a DC offset, (b) output from a passive differentiator circuit with $\tau \gg T$, (c) output from a passive differentiator circuit with $\tau \ll T$, (d) output from a passive differentiator circuit with $\tau > T$. Shown in the output voltage plots are the voltages obtained after a large number of cycles, i.e., when each cycle has the same response.

Cycle 1 ($0 < t < T$): For the first portion of this cycle, $t_o = 0$, $v_{in}(t_o^-) = 0$ and $dv_{in}/dt = v_{in}^o \delta(t)$. Substitution into Eq. (22.72) gives

$$v_{out}(t) = v_{in}^o e^{-t/\tau}, \quad 0 \le t < t_1. \tag{22.74}$$

For the second portion of this cycle, $t_o = t_1$ and $dv_{in}/dt = -v_{in}^o \delta(t - t_1)$. Then from Eq. (22.72) one finds

$$v_{out}(t) = -v_{in}^o \left[1 - e^{-t_1/\tau}\right] e^{-(t-t_1)/\tau}, \quad t_1 \le t < T. \tag{22.75}$$

More compactly the response in the first cycle can be written as

$$\frac{v_{out}(t)}{v_{in}^o} = \begin{cases} e^{-t/\tau}, & 0 \le t < t_1 \\ -\beta e^{-(t-t_1)/\tau}, & t_1 \le t < T \end{cases}, \tag{22.76}$$

where $\beta \equiv 1 - e^{-t_1/\tau} < 1$.

Cycle 2 ($T < t < 2T$): For the first portion of this cycle $t_o = T$, $v_{out}(t_o^-) = -v_{in}^o \beta e^{-(T-t_1)/\tau}$, and $dv_{in}/dt = v_{in}^o \delta(t - T)$. For the second portion of the cycle, $t_o = T + t_1$, $v_{in}(t_o)$ equals the final voltage of the first portion, and $dv_{in}/dt = -v_{in}^o \delta(t - T - t_1)$. Expressions for the transients in both portions of the cycle are obtained from Eq. (22.72).

$$\frac{v_{out}(t)}{v_{in}^o} = \begin{cases} [\alpha + e^{-T/\tau}]e^{-(t-T)/\tau}, & T \le t < T + t_1 \\ -[\beta + \beta e^{-T/\tau}]e^{-(t-T-t_1)/\tau}, & T + t_1 \le t < 2T \end{cases}, \tag{22.77}$$

where $\alpha \equiv 1 - e^{-t_2/\tau} < 1$.

Cycle n ($(n-1)T < t < nT$): One can continue in this fashion cycle after cycle to obtain $v_{in}(t)$ in each successive cycle. By induction, the response for the nth cycle, which begins at $t = t_{n-1} \equiv (n-1)T$, is

$$\frac{v_{out}(t)}{v_{in}^o} = \begin{cases} \left[\alpha \sum_{k=1}^{n-1} e^{-(k-1)T/\tau} + e^{-(n-1)T/\tau}\right] e^{-(t-(n-1)T)/\tau}, & t_{n-1} \le t < t_{n-1} + t_1 \\ -\left[\beta \sum_{k=1}^{n} e^{-(k-1)T/\tau}\right] e^{-(t-(n-1)T-t_1)/\tau}, & t_{n-1} + t_1 \le t < nT \end{cases}. \tag{22.78}$$

Large n: The sums in Eq. (22.78) are a geometric progression

$$S_n = \sum_{k=1}^{n} e^{-(k-1)T/\tau} = 1 + \gamma + \gamma^2 + \ldots \gamma^{n-1} = \frac{1-\gamma^{n-1}}{1-\gamma}, \tag{22.79}$$

where $\gamma \equiv e^{-T/\tau} < 1$. As $n \to \infty$, $S_n \to 1/(1-\gamma)$ and from Eq. (22.78), in the limit of large n,

$$\frac{v_{out}(t)}{v_{in}^o} = \begin{cases} \frac{\alpha}{1-\gamma} e^{-t/\tau}, & 0 \le t < t_1 \\ \frac{-\beta}{1-\gamma} e^{-(t-t_1)/\tau}, & t_1 \le t < T \end{cases}, \tag{22.80}$$

where t is now measured from the start of a cycle.

The area A_1 under $v_{in}(t)$ during the rectangular pulse is

$$A_1 = v_{in}^o \frac{\alpha}{1-\gamma} \int_0^{t_1} e^{-t/\tau}\, dt = \frac{v_{in}^o \alpha\tau}{1-\gamma}(1 - e^{-t_1/\tau}) = \frac{v_{in}^o \alpha\beta\tau}{1-\gamma}. \tag{22.81}$$

Likewise the area A_2 under $v_{in}(t)$ between t_1 and $t_1 + t_2$ is

$$A_2 = -v_{in}^o \frac{\beta}{1-\gamma} e^{t_1/\tau} \int_{t_1}^{t_1+t_2} e^{-t/\tau}\, dt = -\frac{v_{in}^o \beta\tau}{1-\gamma}(1 - e^{-t_2/\tau}) = -\frac{v_{in}^o \alpha\beta\tau}{1-\gamma}. \tag{22.82}$$

Thus, it is seen that $A_1 = -A_2$, i.e., the two shaded areas in Fig. 22.8 are equal. This equality of the areas for the positive and negative transients is true only for the equilibrium cycle ($n \to \infty$) of Eq. (22.80). For the cycles building up to the equilibrium cycle, $A_1 > -A_2$.

Effect of the Time Constant τ The output response is greatly influenced by the circuit time constant τ. Regardless of the average input voltage (the DC offset), the output signal is averaged about zero (no DC component). For an equilibrium cycle (large n), the maximum positive output from Eq. (22.80) is

$$v_{1max} = \frac{\alpha}{1-\gamma} v_{in}^o \tag{22.83}$$

and the minimum positive voltage is

$$v_{1min} = \frac{\alpha}{1-\gamma} v_{in}^o e^{-t_1/\tau} = v_{1max} e^{-t_1/\tau}. \tag{22.84}$$

At t_{p1}, the polarity switches by $-v_i^o$ (as it does for all cycles), producing the maximum negative voltage

$$v_{2max} = \frac{\beta}{1-\gamma} v_{in}^o \tag{22.85}$$

The negative voltage decays towards zero and reaches a minimum negative voltage at $t = T$ of

$$v_{2min} = \frac{\beta}{1-\gamma} v_{in}^o e^{-(t_2-t_1)/\tau} = |v_{2max}| e^{-(t_2-t_1)/\tau}. \tag{22.86}$$

Then, as expected,

$$v_{1min} + v_{2max} = v_{in}^o \left[\frac{\alpha}{1-\gamma} e^{-t_1/\tau} + \frac{\beta}{1-\gamma} \right] = v_{in}^o \frac{[(1-e^{-t_2/\tau})e^{-t_1/\tau} + (1-e^{-t_1/\tau})]}{1-e^{-(t_1+t_2)/\tau}} = v_{in}^o. \tag{22.87}$$

For small $\tau \ll T$, the output rapidly decays to zero before t_p is reached, causing a flip in polarity equal to v_{in}^o at either t_1 (goes negative) or t_2 (goes positive). Thus a small time constant produces spike peaks of equal magnitudes alternating in polarity, as depicted in Fig. 22.8(c). For long $\tau \gg T$, there is hardly any decay, and the output as a rectangular wave that mimics the input, but one with an average of zero (no DC component). This case is depicted in Fig. 22.8(b). The expected result for cases between these extremes is depicted in Fig. 22.8(d).

Exponential Input to Passive High-Pass Filter

Consider an exponential input to a high-pass filter with the form[7]

$$v_{in}(t) = v_{in}^o \left(1 - e^{-t/\tau_{in}}\right) \tag{22.88}$$

where v_{in}^o is the maximum voltage reached as $t \to \infty$ and τ_{in} is the time constant of the input signal. Substitution of Eq. (22.88) into Eq. (22.62) gives the differential equation

$$\frac{v_{in}^o}{\tau_{in}} e^{-t/\tau_{in}} = \frac{dv_{out}}{dt} + \frac{v_{out}}{\tau}, \tag{22.89}$$

where τ is the time constant RC of the high-pass filter. The solution of this equation is

$$v_{out}(t) = \begin{cases} v_{in}^o \dfrac{\tau}{\tau - \tau_{in}} \left(e^{-t/\tau} - e^{-t/\tau_{in}}\right), & \tau \neq \tau_{in} \\ v_{in}^o \dfrac{t}{\tau} e^{-t/\tau}, & \tau = \tau_{in} \end{cases}. \tag{22.90}$$

Now consider two limiting cases of Eq. (22.90), $\tau \ll \tau_{in}$ and $\tau \gg \tau_{in}$. In the former case, one obtains

$$v_{out}(t) \approx v_{in}^o \frac{\tau}{\tau_{in}} e^{-t/\tau_{in}}. \tag{22.91}$$

Here it is seen that as τ decreases with respect to τ_{in}, the maximum output pulse height diminishes and decays according to the input time constant τ_{in}. Hence, the condition where τ is much smaller than the time constant of the input pulse τ_{in} causes a ballistic deficit with a ratio τ/τ_{in}. For the case in which $\tau \gg \tau_{in}$, Eq. (22.90) reduces to

$$v_{out}(t) \approx v_{in}^o e^{-t/\tau}. \tag{22.92}$$

Note that the dependence of $v_{out}(t)$ on τ_{in} vanishes and the output depends only on the network decay constant and the maximum output voltage equal to v_{in}^o. Actual real high-pass networks are somewhere between the limiting cases of Eqs. (22.91) and (22.92). Generally, the amount of ballistic deficit increases as τ decreases.

[7]Compare this signal input to Eq. (22.48), the output solution of a passive low-pass filter.

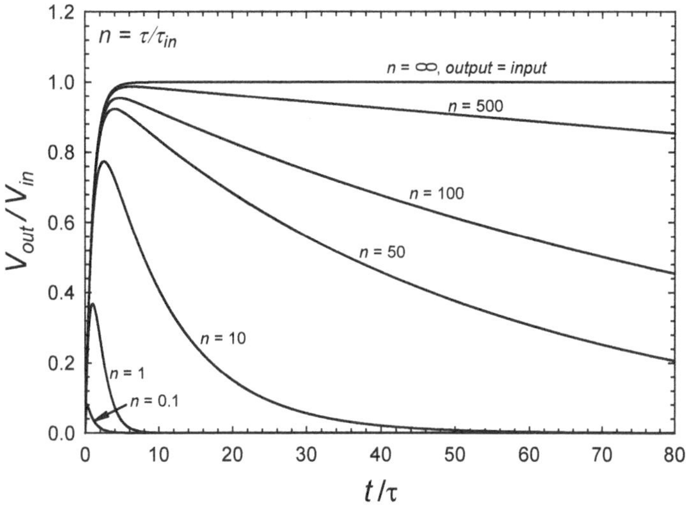

Figure 22.9. Output from a passive high-pass filter from an exponential input described by Eq. (22.89). Shown are normalized outputs to the initial input (v_{out}/v_{in}) as a function of t/τ and τ/τ_{in}. The decrease in the maximum pulse height from unity is the ballistic deficit.

The time to the pulse height maxima is

$$t_{max} = \frac{\tau\tau_{in}}{\tau - \tau_{in}} \ln\left(\frac{\tau}{\tau_{in}}\right). \tag{22.93}$$

This result can then be used to calculate the expected maximum in the output pulse height $v_{out}(t_{max})$. The ballistic deficit is defined as

$$v_{bd} = v_{in}^o - v_{out}(t_{max}). \tag{22.94}$$

Examples of normalized output pulses with varying time constants are shown in Fig. 22.9. For large values of τ/τ_{in}, the pulse height maxima is relatively large and approaches the input value of V_{in}^o. However, as τ decreases, so does the output pulse height maxima, the difference from unity being the ballistic deficit.

22.2.7 Active High-Pass Filter

Simple *active* high-pass filters (inverting configuration) are depicted in Fig. 22.10. Because terminal 1 is a virtual ground, $V_A = 0$ and the output voltage v_{out} equals the negative of the voltage across the resistor R_2 by the current i_2, i.e.,

$$v_{out}(t) = v_1(t) = -i_2(t)R_2. \tag{22.95}$$

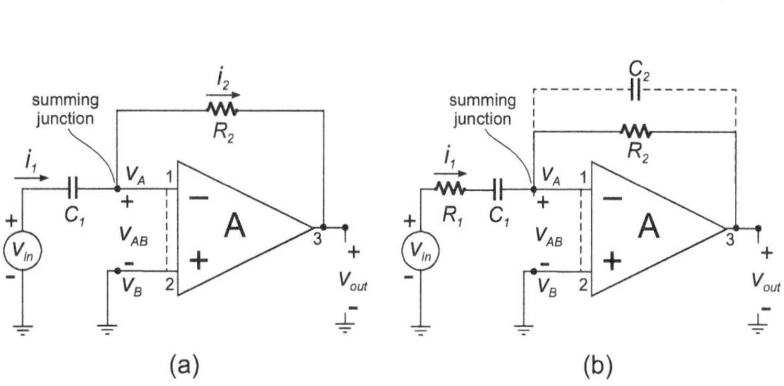

Figure 22.10. Simple high-pass amplification circuits depicting the (a) ideal differentiator and the (b) practical differentiator.

Also because of the high input impedance at terminal 1, the current flowing to the capacitor C_1 must be

$$i_2(t) = i_1(t) = \frac{dQ}{dt} = C_1 \frac{dv_{in}(t)}{dt} \tag{22.96}$$

Substitution of Eq. (22.95) into Eq. (22.96) yields,

$$v_{out}(t) = -R_2 C_1 \frac{dv_{in}(t)}{dt}. \tag{22.97}$$

Hence the output $v_{out}(t)$ is the derivative of the input $v_{in}(t)$ so the circuit in Fig. 22.10 is, understandably, called a *differentiator* circuit.

The closed-loop transfer function for the circuit in Fig. 22.10(a) is readily obtained by taking the Laplace transform of Eq. (22.97) to obtain

$$V_{out}(s) = -R_2 C_1 [sV_{in}(s) - v_{in}(0)]. \tag{22.98}$$

If the initial condition is $v_{in}(0) = 0$, then the transfer function is simply

$$H(s) = \frac{V_{out}(s)}{V_{in}(s)} = -sR_2C_1 \qquad \text{with gain} \qquad G(\omega) = |H(j\omega)| = \omega R_2 C_1. \tag{22.99}$$

The simple differentiator circuit in Fig. 22.10(a) tends to magnify noise and is unstable; hence, this simple active differentiator circuit is relatively unpopular and seldom used, mainly due to sharp spikes produced at the output when there is a shape change with the input voltage [Sedra and Smith 1982]. To remedy these problems, there is an improved version of the differentiator, Fig. 22.10(b) where a resistor R_1 has been added in series with C_1, called the *practical differentiator.* The transfer function is now

$$H(s) = -\frac{Z_2}{Z_1} = \frac{-R_2}{R_1 + \dfrac{1}{sC_1}} = \frac{-sR_2C_1}{1 + sR_1C_1} \qquad \text{with gain} \qquad G(\omega) = |H(j\omega)| = \left[\frac{(\omega R_2 C_1)^2}{1 + (\omega R_1 C_1)^2}\right]^{1/2}. \tag{22.100}$$

At low frequency, the circuit of Fig. 22.10(b) becomes a high-pass filter, while at high frequency the circuit becomes an amplifier, with improved noise rejection, with a gain of R_2/R_1. An additional improvement has a parallel feedback capacitor across R_2, which serves to produce a low-pass filter. Ultimately, the circuit can perform as a *bandpass* filter, with transfer function

$$H(s) = -\frac{Z_2}{Z_1} = \frac{-sR_2C_1}{(1 + sR_1C_1)(1 + sR_2C_2)}. \tag{22.101}$$

From Eq. (22.101), there is a zero at $\omega = 0$, and two poles, one at $\omega_1 = (R_1C_1)^{-1}$ (high-pass differentiator circuit), and one at $\omega_2 = (R_2C_2)^{-1}$ (low-pass integrator circuit).

22.2.8 CR-RC Network

The high-pass filters shown in Fig. 22.7 and Fig. 22.10(a) are far from ideal and seldom used alone. They produce sharp output peaks that are difficult to process and, because they pass high frequencies they allow high-frequency noise to pass through the network. These two problems can be effectively mitigated by adding an integration circuit after the operational amplifier, as shown in Fig. 22.11. The differentiator and integrator networks are isolated by the operational amplifier that has a gain of unity at high frequencies.

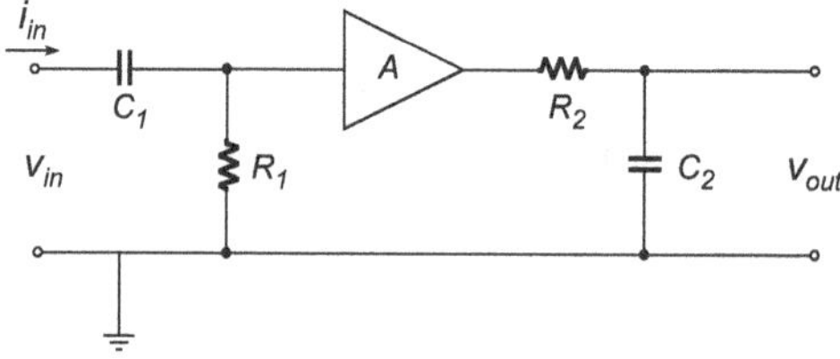

Figure 22.11. Depiction of a *CR-RC* pulse shaping circuit.

In Fig. 22.11 define the time constants $\tau_1 = R_1C_1$ and $\tau_2 = R_2C_2$. Consider the case of a step input $v_{in}(t) = v_{in}^o$ for $t > 0$ into this circuit for which $v_{in} = v_{out} = 0$ for $t \leq 0$. The input to the op amp is described by the output of the high-pass circuit defined by Eq. (22.66), while the output of the low-pass circuit after the op amp is defined by Eq. (22.46). Let the input to the op amp be $v_s(t)$, which is also the output of the op amp because the gain of the op amp is unity for high frequencies. Then

$$\frac{v_s(t)}{\tau_2} = \frac{v_{in}^o}{\tau_2}e^{-t/\tau_1} = \frac{dv_{out}(t)}{dt} + \frac{v_{out}(t)}{\tau_2}. \tag{22.102}$$

Multiply Eq. (22.102) by e^{t/τ_2} to find

$$\frac{v_s(t)e^{t/\tau_2}}{\tau_2} = \frac{dv_{out}(t)}{dt}e^{t/\tau_2} + \frac{v_{out}(t)}{\tau_2}e^{t/\tau_2} = \frac{d}{dt}\left(v_{out}(t)e^{t/\tau_2}\right). \tag{22.103}$$

Substitution of Eq. (22.103) back into Eq. (22.102) leads to

$$\frac{d}{dt}\left(v_{out}(t)e^{t/\tau_2}\right) = \frac{v_{in}^o}{\tau_2}e^{t(1/\tau_2-1/\tau_1)}. \tag{22.104}$$

Integration of Eq. (22.104) from $(0, t)$ yields

$$v_{out}(t)e^{t/\tau_2} - v_{out}(0) = \frac{v_{in}^o}{\tau_2}\int_0^t \exp\left[t\left(\frac{\tau_1-\tau_2}{\tau_2\tau_1}\right)\right] dt, \tag{22.105}$$

which, with $v_{out}(0) = 0$, gives

$$\frac{v_{out}}{v_{in}} = \frac{\tau_1}{\tau_1-\tau_2}\left(e^{-t/\tau_1} - e^{-t/\tau_2}\right). \tag{22.106}$$

For the case that $\tau_2 \to \tau_1$ Eq. (22.106) becomes indeterminate. But with l'Hôpital's rule one obtains

$$\frac{v_{out}(t)}{v_{in}^o} = \lim_{\tau_2\to\tau_1} \frac{\tau_1 \dfrac{d}{d\tau_2}\left(\exp[-t/\tau_1] - \exp[-t/\tau_2]\right)}{\dfrac{d}{d\tau_2}(\tau_1-\tau_2)} = \frac{t}{\tau}e^{-t/\tau}, \tag{22.107}$$

where $\tau_1 = \tau_2 = \tau$. It is notable that the Laplace transform transfer function of the circuit in Fig. 22.11 can be found by simply multiplying the Laplace transform of each stage,

$$H(s) = \frac{\mathcal{L}[V_{out}(t)]}{\mathcal{L}[V_{in}(t)]} = \frac{s\tau_1}{(1+s\tau_1)(1+s\tau_2)}. \tag{22.108}$$

The proof that the inverse transform of Eq. (22.108) yields Eq. (22.106) is left as an exercise.

Example results for the shaping network of Fig. 22.11 are shown in Fig. 22.12, from which a few trends can be seen. First, as τ_1 increases, more time is available to measure the induced charge coming from the detector; hence, the total pulse height increases. Second, as τ_2 increases, the pulse height decreases. Third, as τ_2 decreases, the pulse height increases and also returns to

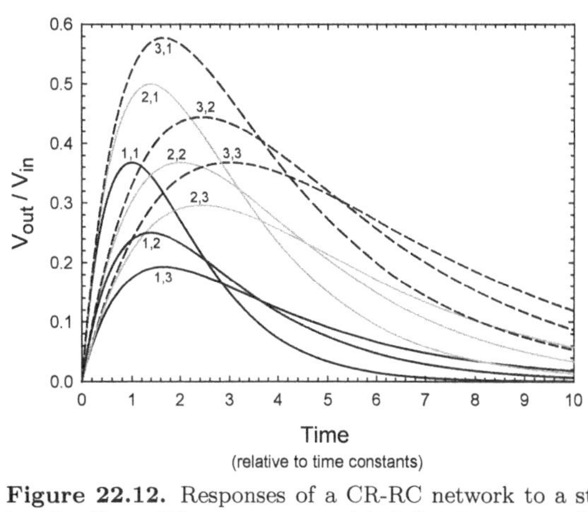

Figure 22.12. Responses of a CR-RC network to a step input voltage. The responses are labeled as τ_1, τ_2 according to Eqs. (22.106) and (22.107).

the baseline faster, thereby producing a faster pulse. To limit problems with pulse pile-up and dead time, these time constants are usually kept short, on the order of a few microseconds.

However, if these time constants are made so short that they are comparable to the rise time of the input signal from the preamplifier, similar to the charge collection time of the detector, then the input to the CR-RC network no longer resembles a step function. Instead it appears as an impulse input with a much different output than that depicted in Fig. 22.7(c). For instance, a square pulse input produces a truncated exponential output that turns negative at the end of the pulse. This unfortunate outcome produces a lower output pulse as a consequence of ballistic deficit. Hence, these time constants are generally designed to be longer than the detector charge collection time. If, however, the time constants are too long, then problems with electronic noise and pulse pile-up arise. As it turns out, for best performance, it is best to optimize the operating system time constants according to the detector type and the radiation measurement environment.

22.2.9 (CR)^2-RC Network

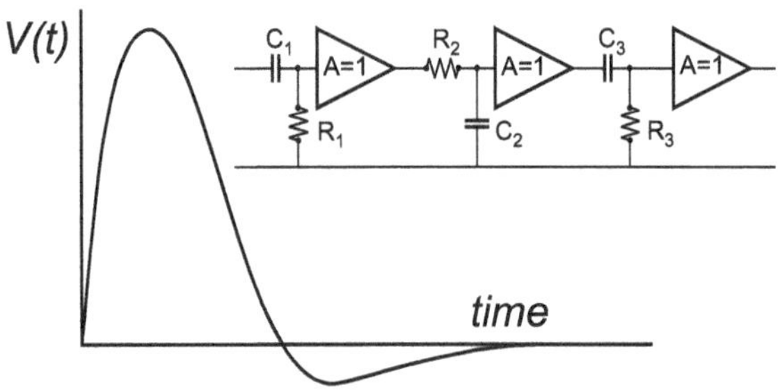

Figure 22.13. Depiction of a (CR)^2-RC circuit and the resulting bi-polar output pulse. Here $\tau_1 = R_1C_1$, $\tau_2 = R_2C_2$, and $\tau_3 = R_3C_3$.

Adding a second differentiator to a CR-RC network, or a double RC differentiator, forms a bipolar output pulse of the shape depicted in Fig. 22.13. The Laplace transform transfer function of this circuit is,

$$H(s) = \frac{s^2\tau_1\tau_3}{(1+s\tau_1)(1+s\tau_2)(1+s\tau_3)}. \tag{22.109}$$

For the fastest return to the baseline, a double differentiator circuit is usually designed with $\tau_1 = \tau_2 = \tau_3$. The advantage of the double differentiator is a reduction in baseline undershoot compared to a unipolar response. However, the bipolar pulse shaper usually has a reduced signal-to-noise ratio and a higher ballistic deficit than those of its unipolar counterpart [Nicholson 1974].

22.2.10 CR-(RC)^n Network

A Gaussian output pulse should theoretically have a shape defined by $e^{-(t/\tau)^2}$ with a nearly optimum signal-to-noise ratio from the symmetric shape. However, such an output is impractical, mainly because the pulse shape would extend to infinity on both sides and, thus, have an infinite time response [Nicholson 1974]. Instead, a network can be devised to approximate a Gaussian output, typically one having several integrator stages beyond the CR stage. The larger the number of integrator stages beyond the high-pass differentiator, the closer the output is to an ideal Gaussian pulse. The Laplace transform transfer function is

$$H(s) = \frac{s\tau_d}{(1+s\tau_d)(1+s\tau_1)(1+s\tau_2)\ldots(1+s\tau_n)}, \tag{22.110}$$

where τ_d is the differentiator (high-pass) time constant and n is the number of integrator (low-pass) stages. If all integrator time constants are equal to τ_i, then

$$H(s) = \frac{s\tau_d}{(1+s\tau_d)(1+s\tau_i)^n}. \tag{22.111}$$

In practice, four integrator stages provide an adequately Gaussian pulse shape [Nicholson 1974]. The output is defined by,

$$V_{out}(t) = \frac{V_{in}}{n!}\left(\frac{t}{\tau}\right)^n e^{-t/\tau}, \tag{22.112}$$

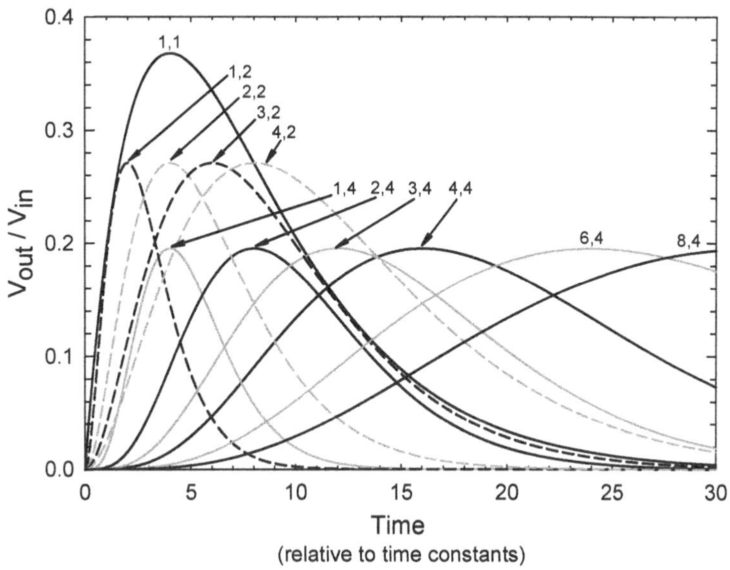

Figure 22.14. Responses of a *CR*-(*RC*)^n network to a step input voltage. The responses are labeled as τ, n according to Eq. (22.112).

where n is the number of integrators beyond the low-pass filter. The additional shaping networks prolong the pulse formation by the number n, and for equal values of τ the CR-(RC)^n network is n times longer than a *CR-RC* network. For instance, given equal values of τ, the CR-(RC)^4 network has a pulse duration 4 times that of a CR-RC network. Also, because the shape is not a true Gaussian, the output pulse still has slowly decaying tail. Consequently, pulse pile-up becomes more of an issue than with a CR-RC network. However, if the time constants are adjusted such that the peaking time matches that of an equivalent CR-RC network, the symmetry of the semi-Gaussian pulse enables it to return to the baseline more quickly and, thereby to reduce pulse pile-up at high count rates. Example pulses calculated with Eq. (22.112) are shown in Fig. 22.14. Notably, a practical Gaussian filter design usually incorporates active components (diodes, transistors, op amps) rather than *RC* filters to produce the desired result [Ortec 2009a].

22.2.11 Delay Line Pulse Shaping

Another method of shaping a voltage pulse is to use two separate signal propagation paths for the input signal. The outputs from each path arrive at different times and, when summed, change the signal. One method used to accomplish this task is depicted in Fig. 22.15. At time t_0, the signal is split between two paths. In the first path, the original signal propagates along the signal cable, such as a coaxial cable, and arrives at the input of an amplifier at some time t_1. Also at time t_0, the signal is diverted through a inverting amplifier with gain of unity,[8] and then propagates along another cable, arriving at the input of the second amplifier at time t_2 slightly delayed behind the other signal. Hence, at time t_2, the second signal is essentially subtracted from the first signal with delay $\Delta t = t_2 - t_1$. The pulse speeds v and arrival times are determined by the cable characteristic properties, where $v = 1/\sqrt{LC}$ (see Sec. 22.6 later in this chapter). The signal delay can be determined by the difference in propagation times of the two transmission line cables.

[8]This gain is at times referred to as a gain of -1. Throughout this text, the gain is the magnitude of the output signal to the input signal, although the sign (negative or positive) of the output may change with respect to the input. Hence, the gain here is 1, while the sign of the output is reversed with respect to the input.

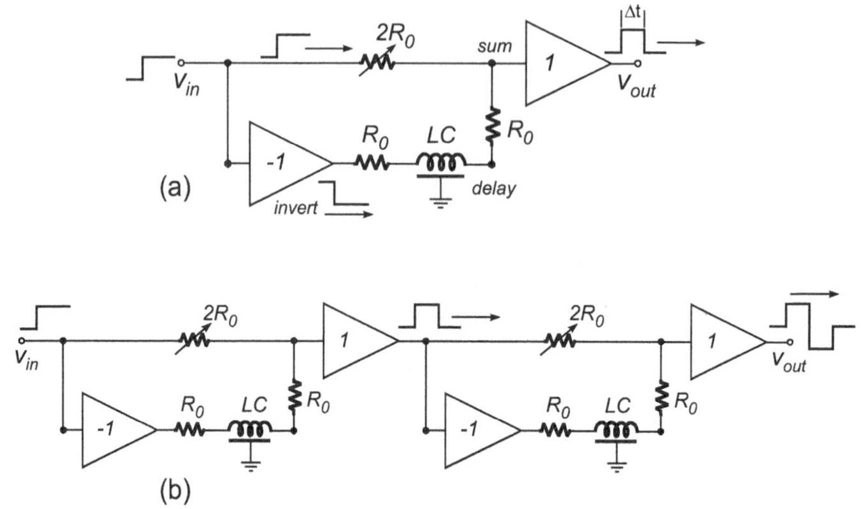

Figure 22.15. Depicted are (a) a single delay line and a (b) double delay line for shaping an input pulse. After Nicholson [1974].

Also, delay line amplifiers are available commercially as nuclear instrument modules (NIM) with a variety of settings and delays. These instruments allow the operator to provide a signal to the module and change the output by simply selecting different delay times and attenuations. The advantage of delay line shaping is the rapid recovery or return to the baseline which cannot be realized with normal Gaussian shaping.

Suppose a step function is applied to the circuit of Fig. 22.15(a) at $t_0 = 0$. One of the signals propagates unaltered, except for attenuation from the cable resistance R_0. The other signal is inverted, attenuated also by $2R_0$, and delayed. When the inverted signal arrives at the summing junction at t_2, with delay of Δt, it is added to the first signal. Because the delayed signal is negative, it causes the step pulse to become a rectangular pulse of width Δt, thereby returning the output pulse to the baseline.

In practice, there is usually some additional attenuation in the delay line, and the added negative signal may not completely restore the output pulse back to the baseline. This effect is especially pronounced for exponential input signals. However, by making the value of $2R_0$ in the first cable adjustable, the signals can be matched to eliminate this problem and properly return the shaped signal back to the baseline.

A bipolar shape can be formed by adding another delay line after the first, as depicted in Fig. 22.15(b). Basically, it adds an inverted shaped pulse directly behind the original shaped pulse. With proper tuning of $2R_0$, this method can also be used for exponential input signals. A disadvantage of the double delay line is that the overall output pulse width is increased.

22.2.12 Pole-Zero Cancellation

The effects of poles and zeros in the transfer function of a circuit can cause the appearance of non-ideal pulse shapes. For instance, the input to a common CR-RC filter usually has an associated input capacitance C_i. The impedance across this capacitance is often regarded as infinite but usually is just very large. Consequently, the input to the filter becomes

$$v_{in}(t) = \frac{Q}{C_i} e^{-t/\tau_i} \tag{22.113}$$

rather than the desired step input.

The equivalent circuit is shown in Fig. 22.16, which has the transfer function

$$H(s) = \frac{s\tau_i\tau_1}{C_i(1+s\tau_i)(1+s\tau_1)(1+s\tau_2)}. \quad (22.114)$$

Consequently, this additional time constant $\tau_i = R_iC_i$ produces an undershoot in the output pulse.

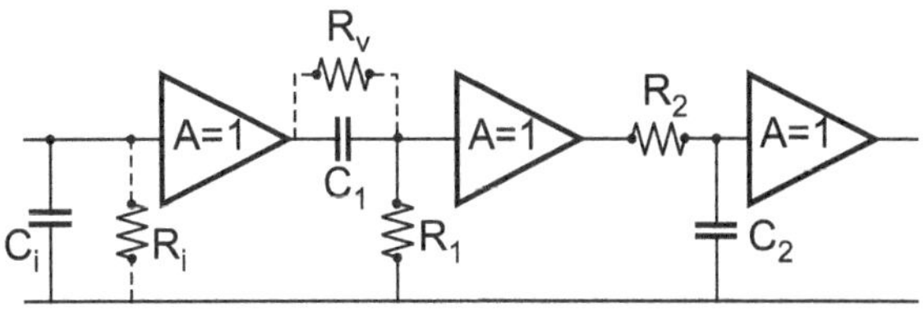

Figure 22.16. Equivalent circuit for a CR-RC filter with input capacitance C_i and pole zero compensation.

At least one method used to counter the effect is to place a variable resistor R_v across the C_1R_1 stage capacitance C_1 [Nowlin and Blankenship 1965]. The new transfer function becomes,

$$H(s) = \frac{\tau_i}{C_i(1+s\tau_i)(1+s\tau_2)} \frac{1}{\left[\dfrac{R_v+R_1}{C_1R_vR_1}+s\right]} \left[\frac{1}{R_vC_1}+s\right]. \quad (22.115)$$

Equation (22.115) is simplified with the substitutions

$$\tau_a = \frac{C_1R_vR_1}{R_v+R_1} \quad \text{and} \quad \tau_b = R_vC_1, \quad (22.116)$$

in which $\tau_b > \tau_a$, so that

$$H(s) = \frac{\tau_i}{C_i(1+s\tau_i)(1+s\tau_2)} \frac{1}{\left[\dfrac{1}{\tau_a}+s\right]} \left[\frac{1}{\tau_b}+s\right]. \quad (22.117)$$

By adjusting the values of R_v, R_1, and C_1 such that $\tau_a = \tau_1$ and $\tau_b = \tau_i$, Eq. (22.117) reduces to

$$H(s) = \frac{\tau_1}{C_i(1+s\tau_1)(1+s\tau_2)}, \quad (22.118)$$

thereby eliminating the undershooting tail. It is common that linear amplifiers come equipped with a pole-zero potentiometer in order to reduce or eliminate undershoot (or overshoot) of the shaped pulse. Note that this compensated result is also achieved with $R_i = R_v = \infty$. Other pole-zero compensation methods for more complicated input pulses are discussed by Blankenship and Nowlin [1966] and Strauss et al. [1967].

22.2.13 Base-Line Shift and Restoration

Amplifiers with AC coupled inputs often produce an output pulse with negative undershoot so the output pulses are not unipolar, but rather have a bipolar characteristic. This effect is described earlier in Sec. 22.2.6 for a series of rectangular wave inputs to a passive high-pass filter. An example of the problem is depicted in Fig. 22.8(b) and (d), in which the RC time constant τ is long compared to the cycle duration T.

The consequence of this undershoot problem is a shift in the base-line of the output pulse. Because the output pulse is measured with respect to the actual zero, the measured amplitude is reduced from its actual value. The problem is worsened in a real counting system because the input pulses arrive at random intervals with random amplitudes. Although a simple counting system can suffer such a difficulty without significant problems, a spectroscopy system suffers reduced energy resolution from the added deviation in pulse heights. The base-line shifting problem increases with count rate, in which T is no longer much greater than τ.

Methods to counteract the base-line shift are described in the literature, including bipolar output, diode clamping, and gated restoration [Nicholson 1974]. The bipolar method uses a circuit fashioned, in general, after the concept shown in Fig. 22.13. Bipolar shapers have positive and negative lobes, and, hence, counteract the problem associated with base-line shifts. If the positive and negative lobes are of equal value, then the average value is zero and the output baseline remains unaltered. The consequence of bipolar shaping is increased electronic noise and a high ballistic deficit. Many modern pulse shape amplifiers have monopolar and bipolar shaping options, allowing the user to optimize the performance of the system by choosing monopolar shaping for low count rates and bipolar shaping for high count rates.

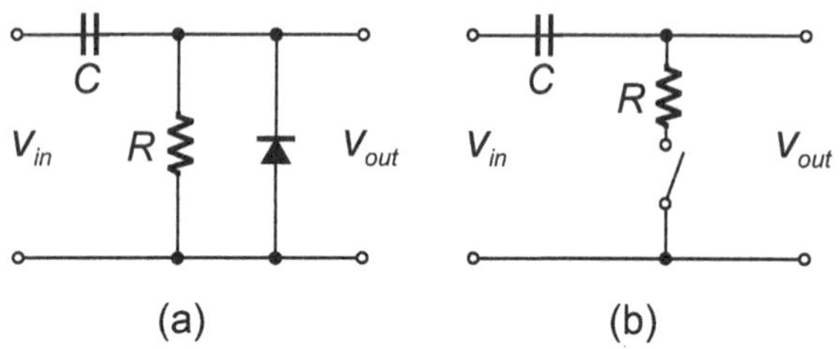

Figure 22.17. Simple base-line restoration circuits with (a) a diode clamp and (b) a gated switch.

The use of a diode in parallel with the resistor, shown in Fig. 22.17(a), acts as a base-line restoration circuit. If at the time the base-line goes negative, t_1 in Fig. 22.8(b), the diode is connected into the circuit and current flows to ground. This flow causes the capacitor to charge up opposite of the held charge and forces the output to return to zero. In reality, the diode has a small forward resistance as does the signal source, both of which combine to produce a small resistance R_s so that complete restoration does not occur. The small negative pulse discharges with time constant RC, and recharge with the next pulse by time constant R_sC. However, if the time constant RC is much greater than T, then the base-line is essentially clamped to zero. In reality, the single diode design is capable of restoring base-line shifts in only one direction. Robinson [1961] and Fairstein [1975] solve this problem by placing two diodes in the circuit to adjust base-line shifts in either direction. Various base-line restoration designs with diode clamps are discussed by Nicholson [1974].

Another method used to restore the base-line is with a gated switch, depicted in Fig. 22.17(b). This configuration eliminates the base-line shift from changing count rates. The switch is open during the input pulse and closed when the output pulse turns negative. By doing so, the CR circuit is active only between pulses, when the voltage is restored to zero. The switch is usually a combination of diodes [Robinson 1961], or more complex circuitry (see references in Nicholson [1974], Karlovac and Blalock [1975] and Fairstein [1975]). The stability of base-line restoration at high counting rates depends on the ability of the gating control circuits to distinguish between the incoming pulses and the base-line. This discrimination can be achieved with a manually adjusted discriminator set slightly above the electronic noise straddling the base-line. Such sophisticated amplifiers include automatic noise discriminators and complicated pulse detection techniques that effectively perform this task [Karlovac and Blalock 1975; Fairstein 1975].

22.3 Components

The main electronic components commonly used in a radiation detection system were introduced in Chapter 2, primarily because many readers using this book to learn about radiation detectors might also need to use some of these components in a laboratory setting as they gain knowledge about the operation of radiation detectors. In the following sections, additional information about these components is provided.

22.3.1 Preamplifiers

A preamplifier unit has two basic purposes, namely (1) to provide a low-noise coupling of the typically high impedance of the detector to a low impedance capable of driving the string of amplifier and readout electronics, and (2) to produce a first stage of signal amplification. This first stage of amplification is usually small. Preamplifiers are designed to be either voltage sensitive, current sensitive, or charge sensitive. Optimum detector performance is obtained when the type of preamplifier is properly matched to the type and characteristics of the detector and the analysis electronics.

Voltage-Sensitive Preamplifiers

Consider the simple preamplifier shown in Fig. 22.18. A resistor R_1 is connected between the input voltage and terminal 1 while terminal 2 is grounded. Also, a feedback resistor R_2 is connected between terminal 1 and terminal 3. The current flowing through R_1 is described by,

$$i_1 = \frac{v_{in} - v_A}{R_1} \tag{22.119}$$

and the current flowing through R_2 is

$$i_2 = \frac{v_A - v_{out}}{R_2}. \tag{22.120}$$

The small voltage appearing between terminals 1 and 2 is,

$$v_{AB} = v_B - v_A, \tag{22.121}$$

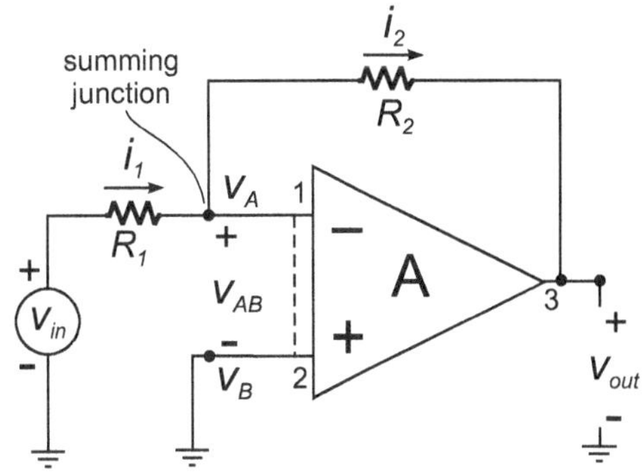

Figure 22.18. Simple voltage sensitive preamplifier with the inverted configuration.

but because terminal 2 is grounded $V_B = 0$ so that

$$v_{AB} = -v_A. \tag{22.122}$$

Because the input impedance into the op amp is large, only an insignificant current flows into terminal 1. Hence, the current flowing through R_1 also flows through R_2, i.e., $i_1 = i_2$, so equating Eqs. (22.119) and (22.120) gives

$$\frac{v_{in} - v_A}{R_1} = \frac{v_A - v_{out}}{R_2}, \tag{22.123}$$

or, upon rearrangement,

$$v_{out}R_1 - v_A(R_1 + R_2) = -v_{in}R_2. \tag{22.124}$$

Equation (22.36) for the present analysis becomes

$$\frac{v_{out}}{A} = v_B - v_A = -v_A \tag{22.125}$$

because here $v_B = 0$. By combining these last two equations, the transfer function of this preamplifier is

$$\frac{v_{out}}{v_{in}} = -\frac{AR_2}{[R_1(A+1) + R_2]}. \tag{22.126}$$

This result can be significantly simplified by recognizing that the open loop gain A is very much larger than unity so that $R_1(A+1) \gg R_2$ and, consequently, Eq. (22.126) reduces to,

$$\frac{v_{out}}{v_{in}} \simeq -\frac{R_2}{R_1}. \tag{22.127}$$

This simple result can also be obtained as follows. Conceptually, one can think of terminal 1 being at virtual ground and causing the input voltage at terminal 1 to also be zero. This means that the current flowing through R_1 must be equal to $i_1 = v_{in}/R_1$. However, the input impedance of the op amp prevents current flow into the device. Instead, the current i_1 must flow through R_2, thereby producing a voltage equivalent to $v_{in}R_2/R_1$. At terminal 3, the output voltage v_{out} is

$$v_{out} = v_1 - i_1R_2 = 0 - v_{in}\frac{R_2}{R_1}, \tag{22.128}$$

which yields the same transfer function as Eq. (22.127).

The gain of Eq. (22.127) is called the *closed loop gain.* First, note that ideally the open loop gain A has no bearing on the closed loop gain.[9] Second, the gain produced has the opposite or inverse sign of the input, hence the name for this circuit is the *inverting configuration.* The inverting configuration is a common circuit used as a voltage sensitive preamplifier, so named because they tend to respond to and linearly amplify the voltage input from the detector into the preamplifier circuit.

As discussed in Chapter 8, the voltage input v_{in} into the preamplifier from the detector is generally an inverse function of the detector capacitance, $v_{in} \approx Q_i/C_d$, where Q_i is the charge induced by the motion of electrons and ions excited in the detector by a radiation interaction event and C_d is the combined capacitance of the detector and coupling cables. Hence, to reduce the capacitance, the preamplifier circuit is usually placed as close as possible to the detector without causing environmental damage to the circuit.

For many detectors, the capacitance remains constant for different applied operating voltages V_o; hence the preamplifier output v_{out} is mostly linear with respect to the charge excited in the detector regardless of the voltage applied to the detector. Examples include gas-filled detectors and scintillation detectors that are coupled to photo-multiplier tubes.

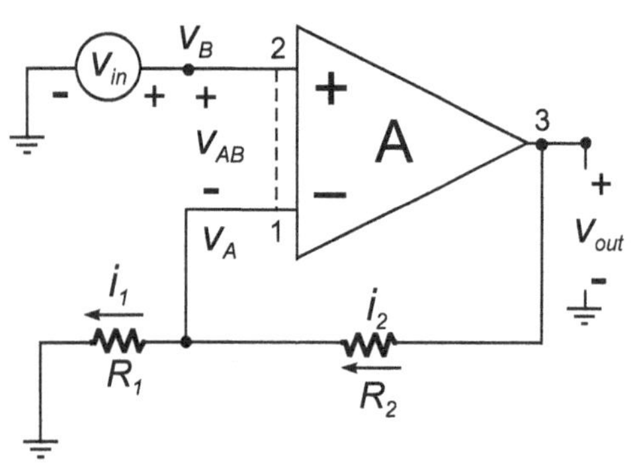

Figure 22.19. Preamplifier in the non-inverting configuration.

Now consider the circuit of Fig. 22.19, referred to as the non-inverting preamplifier configuration. It looks similar to the inverting configuration, except that the voltage input is now at the positive terminal, denoted with the present convention as input 2. As before, the output at terminal 3 is

$$v_{out} = A(v_B - v_A). \tag{22.129}$$

The voltage at input 2 is $v_B = v_{in}$ so that Eq. (22.129) gives

$$\frac{v_{out} + Av_A}{A} = v_{in}. \tag{22.130}$$

From Fig. 22.19 it is seen that

$$v_A = v_{out}\left(\frac{R_1}{R_1 + R_2}\right). \tag{22.131}$$

Substitution of Eq. (22.131) into Eq. (22.130) and rearrangement of terms yields

$$\text{Gain} = \frac{v_{out}}{v_{in}} = \frac{A(R_1 + R_2)}{(A+1)R_1 + R_2}. \tag{22.132}$$

Notice that the output voltage v_{out} retains the same polarity as the input voltage v_{in}. The gain of Eq. (22.132) is often simplified with the assumption that A is very large, so that Eq. (22.132) reduces to

$$\text{Gain} = \frac{v_{out}}{v_{in}} \approx \frac{R_1 + R_2}{R_1} = \frac{R_2}{R_1} + 1. \tag{22.133}$$

However, if the $A+1$ is not much larger than the ratio $(R_2 + R_1)/R_1$, then considerable error can occur if the approximation of Eq. (22.133) is used.

[9] In reality, the rolloff of the open loop gain has a significant effect on the signal gain of the amplifier. See Carter and Brown [2016], Carter and Mancini [2017], Jung [2005], and Zumbahlen [2008] for more details.

Current-Sensitive Preamplifiers

A current-sensitive preamplifier is designed to measure the instantaneous current flowing from a detector. The practical implementation of the topologies can be identical for charge sensitive and current-sensitive preamplifiers, but they are used for different purposes. The component values are optimized to extract different information. This current sensitive preamplifier is seldom used for most detector applications; however, on rare occasion a preamplifier with a fast rise time for timing applications is needed, and a current sensitive preamplifier is required. Examples in which a current sensitive preamplifier is needed include timing applications with fast photomultipliers and microchannel plates. The current sensitive configuration requires that the input impedance be low and the preamplifier gain A be very large. These conditions can be realized by using the general design of a voltage sensitive preamplifier and reducing the load resistance at the input of the preamplifier.

A simplified current sensitive preamplifier is depicted in Fig. 22.20.[10] The two resistors form a parallel circuit so that the current flowing in R_1 is

$$i_1 = i_{in}\frac{R_2}{R_2 + R_1}. \tag{22.134}$$

If R_1 is much smaller than R_2, then Eq. (22.134) indicates that $i_1 \simeq i_{in}$ so that $v_{out} \simeq -Av_A = -Ai_1R_1$. If the input impedance is 50 Ω, then it can be impedance matched to a common 50 Ω transmission line, thereby eliminating reflection losses. The current sensitive preamplifier converts the input current i_{in} to an output voltage v_{out} described by $v_{out} \simeq Ai_{in}(50\Omega)$.

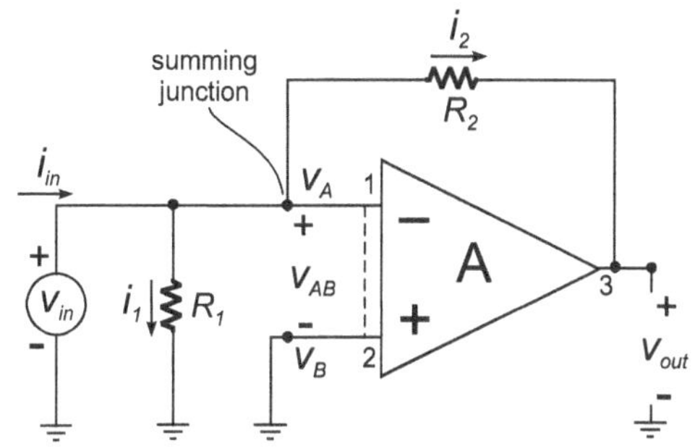

Figure 22.20. Simple current sensitive preamplifier with the inverted configuration.

Because current sensitive preamplifiers are designed to track the detector signal for timing purposes, noise added to the signal causes uncertainty in the actual arrival time of a pulse. This uncertainty is called *jitter*. A preamplifier rise time that is faster than the detector signal really does not help the situation. First, the output signal cannot be faster than the input signal, so a preamplifier rise time that is much faster than the input signal offers no real advantage. Instead, more noise is accepted into the system, thereby causing more jitter. By contrast, if the preamplifier signal is much slower than the detector signal, then the timing resolution suffers. To minimize the effect of jitter, Cova et al. [1991] recommend that the preamplifier rise time be similar to that of the detector input. For practical applications, it is recommended that the preamplifier rise time be within a factor 2 of the detector rise time [Ortec 2009b]. Note that the *Johnson noise* can be high unless the feedback resistor in the preamplifier circuit is $R_2 \geq 10^9$ Ω [Nicholson 1974].

Charge-Sensitive Preamplifiers

There are some detectors that change capacitance as voltage is applied, particularly semiconductor diode detectors. These semiconductor detectors have depletion regions (Chapter 15) that increase with applied operating voltage, which, in turn, causes the capacitance C_d to decrease with increasing voltage. As a result, the preamplifier output changes with a varying operating voltage for identical induced charges q_i. By redesigning the preamplifier feedback circuit, the capacitive component of the detector C_d can be minimized compared to the feedback capacitance C_f of the preamplifier, thereby effectively rendering the preamplifier output almost entirely dependent upon the induced charge q_i.

Consider the amplifier circuit shown in Fig. 22.21. There is a capacitor C_1 connected to input terminal 1 and another capacitor C_2 connected across terminals 1 and 3. The capacitor C_2 is referred to as the feedback

[10] In some cases, the feedback resistor R_2 is missing from an idealized current sensitive preamplifier diagram, implying that it has an infinite feedback resistance.

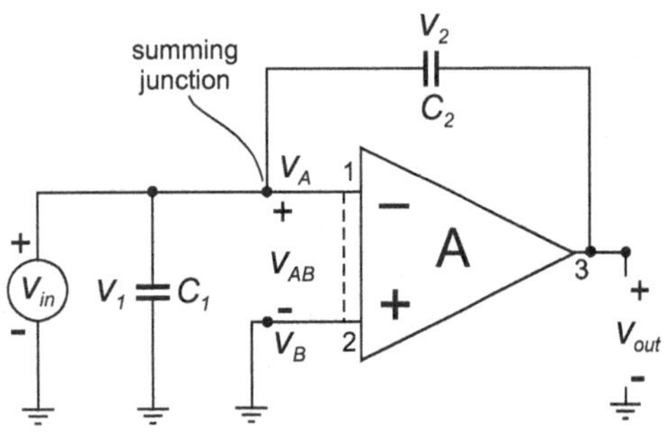

Figure 22.21. Simple charge sensitive preamplifier with the inverted configuration.

capacitor, often denoted C_f. The voltage v_{in} stored on C_1 appears at the summing junction. The voltage at the summing junction is v_{in} and must equal the voltage across C_2 plus v_{out}, i.e.,

$$v_{in} = v_{out} + v_2. \tag{22.135}$$

For an ideal op amp $v_{out} = -Av_{in}$ so that Eq. (22.135) can be written as

$$v_2 = (A+1)v_{in} = -\frac{(A+1)}{A}v_{out}. \tag{22.136}$$

Because $v_2 = q_2/C_2$

$$\frac{q_2}{C_2} = -\frac{(A+1)}{A}v_{out}, \tag{22.137}$$

which upon rearrangement gives

$$v_{out} = -\frac{A}{(A+1)}\frac{q_2}{C_2}. \tag{22.138}$$

Because A is very large and q_2 is the induced charge q_{in} from the detector

$$v_{out} \simeq -\frac{q_{in}}{C_2}. \tag{22.139}$$

The important outcome of Eq. (22.139) is that the output voltage is a function of the feedback capacitor C_2 and the induced change q_i and does not depend on the detector capacitance. Hence, for semiconductor diode detectors, the voltage output v_{out} is largely dependent only upon the induced charge q_i and feedback capacitance C_2, where $v_{out} = -Av_{in} \approx -q_i/C_2$, where A is the preamplifier gain. The *charge gain* is then

$$\frac{v_{out}}{q_{in}} = -\frac{1}{C_2}. \tag{22.140}$$

The goal of the detector system is to measure all of the charge generated in the detector. However, some charge remains on the sensor capacitance, leading to the following ratio for the input charge to the amplifier to the charge remaining on the detector [Spieler 2005],

$$\frac{q_{in}}{q_{in}+q_{det}} = \frac{v_{in}C_{in}}{v_{in}(C_{in}+C_{det})} = \frac{1}{1+\dfrac{C_{det}}{C_{in}}}. \tag{22.141}$$

where C_{in} is the dynamic input capacitance of the preamplifier. For $C_{in} \gg C_{det}$, the ratio $Q_{in}/(Q_{in}+Q_{det})$ approaches 100%.

Commercial charge-sensitive preamplifiers are designed for optimum performance when matched to specific ranges of detector capacitance. Hence, it is advised that the user consult the preamplifier specification sheets to properly match detectors to preamplifiers.

Example 22.5: Consider a silicon surface barrier detector connected to a commercial preamplifier with input capacitance of 10 nF. The detector is circular with diameter of 15 mm. The background doping concentration is 10^{13} cm^{-3} with $V_{bi} = 0.3$ volts. What is the charge ratio measured at self-bias volts? What is it at reverse bias of 150 volts?

Solution:
From Chapter 15, one uses the depletion approximation for the detector self-bias capacitance,

$$C_{det} = \frac{\kappa\epsilon_0 A}{L} = \left[\frac{2\kappa\epsilon_0(V_{bi}-V)}{qN_b}\right]^{-1/2}\kappa\epsilon_0\pi r^2 = \pi r^2\left[\frac{qN_b\kappa\epsilon_0}{2(V_{bi}-V)}\right]^{1/2}$$

$$= \frac{\pi}{4}(1.5\text{ cm})^2\left[\frac{(1.6\times10^{-19}\text{C})(10^{13}\text{ cm}^{-3})(11.9)(8.854\times10^{-14}\text{ F cm}^{-1}}{2(0.3\text{ V})}\right]^{1/2}$$

$$= 2.962\text{ nF}.$$

Recalling that the depletion width increases with $\sqrt{V}$, the capacitance at 150 volts reverse bias is,

$$C_{det} = 2.962\text{ pF}\sqrt{\frac{V_{bi}}{V+V_{bi}}} = 2.962\text{ pF}\sqrt{\frac{0.3}{150.3}} = 132.34\text{ pF}.$$

For the self-biased case,

$$\frac{Q_{in}}{Q_{in}+Q_{det}} = \frac{1}{1+\dfrac{C_{det}}{C_{in}}} = \frac{1}{1+\dfrac{2.962\text{ nF}}{10\text{ nF}}} = 0.7715\text{ or }77.15\%.$$

By the way, the resulting depletion layer width is only 6.29 microns, relatively small compared to alpha particle ranges from most check sources, meaning that the depletion region would be too small to absorb the total particle energy. Consequently, the charge that is deposited in the depletion capacitance is less than the total amount liberated, further diminishing the measured charge.

For the case where the detector is reverse biased at 150 volts,

$$\frac{Q_{in}}{Q_{in}+Q_{det}} = \frac{1}{1+\dfrac{C_{det}}{C_{in}}} = \frac{1}{1+\dfrac{132.34\text{ pF}}{10\text{ nF}}} = 0.9869\text{ or }98.69\%.$$

Also, the depletion depth would be 140.7 microns, wide enough to absorb alpha particle energies up to approximately 15 MeV.

Charge Sensitive Preamplifier with Resistive Feedback

The charge stored across the feedback capacitor in the charge-sensitive configuration shown in Fig. 22.21 accumulates with successive inputs, and eventually the charge reaches the limiting storage capacity. In order to discharge the capacitor between input pulses, a resistor is attached in parallel across the feedback capacitor so the capacitor discharges exponentially with a decay time constant of $\tau = RC$.[11] These time constants can be relatively long and are often about 50 μs [Ortec 2009a].

Pulse pile-up, depicted in Fig. 22.1, is due to subsequent pulses arriving before a prior pulse has returned to the base-line. Although the circuit ideally does not suffer from dead time losses, the output pulses are distorted. If the pulse decay tail is positive, then the subsequent pulse magnitude is greater than it should be, as shown in Fig. 22.1(a). If the pulse decay tail suffers undershoot, then the subsequent pulse is less than it should be, as shown in Fig. 22.1(b). In either case, the result is a broadening of the pulse height spectrum which reduces the energy resolution of the spectrometer. Further, at high count rates, multiple pulses

[11] Note that the value selected for this resistor is critical to stability of the preamplifier.

contribute to the pile-up problem and the total magnitude of the stored charge can exceed the preamplifier limit to produce a condition called "saturation overload," as shown in Fig. 22.22.

Because preamplifier output pulses are usually sent to the input of a shaping amplifier, the amplifier output can be affected by these pulse height distortions created by the preamplifier. There are several amplifiers designed to address pulse pile-up distortions, a few of which are described in latter sections.

Pulsed-Reset Preamplifiers

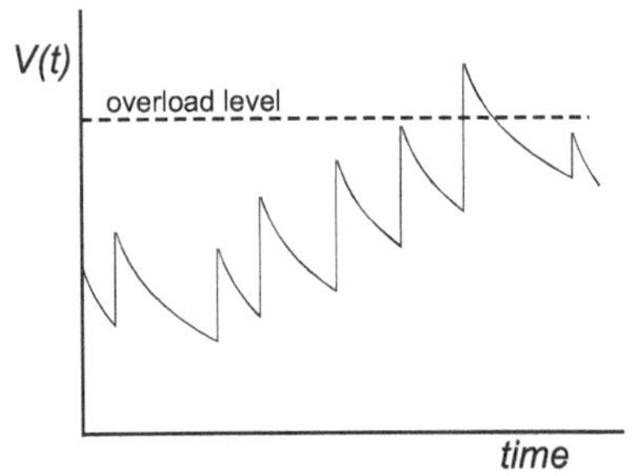

Figure 22.22. Preamplifier pulse pile-up output at high counting rates. Pulses can become distorted if they exceed the preamplifier overload level.

The pulsed-reset preamplifier was designed to reduce noise contributions from the feedback resistor by eliminating it altogether. In other words, R can be regarded as being infinite. Consequently, the decay constant of the feedback capacitor is essentially infinite so the decay tail now becomes a flat step (see Fig. 22.23(a)). Each input to the preamplifier produces a step increase in voltage proportional to the input charge of each pulse. During charge integration, the system does not suffer from dead time, mainly because the preamplifier continues to integrate the input charges that produce the stair-step output. The step inputs are shaped by a shaping amplifier, as described in previous sections. However, if two input pulses occur too close together, a form of pile-up occurs at the shaping amplifier which can cause pulse-height distortions.

A consequence of this preamplifier design is that the feedback capacitance does not discharge the accumulated charge as would occur in a resistive feedback preamplifier. To remedy this problem, an active reset method is incorporated. When the charge accumulated on the feedback resistor reaches a predetermined voltage, labeled V_{ramp}, a reset sequence begins. The design generally replaces a resistor in the feedback loop with either a light emitting diode (LED) or a transistor [Landis et al. 1971; Landis et al. 1982], either one being activated at the end of T_2 to reset the feedback capacitor. First, a short time duration is allowed that passes the last valid input pulse, labeled T_2 in Fig. 22.23(a). The capacitor is rapidly discharged during T_3 using an optical or transistor feedback method. In the case of optical feedback, a light-emitting diode (LED) is triggered that illuminates an optical gate such as a field effect transistor (FET), which opens and discharges the feedback capacitor. Optical feedback has been largely replaced by active FET discharging, in which the gate on the FET is opened to discharge the feedback capacitor when the reset voltage is exceeded.

In either case, the rapid discharge can produce a negative output from the shaping preamplifier. To prevent a negative output, the preamplifier remains disabled for a preset time, designated T_{aol}. During this disabled period, a blocking signal can be supplied to the shaping amplifier for a time T_{inh} to prevent the output of erroneous signals. Although the fractional dead time of the preamplifier would ideally be expressed by the ratio of T_3/T_1 [Britton et al. 1984], the actual fractional deadtime is determined by the time the amplifier cannot accept pulses. Hence, the fractional system deadtime is represented by

$$DT_f = \frac{T_{inh}}{T_1} = \frac{T_{inh}}{\dfrac{V_{ramp}}{(ER)(CG)} + T_2 + T_3}, \tag{22.142}$$

where ER is the input rate of energy into the preamplifier and CG is the conversion gain of the energy to voltage pulse height.

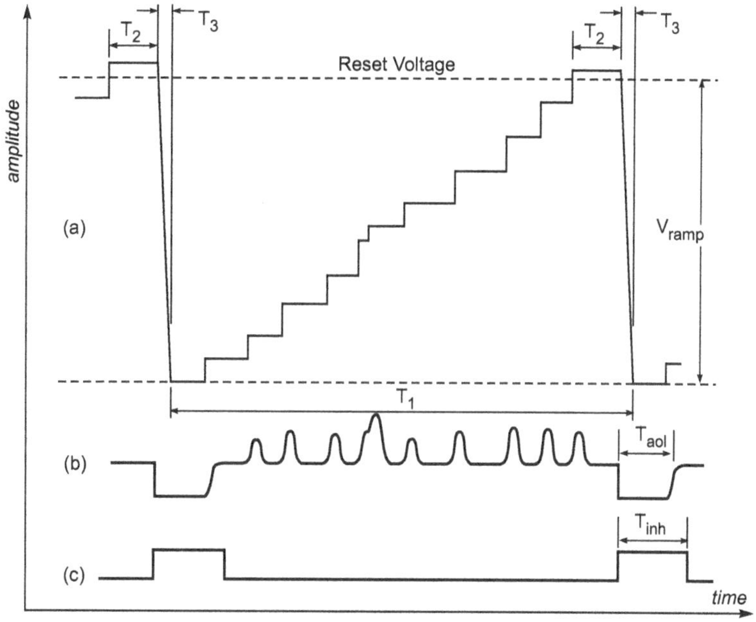

Figure 22.23. Output waveforms from a pulsed-reset preamplifier, where (a) is the output from the preamplifier, (b) are shaped pulses from a Gaussian amplifier, and (c) is the inhibit signal activated to stop data collection during the reset time. After Britton et al. [1984].

22.3.2 Amplifiers

The output pulse from a preamplifier is usually designed to have a short rise time τ_{rise} and a long fall time τ_{fall} so that $\tau_{rise} \ll \tau_{fall}$. Often another radiation interaction occurring in the detector arrives before the tail pulse from the previous radiation interaction is completely discharged. Consequently, subsequent pulses ride atop the tail of prior pulses, as shown in Fig. 22.22. The size of the step increase from a tail pulse contains information about the energy deposited in the detector because it is proportional to charge produced by an interaction and subsequently measured by the preamplifier. Hence, one purpose of the amplifier chain is to measure that signal increment and shape the resulting measured signal into a much shorter pulse signal. Also, because the gain of a preamplifier is usually small in order to prevent overload, the step change is difficult to measure accurately and the small signal can be attenuated significantly over relatively short cable distances. The purpose of the amplifier is to address both of these issues. First, to shape the charge by producing a significantly faster output pulse that still preserves the differential step change of each pulse. Second, to provide additional amplification of the signal. In some cases, fast timing amplifiers allow the preservation of information about the arrival time of different input pulses.

Linear Amplifier

For spectroscopy systems, the preferred amplifier choice is a linear pulse shaping amplifier. The amplifier serves to shape and amplify the input pulses from the preamplifier and produces an output whose amplitude is usually restricted by the maximum 10-volt NIM standard. Thus amplification of a pulse over the maximum limit results in a clipped output with the shaped pulse output being truncated at 10 volts in case of the NIM standard. Users are well advised to adjust the gain such that output pulses for the radiation energy under inspection is well below 10 volts.

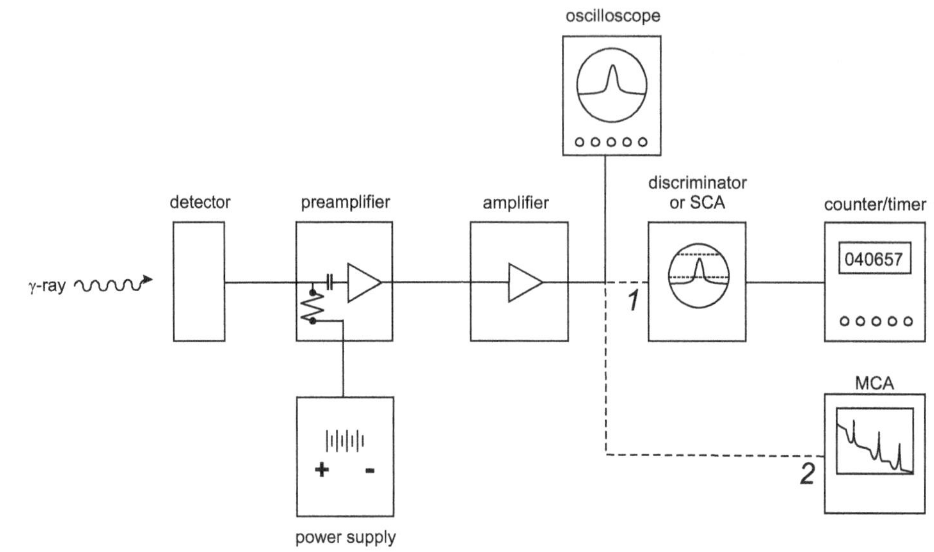

Figure 22.24. Detection system and components for either pulse counting (with SCA) or energy spectroscopy (with MCA).

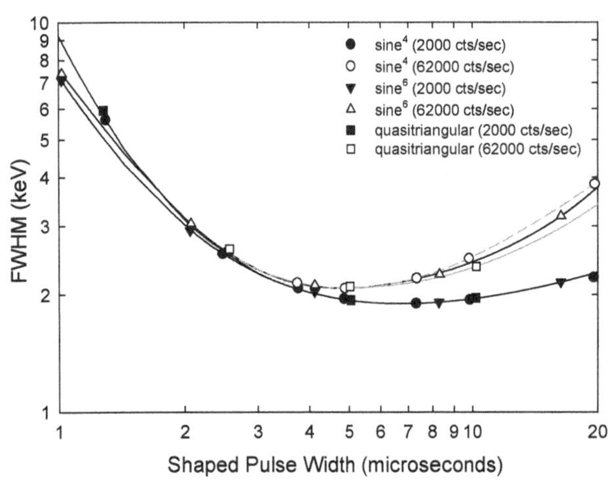

Figure 22.25. The energy resolution as a function of the pulse width for linear amplifiers with three different shaping networks. Also shown are the results from two different counting rates (2000 cps and 62000 cps). Data are from Fairstein [1985].

Most linear amplifiers allow the user to adjust the shaping times, with research units having adjustable independent rise and fall times. This feature allows the user to optimize the energy resolution performance of the system. Gaussian shaping is common with linear shaping amplifiers, although some units offer different shaping outputs such as triangular or quasitriangular as well as variations of Gaussian shaping. Shown in Fig. 22.25 are data obtained for different shaping networks at different count rates for an HPGe detector. As can be seen, the energy resolution changes with shaping time. The minimum FWHM also changes as the count rate changes. From results such as those shown in Fig. 22.25, the shaping times of a linear pulse amplifier can be adjusted to give the best energy resolution. If the shaping times selected are too short, then ballistic deficit losses worsen the energy resolution and the FWHM widens. Also, the contribution from series noise (see Sec. 22.7) increases and consequently decreases the energy resolution. If the shaping time is too long, then parallel noise contributions increase, which also compromises the energy resolution. As the count rate increases, the effects of pulse pile-up also begin to affect detrimentally the energy resolution if long shaping times are used. Consequently, the shaping time should be adjusted for optimum performance at the pulse rate under inspection.

For high-resolution HPGe detectors, the shape of the detector pulse varies depending on the location of the radiation interaction. The induced current entering the preamplifier changes as a function of the weighting field in the detector, which is a function of position in the detector, so that the rise time of the pulse is also a function of the interaction location (see Sec. 16.4). As a consequence, less charge is integrated per unit time for long rise times, while the opposite is true for short rise times. If the shaping time is adequately long, then this effect is minimal because most of the charge is integrated regardless of the rise time. However, if the shaping time is short relative to the rise times, then there is detrimental increase in the variance of the output pulses from the preamplifier for monoenergetic events. This problem is called the *ballistic deficit* (see Sec. 22.2).[12] For low count rates, the shaping time can be set relatively long ($> 6\ \mu s$) to reduce the effects of ballistic deficit. At high count rates, short shaping times are preferred in order to reduce pulse pile-up. However, shaping times shorter than $2\ \mu s$ can cause energy resolution to degrade in high-resolution HPGe detectors [Becker et al. 1981]. Commercially available spectroscopic linear amplifiers quote the spectral broadening for high radiation interaction rates in the detector.[13]

Gated Integrator

Karlovac and Blalock [1975] describe a method for preserving the energy resolution despite a ballistic deficit caused by short shaping times needed for high interaction rates in the detector. The method uses a time-variant amplifier, or "gated integrator," to process the pulse [Radeka 1972]. An output pulse from a semiconductor detector, typically an HPGe detector, is introduced through a preamplifier into a Gaussian shaper. This Gaussian shaper with a short time constant rapidly shapes the input from the preamplifier and delivers the shaped pulse to the gated integrator. The charge leaving the Gaussian shaper is stored on a feedback capacitor across the gated integrator. This charge is the integral (area) of the pulse. At a preset collection time, a shorting gate closes across the feedback capacitor to fully discharge it to ground, while another gate is opened to disconnect the Gaussian shaper from the gated integrator during the discharge process. The time that the gated integrator is active is much longer than the Gaussian shaping time, usually on the order of 10 microseconds. The Gaussian shaper must remain active long enough to measure all charge from the detector.

This gating technique is demonstrated in the modeled results of Fig. 22.26. Here the inputs of three identical charge packets are applied to a fast Gaussian shaper with $0.7\ \mu s$ shaping time. In one case, the induced current is high from fast charge collection (10 ns) and the Gaussian shaper develops the full Gaussian output with no ballistic deficit. In a slightly slower case, the shaping time and the charge collection time ($1\ \mu s$) are comparable. Hence, a small amount of ballistic deficit is observed, but the overall integrated output reaches the same level as the faster pulse. In the third case, the charge collection time ($10\ \mu s$) is much slower than the shaping time. After a short rise, the output fall time causes the input signal to match the output signal, forming a flat top pulse. However, the integral of this slower pulse also reaches the same output as the faster pulses, albeit at a much later time. The integrated charge, shown for all inputs, eventually reaches the same value, despite the rise time differences. The total charge is the same in all cases and is stored on the gated integrator capacitance. Consequently, the output is no longer based on the pulse height as with time-invariant linear amplifiers, but instead the total integrated charge in the pulse. The output is therefore independent of the ballistic deficit perturbations and no longer affects the gated integrator output pulse magnitude, provided that all charge is collected during the processing time and there is no pulse pile-up

[12]Note that ballistic deficit is a consequence of using too short of a shaping time, which results in lost charge integration. For short shaping times, the ballistic deficit can also be affected by specific current induction, which varies as a function of the interaction location, carrier velocity, and the weighting field in the detector [Radeka 1972].

[13]The spectral broadening in terms of percent at a specific high "count rate" is used for this metric. However, this description is somewhat imprecise, because the count rate is the observed response from a system; whereas, it is the input pulse rate that causes the problem. A high input pulse rate is produced by a high radiation interaction rate in the detector and this leads to pulse pile-up. The ballistic deficit is thus exacerbated by short pulse shaping times.

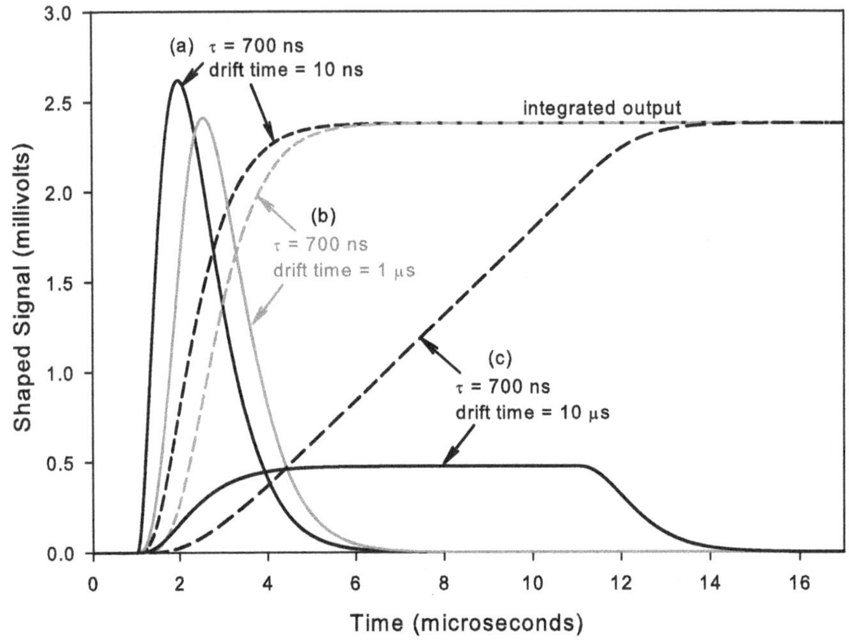

Figure 22.26. Modeled pulses from a Gaussian shaper with $\tau = 0.7\mu s$, showing the effects of different carrier drift times for the same amount of total charge. Also shown is the integrated output, with gain of 500,000, obtained from the shaped signals. In (a), the drift time is 10 ns, yielding a semi-Gaussian output with negligible ballistic deficit. In (b) the drift time is 1 μs and in (c) the drift time is 10 μs. Although the Gaussian peak voltage is different for all cases, the integral of the signals is the same for all cases. Hence, provided that all charge is collected during the processing time and there is no signal pile-up, the integral can be used to preserve the detector spectroscopic information.

during the signal shaping and integration process. Because the circuit is designed to operate under high interaction rates, pile-up rejection is used to reject overlapping pulses.

Two advantages of the gated integrator amplifier circuit are apparent. First, the energy resolution is preserved because ballistic deficit from different rise times is no longer a problem. Second, a shorter time constant allows faster pulse processing; hence more counts are recorded at high rates of gamma-ray interactions. Because the system dead time is largely determined by the processing times of the gated integrator and that of the multichannel analog-to-digital processor, changes in the observed count rate do not necessarily track the interaction rate in the detector [Simpson et al. 1991]. The example spectra in the work of Karlovac and Blalock [1975] shows good energy resolution for an interaction rate of 150,000 gamma-rays per second, but records only (approximately) twice as many counts as observed for a lower count rate of only 5000 interactions per second.[14]

Details for the Common Linear Amplifier

Ideally, a linear pulse amplifier should have the following properties. The output of the amplifier should scale proportionally to the energy input, a condition requiring that any integral and differential non-linearities

[14]The literature states the that input is 150,000 counts per second (150 kcps) [Karlovac and Blalock 1975], but this is a semantics problem. Strictly, the observed count rate is the final tally allowed by the system, which is seriously lower than the gamma-ray interaction rate in the detector. The authors of this text believe that interaction rate, not count rate, is what Karlovac and Blalock actually meant.

should be small. For a linear preamplifier, the minimum output is 0 volts, while the maximum output (NIM rules) is 10 volts. Hence, the minimum input should result in 0 volts output, while the maximum allowed input yields 10 volts output. Ideally the output voltage scales linearly with the input voltage, as depicted in Fig. 22.27. However, there usually is some non-linearity, i.e.,

$$\frac{v_{out}}{v_{in}} = A\left(1 + \epsilon_{int}\right), \tag{22.143}$$

or

$$\epsilon_{int} = \frac{v_{out} - Av_{in}}{Av_{in}}, \tag{22.144}$$

where A is the gain and ϵ_{int} is the integral non-linearity [Kowalski 1970]. The differential non-linearity ϵ_{dif} is defined as

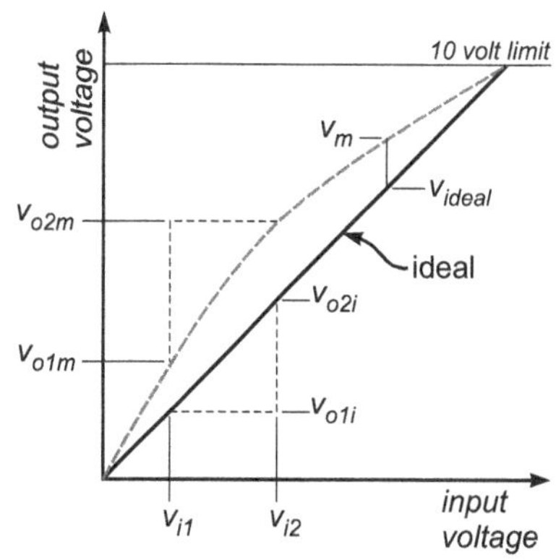

Figure 22.27. Examples of integral and differential non-linearity in an amplifier output.

$$\frac{dv_{out}}{dv_{in}} = A\left(1 + \epsilon_{dif}\right), \tag{22.145}$$

or

$$\epsilon_{dif} = \frac{dv_{out}}{dv_{in}}\frac{1}{A} - 1, \tag{22.146}$$

which describes non-linearity between small differences in pulse heights.

Both non-linearities can be measured from a plot of v_{out} versus v_{in} such as shown in Fig. 22.27. For any given input v_{in} the integral non-linearity of Eq. (22.144) is

$$\epsilon_{int} = \frac{v_m - v_{ideal}}{v_{ideal}} \tag{22.147}$$

where v_m is the measured output and v_{ideal} is the predicted linear output.[15] The differential non-linearity can be determined from the ratio of slopes of the measured outputs to linearly predicted outputs at any input voltage, i.e.,

$$\epsilon_{dif} = \left[\lim_{v_{i2}\to v_{i1}}\left(\frac{v_{o2m} - v_{o1m}}{v_{i2} - v_{i1}}\right)\left(\frac{v_{o2i} - v_{o1i}}{v_{i2} - v_{i1}}\right)^{-1}\right] - 1 = \left[\lim_{v_{i2}\to v_{i1}}\left(\frac{v_{o2m} - v_{o1m}}{v_{i2} - v_{i1}}\right)\left(\frac{A\left(v_{i2} - v_{i1}\right)}{v_{i2} - v_{i1}}\right)^{-1}\right] - 1$$

$$= \left[\lim_{v_{i2}\to v_{i1}}\frac{v_{o2m} - v_{o1m}}{A(v_{i2} - v_{i1})}\right] - 1, \tag{22.149}$$

where v_{o2m} and v_{o1m} are measured outputs and v_{o2i} and v_{o1i} are ideal outputs, both corresponding to inputs v_{i2} and v_{i1}, respectively. Usually, these metrics are reported for the highest observed non-linear behavior over the output voltage range. Commercial spectroscopy linear amplifiers usually have an integral non-linearity $\leq \pm 0.025\%$ between 0 to 10 volts, and a differential non-linearity $\leq \pm 0.012\%$ between -9 to 9 volts.

Another problem that can occur with linear amplifiers is in the amplification of small pulses in the presence of large pulses. If sufficient gain is used to distinguish small pulses from large pulses, the processing of a

[15]To provide a uniform measure over the amplifier range, integral non-linearity for commercial units is often reported as,

$$\epsilon_{int} = \frac{v_m - v_{ideal}}{v_{max}}, \tag{22.148}$$

where v_{max} is the 10 volt NIM limit. Usually it is the highest measured value of ϵ_{int} over the amplifier range that is reported.

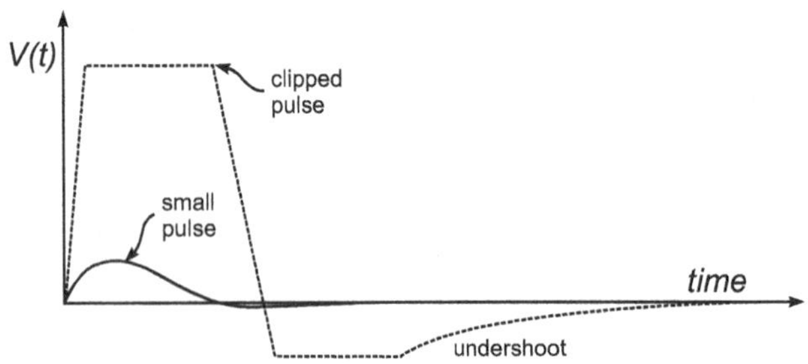

Figure 22.28. Overload pulses can have large undershoot and long recovery time. Small pulses arriving during the recovery time of an overload pulse are distorted.

large pulse may result in a clipped output in which the pulse exceeds the overload setting. Consequently, the pulse undershoot, as shown in Fig. 22.28, of this overload pulse is also large, and the recovery takes excessive time to return to the base-line. During the recovery time, small pulses that arrive are distorted by the negative tail of the overloaded pulse. Modern systems have overload resistant circuits that minimize this effect. Commercial spectroscopy amplifiers usually have the overload recovery specifications reported for maximum gain on the performance sheet.

Thermal stability is important, and the gain should not drift with temperature changes. Commercial units usually have thermal stability of $\leq \pm 0.0075\%$ per °C within a specified temperature range (typically between 0 to 50°C). Also electronic noise must be relatively low to achieve high energy resolution. Hence, the signal-to-noise ratio must be optimized for the device by adjusting the shaping times to coincide with the highest energy resolution (see, for example, Fig. 22.25). The amplifier should respond to high pulse input rates with negligible pile-up effects. Finally, there are some cases where the time correlation of the input pulse must be preserved, and the time trigger method becomes important as explained in the next section.

Amplitude adjustment includes coarse increment changes from 5× up to a maximum gain (usually 2000×). Most units also have a fine adjustment feature, with a continuously adjustable scaling factor between 0.5× to 1.5×. These features allow the user to adjust a pulser input to precise amplitudes for calibration purposes, or to adjust the shaped detector input to fall within the desired range of pulse heights. Detection systems and components for common counting and spectroscopy applications are shown in Fig. 22.24. Other features common to a linear spectroscopy amplifier are pole zero adjustment, automatic base-line restoration, pulse pile-up rejection, and choice of unipolar and bipolar outputs.

Timing Amplifier

For measurements where time correlation is important, the electronic pulse is sent to a timing pick-off gate that uses a threshold method to trigger a time stamp on the pulse. If the pulse height coming from the detector or preamplifier is too low to surmount this threshold, some amplification is then required that does not alter the time signal of the pulse. Common linear amplifiers are not well suited for this purpose, but are instead designed to eliminate high frequency noise while preserving the energy information provided by the detector. Hence, a fast timing preamplifier is needed that mimics the input signal with the added benefit of gain. Timing amplifiers may have rise times in the range of nanoseconds or smaller. Unfortunately, such fast rise times can also detrimentally affect linearity, thermal stability, and electronic noise. Because such amplifiers were designed to process negative output pulses from PMT anodes, the pulse polarity output of commercial units is usually negative to remain compatible with fast timing discriminators.

Earlier it was shown that AC coupling could be used as a DC filter while passing higher frequency signals. Hence, a *wideband amplifier* may be DC coupled to extend the frequency response to zero while still responding to high frequencies. Wideband amplifiers offer no control over the rise time or the decay time of the signal. Also, wideband amplifiers are best used with photomultiplier tube output signals and with silicon diode detectors to preserve timing information [Ortec 2009a]. Pulse shaping adjustment controls are provided with *timing filter amplifiers*, and they also typically have base-line restoration electronics. This type of timing amplifier is generally useful for timing applications with high-resolution HPGe detectors. For timing applications with either type of amplifier, the rise time should be selected to be less than the inherent

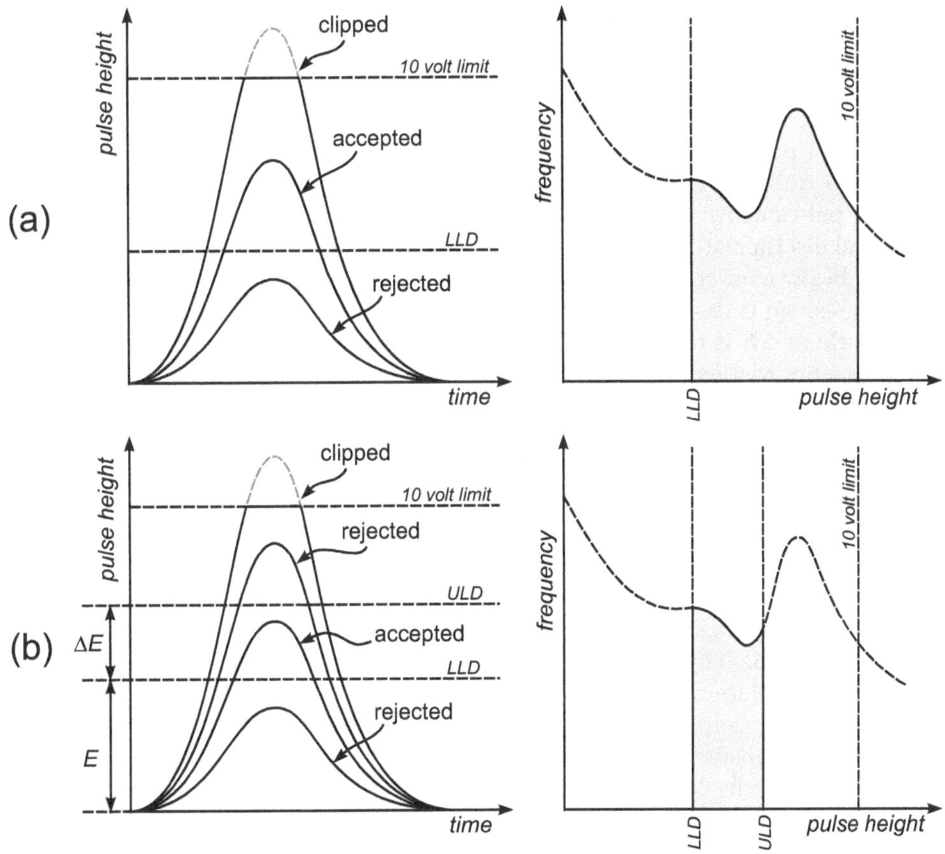

Figure 22.29. The function of an (a) integral discriminator and a (b) single channel analyzer.

rise time of the preamplifier to reduce or eliminate degradation of the signal rise time. As already noted in the discussion on current-sensitive preamplifiers, the rise time should be faster than the input signal rise time, yet not too much faster or electronic noise becomes a problem. Consequently, excessively fast amplifier rise times should be avoided, because they introduce more noise and provide no additional improvement in the signal rise time. The differentiator time constant should be set just large enough to avoid significant loss of the signal amplitude.

22.3.3 Integral Discriminators and Single Channel Analyzers

The electronic pulses emerging from a shaping amplifier are indicative of the energy absorbed within the detector. The amplitude of the pulse varies linearly with the absorbed energy in a well-designed system. Hence, once calibrated, the final signal pulse height is a measure of the energy absorbed in the detector. The main function of both discriminators and single channel analyzers is to reject shaped pulses outside established pulse-amplitude boundaries, thereby passing only pulses whose amplitude voltages are of interest. These systems convert a shaped input pulse into a logic output pulse, usually triggered when the leading edge of the shaped pulse surpasses a preset voltage threshold. Because the output is a logic pulse, energy information is preserved by the electronic boundaries set by the user and not the output pulse amplitude.

Logic pulses are characterized by a binary system, consisting of lows and highs, or alternatively as 0s and 1s. For instance, the Nuclear Instrument Module (NIM) system logic pulse is low (0) at 0 volts and high

(1) for a current of -16 mA through a 50Ω resistor, or a voltage of -0.8 volts.[16] The integral discriminator allows the user to select a signal voltage or energy E defined by a lower level discriminator (LLD) threshold, below which all pulses are rejected (depicted in Fig. 22.29(a)). In most models, any signal above the LLD is passed. However, some systems reject pulses that exceed the nuclear instrument module standard 10-volt upper limit. Because all pulses above the setting E are counted, 10 volt-limit notwithstanding, this device essentially passes all signals above the threshold. Hence, the accumulated pulses during the counting time is the integral of the pulses above E.

Unlike the integral discriminator, the single channel analyzer (SCA) is designed to filter out pulses whose amplitudes are both below a lower level discriminator (LLD) and above an upper level discriminator (ULD) setting. The LLD threshold is designated by E, a relative pulse height energy, while the ULD threshold is designated $E + \Delta E$ where ΔE is the width of the energy channel. Hence, the SCA allows the user to pick a voltage window (or energy window) of interest defined by lower and upper energy thresholds, and pass only those signal pulses that fall within those boundaries, as depicted in Fig. 22.29(b).

One method to accomplish this filtering task is to set two adjustable upper threshold levels in the instrument. The shaped signal input is split and routed through the two threshold circuits. If the signal pulse height is of adequate height to pass through E, then it is routed to an anticoincidence circuit. If the signal is also above the second threshold, it too passes to the anticoincidence circuit, where, upon arriving within a certain time window Δt, the anticoincidence circuit blocks these signals from passing and no logic output pulse is produced. However, if the signal passes above the LLD, but not the ULD, the anticoincidence circuit is not activated, and a logic output pulse is produced. Hence, if the input pulse does not fall within E and $E + \Delta E$ it is rejected. The number of passed signals equals the number of observed counts over a range of energies ΔE to produce a differential result.

Many modern SCA units can be operated as either an integral discriminator or SCA. In integral mode, the device window ΔE is completely opened while the LLD sets the discrimination threshold. When operated as an integral discriminator, the NIM upper 10-volt limit usually applies. Also, many SCA units have two scales, adjusted with a simple switch, with one window that opens to the maximum of 10 volts, and a smaller window with finer resolution that is limited to 1 volt maximum. The reason for the finer window resolution is to allow the experimenter to narrow the energy window with greater precision. A differential pulse height spectrum can be formed by first setting the ΔE window at a desired channel resolution, and subsequently taking measurements at increasing values of E in increments of ΔE from 0 up to 10 volts. Although an educational exercise, the multichannel analyzer (latter section) has made this method of producing a differential pulse height spectrum obsolete.

For most SCA units, the timing information is not accurately correlated to the input time of the pulse. This timing discrepancy occurs, in part, because of the leading-edge timing method usually used to determine when the signal pulse height surmounts the threshold. Unfortunately, this method suffers from so-called "time walk," because the pulse recognition time is dependent on the slope of the input pulse rise time. There are special timing SCAs that use constant-fraction timing, a method that compensates for varying amplitudes and ameliorates this timing shift and gives consistently better timing results. Timing SCAs are designed to minimize timing shift and dead time for a variety of detectors, including NaI:Tl, Si, and HPGe detectors, thereby allowing better system time resolution and higher counting rates.

[16]Other logic systems include resistor-transistor logic (RTL), transistor-resistor logic (TRL), transistor-transistor logic (TTL), complementary metal-oxide-semiconductor (CMOS), and emitter-coupled logic (ECL), all having specific voltage ranges for lows and highs.

22.3.4 Counters (Scalers) and Timers

A counter is used to tally the pulses that emerge from the discriminator or SCA unit. Historically, these units were called "scalers", mainly because their operation required a scaling ratio to tally the number of pulses entering the device. The reason for this operation was that, at one time, electromechanical registers could not keep up with the high rate of delivery from a counting system, capable of recording on average only 60 counts per second [Price 1964]. Instead, a "scale-of-2" was used where one output appears for two inputs. A series of n units gives a dividing factor of 2^n; hence the instrument scales down the number of pulses to manageable levels for the register. The result is an accurate tallying of the counts, giving the scaler the ability to indirectly record thousands of counts per second. The resolving power of these systems determined the count loss rate, manifested as another source of dead time. Other systems with "scale-of-10" were also developed. The period of operation of a scaler was controlled with a separate timer unit. Overall, the name "scaler" is historical, and modern digital electronic registers no longer use this scaling method to record pulses. Instead, the more appropriate name is "counter", and has been adopted as common nomenclature for modern pulse recording units.

High speed counters with rates exceeding 100 MHz are commercially available. These units typically operate with a string of decade counters. Each decade counter can record up to 10 inputs (in binary the number is 1010). Upon reaching the binary number 1010, the counter is reset to zero while a pulse is simultaneously delivered to the next decade counter in the chain, increasing the count by 1. On the second decade counter, a binary number 1010 is then equivalent to 10^2 counts. Hence, a string of n decade counters can register up to 10^n counts before resetting. Commercial units commonly have 8 sequential decade counters (an 8-decade counter).

The timer is used to start and stop the counter at preset time lengths. Usually the timer can be set to count for a certain length of *real* time, where the counter only functions during the preset time interval. Another timing mode feature allows a certain number of counts to be recorded, stopping to show the lapsed time when a preset number of counts is reached. Many timers can be set to scroll, where after a measurement period ends, there is a short delay before the tally is zeroed and another measurement automatically commences.

Timing systems that use the power-line AC frequency as the clock work adequately well and are relatively inexpensive, but they are best suited for use in countries that maintain power-line frequency corrections. However, the actual AC frequency is averaged over a relatively long period of operating time, meaning that short-term timing operations may suffer some inaccuracy. For short measurement time periods, seconds or minutes for instance, a timer based on a crystal-controlled oscillator is preferred.

Depending on the application, counters can be acquired either as independent units, sometimes with more than one counter in a single unit, or as combined counter/timer units having a built-in timer. Display counters have a display for the counting operation, often an 8-digit digital display capable of handling the maximum output from an 8-decade counter. The obvious convenience of a display counter is that the operator can visually monitor the counting process during the operation. Blind counters do not have a counting display, but instead can be interfaced with another display device such as a computer to log the number of counts. In some counter/timer units, there is a display counter, but the timer is blind. Printing counters, as the name implies, have an added capability that they can be connected to a printer to provide an output. Overall, counters and timers can be acquired for various applications, and specification sheets provided by manufacturers supply necessary information on maximum counting rates, display type, and output type.

22.3.5 Ratemeter

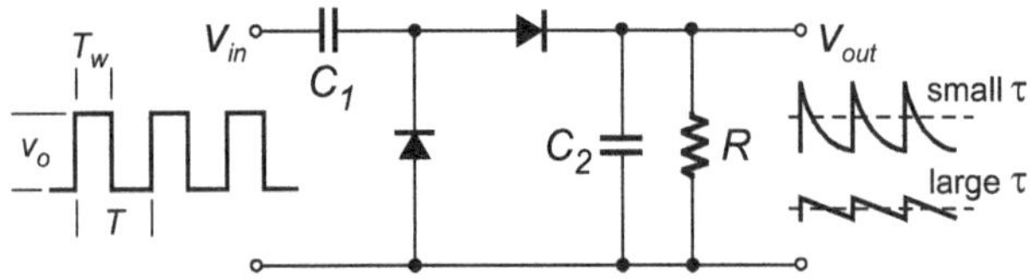

Figure 22.30. The basic diode charge pump ratemeter circuit.

A ratemeter is another type of radiation pulse recording device that produces an output charge (or voltage) proportional to the rate of radiation interactions in the detector. This type of detector was introduced by Gingrich et al. [1936] (see also Evans and Alder [1939]). A ratemeter design in common use is a diode charge pump circuit [Vincent 1960; Vincent and Parker 1970] and is depicted in Fig. 22.30. Consider the input pulse train of square waves as shown in Fig. 22.30. Each pulse has voltage magnitude v_o with width T_w and period T. The capacitor C_1 stores charge q_1 with voltage $v_1 = q_1/C_1$. The pulse input to the diode charge pump causes the capacitor C_2 to accumulate charge q_2, which discharges through the resistor R with a time constant $RC = \tau$. If the capacitance $C_1 \ll C_2$, then the charge passed to C_2 is equivalent to v_1C_1. The instantaneous output on the meter fluctuates in time, although it can be assumed that there is an average input value. The amount of fluctuation about an average is a function of the time constant τ, where a large value of τ decreases the fluctuation (less discharge) while a small τ has greater fluctuation (more discharge).

In reality, radiation pulses arrive at sporadic intervals with varying amplitudes. Instead of a wave train, what is actually observed are fluctuations in stochastic arrivals which contribute to the uncertainty in the measurement. The total charge measured on C_2 is q_2 while each pulse produces an average charge of $\bar{q}$. For instance, recall that a Geiger-Müller counter produces pulses of nearly the same amplitude regardless of the original particle type or energy. Given a fixed measurement time interval, the number of interacting particles within that time period can be found from

$$N = \frac{q_2}{\bar{q}}, \tag{22.150}$$

with fractional standard deviation,

$$\frac{\sigma_N}{N} = \frac{\sqrt{N}}{N} = \frac{1}{\sqrt{N}} = \sqrt{\frac{\bar{q}}{q_2}}. \tag{22.151}$$

The standard deviation of the measured q_2 is thus

$$\sigma_{q_2} = q_2\frac{\sigma_N}{N} = \sqrt{\bar{q}q_2}. \tag{22.152}$$

The amount of charge on the capacitor C_2 at t_0 is the remaining amount after the prior charge input. For an average charge $\bar{q}rdt$ produced between times t and $t + dt$, the amount of remaining charge on C_2 is

$$q_2(t)dt = \bar{q}r \exp\left[-\frac{t_0 - t}{RC_2}\right] dt \tag{22.153}$$

where r is the average rate of particle interactions. The standard deviation of the charge is

$$\sigma_{q_2} = \bar{q}\sqrt{rdt}, \tag{22.154}$$

and, therefore, the contribution to the standard deviation at t_0 is

$$\sigma_{q_2} = \bar{q}\sqrt{rdt}\ \exp\left[-\frac{t_0 - t}{RC_2}\right]. \tag{22.155}$$

For all charge packages delivered to the circuit from the detector, the separate variances can be summed to determine the total standard deviation

$$\sigma_{q_2}^2 = \int_{-\infty}^{t_0} \bar{q}^2 r \, \exp\left[\frac{-2(t_0 - t)}{RC_2}\right] dt = \frac{\bar{q}^2 r RC_2}{2}. \tag{22.156}$$

The average output current from the device is

$$\bar{i} = \bar{q}r, \tag{22.157}$$

while the instantaneous output current is

$$i(t) = \frac{q_2(t)}{RC_2}. \tag{22.158}$$

Hence, with error propagation on Eq. (22.158) and substitution of Eq. (22.157) into Eq. (22.156), the variance of the current is

$$\sigma_i = \frac{\sigma_{q_2}}{RC_2} = \bar{q}\sqrt{\frac{r}{2RC_2}}, \tag{22.159}$$

or the fractional standard deviation with respect to the average current (and also the counting rate) is

$$\frac{\sigma_i}{\bar{i}} = \frac{\sigma_r}{r} = \sqrt{\frac{1}{2rRC_2}}. \tag{22.160}$$

The average output signal is a function of the RC_2 time constant. To understand this outcome, consider the net rate of change on the capacitor C_2, where $r\bar{q}$ is the charge rate and q_2/RC is the discharge rate

$$\frac{dq_2(t)}{dt} = \bar{q}r - \frac{q_2(t)}{RC_2}. \tag{22.161}$$

Assuming at $t_0 = 0$ there is no charge on C_2, then

$$q_2(t) = r\bar{q}RC_2\left[1 - \exp\left(\frac{-t}{RC_2}\right)\right]. \tag{22.162}$$

Notice that as t becomes large ($t \gg RC_2$)

$$q_2(t) \simeq r\bar{q}RC_2, \tag{22.163}$$

which occurs at equilibrium. Hence, Eq. (22.156) and Eq. (22.163) are combined to find the average charge output with standard error,

$$q_2 \simeq \bar{q}rRC_2 \pm \bar{q}\sqrt{\frac{rRC_2}{2}} \tag{22.164}$$

or in terms of voltage

$$v_{out} = \frac{q_2}{C_2} \simeq \bar{q}rR \pm \bar{q}\sqrt{\frac{rR}{2C_2}}. \tag{22.165}$$

The output from a ratemeter can be altered by changing the RC_2 time constant. A long time constant RC_2 provides a relatively steady output reading. However, the device takes longer to reach the equilibrium condition and does not respond quickly to rapid changes in the count rate. Ratemeters accommodate different count rates by making R adjustable. Modern instruments often have automatic meter ranging. To handle large ranges of count rates, some ratemeters produce an output proportional to the log of the interaction rate [for example, see Cooke-Yarborough and Pulsford 1951; Fraser 1974a, 1974b].

22.3.6 Multichannel Analyzers

A multichannel analyzer (MCA) sorts shaped pulses from a linear amplifier according to their amplitudes. The instrument operates by changing the analog signals from the amplifier into digitized outputs. By digitizing and binning the pulse magnitudes, the resulting pulse height histogram produces a digitized image of the differential pulse height spectrum. If analog signals are supplied to the MCA, then the instrument must also incorporate an analog-to-digital converter (ADC). The analog-to-digital conversion architecture affects both the speed and linearity of the system. Some types of ADCs are discussed below.

The ADC outputs are sorted by magnitude and assigned to channels divided equally by 2^n over specific voltage range, usually limited by the NIM maximum standard of 10 volts. For instance, the 10 volts may be divided by 2^9 channels, yielding 512 consecutive channels each 19.53 mV wide. Most modern multichannel analyzers allow the user to change the number of channels 2^n from 256 up to 8192 by changing the value of n. Hence, the multichannel analyzer[17] operates, in some respect, as several SCAs sequentially scaled and operating concurrently.

The MCA evolved over the last 70 years by incorporating many improvements. Early models used vacuum tubes and, consequently, were quite large and were usually floor models. With solid-state electronics, these devices were reduced to table top models with cathode ray tube screens and a pin-fed dot-matrix printer readout. In the 1990s, the table top models were largely replaced by computer cards that would plug into a PC bus. Although the computer card MCAs are still available through resale outfits, they have been largely replaced by compact stand-alone units. Today, MCA instruments can be acquired as nuclear instrument modules that insert into a NIM bin, or as independent portable units. User interface software allows the operator to control the MCA and display the output on a personal computer or other electronic graphical user interface.

The Analog-to-Digital Conversion

An analog-to-digital converter (ADC) is an electronic system that converts a continuous time and amplitude signal into a quantized digital signal. For an MCA, the ADC converts an input analog voltage into a digital output representative of the magnitude of the input pulse. An ADC system is limited by its bandwidth, i.e., by the maximum number of conversions per unit time that can be achieved. This limitation is a consequence of the small time an ADC requires to process a conversion, a time in which additional signals cannot be processed. Factors that affect ADC performance include sampling rate, aliasing, signal-to-noise ratio, jitter, dither, linearity, and resolution, each of which is briefly described here.

The ADC resolution is determined by the number of discrete divisions spread over the analog voltage range (or voltage span). These divisions are determined by the number of bits n. For example, an ADC with n bit resolution over voltage span of V_m has channel resolution of,

$$\Delta E = \frac{V_m}{2^n}. \tag{22.166}$$

The actual resolution is limited by the signal-to-noise ratio, in which the noise consists of noise on the analog signal and quantization noise. Electronic noise appearing with the input signal is also incorporated into the digitized signal. Quantization noise appears as a consequence of the roundoff error between the analog signal and the digitized result.

Linearity is important in spectroscopy measurements for accurate energy identification. The non-linearity of an ADC is defined by either differential non-linearity or integral non-linearity. Differential non-linearity

[17]At one time, multichannel analyzers were called "kick-sorters", a term dating back to an electro-mechanical version of the device that would project small spheres by way of an electromechanical plunger. The magnitude of the kick was proportional to the pulse input, and the balls would land in a linear array of adjacent grooves to produce a histogram as the balls were stacked. The device of Frank et al. [1951] had 30 channels (grooves), each capable of holding 100 balls.

refers to the uniformity of channel widths ΔE defined by Eq. (22.166). The deviation in the width of MCA channels is usually less than $\pm 1\%$ differential non-linearity for modern commercial instruments. Using a linear fit for the channel assignment, the integral non-linearity refers to the maximum difference between the original analog signal pulse height and the assigned digital channel as measured over the entire signal range. For modern MCAs, integral non-linearity is usually on the order of or less than $\pm 0.25\%$ over the top 99% of the dynamic range.

Dither refers to the process of adding white noise to the analog signal before digitization. The reason for adding white noise is to improve the randomization of the quantization error, although such addition increases the system noise. This method slightly extends the range of the ADC while smoothing the quantization error over several noise values. Consequently, hard changes in digitized signals become smoother. Dither is widely used in digital processing of audio and image data.

In many cases, the analog signal is continuous. Consequently, a method is required to measure the signal and hold that specific value constant for a sufficient time to allow the ADC to produce a digitized output. The process of holding the sampled signal constant is performed with a sample-and-hold circuit, in which the sampled signal is used to charge a capacitor to store the charge. This capacitor is then temporarily disconnected from the input while the signal is being digitized. The *sampling rate* is the average speed at which an ADC can measure an analog signal (called sampling) to convert it into a digital output. This speed is usually reported as "samples per second."

To measure a periodic function, it is important that the sampling rate should be high enough to reproduce an accurate representation of the original signal. Typically the sampling rate should be at least greater than twice the analog signal bandwidth. A sampling rate that is too small can result in a problem called *aliasing* in which the reconstructed input is no longer representative of the original periodic input frequency. Usually a lower frequency digitized output is produced. The time between radiation pulses is stochastic, and the task of the ADC is to measure and digitize the peak voltage of the shaped input signal. Related to sampling time is the conversion time, defined as the time that it takes the ADC to sample an analog signal and process a digital output. The ADC is essentially inactive to additional incoming pulses during the conversion process. Hence, it is possible that these shaped pulses arrive at the ADC input faster than the ADC can process them, which ultimately contributes to lost counts.

Jitter refers to the temporal uncertainty in significant events in the ADC process and is observed as a deviation from the ideal timing interval between sampled signals. When the sample-and-hold circuit is opened, the capacitor is connected to the input circuit and stores the signal. The clock closes the circuit after some time Δt, while the ADC converts the stored voltage into a digital output. Jitter is the variation in the time at which the switch is opened. Jitter contributes to uncertainty in the sampled voltage, and is proportional to the amplitude and the rising edge of the input signal. Jitter problems increase with frequency and pulse amplitude.

Stacked Discriminator

An ADC can be realized by combining a series of simultaneously operated SCAs. In this architecture, multiple SCAs are arranged in parallel, each simultaneously accepting the shaped input pulse. Each separate SCA defines a range ΔE, sequentially stacked over the entire acceptable input voltage range. This stack of SCAs feeds a logic circuit that generates a code for each input channel. Ultimately, the stacked discriminator is a fast method to digitize pulses, mainly because the output is routed to the correct accumulator in a single step. Consequently, these ADCs are also called flash ADCs for their high speed capability. However, this architecture is too expensive for modern MCAs with large numbers of channels. For instance, the number of ADCs is equal to the digital resolution, which for an MCA with n-bit resolution requires 2^n comparators. Flash ADCs are usually restricted to 8 bit or to MCAs with only 256 channels. Also, the differential non-linearity is sometimes a problem because the independent threshold values of voltage for each channel have their own associated error and, thus, add variance to the channel widths. Other problems include higher input capacitance than other architectures, higher power consumption, and generally lower energy resolution because of the restricted number of channels.

Wilkinson ADC

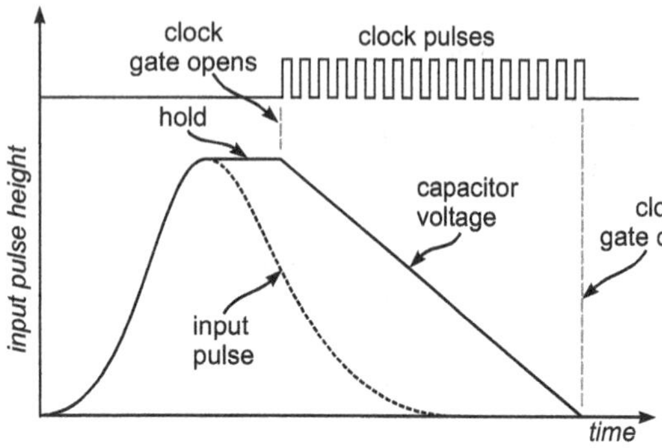

Figure 22.31. Principle of the Wilkinson ADC operation. After Nicholson [1974].

Introduced in 1950 [Wilkinson 1950], the Wilkinson ADC is still one of the most popular architectures used in MCAs. An input voltage into the Wilkinson ADC charges a capacitor while a voltage comparator is used to monitor the charge. When the input capacitor voltage is equivalent to the magnitude of the input voltage, the capacitor is discharged at constant current through a linear network. Hence, the capacitor voltage decreases in time with a constant slope as shown in Fig. 22.31. By measuring the time it takes to fully discharge the capacitor, a relative measure of the original voltage is found. This discharge time is determined by introducing a gate pulse when the capacitor begins to discharge. The gate pulse opens a linear gate that passes pulses produced by a precision oscillator to act as a clocking timer. The gate closes when the capacitor is fully discharged. While the gate is open, a discrete number of pulses is allowed to pass; hence the number of clock cycles that pass through the gate is a digitized indication of the initial input pulse. These clock cycles are counted on a digital address register. To prevent ambiguities in channel assignment, steps are taken to prevent the discharge process from starting in the middle of a clock cycle [Nicholson 1974]. Because the ADC method relies on a time-correlation, it becomes clear that large input pulses take longer to process than small input pulses. Further, the processing time, which depends on the clock frequency, becomes smaller as the clock frequency is increased. Commercial MCAs can be acquired that operate linearly with clock frequencies on the order of 100 MHz.

Successive Approximation ADC

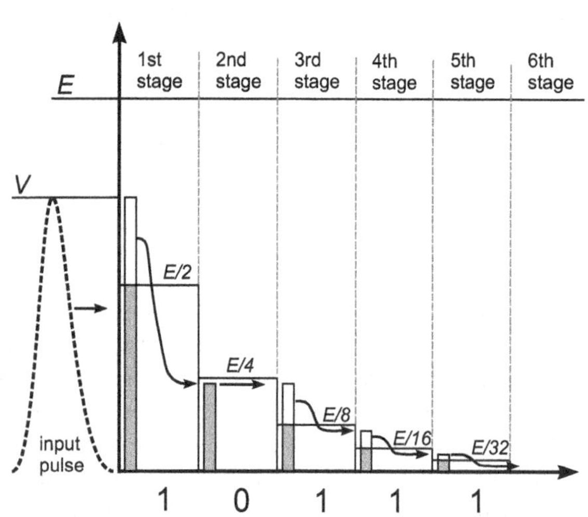

Figure 22.32. Principle of the successive approximation ADC operation. After Nicholson [1974].

A successive approximation ADC uses a comparator and a successive binary approximation method (see Fig. 22.32). An input shaped voltage pulse is initially compared to a range determined by $E/2$, where E is the maximum range of interest. For example, E might correspond to the NIM standard of 10 volts. If the input voltage V is larger than $E/2$, then a '1' is generated in the output, and the difference $V - E/2 = V_1$ is passed to the next stage. If, instead, V is less than $E/2$, then a '0' is assigned and $V = V_1$ is passed to the second stage unaltered. At the second stage, the process is repeated, except the comparison is now performed between V_1 and $E/4$. Hence, if V_1 is greater than $E/4$, then another '1' is assigned and $V_1 - E/4$ is passed. If V_1 is instead less than $E/4$, a '0' is assigned and the unaltered signal is passed. This process continues through a series of binary comparisons until the channel limit is reached. Hence, with 2^n channels for the MCA there are n comparison stages. Each result is stored in a successive approximation register. The final output is the binary number representative of the digital conversion of the analog signal. The main advantage of the successive approximation register is that each input pulse takes essentially the same amount of time to process, which is usually less than that required for large pulses in a Wilkinson ADC in an MCA having a large number of channels. Hence, the successive approximation ADC is potentially faster than the Wilkinson ADC.

Ramp-Compare ADC

The ramp-compare or digital ramp ADC introduces a saw-tooth ramp input signal that is incremented by a counter. The ramp is produced by initiating a digital counter that has its output tied to the input of a digital-to-analog converter (DAC). This DAC ramp output is compared sequentially at each step to the amplitude of the shaped input voltage. The first comparison is performed with the ramp voltage at zero. If the input voltage is higher than the DAC signal, then the comparator produces a '1' which signals the counter to increment by one. The DAC output is therefore also incremented by one voltage division and the comparison is repeated. When the DAC output signal matches the shaped input signal, within a set division tolerance, the counter transfers the binary number to a shift register, and the comparator indicates a match and resets the counter to '0'. The shift register transfers the binary count to the ADC circuit output. Unfortunately, these ADCs are relatively slow and sensitive to the thermal changes. Further, their output speed is a function of the initial input pulse height because it takes longer to process the signal as the pulse height increases.

Sliding Scale Smoothing

A problem arises from the variance in channel widths of an MCA, where some channels are given more weight than others. This variation in channel width ultimately skews the digitized pulses that accumulate in each channel. For instance, pulses of a particular range falls between two levels that mark a channel width. If that channel width is wider or narrower than the average, then the number of pulses assigned to that channel is greater or smaller than the average. The variance and linearity of an ADC can be improved by a technique introduced by Cottini et al. [1963], known as *sliding scale smoothing*. The method alters the shaped input pulse before digitization, returning to its former value afterwards by the following sequence. The analog input pulse V has an analog value E_i added to it. After running through the ADC, the digital value of E_i is subtracted, ultimately returning to the digital value of V. However, the value of E_i is randomized over various i, meaning that the assigned values after digitization also vary. With subtraction of E_i, the process produces an average value spread over a few channels and, thereby it ameliorates the influence of channel width variance.

22.3.7 Pulse Generators

An important electronic component that assists with calibration of the radiation counting system is the pulse generator, which is commonly referred to as a *pulser*. Pulsers provide accurate voltage test pulses into the radiation detection system. The pulses can be set to have different shapes (a tail pulse being the most common), decay times, pulse frequencies, and amplitudes in order to calibrate an amplifier, MCA, SCA, or counter/timer. The test pulses can also be used to help determine system dead times and system electronic noise. Although not necessary to operate a radiation counting system, they are quite valuable when first configuring the system and calibrating it for radiation measurements. Further, a pulser is usually operated during a spectroscopic measurement in order to have a comparison peak with which the system noise can be determined. Typically the pulser peak is set to an MCA channel that does not interfere with the accumulated radiation spectrum. Some vendors offer digital pulsers with multiple pulser outputs.

22.3.8 Power Supplies

Detectors that operate with the NIM system typically produce an electronic output and, therefore, require a power supply of their own. Traditionally, the majority of such NIM power supplies for detectors apply a high voltage to the detectors. Stability of the supply voltage is of particular importance in high-resolution spectroscopic systems. High-voltage power supplies are used to operate detectors that do not suffer changes in capacitance as a function of voltage, such as gas-filled and scintillation detectors. High-voltage bias power supplies are usually used for semiconductor detectors, in which the operating voltage can affect the detector capacitance.

High-Voltage Power Supply

The high-voltage power supply, as the name implies, supplies high voltage to the detection device, and can supply it with either positive or negative polarity. The voltage is applied to the detector across a load resistor with the connection routed through the preamplifier unit. Typically, a power supply has a meter indicating the applied voltage and it is reasonably accurate because the detector resistance is usually much greater than the preamplifier load resistance. HV power supplies are available with the maximum voltages exceeding 3.5 kV. Such units are commonly used with PMTs and gas-filled detectors, neither having capacitance changes as the applied voltage is varied. As discussed in Sec. 14.1.6, a small change in output voltage can produce a large change in a PMT output. Consequently, power supplies used for scintillation spectrometers require high-voltage regulation to prevent voltage drift. Commercial units quote stability specifications as usually less than 0.03% variation over a 24-hour period. Also, because PMTs require high voltage to operate and often with current in the mA range, a HV power supply is usually connected directly into an AC receptacle. For convenience, many models have rails that allow the unit to slip into a NIM bin compartment.

High-Voltage Bias Supply

For semiconductor detectors, the power supply is generally referred to as a "bias supply." As with a HV power supply, the bias supply provides a choice of positive or negative high voltage to the detector. However, this type of power supply usually does not have a meter, but rather only a dial indicator indicating the total applied voltage to the system. Because the system includes load resistances in series with the detector, there is a voltage drop across this resistance, which reduces that applied to the detector. Thus, a meter indicating total voltage would be misleading. Instead jacks are often provided that allow determination of the bias current. Because the series' load resistance is known, the voltage applied to the detector, which is less than the total, can also be determined by subtracting iR from the overall applied voltage. Bias supplies are available for which the maximum allowable voltages can exceed 5 kV. Because typical detectors are highly resistive and draw little current, the power requirement of the bias supply is low and, consequently, most bias supplies are connected directly into the NIM bus for power, unlike HV power supplies. Stability in commercial units is commonly quoted as less than 0.1% variation per hour.

NIM Power Supply

A power supply is attached behind a NIM bin to supply power to the modules. This unit provides ± 6 volts DC, ± 12 volts DC, and ± 24 volts DC to the NIM sockets and voltage jacks,[18] as well as 117 volts AC. Power is commonly rated between 150 and 160 W for the larger power supplies, although lower power models rated at about 95 W are available (usually for models that do not offer the ± 6 volt DC option). The NIM bin sockets and pin assignments conform to the specifications outlined in DOE/ER-0457T. Test jacks in the front of the NIM bin provide convenient access to monitor the DC voltage, and can also be used to conveniently supply power to custom electronics. Modern power supplies are filtered from electromagnetic interference (EMI), are short-circuit proofed, thermally protected, and provide stable power required for

[18]Some models do not provide ± 6 VDC.

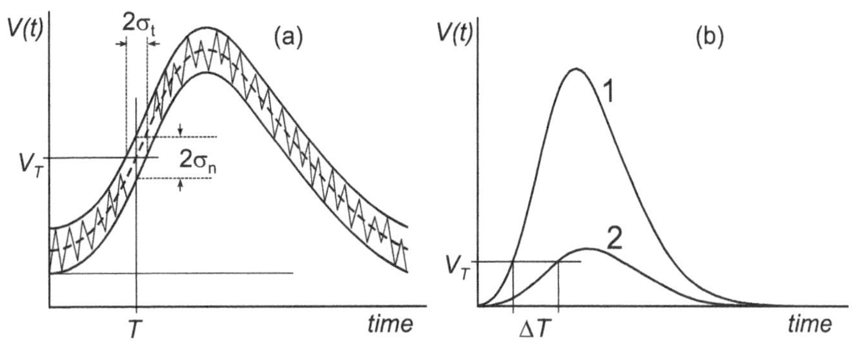

Figure 22.33. The effects of (a) time jitter and (b) time walk on the timing uncertainty, where V_T is the timing trigger threshold.

NIM components. Usually NIM bins have warning indicators that inform the operator if the power supply is approaching the temperature limit.

22.4 Timing

Sometimes it is important to know the time a radiation pulse occurs. Examples where such timing information is needed include coincidence measurements, anticoincidence measurements, and time-of-flight measurements. Timing systems use comparator circuits to determine the time at which a pulse amplitude reaches a preset threshold and then produce a logic pulse at the moment the pulse crosses the threshold. This logic pulse can then be used as a synchronization signal for timing measurements.

22.4.1 Jitter and Time Walk

Jitter refers to noise variations on the input signal as shown in Fig. 22.33(a). Consider a threshold voltage denoted V_T used to trigger a time stamp on an incoming pulse. The average value of an input pulse $v(t)$ is depicted as the dotted line in Fig. 22.33(a). However, electronic noise causes a variance in the time dependent input signal at V_T. This variance is quantified by the standard deviation σ_n of the noise. This variance in noise causes a variance in the actual time that the signal passes the threshold V_T and is quantified by the standard deviation σ_t. Because of these variances, there is uncertainly in the time stamp due to the jitter. Note that σ_n is simply the voltage amplitude of the noise. The contribution of the noise to the timing jitter is represented by [Spieler 2005]

$$\sigma_t = \sigma_n \bigg/ \left. \frac{dV}{dt} \right|_{V_T} . \tag{22.167}$$

If the noise cannot be reduced, the jitter can still be minimized by setting the threshold V_T at the point of the maximum slope of the input pulse.

For some timing methods, the amplitude of the input pulse causes changes in the trigger time. Consider the leading edge of two analog pulses of different amplitude as depicted in Fig. 22.33(b). Both pulses start at the same time and are triggered at the same threshold V_T. For the larger amplitude pulse (1), the pulse crosses V_T before the smaller pulse (2). The difference ΔT in the trigger time is called *time walk.* The trigger timing of pulses shifts with pulse height and, thus, contributes to the uncertainty of the trigger time. Time walk can also appear as a consequence of different rise times, even if the pulse amplitudes are identical. For instance, the pulse output is a function of the location of radiation interaction in a HPGe detector (see Sec. 16.4.2), and these different rise times cause the trigger time to change. Time walk exacerbates the effects of jitter on the timing uncertainty. Some systems may partially compensate for time walk by making appropriate corrections from the slope of the pulse. However, such methods reportedly fail if the input pulses suffer time walk from both amplitude and rise time variances [Spieler 2005]. Otherwise, time walk can be reduced by setting V_T at the lowest possible setting without being triggered by noise. Unfortunately, the optimum settings for time walk and jitter are usually not at the same V_T.

22.4.2 Common Timing Methods

There are many methods for recording the arrival time of pulses, and choosing the appropriate method should include consideration of the required timing resolution and type of detector in use. Briefly outlined in the following sections are a few popular methods of recording pulse arrival times.

Leading-Edge Timing

The leading-edge triggering method is relatively straightforward, and used when timing information is needed with somewhat relaxed timing resolution. The operator can set a threshold V_T and a comparator circuit identifies when the time dependent input pulse matches this preset threshold, at which point a logic pulse is generated. Some units provide a preset logic pulse length, while other units may terminate the logic pulse when the falling slope of the pulse once again matches V_T. However, leading edge timing is susceptible to resolution degradation from time walk and jitter. Nicholson [1974] reviews a few select designs for leading edge timing trigger circuits.

Zero-Crossover Timing

If the input signal is converted into a bipolar pulse, then the zero point crossing of pulses with varied amplitudes is time-invariant as shown in Fig. 22.34. Consider the result of a *CR-RC* shaping network from Eq. (22.107). Adding a differentiator (as in Fig. 22.13) produces a bipolar output

$$V_{bp} = \frac{d}{dt}V_{out} = V_{in}\frac{d}{dt}\frac{t}{\tau}e^{-t/\tau} = V_{in}\left[\frac{\tau - t}{\tau^2}\right]e^{-t/\tau}. \tag{22.168}$$

From this result it is seen that the output becomes zero when $t = \tau$. Note that the zero crossing is only a function of the shaping time and not the amplitude. Zero-crossover timing is best used for detectors that produce a wide range of pulse amplitudes.

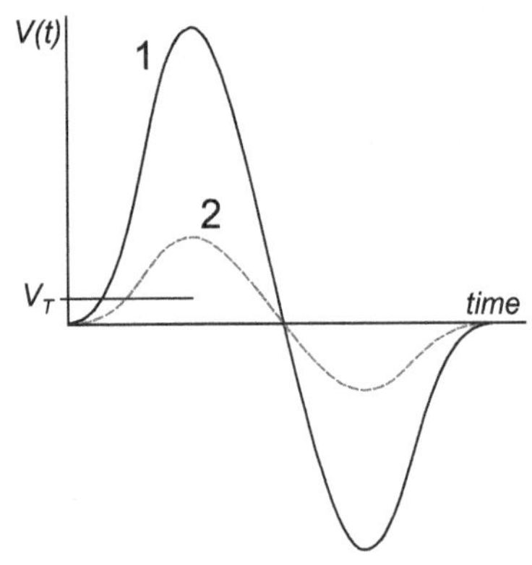

Figure 22.34. The zero-crossover timing method.

Such a system can be set to trigger when the zero-crossover condition is met, ideally eliminating problems with time walk. Noise can also trigger the timing circuit when noise pulses cross the zero point. To prevent confusion between noise and radiation pulses, a leading edge threshold comparator is often included to discriminate against noise. The original pulse is split and passed through both the threshold comparator and the zero crossover network. If the original pulse triggers the threshold comparator, indicating that the pulse is real and not noise, a signal is sent to an AND gate[19] that temporarily allows signals to pass. If both signals arrive in coincidence at the output gate, then a logic pulse is generated. Otherwise, the zero crossover pulse is blocked. The circuits are generally engineered such that the discriminator comparator signal is stretched and arrives at the gate before the zero crossover signal, thereby synchronizing the time signal with the arrival of the zero crossover time signal.

There are a few problems associated with the zero crossover method. First, with any timing circuit, there is a small delay from the time that the trigger condition is met and

[19]An AND gate is a digital logic gate that produces a "high" output (or 1) only when all inputs are also "high". Otherwise, the output is "low" (0).

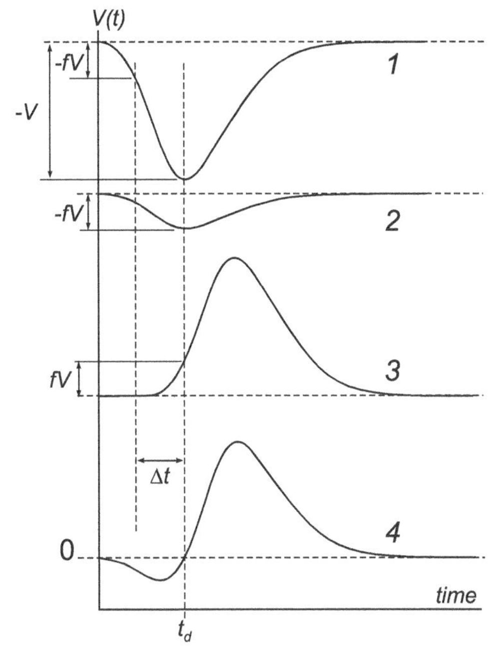

Figure 22.35. Constant fraction timing method. The inverted negative input (1) is reduced by fraction f (2). The input pulse is inverted and delayed by Δt such that a fraction of the input pulse fV equals the attenuated pulse maximum (3). Adding (2) and (3) together causes (4) the summed pulse to cross zero at the delay time t_d.

when the comparator initiates the trigger. Hence, there is a small amount of time walk. Second, the zero crossover point can shift with pulses of different rise times and shapes, such as those produced by HPGe detectors. Finally, jitter can be worse for zero crossover timing than leading-edge timing [Nicholson 1974]. For detector measurements where the pulse amplitudes are similar, the time walk may be less of an issue than jitter, and leading-edge timing may be a better choice.

Constant Fraction Timing

Another method developed to reduce the effect of pulse height time walk is constant fraction timing, which also uses a zero crossover method [Gedcke and McDonald 1967; 1968]. The input pulse V is split and one branch is attenuated by a fraction f of the pulse amplitude, while the other branch is inverted and delayed. The fraction f can be chosen to minimize jitter.

The delay time Δt is chosen to shift the rising edge of the inverted input pulse, the location at fV, to align with the maximum of the attenuated pulse, which is also fV (see Fig. 22.35). The delayed inverted pulse is added to the attenuated pulse, which produces a bipolar pulse that crosses zero at the time denoted t_d. A logic pulse is generated when the bipolar pulse matches the zero condition. The end result is a timing method that no longer suffers from amplitude time walk. This method is also less affected by time walk due to variations in rise time than is the leading edge method [Ortec 2009c]. Further, for relatively low values of f (less than 0.4), the timing resolution remains relatively constant.

ARC Timing

The amplitude and rise time compensated (ARC) timing method was developed for detectors that produce pulses that vary in rise time and pulse amplitude, which is particularly true for HPGe detectors [Chase 1968;

Cho and Chase 1972]. From the explanation for time-dependent pulse height formation in a planar HPGe detector given in Sec. 16.4.2 and shown in Fig. 16.31, the pulse rise takes on various shapes as also shown in Fig. 22.36.

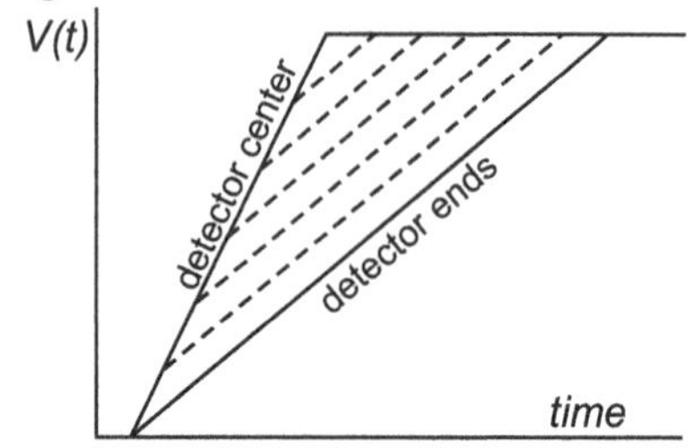

Figure 22.36. Position-dependent pulse shapes for a planar semiconductor detector. If the charge speeds are equal, the shortest rise time is for events occurring at the center of the detector because both electrons and holes are collected at the same time. The longest rise times occur when events take place at the ends of the detector.

The ARC timing method has some similarities to the constant fraction timing methods. The input pulse V is split, where one branch is inverted and attenuated by a fraction f of the pulse amplitude, and the other branch is delayed by Δt and then added to the attenuated pulse. The summed signals produce a bipolar pulse with a zero crossover that triggers a logic pulse. The main difference from constant fraction timing is that the delay Δt is chosen to be small. From Fig. 22.36, pulses from a planar semiconductor detector are initially linear. Once one type of charge carrier is collected (holes or electrons), the pulse shape changes to have a slower rise. ARC timing has time walk if the slope changes before the zero time crossing. By choosing a small Δt to delay the pulse, the zero crossover occurs before the pulse becomes non-linear, and consequently is independent of the overall rise time and pulse amplitude (see Fig. 22.37). As with the bipolar and constant fraction timing methods, a threshold comparator is used to remove zero crossings caused by noise pulses.

The ARC timing method has been shown to work well with coaxial HPGe detectors [Cho and Chase 1972]. However, the slope of pulse shapes from coaxial detectors is non-linear, changing between a mix of convex and concave shapes depending on interaction location (see Fig. 16.32). Consequently, ARC timing does not completely eliminate time walk with coaxial detectors, although it does show better performance than most other timing methods.

Some coaxial HPGe detector pulses can have relatively long rise times, generally caused by the excitation of electron-hole pairs in a region with low electric field, thereby initially producing low carrier speeds. If the discriminating comparator fires relatively late because of a slow rising pulse, then the gate is closed when the zero crossover signal arrives. Instead, the gate is opened late, corresponding to the arrival of the leading edge trigger instead of the zero crossover time. Consequently, these pulses can cause time walk that severely distorts the time spectrum. To counter the effect, slow rise time (SRT) rejection can be used to block the timing output if the zero crossover signal arrives before that of the leading edge discriminator signal [Bedwell and Paulus 1976].

22.5 Coincidence and Anti-Coincidence

There are many applications that require the measurement of coinciding events that occur in two or more separate detectors within a specified time interval. Examples include identification of the simultaneous arrival of photon-photon or photon-particle events, applications involving positron decay, and determination of radioactive decay schemes. Such measurements are commonly referred to as *coincidence measurements*, in which two or more events are considered coincident if they arrive at the coincidence discriminating unit within a preset time window of width τ. Accuracy of these measurements depends on the time correlation of the events that depends on a time window which must be set large enough to identify coincident events while being narrow enough to exclude uncorrelated events.

For example, positron emission tomography (PET) seeks to correlate annihilation photons from a single positron decay event and usually is based on the photons entering separate pairs of detectors arranged 180° about the source. Coincidence measurements can also be used to characterize light yield in scintillators, as explored by Rooney and Valentine [1996]. Coincidence counting is also used to distinguish between radiation

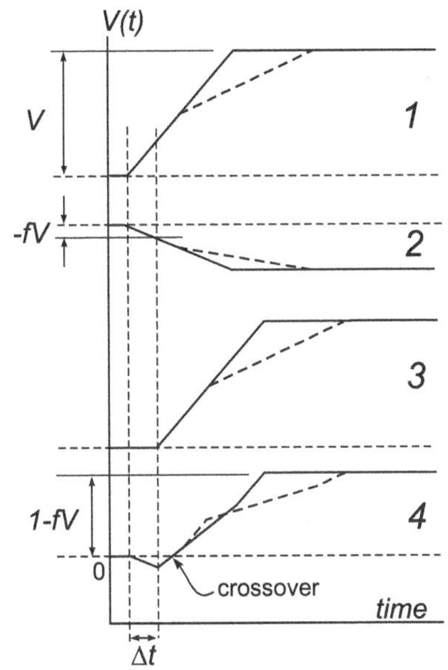

Figure 22.37. The ARC timing method. The input pulses (1) may have different rise times, depending on the interaction location. An input pulse is split into two signals, where one input pulse is inverted and reduced by fraction f (2). The other input pulse (3) is delayed by Δt, starting at only a small fraction of the rising pulse. Adding (2) and (3) together causes the summed pulse (4) to cross zero at the same delay time, regardless of rise time or pulse height.

types, for example the selective imaging of beta particles that are coincident with gamma-ray emissions [NCRP 1985; Dewaraja et al. 1994].

By contrast, there are some measurements that require the opposite condition, i.e., when two radiation events are recorded within time interval τ are rejected. Such measurements are referred to as *anticoincidence* measurements. For example, Compton suppression spectroscopy is a measurement method that uses the technique of anti-coincidence [Lindblad 1978; Bender et al. 2015] to reduce the Compton plateau in a pulse-height spectrum.

22.5.1 Coincidence Analyzers

The main function of a coincidence analyzer is to discern the arrival of two or more radiation particles in separate detectors within some preset time interval τ. In practice, it is not possible to analyze coincidence events with 100% confidence due to the uncertainties associated with the statistical nature of radiation emissions. Statistical timing errors may occur from the detection process and uncertainties in the electronics resulting from timing jitter, amplitude walk and noise, all of which lead to statistically variable time delays between processed events.

A coincidence analyzer accepts the time stamped inputs from two or more detectors, each connected to a separate port. Input signals are converted into stretched logic pulses, where the time duration is set to τ. A simple coincidence circuit solves this problem by essentially summing the two input pulses, passing the resultant sum pulse through a discriminator level, and generating an output pulse when the summed input pulses exceed the discriminator setting.

The period of time in which the two input pulses can be accepted is defined as the coincidence resolving time τ. Commercial coincidence modules allow the user to set τ within limits, usually between tens of ns up to a few μs. A time-stamped pulse enters one port of the unit. If during the set resolving time another pulse enters another port of the unit, it is considered coincident. It does not matter which signal arrives first. If both signals overlap, then the coincidence analyzer produces an output logic pulse. Commercial units also often have an anti-coincidence feature in which, by either flipping a switch or connecting an input to an anti-coincidence port, logic pulses are generated at the output only if pulses are not coincident.

Suppose an input pulse arrives at a random time t_1. It is possible that a pulse already arrived at port 2 between time $t_1 - \tau$, causing a coincidence. It is also possible that a pulse arrives at port 2 between time $t_1 + \tau$, also capable of producing a coincidence signal. Consequently, the actual window for possible coincidence is the sum of these two intervals, namely 2τ. Because detector events occur at random times, accidental coincidences can occur between two pulses which produce background in the coincidence counting. Suppose that two counting rates are recorded by two separate detectors. The experimenter would observe, on average, g_1 cps from detector 1, about which for any single count a coincidence may arrive within $\pm\tau$. The average number of counts recorded from detector 2 within the time window is $g_2(2\tau)$ so that the average rate of accidental or random coincidences is

$$g_a = g_1 g_2 (2\tau). \tag{22.169}$$

For N-fold detectors, the accidental coincidence rate is determined as [Jánossy 1950],

$$g_a = N g_1 g_2 \ldots g_N \tau^{N-1} + \text{higher order terms}. \tag{22.170}$$

For the case where $n_i \ll \tau$, Jánossy [1950] shows that the higher order terms are negligible, yielding [Fenyves and Haiman 1969],

$$g_a \simeq N g_1 g_2 \ldots g_N \tau^{N-1}. \tag{22.171}$$

Consider a two-fold coincidence system where detectors 1 and 2 have counting rates g_1 and g_2, respectively. If the source activity S is known and the source is relatively small, the count rates recorded by each detector are

$$g_i = S \epsilon_i \Omega_{fi} B_i \qquad i = 1, 2, \tag{22.172}$$

where for the ith detector B_i represents correction factors (such as the source branching ratio), Ω_{fi} is the fractional solid angle subtended by the detector from the source, and ϵ_i is the detector counting efficiency for the radiation of interest and source location. The rate of true coincidences depends upon the source activity S and the probability that concurrent emissions arrive at the detectors with respect to the source-to-detector solid angles,

$$g_t = S(\epsilon_1 \Omega_{f1} B_1)(\epsilon_2 \Omega_{f2} B_2). \tag{22.173}$$

The accidental coincidence rate is the detection rate of uncorrelated emissions. From Eq. (22.171), the accidental coincidence rate is

$$g_a = 2\tau S^2 (\epsilon_1 \Omega_{f1} B_1)(\epsilon_2 \Omega_{f2} B_2). \tag{22.174}$$

The ratio of Eq. (22.174) to Eq. (22.173) yields

$$\frac{g_a}{g_t} = 2\tau S, \tag{22.175}$$

which shows that the ratio of accidental to true coincidences increases linearly with source activity and resolving time. Note also that the source activity can be determined by dividing Eq. (22.175) by the value of 2τ.

The number of counts recorded by the detectors depends upon the experimental geometry and the detectors used. A good method to reduce accidental coincidences is to make the resolving time as small as possible. However, the resolving time cannot be better than the system time resolution, which is affected by jitter and time walk. Consequently, the timing methods and detector types together determine the limitations on the resolving time. The limited resolution is depicted in Fig. 22.38(top), where the coincidence count rate is shown as a function of count rate (a differential time spectrum). Given identical equipment with identical delay times between the detectors and the coincidence analyzer inputs, both pulses should arrive at the coincidence analyzer at identical times. In other words, if one pulse arrives at t_1, then the other pulse should also arrive at t_1, although the arrival times may instead be slightly different because of jitter, time walk, and perhaps other effects. Without this deviation, the coincidence peak would be represented by a single channel; however, in reality it is statistically spread out over several channels.

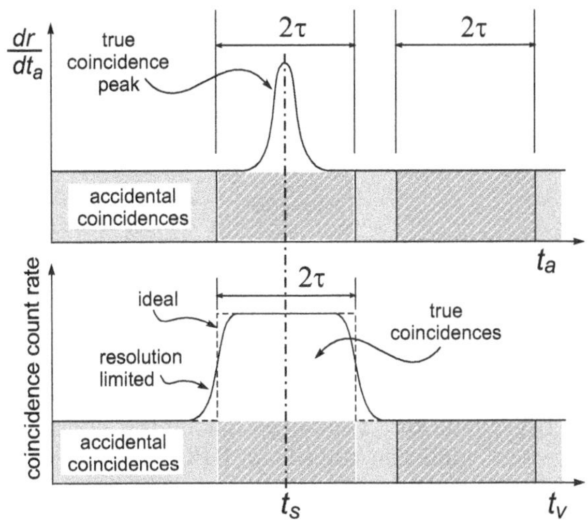

Figure 22.38. (top) The expected differential time spectrum for coincidences, where r is the coincidence count rate, t_s is the set arrival time at one input, and t_a is the variable arrival time at the other input. (bottom) The coincidence count rate observed from a coincidence analyzer with resolving time τ and variable delay input t_v.

Typically, the detectors and electronics for each detector branch are different. Suppose that a resolving time τ is selected that is wider than the actual coincidence peak, as depicted in Fig. 22.38. If the arrival time of one coincidence pulse arrives at time t_s, then a coincidence from the other pulse is logged provided that it arrives within $t_s \pm \tau$. A delay line on each branch of the circuit, depicted in Fig. 22.40, allows the user to set one delay time t_s while also varying the other delay time, denoted by t_v. The operator can then take coincidence measurements at different values of t_v, thereby producing an output response as shown in Fig. 22.38(bottom). If the delay is too small, such that $t_v < t_s - \tau$, or if it is too large, such that $t_v > t_s + \tau$, then any coincidence measured is accidental. If the delayed pulse arrives such that $t_s - \tau < t_v < t_s + \tau$, then real coincidences are tallied along with accidental coincidences. Hence, by setting the variable time delay t_v in the center of the delay curve spectrum, the operator measures the sum of true and accidental coincidences. The true coincidence rate is determined by subtracting the accidental count rate of Eq. (22.169).

The coincidence count rate increases as τ is increased, mainly because the accidental coincidence rate increases. The coincidence analyzer output can be optimized by measuring the delay curve response as a function of both τ and t_v, as depicted in Fig. 22.39 [Nicholson 1974]. If τ is longer than needed, the delay curve has a flat top response, shown as (1) and (2) in Fig. 22.39. If τ is too narrow, true coincidences are rejected and the count rate maximum incorrectly reports the coincidence rate, depicted as (4) in Fig. 22.39. However, at the optimized τ setting, there is no flat top to the delay curve ((3) in Fig. 22.39), and the peak location correctly indicates the sum of the true and accidental coincidence rates. Notably, the delay t_v must equal t_s for the optimized condition. The advantage of narrowing τ to the optimum value is the reduction of background accidental coincidences.

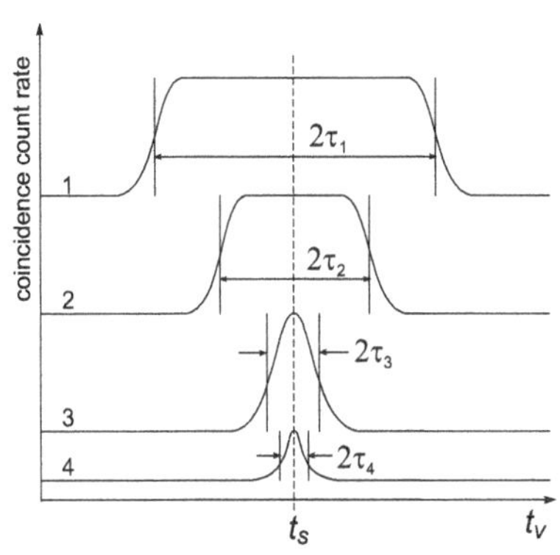

Figure 22.39. Depiction of delay curves for various values of the coincidence resolving time τ, showing the sum of real and accidental coincidences.

Consider the timing analysis system depicted in Fig. 22.40. Two detectors are positioned to measure coincident events from a radiation source. These coincidences may include, for example, positron annihi-

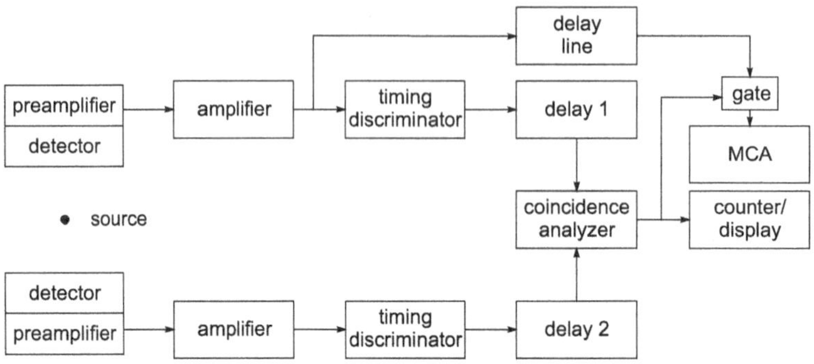

Figure 22.40. A method of using a coincidence analyzer for measuring coincident events.

lation photons, dual gamma-ray emissions such as those from ^{60}Co, and beta particle/gamma-ray coincident events. The selected detectors should detect the radiation emissions of interest with good efficiency and/or energy resolution. They need not be the same type of detector. For instance one detector may be a NaI:Tl detector. while the other is an HPGe detector. The HPGe detector can be used for its excellent energy resolution while the highly efficient scintillator provides the coincidence signal. The signals are passed through preamplifiers and amplifiers compatible with the timing requirements. The time stamps for pulses from either detector are produced by a timing discriminator that uses one of the timing measurement methods previously discussed. The pulse processing chain from the separate detectors may cause coincident events to emerge from the timing discriminators at different times. If so, this problem is remedied by placing a variable delay instrument after the timing discriminator. Although Fig. 22.40 depicts a delay after each timing discriminator, often one delay line or delay amplifier placed on the faster processing chain is adequate to synchronize the pulses as they converge on the coincidence analyzer. If counting coincident events is all that is required, the signals from the coincident analyzer can be routed to a counting system. If the goal is to produce a pulse height spectrum from the high resolution detector only of the coincident events, then the signal from the amplifier can be split and routed to the coincidence analyzer and also into another delay module. If a coincidence is triggered, the logic pulse opens a gate that allows the high-resolution detector pulse to be logged by a multichannel analyzer. The delay on the high resolution pulse is set to arrive within the time that gate is open.

22.5.2 Time-to-Amplitude Converter

The time-to-amplitude converter (TAC) is a measurement module that converts the difference in arrival times between events into a histogram, much like a pulse height spectrum. However, the abscissa is in units of time rather than energy. The timing resolution is defined as the full width at half maximum (FWHM) of the pulse height spectrum. The device operates by accepting a start signal and a small time later accepting a stop signal. Unlike the coincidence analyzer, the order that pulses arrive at the input does matter. A delay is at times employed to ensure that the stop signal arrives at some time Δt behind the start signal. These signals can be provided by separate detectors whose pulses have passed through a timing discriminator, perhaps timed with one of the methods previously described.

The circuit is arranged with a constant current source supply that is routed through start and stop switches. These switches are initially in the closed position, which normally routes the current to ground. The start switch is also in parallel with a converter capacitor that is tied to an amplifier input. When the leading edge of the start signal arrives, the start switch is opened which then allows the constant current source to charge the converter capacitor. When the leading edge of the stop signal arrives, the stop switch is opened to disconnect the constant current source from the circuit. The charging current produces the voltage

$$V = \frac{i\Delta t}{C}, \tag{22.176}$$

where i is the constant current, Δt is the time spacing between the start switch opening and the stop switch opening, and C is the converter capacitance. Hence, the stored voltage is a linear measurement of Δt. Shortly after the stop switch opens, the voltage in the form of a square wave is passed through a linear gate to an amplifier and to the TAC output. The amplitude of the output is indicative of time Δt, which can be supplied to a tallying unit such as a multichannel analyzer. Shortly thereafter, the switches are closed and the converter capacitor is discharged to ground, thereby resetting the circuit for the next timing measurement.

The use of a TAC can greatly simplify a coincidence measurement system. The TAC timing method eliminates the need to produce a delay curve, and there is no need to set a resolving time. The logic signal amplitudes produce a pulse height spectrum in which the pulse height is a function of Δt. Hence, the spectrum directly yields the time resolution of the system. The channels have units of time so the channel resolution (time per channel) can be changed by varying the conversion gain of the MCA. A TAC coincidence spectrum is depicted in Fig. 22.41. The time resolution is measured by the FWHM of the true coincidence peak. Shown is a window about the coincidence spectrum where the total counts in the window are the sum of true and accidental coincidences. The net true counts can be determined by setting a window of equal width (channels) away from the peak, and subtracting the counts in the second window from the counts in the first window.

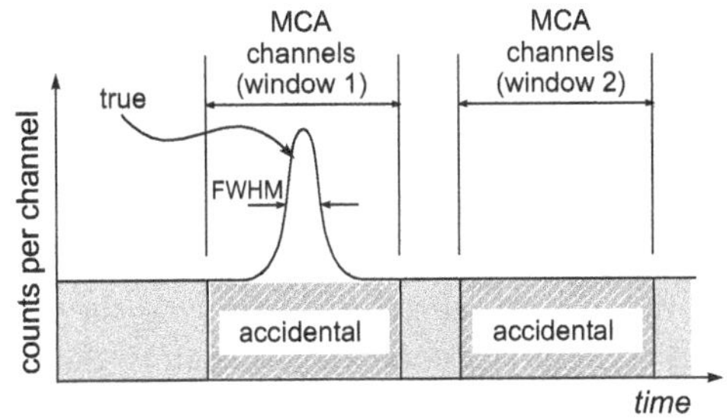

Figure 22.41. Depiction of an MCA spectrum provided by a TAC coincidence system.

One possible arrangement using a TAC is depicted in Fig. 22.42. The system is fundamentally similar to the system of Fig. 22.40, but a single delay line is used to ensure that the stop signal arrives after the start signal. The output of the TAC is a function of the time difference, thereby delivering a differential time spectrum to the MCA. Because the logic pulse amplitudes from the TAC are a function of the Δt, the difference between the start and stop times, the MCA produces a histogram of pulse height frequency as a function of Δt.

Without noise or jitter, the TAC output would deliver a Δt output that is exactly the same for each true coincident event. Hence, the differential time spectrum would fall into a single channel. Yet, this ideal behavior does not occur. Because of timing variations due to jitter and time walk, there is an inherent

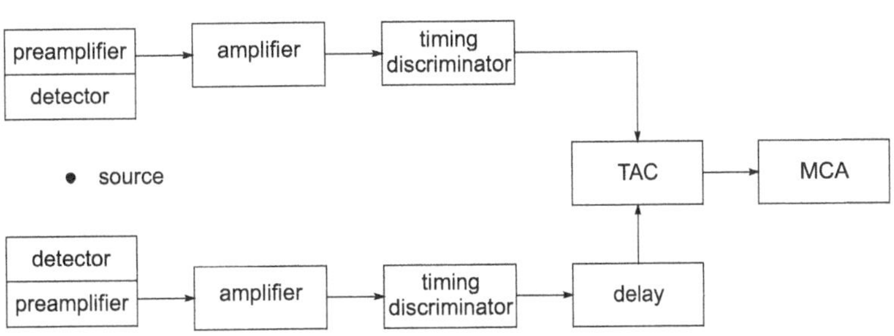

Figure 22.42. A method of using a TAC for measuring coincident events.

limitation to the time resolution. This resolution can be measured by connecting the TAC inputs to a single detector or to a pulse generator. If the signal is split and routed through the two system branches and into the TAC, it is expected that the output would have constant amplitude at some time T. The distribution of pulses about this value of T renders a spectrum that includes the variance in Δt about T. The FWHM of this differential time spectrum is commonly called the *timing resolution*. It is worth noting that the delay shown in Fig. 22.42 is important. Without the delay, the true coincidence spectrum would be centered about T equal to zero. pulses occurring at $T - \Delta t$ would yield a negative output and thus not be counted.

22.6 Instrumentation Standards

Early nuclear instruments used vacuum tube circuits within a standardized 19-inch wide chassis. The components each had their own DC power supply connected directly to AC power. These components, which were easily installed in a standard 19-inch wide instrument rack, could be easily interchanged because each unit operated independently.

The invention of the transistor in 1948 led to more compact electronics, which generated far less heat than vacuum-tube components. Yet, these early transistorized nuclear instruments were built in a similar fashion to their vacuum-tube counterparts and were still designed for large 19-inch wide racks. It was eventually realized, however, that these transistorized electronic components could be built as smaller and more efficient modular units, thereby eliminating the need for a single 19-inch wide chassis. Many independent laboratories and commercial companies began to design and construct systems of various modular electronic components.

Two such systems, advanced at the time, were the systems developed by the European Organization for Nuclear Research (CERN) and the United Kingdom Atomic Energy Research Establishment. The European Standards community on Nuclear Electronics (ESONE), founded in 1961, also began development of a new standardized modular system. Along with these systems, numerous commercial companies began working on various systems of modular nuclear electronics. Further, the independent U.S. national laboratories began developing modular nuclear electronics systems for their own purposes. As a result, numerous different nuclear electronics systems were under development, none of which was compatible with any of the other systems. Overall, research at nuclear laboratories was seriously limited by the compatibility of instruments, and researchers were forced to invest in numerous different modular nuclear components.

22.6.1 NIM Standard

To remedy this divergent development of nuclear modules, the U.S. National Bureau of Standards (NBS),[20] in a 1963 report to the U.S. Atomic Energy Commission (AEC), recommended that the U.S. national laboratories develop a modular system to be duplicated by commercial manufacturers. The U.S. AEC convened a meeting of representatives from several national laboratories on Feb. 26, 1964 to determine if such standardization was of interest to the laboratories. In agreement, the AEC Committee on Nuclear Instrument Modules (NIM) was established and composed of members from all AEC national laboratories and a few other prominent nuclear laboratories.

The first NIM Committee meeting was held on March 17, 1964, with follow-up meetings over several ensuing months. The committee studied the advantages and shortcomings of the existing modular systems, and a system similar to the CERN design was eventually developed. The prototype bins were manufactured by Oak Ridge National Laboratory (ORNL), Lawrence Radiation Laboratory in Berkeley, and Lawrence Radiation Laboratory in Livermore.[21] These early prototypes were studied at later meetings of the NIM Committee and eventually led to a recommended design. Final details were resolved by the NIM Executive

[20]The National Bureau of Standards is now the National Institute of Standards and Technology (NIST).

[21]The Lawrence Radiation Laboratory in Berkeley is now the Lawrence-Berkeley National Laboratory and the Lawrence Radiation Laboratory in Livermore is now the Lawrence-Livermore National Laboratory.

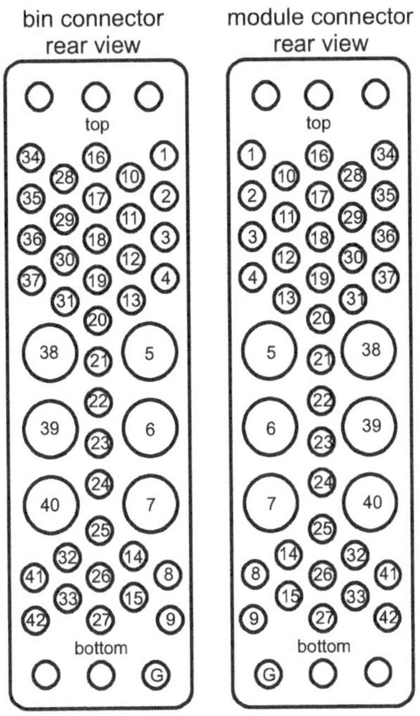

Figure 22.43. Pinout designations for the connectors of nuclear instrument modules and bins. From DOE/ER-0457T.

Committee with representatives from the U.S. NBS, ORNL, Lawrence Radiation Laboratory in Berkeley, and with later added members from Princeton-Pennsylvania Accelerator and Brookhaven National Laboratory (BNL). The final specifications were published in July 1964, and later updated versions of the specifications were published in 1966, 1968, 1969, and 1974. The standard was well received, with the first commercial NIM components made available by November 1964 (see AEC report TID-20893 [1964]). During 1965, numerous other NIM instruments became available, and NIM components accounted for over 95% of all modular nuclear instruments produced in the U.S. in the immediate years following 1967. The NIM standard was again revised in 1990, to make the standard compatible with existent technology and manufacturing processes, as described in Department of Energy (DOE) document DOE/ER-0457T, where detailed information on power, signal, connector, and size dimensions are provided in detail. A few of these specifications are outlined below. Additional information is provided in ORTEC Technical Note 010308 [2008].

Dimensions

A Nuclear Instrument Module (NIM) conforms to a nominal 8.75 inches (22.23 cm) panel height and a basic width such that a standard size NIM Bin accommodates 12 single-width NIMs or any combination of single- and multiple-width NIMs with a total of 12 or less unit widths. A unit width for a single Module is 1.32 inches (3.35 cm).[22] The Module design permits vertical air flow past the components in a NIM Bin which is required to provide an open area (reasonably distributed) of not less than 10% of the total horizontal projection of the Module. Ventilation holes less than 7/64 inch (2.8 mm) diameter are not considered to abide by the recommended 10% condition and do not apply. The connector pin assignments are designated as shown in Fig. 22.43 and listed in Table 22.2.

[22]A limited number of Modules with nominal 5.25-inch panel height have been constructed, along with accompanying NIM Bins with similar reduced dimensions. They are not very popular.

NIM Bins mount in standard EIA or EC 19-inch racks and are designed to accommodate the size and pinout of NIMs. Each standard Bin includes 12 sets of guides and 12 Bin Connectors so as to accommodate up to 12 single-width Modules or any combination of single-width and/or multiple-width Modules with a total of 12 or less unit widths.[23] NIM Bins provide power to the Modules through bussed Bin Connectors that mate with the Module Connectors. Ample air flow through Bins and power supplies is essential to provide adequate cooling for both single Bin and stacked Bin systems. This may necessitate the use of fans, spacers, deflectors, or similar devices. Power at the standard voltages of ±6.00, ±12.00, and ±24.00 DC as well as 117 volts AC is provided at the Bin busses, as shown in Fig. 22.43 and Table 22.2. The ±6 volt option is not available on some NIM bins.

Table 22.2. Pinout designations for NIM components. From DOE/ER-0457T.

Pin	Function	Pin	Function	Pin	Function
1	reserved	15	reserved	29	−24 volts
2	reserved	16	+12 volts	30	spare
3	spare	17	−12 volts	31	spare
4	reserved	18	spare	32	spare
5		19	reserved	33	117 V AC (hot)
6		20	spare	34	power return ground
7		21	spare	35	reset
8	+200 V DC	22	reserved	36	gate
9	spare	23	reserved	37	spare
10	+6 volts	24	reserved	38	
11	−6 volts	25	reserved	39	
12	reserved	26	spare	40	
13	spare	27	spare	41	117 V AC (neutral)
14	spare	28	+24 volts	42	high quality ground
				G	ground guide pin

Coaxial Connectors

Signal connectors shall be either of two standards. One standard is 50 Ω BNC type in accordance with the American National Standards Institute (ANSI) Standard N544, "Signal Connectors for Nuclear Instruments." This standard is also defined in the International Electrotechnical Commission (IEC) Publication 313, "Coaxial Cable Connectors used in Nuclear Instrumentation." The other allowed standard is type 50CM in accordance with Section 4.2.5 of ANSI/IEEE Std 583-1982, "Modular Instrumentation and Digital Interface System (CAMAC)." Connectors for high-voltage applications up to 5 kV shall be the SHV type in accordance with ANSI Standard N42.4, "High Voltage Connectors for Nuclear Instruments," also defined as Type B Connector in IEC Publication 498, "High-Voltage Coaxial Connectors used in Nuclear Instrumentation." Signal and high-voltage connectors may be mounted on the front and rear of the Nuclear Instrument Modules, with stipulation that they shall not be mounted on the bottom 3.00 inches (7.62 cm) of the rear of the Modules.

Logic and Analog Signals

The NIM standard states that the reset and slow gate signals, including those on pins 35 and 36 (see Fig. 22.43 and Table 22.2), shall conform to the levels specified by Table 22.3. In addition, the rise and fall

[23] There are also mini NIM Bins that are designed to hold 4 single-width bins and that, at times, can be quite useful. Also the Bin frame around the NIM connectors is inconveniently oversized and blocks the AC power cords and heat sinks on some HV power supplies.

Table 22.3. NIM standard logic levels for transmission of digital data.

	Nominal Signal Level	Output (shall deliver)	Input (shall respond to)
Logic 1	+4 V	+4.0 V Min	+3.0 V Min
Logic 0	0 V	+1.0 V Max	+1.5 V Max

Table 22.4. NIM standard fast logic levels and characteristics (preferred practice).

	Ouput Driver Current mA into 50 Ω	Receiver Input Voltage Response
Logic 1	−14 to −18	−0.6 max to −1.8 min (or −12 mA to −36 mA)
Logic 0	−1.0 to +1.0	−0.2 min to +1.0 max (or −4 mA to +20 mA)

times of signals on all busses shall be limited so as not to produce excessive cross-talk. Fast signals that can cause cross-talk problems shall not be applied to these busses but may be routed to or from the Modules by external cables.

In Table 22.3, the term "shall deliver" means shall deliver to any load impedance of 1 kΩ or greater. The term "shall respond to" means shall respond fully within specifications. To avoid damage to the circuitry, the NIM standard states that no positive signal (logic level) voltage shall exceed 12 volts and no negative signal voltage shall exceed 2 volts.

According to the NIM-standard, a fast negative logic signal is normally used when rise time or repetition rate requirements exceed the capability of the positive logic pulse standard. The NIM provisions define this signal as one that is furnished into a 50 Ω impedance with the characteristics described in Table 22.4. In Table 22.4, the receiver response to input voltages more positive than +1.0 V or more negative than −1.8 V is not specified. Further, the "Response" means shall respond fully within specifications to any voltage within this range. Circuit designers are alerted to the overdrive required in many receiver circuits to assure "full specification performance," and the need, therefore, to set the receiver trigger threshold accordingly.

For microsecond analog signals, the preferred open-circuit output range is 0 to +10 volts. For fast (submicrosecond) signals, the preferred amplitude ranges are 0 to −1 volt and 0 to −5 volts into a 50 Ω load.

22.6.2 CAMAC Standard

For complex detection systems, such as those encountered at large research facilities and high-energy accelerator centers, there is another international standard for modularized electronics, namely, Computer Automated Measurement And Control, or CAMAC, which defines a standard bus for data acquisition and control [Leo 1994]. The interface system, upon which this standard is based, was developed by the ESONE Committee of European Laboratories with the collaboration of the NIM Committee of the US Department of Energy (DOE) [ESONE 1964; CAMAC 1972]. This standard is based on ERDA Reports TID-25875, July 1972 (corresponding to ESONE Report EUR 4100e) and TID-25877. The standard defines the mechanical construction and electrical dataway for the modules. The CAMAC also includes the IEEE standards of 583 (the basic CAMAC design), 595 (serial highway system), 596 (parallel branch highway system), 675 (auxiliary crate controller), 683 (block transfer specifications), 726 (real-time BASIC computer language subroutines) and 758 (FORTRAN computer language subroutines for CAMAC).

The container for instrumentation modules is referred to as a 'crate' or 'CAMAC crate.' The CAMAC bus allows data exchange between plug-in modules (up to 24 in a single crate) and a crate controller, which then interfaces to a PC or to a VME-CAMAC interface. The original standard was capable of one 24-bit data transfer every μs. Later a revision to the standard was released to support shorter cycles which allow a datum transfer every 450 ns. A follow-on upwardly compatible standard, Fast CAMAC, allows the crate cycle time to be tuned to the capabilities of the modules in each slot. Typically, CAMAC components are significantly more costly than NIM components, and are best used when computer automation is necessary.

Table 22.5. Pinout designations for normal CAMAC modules, from top to bottom viewed from the front of the crate. Here BL stands for "bus-line" and PBL for "power bus-line". From IEEE [1982a].

Description / Function	Pin	Pin	Description / Function
BL / free BL	P1	B	BL / busy
BL / free BL	P2	F16	BL / function
indiv. patch contact	P3	F8	BL / function
indiv. patch contact	P4	F4	BL / function
indiv. patch contact	P5	F2	BL / function
BL / command accepted	X	F1	BL / function
BL / inhibit	I	A8	BL / sub-address
BL / clear	C	A4	BL / sub-address
indiv. line / station no.	N	A2	BL / sub-address
indiv. line / look-at-me	L	A1	BL / sub-address
BL / strobe 1	S1	Z	BL / initialize
BL / strobe 2	S2	Q	BL / response
there are 24 write BLs			
most sign. bit	W24	W23	next most sign. bit
	W22	W21	
⋮	⋮	⋮	⋮
	W4	W3	
2nd least sign. bit	W2	W1	least sign. bit
there are 24 read BLs			
most sign. bit	R24	R23	next most sign. bit
	R22	R21	
⋮	⋮	⋮	⋮
	R4	R3	
2nd least sign. bit	R2	R1	least sign. bit
PBL / −12V DC	−12	−24	PBL / −24V DC
PBL / reserved	−*	−6	PBL / −6V DC
PBL / 117V AC Live	ACL	ACN	PBL / 117 V AC neutral
PBL / suppl. −6 V	Y1	E	PBL / clean earth
PBL / +12V DC	+12	+24	PBL / +24 V DC
PBL / suppl. +6 V	Y2	+6	PBL / +6 V DC
PBL / 0 V (power return)	0	0	PBL / 0 V (power return)

*previously reserved for +200 DC volts.

Dimensions

The basic single-width CAMAC module conforms to a nominal 8.72 inches (22.15 cm) panel height with depth of 12 inches (30.5 cm) [Leo 1994]. The width of a single-width CAMAC module is half that of a NIM, or 0.67 inches (1.70 cm). CAMAC modules may have widths corresponding to integer multiples of the single-width standard. A CAMAC crate is typically 19 inches (48.3 cm) wide with 25 module slots, usually referred to as stations. Of those stations, 24 are considered 'normal stations' while the 25th is a 'control station'.

A CAMAC module plugs into the CAMAC crate by way of an 86-pin PC card connector, with the pins arranged in parallel as 43 pins on each side. It is noted that these connectors are not as robust as the NIM pin connectors, and a greater amount of caution is recommended when inserting and removing CAMAC modules from the crate [Leo 1994]. The connector pin assignments are arranged in a double row and are designated as listed in Table 22.5.

Electrical and Communications

The CAMAC crate is connected to a host computer, and the computer communicates with CAMAC modules through a crate control module. The CAMAC crate must have this control module in order to operate and is inserted into slot 25 (furthest to the right). Usually the CAMAC control module is double-wide so the practical number of available CAMAC module slots is reduced to only 23. The CAMAC crate controller module acts as the crate communication center and manages all information signals on the crate dataway. Commands originating at the computer or at a module in the crate must pass through the crate controller.[24] Ultimately, the crate controller is the *master* and the CAMAC modules are *slaves.* If necessary, multiple crates can be connected to the computer through a branch highway, but each crate must still have a crate controller. Multiple CAMAC crates may be connected in series or parallel configurations. Further, if connected in parallel, the data is managed with a branch driver to control the information flow between the CAMAC crates and the computer.

Table 22.6. CAMAC dataway signal levels.

	Logic 0	Logic 1
Input must accept	+2.0 to 5.5 V	0 to +0.8 V
Output must generate	+3.5 to 5.5 V	0 to +0.5 V

Signal transmission along the dataway is performed with TTL logic with the signals requirements listed in Table 22.6. Notice that, opposite NIM signals, the lower voltages produce a 1 and the higher voltages produce a 0. Communications between the modules are conducted over the dataway through the controller. The dataway is a network of parallel wires that connect the pins at the slots and facilitate communication between the modules and controller. Power lines are designated for the bottom 14 pins, which include ±6 VDC, ±12 VDC, ±24 VDC, as well as 117 VAC and, in some systems, 200 VDC lines. Bussed signals constitute most of pins and are used to transfer data, commands, addresses, and other control signals. Two module pins are reserved for point to point lines that connect the modules to the crate controller, mainly the N (address) and L (look-at-me or LAM) lines. Each bus line has a specific function, as outlined by IEEE [1982a], at locations indicated in Table 22.5. Although most signal transfers are conducted over the dataway, there are some operations that may still require external input. For these external inputs, the IEEE standard [1982a] recommends type 50CM subminiature coaxial connectors. A brief explanation of the listed connector functions is given in Table 22.7.

Three common control signals are available at all stations, namely, one to Initialize (Z) all units (typically after switch-on) in the dataway, one to Clear (C) data registers, and one to Inhibit (I) features such as data recording. The Z signal has priority over all other signals and is always accompanied by signals from I, S2, and B. The same conditions hold for the C signals that are used to clear registers and bistables. The module designer is allowed to determine which registers and bistables are cleared with the C signal, but the signal must also initiate a sequence that includes the S2 and B signals. The I signal inhibits the feature to which it is connected in a module.

[24]There are stand-alone units where the computer microprocessor is incorporated in the crate controller.

Table 22.7. A brief description of the functions for the connector pins of the CAMAC module listed in Table 22.5.

Pin	Description
Ai	($i = 1, 2, 4, 8$): Subaddress, selects a specific section of the module.
B	Busy, indicates that a dataway operation is in progress.
C	Clear, clears registers, accompanied by S2 and B.
E	Clean earth, reference for circuits requiring a clean earth.
Fi	($i = 1, 2, 4, 8, 16$): Function, defines the command function code to be executed by the module.
I	Inhibit, disables features for the duration of the signal.
L	Look-at-me (LAM), indicates request for service. Dedicated point-to-point line from a module to the crate controller.
N	Station number, dedicated point-to-point line that selects a module in the crate for communication.
Pi	($i = 1, 2$): Free bus lines, for unspecified uses.
Pi	($i = 3, 4, 5$): Patch contacts, for unspecified interconnections, no dataway lines.
Q	Response, Indicates status of feature selected by command.
Ri	$i = 1, ..., 24$: Read, i.e., take information from the module.
S1	Strobe 1, controls the first phase of a dataway operation. Dataway signals must not change.
S2	Strobe 2, controls the second phase of a dataway operation. Dataway signals may change.
Wi	$i = 1, ...24$: Write, i.e., bring information to the module.
X	Command accepted, indicates that module is able to perform action required by the command.
Y1	Supplementary, provides supplementary −6 volts.
Y2	Supplementary, provides supplementary +6 volts.
Z	Initialize, sets module to a defined initial state, accompanied by S2 and B.

Status signals come from locations B, L, Q, and X. The busy signal (B) indicates that a dataway operation is in progress. The Busy (B) signal is used to interlock various aspects of a system that can compete for the use of the dataway. The signal $B = 1$ indicates to all units that an operation is in progress. The Look-at-Me (L) signal is a request from a module indicating the need for attention. The Response (Q) signal is a no/yes answer to command operations, where an output of 0 indicates "no" and an output of 1 indicates "yes". Whenever a module is addressed during a command operation, it generates a signal of 1 on the Command Accepted (X) output if it recognizes the command as valid for the instrument, either within the module or in association with external equipment. If the output from X is 0, this indicates a serious malfunction. Examples for a 0 output may include a missing module, no power, bad connections, or improper equipment for the function. The controller may indicate the need for intervention by the operator if a 0 is generated.

Data operations are conducted on the Read (R) and Write (W) lines. Up to 24 bits may be transferred in parallel between the controller and the selected module. Independent lines are provided for the read and write directions of transfer. Read and write lines are activated by command operations.

The state of the signals on the Station Number (N) lines, the Subaddress lines (A), and the Function (F) lines constitute various commands. These command signals are maintained for the full duration of a dataway operation. They are accompanied by a signal on the busy (B) line to indicate that an operation is underway. The Station Number (N) is used to select a module in the crate. The stations are labeled from the front of the CAMAC crate starting with N1 from the left-hand side. The signals for N originate from the controller module. Different sections of a module are addressed by four subaddresses (A1, A2, A4, A8) that specify a subsection of the module. With four locations, there are 16 possible binary codes that are identified as A(0) to A(15). The subaddress codes are defined as A(i) to distinguish them from the subaddresses Ai. Function signals define the function code to be performed at the five F subaddresses (F1, F2, F4, F8, F16) of a module. With five locations, there are 32 possible binary codes, identified and numbered from F(0) to F(31). Note that the type of code F(i) should not be confused with subaddress Fi. The details of each specific F function are provided in the IEEE Std 583-1982.

During each command operation, the controller generates two Strobe signals S1 and S2 in sequence on separate bus lines. In response to these timing signals, the modules initiate various actions appropriate to the dataway command. Only Strobe S2 is mandatory during unaddressed operations, but S1 may also be generated. The timing signals S1 and S2 are used to synchronize events during an operation.

Example 22.6: For the input subaddresses A, what code is selected for A1 = 1, A2 = 0, A4 = 1, and A8 = 0? Also, determine the code for the Function input where F1 = 1, F2 = 0, F4 = 1, F8 = 0, and F16 = 1.

Solution:

From the binary code, the A input is 0101,

$$0101 = 1(1) + 0(2) + 1(4) + 0(8) = 5$$

which is equivalent to 5 in base ten numbers; hence the identifying code is A(5).

From the binary code, the F input is 10101,

$$10101 = 1(1) + 0(2) + 1(4) + 0(8) + 1(16) = 21$$

which is equivalent to 21 in base ten numbers; hence the identifying code is F(21).

22.6.3 VMEbus Standard

The Versa Module Europa bus (VMEbus) is a computer bus standard described in IEEE 1014-1987. The system was originally created in 1971 by Kister and Black for the Motorola MC68000 CPU, receiving upgrades from a larger engineering team in the following few years [Wiener, undated]. It was released in 1981 with standard described in the IEC 821 document and now by ANSI/IEEE 1014-1987. Upgrade standards to 64-bit data paths can be found in ANSI/VITA 1-1994 (or VME64), and also the superset to the VME64 standard called the VME64 Extensions (or VME64x). In both upgrades, the data handling capabilities were greatly improved and many other features were enhanced. Overall, the VMEbus is a scalable and flexible computer backplane bus interface intended to support computer intensive operations [Wiener, undated].

The bus includes four sub-buses on the backplane, identified as the arbitration bus, the data transfer bus, the priority interrupt bus, and the utility bus [IEEE 1987]. The arbitration bus allows an arbiter module and several requester modules to coordinate use of the data transfer bus. The data transfer bus allows master modules to direct the transfer of binary data between themselves and slaves. The data transfer bus supports 8-, 16-, and 32-bit transfers over a non-multiplexed 32-bit data and address highway. The priority interrupt bus allows interrupter modules to send interrupt requests to the various interrupt handlers. The utility bus includes signals that provide periodic timing and coordinates the power-up and power-down of VMEbus systems.

The mechanical specifications of boards, backplanes, subracks, and enclosures are based on the Eurocard form factor, although it has a different signaling system. Each VME card is 20.3 mm wide and 21 of these cards can fit into a 19-inch rack mounted VME crate. The connection between the module and the backplane is made by two 96-pin DIN 41612 connectors. The number of connectors used determines the address space of the card and the card size. Uncommitted pins on the connectors provide support for application specific busses and rear transition modules. 9-U modules[25] may provide a third connector for application specific use.

There are three sizes of VME cards, namely the 3U × 160 mm cards, 6U × 160 mm cards (the most common), and the larger 9U × 400 mm cards. The smaller 3U × 16.0 cm cards have only one backplane connector, labeled J1, with a 24-bit/16-bit address space. Both the 6U × 16.0 cm cards and 9U × 40.0 cm cards have two backplane connectors, labeled J1 and J2, with a 32-bit address/data space. The VME cards are inserted into the slots of a VME crate, which serves as the power supply for the backplane. The actual circuit board should not be greater than 1.8 mm thick in order to slide into the slot. Standard VME voltages are 5 V and ±12 V. The crate is outfitted with a cooling fan or a fan tray.

The VME bus has 31 address lines. The first 23 lines are present on J1 and the remainder being on J2. The lowest address bit (A0) is implied by the transfer cycle and is not present on the backplane. The VME bus has 32 data lines. The low order 16 data lines are on J1, the high order 16 on J2. Hence, the smaller VME modules with only J1 can only perform 16-bit transfers, while larger modules can perform 32-bit transfers. The communications are handled with TTL logic, with the highs between 2.4 to 5 V and lows between 0 to 0.6 V.

Data handling was increased by the newer VME64 and VME64x standards, allowing 32-bit data and 40-bit addressing in the 3U modules, and 64-bit data and addressing on the 6U modules. VME64 can multiplex address and data lines allowing 31 address bits on J1 and an additional 32 address bits on J2. VME64 also has an additional 3.3 volt supply. The bandwidth was increased by the newer standards. The VME64x also describes different connector pins than the VMEbus. Regardless, the older VME modules are compatible with the newer standards, meaning that older modules can be inserted into the newer system. Further, modules designed to the VME64x standard are backward compatible with older backplanes and subracks.

Similar to the CAMAC system, the VMEbus has a controller module, called a crate manager, but it is placed in the left-most crate slot. The VMEbus system also has various modules that perform specific functions, much like the NIM and CAMAC systems.

There are commercial modules available that can perform the familiar standard functions of NIM or CAMAC modules. Modules perform as master, slave, interrupter, interrupter handler, or arbiter. The master can initiate transfers, while the slave module(s) responds to these commands. The bus allows multiple masters and, to enable proper operation, it contains a powerful interrupt scheme which is managed by a crate manager. This managing arbiter is located in slot 1 of the crate. The slave can send an interrupt signal to request an action, for example a signal that data has arrived or an error notification. The interrupt handler is a module that can receive and process interrupt signals, usually a slave module. Data transfer on the VME Bus is asynchronous, thereby enabling modules with a broad variety of response times to be supported.

Module functions are controlled by a computer rather than turn knobs, although some modules may allow direct input lines. Any of the modules can act as a master module or slave module. Master actions send communications through the bus to the crate manager module, which arbitrates the signals and allows access to the bus one module at a time. Once a module assumes the identity as master, it drives the bus

[25]The notation 'U' is defined as a *rack unit*, with a single U being 1.75 inches (4.45 cm) high. Cards that are fit into a slot should be 1.313 inches (3.335 cm) shorter than the slot height. Hence, a 3U card is 10.0 cm high, a 6U card is 23.34 cm high, and a 9U card is 36.67 cm high.

lines to identify a slave module and data transfer to take place. The slave then responds over the data bus lines. After the data transfer, the master relinquishes the bus. Note that slave units, through an interrupt communication over one of the seven interrupt bus lines, can also signal a master module.

Overall, the VMEbus is able to provide high speed of operation combined with a flexible modular system. It is also possible to develop specially designed hardware to operate with the VMEbus for specific functions and requirements.

22.7 Electronic Noise

A biased radiation detector connected to output electronics, such as a preamplifier, randomly produces output signals even though there are no radiation interactions occurring in the device. This noise is generated at the output even when there is no input signal at all. There are numerous sources for these random signals that produce an output voltage [Ambrózy 1982], the important ones of which are described in this section. It is important to understand this electronic noise as it tends to increase the minimum discrimination level as well as increase the variance in the output signal.

Typically the average noise signal is centered about zero. For instance, the average output noise voltage sampled over a long time period is

$$\langle v_{out} \rangle = \lim_{T\to\infty} \frac{1}{T} \int_0^T v_{out}(t)dt = 0. \tag{22.177}$$

This means that the average magnitude of positive pulses equals the average magnitude of negative pulses. The average of the noise provides little information about the noise spectrum because the outcome is (nearly) always zero, regardless of the circuit. To obtain a more meaningful description of the noise, it is the average square of the noise voltage that is usually reported, i.e.,

$$\langle v_{out}^2 \rangle = \lim_{T\to\infty} \frac{1}{T} \int_0^T v_{out}^2(t)dt, \tag{22.178}$$

or, equivalently, the *root mean square* or *rms* of the noise deviation

$$\sigma_n = \langle v_{rms} \rangle = \sqrt{\langle v_{out}^2 \rangle}. \tag{22.179}$$

Electronic noise is superimposed upon radiation induced signals, thereby adding to the total signal variance. It is common to represent the noise signal in terms of the average charge from the detector to produce the same signal, or the current needed at the input of the amplifier to produce an output equivalent to the rms noise voltage. This is called the *input referred noise* and is very valuable in comparing different processing methods as it removes gain and bandwidth effects. The equivalent noise charge q_n can be described in units of coulombs, but it is more commonly normalized by the unit charge q_e and described by the number of equivalent electrons. The equivalent noise charge q_n can then be directly compared to the average number of charges produced by a radiation event. The signal-to-noise ratio is defined as the ratio of the mean signal pulse height v_m to that of the RMS noise voltage

$$\mathrm{SNR} = \frac{v_m}{\langle v_{rms} \rangle}. \tag{22.180}$$

For a given radiation quantum of energy E, the SNR is

$$\mathrm{SNR} = \frac{E}{wq_n} = \frac{Q}{q_n}, \tag{22.181}$$

where w is the average energy required to produce a charge pair. One important goal of the measurement electronics is to reduce the rms electronic noise to values well below the standard deviation of radiation induced signals.

Electronic noise affects the performance of both counters and spectrometers. For counters, q_n greatly affects where the LLD must be set to discriminate between radiation counts and electronic noise. If the LLD is set too low, then noise pulses are included with actual radiation induced pulses, ultimately leading to an enhanced count rate and, often enough, serious system dead time. If the pulses produced in a counter are large, the LLD can be set above the noise and the count rate remains unaffected. However, if the detector pulses are of similar magnitude as the rms noise, then setting the discriminator to remove noise also removes valid radiation counts. If the LLD is set too low, noise is included along with counts and, thus, falsely increases the count rate while also increasing the detector system dead time. The same problem holds true for spectrometers, along with the added inconvenience of spectral broadening manifested as poorer energy resolution.

As an example, consider a planar semiconductor with contacts at the ends. The average density of free charges is a function of temperature. If a bias is applied to the semiconductor, then some current flows

$$i = \frac{q_e N v_c}{d}, \tag{22.182}$$

where q_e is the unit charge, N is the number of charges in motion, v_c is the charge speed, and W_d is the width of the detector between the contacts. The average differential change in current due to fluctuations in the number of charge carriers produced is

$$di_N = \frac{q_e v_c}{W_d}\langle dN\rangle, \tag{22.183}$$

and due to fluctuations in the charge carrier speed is

$$di_v = \frac{q_e N}{W_d}\langle dv_c\rangle. \tag{22.184}$$

The total current fluctuation, or variance, is thus

$$\sigma_n^2 = \langle di\rangle^2 = \left[\left(\frac{q_e v_v}{W_d}\langle dN\rangle\right)^2 + \left(\frac{q_e N}{W_d}\langle dv_c\rangle\right)^2\right]. \tag{22.185}$$

Speed fluctuations arise from the thermal motion of charge carriers. Number fluctuations can result from various sources, including the statistical spread in the average number of charge pairs in motion, thermionic emission past contact barriers, and recombination and generation current.

The variance in the noise signal σ_n^2 must be added to the detector signal variance σ_d^2 to yield the observed standard deviation. For a Gaussian pulse height spectrum, the standard deviation of a full energy peak can be estimated as

$$\sigma = \sqrt{\sigma_d^2 + \sigma_n^2}, \tag{22.186}$$

as depicted in Fig. 22.44.

Note that detectors with severe trapping do not produce Gaussian distributions, but instead can be severely skewed, usually towards the low energy portion of the spectrum. Consequently, the simplicity of Eq. (22.186) yields inaccurate results.

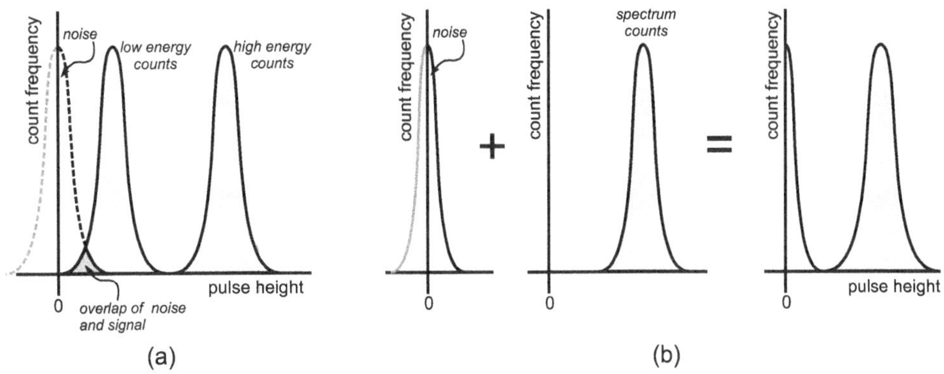

Figure 22.44. The effects of electronic noise: (a) noise floor affects where the LLD must be set to eliminate false counts, and (b) noise fluctuations broaden the energy resolution.

22.7.1 Thermal or Johnson Noise

Thermal or Johnson noise is generated by the thermal motion of charge carriers in a conductive medium, usually electrons in electrical circuits. Thermal agitation causes local fluctuations in the current and also fluctuations in voltage at the conductor contacts. These fluctuations are present even when the component is not connected in a circuit and without applied voltage. If measured over an infinite time period, both the current and voltage must average to zero.

The so-called *spectral power density* (actually power per unit frequency) of the thermal noise is an oft quoted value used to describe noise [Gillespie 1953]

$$\frac{dP}{df} = 4kT, \tag{22.187}$$

where f is the frequency. However, it is usually voltage or current that is being measured; hence, it is convenient to quote noise in terms of voltage or current. From

$$P = iv = \frac{v^2}{R} = i^2R, \tag{22.188}$$

the noise spectral density in terms of voltage and current, expressed as noise per unit bandwidth, are

$$\frac{d\langle v_j^2\rangle}{df} = 4kTR \qquad \text{and} \qquad \frac{d\langle i_j^2\rangle}{df} = \frac{4kT}{R}. \tag{22.189}$$

Of note, for a constant value of R in Eqs. (22.189), the spectral power density is independent of the frequency. To determine the total voltage or current noise contribution, Eqs. (22.189) must be multiplied by the noise bandwidth[26] Δf to give

$$v_j = \sqrt{4kTR(\Delta f)} \qquad \text{and} \qquad i_j = \sqrt{\frac{4kT(\Delta f)}{R}}. \tag{22.190}$$

[26] A common approximation for the noise bandwidth is $\pi/2$ times the system 3 db frequency.

22.7.2 Shot Noise

Pure shot noise is a consequence of fluctuations in the number of charge carriers comprising a current. In other words, there is a variance in the number of charges flowing in a circuit at any time, which affects the current in metals, semiconductors, gases, and through contacts and barriers. There are multiple sources of this form of noise, including thermionic emission, generation-recombination current, and charge carrier trapping. The spectral noise density is [Gillespie 1953]

$$\frac{d\langle i_s^2\rangle}{df} = 2q\langle i_{dc}\rangle, \tag{22.191}$$

where $\langle i_{dc}\rangle$ is the average current. As with Johnson noise, the spectral noise density of Shot noise is also independent of frequency. For these reasons, both Johnson and Shot noise are often called *white noise*,[27] because the power spectral density is nearly constant over a wide frequency spectrum. The total noise contribution is determined by multiplying Eq. (22.191) by the noise bandwidth

$$i_s = \sqrt{2q\langle i_{dc}\rangle(\Delta f)}. \tag{22.192}$$

22.7.3 Flicker or 1/f Noise

Flicker noise is believed to be a consequence of surface effects and structural defect changes on electrodes. It is most important at low frequencies. Although flicker noise is not completely understood, the properties are well documented [Gillespie 1953; Nicholson 1974], and as described by [Kowalski 1970]

$$\frac{d\langle v_F^2\rangle}{df} = \frac{A_F}{f^m}, \tag{22.193}$$

where the parameter m ranges between 0.8 to 1.5 and A_F is a proportionality constant. If m is unity, then the flicker noise can be described by [Spieler 2005],

$$\langle v_F^2\rangle = \int_{f_1}^{f_2} \frac{A_F}{f} df = A\ln\left(\frac{f_2}{f_1}\right). \tag{22.194}$$

where f_1 and f_2 define the low to high frequency range. Consequently, flicker noise is a function of the ratio of the system high and low cutoff frequencies.

Example 22.7: Consider the non-inverting preamplifier configuration of Fig. 22.19 with a bandwidth of 250 kHz operated at room temperature and for which $R_1 = 10$ kΩ, $R_2 = 100$ kΩ, $A = 6000$, and $\langle I_{dc}\rangle = 130$ nA. Determine the zero input electronic noise contribution at the amplifier output. Assume that flicker noise is minimal. The input voltage is terminated to ground with a 5 kΩ resistor.

Solution:
The shot noise is found from Eq. (22.192) as

$$i_s = \sqrt{2q\langle i_{dc}\rangle(\Delta f)} = \sqrt{2(1.6\times10^{-19}\text{ C})(130\times10^{-9}\text{ A})(2.5\times10^{5}\text{ Hz})} = 102\text{ pA rms}.$$

This shot noise current appears on the feedback resistor R_2 to produce,

$$v_{so} = i_s R_2 = (102\text{ pA rms})(10^5\ \Omega) = 10.2\ \mu\text{V rms}.$$

[27] Named so, because much like white light is made up of the different color frequencies, white noise is a combination of multiple noise frequencies each of equal intensity.

The Johnson noise at the input is found from Eq. (22.190),

$$v_j = \sqrt{4kTR(\Delta f)} = \sqrt{4(1.38 \times 10^{-23}\ \text{J/K})(293\ \text{K})(5000\ \Omega)(2.5 \times 10^5\ \text{Hz})} = 4.5\ \mu\text{V rms}.$$

The resulting Johnson noise is amplified at the output to produce,

$$v_{jo} = V_j \left[\frac{R_2}{R_1} + 1\right] = 4.5\ \mu\text{V rms} \left[\frac{10^5\ \Omega}{10^4\ \Omega} + 1\right] = 49.46\ \mu\text{V rms}.$$

The shot and Johnson noise contributions are combined to determine the zero input rms noise voltage from the preamplifier output, namely,

$$v_{no} = \sqrt{v_{so}^2 + v_{jo}^2} = \sqrt{(10.2\ \mu\text{V rms})^2 + (49.46\ \mu\text{V rms})^2} = 50.5\ \mu\text{V rms}.$$

22.7.4 Detector Performance

In radiation spectroscopy, the system noise can be of particular concern. The noise originates partly in the detector and partly in the first stage of a preamplifier circuit. Generally, the optimal performance of the system is obtained by reducing the input capacitance C_i to a minimum while increasing the input resistance R to be as high as is practical. In this way, the input can be approximated by the step input voltage of Q/C_i. The input noise takes the form [Baldinger and Franzen 1956],

$$d\langle v^2 \rangle = \left[2\pi a^2 + \frac{b^2}{2\pi f^2} + \frac{c^2}{f}\right] df = \left[a^2 + \frac{b^2}{\omega^2} + \frac{c^2}{\omega}\right] d\omega. \tag{22.195}$$

The term proportional to a^2 is called 'A noise' or *series noise*, the term proportional to b^2 is called 'B noise' or *parallel noise*, and the term proportional to c^2 represents flicker noise. Series noise is usually a consequence of shot noise in the preamplifier FET. The parallel noise is usually thermal or Johnson noise from the detector leakage current or the current flowing in the amplifier feedback resistor.

As an example, consider a simple *CR-RC* network with a transfer function described by Eq. (22.108). The mean square noise voltage is

$$\langle v_n^2 \rangle = \int_0^\infty \left(a^2 + \frac{b^2}{\omega^2} + \frac{c^2}{\omega}\right)\left(\frac{\omega^2 \tau_1^2}{(1+\omega^2\tau_1^2)(1+\omega^2\tau_2^2)}\right) d\omega, \tag{22.196}$$

which upon integration gives [Nicholson 1974]

$$\langle v_n^2 \rangle = \frac{\pi a^2}{2K_1(K_1+1)\tau_1} + \frac{\pi b^2 \tau_1}{2(K_1+1)} + \frac{c^2}{K_1^2 - 1} \ln K_1, \tag{22.197}$$

where $K_1 = \tau_2/\tau_1$. Baldinger and Franzen [1956] show that the lowest noise is achieved when $K_1 = 1$, for which Eq. (22.197) reduces to

$$\langle v_n^2 \rangle = \frac{\pi a^2}{4\tau} + \frac{\pi b^2 \tau}{4} + \frac{c^2}{2}. \tag{22.198}$$

From the result of Eq. (22.107) for a *CR-RC* network, the maximum pulse voltage from a radiation quantum is seen to occur at $t = \tau$ so that

$$v_{max} = v_{in} e^{-1} = \frac{E}{wC_i} e^{-1} = \frac{Q}{C_i} e^{-1}, \tag{22.199}$$

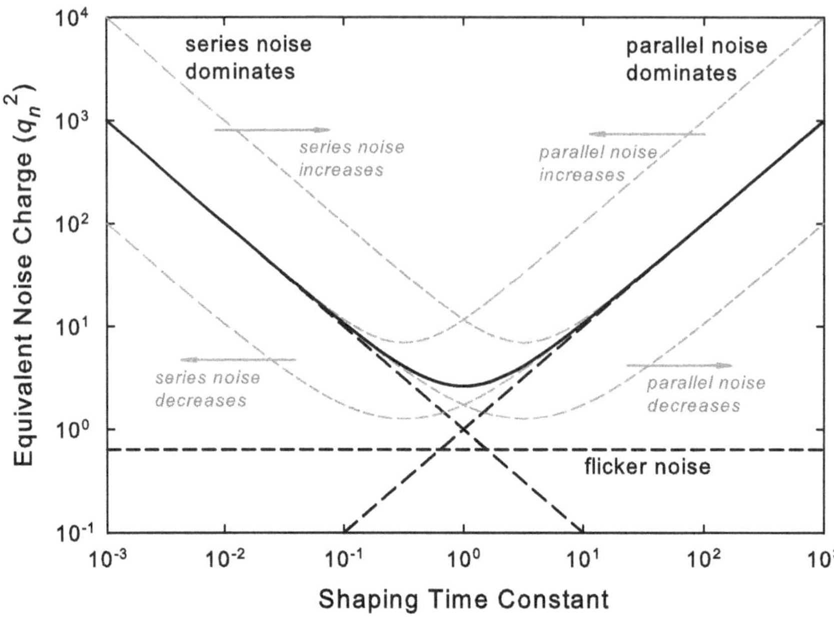

Figure 22.45. Dependence of q_n^2 on the shaping time constant for an *CR-RC* circuit in which $\tau_1 = \tau_2$. As described in the text, the conditions shown here are for $a = b = c = 1$ and $e^2C_i^2 = 4/\pi$.

where E is the radiation particle energy, w is the average energy required to produce an ion pair, and C_i is the input capacitance. Hence, the equivalent noise charge q_n is represented by,

$$q_n^2 = e^2C_i^2\left(\frac{\pi a^2}{4\tau} + \frac{\pi b^2\tau}{4} + \frac{c^2}{2}\right). \tag{22.200}$$

Consider the result of Eq. (22.200), where the series noise has a τ^{-1} dependence, parallel noise has a τ dependence, and flicker noise has no dependence on the shaping time constant. To illustrate the dependence of q_n on the shaping time, several simplifications are introduced. First assume a special case in which $a = b = c = 1$ is considered. Then assume the unlikely happenstance that $e^2C_i^2 = 4/\pi$. For such simplifying assumptions Eq. (22.200) reduces to

$$q_n^2 = \left(\frac{1}{\tau} + \tau + \frac{2}{\pi}\right), \tag{22.201}$$

and the results of this idealized case are shown in Fig. 22.45. At relatively small time constants, the series noise is dominant, while at relatively long shaping times the parallel noise is dominant. In this example there is an optimum τ where the noise equivalent charge is minimized, which turns out to be the general optimizing condition for a semi-Gaussian pulse shaper. Hence, for a spectroscopic measurement, one should adjust the shaping time constant to minimize the electronic noise. If the series noise is decreased, the optimized time constant reduces as does the minimum noise level. If instead, the parallel noise is decreased, the optimized time constant increases while the total noise decreases. The opposite effects are found if either the series or parallel noise contributions are increased.

For high-resolution spectroscopy systems, such as with HPGe detectors, performance is improved by decreasing the coupling capacitance while increasing the input resistance. Typically, the first stage field effect transistor (FET) in the preamplifier is placed close to the detector to reduce the total input capacitance. The sources of noise include shot noise from the detector leakage current, Johnson noise from both parallel and series resistances, and noise from the input FET. Van der Ziel [1962; 1963] shows that the electronic noise for a FET can be described by

$$\frac{d\langle i_{F1}^2\rangle}{df} \simeq \frac{2}{3}\,4kTg_m, \tag{22.202}$$

where g_m is the transconductance (or mutual conductance), with $g_m = dI_{out}/dV_{in}$ of the FET. Hence, the transconductance has units of inverse ohms, or mhos ($1/\Omega$). The output current is the *drain* current, and the input voltage is applied across the *gate* to the *source*.[28] The channel current in the FET is another source of noise, which causes fluctuations in the current across the gate to the source capacitance C_{gs} and can be described by

$$\frac{d\langle i_{F2}^2\rangle}{df} = \frac{kT\omega^2 C_{gs}^2}{g_m}. \tag{22.203}$$

The gate leakage current I_g also produces shot noise as

$$\frac{d\langle i_{F3}^2\rangle}{df} = 2qI_g. \tag{22.204}$$

Leakage current through the detector also contributes to shot noise

$$\frac{d\langle i_{D1}^2\rangle}{df} = 2qI_d, \tag{22.205}$$

while the load resistance contributes to the Johnson noise

$$\frac{d\langle i_{D2}^2\rangle}{df} = \frac{4kT}{R_L}. \tag{22.206}$$

Usually, the load resistor is very high, on the order of 10^9 Ω, so that the noise contribution of Eq. (22.206) is negligible.

The total input impedance Z_L is

$$Z_L = \frac{R_S}{1 + j\omega C_i R_S}, \tag{22.207}$$

where R_S is the sum of all resistances at the input, and has the magnitude

$$|Z_L| = \left[\frac{R^2}{1 + \omega^2 C_i^2 R_S^2}\right]^{1/2}. \tag{22.208}$$

If R is large, which is usually the case required to reduce thermal noise at the input, then the impedance can be approximated by

$$|Z_L| \approx \frac{1}{\omega C_i}. \tag{22.209}$$

The series noise is formed by the combination of Eq. (22.202) and Eq. (22.203),

$$\frac{d\langle i_S^2\rangle}{df} = \frac{2}{3}\,4kTg_m + \frac{kT\omega^2 C_{gs}^2}{g_m}. \tag{22.210}$$

[28]The authors refer to Sze [1981] for a detailed discussion on FET operation.

Equation (22.209) can be rewritten as voltage noise by multiplying by the resistance, namely by selectively applying g_m^{-1} and Z_L (Eq. (22.209)),

$$\frac{d\langle v_S^2\rangle}{df} = \frac{8kT}{3g_m} + \frac{kTC_{gs}^2}{g_m C_i^2}, \qquad (22.211)$$

which is the term $2\pi a^2$ in Eq. (22.200). Substitution into Eq. (22.200) produces the series equivalent noise charge contribution to a *CR-RC* pulse shaping circuit, where the series equivalent noise charge q_{sn} becomes,

$$q_{sn}^2 = e^2 C_i^2 \frac{1}{8\tau}\left(\frac{8kT}{3g_m} + \frac{kTC_{gs}^2}{g_m C_i^2}\right) = e^2 \frac{kT}{g_m \tau}\left(\frac{C_i^2}{3} + \frac{C_{gs}^2}{8}\right). \qquad (22.212)$$

A similar process can be used to describe the parallel noise

$$\frac{d\langle v_p^2\rangle}{df} = Z_L^2\left(2qI_d + 2qI_g + \frac{4kT}{R_L}\right) = \frac{1}{\omega^2 C_i^2}\left(2qI_d + 2qI_g + \frac{4kT}{R_L}\right), \qquad (22.213)$$

and is the term $2\pi b^2/\omega^2$ in Eq. (22.200). Substitution into Eq. (22.200) produces the parallel equivalent noise charge q_{pn}

$$q_{pn}^2 = e^2 C_i^2 \frac{\omega^2 \tau}{8}\frac{1}{\omega^2 C_i^2}\left(2qI_d + 2qI_g + \frac{4kT}{R_L}\right) = \frac{e^2\tau}{8}\left(2q(I_d + I_g) + \frac{4kT}{R_L}\right). \qquad (22.214)$$

Finally, the flicker noise is described by

$$\frac{d\langle i_{F1}^2\rangle}{df} = \frac{A_f}{f}, \qquad (22.215)$$

where A_f is c^2 in Eq. (22.200).

From Eq. (22.212) the input capacitance and the FET capacitance increase the series noise equivalent charge q_{sn}. Consequently, it is important to reduce the input capacitance as much as possible. From Eq. (22.214), parallel noise equivalent charge q_{pn} is not affected by the input capacitance. Note also that both the series and parallel noise can be decreased by cooling the system. The detector shot and Johnson noise decrease as the temperature is lowered, and the series noise in the FET also decreases with lower temperature. Hence, high-resolution detectors are often cooled to reduce noise and improve energy resolution. Finally, the load resistance R_L should be large to reduce the parallel Johnson noise.

22.8 Coaxial Cables

Cables and connectors were introduced in Chapter 2, with descriptions on types and uses. Presented here, with a modest increase in detail, are some of the more common cables used in radiation detection and measurement.

Perhaps the most used cable configuration, in some form or other, is the coaxial cable, invented in 1880 by Oliver Heaviside [Nahin 2002]. The cable consists of a central core electrical conductor surrounded by a concentric insulating layer that is further surrounded by another conducting layer shield as shown in Fig. 22.46. Usually coaxial cables have an outer insulating layer covering the conducting shield. The inner conductive core is usually held at potential with the conducting shield connected to ground. The potential may be applied by the user or may be a signal voltage transmitted by the detector or electronics. This same conducting shield operates as a Faraday cage to divert electromagnetic noise to ground, thereby effectively reducing or eliminating noise pickup.

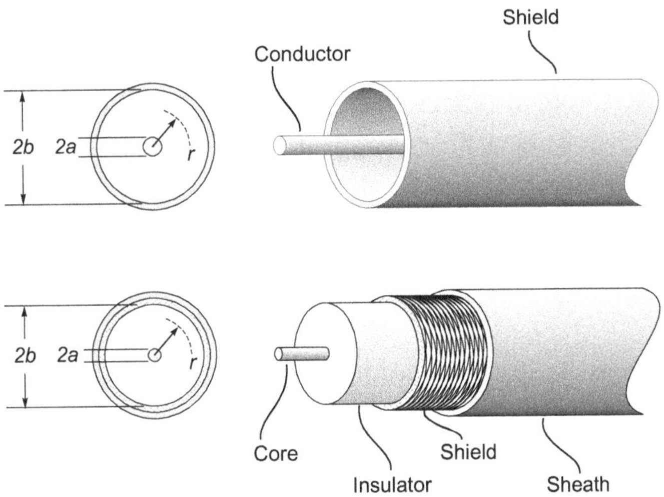

Figure 22.46. (top) Basic coaxial conductor. (bottom) Components of a flexible coaxial cable.

22.8.1 Basic Characteristics

With the notation of Fig. 22.46 and the results derived in Sections 10.2 and 10.3, the electric field in a coaxial device is

$$E(r) = \frac{V_0}{r}\left[\ln\left(\frac{b}{a}\right)\right]^{-1}, \tag{22.216}$$

where V_0 is the applied voltage on the conductive core. The capacitance, per unit length (farads per meter), is

$$C = 2\pi\epsilon_0\kappa\left[\ln\left(\frac{b}{a}\right)\right]^{-1}, \tag{22.217}$$

where ϵ_0 and κ have their usual meanings. The inductance, per unit length (henrys per meter), is

$$L = \frac{\epsilon_0\kappa\mu_0\mu_r}{C} = \frac{\mu_0\mu_r}{2\pi}\ln\left(\frac{b}{a}\right), \tag{22.218}$$

where μ_0 is the permeability of free space ($4\pi \times 10^{-7}$ H m^{-1}) and μ_r is the relative permeability (μ/μ_0) of the insulating material. The conductance, per unit length (siemens per meter), is

$$G = \frac{\sigma}{\epsilon_0\kappa\mu_0\mu_r}C = 2\pi\sigma\left[\ln\left(\frac{b}{a}\right)\right]^{-1}, \tag{22.219}$$

where σ is the usual symbol for electrical conductivity, and $1/\sigma = \rho$, the resistivity (SI unit Ω m).

The complex frequency dependent characteristic impedance of a coaxial cable is defined by

$$Z(j\omega) = \left[\frac{R + j\omega L}{G + j\omega C}\right]^{1/2}. \tag{22.220}$$

At low frequency, the characteristic impedance reduces to

$$Z(0) = \left[\frac{R}{G}\right]^{1/2} = R. \tag{22.221}$$

At high frequency, the characteristic impedance becomes

$$Z_0 \approx \left[\frac{j\omega L}{j\omega C}\right]^{1/2} = \frac{1}{2\pi}\left[\frac{\mu_0 \mu_r}{\epsilon_0 \kappa}\right]^{1/2} \ln\left(\frac{b}{a}\right). \tag{22.222}$$

It is this impedance that is usually quoted for a coaxial cable. Most quality insulators have a relative permeability of 1, so Eq. (22.222) reduces to

$$Z_0 \approx 60\Omega \left[\frac{1}{\kappa}\right]^{1/2} \ln\left(\frac{b}{a}\right). \tag{22.223}$$

The signal velocity in the coaxial cable is defined by,

$$v_s = \frac{1}{\sqrt{\mu_0 \mu_r \epsilon_0 \kappa}} = \frac{c}{\sqrt{\mu_r \kappa}}. \tag{22.224}$$

Example 22.8: Given a Cu core diameter of 0.5 mm and insulator diameter of 3.0 mm, what is the characteristic impedance of a cable if the insulator is HDPE? Also, what is the signal velocity?

Solution:
The relative permittivity κ for HDPE is 2.26 and the relative permeability is approximately equal to 1. From Eq. (22.223), the characteristic impedance is,

$$Z_0 \approx 60\Omega \left[\frac{1}{\kappa}\right]^{1/2} \ln\left(\frac{b}{a}\right) = 60\Omega \left[\frac{1}{2.26}\right]^{1/2} \ln\left(\frac{3}{0.5}\right) = 71.5\ \Omega.$$

The signal velocity is found from Eq. (22.224),

$$v_s = \frac{c}{\sqrt{\mu_r \kappa}} = \frac{2.9979 \times 10^{10}\ \text{cm s}^{-1}}{\sqrt{2.26}} = 1.9942 \times 10^{10} \text{cm s}^{-1} = 0.6652\ c.$$

Characteristics of a few of the hundreds of commercially available coaxial cables are given in Table 22.8. The maximum voltage that can be applied is a function of the insulator material and not necessarily determined by the cable type. For example, RG-58/U cable, a common choice for radiation detection measurements, may have a limiting voltage of 300 volts with a foam polyethylene insulator, but can be increased to 1400 volts with a solid polyethylene insulator. This example shows that the specific properties of coaxial cables must be known for a particular application even if the cables are of the same type.

The optimum ratio of b/a for the maximum operating voltage of a cable can be found from Eq. (22.216) by setting the derivative to zero, i.e.,

$$\frac{dV_{max}}{da} = \frac{d}{da} E_{max} a \left[\ln\left(\frac{b}{a}\right)\right] = E_{max}\left[\ln\left(\frac{b}{a}\right)\right] - E_{max} = 0. \tag{22.225}$$

Thus the derivative vanishes only if $\ln(b/a) = 1$. For an air-gap coaxial cable, Eq. (22.223) indicates that the characteristic impedance that can hold the highest peak voltage is 60 Ω. The maximized ratio $\ln(b/a) = 1$ holds for all coaxial dielectrics, so that the voltage-maximized characteristic impedance Z_0 decreases with $\kappa^{-1/2}$.

Table 22.8. A few examples of coaxial cable properties. All values from Belden catalogue.

Cable Type	Belden Part Number	Charac. Impedance (ohms)	Core		Dielectric		Shield[c]	Sheath		Speed (v/c)	Delay (ns/m)	Atten.		Capac. (pF/m)	Max* RMS Volts
			Type[a]	Diam. (mm)	Type[b]	Diam. (mm)		Type[d]	Diam (mm)			MHz	(dB/m)		
RG-6/U	3131A	75	BCCS	1.02	GIFPE	4.57	DBIV	PVC	7.65	0.82	3.94	400	0.133	53.15	300
RG-6/U	8215	75	BCCS	0.71	PE	4.7	BrC	PE	8.43	0.66	5.05	400	0.193	67.26	2700
RG-8/U	8214	50	BaC	2.74	FPE	7.24	BrC	PVC	10.24	0.78	4.27	400	0.128	85.3	300
RG-8/U	8237	50	BaC	2.16	PE	7.24	BrC	PVC	10.24	0.66	5.05	400	0.139	93.5	3700
RG-11/U	1617A	75	BCCS	1.63	GIFPE	7.11	DBIV	PVC	10.33	0.83	3.94	400	0.085	53.15	300
RG-11/U	8261	75	TC	1.22	PE	7.24	BaC	PVCNC	10.29	0.66	5.05	400	0.138	67.26	3700
RG-58/U	7806A	50	BaC	0.94	GIFHDPE	2.79	DF/TC	PE	4.95	0.77	4.33	450	0.233	79.72	300
RG-58/U	9310	50	BaC	0.84	PE	2.9	DF/TC	PVC	4.9	0.66	5.05	400	0.259	101.05	1400
RG-58A/U	8219	53.5	TC	0.94	FPE	2.95	BrTC	PVC	4.9	0.73	4.56	400	0.328	86.94	300
RG-58A/U	8259	50	TC	0.89	PE	2.97	BrTC	PVC	4.9	0.66	5.05	400	0.407	101.04	1400
RG-58C/U	8262	50	TC	0.89	PE	2.92	BrTC	PVC	4.95	0.66	5.05	400	0.377	101.05	1400
RG-59/U	1186A	75	BCCS	0.81	GIFHDPE	3.68	DBIV	PVC	6.73	0.83	3.28	400	0.16	53.15	300
RG-59/U	8241	75	BCCS	0.58	PE	3.71	BaC	PVC	6.09	0.66	5.05	400	0.23	67.26	300
RG-59/U	8279	75	BCC	0.58	PE	3.71	BrTC	PE	5.59	0.66	5.05	360	0.263	68.9	2300
RG-59/U	8281	75	BaC	0.79	PE	5.03	BrTC	PE	7.75	0.66	5.05	360	0.167	68.9	2900
RG-62/U	8254	93	BCCS	0.635	SSPE	3.71	BrC	PVC	6.05	0.84	3.94	400	0.174	44.29	750
RG-62/U	82262	93	BCCS	0.635	FFEP	3.71	BrC	LSPVC	5.18	0.85	3.94	400	0.174	41.01	300
RG-174/U	7805	50	BaC	0.46	PE	1.55	BFTC	PVC	2.79	0.66	5.05	450	0.499	102.36	1100
RG-178B/U	83265	50	SPCCS	0.305	TFE	0.84	SPC	FEP	1.8	0.695	4.79	400	0.919	95.14	750
RG-179B	83264	75	SPCCS	0.305	TFE	1.6	SPC	FEP	2.54	0.695	4.79	400	0.525	63.98	900
RG-213/U	8267 (mil-spec)	50	BaC	2.26	PE	7.24	BrC	PVCNC	10.29	0.66	5.05	400	0.135	101.05	3700

*Varies according to cable construction, mainly changing with the insulating material.

[a]BaC bare copper; BCC bare compacted copper; BCCS bare copper covered steel; SPCCS silver plated copper covered steel; TC tinned copper

[b]FFEP foam fluorinated ethylene propylene; FFEP foam fluorinated ethylene propylene; GIFHDPE gas-injected foam high density polyethylene GIFPE gas-injected foam polyethylene; PE polyethylene; SSPE semi-solid polyethylene; TFE Teflon®

[c]BaC bare copper; BFTC Beldfoil®+tinned copper; BrC braided copper; BrTC braided tinned copper; DBIV 4 layer shield; DF/TC Duofoil®/tinned copper SPC braided Ag plated Cu

[d]FEP fluorinated ethylene propylene; LSPVC low smoke PVC; PE polyethylene; PVC polyvinyl chloride; PVCNC PVC-noncontaminating

Table 22.9. Select properties of a few dielectrics.

Material	Dielectric Correction Factor κ	Dielectric Strength (kV/mm)
Dry Air	1.0006	1.5–3
Alumina	5.3–5.5	13.4
Fused Quartz	3.8	25–40
Neoprene	6–9	15.7–27.6
Nylon	3.5	14
Polyethylene	2.2–2.3	20–30
PMMA	2.8 2.9	10–30
Polypropylene	2.2–2.3	23–25
PVC	2.9–3.1	14–20
Pyrex®	5.1	13–14
Rubber (natural)	3	100–215
Teflon®	2–2.1	60–70

The maximum *peak power* that a cable can handle is found from

$$P_{max} = \frac{V_{max}^2}{2Z_0}. \tag{22.226}$$

where V_{max} is located at the conductor surface a. Substitution of Eq. (22.216) and Eq. (22.223) into Eq. (22.226) yields

$$P_{max} = \frac{\left(E_{max}a \ln\left[\frac{b}{a}\right]\right)^2 \kappa^{1/2}}{2(60\ \Omega)\ln\left[\frac{b}{a}\right]} = \frac{(E_{max}a)^2 \ln\left[\frac{b}{a}\right] \kappa^{1/2}}{120}. \tag{22.227}$$

It is easily shown from Eq. (22.227) that the maximum power is obtained when the ratio $b/a = \sqrt{e}$. From Eq. (22.223), an air gap coaxial cable transmits maximum power when $Z_0 = 30\ \Omega$.

The values of κ and dielectric strength for several common dielectrics are listed in Table 22.9. The dielectric strength of dry air is approximately 3 kV/mm, but the value diminishes in a humid environment, and a value of 1 kV/mm is probably a safer number to use. Stronger dielectrics, such as HDPE and TFE have substantially higher dielectric strengths, thereby allowing high voltages and power ratings. Hence, many coaxial cables have much higher maximum voltage ratings than others (see Table 22.8). Although the maximum peak voltage rating ($V_{max} \approx V_{RMS}/0.71$) may be substantially high, the weakest location in the cable connections determines the actual maximum allowed voltage and power. Hence, if there is an air gap between connections of a coaxial cable and instrumentation, where air gap appears instead of the dielectric, it is actually the geometry of the connection at that air gap location that determines the breakdown voltage. Excessive voltage can cause arcing, and operators should be aware of this potential problem.

Wave Propagation

Signals propagated along a cable to an instrument are usually partially reflected at the cable/instrument interface. To minimize such reflections between components connected by a cable, all components should have, nominally, the same impedance. In general, a transmission line length should be greater than 1/8 of the propagation wavelength λ. For instance, a signal with a frequency of 400 MHz has a wavelength $\lambda = c/f = 750$ mm, so that a coaxial transmission line should be at least 94 mm (3.7 in) long. When a wave propagating through a transmission line arrives at the load with impedance Z_L, the amount of reflection is determined by the standing wave ratio (SWR) and the reflection coefficient (Γ).

Consider a transmission line connected to a load Z_L, where the load is at the origin $z = 0$. The impedance at location $z = -l$ is $Z(-l)$ for a transmission line of characteristic impedance Z_0. The voltage along the line is written as

$$V(z) = V_+ e^{-jkz} + V_- e^{jkz} \tag{22.228}$$

or

$$V(z) = V_+ \left(e^{-jkz} + \Gamma e^{jkz}\right), \tag{22.229}$$

where V_+ is the incident voltage amplitude for a wave traveling in the positive z direction, V_- is the reflected voltage amplitude for a wave traveling in the negative z direction, Γ is the reflection coefficient, and

$$k = \omega\sqrt{LC} = 2\pi f\sqrt{LC}. \tag{22.230}$$

The *generalized* reflection coefficient, as a function of position, is

$$\Gamma(z) = \frac{V_- e^{jkz}}{V_+ e^{-jkz}} = \Gamma e^{2jkz}. \tag{22.231}$$

Equation (22.231) indicates that the reflection coefficient has real and imaginary components in the complex plane. Substitution of Eq. (22.231) into Eq. (22.229) gives voltage along the cable

$$V(z) = V_+ e^{-jkz}\left[1 + \Gamma(z)\right]. \tag{22.232}$$

If the amplitude V_+ is unity, then the magnitude of Eq. (22.232) becomes

$$|V(z)| = |1 + \Gamma(z)| \tag{22.233}$$

It should be noted that $\Gamma(z)$ can have a positive or negative imaginary component, so that voltage along the transmission line has maximum and minimum voltages where,

$$V_{max} = 1 + |\Gamma| \qquad \text{and} \qquad V_{min} = 1 - |\Gamma|. \tag{22.234}$$

The ratio of the maximum to minimum voltages is an important parameter known as the *voltage standing wave ratio*

$$\text{VSWR} = \frac{V_{max}}{V_{min}} = \frac{1 + |\Gamma|}{1 - |\Gamma|}. \tag{22.235}$$

Rearrangement of this result gives

$$|\Gamma| = \frac{\text{VSWR} - 1}{\text{VSWR} + 1}. \tag{22.236}$$

Also, the VSWR can be defined as

$$\text{VSWR} = \frac{Z_L}{Z_0}, \tag{22.237}$$

which, upon substitution into Eq. (22.236), yields

$$\Gamma = \frac{Z_L - Z_0}{Z_L + Z_0}. \tag{22.238}$$

Notice that if $Z_L < Z_0$, then the reflected voltage signal changes polarity.

Following the result of Eq. (22.226), the power transmitted in the line is

$$P_+ = \frac{|V_+|^2}{2Z_0} \tag{22.239}$$

while the reflected power is

$$P_- = \frac{|V_-|^2}{2Z_0}. \tag{22.240}$$

The ratio of Eq. (22.240) to Eq. (22.239) gives the reflected power

$$R_L \equiv \frac{P_-}{P_+} = \frac{|V_-|^2}{|V_+|^2} = |\Gamma|^2. \tag{22.241}$$

The reflected power, or return loss R_L, is often represented in units of decibels

$$R_L = 20\log_{10}\left|\frac{V_-}{V_+}\right| = 20\log_{10}|\Gamma|. \tag{22.242}$$

Thus, another factor to consider in cable selection is the line loss by attenuation. The line loss reported in decibels per unit length is usually given by cable manufacturers on data specification sheets (see Table 22.8). The following example shows three important cases.

Example 22.9: Consider three cases for a 50 Ω coaxial transmission line. In the first case, the line is shorted at the end, the second case the line is open at the end, and in the third case the line is terminated with a 50 Ω load. What is the condition and percent power reflection for these three cases?

Solution:
In the first case, the line is terminated with a short, so $Z_L = 0$,

$$\Gamma = \frac{Z_L - Z_0}{Z_L + Z_0} = -1 \qquad \text{so} \qquad |\Gamma|^2 = \frac{|Z_L - Z_0|^2}{|Z_L + Z_0)|^2} = 1,$$

which indicates that the signal changes sign and is completely reflected. In the second case, the line is open and $Z_L = \infty$ and

$$|\Gamma|^2 = \frac{|Z_L - Z_0|^2}{|Z_L + Z_0)|^2} = 1,$$

which indicates that the wave is also completely reflected but does not change sign. In the third case, $Z_L = Z_0$,

$$|\Gamma|^2 = \frac{|Z_L - Z_0|^2}{|Z_L + Z_0)|^2} = 0,$$

which shows that the wave is completely transmitted with no reflection losses. The return loss in the first two cases is 0 dB, while the return loss in the last case is $-\infty$ dB.

Cables and connectors are usually impedance matched to the load to reduce reflections so as to transfer the maximum amount of power. The calculations for this operation are usually conducted with the use of a *Smith chart*, an exercise beyond the scope of this text. However, there are numerous books that cover the method of a Smith chart in detail, including Adler et al. [1960], Shen and Kong [1983], and Kosow [1963], and modern electronic equipment (such as a network analyzer) has the capability of performing such calculations for the user.

Delay Lines

Under some circumstances it is necessary to delay the signal from a detector before it enters the analyzing electronics. For instance, a coincidence measurement may be desired for two different types of detectors, but the signal processing times are different for the two devices. A delay line can be used to retard the progression of the faster signal, thereby allowing coincident events to arrive at the same time. A simple delay line consists of a long coaxial cable where the propagation time per unit length of cable is

$$t_d = \frac{1}{v} = \frac{\sqrt{\mu_r \kappa}}{c} = \sqrt{\frac{LC}{d^2}}, \tag{22.243}$$

where v_s is the signal propagation speed, d is the cable length, L is the cable inductance, and C cable capacitance. For typical coaxial cables that have polyethylene as the insulating filling ($\epsilon_r \simeq 2.25$), the delay time is approximately 5 nanoseconds per meter. There are special cables designed as delay lines that have spiral conductors to increase the time delay per unit length, thereby shortening the required cable.

Delay lines not only cause a delay in the arrival time of a signal, but also can attenuate the signal as it travels down the cable. If the delay line is purposely shorted at the output terminal, then the signal can be reflected back, inverted in form, towards the input terminal. As a result, the combined attenuation and reflection of the original signal can be used to alter the final output electronic signal to a desirable outcome, referred to as single delay line pulse shaping, which is described in Sec. 22.1.2 of this chapter.

PROBLEMS

1. Find the Laplace transform of the following functions:

$$\text{(a) } f(t) = \cos at \quad \text{(b) } f(t) = t^n \quad \text{and} \quad \text{(c) } f(t) = \begin{cases} at, & \text{if } 0 \le t \le t_{end} \\ 0, & \text{otherwise} \end{cases}.$$

2. Derive the results of Eq. (22.43).

3. Solve Eq. (22.46) for $v_{in}(t) = v_o H(t)$, where $H(t)$ is the unit step function, by (a) finding a particular solution and the homogeneous solution of the corresponding ordinary differential equation, and (b) by taking the Laplace transform of Eq. (22.46) to find $V_{out}(s)$ and inverting this result to obtain $v_{out}(t)$.

4. Derive the solution to Eq. (22.89) to obtain the results of Eqs. (22.90).

5. Prove that the inverse Laplace transform of Eq. (22.108) yields the result of Eq. (22.106).

6. Given a Cu core diameter of .8 mm with HDPE insulator ($\kappa = 2.26$), what diameter of insulator is required to yield the maximum operating voltage? What is the maximized characteristic impedance?

7. Prove that the maximum power of a coaxial cable is found when $b/a = \sqrt{e}$.

8. For a coaxial cable with inner conductor diameter 0.9 mm and outer HDPE insulator diameter of 3 mm, what is the expected peak breakdown voltage and maximum power? The breakdown field for HDPE is approximately 20 kV/mm.

9. After looking around the lab, all you can find is an RG59/U cable to connect your detector preamplifier to the subsequent electronics. However, the input to your amplifier has an impedance of 50 Ω. What is the expected power loss from this arrangement?

REFERENCES

ADLER, R.B., L.J. CHU, AND R.M. FANO, *Electromagnetic Energy Transmission and Radiation*, New York: Wiley, 1960.

AMBRÓZY, A., *Electronic Noise*, New York: McGraw-Hill, 1982.

American National Standard for Signal Connectors for Nuclear Instruments, ANSI N3.3-1968, New York: American National Standards Institute, 1968.

American National Standard Nomenclature and Dimensions for Panel Mounting Racks, Panels, and Associated Equipment, ANSI N83.9-1968, New York: American National Standards Institute, 1968.

BALDINGER, E. AND W. FRANZEN, "Amplitude and Time Measurement in Nuclear Physics," *Advances in Electronic and Electron Physics*, **8**, 255–315, (1956).

BATEMAN, H., *Tables of Integral Transforms*, Vols. I and II, New York: McGraw-Hill, 1954.

BECKER, T.H., E.E. GROSS, AND R.C. TRAMMELL, "Characteristics of High-Rate Energy Spectroscopy Systems with Time-Invariant Filters," *IEEE Trans. Nucl. Sci.*, **NS-28**, 598–602, (1981).

BEDWELL, M.O. AND T.J. PAULUS, "A New Constant Fraction Timing System with Improved Time Derivation Characteristics," *NS-23*, 234–243, (1976).

BENDER, S., B. HEINRICH, AND K. ÜNLÜ, "Compton Suppressed $LaBr_3$ Detection System for use in Nondestructive Spent Fuel Essay," *Nucl. Instrum. Meth.*, **A784**, 474–481, (2015).

BLANKENSHIP, J.L. AND C.H. NOWLIN, "New Concepts in Nuclear Pulse Amplifier Design," *IEEE Trans. Nucl. Sci.*, **NS-13**, 495–507, (1966).

BRITTON, C.L., T.H. BECKER, T.J. PAULUS, AND R.C. TRAMMELL, "Characteristics of High-Rate Energy Spectroscopy Systems Using HPGe Coaxial Detectors and Time-Variant Filters," *IEEE Trans. Nucl. Sci.*, **NS-31**, 455–460, (1984).

CAMAC, A Modular Instrumentation System for Data Handling - Description and Specification, European Atomic Energy Community, EURATOM Report EUR 4100e, Luxembourg: Office Central de Vente des Publications des Communautes Europeennes, (1972).

CARTER, B. AND T.R. BROWN, *Handbook of Operational Amplifier Applications*, Application Report SBOA092B, Texas Instruments, 2016.

CARTER, B. AND R. MANCINI, *Op Amps for Everyone*, 5th Ed., Oxford: Newnes, 2017.

CHASE, R.L., "Pulse Timing System for Use with Gamma Rays on Ge(Li) Detectors," *Rev. Sci. Instrum.*, **39**, 1318–1326, (1968).

CHIANG, H.H., *Basic Nuclear Electronics*, New York: Wiley, 1969.

CHO, Z.H. AND R.L. CHASE, "Comparative Study of the Timing Techniques Currently Employed with Ge Detectors," *Nucl. Instrum. Meth.*, **98**, 335-347, (1972).

Coaxial Cable Connectors Used in Nuclear Instrumentation, International Electrotechnical Commission Publication 313, 1st. Ed., Switzerland: International Electrotechnical Commission, Geneva, 1969.

COOKE-YARBOROUGH, E.H. AND E.W. PULSFORD, "An Accurate Logarithmic Counting-Rate Meter Covering a Wide Range," *Proc. IEE*, **98**, 323–324, (1951).

COTTINI, C., E. GATTI, AND V. SVELTO, "A New Method for Analog to Digitial Conversion," *Nucl. Instrum. Meth.*, **24**, 241–242, 1963.

COVA, S., M. GHIONI, AND F. ZAPPA, "Optimum Amplification of Microchannel-Plate Photomultiplier Pulses for Picosecond Photon Timing," *Rev. Sci. Instrum.*, **62**, 2596–2601, (1991).

DEWARAJA, Y.K., Z. HE, R.F. FLEMING, D.K. WEHE, S.V. GURU, J.C. FERREIRA, AND R.H. FLEMING, "A Position Sensitive β-γ Coincidence Technique for Multiplexed Gamma Spectroscopy of Many Small Samples," *Nucl. Instrum. Meth.*, **A353**, 588–592, (1994).

ESONE System of Nuclear Electronics, European Atomic Energy Community, EURATOM, Report EUR 1831e, Luxembourg: Office Central de Vente des Publications des Communautes Europeennes, 1964.

EVANS, R.D. AND R.L. ALDER, "Improved Counting Rate Meter," *Rev. Sci. Instrum.*, **10**, 332–336, (1939).

FAIRSTEIN, R., "Gated Baseline Restorer with Adjustable Asymmetry," *IEEE Trans. Nucl. Sci.*, **NS-22**, 463–466, (1975).

FAIRSTEIN, E., "Amplifier Test Standard: Detector and Multichannel Analyzer or Staircase Generator and Oscilloscope," *IEEE Trans. Nucl. Sci.*, **NS-32**, 31–35, (1985).

FENYVES, E. AND O. HAIMAN, *The Physical Principles of Nuclear Radiation Measurements*, New York: Academic Press, 1969.

FRANK, S.G.F., O.R. FRISCH, AND G.G. SCARROTT, "A Mechanical Kick-Sorter (Pulse Height Analyser)," *Phil. Mag.*, **42**, 603–611, (1951).

FRASER, H.J., "A Portable Four-Decade Logarithmic G.M. Survey Meter," *Nucl. Instrum. Meth.*, **118**, 263–267, (1974a).

FRASER, H.J., "An Accurate Logarithmic Ratemeter for Random Pulses," *IEEE Trans. Nucl. Sci.*, **NS-21**, 31–38, (1974b).

GEDCKE, D.A. AND W.J. MCDONALD, "A Constant Fraction of Pulse Height Trigger for Optimum Time Resolution," *Nucl. Instrum. Meth.*, **55**, 377–380, (1967).

GEDCKE, D.A. AND W.J. MCDONALD, "Design of the Constant Fraction of Pulse Height Trigger for Optimum Time Resolution," *Nucl. Instrum. Meth.*, **58**, 253-260, (1968).

GILLESPIE, A.B., *Signal, Noise and Resolution in in Nuclear Counter Amplifiers*, New York: Pergamon, 1953.

GINGRICH, N.S., R.D. EVANS, AND H.E. EDGERTON, "A Direct-Reading Counting Rate Meter for Random Pulses," *Rev. Sci. Instrum.*, **7**, 450–456, (1936).

IEEE STANDARD, *683-1976 IEEE Recommended Practice for Block Transfers in CAMAC Systems*, IEEE, 1976.

IEEE STANDARD, *758-1979 IEEE Standard Subroutines for Computer Automated Measurement and Control (CAMAC)*, IEEE, 1979.

IEEE STANDARD, *583-1982 IEEE Standard Modular Instrumentation and Digital Interface System (CAMAC)*, IEEE, 1982a.

IEEE STANDARD, *595-1982 IEEE Standard Serial Highway Interface System (CAMAC)*, IEEE, 1982b.

IEEE STANDARD, *596-1982 IEEE Standard Parallel Highway Interface System (CAMAC)*, IEEE, 1982c.

IEEE STANDARD, *675-1976 IEEE Standard Multiple Controllers in a CAMAC Crate*, IEEE, 1982d.

IEEE STANDARD, *726-1982 IEEE Standard Real-Time BASIC for CAMAC*, IEEE, 1982e.

IEEE STANDARD, *IEEE Standard for A Versatile Backplane Bus: VMEbus*, IEEE Std 1014, R2008, 1987.

IRVINE, R.G., *Operational Amplifier Characteristics and Applications*, Englewood Cliffs: Prentice-Hall, 1981.

JÁNOSSY, L., *Cosmic Rays*, Oxford: Oxford Press, 1950.

JUNG, W., Ed., *Op Amp Applications Handbook*, Amsterdam: Newnes, 2005

KARLOVAC, N. AND T.V. BLALOCK, "An Investigation of the Count Rate Performance of Baseline Restorers," *IEEE Trans. Nucl. Sci.*, **NS-22**, 457–462, (1975).

KOSOW, I.L., Ed., *Microwave Theory and Measurement*, Englewood Cliffs: Prentice-Hall, 1963.

KOWALSKI, F., *Nuclear Electronics*, New York: Springer-Verlag, 1970.

LANDIS, D.A., F.S. GOULDING, R.H. PEHL, AND J.T. WALTON, "Pulsed Reset Techniques for Semiconductor Detector Radiation Spectrometers," *IEEE Trans. Nucl. Sci.*, **NS-18**, 115–124, (1971).

LANDIS, D.A., C.P. CORK, N.W. MADDEN, AND F.S. GOULDING, "Transistor Reset Preamplifier for High Rate High Resolution Spectroscopy," *IEEE Trans. Nucl. Sci.*, **NS-29**, 619–624, (1982).

LEO, W.R., *Techniques for Nuclear and Particle Physics Experiments*, Berlin: Springer-Verlag, 1994.

LINDBLAD, T., "Design and Performance of a Ge(Li)-NaI(Tl) Compton-Suppression Spectrometer System for In-Beam Experiments," *Nucl. Instrum. Meth.*, **154**, 53–60, (1978).

NCRP, *A Handbook of Radioactivity Measurements Procedures*, NCRP Report No. 58, Bethesda: NCRP, 1985.

NAHIN, P.J., *Oliver Heaviside: The Life, Work, and Times of an Electrical Genius of the Victorian Age*, Baltimore: Johns Hopkins University Press, 2002.

NICHOLSON, P.W., *Nuclear Electronics*, London: Wiley, 1974.

NOWLIN, C.H. AND J.L. BLANKENSHIP, "Elimination of Undesirable Undershoot in the Operation and Testing of Nuclear Pulse Amplifiers," *Rev. Sci. Instrum.*, **36**, 1830–1839, (1965).

ORTEC, *Electronics Standards and Definitions*, Application Note 010308, Oak Ridge: Ortec, 2008.

ORTEC, *Introduction to Amplifiers*, Application Note 082609, Oak Ridge: Ortec, 2009a.

ORTEC, *Preamplifier Introduction*, Application Note 082709, Oak Ridge: Ortec, 2009b.

ORTEC, *Fast-Timing Discriminator Introduction*, Application Note 082709, Oak Ridge: Ortec, 2009c.

PRICE, W.J., *Nuclear Radiation Detection*, 2nd Ed., New York: McGraw-Hill, 1964.

RADEKA, V., "Trapezoidal Filtering of Signals from Large Germanium Detectors at High Rates," *Nucl. Instrum. Meth.*, **99**, 525–539, (1972).

ROBINSON, L.B., "Reduction of Baseline Shift in Pulse-Amplitude Measurements," *Rev. Sci. Instrum.*, **32**, 1057, (1961).

ROONEY, B.D. AND J.D. VALENTINE, "Benchmarking the Compton Coincidence Technique for Measuring Electron Response Non-Proportionality in Inorganic Scintillators," *IEEE Trans. Nucl. Sci.*, **43**, 1271–1276, (1996).

SEDRA, A.S. AND K.C. SMITH, *Microelectronic Circuits*, New York: HRW, 1982.

SHEN, L.C. AND J.A. KONG, *Applied Electromagnetism*, Boston: PWS Publishers, 1983.

SIMPSON, M.L., T.H. BECKER, R.D. BINGHAM, AND R.C. TRAMMELL, "An Ultra-High-Throughput, High-Resolution, Gamma-Ray Spectroscopy System," *IEEE Trans. Nucl. Sci.*, **38**, 89–96, (1991).

SPIELER, H., *Semiconductor Detector Systems*, Oxford: Oxford University Press, 2005.

Standard Nuclear Instrument Modules, U.S. AEC Report TID-20893, Washington DC: U.S. Gov. Printing Office, (1964); superseded in 1974 by TID-20893, Rev. 4.

STRAUSS, M.G., I.S. SHERMAN, R. BRENNER, S.J. BUDNICK, R.N. LARSEN, AND H.M. MANN, "High Resolution Ge(Li) Spectrometer for High Input Rates," *Rev. Sci. Instrum.*, **38**, 725–730, (1967).

SZE, S.M., *Physics of Semiconductor Devices, 2nd Ed.*, New York: Wiley, 1981.

U.S. DEPT. OF ENERGY, *Standard NIM Instrumentation System*, DOE/ER-457T, L. COSTRELL, Chairman, Office of Energy Research, Washington DC, (1990).

VAN DER ZIEL, A., "Thermal Noise in Field-Effect Transistors," *Proc. IRE*, **50**, 1808–1812, (1962).

VAN DER ZIEL, A., "Thermal Noise in Field-Effect Transistors," *Proc. IRE*, **51**, 461–467, (1963).

VINCENT, C.H., "The Rapid Recognition of a Significant Increase in the Rate of Arrival of Pulses Occurring at Random," *Nucl. Instrum. Meth.*, **9**, 181–194, (1960).

VINCENT, C.H. AND J.B. PARKER, "Scale-of-2 Input Circuit to a Saturating Diode-Pump Ratemeter," *Nucl. Instrum. Meth.*, **86**, 169–170, (1970).

WIENER, "Introduction to VME/VXI/VXS Standards," Wiener, Phoenix Mecano Co., Technical Note, undated.

WILKINSON, D.H., "A Stable Ninety-Nine Channel Pulse Amplitude Analyser for Slow Counting," *Proc. Cambridge Phil. Soc.*, **46**, 508–518, (1950).

ZUMBAHLEN, H., Ed., *Linear Circuit Handook*, Amsterdam: Newnes, 2008.

Appendix A

Fundamental Physical Data and Conversion Factors

A.1 Fundamental Physical Constants

Although there are many physical constants, which determine the nature of our universe, the following values are of particular importance when dealing with atomic and nuclear phenomena.

Table A.1. Some important physical constants as internationally recommended in 2018 by CODATA. These and other constants can be obtained through the web from `http://physics.nist.gov/cuu/Constants/index.html`

Constant	Symbol	Value
Speed of light (in vacuum)	c	$2.997\,924\,58 \times 10^{8}$ m s^{-1} (exact)
Electron charge	e	$1.602\,176\,634 \times 10^{-19}$ C (exact)
Atomic mass unit	u	$1.660\,539\,066\,60(50) \times 10^{-27}$ kg ($931.494\,102\,42(28)$ MeV/c^2)
Electron rest mass	m_e	$9.109\,383\,7015(28) \times 10^{-31}$ kg ($0.510\,998\,950\,00(15)$ MeV/c^2) ($5.485\,799\,090\,65(16) \times 10^{-4}$ u)
Proton rest mass	m_p	$1.672\,621\,923\,69(51) \times 10^{-27}$ kg ($938.272\,088\,16(29)$ MeV/c^2) ($1.007\,276\,466\,621(53)$ u)
Neutron rest mass	m_n	$1.674\,927\,498\,04(95) \times 10^{-27}$ kg ($939.565\,420\,52(54)$ MeV/c^2) ($1.008\,664\,915\,95(49)$ u)
Planck constant	h	$6.626\,070\,15 \times 10^{-34}$ J s (exact) $4.135\,667\,696 \times 10^{-15}$ eV s (exact)
Avogadro constant	N_a	$6.022\,140\,76 \times 10^{23}$ mol^{-1} (exact)
Boltzmann constant	k	$1.380\,649 \times 10^{-23}$ J K^{-1} (exact) ($8.617\,333\,262\ldots \times 10^{-5}$ eV K^{-1})
Molar gas constant (STP)	R	$8.314\,462\,618\ldots$ J mol^{-1} K^{-1}
Vacuum electric permittivity	ϵ_o	$8.854\,187\,8128(13) \times 10^{-12}$ F m^{-1}
Vacuum magnetic permeability	μ_o	$1.256\,637\,062\,12(19)$ N A^{-2}

In 2017 the International Bureau of Weights and Measures defined the Avogadro constant N_a to be exactly the very large integer $6.02214076 \times 10^{23}$ entities per mol. This change redefined the mol as the amount of a substance containing N_a entities (atoms, molecules, etc.). As a consequence, the molar mass constant M_u is no longer 1 g/mol but now is 0.999 999 999 65(30) $\times 10^{-3}$ kg mol^{-1}. It also means that 1 mol of ^{12}C atoms no longer has a mass of 0.001 kg. However, the atomic mass unit remains unchanged as 1/12 the mass of a ^{12}C atom.

On the 144th anniversary of the Metre Convention, 20 May 2019, the redefinition of SI base unit was affirmed and came into force. In the redefinition, four of the seven SI base units—the kilogram, ampere, kelvin, and mole—were redefined by setting exact numerical values for the Planck constant h, the elementary electric charge e (q_e as used in this book), the Boltzmann constant k, and the Avogadro constant N_a, respectively.

A.2 The Periodic Table

The periodic table presented in Table A.2 uses the new group format numbers from 1 to 18 approved by the International Union of Pure and Applied Chemistry (IUPAC), while the numbering system used by the Chemical Abstract Service is given in parentheses at the top of each column. In the left of each elemental square, the number of electrons in each of the various electron shells are given. In the upper right of each square, the melting point (MP), boiling point (BP), and critical point (CP) temperatures are given in degrees Celsius. Sublimation and critical temperatures are indicated by s and t. In the center of each elemental square the oxidation states, atomic weight, and natural abundance is given for each element. For elements that do not exist naturally, (e.g., the transuranics), the mass number of the longest-lived isotope is given in square brackets. The abundances are based on meteorite and solar wind data.

The table is from Firestone et al. [1999] and may be found on the web at `http://isotopes.lbl.gov/isotopes/toi.html`. Data for the table are from Lide [1997], Leigh [1990], Anders and Grevesse [1989], and CE News [1985]. Since this table was published, research at several accelerator facilities around the world discovered and added ten new elements to the periodic table and, thereby completed the 7th period (or row) of the periodic table. These new elements officially recognized by IUPAC are listed in Table A.2. The newest elements (Nh, Mc, Ts and Og were recognized in 2016). Often only a few atoms of these short lived radionuclides were produced and little information about their physical properties has been obtained other than that suggested by theory and the elements' location (or group number) in the table.

Table A.2. The "new" elements discovered and officially named since 2000. All are radioactive and the atomic mass number A of the longest lived isotope is given in [ ].

Meit-erium	Darn-stadtium	Roent-genium	Coper-nicium	Nihon-ium	Flero-vium	Moscov-ium	Liver-morium	Tenn-essine	Oga-nesson
$_{109}$**Mt**	$_{110}$**Dd**	$_{111}$**Rg**	$_{112}$**Cn**	$_{113}$**Nh**	$_{114}$**Fl**	$_{115}$**Mc**	$_{116}$**Lv**	$_{117}$**Ts**	$_{118}$**Og**
[278]	[281]	[282]	[285]	[286]	[289]	[290]	[290]	[294]	[294]

A.3 Physical Properties and Abundances of Elements

Some of the important physical properties of the naturally occurring elements are given in Table A.3. In this table the atomic weights, densities, melting points, and boiling points.

The atomic weights are for the elements as they exist naturally on earth, or, in the cases of thorium and protactinium, to the isotopes which have the longest half-lives. For elements whose isotopes are all radioactive, the mass number of the longest lived isotope is given in square brackets.

Mass densities for solids and liquids are given at 25°C, unless otherwise indicated by a superscript temperature (in °C). Densities for normally gaseous elements are for the liquids at their boiling points. The melting and boiling points at normal pressures are in degrees Celsius. Superscripts "t" and "s" are the critical and sublimation temperature (in degrees Celsius).

Data are from the 78th edition of the *Handbook of Chemistry and Physics* [Lide 1997] and have been extracted from tables on the web at `http://isotopes.lbl.gov/isotopes/toi.html`.

A.4 SI Units

With only a few exceptions, units used in radiation detection and nuclear science are those defined by the SI system of metric units. This system is known as the "International System of Units" with the abbreviation SI taken from the French "Le Système International d'Unités." In this system, there are four categories of units: (1) base units of which there are seven (m, kg, s, A, K, cd, and mol), (2) derived units which are combinations of the base units (e.g., N $\equiv$ kg m s^{-2}), (3) supplementary units (rad and sr), and (4) "temporary" units (e.g., b, Ci, R, Gy, and Sv) which are in widespread use for special applications. To accommodate very small and large quantities, the SI units and their symbols are scaled by using the SI prefixes given in Table A.5.

There are several units outside the SI which are in wide use. These include the time units day (d), hour (h), and minute (min); the liter (L or ℓ); plane angle degree (°), minute ($'$), and second ($''$); and, of great use in nuclear and atomic physics, the electron volt (eV) and the atomic mass unit (u). Conversion factors to convert some non-SI units to their SI equivalent are given in Table A.5.

A.5 Internet Data Sources

The data presented in this appendix are sufficient to allow you to do most assigned problems. All of these data have been taken from the web, and you are encouraged to become familiar with these important resources. Some sites, which have many links to various sets of nuclear and atomic data, are

```
http://www.nndc.bnl.gov/
http://physics.nist.gov/PhysRefData/contents.html
http://isotopes.lbl.gov/
http://wwwndc.jaea.go.jp/
http://nucleardata.nuclear.lu.se/nucleardata/toi/
http://atom.kaeri.re.kr/
```

REFERENCES

ANDERS, E. AND N. GREVESSE, "Abundances of the Elements: Meteoritic and Solar," *Geochimica et Cosmochimica Acta*, **53**, 197, (1989).

Chemical and Engineering News, "Group Notation Revised in Periodic Table," **63** (5), 26–27, (1985).

DEBIEVRE, P. AND P.D.P. TAYLOR, *Int. J. Mass Spectrom. Ion Phys.*, **123**, 149, (1993).

FIRESTONE, R.B., C.M. BAGLIN, and S.Y.F. CHU, *Table of Isotopes (1999)*, New York: Wiley, 1999.

LEIGH, G.J., *Nomenclature of Inorganic Chemistry*, Oxford: Blackwells Scientific Publ., 1990.

LIDE, D.R., Ed., *Handbook of Chemistry and Physics*, 77th Ed., Boca Raton, FL: CRC Press, 1996.

Key to Table

Group
Element
K L M N O P Q (shells) — E_Z — M.P.° B.P.° C.P.°
Ox.States
At.Weight
Abundance%

Main table

Group	Element	Symbol	Z	Shells (K L M N O P Q)	M.P.°	B.P.°	C.P.°	Ox. States	At. Weight	Abundance%
1 (IA)	Hydrogen	H	1	1	-259.34°	-252.87°	-240.18°	+1-1	1.00794	91.0%
1 (IA)	Lithium	Li	3	2 1	180.5°	1342°		+1	6.941	1.86×10^{-7}%
1 (IA)	Sodium	Na	11	2 8 1	97.80°	883°		+1	22.989770	0.000187%
1 (IA)	Potassium	K	19	2 8 8 1	63.38°	759°		+1	39.0983	0.0000123%
1 (IA)	Rubidium	Rb	37	2 8 18 8 1	39.31°	688°		+1	85.4678	2.31×10^{-8}%
1 (IA)	Cesium	Cs	55	2 8 18 18 8 1	28.44°	671°		+1	132.90545	1.21×10^{-9}%
1 (IA)	Francium	Fr	87	2 8 18 32 18 8 1	27°			+1	[223]	
2 (IIA)	Beryllium	Be	4	2 2	1287°	2471°		+2	9.012182	2.38×10^{-9}%
2 (IIA)	Magnesium	Mg	12	2 8 2	650°	1090°		+2	24.3050	0.00350%
2 (IIA)	Calcium	Ca	20	2 8 8 2	842°	1484°		+2	40.078	0.000199%
2 (IIA)	Strontium	Sr	38	2 8 18 8 2	777°	1382°		+2	87.62	7.7×10^{-8}%
2 (IIA)	Barium	Ba	56	2 8 18 18 8 2	727°	1897°		+2	137.327	1.46×10^{-8}%
2 (IIA)	Radium	Ra	88	2 8 18 32 18 8 2	700°			+2	[226]	
3 (IIIB)	Scandium	Sc	21	2 8 9 2	1541°	2836°		+3	44.955910	1.12×10^{-7}%
3 (IIIB)	Yttrium	Y	39	2 8 18 9 2	1522°	3345°		+3	88.90585	1.51×10^{-8}%
3 (IIIB)	Lanthanum [2]	La	57	2 8 18 18 9 2	918°	3464°		+3	138.9055	1.45×10^{-9}%
3 (IIIB)	Actinium [3]	Ac	89	2 8 18 32 18 9 2	1051°	3198°		+3	[227]	
4 (IVB)	Titanium	Ti	22	2 8 10 2	1668°	3287°		+2+3+4	47.867	7.8×10^{-6}%
4 (IVB)	Zirconium	Zr	40	2 8 18 10 2	1855°	4409°		+4	91.224	3.72×10^{-8}%
4 (IVB)	Hafnium	Hf	72	2 8 18 32 10 2	2233°	4603°		+4	178.49	5.02×10^{-10}%
4 (IVB)	Rutherfordium	Rf	104	2 8 18 32 32 10 2				+4	[261]	
5 (VB)	Vanadium	V	23	2 8 11 2	1910°	3407°		+2+3+4+5	50.9415	9.6×10^{-7}%
5 (VB)	Niobium	Nb	41	2 8 18 12 1	2477°	4744°		+3+5	92.90638	2.28×10^{-9}%
5 (VB)	Tantalum	Ta	73	2 8 18 32 11 2	3017°	5458°		+5	180.9479	6.75×10^{-11}%
5 (VB)	Hahnium	Ha	105	2 8 18 32 32 11 2					[262]	
6 (VIB)	Chromium	Cr	24	2 8 13 1	1907°	2671°		+2+3+6	51.9961	0.000044%
6 (VIB)	Molybdenum	Mo	42	2 8 18 13 1	2623°	4639°		+6	95.94	8.3×10^{-9}%
6 (VIB)	Tungsten	W	74	2 8 18 32 12 2	3422°	5555°		+6	183.84	4.34×10^{-10}%
6 (VIB)	Seaborgium	Sg	106	2 8 18 32 32 12 2					[266]	
7 (VIIB)	Manganese	Mn	25	2 8 13 2	1246°	2061°		+2+3+4+7	54.938049	0.000031%
7 (VIIB)	Technetium	Tc	43	2 8 18 13 2	2157°	4265°		+4+6+7	[98]	
7 (VIIB)	Rhenium	Re	75	2 8 18 32 13 2	3186°	5596°		+4+6+7	186.207	1.69×10^{-10}%
7 (VIIB)	Nielsbohrium	Ns	107	2 8 18 32 32 13 2					[264]	
8 (VIII)	Iron	Fe	26	2 8 14 2	1538°	2861°		+2+3	55.845	0.00294%
8 (VIII)	Ruthenium	Ru	44	2 8 18 15 1	2334°	4150°		+3	101.07	6.1×10^{-9}%
8 (VIII)	Osmium	Os	76	2 8 18 32 14 2	3033°	5012°		+3+4	190.23	2.20×10^{-9}%
8 (VIII)	Hassium	Hs	108	2 8 18 32 32 14 2					[269]	
9 (VIII)	Cobalt	Co	27	2 8 15 2	1495°	2927°		+2+3	58.933200	7.3×10^{-6}%
9 (VIII)	Rhodium	Rh	45	2 8 18 16 1	1964°	3695°		+3	102.90550	1.12×10^{-9}%
9 (VIII)	Iridium	Ir	77	2 8 18 32 15 2	2446°	4428°		+3+4	192.217	2.16×10^{-9}%
9 (VIII)	Meitnerium	Mt	109	2 8 18 32 32 15 2					[268]	

[2] Lanthanides

Element	Symbol	Z	Shells (K L M N O P)	M.P.°	B.P.°	Ox. States	At. Weight	Abundance%
Cerium	Ce	58	2 8 18 19 9 2	798°	3443°	+3+4	140.116	3.70×10^{-9}%
Praseodymium	Pr	59	2 8 18 21 8 2	931°	3520°	+3	140.90765	5.44×10^{-10}%
Neodymium	Nd	60	2 8 18 22 8 2	1021°	3074°	+3	144.24	2.70×10^{-9}%
Promethium	Pm	61	2 8 18 23 8 2	1042°	3000°	+3	[145]	
Samarium	Sm	62	2 8 18 24 8 2	1074°	1794°	+2+3	150.36	8.42×10^{-10}%
Europium	Eu	63	2 8 18 25 8 2	822°	1596°	+2+3	151.964	3.17×10^{-10}%
Gadolinium	Gd	64	2 8 18 25 9 2	1313°	3273°	+3	157.25	1.076×10^{-9}%

[3] Actinides

Element	Symbol	Z	Shells (K L M N O P Q)	M.P.°	B.P.°	Ox. States	At. Weight	Abundance%
Thorium	Th	90	2 8 18 32 18 10 2	1750°	4788°	+4	232.0381	1.09×10^{-10}%
Protactinium	Pa	91	2 8 18 32 20 9 2	1572°		+5+4	231.03588	
Uranium	U	92	2 8 18 32 21 9 2	1135°	4131°	+3+4+5+6	238.0289	2.94×10^{-11}%
Neptunium	Np	93	2 8 18 32 22 9 2	644°		+3+4+5+6	[237]	
Plutonium	Pu	94	2 8 18 32 24 8 2	640°	3228°	+3+4+5+6	[244]	
Americium	Am	95	2 8 18 32 25 8 2	1176°	2011°	+3+4+5+6	[243]	
Curium	Cm	96	2 8 18 32 25 9 2	1345°		+3	[247]	

Table A.3. The periodic table of the elements. The new IUPAC group format numbers are from 1 to 18, while the numbering system used by the Chemical Abstract Service is given in parentheses. For elements that are not naturally abundant (e.g., the transuranics), the mass number of the longest-lived isotope is given in square brackets. The abundances are based on meteorite and solar wind data. The melting point (MP), boiling point (BP), and critical point (CP) temperatures are given in degrees Celsius. *Source*: Firestone, Baglin, and Chu [1999].

	18 (VIIIA)
	Helium He_2 (2) -272.2° -268.93° -267.96° 0 4.002602 8.9%

13 (IIIA)	14 (IVA)	15 (VA)	16 (VIA)	17 (VIIA)	18 (VIIIA)
Boron B_5 (2 3) 2075° 4000° +3 10.811 6.9×10^{-8}%	**Carbon** C_6 (2 4) 4492t° 3642s° +2+4-4 12.0107 0.033%	**Nitrogen** N_7 (2 5) -210.00° -195.79° -146.94° ±1±2±3+4+5 14.00674 0.0102%	**Oxygen** O_8 (2 6) -218.79° -182.95° -118.56° -2 15.9994 0.078%	**Fluorine** F_9 (2 7) -219.62° -188.12° -129.02° -1 18.9984032 2.7×10^{-6}%	**Neon** Ne_{10} (2 8) -248.59° -246.08° -228.7° 0 20.1797 0.0112%
Aluminum Al_{13} (2 8 3) 660.32° 2519° +3 26.981538 0.000277%	**Silicon** Si_{14} (2 8 4) 1414° 3265° +2+4-4 28.0855 0.00326%	**Phosphorus** P_{15} (2 8 5) 44.15° 280.5° 721° +3+5-3 30.973761 0.000034%	**Sulfur** S_{16} (2 8 6) 115.21° 444.60° 1041° +4+6-2 32.066 0.00168%	**Chlorine** Cl_{17} (2 8 7) -101.5° -34.04° 143.8° +1+5+7-1 35.4527 0.000017%	**Argon** Ar_{18} (2 8 8) -189.35° -185.85° -122.28° 0 39.948 0.000329%

10 (VIII)	11 (IB)	12 (IIB)	13 (IIIA)	14 (IVA)	15 (VA)	16 (VIA)	17 (VIIA)	18 (VIIIA)
Nickel Ni_{28} (2 8 16 2) 1455° 2913° +2+3 58.6934 0.000161%	**Copper** Cu_{29} (2 8 18 1) 1084.62° 2562° +1+2 63.546 1.70×10^{-6}%	**Zinc** Zn_{30} (2 8 18 2) 419.53° 907° +2 65.39 4.11×10^{-6}%	**Gallium** Ga_{31} (2 8 18 3) 29.76° 2204° +3 69.723 1.23×10^{-7}%	**Germanium** Ge_{32} (2 8 18 4) 938.25° 2833° +2+4 72.61 3.9×10^{-7}%	**Arsenic** As_{33} (2 8 18 5) 817t° 614s° 1400° +3+5-3 74.92160 2.1×10^{-8}%	**Selenium** Se_{34} (2 8 18 6) 221° 685° 1493° +4+6-2 78.96 2.03×10^{-7}%	**Bromine** Br_{35} (2 8 18 7) -7.2° 58.8° 315° +1+5-1 79.904 3.8×10^{-8}%	**Krypton** Kr_{36} (2 8 18 8) -157.36° -153.22° -63.74° 0 83.80 1.5×10^{-7}%
Palladium Pd_{46} (2 8 18 18 0) 1554.9° 2963° +2+4 106.42 4.5×10^{-9}%	**Silver** Ag_{47} (2 8 18 18 1) 961.78° 2162° +1 107.8682 1.58×10^{-9}%	**Cadmium** Cd_{48} (2 8 18 18 2) 321.07° 767° +2 112.411 5.3×10^{-9}%	**Indium** In_{49} (2 8 18 18 3) 156.60° 2072° +3 114.818 6.0×10^{-10}%	**Tin** Sn_{50} (2 8 18 18 4) 231.93° 2602° +2+4 118.710 1.25×10^{-8}%	**Antimony** Sb_{51} (2 8 18 18 5) 630.63° 1587° +3+5-3 121.760 1.01×10^{-9}%	**Tellurium** Te_{52} (2 8 18 18 6) 449.51° 988° +4+6-2 127.60 1.57×10^{-8}%	**Iodine** I_{53} (2 8 18 18 7) 113.7° 184.4° 546° +1+5+7-1 126.90447 2.9×10^{-9}%	**Xenon** Xe_{54} (2 8 18 18 8) -111.75° -108.04° 16.58° 0 131.29 1.5×10^{-8}%
Platinum Pt_{78} (2 8 18 32 16 2) 1768.4° 3825° +2+4 195.078 4.4×10^{-9}%	**Gold** Au_{79} (2 8 18 32 18 1) 1064.18° 2856° +1+3 196.96655 6.1×10^{-10}%	**Mercury** Hg_{80} (2 8 18 32 18 2) -38.83° 356.73° 1477° +1+2 200.59 1.11×10^{-9}%	**Thallium** Tl_{81} (2 8 18 32 18 3) 304° 1473° +1+3 204.3833 6.0×10^{-10}%	**Lead** Pb_{82} (2 8 18 32 18 4) 327.46° 1749° +2+4 207.2 1.03×10^{-8}%	**Bismuth** Bi_{83} (2 8 18 32 18 5) 271.40° 1564° +3+5 208.98038 4.7×10^{-10}%	**Polonium** Po_{84} (2 8 18 32 18 6) 254° 962° +2+4 [209]	**Astatine** At_{85} (2 8 18 32 18 7) 302° [210]	**Radon** Rn_{86} (2 8 18 32 18 8) -71° -61.7° 104° 0 [222]
Element-110 110_{110} (2 8 18 32 32 16 2) [271]	**Element-111** 111_{111} (2 8 18 32 32 17 2) [272]	**Element-112** 112_{112} (2 8 18 32 32 18 2) [277]						

Terbium Tb_{65} (2 8 18 27 8 2) 1356° 3230° +3 158.92534 1.97×10^{-10}%	**Dysprosium** Dy_{66} (2 8 18 28 8 2) 1412° 2567° +3 162.50 1.286×10^{-9}%	**Holmium** Ho_{67} (2 8 18 29 8 2) 1474° 2700° +3 164.93032 2.90×10^{-10}%	**Erbium** Er_{68} (2 8 18 30 8 2) 1529° 2868° +3 167.26 8.18×10^{-10}%	**Thulium** Tm_{69} (2 8 18 31 8 2) 1545° 1950° +3 168.93421 1.23×10^{-10}%	**Ytterbium** Yb_{70} (2 8 18 32 8 2) 819° 1196° +2+3 173.04 8.08×10^{-10}%	**Lutetium** Lu_{71} (2 8 18 32 9 2) 1663° 3402° +3 174.967 1.197×10^{-10}%
Berkelium Bk_{97} (2 8 18 32 27 8 2) 1050° +3+4 [247]	**Californium** Cf_{98} (2 8 18 32 28 8 2) 900° +3 [251]	**Einsteinium** Es_{99} (2 8 18 32 29 8 2) 860° +3 [252]	**Fermium** Fm_{100} (2 8 18 32 30 8 2) 1527° +3 [257]	**Mendelevium** Md_{101} (2 8 18 32 31 8 2) 827° +2+3 [258]	**Nobelium** No_{102} (2 8 18 32 32 8 2) 827° +2+3 [259]	**Lawrencium** Lr_{103} (2 8 18 32 32 9 2) 1627° +3 [262]

Table A.3. *(cont.)* The periodic table of the elements. The new IUPAC group format numbers are from 1 to 18, while the numbering system used by the Chemical Abstract Service is given in parentheses. For elements that are not naturally abundant (e.g., the transuranics), the mass number of the longest-lived isotope is given in square brackets. The abundances are based on meteorite and solar wind data. The melting point (MP), boiling point (BP), and critical point (CP) temperatures are given in degrees Celsius. *Source*: Firestone, Baglin, and Chu [1999].

Table A.4. Some physical properties of the elements.

Z	El	Atomic Weight	Mass Density (g/cm^3)	Melting Point (°C)	Boiling Point (°C)	Z	El	Atomic Weight	Mass Density (g/cm^3)	Melting Point (°C)	Boiling Point (°C)
1	H	1.00794	0.0708	-259.34	-252.87	47	Ag	107.8682	10.501	961.78	2162
2	He	4.002602	0.124901	-272.2	-268.93	48	Cd	112.411	8.69	321.07	767
3	Li	6.941	0.534	180.5	1342	49	In	114.818	7.31	156.60	2072
4	Be	9.012182	1.85	1287	2471	50	Sn	118.710	7.287^{26°	231.93	2602
5	B	10.811	2.37	2075	4000	51	Sb	121.760	6.685^{26°	630.63	1687
6	C^a	12.0107	2.2670^{15°	4492^t	3842^s	52	Te	127.60	6.232	449.51	988
7	N	14.00674	0.807	-210.00	-195.79	53	I	126.90447	4.93^{20°	113.7	184
8	O	15.9994	1.141	-218.79	-182.95	54	Xe	131.29	2.953	-111.75	-108.04
9	F	18.9984032	1.50	-219.62	-188.12	55	Cs	132.90545	1.93	28.44	671
10	Ne	20.1797	1.204	-248.59	-246.08	56	Ba	137.327	3.62	727	1897
11	Na	22.989770	0.97	97.80	883	57	La	138.9055	6.15	918	3464
12	Mg	24.3050	0.74	650	1090	58	Ce	140.116	8.16	798	3443
13	Al	26.981538	2.70	660.32	2519	59	Pr	140.90765	6.77	931	3520
14	Si	28.0855	2.3296	1414	3265	60	Nd	144.24	7.01	1021	3074
15	P	30.973761	1.82	44.15	280.5	61	Pm	[145]	7.26	1042	3000
16	S	32.066	2.067	115.21	444.60	62	Sm	150.36	7.52	1074	1794
17	Cl	35.4527	1.56	-101.5	-34.04	63	Eu	151.964	5.24	822	1596
18	Ar	39.948	1.396	-189.35	-185.85	64	Gd	157.25	7.90	1313	3273
19	K	39.0983	0.89	63.38	759	65	Tb	158.92534	8.23	1356	3230
20	Ca	40.078	1.54	842	1484	66	Dy	162.50	8.55	1412	2567
21	Sc	44.955910	2.99	1541	2836	67	Ho	164.93032	8.80	1474	2700
22	Ti	47.867	4.5	1668	3287	68	Er	167.26	9.07	1529	2868
23	V	50.9415	6.0	1910	3407	69	Tm	168.93421	9.32	1545	1950
24	Cr	51.9961	7.15	1907	2671	70	Yb	173.04	6.90	819	1196
25	Mn	54.938049	7.3	1246	2061	71	Lu	174.967	9.84	1663	3402
26	Fe	55.845	7.875	1538	2861	72	Hf	178.49	13.3	2233	4603
27	Co	58.933200	8.86	1495	2927	73	Ta	180.9479	16.4	3017	5458
28	Ni	58.6934	8.912	1455	2913	74	W	183.84	19.3	3422	5555
29	Cu	63.546	8.933	1084.62	2562	75	Re	186.207	20.8	3186	5596
30	Zn	65.39	7.134	419.53	907	76	Os	190.23	22.5	3033	5012
31	Ga	69.723	5.91	29.76	2204	77	Ir	192.217	22.5	2446	4428
32	Ge	72.61	5.323	938.25	2833	78	Pt	195.078	21.46	1768.4	3825
33	As	74.92160	5.776^{26°	817^t	614^s	79	Au	196.96655	19.282	1064.18	2856
34	Se	78.96	4.809^{26°	221	685	80	Hg	200.59	13.5336	-38.83	356.73
35	Br	79.904	3.11	-7.2	58.8	81	Tl	204.3833	11.8	304	1473
36	Kr	83.80	2.418	-157.36	-153.22	82	Pb	207.2	11.342	327.46	1749
37	Rb	85.4678	1.53	39.31	688	83	Bi	208.98038	9.807	271.40	1564
38	Sr	87.62	2.64	777	1382	84	Po	[209]	9.32	254	962
39	Y	88.90585	4.47	1522	3345	85	At	[210]		302	
40	Zr	91.224	6.52	1855	4409	86	Rn	[222]	4.4	-71	-61.7
41	Nb	92.90638	8.57	2477	4744	87	Fr	[223]		27	
42	Mo	95.94	10.2	2623	4639	88	Ra	[226]	5	700	
43	Tc	[98]	11	2157	4265	89	Ac	[227]	10.07	1051	3198
44	Ru	101.07	12.1	2334	4150	90	Th	232.0381	11.72	1750	4788
45	Rh	102.90550	12.4	1964	3695	91	Pa	231.03588	15.37	1572	
46	Pd	106.42	12.0	1554.9	2963	92	U	238.0289	18.95	1135	4131

[a]graphite [t]critical temperature [s]sublimation temperature

Table A.5. SI prefixes.

Factor	Prefix	Symbol
10^{30}	quecca*	Q
10^{27}	ronna*	R
10^{24}	yotta	Y
10^{21}	zetta	Z
10^{18}	exa	E
10^{15}	peta	P
10^{12}	tera	T
10^{9}	giga	G
10^{6}	mega	M
10^{3}	kilo	k
10^{2}	hecto	h
10^{1}	deca	da
10^{-1}	deci	d
10^{-2}	centi	c
10^{-3}	milli	m
10^{-6}	micro	μ
10^{-9}	nano	n
10^{-12}	pico	p
10^{-15}	femto	f
10^{-18}	atto	a
10^{-21}	zepto	z
10^{-24}	yocto	y
10^{-27}	ronto*	r
10^{-30}	quecto*	q

*proposed 2019

Table A.6. Conversion factors.

Property	Unit	SI equivalent
Length	in.	2.54×10^{-2} m[a]
	ft	3.048×10^{-1} m[a]
	mile (int'l)	$1.609\,344 \times 10^{3}$ m[a]
Area	in^2	6.4516×10^{-4} m^2[a]
	ft^2	$9.290\,304 \times 10^{-2}$ m^2[a]
	acre	$4.046\,873 \times 10^{3}$ m^2
	square mile (int'l)	$2.589\,988 \times 10^{6}$ m^2
	hectare	1×10^{4} m^2
Volume	oz (U.S. liquid)	$2.957\,353 \times 10^{-5}$ m^3
	in^3	$1.638\,706 \times 10^{-5}$ m^3
	gallon (U.S.)	$3.785\,412 \times 10^{-3}$ m^3
	ft^3	$2.831\,685 \times 10^{-2}$ m^3
Mass	oz (avdp.)	$2.834\,952 \times 10^{-2}$ kg
	lb	$4.535\,924 \times 10^{-1}$ kg
	ton (short)	$9.071\,847 \times 10^{2}$ kg
Force	kg$_f$	9.806 650 N [a]
	lb$_f$	4.448 222 N
	ton	$8.896\,444 \times 10^{3}$ N
Pressure	lb$_f$/in^2 (psi)	$6.894\,757 \times 10^{3}$ Pa
	lb$_f$/ft^2	$4.788\,026 \times 10^{1}$ Pa
	atm (standard)	$1.013\,250 \times 10^{5}$ Pa[a]
	in. H_2O (@ 4 C)	$2.490\,82 \times 10^{2}$ Pa
	in. Hg (@ 0 C)	$3.386\,39 \times 10^{3}$ Pa
	mm Hg (@ 0 C)	$1.333\,22 \times 10^{2}$ Pa
	bar	1×10^{5} Pa[a]
Energy	eV	$1.602\,19 \times 10^{-19}$ J
	cal	4.184 J[a]
	Btu	$1.054\,350 \times 10^{3}$ J
	kWh	3.6×10^{6} J[a]
	MWd	8.64×10^{10} J[a]

[a]Exact conversion factor.

Source: Standards for Metric Practice, ANSI/ASTM E380-76, American National Standards Institute, New York, 1976.

Appendix B

Cross Sections and Related Data

The tables in this appendix present basic interaction data for thermal neutrons and for photons. These tables give but a very small fraction of the data contained in comprehensive data libraries and other voluminous compilations. The data selected for these tables are for important interactions encountered in many nuclear and atomic analyses. However, for many calculations, it is necessary to obtain interaction data directly from data libraries maintained at several national nuclear data centers around the world. In the United States, the National Nuclear Data Center at Brookhaven National Laboratory (`www.nndc.bnl.gov`) and the Radiation Safety Information and Computational Center (RSICC) at Oak Ridge National Laboratory (`rsicc.ornl.gov`) provide access to extensive nuclear data libraries. The Nuclear Data Service of the International Atomic Energy Agency (`www-nds.iaea.org`) and the Japan Atomic Energy Research Institute (`wwwndc.jaea.go.jp`) also provide nuclear data to the international radiation community.

B.1 Data Tables

The following data are provided in this appendix.

Table B.1. 2200-m/s macroscopic cross sections for the elements.
Table B.2. Thermal microscopic neutron cross sections for special isotopes.
Table B.3. Activation radionuclides produced from thermal neutron capture.
Table B.4. Photon yields for thermal neutron capture.
Table B.5. Prompt gamma-ray energies used for elemental identification.
Table B.6. Photon interaction coefficients for air, water, concrete, iron, and lead.
Table B.7. μ_t/ρ and μ_{t-coh}/ρ for the light elements.

Data for Tables B.1–B.7 have been extracted from the sources provided in the tables. In all these tables, entries such as $1.234 - 5$ should be read as 1.234×10^{-5}.

B.1.1 Thermal Neutron Interactions

The thermal neutron integration rate density for the ith type of interaction is given by Eq. (4.97) for absorption and by Eq. (4.102) for elastic scattering. The total reaction rate density is the sum of all absorption and scattering reactions. Thus

$$F_j = g_j(T)\Sigma_j(E_o)\phi_o, \qquad j = t, s, c, f, \alpha, \ldots \tag{B.1}$$

Here the subscripts t, s, c, f, α refer to total, (elastic) scatter, (n,γ), fission, (n,α) reactions, respectively. In many radiation detection applications, the thermal neutron reaction rates in a natural element are needed.

Cross section data bases are almost always of microscopic cross sections for individual isotopes and, thus, to obtain thermal macroscopic cross sections for an element, it is necessary to average cross section values and the Westcott g-factors over all stable isotopes of an element, namely

$$\Sigma_j(E_o) = \frac{\rho N_a}{\mathcal{A}} \sum_i f_i \sigma_j^i(E_o) \tag{B.2}$$

and

$$g_j(T) = \sum_i f_i g_j^i(T)\sigma_j^i(E_o) \Big/ \sum_i f_i \sigma_j^i(E_o). \tag{B.3}$$

Here i refers to the stable isotopes and f_i is the isotopic abundance. The 2200-m/s macroscopic cross sections for Σ_t, Σ_s, and Σ_c are tabulated for most elements in Table B.1. In this table densities are at 25°C except for elements that are normally gaseous and their densities are at their liquid's boiling point.

B.1.2 Photon Interactions

In Tables B.6 and B.7, listing photon interaction coefficients, the following notation is used: coherent scattering coefficient with electron binding, μ_{coh}; incoherent scattering coefficient with electron binding, μ_c; photoelectric effect coefficient, μ_{ph}; pair-production coefficient, μ_{pp}; total interaction coefficient $\mu_t \equiv \mu_c + \mu_{ph} + \mu_{pp}$; and the total interaction coefficient less the coherent coefficient $\mu_{t-coh} \equiv \mu_{coh} + \mu_c + \mu_{ph} + \mu_{pp}$.

REFERENCES

HUBBELL, J.H. AND S.M. SELTZER, *Tables of X-Ray Attenuation Coefficients 1 keV to 20 MeV for Elements Z = 1 to 92 and 48 Additional Substances of Dosimetric Interest*, Report NISTIR 5632, Gaithersburg, MD: National Institute of Standards and Technology, 1995.

LAMARSH, J.R., *Introduction to Nuclear Reactor Theory*, Reading, MA: Addison-Wesley, 1966.

LONE, M.A., R.A. LEAVITT, AND D.A. HARRISON, "Prompt Gamma Rays from Thermal Neutron Capture," *At. Data Nucl. Data Tables*, **26**, No. 6, 511–559, (1981). [Data files released as Package DLC-140 THERMGAM by the Radiation Shielding Information Center, Oak Ridge National Laboratory, Oak Ridge, TN.]

ORPHAN, V.J., N.C. RASMUSSEN, AND T.L. HARPER, " Line and Continuum Gamma-Ray Yields from Thermal Neutron Capture in 75 Elements," Report GA-10248, San Diego, CA: Gulf General Atomic, Inc., 1970.

RÉVAY, Z., R.B. FIRESTONE, T. BELGYA, AND G.L. MOLNÁR, "Prompt Gamma-Ray Spectrum Catalogue," in *Handbook of Prompt Gamma Ray Activation Analysis in Neutron Beams*, G. MOLNÁR (Ed.), Dordrecht, The Netherlands: Kluwer Acad. Pub., 2004.

SELTZER, S.M., "Calculation of Photon Mass Energy-Transfer and Mass Energy-Absorption Coefficients," *Radiat. Res.*, **136**, 147–179, (1993).

SHIBATA, K., O. IWAMOTO, T. Nakagawa, N. IWAMOTO, A. ICHIHARA, S. KUNIEDA, S. CHIBA, K. FURUTAKA, N. Otuka, T. OHSAWA, T. MURATA, H. MATSUNOBY, A. ZUKERAN, S. KAMADA, AND J. KATAKURA: "JENDL-4.0: A New Library for Nuclear Science and Engineering," *J. Nucl. Sci. Technol.*, **48(1)**, 1-30, (2011).

SHLEIEB, B. (Ed.), *The Health Physics and Radiological Health Handbook*, Silver Spring, MD: Scinta, 1992.

Table B.1. Thermal neutron (2200 m/s) cross sections and Westcott correction factors for most elements. Calculated from data extracted from Jendl 4.0 [Shibata et al. [2011].

Z	El.	Density (g cm^{-3})	$\Sigma_t(E_o)$ (cm^{-1})	$g_t(T)$	$\Sigma_s(E_o)$ (cm^{-1})	$g_s(T)$	$\Sigma_c(E_o)$ (cm^{-1})	$g_c(T)$
1	H	0.0708	1.294	1.083	1.280	1.084	1.404-2	1.001
2	He	0.1249	0.01624	1.121	0.01610	1.122	1.391-12	1.001
3	Li	0.534	3.356	1.003	0.04746	1.127	2.079-3	1.001
4	Be	1.85	0.8049	1.127	0.8039	1.127	1.050-3	1.001
5	B	2.37	101.4	1.002	0.6177	1.128	1.367-2	1.001
6	C	2.267	.5623	1.128	0.5617	1.128	4.388-4	1.001
7	N	0.807	.4293	1.108	0.3591	1.128	2.593-3	1.001
8	O	1.141	.1703	1.129	0.1703	1.129	8.156-6	1.001
9	F	1.5	.1783	1.128	0.1779	1.129	4.550-4	1.001
11	Na	0.97	0.09206	1.110	0.07854	1.129	1.350-2	1.001
12	Mg	0.74	0.06729	1.126	0.06614	1.129	1.155-3	1.001
13	Al	2.7	0.1007	1.110	0.08684	1.129	1.389-2	0.993
14	Si	2.330	0.1100	1.119	0.1018	1.129	8.236-3	1.001
15	P	1.82	0.1186	1.123	0.1127	1.129	5.881-3	1.001
16	S	2.067	0.06070	1.087	0.04046	1.129	1.994-2	1.001
17	Cl	1.56	1.316	1.042	0.4272	1.129	0.8783	1.000
18	Ar	1.396	0.02767	1.065	0.01379	1.129	1.389-2	1.001
19	K	0.89	0.05740	1.066	0.02911	1.129	2.822-2	1.001
20	Ca	1.54	0.08072	1.113	0.0708	1.129	9.916-3	1.001
21	Sc	2.99	1.998	1.059	0.9108	1.129	1.087	1.000
22	Ti	4.5	0.5717	1.048	0.2085	1.129	0.3634	1.001
23	V	6.0	0.7088	1.064	0.3529	1.129	0.3560	1.000
24	Cr	7.15	0.5498	1.068	0.2894	1.129	0.2604	1.000
25	Mn	7.3	1.232	1.018	0.1694	1.129	1.063	1.001
26	Fe	7.875	1.180	1.106	0.9627	1.129	0.2178	1.001
27	Co	8.86	0.4132	1.019	0.05795	1.129	0.3552	1.001
28	Ni	8.912	2.041	1.103	1.633	1.129	0.4083	1.000
29	Cu	8.933	0.9919	1.088	0.6713	1.129	0.3205	1.000
30	Zn	7.134	0.3494	1.104	0.2796	1.129	6.979-2	1.000
31	Ga	5.91	0.5027	1.093	0.3597	1.129	0.1431	1.001
32	Ge	5.323	0.4947	1.103	0.3969	1.129	9.786-2	1.000
33	As	5.776	0.4466	1.074	0.2538	1.129	0.1928	1.001
34	Se	4.809	0.7455	1.053	0.3068	1.129	0.4389	1.000
35	Br	3.11	0.2391	1.044	0.08115	1.129	0.1579	1.000
36	Kr	2.418	0.5874	1.027	0.1390	1.129	0.4482	0.9956
37	Rb	1.53	0.06308	1.121	0.05898	1.129	4.096-3	1.000
38	Sr	2.64	0.1251	1.106	0.1015	1.129	0.02367	1.006
39	Y	4.47	0.2707	1.111	0.2318	1.129	3.888-2	1.000
40	Zr	6.52	0.2846	1.125	0.2762	1.129	8.421-3	1.000

(cont.)

Table B.1. *(cont.)* Thermal neutron (2200 m/s) cross sections and Westcott correction factors for most elements. Calculated from data extracted from Jendl 4.0 [Shibata et al. [2011].

Z	El.	Density (g cm^{-3})	$\Sigma_t(E_o)$ (cm^{-1})	$g_t(T)$	$\Sigma_s(E_o)$ (cm^{-1})	$g_s(T)$	$\Sigma_c(E_o)$ (cm^{-1})	$g_c(T)$
41	Nb	8.57	0.4165	1.109	0.3530	1.129	6.344-2	1.000
42	Mo	10.2	0.5361	1.090	0.3717	1.129	0.1644	1.000
44	Ru	12.1	0.7148	1.088	0.4837	1.129	0.2310	1.001
45	Rh	12.4	9.898	1.026	0.2377	1.123	9.667	1.023
46	Pd	12,0	0.7474	1.048	0.2792	1.129	0.4682	1.000
47	Ag	10.50	3.993	1.013	0.3003	1.127	3.693	1.004
48	Cd	8.69	115.2	1.335	0.3513	1.581	114.9	1.335
49	In	7.31	7.501	1.021	0.09922	1.120	7.403	1.020
50	Sn	7.287	0.1962	1.114	0.1737	1.129	2.249-2	0.9985
51	Sb	6.685	0.2962	1.056	0.1236	1.129	0.1726	1.003
52	Te	6.232	0.2413	1.064	0.1105	1.129	0.1308	1.010
53	I	4.93	0.2259	1.044	0.07615	1.129	0.1498	1.001
54	Xe	2.953	0.5159	1.037	0.1430	1.129	0.3728	1.002
55	Cs	1.93	0.2876	1.018	0.03485	1.126	0.2527	1.003
56	Ba	3.62	0.1011	1.105	0.08229	1.129	1.877-2	1.001
57	La	6.15	0.5081	1.068	0.2686	1.128	0.2395	1.000
58	Ce	8.16	0.1229	1.107	0.1013	1.129	2.156-2	1.001
59	Pr	6.77	0.4111	1.024	0.07899	1.128	0.3321	1.000
60	Nd	7.01	1.945	1.028	0.4765	1.123	1.469	0.9969
62	Sm	7.52	171.9	1.680	0.8999	2.158	171.0	1.677
63	Eu	5.24	94.56	.9043	0.1441	1.077	94.42	0.9041
64	Gd	7.9	1475.5	.8473	4.168	0.9254	1471.4	0.8472
65	Tb	8.23	0.9365	1.031	0.2152	1.128	0.7213	1.002
66	Dy	8.55	32.88	1.001	3.020	1.105	29.86	0.9901
68	Er	9.07	5.375	1.069	0.2929	1.127	5.082	1.066
69	Tm	9.32	3.704	1.012	0.2120	1.124	3.492	1.005
70	Yb	6.9	1.400	1.054	0.5622	1.128	0.8380	1.005
72	Hf	13.3	5.125	1.024	0.4481	1.127	4.677	1.014
73	Ta	16.4	1.438	1.030	0.3092	1.127	1.129	1.003
74	W	19.3	1.463	1.029	0.3154	1.128	1.148	1.002
76	Os	22.5	2.280	1.052	1.025	1.125	1.255	0.9916
79	Au	19.28	6.284	1.015	0.4670	1.127	5.816	1.006
80	Hg	13.53	16.04	.9975	0.9896	1.121	15.05	0.9891
82	Pb	11.34	0.3787	1.127	0.3737	1.129	5.057-3	1.000
83	Bi	9.807	0.2663	1.129	0.2654	1.129	9.668-4	1.000
90	Th	11.72	0.6199	1.080	0.3966	1.128	0.2232	0.9950
92	U	18.95	0.8118	1.065	0.4479	1.129	0.1620	1.000

Table B.2. Thermal neutron (2200 m s^{-1} or 0.0253 eV) cross sections for some special isotopes at 300 K. Cross-section notation: σ_γ for (n,γ); σ_s for elastic scattering; σ_t for total; σ_f for fission; σ_α for (n,α); and σ_p for (n,p).

Isotope	Abundance (atom %)	Half-life	Reaction cross sections (b)		
^{1}H	99.985		$\sigma_\gamma = 333$ mb	$\sigma_s = 30.5$	$\sigma_t = 30.9$
^{2}H	0.015		$\sigma_\gamma = 506$ μb	$\sigma_s = 4.26$	$\sigma_t = 4.30$
^{3}H		12.33 y	$\sigma_\gamma = 6$ μb	$\sigma_s = 1.53$	$\sigma_t = 1.53$
^{6}Li	7.5		$\sigma_\alpha = 941$	$\sigma_\gamma = 38.6$ mb	$\sigma_t = 943$
^{7}Li	92.5		$\sigma_\gamma = 45.7$ mb	$\sigma_s = 1.04$	$\sigma_t = 1.09$
^{10}B	19.6		$\sigma_\alpha = 3840$	$\sigma_\gamma = 0.50$	$\sigma_t = 3847$
^{11}B	80.4		$\sigma_\gamma = 5.53$ mb	$\sigma_s = 5.08$	$\sigma_t = 5.08$
^{12}C	98.89		$\sigma_s = 4.74$	$\sigma_\gamma = 3.4$ mb	$\sigma_t = 4.74$
^{13}C	1.11		$\sigma_\gamma = 1.37$ mb	$\sigma_t = 4.19$	
^{14}C		5736 y	$\sigma_\gamma = 1.0$ μb		
^{14}N	99.64		$\sigma_p = 1.83$	$\sigma_\gamma = 75$ mb	$\sigma_t = 12.2$
^{15}N	0.36		$\sigma_\gamma = 24$ μb	$\sigma_s = 4.58$	$\sigma_t = 4.58$
^{16}O	99.756		$\sigma_\gamma = 190$ μb	$\sigma_s = 4.03$	$\sigma_t = 4.03$
^{17}O	0.039		$\sigma_\alpha = 235$ mb	$\sigma_\gamma = 3.84$ mb	$\sigma_t = 4.17$
^{18}O	0.205		$\sigma_\gamma = 160$ μb		
^{232}Th	100	1.405×10^{10} y	$\sigma_f = 2.5$ μb	$\sigma_\gamma = 5.13$	$\sigma_t = 20.4$
^{233}Th		22.3 m	$\sigma_f = 15$	$\sigma_\gamma = 1450$	$\sigma_t = 1478$
^{233}U		1.592×10^{5} y	$\sigma_f = 529$	$\sigma_\gamma = 46.0$	$\sigma_t = 588$
^{234}U	0.0055	2.455×10^{5} y	$\sigma_f = 0.465$	$\sigma_\gamma = 103$	$\sigma_t = 116$
^{235}U	0.7200	7.038×10^{8} y	$\sigma_f = 587$	$\sigma_\gamma = 99.3$	$\sigma_t = 700$
^{236}U		2.342×10^{7} y	$\sigma_f = 47$ mb	$\sigma_\gamma = 5.14$	$\sigma_t = 14.1$
^{238}U	99.2745	4.468×10^{9} y	$\sigma_f = 11.8$ μb	$\sigma_\gamma = 2.73$	$\sigma_t = 12.2$
^{239}U		23.45 m	$\sigma_f = 14$	$\sigma_\gamma = 22$	
^{239}Pu		24110 y	$\sigma_f = 749$	$\sigma_\gamma = 271$	$\sigma_t = 1028$
^{240}Pu		6564 y	$\sigma_f = 64$ mb	$\sigma_\gamma = 289$	$\sigma_t = 290$
^{241}Pu		14.35 y	$\sigma_f = 1015$	$\sigma_\gamma = 363$	$\sigma_t = 1389$
^{242}Pu		3.733×10^{5} y	$\sigma_f = 1.0$ mb	$\sigma_\gamma = 19.3$	$\sigma_t = 27.0$

Source: ENDF/B-VI and other data extracted from the National Nuclear Data Center Online Service, Brookhaven National Laboratory, Upton, NY, January 1995.

Table B.3. Activation radionuclides resulting from thermal neutron (2200 m s^{-1} or 0.0253 eV) capture. Activation cross sections include production of short-lived metastable daughter nuclides except for ^{59}Co. The gamma photon energies and frequencies for the decay of the activation nuclide can be found at the NUDAT site `www.nndc.bnl.gov\nudat2\indx_dec.jsp`

Activated nuclide		Parent nuclide		
Nuclide	Half-life	Nuclide	Abundance (%) or half-life	Activation cross section (b)
^{12}B	0.0202 s	^{11}B	80.4	5.5 mb
^{16}N	7.13 s	^{15}N	0.38	24 μb
^{17}N[a]	4.173 s	^{17}O	0.039	5.2 μb
^{19}O	26.91 s	^{18}O	0.205	160 μb
^{24}Na	14.96 h	^{23}Na	100	0.530
^{28}Al	2.241 m	^{27}Al	100	0.231
^{31}Si	2.62 h	^{30}Si	3.12	0.107
^{38}Cl	37.24 m	^{37}Cl	24.23	0.433
^{41}Ar	1.822 h	^{40}Ar	99.59	0.660
^{42}K	12.36 h	^{41}K	6.7	1.46
^{46}Sc	83.81 d	^{45}Sc	100	27.2
^{47}Ca	4.536 d	^{46}Ca	0.0033	0.74
^{49}Ca	8.715 m	^{48}Ca	0.185	1.09
^{51}Cr	27.70 d	^{50}Cr	4.35	15.9
^{56}Mn	2.579 h	^{55}Mn	100	13.3
^{59}Fe	44.50 d	^{58}Fe	0.31	1.28
^{60m}Co	10.47 m	^{59}Co	100	20.4
^{60}Co	5.271 y	^{59}Co	100	16.8
		^{60m}Co	10.47 m	
^{75}Se	119.8 d	^{74}Se	0.87	51.8
^{76}As	1.078 d	^{75}As	100	4.5
^{95}Zr	64.02 d	^{94}Zr	17.5	49.9 mb
^{97}Zr	16.91 h	^{96}Zr	2.8	22.9 mb
^{99}Mo	65.94 h	^{98}Mo	23.4	0.130
^{99m}Tc	6.01 h	^{99}Mo	65.94 h	
^{108}Ag	2.37 m	^{107}Ag	51.83	37.6
^{116m}In	54.4 m	^{115}In	95.7	202
^{198}Au	2.6952 d	^{197}Au	100	98.65
^{203}Hg	46.61 d	^{202}Hg	29.8	4.89
^{233}Th	22.3 m	^{232}Th	100	7.37
^{233}Pa	26.97 d	^{233}Th	22.3 m	1500
^{239}U	23.45 m	^{238}U	99.2745	2.68
^{239}Np	2.357 d	^{239}U	23.45 m	22.0

[a]Decays by neutron emission.

Source: Data extracted from the National Nuclear Data Center Online Service, Brookhaven National Laboratory, Upton, NY, January 1995.

Table B.4. Radiative capture cross sections σ_γ and the number of capture gamma rays produced in common elements with natural isotopic abundances. The thermal capture cross sections are for 2200 m s^{-1} (0.0253 eV) neutrons in units of the barn (10^{-24} cm^2). Listed are the numbers of gamma rays produced, per neutron capture, in each of 11 energy groups. Unless otherwise noted, yields are for both line (discrete) and continuum (or unresolved) capture photons.

		Energy group (MeV)										
	σ_γ (b)	0–1	1–2	2–3	3–4	4–5	5–6	6–7	7–8	8–9	9–10	10–11
H	3.32E−1	0.0000	0.0000	1.0000	0.0000	0.0000	0.0000	0.0000	0.0000	0.0000	0.0000	0.0000
Li	3.63E−2	0.1242	0.0491	0.8933	0.0000	0.0000	0.0000	0.0107	0.0402	0.0000	0.0000	0.0000
Be	9.20E−3	0.2641	0.0000	0.2356	0.4530	0.0000	0.0175	0.6375	0.0000	0.0000	0.0000	0.0000
B	1.03E−1	0.0000	0.0000	0.0000	0.0000	1.1014	0.0000	0.3950	0.4785	0.0000	0.0000	0.0000
C	3.37E−3	0.0000	0.2953	0.0000	0.3210	0.6764	0.0000	0.0000	0.0000	0.0000	0.0000	0.0000
N	7.47E−2	0.2632	0.3716	0.2450	0.2819	0.1578	0.8031	0.2064	0.1019	0.0397	0.0222	0.1412
O	2.70E−4	1.0000	0.8200	0.8200	0.1800	0.0000	0.0000	0.0000	0.0000	0.0000	0.0000	0.0000
Na	4.00E−1	0.9266	0.2046	0.7264	0.6537	0.0323	0.0633	0.2244	0.0000	0.0000	0.0000	0.0000
Mg	6.30E−2	0.5963	0.6876	0.6404	0.9584	0.0662	0.1078	0.1156	0.0372	0.0474	0.0075	0.0000
Al	2.30E−1	0.2751	0.0855	0.3225	0.3057	0.3499	0.0863	0.0778	0.3235	0.0000	0.0000	0.0000
Si	1.60E−1	0.1233	0.3403	0.3362	0.8046	0.6467	0.0634	0.1635	0.0903	0.0254	0.0000	0.0040
P	1.80E−1	0.4065	0.5411	0.5212	0.5446	0.2689	0.1290	0.1809	0.0788	0.0000	0.0000	0.0000
S	5.20E−1	0.7555	0.0000	0.7716	0.3642	0.1794	0.6348	0.0000	0.0391	0.0266	0.0000	0.0000
Cl	3.32E+1	0.4818	0.7536	0.4286	0.2597	0.1992	0.1516	0.3858	0.1966	0.0291	0.0000	0.0000
K	2.10E+0	0.7491	0.5656	0.7044	0.4329	0.2654	0.3638	0.0344	0.0670	0.0000	0.0000	0.0000
Ca	4.30E−1	0.2400	0.9349	0.5187	0.1712	0.2303	0.1256	0.4383	0.0216	0.0000	0.0000	0.0000
Ti	6.10E+0	0.3213	0.9772	0.0832	0.1221	0.1187	0.0283	0.6089	0.0109	0.0043	0.0022	0.0003
V	5.04E+0	0.3837	0.2486	0.1335	0.0591	0.0877	0.3158	0.3947	0.1972	0.0000	0.0000	0.0000
Cr	3.10E+0	0.4051	0.1608	0.2067	0.0921	0.0421	0.1103	0.1189	0.2461	0.3766	0.1097	0.0000
Mn	1.33E+1	0.7128	0.1242	0.3838	0.2199	0.2049	0.2981	0.0949	0.3257	0.0000	0.0000	0.0000
Fe	2.55E+0	0.2781	0.2383	0.1018	0.1328	0.1137	0.1097	0.1045	0.5865	0.0087	0.0415	0.0011
Co	3.72E+1	0.9375	0.1737	0.0794	0.0920	0.1121	0.2991	0.2893	0.0980	0.0000	0.0000	0.0000
Ni	4.43E+0	0.2616	0.0658	0.0604	0.0364	0.0371	0.0746	0.1703	0.1404	0.5898	0.0000	0.0000
Cu	3.79E+0	0.8176	0.0602	0.0458	0.0588	0.0917	0.1018	0.1621	0.6488	0.0000	0.0000	0.0000
Zn	1.10E+0	0.1598	0.3949	0.0943	0.0420	0.0596	0.0985	0.1200	0.1429	0.0071	0.0109	0.0000
Ge	2.30E+0	0.9587	0.1459	0.0360	0.4219	0.0576	0.0697	0.1096	0.0266	0.0113	0.0000	0.0000
Br	6.80E+0	0.6347	0.0666	0.0101	0.0030	0.0044	0.0187	0.0318	0.0227	0.0000	0.0000	0.0000
Zr	1.85E−1	0.8081	0.3048	0.2119	0.1361	0.0847	0.0820	0.1745	0.0042	0.0082	0.0000	0.0000
Mo	2.65E+0	0.8097	0.2000	0.0816	0.0416	0.0590	0.0542	0.0611	0.0074	0.0054	0.0000	0.0000
Ag	6.36E+1	0.6831	0.0166	0.0105	0.0102	0.0312	0.0877	0.0266	0.0112	0.0000	0.0000	0.0000
Cd	2.45E+3	1.0399	0.2239	0.1895	0.0736	0.0410	0.0957	0.0129	0.0073	0.0036	0.0025	0.0000
In	1.94E+2	0.3362	0.3534	0.1365	0.0311	0.0381	0.0328	0.0029	0.0000	0.0000	0.0000	0.0000
Sn	6.30E−1	0.1411	0.3611	0.1246	0.0715	0.0307	0.0335	0.0168	0.0035	0.0000	0.0034	0.0000
Ba	1.20E+0	0.3751	0.3971	0.2427	0.0910	0.3210	0.0848	0.0332	0.0088	0.0034	0.0065	0.0000
Ta	2.11E+1	0.4396	0.0090	0.0010	0.0292	0.0771	0.0421	0.0054	0.0000	0.0000	0.0000	0.0000
W	1.85E+1	0.7986	0.0557	0.0520	0.0887	0.1072	0.0987	0.0725	0.0067	0.0000	0.0000	0.0000
Hg	3.76E+2	0.9153	0.3383	0.2106	0.1322	0.2744	0.3008	0.0698	0.0000	0.0000	0.0000	0.0000
Pb	1.70E−1	0.0000	0.0000	0.0000	0.0000	0.0000	0.0000	0.0504	0.9406	0.0000	0.0000	0.0000
Bi	3.30E−2	0.0000	0.0000	0.0000	0.0000	1.1170	0.0000	0.0000	0.0000	0.0000	0.0000	0.0000

Source: M.A. Lone, R.A. Leavitt, and D.A. Harrison, "Prompt Gamma Rays from Thermal Neutron Capture," *At. Data Nucl. Data Tables*, **26**, No. 6, 511–559 (1981). [Data files released as Package DLC-140 THERMGAM by the Radiation Shielding Information Center, Oak Ridge National Laboratory, Oak Ridge, TN.]

Table B.5. Listed are prompt gamma rays from thermal neutron absorption recommended for use in element identification. The *partial gamma ray production cross section* is $f_i P_\gamma \sigma_\gamma^i$ where f_i is the fractional isotopic abundance of the ith isotope, P_γ is the probability the gamma ray is emitted, and σ_γ^i is the thermal neutron capture cross section for the i isotope. Data are from Révay et al. [2004].

E_γ (MeV)	$f_i P_\gamma \sigma_\gamma^i$ (barns)	E_γ (MeV)	$f_i P_\gamma \sigma_\gamma^i$ (barns)	E_γ (MeV)	$f_i P_\gamma \sigma_\gamma^i$ (barns)	E_γ (MeV)	$f_i P_\gamma \sigma_\gamma^i$ (barns)
$_{1}$H $\sigma_\gamma = 0.3326$ b		$_{11}$Na $\sigma_\gamma = 0.53$ b		$_{19}$K $\sigma_\gamma = 2.06$ b		$_{27}$Co $\sigma_\gamma = 37.18$ b	
2.22325	0.3326	0.09099	0.235	0.77031	0.903	0.22988	7.18
		0.47220	0.478	1.15889	0.1600	0.27716	6.77
$_{3}$Li $\sigma_\gamma = 45$ mb		0.87438	0.0760	7.76892	0.117	6.87716	3.02
0.98053	0.00415						
1.05190	0.00414	$_{12}$Mg $\sigma_\gamma = 66.6$ mb		$_{20}$Ca $\sigma_\gamma = 431$ mb		$_{28}$Ni $\sigma_\gamma = 4.39$ b	
2.03230	0.0381	0.58500	0.0314	1.94267	0.352	0.46498	0.843
		2.82817	0.0240	4.41852	0.0708	8.53351	0.721
$_{4}$Be $\sigma_\gamma = 8.8$ mb		3.91684	0.0320	6.41959	0.176	8.99841	1.49
0.85363	0.00208						
3.36745	0.00285	$_{13}$Al $\sigma_\gamma = 231$ mb		$_{21}$Sc $\sigma_\gamma = 27.2$ b		$_{29}$Cu $\sigma_\gamma = 3.795$ b	
6.80961	0.0058	1.77892	0.232	0.14253	4.88	0.27825	0.893
		3.03390	0.0179	0.14701	6.08	7.63740	0.54
$_{5}$B $\sigma_\gamma = 104$ mb		7.72403	0.0493	8.17518	1.80	7.91562	0.869
0.44760	716.						
		$_{14}$Si $\sigma_\gamma = 163$ mb		$_{22}$Ti $\sigma_\gamma = 6.08$ b		$_{30}$Zn $\sigma_\gamma = 1.30$ b	
$_{6}$C $\sigma_\gamma = 3.51$ mb		1.27335	0.0289	0.34171	1.840	0.11522	0.167
1.26177	0.00124	3.53897	0.1190	1.38174	5.18	1.07733	0.356
3.68392	0.00122	4.93389	0.1120	6.76008	2.97	7.83655	0.1410
4.49530	0.00261						
		$_{15}$P $\sigma_\gamma = 172$ mb		$_{23}$V $\sigma_\gamma = 4.96$ b		$_{31}$Ga $\sigma_\gamma = 2.90$ b	
$_{7}$N $\sigma_\gamma = 79.5$ mb		0.63666	0.0311	0.12508	1.61	0.14514	0.466
1.88482	0.0147	3.89989	0.0294	1.43410	4.81	0.69094	0.305
5.26916	0.02367	6.78550	0.0267	6.51728	0.78	0.83408	1.65
10.82912	0.01136						
		$_{16}$S $\sigma_\gamma = 534$ mb		$_{24}$Cr $\sigma_\gamma = 3.07$ b		$_{32}$Ge $\sigma_\gamma = 2.30$ b	
$_{8}$O $\sigma_\gamma = 0.190$ mb		0.84099	0.347	0.74909	0.569	0.59585	1.100
0.87068	0.000177	2.37966	0.208	0.83484	1.38	0.60835	0.250
1.08775	0.000158	5.42057	0.308	8.88436	0.78	0.86790	0.553
2.18442	0.000164						
		$_{17}$Cl $\sigma_\gamma = 33.1$ b		$_{25}$Mn $\sigma_\gamma = 13.36$ b		$_{33}$As $\sigma_\gamma = 4.23$ b	
$_{9}$F $\sigma_\gamma = 9.6$ mb		0.51707	7.58	0.21204	2.13	0.16505	0.996
0.58356	0.00356	1.95114	6.33	0.84675	13.10	0.47100	0.203
1.63353	0.0096	1.95937	4.10	7.24352	1.36	0.55910	2.00
$_{10}$Ne $\sigma_\gamma = 39$ mb		$_{18}$Ar $\sigma_\gamma = 675$ mb		$_{26}$Fe $\sigma_\gamma = 2.56$ b		$_{34}$Se $\sigma_\gamma = 12.0$ b	
0.35072	0.0198	0.16730	0.53	0.35235	0.273	0.13900	2.06
2.03567	0.0245	1.1866	0.34	7.63114	0.653	0.61372	2.14
4.37413	0.0190	4.7453	0.36	7.64554	0.549	6.60069	0.623

(cont.)

Table B.5. *(cont.)* Listed are prompt gamma rays from thermal neutron absorption recommended for use in element identification. The *partial gamma ray production cross section* is $f_i P_\gamma \sigma_\gamma^i$ where f_i is the fractional isotopic abundance of the *i*th isotope, P_γ is the probability the gamma ray is emitted, and σ_γ^i is the thermal neutron capture cross section for the *i* isotope. Data are from Révay et al. [2004].

E_γ (MeV)	$f_i P_\gamma \sigma_\gamma^i$ (barns)	E_γ (MeV)	$f_i P_\gamma \sigma_\gamma^i$ (barns)	E_γ (MeV)	$f_i P_\gamma \sigma_\gamma^i$ (barns)	E_γ (MeV)	$f_i P_\gamma \sigma_\gamma^i$ (barns)
$_{35}$Br $\sigma_\gamma = 6.39$ b		$_{43}$Tc $\sigma_\gamma = 24.3$ b		$_{51}$Sb $\sigma_\gamma = 5.13$ b		$_{59}$Pr $\sigma_\gamma = 11.3$ b	
0.195.60	0.434	0.17202	16.60	0.12150	0.40	0.12684	0.307
0.24520	0.80	0.22334	1.490	0.28265	0.274	0.17686	1.06
0.27137	0.462	0.55077	1.296	0.56424	2.700	5.66617	0.379
$_{36}$Kr $\sigma_\gamma = 25.8$ b		$_{44}$Ru $\sigma_\gamma = 2.75$ b		$_{52}$Te $\sigma_\gamma = 4.6$ b		$_{60}$Nd $\sigma_\gamma = 49.5$ b	
0.88174	20.8	0.47509	0.98	0.60273	2.46	0.61806	13.4
1.21342	8.28	0.53854	1.53	0.64582	0.263	0.69650	33.3
1.46386	7.10	0.68691	0.52	0.72277	0.52	0.81412	4.98
$_{37}$Rb $\sigma_\gamma = 0.38$ b		$_{45}$Rh $\sigma_\gamma = 145$ b		$_{53}$I $\sigma_\gamma = 6.2$ b		$_{62}$Sm $\sigma_\gamma = 5621$ b	
0.48789	0.0494	0.09714	19.5	0.13361	1.42	0.33397	4790
0.55561	0.0407	0.18087	22.6	0.30191	0.17	0.43940	2860
0.55682	0.0913	0.55581	3.14	0.44290	0.600	0.73744	597
$_{38}$Sr $\sigma_\gamma = 1.30$ b		$_{46}$Pd $\sigma_\gamma = 6.9$ b		$_{54}$Xe $\sigma_\gamma = 24$ b		$_{63}$Eu $\sigma_\gamma = 4560$ b	
0.85066	0.275	0.51184	4.00	0.53617	1.71	0.08985	1430
0.89806	0.702	0.61619	0.629	0.66779	6.7	0.22130	73
1.83607	1.030	0.71736	0.777	6.46709	1.33	0.96339	183
$_{39}$Y $\sigma_\gamma = 1.280$ b		$_{47}$Ag $\sigma_\gamma = 63.3$ b		$_{55}$Cs $\sigma_\gamma = 30.3$ b		$_{64}$Gd $\sigma_\gamma = 48770$ b	
0.20253	0.289	0.11745	3.85	0.17640	2.47	0.07951	4010
0.77661	0.659	0.19872	7.75	0.20561	1.560	0.18193	7200
6.08017	0.76	0.65750	1.86	0.30701	1.45	0.94417	3090
$_{40}$Zr $\sigma_\gamma = 0.19$ b		$_{48}$Cd $\sigma_\gamma = 2522$ b		$_{56}$Ba $\sigma_\gamma = 1.18$ b		$_{65}$Tb $\sigma_\gamma = 23.3$ b	
0.56096	0.0285	0.55832	1860	0.62729	0.294	0.09750	0.50
0.93446	0.125	0.65119	358	0.81851	0.212	0.15369	0.44
1.40516	0.0301	0.80585	134.0	4.09584	0.155	0.19143	0.37
$_{41}$Nb $\sigma_\gamma = 1.15$ b		$_{49}$In $\sigma_\gamma = 272$ b		$_{57}$La $\sigma_\gamma = 9.086$ b		$_{66}$Dy $\sigma_\gamma = 944$ b	
0.099407	0.196	0.18621	26.6	0.21822	0.78	0.18425	146
0.25311	0.1320	0.27297	33.1	0.28825	0.73	0.49693	44.9
0.25593	0.176	1.29354	131	5.09773	0.68	0.53861	69.2
$_{42}$Mo $\sigma_\gamma = 2.51$ b		$_{50}$Sn $\sigma_\gamma = 0.54$ b		$_{58}$Ce $\sigma_\gamma = 0.635$ b		$_{67}$Ho $\sigma_\gamma = 64.7$ b	
0.77822	2.02	1.17128	0.0879	0.47504	0.082	0.11684	8.1
0.84760	0.324	1.22964	0.0673	0.66199	0.241	0.13666	14.5
0.84985	0.43	1.29359	0.1340	4.76610	0.113	0.42601	2.88

(cont.)

Table B.5. *(cont.)* Listed are prompt gamma rays from thermal neutron absorption recommended for use in element identification. The *partial gamma ray production cross section* is $f_i P_\gamma \sigma_\gamma^i$ where f_i is the fractional isotopic abundance of the ith isotope, P_γ is the probability the gamma ray is emitted, and σ_γ^i is the thermal neutron capture cross section for the i isotope. Data are from Révay et al. [2004].

E_γ (MeV)	$f_i P_\gamma \sigma_\gamma^i$ (barns)	E_γ (MeV)	$f_i P_\gamma \sigma_\gamma^i$ (barns)	E_γ (MeV)	$f_i P_\gamma \sigma_\gamma^i$ (barns)	E_γ (MeV)	$f_i P_\gamma \sigma_\gamma^i$ (barns)
$_{68}$Er $\sigma_\gamma = 1.568$ b		$_{73}$Ta $\sigma_\gamma = 20.6$ b		$_{78}$Pt $\sigma_\gamma = 10.3$ b		$_{83}$Bi $\sigma_\gamma = 33.8$ mb	
0.184280	56	0.17320	1.210	0.33298	2.580	0.31978	0.0115
0.28466	13.7	0.27040	2.60	0.35568	6.17	4.05457	0.0137
0.81599	42.5	0.40262	1.180	0.52116	0.338	4.17105	0.0171
$_{69}$Tm $\sigma_\gamma = 105$ b		$_{74}$W $\sigma_\gamma = 18.39$ b		$_{79}$Au $\sigma_\gamma = 98.65$ b		$_{90}$Th $\sigma_\gamma = 7.35$ b	
0.14448	5.96	0.14579	0.970	0.21497	9.0	0.47230	0.165
0.14972	7.11	0.68573	3.24	0.24757	5.56	0.96878	0.132
0.20445	8.72	5.26168	0.86	0.41180	95.58	3.47300	0.057
$_{70}$Yb $\sigma_\gamma = 34.9$ b		$_{75}$Re $\sigma_\gamma = 91.5$ b		$_{80}$Hg $\sigma_\gamma = 384$ b		$_{92}$U $\sigma_\gamma = 3.374$ b	
0.51487	9.0	0.14415	1.8	0.36795	251	0.27760	0.382
0.63926	1.43	0.20785	4.44	1.69330	56.2	4.06035	0.186
5.2663	1.4	0.29066	3.5	5.96702	62.5	6.39516	0.0032
$_{71}$Lu $\sigma_\gamma = 76.6$ b		$_{76}$Os $\sigma_\gamma = 16.0$ b		$_{81}$Tl $\sigma_\gamma = 3.44$ b			
0.15039	13.8	0.15510	1.19	0.34796	0.361		
0.45794	8.3	0.18672	2.08	0.87316	0.168		
0.76156	2.60	0.55798	0.84	5.64157	0.316		
$_{72}$Hf $\sigma_\gamma = 119$ b		$_{77}$Ir $\sigma_\gamma = 425$ b		$_{82}$Pb $\sigma_\gamma = 154$ mb			
0.09318	13.3	0.13612	6.5	6.72938	0.00320		
0.21434	22.0	0.22630	4.0	6.73762	0.00691		
0.30399	3.38	0.35169	10.9	7.36778	0.137		

Table B.6. Mass coefficients (cm^2/g) for dry air near sea level. Composition by weight fraction: N 0.755268, O 0.231781, Ar 0.012827, C 0.000124. Nominal density is 1.205×10^{-3} g cm^{-3}. Data are from Hubbell and Seltzer [1995] and Seltzer [1993].

	Air					
E (MeV)	μ_{coh}/ρ	μ_{inc}/ρ	μ_{ph}/ρ	μ_{pp}/ρ	μ_t/ρ	$\mu_{t\text{-}coh}/\rho$
0.0010	1.362	1.038−2	3.605+3		3.606+3	3.605+3
0.0015	1.247	2.116−2	1.190+3		1.191+3	1.190+3
0.0020	1.116	3.340−2	5.267+2		5.279+2	5.267+2
0.0030	8.634−1	5.748−2	1.616+2		1.625+2	1.616+2
0.00320	8.182−1	6.196−2	1.331+2		1.340+2	1.332+2
0.00320K	8.182−1	6.196−2	1.476+2		1.485+2	1.477+2
0.0040	6.648−1	7.770−2	7.713+1		7.788+1	7.721+1
0.0050	5.218−1	9.331−2	3.966+1		4.027+1	3.975+1
0.0060	4.207−1	1.051−1	2.288+1		2.341+1	2.299+1
0.0080	2.946−1	1.213−1	9.505+0		9.921+0	9.626+0
0.010	2.223−1	1.316−1	4.766+0		5.120+0	4.897+0
0.015	1.314−1	1.471−1	1.335+0		1.614+0	1.482+0
0.020	8.752−2	1.556−1	5.347−1		7.779−1	6.904−1
0.030	4.619−2	1.625−1	1.451−1		3.538−1	3.076−1
0.040	2.828−2	1.631−1	5.704−2		2.485−1	2.202−1
0.050	1.905−2	1.613−1	2.755−2		2.080−1	1.889−1
0.060	1.370−2	1.586−1	1.517−2		1.875−1	1.738−1
0.080	8.028−3	1.523−1	5.912−3		1.662−1	1.582−1
0.10	5.254−3	1.460−1	2.847−3		1.541−1	1.489−1
0.15	2.396−3	1.324−1	7.602−4		1.356−1	1.332−1
0.20	1.361−3	1.217−1	3.026−4		1.233−1	1.220−1
0.30	6.096−4	1.061−1	8.604−5		1.067−1	1.061−1
0.40	3.438−4	9.511−2	3.698−5		9.549−2	9.514−2
0.50	2.203−4	8.687−2	1.998−5		8.712−2	8.689−2
0.60	1.531−4	8.039−2	1.246−5		8.055−2	8.040−2
0.80	8.618−5	7.064−2	6.296−6		7.074−2	7.065−2
1.00	5.517−5	6.352−2	3.914−6		6.358−2	6.353−2
1.25	3.533−5	5.682−2	2.545−6	1.781−5	5.687−2	5.684−2
1.50	2.453−5	5.162−2	1.798−6	9.848−5	5.175−2	5.172−2
2.00	1.380−5	4.407−2	1.128−6	3.918−4	4.447−2	4.446−2
3.00	6.133−6	3.467−2	6.276−7	1.132−3	3.581−2	3.580−2
4.00	3.451−6	2.892−2	4.297−7	1.866−3	3.079−2	3.079−2
5.00	2.208−6	2.497−2	3.252−7	2.536−3	2.751−2	2.751−2
6.00	1.534−6	2.207−2	2.611−7	3.147−3	2.522−2	2.522−2
8.00	8.628−7	1.806−2	1.869−7	4.196−3	2.225−2	2.225−2
10.0	5.522−7	1.538−2	1.453−7	5.067−3	2.045−2	2.045−2
15.0	2.454−7	1.138−2	9.323−8	6.717−3	1.810−2	1.810−2
20.0	1.380−7	9.134−3	6.859−8	7.920−3	1.705−2	1.705−2
30.0	6.136−8	6.652−3	4.483−8	9.629−3	1.628−2	1.628−2
40.0	3.451−8	5.286−3	3.329−8	1.082−2	1.610−2	1.610−2
50.0	2.208−8	4.411−3	2.647−8	1.173−2	1.614−2	1.614−2
60.0	1.534−8	3.801−3	2.197−8	1.245−2	1.625−2	1.625−2
80.0	8.625−9	2.998−3	1.640−8	1.354−2	1.654−2	1.654−2
100.0	5.522−9	2.488−3	1.308−8	1.435−2	1.683−2	1.683−2

(cont.)

Table B.6. *(cont.)* Mass coefficients (cm^2/g) for water with density of 1 g/cm^3. Data are from Hubbell and Seltzer [1995] and Seltzer [1993].

	Water					
E (MeV)	μ_{coh}/ρ	μ_{inc}/ρ	μ_{ph}/ρ	μ_{pp}/ρ	μ_t/ρ	$\mu_{t\text{-}coh}/\rho$
0.0010	1.372	1.319−2	4.076+3		4.078+3	4.076+3
0.0015	1.269	2.673−2	1.375+3		1.376+3	1.375+3
0.0020	1.150	4.184−2	6.161+2		6.173+2	6.161+2
0.0030	9.087−1	7.075−2	1.919+2		1.929+2	1.919+2
0.0040	7.082−1	9.430−2	8.198+1		8.278+1	8.207+1
0.0050	5.578−1	1.123−1	4.191+1		4.258+1	4.203+1
0.0060	4.489−1	1.259−1	2.407+1		2.464+1	2.419+1
0.0080	3.102−1	1.440−1	9.919+0		1.037+1	1.006+1
0.0100	2.305−1	1.550−1	4.943+0		5.329	5.098
0.015	1.333−1	1.699−1	1.369+0		1.673	1.539
0.020	8.856−2	1.774−1	5.437−1		8.096−1	7.211−1
0.030	4.694−2	1.829−1	1.457−1		3.756−1	3.286−1
0.040	2.874−2	1.827−1	5.680−2		2.683−1	2.395−1
0.050	1.936−2	1.803−1	2.725−2		2.269−1	2.076−1
0.060	1.392−2	1.770−1	1.493−2		2.059−1	1.920−1
0.080	8.164−3	1.697−1	5.770−3		1.837−1	1.755−1
0.10	5.349−3	1.626−1	2.762−3		1.707−1	1.654−1
0.15	2.442−3	1.473−1	7.307−4		1.505−1	1.481−1
0.20	1.388−3	1.353−1	2.887−4		1.370−1	1.356−1
0.30	6.215−4	1.179−1	8.162−5		1.186−1	1.180−1
0.40	3.506−4	1.058−1	3.495−5		1.061−1	1.058−1
0.50	2.247−4	9.663−2	1.884−5		9.687−2	9.664−2
0.60	1.561−4	8.939−2	1.173−5		8.956−2	8.940−2
0.80	8.790−5	7.856−2	5.920−6		7.865−2	7.857−2
1.00	5.627−5	7.066−2	3.680−6		7.072−2	7.066−2
1.25	3.603−5	6.318−2	2.394−6	1.777−5	6.323−2	6.320−2
1.50	2.501−5	5.741−2	1.689−6	9.820−5	5.754−2	5.751−2
2.00	1.407−5	4.901−2	1.063−6	3.908−4	4.942−2	4.940−2
3.00	6.255−6	3.855−2	5.937−7	1.131−3	3.969−2	3.968−2
4.00	3.519−6	3.216−2	4.075−7	1.867−3	3.403−2	3.402−2
5.00	2.252−6	2.777−2	3.089−7	2.540−3	3.031−2	3.031−2
6.00	1.564−6	2.454−2	2.484−7	3.155−3	2.770−2	2.770−2
8.00	8.796−7	2.008−2	1.780−7	4.211−3	2.429−2	2.429−2
10.0	5.630−7	1.710−2	1.386−7	5.090−3	2.219−2	2.219−2
15.0	2.502−7	1.266−2	8.905−8	6.754−3	1.941−2	1.941−2
20.0	1.407−7	1.016−2	6.555−8	7.974−3	1.813−2	1.813−2
30.0	6.255−8	7.395−3	4.289−8	9.709−3	1.710−2	1.710−2
40.0	3.519−8	5.875−3	3.186−8	1.091−2	1.679−2	1.679−2
50.0	2.252−8	4.906−3	2.534−8	1.183−2	1.674−2	1.674−2
60.0	1.564−8	4.225−3	2.104−8	1.256−2	1.679−2	1.679−2
80.0	8.796−9	3.333−3	1.570−8	1.368−2	1.701−2	1.701−2
100.0	5.630−9	2.767−3	1.253−8	1.450−2	1.727−2	1.727−2

(cont.)

Table B.6. *(cont.)* Mass coefficients (cm^2/g) for ANSI/ANS-6.4.3 standard concrete of density of 2.3 g/cm^3. Composition by weight fraction: H 0.005599, O 0.498250, Na 0.017098, Mg 0.002400, Al 0.045595, Si 0.315768, S 0.001200, K 0.019198, Ca 0.082592, Fe 0.012299. Data are from Hubbell and Seltzer [1995] and Seltzer [1993].

	Concrete					
E (MeV)	μ_{coh}/ρ	μ_{inc}/ρ	μ_{ph}/ρ	μ_{pp}/ρ	μ_t/ρ	$\mu_{t\text{-}coh}/\rho$
0.00100	2.109	1.117−2	3.443+3		3.445+3	3.443+3
0.00104	1.929	2.098−2	3.144+3		3.146+3	3.144+3
0.00107	2.081	1.249−2	2.871+3		2.873+3	2.871+3
0.00107K	2.081	1.249−2	2.972+3		2.974+3	2.972+3
0.00118	1.744	3.115−2	2.300+3		2.302+3	2.300+3
0.00131	1.998	1.701−2	1.775+3		1.777+3	1.775+3
0.00131K	1.998	1.701−2	1.787+3		1.789+3	1.787+3
0.00150	1.929	2.098−2	1.233+3		1.235+3	1.233+3
0.00156	1.907	2.220−2	1.110+3		1.112+3	1.110+3
0.00156K	1.907	2.220−2	1.274+3		1.276+3	1.274+3
0.00169	1.416	5.019−2	1.025+3		1.026+3	1.025+3
0.00184	1.803	2.790−2	8.223+2		8.241+2	8.223+2
0.00184K	1.803	2.790−2	1.733+3		1.734+3	1.733+3
0.0020	1.744	3.115−2	1.454+3		1.456+3	1.454+3
0.00247	1.580	4.044−2	8.479+2		8.495+2	8.479+2
0.00247K	1.580	4.044−2	8.501+2		8.517+2	8.501+2
0.0030	1.416	5.019−2	4.966+2		4.981+2	4.966+2
0.00361	1.252	6.036−2	2.998+2		3.012+2	2.999+2
0.00361K	1.252	6.036−2	3.203+2		3.217+2	3.204+2
0.0040	1.160	6.635−2	2.410+2		2.422+2	2.410+2
0.00404	1.151	6.691−2	2.347+2		2.359+2	2.348+2
0.00404K	1.151	6.691−2	3.094+2		3.107+2	3.095+2
0.0050	9.645−1	7.972−2	1.733+2		1.744+2	1.734+2
0.0060	8.146−1	9.069−2	1.043+2		1.052+2	1.044+2
0.00711	6.852−1	1.007−1	6.448+1		6.527+1	6.458+1
0.00711K	6.852−1	1.007−1	6.884+1		6.962+1	6.894+1
0.0080	6.027−1	1.073−1	4.937+1		5.008+1	4.948+1
0.010	4.640−1	1.190−1	2.599+1		2.657+1	2.611+1
0.015	2.731−1	1.372−1	7.891		8.301	8.028
0.020	1.807−1	1.470−1	3.326		3.654	3.473
0.030	9.677−2	1.557−1	9.629−1		1.215	1.119
0.040	6.047−2	1.579−1	3.942−1		6.126−1	5.521−1
0.050	4.135−2	1.573−1	1.959−1		3.946−1	3.533−1
0.060	3.003−2	1.555−1	1.103−1		2.958−1	2.658−1
0.080	1.785−2	1.503−1	4.432−2		2.125−1	1.947−1
0.10	1.180−2	1.447−1	2.181−2		1.783−1	1.665−1
0.15	5.465−3	1.319−1	6.028−3		1.434−1	1.379−1
0.20	3.132−3	1.214−1	2.447−3		1.270−1	1.239−1
0.30	1.414−3	1.060−1	7.120−4		1.082−1	1.068−1
0.40	8.005−4	9.517−2	3.100−4		9.628−2	9.548−2
0.50	5.139−4	8.699−2	1.687−4		8.768−2	8.716−2
0.60	3.574−4	8.051−2	1.057−4		8.098−2	8.062−2
0.80	2.014−4	7.077−2	5.359−5		7.103−2	7.083−2
1.00	1.290−4	6.366−2	3.332−5		6.382−2	6.369−2
1.25	8.261−5	5.693−2	2.163−5	2.904−5	5.706−2	5.698−2
1.50	5.738−5	5.174−2	1.542−5	1.575−4	5.197−2	5.191−2
2.00	3.228−5	4.416−2	9.586−6	6.212−4	4.483−2	4.480−2
3.00	1.435−5	3.474−2	5.267−6	1.775−3	3.654−2	3.652−2
4.00	8.074−6	2.898−2	3.576−6	2.899−3	3.189−2	3.189−2
5.00	5.167−6	2.503−2	2.692−6	3.915−3	2.895−2	2.895−2
6.00	3.588−6	2.212−2	2.153−6	4.833−3	2.696−2	2.696−2
8.00	2.018−6	1.810−2	1.533−6	6.400−3	2.450−2	2.450−2
10.0	1.292−6	1.541−2	1.187−6	7.696−3	2.311−2	2.311−2
15.0	5.741−7	1.141−2	7.581−7	1.013−2	2.153−2	2.153−2
20.0	3.230−7	9.156−3	5.561−7	1.189−2	2.105−2	2.105−2
30.0	1.435−7	6.665−3	3.625−7	1.438−2	2.105−2	2.105−2
40.0	8.075−8	5.296−3	2.689−7	1.611−2	2.141−2	2.141−2
50.0	5.167−8	4.422−3	2.136−7	1.742−2	2.184−2	2.184−2
60.0	3.588−8	3.808−3	1.772−7	1.845−2	2.226−2	2.226−2
80.0	2.018−8	3.004−3	1.321−7	2.000−2	2.300−2	2.300−2
100.0	1.292−8	2.494−3	1.053−7	2.112−2	2.361−2	2.361−2

(cont.)

Table B.6. *(cont.)* Mass coefficients (cm^2/g) for natural iron with a density of 7.874 g/cm^3. Data are from Hubbell and Seltzer [1995] and Seltzer [1993].

	Iron ($Z = 26$)					
E (MeV)	μ_{coh}/ρ	μ_{inc}/ρ	μ_{ph}/ρ	μ_{pp}/ρ	μ_t/ρ	$\mu_{t\text{-}coh}/\rho$
0.0010	4.536	8.776−3	9.080+3		9.085+3	9.080+3
0.0015	4.241	1.530−2	3.395+3		3.399+3	3.395+3
0.0020	3.933	2.124−2	1.622+3		1.626+3	1.622+3
0.0030	3.354	3.206−2	5.542+2		5.576+2	5.543+2
0.0040	2.849	4.212−2	2.538+2		2.567+2	2.539+2
0.0050	2.421	5.133−2	1.373+2		1.398+2	1.374+2
0.0060	2.065	5.966−2	8.272+1		8.484+1	8.278+1
0.00711	1.747	6.798−2	5.138+1		5.319+1	5.144+1
0.00711K	1.747	6.798−2	4.058+2		4.076+2	4.059+2
0.0080	1.542	7.395−2	3.040+2		3.056+2	3.040+2
0.010	1.201	8.541−2	1.693+2		1.706+2	1.694+2
0.015	7.458−1	1.047−1	5.623+1		5.708+1	5.633+1
0.020	5.169−1	1.162−1	2.505+1		2.568+1	2.516+1
0.030	2.849−1	1.286−1	7.762		8.176	7.891
0.040	1.795−1	1.338−1	3.316		3.629	3.450
0.050	1.244−1	1.355−1	1.698		1.958	1.833
0.060	9.178−2	1.355−1	9.776−1		1.205	1.113
0.080	5.604−2	1.332−1	4.060−1		5.952−1	5.391−1
0.10	3.768−2	1.296−1	2.044−1		3.717−1	3.340−1
0.15	1.781−2	1.200−1	5.861−2		1.964−1	1.786−1
0.20	1.033−2	1.114−1	2.432−2		1.460−1	1.357−1
0.30	4.728−3	9.788−2	7.265−3		1.099−1	1.051−1
0.40	2.694−3	8.810−2	3.209−3		9.400−2	9.131−2
0.50	1.736−3	8.065−2	1.762−3		8.414−2	8.241−2
0.60	1.210−3	7.472−2	1.109−3		7.704−2	7.583−2
0.80	6.833−4	6.574−2	5.650−4		6.699−2	6.631−2
1.00	4.381−4	5.916−2	3.515−4		5.995−2	5.951−2
1.25	2.808−4	5.292−2	2.277−4	7.031−5	5.350−2	5.322−2
1.50	1.951−4	4.811−2	1.627−4	3.580−4	4.883−2	4.863−2
2.00	1.099−4	4.107−2	1.003−4	1.364−3	4.265−2	4.254−2
3.00	4.884−5	3.232−2	5.448−5	3.792−3	3.621−2	3.616−2
4.00	2.748−5	2.697−2	3.669−5	6.085−3	3.312−2	3.309−2
5.00	1.759−5	2.329−2	2.746−5	8.124−3	3.146−2	3.144−2
6.00	1.222−5	2.058−2	2.188−5	9.951−3	3.057−2	3.056−2
8.00	6.870−6	1.684−2	1.548−5	1.305−2	2.991−2	2.991−2
10.00	4.396−6	1.434−2	1.196−5	1.559−2	2.994−2	2.994−2
15.0	1.954−6	1.061−2	7.592−6	2.030−2	3.092−2	3.092−2
20.0	1.099−6	8.520−3	5.554−6	2.371−2	3.224−2	3.223−2
30.0	4.886−7	6.202−3	3.610−6	2.848−2	3.469−2	3.469−2
40.0	2.749−7	4.929−3	2.673−6	3.173−2	3.666−2	3.666−2
50.0	1.759−7	4.114−3	2.122−6	3.416−2	3.828−2	3.828−2
60.0	1.222−7	3.544−3	1.759−6	3.607−2	3.961−2	3.961−2
80.0	6.870−8	2.796−3	1.311−6	3.892−2	4.172−2	4.172−2
100.0	4.396−8	2.321−3	1.045−6	4.097−2	4.329−2	4.329−2

(cont.)

Table B.6. *(cont.)* Mass coefficients (cm^2/g) for natural lead with a density of 11.35 g/cm^3. Data are from Hubbell and Seltzer [1995] and Seltzer [1993].

	Lead ($Z = 82$)					
E (MeV)	μ_{coh}/ρ	μ_{inc}/ρ	μ_{ph}/ρ	μ_{pp}/ρ	μ_t/ρ	$\mu_{t\text{-}coh}/\rho$
0.0010	1.251+1	3.586−3	5.198+3		5.210+3	5.198+3
0.0015	1.201+1	6.600−3	2.344+3		2.356+3	2.344+3
0.0020	1.144+1	9.620−3	1.274+3		1.285+3	1.274+3
0.00248	1.087+1	1.241−2	7.898+2		8.006+2	7.898+2
0.00248M5	1.087+1	1.241−2	1.386+3		1.397+3	1.386+3
0.00253	1.201+1	6.600−3	1.714+3		1.726+3	1.714+3
0.00259	1.075+1	1.298−2	1.933+3		1.944+3	1.933+3
0.00259M4	1.075+1	1.298−2	2.447+3		2.458+3	2.447+3
0.0030	1.027+1	1.525−2	1.954+3		1.965+3	1.955+3
0.00307	1.019+1	1.560−2	1.847+3		1.857+3	1.847+3
0.00307M3	1.019+1	1.560−2	2.136+3		2.146+3	2.136+3
0.00330	9.929	1.684−2	1.784+3		1.794+3	1.784+3
0.00355	9.651	1.814−2	1.486+3		1.496+3	1.486+3
0.00355M2	9.651	1.814−2	1.575+3		1.585+3	1.575+3
0.00370	9.494	1.887−2	1.431+3		1.440+3	1.431+3
0.00385	9.334	1.963−2	1.302+3		1.311+3	1.302+3
0.00385M1	9.334	1.963−2	1.359+3		1.368+3	1.359+3
0.0040	9.178	2.037−2	1.242+3		1.251+3	1.242+3
0.0050	8.208	2.515−2	7.222+2		7.304+2	7.222+2
0.0060	7.362	2.970−2	4.599+2		4.672+2	4.599+2
0.0080	6.002	3.807−2	2.226+2		2.287+2	2.227+2
0.0100	4.982	4.540−2	1.256+2		1.306+2	1.257+2
0.01304	3.854	5.441−2	6.311+1		6.701+1	6.316+1
0.01304L3	3.854	5.441−2	1.582+2		1.621+2	1.582+2
0.0150	3.307	5.920−2	1.082+2		1.116+2	1.083+2
0.01520	3.258	5.965−2	1.045+2		1.078+2	1.045+2
0.01520L2	3.258	5.965−2	1.451+2		1.485+2	1.452+2
0.01553	3.179	6.038−2	1.384+2		1.416+2	1.384+2
0.01586	3.102	6.110−2	1.312+2		1.344+2	1.313+2
0.01586L1	3.102	6.110−2	1.517+2		1.548+2	1.517+2
0.0200	2.338	6.897−2	8.395+1		8.636+1	8.402+1
0.0300	1.377	8.228−2	2.886+1		3.032+1	2.894+1
0.0400	9.202−1	9.019−2	1.335+1		1.436+1	1.344+1
0.0500	6.545−1	9.478−2	7.291		8.041	7.386
0.0600	4.900−1	9.734−2	4.433		5.021	4.531
0.0800	3.078−1	9.922−2	2.012		2.419	2.112
0.08800	2.632−1	9.928−2	1.547		1.910	1.647
0.08800K	2.632−1	9.928−2	7.320		7.683	7.420
0.100	2.127−1	9.893−2	5.238		5.549	5.337
0.150	1.049−1	9.484−2	1.815		2.014	1.910
0.200	6.260−2	8.966−2	8.463−1		9.985−1	9.359−1
0.300	2.988−2	8.036−2	2.928−1		4.031−1	3.732−1
0.400	1.746−2	7.310−2	1.417−1		2.323−1	2.148−1
0.500	1.143−2	6.734−2	8.258−2		1.614−1	1.499−1
0.600	8.059−3	6.263−2	5.407−2		1.248−1	1.167−1
0.800	4.621−3	5.537−2	2.871−2		8.870−2	8.408−2
1.00	2.991−3	4.993−2	1.809−2		7.102−2	6.803−2
1.25	1.930−3	4.476−2	1.169−2	3.781−4	5.876−2	5.683−2
1.50	1.347−3	4.075−2	8.321−3	1.806−3	5.222−2	5.087−2
2.00	7.626−4	3.482−2	5.034−3	5.449−3	4.606−2	4.530−2
3.00	3.406−4	2.744−2	2.631−3	1.193−2	4.234−2	4.200−2
4.00	1.919−4	2.290−2	1.723−3	1.716−2	4.197−2	4.178−2
5.00	1.229−4	1.978−2	1.263−3	2.155−2	4.272−2	4.260−2
6.00	8.542−5	1.749−2	9.893−4	2.535−2	4.391−2	4.382−2
8.00	4.807−5	1.431−2	6.845−4	3.171−2	4.675−2	4.670−2
10.0	3.078−5	1.219−2	5.202−4	3.698−2	4.972−2	4.969−2
15.0	1.368−5	9.021−3	3.229−4	4.722−2	5.658−2	5.656−2
20.0	7.696−6	7.243−3	2.334−4	5.457−2	6.206−2	6.205−2
30.0	3.421−6	5.272−3	1.497−4	6.479−2	7.022−2	7.022−2
40.0	1.924−6	4.188−3	1.101−4	7.180−2	7.610−2	7.610−2
50.0	1.231−6	3.496−3	8.699−5	7.697−2	8.056−2	8.056−2
60.0	8.551−7	3.014−3	7.190−5	8.099−2	8.408−2	8.408−2
80.0	4.810−7	2.376−3	5.339−5	8.691−2	8.934−2	8.934−2
100.0	3.078−7	1.972−3	4.243−5	9.108−2	9.310−2	9.310−2

Table B.7. Mass interaction μ_{t-coh}/ρ (total−coherent) and total μ_t/ρ coefficients (cm^2/g) for various elements. Data source: Seltzer [1993].

E (MeV)	H ($Z=1$)		He ($Z=2$)		Li ($Z=3$)		Be ($Z=4$)		B ($Z=5$)		C ($Z=6$)	
	μ_t/ρ	$\mu_{t\text{-}coh}/\rho$	μ_t/ρ	$\mu_{t\text{-}coh}/\rho$	μ_t/ρ	$\mu_{t\text{-}coh}/\rho$	μ_t/ρ	$\mu_{t\text{-}coh}/\rho$	μ_t/ρ	$\mu_{t\text{-}coh}/\rho$	μ_t/ρ	$\mu_{t\text{-}coh}/\rho$
0.01	3.854−1	3.608−1	2.476−1	1.885−1	3.395−1	2.647−1	6.465−1	5.491−1	1.255	1.132	2.373	2.211
0.015	3.764−1	3.648−1	2.092−1	1.796−1	2.176−1	1.774−1	3.070−1	2.512−1	4.827−1	4.093−1	8.073−1	7.094−1
0.02	3.695−1	3.628−1	1.960−1	1.785−1	1.856−1	1.608−1	2.251−1	1.897−1	3.014−1	2.537−1	4.419−1	3.772−1
0.03	3.570−1	3.540−1	1.838−1	1.757−1	1.644−1	1.524−1	1.792−1	1.615−1	2.064−1	1.820−1	2.562−1	2.225−1
0.04	3.458−1	3.441−1	1.763−1	1.716−1	1.551−1	1.481−1	1.640−1	1.534−1	1.794−1	1.647−1	2.076−1	1.872−1
0.05	3.355−1	3.344−1	1.703−1	1.673−1	1.488−1	1.443−1	1.554−1	1.485−1	1.665−1	1.567−1	1.871−1	1.734−1
0.06	3.260−1	3.253−1	1.651−1	1.630−1	1.438−1	1.406−1	1.493−1	1.444−1	1.584−1	1.514−1	1.753−1	1.655−1
0.08	3.091−1	3.087−1	1.562−1	1.550−1	1.356−1	1.337−1	1.401−1	1.372−1	1.472−1	1.432−1	1.610−1	1.553−1
0.10	2.944−1	2.941−1	1.486−1	1.478−1	1.289−1	1.277−1	1.328−1	1.310−1	1.391−1	1.365−1	1.514−1	1.476−1
0.15	2.651−1	2.650−1	1.336−1	1.333−1	1.158−1	1.152−1	1.190−1	1.182−1	1.244−1	1.232−1	1.347−1	1.330−1
0.20	2.429−1	2.428−1	1.224−1	1.222−1	1.060−1	1.057−1	1.089−1	1.085−1	1.136−1	1.130−1	1.229−1	1.220−1
0.30	2.112−1	2.112−1	1.064−1	1.063−1	9.210−2	9.197−2	9.463−2	9.442−2	9.864−2	9.834−2	1.066−1	1.062−1
0.40	1.893−1	1.893−1	9.535−2	9.530−2	8.249−2	8.241−2	8.471−2	8.460−2	8.836−2	8.819−2	9.546−2	9.522−2
0.50	1.729−1	1.729−1	8.707−2	8.704−2	7.532−2	7.527−2	7.739−2	7.731−2	8.066−2	8.056−2	8.715−2	8.700−2
0.60	1.599−1	1.599−1	8.054−2	8.052−2	6.968−2	6.964−2	7.155−2	7.150−2	7.461−2	7.454−2	8.058−2	8.048−2
0.80	1.405−1	1.405−1	7.076−2	7.074−2	6.121−2	6.119−2	6.286−2	6.283−2	6.550−2	6.546−2	7.076−2	7.070−2
1.00	1.263−1	1.263−1	6.362−2	6.361−2	5.503−2	5.501−2	5.652−2	5.650−2	5.891−2	5.888−2	6.361−2	6.358−2
1.25	1.129−1	1.129−1	5.688−2	5.688−2	4.921−2	4.920−2	5.054−2	5.053−2	5.267−2	5.266−2	5.690−2	5.687−2
1.50	1.027−1	1.027−1	5.173−2	5.172−2	4.476−2	4.475−2	4.597−2	4.597−2	4.792−2	4.791−2	5.179−2	5.177−2
2.00	8.770−2	8.770−2	4.422−2	4.422−2	3.830−2	3.829−2	3.938−2	3.937−2	4.109−2	4.108−2	4.442−2	4.442−2
3.00	6.922−2	6.922−2	3.503−2	3.503−2	3.043−2	3.042−2	3.138−2	3.138−2	3.284−2	3.284−2	3.562−2	3.562−2
4.00	5.807−2	5.807−2	2.949−2	2.949−2	2.572−2	2.572−2	2.664−2	2.663−2	2.799−2	2.798−2	3.047−2	3.047−2
5.00	5.049−2	5.049−2	2.577−2	2.577−2	2.257−2	2.257−2	2.347−2	2.347−2	2.477−2	2.477−2	2.708−2	2.708−2
6.00	4.498−2	4.498−2	2.307−2	2.307−2	2.030−2	2.030−2	2.121−2	2.121−2	2.248−2	2.248−2	2.469−2	2.469−2
8.00	3.746−2	3.746−2	1.940−2	1.940−2	1.725−2	1.725−2	1.819−2	1.819−2	1.945−2	1.945−2	2.154−2	2.154−2
10.00	3.254−2	3.254−2	1.703−2	1.703−2	1.529−2	1.529−2	1.627−2	1.627−2	1.755−2	1.755−2	1.959−2	1.959−2
15.00	2.539−2	2.539−2	1.363−2	1.363−2	1.252−2	1.252−2	1.361−2	1.361−2	1.495−2	1.495−2	1.698−2	1.698−2
20.00	2.153−2	2.153−2	1.183−2	1.183−2	1.109−2	1.109−2	1.227−2	1.227−2	1.368−2	1.368−2	1.575−2	1.575−2
30.00	1.748−2	1.748−2	1.001−2	1.001−2	9.687−3	9.687−3	1.100−2	1.100−2	1.255−2	1.255−2	1.472−2	1.472−2
40.00	1.542−2	1.542−2	9.114−3	9.114−3	9.034−3	9.034−3	1.045−2	1.045−2	1.210−2	1.210−2	1.437−2	1.437−2
50.00	1.419−2	1.419−2	8.608−3	8.608−3	8.683−3	8.683−3	1.018−2	1.018−2	1.191−2	1.191−2	1.426−2	1.426−2
60.00	1.339−2	1.339−2	8.296−3	8.296−3	8.481−3	8.481−3	1.004−2	1.004−2	1.184−2	1.184−2	1.426−2	1.426−2
80.00	1.245−2	1.245−2	7.950−3	7.950−3	8.289−3	8.289−3	9.942−3	9.942−3	1.184−2	1.184−2	1.439−2	1.439−2
100.0	1.194−2	1.194−2	7.784−3	7.784−3	8.214−3	8.214−3	9.940−3	9.940−3	1.192−2	1.192−2	1.457−2	1.457−2

(cont.)

Table B.7. *(cont.)* Mass interaction μ_{t-coh}/ρ (total–coherent) and total μ_t/ρ coefficients (cm^2/g) for various elements. Data source: Seltzer [1993].

E (MeV)	N ($Z=7$)		O ($Z=8$)		F ($Z=9$)		Na ($Z=11$)		Mg ($Z=12$)		Al ($Z=13$)	
	μ_t/ρ	$\mu_{t\text{-}coh}/\rho$	μ_t/ρ	$\mu_{t\text{-}coh}/\rho$	μ_t/ρ	$\mu_{t\text{-}coh}/\rho$	μ_t/ρ	$\mu_{t\text{-}coh}/\rho$	μ_t/ρ	$\mu_{t\text{-}coh}/\rho$	μ_t/ρ	$\mu_{t\text{-}coh}/\rho$
0.01	3.879	3.676	5.951	5.695	8.206	7.900	1.557+1	1.513+1	9.340+1	9.251+1	2.622+1	2.567+1
0.015	1.236	1.116	1.836	1.687	2.492	2.319	4.694	4.447	2.979+1	2.922+1	7.955	7.642
0.02	6.178−1	5.374−1	8.653−1	7.664−1	1.133	1.018	2.057	1.896	1.306+1	1.268+1	3.441	3.236
0.03	3.066−1	2.643−1	3.779−1	3.254−1	4.487−1	3.879−1	7.198−1	6.339−1	4.079	3.873	1.128	1.019
0.04	2.288−1	2.029−1	2.585−1	2.264−1	2.828−1	2.454−1	3.969−1	3.437−1	1.830	1.700	5.685−1	4.999−1
0.05	1.980−1	1.806−1	2.132−1	1.916−1	2.214−1	1.962−1	2.804−1	2.443−1	1.019	9.291−1	3.681−1	3.214−1
0.06	1.817−1	1.693−1	1.907−1	1.752−1	1.920−1	1.738−1	2.268−1	2.008−1	6.578−1	5.913−1	2.778−1	2.440−1
0.08	1.639−1	1.566−1	1.678−1	1.587−1	1.639−1	1.532−1	1.796−1	1.643−1	3.655−1	3.254−1	2.018−1	1.817−1
0.10	1.529−1	1.482−1	1.551−1	1.492−1	1.496−1	1.426−1	1.585−1	1.484−1	2.571−1	2.304−1	1.704−1	1.572−1
0.15	1.353−1	1.331−1	1.361−1	1.334−1	1.298−1	1.266−1	1.335−1	1.289−1	1.674−1	1.548−1	1.378−1	1.317−1
0.20	1.233−1	1.220−1	1.237−1	1.221−1	1.176−1	1.158−1	1.199−1	1.173−1	1.376−1	1.303−1	1.223−1	1.188−1
0.30	1.068−1	1.062−1	1.070−1	1.063−1	1.015−1	1.007−1	1.029−1	1.018−1	1.116−1	1.083−1	1.042−1	1.026−1
0.40	9.557−2	9.526−2	9.566−2	9.527−2	9.073−2	9.027−2	9.185−2	9.118−2	9.782−2	9.595−2	9.276−2	9.187−2
0.50	8.719−2	8.699−2	8.729−2	8.704−2	8.274−2	8.245−2	8.372−2	8.329−2	8.851−2	8.731−2	8.445−2	8.388−2
0.60	8.063−2	8.049−2	8.070−2	8.052−2	7.649−2	7.628−2	7.736−2	7.706−2	8.147−2	8.064−2	7.802−2	7.762−2
0.80	7.081−2	7.073−2	7.087−2	7.077−2	6.717−2	6.705−2	6.788−2	6.771−2	7.121−2	7.074−2	6.841−2	6.818−2
1.00	6.364−2	6.359−2	6.372−2	6.365−2	6.037−2	6.030−2	6.100−2	6.089−2	6.388−2	6.358−2	6.146−2	6.131−2
1.25	5.693−2	5.690−2	5.697−2	5.693−2	5.399−2	5.394−2	5.454−2	5.447−2	5.709−2	5.689−2	5.496−2	5.486−2
1.50	5.180−2	5.177−2	5.185−2	5.183−2	4.915−2	4.912−2	4.968−2	4.963−2	5.206−2	5.193−2	5.006−2	5.000−2
2.00	4.450−2	4.449−2	4.459−2	4.458−2	4.228−2	4.227−2	4.282−2	4.279−2	4.524−2	4.516−2	4.324−2	4.320−2
3.00	3.579−2	3.579−2	3.597−2	3.596−2	3.422−2	3.421−2	3.487−2	3.486−2	3.780−2	3.777−2	3.541−2	3.539−2
4.00	3.073−2	3.073−2	3.100−2	3.100−2	2.960−2	2.960−2	3.037−2	3.037−2	3.395−2	3.393−2	3.106−2	3.105−2
5.00	2.742−2	2.742−2	2.777−2	2.777−2	2.663−2	2.662−2	2.753−2	2.752−2	3.170−2	3.169−2	2.836−2	2.836−2
6.00	2.511−2	2.510−2	2.552−2	2.552−2	2.457−2	2.457−2	2.559−2	2.559−2	3.034−2	3.034−2	2.655−2	2.655−2
8.00	2.209−2	2.209−2	2.263−2	2.263−2	2.195−2	2.195−2	2.319−2	2.318−2	2.892−2	2.892−2	2.437−2	2.437−2
10.00	2.024−2	2.024−2	2.089−2	2.089−2	2.039−2	2.039−2	2.181−2	2.181−2	2.839−2	2.839−2	2.318−2	2.318−2
15.00	1.782−2	1.782−2	1.866−2	1.866−2	1.846−2	1.846−2	2.023−2	2.022−2	2.838−2	2.838−2	2.195−2	2.195−2
20.00	1.673−2	1.673−2	1.770−2	1.770−2	1.769−2	1.769−2	1.970−2	1.970−2	2.903−2	2.903−2	2.168−2	2.168−2
30.00	1.588−2	1.588−2	1.706−2	1.706−2	1.725−2	1.725−2	1.962−2	1.962−2	3.059−2	3.059−2	2.196−2	2.196−2
40.00	1.566−2	1.566−2	1.696−2	1.696−2	1.729−2	1.729−2	1.992−2	1.992−2	3.201−2	3.201−2	2.251−2	2.251−2
50.00	1.566−2	1.566−2	1.706−2	1.706−2	1.748−2	1.748−2	2.029−2	2.029−2	3.322−2	3.322−2	2.306−2	2.306−2
60.00	1.574−2	1.574−2	1.722−2	1.722−2	1.771−2	1.771−2	2.067−2	2.067−2	3.424−2	3.424−2	2.358−2	2.358−2
80.00	1.599−2	1.599−2	1.759−2	1.759−2	1.816−2	1.816−2	2.135−2	2.135−2	3.588−2	3.588−2	2.447−2	2.447−2
100.0	1.626−2	1.626−2	1.794−2	1.794−2	1.858−2	1.858−2	2.191−2	2.191−2	3.712−2	3.712−2	2.517−2	2.517−2

(cont.)

Table B.7. *(cont.)* Mass interaction μ_{t-coh}/ρ (total—coherent) and total μ_t/ρ coefficients (cm^2/g) for various elements. Data source: Seltzer [1993].

E (MeV)	Si ($Z=14$)		P ($Z=15$)		S ($Z=16$)		Cl ($Z=17$)		Ar ($Z=18$)		K ($Z=19$)	
	μ_t/ρ	$\mu_{t\text{-}coh}/\rho$	μ_t/ρ	$\mu_{t\text{-}coh}/\rho$	μ_t/ρ	$\mu_{t\text{-}coh}/\rho$	μ_t/ρ	$\mu_{t\text{-}coh}/\rho$	μ_t/ρ	$\mu_{t\text{-}coh}/\rho$	μ_t/ρ	$\mu_{t\text{-}coh}/\rho$
0.01	3.388+1	3.326+1	4.035+1	3.970+1	5.014+1	4.941+1	5.725+1	5.651+1	6.315+1	6.241+1	7.907+1	7.824+1
0.015	1.034+1	9.977	1.239+1	1.201+1	1.550+1	1.507+1	1.784+1	1.739+1	1.983+1	1.937+1	2.503+1	2.451+1
0.02	4.464	4.230	5.353	5.103	6.709	6.427	7.740	7.444	8.630	8.328	1.093+1	1.058+1
0.03	1.436	1.311	1.700	1.566	2.113	1.962	2.426	2.268	2.697	2.535	3.413	3.225
0.04	7.011−1	6.222−1	8.096−1	7.251−1	9.874−1	8.919−1	1.117	1.017	1.228	1.126	1.541	1.422
0.05	4.385−1	3.845−1	4.917−1	4.336−1	5.850−1	5.192−1	6.483−1	5.792−1	7.012−1	6.305−1	8.678−1	7.857−1
0.06	3.207−1	2.815−1	3.494−1	3.072−1	4.054−1	3.574−1	4.395−1	3.890−1	4.663−1	4.146−1	5.679−1	5.076−1
0.08	2.228−1	1.995−1	2.324−1	2.073−1	2.586−1	2.300−1	2.696−1	2.394−1	2.760−1	2.449−1	3.251−1	2.889−1
0.10	1.835−1	1.682−1	1.865−1	1.699−1	2.020−1	1.831−1	2.050−1	1.850−1	2.043−1	1.837−1	2.345−1	2.103−1
0.15	1.448−1	1.377−1	1.432−1	1.355−1	1.506−1	1.418−1	1.480−1	1.387−1	1.427−1	1.330−1	1.582−1	1.469−1
0.20	1.275−1	1.235−1	1.250−1	1.206−1	1.302−1	1.251−1	1.266−1	1.212−1	1.205−1	1.150−1	1.319−1	1.253−1
0.30	1.082−1	1.063−1	1.055−1	1.035−1	1.091−1	1.068−1	1.054−1	1.029−1	9.953−2	9.701−2	1.080−1	1.050−1
0.40	9.614−2	9.509−2	9.359−2	9.245−2	9.667−2	9.537−2	9.311−2	9.173−2	8.776−2	8.633−2	9.495−2	9.327−2
0.50	8.748−2	8.681−2	8.511−2	8.438−2	8.783−2	8.699−2	8.453−2	8.364−2	7.958−2	7.866−2	8.600−2	8.492−2
0.60	8.077−2	8.031−2	7.854−2	7.803−2	8.104−2	8.046−2	7.795−2	7.733−2	7.335−2	7.271−2	7.922−2	7.847−2
0.80	7.082−2	7.056−2	6.884−2	6.855−2	7.100−2	7.067−2	6.826−2	6.791−2	6.419−2	6.383−2	6.929−2	6.886−2
1.00	6.361−2	6.344−2	6.182−2	6.164−2	6.374−2	6.353−2	6.128−2	6.105−2	5.762−2	5.738−2	6.216−2	6.189−2
1.25	5.688−2	5.677−2	5.526−2	5.515−2	5.698−2	5.684−2	5.477−2	5.463−2	5.150−2	5.135−2	5.556−2	5.539−2
1.50	5.183−2	5.175−2	5.039−2	5.031−2	5.194−2	5.185−2	4.994−2	4.984−2	4.695−2	4.685−2	5.068−2	5.056−2
2.00	4.480−2	4.476−2	4.358−2	4.353−2	4.498−2	4.493−2	4.328−2	4.322−2	4.074−2	4.068−2	4.399−2	4.393−2
3.00	3.678−2	3.676−2	3.590−2	3.588−2	3.716−2	3.713−2	3.585−2	3.582−2	3.384−2	3.382−2	3.666−2	3.663−2
4.00	3.240−2	3.239−2	3.172−2	3.171−2	3.294−2	3.292−2	3.188−2	3.187−2	3.019−2	3.017−2	3.282−2	3.280−2
5.00	2.967−2	2.967−2	2.915−2	2.914−2	3.037−2	3.036−2	2.950−2	2.949−2	2.802−2	2.801−2	3.054−2	3.053−2
6.00	2.788−2	2.787−2	2.747−2	2.746−2	2.872−2	2.872−2	2.798−2	2.797−2	2.667−2	2.666−2	2.915−2	2.914−2
8.00	2.574−2	2.574−2	2.552−2	2.552−2	2.683−2	2.682−2	2.628−2	2.628−2	2.517−2	2.517−2	2.766−2	2.765−2
10.00	2.462−2	2.461−2	2.452−2	2.452−2	2.590−2	2.590−2	2.548−2	2.548−2	2.451−2	2.451−2	2.704−2	2.704−2
15.00	2.352−2	2.352−2	2.364−2	2.364−2	2.517−2	2.517−2	2.496−2	2.496−2	2.418−2	2.418−2	2.687−2	2.686−2
20.00	2.338−2	2.338−2	2.363−2	2.363−2	2.529−2	2.529−2	2.520−2	2.520−2	2.453−2	2.453−2	2.737−2	2.737−2
30.00	2.385−2	2.385−2	2.426−2	2.426−2	2.613−2	2.613−2	2.618−2	2.618−2	2.561−2	2.561−2	2.870−2	2.870−2
40.00	2.454−2	2.454−2	2.507−2	2.507−2	2.708−2	2.708−2	2.721−2	2.721−2	2.669−2	2.669−2	2.998−2	2.998−2
50.00	2.522−2	2.522−2	2.581−2	2.581−2	2.794−2	2.794−2	2.813−2	2.813−2	2.763−2	2.763−2	3.108−2	3.108−2
60.00	2.583−2	2.583−2	2.647−2	2.647−2	2.869−2	2.869−2	2.891−2	2.891−2	2.842−2	2.842−2	3.201−2	3.201−2
80.00	2.685−2	2.685−2	2.755−2	2.755−2	2.990−2	2.990−2	3.016−2	3.016−2	2.970−2	2.970−2	3.349−2	3.349−2
100.0	2.764−2	2.764−2	2.840−2	2.840−2	3.084−2	3.084−2	3.114−2	3.114−2	3.069−2	3.069−2	3.463−2	3.463−2

(cont.)

Table B.7. *(cont.)* Mass interaction μ_{t-coh}/ρ (total−coherent) and total μ_t/ρ coefficients (cm^2/g) for various elements. Data source: Seltzer [1993].

E (MeV)	Ca ($Z = 20$)		Ti ($Z = 22$)		Cr ($Z = 24$)		Mn ($Z = 25$)		Fe ($Z = 26$)		Ni ($Z = 28$)	
	μ_t/ρ	$\mu_{t\text{-}coh}/\rho$	μ_t/ρ	$\mu_{t\text{-}coh}/\rho$	μ_t/ρ	$\mu_{t\text{-}coh}/\rho$	μ_t/ρ	$\mu_{t\text{-}coh}/\rho$	μ_t/ρ	$\mu_{t\text{-}coh}/\rho$	μ_t/ρ	$\mu_{t\text{-}coh}/\rho$
0.01	9.340+1	9.251+1	1.107+2	1.098+2	1.387+2	1.376+2	1.514+2	1.503+2	1.707+2	1.695+2	2.090+2	2.076+2
0.015	2.979+1	2.922+1	3.587+1	3.528+1	4.571+1	4.505+1	5.027+1	4.958+1	5.708+1	5.633+1	7.081+1	6.995+1
0.02	1.306+1	1.268+1	1.585+1	1.545+1	2.038+1	1.992+1	2.253+1	2.205+1	2.568+1	2.516+1	3.220+1	3.160+1
0.03	4.079	3.873	4.971	4.754	6.434	6.185	7.141	6.880	8.176	7.891	1.034+1	1.001+1
0.04	1.830	1.700	2.213	2.076	2.856	2.699	3.169	3.004	3.629	3.450	4.600	4.392
0.05	1.019	9.291−1	1.213	1.118	1.551	1.442	1.714	1.600	1.957	1.833	2.474	2.330
0.06	6.578−1	5.913−1	7.661−1	6.958−1	9.640−1	8.834−1	1.060	9.753−1	1.205	1.113	1.512	1.405
0.08	3.655−1	3.254−1	4.052−1	3.626−1	4.905−1	4.414−1	5.306−1	4.791−1	5.952−1	5.392−1	7.306−1	6.658−1
0.10	2.571−1	2.304−1	2.721−1	2.436−1	3.167−1	2.838−1	3.367−1	3.021−1	3.717−1	3.340−1	4.440−1	4.003−1
0.15	1.674−1	1.548−1	1.649−1	1.515−1	1.788−1	1.633−1	1.838−1	1.675−1	1.964−1	1.786−1	2.208−1	2.001−1
0.20	1.376−1	1.303−1	1.314−1	1.236−1	1.378−1	1.288−1	1.391−1	1.297−1	1.460−1	1.357−1	1.582−1	1.462−1
0.30	1.116−1	1.083−1	1.043−1	1.008−1	1.067−1	1.026−1	1.062−1	1.019−1	1.099−1	1.051−1	1.154−1	1.099−1
0.40	9.782−2	9.595−2	9.081−2	8.880−2	9.213−2	8.979−2	9.133−2	8.887−2	9.400−2	9.131−2	9.765−2	9.450−2
0.50	8.851−2	8.731−2	8.191−2	8.062−2	8.281−2	8.131−2	8.192−2	8.034−2	8.414−2	8.241−2	8.698−2	8.495−2
0.60	8.147−2	8.064−2	7.529−2	7.439−2	7.598−2	7.493−2	7.509−2	7.399−2	7.704−2	7.583−2	7.944−2	7.803−2
0.80	7.121−2	7.074−2	6.572−2	6.521−2	6.620−2	6.561−2	6.537−2	6.475−2	6.699−2	6.631−2	6.891−2	6.811−2
1.00	6.388−2	6.358−2	5.891−2	5.858−2	5.930−2	5.892−2	5.852−2	5.812−2	5.995−2	5.951−2	6.160−2	6.108−2
1.25	5.709−2	5.689−2	5.263−2	5.242−2	5.295−2	5.271−2	5.224−2	5.198−2	5.350−2	5.322−2	5.494−2	5.461−2
1.50	5.206−2	5.193−2	4.801−2	4.787−2	4.832−2	4.815−2	4.769−2	4.751−2	4.883−2	4.863−2	5.015−2	4.992−2
2.00	4.524−2	4.516−2	4.180−2	4.171−2	4.213−2	4.204−2	4.162−2	4.152−2	4.265−2	4.254−2	4.387−2	4.374−2
3.00	3.780−2	3.777−2	3.512−2	3.508−2	3.559−2	3.554−2	3.524−2	3.519−2	3.621−2	3.616−2	3.745−2	3.739−2
4.00	3.395−2	3.393−2	3.173−2	3.171−2	3.235−2	3.232−2	3.213−2	3.211−2	3.312−2	3.309−2	3.444−2	3.440−2
5.00	3.170−2	3.169−2	2.982−2	2.980−2	3.057−2	3.056−2	3.045−2	3.043−2	3.146−2	3.144−2	3.289−2	3.286−2
6.00	3.034−2	3.034−2	2.868−2	2.867−2	2.956−2	2.955−2	2.952−2	2.950−2	3.057−2	3.056−2	3.210−2	3.208−2
8.00	2.892−2	2.892−2	2.759−2	2.759−2	2.869−2	2.868−2	2.875−2	2.875−2	2.991−2	2.991−2	3.164−2	3.163−2
10.00	2.839−2	2.839−2	2.727−2	2.727−2	2.855−2	2.855−2	2.871−2	2.871−2	2.994−2	2.994−2	3.185−2	3.184−2
15.00	2.838−2	2.838−2	2.762−2	2.761−2	2.920−2	2.920−2	2.951−2	2.951−2	3.092−2	3.092−2	3.320−2	3.319−2
20.00	2.903−2	2.903−2	2.844−2	2.844−2	3.026−2	3.026−2	3.068−2	3.068−2	3.224−2	3.223−2	3.476−2	3.476−2
30.00	3.059−2	3.059−2	3.020−2	3.020−2	3.236−2	3.236−2	3.290−2	3.290−2	3.469−2	3.469−2	3.759−2	3.759−2
40.00	3.201−2	3.201−2	3.173−2	3.173−2	3.411−2	3.411−2	3.473−2	3.473−2	3.666−2	3.666−2	3.984−2	3.984−2
50.00	3.322−2	3.322−2	3.300−2	3.300−2	3.556−2	3.556−2	3.625−2	3.625−2	3.828−2	3.828−2	4.167−2	4.167−2
60.00	3.424−2	3.424−2	3.407−2	3.407−2	3.677−2	3.677−2	3.749−2	3.749−2	3.961−2	3.961−2	4.318−2	4.318−2
80.00	3.588−2	3.588−2	3.577−2	3.577−2	3.867−2	3.867−2	3.945−2	3.945−2	4.172−2	4.172−2	4.552−2	4.552−2
100.0	3.712−2	3.712−2	3.706−2	3.706−2	4.008−2	4.008−2	4.092−2	4.092−2	4.329−2	4.329−2	4.726−2	4.726−2

Index

A

N

Q

R

S

T

U

V

W

X

Y

Z

For Product Safety Concerns and Information,
please contact our EU representative GPSR@taylorandfrancis.com
Taylor & Francis Verlag GmbH, Kaufingerstraße 24,
80331 München, Germany

Printed by Integrated Books International,
United States of America